GRAPHENE SCIENCE HANDBOOK

Electrical and Optical Properties

GRAPHENE SCIENCE HANDBOOK

Electrical and Optical Properties

EDITED BY
Mahmood Aliofkhazraei • Nasar Ali
William I. Milne • Cengiz S. Ozkan
Stanislaw Mitura • Juana L. Gervasoni

CRC Press
Taylor & Francis Group
Boca Raton London New York

CRC Press is an imprint of the
Taylor & Francis Group, an **informa** business

CRC Press
Taylor & Francis Group
6000 Broken Sound Parkway NW, Suite 300
Boca Raton, FL 33487-2742

Printed on acid-free paper
Version Date: 20151124

International Standard Book Number-13: 978-1-4665-9131-8 (Hardback)

Library of Congress Cataloging-in-Publication Data

Names: Aliofkhazraei, Mahmood, editor. | Ali, Nasar, editor. | Milne, W. I. (William I.), editor. | Ozkan, Cengiz S., editor. | Mitura, Stanislaw, 1951- editor. | Gervasoni, Juana L., editor.
Title: Graphene science handbook. Electrical and optical properties / edited by Mahmood Aliofkhazraei, Nasar Ali, William I. Milne, Cengiz S. Ozkan, Stanislaw Mitura, and Juana L. Gervasoni Taylor & Francis.
Other titles: Electrical and optical properties
Description: Boca Raton, FL : CRC Press, Taylor & Francis Group, 2016. | "2016 | Includes bibliographical references index.
Identifiers: LCCN 2015042986| ISBN 9781466591318 (hardcover ; alk. paper) |
ISBN 1466591315 (hardcover ; alk. paper)
Subjects: LCSH: Graphene--Handbooks, manuals, etc. | Graphene--Electric properties--Handbooks, manuals, etc. | Graphene--Optical properties--Handbooks, manuals, etc.
Classification: LCC QD341.H9 G684 2016 | DDC 547/.61--dc23
LC record available at http://lccn.loc.gov/2015042986

Visit the Taylor & Francis Web site at
http://www.taylorandfrancis.com

and the CRC Press Web site at
http://www.crcpress.com

Contents

SECTION II Optical Properties

SECTION III Nanocomposites and Applications

Preface

The theory behind "graphene" was first explored by the physicist Philip Wallace in 1947. However, the name "graphene" was not actually coined until 40 years later, where it was used to describe single sheets of graphite. Ultimately, Professor Geim's group in Manchester (UK) was able to manufacture and see individual atomic layers of graphene in 2004. Since then, much more research has been carried out on the material, and scientists have found that graphene has unique and extraordinary properties. Some say that it will literally change our lives in the twenty-first century. Not only is graphene the thinnest possible material, but it is also about 200 times stronger than steel and conducts electricity better than any other material at room temperature. This material has created huge interest in the electronics industry, and Konstantin Novoselov and Andre Geim were awarded the 2010 Nobel Prize in Physics for their groundbreaking experiments on graphene.

Graphene and its derivatives (such as graphene oxide) have the potential to be produced and used on a commercial scale, and research has shown that corporate interest in the discovery and exploitation of graphene has grown dramatically in the leading countries in recent decades. In order to understand how this activity is unfolding in the graphene domain, publication counts have been plotted in Figure P.1. Research and commercialization of graphene are both still at early stages, but policy in the United States as well as in other key countries is trying to foster the concurrent processes of research and commercialization in the nanotechnology domain.

Graphene can be produced in a multitude of ways. Initially, Novoselov and Geim employed mechanical exfoliation by using a Scotch tape technique to produce monolayers of the material. Liquid-phase exfoliation has also been utilized. Several bottom-up or synthesis techniques developed for graphene include chemical vapor deposition, molecular beam epitaxy, arc discharge, sublimation of silicon carbide, and epitaxy on silicon carbide.

The *first volume* of this handbook concerns the fabrication methods of graphene. It is divided into four sections: (1) fabrication methods and strategies, (2) chemical-based methods, (3) nonchemical methods, and (4) advances of fabrication methods.

Carbon is the sixth most abundant element in nature and is an essential element of human life. It has different structures called carbon allotropes. The most common crystalline forms of carbon are graphite and diamond. Graphite is a three-dimensional allotrope of carbon with a layered structure in which tetravalent atoms of carbon are connected to three other carbon atoms by three covalent bonds and form a hexagonal network structure. Each one of these aforementioned layers is called a graphene layer or sheet. Each sheet is placed in parallel on other sheets. Hence, the fourth valence electron connects the sheets to each other via van der Waals bonding. The covalent bond length is 0.142 nm. The bonds that are formed by carbon atoms between layers are weak; therefore, the sheets can slide easily over each other. The distance between layers is 0.335 nm. Due to its unique structure and geometry, graphene possesses remarkable physical–chemical properties, including a high Young's modulus, high fracture strength, excellent electrical and thermal conductivity, high charge carrier mobility, large specific surface area, and biocompatibility.

These properties enable graphene to be considered as an ideal material for a broad range of applications, ranging from quantum physics, nanoelectronics, energy research, catalysis, and engineering of nanocomposites and biomaterials. In this context, graphene and its composites have emerged as a new biomaterial, which provides exciting opportunities for the development of a broad range of applications, such as nanocarriers for drug delivery. The building block of graphene is completely different from other graphite materials and three-dimensional geometric shapes of carbon, such as zero-dimensional spherical fullerenes and one-dimensional carbon nanotubes.

The *second volume* of this handbook is predominantly about the nanostructure and atomic arrangement of graphene. The chapters in this volume focus on atomic arrangement and defects, modified graphene, characterization of graphene and its nanostructure, and also recent advances in graphene nanostructures. The planar structure of graphene provides an excellent opportunity to immobilize a large number of substances, including biomolecules and metals. Therefore, it is not surprising that graphene has generated great interest for its nanosheets, which nowadays can serve as an excellent platform for antibacterial applications, cell culture, tissue engineering, and drug delivery.

It is possible to produce composites reinforced with graphene on a commercial scale and low cost. In these composites, the existence of graphene leads to an increase in conductivity and strength of various three-dimensional materials. In addition, it is possible to use cheaply manufactured graphene in these composites. For example, exfoliation of graphite is one of the cheapest graphene production techniques. The behavior of many two-dimensional materials and their equivalent three-dimensional forms are completely different. The origin of the aforementioned differences in the behavior of these materials is associated with the weak forces that hold a large number of single layers together to create a bulk material. Graphene can be used in nanocomposites. Currently, researchers have been able to produce several tough and light materials by adding small amounts of graphene to metals, polymers, and ceramics. The composite materials usually show better electrical conductivity characteristics compared with pure bulk materials, and they are also more resistant against heat.

The *third volume* describes graphene's electrical and optical properties and also focuses on nanocomposites and their applications. The *fourth volume* relates to the mechanical and chemical properties of graphene and cites recent developments. The *fifth volume* presents other topics, such as size

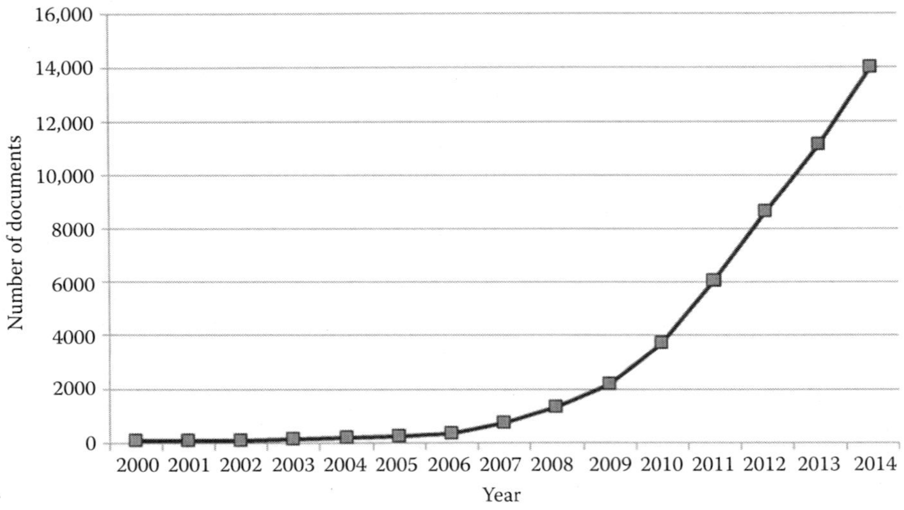

FIGURE P.1 Number of documents published around graphene during recent years, extracted from Scopus search engine by searching "graphene" in title + keywords + abstract.

effects in graphene, characterization, and applications based on size-affected properties. In recent years, scientists have produced advanced composites using graphene, which are excellent from the point of view of mechanical and thermal properties. However, in some of these composites, high electrical conductivity only is desirable. For example, the Chinese Academy of Sciences (IMR, CAS) has created a polymer matrix composite reinforced with graphene, which has a high electrical conductivity. In this composite, a flexible network of graphene has been added to a polydimethylsiloxane matrix (of the silicon family).

Investigation of early corporate trajectories for graphene has led to three major observations. First, the discovery-to-application cycle for graphene seems to be accelerated, for example, compared to fullerene. Even though the discovery of graphene is relatively new, large and small firms have contributed to an upsurge in early corporate activities. Second, a rapid globalization has occurred by companies in the United States, Europe, Japan, South Korea, and other developed economies, which were involved in early graphene activities. Chinese companies are currently starting to enter the graphene domain, resulting in the expansion of research capability of nanotechnology. Nevertheless, science alone does not guarantee commercial exploitation. To clarify the issue, the level of corporate patenting in the United Kingdom, which is a pioneer in graphene research, is slightly ahead of Canada and Germany; however, it is dramatically lower than in the United States, Japan, and South Korea. Third, the potential applications of graphene are rapidly expanding. Corporate patenting trends are indicative of their enthusiasm to utilize the features of graphene in various areas, including transistors, electronic memory and circuits, capacitors, displays, solar cells, batteries, coatings, advanced materials, sensors, and biomedical devices. Although graphene was initially proposed as an alternative to silicon, its initial applications have been in electronic inks and additives to resins and coatings. We have identified six areas of emerging applications for graphene, including displays/screens, memory chips, biomedical devices, batteries/fuel cells, coatings and inks, and materials. In the investigation of the corporate engagement in graphene, we sought to understand early corporate activity patterns related to broader research and invention trends. In traditional innovation models, a lag between research publication and patenting is consistent with the linear model. However, more recent innovation models are stressing concurrent launch, open innovation, and strategic property management.

The *sixth volume* of this handbook is about the application and industrialization of graphene, starting with chapters about biomaterials and continues onto nanocomposites, electrical/sensor devices, and also new and novel applications.

The editorial team would like to thank all contributors for their excellent chapters contributed to the creation of this handbook and for their hard work and patience during its preparation and production. We sincerely hope that the publication of this handbook will help people, especially those working with graphene, and benefit them from the knowledge contained in the published chapters.

Mahmood Aliofkhazraei
Nasar Ali
William I. Milne
Cengiz S. Ozkan
Stanislaw Mitura
Juana L. Gervasoni
Summer 2015

Editors

Mahmood Aliofkhazraei is an assistant professor in the Materials Engineering Department at Tarbiat Modares University. Dr. Aliofkhazraei's research interests include nanotechnology and its use in surface and corrosion science. One of his main interests is plasma electrolysis, and he has published more than 40 papers and a book in this area. Overall he has published more than 12 books and 90 journal articles. He has delivered invited talks, including keynote addresses in several countries. Aliofkhazraei has received numerous awards, including the Khwarizmi award, IMES medal, INIC award, best-thesis award (multiple times), best-book award (multiple times), and the best young nanotechnologist award of Iran (twice). He is on the advisory editorial board of several materials science and nanotechnology journals.

Nasar Ali is a visiting professor at Meliksah University in Turkey. Earlier he held the post of chief scientific officer at CNC Coatings Company based in Rochdale, UK. Prior to this Dr. Ali was a faculty member (assistant professor) at the University of Aveiro in Portugal where he founded and led the Surface Engineering and Nanotechnology group. Dr. Ali has extensive research experience in hard carbon-coating materials, including nanosized diamond coatings and CNTs deposited using CVD methods. He has over 120 international refereed research publications, including a number of book chapters. Dr. Ali serves on a number of committees for international conferences based on nanomaterials, thin films, and emerging technologies (nanotechnology), and he chairs the highly successful NANOSMAT congress. He served as the fellow of the Institute of Nanotechnology for 2 years on invitation. Dr. Ali has authored and edited several books on surface coatings, thin films, and nanotechnology for leading publishers, and he was also the founder of the *Journal of Nano Research*. Dr. Ali was the recipient of the Bunshah prize for presenting his work on time-modulated CVD at the ICMCTF-2002 Conference in San Diego, California.

William I. Milne, FREng, FIET, FIMMM, was head of electrical engineering at the Cambridge University from 1999 until 2014 and has been director of the Centre for Advanced Photonics and Electronics (CAPE) since 2004. He earned a BSc at St. Andrews University in Scotland in 1970 and later earned a PhD in electronic materials at the Imperial College London. In 2003 he was awarded a DEng (honoris causa) by the University of Waterloo, Canada, and he was elected as Fellow of the Royal Academy of Engineering in 2006. He received the JJ Thomson medal from the Institution of Engineering and Technology in 2008 for achievement in electronics and the NANOSMAT prize in 2010. He is a distinguished visiting professor at Tokyo Institute of Technology, Japan, and a distinguished visiting professor at Southeast University in Nanjing, China, and at Shizuoka University, Japan. He is also a distinguished visiting scholar at KyungHee University, Seoul and a high-end foreign expert for the Changchun University of Science and Technology in China. In 2015, he was elected to an Erskine Fellowship to visit the University of Canterbury, New Zealand. His research interests include large area silicon- and carbon-based electronics, thin film materials, and, most recently, MEMS and carbon nanotubes, graphene, and other 1-D and 2-D structures for electronic applications, especially for field emission. He has published/presented approximately 800 papers, of which around 200 were invited/keynote/plenary talks—his "h" index is currently 57 (Web of Science).

Cengiz S. Ozkan has been a professor of mechanical engineering and materials science at the University of California, Riverside, since 2009. He was an associate professor from 2006 to 2009 and an assistant professor from 2001 to 2006. Between 2000 and 2001 he was a consulting professor at Stanford University. He earned a PhD in materials science and engineering at Stanford University in 1997. Dr. Ozkan's areas of expertise include nanomaterials for energy storage; synthesis/processing including graphene, III–V, and II–VI materials; novel battery and supercapacitor architectures; nanoelectronics; biochemical sensors; and nanopatterning for beyond CMOS (complementary metal-oxide semiconductor). He organized and chaired 20 scientific and international conferences. He has written more than 200 technical publications, including journal papers, conference proceedings, and book chapters. He holds over 50 patent disclosures, has given more than 100 presentations worldwide, and is the recipient of more than 30 honors and awards. His important contributions include growth of hierarchical three-dimensional graphene nanostructures; development of a unique high-throughput metrology method for large-area CVD-grown graphene sheets; doping and functionalization of CVD-grown and pristine graphene layers; study of digital data transmission in graphene and InSb materials; memory devices based on inorganic/organic nanocomposites, novel lithium-ion batteries based on nanosilicon from beach sand and silicon dioxide nanotubes; fast-charging lithium-ion batteries based on silicon-decorated three-dimensional nano-carbon architectures; and high-performance supercapacitors based on three-dimensional graphene foam architectures.

Stanislaw Mitura has been a professor in biomedical engineering at the Koszalin University of Technology from 2011. He is a visiting professor at the Technical University (TU) of Liberec and was awarded a doctor honoris causa from TU Liberec. He was a professor of materials science at Lodz University of Technology from 2001 to 2014. He earned an MSc in physics at the University of Lodz in 1974, a PhD in mechanical engineering at the Lodz University of Technology (1985), and a DSc in materials science at the Warsaw University of Technology in 1993. Professor Mitura's most prominent cognitive achievements comprise the following: from the concept of nucleation of diamond powder particles to the synthesis of nanocrystalline diamond coatings (NDC); discovery of diamond bioactivity; a concept of the gradient transition from carbide forming metal to diamond film; and technology development of nanocrystalline diamond coatings for medical purposes. Professor Mitura has published over 200 peer-reviewed articles, communications, and proceedings, over 50 invited talks, and contributed to 7 books and proceedings, including *Nanotechnology for Materials Science* (Pergamon, Elsevier, 2000) and *Nanodiam* (PWN, 2006). He organized and co-organized several conferences focused on materials science and engineering, especially diamond synthesis under reduced pressure. He is an elected member of the Academy of Engineering in Poland, guest editor in few international journals, including *Journal of Nanoscience and Nanotechnology*, *Journal of Superhards Materials* and also a member of the editorial boards of several journals and an elected Fellow of various foreign scientific societies.

Juana L. Gervasoni earned her doctorate in physics at the Instituto Balseiro, Bariloche, Argentina, in 1992. She has been the head of the Department of Metal Materials and Nanostructures, Applied Research of Centro Atomico Bariloche (CAB), National Atomic Energy Commission (CNEA), since 2012. She has been a member of the Coordinating Committee of the CNEA Controlled Fusion Program since 2013. Her area of scientific research involves the interactions of atomic particles of matter, electronic excitations in solids, surfaces, and nanosystems, the absorption of hydrogen in metals, and study of new materials under irradiation. Gervasoni is a researcher at the National Atomic Energy Commission of Argentina and the National Council of Scientific and Technological Research (CONICET, Argentina). She teaches at the Instituto Balseiro and is involved in directing graduate students and postdoctorates. She has published over 100 articles in international journals, some of which have a high impact factor, and she has attended many international conferences. Gervasoni has been a member of the Executive Committee and/or the International Scientific Advisory Board of the International Conference on Surfaces Coatings and Nanostructured Materials (Nanosmat) since 2010, Latin American Conference on Hydrogen and Sustainable Energy Sources (Hyfusen), and the International Conference on Clean Energy (International Conference on Clean Energy, ICCE-2010) and guest editor of the *International Journal of Hydrogen Energy* (Elsevier). Recently she has focused her research on the study of hydrogen storage in carbon nanotubes. Along with her academic and research work, Gervasoni is heavily involved in gender issues in the scientific community, especially in Argentina and Latin America. She is a member of the Third World Organization for Women in Science (TWOWS), branch of the Third World Academy of Science (TWAS), Trieste, Italy, since 2010, as well as of Women in Nuclear (WiN) since 2013.

Contributors

Andrei Alaferdov
Center for Semiconductor Components—CCS
UNICAMP
São Paulo, Brazil

M. Sahabul Alam
Chemical Engineering Department
King Abdullah Institute for Nanotechnology
King Saud University
Riyadh, Kingdom of Saudi Arabia

Luísa Andrade
Laboratory for Process Engineering, Environment, Biotechnology and Energy
Department of Chemical Engineering at the Faculty of Engineering
University of Porto
Porto, Portugal

G. G. N. Angilella
Department of Physics and Astronomy
University of Catania
and
CNISM
UdR di Catania
and
School of Catania
University of Catania
and
INFN
Section of Catania
and
CNR-IMM
Catania, Italy

P. Banerji
Materials Science Centre
Indian Institute of Technology
West Bengal, India

Hui-Ming Cheng
Shenyang National Laboratory for Materials Science
Institute of Metal Research
Chinese Academy of Sciences
Shenyang, China

Gennaro Chiarello
Department of Physics
University of Calabria
Rende (CS), Italy

Y. H. Chiu
Department of Applied Physics
National Pingtung University
Pingtung, Taiwan

Farzana A. Chowdhury
Experimental Physics Division
Atomic Energy Centre
Dhaka, Bangladesh

Rui Cruz
Laboratory for Process Engineering, Environment, Biotechnology and Energy
Department of Chemical Engineering at the Faculty of Engineering
University of Porto
Porto, Portugal

A. G. D'Aloia
Department of Astronautics, Electrical, and Energetic Engineering
Sapienza University of Rome
Rome, Italy

G. De Bellis
Department of Astronautics, Electrical, and Energetic Engineering
Sapienza University of Rome
Rome, Italy

Mircea V. Diudea
Department of Chemistry
"Babes-Bolyai" University
Cluj, Romania

Dirk Englund
Department of Electrical Engineering and Computer Science
Massachusetts Institute of Technology
Cambridge, Massachusetts

Victor Ermakov
Center for Semiconductor Components—CCS
UNICAMP
São Paulo, Brazil

Xiaofeng Fan
College of Materials Science and Engineering
Jilin University
Changchun, China

Yuan Ping Feng
Department of Physics
National University of Singapore
Singapore

C. Galiotis
Foundation of Research and Technology Hellas
Institute of Chemical Engineering and High Temperature Processes
and
Department of Chemical Engineering
University of Patras
Patras, Greece

Xuetao Gan
School of Science
Northwestern Polytechnical University
Xi'an, China

and

Department of Electrical Engineering
Columbia University
New York, New York

Bo Gao
Department of Physics and Astronomy
Hunter College at the City University of New York
New York, New York

Xueyun Gao
Institute of High Energy Physics
Chinese Academy of Sciences
Beijing, China

H. V. Grushevskaya
Physics Department
Belarusan State University
Minsk, Belarus

Godfrey Gumbs
Department of Physics and Astronomy
Hunter College at the City University of New York
New York, New York

and

Donostia International Physics Center
Donostia-San Sebastian, Spain

Xuefei Guo
Department of Chemistry
University of Cincinnati
Cincinnati, Ohio

R. M. Abdel Hameed
Chemistry Department
Cairo University
Giza, Egypt

Linxiang He
Department of Physics and Materials Science
City University of Hong Kong
Hong Kong, China

M. J. Hu
Department of Physics and Materials Science
City University of Hong Kong
Hong Kong, China

Danhong Huang
Air Force Research Laboratory
Space Vehicles Directorate
Kirtland Air Force Base
Albuquerque, New Mexico

Andrii Iurov
Department of Physics and Astronomy
Hunter College at the City University of New York
New York, New York

K. Jagannadham
Materials Science and Engineering
North Carolina State University
Raleigh, North Carolina

Neeraj K. Jaiswal
Discipline of Physics
Indian Institute of Information Technology, Design and Manufacturing
Madhya Pradesh, India

S. Mahaboob Jilani
Materials Science Centre
Indian Institute of Technology
West Bengal, India

Long Jing
Institute of High Energy Physics
Chinese Academy of Sciences
Beijing, China

Saumil Joshi
Department of Electrical Engineering
University of Colorado
Boulder, Colorado

Golap Kalita
Center for Fostering Young and Innovative Researchers
and
Department of Frontier Materials
Nagoya Institute of Technology
Nagoya, Japan

Anupama B. Kaul
Department of Metallurgical and Materials Engineering
Department of Electrical and Computer Engineering (Joint)
University of Texas, El Paso
El Paso, Texas

S. C. Kim
Physics Department
College of Science
Korea University
Seoul, Korea

O. Kon'kov
Division of Solid State Physics
Ioffe Institute
St. Petersburg, Russia

G. G. Krylov
Physics Department
Belarusan State University
Minsk, Belarus

S. S. Kubakaddi
Department of Physics
Karnatak University
Karnataka, India

Chun-Sing Lee
Department of Physics and Materials Science
City University of Hong Kong
Hong Kong, China

and

City University of Hong Kong Shenzhen Research Institute
Shenzhen, China

Woo Hyoung Lee
Department of Civil, Environmental, and Construction Engineering
University of Central Florida
Orlando, Florida

Feng Li
Shenyang National Laboratory of Materials Science
Institute of Metal Research
Chinese Academy of Sciences
Shenyang, China

M. F. Lin
Department of Physics
National Cheng Kung University
Tainan, Taiwan

Sefer Bora Lisesivdin
Department of Physics
Gazi University
Ankara, Turkey

Zhi-Bo Liu
The Key Laboratory of Weak Light Nonlinear Photonics, Ministry of Education, Teda Applied Physics School, and School of Physics
Nankai University
Tianjin, China

Ming-Fai Lo
Department of Physics and Materials Science
City University of Hong Kong
Hong Kong, China

and

City University of Hong Kong Shenzhen Research Institute
Shenzhen, China

José Maçaira
Laboratory for Process Engineering, and Environment, Biotechnology and Energy
Department of Chemical Engineering at the Faculty of Engineering
University of Porto
Porto, Portugal

F. Marra
Department of Astronautics, Electrical, and Energetic Engineering
Sapienza University of Rome
Rome, Italy

Adélio Mendes
Laboratory for Process Engineering, Environment, Biotechnology and Energy
Department of Chemical Engineering at the Faculty of Engineering
University of Porto
Porto, Portugal

Adam Mock
School of Engineering and Technology
Central Michigan University
Mount Pleasant, Michigan

Garret Moddel
Department of Electrical, Computer, and Energy Engineering
University of Colorado
Boulder, Colorado

Stanislav Moshkalev
Center for Semiconductor Components—CCS
UNICAMP
São Paulo, Brazil

Weihua Mu
Department of Physics
Brown University
Providence, Rhode Island

B. G. Mulimani
Department of Physics
B.L.D.E. University
Karnataka, India

Enrique Muñoz
Physics Institute
Pontificia Universidad Católica de Chile, Vicuña Mackenna
Santiago, Chile

Masayoshi Nakano
Department of Materials Engineering Science
Graduate School of Engineering Science
Osaka University
Osaka, Japan

Polina M. Nedorezova
N. N. Semenov Institute of Chemical Physics
Russian Academy of Sciences
Moscow, Russia

Tsz-Wai Ng
Department of Physics and Materials Science
City University of Hong Kong
Hong Kong, China

and

City University of Hong Kong Shenzhen Research Institute
Shenzhen, China

Ottorino Ori
Laboratory of Computational and Structural
Physical Chemistry for Nanosciences and QSAR, Biology-Chemistry Department
West University of Timişoara
Timişoara, Romania

and

Actinium Chemical Research
Rome, Italy

Joe Otsuki
College of Science and Technology
Nihon University
Tokyo, Japan

Y. C. Ou
Department of Physics
National Cheng Kung University
Tainan, Taiwan

Ekmel Ozbay
Nanotechnology Research Center
and
Department of Physics
and
Department of Electrical and Electronics Engineering
Bilkent University
Ankara, Turkey

Alexander N. Ozerin
N. S. Enikolopov Institute of Synthetic Polymer Materials
Russian Academy of Sciences
Moscow, Russia

K. Papagelis
Foundation of Research and Technology Hellas
Institute of Chemical Engineering and High Temperature Processes
and
Department of Materials Science
University of Patras
Patras, Greece

F. M. D. Pellegrino
Department of Physics and Astronomy
University of Catania
and
CNISM
UdR di Catania
Catania, Italy

and

Scuola Normale Superiore
Pisa, Italy

Bradley Pelz
Department of Electrical, Computer, and Energy Engineering
University of Colorado
Boulder, Colorado

A. J. Pérez-Jiménez
Department of Physical Chemistry
University of Alicante
Alicante, Spain

Antonio Politano
Department of Physics
University of Calabria
Rende (CS), Italy

A. Prikhod'ko
Institute of Physics, Nanotechnology and Telecommunications
Peter the Great St. Petersburg Polytechnic University
St. Petersburg, Russia

R. Pucci
Department of Physics and Astronomy
University of Catania
and
CNISM
UdR di Catania
and
CNR-IMM
Catania, Italy

Mihai V. Putz
Laboratory of Computational and Structural Physical Chemistry for Nanosciences and QSAR, Biology-Chemistry Department
West University of Timişoara
and
Laboratory of Renewable Energies—Photovoltaics
R&D National Institute for Electrochemistry and Condensed Matter
Timisoara, Romania

B. S. Razbirin
Ioffe Physical-Technical Institute, RAS
St. Petersburg, Russia

Jeremy T. Robinson
Naval Research Laboratory
Washington, DC

N. N. Rozhkova
Institute of Geology Karelian Research Centre RAS
Petrozavodsk, Russia

J. C. Sancho-García
Department of Physical Chemistry
University of Alicante
Alicante, Spain

N. S. Sankeshwar
Department of Physics
Christ University
Karnataka, India

Beyza Sarikavak-Lisesivdin
Department of Physics
Gazi University
Ankara, Turkey

M. S. Sarto
Department of Astronautics, Electrical, and Energetic Engineering
Sapienza University of Rome
Rome, Italy

G. C. Shan
Department of Physics and Materials Science
City University of Hong Kong
Hong Kong, China

and

Fu Foundation School of Engineering and Applied Science
Columbia University
New York, New York

C. H. Shek
Department of Physics and Materials Science
City University of Hong Kong
Hong Kong, China

E. F. Sheka
People's Friendship University of Russia
Moscow, Russia

Vitaliy G. Shevchenko
N. S. Enikolopov Institute of Synthetic Polymer Materials
Russian Academy of Sciences
Moscow, Russia

Ren-Jye Shiue
Department of Electrical Engineering and Computer Science
Massachusetts Institute of Technology
Cambridge, Massachusetts

Pankaj Srivastava
Nanomaterials Research Group
ABV-Indian Institute of Information Technology and Management
Madhya Pradesh, India

A. Tamburrano
Department of Astronautics, Electrical, and Energetic Engineering
Sapienza University of Rome
Rome, Italy

Masaki Tanemura
Department of Frontier Materials
Nagoya Institute of Technology
Nagoya, Japan

D. Tasis
Foundation of Research and Technology Hellas
Institute of Chemical Engineering and High Temperature Processes
Patras, Greece

and

Department of Chemistry
University of Ioannina
Ioannina, Greece

John Texter
School of Engineering Technology
Eastern Michigan University
Ypsilanti, Michigan

Jian-Guo Tian
Key Laboratory of Weak Light Nonlinear Photonics, Ministry of Education, Teda Applied Physics School, and School of Physics
Nankai University
Tianjin, China

Sie Chin Tjong
Department of Physics and Materials Science
City University of Hong Kong
Hong Kong, China

Masayoshi Umeno
Department of Electronics and Information Engineering
Chubu University
Kasugai, Japan

Alfredo Rodrigues Vaz
Center for Semiconductor Components—CCS
UNICAMP
São Paulo, Brazil

Bing Wang
School of Physics
Huazhong University of Science and Technology
Wuhan, China

Shi Jie Wang
Institute of Material Research and Engineering
Agency for Science Technology and Research (A*STAR)
Singapore

Xiao-Qing Yan
Key Laboratory of Weak Light Nonlinear Photonics, Ministry of Education, Teda Applied Physics School, and School of Physics
Nankai University
Tianjin, China

Ming Yang
Department of Physics
National University of Singapore
Singapore

Qing-Dan Yang
Department of Physics and Materials Science
City University of Hong Kong
Hong Kong, China

and

City University of Hong Kong Shenzhen Research Institute
Shenzhen, China

S.-R. Eric Yang
Physics Department
College of Science
Korea University
Seoul, Korea

Kyohei Yoneda
Department of Chemical Engineering
National Institute of Technology
Nara College
Nara, Japan

Xiang Zhang
Department of Physics
Wuhan University
Wuhan, China

Guangmin Zhou
Shenyang National Laboratory of Materials Science
Institute of Metal Research
Chinese Academy of Sciences
Shenyang, China

Zixu Zhu
University of Colorado
Boulder, Colorado

Klaus Ziegler
Department of Physics
University of Augsburg
Augsburg, Germany

Section I

Electrical Properties

1 Graphene and Graphene Nanoribbons
Properties, Synthesis, and Electronic Applications

Anupama B. Kaul and Jeremy T. Robinson

CONTENTS

ABSTRACT

Research in two-dimensional graphene since it was mechanically exfoliated in 2004 has resulted in significant advances in unveiling its remarkable properties. These remarkable properties arise from graphene's unique hexagonal honeycomb crystal structure and the concominant sp^2 bonding that is also common to its other nanocarbon relatives, one-dimensional carbon nanotubes, and zero-dimensional buckminsterfullerene spheres. Over the past decade, research in harnessing graphene's remarkable material properties has spawned a number of exciting directions for device-related activities where graphene plays a key role. In this chapter, we present a review of the current advances that have taken place in using graphene and graphene nanoribbons for high-frequency transistors, energy-efficient electronics, as well as photonic devices. Given its remarkable mechanical properties, graphene is also very attractive for flexible electronics, as well as nano-electromechanical systems for mechanical resonators and mass-sensing devices, which are also reviewed in this chapter. In addition, techniques for synthesizing electronic-grade graphene using bottom-up techniques such as chemical vapor deposition, as well as the realization of epitaxial graphene through Si sublimation on SiC substrates, are also discussed.

1.1 INTRODUCTION

Since the mechanical exfoliation of graphene in 2004 [1], there has been a surge in research for unveiling the remarkable properties of this single sheet of carbon atoms and applying these properties to a wide range of devices and technological applications. As early as 1962, Boehm et al. [2] reported on the synthesis of single and multilayer graphene through the reduction of graphene oxide (GO) and this technique has gained renewed interest recently. In addition, graphene was also suggested to form on transition metals such as Pt (100) and Ni (111) by hydrocarbon dissociation [3–5], and other techniques such as metal intercalation [6]. While these surface science investigations of graphene on metals [7] occurred several decades ago, the exploration of fundamental properties, such as the quantum Hall effect and other interesting phenomena [8–10], as well as the ensuing applications of graphene, did not occur until after 2004 [1] following the seminal experiments of Novoselov and Geim on mechanically exfoliated graphene that earned them the 2010 Nobel Prize in Physics.

Up until recently, two-dimensional (2D) atomic monolayers were presumed to generally exist only as parts of larger three-dimensional (3D) structures, such as epitaxially grown crystals on lattice-matched substrates [11]. The existence of free-standing graphene was dismissed for some time as not being thermodynamically stable by Landau [12] and then several decades later by Mermin [13]. At the same time, the electronic structure and properties of graphene were analyzed theoretically by Wallace as early as 1947 [14] who showed that a single sheet of sp^2-hybridized carbon would have a linear energy dispersion relation at the *K-point* of the Brillouin zone.

Graphitic carbon nanomaterials such as zero-dimensional (0D) fullerenes, one-dimensional (1D) single-walled carbon nanotubes (SWCNTs), quasi-1D multiwalled (MW) CNTs and carbon nanofibers (CNFs), and 3D graphite are all derivatives of graphene's 2D honeycomb lattice. The sp^2 bonding of carbon atoms in graphitic carbon nanomaterials generally affords them exceptional material properties, such as a high charge carrier mobility, high thermal conductivity, and excellent mechanical flexibility, which make them attractive for a range of applications [15–18]. It is interesting to note that fullerenes—the lowest dimensional form of carbon nanomaterials—were the first to be discovered among the graphitic carbon nanomaterial family by Kroto et al. [19] in 1985, followed by carbon nanotubes in 1991 by Iijima [20], while graphene remained elusive until recently despite being the basis for the other forms of carbon nanomaterials.

The charged carriers in graphene interact with the honeycomb lattice in a remarkable way and lead to unique electron transport properties. Graphene's charge carriers are referred to as massless Dirac fermions, where the electrons move relativistically at speeds of ~10^6 ms^{-1} instead of 3×10^8 ms^{-1} [14]. These unique physical attributes make graphene a remarkable platform in which to also examine relativistic and interesting low-dimensional physics effects, besides the obvious potential it has for technological applications [15]. The massless carriers and minimal scattering over micron-size crystallites results in quantum mechanical effects that can survive even at room temperature. Graphene also exhibits ballistic room-temperature electron mobility of 2.5×10^5 $cm^2/V/s$ [21], the ability to carry ultra-high-current densities of ~10^9 A/cm^2 [22], an ultrahigh Young's modulus of 1 TPa with a breaking strength of ~40 N/m [23] and excellent elasticity for accommodating strains of up to 20% without breaking, a very high thermal conductivity of ~5000 W/m/K [24,25], and optical absorption of 2.3% [26]. Graphene also appears to be an excellent barrier for gases [27].

Owing to its flexibility, strength, high conductivity, transparency, and low cost, graphene has been proposed for a number of applications reviewed recently [28], such as a replacement for indium tin oxide (ITO) [29]. It also appears attractive for organic light-emitting diodes (OLEDs), as well as displays, touch screens [30], and solar cells [31]. The large surface-area-to-volume ratio of graphene suggests that it also has promise in ultracapacitor applications [32] or chemical sensors [33,34], and coupled with its remarkable electronic properties, it is a strong contender for next-generation electronic devices in nanoelectronics [35,36]. Graphene and graphene-like materials have also been applied to composite materials applications [15,37,38].

Although electronic transport in graphene can be ballistic, its predominantly metallic character is somewhat limiting for digital electronics applications. Approaches have been devised where band gaps are induced in graphene through the formation of, for example, graphene nanoribbons (GNRs) [39] with linear widths of <10 nm, which differs from SWCNTs, where 2/3 are semiconducting due to the chemical vapor deposition (CVD) synthesis process. In Section 1.2 of this chapter, an overview of the synthesis techniques for forming graphene and GNRs will be presented, followed in Section 1.3 by the applications of graphene in electronics, photonics, flexible electronics, and nano-electromechanical systems (NEMS). Finally, Section 1.4 provides closing remarks and an outlook for future potential opportunities for graphene.

1.2 SYNTHESIS OF GRAPHENE AND GNRs

1.2.1 Overview of Techniques

As discussed earlier, the formation of graphene can be traced back to the 1960s [2–4] where its synthesis on metal surfaces was explored by surface scientists. Synthesis of graphene has benefitted tremendously from surface science characterization techniques such as low energy electron diffraction (LEED), scanning tunneling microscopy (STM), x-ray photoemission spectroscopy (XPS), and angle-resolved photoemission spectroscopy (ARPES). In addition, Raman spectroscopy has played a central role in the characterization of graphene to evaluate quality and layer number [40], as summarized in several articles recently by Dresselhaus et al. [41,42].

These characterization tools have been instrumental in the examination of mechanically exfoliated graphene using the so-called "scotch-tape" technique for isolating single layers of graphene from 3D bulk graphite [1], which triggered the surge of activities to unveil its remarkable properties [43]. Besides the scotch-tape technique, mechanical exfoliation of highly oriented pyrolytic graphite (HOPG) has also been automated to yield an industrial process [44]. However, given that this technique is of a low yield, it is possible to increase production rates through an ultrasonic cleavage process [45] that can also be aided by chemical techniques such as intercalation to expand the separation between the atomic planes in graphene, further accelerating the sonication process [46]. Such techniques have resulted in polycrystalline films that appear to be of acceptable quality for some applications such as composite materials [45,46].

Besides mechanical exfoliation and its derivatives, numerous other techniques have recently emerged to produce monolayer graphene. This includes CVD [47–49], chemically driven bottom-up approaches for the synthesis of high-quality GNRs and other morphologies such as T and y-shaped connections [50], solid-state carbon source deposition [51,52], and solution exfoliation [53]. The reduction of the oxide on exfoliated graphite oxide has also been revisited recently for the synthesis of graphene [54–56], as well as epitaxial growth techniques involving the vacuum graphitization of SiC substrates [57–59], and molecular-beam epitaxy for synthesizing chemically pure graphene [60]. Other less-explored approaches include laser ablation that has the advantage of enabling graphene growth on arbitrary substrates [61]. Given the wide range of growth and synthesis methods used for forming graphene of varying purity and quality, the choice of the technique to be used is largely dictated by the end-use application.

For electronic and photonic applications, a particularly important parameter is the ability to dope graphene to tailor its physical and chemical properties. Substitutional doping of graphene with elements such as boron and nitrogen adjacent to carbon on the periodic table has been explored theoretically [62]. In the case of boron, a hole is introduced due to its p-type doping character, while nitrogen donates an electron and thus acts as an *n*-type dopant. To introduce the boron dopant, diborane (B_2H_6) is used as a precursor [63] and doping levels of up to 3.1% have been accomplished. On the other hand, nitrogen doping has been achieved with the use of ammonia [64] or pyridine [65] during the CVD growth of graphene, with nitrogen-doping levels approaching 9%, along with reports that demonstrate nitrogen doping in monolayer graphene sheets [66]. Edge doping of graphene ribbons has also been reported with the use of ammonia [67] and subsequent annealing of the ribbons. Both in-plane substitutional doping and edge doping have shown *n*-type behavior, and a band gap in *n*-doped graphene has also been reported [64]. In addition, charge-transfer doping [68], for example with alkali atoms, can also be used to *n*-type dope graphene given the ease with which these alkali elements give up their electrons. Table 1.1 provides a summary of the common synthesis and doping techniques for graphene.

For electronics and photonics applications, due to the stringent requirements on material quality and for minimizing defects, the most promising techniques appear to be CVD and the epitaxial growth of graphene on SiC substrates. These two techniques are discussed in more detail in the next section.

TABLE 1.1
Common Synthesis and Doping Techniques for Graphene

	Synthesis Techniques		
CVD [47–49]	Chemically driven bottom-up approaches [50]	Solid-state carbon source deposition [51,52]	Solution–exfoliation [53]
	Doping		
	Boron	Nitrogen	
	p-type (hole conduction)	*n*-type (electron conduction)	
Precursor	B_2H_6 [63]	Ammonia [64] Pyridine [65]	
Doping level	3.10%	9%	

1.2.2 Techniques for Synthesizing Electronic-Grade Graphene

1.2.2.1 CVD Graphene

Standard CVD is the most investigated method to generate high-quality, large-area graphene films from gaseous carbon sources such as methane using copper or nickel substrates [47–49]. Other metal substrates have also been employed to synthesize graphene by the catalytic decomposition of hydrocarbons on metals [69,70] such as Co (0001), Ni (111), Pt (111), Pd (111), Ru (0001), or Ir (111). In particular, very low defect densities have been observed when low-pressure (LP) CVD of ethylene was used to synthesize monolayer graphene on Ir (111) substrates at temperatures in excess of 800°C [70]. Once formed, Raman spectroscopy is routinely used to characterize graphene's crystalline quality by identifying the position and intensity of the "G peak" that occurs at approximately 1580 wavenumbers (cm^{-1}).

While the CVD synthesis of graphene is typically conducted on metal substrates such as Cu, conventional semiconductor device fabrication requires the films to be transferred onto dielectric surfaces such as SiO_2/Si substrates. Large-scale quantities of graphene-spanning square meters have already been produced on Cu substrates, and transferred to transparent and flexible polyethylene terephthalate (PET) substrates for flexible electronics applications [71]. In general, the future progress of CVD-synthesized graphene will depend on the ability to grow graphene on ultrathin films of metals or directly on dielectrics, to control the doping level, layer number, and grain size, and to minimize defects such as wrinkle formation, in addition to reducing the synthesis temperatures. Besides this, the transfer process also increases the complexity and more work is necessary to minimize defects and damage occurring to the graphene films as a result of the transfer process.

At the same time, there are a number of applications where the transfer process is not required at all. This includes applications such as interconnects [72] that benefit from the high thermal and electrical conductivity of graphene compared to Cu interconnects, for example, where conformal growth of graphene on the surface of the metal is sufficient. Other applications include graphene coatings that appear to be an excellent barrier for gases where the graphene can grow conformally over the surface of choice to protect against corrosion of the underlying metal [27] when it is defect free.

1.2.2.2 Epitaxial Graphene on SiC

In addition to CVD synthesis, high-quality graphene can be formed on SiC substrates. In this technique, the graphitic layers are grown on either the silicon-face or the carbon-face SiC surface, where the Si atoms sublime and leave behind a graphitized carbon film [73]. The formation of graphene on SiC, commonly referred to as epitaxial graphene or "Epi-graphene," was reported as early as 1975 by van Bommel et al. [74], and recently, de Heer and coworkers developed a technique for forming planar graphene layers on SiC substrates by heating SiC wafers to above 1300°C. This was motivated by the use of graphene in 2D electronics where de Heer and coworkers experimentally demonstrated the 2D electron gas properties of graphene's charge carriers in an electric field [75]. Several reviews on the surface science studies of graphene on SiC have also emerged recently [76–78].

For practical applications, SiC is a commonly used material for the high-power electronics industry. Initially, the carbon-terminated face of SiC was used to grow multilayer, randomly oriented polycrystalline films [57], but reports also indicated that the number of graphene layers grown can be controlled [79]. The graphene layers synthesized using this technique are of exceptional quality and exhibit minimal defects, with the crystallites being hundreds of microns in size [80]. High-frequency transistors based on SiC-grown graphene [81] may supersede the existing technology that is based on III–V materials such as InGaAs and GaN that may approach a limit at about 1 THz. Given the high cost of the SiC wafers and the high-synthesis temperatures (exceeding 1000°C), that are somewhat incompatible with the Si-based semiconductor process technology, it is likely that the applications arising from graphene on SiC will be restricted to narrowly focused niche areas such as high-frequency transistors, which are currently dominated by III–V materials.

1.2.3 Band-Gap Engineering: GNRs

Although electronic transport in graphene can be ballistic, its predominantly metallic character is somewhat limiting for its application as, for example, transistor elements in digital logic gates. Engineering semiconducting energy gaps in graphene is possible through several techniques such as the application of an electric field in bilayer graphene [79,82], chemical functionalization [83–85], GNR formation [39,50], as well as single-electron transistors [86,87]. Applying a perpendicular electric field across bilayer graphene can result in a band gap of ~10 meV. In general, the band gaps induced in this manner appear to be relatively small, preventing their applicability for room-temperature field-effect transistors (FETs) [82]. The small band gap limits the ON/OFF ratio to less than 10^3, whereas the requirement for practical digital electronics applications is an ON/OFF ratio of ~10^6. Larger gaps are possible through chemical functionalization of graphene with elements such as hydrogen and fluorine. This yields graphane, fluorographene, and related materials, but the mobilities [88] are reduced in these materials, and there are practical issues particularly in the case of hydrogenated graphane for example, where the hydrogen can desorb easily at moderate temperatures [83,89,90]. Interestingly, band gaps arising from other types of mechanisms have also been observed in graphene, such as those associated with the Josephson effect [91], where supercurrents supported by graphene were carried by Dirac fermions [36].

When charge carriers in graphene are confined in the lateral dimension by the formation of quasi-1D GNRs, numerous theoretical reports suggest that a band gap should emerge that is predicted to depend on the width, as well as the specific crystallographic orientation of the ribbon [92–96]. GNRs with zigzag-shaped edges are presumed to have direct band gaps, where the band gap decreases with increasing width, while armchair-shaped edges can be either metallic or semiconducting depending on their widths. Opening band gaps in GNRs is thus very important for transistor applications, and the planar configuration of graphene makes it compatible with the existing device Si-integrated circuit (IC) process technology.

To form band gaps that are large enough for room-temperature operation, the widths of the GNRs need to be less than ~10 nm. For ribbon widths below ~20 nm, the energy gap ΔE in a GNR is expected to scale with ribbon width W, according to $\Delta E = 2\pi\hbar v_F/3W$ [92,97]. The experimentally observed scaling of ΔE with W was confirmed by Han et al., who demonstrated $\Delta E \sim 200$ meV for GNRs with $W \sim 15$ nm [39]. While theory predicts that ΔE depends sensitively on the boundary conditions at the edges (e.g., if the edges are zigzag or armchair), such a dependence on crystallographic orientation has not been observed experimentally. Some of the techniques used to form GNRs include cutting graphene with an STM [98], an atomic force microscope [99], or using lithographic methods [100,101]. There are already reports where GNRs with widths as low as a few nanometers have been fabricated using top-down techniques [102,103].

GNRs have been formed through direct lithography where atomic oxygen and hydrogen chemisorb on the surface of graphene and, at elevated temperatures, complete oxidation of graphene can occur. Edges of graphene are much more susceptible to chemical attack allowing for selective etching of graphene edges by hydrogen or oxygen at optimized conditions. Nanowires have also been used as masking layers to form GNRs [104,105], specifically copper oxide (CuO) nanowires [106], which are easily synthesized by annealing a copper foil in air [107].

A critical challenge in forming GNRs lies in the ability to achieve precise edges [39] since roughnesses of less than 5 nm are necessary to minimize scattering that ultimately limits the ON/OFF ratios in the transistor applications of such devices [103]. Techniques need to be devised where etching can selectively control the crystallographic orientation of the GNRs at the edges; for example, producing zigzag or armchair-terminated edges on demand. To this end, GNRs have also been patterned recently using Ni [108,109] or Fe [110] metal nanoparticles that can etch graphene if heated in a hydrogen atmosphere. Unlike conventional nanolithography techniques that have no selectivity toward the crystallographic orientation of the graphene lattice, thermally activated (~900°C) Fe or Ni etching appears to be an interesting technique for differentiating between a zigzag and an armchair edge, as depicted in Figure 1.1, to yield ultrasmooth (~0.5 nm) edges.

In addition, GNRs have also been formed from the longitudinal unzipping of nanotubes [111] where almost all of the MWCNTs are converted into GNRs, and thin-film transistor and single GNR FET devices were demonstrated. In recent

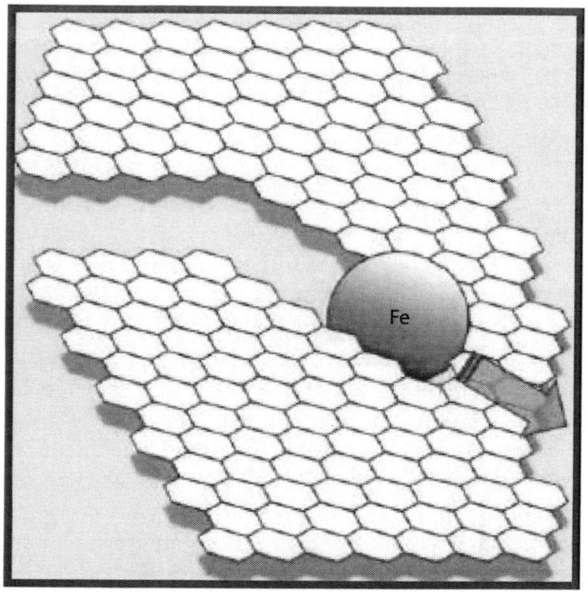

FIGURE 1.1 Schematic of the etching process in graphene or few-layer graphene (FLG) using thermally activated Fe nanoparticles. With this technique, the Fe nanoparticles etch graphene along crystallographic orientations that is useful to maintain the integrity of the edges for electronic device applications. (Reprinted with permission from S. S. Datta et al., *Nano Lett.* **8**, 1912. Copyright 2008 American Chemical Society.)

work [112], a layer-by-layer assembly method was reported for forming GNR films from which early transistor devices were demonstrated.

1.3 APPLICATION OF GRAPHENE AND GNRs IN ELECTRONICS, PHOTONICS, FLEXIBLE ELECTRONICS, AND NEMS

1.3.1 Electronic and Photonic Applications

1.3.1.1 Transistors and Energy-Efficient Tunneling Devices

Over the past several decades, semiconductor scaling based on Moore's Law [113] has increased the density of transistors for IC applications, but the eventual end to continue scaling appears to be in sight by 2020 for conventional Si-based complementary metal–oxide–semiconductors (CMOS). In light of this, new materials and devices are constantly being sought to overcome the fundamental challenges of nanoscale Si transistors given their energy-inefficient switching characteristics. The ON/OFF transition in conventional CMOS is based on the thermionic emission process that leads to a nonabrupt switching transition, and at nanoscale transistor dimensions, this causes excessive power dissipation and increased leakage currents making such devices energy inefficient.

At the same time, the high mobility and ballistic transport in graphene have been attractive features in considering this material for transistor applications and high-frequency electronics. However, to overcome the largely metallic characteristic of graphene, it is important that graphene exhibits a finite band gap for it to be useful for digital electronics and transistor applications. Through the formation of GNRs, FET operation was demonstrated in graphene where the first GNR–FETs utilized e-beam lithography and oxygen plasma etching to pattern the monolayer-thick graphene into ribbon widths between 15 and 100 nm [39]. Transistor ON/OFF ratios of ~10^4 were measured at 4 K and the energy gap for the narrowest ribbons with W ~ 15 nm was found to be ~200 meV [103] but energy gaps as high as ~400 meV have been measured from experimental data.

TABLE 1.2
Device Metrics: Graphene FETs

	Metrics
GNR FETs	ON/OFF ~104 (4 K), E_g ~ 200–400 meV [103]
FETs on SiO_2	Mobility ~ 10,000 $cm^2/V/s$ [1]
Suspended FETs	Mobility ~ 230,000 $cm^2/V/s$ [116]

In addition, the highest measured carrier mobilities of graphene placed on SiO_2 substrates are typically around 10,000 $cm^2/V/s$ [1]. Scattering from external sources, such as scattering from the graphene–substrate interface is the primary mechanism that limits electronic transport and impacts device performance in unsuspended graphene samples. Prior attempts have been made to remove the substrate to yield free-standing graphene over a trench [114,115]. Bolotin et al. [116] recently fabricated suspended graphene sheets and upon current annealing, demonstrated device mobilities of ~230,000 $cm^2/V/s$, which is indicative of ballistic transport over micron-length scales [82]; Table 1.2 provides a summary of the characteristic device metrics enabled by graphene FETs. A general consequence of graphene's atomic thinness is that its transport behavior can be significantly affected by the supporting substrate. Given its pristine surface and low-charged impurities, hexagonal boron nitride (h-BN) appears to be a promising, high-quality substrate for graphene; in addition, other materials have also been considered such as surface-assembled-monolayer (SAM) coatings on SiO_2 in which both *n*- and p-type graphene devices have been fabricated [117].

Besides FET devices, GNRs have also been considered as attractive candidates for tunneling transistors for energy-efficient electronics [118]. In tunneling transistors, the steep subthreshold swing arises from interband tunneling that promises higher drive currents compared to metal–oxide–semiconductor field-effect transistors (MOSFETs) as supply voltages approach 0.1 V. GNR devices appear to be very promising for tunneling FET applications, given their ultrathin body, low tunneling mass, direct band gap, and compatibility with planar processing resulting in high-speed devices with low-energy dissipation [119–121].

1.3.1.2 RF and High-Frequency Applications

While in digital electronics applications, a large ON/OFF ratio is desired for high-performance digital logic devices, for many radio frequency (RF) analog applications, such as amplifiers or mixers, a high ON/OFF ratio is not a critical requirement. Such RF applications would benefit greatly from the high mobility of the charge carriers in graphene. In typical metals, impurities in the metal scatter the electrons, which lead to energy dissipation and the onset of electrical resistance. In contrast, graphene's electrical conductivity is not affected by impurities or chemical dopants [33], which enable the charge carriers in graphene to have long mean-free paths before encountering a scattering event. Such exceptional electronic transport characteristics make graphene a promising material for RF transistors that are the backbone for wireless electronic systems.

The RF and high-frequency electronics area is currently dominated by GaAs-based devices that have been utilized for high-electron mobility transistors (HEMTs) that are widely used in communication systems. The transit time for charge to travel between the source and drain contacts in graphene is ~0.1 ps for channel lengths of ~100 nm assuming ballistic transport; such ultrashort transit times offer the possibility to extend the operational frequency of HEMTs based on graphene into the terahertz frequency regime and beyond. Currently, researchers have demonstrated a graphene RF FET with cutoff frequency f_T of 300 GHz [122,123] at a 40-nm channel length. Gate electrodes can be placed as close as several nanometers above graphene, which would allow even shorter channel lengths and even higher operational frequencies. Although the cutoff frequency of graphene FETs appears to be attractive, more work is required to increase the maximum oscillation frequency or f_{max}, as well as the power gain.

The graphene–substrate interface and the graphene–top-gate dielectric interface are the two dominant sources of scattering in monolayer-thick graphene devices. Practical solutions to achieving high-quality interfaces must be adopted to maintain high transistor performance. Graphene RF FETs have been fabricated using both transferred CVD-grown graphene on nonpolar and nonhydrophilic [25,124] diamond-like carbon substrates, and epi-graphene on SiC [81,122,124]. These reports highlight the rapid progress recently made for enabling practical applications using industrially relevant synthesis techniques, which are a necessary advancement from the very first mechanically exfoliated graphene transistors.

Graphene devices have also been integrated to form wafer-scale graphene-ICs, comprising of FETs and inductors, which were monolithically integrated on a single SiC wafer to yield a broadband RF mixer at frequencies up to 10 GHz [125]. In other work, graphene-based mixers and circuits were also fabricated on glass substrates, where the use of an insulating substrate minimizes the parasitic capacitance, and makes such devices attractive for frequency doubling or mixing circuits operational in the 110-GHz regime [126].

1.3.1.3 Photonic Device Applications

Besides being promising in RF electronics applications, graphene is also attractive for photonic applications [127]. Carriers in graphene behave as massless 2D particles, which enable absorption of normally incident light to occur independently of wavelength below about 3 eV. Mono- and bilayer graphene is also completely transparent owing to the Pauli blocking principle, if the optical energy is smaller than double the Fermi level [128], making graphene suitable for photonic devices.

Among photonic devices, graphene-based photodetectors [129,130] are being explored extensively where the high carrier mobility of graphene enables ultrafast extraction of the photogenerated carriers over a broad spectral range from the ultraviolet (UV) to the infrared (IR); in contrast, semiconducting photodetectors have a limited spectral width. For example, the maximum bandwidth of InGaAs and Ge photodetctors is limited to 150 GHz [131] and 80 GHz [132], respectively, where the bandwidth is determined by the respective carrier transit time. The transit-time-limited bandwidth of graphene photodetectors is calculated to be 1.5 THz [133] and due to the absence of a band gap, graphene-based photodetectors rely on a different carrier extraction model compared to conventional semiconducting photodetectors. There are several possible ways to improve the sensitivity of graphene photodetectors that include the use of plasmonic structures to amplify the local field [134] and integrating waveguides with the device to increase the light–graphene interaction lengths [135]. The high operating bandwidth and ultrafast response time of graphene also makes it suitable for other photonics applications, such as optical modulators [136], and mode-locked lasers and terahertz generators [137], for example.

1.3.2 Flexible Electronics Applications

Both theoretical and experimental work suggest that graphene may be a superior material to commercially available ITO for transparent electrodes [138]. Transparent conductive coatings are widely used in electronic products such as touch screen displays, e-paper, OLEDs that require a low sheet resistance (<30 Ω/square), as well as a transmittance exceeding 90% depending on the particular application. Graphene only absorbs 2.3% of the light intensity per monolayer over a wide spectral range from the IR to the visible [26,139]. Although currently the undoped sheet resistance of graphene may still be higher than ITO, the cost of ITO continues to rise with the growing scarcity of indium. By using various doping methods [71,140,141], monolayer graphene doped by nitric acid exhibits a sheet resistance ~125 Ω/square with a transmittance of 97%, while a four-layer-doped graphene-transparent electrode has been reported to exhibit a sheet resistance of ~30 Ω/square with a transmittance of ~90% that appears competitive with ITO. However, while doping graphene reduces its sheet resistance, more work is necessary to increase the long-term stability of such films.

Unlike ITO, graphene possesses outstanding mechanical flexibility and chemical durability, which are very important characteristics for flexible electronic devices [71,142]. Graphene's fracture strain is an order-of-magnitude higher than ITO, making it attractive for flexible, bendable, and rollable devices such as rollable e-paper, and when coupled with graphene's uniform absorption across the visible spectrum [26], it is suitable for color e-paper. However, some issues remain, such as minimizing the contact resistance [143] between the graphene electrode and the metal lines in the driving circuitry. Transparent conductive coatings and other related applications have less stringent defect requirements compared to semiconductor electronics and photonics applications, allowing the use of inexpensive solution-processed techniques, in contrast to more costly vacuum-based approaches.

Finally, OLEDs require a low sheet resistance (<30 Ω/square) for which graphene appears to be well suited [138]; other crucial parameters for OLEDs are the work function requirement between the electrode and the semiconductor, as well as the electrode surface roughness, since these impact device performance. The tunability of graphene's work function appears to be attractive for OLEDs and its atomically flat surface should help minimize electrical shorts and leakage currents and, in fact, graphene electrodes have already been demonstrated in OLED test cells [144]. In general, given its electronic and photonic transport properties, coupled with its mechanical strength and flexibility, graphene appears to be well poised to play a dominant role in flexible electronics and transparent-conducting electrode applications in the coming years.

1.3.3 NEMS Applications

Mechanical resonators show promise as elements in high-performance sensing [145], miniaturized RF components [146], and computing [147]. Evolving from micromechanical cantilevers developed in the 1980s for force sensing [148], nanomechanical resonators today can achieve impressive sensitivity for force, mass, charge, or displacement measurements. Importantly, improving upon these state-of-the-art performance levels is possible through engineering details of the resonator itself (e.g., physical dimensions and/or material properties). The dimensions of NEMS resonators are closely related to lithographic-processing capabilities and/or the bottom-up synthesis of well-defined nanomaterials. Using conventional thin-film semiconductor technology, it is straightforward to form resonators with at least one dimension—typically thickness—less than 100 nm [149,150]. As NEMS shrink below 100 nm, they begin to achieve high operating frequencies (up to 10^9 Hz) with extreme sensitivities. With the emergence of "2D atomic materials" [15,151], we can now envision combining conventional thin-film technology with the ultimate limit in film thickness—down to a single atomic layer. Within this realm, graphene is the "king," as it has ideal physical properties for NEMS devices/structures (e.g., electrically conducting, low mass, high stiffness, and atomically thin) that will likely define the ultimate performance limits for 2D planar NEMS. Here, we discuss the important elements driving the rapidly advancing and exciting research area of graphene-based NEMS.

From a materials perspective, the critical variables that define the performance of nano-electromechanical structures include strength, stiffness (or Young's modulus), built-in stress, and conductivity. Young's modulus (Y) and density (ρ) dictate the speed of sound in a material (c), which is directly proportional to the resonator frequency (f) as $f \propto c = (Y/\rho)^{1/2}$. Graphene's in-plane stiffness has been experimentally measured at 1 TPa [23,115]—the same as that for diamond—while having an intrinsic resistivity on the order of 10^3–10^4 Ω/square. This implies that for a given resonator geometry, graphene will have a higher operating frequency than, for example, silicon ($Y_{Si} \approx 170$ GPa). Furthermore, graphene's breaking strength is the highest of any material (~130 GPa) [23], allowing the incorporation of high tensile stress. In this regard, stress engineering can enhance the quality factor (Q) of nanomechanical resonators by orders of magnitude [152] leading to improved mass sensitivity or frequency selectivity. The quality factor is a measure of the energy loss within the system defined by $Q = 2\pi(E_{total}/\Delta E_{cycle}) = f/\Delta f$, where E_{total} is the energy stored, ΔE_{cycle} is the energy lost per cycle, and Δf is the full width half maximum of the resonance response. Finally, since all atoms in graphene are surface atoms, the addition and removal of adsorbates has a profound impact on graphene's lattice spacing and bond stiffness [153]. Control over both stress and stiffness within a graphene layer (intraplatelet mechanics), combined with a tunable extent of cross-linking between graphene layers (interplatelet mechanics), are unique for graphene and can provide a degree of freedom unavailable in traditional NEMS materials [153].

1.3.3.1 Materials, Device Structures, and Characterization

Transitioning graphene NEMS from laboratory interests to applications requires film deposition techniques that are large-area, scalable, and IC compatible. To date, two of the most promising approaches include (i) graphene grown on copper [48] or (ii) solution-exfoliated graphene-based materials such as GO [54]. Experimental details for these growth techniques have been reviewed elsewhere [56,154,155], including some used specifically for graphene-based NEMS [156]; so here we emphasize their benefits. (I) *Room-temperature deposition*: In both cases, the graphene materials can be formed separately from their final substrate, which may include extreme temperatures (e.g., >1000°C) or harsh chemical environments (e.g., boiling acids). Then, a room-temperature transfer or deposition step places the film on the target surface [48,157], which allows the use of temperature-sensitive substrates. (II) *Film thickness*: Graphene films grown on copper can be reproducibly formed as single layers on any size Cu-foil substrate, then transferred, and stacked to form films ranging up to a few atomic-layers thick. Solution-based GO films can be spun cast to form films ranging in thickness from nanometers to tens of nanometers on large-area wafers. (III) *Scalability*: The 2D nature and large-area film deposition allows conventional photolithography-processing steps, for example, film etching, contact deposition, and resonator release step, to form ordered arrays of devices (e.g., Reference 158).

The transduction of graphene-based electromechanical devices is one of the most critical components for achieving high-performance applications. Electrical detection schemes rely on a source (S), drain (D), and gate electrode that interface with the mechanical resonance element in the form of capacitive actuation. Such device structures are generally formed by undercutting a sacrificial layer beneath graphene [153,159], or by transferring graphene to prepatterned substrates [157,158,160]. The first graphene resonators were formed by mechanically exfoliating graphene flakes onto prepatterned trenches or wells [115,161], then were formed by building devices from exfoliated flakes alone [159], and finally, working toward large-area applications using CVD [158] or chemically modified graphene films [153]. Beyond the scaling limitations presented by mechanical exfoliation, exfoliated flakes are also randomly shaped and the resulting resonators typically have free (i.e., unclamped) edges that lead to spurious edge modes, increased dissipation, and lower Q-values [162]. These "lossy" modes can be eliminated by shaping circular drum resonators from large-area films, such that the resonator edges are fully clamped [153,157,163].

An important challenge surrounding graphene NEMS is the robustness of an atomically thin layer to conventional processing. As such, there are advantages in identifying the process order, where the final processing step most likely involves undercutting a sacrificial layer to release the graphene resonator. Most often, SiO_2/Si substrates are used, where SiO_2 is implemented as both a sacrificial layer and the dielectric spacer separating the S–D contacts from the bottom silicon gate electrode. Lithography defines the contacts (e.g., Ti/Au metallization) and should generally be executed before the graphene is suspended. After metallization, the SiO_2 layer is etched using hydrofluoric (HF) acid (e.g., aqueous or vapor form) and depending on the size of the suspended region, critical point drying (CPD) may be necessary to avoid capillary forces from crashing the resonator. For smaller structures (less than a few microns), drying from an isopropanol bath can be sufficiently gentle so as not to destroy the suspended resonator. To avoid issues with liquid surface tensions and capillary forces all together, there are advantages in selecting a sacrificial layer that can be dry etched. One such option uses xenon difluoride (XeF_2) gas, which is an isotropic etchant for silicon and has been used for over a decade in forming CMOS-compatible microelectromechanical system (MEMS) structures [164,165]. Exposing graphene to XeF_2 does lead to fluorine functionalization [85], such that the graphene film should be protected during the etch step, or a post annealing step is introduced to remove any bonded fluorine.

To date, two primary designs have been implemented to drive and read out graphene's mechanical motion. In one scheme, a high-frequency mixing approach [159,166,167] uses an RF gate voltage to capacitively drive the resonator, while a second RF voltage at a slightly different frequency is applied to the source. Motion is then detected as a mix-down current at the slightly offset frequency, with acquisition times on the order of Δf (~1 kHz). In addition, a direct current (DC) bias can be applied to electrostatically add tension. In the second approach, the mechanical motion is actuated and detected directly by using a vector network analyzer, employing a local gate to minimize parasitic capacitance [168]. An RF drive and DC voltage are combined and applied to the gate, while the output current is split into DC and RF components, the latter being input to the network analyzer. This approach has significantly faster operating frequencies (~1 MHz) [168]. To improve coupling efficiency in either scheme, the gap (z) between the suspended graphene and gate should be minimized since the signal current is proportional to $1/(z^3)$ [163,168]. However, the critical pull-in instability distance has been measured at approximately 10–20 nm [169], which sets a practical lower bound for the gap distance.

Using these electrical drive/read-out techniques, graphene NEMS devices typically operate in the RF regime on the order of 1–100 MHz, which is dependent on resonator geometry and resonator tension. Most graphene NEMS have "built-in" tension (T_o), likely arising from film deposition or transfer, and so behave as tensioned membranes ($f \propto \sqrt{T_o}$) versus plates ($f \propto \sqrt{Y}$). As mentioned earlier, incorporating tension is an effective means to improve the resonance Q-values, as well as increasing the resonance frequency. Depending on the resonator geometry and this built-in tension, the frequency output can either increase *or* decrease with applied gate voltage and/or operating temperature. The frequency output varies with gate voltage due to electrostatic tension (T_e) changing the effective spring constant of the resonator, which varies as $f \propto \sqrt{T_e}(V)$ [159]. In addition, graphene's small mass means the relative thermal expansion/contraction between graphene and its clamping support leads to the addition of extrinsic tension that is temperature dependent [170]. Such tension can be further tuned using polymer supports (e.g., SU-8 [163,171]), or intrinsically through manipulating the local crystal structure of the resonator itself [153,172]. Using intrinsic stress engineering, graphene-based resonators have shown frequency tunability over 500% and improvements in the quality factor of over 5× [153]. Similar to other NEMS, the dissipation (Q^{-1}) has a temperature (T) dependence. Here, the scaling behavior goes as T^3 down to ~100 K, and as $T^{0.3}$ from approximately 100 K to 5 K [159]. The highest Q-values for single-layer graphene have been measured at $Q \approx 14{,}000$ at 5 K [159], and for multilayer films with significant built-in tension, up to $Q \approx 30{,}000$ at room temperature [153].

1.3.3.2 Applications

While most targeted graphene applications are centered on electronic transport, many technologies can exploit graphene's mechanical properties. Some proposed applications include mass [21], force, or charge sensing, mechanical switches [173–176], separation (e.g., desalination [177], deoxyribonucleic acid [DNA] sequencing [178]), microscope support structure [179–181], gas/liquid encapsulation [27], and electromechanical filtering [182]. Graphene's atomic thinness and 2D "capture area" immediately brings to mind its use in mass sensing. The minimum mass sensitivity (δm_{min}) of a resonant mass sensor can be approximated as $\delta m_{min} \propto 2(M/Q)10^{-DR/20}$, where M is the resonator mass and DR is the dynamic range. Per- micron-square area of graphene, $\delta m_{min} \approx 1.5 \times 10^{-18}/Q$ (using $DR \approx 60$ dB [115]). Maximizing graphene's effectiveness as a mass sensor is best achieved by incorporating tensile stress, which both enhances Q (thus reducing δm_{min}), and reduces spurious effects from unwanted adsorbate–graphene interactions [159]. Prototype graphene NEMS structures have already achieved zeptogram mass sensitivity without significant optimization [115,159].

Similar to other thin-metal films, graphene is being explored as a mechanical switch [176] with the goal of achieving large ON/OFF ratios without leakage. Graphene's sp^2-carbon backbone makes it chemically inert so that van der Waals forces are the principal interacting force with neighboring surfaces [169], and its low mass and large Y can lead to nanosecond switching times [175], significantly faster than conventional MEMS switches. Moreover, graphene NEMS switches can have lower pull-in voltages (e.g., 1.8 V [175]) in-line with CMOS circuit requirements, unlike other NEMS switches that typically have pull-in voltages above 5–10 V [183]. As with all mechanical-type switches, it remains to be seen how well graphene will withstand common failure modes such as stiction, electrical discharge damage, and fatigue/fracture [176]. To date, graphene NEMS switches have achieved lifetimes of ~500 cycles before failure [174]. As discussed earlier, for RF electronics, graphene is already making headway when used as the active element for in-plane circuitry such as ambipolar nonlinear electronics (e.g., frequency doubling) or low noise amplifiers [184]. Given the general advantages that MEMS/NEMS provide for RF applications (e.g., higher quality factors, frequency selectivity, and small sizes), graphene too has distinct qualities, making it worthwhile as the active NEMS material. A primary limitation of conventional RF MEMS/NEMS structures is their large equivalent motional resistance that leads to high insertion losses. The motional resistance (R_m) depends on physical parameters such as resonator mass (M), spring constant (k), Q-values, and electromechanical coupling factors (η) and can be approximated as $R_m \approx \sqrt{kM}/Q\eta$ [182]. Graphene's exceptionally low mass and demonstrated high Q-values lead to estimated motional resistances on the order of 10^2–10^3 Ω for moderate-frequency ranges. The research field is only now at the cusp of implementing graphene NEMS into prototype RF filters.

1.4 SUMMARY AND FUTURE OUTLOOK

The advent of graphene over the past decade has brought with it a wealth of new fundamental science, as well as great promise for fundamentally new applications and for advancing modern state-of-the-art devices. This fast-paced research field has generated a large body of knowledge and understanding, as exemplified by almost 9000 scientific articles published in 2012 alone. These efforts have provided insights into graphene's novel properties and related structures such as nanoribbons, the synthesis of wafer-scale films, the development of characterization tools to understand the quality and the role of imperfections in graphene, and the development of processing methods to integrate graphene into novel device architectures. Graphene's remarkable properties such as high charge carrier mobility, high optical transparency, and mechanical strength and flexibility makes it a promising material for nanoelectronic devices such as energy-efficient transistors and RF electronics, photodetectors and other photonic device applications, flexible electronics and transparent electrodes, and NEMS.

While the transfer of graphene from metal growth substrates has become common practice for laboratory device fabrication, more work is necessary to achieve reproducibility and address issues of defect introduction. As an example, when clean interfaces are achieved between two transferred graphene layers, hybridized properties emerge [185] that open new scientific and technological opportunities. It is also critical to grow graphene directly on high-quality dielectrics such as h-BN, which is free of dangling bonds and charge inhomogeneities, unlike SiO_2. This is crucial to maintain graphene's high charge carrier mobility and, in this regard, all future heterostructure architectures with active graphene elements.

Similarly, the development of novel characterization tools and the ability to manipulate monolayer-thin graphene has opened possibilities to rapidly explore other monolayer-thin 2D crystals [151]. Such 2D atomic crystals appear to have distinct and exploitable properties complementing those of graphene, where the primary focus so far has been on inorganic 2D materials. Examples of such 2D-layered materials include h-BN, transition metal dichalcogenides, transition metal oxides, tertiary compounds of carbonitrides, and topological insulators [186–188]. In addition, there is a whole range of compositions between boron–carbon–nitrogen [189], as well as other complex multicomponent systems such as mica, and talc derived from alumina-silicate layers, which also exhibit a graphite-like layered structure. Being part of such a large and diverse family of 2D, layered materials and their heterostructures should increase the possibilities for graphene-based applications and improve their potential for commercial success.

ACKNOWLEDGMENTS

A.B.K. wishes to acknowledge support for this chapter through the NSF IRD program. J.T.R. acknowledges support in part from ONR and NRL Base Programs.

Any opinion, findings, and conclusions or recommendations expressed in this chapter are those of the author and do not necessarily reflect the views of the National Science Foundation.

REFERENCES

1. K. S. Novoselov, A. K. Geim, S. V. Morozov, D. Jiang, Y. Zhang, S. V. Dubonos, I. V. Grigorieva, and A. A. Firsov, *Science* **306**, 666, 2004.
2. H. P. Boehm, A. Clauss, G. O. Fischer, and U. Hoffmann, *Z. Naturforsch. B* **17**, 150, 1962.
3. A. E. Morgan, and G. A. Somarjai, *Surf. Sci.* **12**, 405, 1968.
4. J. W. May, *Surf. Sci.* **17**, 267, 1969.
5. Y. Gamo, A. Nagashima, M. Wakabayashi, M. Terai, and C. Oshima, *Surf. Sci.* **374**, 61, 1997.
6. Y. S. Dedkov, A. M. Shikin, V. K. Adamchuk, S. L. Molodtsov, C. Laubschat, A. Bauer, and G. Kaindl, *Phys. Rev. B* **64**, 035405, 2001.

7. M. Batzill, *Surf. Sci. Rep.* **67**, 83, 2012.
8. Y. B. Zhang, Y. W. Tan, H. L. Stormer, and P. Kim, *Nature* **438**, 201, 2005.
9. K. S. Novoselov, A. K. Geim, S. V. Morozov, D. Jiang, M. I. Katsnelson, I. V. Grigorieva, S. V. Dubonos, and A. A. Firsov, *Nature* **438**, 197, 2005.
10. D. Zhan, J. Yan, L. Lai, Z. Ni, L. Liu, and Z. Shen, *Adv. Mat.* **24**, 4055, 2012.
11. J. W. Evans, P. A. Thiel, and M. C. Bartelt, *Surf. Sci. Rep.* **61**, 1, 2006.
12. L. D. Landau, *Phys. Z. Sowjetunion* **11**, 26, 1937.
13. N. D. Mermin, *Phys. Rev.* **176**, 250, 1968.
14. P. R. Wallace, *Phys. Rev.* **71**, 622, 1947.
15. A. K. Geim, and K. S. Novoselov, *Nat. Mater.* **6**, 183, 2007.
16. L. Dai, *Carbon Nanotechnology: Recent Developments in Chemistry, Physics, Materials Science and Device Applications* (Ed: L. Dai), Elsevier, Oxford, UK, 2006.
17. H. O. Pierson, *Handbook of Carbon, Graphite, Diamond, and Fullerenes: Properties, Processing and Applications*, Noyes Publications, Park Ridge, NJ, 1993.
18. A. Krüger, *Carbon Materials and Nanotechnology*, Wiley-VCH, Weinheim, Germany, 2010.
19. H. W. Kroto, J. R. Heath, S. C. O'Brien, R. F. Curl, and R. E. Smalley, *Nature* **318**, 162, 1985.
20. S. Iijima, *Nature* **354**, 56, 1991.
21. A. S. Mayorov, R. V. Gorbachev, S. V. Morozov, L. Britnell et al., *Nano Lett.* **11**, 2396, 2011.
22. J. Moser, A. Barreiro, and A. Bachtold, *Appl. Phys. Lett.* **91**, 163513, 2007.
23. C. Lee, X. D. Wei, J. W. Kysar, and J. Hone, *Science* **321**, 385, 2008.
24. A. A. Balandin, S. Ghosh, W. Z. Bao, I. Calizo, D. Teweldebrhan, F. Miao, and C. N. Lau, *Nano Lett.* **8**, 902, 2008.
25. A. A. Balandin, *Nat. Mater.* **10**, 569, 2011.
26. R. R. Nair, P. Blake, A. N. Grigorenko, K. S. Novoselov, T. J. Booth, T. Stauber, N. M. R. Peres, and A. K. Geim, *Science* **320**, 1308, 2008.
27. J. S. Bunch, S. S. Verbridge, J. S. Alden, A. M. van der Zande, J. M. Parpia, H. G. Craighead, and P. L. McEuen, *Nano Lett.* **8**, 2458, 2008.
28. K. S. Novoselov, V. I. Fal'ko, L. Colombo, P. R. Gellert, M. G. Schwab, and K. Kim, *Nature* **490**, 192, 2012.
29. X. Wang, Z. Zhi, and K. Mullen, *Nano Lett.* **8**, 323, 2009.
30. P. Matyba, H. Yamaguchi, G. Eda, M. Chhowalla, L. Edman, and N. D. Robinson, *ACS Nano* **4**, 637, 2010.
31. X. Miao, S. Tongay, M. K. Petterson, K. Berke, A. G. Rinzler, B. R. Appleton, and A. F. Hebard, *Nano Lett.* **12**, 2745, 2012.
32. M. D. Stoller, S. Park, Y. Zhu, J. An, and R. S. Ruoff, *Nano Lett.* **8**, 3498, 2008.
33. F. Schedin, A. K. Geim, S. V. Morozov, E. W. Hill, P. Blake, M. I. Katsnelson, and K. S. Novoselov, *Nat. Mater.* **6**, 652, 2007.
34. J. T. Robinson, F. K. Perkins, E. S. Snow, Z. Wei, and P. E. Sheehan, *Nano Lett.* **8**, 3137, 2008.
35. E. W. Hill, A. K. Geim, K. Novoselov, F. Schedin, and P. Blake, *IEEE Trans. Mag.* **42**, 2694, 2006.
36. H. B. Heersche, P. Jarillo-Herrero, J. B. Oostinga, L. M. K. Vandersypen, and A. F. Morpurgo, *Nature* **446**, 56, 2007.
37. S. Stankovich, R. D. Piner, X. Chen, N. Wu, S. T. Nguyen, and R. S. Ruoff, *J. Mater. Chem.* **16**, 155, 2006.
38. S. Stankovich, D. A. Dikin, G. H. B. Dommett, K. M. Kohlhaas, E. J. Zimney, E. A. Stach, R. D. Piner, S. T. Nguyen, and R. S. Ruoff, *Nature* **442**, 282, 2006.
39. M. Y. Han, B. Ozyilmaz, Y. B. Zhang, and P. Kim, *Phys. Rev. Lett.* **98**, 206805, 2007.
40. A. C. Ferrari, J. C. Meyer, V. Scardaci, C. Casiraghi et al., *Phys. Rev. Lett.* **97**, 187401, 2006.
41. M. S. Dresselhaus, A. Jorio, A. G. Souza, and R. Saito, *Phil. Trans. R. Soc. A*, **368**, 5355, 2010.
42. M. S. Dresselhaus, A. Jorio, and R. Saito, *Annu. Rev. Condens. Matter Phys.* **1**, 89, 2010.
43. K. S. Novoselov, E. McCann, S. V. Morozov, V. I. Falko, M. I. Katsnelson, U. Zeitler, D. Jiang, F. Schedin, and A. K. Geim, *Nat. Phys.* **2**, 177, 2006.
44. http://www.grapheneindustries.com.
45. Y. Hernandez, V. Nicolosi, M. Lotya, F. M. Blighe et al., *Nat. Nanotechnol.* **3**, 563, 2008.
46. D. A. Dikin, S. Stankovich, E. J. Zimney, R. D. Piner et al., *Nature* **448**, 457, 2007.
47. K. S. Kim, Y. Zhao, H. Jang, S. Y. Lee, J. M. Kim, K. S. Kim, J. H. Ahn, P. Kim, J. Y. Choi, and B. H. Hong, *Nature* **457**, 706, 2009.
48. X. Li, W. Cai, J. An, S. Kim et al., *Science* **324**, 1312, 2009.
49. A. Rein, X. Jia, J. Ho, D. Nezich, H. Son, V. Bulovic, M. S. Dresselhaus, and J. Kong, *Nano Lett.* **9**, 30, 2008.
50. J. M. Cai, P. Ruffleux, R. Jaafar, M. Bieri et al., *Nature* **466**, 470, 2010.
51. Z. Sun, Z. Yan, J. Yao, E. Beitler, Y. Zhu, and J. M. Tour, *Nature* **468**, 549, 2010.
52. Z. Yan, Z. Peng, Z. Sun, J. Yao, Y. Zhu, Z. Liu, P. M. Ajayan, and J. M. Tour, *ACS Nano* **5**, 8187, 2011.
53. N. Behabtu, J. R. Lomeda, M. J. Green, A. L. Higginbotham et al., *Nat. Nanotechnol.* **5**, 406, 2010.
54. S. Stankovich, D. A. Dikin, R. D. Piner, K. A. Kohnhass, A. Kleinhammes, Y. Jia, Y. Wu, S. T. Nguyen, and R. S. Ruoff, *Carbon* **45**, 1558, 2007.
55. K. N. Kudin, B. Ozbas, H. C. Schniepp, R. K. Prudhomme, I. A. Aksay, and R. Car, *Nano Lett.* **8**, 36, 2008.
56. D. R. Dreyer, S. Park, C. W. Bielawski, and R. S. Ruoff, *Chem. Soc. Rev.* **39**, 228, 2010.
57. C. Berger, Z. Song, X. Li, X. Wu et al., *Science* **312**, 1191, 2006.
58. J. Hass, R. Feng, T. Li, X. Li, Z. Zong, W. A. de Heer, P. N First, E. H. Conrad, C. A. Jeffery, and C. Berger, *Appl. Phys. Lett.* **89**, 143106, 2006.
59. G. M. Rutter, J. N. Crain, N. P. Guisinger, T. Li, P. N. First, and J. A. Stroscio, *Science* **317**, 219, 2007.
60. J. Hackley, D. Ali, J. DiPasquale, J. D. Demaree, and C. J. K. Richardson, *Appl. Phys. Lett.* **95**, 133114, 2009.
61. S. Dhar, A. R. Barman, G. X. Ni, X. Wang et al., *AIP Adv.* **1**, 022109, 2011.
62. A. Lherbier, X. Blase, Y. M. Niquet, F. Triozon, and S. Roche, *Phys. Rev. Lett.* **101**, 036808, 2008.
63. L. S. Panchakarla, K. S. Subrahmanyam, S. K. Saha, A. Govindaraj, H. R. Krishnamurthy, U. V. Waghmare, and C. N. R. Rao, *Adv. Mater.* **21**, 4726, 2009.
64. D. Wei, Y. Liu, Y. Wang, H. Zhang, L. Huang, and G. Yu, *Nano Lett.* **9**, 1752, 2009.
65. G. Imamura, and K. Saiki, *J. Phys. Chem. C* **115**, 10000, 2011.
66. Z. Jin, J. Yao, C. Kittrell, and J. M. Tour, *ACS Nano* **5**, 4112, 2011.
67. X. Wang, X. Li, L. Zhang, Y. Yoon, P. K. Weber, H. Wang, J. Guo, and H. Dai, *Science* **324**, 768, 2009.
68. W. Chen, D. Qi, X. Gao, and A. T. S. Wee, *Prog. Surf. Sci.* **84**, 279, 2009.
69. J. Coraux, A. T. NDiaye, C. Busse, and T. Michely, *Nano Lett.* **8**, 565, 2008.
70. S. Marchini, S. Gunther, and J. Wintterlin, *J. Phys. Rev. B* **76**, 075429, 2007.

71. S. Bae, H. Kim, Y. Lee, X. Xu et al., *Nat. Nanotechnol.* **5**, 574, 2010.
72. Y. Khatami, H. Li, C. Xu, and K. Banerjee, *IEEE Trans. Electron. Devices* **59**, 2444, 2012.
73. I. Forbeaux, J. M. Themlin, and J. M. Debever, *Phys. Rev. B* **58**, 16396, 1998.
74. A. J. Van Bommel, J. E. Crombeen, and A. Van Tooren, *Surf. Sci.* **48**, 463, 1975.
75. C. Berger, Z. Song, T. Li, X. Li et al., *J. Phys. Chem. B* **108**, 19912, 2004.
76. J. Hass, W. A. de Heer, and E. H. Conrad, *J. Phys. Condens. Matter* **20**, 323202, 2008.
77. A. Bostwick, J. McChesney, T. Ohta, E. Rotenberg, T. Seyller, and K. Horn, *Prog. Surf. Sci.* **84**, 380, 2009.
78. C. Riedl, C. Coletti, and U. Starke, *J. Phys. D: Appl. Phys.* **43**, 374009, 2010.
79. T. Ohta, A. Bostwick, T. Seyller, K. Horn, and E. Rotenberg, *Science* **313**, 951, 2006.
80. C. Vironjanadara, *Phys. Rev. B* **78**, 245403, 2008.
81. Y. M. Lin, C. Dimitrakopoulos, K. A. Jenkins, D. B. Farmer, H. Y. Chiu, A. Grill, and P. Avouris, *Science* **327**, 662, 2010.
82. J. B. Oostinga, H. B. Heersche, X. L. Liu, A. F. Morpurgo, and L. M. K. Vandersypen, *Nat. Mater.* **7**, 151, 2008.
83. D. C. Elias, R. R. Nair, T. M. G. Mohiuddin, S. V. Morozov et al., *Science* **323**, 610, 2009.
84. R. R. Nair, W. C. Ren, R. Jalil, I. Riaz et al., *Small* **6**, 2877, 2010.
85. J. T. Robinson, J. S. Burgess, C. E. Junkermeier, S. C. Badescu et al., *Nano Lett.* **10**, 3001, 2010.
86. L. A. Ponomarenko, F. Schedin, M. I. Katsnelson, R. Yang, E. W. Hill, K. S. Novoselov, and A. K. Geim, *Science* **320**, 356, 2008.
87. C. Stampfer, E. Schurtenberger, F. Molitor, J. Guttinger, T. Ihn, and K. Ensslin, *Nano Lett.* **8**, 2378, 2008.
88. F. Schwierz, *Nat. Nanotechnol.* **5**, 487, 2010.
89. J. O. Sofo, A. S. Chaudhari, and G. D. Barber, *Phys. Rev. B* **75**, 153401, 2007.
90. J. Zhou, M. M. Wu, X. Zhou, and Q. Sun, *Appl. Phys. Lett.* **95**, 103108, 2009.
91. B. D. Josephson, *Phys. Lett.* **1**, 251, 1962.
92. S. V. Morozov, K. S. Novoselov, F. Schedin, D. Jiang, A. A. Firsov, and A. K. Geim, *Phys. Rev. B* **72**, 201401, 2005.
93. M. Ezawa, *Phys. Rev. B* **73**, 045432, 2006.
94. L. Brey, and H. A. Fertig, *Phys. Rev. B* **73**, 235411, 2006.
95. P. Shemella, Y. Zhang, M. Mailman, P. M. Ajayan, and S. K. Nayak, *Appl. Phys. Lett.* **91**, 042101, 2007.
96. Y. W. Son, M. L. Cohen, and S. G. Louie, *Nature* **444**, 347, 2006.
97. V. Barone, O. Hod, and G. E. Scuseria, *Nano Lett.* **6**, 2748, 2006.
98. L. Tapaszto, G. Dobrik, P. Lambin, and L. P. Biro, *Nat. Nanotechnol.* **3**, 397, 2008.
99. S. Fujii, and T. Enoki, *J. Am. Chem. Soc.* **132**, 10034, 2010.
100. A. Sinitskii, and J. M. Tour, *J. Am. Chem. Soc.* **132** 14730, 2010.
101. A. Dimiev, D. V. Kosynkin, A. Sinitskii, A. Slesarev, Z. Sun, and J. M. Tour, *Science* **331**, 1168, 2011.
102. X. Li, X. Wang, L. Zhang, S. Lee, and H. Dai, *Science* **319**, 1229, 2008.
103. X. R. Wang, Y. Ouyang, X. Li, H. Wang, J. Guo, and H. Dai, *Phys. Rev. Lett.* **100**, 206803, 2008.
104. L. Liao, J. W. Bai, R. Cheng, Y. C. Lin, S. Jiang, Y. Huang, and X. F. Duan, *Nano Lett.* **10**, 1917, 2010.
105. L. Liao, J. W. Bai, Y. C. Lin, Y. Q. Qu, Y. Huang, and X. F. Duan, *Adv. Mater.* **22**, 1941, 2010.
106. A. Sinitskii, and J. M. Tour, *Appl. Phys. Lett.* **100**, 103106, 2012.
107. X. C. Jiang, T. Herricks, and Y. N. Xia, *Nano Lett.* **2**, 1333, 2002.
108. L. C. Campos, V. R. Manfrinato, J. D. Sanchez-Yamagishi, J. Kong, and P. Jarillo-Herrero, *Nano Lett.* **9**, 2600, 2009.
109. L. Ci, Z. Xu, L. Wang, W. Gao, F. Ding, K. F. Kelly, B. I. Yakobson, and P. M. Ajayan, *Nano Res.* **1**, 116, 2008.
110. S. S. Datta, D. R. Strachan, S. M. Khamis, and A. T. C. Johnson, *Nano Lett.* **8**, 1912, 2008.
111. D. V. Kosynkin, A. L. Higginbotham, A. Sinitskii, J. R. Lomeda, A. Dimiev, B. K. Price, and J. M. Tour, *Nature* **458**, 872, 2009.
112. Y. Zhu, and J. M. Tour, *Nano Lett.* **10**, 4356, 2010.
113. G. E. Moore, *Electronics* **38**, 114, 1965.
114. J. C. Meyer, A. K. Geim, M. I. Katsnelson, K. S. Novoselov, T. J. Booth, and S. Roth, *Nature* **446**, 60, 2007.
115. J. S. Bunch, A. M. van der Zande, S. S. Verbridge, I. W. Grank, D. M. Tanenbaum, J. M. Parpia, H. G. Craighead, and P. L. McEuen, *Science* **315**, 490, 2007.
116. K. I. Bolotin, K. J. Sikes, Z. Jiang, M. Klima, G. Fudenberg, J. Hone, P. Kim, and H. L. Stormer, *Solid State Commun.* **146**, 351, 2008.
117. Z. Yan, Z. Sun, W. Lu, J. Yao, Y. Zhu, and J. M. Tour, *ACS Nano* **5**, 1535, 2011.
118. A. Seabaugh, and Q. Zhang, *Proc. IEEE* **98**, 2095, 2010.
119. G. Fiori, and G. Iannaccone, *IEEE Electron. Dev. Lett.* **28**, 760, 2007.
120. Q. Zhang, T. Fang, H. Xing, A. Seabaugh, and D. Jena, *IEEE Electron. Dev. Lett.* **29**, 1344, 2008.
121. H. Da, K.-T. Lam, G. Samudra, S.-K. Chin, and G. Liang, *IEEE Trans. Electron. Dev.* **59**, 1454, 2012.
122. Y. Wu, K. A. Jenkins, A. Valdes-Garcia, D. B. Farmer et al., *Nano Lett.* **12**, 3062, 2012.
123. L. Liao, Y. C. Lin, M. Q. Bao, R. Cheng et al., *Nature* **467**, 305, 2010.
124. Y. Q. Wu, Y. M. Lin, A. A. Bol, K. A. Jenkins et al., *Nature* **472**, 74, 2011.
125. Y. M. Lin, A. Valdes-Garcia, S. J. Han, D. B. Farmer et al., *Science* **332**, 1294, 2011.
126. L. Liao, J. Bai, R. Cheng, H. Zhou, L. Liu, Y. Liu, Y. Huang, and X. Duan, *Nano Lett.* **12**, 2653, 2011.
127. P. Avouris, and F. Xia, *MRS Bull.* **37**, 1225–1234, 2012.
128. Z. Q. Li, E. A. Henriksen, Z. Jiang, Z. Hao, M. C. Martin, P. Kim, H. L Stormer, and D. N. Basov, *Nat. Phys.* **4**, 532, 2008.
129. T. Mueller, F. N. A. Xia, and P. Avouris, *Nat. Photon.* **4**, 297, 2010.
130. F. N. Xia, T. Mueller, R. Golizadeh-Mojarad, M. Freitag et al., *Nano Lett.* **9**, 1039, 2009.
131. T. Ishibashi et al., *IEICE Trans. Electron. E* **83C**, 938, 2000.
132. Y. Ishikawa, and K. Wada, *IEEE Photon. J.* **2**, 306, 2010.
133. F. N. Xia, T. Mueller, Y. M. Lin, A. Valdes-Garcia, and P. Avouris, *Nat. Nanotechnol.* **4**, 839, 2009.
134. T. J. Echtermeyer, L. Britnell, P. K. Jasnos, A. Lombardo et al., *Nat. Commun.* **2**, 458, 2011.
135. K. Kim, J. Y. Choi, T. Kim, S. H. Cho, and H. J. Chung, *Nature* **479**, 338, 2011.
136. G. T. Reed, G. Mashanovich, F. Y. Gardes, and D. J. Thomson, *Nat. Photon.* **4**, 518, 2010.
137. F. Rana, *IEEE Trans. Nanotechnol.* **7**, 91, 2008.
138. J.-H. Chen, C. Jang, S. Xiao, M. Ishigami, and M. S. Fuhrer, *Nat. Nanotechnol.* **3**, 206, 2008.
139. A. B. Kuzmenko, E. van Heumen, F. Carbone, and D. van der Marel, *Phys. Rev. Lett.* **100**, 117401, 2008.
140. P. Blake, P. D. Brimicombe, R. R. Nair, T. J. Booth et al., *Nano Lett.* **8**, 1704, 2008.

141. K. K. Kim, A. Reina, Y. Shi, H. Park, L.-J. Li, Y. H. Lee, and J. Kong, *Nanotechnology* **21**, 285205, 2010.
142. A. Nathan, A. Ahnood, M. T. Cole, S. Lee et al., *Proc. IEEE* **100**, 1486, 2012.
143. J. A. Robinson, M. LaBella, M. Zhu, M. Hollander et al., *Appl. Phys. Lett.* **98**, 053103, 2011.
144. T. H. Han, Y. Lee, M. R. Choi, S. H Woo et al., *Nature Photon.* **6**, 105, 2012.
145. K. L. Ekinci, Y. T. Yang, and M. L. Roukes, *J. Appl. Phys.* **95**, 2682, 2004.
146. C. T. C. Nguyen, *Microw. Theory Tech., IEEE Trans.* **47**, 1486, 1999.
147. M. Freeman, and W. Hiebert, *Nat. Nano* **3**, 251, 2008.
148. G. Binnig, C. F. Quate, and C. Gerber, *Phys. Rev. Lett.* **56**, 930, 1986.
149. H. G. Craighead, *Science* **290**, 1532, 2000.
150. M. Roukes, *Phys. World* **14**, 25, 2001.
151. K. S. Novoselov, D. Jiang, F. Schedin, T. J. Booth, V. V. Khotkevich, S. V. Morozov, and A. K. Geim, *Proc. Natl. Acad. Sci. USA* **102**, 10451, 2005.
152. S. S. Verbridge, J. M. Parpia, R. B. Reichenbach, L. M. Bellan, and H. G. Craighead, *J. Appl. Phys.* **99**, 124304, 2006.
153. M. K. Zalalutdinov, J. T. Robinson, C. E. Junkermeier, J. C. Culbertson, T. L. Reinecke, R. Stine, P. E. Sheehan, B. H. Houston, and E. S. Snow, *Nano Lett.* **12**, 4212, 2012.
154. C. Mattevi, H. Kim, and M. Chhowalla, *J. Mater. Chem.* **21**, 3324, 2011.
155. Y. Zhu, S. Murali, W. Cai, X. Li, J. W. Suk, J. R. Potts, and R. S. Ruoff, *Adv. Mater.* **22**, 3906, 2010.
156. R. A. Barton, J. Parpia, and H. G. Craighead, *J. Vac. Sci. Technol. B: Microelectron. Nanometer Struct.* **29**, 050801, 2011.
157. J. T. Robinson, M. Zalalutdinov, J. W. Baldwin, E. S. Snow, Z. Wei, P. Sheehan, and B. H. Houston, *Nano Lett.* **8**, 3441, 2008.
158. A. M. v. d. Zande, R. A. Barton, J. S. Alden, C. S. Ruiz-Vargas, W. S. Whitney, P. H. Q. Pham, J. Park, J. M. Parpia, H. G. Craighead, and P. L. McEuen, *Nano Lett.* **10**, 4869, 2010.
159. C. Chen, S. Rosenblatt, K. I. Bolotin, W. Kalb, P. Kim, I. Kymissis, H. L. Stormer, T. F. Heinz, and J. Hone, *Nat. Nano* **4**, 861, 2009.
160. X. Song, M. Oksanen, M. A. Sillanpää, H. G. Craighead, J. M. Parpia, and P. J. Hakonen, *Nano Lett.* **12**, 198, 2011.
161. I. W. Frank, D. M. Tanenbaum, A. M. v. d. Zande, and P. L. McEuen, *J. Vac. Sci. Technol. B* **25**, 2558, 2007.
162. S. Y. Kim, and H. S. Park, *Nano Lett.* **9**, 969, 2009.
163. S. Lee, C. Chen, V. V. Deshpande, G.-H. Lee et al., *Appl. Phys. Lett.* **102**, 153101, 2013.
164. F. I. Chang, R. Yeh, G. Lin, P. B. Chu, E. G. Hoffman, E. J. Kruglick, K. S. J. Pister, and M. H. Hecht, *Proc. SPIE 2641, Microelectron. Struct. Microelectromech. Dev. Opt. Process. Multimed. Appl.* **117**, 1995, doi: 10.1117/12.220933.
165. N. H. Tea, V. Milanovic, C. A. Zincke, J. S. Suehle, M. Gaitan, M. E. Zaghloul, and J. Geist, *Microelectromech. Syst. J.* **6**, 363, 1997.
166. R. G. Knobel, and A. N. Cleland, *Nature* **424**, 291, 2003.
167. V. Sazonova, Y. Yaish, H. Ustunel, D. Roundy, T. A. Arias, and P. L. McEuen, *Nature* **431**, 284, 2004.
168. Y. Xu, C. Chen, V. V. Deshpande, F. A. DiRenno, A. Gondarenko, D. B. Heinz, S. Liu, P. Kim, and J. Hone, *Appl. Phys. Lett.* **97**, 243111, 2010.
169. X. Liu, N. G. Boddeti, M. R. Szpunar, L. Wang, M. A. Rodriguez, R. Long, J. Xiao, M. L. Dunn, and J. S. Bunch, *Nano Lett.* **13**, 2309, 2013.
170. W. Bao, F. Miao, Z. Chen, H. Zhang, W. Jang, C. Dames, and C. N. Lau, *Nat. Nano* **4**, 562, 2009.
171. Y. Oshidari, T. Hatakeyama, R. Kometani, S. I. Warisawa, and S. Ishihara, *Appl. Phys. Express.* **5**, 117201, 2012.
172. J. T. Robinson, M. K. Zalalutdinov, C. E. Junkermeier, J. C. Culbertson, T. L. Reinecke, R. Stine, P. E. Sheehan, B. H. Houston, and E. S. Snow, *Solid State Commun.* **152**, 1990, 2012.
173. K. M. Milaninia, M. A. Baldo, A. Reina, and J. Kong, *Appl. Phys. Lett.* **95**, 183105, 2009.
174. Z. Shi, H. Lu, L. Zhang, R. Yang et al., *Nano Res.* **5**, 82, 2012.
175. S. M. Kim, E. B. Song, S. Lee, S. Seo, D. H. Seo, Y. Hwang, R. Candler, and K. L. Wang, *Appl. Phys. Lett.* **99**, 023103, 2011.
176. O. Y. Loh, and H. D. Espinosa, *Nat. Nano* **7**, 283, 2012.
177. D. Cohen-Tanugi, and J. C. Grossman, *Nano Lett.* **12**, 3602, 2012.
178. G. G. F. Schneider, S. W. Kowalczyk, V. E. Calado, G. G. Pandraud, H. W. Zandbergen, L. M. K. Vandersypen, and C. Dekker, *Nano Lett.* **10**, 3163, 2010.
179. J. C. Meyer, C. O. Girit, M. F. Crommie, and A. Zettl, *Nature* **454**, 319, 2008.
180. J. Y. Mutus, L. Livadaru, J. T. Robinson, R. Urban, M. H. Salomons, M. Cloutier, and R. A. Wolkow, *New J. Phys.* **13**, 063011, 2011.
181. Z. Lee, K.-J. Jeon, A. Dato, R. Erni, T. J. Richardson, M Frenklach, and V. Radmilovic, *Nano Lett.* **9**, 3365, 2009.
182. F. D. Bannon, J. R. Clark, and C. T. C. Nguyen, *Solid-State Circuits, IEEE J.* **35**, 512, 2000.
183. W. W. Jang, J. O. Lee, J.-B. Yoon, M.-S. Kim, J.-M. Lee, S.-M. Kim, K.-H. Cho, D.-W. Kim, D. Park, and W.-S. Lee, *Appl. Phys. Lett.* **92**, 103110, 2008.
184. T. Palacios, A. Hsu, and H. Wang, *Commun. Mag. IEEE* **48**, 122, 2010.
185. J. T. Robinson, S. W. Schmucker, C. B. Diaconescu, J. P. Long et al., *ACS Nano* **7**, 637, 2013.
186. Q. H. Wang, K. Kalantar-Zadeh, A. Kis, J. N. Coleman, and M. S. Strano, *Nat. Nano* **7**, 699, 2012.
187. http://nsf2dworkshop.rice.edu/home/
188. A. B. Kaul, *J. Mat. Res.* **29**, 348, 2014.
189. L. Song, Z. Liu, A. L. Reddy, N. T. Narayanan, J. Taha-Tijerina, J. Peng, G. Gao, J. Lou, R. Vajtai, and P. M. Ajayan, *Adv. Mat.* **24**, 4878, 2012.

2 Interface between Graphene and High-κ Dielectrics

Ming Yang, Yuan Ping Feng, and Shi Jie Wang

CONTENTS

ABSTRACT

The interfacial interaction between high-k dielectrics and graphene plays an important role in determining the growth kinetics of the high-k dielectric thin films on graphene and also the related electrical properties. In this chapter, we report a recent research progress of integrating high-k dielectrics on graphene, various types of interfacial interactions, and important interfacial properties such as band alignment between graphene and high-k dielectrics. These results provide a detailed understanding of the interface between graphene and high-k dielectrics, and might shed light on tailoring the interfacial properties of graphene-based electronic devices.

For the past decades, the performance of conventional silicon (Si)-based transistors has been improving at a dramatic rate by the scaling down of the size of transistor components. Currently, the size has been decreased approaching its physical limit, especially for the thickness of the gate dielectric (SiO_2) due to the relatively small dielectric constant of SiO_2 ($\kappa \sim 3.9$), which may lead to serious issues for the devices such as large leakage current. Thus, other dielectric materials with high dielectric constant have been proposed to replace SiO_2 as the gate dielectric, of which HfO_2 ($\kappa \sim 25$) has been a dominant high-κ dielectric for current transistors [1].

At the same time, many efforts have been made to explore novel channel materials to substitute Si. More recently, graphene, an atomic layer of graphite, has attracted much attention because of its excellent electronic properties [2,3]. For example, graphene has very high electron mobility. The mobility of free standing graphene at room temperature is up to 200,000 $cm^2\ V^{-1}\ s^{-1}$ [4], which is much higher than that of Si (1400 $cm^2\ V^{-1}\ s^{-1}$). The electrical performance of a transistor is affected by many factors, including the interface between the gate dielectric and channel materials. The interface between conventional high-κ oxides and Si has been extensively studied [1]. However, unlike Si surface, graphene is much more inert due to the strong sp^2 C bonds, which indicates that the interaction between graphene and high-κ dielectrics is quite different. In this chapter, we will focus on the interface of graphene/high-κ dielectrics.

2.1 GROWTH OF HIGH-κ DIELECTRICS ON GRAPHENE

Although graphene has various exceptional electronic, mechanical, and optical properties, there are many challenges to apply graphene as a channel material in complementary metal-oxide-semiconductor (CMOS) devices. One of the main issues is that graphene needs to open a gap in order to realize high and reliable on/off switching [5,6]. Another issue is to integrate a high-κ dielectric on graphene with superior performance. In Si-based devices, due to the ionic bond between Si and oxide dielectric, good quality gate oxides can be grown on Si surface via various growth techniques such as thermal oxidation (SiO_2), sputtering, atomic layer deposition (ALD), chemical vapor deposition (CVD), or pulsed laser deposition (PLD) [1]. In contrast, the graphene surface is very chemically inert, which makes growing high-κ dielectrics on graphene much more difficult, and also causes high density of interfacial dangling bonds and defects. Thus, the integration of high-κ dielectrics on graphene is quite different from that on Si surface, and also much more challenging.

2.1.1 Sputtering Growth of High-κ Oxides on Graphene

Due to its relative high growth efficiency, lower cost, and the ability to grow various oxide dielectrics, sputtering has been widely applied on current Si-based electronic devices. Recently, sputtering also attracts some attention for the deposition of oxide dielectrics on graphene [7,8]. The typical processes for depositing high-κ oxides on graphene are the following: graphene was first thermally annealed to remove surface contaminants, and then the oxide material was deposited by using Ar flow at high vacuum. Due to the hydrophobic and chemically inert nature of the basal plane of the graphene surface, it was found that the interfacial quality of the directly deposited high-κ oxides on graphene is not good [7,9]. This problem can be partially solved by improving the deposition process, in which thin layers of metal were first deposited on the graphene surface, followed with oxidation process using oxygen plasma or atomic oxygen flow [8,9]. After the deposition, the quality of the oxide thin film and interfacial properties can be further improved by postthermal annealing [8]. Figure 2.1 is the x-ray photoemission spectroscopy (XPS) of deposited HfO_2 on graphene with/without thermal annealing. It can be seen that after annealing the core-level peak of Hf 4*f* was shifted toward higher binding energy, indicating the reduction of intrinsic defects in the HfO_2 thin films.

However, the sputtering process can introduce higher defect density in graphene because a high power of Ar source is usually applied, which might cause significant damage, typically carbon vacancies, to graphene. For instance, compared with free standing graphene, the graphene after SiO_2 thin film deposition shows a much larger defect band in the Raman spectra [7,9]. This may cause a remarkable reduction of carrier mobility. The carrier mobilities of graphene electronic devices were degraded to 710 $cm^2\ V^{-1}\ s^{-1}$ (for holes) and 530 $cm^2\ V^{-1}\ s^{-1}$ (for electrons, respectively, after the deposition of SiO_2 as the gate dielectric [7]).

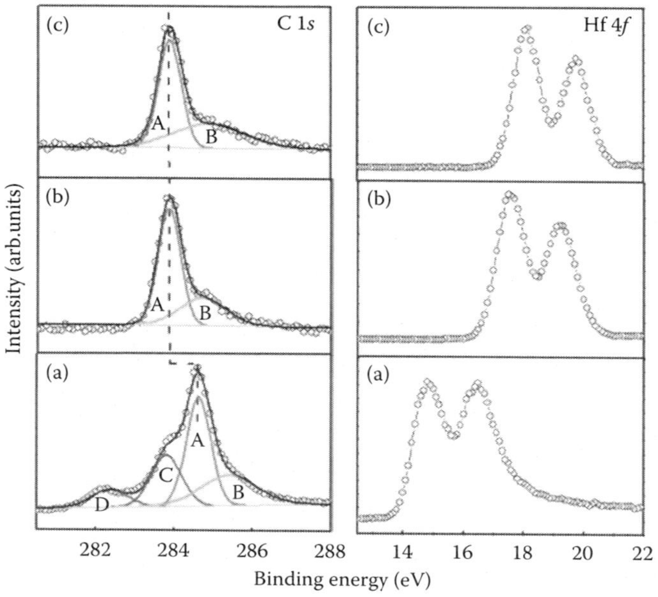

FIGURE 2.1 XPS spectra of C 1*s* and Hf 4*f* for the (a) as deposited, (b) oxidized, and (c) HfO_2 thin films annealed at 650°C on graphene. (Adapted from Chen, Q. et al., 2010. *Applied Physics Letters* 96: 072111–072113.)

2.1.2 ALD Growth of High-κ Oxides on Graphene

ALD is a substrate surface controlled layer-by-layer deposition technique based on a sequential application of gas phase process that results in the deposition of thin film with one atomic layer thickness at a time. Because of this self-limiting process, ALD can produce various thin films with a precise thickness, excellent uniformity, and without pinholes, which makes ALD widely used for the growth of high-κ oxides, carbide, nitride, metal, sulfides, fluorides, or even biomaterials.

For high-quality thin film grown by using ALD, the substrate surface plays an important role. The substrate surface has to be chemically reactive to effectively adsorb precursors, which is crucial for next step reaction [10]. However, unlike conventional semiconductor surfaces that have dangling bonds, which make the surface reactive, for pristine graphene surface, it is hydrophobic, and does not have dangling bonds, suggesting that the graphene surface is chemically inert. It is difficult to use ALD to deposit gate dielectric thin films on graphene directly because the precursors that are needed for the ALD process cannot be adsorbed on graphene effectively [11,12]. There were some attempts to grow Al_2O_3 layers on graphite surface directly using ALD, but it was found that the Al_2O_3 thin films cannot be coated on graphite surface [13].

Thus, in order to use ALD to grow good quality high-κ dielectrics on graphene, the graphene surface has to be functionalized to be chemically reactive. It was found recently that depositing a thin layer of trimethylaluminum (NO_2-TMA) on the graphene surface in the ALD process before the deposition of Al_2O_3 is essential for the later formation of uniform thin Al_2O_3 films [14]. The graphene transistor with the Al_2O_3 as the top gate dielectric based on this process shows inferior electrical properties and low carrier mobilities because of charged impurity scattering associated with NO_2-TMA layer and interfacial phonon scattering [15]. Other functional groups such as O_3 [16] or the intrinsic impurities may also facilitate the ALD to form oxide layers on graphene [17], but the related carrier mobilities are also not high. This problem can be partially resolved by using buffer layers. Dai et al. [18] showed that if graphene was pretreated in 3, 4, 9, 10-perylene tetracarboxylic acid (PTCA) solution for around 30 min, the Al_2O_3 thin films can be formed uniformly on graphene by the subsequent ALD process. The PTCA buffer layer does not introduce so many defects to graphene as the interaction between graphene and PTCA is not strong, and thus higher carrier mobility can be expected for transistors with the Al_2O_3 as the gate dielectric based on this treatment process. Another potential buffer layer is metal oxide [19]. In this method, thin metal layers were first deposited on graphene and then oxidized thoroughly to form thin metal oxide layers. These metal oxide layers serve as the nucleation layers for subsequent standard ALD deposition process to further grow high-κ oxides. Good quality of high-κ oxide can be achieved on graphene. A high mobility up to 8600 cm^2 V^{-1} s^{-1} has been reported for the transistor using ALD deposited Al_2O_3 as the top gate dielectric when a thin Al_2O_3 layer is used as the buffer layer. More recently, low-κ polymer has also attracted much attention as the buffer layer for ALD growth of dielectrics on graphene [20]. Similar to the process of using a molecular buffer layer, the low-κ polymer (NFC 1400-3CP) was diluted in propylene glycol monomethyl ether acetate (PGMEA) and then coated on graphene, in which methyl and hydroxyl are the functional groups. To deposit HfO_2, tetrakis-hafnium and water are usually used as the precursors and then HfO_2 is grown by using ALD at a relatively low temperature (about 125°C). Through carefully controlling the process, uniform buffer layer and gate dielectric thin films can be coated on graphene. More interestingly, it was found that the carrier mobilities of the top gated devices based on this polymer coating process does not degrade significantly. Compared with other types of buffer layers mentioned above, the device using low-κ polymer as the buffer layer shows better electric performance and higher carrier mobilities, suggesting promising applications on graphene-based electronic devices. Nevertheless, in the real semiconductor process, the low dielectric constant and thermal instability of the polymer might not be desirable, nor may the liquid process.

In addition, Duan et al. [21,22] developed a novel method to integrate high-κ dielectric on the graphene surface, in which high-κ oxide nanowires or nanoribbons were first synthesized separately at high temperature with nearly perfect crystalline structures, and then transferred on to the graphene surface at room temperature. This top gated graphene device shows excellent electrical and electronic performance, and the carrier mobility is high up to 23,600 cm^2 V^{-1} s^{-1}. However, this assembly process is relatively more complicated and not compatible with current fabrication process.

2.1.3 Growth of Si_3N_4 Dielectric on Graphene

Silicon nitride (Si_3N_4) has a high dielectric constant (~8), large band gap (~5 eV), high thermal stability, and excellent insulating and mechanical properties, thus Si_3N_4 has been extensively applied on current Si-based semiconductor technology as the gate dielectric or chemical barrier [1,23–25]. For graphene-based electronic devices, Si_3N_4 is also an attractive candidate as the gate dielectric, not only due to the excellent properties mentioned above, but also due to the relative simple fabrication process that is compatible to current Si-based fabrication process. In addition, the surface lattice constant and symmetry of β-Si_3N_4 (the most stable phase of Si_3N_4) match well with the graphene (3 × 3) supercell [26].

Various attempts have been carried out to grow Si_3N_4 on graphene. First, the IBM group reported an application of plasma-enhanced chemical vapor deposition (PECVD) to deposit Si_3N_4 on graphene [27]. PECVD deposition is not based on surface controlled reactions such as ALD, but on gas phase reactions by decomposing precursors initiated in the plasma that are not sensitive to substrate surface. In the process, NH_3, SiH_4, and N_2 were used as the precursors, and Si_3N_4 was deposited with HFRF plasma of 40 W at 400°C. NH_3 may react with graphene at the initial stage to form an effective seed layer, and thus uniform Si_3N_4 thin films were formed on graphene. A relatively high dielectric constant (~6.6) and high breakdown field (11.5 MeV/cm) were found for the Si_3N_4 and graphene-based MOS structure. It was also found that the mobility of graphene was well preserved after the Si_3N_4 gate dielectric deposition, and for the bilayer graphene the mobility was even slightly higher than that of bilayer graphene before deposition due to the more effectively screened charged impurity scattering from the underlying SiO_2 substrate by the higher dielectric constant of Si_3N_4.

Sputtering was also used to deposit amorphous Si_3N_4 on the graphene surface with the atomic nitrogen source assisted [28]. For the atomic nitrogen source (see Figure 2.2 for the

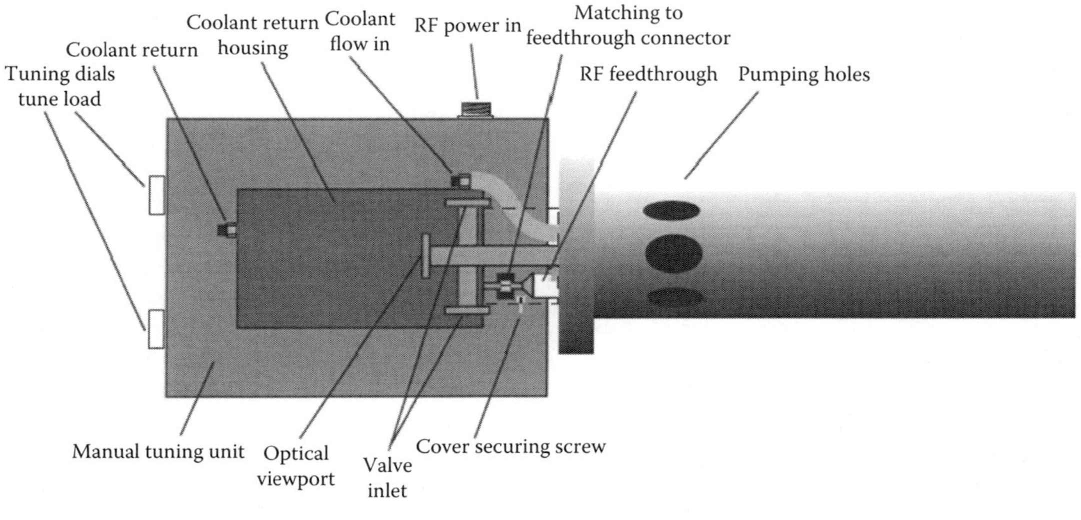

FIGURE 2.2 Schematic diagram of an atomic nitrogen source. (From Yang, M. 2010. Surface passivation and high-K dielectric integration for Ge based FETs. PhD diss., National University of Singapore.)

schematic diagram), the source nitrogen gas is introduced into an all-ceramic cavity (discharge zone). Plasma is induced in the discharge zone by applying inductively coupled radio frequency (RF) excitation. The plasma dissociates the feed gas into ions and neutral reactive atoms. Charged particles are retained within the plasma, and the latter species effuse through an aperture and plasma-confinement plate into the process chamber. This effused atomic oxygen or nitrogen is chemically reactive, and can react with substrates effectively.

Before the deposition of Si_3N_4, the graphene sample was prepared by standard solid graphitization process. Then, Si_3N_4 were deposited on graphene by DC sputtering Si target with 15 W in the atomic nitrogen environment (3.0×10^{-5} torr, 5 μA/cm^2, and a distance of 15 cm between target and gas shutter). During the growth process, the substrate temperature was maintained at 750°C, and the thickness of Si_3N_4 was controlled by varying the deposition time. The XPS spectra of Si 2*p* and N 1*s* of the deposited Si_3N_4 are shown in Figure 2.3a and b, respectively. The fitted Si 2*p* peak shows two components: the peak at 101.9 ± 0.2 eV is due to Si−N bonds in Si_3N_4, and the other peak at 100.6 ± 0.2 eV results from the Si−C bonds in the Si−C substrate. The chemical composition of the grown silicon nitride can be estimated to be near perfectly stoichiometric

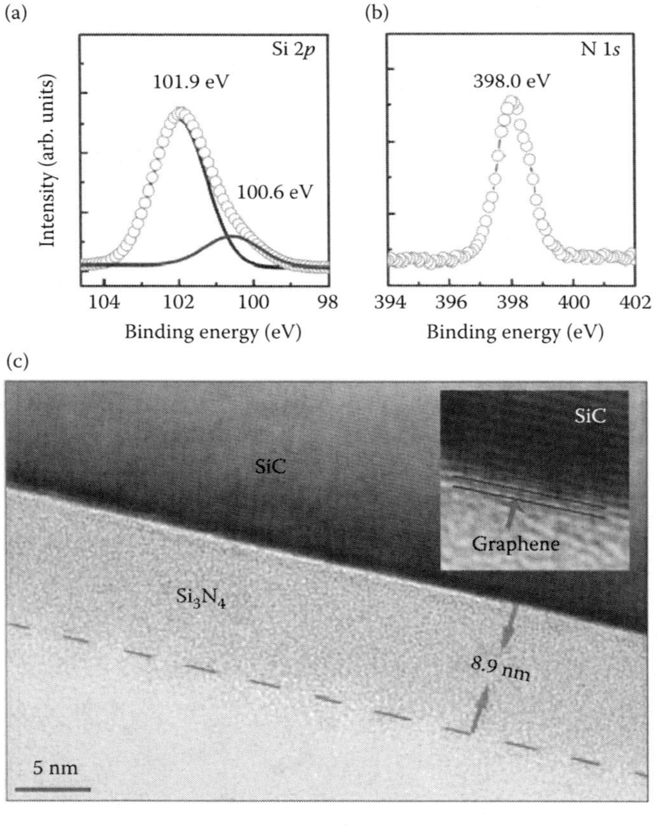

FIGURE 2.3 The original and fitted XPS spectra for (a) Si 2*p* and (b) N1*s*. (c) The HRTEM image of amorphous Si_3N_4 thin films on graphene/SiC. (From Yang, M. et al., 2012. *Journal of Physical Chemistry C* 42: 22315–22318.)

Si_3N_4 as the ratio between N and Si is about 1.32. Figure 2.3c is the corresponding HRTEM image at the scale bar of 5 nm, which clearly shows that the DC sputtered Si_3N_4 thin films are amorphous, and the coverage for Si_3N_4 on graphene is uniform. The Si_3N_4 thickness is about 8.9 nm for the deposition time of 5 min, and without obvious pinholes, suggesting excellent quality of the DC deposited Si_3N_4 thin film assisted by the atomic nitrogen source under high substrate temperature and low power of the DC plasma source. The low power sputtering process may minimize the damage to the graphene surface, and thus high carrier mobility can be expected for the graphene devices top gated with these grown Si_3N_4 thin films [27]. In addition, this sputtering process is simple and compatible with the current semiconductor fabrication process, suggesting promising application in future graphene-based electronic devices.

2.2 INTERFACE BETWEEN GRAPHENE AND HIGH-κ DIELECTRICS

Graphene is only one-atom thick and its physical properties can be easily affected by external influences such as the presence of a high-κ dielectric [6]. From the above discussion, we can see that the interaction between graphene and the gate dielectric is crucial for the growth of gate dielectric on graphene. In addition, this interaction also affects the electronic and electrical properties of graphene-based electronic devices. The associated electronic structure, charge transfer, band offsets, and electrical properties are significantly affected by this interaction. In next subchapters, we will show how the interactions between graphene and different high-κ dielectrics such as HfO_2/graphene, Si_3N_4/graphene, $SrTiO_3$/graphene, and single layer Y_2O_3/graphene altered the related electronic and electric properties.

2.2.1 Interface between Graphene and HfO_2 Dielectric

The intrinsic mobility of free standing graphene is limited by the electron–phonon interaction, which gives a maximum mobility of graphene about 2.0×10^5 cm^2 V^{-1} s^{-1} [4]. However, due to external influences, the mobility of graphene may be reduced, in which the substrate introduced remote interfacial phonon scattering is long-range potential, and reduced the mobility to 4.0×10^4 cm^2 V^{-1} s^{-1} (for SiO_2 substrate). This effect is similar to that of charged impurity scattering. Besides, the mobility of graphene is affected by the disorders in graphene such as ripple or electron–hole puddle (charge inhomogeneity) [4,30]. It is noted that both the remote interfacial phonon scattering and charged impurity scattering are long-term Coulombic interaction, the effects of which can be minimized by replacing the substrate with high-κ dielectric [4,31–33]. HfO_2 is the most promising high-κ dielectric to replace SiO_2 as the gate dielectric in Si-based transistors for post-60 nm technologies, due to its large band gap (~5.9 eV), high dielectric constant (~25), and good thermal stability. The integration of high-κ dielectric on graphene is highly desirable as it can screen the long range scattering from Coulombic impurities, and thus enhance the related electric properties [34].

For the interface between graphene and crystalline HfO_2, the (3 × 3) graphene supercell matches well with the (2 × 2) HfO_2 (111) surface [35]. The most stable interfacial configuration is the structure with maximum surface O atoms of HfO_2 on top of C atoms in graphene, which has the shortest interfacial distance of 3.05 Å. Compared with C−O single (1.42 Å) or double bonds (1.21 Å), and Hf−C bond (2.32 Å) in HfC, which are covalent bonds, the large distance between graphene and HfO_2 is obviously not within these possible covalent bonding ranges. In contrast to the large interfacial distance, the calculated binding energy for graphene/HfO_2 interface is high, −112 meV per C atom, indicating that interaction between graphene and HfO_2 is not dominated by van der Waals [36], but other interaction mechanisms such as orbital hybridization.

Due to the weak interaction between graphene and HfO_2, the linear dispersion bands of graphene are still preserved at Gamma point, but this weak interaction breaks their degeneracy and splits them into four branches that form a tiny gap of about 10 meV. From the projected wave functions (see Figure 2.4c), it was found that the four branches of the linear dispersion bands are actually hybrid states, in which the dominant contribution was from the p_z orbital of C atom in graphene, and other contributions such as Hf 4*d* and O 2*p* states can also be seen. These hybridized states will reduce the electron mobility for graphene integrated with HfO_2 because localized Hf 4*d* states may act as scattering centers for the fast π electrons in graphene. In addition, although the net charge transfer between graphene and HfO_2 is too weak to be noticeable, the interaction still causes the charge inhomogeneity in graphene. As shown in Figure 2.4d, with the presence of HfO_2, the charge density is not neutral in graphene anymore. The electron and hole charge density is unbalanced and forms electron–hole puddles. These electron–hole puddles can further decrease the carrier mobility of graphene.

The interaction between graphene and HfO_2 also determines the band offsets. The band offsets between the semiconductor and gate dielectric is important. In order to minimize tunneling carriers, the height of the band offsets is required to be larger than 1 eV [1]. XPS has been widely used to determine the band offsets at semiconductor/semiconductor or semiconductor/insulator interface, which is based on the assumption that the energy difference between the

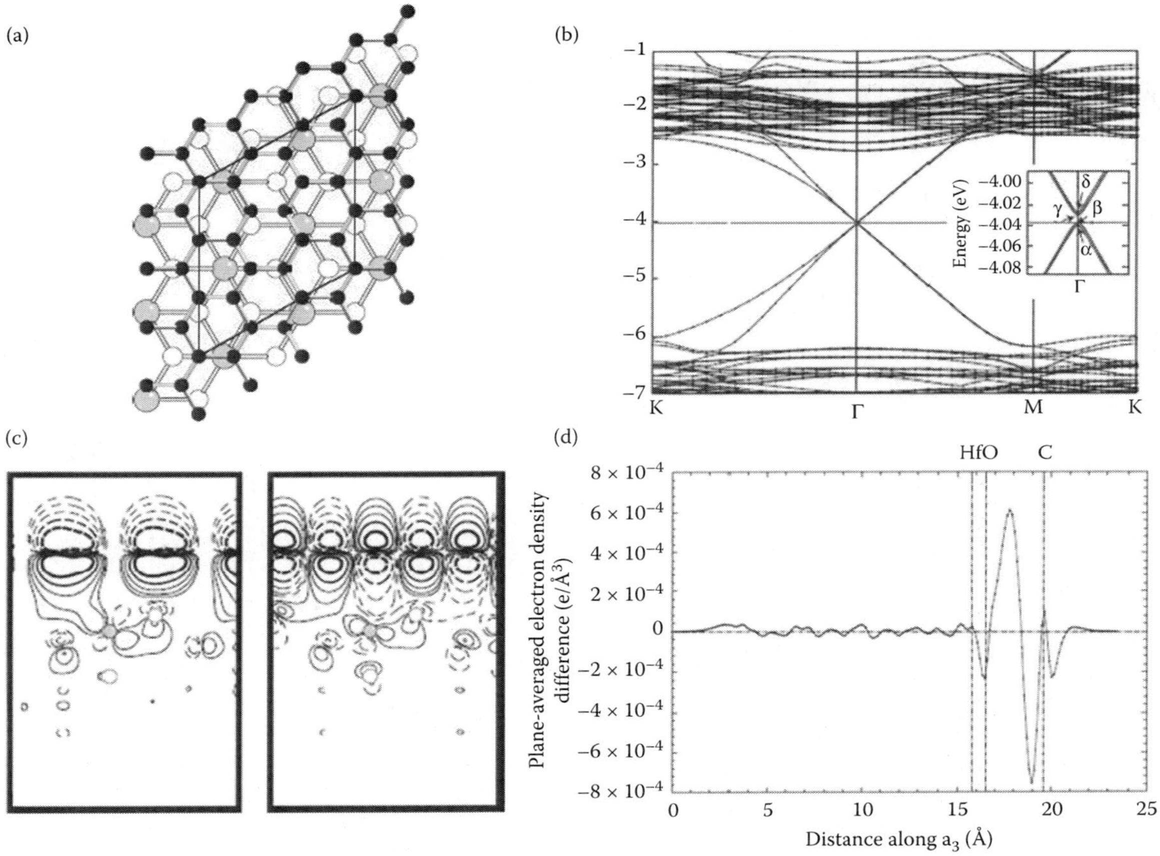

FIGURE 2.4 (a) Top view of graphene on HfO_2. Black, gray, and white circles denote C, Hf, and O atoms, respectively. (b) The band structure of graphene on HfO_2. (c) Contour maps of α (left panel) and β (right panel) states near Fermi level at the Gamma point. The solid and dotted lines denotes the positive and negative values, and the two O atoms and one Hf atom are shown by white and gray circles, respectively. (d) The in-plane averaged charge density difference at the interface. (Adapted from Kamiya, K., Umezawa, N., and Okada, S. 2011. *Physical Review B* 83: 153413–153416.)

valence edge and the core-level peak of the substrate remains constant with/without the growth of the overlayer [37]. The band offsets between graphene and HfO_2 can be measured by the XPS spectra [8], as shown in Figure 2.5. With the reference of the Si 2*p* peak (101.55 eV), the valence band edge of pretreatment annealed graphene/SiC (0001) is at 2.34 eV (Figure 2.5a), and the energy difference between the valence band edge and Si 2*p* core level peak is thus about 99.21 eV. After alignment, the valence band edge of HfO_2/graphene is at 3.71 eV (Figure 2.5c). By direct subtraction, the valence band offset between graphene and HfO_2 is measured to be 1.37 eV. Considering the small gap of graphene on HfO_2 and large band gap of HfO_2 (~5.8 eV) [38], the conduction band offset between graphene and HfO_2 should also be larger than 1 eV, both of which are high enough to minimize the possible carrier tunneling. In addition, it is noted that the valence band edge of HfO_2/graphene is increased to 4.03 eV (Figure 2.5d) after thermal annealing at 650°C, due to improved thin film quality of HfO_2 after annealing.

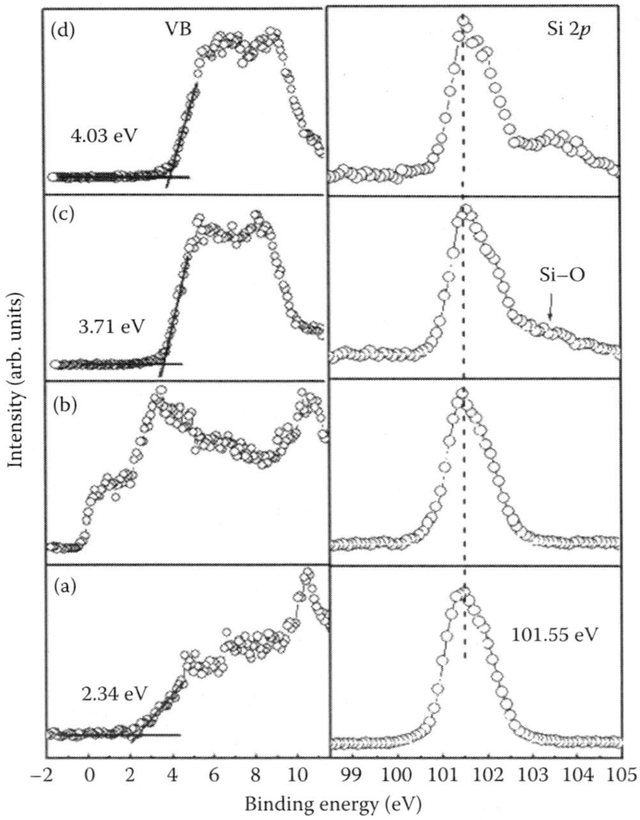

FIGURE 2.5 The valence band edge and Si 2*p* photoemission spectra of (a) pretreatment annealed graphene/4H−SiC (0001) at 500°C, (b) as-deposited, (c) oxidized Hf on graphene/4H−SiC (0001), and (d) thin HfO_2 film on graphene/4H−SiC (000) after annealing at 650°C. (Adapted from Chen, Q. et al., 2010. *Applied Physics Letters* 96: 072111–072113.)

2.2.2 Interface between Graphene and $SrTiO_3$

$SrTiO_3$ (STO) has a relatively small band gap (~3.25 eV) [39], but its dielectric constant is high to 400. Due to the smaller band gap and much larger dielectric constant, the interaction between graphene and STO might not be the same as graphene with other high-κ oxides such as HfO_2. Thus, it is also desirable to study the interface properties of graphene/STO. To model the interface of graphene on a STO substrate, a $(\sqrt{3}\times 3)$ graphene supercell was strained and placed on a (1×2) STO (001) surface with TiO_2 termination and five layers of thickness, in which 7.14% compressive and 7.19% tensile strains were applied along graphene **a** and **b** lattice directions, respectively, to match the lattice constants of the STO substrate. The optimized and most energetically favorable interface structure is shown in Figure 2.6c, where the shortest distance between graphene and STO surface is about 2.83 Å. This distance is much larger than the bond length of the potential interfacial covalent C−O bond (~1.42 Å for single bond in CO) or ionic Ti−C bond (~2.18 Å for bulk TiC). Moreover, the charge transfer at the interface was found to be weak. The calculation of Bader charge redistribution shows that only about 0.002 electrons were transferred from C atom into O atom at the surface, which is negligible. However, the calculated binding energy for this interface structure is high at −475 meV per C atom, indicating that the interfacial interaction is not likely dominated by the van der Waals effect. Large interfacial distance, weak charge transfer, and high binding energy suggest that other mechanisms such as orbital hybridization might be dominant at the interface between graphene and STO.

Figure 2.6d is the PDOS of p_z orbital of C atom in graphene, and p_y and p_z orbital of O atom and *d* orbital of Ti atom at STO surface, respectively, which clearly shows that orbital hybridization is mainly from p_z orbital of C atom in graphene and p_z orbital of O atom at the STO surface, especially at the energy range from −2 to −1 eV, while other orbital hybridization contributions such as C p_z orbital with O p_y or Ti *d* orbital are weak. Because of this orbital hybridization, bands originally only occupied by O p_z electrons can now also be occupied by C p_z electrons. The valence bands of graphene near the Gamma point were lifted upward (Figure 2.6c), compared with the band structure of pristine graphene (Figure 2.6a). In addition, the orbital hybridization also pushed the conduction bands of graphene near the Gamma point slightly downward. It should be noted that this band hybridization is not likely due to the strain effect in graphene. As shown in the band structure of the isolated graphene under the same strain (Figure 2.6b), the strain actually pushes the corresponding valence bands slightly downward. Thus, the hybridized O p_z and C p_z orbital cause the C p_z electrons to appear in some orbital, which are not observed in pristine graphene. These new states of C p_z orbitals due to the orbital hybridization causes a much narrower band width between flat C valence and conduction bands, compared with pristine graphene, which might become new optical excited levels.

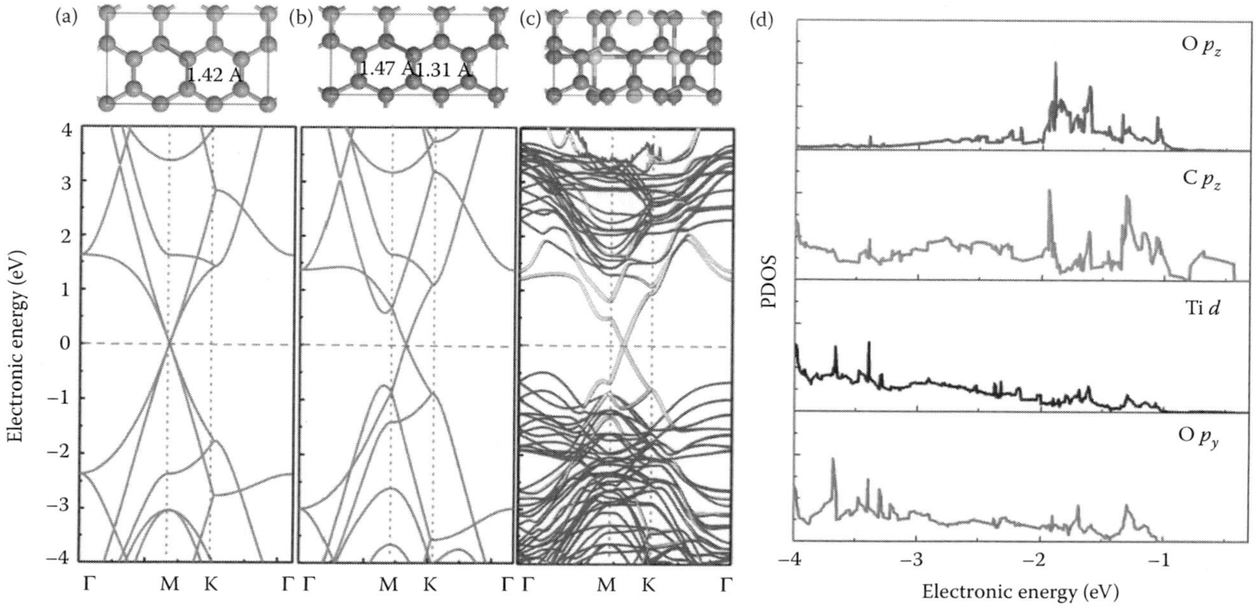

FIGURE 2.6 The atomic structure and band structure of (a) free-standing graphene, (b) strained graphene, and (c) strained graphene on STO. (d) The PDOS of O p_z and p_y, C p_z, and Ti d orbital of stained graphene on STO.

2.2.3 Interface between Graphene and Monolayer Y_2O_3

Very recently, monolayer Y_2O_3 dielectric has been successfully deposited on graphene supported on a Pt (111) surface, and it was found that the Fermi level of graphene had shifted, suggesting charging doping in graphene [40]. The interaction between monolayer high-κ dielectric and graphene is interesting because when the dielectric thickness is reduced to a monolayer, its properties are very different from those of related bulk or thin films.

First-calculations show that the most stable interfacial configuration of unreconstructed monolayer Y_2O_3 on graphene is the structure with maximum O atoms below the bridge site of graphene. After relaxation, a C−O bond with a bond length of 1.45 Å is formed at the interface. This C−O bond is covalent because its length is within the covalent bond length in CO molecules. Moreover, the shortest distance between the Y and C atom is about 2.82 Å. Both these distances are much smaller than that of other dielectric thin films, such as HfO_2 and SiO_2, on graphene, where the interaction is much weaker, just via van der Waals interaction (SiO_2/graphene) or orbital hybridization (HfO_2/graphene), and does not form strong covalent bonds. The strong interaction between monolayer Y_2O_3 and graphene is also confirmed by the high binding energy (−264 meV per C atom). It was found that about 0.82 electrons transferred from Y atoms in monolayer Y_2O_3 to graphene, resulting in electron doping in graphene. The strong interaction also significantly altered the electronic structure of graphene. The linear dispersion disappears and it opens a sizable band gap (about 0.53 eV) in the monolayer Y_2O_3/graphene hybrid system, as shown in the band structure of Figure 2.7a. This large band gap is highly desired for graphene-based field-effect transistors in order to realize reliable on/off switching.

The interface of Y_2O_3 thin films on graphene was also studied. It was noted that the optimized interfacial distance is about 2.51 Å, which is much larger than that of the monolayer Y_2O_3/graphene interface, and its binding energy is also smaller (−65.6 meV), suggesting a much weaker interfacial interaction. The calculated charge density difference is shown in Figure 2.7b. Compared with that of monolayer Y_2O_3/graphene interface, the charge transfer at Y_2O_3 thin films/graphene interface is much smaller. These results suggest that the thickness and size of the high-κ dielectric play are critical to determine the related adsorption, charge transfer, and electronic structure.

2.2.4 Interface between Graphene and Si_3N_4

The interfacial properties of graphene/Si_3N_4 have also been studied by first-principle calculations. It was found that the lattice constant of β-Si_3N_4, the most stable phase of Si_3N_4, matches well with that of a (3 × 3) graphene supercell, in which the lattice mismatch is only about 3.3%. For the hybrid structure of graphene and the Si_3N_4 (0001), the most stable configuration is that of four N atoms of Si_3N_4 located right below C atoms (top site), while three Si atoms are at the site below the C ring center (hollow site), as shown in Figure 2.8a [24]. The shortest interfacial distance between graphene and β-Si_3N_4 is about 3.26 Å. This distance is far beyond the possible interfacial covalent bond length of either Si−C bond (1.89 Å) in SiC or C−N bond (1.47 Å) in C_3N_4. After relaxation, the graphene plane remains planar, and all C atoms are nearly at their original positions. It is also noted that the binding energy

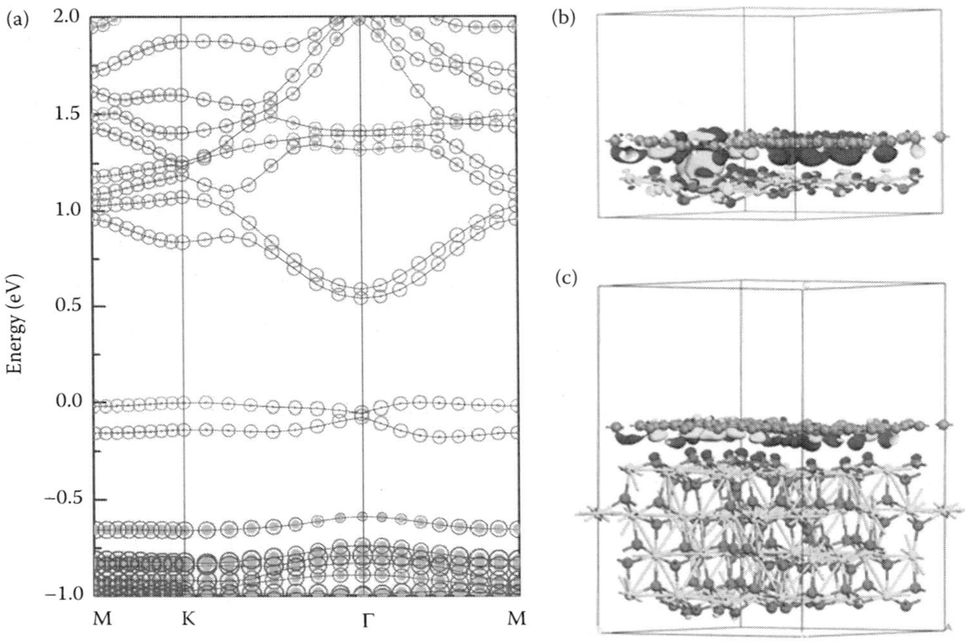

FIGURE 2.7 (a) The band structure of monolayer Y_2O_3 on graphene. The open circle denotes C p_z orbital, and the Fermi level is shifted to 0 eV. The charge density difference of monolayer Y_2O_3 on graphene (b) and Y_2O_3 thin films on graphene (c), in which the black color denotes excess charge density, and gray color is for depleted charge density in isosurface value of 6.0×1.0^{-3} eV/Å.

for graphene on Si_3N_4 is weak. This suggests that the interaction between graphene and Si_3N_4 is dominated by the van der Waals effect. Only about 0.002 electrons are accumulated at the graphene plane based on Bader charge calculations, which shows a very weak charge transfer at the interface. The weak interfacial interaction causes the electronic structure of graphene on Si_3N_4 to be almost identical to that of pristine graphene. As shown in Figure 2.8b, the linear dispersion of the band structure at the Dirac point is well preserved for graphene on Si_3N_4 and there is no band gap opening in graphene.

The band alignments at the interface of graphene/Si_3N_4 interface were studied by XPS [28]. Valence band edge and core-level spectra of Si 2*p* peak from graphene/SiC were used as references to align the grown overlayer to determine the barrier height at the Si_3N_4 and graphene/SiC interface. The aligned core-level spectra and valence band edge of thin and thick Si_3N_4 on graphene/SiC are shown in Figure 2.9a through c. Without the growth of Si_3N_4 overlayer, the core-level spectra of Si 2*p* in graphene/SiC is measured to be 100.6 eV, and the related leading valence band edge is at 2.13 eV (as shown in Figure 2.9a) determined by the intersection between the line segment regression of the edge and the flat energy distribution in the energy gap region. After aligning to the Si 2*p* core level spectra, the leading band edge of thick Si_3N_4 is found to be 2.73 eV, as shown in Figure 2.9c, which gives a barrier height of 0.6 eV for the interface of Si_3N_4 and graphene/SiC (0001).

When the valence band edge of Si_3N_4 on the graphene/6H−SiC substrate was measured at a smaller take-off angle 20°C, the real valence band edge of graphene could be resolved from that of the SiC, as shown in Figure 2.7d, which is 2.1 eV lower than that of the SiC. This suggests that previously

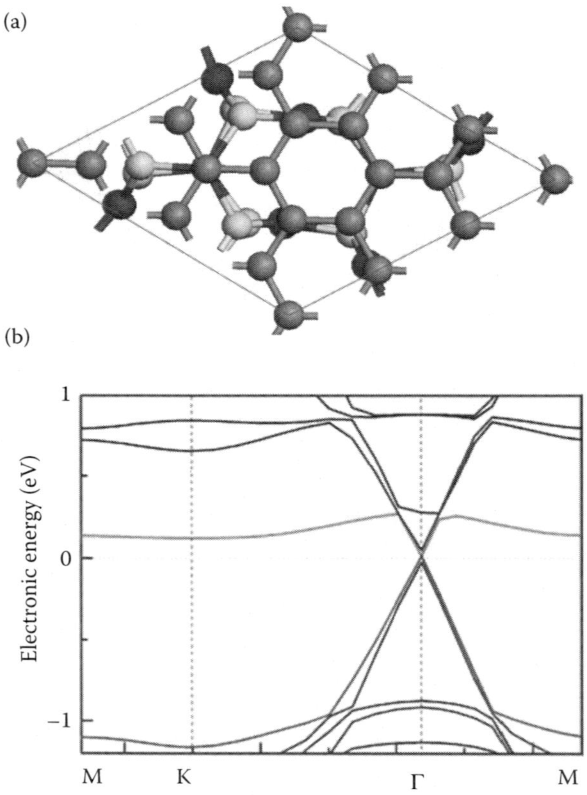

FIGURE 2.8 (a) The relaxed most stable graphene/Si_3N_4 structure. The gray, dark gray, and black balls denotes Si, N, and C atoms, respectively. (b) The band structure of graphene on β-Si_3N_4 (0001) surface. (Adapted from Yang, M. et al., 2009. *Journal of Applied Physics* 105: 024108–024112.)

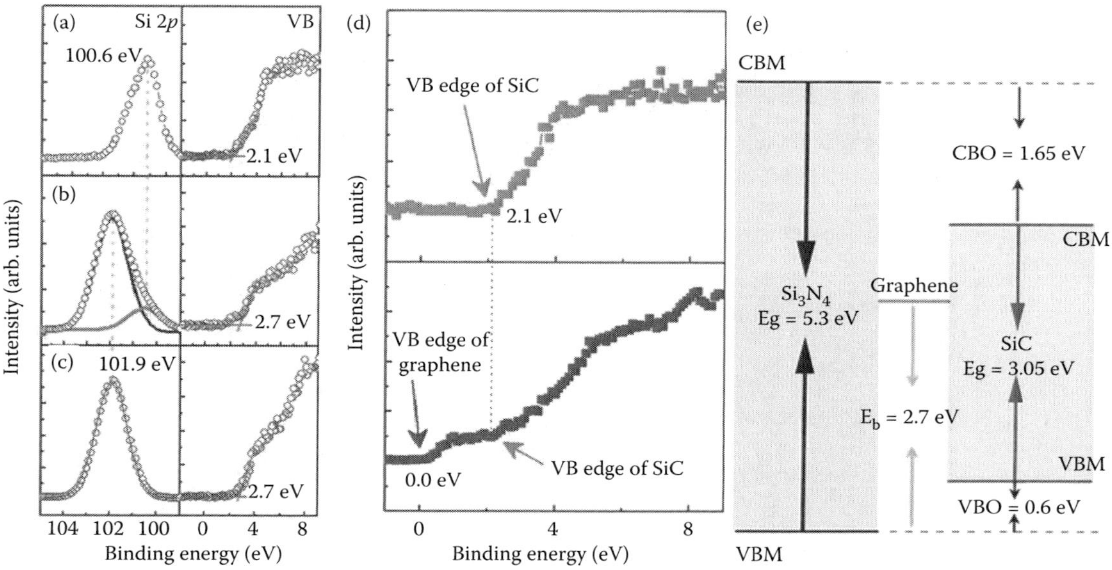

FIGURE 2.9 The core-level Si 2*p* and valence band edge of (a) epitaxial graphene on 6H−SiC, (b) thin Si_3N_4 film on graphene/SiC, and (c) thicker Si_3N_4 film on graphene/6H−SiC. (d) Angle resolved valence band of Si_3N_4 film on graphene/6H−SiC. (e) Band diagram of Si_3N_4 on graphene/6H−SiC substrate. (Adapted from Yang, M. et al., 2012. *Journal of Physical Chemistry C* 42: 22315–22318.)

measured valence band offset 0.6 eV is actually for the Si_3N_4/SiC interface, in good agreement with other results [37]. Thus, the real valence band offset for Si_3N_4/graphene interface is about 2.7 eV, large enough for electronic applications. Given the band gap of Si_3N_4 (~5.3 eV) and 6H−SiC (~3.05 eV), the band diagram for Si_3N_4/graphene/6H−SiC can be determined as Figure 2.7e shows. The VBM of graphene is almost at the gap center of Si_3N_4, indicating that Si_3N_4 can also provide a large enough conduction band offset for graphene even if graphene is engineered with a band gap. It is noted that there are six dangling bonds (three from Si atoms and the other three from N atoms) at the â-Si_3N_4 (0001) surface, and they may be charge scattering centers that will cause the reduction of carrier mobility in graphene. But these dangling bonds can be passivated by H atoms through hydrogenation process. It has been reported that graphene on a hydrogenated Si (001) surface shows high electron mobility [40,41]. A similar hydrogenation process is expected to work on a â-Si_3N_4 (0001) surface also, but further studies are needed.

The weak interaction, large band offsets, and high thermal stability for graphene on Si_3N_4 substrate, together with high dielectric constant, large band gap, and excellent insulating and mechanical properties of Si_3N_4, suggest that Si_3N_4 is a promising gate dielectric or substrate support for future graphene-based electronic devices.

2.3 CONCLUSION

In short, in this chapter, the growth processes of various high-κ dielectrics on graphene have been introduced. Various interfacial interactions such as orbital hybridization type (HfO_2/graphene and STO/graphene), covalent bonding type (unreconstructed monolayer Y_2O_3/graphene), and van der Waals interaction (Si_3N_4/graphene) were discussed, as well as the related electronic structures. The electronic properties of graphene are quite different due to these different interfacial interactions. These studies may pave a way to understand the interface between graphene and high-κ dielectrics and thus to improve the electrical performance of graphene-based electronic devices. Besides, the study of these different interactions between graphene and high-κ dielectrics may be beneficial also for exploring new functionalities of these graphene/high-κ dielectrics hybrid systems.

REFERENCES

1. Wilk, G. D., Wallace, R. M., and Anthony, J. M. 2001. High-kappa gate dielectrics: Current status and materials properties considerations. *Journal of Applied Physics* 89: 5243–5275.
2. Novoselov, K. S., Geim, A. K., Morozov, S. V., Jiang, D., Katsnelson, M. I., Grigorieva, I. V., Dubonos, S. V., and Firsov, A. A. 2005. Two-dimensional gas of massless Dirac fermions in graphene. *Nature* 438: 197–200.
3. Novoselov, K. S., Geim, A. K., Morozov, S. V., Jiang, D., Katsnelson, M. I., Grigorieva, I. V., Dubonos, S. V., and Firsov, A. A. 2004. Electric field effect in atomically thin carbon films. *Science* 306: 666–669.
4. Chen, J. H., Jang, C., Xiao, S. D., Ishigami, M., and Fuhrer, M. S. 2008. Intrinsic and extrinsic performance limits of graphene devices on SiO_2. *Nature Nanotechnology* 3: 206–209.
5. Geim, A. K., and Novoselov, K. S. 2007. The rise of graphene. *Nature Materials* 6: 183–191.
6. Castro Neto, A. H., Guinea, F., Peres, N. M. R., Novoselov, K. S., and Geim, A. K. 2009. The electronic properties of graphene. *Review of Modern Physics* 1: 109–162.
7. Lemme, M. C., Echtermeyer, T. J., Baus, M., and Kurz, H. 2007. A graphene field-effect device. *IEEE Electron Device Letters* 28: 282–284.

8. Chen, Q., Huang, H., Chen, W., Wee, A. T. S., Feng, Y. P., Chai, J. W., Zhang, Z., Pan, J. S., and Wang, S. J. 2010. *In situ* photoemission spectroscopy study on formation of HfO_2 dielectrics on epitaxial graphene on SiC substrate. *Applied Physics Letters* 96: 072111–072113.
9. Ni, Z. H., Wang, H. M., Ma, Y., Kasim, J., Wu, Y. H., and Shen, Z. X. 2008. Tunable stress and controlled thickness modification in graphene by annealing. *ACS Nano* 2: 1033–1039.
10. Yang, F. H., and Yang, R. T. 2002. *Ab initio* molecular orbital study of the mechanism of SO_2 oxidation catalyzed by carbon. *Carbon* 40: 437–444.
11. Wu, Y. Q., Ye, P. D., Capano, M. A., Xuan, Y., Sui, Y., Qi, M., Cooper, J. A. et al. 2008. Top-gated graphene field-effect-transistors formed by decomposition of SiC. *Applied Physics Letters* 92: 092102–092104.
12. Farmer, D. B., and Gordon, R. G. 2006. Atomic layer deposition on suspended single-walled carbon nanotubes via gas-phase noncovalent functionalization. *Nano Letters* 6: 699–703.
13. Hashimoto, A., Suenaga, K., Gloter, A., Urita, K., and Iijima, S. 2004. Direct evidence for atomic defects in graphene layers. *Nature* 430: 870–873.
14. Williams, J. R., DiCarlo, L., and Marcus, C. M. 2007. Quantum Hall effect in a gate-controlled p-n junction of graphene. *Science* 317: 638–641.
15. Lin, Y. M., Jenkins, K. A., Valdes-Garcia, A., Farmer, D. B., and Avouris, A. 2009. Operation of graphene transistors at gigahertz frequencies. *Nano Letters* 9: 422–426.
16. Lee, B. K., Park, S. Y., Kim, H. C., Cho, K., Vogel, E. M., Wallace, R. M., and Kim, J. Y. 2008. Conformal Al_2O_3 dielectric layer deposited by atomic layer deposition for graphene-based nanoelectronics. *Applied Physics Letters* 92: 203102–203104.
17. Meric, I., Han, M. Y., Young, A. F., Ozyilmaz, B., Kim, P., and Stepard, K. L. 2008. Current saturation in zero-bandgap, topgated graphene field-effect transistors. *Nature Nanotechnology* 3: 654–659.
18. Wang, X. R., Tabakman, S. M., and Dai, H. J. 2008. Atomic layer deposition of metal oxides on pristine and functionalized graphene. *The American Chemical Society* 130: 8152–8153.
19. Kim, S., Nah, J., Shahrjerdi, I. J. D., Colombo, L., Yao, Z., Tutuc, E., and Banerjee, S. K. 2009. Realization of a high mobility dual-gated graphene field-effect transistor with Al_2O_3 dielectric. *Applied Physics Letters* 94: 062107–062109.
20. Farmer, D. B., Chiu, H. Y., Lim, Y. M., Jenkins, K. A., Xia, F. N., and Avouris, P. 2009. Utilization of a buffered dielectric to achieve high field-effect carrier mobility in graphene transistors. *Nano Letters* 9: 4474–4478.
21. Liao, L., Lin, Y. C., Bao, M. Q., Cheng, R., Bai, J. W., Liu, Y. A., Qu, Y. Q., Wang, K. L., Huang, Y., and Duan, X. F. 2010. High-speed graphene transistors with a self-aligned nanowire gate. *Nature* 467: 305–308.
22. Liao, L., and Duan, X. F. 2010. Graphene-dielectric integration for graphene transistors. *Materials Science and Engineering B* R70: 354–370.
23. Wang, X. S., Zhai, G. J., Yang, J. S., and Cue, N. S. 1998. Crystalline Si_3N_4 thin films on Si (111) and the 4 × 4 reconstruction on Si_3N_4 (0001). *Physical Review B* 60: R2146–R2149.
24. Yang, M., Wu, R. Q., Deng, W. S., Shen, L., Sha, Z. D., Cai, Y. Q., Wang, S. J., and Feng, Y. P. 2009. Electronic structures of beta-Si_3N_4 (0001)/Si(111) interfaces: Perfect bonding and dangling bond effects. *Journal of Applied Physics* 105: 024108–024112.
25. Ma, C. T. 1998. Making silicon nitride film a viable gate dielectric. *IEEE Transaction on Electronic Devices* 45: 680–690.
26. Yang, M., Zhang, C., Wang, S. J., Feng, Y. P., and Ariando. 2011. Graphene on beta-Si_3N_4: An ideal system for graphene-based electronics. *AIP Advances* 1: 032111–032115.
27. Zhu, W. J., Neumayer, D., Perebeinos, V., and Avouris, P. 2010. Silicon nitride gate dielectric and band gap engineering in graphene layers. *Nano Letters* 9: 3572–3576.
28. Yang, M., Chai, J. W., Wang, Y. Z., Wang, S. J., and Feng, Y. P. 2012. Interfacial properties of silicon nitride grown on epitaxial graphene on 6H-SiC substrate. *Journal of Physical Chemistry C* 42: 22315–22318.
29. Yang, M. 2010. Surface passivation and high-K dielectric integration for Ge based FETs. PhD diss., National University of Singapore.
30. Das Sarma, S., Adam, S., Hwang, E. H., and Rossi, E. 2011. Electronic transport in two-dimensional graphene. *Reviews of Modern Physics* 83: 407–470.
31. Jena, D., and Kona, A. 2007. Enhancement of carrier mobility in semiconductor nanostructures by dielectric engineering. *Physical Review Letters* 98: 138805–136808.
32. Chen, F., Xia, J. L., Ferry, D. K., and Tao, N. J. 2009. Dielectric screening enhanced performance in graphene FET. *Nano Letters* 9: 2571–2574.
33. Chen, F., Xia, J. L., and Tao, N. J. 2009. Ionic screening of charged-impurity scattering in graphene. *Nano Letters* 9: 1621–1625.
34. Chen, J. H., Jang, C., Adam, S., Fuhrer, M. S., William, E. D., and Ishigami, M. 2009. Charged-impurity scattering in graphene. *Nature Physics* 4: 377–381.
35. Kamiya, K., Umezawa, N., and Okada, S. 2011. Energetics and electronic structure of graphene adsorbed on HfO_2 (111): Density functional theory calculations. *Physical Review B* 83: 153413–153416.
36. Cuong, N. T., Otani, M., and Okada, S. 2011. Semiconducting electronic property of graphene adsorbed on (0001) surfaces of SiO_2. *Physical Review Letters* 106: 106801–106804.
37. Kraut, E. A., Grant, R. W., Waldrop, J. R., and Kowalczyk, S. P. 1980. Precise determination of the valence-band edge in x-ray photoemission spectra of semiconductor interface potentials. *Physical Review Letters* 44: 1620–1623.
38. Robertson, J., and Falabretti, B. 2006. Band offsets of high K gate oxides on III-V semiconductors. *Journal of Applied Physics* 100: 014111–014116.
39. Pontes, F. M., Lee, E. J. H., Leite, E. R., and Longo, E. 2000. High dielectric constant of $SrTiO_3$ thin films prepared by chemical process. *Journal of Materials Science* 35 4783–4787.
40. Addou, R., Dahal, A., and Batzill, M. 2013. Growth of a two-dimensional dielectric monolayer on quasi-freestanding graphene. *Nature Nanotechnology* 8: 41–45.
41. Xu, Y., He, K. T., Schmucker, S. W., Guo, Z., Koepke, J. C., Wood, J. D., Lyding, J. W., and Aluru, N. R. 2011. Inducing electronic changes in graphene through silicon (100) substrate modification. *Nano Letters* 11: 2735–2742.

3 Conventional and Laser Annealing to Improve Electrical and Thermal Contacts between Few-Layer or Multilayer Graphene and Metals

Alfredo Rodrigues Vaz, Andrei Alaferdov, Victor Ermakov, and Stanislav Moshkalev

CONTENTS

ABSTRACT

Formation of high-quality electrical and thermal contacts between nanostructured materials such as carbon nanotubes or graphene and metal electrodes is one of the fundamental issues in modern nanoelectronics. The quality of contact can be improved by nonlocal laser annealing at increased power. The improvement of thermal contacts to initially rough metal electrodes is attributed to local melting of the metal surface under laser heating, and increased area of real metal–graphene contact. The accuracy of thermal conductivity measurements for suspended multilayer graphene (MLG) flakes by micro-Raman technique was shown to depend critically on the quality of thermal contacts between the flakes and metal electrodes used as a heat sink. Improvement of the thermal contacts was observed also between MLG and silicon oxide surface, with more efficient heat transfer from graphene as compared with the graphene–metal case.

3.1 INTRODUCTION

The extraordinary thermal properties of graphene motivated numerous studies aiming at its applications for heat management in micro- and nanoelectronics [1–4]. However, on the way to successful applications, many important issues have yet to be addressed, in particular, in the formation of high-quality metal–graphene thermal contacts [2,5–10]. In many cases, poor contacts may affect not only the electrical transport but also efficiency of heat dissipation in sub-micrometer scale graphene-based devices [6]. Thus, high graphene thermal conductivity and high quality of thermal contacts with metal electrodes are equally important for heat management in such devices. Note that experimental works focused on the thermal properties of graphene–metal contacts are still scarce when compared with theoretical studies [11]. A number of techniques were reported in the literature to measure thermal conductivity of graphene or reduced graphene oxide, including the opto-thermal micro-Raman method [9,10,12–14] that is used in the present research, as well as the thermal bridge [15] and four point probe [16] methods, where sample heating and local temperature measurements are realized indirectly by either optical or electrical means.

Methods to prepare graphene flakes of different thicknesses (single-layer graphene [SLG], few-layer graphene [FLG], and multilayer graphene [MLG]) are summarized in Table 3.1, and the results of recent experimental studies on thermal conductivity in graphene are presented in Table 3.2.

Here, for studies of thermal contacts between FLG and MLG, we used the confocal micro-Raman method at low laser power. The graphene flakes were prepared for natural graphite by sonication and deposited onto metallic electrodes by AC di-electrophoresis technique. By applying increased laser power, it was possible to observe the formation of improved thermal contacts resulting in significantly decreased local temperature of graphene under laser heating.

TABLE 3.1
Methods Used for Fabrication of Graphene Sheets and Flakes

Method	SLG, FLG, or MLG	Advantages	Process Conditions	Reference
CVD	SLG, FLG	Large-area sheets, possibility to transfer to large variety of substrates	Hot wall reactor (1000°C)	[17]
Super-short-pulsed laser-induced deposition	FLG, MLG (up to 8 nm thick)	Freestanding; largely covers the whole substrates	Hot wall reactor (1100–1300°C)	[18]
Reduction of MLG oxide	SLG, FLG, or MLG, depending on the starting material	Thickness depends on the starting material	Rapid thermal expansion in hot wall reactor (1050°C)	[19]
Exfoliation through thermal shock	Isolated graphite nanoribbons (MLG)	Mean aspect ratio ~250	Vacuum oven, use of sulfuric and fuming nitric acids	[20]
Acid treatment of expandable graphite	SLG or FLG nanoribbons MLG	Sub-10-nm width; ultrasmooth edges	Heating in forming gas to 1000°C	[21]
Liquid-phase exfoliation		Freestanding, possesses quality of initial bulk graphite	Need to use solvents	[22]

TABLE 3.2
Methods to Measure Thermal Conductivity in Graphene (SLG, FLG, MLG—Single-Layer, Few-Layer, and Multilayer Graphene)

Graphene Type/ Fabrication Method	Method of Measurements	Configuration of Measurements	Thermal Conductance ($W\ m^{-1}\ K^{-1}$)	Raman Shift/Temperature Conversion Coefficient	Reference
SLG/CVD	Microelectrothermal system	Suspended samples	1500–1800 (300 K) (increases with length)		[23]
SLG/CVD Isotropically modified	Raman shift (2D)	Circle	2197–4419 (300 K)	$6.98–7.31 \times 10^{-2}$ cm^{-1}/K	[24]
SLG/CVD	Raman shift (2D)	Circle	$2.6–3.1 \times 10^3$ (350 K)	7.2×10^{-2} (2D), 4.4×10^{-2} cm^{-1}/K (G)	[25]
SLG/exfoliation	Raman shift (2D)	Circle	1800 (325 K) 710 (500 K)	0.072 cm^{-1}/K	[14]
SLG/CVD	Raman shift (G)	Circle	2500 (350 K) 1400 (500 K)	0.0405 cm^{-1}/K	[9]
SLG/CVD	Raman stokes/anti stokes ratio	Circle	600 (295 K)		[10]
SLG/exfoliation	Raman shift (G)	Trench	4840–5300 (300 K)	0.016 cm^{-1}/K	[8]
FLG/exfoliation	Microelectrothermal system	Thermal bridge	1400–1495 (300 K)		[26]
FLG/exfoliation	Thermometer	Trench	560–620 (300 K)		[15]
Nanoribbons of FLG	Electrical	Trench	1100 (300 K)		[27]
MLG/exfoliation	Thermal flash	Platelets	776–2275 (300 K)		[28]

3.2 PREPARATION AND CHARACTERIZATION OF FLG AND MLG

Methods of preparation of FLG and MLG flakes (Table 3.1, see also [6,7,29]) include chemical vapor deposition [17], super-short-pulse laser produced plasma deposition [18], reduction of MLG oxide [19], exfoliation through thermal shock [20], acid treatment of expandable graphite [21], etc. All these methods have certain merits and disadvantages. Some of them are toxic, others are energy-intensive and time consuming, can produce defects in graphene, or cannot be used for large-scale production. In this work, the conventional low-cost liquid-phase exfoliation method [30–36] was employed to prepare FLG and MLG. For this, we used natural graphite powder (size of polycrystals 1–3 mm) from Nacional de Grafite Ltda (Brazil). Two types of solvents were used: *N,N*-dimethylformamide (DMF) and isopropyl alcohol (IPA), both analytical grades. FLG/MLG dispersion was prepared from a mixture of natural graphite flakes (1 mg) in DMF or IPA (1 mL) using ultrasound processing followed by centrifugation. It is known that the surface energy of solvents and solute (Hildebrand solubility parameter) must be very close to each other [22,33,37] for effective process of exfoliation. In our case, Hildebrand parameters for DMF and IPA are 24.86 and 23.58 $(MPa)^{1/2}$, respectively, and ~23 $(MPa)^{1/2}$ for graphene [38]. After centrifugation, the supernatant high-density solution (~0.4 mL), containing graphene flakes with lateral sizes from 0.5 to 10 μm (with lateral size-to-thickness ratios up to 200–300 as measured using SEM), were carefully removed by pipette and retained for analysis and further use. Exfoliation of graphite by sonication in a liquid phase is believed to be the result of the action of shock waves and microjets generated in liquid sonication [39–41].

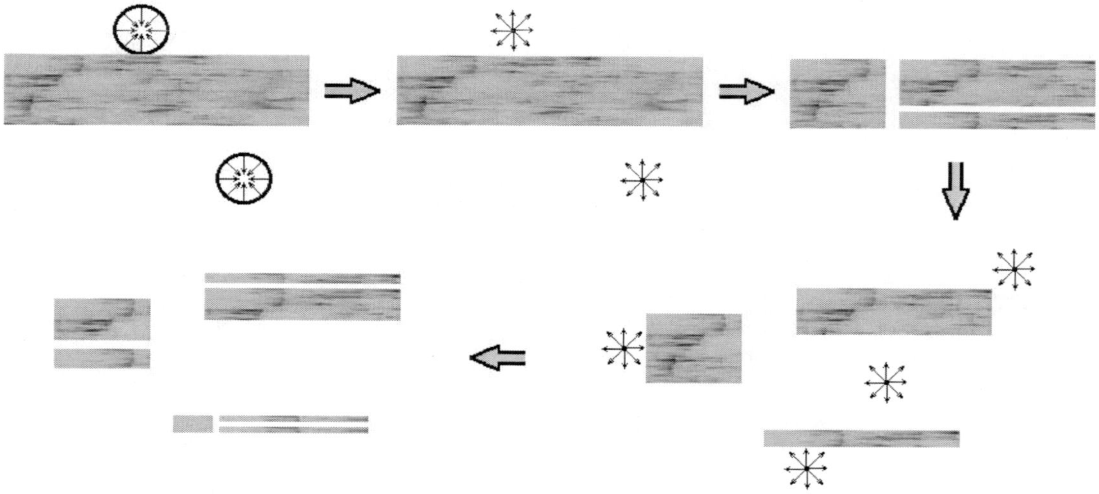

FIGURE 3.1 A schematic overview of the process of exfoliation of natural graphite in liquid phase. During sonication, bubbles and shock waves are produced, followed by cavitation with delamination graphite.

A schematic of the exfoliation process is shown in Figure 3.1. Cavitation is a nonlinear phenomenon that concentrates and transforms the low-density elastic wave energy into higher energy densities through rapid formation and collapse of gas bubbles in the liquid. Collapse of bubbles on the solid graphite surface will cause breaking of the solid by the produced shock wave, whereas collapse in the liquid close to the surface causes a microjet of solvent that can hit the solid with a great kinetic energy [42,43]. The energy released is sufficient to remove top graphitic layers after disrupting weak molecular interactions between them, leading to gradual delamination and dispersion of the initial graphite flakes. It is usually assumed that polycrystalline natural graphite starts to exfoliate in the areas containing defects and next at the grain boundaries. Due to complexity of the cavitation process in liquids, many parameters can affect the resulting dimensions of flakes (lateral size and thickness), including the bath temperature, sonication, and centrifugation time, see Reference 39 for more details. Suspensions of graphene sheets have been fabricated in DMF and IPA with different time of sonication (Figure 3.2).

For the size characterization, graphene flakes were deposited over holey carbon transmission electrons microscopy (TEM) grids with the mean size of holes near 1 μm. The statistical analysis for large number (hundreds) of sheets was performed to characterize the lateral graphene size distribution. Measured distributions were found to follow the so-called log normal size distribution, that is, the statistics for the particle size logarithms is normal (Gaussian), as first reported by Kolmogorov in 1941 [44], for the case of gold particles obtained by mechanical processing (attrition) of gold-containing rocks (Equation 3.1):

$$f(x) = \frac{1}{x\sigma\sqrt{2\pi}} \exp\left\{-\frac{1}{2}\left(\frac{\ln x - \mu}{\sigma}\right)^2\right\} \tag{3.1}$$

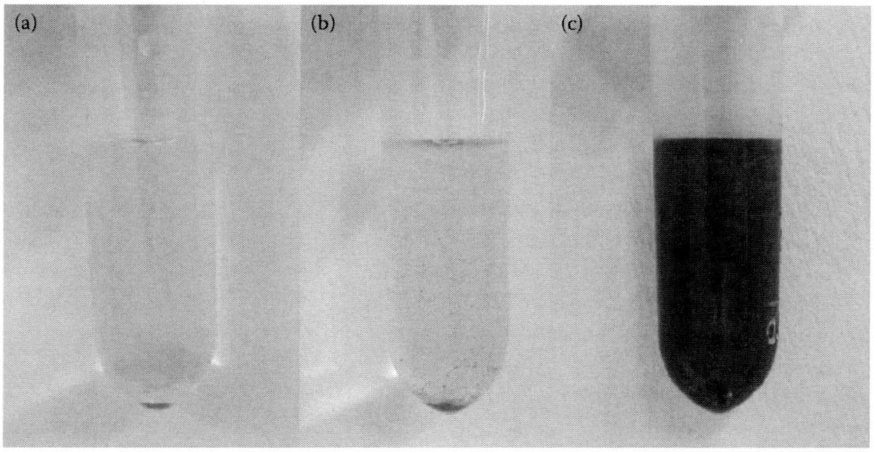

FIGURE 3.2 Solutions with FLG in DMF. (a) Mixture of DMF and natural graphite, (b) suspension after 2 min of sonication, and (c) suspension after 240 min of sonication.

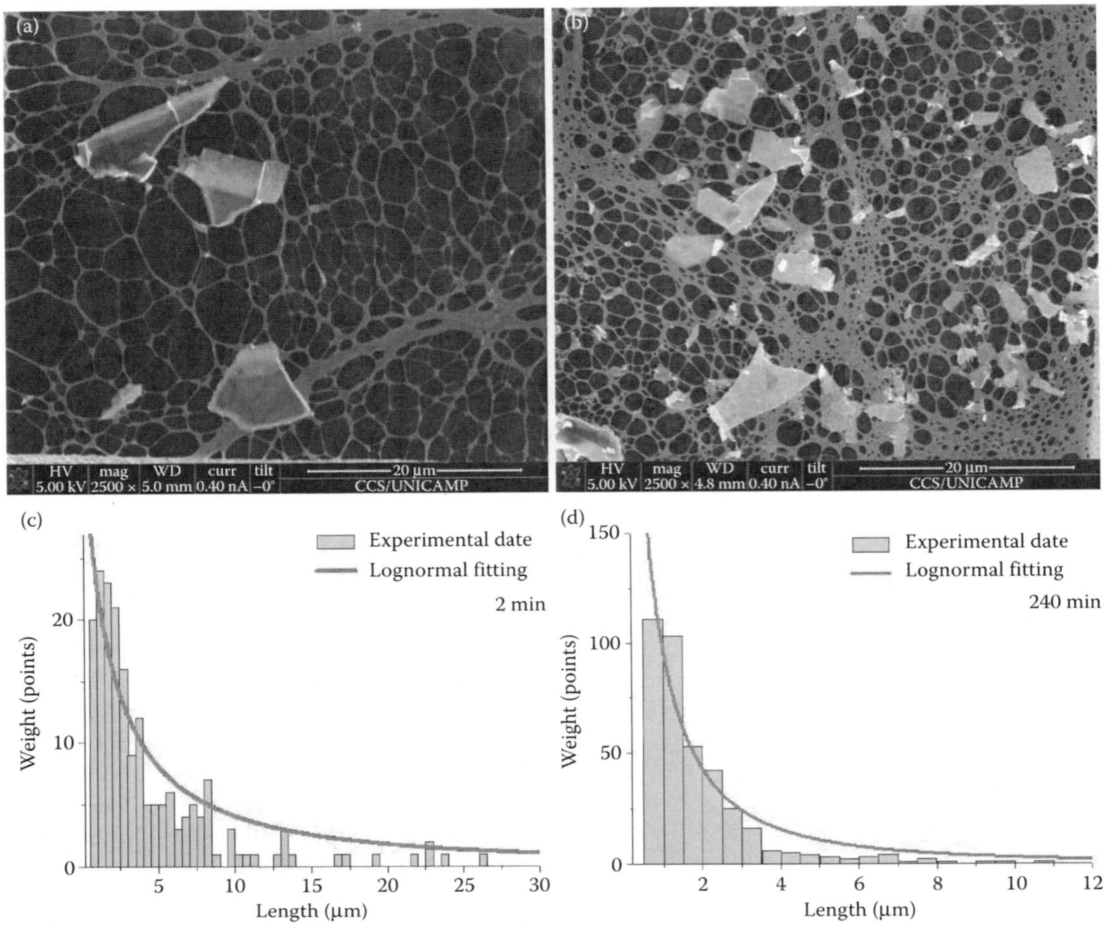

FIGURE 3.3 SEM images (a,b) of FLG deposited from DMF solution on holey carbon grids, for different sonication times (2 and 240 min, respectively) and (c,d) statistical distribution of graphene lateral sizes and lognormal fitting to the histogram for 2 and 240 min.

where

$\exp(\mu) = L_c$ (median size)

$\langle L_c \rangle = \exp(\mu + \sigma^2/2)$ (expected value)

$w = \exp(\mu + (\sigma^2/2))\sqrt{\exp(\sigma^2) - 1}$ (standard deviation)

This kind of distribution was observed later in many different situations, and can be expected to be valid in our case. To analyze the size distribution, histograms of lateral sizes for graphene sheets deposited over holey TEM grids were first obtained, with the number of measurements always being high enough (>100), see Figure 3.3 for DMF, similar results were obtained for IPA, not shown.

With increasing sonication time, the mean size was found to decrease from ~5 μm at 2 min to ~2 μm for 240 min, see Figure 3.4. The thickness of samples was found to vary from less than 2 to 30 nm (decreasing with sonication time), most of the samples being MLG or thin graphite nanoplatelets [39]. The aspect ratio (lateral size/thickness) was estimated roughly using SEM images to vary within a range 50–300 for all cases studied here, depending slightly on the sonication and centrifugation time and the type of solvent. Raman analysis with a micro-Raman spectrometer in confocal configuration ($\lambda = 473$ nm and 633 nm), was performed for flakes deposited on the TEM-grids to confirm high quality of graphene, with small full width at half maximum values, usually near 16–18 cm^{-1} (Figure 3.5). The D/G band ratio, widely used to evaluate the quality of the flakes, was found to vary from near 0 to 0.2 (tending to increase with sonication time), being consistent with small defect-free flakes.

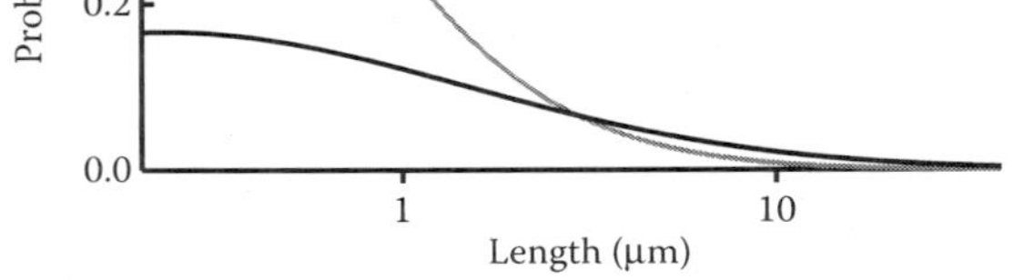

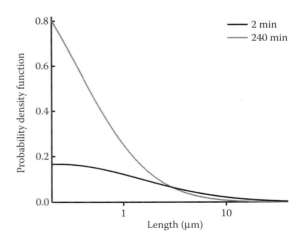

FIGURE 3.4 Distribution function (after fitting) of graphene sizes for 2 and 240 min of sonication in DMF.

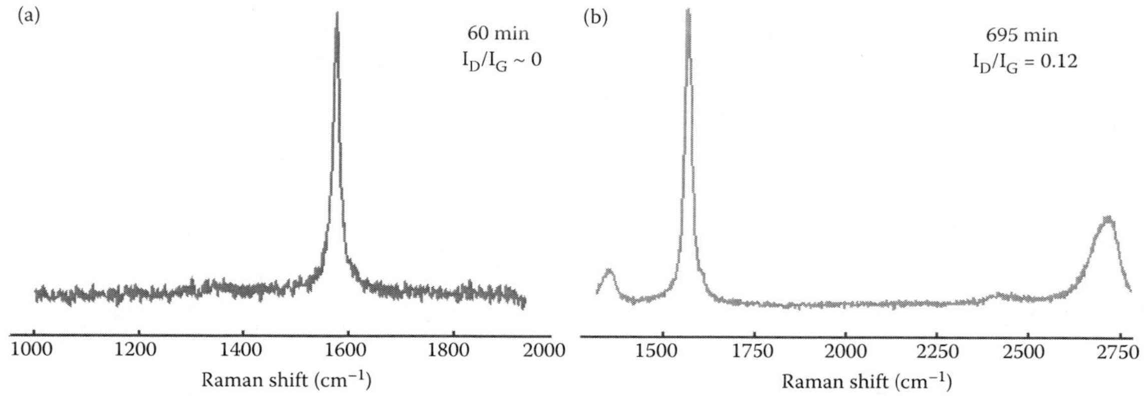

FIGURE 3.5 Raman spectra of graphene flakes with (a) 60 and (b) 695 min of sonication.

3.3 DEPOSITION OF MLG

For the experimental study of the MLG/metal contact formation, linear test structures of W, Ti, and Au (100 nm thick) were prepared by a conventional lift-off technique and metal sputtering (Figure 3.6) [45]. Then, gaps between electrodes (~1 μm wide, ~5 μm deep) were cut by FIB milling to prepare a pair of metallic electrodes [46]. MLG flakes were deposited from liquid solutions between electrodes using an AC-dielectrophoresis method (DEP) with normal process parameters: frequency of 40 MHz, 1–3 V peak-to-peak voltages, and 30–180 s deposition time [47,48]. Electrical contacts between MLG and metal electrodes were further improved with thermal annealing under high-vacuum conditions (10^{-6} Torr) and the electrical resistances were measured, using a two terminals method. The results of two terminals measurements show that after annealing the contact resistances reduced by orders of magnitude, with values in the range from 0.1 to 10 kΩ μm² [49,50].

Examples of samples obtained after DEP depositions over Ti and W electrodes followed by annealing in high vacuum (700°C, 1 h), as well as cross sectioning of the deposited flakes using FIB can be seen in Figures 3.7 and 3.8. For cross sectioning, the MLG flakes deposited over electrodes were protected with a Pt layer deposited by focused electron beam (FEB) (Figure 3.7d), and cross sectioning of contacts between MLG and metal were made by FIB to expose the structure of contacts (Figure 3.7e). The results for Ti indicate very tight contact apparently without diffusion of titanium inside the graphene flake (Figure 3.7f and g). Stronger metal–MLG interaction (due to annealing) can be observed in the case of W electrodes (Figure 3.8b), where an area with intermediate contrast at the metal–graphene interface (20–30 nm thick) is clearly seen, probably due to possible formation of tungsten carbide during high-temperature annealing. After annealing, some samples with one or two MLG flakes deposited between electrodes were selected for contact resistance measurements. For this, the total two terminal resistances were measured, and the resistances of contact leads and MLGs were estimated from the flake dimensions and known electrical resistances for electrode material and graphite. Then contact resistance (*R*) and resistivity (*R.A*) values were calculated, where *A* is the contact area. Values of contact resistance and resistivity for W electrodes (similar data were obtained for Ti) are shown in the Table 3.3 and Figure 3.9. The data indicate a strong reduction of contact resistance/resistivity with increasing annealing temperature.

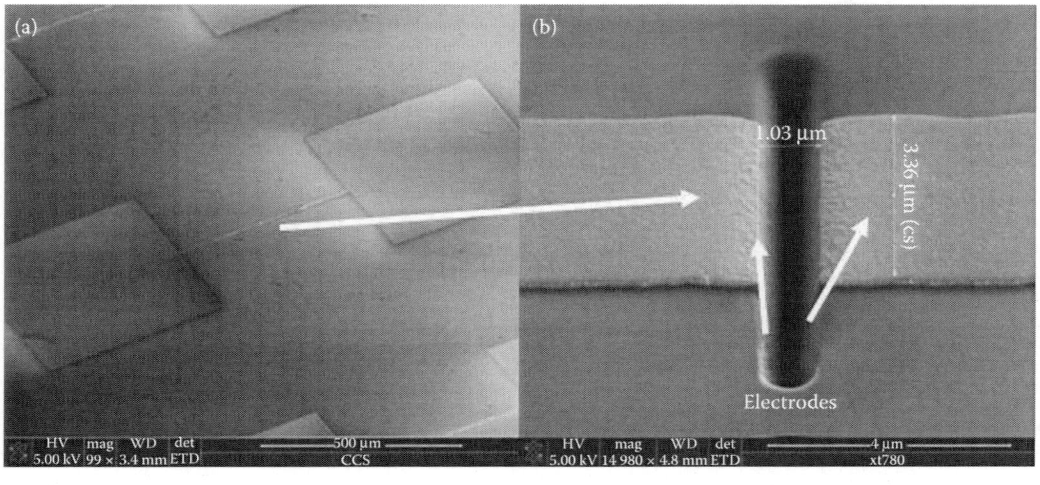

FIGURE 3.6 Test structures prepared by conventional lithography (a) and electrodes for FLG deposition prepared by FIB (b).

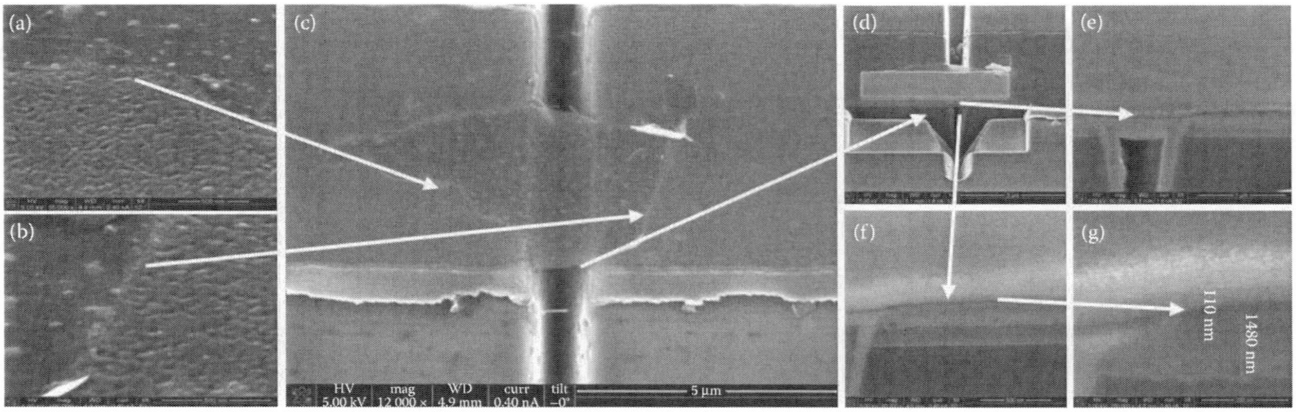

FIGURE 3.7 SEM images of MLG and details of border contact (a,b,c). Flakes were protected with an FEB-deposited Pt layer followed by FIB milling, and cross sectioning of MLG/Ti contacts were further performed to expose the structure of contacts (d,e). (f,g) Cross section with larger magnification in the supported part of MLG.

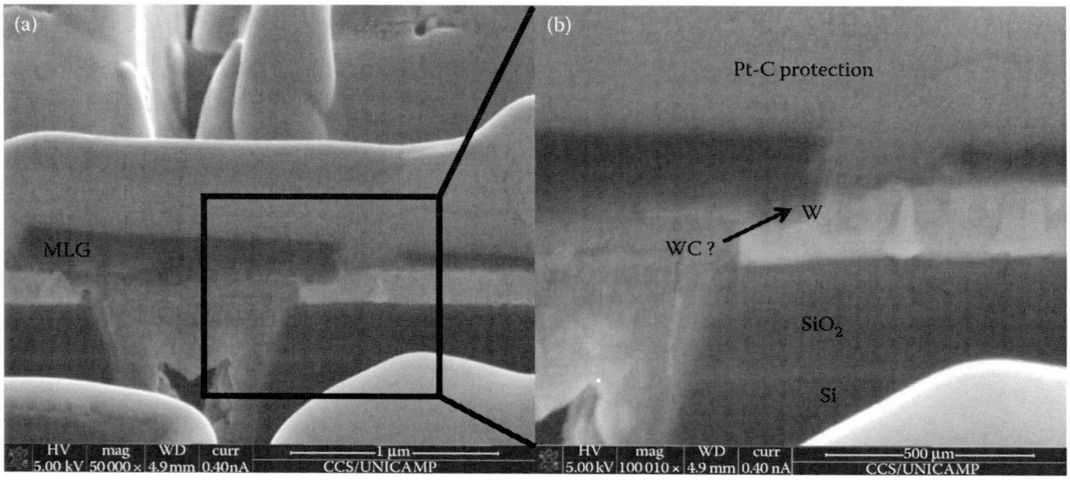

FIGURE 3.8 SEM images of MLG/W-electrode contact after annealing. (a) Flakes were protected with an FEB-deposited Pt layer followed by cross sectioning of contacts to expose the interface structure. (b) The results indicate strong diffusion for contacts improved with thermal treatment.

TABLE 3.3
Typical Values of Contact Resistivity for W Electrodes versus Annealing Temperature

Temperature of Annealing (°C)	Contact Resistivity (kΩ μm²)
500	7.5 ± 2.5
600	2.3 ± 1.0
700	0.4 ± 0.2

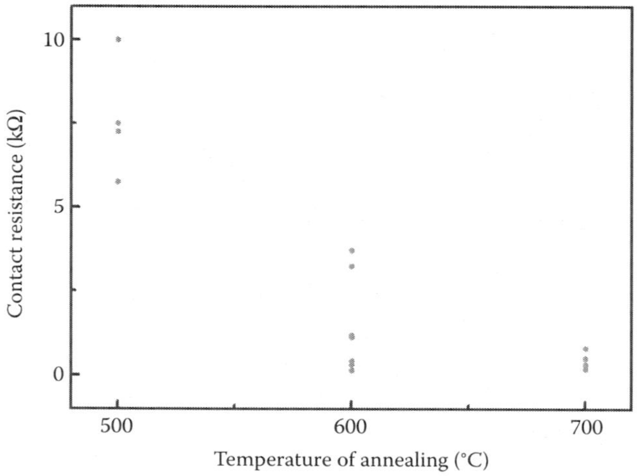

FIGURE 3.9 Resistance of contact area as a function of annealing temperature.

3.4 MICRO-RAMAN METHOD TO MEASURE THERMAL CONDUCTIVITY OF MLG

Potential applications of graphene for thermal management in electronic devices are very promising as in-plane thermal conductivity of SLG is extremely high, up to 5.3×10^3 W m^{-1} K^{-1} [29]. With increasing number N of layers in FLG/MLG, thermal conductivity K of graphene was shown to decrease rapidly, reaching the value for bulk graphite (near 1000 W m^{-1} K^{-1} at room temperature) for $N > 5$ [12]. The theoretical studies predict K value for ideal infinite SLG as high as 10^4 W m^{-1} K^{-1} [51], however reported experimental values vary widely, from 600 to 5000 W m^{-1} K^{-1}. This variation can be attributed in part to the difficulties to prepare and transfer high-quality (defect-less) graphene with number of layers known, as well as to uncertainties in the laser absorbed power, finite size of the laser spot, contribution of impurities, and edge effects [24,27]. One more source of uncertainties in K measurements is related to poor thermal contacts between graphene and metal electrodes. Heat transfer through contacts and interfaces between nanostructured materials is an important issue in nanoelectronics, in studies of nanocomposites and thermal interface materials, but it is still poorly understood and the existing models of electrical and thermal (macro) contacts are often inapplicable at the nanoscale [52,53]. In conventional electronic devices, interfaces between two thermally dissimilar materials like metals and semiconductors are assumed to be planar and perfectly matched. In reality, the contact thermal resistance between nanostructured materials is determined by two factors: the overlap of phonon states for two solids and the properties of the interface itself (roughness, adhesion between materials). The heat transfer between two solids in a perfect contact (strong bond) can be calculated using diffusive mismatch model (DMM) or acoustic mismatch model (AMM) [24,54]. The efforts to improve the thermal interface models continue [55–57], however, the existing models are based on various simplifying assumptions and often fail to describe real nanoscale systems composed by materials with large differences in mechanical and thermal properties. In particular, graphene materials are characterized by extremely high anisotropy (in-plane/cross-plane) and large vibrational mismatch with most solid materials used in nanoelectronics. The maximum vibrational frequencies in graphene are much higher than those in metals [58,59] (45 THz compared to 7–10 THz in metals), with the difference being smaller for oxides [60,61]. The sp^2 bonding in graphene is responsible for high elastic stiffness in a basal plane, while rough metals surfaces can suffer plastic deformation at short length scale (at point contacts) under heating and pressure [62]. The existing models have difficulties also in describing weakly interacting solids with small adhesion energy. This is the case of graphene supported over most metals or oxides like SiO_2 that do not form chemical bonds with carbon. Adhesion energy of graphene to amorphous SiO_2 has been recently measured to be 0.45 and 0.31 J m^{-2} for SLG and FLG, respectively [63], being slightly higher for metals like Cu—0.72 J m^{-2} [64] and Ni (fcc)—0.81 J m^{-2} [65], respectively. Adhesion between graphene and metals can be calculated using molecular dynamics (MD) simulations [53,65]. MD simulations were shown to be useful also for analyses of the effects of atomistic changes (impurities, defects, strain) on thermal properties of materials and interfaces [66]. Using the modified AMM method, Prasher [56] has shown that reduced adhesion implies significantly lower heat transfer between two materials. It should be noted however, that even low adhesion between two surfaces produces some pressure squeezing the materials against each other, and this is especially important when plastic deformation of contacting materials is possible. The real contact area between rough materials (or between rough and plane materials) can be just a small fraction of the nominal contact area. Then the local pressure due to adhesion at the isolated contact points under heat flow can eventually produce a kind of "welding" between two solid surfaces resulting in a significant increase of the real contact area and decrease of the thermal contact resistance [67,68]. An MLG is essentially a rigid structure with atomically flat surface that cannot conform to rough metal surfaces. Therefore, for the MLG–metal contact improvement, the metal surface must be smoothened. Alternatively, for metals that can interact chemically with carbon forming carbides (like tungsten) stronger thermal contact can be established at high enough annealing temperature [69]. The adhesion energy can be also increased through functionalization of contacting surfaces, substituting a weak *van der Waals* interaction by stronger covalent bonding [70].

The quality of thermal contacts is characterized by thermal contact resistance R_{th} (or thermal boundary conductance G) [66,68] that can be calculated by using the equation:

$$G = (R_{th})^{-1} = \frac{Q}{(S\Delta T)} \tag{3.2}$$

where Q is the heat flow through the interface (absorbed laser power), S is the contact area, and ΔT is the temperature drop at the interface. Schmidt et al. [71] obtained experimental values of metal–graphite (HOPG) thermal boundary conductance (TBC) at room temperature (RT) of ~3×10^7 W m^{-2} K^{-1} for Au, ~5×10^7 W m^{-2} K^{-1} for Cr and Al, and ~10^8 W m^{-2} K^{-1} for Ti. Similar data for Au–HOPG interface were reported by Norris et al. [55]: near 3×10^7 W m^{-2} K^{-1} at RT for as-cleaved HOPG surfaces. Much higher values were obtained by Hirotani et al. [72], in a study of thermal interfaces between Au and graphitic layers (c-axis) in ~100 nm diameter multiwalled carbon nanotubes (MWCNTs), varying from 8.6×10^7 W m^{-2} K^{-1} to 2.2×10^8 W m^{-2} K^{-1}.

TBC between SLG or FLG (1.2–3.0 nm thick) and amorphous SiO_2 was measured to range from ~0.8×10^8 to ~2×10^8 W m^{-2} K^{-1}, with no clear dependence on the number of layers by Chen et al. [73]. Different models give values varying from 6×10^7 W m^{-2} K^{-1} [40] to 3×10^8 W m^{-2} K^{-1} [67] for perfectly flat graphene–amorphous SiO_2 interface at RT. Even higher TBC can be expected for graphene interfaces with crystalline oxide materials such as sapphire [74].

For graphene (c-axis) interfaces with metals, the TBC values measured in most studies vary from 3×10^7 W m^{-2} K^{-1} to 1×10^8 W m^{-2} K^{-1} at room temperature, whereas higher values are measured for silicon oxide, up to 2×10^8 W m^{-2} K^{-1}. It should be emphasized also that experimental methods may give reduced TBC values in particular for metals, as real contact areas for rough metal surfaces are lower than nominal graphene/metal contact areas, while less difference can be expected for graphene/silicon oxide (smooth) surfaces. In practice, the presence of a thermal contact resistance together with a small contact area means that a considerable temperature drop (tens to hundreds *K*) can appear at interfaces between two dissimilar materials under heat flow. Such high temperatures can result in graphene burning and failure of the nanodevices, and also can affect the accuracy of *K* measurements.

In principle, high-quality thermal contacts between graphene and metals can be obtained using conventional high-temperature annealing in vacuum. However, as shown in our study [50], this can induce strain in MLG due to the formation of tight mechanical contacts between the graphene and metal electrodes followed by the shrinkage of the electrodes, thus stretching the graphene flake. Therefore, it is desirable to perform annealing by heating the graphene flake and the area of contacts locally, without substantial heating of the electrode bodies. Here, it is proposed to do such annealing by using the same laser resulting in mild "welding" of the graphene to the initially rough metal surface. The micro-Raman method was used here for determination of thermal conductivity along suspended MLG flakes, and the dramatic effect of localized laser thermal annealing for improving of thermal contacts between graphene and metals or silicon oxide, as well on the results of *K* measurements, was shown.

Experimentally, suspended flakes of graphene were used to study the properties of thermal conductivity in MLG. A 473 nm laser focused on the sample with 100× objective (~400 nm diameter laser spot, max laser power at the sample of 10.5 MW) was used for monitoring of the sample's temperature, using the optothermal micro-Raman method [29] based on measurements of the frequency downshift $\Delta\omega_G$ of a graphene G-line with increasing temperature, with the same laser used for local heating of suspended graphene samples (Figure 3.10) [75]. From the measured $\Delta\omega_G$ value, the local graphene temperature rise ΔT above room temperature T_{RT} can be estimated, using conversion coefficient $g = \Delta\omega_G/\Delta T$ [cm^{-1} K^{-1}], obtained in calibration experiments. The coefficients were measured to be −0.016, −0.015, and −0.011 cm^{-1} K^{-1} for SLG, bilayer graphene, and MLG, respectively [76]. In other studies, values from −0.011 to −0.024 cm^{-1} K^{-1} for MLG (or HOPG) were reported [77,78]. The reasons for such a discrepancy are not clear at the moment. This may be attributed to different quality of samples, not perfect thermal contacts between the graphene and heat sink, and different geometries of the experiment. For example, a finite size of the laser spot on a sample can affect the measurement accuracy if it is comparable with the distance between the heating point and heat sink. Here, we accepted the value $\Delta\omega_G = -0.011$ cm^{-1} K^{-1}. Resuming, from the measured G-peak downshift $\Delta\omega_G$, the local MLG temperature rise ΔT above room temperature T_{RT} was estimated, using a relation $\Delta T(\mathrm{K}) = -\Delta\omega_G(\mathrm{cm}^{-1})/0.011$ cm^{-1} K^{-1}. Assuming that a perfect thermal contact is established between a graphene and a heat sink (i.e., room temperature T_{RT} is achieved at the contact point), the temperature gradient $\Delta T/L = (T_h - T_{RT})/L$ is estimated, where L is the distance between the heating point and heat sink, and T_h is the temperature measured at the point of laser heating. As a heat sink, metal electrodes known as good heat conductors are commonly used. Further, using the Fourier equation the *K* value can be calculated:

$$K = -\frac{P_{abs}L}{(2A\Delta T)} \tag{3.3}$$

where P_{abs} is the laser power absorbed on the sample and A is the sample cross-sectional area. In Equation 3.3, we consider that heating occurs in the center of the graphene flake and both electrodes contribute equally to heat absorption. Note that heat dissipation onto the underlying substrate occurs within the length scale given by the so-called thermal healing length L_H, estimated to be ~0.1–0.2 μm [66]. The L_H value should be added to L in Equation 3.3 while calculating *K* for short samples.

It is generally assumed that 2.3% of laser power is absorbed by a single graphene layer in a visible spectral range [79], however in the UV region the absorption can be higher [80], and in micro-Raman experiments done by Cai et al., a value of 3.3 ± 1.1% was reported for 532 nm laser wavelength [2]. Uncertainties in optical absorption in graphene (due to defects, impurities, varying number of layers) obviously contribute to errors in estimations of *K* using the micro-Raman technique. Further, it should be noted that the G-line downshift may depend on the heating mechanism, and heating by electrical current or by laser may result in a partial thermal equilibrium (not all phonon modes are equally excited), in contrast to the full thermal equilibrium reached in hot cells [81]. Therefore, the use of conversion coefficients determined under thermal equilibrium conditions for evaluation of local temperatures for the laser heated graphene may result in additional errors.

Nair et al. [79] has shown that the reflectance of SLG is less than 0.1%. No data are available for reflectance from an FLG, however, the reflectance of bulk graphite at the range of 500 nm was found to be as high as ~30%. For FLG graphene, the reflectance is probably lower, but should approach to 30%

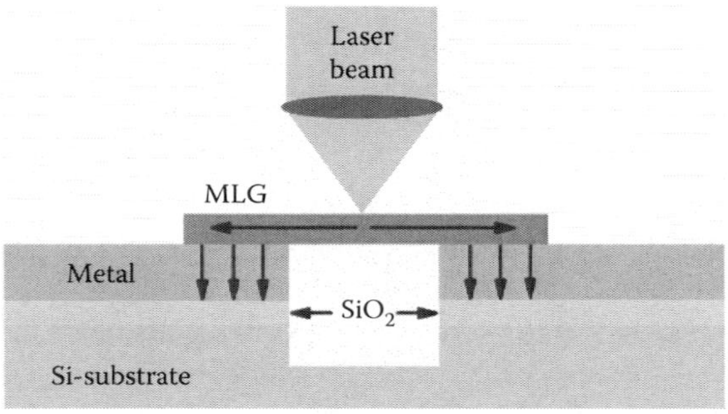

FIGURE 3.10 Schematic setup of the experiment. (The source of the material, V. A. Ermakov et al., Nonlocal laser annealing to improve thermal contacts between multi-layer graphene and metals, *Nanotechnology*, vol. 24, no. 15, p. 155301. Copyright 2013, IOP Publishing Ltd. is acknowledged.)

for thick samples (>20 nm). We used here the 30% reflectance value for our estimates of absorbed laser power.

Poor electrical or thermal contacts between the graphene surface and substrate could be also due to the presence of the so-called dead layer [1] (the graphene-substrate gap filled by air, water, or other impurities), and also can depend on the way of sample preparation and the roughness of the substrate surface especially in the case of metallic contacts deposited by sputtering. In contrast, the surface of the silicon dioxide has a smooth surface that in some cases (big contact area) can result in a small thermal contact resistance [12,29].

Other sources of uncertainties are related to possible deviations of temperature distribution from the idealized scheme shown in Figure 3.10, due to poor thermal contacts between the graphene flake and metal electrodes. In this case, the real values of ΔT may be lower and L may be larger than shown on Figure 3.10, resulting in underestimated K values.

3.5 EXPERIMENTAL DETERMINATION OF MLG THERMAL CONDUCTIVITY AND THERMAL CONTACT RESISTIVITY

Examples of MLG deposited over various metal (W, Au, Ti) electrodes forming side contacts with the metal are shown in Figure 3.11a through f. As the cuts made by FIB were longer than metal electrodes, in a few cases we also found graphene flakes deposited over the gaps outside the electrodes (over the silicon oxide surface) that allowed us to study the effects of annealing between graphene and silicon oxide (Figure 3.11e). Cuts between electrodes were made with width of 1 μm (±10%), and a tilted image of the cut is shown in Figure 3.11d, where nonvertical cut walls can be seen.

Experiments were performed in air, using a 473 nm wavelength laser beam with a spot of ~400 nm and the maximum power up to 10 MW that was focused at the center of suspended MLG platelets. Most of experiments were performed with 20–30 nm thick (60–90 monolayers) samples, so that practically full laser power (with a correction for reflection) was absorbed by the samples. In other words, in such experiments there is no uncertainty in absorbed laser power. The G-line position was found first to shift linearly with laser power (see Figure 3.12a, black squares) indicating gradual heating of the sample proportional to the absorbed power, as expected. However, with increasing power the shift was found to saturate or even reduce. In the case shown in Figure 3.12a, the saturation occurs at -22 cm^{-1}, in other cases the maximum shift was found to vary between 7 and 23 cm^{-1} for laser power up to 10.5 MW, depending on the flake dimensions and quality of contact with electrodes, the smallest downshifts were obtained for Au electrodes. Using the conversion coefficient $g = -0.011$ cm^{-1} K^{-1}, very high local graphene temperatures (from 600°C to 2000°C) can be estimated. The saturation can be attributed to two different mechanisms. First, improvement of the thermal contact between the MLG and metal electrode can occur during sample heating, then heat losses to the electrodes become stronger and the sample temperature drops. Second, gradual sample thinning due to graphene etching in air and thus reduced laser absorption during the heating process, as the increasing part of the laser power passes through the flake without absorption. Obviously, such strong heating should inevitably result in gradual layer-by-layer graphene etching and thinning by oxygen present in the air [82,83]. However, our results proved this process to be relatively slow, with a maximum rate much less than one monolayer per second for $T_{max} \sim 2000$ K, as revealed by special thinning experiments, see below. This means that a considerable number of spectra can be acquired during the thinning process before the moment when MLG becomes thin enough to reduce absorption of laser light in the sample. Furthermore, care was taken here to keep the sample expositions at minimum required for the spectra acquisition (typically, ~1 s at T_{max}) leading to minimum sample changes during the exposition.

In the thinning experiments, designed to determine the MLG etching rate at the maximum laser power, we deposited MLG flakes onto a crystalline Si substrate covered by

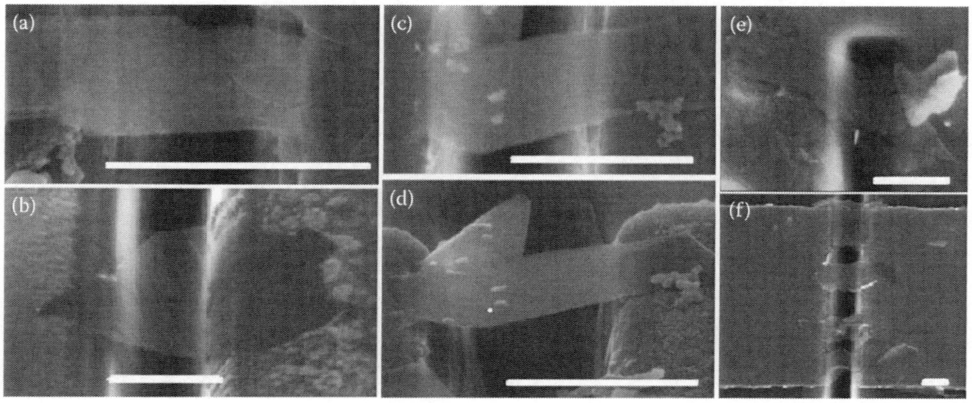

FIGURE 3.11 MLG flakes deposited over metal electrodes: W (a), Au (b), Ti (c), and SiO_2 layer (e). (f) An overview of electrodes with several FLG flakes deposited over the gap, (a,c,e,f) are top images and (d) is a tilted image (52°). The scale bar corresponds to 1 μm. (The source of the material, V. A. Ermakov et al., Nonlocal laser annealing to improve thermal contacts between multi-layer graphene and metals, *Nanotechnology*, vol. 24, no. 15, p. 155301. Copyright 2013, IOP Publishing Ltd. is acknowledged.)

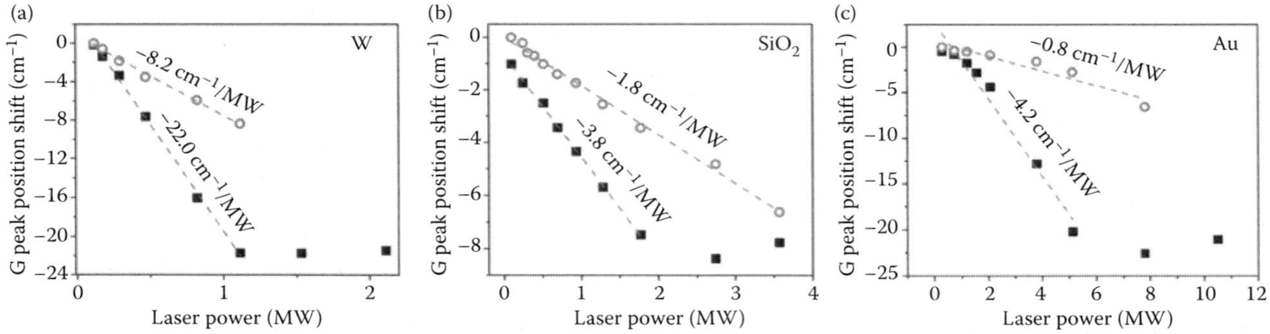

FIGURE 3.12 G peak position versus absorbed laser power during the first irradiation (initial laser annealing, black squares) and after annealing (gray open circles) for the same FLG flake on W contacts (a), SiO_2 (b), and Au contacts (c). (The source of the material, V. A. Ermakov et al., Nonlocal laser annealing to improve thermal contacts between multi-layer graphene and metals, *Nanotechnology*, vol. 24, no. 15, p. 155301. Copyright 2013, IOP Publishing Ltd. is acknowledged.)

thermally grown 300 nm thick SiO_2 layers. We applied a method to measure the thickness of FLG over Si, recently proposed by Han et al. [82] and Koh et al. [83], to monitor the changes in the flake thickness during the sample annealing in air. The method is based on comparison between the integrated intensities of the Raman lines of graphene (G-line) and first-order optical phonon peak from underlying Si (520 cm^{-1}). The MLG thickness, before and after the thinning experiment, was also measured using atomic force microscopy that allowed an independent evaluation of the thickness changes. Due to much lower cross-layer heat conductivity of graphitic materials [66,80], absorption of the laser light in the flake leads to strong heating mostly of the upper layers and their layer-by-layer etching due to interaction with oxygen and the formation of volatile products such as CO or CO_2. The MLG flakes chosen for such experiments were thin enough (usually, thinner than 10 nm) to detect the characteristic Si-line (520 cm^{-1}) from the substrate. Continuous detection of the I(G)/I(Si) ratio allowed us to track the flake thinning while the G-line downshift provided the information about the local temperature. Note that the method is suitable for reliable measurements of the number of layers N up to ~10. For thicker flakes the relation between the I(G)/I(Si) ratio and the N number becomes nonlinear affecting the measurements accuracy, however semiquantitative information on the sample thickness evolution still can be obtained. For these measurements, only flakes with lateral dimensions exceeding 1.5–2.0 μm were used, to avoid a parasitic contribution from the bulk Si signal collected from the area around the flake (i.e., without absorption by the flake). Typical temporal evolution of the I(G)/I(Si) ratio and temperature is shown in Figure 3.13. The process of thinning (reduction of the I(G)/I(Si) ratio) was accompanied by gradual decrease of the sample temperature (G-line downshift) due to decreasing absorption in the flake. In Figure 3.13, the ratio changes from 0.8 to 0.25, that translates in evolution of N from ~10 to ~4, following the calibration given in Reference 83 for approximately the same conditions. The thinning process was relatively fast for T as high

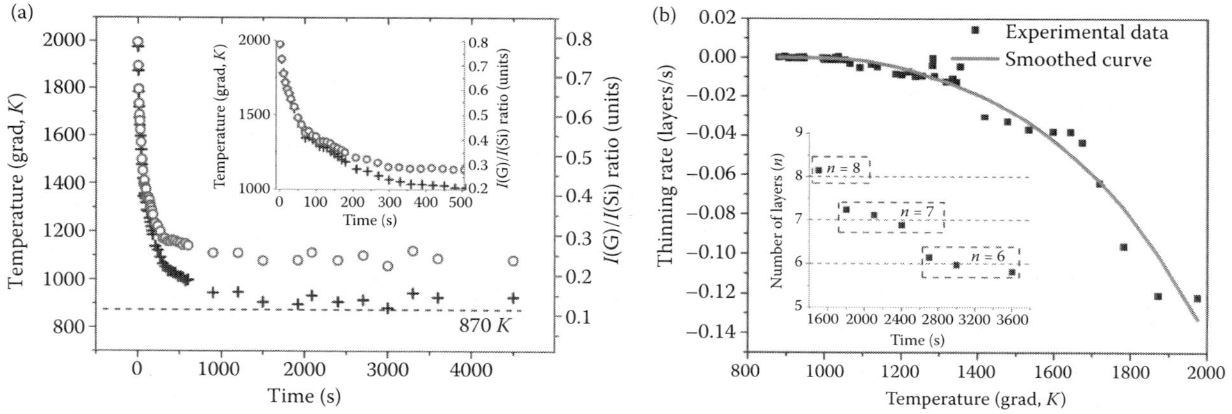

FIGURE 3.13 (a) Layer-by-layer thinning process for FLG deposited onto oxidized Si substrate—I(G)/I(Si) ratio and temperature of the flake, +—temperature, open circles—I(G)/I(Si) ratio. The inset shows the signals evolution during first 500 s. (b) Dependence of thinning rate versus temperature (dots—derived from experimental data, line—smoothed curve). The inset in (b) shows evolution of the number of layers at the final stage of thinning. (The source of the material, V. A. Ermakov et al., Nonlocal laser annealing to improve thermal contacts between multi-layer graphene and metals, *Nanotechnology*, vol. 24, no. 15, p. 155301. Copyright 2013, IOP Publishing Ltd. is acknowledged.)

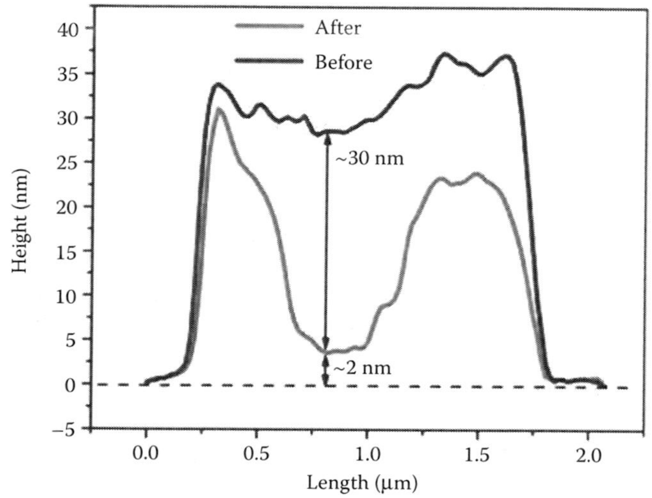

FIGURE 3.14 AFM measurements of MLG thickness before and after laser thinning. Note that the minimum thickness of the flake after the thinning process is ~2 nm. (The source of the material, V. A. Ermakov et al., Nonlocal laser annealing to improve thermal contacts between multi-layer graphene and metals, *Nanotechnology*, vol. 24, no. 15, p. 155301. Copyright 2013, IOP Publishing Ltd. is acknowledged.)

as 1500–2000 K, when ~4 monolayers were etched out during the first 60 s under laser irradiation with power of 10 MW. Further, under the present conditions the thinning process was always found to stop when the temperature reduced to ~850–900 K, when monolayer graphene was not reached yet, see Figure 3.13. AFM measurements were performed to find that the minimum thickness obtained in our experiments was limited to ~1.5–2.0 nm, compatible with the estimates of N_{min} near five layers (see an example in Figure 3.14). The temperature when the graphene etching stops (near 600°C) is comparable with the results of TGA analysis by Han et al. [82] for graphene and Pang et al. [84] for nanotubes (slightly lower T values were obtained for graphite). Interestingly, at the final stage of thinning ($T < 1000$ K) it was possible to observe step-like changes of the I(G)/I(Si) ratio and thus the number of layers (see inset for Figure 3.13b), indicating basically a layer-by-layer character of the MLG thinning, where a fast layer removal is followed by very long periods (up to 700–1000 s) of small changes in the I(G)/I(Si) ratio.

Further, the quality of graphene flakes apparently did not change during annealing, as G-line FWHM and I(D)/I(G) ratios remained at the same level (in some case, even a small reduction of the ratio was detected). In Figure 3.15a, one example of Raman spectra taken before and after annealing in the thinning experiment is shown for the flake with the G-line FWHM and I(D)/I(G) ratio of ~15 cm^{-1} and ~0.05, respectively. In general, the I(D)/I(G) ratio was found to vary from ~0.01 to 0.05 for different samples (in Figure 3.15b, a spectrum with a small D line is shown; such samples were mostly used for K measurements). This is much smaller than observed in other studies with narrower graphene nanoribbons [66,85], indicating very low density of defects and relatively large average size of graphene crystallites in our samples. Following the approach developed by Cançado et al [86] (see also [13,85]), the average crystallite size L_a in graphene can be found using the formula: $L_a \sim 560\ [I(D)/I(G)]^{-1}/E_L^4$, where E_L is the laser photon energy in eV. For $I(D)/I(G) = 0.05$, $L_a \sim 280$ nm can be estimated, being higher for smaller ratios. The fact that an increase of I(D)/I(G) ratio was not observed during annealing indicates that no additional defects (or average crystallites size reduction) were induced by laser processing under the present conditions, characterized by very short exposures (<1 s) by laser light.

The saturation of the G-line position downshift with increasing power (Figure 3.12) therefore can be attributed mostly to the gradual annealing of the graphene–metal contact, resulting in improving heat transfer through the contact and thus reduced graphene heating. This is confirmed by results obtained in the second run (performed immediately after the first one), when much lower G-line downshifts are detected for the same laser powers, with the curve slope of

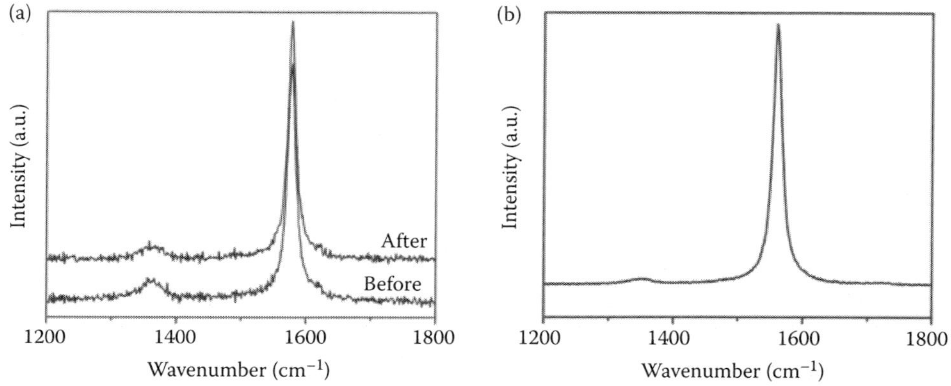

FIGURE 3.15 Raman spectra of different FLG: (a) the same sample before and after laser annealing, at laser power of 0.29 MW, 1 s; (b) another sample, spectrum obtained at laser power of 2 MW, 5 s. (The source of the material, V. A. Ermakov et al., Nonlocal laser annealing to improve thermal contacts between multi-layer graphene and metals, *Nanotechnology*, vol. 24, no. 15, p. 155301. Copyright 2013, IOP Publishing Ltd. is acknowledged.)

-8×10^3 cm^{-1} W^{-1}, as compared with -2.2×10^4 cm^{-1} W^{-1} for the first irradiation (Figure 3.12a). Heat losses from the graphene surface due to thermal contact with air (~10^5 W m^{-2} K^{-1}) [87] and by radiation are estimated to be much smaller than the absorbed laser power (~0.1 and 10^{-8} MW, respectively). Thus, under the present conditions the main mechanism of heat losses is due to heat transfer through the graphene–metal contacts, and the reduced sample heating during the second run clearly indicates the improved thermal conductivity of the graphene–metal contact after annealing occurred during the first irradiation. The thermal contact improvement is likely due to the partial melting of the metal surface in contact with the graphene, reducing of the initially high metal surface roughness under the graphene, and corresponding increase of the contact area and adhesion between surfaces. This is possible because of distinctly different mechanical properties of two contacting materials: (i) MLG with a perfectly flat surface, characterized by high in-plane elastic stiffness and high cross-plane hardness, and (ii) rough metal surface that is subject to plastic deformation at nanoscale length under pressure (due to adhesion) and heat flow.

Interestingly, much weaker graphene heating was observed for samples deposited over SiO_2 (Figure 3.11d), both before and after the annealing (G-line downshift -1.6×10^3 cm^{-1} W^{-1} and -8×10^2 cm^{-1} W^{-1}, respectively). Note that the contact annealing in this case was possible only for higher powers (>5 MW) and lower G-line downshift (−7 cm^{-1}) corresponding to T_{max} ~ 600°C, see Figure 3.12b. This is a clear indication of initially better thermal contacts between the graphene and thermal oxide surface that is much flatter (roughness ~0.1 nm) compared with metals (W and Au). Note that for the oxide substrate, the maximum sample heating does not exceed ~600°C, when no sample burning and thinning can be expected, thus the significant reduction of the sample heating can be attributed only to the improvement of thermal contacts.

Note that the Raman measurements provide only local values of temperature (ΔT_1 and ΔT_2), before and after annealing (Figure 3.16) while temperature drops at the graphene–substrate interface (ΔT_{i1} and ΔT_{i2}), are unknown. This leads to underestimated values of K as real values of the temperature change along the graphene (ΔT_g) are smaller than those measured by the micro-Raman method. Assuming that annealing changes only the interface properties (reducing ΔT_{i1} due to the contact area increase) and the temperature change along the MLG remains the same (ΔT_g), one can get: $\Delta T_1 = \Delta T_g + \Delta T_{i1}$, $\Delta T_2 = \Delta T_g + \Delta T_{i2}$, where $\Delta T_{i1} = P_{abs}/(S_1 G)$ and $\Delta T_{i2} = P_{abs}/(S_2 G)$, S_1 and S_2 are real contact areas before and after annealing, and ΔT_g is determined by Equation 3.2: $\Delta T_g = -P_{abs}L/(2KA)$. Further, assuming that after annealing the contact area becomes close to the nominal contact area S_{nom} ($S_1 \ll S_2 \approx S_{nom}$), estimates of the temperature drop at the interface after annealing can be made using the literature data on the thermal boundary conductance G: $\Delta T_{i2} = P_{abs}/(S_{nom}G)$. Further, the temperature change along the graphene can be calculated using the results after annealing: $\Delta T_g = \Delta T_2 - \Delta T_{i2} \approx \Delta T_2 - P_{abs}/(S_{nom}G)$. Then, the temperature drop at the interface before annealing and the contact area increase due to annealing can be estimated: $\Delta T_{i1} = \Delta T_1 - \Delta T_g \approx \Delta T_1 - \Delta T_2 + \Delta T_{i2} = P_{abs}/(S_1 G)$ and $S_2/S_1 = \Delta T_{i1}/\Delta T_{i2}$, respectively. Finally, estimates were made to show that corrections to ΔT_g in our experiments after annealing (due to ΔT_{i2}) can vary from ~10 K (for SiO_2) to 100–150 K (for metals). These corrections result in considerable increase (that can reach 30%–40% or even more) of the calculated K values for metal electrodes, with smaller corrections for SiO_2. Note also that if the annealing is not complete ($S_{nom} > S_2$), the ΔT_{i2} is in fact higher and the ΔT_g value is then overestimated, thus K is still underestimated.

Using the approach described here and determined above ΔT_{i1} values, estimations gave G near (2.5–3) × 10^6 W m^{-2} K^{-1} for W and Au and 6×10^7 W m^{-2} K^{-1} for SiO_2 before annealing, and contact areas as low as 6%–8% for metals and 30%

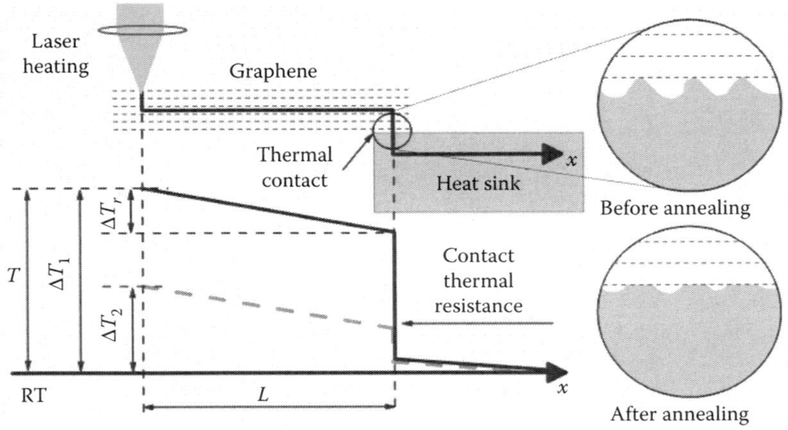

FIGURE 3.16 Scheme showing improvement of thermal contacts between FLG and metal contacts during laser annealing, and the resulting changes of temperature distribution over the sample. (The source of the material, V. A. Ermakov et al., Nonlocal laser annealing to improve thermal contacts between multi-layer graphene and metals, *Nanotechnology*, vol. 24, no. 15, p. 155301. Copyright 2013, IOP Publishing Ltd. is acknowledged.)

for SiO_2 (see Table 3.4). The thermal boundary conductance strongly increased as a result of laser annealing, and the highest TBC was observed for SiO_2. This finding proves that heat transfer by phonons from graphene to silicon oxide can be even more efficient than to metals, see also [71,73]. Based on these observations, two important conclusions can be made: (i) high-temperature ($T \geq 1000°C$) local annealing is necessary to achieve reasonable quality of contacts with metal surfaces that are usually relatively rough (although the degree of roughness may depend on the metal deposition method), (ii) high-quality thermal contacts between two nearly atomically flat surfaces (MLG and silicon oxide) can be achieved at lower temperatures ($T \sim 600°C$). The improvement of contact in the latter case can be attributed to removal of air pockets, solvent traces, and other possible impurities.

Further, using Equation 3.3 and corrections as discussed above, the thermal conductivity values were estimated for a number of MLG samples before and after annealing. For example, the values obtained were $K \sim 340$ W m^{-1} K^{-1} and 560 W m^{-1} K^{-1} for the samples shown in Figure 3.11a,b (400 nm and 1 μm wide, respectively, both ~20 nm thick), near room temperature. In calculations of K using Equation 3.3, thermal healing length L_H of 0.1 μm was added to L value, as discussed above. It is important to note that much lower values (not exceeding 150 W m^{-1} K^{-1}) were obtained for samples before laser annealing. The obtained data (after annealing) are comparable with other reported results for narrow graphene samples where K is reduced due to a strong effect of phonon scattering at the edges [66,84,88]. For instance, for narrow SLG nanoribbons supported over SiO_2 [50] K median values were measured to be ~80 W m^{-1} K^{-1} for samples 20–60 nm wide being much smaller than measured for graphene of larger size (≫0.3 μm) under the same conditions. The same trend was observed for FLG (3–5 layers) by Wang et al. [88], where K values were found to be much smaller for narrow samples, for both supported and suspended graphene, decreasing from 1250 W m^{-1} K^{-1} for 5 μm wide sample to 150 W m^{-1} K^{-1} (supported) or 170 W m^{-1} K^{-1} (suspended) for 1 μm wide samples. These data show the strong effect of the sample width on graphene thermal conductivity.

TABLE 3.4
Estimates of Thermal Boundary Conductance between MLG and Metals (W, Au) and SiO_2

	G (W m^{-2} K^{-1})		
Conditions	**W**	**Au**	**SiO_2**
Before annealing	3×10^6	2.5×10^6	6×10^7
	(6%)	(8%)	(30%)
After annealing[a]	5×10^7	3×10^7	2×10^8

Source: The source of the material, R. R. Nair et al., Fine structure constant defines visual transparency of graphene, *Science*, vol. 320, no. 5881, p. 1308, 2008. Copyright 2013 IOP Publishing Ltd. is acknowledged.

[a] Assuming perfect contact after annealing, data from References 71 and 73.

It is important to emphasize again that the main point of this study is to show the importance of good thermal contacts on the K measurements accuracy (that is frequently overlooked in the experiments).

In addition, we performed similar experiments with MWCNTs (diameters in the range of 20–30 nm, where the number of graphitic layers is about the same as in the experiments with MLG described above) deposited under the same conditions over metal (Au, W) electrodes with 1 μm gaps, followed by the same laser annealing procedure. Similar effect of thermal contacts improvement was observed, and K values up to ~1000 W m^{-1} K^{-1} were estimated after annealing, in good agreement with data reported for MWCNTs of 20–30 nm diameter by Fujii et al. [89], where K was found to decrease fast for increasing nanotube diameter, from $K \sim 2000$ W m^{-1} K^{-1} to $K \sim 500$ W m^{-1} K^{-1} at 10 nm and 28 nm diameter, respectively. The main difference between nanotubes and MLG in such measurements consists in the absence of edges for nanotubes where enhanced phonon scattering occurs. This comparison strengthens the hypothesis that the main reason for relatively small K values obtained here can be attributed mostly to the small size of flakes, rather than to other factors such as contaminants and graphene quality.

Finally, electrical two-terminal measurements performed before and after laser annealing, showed some reduction of the resistance for most of samples, with lowest electrical contact resistivity estimated to be at the order of ~1 kΩ μm^2 or higher after annealing [50]. However, in most cases this electrical contact reduction was much smaller as compared with the improvement obtained here for thermal contact resistance (up to an order of magnitude). This can be probably attributed to the fact that annealing was performed in air, resulting likely in the notable oxidation of metal (W and Ti) electrodes.

3.6 CONCLUSIONS

The formation of thermal contacts between MLG flakes and different metals (W, Ti, and Au) as well as with silicon oxide surfaces was studied here using confocal Raman spectroscopy. It was demonstrated that the accuracy of thermal conductivity measurements for suspended MLG flakes by an optothermal micro-Raman technique depends critically on the quality of thermal contacts between the flake and initially rough metal electrode surface. The same laser at higher power was used for the contacts improvement by annealing that occurs usually at temperatures above 900°C. After very short laser exposures (~1 s, with just a few mJ of laser energy deposited at a spot), compared with much longer periods of conventional thermal annealing (~1 h), considerable reduction of the sample heating by laser was observed. After improvement of the thermal contacts achieved by the laser annealing, we were able to perform more precise measurements of MLG thermal conductivity. The maximum values obtained for narrow MLG samples used here are close to 600 W m^{-1} K^{-1} (consistent with the literature data for flakes with submicron lateral dimensions) compared with much lower values <150 W m^{-1} K^{-1} obtained without annealing. This finding shows the importance of establishing good thermal contacts between graphene and metals. The improvement of thermal contacts to initially rough metal electrodes was attributed to local melting of the metal surface under laser heating (heat flow through the contact area) and increasing area of real contact and enhanced adhesion. Note that for the graphene–metal contact improvement, the localized heating by laser occurs in another (suspended) part of graphene, so that the contact annealing can be called remote or nonlocal. It is also interesting to note that improvement of thermal contacts was observed also between MLG and FLG with silicon oxide surface, for the latter the graphene–substrate heat transfer was found to be much higher, with the thermal boundary conductance estimated to be ~200 MWm^{-2} K^{-1} as compared to the metals such as Au and W used in the present work (~30–50 MW m^{-2} K^{-1}).

ACKNOWLEDGMENTS

The authors thank CNPq, FAPESP, and INCT NAMITEC for financial support, CCS-UNICAMP staff for technical assistance, and CTI Renato Archer, R. Savu (CCS-UNICAMP) and A. L. Gobbi (LNNano) for help in the preparation of samples.

REFERENCES

1. K. S. Novoselov, A. K. Geim, S. V. Morozov, D. Jiang, Y. Zhang, S. V. Dubonos, I. V. Grigorieva, and A. A. Firsov, Electric field effect in atomically thin carbon films, *Science*, vol. 306, no. 5696, pp. 666–669, 2004.
2. A. K. Geim, and K. S. Novoselov, The rise of graphene, *Nat. Mater.*, vol. 6, no. 3, pp. 183–191, 2007.
3. K. S. Novoselov, A. K. Geim, S. V. Morozov, D. Jiang, M. I. Katsnelson, I. V. Grigorieva, S. V. Dubonos, and A. A. Firsov, Two-dimensional gas of massless Dirac fermions in graphene, *Nature*, vol. 438, no. 7065, pp. 197–200, 2005.
4. M. H. Gass, U. Bangert, A. L. Bleloch, P. Wang, R. R. Nair, and A. K. Geim, Free-standing graphene at atomic resolution, *Nat. Nano*, vol. 3, no. 11, pp. 676–681, 2008.
5. F. Xia, V. Perebeinos, Y. Lin, Y. Wu, and P. Avouris, The origins and limits of metal-graphene junction resistance, *Nat. Nanotechnol.*, vol. 6, no. 3, pp. 179–184, 2011.
6. Y. K. Koh, M.-H. Bae, D. G. Cahill, and E. Pop, Heat conduction across monolayer and few-layer graphenes, *Nano Lett.*, vol. 10, no. 11, pp. 4363–4368, 2010.
7. K. L. Grosse, M.-H. Bae, F. Lian, E. Pop, and W. P. King, Nanoscale joule heating, peltier cooling and current crowding at graphene-metal contacts, *Nat. Nano*, vol. 6, no. 5, pp. 287–290, 2011.
8. A. A. Balandin, S. Ghosh, W. Bao, I. Calizo, D. Teweldebrhan, F. Miao, and C. N. Lau, Superior thermal conductivity of single-layer graphene, *Nano Lett.*, vol. 8, no. 3, pp. 902–907, 2008.
9. W. Cai, A. L. Moore, Y. Zhu, X. Li, S. Chen, L. Shi, and R. S. Ruoff, Thermal transport in suspended and supported monolayer graphene grown by chemical vapor deposition, *Nano Lett.*, vol. 10, no. 5, pp. 1645–1651, 2010.
10. C. Faugeras, B. Faugeras, M. Orlita, M. Potemski, R. R. Nair, and A. K. Geim, Thermal conductivity of graphene in corbino membrane geometry, *ACS Nano*, vol. 4, no. 4, pp. 1889–1892, 2010.
11. F. Xia, V. Perebeinos, Y. Lin, Y. Wu, and P. Avouris, The origins and limits of metal-graphene junction resistance, *Nat. Nanotechnol.*, vol. 6, no. 3, pp. 179–184, 2011.
12. S. Ghosh, W. Bao, D. L. Nika, S. Subrina, E. P. Pokatilov, C. N. Lau, and A. A. Balandin, Dimensional crossover of thermal transport in few-layer graphene, *Nat. Mater.*, vol. 9, no. 7, pp. 555–558, 2010.
13. I. Vlassiouk, S. Smirnov, I. Ivanov, P. F. Fulvio, S. Dai, H. Meyer, M. Chi, D. Hensley, P. Datskos, and N. V. Lavrik, Electrical and thermal conductivity of low temperature CVD graphene: The effect of disorder, *Nanotechnology*, vol. 22, no. 27, p. 275716, 2011.
14. J.-U. Lee, D. Yoon, H. Kim, S. W. Lee, and H. Cheong, Thermal conductivity of suspended pristine graphene measured by Raman spectroscopy, *Phys. Rev. B*, vol. 83, no. 8, p. 81419, 2011.
15. M. T. Pettes, I. Jo, Z. Yao, and L. Shi, Influence of polymeric residue on the thermal conductivity of suspended bilayer graphene, *Nano Lett.*, vol. 11, no. 3, pp. 1195–1200, 2011.
16. T. Schwamb, B. R. Burg, N. C. Schirmer, and D. Poulikakos, An electrical method for the measurement of the thermal and electrical conductivity of reduced graphene oxide nanostructures, *Nanotechnology*, vol. 20, no. 40, p. 405704, 2009.
17. A. Reina, X. Jia, J. Ho, D. Nezich, H. Son, V. Bulovic, M. S. Dresselhaus, and J. Kong, Large area, few-layer graphene films on arbitrary substrates by chemical vapor deposition, *Nano Lett.*, vol. 9, no. 1, pp. 30–35, 2008.
18. H. Zhang, and P. X. Feng, Fabrication and characterization of few-layer graphene, *Carbon N. Y.*, vol. 48, no. 2, pp. 359–364, 2010.
19. Z.-S. Wu, W. Ren, L. Gao, B. Liu, C. Jiang, and H.-M. Cheng, Synthesis of high-quality graphene with a pre-determined number of layers, *Carbon N. Y.*, vol. 47, no. 2, pp. 493–499, 2009.
20. G. Chen, W. Weng, D. Wu, C. Wu, J. Lu, P. Wang, and X. Chen, Preparation and characterization of graphite nanosheets from ultrasonic powdering technique, *Carbon N. Y.*, vol. 42, no. 4, pp. 753–759, 2004.
21. X. Li, X. Wang, L. Zhang, S. Lee, and H. Dai, Chemically derived, ultrasmooth graphene nanoribbon semiconductors, *Science*, vol. 319, no. 5867, pp. 1229–1232, 2008.
22. J. N. Coleman, Liquid-phase exfoliation of nanotubes and graphene, *Adv. Funct. Mater.*, vol. 19, no. 23, pp. 3680–3695, 2009.
23. X. Xu, L. F. C. Pereira, Y. Wang, J. Wu, K. Zhang, X. Zhao, S. Bae et al., Length-dependent thermal conductivity in suspended single-layer graphene, *Nat. Commun.*, vol. 5, p. 3689, 2014.
24. S. Chen, Q. Wu, C. Mishra, J. Kang, H. Zhang, K. Cho, W. Cai, A. A. Balandin, and R. S. Ruoff, Thermal conductivity of isotopically modified graphene, *Nat. Mater.*, vol. 11, no. 3, pp. 203–207, 2012.
25. S. Chen, A. L. Moore, W. Cai, J. W. Suk, J. An, C. Mishra, C. Amos et al., Raman measurements of thermal transport in suspended monolayer graphene of variable sizes in vacuum and gaseous environments, *ACS Nano*, vol. 5, no. 1, pp. 321–328, 2011.
26. J. Wang, L. Zhu, J. Chen, B. Li, and J. T. L. Thong, Suppressing thermal conductivity of suspended tri-layer graphene by gold deposition, *Adv. Mater.*, vol. 25, no. 47, pp. 6884–6888, 2013.
27. R. Murali, Y. Yang, K. Brenner, T. Beck, and J. D. Meindl, Breakdown current density of graphene nanoribbons, *Appl. Phys. Lett.*, vol. 94, no. 24, pp. 243113–243114, 2009.

28. N. K. Mahanta, and A. R. Abramson, Thermal conductivity of graphene and graphene oxide nanoplatelets, in *13th InterSociety Conference on Thermal and Thermomechanical Phenomena in Electronic Systems*, Sydney, Australia, 2012, pp. 1–6.
29. A. A. Balandin, S. Ghosh, W. Bao, I. Calizo, D. Teweldebrhan, F. Miao, and C. N. Lau, Superior thermal conductivity of single-layer graphene, *Nano Lett.*, vol. 8, no. 3, pp. 902–907, 2008.
30. Y. Hernandez, V. Nicolosi, M. Lotya, F. M. Blighe, Z. Sun, S. De, I. T. McGovern et al., High-yield production of graphene by liquid-phase exfoliation of graphite, *Nat. Nanotechnol.*, vol. 3, no. 9, pp. 563–568, 2008.
31. A. A. Green, and M. C. Hersam, Solution phase production of graphene with controlled thickness via density differentiation, *Nano Lett.*, vol. 9, no. 12, pp. 4031–4036, 2009.
32. M. Lotya, Y. Hernandez, P. J. King, R. J. Smith, V. Nicolosi, L. S. Karlsson, F. M. Blighe et al., Liquid phase production of graphene by exfoliation of graphite in surfactant/water solutions, *J. Am. Chem. Soc.*, vol. 131, no. 10, pp. 3611–3620, 2009.
33. Y. Hernandez, M. Lotya, D. Rickard, S. D. Bergin, and J. N. Coleman, Measurement of multicomponent solubility parameters for graphene facilitates solvent discovery, *Langmuir*, vol. 26, no. 5, pp. 3208–3213, 2009.
34. U. Khan, A. O'Neill, M. Lotya, S. De, and J. N. Coleman, High-concentration solvent exfoliation of graphene, *Small*, vol. 6, no. 7, pp. 864–871, 2010.
35. U. Khan, A. O'Neill, H. Porwal, P. May, K. Nawaz, and J. N. Coleman, Size selection of dispersed, exfoliated graphene flakes by controlled centrifugation, *Carbon N. Y.*, vol. 50, no. 2, pp. 470–475, 2012.
36. M. Yi, Z. Shen, S. Ma, and X. Zhang, A mixed-solvent strategy for facile and green preparation of graphene by liquid-phase exfoliation of graphite, *J. Nanoparticle Res.*, vol. 14, no. 8, pp. 1–9, 2012.
37. A. O'Neill, U. Khan, P. N. Nirmalraj, J. Boland, and J. N. Coleman, Graphene dispersion and exfoliation in low boiling point solvents, *J. Phys. Chem. C*, vol. 115, no. 13, pp. 5422–5428, 2011.
38. C. Panayiotou, Redefining solubility parameters: The partial solvation parameters, *Phys. Chem. Chem. Phys.*, vol. 14, no. 11, pp. 3882–3908, 2012.
39. A. V. Alaferdov, A. Gholamipour-Shirazi, M. A. Canesqui, Y. A. Danilov, and S. A. Moshkalev, Size-controlled synthesis of graphite nanoflakes and multi-layer graphene by liquid phase exfoliation of natural graphite, *Carbon N. Y.*, vol. 69, pp. 525–535, 2014.
40. S. E. Skrabalak, Ultrasound-assisted synthesis of carbon materials, *Phys. Chem. Chem. Phys.*, vol. 11, no. 25, pp. 4930–4942, 2009.
41. G. Cravotto, and P. Cintas, Sonication-assisted fabrication and post-synthetic modifications of graphene-like materials, *Chem. Eur. J.*, vol. 16, no. 18, pp. 5246–5259, 2010.
42. K. S. Suslick, and D. J. Flannigan, Inside a collapsing bubble: Sonoluminescence and the conditions during cavitation, *Annu. Rev. Phys. Chem.*, vol. 59, pp. 659–683, 2008.
43. T. J. Mason, *Practical Sonochemistry: User's Guide to Applications in Chemistry and Chemical Engineering*. New York; London; Toronto [etc.]: Ellis Horwood, 1991.
44. A. N. Kolmogorov, No Title, in *The Proceedings of the USSR Academy of Sciences*, vol. 31, no. 2, pp. 99–101, 1941.
45. M. J. Madou, *Fundamentals of Microfabrication: The Science of Miniaturization*, vol. 29, no. 14. CRC Press, Boca Raton, Florida, 2002, p. 723.
46. I. Utke, S. A. Moshkalev, and P. Russell, *Nanofabrication Using Focused Ion and Electron Beams: Principles and Applications*, Oxford University Press, New York, New York, 2012.
47. R. Krupke, F. Hennrich, H. B. Weber, M. M. Kappes, H. V. Löhneysen, and H. Lo, Simultaneous deposition of metallic bundles of single-walled carbon nanotubes using ac-dielectrophoresis, *Nano Lett.*, vol. 3, no. 8, pp. 1019–1023, 2003.
48. S. A. Moshkalev, J. Leon, C. Veríssimo, A. R. Vaz, A. Flacker, M. B. De Moraes, and J. W. Swart, Controlled deposition and electrical characterization of multi-wall carbon nanotubes, *J. Nano Res.*, vol. 3, no. 3, pp. 25–32, 2008.
49. K. Nagashio, T. Nishimura, K. Kita, and A. Toriumi, Contact resistivity and current flow path at metal/graphene contact, *Appl. Phys. Lett.*, vol. 97, no. 14, pp. 143513–143514, 2010.
50. F. P. Rouxinol, R. V. R. V. Gelamo, R. G. Amici, A. R. Vaz, and S. A. Moshkalev, Low contact resistivity and strain in suspended multilayer graphene, *Appl. Phys. Lett.*, vol. 97, no. 25, p. 253104, 2010.
51. W. J. Evans, L. Hu, and P. Keblinski, Thermal conductivity of graphene ribbons from equilibrium molecular dynamics: Effect of ribbon width, edge roughness, and hydrogen termination, *Appl. Phys. Lett.*, vol. 96, no. 20, pp. 203112–203113, 2010.
52. F. Léonard, A. A. Talin, and F. Leonard, Electrical contacts to one- and two-dimensional nanomaterials, *Nat. Nano*, vol. 6, no. 12, pp. 773–783, 2011.
53. G. Chen, *Nanoscale Energy Transport and Conversion*. Oxford University Press, New York, New York, 2005, p. 531.
54. E. T. Swartz, and R. O. Pohl, Thermal boundary resistance, *Rev. Mod. Phys.*, vol. 61, no. 3, pp. 605–668, 1989.
55. P. M. Norris, J. L. Smoyer, J. C. Duda, and P. E. Hopkins, Prediction and measurement of thermal transport across interfaces between isotropic solids and graphitic materials, *J. Heat Transfer*, vol. 134, no. 2, p. 20910, 2011.
56. R. Prasher, Acoustic mismatch model for thermal contact resistance of van der Waals contacts, *Appl. Phys. Lett.*, vol. 94, no. 4, pp. 41903–41905, 2009.
57. T. Beechem, J. C. Duda, P. E. Hopkins, and P. M. Norris, Contribution of optical phonons to thermal boundary conductance, *Appl. Phys. Lett.*, vol. 97, no. 6, pp. 61903–61907, 2010.
58. R. Nicklow, N. Wakabayashi, and H. G. Smith, Lattice dynamics of pyrolytic graphite, *Phys. Rev. B*, vol. 5, no. 12, pp. 4951–4962, 1972.
59. C. V. Pandya, P. R. Vyas, T. C. Pandya, N. Rani, and V. B. Gohel, An improved lattice mechanical model for FCC transition metals, *Phys. B Condens. Matter*, vol. 307, no. 1–4, pp. 138–149, 2001.
60. R. Sikora, Ab initio study of phonons in the rutile structure of TiO_2, *J. Phys. Chem. Solids*, vol. 66, no. 6, pp. 1069–1073, 2005.
61. M. E. Striefler, and G. R. Barsch, Lattice dynamics of α-quartz, *Phys. Rev. B*, vol. 12, no. 10, pp. 4553–4566, 1975.
62. R. L. Plackett, Karl Pearson and the Chi-Squared test, *Int. Stat. Rev./Rev. Int. Stat.*, vol. 51, no. 1, pp. 59–72, 1983.
63. S. P. Koenig, N. G. Boddeti, M. L. Dunn, and J. S. Bunch, Ultrastrong adhesion of graphene membranes, *Nat. Nano*, vol. 6, no. 9, pp. 543–546, 2011.
64. T. Yoon, W. C. Shin, T. Y. Kim, J. H. Mun, T.-S. Kim, and B. J. Cho, Direct measurement of adhesion energy of monolayer graphene as-grown on copper and its application to renewable transfer process, *Nano Lett.*, vol. 12, no. 3, pp. 1448–1452, 2012.

65. J. Lahiri, T. S. Miller, A. J. Ross, L. Adamska, I. I. Oleynik, and M. Batzill, Graphene growth and stability at nickel surfaces, *New J. Phys.*, vol. 13, no. 2, p. 25001, 2011.
66. E. Pop, V. Varshney, and A. K. Roy, Thermal properties of graphene: Fundamentals and applications, *MRS Bull.*, vol. 37, no. 12, pp. 1273–1281, 2012.
67. B. N. J. Persson, A. I. Volokitin, and H. Ueba, Phononic heat transfer across an interface: Thermal boundary resistance, *J. Phys. Condens. Matter*, vol. 23, no. 4, p. 45009, 2011.
68. B. N. J. Persson, B. Lorenz, and A. I. Volokitin, Heat transfer between elastic solids with randomly rough surfaces, *Eur. Phys. J. E*, vol. 31, no. 1, pp. 3–24, 2010.
69. J. Luthin, and C. Linsmeier, Carbon films and carbide formation on tungsten, *Surf. Sci.*, vol. 454–456, no. 0, pp. 78–82, 2000.
70. P. E. Hopkins, M. Baraket, E. V. Barnat, T. E. Beechem, S. P. Kearney, J. C. Duda, J. T. Robinson, and S. G. Walton, Manipulating thermal conductance at metal–graphene contacts via chemical functionalization, *Nano Lett.*, vol. 12, no. 2, pp. 590–595, 2012.
71. A. J. Schmidt, K. C. Collins, A. J. Minnich, and G. Chen, Thermal conductance and phonon transmissivity of metal–graphite interfaces, *J. Appl. Phys.*, vol. 107, no. 10, p. 104907, 2010.
72. J. Hirotani, T. Ikuta, T. Nishiyama, and K. Takahashi, Thermal boundary resistance between the end of an individual carbon nanotube and a Au surface, *Nanotechnology*, vol. 22, no. 31, p. 315702, 2011.
73. Z. Chen, W. Jang, W. Bao, C. N. Lau, and C. Dames, Thermal contact resistance between graphene and silicon dioxide, *Appl. Phys. Lett.*, vol. 95, no. 16, p. 161910, 2009.
74. H. Maune, H.-Y. Chiu, and M. Bockrath, Thermal resistance of the nanoscale constrictions between carbon nanotubes and solid substrates, *Appl. Phys. Lett.*, vol. 89, no. 1, pp. 13103–13109, 2006.
75. V. A. Ermakov, A. V. Alaferdov, A. R. Vaz, A. V. Baranov, and S. A. Moshkalev, Nonlocal laser annealing to improve thermal contacts between multi-layer graphene and metals, *Nanotechnology*, vol. 24, no. 15, p. 155301, 2013.
76. I. Calizo, F. Miao, W. Bao, C. N. Lau, and A. A. Balandin, Variable temperature Raman microscopy as a nanometrology tool for graphene layers and graphene-based devices, *Appl. Phys. Lett.*, vol. 91, no. 7, p. 71913, 2007.
77. S. Osswald, E. Flahaut, H. Ye, and Y. Gogotsi, Elimination of D-band in Raman spectra of double-wall carbon nanotubes by oxidation, *Chem. Phys. Lett.*, vol. 402, no. 4–6, pp. 422–427, 2005.
78. P. Tan, Y. Deng, Q. Zhao, and W. Cheng, The intrinsic temperature effect of the Raman spectra of graphite, *Appl. Phys. Lett.*, vol. 74, no. 13, pp. 1818–1820, 1999.
79. R. R. Nair, P. Blake, A. N. Grigorenko, K. S. Novoselov, T. J. Booth, T. Stauber, N. M. R. Peres, and A. K. Geim, Fine structure constant defines visual transparency of graphene, *Science*, vol. 320, no. 5881, p. 1308, 2008.
80. A. A. Balandin, Thermal properties of graphene and nanostructured carbon materials, *Nat. Mater.*, vol. 10, no. 8, pp. 569–581, 2011.
81. C.-L. Tsai, A. Liao, E. Pop, and M. Shim, Electrical power dissipation in semiconducting carbon nanotubes on single crystal quartz and amorphous SiO_2, *Appl. Phys. Lett.*, vol. 99, no. 5, p. 53120, 2011.
82. G. H. Han, S. J. Chae, E. S. Kim, F. Gu¨nes¸, I. H. Lee, S. W. Lee, S. Y. Lee et al., Laser thinning for monolayer graphene formation: Heat sink and interference effect, *ACS Nano*, vol. 5, no. 1, pp. 263–268, 2010.
83. Y. K. Koh, M.-H. Bae, D. G. Cahill, and E. Pop, Reliably counting atomic planes of few-layer graphene ($n > 4$), *ACS Nano*, vol. 5, no. 1, pp. 269–274, 2010.
84. L. S. K. Pang, J. D. Saxby, and S. P. Chatfield, Thermogravimetric analysis of carbon nanotubes and nanoparticles, *J. Phys. Chem.*, vol. 97, no. 27, pp. 6941–6942, 1993.
85. A. D. Liao, J. Z. Wu, X. Wang, K. Tahy, D. Jena, H. Dai, and E. Pop, Thermally limited current carrying ability of graphene nanoribbons, *Phys. Rev. Lett.*, vol. 106, no. 25, p. 256801, 2011.
86. L. G. Cancado, K. Takai, T. Enoki, M. Endo, Y. A. Kim, H. Mizusaki, A. Jorio, L. N. Coelho, R. Magalhaes-Paniago, and M. A. Pimenta, General equation for the determination of the crystallite size L[sub a] of nanographite by Raman spectroscopy, *Appl. Phys. Lett.*, vol. 88, no. 16, pp. 163103–163106, 2006.
87. I.-K. Hsu, M. T. Pettes, M. Aykol, L. Shi, and S. B. Cronin, The effect of gas environment on electrical heating in suspended carbon nanotubes, *J. Appl. Phys.*, vol. 108, no. 8, p. 84307, 2010.
88. Z. Wang, R. Xie, C. T. Bui, D. Liu, X. Ni, B. Li, and J. T. L. Thong, Thermal transport in suspended and supported few-layer graphene, *Nano Lett.*, vol. 11, no. 1, pp. 113–118, 2010.
89. M. Fujii, X. Zhang, H. Xie, H. Ago, K. Takahashi, T. Ikuta, H. Abe, and T. Shimizu, Measuring the thermal conductivity of a single carbon nanotube, *Phys. Rev. Lett.*, vol. 95, no. 6, p. 65502, 2005.

4 Strain Effect on the Electronic Spectrum of Graphene
Beyond Two Dimensionality

F. M. D. Pellegrino, G. G. N. Angilella, and R. Pucci

CONTENTS

ABSTRACT

We study the dependence of the plasmon dispersion relation of graphene on applied uniaxial strain. Besides electron correlation at the random-phase approximation level, we also include local field effects specific for the honeycomb lattice. As a consequence of the two-band character of the electronic band structure, we find two distinct plasmon branches. We recover the square-root behavior of the low-energy branch, and find a nonmonotonic dependence of the strain-induced modification of its stiffness, as a function of the wave-vector orientation with respect to applied strain. We also take into account the full three-dimensional representation for the wave functions of single-particle excitations. This induces quantitative changes in the plasmon dispersion relation.

4.1 BACKGROUND AND OUTLINE

Graphene is an atomically thick single layer of carbon atoms arranged according to a honeycomb lattice. Its quite recent discovery and the realization of sufficiently large graphene flakes in the laboratory have stimulated an enormous outburst of both experimental and theoretical investigation [1,2]. Most of its remarkable electronic and transport properties can be related to its two-dimensional (2D) character, as well as to its lattice symmetry. These include a linear low-energy dispersion relation [3], a minimal, finite conductivity in the clean limit at zero temperature [3], and a nearly constant optical conductivity over a large interval of frequencies [4,5].

Graphene is also notable for its remarkable mechanical properties. In particular, recent *ab initio* calculations [6] as well as experiments [7] have demonstrated that graphene single layers can reversibly sustain elastic deformations as large as 20%. In microelectronics, the effect of strain is often used to modify the electronic and transport properties of materials to improve the performance of the devices [8]. In graphene, the application of strain (e.g., by stretching [9] or bending [7]) allows the tuning of its electronic properties [10,11].

Recently, there has been a great interest toward the study and realization of graphene-based electronic devices designed by a suitable tailoring of the electronic structure exploiting not only the electric field effect but also the applied strain. Both these techniques allow to tune the electronic properties of graphene in a reversible and clean way, that is, without adding any source of disorder.

Therefore, an in-depth knowledge of the effects of strain on graphene could be exploited to improve graphene-based devices. In this chapter, we study in detail the influence that applied strain can have on the plasmonic properties of graphene. Owing to its low dimensionality and the large mean-free path, graphene is an interesting material for applications in plasmonics [12]. Hence, the application of strain for tuning the plasmonic spectrum could have worthwhile technological implications.

Starting from the tight-binding model of graphene under applied uniaxial strain [13,14], we focus on the electronic properties at low energy [14,15]. In particular, a general correspondence has been derived between linear response correlation functions in graphene with and without applied uniaxial strain. We have analytically studied the dependence on the strain modulus and direction of selected electronic properties, such as the plasmonic dispersion relation, the optical conductivity, as well as the static, magnetic, and electric susceptibilities. Specifically, we have considered how the uniaxial strain can change the dispersion of the recently predicted transverse collective excitation that exhibits an anisotropic deviation from the photonic behavior [16], thus facilitating its experimental detection in strained graphene samples.

4.2 MAIN PHYSICAL PROPERTIES OF GRAPHENE

The enormous outburst of both experimental and theoretical investigation of graphene has been fueled, mainly, by numerous remarkable properties that make graphene an ideal candidate for applications in nanotechnologies. In this section, we list several properties that make graphene such an attractive material.

4.2.1 Transport Properties

Owing to its lattice symmetry, graphene is a zero-gap semiconductor and is characterized by a low-energy linear dispersion relation [3]. In other words, the low-energy quasiparticles are massless and are characterized by an energy-independent effective velocity. The dynamics of the low-energy excitations is described by the Dirac–Weyl equation, which is used for massless fermions [17].

Intrinsic graphene disposes of no charge carriers. However, using electric doping, it is possible to have either electrons or holes as charge carriers. Exploiting the electric field effect, a gate electrode can continuously change both the carrier density and type [1].

An attractive feature of graphene is its high carrier mobility at room temperature. Mobility in graphene on SiO_2 is generally of the order of $10{,}000 \div 15{,}000$ cm^2 V^{-1} s^{-1} [18]. Moreover, mobility on SiO_2 is almost constant at low temperature ($T = 300$ K) and is limited by disorder. Disorder in exfoliated graphene on SiO_2 is mainly due to the charges trapped on the surface of the substrate or adsorbed on graphene. At high temperature $T \approx 300$ K, most properties of graphene are strongly dependent on temperature because of the optical phonons of the substrate [19].

In the case of suspended graphene, and removing the substrate, one gets rid of an extrinsic disorder and so, mobility increases of an order of magnitude with respect to graphene on the substrate. In suspended graphene, it is possible to measure a mobility as high as 250,000 cm^2 V^{-1} s^{-1} at low temperature and 120,000 cm^2 V^{-1} s^{-1} at 240 K [20,21]. Unlike in graphene on the substrate, the mobility in suspended graphene is strongly temperature dependent also at low temperature [20]. These large values of the mobility mean that in suspended graphene, the mean-free path is of the order of 1 μm, which is comparable to the dimension of a typical device [20,22].

For comparison, we consider the modulation-doped field-effect transistor (MODFET). It is based on an heterostructure (e.g., AlGaAs and GaAs), where the wide energy gap material (e.g., AlGaAs) is doped and carriers diffuse to the intrinsic narrow bandgap layer (e.g., GaAs), at whose interface, a 2D electron gas (channel) is formed [23]. The physical distance from the channel and dopants allows to obtain a high mobility. This methodology is called modulation doping and it was invented by Horst Stormer at Bell Labs [24]. Modulation doping represents the best technique to obtain a very large value of electron mobility in a bulk system. At cryogenic temperatures, it is possible to reach a mobility of the order of 10^6 cm^2 V^{-1} s^{-1}, but at temperatures above $T \approx 80$ K, the mobility of these systems drops to values of the order of 10^4 cm^2 V^{-1} s^{-1} [23]. Comparing the mobility as a function of the temperature for graphene on SiO_2 [25] and for a MODFET [23], one can see that at room temperature, the mobility of graphene can be larger than that of a MODFET.

Owing to the gapless energy spectrum, low carrier density, and atomic thickness, it is possible to modify the profile of carriers along a graphene layer exploiting the electric field effect. For example, it is possible to realize a *p–n* or a *p–n–p* multipolar configuration by electrostatic gates [26]. Nam et al. [27] have realized such a *p–n–p* structure, where the back gate (V_{BG}) is responsible for the electric doping in the *p*-regions whereas the local gate (V_{LG}) is responsible for the electric doping in the *n*-region. In the experiment by Nam et al. [27], the device is made of a highly doped Si wafer (back gate), an insulating SiO_2 layer where embedded inside, there is a polysilicon layer (local gate) that is conductive by implantation of phosphorus ions. Finally, above the substrate, there is a graphene monolayer with metal electrodes (Ti/Au). The local gate has a width of 130 nm, which is comparable to the mean-free path of the sample. Indeed, Nam et al. have found ballistic and phase-coherent carrier transport.

Looking at the conductivity as a function of the local-gate V_{LG} at a fixed back-gate V_{BG}, one observes an oscillating behavior due to the Fabry–Pérot interference between the two *p–n* interfaces. Moreover, there is an important interest toward the theoretical study of the spectra and the electronic transport through differently doped regions, whose behavior differs from that of conventional 2D electron gases [28,29]. Recently, Rossi et al. [30] used a microscopic model where the disorder is dominated by charge impurities and transport properties are obtained fully quantum mechanically. In particular, they have studied the effects of disorder on transport through p–n–p junctions. The crossover from the ballistic transport governed by Klein tunneling, to the disordered diffusive transport is found to take place as the mean-free path becomes of the order of the distance between the two p–n interfaces consistent with the experiments. These results demonstrate that the signatures of coherent transport are observable for impurity densities as high as 10^{12} cm^{-2}, and then, the quantum transport properties are sufficiently robust in graphene [31].

4.2.2 Optical Properties

In addition to a high charge mobility, graphene is characterized by excellent optical transparency. Such properties make graphene an attractive material for photonics and optoelectronic devices such as displays, touch screen, light-emitting diodes (LEDs), and solar cells [32].

Current transparent conductors are semiconductor based: doped indium oxide (In_2O_3), zinc oxide (ZnO), tin oxide (SnO_2), or ternary compounds based on their combinations [32]. The most widely used material is indium tin oxide (ITO), a doped n-type semiconductor composed of ≈90% In_2O_3 and ≈10% SnO_2 [33]. ITO is commercially available with transmittance $T \approx 80\%$. Moreover, ITO is brittle for applications involving bending, such as touch screens and flexible displays. For this reason, there is interest in the research of new transparent conductor materials with improved performance. Metal grids, metallic nanowires, or other metal oxides have been explored as alternatives. Nanotubes and graphene are promising materials. Usually, graphene films have a higher trasmittance over a wider wavelength range than single-walled carbon nanotube (SWNT) films, thin metallic films, and ITO [32].

Despite its thickness, Nair et al. have observed that graphene absorbs a significant fraction ($\pi\alpha \approx 2.3\%$, where α is the fine structure constant) of incident light, from the near infrared (IR) to violet [34]. Within this experimental range, Nair et al. [34] find a constant value for the transmittance ($\approx 1 - \pi\alpha$).

4.2.3 Plasmonic Properties

It has been already said that it is possible to modify the type and density of charge carriers in graphene using an external voltage. This feature can be effectively exploited in plasmonics. A doped graphene monolayer can sustain low-energy plasmons that are tunable by means of the electric field effect. In particular, plasmons in doped graphene enable low losses and significant wave localization of the light in the terahertz (THz) and IR domains [35]. These properties make graphene relevant for possible applications in plasmonics.

The recent attraction toward plasmonics is immediately motivated by the constant effort toward improving the performance of devices. A limitation to an improvement of the speed of digital circuits is due to electronic interconnections. A possible solution is offered by photonics by implementing faster communication systems based on optical fibers and photonic circuits [36]. However, the replacement of electric circuits by photonic ones is hindered by the low level of integration and miniaturization of the photonic components. The wavelength of light used in photonic circuits is of the order of 1000 nm; hence, it is larger than the typical dimensions of an electronic circuit. Thus, if the dimensions of the optical components should be reduced further and become comparable with the wavelength of light, propagation would be obstructed by optical diffraction. One way to avoid this obstacle is suggested by plasmonics. Surface plasmons enable the confining of light within very small dimensions, as electromagnetic waves can be trapped near the surface due to their interaction with the electron plasma. Hence, the idea is to use plasmonic guides instead of optical fibers.

To date, the noble metals are the materials mainly investigated for developments in plasmonics, but they are hardly tunable and have large ohmic losses that limit their applicability. In graphene, both characteristics are improved, and the confinement of plasmons is much stronger than that of surface plasmons in metals due to the 2D nature of graphene. In particular, graphene plasmons are confined to volumes $\approx 10^6$ times smaller than the diffraction limit, thus facilitating strong light–matter interactions [12].

In aluminum, which is a relatively absorbing metal, the propagation length is 2 mm at a wavelength of 500 nm, whereas in silver, which is a low-loss metal, at the same wavelength, the propagation length is 20 mm. For slightly longer wavelengths, such as 1.55 mm, the propagation length is around 1 mm [37], whereas in graphene, the propagation distance can reach values well above 100-plasmon wavelengths [12,35].

4.2.4 Mechanical Properties

In addition to its electronic properties, graphene is a remarkable material also for its mechanical properties. Generally, carbon nanostructures characterized by sp^2 bonds, such as carbon nanotubes, show an exceptional resistance to mechanical stress, notwithstanding low dimensionality [38]. Lee et al. have measured the mechanical properties of graphene using atomic force microscope (AFM) nanoindentation, and this technique has been used to study a single layer suspended over an aperture of a substrate [39]. The experimental apparatus consists of an array of circular holes in a substrate, on top of which graphene monolayers have been deposited. Once a graphene sample placed over a hole was detected, the mechanical properties of the suspended membrane have been measured by indenting with AFM. The measured breaking strength of graphene is 42 N m^{-1}. To compare the mechanical properties of graphene with those of other three-dimensional materials, Lee et al. considered a graphene sheet as a three-dimensional slab having an effective height equal to the distance between two adjacent graphene planes in graphite ($h = 3.35$ Å). Thus, Young's modulus of graphene is $E = 1.0$ TPa and the third-order elastic stiffness is $D = -2.0$ TPa. These values allow the listing of graphene among the strongest materials ever measured [39].

Another methodology to study graphene under strain is to transfer it onto a flexible substrate, so that one can apply controllable (uniaxial or biaxial) strain to graphene by applying stress on the supporting substrate [7,40]. Exploiting this technique, Kim et al. have found that graphene can reversibly sustain elastic deformations as large as 20% [7]. Theoretical results are in agreement with these measurements. Indeed, according to *ab initio* calculations, the graphene lattice is stable with respect to uniaxial deformations up to around 20% [6].

Graphene behaves as an impermeable membrane and can support pressure differences larger than 1 atm. Exploiting this property, it is possible to deform graphene [41,42]. In the experimental setup, the impermeability of graphene is used. There is a graphene layer suspended over a well in an SiO_2 substrate. The graphene membrane is clamped to the substrate through the Van der Waals interaction. Inside the well, there is a gas at pressure P_{int}, whereas outside the pressure, P_{ext} is different; so, the difference of pressure ($\Delta P = P_{int} - P_{ext}$) allows to have a controlled deformation of graphene. A similar methodology to have a desired amount of strain in graphene is obtained by using a gate. In this case, an electric field induces an electron concentration in graphene and exerts on it a pressure of electrostatic nature [43].

Uniaxial or biaxial strain induces modifications not only in the phonon spectrum but also in the electronic spectrum, and can be measured directly using Raman spectroscopy [40,42]. In graphene, Raman measurements give information not only about phonons but also about the electronic properties since graphene is a nonpolar crystal, and so, Raman scattering involves electronic excitations as intermediate states [44].

Material science teaches that the presence of strain can significantly affect device performance. Indeed, sometimes, strain is intentionally applied to improve mobility, as in the strained silicon technology, which is used in modern microelectronics [45]. Recently, an appealing challenge is to exploit the modifications of the electronic structure due to the strain to realize an all-graphene circuit where all the elements are made of graphene with different amounts and types of strain [42,46]. Further methodologies to accomplish this challenge and to have a controlled strain profile in a graphene sample are obtained by means of an appropriate geometrical pattern in a homogeneous substrate, or by means of a heterogeneous substrate so that each region interacts with graphene in a different way [47–49]. Among the experimental methodologies to realize strain superstructures in graphene, one is based on the relatively large and negative thermal coefficient of graphene (which is around 5 times larger than that of bulk graphite in the basal plane). Bao et al. [50] have experimentally realized a strain superlattice in graphene, and it is possible to manipulate the orientation and dimensions of ripples exploiting the boundary conditions and the difference in the thermal expansion coefficients between graphene and the substrate. The graphene membrane is annealed up to 700 K; so, any preexisting ripple disappears. After this phase, the sample is cooled and the graphene layer exhibits ordered ripples, whose geometry depends on the boundary conditions. In particular, the necessary tension to produce this structure is due to the different sign of the thermal coefficients of graphene and the substrate [50].

Such recent ideas to exploit mechanical modifications to realize an all-graphene device are attractive as strain engineering would allow the tailoring of electronic properties, in a controlled manner, without the introduction of disorder [46].

4.3 STRAIN EFFECT ON THE PLASMONIC SPECTRUM

Most of the electronic properties of graphene are encoded in the electron polarization, which has been studied within the Dirac cone approximation at zero [51] and finite temperature [52] for pristine graphene, as well as for doped graphene [53,54]. These results have been recently extended beyond the Dirac cone approximation [55].

In this chapter, we are concerned with the dynamical polarization of graphene within the full first Brillouin zone (1BZ) of the honeycomb lattice. While electron correlations are treated at the level of the random-phase approximation (RPA), we explicitly include local field effects (LFEs) [56], which are characteristic of the lattice structure of graphene. The importance of LFE has been shown to be more important in graphene than in bulk semiconductors, in connection with the static dielectric properties of graphene [57,58]. By discussing the singularities of the polarization, we can identify the longitudinal collective modes of the correlated electron liquid. We are mainly interested in the plasmon modes, which dominate the long-wavelength charge density fluctuations. The role of electron–plasmon interaction in renormalizing the (especially low-energy) quasiparticle dispersion relation has been emphasized [59,60], and plasmons in graphene are potentially interesting for applications in nanophotonics [35].

Specifically, we are interested in the dependence of the plasmon modes on applied uniaxial strain. To this aim, we use the tight-binding model modified under strain, which we have presented in References 13 and 14. Despite its simplicity, the tight-binding model is successful because it is closely related to the symmetry properties of graphene. In particular, the tight-binding approximation allows to include important features of the electronic band dispersion, such as a finite bandwidth and the occurrence of Van Hove singularities These features play an essential role in deriving some of the characteristics of the plasmon dispersion.

4.4 LFEs ON THE ELECTRON POLARIZATION

Within linear response theory, plasmon modes can be described as poles of the density–density correlation function, that is, the polarization. The RPA is then the simplest, infinite-order, diagrammatic procedure to include electron correlations in the dielectric screening giving rise to the polarization [61]. Besides electron–electron correlations, another source of k-space dependence of the dielectric function is provided by LFEs [62]. This is due to the generally atomic consistence of matter and, in the case of solids, to the periodicity of the crystalline lattice. An account of the LFE on the dielectric function of crystalline solids dates back at least to the original paper of Adler [56,63,64], and is generalized below to the case of graphene, including both valence and conduction bands.

We start by considering the polarization, which for a noninteracting system at finite temperature T reads

$$\begin{aligned}\Pi^0_{\rho\rho}(\mathrm{x},\mathrm{x}',i\omega_m)\\ &= \frac{1}{\hbar^2\beta}\sum_{i\omega_n}\sum_{\mathrm{k}\lambda\mathrm{k}'\lambda'}\psi^*_{\mathrm{k}\lambda}(\mathrm{x}')\mathcal{G}^0_\lambda(\mathrm{k},i\omega_n)\psi_{\mathrm{k}\lambda}(\mathrm{x})\\ &\times\psi^*_{\mathrm{k}'\lambda'}(\mathrm{x})\mathcal{G}^0_{\lambda'}(\mathrm{k}',i\omega_n+i\omega_m)\psi_{\mathrm{k}'\lambda'}(\mathrm{x}')\end{aligned} \quad (4.1)$$

where $\psi_{\mathrm{k}\lambda}(\mathrm{x})$ is the 2D eigenfunction, $\mathcal{G}^0_\lambda(\mathrm{k},i\omega_n)=(i\omega_n-\xi_{\mathrm{k}\lambda}/\hbar)^{-1}$ is Green's function for the noninteracting system, and $\hbar\omega_n=(2n+1)\pi k_B T$ $[\hbar\omega_m=2m\pi\ k_B T]$ denote the fermionic (bosonic) Matsubara frequencies at temperature T, with $\hbar$ Planck's constant and k_B Boltzmann's constant. In treating systems at finite temperatures, it is convenient to use the grand canonical ensemble [65]. Hence, we use the single-particle energy $\xi_{\mathrm{k}\lambda}$, as a natural variable, which is defined as $\xi_{\mathrm{k}\lambda}=E_{\mathrm{k}\lambda}-\mu$, where μ is the chemical potential, and $E_{\mathrm{k}\lambda}$ is the electronic dispersion relation where $\lambda=1$ refers to the valence band and $\lambda=2$ refers to the conduction band. Fourier transforming into momentum space (Equation 4.1), and performing the summation over the Matsubara frequencies, one finds

$$\begin{aligned}\Pi^0_{\rho\rho}(\mathrm{q}+\mathrm{G},-\mathrm{q}'-\mathrm{G}',i\omega_m)\\ &=(2\pi)^2A^{-1}_{cell}\delta(\mathrm{q}-\mathrm{q}')\times\frac{1}{N}\sum_{\mathrm{k}\lambda\lambda'}T_{\mathrm{k}\lambda,\mathrm{k}-\mathrm{q}\lambda'}(i\omega_m)\\ &\times\langle\mathrm{k}-\mathrm{q}\lambda'|e^{-i(\mathrm{q}+\mathrm{G})\cdot\hat{\mathrm{r}}}|\mathrm{k}\lambda\rangle\langle\mathrm{k}\lambda|e^{i(\mathrm{q}+\mathrm{G}')\cdot\hat{\mathrm{r}}}|\mathrm{k}-\mathrm{q}\lambda'\rangle\end{aligned} \quad (4.2)$$

where

$$T_{\mathrm{k}\lambda,\mathrm{k}-\mathrm{q}\lambda'}(i\omega_m)=\frac{n_F(\xi_{\mathrm{k}-\mathrm{q}\lambda'})-n_F(\xi_{\mathrm{k}\lambda})}{i\hbar\omega_m+\xi_{\mathrm{k}-\mathrm{q}\lambda'}-\xi_{\mathrm{k}\lambda}} \quad (4.3)$$

Here, $n_F(\omega)$ is the Fermi function, $A_{cell}=3\sqrt{3}a^2/2$ is the area of the unit cell, **q**, **q′** belong to the 1BZ, **G**, **G′** are vectors of the reciprocal lattice, and LFE is embedded in Adler's weights [56]

$$\begin{aligned}\langle\mathrm{k}-\mathrm{q}\lambda'|e^{-i(\mathrm{q}+\mathrm{G})\cdot\hat{\mathrm{r}}}|\mathrm{k}\lambda\rangle &= \int d^2\mathrm{x}\,e^{-i(\mathrm{q}+\mathrm{G})\cdot\mathrm{x}}\psi_{\mathrm{k}\lambda}(\mathrm{x})\psi^*_{\mathrm{k}-\mathrm{q}\lambda'}(\mathrm{x})\\ &\simeq\frac{1}{2}\left[(-1)^{\lambda-\lambda'}+e^{i(\theta_{\mathrm{k}-\mathrm{q}}-\theta_{\mathrm{k}})-i\mathrm{G}\cdot\delta_3}\right]e^{-\sigma_g^2|\mathrm{q}+\mathrm{G}|^2/4}\end{aligned} \quad (4.4)$$

where in the last line, only the on-site overlap between pairs of atomic orbitals [13,66], centered on either sublattices, has been retained; on account of their localized character, we have retained only the lowest (zeroth) order contributions in the overlap function g_{k}, and $e^{i\theta_{\mathrm{k}}}=-f_{\mathrm{k}}/|f_{\mathrm{k}}|$. Using a more compact notation, one may also write

$$\begin{aligned}\Pi^0_{\rho\rho}(\mathbf{q}+\mathbf{G},-\mathbf{q}'-\mathbf{G}',i\omega_m)\\ &=(2\pi)^2A^{-1}_{cell}\delta(\mathbf{q}-\mathbf{q}')\sum_{\alpha\beta}\rho_{\mathrm{q}\alpha}(\mathbf{G})Q^0_{\alpha\beta}(\mathbf{q},i\omega_m)\rho^*_{\mathrm{q}\beta}(\mathbf{G}')\end{aligned} \quad (4.5)$$

where

$$Q^0_{\alpha\beta}(\mathbf{q},i\omega_m)=\frac{1}{N}\sum_{\mathrm{k}\lambda\lambda'}u^\alpha_{\mathrm{k}\lambda}u^{\beta*}_{\mathrm{k}\lambda}u^{\alpha*}_{\mathrm{k}-q\lambda'}u^\beta_{\mathrm{k}-q\lambda'}T_{\mathrm{k}\lambda,\mathrm{k}-\mathrm{q}\lambda'}(i\omega_m) \quad (4.6)$$

with $u^\alpha_{\mathrm{k}\lambda}$, the components of $u_{\mathrm{k}\lambda}$, which are solutions of the generalized eigenvalue problem for the tight-binding Hamiltonian [13], and

$$\rho_{\mathrm{q}\alpha}(\mathbf{G})=\exp(-i\mathbf{G}\cdot\delta_\alpha-\sigma_g^2|\mathrm{q}+\mathbf{G}|^2/4) \quad (4.7)$$

are the LFE weights. The indices α and β refer to the pseudospin space ($\alpha,\beta=A,B$), whereas the indices λ and λ' refer to the conduction and valence bands ($\lambda,\lambda'=1,2$). Moreover, we also set $\delta_A=0$ and $\delta_B=\delta_3$. The continuum limit is recovered when $\mathbf{G}=\mathbf{G}'=0$.

Many-body correlations are then included within RPA, yielding a renormalized polarization

$$\begin{aligned}\Pi_{\rho\rho}(\mathbf{q}+\mathbf{G},-\mathbf{q}'-\mathbf{G}',i\omega_m)\\ &=(2\pi)^2A^{-1}_{cell}\delta(\mathbf{q}-\mathbf{q}')\sum_{\alpha\beta}\rho_{\mathrm{q}\alpha}(\mathbf{G})Q_{\alpha\beta}(\mathbf{q},i\omega_m)\rho^*_{\mathrm{q}\beta}(\mathbf{G}')\end{aligned} \quad (4.8)$$

where now

$$Q(\mathbf{q},i\omega_m)=g_sQ^0(\mathbf{q},i\omega_m)\left[1-g_sA^{-1}_{cell}V(\mathbf{q})Q^0(\mathbf{q},i\omega_m)\right]^{-1} \quad (4.9)$$

where matrix products are being understood, $g_s=2$ is a factor for spin degeneracy, and

$$V_{\alpha\beta}(\mathbf{q})=\sum_{\mathrm{G}''}\rho^*_{\mathrm{q}\alpha}(\mathbf{G}'')V_0(\mathbf{q}+\mathbf{G}'')\rho_{\mathrm{q}\beta}(\mathbf{G}'') \quad (4.10)$$

is the renormalized Coulomb potential; $V_0(\mathrm{q})=e^2/(2\varepsilon_0\varepsilon_r q)$, is now a matrix over band indices. Here, $\varepsilon_r=(\varepsilon_{r1}+\varepsilon_{r2})/2$ denotes the average relative dielectric constants of the two media surrounding the graphene layer, namely, air for suspended graphene ($\varepsilon_{r1}=\varepsilon_{r2}=\varepsilon_r=1$). In the case of a stronger dielectric substrate, we therefore expect a softening of the correlation effects on the plasmon frequency. It is relevant to note that the renormalized potential already includes LFE. Finally, the approximation used to obtain the electron polarization in Equation 4.8 is shown diagrammatically in Figure 4.1.

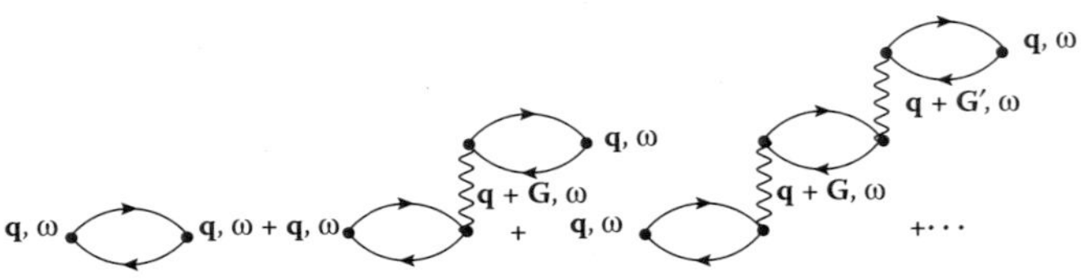

FIGURE 4.1 Diagrammatic representation of the RPA, including LFEs, for the electron polarization. The exchange momentum in the interaction terms can be outside the 1BZ. Indeed, **q** belongs to the 1BZ, whereas **G**, **G′** are vectors of the reciprocal lattice.

4.4.1 Plasmons

Plasmons are defined as collective excitations of the electron liquid corresponding to poles of the retarded polarization

$$\Pi_{\rho\rho}(\mathbf{q},\omega) \equiv \Pi_{\rho\rho}(\mathbf{q},-\mathbf{q},i\omega_m \to \omega + i0^+) \quad (4.11)$$

where $\mathbf{q} \in$ 1BZ. Here and in what follows, we shall restrict to the case $\mathbf{G} = \mathbf{G}' = 0$. Indeed, it is apparent from the definition of $\Pi_{\rho\rho}(\mathbf{q},\omega)$ that its poles can only arise from the vanishing of $\det[1 - V(\mathbf{q})Q^0(\mathbf{q},\omega)]$ in Equation 4.9, which already contains LFE via the renormalized Coulomb potential, Equation 4.10. We therefore define the dispersion relation $\omega_\ell(\mathbf{q})$ of the ℓth plasmon branch as

$$\Pi^{-1}_{\rho\rho}(\mathbf{q},\omega_\ell(\mathbf{q})) = 0 \quad (4.12)$$

This clearly involves vanishing of both real and imaginary parts of the inverse polarization. It will be useful to define the dispersion relation $\tilde{\omega}_\ell(\mathbf{q})$ of damped plasmons through

$$\mathrm{Re}[\Pi^{-1}_{\rho\rho}(\mathbf{q},\tilde{\omega}_\ell(\mathbf{q}))] = 0 \quad (4.13)$$

Correspondingly, the inverse lifetime $\tau^{-1}(\mathbf{q},\omega)$ of such damped plasmons is proportional to $-\mathrm{Im}\Pi_{\rho\rho}(\mathbf{q},\omega)$, for $\omega = \tilde{\omega}_\ell(\mathbf{q})$.

Figure 4.2 shows our numerical results for the plasmon dispersion relation in doped suspended graphene ($\mu = 1$ eV, $\varepsilon_{r1} = \varepsilon_{r2} = 1$, and $\mu = 1$ eV) at finite temperature ($T = 3$ K) along a symmetry contour in the 1BZ, without LFE ($\mathbf{G}'' = 0$ in Equation 4.10, left panel) and including LFE (right panel). At small wavevectors and low frequencies, one recognizes a square-root plasmon mode $\omega_1(\mathbf{q}) \sim \sqrt{q}$, typical of a 2D system [61]. This is in agreement with earlier studies of the dynamical screening effects in graphene at RPA level, employing an approximate conic dispersion relation for electrons around the Dirac points [53,54]. Such a result has also been confirmed for a tight-binding band [55,67], and is generalized here with the inclusion of LFE.

The high-energy (5–20 eV) pseudoplasmon mode, extending throughout the whole 1BZ, is rather associated with a logarithmic singularity of the bare polarization $Q^0(\mathbf{q},\omega)$ in Equation 4.9, and therefore does not correspond to a true pole of the polarization. This collective mode can be related to an interband transition between the Van Hove singularities in the valence and conduction bands of graphene, and has been identified with a $\pi \to \pi^*$ transition [55,68].

At large wavevectors, specifically along the zone boundary between the M and the K (Dirac) points, full inclusion of LFE determines the appearance of a second, high-frequency (20–25 eV), optical-like plasmon mode $\omega_2(\mathbf{q})$, weakly dispersing as $q \to 0$.

Multiple plasmon modes are a generic consequence of the possibility of interband transitions, whenever several such bands are available. This is, for example, the case of quasi-2D quantum wells (2DQWs), whose energy spectrum is characterized by quantized levels in the direction perpendicular to the plane of the well, while electrons can roam freely within the plane [61]. In this case, collective modes arise as zeroes of the determinant of the dielectric function. At low temperatures, at most, the two lowest subbands need to be considered. One usually obtains an "acoustic" mode associated to intrasubband coupling, and an "optical" mode associated to intersubband coupling [69]. Here, such a situation is paralleled by the

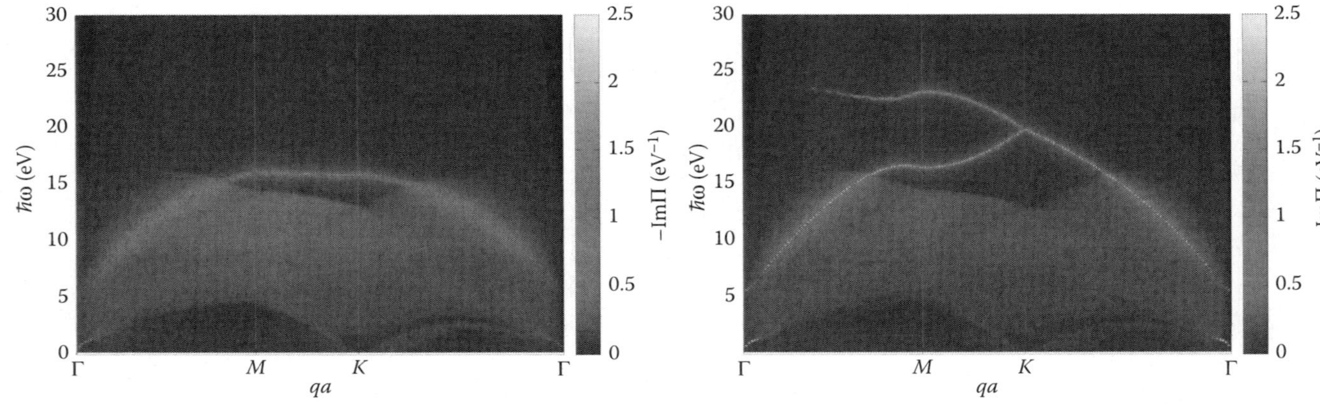

FIGURE 4.2 Plasmon dispersion relation for suspended doped graphene ($\mu = 1$ eV, $\varepsilon_{r1} = \varepsilon_{r2} = 1$) at finite temperature ($T = 3$ K), not including (left panel) and including (right panel) LFE. Results are shown along a symmetry contour in the 1BZ, with $\Gamma = (0,0)$, $M = (2\pi/3a,0)$, and $K = (2\pi/3a, 2\pi/3\sqrt{3}a)$. Energies $\hbar\omega$ are in electron volts. The shaded background is a contour plot of $-\mathrm{Im}\Pi_{\rho\rho}(\mathbf{q},\omega)$ (arbitrary scale), while continuous lines are the dispersion relation of damped plasmons, $\tilde{\omega}_\ell(\mathbf{q})$, Equation 4.13, which is shown as a dotted line.

case of graphene, and the role of the two subbands of 2DQW being played by the valence and conduction bands, touching at the Dirac points in the neutral material. It should be noted that the plasmon mode due to interband coupling is suppressed when LFE is neglected. In 2DQW, the discrete nature of the electronic subbands is due to the real-space confinement of the electron liquid in the direction perpendicular to the plane, that is, to the *quasi*-2D character of the quantum well. In graphene, the origin of the two bands ultimately lies in the specific lattice structure of this material. Therefore, the high-energy, "optical" plasmon mode disappears in the absence of LFE (Figure 4.2, top panel), as expected whenever the lattice structure of graphene is neglected. In other words, while in the absence of LFE, only scattering processes with momenta within the 1BZ are considered, LFE allows the inclusion of all scattering processes with arbitrarily low wavelengths, thereby taking into account the discrete nature of the crystalline lattice. Such a structure need not be considered in the case of a 2DQW. Our finding of a high-energy "optical" plasmon branch, as a generic consequence of the two-band electronic structure of graphene, should stimulate further investigation of the electronic collective modes in graphene [67,70], in view of the role of electron–electron correlations in interpreting the results of electron spectroscopy for interband transitions [71].

4.4.1.1 Beyond Two Dimensionality

Usually, the electronic system in graphene is considered as a 2D electron gas. Here, we take into account a full three-dimensional representation for the wave functions of the single particles. The generic electron wave function, corresponding to a λ band, is written as

$$\Psi_{\mathbf{k}\lambda}(\mathbf{r}) = \psi_{\mathbf{k}\lambda}(x)\Phi_{\mathbf{k}\lambda}(z) \tag{4.14}$$

where $\psi_{\mathbf{k}\lambda}(x)$ is the 2D eigenfunction, which has been considered previously, and $\Phi_{\mathbf{k}\lambda}(z)$ describes the z-dependence of the electron wave function.

Neglecting the z-dimension is equivalent to approximate the $\Phi_{\mathbf{k}\lambda}(z)$ so that its square modulus is a Dirac delta function

$$|\Phi_{\mathbf{k}\lambda}(z)|^2 \approx \delta(z) \tag{4.15}$$

A simple approximation to describe the finite extension of the electron wave function along the z-direction is an exponential function

$$\Phi_{\mathbf{k}\lambda}(z) = \mathrm{sgn}(z)\sqrt{\frac{\kappa_z}{2}}e^{-(\kappa_z/2)|z|} \tag{4.16}$$

where we consider κ_z as a constant. In particular, we set $\kappa_z = 3$ Å^{-1}, which is in good agreement with *ab initio* calculations [72]. In limit $\kappa_z \to \infty$, the square modulus of $\Phi_{\mathbf{k}\lambda}(z)$ becomes the Dirac delta function. In addition, one can see that the electron wave functions are odd under reflection symmetry with respect to the basal plane, like the p_z wave functions.

Taking into account the finite extension of the electron wave function along the z-direction, the renormalized polarization maintains the expression in Equation 4.8, but the 2D Coulomb potential $V_0(\mathbf{q}) = e^2/[\varepsilon_0(\varepsilon_{r1} + \varepsilon_{r2})q]$ is replaced by a more complex formula

$$V_0(\mathbf{q}) = \frac{e^2}{\varepsilon_0(\varepsilon_{r1}+\varepsilon_{r2})q}\left[\frac{\kappa_z(2\kappa_z+q)}{2(\kappa_z+q)^2} + \frac{(\varepsilon_{r1}-\varepsilon_{r2})^2}{8\varepsilon_{r1}\varepsilon_{r2}}\frac{\kappa_z q}{(\kappa_z+q)^2}\right] \tag{4.17}$$

where ε_{r1} and ε_{r2} denote the relative dielectric constants of the two media surrounding the graphene layer. The correction to the Coulomb potential in Equation 4.17 is negligible at a small momenta, $q \ll \kappa_z$, whereas its contribution is sizable for a large momenta, $q \gg \kappa_z$. Hence, the scattering processes with a large exchange momentum are particularly interested by the correction in Equation 4.17.

Figure 4.3 shows our numerical results for the plasmon dispersion relation in suspended doped graphene ($\mu = 1$ eV,

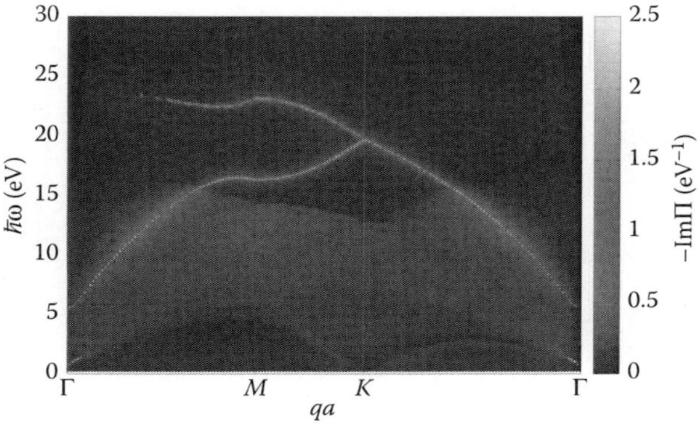

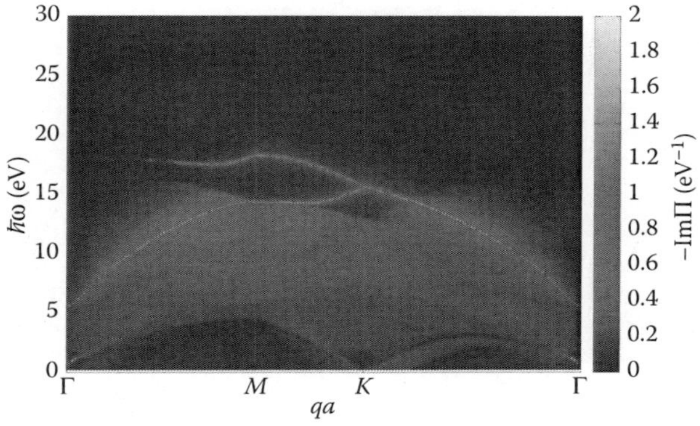

FIGURE 4.3 Plasmon dispersion relation for suspended doped graphene ($\mu = 1$ eV, $\varepsilon_{r1} = \varepsilon_{r2} = 1$) at finite temperature ($T = 3$ K), not including (left panel) and including (right panel) the z-extension of the electron wave functions, considering the LFE. Results are shown along a symmetry contour in the 1BZ, with $\Gamma = (0,0)$, $M = (2\pi/3a,0)$, and $K = (2\pi/3a, 2\pi/3\sqrt{3}a)$. Energies $\hbar\omega$ are in electron volts. The shaded background is a contour plot of $-\mathrm{Im}\Pi_{\rho\rho}(\mathbf{q},\omega)$ (arbitrary scale), while continuous lines are the dispersion relation of damped plasmons, $\tilde{\omega}_\ell(\mathbf{q})$, Equation 4.13, which is shown as a dotted line.

$\varepsilon_{r1} = \varepsilon_{r2} = 1$) at finite temperature ($T = 3$ K) along a symmetry contour in the 1BZ, without (left panel) and with (right panel) the z-extension of the electron wave functions, and both include the LFE. By a comparison of both panels in Figure 4.3, at low energies and small wavevectors, the contribution due to the z-extension of the electron wave functions has no appreciable effect. On the other hand, at high energies, there is a quantitative, but not qualitative, modification of the plasmon dispersion relation due to the z-extension of the electron wave functions. In particular, there is an energy lowering of the "optical" plasmon branch because of the correction on the Coulomb potential in Equation 4.17. Moreover, one may observe that the high-energy plasmon branch maintains the same form, since the features of this collective excitation are related to the LFE, and more generally to the lattice symmetry.

Moreover, a quantitative improvement for the description of the high-energy collective excitations could be obtained adding in the tight-binding model, the next-neighbor terms, or further terms. However, these terms would make the model more complicated without adding new features of the π electronic structure. Indeed, our tight-binding model contains all principal properties of the π electronic structure, that is, the Dirac cones and the saddle points.

Finally, a qualitative improvement for the description of the high-energy plasmon branch could be obtained taking into account the σ electrons beyond the π electrons. In particular, these further electronic bands could heavily change the structure of the high-energy plasmon dispersion relation, and they could induce a finite lifetime to these collective excitations because of the further promotion of electrons from the valence band into the higher (σ*) energy band.

4.4.2 Asymptotic Behaviors

In certain limiting regimes, one may derive the asymptotic behavior of the polarization in a closed form. At low energies ($\hbar\omega \lesssim |t|$) and small wavevectors ($q \to 0$, i.e., $qa \ll 1$), LFE and z-extension can be neglected. The matrix product entering the definition of the polarization through Equation 4.9 then reduces to

$$g_s A_{cell}^{-1} V(\mathbf{q}) Q^0(\mathbf{q},\omega) = g_s A_{cell}^{-1} V_0(\mathbf{q}) \sum_{\alpha\beta} Q^0_{\alpha\beta}(\mathbf{q},\omega)$$

$$= \frac{\tilde{V}_0}{qa}\frac{1}{N}\sum_{\mathbf{k}\lambda}\delta_T(\xi_{\mathbf{k}\lambda})\left(\frac{\mathbf{q}\cdot\nabla_{\mathbf{k}}E_{\mathbf{k}\lambda}}{\hbar\omega}\right)^2 \quad (4.18)$$

where $\tilde{V}_0 = g_s(8\pi/3\sqrt{3})(a_0/a)$ Ry, a_0 being Bohr's radius, and $\delta_T(\varepsilon) \equiv -(\partial n_F(\varepsilon))/(\partial\varepsilon) \to \delta(\varepsilon)$, as $T \to 0$. In the latter limit, the δ-function effectively restricts the integration over wavevectors along the Fermi line. Whenever the cone approximation holds (i.e., for a sufficiently low chemical potential and strain), this can be taken as a constant energy ellipse [13]. The $\mathbf{k}$-integration in Equation 4.18 can then be performed analytically, and the retarded polarization, Equation 4.11, then reads

$$\Pi_{\rho\rho}(\mathbf{q},\omega) \approx \frac{g_s A_{cell}^{-1}\tilde{V}_0^{-1}\tilde{\omega}_1^2 q^2 a^2}{\hbar^2\omega^{+2} - \hbar^2\omega_1^2(\mathbf{q})} \quad (4.19)$$

where $\omega^+ \equiv \omega + i0^+$, and

$$\hbar\tilde{\omega}_1 = (\tilde{V}_0\rho(\mu))^{1/2}\,|\nabla_{\mathbf{q}}E_{\mathbf{q}2}/a\,| \quad (4.20)$$

with $\rho(\mu)$, the density of states (DOS) at the Fermi level. To a leading order in qa from Equation 4.19, one thus obtains

$$\omega_1(\mathbf{q}) \approx \tilde{\omega}_1\sqrt{qa} \quad (4.21)$$

for the acoustic-like plasmon dispersion relation. One thus recovers the square-root behavior of the plasmon dispersion relation, as it is typical in 2D electron systems [61]. Moreover, one recovers the dependence of the coefficient $\tilde{\omega}_1 \sim n^{1/4}$ on the carrier density n, rather than $\sim n^{1/2}$, as is the case for a parabolic dispersion relation of the quasiparticles [53,73]. The acoustic-like plasmon mode may be related to the Drude weight [74], thus enabling the observation of strain effects from optical measurements [14]. In the case of graphene on a dielectric substrate ($\varepsilon_r > 1$), one has a reduction of $\tilde{\omega}_1$, and thus, a softening of the plasmon mode. From Equation 4.19, one may also read off the imaginary part of the retarded polarization, which close to the "acoustic" plasmon mode [$\omega \sim \omega_1(\mathbf{q})$] reads

$$\mathrm{Im}\Pi_{\rho\rho}(\mathbf{q},\omega^+) \approx -\frac{\pi}{2} g_s A_{cell}^{-1}\tilde{V}_0^{-1/2}\tilde{\omega}_1(qa)^{3/2}\delta(\omega - \omega_1(\mathbf{q})) \quad (4.22)$$

We now turn to the asymptotic behavior of the second branch of the plasmonic spectrum, $\omega_2(\mathbf{q})$. We have already established that it displays an optical-like character, with $\omega_2(\mathbf{q}) \to \omega_2(0)$, as $q \to 0$. Here, $\omega_2(0)$ is greater than the distance between the top of the conduction band and the bottom of the valence band. At small wavevectors, it is useful to consider the expansions of the relevant terms in Equation 4.9, which to a leading order in q_i ($i = x,y$) read

$$Q^0_{AA}(\mathbf{q},\omega) \approx Q_{AA}(0,\omega) + \sum_{ij} q_i y_{ij}(\omega) q_j \quad (4.23)$$

$$Q^0_{AB}(\mathbf{q},\omega) \approx -Q_{AA}(0,\omega) + \sum_{ij} q_i z_{ij}(\omega) q_j \quad (4.24)$$

where $y_{ij}(\omega)$, and $z_{ij}(\omega)$ are real-valued functions of the frequency ω, and

$$Q^0_{AA}(0,\omega_2(0)) = \frac{1}{4N}\sum_{\mathbf{k}\lambda}\frac{n_F(\xi_{\mathbf{k}\bar{\lambda}}) - n_F(\xi_{\mathbf{k}\lambda})}{\omega_2(0) + \xi_{\mathbf{k}\bar{\lambda}} - \xi_{\mathbf{k}\lambda}} \quad (4.25)$$

The asymptotically constant value of the optical-like plasmon frequency is then implicitly given by

$$1-4Q_{AA}^{0}\left(0,\omega_2(0)\right)g_sA_{cell}^{-1}\sum_{\mathbf{G}}V_0(\mathbf{G})\sin^2\left(\frac{1}{2}\mathbf{G}\cdot\delta_3\right)=0 \quad (4.26)$$

whereas the imaginary part of the retarded polarization, close to the second plasmon branch [$\omega \sim \omega_2(0)$], to a leading order in q, reads

$$\text{Im}\Pi_{\text{pp}}(\text{q},\omega^+)\approx -\pi g_sA_{cell}^{-1}\left|\frac{1}{4N}\sum_{\mathbf{k}\lambda}\frac{n_F(\xi_{\mathbf{k}\bar{\lambda}})-n_F(\xi_{\mathbf{k}\lambda})}{(\omega_2(0)+\xi_{\mathbf{k}\bar{\lambda}}-\xi_{\mathbf{k}\lambda})^2}\right|^{-1}$$
$$\times\sum_{ijhk}q_iq_h(z_{ij}-y_{ij})(z_{hk}+y_{hk})q_jq_k\delta(\omega-\omega_2(0)) \quad (4.27)$$

In particular, it follows that the spectral weight of $\text{Im}\Pi_{\text{pp}}$ close to $\omega_2(0)$ decreases as $\sim q^4$, as $q\to 0$, rather than as $\sim q^{3/2}$, as is the case for the acoustic-like plasmon mode, Equation 4.22. This justifies the reduced spectral weight associated with the second plasmon branch at a small wavevector in Figure 4.3.

In the case of graphene on a dielectric substrate ($\varepsilon_r > 1$), inspection of Equations 4.26 and 4.27 yields a reduction of $\omega_2(0)$.

Usually, experimental methodologies to detect plasmon dispersion relation, such as electron energy loss spectroscopy (EELS), measure the collective excitation at a small wavevector limit ($q\to 0$) [70]. In graphene, to date, there are measurements about the low-energy plasmon [75,76] and the pseudoplasmon excitation [70], whereas there is no clear experimental evidence about the high-energy plasmon. The detection of the high-energy branch at a small wavevector could be difficult not only because of the reduced spectral weight associated with the high-energy branch, but also because these plasmons could be damped by the further promotions of electrons from the valence band into the higher (σ^*) energy band. Here, we have not considered the electronic bands due to the σ electrons, and this possible correction will be the subject of future work.

4.5 EFFECT OF STRAIN ON THE PLASMON DISPERSION RELATION

We now turn to consider the effect of strain on the plasmon dispersion relation. As in References 10 and 13, applied uniaxial strain can be modeled by explicitly considering the dependence on the strain tensor ε of the tight-binding parameters $t_l = t(\delta_l)$ through the vectors δ_l connecting two NN sites ($l = 1,2,3$). A linear dependence of δ_l on ε is justified in the elastic limit. Such an assumption is however quite robust, due to the extreme rigidity of graphene [9], and is supported by *ab initio* calculations [77,78].

Below, the strain tensor ε will be parametrized by a strain modulus ε, and by the angle θ between the direction of applied strain and the x axis in the lattice coordinate system. Specifically, one has $\theta = 0$ (resp., $\theta = \pi/6$) for strain applied along the armchair (resp., zigzag) direction.

Figure 4.4 shows the dispersion relation of the plasmon branches studied in Section 4.4.1, including LFE and z-extension, along a symmetry contour of the 1BZ, for strain applied along the armchair direction ($\theta = 0$), with increasing strain modulus ($\varepsilon = 0 - 0.275$). The low-frequency, "acoustic" plasmon mode $\omega_1(\mathbf{q})$ is not qualitatively affected by the applied strain. In particular, the dominant square-root behavior is independent with respect to the opening of a gap. On the other hand, one observes an increase of spectral weight associated with the high-frequency, "optical" plasmon mode $\omega_2(\mathbf{q})$ at small wavevectors. The overall flattening of the second plasmon branch over the symmetry contour under consideration can be traced back to the strain-induced shrinking of both valence and conduction bands.

A qualitatively similar analysis applies to the case of strain applied along the zigzag direction ($\theta = \pi/6$, Figure 4.5), and for strain applied along a generic direction ($\theta = \pi/4$, Figure 4.6), with $\omega_2(\mathbf{q})$ dispersing more weakly as the strain increases.

Finally, we turn to study the $\mathbf{q}$-dependence of the low-frequency, "acoustic" mode $\omega_1(\mathbf{q}) \equiv \omega_1(q, \varphi_q)$ under applied strain, where $q = |\mathbf{q}|$ and φ_q denotes the angle between $\mathbf{q}$ and the $\hat{x}$ axis. Figure 4.7 then shows the dispersion relation of the lower plasmon branch as a function of q for several values of φ_q, for increasing strain applied along the armchair direction ($\theta = 0$). While the overall square-root shape $\omega_1 \approx \tilde{\omega}_1\sqrt{qa}$, Equation 4.21, is maintained in all cases, one observes a stiffening of such a plasmonic mode with increasing strain and a maximum of the coefficient $\tilde{\omega}_1$, Equation 4.20, when $\varphi_q - \theta \approx \pi/2$. The same description also qualitatively applies to the cases of strain applied along the armchair ($\theta = \pi/6$), and along a generic ($\theta = \pi/4$) direction. Such a behavior can be justified analytically in the limit of no LFE (cf. Section 4.4.2), and corresponds to the strain dependence obtained for the optical conductivity [13]. Indeed, from Equation 4.20, one may note that all the strain dependence is contained in the modulus square of the quasiparticle dispersion relation of the conduction band at the Fermi level, $|\nabla_q E_{k2}/a|$. One finds

$$\tilde{\omega}_1 \propto |\nabla_{\mathbf{q}}E_{\mathbf{q}2}| = \left(\frac{\cos^2(\varphi_{\mathbf{q}}-\eta)}{A^2}+\frac{\sin^2(\varphi_{\mathbf{q}}-\eta)}{B^2}\right)^{1/2} \quad (4.28)$$

where A and B denote the semiaxes of the constant energy ellipse, whereas the angle η is a function of the hopping parameters [13]. It follows that $\tilde{\omega}_1$ attains its maximum values whenever $\varphi_{\mathbf{q}} - \eta = \pi/2$ (modulo π), and its minimum values whenever $\varphi_{\mathbf{q}} - \eta = 0$ (modulo π). It turns out that $\eta = \theta$ in the zigzag and armchair cases (cf. Figure 4.7), whereas η ; θ in the generic case.

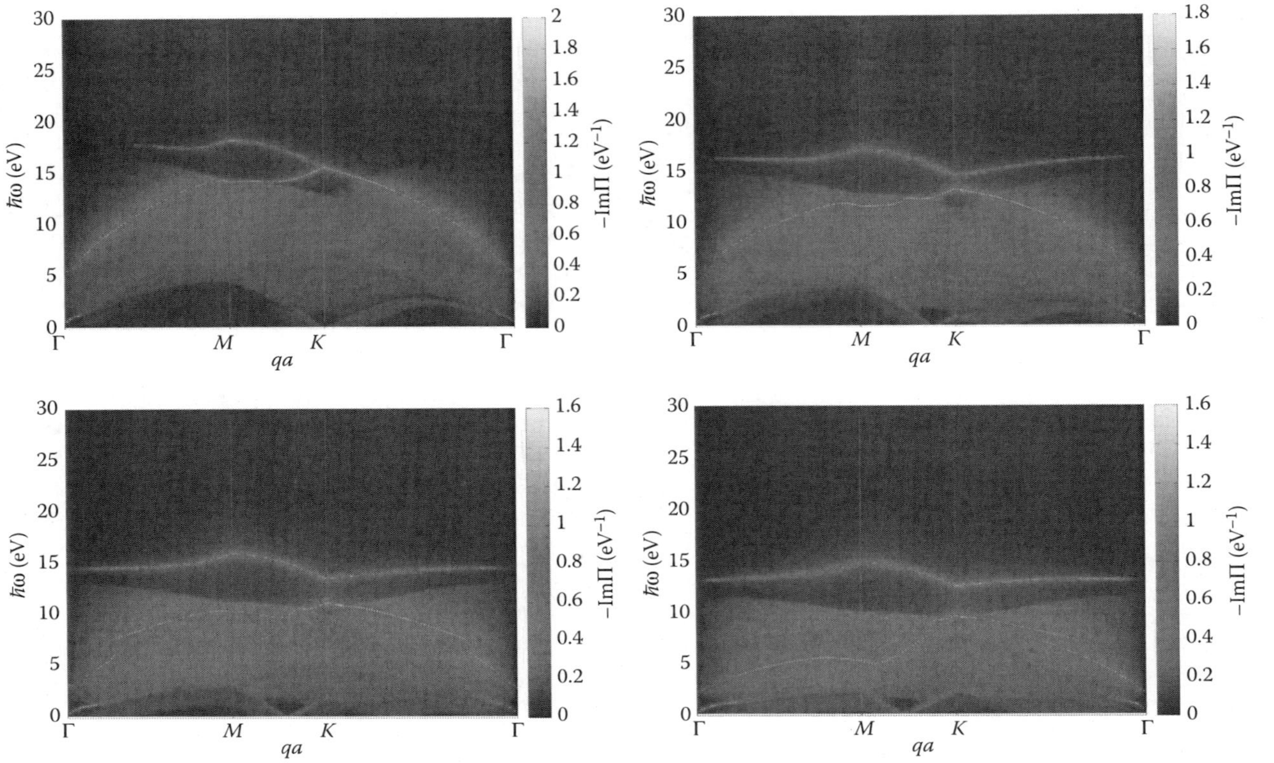

FIGURE 4.4 Plasmon dispersion relation for suspended doped graphene ($\mu = 1$ eV, $\varepsilon_{r1} = \varepsilon_{r2} = 1$), including LFE and z-extension, with strain applied along the $\theta = 0$ (armchair) direction. Strain increases (from left to right, from top to bottom) as $\varepsilon = 0$, 0.075, 0.175, and 0.275.

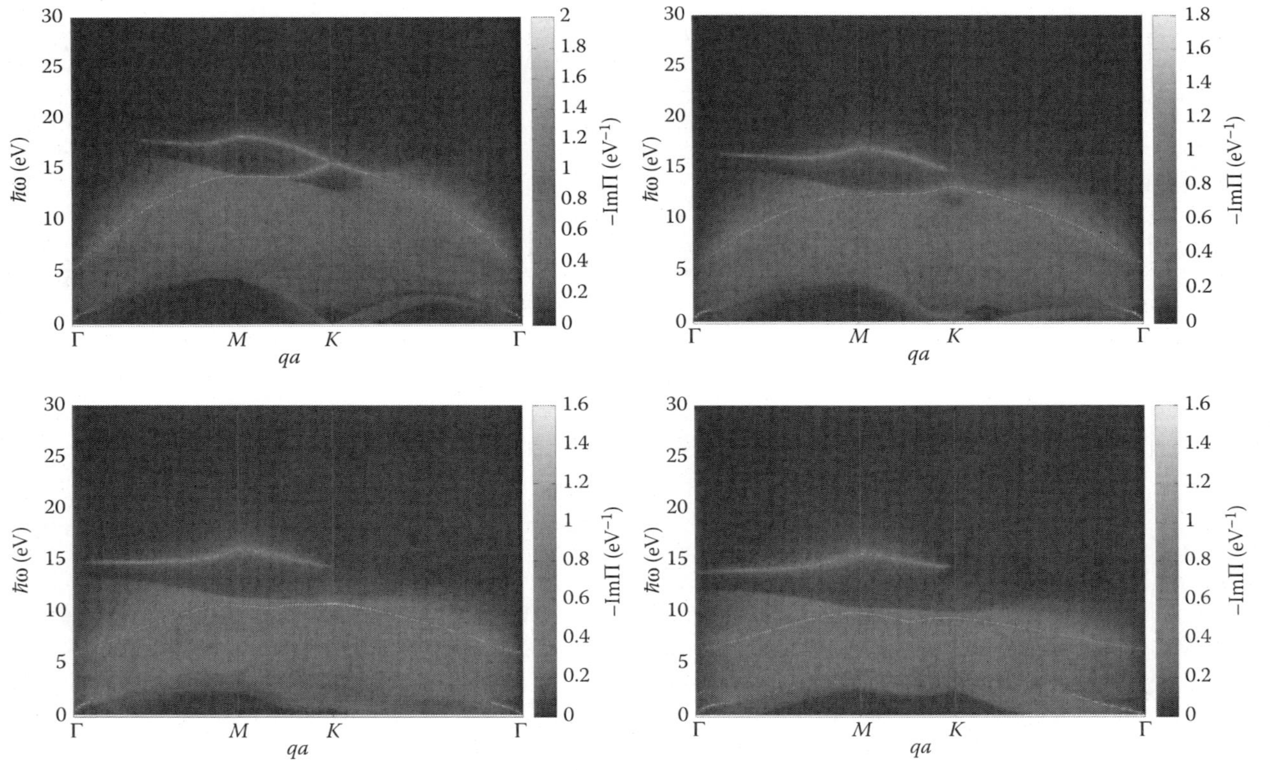

FIGURE 4.5 Plasmon dispersion relation for suspended doped graphene ($\mu = 1$ eV, $\varepsilon_{r1} = \varepsilon_{r2} = 1$), including LFE and z-extension, with strain applied along the $\theta = \pi/6$ (zigzag) direction. Strain increases (from left to right, from top to bottom) as $\varepsilon = 0$, 0.075, 0.175, and 0.275.

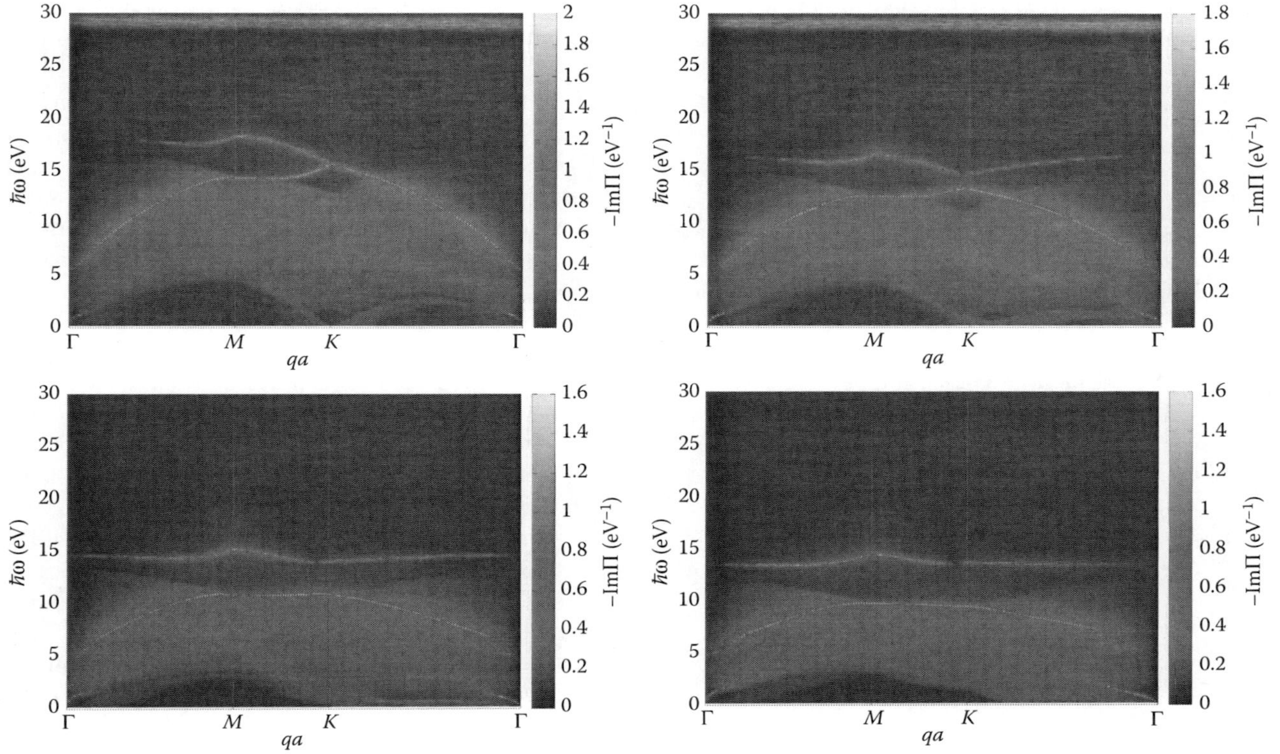

FIGURE 4.6 Plasmon dispersion relation for suspended doped graphene ($\mu = 1$ eV, $\varepsilon_{r1} = \varepsilon_{r2} = 1$), including LFE and z-extension, with strain applied along the $\theta = \pi/4$ (generic) direction. Strain increases (from left to right, from top to bottom) as $\varepsilon = 0$, 0.075, 0.175, and 0.275.

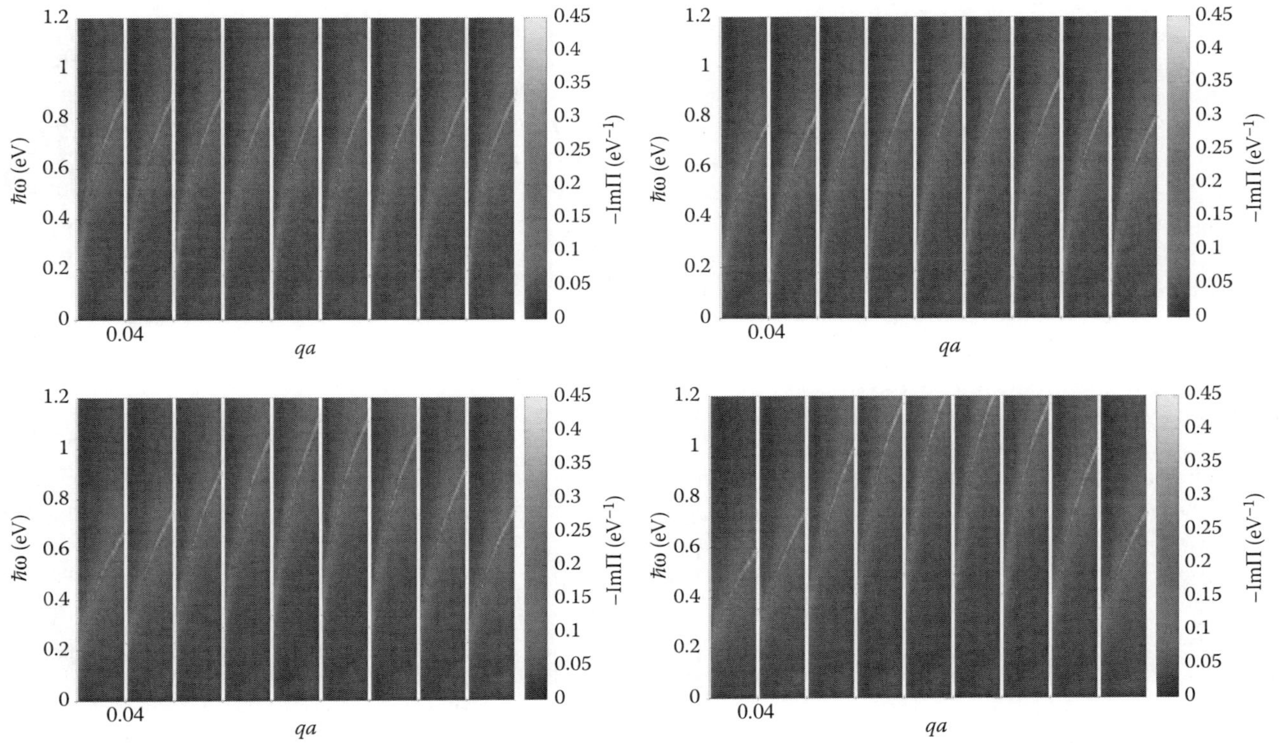

FIGURE 4.7 Plasmon dispersion relation for suspended doped graphene ($\mu = 1$ eV, $\varepsilon_{r1} = \varepsilon_{r2} = 1$), with strain applied along the $\theta = 0$ (armchair) direction. Strain increases (from left to right, from top to bottom) as $\varepsilon = 0$, 0.075, 0.175, and 0.275. In each graph, different panels refer to $\omega_1(\mathbf{q}) = \omega_1(q, \varphi_q)$, with $\varphi_q = 0°, 20°, 40°, \ldots 160°$.

The Drude peak appears in the optical conductivity for $\omega \to 0$ in doped graphene as

$$\sigma(\omega \to 0) = D\delta(\omega) \tag{4.29}$$

where D is called Drude weight. The Drude weight can be connected by means of an effective f-sum rule to the dispersion relation of plasmons [74], which has also been studied under applied strain [14,79].

Through the continuity equation, one obtains

$$\sigma_{\phi\phi}(\omega) = \frac{ie^2}{\omega} \lim_{q \to 0} \frac{\omega^2}{q^2} \Pi_{\rho\rho}(\mathbf{q}, -\mathbf{q}, \omega) \tag{4.30}$$

Letting $\omega \to \omega + i0^+$, and extracting the real part, one recognizes the Drude weight as

$$D_\phi = \pi e^2 \lim_{\omega \to 0} \lim_{q \to 0} \frac{\omega^2}{q^2} \mathrm{Re}\Pi_{\rho\rho}(\mathbf{q}, -\mathbf{q}, \omega) \tag{4.31}$$

Using the asymptotic limit of polarization (4.19), one finds

$$D_\phi = 4\mu\sigma_0 \left[\pi A_{cell}^{-1} \rho_1 \left(\frac{\cos^2(\phi - \eta)}{A^2} + \frac{\sin^2(\phi - \eta)}{B^2} \right) \right] \tag{4.32}$$

where $\sigma_0 = \pi e^2/2h$ is the so-called universal interband electrical conductivity of neutral graphene [34], $A_{cell} = (1+\varepsilon)(1-\nu\varepsilon)3\sqrt{3}a^2/2$ is the area of the unit cell, ρ_1 is the strain-dependent prefactor in the linear dependence of the DOS on the chemical potential at low energy, $\rho(\mu) = \rho_1|\mu|$ [13], and ϕ is the direction of the normally incident monochromatic electric field.

In Equation 4.32, the quantity between square brackets goes to unity in the limit $\varepsilon \to 0$, where in particular $\rho_1 = 4/(\pi\sqrt{3}t^2)$. From Equation 4.32, it follows that D_ϕ attains its maximum values whenever $\phi - \eta = \pi/2$ (modulo π), and its minimum values whenever $\phi - \eta = 0$ (modulo π). The ellipse semiaxes A and B depend on strain, whose role is that of increasing the ellipse anisotropy. In the unstrained limit, $\varepsilon = 0$, one has $A = B$; hence, one recovers that the Drude weight is independent of polarization of the incident electromagnetic field. Such a dependence of the Drude weight on applied uniaxial strain is amenable to experimental verification. Finally, the prefactor $\tilde{\omega}_1$ in the long-wavelength dispersion relation of low-energy plasmons in graphene [14], $\omega_\mathbf{q} = \tilde{\omega}_1\sqrt{qa}$, is related to the Drude weight through [74]

$$\frac{D_\phi}{(\hbar\tilde{\omega}_1)^2 a} = \frac{2\pi\varepsilon_0\varepsilon_r}{\hbar} \tag{4.33}$$

4.6 CONCLUSIONS

We studied the strain-induced modifications of plasmons in graphene. By studying the electronic polarization, we have derived the dispersion relation of the plasmon modes in graphene. Besides including electron–electron correlation at the RPA level, we have considered LFEs, which are specific to the peculiar lattice structure under study, and we have also taken into account the z-extension of the electron wave functions, without spoiling the inherent 2D character of graphene. Both terms are sizable in electron–electron scattering processes with a large exchange momentum ($q \sim \pi/a$). As a consequence of the two-band character of the electronic band structure of graphene, in general, we have found two plasmonic branches: (1) a low-energy branch, with a square-root behavior at small wavevectors, and (2) a high-energy branch, weakly dispersing at small wavevectors. In particular, we have found that the high-energy plasmon mode disappears neglecting LFE. While in the absence of LFE, only scattering processes with momenta within the 1BZ are considered, LFE allows the inclusion of all scattering processes with arbitrarily low wavelengths, thereby taking into account the discrete nature of the crystalline lattice. Hence, the Umklapp electron–electron scattering processes have a fundamental role in the system sustaining the high-energy plasmon mode. Moreover, we have found an intermediate-energy pseudoplasmon mode, associated with a logarithmic divergence of the polarization, which can be related to an interband transition between the Van Hove singularities in the valence and conduction bands of graphene, and it can be identified with a $\pi \to \pi^*$ transition. In graphene, to date, there are measurements about the low-energy plasmon [75,76] and the pseudoplasmon excitation [70], whereas there is no clear experimental evidence about the high-energy plasmon. Usually, experimental methodologies to detect plasmon dispersion relation, such as EELS, measure the collective excitation at a small wavevector limit ($q \to 0$) [70]. The detection of the high-energy branch at a small wavevector could be difficult, first of all, because of the reduced spectral weight associated with the high-energy branch, and also because these plasmons could be damped by the promotions of electrons from the valence band into the higher (σ^*) energy band. Here, we have not considered the electronic bands due to the σ electrons, and this possible correction will be the subject of future investigation.

REFERENCES

1. K. S. Novoselov, A. K. Geim, S. V. Morozov, D. Jiang, Y. Zhang, S. V. Dubonos, I. V. Grigorieva, and A. A. Firsov, Electric field effect in atomically thin carbon films. *Science* **306**, 666, 2004.
2. K. S. Novoselov, A. K. Geim, S. V. Morozov, D. Jiang, M. I. Katsnelson, I. V. Grigorieva, S. V. Dubonos, and A. A. Firsov, Two-dimensional gas of massless Dirac fermions in graphene. *Nature* **438**, 197, 2005.
3. A. H. Castro Neto, F. Guinea, N. M. R. Peres, K. S. Novoselov, and A. K. Geim, The electronic properties of graphene. *Rev. Mod. Phys.* **81**, 000109, 2009.
4. V. P. Gusynin, and S. G. Sharapov, Transport of Dirac quasiparticles in graphene: Hall and optical conductivities. *Phys. Rev. B* **73**, 245411, 2006.
5. T. Stauber, N. M. R. Peres, and A. K. Geim, Optical conductivity of graphene in the visible region of the spectrum. *Phys Rev. B* **78**, 085432, 2008.

6. F. Liu, P. Ming, and J. Li, *Phys. Rev. B* **76**, 064120, 2007.
7. K. S. Kim, Y. Zhao, H. Jang, S. Y. Lee, J. M. Kim, K. S. Kim, J. H. Ahn, P. Kim, J. Choi, and B. H. Hong, Large-scale pattern growth of graphene films for stretchable transparent electrodes. *Nature* **457**, 706, 2009.
8. M. Hytch, F. Houdellier, F. Hue, and E. Snoeck, Nanoscale holographic interferometry for strain measurements in electronic devices. *Nature* **453**, 1086, 2008.
9. T. J. Booth, P. Blake, R. R. Nair, D. Jiang, E. W. Hill, U. Bangert, A. Bleloch et al., Macroscopic graphene membranes and their extraordinary stiffness. *Nano Lett.* **8**, 2442, 2008.
10. V. M. Pereira, A. H. Castro Neto, and N. M. R. Peres, Tight-binding approach to uniaxial strain in graphene. *Phys. Rev. B* **80**, 045401, 2009.
11. N. M. R. Peres, Colloquium: The transport properties of graphene: An introduction. *Rev. Mod. Phys.* **82**, 2673, 2010.
12. F. H. Koppens, D. E. Chang, and J. G. de Abajo, Graphene plasmonics: A platform for strong light–matter interactions. *Nano Lett.* **11**, 3370, 2011.
13. F. M. D. Pellegrino, G. G. N. Angilella, and R. Pucci, Strain effect on the optical conductivity of graphene. *Phys. Rev. B* **81**, 035411, 2010.
14. F. M. D. Pellegrino, G. G. N. Angilella, and R. Pucci, Dynamical polarization of graphene under strain. *Phys. Rev. B* **82**, 115434, 2010.
15. F. M. D. Pellegrino, G. G. N. Angilella, and R. Pucci, Linear response correlation functions in strained graphene. *Phys. Rev. B* **84**, 195407, 2011.
16. S. A. Mikhailov, and K. Ziegler, New electromagnetic mode in graphene. *Phys. Rev. Lett.* **99**, 016803, 2007.
17. G. W. Semenoff, Condensed-matter simulation of a three-dimensional anomaly. *Phys. Rev. Lett.* **53**, 2449, 1984.
18. F. Schwierz, Graphene transistors. *Nat. Nanotechnol.* **5**, 487, 2010.
19. S. V. Morozov, K. S. Novoselov, M. I. Katsnelson, F. Schedin, D. C. Elias, J. A. Jaszczak, and A. K. Geim, Giant intrinsic carrier mobilities in graphene and its bilayer. *Phys. Rev. Lett.* **100**, 016602, 2008.
20. K. I. Bolotin, K. J. Sikes, J. Hone, H. L. Stormer, and P. Kim, Temperature-dependent transport in suspended graphene. *Phys. Rev. Lett.* **101**, 096802, 2008.
21. K. I. Bolotin, K. J. Sikes, Z. Jiang, M. Klima, G. Fudenberg, J. Hone, P. Kim, and H. L. Stormer, Ultrahigh electron mobility in suspended graphene. *Solid State Commun.* **146**, 351, 2008.
22. X. Du, I. Skachko, A. Barker, and E. Y. Andrei, Approaching ballistic transport in suspended graphene. *Nat. Nanotech.* **3**, 491, 2008.
23. S. M. Sze, and K. K. Ng, *Physics of Semiconductor Devices* (John Wiley & Sons, Hoboken, New Jersey, 2007).
24. H. L. Störmer, R. Dingle, A. C. Gossard, W. Wiegmann, and M. D. Sturge, Two-dimensional electron gas at a semiconductor-semiconductor interface. *Solid State Commun.* **29**, 705, 1979.
25. M. S. Fuhrer, C. N. Lau, and A. H. MacDonald, Graphene: materially better carbon. *MRS Bull.* **35**, 289, 2010.
26. A. F. Young, and P. Kim, Quantum interference and Klein tunnelling in graphene heterojunctions. *Nat. Phys.* **5**, 222, 2009.
27. S. G. Nam, D. K. Ki, J. W. Park, Y. Kim, J. S. Kim, and H. J. Lee, Ballistic transport of graphene pnp junctions with embedded local gates. *Nanotechnology* **22**, 415203, 2011.
28. D. P. Arovas, L. Brey, H. A. Fertig, E. A. Kim, and K. Ziegler, Dirac spectrum in piecewise constant one-dimensional (1D) potentials. *New J. Phys.* **12**, 123020, 2010.
29. C. W. J. Beenakker, Colloquium: Andreev reflection and Klein tunneling in graphene. *Rev. Mod. Phys.* **80**, 1337, 2008.
30. E. Rossi, J. H. Bardarson, P. W. Brouwer, and S. Das Sarma, Signatures of Klein tunneling in disordered graphene p–n–p junctions. *Phys. Rev. B* **81**, 121408, 2010.
31. S. Das Sarma, S. Adam, E. H. Hwang, and E. Rossi, Electronic transport in two-dimensional graphene. *Rev. Mod. Phys.* **83**, 407, 2011.
32. F. Bonaccorso, Z. Sun, T. Hasan, and A. C. Ferrari, Graphene photonics and optoelectronics. *Nat. Photon.* **4**, 611, 2010.
33. I. Hamberg, and C. G. Granqvist, Evaporated Sn-doped In2O3 films: Basic optical properties and applications to energy-efficient windows. *J. Appl. Phys.* **60**, R123, 1986.
34. R. R. Nair, P. Blake, A. N. Grigorenko, K. S. Novoselov, T. J. Booth, T. Stauber, N. M. R. Peres, and A. K. Geim, Fine structure constant defines visual transparency of graphene. *Science* **320**, 1308, 2008.
35. M. Jablan, H. Buljan, and M. Soljačić, Plasmonics in graphene at infrared frequencies. *Phys. Rev. B* **80**, 245435, 2009.
36. E. Ozbay, Plasmonics: Merging photonics and electronics at nanoscale dimensions. *Science* **311**, 189, 2006.
37. W. L. Barnes, A. Dereux, and T. W. Ebbesen, Surface plasmon subwavelength optics. *Nature* **424**, 824, 2003.
38. M. F. Yu, O. Lourie, M. J. Dyer, K. Moloni, T. F. Kelly, and R. S. Ruoff, Strength and breaking mechanism of multiwalled carbon nanotubes under tensile load. *Science* **287**, 637, 2000.
39. C. Lee, X. Wei, J. W. Kysar, and J. Hone, Measurement of the elastic properties and intrinsic strength of monolayer graphene. *Science* **321**, 385, 2008.
40. T. M. G. Mohiuddin, A. Lombardo, R. R. Nair, A. Bonetti, G. Savini, R. Jalil, N. Bonini et al., Uniaxial strain in graphene by Raman spectroscopy: G peak splitting, Grüneisen parameters, and sample orientation. *Phys. Rev. B* **79**, 205433, 2009.
41. J. S. Bunch, S. S. Verbridge, J. S. Alden, A. M. van der Zande, J. M. Parpia, H. G. Craighead, and P. L. McEuen, Impermeable atomic membranes from graphene sheets. *Nano Lett.* **8**, 2458, 2008.
42. J. Zabel, R. R. Nair, A. Ott, T. Georgiou, A. K. Geim, K. S. Novoselov, and C. Casiraghi, Raman spectroscopy of graphene and bilayer under biaxial strain: Bubbles and balloons. *Nano Lett.* **12**, 617, 2012.
43. M. M. Fogler, F. Guinea, and M. I. Katsnelson, Pseudomagnetic fields and ballistic transport in a suspended graphene sheet. *Phys. Rev. Lett.* **101**, 226804, 2008.
44. D. M. Basko, Resonant low-energy electron scattering on short-range impurities in graphene. *Phys. Rev. B* **78**, 115432, 2008.
45. H. M. Lee, S. B. Suh, P. Tarakeshwar, and K. S. Kim, Origin of the magic numbers of water clusters with an excess electron. *J. Chem. Phys.* **122**, 044309, 2005.
46. V. M. Pereira, and A. H. Castro Neto, Strain engineering of graphene's electronic structure. *Phys. Rev. Lett.* **103**, 046801, 2009.
47. T. Low, V. Perebeinos, J. Tersoff, and Ph. Avouris, Deformation and scattering in graphene over substrate steps. *Phys. Rev. Lett.* **108**, 096601, 2012.
48. I. Pletikosić, M. Kralj, P. Pervan, R. Brako, J. Coraux, A. T. N'Diaye, C. Busse, and T. Michely, Dirac cones and minigaps for graphene on Ir(111). *Phys. Rev. Lett.* **102**, 056808, 2009.
49. A. L. Vázquez de Parga, F. Calleja, B. Borca, M. C. G. Passeggi, J. J. Hinarejos, F. Guinea, and R. Miranda, Periodically rippled graphene: Growth and spatially resolved electronic structure. *Phys. Rev. Lett.* **100**, 056807, 2008.
50. W. Bao, F. Miao, Z. Chen, H. Zhang, W. Jang, C. Dames, and C. N. Lau, Controlled ripple texturing of suspended graphene and ultrathin graphite membranes. *Nat. Nanotechnol.* **5**, 562, 2009.

51. J. González, F. Guinea, and M. A. H. Vozmediano, Marginal-fermi-liquid behavior from two-dimensional Coulomb interaction. *Phys. Rev. B* **59**, R2474, 1999.
52. O. Vafek, Thermoplasma polariton within scaling theory of single-layer graphene. *Phys. Rev. Lett.* **97**, 266406, 2006.
53. E. H. Hwang, and S. Das Sarma, Dielectric function, screening, and plasmons in two-dimensional graphene. *Phys. Rev. B* **75**, 205418, 2007.
54. B. Wunsch, T. Stauber, F. Sols, and F. Guinea, Dynamical polarization of graphene at finite doping. *New J. Phys.* **8**, 318, 2006.
55. T. Stauber, J. Schliemann, and N. M. R. Peres, Dynamical polarizability of graphene beyond the Dirac cone approximation. *Phys. Rev. B* **81**, 085409, 2010.
56. S. L. Adler, Quantum theory of the dielectric constant in real solids. *Phys. Rev.* **126**, 413, 1962.
57. P. E. Trevisanutto, M. Holzmann, M. Côté, and V. Olevano, Ab initio high-energy excitonic effects in graphite and graphene. *Phys. Rev. B* **81**, 121405, 2010.
58. M. van Schilfgaarde, and M. I. Katsnelson, First-principles theory of nonlocal screening in graphene. *Phys. Rev. B* **83**, 081409, 2011.
59. A. Bostwick, T. Ohta, T. Seyller, K. Horn, and E. Rotenberg, Quasiparticle dynamics in graphene. *Nat. Phys.* **3**, 36, 2007.
60. V. W. Brar, S. Wickenburg, M. Panlasigui, C.-H. Park, T. O. Wehling, Y. Zhang, R. Decker et al., Observation of carrier-density-dependent many-body effects in graphene via tunneling spectroscopy. *Phys. Rev. Lett.* **104**, 036805, 2010.
61. G. Giuliani, and G. Vignale, *Quantum Theory of the Electron Liquid* (Cambridge University Press, Cambridge, 2005).
62. W. Schattke, in *Encyclopedia of Condensed Matter Physics*, edited by F. Bassani, G. L. Liedl, and P. Wyder (Elsevier, Amsterdam, 2005), Vol. 1, p. 145.
63. W. Hanke, and L. J. Sham, Dielectric response in the wannier representation: Application to the optical spectrum of diamond. *Phys. Rev. Lett.* **33**, 582, 1974.
64. W. Hanke, and L. J. Sham, Local-field and excitonic effects in the optical spectrum of a covalent crystal. *Phys. Rev. B* **12**, 4501, 1975.
65. A. L. Fetter, and J. D. Walecka, *Quantum Theory of Many-Particle Systems* (Dover, New York, 2004).
66. F. M. D. Pellegrino, G. G. N. Angilella, and R. Pucci, Effect of impurities in high-symmetry lattice positions on the local density of states and conductivity of graphene. *Phys. Rev. B* **80**, 094203, 2009.
67. A. Hill, S. A. Mikhailov, and K. Ziegler, Dielectric function and plasmons in graphene. *EPL (Europhys. Lett.),* **87**, 27005, 2009.
68. Mh. H. Gass, U. Bangert, A. L. Bleloch, P. Wang, R. R. Nair, and A. K. Geim, Free-standing graphene at atomic resolution. *Nat. Nanotech.* **3**, 676, 2008.
69. C. A. Ullrich, and G. Vignale, Time-dependent current-density-functional theory for the linear response of weakly disordered systems. *Phys. Rev. B* **65**, 245102, 2002.
70. T. Eberlein, U. Bangert, R. R. Nair, R. Jones, M. Gass, A. L. Bleloch, K. S. Novoselov, A. Geim, and P. R. Briddon, Plasmon spectroscopy of free-standing graphene films. *Phys. Rev. B* **77**, 233406, 2008.
71. M. Polini, R. Asgari, G. Borghi, Y. Barlas, T. Pereg-Barnea, and A. H. MacDonald, Plasmons and the spectral function of graphene. *Phys. Rev. B* **77**, 081411(R), 2008.
72. T. O. Wehling, K. S. Novoselov, S. V. Morozov, E. E. Vdovin, M. I. Katsnelson, A. K. Geim, and A. I. Lichtenstein, Molecular doping of graphene. *Nanoletters* **8**, 173, 2008.
73. S. Das Sarma, and E. H. Hwang, Collective modes of the massless dirac plasma. *Phys. Rev. Lett.* **102**, 206412, 2009.
74. S. H. Abedinpour, G. Vignale, A. Principi, M. Polini, W. K. Tse, and A. H. MacDonald, Drude weight, plasmon dispersion, and ac conductivity in doped graphene sheets. *Phys. Rev. B* **84**, 045429, 2011.
75. J. Chen, M. Badioli, P. Alonso-Gonzalez, S. Thongrattanasiri, F. Huth, J. Osmond, M. Spasenovic et al., Optical nano-imaging of gate-tunable graphene plasmons. *Nature* **487**, 77, 2012.
76. A. N. Grigorenko, M. Polini, and K. S. Novoselov, Graphene plasmonics. *Nat. Photon.* **6**, 749, 2012.
77. E. Cadelano, P. L. Palla, S. Giordano, and L. Colombo, nonlinear elasticity of monolayer graphene. *Phys. Rev. Lett.* **102**, 235502, 2009.
78. J.-W. Jiang, J.-S. Wang, and B. Li, Elastic and nonlinear stiffness of graphene: A simple approach. *Phys. Rev. B* **81**, 073405, 2010.
79. F. M. D. Pellegrino, G. G. N. Angilella, and R. Pucci, Effect of uniaxial strain on the Drude weight of graphene. *High Press Res.* **31**, 98, 2011.

5 Bondonic Electronic Properties of 2D Graphenic Lattices with Structural Defects

Mihai V. Putz, Ottorino Ori, and Mircea V. Diudea

CONTENTS

ABSTRACT

We propose an extensive simulation of electronic features in sp^2 carbon layers in the presence of structural defects. Among the variety of possible defects, we focus on Stone–Wales (SW) defects. Accordingly, striped-hexagons (or nodes) in pristine graphene G are suitable to undertake both, the central bond rotation producing single SW defect 5/7/7/5 divacancy reconstruction. Based on quantum calculations and topological methods, the original bondonic model is extended to describe the electronic fingerprints of these defective structures under the progressive variation and defect densities. Along surveying the physical origin of bondon, its chemical features, and the paradigmatic homopolar (Heitler–London) chemical bonding phenomenology in a generalized manner, this chapter showcases the novel idea of considering the topological indices with energetic relevance or correspondence (as recently was proven for Wiener index) as working potential in the physical analytical problems when the particle-bondon feels the topo-energetical action of the entire structure it encompasses. Our alternative quantum mechanical descriptions of total energy given by Bohmian theory evidences an important regime of nonlocal phenomena with the prediction of transition effects of typical localization scale and defect densities through identifying the origin of defects in topological tension at the pristine level as well as providing a stationary propagation of defects upon certain defective phase transition, as revealed by bondonic information contained in the caloric capacity at critical temperature.

5.1 INTRODUCTION: GRAPHENE AND STONE–WALES DEFECTS

Graphene monolayer is made of sp^2-hybridized carbon atoms forming a regular hexagonal (honeycomb) network. In this genuine bidimensional system, conductance π-electrons behave like massless Dirac carriers exhibiting scientific wonders like quantum Hall effect, the highest mobility recorded so far for any known material (200,000 cm^2/Vs at room temperature), a thermal conductivity 20 times bigger than copper, making graphene the candidate for the next-generation electronics. Graphene properties largely depend on the presence of *topological defects* created by bond rotation to generate different polygons such as pentagons, heptagons, octagons, etc. The Stone–Wales (SW) transformation produces a 5|7 double pair that plays an important role in influencing the properties of such an hexagonal system. Iterated SW transformations lead to a (reversible) *propagation wave-like mechanism* of the 5|7 rings, creating large dislocation dipole whose topological and structural peculiarities are described in this section.

5.1.1 Graphene as Extended Nanostructures

Graphene is a graphite sheet consisting of only sp^2 carbon atoms disposed in hexagonal faces. Single- or multiple-sheet graphene can be exfoliated from the bulk graphite by forming graphene oxide (Okamoto and Miyamoto 2001; Lee et al. 2009; Tylianakis et al. 2010) or via chemical functionalization (Paci et al. 2007; Denis 2009; Ueta et al. 2010). Cataldo and coworkers (Cataldo et al. 2012) recently synthesized graphene nanoribbons (GNRs) by oxidative unzipping of single-wall carbon nanotubes (SWCNTs). Morphological studies conducted by transmission electron microscopy (TEM) and atomic force microscopy confirm that these carbon nanosystems possess the characteristic structure of graphenic nanoribbon. Updates on graphene production methods may be found in recent reviews (Paton et al. 2014; Seaha et al. 2014). New carbon allotropes of graphene with honeycomb structures such as graphyne and graphdiyne (Peng et al. 2014), nowadays attract the interest of researchers on the next generation of bidimensional systems with exceptional technological properties.

Graphene has attracted considerable interest in the research community because of its outstanding properties. Graphene is in fact considered to be a zero-bandgap semiconductor material with the possibility of multiple applications in electronics, spintronics, hydrogen storage (Okamoto and Miyamoto 2001; Tylianakis et al. 2010), electrical batteries (Seger and Kamat, 2009; Abouimrane et al. 2010), capacitors (Wang et al. 2009a,b; Yu and Dai 2010), etc. Interactions with oxygen, fluorine, sulfur, and other chemical species permit various chemical modifications of the graphene (Paci et al. 2007; Denis 2009; Lee et al. 2009). Composites of graphene oxide and titanium oxide GO–TiO_2 exhibit improved photocatalytic activity toward mineralization of organic pollutants (Ng et al. 2010). Decoration of graphene with metal nanoparticles (Goncalves et al. 2009) or proteins is also possible (Liu et al. 2010). It is nevertheless hoped that the future microprocessors could be fabricated by etching graphene wafers into desired device architectures (Geim and Novoselov 2007) while closely packed graphene sheets could be employed in applications such as transparent conductors (Watcharotone et al. 2007; Green and Hersam 2009), field emission displays (Novoselov et al. 2004; Eda et al. 2008), and various composite materials (Stankovich et al. 2006).

Polymer composites of carbon nanotubes (CNTs) are known to exhibit high mechanical strength. Similar studies on graphene–polymer composites have shown promising mechanical properties. It has been shown that mechanical properties of poly (vinyl alcohol) and poly (methyl methacrylate) composites reinforced with small quantities of few-layer graphene (FLG) lead to a significant increase in both elastic modulus and hardness (Das et al. 2009; Prasad et al. 2009).

There are studies about the formation of carbon nanoscrolls (CNSs) from graphene. The scrolls can be prepared by ultrasonication of potassium intercalated graphite (Viculis et al. 2003; Wang et al. 2009a,b) or by simply dipping the single-layer graphene deposited on a substrate in isopropylalcohol (Xie et al. 2009). Because of the novel scroll topology, properties of CNSs differ from those of single-walled nanotubes (SWNTs) and multiwalled nanotubes (MWNTs). CNSs provide interlayer galleries that can be intercalated with donors and acceptors and may be valuable in energy storage, in supercapacitors or batteries. Structural modifications of graphene can be performed by including ring defects, and such modified graphenes are no more planar. In this respect, we recall the "CorSu" net $\{[6:6_6],[6:(5,6)_3)]_6\}$, theoretically described (Diudea and Ilić 2009; Diudea 2010, Saheli et al. 2010) as a corrugated surface tessellated with alternating Coronene $[6:6_6]$, and Sumanene $[6:(5,6)_3]$ patches (Figure 5.1). The number of atoms in a hexagonal domain of the lattice is $v = 96 + 273k + 243k^2$ while the number of edges is calculated by the formula: $e = 138 + 402k + 378k^2$.

Generalized cone structures (Krishnan et al. 1997; Ebbesen 1998) are also possible (Figure 5.2). Denote a cone by $C(a,n)$, with a = number of apex polygon and n = number of hexagon circles in the cone. The number of vertices and edges in the cone $C(a,n)$ are given by the formulas (Ilić et al. 2010): $v(C(a,n)) = a(n + 1)^2$; $e(C(a,n)) = (a/2)(3n^2 + 5n + 2)$. Structural modifications may also include heteroatoms (N, P, etc.). Topological invariants for pentagonal nanocones have been originally computed in Cataldo et al. (2010) taking into consideration the infinite lattice limit; those theoretical investigations demonstrate the relative chemical stability of fullerenic fragments (nanocones) which exhibit a larger topological efficiency when compared to graphenic fragments.

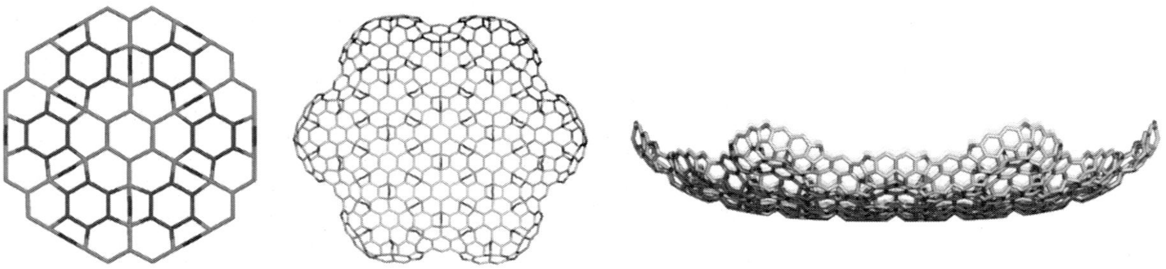

FIGURE 5.1 CorSu lattice, hexagonal domains: $k = 0$; $v = 96$; $v = 138$ (left); $k = 1$; $v = 612$; $e = 918$ (middle); $k = 3$; $v = 1614$; and $e = 2454$ (right). (Rearranged from Diudea, M.V., A. Ilić. 2009. *Studia Univ. "Babes-Bolyai"—Chemia* 54:171–177; Saheli, M. et al. 2010. *Croat. Chem. Acta* 83:395–401.)

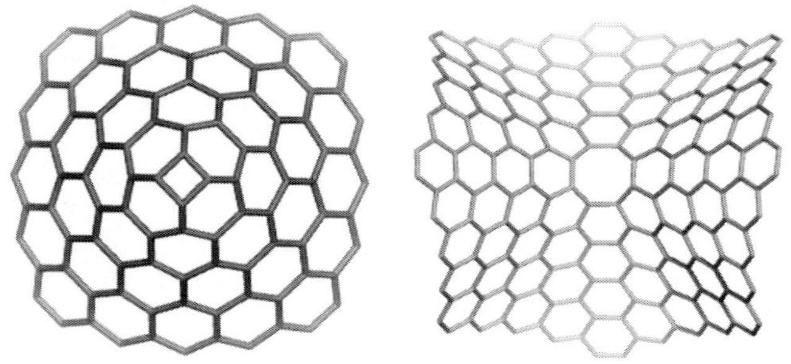

FIGURE 5.2 Structural modifications of graphene sheet. Conical domains (C(4,4) in left and C(8,4) in right). (Rearranged from Ilić, A. et al. 2010. In *Novel Molecular Structure Descriptors—Theory and Applications*, Vol. I, ed. I. Gutman, B. Furtula, 217–226. Kragujevac: University of Kragujevac.)

5.1.2 Graphene as Bondonic Environment*

However, besides the different experimental approaches to graphene synthesis, graphenic structures represent the recognized new frontier in physics of electronic systems with reduced dimensionality. Quasi-two-dimensional systems are well known for outstanding behaviors in the condensed matter physics as the quantum Hall effect and the high-temperature copper-oxide superconductors. According to the literature, what makes graphene "qualitatively new" (Evans 2000) is its semimetallic nature with low-energy quasiparticles behaving as "relativistic" Dirac spinors in the conducting band. Within this context, the very recent identification of the "bondon" as quantum particle of the chemical bond, rooting either in Bohm quantum mechanics as well in Dirac relativistic quantum theory (Putz 2010a), may provide a comprehensive tool for modeling physical properties of extended nanosystems in general (Putz 2012a,b,c; Putz 2013), and those of graphite layer (Putz and Ori 2012), or group-IV hexagonal nanoribbon (Putz and Ori 2014), as linking the microscopic behavior of electrons in condensed systems with macroscopic observable quantities described by quantum statistics.

5.1.3 Stone–Wales Waves Defects in Graphene†

Topological defects play an indubitable important role in influencing chemical, mechanical, and electronic properties of carbon systems arranged in hexagonal networks; many reviews have been published on the subject, see the exhaustive article by Terrones at al. (2010). The importance of topological mechanisms has been recently confirmed by a study (Ori and Putz 2014) on the formation of 5|8|5 defects without vacancies. The isolated pentagon–heptagon *single* pair, also called 5|7 dislocation or *pearshaped polygon* (Liu and Yakobson 2010), the pentagon–heptagon *double* pair 5/7/7/5, the dislocation dipole 5/7{6|6}7/5—in which a zip made of hexagon–hexagon pairs connects the 5|7 pairs—are just relevant examples of topological structural defects commonly affecting graphene layers, GNR's, and CNT's. As it has been reported in the study (Ori et al. 2011), a large number of these defects emerge from the celebrated SW isomeric transformation or rotation (Stone and Wales 1986) including the well-known 5|8|5 defects normally attributed to a divacancy alteration of the graphenic pristine plane, as the paper (Ori and Putz 2014) clearly demonstrates. Figure 5.3 depicts the general SW transformation, $SW_{q/r}$ (Figure 5.3a), associated to the most studied variants, the $SW_{6/6}$ in graphene (Figure 5.3b) often called Stone–Thrower–Wales rotation, and the $SW_{5/6}$ in fullerenes (Figure 5.3c) the so-called pyracylene rearrangement.

Another interesting topological effect has been introduced, consisting in the diffusion of a 5|7 pair in the hexagonal network as a consequence of iterated SW rotations; this topology-based mechanism *that produces a linear rearrangement of the hexagonal mesh* is called here the *Stone–Wales wave* (SWw). Whereas mechanically exfoliated monolayer graphene is structurally (almost) perfect in atomic scale (Terrones et al. 2010), graphene layers produced by chemical vapor deposition (CVD) techniques present a parade of defects, induced by the presence of irregularities on the surface of the substrate. New stable carbon allotropes have been, therefore, proposed (Lusk and Carr 2008) by considering the presence of periodical arrangements of defective building blocks such as SW defects, inverse SW defects, vacancy defects, and other structural modifications of the pristine hexagonal plane.

The first experimental observation of a particular type of linear topological defects is reported in Lahiri et al. (2010) where extended chains of octagonal and pentagonal sp^2-hybridized carbon rings, detected by scanning tunneling microscopy (STM) images, function as a

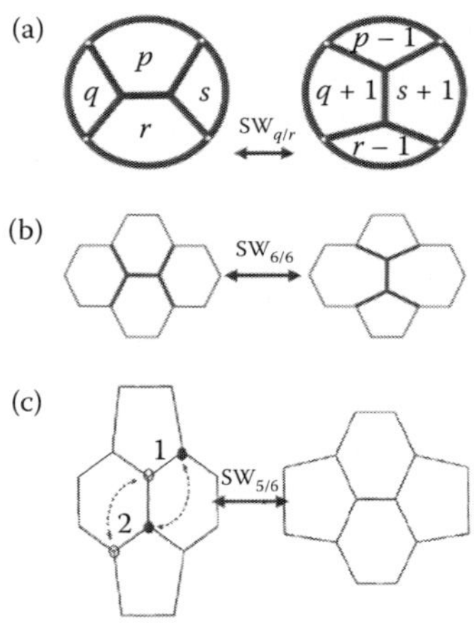

FIGURE 5.3 (a) Local transformation $SW_{q/r}$ changes a group of four proximal faces with *p*, *q*, *r*, and *s* atoms in four new rings with *p*–1, *q* + 1, *r*–1, and *s* + 1 atoms; (b) on the graphene layer ($p = q = r = s = 6$), $SW_{6/6}$ reversibly flips four hexagons in a 5|7 double pair; (c) $SW_{5/6}$ reversible flip on the fullerene surface. (From Ori, O., F. Cataldo, M.V. Putz. 2011. *Int. J. Mol. Sci.* 12:7934–7949.)

* Section 5.1.2 is excerpted from Putz, M.V., O. Ori. 2012. Bondonic characterization of extended nanosystems: Application to graphene's nanoribbons. *Chem. Phys. Lett.* 548:95–100.

† Section 5.1.3 is excerpted from Ori, O., F. Cataldo, M.V. Putz. 2011. Topological anisotropy of Stone-Wales waves in graphenic fragments. *Int. J. Mol. Sci.* 12:7934–7949.

quasi-one-dimensional metallic wire and may be the building blocks for new all-carbon electronic devices. This important experimental finding enforces meanwhile the theoretical role of the SWw, that are in principle structurally simpler than the pentagons–octagons chain reported in Lahiri et al. (2010), as a possible *hexagonal intergrain spacing*, see Ori et al. (2011) between graphenic fragments. Molecular mechanics simulations show that in graphene the presence of cylindrical curvature energetically facilitates such a split of the 5/7/7/5 SW dislocation dipole (Samsonidze et al. 2002), assigning to this class wave-like atomic-scale rearrangements a fundamental role in nanoengineering of graphenic lattices. One has however to notice that other TEM detailed measurements point out (Meyer et al. 2008; Kotakoski et al. 2011a,b) that the migration and the separation of the pentagon–heptagon pairs does not happen on planar graphene membranes where the 5|7 defects relax back reconstructing the original graphene lattice. These experiments indicate that extended *dislocation dipole*, favored by the presence of structural strain, preferably appear in curved graphitic structures or systems like CNT or fullerene molecules. In epitaxial graphene grown at high temperatures on mechanically polished SiC(0001), a characteristic sixfold "flower" defect results from STM measures (Rutter et al. 2007; Meyer et al. 2011). We underline that the observed rotational grain boundaries are conveniently describable as a *radial* type of the *SWw* suggesting the general applicability of the wave-like graph-theoretical mechanism presented here.

The SW rotations applied in the present studies derive from the general (Figure 5.3a) SW local and isomeric transformation $SW_{p/r}$ varying the internal connectivity of four generic carbon rings made of p, q, r, and s atoms to produce four new adjacent rings with $p-1$, $q+1$, $r-1$, $s+1$ atoms without changing the network of the surrounding lattice. $SW_{p/r}$ reversibly rotates the bond shared by the two rings p and r, preserving both, the total number of carbon atoms $v = p + q + r + s - 8$ and the total number of carbon–carbon bonds $e = v + 3$. On the graphene ideal surface, made only of hexagonal faces, the $SW_{6/6}$ rotation transforms four hexagons in two 5|7 adjacent pairs (Figure 5.3b) symbolized in literature (Samsonidze et al. 2002; Carpio et al. 2008) as 5/7/7/5 defect and also quoted as the *SW defect* or the *dislocation dipole*. We remember here that the SW rotations play an important role in connecting the isomers of a given C_n fullerene with different symmetries. In the crucial case of the C_{60} fullerene, its 1812 isomers are grouped by the *pyracylene* rearrangements $SW_{5/6}$ (Figure 5.3c) in 13 inequivalent sets (the larger one consisting of 1709 cages) connected to the buckminsterfullerene (C_{60}-I_h) through one or more SW transformations (Kumeda and Wales 2003), leaving 31 isomers unconnected to any of these sets. This limitation has been overcome by the introduction of nonlocal generalized SW transformations (Babic et al. 1995) to generate the whole C_{60} isomeric space starting from just one C_{60} isomer.

Theoretical investigations based on plane-wave density-functional methods (Kumeda and Wales 2003) set to no less than 6.30 eV the uphill energy barrier dividing the buckminsterfullerene from the SW connected isomer with C_{2v} symmetry; this barrier reaches 9 eV for hexagonal systems such as nanotubes or large graphene portions. Using the extended Hückel method, enlarging the relaxation region around the SW defect, it can be found that the formation energy of an SW defect considerably decreases to 6.02 eV for a flat graphene fragment case. This result has been verified by using *ab initio* pseudopotential (Zhoua and Shib 2003). This result seems to preclude the formation of any SW 5/7/7/5 defect in nature, but as it has been reported (Ewels et al. 2002; Collins 2011) that this barrier drops rapidly, reducing to 2.29 eV the creation barrier of SW rotations due to the catalyzing action of interstitials defects or ad-atoms present in the hexagonal networks. Pentagon–heptagon pairs have been predicted to be stable defects also in important theoretical articles (Nordlund et al. 1996; Krasheninnikov et al. 2001) showing that energetic particles, as electrons and ions, generate 5|7 pairs in graphite layers or CNT's as a result of knock-on atom displacements. On the experimental side, accurate high-resolution TEM studies made on SWCNTs (Hashimoto et al. 2004) or electron-irradiated pristine graphene (Kotakoski et al. 2011a,b) document *in situ* formation of SW dislocation dipoles. TEM measurements also evidenced (Chuvilin et al. 2009) stable grain boundaries with alternating sequence of pentagons and heptagons that show the relevance of wave-like defects during graphene edge reconstruction.

Extended theoretical investigations (Jeong et al. 2008) by means of first-principles density-functional computations demonstrate that, on graphene layers the dislocation dipole 5/7/7/5 defects become particularly stable—in comparison to other possible local defective structures as haeckelite units with three pentagons and three heptagons—when the two 5|7 pairs are separated by lattice vacancies in the number of 10 or over. Moreover, recent literature (Terrones et al. 2010) on GNR's constructed from haeckelites considers systems with SW defects as new hypothetical nanoarchitectures with fascinating applications in electronics. Isolated 5|7 pairs could also appear at the grain boundary in graphene fragments, changing their edge termination and electronic properties, forming *hybrid* GNR's. These hybrids exhibit half metallicity in the absence of an electric field, and could be used to transport spin-polarized electrons, which could be a step forward in new spintronic devices.

Figure 5.4 shows the fundamental topological operations for the generation and the propagation of SWw in the graphene lattice. The first rotation $SW_{6/6}$ (Figure 5.4a) of the chemical bond (arrowed) shared by the two hexagons creates the two 5|7 pairs (the SW defect 5/7/7/5). The second operator $SW_{6/7}$ turns the bond between the heptagon and the nearby shaded hexagon and inserts the 6|6 couple of shaded hexagons between the two original 5|7 pairs (this topological defect is also referenced in Samsonidze et al. (2002) as 5/7/6/6/7/5), leading to the overall structural effect of initiating the propagation of the SWw along the dotted direction (Figure 5.4b). Iterated transformations $SW_{6/7}$ successively drifts the 5|7 pairs in the lattice (along the dotted directions in Figure 5.4b, producing the topological SWw.

SWw mechanism provides theoretical support to recent studies on graphenic structures. Some authors (Liu and

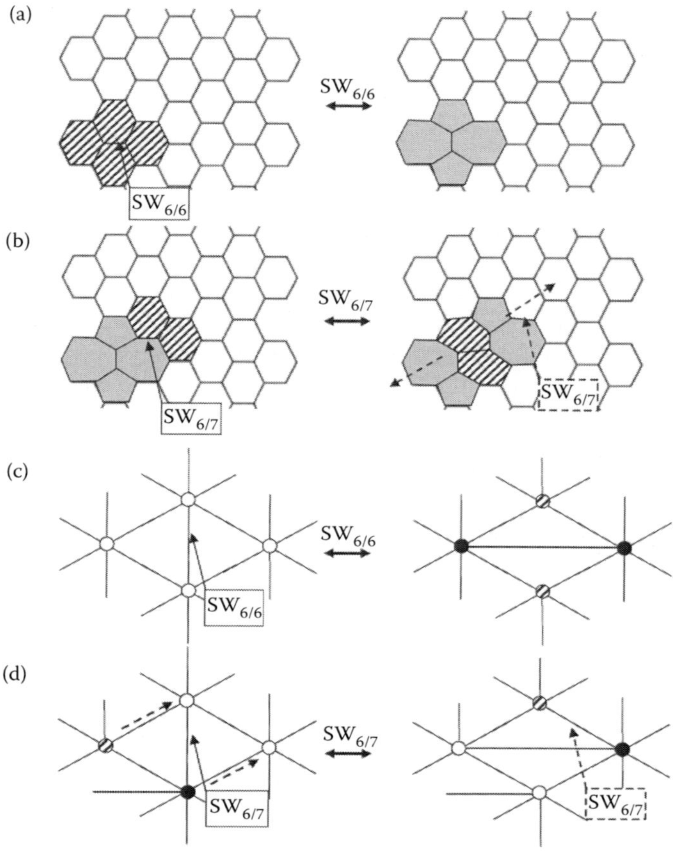

FIGURE 5.4 (a) $SW_{6/6}$ originates two 5|7 pairs (in gray); (b) $SW_{6/7}$ splits the pairs by swapping one of them with two nearby hexagons (shaded). Dotted $SW_{6/7}$ pushes the SWw in the dashed direction; (c–d) Mechanisms (a,b) in the graphene dual plane. Hexagons, pentagons, and heptagons are represented by white, shaded, and black circles, respectively. (From Ori, O., F. Cataldo, M.V. Putz. 2011. *Int. J. Mol. Sci.* 12:7934–7949.)

Yakobson 2010) emphasize the importance of 5|7 dislocations monopole at the grain boundaries of polycrystalline graphene, stating that these defects cannot be annealed by any local reorganization of the lattice. SWw allow 5|7 dislocations also to anneal by just involving surrounding 6|6 pairs and moving *backward*, being all transformations in Figure 5.4 completely *reversible*. Theoretically, SW defects and isolated 5|7 pairs have been extensively investigated in Carpio et al. (2008) where *ab initio* simulations of the electronic properties are reported; authors conclude that a single heptagon–pentagon dislocation is a stable defect whereas the SW adjacent pairs are dynamically unstable. These two conformations may easily find a unified description considering that these lattice defects correspond to different propagation steps of the same SWw.

Considering the very rich variety and complexity of all possible paths that SWw may describe on the graphenic surface, involving a variable numbers of 5|7 pairs, this chapter just focuses on the topological properties exhibited by the linear propagation of the basic SW defect, the 5|7 double pair (Figure 5.3b). This choice limits the $SW_{p/r}$ rotations to just to the operators $SW_{6/6}$ and $SW_{6/7}$. In spite of the apparent simplicity of our model, *SWw present* an evident and *marked topological anisotropy* immediately signaled by the Wiener index (Todeschini and Consonni 2000) *W*(*N*) of the graphenic system under study (graphene fragments, CNT's, and GNR's).

It is really important to note that, more and more, various anisotropic effects are evidenced in literature (Samsonidze et al. 2002; Dinadayalane and Leszczynski 2007; Huang et al. 2009; Bhowmick and Waghmare 2010; Terrones et al. 2010; Zeng et al. 2011) by applying first-principle techniques to the determination of the energy–stress behaviors of different configurations of SW defects on graphene nanotubes and nanoribbons. Similar effects appear in the theoretical distribution of magnetic dipoles in defective carbon metallic nanotubes (Im et al. 2011).

The remaining of the chapter is organized as follows. First, a survey of the bondonic approach to the chemical bonding is presented in certain details. Then, the main characteristics (e.g., bosonic mass, charge, velocity, and lifetime) of the quasiparticle corresponding to the chemical bonding wave function are emphasized. These characteristics deeply influence the type of chemical bonding itself. A worthy illustration of how the bosonization of electrons in chemical bonding affects the chemical bonding levels toward the associate bonding condensate is also presented in the framework of paradigmatic Heitler–London bindings. The bondonic formalism is further applied to extended graphenic systems, for the pristine case as well as for a moving SW defects in a certain periodic supercell so that the phase transition models and macroscopic statistical parameters are provided by the aid of path-integral formalism.

5.2 NATURE OF CHEMICAL BOND: BONDONS*

"A chemical bond is not real thing: it does not exist; no-one has ever seen it: no-one ever can. It is a figment of our own imagination"—so appeared in 1951 the "Coulson's dream;" such mixed feeling about chemical bonding is legitimate due to the plethora of unicorns in the world of physical-chemistry of atoms in molecules, rooted in three interrelated directions: (i) One is due to the Lewis's (1916) cornerstone paper advancing the model of cubic atoms in molecule (despite the orbital model of Bohr available from 1913) according which *the Coulomb law changes its nature in electronic pairs of chemical bonding*, that is, at very short distances between electrons, in some way anticipating the quantum Pauli (spin) repulsion; (ii) on other side, since the ionic and covalent (overlapping) characters should be also equilibrated in chemical gauge bonding ($A - B \leftrightarrow A^- : B^+ \leftrightarrow A^+ : B^-$) the concept of resonance structures using localized pairs modeling valence bond first emerged by the Pauling (1939) insight, nowadays arriving to the message that by mimicking the covalent bonding (as in H_2) the *ionic-covalent bonds* (as in F_2) *may be viewed as a realization of distance interaction of parallel spins in triplet (valence) states* (Shaik et al. 2009); (iii) the "polyphonic culture" of chemical bonding is completed by the molecular orbital (MO) approach initiated by the seminal work of Heitler and London (1927) with H_2 as a paradigmatic covalent system while triggering the nowadays, *molecular orbitals' delocalization models which for the size increasingly systems loose the chemical bonding intuition* (Malrieu et al. 2007). In this line, our chapter proceeds in advancing a recent frontier idea of chemical bonding while explaining why, till now, it was not an observable quantum object: as simple as one cannot observe a wave function since not having a reality observable nature (Scerri 2000) instead, one should consider the complete "circle of chemical bonding": electrons → wave functions → interference (resonance, exchange, and correlation) resulting in *chemical bonding wave function → the bondon* (Putz 2010a) as *associated quantum quasiparticle*! This idea is in the background of the actual endeavor and will be jointly unfolded on the fundamental, and applied on topological graphene environment in the present section and of its sequel.

5.2.1 Physical Origins of Bondons

In modern quantum chemistry, while passing from the Lewis point-like ansatz to the undulatory modeling of electrons in bonding, the reverse passage was still missing in an analytical formulation. Only recently the first attempt was formulated, based on the broken-symmetry approach of the Schrödinger Lagrangian with the electronegativity-chemical hardness parabolic energy dependency, showing that a systematical quest for the creation of particles from the chemical bonding fields is possible (Putz 2008a, 2010a,b,c).

Following this line, one can make a step forward and considering the gauge transformation of the electronic wave function and spinor over the de Broglie–Bohm augmented nonrelativistic and relativistic quantum pictures of the Schrödinger and Dirac electronic (chemical) fields, respectively. As a consequence, the reality of the chemical field in bonding may be proven in either framework while providing the corresponding bondonic particle with the associate mass and velocity in a full quantization form. In fact, the Dirac bondon stays as a natural generalization of the Schrödinger one while supplementing it with its anti-bondon particle (66) for the positron existence in the Dirac Sea (Putz 2010a,b).

The bondon is the quantum particle corresponding to the superimposed electronic pairing effects or distribution in chemical bond; accordingly, through the values of its mass and velocity it may be possible to indicate the type of bonding (in particular) and the characterization of electronic behavior in bonding (in general).

However, one of the most important consequences of bondonic existence is that the chemical bonding may be described in a more complex manner than relaying only on the electrons, but eventually employing the fermionic (electronic)–bosonic (bondonic) mixture: the first preeminent application is currently on progress, that is, exploring the effect that the Bose–Einstein condensation (BEC) has on chemical bonding modeling (Putz 2011a,b; Putz 2012d). Yet, such a possibility arises due to the fact that whether the Pauli principle is an independent axiom of quantum mechanics or whether it depends on other quantum description of matter is still under question (Kaplan 2002), as is the actual case of involving hidden variables and the entanglement or nonlocalization phenomenology that may be eventually mapped onto the delocalization and fractional charge provided by quantum chemistry over and on atomic centers of a molecular complex/chemical bond respectively.

However, the search for the bondons employs the general Bohmian project, resumed here.

Considering the de Broglie–Bohm electronic wave function/spinor Ψ_0 formulation of the associated quantum

* Section 5.2 is excerpted from Putz, M.V. 2010a. The bondons: The quantum particles of the chemical bond. *Int. J. Mol. Sci.* 11:4227–4256; Putz, M.V. 2010b. Beyond quantum nonlocality: Chemical bonding field. *Int. J. Env. Sci.* 1:25–31; Putz, M.V. 2011a. Hidden side of chemical bond: The bosonic condensate. In *Advances in Chemistry Research*, Vol. 10, ed. J.C. Taylor, 261–298. New York: NOVA Science Publishers; Putz, M.V. 2011b. Conceptual density functional theory: From inhomogeneous electronic gas to Bose–Einstein condensates. In *Chemical Information and Computational Challenges in 21st Century*, ed. M.V. Putz, 1–60. New York: NOVA Science Publishers; Putz, M.V. 2012a. Density functional theory of Bose–Einstein condensation: Road to chemical bonding quantum condensate. *Struct. Bond* 149:1–50; Putz, M.V. 2012b. Nanoroots of quantum chemistry: Atomic radii, periodic behavior, and bondons. In *Nanoscience and Advancing Computational Methods in Chemistry: Research Progress*, ed. E.A. Castro, A.K. Haghi, 103–143. Hershey: IGI Global (formerly Idea Group Inc.); Putz, M.V. 2012c. *Chemical Orthogonal Spaces. Mathematical Chemistry Monographs*, Vol. 14, Kragujevac: University of Kragujevac—Faculty of Sciences and Mathematics.

Schrödinger/Dirac equation of motion (de Broglie and Vigier 1953; Bohm and Vigier 1954)

$$\Psi_0(t,x) = R(t,x)\exp\left(i\frac{S(t,x)}{\hbar}\right) \quad (5.1)$$

with the R-amplitude and S-phase action factors given, respectively, as

$$R(t,x) = \sqrt{\Psi_0(t,x)^2} = \rho^{1/2}(x) \quad (5.2)$$

$$S(t,x) = px - Et \quad (5.3)$$

in terms of electronic density ρ, momentum p, total energy E, and time-space (t, x) coordinates, without the spin. On the other side, written within the de Broglie–Bohm framework, the spinor solution looks like (Dirac, 1944; Putz, 2011c)

$$\begin{aligned}\breve{\Psi}_0 &= \frac{1}{\sqrt{2}}R(t,x)\begin{bmatrix}\varphi\\ \phi\end{bmatrix}\\ &= \frac{1}{\sqrt{2}}R(t,x)\begin{bmatrix}\exp\left\{\frac{i}{\hbar}[S(t,x)+s]\right\}\\ \exp\left\{-\frac{i}{\hbar}[S(t,x)+s]\right\}\end{bmatrix}, \quad s = \pm\frac{1}{2}\end{aligned} \quad (5.4)$$

that from the consistency it satisfies the necessary electronic density condition

$$\breve{\Psi}_0^*\breve{\Psi}_0 = R^*R = \rho \quad (5.5)$$

In these conditions, since one perfumes the wave function partial derivatives respecting space and time,

$$\frac{\partial^2\Psi_{BB}}{\partial x^2} = \left[\frac{\partial^2 R}{\partial x^2} + 2\frac{i}{\hbar}\frac{\partial R}{\partial x}\frac{\partial S}{\partial x} + \frac{i}{\hbar}R\frac{\partial^2 S}{\partial x^2} - \frac{R}{\hbar^2}\left(\frac{\partial S}{\partial x}\right)^2\right]\exp\left(\frac{i}{\hbar}S\right) \quad (5.6)$$

$$\frac{\partial\Psi_{BB}}{\partial t} = \left[\frac{\partial R}{\partial t} + \frac{i}{\hbar}R\frac{\partial S}{\partial t}\right]\exp\left(\frac{i}{\hbar}S\right) \quad (5.7)$$

the conventional Schrödinger equation (Schrodinger 1926)

$$i\hbar\frac{\partial\Psi_{BB}}{\partial t} = -\frac{\hbar^2}{2m}\frac{\partial^2\Psi_{BB}}{\partial x^2} + V\Psi_{BB} \quad (5.8)$$

takes the real and imaginary forms

$$\frac{\partial R}{\partial t} = -\frac{1}{2m}\left[2\frac{\partial R}{\partial x}\frac{\partial S}{\partial x} + R\frac{\partial^2 S}{\partial x^2}\right] \quad (5.9)$$

$$-R\frac{\partial S}{\partial t} = -\frac{\hbar^2}{2m}\frac{\partial^2 R}{\partial x^2} + \frac{R}{2m}\left(\frac{\partial S}{\partial x}\right)^2 + VR \quad (5.10)$$

that can be further arranged as

$$\frac{\partial R^2}{\partial t} + \frac{\partial}{\partial x}\left[\frac{R^2}{m}\frac{\partial S}{\partial x}\right] = 0 \quad (5.11)$$

$$\frac{\partial S}{\partial t} - \frac{\hbar^2}{2m}\frac{1}{R}\frac{\partial^2 R}{\partial x^2} + \frac{1}{2m}\left(\frac{\partial S}{\partial x}\right)^2 + V = 0 \quad (5.12)$$

Worth noting that the first Equation 5.11 recovers in 3D coordinates the charge current ($\mathbf{j}$) conservation law

$$\frac{\partial\rho}{\partial t} + \nabla\mathbf{j} = 0, \quad \mathbf{j} = (R^2/m)\overline{\nabla S} \quad (5.13)$$

while the second Equation 5.12 in 3D

$$\frac{\partial S}{\partial t} - \frac{\hbar^2}{2m}\frac{\nabla^2 R}{R} + \frac{1}{2m}(\nabla S)^2 + V = 0 \quad (5.14)$$

extends the basic Schrödinger Equation 5.8 to include further quantum complexity. It may be clearly seen since recognizing that

$$(\nabla S)^2 = p^2 \Rightarrow \frac{1}{2m}(\nabla S)^2 = \frac{p^2}{2m} = T; \quad \frac{\partial S}{\partial t} = -E \quad (5.15)$$

one gets from Equation 5.14 the total energy expression

$$E = T + V + V_{qua} \quad (5.16)$$

in terms of newly appeared so-called quantum (or Bohm) potential

$$V_{qua} = -\frac{\hbar^2}{2m}\frac{\nabla^2 R}{R} \quad (5.17)$$

The existence of the Bohm potential (5.17) enriches the electronic quantum behavior with new interesting features. We will survey some of them in the following part of the section.

From definition (5.17), one can easily see that the Bohm potential comprises both classical and quantum characters:

$$V_{qua} = \begin{cases}\infty, m \to 0, \ldots \text{small quantum particles}\\ 0, \dfrac{\hbar}{m} \to 0, \ldots \text{classical particles and limit}\end{cases} \quad (5.18)$$

Although applicable in the case of stationary and localized quantum systems, potential (5.17) inherently comprises nonlocality features as well. For instance, one could observe that the quantum field is independent of the intensity of the measurement (the magnitude of aR) due to the elision:

$$V_{qua} = -\frac{\hbar^2}{2m}\frac{\nabla^2(aR)}{(aR)} = -\frac{\hbar^2}{2m}\frac{\nabla^2 R}{R} \quad (5.19)$$

Moreover, following the basic properties of the correct wave functions, we have that both the wave function and the amplitude of de Broglie–Bohm wave-packet cancel at "infinitum":

$$\Psi_{BB}(x \to \infty) = 0; \quad R(x \to \infty) = 0 \quad (5.20)$$

the quantum potential (5.17) does not vanish asymptotically, meaning that it does not behave locally, being distributed in a spatial-temporal manner that preserves its nonlocal (or non-separated) entangled action (Boeyens 2005).

An immediate generalization for many-body systems is provided for the generalized Schrödinger equation:

$$i\hbar\frac{\partial}{\partial t}\Psi_{BB}(x_1,...,x_N,t) = \left[-\frac{\hbar^2}{2m}\sum_{i=1}^{N}\nabla_i^2 + V\right]\Psi_{BB}(x_1,...,x_N,t) \quad (5.21)$$

with

$$\Psi_{BB}(x_1,...,x_N,t) = R(x_1,...,x_N,t)\exp\left(i\frac{S(x_1,...,x_N,t)}{\hbar}\right) \quad (5.22)$$

while the N-body quantum potential becomes

$$V_{N-qua} = -\frac{\hbar^2}{2m}\frac{\nabla^2 R(x_1,...,x_N,t)}{R(x_1,...,x_N,t)} \quad (5.23)$$

Giving the N-particles quantum field (5.23) one notices its peculiar behavior respecting ordinary potential shapes; namely, it does not show decreasing feature with the distance while each particle may nonlocally depend on the configuration of all others, no matter how far they may be (Putz 2010b). This way one may question about the more general item regarding the individual objects existence in the entire Universe: are they connected or disconnected? Or, in other terms: how the nonlocal entangled states act in the Universe? This paradox may be solved by the fact not all systems are equally correlated; actually, one can consider them as independent and achieve the total working function by the custom quantum factorization of individual object wave functions:

$$\Psi_{BB}(x_1,...,x_N,t) = \prod_{i=1}^{N}\Psi_{BBi}(x_i,t) \quad (5.24)$$

Now, the *global* quantum potential (5.23) rewrites as the *total* quantum potential summing the individual terms' contributions:

$$V_{N-qua} = \sum_{i=1}^{N} V_{i-qua} = -\frac{\hbar^2}{2m}\sum_{i=1}^{N}\frac{\nabla^2 R(x_i,t)}{R(x_i,t)} \quad (5.25)$$

There, it is clear that now the interaction at the level of each i-subunit may be treated in a traditional (observable) quantum-mechanical way while their interaction behavior is driven by nonlocal or entangled quantum potential. With these last considerations, it is clear that the Bohmian framework considerably enlarges the quantum mechanical concepts and the power of the quantum interpretation to describe the nature in general and the atoms and molecules in special (Putz 2010a, 2012d,e), although the concrete measure of the chemical bonding filed was missing from the Bohmian quantum mechanics till very recently (Putz 2010a). Nevertheless, since relating it with the localization of electronic pairs in a new quantum particle—bondonic reality, paralleling the nonlocality spreading of this bosonic bondon induced by electronic delocalization provides actual metastates picture of chemical bonding. In other words, *the chemical bonding field may be seen as the critical entity that transforms the locality in nonlocality and vice versa*; accordingly, it may be considered as the major concept in evolving Life and observable objects in Nature ... since being at the midway between (nano- to hidden) physical variables and the macroscopic biological reality (Putz 2010b)!

5.2.2 Bondonic Properties and Chemical Bonds*

The effective bondonic algorithm unfolds as follows (Putz 2010a, 2012a,b,c):

One reloads the electronic wave function/spinor under the augmented U(1) or SU(2) group form

$$\Psi_G(t,x) = \Psi_0(t,x)\exp\left(\frac{i}{\hbar}\frac{e}{c}\aleph(t,x)\right) \quad (5.26)$$

with the standard abbreviation $e = e_0^2/4\pi\varepsilon_0$ in terms of the chemical field $\aleph$ considered as the inverse of the fine-structure order

$$\aleph_0 = \frac{\hbar c}{e} \sim 137.03599976\left[\frac{\text{joule} \times \text{meter}}{\text{coulomb}}\right] \quad (5.27)$$

since upper bounded, in principle, by the atomic number of the ultimate chemical stable element ($Z = 137$). Although apparently small enough to be neglected in the quantum range, the quantity (5.27) plays a crucial role for chemical bonding

* Section 5.2.2 is excerpted from Putz, M.V. 2010a. The bondons: The quantum particles of the chemical bond. *Int. J. Mol. Sci.* 11:4227–4256. Putz, M.V. 2012c. *Chemical Orthogonal Spaces. Mathematical Chemistry Monographs*, Vol. 14, Kragujevac: University of Kragujevac—Faculty of Sciences and Mathematics.

where the energies involved are around the order of 10^{-19} J (eV)! Nevertheless, for establishing the physical significance of such chemical bonding quanta, one can proceed with the chain equivalences (Putz 2010b)

$$\aleph_{B} \sim \frac{\text{energy} \times \text{distance}}{\text{charge}} \sim \frac{\text{charge} \times \text{potential difference} \times \text{distance}}{\text{charge}} \sim \text{potential difference} \times \text{distance} \tag{5.28}$$

revealing that the chemical bonding field carries *bondons* with unit quanta $\hbar c/e$ along the distance of bonding within the potential gap of stability or by tunneling the potential barrier of encountered bonding attractors.

Then one rewrites the quantum wave function/spinor equation with the group object Ψ_G while separating the terms containing the real and imaginary $\aleph$ chemical field contributions; the next operations follows:

- Identifying the chemical field charge current and term within the actual group transformation context
- Establishing the global/local gauge transformations that resemble the de Broglie–Bohm wave function/spinor ansatz Ψ_0 of Equation 5.26
- Imposing invariant conditions for Ψ_G wave function on pattern quantum equation respecting the Ψ_0 wave function/spinor action of Equation 5.26
- Establishing the chemical field $\aleph$ specific equations
- Solving the system of chemical field $\aleph$ equations
- Assessing the stationary chemical field

$$\frac{\partial \aleph}{\partial t} \equiv \partial_t \aleph = 0 \tag{5.29}$$

that is the case in chemical bonds at equilibrium (ground state condition) to simplify the quest for the solution of the chemical field $\aleph$

- The manifested bondonic chemical field $\aleph_{bondon}$ is eventually identified along the bonding distance (or space)
- Checking the eventual charge flux condition of Bader within the vanishing chemical bonding field (Bader 1990)

$$\aleph_{B} = 0 \Leftrightarrow \nabla \rho = 0 \tag{5.30}$$

- Employing the Heisenberg time-energy relaxation-saturation relationship through the kinetic energy of electrons in bonding

$$v = \sqrt{\frac{2T}{m}} \sim \sqrt{\frac{2}{m}\frac{\hbar}{t}} \tag{5.31}$$

- Equate the bondonic chemical bond field with the chemical field quanta (5.27) to get the bondons' mass

$$\aleph_{B}(m_{B}) = \aleph_0 \tag{5.32}$$

This algorithm may be implemented in both for nonrelativistic as well as for relativistic electronic motion to quest upon the bondonic existence, eventually emphasizing their difference in anti-bondons' manifestations. Accordingly, through performing the above algorithm, either by Schrodinger or Dirac quantum level analysis, one obtains the working expression for the bondonic mass (Putz 2010a)

$$m_{B} = \frac{\hbar^2}{2}\frac{(2\pi n+1)^2}{E_{bond}X_{bond}^2}, \quad n = 0,1,2\ldots \tag{5.33}$$

that unifies the traditional characterization of bonding types in terms of length and energy of bonding; it may further assume the numerical ground-state ratio form

$$\varsigma_m = \frac{m_{B}}{m_0} = \frac{87.8603}{(E_{bond}[\text{kcal/mol}])\left(X_{bond}[\overset{0}{A}]\right)^2} \tag{5.34}$$

when the available bonding energy and length are considered (as is accustomed for chemical information) in kcal/mol and Angstrom, respectively. Note that having the bondon's mass in terms of bond energy implies the inclusion of the electronic pairing effect in the bondonic existence, without the constraint that the bonding pair may accumulate in the internuclear region (Berlin, 1951).

Equally, the quantified bondonic to light velocity ratio is found to be (Putz 2010a)

$$\frac{v_{B}}{c} = \frac{1}{\sqrt{1+(1/64\pi^2)((\hbar^2c^2(2\pi n+1)^4)/(E_{bond}^2X_{bond}^2))}}, \quad n = 0,1,2\ldots \tag{5.35}$$

Or, numerically, in the bonding ground state, as

$$\varsigma_v = \frac{v_{B}}{c} = \frac{100}{\sqrt{1+((3.27817\times 10^6)/(E_{bond}[\text{kcal/mol}])^2(X_{bond}[\text{Å}])^2)}}[\%] \tag{5.36}$$

while it carries the electrical charge (Putz 2010a) under the quantified form

$$e_{B} = \frac{4\pi\hbar c}{137.036}\frac{1}{\sqrt{1+((64\pi^2E_{bond}^2X_{bond}^2)/(\hbar^2c^2(2\pi n+1)^4))}}, \quad n = 0,1,2\ldots \tag{5.37}$$

TABLE 5.1
Predicted Basic Values for Bonding Energy and Length, along the Associated Bondonic Lifetime and Velocity Fraction from the Light Velocity for a System Featuring Unity Ratios of Bondonic Mass and Charge, Respecting the Electron Values, through Employing the Basic Formulas 5.33 through 5.39

X_{bond} (Å)	E_{bond} (kcal/mol)	$t_{\mathcal{B}}[\times 10^{15}]$ (s)	$\varsigma_v = v_{\mathcal{B}}/c$ (%)	$\varsigma_m = m_{\mathcal{B}}/m_0$	$\varsigma_e = e_{\mathcal{B}}/e_0$
1	87.86	10.966	4.84691	1	0.4827×10^{-3}
1	182019	53.376	99.9951	4.82699×10^{-4}	1
10	18201.9	533.76	99.9951	4.82699×10^{-5}	1
100	1820.19	5337.56	99.9951	4.82699×10^{-6}	1

Source: Adapted from Putz, M.V. 2010a. *Int. J. Mol. Sci.* 11:4227–4256.

Note: The first row corresponds to equal electron–bondonic mass implying low-velocity propagation yet, the bondon carrying much less electronic charge within the shorter time for the low energy of bonding. The other rows consider, for increasing bondonic distances and lifetimes, equal bondon–electronic charge, and comparable bondon–photonic propagation velocity while decreasing the bond energy and the bondon–photonic mass ratio.

employed toward the working ratio between the bondonic and electronic charges in the ground state of bonding—revealed as

$$\varsigma_e = \frac{e_{\mathcal{B}}}{e} \sim \frac{1}{32\pi} \frac{\left(E_{bond}[\text{kcal/mol}]\right)\left(X_{bond}[\text{Å}]\right)}{\sqrt{3.27817} \times 10^3} \tag{5.38}$$

With these quantities, the predicted lifetime of corresponding bondons is obtained from the bondonic mass and velocity working expressions throughout the basic time-energy Heisenberg relationship—here, restrained at the level of kinetic energy only for the bondonic particle; this way one yields the successive analytical forms

$$t_{\mathcal{B}} = \frac{\hbar}{T_{\mathcal{B}}} = \frac{2\hbar}{m_{\mathcal{B}} v_{\mathcal{B}}^2} = \frac{2\hbar}{(m_0 \varsigma_m)(c\varsigma_v \cdot 10^{-2})^2}$$
$$= \frac{\hbar}{m_0 c^2} \frac{2 \cdot 10^4}{\varsigma_m \varsigma_v^2} = \frac{0.0257618}{\varsigma_m \varsigma_v^2} \times 10^{-15} [s]_{SI} \tag{5.39}$$

Note that defining the bondonic lifetime by last equation is appropriate, since it involves the basic bondonic (particle!) information, mass, and velocity; instead, when directly evaluating the bondonic lifetime by only the bonding energy one deals with the working formula

$$t_{bond} = \frac{\hbar}{E_{bond}} = \frac{1.51787}{E_{bond}[\text{kcal/mol}]} \times 10^{-14} [s]_{SI} \tag{5.40}$$

In this way, the bondon is affirmed as a special particle of Nature, in that when behaving like an electron in charge it is behaving like a photon in velocity and like neutrino in mass while having an observable (at least as femtosecond) lifetime for nanosystems having chemical bonding in the range of hundred of Angstroms and thousands of kcal/mol (see Table 5.1 and Figure 5.5)!

On the other side, the mass, velocity, charge, and lifetime properties of the bondons were employed for analyzing some typical chemical bonds, see Table 5.2, this way revealing a sort of fuzzy classification of chemical bonding types in terms of the bondonic-to-electronic mass and charge ratios ς_m and ς_e, and of the bondonic-to-light velocity percent ratio ς_v, along the bondonic observable lifetime, $t_{\mathcal{B}}$, respectively—summarized in Table 5.3.

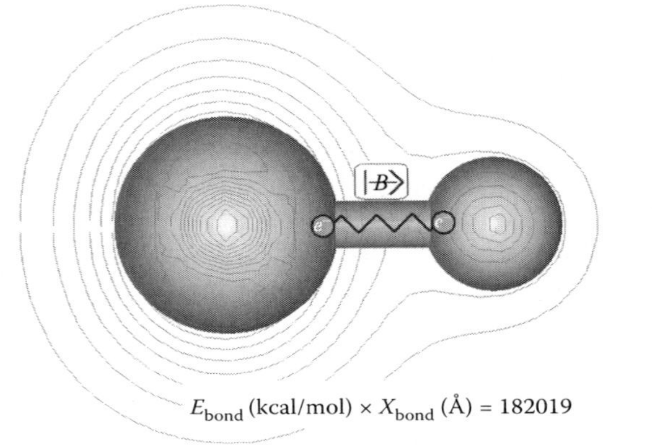

FIGURE 5.5 The bondon as the quantum particle of chemical bonding interaction, alongside its Heisenberg-like relationship, abstracted from Table 5.1.

5.2.3 Chemical Bond as Bondonic Condensates*

Since December 22, 1995, when the *Science* magazine declared the Bose condensate as the "molecule of the year," the Bose–Einstein condensation (BEC, in short), basically, viewed as *the macroscopic occupation of the same single-particle state in a many-body systems of bosons*, had received new impetus both at theoretical and experimental levels in searching and comprehending new states of matter (Anderson et al. 1995; Ketterle 2002). However, with the ever-increasing number of experiments revealing quantum phase transitions

* Section 5.2.3 is excerpted from Putz, M.V. 2010c. Developing density functional theory for Bose–Einstein condensates. The case of chemical bonding. Preprint for *AIP Conference Proceedings of 8th International Conference of Computational Methods in Sciences and Engineering (ICCMSE)*, Kos: Greece, October 3–8, 2010.

TABLE 5.2
Ratios for Bondonic-to-Electronic Mass and Charge and for Bondon-to-Light Velocity, along the Associated Bondonic Lifetime for Typical Chemical Bonds in Terms of Their Basic Characteristics Such as Bond Length and Energy (Findlay 1955; Oelke 1969) through Employing Basic Formulas 5.34 through 5.39 for Ground States, Respectively

Bond Type	X_{bond}(Å)	E_{bond}(kcal/mol)	$\varsigma_m = m_B/m_0$	$\varsigma_v = v_B/c(\%)$	$\varsigma_e = e_B/e[\times 10^3]$	$t_B[\times 10^{15}]$ (s)
H–H	0.60	104.2	2.34219	3.451	0.3435	9.236
C–C	1.54	81.2	0.45624	6.890	0.687	11.894
C–C[a]	1.54	170.9	0.21678	14.385	1.446	5.743
C=C	1.34	147	0.33286	10.816	1.082	6.616
C≡C	1.20	194	0.31451	12.753	1.279	5.037
N≡N	1.10	225	0.32272	13.544	1.36	4.352
O=O	1.10	118.4	0.61327	7.175	0.716	8.160
F–F	1.28	37.6	1.42621	2.657	0.264	25.582
Cl–Cl	1.98	58	0.3864	6.330	0.631	16.639
I–I	2.66	36.1	0.3440	5.296	0.528	26.701
C–H	1.09	99.2	0.7455	5.961	0.594	9.724
N–H	1.02	93.4	0.9042	5.254	0.523	10.32
O–H	0.96	110.6	0.8620	5.854	0.583	8.721
C–O	1.42	82	0.5314	6.418	0.64	11.771
C=O[b]	1.21	166	0.3615	11.026	1.104	5.862
C=O[c]	1.15	191.6	0.3467	12.081	1.211	5.091
C–Cl	1.76	78	0.3636	7.560	0.754	12.394
C–Br	1.91	68	0.3542	7.155	0.714	14.208
C–I	2.10	51	0.3906	5.905	0.588	18.9131

Source: Putz, M.V. 2010a. *Int. J. Mol. Sci.* 11:4227–4256.
[a] In diamond.
[b] In CH_2O.
[c] In O=C=O.

at atomic scales (Yukalov 2004), the need for accurate models for this new state of matter became imperative. Yet, although powerful variational and perturbation methods are available (Kleinert et al. 2004), a basic approach, centered on the key object of BEC—the bosonic gas density ρ—it is not yet systematically developed and implemented (Vetter 1997).

TABLE 5.3
Phenomenological Classification of Chemical Bonding Types by Bondonic (Mass, Velocity, Charge, and Lifetime) Properties Abstracted from Table 5.2

Property Chemical Bond	ς_m	ς_v	ς_e	t_B
Covalence	≫	≪	≪	≫
Multiple bonds	<	>	>	<
Metallic	≪	>	>	<
Ionic	~>	~	~	~>

Source: Putz, M.V. 2010a. *Int. J. Mol. Sci.* 11:4227–4256.
Note: The used symbols are: > and ≫ for "high" and "very high" values; < and ≪ for "low" and "very low" values; ~ and ~> for "moderate" and "moderate high and almost equal" values in their class of bonding.

Fortunately, the celebrated density functional theory (DFT) of atoms and molecules may provide a suitable framework for BEC modeling by exploiting DFT main features (Dreizler and Gross 1995).

It is a many (N)-body theory whose *main vehicle is the density itself*

$$\rho(\mathbf{r}) = N\int d\mathbf{r}_2\int d\mathbf{r}_3\ldots\int d\mathbf{r}_N\Psi^*(\mathbf{r},\mathbf{r}_2\ldots\mathbf{r}_N)\Psi(\mathbf{r},\mathbf{r}_2\ldots\mathbf{r}_N) \tag{5.41}$$

fulfilling the normalization condition

$$N = \int d\mathbf{r}\rho(\mathbf{r}) \tag{5.42}$$

matching in this way the Gross–Pitaevsky uni-particle field ψ, which describes the dilute bosonic gases through the nonlinear Schrödinger equation having chemical potential μ as the eigenvalue (Pitaevskii and Stringari 2003):

$$\left[-\frac{\hbar^2}{2m_B}\nabla^2 + V(\mathbf{r}) + g|\psi|^2\right]\psi = \mu\psi \tag{5.43}$$

Equation 5.43 accounts for the bosonic condensation through the self-interaction coupling:

$$g = \left.\frac{4\pi a_B \hbar^2}{m_B}\right|_{a_B \to R_{bond}} = 8\pi E_{bond} R_{bond}^3 = \frac{8\pi E_{bond}}{|\langle \psi(\mathbf{r})\rangle|^2} \quad (5.44)$$

by considering the mass of the bondons—the quantum bosonic particles of the chemical bonding (Putz 2010a) defined as

$$m_B = \frac{\hbar^2}{2} \frac{1}{E_{bond} R_{bond}^2} \quad (5.45)$$

representing E_{bond} the bonding equilibrium energy. Generalizing the s-scattering length a_B to the bonding length R_{bond}, the uni-bondonic volume is obtained:

$$R_{bond}^3 = V_{bond} = \frac{1}{\underset{(bosons)}{\rho_{bondons}(\mathbf{r})}} = \frac{1}{|\langle \psi(\mathbf{r})\rangle|^2} \quad (5.46)$$

on the basis of mean-field approximation $\rho \sim \langle \psi \rangle^2$, attributing to the condensate density as the role of order parameter.

DFT *supports the variational principle* (Dreizler and Gross, 1995):

$$\delta\left(E[\rho] - \mu \int d\mathbf{r} \rho(\mathbf{r}) \right) = 0 \quad (5.47)$$

in minimizing the energy functional

$$E[\rho] = C_A[\rho] + F_{HK}[\rho] \quad (5.48)$$

by means of the Lagrangian method in terms of a chemical potential parameter. Note that Equation 5.48 expresses the effects of the chemical action $C_A[\rho] = \int d\mathbf{r} V(\mathbf{r})\rho(\mathbf{r})$, directly related to chemical bonding and reactivity (Putz 2009), added to the universal Hohenberg–Kohn functional $F_{HK}[\rho]$ that, by containing the kinetic effects, due to Coulombic and exchange-correlation contributions (Hohenberg and Kohn 1964), it is, therefore, susceptible in being related to the bosonic condensation phenomenology.

Overall, by originally employing DFT to model the BEC, the analytical DFT–BEC relationships here are primarily reported along their consequences in modeling the chemical bonding. The last item is also motivated by the intrigued case of forbidden He–He bonding within ordinary molecular orbital theory, albeit largely allowed through supefluidic (i.e., within nonlinear Schrodinger) treatment (London 1938); the hidden BEC side of the fermionic wave functions superposition in general and of chemical bonding in special, it is, therefore, worth to explore.

The practical implementation of BEC and of its mean field approximation usually makes use of thermodynamically limited constraint which enables the usage of the so-called Thomas–Fermi approximation of DFT (Parr and Yang 1989) for systems with many-to-infinite number of particles ($N \to \infty$), since in condensation phenomenon "infinite more is the same" (Kadanoff 2009).

By considering that the potential and the interaction energies are larger than the kinetic energy, the kinetic term can be neglected in the Gross–Pitaevsky Equation 5.43, reducing it to the algebraic form:

$$|\psi|^2 \cong \frac{1}{g}[\mu - V(\mathbf{r})] \quad (5.49)$$

On the other side, the variational DFT principle (5.47) provides the working relationship among the chemical potential, external potential, and the Hohenberg–Kohn functional:

$$\mu = V(\mathbf{r}) + \frac{\delta F_{HK}[\rho]}{\delta\rho(\mathbf{r})} \quad (5.50)$$

Combining equations 5.49 and 5.50 the *second DFT–BEC connection* is directly obtained:

$$|\psi|^2 \cong \frac{1}{g}\left[\frac{\delta F_{HK}[\rho]}{\delta\rho(\mathbf{r})}\right]_{V(r)} \quad (5.51)$$

However, the *third working DFT–BEC connection* can be established by employing the interbosonic quantum average (Putz 2011a):

$$\int d\mathbf{r} |\psi(\mathbf{r})|^4 \cong \frac{1}{g^2} \int d\mathbf{r} [\mu - V(\mathbf{r})]^2 \cong -\frac{F_{HK}\eta}{g^2} \quad (5.52)$$

where the chemical hardness η was recognized through its local integrated version (Baekelandt et al. 1995):

$$\eta = \int d\mathbf{r} \eta(\mathbf{r}) \quad (5.53)$$

$$\eta(\mathbf{r}) = \frac{\delta\mu}{\delta\rho(\mathbf{r})} = \frac{\delta^2 F_{HK}[\rho]}{\delta\rho^2(\mathbf{r})} \quad (5.54)$$

involving the Hohenberg–Kohn functional. In this way, the chemical hardness manifests its straight involvement in bosonic condensates, in general; its particular contribution in chemical bonding are treated next.

Based on previously emphasized fermionic-bosonic (DFT–BEC) relationships, one can consider the *build-in-bondonic* (BB) superposition of fermionic (real) wave functions coming from the two atoms labeled as "1" and "2":

$$\underbrace{\psi_{BB}(\mathbf{r})}_{BONDONIC} = \underbrace{c_1\psi_1(\mathbf{r}) + c_2\psi_2(\mathbf{r})}_{FERMIONIC\ SUPERPOSITION}, c_1, c_2 \in \Re \quad (5.55)$$

in providing the bonding energy of the homopolar system "1 and 2" (i.e., a system in which all atoms are of the same type with different wave functions contributing to the bond). The corresponding eigenvalue is written as the expectation value of the Hamiltonian of Equation 5.43 emphasizing on Schrodinger (Sch) and nonlinear (BB) entries:

$$E_{bond} = \frac{\left[\begin{array}{l}\int d\mathbf{r}\left(c_1\psi_1(\mathbf{r})+c_2\psi_2(\mathbf{r})\right)^* H_{Sch}\left(c_1\psi_1(\mathbf{r})+c_2\psi_2(\mathbf{r})\right)\\ +g\int d\mathbf{r}\left|\psi_{BB}(\mathbf{r})\right|^4\end{array}\right]}{\int d\mathbf{r}\left(c_1\psi_1(\mathbf{r})+c_2\psi_2(\mathbf{r})\right)^*\left(c_1\psi_1(\mathbf{r})+c_2\psi_2(\mathbf{r})\right)}$$

$$\cong \frac{\left(c_1^2+c_2^2\right)\left(H_{11}+(\eta_{Molec}/g)V_{11}\right) + 2c_1c_2\left(H_{12}+(\eta_{Molec}/g)V_{12}\right)}{c_1^2+c_2^2+2c_1c_2S+(\eta_{Molec}/g)} \quad (5.56)$$

The notations in Equation 5.56 represent the integrals of the intraatomic $H_{11} = \int\psi_{1/2}H_{Sch}\psi_{1/2}$, interatomic $H_{12} = \int\psi_{1/2}H_{Sch}\psi_{2/1}$, and overlapping $S = \int[\psi_{1/2}]^2$ contributions, along the potential integrals paralleling the Coulombic and exchange contributions for electrons in bonding, namely $V_{11} = \int\psi_{1/2}V\psi_{1/2}$ and $V_{12} = \int\psi_{1/2}V\psi_{2/1}$, respectively. Also, note that in deriving the result of Equation 5.56 the universal Hohenberg–Kohn functional was implemented with the expression

$$F_{HK} = E_{bond} - C_A = E_{bond} - \left(c_1^2+c_2^2\right)V_{11} - 2c_1c_2V_{12} \quad (5.57)$$

Now, the variational principle respecting the MO-coefficients of Equation 5.55 upon the trial energy (5.56) yields that the *bonding* (+) and *antibonding* (–) wave functions display the same forms as in the classical Heitler–London homopolar model (Heitler and London 1927) while the corresponding energies are corrected by the *bondonic* BEC contribution

$$E_{bond}^{\pm} = \frac{H_{11}\pm H_{12}}{1\pm S} + \left|\langle\psi(\mathbf{r})\rangle\right|^2 \frac{\eta_{Molec}}{8\pi E_{bond}^{\pm}}\frac{V_{11}\pm V_{12}}{1\pm S} \quad (5.58)$$

In Equation 5.58 one recognizes the inverse of the bosonic coupling parameter (1/*g*) considered with its bondonic-order parameter form of Equation 5.44; nevertheless, unlike the classical treatment, variational result (5.58) shapes as an additional quadratic equation of type $x = a + b/x$ of which the basic solutions $x_{1,2} = 1/2[a \pm \sqrt{(a^2 + 4b)}]$ provides in the limit of small bosonic interaction, that is, $b \sim \langle\psi\rangle \to 0$, as is the case of bondons in chemical bonding, the actual BEC–DFT twofold bonding and antibonding energies:

$$E_{bond-BEC-I}^{\pm} = -\left|\langle\psi(\mathbf{r})\rangle\right|^2 \frac{\eta_{Molec}}{8\pi}\frac{V_{11}\pm V_{12}}{H_{11}\pm H_{12}} \quad (5.59)$$

$$E_{bond-BEC-II}^{\pm} = \frac{H_{11}\pm H_{12}}{1\pm S} + \left|\langle\psi(\mathbf{r})\rangle\right|^2 \frac{\eta_{Molec}}{8\pi}\frac{V_{11}\pm V_{12}}{H_{11}\pm H_{12}} \quad (5.60)$$

The results of Equations 5.59 and 5.60 considerably enlarge (and enrich) the chemical bonding paradigm by including condensation and ordering phenomena, especially those associated with ^4He or alkali gases (Leggett 2001).

The DFT formulation of BEC is currently a very vital topic due to innumerous recent theoretical advances in describing both fermionic and bosonic matter, based on the main DFT and BEC concepts, namely the nonlinear Schrödinger equation, the condensate density, the Hohenberg–Kohn universal functional, the Thomas–Fermi limit, the self-interacting bosonic strength, and chemical hardness, enabling the bosonic (bondonic)–fermionic mixture formulation of the chemical bonding as a direct conceptual consequence. The results show that the bosonic nature of the chemical bonding primarily depends on the degree with which the atoms involved in bonding display the bosonic condensation when considered as single gas. As such, when the small nonzero order parameter ($\langle\psi\rangle \neq 0$, $\langle\psi\rangle \to 0$) is assumed upon the quantum particles of the chemical bonding—the bondons, see Equations 5.44 through 5.46, the BEC influence is manifested over Heitler–London bonding and antibonding energetic terms in Equations 5.59 and 5.60. Equally relevant, the presence of the molecular chemical hardness η_{Molec} on the nominator of Equations 5.59 and 5.60 further supports the condensation phenomenology by its consecrated maximum hardness principle for a stable molecule (Chattaraj et al. 1991, 1995; Ayers and Parr 2000; Putz 2008b,c): *the higher the hardness the higher the stability of the chemical system.*

The model is conceptually verified by the prediction of interdiction of the He–He chemical bond, even through the bondonic particle, specifying instead new bonding states for the H_2 molecule suitable to be experimentally detected in condition of BEC (temperatures of nanoKelvin degree), see Figure 5.6.

5.3 PROPAGATING SW DEFECTS IN GRAPHENE: PHASE TRANSITIONS BY BONDONS

5.3.1 Topological Representation of SW Propagations*

Before modeling the SWw propagation, it is worth introducing the graphic tool used to generate this kind of defects on the hexagonal structures. The effectiveness of such an algorithm derives from the choice to operate in the *dual topological representation* of the graphenic layers as shown in Figure 5.7. Also, the topological modeling will be conducted in the dual space.

The generation of the SW rotations is greatly facilitated by considering the dual representation of the graphene layer

* Section 5.3.1 is excerpted from Ori, O., F. Cataldo, M.V. Putz. 2011. Topological anisotropy of Stone-Wales Waves in graphenic fragments. *Int. J. Mol. Sci.* 12:7934–7949.

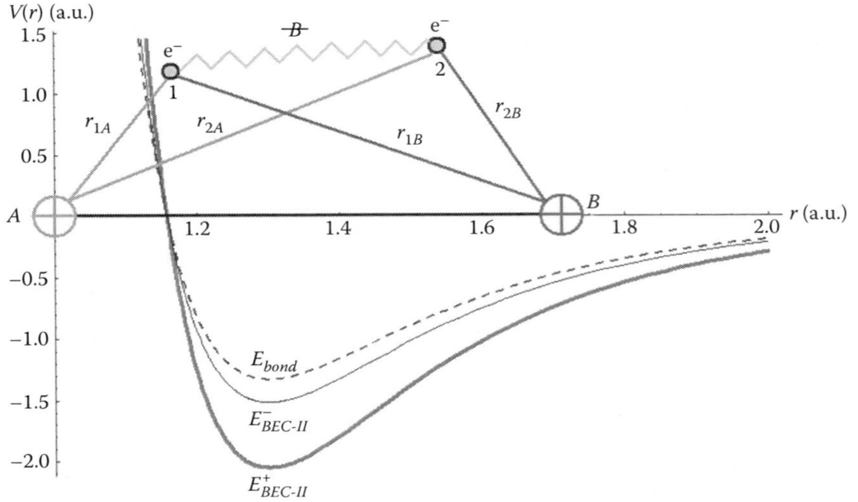

FIGURE 5.6 Illustration of the biatomic chemical bond through the bondonic–bosonic contribution at interelectronic interaction; the model is overlapped with the states of the molecular hydrogen thus treated, with the highlighting of the BEC levels of bonding (thick line) and antibonding (thin line), both placed on the level of "normal" bonding of the molecule in gaseous state. (Adapted from Putz, M.V. 2012a. *Struct. Bond.* 149:1–50.)

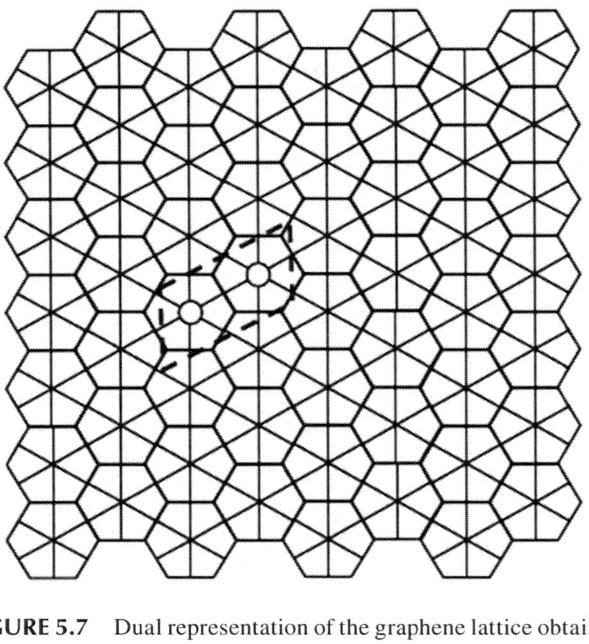

FIGURE 5.7 Dual representation of the graphene lattice obtained by replacing each hexagonal face by the central six-connected graph node. Graphene plane is then equivalently tiled by hexagons (direct space) or by starred nodes (dual space). The x-periodic (y-periodic) direct GNR has the armchair (zigzag) orientation. The framed unit cell has been used to build this 4×7 graphenic fragment. (From Ori, O., F. Cataldo, M.V. Putz. 2011. *Int. J. Mol. Sci.* 12:7934–7949.)

by assigning to each hexagonal face the corresponding six-connected (starred) vertex. Figure 5.7 visually overlaps both direct and dual graphene representations showing their topological equivalency: each pair of adjacent faces in the direct lattice corresponds in fact to a pair of bonded nodes in the dual graph and vice versa. The graphene fragment is taken in its *armchair* orientation along x and Figure 5.7 evidences a 4×7 dual lattice and its unit cell. In the dual lattice, the generic $SW_{p/r}$ *simply rotates* the internal edge between the p- and the r-connected nodes, making the study of the SW rearrangement very simple and suitable for automatic procedures.

On the graphene dual layer the $SW_{6/6}$ rotation (Figure 5.4c) then changes four six-connected nodes (white circles) into two five-connected (shaded circles) and two seven-connected (black circles) vertices, matching the standard transformation in the direct lattice of four hexagons in two pentagons and two heptagons (Figure 5.4a). Moreover 5|7 pairs may also *migrate* in the graphene lattice, pushed by consecutive SW transformations of $SW_{6/7}$ type that rotate, in the dual space, the *vertical* edge between the six-, and the seven-connected vertices, driving the *diagonal* diffusion of a 5|7 pair in the graphene lattice. Figure 5.2d gives more details about the swapping mechanism between the 5|7 and the 6|6 couples. The repeated action of the $SW_{6/7}$ operator originates the topological SWw in both lattice representations.

Figure 5.8a represents the diagonal diffusion of the SWw (dislocation dipole) after four $SW_{6/7}$ rearrangements, evidencing with the dashed arrows the increasing distance between the two 5|7 pairs of the original $SW_{6/6}$ dislocation. At each step, the pentagon (shaded circle) and the heptagon (black circle) interchange their locations with those of two hexagons (white circles) producing the *diagonal SWw*, a large dislocation dipole that modifies the landscape of direct and dual lattices (Figure 5.8a and b). Being η the size of the dislocations (e.g., η equals the number of 6|6 pairs included between the two 5|7 pairs) both examples in Figure 5.8 have size $\eta = 4$, assuming size $\eta = 0$ for the basic $SW_{6/6}$ rotation of Figure 5.3b. Equivalently, η equals the number of $SW_{6/7}$ rearrangements used to generate the dislocations in both spaces.

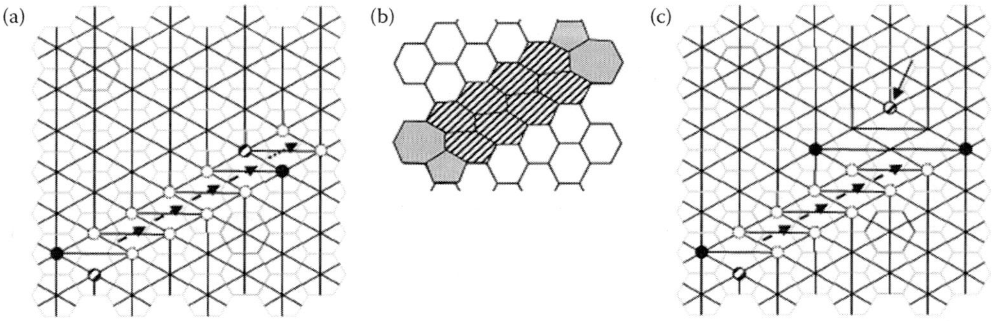

FIGURE 5.8 (a) Diagonal SWw (dislocation dipole) in the dual graphene layer after the generation and four propagation steps (size $\eta = 4$); at each step (dashed arrows), $SW_{6/7}$ swaps the pair made by one pentagon (dashed circle) and one heptagon (black circle) with two connected hexagons (dotted circles); dotted arrow indicates the next available translation of the 5|7 pair; (b) the topological modification (a) originates, in the direct lattice, a hexagonal inter-grain spacing (dashed rings); (c) after a few more SW rotations, an isolated pentagon (arrowed), forming a small nanocone, is generated. (From Ori, O., F. Cataldo, M.V. Putz. 2011. *Int. J. Mol. Sci.* 12:7934–7949.)

An SWw produces (Figure 5.8b) a characteristic *hexagonal inter-grain spacing*, isomeric to the pristine graphene layer that represents, therefore, a good theoretical model for the boundary between graphenic fragments.

The dual space represents the natural arena for studying all sorts of SW flips, avoiding the graphical difficulties that one usually encounters in redistributing the carbon atoms and bonds in the direct lattice. One easily generates the *vertical* SWw by applying, in fact, our graphical algorithm to the *diagonal* edges of the graphene dual lattice (Figure 5.9). SW transformations also produce very complex rearrangements of the graphenic layer including isolated pentagonal nanocones, as the very little one on top of the diagonal SWw in Figure 5.8c, creating fullerenic-like regions in the graphenic plane (Cataldo et al. 2010). The proposed dual space graphical algorithm appears, therefore, capable to handle complex combinations of general SW rotations $SW_{p/r}$ to create novel classes of isomeric rearrangements, with rings made of various numbers of atoms, of fullerene (dimensionality $D = 0$), nanotubes ($D = 1$), graphenic structures ($D = 2$), or crystals ($D = 3$) as schwarzites or zeolites.

The somehow arbitrary, definition of *diagonal* or *vertical* direction assigned to the SWw on closed surfaces of graphenic fragments, nanoribbons, and nanotubes, considers the *armchair* graphene orientation selected in Figure 5.7. Topologically, the extension of the region interested by the dislocation dipole may be arbitrarily enlarged by applying more and more $SW_{6/7}$ rotations. In summary, on the armchair-oriented graphene (Figure 5.7), the simplest propagation mechanisms available for the 5|7 pairs are

- *Diagonal* SWw, Figure 5.8: $SW_{6/7}$ rotates the *vertical* bond of the graphene dual lattice between the six-connected node and the seven-connected node of the diffusing 5|7 pair, causing the *diagonal* drift of the pair and the creation of a new horizontal hexagon–hexagon bond (Figure 5.4d gives some more details).
- *Vertical* SWw, Figure 5.9: $SW_{6/7}$ rotates the *diagonal* bond of the graphene dual lattice between the six-connected node and the seven-connected node of the diffusing 5|7 pair, with the overall effect to *vertically* shift the pair, generating a new *antidiagonal* hexagon–hexagon bond.

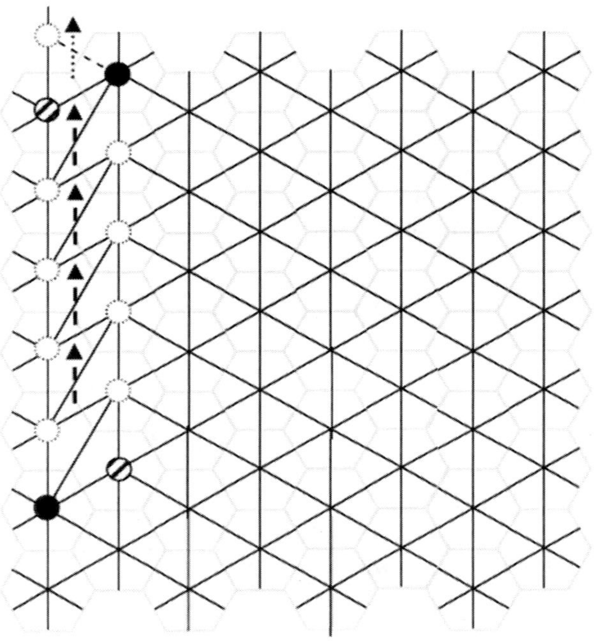

FIGURE 5.9 SW vertical wave in the dual graphene layer after four propagation steps (dashed arrows); $SW_{6/7}$ swaps the pentagon (dashed circle) heptagon (black circle) pair with two hexagons (dotted circles); dotted arrow indicates the next possible translation of the 5|7 pair, induced by an $SW_{6/7}$ rotation of the hexagon–heptagon diagonal dashed bond. The SWw generates antidiagonal hexagon–hexagon bonds with respect to the unrotated one. (From Ori, O., F. Cataldo, M.V. Putz. 2011. *Int. J. Mol. Sci.* 12:7934–7949.)

The diffusion processes mentioned above apply to an isolated 5|7 dislocation monopole as well to the 5|7 double pair arising from a $SW_{6/6}$ rearrangement. In the following, we mainly study the latter case, focusing on the mechanisms leading to the creation of diagonal or vertical *extended dislocation dipoles* in the graphene lattice. It is worth noting that similar topological tools are used in other disciplines, for example, in biology where wave-like diffusion mechanisms model cells proliferation processes (Pyshnov 1980).

In concluding this section, we observe that from the pure topological point of view one may consider each lattice configuration illustrated in this work as the result of an *instantaneous* transformation caused by a *single, nonlocal* SW rotation. This new class of transformations represents a further generalization, potentially infinite, of the nonlocal rearrangements early proposed (Babic et al. 1995) to generate the entire isomeric space of a given C_n fullerene starting for a limited number of inequivalent cages.

5.3.2 Nanoribbon Phase Transition by Bondonic Path Integral*

In order making further bondonic connection with extended systems, graphene in particular and of their propagating defects, one considers a particle (the bondon) with mass M moving between the space points x_a and x_b under the potential $V(\bar{x})$ (the molecular or molecular net topological potential, implemented here—see the following) with $\bar{x} = (x_a + x_b)/2$ the associate quantum statistical propagator may be expanded up to the second order in $\Delta x = x_b - x_a$ (Kleinert 2004)

$$(x_b\hbar\beta; x_a 0) = \sqrt{\frac{M}{2\pi\hbar^2\beta}}\exp\left\{-\frac{M}{2\hbar^2\beta}(\Delta x)^2 - \beta V(\bar{x})\right\}$$

$$\times\left\{1 - \frac{\hbar^2\beta^2}{12M}\nabla^2 V(\bar{x}) - \frac{\beta}{24}(\Delta x\nabla)^2 V(\bar{x}) + \frac{\hbar^2\beta^3}{24M}\left[\nabla V(\bar{x})\right]^2\right\} \tag{5.61}$$

with β being the inverse of the thermal energy $k_B T$ and $\hbar$ the reduced Planck constant. Upon close integration over the classical or average path, it provides the immediate partition function

$$z(\beta) = \int (x_b\hbar\beta; x_a 0)\Big|_{x_a = x_b = \bar{x}} d\bar{x}$$

$$= \sqrt{\frac{M}{2\pi\hbar^2\beta}}\int \exp\left[-\beta V(\bar{x}) - \frac{\hbar^2\beta^2}{24M}\nabla^2 V(\bar{x})\right] d\bar{x} \tag{5.62}$$

that further specializes for the bondonic evolution and mass (5.45) to the working form

$$z_{\mathcal{B}}(\beta, X_{bond}, E_{bond}) = \frac{1}{2X_{bond}}\sqrt{\frac{1}{\pi E_{bond}\beta}}$$

$$\int \exp\left[-\beta V(\bar{x}) - \frac{\beta^2}{12}E_{bond}\bar{x}^2\nabla^2 V(\bar{x})\right] d\bar{x} \tag{5.63}$$

where, for consistency, X_{bond} was considered as $\bar{x}$ under spatial integration.

The connection with topological properties of the periodic nets is achieved through considering the correspondence of the driving potential with a proper topological descriptor for the periodic cell $V(\bar{x}) \to \Xi$ so that the extension to superior orders $\nabla^2 V(\bar{x}) \to \Xi^{[2]}$ also holds. Two basal properties of *distance-based* topological descriptors perfectly fit in describing bondons properties: (i) topological potentials Ξ behave as long-range interatomic potentials connecting all atoms pairs in the system; (ii) topological invariants exhibit, also in case of periodic structures, peculiar polynomial behaviors with leading coefficients only depending from the dimensionality D of the system—see the recent review on topological modeling methods and results (Iranmanesh et al. 2012).

This causes the immediate consequence of fixing the energy-length realm of the bondon in analytical manner: it starts with considering the partition function (5.63) to its 0th order, specific for motion under thermal and topological energy, further restrained to the bondonic existence, for example, over the periodic cell or elementary bonding length in a system (Putz and Ori 2012)

$$z_{\mathcal{B}}^{[0]}(\beta, E_{bond}) = \frac{1}{2X_{bond}}\sqrt{\frac{1}{\pi E_{bond}\beta}}\int_0^{X_{bond}} \exp[-\beta\Xi] d\bar{x}$$

$$= \frac{1}{2}\sqrt{\frac{1}{\pi E_{bond}\beta}}\exp[-\beta\Xi] \tag{5.64}$$

to get the averaging bonding energy as a realization of the allied statistical internal energy

$$\langle E_{Bond}\rangle[\text{kcal/mol}] = -\frac{\partial \ln z_{\mathcal{B}}^{[0]}(\beta, E_{bond})}{\partial\beta} = \frac{1}{2\beta} + \Xi$$

$$= \begin{cases}\Xi, \ldots \beta \to \infty (T \to 0K) \\ \infty, \ldots \beta \to 0 (T \to \infty K)\end{cases} \tag{5.65}$$

Equation 5.65 generally expresses the "total" bonding energy as a sum of the thermal energy as a representation of kinetic motion of the bondon with the topological contribution carrying the physical information of the nanostructure; it naturally resumes to the last quantity for the low temperature stabilization while having no superior bounding for high temperature limit—a specific manifestation of the ergodic character of the bondonic (Bohmian based) representation contained in *X–E* equation of Figure 5.6. Moreover, the expressions (5.73) and (5.66) fixed the averaged bonding length to be (Putz and Ori 2012)

$$\langle X_{Bond}\rangle[\text{Å}] = \frac{2a\beta}{1 + 2\beta\Xi} = \begin{cases}\frac{a}{\Xi}, \ldots \beta \to \infty (T \to 0\text{K}) \\ 0, \ldots \beta \to 0 (T \to \infty\text{K})\end{cases} \tag{5.66}$$

* Section 5.3.2 is excerpted from Putz, M.V., O. Ori. 2012. Bondonic characterization of extended nanosystems: Application to graphene's nanoribbons. *Chem. Phys. Lett.* 548:95–100.

This result has the rationale in predicting no bonding (length) for indefinitely heated nanostructure (when all periodicity and bonding structures are destroyed), yet with a finite β-dependency for stable periodic nets with a ground state limit in terms of topological index Ξ as an energetic measure for the bonding in the system, consistent with *X–E* equation of Figure 5.6.

The bondonic length of Equation 5.66 may be further employed to the full partition expression of Equation 5.63 for modeling the transition between two phases of a nanosystem, for example, an ideal and a modified (with defects) ones, as follows: let's consider the topological index computed for ideal net system Ξ_0 as well as for the net with (propagating) defects Ξ_D alike, the so-called critical temperature $1/\beta_{CRITIC}$ for the phase transition between these two systems may be carried out through the caloric capacity equation

$$C_{\mathcal{B}}(\beta_{CRITIC},\Xi_0)\big|_{IDEAL} = C_{\mathcal{B}}(\beta_{CRITIC},\Xi_D)\big|_{DEFECTS} \tag{5.67}$$

Analytically, for a net with *N*-periodic cells, one gets first the bondonic energy per cell, under grand canonical conditions, as (Putz and Ori 2012)

$$N^{-1}\langle E_{Bond}^{N}\rangle[\text{kcal/mol}] = -\frac{\partial}{\partial\beta}\ln\left\{\frac{z_{\mathcal{B}}^{N}(\beta)_{\langle X_{bond}\rangle,\Xi,\Xi^{[2]}}}{N!}\right\} = \frac{3}{2\beta}+\Xi+\frac{1}{\beta+2\beta^2\Xi} \tag{5.68}$$

having the same thermodynamic limits as in Equation 5.65 yet with 3/2 correction on the kinetic contribution as well as with more complexity on the topological dependency for finite temperature regime.

In the same grand canonical framework, the bondonic caloric capacity writes as

$$C_{\mathcal{B}}(N,\beta) = k_B\beta^2\frac{\partial^2}{\partial\beta^2}\ln\left\{\frac{z_{\mathcal{B}}^{N}(\beta)_{\langle X_{bond}\rangle,\Xi,\Xi^{[2]}}}{N!}\right\} = \frac{5+4\beta\Xi(5+3\beta\Xi)}{2(1+2\beta\Xi)^2}Nk_B \tag{5.69}$$

Remarkably, the expression (5.69) recovers the Debye energy limit $5Nk_B/2$ for conducting electrons in solids either for high temperature limit and or for the no topological structure ($\beta\vee\Xi\to 0$) while having minimal limit as $3Nk_B/2$ energy for the bondons in most stable state or with maximum topological potential ($\beta\vee\Xi\to\infty$) corresponding with their freely spanning over the net considerate as a quantum box with infinite potential.

Employing the form (5.69) to phase transition modeled by Equation 5.67 the absolute finite critical temperature is predicted with the form (Putz and Ori 2012)

$$|\beta_{CRITIC}| = \frac{\Xi_0+\Xi_D}{4\Xi_0\Xi_D} = \begin{cases}\dfrac{1}{2\Xi_0} & \ldots\Xi_0=\Xi_D\\ \infty & \ldots\ \Xi_0\to 0\ \vee\ \Xi_D\to 0\end{cases} \tag{5.70}$$

From the limits of Equation 5.70 one observes the major role the topological index plays in predicting the characteristic temperature $T = 2\Xi/k_B$ at which a given nanosystem (topology) is stabilized (either as ideal or with defects) by the bondonic motion through it; otherwise, for an absent topology the bondonic motion is purely entropic at $T = 0$ K, without observable character, according with Equation 5.70 for the indefinite limit $\beta\Xi\to\infty\times 0$.

Worth noting that, through introducing the relative temperature and topological parameters respecting the critical and the defect ones, $\delta = \beta/\beta_{CRITIC}$ and $\sigma = \Xi_0/\Xi_D$, respectively, in Equation 5.70 and then in Equation 5.69 one obtains the predicted normalized bondonic caloric capacity per periodic cell (Putz and Ori 2012)

$$\frac{C_{\mathcal{B}}(\sigma,\delta)}{Nk_B} = \frac{3}{2}+\frac{1+\delta\sigma(1+\sigma)}{(1+0.5\delta\sigma(1+\sigma))^2}\xrightarrow{\delta\&\sigma\to 1}\frac{43}{18}\cong 2.38889 \tag{5.71}$$

representing a sort of universal value for twofold critical capacity of nanosystems, that is, at critical temperature and coexistence of ideal-with-defect structures. However, in the light of above discussion, the closer value of Equation 5.71 with Debye value of Equation 5.69, $C_{\mathcal{B}}(N,\beta\to 0)/Nk_B\to 2.5$, indicates that at critical regime ($\delta\&\sigma\to 1$) the bondons behave like a Debye conducting electronic gas in solids at high temperature.

5.3.3 Case of SW Moving Defects in Graphene

Our theoretical considerations are based on the calculation and minimization of certain distance-based topological invariants Ξ of the atomic network representing the graphenic fragment under present study. In our approach, long-range topological invariants Ξ in fact work like the effective potential energy of the electronic system (*the topological potential energy* in effect) and they are, therefore, subject of a calibration phase obtained by fitting the Ξ curve with potential energy values obtained by *ab initio* computations on some significant lattice configurations. An introductory exposure of topological modeling methods is provided in (Iranmanesh et al. 2012).

Bondons quite naturally match with the topological descriptions of a given chemical system, since both these newly predicted bosonian particles and the topological potentials possess inherent nonlocal characters receiving contributions from all the atoms of the system. In this section, the suitable topological descriptors for the subsequent treatment of the quantum statistical behavior of the bondons are derived.

To reach that goal, a peculiar set of defective structures has been taken into consideration in order to effectively describe the influence on the topological properties of the hexagonal

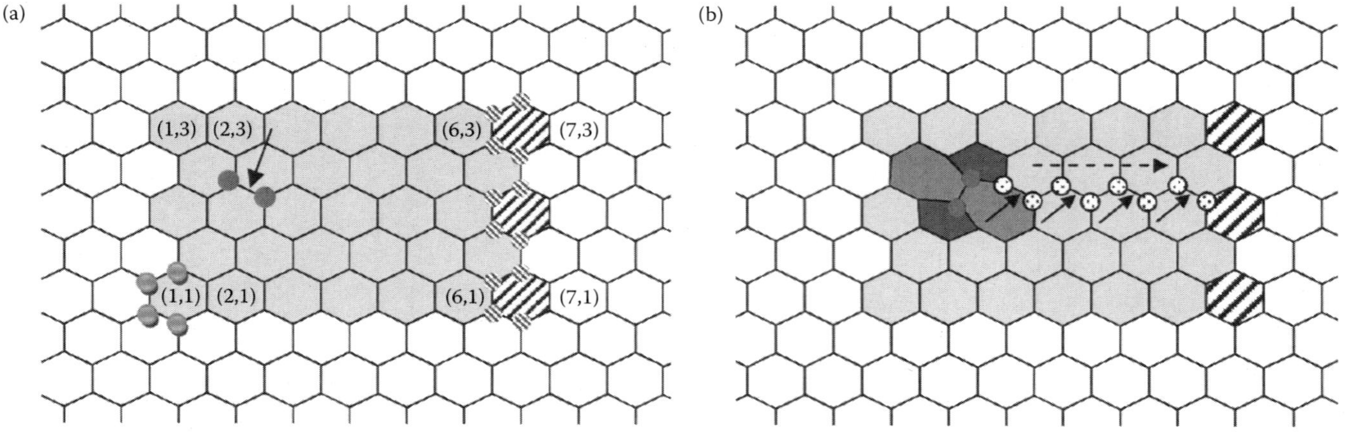

FIGURE 5.10 (a) The colored hexagons constitute the $\boldsymbol{E}$ graphenic supercell made of $n = 84$ carbon atoms. This supercell in its pristine form gets reproduced by translating the four colored carbon atoms shown in the bottom-left (1,1) unit block. The $\boldsymbol{E}$ supercell shares, on the right side, the stripped hexagons with the lattice; (b) generation of the SW defects 5/7/7/5, which may propagate along the dotted line by rotating the arrowed bonds of the pristine lattice.

network exerted by the presence of a SW dislocation which propagates in the lattice.

Figure 5.10 depicts the supercell $\boldsymbol{E}$ with $n = 84$ carbon atoms used in this article which is obtained by translating the unitary block made of four colored carbon atoms shown by the (1,1) unit in the bottom-left of the supercell. The $n = 84$ carbon atoms are grouped in $h = 21$ unitary blocks labeled from (1,1) to (7,3) in Figure 5.10a. The structure of $\boldsymbol{E}$ is such that it shares with the surrounding lattice the right-most units, which are represented as stripped hexagons in Figure 5.10a. The arrow indicates the bond rotated by the SW rotation. This isomeric rotation originates the well-known Stone–Thrower–Wales topological defect 5/7/7/5, by transforming four touching hexagons in the pentagon–heptagon *double pair*. A peculiar topological feature of this defect is represented by the fact that it has the ability to migrate in the lattice by rotating a variable number η of bonds ($\eta = 1$ corresponds to the initial SW rotation), inserting between the two 5|7 pairs a number $\eta - 1$ of pairs of hexagons which split the initial pentagon–heptagon couples; in Figure 5.10b some bonds available for subsequent rotations are indicated by arrows, whereas Figure 5.11 shows that the migration of the two pairs of hexagons have been interposed between the 5|7 pairs, the size of the SW dislocation dipole being in this case $\eta = 3$. This purely topological property has been originally reported by the authors (Ori et al. 2011), see above sections, as a pure topological explanation of the linear rearrangement of the hexagonal mesh which in literature is often called the SW dislocation dipole or, following the authors, the SWw.

The SWw topological rearrangement of the graphenic supercell $\boldsymbol{E}$ after $\eta = 3$ migration steps is shown in Figure 5.11, the structure with $\eta = 0$ and $\eta = 1$ corresponding respectively to the pristine graphene lattice and to the initial SW double-pair given in Figure 5.10b.

In the topological approach currently adopted for the chemical systems—the graphenic lattices—are simply described as graphs made by n three-connected nodes connected in hexagonal meshes, being the physic-chemical

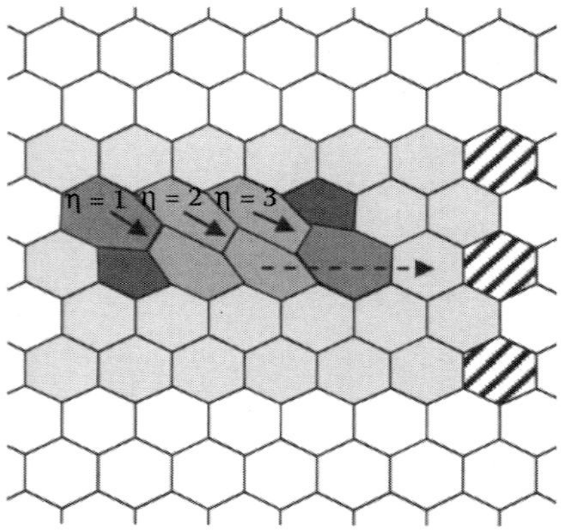

FIGURE 5.11 The propagation of SWw defect along the dotted line after its size reaches the $\eta = 3$; η value equals the number of bonds rotated to generate the configuration; $\eta - 1$ is the number hexagons pairs which split the two pentagon–heptagon flat regions.

information stored in the long-range coordination shells of each atom. In practice, topological approximation found its nontrivial predictive power by looking at *each atom in interaction with all the remaining atoms* of the chemical network. Topological modeling methods applied for the determination of the form of the topological invariants Ξ typically considers the distance d_{ij} between two nodes i and j in the lattice (the chemical distances), a pure topological quantity which corresponds to the number of bonds connecting the two atoms following the *shortest path* in the graph network; clearly when $d_{ij} = k$, atom j belongs to the k-th coordination shell of atom i and vice versa. Indicating with M the length of the longest path of the graph—the integer M corresponds to the graph *diameter*—and with b_{ik} the number of k-neighbors of atom i, the effect of *long-range* lattice connectivity on this atom is effectively summarized by topological invariant w_i

$$w_i = ½\, \Sigma_k k b_{ik} \quad \text{with} \quad k = 1,2,\ldots,M-1,M \tag{5.72}$$

where $n - 1 = \Sigma_k b_{ik}$ and $b_{i1} = 3$ for atoms of graphenic structures without vacancies.

Symbol $\underline{w}(\underline{\underline{w}})$ indicates the smallest (largest) in the set of integers $\{w_i\}$. Nodes having $w_i = \underline{w}$ are the so-called *minimal vertices* of the graph. Set $\{b_{ik}\}$ identify the Wiener-weights (WW) of vertex *i*. Wiener index *W*, the oldest among the molecular invariants (Iranmanesh et al. 2012), derives from the half-summation of the chemical distances d_{ij} of the graph with *n* nodes

$$W(n) = \tfrac{1}{2}\Sigma_{ij}d_{ij} = \Sigma_i w_i \quad \text{with } i,j = 1,2,\ldots,n-1,n \tag{5.73}$$

It measures the overall compactness of the chemical structure.

A… higher stability is assigned to lattices with low W(n)

This strong approximation holds when similar structures are compared. The present computational method relates, therefore, the topological potential Ξ to the Wiener index by introducing a suitable constant α to restore the correct energy dimension

$$\Xi = \alpha W \tag{5.74}$$

We end the present section by determining the exact expression for the Wiener index when the propagation parameter η is taken into consideration. First of all, let's select the reference 3×3 lattice in which the SWw defect migrates. To avoid marked self-interaction topological effects, the supercell ***E*** is isolated by a frame of seven equal cells. Then, the generation of the SW defects and the subsequent separation of the 5|7 pairs takes place only in the central supercell ***E***. The elegant polynomial dependence of *W* and η derived for such a lattice is

$$\begin{cases} W(\eta) = W_0 + \dfrac{3433}{60}\eta^5 - 732\eta^4 + \dfrac{37151}{12}\eta^3 - 3776\eta^2 - \dfrac{92777}{15}\eta \\ W_0 = W(\eta = 0) = 4467960 \end{cases} \tag{5.75}$$

The appropriate α values are determined, for the current defective supercell ***E***, by a specific interpolation process that is described in the following.

One consider various energetic quantities, among which total energy, electronic energy or binding energy, alongside the so-called parabolic pi-energy computed upon combining the electronegativity and chemical hardness orbital frontier (based on highest occupied molecular orbital [HOMO] and lowest unoccupied molecular orbital [LUMO]) definitions with the number of pi-electrons engaged in the molecular reactivity, upon the consecrated formula (Putz 2011d,e)

$$E_\pi = -\underbrace{\left(-\frac{\varepsilon_{LUMO} + \varepsilon_{HOMO}}{2}\right)}_{\text{Electronegativity}} N_\pi + \frac{1}{2}\underbrace{\left(\frac{\varepsilon_{LUMO} - \varepsilon_{HOMO}}{2}\right)}_{\text{Chemical hardness}} N_\pi^2 \tag{5.76}$$

It may also be equivalently considered under the ionization potential (IP) and electronic affinity (EA) by rewriting (5.76) within the frozen core approximation or by Koopmans' theorem (Putz 2011d,e)

$$E_\pi = -\frac{IP + EA}{2}N_\pi + \frac{IP - EA}{4}N_\pi^2 \tag{5.77}$$

Table 5.4 presents that the numerical values of these energies are reported paralleling the molecular fragments carrying the graphenic SW topological defects with the defect-step evolution and associate Wiener values computed from Equation 5.75. Next, the "best" correlation of such energetic frameworks is selected for performing the topo-energetic calibration according with the Equation 5.74. However, one notes from the results of Table 5.4 the anomalous *binding energy* behavior around the defect propagation in the stage $\eta = 3$, although having this energy an overall superior correlation factor respecting other energetic variants. Nevertheless, one observes the second best correlation as the parabolic energy and not the total one while the electronic energy alone does not match well the Wiener index curve of the propagating SW dislocation defect. Yet, in order to differently characterize the pristine and defective graphenic 2D lattices, one should produce two kinds of polynomial forms out of the topological one (5.75) even energetically rescaled; in this way, one consider two regimes of Equation 5.75.

- One free of defects (the "0" structure) for which the polynomial (5.75) is equally specialized within [0,1] range of the η "step" parameter producing the *W* values for six equally distanced step-points, i.e., taking the values for $W(\eta = 0)$, $W(\eta = 0.2)$, $W(\eta = 0.4)$, $W(\eta = 0.6)$, $W(\eta = 0.8)$, and $W(\eta = 1)$, thereafter interpolated to produce the working polynomial of the fifth degree; being it corrected by the binding energy scale factor of Table 5.4 and then by the conversion factor to [kcal/mol] units one obtains the final pristine polynomial form

$$\Xi_0(\text{kcal/mol}) = 135849 - 26.0167\eta - 7.06913\eta^2 + 0.900852\eta^3 - 0.038385\eta^4 + 0.000556568\eta^5 \tag{5.78}$$

- The defective polynomial is obtained by specializing Equation 5.75 with the integer step over the set of six points, specifically the values $W(\{0,1,2,3,4\})$ + that one obtained by extrapolating Equation 5.75 to the value of $W(\eta = 5)$ so that maintaining the fifth order of the working potential; they are then employed toward interpolation while, after energetic calibration and kcal/mol conversion one yields the unfolded form

$$\Xi_D(\text{kcal/mol}) = 135772 + 421.581\eta - 548.013\eta^2 + 200.508\eta^3 - 30.9477\eta^4 + 1.73927\eta^5 \tag{5.79}$$

TABLE 5.4
Synopsis of the Topo-Reactive Parameters for the Defects Instances (Including Pristine Net at the Step $\eta = 0$) from SW Propagations in Graphene Sheet as Described in Figures 5.10 and 5.11

Defect Instant Structure	Defect Step, $W(\eta) = W_\eta$	Electronic Energy (eV)	Total Energy (eV)	Binding Energy (eV)	Parabolic Energy (eV)
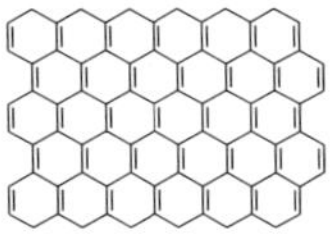	$\eta = 0$ $W_0 = 4{,}467{,}960$	2,858.69979	2,595.306	7,308.17	13,063.1207
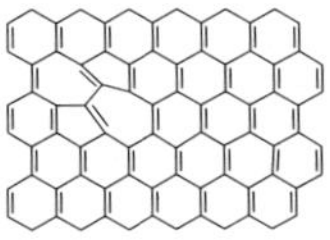	$\eta = 1$ $W_1 = 4{,}460{,}420$	2,425.90314	2,409.47	7,494.0069	102,344.109
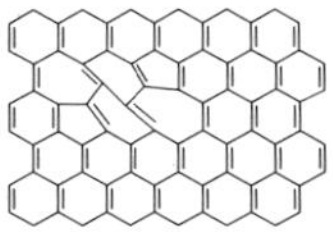	$\eta = 2$ $W_2 = 4{,}455{,}370$	2,641.35644	2,410.023	7,493.4534	100,091.3384
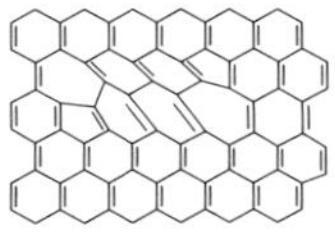	$\eta = 3$ $W_3 = 4{,}453{,}620$	90,063.404	10,331.43	−427.9522	6,770.427006
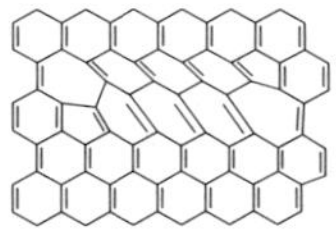	$\eta = 4$ $W_4 = 4{,}452{,}140$	2,428.23769	2,408.129	7,495.3472	102,353.338
Correlation slope, α in Equation 5.74		0.0045014	0.000904	0.0013177	0.01455902
Correlation factor R^2		0.24739544	0.620242	0.776828	0.675377171

Note: The Wiener index abstracted from the polynomial form of Equation 5.75, electronic, total, binding, and parabolic energy—the last one computed from Equations 5.76 and/or 5.77 with the total number of pi-electrons $N_\pi = 82$, in each case, within the semiempirical AM1 framework (Hypercube, 2002), respectively; the bottom of the table reports the correlation slope of Equation 5.74 and the associate correlation factor for each set of structural energies respecting the topological Wiener one toward providing the actual working calibration recipe.

These polynomials are now implemented in the analytical results of the previous section for modeling the critical regime of pristine-to-defective graphenic electronic properties by the aid of bondonic treatment of bonding. The respective results are presented by the draws of Figure 5.12 while revealing interesting features as

- One notes that in spite of the apparent monotonic decrease of the topo-energetic Wiener potentials (Ξ_0 and Ξ_D), either for pristine or defective graphenic nanoribbons, their difference turns into a nonlinear shape, with two inflections points, which is then preserved to all the remaining physical–chemical parameters around the critical regime.
- The inverse of critical thermal energy β_{CRIT} computed upon Equation 5.70, despite being a combination of the two Wiener topo-energetic potentials shown above, already displays a nonlinear attractive-repulsive shape which is further saturated under the "sigmoid" form for its difference respecting the pristine graphenic configuration as well as for D-to-0 difference, as already predicted by the difference of the topo-potentials themselves as above. Remarkable, the bondonic range of action for the pristine graphene (X_0) follows the same trend as the critical inverse of thermal energy, which is more pronounced for defective form (X_D), while their difference recovers the sigmoid shape with two inflexion points: one placed at the mid phenomenological region between the pristine and the first defective step while the second point corresponds with the step $\eta = 3$ which already marked the binding anomaly in the graphenic structure as recorded in Table 5.4. Moreover, the bondonic scale results of the size of the aromatic ring in accordance with the bonds' rearrangement complex mechanisms which takes place during the isomeric topological transformations of the lattice due to SW defects formations and propagation; instead, the total energy analysis provides a higher bondonic range in order to improve bonding stability. Worth noting that in accordance with the maximum slope recorded in the hierarchy of Table 5.4 the parabolic energy framework will provide the bondonic range of action to be on the

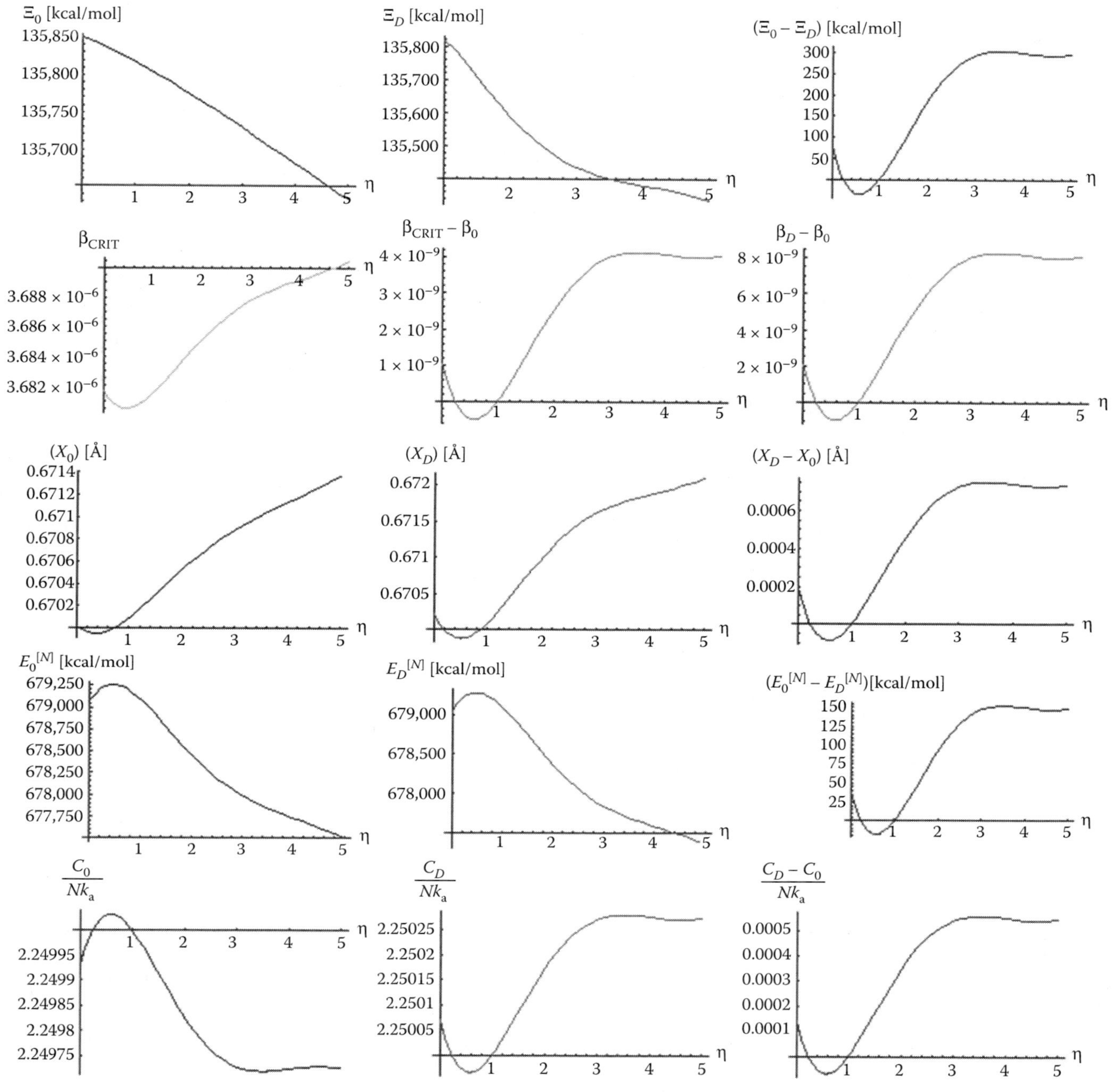

FIGURE 5.12 From top to bottom and left to right: the Wiener-based topological potentials of Equations 5.78 and 5.79 and their pristine (0) to SW defect (D) difference, the critical inverse of thermal energy (5.70) along its pristine-defect-induced differences, the distance of bondonic actions upon Equation 5.66 and their D-to-0 difference, the bondonic N-normalized grand canonical internal energies of Equation 5.68 and the 0-to-D difference, as well as the absolute caloric capacity from Equation 5.69 and the D-to-0 difference.

lowest scale (about 0.06 Å), that is, its action will be restrained about the atomic frontier levels of the carbon atoms in graphenic network; nevertheless, this is in accordance with the reactivity nature of the approach as imposed by the electronegativity and chemical hardness presence in Equations 5.76 and 5.77.

- Going to the energetics of bondons ($E_0^{[N]}$ and $\Xi_D^{[N]}$) one notes the natural inverse behavior respecting the bondonic range of action, due to the inverse X–E dependency equation of Figure 5.6. However, their difference lays in the range of chemical bonding formation, with a slightly increased value respecting the total energy analysis, and with one order less respecting the parabolic energetic case when the defective forms of graphene induce much tension to the pi-electronic bonding condensate so that promoting it to higher excited states; overall the energetic hierarchy is well respected: total energy (aka ground state) < binding energy (aka valence shell) < parabolic energy (aka reactive states) also at the level of bondonic analysis—such that appearing as a reliable tool in characterizing chemical bonding for extended molecular systems; yet, one should also notice the specific Morse-like shape of the 0-to-D difference in bondonic energy so that presenting both the attractive and the repulsive parts of the custom molecular bonding potential nevertheless, here corresponding to the pristine and the defect propagation to saturation, respectively.

- Finally, the bondonic caloric capacities (C_0 and C_D) display both the same range, shape, and D-to-0 difference for all total, binding, and parabolic energetic analyses, thus, proving its universal mean in assessing phase transition. Accordingly, from the sigmoid shapes of Figure 5.12 one concludes upon the existence of the defective information even in the vicinity (the bondonic vibrations) of the pristine graphene with the useful insight upon the natural creation of defects (almost spontaneously) based only on the topological potential (in this case, the Wiener index) even before its macroscopic manifestation (i.e., before the completion of the first step $\eta = 1$). The second inflexion point around the step $\eta = 3$, apart from being consistent with the binding sign reversal in Table 5.4—a feature not recorded to the other energetic frameworks despite the similar final bondonic sigmoid shapes in the D/to/0 structures' differences, actually records the graphenic entering in a plateau/saturation of quantum-statistical information—a situation specific only to caloric capacities C_0 and C_D, either in minimum or maximum forms, respectively. Since for all other physical–chemical quantities computed, such feature was manifested only to D/to/0 structures' differences but not to the separate structures themselves; this means that in terms of thermal actions on the long-range graphenic 2D nanosystems, the pristine phase goes to minimum constant thermal energy C_0 while the defective phase to the maximum-stable thermal information C_D corresponding with *stationary propagation of defects* after the step $\eta = 3$, also in accord with the long-range nature of bondonic range of action. The present results nicely accord also with the present computational restrained dimensionality since all relevant effects appears inside the supercell E of Figure 5.10 while further extension in either direction by increasing the size of the supercell or by the evolution of the topological isomerization of the SW dipole toward other possible defects in graphene, for instance, 5|8|5 bonding rotations and waves (Ori and Putz 2013), are also possible and are envisaged in the near future communications.

5.4 CONCLUSIONS

Exposure of graphene to electron- or ion-beam led to atom dislocation and ring shrinking, with the local apparition of conical domains up to cage closure. Such local defects influence the properties of structurally modified graphenes; the reactivity included. In the future, it is expected that theoretical studies on graphenes will elucidate important results, from design to a possible mechanism of experimental realization of more complex structures.

In this chapter, a possible advanced physical framework for treating graphenic structures and of their electronic and, thus, bonding defects, here in topological (isomeric) sense, is exposed: the Bohmian approach of the quantum description of the matter allows the quantification of the chemical bond by the quasiparticle bondon, able to explain through the mass, the energy and the proper length of action the chemical bonding of the extended systems (for graphene, in area 15–30 Å), and of the nanostructures and of their phase transformations. The applications on graphene give a valuable premise for the further projections of the design of reactive supports for the pharmaceutical and cosmetic compounds, due to the conductor properties, magnetic and of flexibility and chemical saturation—unique for these new types of materials (as recognized also by the Nobel Prize in Physics in 2010). Other utilities and promising research directions, based on the bondonic model of nanostructures contain the description of the optical and acoustic branches through the bondon–phonon interaction, the description of the transitions of phase by the bondonic propagators of fourth order (equal to the maximum of the order of bond in chemical systems), the study of the transitions of phase on defective bases (of order-disorder type), the modeling of nanostructures of "honeycomb" type through the chemical field associated and the specific gauge transformations, the remote quantification of Coulomb's action, the bondonic identification in the IR, and Raman spectra of the chemical compounds, as a size of their reactivity—base of the activity and toxicity bio-, eco-, and pharmaco-logical.

Worthy to notice that the nonlinear character of the pristine-to-defective shapes of all analyzed physical chemical characteristics of 2D graphenic systems such as the Wiener topo-energetic potential, critical inverse of thermal energy relating critical temperature of phase transition, the bondonic range of action and it energy, and along the caloric capacities, were essentially rooted in considering the bondonic–bosonic condensation approach of the chemical bonding, through the generalization of the Schrödinger equation by the so-called Gross-Pitaevsky equation; it further allows the connection of DFT with BEC through the Thomas–Fermi approximation; it was further used for the generalization of the classical model Heitler–London of the chemical bond, here for the homopolar bond, but consistent enough to be considered a premise also for the approach of the cyclic systems, aromatic and conjugated systems. This treatment was centered on the new quantum particle responsible for the chemical bond—the *boson bondon*—having as a consequence its condensation in the chemical bond, thus explaining the orbitalic paradigm of the bonding-antibonding levels with new BEC specific levels, not yet explicitly explored in an experimental way. There is nevertheless hope that before the individual bondonic existence will be observed, the graphenic systems, by their inherent extended molecular structure (in between the molecular and solid state forms) be the first appropriate physicochemical environment in which the long-range properties of bondonic information may be revealed at macroscopic scale.

ACKNOWLEDGMENTS AND COPYRIGHT NOTES

MVP thanks the support by CNCS-UEFISCDI, project number PN II-RU TE16/2010-2013. MVP and OO are grateful to Elsevier for allowing to reuse extended excerpts from their recent work (Putz and Ori 2012) in Sections 5.1.2 and 5.3.2; Section 5.2.3 reconsiders the AIP Conference Proceedings preprint of Putz (2010c) under nonexclusive licensee to author; Sections 5.1.3 and 5.3.1 are essentially based on the original work of Ori et al. (2011), while Sections 5.2.1 and 5.2.2 are essentially based on the original work of Putz (2010a); in all cases, authors' copyright are licensed.

REFERENCES

Abouimrane, A., O.C. Compton, K. Amine, S.T. Nguyen. 2010. Non-annealed graphene paper as a binder-free anode for lithium-ion batteries. *J. Phys. Chem. C* 114:12800–12804.

Anderson, M.H., J.R. Ensher, M.R. Matthews, C.E. Wieman, E.A. Cornell. 1995. Observation of Bose–Einstein condensation in a dilute atomic vapor. *Science* 269:198–201.

Ayers, P.W., R.G. Parr. 2000. Variational principles for describing chemical reactions: The Fukui function and chemical hardness revisited. *J. Am. Chem. Soc.* 122:2010–2018.

Babic, D., S. Bassoli, M. Casartelli, F. Cataldo, A. Graovac, O. Ori, B. York. 1995. Generalized Stone-Wales transformations. *Mol. Simul.* 14:395–401.

Bader, R.F.W. 1990. *Atoms in Molecules—A Quantum Theory.* Oxford: Oxford University Press.

Baekelandt, B.G., A. Cedillo, R.G. Parr. 1995. Reactivity indices and fluctuation formulas in density functional theory: Isomorphic ensembles and a new measure of local hardness. *J. Chem. Phys.* 103:8548–8556.

Berlin, T. 1951. Binding regions in diatomic molecules. *J. Chem. Phys.* 19:208–213.

Bhowmick, S., U.V. Waghmare. 2010. Anisotropy of the Stone-Wales defect and warping of graphene nano-ribbons: A first-principles analysis. *Phys. Rev. B.* 81:155416/1–7.

Boeyens, J.C.A. 2005. *New Theories for Chemistry*: New York: Elsevier.

Bohm, D., J.P. Vigier. 1954. Model of the causal interpretation of quantum theory in terms of a fluid with irregular fluctuations. *Phys. Rev.* 96:208–216.

Carpio, A., L.L. Bonilla, F. de Juan, M.A.H. Vozmediano. 2008. Dislocations in graphene. *New J. Phys.* 10:053021/1–13.

Cataldo F., G. Compagnini, G. Patané, O. Ursini, G. Angelini, P.R. Ribic, G. Margaritondo, A. Cricenti, G. Palleschi, F. Valentini. 2012. Graphene nanoribbons produced by the oxidative unzipping of single-wall carbon nanotubes. *Carbon* 48:2596–2602.

Cataldo, F., O. Ori, S. Iglesias-Groth. 2010. Topological lattice descriptors of graphene sheets with fullerene-like nanostructures. *Mol. Simul.* 36:341–353.

Chattaraj, P.K., G.H. Liu, R.G. Parr. 1995. The maximum hardness principle in the Gyftpoulos-Hatsopoulos three-level model for an atomic or molecular species and its positive and negative ions. *Chem. Phys. Lett.* 237:171–176.

Chattaraj, P.K., H. Lee, R.G. Parr. 1991. Principle of maximum hardness. *J. Am. Chem. Soc.* 113:1854–1855.

Chuvilin, A., J.C. Meyer, G. Algara-Siller, U. Kaiser. 2009. From graphene constrictions to single carbon chains. *New J. Phys.* 11:083019/1–10.

Collins, P.G. 2011. Defects and disorder in carbon nanotubes. In *Oxford Handbook of Nanoscience and Technology: Frontiers and Advances*, ed. A.V. Narlikar, Y.Y. Fu. Oxford: Oxford University Press.

Das, B., K. Eswar, U. Ramamurty, C.N.R. Rao. 2009. Nano-indentation studies on polymer matrix composites reinforced by few-layer graphene. *Nanotechnology* 20:125705.

de Broglie, L., M.J.P. Vigier. 1953. *La Physique Quantique Restera-t-Elle Indéterministe?* Paris: Gauthier-Villars.

Denis, P.A. 2009. Density functional investigation of thioepoxidated and thiolated graphene. *J. Phys. Chem. C* 113:5612–5619.

Dinadayalane, T.C., J. Leszczynski. 2007. Stone–Wales defects with two different orientations in (5, 5) single-walled carbon nanotubes: A theoretical study. *Chem. Phys. Lett.* 434:86–91.

Dirac, P.A.M. 1944. *The Principles of Quantum Mechanics.* Oxford: Oxford University Press.

Diudea, M.V. 2010. Omega polynomial: Composition rules in CorSu lattice. *Int. J. Chem. Model.* 2:1–6.

Diudea, M.V., A. Ilić. 2009. CorSu network—A new graphene design. *Studia Univ. "Babes-Bolyai"—Chemia* 54:171–177.

Dreizler, R., E.K.U. Gross. 1995. *Density Functional Theory.* New York: Plenum Press.

Ebbesen, T.W. 1998. Cones and tubes: Geometry in the chemistry of carbon. *Acc. Chem. Res.* 31:558–566.

Eda, G., H.E. Unalan, N. Rupesinghe, G.A.J. Amaratunga, M. Chhowalla. 2008. Field emission from graphene based composite thin films. *Appl. Phys. Lett.* 93:233502.

Evans, M.R. 2000. Phase transitions in one-dimensional nonequilibrium systems. *Brazil J. Phys.* 30: 42–57.

Ewels, C.P., M.I. Heggie, P.R. Briddon. 2002. Adatoms and nanoengineering of carbon. *Chem. Phys. Lett.* 351:178–182.

Findlay, A. 1955. *Practical Physical Chemistry.* London: Longmans.

Geim, A.K., K.S. Novoselov. 2007. The rise of graphene. *Nature Mat.* 6:183–191.

Goncalves, G., P.A.A.P. Marques, C.M. Granadeiro, H.I.S. Nogueira, M.K. Singh, J. Grácio. 2009. Surface modification of graphene nanosheet with gold nanoparticles: The role of oxygen moieties at the graphene surface on gold nucleation and growth. *Chem. Mat.* 21:4796–4802

Green, A.A., M.C. Hersam. 2009. Solution phase production of graphene with controlled thickness via density differentiation. *Nano Lett.* 9:4031–4036.

Hashimoto, A., K. Suenaga, A. Gloter, K. Urita, S. Iijima. 2004. Direct evidence for atomic defects in graphene layers. *Nature* 430:870–873.

Heitler, W., F. London. 1927. Wechselwirkung neutraler Atome und homöopolare Bindung nach der Quantenmechanik. *Z. Phys.* 44:455–472.

Heitler, W., F. London. 1927. Wechselwirkung neutraler atome und homöopolare bindung nach der quantenmechanik. *Z. Phys.* 44:455–472.

Hohenberg, P., W. Kohn. 1964. Inhomogeneous electron gas. *Phys. Rev.* 136: B864–B871.

Huang, B., M. Liu, N. Su, J. Wu, W. Duan, B.-L. Gu, F. Liu. 2009. Quantum manifestations of graphene edge stress and edge instability: A first-principles study. *Phys. Rev. Lett.* 102:166404/1–4.

Hypercube, Inc. 2002. *HyperChem 7.01 [Program Package]*. Gainesville, FL, USA: Hypercube, Inc.

Ilić, A., M.V. Diudea, F. Gholami-Nezhaad, A.R. Ashrafi. 2010. Topological indices in nanocones. In *Novel Molecular Structure Descriptors—Theory and Applications*, Vol. I, ed. I. Gutman, B. Furtula, 217–226. Kragujevac: University of Kragujevac.

Im, J., Y. Kim, C.-K. Lee, M. Kim, J. Ihm, H.J. Choi. 2011. Nanometer-scale loop currents and induced magnetic dipoles in carbon nanotubes with defects. *Nano Lett.* 11:1418–1422.

Iranmanesh, A., A.R. Ashrafi, A. Graovac, F. Cataldo, O. Ori. 2012. Wiener index. Role in topological modeling of hexagonal systems. From fullerenes to graphene. In *Distance in Molecular Graphs—Applications. Mathematical Chemistry Monographs*, Vol. 13, ed. I. Gutman, B. Furtula, 135–155. Kragujevac: University of Kragujevac.

Jeong, B.W., J. Ihm, G.-D. Lee. 2008. Stability of dislocation defect with two pentagon-heptagon pairs in graphene. *Phys. Rev. B.* 78:165403/1–5.

Kadanoff, L.P. 2009. More is the same; mean field theory and phase transitions. *J. Stat. Phys.* 137:777–797.

Kaplan, I.G. 2002. Is the Pauli exclusive principle an independent quantum mechanical postulate? *Int. J. Quantum Chem.* 89:268–276.

Ketterle, W. 2002. Nobel Lecture: When atoms behave as waves: Bose–Einstein condensation and the atom laser. *Rev. Mod. Phys.* 74: 1131–1151.

Kleinert, H. 2004. *Path Integrals in Quantum Mechanics, Statistics, Polymer Physics, and Financial Markets*. 3rd ed. Singapore: World Scientific.

Kleinert, H., S. Schmidt, A. Pelster. 2004. Reentrant phenomenon in the quantum phase transitions of a gas of bosons trapped in an optical lattice. *Phys Rev. Lett.* 93: 160402.

Kotakoski, J., A.V. Krasheninnikov, U. Kaiser, J.C. Meyer. 2011a. From point defects in graphene to two-dimensional amorphous carbon. *Phys. Rev. Lett.* 106:105505/1–4.

Kotakoski, J., J.C. Meyer, S. Kurasch, D. Santos-Cottin, U. Kaiser, A.V. Krasheninnikov. 2011b. Stone-Wales-type transformations in carbon nanostructures driven by electron irradiation. *Phys. Rev. B* 83:245420/1–6.

Krasheninnikov, A.V., K. Nordlund, M. Sirviö, E. Salonen, J. Keinonen. 2001. Formation of ion-irradiation-induced atomic-scale defects on walls of carbon nanotubes. *Phys. Rev. B* 63: 245405/1–6.

Krishnan, A., E. Dujardin, M.M.J. Treacy, J. Hugdahl, S. Lynum, T.W. Ebbesen. 1997. Graphitic cones and the nucleation of curved carbon surfaces. *Nature* 388:451–454.

Kumeda, Y., D.J. Wales. 2003. *Ab initio* study of rearrangements between C_{60} fullerenes. *Chem. Phys. Lett.* 374:125–131.

Lahiri, J., Y. Lin, P. Bozkurt, I.I. Oleynik, M. Batzill. 2010. An extended defect in graphene as a metallic wire. *Nat. Nano* 5:326–329.

Lee, G., B. Lee, J. Kim, K. Cho. 2009. Ozone adsorption on graphene: *Ab initio* study and experimental validation. *J. Phys. Chem. C* 113:14225–14229.

Leggett, A.J. 2001. Bose–Einstein condensation in the alkali gases: Some fundamental concepts. *Rev. Mod. Phys.* 73:307–356.

Lewis, G.N. 1916. The atom and the molecule. *J. Am. Chem. Soc.* 38:762–785.

Liu, J., S. Fu, B. Yuan, Y. Li, Zh. Deng. 2010. Toward a universal "adhesive nanosheet" for the assembly of multiple nanoparticles based on a protein-induced reduction/decoration of graphene oxide. *J. Am. Chem. Soc.* 132:7279–7281.

Liu, Y., B.I. Yakobson. 2010. Cones, pringles, and grain boundary landscapes in graphene topology. *Nano Lett.* 10: 2178–2183.

London, F. 1938. On the Bose–Einstein condensation. *Phys. Rev.* 54:947–954.

Lusk, M.T., L.D. Carr. 2008. Nanoengineering defect structures on graphene. *Phys. Rev. Lett.* 100:175503/1–4.

Malrieu, J.P., N. Guihéry, C. Jiménez-Calzado, C. Angeli. 2007 Bond electron pair: Its relevance and analysis from the quantum chemistry point of view. *J. Comput. Chem.* 28:35–50.

Meyer, J.C., C. Kisielowski, R. Erni, M.D. Rossell, M.F. Crommie, A. Zettl. 2008. Direct imaging of lattice atoms and topological defects in graphene membranes. *Nano Lett.* 8:3582–3586.

Meyer, J.C., S. Kurasch, H.J. Park, V. Skakalova, D. Künzel, A. Groß et al. 2011. Experimental analysis of charge redistribution due to chemical bonding by high-resolution transmission electron microscopy. *Nat. Mater.* 10:209–215.

Ng, Y.H., I.V. Lightcap, K. Goodwin, M. Matsumura, P.V. Kamat. 2010. To what extent do graphene scaffolds improve the photovoltaic and photocatalytic response of TiO_2 nanostructured films? *J. Phys. Chem. Lett.* 1:2222–2227.

Nordlund, K., J. Keinonen, T. Mattila. 1996. Formation of ion irradiation induced small-scale defects on graphite surfaces. *Phys. Rev. Lett.* 77:699–702.

Novoselov, K.S., A.K. Geim, S.V. Morozov, D. Jiang, Y. Zhang, S.V. Dubonos et al. 2004. Electric field effect in atomically thin carbon films. *Science* 306:666–669.

Oelke, W.C. 1969. *Laboratory Physical Chemistry*. New York: Van Nostrand Reinhold Company.

Okamoto, Y., Y. Miyamoto. 2001. *Ab initio* investigation of physisorption of molecular hydrogen on planar and curved graphenes. *J. Phys. Chem. B* 105:3470–3474.

Ori, O., F. Cataldo, M.V. Putz. 2011. Topological anisotropy of Stone-Wales waves in graphenic fragments. *Int. J. Mol. Sci.* 12:7934–7949.

Ori, O., M.V. Putz. 2014. Isomeric formation of 51815 defects in graphenic systems. *Fuller. Nanotub. Carbon Nanostruct.* 22:887–900.

Paci, J.T., T. Belytschko, G.C. Schatz. 2007. Computational studies of the structure, behavior upon heating, and mechanical properties of graphite oxide. *J. Phys. Chem. C* 111:18099–18111.

Parr, R.G., W. Yang. 1989. *Density-Functional Theory of Atoms and Molecules*. New York: Oxford University Press.

Paton, K.R., E. Varrla, C. Backes, R.J. Smith, U. Khan, A. O'Neill et al. 2014. Scalable production of large quantities of defect-free few-layer graphene by shear exfoliation in liquids. *Nat. Mater.* doi:10.1038/nmat3944.

Pauling, L. 1939. *The Nature of the Chemical Bond and the Structure of Molecules and Crystals*. Ithaca: Cornell University Press.

Peng, Q., A.K. Dearden, J. Crean, L. Han, S. Liu, X. Wen, S. De. 2014. New materials graphyne, graphdiyne, graphone, and graphane: Review of properties, synthesis, and application in nanotechnology. *Nanotechnol. Sci. Appl.* 7:1–29.

Pitaevskii, L., S. Stringari. 2003. *Bose–Einstein Condensation*. Oxford: Clarendon Press.

Prasad, K.E., B. Das, U. Maitra, U. Ramamurty, C.N.R. Rao. 2009. Extraordinary synergy in the mechanical properties of polymer matrix composites reinforced with 2 nanocarbons. *Proc. Natl. Acad. Sci. USA* 106:13186–13189.

Putz, M.V. 2008a. The chemical bond: Spontaneous symmetry–breaking approach. *Symmetr. Cult. Sci.* 19:249–262.

Putz, M.V. 2008b. *Absolute and Chemical Electronegativity and Hardness*. New York: NOVA Science Publishers.

Putz, M.V. 2008c. Maximum hardness index of quantum acid-base bonding. *MATCH Commun. Math. Comput. Chem.* 60:845–868.

Putz, M.V. 2009. Chemical action and chemical bonding. *J. Mol. Struct. THEOCHEM* 900: 64–70.

Putz, M.V. 2010a. The bondons: The quantum particles of the chemical bond. *Int. J. Mol. Sci.* 11:4227–4256.

Putz, M.V. 2010b. Beyond quantum nonlocality: Chemical bonding field. *Int. J. Env. Sci.* 1:25–31.

Putz, M.V. 2010c. Developing density functional theory for Bose–Einstein condensates. The case of chemical bonding. Preprint for *AIP Conference Proceedings of 8th International Conference of Computational Methods in Sciences and Engineering (ICCMSE)*, Kos: Greece, October 3–8, 2010.

Putz, M.V. 2011a. Hidden side of chemical bond: The bosonic condensate. In *Advances in Chemistry Research*, Vol. 10, ed. J.C. Taylor, 261–298. New York: NOVA Science Publishers.

Putz, M.V. 2011b. Conceptual density functional theory: From inhomogeneous electronic gas to Bose–Einstein condensates. In *Chemical Information and Computational Challenges in 21st Century*, ed. M.V. Putz, 1–60. New York: NOVA Science Publishers.

Putz, M.V. 2011c. Fulfilling the Dirac promises on quantum chemical bond. In *Quantum Frontiers of Atoms and Molecules*, ed. M.V. Putz, 1–20. New York: NOVA Science Publishers.

Putz, M.V. 2011d. Electronegativity and chemical hardness: Different patterns in quantum chemistry. *Curr. Phys. Chem.* 1:111–139.

Putz, M.V. 2011e. Quantum parabolic effects of electronegativity and chemical hardness on carbon π-systems. In *Carbon Bonding and Structures: Advances in Physics and Chemistry, Series Carbon Materials: Chemistry and Physics*, Vol. V, ed. M.V. Putz, 1–32. London: Springer Verlag.

Putz, M.V. 2012a. Density functional theory of Bose–Einstein condensation: Road to chemical bonding quantum condensate. *Struct. Bond* 149:1–50.

Putz, M.V. 2012b. Nanoroots of quantum chemistry: Atomic radii, periodic behavior, and bondons. In *Nanoscience and Advancing Computational Methods in Chemistry: Research Progress*, ed. E.A. Castro, A.K. Haghi, 103–143. Hershey: IGI Global (formerly Idea Group Inc.).

Putz, M.V. 2012c. *Chemical Orthogonal Spaces. Mathematical Chemistry Monographs*, Vol. 14, Kragujevac: University of Kragujevac—Faculty of Sciences and Mathematics.

Putz, M.V. 2012d. *Quantum Theory: Density, Condensation, and Bonding*. Ontario: Apple Academics Press.

Putz, M.V. 2012e Valence atom with Bohmian quantum potential: The golden ratio approach. *Chem. Central J.* 6:135 (16 pages).

Putz, M.V. 2013. Chemical orthogonal spaces (COSs): From structure to reactivity to biological activity. *Int. J. Chem. Model* 5:000–000.

Putz, M.V., O. Ori. 2012. Bondonic characterization of extended nanosystems: application to graphene's nanoribbons. *Chem. Phys. Lett.* 548:95–100.

Putz, M.V., O. Ori. 2014. Bondonic effects in group-IV honeycomb nanoribbons with Stone-Wales topological defects. *Molecules* 19:4157–4188.

Pyshnov, M.B. 1980. Topological solution for cell proliferation in intestinal crypt. *J. Theor. Biol.* 87:189–200.

Rutter, G.M., J.N. Crain, N.P. Guisinger, T. Li, P.N. First, J.A. Stroscio. 2007. Scattering and interference in epitaxial graphene. *Science* 317:219–222.

Saheli, M., M. Neamati, A. Ilić, M.V. Diudea. 2010. Omega polynomial in a combined coronene-sumanene covering. *Croat. Chem. Acta* 83:395–401.

Samsonidze, G.G., G.G. Samsonidze, B.I. Yakobson. 2002. Energetics of Stone–Wales defects in deformations of monoatomic hexagonal layers. *Comput. Mater. Sci.* 23:62–72.

Scerri, E.R. 2000. Have orbitals really been observed? *J. Chem. Edu.* 77:1492–1494.

Schrodinger, E. 1926. An undulatory theory of the mechanics of atoms and molecules. *Phys. Rev.* 28:1049–1070.

Seaha, C.-M., S.-P. Chaib, A.R. Mohameda. 2014. Mechanisms of graphene growth by chemical vapour deposition on transition metals. *Carbon* 70:1–21.

Seger, B., P.V. Kamat. 2009. Electrocatalytically active graphene-platinum nanocomposites. Role of 2-D carbon support in PEM fuel cells. *J. Phys. Chem. C* 113:7990–7995.

Shaik, S., D. Danovich, W. Wu, P. Hiberty. 2009. Charge-shift and its manifestations in chemistry. *Nat. Chem.* 1:443–449.

Stankovich, S., D.A. Dikin, G.H.B. Dommett, K.M. Kohlhaas, E.J. Zimney, E.A. Stach et al. 2006. Graphene-based composite materials. *Nature* 442:282–286.

Stone, A.J., D.J. Wales. 1986. Theoretical studies of icosahedral C_{60} and some related species. *Chem. Phys. Lett.* 128:501–503.

Terrones, M., A.R. Botello-Mendez, J. Campos-Delgado, F. Lopez-Urias, Y.I. Vega-Cantu, F.J. Rodriguez-Macias et al. 2010. Graphene and graphite nanoribbons: Morphology, properties, synthesis, defects and applications. *Nano Today* 5:351–372.

Todeschini, R., V. Consonni. 2000. *Handbook of Molecular Descriptors*. Weinheim: Wiley-VCH.

Tylianakis, E., G.M. Psofogiannakis, G.E. Froudakis. 2010. Li-doped pillared graphene oxide: A graphene-based nanostructured material for hydrogen storage. *J. Phys. Chem. Lett.* 1:2459–2464.

Ueta, A., Y. Tanimura, O.V. Prezhdo. 2010. Distinct infrared spectral signatures of the 1,2- and 1,4-fluorinated single-walled carbon nanotubes: A molecular dynamics study. *J. Phys. Chem. Lett.* 1:1307–1311.

Vetter, A. 1997. *Density Functional Theory for BEC*. Graduate thesis, in German. Würzburg: Institute of Theoretical Physics, Bayerische Julius-Maximilians University.

Viculis, L.M., J.J. Mack, R.B. Kaner. 2003. A chemical route to carbon nanoscrolls. *Science* 99:1361–1361.

Wang, S., L.Al. Tang, Q. Bao, M. Lin, S. Deng, B.M. Goh, K.P. Loh. 2009a. Room-temperature synthesis of soluble carbon nanotubes by the sonication of graphene oxide nanosheets. *J. Am. Chem. Soc.* 131:16832–16837.

Wang, Y., Z.Q. Shi, Y. Huang, Y.F. Ma, C.Y. Wang, M.M. Chen, Y.S. Chen. 2009b. Supercapacitor devices based on graphene materials. *J. Phys. Chem. C* 113:13103–13107.

Watcharotone, S., D.A. Dikin, S. Stankovich, R. Piner, I. Jung, G.H.B. Dommett et al. 2007. Graphene—silica composite thin films as transparent conductors. *Nano Lett.* 7:1888–1892.

Xie, X., L. Ju, X. Feng, Y. Sun, R. Zhou, K. Liu et al. 2009. Controlled fabrication of high-quality carbon nanoscrolls from monolayer graphene. *Nano Lett.* 9:2565–2570.

Yu, D.S., L.M. Dai. 2010. Self-assembled graphene/carbon nanotube hybrid films for supercapacitors. *J. Phys. Chem. Lett.* 1:467–470.

Yukalov, V.I. 2004. Principal problems in Bose–Einstein condensation of dilute gases. *Laser Phys. Lett.* 1:435–461.

Zeng, H., J.P. Leburton, Y. Xu, J. Wei. 2011. Defect symmetry influence on electronic transport of zigzag nanoribbons. *Nanoscale Res. Lett.* 6:254/1–6.

Zhoua, L.G., S.Q. Shib. 2003. Formation energy of Stone–Wales defects in carbon nanotubes. *Appl. Phys. Lett.* 83:1222–1224.

6 Electric Lens in Graphene

Weihua Mu

CONTENTS

ABSTRACT

The theory of negative refraction proposed by Veselago in 1968 is the foundation of making a "perfect" optical lens with negative refraction medium to focus the light into a fine point. Recently, it was suggested that an electron equivalent of the negative refraction can be realized in a mono-layered graphene to fabricate the electric lens in graphene. The long electron mean free path, ballistic electronic transport, and high current density of graphene make graphene a good candidate for new devices based on the electric lens effect. In this chapter, we review the principles of this amazing physical phenomenon, as well as the potential applications based on it.

6.1 INTRODUCTION

In classical theory, graphene, the two-dimensional honeycomb structure consisting of a single atomic layer of carbon atoms would be unstable in reality, since the thermal fluctuations violate its long-range crystalline order at finite temperature (Perierls 1935, Landau 1937, Mermin 1968). However, in 2004, Geim and Novoselov first prepared the monolayer graphene by simply using a scotch tape to mechanically exfoliate it from graphite (Novoselov et al. 2004, 2005). Since then, a variety of phenomena of physics have been explored. The excellent properties of graphene have aroused great interest in scientists to study the basic physics of graphene (Bunch et al. 2007, Meyer et al. 2007, Novoselov et al. 2007, Garcia-Sanchez et al. 2008, Gorbachev et al. 2008, Chen et al. 2009, Geim 2009).

Graphene is an amazing material with potential electronic (Lin et al. 2010), mechanical (Bunch et al. 2007), photonic (Li et al. 2009, Xia et al. 2009), thermal (Hu et al. 2009), chemical (Ling et al. 2010), and acoustic applications (Tian et al. 2011), due to its distinctive electronic mobility (Geim and Novoselov 2007), high thermal conductivity (Balandin et al. 2008, Wang et al. 2011), and high mechanical strength (Frank et al. 2007).

In this chapter, we will briefly review the basic electronic properties of graphene, including the electronic structure and related electrical transport properties. We emphasize a novel phenomenon in monolayer graphene, namely, the electric lens, which is a type of negative refractory effect, as well as discuss its physical fundamentals and possible applications in detail. Some recent progresses in the electric lens based on graphene-like materials are also addressed in this chapter.

6.2 DIRAC FERMIONS IN GRAPHENE

6.2.1 Electronic Structure of Graphene

Graphene is the thinnest material with only one-atomic thickness. It is a two-dimensional hexagonal lattice formed by sp^2-bonded carbon atoms, as illustrated in Figure 6.1. Graphene can be thought of as the building block for other carbon-based materials, such as fullerene, multiwalled carbon nanotube, and buck graphite.

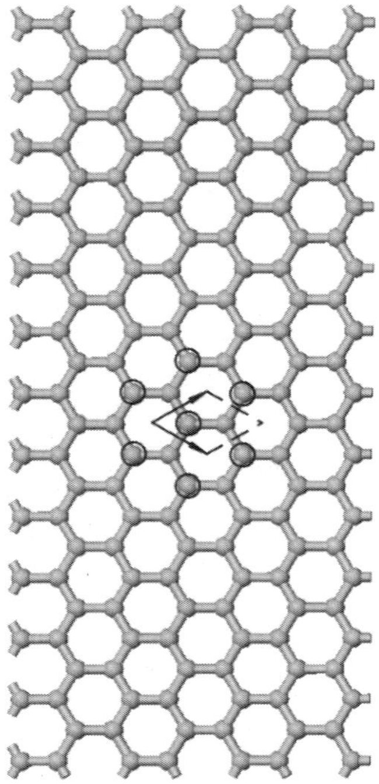

FIGURE 6.1 Honeycomb lattice of monolayer graphene, with lattice constant $a = 1.42$ Å. The carbon atoms are connected by covalent bonds. There are two sublattices in the honeycomb lattice, A (labeled by the circle) and B. There are two carbon atoms at each unit cell.

There are two types of atoms in the honeycomb lattice of graphene. In each unit cell, there are two carbon atoms, and the lattice vectors in real space are

$$\vec{a}_1 = \sqrt{3}a_0\left(\frac{\sqrt{3}}{2},1\right) \equiv a\left(\frac{\sqrt{3}}{2},1\right)$$
$$\vec{a}_2 = \sqrt{3}a_0\left(\frac{\sqrt{3}}{2},-1\right) \equiv a\left(\frac{\sqrt{3}}{2},-1\right) \tag{6.1}$$

Here, $a_0 = 1.42$ Å, is the bond length of the carbon–carbon covalent bond, $a = \sqrt{3}a_0$. The corresponding lattice vectors for the reciprocal lattice are

$$\vec{b}_1 = \frac{2\pi}{a}\left(\frac{1}{\sqrt{3}},1\right),\quad \vec{b}_2 = \frac{2\pi}{a}\left(\frac{1}{\sqrt{3}},-1\right) \tag{6.2}$$

The high symmetric points in the first Brillouin zone include two types of the corners, K and K' points, and M point, the midpoint of K and K'

$$\vec{K} = \left(\frac{4\pi}{3a},0\right),\quad \vec{K}' = \left(-\frac{4\pi}{3a},0\right),\quad \vec{M} = \left(0,-\frac{2\pi}{\sqrt{3}a}\right) \tag{6.3}$$

The electronic structures of graphene and graphite have been studied extensively (Wallace 1947, McClure 1957, Slonczewski and Weiss 1958). In 1947, Wallace obtained the energy bands with the concerning of only π-bonds in graphene by the tight-binding approximate (TBA) method. We use a similar method to illustrate the basic physics by the simple model. The wave function is the superposition of wave functions of the sublattices A and B

$$\Psi^k(\vec{r}) = c_A(\vec{k})\psi_A^k(\vec{r}) + c_B(\vec{k})\psi_B^k(\vec{k})$$
$$= \frac{1}{\sqrt{N}}\sum_{j=1}^{N}[e^{i\vec{k}\cdot\vec{R}_j^A}c_A(\vec{k})\phi(\vec{r}-\vec{R}_j^A) + e^{i\vec{k}\cdot\vec{R}_j^B}c_B(\vec{k})\phi(\vec{r}-\vec{R}_j^B)] \tag{6.4}$$

Here, N is the number the two-atom unit cells in graphene, and the function $\phi(\vec{r}-\vec{R}_j)$ denotes the local wave function of $2p_z$ orbit of a carbon atom to form π-bond. The positions of carbon atoms are described by vectors $\vec{R}_j^A$ and $\vec{R}_j^B$, respectively. A type "A" carbon atom has three nearest-neighboring-type B atoms, with the relation $\vec{R}_j^B = \vec{R}_j^A + \vec{\delta}_i$, $i = 1, 2, 3$, and $\vec{\delta}_1 = (a/\sqrt{3})(0,1)$, $\vec{\delta}_2 = (a/2)(1,-(1/\sqrt{3}))$, $\vec{\delta}_3 = (-1,-(1/\sqrt{3}))$.

The tight-binding Hamiltonian with the only nearest-neighboring interaction can be described by a Hamiltonian (Bena and Montambaux 2009)

$$\hat{H} = -t\sum_{\langle ij\rangle} |\phi_A(\vec{r}-\vec{R}_i^A)\rangle\langle\phi_B(\vec{r}-\vec{R}_j^B)| + H.c. \tag{6.5}$$

Substituting Equations 6.4 and 6.5, we get

$$\varepsilon(\vec{k})c_A(\vec{k}) = -t(e^{-i\vec{k}\cdot\vec{\delta}_1} + e^{-i\vec{k}\cdot\vec{\delta}_2} + e^{-i\vec{k}\cdot\vec{\delta}_3})c_B(\vec{k}) \equiv f(\vec{k})c_B(\vec{k})$$
$$\varepsilon(\vec{k})c_B(\vec{k}) = -t(e^{i\vec{k}\cdot\vec{\delta}_1} + e^{i\vec{k}\cdot\vec{\delta}_2} + e^{i\vec{k}\cdot\vec{\delta}_3})c_A(\vec{k}) \equiv f^*(\vec{k})c_A(\vec{k}) \tag{6.6}$$

Thus, the Hamiltonian matrix is reduced to

$$H(\vec{k}) = \begin{pmatrix} 0 & f(\vec{k}) \\ f^*(\vec{k}) & 0 \end{pmatrix} \equiv |f(\vec{k})| \begin{pmatrix} 0 & e^{i\theta_1(\vec{k})} \\ e^{-i\theta_1(\vec{k})} & 0 \end{pmatrix} \tag{6.7}$$

with $\theta_1(\vec{k})$, the phase of $f(\vec{k})$. The energy dispersion of graphene can be obtained by diagonalizing Equation 6.7

$$E(k_x,k_y) = \pm t\sqrt{1 + 4\cos\left(\frac{\sqrt{3}k_y a}{2}\right)\cos\left(\frac{k_x a}{2}\right) + 4\cos^2\left(\frac{k_x a}{2}\right)} \tag{6.8}$$

where k_x and k_y are the x and y components of the wave vector $\vec{k}$, and $t = -2.75$ eV is the hopping parameter presented in the TBA Hamiltonian. The electronic properties of graphene are characterized by its symmetric conduction and valence energy bands (measured from Fermi energy), as shown in Figure 6.2.

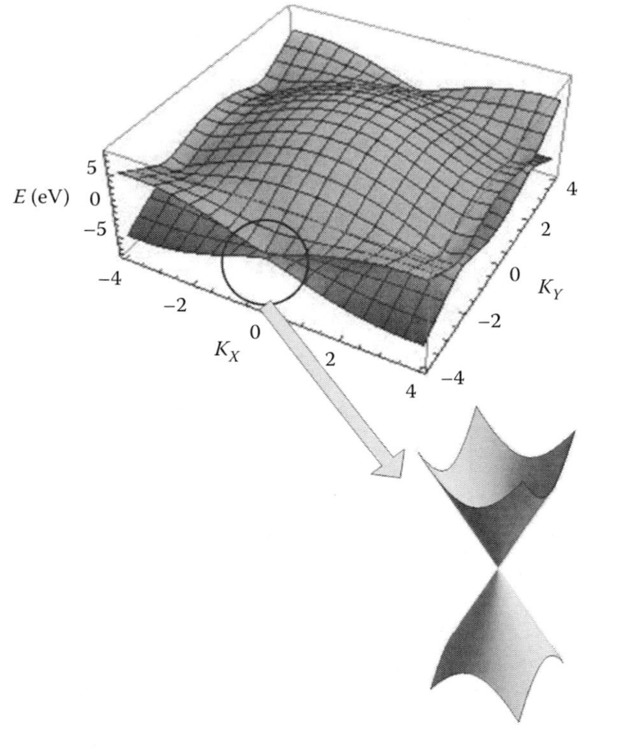

FIGURE 6.2 The $2p_z$ electrons of a carbon atom form a delocalized $\pi(\pi^*)$ band, which are the valence (conduction) band of graphene, respectively. The symmetric conduction and covalent electron energy bands of graphene are calculated by the nearest-neighboring tight-binding approximation, with the hopping parameter $t = -2.75$ eV. The two bands touch each other at six corners in the hexagonal Brillouin zone. The Fermi surface is degenerated to six points (inset). The conic shape energy band near the high symmetric K (K') points, which are the six corners of the first Brillouin zone, implies low-energy excitation near these corners, $E_{c(\upsilon)} = \pm\hbar\upsilon\,|\,\vec{k}\,|$. The low-energy electron (hole) of graphene behaves like a massless relativistic particle, which is the physical origin of the electric lens in graphene.

The valence and conduction bands of graphene touch at six points, which are the corners of the first Brillouin zone. The two inequivalent corners are labeled as K and K' valleys. The Fermi surface of graphene is degenerated to six Fermi points in its first Brillouin zone. At the Fermi points, the energy gap vanishes; thus, graphene is called a "semimetal material."

The eigenfunction of this Hamiltonian Equation 6.5 is

$$\psi_{\vec{k}}(\vec{r}) = \frac{1}{\sqrt{2N}}\left[\sum_{j=1}^{N} e^{i\vec{k}\cdot\vec{R}_j^A}\phi(\vec{r}-\vec{R}_j^A) \pm \sum_{j=1}^{N} e^{i\vec{k}\cdot\vec{R}_j^B} e^{-i\theta(\vec{k})}\phi(\vec{r}-\vec{R}_j^B)\right] \tag{6.9}$$

Here, the "+" is for the eigenfunction of the conduction band, and "–" is for that of the covalent band. As mentioned before, the two bands touch at the K and K' points in the first Brillouin zone, and the band structure near the K and K' points is important in the low-energy electronic excitation of graphene. At the vicinity of K and K' points, $\vec{k}_{\pm} = (\pm 4\pi/3a, 0) + \vec{q}$ with "+" for K valley, and "–" for K' valley. Here, $f_{K(K')}(\vec{q}) \approx \pm\upsilon(q_x - iq_y)$ with $\upsilon \equiv \sqrt{3}t/2a$. Thus, the low-energy effective Hamiltonian for graphene is

$$\hat{H}(\vec{q}) = \begin{pmatrix} 0 & f_K & 0 & 0 \\ f_K^* & 0 & 0 & 0 \\ 0 & 0 & 0 & f_{K'} \\ 0 & 0 & f_{K'}^* & 0 \end{pmatrix} = \upsilon\tau_z \otimes (q_x\sigma_x + q_y\sigma_y)$$

$$= \upsilon\,|\,\vec{q}\,| \begin{pmatrix} 0 & e^{-i\theta(\vec{q})} & 0 & 0 \\ e^{i\theta(\vec{q})} & 0 & 0 & 0 \\ 0 & 0 & 0 & -e^{-i\theta(\vec{q})} \\ 0 & 0 & -e^{i\theta(\vec{q})} & 0 \end{pmatrix} \tag{6.10}$$

Here, $\tan\theta(\vec{q}) \equiv q_y/q_x$. The electronic properties of graphene are invariant by interchanging the K and K' states, which implies that the two valleys are related by time-reversal symmetry.

Near K point, $\vec{q} = \vec{k} - \vec{K}$, the 2×2 effective Hamiltonian is written as (we add $\hbar$ here)

$$\hat{H}_K = \hbar\upsilon\begin{pmatrix} 0 & q_x - iq_y \\ q_x + iq_y & 0 \end{pmatrix} \tag{6.11}$$

which leads to linear energy dispersion, $E(\vec{q}) = \pm\hbar\upsilon\,|\,\vec{q}\,|$, with $\upsilon = c/300$, and c is the velocity of light. Thus, near the Fermi points, the conduction and valence bands show the conical shape, which is much different from the quadratic energy–momentum relation of electrons at the band edges in conventional semiconductors.

The analog of the linear energy dispersion of graphene and the energy–momentum relation of a massless relativistic particle, suggests that the charge carriers in graphene behave as the Dirac fermions with the effective Fermi velocity approximately 1/300 of the speed of light. The massless relativistic low-energy excitations in graphene make graphene suitable to be used to investigate quantum electrodynamic phenomena, which previously could only be observed in high-energy physics processes.

6.2.2 Dirac Fermion Description

There are a variety of interesting phenomena related to the Dirac relativistic particles in graphene, such as the absence of localization effects (Novoselov et al. 2005, Katsnelson and Novoselov 2007), and half-integer quantum Hall effect (Zhang et al. 2005, Novoselov et al. 2007). The Dirac fermion of graphene is a spinor with a degree of freedom of pseudospin and valley (Castro Neto et al. 2009). The terminology spinor means that the phase of the electron's wave function changes π when it rotates 2π in momentum space.

The pseudospin quantum number refers to the A and B sublattices in graphene.

In the two-component spinor of $K(K')$ valley, the two components describe the distribution of the electron in two types of sublattices. This degree of freedom is called "pseudospin" to distinguish it from the degree of freedom for real spin in Dirac spinors for quantum relativistic particles. Here, the "up" and "down" of the pseudospin correspond to the electron distribution at the A or B sublattice.

The plane wave eigenfunction of the low-energy effective Hamiltonian Equation 6.11 is found to be (Katsnelson 2012)

$$\psi_{K}^{e,h} = \frac{1}{\sqrt{2}}\begin{pmatrix} e^{-i\theta(\vec{k})/2} \\ \pm e^{i\theta(\vec{k})/2} \end{pmatrix} \tag{6.12}$$

and

$$\psi_{K'}^{e,h} = \frac{1}{\sqrt{2}}\begin{pmatrix} e^{i\theta(\vec{k})/2} \\ \pm e^{-i\theta(\vec{k})/2} \end{pmatrix} \tag{6.13}$$

It is easy to verify that $E_{e,h}(k) = \pm\hbar\upsilon k$.

It is convenient to define a quantity, chirality, or helicity of the present Dirac spinor, which is the projection of the pseudospin in the direction of momentum of the electron (hole) in graphene

$$\hat{h} \equiv \frac{\vec{\sigma}\cdot\vec{k}}{|\vec{k}|} \tag{6.14}$$

The electrons and holes of graphene have definite chiralities

$$\hat{h}\psi^{e,h} = \pm\psi^{e,h} \tag{6.15}$$

with "+" for electrons, and "−" for holes.

If there is no interaction to change the pseudospin of the wave function of the electron (hole), its chirality is conserved, which plays an important role in the electric transport properties of graphene, such as Klein tunneling (Katsnelson et al. 2006) and electric lens effect (Cheianov et al. 2007).

6.3 ELECTRIC LENS IN GRAPHENE

6.3.1 Klein Tunneling

Since the low-energy excitation of graphene is similar to the quantum relativistic particles, the electron of graphene can cross a potential barrier with arbitrary height and width, namely, the phenomenon related to "Klein paradox." It was predicted that the incoming electrons with a well-defined momentum and pseudospin can propagate through the energy barrier as if it does not exist. The physics behind this "Klein paradox" is that the electrons occupy the hole states with opposite momentum and equivalent pseudospin in the region of energy barriers, which serve as the bridge across the barrier for the transport of electrons. The energy barrier can be introduced and adjusted by electrostatic gates, and the electronic transport across this type of a sharp potential barrier has been studied extensively (Huard et al. 2007, Özyilmaz et al. 2007, Williams et al. 2007, Liu et al. 2008, Oostinga et al. 2008, Stander et al. 2009).

The Klein tunneling of monolayer graphene shows perfect transmission ($T = 1$) for normal incidence, which is the consequence of conservation of chirality, since there is no interaction with a small character-length scale that can distinguish two types of sublattices (Ando et al. 1998, Mceuen et al. 1999, Katsnelson et al. 2006). The details of Klein tunneling have been shown in Katsnelson et al. (2006) and Katsnelson (2012).

6.3.2 Veselago's Lens

In 1968, Veselago discussed the theory of negative refraction for the first time (Veselago 1968). In his original work, Veselago suggested the lens based on the electromagnetic material with a simultaneously negative dielectric constant and magnetic permeability, $\varepsilon < 0$, and $\mu < 0$, as shown in Figure 6.3.

The negative refraction phenomenon involving the left-handed material can be used to make the so-called "perfect optical lenses," which can focus light rays to an extremely fine point (Pendry 2000).

6.3.3 Negative Refractory in Graphene

In 2007, inspired by the analogy between ballistic electron transport in graphene and light rays in a dielectric medium, Cheianov et al. proposed that an electron equivalent of the negative refractory can be realized in a monolayered graphene (Cheianov et al. 2007). In this novel designing, the wave vectors of the electron (hole) in graphene are adjusted by controlling the gate voltage applied to graphene. The positive gate electrode makes the zero-energy charge carriers in this area being electrons, while the negative gate electrode ensures holes being charge carriers.

By fine tuning the density of charge carriers on both sides of PN junction (PNJ) to be equal values, the electron flow radiated from a point-like electric current source (divergent beams) can focus exactly on the point on the other side of PNJ due to the

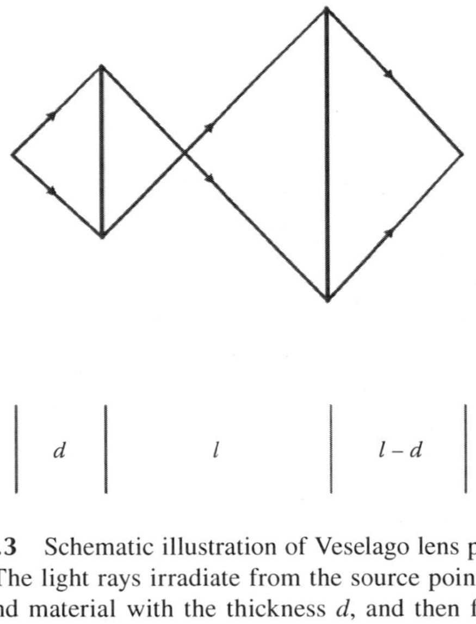

FIGURE 6.3 Schematic illustration of Veselago lens proposed by Veselago. The light rays irradiate from the source point, then pass the left-hand material with the thickness d, and then focus at the drain point. The negative refractive index is $n = -1$. (The source of the material, Weihua, M. et al., Electric current focusing efficiency in a graphene electric lens. *J. Phys. Condens. Matter* 23:495302. Copyright 2011, IOP Publishing is acknowledged.)

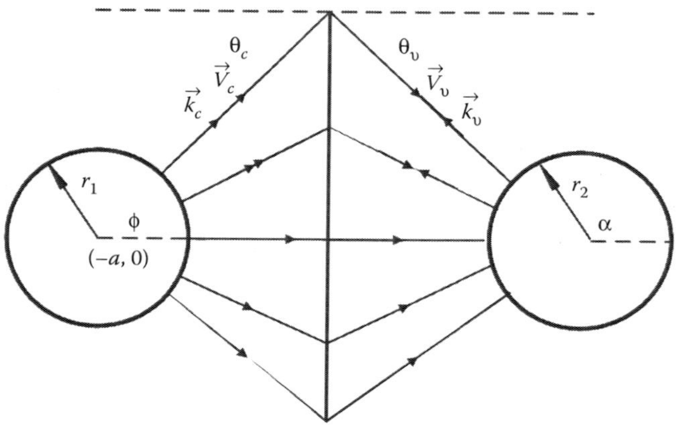

FIGURE 6.4 Schematic illustration of the negative refractory in graphene PNJ (left: N region, right: P region). The N and P regions are controlled by a positive and negative gate voltage. In N region, the conduction band electron flow with $\vec{V}_c$ and $\vec{k}_c$ propagates toward the interface of NP junction, and then enters P region after negative refraction. At P region, the electron flow of valence band electrons with $\vec{V}_\upsilon$ and $\vec{K}_\upsilon$ focus toward a point. The momentum conservation in y direction requires that $k_c \sin\theta_c = -k_\upsilon \sin\theta_\upsilon$, which implies the relation between the angle of refraction θ_υ, and the incident angle θ_c, similar to Snell's law in optical refractory, $\sin\theta_c/\sin\theta_\upsilon = -k_\upsilon/k_c \equiv n$. (The source of the material, Weihua, M. et al., Electric current focusing efficiency in a graphene electric lens. *J. Phys. Condens. Matter* 23:495302. Copyright 2011, from IOP Publishing is acknowledged.)

negative refractory of the electron wave propagating from the conduction band to the valence band, as shown in Figure 6.4.

The theory of negative refractory of electron flow in graphene can be understood by the group velocities of electrons and holes in graphene. The energy dispersion $E_{e,h} = \pm\hbar\upsilon|\vec{k}|$ implies that the group velocities of electrons are $\vec{V}_{c(\upsilon)} \equiv \partial E(\vec{k})/\partial(\hbar\vec{k}) = \pm\upsilon\vec{k}/k$ with "+" for the conduction band electron, and "–" for the valence band electron. The negative refractory index phenomenon ($\theta_r = -\theta_i$) results from a different sign in the expression of group velocity for the conduction and valence electrons, which are separated by PNJ. The continuity of electronic current requires the equality of the velocity of the conduction band electron in N region and velocity of the valence band electron on the other side of the junction, which determines that $k_x^{left} = -k_x^{right}$, whereas the conservation of wave vector component in the direction perpendicular to the direction that the potential varies, requires that $k_y^{left} = k_y^{right}$, thus $\theta_\upsilon = -\theta_c$. Here, $k_{\upsilon,c}^x = k_{\upsilon,c}\cos\theta_{\upsilon,c}$ and $k_{\upsilon,c}^y = k_{\upsilon,c}\sin\theta_{\upsilon,c}$.

Without concerning the intervalley scattering and the degree of freedom for real spin, a two-component spinor is a good description of electrons and holes for the negative refractory in graphene, and the low-energy effective Hamiltonian $\hat{H} = \hbar\upsilon\vec{\sigma}\cdot\vec{\nabla}$ with $\upsilon \approx 10^8$ cm · s^{-1}, $\vec{\sigma} = (\sigma_x, \sigma_y)$.

6.3.4 Theory of Electronic Transport in Graphene PNJ

In the graphene-based Veselago lens, the source electrode of electron flow is a two-dimensional current source electrode with circular symmetry, which is placed at the left N region with its center located at $(-a, 0)$. The radius of the electrode should be much smaller than the size of the whole graphene sheet, but it still satisfies the "far field" condition $k_cR = 2\pi R/\lambda_F \gg 1$, with λ_F the Fermi wavelength of the electron wave near the source electrode.

For convenience, the injecting electrons with zero energy are studied. The wave function of zero-energy electrons closely near the source electrode has the explicit expression in the form of a two-component spinor

$$\psi_1(x,y) = A\begin{pmatrix} H_0^{(1)}(k_c r_0) \\ iH_1^{(1)}(k_c r_0)e^{i\phi(x,y)} \end{pmatrix} \approx A\sqrt{\frac{2}{\pi k_c r_0}}\begin{pmatrix} 1 \\ e^{i\phi(x,y)} \end{pmatrix} e^{i(k_c r_0 - (\pi/4))} \tag{6.16}$$

which is one of the cylindrical function solutions of Dirac equation

$$(-i\sigma_x\partial_x - i\sigma_y\partial_y - k_c)\psi_1(x,y) = 0 \tag{6.17}$$

where A is a constant to be determined by a certain boundary condition, and $H_n^{(1)}(z)$ is the Hankel function of the first kind. The $\tan\phi = y/(x+1), r_0 \equiv \sqrt{(x+a)^2 + y^2}$.

Here, the lowest-order asymptotic expansion of $H_n^{(1)}(z)$ is good enough, due to the far-field condition $k_c r_0 \gg 1$. It is easy to verify that only the current density of this type of wave function is isotropic near the edge of the source electrode

$$\vec{j}_{in} = e\upsilon_F\psi_1^\dagger\vec{\sigma}\psi_1 = \frac{2e\upsilon_F A^2}{(\pi k_c R)(\cos\phi, \sin\phi)} \tag{6.18}$$

while all other cylindrical function solutions lead to $\vec{j}_{in} \propto (\cos n\phi, \sin n\phi)$, $n \geq 2$ that has C_n, $n \geq 2$ symmetry instead of circular symmetry. The constant in Equation 6.16 can be determined by the incoming electron flow. Given the electric current I_0 inputted into the source electrode, the continuity of electric current at the boundary of the source electrode implies that $I_0 = 2\pi R|\vec{j}_{in}|$ that determines the constant $A = \sqrt{k_c I_0/(8e\upsilon_F)}$.

The negative refractory phenomenon of graphene, fundamentally, is a penetration problem of the incoming electron wave. The solutions of the penetration problem of an incident plane wave of an electron at interfaces of graphene PNJ have been solved (Katsnelson et al. 2006). To use these fruits in graphene Veselago's lens involving a cylindrical incident wave, the key point is to find the expansion of a cylindrical wave by a set of plane wave basis, that is, describe the cylindrical wave as the superposition of a series of plane waves (Gerlach 2010). After decomposition, it is easy to find the refractory (for PNJ) and transmission wave (for NPN junction) of each plane wave components of an incident wave. The whole refractive (transmitted) electron wave can be obtained by adding all the components together.

The expansion of an incident cylindrical electron wave as given in Equation 6.16 can be written as

$$\psi_1(x,y) = \frac{A}{\pi}\int_{-\varepsilon+i\infty}^{\varepsilon-i\infty} e^{ik_c a\cos\theta_c}\begin{pmatrix}1\\ e^{i\theta_c}\end{pmatrix} e^{ik_c(x\cos\theta_c + y\sin\theta_c)}d\theta_c \quad (6.19)$$

In general, the contour of the integral is shown in Figure 6.4a. Here, $\vec{k} \equiv (k_c\cos\theta_c,\ k_c\sin\theta_c)$ is the wave vector of an incident plane electron wave component with an incident angle θ_c.

The reflection and refraction of incident electron flow spreading from a circular source electrode in graphene PNJ are shown in the schematic illustration of Figure 6.5.

The plane wave with the angle of incidence θ_c has the form (Castro Neto 2009)

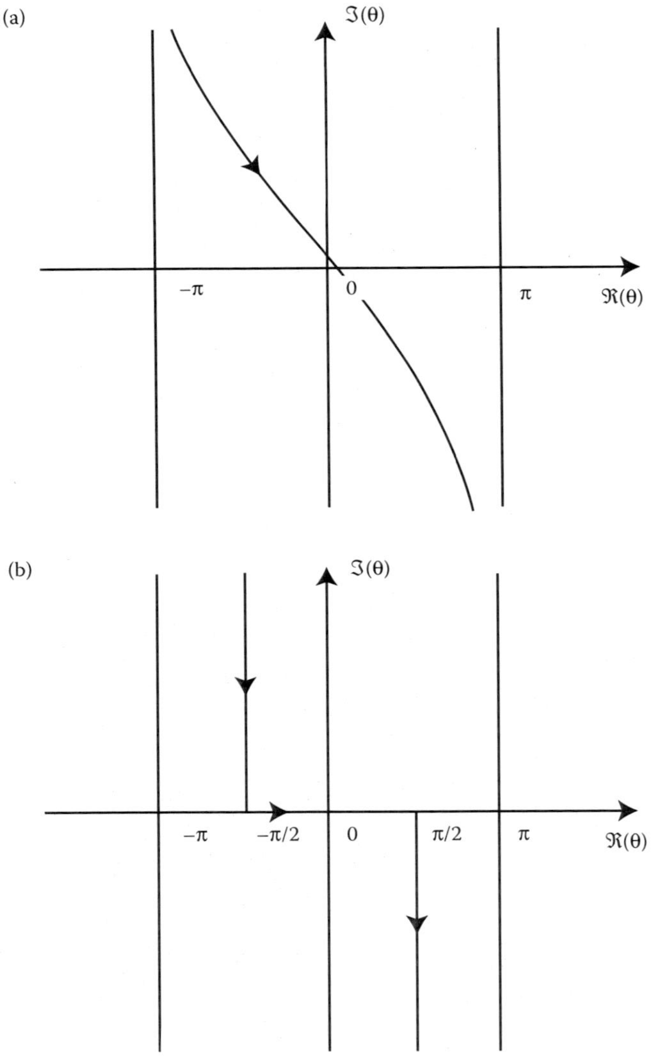

FIGURE 6.5 (a) Integral contour for θ_c at the complex plane. It is used in the integral representation of injecting the electron wavefunction near the source electrode, which is described by Hankel functions. (b) The integral contour of the complex variable θ_c is shifted to obtain an analytical expression of the transmission wave of a symmetric PNJ.

$$\psi_{plane\text{-}in}(x,y) = \frac{1}{\sqrt{2}}\begin{pmatrix}1\\ e^{i\theta_c}\end{pmatrix} e^{ik_c(x\cos\theta_c + y\sin\theta_c)} \quad (6.20)$$

At the interface of PNJ, the reflective and refractive plain waves are written as

$$\psi_{plane\text{-}refl}(x,y) = \frac{1}{\sqrt{2}}\begin{pmatrix}1\\ e^{i(\pi-\theta_c)}\end{pmatrix} e^{ik_c(-x\cos\theta_c + y\sin\theta_c)} \quad (6.21)$$

and

$$\psi_{plane\text{-}refr}(x,y) = \frac{1}{\sqrt{2}}\begin{pmatrix}1\\ e^{i\theta_\upsilon}\end{pmatrix} e^{-ik_\upsilon(x\cos\theta_\upsilon + y\sin\theta_\upsilon)} \quad (6.22)$$

respectively. Here, the three wave functions $\psi_{plane\text{-}in}$, $\psi_{plane\text{-}refl}$, and $\psi_{plane\text{-}refr}$ are the plane wave solutions of Dirac equations

$$\begin{cases}(-i\sigma_x\partial_y - i\sigma_y\partial_y + k_c)\psi(x,y) = 0, & x<0\\ (-i\sigma_x\partial_x - i\sigma_y\partial_y - k_\upsilon)\psi(x,y) = 0, & x>0\end{cases} \quad (6.23)$$

The factor $e^{i\pi}$ in the reflective wave function $\psi_{plane\text{-}refl}(x, y)$ is related to Berry phase (Castro Neto et al. 2009).

At the interface of PNJ, the continuity of the plane wave function in P and N regions

$$\psi_{plane\text{-}in}(0,y) + r\psi_{plane\text{-}refr}(0,y) = t\psi_{plane\text{-}refl}(0,y) \quad (6.24)$$

determines the coefficient of the refraction of the plane wave $t = 2\cos\theta_c/(e^{i\theta_\upsilon} + e^{-i\theta_c})$, and the coefficient of the reflection $r = 1 - t$.

Applying the results of plane waves, the refractive wave function of the cylindrical wave is obtained as

$$\psi_{refr}(x,y) = \frac{A}{\pi}\int_{-\varepsilon+i\infty}^{\varepsilon-i\infty} e^{ik_c a\cos\theta_c} \times \frac{2\cos\theta_c}{e^{i\theta_\upsilon} + e^{-i\theta_c}}\begin{pmatrix}1\\ e^{i\theta_\upsilon}\end{pmatrix} e^{-ik_\upsilon\cos\theta_\upsilon x - ik_\upsilon\sin\theta_\upsilon y}d\theta_c, \quad x>0 \quad (6.25)$$

The refractory for a symmetric PNJ is much simple, with $k_\upsilon = k_c$, $n = -1$, and $\theta_\upsilon = -\theta_c$. The refractive wave at P region has the analytical form

$$\psi_{refr}(r_1,\alpha) = \frac{A}{\pi}\int_{-(\pi/2)+i\infty}^{(\pi/2)-i\infty} d\theta_c \frac{1}{\sqrt{2}}\cos\theta_c \begin{pmatrix} e^{i\theta_c}\\ 1\end{pmatrix} e^{ik_c\hat{r}_1\cos(\theta_c-(\pi-\alpha))}$$

$$\approx \begin{cases} A\sqrt{\dfrac{2}{\pi k_c r_1}}e^{-i(k_c\hat{r}_1 - \pi/4)}\begin{pmatrix} e^{-i\alpha}\\ 1\end{pmatrix}\cos\alpha, & \alpha \in \left[-\dfrac{\pi}{2}, \dfrac{\pi}{2}\right]\\ A\sqrt{\dfrac{2}{\pi k_c r_1}}e^{i(k_c\hat{r}_1 - \pi/4)}\begin{pmatrix} e^{-i\alpha}\\ -1\end{pmatrix}\cos\alpha, & \alpha \in \left[\dfrac{\pi}{2}, \dfrac{3\pi}{2}\right]\end{cases} \quad (6.26)$$

Here, the integral contour is shifted for convenience, as shown in Figure 6.4b.

We have used the steepest descent method with the inclusion of only one leading term, which implies that the approximate refractive wave function in Equation 6.26 is valid for $k_\upsilon r_1 \gg 1$. The current density $\vec{j} = e\upsilon_F \psi^\dagger_{refr}\vec{\sigma}\psi_{refr}$ is therefore

$$\vec{j}_1(r_1,\alpha)(k_\upsilon R) = \begin{cases} \dfrac{I_0}{2\pi r_1}\cos^2\alpha(\cos\alpha,\sin\alpha), & \alpha \in \left[-\dfrac{\pi}{2},\dfrac{\pi}{2}\right) \\ -\dfrac{I_0}{2\pi r_1}\cos^2\alpha(\cos\alpha,\sin\alpha), & \alpha \in \left[\dfrac{\pi}{2},\dfrac{3\pi}{2}\right) \end{cases} \quad (6.27)$$

The divergent electrons wave at N region become convergent at P region, and tends to focus on the point $(a, 0)$, which is symmetrical to the center of the source electrode.

The intensity of current density around but not too close to the focus $(a, 0)$ has an expression, $j_1 \infty \cos^2\alpha/r_1 = (x-a)^2/[(x-a)^2+y^2]^{3/2}$, which is exactly the result numerically obtained by Cheianov et al.

6.3.5 Electric Current Collection Efficiency of Graphene PNJ

Only a part of electric current from the source electrode at N region can be collected by the drain electrode placed at the focus at P region. The electric collection efficiency is an important parameter for graphene electric lens-based nanoelectronic devices. If a detecting circular electrode with radius R is placed at $(a, 0)$, the maximal electric current collected by it can be written as

$$I_1 = R\int_{\pi/2}^{3\pi/2} \vec{j}(R,\alpha)\cdot(\cos\alpha,\sin\alpha) = \frac{I_0}{4} \quad (6.28)$$

Thus, only 1/4 of the total current from the source electrode can be collected by the detecting drain electrode, considering the absorption boundary condition of the graphene sheet. A device of graphene electronic lens should be designed sophisticatedly for high efficiency.

In general, an asymmetric PNJ leads to a refractory index $n \neq 1$, and the refractive wave function does not have the simple analytical form. The integral in Equation 6.25 can be numerically calculated by the steepest descent method. The electron flow transporting across an asymmetric PNJ forms caustics, with the cusp located at $(|n|a,0)$ (Cheianov et al. 2007, Cserti et al. 2007).

6.3.6 Electric Lens Based on Graphene NPN Junction

In the original prediction of an optical Veselago's lens, the electromagnetic wave from a point source focuses on a point after transmitting across the left-hand material, which is in close analogy with the focusing of electron flow at the second N region in graphene NPN junction.

Using a similar method for the refractive wave of PNJ, it is not difficult to find each plane electron wave components of the transmission wave, and then integrate all the components together.

The plane wave's transmission across the P region between two N regions can be solved by a standard method for an electron wave propagating across a square energy barrier (Katsnelson 2007). The two-dimensional space in graphene NPNJ was divided into three regions $x \le 0$, $0 \le x \le d$, and $x > d$, which are labeled by I, II, and III, respectively. The Dirac equations for three regions of NPNJ are

$$\begin{cases} (-i\sigma_x\partial_x - i\sigma_y\partial_y + k_c)\psi(x,y) = 0, & x \le 0 \text{ or } x \ge d \\ (-i\sigma_x\partial_y - i\sigma_y\partial_y - k_\upsilon)\psi(x,y) = 0, & 0 \le d \end{cases} \quad (6.29)$$

The plane wave function at each region is the solution of Equation 6.29, which can be written as

$$\begin{aligned} \psi_{\rm I}(x,y) &= \frac{1}{\sqrt{2}}\begin{pmatrix}1\\ e^{i\theta_c}\end{pmatrix} e^{ik_c\cos\theta_c x + ik_c\sin\theta_c y} \\ &\quad + \frac{r}{\sqrt{2}}\begin{pmatrix}1\\ -e^{-i\theta_c}\end{pmatrix} e^{-ik_c\cos\theta_c x + ik_c\sin\theta_c y}, \quad x<0 \\ \psi_{\rm II}(x,y) &= \frac{a}{\sqrt{2}}\begin{pmatrix}1\\ -e^{-i\theta_\upsilon}\end{pmatrix} e^{-ik_\upsilon\cos\theta_\upsilon x + ik_\upsilon\sin\theta_\upsilon y} \\ &\quad + \frac{b}{\sqrt{2}}\begin{pmatrix}1\\ e^{i\theta_\upsilon}\end{pmatrix} e^{ik_\upsilon\cos\theta_\upsilon x - ik_c\sin\theta_\upsilon y}, \quad 0<x<d \\ \psi_{\rm III}(x,y) &= \frac{t}{\sqrt{2}}\begin{pmatrix}1\\ e^{i\theta_c}\end{pmatrix} e^{ik_c\cos\theta_c x + ik_c\sin\theta_c y}, \quad x>d \end{aligned} \quad (6.30)$$

The coefficients of transmission and reflection, as well as the coefficients a and b, are obtained by the continuity of wave functions at two interfaces $x = 0$, and $x = d$

$$\psi_{\rm I}(0,y) = \psi_{\rm II}(0,y), \quad \psi_{\rm II}(d,y) = \psi_{\rm III}(d,y) \quad (6.31)$$

For a symmetric NPNJ, $n = -1$, $\theta_\upsilon = -\theta_c$, and $k_\upsilon = k_c$, the coefficient of transmission gets

$$t = \frac{\cos^2\theta_c}{e^{2ik_c d\cos\theta_c} - \sin^2\theta_c} \approx \cos^2\theta_c e^{-2ik_c d\cos\theta_c} \quad (6.32)$$

Thus, the transmission wave at the second N region can be written as

$$\begin{aligned} \psi_{transmission}(x,y) &\equiv \frac{A}{\pi}\int_{-\pi/2+i\infty}^{\pi/2-i\infty} \cos^2\theta_c e^{ik_c[x-(2d-a)]\cos\theta_c} \\ &\quad \times \begin{pmatrix}1\\ e^{i\theta_c}\end{pmatrix} e^{ik_c y\sin\theta_c} d\theta_c \\ &= \frac{A}{\pi}\int_{-\pi/2+i\infty}^{\pi/2-i\infty} \cos^2\theta_c e^{ik_c e^{ik_c\tau 2}\cos(\theta_c-\beta)}. \end{aligned} \quad (6.33)$$

Here, $r_2 = \sqrt{[x-(2d-a)]^2 + y^2}$ and $\tan\beta = y/[x-(2d-a)]$. Using the steepest descent method, the approximate expression of the transmitting electron wave is written as

$$\psi_{transmission}(x,y) \approx \begin{cases} \sqrt{\dfrac{2}{\pi k_c r_2}}\cos^2\beta\begin{pmatrix}1\\ e^{i\beta}\end{pmatrix}e^{i(k_c r_2-\pi/4)}, & \beta\in\left[-\dfrac{\pi}{2},\dfrac{\pi}{2}\right] \\ \sqrt{\dfrac{2}{\pi k_c r_2}}\cos^2\beta\begin{pmatrix}1\\ -e^{i\beta}\end{pmatrix}e^{-i(k_c r_2-\pi/4)}, & \beta\in\left(\dfrac{\pi}{2},\dfrac{3\pi}{2}\right] \end{cases} \quad (6.34)$$

Similar to the discussion for the refractive wave function in PNJ, at the second N region (Region III), the electric current density around the point $(a_1, 0)$ is

$$\vec{j}(r_2,\beta) = \pm\frac{I_0\cos^4\beta}{2\pi r_2}(\cos\beta,\sin\beta), \quad k_c r_2 \gg 1 \quad (6.35)$$

with "+" for $\beta \in [-\pi/2, \pi/2]$, and "−" for $\beta \in (\pi/2, 3\pi/2]$, similar to the definitions used in Equation 6.27.

Around but not too close to the focus, the intensity of current density is

$$j_2 = |\vec{j}(x,y)| = \frac{I_0}{2\pi}\frac{(x-a_1)^4}{[(x-a_1)^2+y^2]^{5/2}} \quad (6.36)$$

The maximal current that can be collected by a drain electrode at focus $(a_1, 0)$ is $I_0\int_{\pi/2}^{3\pi/2}\cos^4\beta/(2\pi)d\beta = 3/16 I_0$, which is nearly one-fourth of the total electric current spread from the source electrode. In the above analytical expressions, only the leading term in the results obtained by the deepest descent method is kept, which is reasonable for the wave at the positions $r_1 \gg \lambda_F$ (NPJ) and $r_2 \gg \lambda_F$ (NPNJ). The low-energy effective Hamiltonian of graphene is valid for energy less than 1 eV; thus, λ_F is at most several nanometers.

In the experiments of graphene-based devices, the source/drain (detective) electrode's size is roughly tens of nanometers, and the far-field condition $k_\upsilon r_1(k_c r_2) \gg 1$ is reasonable. In fact, due to the short wavelength of the electron wave, high resolutions of one wavelength or even subwavelength are less meaningful than that for photons (Pendry 2007).

6.3.7 Applications of Electric Lens in Graphene

Cheianov et al. suggested two potential applications of graphene-based electric lens. One is the gate voltage-controlled transistor, and the other is the splitter. Consider graphene NPN junction, which can be prepared by a pair of split gates. The electron flow from a fixed source electrode at the left N region becomes converge beams at the right P region. The electric current captured by the drain electrode located at the focus in P region is sensitive to the gate voltage. It is possible to realize "0" and "1" states by changing the gate voltage.

By the analog of the optical lens, the electric lens can split one electric beam into two beams. Unlike the graphene transistor achieved by the rectangle-shaped top gate electrode, the beam splitter can be made using prism-shaped top gate, which can be used to fabricate logical gates and interconnects in nanoelectronics, or in the studies of fundamental physics, such as Einstein–Podolsky–Rosen (EPR) paradox, suggested by Cheianov et al. (2007).

6.3.8 Discussion on the Sharp PNJ/NPN Junction Assumption

The idea of graphene electric lens is attractive, but it is difficult to realize experimentally. In Cheianov et al.'s original work, the steep electrostatic potential was adopted, which implies the abrupt PNJ interface, and it is not realistic. In fact, the gate electrodes provide a smooth electrostatic potential landscape, varying slowly on the scale of the electron wavelength. Since the transmitting probability for Klein tunneling necessary to Veselago's lens effect decreases exponentially, as ~$\exp(-d/\lambda_F)$, with d the interface width and λ_F the Fermi wavelength. In addition, the collimation effect should also be considered for transmission across the smooth junction. However, it has been demonstrated that by using the bottom/top gate, it can make the electrostatic potential change at the interface more abruptly than that of split gates system. Moreover, although the smooth junction decreases the resolution of an electronic lens in graphene, the theoretical prediction based on a sharp junction assumption can still provide useful results for Veselago-type devices of graphene. The discussions in this section based on the sharp junction, such as the electric current collection efficiency can be understood as providing its theoretical upper bound for realistic devices.

6.3.9 Recent Progresses in Electric Lens

The investigation of the electric lens concept has been extended to the systems beyond a monolayer graphene sheet, that is, the original design of Cheianov et al.'s, for example, circular quantum dots of graphene (Cserti et al. 2007). Moreover, a recent progress on the flat-lens focusing of electrons on the surface of a topological insulator suggests a high efficient electric lens effect based on the topological insulator. In this electric lens, only conduction band electrons are used, which avoids high interface resistance in the PNJ (Hassler et al. 2010).

6.4 CONCLUSION

In this chapter, we have introduced the general properties of graphene, as well as the basic electronic structure of graphene and the Dirac spinor description of low-energy electrons and holes. The electric lens phenomena in graphene PNJ and NPN junction based on chiral tunneling of electrons have been discussed, theoretically, and also the analytical expressions for the current density distribution in PNJ and NPN junction have been shown. The maximal possible current that can be collected by the drain electrode in a symmetric PNJ and NPN

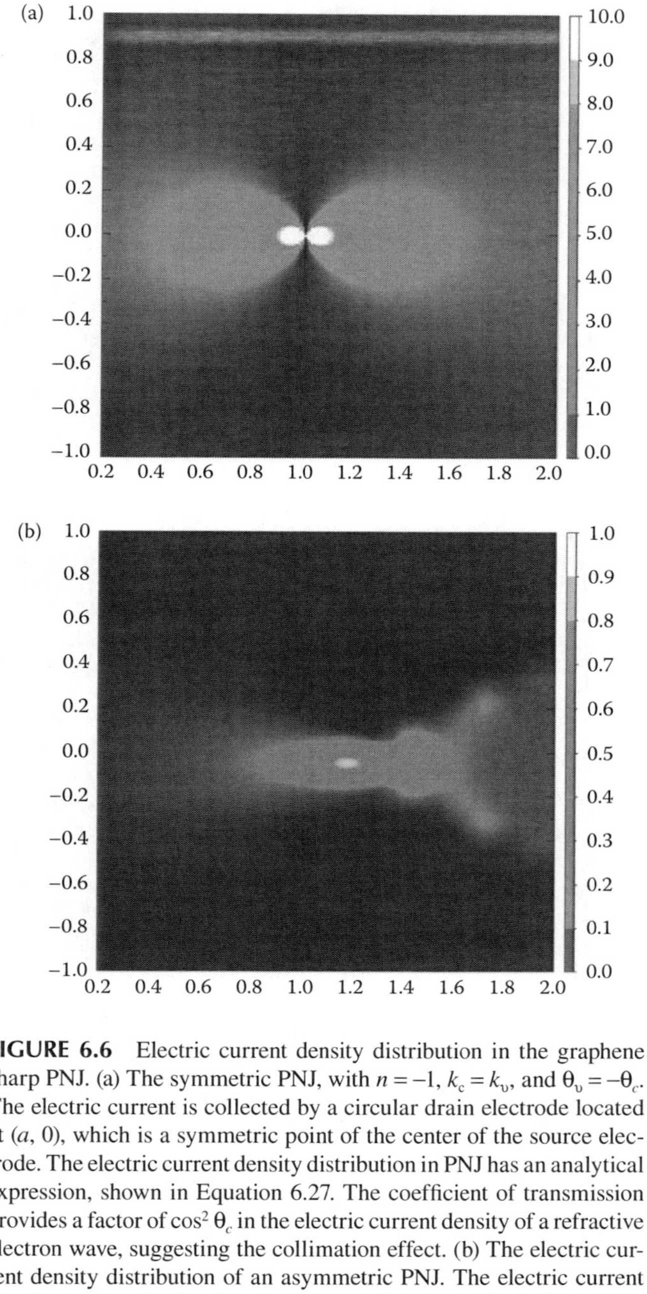

FIGURE 6.6 Electric current density distribution in the graphene sharp PNJ. (a) The symmetric PNJ, with $n = -1$, $k_c = k_\upsilon$, and $\theta_\upsilon = -\theta_c$. The electric current is collected by a circular drain electrode located at $(a, 0)$, which is a symmetric point of the center of the source electrode. The electric current density distribution in PNJ has an analytical expression, shown in Equation 6.27. The coefficient of transmission provides a factor of $\cos^2 \theta_c$ in the electric current density of a refractive electron wave, suggesting the collimation effect. (b) The electric current density distribution of an asymmetric PNJ. The electric current distribution is calculated by the steepest descent method, since it has no analytical form expression. The electric density concentrates near the "cusp," $(|n|a, 0)$, which is one of the caustic phenomenon.

junction that can be used in the design of devices, such as logical gates and interconnects based on graphene Veselago lens have also been presented. Also, some problems for the realization of graphene electric lens have been discussed, and some recent progresses on the electric lens have been addressed (Figure 6.6).

ACKNOWLEDGMENT

The authors gratefully acknowledge the National Science Foundation of China (NSFC) under Grant No. 11074259 and the Major Research Plan for the National Natural Science Foundation of China (Grant No. 91027045) for the support of this chapter.

REFERENCES

Ando, T., T. Nakanishi, and R. Saito. 1998. Berry's phase and absence of back scattering in carbon nanotubes. *J. Phys. Soc. Japan* 67:2857.

Balandin, A. A., S. Ghosh, W. Bao, I. Calizo, D. Teweldebrhan, F. Miao, and C. N. Lau. 2008. Superior thermal conductivity of single-layer graphene. *Nano Lett.* 8:902.

Bena, C., and G. Montambaux. 2009. Remarks on the tight-binding model of graphene. *New J. Phys.* 11:095003.

Bunch, J. S., A. M. van der Zande, S. S. Verbridge, I. W. Frank, D. M. Tanenbaum, J. M. Parpia, H. G. Craighead, and P. L. McEuen. 2007. Electromechanical resonators from graphene sheets. *Science* 315:490.

Castro Neto, A. H., F. Guinea, N. M. R. Peres, K. S. Novoselov, and A. K. Geim. 2009. The electronic properties of graphene. *Rev. Mod. Phys.* 81:109.

Cheianov, V. V., V. I. Fal'ko, and B. L. Altshuler. 2007. The focusing of electron flow and a Veselago lens in graphene p–n junctions. *Science* 315:1252.

Chen, C., S. Rosenblatt, K. I. Bolotin, W. Kalb, P. Kim, I. Kymissis, H. L. Stormer, T. F. Heinz, and J. Hone. 2009. Performance of monolayer graphene nanomechanical resonators with electrical readout. *Nat. Nanotechnol.* 4:861.

Cserti, J., A. Pályi, and C. Péterfalvi. 2007. Caustics due to a negative refractive index in circular graphene p–n junctions. *Phys. Rev. Lett.* 99:246801.

Frank, I. W., D. M. Tanenbaum, A. M. van der Zande, and P. L. McEuen. 2007. Mechanical properties of suspended graphene sheets. *J. Vac. Sci. Technol. B* 25:2558.

Garcia-Sanchez, D., A. M. van der Zande, A. S. Paulo, B. Lassagne, P. L. McEuen, and A. Bachtold. 2008. Imaging mechanical vibrations in suspended graphene sheets. *Nano Lett.* 8:1399.

Geim, A. K. 2009. Graphene: Status and prospects. *Science* 324:1530.

Geim, A. K., and K. S. Novoselov. 2007. The rise of graphene. *Nat. Mater.* 6:183.

Gerlach, U. H. 2010. Linear Mathematicas in in_nite dimensions signals boundary value problems and special functions. http://www.math.osu.edu/gerlach/math/BVtypset/node122.html.

Gorbachev, R. V., A. S. Mayorov, A. K. Savchenko, D. W. Horsell, and F. Guinea. 2008. Conductance of p–n–p graphene structures with "air-bridge" top gates. *Nano Lett.* 8:1995.

Hassler, F., A. R. Akhmerov, and C. W. J. Beenakker. 2010. Flat-lens focusing of electrons on the surface of a topological insulator. *Phys. Rev. B* 82:125423.

Hu, J., X. Ruan, and Y. P. Chen. 2009. Thermal conductivity and thermal rectification in graphene nanoribbons: A molecular dynamics study. *Nano Lett.* 9:730.

Huard, B., J. A. Sulpizio, N. Stander, K. Todd, B. Yang, and D. Goldhaber-Gordon. 2007. Transport measurements across a tunable potential barrier in graphene. *Phys. Rev. Lett.* 98:236803.

Katsnelson, M. I. 2012. *Graphene: Carbon in Two Dimensions*. Cambridge University Press (New York).

Katsnelson, M. I., and K. S. Novoselov. 2007. Graphene: New bridge between condensed matter physics and quantum electrodynamics. *Solid State Commun.* 143:3.

Katsnelson, M. I., K. S. Novoselov, and A. K. Geim. 2006. Chiral tunnelling and the Klein paradox in graphene. *Nat. Phys.* 2:620.

Landau, L. 1937. Zur Theorei der phasenumwandlugen II. *Physikalische Zeitschrift Sowjetunion* 11:26.
Li, X. S., Y. W. Zhu, W. W. Cai, M. Borysiak, B. Y. Han, D. Chen, R. D. Piner, L. Colombo, and R. S. Ruoff. 2009. Transfer of large-area graphene films for high-performance transparent conductive electrodes. *Nano Lett.* 9:4359.
Lin, Y. M., C. Dimitrakopoulos, K. A. Jenkins, D. B. Farmer, H. Y. Chiu, A. Grill, and P. Avouris. 2010. 100-GHz transistors from wafer-scale epitaxial graphene. *Science* 327:662.
Ling, X., L. Xie, Y. Fang, H. Xu, H. Zhang, J. Kong, M. S. Dresselhaus, J. Zhang, and Z. Liu. 2010. Can graphene be used as a substrate for Raman enhancement? *Nano Lett.* 10:553.
Liu, G., J. V. Jr., W. Bao, and C. N. Lau. 2008. Fabrication of graphene p–n–p junctions with contactless top gates. *Appl. Phys. Lett.* 92:203103.
McClure, J. W. 1957. Band structure of graphite and de Haas-van Alphen effect. *Phys. Rev.* 108:612.
McEuen, P. L., M. Bockrath, D. H. Cobden, Y. G. Yoon, and S. G. Louie. 1999. Disorder, pseudospins, and backscattering in carbon nanotubes. *Phys. Rev. Lett.* 83:5098.
Mermin, N. D. 1968. Crystalline order in two dimensions. *Phys. Rev.* 176:250.
Meyer, J. C., A. K. Geim, M. I. Katsnelson, K. S. Novoselov, T. J. Booth, and S. Roth. 2007. The structure of suspended graphene sheets. *Nature* 446:60.
Noveselov, K. S., A. K. Geim, S. V. Morozov, D. Jiang, M. I. Katsnelson, I. V. Grigorieva, S. V. Dubonos, and A. A. Firsov. 2005. Low-dimensional gas of massless Dirac fermions in graphene. *Nature* 438:197.
Novoselov, K. S., A. K. Geim, S. V. Morozov, D. Jiang, Y. Zhang, S. V. Dubonos, I. V. Grigorieva, and A. A. Firsov. 2004. Electric field effect in atomically thin carbon films. *Science* 306:666.
Novoselov, K. S., Z. Jiang, Y. Zhang, S. V. Morozovl, H. L. Stormer, U. Zeitler, J. C. Maan, G. S. Boebinger, P. Kim, and A. K. Geim. 2007. Room-temperature quantum Hall effect in graphene. *Science* 315:1379.
Oostinga, J. B., H. B. Heersche, X. Liu, A. F. Morpurgo, and L. M. K. Vandersypen. 2008. Gate-induced insulating state in bilayer graphene devices. *Nat. Mater.* 7:151.
Özyilmaz, B., P. Jarillo-Herrero, D. Efetov, D. A. Abanin, L. S. Levitov, and P. Kim. 2007. Electronic transport and quantum Hall effect in bipolar graphene p–n–p junctions. *Phys. Rev. Lett.* 99:166804.
Pendry, J. B. 2000. Negative refraction makes a perfect lens. *Phys. Rev. Lett.* 85:3966.
Pendry, J. B. 2007. Negative refraction for electrons? *Science* 315:1226.
Perierls, R. 1935. Quelques properties typiques des corpes solides. *Annales d'Institut Henri Poincare* 5:177.
Slonczewski, J. C., and P. R. Weiss. 1958. Band structure of graphite. *Phys. Rev.* 109:272.
Stander, N., B. Huard, and D. Goldhaber-Gordon. 2009. Evidence for Klein tunneling in graphene p–n junctions. *Phys. Rev. Lett.* 102:026807.
Tian, H., T.-L. Ren, D. Xie, Y.-F. Wang, C.-J. Zhou, T.-T. Feng, D. Fu et al. 2011. Graphene-on-paper sound source devices. *ACS Nano* 5:4878.
Veselago, V. G. 1968. The electrodynamics of substances with simultaneously negative values of ε and μ. *Sov. Phys. Usp.* 10:509.
Wallace, P. R. 1947. The band theory of graphite. *Phys. Rev.* 71:622.
Wang, Z., R. Xie, C. T. Bui, D. Liu, X. Ni, B. Li, and J. T. L. Thong. 2011. Thermal transport in suspended and supported few-layer graphene. *Nano Lett.* 11:113.
Weihua, M., G. Zhang, Y. Tang, W. Wang, and Z. Ou-Yang. 2011. Electric current focusing efficiency in a graphene electric lens. *J. Phys. Condens. Matter* 23:495302.
Williams, J. R., L. DiCarlo, and C. M. Marcus. 2007. Quantum Hall effect in a gate-controlled p–n junction of graphene. *Science* 317:638.
Xia, F. N., T. Mueller, Y. M. Lin, A. V. Garcia, and P. Avouris. 2009. Ultrafast graphene photodetector. *Nat. Nanotechnol.* 4:839.
Zhang, Y., Y.-W. Tan, H. L. Stormer, and P. Kim. 2005. Experimental observation of the quantum Hall effect and Berry's phase in graphene. *Nature* 438:201.

7 Electronic Properties and Transport in Finite-Size Two-Dimensional Carbons

J. C. Sancho-García and A. J. Pérez-Jiménez

CONTENTS

ABSTRACT

By quantum-chemical calculations, here, we ascertain the intrinsic electronic and conducting properties of molecular material prototypes for graphene-based nanostructures, and how these properties evolve as a function of their size with an increased dimensionality (from quasi-one dimensional in the case of isolated nanoribbons to three-dimensional arrangements when they self-organize in common samples). As it is expected in regular devices architectures, whether they are purely organic thin films allowing for the diffusive transport of the charge carrier upon its injection from an external source, or electrode–molecule(s)–electrode nanojunctions targeted to study charge transport in the coherent regime, the relative positions of the interacting molecules can significantly alter the conclusions found for an isolated molecule. Since the supramolecular ordering of the organic layers is intimately related to the existing noncovalent interactions driving the specific mode of packing, we also investigate how to efficiently incorporate these effects into state-of-the-art calculations without giving up the computational cost-effectiveness that allow the tackling of longer and more complicated systems. Thus, owing to a clear understanding of structure–property relationships for isolated and packed molecules in both regimes, we can confirm two-dimensional nanographene-based materials as promising candidates for organic and molecular electronics.

7.1 INTRODUCTION

The graphene revolution is on. Everything seems to indicate that from now on, in the souvenir room of the Nobel foundation, a tiny piece of transparent scotch tape will be accompanied by a pair of electronic transistors and a sample of graphite. This small yet representative collection is the kind gift of Nobel Laureates Sir A. Geim and Sir K. Novoselov, who were awarded the Nobel Prize for Physics in 2010 [1]. These objects are the symbols of a major breakthrough in the field of materials science within the last few years: graphene-based materials and their envisioned applications. As a matter of illustration, the European Union will devote 1000 million euros in the next 10 years to move graphene, and related layered or two-dimensional (2D) materials, from academic laboratories out into society, in an attempt to revolutionize multiple industries creating a robust platform for economic growth and a new job market in Europe [2]. And this is so because the expected applications in materials science, not yet even entirely contemplated, might largely impact most existing technological domains, including electronic and optical devices, consumer electronics and screens, advanced batteries, medical applications and aids, lighter and more efficient aeronautic materials, clean energy devices, sensors and membranes, and light-harvesting antennae, among others. A recent analysis [3] by a specialized company revealed an intense patent landscape in the covered period (2007–2013), with international companies and academic centers sharing thousands of

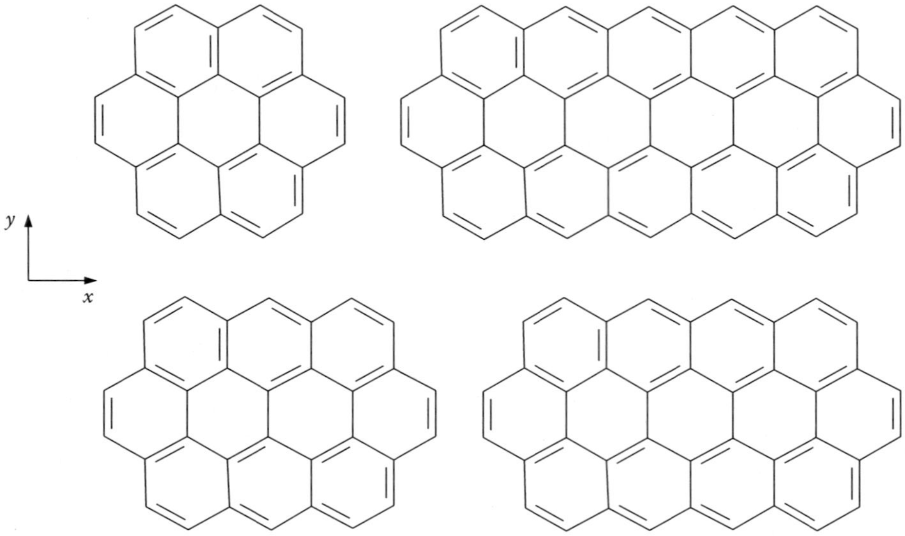

FIGURE 7.1 Chemical structure of circum(oligo)acenes: circumbenzene (cB) or coronene, cN or ovalene, and cA and cT (counterclockwise). The peripheral hydrogen atoms and corresponding C–H bonds have been omitted for clarity.

patent ownerships or portfolios. Briefly speaking, graphene is a ubiquitous, versatile, and extremely flat material and, actually, no other known compound combines these features in a manner so greatly promising. We will mostly concentrate on the latter characteristics, flatness, and thinness, since reducing the size of the original 2D graphene sheet, and indeed thus dealing with nanographenes, might also bring new facilities to control the chemical synthesis of high purity and associated versatility [4], although the price to be paid for it deals with new supramolecular issues (*vide infra*) that are no longer affecting isolated sheets. For this purpose, we have selected the set of compounds presented in Figure 7.1. Note that these are commonly dubbed as circum(oligo)acenes by analogy to (oligo)acenes (a subset of polycyclic aromatic hydrocarbons known some time ago) and that the entire family of compounds has been successfully synthesized in recent years.

For instance, the peripheral functionalization of these systems with electroactive substituents might serve to fine tune their electrical and optical properties [5,6]. Furthermore, these nanographenes may also be converted into (quasi-one-dimensional [1D]) nanoribbons by enlarging the backbone while asymmetrically reducing their size, which would efficiently introduce a nonzero band gap upon constraint of the dimensions, thus representing an alternative to other techniques to induce it such as multilayered or chemically doped graphene, to name just a few [7]. However, these nanographenes concomitantly self-assemble in crystalline samples, thus necessarily introducing a larger dimensionality in defect-free supramolecular arrays [8]. Here, our aim is to theoretically disentangle these challenging issues [9], and to correspondingly analyze the *pros* and *cons* of the engineered finite-size nanographenes presented in Figure 7.1, after simultaneously addressing intramolecular (single-molecule) magnitudes related to charge transport as well as the expected intermolecular (supramolecular) influence of the packing driven by subtle and weak noncovalent interactions. Furthermore, our theoretical tools have allowed the analysis of charge transport in two extreme yet related situations, either in a diffusive regime after charge injection and transport across the organic layer, or in a ballistic-like regime when the array of molecules is connected to a metallic electrode acting as a reservoir [10–12]. Note that although the two mentioned regimes shared the agents, charge carriers, and some array of molecules, involved in charge transport, they differ in the underlying interactions defining the transport regime because the molecular arrangements used on each one are completely different. Note that we will avoid referring to technical details as much as possible in the following; however, the interested reader may consult the original references quoted in the chapter.

7.2 PREVIOUS CONSIDERATIONS

7.2.1 Theoretical Methods

We choose Hamiltonian, the Fock operator (F) as a model typical of density functional theory (DFT) in various flavors, whether they are of hybrid or double-hybrid forms. This operator acts self-consistently on a set of eigenvectors (φ) providing the corresponding eigenvalues (ε) upon convergence of the one-electron Kohn–Sham equations. The form of this operator is given by a kinetic term together with the external potential (v_s) felt by the particles and that includes many-body interactions such as Coulomb and exchange-correlation (v_{xc}) effects. The approximation scheme to calculate physical observables relies on accurate expressions for the exchange-correlation density functional $E_{xc}(\rho)$, which will thus provide the corresponding v_{xc} potential through the associated functional derivative, in the following sense: (i) a hybrid approximation mixes an exact-like (i.e., Hartree–Fock) exchange and an exchange energy density functional, together with some functional form for correlation effects, with (possibly) different weights (e.g., B3LYP model) [13]; and (ii) a double-hybrid approximation also mixes the second-order perturbation theory and a correlation energy density functional again with variable weights (e.g., B2-PLYP model) [14]. In the past, we have intensively used these two schemes to calculate the key magnitudes needed to address charge transport issues in both regimes. However, the conclusions did not differ too much (in this framework) regarding the choice of the functional, which allows us to draw the conclusion on a much firmer basis. All the calculations presented in this section were always performed with large basis sets (e.g., cc-pVTZ) and dense integration grids, to avoid numerical errors as much as possible, and with the ORCA [15] quantum-chemical package.

7.2.2 Isolated Molecules

First, we note that the set of compounds chosen should allow us to systematically ascertain the convergence (if any) of key properties with the system size, before trying to analyze the influence of more real environments in depth, such as molecular packing or nanojunctions, in the second step. It is well known that single events in the diffusive regime for charge transport will consist of a sequence of steps involving the charge/uncharge of an isolated molecule, roughly neglecting the polarization of the adjacent molecules. The energy needed to charge/uncharge the molecule, thanks to the coupling (or trapping) between the arrival charge and the rigid molecular backbone, is called the reorganization energy (λ) and can be adequately calculated by theoretical methods that are able to deal with subtle exchange-correlation effects of a truly quantum nature [16]. Evidently, the lower the reorganization energy, the lower the probability of trap states, and then the higher the probability to successfully transfer the charge from one molecule to the neighboring one.

Figure 7.2 shows how this reorganization energy evolves as a function of the size of the molecule for circum(oligo)acenes, and how a much more pronounced slope is observed when

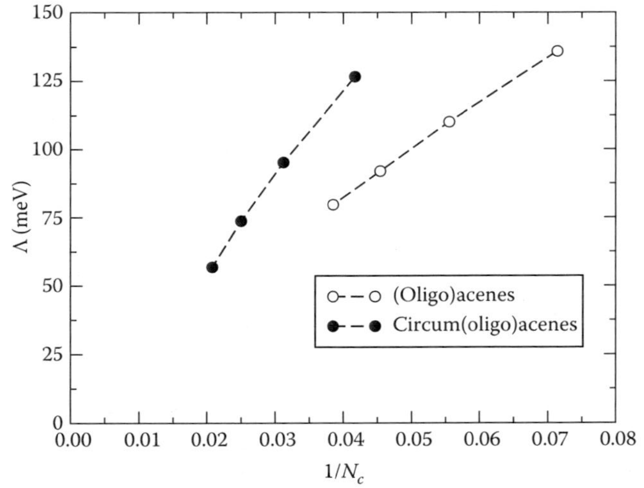

FIGURE 7.2 Evolution of reorganization energies of (oligo)acenes and circum(oligo)acenes as a function of the inverse number of carbon atoms (N_c).

compared with the smaller corresponding (oligo)acenes. On the other hand, when these candidates directly interact with metallic electrode(s), thus forming a nanojunction or a nanobridge, the efficiency of charge transport depends, besides details affecting the metal-contact geometry, on the alignment between the energy (ε) of the highest-occupied molecular orbital (HOMO) with the metal state, as well as the corresponding energy gap between the frontier-occupied and frontier-unoccupied orbitals ($\Delta\varepsilon$). While the value of ε (HOMO) decreases as expected with the system size, 4.94, 4.60, and 4.35 eV for circumnaphthalene (cN), circumanthracene (cA), and circumtetracene (cT), respectively, the evolution of $\Delta\varepsilon$ is more marked now, having values of 2.92, 2.15, and 1.60 eV, highlighting the strong relationships between structural and electronic properties, and allowing the choice of the metal used to build the nanobridges thanks to this band gap engineering.

7.2.3 Interacting Molecules

If one aims to describe the pristine self-assembly of these molecules driven by the underlying subtle forces between adjacent molecules (i.e., intermolecular interactions), it is known that large computational efforts are needed to accurately tackle these issues [17]. The solid-state structures are driven by through-space interactions and, despite the weakness of these forces, they really matter since any chemical functionalization of the molecular backbone might alter the mode of packing and thus all the expected supramolecular properties. These graphene nanoribbons are known to self-assemble in a herringbone-like pattern, as shown in Figure 7.3 in the simplest case of coronene taken as an example, for which the association energy for every dimer is defined as $\Delta E = E(\text{dimer}) - 2\,E(\text{monomer})$, taking into account all the possible orientations within the unit cell. Note that a negative value ($\Delta E < 0$) thus implies the existence of the complex with respect to the pair of isolated monomers. As a matter of example, we focus on the ΔE energy of the (0, *b*, 0) pair, which actually drives the herringbone-like growing of coronene molecules and contributes the most to the lattice (or cohesive) energy. Note that the dimer is predicted to be unbound without adding damped atom-pairwise long-range C_6R^{-6} energy corrections [18], and that the association energy is estimated to be in the order of −16 to −20 kcal/mol, depending on certain technical details, after this correction. Also note that some kind of π-stacking is also expected if the molecules are left to freely interact with metal tips to form the nanobridges.

Since the largest overlap between electronic clouds is expected for fully cofacial arrangements of these molecules, we have also calculated (Table 7.1) the association energy for cofacial dimers of circum(oligo)acenes, separated by a fixed distance of 3.5 Å between the center of mass of the molecules.

Strikingly, it is seen how the uncorrected DFT methods make an error of about 1.5 kcal/mol per atom, and that they slightly increase with system size which discourages the blind use of DFT for the treatment of increasingly

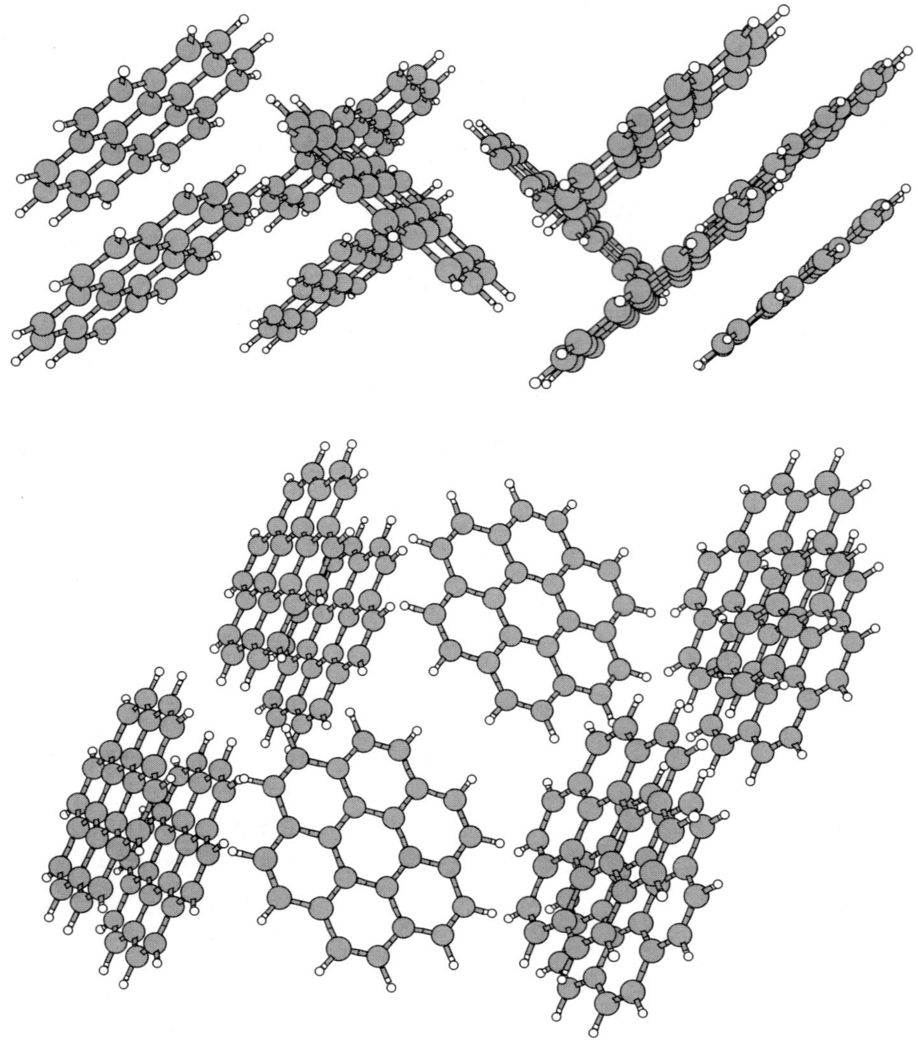

FIGURE 7.3 Interacting coronene molecules extracted from the crystalline structure (unit cell) in two views: along the *c* (top) and *b* (bottom) crystallographic axes.

TABLE 7.1
Association Energies (kcal/mol) of Circum(oligo)acene Dimers as Calculated by DFT-Based Methods, with (-D) and without Inclusion of Dispersion Interactions

Method	cB	cN	cA	cT
DFT	16	21	25	30
DFT-D	−16	−23	−31	−38

larger carbon nanostructures if they are expected to weakly interact. However, the results are dramatically altered in the right direction upon adding the aforesaid dispersive-like correction (DFT-D) for noncovalent interactions. It is easily seen again how the interaction energies increase with system size, which would predict a strong-enough attraction between graphene nanoribbons independently of their size and/or orientation. Furthermore, these methods would also allow the exploration of unknown potential energy hypersurfaces of related compounds to disclosure supramolecular order, polymorphism, and other related issues [19].

Concerning the pervasive competition between slipped face-to-face and *T*-shaped orientations, which can be viewed as the most commonly found patterns in this kind of sample, next, we will use the electrostatic potential (ESP) as an interpretative tool [20] to rationalize the origin and strength of the underlying interactions driving these two motifs. This potential, defined at every position $\boldsymbol{r}$ in the space, $V(\boldsymbol{r})$, is composed of two terms: the first term corresponds to the classical ESP of all the nuclei, and the second term corresponds to the quantum-mechanical ESP of the electrons at all other positions $\boldsymbol{r}'$ inducing the ESP at any point $\boldsymbol{r}$. Figure 7.4 displays the ESP values, mapped onto electron density isosurfaces (0.01 e/au^3), for tetracene, 5,11-disubstitutedtetracene, and 5,6,11,12-tetrasubstitutetetracene, being the

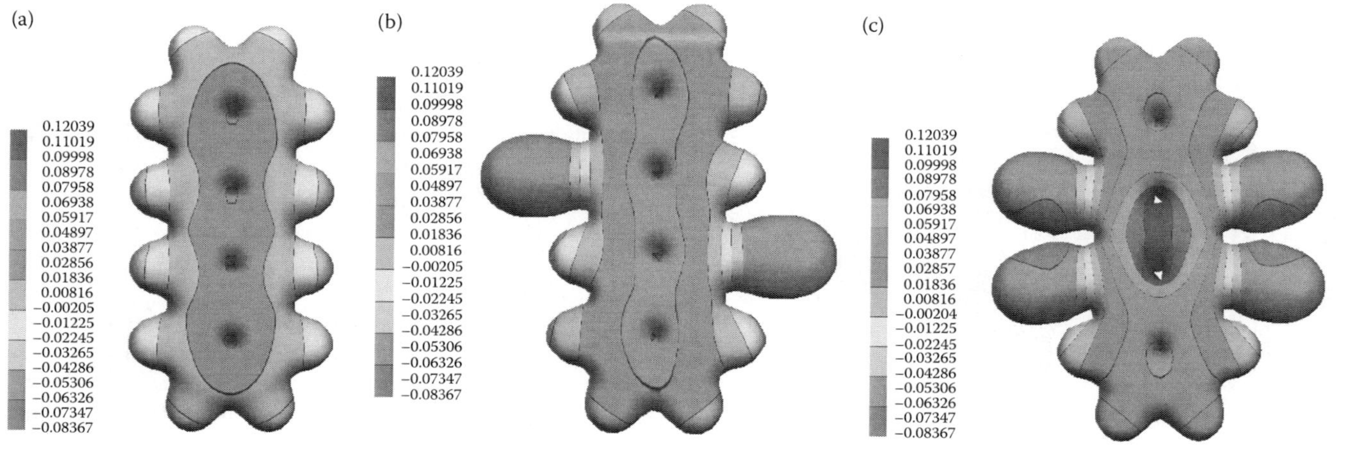

FIGURE 7.4 Isocontour plots of the ESP for (from (a) to (c)) tetracene, 5,11-disubstitutedtetracene, and 5,6,11,12-tetrasubstitutedtetracene.

substituents of an electro-withdrawing group in all the cases studied.

It is easily seen how the shape and values of the ESP markedly differ among the three compounds, which helps to understand the different packing motifs mainly in terms of the interactions between the electron-rich and electron-poor regions, favoring edge-to-face motifs for neutral and unsubstituted aromatic compounds and parallel-displaced orientations when they are substituted with electroactive groups. If the same reasoning applies to substituted graphene nanoribbons of increasing size, it is still a matter of current research in our group.

7.3 CHARGE TRANSPORT IN THE HOPPING REGIME

In this regime, the ease of charge injection from source electrodes together with the intrinsic ability of the material to transport the generated carriers, often determines the *p*- and/or *n*-type behavior of the material. For instance, efficient injection of holes (h^+) into a layer made of organic molecules relies on a close match between its ionization potential (I_p) and the work function of the inorganic electrode material used. Then, the charge transport rate (k^{CT}) can be estimated in the first approximation [21] by using expressions normally derived from Fermi's golden rule, where $k^{CT} \alpha V^2_{if} \rho(E_i - E_f)$, V_{if} being the electronic coupling between the two interacting initial (particle on donor) and final (particle on acceptor) states, and $\rho(E_i - E_f)$ the corresponding density of states depending on the nuclear deformations associated with the charge transfer process. While the electronic coupling can be expressed as a function of the wavefunctions of the initial and final states, $V_{if} = \langle \psi_i | F | \psi_f \rangle$, the density of states depends on certain semi-classical assumptions [22] and for a homogeneous distribution of identical molecules, it mainly depends on the reorganization energy (λ). The associated charge carrier mobility (μ) can be estimated from these transfer rates (in a random walk for multiple paths) considering all the nonequivalent orientation and distances between neighboring molecules in the crystal, since the magnitude of the electronic coupling is known to severely depend on the electronic overlap [22], either if it is experimentally known or if it needs to be computationally obtained by the methods stated above. Once the charges are injected or as the electron and hole separate from each other, note that the carrier–carrier and carrier–medium interactions are usually neglected, their motion (migration) can be regarded as diffusive on a large-length scale. In the past, we have tackled the computational evaluation of these two key quantities by modern yet cost-effective DFT-based methods, and the results will be summarized next. Remarkably, the reorganization energies for the largest circum(oligo)acenes considered here, cA and cT, are one of the lowest reported to date, and compare very favorably with the best-suited hole-transporting materials currently in use [22]. It is also interesting to note that *per*-functionalization of these molecules, which implies the full substitution of the hydrogen atoms by an electroactive group, might become an efficient alternative to tune their behavior from *p*-type to *n*-type, or even to have an ambipolar material with balanced hole and electron transport [23]. As a matter of example, in the case of coronene, while the ratio between μ values for hole and electron transport, respectively, is 0.76, this ratio closes to unity (0.98) in the case of *per*-fluorination. However, selective *di*-functionalization at specific positions is able to switch from these two extremes, ranging from 0.85 to 0.93 depending on their relative position, and show again the possibilities for further molecular engineering [24]. Actually, one can also estimate the ratio of the corresponding room-temperature charge mobilities for the two materials, circum(oligo)acenes and (oligo)acenes of the same width, in an idealized cofacial packing motif, after feeding the results of the calculations, λ and V_{if}, into the transfer rates and also neglecting (energetic or position) disorder effects. We obtain values of 1.2, 1.6, and 1.9 for circumbenzene/anthracene, cN/tetracene, cA/pentacene (P), and cT/hexacene ratios.

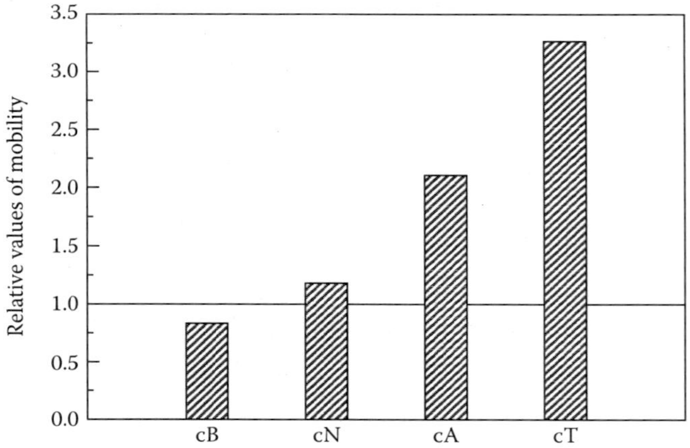

FIGURE 7.5 Estimates of relative values of mobility for circum(oligo)acenes with respect to tetracene at the DFT level. The solid line is a guide to the eye.

Interestingly, we note that in all cases, the expected mobilities for circum(oligo)acenes are larger than those of (oligo)acenes, also taking into account that tetracene and P are universally considered state-of-the-art materials for organic electronics [25]. It is thus worthwhile to now compare the mobility ratios of circum(oligo)acenes with respect to tetracene, Figure 7.5, which is known to possess a decent mobility in the range of 1–5 $cm^2/V/s$ for an herringbone solid-state packing. If all the external conditions, including the sample quality, remain approximately the same, we can anticipate larger mobilities for this family of compounds, which renders circum(oligo)acenes of a large size as potential synthetic targets that are useful in molecular electronics application.

7.4 CHARGE TRANSPORT IN THE COHERENT REGIME

In the preceding paragraphs, we have mainly focused on the so-called "hopping" charge transport regime, characterized by a *transfer rate constant*, k^{CT}, which measures the speed at which the charge is transferred from an initial state, where it remains localized in one molecule, to a final state, where it localizes in an adjacent molecule. But, circum(oligo)acenes can also be used to fabricate molecular nanobridges, where they are "sandwiched" between two metallic contacts. In this case, charge transport frequently occurs within the coherent regime, where the molecule can be considered as a coherent scatterer, and the key quantity is the *transmission probability*, T, of an incoming delocalized charge in the metal lead to traverse the energy barrier of the molecule–metal nanobridge [26]. In fact, at low bias, the conductance, G, of such a molecular nanojunction, is related to T via the well-known Landauer's formula [26], $G = G_0T(E_F)$, where $G_0 = 2e^2/h$ is the so-called "quantum of conductance" and $T(E_F)$ represents the transmission probability evaluated at the Fermi energy of the system.

The value of $T(E_F)$ can be computed by formally dividing the system into three parts: two semi-infinite leads and a finite-scattering region composed of the molecule and the metal atoms near it, which hereafter will be termed as a "cluster." The transmission probability can henceforth be expressed in terms of Green's functions of the cluster, which take into account the electron propagation through this portion of the nanodevice, and the coupling matrices with the left and right electrodes, which describe the interaction between the cluster and the metallic leads [26].

The great advantage of using the above Green's function formalism is that it allows us to replace, without introducing any approximation, the infinite open system composed of two semi-infinite leads connected through the molecular conductor (the cluster), by an equivalent but finite one where the effect of the semi-infinite leads on the cluster is incorporated through the corresponding self-matrices [26].

What differentiates the algorithms that compute T using the above formalism are the following three key points: (i) the model used to compute the self-energies of the leads, (ii) the model used to compute the Hamiltonian of the cluster, and (iii) the procedure used to determine the Fermi energy. The calculations reported here follow the method proposed by Palacios et al., which has proved to be both efficient and reliable for a number of nanojunctions [27–32], and is based on the following choices for each one of the three aforementioned topics. First, a Bethe lattice model is employed to describe the leads, because aside from its simplicity and computational convenience, it reproduces the short-range order and the bulk density of states of the metallic leads quite well, being a sound choice if one tries to model the polycrystalline structures of the metallic electrodes present in real nanobridges [27–30]. Second, a Kohn–Sham DFT Hamiltonian is used to approximate the electronic structure of the cluster, where the B3LYP exchange-correlation functional is normally chosen, because it yields reasonable values for both the molecule–metal nanocontact geometry and the HOMO–lowest-unoccupied molecular orbital (LUMO) gap (see, e.g., Reference 11 and the references therein), both magnitudes being of primary importance to obtain accurate enough conductance values, as will be shown below. Finally, the Fermi energy is determined self-consistently by forcing charge neutrality in the cluster region once it is attached to the semi-infinite leads, guaranteeing the correct alignment between the molecular resonances and the Fermi energy, which proves to be the key to obtain meaningful results [30].

In the next two sections, we will see that the above-mentioned formalism can be applied to calculate and analyze the zero-bias conductance of several gold–circum(oligo)acene–gold molecular nanobridges. In the first section, we focus on the role that some structural and electronic properties play in the relationship between both transport regimes (hopping and coherent) for this kind of "nanographenes." On the basis of the information gathered in that section, the second section will be devoted to analyze the effect of "piling up" several molecules on the conductance of the nanodevice.

7.5 SIZE OF NANOGRAPHENES IS KEY TO BOTH COHERENT AND HOPPING TRANSPORT REGIMES

In this section, we analyze the evolution in the charge transport capabilities of several (oligo)acene and circum(oligo) acene molecules of varying sizes, both in the coherent and in the hopping transport regimes. The three molecules considered here are cN, P, and cA.

Their respective lengths along the longitudinal axis (see Figure 7.6 and Table 7.2) are 9.8, 12.1, and 12.3 Å, as obtained from DFT geometry optimizations performed with the Gaussian03 package [33] on the isolated molecules, using the B3LYP exchange-correlation functional and the 6–31G* basis set.

The values of k^{CT} for cN, P, and cA using the methodology described in Section 7.4, and listed in Table 7.2, are, respectively, 21×10^{14}, 26×10^{14}, and 34×10^{14} s^{-1} [12]; clearly increasing with the size of the molecule. This can be qualitatively justified by noting, as suggested by Nitzan and coworkers [34], that k^{CT} scales with the length of the localized state of the charge, which to a first approximation can be related to the length of the molecule it resides on. It is also found that cA yields larger rate transfer rates than expected solely from its length, which points to it as a good candidate to fabricate efficient devices for organic electronics.

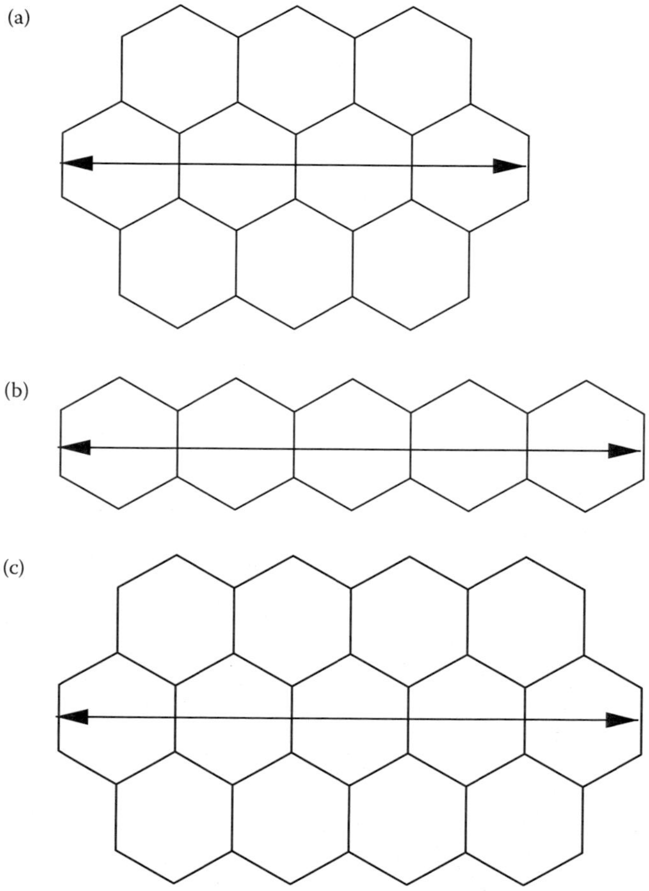

FIGURE 7.6 Carbon backbone of P (a), cN (b), and cA (c). The molecular length along the longitudinal axis is depicted in each case.

TABLE 7.2
Relationship between Geometrical, Electronic, and Charge Transport Parameters in the Hopping and Coherent Regime for cN, P, and cA

Molecule	L (Å)	HOMO–LUMO Gap (eV)	$K^{CT} \times 10^{-14}$ (s^{-1})	G (μS) *OS*	G (μS) *SS*
cN	9.8	2.94	21	23	0.77
P	12.24	2.22	26	36	1.24
cA	12.27	2.13	34	48	3.41

Note: See the text for the meaning of the symbols.

We have also studied whether the relative ordering found above for the transport in the hopping regime would also be found in the coherent regime. With this goal in mind, we perform transport calculations using the methodology outlined in Section 7.5 on several gold–molecule–gold nanobridges using all three molecules.

However, what determines the conductance in the coherent transport regime is the relative positioning of the Fermi level with the resonances derived from the molecular orbitals, once they couple with the semi-infinite electrodes: the so-called "band lineup" problem [35]. This, in turn, is largely influenced by the molecule–metal detailed structure, which can vary quite uncontrollably from experiment to experiment. To account for such variability in the calculations while keeping their number at an affordable value, we study two sets of metal–molecule structures, representative of the two most common types of geometries usually found in molecular nanojunctions. In the first one, see Figure 7.7 (top), the two metallic contacts are placed on opposite sides of the molecular plane (from now on, they will be termed *OS* structures), which are typical of nanobridges built with a scanning tunneling microscope. In the second one, see Figure 7.7 (bottom), the two metallic leads are placed on the same side of the molecular plane (in the following, these will be labeled *SS* structures), representing the situation found when the nanodevice is built using a mechanically controllable break junction.

Figure 7.7 depicts the structures obtained from B3LYP constrained geometry optimizations for both kinds of nanodevices on P and cN (the corresponding ones obtained for cA are similar, but are not shown here for simplicity). The calculations were done using the 6–31G* basis set for C and H, and the Christiansen and coworkers [36] pseudopotential and basis set for Au. In both cases, the relative positions of the gold atoms representing fcc [111] tips are kept fixed, although in the *OS* case, the distance between both tips is allowed to vary. Interestingly, the gold atoms attach to the edge C–C bonds

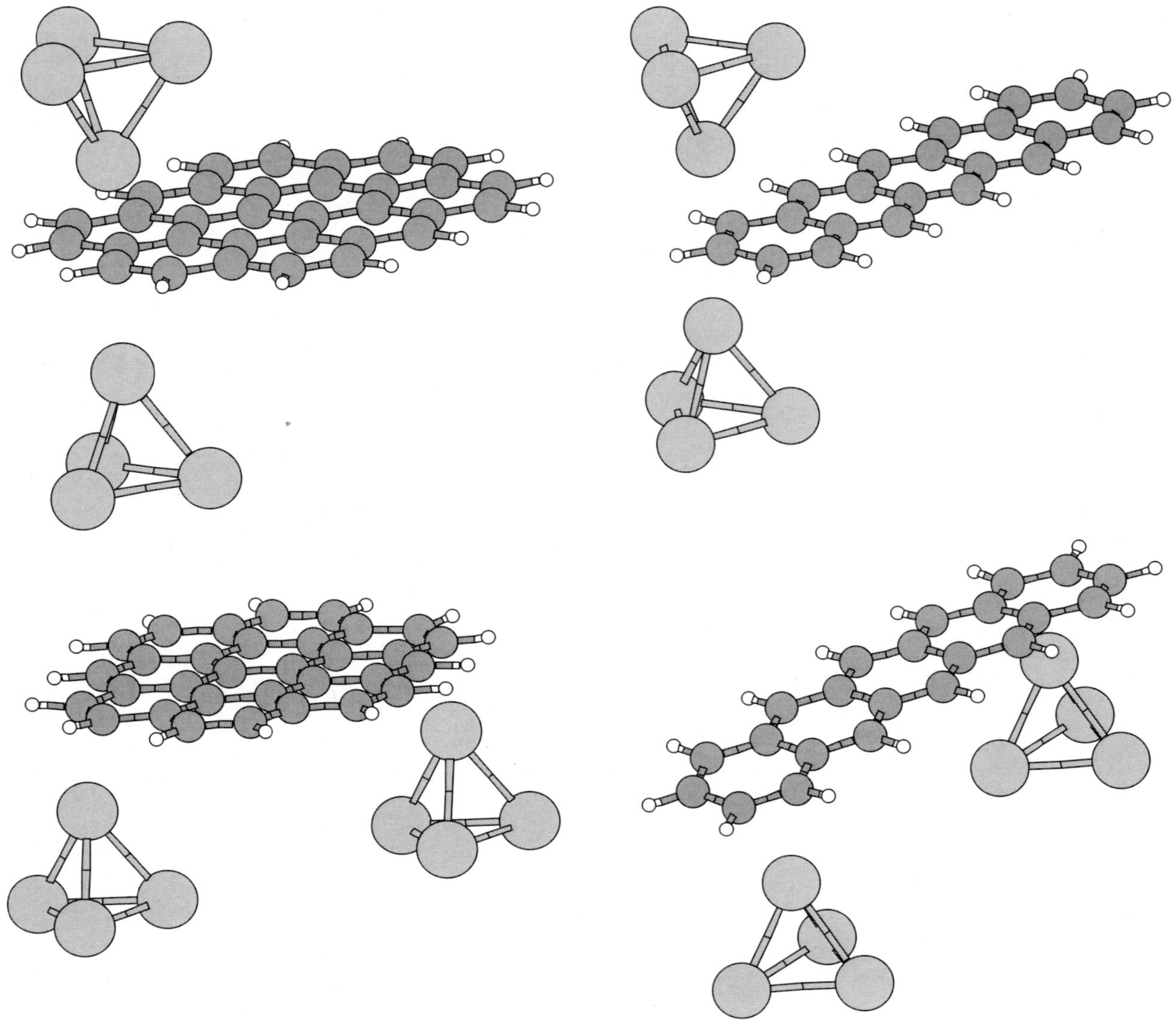

FIGURE 7.7 Optimized geometries for gold–molecule–gold nanobridges. Top: *OS* structures of P (right) and cN (left). Bottom: *SS* structures for the same molecules.

lying furthest from the center of the molecule, and agree well with the fact that graphene sheets have their highest electron density and conductivity at their edges [12,37].

The conductances calculated in the coherent regime for the aforementioned nanodevices using the methodology introduced in Section 7.4 are listed in Table 7.2. It is found that the relative ordering obtained in the hopping regime: cN < P < cA, is also maintained in the coherent regime, for both the *OS* and *SS* types of structures. This, to a first approximation, is a consequence of the increasing length of the molecules considered because, as it can be seen in Table 7.2, the greater the length of the circum(oligo)acene molecule, the lower the HOMO–LUMO gap becomes, which in turn helps in the band lineup between the HOMO resonance and the Fermi level of the nanodevice [12], increasing the conductance.

On the other hand, we must not overlook the influence that nanodevice geometry has on its conductance. Regarding this, we can check, from the values listed in Table 7.2 that the conductance also depends to a large extent on the relative positioning between the metal and the molecule atoms: the *OS* structures yielding much larger values than the *SS* structures. The different geometry of the metal–molecule structure induces changes in the interaction between the metal atoms in the two tips, as well as between them and the molecule atoms [12]. In the *OS* structures, both the amount of charge transfer and the stronger tip–tip interaction increase the conductance at the Fermi level, yielding values of G that are almost an order of magnitude larger than those calculated for the *SS* structures [12].

However, the relative ordering of the conductance found for the three molecules considered *within the same molecule–metal-contact structure* is independent of the geometry, and can be rationalized using the above-mentioned relationship between the molecular length and the HOMO–LUMO gap.

As regards this, we want to stress in passing that this interplay has also been noted for finite graphene sheets [37], and is also found between the width and the band gap of graphene nanoribbons [38]. In fact, the latter can be considered (at least formally) as derivatives of infinitely long oligoacenes, where the additional layers of fused benzene rings are also of finite length.

Therefore, the calculations indicate that, *a priori*, larger circum(oligo)acenes are promising candidates to improve *both* the transport properties of organic electronics *and* molecular electronics, because in both cases, the underlying transport mechanism is favored by an increase of the molecular size: in the former case, the localized charge typical of the hopping transport can be distributed over a larger area, while in the latter case, the corresponding reduced HOMO–LUMO gap improves the band lineup between the Fermi level and the molecular resonances, which is a key factor within the coherent transport regime.

In the next section, it will be shown that the HOMO–LUMO gap of circum(oligo)acenes can also be altered by molecular π-stacking, opening an alternative way to increase the conductance of molecular electronic nanodevices.

7.6 INCREASING CONDUCTANCE BY Π-STACKING NANOGRAPHENES IN MOLECULAR NANOBRIDGES

As already commented in the previous section, the conductance at zero bias in gold–circum(oligo)acene–gold nanobridges is largely influenced by the HOMO–LUMO gap of the molecule "sandwiched" between the two metal contacts. More precisely, $G(E_F)$ is controlled by the width and alignment of a HOMO-derived resonance with respect to the Fermi level, because E_F always lies slightly above the former [11]. On the other hand, it has been demonstrated that the energy of the Kohn–Sham HOMO orbital obtained with the *exact* but unknown exchange-correlation functional is equal to minus the ionization potential of the corresponding molecule [39,40], although for the functionals proposed in the literature, such equality holds only approximately.

Therefore, and to a first approximation, the lower the ionization potential of the molecule is, the higher the conductance it yields. Since, in turn, the ionization potential and the HOMO–LUMO gap in circum(oligo)acenes are inversely proportional to its length, one can, in principle, control the conductance of molecular nanojunctions with this kind of molecule by using acenes of increasing size. Here, we prove this finding by adding a molecule of increased size, namely cT, to the set of three molecules analyzed in Section 7.5, and calculating its conductance at zero bias for the same type of metal–molecule contact geometries.

The results are plotted in Figure 7.8, where the trend between G and the molecular size, already observed in Reference 12 and commented on in the previous section can be graphically visualized again, and which can be put into correspondence with a diminishing ionization potential for the four molecules considered: 4.9 eV (cN) < 4.6 eV (P and cA) < 4.4 eV (cT).

The quite nonlinear increase in the conductance with the molecular size calculated for *SS* structures compared to those obtained for *OS* ones is due to the Lorentzian form of the conductance peaks [11]: in the *OS* structures, the lineup between the HOMO resonance and the Fermi level is such that E_F lies much closer to the peak maximum than in the *SS* structures, thus making the latter structures quite sensitive to small changes in such a lineup. In fact, charging the *SS* gold–cT–gold nanojunctions by a single additional electron, its conductance varies by about three orders of magnitude: from $0.3G_0$ to $0.0004G_0$, which makes this kind of molecular nanodevice an interesting candidate to build field-effect transistors.

Figure 7.8 also plots the conductance calculated when several acene molecules pile up in a perfect stack (for simplicity, only the stacked structures of P and cN have been depicted: see Figures 7.9 and 7.10), and shows that the trend observed previously for single-molecule nanobridges is still valid.

However, while there is not much difference in the conductance between single-molecule and two-molecule *OS*

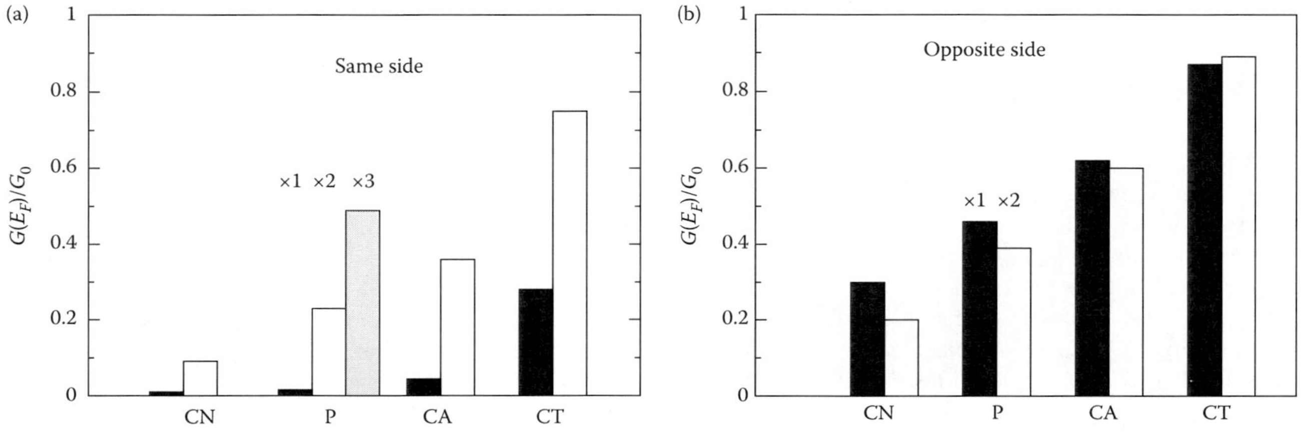

FIGURE 7.8 Conductance at zero bias of gold–molecule–gold nanobridges for P, cN, cA, and cT. (a) Cofacial *SS* nanojunctions. (b) Cofacial *OS* nanojunctions.

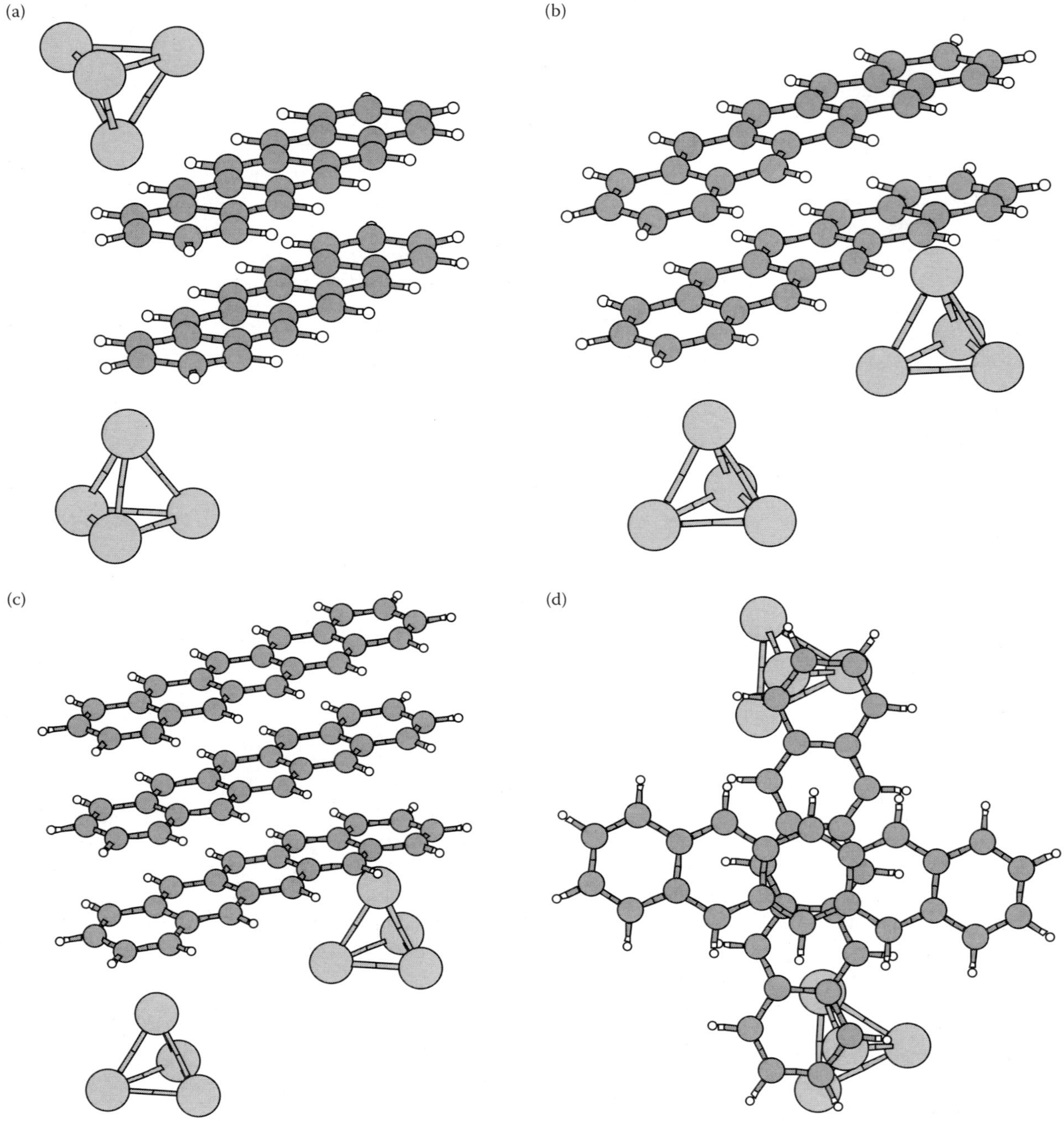

FIGURE 7.9 Structures of several π-stacked gold–P–gold nanobridges. (a) $OS \times 2$. (b) $SS \times 3$. (c) $SS \times 2$. (d) $SS \times 2$ rotated.

structures, a significant increase in conductance is found for the *SS* structures.

This can be traced back, again, to the effect that the ionization potential of the molecular system has on the alignment between the Fermi energy and the molecular resonances, because, as Figure 7.11 shows, both the ionization potential and the HOMO–LUMO gap vary linearly with the inverse of the number of acene molecules in the stack, a fact previously noted for thiophene complexes [41].

Therefore, as the number of molecules in the stack increases, the Fermi energy progressively moves toward the peak maximum of the HOMO-derived resonance of the acene molecule. Since, for the *OS* structures, E_F is already close to the peak maximum, this effect does not change the conductance much; but for the *SS* structures, it has a dramatic effect, with a significant increase in conductance for the four molecules considered, as Figure 7.8 clearly shows.

Finally, we want to point out that the conductance of π-stacks also depends on the amount of overlap between the π-electronic clouds of the molecules within the stack: for instance, the conductance of the structure with two rotated Ps, depicted at the bottom right of Figure 7.9, is

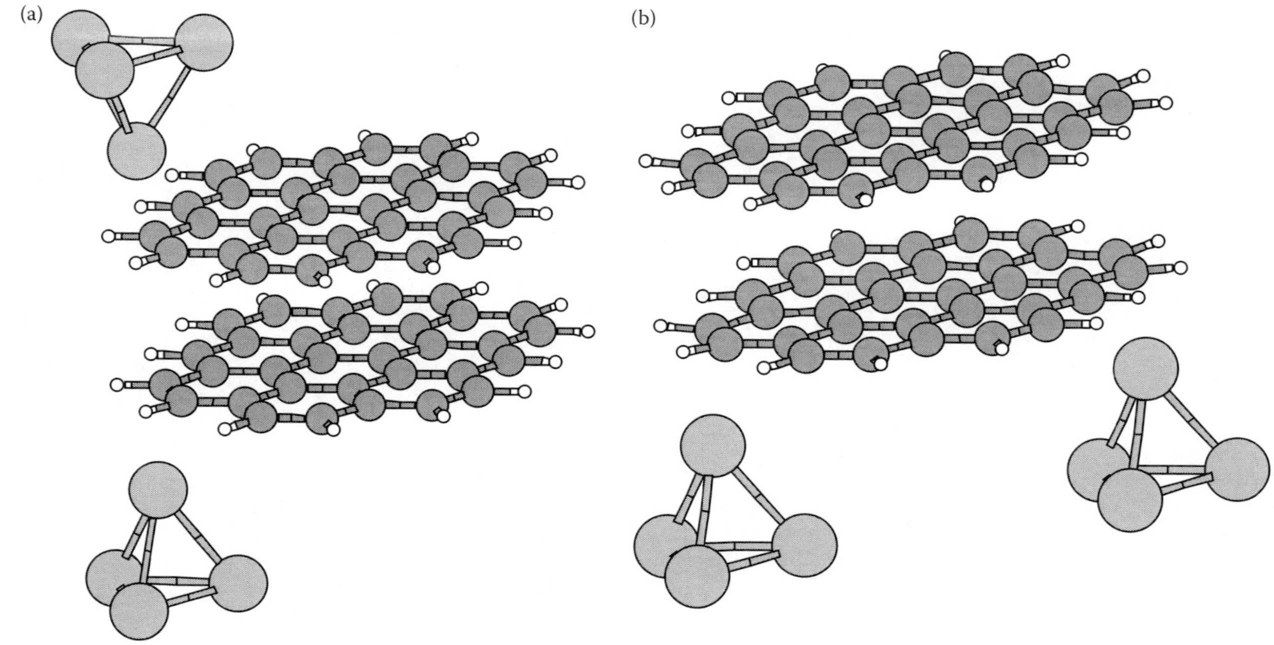

FIGURE 7.10 Structures of several π-stacked gold–cN–gold nanobridges. (a) $OS \times 2$. (b) $SS \times 2$.

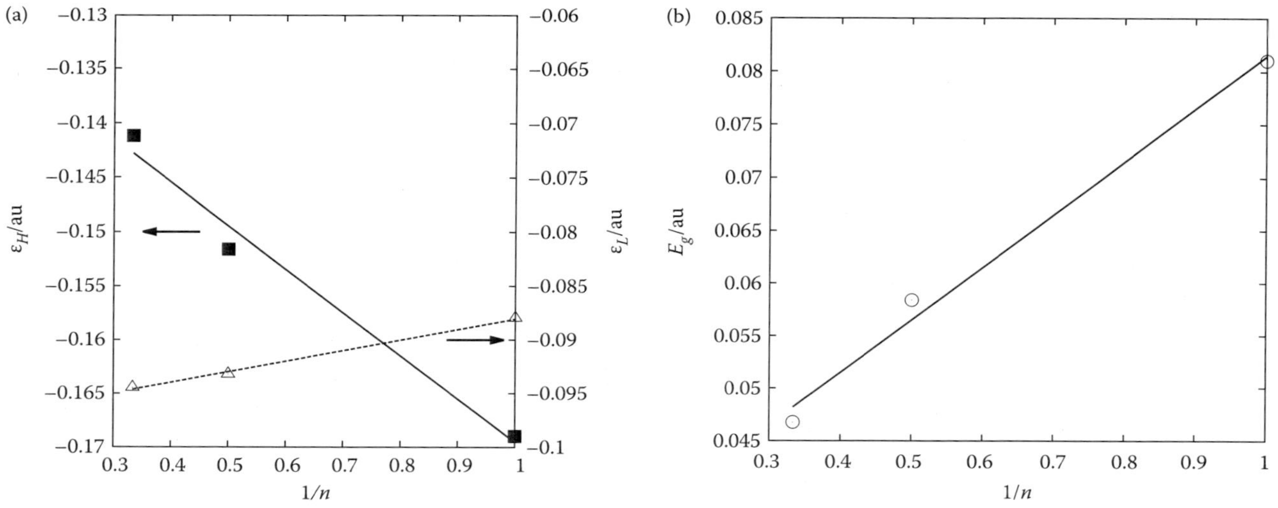

FIGURE 7.11 (a) HOMO energy (squares; left axis) and LUMO energy (triangles; right axis) versus the inverse of the number of π-stacked molecules (n) in SS gold–P–gold nanobridges. (b) Evolution of the corresponding HOMO–LUMO gap for the same systems. The lines represent linear fits to the data.

almost one-third of that obtained for the perfectly cofacial structure [11].

7.7 CONCLUSIONS

The applications shown here for some selected graphene nanoribbons are intended to pave the way toward further engineering of these molecules in the search of more efficient and relevant devices. Despite state-of-the-art-available theoretical tools, accurate enough as those discussed here, a number of fundamental aspects governing charge transport are not yet fully understood, and more investigations are still needed. Structure–property relationships have provided some clarity, especially concerning the dependence of the results with the system size and/or the way of packing, which are the key concepts to be explored when screening potential candidates. The chemical morphology of the samples, since a new dimension is introduced when the graphene sheet is reduced to nanographenes of a controlled shape, is judged to be most important for device performance. Furthermore, our research efforts always aim at deriving more accurate cost-effective theoretical methods for dealing with these issues, while at the same time providing further insights into charge transport phenomena in the case of graphene nanoribbons.

ACKNOWLEDGMENTS

The work on this chapter has been supported over the last few years by the "Ministerio de Economía y competitivided" of Spain and the "European Regional Development Fund" through projects CTQ2011-27253 and CTQ2014-55073P.

REFERENCES

1. http://www.nobelprize.org/nobel_prizes/physics/laureates/2010/ (accessed June 12, 2014).
2. http://www.graphene-flagship.eu/GF/index.php (accessed June 12, 2014).
3. http://www.cambridgeip.com/industries/nanotechnology/graphene/ (accessed June 12, 2014).
4. Loh, K.-P., Bao, Q., Kailian, P., and Yang, J. 2010. The chemistry of graphene. *J. Mater. Chem.* 20: 2277–2289.
5. Wan, X., Long, G., Huang, L., and Chen, Y. 2011. Graphene—A promising material for organic photovoltaic cells. *Adv. Mater.* 23: 5342–5358.
6. Stolar, M., and Baumgartner, T. 2013. Organic *n*-type materials for charge transport and charge storage applications. *Phys. Chem. Chem. Phys.* 15: 9007–9024.
7. Burghard, M., Klauk, H., and Kern, K. 2009. Carbon-based field-effect transistors for nanoelectronics. *Adv. Mater.* 21: 2586–2600.
8. Roncali, J., Leriche, P., and Cravino, A. 2007. From one- to three-dimensional organic semiconductors: In search of the organic silicon? *Adv. Mater.* 19: 2045–2060.
9. Baumeier, B., Lennartz, C., and Andrienko, D. 2012. Challenges for *in silico* design of organic semiconductors. *J. Mater. Chem.* 22: 20971–20976.
10. Sancho-García, J. C., and Pérez-Jiménez, A. J. 2009. Charge-transport properties of prototype molecular materials for organic electronics based on graphene nanoribbons. *Phys. Chem. Chem. Phys.* 11: 2741–2746.
11. Pérez-Jiménez, A. J., and Sancho-García, J. C. 2009. Conductance enhancement in nanographene–gold junctions by molecular π-stacking. *J. Am. Chem. Soc.* 131: 14857–14867.
12. Pérez-Jiménez, A. J., and Sancho-García, J. C. 2009. Using circumacenes to improve organic electronics and molecular electronics: Design clues. *Nanotechnology* 20: 475201.
13. Becke, A. D. 1993. Density-functional thermochemistry. III. The role of exact exchange. *J. Chem. Phys.* 98: 5648–5652.
14. Grimme, S. 2006. Semiempirical hybrid density functional with perturbative second-order correlation. *J. Chem. Phys.* 124: 034108.
15. Neese, F. 2012. The ORCA program system. *WIREs Comput. Mol. Sci.* 2: 73–78.
16. Sancho-García, J. C. 2007. Assessment of density-functional models for organic molecular semiconductors: The role of Hartree–Fock exchange in charge-transfer processes. *Chem. Phys.* 331: 321–331.
17. Sancho-García, J. C., and Olivier, Y. 2012. Reliable DFT-based estimates of cohesive energies of organic solids: The anthracene crystal. *J. Chem. Phys.* 137: 194311.
18. Grimme, S. 2006. Semiempirical GGA-type density functional constructed with a long-range dispersion correction. *J. Comput. Chem.* 27: 1787–1799.
19. Sancho-García, J. C., Pérez-Jiménez, A. J., Olivier, Y., and Cornil, J. 2010. Molecular packing and charge transport parameters in crystalline organic semiconductors from first-principles calculations. *Phys. Chem. Chem. Phys.* 12: 9381–9388.
20. Sancho-García, J. C., and Pérez-Jiménez, A. J. 2010. A theoretical study of π-stacking tetracene derivatives as promising organic molecular semiconductors. *Chem. Phys. Lett.* 499: 146–151.
21. Nitzan, A. 2001. Electron transmission through molecules and molecular interfaces. *Annu. Rev. Phys. Chem.* 52: 681–750.
22. Coropceanu, V., Cornil, J., da Silva Filho, D. A., Olivier, Y., Silbey, R., and Brédas, J.-L. 2007. Charge transport in organic semiconductors. *Chem. Rev.* 107: 926–952.
23. Sancho-García, J. C. 2012. Application of double-hybrid density functionals to charge transfer in *N*-substituted pentacenequinones. *J. Chem. Phys.* 136: 174703.
24. Sancho-García, J. C., and Pérez-Jiménez, A. J. 2014. Theoretical study of stability and charge-transport properties of coronene molecule and some of its halogenated derivatives: A path to ambipolar organic-based materials. *J. Chem. Phys.* 141: 134708.
25. Chen, F., and Tao, N. J. 2009. Electron transport in single molecules: From benzene to graphene. *Acc. Chem. Res.* 42: 429–438.
26. Datta, S. 1995. *Electronic Transport in Mesoscopic Systems.* Cambridge University Press (Cambridge).
27. Palacios, J. J., Pérez-Jiménez, A. J., Louis, E., and Vergés, J. A. 2001. Fullerene-based molecular nanobridges: A first-principles study. *Phys. Rev. B* 64: 115411.
28. Palacios, J. J., Pérez-Jiménez, A. J., Louis, E., SanFabián, E., and Vergés, J. A. 2002. First-principles approach to electrical transport in atomic-scale nanostructures. *Phys. Rev. B* 66: 035322.
29. Louis, E., Vergés, J. A., Palacios, J. J., Pérez-Jiménez, A. J., and SanFabián, E. 2003. Implementing the Keldysh formalism into *ab initio* methods for the calculation of quantum transport: Application to metallic nanocontacts. *Phys. Rev. B* 67: 155321.
30. Palacios, J. J., Pérez-Jiménez, A. J., Louis, E., SanFabián E., Vergés, J. A., and García, Y. 2005. Molecular electronics with Gaussian98/03. *Computational Chemistry: Reviews of Current Trends,* vol. 9, ed. Leszczynski, J., World Scientific (Singapore).
31. Palacios, J. J., Pérez-Jiménez, A. J., Louis, E., SanFabián, E., and Vergés, J. A. 2003. First-principles phase-coherent transport in metallic nanotubes with realistic contacts. *Phys. Rev. Lett.* 90: 106801.
32. Pérez-Jiménez, A. J., Palacios, J. J., Louis, E., SanFabián, E., and Vergés, J. A. 2003. Analysis of scanning tunneling spectroscopy experiments from first principles: The test case of C_{60} adsorbed on Au(111). *Chem. Phys. Chem.* 4: 388–392.
33. Frisch, M. J. et al. 2003. *GAUSSIAN03, Revision B.01.* Gaussian, Inc. (Pittsburgh, PA).
34. Segal, D., Nitzan, A., Davis, W. B., Wasielewski, M. R., and Ratner, M. A. 2000. Electron transfer rates in bridged molecular systems 2. A steady-state analysis of coherent tunneling and thermal transitions. *J. Phys. Chem. B* 104: 3817.
35. Lindsay, S. M., and Ratner, M. A. 2007. Molecular transport junctions: Clearing mists. *Adv. Mater.* 19: 23–31.

36. Roos, R. B., Powers, J. M., Atashroo, T., Ermler, W. C., LaJohn, L. A., and Christiansen, P. A. 1990. Ab initio relativistic effective potentials with spin–orbit operators. IV. Cs through Rn. *J. Chem. Phys.* 93: 6654–6670.
37. Banerjee, S., and Bhattacharyya, D. 2008. Electronic properties of nano-graphene sheets calculated using quantum chemical DFT. *Comput. Mater. Sci.* 44: 41–45.
38. Son, Y.-W., Cohen, M. L., and Louie, S. G. 2006. Energy gaps in graphene nanoribbons. *Phys. Rev. Lett.* 97: 216803.
39. Perdew, J. P., Parr, R. G., Levy, M., and Balduz, J. L. Jr. 1982. Density-functional theory for fractional particle number: Derivative discontinuities of the energy. *Phys. Rev. Lett.* 49: 1691–1694.
40. Levy, M., Perdew, J. P., and Sahni, V. 1984. Exact differential equation for the density and ionization energy of a many-particle system. *Phys. Rev. A* 30: 2745–2748.
41. Rodríguez-Ropero, F., Casanovas, J., and Alemán, C. 2007. *Ab initio* calculations on π-stacked thiophene dimer, trimer and tetramer: Structure, interactions energy, cooperative effects, and intermolecular electronic parameters. *J. Comput. Chem.* 29: 69–78.

8 Electronic Properties of Graphene Nanoribbons with Transition Metal Impurities

Neeraj K. Jaiswal and Pankaj Srivastava

CONTENTS

ABSTRACT

This chapter begins with a brief introduction of graphene/graphene nanoribbons (GNR). Thereafter, the experimental and the theoretical advancements have been discussed in the area of graphene/GNR research. Review on the theoretical work includes the exploration of exotic electronic/transport properties of GNR and the methods adopted to tailor the electronic properties of GNR via chemical functionalization, adsorption of impurity adatoms, or the substitutional doping of foreign atoms. The chapter also describes how the electronic/transport properties of GNR can be tailored via transition metal (TM) impurities. The impurities in GNR can be present either in the form of adsorbed adatoms or as a substitutional dopant. It is revealed that both types of impurities result in different bonding with host carbon atoms and stability in GNR. The electronic properties of GNR are found sensitive to the presence of TM impurities. This indicates toward a viable way to alter the electronic properties of GNR through TM impurities.

8.1 INTRODUCTION

As the conventional complementary metal-oxide semiconductor (CMOS) technology is approaching its physical limits, low dimensional materials have become a subject of great importance. Among these, C-based nanostructures have attracted special attention. Owing to its unique bonding capability, C exhibits a number of allotropes. Whenever we write with a pencil, we leave a number of graphene layers stacked on the paper. If we peel off these samples layer by layer, a monolayer graphene could be achieved. This is what Novoselov et al. [1] did to demonstrate the existence of monolayer graphene. They obtained samples of monolayer graphene through mechanical exfoliation of highly oriented pyrolytic graphite [1]. However, in this method graphene samples contained a number of few-layered graphene (FLG) as well.

The existence of graphene was predicted as a theoretical model [2] many years before it could be verified experimentally [1]. The reason for a long delay between its theoretical prediction and the experimental realization was not actually the synthesis part. The main obstacle was to detect the presence of freestanding monolayer graphene sheet. Since monolayer graphene absorbs only 2.3% of the incident light [3], some special conditions were required for its detection. Finally, the presence of monolayer graphene could only be verified on the top of a 300 nm thick substrate of SiO_2. Earlier, graphene was assumed not to exist in the stable state and, therefore, regarded only as an academic material [4]. Peierls and Landau separately predicted the nonexistence of purely two-dimensional (2-D) crystals due to thermal phonon fluctuations at arbitrarily low temperature [5]. In reality, it was observed that graphene crumples in the direction perpendicular to its plane. This phenomenon is energetically favorable for the stability of its purely flat 2-D structure. Owing to unique confinement of electronic states, charge carriers in graphene mimic relativistic particles. The properties of charge carriers in graphene are similar to massless Dirac particles, and their behavior can be more accurately described by the Dirac equation instead of the Schrodinger equation. Wallace [2] was the first who calculated the E–K relationship for the band structure of graphite. A linear E–K relationship was found at low energies near the vertices of the hexagonal Brillouin zone (BZ).

The valence and conduction bands are in conical form and touch each other at the Fermi level. The points where these bands touch exists at the corners of the hexagonal BZ. These points are formally known as the Dirac cones and denoted by K. Since the valence and conduction bands are just touching the Fermi level, the density of states (DOS) is zero at Fermi level and, consequently, 2-D graphene is a zero bandgap semimetal [1]. This intrinsic semimetallicity is, in fact, a major obstacle in the path of graphene toward its application in various semiconducting devices. Owing to this semimetallicity, using graphene in field effect transistors (FETs) causes a finite resistivity even at vanishing gate voltage. Thus, it is not possible to turn off the graphene-based FETs even at the point of zero charge carrier density. However, researchers have revealed that the limitations of graphene could be overcome by using nanoribbons.

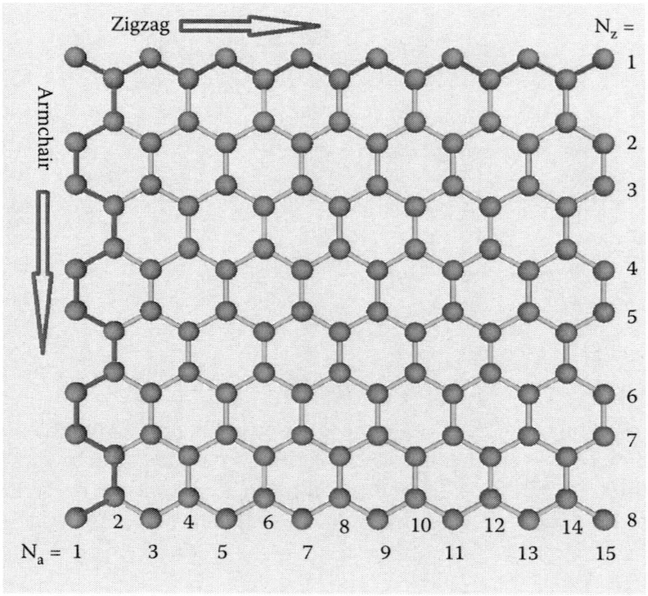

FIGURE 8.1 Directions of cutting the graphene sheet to obtain AGNR and ZGNR along with the convention of defining the ribbon width for both types of GNR.

8.1.1 Nanoribbons

Use of graphene in various electronic devices demands its thin strips of different shapes and sizes. These thin strips are formally known as graphene nanoribbons (GNR). Unlike 2-D graphene, GNR are confined in two directions while periodic along a third direction and, therefore, regarded as quasi one-dimensional (1-D) structures. Due to unique confinement of electronic states along 1-D, nanoribbons exhibit electronic properties that are different as compared to its 2-D counterpart, that is, graphene. GNR can be produced by controlled cutting of 2-D graphene. Depending upon the cutting direction, nanoribbons have either armchair or zigzag edges and accordingly known as the armchair graphene nanoribbons (AGNR) or zigzag graphene nanoribbons (ZGNR). The electronic properties of GNR are highly dependent upon the ribbon width, and the shape of the ribbon edges. Figure 8.1 depicts direction of cutting the graphene sheet to produce AGNR or the ZGNR. The widths of the nanoribbons are defined in terms of a number of C atoms across the ribbon edge as shown in Figure 8.1 for zigzag (N_z) and armchair (N_a) ribbons. Besides AGNR and ZGNR, there also exists another family of GNR called the chiral nanoribbons. These nanoribbons do not have a particular cutting angle and their electronic properties are influenced by the angle of cutting direction. The edges of chiral nanoribbons consist of a combination of both, the zigzag and the armchair shapes. Figure 8.2 shows the schematics of ZGNR, AGNR, and the chiral nanoribbons.

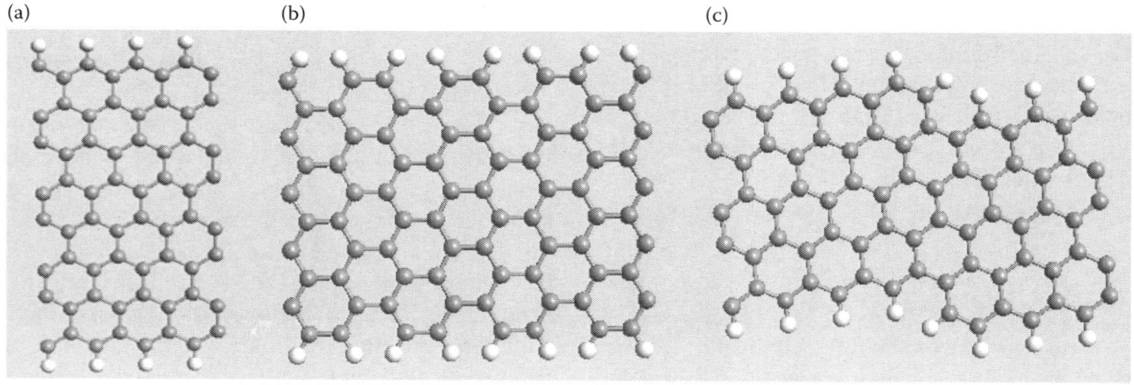

FIGURE 8.2 Schematics of (a) zigzag, (b) armchair, and (c) chiral nanoribbons.

8.2 WHY NANORIBBONS?

GNR emerged as an alternative to overcome the drawback of semimetallicity of graphene. In nanoribbons, the edge C atom is attached with only two nearest neighboring C atoms (which in case of graphene is three). Therefore, an additional unsaturated chemical bond (dangling bond) is associated with each C atom at both the edges of the ribbons. The presence of these dangling bonds at the ribbon edges provides a convenient way to alter the electronic properties and the stability of GNR through termination or the chemical edge modification [6]. Various elements as well as functional groups such as $-NH_2$, –OH, –O, etc. have been used for the functionalization of graphene and GNR [7–9].

Graphene is the most promising member in the family of organic nanomaterials with remarkable physical properties. Even more striking are the GNR, which have been found as a better candidate for making electrodes of Li-ion batteries as compared to carbon nanotubes (CNT) [10]. Moreover, GNR can get rid of overheating in the path of nanoelectronics [11]. Various other devices have also been realized using GNR [12,13]. The bandgap and, hence, the electronic properties of the ribbons are sensitive to their width or the shape of edges. All these properties make GNR an ideal material for experimental as well as theoretical research. Being a vital catalyst for the synthesis of organic nanomaterial such as fullerenes, nanotubes, graphene, etc., TM are most likely to be found as an impurity in organic nanomaterial. Therefore, TM can serve as a natural dopant in GNR. Since the properties of GNR are sensitive to the presence of impurities, it is important to understand the electronic properties of GNR with TM impurities.

8.3 SYNTHESIS OF GRAPHENE

First successful synthesis of monolayer graphene came into existence through mechanical exfoliation of highly oriented pyrolytic graphite [1]. This method provides an important route to single layered graphene as it gives the sample of the high quality. Owing to its simplicity, mechanical cleavage is a popular technique, especially for academic purpose. The samples produced in this way consist of graphene flakes of several μm. In spite of its simplicity, this method could not be applied for the mass production of graphene as required for industrial applications. The serious problem associated with this technique is the coexistence of monolayered graphene among innumerable samples of FLG. Therefore, one has to search for a monolayer flake with an optical microscope that itself is a tedious task.

The commercial applications of graphene demands for its mass production and large-area synthesis techniques. Therefore, extensive efforts are going on to explore novel, alternative ways of producing monolayered and few layered graphene sheets. As an outcome of these efforts, graphene has been grown epitaxially on SiC by vacuum graphitization [14,15]. It was observed that first layer on the substrate is electron doped, and further layers are undoped. In such graphene samples, the characteristic quantum Hall effect as observed in epitaxial graphene is missed. This separates the epitaxial graphene from the one obtained through exfoliation.

Osvath et al. [16] demonstrated the synthesis of graphene layers via thermal oxidation of exfoliated graphite plates. They were able to produce samples with desired number of graphene layers (including monolayer graphene) through controlling the thermal oxidation. Wang et al. [17] used chemical synthetic route consisting of graphite oxidation, ultrasonic exfoliation, and then a chemical reduction to produce graphene in a large quantity. Monolayer graphene sheets were also synthesized through pyrolytic cleavage of C_2H_2 on the single crystal Ir(111) [18]. The identification and transferring of single and FLG from SiO_2/Si substrates to some other desired substrates were also achieved by the researchers [19]. This method is of particular importance as it allows the fabrication of the device on a substrate other than SiO_2 or Si. Epitaxial growth techniques have also attracted the enormous attention of research community as it pledges for the growth of large-area graphene sheets with uniform thickness. The epitaxial graphene, grown on SiC was used to produce freestanding samples [20].

Lotya et al. [21] came with a novel idea and demonstrated the production of graphene in the liquid phase. They dispersed and exfoliate graphite to produce graphene samples in water-surfactant solution. It was an efficient path for production of FLG sheets. Almost 40% samples obtained in this method consisted of less than five graphene layers and monolayer graphene flakes contributed about 3%. The chemical vapor deposition (CVD) is another popular method for synthesizing high-quality graphene samples on the surface of TM [22–25]. Among these metals, Cu has attracted special attention due to the low solubility of C in Cu [23,26–28]. Once the synthesis process is completed, the Cu substrate can be evaporated to obtain the monolayer graphene for transferring on any insulating substrate. Partial oxidation approach was also demonstrated to realize the large area synthesis of graphene [29]. Recently, ultrasmall graphene structures have been demonstrated in a scalable quantity with atomic precision enabling us to fabricate various devices at the molecule level [30].

8.4 SYNTHESIS OF GNR

GNR can be achieved through precise cutting of 2-D graphene sheet. Datta et al. [31] used thermally activated metal nanoparticle for the etching of FLG samples on the insulating substrate. The method produces long (>1 μm) crystallographic edges of GNR and can be a potential way for atomically precise graphene device fabrication. In a similar manner, Ci et al. [32] used Ni nanoparticle as a knife to cut the graphene sheet with atomic precision. Nanowire lithography technique has also been demonstrated by Fasoli et al. [33] for the fabrication of narrow GNR. They used Si NW as selective etch masks lying over the graphene sheet. The etching was done by oxygen plasma. The GNR produced by this technique have the width equal to that of the NW diameter [33]. Jia et al. [34] demonstrated the formation of GNR with sharp zigzag or armchair edges. Kosynkin et al. [35] came with a novel idea for the fabrication of high yield GNR. In their experiment, the solution-based oxidative process was adopted for the unzipping of CNT. The bottom-up fabrication method is also capable of producing GNR of different topologies and widths [36]. The facile approach was demonstrated by Wu et al. [37] to synthesize narrow and long GNR by sonochemical cutting of chemically derived graphene sheets. The results show that GNR were produced in a yield of ~5 wt%. The site and alignment controlled synthesis of GNR has also been achieved using Ni nanobars [38]. Thin ~23 nm nanoribbons were grown directly as the channel of an FET without transfer, lithography, or any other treatments.

8.5 CHARACTERIZATION TECHNIQUES

From the above section, it is clear that graphene/GNR have witnessed considerable progress in the research. These efforts results in various techniques for the growth of monolayer/FLG sheets on various substrates. This enables the development of characterizing techniques for the identification of monolayer graphene and the sharp edges of the nanoribbons. Scanning electron microscopy (SEM), tunneling electron microscopy (TEM), atomic force microscopy (AFM), and Raman spectra play a key role in the characterization of graphene/GNR. Attempts have also been made to distinguish between mono- and multilayered graphene films on the basis of unconventional quantum Hall effect [39,40]. Recently, Raman spectra have evolved as a powerful tool for the identification and characterization of mono, bi, and multilayered graphene. Ferrari et al. [41] studied Raman spectra of graphene samples on Si, SiO_2, and also in the freestanding form. The study revealed that electronic structure of graphene is completely captured in Raman Spectra. Pristine 2-D graphene exhibit two prominent peaks, the G peak at ~1580 cm^{-1} and the G′ peak at ~2700 cm^{-1}. Further, the position of these peaks is sensitive to the number of layers and the presence of defects in a graphene sheet. Researchers have also studied SEM of mono- and bilayered graphene sheet grown epitaxially on SiC(0001) surface [42]. Ishigami et al. [43] employed scanning probe microscopy to reveal atomic structures and nanoscale morphology of graphene-based electronic devices. One of the drawbacks that limit the use of AFM/STM, etc. while observing the atomic scale resolution of surfaces or graphitic edges is the slow scanning rate. The typical scanning rate may be of the order of several minutes or even in hours. Second, the stability and dynamics are examined in the range of cryogenic temperature where the dynamics may be ceased. This difficulty was resolved by TEM. Wang et al. [44] used high resolution TEM observations and selected area electron diffraction to confirm the ordered graphite crystal structure of graphene produced via soft chemistry synthetic route involving graphite oxidation, exfoliation, and chemical reduction. The movies of the dynamics of carbon atoms at the edge of a defected and suspended, single layered graphene were successfully recorded by TEM with a temporal resolution of 1 s [45]. Thin nanoribbons were also characterized by AFM, SEM, TEM, and Raman spectroscopy by Wu et al. [37].

8.6 FUNCTIONALIZATION OF GNR

Owing to its perfect 2-D structure, graphene presents a unique confinement of charge carriers that behave as massless Dirac fermions [46]. Graphene has astounded the intellectual community with its exotic physical properties [40,47]. Even more interesting are the nanoribbons exhibiting finite bandgap depending upon the ribbon width [48]. The properties of GNR are much influenced by the shape of the ribbon edges and the ribbon widths [49,50]. Previous theoretical calculations revealed that ZGNR exhibit magnetic ground state [51] whereas AGNR are nonmagnetic semiconductors whose bandgap depends upon the ribbon width [52]. Research on graphene/GNR is also going on toward experimental advancements [16,53] as well as for the theoretical understanding [49,54,55]. The stable 2-D geometry and ultrahigh mobility of charge carriers promise toward the use of graphene in future device applications [56,57].

To incorporate graphene in various future devices, it is necessary to understand the electronic properties of GNR with various shapes and sizes. Previous investigations show that electronic properties of organic nanomaterials viz. CNTs, graphene/GNR, etc. are sensitive to the imperfections either in the form of vacancy or the presence of foreign atom [54,58–61]. Motivated from this, various attempts have been made to tailor the electronic properties of GNR via chemical functionalization [6,8,9,62–65] or through termination with different elements [7,66]. Doping/adsorption of impurity have also been used to tune the electronic properties of GNR [67,68]. Experimental efforts are also being made to tailor the electronic structure of graphene [69,70]. There exists plenty of literature studying nonmetallic impurities (particularly B/N) in CNT, graphene/GNR [8,64,70–73]. On the other hand, not much work is done with the metallic impurities in graphene [74,75]. The experimental verification that Au and Pt impurities have stable binding in graphene [76] suggested toward the use of metallic impurities as a potential dopant in graphene. Attempts have been started to investigate metallic impurities in graphene [77–79]. Gierz et al. [78] reported that gold doping is analogous to the hole doping of graphene. Zhang et al. [79] investigated the migration of Au atoms attached to the monovacancy at the edge of GNR. Amongst other metallic impurities, TM impurities have attracted special attention of researchers due to their magnetic behavior [80–82] and stable bonding with graphene/GNR [77,83–88]. Moreover, TM impurities have a significant impact on the electronic, magnetic, and transport properties of CNT, graphene/GNR [85–91]. In an atomic resolution electron microscopy study, Ni has been found to take in-plane position in the graphitic network, which was further supported by first-principles calculations [83]. This indicates the possibility of bandgap engineering of GNR through TM impurities.

8.7 GNR INTERACTION WITH TM IMPURITIES

Doping of impurities is a preferred way to tailor the bandgap of GNR for various applications. B/N are the most common and extensively studied dopants in GNR [64,71,72]. In the past few years, the trend has been changed and now attempts are being made to study metallic impurities in GNR, which has shown the significant possibility for sensors, H storage, and various device applications. TM are used as catalysts for the synthesis of graphene/GNR, which make them most probable to be found as impurities. This may ultimately cause a difference in the electronic and transport properties of graphene from that of the pristine ones. Other properties (optical, mechanical, etc.) will also change but to be concise, here our concern is limited to electronic properties only.

TM impurities can be present either in the form of absorbed atoms or as a substitutional dopant. Duffy and Blackman [74] investigated magnetic properties of TM atoms adsorbed on the surface of graphite. They considered Sc through Ni in the form of monomers and dimers. It was revealed that Sc to Mn, monomers prefer atop position (i.e., above the C atom), whereas Fe, Co, and Ni prefer to adsorbed on the hole site (center of hexagon). Kan et al. [84] studied the adsorption of Ti chain on the semiconducting (armchair) nanoribbons to predict half-metallicity in a narrow (width less than 2.1 nm) ribbons. Ti atoms prefer hole site to settled in an FM ground state. Adsorbed Co impurity also stabilizes at the hole site in ZGNR at an equilibrium distance of 1.5 Å out of the plane with a C–Co bond length between 2 and 2.15 Å [90]. Pd atoms and dimers adsorbed on graphene were found stable with binding energy (BE), ranging from −0.986 to −1.135 eV, exhibit spin polarization and higher (up to 1.99 μ_B) magnetic moment [88]. Another theoretical group reported structural and magnetic properties of atomic TM impurities in monovacancy and divacancy in a graphite sheet [86]. In another study, substitutional Co impurity was investigated by Santos et al. [82]. It was observed that Co impurity induces a spin polarization in graphene unlike Fe and Ni impurities [86]. Further, the induced spin polarization was sensitive to the position of Co atom at A or B sublattice in the graphene. Further, the results indicate that substitutional Co doping is analogous to single π defects in graphene. Interestingly, substitutional Fe impurity in ZGNR exhibit a finite magnetic moment [87] in contrast to that in 2-D graphene [86]. Stability analysis shows that substitutional Fe impurities with a BE of 4.05 eV are more stable in the graphitic network as compared to adsorbed Fe impurity 2 eV [74].

We also have investigated the GNR–TM interaction by considering TM as a terminating element as well as a substitutional dopant at various sites of GNR [91–96]. Four representative TM impurities Fe, Co, Ni, and Cu were selected for the investigations as they are the widely used metals for the synthesis of organic nanomaterials such as CNTs, graphene/GNR. A total of five configurations were examined which include one-edge TM-termination, both-edges TM-termination, one-edge TM-doping, both-edges TM-doping, and the center TM-doping. Further, ribbons with zigzag as well armchair edges were investigated to account the effect of edge geometry on the electronic properties of the ribbons. The size effect was also taken into account by considering ribbon with widths 4–9 for AGNR (N_a) [92] and 3–8 for ZGNR (N_z) [94].

8.7.1 Structural Stability

8.7.1.1 Armchair Graphene Nanoribbons

In order to probe the strength of bonding between TM and GNR, the stability of TM impurities (Fe to Cu) in AGNR and ZGNR, is discussed in terms of per atom BE of the doped ribbons. The BE is calculated as

$$E_b = E(TM-GNR) - E(GNR) - nE(TM) \qquad (8.1)$$

where E(TM–GNR) is the total energy of GNR with TM impurity, and E(GNR) is the total energy of the reconstructed GNR with a vacancy at the site of TM impurity [86]. E(TM) is the total energy of a single isolated TM atom and n is the number of impurity (TM) atoms in the supercell of the ribbon. The comparison of BE for all the considered TM impurities in AGNR ($N_a = 9$) is depicted in Table 8.1. The results indicate toward the stable bonding between TM–AGNR except for both-edges Cu-termination and both-edges Cu-doping of AGNR, which exhibit a weak binding as evident from positive BE of these structures (Table 8.1). Further, larger negative BE and hence more stable TM–AGNR bonding is noticed as we go from Cu to Fe. This is because the difference between atomic radii of C and impurity atom reduces Cu to Fe. Therefore, Fe exhibit strongest binding with BE value of −9.83 eV and for Cu its only −2.12 eV. Further, it is noticed that center of the ribbon is the most favorable doping site for the substitutional TM impurity in AGNR due to a maximum number of nearest neighbors available for bonding.

8.7.1.2 Zigzag Graphene Nanoribbons

The computed BE values for ZGNR interacting with TM have been presented in Table 8.2. Co possesses the strongest bonding as compared to other TM impurities. Next is Fe with BE difference of few meV. Moreover, the edge is the most favorable doping site for substitutional TM in ZGNR in consistent with previous results [76,83,86]. To sum up, the substitutional TM impurities exhibit enough stability in GNR making considered structures thermodynamically stable. Moreover, it is

TABLE 8.1
Calculated Binding Energy for AGNR ($N_a = 9$) Interacting with TM

TM	One-Edge Terminated	Both-Edges Terminated	One-Edge Doped	Both-Edges Doped	Center Doped
Cu	−2.12	5.74	−2.75	4.52	−1.51
Ni	−9.01	−3.60	−1.50	−1.50	−2.80
Co	−9.42	−3.96	−2.96	−2.95	−3.29
Fe	−9.83	−4.23	−3.54	−3.56	−3.95

TABLE 8.2
Calculated Binding Energy for ZGNR ($N_z = 8$) Interacting with TM

TM	One-Edge Terminated	Both-Edges Terminated	One-Edge Doped	Both-Edges Doped	Center Doped
Cu	−8.93	−2.74	−3.23	−3.20	−1.78
Ni	−10.40	−4.12	−5.22	−5.21	−4.27
Co	−10.61	−4.43	−6.92	−6.87	−5.50
Fe	−10.44	−4.29	−6.34	−6.31	−4.83

revealed through BE analysis that substitutional doping results in greater stability than the adsorption of the same impurity [80,93]. The stability of TM–C bonding is also affected by the geometry of ribbon edges. It is revealed that for AGNR, center doping is most favorable configuration, whereas for ZGNR, ribbon edges are more suitable for the substitutional doping of TM impurity.

8.7.2 Electronic and Magnetic Properties

Electronic properties of TM interacting GNRs have been investigated in two ways, that is, considering TM as an adsorbed atom and considering it as a substitutional dopant in GNR. Both types of TM impurities (adsorbed and dopant) affect the electronic properties of GNR in a different manner. However, a clear comparison of both types of impurities is yet to be explored. The adsorption of various TM adatoms and their dimmers on graphite was studied by Duffy and Blackman [74]. They investigated Sc through Ni to reveal the magnetic behavior of TM impurities. It was reported that Sc to Mn like to site at the atop site (right above the C atom) while Fe to Ni tend to position themselves at the hole position (above the center of hexagon). Further, the magnetic moments of adatoms with atop (hole) as the most preferred site were found higher (lower) as compared to their free atom values [74]. Kan et al. [84] reported that adsorption of Ti atomic chains on the AGNR results in a magnetic ground state of the system. Moreover, a high metallicity was observed in the Ti-adsorbed system. The observed magnetic moment varies between 0.49 and 2.05 μ_B. Zanella et al. [80] investigated the adsorption of Ti and Fe on graphene, considering the three important sites hole, atop, and bridge (just above the C–C bond). Hole sites were found to be the most stable sites, however, other sites also revealed stable bonding between graphene and adatoms except for Ti at bridge position. Interestingly, highest magnetic moment (3.26 μ_B) was obtained for the longest C–TM bond length, that is, for Ti at hole site [80]. Rigo et al. [85] investigated the adsorption of Ni atoms on ZGNR at various possible sites as suggested by Zanella et al. [80]. The optimized C–Ni bond length varied from 1.9 to 2.2 Å depending upon the adsorption site. A nonzero magnetic moment (0.46–0.90°μ_B) was observed for Ni with the highest value for atop site. Further, it was revealed that Ni impurity could induce the metallicity with the possibility of spin-dependent electronic transport. Various adsorption cited were investigated for Co by Cocchi et al. [90] and a magnetic moment of 1 μ_B was reported irrespective of the adsorption site. On the other hand, the adsorption energy was sensitive to the site of Co atom and varies from −1.05 to −1.39 eV. Their observation of electronic structure revealed the broken spin degeneracy that was otherwise preserved for ideal ZGNR. The results suggest that Co adsorption enhances the spin rectifying property of the ribbons when present at the edges; however, a finite net spin polarization exists for all the considered sites. Adsorption of coin metals (Cu, Ag, and Au) on ZGNR was studied by Wei et al. [97]. It was reported that coin metals like to sit at the atop position instead of the hole or bridge position. A small magnetic moment was observed for all these coin metals. The results show that Cu adsorbed ZGNR are semiconducting in nature whereas Ag and Au turn them half-metallic. Study of adsorption of Pd atoms and its dimmers on graphene has also been done recently [88]. It is observed that Pd adatoms induce no magnetic moment whereas Pd dimmers can induce the magnetism in the system. The electronic band structure of graphene was modified by the adatom impurities and a spin polarization up to 68% was noticed for Pd dimer adsorption. Amongst substitutional doping of TM in the graphitic network, Krasheninnikov et al. [86] considered Sc to Zn, Pt, and Au impurities in graphene at mono- and divacancies. TM impurities were found to have stable bonding with graphene. The considered TM exhibits a finite magnetic moment except for Sc, Ti, Ni, Zn, and Pt, which revealed zero magnetic moments with monovacancy as well as divacancy. In another study of substitutional TM impurities in graphene, it was noticed that only Co induces a spin polarization among other traditional elements (Fe, Co, and Ni) [82]. The electronic band structure of graphene with substitutional Co impurity represents an impurity level at the Fermi level that is equivalent to the realization of single π vacancies. Longo et al. [87] investigated magnetic properties of substitutional Fe impurities in ZGNR and revealed a finite magnetism in contrast to the nonmagnetic behavior of Fe when doped in graphene.

From the previous literature, it is revealed that electronic structure of GNR with substitutional TM impurities is highly dependent on the impurity site and type of impurity [80,90,92,94]. The comparison of electronic behavior for various substitutional TM impurities has been depicted in Tables 8.3 and 8.4 for AGNR and ZGNR, respectively. It is observed that presence of TM impurity (Fe through Cu) causes metallicity in AGNR except for Cu-terminated ribbons that show semiconducting behavior.

Interestingly, half-metallicity is observed for one-edge Cu-terminated ZGNR, which disappears in both-edges Cu-terminated ZGNR leaving the structure semimetallic in nature. Rests of other Cu–ZGNR are metallic irrespective of doping site. Ni-terminated ZGNR are purely metallic whereas substitutional Ni-doping makes the structures semimetallic. Ni impurities caused metallicity in each configuration

TABLE 8.3
Electronic Behavior of AGNR ($N_a = 9$) Interacting with TM

TM	One-Edge Terminated	Both-Edges Terminated	One-Edge Doped	Both-Edges Doped	Center Doped
Cu	SC	SC	M	M	M
Ni	M	M	M	M	M
Co	M	M	M	M	M
Fe	M	SM	M	M	M

Note: SC = semiconducting; SM = semimetallic; M = metallic.

TABLE 8.4
Electronic Behavior of ZGNR ($N_z = 8$) Interacting with TM

TM	One-Edge Terminated	Both-Edges Terminated	One-Edge Doped	Both-Edges Doped	Center Doped
Cu	HM	SM	M	M	M
Ni	M	M	SM	SM	SM
Co	M	M	M	M	M
Fe	SM	M	M	M	M

Note: HM = half-metallic; SM = semimetallic; M = metallic.

for ZGNR as well as for AGNR. The similar behavior was observed for Co impurities in AGNR. Fe impurity induces enhanced metallicity and spin polarization in ZGNR for termination as well as substitutional doping [98]. Thus, one can conclude that TM impurities have a profound effect on the electronic properties of GNR. By selecting the proper TM, it is possible to obtain semiconducting, semimetallic, purely metallic, and even half-metallic nanoribbons. The position of metallic impurity can be controlled dynamically. This can be done in two steps: (i) removing a predefined C atom to create a vacancy with atomic preciseness using TEM with focused electron beam [86,99] and (ii) after creating the vacancy, desired metallic impurities can be pinned into the vacant sites by thermal excitation. Therefore, it could lead to a practical method to alter the electronic properties of GNR via TM impurities. The feasibility of substitutional TM impurities has been recently verified by Wang et al. [99] who successfully achieved substitutional doping of TM atoms in monolayer graphene. Conclusively, the findings of previous studies of GNR interaction with TM impurities have been summarized in Table 8.5.

8.8 GRAPHENE/GNR FOR TECHNOLOGICAL APPLICATIONS

Graphene and GNR pledge for their use in various technological applications. First-principles calculation-based investigations have predicted that GNR containing TM atoms could be potentially used for interconnects and spintronic applications [100]. The theoretical predictions of graphene-based devices are further supported by the realization of p–n junctions [101], FETs [102]. High electron mobility [56,57] in graphene can play a key role for the electrical applications. Ballistic transport of charge carrier was maintained in ZGNR and AGNR in the presence of a substrate-induced disorder [103]. Therefore, attempts have been made to study GNR/CNT as an interconnect material [100,104,105]. Research is going on for the realization and investigation of various devices made from graphene/GNR [13,106–108]. All the atoms on graphene surface are prone to the environment, which makes it a promising material for sensor applications. Schedin et al. [109] reported that graphene-based chemical sensors are capable of detecting the individual gas molecules. Inspired from this, various sensors are investigated using graphene [110–113]. Further, GNR can serve as a potential material for H storage [114–118]. It was also observed that strain effects can affect the H storage capability of GNR [119]. 100% spin polarization was achieved in CNT with adsorbed TM impurities [120]. Later on, the half-metallicity has also been observed in GNR in the presence of external transverse electric field [121]. This finding is important and has accelerated the research to obtain high spin polarization in GNR, which is useful for spintronic applications [8,65,122–124]. The phenomenon of half-metallicity is particularly useful for designing the logic circuits [125] and for the memory applications [13,101]. Graphene devices are also being used for biological applications. Min et al. [126] reported that graphene-based nanochannel device could be used for fast DNA sequencing. Recently, the ballistic transport has been predicted in the atomic chain of C atoms coupled with graphene electrodes with zigzag and armchair edges [127]. This suggests that GNR can be used as contact material in organic electronic devices. Recently, Ren et al. [128] have demonstrated that prelithiated graphene nanosheets can serve as a better material for the negative electrode to provide high energy and power density. Ni/graphene hybrid nanostructure has also been used for enhanced electrochemical hydrogen storage [129]. Rapidly evolving area of smart sensors uses graphene oxide–DNA-based sensors with DNA as recognition element [130]. It is also reported that the efficiency of TiO_2-based solar cells can be improved with the help of graphene oxide [131]. The combination of spintronics and thermoelectric, that is, spin caloritronics effect was studied in a quantum dot between graphene electrodes [132]. The ballistic transport properties were reported in doped graphene sandwiched between pristine graphene leads [133], which could of particular importance for novel electronic device applications. The presence of relativistic tunneling features in the I–V characteristic of a graphene tunnel diode pledges for its use in future tunneling FET (TFET) devices [134]. Further, Kang et al. [135] has also presented a simplified model for a nanoelectromechanical device based on graphene nanoflakes [135]. Therefore, it can be concluded that there are many opportunities toward exploring the hidden potential of this wonderful material "graphene."

TABLE 8.5
Summary of Previous Findings

TM	Type of Impurity	Host System	C–TM Bond Length (Å)	Binding/Adsorption Energy (eV)	Magnetic Moment (μ_B)	Out-of-Plane Height (Å)
[74]Sc, Ti, V, Cr, Mn, Fe, Co, Ni	Adsorption	Graphite	NA	Sc (1.9), Ti (2.1), V (1.6), Cr (1.0), Mn (1.0), Fe (2.0), Co (2.4), Ni (2.5)	Sc (2.0), Ti (3.0), V (4.0), Cr (5.0), Mn (6.0), Fe (2.0), Co (1.0), Ni (0)	Sc (2.18), Ti (2.10), V (2.08), Cr (2.13), Mn (2.14), Fe (1.52), Co (1.52), Ni (1.53)
[75]Fe, Co, Ni, Cu, Ag, Au	Terminations	ZGNR	Fe (1.87), Co (1.82), Ni (1.80), Cu (2.07), Ag (2.15), Au (2.07)	Fe (4.13), Co (4.26), Ni (4.44), Cu (3.42), Ag (2.38), Au (3.00)	Fe (2.19), Co (1.17), Ni (0.18), Cu (0.26), Ag (0.28), Au (0.26)	NA
[77]Fe, Pd, Au	Adsorption	Graphene	Fe (2.11–2.35), Pd (2.10–2.46), Au (2.55–3.80)	Fe (0.15–0.75), Pd (0.85–1.08), Au (0.09–0.1)	NA	Fe (1.53–2.22), Pd (2.03–2.21), Au (2.69–3.53)
[80]Ti, Fe	Adsorption	Graphene	Ti (2.12–2.26), Fe (2.13–2.24)	Ti (2.55–3.26), Fe (0.58–1.10)	Ti (2.71–2.88), Fe (2.00–4.00)	NA
[81]Fe, Mn, Co	Adsorption	Graphene	Fe (2.08), Mn (2.52), Co (2.09)	Fe (−0.25 to −1.42), Mn (−0.76 to −0.93), Co (−0.58 to −1.32)	Fe (2.20), Mn (5.62), Co (1.10)	Fe (1.47), Mn (2.05), Co (1.56)
[82]Co	Substitution	Graphene	1.77	7.6	1.00	0.92
[83]Ni	Substitution	Graphite	2.10	−4.70	NA	1.0
[84]Ti	Adsorption	AGNR	NA	0.56–2.04	0.51–2.05	NA
[85]Ni	Adsorption	GNR	1.90–2.20	−1.89 to −2.21	0.43–0.97	NA
[86]Sc, Ti, V, Cr, Mn, Fe, Co, Ni, Cu, Zn, Pt, Au	Substitution	Graphene	Sc (2.06), Ti (1.94), V (1.88), Cr (1.86), Mn (1.84), Fe (1.76), Co (1.76), Ni (1.78), Cu (1.88), Zn (1.88), Pt (1.94), Au (2.09)	Sc (−7), Ti (−8), V (−7.5), Cr (−6.4), Mn (−6.4), Fe (−7.4), Co (−7.8), Ni (−6.8), Cu (−3.2), Zn (−1.5), Pt (−7.1), Au (−2.2)	Sc (0), Ti (0), V (1.0), Cr (2.0), Mn (2.7), Fe (0), Co (1.0), Ni (0), Cu (1.5), Zn (0), Pt (0), Au (1.0)	Sc (2.2), Ti (1.9), V (1.8), Cr (1.8), Mn (1.7), Fe (1.3), Co (1.4), Ni (1.2), Cu (1.7), Zn (1.8), Pt (1.7), Au (1.8)
[87]Fe	Substitution	ZGNR	1.78	4.05	0.31	1.29
[88]Pd	Adsorption	Graphene	2.088–2.425	−0.986 to −1.135	NA	0
[90]Co	Adsorption	ZGNR	2–2.16	1.05–1.33	1.0	1.5
[91]Ni	Substitution	AGNR	1.80–1.85	−1.46 to −3.60	NA	NA
[92]Cu	Substitution	AGNR	1.87	−1.31 to 5.38	NA	NA
[93]Fe	Adsoption[a]/ substitution[b]	ZGNR	1.99[a], (1.81–1.91)[b]	(−1.83 to −1.94)[a], (−4.93 to −6.04)[b]	NA	NA
[94]Ni	Substitution	ZGNR	1.78–1.79	−4.07 to −10.35	NA	NA
[95]Co	Substitution	ZGNR	1.79–1.90	−5.34 to −6.92	NA	NA
[97]Cu, Ag, Au	Adsorption	ZGNR	Cu (1.99), Ag (2.39), Au (2.50)	Cu (−1.24 to −1.92), Ag (−1.29 to −1.65), Au (−1.25 to −1.61)	Cu (0.87), Ag (0.926), Au (0.947)	Cu (1.99), Ag (2.39), Au (2.50)
[99]Pt, Co, In	Substitution	Graphene	NA	Pt (4.40–6.80)	Co (1.0–1.56)	Co (1.00)
[100]Fe	Substitution	AGNR	1.83	−2.32 to −9.83	0–2.90	NA

Note: NA = not available.
[a] Adsorption.
[b] Substitution.

REFERENCES

1. Novoselov, K. S., Geim, A. K., Morozov, S. V., Jiang, D., Zhang, Y., Dubonos, S. V., Grigorieva, I. V., and Firsov, A. A. 2004. Electric field effect in atomically thin carbon films. *Science* 306:666–69.
2. Wallace, P. R. 1947. The band theory of graphite. *Phys. Rev.* 71:622–34.
3. Nair, R. R., Blake, P., Grigorenko, A. N., Novoselov, K. S. Booth, T. J., Stauber, T., Peres, N. M. R., and Geim, A. K. 2008. Fine structure constant defines visual transparency of graphene. *Science* 320:1308.
4. Fradkin, E. 1986. Critical behavior of disordered degenerate semiconductors. II. Spectrum and transport properties in mean-field theory. *Phys. Rev. B* 33:3263–68.

5. Landau, L. D. 1937. Zur theorie der phasenumwandlungen. *Phys. Z* 11:26–35.
6. Wang, Z. F., Li, Q., Zheng, H., Ren, H., Su, H., Shi, Q. W., and Chen, J. 2007. Tuning the electronic structure of graphene nanoribbons through chemical edge modification: A theoretical study. *Phys. Rev. B* 75:113406-1–4.
7. Ramasubramaniam, A. 2010. Electronic structure of oxygen-terminated zigzag graphene nanoribbons: A hybrid density functional theory study. *Phys. Rev. B* 81:245413-1–6.
8. Kan, E.-J., Li, Z., Yang, J., and Hou, J. G. 2008. Half metallicity in edge modified zigzag graphene nanoribbons. *J. Am. Chem. Soc.* 130:4224–25.
9. Zheng, H., and Duley, W. 2008. First principles study of edge chemical modifications in graphene nanodots. *Phys. Rev. B* 78:045421-1–5.
10. Bhardwaj, T., Antic, A., Pavan, B., Barone, V., and Fahlman, B. D. 2010. Enhanced electrochemical lithium storage by graphene nanoribbons. *J. Am. Chem. Soc.* 132:12556–58.
11. Seol, J. H., Jo, I., Moore, A. L., Lindsay, L., Aitken, Z. H., Pettes, M. T., Li, X. et al. 2010. Two-dimensional phonon transport in supported graphene. *Science* 328:213–16.
12. Abanin, D. A., and Levitov, L. S. 2007. Quantized transport in graphene p-n junctions in magnetic field. *Science* 317:641–43.
13. Standley, B., Bao, W., Zhang, H., Bruck, J., Lau, C. N., and Bockrath, M. 2008. Graphene-based atomic-scale switches. *Nano Lett.* 8:3345–49.
14. Berger, C., Song, Z., Li, X., Wu, X., Brown, N., Naud, C., Mayou, D. et al. 2006. Electronic confinement and coherence in patterned epitaxial graphene. *Science* 312:1191–96.
15. de Heer, W. A., Berger, C., Wu, X., First, P. N., Conrad, E. H., Li, X., Li, T. et al. 2007. Epitaxial graphene. *Solid State Commun.* 143:92–100.
16. Osvath, Z., Darabont, A., Incze, P. N., Horvath, E., Horvath, Z. E., and Biro, L. P. 2007. Graphene layers from thermal oxidation of exfoliated graphite plates. *Carbon* 45:3022–26.
17. Wang, G., Yang, J., Park, J., Gou, X., Wang, B., Liu, H., and Yao, J. 2008. Facile synthesis and characterization of graphene nanosheets. *J. Phys. Chem. C* 112:8192–95.
18. Diaye, A. T. N'. Coraux, J., Plasa, T. N., Busse, C., and Michely, T. 2008. Structure of epitaxial graphene on Ir(111). *N. J. Phys.* 10:043033-1–16.
19. Reina, A., Son, H., Jiao, L., Fan, B., Dresselhaus, M. S., Liu, Z. F., and Kong, J. 2008. Transferring and identification of single and few-layer graphene on arbitrary substrates. *J. Phys. Chem. C* 112:17741–44.
20. Shivaraman, S., Barton, R. A., Yu, X., Alden, J., Herman, L., Chandrashekhar, M. V. S., Park, J. et al. 2009. Free-standing epitaxial graphene. *Nano Lett.* 9:3100–05.
21. Lotya, M., Hernandez, Y., King, P. J., Smith, R. J., Nicolosi, V., Karlsson, L. S., Blighe, F. M. et al. 2009. Liquid phase production of graphene by exfoliation of graphite in surfactant/water solutions. *J. Am. Chem. Soc.* 131:3611–20.
22. Kim, K. S., Zhao, Y., Jang, H., Lee, S. Y., Kim, J. M., Kim, K. S., Ahn, J.-H., Kim, P., Choi, J.-Y., and Hong, B. H. 2009. Large-scale pattern growth of graphene films for stretchable transparent electrodes. *Nat. Lett.* 457:706–10.
23. Li, X., Cai, W., An, J., Kim, S., Nah, J., Yang, D., Piner, R. et al. 2009. Large-area synthesis of high-quality and uniform graphene films on copper foils. *Science* 324:1312–14.
24. Kwon, S.-Y., Ciobanu, C. V., Petrova, V., Shenoy, V. B., Baren˜o, J., Gambin, V., Petrov, I., and Kodambaka, S. 2009. Growth of semiconducting graphene on palladium. *Nano Lett.* 9:3985–90.
25. Cai, W., Zhu, Y., Li, X., Piner, R. D., and Rouff, R. S. 2009. Large area few-layer graphene/graphite films as transparent thin conducting electrodes. *Appl. Phys. Lett.* 95:123115-1–3.
26. Srivastava, A., Galande, C., Ci, L. J., Song, L., Rai, C., Jariwala, D., Kelly, K. F., and Ajayan, P. M. 2010. Novel liquid precursor-based facile synthesis of large area continuous, single, and few-layer graphene films. *Chem. Mater.* 22:3457–61.
27. Bae, S., Kim, H., Lee, Y., Xu, X., Park, J.-S., Zheng, Y., Balakrishnan, J. et al. 2010. Roll-to-roll production of 30-inch graphene films for transparent electrodes. *Nat. Nanotech.* 5:574–78.
28. Cao, H., Yu, Q., Jauregui, L. A., Tian, J., Wu, W., Liu, Z., Jalilian, R. et al. 2010. Electronic transport in chemical vapor deposited graphene synthesized on Cu: Quantum Hall effect and weak localization. *Appl. Phys. Lett.* 96:122106-1–3.
29. Eda, G., Ball, J., Mattevi, C., Acik, M., Artiglia, L., Granozzi, G., Chabal, Y., Anthopoulos, T. D., and Chhowalla, M. 2011. Partially oxidized graphene as a precursor to graphene. *J. Mat. Chem.* 21:11217–23.
30. Christopher, J. R., and Golovchenko, J. A. 2012. Atom by atom nucleation and growth of graphene noanopores. *PNAS* 109:5953–57.
31. Datta, S. S., Strachan, D. R., Khamis, S. M., and Johnson, A. T. C. 2008. Crystallographic etching of few-layer graphene. *Nano Lett.* 8:1912–15.
32. Ci, L., Xu, Z., Wang, L., Gao, W., Ding, F., Kelly, K. F., Yakobson, B. I., and Ajayan, P. M. 2008. Controlled nanocutting of graphene. *Nano Res.* 1:116–22.
33. Fasoli, A., Colli, A., Lombardo, A., and Ferrari, A. C. 2009. Fabrication of graphene nanoribbons via nanowire lithography. *Phys. Status Solidi (B)* 246:2514–17.
34. Jia, X., Hofmann, M., Meunier, V., Sumpter, B. G., Delgado, J. C., Herrera, J. M. R., Son, H. et al. 2009. Controlled formation of sharp zigzag and armchair edges in graphitic nanoribbons. *Science* 323:1701–05.
35. Kosynkin, D. V., Higginbotham, A. L., Sinitskii, A., Lomeda, J. R., Dimiev, A., Price, B. K., and Tour, J. M. 2009. Longitudinal unzipping of carbon nanotubes to form graphene nanoribbons. *Nature* 458:872–76.
36. Cai, J., Ruffieux, P., Jaafar, R., Bieri, M., Braun, T., Blankenburg, S., Muoth, M. et al. 2010. Atomically precise bottom-up fabrication of graphene nanoribbons. *Nature* 466:470–73.
37. Wu, Z. S., Ren, W., Gao, L., Liu, B., Zhao, J., and Cheng, H. M. 2010. Efficient synthesis of graphene nanoribbons sonochemically cut from graphene sheets. *Nano Res.* 3:16–22.
38. Kato, T., and Hatakeyama, R. 2012. Site and alignment-controlled growth of graphene nanoribbons from nickel nanobars. *Nat. Nanotech.* 7:651–56.
39. Zhang, Y., Tan, Y. W., Stormer, H. L., and Kim, P. 2005. Experimental observation of the quantum Hall effect and Berry's phase in graphene. *Nature* 438:201–05.
40. Novoselov, K. S., McCann, E., Morozov, S. V., Fal'ko, V. I., Katsnelson, M. I., Zeitler, U., Jiang, D., Schedin, F., and Geim, A. K. 2006. Unconventional quantum Hall effect and Berry's phase of 2π in bilayer graphene. *Nat. Phys.* 2:177–80.
41. Ferrari, A. C., Meyer, J. C., Scardaci, V., Casiraghi, C., Lazzeri, M., Mauri, F., Piscanec, S. et al. 2006. Raman spectrum of graphene and graphene layers. *Phys. Rev. Lett.* 97:187401-1–4.
42. Brar, V. W., Zhang, Y., Yayon, Y., Ohta, T., McChesney, J. L., Bostwick, A., Rotenberg, E., Horn, K., and Crommie, M. F. 2007. Scanning tunneling spectroscopy of inhomogeneous electronic structure in monolayer and bilayer graphene on SiC. *Appl. Phys. Lett.* 91:122102-1–3.

43. Ishigami, M., Chen, J. H., Cullen, W. G., Fuhrer, M. S., and Williams, E. D 2007. Atomic structure of graphene on SiO_2. *Nano Lett.* 7:1643–48.
44. Wang, G., Yang, J., Park, J., Gou, X., Wang, B., Liu, H., and Yao, J. 2008. Facile synthesis and characterization of graphene nanosheets. *J. Phys. Chem. C* 112:8192–95.
45. Girit, C. O., Meyer, J. C., Erni, R., Rossell, M. D., Kisielowski, C., Yang, L., Park, C. H. et al. 2009. Graphene at the edge: Stability and dynamics. *Science* 323:1705–08.
46. Novoselov, K. S., Geim, A. K., Morozov, S. V., Jiang, D., Katsnelson, M. I., Grigorieva, I. V., and Dubonos., S. V. 2005. Two-dimensional gas of massless Dirac fermions in graphene. *Nature* 438:197–200.
47. Toke, C., Lammert, P. E., Crespi, V. H., and Jain, J. K. 2006. Fractional quantum Hall effect in graphene. *Phys. Rev. B* 74:235417-1–5.
48. Son, Y. W., Cohen, M. L., and Louie, S. G. 2006. Energy gaps in graphene nanoribbons. *Phys. Rev. Lett.* 97:216803-1–4.
49. Nakada, K., Fujita, M., Dresselhaus, G., and Dresselhaus, M. S. 1996. Edge state in graphene ribbons: Nanometer size effect and edge shape dependence. *Phys. Rev. B* 54:17954–61.
50. Zheng, H., Wang, Z. F., Luo, T., Shi, Q. W., and Chen, J. 2007. Analytical study of electronic structure in armchair graphene nanoribbons. *Phys. Rev. B* 75:165414-1–6.
51. Pisani, L., Chan, J. A., Montanari, B., and Harrison, N. M. 2007. Electronic structure and magnetic properties of graphitic ribbons. *Phys. Rev. B* 75:064418-1–9.
52. Fujita, M., Wakabayashi, K., Nakada, K., and Kusakabe, K. 1996. Peculiar localized state at zigzag graphite edge. *J. Phys. Soc. Jpn* 65:1920–23.
53. Jiang, Z., Zhang, Y., Tan, Y., Stormer, H. L., and Kim, P. 2007. Quantum Hall effect in graphene. *Solid State Commun.* 143:14–19.
54. Topsakal, M., Akturk, E., Sevin¸cli, H., and Ciraci, S. 2008. First-principles approach to monitoring the band gap and magnetic state of a graphene nanoribbon via its vacancies. *Phys. Rev. B* 78:235435-1–6.
55. Sinitsyn, N. A., Hill, J. E., Min, H., Sinova, J., and MacDonald, A. H. 2006. Charge and spin Hall conductivity in metallic graphene. *Phys. Rev. Lett.* 97:106804-1–4.
56. Chen, Z., Lin, Y.-M., Rooks, M. J., and Avouris, P. 2007. Graphene nano-ribbon electronics. *Physica E* 40:228–32.
57. Bolotin, K. I., Sikes, K. J., Jiang, Z., Klima, M., Fudenberg, G., Hone, J., Kim, P., and Stormer, H. L. 2008. Ultrahigh electron mobility in suspended graphene. *Solid State Commun.* 146:351–55.
58. El-Barbary, A. A., Telling, R. H., Ewels, C. P., Heggie, M. I., and Briddon, P. R. 2003. Structure and energetics of vacancy in graphene. *Phys. Rev. B* 68:144107-1–7.
59. Peres, N. M. R., Guinea, F., and Neto, A. H. C. 2006. Electronic properties of disordered two-dimensional carbon. *Phys. Rev. B* 73:125411-1–23.
60. Boukhvalov, D. W., Katsnelson, M. I., and Lichtenstein, A. I. 2008. Hydrogen on graphene: Electronic structure, total energy, structural distortions and magnetism from first-principles calculations. *Phys. Rev. B* 77:035427-1–7.
61. Jafri, S. H. M., Carva, K., Widenkvist, E., Blom, T., Sanyal, B., Fransson, J., Eriksson, O. et al. 2010. Conductivity engineering of graphene by defect formation. *J. Phys. D. Appl. Phys.* 43:045404-1–8.
62. Sodi, F. C., Csanyi, G., Piscanec, S., and Ferrari, A. C. 2008. Edge functionalized and substitutionally doped graphene nanoribbons: Electronic and spin properties. *Phys. Rev. B* 77:165427-1–13.
63. Gorjizadeh, N., Farajian, A. A., Esfarjani, K., and Kawazoe, Y. 2008. Spin and band-gap engineering in doped graphene nanoribbons. *Phys. Rev. B* 78:155427-1–6.
64. Song, L. L., Zheng, X. H., Wang, R. L., and Zeng, Z. 2010. Dangling bond states, edge magnetism, and edge reconstruction in pristine and B/N-terminated zigzag graphene nanoribbons. *J. Phys. Chem. C* 114:12145–50.
65. Li, L., Qin, R., Li, H., Yu, L., Liu, Q., Luo, G., Gao, Z., and Lu, J. 2011. Functionalized graphene for high performance two-dimensional spintronic devices. *ACS Nano* 5:2601–10.
66. Miyamoto, Y., Nakada, K., and Fujita, M. 1999. First principles study of edge states of H terminated graphitic ribbons. *Phys. Rev. B* 59:9858–61.
67. Wehling, T. O., Novoselov, K. S., Morozov, S. V., Vdovin, E. E., Katsnelson, M. I., Geim, A. K., and Lichtenstein, A. I. 2008. Molecular doping of graphene. *Nano Lett.* 8:173–77.
68. Yu, S. S., Zheng, W. T., and Jiang, Q. 2010. Electronic properties of nitrogen-/boron-doped graphene nanoribbons with armchair edges. *IEEE Trans. Nanotech.* 9:78–81.
69. Tapaszto, L., Dobrik, G., Lambin, P., and Bir´o, L. P. 2008. Tailoring the atomic structure of graphene nanoribbons by scanning tunneling microscope lithography. *Nat. Nanotech.* 3:397–401.
70. Wang, X., Li, X., Zhang, L., Yoon, Y., Weber, P. K., Wang, H., Guo, J., and Dai, H. 2009. N doping of graphene through electrothermal reactions with ammonia. *Science* 324:768–71.
71. Kotakoski, J., Krasheninnikov, A. V., Ma, Y., Foster, A. S., Nordlund, K., and Nieminen, R. M. 2005. B and N ion implantation into carbon nanotubes: Insight from atomistic simulations. *Phys. Rev. B* 71:205408-1–6.
72. Yu, S. S., Zheng, W. T., Wen, Q. B., and Jiang, Q. 2008. First principle calculations of the electronic properties of nitrogen doped carbon nanoribbons with zigzag edges. *Carbon* 46:537–43.
73. Biel, B., Blase, X., Triozon, F., and Roche, S. 2009. Anomalous doping effects on charge transport in graphene nanoribbons. *Phys. Rev. Lett.* 102:096803-1–4.
74. Duffy, D. M., and Blackman, J. A. 1998. Magnetism of 3D transition-metal adatoms and dimers on graphite. *Phys. Rev. B* 58:7443–49.
75. Wang, Y., Cao, C., and Cheng, H. P. 2010. Role of edge states in metal-terminated graphene nanoribbons. *Phys. Rev. B* 82:205429-1–5.
76. Gan, Y., Sun, L., and Banhart, F. 2008. One-and two-dimensional diffusion of metal atoms in graphene. *Small* 4:587–91.
77. Chan, K. T., Neaton, J. B., and Cohen, M. L. 2008. First-principles study of metal adatom adsorption on graphene. *Phys. Rev. B* 77:235430-1–12.
78. Gierz, I., Riedl, C., Starke, U., Ast, C. R., and Kern, K. 2008. Atomic hole doping of graphene. *Nano Lett.* 8:4603–07.
79. Zhang, W., Sun, L., Xu, Z., Krasheninnikov, A. V., Huai, P., Zhu, Z., and Banhart, F. 2010. Migration of gold atoms in graphene ribbons: Role of the edges. *Phys. Rev. B* 81:125425-1–5.
80. Zanella, I., Fagan, S. B., Mota, R., and Fazzio, A. 2008. Electronic and magnetic properties of Ti and Fe on graphene. *J. Phys. Chem. C* 112:9163–67.
81. Mao, Y., Yuan, J., and Zhong, J. 2008. Density functional calculation of transition metal adatom adsorption on graphene. *J. Phys.: Condens. Matter.* 20:115209-1–6.
82. Santos, E. J. G., Portal, D. S., and Ayuela, A. 2010. Magnetism of substitutional Co impurities in graphene: Realization of single π vacancies. *Phys. Rev. B* 81:125433-1–6.

83. Banhart, F., Charlier, J., and Ajayan, P. M. 2000. Dynamic behavior of nickel atoms in graphitic networks. *Phys. Rev. Lett.* 84:686–89.
84. Kan, E.-J., Xiang, H. J., Yang, J., and Hou, J. G. 2007. Electronic structure of atomic Ti chains on semiconducting graphene nanoribbons: A first-principles study. *J. Chem. Phys.* 127:164706-1–5.
85. Rigo, V. A., Martins, T. B., da Silva, A. J. R., Fazzio, A., and Miwa, R. H. 2009. Electronic, structural and transport properties of Ni-doped graphene nanoribbons. *Phys. Rev. B* 79:075435-1–9.
86. Krasheninnikov, A. V., Lehtinen, P. O., Foster, A. S., Pyykko, P., and Nieminen, R. M. 2009. Embedding transition-metal atoms in graphene: Structure, bonding, and magnetism. *Phys. Rev. Lett.* 102:126807-1–4.
87. Longo, R. C., Carrete, J., and Gallego, L. J. 2011. Magnetism of substitutional Fe impurities in graphene nanoribbons. *J. Chem. Phys.* 134:024704-1–4.
88. Thapa, R., Sen, D., Mitra, M. K., and Chattopadhyay, K. K. 2011. Palladium atoms and its dimmers adsorbed on graphene. *Physica B* 406:368–73.
89. Yang, C-.K., Zhao, J., and Lu, J. P. 2002. Binding energies and electronic structures of adsorbed titanium chains on carbon nanotubes. *Phys. Rev. B* 66:041403-1–4.
90. Cocchi, C., Prezzi, D., Calzolari, A., and Molinari, E. 2010. Spin-transport selectivity upon Co adsorption on antiferromagnetic graphene nanoribbons. *J. Chem. Phys.* 133:124703-1–5.
91. Jaiswal, N. K., and Srivastava, P. 2011. Structural stability and electronic properties of Ni-doped armchair graphene nanoribbons. *Solid State Commun.* 151:1490–95.
92. Jaiswal, N. K., and Srivastava, P. 2011. First principles calculations of armchair graphene nanoribbons interacting with Cu atoms. *Physica E* 44:75–79.
93. Jaiswal, N. K., and Srivastava, P. 2011. First principles study of adsorbed and substitutionally doped Fe atoms in zigzag graphene nanoribbons. *Int. Conf. ICONSET, IEEE* 978-1-4673-0073-5/11, November 28–30, Chennai, India.
94. Jaiswal, N. K., and Srivastava, P. 2012. Ab-initio study of transition metal (Ni) interaction with zigzag graphene nanoribbons. *J. Comput. Theo. Nanosci.* 9:555–59.
95. Jaiswal, N. K., and Srivastava, P. 2012. First principles calculations of cobalt doped zigzag graphene nanoribbons. *Solid State Commun.* 152:1489–92.
96. Jaiswal, N. K., and Srivastava, P. 2013. Tailoring the electronic structure of zigzag graphene nanoribbons via copper impurities. *J. Comput. Theo. Nanosci.* 10:1441–45.
97. Wei, M., Chen, L., Lun, N., Sun, Y., Li, D., and Pan, H. 2011. Electronic and magnetic properties of copper-family-element atom adsorbed graphene nanoribbons with zigzag edges. *Solid State Commun.* 151:1440–43.
98. Jaiswal, N. K., and Srivastava, P. 2013. Enhanced metallicity and spin polarization in zigzag graphene nanoribbons with Fe impurities. *Physica E* 54:103–08.
99. Wang, H., Wang, Q., Cheng, Y., Li, K., Yao, Y., Zhang, Q., Dong, C. et al. 2012. Doping monolayer graphene with single atom substitutions. *Nano Lett.* 12:141–44.
100. Jaiswal, N. K., and Srivastava, P. 2013. Fe-doped armchair graphene nanoribbons for spintronic/interconnect applications. *IEEE Trans. Nanotech.* 12:685–91.
101. Williams, J. R., Dicarlo, L., and Marcus, C. M. 2007. Quantum Hall effect in a gate-controlled p-n junction of graphene. *Science* 317:638–41.
102. Huang, B., Yan, Q., Zhou, G., Wu, J., Gu, B-L., and Duana, W. 2007. Making a field effect transistor on a single graphene nanoribbon by selective doping. *Appl. Phys. Lett.* 91:253122-1–3.
103. Areshkin, D. A., Gunlycke, D., and White, C. T. 2007. Ballistic transport in graphene nanostrips in the presence of disorder: Importance of edge effects. *Nano Lett.* 7:204–10.
104. Xu, C., Li, H., and Banerjee, K. 2009. Modeling, analysis, and design of graphene nano-ribbon interconnects. *IEEE Trans. Elect. Dev.* 56:1567–78.
105. Kulkarni, C., and Khot, A. 2010. Carbon nanotubes as interconnects. *Ind. J. Pure Appl. Phys.* 48:305–10.
106. Bunch, J. S., van der Zande, A. M., Verbridge, S. S., Frank, I. W., Tanenbaum, D. M., Parpia, J. M., Craighead, H. G., and McEuen, P. L. 2007. Electromechanical resonators from graphene sheets. *Science* 315:490–93.
107. Xia, F., Mueller, T., Lin, Y.-M., Garcia, A. V., and Avouris, P. 2009. Ultrafast graphene photodetector. *Nat. Nanotech.* 4:839–43.
108. Russo, S., Craciun, M. F., Yamamoto, M., Tarucha, S., and Morpurgo, A. F. 2009. Double-gated graphene-based devices. *N. J. Phys.* 11:095018-1–11.
109. Schedin, F., Geim, A. K., Morozov, S. V., Hill, E. W., Blake, P., Katsnelson, M. I., and Novoselov, K. S. 2007. Detection of individual gas molecules adsorbed on graphene. *Nat. Mater.* 6:652–55.
110. Huang, B., Li, Z., Liu, Z., Zhou, G., Hao, S., Wu, J., Gu, B-L., and Duan, W. 2008. Adsorption of gas molecules on graphene nanoribbons and its implication for nanoscale molecule sensor. *J. Phys. Chem. C* 112:13442–46.
111. Chowdhury, R., Adhikari, S., Rees, P., Wilks, S. P., and Scarpa, F. 2011. Graphene-based biosensor using transport properties. *Phys. Rev. B* 83:045401-1-8.
112. Vedala, H., Sorescu, D. C., Kotchey, G. P., and Star, A. 2011. Chemical sensitivity of graphene edges decorated with metal nanoparticles. *Nano Lett.* 11:2342–47.
113. Yavari, F., and Koratkar, N. 2012. Graphene-based chemical sensors. *J. Phys. Chem. Lett.* 3:1746–53.
114. Park, N., Hong, S., Kim, G., and Jhi, S-H. 2007. Computational study of hydrogen storage characteristics of covalent-bonded graphenes. *J. Am. Chem. Soc.* 129:8999–9003.
115. Ataca, C., Akturk, E., Ciraci, S., and Ustunel, H. 2008. High-capacity hydrogen storage by metallized graphene. *Appl. Phys. Lett.* 93:043123-1–3.
116. Bhattacharya, A., Bhattacharya, S., Majumder, C., and Das, G. P. 2010. Transition-metal decoration enhanced room-temperature hydrogen storage in a defect-modulated graphene sheet. *J. Phys. Chem. C* 114:10297–301.
117. Lee, H., Ihm, J., Cohen, M. L., and Louie, S. G. 2010. Calcium-decorated graphene-based nanostructures for hydrogen storage. *Nano Lett.* 10:793–98.
118. Pumera, M. 2011. Graphene-based nanomaterials for energy storage. *Energy Environ. Sci.* 4:668–74.
119. Zhou, M., Lu, Y., Zhang, C., and Feng, Y. P. 2010. Strain effects on hydrogen storage capability of metal-decorated graphene: A first-principles study. *Appl. Phys. Lett.* 97:103109-1–3.
120. Yang, C. K., Zhao, J., and Lu, J. P. 2004. Complete spin polarization for a carbon nanotube with an adsorbed atomic transition-metal chain. *Nano Lett.* 4:561–63.
121. Son, Y. W., Cohen, M. L., and Louie, S. G. 2006. Half-metallic graphene nanoribbons. *Nature* 444:347–49.
122. Dutta, S., Manna, A., and Pati, S. 2009. Intrinsic half-metallicity in modified graphene nanoribbons. *Phys. Rev. Lett.* 102:096601-1–4.

123. Dutta, S., and Pati, S. K. 2008. Half-metallicity in undoped and boron doped graphene nanoribbons in the presence of semilocal exchange-correlation interactions. *J. Phys. Chem. B* 112:1333–35.
124. Da, H., Feng, Y. P., and Liang, G. 2011. Transition-metal-atom-embedded graphane and its spintronic device applications. *J. Phys. Chem. C* 115:22701–06.
125. Zeng, M., Shen, L., Su, H., Zhang, C., and Feng, Y. 2011. Graphene-based spin logic gates. *Appl. Phys. Lett.* 98:092110-1–3.
126. Min, S. K., Kim, W. Y., Cho, Y., and Kim, K. S. 2011. Fast DNA sequencing with a graphene-based nanochannel device. *Nat. Nanotech.* 6:162–65.
127. Ambavale, S. K., and Sharma, A. C. 2012. *Ab initio* investigations on ballistic transport through carbon atomic chains attached to armchair and zigzag edged graphene electrodes. *Phys. Stat. Sol. B* 249:107–12.
128. Ren, J. J., Su, L. W., Qin, X., Yang, M., Wei, J. P., Zhou, Z., and Shen, P. W. 2014. Pre-lithiated graphene nanosheets as negative electrode material for Li-ion capacitors with high power and energy density. *J. Power Sources* 264:108–13.
129. Choi, M.-H., Min, Y.-J., Gwak, G.-H., Paek, S.-M., and Oh, J.-M. 2014. A nanostructured Ni/graphene hybrid for enhanced electrochemical hydrogen storage. *J. Alloys Comp.* 610:231–35.
130. Gao, L., Lian, C., Zhou, Y., Li, Q., Zhang, C., Chen, L., and Chen, K. 2014. Graphene-oxide-DNA based sensors. *Biosens. Bioelectron.* 60:22–29.
131. Al-Ghamdi, A. A., Gupta, R. K., Kahol, P. K., Wageh, S., Al-Turki, Y. A., Shirbeeny, W. E., and Yakuphanoglu, F. 2014. Improved solar efficiency by introducing graphene oxide in purple cabbage dye sensitized TiO_2 based solar cells. *Solid State Commun.* 183:56–59.
132. Frota, H. O., and Ghosh, A. 2014. Spin caloritronics in graphene. *Solid State Commun.* 191:30–34.
133. Ardenghi, J. S., Bechthold, P., Gonzalez, E., Jasen, P., and Juan, A. 2014. Ballistic transport properties in pristine/doped/pristine graphene junctions. *Superlattices Microstruc.* 72:325–35.
134. Jimenez, M. S., and Dartora, C. A. 2014. The I-V characteristics of a graphene tunnel diode. *Physica E* 59:1–5.
135. Kang, J. W., Park, J., and Kwon, O. K. 2014. Developing a nanoelectromechanical shuttle graphene-nanoflake device. *Physica E* 58:88–93.

9 Electronic Structure and Transport in Graphene

QuasiRelativistic Dirac–Hartree–Fock Self-Consistent Field Approximation

H. V. Grushevskaya and G. G. Krylov

CONTENTS

ABSTRACT

The secondary-quantized self-consistent Dirac–Hartree–Fock approach has been used to consider the electronic properties of monolayer graphene with an accounting of spin-polarized states. The approach allows the coherent explanation of experimental results on energy band minigaps and charge carrier asymmetry in graphene and the proposal of a description of valent and conduction zone shifts and gives a theoretical estimation of electron and hole masses that is in accordance with known experimental data.

9.1 INTRODUCTION

Graphene and graphene-like materials are considered today as prominent candidates to be used in new devices with functionality based on quantum effects and (or) spin-dependent phenomena in low-dimensional systems (see, e.g., References 1–7). Technically, the main obstacle to such devices implementation is the lack of methods that provide minimization of distortion of these material unique properties in bulk nanoheterostructures [8,9]. Theoretical approaches and computer simulation play an important role in a systematic search of these kinds of nanostructures for nanoelectronic applications. The main approaches to simulation of the band structure in graphene are represented in Table 9.1. A significant part of the theoretical consideration of graphene-like materials as well as modern model representations of charge transport in these systems on a two-dimensional (2D) pseudo-Dirac massless fermion model, are originally based on the tight-binding approximation and applied to the description of graphite [10–13], which is a bulk three-dimensional (3D) material. But, as one can see from Table 9.1, for graphene, being a genuine 2D material in fact has no pure 2D approaches for its description. 2D theoretical models are used for fermionic excitations in graphene whereas interaction between them (primarily electromagnetic) has to be considered in covering 3D space (or [3 + 1] for time-dependent problems). As for numeric simulations, it is typical to use a supercell approach when the 2D graphene plane is simply changed to a periodic 3D structure with a large enough distance (~15 Å, see, e.g., Reference 21) between planes.

According to the approach [11,13,22] utilizing 2D motion equations, $\pi(p_z)$-electrons in graphene are massless fermion-type quasiparticle excitations moving with Fermi velocity. While the approach has been seriously developed [14,23–27] and successfully applied to a number of experimental situations, we mention only a few of them [28–31]; some review papers on the topics and further references may be found in References 9, 17, and 32–36.

There are a few known key points where (to our knowledge) one could expect the necessity of some generalized consideration. The first one is the cyclotron mass dependence upon the carriers' concentration. Owing to weakness of the

TABLE 9.1
Models of Graphene and Interactions in It

Model Name and Motion Equation	Interaction Model	References
I. Pseudo-electrodynamics, massless pseudofermion 2D-motion equation	Hartree–Fock and random-phase approximations (RPAs) for 3D Coulomb interactions	[14–17]
II. Gauge field theory, physical flowers: pseudospin, pseudocharge	3D Dyson–Schwinger equation, RPA	[18]
III. *Ab initio* calculations, 3D Schrödinger equation with quasirelativistic corrections	3D Poisson's equation with gradient or muffin-tin atomic ellipsoids approximations for electron density for a monoatomic layer in vacuum, RPA	[19,20]

signal, modern experimental techniques can register cyclotron mass of charge carriers that is just a little smaller than 0.02 of a free-electron mass [22,37–39]. Assessments are absent whether this mechanism of conductivity prevails in the region of very small values of charge carriers' concentration.

Another point is the experimentally observable carrier asymmetry in graphene. According to modern theoretical concepts, the bands for pseudorelativistic electrons and holes in graphene must be symmetrical. With this in mind, in the paper [19] in the generalized gradient approximation, the hexagonal Si and Ge were modeled, with the same structure as in graphene. But the electrons and holes band near the Dirac points $K(K')$ in the Brillouin zone turned out to be a strongly asymmetric one for both cases. First, a Dirac cone deformation takes place far away from a circular shape, as for the second, the Dirac velocities for the valent and conduction bands are different [19]. Therefore, one can assume the existence of some asymmetry in the behavior of pseudorelativistic electrons and holes of graphene as well. Since the value of the asymmetry seems to be very small, for its experimental observation, one should use a highly sensitive method, such as, for example, based on the measurement of noises [40]. For graphene, such a method is based on large amplitudes of nonuniversal fluctuations of charge carriers current in the form of nonmonotonic $1/f$ noise in the crossover region of the scattering [41]. At high charge densities, the contribution to the resistance of clean graphene basically gives the scattering of charge carriers on long-range impurities at ordinary (symplectic) diffusion. The regime of pseudodiffusion with charge carriers scattering on short-range impurities is realized in the vicinity of the Dirac points of the Brillouin zone. In the papers [42,43], measurements of quantum interference noise in a crossover between a pseudodiffusive and symplectic regime and magnetoresistance measurements in graphene p–n junctions have been performed, which established the asymmetric behavior of pseudorelativistic electrons and holes based on an asymmetric form of nonmonotonic dependence of noise and magnetoresistance. And, the third-known key point-needing theoretical explanation is the replica's existence. A weakly interacting epitaxial graphene on the surface of Ir(111) has a nondistorted hexagonal symmetry due to the weakness of the interaction with the substrate in the temperature range up to room temperature. Therefore, in angle-resolved photoelectron spectroscopy (ARPES) spectra, the perturbation of the band structure is manifested in the form of a replica of the inverted Dirac cone and minigaps in places of quasi-crossings of replicas and the Dirac cone [44]. The authors of the paper [44] proposed the explanation of the replica's existence such that replicas were produced only in the areas of convergence of C and Ir atoms. The last allows the explanation of the weak intensity of photoelectron emission replicas and brightness of the main cone. However, besides different intensities, the asymmetry of photoelectron emission spectra also manifests itself in the fact that if the corners of oppositely disposed replicas are in the neighborhood $E - E_F \approx 0$, then, the Dirac cone in the ARPES spectrum is located lower than the corners. Together with this, there exists a usual experimental situation when replicas oppositely arranged on the hexagon and having equal maxima are lower than zero maximum ($E - E_F \approx 0$) of the Dirac cone. The described above is possible if the axes of the Dirac cone and its replicas are not parallel. It means that the top of the replica does not correspond to the corners of the hexagonal mini-Brillouin zone, centered at the Dirac cone corner for the epitaxial graphene on the surface of Ir(111). It has also been demonstrated that epitaxial graphene on SiC(0001) holds a hexagonal mini-Brillouin zone near the Dirac points [45,46].

And finally, a somewhat more philosophical but also important comment. The majority of modern software for *ab initio* band structure simulations [47–50] uses models that are some variant of the Dirac equation or at least take into account the known relativistic corrections to the Schrödinger equation when attacking the problem [51,52]. In this connection, the quasi-Dirac massless fermion approach purely based on a tight-binding nonrelativistic Hamiltonian seems to be oversimplified and hardly extendable.

With the goal to investigate the balance of exchange and correlation interactions, the *ab initio* band structure simulations of loosely packed solids (it was used to develop the generalization of the linear muffin-tin orbital [LMTO] method [53,54]) have been performed and demonstrated that the strong exchange leads to the appearance of an energy gap in the spectrum whereas strong correlation interactions lead to a tightening of this gap [20]. The spin nonpolarized *ab initio* simulations of partial electron densities of 2D graphite have shown that the material is a semiconductor. Interlayer correlations tighten the energy gap that results in semimetal behavior of 3D graphite [20]. This means that in the absence of correlation holes, the correlation interaction in a monoatomic carbon layer (monolayer) is weak in comparison with the exchange. This theoretical prediction for spin-nonpolarized graphene was experimentally confirmed in References 45 and 46, where it demonstrated a band gap in bilayer graphene on SiC(0001) and its diminishing up to vanishing in multilayer graphene.

By the way, in a monolayer graphene on SiC(0001), one observes the dispersion of the Dirac cone apexes [45,46].

Experimentally manufactured quasi-2D systems, such as graphene, carbon armchair nanotubes and ribbons, as well as some types of zigzag carbon nanotubes manifest metallic properties (see, e.g., References 32 and 55). In this regard, there are discrepancies between theoretical predictions and experimental data.

Therefore, based on the results of *ab initio* spin-nonpolarized simulations of 2D and 3D graphite [20], we can make the following assumption. Carbon low-dimensional systems having properties similar to graphite-like materials should possess spin-polarized electronic states with the correlation holes (a magnetic ordering). Enhancing of the correlation interaction due to correlation-hole contribution leads to tightening of the energy gap and, as a consequence, the emergence of semimetal conductivity.

The approach we use has been developed earlier in Reference 56 and applied for graphene-like material in Reference 57.

The goal of this chapter is to represent a Dirac–Hartree–Fock self-consistent field quasirelativistic approximation for quasi-2D systems and to describe the origin of asymmetry of electron– correlation hole carriers in graphene-like materials.

9.2 GRAPHENE MODEL BANDS WITH CORRELATION HOLES

A single atomic layer of carbon atoms (2D graphite) is called monolayer graphene. Its hexagonal structure can be represented by two triangular sublattices A, B [36,58]. The primitive unit cell of the graphene contains two carbon atoms C_A and C_B belonging to the sublattices A and B respectively.

The carbon atom has four valent electrons s, p_x, p_y, and p_z. Electrons s, p_x, and p_y are hybridized in the plane of the monolayer, and p_z-electron orbitals form a half-filled band of π-electron orbitals on a hexagonal lattice.

Electrons with spin $+\sigma$ "down" ("up") are placed on the sublattice A (B), and the electrons with spin $-\sigma$ "up" ("down") are placed on the sublattice B (A), as shown in Figure 9.1. With such symmetry of the problem, all the relevant bands of sublattices are half-filled and are formed due to correlation holes. In the representation of secondary quantization and Hartree–Fock self-consistent field approximation, when not accounting for the electron density fluctuation correlation, the hole energy $\varepsilon(k)$ is simply added to the electron energy $\varepsilon_m(0)$ [59]:

$$\left[H(\vec{r})+\hat{V}^{sc}(kr)-\hat{\Sigma}^{x}(kr)\right]\psi_m(kr)=\left(\varepsilon_m(0)-\sum_{j=1}^{n}\hat{\varepsilon}^{\dagger}P_j\right)\psi_m(kr) \tag{9.1}$$

$$\hat{\varepsilon}^{\dagger}=\varepsilon(k)\hat{I} \tag{9.2}$$

because the sum $\Sigma_j P_j$ of projection operators P_j in parentheses equals to the identity operator $\hat{I}:\Sigma_j P_j=\hat{I}$.

We denote spinor wave functions of the valent electrons of graphene as $\hat{\chi}^{\dagger}_{\sigma}(\vec{r}_A)|0,+\sigma\rangle$ and $\hat{\chi}^{\dagger}_{-\sigma}(\vec{r}_B)|0,-\sigma\rangle$. From Figure 9.1, it follows that the spinor quantum fields $\hat{\chi}^{\dagger}_{\sigma}(\vec{r}_A)$ and $\hat{\chi}^{\dagger}_{-\sigma}(\vec{r}_B)$ are transformed into each other under the mirror reflection

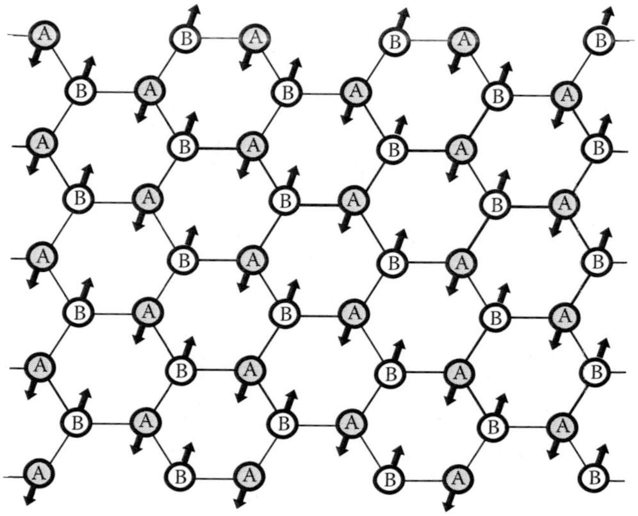

FIGURE 9.1 Hexagonal lattice of carbon monolayer with spin-ordering sublattices A, B.

$A\rightarrow B$. Therefore, a quasiparticle excitation in the proposed model of graphene is a pair of an electron and a correlation hole. As electron–hole pairs at the same time represent themselves on their proper antiparticle, the wave functions belong to the space of Majorana bispinors Ψ', and the upper and lower spin components ψ', $\dot{\psi}'$ are transformed via different representations of the Lorenz group

$$\Psi'=\begin{pmatrix}\psi'_{\sigma}\\ \dot{\psi}'_{-\sigma}\end{pmatrix}=\begin{pmatrix}e^{(\kappa/2)\vec{\sigma}\cdot\vec{n}}\psi_{\sigma}\\ e^{(\kappa/2)(-\vec{\sigma})\cdot\vec{n}}\dot{\psi}_{-\sigma}\end{pmatrix} \tag{9.3}$$

It means that $\hat{\chi}^{\dagger}_{\sigma}(\vec{r}_A)|0,+\sigma\rangle$ behaves as a component ψ_{σ}, and $\hat{\chi}^{\dagger}_{-\sigma}(\vec{r}_B)|0,-\sigma\rangle$—behaves as a component $\dot{\psi}_{-\sigma}$ of bispinor (9.3). Using expression (9.3) and properties of these operators:

$$\hat{\chi}^{\dagger}_{-\sigma_A}(\vec{r})|0,+\sigma\rangle\overset{def}{=\!=}0,\quad \hat{\chi}^{\dagger}_{+\sigma_B}(\vec{r})|0,-\sigma\rangle\overset{def}{=\!=}0$$

one gets the following expression for the bispinor wave function $|\Psi\rangle$ of an electron in graphene:

$$|\Psi\rangle=\begin{pmatrix}\hat{\chi}^{\dagger}_{-\sigma_A}(\vec{r})|0,-\sigma\rangle\\ \hat{\chi}^{\dagger}_{\sigma_B}(\vec{r})|0,+\sigma\rangle\end{pmatrix}=\begin{pmatrix}|0,+\sigma\rangle\hat{\chi}^{\dagger}_{-\sigma_A}(\vec{r})|0,-\sigma\rangle\\ |0,-\sigma\rangle\hat{\chi}^{\dagger}_{\sigma_B}(\vec{r})|0,+\sigma\rangle\end{pmatrix}$$
$$=\begin{pmatrix}|0,+\sigma\rangle\hat{\chi}^{\dagger}_{-\sigma_A}(\vec{r})|0,-\sigma\rangle+|0,-\sigma\rangle\hat{\chi}^{\dagger}_{-\sigma_A}(\vec{r})|0,+\sigma\rangle\\ |0,-\sigma\rangle\hat{\chi}^{\dagger}_{\sigma_B}(\vec{r})|0,+\sigma\rangle+|0,+\sigma\rangle\hat{\chi}^{\dagger}_{\sigma_B}(\vec{r})|0,-\sigma\rangle\end{pmatrix}$$
$$=\begin{pmatrix}\hat{\chi}^{\dagger}_{-\sigma_A}(\vec{r})|0,-\sigma\rangle|0,+\sigma\rangle\\ \hat{\chi}^{\dagger}_{\sigma_B}(\vec{r})|0,+\sigma\rangle|0,-\sigma\rangle\end{pmatrix}=\begin{pmatrix}\hat{\chi}^{\dagger}_{-\sigma_A}(\vec{r})\\ \hat{\chi}^{\dagger}_{\sigma_B}(\vec{r})\end{pmatrix}|0\rangle \tag{9.4}$$

where $|0\rangle$ is a vacuum vector that consists of uncorrelated vacuum states with spin "down" $|0,-\sigma\rangle$ and "up" $|0,+\sigma\rangle$: $|0\rangle=|0,+\sigma\rangle\,|0,-\sigma\rangle$.

The density matrix $\hat{\rho}_{rr'}$ is expressed through the components of bispinor (9.4) as

$$\hat{\rho}^{AB}_{rr'} = \begin{pmatrix} \hat{\chi}^{\dagger}_{-\sigma_A}(\vec{r})\hat{\chi}_{\sigma_A}(\vec{r}) & \hat{\chi}^{\dagger}_{-\sigma_A}(\vec{r})\hat{\chi}_{-\sigma_B}(\vec{r}') \\ \hat{\chi}^{\dagger}_{\sigma_B}(\vec{r}')\hat{\chi}_{\sigma_A}(\vec{r}) & \hat{\chi}^{\dagger}_{\sigma_B}(\vec{r}')\hat{\chi}_{-\sigma_B}(\vec{r}') \end{pmatrix} \quad (9.5)$$

9.3 EQUATION FOR THE DENSITY MATRIX

In description, we will consider only valent electrons. We denote the number of atoms in two sublattices by N. For valent electrons, the Dirac Hamiltonian H_D has the following form:

$$H_D = \sum_{L=A,B}\sum_{i=1}^{N/2}\sum_{\nu=1}^{4}\left\{ c\vec{\alpha}\cdot\vec{p}_{i_L^{\nu}} + \beta m_e c^2 - \sum_{k=1}^{N}\frac{Ze^2}{|\vec{r}_{i_L^{\nu}} - \vec{R}_k|} + \sum_{L<L'=A,B}\sum_{i<j=1}^{N/2}\sum_{\nu'=1}^{4}\frac{e^2}{|\vec{r}_{i_L^{\nu}} - \vec{r}_{j_{L'}^{\nu'}}|}\right\} \quad (9.6)$$

$$\vec{p} = -\imath\hbar\vec{\nabla},\ \vec{\alpha} = \begin{pmatrix} 0 & \vec{\sigma} \\ \vec{\sigma} & 0 \end{pmatrix},\ \beta = \begin{pmatrix} 1 & 0 \\ 0 & -1 \end{pmatrix} \quad (9.7)$$

Here, $\vec{\alpha}$, β is a set 4×4 of Dirac matrices, $\vec{\sigma}$ is the set of 2×2 Pauli matrices, indices ν and ν' enumerate s-, p_x-, p_y-, and p_z-electron orbitals, indices L, i and L', and j enumerate sublattices and atoms within them respectively, $\vec{r}_{i_L^{\nu}}$ is the electron radius vector, $\vec{R}_k$ is the radius vector of kth carbon atom without valent electrons (atomic core), $-Ze = -4e$ is the charge of the atomic core, e is the electron charge, m_e is the free-electron mass, and c is the speed of light.

The operator (matrix) of the electron density $\hat{\rho}_{nn';rr'}$ in the mean field approximation when neglecting correlation interactions between electrons, satisfies the equation [59,60]

$$\left(\imath\frac{\partial}{\partial t} - (\hat{h} + \Sigma^x + V^{sc})\right)\sum_n \hat{\rho}_{nn';rr'} = (-\varepsilon_n(0))N_\nu N\delta_{rr'}\delta(t-t') \quad (9.8)$$

where $\hat{h}$ is the kinetic energy operator for a single-particle state, V^{sc} is the self-consistent potential, Σ^x is the exchange interaction, $\varepsilon_n(0)(\varepsilon_n(0) < 0)$ is the eigenvalue of a nonexcited single-particle state (energy of an electron orbital for an isolated atom), and N_ν, $N_\nu = 4$ is the number of valent electrons. At $N\to\infty$, Equation 9.8 can be considered as the equation for the Green function of the quasiparticle excitations [59,60]:

$$\left(\imath\frac{\partial}{\partial t} - (\hat{h} + \Sigma^x + V^{sc})\right)\sum_{n'} \hat{\rho}_{nn';rr'} = \delta(\vec{r} - \vec{r}')\delta(t-t') \quad (9.9)$$

Further, the "electron" will be used in the sense of a quasiparticle.

A Dirac–Hartree–Fock Hamiltonian H_{DFH} for quasiparticle excitations in graphene can be obtained by the procedure of secondary quantization of the Dirac Hamiltonian H_D (9.6):

$$\frac{1}{N_\nu N}\left(\imath\frac{\partial}{\partial t} - \hat{h}_{DFH}\right)\sum_n \hat{\rho}^{AB}_{nn';rr'} = (-\varepsilon_n(0))\delta_{rr'}\delta(t-t') \quad (9.10)$$

$$H_{DFH} = \hat{h}_D + \Sigma^x_{rel} + V^{sc}_{rel} \quad (9.11)$$

where $\hat{h}_D$, Σ^x_{rel}, and V^{sc}_{rel} are relativistic analogs of operators $\hat{h}$, Σ^x, and V^{sc}. Equation 9.10 can be rewritten for the quasiparticle field $\chi(\vec{r},t)$ as

$$\left[E(p) - (\hat{h}_D + \Sigma^x_{rel} + V^{sc}_{rel})\right]\chi(\vec{r},t) = 0 \quad (9.12)$$

Here, $E(p)$ is the energy of the quasiparticle excitation.

Now, one can write the relativistic Equation 9.12 in an explicit form:

$$\begin{pmatrix} m_e c^2 - \sum_{k=1}^{N}\frac{Ze^2}{|\vec{r} - \vec{R}_k|} & c\vec{\sigma}\cdot\vec{p} \\ c\vec{\sigma}\cdot\vec{p} & -m_e c^2 - \sum_{k=1}^{N}\frac{Ze^2}{|\vec{r} - \vec{R}_k|} \end{pmatrix} \times \begin{pmatrix} \hat{\chi}^{\dagger}_{-\sigma_A}(\vec{r}) \\ \hat{\chi}^{\dagger}_{\sigma_B}(\vec{r}) \end{pmatrix} |0,-\sigma\rangle|0,\sigma\rangle + \frac{1}{N_\nu N V}\sum_{i,i'=1}^{N_\nu N}\iint d\vec{r}_i d\vec{r}_{i'}$$

$$\times \begin{pmatrix} \langle 0,-\sigma_i|\hat{\chi}^{\dagger}_{-\sigma_i^A}(\vec{r}_i)V_{r_i,r}\hat{\chi}_{\sigma_i^A}(\vec{r}_i)|0,-\sigma_i\rangle & \langle 0,-\sigma_i|\hat{\chi}^{\dagger}_{-\sigma_i^A}(\vec{r}_i)V_{r_i,r}\hat{\chi}_{-\sigma_{i'}^B}(\vec{r}_i)|0,-\sigma_{i'}\rangle \\ \langle 0,\sigma_{i'}|\hat{\chi}^{\dagger}_{\sigma_{i'}^B}(\vec{r}_{i'})V_{r_{i'},r}\hat{\chi}_{\sigma_i^A}(\vec{r}_{i'})|0,\sigma_i\rangle & \langle 0,\sigma_{i'}|\hat{\chi}^{\dagger}_{\sigma_{i'}^B}(\vec{r}_{i'})V_{r_{i'},r}\hat{\chi}_{-\sigma_{i'}^B}(\vec{r}_{i'})|0,\sigma_{i'}\rangle \end{pmatrix}$$

$$\times \begin{pmatrix} \hat{\chi}^{\dagger}_{-\sigma_i^A}(\vec{r}) \\ \hat{\chi}^{\dagger}_{\sigma_{i'}^B}(\vec{r}) \end{pmatrix}|0,-\sigma\rangle|0,\sigma\rangle = E(p)I\begin{pmatrix} \hat{\chi}^{\dagger}_{-\sigma_A}(\vec{r}) \\ \hat{\chi}^{\dagger}_{\sigma_B}(\vec{r}) \end{pmatrix}|0,-\sigma\rangle|0,\sigma\rangle \quad (9.13)$$

$$\hat{\chi}^{\dagger}_{-\sigma_A}(\vec{r}) = \frac{1}{N_\nu N}\sum_{i'=1}^{N_\nu N}\chi^{\dagger}_{-\sigma_{i'}^A}(\vec{r}),\quad \hat{\chi}^{\dagger}_{\sigma_B}(\vec{r}) = \frac{1}{N_\nu N}\sum_{i=1}^{N_\nu N}\chi^{\dagger}_{\sigma_i^B}(\vec{r}) \quad (9.14)$$

where $V_{r_{i(i')},r} \equiv V(r_{i(i')} - r)$, I is the identity matrix, and index "i" ("i'") enumerates all valent electrons of graphene. From Equations 9.13 and 9.14, the expressions follow for the relativistic self-consistent Coulomb potential V^{sc}_{rel}

$$V_{rel}^{sc}\begin{pmatrix}\hat{\chi}^{\dagger}_{-\sigma_A}(\vec{r})\\ \hat{\chi}^{\dagger}_{\sigma_B}(\vec{r})\end{pmatrix}|0,-\sigma\rangle\,|0,\sigma\rangle=\begin{pmatrix}(V_{rel}^{sc})_{AA} & 0\\ 0 & (V_{rel}^{sc})_{BB}\end{pmatrix}$$

$$\times\begin{pmatrix}\hat{\chi}^{\dagger}_{-\sigma_A}(\vec{r})\\ \hat{\chi}^{\dagger}_{\sigma_B}(\vec{r})\end{pmatrix}|0,-\sigma\rangle\,|0,\sigma\rangle \tag{9.15}$$

$$(V_{rel}^{sc})_{AA}=\sum_{i=1}^{N_v N}\int d\vec{r}_i\langle 0,-\sigma_i\,|\,\hat{\chi}^{\dagger}_{-\sigma_i^A}(\vec{r}_i)V(\vec{r}_i-\vec{r})\hat{\chi}_{\sigma_i^A}(\vec{r}_i)\,|\,0,-\sigma_i\rangle \tag{9.16}$$

$$(V_{rel}^{sc})_{BB}=\sum_{i'=1}^{N_v N}\int d\vec{r}_{i'}\langle 0,\sigma_{i'}\,|\,\hat{\chi}^{\dagger}_{\sigma_{i'}^B}(\vec{r}_{i'})V(\vec{r}_{i'}-\vec{r})\hat{\chi}_{-\sigma_{i'}^B}(\vec{r}_{i'})\,|\,0,\sigma_{i'}\rangle \tag{9.17}$$

and exchange interaction term Σ_{rel}^{x} [56]

$$\Sigma_{rel}^{x}\begin{pmatrix}\hat{\chi}^{\dagger}_{-\sigma_A}(\vec{r})\\ \hat{\chi}^{\dagger}_{\sigma_B}(\vec{r})\end{pmatrix}|0,-\sigma\rangle\,|0,\sigma\rangle=\begin{pmatrix}0 & (\Sigma_{rel}^{x})_{AB}\\ (\Sigma_{rel}^{x})_{BA} & 0\end{pmatrix}$$

$$\times\begin{pmatrix}\hat{\chi}^{\dagger}_{-\sigma_A}(\vec{r})\\ \hat{\chi}^{\dagger}_{\sigma_B}(\vec{r})\end{pmatrix}|0,-\sigma\rangle\,|0,\sigma\rangle \tag{9.18}$$

$$(\Sigma_{rel}^{x})_{AB}\hat{\chi}^{\dagger}_{\sigma_B}(\vec{r})\,|0,\sigma\rangle=\sum_{i=1}^{N_v N}\int d\vec{r}_i\hat{\chi}^{\dagger}_{\sigma_i^B}(\vec{r})\,|0,\sigma\rangle$$

$$\times\langle 0,-\sigma_i|\,\hat{\chi}^{\dagger}_{-\sigma_i^A}(\vec{r}_i)V(\vec{r}_i-\vec{r})\hat{\chi}_{-\sigma_B}(\vec{r}_i)\,|0,-\sigma_{i'}\rangle \tag{9.19}$$

$$(\Sigma_{rel}^{x})_{BA}\hat{\chi}^{\dagger}_{-\sigma_A}(\vec{r})\,|0,-\sigma\rangle=\sum_{i'=1}^{N_v N}\int d\vec{r}_{i'}\hat{\chi}^{\dagger}_{-\sigma_{i'}^A}(\vec{r})\,|0,-\sigma\rangle$$

$$\times\langle 0,\sigma_{i'}|\,\hat{\chi}^{\dagger}_{\sigma_{i'}^B}(\vec{r}_{i'})V(\vec{r}_{i'}-\vec{r})\hat{\chi}_{\sigma_A}(\vec{r}_{i'})\,|0,\sigma_i\rangle \tag{9.20}$$

Substitution of expressions (9.15) and (9.18) into Equation 9.13 gives

$$\left[\begin{pmatrix} m_e c^2-\sum_{k=1}^{N}\frac{Ze^2}{|\vec{r}-\vec{R}_k|}+(V_{rel}^{sc})_{AA} & c\vec{\sigma}\cdot\vec{p}+(\Sigma_{rel}^{x})_{AB}\\ c\vec{\sigma}\cdot\vec{p}+(\Sigma_{rel}^{x})_{BA} & -m_e c^2-\sum_{k=1}^{N}\frac{Ze^2}{|\vec{r}-\vec{R}_k|}+(V_{rel}^{sc})_{BB}\end{pmatrix}-E(p)I\right]$$

$$\times\begin{pmatrix}\hat{\chi}^{\dagger}_{-\sigma_A}(\vec{r})\,|0,-\sigma\rangle\\ \hat{\chi}^{\dagger}_{\sigma_B}(\vec{r})\,|0,\sigma\rangle\end{pmatrix}=0 \tag{9.21}$$

Let us perform a variable change $E\rightarrow E+m_e c^2$ and write down system (9.21) in components

$$\left[-\sum_{k=1}^{N}\frac{Ze^2}{|\vec{r}-\vec{R}_k|}+(V_{rel}^{sc})_{AA}-E(p)\right]\hat{\chi}^{\dagger}_{-\sigma_A}(\vec{r})\,|0,-\sigma\rangle$$

$$+[c\vec{\sigma}\cdot\vec{p}+(\Sigma_{rel}^{x})_{AB}]\hat{\chi}^{\dagger}_{\sigma_B}(\vec{r})\,|0,\sigma\rangle=0 \tag{9.22}$$

$$[c\vec{\sigma}\cdot\vec{p}+(\Sigma_{rel}^{x})_{BA}]\hat{\chi}^{\dagger}_{-\sigma_A}(\vec{r})\,|0,-\sigma\rangle$$

$$+\left[-2m_e c^2-\sum_{k=1}^{N}\frac{Ze^2}{|\vec{r}-\vec{R}_k|}+(V_{rel}^{sc})_{BB}-E(p)\right]\hat{\chi}^{\dagger}_{\sigma_B}(\vec{r})\,|0,\sigma\rangle=0 \tag{9.23}$$

From the last equation of systems (9.22) and (9.23), we find the equation for the component $\hat{\chi}^{\dagger}_{\sigma_B}(\vec{r})\,|0,\sigma\rangle$

$$\hat{\chi}^{\dagger}_{\sigma_B}(\vec{r})\,|0,\sigma\rangle=\frac{1}{2m_e c^2}\left\{1+\left[\frac{\sum_{k=1}^{N}\frac{Ze^2}{|\vec{r}-\vec{R}_k|}-(V_{rel}^{sc})_{BB}+E(p)}{2m_e c^2}\right]\right\}^{-1}$$

$$\times[c\vec{\sigma}\cdot\vec{p}+(\Sigma_{rel}^{x})_{BA}]\hat{\chi}^{\dagger}_{-\sigma_A}(\vec{r})\,|0,-\sigma\rangle \tag{9.24}$$

9.4 QUASIRELATIVISTIC CORRECTIONS

In the quasirelativistic limit $c\rightarrow\infty$, it is possible to neglect the lower components of a bispinor with respect to the upper components, as the components of the bispinor $\hat{\chi}^{\dagger}_{\sigma_B}(\vec{r})\,|0,\sigma\rangle$ have an order of $O(c^{-1})$. So, it is sufficient to find the upper components to describe the behavior of the system. With this in mind, we eliminate the small lower components in Equation 9.22, expressing small components through large ones with the help of Equation 9.24:

$$\left[-\sum_{k=1}^{N}\frac{Ze^2}{|\vec{r}-\vec{R}_k|}+(V_{rel}^{sc})_{AA}-E(p)\right]\hat{\chi}^{\dagger}_{-\sigma_A}(\vec{r})\,|0,-\sigma\rangle$$

$$+\frac{1}{2m_e c^2}[c\vec{\sigma}\cdot\vec{p}+(\Sigma_{rel}^{x})_{AB}]$$

$$\times\left\{1-\left[\frac{-\sum_{k=1}^{N}\frac{Ze^2}{|\vec{r}-\vec{R}_k|}+(V_{rel}^{sc})_{BB}-E(p)}{2m_e c^2}\right]\right\}^{-1}$$

$$\times[c\vec{\sigma}\cdot\vec{p}+(\Sigma_{rel}^{x})_{BA}]\hat{\chi}^{\dagger}_{-\sigma_A}(\vec{r})\,|0,-\sigma\rangle=0 \tag{9.25}$$

Expanding the factor in curly brackets in a power series on a small parameter

$$\left|\frac{(-Ze^2/r)+(V_{rel}^{sc})_{BB}-E(p)}{2m_ec^2}\right| \ll 1$$

we obtain the quasirelativistic Dirac–Hartree–Fock approximation for graphene:

$$\left[-\sum_{k=1}^{N}\frac{Ze^2}{|\vec{r}-\vec{R}_k|}+(V_{rel}^{sc})_{AA}-E(p)\right]\hat{\chi}^{\dagger}_{-\sigma_A}(\vec{r})\,|0,-\sigma\rangle$$
$$+\frac{1}{2m_ec^2}[c\vec{\sigma}\cdot\vec{p}+(\Sigma_{rel}^{x})_{AB}]$$
$$\times[c\vec{\sigma}\cdot\vec{p}+(\Sigma_{rel}^{x})_{BA}]\hat{\chi}^{\dagger}_{-\sigma_A}(\vec{r})\,|0,-\sigma\rangle$$
$$+\frac{1}{2m_ec^2}[c\vec{\sigma}\cdot\vec{p}+(\Sigma_{rel}^{x})_{AB}]\left[\frac{-\sum_{k=1}^{N}\frac{Ze^2}{|\vec{r}-\vec{R}_k|}+(V_{rel}^{sc})_{BB}-E(p)}{2m_ec^2}\right]$$
$$\times[c\vec{\sigma}\cdot\vec{p}+(\Sigma_{rel}^{x})_{BA}]\hat{\chi}^{\dagger}_{-\sigma_A}(\vec{r})\,|0,-\sigma\rangle=0 \tag{9.26}$$

Let us find the nonrelativistic limit. With this goal in Equation 9.26, we write down

$$(\Sigma_{rel}^{x})_{AB}\hat{\chi}^{\dagger}_{-\sigma_A}(\vec{r})\,|0,-\sigma\rangle \rightarrow (\Sigma_{rel}^{x})_{AB}\hat{\chi}^{\dagger}_{\sigma_B}(\vec{r})\,|0,\sigma\rangle \tag{9.27}$$

and leave only the first-order terms on $(c^2)^{-1}$:

$$\left[-\sum_{j=1}^{N}\frac{Ze^2}{|\vec{r}-\vec{R}_k|}+(V_{rel}^{sc})_{AA}-E(p)\right]\hat{\chi}^{\dagger}_{-\sigma_A}(\vec{r})\,|0,-\sigma\rangle$$
$$+\frac{1}{2m_ec^2}[c\vec{\sigma}\cdot\vec{p}\;c\vec{\sigma}\cdot\vec{p}+c\vec{\sigma}\cdot\vec{p}(\Sigma_{rel}^{x})_{BA}$$
$$+(\Sigma_{rel}^{x})_{AB}(\Sigma_{rel}^{x})_{BA}]\,\hat{\chi}^{\dagger}_{-\sigma_A}(\vec{r})\,|0,-\sigma\rangle$$
$$+\frac{1}{2m_ec^2}(\Sigma_{rel}^{x})_{AB}c\vec{\sigma}\cdot\vec{p}\hat{\chi}^{\dagger}_{\sigma_B}(\vec{r})\,|0,\sigma\rangle=0 \tag{9.28}$$

After some elementary algebra, we transform Equation 9.28 to the form

$$\left[\frac{\vec{p}^2}{2m_e}-\sum_{k=1}^{N}\frac{Ze^2}{|\vec{r}-\vec{R}_k|}+(V_{rel}^{sc})_{AA}-E(p)\right]\hat{\chi}^{\dagger}_{-\sigma_A}(\vec{r})\,|0,-\sigma\rangle$$
$$+\frac{1}{2}\left[(\Sigma_{rel}^{x})_{BA}\,\hat{\chi}^{\dagger}_{-\sigma_A}(\vec{r})\,|0,-\sigma\rangle+(\Sigma_{rel}^{x})_{AB}\,\hat{\chi}^{\dagger}_{\sigma_B}(\vec{r})\,|0,\sigma\rangle\right]$$
$$+\frac{1}{2m_ec^2}(\Sigma_{rel}^{x})_{AB}(\Sigma_{rel}^{x})_{BA}\,\hat{\chi}^{\dagger}_{-\sigma_A}(\vec{r})\,|0,-\sigma\rangle=0 \tag{9.29}$$

Since at replacements $A\leftrightarrow B$ and $\sigma_A\leftrightarrow-\sigma_B$, the first two terms do not change the form of the equation, and then, they give the nonrelativistic contributions. Quadratic summand

$$\left(\Sigma_{rel}^{x}\right)_{AB}\left(\Sigma_{rel}^{x}\right)_{BA} \tag{9.30}$$

is a quasirelativistic correction, because its form is sensitive to the above-mentioned change. Since in the nonrelativistic limit the quasirelativistic quadratic correction (9.30) should be omitted, the substitution of expressions (9.15) and (9.18) into Equation 9.29 leads to the nonrelativistic equation [61]

$$\left[\frac{\vec{p}^2}{2m_e}-\sum_{k=1}^{N}\frac{Ze^2}{|\vec{r}-\vec{R}_k|}\right.$$
$$\left.+\sum_{i=1}^{N_vN}\int d\vec{r}_i\langle 0,-\sigma_i|\,\hat{\chi}^{\dagger}_{-\sigma_i^A}(\vec{r}_i)V_{\vec{r}_i,\vec{r}}\hat{\chi}_{\sigma_i^A}(\vec{r}_i)\,|0,-\sigma_i\rangle-E(p)\right]$$
$$\times\chi^{\dagger}_{-\sigma_A}(\vec{r})\,|0,-\sigma\rangle+\frac{1}{2}\left[\sum_{i'=1}^{N_vN}\int d\vec{r}_{i'}\hat{\chi}^{\dagger}_{-\sigma_{i'}^A}(\vec{r})\,|0,-\sigma\rangle\right.$$
$$\times\langle 0,\sigma_{i'}|\,\chi^{\dagger}_{\sigma_{i'}^B}(\vec{r}_{i'})V_{\vec{r}_{i'},\vec{r}}\hat{\chi}_{\sigma_A}(\vec{r}_{i'})\,|0,\sigma_i\rangle$$
$$\left.+\sum_{i=1}^{N_vN}\int d\vec{r}_i\hat{\chi}^{\dagger}_{\sigma_i^B}(\vec{r})\,|0,\sigma\rangle\langle 0,-\sigma_i|\,\hat{\chi}^{\dagger}_{-\sigma_i^A}(\vec{r}_i)V_{\vec{r}_i,\vec{r}}\hat{\chi}_{-\sigma_B}(\vec{r}_i)\,|0,-\sigma_{i'}\rangle\right]=0 \tag{9.31}$$

Presenting $E(p)$ as a difference of mth energy eigenvalue for one-electron nonexcited state $\varepsilon_m^{(0)}$ and the energy eigenvalue for the hole $\varepsilon^{\dagger}=\varepsilon(p)I : E(p)=\varepsilon_m^{(0)}-\varepsilon(p)$ and taking into account the chain of equalities

$$\sum_{i'=1}^{N_vN}\int d\vec{r}_{i'}\hat{\chi}^{\dagger}_{-\sigma_{i'}^A}(\vec{r})\,|0,-\sigma\rangle\langle 0,\sigma_{i'}|\,\hat{\chi}^{\dagger}_{\sigma_{i'}^B}(\vec{r}_{i'})V(\vec{r}_{i'}-\vec{r})\hat{\chi}^{\dagger}_{\sigma_A}(\vec{r}_{i'})\,|0,\sigma_i\rangle$$
$$\equiv\sum_{i'=1}^{N_vN}\int d\vec{r}_{i'}\hat{\chi}^{\dagger}_{-\sigma_{i'}^A}(\vec{r})\,|0,-\sigma\rangle\langle 0,\sigma_{i'}|\,\hat{\chi}^{\dagger}_{\sigma_{i'}^B}(\vec{r}_{i'})V(\vec{r}_{i'}-\vec{r})\hat{\chi}^{\dagger}_{-\sigma_B}(\vec{r}_{i'})\,|0,-\sigma_{i'}\rangle$$
$$=\sum_{i'=1}^{N_vN}\int d\vec{r}_{i'}\hat{\chi}^{\dagger}_{\sigma_{i'}^B}(\vec{r})\,|0,-\sigma\rangle\langle 0,-\sigma_{i'}|\,\hat{\chi}^{\dagger}_{-\sigma_{i'}^A}(\vec{r}_{i'})V(\vec{r}_{i'}-\vec{r})\hat{\chi}^{\dagger}_{-\sigma_B}(\vec{r}_{i'})\,|0,-\sigma_{i'}\rangle \tag{9.32}$$

one can rewrite Equation 9.31 as

$$\left[\frac{\vec{p}^2}{2m_e}-\sum_{k=1}^{N}\frac{Ze^2}{|\vec{r}-\vec{R}_k|}\right.$$
$$\left.+\sum_{i=1}^{N_vN}\int d\vec{r}_i\langle 0,-\sigma_i|\,\hat{\chi}^{\dagger}_{-\sigma_i^A}(\vec{r}_i)V(\vec{r}_i-\vec{r})\hat{\chi}_{\sigma_i^A}(\vec{r}_i)\,|0,-\sigma_i\rangle-(\varepsilon_m^{(0)}-\varepsilon(p))\right]$$
$$\times\hat{\chi}^{\dagger}_{-\sigma_A}(\vec{r})\,|0,-\sigma\rangle+\sum_{i=1}^{N_vN}\int d\vec{r}_i$$
$$\times\hat{\chi}^{\dagger}_{\sigma_i^B}(\vec{r})\,|0,-\sigma\rangle\langle 0,-\sigma_i|\,\hat{\chi}^{\dagger}_{-\sigma_i^A}(\vec{r}_i)V(\vec{r}_i-\vec{r})\chi_{-\sigma_B}(\vec{r}_i)\,|0,-\sigma_i\rangle=0 \tag{9.33}$$

As mentioned above, in the nonrelativistic limit, the indices A, B can be omitted and Equation 9.33 can be written in a final form

$$\varepsilon_m^{(0)}\hat{\psi}_{\sigma_m}^{\dagger}(\vec{r}_m)|0,\sigma_m\rangle - \langle 0,-\sigma_i|\hat{\varepsilon}^{\dagger}\hat{\mathrm{I}}|0,-\sigma_i\rangle\hat{\psi}_{\sigma_m}^{\dagger}(\vec{r}_m)|0,\sigma_m\rangle$$
$$= \hat{h}(\vec{r}_m)\hat{\psi}_{\sigma_m}(\vec{r}_m)|0,\sigma_m\rangle$$
$$-\sum_{i=1}^{n}\int d\vec{r}_i\hat{\psi}_{\sigma_i}^{\dagger}(\vec{r}_m)|0,\sigma_m\rangle V(\vec{r}_i-\vec{r}_m)$$
$$\times\langle 0,-\sigma_i|\hat{\psi}_{-\sigma_i}^{\dagger}(\vec{r}_i)\hat{\psi}_{\sigma_m}(\vec{r}_i)|0,-\sigma_i\rangle$$
$$+\sum_{i=1}^{n}\int d\vec{r}_i\hat{\psi}_{\sigma_m}^{\dagger}(\vec{r}_m)|0,\sigma_m\rangle V(\vec{r}_i-\vec{r}_m)$$
$$\times\langle 0,-\sigma_i|\hat{\psi}_{-\sigma_i}^{\dagger}(\vec{r}_i)\hat{\psi}_{\sigma_i}(\vec{r}_i)|0,-\sigma_i\rangle \tag{9.34}$$

where $\sigma_m \equiv -\sigma$, $\vec{r}_m \equiv \vec{r}$

$$\hat{h}(\vec{r}_m) = \frac{\vec{p}^2}{2m_e} - \sum_{k=1}^{N}\frac{Ze^2}{|\vec{r}_m - \vec{R}_k|} \tag{9.35}$$

Formula (9.34) precisely represents the Hartree–Fock equation for the spin electron density as it was shown in Reference 56.

Thus, distinctions of spinor wave functions of the electrons belonging to different sublattices are manifested through the interaction of the sublattices. The spin dependence of Dirac cones is manifested in the first order in $(c^2)^{-1}$ when one cannot neglect the lower components of the bispinor. When neglecting the small lower spinor components, the description becomes nonrelativistic, and therefore does not allow the description of spin-dependent polarization of the band structure of graphene and graphene-like material.

Now, it is possible to consider the energy band structure of graphene with the second quantized Hamiltonian (9.11). We choose the Bloch functions

$$\chi_n(\vec{k},\vec{r}) = e^{i\vec{k}\cdot\vec{r}}u_n(\vec{r}) \tag{9.36}$$

as a basis to describe a wave function of quasiparticles in graphene. The Bloch function with a wave vector $\vec{k}$ at a point with radius vector $\vec{r}$ has the form

$$\chi_n(\vec{k},\vec{r}) = \frac{1}{(2\pi)^{3/2}\sqrt{2N}}\sum_{\vec{R}_\iota} e^{i\vec{k}\cdot\vec{R}_\iota}\psi_{\{n\}}(\vec{r}-\vec{R}_\iota) \tag{9.37}$$

where $\psi_{\{n\}}$ is the atomic orbital with a set of quantum numbers $\{n\}$.

A wave function $\Psi(\vec{k},\vec{r})$ of an electron in graphene has the form

$$\Psi(\vec{k},\vec{r}) = \frac{1}{\sqrt{2}}\sum_n\left[c_A^n\chi_n^{(A)}(\vec{k},\vec{r}) + c_B^n\chi_n^{(B)}(\vec{k},\vec{r})\right] \tag{9.38}$$

with the normalization condition given by

$$\int|\Psi(\vec{k},\vec{r})|^2 d\vec{r} = \sum_n[(c_A^n)^2 + (c_B^n)^2] = 1$$

As a zero-order approximation $\psi_m^{(0)}(\kappa r)$, $\kappa r \equiv \vec{\kappa}\cdot\vec{r}$ for functions $\psi_{\{m\}}$, we adopt the solution of a single-electron problem for an isolated atom. As the number of electrons for C atom is even, there are no pseudopotential terms in the self-consistent Hartree–Fock equation [62]:

$$\left[h(\vec{r}) + \hat{V}^{sc}(\kappa r) - \hat{\Sigma}^{x}(\kappa r)\right]\psi_m^{(0)}(\kappa r) = \varepsilon_m^{(0)}\psi_m^{(0)}(\kappa r) \tag{9.39}$$

where $\varepsilon_m^{(0)}$ is the mth eigenvalue of the Hamilton operator for a single-electron state of the isolated atom C.

9.5 BRILLOUIN ZONE CORNER APPROXIMATION

Let us consider peculiar points in momentum space for graphene. These points correspond to the diffraction peaks, the so-called reflexes of the diffraction pattern. These are the corners K and K' of the graphene Brillouin hexagonal zone, which we designate as K_A and K_B, respectively. Their positions are given by [32]

$$\vec{K}_A = \left(\frac{2\pi}{3a}, \frac{2\pi}{3\sqrt{3}a}, 0\right), \quad \vec{K}_B = \left(\frac{2\pi}{3a}, -\frac{2\pi}{3\sqrt{3}a}, 0\right) \tag{9.40}$$

Here, $a \approx 1.44$ Å is the carbon–carbon distance.

The basis Bloch function $\chi_n^{(i)}(\vec{k}^{(i)},\vec{r})$, $i = A$, B for the description of an electron in one of the sublattices has the form

$$\chi_n^{(i)}(\vec{k}^{(i)},\vec{r}) = \frac{1}{(2\pi)^{3/2}\sqrt{N/2}}\sum_{\vec{R}_l^{(i)}}\exp\{i\vec{k}^{(i)}\cdot\vec{R}_l^{(i)}\}\psi_{\{n\}}(\vec{r}-\vec{R}_l^{(i)}) \tag{9.41}$$

Let us make a variables change

$$\vec{k}^{(i)} = \vec{K}_i - (\vec{K}_i - \vec{k}^{(i)}) \equiv \vec{K}_i - \vec{q} \tag{9.42}$$

Taking into account the change (9.42), the wave functions (9.41) in the primitive subcell of the graphene space are approximately described as

$$\chi_n^{(i)}(\vec{K}_i - \vec{q},\vec{r}) = \frac{1}{(2\pi)^{3/2}\sqrt{N/2}}$$
$$\times\sum_{\vec{R}_l^{(i)}}\exp\{i[\vec{K}_i-\vec{q}]\cdot\vec{R}_l^{(i)}\}\psi_{\{n\}}(\vec{r}-\vec{R}_l^{(i)})$$
$$= \exp\{i[\vec{K}_i-\vec{q}]\cdot\vec{r}\}\frac{1}{(2\pi)^{3/2}\sqrt{N/2}}$$
$$\times\sum_{\vec{R}_l^{(i)}}\exp\{i[\vec{K}_i-\vec{q}]\cdot[\vec{R}_l^{(i)}-\vec{r}]\}$$
$$\times\psi_{\{n\}}(\vec{r}-\vec{R}_l^{(i)}) \equiv \exp\{i[\vec{K}_i-\vec{q}]\cdot\vec{r}\}\Psi_n^{(i)}(\vec{r}) \tag{9.43}$$

9.6 SECONDARY-QUANTIZED HAMILTONIAN OF QUASI-2D GRAPHENE

Now, we construct the secondary-quantized Hamiltonian H_{qu} in this approximation. By left multiplying Equation 9.28 on the Dirac bra-vector $\langle 0,\sigma|\,\hat{\chi}_{-\sigma_B}(\vec{r})$, we find that

$$\langle 0,\sigma|\,\hat{\chi}_{-\sigma_B}(\vec{r})\left[-\sum_{j=1}^{N}\frac{Ze^2}{|\vec{r}-\vec{R}_j|}+(V_{rel}^{sc})_{AA}-E(p)\right]\hat{\chi}^{\dagger}_{-\sigma_A}(\vec{r})\,|0,-\sigma\rangle$$
$$+\langle 0,\sigma|\,\hat{\chi}_{-\sigma_B}(\vec{r})\frac{1}{2m_ec^2}[c\vec{\sigma}\cdot\vec{p}c\vec{\sigma}\cdot\vec{p}+c\vec{\sigma}\cdot\vec{p}(\Sigma_{rel}^{x})_{BA}$$
$$+(\Sigma_{rel}^{x})_{AB}(\Sigma_{rel}^{x})_{BA}]\hat{\chi}^{\dagger}_{-\sigma_A}(\vec{r})\,|0,-\sigma\rangle+\langle 0,\sigma|\,\hat{\chi}_{-\sigma_B}(\vec{r})$$
$$\times\frac{1}{2m_ec^2}(\Sigma_{rel}^{x})_{AB}\,c\vec{\sigma}\cdot\vec{p}\hat{\chi}^{\dagger}_{\sigma_B}(\vec{r})\,|0,\sigma\rangle=0 \quad (9.44)$$

Let us consider the quasi-2D model of graphene, when the radius vectors $\vec{r}$ deviate slightly from the plane of the monolayer. Therefore, the values of q are small: $q < 1$, and one can omit terms of order q^2. In this case, the use of Equation 9.43 allows the transformation of the multiplier

$$\left[\langle 0,\sigma|\,\hat{\chi}_{-\sigma_B}(\vec{r})\vec{\sigma}\cdot\vec{\nabla}\right]\left[\vec{\sigma}\cdot\vec{\nabla}\hat{\chi}^{\dagger}_{-\sigma_A}(\vec{r})\,|0,-\sigma\rangle\right]$$

in Equation 9.44 to the following form:

$$-\vec{\sigma}\cdot(\vec{q}-\vec{K}_B)\vec{\sigma}\cdot(\vec{K}_A-\vec{q})\approx(\vec{\sigma}\cdot\vec{K}_A)(\vec{\sigma}\cdot\vec{K}_B)-(\vec{\sigma}\cdot\vec{q})(\vec{\sigma}\cdot(\vec{K}_A+\vec{K}_B)) \quad (9.45)$$

Let us renormalize the energy as follows:

$$E\rightarrow\tilde{E}-\frac{\hbar^2}{2m_e}(\vec{\sigma}\cdot\vec{K}_A)(\vec{\sigma}\cdot\vec{K}_B) \quad (9.46)$$

Substitution of Equations 9.45 and 9.46 into Equation 9.44 gives the following equation:

$$\langle 0,\sigma|\,\hat{\chi}_{-\sigma_B}(\vec{r})\left[-\sum_{j=1}^{N}\frac{Ze^2}{|\vec{r}-\vec{R}_j|}+(V_{rel}^{sc})_{AA}\right]\hat{\chi}^{\dagger}_{-\sigma_A}(\vec{r})\,|0,-\sigma\rangle$$
$$+\langle 0,\sigma|\,\hat{\chi}_{-\sigma_B}(\vec{r})\frac{1}{2m_ec^2}\left[c^2\hbar^2(\vec{\sigma}\cdot\vec{q})(\vec{\sigma}\cdot(\vec{K}_A+\vec{K}_B)\right.$$
$$+c\vec{\sigma}\cdot\vec{p}(\Sigma_{rel}^{x})_{BA}+(\Sigma_{rel}^{x})_{AB}(\Sigma_{rel}^{x})_{BA}]\hat{\chi}^{\dagger}_{-\sigma_A}(\vec{r})\,|0,-\sigma\rangle$$
$$+\frac{1}{2m_ec^2}\langle 0,\sigma|\,\hat{\chi}_{-\sigma_B}(\vec{r})(\Sigma_{rel}^{x})_{AB}c\vec{\sigma}\cdot\vec{p}\hat{\chi}^{\dagger}_{\sigma_B}(\vec{r})\,|0,\sigma\rangle$$
$$=\langle 0,\sigma|\,\hat{\chi}_{-\sigma_B}(\vec{r})\tilde{E}(p)\hat{\chi}^{\dagger}_{-\sigma_A}(\vec{r})\,|0,-\sigma\rangle \quad (9.47)$$

Since

$$\hbar(\vec{\sigma}\cdot\vec{q})\hat{\chi}^{\dagger}_{-\sigma_A}(\vec{r})\,|0,-\sigma\rangle=-i\hbar(\vec{\sigma}\cdot\vec{\nabla})\hat{\chi}^{\dagger}_{-\sigma_A}(\vec{r})\,|0,-\sigma\rangle \quad (9.48)$$

Equation 9.47 can be transformed to the form

$$\langle 0,\sigma|\,\hat{\chi}_{-\sigma_B}(\vec{r})\left[-\sum_{j=1}^{N}\frac{Ze^2}{|\vec{r}-\vec{R}_j|}+(V_{rel}^{sc})_{AA}\right]\hat{\chi}^{\dagger}_{-\sigma_A}(\vec{r})\,|0,-\sigma\rangle,$$
$$+\langle 0,\sigma|\,\hat{\chi}_{-\sigma_B}(\vec{r})\frac{1}{2m_ec^2}\left[c^2\hbar(-i\hbar(\vec{\sigma}\cdot\vec{\nabla}))(\vec{\sigma}\cdot(\vec{K}_A+\vec{K}_B)\right.$$
$$+c\vec{\sigma}\cdot\vec{p}(\Sigma_{rel}^{x})_{BA}+(\Sigma_{rel}^{x})_{AB}(\Sigma_{rel}^{x})_{BA}]\hat{\chi}^{\dagger}_{-\sigma_A}(\vec{r})\,|0,-\sigma\rangle$$
$$+\frac{1}{2m_ec^2}\langle 0,\sigma|\,\hat{\chi}_{-\sigma_B}(\vec{r})(\Sigma_{rel}^{x})_{AB}c\vec{\sigma}\cdot\vec{p}\hat{\chi}^{\dagger}_{\sigma_B}(\vec{r})\,|0,\sigma\rangle$$
$$=\langle 0,\sigma|\hat{\chi}_{-\sigma_B}(\vec{r})\tilde{E}(p)\hat{\chi}^{\dagger}_{-\sigma_A}(\vec{r})\,|0,-\sigma\rangle \quad (9.49)$$

The overlap integrals for the same sublattices are much smaller than that for different sublattices. Furthermore, for the quasi-2D graphene, the first term on the left-hand side of Equation 9.49 describing the screening is also small. Therefore, we can neglect the first and the last terms on the left-hand side of Equation 9.49:

$$\langle 0,\sigma|\,\hat{\chi}_{-\sigma_B}(\vec{r})\frac{1}{2m_ec^2}\left\{c\vec{\sigma}\cdot\vec{p}\left[(\Sigma_{rel}^{x})_{BA}+c\hbar\vec{\sigma}\cdot(\vec{K}_A+\vec{K}_B)\right]\right.$$
$$\left.+(\Sigma_{rel}^{x})_{AB}(\Sigma_{rel}^{x})_{BA}\right\}\hat{\chi}^{\dagger}_{-\sigma_A}(\vec{r})\,|0,-\sigma\rangle$$
$$=\langle 0,\sigma|\,\hat{\chi}_{-\sigma_B}(\vec{r})\tilde{E}(p)\hat{\chi}^{\dagger}_{-\sigma_A}(\vec{r})\,|0,-\sigma\rangle \quad (9.50)$$

The left-hand side of Equation 9.50 represents matrix elements of the secondary-quantized Hamiltonian H_{qu} of quasi-2D graphene, in which the motion of quasiparticle excitations of the electronic subsystem is described by the equation of the form

$$\left\{\vec{\sigma}\cdot\vec{p}\left[(\Sigma_{rel}^{x})_{BA}+c\hbar\vec{\sigma}\cdot(\vec{K}_A+\vec{K}_B)\right]\right.$$
$$\left.+\frac{1}{c}(\Sigma_{rel}^{x})_{AB}(\Sigma_{rel}^{x})_{BA}\right\}\hat{\chi}^{\dagger}_{-\sigma_A}(\vec{r})\,|\,0,-\sigma\rangle$$
$$=E_{qu}(p)\hat{\chi}^{\dagger}_{-\sigma_A}(\vec{r})\,|\,0,-\sigma\rangle \quad (9.51)$$

and $E_{qu}(p)$ is defined by the expression

$$E_{qu}(p)=2m_ec\tilde{E}(p) \quad (9.52)$$

Equation 9.51 is nothing but an equation that describes the motion of the Dirac charge carrier in the quasi-2D graphene:

$$\left\{\vec{\sigma}\cdot\vec{p}\hat{v}_F^{qu}-\frac{1}{c}(i\Sigma_{rel}^{x})_{AB}(i\Sigma_{rel}^{x})_{BA}\right\}\hat{\chi}^{\dagger}_{-\sigma_A}(\vec{r})\,|0,-\sigma\rangle$$
$$=E_{qu}(p)\hat{\chi}^{\dagger}_{-\sigma_A}(\vec{r})\,|0,-\sigma\rangle \quad (9.53)$$

where operator $\hat{v}_F^{qu}$ is defined as

$$\hat{v}_F^{qu}=\left[(\Sigma_{rel}^{x})_{BA}+c\hbar\vec{\sigma}\cdot(\vec{K}_A+\vec{K}_B)\right] \quad (9.54)$$

The physical meaning of quasirelativistic corrections (9.30), entered in Equation 9.53, is in appearance of a pseudomass m_{pseudo} for the charge carriers in graphene:

$$m_- = \frac{1}{c}(i\Sigma_{rel}^x)_{AB}(i\Sigma_{rel}^x)_{BA} \tag{9.55}$$

Since quasirelativistic correction (9.30) is included with a small factor c^{-1}, then, it may be neglected and one obtains the equation of motion for a massless quasiparticle charge carrier in quasi-2D graphene:

$$\{\vec{\sigma}\cdot\vec{p}\hat{v}_F^{qu}\}\hat{\chi}_{-\sigma_A}^{\dagger}(\vec{r})\,|0,-\sigma\rangle = E_{qu}(p)\hat{\chi}_{-\sigma_A}^{\dagger}(\vec{r})\,|0,-\sigma\rangle \tag{9.56}$$

According to Equation 9.56, the massless charge carrier moves with the Fermi velocity operator $\hat{v}_F^{qu}$ (9.54).

By transforming the Fermi velocity operator $\hat{v}_F^{qu}$ (9.54) into a matrix form, one arrives at different values of Fermi velocity in different directions.

9.7 CHARGE CARRIERS ASYMMETRY

The operator of pseudomass (9.55) is not invariant in respect to transformation $A \to B$:

$$m_-\hat{\chi}_{-\sigma_A}^{\dagger}(\vec{r})\,|0,-\sigma\rangle$$
$$\neq \frac{1}{c}(i\Sigma_{rel}^x)_{BA}(i\Sigma_{rel}^x)_{AB}\hat{\chi}_{\sigma_B}^{\dagger}(\vec{r})\,|0,\sigma\rangle \stackrel{def}{=} m_+\hat{\chi}_{\sigma_B}^{\dagger}(\vec{r})\,|0,\sigma\rangle \tag{9.57}$$

where m_+ is a hole mass in graphene. Owing to the factor c^{-1}, pseudomass $m_\pm$ in Equation 9.57 is small.

Energy can be calculated based on the following equation:

$$E_{qu}(p) = \langle 0,\sigma|\,\hat{\chi}_{-\sigma_B}(\vec{r})\vec{\sigma}\cdot\vec{p}\hat{v}_F^{qu}\hat{\chi}_{-\sigma_A}^{\dagger}(\vec{r})\,|0,-\sigma\rangle$$
$$+\frac{1}{c}\langle 0,\sigma|\,\hat{\chi}_{-\sigma_B}(\vec{r})(\Sigma_{rel}^x)_{AB}(\Sigma_{rel}^x)_{BA}\hat{\chi}_{-\sigma_A}^{\dagger}(\vec{r})\,|0,-\sigma\rangle \tag{9.58}$$

From the last equation, one arrives at the energy dispersion law for graphene:

$$E_{qu}^2(p) = \langle 0,\sigma|\,\hat{\chi}_{-\sigma_B}(\vec{r})p\hat{v}_F^{qu}\hat{\chi}_{-\sigma_A}^{\dagger}(\vec{r})\,|0,-\sigma\rangle^2$$
$$+\frac{1}{c^2}\langle 0,\sigma|\,\hat{\chi}_{-\sigma_B}(\vec{r})(\Sigma_{rel}^x)_{AB}(\Sigma_{rel}^x)_{BA}\hat{\chi}_{-\sigma_A}^{\dagger}(\vec{r})\,|0,-\sigma\rangle^2 \tag{9.59}$$

Let us represent dispersion law (9.59) in a dimensionless form

$$\frac{E_-}{m_e v_F^2} = \frac{1}{v_F^2}\sqrt{\frac{v_F^2p^2}{m_e^2}+\langle 0,\sigma|\,\hat{\chi}_{-\sigma_B}(\vec{r})(\hat{v}_F^{qu})^4\hat{\chi}_{-\sigma_A}^{\dagger}(\vec{r})\,|0,-\sigma\rangle^2} \tag{9.60}$$

A similar expression can be written for holes.

Performing the series expansion of Equation 9.60, one arrives at

$$\frac{E_\mp}{m_e v_F^2} \approx \frac{\hbar}{v_F^2}\left[\frac{v_F k}{m_e} \pm \frac{1}{2}\frac{m_e\langle 0,\sigma|\,\hat{\chi}_{-\sigma_B}(\vec{r})(\hat{v}_F^{qu})^4\hat{\chi}_{-\sigma_A}^{\dagger}(\vec{r})\,|0,-\sigma\rangle^2}{\hbar^2 v_F k}\right] \tag{9.61}$$

Next, utilizing the expression for wave function (9.43), we estimate the matrix element

$$\langle 0,\sigma|\,\hat{\chi}_{-\sigma_B}(\vec{r})(\hat{v}_F^{qu})^4\hat{\chi}_{-\sigma_A}^{\dagger}(\vec{r})\,|0,-\sigma\rangle$$

in the tight-binding approximation:

$$\langle 0,\sigma|\,\hat{\chi}_{-\sigma_B}(\vec{r})(\hat{v}_F^{qu})^4\hat{\chi}_{-\sigma_A}^{\dagger}(\vec{r})\,|0,-\sigma\rangle = \frac{1}{(2\pi)^{3/2}\sqrt{N/2}}\int d^2r(\chi_n^B)^{\dagger}$$
$$\times\left(\hat{v}_F^{qu}\right)^4\sum_{\vec{R}_l^A}\exp\{i[\vec{K}_A-\vec{k}]\cdot\vec{R}_l^A\}\psi_{\{n\}}(\vec{r}-\vec{R}_l^A)$$
$$\approx\frac{N[V_{WS}/2]^{-1}}{(2\pi)^{3/2}\sqrt{N/2}}\int_{V_{WS}/2} d^2r\int_{V_{BZ}} d^2k'(\chi_n^B)^{\dagger}(\hat{v}_F^{qu})^4$$
$$\times\sum_{\vec{\delta}_l}\exp\{i[\vec{K}_A-\vec{k}+\vec{k}']\cdot\vec{\delta}_l\}\psi_{\{n\}}(\vec{\tilde{k}}-\vec{K}_A)e^{i(\vec{\tilde{k}}-\vec{K}_A)\cdot\vec{r}} \tag{9.62}$$

where $\vec{\delta}_1 = (a/2)(1,\sqrt{3})$, $\vec{\delta}_2 = (a/2)(1,\sqrt{3})$, and $\vec{\delta}_3 = -a(1,0)$ are three nearest-neighbor vectors in real lattice space whose length δ_i, $i = 1, 2, 3$ is given by

$$\delta_1 = \delta_2 = \delta_3 = a \tag{9.63}$$

V_{WS} is a volume of the Wigner–Seitz 2D-cell, and V_{BZ} is a volume of the Brillouin 2D-zone, $\vec{\tilde{k}}-\vec{K}_A = \vec{k}'$. Using the facts that $\sum_{\vec{\delta}_i} e^{i\vec{K}_A\cdot\vec{\delta}_i} = 0$ and

$$e^{i(\vec{K}_A-\vec{k}+\vec{k}')\cdot\vec{\delta}_i} \simeq (1+ik'\delta_i)[e^{i\vec{K}_A\cdot\vec{\delta}_i} - i\sin(\vec{k}\cdot\vec{\delta}_i)\cos(\vec{K}_A\cdot\vec{\delta}_i)]$$
$$\simeq i\vec{k}'\cdot\vec{\delta}_i e^{i\vec{K}_A\cdot\vec{\delta}_i} - iak + a^2kk' + e^{i\vec{K}_A\cdot\vec{\delta}_i}$$

for $k \ll 1$ and $a \to 0$, performing inverse Fourier transformation for the function $\psi_{\{n\}}(\vec{\tilde{k}}+\vec{K}_A)$ and integrating over d^2r, one derives that the following expression takes place in the vicinity of the Dirac point:

$$\langle 0,\sigma\,|\,\hat{\chi}_{-\sigma_B}(\vec{r})(\hat{v}_F^{qu})^4\hat{\chi}_{-\sigma_A}^{\dagger}(\vec{r})\,|\,0,-\sigma\rangle \approx \frac{N[V_{WS}/2]^{-1}}{(2\pi)^{3/2}\sqrt{N/2}}\delta(\vec{\tilde{k}}-\vec{K}_A)$$
$$\times\int d^2r'(\chi_n^B)^{\dagger}(\hat{v}_F^{qu})^4\sum_{l=1}^{3}\int_{-\vec{\tilde{k}}}^{\vec{\tilde{k}}} i[k'\delta_l(\cos\theta_{\vec{k}',\vec{\delta}_l})-ak]dk'\,\psi_{\{n\}}(\vec{r}') \tag{9.64}$$

Now, let us expand operators $\hat{\chi}, \hat{\chi}^{\dagger}$ entering into $(\Sigma^x_{rel})_{BA}$ (9.20), and $(\Sigma^x_{rel})_{AB}$ (9.19), respectively, up to a linear order in δ_i. We act by the operator expansions $(\Sigma^x_{rel})_{B(A)A(B)}$, which enter in $(\hat{v}_F^{qu})^4$, on appropriate wave functions $\psi_{\{n\}}$, and use an explicit form of the wave function χ^B_n, analogous to Equation 9.43. As a result, we get

$$\langle 0,\sigma \mid \hat{\chi}_{-\sigma_B}(\vec{r})(\hat{v}_F^{qu})^4\hat{\chi}^{\dagger}_{-\sigma_A}(\vec{r}) \mid 0,-\sigma\rangle \approx \frac{\delta_{\vec{k},\vec{K}_A}}{V_{WS}}$$
$$\times \sum_{\vec{R}^A_j,\vec{R}^B_j} \exp\{-i\vec{k}\cdot(\vec{R}^A_j - \vec{R}^B_j)\}$$
$$\times \left[\int d^2r' \mid \psi_n \mid^2 (\vec{r}')\right]\left[\frac{V_{WS}}{2}\right]^4 \left[\int_{V_{WS/2}} \psi^{\dagger}_n(\vec{r}-\delta_i)V(r)\psi_n(\vec{r})d\vec{r}\right]^4$$
$$\times \sum_{l,n,m,p=1}^{3} \frac{1}{(2\pi)^3}\int_{-\vec{k}}^{\vec{k}} [k'^5\delta_l^4(\cos\theta_{\vec{k}',\vec{\delta}_l})^4 - k'^4\delta_l^3(\cos\theta_{\vec{k}',\vec{\delta}_l})^3 ak]dk' \tag{9.65}$$

Since the integral of the odd function $\int_{-\vec{k}}^{\vec{k}}[k'^5\delta_l^4(\cos\theta_{\vec{k}',\vec{\delta}_l})^4]dk'$ entering into Equation 9.65 vanishes and $\int \mid \psi_n \mid^2 (\vec{r}')d^2r' = 1$, we find the desired matrix element

$$\langle 0,\sigma \mid \hat{\chi}_{-\sigma_B}(\vec{r})(\hat{v}_F^{qu})^4\hat{\chi}^{\dagger}_{-\sigma_A}(\vec{r}) \mid 0,-\sigma\rangle \approx \delta_{\vec{k},\vec{K}_A}\left[\frac{3V_{WS}}{2}\right]^4 ak(\vec{k})^5$$
$$\times \left[\int_{V_{WS}/2} \psi^{\dagger}_n(\vec{r}-\delta_i)V(r)\psi_n(\vec{r})d\vec{r}\right]^4 \frac{3a^3}{V_{WS}} \approx -3(v_F)^4 ak \tag{9.66}$$

where $v_F = (3a/2)\int_{V_{WS}/2}\psi^{\dagger}_n(\vec{r}-\delta_i)V(r)\psi_n(\vec{r})d\vec{r}$, $V_{WS} \approx a^2$ Substituting Equation 9.66 into Equation 9.61, we obtain the energy dispersion law in graphene

$$\frac{E_{\mp}}{m_e v_F^2} = \frac{\hbar}{m_e}\left(\frac{k}{v_F} \mp \frac{1}{2}3a\frac{m_e^2 v_F}{\hbar^2}\right) \tag{9.67}$$

The discrepancy between the energy dispersion laws of electrons and holes due to their different pseudomasses will lead to a discrepancy, though small, of the cyclotron mass $m^*_{\pm}$ of holes and electrons.

The addition of a small mass term leads to electron–hole asymmetry of the dependence of cyclotron mass upon charge carrier concentration. Let us estimate this asymmetry based on the concentration dependence of cyclotron mass with respect to the free-electron mass m_e. Within the semiclassical approximation [63], the cyclotron mass is defined by the formula

$$\frac{m*}{m_e} = \frac{1}{2\pi m_e}\left[\frac{\partial S(E)}{\partial E}\right]_{E=E_F} = \frac{E_F}{m_e v_F^2} \tag{9.68}$$

where the area $S(E)$ in momentum space enclosed by the orbit is given as $S(E) = \pi q^2(E) = \pi E^2/v_F^2$, where v_F is an ordinary scalar Fermi velocity in graphene: $v_p \simeq 10^6$ m/s. With this in mind, let us substitute Equation 9.67 into Equation 9.68 and find the difference of cyclotron masses of charge carriers

$$\frac{\Delta m^*_{theor}}{m_e} \sim 3\frac{\hbar}{m_e}a\frac{m_e^2 v_F}{\hbar^2} \approx 3\frac{1.44\hbar}{m_e} \tag{9.69}$$

where $\Delta m^* = m^*_+ - m^*_-$

Cyclotron mass $m^*_{\pm}$ of charge carriers in graphene has been extracted from the temperature dependence of the Shubnikov–de Haas oscillations in Reference 22. The experimental dependencies of cyclotron masses m^*_+, m^*_- of the hole and electron upon concentration n^{exp}_+ and n^{exp}_- of the charged carriers have been fitted based on a dimensionless formula for massless quasiparticles:

$$\frac{m^*_{\pm}}{m_e} = \frac{E_F^{\pm}}{v_F^2 m_e} = \frac{\hbar k_F^{\pm}}{v_F m_e} = \frac{\hbar\sqrt{\pi}}{v_F m_e}\sqrt{n^{exp}_{\pm}} \tag{9.70}$$

when accounting the relation between the surface density $n^{exp}_{\pm}$ and the Fermi momentum $k_F^{\pm}$ in the form $(k_F^{\pm})^2/\pi = n^{exp}_{\pm}$. Taking into account the small difference of the pseudomasses $m_{\pm}$, an estimation of the experimental data based on formula (9.70) gives the value Δm^*_{exp} of the asymmetry in a form

$$\Delta m^*_{exp}/m_e = \frac{\hbar\sqrt{\pi}}{v_F m_e}\left(\sqrt{n^{exp}_+} - \sqrt{n^{exp}_-}\right) = \frac{\hbar\sqrt{\pi}}{v_F m_e}\left(\sqrt{n^{exp}_- + \Delta n} - \sqrt{n^{exp}_-}\right)$$
$$\simeq \frac{\hbar\sqrt{\pi}}{v_F m_e}\frac{\Delta n}{2\sqrt{n^{exp}_-}} \simeq 3\frac{\hbar\sqrt{\pi}}{m_e} \tag{9.71}$$

where Δn is the experimental value.

The practical coincidence of experimental (9.71) and theoretical (9.69) estimates of electron–hole asymmetry demonstrates the fruitfullness of the developed approach and the valuable argument to support the assumption that charge carriers in graphene are pseudo-Dirac quasiparticles having a small but finite pseudomass.

So, the quasirelativistic correction perturbs the system and removes double degeneracy of atomic wave functions for C_A and C_B. Besides, due to the hexagonal lattice symmetry, one of these Dirac cones is a replica repeated 6 times. The replicas form a hexagonal mini-Brillouin zone around the primary Dirac cone.

In what follows, we estimate the effects of Dirac cone splitting on the primary Dirac cone and its replicas. The influence of an operator of the Fermi velocity $\hat{v}_F^{qu}$ introduced in Section 9.6 on the form of a dispersion law in graphene will also be considered.

9.8 2D GRAPHENE

Consider a model of graphene, when the radius vectors $\vec{r}$ practically do not leave the plane of the monolayer. Such a mode is applicable for a very small $\vec{q}, q \ll 1$. Therefore, for 2D graphene in the vicinity of corners K_i, one can set the z-component of vectors equal to zero $\vec{r} = (x, y, 0)$, $\vec{q} = (q_x, q_y, 0)$, $q \ll 1$. Since the vectors lie in one plane, near the corners K_i it is possible to transform expression (9.45) to the following form:

$$(\vec{\sigma}\cdot\vec{q})(\vec{\sigma}\cdot(\vec{K}_A + \vec{K}_B)) = \vec{q}\cdot(\vec{K}_A + \vec{K}_B) + i\vec{\sigma}\cdot[(\vec{K}_A + \vec{K}_B)\times\vec{q}]$$
$$= \vec{q}\cdot(\vec{K}_A + \vec{K}_B) + i\sigma_z[(\vec{K}_A + \vec{K}_B)\times\vec{q}]_z, \quad q \ll 1 \tag{9.72}$$

Accounting of Equation 9.72 and neglecting the quasirelativistic correction (9.30), it is possible to simplify Equation 9.56 in the following way:

$$\left\{\vec{\sigma}_{2D}\cdot\hat{p}\hat{v}_F + \vec{q}\cdot(\vec{K}_A + \vec{K}_B) + i\vec{\sigma}_z[(\vec{K}_A + \vec{K}_B)\times\vec{q}]_z\right\}\hat{\chi}^{\dagger}_{-\sigma_A}(\vec{r})|0,-\sigma\rangle$$
$$= E_{qu}(p)\hat{\chi}^{\dagger}_{-\sigma_A}(\vec{r})|0,-\sigma\rangle, \quad q \ll 1 \tag{9.73}$$

where $\vec{\sigma}_{2D}$ is the 2D vector of the Pauli matrices: $\vec{\sigma}_{2D} = (\sigma_x, \sigma_y)$, and the operator of 2D Fermi velocity $\hat{v}_F$ is defined by the expression

$$\hat{v}_F = (\Sigma^x_{rel})_{BA} \tag{9.74}$$

Let us demonstrate that the presence of operator $\hat{v}_F$ rather than scalar v_F, leads to a rotation of the Dirac cone with respect to replicas. To do this, we perform the following nonunitary transformation of the wave function for graphene:

$$\hat{\tilde{\chi}}^{\dagger}_{-\sigma_A}|0,-\sigma\rangle = (\Sigma^x_{rel})_{BA}\hat{\chi}^{\dagger}_{-\sigma_A}|0,-\sigma\rangle \tag{9.75}$$

After this transformation, Equation 9.73 takes a form similar to the pseudo-Dirac approximation of 2D graphene:

$$\left\{\vec{\sigma}^{AB}_{2D}\cdot\vec{p}_{BA} + \vec{q}\cdot(\vec{K}_A + \vec{K}_B)\hat{v}_F^{-1} + i\sigma_z[(\vec{K}_A + \vec{K}_B)\times\vec{q}]_z\hat{v}_F^{-1}\right\}$$
$$\times\hat{\tilde{\chi}}^{\dagger}_{-\sigma_A}(\vec{r})|0,-\sigma\rangle$$
$$= \tilde{E}_{qu}(p)\hat{\tilde{\chi}}^{\dagger}_{-\sigma_A}(\vec{r})|0,-\sigma\rangle, \quad q \ll 1 \tag{9.76}$$

where $\vec{\sigma}^{AB}_{2D} = (\Sigma^x_{rel})_{BA}\vec{\sigma}_{2D}(\Sigma^x_{rel})^{-1}_{BA}$, $\vec{p}_{BA}\hat{\tilde{\chi}}^{\dagger}_{-\sigma_A} = (\Sigma^x_{rel})_{BA}\vec{p}(\Sigma^x_{rel})^{-1}_{BA}\hat{\tilde{\chi}}^{\dagger}_{-\sigma_A}$, $\tilde{E}_{qu} = E_{qu}\hat{v}_F^{-1}$. Owing to the fact that $(\Sigma^x_{rel})_{BA} \neq (\Sigma^x_{rel})_{AB}$, the vector $\vec{p}_{BA}$ of the Dirac cone axis is somehow rotated in respect to the vector $\vec{p}_{AB}$ of its replica as qualitatively shown in Figure 9.2. We investigate this in detail a bit later.

The term with a scalar product $\vec{q}\cdot(\vec{K}_A + \vec{K}_B)$ in Equation 9.76 equally shifts the bands of quasiparticles, whereas the term with the vector product leads to a different sign of shift for bands of electrons and holes due to the presence of σ_z. The

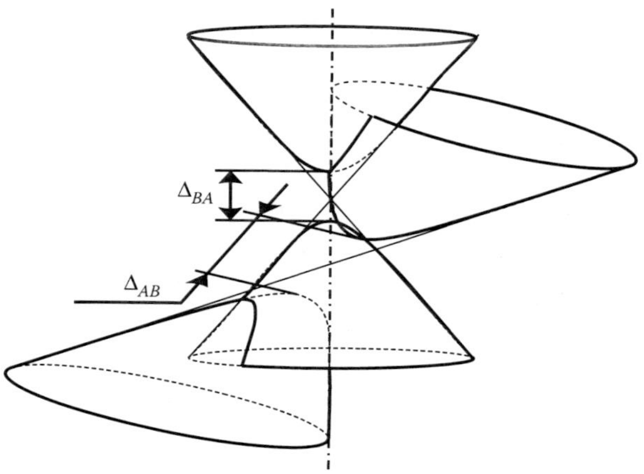

FIGURE 9.2 Qualitative representation of splitting for the double-degenerated Dirac cone.

pseudomass m_- (9.55) with corrections on these shifts leads to the appearance of an energy gap Δ_{BA} between valent and conduction Dirac zones (Figure 9.2). By analogy, the pseudomass m_+ (9.57) with corrections on these shifts leads to the appearance of an energy gap Δ_{AB}, $\Delta_{AB} > \Delta_{BA}$ between valent and conduction zones of the Dirac replica (see Figure 9.2). As one can see from Figure 9.2, the conduction zone of the replica leads to the gap between valent and conduction zones of the primary Dirac cone. Therefore, in spite of the gap, the monolayer graphene is semimetal. This theoretically predicted effect can qualitatively explain the high-intensity narrow strip connecting the valence and the conduction band of graphene in the vicinity of the Dirac point, which has been experimentally observed in ARPES spectra of the monolayer epitaxial graphene on SiC(0001) [45,46].

Besides, due to diagonality of the Pauli matrix σ_z, a mixing of waves Ψ_{K_A} and Ψ_{K_B} takes place. In real experiments investigating the electron motion in the vicinity of the corners K_A, K_b of the graphene Brillouin zone, the value of q is not infinitely small but gets finite though small values. Therefore, the effect of solutions mixing in the vicinity of points K_A, K_B is always presented though it could be small enough. According to results (9.54) and (9.56) of the previous section, the effect of the mixing is a manifestation of the graphene lattice anisotropy. To existing physical systems whose properties are strongly dependent on the presence of graphene lattice anisotropy, one can refer to graphene nanoribbons with zigzag and armchair edges, including carbon nanotubes [64–66]. It is stipulated by the fact that, due to the finite width, these systems effectively represent quasi-one-dimensional configurational ones [59,60]. The maximal mixing effect should be expected for charge carriers motion in graphene nanoribbons with zigzag and armchair edges.

Thus, the proposed graphene model can qualitatively explain an experimentally observed different electrical conductivity of zigzag and armchair graphene nanoribbons as a result of the electron–holes asymmetry of graphene.

Now, let us estimate the effects stipulated by the Fermi velocity operator $\hat{v}_F$.

9.9 APPROXIMATION OF π (p_z) ELECTRONS

We consider electrons in monolayer graphene in a commonly used assumption on the absence of mixing of states for Dirac points K_A and K_B. Then, Equation 9.76 and an equation

$$\vec{\sigma}_{2D}^{BA}\cdot\vec{p}_{AB}\hat{\tilde{\chi}}^{\dagger}_{+\sigma_B}(\vec{r})|0,-\sigma\rangle=\tilde{E}_{qu}(p)\hat{\tilde{\chi}}^{\dagger}_{+\sigma_B}(\vec{r})|0,-\sigma\rangle,\quad q\ll 1 \tag{9.77}$$

describe the delocalized π-electron on a Dirac cone and its replicas. Here, $\vec{\sigma}_{2D}^{BA}=(\Sigma^x_{rel})_{AB}\vec{\sigma}_{2D}(\Sigma^x_{rel})^{-1}_{AB}$, $\vec{p}_{AB}\hat{\tilde{\chi}}^{\dagger}_{+\sigma_B}=(\Sigma^x_{rel})_{AB}\vec{p}(\Sigma^x_{rel})^{-1}_{AB}\hat{\tilde{\chi}}^{\dagger}_{+\sigma_B}$, $\hat{\tilde{\chi}}^{\dagger}_{+\sigma_B}(\vec{r})|0,\sigma\rangle=(\Sigma^x_{rel})_{AB}\hat{\chi}^{\dagger}_{+\sigma_B}(\vec{r})|0,\sigma\rangle$.

We write down wave functions $\phi_\uparrow$ and $\phi_\downarrow$ with spin "up" and "down" appropriately in the form

$$\phi_\uparrow=\frac{1}{\sqrt{2}}\begin{pmatrix}\phi_1\\0\end{pmatrix},\quad \phi_\downarrow\frac{1}{\sqrt{2}}\begin{pmatrix}0\\\phi_2\end{pmatrix} \tag{9.78}$$

To obtain a numerical estimate but without full-scale *ab initio* simulations, we restrict ourselves by the consideration of a non-self-consistent problem. With this in mind, the bispinor wave functions of quasiparticles (in the vicinity of the Dirac cone) are presented as an (almost) free massless Dirac field (p_z-electrons):

$$\begin{pmatrix}\hat{\chi}^{\dagger}_{-\sigma_{A(B)}}(\vec{r})|0,-\sigma\rangle\\ \hat{\chi}^{\dagger}_{\sigma_{B(A)}}(\vec{r})|0,\sigma\rangle\end{pmatrix}=\frac{e^{-\imath(\vec{K}_{A(B)}-\vec{q})\cdot\vec{r}}}{\sqrt{2}}\begin{pmatrix}\exp\{-\imath\theta_{k_{A(B)}}\}\phi_1\\ \exp\{-\imath\theta_{k_{A(B)}}\}\phi_2\\ -\exp\{\imath\theta_{k_{B(A)}}\}\phi_2\\ \exp\{\imath\theta_{k_{B(A)}}\}\phi_1\end{pmatrix}$$

$$\equiv\begin{pmatrix}\chi_{-\sigma_{A(B)}}\\ \chi_{+\sigma_{B(A)}}\end{pmatrix} \tag{9.79}$$

$$\begin{pmatrix}\hat{\chi}^{\dagger}_{+\sigma_{A(B)}}(\vec{r})|0,\sigma\rangle\\ \hat{\chi}^{\dagger}_{-\sigma_{B(A)}}(\vec{r})|0,-\sigma\rangle\end{pmatrix}=\frac{e^{-\imath(\vec{K}_{A(B)}-\vec{q})\cdot\vec{r}}}{\sqrt{2}}\begin{pmatrix}-\exp\{\imath\theta_{k_{A(B)}}\}\phi_2\\ \exp\{\imath\theta_{k_{A(B)}}\}\phi_1\\ \exp\{-\imath\theta_{k_{B(A)}}\}\phi_1\\ \exp\{-\imath\theta_{k_{B(A)}}\}\phi_2\end{pmatrix}$$

$$\equiv\begin{pmatrix}\chi_{+\sigma_{A(B)}}\\ \chi_{-\sigma_{B(A)}}\end{pmatrix} \tag{9.80}$$

where

$$\phi_i=\frac{1}{(2\pi)^{3/2}\sqrt{N/2}}\sum_{\vec{R}_l^{A(B)}}\exp\{\imath[\vec{K}_{A(B)}-\vec{q}]\cdot[\vec{R}_l^{A(B)}-\vec{r}]\}\psi_{\{n_i\}}(\vec{r}-\vec{R}_l^{A(B)}) \tag{9.81}$$

This form is coherent with known scattering problem considerations [23] when all ϕ_i should be placed as unity. Then, in accordance with formulas (9.79) and (9.80), the expression for wave function (9.14) is transformed into the following:

$$\hat{\chi}^{\dagger}_{-\sigma_{i'}A}(\vec{r})\equiv\hat{\chi}^{\dagger}_{-\sigma_A}(\vec{r}),\quad \hat{\chi}^{\dagger}_{\sigma_iB}(\vec{r})\equiv\hat{\chi}^{\dagger}_{\sigma_B}(\vec{r})\quad\text{for }\forall i,i' \tag{9.82}$$

Now, it is possible to write down matrices $(\Sigma^x_{rel})_{AB}\approx\Sigma_{AB}$ and $(\Sigma^x_{rel})_{BA}\approx\Sigma_{BA}$ without self-action, for example, for Σ_{AB} as

$$\begin{aligned}(\Sigma^x_{rel})_{AB}\hat{\chi}^{\dagger}_{+\sigma_B}(\vec{r})|0,\sigma\rangle &\approx\Sigma_{AB}\,\chi_{\sigma_B}=\sum_{i=1}^{N_\nu N-1}\int d\vec{r}_i\\ &\times V(\vec{r}_i-\vec{r})[\chi_{-\sigma_A}(\vec{r}_i)\cdot\chi^*_{-\sigma_B}(\vec{r}_i)]\chi_{+\sigma_B}(\vec{r})\\ &=\frac{1}{2^{3/2}}\sum_{i=1}^{N_\nu N-1}\int d\vec{r}_i\,V(\vec{r}_i-\vec{r})\\ &\times\begin{bmatrix}e^{-\imath\theta_{kA}}\phi_1(\vec{r}_i)e^{\imath\theta_{kB}}\phi_1^*(\vec{r}_i) & e^{-\imath\theta_{kA}}\phi_1(\vec{r}_i)e^{\imath\theta_{kB}}\phi_2^*(\vec{r}_i)\\ e^{-\imath\theta_{kA}}\phi_2(\vec{r}_i)e^{\imath\theta_{kB}}\phi_1^*(\vec{r}_i) & e^{-\imath\theta_{kA}}\phi_2(\vec{r}_i)e^{\imath\theta_{kB}}\phi_2^*(\vec{r}_i)\end{bmatrix}\\ &\times\begin{bmatrix}-e^{-\imath[(\vec{K}_A-\vec{q})\cdot\vec{r}-\theta_{kB}]}\phi_2(\vec{r})\\ e^{-\imath[(\vec{K}_A-\vec{q})\cdot\vec{r}-\theta_{kB}]}\phi_1(\vec{r})\end{bmatrix}\end{aligned} \tag{9.83}$$

and a similar expression for Σ_{BA}.

Now, we use the tight-binding approximation for further simplification of Σ_{AB} and Σ_{BA}. Accounting only nearest neighbors and choosing $\psi_{\{n_2\}}$, $\psi_{\{n_1\}}$ as orbitals of the π-electron

$$\psi_{\{n_2\}}=c_1\psi_{p_z}(\vec{r}\pm\vec{\delta}_i)+c_2\psi_{p_z}(\vec{r}),\quad\sum_{i=1}^{2}c_i=1;\quad \psi_{\{n_1\}}=\psi_{p_z}(\vec{r})$$

after some algebra, we get

$$\begin{aligned}\Sigma_{AB}&=\frac{1}{\sqrt{2}(2\pi)^3}e^{-\imath(\theta_{kA}-\theta_{kA})}\sum_{i=1}^{3}\exp\{\imath[\vec{K}_A-\vec{q}]\cdot\vec{\delta}_i\}\int V(\vec{r})d\vec{r}\\ &\times\begin{pmatrix}\sqrt{2}\psi_{p_z}(\vec{r})\psi^*_{p_z,-\vec{\delta}_i}(\vec{r}) & \psi_{p_z}(\vec{r})[\psi^*_{p_z}(\vec{r})+\psi^*_{p_z,-\vec{\delta}_i}(\vec{r})]\\ \psi^*_{p_z,-\vec{\delta}_i}(\vec{r})[\psi_{p_z,\vec{\delta}_i}(\vec{r})+\psi_{p_z}(\vec{r})] & \dfrac{[\psi_{p_z,\vec{\delta}_i}(\vec{r})+\psi_{p_z}(\vec{r})][\psi^*_{p_z}(\vec{r})+\psi^*_{p_z,-\vec{\delta}_i}(\vec{r})]}{\sqrt{2}}\end{pmatrix}\end{aligned} \tag{9.84}$$

$$\begin{aligned}\Sigma_{BA}&=\frac{1}{\sqrt{2}(2\pi)^3}e^{-\imath(\theta_{kA}-\theta_{kB})}\sum_{i=1}^{3}\exp\{\imath[\vec{K}_A-\vec{q}]\cdot\vec{\delta}_i\}\int V(\vec{r})d\vec{r}\\ &\times\begin{pmatrix}\dfrac{[\psi_{p_z,\vec{\delta}_i}(\vec{r})+\psi_{p_z}(\vec{r})][\psi^*_{p_z}(\vec{r})+\psi^*_{p_z,-\vec{\delta}_i}(\vec{r})]}{\sqrt{2}} & -\psi^*_{p_z,-\vec{\delta}_i}(\vec{r})[\psi_{p_z,\vec{\delta}_i}(\vec{r})+\psi_{p_z}(\vec{r})]\\ -\psi_{p_z}(\vec{r})[\psi^*_{p_z}(\vec{r})+\psi^*_{p_z,-\vec{\delta}_i}(\vec{r})] & \sqrt{2}\psi_{p_z}(\vec{r})\psi^*_{p_z,-\vec{\delta}_i}(\vec{r})\end{pmatrix}\end{aligned} \tag{9.85}$$

Here, $c_1 = c_2 = 1/\sqrt{2}$, we choose the upper sign for π-orbital $\psi_{\{n_2\}}$ and introduce a notion $\psi_{p_z,\pm\vec{\delta}_i}(\vec{r}_{2D}) = \psi_{p_z}(\vec{r}_{2D} \pm \vec{\delta}_i)$.

Substituting a known expression for the eigenfunctions of a hydrogen-like atom, and evaluating integrals, we obtain a rather lengthy $(\vec{q})$-dependent invertible matrices; for $q = 0$, they are purely numeric and up to a common scalar prefactor read

$$\Sigma_{AB}(\vec{K}_A) = \begin{pmatrix} 0.062 & 0.54 \\ 0.045 & 0.42 \end{pmatrix} \quad (9.86)$$

$$\Sigma_{BA}(\vec{K}_A) = \begin{pmatrix} 0.42 & -0.045 \\ -0.54 & 0.062 \end{pmatrix} \quad (9.87)$$

The most interesting thing is that when using the last two expressions for Σ_{AB}, Σ_{BA}, the eigenvalue problem (9.77) precisely gives the known dispersion law $E(\vec{q}) = \pm\sqrt{q_x^2 + q_y^2}$ as for a massless Dirac fermion in graphene, that is, the problem is persistently linear in $|\vec{q}|$ variations. One can estimate the dynamical mass (9.57) (in eV) utilizing expressions (9.86) and (9.87):

$$m_\pm \sim \frac{\Sigma_{AB}\Sigma_{BA}v_F}{c} \sim 10^{-4} \quad (9.88)$$

Now, one can see that from Table 9.2, the obtained estimate is the most acceptable one for the description of existing experimental data.

TABLE 9.2
Estimations of Energy Gap E_g for Graphene

Model or Experiment	E_g (eV)	References
Spin-orbital couplings	$<10^{-4}$	[67]
Dynamical mass in Gauge field theory	$\sim 10^{-6}$	[18]
Nanostructures with photon-dressed ground state	$\sim 10^{-2}$	[68]
Spin nonpolarized *ab initio* simulation	0.85	[20]
Experiment	$<10^{-3}$	[69]
Dynamical mass in Dirac–Hartree–Fock self-consistent field approximation	$\sim 10^{-4}$	

Now, we take into account a higher order in $|\vec{q}|$ terms when evaluating Σ_{AB}, Σ_{BA}. The spectrum corresponding to Equation 9.77 deviates from the conic form, which we demonstrate by $E(\vec{q})$ surface sections for few q_y in Figure 9.3. Figure 9.3a emphasizes that in the vicinity of the Dirac point, the cone is persistent due to symmetry (the section crosses the original cone and its replicas simultaneously); at higher values of q, higher-order corrections start to contribute. When the section crosses only the original cone (Figure 9.3b), we find higher-order corrections to charge the carrier asymmetry. In Figure 9.3c, we can observe that the dispersion curve corresponded to the section crossing the original Dirac cone and one of its replicas. We can estimate that the q-distance between the K_A point and one of its reflexes in ARPES experiments

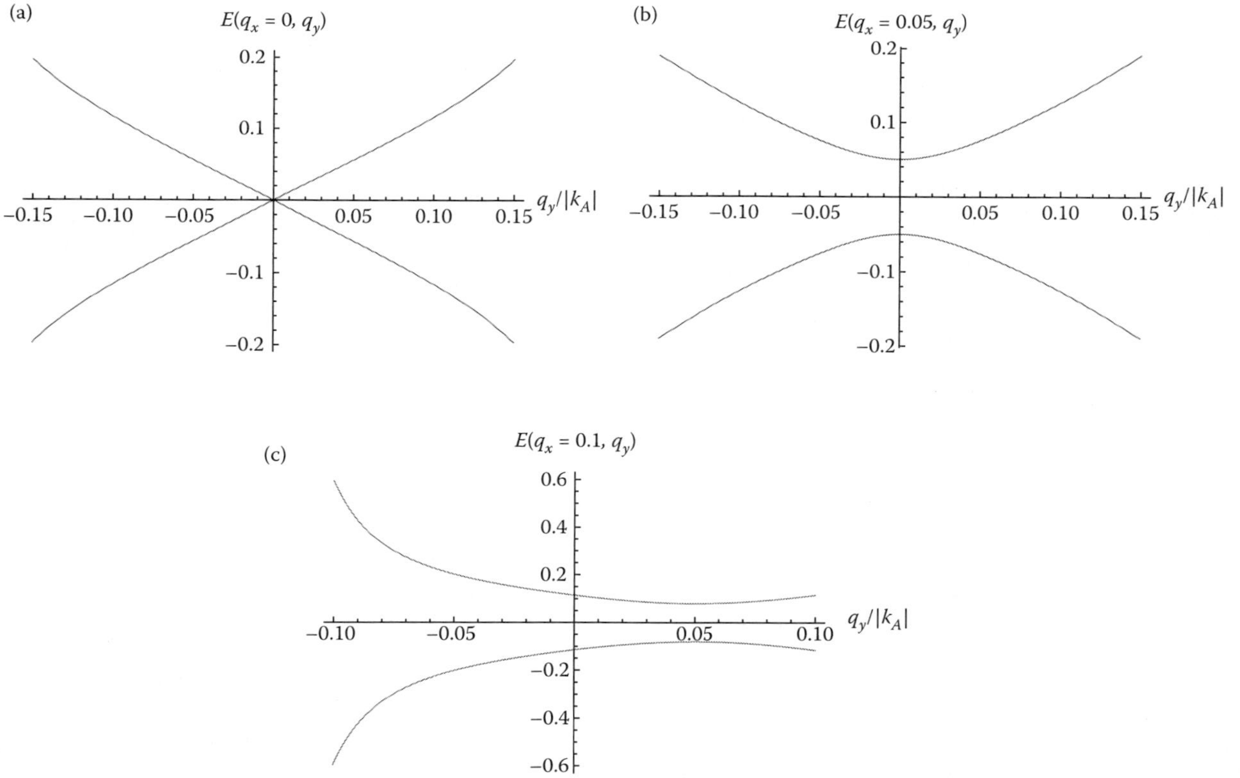

FIGURE 9.3 Dependence $E(q_x, q_y)$ for several different values of $q(x)$: (a) $-q_x/|k_A| = 0$, (b) $-q_x/|k_A| = 0.05$, and (c) $-q_x/|k_A| = 0.1$.

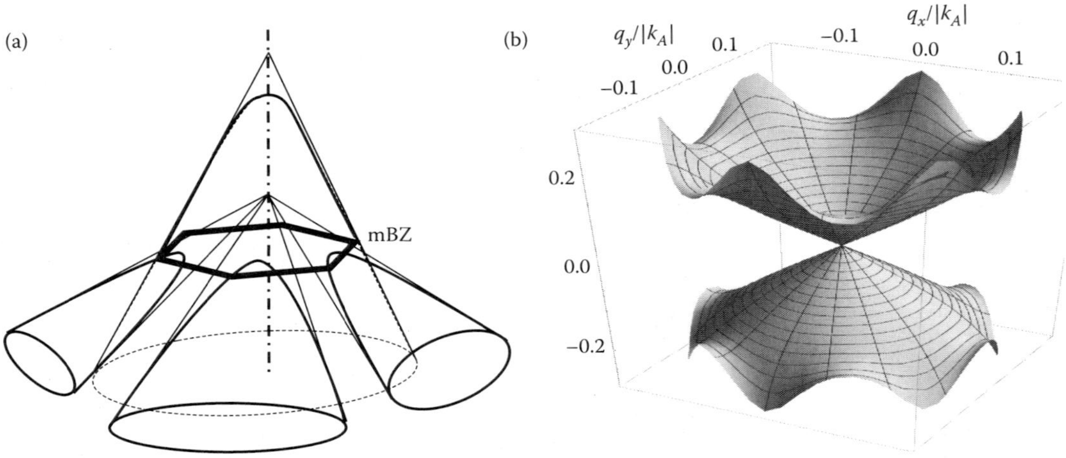

FIGURE 9.4 Mini-Brillouin zone (mBZ) formed from the Dirac cone replicas in the vicinity of the primary Dirac cone; three of six replicas are shown on a sketch (a), and simulation (b).

as a distance from the origin to q_x value corresponded to the local curve minimum near $q_x/k_F = 0.27$ (and minigap $E_{mg} \approx 0.02k_F v_F$), but of course, the approximations we made were too rough to obtain a credible result, and it should be considered as qualitative only.

As it has been shown above, the divergence of the Dirac cone and its replicas relative to each other leads to the electron–hole asymmetry. Then, the sixfold rotational symmetry of graphene near the Dirac point energy breaks. The displacement of replica points in the graphene Brillouin zone with respect to the primary Dirac cone points occurs at a distance

$$|\Delta\vec{q}_{AB}| = |\vec{q}_{AB} - \vec{q}_{BA}| \quad (9.89)$$

Since in the neighborhood of the top of the Dirac cone $q \to 0$, and (as was shown above), the cone then persists in accordance with Equation 9.89 in the graphene Brillouin zone and all points of the Dirac cone replicas shift, except their tops (see Figure 9.4).

To understand what numerical value it could correspond to, we choose $\vec{q} \propto \vec{K}_A$ and find $\Delta\vec{q}_{AB}$ based on Equation 9.89 and expressions (9.86) and (9.87) for Σ_{BA}, Σ_{AB}. Then, the rotation angle $\alpha = \arccos(\vec{q}_{AB} \cdot \vec{q}_{BA} / |\vec{q}_{AB}||\vec{q}_{BA}|) \approx \pi/2$; so, such a rotation could be large enough for some points in momentum space.

Thus, the removal of the degeneracy leads to the appearance of the hexagonal mini-Brillouin zone in the vicinity of the Dirac point, such that the corners of the cones lie on the same line, whereas the replicas are rotated with respect to the Dirac cone (Figure 9.4). The electron density distribution is unstable at the intersection of the cones of the mini-Brillouin bands with the Dirac cone of the graphene Brillouin zone. Therefore, in these places, the Dirac cone and its replicas can only quasi-cross to form energy minigaps in the ARPES band as shown in Figure 9.5. Since $\Delta_{BA} < \Delta_{AB}$, the probability of transitions for replica photoelectrons is lower than for photoelectrons near the Dirac cone. This fact is represented in Figure 9.5 by lines at the intersection of the valent zones with the ARPES experimental plane that have different intensities (shown in the figure by different thicknesses). The more intensive line (thick) represents the ARPES band for photoelectrons near the Dirac cone.

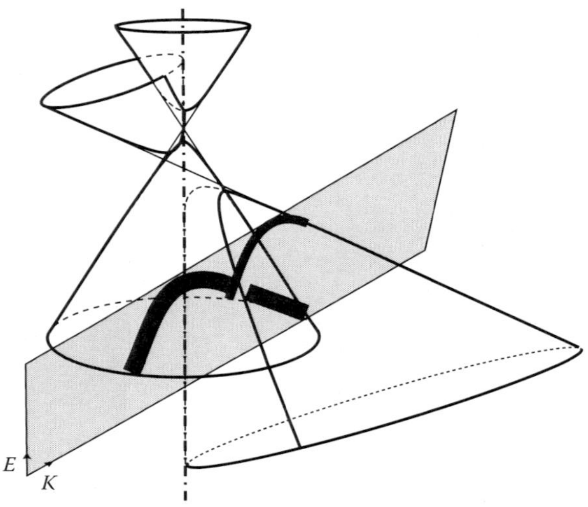

FIGURE 9.5 ARPES experimental plane (shaded) cuts the valent zone and its replica in the vicinity of the Dirac point $K_{A(B)}$. Lines at the intersection mark the ARPES band (thick line) and its replica (less thick line).

Minigaps in the vicinity of the Dirac point have been experimentally observed as minigaps (<0.2 eV) in asymmetrical photoemission intensity of ARPES spectra for weakly interacting graphene on iridium support [44].

9.10 CONCLUSION

To summarize, the application of secondary-quantized self-consistent Dirac–Hartree–Fock approach to consider electronic properties of monolayer graphene with accounting of spin-polarized states allows the coherent explanation of experimental results on energy band minigaps and charge carrier asymmetry in graphene, proposes a description of valent and conduction zone shifts, and gives a nice theoretical estimation of electron and hole masses that is in accordance with known experimental data.

REFERENCES

1. Kane, C.L. and Mele, E.J. 2005. Quantum spin Hall effect in graphene. *Phys. Rev. Lett.* 95:226801.
2. Huertas-Hernando, D., Guinea, F., and Brataas, A. 2006. Spinorbit coupling in curved graphene, fullerenes, nanotubes, and nanotube caps. *Phys. Rev. B* 74:155426.
3. Min, H. et al. 2006. Intrinsic and Rashba spinorbit interactions in graphene sheets. *Phys. Rev. B* 74:165310.
4. Han, W. et al. 2009. Electron–hole asymmetry of spin injection and transport in single-layer graphene. *Phys. Rev. Lett.* 102:137205.
5. Han, W. et al. 2012. Spin transport and relaxation in graphene. *J. Magnet. Magnet. Mat.* 324:369.
6. Pesin, D. and MacDonald, A.H. 2012. Spintronics and pseudospintronics in graphene and topological insulators. *Nat. Mat.* 11:409.
7. Ferreira, A. et al. 2014. Extrinsic spin Hall effect induced by resonant skew scattering in graphene. *Phys. Rev. Lett.* 112:066601.
8. Zhang, T. et al. 2012. Theoretical approaches to graphene and graphene-based materials. *Nano Today* 7:180.
9. Cooper, D.R. et al. 2012. Experimental review of graphene. *ISRN Condensed Matter Phys.* 2012, Article ID 501686.
10. Wallace, P.R. 1947. The band theory of graphite. *Phys. Rev.* 71:622.
11. Semenoff, G.W. 1984. Condensed-matter simulation of a three-dimensional anomaly. *Phys. Rev. Lett.* 53:2449.
12. Saito, R.G. et al. 1998. *Physical Properties of Carbon Nanotubes*. London: Imperial.
13. Reich, S. et al. 2002. Tight-binding description of graphene. *Phys. Rev. B* 66:035412.
14. Gusynin, V.P., Sharapov, S.G., and Carbotte, J.P. 2007. AC conductivity of graphene: From tight-binding model to 2 + 1-dimensional quantum electrodynamics. *Int. J. Mod. Phys. B* 21:4611.
15. Falkovsky, L.A. 2011. Optics of semiconductors with a linear electron spectrum. *Low Temp. Phys.* 37:480.
16. Falkovsky, L.A. and Varlamov, A.A. 2007. Space–time dispersion of graphene conductivity. *Eur. Phys. J. B* 56:281.
17. Falkovsky, L.A. 2008. Optical properties of graphene and IV–VI semiconductors. *Physics—Uspekhi* 51:887.
18. Jing-Rong, W. and Guo-Zhu, L. 2011. Eliashberg theory of excitonic insulating transition in graphene. *J. Phys. Cond. Matt.* 23:155602.
19. Rojas-Cuervo, A.M. and Rey-González, R.R. 2013. Asymmetric Dirac cones in monatomic hexagonal lattices. arXiv: 1304.4576v1 [cond-mat.mes-hall] April 16 2013.
20. Grushevskaya, G.V. et al. 1998. Exchange and correlation interactions and band structure of non-close-packed solids. *Phys. Solid State* 40:1802.
21. Deslippe, J. et al. 2012. BerkeleyGW: A massively parallel computer package for the calculation of the quasiparticle and optical properties of materials and nanostructures. *Comp. Phys. Comm.* 183:1269
22. Novoselov, K.S. et al. 2005. Two-dimensional gas of massles Dirac fermions in graphene. *Nature* 438:197.
23. Katsnelson, M.I. et al. 2006. Chiral tunneling and the Klein paradox in graphene. *Nat. Phys.* 2:620.
24. Katsnelson, M.I. 2007. Graphene: Carbon in two dimensions. *Mater. Today* 10:20.
25. Katsnelson, M.I. and Novoselov, K.S. 2007. Graphene: New bridge between condensed matter physics and quantum electrodynamics. *Solid State Commum.* 143:3.
26. Katsnelson, M.I. 2006. Zitterbewegung, chirality, and minimal conductivity in graphene. *Eur. Phys. J. B* 51:157.
27. Drut, J.E. and Lähde, T.A. 2009. Lattice field theory simulations of graphene. *Phys. Rev. B* 79:165425.
28. Young-Woo, S. et al. 2006. Half-metallic graphene nanoribbons. *Nat. Lett.* 444:347.
29. Miao, F. et al. 2007. Phase-coherent transport in graphene quantum billiards. *Science* 317:1530.
30. Bolotin, K.I. et al. 2008. Ultrahigh electron mobility in suspended graphene. *Solid State Comm.* 146:351.
31. Bardarson, J.H. et al. 2007. One-parameter scaling at the Dirac point in graphene. *Phys. Rev. Lett.* 99:106801.
32. Castro Neto, A.H. et al. 2009. The electronic properties of graphene. *Rev. Mod. Phys.* 81:109.
33. Geim, A.K. and Novoselov, K.S. 2007. The rise of graphene. *Nat. Mat.* 6:183.
34. Sarma, S.D. et al. 2011. Electronic transport in two-dimensional graphene. *Rev. Mod. Phys.* 83:407.
35. Hancock, Y. 2011. The 2010 Nobel Prize in physics groundbreaking experiments on graphene. *J. Phys. D* 44:473001.
36. Peres, N.M.R. 2009. The transport properties of graphene. *J. Phys. Condens. Matt.* 21:323201.
37. Zhang, Y. et al. 2005. Experimental observation of the quantum Hall effect and Berry's phase in graphene. *Nature* 438:201.
38. Deacon, R.S. et al. 2007. Cyclotron resonance study of the electron and hole velocity in graphene monolayers. Cyclotron resonance study of the electron and hole velocity in graphene monolayers. *Phys. Rev. B* 76:081406(R).
39. Jiang, Z. et al. 2007. Infrared spectroscopy of Landau levels of graphene. *Phys. Rev. Lett.* 98:197403.
40. Altshuler, B.L. and Spivak, B.Z. 1985. Change of random potential realization and conductivity of small size sample. *JETP Lett.* 42:447.
41. Rossi, E. et al. 2012. Universal conductance fluctuations in Dirac materials in the presence of long-range disorder. *Phys. Rev. Lett.* 109:096801.
42. Rahman, A. et al. 2013. Asymmetric scattering of Dirac electrons and holes in graphene. arXiv:1304.6318v1 [cond-mat.mes-hall] April 23 2013.
43. Rahman, A. et al. 2013. Direct evidence of angle-selective transmission of Dirac electrons in graphene p–n junctions. arXiv:1304.5533v1 [cond-mat.mes-hall] April 19 2013.
44. Pletikosić, I. et al. 2009. Dirac cones and minigaps for graphene on Ir(111). *Phys. Rev. Lett.* 102:056808, arXiv: 0807.2770v2 [cond-mat.mtrl-sci] February 13 2009.
45. Zhou, S.Y. et al. 2008. Origin of the energy bandgap in epitaxial graphene. *Nat. Mater.* 7:259.
46. Zhou, S.Y. et al. 2007. Substrate-induced band gap opening in epitaxial graphene. *Nat. Mat.* 6:770, preprint at arXiv: 0709.1706v2.
47. Gonze, X. et al. 2002. First-principles computation of material properties: The ABINIT software project. *Comput. Mat. Sci.* 25:478.
48. Blaha, P. et al. 1990. Full-potential, linearized augmented plane wave programs for crystalline systems. *Comput. Phys. Commun.* 59:399; see also WIEN2k package home site at http://www.wien2k.at/index.html.
49. Kresse, G. and Hafner, J. 1993. *Ab initio* molecular dynamics for liquid metals. *Phys. Rev. B* 47:558; see also *Vienna Ab initio Simulation Package* at http://www.vasp.at/index.php.

50. Methfessel, M., Rodriguez, C.O., and Andersen, O.K. 1989. Fast full-potential calculations with a converged basis of atom-centered linear muffin-tin orbitals: Structural and dynamic properties of silicon. *Phys. Rev. B* 40:2009(R).
51. Eschrig, H., Richter, M., and Opahle, I. 2004. Relativistic solid state calculations, in *Relativistic Electronic Structure Theory*, Part 2. *Applications* (ed. P. Schwerdtfeger, Elsevier), *Theor. Comput. Chem.* 13:723.
52. Opahle, I. et al. 2008. *Relativistic Code*, at http://www.fplo.de/workshop/ws2008/rfplo08.pdf
53. Andersen, O.K. 1973. Simple approach to the band-structure problem. *Solid State Commun.* 13:133.
54. Skriver, H.L. 1984. *The LMTO Method*. New York: Springer.
55. Reich, S. et al. 2002. Electronic band structure of isolated and bundled carbon nanotubes. *Phys. Rev. B* 65:155411.
56. Krylov, G.G., Krylova, H.V., and Belov, M.A. 2011. Electron transport in low-dimensional systems: Localization effects. In *Dynamical Phenomena in Complex Systems,* ed. A.V. Mokshin et al. 161–180. Kazan: Publishing MOiN RT.
57. Grushevskaya, G.V. and Krylov, G.G. 2013. Charge carriers asymmetry and energy minigaps in monolayer graphene: Dirac Hartree Fock approach. *Int. J. Nonlin. Phen. Compl. Sys.* 16:189.
58. Kelly, B.T. 1981. *Physics of Graphite*. London: Applied Science Publishing.
59. Krylova, H. and Hursky, L. 2013. *Spin Polarization in Strong-Correlated Nanosystems.* Germany: LAP LAMBERT Academic Publishing, AV Akademikerverlag GmbH & Co. (in Russian).
60. Krylova, H.V. 2008. *Electric Charge Transport and Nonlinear Polarization of Periodically Packed Structures in Strong Electromagnetic Fields.* Minsk: Publishing Center of BSU. (in Russian).
61. Grushevskaya, G.V. et al. 2007. Effects of spatial distribution of the electron density functional in crystals. *Bull. Lebedev Phys. Inst.* 34:127.
62. Fock, V.A. 1931. *Foundations of Quantum Mechamos.* Moscow: Science Publisher, 1976 (in Russian).
63. Ashcroft, N.W. and Mermin, N.D. 1976. *Solid State Physics.* Philadelphia: Sounders College.
64. Brey, L. and Fertig, H. 2006. Edge states and the quantized Hall effect in graphene. *Phys. Rev. B* 73:195408; Brey, L. and Fertig, H. 2006. Electronic states of graphene nanoribbons studied with the Dirac equation. *Phys. Rev. B* 73:235411.
65. Nakada, K. et al. 1996. Edge state in graphene ribbons: Nanometer size effect and edge shape dependence. *Phys. Rev. B* 54:17954.
66. Maksimenko, S.A. and Slepyan, G.Y. 2003. Electromagnetics of carbon nanotubes. In *Introduction to Complex Medium for Optics and Electromagnetics*, eds. W.S. Wieglhofer and A. Lakhtakia, 507–545. Bellingham: SPIE Press.
67. Gmitra, M. et al. 2009. Band-structure topologies of graphene: Spinorbit coupling effects from first principles. *Phys. Rev. B* 80:235431.
68. Kibis, O.V. 2011. Dissipationless electron transport in photon-dressed nanostructures. *Phys. Rev. Lett.* 107:106802.
69. Elias, D.C. et al. 2012. Dirac cones reshaped by interaction effects in suspended graphene. *Nat. Phys.* 8:172.

10 Graphene and Its Hybrids as Electrode Materials for High-Performance Lithium-Ion Batteries

Guangmin Zhou, Feng Li, and Hui-Ming Cheng

CONTENTS

ABSTRACT

In this chapter, we will present an overview on electrochemical properties of graphene by summarizing the recent scientific researches and developments of graphene and its hybrids for lithium ion batteries. Graphene-based hybrids exhibited exciting properties which exceed the intrinsic properties of each component and/or introduced some new properties and functions such as improved electrical conductivity, larger capacity, higher rate capability, and excellent cycling stability. The beneficial role of graphene in the hybrids is due to its unique structures and properties such as high surface area and electrical conductivity, which can be used as an ideal support or construct a 3D conductive porous network for achieving excellent properties of the hybrids. The proposed interaction mechanisms for the enhanced electrochemical performance, such as synergistic effects, are highlighted. Finally, the challenges and perspective on future research directions in this field are also discussed.

10.1 INTRODUCTION

Lithium-ion batteries (LIBs), as environmentally friendly energy-storage devices, are widely used for portable devices and are the future option for electric/hybrid vehicles.[1–3] There are growing demands for next-generation LIBs with high energy/power density and long life to meet the challenges of large-scale electric energy storage. The current commercialized electrode materials suffer from a limited theoretical capacity that cannot meet the requirements of high-energy density, power density, and longer cycling life.[4] Therefore, efforts have been made for developing new materials to obtain better performance. Graphene is a promising material in the electrochemical energy-storage field due to its unique properties including a large surface area, good chemical and thermal stability, wide potential windows, rich surface chemistry, high electrical conductivity, and flexible structure (Table 10.1).[5] Low-cost, large-scale production with various processing methods promotes the available utilization of graphene and its hybrids as high-performance electrode materials in energy storage.[6] The easy functionalization of graphene endows graphene-based hybrids with many unique properties in the electrochemical process through interaction. In this chapter, we will present an overview on electrochemical properties of graphene by summarizing recent scientific research and development of graphene and its hybrids for LIBs. The effects of nanostructured graphene-based anode/cathode materials and graphene as a conductive additive on improving the capacity, rate capability, and cycling stability of the LIBs are presented. The proposed interaction mechanisms for enhanced electrochemical performance, such as synergistic effects, are highlighted. Finally, the challenges and perspectives on future research directions in this field are also discussed.

10.2 GRAPHENE AS ELECTRODE MATERIALS IN LIBs

Graphene has many merits compared with graphite in LIBs, such as a high specific surface area with more reaction-active sites for lithium storage. Another characterization of graphene is the open-porous systems that exhibit a great advantage in

TABLE 10.1
Properties of Graphene

Electrical Conductivity (S cm^{-1})	Thermal Conductivity (W m^{-1} K^{-1})	Specific Surface Area (m^2 g^{-1})	Strength (GPa)	Young's Modulus (TPa)	Electron Mobility (cm^2 V^{-1} s^{-1})
2000	4840~5300	2620	130	~1.0	~10^6

fast ion transport, enabling the high rate capability,[7,8] which is a bottleneck in graphite with its long bulk diffusion distance. Therefore, graphene has been proposed as a candidate to replace graphite-based anodes. The lithium storage performance of graphene was first reported by Honma and coworkers[9] that the specific capacity could reach 540 mAh g^{-1}, much larger than that of graphite. It was also demonstrated that the lithium storage properties are affected by the layer spacing between graphene. By controlling the layer spacing between the graphene nanosheets (GNSs), higher capacity was obtained by embedding carbon nanotubes (CNTs, 730 mAh g^{-1}) or fullerene macromolecules (784 mAh g^{-1}) into graphene layers to increase the interlayer distance (Figure 10.1a), which may provide additional sites for the accommodation of lithium ions.[9] Similar ideas were reported by growing CNTs[10] or carbon nanofibers[11] on GNSs to prevent their restacking and provide extra space for lithium ions storage. Higher capacity means different lithium storage mechanisms and Wang et al.[12] claimed that lithium ions could not only adsorb on both the surfaces of graphene (forming Li_2C_6 structure), but could also be stored on the edges and covalent sites. A large number of functional groups and the micropores and/or defects in reduced graphene oxide (RGO) were demonstrated to contribute toward a higher capacity.[13] High-quality graphene with a large specific surface area shows an initial capacity as high as 1264 mAh g^{-1} at 100 mA g^{-1}, which is attributed to the few layers of graphene with a curled morphology and disordered structure, many edge-type sites or nanopores, the reaction of lithium ions with the residual H, and a broad electrochemical window (0.01–3.5 V).[14] Other parameters such as the degree of disorder (intensity ratio of the Raman D band to the G band) prepared by different reduction methods,[15] thickness, and crystallinity of graphene tuned from the graphite oxide with different degrees of oxidation are also the key structural factors influencing lithium storage properties[16] (Figure 10.1b).

Heteroatom doping and electrode structure are important factors affecting the energy-storage capacity of LIBs. Wu et al.[17] demonstrated that nitrogen- or boron-doped graphene can be used as a promising anode for high-energy LIBs under high rate charge and discharge conditions. The doped graphene can be charged/discharged in a very short time down to several tens of seconds with excellent long-term cyclability. The unique two-dimensional structure, disordered surface morphology, heteroatomic defects, and better electrode/electrolyte wettability of the doped graphene are beneficial to rapid surface Li^+ absorption and diffusion, contributing to superior performance.[17] Yin et al. reported a scalable self-assembly strategy to create hierarchical structures composed of functionalized graphene to work as anodes of LIBs. Owing to high-conductivity, three-dimensional (3D) porous structures with an interconnected graphene network, the material exhibits an optimized ion transport path and a

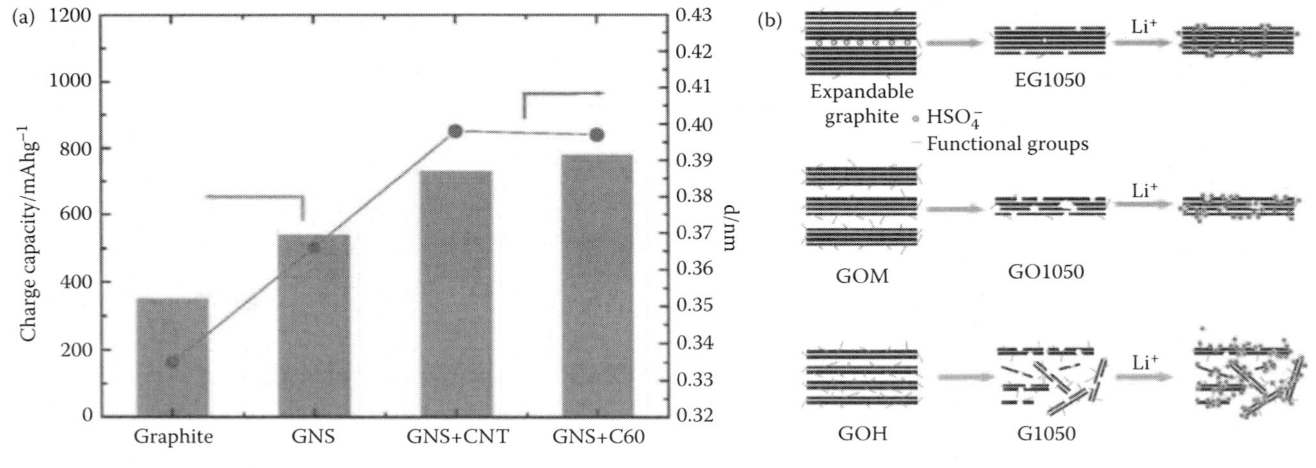

FIGURE 10.1 (a) Relationship between the d-spacing and the charge capacity of GNS families and graphite. (Reprinted with permission from Yoo, E. et al., 2008. Large reversible Li storage of graphene nanosheet families for use in rechargeable lithium ion batteries. *Nano Lett.* 8:2277–2282. Copyright 2008, American Chemical Society.) (b) An illustration of the lithium storage mechanisms of various GNSs. (Xiang, H. F.; Li, Z. D.; Xie, K.; Jiang, J. Z.; Chen, J. J.; Lian, P. C.; Wu, J. S.; Yu, Y.; Wang, H. H. 2012. Graphene sheets as anode materials for Li-ion batteries: Preparation, structure, electrochemical properties and mechanism for lithium storage. *RSC Adv.* 2:6792–6799. Reproduced by permission of The Royal Society of Chemistry.)

TABLE 10.2
Advantages and Disadvantages of Graphene in LIBs

Advantages	Superior Electrical Conductivity	Large Surface Area	Abundant Surface Functional Groups	Structural Flexibility	Ultrathin Thickness	Broad Electrochemical Window
Disadvantages	Serious agglomeration	Restacking	Large irreversible capacity	Low initial coulombic efficiency	No obvious voltage plateau	Large voltage hysteresis

high reversible capacity of up to 1600 mAh g^{-1} at a current density of 50 mA g^{-1}.[18] Yang et al. constructed an expanded graphene nanostructure with fast Li^+ transport channels, realizing a high capacity and excellent rate performance as an anode material for LIBs. The expanded structure is obtained by employing linear polymers as the spacers in the stacking process of GNSs.[19] Recently, Koratkar and coworkers[20] reported the photoflash and laser-reduced graphene paper as high rate anodes for LIBs. The instantaneous heating and deoxygenation reaction with rapid outgassing create microscale pores, cracks, and intersheet voids in graphene paper, which enable the access of lithium ions to the underlying sheets of graphene and facilitates efficient intercalation kinetics at ultrafast charge/discharge rates of >100°C.[20]

Good lithium storage properties for graphene as electrode materials in LIBs have been achieved and different explanations have been proposed,[21–24] while detailed lithium storage sites and mechanisms are still not clear and are worthy of further investigation and elucidation. Besides, several challenges still remain that impede graphene as an anode material in LIBs. It suffers from a large irreversible capacity, low initial coulombic efficiency, and fast capacity fading, which are mainly due to the restacking of graphene and side reactions between graphene and electrolytes arising from the functional groups and defects. Another disadvantage of graphene is that there is no obvious voltage plateau to provide stable potential outputs, and a large hysteresis (strong polarization) existing between the charge/discharge curves, will be the major drawback for its practical use in commercial LIBs. The advantages and disadvantages of graphene in LIBs are summarized in Table 10.2.

10.3 GRAPHENE-BASED ANODE MATERIALS IN LIBs

In view of the above-mentioned issues remaining for graphene as the sole component in LIB anodes, graphene is considered as an ideal building block in composite materials combined with a variety of semiconductors, metals, and inorganic compounds, such as Si, Sn, metal oxides, and $Li_4Ti_5O_{12}$, to produce exceptional performance in LIB applications. This is because graphene layers are prone to agglomerate and restack during the reduction and drying process due to strong π–π stacking and hydrophobic interactions.[25] Consequently, many unique properties of graphene are unavailable and the electrochemical performance of graphene in LIBs is significantly compromised. The agglomeration and restacking of graphene can be prevented by sandwiching other active electrode materials between it, yielding an increased active surface area and porous structure to achieve high electrochemical activity. Also, graphene as a support matrix can promote uniform dispersion of electroactive materials with a controlled morphology. It facilitates electron transport and lithium-ion diffusion and acts as a structural buffer to alleviate a large volume change of high-capacity electrode materials. Thus, graphene-based anodes are expected to exhibit a high performance in LIBs.

10.4 GRAPHENE–Si/Sn HYBRIDS IN LIBs

The elements Si and Sn are considered as next-generation LIB anode materials due to their low discharge potential and high specific capacity of 4200 and 994 mAh g^{-1} that are much higher than the theoretical value of graphite.[26] However, their application in practical LIBs is significantly hindered by the poor cyclic performance arising from a huge volume expansion and low electronic conductivity.

Chou et al.[27] synthesized the Si/graphene composite by simply mixing a commercially available nanosize Si and graphene. The composite maintains a capacity of 1168 mAh g^{-1} after 30 cycles with an average coulombic efficiency of 93%, which is obviously better than a pure Si nanoparticle with fast capacity fading caused by pulverization. The improved cycling stability is attributed to good electrical contact between graphene and Si nanoparticles, and the high surface area and porous structure of graphene facilitate the penetration of the electrolyte and accommodate the large volume changes. Self-supported silicon nanoparticles/graphene paper composites were prepared by the dispersing Si nanoparticles between graphene oxide sheets through filtration followed by thermal reduction. The composites exhibit high lithium storage capacities of >2200 mAh g^{-1} after 50 cycles and >1500 mAh g^{-1} after 200 cycles with a decrease rate of 0.5% per cycle.[28] It is suggested that the reconstitution of graphene to form a 3D network can provide better electrical connection with Si nanoparticles and serve as a mechanically strong framework to sandwich and trap the Si nanoparticles. A similar idea was demonstrated by Wang et al.[29] using a flexible, free-standing, paper-like graphene–silicon composite as a high-performance anode material. The Si nanoparticles are encapsulated in graphene and show a much higher discharge capacity after 100 cycles (708 mAh g^{-1}) than that of pure graphene (304 mAh g^{-1}). On the basis of graphene paper, Kung and coworkers[30] prepared an in-plane carbon vacancy-enabled high-power Si/graphene composite anode *via* a mild

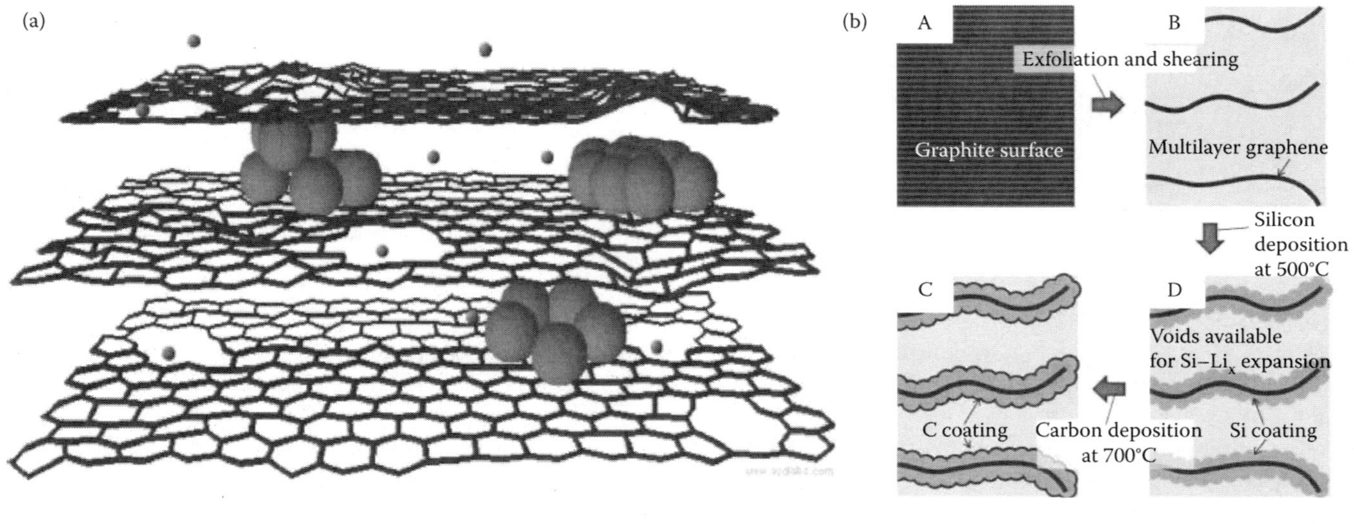

FIGURE 10.2 (a) Schematic drawing of a section of a composite electrode material constructed with a graphene scaffold with in-plane carbon vacancy defects. (Zhao, X. et al.: In-plane vacancy-enabled high-power Si–graphene composite electrode for lithium-ion batteries. *Adv. Energy Mater.* 2011. 1. 1079–1084. Copyright Wiley-VCH Verlag GmbH & Co. KGaA. Reproduced with permission.) (b) Schematic of C–Si–graphene composite formation: (A) natural graphite is transformed into (B) graphene and then (C) coated by Si nanoparticles and (D) a thin carbon layer. (Evanoff, K.; Magasinski, A.; Yang, J. B.; Yushin, G.: Nanosilicon-coated graphene granules as anodes for Li-ion batteries. *Adv. Energy Mater.* 2011. 1. 495–498. Copyright Wiley-VCH Verlag GmbH & Co. KGaA. Reproduced with permission.)

acid etching. The in-plane carbon vacancies enable facile ion transport throughout the electrode, and the good flexibility of graphene accommodates large volume variations during cycling (Figure 10.2a). As a result, this composite electrode exhibited a high reversible capacity of approximately 3200 mAh g^{-1} at 1 A g^{-1} and 83% of its theoretical capacity was maintained after 150 cycles. Even tested at 8 A g^{-1}, a rate equivalent to full discharge in 8 min, a reversible capacity of around 1100 mAh g^{-1} was still achieved. To fabricate an Si/graphene composite with good electronic conduction and structural integrity, Si nanoparticles are covalently bound to the surface of graphene *via* aromatic linkers through diazonium chemistry.[31] The composite delivered an initial capacity of 1079 mAh g^{-1} and retained 828 mAh g^{-1} after the 50th cycle at a current density of 300 mA g^{-1}. Nanostructure design is another important strategy to improve the performance of Si/graphene composites. Yushin et al. demonstrated uniform Si and carbon coatings on graphene *via* vapor decomposition routes (Figure 10.2b). The composite exhibited a specific capacity of 2000 mAh g^{-1} at the current density of 140 mA g^{-1} and excellent stability over 150 cycles with an average coulombic efficiency in excess of 99%.[32] Other engineered structures such as a lily-like graphene-wrapped nano-Si composite[33] and 3D porous architecture of Si/graphene nanocomposite[34] synthesized *via* a spray-drying process, silicon nanoparticles inserted in GNSs *via* freeze-drying and thermal reduction,[35] silicon nanoparticles encapsulated in graphene through electrostatic attraction,[36] or capillary-driven assembly route[37] show promising results in constructing high-performance LIB anode materials.

Sn is another high-capacity anode material in LIBs, which can reversibly form alloys with lithium, while it suffers from a severe volume change as that of Si anode during cycling. A 3D composite of nanosize Sn particles and graphene was prepared through $NaBH_4$ reduction of graphene oxide to graphene and Sn^{2+} to Sn in an aqueous solution.[38] The graphene constructs voids to efficiently buffer the volume change of Sn, and Sn nanoparticles act as spacers to effectively separate graphene; thus, the composite electrode shows a reversible capacity of 795 mAh g^{-1} in the second cycle and 508 mAh g^{-1} in the 100th cycle, which is much better compared to bare graphene and an Sn electrode.[38] Ji et al.[39] prepared a 3D multilayered nanostructure by embedding Sn nanopillars between graphene using a film deposition and annealing process. The composite can be directly used as an anode material without adding any polymer binder and conductive additives, which shows a reversible capacity of 734 mAh g^{-1} at 50 mA g^{-1} and high rate performance of 408 mAh g^{-1} at 5 A g^{-1}. The Sn nanopillars prevent the agglomeration of graphene, improve the penetration of Li ions, and provide space to accommodate volume change during Li insertion/extraction. A hierarchical Sn@CNT nanostructure rooted in graphene was prepared by Wang and coworkers,[40] which exhibits a large reversible capacity of 982 mAh g^{-1} at 100 mA g^{-1} after 100 cycles and high rate capability of 594 mAh g^{-1} at 5 A g^{-1}. The excellent electrochemical properties were attributed to the efficient prevention of graphene agglomeration by Sn@CNT decoration and the efficient buffering of Sn by CNT shell protection and graphene support. A similar nanostructure of Sn core/carbon-sheath coaxial nanocable directly rooted onto graphene surface was reported by Luo et al.[41] (Figure 10.3a), which also demonstrates excellent lithium storage performance compared to the control sample of Sn/graphene composite. Recently, graphene-confined Sn nanosheets through glucose-assisted chemical protocol were also developed by this group.[42] Glucose is an important precursor that not only

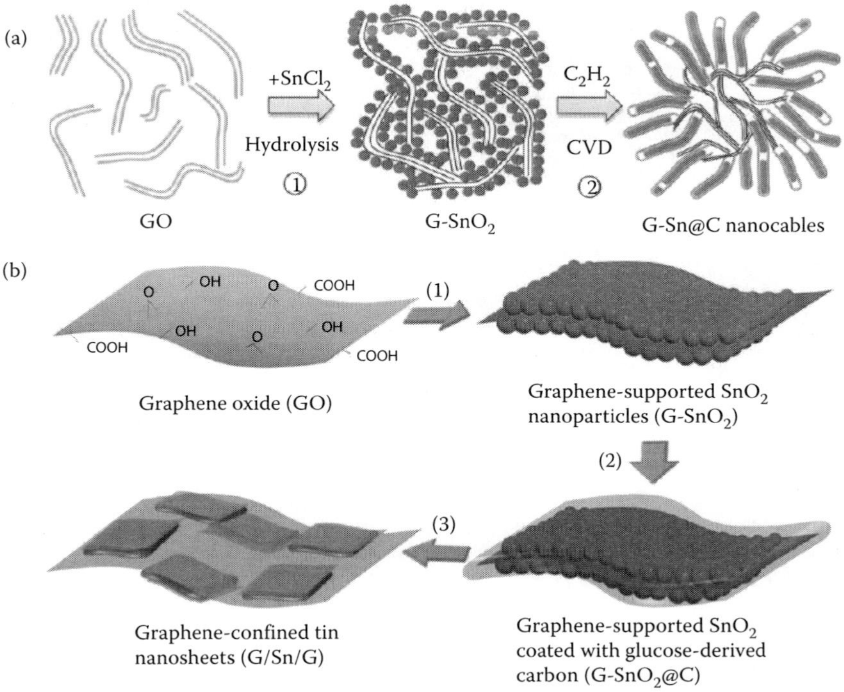

FIGURE 10.3 (a) Schematic of procedures for the synthesis of RGO-supported Sn@C nanocables: ① SnO_2 nanoparticle-decorated RGO (RGO–SnO_2) hybrids are first synthesized by a concurrent hydrolysis and reduction process; ② RGO–Sn@C nanocables are grown from the RGO–SnO_2 hybrids by CVD. (Luo, B. et al.: Reduced graphene oxide-mediated growth of uniform tin-core/carbon-sheath coaxial nanocables with enhanced lithium ion storage properties. *Adv. Mater.* 2012. 24. 1405–1409. Copyright Wiley-VCH Verlag GmbH & Co. KGaA. Reproduced with permission.) (b) Procedure for the formation of graphene-confined tin nanosheets: (1) SnO_2 nanoparticles were first decorated on GO nanosheets *via* a hydrolysis process; (2) G-SnO_2 was then coated with glucose-derived carbon through a hydrothermal method followed by a precarbonization process; and (3) G-SnO_2@C was transformed into G/Sn/G nanosheets through a rapid heat treatment. (Luo, B. et al.: Graphene-confined Sn nanosheets with enhanced lithium storage capability. *Adv. Mater.* 2012. 24. 3538–3543. Copyright Wiley-VCH Verlag GmbH & Co. KGaA. Reproduced with permission.)

acts as a solid carbon source, but also directs the expansion of liquid tin on the surface of graphene. The sheet-to-sheet binding between graphene and Sn efficiently accommodates the volume change of tin during the alloying–dealloying reactions and thus prevents pulverization of the electrode (Figure 10.3b). As a result, a high reversible capacity and good cycling stability were achieved.[42]

10.5 GRAPHENE–METAL OXIDE HYBRIDS IN LIBs

Apart from Si and Sn, metal oxides that store lithium through a conversion reaction, have attracted great interest due to their higher capacity that is 2 times larger compared with graphite.[43] However, their poor cyclic performance caused by large volume changes leads to the loss of electrical connection of the active material from current collectors. Graphene can be used as a good conductive support to disperse metal oxides and suppress their volume change and agglomeration, while metal oxides can prevent the restacking of graphene. To integrate the advantages of graphene and metal oxides, various metal oxide/graphene composites have been attempted such as Fe_3O_4,[44–50] Co_3O_4,[51–56] SnO_2,[57–66] Mn_3O_4,[67–71] NiO,[8,72–76] Fe_2O_3,[77–80] etc.

Zhou et al. reported a 3D graphene-wrapped Fe_3O_4 composite as an anode material for high-performance LIBs.[44] The well-organized flexible-interleaved composite of graphene decorated with Fe_3O_4 particles was synthesized through *in situ* reduction of iron hydroxide between graphene. It is suggested that the role of graphene to flexibly wrap Fe_3O_4 particles can effectively accommodate the strain and stress of volume change during cycling (Figure 10.4a). Therefore, the graphene-wrapped Fe_3O_4 composite shows a high reversible specific capacity of 1026 mAh g^{-1} after 30 cycles, which is much higher than that of commercial Fe_3O_4 (475 mAh g^{-1}) and bare Fe_2O_3 (359 mAh g^{-1}) with a rapid capacity fading. The graphene/Fe_3O_4 composite also exhibits a much better rate performance than commercial Fe_3O_4 and bare Fe_2O_3 particles. In particular, when the rate reaches a value as high as 1750 mA g^{-1}, the capacity of the composite is still 520 mAh g^{-1}, 53% of the initial capacity, while the capacities of Fe_3O_4 and Fe_2O_3 electrodes drop dramatically to 10% and 3% of the initial capacity at such a high rate (Figure 10.4b).[44] Microwave irradiation was used to synthesize Fe_3O_4/graphene anode materials and showed a high initial discharge capacity of 1320 mAh g^{-1} and retains 650 mAh g^{-1} after 50 cycles. A high rate performance of 350 mAh g^{-1} at 5°C that is equal to discharge/charge within 12 min was also demonstrated.[45] Graphene-encapsulated

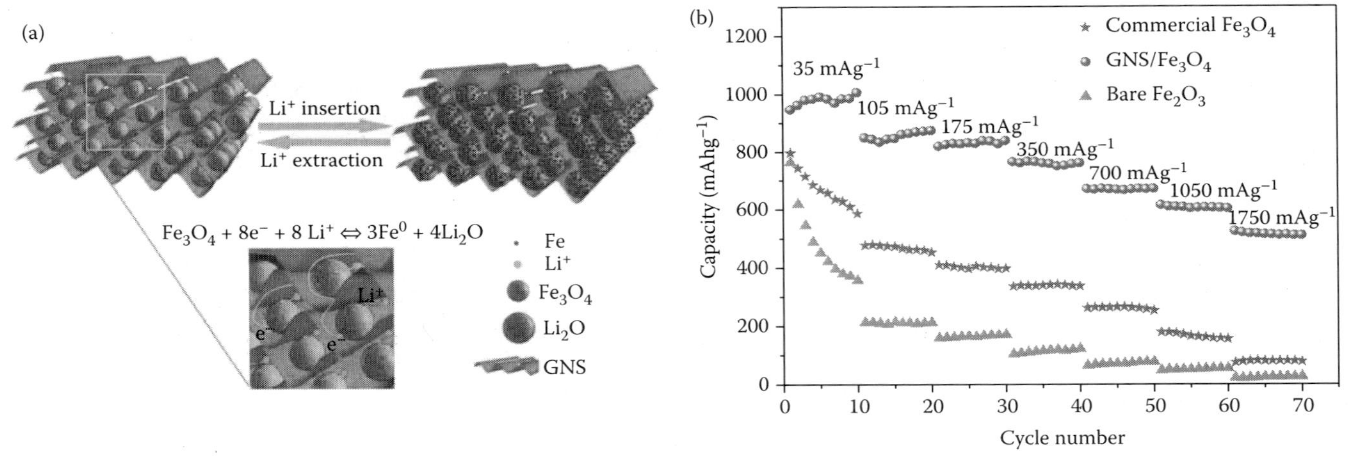

FIGURE 10.4 (a) Schematic of a flexible interleaved structure consisting of graphene and Fe_3O_4 particles. (b) Rate performance of the commercial Fe_3O_4 particles, graphene/Fe_3O_4 composite, and bare Fe_2O_3 particles at different current densities. (Reprinted with permission from Zhou, G. M. et al., 2010. Graphene-wrapped Fe_3O_4 anode material with improved reversible capacity and cyclic stability for lithium ion batteries. *Chem. Mater.* 22:5306–5313. Copyright 2010 American Chemical Society.)

Fe_3O_4 nanoparticles[49] and hollow Fe_3O_4 nanoparticle aggregates[48] were also fabricated and used as anode materials in LIBs, which exhibited good electrochemical performance due to graphene encapsulation and its porous structure.

In another study, Wu et al.[51] reported a chemical *in situ* deposition strategy to synthesize Co_3O_4 nanoparticles anchored on graphene as an anode material for high-performance LIBs. The Co_3O_4 nanoparticles obtained are 10–30 nm in size and homogeneously dispersed on graphene and function as spacers to keep the graphene separated (Figure 10.5a and b). This graphene/Co_3O_4 composite displays a higher capacity, better cyclic stability, and rate performance than Co_3O_4 and graphene, highlighting the importance of anchoring nanoparticles on graphene for a better use of Co_3O_4 and graphene in LIB applications.[51] Graphene can be functionalized to be either positively or negatively charged. Thus, it can form a composite with other charged components. On the basis of these principles, Yang et al.[52] successfully prepared graphene-encapsulated Co_3O_4 by electrostatic coassembly between negatively charged graphene oxide and positively charged Co_3O_4 nanoparticles, followed by chemical reduction (Figure 10.5c and d). The resulting composites exhibit a high reversible capacity of 1100 mAh g^{-1} in the initial cycle, and over 1000 mAh g^{-1} after 130 cycles.[52] Other attempts include an *in situ* reduction process to fabricate graphene/Co_3O_4 hybrid material,[53] microwave-assisted synthesis of a Co_3O_4/graphene sheet-on-sheet nanostructure,[54] and a one-spot solvothermal method to synthesize Co_3O_4 nanorods/graphene nanocomposites,[55] all of which show much better lithium storage capability than the sole component. The superior electrochemical performance of these composites is due to the unique nanostructure that provides a buffer space to accommodate the volume expansion/contraction combined with the good electronic-conductive graphene network.

SnO_2 is considered one of the most promising anode candidates due to its high capacity and safety. Paek et al.[57] reported a 3D nanoporous graphene/SnO_2 composite with a delaminated structure (Figure 10.6a), which displays an initial reversible capacity of 810 mAh g^{-1} and remains at 570 mAh g^{-1} after 30 cycles at a current density of 50 mA g^{-1}, much better than that of a bare SnO_2 nanoparticle without graphene that decreased rapidly to 60 mAh g^{-1} only after 15 cycles. SnO_2 nanosheets hybrided with graphene exhibited improved lithium storage properties as well. Ding et al.[58] developed a hydrothermal method to directly grow SnO_2 nanosheets on graphene, which exhibits greatly improved lithium storage properties compared to pure SnO_2 nanosheets, demonstrating the importance that graphene brings to the composite. Kim et al. controlled surface charge to prepare echinoid-like SnO_2 nanoparticles uniformly decorated on graphene through electrostatic attraction. The SnO_2/graphene composite shows a reversible capacity of 634 mAh g^{-1} with a coulombic efficiency of 98% after 50 cycles, which is much better compared to the commercial SnO_2.[59] Monodispersed SnO_2 nanoparticles on both sides of single-layer graphene sheets,[60] and a carbon-coated SnO_2/graphene composite with a double-protection layer,[61] improve the electrochemical performance of SnO_2 in lithium storage. Recently, SnO_2/*N*-doped graphene sandwich papers using 7,7,8,8-tetracyanoquinodimethane ions as the nitrogen source and complexing agent (Figure 10.6b),[62] and SnO_2 nanocrystals/nitrogen-doped graphene linked by Sn–N bonding,[63] show extremely high capacity, excellent rate capability, and a long cycle life. In addition, interface chemistry engineering was introduced to enhance the capacity and cyclic performance of SnO_2/graphene electrode materials.[64] An SnO_2/graphene composite was coated by a nanothick polydopamine layer and the polydopamine-coated composite was then cross-linked with poly(acrylic acid) that was used as a binder to link the whole anode electrode. The cross-link reaction between poly(acrylic acid) and polydopamine produced a robust network in the anode system to stabilize the whole anode during cycling. As a result, the designed anode exhibits a stable capacity

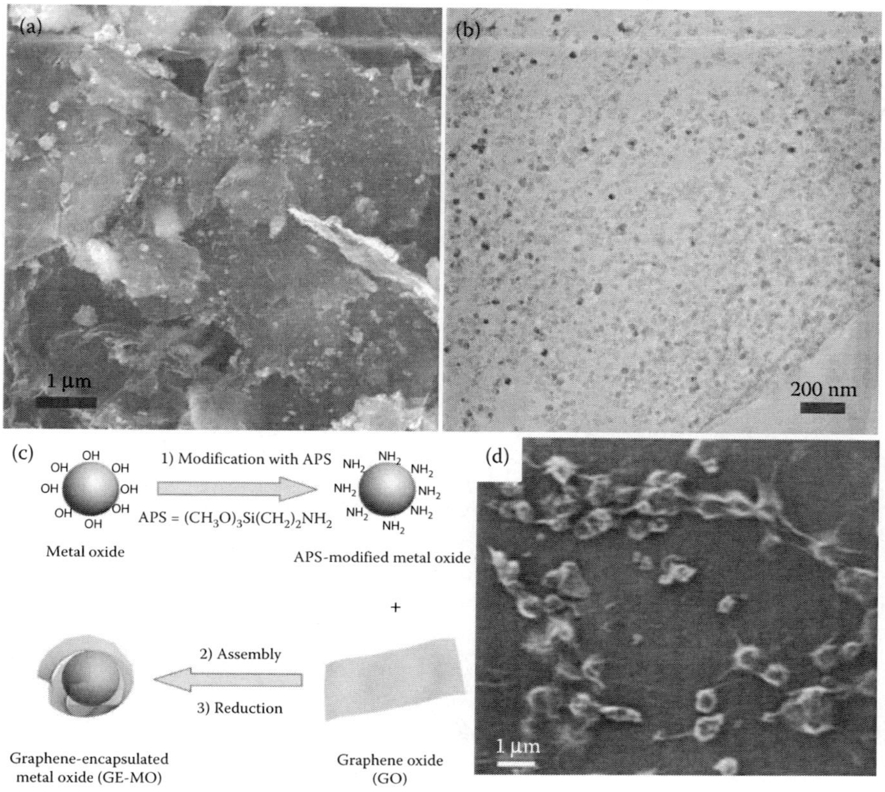

FIGURE 10.5 (a) SEM image of Co_3O_4/graphene composite. (b) Low-magnification TEM image of Co_3O_4/graphene composite. (Reprinted with permission from Wu, Z. S.; Ren, W. C.; Wen, L.; Gao, L. B.; Zhao, J. P.; Chen, Z. P.; Zhou, G. M.; Li, F.; Cheng, H. M. 2010. Graphene anchored with Co_3O_4 nanoparticles as anode of lithium ion batteries with enhanced reversible capacity and cyclic performance. *ACS Nano* 4:3187–3194. Copyright 2010 American Chemical Society.) (c) Fabrication of graphene-encapsulated metal oxide. (d) SEM image of graphene-encapsulated Co_3O_4. (Yang, S. B. et al.: Fabrication of graphene-encapsulated oxide nanoparticles: Towards high-performance anode materials for lithium storage. *Angew. Chem. Int. Edit.* 2010. 49. 8408–8411. Copyright Wiley-VCH Verlag GmbH & Co. KGaA. Reproduced with permission.)

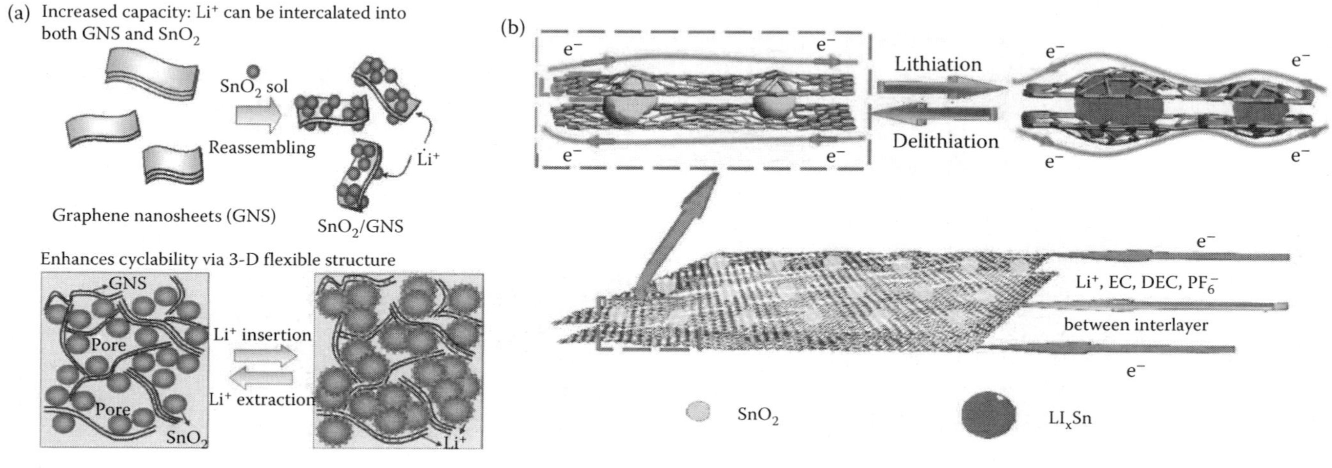

FIGURE 10.6 (a) Schematic of the synthesis and structure of graphene/SnO_2. (Reprinted with permission from Paek, S. M.; Yoo, E.; Honma, I. 2009. Enhanced cyclic performance and lithium storage capacity of SnO_2/graphene nanoporous electrodes with three-dimensionally delaminated flexible structure. *Nano Lett.* 9, 72–75. Copyright 2009 American Chemical Society.) (b) Schematic representation showing paths for lithium ions and electrons in the *N*-doped G-SnO_2 paper. (Wang, X. et al.: *N*-doped graphene–SnO_2 sandwich paper for high-performance lithium-ion batteries. *Adv. Funct. Mater.* 2012. 22. 2682–2690. Copyright Wiley-VCH Verlag GmbH & Co. KGaA. Reproduced with permission.)

of 718 mAh g^{-1} at a current density of 100 mA g^{-1} after 200 cycles.[64] Similar results by introducing graphene to significantly improve the capacity, rate capability, and cyclic stability were also reported for graphene/Mn_3O_4,[67–71] graphene/NiO,[72–76] and graphene/Fe_2O_3[77–80] composites. The superiority of the above composites can be ascribed to the confinement of metal oxides by surrounding graphene that limits the volume expansion upon lithium insertion/extraction, the pores developed between metal oxides and graphene that can be used as a buffer space during charge and discharge, the interconnected graphene framework that suppresses particle aggregation, and the electrically conducting graphene network that ensures good electrical contacts with various metal oxides.

Another lithium storage mechanism of metal oxides is based on a lithium ion intercalation reaction, and TiO_2 is a typical example. The lithium intercalation and extraction process with a small lattice change (smaller than 4%) ensures its structural stability and cycling life. Li et al.[81] prepared mesoporous anatase TiO_2 nanospheres on graphene by a template-free self-assembly process (Figure 10.7a through c). In comparison to the reference TiO_2, the composite shows substantial improvement in specific capacity from 1°C to 50°C. More strikingly, the specific capacity of the composite is as high as 97 mAh g^{-1} at the rate of 50°C, 6 times higher than that of the reference TiO_2 (Figure 10.7d).[81] Wang et al.[82] used anionic sulfate surfactants to assist the stabilization of graphene in aqueous solutions and facilitate the self-assembly of *in situ*-grown nanocrystalline TiO_2, rutile, and anatase, with graphene for increased lithium-ion insertion. Even at a high rate of 30°C, equal to 2 min of charging/discharging, the specific capacity of the hybrid material is 87 mAh g^{-1}, which is more than double the capacity (35 mAh g^{-1}) of the control rutile TiO_2.[82] Yang et al. developed a nanocasting method to prepare sandwich-like, graphene-based mesoporous titania nanosheets. The resulting hybrids possess a thin nature, large aspect ratio, and mesoporous structure, which facilitate the fast diffusion of lithium ions and exhibit a high rate capability and excellent cycle performance.[83] Other strategies that combined graphene with the nanostructure of TiO_2 hollow particles[84] or anatase TiO_2 nanosheets with exposed (001) high-energy facets[85] also show promising advantages.

$Li_4Ti_5O_{12}$ (LTO) also belongs to the lithium ion intercalation electrode, which is a promising candidate anode material for LIBs due to its well-known zero-strain merits. However, the electronic conductivity of LTO is relatively low, which seriously limits its high rate capability. Shi et al.[86] reported a graphene-embedded LTO composite with a nanosized architecture to improve the electrochemical properties of LTO in

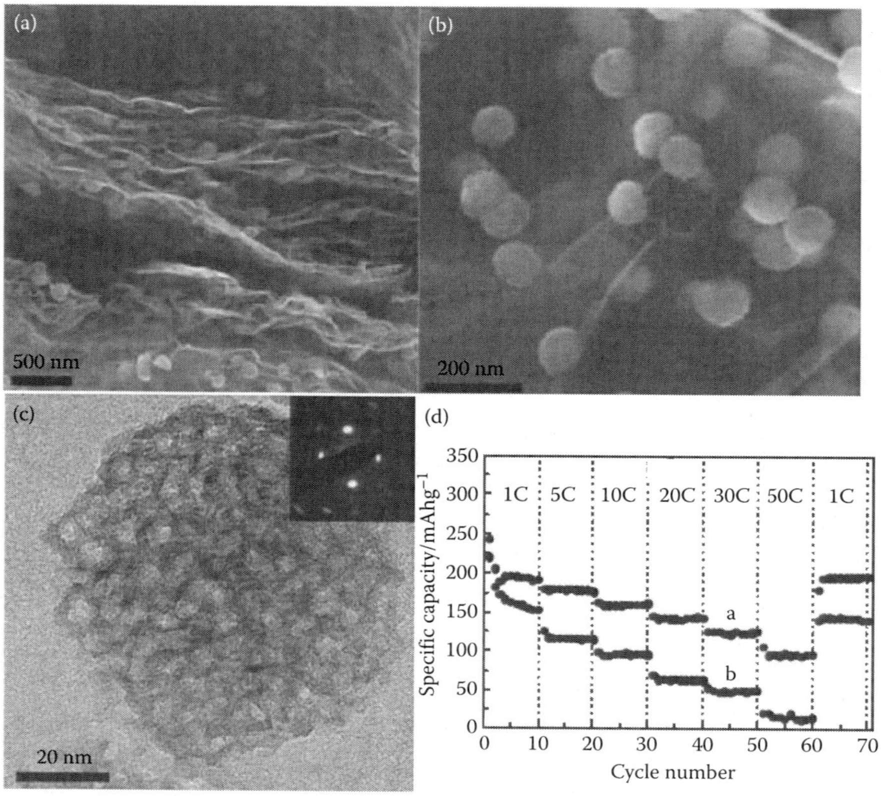

FIGURE 10.7 (a) Side-view and (b) top-view SEM images of mesoporous anatase TiO_2 nanospheres on graphene. (c) TEM image of a single mesoporous anatase nanosphere on graphene. (d) Comparison of the specific capacity at different rates between a: TiO_2/graphene and b: TiO_2 electrodes. (Li, N. et al.: Battery performance and photocatalytic activity of mesoporous anatase TiO(2) nanospheres/graphene composites by template-free self-assembly. *Adv. Funct. Mater.* 2011. 21. 1717–1722. Copyright Wiley-VCH Verlag GmbH & Co. KGaA. Reproduced with permission.)

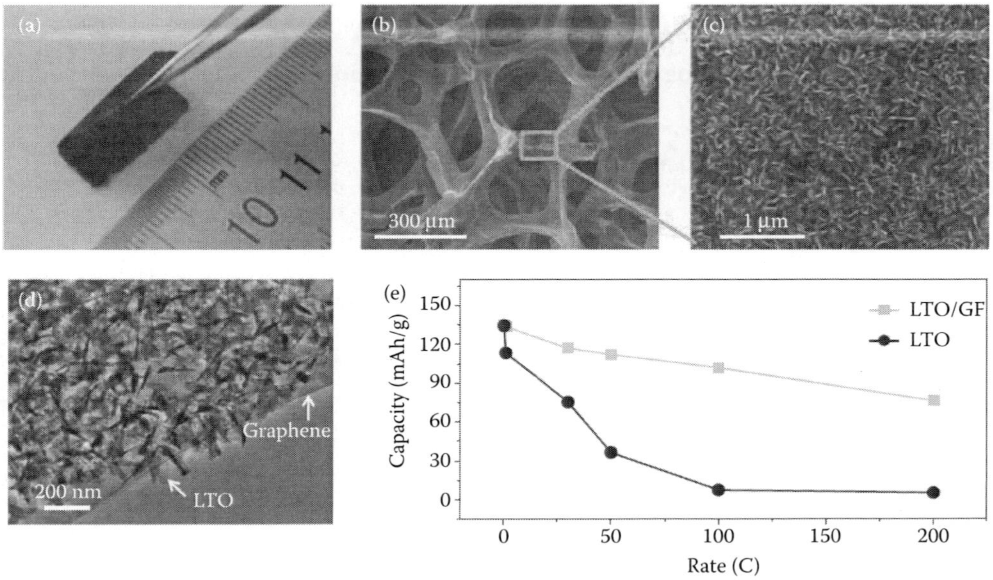

FIGURE 10.8 (a) Photograph of a free-standing flexible LTO/GF being bent. (b, c) SEM images of the LTO/GF. (d) TEM image of the LTO/GF. (e) Specific capacities of the LTO/GF and reference LTO at various charge/discharge rates. (Li, N. et al., 2012. Flexible graphene-based lithium ion batteries with ultrafast charge and discharge rates. *Proc. Natl. Acad. Sci. USA* 109:17360–17365, Copyright 2012 National Academy of Sciences, U.S.A.)

LIBs, in which graphene was chosen as an effective carbon coating to improve the electrical conductivity of the nanocomposites. It was found that with only 5 wt% graphene, the high rate performance can be greatly improved, and the specific capacity can reach 122 mAh g^{-1} when charged/discharged at a rate of 30°C.[86] Li et al.[87] reported a lightweight and flexible LIB anode using graphene foam (GF) as a current collector, loaded with LTO without conducting additives and binders (Figure 10.8a through d). The excellent electrical conductivity and pore structure of the hybrid electrodes enable rapid electron and ion transport; thus, the hybrid shows extremely high charge/discharge rates. For example, even at a charge and discharge rate of 200°C (corresponding to an 18-s full discharge), it still retains a specific capacity of 135 mAh g^{-1}, corresponding to ~80% of the specific capacity at the 1°C rate. In contrast, the reference LTO, which was prepared using the same process but without the presence of GF, shows a capacity of almost zero at 200°C (Figure 10.8e).[87] Other studies such as *in situ* synthesis of LTO/graphene hybrids[88] or two-step microwave-assisted solvothermal reaction followed by heat treatment to fabricate LTO/graphene hybrids,[89] show excellent rate capability and good cycling properties.

These results demonstrate the benefits of using graphene to significantly improve the performance of graphene-based hybrids. Some of the results were briefly summarized in Table 10.3. The important role of graphene in the composites is due to its excellent electrical conductivity, high surface area, ultrathin thickness, and mechanical flexibility. It prevents the active particles from aggregation or cracking upon cycling, provides a large buffer space to accommodate the volume expansion/contraction of high-capacity active materials during the lithium insertion/extraction process, dramatically improves the electrical conductivity of the hybrids, and thus achieves a large capacity, good cycling performance, and high rate capability.

10.6 GRAPHENE-BASED CATHODE MATERIALS IN LIBs

Commonly used cathode materials for LIBs, such as $LiCoO_2$, $LiMn_2O_4$, and $LiFePO_4$, suffer from poor electrical conductivity and impair the rate capability of the electrodes. Owing to the high electrical conductivity and large specific surface area, graphene plays a key role in improving the capacity and rate capability of the cathodes. Ding et al.[90] prepared nano-structured graphene/$LiFePO_4$ composites by the coprecipitation method to improve the electrical conductivity of the composites. The composites with only 1.5 wt% graphene deliver a higher specific capacity of 160 mAh g^{-1} than $LiFePO_4$ (113 mAh g^{-1}) at 0.2°C and retained 110 mAh g^{-1} even at a high rate of 10°C.[90] Constructing a 3D graphene network can not only improve electrical conductivity, but also enables fast Li^+ diffusion. $LiFePO_4$ particles wrapped homogeneously by a graphene network were prepared by spray-drying and annealing processes (Figure 10.9a).[91] The interconnected graphene network facilitates electron migration throughout the whole particles, and the presence of voids was beneficial for Li^+ diffusion. As a result, the composite cathode delivers a high capacity of 70 mAh g^{-1} at 60°C and shows an excellent cycling stability under 10°C charging and 20°C discharging for 1000 cycles (Figure 10.9b).[91] Graphene was also incorporated into the 3D porous hierarchical network by a template-free sol–gel approach to improve the electrical conductivity of the $LiFePO_4$/graphene composite, and it demonstrates a

TABLE 10.3
Examples of Synthesis Methods, Morphology, and Electrochemical Properties of Graphene-Based Anode Materials for LIBs Reported in Literature

Graphene-Based Materials	Synthesis Methods	Morphology	Performance	References
Si/G	Mixing	NPs	A capacity of 1168 mAh g^{-1} after 30 cycles with an average coulombic efficiency of 93%	[27]
Si/G	Vacuum filtration and thermal reduction	NPs	A high lithium storage capacities of >2200 mAh g^{-1} after 50 cycles and >1500 mAh g^{-1} after 200 cycles with a decrease rate of 0.5% per cycle	[28]
Si/G	Vacuum filtration and thermal reduction	NPs	A high reversible capacity of approximately 3200 mAh g^{-1} at 1 A g^{-1} and 83% of its theoretical capacity was maintained after 150 cycles, a reversible capacity of around 1100 mAh g^{-1} at 8 A g^{-1}	[30]
Si/G	Diazonium chemistry	NPs	An initial capacity of 1079 mAh g^{-1} and retained 828 mAh g^{-1} after the 50th cycle at a current density of 300 mA g^{-1}	[31]
Si/G	Vapor decomposition	NPs	A specific capacity of 2000 mAh g^{-1} at the current density of 140 mA g^{-1} and excellent stability over 150 cycles with an average coulombic efficiency in excess of 99%	[32]
Sn/G	$NaBH_4$ reduction of graphene oxide and Sn^{2+}	NPs	A reversible capacity of 795 mAh g^{-1} in the second cycle and 508 mAh g^{-1} in the 100th cycle	[38]
Sn/G	Film deposition and annealing process	Nanopillars	A reversible capacity of 734 mAh g^{-1} at 50 mA g^{-1} and high rate performance of 408 mAh g^{-1} at 5 A g^{-1}	[39]
Sn/G	Hydrolysis and CVD process	Sn core/carbon-sheath coaxial nanocable	Specific capacities higher than 760 mAh g^{-1} in the initial 10 cycles and higher than 630 mAh g^{-1} after the 50th cycle at 50 mA g^{-1}	[41]
Sn/G	Glucose-assisted chemical protocol	Nanosheets	Specific capacities of 435 mAh g^{-1} when cycled at 400 mA g^{-1}, 357 mAh g^{-1} at 800 mA g^{-1}, and 265 mAh g^{-1} at 1600 mA g^{-1}	[42]
Fe_3O_4/G	*In situ* reduction of iron hydroxide	Particles	A high reversible specific capacity approaching 1026 mAh g^{-1} after 30 cycles at 35 mA g^{-1} and 580 mAh g^{-1} after 100 cycles at 700 mA g^{-1} as well as improved cyclic stability and excellent rate capability	[44]
Fe_3O_4/G	Microwave irradiation	NPs	A high initial discharge capacity of 1320 mAh g^{-1} and retains 650 mAh g^{-1} after 50 cycles at 0.1°C and a high rate performance of 350 mAh g^{-1} at 5°C	[45]
Fe_3O_4/G	*In situ* hydrothermal method	NPs	A stable capacity of about 650 mAh g^{-1} with no noticeable fading for up to 100 cycles at 100 mA g^{-1}	[49]
Co_3O_4/G	*In situ* deposition	NPs	Large reversible capacity of 935 mAh g^{-1} after 30 cycles at 50 mA g^{-1}, excellent cyclic performance, and good rate capability	[51]
Co_3O_4/G	Electrostatic coassembly	NPs	A high reversible capacity of 1100 mAh g^{-1} in the initial cycle, and over 1000 mAh g^{-1} after 130 cycles at 74 mA g^{-1}	[52]
Co_3O_4/G	*In situ* reduction	NPs	>800 mAh g^{-1} reversibly at 200 mA g^{-1}, >550 mAh g^{-1} at 1000 mA g^{-1}	[53]
SnO_2/G	Reassembly	NPs	An initial reversible capacity of 810 mAh g^{-1} and remains at 570 mAh g^{-1} after 30 cycles at 50 mA g^{-1}	[57]
SnO_2/G	Hydrothermal method	Nanosheets	A reversible capacity of 518 mAh g^{-1} after 50 cycles at 400 mA g^{-1}	[58]
SnO_2/G	Electrostatic attraction	Echinoid-like NPs	A reversible capacity of 634 mAh g^{-1} with a coulombic efficiency of 98% after 50 cycles	[59]
SnO_2/G	Hydrothermal method	NPs	A stable capacity of 718 mAh g^{-1} at a current density of 100 mA g^{-1} after 200 cycles	[64]
TiO_2/G	Template-free self-assembly	Mesoporous nanospheres	The specific capacity of the composite is as high as 97 mAh g^{-1} at the rate of 50°C	[81]
TiO_2/G	Self-assembly and *in situ* growth	NPs	The specific capacity of the hybrid material is 87 mAh g^{-1} at 30°C	[82]
TiO_2/G	Nanocasting	Nanosheets	The first discharge capacity of 269 mAh g^{-1} is achieved at 0.2°C and a capacity of 202 mAh g^{-1} in the charging process; the reversible capacities are retained at 162 and 123 mAh g^{-1} at 1°C and 10°C	[83]
$Li_4Ti_5O_{12}$/G	Mixed	NPs	A specific capacity of 122 mAh g^{-1} even at a very high charge/discharge rate of 30°C	[86]
$Li_4Ti_5O_{12}$/G	CVD and hydrothermal reaction	Nanosheets	A specific capacity of 135 mAh g^{-1} at a charge and discharge rate of 200°C, corresponding to ~80% of the specific capacity at the 1°C rate	[87]

Note: G, graphene; NPs, nanoparticles.

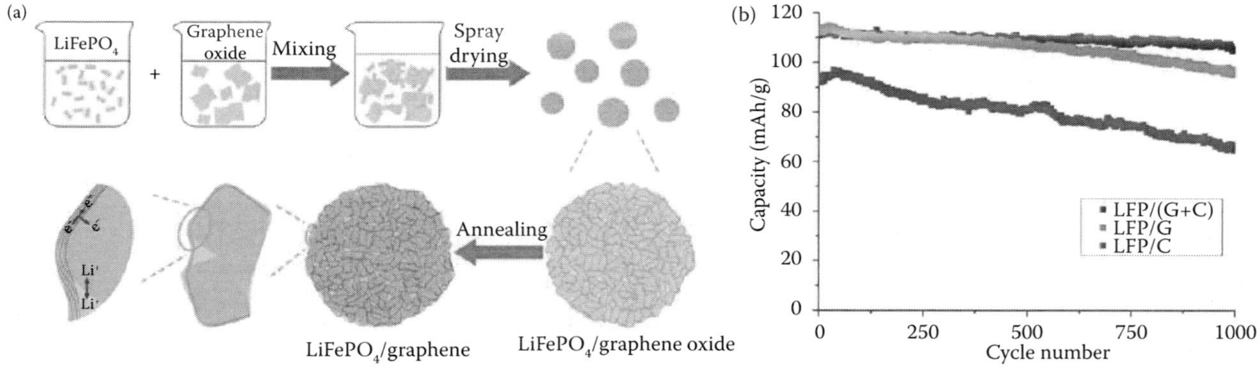

FIGURE 10.9 (a) Illustration of the preparation process and the microscale structure of $LiFePO_4$/graphene composite. (b) Comparative cycling performances of $LiFePO_4$/G, $LiFePO_4$/C, and $LiFePO_4$/(G + C) operated under 10°C charging and 20°C discharging. (Zhou, X. F. et al., 2011. Graphene modified $LiFePO_4$ cathode materials for high power lithium ion batteries. *J. Mater. Chem.* 21:3353–3358. Reproduced by permission of The Royal Society of Chemistry.)

reversible capacity of 146 mAh g^{-1} at 17 mA g^{-1} after 100 cycles and improved Li-ion insertion/extraction kinetics.[92] 3D GF prepared by chemical vapor deposition (CVD) on a 3D porous Ni template followed by template removal was also used as a matrix to load $LiFePO_4$ as a high-performance cathode in LIBs.[93,94] With the assistance of the highly conductive and interconnected porous 3D graphene network, the composite exhibits good rate performance and excellent electrochemical stability. In other studies, an $LiFePO_4$/graphene/carbon composite was prepared by using an *in situ* solvothermal method followed by a carbon-coating process[95] or a one-pot, microwave-assisted hydrothermal method followed by sintering in an H_2/Ar atmosphere.[96] The electrochemical test results indicated that the comodification of $LiFePO_4$ with graphene and carbon could construct an effective conducting network and ion transportation route, which significantly improve the electrochemical activity of $LiFePO_4$ with a high capacity, good rate performance, as well as an excellent cycling stability. Wang et al. reported a two-step approach for the synthesis of $LiMn_{1-x}Fe_xPO_4$ nanorods on RGO to significantly boost the electrical conductivity and low ionic resistance. The resulting hybrid with a total content of 26 wt% conductive carbon showed capacities of 132 and 107 mAh g^{-1} at high discharge rates of 20°C and 50°C, which is 85% and 70% of the capacity at C/2 (155 mAh g^{-1}).[97]

$Li_3V_2(PO_4)_3$ is another phosphate material for use as a cathode material in LIBs, which exhibits better ionic conductivity compared to $LiFePO_4$ because it has a 3D path for lithium-ion diffusion, while $LiFePO_4$ only has a one-dimensional channel. It also demonstrates a higher average voltage (4.0 V) and theoretical capacity (197 mAh g^{-1}).[98] However, the low electronic conductivity of $Li_3V_2(PO_4)_3$ also limits its high rate performance. $Li_3V_2(PO_4)_3$ was selectively enwrapped by a minor amount of graphene (1.14 wt%) to form an electronic conducting network, resulting in excellent high rate capability and cycling stability.[98] Zhang et al.[99] encapsulated $Li_3V_2(PO_4)_3$ particles within carbon shells and anchored them on graphene to greatly improve the electronic conductivity. In addition, the coated carbon can avoid contacts between the electrolyte and $Li_3V_2(PO_4)_3$ particles, and reduce the side reactions. Thus, the excellent rate performance (85 mAh g^{-1} at 50°C) as well as a good cycling performance between 3.0 and 4.8 V was achieved.[99]

Other cathode materials such as $LiNi_{1/3}Mn_{1/3}Co_{1/3}O_2$, $LiMn_2O_4$, and V_2O_5 were also hybrided with graphene to achieve a high rate capability. For example, an electronically conductive 3D network of graphene was introduced into $LiNi_{1/3}Mn_{1/3}Co_{1/3}O_2$ cathode material in a special nano/micro hierarchical structure, which decreases the resistance, promotes the kinetics of the batteries, and leads to high rate performance.[100] A microwave-assisted hydrothermal method was adopted to prepare nanosized $LiMn_2O_4$ dispersing on graphene as a cathode material for high rate LIBs. The hybrid delivered a high specific capacity of 137 mAh g^{-1} at 1°C rate and a remarkable discharge capacity of 117 and 101 mAh g^{-1} even at a fast rate of 50°C and 100°C, which is ascribed to the nanosized particles, high crystallinity of $LiMn_2O_4$, and high electrical conductivity of graphene.[101] Ultra-long single-crystalline V_2O_5 nanowire[102] or porous V_2O_5 spheres[103] supported by graphene were realized and exhibited good cycling stability and excellent rate capability. Notably, ultrathin V_2O_5 nanowires were uniformly incorporated into graphene paper and enabled a decent specific capacity that can be preserved over 100,000 cycles, which are 2–3 orders of magnitude larger than commonly used cathode materials.[104] Recently, a bottom-up approach toward single-crystalline vanadium oxide (VO_2) ribbons with graphene layers was reported to construct 3D architectures (Figure 10.10a) with ultrafast charging and discharging capability (190°C, a full charge/discharge in 20 s) and a long cycling life (more than 1000 cycles, Figure 10.10b through d).[105] The superior performance was ascribed to the high electrical conductivity of the VO_2–graphene network, high active material loading, ultrathin VO_2 ribbons with a short lithium diffusion length, and multichannels enabling rapid diffusion of ions in the electrode.[105]

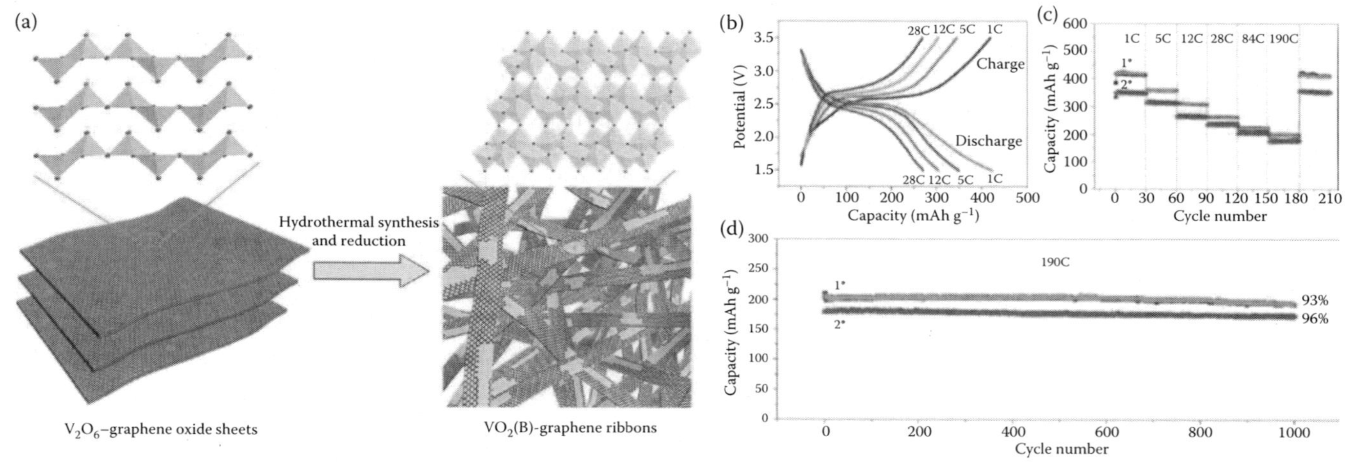

FIGURE 10.10 (a) Fabrication of VO_2–graphene ribbons by a simultaneous hydrothermal synthesis and reduction of layered V_2O_5–graphite oxide composites at 180°C. (b) Representative discharge–charge curves of VO_2–graphene architecture at various C-rates over the potential range of 1.5–3.5 V versus Li^+/Li. (c) Rate capacities of VO_2–graphene architectures with different VO_2 contents, measured for 30 cycles at each selected rate from 1°C to 190°C. (d) Capacity retentions of VO_2–graphene architectures when performing a full discharge–charge at the highest rate of 190°C (37.2 A g^{-1}) for 1000 cycles. 1′ and 2′ are denoted as VO_2–graphene architectures with the VO_2 contents of 78% and 68%. (Reprinted with permission from Yang, S. et al., 2013. Bottom-up approach toward single-crystalline VO_2–graphene ribbons as cathodes for ultrafast lithium storage. *Nano Lett.* 13, 1596–1601. Copyright 2013 American Chemical Society.)

Owing to the low capacity of the currently used lithium-transition metal oxides or lithium–phosphate materials, sulfur is a promising cathode material with a high theoretical specific capacity of 1672 mAh g^{-1}, about 5 times than that of traditional cathode materials.[106,107]

However, the low electrical conductivities of sulfur, the intermediate polysulfide products, and the final Li_2S product affect the utilization of the active sulfur material and the rate capability of the battery.[108] Graphene with a large surface area, rich surface chemistry, and high electrical conductivity has been extensively studied and the detailed discussion can be found in our recent review[109] and other excellent reviews.[110,111] The above-mentioned examples and discussions (some are summarized in Table 10.4) suggest that graphene can not only retain the high overall conductivity of the cathodes, but also greatly improve their electrochemical performance upon cycling, especially at a high rate.

TABLE 10.4
Examples of Synthesis Methods, Morphology, and Electrochemical Properties of Graphene-Based Cathode Materials for LIBs Reported in Literature

Graphene-Based Materials	Synthesis Methods	Morphology	Performance	References
$LiFePO_4$/G	Coprecipitation	NPs	A higher specific capacity of 160 mAh g^{-1} at 0.2°C and retained 110 mAh g^{-1} even at a high rate of 10°C	[90]
$LiFePO_4$/G	Spray-drying and annealing processes	Particles	A high capacity of 70 mAh g^{-1} at 60°C and shows an excellent cycling stability under 10°C charging and 20°C discharging for 1000 cycles	[91]
$LiFePO_4$/G	Template-free sol–gel approach	Particles	A reversible capacity of 146 mAh g^{-1} at 17 mA g^{-1} after 100 cycles and improved Li-ion insertion/extraction kinetics	[92]
$LiMn_{1-x}Fe_xPO_4$/G	Two-step solution-phase reaction	Nanorods	Capacities of 132 and 107 mAh g^{-1} at high discharge rates of 20°C and 50°C, which is 85% and 70% of the capacity at C/2	[97]
$Li_3V_2(PO_4)_3$/G	Modified Pechini method	Particles	Excellent rate performance of 85 mAh g^{-1} at 50°C	[99]
$LiMn_2O_4$/G	Microwave-assisted hydrothermal method	NPs	A high specific capacity of 137 mAh g^{-1} at 1°C rate and a remarkable discharge capacity of 117 and 101 mAh g^{-1} even at a fast rate of 50°C and 100°C	[101]
V_2O_5/G	Solvothermal approach and annealing process	Porous spheres	A reversible capacity of 235 mAh g^{-1} at 0.3°C and high rate performance of 102 mAh g^{-1} at 19°C	[103]
V_2O_5/G	Hydrothermal reaction	Nanowires	A high reversible specific capacity of 94.4 mAh g^{-1} after 100,000 cycles at 10,000 mA g^{-1}	[104]
VO_2/G	Bottom-up approach	Ribbons	Ultrafast charging and discharging capability (190°C, a full charge/discharge in 20 s) and long cycling life (more than 1000 cycles)	[105]

10.7 GRAPHENE AS CONDUCTIVE ADDITIVES IN LIBs

To ensure sufficient electrical conductivity of electrodes during the charge and discharge process, conductive additives such as carbon black and CNTs are necessary to decrease the electrical resistivity of the electrodes. Generally, the amount of conductive carbon additives is below 10 wt% of the total electrode mass because they contribute a much lower capacity than the active materials. Recently, graphene has been implemented as a new electron-conducting additive to improve the electrical conductivity of electrodes, especially for high-power electrode materials, such as $LiFePO_4$[112–116] and LTO,[117,118] which seriously suffer from low electrical conductivity. It is considered that graphene can be used as a conductive carbon material to construct a 3D conductive network through electrodes. For example, Yang's group introduced graphene as a planar conductive additive to construct the conducting network through plane-to-point contact.[112] The graphene-modified $LiFePO_4$ cathode shows better charge/discharge performance compared to other control additives such as CNTs and carbon black (super-P) with similar fractions. The use of graphene to create plane-to-point contact instead of point-to-point contact of carbon black (Figure 10.11) not only reduces the mass fraction of carbon-based additives (2 wt% graphene), but also shows an improved high rate performance.[112] Wang et al.[113] developed a hydrothermal route to prepare the $LiFePO_4$/graphene composite followed by heat treatment, with a 3D mixed electron and ion-conducting network. Such a 3D architecture efficiently bridges graphene and $LiFePO_4$ enabling fast electron transport to improve the rate performance of $LiFePO_4$. The results indicate that the composite with 8 wt% graphene exhibits a highly stable reversible capacity of 81.5 mAh g^{-1} at the current density of 10°C, with only a value of 30.6 mAh g^{-1} for $LiFePO_4$. The morphology of graphene used as a conductive additive also plays important roles in determining the electrochemical performance. Yang et al.[114] investigate the impact of unfolded and stacked graphene as a conducting matrix on utilizing $LiFePO_4$ active materials. Compared with stacked graphene, the unfolded graphene provides better dispersion and good electric contact of $LiFePO_4$ and restricts the particle size at the nanoscale. Thus, the composite with unfolded graphene (1.5 wt%) displayed a discharge capacity of 166.2 mAh g^{-1} in the first cycle, whereas with stacked graphene, a similar carbon content delivers only 77 mAh g^{-1}.[114] Recently, electrochemically exfoliated graphene (2 wt%) was used to modify $LiFePO_4$ and the composite is able to deliver 208 mAh g^{-1} in a specific capacity at 0.1°C that exceeded the theoretical value of 170 mAh g^{-1}.[115] The excess capacity is due to the reversible reduction–oxidation reaction between the lithium ions of the electrolyte and the exfoliated few-layer graphene exhibiting a capacity higher than 2000 mAh g^{-1}. The highly conductive graphene assists the electron migration during cycling, diminishes the irreversible capacity, and leads to ~100% coulombic efficiency at high rates.

Graphene also functions as an efficient conducting agent in anode materials such as $Li_4Ti_5O_{12}$ and graphite. The electrospinning method was adopted to produce a nanosized architecture of $Li_4Ti_5O_{12}$ with a shorted ion and electron transport path, combined with a graphene coating to improve the conductivity of the composites.[117] The composites with only 1 wt% graphene exhibited a specific capacity of around 110 mAh g^{-1} at a high rate of 22°C, which is more than twice the capacity (47 mAh g^{-1}) of the electrospun $Li_4Ti_5O_{12}$ without graphene and 3 times the capacity (34 mAh g^{-1}) of the sol–gel $Li_4Ti_5O_{12}$ powder. Even after being charged/discharged at 1300 cycles at 22°C, the composites retained ~91%

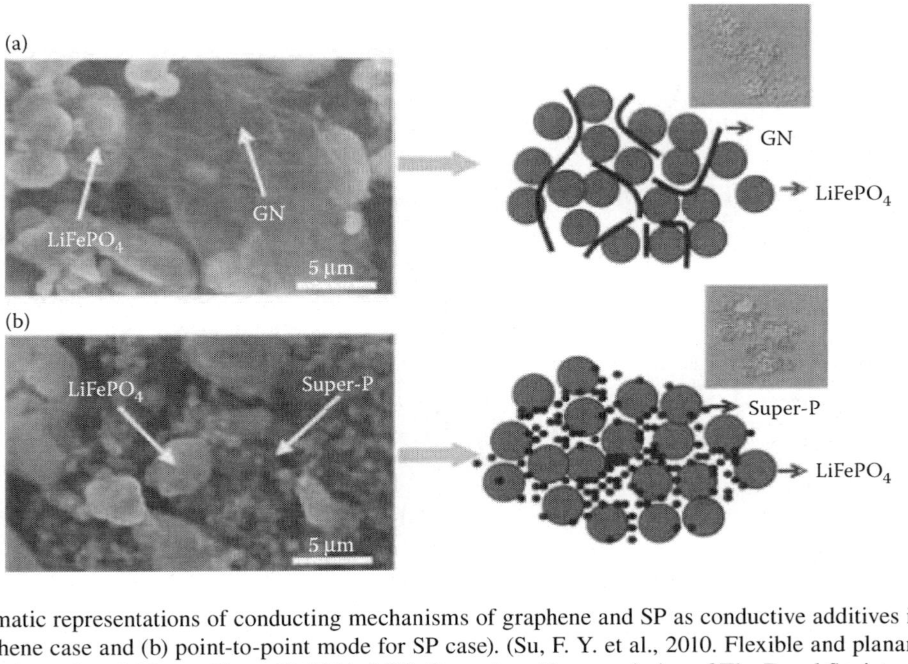

FIGURE 10.11 Schematic representations of conducting mechanisms of graphene and SP as conductive additives in $LiFePO_4$ ((a) plane-to-point mode for graphene case and (b) point-to-point mode for SP case). (Su, F. Y. et al., 2010. Flexible and planar graphene conductive additives for lithium-ion batteries. *J. Mater. Chem.* 20:9644–9650. Reproduced by permission of The Royal Society of Chemistry.)

retention (101 mAh g^{-1}) of the initial capacity demonstrating excellent cycling stability.[117] Zhang et al.[118] systematically investigated the influence of graphene content and pointed out that graphene with a diameter of 46 μm and a thickness of 4.5 nm has a percolation threshold of 1.8 wt%. Combined with the interparticle distance concept, it predicts a percolation threshold of 0.54 wt% for graphene, which is an order of magnitude lower than that for commonly used carbon black particles. The $Li_4Ti_5O_{12}$ anodes with 5 wt% graphene deliver a much better rate capability than those with 15 wt% carbon black, which are ascribed to the low percolation, high aspect ratio, and excellent electrical conductivity of graphene. It also suggests that a higher graphene content (10 wt%) leads to the restacking of graphene, deteriorating the diffusion of Li ions through the graphene layers.[118] Guo et al.[119] loaded graphene in graphite electrode and found that it plays significant roles in improving the rate performance, specific capacity, and cycle performance. Compared with a traditional spherical-conducting agent such as acetylene black with a discontinuous conductive network in connecting active materials, the high aspect ratio of graphene ensures the formation of a continuous conducting network throughout the anode and lowers the amount of conductive additives needed for efficient electronic transport.

10.8 INTERACTION MECHANISMS FOR THE IMPROVED ELECTROCHEMICAL PERFORMANCE

As discussed above, graphene has dramatically improved the capacity, rate capability, and cycling performance of nearly any electrode material for batteries. However, the interaction mechanism between graphene and these electrode materials is not clearly elucidated. Zhou et al.[8] proposed oxygen bridges as a possible origin of the synergistic mechanism between graphene and metal oxides, and took NiO as an example. The existence of oxygen bridges was confirmed by x-ray photoelectron spectroscopy, Fourier transform infrared spectroscopy, Raman spectroscopy, and the first-principles calculations. It was found that NiO nanosheets (NiO NSs) are bonded strongly to graphene through oxygen bridges. The oxygen bridges mainly originate from the pinning of hydroxyl/epoxy groups from graphene on the Ni atoms of NiO NSs. The calculated diffusion barriers of the Ni adatom on the oxygenated graphene surface (2.23 and 1.69 eV for graphene with hydroxyl and epoxy) are much larger than that on graphene (0.19 eV, Figure 10.12a).[8] Therefore, the NiO NS is anchored strongly on graphene through a C–O–Ni bridge, which allows a high reversible capacity, excellent high rate, and cycling performance. The electrochemical test results show that the capacity of the NiO NS/graphene composite is much higher than the total sum of the individual capacities of NiO NSs and graphene, indicating a synergistic effect between these two components (Figure 10.12b).[8] Interfacial interaction was also reported by Zhou et al.[120] between Fe_3O_4 nanoparticles and graphene through the preparation process of direct pyrolysis of $Fe(NO_3)_3$ on graphene. It was found that Fe_3O_4 nanoparticles anchored on graphene through strong covalent bond interactions. The strong covalent links ensure the synergistic improvement of the Fe_3O_4/graphene hybrids on specific capacity, rate property, and cyclic stability. As a result, the hybrid delivered a high capacity of 796 mAh g^{-1} after 200 cycles without capacity loss at a current density of 500 mA g^{-1}. A stable reversible capacity of about 550 mAh g^{-1} for this hybrid remained after 300 cycles at 1000 mA g^{-1}, which dramatically outperformed the electrodes made from mechanical mixing of an Fe_3O_4–graphene hybrid under the same conditions.[120]

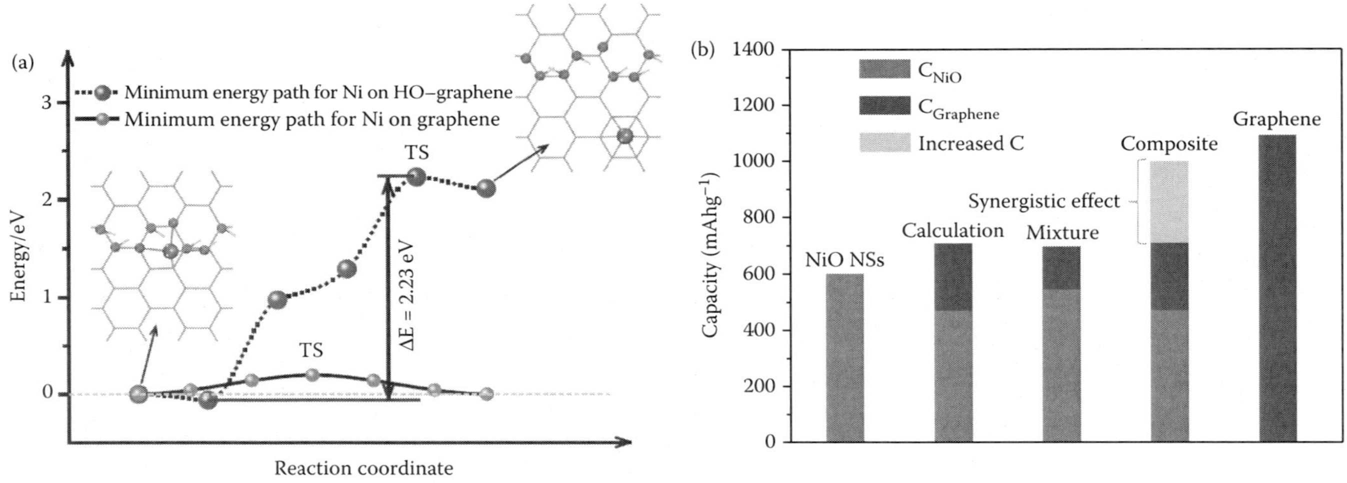

FIGURE 10.12 (a) Minimum energy path for a Ni adatom diffusing on graphene and HO–graphene surface. (b) The first reversible specific capacity of graphene, NiO NSs, NiO NS/graphene composite, NiO NS–graphene mixture, and the calculated specific capacity based on experimental values. The light gray rectangle indicates the increased capacity due to the synergistic effect between graphene and NiO NSs. (Reprinted with permission from Zhou, G. M. et al., 2012. Oxygen bridges between NiO nanosheets and graphene for improvement of lithium storage. *ACS Nano* 6:3214–3223. Copyright 2012 American Chemical Society.)

Miao et al.[121] used first-principles calculations to investigate the synergistic mechanism on the stability, electronic, and Li diffusion performance of the SnO_2 and graphene hybrids. It revealed that the stable interface formed by C–O covalent bonds not only improves the structural stability but also makes the hybrids more conductive than pure SnO_2. The calculated binding energies and diffusion barriers show that Li tends to insert into the interface, and diffuses along the SnO_2 (110) direction with a low barrier. The synergistic mechanism contributes from the interaction between graphene and SnO_2 at the interface, while the rest of the graphene and SnO_2 far away from the interface have little effect and remain in their respective Li storage performance. In a cathode material system, Radich and Kamat[122] discussed the origin of graphene enhancements in electrochemical energy storage for an MnO_2 electrode. It was found that electron storage properties of graphene enable better electrode kinetics, more rapid diffusion of Li^+ to intercalation sites, and a greater capacitance effect during discharge. This is due to graphene that can store electrons in π–π network and act as a kinetic mediator between electrons and Li^+ ions in the electrolyte. These insights give an interpretation for the superior performance of graphene-based materials in LIBs and other energy-related applications, and are important to understand and design other graphene hybrid materials.

10.9 SUMMARY AND PERSPECTIVE

We reviewed recent advances in the applications of graphene and graphene-based materials for LIBs. Graphene-based hybrids exhibited exciting properties that exceed the intrinsic properties of each component and/or introduced some new properties and functions such as improved electrical conductivity, larger capacity, higher rate capability, excellent cycling stability, and so on. The beneficial role of graphene in the hybrids is due to its unique structures and properties such as a high surface area and electrical conductivity, which can be used as an ideal support to construct a 3D conductive porous network for achieving excellent properties of the hybrids. The properties of graphene strongly depend on the routes for their preparation, and a suitable method selected plays a vital role in the final performance. The chemical routes from the oxidation of graphite or liquid-phase exfoliation followed by reduction are able to produce graphene on a large scale with relative low costs, which are the premise for their wide range of practical applications.

Although many remarkable achievements have been obtained in this field, much remains to be done before this rising star can be practically used in LIBs. First, understanding of the relationship between lithium storage and defects, layer numbers, sizes of graphene and oxygen-containing functional groups, fine control of the number of graphene layers, sizes of graphene, and the avoidance of agglomeration for graphene or composites during preparation are crucial for realizing their optimal performance in LIBs. Second, more efforts should be focused on the interactions between graphene and foreign functional building blocks, and a deep understanding of the interaction mechanism is crucial for achieving a well-defined ideal structure and increasing the charge transfer. For example, given that chemically prepared graphene can be negatively charged when dispersed in a solution, a wide range of intriguing graphene-based composite electrodes can be created by grafting or self-assembly with functional molecules and nanoparticles that are positively charged. *In situ* measurement techniques such as *in situ* transmission electron microscopy (TEM) or x-ray diffraction are promising approaches in revealing and understanding the mechanisms and structural evolution beyond electrochemical test results. Third, due to the lightweight and flexible characteristics of graphene, the development of thin, lightweight, flexible, and all-solid-state energy-storage devices for advanced thin and wearable electronics is highly anticipated. Fourth, the design and optimization of the structure and doping or surface modification of the graphene electrode to achieve supercapacitor-like rate capabilities while maintaining battery-like storage capacities are highly desirable yet still very challenging. Last but not the least, to reveal the structure–property relationship in the hybrids is very necessary, and theoretical calculations are an important tool that can be combined for investigating the ion and electron transport behaviors, defects, doping effects, and interfacial interactions of graphene-based hybrids. With further exploitation and multidisciplinary efforts, it is believed that the development of graphene-based materials will lead to many advances in science and technology and play an important role for the development of high-performance energy-storage devices applying in portable electronics, electric vehicles, and various fields.

REFERENCES

1. Tarascon, J. M.; Armand, M. 2001. Issues and challenges facing rechargeable lithium batteries. *Nature* 414:359–367.
2. Armand, M.; Tarascon, J. M. 2008. Building better batteries. *Nature* 451:652–657.
3. Bruce, P. G.; Scrosati, B.; Tarascon, J. M. 2008. Nanomaterials for rechargeable lithium batteries. *Angew. Chem. Int. Edit.* 47:2930–2946.
4. Liu, C.; Li, F.; Ma, L. P.; Cheng, H. M. 2010. Advanced materials for energy storage. *Adv. Mater.* 22:28–62.
5. Pumera, M. 2010. Graphene-based nanomaterials and their electrochemistry. *Chem. Soc. Rev.* 39:4146–4157.
6. Park, S.; Ruoff, R. S. 2009. Chemical methods for the production of graphenes. *Nat. Nanotechnol.* 4:217–224.
7. Lv, W.; Tang, D. M.; He, Y. B.; You, C. H.; Shi, Z. Q.; Chen, X. C.; Chen, C. M.; Hou, P. X.; Liu, C.; Yang, Q. H. 2009. Low-temperature exfoliated graphenes: Vacuum-promoted exfoliation and electrochemical energy storage. *ACS Nano* 3:3730–3736.
8. Zhou, G. M.; Wang, D.-W.; Yin, L.-C.; Li, N.; Li, F.; Cheng, H.-M. 2012. Oxygen bridges between NiO nanosheets and graphene for improvement of lithium storage. *ACS Nano* 6:3214–3223.
9. Yoo, E.; Kim, J.; Hosono, E.; Zhou, H.; Kudo, T.; Honma, I. 2008. Large reversible Li storage of graphene nanosheet families for use in rechargeable lithium ion batteries. *Nano Lett.* 8:2277–2282.

10. Chen, S. Q.; Chen, P.; Wang, Y. 2011. Carbon nanotubes grown *in situ* on graphene nanosheets as superior anodes for Li-ion batteries. *Nanoscale* 3:4323–4329.
11. Fan, Z.-J.; Yan, J.; Wei, T.; Ning, G.-Q.; Zhi, L.-J.; Liu, J.-C.; Cao, D.-X.; Wang, G.-L.; Wei, F. 2011. Nanographene-constructed carbon nanofibers grown on graphene sheets by chemical vapor deposition: High-performance anode materials for lithium ion batteries. *ACS Nano* 5:2787–2794.
12. Wang, G. X.; Shen, X. P.; Yao, J.; Park, J. 2009. Graphene nanosheets for enhanced lithium storage in lithium ion batteries. *Carbon* 47:2049–2053.
13. Guo, P.; Song, H. H.; Chen, X. H. 2009. Electrochemical performance of graphene nanosheets as anode material for lithium-ion batteries. *Electrochem. Commun.* 11:1320–1324.
14. Lian, P. C.; Zhu, X. F.; Liang, S. Z.; Li, Z.; Yang, W. S.; Wang, H. H. 2010. Large reversible capacity of high quality graphene sheets as an anode material for lithium-ion batteries. *Electrochim. Acta* 55:3909–3914.
15. Pan, D. Y.; Wang, S.; Zhao, B.; Wu, M. H.; Zhang, H. J.; Wang, Y.; Jiao, Z. 2009. Li storage properties of disordered graphene nanosheets. *Chem. Mater.* 21:3136–3142.
16. Xiang, H. F.; Li, Z. D.; Xie, K.; Jiang, J. Z.; Chen, J. J.; Lian, P. C.; Wu, J. S.; Yu, Y.; Wang, H. H. 2012. Graphene sheets as anode materials for Li-ion batteries: Preparation, structure, electrochemical properties and mechanism for lithium storage. *RSC Adv.* 2:6792–6799.
17. Wu, Z. S.; Ren, W. C.; Xu, L.; Li, F.; Cheng, H. M. 2011. Doped graphene sheets as anode materials with superhigh rate and large capacity for lithium ion batteries. *ACS Nano* 5:5463–5471.
18. Yin, S. Y.; Zhang, Y. Y.; Kong, J. H.; Zou, C. J.; Li, C. M.; Lu, X. H.; Ma, J.; Boey, F. Y. C.; Chen, X. D. 2011. Assembly of graphene sheets into hierarchical structures for high-performance energy storage. *ACS Nano* 5:3831–3838.
19. Chen, X. C.; Wei, W.; Lv, W.; Su, F. Y.; He, Y. B.; Li, B. H.; Kang, F. Y.; Yang, Q. H. 2012. A graphene-based nanostructure with expanded ion transport channels for high rate Li-ion batteries. *Chem. Commun.* 48:5904–5906.
20. Mukherjee, R.; Thomas, A. V.; Krishnamurthy, A.; Koratkar, N. 2012. Photothermally reduced graphene as high-power anodes for lithium-ion batteries. *ACS Nano* 6:7867–7878.
21. Zhu, Y. W.; Murali, S.; Cai, W. W.; Li, X. S.; Suk, J. W.; Potts, J. R.; Ruoff, R. S. 2010. Graphene and graphene oxide: Synthesis, properties, and applications. *Adv. Mater.* 22:3906–3924.
22. Kaskhedikar, N. A.; Maier, J. 2009. Lithium storage in carbon nanostructures. *Adv. Mater.* 21:2664–2680.
23. Wu, Z.-S.; Zhou, G. M.; Yin, L.-C.; Ren, W.; Li, F.; Cheng, H.-M. 2012. Graphene/metal oxide composite electrode materials for energy storage. *Nano Energy* 1:107–131.
24. Sun, Y. Q.; Wu, Q. O.; Shi, G. Q. 2011. Graphene based new energy materials. *Energy Environ. Sci.* 4:1113–1132.
25. Li, C.; Shi, G. Q. 2012. Three-dimensional graphene architectures. *Nanoscale* 4:5549–5563.
26. Park, C. M.; Kim, J. H.; Kim, H.; Sohn, H. J. 2010. Li-alloy based anode materials for Li secondary batteries. *Chem. Soc. Rev.* 39:3115–3141.
27. Chou, S. L.; Wang, J. Z.; Choucair, M.; Liu, H. K.; Stride, J. A.; Dou, S. X. 2010. Enhanced reversible lithium storage in a nanosize silicon/graphene composite. *Electrochem. Commun.* 12:303–306.
28. Lee, J. K.; Smith, K. B.; Hayner, C. M.; Kung, H. H. 2010. Silicon nanoparticles–graphene paper composites for Li ion battery anodes. *Chem. Commun.* 46:2025–2027.
29. Wang, J. Z.; Zhong, C.; Chou, S. L.; Liu, H. K. 2010. Flexible free-standing graphene–silicon composite film for lithium-ion batteries. *Electrochem. Commun.* 12:1467–1470.
30. Zhao, X.; Hayner, C. M.; Kung, M. C.; Kung, H. H. 2011. In-plane vacancy-enabled high-power Si–graphene composite electrode for lithium-ion batteries. *Adv. Energy Mater.* 1:1079–1084.
31. Yang, S. N.; Li, G. R.; Zhu, Q.; Pan, Q. M. 2012. Covalent binding of Si nanoparticles to graphene sheets and its influence on lithium storage properties of Si negative electrode. *J. Mater. Chem.* 22:3420–3425.
32. Evanoff, K.; Magasinski, A.; Yang, J. B.; Yushin, G. 2011. Nanosilicon-coated graphene granules as anodes for Li-ion batteries. *Adv. Energy Mater.* 1:495–498.
33. He, Y. S.; Gao, P. F.; Chen, J.; Yang, X. W.; Liao, X. Z.; Yang, J.; Ma, Z. F. 2011. A novel bath lily-like graphene sheet-wrapped nano-Si composite as a high performance anode material for Li-ion batteries. *RSC Adv.* 1:958–960.
34. Xin, X.; Zhou, X. F.; Wang, F.; Yao, X. Y.; Xu, X. X.; Zhu, Y. M.; Liu, Z. P. 2012. A 3D porous architecture of Si/graphene nanocomposite as high-performance anode materials for Li-ion batteries. *J. Mater. Chem.* 22:7724–7730.
35. Zhou, X. S.; Yin, Y. X.; Wan, L. J.; Guo, Y. G. 2012. Facile synthesis of silicon nanoparticles inserted into graphene sheets as improved anode materials for lithium-ion batteries. *Chem. Commun.* 48:2198–2200.
36. Zhou, X. S.; Yin, Y. X.; Wan, L. J.; Guo, Y. G. 2012. Self-assembled nanocomposite of silicon nanoparticles encapsulated in graphene through electrostatic attraction for lithium-ion batteries. *Adv. Energy Mater.* 2:1086–1090.
37. Luo, J. Y.; Zhao, X.; Wu, J. S.; Jang, H. D.; Kung, H. H.; Huang, J. X. 2012. Crumpled graphene-encapsulated Si nanoparticles for lithium ion battery anodes. *J. Phys. Chem. Lett.* 3:1824–1829.
38. Wang, G. X.; Wang, B.; Wang, X. L.; Park, J.; Dou, S. X.; Ahn, H.; Kim, K. 2009. Sn/graphene nanocomposite with 3D architecture for enhanced reversible lithium storage in lithium ion batteries. *J. Mater. Chem.* 19:8378–8384.
39. Ji, L. W.; Tan, Z. K.; Kuykendall, T.; An, E. J.; Fu, Y. B.; Battaglia, V.; Zhang, Y. G. 2011. Multilayer nanoassembly of Sn-nanopillar arrays sandwiched between graphene layers for high-capacity lithium storage. *Energy Environ. Sci.* 4:3611–3616.
40. Zou, Y. Q.; Wang, Y. 2011. Sn@CNT nanostructures rooted in graphene with high and fast Li-storage capacities. *ACS Nano* 5:8108–8114.
41. Luo, B.; Wang, B.; Liang, M. H.; Ning, J.; Li, X. L.; Zhi, L. J. 2012. Reduced graphene oxide-mediated growth of uniform tin-core/carbon-sheath coaxial nanocables with enhanced lithium ion storage properties. *Adv. Mater.* 24:1405–1409.
42. Luo, B.; Wang, B.; Li, X. L.; Jia, Y. Y.; Liang, M. H.; Zhi, L. J. 2012. Graphene-confined Sn nanosheets with enhanced lithium storage capability. *Adv. Mater.* 24:3538–3543.
43. Poizot, P.; Laruelle, S.; Grugeon, S.; Dupont, L.; Tarascon, J. M. 2000. Nano-sized transition-metaloxides as negative-electrode materials for lithium-ion batteries. *Nature* 407:496–499.
44. Zhou, G. M.; Wang, D. W.; Li, F.; Zhang, L. L.; Li, N.; Wu, Z. S.; Wen, L.; Lu, G. Q.; Cheng, H. M. 2010. Graphene-wrapped Fe_3O_4 anode material with improved reversible capacity and cyclic stability for lithium ion batteries. *Chem. Mater.* 22:5306–5313.
45. Zhang, M.; Lei, D. N.; Yin, X. M.; Chen, L. B.; Li, Q. H.; Wang, Y. G.; Wang, T. H. 2010. Magnetite/graphene composites: Microwave irradiation synthesis and enhanced cycling and rate performances for lithium ion batteries. *J. Mater. Chem.* 20:5538–5543.

46. Lian, P. C.; Zhu, X. F.; Xiang, H. F.; Li, Z.; Yang, W. S.; Wang, H. H. 2010. Enhanced cycling performance of Fe_3O_4–graphene nanocomposite as an anode material for lithium-ion batteries. *Electrochim. Acta* 56:834–840.
47. Li, B.; Cao, H.; Shao, J.; Qu, M.; Warner, J. H. 2011. Superparamagnetic Fe_3O_4 nanocrystals@graphene composites for energy storage devices. *J. Mater. Chem.* 21:5069–5075.
48. Chen, D.; Ji, G.; Ma, Y.; Lee, J. Y.; Lu, J. 2011. Graphene-encapsulated hollow Fe_3O_4 nanoparticle aggregates as a high-performance anode material for lithium ion batteries. *ACS Appl. Mater. Interfaces* 3:3078–3083.
49. Wang, J. Z.; Zhong, C.; Wexler, D.; Idris, N. H.; Wang, Z. X.; Chen, L. Q.; Liu, H. K. 2011. Graphene-encapsulated Fe_3O_4 nanoparticles with 3D laminated structure as superior anode in lithium ion batteries. *Chem.-Eur. J.* 17:661–667.
50. Chen, W. F.; Li, S. R.; Chen, C. H.; Yan, L. F. 2011. Self-assembly and embedding of nanoparticles by *in situ* reduced graphene for preparation of a 3D graphene/nanoparticle aerogel. *Adv. Mater.* 23:5679–5683.
51. Wu, Z. S.; Ren, W. C.; Wen, L.; Gao, L. B.; Zhao, J. P.; Chen, Z. P.; Zhou, G. M.; Li, F.; Cheng, H. M. 2010. Graphene anchored with Co_3O_4 nanoparticles as anode of lithium ion batteries with enhanced reversible capacity and cyclic performance. *ACS Nano* 4:3187–3194.
52. Yang, S. B.; Feng, X. L.; Ivanovici, S.; Mullen, K. 2010. Fabrication of graphene-encapsulated oxide nanoparticles: Towards high-performance anode materials for lithium storage. *Angew. Chem. Int. Edit.* 49:8408–8411.
53. Kim, H.; Seo, D. H.; Kim, S. W.; Kim, J.; Kang, K. 2011. Highly reversible Co_3O_4/graphene hybrid anode for lithium rechargeable batteries. *Carbon* 49:326–332.
54. Chen, S. Q.; Wang, Y. 2010. Microwave-assisted synthesis of a Co_3O_4–graphene sheet-on-sheet nanocomposite as a superior anode material for Li-ion batteries. *J. Mater. Chem.* 20:9735–9739.
55. Tao, L. Q.; Zai, J. T.; Wang, K. X.; Zhang, H. J.; Xu, M.; Shen, J.; Su, Y. Z.; Qian, X. F. 2012. Co_3O_4 nanorods/graphene nanosheets nanocomposites for lithium ion batteries with improved reversible capacity and cycle stability. *J. Power Sources* 202:230–235.
56. Li, B. J.; Cao, H. Q.; Shao, J.; Li, G. Q.; Qu, M. Z.; Yin, G. 2011. Co_3O_4@graphene composites as anode materials for high-performance lithium ion batteries. *Inorg. Chem.* 50:1628–1632.
57. Paek, S. M.; Yoo, E.; Honma, I. 2009. Enhanced cyclic performance and lithium storage capacity of SnO_2/graphene nanoporous electrodes with three-dimensionally delaminated flexible structure. *Nano Lett.* 9:72–75.
58. Ding, S.; Luan, D.; Boey, F. Y. C.; Chen, J. S.; Lou, X. W. 2011. SnO_2 nanosheets grown on graphene sheets with enhanced lithium storage properties. *Chem. Commun.* 47:7155–7157.
59. Kim, H.; Kim, S. W.; Park, Y. U.; Gwon, H.; Seo, D. H.; Kim, Y.; Kang, K. 2010. SnO_2/graphene composite with high lithium storage capability for lithium rechargeable batteries. *Nano Res.* 3:813–821.
60. Zhang, L. S.; Jiang, L. Y.; Yan, H. J.; Wang, W. D.; Wang, W.; Song, W. G.; Guo, Y. G.; Wan, L. J. 2010. Mono dispersed SnO_2 nanoparticles on both sides of single layer graphene sheets as anode materials in Li-ion batteries. *J. Mater. Chem.* 20:5462–5467.
61. Su, Y. Z.; Li, S.; Wu, D. Q.; Zhang, F.; Liang, H. W.; Gao, P. F.; Cheng, C.; Feng, X. L. 2012. Two-dimensional carbon-coated graphene/metal oxide hybrids for enhanced lithium storage. *ACS Nano* 6:8349–8356.
62. Wang, X.; Cao, X. Q.; Bourgeois, L.; Guan, H.; Chen, S. M.; Zhong, Y. T.; Tang, D. M. et al. 2012. *N*-doped graphene–SnO_2 sandwich paper for high-performance lithium-ion batteries. *Adv. Funct. Mater.* 22:2682–2690.
63. Zhou, X. S.; Wan, L. J.; Guo, Y. G. 2013. Binding SnO_2 nanocrystals in nitrogen-doped graphene sheets as anode materials for lithium-ion batteries. *Adv. Mater.* 25:2152–2157.
64. Wang, L.; Wang, D.; Dong, Z. H.; Zhang, F. X.; Jin, J. 2013. Interface chemistry engineering for stable cycling of reduced GO/SnO_2 nanocomposites for lithium ion battery. *Nano Lett.* 13:1711–1716.
65. Yao, J.; Shen, X. P.; Wang, B.; Liu, H. K.; Wang, G. X. 2009. *In situ* chemical synthesis of SnO_2–graphene nanocomposite as anode materials for lithium-ion batteries. *Electrochem. Commun.* 11:1849–1852.
66. Li, Y. M.; Lv, X. J.; Lu, J.; Li, J. H. 2010. Preparation of SnO_2–nanocrystal/graphene-nanosheets composites and their lithium storage ability. *J. Phys. Chem. C* 114:21770–21774.
67. Wang, H. L.; Cui, L. F.; Yang, Y. A.; Casalongue, H. S.; Robinson, J. T.; Liang, Y. Y.; Cui, Y.; Dai, H. J. 2010. Mn_3O_4–graphene hybrid as a high-capacity anode material for lithium ion batteries. *J. Am. Chem. Soc.* 132:13978–13980.
68. Liu, S. Y.; Xie, J.; Zheng, Y. X.; Cao, G. S.; Zhu, T. J.; Zhao, X. B. 2012. Nanocrystal manganese oxide (Mn_3O_4, MnO) anchored on graphite nanosheet with improved electrochemical Li-storage properties. *Electrochim. Acta* 66:271–278.
69. Lavoie, N.; Malenfant, P. R. L.; Courtel, F. M.; Abu-Lebdeh, Y.; Davidson, I. J. 2012. High gravimetric capacity and long cycle life in Mn_3O_4/graphene platelet/LiCMC composite lithium-ion battery anodes. *J. Power Sources* 213:249–254.
70. Li, L.; Guo, Z. P.; Du, A. J.; Liu, H. K. 2012. Rapid microwave-assisted synthesis of Mn_3O_4–graphene nanocomposite and its lithium storage properties. *J. Mater. Chem.* 22:3600–3605.
71. Mao, S.; Wen, Z. H.; Kim, H.; Lu, G. H.; Hurley, P.; Chen, J. H. 2012. A general approach to one-pot fabrication of crumpled graphene-based nanohybrids for energy applications. *ACS Nano* 6:7505–7513.
72. Zou, Y.; Wang, Y. 2011. NiO nanosheets grown on graphene nanosheets as superior anode materials for Li-ion batteries. *Nanoscale* 3:2615–2620.
73. Kottegoda, I. R. M.; Idris, N. H.; Lu, L.; Wang, J.-Z.; Liu, H.-K. 2011. Synthesis and characterization of graphene–nickel oxide nanostructures for fast charge–discharge application. *Electrochim. Acta* 56:5815–5822.
74. Zhu, X. J.; Hu, J.; Dai, H. L.; Ding, L.; Jiang, L. 2012. Reduced graphene oxide and nanosheet-based nickel oxide microsphere composite as an anode material for lithium ion battery. *Electrochim. Acta* 64:23–28.
75. Mai, Y. J.; Shi, S. J.; Zhang, D.; Lu, Y.; Gu, C. D.; Tu, J. P. 2012. NiO–graphene hybrid as an anode material for lithium ion batteries. *J. Power Sources* 204:155–161.
76. Huang, Y.; Huang, X. L.; Lian, J. S.; Xu, D.; Wang, L. M.; Zhang, X. B. 2012. Self-assembly of ultrathin porous NiO nanosheets/graphene hierarchical structure for high-capacity and high-rate lithium storage. *J. Mater. Chem.* 22:2844–2847.
77. Zhu, X.; Zhu, Y.; Murali, S.; Stoller, M. D.; Ruoff, R. S. 2011. Nanostructured reduced graphene oxide/Fe_2O_3 composite as a high-performance anode material for lithium ion batteries. *ACS Nano* 5:3333–3338.
78. Zou, Y. Q.; Kan, J.; Wang, Y. 2011. Fe_2O_3–graphene rice-on-sheet nanocomposite for high and fast lithium ion storage. *J. Phys. Chem. C* 115:20747–20753.

79. Wang, G.; Liu, T.; Luo, Y. J.; Zhao, Y.; Ren, Z. Y.; Bai, J. B.; Wang, H. 2011. Preparation of Fe_2O_3/graphene composite and its electrochemical performance as an anode material for lithium ion batteries. *J. Alloys Compd.* 509: L216–L220.
80. Zhang, M.; Qu, B. H.; Lei, D. N.; Chen, Y. J.; Yu, X. Z.; Chen, L. B.; Li, Q. H.; Wang, Y. G.; Wang, T. H. 2012. A green and fast strategy for the scalable synthesis of Fe_2O_3/graphene with significantly enhanced Li-ion storage properties. *J. Mater. Chem.* 22:3868–3874.
81. Li, N.; Liu, G.; Zhen, C.; Li, F.; Zhang, L. L.; Cheng, H. M. 2011. Battery performance and photocatalytic activity of mesoporous anatase TiO(2) nanospheres/graphene composites by template-free self-assembly. *Adv. Funct. Mater.* 21:1717–1722.
82. Wang, D. H.; Choi, D. W.; Li, J.; Yang, Z. G.; Nie, Z. M.; Kou, R.; Hu, D. H.et al. 2009. Self-assembled TiO_2–graphene hybrid nanostructures for enhanced Li-ion insertion. *ACS Nano* 3:907–914.
83. Yang, S. B.; Feng, X. L.; Mullen, K. 2011. Sandwich-like, graphene-based titania nanosheets with high surface area for fast lithium storage. *Adv. Mater.* 23:3575–3579.
84. Chen, J. S.; Wang, Z.; Dong, X. C.; Chen, P.; Lou, X. W. 2011. Graphene-wrapped TiO_2 hollow structures with enhanced lithium storage capabilities. *Nanoscale* 3:2158–2161.
85. Ding, S.; Chen, J. S.; Luan, D.; Boey, F. Y. C.; Madhavi, S.; Lou, X. W. 2011. Graphene-supported anatase TiO_2 nanosheets for fast lithium storage. *Chem. Commun.* 47:5780–5782.
86. Shi, Y.; Wen, L.; Li, F.; Cheng, H. M. 2011. Nanosized Li(4) Ti(5)O(12)/graphene hybrid materials with low polarization for high rate lithium ion batteries. *J. Power Sources* 196:8610–8617.
87. Li, N.; Chen, Z. P.; Ren, W. C.; Li, F.; Cheng, H. M. 2012. Flexible graphene-based lithium ion batteries with ultrafast charge and discharge rates. *Proc. Natl. Acad. Sci. USA* 109:17360–17365.
88. Shen, L. F.; Yuan, C. Z.; Luo, H. J.; Zhang, X. G.; Yang, S. D.; Lu, X. J. 2011. *In situ* synthesis of high-loading $Li_4Ti_5O_{12}$–graphene hybrid nanostructures for high rate lithium ion batteries. *Nanoscale* 3:572–574.
89. Kim, H. K.; Bak, S. M.; Kim, K. B. 2010. $Li_4Ti_5O_{12}$/reduced graphite oxide nano-hybrid material for high rate lithium-ion batteries. *Electrochem. Commun.* 12:1768–1771.
90. Ding, Y.; Jiang, Y.; Xu, F.; Yin, J.; Ren, H.; Zhuo, Q.; Long, Z.; Zhang, P. 2010. Preparation of nano-structured $LiFePO_4$/graphene composites by co-precipitation method. *Electrochem. Commun.* 12:10–13.
91. Zhou, X. F.; Wang, F.; Zhu, Y. M.; Liu, Z. P. 2011. Graphene modified $LiFePO_4$ cathode materials for high power lithium ion batteries. *J. Mater. Chem.* 21:3353–3358.
92. Yang, J. L.; Wang, J. J.; Wang, D. N.; Li, X. F.; Geng, D. S.; Liang, G. X.; Gauthier, M.; Li, R. Y.; Sun, X. L. 2012. 3D porous $LiFePO_4$/graphene hybrid cathodes with enhanced performance for Li-ion batteries. *J. Power Sources* 208:340–344.
93. Tang, Y. F.; Huang, F. Q.; Bi, H.; Liu, Z. Q.; Wan, D. Y. 2012. Highly conductive three-dimensional graphene for enhancing the rate performance of $LiFePO_4$ cathode. *J. Power Sources* 203:130–134.
94. Ji, H. X.; Zhang, L. L.; Pettes, M. T.; Li, H. F.; Chen, S. S.; Shi, L.; Piner, R.; Ruoff, R. S. 2012. Ultrathin graphite foam: A three-dimensional conductive network for battery electrodes. *Nano Lett.* 12:2446–2451.
95. Su, C.; Bu, X. D.; Xu, L. H.; Liu, J. L.; Zhang, C. 2012. A novel $LiFePO_4$/graphene/carbon composite as a performance-improved cathode material for lithium-ion batteries. *Electrochim. Acta* 64:190–195.
96. Shi, Y.; Chou, S. L.; Wang, J. Z.; Wexler, D.; Li, H. J.; Liu, H. K.; Wu, Y. P. 2012. Graphene wrapped $LiFePO_4$/C composites as cathode materials for Li-ion batteries with enhanced rate capability. *J. Mater. Chem.* 22:16465–16470.
97. Wang, H. L.; Yang, Y.; Liang, Y. Y.; Cui, L. F.; Casalongue, H. S.; Li, Y. G.; Hong, G. S.; Cui, Y.; Dai, H. J. 2011. $LiMn_{1-x}Fe_xPO_4$ nanorods grown on graphene sheets for ultrahigh-rate-performance lithium ion batteries. *Angew. Chem. Int. Edit.* 50:7364–7368.
98. Liu, H.; Gao, P.; Fang, J.; Yang, G. 2011. $Li_3V_2(PO_4)_3$/graphene nanocomposites as cathode material for lithium ion batteries. *Chem. Commun.* 47:9110–9112.
99. Zhang, L.; Wang, S. Q.; Cai, D. D.; Lian, P. C.; Zhu, X. F.; Yang, W. S.; Wang, H. H. 2013. $Li_3V_2(PO_4)$(3)@C/graphene composite with improved cycling performance as cathode material for lithium-ion batteries. *Electrochim. Acta* 91:108–113.
100. Jiang, K. C.; Xin, S.; Lee, J. S.; Kim, J.; Xiao, X. L.; Guo, Y. G. 2012. Improved kinetics of $LiNi_{1/3}Mn_{1/3}Co_{1/3}O_2$ cathode material through reduced graphene oxide networks. *Phys. Chem. Chem. Phys.* 14:2934–2939.
101. Bak, S. M.; Nam, K. W.; Lee, C. W.; Kim, K. H.; Jung, H. C.; Yang, X. Q.; Kim, K. B. 2011. Spinel $LiMn_2O_4$/reduced graphene oxide hybrid for high rate lithium ion batteries. *J. Mater. Chem.* 21:17309–17315.
102. Liu, H. M.; Yang, W. S. 2011. Ultralong single crystalline V_2O_5 nanowire/graphene composite fabricated by a facile green approach and its lithium storage behavior. *Energy Environ. Sci.* 4:4000–4008.
103. Rui, X. H.; Zhu, J. X.; Sim, D.; Xu, C.; Zeng, Y.; Hng, H. H.; Lim, T. M.; Yan, Q. Y. 2011. Reduced graphene oxide supported highly porous V_2O_5 spheres as a high-power cathode material for lithium ion batteries. *Nanoscale* 3:4752–4758.
104. Lee, J. W.; Lim, S. Y.; Jeong, H. M.; Hwang, T. H.; Kang, J. K.; Choi, J. W. 2012. Extremely stable cycling of ultra-thin V_2O_5 nanowire–graphene electrodes for lithium rechargeable battery cathodes. *Energy Environ. Sci.* 5:9889–9894.
105. Yang, S.; Gong, Y.; Liu, Z.; Zhan, L.; Hashim, D. P.; Ma, L.; Vajtai, R.; Ajayan, P. M. 2013. Bottom-up approach toward single-crystalline VO_2–graphene ribbons as cathodes for ultrafast lithium storage. *Nano Lett.* 13:1596–1601.
106. Ji, X. L.; Lee, K. T.; Nazar, L. F. 2009. A highly ordered nanostructured carbon–sulphur cathode for lithium–sulphur batteries. *Nat. Mater.* 8:500–506.
107. Bruce, P. G.; Freunberger, S. A.; Hardwick, L. J.; Tarascon, J.-M. 2012. Li–O_2 and Li–S batteries with high energy storage. *Nat. Mater.* 11:19–29.
108. Zhou, G. M.; Yin, L.-C.; Wang, D.-W.; Li, L.; Pei, S.; Gentle, I. R.; Li, F.; Cheng, H.-M. 2013. A fibrous hybrid of graphene and sulfur nanocrystals for high performance lithium–sulfur batteries. *ACS Nano* 7(6):5367–5375.
109. Wang, D. W.; Zeng, Q.; Zhou, G. M.; Yin, L.; Li, F.; Cheng, H.-M.; Gentle, I.; Lu, G. Q. 2013. Carbon/sulfur composites for Li–S batteries: Status and prospects. *J. Mater. Chem. A.* 1(33):9382–9394.
110. Manthiram, A.; Fu, Y.; Su, Y.-S. 2012. Challenges and prospects of lithium–sulfur batteries. *Acc. Chem. Res.* 46(5):1125–1134.
111. Yang, Y.; Zheng, G.; Cui, Y. 2013. Nanostructured sulfur cathodes. *Chem. Soc. Rev.* 42:3018–3032.
112. Su, F. Y.; You, C. H.; He, Y. B.; Lv, W.; Cui, W.; Jin, F. M.; Li, B. H.; Yang, Q. H.; Kang, F. Y. 2010. Flexible and planar graphene conductive additives for lithium-ion batteries. *J. Mater. Chem.* 20:9644–9650.

113. Wang, L.; Wang, H.; Liu, Z.; Xiao, C.; Dong, S.; Han, P.; Zhang, Z.; Zhang, X.; Bi, C.; Cui, G. 2010. A facile method of preparing mixed conducting $LiFePO_4$/graphene composites for lithium-ion batteries. *Solid State Ion.* 181:1685–1689.
114. Yang, J. L.; Wang, J. J.; Tang, Y. J.; Wang, D. N.; Li, X. F.; Hu, Y. H.; Li, R. Y.; Liang, G. X.; Sham, T. K.; Sun, X. L. 2013. $LiFePO_4$–graphene as a superior cathode material for rechargeable lithium batteries: Impact of stacked graphene and unfolded graphene. *Energy Environ. Sci.* 6:1521–1528.
115. Lung-Hao Hu, B.; Wu, F.-Y.; Lin, C.-T.; Khlobystov, A. N.; Li, L.-J. 2013. Graphene-modified $LiFePO_4$ cathode for lithium ion battery beyond theoretical capacity. *Nat. Commun.* 4:1687.
116. Bi, H.; Huang, F. Q.; Tang, Y. F.; Liu, Z. Q.; Lin, T. Q.; Chen, J.; Zhao, W. 2013. Study of $LiFePO_4$ cathode modified by graphene sheets for high-performance lithium ion batteries. *Electrochim. Acta* 88:414–420.
117. Zhu, N.; Liu, W.; Xue, M. Q.; Xie, Z. A.; Zhao, D.; Zhang, M. N.; Chen, J. T.; Cao, T. B. 2010. Graphene as a conductive additive to enhance the high-rate capabilities of electrospun $Li_4Ti_5O_{12}$ for lithium-ion batteries. *Electrochim. Acta* 55:5813–5818.
118. Zhang, B.; Yu, Y.; Liu, Y. S.; Huang, Z. D.; He, Y. B.; Kim, J. K. 2013. Percolation threshold of graphene nanosheets as conductive additives in $Li_4Ti_5O_{12}$ anodes of Li-ion batteries. *Nanoscale* 5:2100–2106.
119. Guo, P.; Song, H.; Chen, X.; Ma, L.; Wang, G.; Wang, F. 2011. Effect of graphene nanosheet addition on the electrochemical performance of anode materials for lithium-ion batteries. *Anal. Chim. Acta* 688:146–155.
120. Zhou, J. S.; Song, H. H.; Ma, L. L.; Chen, X. H. 2011. Magnetite/graphene nanosheet composites: Interfacial interaction and its impact on the durable high-rate performance in lithium-ion batteries. *RSC Adv.* 1:782–791.
121. Miao, L.; Wu, J. B.; Jiang, J. J.; Liang, P. 2013. First-principles study on the synergistic mechanism of SnO_2 and graphene as a lithium ion battery anode. *J. Phys. Chem. C* 117:23–27.
122. Radich, J. G.; Kamat, P. V. 2012. Origin of reduced graphene oxide enhancements in electrochemical energy storage. *ACS Catal.* 2:807–816.

11 Graphene Oxide

An Important Derivative of Graphene with Interesting Electrical Properties

S. Mahaboob Jilani and P. Banerji

CONTENTS

It is a profound and necessary truth that the deep things in science are not found because they are useful; they are found because it was possible to find them.

J. Robert Oppenheimer

ABSTRACT

Graphene oxide (GO), a single layer of graphite oxide, emerges as a precursor material for the chemical synthesis of graphene; however, it has some interesting electrical properties that are strongly influenced by the oxygen functional groups arranged on its surface and edges. By adjusting the ratio of carbon to oxygen atoms, a wide range of electrical properties with insulator-to-semimetal characteristics are obtained in GO. In this chapter, we have discussed the electrical properties in GO as an insulator, a semiconductor, and a semimetal. Further, the applications of GO as an active material in resistive-switching nonvolatile memory devices and its composites with ZnO in thin-film transistors have also been discussed.

11.1 INTRODUCTION

Graphene, a monolayer of graphite consisting of a hexagonal arrangement of carbon atoms in a honeycomb structure, was studied as a theoretical entity for a long time until it was experimentally discovered by Novosolov et al. in 2004 [1]. Various methods for the synthesis and some of the remarkable properties of graphene are summarized in Table 11.1.

For large area and flexible applications of graphene, the chemical exfoliation method is the most suitable one. The method involves intercalation or oxygen group functionalization of individual graphene layers in graphite. Therefore, graphene oxide (GO) is defined as a monolayer of graphene with various oxygen-containing functional groups such as hydroxyl and epoxy groups attached on the basal plane and carboxyl, carbonyl, and phenol groups at the edges. The schematic illustration of the synthesis of GO is shown in Figure 11.1.

The oxidation of graphite provides two advantages: (i) the separation between graphene layers in graphite increases due to bonding of oxygen functional groups both on the surface

TABLE 11.1
Different Methods and Properties of Graphene

Various Methods for the Synthesis of Graphene	Remarkable Properties of Graphene [2]
• Mechanical exfoliation [3]	• High carrier mobility (200,000 $cm^2/V/s$ [300 K])
• Chemical vapor deposition [4]	• Superior thermal conductivity (~5000 W/m/K)
• Epitaxial technique [5]	• High Young's modulus (~1.0 TPa)
• Chemical exfoliation [6]	• Optical transmittance (~97.7%)

and at the edges of individual layers, which suppress the van der Waals force between them, and (ii) after oxidation, graphite becomes hydrophilic; it allows graphite oxide to disperse in wide varieties of solvents including deionized water to produce GO. Once the oxygen groups are removed from the surface and the edges of a single-layer GO, a graphene-like material could be obtained that is generally termed as reduced GO. During reduction, the oxygen groups take away some of the carbon atoms from the lattice. Therefore, compared to pristine graphene, reduced GO contains structural defects and consequently, the mobility of charge carriers in reduced GO is moderate. In this chapter, we shall emphasize the electrical properties of GO as an insulator in its pristine form and as a semimetal after its reduction. An application of pristine GO in nonvolatile memory (NVM) will be discussed. An effort will be made to show the efficacy of a GO-based nanocomposite as a conducting channel in thin-film transistors (TFTs).

11.2 BRIEF HISTORY AND SYNTHESIS OF GO

GO, in the form of graphite oxide, is known to researchers for more than a century. The oxidation of graphite was first carried out by the British chemist, B. C. Brodie [7] to calculate the molecular weight of graphite. The process of evolution of

TABLE 11.2
Timeline: Synthesis of Graphite Oxide

Scientist and Year	Synthesis Procedure	Remarks
B. C. Brodie [7] in 1859	Successive oxidation of the slurry of graphite with potassium chlorate ($KClO_3$) in fuming nitric acid (HNO_3)	Time consuming and involves emission of hazardous gases during the reaction
Staudenmaier [8] in 1898	This is an improved Brodie's method by adding sulfuric acid (H_2SO_4) into the mixture of $KClO_3$ and HNO_3 to increase its acidic nature so that the oxidation process can be carried out in a single step	This is also time consuming and involves emission of hazardous gases during the reaction as above
Hummers and Offeman [9] in 1958	Hummers' method involves oxidation of graphite using strong oxidizing agents such as potassium permanganate ($KMnO_4$) and sodium nitrate ($NaNO_3$) in concentrated H_2SO_4	Widely used for the synthesis of graphite oxide in these days

Hummers' method in synthesizing graphite oxide is shown in Table 11.2.

11.3 ELECTRICAL PROPERTIES OF GO

GO is a nonstoichiometric compound [6]; hence, its properties are very much dependent on its synthesis process and ambient conditions. The electrical conductivity of GO is strongly influenced by the oxygen functional groups. Therefore, the band gap of GO could be tuned from an insulator to a semimetal by the selective reduction of GO. Otherwise, one can use it by making a composite with semiconducting materials, say, ZnO to induce electrical activity into it.

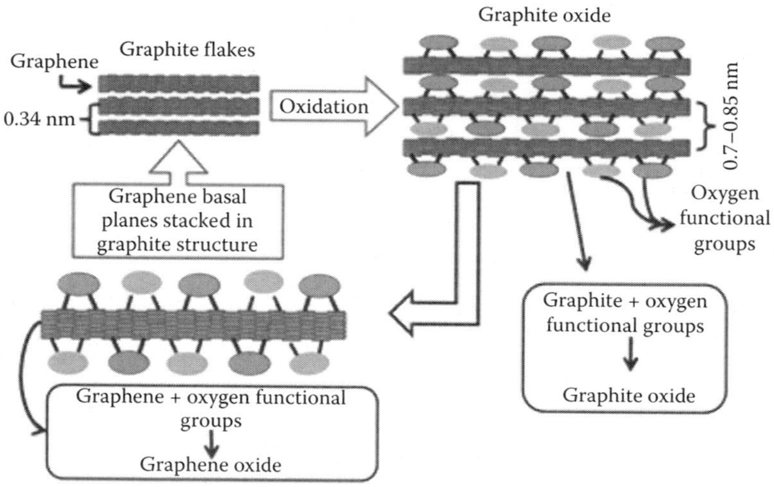

FIGURE 11.1 Schematic illustration of the synthesis of GO.

11.3.1 Effect of Oxygen Functional Groups on Electrical Conductivity

To understand the electrical properties of GO, let us have a look at the structure and electronic properties of pristine graphene [2]. In a two-dimensional (2D) honeycomb structure of carbon atoms (graphene), out of four hybridized electrons ($2s^1$, $2p_x^1$, $2p_y^1$, and $2p_z^1$), three electrons form sp^2 σ-bond with three in-plane carbon atoms and the fourth electron spreads in or out of the graphene surface as sp^2 π-electron cloud. These π-electrons on the surface of a graphene plane move like massless fermions and are responsible for high electron mobility in graphene. However, in case of GO, these π-electrons form a covalent bond with the oxygen functional groups, interrupt the sp^2 π-electron network, and consequently, GO becomes an insulator. Hence, depending on the degree of oxidation (C/O ratio), the bandgap of GO varies from 1.7 to 3.6 eV [10,11]. The surface of GO appears to be like the islands of oxygen functional groups in the π-electrons network. The degree of oxidation, that is, the ratio of π-electrons with that of oxygen functional groups, determines the conductivity in the GO layer with a value ranging from that of an insulator to a semimetal. The electronic band structure of GO is a subject of debate due to the nonstoichiometric distribution of oxygen functional groups on the graphene surface [12]. To restore the π-electron network, several reduction processes are followed to produce reduced GO, a graphene-like semimetal. The different methods are broadly categorized into thermal and chemical reduction processes.

11.3.1.1 Transport Mechanism in Pristine GO

GO has potential applications in electronic and optoelectronic devices due to its unique electrical and optical properties [13]. Venugopal et al. [14] have studied the transport processes in pristine GO deposited onto SiO_2/Si using silver electrodes under van der Pauw configuration in the temperature range 120–300 K. Further, the authors fabricated bottom-gated field-effect transistors (FETs) by depositing Ag source and drain on GO to observe its field-dependent transport properties.

11.3.1.2 Temperature-Dependent Transport Characteristics

Figure 11.2a shows the temperature-dependent resistance in GO thin film. At 300 K, the bulk resistivity of the thin film is found to be 0.442 Ω cm. It is observed that the resistance decreases with increasing temperature indicating the semiconducting nature of GO [14]. Initially, at room temperature, that is, at 300 K, the resistance is found to be 10.4 kΩ that increases gradually to 1.37 MΩ at 120 K. The current–voltage characteristics at different temperatures in the voltage range of −0.8 to 0.8 V are shown in Figure 11.2b that confirm the formation of ohmic contact between Ag and GO. The charge transport in GO is found to follow a variable-range hopping (VRH) model, which describes the successive inelastic tunneling between two localized states. The process, in general, is represented by the relation [14].

$$I = I_0 \exp\left[\left(\frac{-T_0}{T}\right)^{1/n}\right] \quad (11.1)$$

where (n − 1) is the dimensionality of GO.

The temperature-dependent resistance data were fitted with different values of n (2, 3, and 4) and a linear plot was found for lnG versus $T^{-1/4}$ (for n = 4) as shown in Figure 11.2c, where G is the conductance of GO thin film. Therefore, charge transport in multilayer GO follows a three-dimensional VRH mechanism.

11.3.1.3 Field-Effect Transport Properties

To study the field-assisted transport characteristics, TFTs were fabricated with GO as a channel material on SiO_2/Si substrates [14]. The silver source and drain electrodes of width 50 μm and length 60 μm with gold as the back-gate electrode were deposited using thermal evaporation. The transfer characteristics of transistors at drain-to-source voltage, $V_{DS} = 0.15$ V, both in vacuum and atmospheric ambient are shown in Figure 11.2d. From these characteristics, it is observed that the drain current (I_d) is high during the negative sweep of the gate bias, indicating the holes as dominant charge carriers. The threshold voltage and field-effect mobility are found to be, respectively, −14.4 V and 0.25 cm^2/V s in air. However, the threshold voltage shifted to −10.5 V and the field-effect mobility increases to 0.596 cm^2/V s in vacuum. The authors [14] explained the enhancement in conductivity in vacuum. The electrical properties of hygroscopic materials such as GO are strongly influenced by moisture and other adsorbed molecules on the surface of GO. In ambient air, the major reason behind the deterioration of the electrical conductivity in GO is due to the adsorption of oxygen contaminants (since the adsorption of water molecules slightly increases the electrical conductivity [15]). However, in vacuum, these adsorbents are removed from the GO surface, thereby enhancing its electrical conductivity.

Therefore, in its pristine form, the current conduction in GO is due to the hopping of charge carriers through sp^2 islands interrupted by oxygen functional groups. It is also evaluated from the field-effect transfer characteristics that the electrical properties of GO are very much sensitive to the ambient and it behaves as a p-type semiconductor.

11.3.2 GO as Gate Dielectric for FETs

Owing to its unique 2D structures and exceptional carrier mobility, graphene is considered as a suitable active material for future FET applications. Graphene-based FETs are widely reported with different gate dielectrics such as SiO_2, Al_2O_3, HfO_2, and ZrO_2. However, these materials are not suitable as a gate dielectric in flexible TFTs due to the requirement of a high temperature for their processing and other limitations [16]. Recently, GO, an insulating form of graphene, has been reported as a gate dielectric for graphene-based TFTs on flexible substrates. Owing to its transparency in the visible region, the ease of large-area deposition by the solution process, and

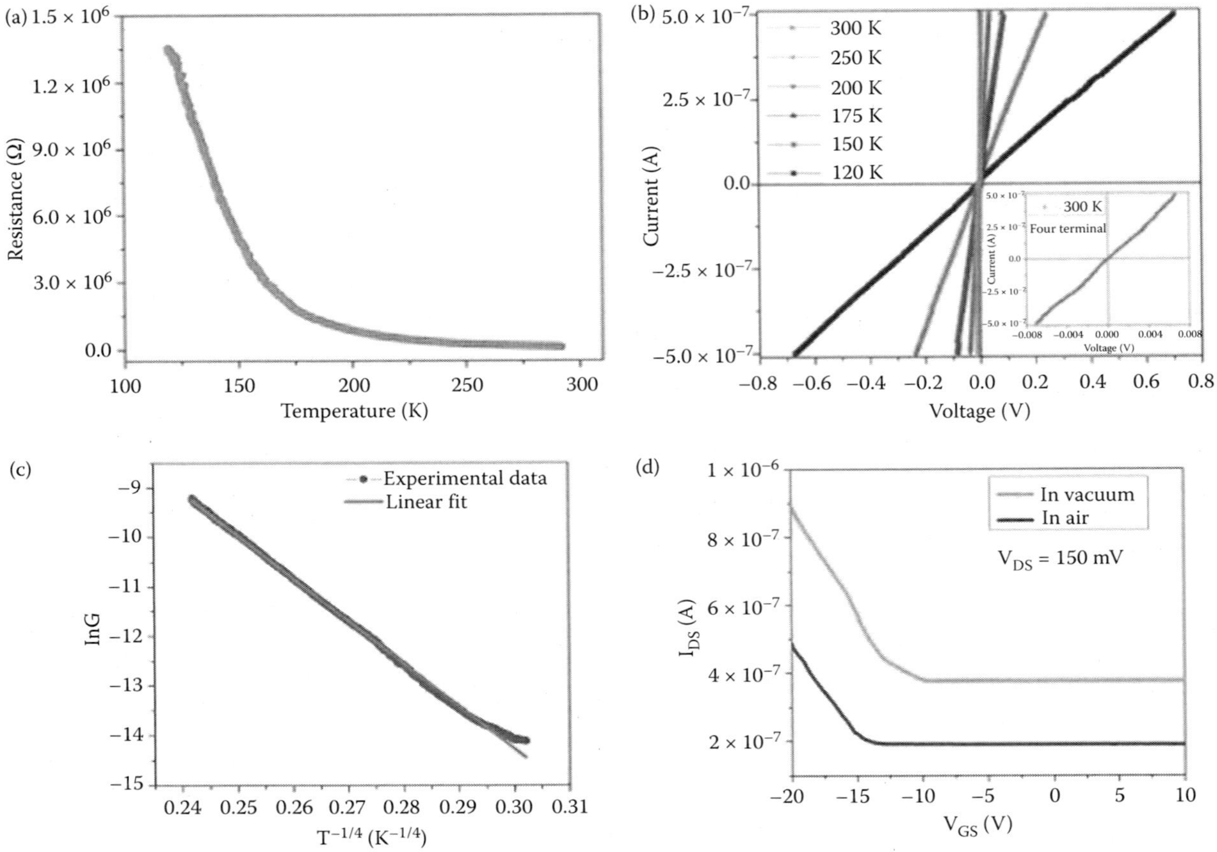

FIGURE 11.2 Temperature-dependent (a) resistance characteristics, (b) two-probe current–voltage characteristics of GO, (c) resistance plot fitted with VRH model, and (d) transfer characteristics of FET with GO as the channel layer. (Reprinted from *Mater. Chem. Phys.* 132, Venugopal, G. et al., An investigation of the electrical transport properties of graphene-oxide thin films, 29–33, Copyright 2012, with permission from Elsevier.)

stretchable mechanical properties, GO is a suitable material for a gate dielectric for flexible TFT applications. Further, the electronic properties of graphene in TFTs are strongly influenced by the interface between the gate dielectric and graphene. Being a derivative of graphene is an added advantage of GO as a dielectric for transistor applications.

TFTs fabricated with all-graphene-based materials has shown promising electronic and optical properties with appreciable mechanical flexibility. The bottom-gated graphene TFTs with GO as the gate dielectric is fabricated in the following process. Single-layer graphene films grown on Cu foil using chemical vapor deposition are transferred onto an oxidized silicon substrate. The gate dielectric of GO (100-nm-thin film) was deposited repeatedly by Langmuir–Blodgett technique onto a patterned graphene gate electrode. Then, monolayer graphene was transferred onto the GO dielectric. Finally, monolithic graphene source and drain electrodes of width 25 μm and length 10 μm were patterned on the active graphene layer (for more details, the reader can refer to Reference 16). This increases the optical transmittance of the device and improves the contact between the source and drain with the active layer. The schematic of a typical all-graphene-based TFT is shown in Figure 11.3a. The transfer and output characteristics of TFT are shown in Figure 11.3b (inset). The device shows ambipolar carrier transport with hole mobility greater than the electron mobility, which is generally observed in graphene FETs on SiO_2/Si. The hole and electron field-effect mobility are found to be 300 and 250 cm^2/Vs at a drain voltage of −0.1 V. This value of the mobility is found to be less compared to that of graphene transistors on SiO_2 dielectric. The resistance of the graphene channel layer with an applied gate bias voltage at temperatures 200, 260, and 300 K is shown in Figure 11.3c. It is observed that the resistance of the channel material decreases with an increase in temperature [16]. Overall, the TFTs fabricated using GO as the gate dielectric have shown reliable performance and are found to be suitable for applications in flexible and transparent electronics.

11.3.3 GO as an Active Material for Resistive Switching

Silicon-based floating-gate memory devices are extensively used as NVM elements in dynamic random access memory. However, further miniaturization of these memory cells poses both technological and fundamental challenges. For this

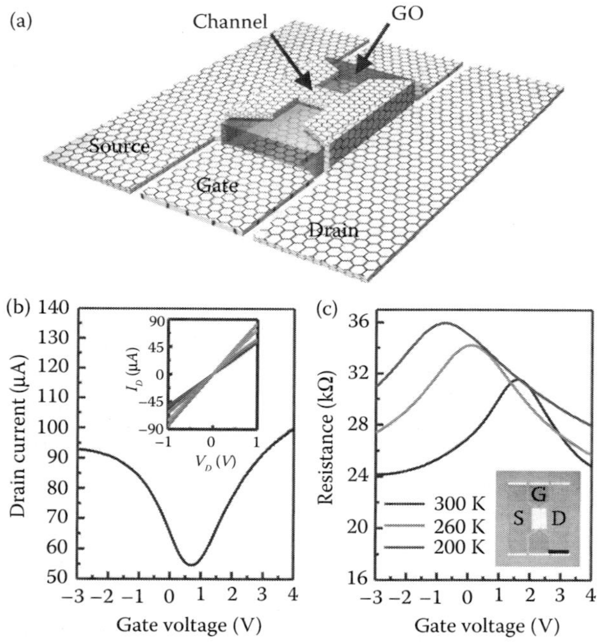

FIGURE 11.3 (a) Schematic representation of all-graphene-based TFTs on flexible substrates. (b) Transfer and output (inset) characteristics of graphene transistors. (c) Resistance versus gate voltage characteristics of TFT at different temperatures. (Reprinted with permission from Lee, S. K. et al. 2012. All graphene-based thin film transistors on flexible plastic substrates. *Nano Lett.* 12, 3472–6. Copyright 2012 American Chemical Society.)

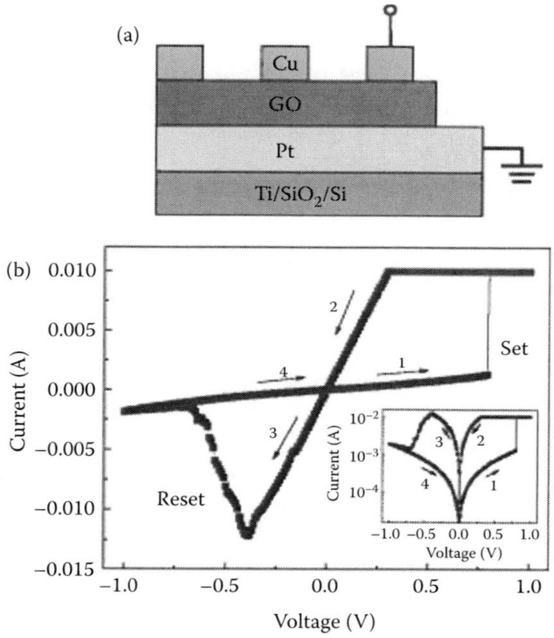

FIGURE 11.4 (a) Schematic illustration of Cu/GO/Pt memory cell on $Ti/SiO_2/Si$ substrate and (b) the current–voltage characteristics of GO-based memory cell. (Reprinted with permission from He, C. L. et al. 2009. Nonvolatile resistive switching in graphene oxide thin films. *Appl. Phys. Lett.* 95: 232101–3. Copyright 2009, American Institute of Physics.)

reason, simple three-layer metal–insulator–metal-structured resistive-switching NVM cells are emerging as an alternative to Si- based floating-gate memory cells. The active insulating layer between the top and bottom electrodes shows a bistability in its resistance state with the applied bias. These are known as high-resistance state (HRS) and low-resistance state (LRS) and are expressed, respectively, as the state "0" and "1" for digital logic. Resistive-switching memory cells are suitable for high-density memory chips as an additional transistor in each memory cell is not required for their functioning. GO, due to its unique 2D structures, is a promising material as an active insulating layer. In the following paragraphs, we discuss the performance of GO-based NVM cells.

The first GO-based NVM device was reported by He et al. [17]. The memory cell is fabricated by depositing 20–40-nm-thick GO thin film onto commercially available $Pt/Ti/SiO_2/Si$ substrates, where Pt acts as the bottom electrode. The top Cu electrode was deposited by room-temperature electron beam evaporation through a shadow mask. The schematic of the memory cell is shown in Figure 11.4a. The current–voltage characteristics were studied using a Keithley 4200 semiconductor parameter analyzer. While taking the measurements, the bias voltage was applied to the top Cu electrode, and the bottom Pt electrode was grounded. The voltage was swept from 0 to 1 V, 1 to 0 V, 0 to −1 V, and then −1 to 0 V with a rate of 0.01 V/s. The current–voltage characteristics are shown in Figure 11.4b. Initially, the device was in HRS, and at about 0.8 V, the cell was switched to LRS. This is called "set" process and the device remained in the same LRS throughout the reverse sweep from 1 to 0 V. During the negative voltage sweep, at about −0.4 V, the memory cell switched back to HRS, called the "reset" process and the cell remains in the same state until the next "set" process exhibits bipolar-resistive characteristics [17].

The authors predicted [17] that the resistive- switching characteristic of GO is due to adsorption/desorption of oxygen functional groups with the applied voltage bias. When a positive potential is applied, the carboxyl, hydroxyl, and epoxide groups are desorbed from the GO surface, restoring sp^2 π-electron network. As a result, the conductivity of GO increases and the memory cell shifts to LRS. At a particular reverse voltage bias, the desorbed oxygen groups at Cu/GO interface are adsorbed on the GO surface forming sp^3-hybridized carbon atoms and consequently, the conductivity of GO decreases and the memory cell switches back to HRS. The resistive switching was also observed with different top electrodes (such as Ti, Au, and Ag); however, the yield and on/off ratio was high for Cu top electrodes compared to Au electrodes. This suggests that the diffusion of metal particles from the top electrode is also responsible for resistive switching in a memory cell.

To find the mechanism of resistive switching in GO, Jeong et al. [13] studied the Al/GO and GO/Al interfaces in Al/GO/Al memory cells using x-ray photoelectron spectroscope (XPS) and x-ray diffraction (XRD).

The XRD patterns of GO/Al and Al/GO/Al are obtained using the coupled-scan method (θ–2θ) as shown in Figure 11.5a. It shows (002) peak of graphite oxide at $2\theta = 11.6°$ corresponding to the d-spacing of 7.6 Å, a value that lies within the d-spacing range of GO layers in graphite oxide. The intensity of (002) peak is found to decrease significantly after the deposition of Al top electrode, indicating that the structure of GO was modified due to the deposition of Al top electrode of thickness 10 nm. The Al/GO/Al structure was further investigated using XRD by fixing $\theta = 2°$. From Figure 11.5b, it is observed that the d-spacing in GO film decreases with increasing the top Al electrode thickness due to the formation of an aluminum oxide layer at Al/GO interface with the loss of oxygen on GO surface [13].

An *in situ* XPS measurement was carried out during deposition of the top Al electrode to study the chemical reaction at the interface between the top Al electrode and GO. In general, the XPS spectrum of as-deposited GO film contains two major peaks. The first one (at 284.6 eV) corresponds to sp^2 carbon–carbon (C–C/C=C) bonding whereas the second major peak (286.5 eV) related to the bonding between sp^3 carbon atoms and oxygen functional groups. From Figure 11.5c, it is found that the second carbon–oxygen sp^3 peak gradually decreases with an increase in Al thickness. Further, the formation of bonding between Al and oxygen is observed from Figure 11.5d and the intensity of Al–O peak increases with Al thickness.

From the results of XRD and XPS, a plausible mechanism of bipolar-resistive switching in Al/GO/Al memory cell in transition from ON (LRS) to OFF (HRS) state is schematically shown in Figure 11.6a. The amorphous insulating layer of Al–O, at Al/GO interface, formed due to the redox reaction of GO, plays a crucial role as an insulating barrier in the HRS of the memory cell [13]. When negative bias is applied at Al top electrode, a conductive filament is formed due to the diffusion of field-induced oxygen ions in the GO film, and the device switches to LRS as shown in Figure 11.6b. This state is maintained throughout the negative bias. When sufficient positive bias is applied, the diffused oxygen ions adsorb on the GO surface and the memory cell switches to HRS. Therefore, the memory cell can be switched from one state to another with applied bias.

The resistive-switching NVM cells discussed so far are based on metal–GO–metal structures. Owing to the heterogeneous structure, the current rectification ratio in metal–GO–metal structures is low (<10). Thus, when memory cells are stacked in a cross bar array, the cross talk between adjacent memory cells cannot be avoided due to sneak current. In general practice, a diode has to be engraved in each memory cell to introduce rectification in current–voltage characteristics so that the sneak current can be blocked. However, the complexity of the metal–GO–metal structures increases due to these additional diodes. To avoid this problem, GO resistive-switching memory cells based on Al–GO–semiconductor structures have been proposed [18]. In metal–insulator–metal-based NVM cells, any material with good electrical conductivity can be used as the bottom electrode. Hence, to avoid the sneak current, an elementary semiconductor such as Si or Ge

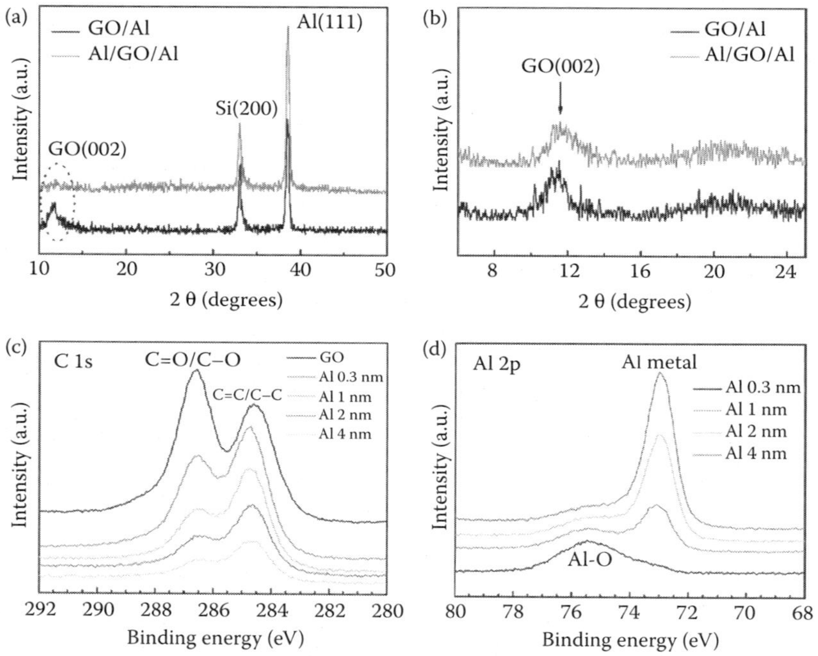

FIGURE 11.5 XRD patterns at GO/Al and Al/GO interface characterized by (a) θ–2θ coupled and (b) fixed θ (2°), method, respectively. XPS spectra of (c) GO C 1s and (d) Al 2p, measured during the deposition of Al top electrode on GO. (Reprinted with permission from Jeong, H. Y. et al. 2010. Graphene oxide thin films for flexible nonvolatile memory applications. *Nano Lett.* 10, 4381–6. Copyright 2010 American Chemical Society.)

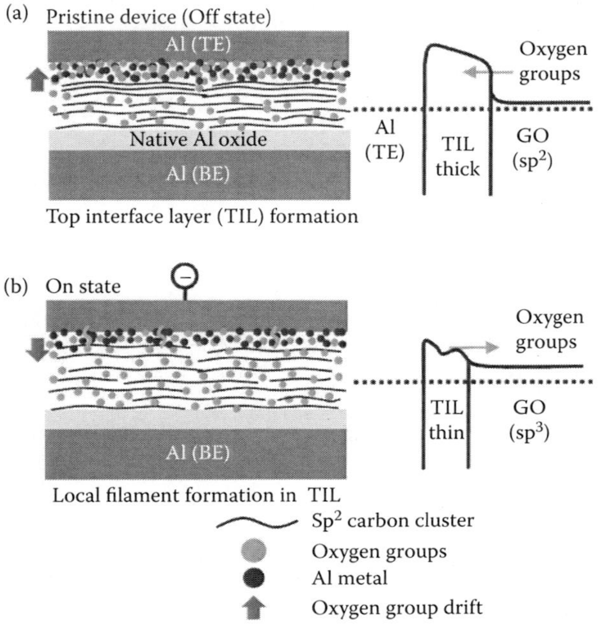

FIGURE 11.6 Schematic of the resistive-switching mechanism in Al/graphene-oxide/Al memory cell in (a) Off state and (b) On state. (Reprinted with permission from Jeong, H. Y. et al. 2010. Graphene oxide thin films for flexible nonvolatile memory applications. *Nano Lett.* 10, 4381–6. Copyright 2010 American Chemical Society.)

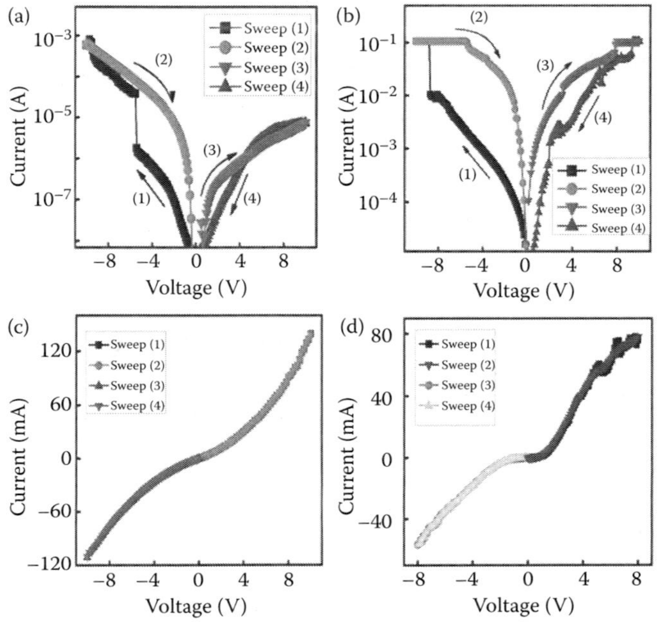

FIGURE 11.7 Current–voltage characteristics of (a) Al/GO/Si, (b) Al/GO/Ge memory cells, (c) Al/Si, and (d) Al/Ge structures. (Reprinted from Carbon, DOI: http://dx.doi.org/10.1016/j.carbon.2013.07.05, Jilani, S. M. et al., Studies on resistive switching characteristics of Al/graphene-oxide/semiconductor memory cells, Copyright 2013, with permission from Elsevier.)

has been used as the bottom electrode. In the following section, the fabrication, performance, and mechanism of resistive switching in Al/GO/Si (or Ge) memory cells are discussed.

11.3.4 Resistive-Switching Characteristics of Al/GO/Si (or Ge) Memory Cells

GO flakes are dispersed in deionized water with a concentration of 1 mg/mL. The solution thus produced is spin coated onto cleaned Si (or Ge) substrates. The substrate was annealed for 10 min and the process was repeated to deposit a GO thin film of thickness ~50 nm. Aluminum electrodes of dimensions 600 × 600 μm are deposited using thermal evaporation on the surface of GO/Si (or Ge) as the top electrode. Large-area Al electrodes are also deposited at the bottom of the Si (or Ge) to form ohmic contacts to the semiconductors. Two more structures, namely, Al/Si/Al and Al/Ge/Al, without GO, are also prepared for comparison [18].

The current–voltage characteristics of Al/GO/Si, Al/GO/Ge, Al/Si/Al, and Al/Ge/Al are measured by grounding the bottom electrode and sweeping the voltage at the top electrode from 0 to −10 V, −10 to 0 V, 0 to 10 V, and 10 to 0 V. From the current– voltage characteristics of Al/GO/Si memory cells (Figure 11.7a), it is observed that during negative sweep from 0 to −10 V (sweep 1), the device was initially in HRS; however, it switches to LRS at −5.4 V. The device remained in the same LRS during sweeping from −10 to 0 V (sweep 2) and switches to the HRS in the positive sweep from 0 to 10 V (sweep 3) [18]. However, in the case of Al/GO/Ge memory cell (Figure 11.7b), the device was set to LRS at −8.7 V and current in the device reaches to compliance current limit (0.1 A in our case) showing breakdown current–voltage characteristics. No such resistive switching was observed in current–voltage characteristics of Al/Si/Al and Al/Ge/Al structures (shown, respectively in Figure 11.7c and d), indicating that the origin of resistive switching lies in GO.

To understand the origin of resistive switching, XPS analysis has been carried out at Al/GO and GO/Si interfaces in Al/GO/Si memory cells. Figure 11.8a shows the spectrum of Al 2p region at Al/GO interface. The peak corresponding to Al–O bond is due to oxidation of Al utilizing the oxygen groups on the surface of GO and consequently, the GO at Al/GO interface reduces. Comparing the spectra of GO at Al (Figure 11.8b) and Si (Figure 11.8c) interface, it is observed that the relative intensity of C–O/C=O peak is low at Al/GO interface due to the reduction of GO. Hence, a conducting filament is formed at Al/GO interface. Generally, the oxide layer on the bottom electrode of a metal–insulator–metal structure plays an important role in preventing the memory cells from breakdown. The reactive metals such as Al or Ag are preferred as the bottom electrode over the nonreactive noble metals such as Au or Pt [13]. In case of Al–GO–semiconductor structures, breakdown in the memory cell was observed in Al/GO/Ge, whereas, no such phenomenon was observed in Al/GO/Si. The native SiO_2 at GO/Si interface (Figure 11.8d), due to its insulating nature, prevents the Al/GO/Si memory cell from breakdown. The existence of SiO_2 is established in XPS measurements [18]. However, breakdown was observed

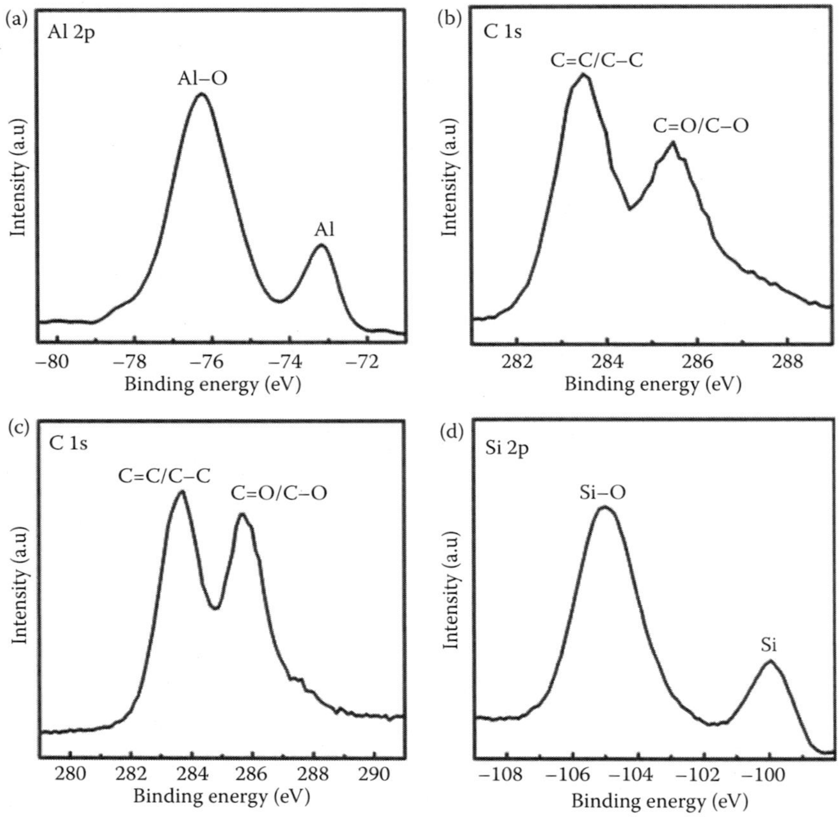

FIGURE 11.8 XPS spectra of (a) Al 2p, (b) GO (C 1s) at Al/GO interface, (c) GO (carbon 1s), and (d) Si at GO/Si interface of Al/GO/Si memory cell. (Reprinted from Carbon, DOI: http://dx.doi.org/10.1016/j.carbon.2013.07.05, Jilani, S. M. et al., Studies on resistive switching characteristics of Al/graphene-oxide/semiconductor memory cells, Copyright 2013, with permission from Elsevier.)

in Al/GO/Ge devices due to the absence of a stable native oxide at GO/Ge interface (such as SiO_2 in Si).

The mechanism of resistive switching in GO memory cells is explained using Figure 11.9a and b. Initially, the memory cell was in the HRS state. During negative bias, the holes from the semiconductor and electrons from the Al electrode are injected into the GO layer. The electrons are trapped by oxygen functional groups and at a sufficient negative potential, the oxygen groups get detached from the GO surface and move toward the GO/Si (or Ge) interface forming the conducting filament between Al and Si (or Ge). Hence, the device switches to LRS. During positive sweep (Figure 11.9b), the charge carriers from Al and Si or Ge (majority charge carriers) move away from the GO layer and hence, a rectification in current–voltage characteristics is observed and the sneak current is minimized. At a sufficient positive bias, the oxygen groups at the GO/Si (or Ge) interface are readsorbed on the GO surface and consequently, the conducting filament is closed and the device switches back to HRS [18]. Therefore, with the application of a semiconductor as the bottom electrode, the sneak current can be minimized without any additional diode structure.

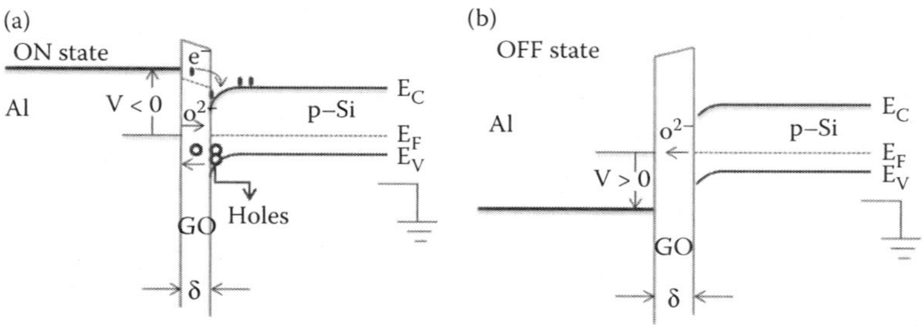

FIGURE 11.9 Schematic representation of resistive switching in Al/GO/Si memory cell. (Reprinted from Carbon, DOI: http://dx.doi.org/10.1016/j.carbon.2013.07.05, Jilani, S. M. et al., Studies on resistive switching characteristics of Al/graphene-oxide/semiconductor memory cells, Copyright 2013, with permission from Elsevier.)

11.3.5 GO–ZnO-Based Nanocomposite as a Channel Material for Transistors

Exploiting the unique property of tunable electrical conductivity, pristine GO-based thin-film FETs have been fabricated by Jin et al. [19]. The author employed partially oxidized (duration of oxidation: 5 min) as well as heavily oxidized (duration of oxidation: 60 min) GO as the channel material. The heavily oxidized GO (band gap 2.1 eV) behaves like an insulator, exhibiting poor transfer characteristics. However, the partially oxidized GO (band gap 1.7 eV) has shown p-type electrical conductivity in argon ambient. So, partially oxidized GO is a suitable candidate as a channel material for FET applications. However, the partially oxidized GO has enough tendency to agglomerate because of the van der Waals forces between consecutive graphene layers [6]. To enhance the electrical conductivity and to prevent the agglomeration of graphene during the reduction process, pristine GO-based nanocomposites can be used for transistors. Metal or metal-oxide-based nanostructures can be anchored onto the GO surface to enhance its electrical conductivity. By anchoring ZnO or TiO_2 nanostructures, GO can be reduced through photocatalytic reaction.

Using GO–ZnO nanocomposite as a channel material, TFTs have been fabricated [21] to enhance the electrical conductivity in GO-based TFTs. The conductivity of the composite was increased by photocatalytic reduction of GO with ZnO nanostructures. The schematic of a GO–ZnO-based transistor on SiO_2/Si with aluminum as the source, drain, and back-gate electrodes is shown in Figure 11.10a. The surface morphology of the GO–ZnO composite channel with ZnO nanostructures is characterized by a field-effect-scanning electron microscope. The ZnO nanostructures of length 140 nm and width 50 nm anchored on the GO surface are shown in Figure 11.10b. For a comparison, TFTs with pristine GO as the channel layer are also fabricated.

The field-effect transfer and output characteristics of GO-based TFTs are shown in Figure 11.11a. It is observed that

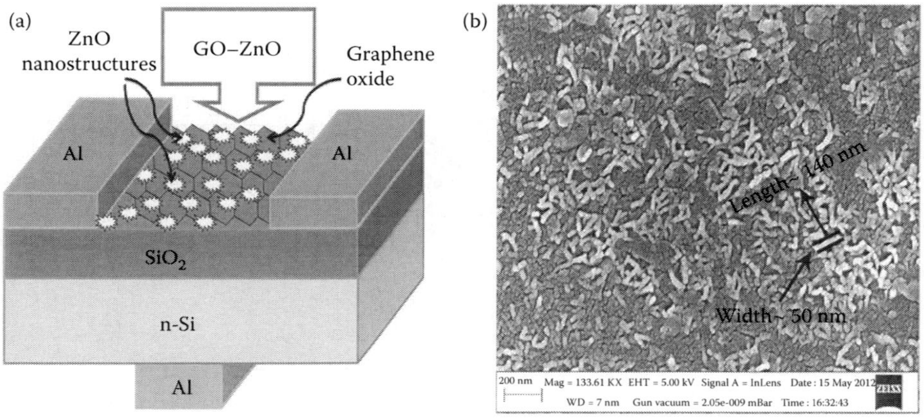

FIGURE 11.10 (a) Schematic representation of GO–ZnO TFTs and (b) the field-effect scanning electron microscope image of GO–ZnO composites. (Reprinted with permission from Jilani, S. M., Gamot, T. D. and Banerji, P. 2012. Thin-film transistors with graphene oxide nanocomposite channel. *Langmuir* 28, 16485–9. Copyright 2012 American Chemical Society.)

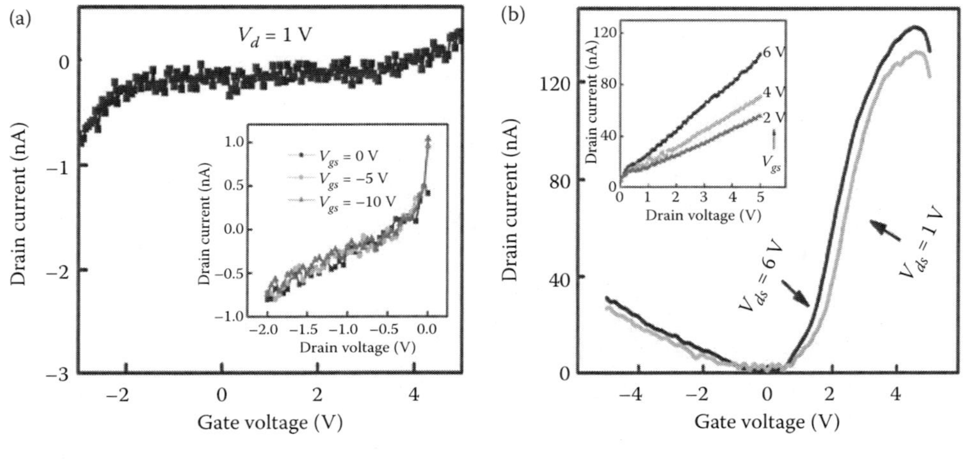

FIGURE 11.11 Transfer and output characteristics (inset) of (a) GO-based TFT and (b) GO–ZnO TFT. (Reprinted with permission from Jilani, S. M., Gamot, T. D. and Banerji, P. 2012. Thin-film transistors with graphene oxide nanocomposite channel. *Langmuir* 28, 16485–9. Copyright 2012 American Chemical Society.)

the conductivity of the transistors remains nearly the same with an applied gate electric field due to the insulating characteristics of the GO channel layer. However, in case of GO–ZnO TFT, the device showed ambipolar electrical conductivity (Figure 11.11b), though the dominant carriers are *n*-type. The *n*-type conductivity of the channel material originates from anchoring of ZnO nanostructures, the as-grown ZnO being *n*-type. The mobility of electrons is found to be 1.94 cm^2/Vs. The transport characteristics are explained by fluctuation-induced charge transport model. Therefore, by anchoring of the metal oxide semiconductor such as ZnO, the field-effect conductivity in pristine GO can be enhanced and thus, the resultant composite can be used as a channel material for the fabrication of TFTs. Yoo et al. [22] reported that the *n*-type conductivity in GO can be enhanced by anchoring metal oxides such as ZnO or TiO_2 on GO surface. The metal oxide nanoparticles (ZnO or TiO_2) generate electron–hole pairs and these electrons are transferred to the graphene lattice enhancing the *n*-type electrical conductivity in GO composites.

11.3.5.1 Electrical Properties of Reduced Graphene Oxide

One of the important applications of GO is as a precursor material for the chemical synthesis of graphene. Once a single layer of GO is obtained, a graphene-like material called reduced graphene oxide (r-GO) can be prepared by removing oxygen functional groups from its surface and edges. r-GO is also a 2D sheet of graphite having some similar properties of graphene such as a high specific area, low sheet resistance 10^2–10^3 Ω per square, and high optical transmittance (80% at 550 nm) [23]. In general, r-GO is not simply pronounced as being "graphene" because the properties of r-GO are not the same as that of pristine graphene due to the presence of defects and residual oxygen moieties even after its reduction. The reduction of GO can be broadly made by (i) chemical, and (ii) thermal methods that are described in the following sections.

11.3.5.2 Chemical Reduction of GO

GO sheets can be chemically reduced using strong reducing agents such as hydrazine, hydrazine hydrate, dimethylhydrazine, and sodium borohydride [23]. The chemical reduction involves the reaction of reducing reagents with GO at room temperature or at moderate temperature below 100°C. Using a simple equipment via the chemical reduction process, mass production of graphene is thus possible at cheaper prices. Here, we represent the chemically r-GO as "chemically modified graphene" (CMG) to indicate the reduction of GO by the chemical route.

11.3.5.3 Electrical Properties of Chemically r-GO

To study the electrical properties of CMG, GO was deposited onto SiO_2/Si substrates using the spray deposition technique [23]. In this technique, the GO dispersions are spray coated onto a preheated SiO_2/Si substrate so that the GO sheets stick onto the SiO_2 surface and the solvents evaporate instantly. While depositing the aqueous dispersions of GO, a small amount of organic solvents such as dimethylformamide, methanol, ethanol, or acetone is used to avoid the agglomeration of GO platelets in deionized water. Compared to other methods such as drop casting or dip coating, a uniform thin film of GO sheets can be obtained using this method.

The reduction of GO on SiO_2/Si by anhydrous hydrazine vapor is explained as follows. The SiO_2/Si substrate is heated to 80°C and hydrazine vapors are passed onto the substrate using helium as the carrier gas. This setup is used to avoid the condensation of hydrazine liquid and moisture onto the deposited GO surface. The source and drain electrodes (Au) of thickness 30 nm are deposited onto CMG sheets as shown in Figure 11.12a. The channel length and width of the electrodes are respectively, 500 nm and 20 μm. Figure 11.12b shows the atomic force microscopic image of source and drain electrodes deposited onto the CMG sheet. The thickness of the deposited thin films of CMG is observed to be 8–9 Å (from Figure 11.12c) representing two CMG sheets. To compare the electrical properties of CMG with GO, necessary electrical contacts are prepared on the GO surface without its reduction.

The sheet resistance of GO and that of CMG are reported to be 4×10^{10} and 4×10^6 Ω per square, respectively. The enhancement in the electrical conductivity in CMG (Figure 11.13) is due to the removal of oxygen groups from

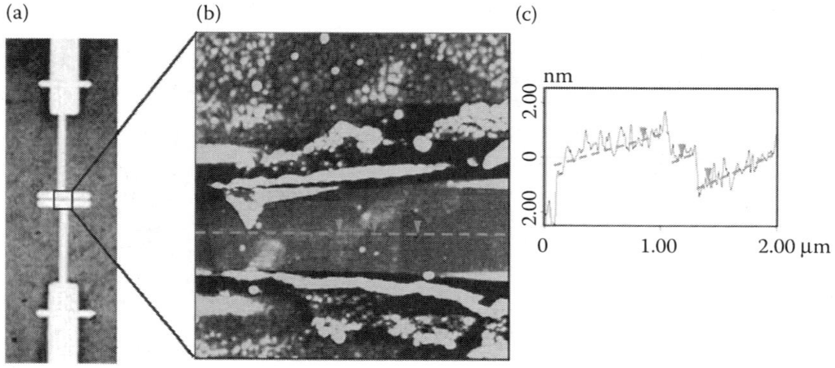

FIGURE 11.12 (a) Optical microscope image of source and drain electrodes onto CMG sheet, (b) the atomic microscope image of CMG sheet between source and drain electrodes, and (c) the thickness of CMG is 8–9 Å. (Reprinted with permission from Gilje, S. et al. 2007. A chemical route to graphene for device applications. *Nano Lett.* 7, 3394–8. Copyright 2007 American Chemical Society.)

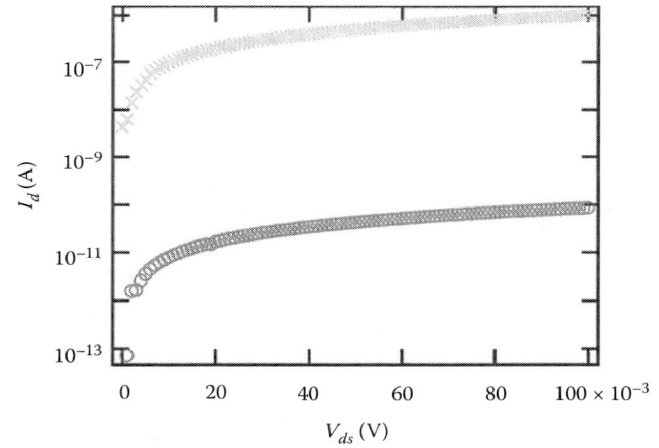

FIGURE 11.13 Drain current before (circles) and after reduction (cross) of GO. (Reprinted with permission from Gilje, S. et al. 2007. A chemical route to graphene for device applications. *Nano Lett.* 7, 3394–8. Copyright 2007 American Chemical Society.)

the surface of GO. The ratio of sheet resistance of GO and CMG, fabricated following the above process, is found to be 10^4–10^5. Therefore, the conductivity of GO can be enhanced by several orders after its reduction.

To study the field-effect transport properties in CMG, a back-gated FET is fabricated by depositing an Au electrode at the back of SiO_2/Si substrates. The gate voltage varies from −15 to 15 V as shown in the inset of Figure 11.14a and the current is found to be increasing due to the lowering of gate voltage such as a p-type semiconductor. The FETs fabricated using pristine graphene as the channel layer show ambipolar electrical conductivity [16] due to the zero band gap of graphene. However, in CMG layers, the band gap is induced due to the presence of residual oxygen moieties and defects formed due to the loss of carbon atoms during reduction in the honeycomb structure.

To find the type of contact between Au and CMG, Gilje et al. [23] measured the drain current at different temperatures and fitted it with the Schottky equation $I = AT^2e^{-q(\Phi-V)/kT}$. The barrier height ($\Phi = 54.8$ meV) between the metal and CMG was determined from the I_d/T^2 versus $1/T$ plot as shown in Figure 11.14b. This further confirms that the CMG exhibits semiconducting characteristics. Therefore, by partially reducing GO, p-type semiconducting CMG can be prepared. Using chemical methods, large-area CMG could be deposited onto different types of substrates at relatively low temperatures below 100°C. Kaiser et al. [24] studied the electrical characteristics of completely reduced GO monolayers and reported that the electrical conduction was dominated by 2D VRH and tunneling (due to an applied electrical field). The authors also showed that the VRH conduction dominated from 40 to 200 K, whereas the field-driven tunneling dominates at low temperatures, down to 2 K.

11.3.5.4 Thermal Reduction of GO

GO can be reduced to a graphene-like material by proper heat treatment. The sheet resistance of GO in its pristine form is 10^{12} Ω per square. However, the sheet resistance drops to 0.37 MΩ per square after its thermal reduction, which is even high compared to the pristine graphene. This is due to the residual oxygen content that is still present in GO after its thermal reduction. In this chapter, we have not discussed the reduction mechanism in detail; however, for more information, the reader may follow the review of Pei and Cheng [25].

11.3.5.5 Low-Temperature Thermal Reduction of GO

The GO sheets are reduced using the step-by-step annealing process at 250°C [26]. The change in the electrical characteristics of GO with variation in annealing temperature and time are discussed in this section. To fabricate the samples, Jung et al. [26] used graphite oxide sheets that are prepared using a modified Hummers' method. Instead of following the conventional sonication process, the graphite oxide is stirred in deionized water for nearly 1 week to obtain large-area (the lateral size being 100 μm) single-layer GO sheets. The sonication process degrades the size of the GO sheets; hence, the stirring process is preferred. The GO sheets are deposited onto SiO_2/Si substrate where the thickness of SiO_2 is 300 nm.

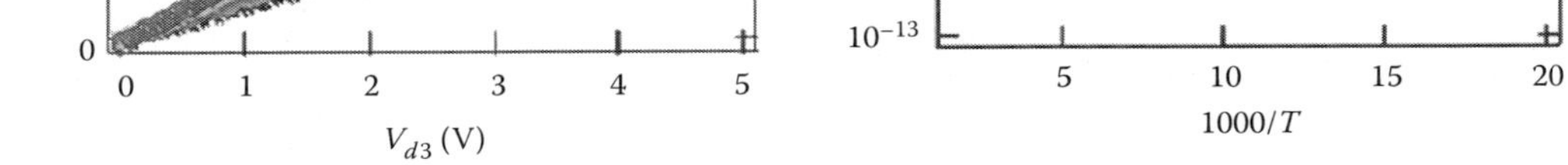

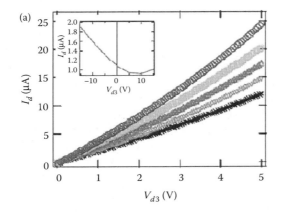

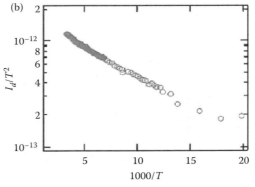

FIGURE 11.14 (a) Transfer (inset) and output characteristics of CMG-based FET and (b) the temperature-dependent resistance data fitted in I_d/T^2 versus $1/T$ plot. (Reprinted with permission from Gilje, S. et al. 2007. A chemical route to graphene for device applications. *Nano Lett.* 7, 3394–8. Copyright, 2007 American Chemical Society.)

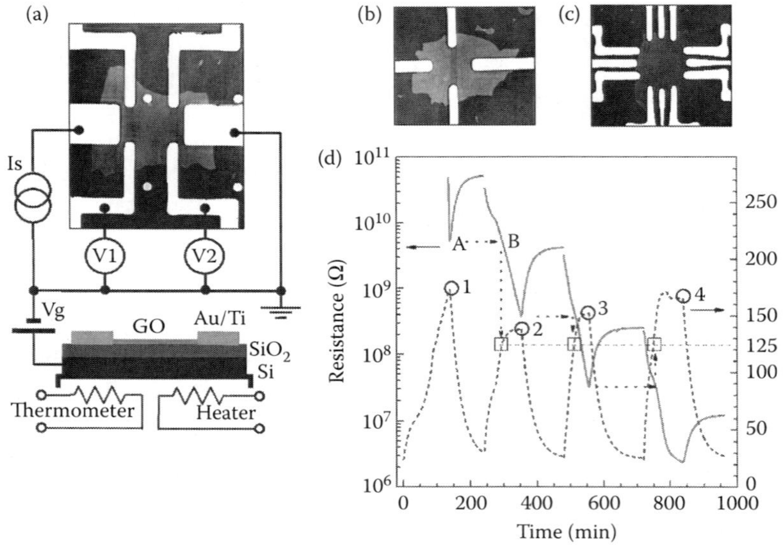

FIGURE 11.15 (a) Electrical circuit diagram for measuring resistivity using van der Pauw configuration and field-effect transport properties; optical microscopic image of (b) Au electrodes in the four-probe arrangement, (c) different combinations of source and drain electrodes, and (d) the resistance of GO at different temperatures and time of annealing. (Reprinted with permission from Pei, S. and Cheng, H. M. 2012. Reduction of graphene oxide. *Carbon* 50, 3210–28. Copyright 2012 American Chemical Society.)

Single-layer and large-area GO sheets of thickness ~1 nm are obtained. For measuring the electrical properties, four Au/Ti electrodes in van der Pauw configuration and different source and drain combinations of channel length ranging from 10 to 50 μm are deposited using electron beam lithography.

The optical microscopic images of different combinations of source and drain electrodes are shown in Figure 11.15a through c. The resistance of r-GO sheets at different annealing temperatures and time is shown in Figure 11.15d. The measurements are carried out in a vacuum chamber (~10^{-5} torr) to avoid the influence of atmospheric conditions. The variation in the chemical composition and structure of GO is observed through the change in the resistance of GO. In Figure 11.15d, the solid line indicates the resistance of the GO sheet and the dashed line represents the temperature with respect to time. The temperature is raised by 1°C/min and interrupted at four different temperatures, namely, 172°C (point 1), 138°C (point 2), 152°C (point 3), and 165°C (point 4). At each interruption, the sample is brought to room temperature and the corresponding resistance is measured. It is observed that a considerable change in resistance of the sample occurred after 125 ± 5°C in each heating and cooling cycle and the reduction process starts at about 160°C. At the fourth cycle of heating and cooling, that is, after 800 min, the resistance of the sample saturates and attains a value of 2 MΩ.

At each interruption, when the sample is cooled to room temperature, its current–voltage characteristics are studied. The following observations are made from the I–V plot as shown in Figure 11.16a: (i) the current in GO varies nonlinearly with applied voltage, (ii) there is asymmetry in I–V curves, and (iii) the resistance of the sample decreases with an increase in applied direct current (dc) bias. The differential conductance is found to vary linearly with the current at a low level of reduction (Figure 11.16b); however, such dependence is not observed at a high level of reduction. As the authors [26] used the four-probe configuration, the effect of parasitic resistance is suppressed and so, the current is not influenced by the contact between the metal and GO. In semiconductors having low mobility, the current is due to space charge-limited conduction (SCLC) when the number of injected carriers through a metal electrode is higher than its intrinsic carrier concentration [27]. In the case of GO, at high electric fields, the injected charge carriers are trapped by the defects and residual oxygen moieties present in r-GO and generate the local electric field, which enhances the conductivity in r-GO. Therefore, the charge transport in GO is dominated by SCLC as r-GO behaves like a p-type semiconductor and thus, the resistance decreases at high electric fields. To study the SCLC conduction in GO, the current–voltage curves are plotted in log (I)–log (V) scale and fitted with the relation $V(I) = R(I) \times I$. For the second (low-level reduction) and fourth (high-level reduction) level of GO reduction, the fitted curves are shown, respectively, in Figure 11.16c and d. For high-level reduction, the current fairly follows the theoretical plot (solid line) in Figure 11.16d; however, a deviation is observed for low-level reduction of GO in Figure 11.16c. Thus, it is found that the experimental data are reasonably consistent with the theory of SCLC conduction in GO samples, where using thermal treatment, most of the oxygen groups have been removed.

Jung et al. [28] studied the electrical conductivity in thermally and chemically reduced GO and found that the conduction in r-GO is influenced by the water molecules in ambient conditions. The conductivity was found to be superior in chemical reduction followed by thermally reduced GO samples compared to either chemically or thermally reduced GO.

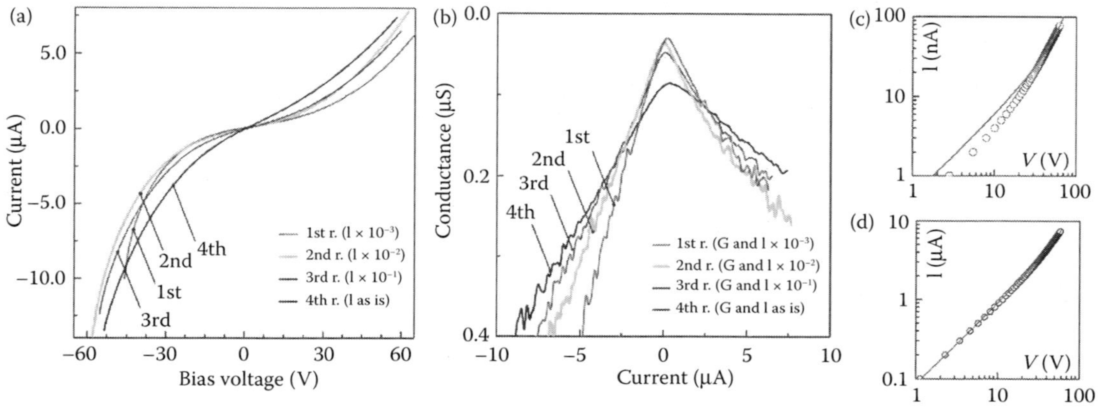

FIGURE 11.16 (a) Current–voltage characteristics of GO at room temperature after each step of reduction, (b) the differential conductance versus current plot, and (c) and (d) log I versus log V plots, respectively, after the second and fourth step of reduction. (Reprinted with permission from Pei, S. and Cheng, H. M. 2012. Reduction of graphene oxide. *Carbon* 50, 3210–28. Copyright 2012 American Chemical Society.)

To further enhance the conductivity in r-GO, Cheng et al. [29] reported the methane plasma restoration approach to restore the lost carbon atoms during the reduction of GO. Owing to its graphene-like structure, moderate electrical conductivity, and optical transparency, r-GO is being used as transparent electrodes to replace the conventional indium tin oxide (ITO) [30].

11.4 CONCLUSION

In this chapter, the interesting electrical properties of GO and its applications in pristine, insulating, and reduced graphene-like semimetallic forms have been discussed. It can be summarized as follows:

1. By adjusting the content of oxygen functional groups in GO, the electrical properties can be tuned from an insulator to a semimetal.
2. Its pristine form behaves like an insulator and the electrical transport can be explained by invoking the VRH model.
3. It is found to be a potential candidate for the gate dielectric in all-graphene-based TFTs for transparent electronic applications.
4. Resistive switching in GO thin films in metal–GO–metal structures has been discussed. From the resistive-switching characteristics of Al/GO/semiconductor memory cells, it is found that the sneak current between the adjacent devices can be minimized by replacing the bottom metal electrode with a semiconductor.
5. The electrical conductivity in GO can be enhanced by anchoring metal oxide, say, ZnO nanostructures on its surface. The field-effect electrical properties of GO–ZnO composites are explained from the TFTs fabricated using GO–ZnO nanocomposite as the channel layer.
6. The low-temperature thermal reduction of GO is discussed and it is observed that in partially reduced GO, the electrical transport is dominated by SCLC.
7. GO can be reduced chemically using hydrazine as the reducing agent. In chemically reduced GO, the transport of charge carriers follows the VRH mechanism from 200 K to 40 K and below 40 K, the conduction is due to the tunneling.

REFERENCES

1. Novoselov, K. S., Geim, A. K., Morozov, S. V., Jiang, D., Zhang, Y., Dubonos, S. V., Grigorieva, I. V. and Firsov, A. A. 2004. Electric field effect in atomically thin carbon films. *Science* 306: 666–69.
2. Geim, A. K. and Novoselov, K. S. 2007. The rise of graphene. *Nat. Mater.* 6: 183–91.
3. Novoselov, K. S., Geim, A. K., Morozov, S. V., Jiang, D., Katsnelson, M. I., Grigorieva, I. V. and Dubonos, S. V. 2005. Two dimensional gas of massless fermions in graphene. *Nature* 438: 197–200.
4. Li, X. and Cai, W. 2009. Large-area synthesis of high quality and uniform graphene films on copper foils. *Science* 324: 1312–4.
5. Hass, J., de Heer, W. A. and Conrad, E. H. 2008. The growth and morphology of epitaxial multilayer graphene. *J. Phys.: Condens. Matter* 20: 323202.
6. Marcano, D. C., Kosynkin, D. V., Berlin, J. M., Sinitskii, A., Sun, Z., Slesarev, A., Alemany, L. B., Lu, W. and Tour, J. M. 2010. Improved synthesis of graphene oxide. *ACS Nano* 4: 4806–14.
7. Brodie, B. C. 1859. On the atomic weight of graphite. *Phil. Trans. R. Soc. Lond.* 149: 249–59.
8. Staudenmaier, L. 1898. Verfahren zur Darstellung der Graphitsäure. *Ber. Dtsch. Chem. Ges.* 31: 1481–7.
9. Hummers, W. S. and Offeman, R. E. 1958. Preparation of graphitic oxide. *J. Am. Chem. Soc.* 80: 1339.
10. Jeong, H. K., Jin, M. H., So, K. P., Lim, S. C. and Lee, Y. H. 2009. Tailoring the characteristics of graphite oxides by different oxidation times. *J. Phys. D: Appl. Phys.* 42: 065418.
11. Mathkar, A., Tozier, D., Cox, P., Ong, P., Galande, C., Balakrishnan, K., Reddy, A. L. M. and Ajayan, P. M. 2012. Controlled, stepwise reduction and bandgap manipulation of graphene oxide. *J. Phys. Chem. Lett.* 3: 986–91.

12. Dreyer, D. R., Park, S., Beilawski, C. W. and Rouff, R. S. 2010. Chemistry of graphene oxide. *Chem. Soc. Rev.* 39: 228–40.
13. Jeong, H. Y., Kim, J. Y., Kim, J. W., Hwang, J. K., Kim, J. E. and Lee, J. Y. 2010. Graphene oxide thin films for flexible nonvolatile memory applications. *Nano Lett.* 10: 4381–6.
14. Venugopal, G., Krishnamoorthy, K., Mohan, R. and Kim, S. J. 2012. An investigation of the electrical transport properties of graphene-oxide thin films. *Mater. Chem. Phys.* 132: 29–33.
15. Yao, Y., Chen, X., Zhu, J., Zeng, B., Wu, Z. and Li, X. 2012. The effect of ambient humidity on the electrical properties of graphene oxide films. *Nanoscale Res. Lett.* 7: 363.
16. Lee, S. K., Jang, H. Y., Jang, S., Choi, E., Hong, B. H., Lee, J., Park, S. and Ahn, J. H. 2012. All graphene-based thin film transistors on flexible plastic substrates. *Nano Lett.* 12: 3472–6.
17. He, C. L., Zhunge, F., Zhou, X. F., Li, M., Zhou, G. C. and Liu, Y. W. 2009. Nonvolatile resistive switching in graphene oxide thin films. *Appl. Phys. Lett.* 95: 232101–3.
18. Jilani, S. M., Gamot, T. D., Banerji, P. and Chakraborty, S. 2013. Studies on resistive switching characteristics of Al/graphene-oxide/semiconductor memory cells. *Carbon*, DOI: http://dx.doi.org/10.1016/j.carbon.2013.07.05
19. Jin, M., Jeong, H. K., Yu, W. J., Bae, D. J., Kang, B. R. and Lee, Y. H. 2009. Graphene oxide thin field effect transistor without reduction. *J. Phys. D: Appl. Phys.* 42: 135109.
20. Kamat, P. V. 2010. Graphene-based nanoarchitectures: Anchoring semiconductor and metal nanoparticles on a two-dimensional carbon support. *J. Phys. Chem. Lett.* 1: 520–7.
21. Jilani, S. M., Gamot, T. D. and Banerji, P. 2012. Thin-film transistors with graphene oxide nanocomposite channel. *Langmuir* 28: 16485–9.
22. Yoo, H., Kim, Y., Lee, J., Lee, M., Yoon, Y., Kim, G. and Lee, H. 2012. *n*-Type reduced graphene oxide field effect transistors (FETs) from photoactive metal oxide. *Chem. Eur. J.* 18: 4923–9.
23. Gilje, S., Han, S., Wang, M., Wang, K. L. and Kaner, R. B. 2007. A chemical route to graphene for device applications. *Nano Lett.* 7: 3394–8.
24. Kaiser, A. B., Navarro, C. G., Sundaram, R. S., Burghard, M. and Kern, K. 2009. Electrical conduction mechanism in chemically derived graphene monolayers. *Nano Lett.* 9: 1787–92.
25. Pei, S. and Cheng, H. M. 2012. Reduction of graphene oxide. *Carbon* 50: 3210–28.
26. Jung, I., Dikin, D. A., Piner, R. D. and Rouff, R. S. 2007. Tunable electrical conductivity of individual graphene oxide sheets reduced at low temperatures. *Nano Lett.* 8: 4283–7.
27. Joung, D., Chunder, A., Zhai, L. and Khondaker, S. 2010. Space charge limited conduction with exponential trap distribution in reduced graphene oxide sheets. *Appl. Phys. Lett.* 97: 093105.
28. Jung, I., Dikin, D., Park, S., Cai, W., Mielke, S. L. and Ruoff, R. S. 2008. Effect of water vapor on electrical properties of individual reduced graphene oxide sheets. *J. Phys. Chem. C* 112: 2026–68.
29. Cheng, M., Yang, R., Zhang, L., Shi, Z., Yang, W., Wang, D., Xie, G., Shi, D. and Zhang, G. 2012. Restoration of graphene from graphene oxide by defect repair. *Carbon* 50: 2581–7.
30. Yin, Z., Sun, S., Salim, T., Wu, S., Haung, X., He, Q., Lam, Y. M. and Zhang, H. 2010. Organic photovoltaic devices using highly flexible reduced graphene oxide films as transparent electrodes. *ACS Nano* 28: 5263–8.

12 Modified Electronic Properties of Graphene

Xiaofeng Fan

CONTENTS

ABSTRACT

Graphene has attracted enormous attention due to its fascinating linear behavior of electronic bands around Dirac points and the special single-layer atomic structure. However, for many practical applications, such as traditional microelectronic devices, the electronic structure of graphene needs to be modified. The modified graphene is expected to have a proper bandgap, or to be a p-type/n-type electronic material. In this chapter, we will review the major progresses in the theoretical part of engineering, the electronic properties of graphene. First, we check the effect of layer stacking to electronic properties of graphene, especially two-layer graphene. Second, we introduce the effect of electric and magnetic fields. Finally, we will review the change of electronic properties induced by chemically modifying the atomic structure of graphene including the change of graphene width, hydrogenation and fluorination of graphene, molecular adsorption on graphene, atom/molecule intercalation in multilayer graphene, and effect of defects and alloying.

12.1 INTRODUCTION

Carbon with the special sp^3 and sp^2 hybridizations can form the two important carbon materials, diamond and graphite. Due to its stable structure, large bandgap, and superhard properties, diamond can be used as the good cutting materials, electrically insulating materials, thermal conductivity materials, and other functional materials. Graphite with the weak van der Waals (vdW) force between layers can be used as pencils and lubricating materials. Due to the renewed combination of graphite's microstructures, the amorphous carbon can be formed and used for the adsorption of different gases and impurities. In addition, graphite can be used to construct the intercalation compounds. With the good conductivity, it can be used as the anode of Li-ion battery, the electrode of supercapacity, and so on. It is also used as a neutron moderator in the nuclear engineering due to the nuclear scattering to a neutron. It is more important that the stable 2D nanostructures, such as small 2D benzene-based molecule, carbon nanotubes,[1] and fullerenes,[2] can be formed with the benzene ring as the basis. Graphene, an isolated single atomic layer with large size, is considered to be impossible to have existed stably for a long time. In 2004, this crystal atomic film was observed on the surface of SiO_2 due to the ingenious optical effect and, has drawn the widespread attention due to its fascinating physical properties (ballistic electronic transport, abnormal quantum Hall effects, and massless Dirac fermions) and the potential applications in different fields.[3–7]

The electronic properties of graphene are controlled by its π electrons.[8] The electronic bands are half filled because each carbon atom just contributes one electron for the π orbitals. It is well known that half-filling of the electronic band has an important impact on the strong correlation of electrons.[9] For an example, in some transition metal oxides, the strong Coulomb repulsion among localized *d* electrons on same sites results in the strong correlation effects which is related to the magnetic and insulating properties. In some early views, the resonant valence bond structure of benzene ring is possible to exist for graphene.[10] However, the experiments support the electronic band structure predicted by the first-principles calculations.[3,5,11] Based on the linear dispersion relation of low-energy electrons under the one-electron approximation, the questions about many-body physics of graphene's electrons such as the electron–electron interactions and the related electron correlation effects have become one important aspect of the study of the physical properties of graphene.[12] Since the low-energy quantum states near Fermi lever satisfy the relativistic Dirac equation, the excited electrons with linear dispersion relation is named as Dirac fermions or massless fermions. In addition, the elementary excitations are with chiral properties due to the hexagonal lattice. Though the Fermi velocity of Dirac fermions is only 1/300 of the speed of light, the characteristic of its massless satisfies the quantum chromodynamics theory.[13] So, it should have the exotic properties of relativistic particles under quantum chromodynamics which can lead to many new physical phenomena such as the singular integer quantum Hall effect under external magnetic field observed especially at room temperature.[5,11,13–17] In addition, with the Klein paradox which means Dirac fermions are not sensitive to the external electrostatic potential, Dirac fermions can go through the classic barrier-restricted area with the probability of 100%.[15] It is also expected that the jittery motion of the electron wave under limited external potential (called as Zitterbewegung phenomenon) is observed on graphene.[14] Since graphene is just with a single atomic layer, the substrate (such as SiO_2 surface), the defects in graphene, and the scattering of the out-of-plane phonons are expected to have obvious effect on its electronic conductivity. The effect of the substrate includes the electrostatic scattering of impurity charges from the substrate surface, the local potential of defects and the coupling between graphene and substrate.[18] However, the conductivity induced by Dirac fermions is possible to be insensitive to these scattering effects.[19] The experiments have confirmed that the electrons in graphene can indeed spread the distances of micron scale without scattering. With the high mobility of electrons and the structural characteristic of the single atomic layer, graphene is expected to be the kernel material of the next-generation microelectronic devices. In addition, with the singular characteristic of Dirac fermions, a lot of other macroscopic quantum effects including the quantum interference caused by weak localization, the universal conductance, AB effect of the electron, and so on, are also expected to observe on graphene.[20–23] In Table 12.1, the special electronic properties of graphene and related phenomena are summarized.

The linear dispersion of energy bands near Dirac points is contributed to the singular physical properties of graphene while the gapless spectrum is conflicted with the basic requirement in the application of traditional microelectronic devices such as field effect transistors. The high on-off ratio of electronic current in the devices requires that there is a gap in the electronic spectrum of graphene.[24–27] In addition, with the miniaturization of electronic devices, the requirement of nanoelectronic devices with less than the size of 50 nm makes graphene an ideal candidate material. The 2D structure with low-concentration impurities can lead to the high-speed ballistic transport of electrons in small size that is consistent with the requirement of the high response speed of the electronic switch. Accordingly, the boundary problem is worth considering when graphene is clipped into the nanosize such as graphene nanoribbons (GNRs).[28–30] Therefore, the electronic structure of graphene needs to be modified to obtain a proper bandgap.[31] Then, the modified graphene is expected to be an excellent *p*-type/*n*-type electronic material. In this chapter, we will review the major progresses in the theoretical part of engineering the electronic properties of graphene.

12.2 STACKING EFFECT ON THE ELECTRONIC PROPERTIES OF GRAPHENE

Graphene is constituted of a single layer of carbon atoms with 2D hexagonal structure. As shown in Figure 12.1, the structure of graphene is a 2D Bravais lattice with the space group P6/mmm. In each unit cell, there are two carbon atoms with the C–C bond length (a_0) of 1.42 Å. Therefore, the lattice constant a is about 2.46 Å. Taken $\mathbf{a}_1 = (\sqrt{3}/2, -1/2)a$ and $\mathbf{a}_2 = (\sqrt{3}/2, 1/2)a$ as the unit vectors, the relative coordinates of two carbon atoms are (0, 0) and (1/3, 1/3). With the relation between lattice vectors and reciprocal vectors, the reciprocal lattice vectors can be written as, $\mathbf{G}_1 = (1/2, -\sqrt{3}/2)2\pi/a$ and $\mathbf{G}_2 = (1/2, \sqrt{3}/2)2\pi/a$.

With honeycomb lattice structure, the three-coordinated carbon atom is with sp^2 hybridization. Two nearest-neighbor carbon atoms form σ bond by the coupling between sp^2-hybridized orbitals. The stable σ bond results in the formation of the stable graphene sheet. Besides the three sp^2-hybridized orbitals, there is a p_z orbital for each carbon atom. Due to the parallel of each p_z orbital, the coupling results in that the split between the energy of π band and that of π* band is possible to be small while the energy gap between σ band σ* band will be large due to the strong coupling of sp^2-hybridized orbitals. Therefore, the electronic properties of graphene are

TABLE 12.1
Electronic Properties of Graphene and Related Phenomena

Properties and Phenomena	Comments	References
Linear dispersion relation	Dirac fermions, high mobility, and zero bandgap	[3,5,8,11]
Chirality	Electrons and holes linked via charged conjugation, edge effects	[7,15,28,29,30]
Many-body effect	Electron–phonon interactions, optical phonons and Raman spectroscopy, electron–electron interactions, and 2D plasmons	[12,17,33,35,36]
Klein paradox	Absence of localization, finite minimum conductivity, and absence of backscattering	[13,15,19,20,24]
Anomalous quantum Hall effect	Half-integer effect, observed at room temperature	[5,11,16,27,32]
Zitterbewegung effect and AB effect	Jittery motion of the electron wave under limited external potential, quantum interference of electronic waves	[14,21,22]

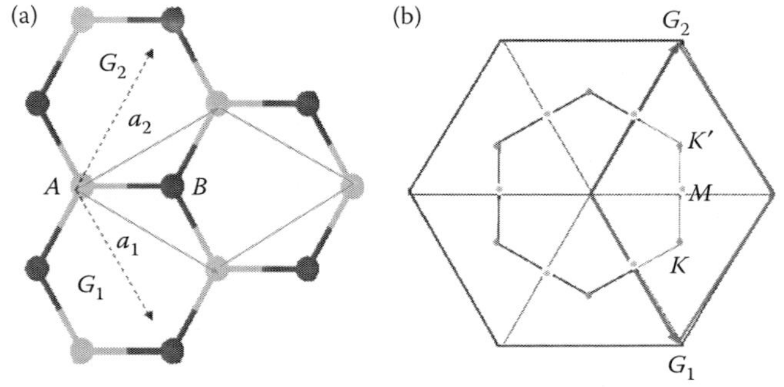

FIGURE 12.1 The lattice structure of graphene (a) and the Brillouin zone in reciprocal space (b).

controlled by the π electron that is decided by the electronic states near Fermi level.

With the tight-binding (TB) method, we consider the electronic structure from the π electrons. With the nearest-neighbor approximation, the parameter t (≈2.8 eV) is for the transition matrix of electron between nearest-neighbor sites $\langle \varphi_A | H | \varphi_B \rangle$, where φ_A and φ_B are the atomic wavefunctions of sublattice site A and B in one unit cell for p_z orbitals, respectively. Under the TB method, the wavefunction that should be consistent with Block theorem can be expressed as[8]

$$\psi(k) = \sum_n e^{ik \cdot R_n} \varphi(r - R_n)$$

with

$$\varphi(r) = c_A \varphi_A(r) + c_B \varphi_B(r) \quad (12.1)$$

With φ_A and φ_B as the basis functions and the approximation $\langle \varphi_A(r) \mid \varphi_B(r - R_n) \rangle = 0$, the Hamiltonian can be expressed as

$$H(k) = \begin{pmatrix} 0 & H_{A,B}(k) \\ H^*_{A,B}(k) & 0 \end{pmatrix} \quad (12.2)$$

where $H_{A,B}(k) = t(1 + e^{-ik \cdot a_1} + e^{-ik \cdot a_2})$, if we assure that the $H_{i,i} = \langle \varphi_i | H | \varphi_i \rangle (i = A, B)$ equals to zero. It is noticed that both $H_{A,B}(K)$ and $H_{A,B}(K')$ equal to zero at the points K and K'. Therefore, the electronic properties of the points near six vertices of the Brillouin zone will be important. With the reciprocal vectors $\mathbf{G}_1$ and $\mathbf{G}_2$, the relative coordinates of K and K' are (1/3, 2/3) and (2/3, 1/3). For the point k near the K, the coordinate can be expressed as $\mathbf{k}(q) = 1/3\mathbf{G}_1 + 2/3\mathbf{G}_2 + \mathbf{q}$, where $\mathbf{q}$ is a small vector ($\mathbf{q} = q_x + iq_y$). Then, the Hamiltonian under Taylor series about $\mathbf{q}$ can be expressed as[7]

$$H_K = \frac{3}{2} t a_0 \begin{pmatrix} 0 & q_x - iq_y \\ q_x + iq_y & 0 \end{pmatrix} \quad (12.3)$$

The Hamiltonian near the K' is similar to that of K and is with the minus. The Fermi velocity υ_F is defined as $3ta_0/2$ and the value of υ_F is about 1×10^6 m/s. Near the Dirac points (K and K'), the electronic band is composed of two cones with top to top. Obviously, the dispersion relation is different from that ($E(q) = q^2/m^*$) of the ordinary semiconductors. This singular dispersion relation results in that the electrons or holes behave as the massless fermions.

With density functional theory under the local density approximation (LDA) or generalized gradient approximation (GGA), the lattice parameter and band structure of graphene can be given reasonably and be consistent with the results of experiments. As shown in Figure 12.2, the bands from the σ electrons are distributed at the relative high energy. The highest filled band is at about −3 eV under Fermi level for Γ point. The π electrons occupy the states near Fermi level. Similar to the result of TB method, a linear dispersion relation near the Dirac point is found. For the total electrons (2s2p), the distribution in real space is around the benzene ring. The 2D electron gas with hexagonal structure is formed. However, the π electrons contributed to the physical properties near Fermi level are distributed mostly on the top of carbon atoms. Therefore, the energy states near Fermi level are sensitive to the external perturbation.

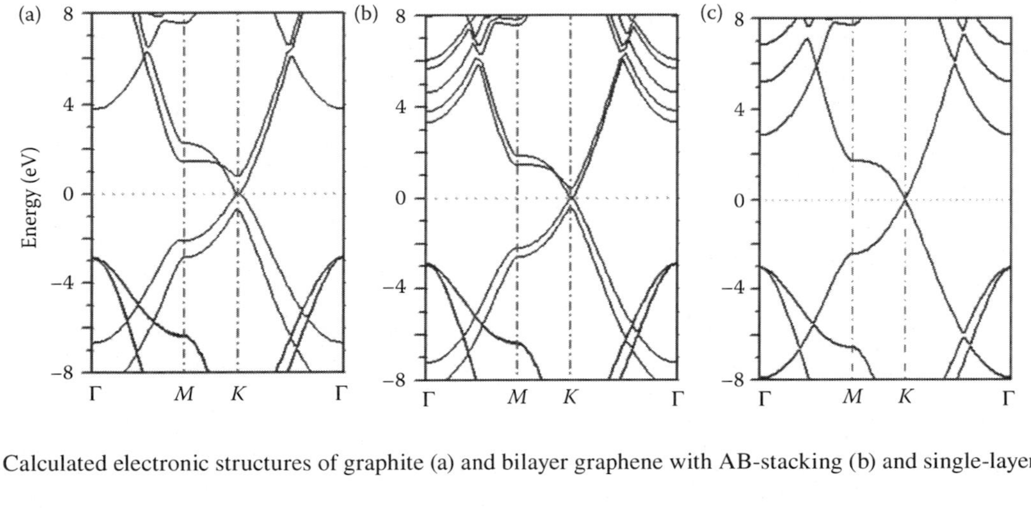

FIGURE 12.2 Calculated electronic structures of graphite (a) and bilayer graphene with AB-stacking (b) and single-layer graphene (c).

It is well known the spin–orbit coupling from the relativistic effect is important for the bands of compounds composed of heavy elements. Though graphene is composed by the light element carbon, the states near Fermi level in bands of graphene is also possible to be affected by this relative effect. The spin–orbit coupling can be expressed by the formula, $V_{SO} = (\hbar/4m_e^2c^2)\sigma\cdot\nabla V \times \mathbf{P}$, where σ is Pauli matrix. With a simple Coulomb interaction $V(r) = e^2/r$, the strength of V_{SO} estimated by Kane et al.[32] equals to $2\pi^2e^2\hbar^2/3m^2c^2a^3$ at the Dirac points. This results in an opening of bandgap about 0.2 meV. With the first-principles calculations, a relatively rigorous assessment about V_{SO} is given by Yao et al.[33] Due to the reflection symmetry, the spin–orbit coupling in the z-direction is just preserved. The spin–orbit coupling is found to result in a bandgap splitting of 10^{-3} meV. With full-electron calculations, the spin–orbit coupling is carried out to assess further by Boettger et al.[34] and a bandgap opening of 0.1 meV is found. In short, the spin–orbit coupling results in the symmetry breaking in the z-direction of the 2D electron gas. A tiny bandgap is opened and need be found at a low temperature about 1 K.

Under the frame of DFT with LDA/GGA, the many-body effects due to the electron–electron interaction cannot be included properly. With the GW approximation, the introduction of electron self-energy can correct the result of the LDA/GGA effectively. The self-energy can be expressed by the single Green's function and the shielded Coulomb potential as

$$\Sigma^{GW}(r_1, r_2, \omega) = i/2\pi \int d\omega' G(r_1, r_2, \omega - \omega')W(r_1, r_2, \omega') \quad (12.4)$$

and

$$W(r_1, r_2, \omega) = \varepsilon^{-1}(\omega)V(r_1, r_2) \quad (12.5)$$

where $\varepsilon(\omega)$ is the dynamic dielectric function in which the many-body screening effect is included and expressed by the polarization function. Under the framework of single quasi-particle, Trevisanutto et al.[35] analyze the many-body effects introduced by electron–electron interaction with DFT–GW method. It is found that the spectral distribution of energy loss function ($-\text{Im}\varepsilon^{-1}(\omega)$) of graphene is composed mainly of three parts including the high-energy plasma polarization mode ($\sigma + \pi$ at about 15 eV), plasma mode of π electrons (about 2 eV), and single-electron excitation mode ($\pi \rightarrow \pi^*$ at about 5 eV).[35] Compared with the results of LDA, many-body effects lead to an enhancement of electron Fermi velocity of about 17%. In addition, the linear dispersion relation is also renormalized. At the points with a distance of about 0.025 Å^{-1} from K point, the GW results lead to a negative correction (about −0.12 eV). Therefore, a kink is introduced at the points, which are at about 0.1 eV under Dirac point.[12]

Besides the electron–electron interaction, the electron–phonon interaction also results in the change of electronic properties such as electron effective mass, Fermi velocity, and the energy band under single electron approximation. In the phonon spectrum of graphene, there are two zones about Kohn singularities that are the E_{2g} mode at Γ and $A_{1'}$ mode at K. At both singular zones, there is the strong interaction between electrons and phonons (such as E_{2g} and $A_{1'}$).[36] The interaction between a phonon mode ν and the electron near Fermi level can be evaluated by the dimensionless parameter about electron–phonon coupling which can be expressed as

$$\lambda_{\nu,q} = \frac{2}{\hbar\omega_{\nu,q}N(\sigma, E_F)} \int_{BZ} \frac{dk}{\Omega} \left| g^{\nu}_{i,k;j,(k+q)} \right|^2 \delta(E_k - E_F)\delta(E_{k+q} - E_F) \quad (12.6)$$

where $\omega_{\nu,q}$ is the phonon frequency, i and j are the indexes of energy bands, $N(\sigma, E_F)$ is the spin density of states (DOS) near Fermi level, and Ω is the area of Brillouin zone. For graphene, the spin DOS near the Fermi level is expressed as $4\pi|E_F|/(\Omega v_F^2)$, where the zero point of energy is set at Dirac point. Scattering matrix element is expressed through the scattering potential[37]

$$g^{\nu}_{i,k;j,(k+q)} = \sqrt{\hbar/(2M\omega_{\nu,q})}\langle k+q, j | \Delta V_{\nu,q} | i, k \rangle \quad (12.7)$$

where $\Delta V_{\nu,q}$ is the Kohn–Sham self-consistent potential change due to the lattice distortion. In graphene, significant scattering is divided into two processes, which are the scattering of electrons near Fermi surface within one Dirac cone with the assistance of E_{2g} phonon at Γ and the electron scattering between two cones with that of E_{2g} phonon near K (K'). The electron–phonon interaction can result in the first-order correction of electron's self-energy. With mean-field approximation, Calandra et al. get the spectral distribution function with the electron's self-energy under the correction of electron-phonon coupling about E_{2g} mode and A_1' mode.[38] They find that there are the band distortions at about −0.16 eV and −0.195 eV which is used to explain the kink at −0.2 eV observed at the experiments of angle resolved photoemission spectroscopy (ARPES). The distortion at −0.195 eV is attributed to the E_{2g} phonon and that at −0.16 eV is due to the A_1' mode. With first-principles calculation, Park et al. state that the band speed of electron depends on the position of Fermi level and the electron-phonon coupling can reduce the band speed by about 4%–8%.[39,40] With the theoretical analysis, Tse et al. get the renormalized electron's self-energy, spectral function, and band speed.[41] They also find that there is a kink about 0.2 eV under Fermi level due to the electron-phonon coupling and the band speed is reduced by about 10%–20% for the doping level of experiments.

Graphite is formed by the periodic AB-stacking of graphene layers. The distance between layers is about 0.335 nm. For bilayer graphene, the AB-stacked structure is considered the most stable structure that is similar to graphite. In addition, a metastable configuration (AA′-stacked structure) is found based on the analysis of the energy surface of the interaction between two layers (in Figure 12.3). Since the π electron is distributed on the outside of graphene plane, the bands of π

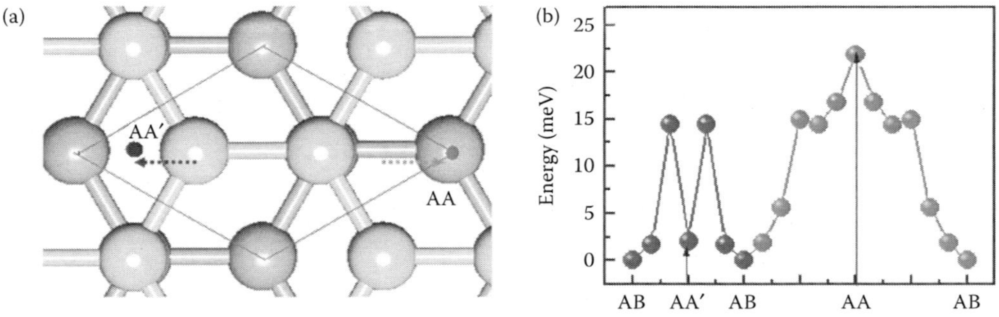

FIGURE 12.3 Schematic diagram of the translation of the second-layer graphene on the first-layer graphene (a) and the energy surface (b) related to the translation for bilayer graphene.

are splitted due to the coupling between graphene layers. This results in the change of dispersion relation between energy and momentum near Dirac point ($E(k) \propto k^2$). For few-layer graphene, there is a similar phenomenon. The liner dispersion near Dirac points disappears due to the coupling of layers. For graphite, along the $M \rightarrow K$ (K′) direction, there is a platform in π^* band at 1.4 eV above Fermi level and a cross at about 1.2 eV above Fermi level where two π^* bands overlap. The significant distortion results in that the energy surface likes a triangle shape at 1.2 eV above Fermi level. Compared with that of graphite, the splitting of bands of few-layer graphene at K (K′) and M is weak. For bilayer graphene, the weaker bands splitting results in that the platform width at M is decreased, and the cross of two π^* bands occurs at about 1.1 eV above Fermi level. Based on TB method, the triangular twist at 1.1 eV is due to the C_3-symmetric transition parameters.

AB-stacking makes the configuration with the lowest energy. However, for few-layer graphene, especially bilayer graphene, the structures with other stacking ways such as AA-stacking and AA′-stacking that are the metastable states of energy surface will be also possible to exist due to the dynamic mechanism in the process of fabrication such as the mechanical folding and catalytic epitaxial growth (e.g., CVD growth). In addition, for bilayer graphene, the configurations with Moire structures are also induced due to the rotation between two layers. In Figure 12.4, there are four typical configurations of bilayer graphene constructed in a supercell with 28 carbon atoms including the AA-stacking, AB-stacking, AA′-stacking, and the AA-R-stacking in which the second layer graphene is rotated 21.787° related to the first layer. With DFT–LDA method, the electronic structures are calculated. It can be found that the different configurations are with the apparently different band structures, as shown in Figure 12.5. For the stable AB-stacking and metastable AA′-stacking, the linear dispersion relation is destroyed due to the stronger interaction between layers. For the AA-stacking which is an unstable configuration, the bands near Dirac point from both graphene layers are translated from K (K') with each other and the linear dispersion relation near Dirac point for each layer still maintains. For the AA-R-stacking, the Dirac electron structure does not change compared with that of the graphene. However, in the relatively high energy states (such as

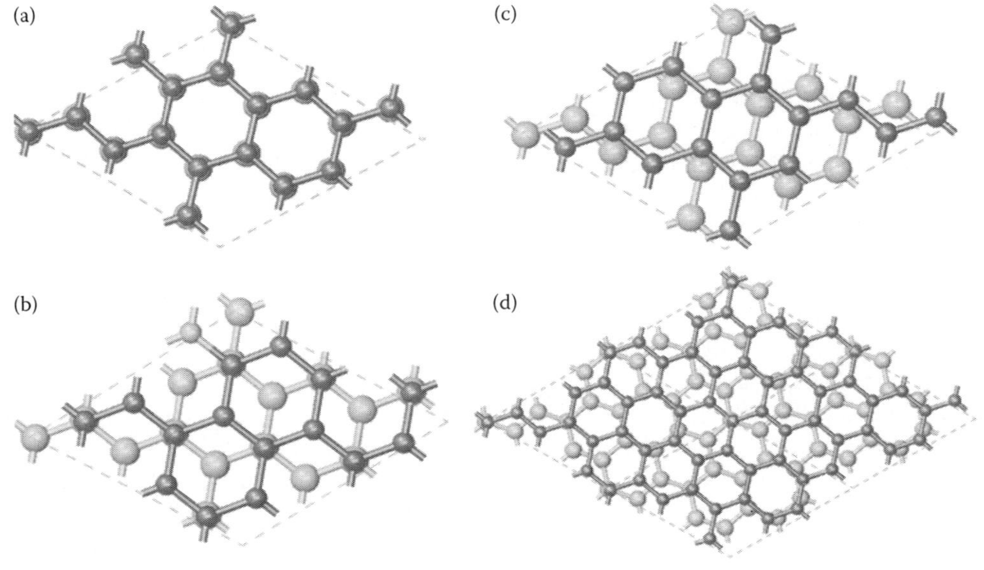

FIGURE 12.4 Different configurations of bilayer graphene in the model with 28 carbon atoms, including AA-stacking (a), AB-stacking (b), AA′-stacking (c), and AA-R-stacking in which the second-layer graphene is rotated 21.787° related to the first layer (d).

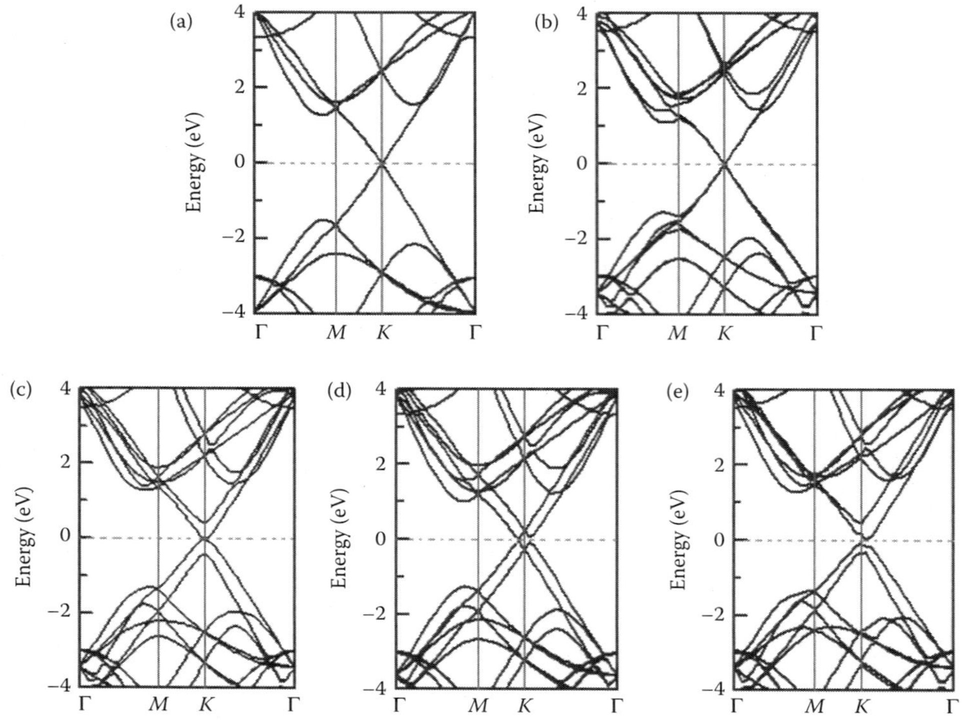

FIGURE 12.5 Calculated electronic structure of single-layer graphene with the model of 14 carbon atoms (a) and that of bilayer graphene with AA-stacking (b), AB-stacking (c), AA′-stacking (d), and AA-R-stacking (e) in the model of 28 carbon atoms.

that near M point), the coupling between both layers results in that the top of valence band moves upward and the bottom of conduction band moves downward. These changes result in the move of Van Hove singularity in the electronic structure of bilayer graphene.[42]

12.3 EFFECT OF ELECTRIC AND MAGNETIC FIELDS AND SUBSTRATE

As the earlier experiment shown, the conductivity of graphene can be as a function of gate voltage and the effect of the ambipolar electric field is observed in graphene.[3] There is the charge transfer between graphene and substrate with the electric field. With positive electric field, the states above Dirac point are occupied by the electrons from substrate and graphene is doped with electrons. With negative electric field, the opposite happens, and holes are introduced into graphene plane. With the increase of the electric field, the increase of the carriers' concentration results in the increase of conductivity. Therefore, with the perpendicular electronic field, the electronic properties of graphene can be controlled. With in-plane homogeneous electric fields, the electronic properties of graphene nanostructure are also modulated. For zigzag graphene nanoribbon (ZGNR), both spin-up and spin-down electrons are separated at the edges due to the symmetry break of A–B sites and the contribution of edge states. Due to the separation of spin-polarized electrons at the edges, there is a bandgap for ZGNR. If the in-plane electric field is applied, the symmetry of both sides is broken, and the energy level shifts of both sides will be with the opposite signs.[29] Since both sides with opposite spin-ordered edge states, one of spins will close its band gap with the increase of external electric field, because the energy shifts of hole state and electron state are also opposite for each spin. This results in the half-metallicity of nanometer-scale GNR. These effects and phenomena related to the electric and magnetic fields and substrates are summarized in Table 12.2.

Due to the high mobility in 2D hexagonal structure, the electronic properties are also expected to be controlled by the magnetic fields as in the 2D electron gas from usual semiconductor heterojunction. In graphene, the quantum Hall effect is observed with the anomalous disappearing of the Hall plateau at Dirac point of pristine graphene. This anomalous Hall effect is even observed at room temperature. In the Dirac equation of K point, the moment $\mathbf{p}$ ($=q_x + iq_y$) is replaced by $\mathbf{p} + e\mathbf{A}/c$, the eigenenergy of each Hall level can be expressed as $E_n = V_F\sqrt{2h|eB|n/c}$, where $n = 0, 1, 2, \ldots$ is a positive integer.[11,43] For the K' point, there is a same spectrum of eigenenergy. Therefore, each Landau level is doubly degenerate. With Laughlin's framework, the Landau levels are broadened with the disorder introduced by the thermal defects and the energy spectrum is constituted by the broadened Landau levels and the localized defect states. When the chemical potential under the control of the magnetic field is changed to the region of localized state, the longitudinal conductivity vanishes which means the appearing of new Hall plateau.[7] In the localized state, the conductivity σ_{xy} can be obtained based on Laughlin's gauge invariance. The change of magnetic flux with a quantum $\Phi_0 = hc/e$ will result in that an integer number of states transfer from one side to another one under the

TABLE 12.2
Modification of Graphene's Properties and Related Phenomena under Electric and Magnetic Fields with Substrates

Fields and Substrates	Comments	References
Perpendicular electric field	Ambipolar effect, charge transfer with substrates	[3,4]
In-plane electric fields	To tune electrons' spins at the edges	[29]
Perpendicular electric field for AB-stacked bilayer graphene	Symmetry breaking in z-direction, bandgap opening	[45]
Perpendicular magnetic field	Anomalous Hall effect, discrete energy levels	[7,11,43]
SiC substrate	High-temperature epitaxial growth, charge transfer, possible buffer layer	[46,47,48,49]
BN substrate	Symmetry breaking of A–B sites, bandgap opening	[54,55]
SiO_2/Si substrate	Light interference, charge transfer, possible chemical bonds	[3,56,57,58,59]
Metal substrates	Shift of Fermi level, modulation of work function	[60,61]

Hall voltage V_H. The current is given by $I = c\Delta E/\Phi_0$, where ΔE is the change of energy due to the transfer of states. For graphene, the degeneracy of the states is four due to the two spins and two Dirac cones (K and K′). Therefore, the energy change is $\pm 4neV_H$, where n is an integer. The conductivity σ_{xy} can be expressed as[7] $I/V_H = \pm 4ne^2/h$. It needs to be noticed that the zero mode in the Landau energy spectrum is shared by two Dirac points. Therefore, the transferred states $4n$ due to the change of integer quantum flux should be replaced by $2(2n + 1)$. The conductivity σ_{xy} can be expressed by the formula $\sigma_{xy} = \pm 2(2n + 1)e^2/h$.

For AB-stacked bilayer graphene, the highest valence band is parallel nearly with the lowest conduction band at Fermi level. With the absence of perpendicular electric field, both bands touch each other at K (K') with a zero bandgap. With the electrical gating, the charges transferred from the substrate to both graphene layers will be different.[44] This induces a net electric displacement fields and results in the breaking of the inversion symmetry of the Bernal-stacked structure. Therefore, a gap Δ_g is introduced between conduction bands and valence bands. With the external electrostatic potential V, the gap can be expressed as,[45] $\Delta_g = \left[e^2V^2t_{\text{in}}^2/(e^2V^2t_{\text{in}}^2 + e^2V^2)\right]^{1/2}$, where t_{in} is the parameter of interplane hopping.

One of the important method to obtain graphene is high-temperature epitaxial growth on (0001) (Si-polarized) and (000-1) (C-polarized) SiC surface. In both directions, there is the structure of Si–C bilayer, as shown in Figure 12.6. In the bilayer, each atom is bonded to three atoms by three chemical bonds while there just is one bond per atom to connect the nearest-neighbor bilayer. This results in that the atomic bilayer on the (0001) or (000-1) surface is easy to vapor at high-temperature. Due to the C–C bond energy (346 kJ/mol) larger than that of C–Si bond (318 kJ/mol) and Si–Si bond (222 kJ/mol), it is possible that the silicon atoms are evaporated and the carbon atoms are deposited on the surface at the appropriate temperature. Due to the lattice mismatch between the deposited carbon layer and SiC surface, the carbon layer can settle stably on the SiC surface with the weak coupling. Since the electronic structure of graphene is very sensitive to the environment, the interaction between the graphene and SiC surface is worth to analyze. With the theoretical and experimental studies, it has been shown that the charge transfer between graphene and the SiC surface has an important impact on the electronic structure of graphene. In addition, it is possible that the 2D carbon layer interact with SiC surface to form a buffer layer composed of carbon atoms on which single-layer or few-layer graphene is formed. There just is a weak coupling between graphene and the buffer layer.

The lattice constant of SiC surface is a little different from that of graphene results in that the 13×13 supercell of graphene matches with the $6\sqrt{3} \times 6\sqrt{3}R30^\circ$ surface, which has been demonstrated by the observing of LEED and Raman spectrum.[46] For the theoretical calculations, a small supercell model of $\sqrt{3} \times \sqrt{3}R30^\circ$ surface with a 2×2 one- or two-layer graphene is constructed to analyze the effect of substrate on the electronic properties of graphene. With such a model, the lattice constant of graphene is expanded by 8%. With the strain of 8%, the electronic properties of graphene near Dirac points are not changed obviously. So, such a model

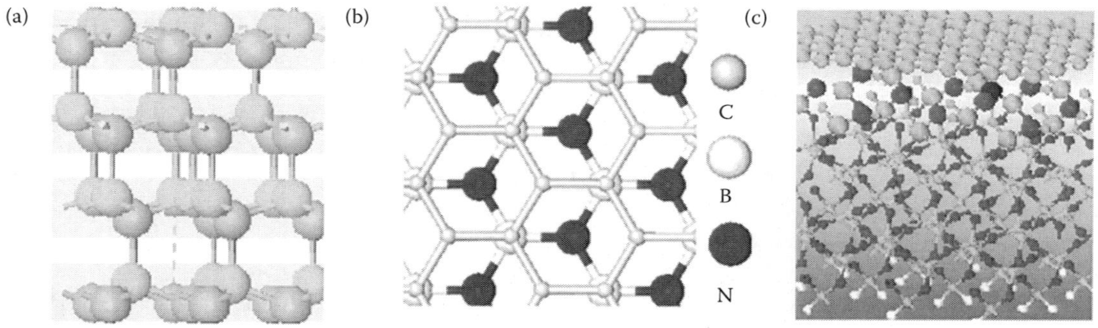

FIGURE 12.6 Schematic diagrams of Si-polarized SiC(0001) surface with Si–C bilayer (a), graphene on BN with AB-stacking (b), and single-layer graphene on SiO_2 surface with one O defect (c).

is considered to be reasonable to deal with the interface between graphene and SiC surface. For either Si-polarized or C-polarized surface of SiC, there are the two in eight carbon atoms in 2×2 cell on the two of three carbon (or silicon) atoms which results in the strong interaction. Therefore, the suspended graphene on the surface will be possible to bond chemically with SiC surface. With the theoretical calculation, the formation energy of chemical bond is found to be larger than the strain induced locally by the quasi-sp^3 bond. Due to the formation of two quasi-sp^3 bonds, the entire low-energy electronic properties are destroyed.[47,48] However, if, with the real model about 13×13 graphene, the low-energy electronic properties are possible to be undestroyed, obviously it is due to the low amount of chemical bonds for the interface. In addition, there still is one atom with a dangling bond on SiC surface. Therefore, near the Fermi level, a localized surface state will be introduced. Since the weak interaction between graphene layers, the second layer graphene will be free and with the ideal low-energy properties. Therefore, the distorted first-layer graphene usually is considered as a buffer layer. If the three atoms in the SiC surface are silicon atoms, due to the charge transfer, the second layer graphene will be doped with electrons. If the three atoms are carbon atoms, there will be no charge transfer and the Fermi level still will be at Dirac point with a localized impurity state from the contribution of SiC surface. Another explanation states that the interaction between buffer layer graphene and SiC surface is weak, and the buffer layer retains its low-energy electronic properties. However, for the effect of substrate on the low-energy properties, things are not just simple. With theoretical calculation, Magaud et al.[49] state that it is possible to have the adsorbed silicon atoms on the C-polarized surface which results in a 2×2 reconstruction of SiC surface. With a model of one-layer 5×5 graphene on a 4×4 SiC surface, it is found that the electronic properties of Dirac fermions are retained. However, with the ARPES experiments, the buffer layer graphene seems to be proved.[50] It is interesting that the second-layer graphene is with a bandgap of 0.26 eV at Dirac point due to the coupling with the buffer layer. It is considered that the buffer layer breaks the lattice symmetry of A–B sites of the second layer graphene, and the bandgap is opened. Peng et al.[51] give another explanation that the bandgap induced is due to the Stone–Wales (SW) impurity states.[51] Ohta et al.[52] consider that the bandgap is due to the presence of electric dipole field that breaks the z-direction symmetry of bilayer graphene. The electric dipole field due to the charge transfer results in the charge depletion of SiC surface and the charge accumulation of the buffer layer.

Similar to graphene, single-layer BN is with the hexagonal structure. Due to the chemical potential difference between A and B site which results in the symmetry breaking of A–B sites, single-layer BN is with a large bandgap of about 5.97 eV.[53] With DFT calculation, it is found that the most stable configuration of graphene on BN is with the Bernal-stacking.[54] As shown in Figure 12.6, the one of two carbon atoms is on the *B* atom of BN. Therefore, the A-site and B-site will have different chemical potential, and the bandgap of graphene can be opened. With DFT–LDA, the bandgap is about 53 meV. The interaction between graphene and cubic BN (*c*-BN) is also studied theoretically.[55] It is found, a bandgap of about 0.13 eV is induced in the N-polarized surface of *c*-BN due to the lattice symmetry breaking. In addition, the spin polarization of surface states has also led to the spin polarization of graphene.

Another important surface on which graphene is deposited is the surface of Si. Due to the oxidation of the surface, the surface is actually constructed by a thin layer of Si_2O with disordered structure. Currently, most of the experimental study about graphene is completed on Si/SiO_2 surfaces. The Si/SiO_2 surface has its advantages such as easy designing of the device with Si and with excellent insulating properties. In addition, when the thickness of SiO_2 surface layer is about 300 nm, the interference between light from the SiO_2/graphene interface and from Si/SiO_2 interface will be enhanced and graphene can be observed just with the optical microscope. The disorder of SiO_2 surface results in a complex surface structure that makes the direct calculation difficult. Typically, the structure of α-SiO_2 can reflect the local characteristic of the disordered SiO_2 structure such as that of the basic unit SiO_4. Therefore, the α-SiO_2 structure can be used to simulate the surface of disordered SiO_2. The results obtained by Shemella et al.[56] show that graphene forms the chemical bonds with oxygen atoms of the O-ended surface that results in the destroying of the π electron structure.[56] While the weak interaction between H-passivated surface and graphene does not change the electronic structure of graphene. Kang et al.[57] simulate the Si-ended surface and find that the Si dangling bond does not associate with the graphene layer with chemical bonding and the graphene maintains its electronic structure. Cuong et al.[58] find that the O-ended surface is possible to be reconstructed which results in the disappearance of the O-dangling bond. Therefore, there is a small opened bandgap on graphene's electronic structure due to the weak interaction with the reconstructed surface. Fan et al.[59] analyze the SiO_2 surface with the surface models of α-SiO_2 and cristobalite. They find that the oxygen defect in a SiO_2 surface can shift the Fermi level of graphene down and result in the hole doping of graphene adsorbed on a SiO_2 observed in many experiments.

Since the electrode is an important part of the electronic device, the interaction between graphene and metal substrate is important.[60,61] The interaction can be divided into two categories: one is the chemical adsorption of graphene on the metals such as Co, Ni, and Pd substrate and the other one is the physical adsorption on the substrates such as Al, Cu, Ag, Au, and Pt. For the chemical adsorption, the low-energy electronic structure of graphene is destroyed, and the mixed electronic properties of metal and graphene are formed with the metallic character. For the physical adsorption, the electronic structure of graphene is maintained and its Fermi level is shifted, following the change of metals' work function. Hence, the p-type and n-type graphene can be formed by the contact with the different metals. The transition between p-type and n-type occurs at the work function of about 5.4 eV that is larger than that of the graphene (about 4.5 eV). This is attributed to the formation of dipole moment and chemical effects due to the charge transfer. Both the effects result in an internal electric field that causes the transition from n-type to p-type occurs at the larger work function.[61]

12.4 CHEMICALLY MODIFYING THE ATOMIC STRUCTURE OF GRAPHENE

12.4.1 Electronic Structure of GNRs

With limited size, the bandgap of graphene is expected to open. There are two basic configurations for the edges, which are called as armchair and zigzag. Therefore, GNR, which is 1D graphene structure exist two basic categories that are armchair graphene nanoribbon (AGNR) and ZGNR. Based on the folding theory of band structure, the band structure of GNR can be obtained by the 1D folding of that of graphene in Figure 12.7. For the lateral dimension of GNR, there should be the periodic boundary which can be written as $\mathbf{k} \cdot \mathbf{c} = 2\pi j$, where $|\mathbf{c}|$ is the width of GNR which results in that k is discrete for the lateral dimension. In order to figure out whether the K (or K') point belongs to the allowed k in the 1D Brillouin zone, the vector $\mathbf{K}$ should satisfy the periodic boundary condition which is $\mathbf{K} \cdot \mathbf{c} = 2\pi j$. This will results in the ZGNR and special AGNR with zero bandgap, since K belongs to the allowed k points in the 1D Brillouin zone. According to the quantum confinement theory, the limited size in lateral dimensions will be possible to induce a finite gap in the energy bands. However, due to the special properties of the hexagonal structure with the characteristic of 1D edges, the simple quantum confinement theory does not explain the band structure of GNR.

For the first-principles calculations, the atomic structures of the edges should be considered first. For the edge of both zigzag and armchair, each carbon atom with a sp^2-hybridized dangling bond will inevitably introduce the localized impurity states at the Fermi level. As shown in Figure 12.8, the electronic structures of armchair-N18 and zigzag-N19 demonstrate the localized states near Fermi level. The dangling bonds will lead to the instability of edge carbon atoms, which results in that the other atoms are easily adsorbed on edges. Typically, a hydrogen atom with a single electron is used to passivate the boundary carbon atoms. Son et al.[30] have made a systematical calculation about the electronic structures of ZGNR and AGNR. For AGNR, the bandgaps of all the different widths will be opened due to the quantum confinement effect. The width of bandgap decreases with the width increase of GNR with the law of $N = 3p + i$ ($i = 0, 1, 2$).[30] For ZGNR, the TB method states that the breaking of A–B symmetry due to the boundary leads to the formation of edge states at Fermi level. This results in the closing of the bandgap, no matter how small the width is. Without the consideration of spin polarization, the result of DFT is similar to that of TB. However, with the break of A–B symmetry, the spin polarized electrons are separated, especially at the edges. The coupling of the spin polarized states results in the opening of the bandgap, as shown in Figure 12.9. The bandgap decreases with the increase of GNR width, in accordance with the quantum confinement effect.

With the 2D electronic structure, the weak Coulomb screening effect will improve the correction of electron self-energy in graphene. Due to the quasi-1D structure of GNR, the electron self-energy correction will be improved further. In addition, the quantum confinement also results in the increase of electron–electron interaction. The increase of the electron–electron interaction may lead to the significant change in bandgap. With GW approximation, Yang et al.[62] consider the effect of electron self-energy correction. For AGNR, the band gaps from GW are larger than that from LDA/GGA. Following the decrease of GNR width, the correction is more obvious. For ZGNR, the correction of bandgap with GW is relatively smaller. This is because the bandgap of ZGNR is decided mainly by edge states. The edge states are distributed mainly at the boundary. This results in the quantum confinement effect and is weak to the coupling of edge states. Yang et al.'s[62] results show that the GW correction of the first bandgap at the $3/4(\pi/a)$ from Brillouin center changes with the variation of width and the correction of the second bandgap at the boundary of Brillouin zone does not

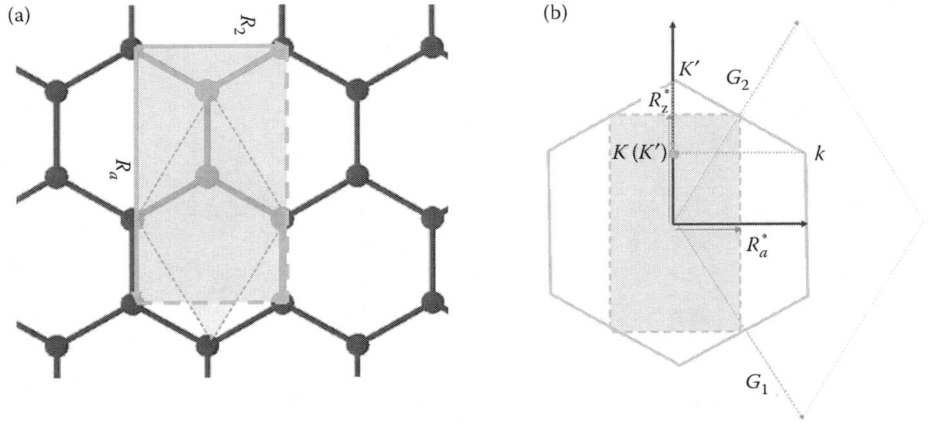

FIGURE 12.7 Schematic diagram of graphene with a rectangular cell for showing ZGNR and AGNR (a) and the folding of Brillouin zone for the 1D Brillouin zones of ZGNR and AGNR (b).

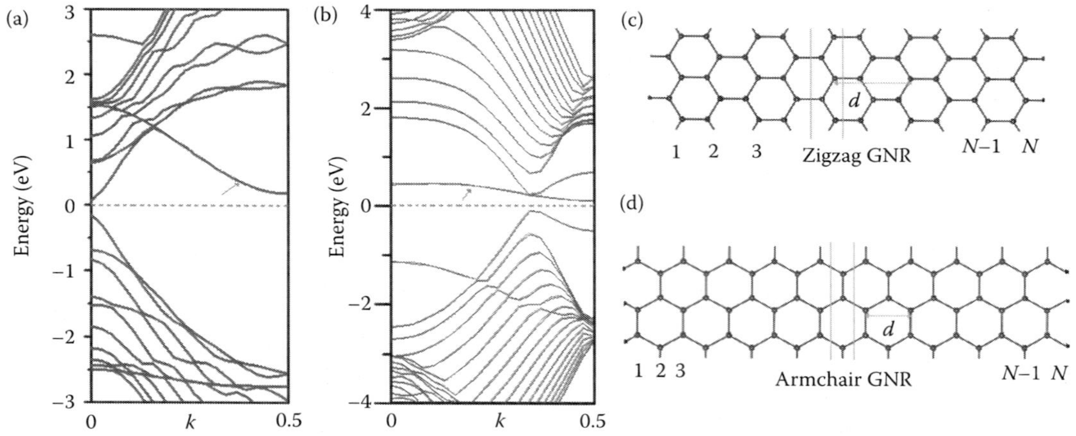

FIGURE 12.8 Electronic structures of armchair-N18 GNR (a) and zigzag-N19 GNR with spin-polarization (b) and schematic structures of ZGNR (c) and AGNR (d). Note the edges are not passivated by hydrogen and the impurity states is introduced near the Fermi level. The widths of ZGNR and AGNR is defined as $w_{zig} = (d/2)(N-1) = 3/2(N-1)a_0$ and $w_{arm} = (d/2)(N-1) = \sqrt{3}/2(N-1)a_0$, respectively.

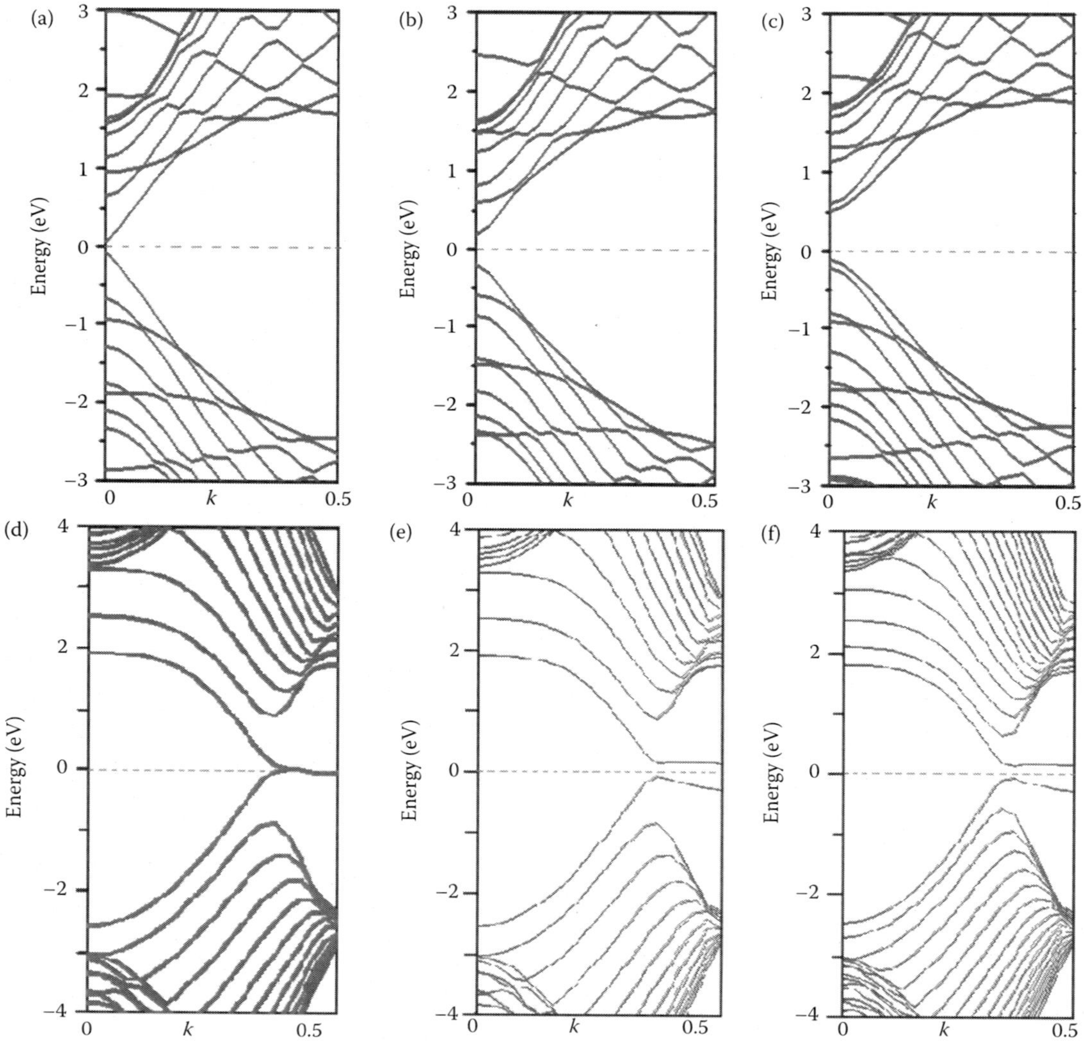

FIGURE 12.9 Electronic structures of AGNR with $N = 17$ (a), $N = 18$ (b), and $N = 19$ (c), and electronic structure of ZGNR with $N = 19$ (d) and spin-polarized electronic structures of ZGNR with $N = 12$ (e) and $N = 19$ (f).

change obviously. With the analysis of charge distribution, the boundary states contributed to the first bandgap decay, slowly. Therefore, the GW correction to quantum confinement changes the bandgap to some extent.

12.4.2 Physical and Chemical Adsorption of Molecules on Graphene

The interaction between graphene and organic molecules is weak. The adsorption of adenine on graphene demonstrates the weak interaction that is considered to be mainly vdW force.[63] However, due to the charge transfer and weak hybridization of orbitals of graphene and molecules, particularly small molecules, the electronic properties of graphene are possible to change significantly. This weak coupling mainly changes the position of Fermi level of graphene and induces a perturbation to the low-energy electron bands. For graphene, the adjusting of Fermi level easily changes its conductivity. Therefore, the conductivity of graphene is very sensitive to the adsorption of small molecules, such as NO_2, H_2O, CO, NH_3, and so on, and is even used to judge the event of single molecule adsorption.[64] Due to the adsorption of small molecules, it is possible to open the bandgap at the Dirac point.

The results of the experimental study show that the adsorption of NO_2 on graphene leads to a transition between metallic and insulating states. With the theoretical calculations, the adsorption of NO_2 monomers on graphene results in that the spin bands of graphene are split with the stable paramagnetic state and the Fermi level is down-shifted about 0.4 eV which leads to a hole-doping.[65] However, the NO_2 usually is easy to form N_2O_4. With the N_2O_4 adsorbed on graphene due to its closed shell of electrons, the spin bands are not split, and the Fermi level of graphene does not also move. Halogens can also be used to adjust the electronic structure of graphene. With the theoretical calculation, the conductivity of graphene can be controlled by the UV light that can result in the transition of Br atoms and Br_2 molecule.[66] The reason is that the adsorption of Br atoms can result in a large shift of Fermi level and the adsorption of Br_2 molecule just results in a small shift. Due to the diffusion barrier, the Br atoms are easy to move on graphene and form a stable state of Br_2. Due to the adsorption on graphene, the transition barrier from Br_2 to Br is deceased largely and be controlled by the light. With the charge transfer, the small organic molecule can also lead to the shift of Fermi level. For an example, tetracyanoethylene ($C_2(CN)_4$) which is a typical receptor of charge can result in the hole doping of graphene.[55] With the aromatic molecules, such as $B_3N_3H_6$ and $C_3N_3H_6$, adsorbed on graphene, the bandgap even can be opened, due to the symmetry break of graphene lattice.[67] Obviously, the metal atoms can be used to modify the electronic properties of graphene. Gierz et al.[68] demonstrate experimentally that bismuth, antimony, and gold atoms can lead to the hole doping of graphene. The strong interaction between polar molecules and graphene also induce a gap in the bands. Theoretical calculations show that CrO_3 as an electron acceptor adsorbed on some sites of graphene can induce a bandgap at Dirac point.[69]

The dangling bonds of small molecular group can be used to modify chemically the structure of graphene, which be possible to destroy the low-energy electronic network of graphene to some extent and lead to the change of electronic properties. Yan et al.[70] consider the structures and electronic properties of graphene oxide by the theoretical calculation combined with the finds of experiments in details. The configurations of graphene oxides are considered to be composed of the combination of epoxy group and hydroxyl group. The bandgap of graphene oxide can be controlled by the degree of oxidation of graphene and the mediation of the ratio of epoxy group and hydroxyl group. Ghaderi et al.[71] analyze the effect of hydroxyl group on the structure of graphene with defects and epoxy group. For the defect-free graphene, the hydroxyl group interacts with carbon atom by a weak chemical bond of about 0.54 eV. This results in a localized magnetic moment. With a reaction barrier of about 0.5 eV, two hydroxyl groups can react with each other to produce a water molecule adsorbed weakly on graphene. In addition, hydrogenation and fluorination of graphene are considered to form the bases for the new 2D materials graphane and fluorographene which are with very large bandgaps.[72–77]

The chemical adsorption of a single atom, such as F, O, N, S, and H, can change the electronic character of graphene obviously. To explore the magnetic properties of graphite, Lehtinen et al.[78] have analyzed the adsorption of single carbon atom on graphene. The stable configuration is the carbon atom adsorbed on the bridge site that disturbs the under graphene structure. For the adsorbed carbon atoms, two electrons participate in the bonding with the below two carbon atoms of graphene, one is for the sp^2-hybridized dangling bond and the last one is shared by the sp^2-hybridized state and the localized π state, which leads to a formation of localized magnetic moment of about 0.5 μB. The adsorption of N, O, and F atoms on graphene was studied by Wu et al.[79] With just one of dangling bonds, the F atom is adsorbed easily on carbon atom while N and O are easy to be adsorbed on the bridge site, which is similar to C atom. The adsorption of N atom can induce a localized spin. The main reason is the adsorbed N atom has a partially occupied orbital. With the exchange splitting of Hund, the spin-up states at the 2.05 eV under Dirac point are occupied fully, and the spin-down state at the 0.39 eV under Dirac point are occupied partially.

Impurities and the edges have a significant impact on the electronic properties. Therefore, the adsorption of molecules on the defected graphene and the edge of graphene will be an interesting phenomenon to study. Huang et al.[80] studied the adsorption of CO, NO, NO_2, O_2, N_2, CO_2, and NH_3 molecules on the edge of AGNR. H-passivated AGNR is with a distinct bandgap, which can be used for the design of microelectronic devices. However, it is inevitable that the defected states, such as hydrogen vacancy (V_H), will be possible to occur. If NO_2 is adsorbed on the $V_{H,}$ which can hybridize with the electronic states of graphene, Fermi level will be pinned in the valence band top, and the defect induced by V_H disappears. If NH_3 is adsorbed on the V_H, an n-type impurity is introduced, and the conductivity of the nanoribbon is enhanced. The adsorption of CO_2 and O_2 will introduce the impurity band in the valence band and GNR will become a p-type semiconductor. The adsorption of NO and CO will introduce the deep impurity level in bandgap. The adsorption of oxygen on B-, N-, Al-, Si-, P-, Cr-, or Mn-doped graphene has been analyzed theoretically.[81] The adsorption on B- and N-doped graphene belongs to the physical adsorption and that on Al-, Si-, P-, Cr-, and Mn-doped graphene is chemical adsorption. This reason is possible to be related to the heave of the dopant, which can be attributed to the longer X–C bond where X is the dopant. The chemical adsorption on defected graphene is also possible to induce the localized spin polarization. The adsorption of the different molecule including H_2O, NH_3, CO, and HF on the small-sized graphene can cause the change of the edge states, has been studied.[82] The effect of the adsorption of H_2O and other molecules on the electronic properties of graphene is similar to that of the defect. With the adsorption, the bandgap can be adjusted, and the magnetic order on the edge can be controlled.

12.4.3 Defects on Graphene and Doping of Graphene

The interest on the origin of defect states comes from the exploration of the mechanism of magnetism on graphite. The experimental work of Esquinazi et al.[83] about the high-energy (2.25 MeV) neutron bombardment of graphite suggests that the defected graphite has the ferromagnetism. It is considered that this phenomenon is associated with the ferromagnetism of carbon at room temperature (RT). Therefore, the RT ferromagnetism of carbon materials may be induced by the nonmagnetic impurities. Theoretically, the different physical mechanisms, in particular, the mechanism of defect states is used to discuss the RT ferromagnetic properties.[78,84–88] The defect states, such as the carbon atom adsorbed on the bridge site, the single vacancy and the complex of carbon vacancy and H, can introduce a localized magnetic moment.[78] With the band theory, the defect states introduce the flat impurity levels near the Fermi level, which leads to the increase of DOS at Fermi level. Therefore, the ferromagnetic state will be possible, to be stabilized by the defects. However, there is no strong interaction between these localized spin states induced by the defects. When the distance between defects in the graphene plane is larger than about 11 Å (i.e., the defect density is less than 1.56%), the ferromagnetic coupling between the local magnetic moments can be ignored.[88] With the theoretical calculations, Fan et al.[88] state that *sp*-hybridized carbon may play an important role in the RT ferromagnetism of carbon. Of course, there are other mechanisms of ferromagnetic coupling proposed such as the ferromagnetism of edge states due to the H-passivation of both zigzag edges with different degrees based on Lieb's AB-site theory about graphene lattice.[89]

By the doping, the electrical properties of the semiconductors can be controlled effectively. Similarly, the electronic properties of graphene can be controlled effectively by adjusting the Fermi level and opening the bandgap at Dirac point. It is shown experimentally that N-substitution in the graphene plane can effectively shift the Fermi level up.[90] At the same time, B-doping can result in the conductivity of hole. With the theoretical calculations, it is found that ZGNR can translate from the metallic state to the semiconductor by the replacing of *B* or *N* at the edges.[91,92] Due to the asymmetry of the replacing at both edges, two spin bands will have different band gaps. This even causes a spin-polarized half-metallic state. The bandgap of AGNR is not affected significantly by the B-doping or N-doping.[93] However, the doping at the different location has an obvious effect on the shift of Fermi level. Small BN domains are possible to be formed due to the phase separation between graphene and BN. With the doping of small BN doping, it is found, theoretically, that the band gap of graphene can be opened effectively.[94]

A unique character of graphene lattice is its local structure can be reconstructed by non-six-membered ring (non-six-ring) structures.[95] A typical example is the SW structural defect, which is a topological distortion. Four neighboring six rings can be transferred to a pair of five rings and seven rings by rotating a C–C bond, as shown in Figure 12.10. The formation energy of SW defect is about 5 eV.[96,97] In the process of the rotation, there is a kinetic barrier about 10 eV.[97] Therefore, this kind of defect is not easy to form. If this defect is formed, it will be very stable since there is a potential barrier of about 5 eV to across in order to repair this defect. With the theoretical analysis, a carbon leaves away its lattice without the assistance of vacancies needs a minimum energy of about 20 eV.[98] In fact, with the high-energy electron beam, many atoms in a localized location can be excited at the same time. Therefore, the SW defect can be formed with some thermal energy that is far below the threshold of the barrier (20 eV).

Vacancy is the popular defect in the semiconductor materials, which has a very important influence on the electronic properties of the materials. In graphene, a single vacancy

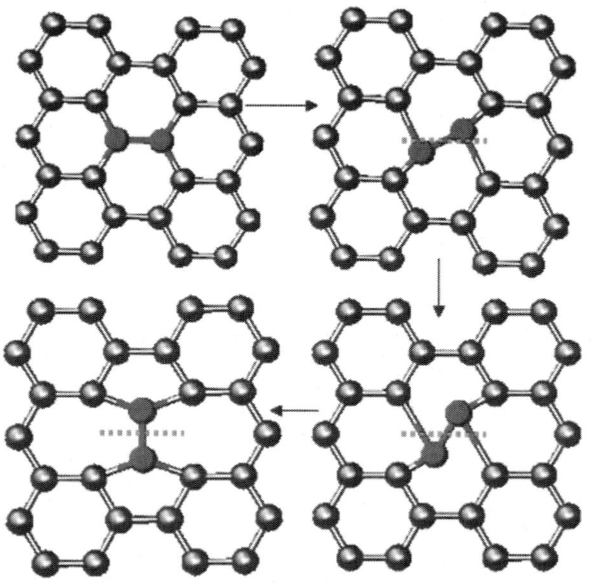

FIGURE 12.10 Schematic diagrams for the formation process of the SW defect (two nearest-neighbor carbon atoms with 90° rotation results in the formation of SW structural defect).

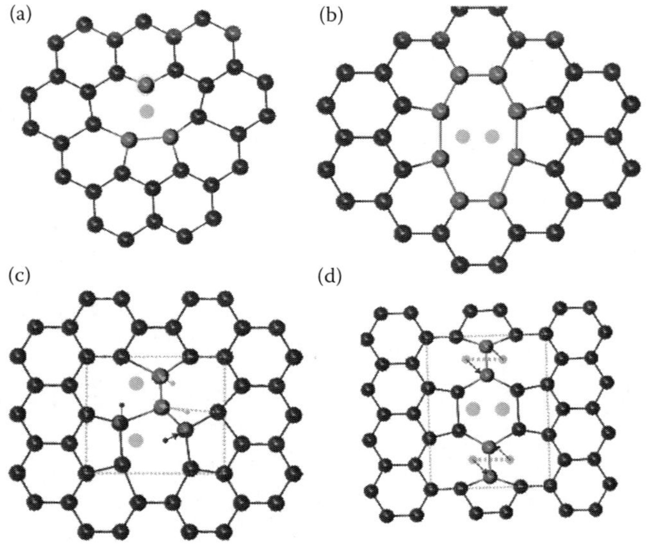

FIGURE 12.11 Schematic structural diagram of single atomic vacancy with Jahn–Teller distortion (a) and schematic structures of three kinds of divacancies, including V_2-5-8-5 (b), V_2-555-777 (c), and V_2-5555-6-7777 (d).

causes each of three carbon atoms in its nearest neighbor has a dangling bond. With the deformation of Jahn–Teller distortion, two dangling bonds reform a new bond results in a local structure of 5–9 ring, as shown in Figure 12.11. However, there still is one carbon atom with a dangling bond in the nine-membered ring that leads to a high localized DOS near Fermi level with a localized magnetic moment. Due to the existence of dangling bonds, the formation energy of single vacancy is very high and about 7.5 eV.[99] However, the migration barrier of single vacancy is only about 1.3 eV.[100] Thus, at low temperatures (200°C), the single vacancy can move in the lattice. Under the irradiation of the high-energy electron beam, the multivacancy defect is also formed due to the local structural distortion. Interestingly, the formation of double-vacancy defects can make the dangling bond disappear due to the topological distortion of the structure. As shown in Figure 12.11, there are three kinds of divacancies, V_2-5-8-5, V_2-555-777, and V_2-5555-6-7777.[95] The formation of defect V_2-5-8-5 is about 8 eV. This means that each carbon atom removed just consumes the energy of about 4 eV. From the thermodynamic point of view, the divacancy is more stable. That of defect V_2-555-777 is 7 eV. The formation energy of V_2-5555-6-7777 is between that of V_2-5-8-5 and V_2-555-777. However, the migration energy of divacancy is about 7 eV. It will be difficult to form a more complex multivacancies by the migration of divacancy.

12.5 SUMMARY

Graphene as a truly 2D material has properties of soft material and can be used to mix with other nanomaterials to form the special composite materials for a lot of applications in different fields such as supercapacitor, fuel cell, sensor, and so on. With its singular electronic properties described by Dirac fermions, graphene also affords an ideal model for the condensed matter research. Graphene, as an unusual semiconductor with zero bandgap, has very high mobility with the long mean free path and is expected to replace the role of silicon in the electronic devices of the next generation. Based on band theory, we have reviewed the fundamental electronic properties of graphene and modified graphene with different ways such as electric and magnetic fields, substrate effect, doping effect, etc. The electric field can shift the Fermi level and tune the carrier concentration of graphene. With high electric field, a bandgap is even opened in double-layer graphene. With magnetic field, the discrete Landau levels are formed with anomalous Hall effect. Substrates are possible to interact with graphene by the formation of chemical bonds to result in the obvious effect on electronic properties. By controlling the stacking ways and rotational angle of layers, the electronic properties are also tuned. With the quantum confinement effect of size and chirality of graphene's edge, graphene quantum dots, and nanoribbons are the promising way to control the electronic properties for the practical applications. With the opening of the bandgap, the modified GNRs make the logic devices with high on/off ratio possible. With the chemical and/or physical adsorption of small molecules/atoms, such as H, F, and Br, the properties of graphene can be modulated effectively. The chemical methods usually induce quasi-sp^3-hybridized carbon and defects of vacancies in graphene plane and result in graphene with the limited size. For example, the defects have dramatically lowered the carrier mobility and conductivity in reduced graphene oxide (RGO). The physical adsorption of aromatic molecules with vdW force may disturb slightly the distribution of π electrons with symmetry breaking and result in the opening of the band gap. In addition, alloying with BN may also be an effective way to open band gap. Obviously, to satisfy the industrial requirement of graphene-based microelectronics, the mobility with the long mean free path and opened bandgap need much more work in modifying the properties of graphene.

REFERENCES

1. Iijima, S. 1991. Helical microtubules of graphitic carbon. *Nature* 354:56–58.
2. Kroto, H. W.; Heath, J. R.; O'Brien, S. C.; Curl, R. F.; Smalley, R. E. 1985. C_{60}: Buckminsterfullerene. *Nature* 318:162–163.
3. Novoselov, K. S.; Geim, A. K.; Morozov, S. V.; Jiang, D.; Zhang, Y.; Dubonos, S. V.; Grigorieva, I. V.; Firsov, A. A. 2004. Electric field effect in atomically thin carbon films. *Science* 306:666–669.
4. Geim, A. K.; Novoselov, K. S. 2007. The rise of graphene. *Nat. Mater.* 6:183–191.
5. Novoselov, K. S.; Geim, A. K.; Morozov, S. V.; Jiang, D.; Katsnelson, M. I.; Grigorieva, I. V.; Dubonos, S. V.; Firsov, A. A. 2005. Two-dimensional gas of massless Dirac fermions in graphene. *Nature* 438:197–200.
6. Geim, A. K. 2009. Graphene: Status and prospects. *Science* 324:1530–1534.
7. Castro Neto, A. H.; Guinea, F.; Peres, N. M. R.; Novoselov, K. S.; Geim, A. K. 2009. The electronic properties of graphene. *Rev. Mod. Phys.* 81:109.

8. Wallace, P. R. 1947. The band theory of graphite. *Phys. Rev.* 71:622.
9. Phillips, P. 2006. Mottness. *Ann. Phys.* 321:1634–1650.
10. Pauling, L. 1960. *Nature of the Chemical Bond*. Cornell University Press: New York.
11. Zhang, Y.; Tan, J. W.; Stormer, H. L.; Kim, P. 2005. Experimental observation of the quantum Hall effect and Berry's phase in graphene. *Nature* 438:201–204.
12. Bostwick, A.; Speck, F.; Seyller, T.; Horn, K.; Polini, M.; Asgari, R.; MacDonald, A. H.; Rotenberg, E. 2010. Observation of plasmarons in quasi-freestanding doped graphene. *Science* 328:999–1002.
13. Peres, N. M. R.; Castro Neto, A. H.; Guinea, F. 2006. Conductance quantization in mesoscopic graphene. *Phys. Rev. B* 73:195411.
14. Katsnelson, M. I. 2006. Zitterbewegung, chirality, and minimal conductivity in graphene. *Eur. Phys. J. B Condens Matter Complex Syst.* 51:157–160.
15. Katsnelson, M. I.; Novoselov, K. S.; Geim, A. K. 2006. Chiral tunneling and the Klein paradox in graphene. *Nat. Phys.* 2:620–625.
16. Gusynin, V. P.; Miransky, V. A.; Sharapov, S. G.; Shovkovy, I. A. 2006. Excitonic gap, phase transition, and quantum Hall effect in graphene. *Phys. Rev. B* 74:195429.
17. Pisana, S.; Lazzeri, M.; Casiraghi, C.; Novoselov, K. S.; Geim, A. K.; Ferrari, A. C.; Mauri, F. 2007. Breakdown of the adiabatic Born-Oppenheimer approximation in graphene. *Nat. Mater.* 6:198–201.
18. Meyer, J. C.; Geim, A. K.; Katsnelson, M. I.; Novoselov, K. S.; Booth, T. J.; Roth, S. 2007. The structure of suspended graphene sheets. *Nature* 446:60–63.
19. Lee, P. A.; Ramakrishnan, T. V. 1985. Disordered electronic systems. *Rev. Mod. Phys.* 57:287.
20. Morozov, S. V.; Novoselov, K. S.; Katsnelson, M. I.; Schedin, F.; Ponomarenko, L. A.; Jiang, D.; Geim, A. K. 2006. Strong suppression of weak localization in graphene. *Phys. Rev. Lett.* 97:016801.
21. Recher, P.; Trauzettel, B.; Rycerz, A.; Blanter, Y. M.; Beenakker, C. W. J.; Morpurgo, A. F. 2007. Aharonov-Bohm effect and broken valley degeneracy in graphene rings. *Phys. Rev. B* 76:235404.
22. Huefner, M.; Molitor, F.; Jacobsen, A.; Pioda, A.; Stampfer, C.; Ensslin, K.; Ihn, T. 2010. The Aharonov–Bohm effect in a side-gated graphene ring. *New J. Phys.* 12:043054.
23. Heersche, H. B.; Jarillo-Herrero, P.; Oostinga, J. B.; Vandersypen, L. M. K.; Morpurgo, A. F. 2007. Bipolar supercurrent in graphene. *Nature* 446:56–59.
24. Cheianov, V. V.; Fal'ko, V. I.; Altshuler, B. L.; Aleiner, I. L. 2007. Random resistor network model of minimal conductivity in graphene. *Phys. Rev. Lett.* 99:176801.
25. Huard, B.; Sulpizio, J. A.; Stander, N.; Todd, K.; Yang, B.; Goldhaber-Gordon, D. 2007. Transport measurements across a tunable potential barrier in graphene. *Phys. Rev. Lett.* 98:236803.
26. Tworzydlstroko, J.; Trauzettel, B.; Titov, M.; Rycerz, A.; Beenakker, C. W. J. 2006. Sub-Poissonian shot noise in graphene. *Phys. Rev. Lett.* 96:246802.
27. Williams, J. R.; DiCarlo, L.; Marcus, C. M. 2007. Quantum Hall effect in a gate-controlled p-n junction of graphene. *Science* 317:638–641.
28. Biel, B.; Blase, X.; Triozon, F.; Roche, S. 2009. Anomalous doping effects on charge transport in graphene nanoribbons. *Phys. Rev. Lett.* 102:096803.
29. Son, Y.-W.; Cohen, M. L.; Louie, S. G. 2006. Half-metallic graphene nanoribbons. *Nature* 444:347–349.
30. Son, Y.-W.; Cohen, M. L.; Louie, S. G. 2006. Energy gaps in graphene nanoribbons. *Phys. Rev. Lett.* 97:216803.
31. Novoselov, K. 2007. Graphene: Mind the gap. *Nat. Mater.* 6:720–721.
32. Kane, C. L.; Mele, E. J. 2005. Quantum spin Hall effect in graphene. *Phys. Rev. Lett.* 95:226801.
33. Yao, Y.; Ye, F.; Qi, X.-L.; Zhang, S.-C.; Fang, Z. 2007. Spin-orbit gap of graphene: First-principles calculations. *Phys. Rev. B* 75:041401.
34. Boettger, J. C.; Trickey, S. B. 2007. First-principles calculation of the spin-orbit splitting in graphene. *Phys. Rev. B* 75:121402.
35. Trevisanutto, P. E.; Giorgetti, C.; Reining, L.; Ladisa, M.; Olevano, V. 2008. Ab initio GW many-body effects in graphene. *Phys. Rev. Lett.* 101:226405.
36. Bonini, N.; Lazzeri, M.; Marzari, N.; Mauri, F. 2007. Phonon anharmonicities in graphite and graphene. *Phys. Rev. Lett.* 99:176802.
37. Calandra, M.; Mauri, F. 2005. Theoretical explanation of superconductivity in C_6Ca. *Phys. Rev. Lett.* 95:237002.
38. Calandra, M.; Mauri, F. 2007. Electron-phonon coupling and electron self-energy in electron-doped graphene: Calculation of angular-resolved photoemission spectra. *Phys. Rev. B* 76:205411.
39. Park, C.-H.; Giustino, F.; Cohen, M. L.; Louie, S. G. 2007. Velocity renormalization and carrier lifetime in graphene from the electron-phonon interaction. *Phys. Rev. Lett.* 99:086804.
40. Park, C.-H.; Giustino, F.; Spataru, C. D.; Cohen, M. L.; Louie, S. G. 2009. First-principles study of electron linewidths in graphene. *Phys. Rev. Lett.* 102:076803.
41. Tse, W.-K.; Das Sarma, S. 2007. Phonon-induced many-body renormalization of the electronic properties of graphene. *Phys. Rev. Lett.* 99:236802.
42. Li, G.; Luican, A.; Lopes dos Santos, J. M. B.; Castro Neto, A. H.; Reina, A.; Kong, J.; Andrei, E. Y. 2010. Observation of Van Hove singularities in twisted graphene layers. *Nat. Phys.* 6:109–113.
43. Zheng, Y.; Ando, T. 2002. Hall conductivity of a two-dimensional graphite system. *Phys. Rev. B* 65:245420.
44. Zhang, Y.; Tang, T.-T.; Girit, C.; Hao, Z.; Martin, M. C.; Zettl, A.; Crommie, M. F.; Shen, Y. R.; Wang, F. 2009. Direct observation of a widely tunable bandgap in bilayer graphene. *Nature* 459:820–823.
45. Castro, E. V.; Novoselov, K. S.; Morozov, S. V.; Peres, N. M. R.; dos Santos, J. M. B. L.; Nilsson, J.; Guinea, F.; Geim, A. K.; Neto, A. H. C. 2007. Biased bilayer graphene: semiconductor with a gap tunable by the electric field effect. *Phys. Rev. Lett.* 99:216802.
46. Ni, Z. H.; Chen, W.; Fan, X. F.; Kuo, J. L.; Yu, T.; Wee, A. T. S.; Shen, Z. X. 2008. Raman spectroscopy of epitaxial graphene on a SiC substrate. *Phys. Rev. B* 77:115416.
47. Mattausch, A.; Pankratov, O. 2007. Ab initio study of graphene on SiC. *Phys. Rev. Lett.* 99:076802.
48. Varchon, F.; Feng, R.; Hass, J.; Li, X.; Nguyen, B. N.; Naud, C.; Mallet, P. et al. 2007. Electronic structure of epitaxial graphene layers on SiC: Effect of the substrate. *Phys. Rev. Lett* 99:126805.
49. Magaud, L.; Hiebel, F.; Varchon, F.; Mallet, P.; Veuillen, J. Y. 2009. Graphene on the C-terminated SiC (000-1) surface: An ab initio study. *Phys. Rev. B* 79:161405.
50. Zhou, S. Y.; Gweon, G. H.; Fedorov, A. V.; First, P. N.; de Heer, W. A.; Lee, D. H. et al. 2007. Substrate-induced bandgap opening in epitaxial graphene. *Nat. Mater.* 6:770–775.
51. Peng, X.; Ahuja, R. 2008. Symmetry breaking induced bandgap in epitaxial graphene layers on SiC. *Nano Lett.* 8:4464–4468.

52. Ohta, T.; Bostwick, A.; Seyller, T.; Horn, K.; Rotenberg, E. 2006. Controlling the electronic structure of bilayer graphene. *Science* 313:951–954.
53. Watanabe, K.; Taniguchi, T.; Kanda, H. 2004. Direct-bandgap properties and evidence for ultraviolet lasing of hexagonal boron nitride single crystal. *Nat. Mater.* 3:404.
54. Giovannetti, G.; Khomyakov, P. A.; Brocks, G.; Kelly, P. J.; van den Brink, J. 2007. Substrate-induced band gap in graphene on hexagonal boron nitride: Ab initio density functional calculations. *Phys. Rev. B* 76:073103.
55. Lu, Y. H.; Chen, W.; Feng, Y. P.; He, P. M. 2008. Tuning the electronic structure of graphene by an organic molecule. *J. Phys. Chem. B* 113:2–5.
56. Shemella, P.; Nayak, S. K. 2009. Electronic structure and band-gap modulation of graphene via substrate surface chemistry. *Appl. Phys. Lett.* 94:032101.
57. Kang, Y.-J.; Kang, J.; Chang, K. J. 2008. Electronic structure of graphene and doping effect on SiO_2. *Phys. Rev. B* 78:115404.
58. Cuong, N. T.; Otani, M.; Okada, S. 2008. Semiconducting electronic property of graphene adsorbed on (0001) surfaces of SiO_2. *Phys. Rev. Lett.* 106:106801.
59. Fan, X. F.; Zheng, W. T.; Viorel, C.; Shen, Z. X.; Jer-Lai, K. 2012. Interaction between graphene and the surface of SiO_2. *J. Phys.: Condens. Matter.* 24:305004.
60. Giovannetti, G.; Khomyakov, P. A.; Brocks, G.; Karpan, V. M.; van den Brink, J.; Kelly, P. J. 2008. Doping graphene with metal contacts. *Phys. Rev. Lett.* 101:026803.
61. Khomyakov, P. A.; Giovannetti, G.; Rusu, P. C.; Brocks, G.; van den Brink, J.; Kelly, P. J. 2009. First-principles study of the interaction and charge transfer between graphene and metals. *Phys. Rev. B* 79:195425.
62. Yang, L.; Park, C.-H.; Son, Y.-W.; Cohen, M. L.; Louie, S. G. 2007. Quasiparticle energies and band gaps in graphene nanoribbons. *Phys. Rev. Lett.* 99:186801.
63. Ortmann, F.; Schmidt, W. G.; Bechstedt, F. 2005. Attracted by long-range electron correlation: Adenine on graphite. *Phys. Rev. Lett.* 95:186101.
64. Schedin, F.; Geim, A. K.; Morozov, S. V.; Hill, E. W.; Blake, P.; Katsnelson, M. I.; Novoselov, K. S. 2007. Detection of individual gas molecules adsorbed on graphene. *Nat. Mater.* 6:652–655.
65. Wehling, T. O.; Novoselov, K. S.; Morozov, S. V.; Vdovin, E. E.; Katsnelson, M. I.; Geim, A. K.; Lichtenstein, A. I. 2007. Molecular doping of graphene. *Nano Lett.* 8:173–177.
66. Fan, X.; Liu, L.; Kuo, J.-L.; Shen, Z. 2010. Functionalizing single- and multi-layer graphene with Br and Br_2. *J. Phys. Chem. C* 114:14939–14945.
67. Chang, C.-H.; Fan, X.; Li, L.-J.; Kuo, J.-L. 2012. Band gap tuning of graphene by adsorption of aromatic molecules. *J. Phys. Chem. C* 116:13788–13794.
68. Gierz, I.; Riedl, C.; Starke, U.; Ast, C. R.; Kern, K. 2008. Atomic hole doping of graphene. *Nano Lett.* 8:4603–4607.
69. Zanella, I.; Guerini, S.; Fagan, S. B.; Mendes Filho, J.; Souza Filho, A. G. 2008. Chemical doping-induced gap opening and spin polarization in graphene. *Phys. Rev. B* 77:073404.
70. Yan, J.-A.; Xian, L.; Chou, M. Y. 2009. Structural and electronic properties of oxidized graphene. *Phys. Rev. Lett.* 103:086802.
71. Ghaderi, N.; Peressi, M. 2010. First-principle study of hydroxyl functional groups on pristine, defected graphene, and graphene epoxide. *J. Phys. Chem. C* 114:21625–21630.
72. Sofo, J. O.; Chaudhari, A. S.; Barber, G. D. 2007. Graphane: A two-dimensional hydrocarbon. *Phys. Rev. B* 75:153401.
73. Leb; egrave; gue, S.; Klintenberg, M.; Eriksson, O.; Katsnelson, M. I. 2009. Accurate electronic band gap of pure and functionalized graphane from GW calculations. *Phys. Rev. B* 79:245117.
74. Nair, R. R.; Ren, W.; Jalil, R.; Riaz, I.; Kravets, V. G.; Britnell, L.; Blake, P. et al. 2010. Fluorographene: A two-dimensional counterpart of teflon. *Small* 6:2877–2884.
75. Klintenberg, M.; Lebègue, S.; Katsnelson, M. I.; Eriksson, O. 2010. Theoretical analysis of the chemical bonding and electronic structure of graphene interacting with Group IA and Group VIIA elements. *Phys. Rev. B* 81:085433.
76. Leenaerts, O.; Peelaers, H.; Hernández-Nieves, A. D.; Partoens, B.; Peeters, F. M. 2010. First-principles investigation of graphene fluoride and graphane. *Phys. Rev. B* 82:195436.
77. Samarakoon, D. K.; Chen, Z.; Nicolas, C.; Wang, X.-Q. 2011. Structural and electronic properties of fluorographene. *Small* 7:965–969.
78. Lehtinen, P. O.; Foster, A. S.; Ayuela, A.; Krasheninnikov, A.; Nordlund, K.; Nieminen, R. M. 2003. Magnetic properties and diffusion of adatoms on a graphene sheet. *Phys. Rev. Lett.* 91:017202.
79. Wu, M.; Liu, E.-Z.; Jianga, J. Z. 2008. Magnetic behavior of graphene absorbed with N, O, and F atoms: A first-principles study. *Appl. Phys. Lett.* 93:082504.
80. Huang, B.; Li, Z.; Liu, Z.; Zhou, G.; Hao, S.; Wu, J.; Gu, B.-L.; Duan, W. 2008. Adsorption of gas molecules on graphene nanoribbons and its implication for nanoscale molecule sensor. *J. Phys. Chem. C* 112:13442–13446.
81. Dai, J.; Yuan, J. 2010. Adsorption of molecular oxygen on doped graphene: Atomic, electronic, and magnetic properties. *Phys. Rev. B* 81:165414.
82. Berashevich, J.; Chakraborty, T. 2009. Tunable band gap and magnetic ordering by adsorption of molecules on graphene. *Phys. Rev. B* 80:033404.
83. Esquinazi, P.; Spemann, D.; Höhne, R.; Setzer, A.; Han, K. H.; Butz, T. 2003. Induced magnetic ordering by proton irradiation in graphite. *Phys. Rev. Lett.* 91:227201.
84. Vozmediano, M. A. H.; López-Sancho, M. P.; Stauber, T.; Guinea, F. 2005. Local defects and ferromagnetism in graphene layers. *Phys. Rev. B* 72:155121.
85. Faccio, R.; Pardo, H.; Denis, P. A.; Oeiras, R. Y.; Araújo-Moreira, F. M.; Veríssimo-Alves, M.; Mombrú, A. W. 2008. Magnetism induced by single carbon vacancies in a three-dimensional graphitic network. *Phys. Rev. B* 77:035416.
86. Uchoa, B.; Kotov, V. N.; Peres, N. M. R.; Neto, A. H. C. 2008. Localized magnetic states in graphene. *Phys. Rev. Lett.* 101:026805.
87. Yazyev, O. V. 2008. Magnetism in disordered graphene and irradiated graphite. *Phys. Rev. Lett.* 101:037203.
88. Fan, X. F.; Liu, L.; Wu, R. Q.; Peng, G. W.; Fan, H. M.; Feng, Y. P.; Kuo, J. L.; Shen, Z. X. 2010. The role of *sp*-hybridized atoms in carbon ferromagnetism: A spin-polarized density functional theory calculation. *J. Phys.: Condens. Matter* 22:046001.
89. Kusakabe, K.; Maruyama, M. 2003. Magnetic nanographite. *Phys. Rev. B* 67:092406.
90. Wei, D.; Liu, Y.; Wang, Y.; Zhang, H.; Huang, L.; Yu, G. 2009. Synthesis of N-doped graphene by chemical vapor deposition and its electrical properties. *Nano Lett.* 9:1752–1758.
91. Martins, T. B.; Miwa, R. H.; da Silva, A. J. R.; Fazzio, A. 2007. Electronic and transport properties of boron-doped graphene nanoribbons. *Phys. Rev. Lett.* 98:196803.
92. Huang, B.; Yan, Q.; Zhou, G.; Wu, J.; Gu, B.-L.; Duan, W.; Liu, F. 2007. Making a field effect transistor on a single graphene nanoribbon by selective doping. *Appl. Phys. Lett.* 91:253122.
93. Cervantes-Sodi, F.; Csányi; G.; Piscanec, S.; Ferrari, A. C. 2008. Edge-functionalized and substitutionally doped graphene nanoribbons: Electronic and spin properties. *Phys. Rev. B* 77:165427.

94. Fan, X.; Shen, Z.; Liu, A. Q.; Kuo, J.-L. 2012. Band gap opening of graphene by doping small boron nitride domains. *Nanoscale* 4:2157–2165.
95. Banhart, F.; Kotakoski, J.; Krasheninnikov, A. V. 2011. Structural defects in graphene. *ACS Nano* 5:26–41.
96. Ma, J.; Alfè, D; Michaelides, A.; Wang, E. 2009. Stone-Wales defects in graphene and other planar sp^2-bonded materials. *Phys. Rev. B* 80:033407.
97. Li, L.; Reich, S.; Robertson, J. 2005. Defect energies of graphite: Density-functional calculations. *Phys. Rev. B* 72:184109.
98. Yazyev, O. V.; Helm, L. 2007. Defect-induced magnetism in graphene. *Phys. Rev. B* 75:125408.
99. Krasheninnikov, A. V.; Lehtinen, P. O.; Foster, A. S.; Nieminen, R. M. 2006. Bending the rules: Contrasting vacancy energetics and migration in graphite and carbon nanotubes. *Chem. Phys. Lett.* 418:132–136.
100. Rossato, J.; Baierle, R. J.; Fazzio, A.; Mota, R. 2004. Vacancy formation process in carbon nanotubes: First-principles approach. *Nano Lett.* 5:197–200.

13 Novel Electronic Properties of a Graphene Antidot, Parabolic Dot, and Armchair Ribbon

S.-R. Eric Yang and S. C. Kim

CONTENTS

ABSTRACT

Graphene nanostructures have great potential for device applications. However, they can exhibit several counterintuitive electronic properties not present in ordinary semiconductor nanostructures. In this chapter, we review several of these graphene nanostructures. A first example is a graphene antidot that possesses boundstates inside the antidot potential in the presence of a magnetic field. As the range of the repulsive potential decreases in comparison to the magnetic length, the effective coupling constant between the potential and electrons becomes more repulsive, and then, it changes the sign and becomes attractive. This is a consequence of a subtle interplay between Klein tunneling and quantization of Landau levels. In this regime, wavefunctions become anomalous with a narrow probability density peak inside the barrier and another broad peak outside the potential barrier with the width comparable to the magnetic length. The second example is a graphene parabolic dot in the presence of a magnetic field. One counterintuitively finds that resonant quasi-bound states of both positive and negative energies exist in the energy spectrum. The presence of resonant quasi-bound states of negative energies is a unique property of massless Dirac fermions. Also, when the strength of the potential increases, resonant and nonresonant states transform into discrete anomalous states with a narrow probability density peak inside the well and another broad peak under the potential barrier with the width comparable to the magnetic length. The last example is a one-dimensional electron gas in the lowest energy conduction subband of graphene armchair ribbons. Bulk magnetic properties of it may sensitively depend on the width of the ribbon. For ribbon widths $L_x = 3Ma_0$, depending on the value of the Fermi energy, a ferromagnetic or paramagnetic state can be stable while for $L_x = (3M + 1)a_0$, the paramagnetic state is stable (M is an integer and a_0 is the length of the unit cell). Ferromagnetic and paramagnetic states are well suited for spintronic applications.

13.1 PART I: GATE INDUCED ANTIDOTS AND DOTS IN A MAGNETIC FIELD

13.1.1 Antidot in Magnetic Fields

We consider antidots fabricated by potential modulation in a graphene sheet (Vázquez de Parga et al., 2008, Meyer et al., 2008, Park et al., 2010, 2011a,b,c, Kim et al., 2012) (our antidots are not created by punching holes in the graphene sheet). Before we review details of solutions of the antidot problem in the presence of a magnetic field, it is useful to consider some general properties of the solutions. Let us consider how Dirac electrons in a magnetic field respond to a localized potential. The Dirac Hamiltonian near $\vec{K}$ point is $H = v_F\vec{\sigma}\cdot(\vec{p}+(e/c)\vec{A})+\tilde{V}(r)I$, where I is a 2×2 unit matrix, v_F is the Fermi velocity, and $\vec{\sigma}=(\sigma_x,\sigma_y)$ are Pauli spin matrices (the Dirac Hamiltonian can be derived from a tight-binding Hamiltonian, see Appendix 13A). The second term $\tilde{V}(r)=(V_0/(R^2\pi))e^{-r^2/R^2}$ represents a regularized two-dimensional (2D) delta-function potential. In the limit where the potential range $R\to 0$, the potential becomes delta-function potential and the wavefunction will diverge at $r = 0$. One way to circumvent this problem is to regularize the delta function with a finite range R. This problem has two length scales: the magnetic length $\ell = \sqrt{(\hbar c/eB)}$ and the range of the potential R. Note that the localized potential scales as $1/r^2$, while the kinetic term scales as $1/r$ (Jackiw, 1991). So, when $R \ll \ell$, the localized potential in graphene dominates over the kinetic term, and the ordinary perturbational approach fails. In graphene, another new feature is present, namely Klein tunneling. The interplay between the localized potential and Klein tunneling exhibits several counterintuitive features, and affects both the width of the localized wavefunction and its energy in a nontrivial way. We will show that the resulting state is *anomalous* with a narrow probability density peak inside the barrier and another broad peak outside the potential barrier with the width comparable to the magnetic length. These unusual states are present in cylindrical (Park et al., 2011a,b,c), parabolic (Kim et al., 2012), and Coulomb (Kim and Yang, 2014a) potentials.

It is also useful to consider a symmetry between antidot and dot potentials $V_a(r)$ and $V_d(r)$ satisfying $V_a(r) + V_d(r) = V$. It can be shown that dot and antidot states with different energies $E_d = V - E_a$ and E_a have the same wavefunctions, see Figure 13.1. This means that the induced density is invariant under the transformation $V_a(r) \to V_d(r)$ (see Figure 13.2). Using this transformation, we can also understand why there is a boundstate inside an antidot (see Figure 13.1c and d).

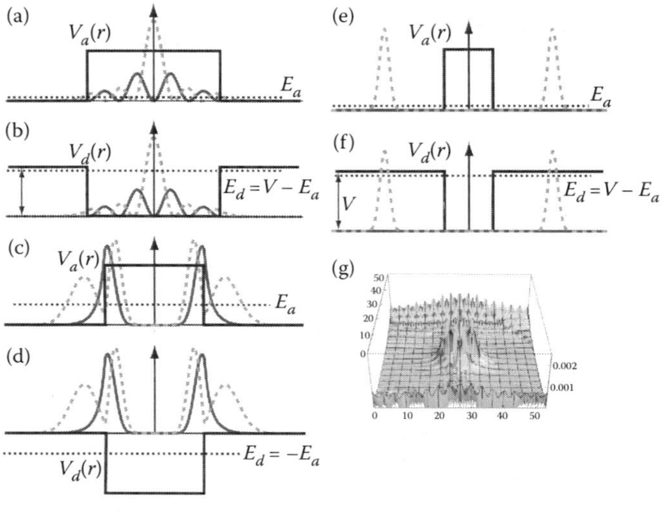

FIGURE 13.1 Schematic drawings of some typical probability wavefunctions of antidots and dots. (a) and (b) display antidot and dot states with the same wavefunction but with different energies $E_d = V - E_a$ and E_a. The dolid (dashed) line represents the probability wavefunction of A (B) component. Naively, one may expect that the energy of the antidot state of (a) should be larger than the dot state by an amount V since the antidot state is in the barrier with the height V, but this reasoning is actually incorrect. (c) and (d) are the same as in (a) and (b) but with $V_a(r)+V_d(r)=0$. Antidot and dot states have energies that are E_a and $E_d = -E_a$. (e) and (f) are the same as (a) and (b) but wavefunctions are localized outside the antidot and dot regions. (g) shows the tight-binding probability density of an antidot state of a zigzag nanoribbon. (Adapted from Park, P. S., S. C. Kim, and S.-R. Eric Yang, 2010, *J. Phys.: Condens. Matter* **22**, 375302.)

It should be noted that in the presence of a potential, density accumulation changes into depletion as the ratio R/ℓ increases, see Figure 13.2a and b. The presence of positive-induced density inside the repulsive barrier may appear to be counterintuitive. This charge accumulation can be understood by the invariance of the induced density of a filled Landau level (LL) under the transformation of a repulsive potential to an attractive potential: $V(r) \to -V(r)$. However, if the perspective of an attractive potential is adopted to explain the charge accumulation, Figure 13.2c, the density depletion, shown in Figure 13.2d, cannot be explained since the transformed

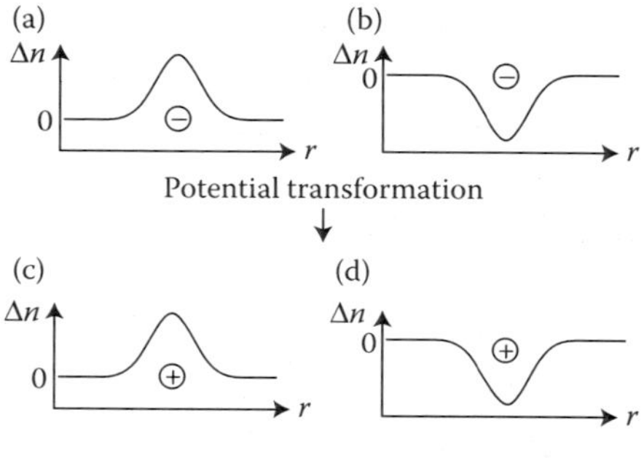

FIGURE 13.2 (a) and (b): Induced densities $\Delta n(r)$ for a repulsive barrier at two different values of R/ℓ. (c) and (d): Same as in (a) and (b) but for an attractive well. Symbols $-$ and $+$ represent the repulsive barrier and attractive well.

attractive potential would accumulate electrons. We will show that the crossover between density accumulation and depletion is highly nontrivial.

13.1.1.1 Effect of Magnetic Field: Eigenenergies and Eigenfunctions

We consider some consequences of LL quantization in the presence of a rotationally invariant impurity potential in graphene. In the absence of a magnetic field, the energy spectrum of a Dirac electron moving near a dot or antidot is continuous with quasi-bound states that have finite trapping times (Hewageegana and Apalkov, 2008, Matulis and Peeters, 2008). However, in the presence of an infinitely small magnetic field, these quasi-bound states become *true bound states* with infinite lifetimes (Giavaras et al., 2009). A Dirac electron can move inside an antidot potential due to Klein. This behavior is in sharp contrast to the electron motion of skipping orbits around an antidot potential of an ordinary semiconductor in a magnetic field. The perturbative method gives qualitatively incorrect results and fails to describe these boundstates inside antidots. Since the characteristics of solutions change qualitatively, the presence of a magnetic field represents a singular perturbation.

Consider a rotationally invariant antidot potential in graphene in the presence of a magnetic field. The model potential is cylindrical

$$\tilde{V}(r) = \begin{cases} V & \text{for } r \le R \\ 0 & \text{for } r > R \end{cases} \tag{13.1}$$

We assume that the strength of the potential is such that energies of impurity states of an LL do not overlap with those of adjacent LLs. For example, $N = 0$ LL states must not overlap with states of $N = \pm 1$ LLs. In the presence of an impurity, potential boundstates will split off from degenerate LLs, see Figure 13.3. The Dirac equations can be solved exactly using

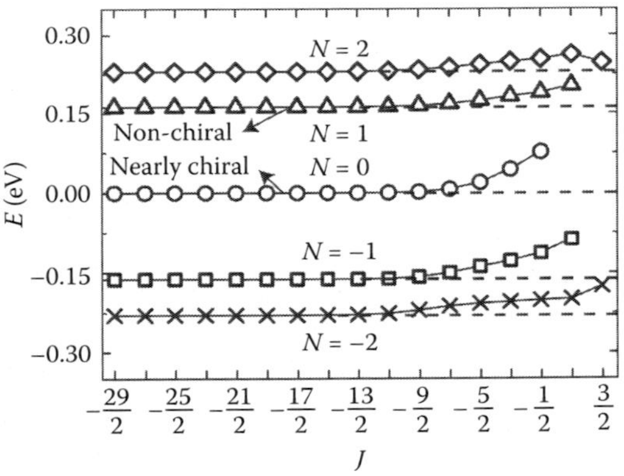

FIGURE 13.3 Single-particle energy spectrum of a massless Dirac fermion as a function of angular momentum quantum number J in the presence of a repulsive potential and magnetic field. Energies of five LLs are shown for $N = -2, -1, 0, 1$, and 2. Dashed lines represent unperturbed energies, and those that deviate from these are energies of boundstates (at $B = 20$ T, $R = 10$ nm, and $V = 0.1$ eV). (Adapted from Park, P. S., S. C. Kim, and S.-R. Eric Yang, 2011a, *Phys. Rev. B* **84**, 085405.)

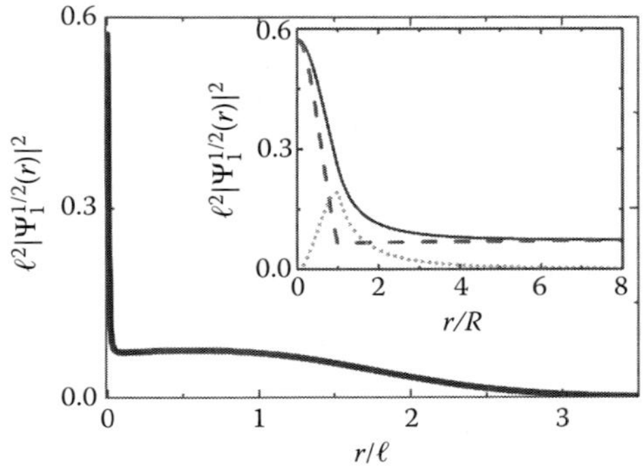

FIGURE 13.4 Probability density of a boundstate with LL index $N = 1$ and angular momentum $J = 1/2$ in the strong-coupling limit for $V/E_c = 1.82(E_c = (\hbar v_F/R))$, $R/\ell = 0.01$, $R = 1$ nm, $V = 1.2$ eV, and $E = 0.009$ eV. Inset: A (B) component is the dashed (dotted) line. (Adapted from Park, P. S., S. C. Kim, and S.-R. Eric Yang, 2011a, *Phys. Rev. B* **84**, 085405.)

confluent hypergeometric functions (Recher et al., 2009). Park et al. (2011a) find that a repulsive potential of graphene in the presence of a magnetic field has bound states that are peaked inside the barrier with tails extending over $\ell(N + 1)$, where N is the LL index. An exact solution is displayed in Figure 13.4. Note the presence of a sharp peak inside the antidot potential.

It is useful to rewrite the Dirac equation into an equation that looks like a Schrödinger equation (see Appendix 13B)

$$\begin{aligned} -\varphi_A''(r) + \tilde{U}(r)\varphi_A(r) &= \varepsilon\varphi_A(r), \\ -\varphi_B''(r) + U(r)\varphi_B(r) &= \varepsilon\varphi_B(r). \end{aligned} \tag{13.2}$$

Here, the wavefunctions $\varphi_A(r)$ and $\varphi_B(r)$ are related to the A- and B-components of the solution of the Dirac Hamiltonian, see Appendix 13B. The effective potentials for $\sigma = A$- and B-components are, respectively, $\ell^2\tilde{U}(r) = \ell^2\tilde{U}_1(r) + \ell^2\tilde{U}_2(r)$ and $\ell^2U(r) = \ell^2U_1(r) + \ell^2U_2(r)$, see Appendix 13B. It can be shown that the effective potentials behave like r^2 for $r \to \infty$ and *no quasi-bound states can exist.* These potentials for the state shown in Figure 13.4 are plotted in Figure 13.5. According to the shapes of these potentials, the A-component of the wavefunction should have a peak near $r = 0$ while the B-component of the wavefunction should be zero at $r = 0$, which is consistent with the result of Figure 13.4.

13.1.1.2 Induced Density of Filled LLs

Park et al. (2011a) have investigated how bound states inside the antidot potential affect the scaling properties of the induced density of *filled* LLs of massless Dirac fermions in different coupling regimes (they define the strong, intermediate, and weak-coupling regimes as $R/\ell \ll 1$, $R/\ell \sim 1$, and $R/\ell \gg 1$, respectively). The electron density of the Nth-split LL is given by

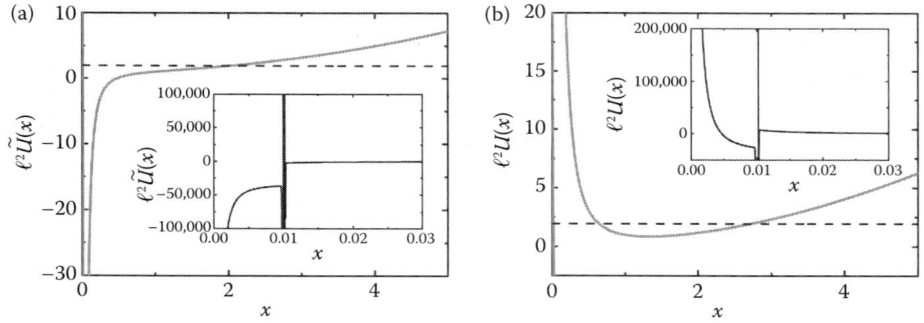

FIGURE 13.5 (a) Effective potential for A-component of the wavefunction shown in Figure 13.4 ($x = r/\ell$ is the dimensionless coordinate). (b) Effective potential for B-component of the wavefunction shown in Figure 13.4. $R/\ell = 0.01$, $B = 0.066$ T, $J = 1/2$, $E = 0.009$ eV, and $(E/E_M)^2 = 1.94$.

$$n_N(r) = \sum_{J \in \text{filled Landau level}} |\Psi_N^J(r)|^2. \tag{13.3}$$

The eigenstates $\Psi_N^J(r)$ and eigenvalues $E_N(J)$ are labeled by LL index N and the half-integer angular momentum quantum number J (here the LL is assumed to be completely filled). This definition is valid as long as energies of impurity states of the LL do not overlap with those of adjacent LLs. Figure 13.3 displays energies of these antidot states. In the absence of a localized potential, the dimensionless density takes the value

$$\ell^2 n_N(r) = \frac{1}{2\pi}. \tag{13.4}$$

It should be noted that the total number of electrons is equal to the LL degeneracy

$$\int d\vec{r} n_N(\vec{r}) = \frac{A}{2\pi\ell^2} = N_\varphi, \tag{13.5}$$

where A is the area of the system. For chiral fermions, Park et al. (2011a) find, in a strong-coupling regime, that the density inside the repulsive potential can be greater than the value in the absence of the potential while in the weak-coupling regime, we find negative-induced density, see Figure 13.6.

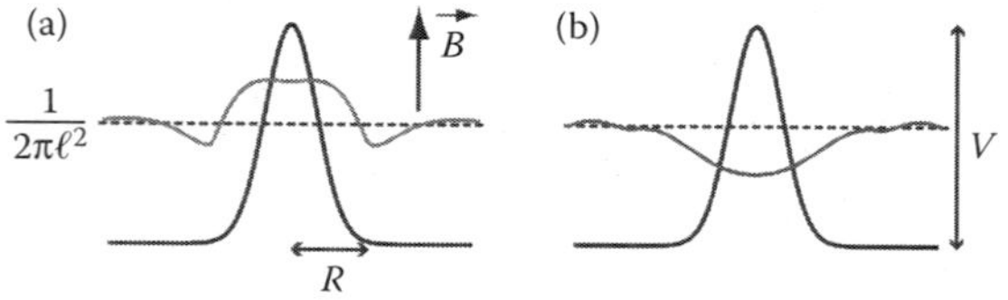

FIGURE 13.6 Schematic plot of electron density near an antidot of size R for a filled LL. The electron density in the absence of the antidot potential is $1/2\pi\ell^2$, represented by the dashed horizontal line. In graphene LLs, under certain conditions, induced density can be positive (a), in contrast to the usual case where it is negative (b). (Adapted from Park, P. S., S. C. Kim, and S.-R. Eric Yang, 2011a, *Phys. Rev. B* **84**, 085405.)

Similar results hold also for nonchiral fermions. As one moves from weak- to strong-coupling regimes, the effective coupling constant between the potential and electrons becomes more repulsive, and then, it changes the sign and becomes attractive. These unusual properties may be probed experimentally through scanning tunneling spectroscopy. The value of the dimensionless electron density at the center of the potential ($r = 0$) gives a good indication of how strong the *effective coupling constant* between the repulsive potential and an electron is. Thus, we will call this dimensionless-induced density at $r = 0$, with sign change, as the effective coupling constant $\alpha = -\ell^2 \Delta n_N(0)$.

Park et al. (2011a) calculate the induced density using second-order perturbation theory. The electron density at the origin of the $N = 0$ LL is given by the value of the probability density of the eigenstate $\Psi_0^{-1/2}(0)$. The eigenstate $\Psi_0^{-1/2}(\vec{r})$ is expanded in the basis states $\psi_N^J(r)$ of graphene LLs (see Appendix 13C)

$$\Psi_0^{-1/2}(\vec{r}) = \sum_N C_N \psi_N^{-1/2}(\vec{r}). \tag{13.6}$$

The density at $r = 0$ is

$$\begin{aligned} n(0) = |\Psi_0^{-1/2}(0)|^2 &= \sum_{N,N'} C_N^* C_{N'} \psi_N^{-1/2}(0)^* \psi_{N'}^{-1/2}(0) \\ &= |C_0|^2 |\psi_0^{-1/2}(0)|^2 \end{aligned} \tag{13.7}$$

In deriving this result, the following equations are used: $\sum_{N\neq 0} C_N \psi_N^{-1/2}(0)^* \psi_0^{-1/2}(0) = 0$ and $\sum_{N,N'\neq 0} C_N^* C_{N'} \psi_N^{-1/2}(0)^* \psi_{N'}^{-1/2}(0) = 0$, Kim and Yang (2012). Note that in the absence of the potential, this value of $n(0)$ is $|\psi_0^{-1/2}(0)|^2$, where $\psi_0^{-1/2}(r) = (0, \phi_{0,0}(r))$. The density at $r = 0$ is thus renormalized by $|C_0|^2$. The coefficient C_0 may be evaluated up to second order in V from the wavefunction renormalization (Baym, 1969)

$$Z = |C_0|^2 = 1 - \sum_{N\neq 0} \frac{\left|\langle \psi_N^{-1/2} \right| V \left| \psi_0^{-1/2} \rangle\right|^2}{\varepsilon_N^2}, \tag{13.8}$$

where LL energy is $\varepsilon_N = E_M\sqrt{2|N|}$ with $E_M = \hbar v_F/\ell$. The density difference at the origin with and without the localized potential is thus

$$\Delta n(0) = |\Psi_0^{-1/2}(0)|^2 - \frac{1}{2\pi\ell^2} = -\frac{1}{2\pi\ell^2}\sum_{N\neq 0}\frac{|\langle\psi_N^{-1/2}|V|\psi_0^{-1/2}\rangle|^2}{\varepsilon_N^2}. \tag{13.9}$$

Here, we use that, when $V = 0$, the density of each filled band is a constant $n(r) = 1/2\pi\ell^2$. This induced density can be written in a dimensionless-scaling form

$$\ell^2\Delta n(0) = \left(\frac{V}{E_c}\right)^2 g\left(\frac{R}{\ell}\right). \tag{13.10}$$

where

$$g\left(\frac{R}{\ell}\right) = -\frac{A_{0,0}}{(R/\ell)^2}\sum_{N\neq 0} A_{|N|,|N|}\frac{1}{2|N|}\int_0^{R/\ell} dx\, x e^{-(x^2/2)} L_{|N|}^0\left(\frac{x^2}{2}\right). \tag{13.11}$$

($E_c = \hbar v_F/R$ and $A_{N,N}$ are normalization constants of LL wavefunctions and are given in Appendix 13C). Figure 13.7 compares this result with the exact result. For large R/ℓ, the agreement is excellent. However, the perturbative result fails dramatically for $R/\ell < 1$ as the sign of its induced density is negative while that of the exact result is positive. These results demonstrate that perturbative methods cannot be applied in the intermediate- and strong-coupling regimes. Note that the sign of the induced density does not change under the transformation $V(r) \rightarrow -V(r)$.

The sign of induced densities for various values of (R/ℓ, V/E_c) is schematically shown in Figures 13.8 and 13.9. The boundaries between positive- and negative-induced densities of chiral and nonchiral fermions are displayed in Figures 13.8 and 13.9, respectively. For nonchiral fermions, the range of R/ℓ, where the induced density is positive, expands, as V/E_c increases. However, for chiral fermions, the critical value of R/ℓ is about 1, independent of V/E_c. However, our single-electron energies are not the renormalized Hartree–Fock (HF) result. This correction will somewhat affect our numerical estimate of the boundary between overlapping LLs of Figures 13.8 and 13.9.

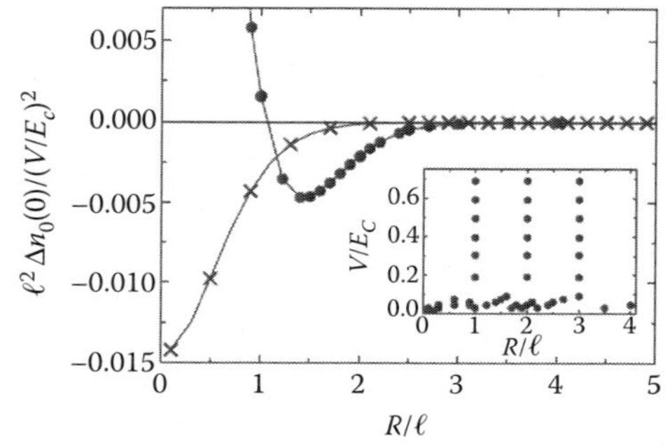

FIGURE 13.7 Crosses represent the results of $g_0(R/\ell)$ obtained by treating $V(r)$ in the second-order perturbation theory. Circles represent exact values of $g_0(R/\ell)$ obtained by data collapse for $N = 0$ and $V/E_c < 0.7$. Values (R/ℓ, V/E_c) used in the data collapse are shown in the inset. Note that the induced density changes the sign near $R/\ell = 1$. (Adapted from Park, P. S., S. C. Kim, and S.-R. Eric Yang, 2011a, *Phys. Rev. B* **84**, 085405.)

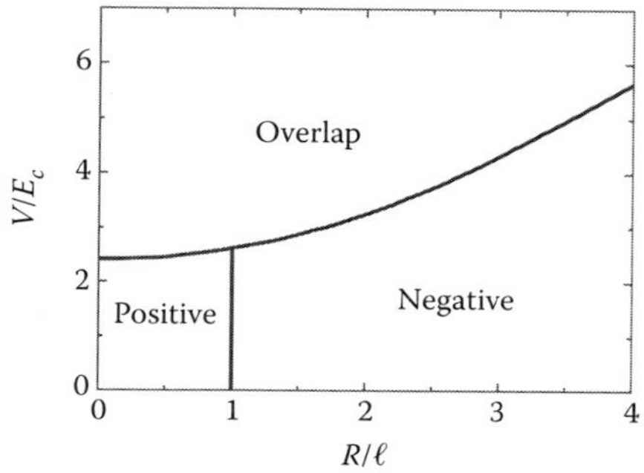

FIGURE 13.8 Sign of induced density at $r = 0$ in the parameter space (R/ℓ, V/E_c). In the region denoted as "overlap," $N = 0$ LL states overlap with $N = \pm 1$ LLs. (Adapted from Park, P. S., S. C. Kim, and S.-R. Eric Yang, 2011a, *Phys. Rev. B* **84**, 085405.)

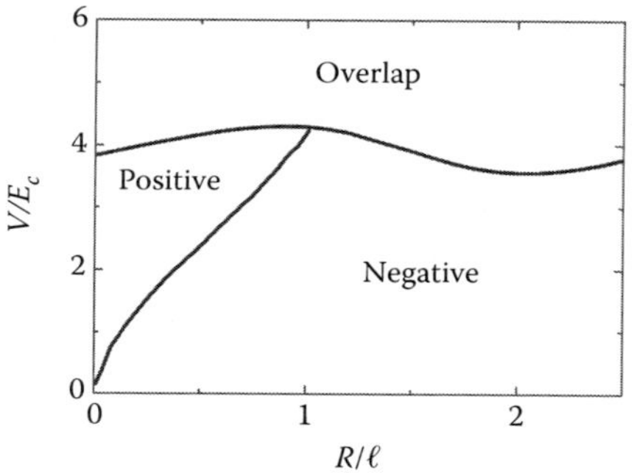

FIGURE 13.9 Sign of induced density at $r = 0$ in the parameter space (R/ℓ, V/E_c). The range of R/ℓ, where the induced density at $r = 0$ is positive increases with increasing V/E_c. In the region denoted as "overlap," $N = 1$ LL states overlap with $N = 0$, 2 LLs. (Adapted from Park, P. S., S. C. Kim, and S.-R. Eric Yang, 2011a, *Phys. Rev. B* **84**, 085405.)

13.1.1.3 Scaling Properties of Induced Density of Chiral and Nonchiral Dirac Fermions in Magnetic Fields

In the presence of the potential, the induced density is defined as the difference between densities with and without the potential

$$\ell^2\Delta n_N(r) = \ell^2 n_N(r) - \frac{1}{2\pi}. \tag{13.12}$$

Here, we have used that each filled LL has electron density $\ell^2/2\pi$. A cylindrical potential in zero-magnetic field possesses quasi-bound states with complex energies. The energy-level spacing of these quasi-bound states is roughly $E_c = \hbar v_F/R$ with finite lifetimes. The ratio between this energy scale and the energy scale of LLs $E_M = \hbar v_F/\ell$ is given by

$$\frac{E_M}{E_c} = \frac{R}{\ell}. \quad (13.13)$$

Park et al. (2011a) find that the correct scaling function of the induced density has the form

$$\ell^2 \Delta n_N(r) = s_N\left(\frac{r}{R}, \frac{V}{E_c}, \frac{R}{\ell}\right). \quad (13.14)$$

It is different from the scaling function of ordinary LLs where one would expect that V/E_M appears instead of V/E_c. In graphene, V/E_M does not contain a nonperturbative effect of the formation of bound states in the barrier, and it cannot be used instead of V/E_c.

The effective coupling constant has the scaling form

$$\alpha_N\left(\frac{V}{E_c}, \frac{R}{\ell}\right) = -\ell^2 \Delta n_N(0) = -s_N\left(0, \frac{V}{E_c}, \frac{R}{\ell}\right). \quad (13.15)$$

The numerical results of Park et al. (2011a) show that as one moves from weak to strong-coupling regimes, the effective coupling constant becomes increasingly more repulsive. Upon further increase, it starts to decrease, passing through zero, and finally becomes attractive, see Figures 13.6 and 13.7. This means that a repulsive potential can effectively attract electrons in the strong-coupling regime. The formation of bound states with sharp peaks of the probability density inside the barrier is responsible for this effect, making the induced density positive (see Figure 13.4). The main physics behind this effect is the competition between localization of the wavefunction in the barrier and Klein tunneling effect. Also, note that the value of the induced density does not change under the transformation $V_a(r) \rightarrow V_d(r)$ (see Figure 13.1).

In the limit $V/E_c \ll 1$, Park et al. (2011a) find that the induced density at $r = 0$ is given by a power law of V/E_c

$$\ell^2 \Delta n_N(0) = g_N\left(\frac{R}{\ell}\right)\left(\frac{V}{E_c}\right)^{\delta_N}, \quad (13.16)$$

with $\delta_N = 2$ for the chiral $N = 0$ LL, but $\delta_N = 1$ for nonchiral $N = 1$ LL. A plot of $g_0(R/\ell)$ for chiral fermions is shown in Figure 13.10. Note that the power law holds even for $V/E_c \sim 1$. It is noteworthy that $g_0(R/\ell)$ takes the minimum value near $R/\ell \approx 1.5$. This implies that the electron density is depleted most strongly for $R/\ell \approx 1.5$. However, further decrease in R/ℓ has the opposite effect of increasing more penetration of electrons into the barrier. Near $R/\ell = 1$, the scaling function $g_0(R/\ell)$ changes the sign. For $R/\ell < 1$, electrons accumulate in the barrier and the density becomes greater than

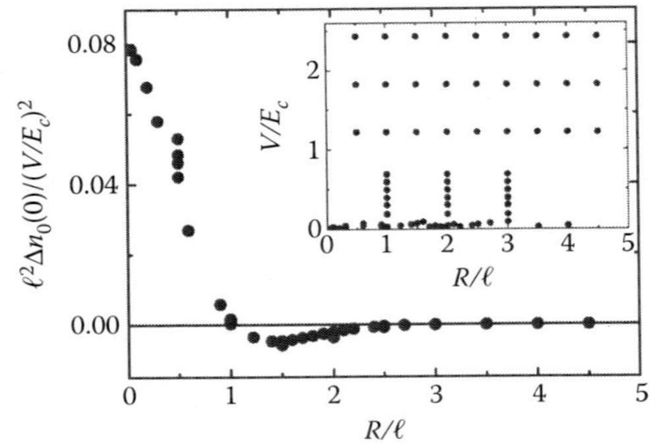

FIGURE 13.10 Approximate data collapse of $g_0(R/\ell)$ is obtained for a larger range of V/E_c than the one used in Figure 13.7. Values of $(R/\ell, V/E_c)$ used are shown in the inset. (Adapted from Park, P. S., S. C. Kim, and S.-R. Eric Yang, 2011a, *Phys. Rev. B* **84**, 085405.)

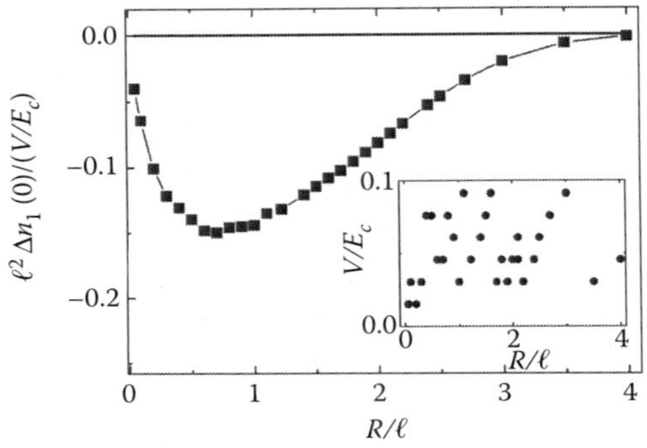

FIGURE 13.11 $g_1(R/\ell)$ is obtained by data collapse for $N = 1$ and $V/E_c \ll 1$. Values of $(R/\ell, V/E_c)$ used are shown in the inset. (Adapted from Park, P. S., S. C. Kim, and S.-R. Eric Yang, 2011a, *Phys. Rev. B* **84**, 085405.)

the density of the unperturbed LL. This dependence on R/ℓ is thus strongly *nonlinear*. Similar behavior is also seen in $g_1(R/\ell)$ for nonchiral fermions, as shown in Figure 13.11. The dependence of the induced density on R/ℓ is again strongly nonlinear: $g_1(R/\ell)$ takes the minimum value near $R/\ell = 0.7$.

The scaling function has two critical points $y_{c,1}$ and $y_{c,2}$

$$\left.\frac{\partial s_N(0,x,y)}{\partial y}\right|_{y=y_{c,1}} = 0, \quad (13.17)$$

$$\left. s_N(0,x,y)\right|_{y=y_{c,2}} = 0, \quad (13.18)$$

where ($x = V/E_c$ and $y = R/\ell$). We find that $y_{c,1} \sim 1$. The induced density function takes the global minimum at $y_{c,1}$ and changes the sign at $y_{c,2}$.

13.1.1.4 Antidot in a Zigzag Ribbon in Magnetic Fields

An antidot in a zigzag ribbon can interact with surface states of zigzag edges. This interaction is nontrivial and can produce several interesting chiral states. The energy spectrum of an antidot is displayed in Figure 13.12 as a function of the magnetic field. Note that there are several states that have nearly

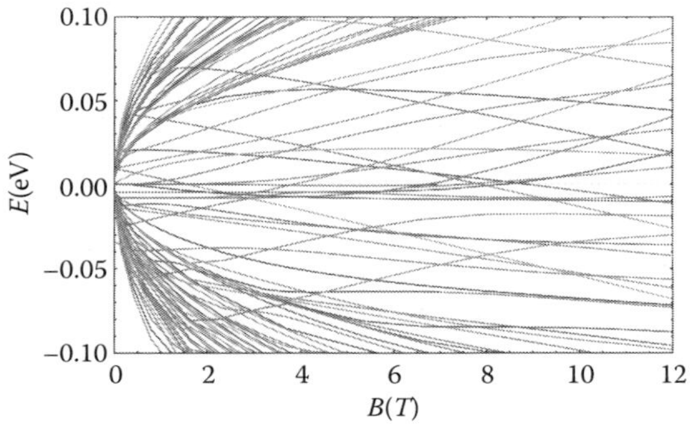

FIGURE 13.12 Dependence of the energy spectrum of an antidot on B. Parameters are $R = 200$ Å, $B = 10.28$ T, $\ell = 80$ Å, $V = 0.26$ eV, and $E_c = 0.033$ eV. Here, $V/E_c \gg 1$; so, it overlaps with adjacent LLs. (Adapted from Park, P. S., S. C. Kim, and S.-R. Eric Yang, 2010, *J. Phys.: Condens. Matter* **22**, 375302.)

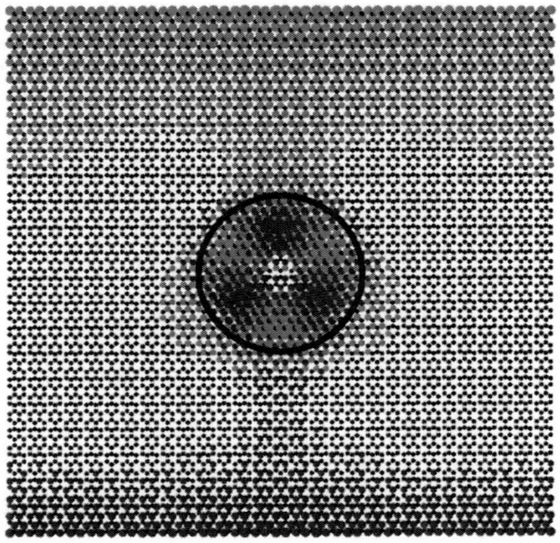

FIGURE 13.13 Occupation probabilities on A and B carbon atoms of a state near the Dirac point with $E = -0.01$ eV. Small black dots indicate lattice sites and gray (black) dots indicate values of the probability wavefunction on A (B) carbon atoms. The circle represents the antidot. Parameters are $B = 20$ T, $V = 1$ eV, $R = 19.68$ Å, $R/\ell = 0.34$, and $V/E_c = 2.99$. (Adapted from Park, P. S., S. C. Kim, and S.-R. Eric Yang, 2010, *J. Phys.: Condens. Matter* **22**, 375302.)

zero energy. Probability density of such a state is shown in Figure 13.13. Different regions of the ribbon display different chiralities. The region with a strong mixed chirality is almost absent. On the other hand, probability densities of eigenstates away from $E = 0$ are not peaked along the zigzag edges, see Figure 13.14.

13.1.1.5 Periodic Antidots in Magnetic Fields

How Klein tunneling changes as a magnetic function can be clearly seen in a periodic antidot array, see Figure 13.15. Experimentalists have succeeded in fabricating graphene antidot arrays by creating a 2D periodic potential consisting of circular dots (Meyer et al., 2008, Vázquez de Parga et al., 2008). Before we review the results of periodic antidots, we first consider the case of a single antidot. The lowest positive energy of the eigenstate $\Psi_0^{-1/2}(r)$ of an antidot is plotted as a function of R/ℓ in Figure 13.16. When $R/\ell < 1$, one is in the quantum regime, where Klein tunneling is possible. In this case, eigenenergies can be nearly zero. When multiple antidots are present, the probability density is localized inside antidots, see Figure 13.17. On the other hand, a semi classical argument (Xiao et al., 2010) can be applied when $R/\ell \gg 1$. In this case, an electron is localized outside the antidot, see Figure 13.18. When there are multiple antidots, an electron with low energy will be in the region between the antidot potentials, see Figure 13.19.

Kim and Yang (2012) have investigated an antidot array on a rectangular graphene sheet using a tight-binding Hamiltonian. Let us give a short summary of their theoretical results of such a system. A probability density is localized inside antidots when $R/\ell < 1$, while it is delocalized outside antidots for $R/\ell \gg 1$, see Figures 13.17 and 13.19. Skipping orbits can form near the edges of antidots in the intermediate regime $R/\ell > 1$. These results suggest that the usual semiclassical results of a strong magnetic field hold in the regime $R/\ell \gg 1$ while Klein tunneling is operative in the regime $R/\ell < 1$. Scattering from multiple antidots is important in localizing the probability in antidots.

13.1.2 Parabolic Dot in Magnetic Fields

Consider an electron moving in an external potential in the presence of a magnetic field. Here, we consider a very simple potential, that is, a parabolic dot. This model of the dot is one of the most widely studied models in ordinary semiconductor heterostructures, both experimentally and theoretically. A parabolic dot can effectively confine electrons, and the number of electrons in it can be controlled using a gate potential. For a massful electron, the Hamiltonian is given by $H = (1/2m)(\vec{p} + (e/c)\vec{A})^2 + (1/2)m\Omega^2 r^2$ with a magnetic field B that is applied perpendicular to the 2D plane (vector potential $\vec{A}$ is given in a symmetric gauge). The characteristic length scale of the problem is given by $\lambda^2 = (\hbar/m\sqrt{4\Omega^2 + \omega_c^2})$, where $\omega_c = eB/mc$ is the cyclotron frequency. It can be solved exactly and all the eigenenergies are *positive* and their spectrum is discrete (Fock, 1928, Darwin, 1930).

The problem has several new features in graphene in comparison to ordinary semiconductors. First, conduction and valence bands are represented by the Dirac cones and they touch each other at the Dirac points. Unlike the case of ordinary semiconductor dots, the parabolic potential in graphene couples conduction and valence band states. This effect can make an electron leak out of the well, see Figure 13.20. The other new feature is the presence of negative energy states under the potential barrier, which will get strongly perturbed by the parabolic potential, as the following argument indicates. The first-order energy correction of a graphene LL state $\psi_{n,m}(r)$ is, for a sufficiently large n

$$\langle \psi_{n,m} | V(r) | \psi_{n,m} \rangle \sim \kappa \langle r^2 \rangle \sim \kappa \ell^2 |n|,$$

where $\kappa = m\,\Omega^2$. This result suggests that a LL state with a large negative energy, $-E_M\sqrt{2|n|}$ with $|n| \gg 1$, corresponding

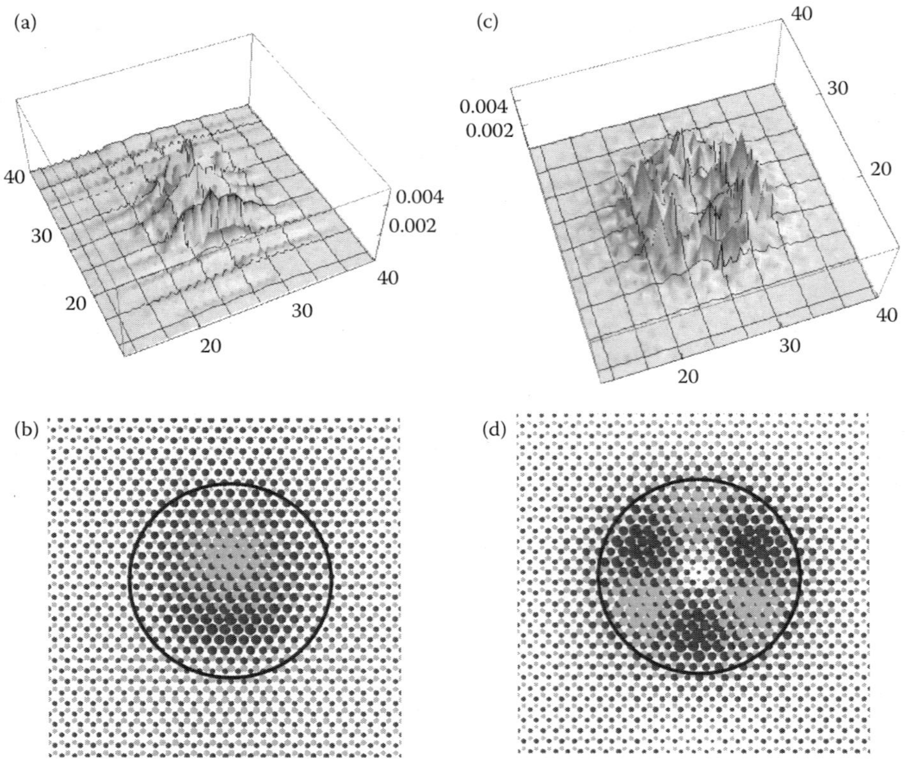

FIGURE 13.14 Occupation probabilities for states with $E = 0.18$ eV (a), and $E = -0.04$ eV (c). Parameters are $B = 20$ T, $V = 1$ eV, $R = 19.68$ Å, $R/\ell = 0.34$, and $V/E_c = 2.99$. Corresponding probability densities on A and B carbon atoms are shown in (b) and (d). The circle represents the antidot. (Adapted from Park, P. S., S. C. Kim, and S.-R. Eric Yang, 2010, *J. Phys.: Condens. Matter* **22**, 375302.)

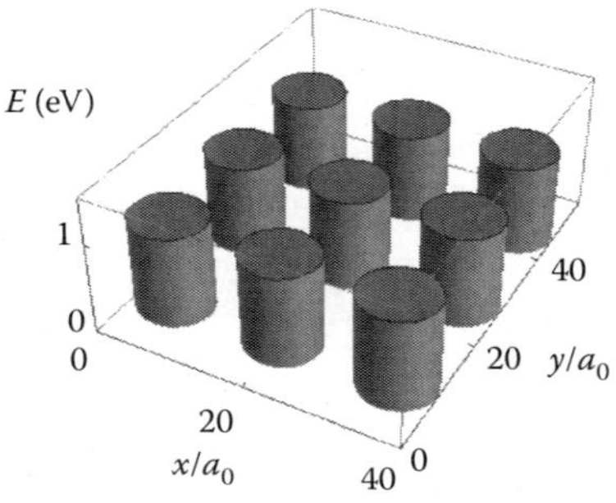

FIGURE 13.15 Cylindrical potentials can produce an antidot array on a rectangular graphene sheet. Length unit is the unit cell length $a_0 = \sqrt{3}a$, where $a = 1.42$ Å is the carbon–carbon distance.

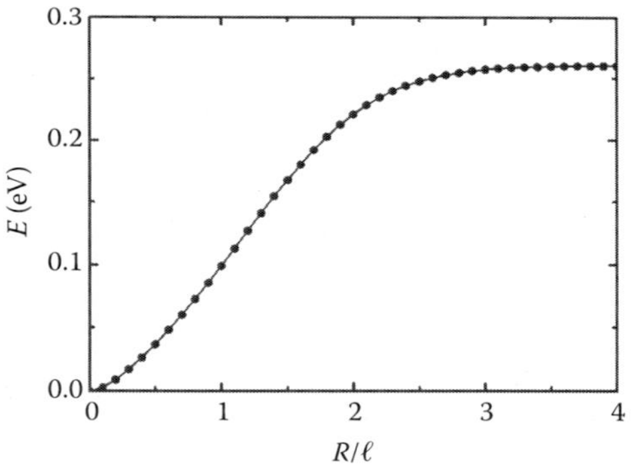

FIGURE 13.16 Lowest positive energy state $\Psi_0^{-1/2}(r)$ of an antidot as a function of R/ℓ. The height of the antidot potential is $V = 0.26$ eV. Here, $V/E_c < 1.98$; so, there is no overlap with adjacent LLs.

to a large average radius $\sqrt{\langle r^2 \rangle}$, acquires a significant positive energy correction, which can make the renormalized energy *positive*. The ratio between the potential strength and the LL energy spacing serves as the coupling constant of this problem $\alpha = \kappa\ell^2/E_M = \kappa\ell^3/\hbar v_F$. In the dimensionless units, the energy correction is $\kappa\ell^2|n|/E_M = \alpha|n|$, which suggests that even for a small value of α, the correction can be significant for $|n| \gg 1$. Another new feature is the presence of anomalous states. Under the transformation $V_a(r) \to V_d(r)$, a finite parabolic well can be considered as a localized potential, see Figure 13.20c, and the results of a cylindrical potential apply qualitatively, see Section 13.1.1. Thus, we expect that anomalous states should exist when the potential is strong. In the strong-coupling limit, the parabolic well has a natural length scale $\xi = (\hbar v_F/\kappa)^{1/3}$, in addition to ℓ. Note that $\xi \ll \ell$.

13.1.2.1 Coupling between Conduction and Valence Band States

A 2D parabolic quantum dot of graphene can be modeled by the Hamiltonian

$$H = v_F\vec{\sigma}\cdot\left(\vec{p}+\frac{e}{c}\vec{A}\right)+\frac{1}{2}\kappa r^2. \tag{13.19}$$

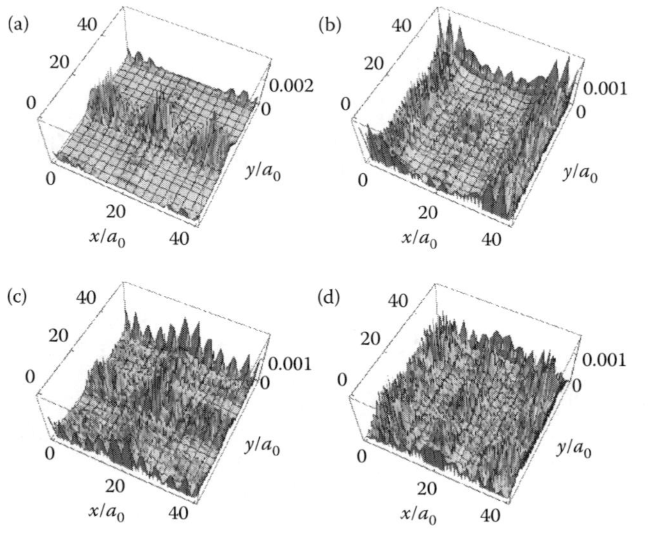

FIGURE 13.17 Some probability densities for the antidot array shown in Figure 13.15 with $R/\ell = 0.48$. $N = 1$ LL energy is 0.36 eV. Eigenenergies are (a) $E = 0.041$ eV, (b) $E = 0.122$ eV, (c) $E = 0.157$ eV, and (d) $E = 0.200$ eV. Probability densities take maximum values inside antidot potentials. (Adapted from Kim, S. C., and S.-R. Eric Yang, 2012, *J. Phys.: Condens. Matter* **24**, 195301.)

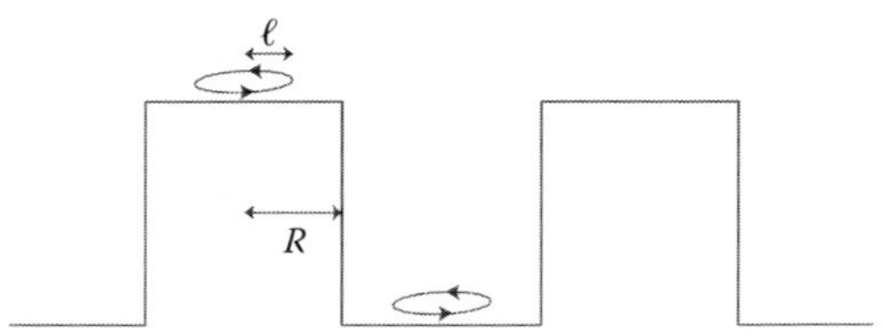

FIGURE 13.18 When $R/\ell > 1$, the lowest positive eigenenergy state $\Psi_0^{-1/2}(r)$ of an antidot is about V.

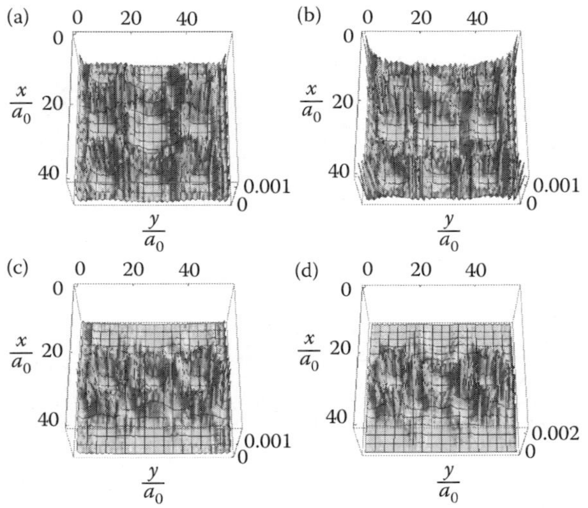

FIGURE 13.19 Some probability densities for the antidot array shown in Figure 13.15 with $R/\ell = 2$. Eigenenergies are (a) $E = 0.087$ eV, (b) $E = 0.104$ eV, (c) $E = 0.153$ eV, and (d) $E = 0.166$ eV. All probability densities are nearly zero inside antidot potentials. (Adapted from Kim, S. C., and S.-R. Eric Yang, 2012, *J. Phys.: Condens. Matter* **24**, 195301.)

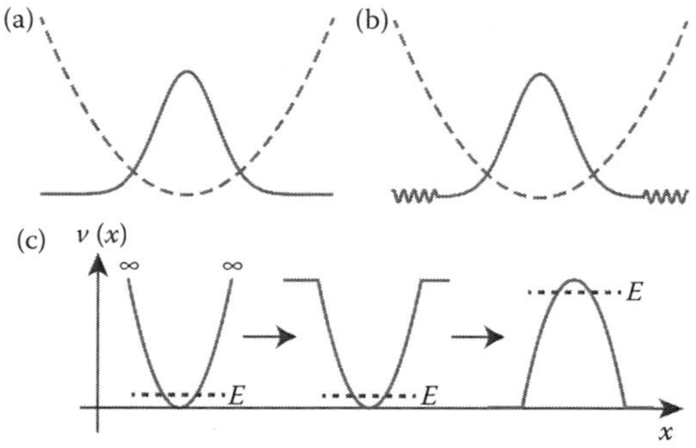

FIGURE 13.20 (a) Coupling between conduction and valence band states by the parabolic potential is ignored. Electrons are confined in the well. (b) Parabolic potential couples conduction and valence band states. Electrons can now leak out of the well. (c) A low-energy level is indicated in an infinite parabolic well. The same energy level is indicated in a finite parabolic well. The corresponding energy level under the transformation $V_a(r) \to V_d(r)$ is indicated (the role of this transformation is explained in Figure 13.1).

No exact solutions of this problem are known and several fundamental properties are unknown. We will see that several new features are present, such as the existence of resonant and nonresonant states. These basic properties may affect the optical spectrum in a profound way. Another new feature is the coupling between conduction and valence band states. Park et al. (2012) show that the value of azimuthal velocity for the nearly chiral LL ($n = 0$) states can be computed correctly only by including *both* positive and negative energy basis states. To demonstrate this point in a simple way, they apply first-order perturbation theory containing three terms, and treat the harmonic potential as a perturbation. The basis states of the Hamiltonian matrix are chosen as LL states of graphene (see Appendix 13C)

$$\psi_{n,m}(r,\theta) = c_n \begin{pmatrix} -\mathrm{sgn}(n) i \phi_{|n|-1,m}(r,\theta) \\ \phi_{|n|,m}(r,\theta) \end{pmatrix}. \tag{13.20}$$

Inter-LL index n and intra-LL index m are related to the angular momentum quantum J through $J = |n| - m - 1/2$. The perturbed wavefunction of a nearly chiral state with $n = 0$ is

$$|\psi_{0,m}^{(1)}\rangle \approx |\psi_{0,m}\rangle + C_1 |\psi_{1,m+1}\rangle + C_{-1} |\psi_{-1,m+1}\rangle. \tag{13.21}$$

All three states of the right-hand side of the equation have the same angular momentum $J = -m - 1/2$. The expansion coefficients are

$$C_{\pm 1} = \frac{\langle \psi_{\pm 1,m+1} |V| \psi_{0,m}\rangle}{-E_{\pm 1}}. \tag{13.22}$$

Since $C_{-1} = -C_1$, the expectation value of the azimuthal velocity, $v_\theta = (-\sin\theta\sigma_x + \cos\theta\sigma_y)v_F$, is *2 times* bigger when both positive and negative energy basis states are included than when only positive energy basis states are included in the perturbation series:

$$\langle \psi^{(1)}_{0,m} | v_\theta | \psi^{(1)}_{0,m} \rangle = 2\sqrt{2} v_F \, \mathrm{Re}\left[C_1 \langle \phi_{0,m} | e^{i\theta} | \phi_{0,m+1} \rangle \right]. \quad (13.23)$$

13.1.2.2 Resonant, Nonresonant, and Anomalous States

In the weak-coupling limit of $\alpha \to 0$, the energy spectrum in each Hilbert subspace of an angular momentum J consists of the discrete energies of the LLs of graphene. However, away from the weak-coupling regime, it is highly nontrivial to find the energy spectrum, and it is unclear how eigenstates evolve from weak to strong parabolic potentials. A simple dimensional analysis suggests that the energy scale of the problem in the strong-coupling limit of $B \to 0$, or $\alpha \to \infty$, is $\kappa^{1/3}(\hbar v_F)^{2/3}$. In units of E_M, this energy scale is $\alpha^{1/3}$. It indicates that the dimensionless energy-level spacing $\Delta E/E_M$ increases from ~1 to ~$\alpha^{1/3}$ as one moves from weak- to strong-coupling regimes. However, studies in ordinary semiconductors suggest that the crossover regime may be nontrivial (MacDonald and Ritchie, 1986). Kim et al. (2012) find in the intermediate-coupling regime that nonresonant states form a closely spaced energy spectrum. In addition, resonant quasi-bound states of both positive and negative energies exist in the spectrum, see Figure 13.21. The presence of resonant quasi-bound states of negative energies is a counterintuitive and unique property of massless Dirac fermions. As the strong-coupling limit is approached, resonant and nonresonant states transform into anomalous states, whose probability densities develop a narrow peak inside the well and another broad peak under the potential barrier. These features are qualitatively different from those of a parabolic quantum dot of semiconductor heterostructures. They may be experimentally investigated by measuring optical transition energies that can be described by a scaling function of the coupling constant.

Kim et al. (2012) use a Hamiltonian matrix approach using LL states as basis states

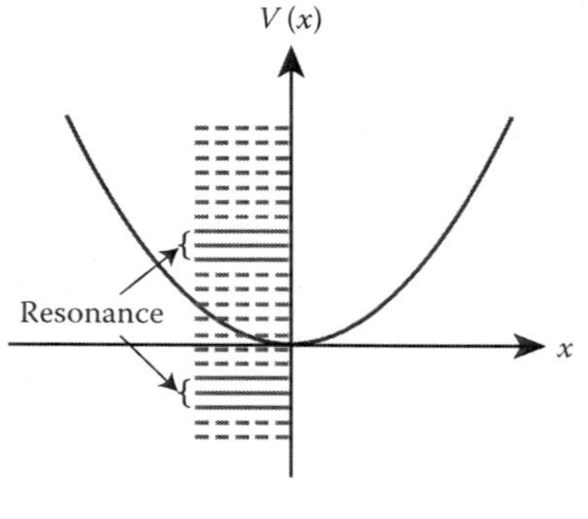

FIGURE 13.21 Schematic energy spectrum of a parabolic dot in the intermediate-coupling regime $\alpha \sim 1$. The energy spectrum is closely spaced, and resonant states of positive and negative energies are present. We diagonalize the Hamiltonian matrix in a Hilbert subspace of angular momentum J. (Adapted from Kim, S. C., J. W. Lee, and S.-R. Eric Yang, 2012, *J. Phys.: Condens. Matter* **24**, 495302.)

$$\psi_{nm}(\vec{r}) = \begin{pmatrix} \chi^A_{nm}(r) e^{i(m-1)\theta} \\ \chi^B_{nm}(r) e^{im\theta} \end{pmatrix}. \quad (13.24)$$

The matrix elements $\langle \psi_{nm} | H | \psi_{n'm'} \rangle$ can be made Hermitian when the simple size R is finite

$$\int (\psi^*_{nm} H \psi_{n'm'} - (H\psi_{nm})^* \psi_{n'm'}) d\vec{r}$$

$$= -2\pi i v_F \left[(\chi^{A*}_{nm} \chi^B_{n'm'} + \chi^{B*}_{nm} \chi^A_{n'm'}) r \right]_0^R = 0 \quad (13.25)$$

From this we see that it is sufficient that only one component, either A or B, vanishes at R, that is, both components of the wavefunction need not vanish in the limit of large R. Hence both confined and deconfined states can be described within this Hamiltonian matrix approach (Giavaras et al., 2009).

Figure 13.22 displays the computed energy spectrum in the weak and intermediate regimes. Resonant and nonresonant

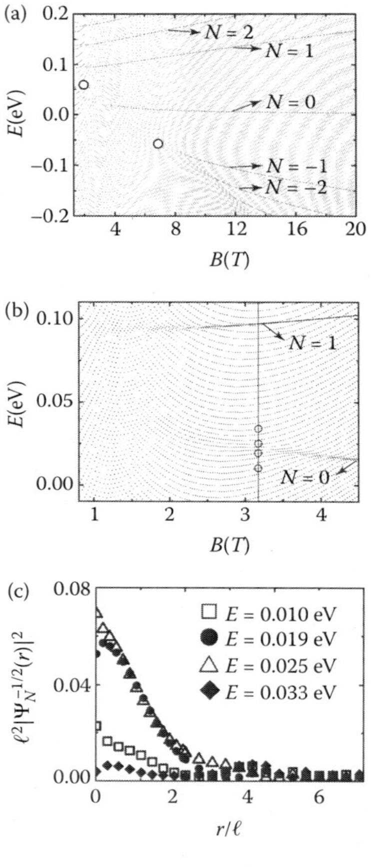

FIGURE 13.22 (a) Eigenenergy spectrum of Hilbert subspace of $J = -1/2$. Lines labeled by N represent resonant quasi-bound states. 2001×2001 matrix is used. (b) Enlarged energy spectrum for $J = -1/2$. Resonant quasi-boundstate $|\Psi_0^{-1/2}\rangle$ anticrosses strongly at $B = 3.14$ T ($\alpha = 0.47$). Four circles represent these coupled states. (c) Probability densities of these four states are displayed. Together, they form a resonance with the approximate resonant energy $\epsilon_0^{-1/2}(B) = 0.025$eV. (Adapted from Kim, S. C., J. W. Lee, and S.-R. Eric Yang, 2012, *J. Phys.: Condens. Matter* **24**, 495302.)

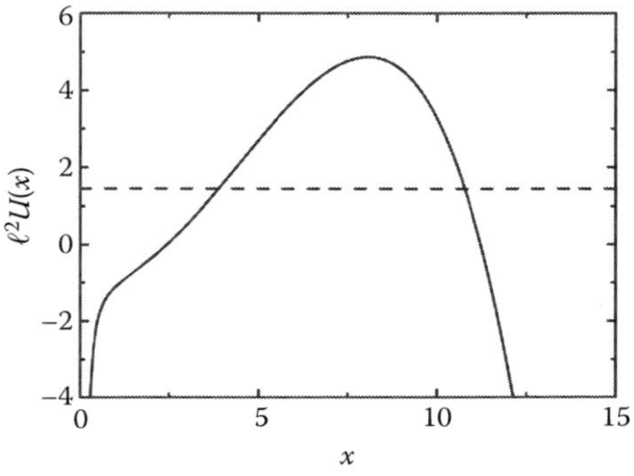

FIGURE 13.23 Effective potential for the B-component of the quasi-boundstate with $J = -1/2$ and negative energy $E = -0.100$ eV at $B = 11$ T. Dimensionless parameters are $\alpha = 0.072$ and $(E/E_M)^2 = 1.45$. The dashed line indicates the value of E.

states are present. Note that resonant states with negative energies are also present. Figure 13.22b and c show the structure of a resonance. It is instructive to plot the effective potential of the quasi-boundstate with $J = -1/2$ and negative energy $E = -0.100$ eV, see Figure 13.22a. The result is consistent with the presence of a resonant state, see Figure 13.23. Consider another quasi-bound state with $J = -1/2$ and negative energy $E = -0.131$ eV, see Figure 13.22a. The effective potential of this state is shown in Figure 13.24. Note that the energy of this state is closer to the top of the barrier than that of the state shown in Figure 13.23. Outside the barrier, the wavefunction should oscillate since $U(x) < E^2$.

We have also computed the probability densities of a nonresonant state indicated as an open circle at $B = 2.04$ T in Figure 13.22a. The probability densities of A- and B-components of this state are displayed in Figure 13.25. Note that a large peak exists at a finite value of r, suggestive of a nonresonant state.

So far, we have investigated states in the regime $\alpha \leq 1$. In the strong-coupling regime $\alpha \gg 1$, resonant and nonresonant states evolve into anomalous states. The probability density

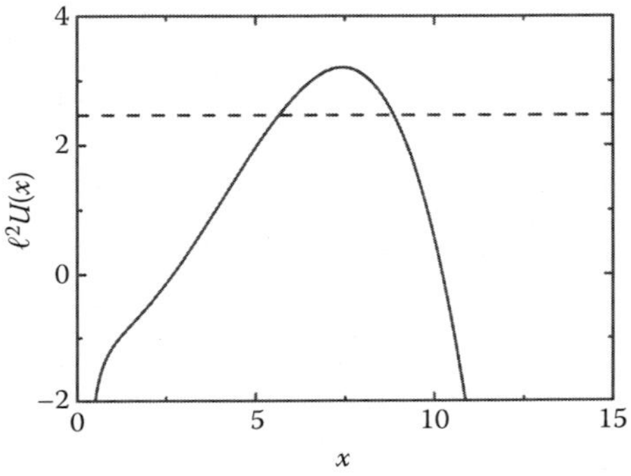

FIGURE 13.24 Effective potential for the B-component of the quasi-boundstate with $J = -1/2$ and negative energy $E = -0.131$ eV at $B = 11$ T. Dimensionless parameters are $\alpha = 0.072$ and $(E/E_M)^2 = 2.47$.

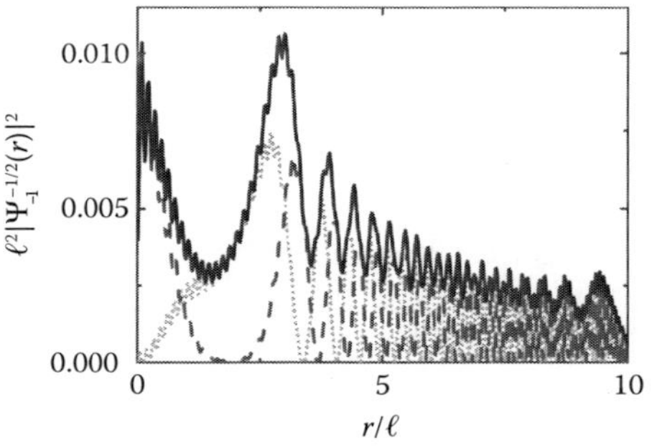

FIGURE 13.25 Example of a nonresonant state (indicated as an open circle in Figure 13.22a). The dotted (dashed) line represents A (B) components of the probability density $|\Psi_N^J\rangle = |\Psi_{-1}^{-1/2}\rangle$ with energy $E = 0.054$ eV and $B = 2.04$ T ($\alpha = 0.9$). As $r \to 0$, the dashed line approaches a finite value while the dotted line goes to zero. 2001×2001 matrix is used. (Adapted from Kim, S. C., J. W. Lee, and S.-R. Eric Yang, 2012, *J. Phys.: Condens. Matter* **24**, 495302.)

of an anomalous state is plotted in Figure 13.26. Note that this state has two length scales $\xi = (\hbar v_F/\kappa)^{1/3}$ and ℓ, and that $\xi \ll \ell$. The effective potential of this anomalous state is displayed in Figure 13.27. Note that it has a very deep well near $r = 0$, giving rise to the anomalous peak.

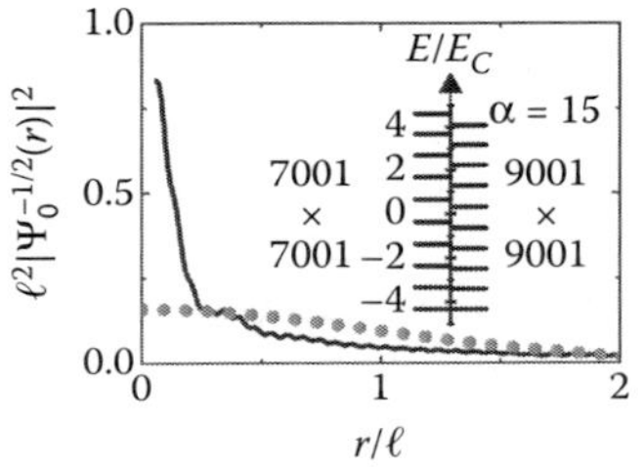

FIGURE 13.26 Total probability density of $|\Psi_0^{-1/2}\rangle$ with energy 0.177 eV at $B = 0.312$ T ($\alpha = 15$) (solid). The corresponding state at $\alpha = 0$ is shown as a dotted line. The eigenstate is obtained by diagonalizing 7001×7001 Hamiltonian matrix. Energy levels are also shown at $B = 0.312$ T ($\alpha = 15$) for 7001×7001 and 9001×9001. (Adapted from Kim, S. C., J. W. Lee, and S.-R. Eric Yang, 2012, *J. Phys.: Condens. Matter* **24**, 495302.)

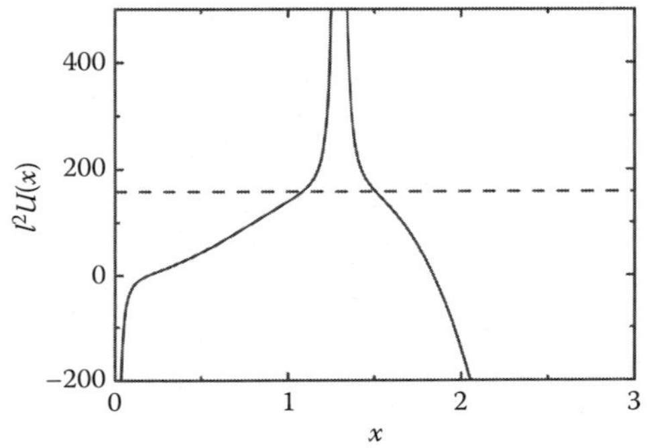

FIGURE 13.27 Effective potential for the B-component of an anomalous state with $J = -1/2$ and energy $E = 0.177$ eV at $B = 0.312$ T. Dimensionless parameters are $\alpha = 15$ and $(E/E_M)^2 = 158.6$.

13.1.2.3 Scaling and Optical Properties

So far, the energy spectrum is computed for each given values of B and κ. These two parameters can be combined into a single dimensionless parameter α. When the excitation energies are measured in units of E_M, the resonant energies follow scaling functions as a function of α, see Figure 13.28.

The absorption selection rules for graphene intraband and interband LL transitions are given by (Kim and Yang, 2013)

$$|\langle n,J|\sigma_x|n',J'\rangle|^2$$
$$= |c_n^* c_{n'}(\mathrm{sgn}(n)\delta_{|n|-1,|n'|}\delta_{J-1,J'} - \mathrm{sgn}(n')\delta_{|n|,|n'|-1}\delta_{J,J'-1})|^2, \quad (13.26)$$

(photons are polarized along x-axis and $|n, J\rangle$ is a graphene LL state). Allowed transitions are $(n, -1/2) \to (n+1, 1/2)$ for $n \geq 0$, and $(n, 1/2) \to (n+1, -1/2)$ for $n < 0$. There are also interband LL transitions with the selection rule $(n, -1/2) \to (|n|+1, 1/2)$ for $n \leq -1$, and $(n, 1/2) \to (|n|-1, -1/2)$ for $n \leq -2$. These selection rules are depicted in Figure 13.29. A parabolic potential will mix different LLs so that the selection rule for Δn must be relaxed. However, the selection rule for ΔJ remains unchanged. Figure 13.30 shows possible transitions for a parabolic dot in magnetic fields. A transition between quasi-bound states with positive energies is shown in (e). Also, transition from a negative energy quasi-bound state to a positive energy quasi-bound state is possible (f). Transitions between nonresonant states are labeled as (a), (b), and (c). For some of these transitions, $\Delta n \neq 1$. The effects of many-body interactions on these impurity cyclotron resonances can be explored using a time-dependent H–F approximation (Kim et al., 2014b).

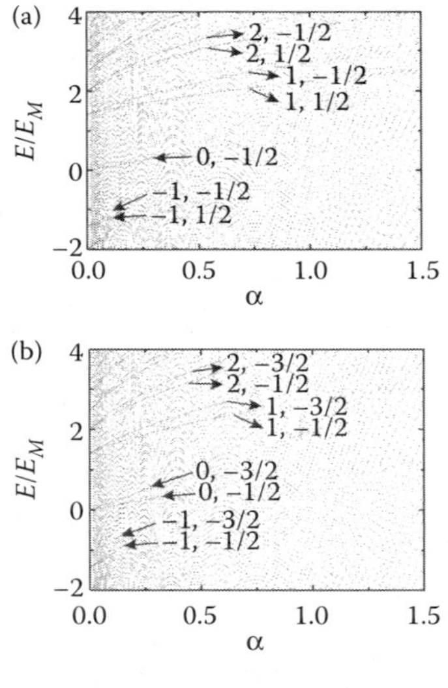

FIGURE 13.28 Energy spectra for $J = -(1/2)$ and $1/2$ plotted together. Matrix sizes N_c are 2001 for $J = -1/2$ and 2000 for $J = 1/2$. (Adapted from Kim, S. C., J. W. Lee, and S.-R. Eric Yang, 2012, *J. Phys.: Condens. Matter* **24**, 495302.)

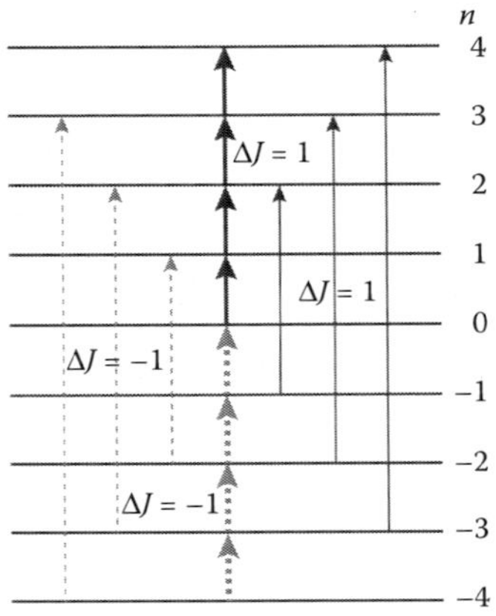

FIGURE 13.29 Absorption transitions in the absence of the impurity potential. Intraband transitions satisfy $\Delta n = 1$ (thick solid and dashed lines). There are two types of interband transitions: $n \to |n|+1$ with $\Delta J = 1$ ($n \leq -1$) (solid lines) and $n \to |n|-1$ with $\Delta J = -1$ ($n \leq -2$) (dashed line).

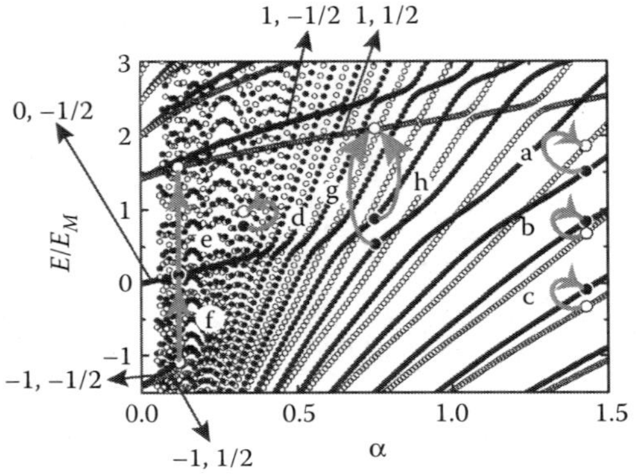

FIGURE 13.30 To display clearly possible optical transitions, we show the energy spectra for a small $N_c = 101$ and 100 for $J = -1/2$ (black dots) and 1/2 (white dots) plotted together. The potential strength is $\kappa = 0.1$ meV/nm^2. Note that the value of the strength of the potential κ is such that $B = 1.5$ T corresponds to $\alpha = 1.424$ and $B = 8$ T corresponds to $\alpha = 0.116$. Quasi-boundstate energies $\epsilon_N^J(\alpha)$ are labeled by (N, J). (Adapted from Kim, S. C., J. W. Lee, and S.-R. Eric Yang, 2012, *J. Phys.: Condens. Matter* **24**, 495302.)

13.2 PART II: DOTS WITH ZIGZAG AND ARMCHAIR EDGES IN A MAGNETIC FIELD

13.2.1 Rectangular Dot in Magnetic Fields

An electron can be confined in a nano-sized island of graphene. Such a system can be cut out of a graphene sheet. Properties of such a dot are expected to sensitively depend on the types of edges that surround the island. Kim et al. (2010) consider a rectangular dot that has two zigzag edges and two armchair edges (Fujita et al., 1996) (see Figure 13.31), and show that some unusual chiral states with nearly zero energy can exist in a perpendicular magnetic field, see Figure 13.32.

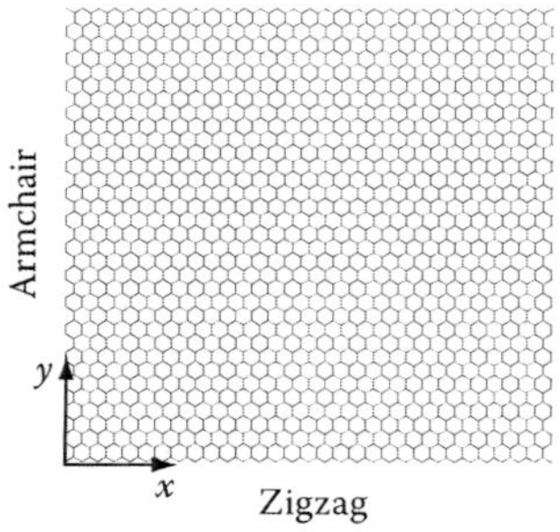

FIGURE 13.31 Armchair edges are along the y-axis and zigzag edges are along the x-axis.

13.2.1.1 Zigzag Edges in $B = 0$

Before we present the properties of a rectangular dot, we first review the basic properties of a zigzag ribbon in the absence of a magnetic field (Brey and Fertig, 2006). One unique feature of a zigzag is the existence of chiral states. One example is surface states of zigzag nano ribbons. If the surface state is of A-type on one zigzag edge, then, it is of B-type on the opposite edge. In addition to the surface states, there are also confined modes.

The Dirac Hamiltonians near $\vec{K}$ and $\vec{K}'$ valleys are derived for the zigzag edge in Appendix 13A. Putting the two Hamiltonians of the two valleys together, we have

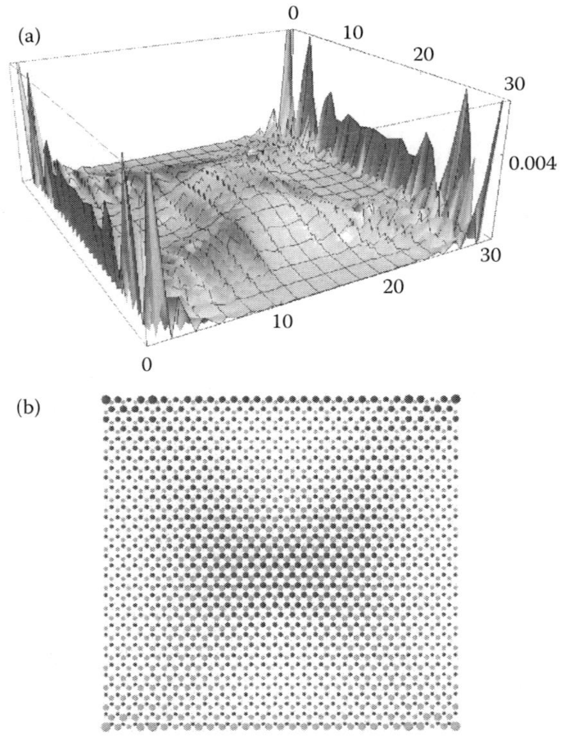

FIGURE 13.32 (a) Probability wavefunction of the eigenstate with energy -2.0×10^{-3} eV at $\ell/L_x = 0.12$. Probability density is peaked at the zigzag edges. (b) Profile of probability densities of A- and B-components of the state shown in (a). (Adapted from Kim, S. C., P. S. Park, and S.-R. Eric Yang, 2010, *Phys. Rev. B* **81**, 085432.)

$$H = \frac{3\gamma_0 a}{2}\begin{pmatrix} 0 & k_x - ik_y & 0 & 0 \\ k_x + ik_y & 0 & 0 & 0 \\ 0 & 0 & 0 & -k_x - ik_y \\ 0 & 0 & -k_x + ik_y & 0 \end{pmatrix}. \tag{13.27}$$

The wavefunction solution for the sublattice A is

$$\Psi_A(\vec{r}) = e^{i\vec{K}\cdot\vec{r}}F_A(\vec{r}) + e^{i\vec{K}'\cdot\vec{r}}F_A'(\vec{r}), \tag{13.28}$$

and for sublattice B

$$\Psi_B(\vec{r}) = e^{i\vec{K}\cdot\vec{r}}F_B(\vec{r}) + e^{i\vec{K}'\cdot\vec{r}}F_B'(\vec{r}), \tag{13.29}$$

where $F_A(\vec{r})$ and $F_B(\vec{r})$ are the effective mass wavefunctions of $\vec{K}$ and $F_A'(\vec{r})$ and $F_B'(\vec{r})$ are of $\vec{K}'$. Assume that the zigzag edges are parallel to the x-axis, see Figure 13.33. In this case, the translational symmetry guarantees that the envelope wave function of $\vec{K}$ can be written as

$$F(\vec{r}) = e^{ik_x x}\begin{pmatrix} \phi_A(y) \\ \phi_B(y) \end{pmatrix}. \tag{13.30}$$

A similar equation holds for the envelope wave function of $\vec{K}'$. For zigzag edges, the boundary conditions are

$$\Psi_B(y = 0) = 0, \Psi_A(y = L) = 0. \tag{13.31}$$

These boundary conditions are satisfied for any x by the choice

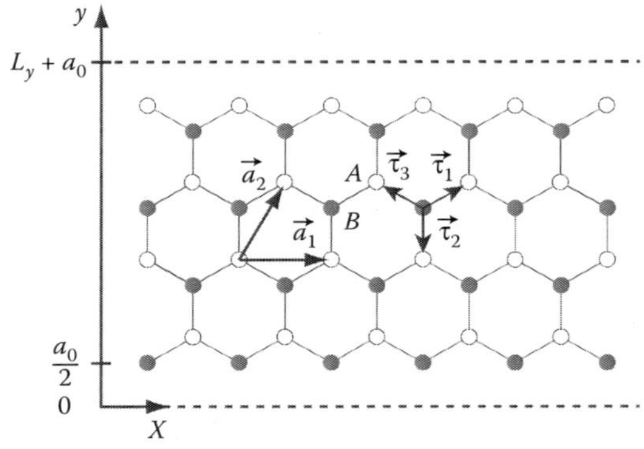

FIGURE 13.33 Zigzag edges are along the x-axis. The hard walls are placed just outside of the ribbon (distance $a_0/2$ from the ribbon). The value of wavefunctions is 0 at the positions of the hard walls $y = 0$ and $y = L$. Note that $L = L_y + a_0$, where L_y is the width of the ribbon. The Dirac points are chosen $\vec{K} = (2\pi/a_0)(1/3, 1/\sqrt{3}) = (K_x, K_y)$ and $\vec{K}' = (2\pi/a_0)(-(1/3), 1/\sqrt{3}) = (-K_x, K_y)$. Vectors from B to A carbon atoms are $\vec{\tau}_1 = a(\sqrt{3}/2, 1/2)$, $\vec{\tau}_2 = -a(0,1)$, and $\vec{\tau}_3 = a(-(\sqrt{3}/2), 1/2)$. Note that A (B) carbon atoms are on the upper (lower) zigzag edge while both types of atoms are on each armchair edge.

$$\phi_B(y=0) = \phi_A(y=L) = 0, \tag{13.32}$$

and

$$\phi'_B(y=0) = \phi'_A(y=L) = 0. \tag{13.33}$$

Note that there are *four* boundary conditions for the envelope wavefunctions, and that $\vec{K}$ and $\vec{K}'$ valley envelope wavefunctions are *decoupled*.

For $\vec{K}$ valley, the envelope function $\phi_B(y)$ satisfies a *second*-order differential equation

$$(-\partial_y^2 + k_x^2)\phi_B = \tilde{\varepsilon}^2\phi_B, \tag{13.34}$$

where $\tilde{\varepsilon} = \varepsilon/v_F$ and ε is the energy eigenvalue. The solution of the surface states has the form

$$\phi_B(y) = Ae^{qy} + Be^{-qy}, \tag{13.35}$$

where $q > 0$. Similar solutions exist for $\vec{K}'$ valley. Using these, we find the four-component envelope wavefunction along the y-axis

$$\begin{pmatrix}\phi_A\\ \phi_B\\ \phi'_A\\ \phi'_B\end{pmatrix} = \begin{pmatrix}\frac{A}{\tilde{\varepsilon}}(k_x+q)(e^{qy}-e^{2qL}e^{-qy})\\ A(e^{qy}-e^{-qy})\\ -\frac{C}{\tilde{\varepsilon}}(k_x+q)(e^{qy}-e^{2qL}e^{-qy})\\ C(e^{qy}-e^{-qy})\end{pmatrix} \tag{13.36}$$

Here, wavevectors q and k_x are related by

$$e^{-2qL} = \frac{k_x - q}{k_x + q}, \tag{13.37}$$

for $\vec{K}$ valley, and

$$e^{-2qL} = \frac{k_x + q}{k_x - q}, \tag{13.38}$$

for $\vec{K}'$ valley. There are real solutions of q for $k_x > 1/L$ with eigenenergy satisfying $\tilde{\varepsilon}^2 = k_x^2 - q^2$. These solutions correspond to surface waves existing near the edges of the graphene ribbon. Their energies are nearly zero. There are also confined solutions with imaginary values of $q = iq_n$. They are

$$\begin{pmatrix}\phi_A\\ \phi_B\\ \phi'_A\\ \phi'_B\end{pmatrix} = \begin{pmatrix}\frac{1}{\tilde{\varepsilon}}(k_n \sin q_n y - q_n \cos q_n y)\\ \sin q_n y\\ -\frac{1}{\tilde{\varepsilon}}(k_n \sin q_n y - q_n \cos q_n y)\\ -\sin q_n y\end{pmatrix}, \tag{13.39}$$

where wavevectors q_n and k_x are related through

$$k_x = \frac{q_n}{\tan(q_n L)}, \tag{13.40}$$

for $\vec{K}$ valley, and

$$k_x = -\frac{q_n}{\tan(q_n L)}, \tag{13.41}$$

for $\vec{K}'$.

13.2.1.2 Rectangular Dot with Two Zigzag Edges and Two Armchair Edges in $B = 0$

We consider a rectangular graphene dot with two zigzag edges and two armchair edges in the absence of a magnetic field. Solutions of such a system are given in Tang et al. (2008), and here, we present their derivation. As shown in Figure 13.33, the zigzag edges are along the x-axis and armchair edges are along the y-axis. The boundary conditions are that both A- and B-components of the wavefunction are zero on the armchair edges

$$\Psi_A(x=0) = \Psi_B(x=0) = \Psi_A(x=L) = \Psi_B(x=L) = 0. \tag{13.42}$$

The envelope wavefunction can be written as

$$F(\vec{r}) = f(y)\begin{pmatrix}\phi_A(x)\\ \phi_B(x)\end{pmatrix}, \tag{13.43}$$

where the wavefunction $f(y)$ is studied in Section 13.3.1. The hard wall boundary conditions (Tang et al., 2008) on each of the total A- and B-components of the wavefunction on the armchair edges $x = 0$ and $x = L$ are

$$\phi_\mu(x=0) + \phi'_\mu(x=0) = 0, \tag{13.44}$$

and

$$e^{iK_xL}\phi_\mu(x=L) + e^{-iK_xL}\phi'_\mu(x=L) = 0, \tag{13.45}$$

with $\mu = A, B$. These boundary conditions are satisfied for any y. It is clear that these boundary conditions *mix* states of $\vec{K}$ and $\vec{K}'$ valleys. Note that there are all together four boundary conditions on the envelope wavefunctions $\phi_A^{(\prime)}$ and $\phi_B^{(\prime)}$. The functions $\phi_A^{(\prime)}$ and $\phi_B^{(\prime)}$ obey the second-order differential equations

$$\begin{aligned}(-\partial_x^2 + k_y^2)\phi_A^{(\prime)} &= \tilde{\varepsilon}^2\phi_A^{(\prime)},\\ (-\partial_x^2 + k_y^2)\phi_B^{(\prime)} &= \tilde{\varepsilon}^2\phi_B^{(\prime)}.\end{aligned} \tag{13.46}$$

The solutions have the form

$$\phi_B = Ae^{ik_n x} + Be^{-ik_n x},$$
$$\phi'_B = Ce^{ik_n x} + De^{-ik_n x}. \quad (13.47)$$

Applying the boundary conditions of Equations 13.44 and 13.45 on the armchair edges, one obtains

$$0 = A + B + C + D,$$
$$0 = Ae^{i(k_n+K_x)L} + De^{-i(k_n+K_x)L} + Be^{-i(k_n-K_x)L} + Ce^{i(k_n-K_x)L}. \quad (13.48)$$

The boundary conditions are satisfied with the choice

$$A = -D, \quad B = C = 0, \quad (13.49)$$

which leads to $\sin[L(k_n + (2\pi/3a_0))] = 0$ (note that this result is valid only when $\vec{K}$ and $\vec{K}'$ valleys are chosen such that their x-components satisfy $K_x = -K'_x$). From this, we find

$$k_n = \frac{n\pi}{L} - \frac{2\pi}{3a_0}. \quad (13.50)$$

The B-component of wavefunctions along the x-axis are $\phi_B(x) = e^{ik_n x}$ and $\phi'_B(x) = -e^{-ik_n x}$.

Combining these solutions along the x-axis with the wavefunctions along the y-axis, given by Equation 13.39, we find the total envelope wavefunctions of the confined modes

$$\begin{pmatrix} F_A \\ F_B \\ F'_A \\ F'_B \end{pmatrix} = C \begin{pmatrix} \frac{1}{\varepsilon_n}(k_n \sin(qy) - q\cos(qy))e^{ik_n x} \\ \sin(qy)e^{ik_n x} \\ -\frac{1}{\varepsilon_n}(k_n \sin(qy) - q\cos(qy))e^{-ik_n x} \\ -\sin(qy)e^{-ik_n x} \end{pmatrix}, \quad (13.51)$$

where C is the normalization constant. To find *surface* states, we can replace wavevector q with iq_n. The wavefunction of the surface states localized at the zigzag edges is $\phi_B(y) = A(e^{q_n y} - e^{-q_n y})$. According to Equations 13.40 and 13.41, the wavevectors q_n and k_x are related

$$k_{n,x} = \frac{q_n}{\tanh q_n L}, \quad (13.52)$$

where the relation $\tan(iq) = i\tanh(q)$ is used (the eigenenergies are $\varepsilon_n = \pm\sqrt{k_{n,x}^2 - q_n^2}$). Let us find out how many surface states there are. The boundary conditions on the armchair edges give allowed values $k_{n,x}$ that are given by Equation 13.50. In addition, $k_{n,x}$ must satisfy

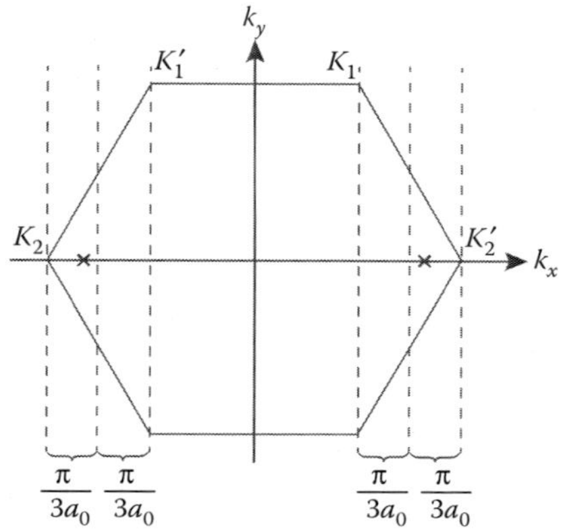

FIGURE 13.34 If $k_{n,x} > \pi/3a_0$ one can find a transformation between equivalent valleys $(\vec{K}_1, \vec{K}'_1) \rightarrow (\vec{K}_2, \vec{K}'_2)$ and a wavevector $\tilde{k}_x < \pi/3a_0$ so that $e^{i(K_{2,x}+\tilde{k}_x)x} = e^{i(K'_{1,x}-k_{n,x})x}$.

$$\frac{1}{L} < k_{n,x} < \frac{\pi}{3a_0}. \quad (13.53)$$

The left inequality originates as follows. As a function of q, the wavevector $k_{n,x}$ is always larger than $1/L$, see Equation 13.52. The values of $k_{n,x}$, when measured from a $\vec{K}$ point, should be less than $\pi/3a_0$. This is because a wavefunction with $k_{n,x}$ larger than $\pi/3a_0$ does not provide a new independent solution, see Figure 13.34.

This gives the right inequality of Equation 13.53. Rewriting these inequalities together, we find

$$\frac{2}{3}\frac{L}{a_0} < n < \frac{L}{a_0}. \quad (13.54)$$

The total number of possible surface states is thus $N_l = 2 \times (1/3)(L/a_0)$ (a factor 2 is due to the presence of valence band states).

13.2.1.3 Rectangular Dot in Magnetic Fields

The confined states of $B = 0$ found in the previous section turn into LL states in finite magnetic fields. One of the unique features of these LL states is that some of them are chiral: the $n = 0$ LL states of $\vec{K}$ valley have only the nonzero B-component (Figure 13.35c) while those of $\vec{K}'$ valley have only the nonzero A-component (Figure 13.35b). The surface states of zigzag nanoribbons also exist in the presence of a magnetic field. If a surface state is of A-type on one zigzag edge, then, it is of B-type on the opposite edge, see Figure 13.35a.

Kim et al. (2010) show that there can be a new type of nearly chiral state in a rectangular dot in magnetic fields. Surface states that are localized on the zigzag edges and LL states that are localized inside the dot are coupled to each other in a rectangular dot with two zigzag edges and two armchair edges. An example of such a state is shown in Figure 13.36. Its probability density displays two localized peaks

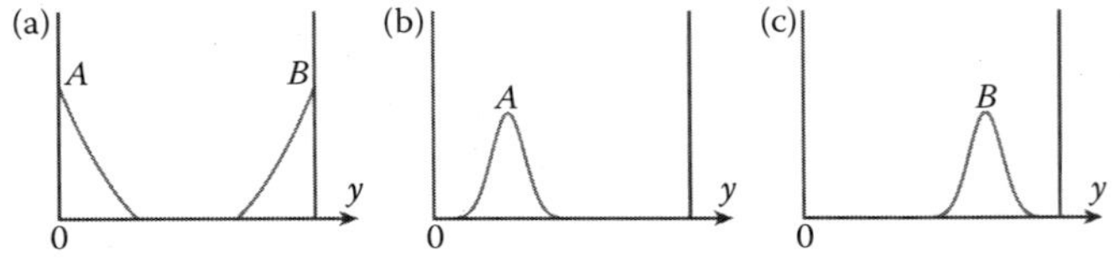

FIGURE 13.35 Cross section of probability wavefunctions of a nanoribbon with infinitely long zigzag edges along the x-axis in the presence of a perpendicular magnetic field. A and B indicate the chirality of wavefunctions. (a) Surface state with nearly zero energy. (b) and (c) depict $n = 0$ LL states of graphene. (Adapted from Park, P. S., S. C. Kim, and S.-R. Eric Yang, 2011b, *J. Nanosci. Nanotechnol.* **11**, 629.)

with different chiralities at the opposing zigzag edges. It also shows a broad peak inside the dot, which displays mixed chirality. These coupled modes have nearly zero energy.

The total number of nearly zero energy states in a magnetic field is then

$$N_T = N_l + N_\phi. \tag{13.55}$$

Here, N_ϕ is the $n = 0$ LL degeneracy of the rectangular sheet and N_l is the number of surface states. These quantities can be estimated from Figure 13.37.

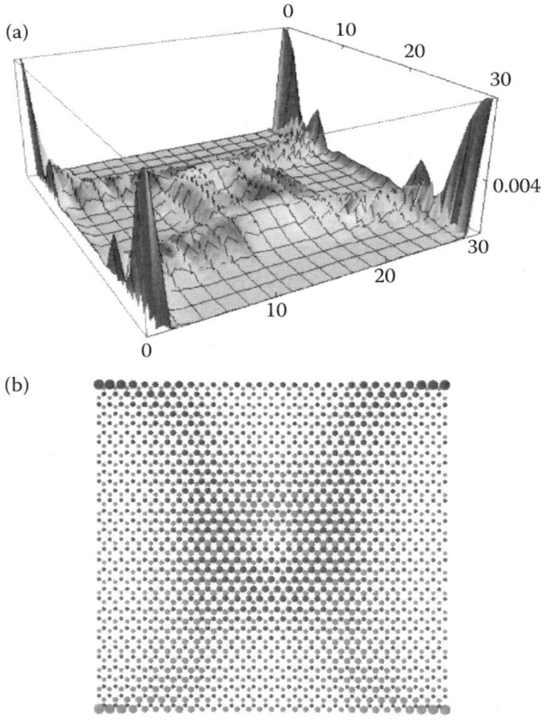

FIGURE 13.36 (a) Probability wavefunction of the state with energy -2.04×10^{-5} eV at $\phi = 0.01$. (b) Profile of z-component of pseudospin of the same state. (Adapted from Kim, S. C., P. S. Park, and S.-R. Eric Yang, 2010, *Phys. Rev. B* **81**, 085432.)

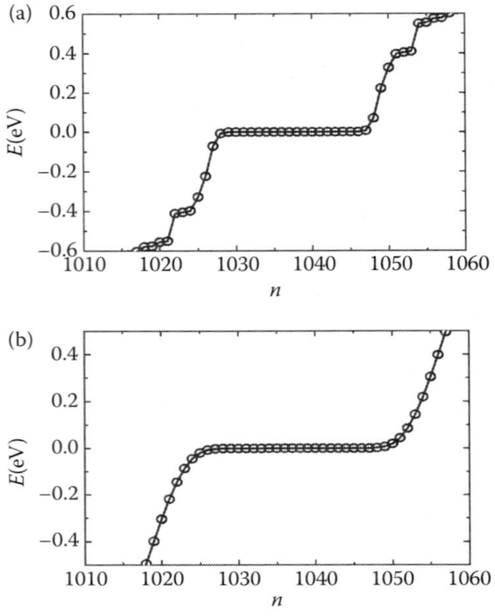

FIGURE 13.37 (a) Eigenenergies at $B = 0$. Quasi-degenerate states are present near zero energy. The size of the dot is 74×71 Å. Quantization energy of order 0.03 eV can be seen as an excitation gap near zero energy. (b) Eigenenergies at a dimensionless magnetic flux $\phi = \Phi/\Phi_0 = 0.01$, where $\Phi_0 = hc/e$ and $\Phi = BA_h\,(A_h = (3\sqrt{3}/2)a^2)$ is the area of a hexagon. The number of nearly zero energy states is N_T, given by Equation 13.55. (Adapted from Kim, S. C., P. S. Park, and S.-R. Eric Yang, 2010, *Phys. Rev. B* **81**, 085432.)

13.3 PART III: ARMCHAIR RIBBONS

13.3.1 Spintronic Properties of One-Dimensional Electron Gas in Graphene Armchair Ribbons

Owing to small spin–orbit coupling and low hyperfine interaction, electrons of graphene can move very long distances without scattering. Lee et al. (2012) have theoretically explored whether armchair nanoribbons can be used to generate spin-polarized currents (Geim and MacDonald, 2007, Schwierz, 2010). Graphene nanoribbons have been fabricated recently (Cai et al., 2010, Kato and Hatakeyama, 2012). These graphene armchair ribbons have a gap between conduction and valence subbands (Brey and Fertig, 2006, Son et al., 2006, Brey and Fertig, 2007a,b, Castro Neto et al., 2009). Such a system also has subbands, and when the system is doped, electrons occupy them and a one-dimensional electron gas is formed. Electrons in these subbands interact via a long-range Coulomb interaction. The ferromagnetic state can be stable when the nearest-neighbor Coulomb interaction is included in the one-dimensional Hubbard model (Strack and Vollhardt, 1955). The other factor that can make a ferromagnetic state stable is that many-body exchange self-energies that are one order of magnitude larger in comparison to the corresponding values of an electron gas of ordinary semiconductors. This is because the value of the dielectric constant of graphene is small ($\varepsilon \sim 1$) in comparison to that of an ordinary semiconductor ($\varepsilon \sim 10$). These factors can make the ferromagnetic

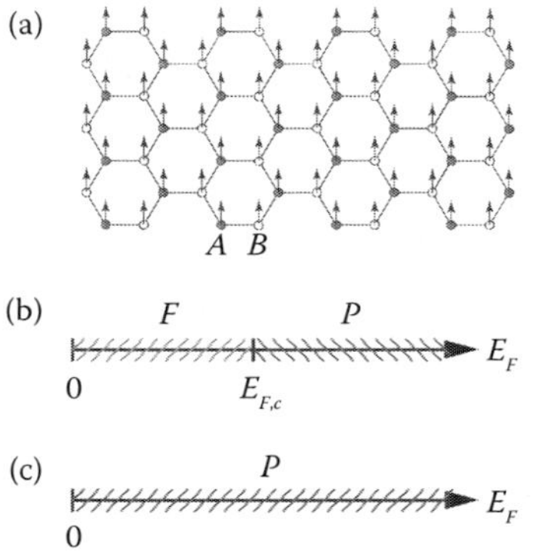

FIGURE 13.38 (a) Bulk ferromagnetic state of one-dimensional electron gas in a graphene armchair ribbon. We assume that only the conduction subband of a graphene armchair ribbon that is closest in energy to $E = 0$ is occupied. The average spin value per carbon site is shown as vertical arrows (only spins of electrons belonging to the lowest conduction subband are shown). (b) Phase diagram as a function of the Fermi energy E_F for the width of ribbon $L_x = 3Ma_0$. E_F is measured from the bottom of the conduction subband. F (P) stands for the ferromagnetic (paramagnetic) state. (c) Phase diagram for $L_x = (3M + 1)a_0$. (Reprinted from *Solid State Commun.* **152**, Lee, J. W., S. C. Kim, and S.-R. Eric Yang, 1929, Copyright 2012, with permission from Elsevier.)

state stable, see Figure 13.38. Lee et al. (2012) investigate how bulk magnetic properties of such a one-dimensional electron gas depend on the width of the armchair ribbon. In this section, we review their results.

13.3.1.1 Formation of One-Dimensional Subbands in Armchair Ribbons

We now consider an armchair nanoribbon with edges along the y direction in the absence of an external potential (Brey and Fertig, 2006). The effective mass approach (EMA), which can be derived from the tight-binding method (Ando and Nakanishi, 1998), can approximately describe the physical properties of graphene. Deriving the effective mass Hamiltonians for $\vec{K}$ and $\vec{K}'$ valleys and putting them together, we find the total Hamiltonian, which is identical to the Hamiltonian given by Equation 13.27. The A- and B-components of the wavefunction are given by Equations 13.28 and 13.29. The total envelope wavefunction Ψ of conduction subbands is

$$\Psi_n(x,y,k_y) = \frac{e^{ik_y y}\theta(x)}{2\sqrt{L}\sqrt{L_y}} \begin{pmatrix} e^{-i\theta_{k_n,k_y}} e^{ik_n x} \\ e^{ik_n x} \\ -e^{-i\theta_{k_n,k_y}} e^{-ik_n x} \\ -e^{-ik_n x} \end{pmatrix}, \quad (13.56)$$

where $\theta(x) = 1$ for $0 \le x \le L$ and 0 otherwise, and $\vartheta_{k_n,k_y} = \mathrm{Arc}\tan(k_y/k_n)$. The eigenenergies are given by

$$\varepsilon_n(k_y) = \gamma a_0 \sqrt{k_y^2 + k_n^2}, \quad (13.57)$$

with $\gamma = (\sqrt{3}/2)t$ and $t = 2.7$ eV. The allowed values of k_n are given by Equation 13.50. Note that envelope wavefunctions along the x-axis are $e^{\pm ik_n x}$ and have the opposite wavevectors k_n and $-k_n$ for $\vec{K}$ and $\vec{K}'$. It is important to note that A- and B-components of $\vec{K}$ and $\vec{K}'$ wavefunctions are identical except for the sign. This property will have important implications for the exchange self-energy. Note that the direction of the current with the components $j_x = v_F\sigma_x \otimes I$ and $j_y = v_F\sigma_y \otimes I$ is the same as the direction of the wavevector (k_n, k_y).

Let us find the index n for the lowest energy subband. Lee et al. (2012) define the dimensionless variable $X_n = k_n L$. In the case the width of the ribbon is $L_x = 3Ma_0$ (note that in the hard wall boundary condition $L = L_x + a_0$), one has $X_n = \pi(n - 2M - 2/3)$, and the conduction subband with the lowest energy has the value

$$X_n = \frac{\pi}{3}, \quad (13.58)$$

corresponding to the value $n = 2M + 1$. When $L_x = 3Ma_0 = 24a_0$, the lowest energy occurs for $n = 17$ and the next lowest energy occurs for $n = 16$. In Figure 13.39, $|X_n|$ is plotted as a function of n for width $L_x = 3Ma_0$.

For $L_x = (3M + 1)a_0$, we have $X_n = \pi(n - 2M - 4/3)$, and the conduction subband with the lowest energy has the value

$$X_n = -\frac{\pi}{3}, \quad (13.59)$$

corresponding to $n = 2M + 1$. When $L_x = (3M + 1)a_0 = 25a_0$, the lowest energy occurs for $n = 17$ and the next lowest energy occurs for $n = 18$. In Figure 13.40, $|X_n|$ is plotted as a function of n for width and $L_x = (3M + 1)a_0$. The lowest energy subbands are shown in Figure 13.41.

These EMA results for the subbands agree with the local density approximation (LDA) result for sufficiently large values of L_x. Figure 13.42 displays both LDA (Son et al., 2006) and EMA values of the gap of armchair ribbons as a function of the width for *undoped* armchair ribbons. LDA is a

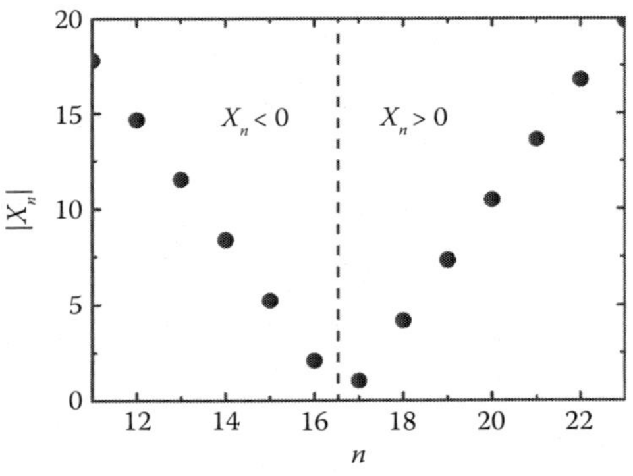

FIGURE 13.39 Plot of $|X_n|$ as a function of n. Here, $L_x = 3Ma_0 = 24a_0$.

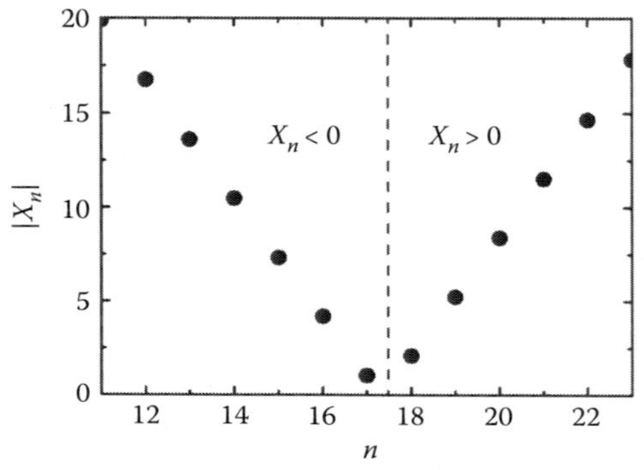

FIGURE 13.40 Plot of $|X_n|$ as a function of n. Here, $L_x = (3M + 1)a_0 = 25a_0$.

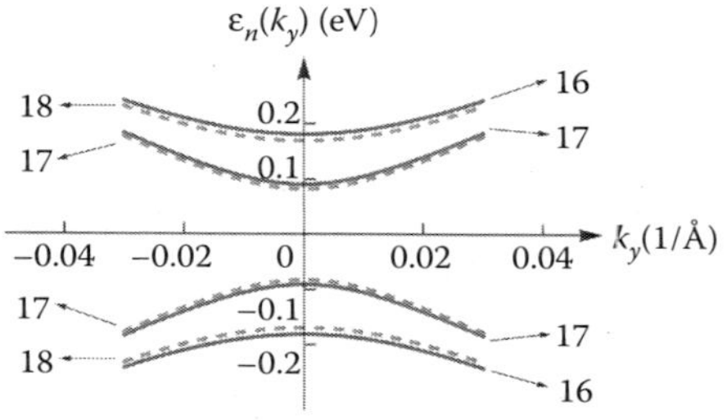

FIGURE 13.41 Conduction and valence subbands with energies that are closest near $E = 0$. Solid lines are for $L_x = 3Ma_0 = 24a_0$. Dashed lines are for $L_x = (3M + 1)a_0 = 25a_0$. The numbers beside the curves indicate values of the subband index n. (Reprinted from *Solid State Commun.* **152**, Lee, J. W., S. C. Kim, and S.-R. Eric Yang, 1929, Copyright 2012, with permission from Elsevier.)

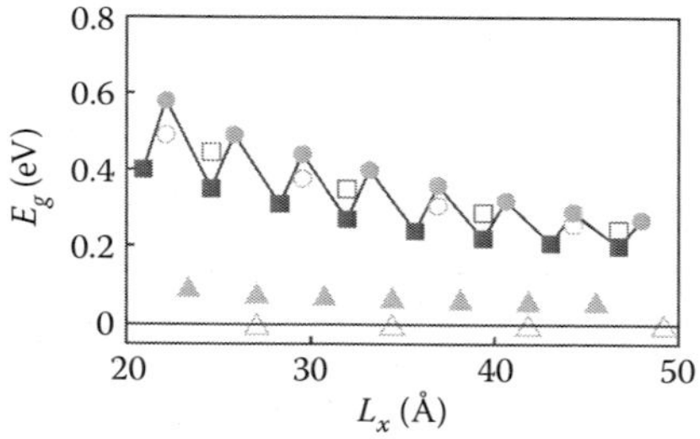

FIGURE 13.42 Size of gap E_g of undoped graphene is shown. LDA results (Son et al., 2006) for widths $L_x = (3M + 1)a_0$ (filled circles), $L_x = 3Ma_0$ (filled squares), and $L_x = (3M + 2)a_0$ (filled triangles). EMA results for the same widths are open squares, circles, and triangles. (Reprinted from *Solid State Commun.* **152**, Lee, J. W., S. C. Kim, and S.-R. Eric Yang, 1929, Copyright 2012, with permission from Elsevier.)

better approximation than EMA since it includes electron–electron interaction effects and goes beyond nearest-neighbor hopping. According to LDA, when the width of an armchair ribbon is $L_x = (3M + 1)a_0$ or $L_x = 3Ma_0$, a gap exists in the energy spectrum (Son et al., 2006) (a rather small gap exists when $L_x = (3M + 2)a_0$, and this case will not be considered here). We see that the LDA values of the magnitude of the gap are in rough agreement with those of the EMA results, and the agreement becomes better for larger values of L_x. For $L_x = (3M + 1)a_0$ or $3Ma_0$, the EMA results approach the LDA results when L_x is larger than 45 Å.

13.3.1.2 Exchange Self-Energy of Doped Armchair Ribbon

Lee et al. (2012) use the EMA to compute the approximate exchange self-energy. They assume that a gap separates conduction and valence subbands, and that only the lowest energy conduction subband is occupied with electrons. They calculate the many-body exchange self-energy Σ_{ex} of a spin-polarized one-dimensional electron gas of such a system for relatively large values of the width $L_x \geq 45$ Å, where the LDA and EMA results agree approximately (see Figure 13.42). This dependence on the width can be understood by examining the wavefunction overlap that appears in the expression for the exchange self-energy. The effect is a consequence of a subtle width-dependent mixture of $\vec{K}$ and $\vec{K}'$ states in the eigenstates. The large difference in the values of dielectric constants of graphene and ordinary semiconductors cannot explain these width-dependent magnetic properties.

Lee et al. (2012) consider spin-polarized electrons in the nth conduction subband with the electron density $n_D = k_F/\pi$, where k_F is the Fermi wavevector. We are in the quantum limit where only the lowest energy conduction subband is occupied. The Coulomb coupling between different subbands is ignored in the calculation of self-energies. Wavefunctions along the y-axis do not change since Coulomb interactions do not break translational invariance. In this case, wavefunctions of Equation 13.56 are self-consistent solutions (Ashcroft and Mermin, 1976). We denote these wavefunctions by $|k_y\rangle$. At $k_y = 0$, the exchange self-energy (Ashcroft and Mermin, 1976) is

$$\Sigma_{n,ex}(0) = -\frac{L_y}{2\pi}\int_{-k_F}^{k_F} dk_y \left\langle 0, k_y \left| \frac{e^2}{\epsilon\,|\mathbf{r}_1 - \mathbf{r}_2|} \right| k_y, 0 \right\rangle, \tag{13.60}$$

which can be written as $\Sigma_{n,ex}(0) = -(e^2/\epsilon L)\Delta(Y_F, X_n)$ with the dimensionless exchange self-energy defined as

$$\Delta(Y_F, X_n) = \frac{1}{2\pi}\int_{-\infty}^{\infty} dX \int_{-Y_F}^{Y_F} dY \left(1 + \frac{X_n}{\sqrt{X_n^2 + Y^2}}\right)\frac{(1 - \cos X)}{\sqrt{X^2 + Y^2}\,X^2}, \tag{13.61}$$

where $A = LL_y$. Here, we have used

$$\left\langle 0, k_y \left| \frac{1}{A}\sum_{\vec{q}} \frac{2\pi}{|\vec{q}|} e^{i\vec{q}\cdot\vec{r}} \right| k_y, 0 \right\rangle = \frac{1}{L_y}\int_{-\infty}^{\infty} dX \left(1 + \frac{X_n}{\sqrt{X_n^2 + Y^2}}\right)\frac{(1 - \cos X)}{\sqrt{X^2 + Y^2}\,X^2}, \tag{13.62}$$

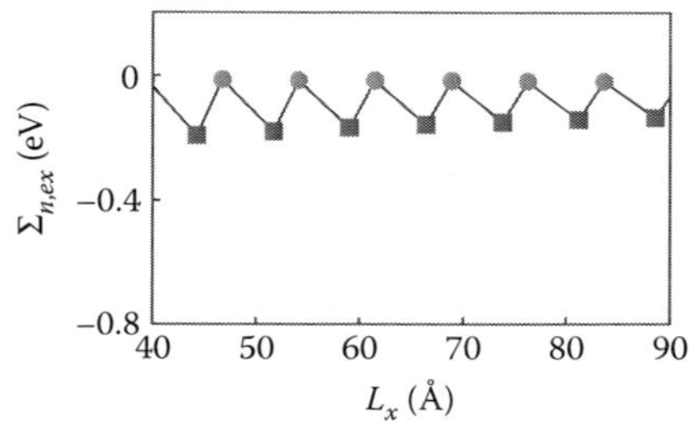

FIGURE 13.43 Exchange self-energies of doped graphene for ribbon widths $L_x = (3M + 1)a_0$ (circles) and $L_x = 3Ma_0$ (squares), where $a_0 = 2.46$ Å and M is an integer. The EMA is used. Here, the dielectric constant is $\epsilon = 3$ and the dimensionless Fermi wavevector is $k_F a_0 = 0.07$. (Reprinted from *Solid State Commun.* **152**, Lee, J. W., S. C. Kim, and S.-R. Eric Yang, 1929, Copyright 2012, with permission from Elsevier.)

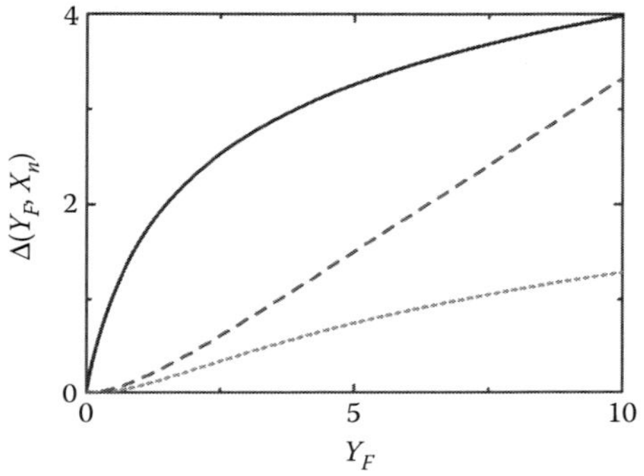

FIGURE 13.44 Magnitude of the dimensionless exchange self-energy $\Delta(Y_F, X_n)$ as a function of Y_F for $X_n = \pi/3$ (solid line) and $-\pi/3$ (dotted line). The dashed line is $E_F/(e^2/\epsilon\gamma a_0)$.

where $Y = k_y L$ and $Y_F = k_F L$. The Hartree self-energy will cancel with the potential of the uniform positive background. Computed exchange self-energies of doped graphene for various ribbon widths are shown in Figure 13.43. In Figure 13.44, $\Delta(Y_F, X_n)$ is plotted as a function of Y_F for $X_n = \pi/3$ and $-\pi/3$.

13.3.1.3 Phase Diagram

When the magnitude of the exchange self-energy is larger than the Fermi energy, spontaneous spin splitting will occur in the lowest conduction subband

$$|\Sigma_{n,ex}(0)| > E_F, \tag{13.63}$$

which is equivalent to $(e^2/\varepsilon\gamma a_0)\Delta(Y_F, X_n) > \sqrt{X_n^2 + Y_F^2} - |X_n|$. When they are equal, a critical point exists. This critical point is schematically shown in Figure 13.38b. At $X_n = -\pi/3$, no critical point exists since $\Delta(Y_F, X_n)$ is too small, see Figure 13.44. In this case, the paramagnetic state is stable. At $X_n = \pi/3$, the inequality is satisfied when $Y_F < 11.309$ for $\epsilon = 1$ and $Y_F < 3.22$ for $\epsilon = 3$. Using these results for the ribbon width $L_x = 3Ma_0$, we find that a critical point separates ferromagnetic and paramagnetic states while the paramagnetic state is stable for $L_x = (3M + 1)a_0$. Figure 13.38 illustrates these results. For consistency of our model, we must require that the exchange self-energy is smaller than the gap between the conduction and valence subbands: $|\Sigma_{n,ex}(0)| < E_g = 2\gamma a_0 |k_n|$, which is equivalent to $(e^2/\epsilon\gamma a_0)\Delta(Y_F, X_n) < 2|X_n| = 2\pi/3$. We find that this condition is satisfied when $Y_F < 0.364$ for $\epsilon = 1$ and $Y_F < 2.439$ for $\epsilon = 3$.

The dependence on the width can be understood by examining the wavefunction overlap that appears in the expression for the exchange self-energy. When $\vec{r}_1 = \vec{r}_2$ it is

$$\psi_n^*(x,y,0)\psi_n(x,y,k_y) \sim \left(e^{-ik_n x}, e^{-ik_n x}, -e^{ik_n x}, -e^{ik_n x}\right) \cdot \begin{pmatrix} e^{-i\theta_{k_n,k_y}} e^{ik_n x} \\ e^{ik_n x} \\ -e^{-i\theta_{k_n,k_y}} e^{-ik_n x} \\ -e^{-ik_n x} \end{pmatrix} \sim 1 + e^{-i\theta_{k_n,k_y}}. \tag{13.64}$$

The absolute square of this value appears in the exchange self-energy

$$|\psi_n^*(x,y,0)\psi_n(x,y,k_y)|^2 \sim (1 + e^{i\theta_{k_n,k_y}})(1 + e^{-i\theta_{k_n,k_y}}) \sim 2 + e^{i\theta_{k_n,k_y}} + e^{-i\theta_{k_n,k_y}} \sim 2(1 + \cos\theta_{k_n,k_y}). \tag{13.65}$$

When $k_n < 0$, the quantity $\cos\theta_{k_n,k_y} = (X_n/\sqrt{X_n^2 + Y^2}) < 0$, which makes the dimensionless self-energy $\Delta(Y_F, X_n)$ smaller, see the integrand in Equation 13.61. This effect is a consequence of the presence of the phase factors $e^{-i\theta_{k_n,k_y}}$ only in certain components of the eigenstate wavefunctions (see Equation 13.56), and that the value of k_n for the lowest conduction subband is *negative* when $L_x = (3M + 1)a_0$. In other words, it is a consequence of a subtle width-dependent mixture of $\vec{K}$ and $\vec{K}'$ states in the eigenstate wavefunctions.

The electron gas in the subband is partially spin polarized when the electron density is such that the exchange self-energy is smaller than the Fermi self-energy, see Figure 13.45a. The electron density in the subband may be controlled by changing the gate voltage, Schwierz (2010). Depending on the size of the Fermi wavevector k_F, which depends on electron density, the magnitude of the exchange spin splitting can vary in the range of 10–100 meV. Hence, if the electron density is lowered, the exchange self-energy can be larger than the Fermi self-energy, see Figure 13.45b. In this case, the electron gas is spin polarized.

APPENDIX 13A: DERIVATION OF THE DIRAC EQUATION FROM TIGHT-BINDING APPROACH

The nearest-neighbor tight-binding Hamiltonian is

$$\begin{aligned} H &= H_0 + H_{imp} \\ &= -\gamma_0 \sum_{i\in A} \sum_{a=1,2,3} ((c_{i+a}^B)^\dagger c_i^A + (c_i^A)^\dagger c_{i+a}^B) \\ &\quad + \sum_{i\in A} \tilde{u}_A(\vec{R}_{A,i})(c_i^A)^\dagger c_i^A + \sum_{i\in B} \tilde{u}_B(\vec{R}_{B,i})(c_i^B)^\dagger c_i^B. \end{aligned} \tag{13A.1}$$

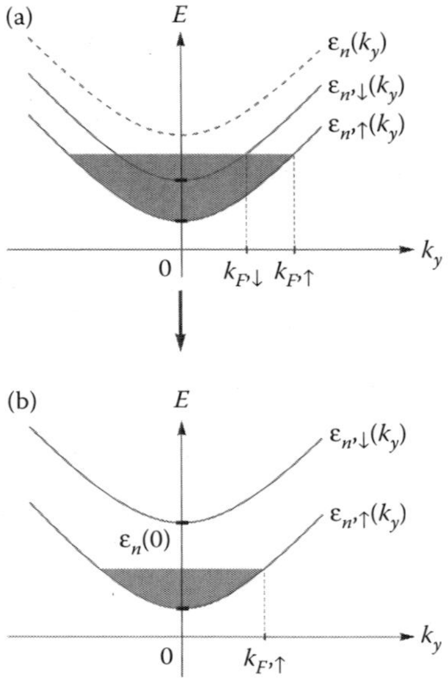

FIGURE 13.45 (a) The dashed line indicates spin-degenerate subband energy in the absence of electron–electron interactions. These degenerate subbands split into spin-up and spin-down subbands when the exchange interaction is turned on. (b) For a smaller electron density, the exchange self-energy can be larger than the Fermi energy and the electron gas is fully spin polarized. (Reprinted from *Solid State Commun.* **152**, Lee, J. W., S. C. Kim, and S.-R. Eric Yang, 1929, Copyright 2012, with permission from Elsevier.)

$(c_i^A)^\dagger/(c_i^B)^\dagger$ creates an electron on site i on the sublattice A/B and $\tilde{u}_{A(B)}(\vec{R}_{A(B),i})$ is the external potential energy at position $\vec{R}_{A(B),i}$. Bloch wavefunctions are

$$|\psi_A^{\vec{k}}\rangle = \sum_{i\in A} \psi_A(\vec{R}_{A,i})(c_i^A)^\dagger\,|0\rangle,$$
$$|\psi_B^{\vec{k}}\rangle = \sum_{i\in B} \psi_B(\vec{R}_{A,i})(c_i^B)^\dagger\,|0\rangle. \tag{13A.2}$$

The tight-binding equations for sites A and B are given by

$$(\varepsilon - \tilde{u}_A(\vec{R}_A))\psi_A(\vec{R}_A) = -\gamma_0 \sum_{a=1,2,3} \psi_B(\vec{R}_A - \vec{\tau}_a),$$
$$(\varepsilon - \tilde{u}_B(\vec{R}_B))\psi_B(\vec{R}_B) = -\gamma_0 \sum_{a=1,2,3} \psi_A(\vec{R}_B + \vec{\tau}_a), \tag{13A.3}$$

where γ_0 is the transfer integral. In the absence of external potentials

$$\begin{pmatrix}\psi_A(\vec{R}_A)\\ \psi_B(\vec{R}_B)\end{pmatrix} = \begin{pmatrix} e^{i\vec{K}\cdot\vec{R}_A}\\ e^{i\vec{K}\cdot\vec{R}_B}\end{pmatrix} \tag{13A.4}$$

is the eigenstate of $\vec{K}$ point with zero energy since

$$-\gamma_0 \sum_{a=1,2,3} \psi_B(\vec{R}_A - \vec{\tau}_a) = 0,$$
$$-\gamma_0 \sum_{a=1,2,3} \psi_A(\vec{R}_B + \vec{\tau}_a) = 0. \tag{13A.5}$$

Similarly, the wavefunction

$$\begin{pmatrix}\psi_A(\vec{R}_A)\\ \psi_B(\vec{R}_B)\end{pmatrix} = \begin{pmatrix} e^{i\vec{K}'\cdot\vec{R}_A}\\ e^{i\vec{K}'\cdot\vec{R}_B}\end{pmatrix}, \tag{13A.6}$$

is the solution of $\vec{K}'$ point with zero energy.

Ando and Nakanishi (1998) derive the Dirac Hamiltonian from Equation 13A.3 in the presence of slowly varying potentials using the effective mass approximation. Here, we present their derivations. Since

$$\begin{pmatrix}\psi_A(\vec{R}_A)\\ \psi_B(\vec{R}_B)\end{pmatrix} = \begin{pmatrix} e^{i\vec{K}\cdot\vec{R}_A}\\ e^{i\vec{K}\cdot\vec{R}_B}\end{pmatrix} \quad \text{and} \quad \begin{pmatrix}\psi_A(\vec{R}_A)\\ \psi_B(\vec{R}_B)\end{pmatrix} = \begin{pmatrix} e^{i\vec{K}'\cdot\vec{R}_A}\\ e^{i\vec{K}'\cdot\vec{R}_B}\end{pmatrix}$$

satisfy the tight-binding equations at $\vec{K}$ and $\vec{K}'$ points, the eigenstates must have the following form in the presence of an external potential:

$$\psi_A(\vec{R}_A) = e^{i\vec{K}\cdot\vec{R}_A}F_A^K(\vec{R}_A) + e^{i\eta}e^{i\vec{K}'\cdot\vec{R}_A}F_A^{K'}(\vec{R}_A),$$
$$\psi_B(\vec{R}_B) = -\omega e^{i\eta}e^{i\vec{K}\cdot\vec{R}_B}F_B^K(\vec{R}_B) + e^{i\vec{K}'\cdot\vec{R}_B}F_B^{K'}(\vec{R}_B), \tag{13A.7}$$

where $\omega = \exp(2\pi i/3)$ and η is the angle between the chiral vector L and the x' direction fixed on the graphene plane (see Figure 13A.1). The factors ω and $e^{i\eta}$ are introduced so that the final form of the $\vec{k}\cdot\vec{p}$ Hamiltonian has a simple form. Envelope wavefunctions $F_A^K, F_B^K, F_A^{K'}$, and $F_B^{K'}$ are slowly varying on the scale of the lattice constant a_0. We use the approximation

$$\psi_B(\vec{R}_A - \vec{\tau}_a) = -\omega e^{i\eta}e^{i\vec{K}\cdot\vec{R}_A}e^{-i\vec{K}\cdot\vec{\tau}_a}(1 - i\vec{\tau}_a\cdot\hat{k}')F_B^K(\vec{R}_A)$$
$$+ e^{i\vec{K}'\cdot\vec{R}_A}e^{-i\vec{K}'\cdot\vec{\tau}_a}(1 - i\vec{\tau}_a\cdot\hat{k}')F_B^{K'}(\vec{R}_A),$$
$$\psi_A(\vec{R}_B + \vec{\tau}_a) = e^{i\vec{K}\cdot\vec{R}_B}e^{i\vec{K}\cdot\vec{\tau}_a}(1 + i\vec{\tau}_a\cdot\hat{k}')F_A^K(\vec{R}_B)$$
$$+ e^{i\eta}e^{i\vec{K}'\cdot\vec{R}_B}e^{i\vec{K}'\cdot\vec{\tau}_a}(1 + i\vec{\tau}_a\cdot\hat{k}')F_A^{K'}(\vec{R}_B), \tag{13A.8}$$

where $\hat{k}' = (\hat{k}_{x'}, \hat{k}_{y'}) = -i(\partial_{x'}, \partial_{y'})$ and $F_B^K(\vec{R}_A - \vec{\tau}_a) \simeq (1 - i\vec{\tau}_a\cdot\hat{k}')F_B^K(\vec{R}_A)$. We insert Equation 13A.7 into the first equation of Equation 13A.3

$$(\varepsilon - \tilde{u}_A(\vec{R}_A))\left(e^{i\vec{K}\cdot\vec{R}_A}F_A^K(\vec{R}_A) + e^{i\eta}e^{i\vec{K}'\cdot\vec{R}_A}F_A^{K'}(\vec{R}_A)\right)$$
$$= e^{i\vec{K}\cdot\vec{R}_A}\gamma e^{i\eta}(\hat{k}'_x - i\hat{k}'_y)F_B^K(\vec{R}_A)$$
$$+ e^{i\vec{K}'\cdot\vec{R}_A}\gamma(\hat{k}'_x + i\hat{k}'_y)F_B^{K'}(\vec{R}_A), \tag{13A.9}$$

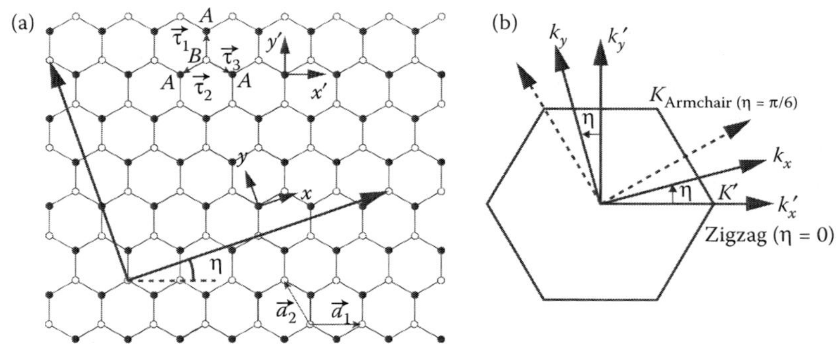

FIGURE 13A.1 (a) For $\eta = 0$, zigzag nanoribbon is realized along the x'-axis and armchair ribbon is realized along the y'-axis. Armchair nanoribbon can also be realized along the x-axis for $\eta = \pi/6$. Basis vectors of the honeycomb lattice are $\vec{a}_1 = a_0(1,0)$ and $\vec{a}_2 = (a_0/2)(-1,\sqrt{3})$, where $a_0 = \sqrt{3}a$ is the length of the unit cell with carbon–carbon distance a. Vectors from B to A carbon atoms are $\vec{\tau}_1 = a(0,1)$, $\vec{\tau}_2 = a(-(\sqrt{3}/2),-(1/2))$, and $\vec{\tau}_3 = a((\sqrt{3}/2),-(1/2))$. (b) Among six corners in the Brillouin zone (BZ), only two points are independent with respect to reciprocal lattice vectors, and they are called $\vec{K}$ and $\vec{K}'$ points. For $\eta = 0$, two possible Dirac points are $\vec{K} = (2\pi/a_0)(1/3,1/\sqrt{3})$ and $\vec{K}' = (2\pi/a_0)(2/3,0)$. (Adapted from Ando, T. and T. Nakanishi, 1998, *J. Phys. Soc. Jap.* **67**, 1704.)

where $\gamma \equiv (\sqrt{3}/2)a\gamma_0$. Here, we have derived $-\omega\Sigma_{a=1}^{3}[e^{i\vec{K}\cdot\vec{\tau}_a}(1-i\vec{\tau}_a\cdot\hat{k}')] = \hat{k}'_x - i\hat{k}'_y$ and $\Sigma_{a=1}^{3}[e^{i\vec{K}'\cdot\vec{\tau}_a}(1-i\vec{\tau}_a\cdot\hat{k}')] = \hat{k}'_x + i\hat{k}'_y$ using $\vec{K} = (2\pi/a_0)(1/3,1/\sqrt{3})$ and $\vec{K}' = (2\pi/a_0)((2/3),0)$. To derive the Dirac equations for the envelope wavefunctions at the $\vec{K}$ point, we rewrite Equation 13A.9 as

$$(\varepsilon - \tilde{u}_A(\vec{R}_A))(F_A^K(\vec{R}_A) + e^{i\eta}e^{i(\vec{K}'-\vec{K})\cdot\vec{R}_A}F_A^{K'}(\vec{R}_A))$$
$$= \gamma e^{i\eta}(\hat{k}'_x - i\hat{k}'_y)F_B^K(\vec{R}_A) + e^{i(\vec{K}'-\vec{K})\cdot\vec{R}_A}\gamma(\hat{k}'_x + i\hat{k}'_y)F_B^{K'}(\vec{R}_A) \quad (13A.10)$$

We introduce a function $g(\vec{R})$ normalized in such a way that $\Sigma_{\vec{R}}g(\vec{R}) = 1$. We assume that $g(\vec{R})$ is real with a range that is a few times of the lattice constant. This means that the spatial variation of envelope functions in this region can be safely neglected. Multiplying both sides of Equation 13A.10 by $g(\vec{R}-\vec{R}_A)$ and summing over $\vec{R}_A$, we get

$$\varepsilon F_A^K(\vec{R}) = \gamma e^{i\eta}(\hat{k}'_x - i\hat{k}'_y)F_B^K(\vec{R}) + u_A(\vec{R})F_A^K(\vec{R}) + e^{i\eta}u'_A(\vec{R})F_A^{K'}(\vec{R}), \quad (13A.11)$$

where

$$u_A(\vec{R}) = \sum_{\vec{R}_A} g(\vec{R}-\vec{R}_A)\tilde{u}_A(\vec{R}_A),$$
$$u'_A(\vec{R}) = \sum_{\vec{R}_A} g(\vec{R}-\vec{R}_A)e^{i(\vec{K}'-\vec{K})\cdot\vec{R}_A}\tilde{u}_A(\vec{R}_A). \quad (13A.12)$$

In deriving these equations, we have ignored terms such as

$$\sum_{\vec{R}_A} e^{i(\vec{K}'-\vec{K})\cdot\vec{R}_A}S(\vec{R}_A)g(\vec{R}-\vec{R}_A) \approx 0, \quad (13A.13)$$

where $S(\vec{R}_A)$ is a slowly varying function over the lattice constant. This is because many phase factors $e^{i(\vec{K}'-\vec{K})\cdot\vec{R}_A}$ with $\vec{R}_A$ in the range of $g(\vec{R}-\vec{R}_A)$ contribute to the sum. If $S(\vec{R}_A)$ has a shorter range, these terms cannot be neglected, see Figure 13A.2.

Similarly, we find the A-component of the envelope function of $\vec{K}'$ point. We rewrite Equation 13A.9 as

$$(\varepsilon - \tilde{u}_A(\vec{R}_A))(e^{i(\vec{K}-\vec{K}')\cdot\vec{R}_A}F_A^{K'}(\vec{R}_A) + e^{i\eta}F_A^K(\vec{R}_A))$$
$$= \gamma e^{i\eta}e^{i(\vec{K}-\vec{K}')\cdot\vec{R}_A}(\hat{k}'_x - i\hat{k}'_y)F_B^K(\vec{R}_A) + \gamma(\hat{k}'_x + i\hat{k}'_y)F_B^{K'}(\vec{R}_A) \quad (13A.14)$$

Multiplying both sides of Equation 13A.14 by $g(\vec{R}-\vec{R}_A)$ and summing over $\vec{R}_A$, we get

$$\varepsilon F_A^{K'}(\vec{R}) = \gamma e^{-i\eta}(\hat{k}'_x + i\hat{k}'_y)F_B^{K'}(\vec{R})$$
$$+ u_A(\vec{R})F_A^{K'}(\vec{R}) - e^{-i\eta}u'_A(\vec{R})^* F_A^{K'}(\vec{R}), \quad (13A.15)$$

Using the second equation of Equation 13A.3, similarly, we find the equations for the B-components

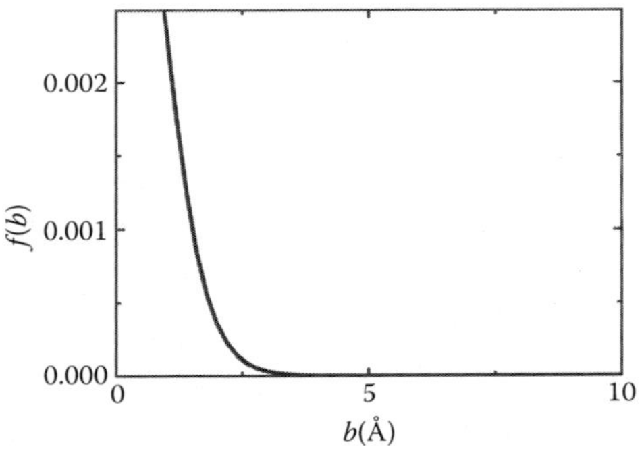

FIGURE 13A.2 Plot of the left-hand side of Equation 13A.13: $f(b) \equiv \Sigma_{\vec{R}_A}e^{i(\vec{K}'-\vec{K})\cdot\vec{R}_A}S(\vec{R}_A)g(\vec{R}-\vec{R}_A)$. We have used $S(\vec{r}) = (1/\pi b^2)e^{-r^2/b^2}$, $g(\vec{r}) = (1/\pi\xi^2)e^{-r^2/\xi^2}$, $\xi = 2a_0$, and $\vec{R} = (0,0)$.

$$\varepsilon F_B^K(\vec{R}) = \gamma e^{-i\eta}(\hat{k}'_x + i\hat{k}'_y)F_A^K(\vec{R}) + u_B(\vec{R})F_B^K(\vec{R}) - \omega^{-1}e^{-i\eta}u'_B(\vec{R})F_B^{K'}(\vec{R}),$$
$$\varepsilon F_B^{K'}(\vec{R}) = \gamma e^{i\eta}(\hat{k}'_x - i\hat{k}'_y)F_A^{K'}(\vec{R}) + u_B(\vec{R})F_B^{K'}(\vec{R}) - \omega e^{i\eta}u'_B(\vec{R})^* F_B^{K'}(\vec{R}), \quad (13A.16)$$

where

$$u_B(\vec{R}) = \sum_{\vec{R}_B} g(\vec{R} - \vec{R}_B)\tilde{u}_B(\vec{R}_B),$$
$$u'_B(\vec{R}) = \sum_{\vec{R}_B} g(\vec{R} - \vec{R}_B)e^{i(\vec{K}'-\vec{K})\cdot\vec{R}_B}\tilde{u}_B(\vec{R}_B) \quad (13A.17)$$

These equations can be written together as a matrix eigenvalue problem

$$\begin{pmatrix} u_A(r) & \gamma(\hat{k}_x - i\hat{k}_y) & e^{i\eta}u'_A(r) & 0 \\ \gamma(\hat{k}_x + i\hat{k}_y) & u_B(r) & 0 & -\omega^{-1}e^{-i\eta}u'_B(r) \\ e^{-i\eta}u'_A(r)^* & 0 & u_A(r) & \gamma(\hat{k}_x + i\hat{k}_y) \\ 0 & -\omega e^{i\eta}u'_B(r)^* & \gamma(\hat{k}_x - i\hat{k}_y) & u_B(r) \end{pmatrix} \times \begin{pmatrix} F_A^K(r) \\ F_B^K(r) \\ F_A^{K'}(r) \\ F_B^{K'}(r) \end{pmatrix} = E \begin{pmatrix} F_A^K(r) \\ F_B^K(r) \\ F_A^{K'}(r) \\ F_B^{K'}(r) \end{pmatrix}, \quad (13A.18)$$

where $(\hat{k}_x \pm i\hat{k}_y) = e^{\mp i\eta}(\hat{k}'_x \pm i\hat{k}'_y)$. Note that $u'_A(r)$ and $u'_B(r)$ vanish when the range of $u_A(r)$ and $u_B(r)$ becomes large. These short-range potentials play an important role in the x-ray edge problem of graphene (Yang and Lee, 2007). If we use, instead of Equation 13A.7, a different ansatz

$$\psi_A(\vec{R}_A) = e^{i\pi/6}(e^{i\vec{K}\cdot\vec{R}_A}F_A^K(\vec{R}_A) + e^{i\vec{K}'\cdot\vec{R}_A}F_A^{K'}(\vec{R}_A)),$$
$$\psi_B(\vec{R}_B) = e^{-i\pi/6}(e^{i\vec{K}\cdot\vec{R}_B}F_B^K(\vec{R}_B) + e^{i\vec{K}'\cdot\vec{R}_B}F_B^{K'}(\vec{R}_B)), \quad (13A.19)$$

we find

$$\begin{pmatrix} u_A(r) & \gamma(\hat{k}_x - i\hat{k}_y) & u'_A(r) & 0 \\ \gamma(\hat{k}_x + i\hat{k}_y) & u_B(r) & 0 & u'_B(r) \\ u'_A(r)^* & 0 & u_A(r) & \gamma(\hat{k}_x + i\hat{k}_y) \\ 0 & u'_B(r)^* & \gamma(\hat{k}_x - i\hat{k}_y) & u_B(r) \end{pmatrix} \times \begin{pmatrix} F_A^K(r) \\ F_B^K(r) \\ F_A^{K'}(r) \\ F_B^{K'}(r) \end{pmatrix} = E \begin{pmatrix} F_A^K(r) \\ F_B^K(r) \\ F_A^{K'}(r) \\ F_B^{K'}(r) \end{pmatrix} \quad (13A.20)$$

The phase factors $e^{\pm i\pi/6}$ are immaterial in the problems treated here and will be ignored. In the absence of the external potentials, the 4×4 Hamiltonian operator separates into two 2×2 Hamiltonian operators

$$H_{\vec{K}} = \begin{pmatrix} 0 & \gamma(\hat{k}_x - i\hat{k}_y) \\ \gamma(\hat{k}_x + i\hat{k}_y) & 0 \end{pmatrix},$$
$$H_{\vec{K}'} = \begin{pmatrix} 0 & \gamma(\hat{k}_x + i\hat{k}_y) \\ \gamma(\hat{k}_x - i\hat{k}_y) & 0 \end{pmatrix}. \quad (13A.21)$$

It should be noted that the effective mass approximation should be used with some caution, especially for graphene ribbons with small widths.

The eigenstates near $\vec{K}$ point are given by

$$\psi_{c,\vec{k}}^K(\vec{r}) = \frac{e^{i\vec{k}\cdot\vec{r}}}{\sqrt{2S}}\begin{pmatrix} 1 \\ e^{i\theta_{\vec{k}}} \end{pmatrix}, \quad \psi_{v,\vec{k}}^K(\vec{r}) = \frac{e^{i\vec{k}\cdot\vec{r}}}{\sqrt{2S}}\begin{pmatrix} 1 \\ -e^{i\theta_{\vec{k}}} \end{pmatrix}. \quad (13A.22)$$

Here, $\theta_{\vec{k}}$ is defined by $(k_x, k_y) \equiv |\vec{k}|(\cos\theta_{\vec{k}}, \sin\theta_{\vec{k}})$. We can write the Hamiltonian matrix as

$$H_{\vec{K}} = \hbar v_F |\vec{k}| \begin{pmatrix} 0 & e^{-i\theta_{\vec{k}}} \\ e^{i\theta_{\vec{k}}} & 0 \end{pmatrix}. \quad (13A.23)$$

The energy eigenvalue of Equation 13A.23 is given by $\pm v_F |\vec{k}|$ and the energy dispersion relation shows a linear dispersion relation near the $\vec{K}$ point, known as the Dirac cone. At the $\vec{K}$ (or $\vec{K}'$) point, the valence and conduction bands touch each other and thus, we call the degenerated point as the Dirac point. The directions of the pseudospins $\langle\vec{T}\rangle$ are parallel to $\vec{k}$ and $-\vec{k}$, respectively. Similarly, the eigenstates near $\vec{K}'$ point are

$$\psi_{c,\vec{k}}^{K'}(\vec{r}) = -\frac{e^{i\vec{k}\cdot\vec{r}}}{\sqrt{2S}}\begin{pmatrix} 1 \\ e^{-i\theta_{\vec{k}}} \end{pmatrix}, \quad \psi_{v,\vec{k}}^{K'}(\vec{r}) = -\frac{e^{i\vec{k}\cdot\vec{r}}}{\sqrt{2S}}\begin{pmatrix} 1 \\ -e^{-i\theta_{\vec{k}}} \end{pmatrix}. \quad (13A.24)$$

The effective-mass Hamiltonian matrix for the $\vec{K}'$ point is given by

$$H_{\vec{K}'} = \hbar v_F \left|\vec{k}\right| \begin{pmatrix} 0 & e^{i\theta_{\vec{k}}} \\ e^{-i\theta_{\vec{k}}} & 0 \end{pmatrix}. \quad (13A.25)$$

The pseudospins $\langle\vec{T}\rangle$ point along $(k_x, -k_y)$ and $(-k_x, k_y)$, respectively.

APPENDIX 13B: DERIVATION OF EFFECTIVE SCHRÖDINGER EQUATION

In the presence of a rotationally invariant potential, eigenfunctions of the Dirac equations can be written as

$$\Psi = \frac{1}{r}\begin{pmatrix} e^{i\phi(J-1/2)} f(r) \\ e^{i\phi(J+1/2)} g(r) \end{pmatrix}. \tag{13B.1}$$

It is useful to rewrite the Dirac equation into an equation that looks like a Schrödinger equation (Gamayun et al., 2009, Giavaras et al., 2009). It will provide information on the properties of solutions of the Dirac equation in the presence of a strongly localized potential, especially the presence of quasi-stationary solutions with complex energy values. The Dirac equations are

$$\begin{aligned} f' - \frac{J+1/2}{r} f - \frac{r}{2\ell^2} f + \frac{E+\xi\Delta(r)-V(r)}{\hbar v_F} g = 0, \\ g' + \frac{J-1/2}{r} g + \frac{r}{2\ell^2} g + \frac{E-\xi\Delta(r)-V(r)}{\hbar v_F} f = 0. \end{aligned} \tag{13B.2}$$

Here, ξ is 1 and −1 for $\vec{K}$ and $\vec{K}'$ valleys (Berry and Mondragon, 1987). A- and B-component wavefunctions $f(r)$ and $g(r)$ satisfy second-order differential equations

$$\begin{aligned} & f'' + \left(\frac{V'-\xi\Delta'}{E+\xi\Delta-V} - \frac{1}{r} \right) f' \\ & + \left[\frac{(V-E)^2-\Delta^2}{(\hbar v_F)^2} - \left(\frac{r}{2\ell^2} + \frac{J}{r} + \frac{1}{2r} \right) \frac{V'-\xi\Delta'}{E+\xi\Delta-V} \right. \\ & \left. - \frac{J^2-J-3/4}{r^2} - \frac{r^2}{4\ell^4} - \frac{J+1/2}{\ell^2} \right] f = 0, \\ & g'' + \left(\frac{V'+\xi\Delta'}{E-\xi\Delta-V} - \frac{1}{r} \right) g' \\ & + \left[\frac{(V-E)^2-\Delta^2}{(\hbar v_F)^2} - \left(\frac{r}{2\ell^2} + \frac{J}{r} - \frac{1}{2r} \right) \frac{V'+\xi\Delta'}{E-\xi\Delta-V} \right. \\ & \left. - \frac{J^2+J-3/4}{r^2} - \frac{r^2}{4\ell^4} - \frac{J-1/2}{\ell^2} \right] g = 0. \end{aligned} \tag{13B.3}$$

The first-order derivatives in Equation 13B.3 are eliminated by putting

$$f(r) = \eta(r) e^{-\int (a(r)/2) dr}, \quad g(r) = \chi(r) e^{-\int (b(r)/2) dr}, \tag{13B.4}$$

where $a(r) = (V' - \xi\Delta')/(E + \xi\Delta - V) - 1/r$ and $b(r) = (V' + \xi\Delta')/(E - \xi\Delta - V) - 1/r$. When $V(r) = 0$, we find $b(r) = -1/r$ and $g(r) = \chi(r) r^{1/2} \sim e^{-r^2/4}$. The functions $\eta(r)$ and $f(r)$ are related to each other by

$$[E + \xi\Delta - V(r)]^{1/2} \mu(r) = \frac{f(r)}{\sqrt{r}}, \tag{13B.5}$$

and the functions $\chi(r)$ and $g(r)$ are related to each other by

$$[E - \xi\Delta - V(r)]^{1/2} \chi(r) = \frac{g(r)}{\sqrt{r}}. \tag{13B.6}$$

The A-component $\eta(r)$ satisfies an equation that looks like a Schrödinger equation

$$-\eta''(r) + \tilde{U}(r)\eta(r) = \varepsilon\eta(r), \tag{13B.7}$$

where the effective potential is $\tilde{U} = \tilde{U}_1 + \tilde{U}_2$ with

$$\tilde{U}_1 = \frac{V(2E-V)}{(\hbar v_F)^2} + \frac{J(J-1)}{r^2} + \frac{r^2}{4\ell^4} + \frac{J+1/2}{\ell^2}, \tag{13B.8}$$

and

$$\begin{aligned} \tilde{U}_2 = \frac{1}{2}\left[\frac{V''-\xi\Delta''}{E+\xi\Delta-V} + \frac{3}{2}\left(\frac{V'-\xi\Delta'}{E+\xi\Delta-V} \right)^2 \right. \\ \left. + \left(\frac{J}{r} + \frac{r}{2\ell^2} \right) \frac{2(V'-\xi\Delta')}{E+\xi\Delta-V} \right] \end{aligned} \tag{13B.9}$$

When $\varepsilon > U(r)$, the wavefunction displays oscillatory behavior while for $\varepsilon < U(r)$, it shows an exponentially decaying behavior. When a resonant barrier forms, a quasi-bound state with a complex energy is possible. A Dirac electron may be confined by using the spatial variation of $\Delta(r)$ (Schnez et al., 2008).

The B-component $\chi(r)$ satisfies an equation that looks like a Schrödinger equation

$$-\chi''(r) + U(r)\chi(r) = \varepsilon\chi(r), \tag{13B.10}$$

where $\varepsilon = (E^2 - \Delta^2)/(\hbar v_F)^2$ and the effective potential $U = U_1 + U_2$, where

$$U_1 = \frac{V(2E-V)}{(\hbar v_F)^2} + \frac{J(J+1)}{r^2} + \frac{r^2}{4\ell^4} + \frac{J-1/2}{\ell^2}, \tag{13B.11}$$

and

$$\begin{aligned} U_2 = \frac{1}{2}\left[\frac{V''+\xi\Delta''}{E-\xi\Delta-V} + \frac{3}{2}\left(\frac{V'+\xi\Delta'}{E-\xi\Delta-V} \right)^2 \right. \\ \left. - \left(\frac{J}{r} + \frac{r}{2\ell^2} \right) \frac{2(V'+\xi\Delta')}{E-\xi\Delta-V} \right]. \end{aligned} \tag{13B.12}$$

For the cylindrical potential $\tilde{V}(r) = V_0\theta(R - r)$, we use the dimensionless variables $x = r/\ell$ and $X = R/\ell$, which gives

$(\tilde{V}(r)/E_M)' = -V_0\delta(X-x)/E_M$ and $(\tilde{V}(r)/E_M)'' = V_0(\delta(X-x)/(X-x))(1/E_M)$, where $E_M = \hbar v_F/\ell$ is the energy scale of LLs. In addition, we use the representation $\delta(x) = \lim_{a\to 0}(1/a\sqrt{\pi})e^{-x^2/a^2}$ to evaluate the effective potentials.

$$\ell^2\tilde{U}_1 = \frac{\tilde{V}}{E_M}\left(\frac{2E}{E_M} - \frac{\tilde{V}}{E_M}\right) + \frac{J(J-1)}{x^2} + \frac{x^2}{4} + J + \frac{1}{2},$$

$$\ell^2\tilde{U}_2 = \frac{1}{2}\left[\frac{(V_0/E_M)\delta(X-x)/(X-x)}{E/E_M + \xi\Delta/E_M - \tilde{V}/E_M} + \frac{3}{2}\left(\frac{(V_0/E_M)\delta(X-x)}{E/E_M + \xi\Delta/E_M - \tilde{V}/E_M}\right)^2 + \left(\frac{J}{x} + \frac{x}{2}\right)\frac{2(V_0/E_M)\delta(X-x)}{E/E_M + \xi\Delta/E_M - \tilde{V}/E_M}\right], \quad (13B.13)$$

$$\ell^2 U_1 = \frac{\tilde{V}}{E_M}\left(\frac{2E}{E_M} - \frac{\tilde{V}}{E_M}\right) + \frac{J(J+1)}{x^2} + \frac{x^2}{4} + J - \frac{1}{2},$$

$$\ell^2 U_2 = \frac{1}{2}\left[\frac{(V_0/E_M)\delta(X-x)/(X-x)}{E/E_M - \xi\Delta/E_M - \tilde{V}/E_M} + \frac{3}{2}\left(\frac{(V_0/E_M)\delta(X-x)}{E/E_M - \xi\Delta/E_M - \tilde{V}/E_M}\right)^2 - \left(\frac{J}{x} + \frac{x}{2}\right)\frac{2(V_0/E_M)\delta(X-x)}{E/E_M - \xi\Delta/E_M - \tilde{V}/E_M}\right] \quad (13B.14)$$

For a parabolic potential, we introduce the dimensionless quantities $x = r/\ell$, $V/E_M = (1/2)\alpha x^2$, $(V/E_M)' = \alpha x$, and $(V/E_M)'' = \alpha$. We find the effective potential $U = U_1 + U_2$ for the B-component of the Dirac wavefunction

$$\ell^2 U_1 = \frac{V}{E_M}\left(\frac{2E}{E_M} - \frac{V}{E_M}\right) + \frac{J(J+1)}{x^2} + \frac{x^2}{4} + J - \frac{1}{2},$$

$$\ell^2 U_2 = \frac{1}{2}\left[\frac{\alpha}{E/E_M - V/E_M} + \frac{3}{2}\left(\frac{\alpha x}{E/E_M - V/E_M}\right)^2 - \left(\frac{J}{x} + \frac{x}{2}\right)\frac{2\alpha x}{E/E_M - V/E_M}\right]. \quad (13B.15)$$

$$\ell^2\tilde{U}_1 = \frac{V}{E_M}\left(\frac{2E}{E_M} - \frac{V}{E_M}\right) + \frac{J(J-1)}{x^2} + \frac{x^2}{4} + J + \frac{1}{2},$$

$$\ell^2\tilde{U}_2 = \frac{1}{2}\left[\frac{\alpha}{E/E_M - V/E_M} + \frac{3}{2}\left(\frac{\alpha x}{E/E_M - V/E_M}\right)^2 + \left(\frac{J}{x} + \frac{x}{2}\right)\frac{2\alpha x}{E/E_M - V/E_M}\right] \quad (13B.16)$$

APPENDIX 13C: GRAPHENE LLS

In the presence of a magnetic field, the Dirac equations for A- and B-components can be combined together and written as the following second-order equations:

$$f'' - \frac{1}{r}f' + \left[\frac{E^2}{(\hbar v_F)^2} - \frac{J^2 - J - 3/4}{r^2} - \frac{r^2}{4\ell^4} - \frac{J+1/2}{\ell^2}\right]f = 0,$$

$$g'' - \frac{1}{r}g' + \left[\frac{E^2}{(\hbar v_F)^2} - \frac{J^2 + J - 3/4}{r^2} - \frac{r^2}{4\ell^4} - \frac{J-1/2}{\ell^2}\right]g = 0 \quad (13C.1)$$

Here, we have used a symmetric gauge (no potentials are present). Setting $\varepsilon = (E/E_M)^2$ and using the dimensionless variable $x = r/\ell$, we find

$$g'' - \frac{1}{x}g' + \left[\varepsilon^2 - \frac{J^2 + J - 3/4}{x^2} - \frac{x^2}{4} - (J - 1/2)\right]g = 0. \quad (13C.2)$$

Substituting $g(x) = xG(x)$ in the above equation

$$4x(G'(x) + xG''(x)) + (4J^2 + (-1 + x^2)^2 + 4J(1 + x^2) + \varepsilon)G(x) = 0. \quad (13C.3)$$

Substituting $x^2/2 \equiv z$ and using

$$\frac{d^2G}{dx^2} = \frac{d}{dx}\left(x\frac{dG}{dz}\right) = \frac{dG}{dz} + x\frac{dz}{dx}\frac{dG}{dz} = \frac{dG}{dz} + x^2\frac{d^2G}{dz^2}, \quad (13C.4)$$

we find

$$(G'(z) + zG''(z)) + \left(-\frac{J-1/2}{2} + \frac{\varepsilon}{2} - \frac{z}{4} - \frac{(J+1/2)^2}{4z}\right)G(z) = 0. \quad (13C.5)$$

The differential equation

$$(K' + zK'') + \left(m + \frac{\alpha + 1}{2} - \frac{z}{4} - \frac{\alpha^2}{4z}\right)K = 0, \quad (13C.6)$$

has a solution (Bateman, 1953)

$$K = e^{-z^2/2} z^{|\alpha|/2} L_m^{|\alpha|}(z). \quad (13C.7)$$

Identifying $\alpha = J + 1/2$, we require

$$\frac{\varepsilon}{2} - \frac{J - 1/2}{2} = m + \frac{(J+1/2)+1}{2},$$

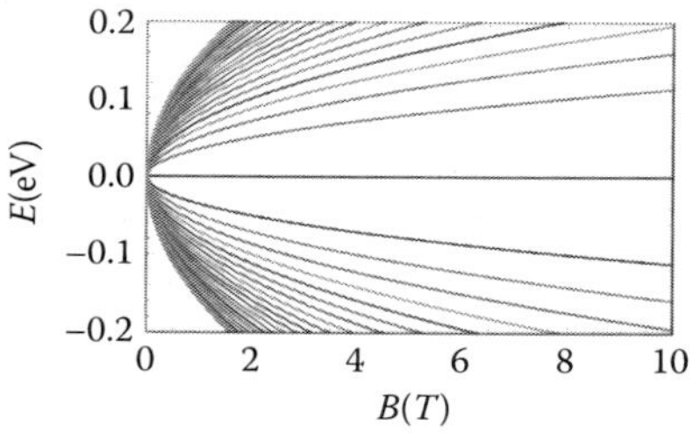

FIGURE 13C.1 Graphene LL energy is plotted as a function of B. Note that the conduction band LL states ($E > 0$) are identical to those of the valence band LLs except for some phase factors, see the text below. Their energies are also identical except for the sign.

or

$$\frac{\varepsilon}{2} = m + \frac{(J+1/2)+1}{2} + \frac{J-1/2}{2} = m + J + \frac{1}{2}. \tag{13C.8}$$

Since m is an integer ($m \geq 0$) and J are half-integers, the quantity $\varepsilon/2 = (E/E_M)^2$ is a positive integer $|n|$. Thus, we find $\varepsilon = 2|n|$. The magnetic field dependence of these energies is plotted in Figure 13C.1 for different values of n. In Table 13C.1, we give possible values of $J = |n| - m - 1/2$ for a given n.

The solutions are

$$G(z) = e^{-z/2} z^{|J+1/2|/2} L_m^{\alpha}(z), \tag{13C.9}$$

where $m = (2|n| - J - 1/2 - |J + 1/2|)$ is an integer. The LL states of graphene are given by

TABLE 13C.1
Eigenstates Labeled by Two Quantum Numbers: Band Index n and Angular Momentum Quantum Number J

n	J
⋮	⋮
2	$\frac{3}{2}, \frac{1}{2}, -\frac{1}{2}, -\frac{3}{2}, -\frac{5}{2}, \ldots$
1	$\frac{1}{2}, -\frac{1}{2}, -\frac{3}{2}, -\frac{5}{2}, \ldots$
0	$-\frac{1}{2}, -\frac{3}{2}, -\frac{5}{2}, \ldots$
−1	$\frac{1}{2}, -\frac{1}{2}, -\frac{3}{2}, -\frac{5}{2}, \ldots$
−2	$\frac{3}{2}, \frac{1}{2}, -\frac{1}{2}, -\frac{3}{2}, -\frac{5}{2}, \ldots$
⋮	⋮

Source: Park, P. S., S. C. Kim, and S.-R. Eric Yang, 2011a, *Phys. Rev. B* **84**, 085405.
Note: Possible values of n and J are listed.

$$\Psi_n^J(r,\theta) = e^{i(J-1/2)\theta} \begin{pmatrix} \chi_A(r) \\ \chi_B(r) e^{i\theta} \end{pmatrix}, \tag{13C.10}$$

where

$$\chi_A(r) = -i\,\mathrm{sgn}(n) c_{|n|-1} A_{|n|-1,J} \exp\left(-\frac{r^2}{4\ell^2}\right)\left(\frac{r}{\ell}\right)^{|J+1/2|} \times L^{|J+1/2|}_{(2|n|-J-5/2-|J+1/2|)/2}\left(\frac{r^2}{2\ell^2}\right),$$

$$\chi_B(r) = c_{|n|} A_{|n|,J} \exp\left(-\frac{r^2}{4\ell^2}\right)\left(\frac{r}{\ell}\right)^{|J+1/2|} \times L^{|J+1/2|}_{(2|n|-J-1/2-|J+1/2|)/2}\left(\frac{r^2}{2\ell^2}\right), \tag{13C.11}$$

$$A_{n,J} = \frac{1}{\ell}\left(2\pi 2^{\alpha} \frac{\Gamma(\beta+\alpha+1)}{\beta!}\right)^{-(1/2)}.$$

Here $\beta = (2|n| - J - 1/2 - \alpha)/2$ and $c_n = 1$ for $n = 0$ and $1/\sqrt{2}$ otherwise. We define $\mathrm{sgn}(n) = -1$, 0, and 1 for $n < 0$, $n = 0$, and $n > 0$.

Introducing an integer m given by $J = |n| - m - 1/2$, LL states of graphene can be written as

$$\psi_{n,m}(r,\theta) = c_n \begin{pmatrix} -\mathrm{sgn}(n) i \phi_{|n|-1,m}(r,\theta) \\ \phi_{|n|,m}(r,\theta) \end{pmatrix}. \tag{13C.12}$$

Here

$$\phi_{n,m}(r,\theta) = A_{n,m} \exp\left(i(n-m)\theta - \frac{r^2}{4\ell^2}\right) \times \left(\frac{r}{\ell}\right)^{|m-n|} \times L^{|m-n|}_{(n+m-|m-n|)/2}\left(\frac{r^2}{2\ell^2}\right), \tag{13C.13}$$

where the normalization constants are

$$A_{n,m} = \frac{1}{\ell}\left(2\pi 2^{a} \frac{\Gamma(b+a+1)}{b!}\right)^{-(1/2)}, \tag{13C.14}$$

with $a = |m - n|$ and $b = (n + m - a)/2$ (Yoshioka, 1998). For $n = 0$, the eigenstate $\psi_{n,m}(r, \theta)$ has only one component and is *chiral* while for $n \neq 0$, eigenstates are *nonchiral.*

ACKNOWLEDGMENTS

This chapter was supported by the Basic Science Research Program through the National Research Foundation of Korea (NRF) funded by the Ministry of Science, ICT, and Future

Planning (MSIP) (NRF-2012R1A1A2001554). In addition, this chapter was supported by a Korea University Grant.

REFERENCES

Ando, T. and T. Nakanishi, 1998, *J. Phys. Soc. Jap.* **67**, 1704.

Ashcroft, N. W. and N. D. Mermin, 1976, *Solid State Physics*, Brooks Cole, United States.

Bateman, H., 1953, *Higher Transcendental Functions*, Vol. II, McGraw-Hill, New York.

Baym, G., 1969, *Lectures on Quantum Mechanics*, W. A. Benjamin, Inc., London.

Berry, M. V. and R. J. Mondragon, 1987, *Proc. R. Soc. Lond. A* **412**, 53.

Brey, L. and H. A. Fertig, 2006, *Phys. Rev. B* **73**, 195408.

Brey, L. and H. A. Fertig, 2007a, *Phys. Rev. B* **73**, 235411.

Brey, L. and H. A. Fertig, 2007b, *Phys. Rev. B* **75**, 125434.

Cai, J., P. Ruffieux, R. Jaafar, M. Bieri, T. Braun, S. Blankenburg, M. Muoth et al., 2010, *Nature* **466**, 470.

Castro Neto, A. H., F. Guinea, N. M. R. Peres, K. S. Novoselov, and A. K. Geim, 2009, *Rev. Mod. Phys.* **81**, 109.

Darwin, C. G., 1930, *Proc. Camb. Philos. Soc.* **27**, 86.

Fock, V., 1928, *Z. Phys.* **47**, 446.

Fujita, M., K. Wakabayashi, K. Nakada, and K. Kusakabe, 1996, *J. Phys. Soc. Japan* **65**, 1920.

Gamayun, O. V., E. V. Gorbar, and V. P. Gusynin, 2009, *Phys. Rev. B* **80**, 165429.

Geim, A. K. and A. H. MacDonald, 2007, *Phys. Today* **60**, 35.

Giavaras, G., P. A. Maksym, and M. Roy, 2009, *J. Phys.: Condens. Matter* **21**, 102201.

Hewageegana, P. and V. Apalkov, 2008, *Phys. Rev. B* **77**, 245426.

Jackiw, R., 1991, Delta function potentials in two- and three-dimensional quantum mechanics, in *M.A.B. Bég Memorial Volume*, eds. A. Ali and P. Hoodbhoy (World Scientific, Singapore).

Kato, T. and R. Hatakeyama, 2012, *Nat. Nanotech.* **7**, 651.

Kim, S. C. and S.-R. Eric Yang, 2012, *J. Phys.: Condens. Matter* **24**, 195301.

Kim, S. C. and S.-R. Eric Yang, 2013, *J. Nanosci. Nanotechnol.* **13**, 6345.

Kim, S. C. and S.-R. Eric Yang, 2014a, *Ann. Phys.* **347**, 21–31.

Kim, S. C., S.-R. Eric Yang, and A. H. MacDonald, 2014b, *J. Phys.: Condens. Matter* **26**, 325302.

Kim, S. C., J. W. Lee, and S.-R. Eric Yang, 2012, *J. Phys.: Condens. Matter* **24**, 495302.

Kim, S. C., P. S. Park, and S.-R. Eric Yang, 2010, *Phys. Rev. B* **81**, 085432.

Lee, J. W., S. C. Kim, and S.-R. Eric Yang, 2012, *Solid State Commun.* **152**, 1929.

MacDonald, A. H. and D. S. Ritchie, 1986, *Phys. Rev. B* **33**, 8336.

Matulis, A. and F. M. Peeters, 2008, *Phys. Rev. B* **77**, 115423.

Meyer, J. C., C. O. Grit, M. F. Crommie, and A. Zettl, 2008, *Appl. Phys. Lett.* **92**, 123110.

Park, P. S., S. C. Kim, and S.-R. Eric Yang, 2010, *J. Phys.: Condens. Matter* **22**, 375302.

Park, P. S., S. C. Kim, and S.-R. Eric Yang, 2011a, *Phys. Rev. B* **84**, 085405.

Park, P. S., S. C. Kim, and S.-R. Eric Yang, 2011b, *J. Nanosci. Nanotechnol.* **11**, 629.

Park, P. S., S. C. Kim, and S.-R. Eric Yang, 2011c, *J. Nanosci. Nanotechnol.* **11**, 6332.

Park, P. S., S. C. Kim, and S.-R. Eric Yang, 2012, *Phys. Rev. Lett.* **108**, 169701.

Recher, P., J. Nilsson, G. Burkard, and B. Trauzettel, 2009, *Phys. Rev. B* **79,** 085407.

Schnez, S., K. Ensslin, M. Sigrist, and T. Ihn, 2008, *Phys. Rev. B* **78,** 195427.

Schwierz, F., 2010, *Nat. Nanotechnol.* **5**, 487.

Son, Y.-W., M. L. Cohen, and S. G. Louie, 2006, *Phys. Rev. Lett.* **97**, 216803.

Strack, R. and D. Vollhardt, 1995, *J. Low. Phys.* **99**, 385.

Tang, T., W. Yan, Y. Zheng, G. Li, and L. Li, 2008, *Nanotechnology* **19**, 435401.

Vázquez de Parga, A. L., F. Calleja, B. Borca, M. C. G. Passeggi, J. J. Hinarejos, F. Guinea, and R. Miranda, 2008, *Phys. Rev. Lett.* **100,** 056807.

Xiao, D., M.-C. Chang, and Q. Niu, 2010, *Rev. Mod. Phys.* **82**, 1959.

Yang, S.-R. Eric, and H. C. Lee, 2007, *Phys. Rev. B* **76**, 245411.

Yoshioka, D., 1998, *The Quantum Hall Effect* (Berlin, Springer).

14 Self-Organized Criticality, Percolation, and Electrical Instability in Graphene Analogs

A. Prikhod'ko and O. Kon'kov

CONTENTS

ABSTRACT

The self-organized criticality mechanism at levels of nano- and microsizes has been investigated using carbon multiwalled and single-walled nanotubes as graphene analogs. The formation of samples in the shape of sand heaps has been described. The specific features of this experiment are electrical resistance measurements conducted for lateral samples at certain decline angles of the surface at which the sample forms, and prebreakdown voltage values. The power law dependency between resistivity of the samples and the amount of material portions has demonstrated the number of the self-organized criticality manifestation. The processes of avalanche formation, percolation, and electrical instability have been studied experimentally using carbon nanotubes as examples. Described investigations are based on comparing electrical conductivity dynamics in classical experiments such as the "sand heap," as well as comparing the two-dimensional grid of resistances with a stochastic node blocking, and the nanosecond percolation in an electrical instability mode in nanotube tangles or granules. Regular patterns and the general concept have been identified. Nanosecond-pulsed voltage–current curves of an array of multi- and single-walled carbon nanotubes have been studied in the presence of electric fields where instabilities with negative differential conductivity can be observed. It is established that the development of electric instability in these structures obeys the classical percolation mechanism. Processes in weak electric fields analogous to processes in the base grid with embedded inhomogeneities have been revealed.

14.1 INTRODUCTION

The different teams investigating the electrical conductivity of physical systems that contain a large number of interacting elements in the years 1984–2002 developed the idea of studying samples similar to graphene. The chronology of these studies—since percolation theory, fractal model of dielectric breakdown, behavior of magnetic vortices in superconductors, and following the concept of self-organized criticality (SOC), through our studies of graphene analogs—are presented in Figure 14.1. Attention must be paid to the fact that all physical systems that define models and were studied earlier contain a large number of statistically distributed interacting elements. We assumed that the SOC concept cannot be applied to such two-dimensional (2D) physical systems like graphenes. It can however be applied to the interaction of a large amount of these elements, such as graphene planes twisted into a nanotube. Thus, we tried to solve the problem of interaction between stranded (deformed) planes within nonoriented arrays based on single-layer nanotubes, and the problem of interaction between stranded planes within a multilayer nanotube. Note that recent studies of narrow graphene strips [1] have revealed their three-dimensional distortions, that is, real graphene planes are distorted. We want to draw the attention of researchers to the fact that real graphene planes and highly

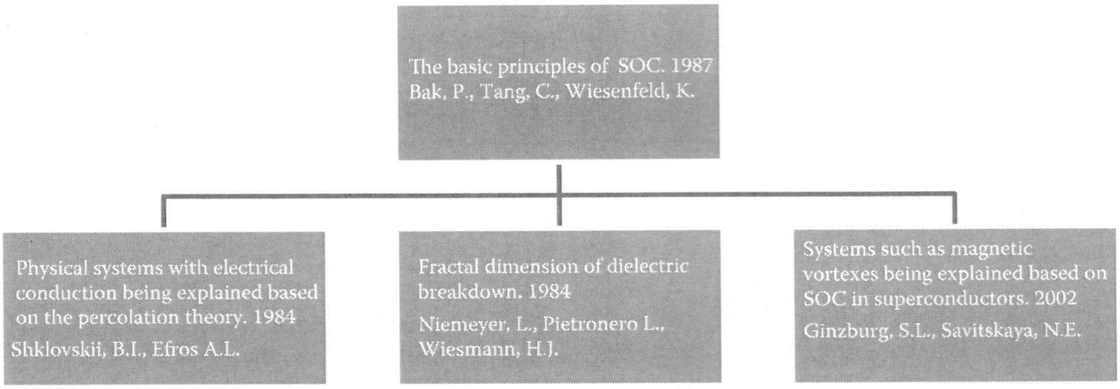

FIGURE 14.1 Chronology studies.

distorted planes such as nanotubes exhibit similar properties within statistically distributed arrays.

According to the basic principles of SOC [2], there are many giant dissipative dynamic systems consisting of a great number of interacting elements that are capable of accumulating small external perturbations. The mathematical criterion of self-organization is power behavior of the probability density for sizes of avalanches. An avalanche descent actually means that the system is in the critical state. For the sand heap experiment, this state corresponds to the avalanche descent at the lateral surface. Systems such as magnetic vortexes in superconductors [3] fit well into power law frameworks of the SOC concept.

The aim of this chapter is to show that nanostructures such as graphene analogs (2D-layered materials): carbon nanotubes (CNTs) and powder polycrystalline graphite can be convenient objects for SOC experimental studies.

Since its discovery, graphene, a 2D array of carbon atoms having the same structure as planes of graphite, has been the subject of much research. Nanoribbons, which are narrow strips of graphene of nanometer length may be the most useful for applications. Calculation of the minimum energy structure of zigzag and armchair graphene and boron nitride nanoribbons, having a single atomic vacancy show that the structures are significantly distorted from two dimensions [1]. Note that the approach known to making graphene nanoribbons (GNRs) by unzipping multiwalled carbon nanotubes (MWCNTs) by plasma etching of nanotubes is partly embedded in a polymer film. Unzipping CNTs with well-defined structures in an array will allow the production of GNRs with controlled widths, edge structures, placement, and alignment in a scalable manner for device integration [4].

Allotropes of carbon, for example, graphene and CNTs, are the subject of intense scientific interest, as an example of 2D and three-dimensional structures with similar properties. Graphene and CNTs, composed of sp^2-bonded carbon atoms, exhibit remarkable electrical and mechanical properties. The growth of CNTs and graphene on metals has been well studied using chemical vapor deposition (CVD). While conventional CVD has emerged as the leading technology for the production of graphene and CNTs, the process can be costly, requires lengthy processing times, and be restricted to a confined synthesis [5]. A simple solution-based oxidative process for producing a nearly 100% yield of nanoribbon structures by lengthwise cutting and unraveling of MWCNT side walls is well known. Subsequent chemical reduction of the nanoribbons from MWCNTs results in restoration of electrical conductivity [6].

It is known that the active impact of strong electric fields in carbon nanostructures can lead not only to S-type electric instability but also to structural transformations of various types, in particular [7] to nanotubes transforming into graphene planes.

It is possible that the study of current with a short duration of the electric field can reveal such nanotubes transforming into graphene planes and, as a consequence, the appearance of additional grids in the percolation array.

An integral property of the dynamics of giant dissipative systems consisting of a great number of interacting elements is the presence of chain reactions: from avalanches in neuron networks and snowy avalanches to the magnetic field avalanches in the Josephson medium. Nanoscale experimental studies of SOC have not been performed earlier. We are not aware of similar SOC studies for graphene-like carbon as well.

Our experiment is based on the similarity between sand heap growth dynamics and conductivity/resistivity dynamics in a preavalanche state. This chapter is devoted to the experimental detection of resistivity power behavior. We attempted to clarify the conditions of controlling the avalanche formation using electric field action at the critical angles related to the avalanche descent.

Physical systems with electrical conduction being explained based on the percolation of the grid of conducting and nonconducting (blocked) links/nodes, is described by a simple power law with index $\pm t$ (where t is a critical electrical conductivity index) [8]. The electrical conductivity model based on the percolation theory has some features that are in common with a model based on the SOC concept [2]. These models describe statistical systems without any regular structure. The goal of this chapter is to define conditions, under which physical systems with the electrical conductivity model based on the percolation theory can be described within the SOC concept framework.

Negative differential conductivity (NDC) of single quasi-one-dimensional contacts in a static electric field has received some attention in view of a possible use of nanodiodes in microwave electronics. In recent years, we have performed first investigations of nanosecond voltage–current (VI) characteristics in the electric instability mode for an array of quasi-one-dimensional contacts in a polymer matrix with CNTs [9] and for arrays of multi- (MWCNTs) and single-walled carbon nanotubes (SWCNTs) [10,11]. It can be expected that the electric field can induce NDC in an inhomogeneous three-dimensional medium consisting of tangled nanotube bundles. It is also possible that electric fields close to the breakdown threshold can induce a process of current redistribution analogous to the mechanism of percolation or SOC [11]. It is well known for disordered macrostructures that, in postbreakdown fields where a low-resistance nanosecond-switching state has been observed, a volume mosaic structure can exist with percolation pathways taking the form of branched channels/filaments [12].

As we have shown in References 11,13, and 14, current filament formation within "entangled" arrays of CNTs is connected with percolation mechanism specifics for an inhomogeneous medium. The array can include inhomogenities as embedded grids with cells of different size. It is known [8] that for a basic grid consisting of such an array, the current mechanism is based both on self-intersections and on the influence of "dead ends." In our opinion, the electrical field creates a percolation threshold equal to the breakdown field for the embedded grid in a real structure consisting of nano tangles and self-intersections of various sizes. This electrical breakdown will happen at different electrical field values for each embedded grid. It could happen that a strongly inhomogeneous three-dimensional medium consisting of nanotube tangles is a percolation cluster that demonstrates fractal properties. We are not aware of experiments where similar objectives have been set for tangled nanotube media. Let us note that known fractal systems such as dielectric breakdown can only be observed at electrical fields above the breakdown [15].

We considered the following obstacles when we planned our experimental study. It is well known that electrical instability specifics can be identified by applying the electrical field actively or passively. This is shown in References 11 and 14 for arrays of CNTs and an active electrical field impact, when thermal instability conditions can be obtained. The same was shown in Reference 12 where an electrical field of nanosecond duration passively applied to amorphous semiconductor structures and current filaments created by an active electrical field impact were studied. An example of such a dual impact can be an experiment with 2-ns pulses of different durations and electrical field values. Experimental studies of this kind, known as transient on characteristic (TONC) [16], were used to investigate VI curves of current filaments in disordered semiconductors. In a nanosecond experiment, thermal overheating of current channels can be managed. Therefore, new patterns of current flow in amorphous structures can be identified by a combination of active and passive electrical field impacts. Considering the above, our methodology is based on impacting samples by active and passive electrical fields. In our opinion, by controlling thermal overheating by a passive electrical field impact, one can observe details of current flow mechanism in relatively weak electrical fields, when threshold effects of the electrical field active impact are suppressed. Current flow mechanism for an array of nondirectional SWCNTs when applying an electric field of nanosecond duration in the active and passive mode is discussed in Section 14.3.4.

14.2 EXPERIMENTAL TRICKS

When selecting the object for the investigation, we adhered to the following: the methods to prepare either samples of CNTs or graphene samples should be the same. The temperature mode could be different for different samples. The CVD method corresponds to this approach. In addition, we used the entangled chains (threads) creation technique to implement various conditions of current flow. We have also created conditions to destroy nanotubes thermally by the strong electric field of nanosecond duration. And finally, we used the current flow study technique before and after destroying nanotubes. Below, we take a closer look at each stage of the study.

The experiments on SOC have been performed using carbon multiwall nanotube (CMWNT) samples. The samples consisted of disoriented CMWNT "Taunit" arrays [17] 20–70 nm in diameter and more than 150 nm in length. Granules based on "Taunit" nanotubes are one-dimensional nanoscale-fiber-like formations with a structure of entangled bunches of multiwall tubes. The method of producing the granules is gas-phase chemical deposition (catalytic pyrolysis [CVD]) of C_xH_y hydrocarbons on Ni/Mg catalyst at the atmosphere pressure and at temperatures of 580–650°C. We have also studied SWCNT samples, which consist of disoriented nanotubes [18] that are 0.7–3 nm in diameter and more than 100 nm in length. The SWCNT-based granules are the one-dimensional filamentary formations with the structure of tangled bundles as well. The MWCNT and SWCNT samples have some features in common. They are filamentary formations consisting of tangled nanoscale carbon chains. It is well known that ideal polymer chains can form tangled bundles of a size proportional to the chain length square root [19]. With the same sizes of the bundle consisting of nanotube chains (which takes place with their identical length), the "entanglement" degree increases as the nanotube diameter decreases. The "entanglement" degree of the bundle is determined by the number of self-intersections of the chains or "shortened" long chains. This approach has allowed us to classify samples under study by an increase in the "entanglement" degree from the MWCNT samples to SWCNT samples. Investigations of microscale objects having no such features, namely, powder-like polycrystalline graphite samples with a granule size of 3–5 μm, are presented for comparison. The samples were formed on a flat surface containing current contacts using the known technique of preparing a sand heap [20]. Figure 14.2 shows the experiment geometry. We measured

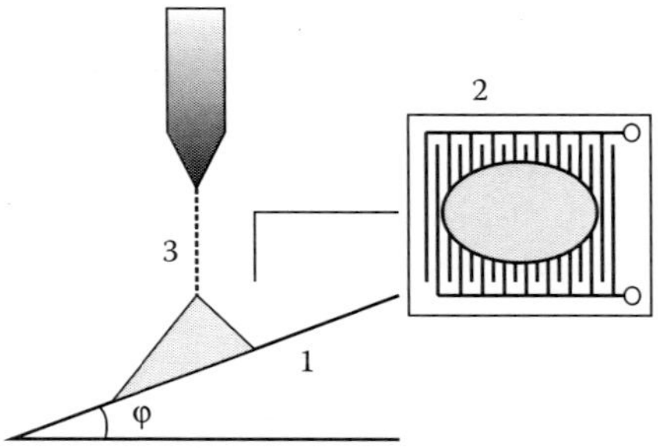

FIGURE 14.2 Geometry of the experiment: (1) plane of sample formation (φ is the plane slope); (2) current electrodes; and (3) powder-feeding direction. A part of the comb of current electrodes is shown inside (top view).

the electrical resistance of sample lateral layers in the current contact plane. The slope φ of the current contact plane can be changed with respect to the horizontal surface. During the experiment, the angle φ was below or equal to φ_c—the critical angle corresponding to the avalanche development. This geometry allows a decrease in the layer thickness and, thus, an increase in the measurement sensitivity. The powder portions of ~10^{-5} cm^3 (n is the number of portions) were fed on the flat surface at which the current contacts were disposed using an electromagnetic dispenser at a frequency of 1 Hz. The distance from the dispenser to the contact plane was 50 mm. The dispenser nozzle diameter was 1 mm. The static VI characteristics of the samples were measured in the current generator mode. Current contacts were in the form of a contrary comb, in which the total contact area S_0 exceeded the area of the powder sample spreading S by a factor of 7. In this case, the total sample resistance was measured. The distance between current contacts was 0.7 mm, and the number of contacts in the comb was 10, and $S_0 = 700$ mm^2. The interrelation between the spreading area S and the current contact plane slope using graphite samples as examples allows us to choose an optimal current contact area S_0 taking into account the total resistance. The optimal value S_0 must exceed 400 mm^2.

To observe the percolation process between nanotube granules/bundles, we used the nanosecond technique described in Reference 21. The generator based on the mercury relay provided the duration of the voltage pulse in a range of 1–20 ns at the pulse repetition rate of 100 Hz. The powder-like sample, which consisted of nanotube granules, was arranged in a holder between two cylindrical electrodes made of glassy carbon, thereby closing the coaxial line. To detect voltage pulse U_i incident onto the sample and that one reflected from U_r, we used a stroboscopic oscilloscope. The voltage and current across the sample were calculated by known formulas [21]:

$$U = U_i + U_r, \quad I = \frac{(U_i - U_r)}{\rho},$$

where $\rho = 50\ \Omega$ is the coaxial line wave resistance.

TONC-adopted nanosecond techniques were used to study VI curves in active and passive impact modes [16]. A distinctive feature of the experiment is that the double nanosecond pulse mode was used to control and modify VI curves. The first pulse, of 3-ns duration, is the control pulse. It was used to create the basic state of the sample. The second pulse, of 15-ns duration, modifies the basic state; therefore, it is the modifying and at the same time, the probe pulse. Basic and modifying VI curves as well as control (U_2) and modifying (U_1) pulses incident on the sample are shown in Figure 14.3. Time stamps ($t_1 = 1$ ns, $t_2 = 7$ ns) indicate time to measure longitudes of incident and reflected pulses, respectively.

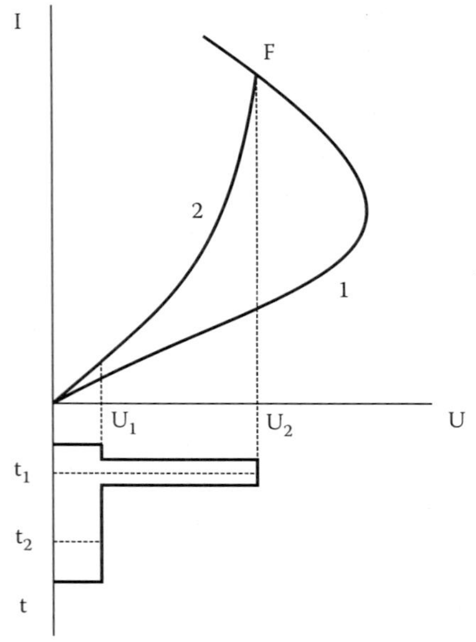

FIGURE 14.3 General view of basic (1) and modifying (2) VI curves, shapes of control (U_2), and probe (U_1) voltage pulses at the sample. t_1 and t_2—measurement points. F—formation point.

The value of the modifying pulse (U_1) was chosen in a way as to not to change the sample basic state. It is found that the basic state was practically not changed at $U_2/U_1 \sim 10$. The duration of the modifying pulse was chosen based on the current filament disappearance condition [22]. The estimate is that for a filament area of about 100 nm, which corresponds to the average length of a single nanotube, the thermal relaxation time (typical time of filament cooling) is of the order of the dielectric relaxation time (10^{-9}–10^{-11}) s. For amorphous semiconductors, for example, for amorphous selenium in the switching mode, for the conductive channel of 10-μm diameter and 1-μm length, and for pulse duration of 100–200 ns, the thermal relaxation time is of the order of 10^{-7} s [23]. The selected interval of control voltage pulse duration (1–8) ns allows creating the conductive state. Control pulse duration variations within (1–8)-ns range were dependent on the equipment limitations and did not change the results significantly.

14.3 WHAT TO DO AND WHAT TO GET

14.3.1 SOC in Graphene Analogs

Following the main SOC concepts, we carried out experiments on the determination of critical angles related to avalanche formations and, as a result, to a sharp decrease in the thickness of the layer above the current contacts. Our results from Reference 11 show the resistance dynamics in graphite powder samples and CMWNT powder measured at the same number of fed powder portions n. Critical angles of avalanche development φ_c are found as angles of a sharp increase in the resistance that corresponds to a sharp decrease in the sample thickness using the sand heap technique. The experiment was carried out in a weak electric field, which corresponds to the ohmic segment of the VI curve.

To study the electric field impact to an avalanche development, we studied static VI curves of the samples in detail. Let us note the existence of the electrical S-type instability among other features. As the current increases, the prebreakdown segment of the VI curve is recorded at voltages U_{max} for all the samples under study. The ohmic behavior is observed at U_{min}. Thus, U_{max} and U_{min} determine the prebreakdown and ohmic segments of the VI characteristic. Note that the S-shaped VI curve observed in the CMWNT samples is typical of multilayer CNTs [9–11]. Figures 14.4 and 14.5 show VI curves of the samples under study measured at incline angles close to critical angles of the avalanche development.

We found typical VI curve patterns when increasing the number of portions n at the precritical angles of the avalanche formation ($\varphi \leq \varphi_c$). In the case of the CMWNT samples, VI curves are practically unchanged with increasing n to 130. VI curves of graphite samples are different: they are not changed with increasing n above 100. At smaller n, VI curves are changed due to a decrease in the resistance, for example, because of thermal effects. On the basis of SOC mathematical criterion expressing the power dependence of sand heap sizes, we assume the following. The avalanche-like dynamics detection in the sand heap model can be related to the electrical resistance dynamics at the slope of a conducting sand heap as the number of powder portions fitted on the electrical contacts increases. Thus, a direct change in the sand heap size can be observed as a change in the number of powder portions n, and the mechanism of SOC can manifest itself as the change in the resistance at the slope (the lateral layer resistance). Figures 14.6 and 14.7 show R(n) dependence for the samples under study measured at slopes φ close to critical angles. For all the samples, the power dependence is observed. Particular emphasis is placed on the following feature: the slope index B substantially depends on the applied voltage. For all figures, the change in B is marked by an arrow. The slope index change depending on the voltage can be indicative of the intergranular current influence on the formation of the sample critical state. In the prebreakdown voltage region, the influence is maximal, and it can suppress the avalanche formation, which is equivalent to a slope change. At precritical slopes φ, $B = -1.52$ (U_{min}) and $B = -0.28$ (U_{max}) in the CMWNT samples and $B = -1.27$ (U_{min}) and $B = -0.45$ (U_{max}) in the graphite samples. The difference in experimental data obtained for CMWNT and

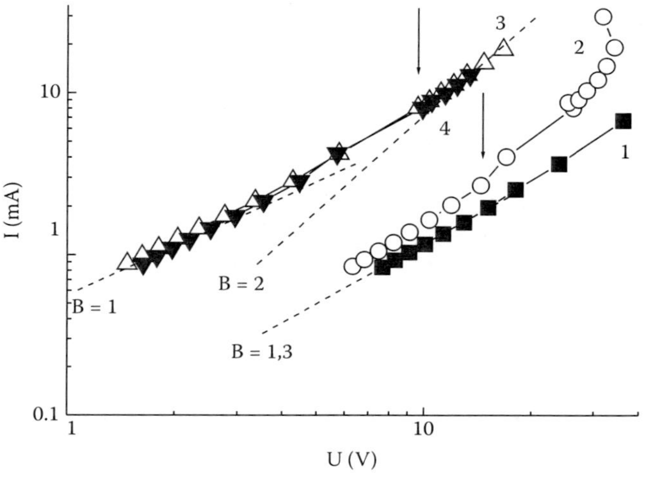

FIGURE 14.5 VI characteristics of the graphite samples measured at the number of portions n = (1) 20, (2) 50, (3) 70, and (4) 100; $\varphi = 35°$.

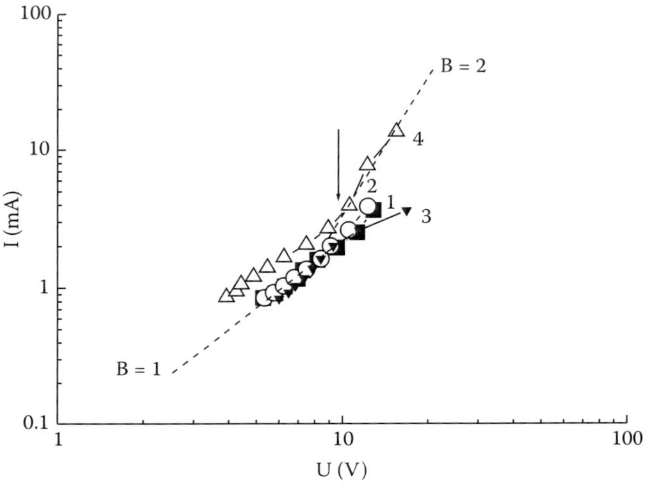

FIGURE 14.4 VI characteristics of the CMWNT samples measured at the number of portions n = (1) 20, (2) 50, (3) 100, and (4) 150; $\varphi = 55°$.

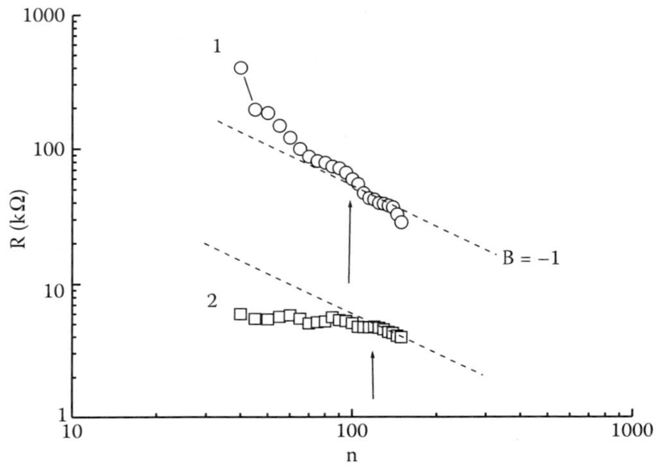

FIGURE 14.6 Dependence of R on the number of portions n for the CMWNT samples measured at (1) U_{max} and (2) U_{min}; $\varphi = 55°$.

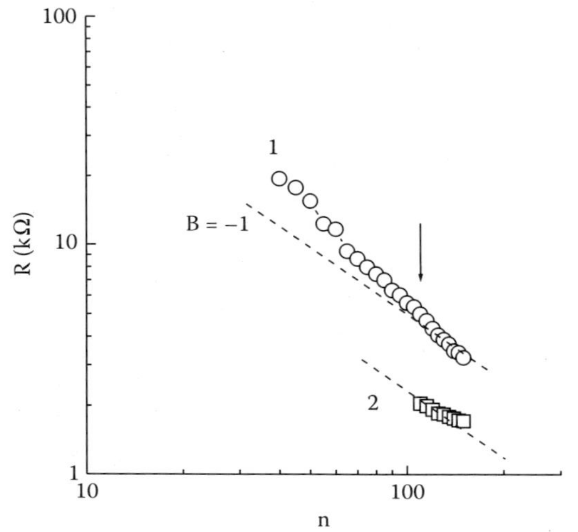

FIGURE 14.7 Dependence of R on the number of portions n for the graphite samples measured at (1) U_{max} and (2) U_{min}; $\varphi = 35°$.

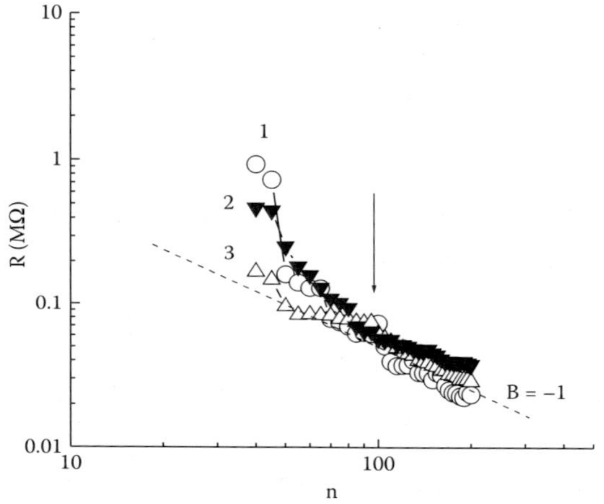

FIGURE 14.8 Dependence of R on the number of portions n for the CMWNT samples measured at φ = (1) 55, (2) 30, and (3) 5°. $U = U_{min}$.

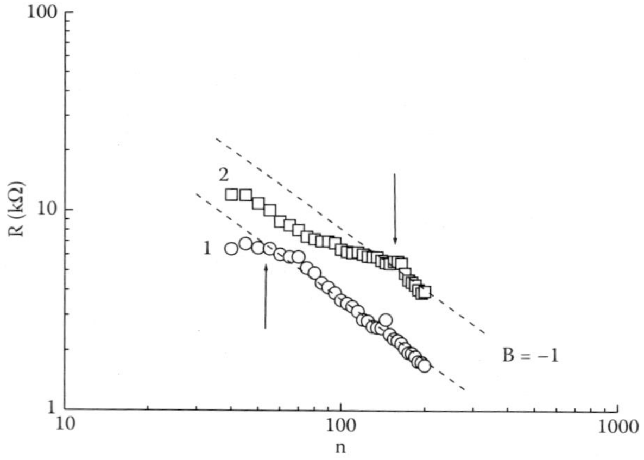

FIGURE 14.9 Dependence of R on the number of portions n for the graphite samples measured at φ = (1) 35 and (2) 5°. $U = U_{min}$.

graphite powder can be caused by the difference of the particle sizes (nm and μm), their geometric characteristics, and the interaction between them. We stated one more specific feature: the slope index in R(n) curves is undependent on the angle φ in the ohmic segment of VI curves (Figures 14.8 and 14.9) at a fairly large n (n > 100).

As the angle φ increases, the power law exponent is changed for all the samples. We can say on similarity of the processes of avalanche formation as the slope of the contact plane increases and as the applied voltage decreases.

14.3.1.1 Conclusions

We have successfully demonstrated that the experimental method of measuring the lateral layer resistivity at certain slopes of the sample-forming surface on the basis of entangled bunches of multiwalled nanotubes or powder polycrystalline graphite can be used to study SOC in the sand heap technique. It has been shown experimentally that the observed self-maintained critical state can be controlled using an external electric action. We found that, in prebreakdown electric fields, likely, an analogy of thermally controlled avalanches appears [24], but the avalanches, unlike hard second-order superconductors, do not cause destruction of the critical state and only change (decrease) the exponent in the SOC law.

14.3.2 Percolation and Electrical Instability in Graphene Analogs

The variation in normalized conductivity R_0/R depending on the number of randomly removed nodes n_i (Figure 14.10) for a 2D graphite grid at $n_i > 10$ is in well accordance with the classical experiment [8], where the critical electrical conductivity index $t = -(1.15 \pm 0.15)$ is determined. In our case (Figure 14.10, curve 1), this corresponds to the slope index B = −1. Curve 1 corresponds to a symmetric grid with

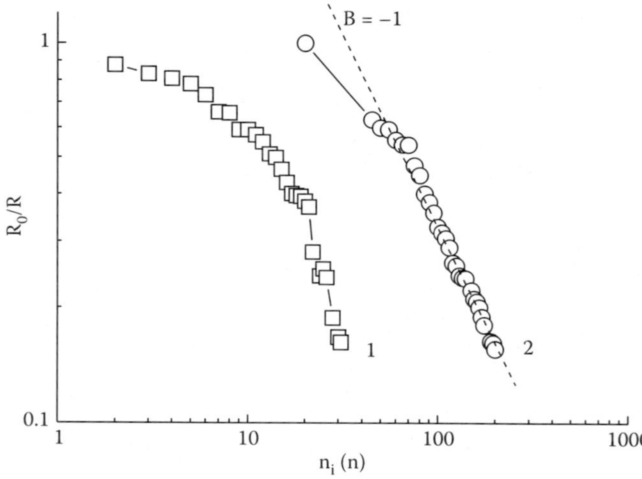

FIGURE 14.10 Dependence $R_0/R(n_i)$ for 2D graphite grids 10 × 10 elements in size (1) and $R_0/R(n)$ for graphite powder on the ohmic portion of the VI characteristic, $\varphi = 35°$(2). R_0 corresponds to the grid with all nodes ($n_i = 0$, n = 20).

identical distances between the nodes. The following classical experiment on the avalanche-like dynamics for graphite powder that we performed in the ohmic voltage region is presented in Figure 14.10, curve 2. Slope index for curves R(n), where n is the number of powder portions, for the slope angle of the current contacts plane relative to the horizontal surface φ equal to the precritical angle of avalanche formation φ = 35°, is almost undistinguishable from B and t. Thus, we can affirm that the conductivity dynamics of lateral layers in the "sand heap" technique is similar to percolation dynamics at precritical angles of avalanche formation in the region of the ohmic behavior of the VI characteristic. Precritical angles of avalanche formation correspond to an abrupt decrease in the sample thickness increment as the number of portions increases, while the random variation/annihilation of the node numbers corresponds to the redistribution of powder portions in the spreading region in terms of the percolation model. According to our experiment [13], this region retains the constant area until the critical angles are approached. It was also established for small angles that a considerable decrease in the absolute value of the slope index is detected up to value B = –0.45 (see Figure 14.10, curve 1). It is likely that the powder redistribution over the bulk occurs at such angles in the actual three-dimensional carbon nanostructure. In this case, the existence of the so-called "dead ends" can be admitted [8]. These "dead ends" are resistance chains that are finished by nonconductive deadlocks. The value of B does not coincide with the classical grid critical electrical conductivity index for this case, but it coincides with the critical density index of the infinite cluster β. According to calculations for the three-dimensional case [8], β = 0.35–0.47, which agrees well with our experiments for small angles. There is one more possibility to implement the percolation mechanism, which is associated with the self-intersection model of long chains doubling short ones [8], or the so-called mechanism of "shortened" long chains. This mechanism can be experimentally confirmed by our experiment on nanosecond transport for the samples of carbon nanostructures differing by the degree of "entanglement," namely, MWCNTs and SWCNTs. It was established previously [13] that general behavior patterns revealed for graphite and reflecting dynamics of avalanche formation are inherent to less-tangled MWCNT samples. It is known that S-type electrical instabilities can occur in powder graphite [13]. Figure 14.11 represents a nanosecond VI characteristic and dependence between current and resistance for such samples. The arrows denote threshold currents. S-type electrical instabilities in the nanosecond range were previously found for MWCNT samples [9] as well as for SWCNT samples. Figures 14.12 through 14.14 represent typical nanosecond VI curves of MWCNT and SWCNT samples in the region of large overvoltages leading to the implementation of S-type electrical instability.

We can affirm that electrical instability in a strongly nonuniform three-dimensional medium consisting of nanotube tangles/granules can be observed in a strong electric field. It seems likely that the process similar to percolation and SOC mechanisms occurs under electric fields exceeding the breakdown field. It is well known that a bulk mosaic with percolation paths in the shape of a branched grid of channels/filaments exists in a low ohmic state of nanosecond switching [12]. The percolation mechanism can also be seen by closeness of slope indices of experimental curves to B ~ –1 and the critical electrical conductivity index t = –(1.15 ± 0.15). It was assumed [25] that the infinite cluster at threshold parameters is not the set of locally one-dimensional "percolation channels." In fact, it involves channels with closed/intersecting paths. The fraction of such clusters does not decrease as the cluster size increases. In this case, the investigation of samples with various "entanglement" degrees can reveal features of such channels. Indeed, the analysis of VI curves for the low ohmic state of samples consisting of single-walled and multiwalled nanotubes has revealed some interesting features. First, the critical region of the low ohmic state exists, for which VI curves for samples of two types coincide. Second, the effect

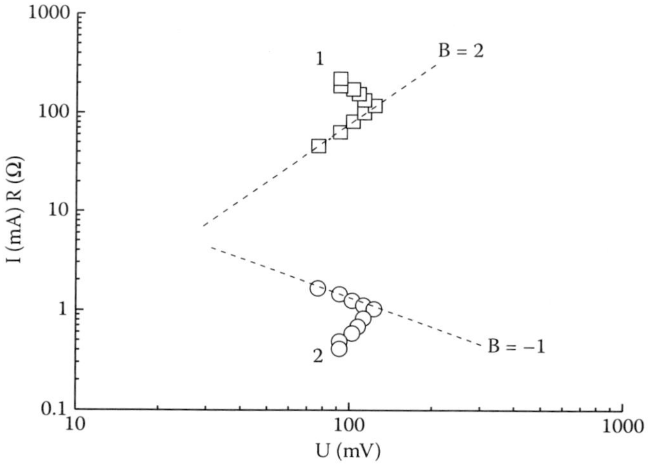

FIGURE 14.11 VI characteristic (1) and R(I) (2) dependence for powder graphite. Pulse duration is 20 ns.

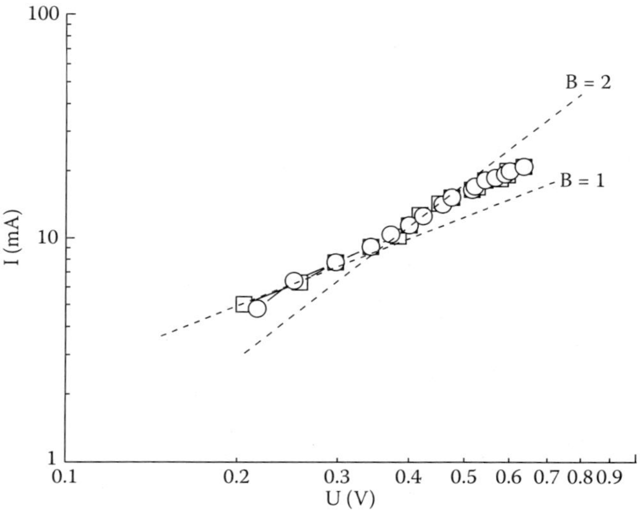

FIGURE 14.12 SWCNT sample prebreakdown VI curve in forward and reverse directions. Pulse duration is 5 ns.

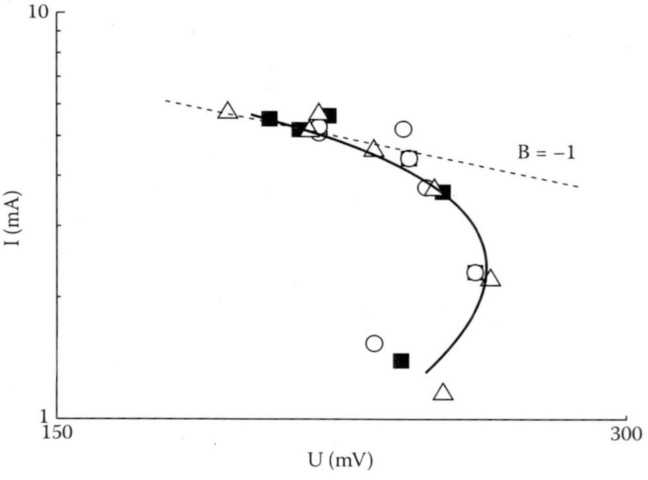

FIGURE 14.13 SWCNT sample postbreakdown VI curve (three consequent passes by voltage). Pulse duration is 5 ns.

of irreversible variation in VI characteristics (forming), which enhances with increasing the pulse duration even for 1 ns, manifests itself for MWCNT samples at currents lower than the critical one. This effect cannot be observed for SWCNT samples at sufficiently prolonged nanosecond voltage pulses (up to 20 ns). The possible cause can be a spatial transformation of multiwalled nanotubes into single-walled ones or into graphene planes, for example, due to the effect of strong electric fields of nanosecond range. Observed results are the variation in the "entanglement" degree, a decrease in the fraction of closed paths, and, as a consequence, effects of irreversible variation in the VI characteristic. We note that the use of the "electrical filament" technique is known for the destruction of multiwalled nanotubes in a static mode of VI characteristics over the entire range of used voltages [7]. In this case, one

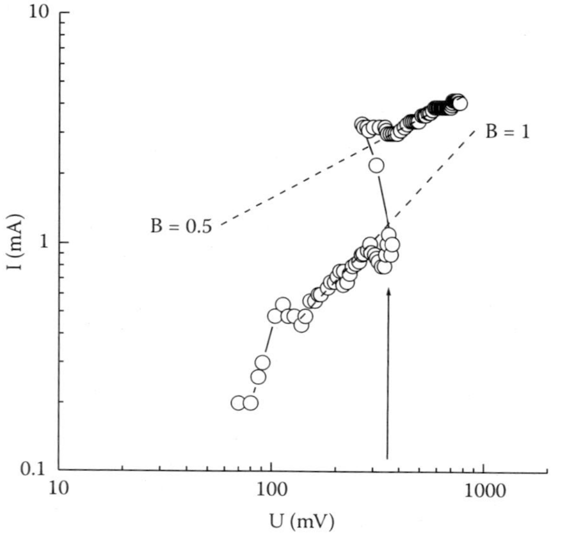

FIGURE 14.14 VI characteristic for MWCNT samples measured at a voltage pulse duration of 20 ns. Breakdown field is marked by an arrow.

can observe smooth degradation of static VI characteristics of separate nanotubes rather than electrical instability (and, correspondingly, the S-type VI curves in the granular structure).

14.3.2.1 Conclusions

As a result of performed investigations, it is established that conductivity dynamics of lateral layers in the course of the classical SOC experiment reflects percolation dynamics in conditions of avalanche-like formation near critical angles of the surface of current electrodes with a weak electric field (the ohmic VI characteristic portion). If the avalanche formation occurs at small angles, percolation can be obtained using the mechanism based on both the self-intersection of long chains and on the effect of "dead ends." This conclusion is related to the role of "tangled" chains in the course of percolation. An increase in tangle size or a decrease in chain diameter, which are associated with the "entanglement" degree, can lead to an increase in the volume of self-intersecting chains and, as a consequence, to a decrease in the effects of forming, degradation, and others, that is, the effects leading to an unstable operation of nanodevices. We can conclude that further increase in the CNT tangle will allow the increase in the reliability of nanotube-based nanodevices. We can also conclude that electrical instability is obtained in a strongly nonuniform three-dimensional medium consisting of "tangled" nanotubes under the effect of a strong electric field, which leads to the transition into a low ohmic state. The process similar to the percolation mechanism or SOC in an irregular statistic system based on tangles apparently occurs at electric fields exceeding the breakdown voltage. This can be indicated by comparing slope values and the critical electrical conductivity index. It seems likely that a new direction in studying the effects of electrical instability in macrosystems/tangles based on nanostructured objects of various topologies such as structures of multiwalled and single-walled nanotubes, clusters, and systems of graphene planes is revealed in the course of performed investigations.

14.3.3 Specific Features of Nanosecond VI Characteristics of an Array of Graphene Analogs

The critical electric conductivity index t according to a percolation model [15] was previously evaluated using a classical experiment with a 2D rectangular graphite grid of 10 × 10 nodes (Figure 14.10). It was established that dependency between normalized conductivity R_0/R (where R_0 is the resistance of an intact grid with all nodes) and the number n_i of randomly missing nodes, representing a line with a slope of $B = -1$, is in good agreement for $n_i > 10$ with a critical index of $t = -1.15 \pm 0.15$. For smaller n_i, the slope decreases down to $B = -0.5$ due to the effect of dead ends [8]. We suggest that dead ends have a decisive effect on the conductivity of a grid with a small number of missing nodes. In this chapter, the VI characteristics of MWCNTs (Figure 14.14) have been studied using nanosecond pulses of moderate length (~20 ns).

It was established that the dependence between relative conductivity (R_0/R) and electric field E (where R_0 is the sample resistance for E = 0) during the formation of an electric instability in supercritical fields $E > E_c$ (where E_c is the breakdown field) corresponds to the behavior of a 2D graphite grid with a randomly varied number of missing nodes. The curve slope upon breakdown is B = –1 and then, it decreases to B = –0.5 at greater fields, in agreement with B values in the percolation experiment [11]. The experimental results showed that the critical electric conductivity index, on one hand, reflects certain changes in the conductivity state with an increasing number of randomly missing nodes of the base grid [8]. On the other hand, this index may reflect an analogous conductivity variation for samples related to percolation development.

A specific feature of the nanosecond-pulsed experiment is that variations in thermal overheating of current channels were controlled. One can expect that, by decreasing the pulse duration, it will be possible to trace the manifestation of the conductivity mechanism in fields below the percolation threshold for the base grid. This can be of interest in cases where cell dimensions of embedded grids (inhomogeneities) are smaller than the ones of the base grid. The results of a computer experiment [26] showed that critical indices of the percolation theory follow variations in the linear size of inhomogeneities. As the cell size of the embedded grid (inhomogeneity) decreases, the slope B equivalent to the electric conductivity critical index must decrease. We cannot exclude that, with decreasing pulse duration, only percolation thresholds of embedded grids with smaller cell sizes will be detected. The structure under consideration represented nanoballs with various dimensions of self-intersection circuits—that is, with various sizes of inhomogeneities. Thus, variation of the applied electric field can reveal percolation thresholds both in the base grid and in those grids corresponding to inhomogeneities. We suggest that a more "tangled" array of SWCNTs (consisting of thin CNTs) [11] represents a more homogeneous percolation network, whereas a less- "tangled" array of MWCNTs (involving thicker CNTs) is less homogeneous and includes several inhomogeneities in the form of built-in grids with various cell sizes. Indeed, Figure 14.15 shows R_0/R versus U plot measured for a voltage pulse duration of 5 ns, which reveals the appearance of one (for SWCNT) or several (for MWCNTs) threshold voltages with a power behavior (and, accordingly a smaller slope B) in the subthreshold region.

As the pulse duration grows, the increasing overheating of current channels leads to manifestation of the base grid with a greater cell size. Thus, an increase in the critical percolation thresholds observed for SWCNT and MWCNT samples does not contradict the conclusion that the percolation threshold increases upon the introduction of inhomogeneities into the grid [26].

14.3.3.1 Conclusions

Thus, the results of our experiments with CNT arrays reflect the dynamics of instability development in the framework of a percolation mechanism for various durations of nanosecond voltage pulses. For the base grid (and supercritical fields), the percolation proceeds according to a mechanism involving both self-intersection of long chains and the effect of

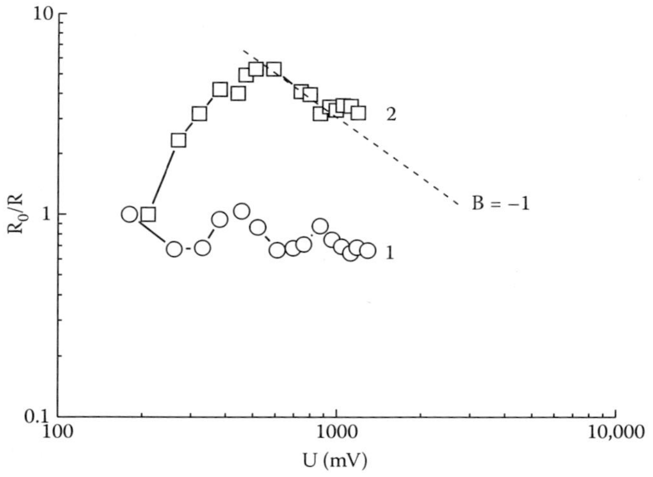

FIGURE 14.15 Plots of R0/R versus U for MWCNT (1) and SWCNT (2) samples measured at a voltage pulse duration of 5 ns.

dead ends. In subthreshold fields, percolation is obtained in the base grid containing embedded ones, which play the role of linear inhomogeneities with various cell sizes. Obtained results demonstrate the universal nature of electric current flows in nanostructures consisting of CNTs. Electric instability development in these structures follows the percolation mechanism. Under the action of an applied field, a percolation threshold is attained (i.e., analogous to the electric breakdown field) and the electric instability is developed according to the percolation mechanism in a strongly inhomogeneous three-dimensional medium consisting of "tangled" CNT balls. The possibility of MWCNT transformation (breakdown) under the action of strong electric fields was previously reported [7]. We cannot exclude that observed features of VI curves measured at small voltage pulse durations are related to this kind of MWCNT transformation and, hence, the appearance of additional grids in the array. It should be noted that appearance of electric instabilities in strongly inhomogeneous random (statistical) structures (e.g., amorphous semiconductors [22]) can also be related to the formation of percolation thresholds. This mechanism is critical with respect to the duration of nanosecond-scale action. At sufficiently large voltage pulse durations, an inhomogeneous statistical structure converts into a virtually homogeneous structure. A strong heating, which increases with the current amplitude, leads to the appearance of a single pinch filling the entire sample. In a prebreakdown state, grids become more homogeneous. This can also be observed in our nanosecond-pulsed experiments, where an increase in the voltage pulse duration makes features on the VI curve almost indistinguishable (Figure 14.16).

It can be confirmed that effects related to the percolation mechanism are absent in an ideal array without nanoballs and in SWCNT arrays. Therefore, the proposed investigation method, based on the formation of electric instabilities during nanosecond voltage pulses and the subsequent analysis in the percolation model framework, can be used for diagnostics of the state of these arrays.

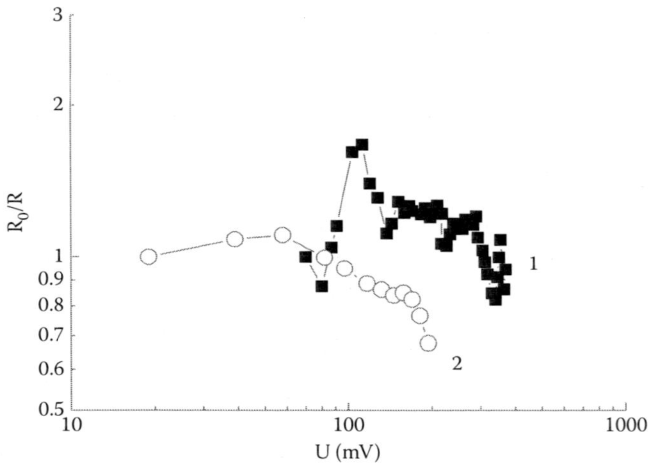

FIGURE 14.16 Plot of R_0/R versus U for MWCNT samples measured at a voltage pulse duration of (1) 20 ns and (2) 10 ns.

14.3.4 Structure Modification and Current Flow in Graphene Analogs

As reported in Reference 14, S-type nanosecond switching has been observed in SWCNT samples. The investigation has shown no delay time in the switching process, and the time of switching depends on voltage (Figure 14.17). It is known that nanosecond-switching development with no delay is caused by thermal processes. Thermal instability can be obtained with no delay within a time of about a conductive cord thermal relaxation [23]. Presented VI and R(U) curves for a control pulse (Figures 14.18 and 14.19) present the appearance of an electrical instability (Figure 14.18—inside) and confirm that the flow threshold value has been achieved.

Let us note that VI curve for SWCNT array differs from the one for multiwalled nanotubes array [14]. The study of modifying U_I curves of SWCNT samples for weaker electrical fields has found some specifics that is in common with managing U_I curves (Figures 14.20 and 14.21).

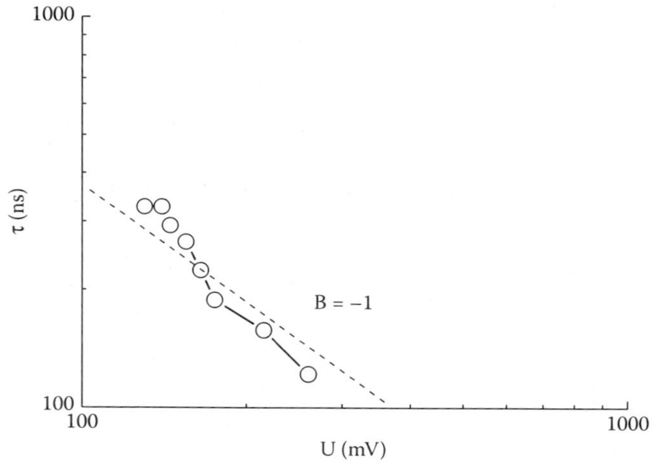

FIGURE 14.17 Time of switching τ versus voltage for SWCNT samples.

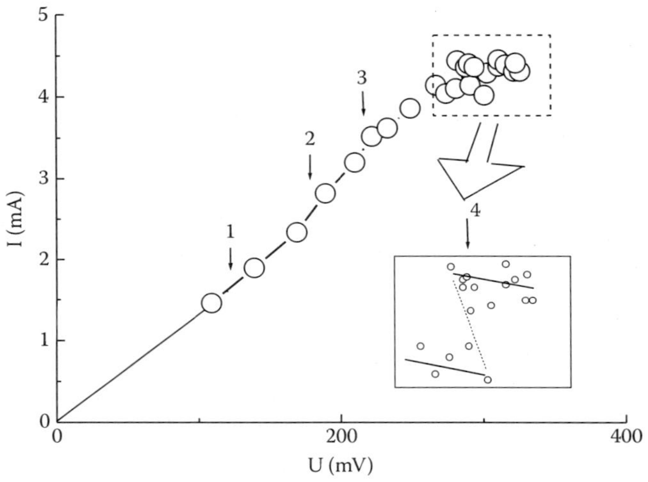

FIGURE 14.18 Managing VI curve of SWCNT sample. Markers 1–4 show fixed amounts of the control pulse for modifying VI curves. Details of the selected area are shown inside.

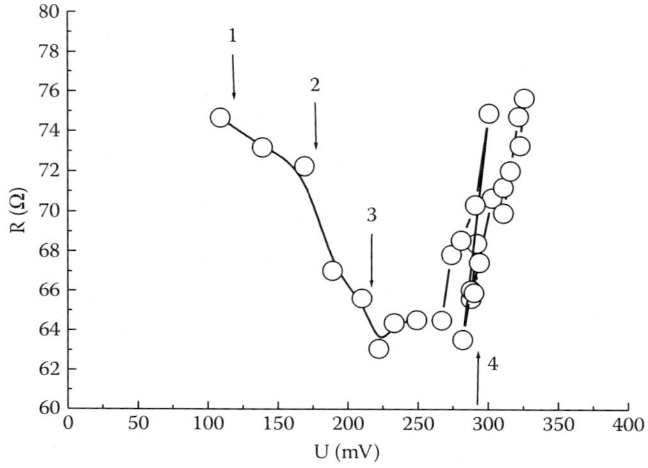

FIGURE 14.19 R(U) dependency of SWCNT sample. Markers 1–4 show fixed amounts of the control pulse for modifying VI curves.

One can clearly indicate the area where flow threshold is achieved with a control pulse of a fixed voltage in Figure 14.20 (fixing the pulse is presented in Figure 14.18, marker no. 4). At the same time, sample resistance decreases as a power law with B = −1 that is typical to the electrical percolation model and is discussed earlier in References 11,13, and 14. Generalized dependencies that are combinations of managing and modifying dependencies (see Figures 14.18 through 14.21), with two clearly visible flow thresholds U_{c1}, U_{c2}, are shown in Figures 14.22 and 14.23.

14.3.4.1 Conclusions

It has been presented that current flows in tangled nanostructures can be characterized in a universal way. An electrical field creates a flow threshold that is equal to a breakdown electrical field; therefore, the flow is carried out in a strongly heterogeneous three-dimensional environment consisting of nanotube

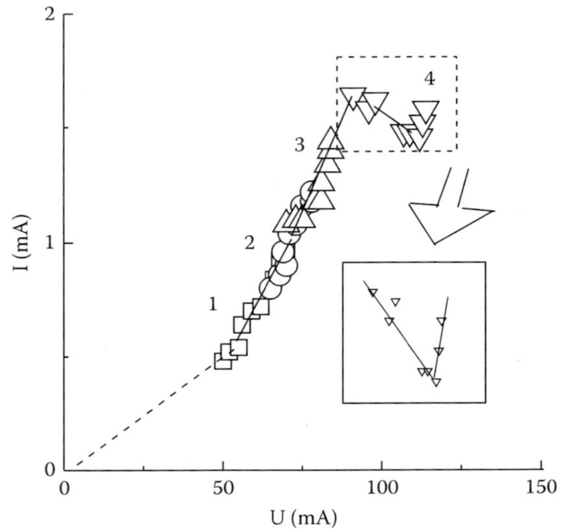

FIGURE 14.20 Modifying VI curves of SWCNT sample. Curves 1–4 correspond to fixed amounts of the control pulse marked as 1–4 in Figure 14.18. Details of the selected area are shown inside.

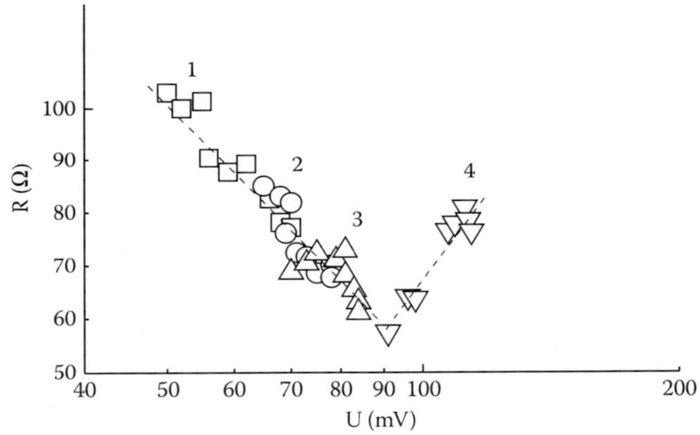

FIGURE 14.21 R(U) dependencies of SWCNT sample. Curves 1–4 correspond to fixed amounts of the control pulse marked as 1–4 in Figure 14.19.

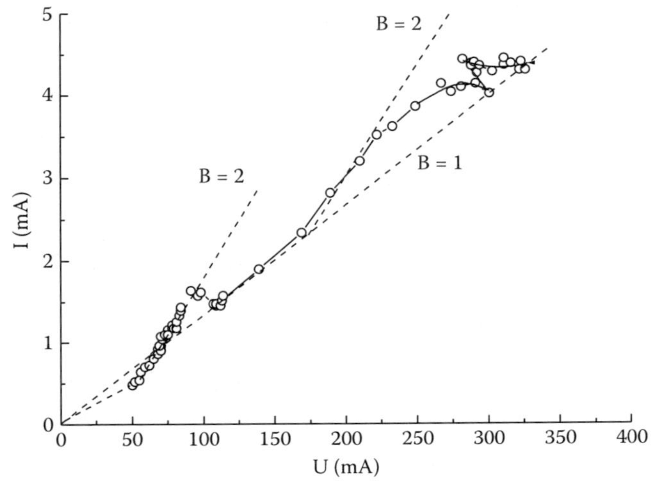

FIGURE 14.22 Generalized VI curve for SWCNT sample.

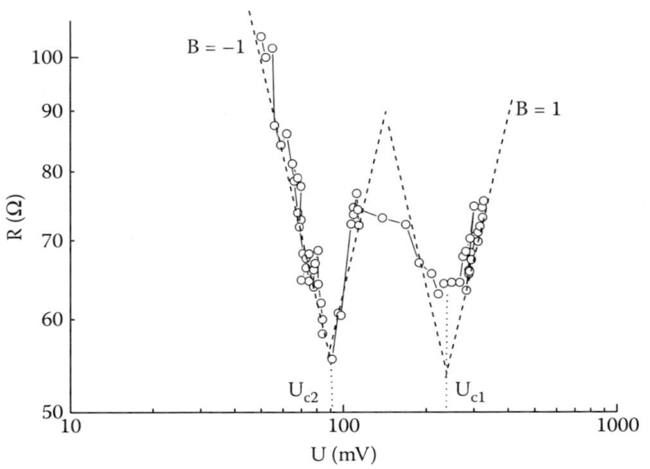

FIGURE 14.23 Generalized R(U) dependency for SWCNT sample.

tangles. In the conditions of our experiment, the flow threshold can be achieved in one of the two ways. The first one is to change the voltage of the control pulse. As the electrical field of the control pulse increases, then, thermal heating becomes sufficient to create the basic conductive grid. The last weakest link, which is the tangled nanotube self-intersection area of the basic grid, disappears since the corresponding voltage U_{c1} is reached (Figure 14.23). The second option is based on creating conditions for the sample conductivity relaxation mode, which is well known earlier as the high-resistance state recovery mode [22]. The low-resistance state should be fixed by the voltage control pulse and the second (modifying) pulse should be applied to "support" conductivity in the given timeframe. In our case, it happens within an area of lower electrical fields corresponding to the modification of an electrical field. As the modifying pulse electrical field increases, then, thermal heating becomes sufficient to create an additional conductive grid. The last weakest link, which is the tangled nanotube self-intersection area of the new grid, disappears since the corresponding voltage U_{c2} is reached (Figure 14.23). Let us note that given an explanation can provide the role of various "tangled" grid chains during the percolation process. Analysis of the generalized dependency (Figure 14.23) has uncovered an interesting feature: the specific ratio of threshold energies $(U_{c1}/U_{c2})^2 \cdot (t_1/t_2) \sim 1$. This fact indicates that transition energy of the percolation mode, which creates a low-resistance state at threshold voltages, can achieve values of $(10^{-11} - 10^{-12})$ J, that is, it is virtually unchanged. An "energetic" criteria of this kind is well known in the switching of amorphous semiconductors. It states that Joule energy released during the delay should approach the critical value of

$$\int_0^t IU\,dt = 3 \cdot 10^{-11}\,\text{J}.$$

In our case, with no delay in switching, we obviously need to consider the time of low-resistance state creation. On the basis of general ideas of nanosecond-switching mechanism [12], the same energy absorbed by spatial structures during the switching also means the same percolation topology for different electrical fields. This would suggest the manifestation of scaling laws for nanostructures based on "tangled" environments.

Note that the increase in tangle size, or decrease in nanotube diameter, which are uniquely associated with growth in "entanglement," could lead to an increase in the volume of self-intersecting chains, and, as a consequence, to reduction in the effects of mold, degradation, and others that lead to an unstable operation of nanodevices.

14.4 SUMMARY

It seems likely that a new direction in studying the effects of electrical instability in macrosystems/tangles based on nanostructured objects of various topologies such as the structures of multiwalled and single-walled nanotubes, clusters, and systems of graphene planes is revealed in the course of performed investigations. During this chapter, the questions arise that, it is possible, will be interesting for experimenters:

1. What is the manner in which the avalanche formation of the electric flux occurs in granulated nanostructures with S type of electrical instabilities? (A possible analogy is the magnetic flux in artificial Josephson junctions.)
2. What are the nucleation mechanisms of the avalanches for nanosecond time intervals and avalanche formation in active nanosized sand heaps, for example, at electrical and magnetic external actions?

Note that using an array of nonoriented graphene planes would be an ideal SOC study. Currently, we are not aware of the creation or study of such an array. We hope that our investigations and their further development will allow a new direction to appear: SOC of 2D objects, graphene, and graphene-like structures.

REFERENCES

1. Miller, M., Owens, F.J. 2013. Defect induced distortion of armchair and zigzag graphene and boron nitride nanoribbons. *Chem. Phys. Lett.*, 570: 42–45.
2. Bak, P., Tang, C., Wiesenfeld, K. 1987. Self-organized criticality: An explanation of the 1/f noise. *Phys. Rev. Lett.*, 59: 381–384.
3. Ginzburg, S.L., Savitskaya, N.E. 2002. Granular superconductors and a sandpile model with intrinsic spatial randomness. *Phys. Rev. E*, 66: 026128.
4. Liying, J., Li, Z., Xinran, W. et al. 2009. Narrow graphene nanoribbons from carbon nanotubes. *Nature*, 458: 877–880.
5. Memon, N.K., Kear, B.H., Tse, S.D. 2013. Transition between graphene-film and carbon-nanotube growth on nickel alloys in open-atmosphere flame synthesis. *Chem. Phys. Lett.*, 570: 90–94.
6. Kosynkin, D.V., Higginbotham, A.L., Sinitskii, A., Lomeda, J.R., Dimiev, A., Katherine Price, B., Tour James, M. 2009. Longitudinal unzipping of carbon nanotubes to form graphene nanoribbons. *Nature*, 458: 872–876.
7. Huang, J.Y., Chen, S., Jo, S.H. et al. 2005. Kink formation and motion in carbon nanotubes. *Phys. Rev. Lett.*, 94: 236802-1-4.
8. Shklovskii, B.I., Efros, A.L. 1984. *Electronic Properties of Doped Semiconductors.* Springer (Title of the original Russian edition: Etekironntye svoistva leglrcvannykh poluprovodnlkov © by "Nauka" Publishing House, Moscow, 1979).
9. Prikhodko, A., Konkov, O., Terukov, E. et al. 2010. Nanosecond S-type electrical instability in carbon-nanotube–polymer composites. *Full. Nanot. Carb. Nanostruct.*, 19: 1–4.
10. Kon'kov, O.I., Prikhod'ko, A.V. 2010. Carbon nanotubes: Specific features of nanosecond current–voltage characteristics of an array of carbon nanotubes. *Proceedings of the 7th International Conference on Amorphous and Microcrystalline Semiconductors*: St. Petersburg [in Russian].
11. Prikhod'ko, A.V., Kon'kov, O.I. 2012. Percolation, self-organized criticality, and electrical instability in carbon nanostructures. *Phys. Solid State*, 54: 2325–2328.
12. Balevichius, S., Gruzhinskis, V., Poshkus, A. 1983. Model of "on"-state created during nanosecond electrical switching in amorphous semiconductor films. *Sov. Phys. Semicond.*, 16: 1934–1937.
13. Prikhod'ko, A.V., Kon'kov, O.I. 2012. Carbon nanostructures as an example of the self-organized criticality. *Phys. Solid State*, 54: 642–646.
14. Prikhod'ko, A.V., Kon'kov, O.I. 2013. Specific features of nanosecond current–voltage characteristics of an array of carbon nanotubes. *Tech. Phys. Lett.*, 39: 302–304.
15. Niemeyer, L., Pietronero, L., Wiesmann, H.J. 1984. Fractal dimension of dielectric breakdown. *Phys. Rev. Lett.*, 52: 1033–1036.
16. Pryor, R.W., Henisch, H.K. 1972. Nature of the on-state in chalcogenide glass threshold switches. *J. Non-Cryst. Solids*, 7: 181–186.
17. Tkachev, A.G. 2009. Equipment and technology of nanostructured carbon materials production. *Proceedings of the NATO Advanced Research Workshop on Using Carbon Nanomaterials in Clean Energy Hydrogen Systems, NATO Science for Peace and Security Series*, Springer, September 22–28, 2007, Sudak, Crimea, Ukraine, 301–306.
18. Knerelman, E.I., Zvereva, G.I., Kislov, M.B. et al. 2010. Characterization of products on the base of single-walled carbon nanotubes by the method of nitrogen adsorption. *Nanotechnol. Russ.*, 5: 786–794.
19. Grosberg, A.Y., Khokhlov, A.R. 1994. *Statistical Physics of Macromolecules.* American Institute of Physics, New York.
20. Held, G.A., Solina, D.H., Keane, D.T. et al. 1990. Experimental study of critical-mass fluctuations in an evolving sandpile. *Phys. Rev. Lett.*, 65: 1120–1123.
21. Jantsch, W., Heinrich, H.R. 1970. A method for subnanosecond pulse measurements of I–V characteristics. *Sci. Instr.*, 41: 228–230.
22. Sandomirskij, V.B., Suchanov, A.A. 1976. The phenomenon of electrical instability (switching) in amorphous semiconductors. *Zarubezhnaja Radioelektronika*, 9: 68–101 [in Russian].
23. Prikhod'ko, A.V., Chesnys, A.A., Bareikis, V.A. 1981. Investigation of microwave noise in amorphous selenium films under monostable switching conditions. *Sov. Phys. Semicond.*, 15: 303–306.
24. Prikhod'ko, A.V., Deksnys, A.P., Chesnys, A.A. 1979. Nanosecond switching in selenium. *Sov. Phys. Semicond.*, 13: 111–112.
25. Kirkpatrick, S. 1973. Percolation and conduction. *Rev. Mod. Phys.*, 45: 574–588.
26. Bagnich, S.A., Konash, F.D. 2001. The influence of inhomogeneous properties of a system on the percolation process in two-dimensional space. *Phys. Solid State*, 43: 2313–2320.

15 Effects of the Interaction of Transition Metals on the Electronic Properties of Graphene Nanosheets and Nanoribbons

Sefer Bora Lisesivdin, Beyza Sarikavak-Lisesivdin, and Ekmel Ozbay

CONTENTS

ABSTRACT

One of the most interesting elements in the periodic table is carbon. Beside diamond and graphite, carbon has other allotropes such as a nanotube and nanoribbons. After its experimental proof, graphene has gained in importance and is studied intensely due to its unique electronic properties. To use graphene in electronics, various kinds of graphene are required. To modify electronic properties, doping is one of the most used methods. There are many theoretical studies on this subject, with both n-type and p-type behaviors observed. For the synthesis of graphene, growth on transition metals is gaining in importance due to the quality and cost of ownership. Therefore, interaction between graphene and transition metals is also important. In this chapter, we compile the important findings on the effects of doping with transition metals on the electronic properties of graphene nanosheets and the effects of both doping and termination with transition metals on the electronic properties of graphene nanoribbons.

15.1 INTRODUCTION

Carbon nanostructures have provided new research areas in highly interesting nanoelectronics [1]. Carbon nanostructures, which include carbon nanotubes, nanowires, and fullerenes, have been at the forefront material-based research for nearly the last two decades. Graphene—a single layer of C atoms arranged in a honeycomb lattice—is known to be a two-dimensional (2D) basic material to form these carbon-based nanostructures and graphite. In addition to being a basic building block of these nanostructures, graphene has also attracted attention since 1947 due to its interesting electronic structure [2,3]. After the synthesis of stable graphene in 2004 [4], unique mechanical and electronic properties—that is, massless Dirac electrons at K-point [5], high mobility even at high temperatures [6], and anomalous quantum Hall effect [5]—there is a tremendous increase in both experimental and theoretical research on this material. These unique features of graphene are mostly caused by its honeycomb lattice structure. Two π bands intersecting linearly at K-point, conclude that the electronic excitations can be described in terms of relativistic Dirac fermions [7]. The Fermi level is normally located at these intersection points called Dirac points. The "mass-less" Dirac fermions near these Dirac points can coherently propagate very large distances that allow graphene to maintain large currents [8]. In addition, controlling the carrier density or the type—electron or hole—of these carriers can be easily done with a simple gating structure, which means endless possibilities of applications [9].

With many application possibilities, the most significant problem about graphene research is how to make high-quality large graphene samples. Several fabrication methods have been reported as thermal decomposition on the (0001) surface of 6H-SiC, using polystyrene–graphene composites and chemical vapor deposition (CVD) on transition metals (TMs) [10–12]. These new fabrication methods, especially the latter one, have successfully resulted in the growth of graphene sheets. By getting high-quality, larger sheets, research is now directed to the device studies that can be done with getting the quasi-one-dimensional (q1D) form, graphene nanoribbons (GNRs), or quasi-zero-dimensional (q0D) form graphene nanodots (GNDs) [13,14]. With a tunable bandgap, these quantum structures are used to fabricate transistors [15,16], sensors [17], p–n junctions [18], spin-valve devices [19], memory devices [20], hydrogen storage [21], and other new device possibilities [22,23].

To functionalize GNR and GND devices, proper doping, adsorption, and termination is needed [24]. Therefore, it is necessary to understand and control the interaction of the atoms or molecules which are used for doping, adsorption or termination, and graphene. This topic has a fundamental importance that is caused by the nature of graphene. By modifying the intrinsic properties of GNR and GND devices, with doping, adsorption, or termination, useful new behaviors can be found or the limits of these devices can be conquered.

As mentioned above, graphene growth on TMs has gained in importance due to the simplicity in finding the proper TM substrate and the ability in gaining high-quality large-area graphene sheets [12,25,26]. The TM atoms with their large sizes can be incorporated into the atomic network as an impurity as well [27], or can be added as an adatom [28]. The interaction of TMs with graphene has received much attention due to the possible usage in applications such as spintronics [29,30], and hydrogen storage [31] in addition to the applications mentioned previously.

The interaction of TMs with graphene and the effect of TM on the electronic properties of graphene are mostly focused on doping with TMs, adsorption of TM adatoms, graphene growth on TM foils, and nanoribbon studies. Most of the studies are theoretical—done with the help of the density functional theory (DFT)—due to a lack of experimental studies. Studies of nanoribbon structures with TM doping, adsorption, or termination seem to be at their first stage. There are a few important experimental findings on the effect of TM interaction on the electronic band structure of graphene using the state-of-the-art angle-resolved photoemission spectroscopy (ARPES) method.

The rest of the chapter is organized as follows: in the next section, we briefly discuss the important findings and parameters of both theoretical and experimental studies of graphene nanosheets. For nanoribbons, we again discuss the important findings in the literature and then summarize some results for the adsorption of period 3d TMs on selected zigzag and armchair GNRs, and draw conclusions.

15.2 INTERACTION OF TMs WITH GRAPHENE NANOSHEETS

With the inspirational experimental study of N'Diaye et al. [32], every type of study about the interactions of TMs with graphene nanosheets have emerged. There are many theoretical studies that are related to the effects of doping, adsorbing of TMs, or growing on TMs. In this chapter, we briefly present the most important milestones of these studies.

15.2.1 Elemental TM Doping

In Figure 15.1a through c, DFT band calculation results for pure graphene that are excessively doped with Au and Ru atoms are shown, as an example for an elemental TM-doping case. In calculation, $9 \times 9 \times 1$ k-point sampling and cutoff

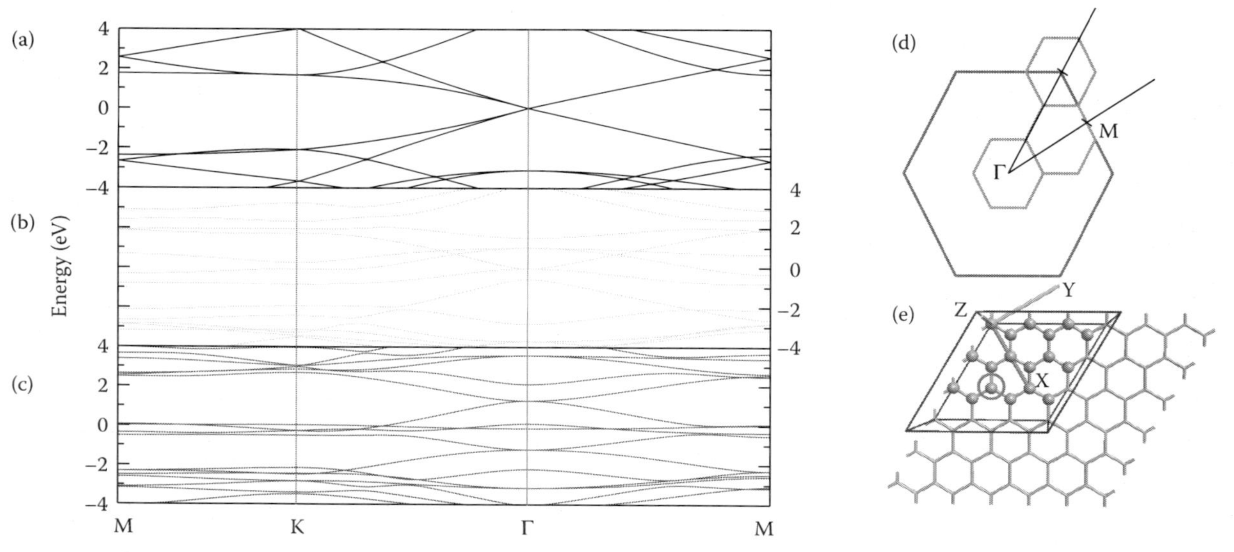

FIGURE 15.1 Band structures of 3×3 (a) pure, (b) Au adsorbed, and (c) Ru-adsorbed graphene nanosheets. (d) Brillouin zone of 1×1 and 3×3 graphene. (e) 3×3 supercell of graphene. The atom in the circle shown in unit cell is used for excessive doping of Au and Ru.

energy of 500 eV is used. Other details about the calculation method are described in Section 15.3.1. For the calculations, a 3×3 supercell is chosen as shown in Figure 15.1e. Because of the K-point in the 1×1 Brillouin zone, it is folded into the Γ-point in the 3×3 supercell Brillouin zone, and the Dirac point is observed at the Γ-point in Figure 15.1a. With the excessive doping of Au, Dirac cones are not observed and instead of linear dispersion, classic parabolic dispersion lines are observed at the Γ-point as shown in Figure 15.1b. With the excessive doping of Ru—even a no-doping atom interacts with other doping atoms—the Dirac point is completely destroyed because of the strong hybridization of graphene's p_z states and the doped TM d-states. Therefore, strong interactions, which can alter and totally destroy graphene's unique electronic band structure, are observed with the elemental doping of TM atoms.

In general, due to the structural stability of the incorporation of TM impurities into the existing vacancies of carbon nanostructures, they are a common phenomenon [33,34]. With a Ni-containing catalysis process, Ushiro et al. [33] showed important amounts of substitutional Ni impurities in carbon. And with TM impurity, these carbon nanostructures and especially graphene nanosheets are showing important electronic structure changes related to gaining a magnetic nature, which can be used in spintronic applications [35], superconductivity [36], or hydrogen storage [37]. With the knowledge of these applications, there are plenty of studies about the doping of graphene nanosheets with TM impurities [28,31,35,37–43].

In one of the early studies, Santos et al. [35] showed that an Ni impurity presents a strong covalent interaction, which stabilizes the 3d orbitals of Ni as a fully occupied shell, with the neighboring carbon atoms in the graphene lattice. Therefore, Ni impurities cannot induce any magnetic moment in the graphene nanosheets [35].

According to Pi et al. [38], the relative work function between the graphene nanosheets and the TM is a crucial parameter to investigate and understand the charge transfer between the graphene and the TM. In their study, they investigated the Ti, Fe, and Pt metals experimentally and found the work functions 4.3, 4.7, and 5.9 eV, respectively [44]. These values are quite compatible with graphene's own work function [45]. As important observations, they found that the doping efficiency is related to a work function at low concentrations. At these concentrations, TM clusters are found to show a different behavior compared to Coulomb scattering. All these metals exhibit an n-type behavior, which is consistent with the theoretical predictions [46]; only Pt-doping can cause p-type behavior with changing the concentration.

Boukhvalov and Katsnelson [39] showed that TMs can cause not only aimed band structure tailoring, but also the destruction of graphene. In their study, they investigated the effect of Fe, Ni, Au, and Co metals with the spin-included DFT method on vacancy formation energies.

$$E_{\text{vacancy}} = E_{\text{graphene+vacancy+TM}} - E_{\text{graphene+TM}} - E_{\text{carbon}}, \quad (15.1)$$

where E_{carbon} is the total energy of the single-carbon atom. The calculation results showed that Fe, Ni, and especially Co drastically reduced the E_{vacancy} values for monovacancies and divacancies. However, Au has almost no effect on the vacancy behavior [39]. Most importantly, with its minimal vacancy formation energy for all cases, Co strongly destabilized the graphene lattice and, therefore can be used to slice graphene sheets into the intended shapes such as GNRs and GNDs. Zan et al. [47] also studied the interaction of Au with graphene and could not find any important interaction between them. In their experimental method—like other experimental observations [48]—they observed Au clusters, which is also an experimental proof of the Au–Au interaction and is much stronger than the Au–graphene interaction.

Another study on the effects of vacancy defects in graphene on TM anchoring was carried out by Kim et al. [37]. With the help of DFT calculations, they calculated cohesive energy per atom and binding energies of the studied Sc, Ti, and V atoms on pristine or defective graphene. They showed that the binding energies of TMs are significantly smaller than their cohesive energy, which results in clustering instead of single-atom covering. However, graphene with vacancies is more likely to bind single atoms, which can be studied as doping. In addition, according to the study, alkaline earth metals are more likely to hold more hydrogen per atom than TM atoms, which is an important result for studies of a suitable catalyst search for hydrogen storage [37].

Bhattacharya et al. [31] also investigated the room- temperature hydrogen storage of graphene with vacancies, which is doped at these vacancy sites with Sc, Fe, Co, Ti, V, and Mn atoms. They investigated two different 3d TM atoms that are connected on either side of a vacancy site with the DFT method. For hydrogen storage, the desorption of the stored hydrogen is also very important [49]. From the desorption point of view, they reported Ti/Fe on either side of a vacancy site of a graphene structure as the best one. They reported that desorption begins at 400 K, and at 600 K, most of the hydrogen was released without any catastrophic effects on the Ti-/Fe-decorated graphene structure. They calculated a hydrogen storage of ~5.1 wt% at room temperature, which is comparable with the previously reported other carbon-based nanostructures [49]. Choi et al. [42] also investigated the hydrogen storage of graphene with vacancies, which is doped at these vacancy sites with Sc, Ti, V, Cr, Mn, Fe, Co, Ni, Cu, and Zn TM atoms.

TMs can be used not only for hydrogen storage, but also as a catalyst for CO oxidation [40]. CO oxidation with a barrier of 0.5 eV is expected to occur at room temperature. According to Lu et al.'s [40] DFT study, with Au-embedded graphene, this value is as low as 0.31 eV and the reaction is likely to proceed at room temperature.

Lim and Wilcox [43] studied the structural and the electronic properties of O_2 adsorption of—not the single atom but—the Pt_{13} clusters on a monovacancy of graphene with the DFT method. They found the monovacancy site of graphene to provide minimal Pt_{13} aggregation on the graphene surface. Their results on O_2 adsorption show similar results of O_2-adsorbed Pt(111) bulk layers [50].

Krasheninnikov et al. [41] studied the interaction of many TMs including Sc, Ti, V, Cr, Mn, Fe, Ni, Co, Au, Pt, Cu, and Zn with the pristine graphene and graphene sheets with monovacancies and divacancies. They also found that nearly all TMs are strongly bound to defective graphene. Also, different defects with different TMs play a role in having different magnetic moments. For example, for V, Cr, and Mn, which have single-filled d-states, bonded to a monovacancy, magnetic properties can be observed. However, with a double-occupied d-state, TMs bonded to a monovacancy can show both magnetic and nonmagnetic behaviors. Elements with an even and odd number of electrons with a double-occupied d-state show nonmagnetic and magnetic behavior, respectively. They also found that, from V to Co, nearly all TM-decorated divacancy structures show magnetic behaviors [41]. They calculated the activation barriers, which is the energy that is required to migrate the TM atom on a monovacancy by rotating one of the TM–graphene bonds, for Au, Pt, Fe, Co, and Ni as 2.1, 3.1, 3.6, 3.2, and 3.1 eV, respectively. And for a divacancy, the value is nearly 5 eV. These values are quite in agreement with the available experimental and theoretical values [48,51]. Similarly, Krasheninnikov and Nieminen [28] also studied W, Cr, and Mo TMs with additional defect structures, which are dominant in a wide temperature range, in addition to monovacancies and divacancies, and they found that these defects also behave like strong trapping centers for TMs.

15.2.2 Elemental TM Adsorption

Because TMs adsorbed on carbon nanotubes have been experimentally shown to form a variety of structures such as clusters and coatings [52], the effects of graphene's TM adsorption on the band structure is also a subject that has gained a lot of attention [53–66]. In addition to TM adsorption on carbon nanotube studies [67,68], one of the first studies about electronic band structures for TMs adsorption on graphene is reported by Chan et al. [53] for TM elements Ti, Pd, Fe, and Au. For the three possible adsorption sites, hollow (H), bridge (B), and top (T), they calculated the preferable site binding energy of the adsorbed atom and some geometrical parameters [53]. In Figure 15.2, three possible states for an adsorption process are shown.

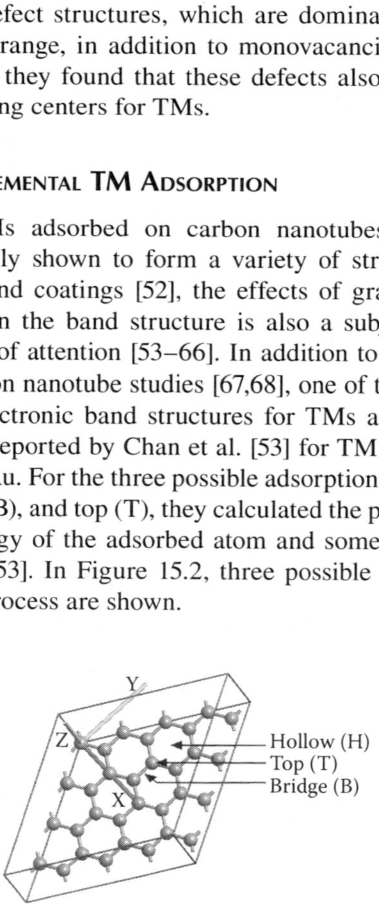

FIGURE 15.2 Three possible adsorption sites; hollow, top, and bridge.

Binding energy of the adsorbed atom is calculated as

$$E_b = E_{\text{graphene+adatom}} - E_{\text{adatom}} - E_{\text{graphene}} \tag{15.2}$$

Here, $E_{\text{graphene+adatom}}$, E_{adatom}, and E_{graphene} are the total energies of graphene with an adatom, adatom itself, and graphene itself, respectively. In addition to their findings, the binding energies of several TMs, which are also listed in the literature, are listed in Table 15.1 [53,55,65,69–72].

According to Chan et al. [53], Ti has the largest (1.301 eV) and Au has the lowest (0.096 eV) binding energy values. Ti, Fe, and Pd are covalently bonded to graphene. Because the covalent bond is directional, the adsorption site is also direction dependent. However, because Au has a single s valance electron, its adsorption energy is small. Instead of covalent bonding, the van der Walls (vdW) forces are mostly involved in the bonding process. They also reported significant differences in the band structure of graphene due to the adsorption of TMs. Owing to the strong hybridization of Ti 3d and the graphene states, the Dirac point is found to be no longer evident. With spin-dependent DFT calculations, they observed a spin-dependent charge transfer from Ti to graphene, which also explains the reduction in the calculated magnetic moment from the isolated TM atom to the adsorbed TM atom. The reduction in the calculated magnetic moment is the largest in Fe, which is dropped from 4.00 μ_B to 2.03 μ_B [53]. In addition to Chan et al.'s study, magnetic moments related with adsorbed TM atom systems are listed in Table 15.2 as a broad review [41,53,55,57,64,72–75].

Uchoa et al. [54] studied the effects of Pd coating on the electronic band structure of graphene. Because of its closed shell of configuration of 4d orbital, elemental Pd is not a magnetic material. However, in bulk, with s and p band hybridization, Pd exhibits strong Pauli paramagnetism, and even ferromagnetism, if the Stoner criterion is satisfied [76]. In their study, Uchoa et al. [54] showed a strong enhancement of the density of states (DOS) at the Fermi level and a strongly polarized Pd bands through Stoner criterion. Therefore, it shows the possibility of inducing a ferromagnetic instability in graphene, which can be used to build spintronic devices, with coating TMs.

Wehling et al. [55] investigated the effects on the electronic band structure of the Cr, Mn, Fe, Co, and Ni atom-adsorbed graphene that is affected by local Coulomb interactions. They found that high-spin configurations are favorable for Cr and Mn adatoms independently when generalized gradient approximation (GGA), GGA + U, or local density approximation (LDA) functionals are used. In addition to magnetic properties, they also studied the hydrogenated Co and Ni adatoms and found the hydrogen-binding energies with GGA + U functional as 2.7 and 3.0 eV for Ni and Co, respectively [55].

In addition to doping studies and Wehling et al.'s study, Zhou et al. [56] also investigated the hydrogen storage capability of TM-decorated graphene. For their studies, they chose Ti. With the applied strain to the system, they showed that the adsorption energies, magnetic moments, and the total hydrogen storage capacity are highly altered. According to

TABLE 15.1
Some Binding Energies for Adsorbed TMs on Graphene

Atom	Bonding Site	Binding Energy (eV)	Notes
		3d TMs	
^{21}Sc	H	0.792 [61]	
		1.59 [69]	
^{22}Ti	H	2.22 [61]	
		1.22 [69]	
^{23}V	H	1.127 [61]	
	T, B, and H	1.301, 1.301, and 0.748 [53]	
^{24}Cr	T	0.34, 0.49 [55]	In Reference 55, first values are calculated with GGA + U, and seconds are calculated with GGA functionals.
	B	0.34, 0.50 [55]	
	H	0.33, 0.47 [55]	
	H	0.334 [61]	
^{25}Mn	T	0.30, 0.35 [55]	In Reference 55, first values are calculated with GGA + U, and seconds are calculated with GGA functionals.
	B	0.28, 0.34 [55]	
	H	0.28, 0.42 [55]	
^{26}Fe	T	0.29, 0.24 [55]	In Reference 55, first values are calculated with GGA + U, and seconds are calculated with GGA functionals.
	B	0.25, 0.2 [55]	
	H	0.27, 0.71 [55]	
	T, B, and H	0.149, 0.231, and 0.852 [53]	
^{27}Co		0.030 [70]	vdW-DF
		0.175 [70]	LDA
	T	0.62, 0.75 [55]	In Reference 55, first values are calculated with GGA + U, and seconds are calculated with GGA functionals.
	B	0.56, 0.88 [55]	
	H	0.41, 1.35 [55]	
^{28}Ni		0.037 [70]	vdW-DF
		0.123 [70]	LDA
	T	0.74, 1.44 [55]	In Reference 55, first values are calculated with GGA + U, seconds are calculated with GGA functionals.
	B	0.82, 1.52 [55]	
	H	0.97, 1.72 [55]	
^{29}Cu		0.038 [70]	vdW-DF
		0.035 [70]	LDA
^{30}Zn		1.07 [71]	
		4d TMs	
^{44}Ru			
^{45}Rh		2.02 [65]	
^{46}Pd		0.039 [70]	
		0.079 [70]	vdW-DF
		1.077 [70]	LDA
	T	1.044 [53]	
	B	1.081 [53]	
	H	0.085 [53]	
^{47}Ag		0.033 [70]	vdW-DF
		0.045 [70]	LDA
		5d TMs	
^{78}Pt		0.043 [70]	vdW-DF
		0.033 [70]	LDA
		1.552 [72]	

(Continued)

TABLE 15.1 (*Continued*)
Some Binding Energies for Adsorbed TMs on Graphene

Atom	Bonding Site	Binding Energy (eV)	Notes
^{79}Au		0.038 [70]	vdW-DF
		0.031 [70]	LDA
		0.075 [72]	
	T	0.096 [53]	
	B	0.089 [53]	

Source: Chan, K. T., J. B. Neaton, and M. L. Cohen. 2008. *Phys. Rev. B* 77(23):235430 (12pp); Wehling, T. O., A. I. Lichtenstein, and M. I. Katsnelson. 2011. *Phys. Rev. B* 84(23):235110 (7pp); Zhou, Q. et al. 2014. *Phys. Rev. Lett.* 110(13):136804; Yazyev, O. V., and A. Pasquarello. 2010. *Phys. Rev. B* 82(4):045407(5pp); Durgun, E., S. Ciraci, and T. Yildirim. 2008. *Phys. Rev. B* 77(8):085405 (9pp); Vanin, M. et al. *Phys. Rev. B* 81(8):081408 (4pp); With kind permission from Springer Science+Business Media: *Graphene Nanoelectronics: Metrology, Synthesis, Properties and Applications*, 2011, Raza, H., Berlin; Tang, Y., Z. Yang, and X. Dai. 2011. *J. Chem. Phys.* 135(22):224704 (7pp).

their results, the adsorption energy of the Ti atom increased as much as 71% with an applied strain of 10%. Meanwhile, the magnetic moment of the system is reduced from 3.4 μ_B to 2.3 μ_B [56]. For the strained structure, the resulting hydrogen storage capacity is calculated as 9.5 wt%, which is higher than the previously calculated values for the systems without strain [49].

Because of their known magnetic properties, Mao et al. [57] investigated the magnetic moments of the adsorbed Mn, Fe, and Co on graphene. They also estimated the bond lengths of the adsorbed atoms. They found the largest magnetic moment (5.62 μ_B), as well as the largest bond (2.52 Å) for Mn. They found the H site to be the most stable site for Mn, Fe, and Co. Unlike the DFT studies on graphene grown on TM substrates—because the adsorbed atom bond length is short—they suggested that the interaction between the TMs and graphene is very far away from a vdW dominant system. Therefore, they pointed out that the equilibrium is mostly determined by the charge redistribution that is a Coulomb interaction [57]. According to spin-polarized band calculations, the conduction band of graphene is highly modified by the coupling between Mn and graphene. Fe- and Co-adsorbed graphene structures are found to represent semi-half-metallic and metallic behaviors, respectively [57].

In their study, Ishii et al. [58] also investigated the adsorption sites with DFT, and suggested two types of adsorption with a migration barrier energy of less than and greater than 0.4 eV. This energy value roughly corresponds to the threshold energy for atomic migration at room temperature. Therefore, according to their study, TMs Sc, Ti, Ni, Cu, Zn, Pd, Ag, Pt, and Au can migrate on other sites even at room temperature. However, TMs V, Cr, Mn, Fe, Co, Mo, and Ru have large migration energies, and so are more stable at room temperature. For the TMs with a large migration barrier energy, the

TABLE 15.2
Some Local and Total Magnetic Moments for TM-Adsorbed Graphene Systems

Atom	Bonding Site	Magnetic Moment (μ_B)	Notes
		3d TMs	
^{21}Sc		2.35 [73]	
		0.0 [74]	Local magnetic moment
^{22}Ti		~3.3 [53,73]	
		0.0 [74]	Local magnetic moment
^{23}V		4.88 [73]	
		1.0 [41,74]	Local magnetic moment
^{24}Cr		6.0 [73]	
		2.0 [41,74]	Local magnetic moment
	T	5.9, 5.9 [55]	In Reference 55, first values are
	B	5.9, 5.8 [55]	calculated with GGA + U, and
	H	5.9, 5.6 [55]	seconds are calculated with GGA functionals.
^{25}Mn		5.62 [57]	
		3.0 [41,74]	Local magnetic moment
	T	5.0, 5.1 [55]	In Reference 55, first values are
	B	5.0, 5.1 [55]	calculated with GGA + U, and
	H	5.0, 5.2 [55]	seconds are calculated with GGA functionals.
^{26}Fe		~2.1 [53,73]	
		2.21 [61]	
		0.0 [74]	Local magnetic moment
	T	3.9, 4.1 [55]	In Reference 55, first values are
	B	3.9, 4.1 [55]	calculated with GGA + U, and
	H	3.6, 2.0 [55]	seconds are calculated with GGA functionals.
^{27}Co		1.1 [57]	
		1.4 [73]	
		1.19 [64]	
		1.0 [41,74]	Local magnetic moment
	T	2.9, 1.1 [55]	In Reference 55, first values are
	B	2.9, 1.1 [55]	calculated with GGA + U,
	H	2.6, 1.1 [55]	seconds are calculated with GGA functionals.
^{28}Ni	T, B,	0.0 [73]	Local magnetic moment
	and H	0.0 [55,64]	In Reference 55, results are zero for both calculations that used GGA + U and GGA functionals.
^{29}Cu		1.0 [73,74]	Local magnetic moment
		0.84 [64]	
^{30}Zn		0.0 [73,74]	Local magnetic moment
		4d TMs	
Atom	**Bonding Site**	**Binding Energy (eV)**	**Notes**
^{42}Mo		0.0 [73]	
^{46}Pd		0.0 [53,72,73]	
^{47}Ag		1.0 [73,74]	Local magnetic moment
		5d TMs	
^{78}Pt		0.0 [72–74]	Local magnetic moment
^{79}Au		−0.15 ~ −0.17 [75]	

(Continued)

TABLE 15.2 (*Continued*)
Some Local and Total Magnetic Moments for TM-Adsorbed Graphene Systems

Atom	Bonding Site	Binding Energy (eV)	Notes
		~1.0 [53,73]	
		0.4 [72]	

Source: Krasheninnikov, A. V. et al. 2009. *Phys. Rev. Lett.* 102(12):126807 (4pp); Chan, K. T., J. B. Neaton, and M. L. Cohen. 2008. *Phys. Rev. B* 77(23):235430 (12pp); Wehling, T. O., A. I. Lichtenstein, and M. I. Katsnelson. 2011. *Phys. Rev. B* 84(23):235110 (7pp); Mao, Y., J. Yuan, and J. Zhong. 2008. *J. Phys. Condens. Matter* 20:115209 (6pp); Cao, C. et al. *Phys. Rev. B* 81(20):205424 (9pp); Tang, Y., Z. Yang, and X. Dai. 2011. *J. Chem. Phys.* 135(22):224704 (7pp); Hu, L. et al. 2010. *Physica B: Cond. Mat.* 405(16):3337–3341; Santos, E. J. G., A. Ayuela, and D. Sánchez-Portal. 2010. *New J. Phys.* 12(5):053012 (31pp); Varns, R., and P. Strange. 2008. *J. Phys. Cond. Mat.* 20(22):225005 (8pp).

stable adsorption site is found to be H. Other TMs may show migration due to thermal energy [58].

Zolyomi et al. [59,60] reported a series of DFT calculations on the adsorption of 4d and 5d TMs on graphene. In their calculations, they found binding energies and magnetization for all 4d and 5d elements except Y, Cd, Lu, and Hg. Calculations for all H, T, and B sites and for low-coverage and high-coverage cases are also presented. In a high-coverage case, TMs interact and present a triangular lattice above the graphene sheet [59]. According to their results, Au and Ag bind weakly and Mo, Hf, Ta, and W bind strongly for low-coverage cases [59].

In their study, Zhou et al. [61] present a detailed study of the combined effects of TM metal adatoms (Sc, Ti, V, and Cr) and Stone–Walles-type defects. They found that adatom–graphene interaction is enhanced in the presence of Stone–Walles defects. Also, they showed that the Sc-adsorbed system exhibits metallicity while the Ti, V, and Cr exhibit a half-metallicity. Eelbo et al. [62] found top site adsorption for Co monomers in their DFT-supported experimental study and an absence of the local magnetic moment in the Ni adatom due to showing a configuration of d^{10}. Also, they pointed a transition from nonmagnetic Ni clusters toward magnetic ones when the Ni cluster size exceeds four atoms. In another study of Eelbo et al. [63], they studied the Co and Ni adatoms on two different (monolayer graphene and quasi-freestanding monolayer graphene) model types of graphene on SiC(0001). In the case of monolayer graphene, adatoms can be adsorbed only on hollow-type sites. However, for quasi-freestanding monolayer graphene, both hollow and top sites are found to be possible.

In addition to studies about the adsorptions of the single TM element, Cao et al. [64] studied the electronic structures and magnetic properties of both single adatoms and dimers adsorbed on graphene for the elements Fe, Co, Ni, and Cu. They found drastic changes in the band structures, in which even the concentration of the TM adsorption was very low (~1%). Fe- and Co-adsorbed systems represented a 0.2-eV-wide DOS opening for the majority spin, but remained continuous for the minority spin. With an unsymmetrical spin choice, these systems can be used as spin-filtering materials as they pointed out [77]. Because of strong binding in the Ni adsorption case, the system showed a small gap, which is caused by the destruction of graphene's delocalized p orbitals, of 0.1 eV at the Fermi level. For Cu, both spins represent a continuous DOS structure at the Fermi level. However, the minority spin is nearly zero and the majority spin has a peak at the Fermi level [64]. For dimer adsorption, they found magnetic moments of 2.99 μ_B and 1.49 μ_B for Fe and Co dimers adsorbed with the help of the spin included in LDA calculations, respectively. For Ni and Cu dimers, they found zero magnetic moments.

Before continuing with the electronic band structure studies about large-area graphene growth on TM substrates, it may be better to look at Yazyev and Pasquarello's study [65]. In their study, instead of using freestanding graphene and a TM adatom, they used a TM adatom-decorated graphene that is also grown on a TM substrate. In their study, they used a Co adatom and Ni(111) and Cu(111) substrates, which are strongly and weakly interacting substrates with monolayer graphene. Because of the moiré pattern, which will be explained in the next section, they determined three different adsorption points on the graphene. With a strongly interacting substrate such as Ni(111), they found that the adsorption energy of the Co adatom is highly dependent on the site places at the template that is induced by the moiré pattern [65].

Sutter et al.'s [66] study also investigated the interactions between the TM adatom/cluster on the graphene system and the underlying TM substrate. In their experimental study, which is also supported with intense calculations, they reported findings about the self-assembly of arrays of Ru clusters on monolayer graphene on Ru(0001). With the same TM being used as a substrate and the clusters formed, the symmetrical geometries and similar interaction strengths on both the graphene sheets are maintained. They found a strong coupling between the moiré pattern and the Ru nucleation on graphene, which is consistent with Yazyev and Pasquarello's study [65,66].

15.2.3 Theoretical Electronic Band Structure Studies about Large-Area Graphene Growth on TM Foils

A strong affinity between carbon, or carbon-containing molecules, and the Pt surface has been a known phenomenon since the 1960s [78]. The deposition of carbon, which can be observed with the low-energy electron diffraction (LEED) method, generated by flowing ethylene onto the hot single-crystal Pt, causes the formation of graphitic layers on the substrate [79]. With this knowledge and Novoselov et al.'s experimental proof of 2D carbon structure, graphene growths on TMs have attracted an amount of interest in both an experimental and theoretical way [26,70,80–84].

Effects of the underlying TM substrate on the electronic band structure of the graphene layer can be observed with an ARPES experiment [26,80–82]. Before continuing with ARPES, we will look at some theoretical studies about the band structure of graphene that are grown on TM substrate [70,83,84].

On the theory side, before 2009, the problem was that DFT studies could only poorly describe the vdW forces. However, these forces play a very important role for graphene/TM interaction. In 2010, Vanin et al. [70] presented a graphene on metal study, which uses DFT with a relatively new van der Walls density functional (vdW-DF). They studied the adsorption of graphene in various TM substrates such as Co, Ni, Pd, Ag, Au, Cu, Pt, and Au. For the Ag, Au, Cu, Pt, and Al substrates, density-functional LDA and vdW-DF showed a weak binding with the graphene layer. For Ni, Co, and Pd, LDA showed a strong binding, which is due to hybridization between the d-states of related TM and graphene. However, vdW-DF showed a weak binding like other TM elements, which appears to conflict with experimental observations.

It is also very important to understand the role of the atomic step edges in a CVD growth of graphene on TM substrates for growing large-area sheets experimentally. In their study, Feng et al. [85] investigated the effect of Au alloying, which is known to decrease the growth temperature, on the stepped Ni(111) surface. They found that the step edges are the most preferable sites for C adsorption. Therefore, the density of step edges and total morphology of the surface highly affect the quality of the grown graphene film. In Au alloying, they found that Au atoms prefer to substitute the Ni atoms at the step edges. This makes the growth of graphene on the stepped surface more uniform [85].

In many situations, the modification of the graphene morphology due to the presence of the substrate can be observed. One of the cases, which has attracted attention, is the moiré pattern appearance in graphene that is grown on Ru(0001) substrate [86]. This subject is intensively investigated in many DFT studies [87–89]. In their studies, Stradi et al. investigated the graphene-on-TM interactions with the help of the functional DFT + D/PBE, like the vdW-DF of the previous study of Vanin et al. [70,83]. However, instead of using a smaller unit cell, they used a unit cell, which was large enough to account for the moiré pattern. With the help of a larger unit cell and including the vdW interaction, they found that the vdW interactions are responsible for the reduction in the corrugation of the moiré pattern, which is in very good agreement with that observed in scanning tunneling microscopy (STM) experiments.

In their studies, Peng and Ahuja [84] also used the big unit cell that includes nearly fitted graphene(12×12) and Ru(11×11) periodic structures with a total of 651 atoms. They found that the graphene monolayer on a Ru substrate is nearly flat rather than strongly corrugated. The moiré structure is present in graphene structures with more than one layer. Because the second layer has a higher carbon density, a small corrugation occurs [84]. With this structure, the electronic states of the first layer showed a 1 eV of shifting down, and the upper layers represented a free graphene-like behavior [81]. In addition to their findings, the Fermi-level shifting related to the band structures of graphene on TMs are listed in Table 15.3 [70,84,90].

TABLE 15.3
Band Structure Shiftings in Graphene on TM Systems

Atom	Fermi-Level Shift (eV)	Notes
	3d TMs	
^{27}Co	−0.20 [70]	
^{28}Ni	0.13 [70]	
^{29}Cu	−0.17 [90]	
	−0.43 [70]	
	4d TMs	
^{44}Ru	~1.0 [84]	First layer
^{46}Pd	0.65 [70]	
^{47}Ag	−0.32 [90]	
	−0.40 [70]	
	5d TMs	
^{78}Pt	0.33 [90]	
	0.66 [70]	
^{79}Au	0.19 [90]	
	0.21 [70]	

Source: Vanin, M. et al. 2010. *Phys. Rev. B* 81(8):081408 (4pp); Peng, X., and R. Ahuja. 2010. *Phys. Rev. B* 82(4): 045425 (5pp); Khomyakov, P. A. et al. 2009. *Phys. Rev. B* 79(19):195425 (12pp).

15.2.4 ARPES Studies

The single-particle spectral function $A(\vec{k},\omega)$ includes the effect of Coulomb and phonon interactions on the energy band properties of an investigated crystal [91]. The ARPES experiment, which measures the single-particle properties directly, is a powerful tool for probing $A(\vec{k},\omega)$ in 2D crystals because it can map momentum (k) space [92]. The quantity determined in ARPES experiments is the single-particle spectral function $A(\vec{k},\omega)$, which depends on the self-energy as follows:

$$A(\vec{k},\omega) = -\frac{1}{\pi}\frac{\sum_2(\vec{k},\omega)}{\left(\omega-\varepsilon_k-\sum_1(\vec{k},\omega)\right)^2+\left(\sum_2(\vec{k},\omega)\right)^2}$$

where ε_k, $\Sigma_1(\vec{k},\omega)$, and $\Sigma_2(\vec{k},\omega)$ represent the bare band structure, and the real and imaginary parts of the self-energy function $\Sigma(\vec{k},\omega)$, respectively.

In Figure 15.3, the experimental layout of the ARPES system is shown [93]. Because ARPES gives information on the direction, speed, and scattering of valence electrons in a sample, both the energy and momentum of an electron can be calculated easily. And with energy and momentum information, detailed information on band dispersion and, therefore, momentum space mapping can be done. Therefore, the ARPES method is also used to understand the nature of the interaction between TM substrates and on-grown graphene [26,80–82,94]. In an earlier study, Wintterlin and Bocquet [95] presented a detailed overview of results of experimental-based studies that appeared in June 2008. In their study, they presented a detailed reference list on graphene-on-TM experimental growths, which includes Co(0001), Ni(111), Ni(100), Ru(0001), Rh(111), Rh(100), Pd(111), Pd(100), Ir(111), Pt(111), Pt(100), and Pt(110) surfaces [95 references in 8–12,22–25,29–39,41,44–52,55–57,59,64,67,68,71,73–78,82–86]. Also, in their review, they analyzed the ARPES studies for graphene grown on Ni(100), Ni(111), Ru(0001), and Ir(111) that appeared in June 2008 [95 references in 31,36,37,39–41,71].

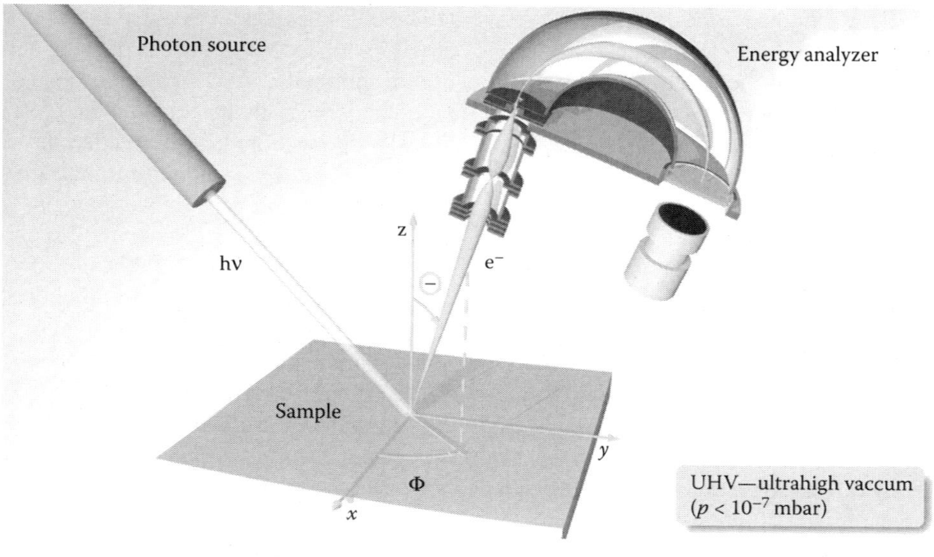

FIGURE 15.3 Experimental setup for the ARPES method. (Adapted from public domain image taken from Wikipedia http://en.wikipedia.org/wiki/Angle-resolved_photoemission_spectroscopy.)

For ARPES studies of graphene, the Brookhaven National Laboratory comes in the first place [26,80–82]. In Sutter et al.'s [26] study that was published in *Applied Physics Letters* in 2009, they found a strong coupling between the single-graphene layer and the underlying Ru substrate. However, in more than one graphene layer, this interaction seemed to be screened by the first graphene layer. Therefore, in bilayer or more-layer graphene sheets, above-graphene layers show the features of freestanding graphene. With micro-ARPES measurements, they observed linear band dispersion near the K-point for a bilayer graphene sheet with a Dirac point 0.5 eV below the Fermi level. This behavior confirms that the top sheet of the investigated graphene sheet is to be decoupled and to leave only the π-states of an isolated top graphene layer near the Fermi level at the K-point [26]. These findings were also explained in detail with new measurements in a study that was published in *Nano Letters* in 2009 [80]. Therefore, with these findings, they show a few-layer graphene structure, which is grown on a Ru substrate, that the outermost layers could be screened from the d-states of the Ru substrate by the first layer of graphene. Therefore, they showed the possibility of growing structures on Ru substrates where they are recovered, required for spin-filtering applications or high-mobility applications [96]. However, for weakly interacting Ir and Pt substrates, Pletikosic et al. and Sutter et al. showed a possibility to observe the massless Dirac fermions also in the first monolayer, respectively [81,97]. To prevent the strong effects of the Ru substrate, Enderlein et al. [94] suggested the use of the intercalation of a single monolayer of an Au layer between the single-graphene layer and the underlying Ru substrate. They found a shift in the Dirac point and a "surprising" energy gap formation in the band structure due to the fact that the symmetry for two carbon sublattices is broken in the graphene/Au/Ru system.

Like Au intercalation, Sutter et al. [82] pointed out a possible atomic and molecular intercalation of metals, semiconductors, or oxygen molecules; it may be possible to build gate electrodes with thin oxides or other metallic source and drain electrodes.

15.3 INTERACTION OF TMs WITH GNRs

Investigating the effects of TMs on GNRs is a relatively less-studied subject with respect to graphene nanosheet studies [98–102]. The electronic properties of GNR are very dependent on the width of the GNR and its shape. Like carbon nanotubes, chirality has an important effect on the electronic properties. In contrast to carbon nanotubes, zigzag graphene nanoribbons (ZGNRs) show metallic and most of the armchair graphene nanoribbons (AGNRs) show semiconductor behavior [98]. It is also known that the adsorption and doping of atoms and molecules on GNRs can alter the electronic band structure, like graphene nanosheets [103–106]. Therefore, all these alterations can be used for successful band engineering. Because of the known effect of TMs on graphene nanosheets, which are briefly discussed in the previous sections, TMs on GNR structures are also gaining attention.

In their study, Gorjizadeh et al. [99] chose an AGNR structure with a width of $N = 8$. Here, N is an integer and an indicator of the width. Even- and odd-numbered widths are corresponding to the antisymmetrical and symmetrical edges. Therefore, Gorjizadeh et al. chose an antisymmetrical GNR for their studies. With spin-polarized DFT, they studied the effects of Ti, Cr, Mn, Fe, and Co atoms that are positioned at the edges of the ribbon. All cases show metallic behavior. Ti- and Cr-terminated cases are found to be more stable in ferromagnetic behavior. However, Mn-, Fe-, and Co-terminated cases are found to be more stable in antiferromagnetic behavior [99]. According to their findings, they reported that the band structure of the spin channels does not present symmetry and, therefore, half- metallic or half-semiconductor behaviors can be observed in TM-doped ribbons. To compare with AGNRs, they also improved the calculations for ZGNRs with $N = 5$. For the doped Fe and Co, binding energies are found to be larger than the AGNRs.

For the effect of the width on the electronic properties, Jaiswal and Srivastava [98,100] reported a series of studies about the DFT calculations of Ni and Cu doped and terminated AGNRs. For Ni, they calculated the AGNRs with $N = 4–9$ including one edge and both edge termination and different doping configurations [98]. For both edge termination cases, conduction and valence bands show that the termination of degeneracy and spin states are separated. Doped Ni is found to always lead to metallic behavior in AGNRs, which is different from the findings of Rigo et al.'s [105] study for the adsorption of Ni, where GNR may show semiconductor behavior. For Cu, they calculated the energy band change with the changing ribbon width [100]. They found $N = 3m + 2$ families with hydrogen terminated or one-edge Cu-terminated ribbons show minimum band gap. Here, m is an integer. Both edge Cu-terminated case $N = 3m$ families show the minimum. Wang et al. [106] studied the Co-/Ni-adsorbed ZGNRs. Their calculations show that the most stable site for Co/Ni atom adsorption is on the edge compared with the middle of the ribbon. Also, they pointed that the interaction between the Ni atom and the ZGNR is stronger than the interaction between the Co atom and the ZGNR.

In their studies, Sarikavak-Lisesivdin et al. [101] investigated the effects of Ru termination and doping on an electronic band structure of $N = 9$ symmetrical AGNR structure. In addition to the expected one-dimensional (1D) DOS behavior, they also observed q0D and q1D behaviors in DOS. For the Ru doping at sites $Na = 1–5$, which correspond to all the sites of $N = 1–9$ because of symmetry, various mini band gaps of 12.4–89.6 meV are observed [101].

Similarly, Kuloglu et al. [102] studied the effects of Pd termination and doping on the electronic band structure of $N = 11$ symmetrical AGNR structure. They found metallic behavior for both one- and both-sided Pd-terminated cases. They also pointed out the q0D and q1D peaks in DOS and suggested that for these structures, a highly localized Pd 4d shell may result in lower-dimensional peaks in the DOS spectra, with the extensively doped Pd density [102]. They also performed width-dependent calculations and showed a drop in binding

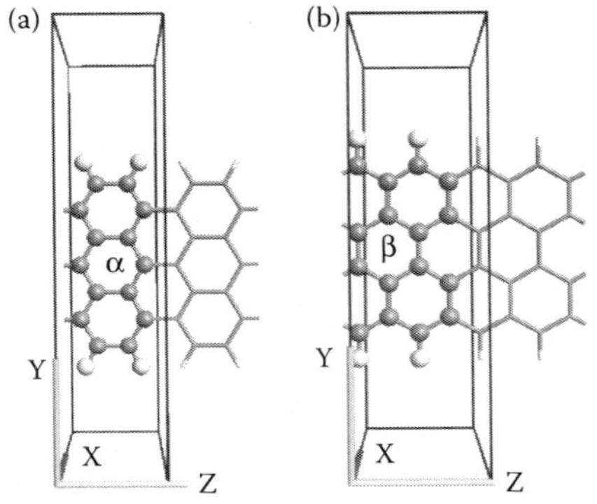

FIGURE 15.4 Used unit cells for (a) symmetrical armchair and (b) zigzag nanoribbon calculations. Sites α and β, are the H sites for adsorption.

energies at N = 11 and N = 13 for one- or both- edge Pd termination cases, respectively. With the drops, binding energies being close to the hydrogenated cases, and with the increasing N, the both edges Pd-terminated case began to have greater binding energies than the one-edge Pd-terminated case. They pointed out that the binding energies of the GNRs are linearly dependent on Young's modulus [107] and, therefore, these characteristic drops also correspond to a change in the stiffness of these AGNRs [102].

15.3.1 Calculation Method

In this study, we used a first-principles band calculation technique based on DFT by using Atomistix Toolkit–Visual NanoLab (ATK–VNL) [108–110]. For the nanoribbon studies, periodic boundary conditions on the z-axis are accepted. For the vacuum confinement region, the interspace between two periodic slabs was kept enough to avoid the interaction between periodic graphene layers. All atoms were fully relaxed and the maximum strains on atoms were accepted to be less than 0.005 eV/Å. We adopted the LDA wherein spin interaction is included. This functional was used in the form of Perdew and Zunger [111] with an energy cutoff value of 500 eV and with a regular Monkhorst–Pack k-point grid of $1 \times 1 \times 100$ [112]. For the nanosheet calculations that are presented in Section 15.2.1, a regular Monkhorst–Pack k-point grid of $9 \times 9 \times 1$ is used. In Figure 15.4a and b, the unit cells for symmetrical armchair and zigzag nanoribbons are shown, respectively. Sites α and β, are the H sites, which are known as minimum energy sites, for the adsorption of the studied TM atoms [58]. In the calculations, 3d TM elements Sc, Ti, V, Cr, Mn, Fe, Co, Ni, and Cu are used as adsorbed elements.

15.3.2 Computational Results for TM-Adsorbed AGNRs and ZGNRs

In Figure 15.5, the band structures of AGNRs of (a) hydrogenated only and (b through j) that additionally adsorb TM atoms are shown. All band structures show nonmagnetic behavior, where all bands for spin-up and spin-down cases are overlapped. For a hydrogenated AGNR, direct semiconductor behavior with ~1.51-eV band gap is observed as expected [101,102]. With a TM atom adsorbed, this band gap tends to be reduced and becomes indirect. The smallest band gap is found to be ~0.65 eV for the Fe-adsorbed case. For the hydrogenated-only and Ni-adsorbed cases, the Fermi level is located nearly at the center of the band gap. However, for Sc, Ti, V, Cr, Mn, and Cu, AGNRs show degenerate n-type behavior and Fe and Co show degenerate p-type behavior.

Figure 15.6 shows the DOS structure of cases that are shown in Figure 15.5. Many 1D DOS peaks, which are mostly symmetrical, are present in the hydrogenated case. With an adsorption of TM atom, the DOS show an increase, and the saddle-shaped peaks can be observed. These peaks are called q0D peaks and they are caused by the strong hybridization of the d-states of TM atoms with carbon's p-states [101,113].

In Figure 15.7, the band structures of the ZGNRs of (a) hydrogenated only and (b through j) that additionally adsorb TM atoms are shown. All structures including the hydrogenated-only case show magnetic behavior. Spin-up and spin-down are shown in Figures 15.7 and 15.8 as black lines and gray lines, respectively. All the structures show metallic behavior. For V, Mn, and Fe, the spin-selective band gap, which may not be classified as half-metallicity, is observed. Figure 15.8 shows the DOS structure of cases that are shown in Figure 15.7. The q0D peaks can also be observed in this figure for both spins. The spin-dependent behavior of these ZGNRs is open for further investigation for different ribbon widths and different adsorption sites that are beyond the scope of this chapter.

15.4 CONCLUDING REMARKS

In this chapter, we briefly report the important findings on the effects of the interaction of TMs on the electronic properties of graphene nanosheets and GNRs. These interactions include doping, adsorption, and grown-on cases and are mostly investigated with DFT studies and some of them include experimental ARPES results. In addition to the band structure, the related parameters which can be found from these calculations, binding energies of the TM atoms, and the change in magnetic moments are also investigated and the brief findings in the literature are summarized as tables.

For the GNRs, studies in the literature are limited. Here, we reported on some band calculation results for hydrogenated-only AGNR and ZGNRs, and the ones with 3d TM elements such as Sc, Ti, V, Cr, Mn, Fe, Co, Ni, and Cu that were adsorbed. The results show that—like nanosheet studies—GNRs also interact with TM atoms, and both semiconductor and metallic or spin-selective metallic behaviors can be observed with TM atom adsorption on both AGNRs and ZGNRs. Further studies, which may include the density of adsorbed atoms, width of GNR, and adsorption site, will result in a better understanding of the subject and will open many spintronic or electronic device applications.

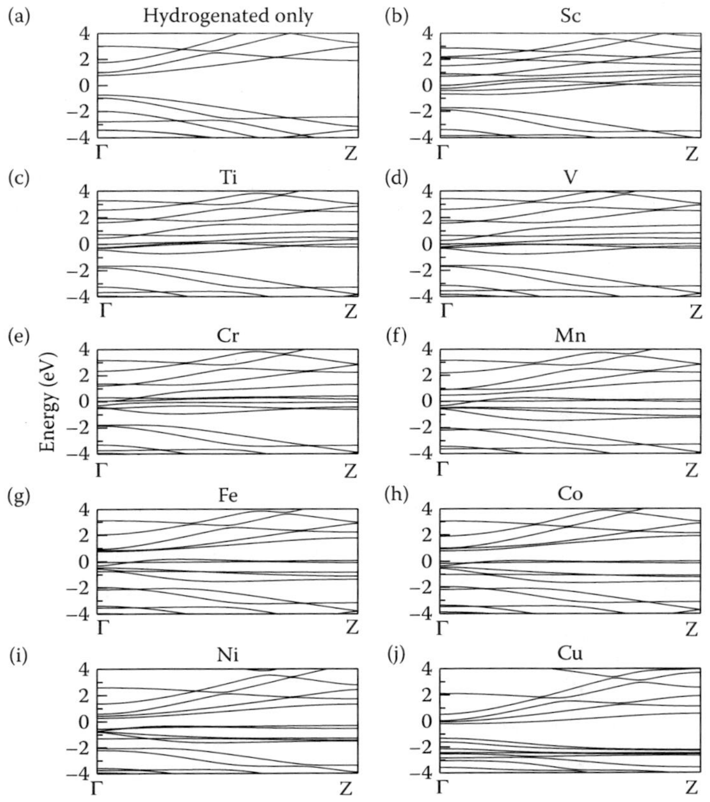

FIGURE 15.5 Band structures of hydrogenated (a), pristine, and (b) through (j) TM-adsorbed symmetrical armchair nanoribbons.

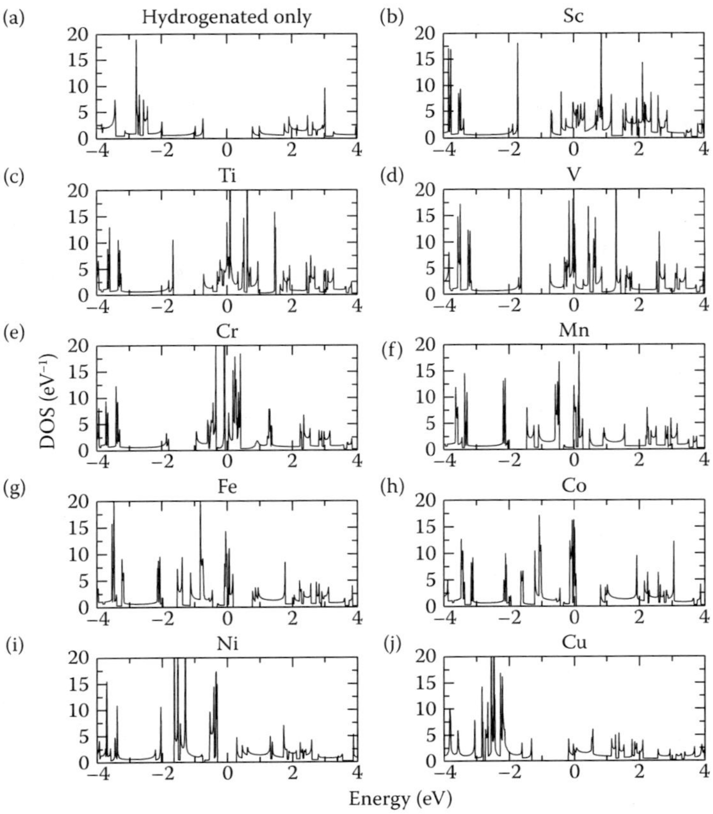

FIGURE 15.6 DOS of hydrogenated (a), pristine, and (b) through (j) TM-adsorbed armchair nanoribbons.

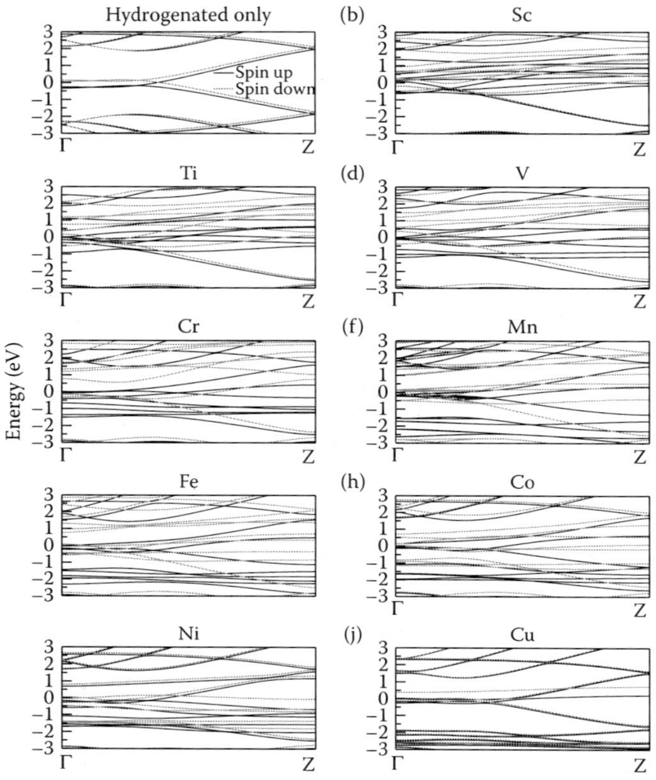

FIGURE 15.7 Band structures of hydrogenated (a), pristine, and (b) through (j) TM-adsorbed zigzag nanoribbons. Black lines and gray lines correspond to up and down spin states, respectively.

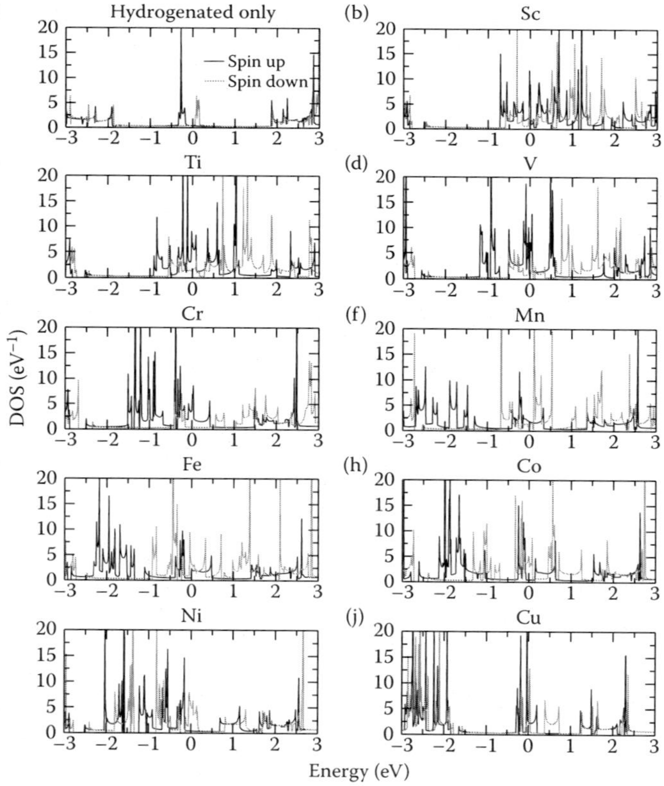

FIGURE 15.8 DOS of hydrogenated (a), pristine, and (b) through (j) TM-adsorbed zigzag nanoribbons. Black lines and gray lines correspond to up and down spin states, respectively.

ACKNOWLEDGMENTS

This chapter was supported by the projects DPT-HAMIT, DPT-FOTON, NATO-SET-193, and TUBITAK under Project Nos. 113F364, 113E331, 109A015, and 109E301. One of the authors (E.O.) also acknowledges partial support from the Turkish Academy of Sciences.

REFERENCES

1. Jorio, A., M. S. Dresselhaus, and G. Dresselhaus. 2008. *Carbon Nanotubes: Advanced Topics in the Synthesis, Structure, Properties and Applications*. Berlin: Springer-Verlag.
2. Wallace, P. R. 1947. The band theory of graphite. *Phys. Rev.* 71(9): 622–634.
3. McClure, J. W. 1956. Diamagnetism of graphite. *Phys. Rev.* 104(3): 666–671.
4. Novoselov, K. S., A. K. Geim, S. V. Morozov, D. Jiang, Y. Zhang, S. V. Dubonos, I. V. Grigorieva, and A. A. Firsov. 2004. Electric field effect in atomically thin carbon films. *Science* 306(5696): 666–669.
5. Yuanbo, Z., Y.-W. Tan, H. L. Stormer, and P. Kim. 2005. Experimental observation of the quantum Hall effect and Berry's phase in graphene. *Nature* 438:201–204.
6. Bolotin, K. I., K. J. Sikes, Z. Jiang, M. Klima, G. Fudenberg, J. Hone, P. Kim, and H. L. Stormer. 2008. Ultrahigh electron mobility in suspended graphene. *Solid State Comm.* 146:351–355.
7. Geim, A. K., and K. S. Novoselov. 2007. The rise of graphene. *Nat. Mater.* 6(3):183–191.
8. Heersche, H. B., P. Jarillo-Herrero, J. B. Oostinga, L. M. K. Vandersypen, and A. F. Morpurgo. 2007. Bipolar supercurrent in graphene. *Nature* 446(7131): 56–59.
9. Zhou, S. Y., G.-H. Gweon, A. V. Fedorov, P. N. First, W. A. de Heer, D.-H. Lee, F. Guinea, A. H. Castro Neto, and A. Lanzara. 2007. Substrate-induced bandgap opening in epitaxial graphene. *Nat. Mater.* 6:770–775.
10. Berger, C., Z. Song, T. Li, X. Li, A. Y. Ogbazghi, R. Feng, Z. Dai et al. 2004. Ultrathin epitaxial graphite: 2D electron gas properties and a route toward graphene-based nanoelectronics. *J. Phys. Chem. B* 108(52):19912–19916.
11. Stankovich, S., D. A. Dikin, G. H. B. Dommett, K. M. Kohlhaas, E. J. Zimney, E. A. Stach, R. D. Piner, S. T. Nguyen, and R. S. Ruoff. 2006. Graphene-based composite materials. *Nature* 442(7100):282–286.
12. De Arco, L. G., Y. Zhang, A. Kumar, and C. Zhou. 2009. Synthesis, transfer, and devices of single- and few-layer graphene by chemical vapor deposition. *Nanotechnol., IEEE Trans.* 8(2):135–138.
13. Wang, X., Y. Ouyang, X. Li, H. Wang, J. Guo, and H. Dai. 2008. Room-temperature all-semiconducting sub-10-nm graphene nanoribbon field-effect transistors. *Phys. Rev. Lett.* 100:206803 (4pp).
14. Hod, O., V. Barone, and G. E. Scuseria. 2008. Half-metallic graphene nanodots: A comprehensive first-principles theoretical study. *Phys. Rev. B* 77:035411 (6pp).
15. Lemme, M. C., T. J. Echtermeyer, M. Baus, and H. Kurz. 2007. A graphene field-effect device. *Electron. Device Lett., IEEE* 28(4):282–284.
16. Liang, X., Z. Fu, and S. Y. Chou. 2007. Graphene transistors fabricated via transfer-printing in device active-areas on large wafer. *Nano Lett.* 7(12):3840–3844.
17. Wehling, T. O., K. S. Novoselov, S. V. Morozov, E. E. Vdovin, M. I. Katsnelson, A. K. Geim, and A. I. Lichtenstein. 2008. Molecular doping of graphene. *Nano Lett.* 8(1):173–177.
18. Williams, J. R., L. DiCarlo, and C. M. Marcus. 2007. Quantum Hall effect in a gate-controlled pn junction of graphene. *Science* 317(5838):638–641.
19. Martins, T. B., R. H. Miwa, A. J. R. da Silva, and A. Fazzio. 2007. Electronic and transport properties of boron-doped graphene nanoribbons. *Phys. Rev. Lett.* 98(19):196803 (4pp).
20. Kim, T.-W., Y. Gao, O. Acton, H.-L. Yip, H. Ma, H. Chen, and A. K.-Y. Jen. 2010. Graphene oxide nanosheets based organic field effect transistor for nonvolatile memory applications. *Appl. Phys. Lett.* 97(2): 023310 (3pp).
21. Dimitrakakis, G. K., E. Tylianakis, and G. E. Froudakis. 2008. Pillared graphene: A new 3-D network nanostructure for enhanced hydrogen storage. *Nano Lett.* 8(10):3166–3170.
22. Blake, P., P. D. Brimicombe, R. R. Nair, T. J. Booth, D. Jiang, F. Schedin, L. A. Ponomarenko et al. 2008. Graphene-based liquid crystal device. *Nano Lett.* 8:1704–1708.
23. Liu, M., X. Yin, E. Ulin-Avila, B. Geng, T. Zentgraf, L. Ju, F. Wang, and X. Zhang. 2011. A graphene-based broadband optical modulator. *Nature* 474:64–67.
24. Huang, B., Q. Yan, G. Zhou, J. Wu, B.-L. Gu, W. Duan, and F. Liu. 2007. Making a field effect transistor on a single graphene nanoribbon by selective doping. *Appl. Phys. Lett.* 91(25):253122 (3pp).
25. Sutter, P. W., J.-I. Flege, and E. A. Sutter. 2008. Epitaxial graphene on ruthenium. *Nat. Mater.* 7(5):406–411.
26. Sutter, E., D. P. Acharya, J. T. Sadowski, and P. Sutter. 2009. Scanning tunneling microscopy on epitaxial bilayer graphene on ruthenium (0001). *Appl. Phys. Lett.* 94(13):133101 (3pp).
27. Yagi, Y., T. M. Briere, M. H. F. Sluiter, V. Kumar, A. A. Farajian, and Y. Kawazoe. 2004. Stable geometries and magnetic properties of single-walled carbon nanotubes doped with 3d transition metals: A first-principles study. *Phys. Rev. B* 69(7):075414 (9pp).
28. Krasheninnikov, A. V., and R. M. Nieminen. 2011. Attractive interaction between transition-metal atom impurities and vacancies in graphene: A first-principles study. *Theor. Chem. Acc.* 129:625–630.
29. Hentschel, M., and F. Guinea. 2007. Orthogonality catastrophe and Kondo effect in graphene. *Phys. Rev. B* 76(11):115407 (7pp).
30. Dóra, B., and P. Thalmeier. 2007. Reentrant Kondo effect in Landau-quantized graphene: Influence of the chemical potential. *Phys. Rev. B* 76(11):115435 (6pp).
31. Bhattacharya, A., S. Bhattacharya, C. Majumder, and G. P. Das. 2010. Transition-metal decoration enhanced room-temperature hydrogen storage in a defect-modulated graphene sheet. *J. Phys. Chem. C* 114(22):10297–10301.
32. N'Diaye, A. T., S. Bleikamp, P. J. Feibelman, and T. Michely. 2006. Two-dimensional Ir cluster lattice on a graphene Moiré on Ir(111). *Phys. Rev. Lett.* 97:215501.
33. Ushiro, M., K. Uno, T. Fujikawa, Y. Sato, K. Tohji, F. Watari, W.-J. Chun, Y. Koike, and K. Asakura. 2006. X-ray absorption fine structure (XAFS) analyses of Ni species trapped in graphene sheet of carbon nanofibers. *Phys. Rev. B* 73(14):144103 (11pp).
34. Banhart, F., J.-C. Charlier, and P. M. Ajayan. 2000. Dynamic behavior of nickel atoms in graphitic networks. *Phys. Rev. Lett.* 84(4):686–689.
35. Santos, E. J. G., A. Ayuela, S. B. Fagan, J. Mendes Filho, D. L. Azevedo, A. G. Souza Filho, and D. Sánchez-Portal. 2008. Switching on magnetism in Ni-doped graphene: Density functional calculations. *Phys. Rev. B* 78:195420.
36. Uchoa, B., and A. H. Castro Neto. 2007. Superconducting states of pure and doped graphene. *Phys. Rev. Lett.* 98(14):146801 (4pp).

37. Kim, G., S.-H. Jhi, S. Lim, and N. Park. 2009. Effect of vacancy defects in graphene on metal anchoring and hydrogen adsorption. *Appl. Phys. Lett.* 94(17):173102 (3pp).
38. Pi, K., K. M. McCreary, W. Bao, W. Han, Y. F. Chiang, Y. Li, S.-W. Tsai, C. N. Lau, and R. K. Kawakami. 2009. Electronic doping and scattering by transition metals on graphene. *Phys. Rev. B* 80(7): 075406 (5pp).
39. Boukhvalov, D. W., and M. I. Katsnelson. 2009. Destruction of graphene by metal adatoms. *Appl. Phys. Lett.* 95(2):023109 (3pp).
40. Lu, Y.-H., M. Zhou, C. Zhang, and Y.-P. Feng. 2009. Metal-embedded graphene: A possible catalyst with high activity. *J. Phys. Chem. Lett.* 113:20156–20160.
41. Krasheninnikov, A. V., P. O. Lehtinen, A. S. Foster, P. Pyykkö, and R. M. Nieminen. 2009. Embedding transition-metal atoms in graphene: Structure, bonding, and magnetism. *Phys. Rev. Lett.* 102(12):126807 (4pp).
42. Choi, W. I., S.-H. Jhi, K. Kim, and Y.-H. Kim. 2010. Divacancy–nitrogen-assisted transition metal dispersion and hydrogen adsorption in defective graphene: A first-principles study. *Phys. Rev. B* 81(8):085441 (5pp).
43. Lim, D.-H., and J. Wilcox. 2011. DFT-based study on oxygen adsorption on defective graphene-supported Pt nanoparticles. *J. Phys. Chem. C* 115:22742–22747.
44. Lide, D. R., ed. 2008. *CRC Handbook of Chemistry and Physics, 89th Edition*. Boca Raton: CRC Press, Taylor & Francis Group.
45. Sque, S. J., R. Jones, and P. R. Briddon. 2007. The transfer doping of graphite and graphene. *Phys. Stat. Sol. (a)* 204(9):3078–3084.
46. Giovannetti, G., P. A. Khomyakov, G. Brocks, V. M. Karpan, J. Van den Brink, and P. J. Kelly. 2008. Doping graphene with metal contacts. *Phys. Rev. Lett.* 101(2):026803 (4pp).
47. Zan, R., U. Bangert, Q. Ramasse, and K. S. Novoselov. Metal–graphene interaction studied via atomic resolution scanning transmission electron microscopy. *Nano Lett.* 11:1087–1092.
48. Gan, Y., L. Sun, and F. Banhart. 2008. One- and two-dimensional diffusion of metal atoms in graphene. *Small* 4(5):587–591.
49. Sun, Q., Q. Wang, and P. Jena. 2005. Storage of molecular hydrogen in BN cage: Energetics and thermal stability. *Nano Lett.* 5(7):1273–1277.
50. Qi, L., X. Qian, and J. Li. 2008. Near neutrality of an oxygen molecule adsorbed on a Pt (111) surface. *Phys. Rev. Lett.* 101(14):146101 (4pp).
51. Malola, S., H. Hakkinen, and P. Koskinen. 2009. Gold in graphene: In-plane adsorption and diffusion. *Appl. Phys. Lett.* 94(4):043106 (3pp).
52. Zhang, Y., N. W. Franklin, R. J. Chen, and H. Dai. 2000. Metal coating on suspended carbon nanotubes and its implication to metal–tube interaction. *Chem. Phys. Lett.* 331(1):35–41.
53. Chan, K. T., J. B. Neaton, and M. L. Cohen. 2008. First-principles study of metal adatom adsorption on graphene. *Phys. Rev. B* 77(23):235430 (12pp).
54. Uchoa, B., C.-Y. Lin, and A. H. Castro Neto. 2008. Tailoring graphene with metals on top. *Phys. Rev. B* 77(3):035420 (5pp).
55. Wehling, T. O., A. I. Lichtenstein, and M. I. Katsnelson. 2011. Transition-metal adatoms on graphene: Influence of local Coulomb interactions on chemical bonding and magnetic moments. *Phys. Rev. B* 84(23):235110 (7pp).
56. Zhou, M., Y. Lu, C. Zhang, and Y. P. Feng. 2010. Strain effects on hydrogen storage capability of metal-decorated graphene: A first-principles study. *Appl. Phys. Lett.* 97(10):103109 (3pp).
57. Mao, Y., J. Yuan, and J. Zhong. 2008. Density functional calculation of transition metal adatom adsorption on graphene. *J. Phys. Condens. Matter* 20:115209 (6pp).
58. Ishii, A., M. Yamamoto, H. Asano, and K. Fujiwara. 2008. DFT calculation for adatom adsorption on graphene sheet as a prototype of carbon nanotube functionalization. *J. Phys. Conf. Series* 100:052087 (4pp).
59. Zolyomi, V., A. Rusznyak, J. Koltai, J. Kürti, and C. J. Lambert. 2010. Functionalization of graphene with transition metals. *Phys. Stat. Sol. B* 247(11–12): 2920–2923.
60. Zolyomi, V., A. Rusznyak, J. Kürti, and C. J. Lambert. 2010. First principles study of the binding of 4d and 5d transition metals to graphene. *J. Phys. Chem. C* 114:18548–18552.
61. Zhou, Q., Y. Tang, C. Wang, Z. Fu, and H. Zhang. 2014. Electronic and magnetic properties of transition-metal atoms absorbed on Stone–Wales defected graphene sheet: A theory study. *Comput. Mater. Sci.* 81: 348–352.
62. Eelbo, T., M. Waśniowska, P. Thakur, M. Gyamfi, B. Sachs, T. O. Wehling, S. Forti et al. 2013. Adatoms and clusters of 3 d transition metals on graphene: Electronic and magnetic configurations. *Phys. Rev. Lett.* 110(13):136804.
63. Eelbo, T., M. Waśniowska, M. Gyamfi, S. Forti, U. Starke, and R. Wiesendanger. 2013. Influence of the degree of decoupling of graphene on the properties of transition metal adatoms. *Phys. Rev. B* 87(20):205443.
64. Cao, C., M. Wu, J. Jiang, and H.-P. Cheng. 2010. Transition metal adatom and dimer adsorbed on graphene: Induced magnetization and electronic structure. *Phys. Rev. B* 81(20):205424 (9pp).
65. Yazyev, O. V., and A. Pasquarello. 2010. Metal adatoms on graphene and hexagonal boron nitride: Towards rational design of self-assembly templates. *Phys. Rev. B* 82(4):045407 (5pp).
66. Sutter, E., P. Albrecht, B. Wang, M.-L. Bocquet, L. Wu, Y. Zhu, and P. Sutter. 2011. Arrays of Ru nanoclusters with narrow size distribution templated by monolayer graphene on Ru. *Surf. Sci.* 605(17):1676–1684.
67. Caragiu, M., and S. Finberg. 2005. Alkali metal adsorption on graphite: A review. *J. Phys. Cond. Mat.* 17(35):R995–R1024.
68. Durgun, E., S. Dag, S. Ciraci, and O. Gülseren. 2004. Energetics and electronic structures of individual atoms adsorbed on carbon nanotubes. *J. Phys. Chem. B* 108(2): 575–582.
69. Durgun, E., S. Ciraci, and T. Yildirim. 2008. Functionalization of carbon-based nanostructures with light transition-metal atoms for hydrogen storage. *Phys. Rev. B* 77(8):085405 (9pp).
70. Vanin, M., J. J. Mortensen, A. K. Kelkkanen, J. M. Garcia-Lastra, K. S. Thygesen, and K. W. Jacobsen. 2010. Graphene on metals: A van der Waals density functional study. *Phys. Rev. B* 81(8):081408 (4pp).
71. Raza, H., ed. 2011. *Graphene Nanoelectronics: Metrology, Synthesis, Properties and Applications*. Berlin: Springer.
72. Tang, Y., Z. Yang, and X. Dai. 2011. Trapping of metal atoms in the defects on graphene. *J. Chem. Phys.* 135(22):224704 (7pp).
73. Hu, L., X. Hu, X. Wu, C. Du, Y. Dai, and J. Deng. 2010. Density functional calculation of transition metal adatom adsorption on graphene. *Physica B: Cond. Mat.* 405(16):3337–3341.
74. Santos, E. J. G., A. Ayuela, and D. Sánchez-Portal. 2010. First-principles study of substitutional metal impurities in graphene: Structural, electronic and magnetic properties. *New J. Phys.* 12(5):053012 (31pp).
75. Varns, R., and P. Strange. 2008. Stability of gold atoms and dimers adsorbed on graphene. *J. Phys. Cond. Mat.* 20(22):225005 (8pp).

76. Sampedro, B., P. Crespo, A. Hernando, R. Litrán, J. C. Sánchez López, C. López Cartes, A. Fernandez, J. Ramírez, J. González Calbet, and M. Vallet. 2003. Ferromagnetism in fcc twined 2.4 nm size Pd nanoparticles. *Phys. Rev. Lett.* 91(23):237203 (4pp).
77. Józsa, C., M. Popinciuc, N. Tombros, H. T. Jonkman, and B. J. Van Wees. 2009. Controlling the efficiency of spin injection into graphene by carrier drift. *Phys. Rev. B* 79(8):081402 (4pp).
78. Morgan, A. E., and G. A. Somorjai. 1968. Low energy electron diffraction studies of gas adsorption on the platinum (100) single crystal surface. *Surf. Sci.* 12(3):405–425.
79. Lang, B. 1975. A LEED study of the deposition of carbon on platinum crystal surfaces. *Surf. Sci.* 53(1):317–329.
80. Sutter, P., M. S. Hybertsen, J. T. Sadowski, and E. Sutter. 2009. Electronic structure of few-layer epitaxial graphene on Ru (0001). *Nano Lett.* 9(7):2654–2660.
81. Sutter, P., J. T. Sadowski, and E. Sutter. 2009. Graphene on Pt (111): Growth and substrate interaction. *Phys. Rev. B* 80(24):245411 (10pp).
82. Sutter, P., J. T. Sadowski, and E. A. Sutter. 2010. Chemistry under cover: Tuning metal–graphene interaction by reactive intercalation. *J. Am. Chem. Soc.* 132(23):8175–8179.
83. Stradi, D., S. Barja, C. Diaz, M. Garnica, B. Borca, J. J. Hinarejos, D. Sánchez-Portal et al. 2011. Role of dispersion forces in the structure of graphene monolayers on Ru surfaces. *Phys. Rev. Lett.* 106(18):186102 (4pp).
84. Peng, X., and R. Ahuja. 2010. Epitaxial graphene monolayer and bilayers on Ru (0001): *Ab initio* calculations. *Phys. Rev. B* 82(4): 045425 (5pp).
85. Feng, Y., X. Yao, Z. Hu, J.-J. Xu, and L. Zhang. 2013. Passivating a transition-metal surface for more uniform growth of graphene: Effect of Au alloying on Ni (111). *Phys. Rev. B* 87(19):195421.
86. Marchini, S., S. Gunther, and J. Wintterlin. 2007. Scanning tunneling microscopy of graphene on Ru(0001). *Phys. Rev. B* 76:075429 (9pp).
87. Moritz, W., B. Wang, M.-L. Bocquet, T. Brugger, T. Greber, J. Wintterlin, and S. Günther. 2010. Structure determination of the coincidence phase of graphene on Ru (0001). *Phys. Rev. Lett.* 104(13): 136102 (4pp).
88. Wang, B., M.-L. Bocquet, S. Marchini, S. Günther, and J. Wintterlin. 2008. Chemical origin of a graphene moiré overlayer on Ru(0001). *Phys. Chem. Chem. Phys.* 10(24): 3530–3534.
89. Jiang, D., M.-H. Du, and S. Dai. 2009. First principles study of the graphene/Ru(0001) interface. *J. Chem. Phys.* 130(7):074705 (5pp).
90. Khomyakov, P. A., G. Giovannetti, P. C. Rusu, G. Brocks, J. Van den Brink, and P. J. Kelly. 2009. First-principles study of the interaction and charge transfer between graphene and metals. *Phys. Rev. B* 79(19):195425 (12 pp).
91. Giuliani, G. F., and G. Vignale. 2005. *Quantum Theory of the Electron Liquid*. Cambridge: Cambridge University Press.
92. Damascelli, A., Z. Hussain, and Z.-X. Shen. 2003. Angle-resolved photoemission studies of the cuprate superconductors. *Rev. Mod. Phys.* 75(2):473–541.
93. Public domain image is taken from Wikipedia http://en.wikipedia.org/wiki/Angle-resolved_photoemission_spectroscopy.
94. Enderlein, C., Y. S. Kim, A. Bostwick, E. Rotenberg, and K. Horn. 2010. The formation of an energy gap in graphene on ruthenium by controlling the interface. *New J. Phys.* 12:033014 (9pp).
95. Wintterlin, J., and M.-L. Bocquet. 2009. Graphene on metal surfaces. *Surf. Sci.* 603:1841–1852.
96. Karpan, V. M., G. Giovannetti, P. A. Khomyakov, M. Talanana, A. A. Starikov, M. Zwierzycki, J. Van Den Brink, G. Brocks, and P. J. Kelly. 2007. Graphite and graphene as perfect spin filters. *Phys. Rev. Lett.* 99(17): 176602 (4pp).
97. Pletikosić, I., M. Kralj, P. Pervan, R. Brako, J. Coraux, A. T. N'diaye, C. Busse, and T. Michely. 2009. Dirac cones and minigaps for graphene on Ir (111). *Phys. Rev. Lett.* 102(5):056808 (4pp).
98. Jaiswal, N. K., and P. Srivastava. 2011. Structural stability and electronic properties of Ni-doped armchair graphene nanoribbons. *Solid State Commun.* 151(20): 1490–1495.
99. Gorjizadeh, N., A. A. Farajian, K. Esfarjani, and Y. Kawazoe. 2008. Spin and band-gap engineering in doped graphene nanoribbons. *Phys. Rev. B* 78(15): 155427 (6pp).
100. Jaiswal, N. K., and P. Srivastava. 2011. First principles calculations of armchair graphene nanoribbons interacting with Cu atoms. *Physica E: Low-Dimens. Syst. Nanostruct.* 44(1):75–79.
101. Sarikavak-Lisesivdin, B., S. B. Lisesivdin, and E. Ozbay. 2012. *Ab initio* study of Ru-terminated and Ru-doped armchair graphene nanoribbons. *Mol. Phys.* 110(18): 2295–2300.
102. Kuloglu, A. F., B. Sarikavak-Lisesivdin, S. B. Lisesivdin, and E. Ozbay. 2013. First-principles calculations of Pd-terminated symmetrical armchair graphene nanoribbons. *Comput. Mater. Sci.* 68:18–22.
103. Kan, E.-J., H. J. Xiang, J. Yang, and J. G. Hou. 2007. Electronic structure of atomic Ti chains on semiconducting graphene nanoribbons: A first-principles study. *J. Chem. Phys.* 127:164706 (5pp).
104. Sevincli, H., M. Topsakal, E. Durgun, and S. Ciraci. 2008. Electronic and magnetic properties of 3d transition-metal atom adsorbed graphene and graphene nanoribbons. *Phys. Rev. B* 77(19):195434 (7pp).
105. Rigo, V. A., T. B. Martins, A. J. R. da Silva, A. Fazzio, and R. H. Miwa. 2009. Electronic, structural, and transport properties of Ni-doped graphene nanoribbons. *Phys. Rev. B* 79(7):075435 (9pp).
106. Wang, Z., J. Xiao, and M. Li. 2013. Adsorption of transition metal atoms (Co and Ni) on zigzag graphene nanoribbon. *Appl. Phys. A* 110(1):235–239.
107. Zeinalipour-Yazdi, C. D., and C. Christofides. 2009. Linear correlation between binding energy and Young's modulus in graphene nanoribbons. *J. Appl. Phys.* 106(5):054318 (5pp).
108. Version 12.8.2 QuantumWise A/S (http://www.quantumwise.com).
109. Brandbyge, M., J.-L. Mozos, P. Ordejón, J. Taylor, and K. Stokbro. 2002. Density-functional method for nonequilibrium electron transport. *Phys. Rev. B* 65(16):165401 (17pp).
110. Soler, J. M., E. Artacho, J. D. Gale, A. García, J. Junquera, P. Ordejón, and D. Sánchez-Portal. 2002. The SIESTA method for *ab initio* order-N materials simulation. *J. Phys. Condens. Matt.* 14(11):2745–2780.
111. Perdew, J. P., and A. Zunger. 1981. Self-interaction correction to density-functional approximations for many-electron systems. *Phys. Rev. B* 23:5048–5079.
112. Monkhorst, H. J., and J. D. Pack. 1976. Special points for Brillouin-zone integrations. *Phys. Rev. B* 13:5188–5192.
113. Cho, H., and P. R. Prucnal. 1989. Density of states of quasi-two, -one, and -zero dimensional superlattices. *J. Vac. Sci. Technol. B* 7:1363–1367.

16 Electric Properties of Graphene and Its Chemisorption Derivatives

Long Jing and Xueyun Gao

CONTENTS

16.1 ELECTRONIC PROPERTIES OF INTRINSIC GRAPHENE

Graphene is the first material of a truly two-dimensional (2D) crystal that was extracted from three-dimensional (3D) graphite using a simple technique, called micromechanical cleavage, in 2004.[1,2] After fine-tuning, this technique now provides high-quality graphene crystallites of enough size for most research purposes. Before this, it was commonly believed that strictly 2D crystals were thermodynamically unstable and could not exist in the free state.[3] Besides serving as a basic 2D model system for material science and condensed matter physics interest, it is the unique electronic properties including the ultra-high mobility,[4] ballistic transport,[5] the thinness and stability, quantum Hall effect (QHE), and so on that inspire the graphene upsurge which aims to make it promising for next generation nanoscale electronic and magnetic applications in real devices.[6]

The intrinsic graphene is a unique semimetal or zero-gap semiconductor. The outstanding electronic properties of graphene ascribe to the hybrid states of carbon and its unique topological structure. Originally, carbon has two 2s and two 2p electrons. These four electrons produce different kinds of sp-hybridized orbitals. In the graphene, the carbon atoms arrange in the unique one-atom-thick honeycomb lattice structure so that each carbon atom is surrounded by three neighboring ones. In fact, the honeycomb lattice can be viewed as two equivalent triangular sublattices (labeled as A and B) that interpenetrate with each other, which means it is a compound lattice. Thus, the primitive cell of the honeycomb lattice is a rhombus containing two inequivalent carbon atoms, as shown in Figure 16.1a. All the carbon atoms in graphene sp^2-hybridize, where three sp^2 orbits of each carbon form σ bonds to form the framework of XY plane with a 120° angle between either two bonds and the p–p orbit overlaps along Z-axis form the conjugated π bonding network over the both sides of the whole graphene plane. These σ and π electrons occupy three filled σ bands and two degenerate half-filled π bands near the Fermi level when considering the intrinsic graphene's band structure.

As a basic 2D crystal and honeycomb lattice symmetry model in condensed matter physics, the electronic structure and transport properties of graphene were predicted, theoretically, for many years. Using density functional theory (DFT) and tight binding methods adequately describe the shape of π bands and σ bands of graphene. At the Γ point of the 2D Brillouin zone, energy gaps of 34 and 20 eV, respectively, separate the σ- and π-bonding from the corresponding antibonding bands.[7] However, at the K and K′ points, the linear π bands intersect at the point-like Fermi surface, the so-called Dirac cone.[8] Unlike most of 3D condensed matter systems that describe the electronic states using the Schrödinger equation, graphene's charge in the Dirac cone carriers mimic relativistic particles and are more suitably described with the Dirac equation in quantum electrodynamics.[8] The interactions between the relative free π electron wave and the periodic potential of graphene's honeycomb lattice give rise to new quasiparticles in reciprocal space, those at low energies E are well described by the (2 + 1)-dimensional Dirac equation with an effective speed of light $v_F = 10\ m^{-1}\ s^{-1}$. These quasiparticles, named as massless Dirac fermions, can be seen as electrons that have lost their rest mass m_0.[8] According to the description of the Dirac equation we can summarize some properties of graphene's electronic structure. First, the fully filled valence band and the totally empty conduction band connect with each other at the K and K′ points of the Brillouin zone, thus the band gap of graphene vanishes at these points, as shown in Figure 16.1b.[8] Without a band gap, graphene can be continuously tuned from p-type to n-type doping by an external electrostatic field, resulting in the bipolar field effect.[1] Second, near the Dirac cone, the dispersion relation model of the energy band is linear in the low excitation regime, leading

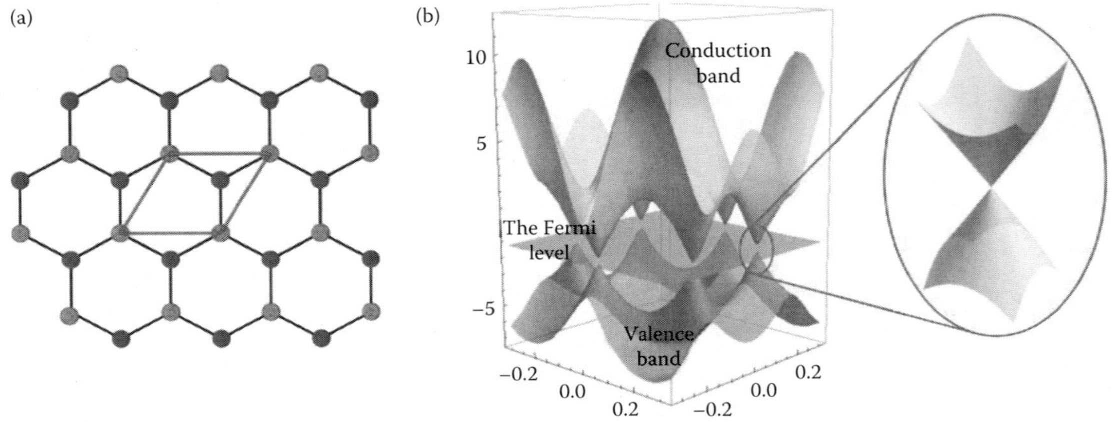

FIGURE 16.1 (a) Lattice structure of graphene; also illustrated is the primitive cell. (b) Electronic band structure of graphene. Zoom in is the linear dispersion relation near the Fermi level at the K point of the Brillouin zone.

to massless Dirac fermions. Third, the sublattice symmetry introduces the pseudospin as an extra quantum number into graphene wave functions to specify which sublattice the electron belongs to. This property is called the chirality and manifests itself in quantum transport processes such as the half-integer QHE[9] and unusual weak localization.

16.2 BRIEF VIEW OF TUNING THE ELECTRONIC STRUCTURE OF GRAPHENE

On the other hand, however, the zero bandgap and low density of electronic states features limit the direct application of graphene in electronic and photonic devices.[10] Thus, various approaches, both experimental and theoretical, have been devoted to introducing a finite band gap in graphene (e.g., 1–3 eV).

Using physical strategies such as realized by the substrate–graphene interaction,[11,12] applying an external electric field on bilayer graphene[13] or uniaxial strain,[14] cutting graphene into nanoribbons, and physical adsorption of molecules (e.g., H_2SO_4, H_2O, NH_3, and CrO_3) on the graphene surface will open a small but tunable band gap up to 250 meV.[15,16]

Chemisorbing a chemical functional group (e.g., H, F, O, OH, and nitrophenyl) on graphene is a direct way to control the electronic structure and magnetic properties of graphene-based materials.[17–20] The main mechanism of chemisorption on graphene, which can also be regarded as covalent modification of graphene, transit the hybridized state of the carbon host atom from sp^2 to sp^3 to break a π bond and produce an additional σ bond, thus the sp^3 hybridized carbon connect other four neighbor atoms via four σ bonds. This sp^3 hybridized carbon can be viewed as a single-atom defect on a fine graphene crystal because it will reduce the π electron to form a σ bond. It can be inferred that appropriate coverage of chemical absorptions introduce a barrier to the p electronic flow and induces the opening of a band gap. For example, patterned absorption of atomic hydrogen onto the moiré superlattice of epitaxial graphene results in a band gap of 450 meV.[21]

Whereas the ideal graphene are known to be nonmagnetic, chemisorption on graphene may create the magnetic moment on graphene-based materials, which is discovered in hydrogenated graphene and nitrophenyl diazonium functionalized graphene (NP-graphene).[22–24] At the adsorption site, when an unpaired single-atom defect (sp^3 hybridized carbon) is created on sublattice A (B), the p_z orbital associated with carbon atoms in the sublattice B (A) are unpaired. The unpaired p electrons contribute to quasi-localized states near the Fermi level, which extend over several nanometers around the defects.[23] This suggests that the itinerant magnetic exchange coupling can be attributed to the extended p electron interactions, which leads to a high Curie temperature (T_C) predicted by the Stroner ferromagnetism for sp electron systems.[25] While on the other hand, the Ruderman–Kittel–Kasuya–Yoshida (RKKY) coupling of localized magnetic moments in graphene is too weak to result in such a high T_C and the RKKY type interactions in graphene are proved to be suppressed.[23,26,27]

16.3 STRUCTURE, MAGNETISM, AND ELECTRONIC PROPERTIES OF HYDROGENATED GRAPHENE

Hydrogenated graphene, which attach the H atom on the graphene plane, is an attractive object for hydrogen storage and electronic structure engineering. It is commonly accepted now that the adsorption of a hydrogen atom on graphene is a process of chemisorption and a promise to the appearance of magnetic moments and bandgaps in the system. In this section, we introduce some experimental and theoretical work to explore the structure, magnetism, and electronic properties of various hydrogenated graphenes such as the chemisorption of single hydrogen atom and hydrogen dimmer on graphene, fully hydrogenated graphene (namely graphane), fully hydrogenate by one side of graphene (namely graphone), partial hydrogenated graphene and so on.

The theoretical studies using the DFT calculations give us a convenient and clear atomic image about the microstructure and the adsorption mechanisms of the very low concentrations of hydrogenated single-layer graphene. And then, these results are proved by a variety of experimental techniques including scanning tunneling microscope (STM),[21,28,29] Raman spectroscopy,[30] transmission electron microscopy (TEM),[30,31] and angle-resolved photoemission spectroscopy (ARPES).[21]

Boukhvalov et al.[17] and Ferro et al.[24] carry on the molecular and periodic supercell DFT approaches using the Perdew, Burke, and Ernzerhof gradient corrected generalized gradient approximation (GGA) functional for exchange and correlation to investigate the basics of single H atom and H dimer adsorption on single-layer and bilayer graphene. After the geometry optimized, they found that when a single H adsorbed on an infinite graphene surface (at the sublattice A for example), the hydrogen–carbon bond length is about 1.1 Å, and shifts the carbon atom bonded with the hydrogen atom about 0.3 Å along Z direction. The amplitude of the modulation of the graphene sheet in the perpendicular direction around the hydrogen atom was estimated as 0.4 Å, which is comparable with the height of intrinsic ripples on graphene of the order of 0.7 Å found in atomistic simulations. The radius of the distorted region around the hydrogen atom turned out to be about 3.8 Å. The C–C bond length test is the direct proof to determine the hybridization of the carbon. A typical bond length for sp^2 C–C bonds is 1.42 Å for graphene, and the standard bond angle is 120°. For sp^3 hybridization, the standard value of C–C bond length is 1.54 Å, and the angle is 109.5°. A typical value for the single C–H bond length is 1.086 Å. One can see that for the single hydrogen atom, the C–H bond length is close to the standard value, but the bond angles of C–C–H and C–C–C are intermediate between 90° and 109.5° and 120° and 109.5°, respectively. In addition, the length of the C–C bond is in between 1.42 and 1.54 Å. This means an intermediate character of the hybridization between sp^2 and sp^3.

Figure 16.2 shows the spin-density mapping around the adsorption site induces a magnetic defect on the graphene. As we mentioned above, when an H atom adsorbs on the sublattice A, one aromatic π bond breaks and one p electron is unpaired on the sublattice B. As a result, the graphene becomes magnetic, and the carbon atom bonded to the hydrogen atom carries the magnetic moment of 1 μB. The spin density is borne by all the carbon atoms in lattice B surrounding the adsorption site.

Once the single H atom is adsorbed, it can be seen as a radical delocalized at sublattice B to form a triplet spin conformation. If another H atom adsorbs on the sublattice B and at ortho and para sites of the first H atom adsorption site on the same aromatic ring, the unpaired electron forms the covalent C–H bonds. And then, the system is in a nonmagnetic singlet electronic conformation when in the thermodynamic ground state. If the two H adsorption sites are in the meta position, which means that they are bonded by carbon atoms from the same sublattice, the formation of a chemical C–H bond implies breaking another aromatic π bond. As in the case of the single H atom adsorption, this conformation is known as non-Kekule with the ground state in a triplet spin conformation too. Boukhvalov et al. calculated the H dimmer model and shown that the most thermodynamic stable conformation of low hydrogenated graphene layer corresponds to the nonmagnetic pair of hydrogen atoms attached to the different sublattices of graphene from the different sides (Figure 16.3).

Actually, the fully hydrogenated graphene (namely, graphane) was theoretically investigated by Sofo et al.[18] The graphane has two favorable conformers: a chair-like conformer with the hydrogen atoms alternating on both sides of the plane with the space group of *P-3m*1 (Figure 16.4a) and a boat-like conformer with the hydrogen atoms alternating in pairs with the space group of *Pmmn*. They found that graphane is among the most stable hydrocarbons, and it is the most stable for its hydrogen concentration. It is also more stable than mixtures of cyclohexene and graphite. The electronic band structures of the two conformers are very similar. They have a direct bandgap at the Γ point and with the large bandgap of 5.4 eV for the chair conformation (see Figure 16.4b) and 4.9 eV for the boat conformation.[32] It suggests that a variable electron bandgap between that of graphene (0 eV) and its 100% hydrogenated

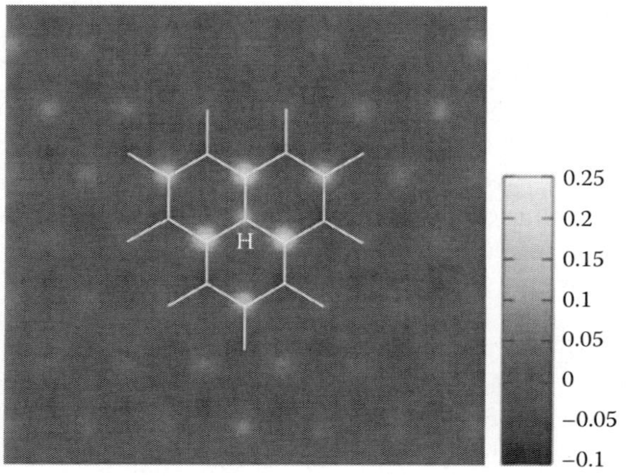

FIGURE 16.2 Single H adsorption on graphene. The projected spin distribution on the graphene plane.

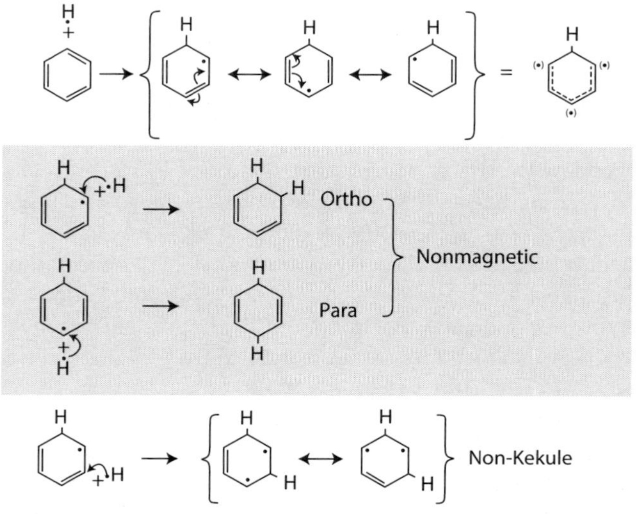

FIGURE 16.3 Double H adsorptions on an aromatic ring. Mesomeric effect and non-Kekule structures.

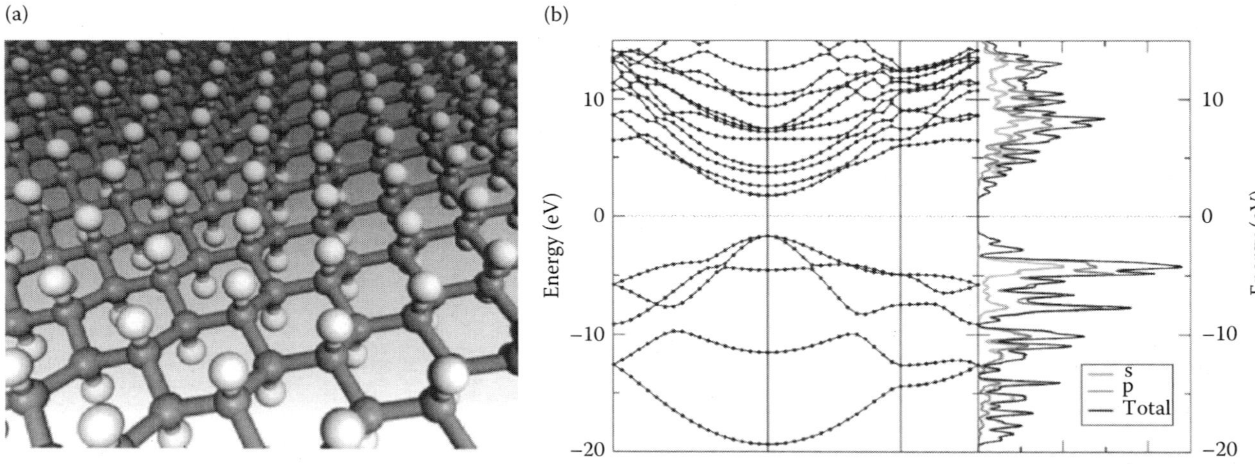

FIGURE 16.4 (a) Structure of graphane in the chair conformation. The carbon atoms are shown in gray and the hydrogen atoms in white. The figure shows the hexagonal network with carbon in the sp^3 hybridization. (b) Band structure (left) and DOS (right) of the chair conformer. The decomposition into states with s and p (gray) symmetries is shown together with the total DOS (black).

counterpart graphane (5 eV) may be achieved by the partial hydrogenation of graphene.

The relation between the band gap and hydrogen coverage was investigated by Gao et al.[33] using the random model, pattern model, and DFT calculations. It was known that the structural defect (like H adsorption) on graphene or (H vacancy) on graphane might induce impurity states in the middle of the gap region (called mid-states). For example, as mentioned above, removal of one H atom from graphane introduces an sp^2 hybridized C atom into the sp^3 carbon network. Each sp^2 carbon atom creates one unpaired local π electron (or radical), which result in the mid-states in the band structure. Based on this physical picture, they tested three types of hydrogenated graphene conformations as type one, randomly removing H pairs from fully hydrogenated graphene; type two, randomly removing individual H atoms from fully hydrogenated graphene; and type three, creating paired H vacancies according to some ordered pattern.

Based on the DFT–GGA calculations, they concluded: first, among the three types of configurations considered, the most stable configurations are those generated by random pairwise hydrogenation of the perfect graphene, suggesting that the clustering of H vacancy pairs stabilizes the structure; second, within the coverage range of 66.7%–100%, the hydrogen binding energy of the hydrogenated graphene is higher than that of the H_2 molecule; and third, the hydrogenated graphene with paired H vacancies presents an energy gap without mid-states due to the formation of the C=C double bond (or removal of radicals). For randomly hydrogenated graphene, the GGA bandgap reduces from 4.66 to 0 eV as the H coverage varies from 100% to 66.7%.

In the experiment, Elias et al.[30] reported reversible hydrogenation of graphene and showed that the electronic properties could be controlled by H adsorption and desorption. In the following, we will describe this work in detail. To obtain the large area graphene crystals, they prepared the samples by the use of micromechanical cleavage of graphite. After the initial characterization, the hydrogenated graphene were made by exposing the graphene to a cold hydrogen plasma. Under these conditions, it typically required 2 h of plasma treatment to reach the saturation in measured characteristics.

Before being hydrogenated, the graphene sample exhibited the standard ambipolar field effect with the neutrality point (NP) near zero gate voltage[1] and weak temperature dependence of its resistivity at all gate voltages. The metallic dependence of the sample was close to the NP below 50 K and the half-integer QHE at cryogenic temperatures (Figure 16.5), both of which are the characters of single-layer graphene.

These properties changed completely after the devices were hydrogenated. The devices exhibited an insulating behavior such that the resistivity ρ grew by two orders of magnitude with decreasing temperature T from 300 to 4 K (Figure 16.5c). Carrier mobility decreased at liquid-helium temperatures down to values of ~10 $cm^2\ V^{-1}\ s^{-1}$ for typical carrier concentrations n on the order of $10^{12}\ cm^{-2}$. The quantum Hall plateaus, so marked in the original devices, completely disappeared, with only weak signatures of Shubnikov–de Haas oscillations remaining in magnetic field B of 14 T. In addition, a shift of NP to gate voltages $V_g \approx +50$ V, which showed that graphene became p-doped in concentration of $\approx 3 \times 10^{12}\ cm^{-2}$ (probably due to adsorbed water). At carrier concentrations of less than $3 \times 10^{12}\ cm^{-2}$, the observed temperature dependences r(T) can be well fitted by $\exp[(T_0/T)^{1/3}]$, (T_0 is the parameter that depends on V_g) (Figure 16.6), which is a signature of variable-range hopping in two dimensions.[34] T_0 exhibits a maximum at NP of ~250 K and strongly decreases away from NP (Figure 16.6b). At $n > 4 \times 10^{12}\ cm^{-2}$ (for both electrons and holes), changes in r with T became small (similar to those in pristine graphene), which indicates a transition from the insulating to the metallic regime.

How dramatic it is that the original metallic state of graphene could be restored after annealing. The hydrogenated samples were stable at room T for many days and showed the same characteristics during repeated measurements. After the

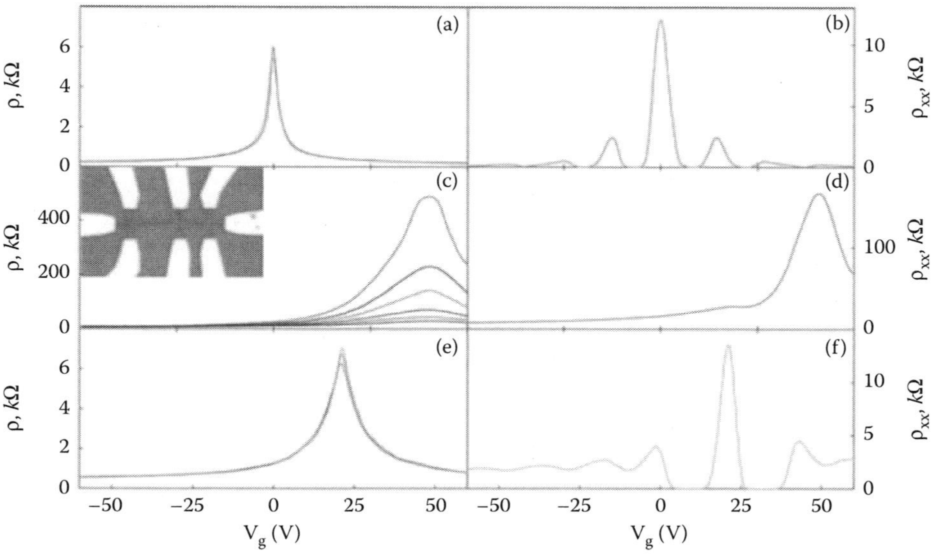

FIGURE 16.5 Control of the electronic properties of graphene by hydrogenation. The electric field effect for one of the samples at zero B at various temperatures T (left column) and in B = 14 T at 4 K (right). (a and b) Sample before its exposure to atomic hydrogen; curves in (a) for three temperatures (40, 80, and 160 K) practically coincide. (c and d) After atomic hydrogen treatment. In (c), temperature increases from the top; T = 4, 10, 20, 40, 80, and 160 K. (e and f) Same sample after annealing. (e) T = 40, 80, and 160 K, from top to bottom. (Inset) Optical micrograph of a typical Hall bar device. The scale is given by its width of 1 mm.

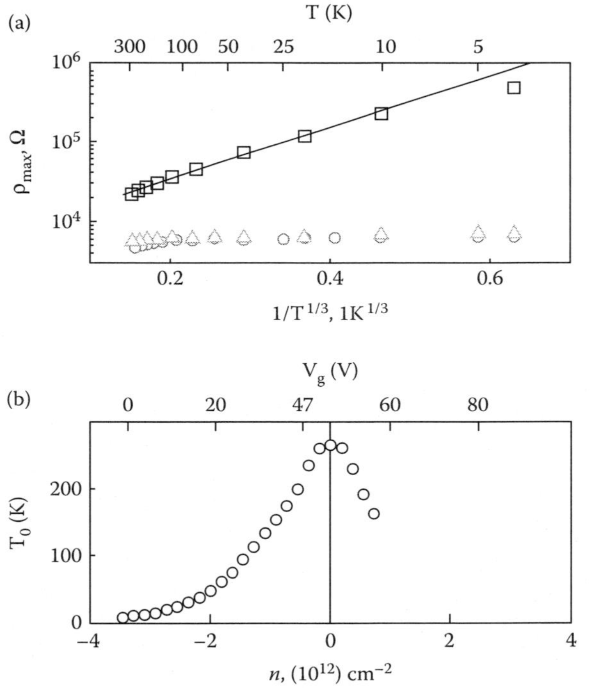

FIGURE 16.6 Metal–insulator transition in hydrogenated graphene. (a) Temperature dependence of graphene's resistivity at NP for the sample shown in Figure 16.5. Circles, squares, and triangles are for pristine, hydrogenated, and annealed graphene, respectively. The solid line is a fit by the variable-range hopping dependence $\exp[(T_0/T)^{1/3}]$. (b) Characteristic exponents T_0 found from this fitting at different carrier concentrations.

annealing, the samples practically returned to the same state as before hydrogenation: ρ as a function of V_g again reached a maximum value of $\approx h/4e^2$, where h is Planck's constant and e is the electron charge and became only weakly T dependent. Also, μ recovered to ~3500 $cm^2\ V^{-1}\ s^{-1}$, and the QHE reappeared. However, the recovery was not complete: Graphene remained p-doped, the QHE did not restore at filling factors ν larger than ±2, and zero B field conductivity σ(=1/ρ) became a sublinear function of n, which indicates an increased number of short-range scatterers.[4] The remnant properties attribute to vacancies induced by plasma damage or residual oxygen during annealing. To this end, after annealing, the distance (as a function of V_g) between the peaks in ρ_{xx} at n = 0 and ±4 became notably greater (~40%) than that between all the other peaks for both annealed and original samples. The distance indicates the mid-states[35] that were caused during the processing, which was in agreement with the observed sublinear behavior of the conductivity. The extra charge required to fill these states[35] yields their density of about 1×10^{12} cm^{-2}.

The changes in Raman spectra are in agreement with the transport measurements. The main features in the Raman spectra of pristine graphene are the G and 2D peaks that lie at around 1580 and 2680 cm^{-1}, respectively. The G peak corresponds to optical E_{2g} phonons at the Brillouin zone center. The 2D peak is the sum of two phonons with opposite momentum, and its shape identifies monolayer graphene.[36] However, a D peak at 1350 cm^{-1} is caused by and requires a defect for its activation via an intervalley double-resonance Raman process, and its intensity provides a convenient measure of the amount of disorder in graphene.[36,37] There is also a

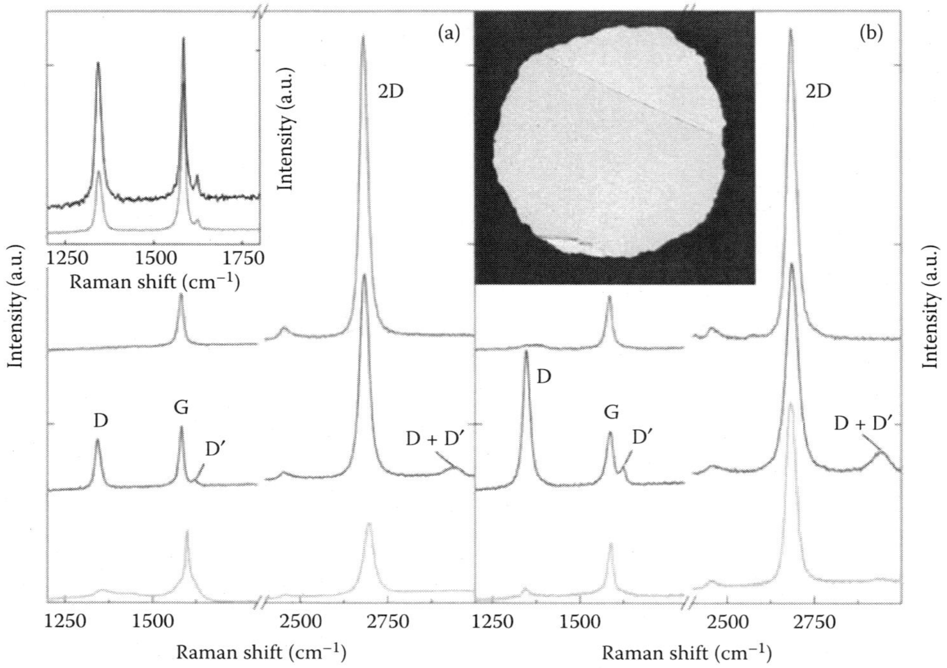

FIGURE 16.7 Changes in Raman spectra of graphene caused by hydrogenation. The spectra are normalized to have a similar intensity of the G peak. (a) Graphene on SiO_2. (b) Free-standing graphene. Curves (top to bottom) correspond to pristine, hydrogenated, and annealed samples, respectively.

peak at ~1620 cm^{-1}, called D′, which occurs via an intravalley double-resonance process in the presence of defects. Figure 16.7 shows the evolution of the Raman spectra for graphene crystals that are hydrogenated and annealed simultaneously with the sample. Hydrogenation resulted in the appearance of sharp D and D′ peaks, slight broadening and suppressing the height of the 2D peak relative to the G peak, and the onset of a combination mode (D + D′) around 2950 cm^{-1}, which, unlike the 2D and 2D′ bands, requires a defect for its activation because it is a combination of two phonons with different momentum. A sharp D peak in hydrogenated graphene is observed at 1342 cm^{-1}, as compared with that in disordered or nanostructured carbon-based materials.[38] This sharp D peak of the hydrogenated sample breaks off the translational symmetry of C–C sp^2 bonds after the formation of C–H sp^3 bonds. After annealing, the Raman spectrum recovered to almost its original shape, and all of the defect-related peaks (D, D′, and D + D′) were strongly suppressed. However, two broad low-intensity bands appeared, overlapping a sharper G and residual D peaks. These bands are indicative of some residual structural disorder. The 2D peak remained relatively small with respect to the G peak when compared with the same ratio in the pristine sample, and both shifted to higher energies, indicating that the annealed graphene is p-doped, as shown in Figure 16.7.[39]

Balog et al.'s[21] experiment realizes patterned H adsorption on the moiré superlattice positions of graphene and gains a bandgap. The sample was prepared using epitaxial graphene with the moiré pattern on Ir(111) as a template. After increasing the exposure to atomic hydrogen, the STM image illustrates that hydrogen atoms adsorbed hydrogen atoms, followed the bright parts of the moiré pattern, and then formed the moiré periodicity. Figure 16.8a shows the clean graphene surface. It reveals the characteristic moiré pattern. In Figure 16.8b, graphene exposed to a very low dose of atomic hydrogen is shown. Some protrusions appear on the surface, and these are observed to be located at the bright parts of the moiré pattern. Increased hydrogen exposure leads to the formation of ring-like structures along the moiré superlattice, as shown in Figure 16.1c. At even higher hydrogen doses, the ring-like structures merge and form elongated structures, see Figure 16.8d and e. Figure 16.8f shows a Fourier transform of the STM image in Figure 16.8e, revealing that even at the high coverage the adsorbed hydrogen follows the moiré superlattice periodicity.

ARPES demonstrates the size of the overall band gap. Figure 16.8a shows the photoemission intensity near the Fermi level E_F for a pristine layer of epitaxial graphene on Ir(111). The data follows the path of A–K–A′ in the Brillouin zone, as shown in the inset of Figure 16.9a. One clearly identifies the π-band dispersion associated with graphene and connects the π-band at the Dirac point near the Fermi lever. The mini-gaps visible in the bare graphene π-band at a binding energy of about 0.7 eV are related to the moiré superstructure visible in low energy electron diffraction and STM images. Their observation in ARPES is indicative of the excellent long-range ordering in the graphene layer. Figure 16.9b and c illustrates the evolution of the π-band for increasing exposures to atomic hydrogen. A comparison of the ARPES results for the clean and hydrogenated graphene layer gives rise to three conclusions. First, hydrogen adsorption induces a significant

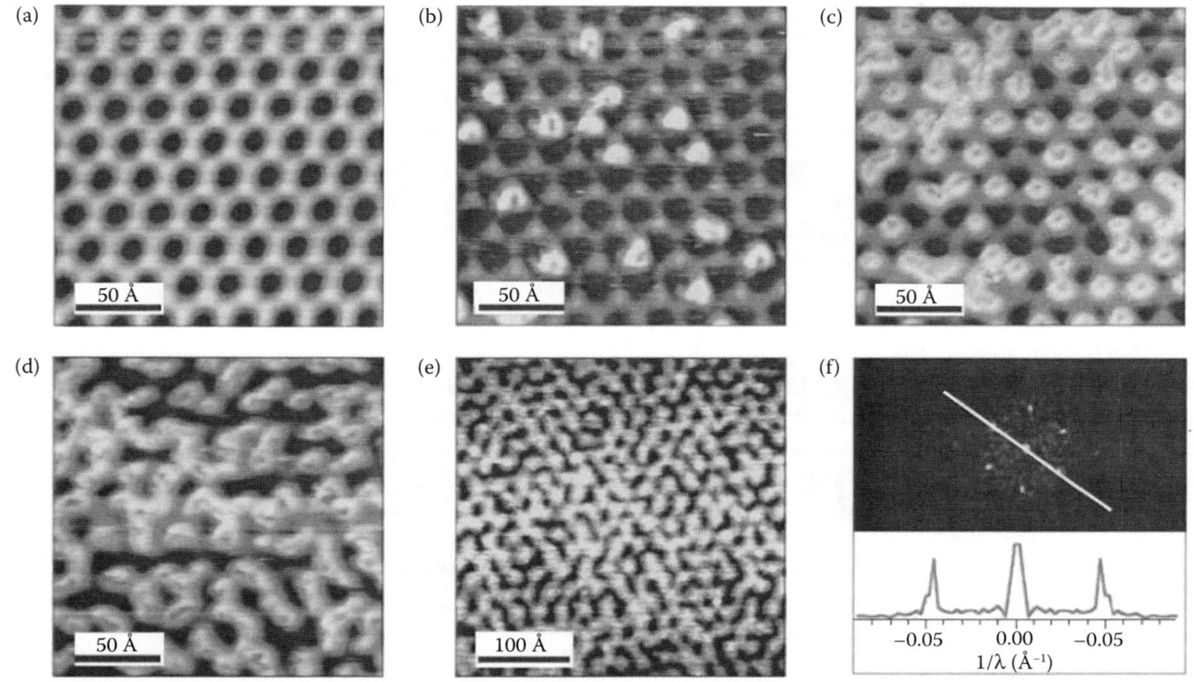

FIGURE 16.8 STM images of hydrogenate structures following and preserving the moiré pattern of graphene on Ir(111). (a) Moiré pattern of clean graphene on Ir(111) with the superlattice cell indicated. (b) through (e) Graphene exposed to atomic hydrogen for very low dose, 15, 30, and 50 s, respectively. (f) Fourier transform of the image in (e) illustrating that hydrogen adsorbate structures preserve the moiré periodicity.

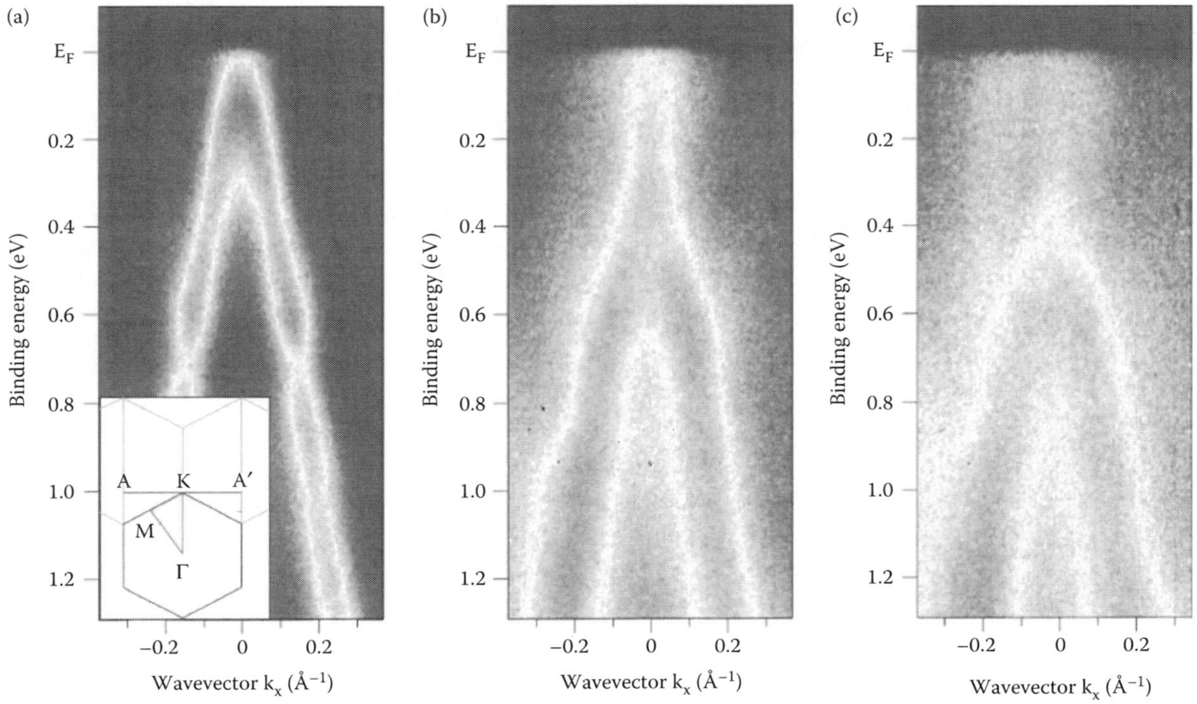

FIGURE 16.9 Observation of a gap opening in hydrogenated graphene (a through c). Photoemission intensity along the A–K–A′ direction of the Brillouin zone for clean graphene on Ir(111) (a), graphene exposed to a 30 s dose of atomic hydrogen (b), and graphene exposed to a 50 s dose of atomic hydrogen (c).

gap opening at the Fermi level. For the intermediate dose of 30 s in Figure 16.9b, the size of the gap is not well defined. However, for the 50 s dose in Figure 16.9c, one can observe a well-defined gap between the top of the observed π-band 450 meV below the Fermi level and, hence, the gap must have at least this width. As ARPES probes only occupied electronic states, a determination of the absolute gap size is not possible. Second, the π-band broadens with increasing hydrogen exposure and, third, the overall π-band signal weakens.

The superlattice structure and band gap properties are also investigated by DFT and DFT-based tight-binding (DFTB) calculations. The calculations illustrate two conclusions. First, structures with hydrogen atoms in stable dimer positions on ortho sites or opposite sides of para sites, which agree with Boukhvalov et al. and Ferro et al.'s work. Second, a clear band gap is observed at the K point and at the 23% coverage, a band gap opening as large as 0.73 eV is obtained (see Figure 16.4). Thus, the band gap opening by patterned hydrogen adsorption is not dependent on the formation of graphane-like islands, but could also be realized in other graphene systems. However, in general, partial hydrogenation is a feasible way to tune the band structure of graphene in applications.

16.4 STRUCTURE, ELECTRONIC, AND TRANSPORT PROPERTIES OF NITROPHENYL DIAZONIUM FUNCTIONALIZED GRAPHENE

In experimental studies, one verified way to create the single-atom chemisorption defects on graphene is covalent functionalization by diazonium salts. Nitrophenyl diazonium salts were widely used to functionalize amorphous graphite,[40] graphite crystals,[41] carbon nanotubes,[42] and graphene {Bekyarova, 2009 #43; Bekyarova, 2012 #44; Bekyarova, 2009 #43; Huang, 2013 #45}. This reaction mechanism is illustrated in Scheme 16.1. In the reaction, the nitrophenyl diazonium salts accept one electron from the carbon surface and release a nitrogen molecule to a nitrophenyl radical. The nitrophenyl radical is highly reactive so that it can then covalently bond to the carbon surface that provides the electron in the former step. The entire process is quick and dramatic. Its advantages are the solution-phase reaction nature, so that is easy to realize and control. Since the big size of the nitrophenyl group induces the stereo-hindrance effect, the coverage of NP-graphene on graphene can not reach as high as that of the chemisorption of the monatomic functional groups (compared with H and F) on graphene. Due to the size of nitrophenyl, the van der Waals effect and the stereo-hindrance effect are two competitive factors in achieving stability of high coverage NP-graphene in NP-graphene structures.

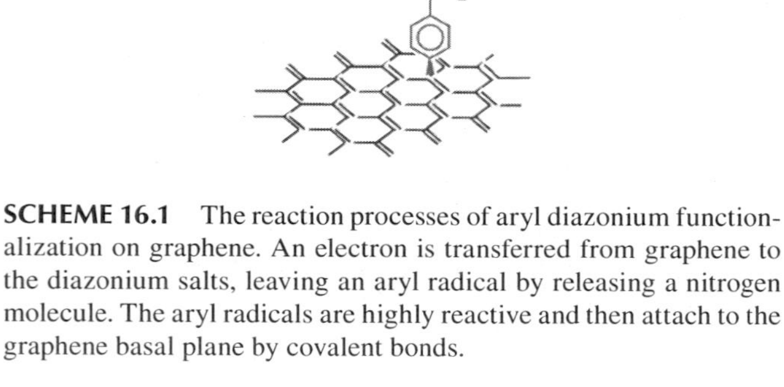

SCHEME 16.1 The reaction processes of aryl diazonium functionalization on graphene. An electron is transferred from graphene to the diazonium salts, leaving an aryl radical by releasing a nitrogen molecule. The aryl radicals are highly reactive and then attach to the graphene basal plane by covalent bonds.

In this chapter, we introduce a systematic investigation of nitrophenyl diazonium functionalized graphene (NP-graphene) including microstructures, electronic structures, and transport properties of random distribution, and long-range order structures, to provide a deep insight into the structure–property relation of this system.

16.4.1 MICROSTRUCTURES OF NP-GRAPHENE

The structure of NP-graphene can be characterized via atomic force microscope (AFM), Raman spectroscopy, and TEM, respectively. For the local chemical configurations, the AFM measurements revealed a height difference of 0.686 nm between pristine and functionalized graphene, which was consistent with the DFT calculations adopting the mode of the nitrophenyl groups (NPs) connecting to one side of the graphene basal plane perpendicularly.[43] The Raman spectra of the exhibited the D mode (at ~1350 cm^{-1}) and peaks deriving from the NPs (C–N symmetric stretching at 1108 cm^{-1}; N–O antisymmetric stretching at 1389 cm^{-1}, 1438 cm^{-1}, and 1515 cm^{-1}, respectively), besides the G and the 2D mode of pristine graphene, shown in Figure 16.10.[37] The D mode represents the defect-involving double resonant scattering process and the characteristics of the covalent bonds formed between graphene carbon atoms and NPs.[44] The simultaneous appearance of the D mode and NPs' peaks directly demonstrate that the NPs connected to the graphene basal plane via covalent bonds.

2D Raman maps could provide a global mapping of the spatial distribution of the NPs on graphene, and one such example is shown in Figure 16.10b. The intensity map of the G mode indicates the distribution of the stack layers of graphene, that is, more layers exhibit stronger G mode intensity. Also, the intensity maps of the D mode and the NPs' peak indicate the distribution of the covalent bonds and the NPs, respectively, which were nearly the same. Thus, the D and NPs maps both indicate the distribution of the covalently bonded NPs on graphene, which can be found quite inhomogeneous. Both these can be indicating that the more the layers stack, the fewer the NPs covered and more NPs intend to attach to the boundary areas, including the graphene sheet's edges as well as the boundaries between domains with different stack layers, than the bulk parts. This can be attributed to

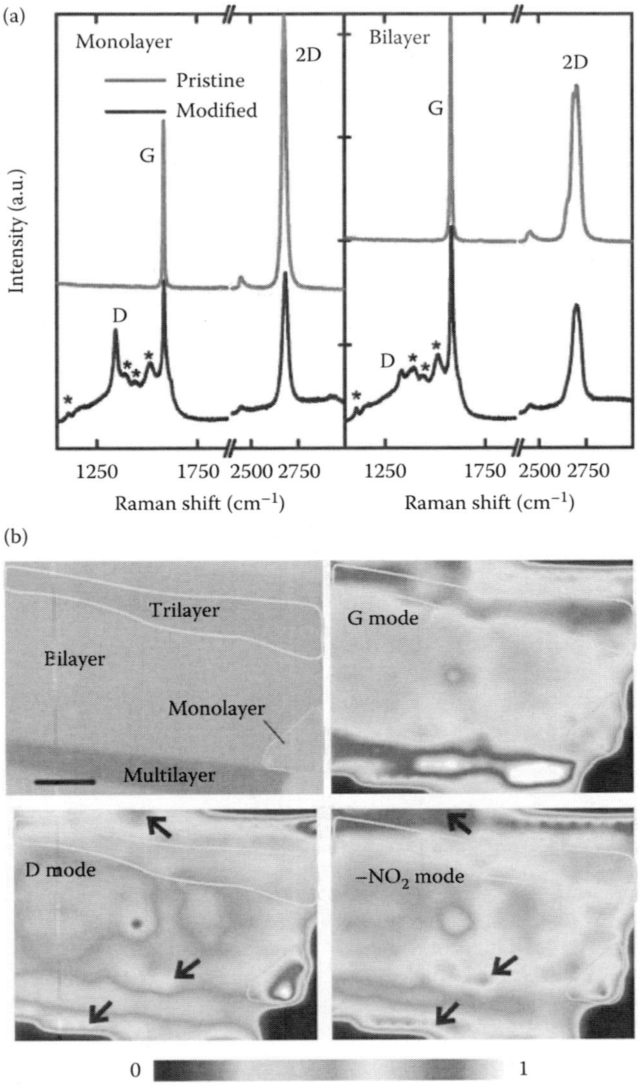

FIGURE 16.10 (a) Raman spectra of pristine and modified mono- and bilayer graphene. Peaks introduced by nitrophenyl groups are asterisked (*). (b) (Top left panel) Optical image of the investigated graphene sheet, where the scale bar is 2.5 μm and domains with different numbers of stack layers is outlined. The 2D Raman maps of the G mode, D mode, and N–O vibration mode.

the differences in the reactive ability, that is, more stack layers lead to stronger π–π interaction between graphene layers which enhances the reaction barriers {Koehler, 2010 #5083} and, on the other hand, the unsaturated dangling bonds in the boundary areas make them more reactive {Sharma, 2010 #5085}.

The inhomogeneous distribution of NPs on graphene reflects the random feature of the diazonium functionalization. The NPs can be viewed to form a random lattice with no symmetry elements. Thus, the combination of the random NP lattice and the graphene basal plane is expected to exhibit the global symmetry of graphene. Selected area electron diffraction (SAED) investigations gave direct and more detailed information on the lattice structures of the NP-graphene. The SAED patterns of the NP-graphene all revealed the graphene hexagonal symmetry, indicating the preserving of the graphene honeycomb lattice, as shown in Figure 16.11c.

16.4.2 Long-Range Ordering of NP-Graphene

Zhu et al.[43] reported that, for the first time, they obtained the long-range ordering of NP-graphene with approximately twice that of pristine graphene's crystal lattice constant. Is it possible for the NPs to form fine long-range ordered patterns on the graphene basal plane like this? Generally, the diazonium functionalization reaction may be under kinetic rather than thermodynamic control due to its high rate reaction mechanism. Thus, unlike small radicals such as H that easily arrives at the thermodynamic ground state, the NPs take random distribution since the equilibrium state of long-range ordered arrangement cannot be achieved in NP-graphene in a global view. However, fluctuations may lead to long-range ordering in a local domain, even if the possibility is ultrasmall. There are many factors that could induce fluctuations of the reaction, such as local thermal excitation, inhomogeneity of the reagent solution, corrugation of the graphene sheet, and so on. Figure 16.12a shows the SAED pattern of an NP-graphene that exhibits superlattice structure. This superlattice structure implies that 1/8 of the carbon atoms of the graphene were transformed from sp^2 to sp^3 via covalently bound, and the equivalence of two sublattices of pristine graphene was broken as illustrated in Figure 16.12. This configuration is a non-Kekule structure, which means it is commonly thermodynamically unstable, as mentioned above. This structure and its allotropes with the same coverage were carried using the DFT calculation {Jing, 2013 #20}. But the GGA method, which neglects the long-range van der Waals interaction among nitrophenyl groups in theoretical studies, reveals a positive increment in the total energy after the formation of all these long-range ordered structures, implying they are thermodynamically unstable. When using DFT-D2[45] and the DFT + LAP methods[46] that count the van der Waals interactions to estimate the binding energies of NPs adding to the graphene basal plane, thermodynamically favorable binding energies were obtained. It is predicted, however, that the configuration of NPs on two different sublattices is more stable than the configuration of NPs on the identical sublattice, which is also demonstrated by our own calculation results for the former which has a lower binding energy than the latter. Since we speculate that the observed superlattice structure arises from the fluctuation during the reaction process, and is stable in energy, it should situate in a potential well in the energy spectrum, that is, its energy is a local minimum, and as a result, the potential barriers around prevent it from relaxing to the local/global minimums of other configurations. And, the van der Waals interactions between NPs and steric hindrance effects would inhibit the mesomeric effect of the nitrophenyl-radical functionalization to reach the energy minimum, which infers that NP-graphene is inclined to form multiple structures, especially some non-Kekule configurations.

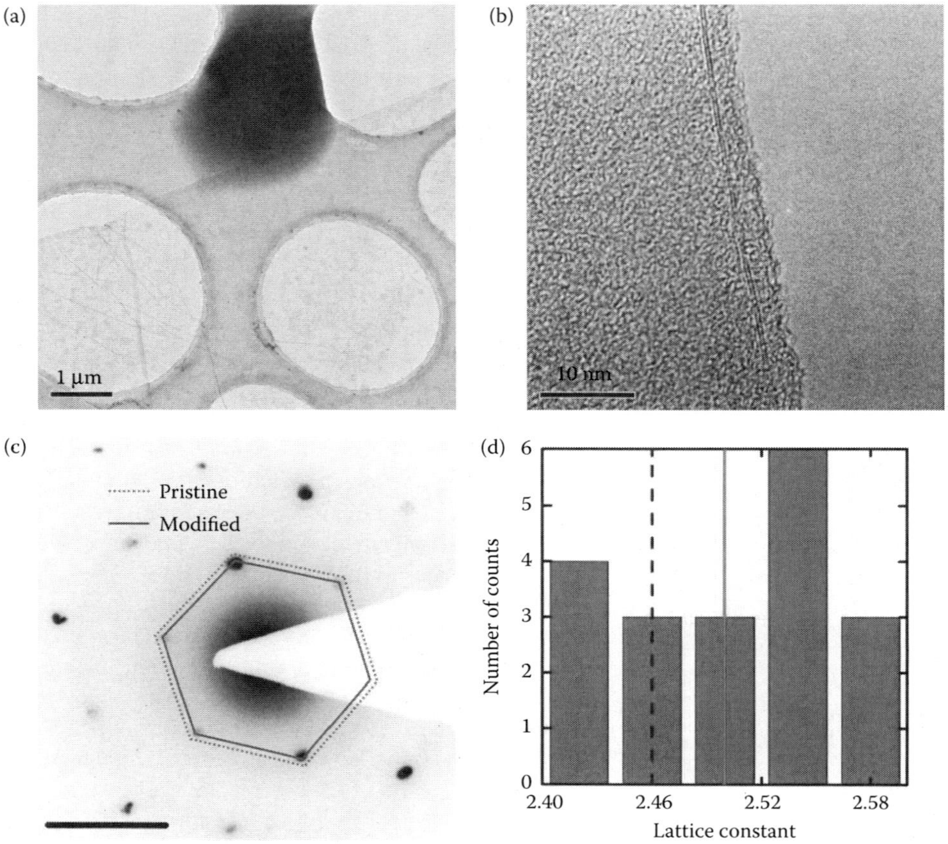

FIGURE 16.11 (a) Low-magnification overview image of a suspended modified graphene sheet on lacey support foil. (b) High-resolution image of a modified graphene sheet near its edge. Note that the exhibition of two dark edges indicates a bilayer domain. (c) SAED image of a modified graphene membrane. The inner and outer hexagons indicate the diffraction spots of modified and pristine graphene, respectively. Scale bar, 5 nm^{-1}. (d) Distribution of lattice constant measured in our modified graphene samples. The dashed and solid lines indicate the lattice constant of pristine and modified graphene, respectively.

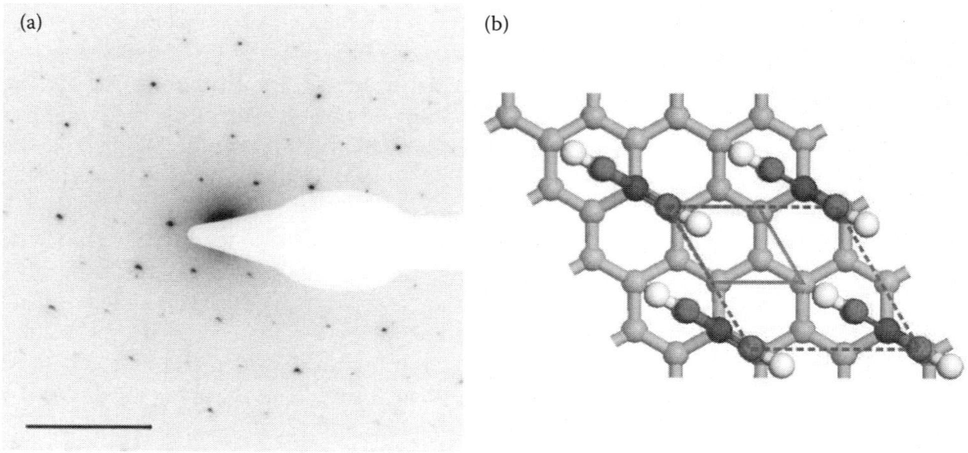

FIGURE 16.12 (a) SAED pattern of diazonium functionalized graphene exhibiting superlattice structure. Scale bar, 5 nm^{-1}. (b) Illustration of the primitive cell in which the nitrophenyl groups form graphene honeycomb superlattice with twice the lattice constant of graphene. The inner and outer diamonds are a guide for the eyes and indicate primitive cells of pristine and modified graphene, respectively.

The electronic structure of the superlattice was investigated in detail. The band structure shown in Figure 16.13a represents a direct bandgap semiconductor with valence band maximum located at the G point on spin-up bands, and conduction band minimum at the G point on spin-down bands. The valence band and conduction band are narrow bands mainly formed by the quasi-localized electronic states of the defects. These states (as shown in Figure 16.13a) are the narrow peaks of the DOS near the Fermi level and belong to the unpaired p_z orbitals in graphene. The width of the defects' narrow peak is less than 0.2 eV, which are necessary for the stability of magnetic ordering at high temperatures.[25] The energy bandgap was found to be 0.50 eV from the band structure. This bandgap is ascribed to the defect state exchange splitting, which is defined as the difference between the corresponding majority spin and minority spin band maxima.

The ortho-configured NP superlattice leads to spin polarization. The spin-density distribution over the graphene basal plane is plotted in Figure 16.13b. This indicates that the C atoms belonging to sublattice B induced large positive magnetic moments while the C atoms in sublattice A attached by NPs carried vanishing magnetic moments while the other C atoms in sublattice A carried small negative magnetic moments (less than 0.1 μB). The long-range ordering of the absorbent NPs creates the defects that introduce vacant sites of spin density on graphene. The defect state is distributed over the sites of the sublattice complementary to the one in which the defect was created (i.e., over the odd nearest neighbors). The major positive contribution to the electron spin density originates from the exchange splitting of the defect states. In addition, the exchange spin-polarization effect (i.e., the response of the fully populated valence bands to the magnetization of the defect states) results in a negative spin density on the even-nearest-neighbor sites and in the enhancement of a positive spin density on the odd-nearest-neighbor sites.[23] The defect sites on the graphene basal plane clearly exhibit triangular symmetry, and the vacant sites of spin density in the sublattice A leave unpair p_z spin in the sublattice B, leading

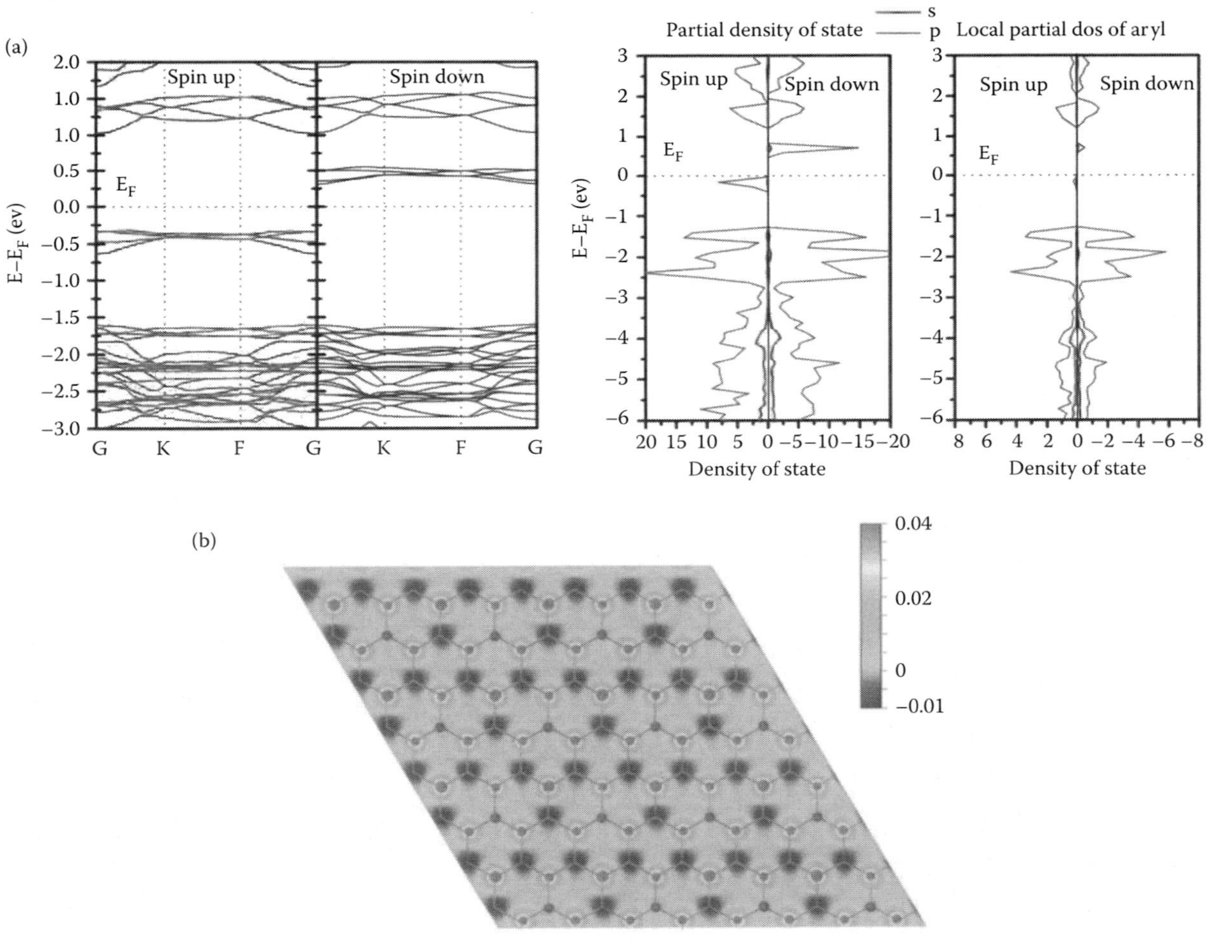

FIGURE 16.13 (a) Band structures and the project density of states (PDOS) of NP-formed superlattice configuration. The k path of the band structures follows G–K–F–G, where G = (0,0,0), K = (−1/3,2/3,0) represent the Dirac point of graphene, and F = (0,1/2,0). In the band structures, the left curves and right curves represent the spin-up bands and spin-down bands, respectively. (b) Calculated spin density (ρ↑−ρ↓) of the NP-formed superlattice configuration (8–1). The cycle dot on the honeycomb lattice of graphene indicates the position of the carbon atoms modified by nitrophenyl groups.

to the ferromagnetic case. Thus, three interesting domains can be distinguished on the graphene basal plane. First, vanishing spin density is the background of the whole graphene plane including those carbon atoms bonding with nitrophenyl groups. Second, large spin-up density can be found around the carbon atoms in the complementary sublattice of the bonded one with the triangular lattice. Third, the small spin-down density forms the kagome lattice. As the valence electrons in p-states are quasi-localized defect states that are more delocalized than those in d- or f-electrons, they have a much larger spatial extension that promotes long-range exchange interaction. Therefore, the long-range magnetic coupling between 2p moments can be attributed to the extended p–p interactions. As a result, the total spin density will be in the spin-up direction to form a spin-polarized configuration and, hence, macro ferromagnetism may be demonstrated in this structure.

16.4.3 Transport Properties of Random Distributions of NP-Graphene

The transport properties of graphene are ascribed to the conjugated π electronic network. Appropriate chemical absorptions on graphene induce defects to break the π electronic network and, hence, change the electronic structure and conductivity properties. The random distribution of NPs on the graphene basal plane induces two competitive effects to its electronic properties: one is the charge transfer from graphene to NPs and the other is the decrease of the carrier mobility by enhanced scattering. With a para nitryl connecting on the benzene, the NPs are typical electron-withdrawing groups. When NPs covalently bond with the graphene, the electron density between NPs and graphene carbon atoms is more inclined to the NPs, thus the NP-graphene becomes p-type doping. In a certain range, which depends on the specific configuration of the NPs on graphene, the charge transfer-induced holes density increases with the concentration of the NPs. The increase in the hole density induces enhancement of the electric conductivity. On the other hand, the covalent bonds of NPs with graphene introduce defects into the system, which increase the scattering probability of the charge carriers propagating in the NP-graphene and, hence, suppress the electric conductivity.

Unlike the full chemisorptions of small atoms on graphene, due to the big size of the nitrophenyl groups, however, their obvious steric hindrance effect limits their coverage on graphene. As mentioned above, highly hydrogenated graphene would be an insulator and the temperature dependence of its resistivity can be described by 2D variable ranges hopping, which is the transport mechanism for strongly localized systems.[30] In fact, as more and more graphene carbon atoms change from sp^2 to sp^3 hybridization by covalent functionalization, the conjugated π electrons will vanish after a certain critical point at which the 2D percolation phase transition occurs.[33] In this situation, free electrons that can conduct current in graphene are very rare. If the H atoms are randomly distributed (the situations of high but not full functionalization), the strong scattering makes the electrons localized; if the H or F atoms bond to every carbon atom with full reaction, a wide gap will be opened in the energy band, as predicted by theoretical calculations.[32] Both situations will result in insulating graphene.

The transport mechanism of the NP-graphene can be derived from further investigations. Let us take one sample as an example, shown in Figure 16.14. The ρ–T data can be well fitted by the Bloch–Grüneisen's law that describes the temperature dependence of the resistivity of monovalent metals as well as 2D electron gases[47]

$$\rho(T) = \rho_0 + \alpha\left(\frac{T}{\Theta_D}\right)^5 + \int_0^{\Theta_D/T} \frac{z^5}{(e^z - 1)(1 - e^{-z})} dz$$

where ρ_0 is the residual resistivity, α is a constant, and Θ_D is the Debye temperature below which only the long wave phonons (i.e., lattice oscillation modes) of the crystal are excited. The data revealed the Θ_D to be ~1840 K, slightly lower than that of pristine graphene of ~2000 K, which indicate that the origin of the NP-graphene's resistance is mainly from the

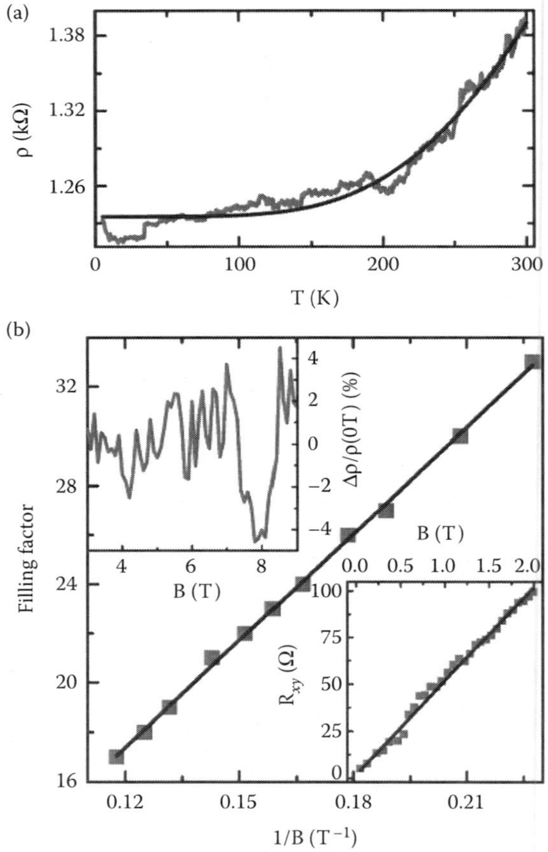

FIGURE 16.14 (a) ρ–T characteristics of Diazonium functionalized graphene (DFG) (gray line); fitting curve to Equation 16.1 is also shown (black line). (b) Landau fan diagram of the DFG at each SdH peaks and its linear fitting at 5 K, revealing an SdH oscillation frequency of BF ~ 146 T. Note that the intersection is zero, which means a Berry's phase of π. Top inset: the SdH oscillation of the DFG at 5 K. Bottom inset: Hall effect of the DFG.

phonon scattering. Unlike the semiconductive NP-graphene, the resistivity of which increased monotonically with decreasing temperature, implying thermal excitation of the charge carriers is the dominant factor. The high intrinsic hole concentration and phonon scattering-induced resistance indicate that the NP-graphene behaved as 2D hole gases (2DHGs).

The bonding of the NPs will locally break the honeycomb lattice symmetry of graphene so that the linear band structure will become nonlinear. As a result of this, the effective mass of the electrons in the NP-graphene will be finite rather than vanishing as in pristine graphene with linear dispersion relation. The effective mass of electrons can be estimated from the temperature evolution of the amplitude of Shubnikov–de Haas (SdH) oscillation according to the following equation[47]:

$$\frac{\Delta\rho}{\rho_0} = \frac{2\pi^2 m^* k_B T/\hbar eB}{\sinh(2\pi^2 m^* k_B T/\hbar eB)}$$

where $\Delta\rho$ is the amplitude of SdH oscillation, ρ_0 the resistivity without a magnetic field, m^* the effective mass of electrons, k_B the Boltzmann constant, and $\hbar$ the Planck's constant. The SdH oscillation was observed in our magnetoresistance (MR) measurements taken at 5, 10, and 50 K, respectively, which can be clearly extracted by subtracting a linear and nearly temperature-independent MR background, as shown in Figure 16.4. The temperature evolution of the peaks at B = 6.3 and 7.0 T were fitted to Equation 16.2 and the results revealed 0.574me and 0.565me, respectively (me is the mass of free electron), while experimentally derived effective mass of pristine graphene was ~0.02me.[47]

The effective mass determines the electrons' response to the external electric field and virtually reflects the effect of the lattice periodic potential, which give more detailed information about the band structure of the NP-graphene. The density of states (DOS) can be estimated as $g(E) = g_s g_v m^*/2\pi\hbar$ ~4.8×10^{11} $cm^{-2} \cdot meV^{-1}$ (where g_s and g_v are the spin and valley degeneracy, respectively), and such a giant DOS is the origin of its high intrinsic hole concentration. The Fermi velocity $v_F = \hbar k_F/m^*$ was then ~1.4×10^5 m s^{-1} (where $k_F = \sqrt{4\pi n/g_s g_v}$ is the Fermi wave vector), one magnitude lower than that of pristine graphene (which is 1×10^6 m s^{-1}). This means that the local band structure of the NP-graphene is indeed like a 2D hole gas.

On transport parameters, the mean free path l of the electrons determined by elastic scattering is $l = h/2e^2 k_F \rho$ ~10 nm. Meanwhile, the defect domain size of graphene can be empirically estimated from Raman spectrum via $L = 2.4 \times 20^{-10}$ $(nm^{-3}) \cdot \lambda^4(nm^4) \cdot I_G/I_D$, where λ is the excitation wavelength and I_G and I_D is the intensity of the G and the D mode, respectively.[48] The Raman spectrum of the NP-graphene revealed a domain scale of L ~8.9 nm, consistent with the mean free path, implying the defects induced by the covalent bonding with NPs was the origin of the enhanced elastic scattering in the NP-graphene. Since the carrier mobility μ is positively dependent on the mean free path, it is the significant suppression of l that leads to the suppression of μ.

Although NP-graphene can hardly lead to thorough localization of the charge carriers due to the limitation in the coverage, clear weak localization phenomenon was still observed in the NP-graphene. Weak localization was arising from the constructive interference of electron waves propagating along two time-reversal symmetric closed paths, that is, clockwise and counter-clockwise around a closed path, as illustrated in Figure 16.15. The constructive interference enhances the possibility of electron waves propagating to the origin so that adds a positive correction to the semiclassical resistivity. Weak localization manifests itself as a peak in an MR plot around zero fields since the phase coherence of electrons can be destroyed by an external magnetic field, as shown in Figure 16.15. Weak localization in NP-graphene needs more investigations because it is the middle status between ideal defect-free graphene and totally disordered functionalized graphene (such as hydrogenated graphene). The knowledge of this status will significantly benefit the understanding of how graphene transforms from semimetal to insulator, especially the breakdown process of the conjugated π-bond.[49]

Recently, several improvement methods have been employed to obtain the controllable coverage NP-graphene. Bissett et al.[50] transferred single-layer graphene onto a flexible substrate and investigated the functionalization using different aryl diazonium molecules while applying mechanical strain. They found that mechanical strain can alter the structure of graphene, and increase the reaction rate of NP-graphene, as well as increase the final degree of functionalization. Furthermore, they demonstrated that mechanical strain enables functionalization of graphene for both p- and n-type dopants, where unstrained graphene showed negligible reactivity. Shih et al.[51] used an electrochemical approach involving aryl diazonium salts to systematically probe electronic modification in SiO_2 supported monolayer and bilayer graphene with increasing functional coverage. They found that both monolayer and bilayer graphene retained their relatively high conductivity after functionalization even at high

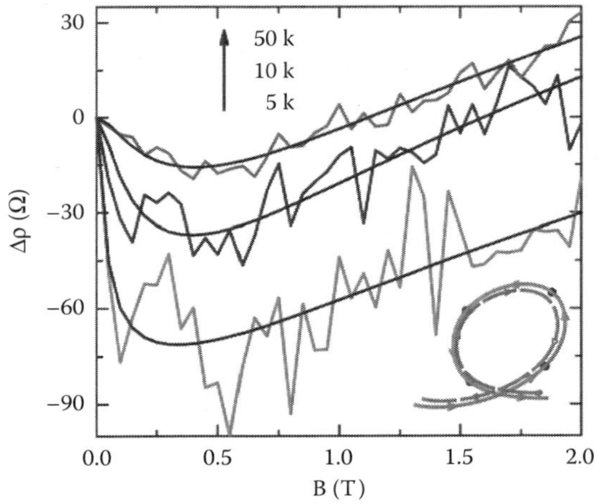

FIGURE 16.15 Weak localization of diazonium functionalized graphene in 5, 10, and 50 K. Inset: illustration of the interference of time-reversal paths that induces weak localization.

conversion, as mobility losses are offset by increases in carrier concentration. For monolayer graphene, they found that the bandgap opening was below 0.1 meV in magnitude for SiO_2 supported graphene. On the other hand, heavily functionalized bilayer graphene retains a signature dual-gated bandgap opening due to electric-field symmetry breaking. They found a notable asymmetric deflection of the charge neutrality point (CNP) under positive bias which increases the apparent on/off current ratio by 50%, suggesting that synergy between symmetry breaking, disorder, and quantum interference may allow the observation of new transistor phenomena. In summary, these new mechanisms of covalent functionalized NP-graphene may be expanded to a wide variety of potential graphene-based devices.

REFERENCES

1. Novoselov, K. S.; Geim, A. K.; Morozov, S. V.; Jiang, D.; Zhang, Y.; Dubonos, S. V.; Grigorieva, V.; Firsov, A. A., Electric field effect in atomically thin carbon films. *Science* 2004, 306, 666–669.
2. Geim, A. K.; Novoselov, K. S., The rise of graphene. *Nat. Mater.* 2007, 6, 183–191.
3. Mermin, N. D., Crystalline order in 2 dimensions. *Phys. Rev.* 1968, 176, 250–254.
4. Morozov, S. V.; Novoselov, K. S.; Katsnelson, M. I.; Schedin, F.; Elias, D. C.; Jaszczak, J. A.; Geim, A. K., Giant intrinsic carrier mobilities in graphene and its bilayer. *Phys. Rev. Lett.* 2008, 100, 106602.
5. Miao, F.; Wijeratne, S.; Zhang, Y.; Coskun, U. C.; Bao, W.; Lau, C. N., Phase-coherent transport in graphene quantum billiards. *Science* 2007, 317, 1530–1533.
6. Avouris, P., Graphene: Electronic and photonic properties and devices. *Nano Lett.* 2010, 10, 4285–4294.
7. Feng, M.; Zhao, J.; Huang, T.; Zhu, X. Y.; Petek, H., The electronic properties of superatom states of hollow molecules. *Acc. Chem. Res.* 2011, 44, 360–368.
8. Geim, A. K., Graphene: Status and prospects. *Science* 2009, 324, 1530–1534.
9. Allor, D.; Cohen, T. D.; McGady, D. A., Schwinger mechanism and graphene. *Phys. Rev. D* 2008, 78, 096009.
10. Novoselov, K., Graphene: Mind the gap. *Nat. Mater* 2007, 6, 720–721.
11. Giovannetti, G.; Khomyakov, P. A.; Brocks, G.; Kelly, P. J.; van den Brink, J., Substrate-induced band gap in graphene on hexagonal boron nitride: Ab initio density functional calculations. *Phys. Rev. B* 2007, 76, 073103.
12. Zhou, S. Y.; Gweon, G. H.; Fedorov, A. V.; First, P. N.; de Heer, W. A.; Lee, D. H.; Guinea, F.; Castro Neto, A. H.; Lanzara, A., Substrate-induced bandgap opening in epitaxial graphene. *Nat. Mater.* 2007, 6, 770–775.
13. Zhang, Y.; Tang, T.-T.; Girit, C.; Hao, Z.; Martin, M. C.; Zettl, A.; Crommie, M. F.; Shen, Y. R.; Wang, F., Direct observation of a widely tunable bandgap in bilayer graphene. *Nature* 2009, 459, 820–823.
14. Gui, G.; Li, J.; Zhong, J., Band structure engineering of graphene by strain: First-principles calculations. *Phys. Rev. B* 2008, 78, 075435.
15. Lu, Y. H.; Chen, W.; Feng, Y. P.; He, P. M., Tuning the electronic structure of graphene by an organic molecule. *J. Phys. Chem. B* 2008, 113, 2–5.
16. Ribeiro, R. M.; Peres, N. M. R.; Coutinho, J.; Briddon, P. R., Inducing energy gaps in monolayer and bilayer graphene: Local density approximation calculations. *Phys. Rev. B* 2008, 78, 075442.
17. Boukhvalov, D. W.; Katsnelson, M. I.; Lichtenstein, A. I., Hydrogen on graphene: Electronic structure, total energy, structural distortions and magnetism from first-principles calculations. *Phys. Rev. B* 2008, 77, 035427.
18. Sofo, J. O.; Chaudhari, A. S.; Barber, G. D., Graphane: A two-dimensional hydrocarbon. *Phys. Rev. B* 2007, 75, 153401.
19. Yan, J. A.; Xian, L. D.; Chou, M. Y., Structural and electronic properties of oxidized graphene. *Phys. Rev. Lett.* 2009, 103.
20. Gao, B. L.; Xu, Q. Q.; Ke, S. H.; Xu, N.; Hu, G.; Wang, Y. Z.; Liang, F.; Tang, Y. L.; Xiong, S. J., Band-gap modulation of graphane-like SiC nanoribbons under uniaxial elastic strain. *Phys. Lett. A* 2014, 378, 565–569.
21. Balog, R.; Jørgensen, B.; Nilsson, L.; Andersen, M.; Rienks, E.; Bianchi, M.; Fanetti, M. et al. Bandgap opening in graphene induced by patterned hydrogen adsorption. *Nat. Mater.* 2010, 9, 315–319.
22. Hong, J.; Niyog, S.; Bekyarova, E.; Itkis, M. E.; Ramesh, P.; Amos, N.; Litvinov, D. et al. Effect of nitrophenyl functionalization on the magnetic properties of epitaxial graphene. *Small* 2011, 9, 1175–1180.
23. Yazyev, O.; Helm, L., Defect-induced magnetism in graphene. *Phys. Rev. B* 2007, 75, 125408.
24. Ferro, Y.; Teillet-Billy, D.; Rougeau, N.; Sidis, V.; Morisset, S.; Allouche, A., Stability and magnetism of hydrogen dimers on graphene. *Phys. Rev. B* 2008, 78, 085417.
25. Edwards, D. M.; Katsnelson, M. I., High-temperature ferromagnetism of sp electrons in narrow impurity bands: Application to CaB(6). *J. Phys.-Condens. Mater.* 2006, 18, 7209–7225.
26. Vozmediano, M. A. H.; López-Sancho, M. P.; Stauber, T.; Guinea, F., Local defects and ferromagnetism in graphene layers. *Phys. Rev. B* 2005, 72, 155121.
27. Dugaev, V. K.; Litvinov, V. I.; Barnas, J., *Phys. Rev. B* 2006, 74, 224438.
28. Hornekær, L.; Šljivančanin, Ž.; Xu, W.; Otero, R.; Rauls, E.; Stensgaard, I.; Lægsgaard, E.; Hammer, B.; Besenbacher, F., Metastable structures and recombination pathways for atomic hydrogen on the graphite (0001) surface. *Phys. Rev. Lett.* 2006, 96, 156104.
29. Balog, R.; Jørgensen, B.; Wells, J.; Lægsgaard, E.; Hofmann, P.; Besenbacher, F.; Hornekær, L., Atomic hydrogen adsorbate structures on graphene. *J. Am. Chem. Soc.* 2009, 131, 8744–8745.
30. Elias, D. C.; Nair, R. R.; Mohiuddin, T. M. G.; Morozov, S. V.; Blake, P.; Halsall, M. P.; Ferrari, A. C. et al. Control of graphene's properties by reversible hydrogenation: Evidence for graphane. *Science* 2009, 323, 610–613.
31. Meyer, J. C.; Geim, A. K.; Katsnelson, M. I.; Novoselov, K. S.; Booth, T. J.; Roth, S., The structure of suspended graphene sheets. *Nature* 2007, 446, 60–63.
32. Lebegue, S.; Klintenberg, M.; Eriksson, O.; Katsnelson, M. I., Accurate electronic band gap of pure and functionalized graphane from GW calculations. *Phys. Rev. B* 2009, 79.
33. Gao, H.; Wang, L.; Zhao, J.; Ding, F.; Lu, J., Band gap tuning of hydrogenated graphene: H coverage and configuration dependence. *J. Phys. Chem. C* 2011, 115, 3236–3242.
34. Mott, N. F., Conduction in non-crystalline materials. *Philos. Mag.* 1969, 19, 835–852.
35. Wehling, T. O.; Novoselov, K. S.; Morozov, S. V.; Vdovin, E. E.; Katsnelson, M. I.; Geim, A. K.; Lichtenstein, A. I., Molecular doping of graphene. *Nano Lett.* 2008, 8, 173–177.

36. Ferrari, A. C.; Meyer, J. C.; Scardaci, V.; Casiraghi, C.; Lazzeri, M.; Mauri, F.; Piscanec, S. et al. Raman spectrum of graphene and graphene layers. *Phys. Rev. Lett.* 2006, 97.
37. Ferrari, A. C., Raman spectroscopy of graphene and graphite: Disorder, electron-phonon coupling, doping and nonadiabatic effects. *Solid State Commun.* 2007, 143, 47–57.
38. Ferrari, A. C.; Robertson, J., Interpretation of Raman spectra of disordered and amorphous carbon. *Phys. Rev. B* 2000, 61, 14095–14107.
39. Das, A.; Pisana, S.; Chakraborty, B.; Piscanec, S.; Saha, S. K.; Waghmare, U. V.; Novoselov, K. S. et al. Monitoring dopants by Raman scattering in an electrochemically top-gated graphene transistor. *Nat. Nanotechnol.* 2008, 3, 210–215.
40. Liu, Y. C.; Mccreery, R. L., Reactions of organic monolayers on carbon surfaces observed with unenhanced Raman spectroscopy. *J. Am. Chem. Soc.* 1995, 117, 11254–11259.
41. Allongue, P.; Delamar, M.; Desbat, B.; Fagebaume, O.; Hitmi, R.; Pinson, J.; Saveant, J. M., Covalent modification of carbon surfaces by aryl radicals generated from the electrochemical reduction of diazonium salts. *J. Am. Chem. Soc.* 1997, 119, 201–207.
42. Bahr, J. L.; Yang, J. P.; Kosynkin, D. V.; Bronikowski, M. J.; Smalley, R. E.; Tour, J. M., Functionalization of carbon nanotubes by electrochemical reduction of aryl diazonium salts: A bucky paper electrode. *J. Am. Chem. Soc.* 2001, 123, 6536–6542.
43. Zhu, H. R.; Huang, P.; Jing, L.; Zuo, T. S.; Zhao, Y. L.; Gao, X. Y., Microstructure evolution of diazonium functionalized graphene: A potential approach to change graphene electronic structure. *J. Mater. Chem.* 2012, 22, 2063–2068.
44. Niyogi, S.; Bekyarova, E.; Itkis, M. E.; Zhang, H.; Shepperd, K.; Hicks, J.; Sprinkle, M. et al. Spectroscopy of covalently functionalized graphene. *Nano Lett.* 2010, 10, 4061–4066.
45. Grimme, S., Accurate description of van der Waals complexes by density functional theory including empirical corrections. *J. Comput. Chem.* 2004, 25, 1463–1473.
46. Sun, Y. Y.; Kim, Y.; Lee, K.; Zhang, S. B., Accurate and efficient calculation of van der Waals interactions within density functional theory by local atomic potential approach. *J. Chem. Phys.* 2008, 129, 8.
47. Novoselov, K. S.; Geim, A. K.; Morozov, S. V.; Jiang, D.; Katsnelson, M. I.; Grigorieva, I. V.; Dubonos, S. V.; Firsov, A. A., Two-dimensional gas of massless Dirac fermions in graphene. *Nature* 2005, 438, 197–200.
48. Cancado, L. G.; Takai, K.; Enoki, T.; Endo, M.; Kim, Y. A.; Mizusaki, H.; Jorio, A.; Coelho, L. N.; Magalhaes-Paniago, R.; Pimenta, M. A., General equation for the determination of the crystallite size L-a of nanographite by Raman spectroscopy. *Appl. Phys. Lett.* 2006, 88.
49. Moser, J.; Tao, H.; Roche, S.; Alzina, F.; Torres, C. M. S.; Bachtold, A., Magnetotransport in disordered graphene exposed to ozone: From weak to strong localization. *Phys. Rev. B* 2010, 81.
50. Bissett, M. A.; Konabe, S.; Okada, S.; Tsuji, M.; Ago, H., Enhanced chemical reactivity of graphene induced by mechanical strain. *ACS Nano* 2013, 7, 10335–10343.
51. Shih, C.-J.; Wang, Q. H.; Jin, Z.; Paulus, G. L. C.; Blankschtein, D.; Jarillo-Herrero, P.; Strano, M. S., Disorder imposed limits of mono- and bilayer graphene electronic modification using covalent chemistry. *Nano Lett.* 2013, 13, 809–817.

17 Thermal and Thermoelectric Transport in Graphene

The Role of Electron–Phonon Interactions

Enrique Muñoz

CONTENTS

ABSTRACT

Single-layer graphene possesses remarkable mechanical and electronic properties. Its great flexibility, surprisingly, goes along with the highest Young's modulus known for any material up to date. These mechanical properties, combined with the two-dimensional topology of graphene, determine its phonon spectrum. On the other hand, the singular features of the electronic spectrum determine its semimetallic behavior in the vicinity of the so-called Dirac points, where conduction electrons exhibit relativistic dynamics as chiral massless fermions in two dimensions. Transport properties at nonzero temperature are strongly affected by the interplay between mechanics and electronics: the interaction between electrons and phonons. Along this chapter, thermal and thermoelectric transport properties of graphene will be described and discussed. Different scattering mechanisms affecting transport coefficients will be analyzed, with emphasis on the role of electron–phonon interactions and mechanical stress. Theoretical approaches to include these scattering mechanisms in the modeling of thermal and thermoelectric transport will be covered in detail, and predictions will be compared with experimental values. Finally, we will discuss the possibilities for engineering thermal and thermoelectric transport coefficients in graphene by controlling electron–phonon interactions.

17.1 INTRODUCTION

Theoretical studies on the electronic spectrum and conductivity of "single-layer graphite" go back to the seminal work by Wallace [1]. This monolayer of carbon atoms, arranged in a two-dimensional hexagonal lattice, was then called graphene. For long, it was believed that graphene was an unstable structure, impossible to find in nature as a flat membrane [2], except for its folded related structures such as carbon nanotubes and fullerenes. The experimental discovery of graphene by Novoselov [3] dramatically changed this picture, and since then significant progress has been made in developing efficient methods for the synthesis and functionalization of graphene and related nanostructures [4–12].

The experimental confirmation by Novoselov et al. [13] that charge carriers in graphene indeed behave like a two-dimensional gas of massless chiral Dirac fermions [14], opened a new venue for fundamental scientific research and potential technological [15–21] and even biotechnological [15,22] applications, with graphene nanoribbons and related structures being serious candidates for a next generation of quasi-ballistic nanoelectronics [23,24]. The electronic properties of graphene in the presence of external magnetic fields [25,26] are also fascinating, particularly the anomalies observed in the Integer Quantum Hall effect [13,27].

Single-layer graphene also exhibits remarkable mechanical properties, with the highest Young's modulus ever measured for any known material [28]. The mechanical properties of graphene, which determine its phonon spectrum, are sensitive to the presence of structural defects [29]. The phonon spectrum of graphene possesses three acoustic branches, reflecting the bipartite character of the hexagonal lattice, and the fact that atoms in the lattice can experience out-of-plane motion. This, in turn, determines the quadratic dispersion of the ZA or flexural mode. The divergence in the number of flexural phonons at any finite temperature in a soft membrane [2] lead in the beginning to the general belief that graphene could not exist in nature as a stable material. The structure, however, exists, and it is in practice stabilized by the presence of a substrate or scaffold, which nevertheless may create static "ripples." Such ripples and the associated local curvature, combined with other types of strain, determine complex electron–phonon interactions [30]. Within the framework of the Dirac fermion theory for charge carriers in graphene, electron–phonon interactions are described by the presence of gauge fields, representing a nonuniform pseudo-magnetic field. Such electron–phonon interactions have important consequences in determining the transport coefficients in graphene, which is the main subject to be covered by this chapter.

The chapter is divided into four main sections. In the first section, the phonon spectrum of single-layer graphene, as obtained from the mechanical properties of this material is described. Theoretical estimations for the phonon contribution to the thermal conductivity, which is currently believed to be dominated by a nearly ballistic transport mechanism [31], is presented and compared with experiments reported in the recent literature. In the second section, we briefly discuss the electronic spectrum of single-layer graphene, and the Dirac fermion representation is obtained from a tight-binding atomistic model. The third section presents electron–phonon interactions, and their quantum-mechanical description in terms of gauge fields [32,33]. In particular, a second quantized representation for the deformation potential type of interaction is formulated [34]. The fourth and final section of this chapter presents theoretical calculations, based on analytical solutions of the Boltzmann equation, for the different transport coefficients limited by electron–phonon scattering processes. Such calculations are found to be in excellent agreement with experimental data in the recent literature [34–36].

17.2 PHONON SPECTRUM OF SINGLE-LAYER GRAPHENE

Phonons constitute quantized normal modes of the crystal lattice vibrations. Therefore, within the Born approximation, the phonon spectrum can be obtained from the classical equations of motion for the ions in the lattice. Let us formulate the classical lattice Hamiltonian by first defining the equilibrium positions of the ions by the set of vectors $\boldsymbol{R}_{n\alpha} = \boldsymbol{R}_n + \boldsymbol{R}_\alpha$. Here, $\boldsymbol{R}_n$ denotes the position of a reference point in the nth unit cell, whereas $\boldsymbol{R}_\alpha$ refers to the position of the αth basis ion within the unit cell. In graphene $\alpha = (A, B)$, taking into account the

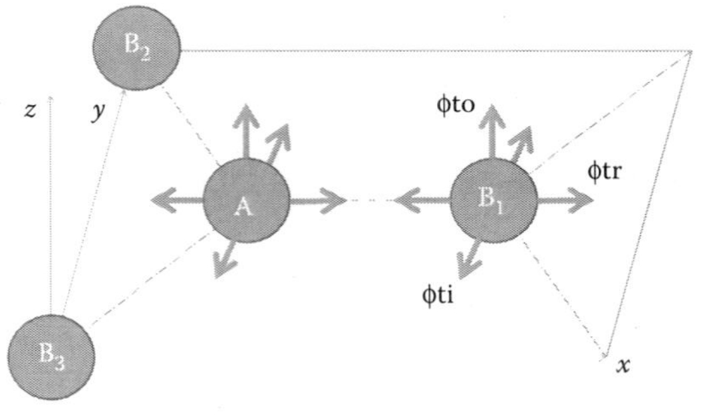

FIGURE 17.1 Displacements of the carbon atoms in the hexagonal lattice, showing the interpretation of the atomic force constants. (From Jishi, R.A. et al. *Chemical Physics Letters*, 1993. **209**(1–2): p. 77–82; Saito, R., Dresselhaus, G., and Dresselhaus, M.H., *Physical Properties of Carbon Nanotubes*. 1998, Imperial College, London: Imperial College Press.)

A and B carbon atoms of the basis (see Figure 17.1). The time-dependent ionic displacements with respect to the equilibrium positions are defined by the vectors $\boldsymbol{s}_{n\alpha}(t)$. We then expand the interatomic potential of the lattice with respect to its value at the equilibrium position, up to second order in the displacement coordinates [37]

$$V(\{\boldsymbol{s}_{n\alpha}(t)\}) = V(\{\boldsymbol{R}_{n\alpha}\}) + \frac{1}{2}\sum_{n,n',\alpha,\alpha',i,i'} s_{n\alpha i}\Phi_{n\alpha i}^{n'\alpha' i'} s_{n'\alpha' i'} \quad (17.1)$$

where the $\Phi_{n\alpha i}^{n'\alpha' i'} = \partial^2 V/(\partial R_{n\alpha i}\partial R_{n'\alpha' i'})$ are the so-called atomic force constants [37]. The classical Hamiltonian for the lattice displacements is then

$$H_{ion}^{Class} = \sum_{n,\alpha,i}\frac{M}{2}(\dot{s}_{n\alpha i})^2 + \frac{1}{2}\sum_{n,n',\alpha,\alpha',i,i'} s_{n\alpha i}\Phi_{n\alpha i}^{n'\alpha' i'} s_{n'\alpha' i'} \quad (17.2)$$

The classical equations of motion arising from this Hamiltonian correspond to a system of coupled harmonic oscillators.

$$M\ddot{s}_{n\alpha i} = -\sum_{n',\alpha',i'}\Phi_{n\alpha i}^{n'\alpha' i'} s_{n'\alpha' i'} \quad (17.3)$$

We look for solutions of the equations of motion that are periodic in time and in the spatial lattice coordinates, using the ansatz [37]

$$s_{n\alpha i}(t) = \frac{1}{\sqrt{M}} c_{\alpha i} e^{i(\boldsymbol{q}\cdot\boldsymbol{R}_n - \omega t)} \quad (17.4)$$

By taking into account the translational symmetry of the lattice, the atomic force constants must satisfy [37] $\Phi_{n\alpha i}^{n'\alpha' i'} = \Phi_{\alpha i}^{\alpha' i'}(n' - n)$. Therefore, the equations of motion for the ionic displacements become

TABLE 17.1
Atomic Force Constants for Graphene, Up to Fourth Nearest Neighbors

Nearest Neighbor Order i	$\phi_r^{(i)}$ 10^4 (dyn/cm)	$\phi_{to}^{(i)}$ 10^4 (dyn/cm)	$\phi_{ti}^{(i)}$ 10^4 (dyn/cm)
1	36.5	9.82	24.5
2	8.80	−0.40	−3.23
3	3.00	0.15	−5.25
4	−1.92	−0.58	2.29

Source: Saito, R., Dresselhaus, G., and Dresselhaus, M.H., *Physical Properties of Carbon Nanotubes*. 1998, Imperial College, London: Imperial College Press.
Note: The geometrical interpretation is shown in Figure 17.1.

$$\omega^2 c_{\alpha i} = \sum_{\alpha',i'} D_{\alpha i}^{\alpha' i'}(\boldsymbol{q}) c_{\alpha' i'}, \tag{17.5}$$

where we have defined $D_{\alpha i}^{\alpha' i'}(\boldsymbol{q}) = (1/M)\sum_n \Phi_{\alpha i}^{\alpha' i'}(n)\exp[i\boldsymbol{q}\cdot\boldsymbol{R}_n]$. The solutions to the eigenvalue Equation 17.5 provide the phonon spectrum. In general, for a lattice with r atoms in the basis, one finds $3r$ independent branches from Equation 17.5. In graphene $r = 2$, which determines three acoustic and three optical phonon branches. The numerical values for the elastic constants [38,39], up to fourth nearest neighbors, are displayed in Table 17.1.

The geometrical interpretation of the displacements and associated elastic constants is shown in Figure 17.1.

The calculated phonon spectrum of graphene, as obtained from the solution of Equation 17.5, is depicted in Figure 17.2. It shows $3 \times 2 = 6$ branches, with three of them representing acoustic modes: longitudinal acoustic (LA), transverse acoustic (TA), and an out-of-plane flexural mode (ZA). Near the Γ point ($\boldsymbol{q} = 0$), the LA and TA branches display the usual linear dispersion of acoustic phonons. On the other hand, the ZA mode possesses a quadratic rather than linear dispersion.

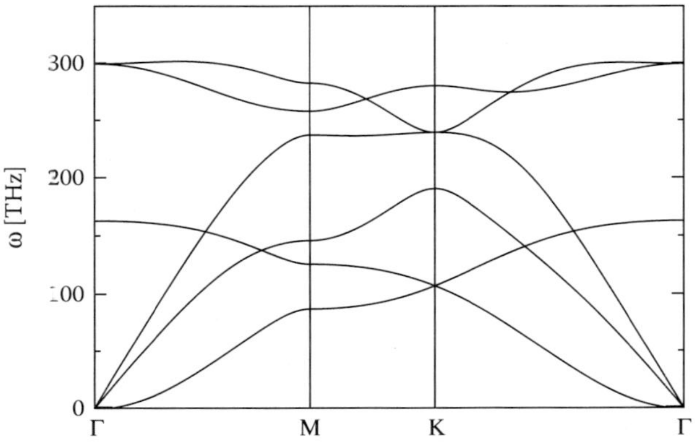

FIGURE 17.2 Calculated phonon spectrum of graphene, along the main symmetry directions, by solving the eigenvalue Equation 17.5 up to four nearest neighbors. The numerical values used for the atomic force constants are shown in Table 17.1.

The dispersion relations of the three acoustic branches can be obtained analytically, at low frequencies, from a continuum model [40–42] representing the dynamics of elastic deformations of a thin plate

$$\frac{\partial^2 u}{\partial \tau^2} = \frac{C}{\rho_s(1-\nu^2)}\left\{\frac{\partial^2 u}{\partial x^2} + \frac{1-\nu}{2}\frac{\partial^2 u}{\partial y^2} + \frac{1+\nu}{2}\frac{\partial^2 v}{\partial x\,\partial y}\right\},$$
$$\frac{\partial^2 v}{\partial \tau^2} = \frac{C}{\rho_s(1-\nu^2)}\left\{\frac{\partial^2 v}{\partial y^2} + \frac{1-\nu}{2}\frac{\partial^2 v}{\partial x^2} + \frac{1+\nu}{2}\frac{\partial^2 u}{\partial x\,\partial y}\right\},$$
$$\frac{\partial^2 w}{\partial \tau^2} = -\frac{D}{\rho_s}\left\{\frac{\partial^4 w}{\partial x^4} + 2\frac{\partial^4 w}{\partial x^2\,\partial y^2} + \frac{\partial^4 w}{\partial y^4}\right\}. \tag{17.6}$$

Here, (u, v, w) are the displacements along the parallel (x, y), and perpendicular (z) directions with respect to the reference plane representing the flat surface. The elastic parameters involved in this continuum model are the surface mass density ρ_s, the Poisson ratio ν, the flexural rigidity $D = Yt^3/[12(1-\nu^2)]$, and the in-plane stiffness $C = Yt$. In these equations, Y is the Young modulus and $t \to 0$ the plate thickness. If one considers harmonic solutions to the system equation (17.6) of the form

$$\begin{pmatrix} u \\ v \\ w \end{pmatrix} = \begin{pmatrix} A_x \\ A_y \\ A_z \end{pmatrix} e^{i(q_l x + q_w y - \omega\tau)} \tag{17.7}$$

then Equation 17.6 transforms into the eigenvalue equation $[\boldsymbol{M}(\boldsymbol{q}) - \omega^2 \boldsymbol{I}]\boldsymbol{A} = 0$, where we have defined

$$\boldsymbol{M}(\boldsymbol{q}) = \begin{bmatrix} \frac{C}{\rho_s(1-\nu^2)}\left(q_l^2 + \frac{1-\nu}{2}q_w^2\right) & \frac{C q_l q_w}{2\rho_s(1-\nu)} & 0 \\ \frac{C q_l q_w}{2\rho_s(1-\nu)} & \frac{C}{\rho_s(1-\nu^2)}\left(q_w^2 + \frac{1-\nu}{2}q_l^2\right) & 0 \\ 0 & 0 & \frac{D}{\rho_s}(q_l^2 + q_w^2)^2 \end{bmatrix} \tag{17.8}$$

The frequency eigenvalues, obtained as solutions of the polynomial equation $det[\boldsymbol{M}(\boldsymbol{q}) - \omega^2\mathbf{I}] = 0$, are given by the analytical expressions

$$\omega_{ZA}(\boldsymbol{q}) = \sqrt{\frac{D}{\rho_s}}\boldsymbol{q}^2, \quad \text{(ZA)}$$
$$\omega_{LA}(\boldsymbol{q}) = \sqrt{\frac{C}{\rho_s(1-\nu^2)}}q, \quad \text{(LA)} \tag{17.9}$$
$$\omega_{TA}(\boldsymbol{q}) = \sqrt{\frac{C}{\rho_s(1+\nu)}}q, \quad \text{(TA)}$$

where we have defined $q = \sqrt{q_l^2 + q_w^2} = \|\boldsymbol{q}\|$.

TABLE 17.2
Elastic Parameters for Graphene, Calculated from First Principles [43, 44], and Speed of Sound for the LA and TA Modes according to Equation 17.9

ρ_s (mg/m²)	C (J/m²)	D (eV)	ν	t (Å)	c_{LA} (m/s)	c_{TA} (m/s)
0.75	345	1.46	0.15	0.88	2.17×10^4	1.42×10^4

Numerical values for the elastic parameters involved in the model have been obtained from first-principles calculations [43,44]. The values used in [40] are summarized in Table 17.2.

From the parameters displayed in Table 17.2 combined with the analytical dispersion Equation 17.9, the speed of sound for the LA and TA modes are given by $c_{LA} = 2.17 \times 10^4$ m/s and $c_{TA} = 1.42 \times 10^4$ m/s, as listed in Table 17.2. This analytical result is in agreement with the values extracted from the numerical dynamical matrix calculation displayed in Figure 17.2.

Flexural phonons (ZA) possess a quadratic ($\sim q^2$) dispersion, and hence they constitute the dominant mode at low temperatures. From the theory of continuum elasticity [42,45,46], the graphene surface can be described in terms of its deviation from the planar shape $z = 0$, by defining a local height function $\phi(\boldsymbol{r})$, with $\boldsymbol{r} = (x, y)$ the position of a vector in the plane. Therefore, a point on the surface is associated with the coordinates $\mathcal{R} = (\boldsymbol{r}, \phi(\boldsymbol{r}))$, and the equation of the surface is $\mathcal{S}(x, y, z) = z - \phi(\boldsymbol{r}) = 0$. The normal to the surface is then $\hat{n} = \nabla\mathcal{S}/\|\nabla\mathcal{S}\| = (1, -\nabla\phi)/\sqrt{1+(\nabla\phi)^2}$. The elastic energy associated with deviations from the flat geometry (i.e., when $\nabla \cdot \hat{n} = 0$) is described by the term [45,46] $\nabla \cdot \hat{n} \approx \nabla^2\phi$ (for $\|\nabla\phi\| \ll 1$)

$$E_0 = \frac{D}{2}\int d^2r(\nabla^2\phi)^2 \tag{17.10}$$

where $D \sim 1.46$ eV (see Table 17.2) is the bending rigidity.

Under tension, which breaks down the rotational symmetry of the graphene sheet ($\nabla\phi \neq 0$), there is an additional energy term [45,46]

$$E_T = \frac{\gamma}{2}\int d^2r(\nabla\phi)^2, \tag{17.11}$$

where γ is the interfacial stiffness. Finally, for a supported graphene layer on top of a substrate, the local topology of the substrate surface (represented by the height function $\chi(\boldsymbol{r})$) will determine complex interactions, like van der Waals for instance, which depend on the relative distance $|\chi(\boldsymbol{r}) - \phi(\boldsymbol{r})|$. The energy cost of the graphene–surface interactions can be modeled in the continuum limit by the term [47]

$$E_S = \frac{v}{2}\int d^2r[\chi(\boldsymbol{r}) - \phi(\boldsymbol{r})]^2 \tag{17.12}$$

By introducing the spatial Fourier transform of the local field $\phi_q = \int d^2re^{iq\cdot r}\phi(\boldsymbol{r})$, the elastic energy can be expressed as

$$V[\phi] = E_0 + E_T = \sum_q \left\{\frac{D}{2}q^4 + \frac{\gamma}{2}q^2\right\}\phi_q\phi_{-q} \tag{17.13}$$

To quantize this model, let us introduce a set of momentum operators Π_q, which satisfy canonical commutation relations $[\phi_q, \Pi_{q'}] = i\delta_{q,q'}$. The "kinetic" term in the fields is then given by $K[\Pi] = (1/2\rho_s)\Sigma_q \Pi_q\Pi_{-q}$, with ρ_s the surface mass density of graphene (see Table 17.2).

The resulting quantized Hamiltonian operator is [46]

$$H = K[\Pi] + V[\phi] = \sum_q \left\{\frac{1}{2\rho_s}\Pi_q\Pi_{-q} + \left(\frac{D}{2}q^4 + \frac{\gamma}{2}q^2\right)\phi_q\phi_{-q}\right\} \tag{17.14}$$

The equations of motion for the quantized fields are obtained through commutation relations with the Hamiltonian Equation 17.14

$$i\dot{\phi}_q = [\phi_q, H] = i\rho_s^{-1}\Pi_q \tag{17.15}$$

$$i\dot{\Pi}_q = [\Pi_q, H] = -i(Dq^4 + \gamma q^2)\phi_{-q} \tag{17.16}$$

Taking the time derivative of Equation 17.15, and substituting into Equation 17.16, one obtains

$$\ddot{\phi}_q = -\rho_s^{-1}(Dq^4 + \gamma q^2)\phi_q \tag{17.17}$$

From Equation 17.17, it follows that the vibrational frequency of the flexural modes is given by the expression

$$\omega_{Flex}(q) = \sqrt{\frac{D}{\rho_s}}\left(q^2\frac{\gamma}{D}\right)^{1/2} \tag{17.18}$$

In the absence of tension $\gamma \to 0$, one finds $\omega_{Flex}(q) \to \sqrt{D/\rho_s}q^2 = \omega_{ZA}(q)$, which is precisely the long-wavelength approximation for the ZA mode obtained in Equation 17.9. It is interesting to remark that at low temperatures, it is precisely these flexural modes that dominate among the total population of thermally excited phonons. In thermal equilibrium, these modes are excited according to the Bose–Einstein distribution $n_B(\omega) = (\exp(\hbar\omega/k_BT) - 1)^{-1}$, and then the total number of flexural phonons (ZA) per unit area is

$$N_{ZA} = \int \frac{d^2q}{(2\pi)^2} n_B(\omega_{ZA}(q)) = \frac{k_BT}{4\pi\hbar\sqrt{D/\rho_s}} \int_{x_{\min}}^{\infty} \frac{e^{-x}dx}{1-e^{-x}}$$

$$= \frac{k_BT}{4\pi\hbar\sqrt{D/\rho_s}} \ln\left(\frac{1}{1-e^{-x_{\min}}}\right) \tag{17.19}$$

where the change of variables $x = q^2\hbar\sqrt{D/\rho_s}/(k_BT)$ was made. Unless one introduces a lower cutoff $x_{\min} > 0$, this integral is divergent. As argued in Reference 46, for a finite sample of linear dimension L, a natural cutoff is given by $q_{\min} \sim 2\pi/L$. If one assumes that the size of the sample is much larger than the thermal wavelength of these flexural modes $L_T = 2\pi(D/\rho_s)^{1/4}(k_BT)^{-1/2}$, that is, $L \gg L_T$, then after Equation 17.19 the total number of ZA phonons diverge logarithmically with the system size [46]

$$N_{ZA} \approx \frac{2\pi}{L_T} \ln\left(\frac{L}{L_T}\right) \tag{17.20}$$

This indicates that if bending was the sole mechanical deformation involved, then the suspended graphene layer would become unstable at any finite temperature, in agreement with the usual statement that long-range structural order is absent in two-dimensional membranes [48]. This infrared "catastrophe" is saved by the coupling between flexural and in-plane modes, the existence of topological defects, and the application of tension to the membrane [47]. On the other hand, if graphene is supported on top of a substrate, then flexural modes become partially frozen [49] due to interactions with the substrate surface. Moreover, static ripples [49] can be induced and observed under these conditions.

17.2.1 Phonon Thermal Conductance

The thermal conductivity of single-layer graphene, both suspended and supported over a solid-state substrate such as SiO_2, has been measured by different methods, from Raman optothermal to electrical heating [50–59]. As shown in Table 17.3 for comparison, the different experimental methods report values for the thermal conductivity of single-layer graphene (near room temperature) in the range $\kappa \sim 600$–5000 W m^{-1} K^{-1}, which by far exceeds that of copper (~400 W m^{-1} K^{-1}). Remarkably, these measured thermal conductivity values are not only sensible to the method, but also to the growth conditions (i.e., exfoliated or chemical vapor deposition [CVD]) and supporting conditions (i.e., suspended or supported on a substrate). The higher values reported for suspended conditions [50,52–54,56] suggest that the role of

TABLE 17.3
Thermal Conductivity of Single-Layer Graphene and Related Nanostructures Near Room Temperature

Experimental Method	Comments	Thermal Conductivity (W m^{-1} K^{-1})
Raman, optothermal	Suspended, exfoliated	2000–5000 [50,52]
	Suspended, CVD	1500–5000 [53,54,61]
Electrical	Suspended, exfoliated (660 K)	~600 [55]
	Suspended, CVD, exfoliated	~2500 [56]
	Supported, exfoliated	~600 [57]
	Encased within SiO_2	<160 [58]
	Three layers	~1250 [59]
Theoretical Method		
Boltzmann transport equation, valence force field	Width dependence	2000–2500 [62]
Boltzmann transport equation	Length dependence	1400–2400 [63]
	Non-monotonic length dependence	1000–8000 [64]
Relaxation time approx.	Size dependence	1000–5000 [60]
Molecular dynamics	Tersoff pot., size dependence	8000–10,000 [65]

flexural modes is very important for thermal transport in single-layer graphene, in agreement with the analysis presented in the previous section. The ballistic thermal conductance due to phonons in graphene nanoribbons and extended layers has been analytically calculated in References 40 and 41. The model assumes that a ribbon of finite width w and length $L \gg w$ is connected in between two macroscopic thermal reservoirs, which are maintained at temperatures $T_L = T + \Delta T/2$ (left) and $T_R = T - \Delta T/2$ (right), respectively. Phonons distributed according to the Bose–Einstein distribution at the temperatures $T_{L(R)}$ are emitted from each thermal reservoir, and assumed to counter-propagate ballistically (i.e., with negligible scattering) along the ribbon, such that the heat flow (energy/time) across the structure is given by the expression [40,41]

$$\dot{Q} = \int_0^{q_D} \frac{dq_l}{2\pi} \sum_{p=LA,TA,ZA,\tau} \sum_{<l=0>}^{(w/a)-1} \hbar\omega_p(\boldsymbol{q})v_{p,l}(\boldsymbol{q})[n_B(\omega_p(\boldsymbol{q}),T+\Delta T/2) - n_B(\omega_p(\boldsymbol{q}),T-\Delta T/2)] \tag{17.21}$$

For a thin ribbon, in addition to the phonon modes presented in Equation 17.9, a torsional mode with dispersion [40,41]

$$\omega_\tau(\boldsymbol{q}) = \sqrt{8(1-\nu)/\rho_s w^2}\, q_l \equiv c_\tau q_l \tag{17.22}$$

must be taken into account. Note from Equation 17.22 that the speed of sound for this torsional mode scales as $c_\tau \sim w^{-1}$, and hence its contribution to thermal transport vanishes in the limit of a wide graphene sheet. The bracket in the summation index in Equation 17.21 indicates that the term $l = 0$ is to be included only for the torsional mode. All four modes split into separate branches, due to the discreteness of the transverse component of the crystal wavevector $q_w = l\pi/w$, with $0 \le l \le w/a$ and a being the primitive cell size. In the spirit of the Debye model, a wavevector cutoff $q_D = 3\pi n_s/((3/a) + (1/w))$ was introduced, for $n_s = N/A$, in order to account for the $3N - 6 \sim 3N$ atomic degrees of freedom. The ballistic thermal conductance due to phonons is obtained from Equation 17.21 by the definition $\sigma_B(T) = \lim_{\Delta T \to 0} \dot{Q}/\Delta T$, which yields the analytical expression

$$\sigma_B(T) = \frac{k_B^2 T}{h}\left\{ \sum_{p=(ZA,LA,TA,\tau)} f_2\left(\frac{\theta_p}{T}\right) + \sum_{l=1}^{(w/a)-1}\left\{ \sum_{p=(LA,TA)}\left[f_2\left(\frac{\theta_p\sqrt{1+(l\pi/q_D w)^2}}{T}\right) - f_2\left(\frac{\theta_p \pi l}{q_D w T}\right)\right]\right\} + f_2\left(\frac{\theta_{ZA}[1+(l\pi/q_D w)^2]}{T}\right) - f_2\left(\frac{\theta_{ZA}[l\pi/q_D w]^2}{T}\right)\right\} \quad (17.23)$$

Here, we have defined the special functions

$$f_n(z) = \int_0^z dx x^n \frac{e^x}{(e^x - 1)^2} \quad (17.24)$$

In this model, each phonon mode has its own Debye temperature: $\theta_p = \mathrm{h}c_p q_D/k_B$ for p = LA, TA, and τ, and $\theta_{ZA} = \hbar c_{ZA} q_D^2/k_B$. It can be shown [40,41] that Equation 17.23 predicts a low-T behavior with different exponents depending on the width of the ribbon, ranging from $\sigma_B(T) \sim (4(\pi^2 k_B^2/3h))T$ for a very narrow ribbon, toward $\sigma_B(T)/w \sim 0.23(k_B^5 c_{LA}^{-1}\hbar^{-3})^{1/2} T^{3/2}$ for an extended ($w \to \infty$) graphene sheet [40,41]. The transition between the different width-dependent exponents is depicted in Figure 17.3.

The phonon contribution to the thermal conductivity can be estimated by combining the ballistic thermal conductance calculated in Equation 17.23, via Matthiessen's rule, with diffusive scattering effects represented by an overall phonon mean free path λ. This yields the expression [40]

$$\kappa_{ph}(L) = \frac{\sigma_B/(tw)}{1 + L/\lambda} L \quad (17.25)$$

with L the (finite) length of the ribbon and t = 3.4 Å its "thickness," defined by convention as the interlayer distance in bulk graphite. At room temperature, the mean free path was estimated from molecular dynamics simulations [40] and experiments [50,52,60] as $\lambda \sim$ 750 nm. When this value is substituted into Equation 17.25, in combination with the analytical value for the ballistic thermal conductance σ_B at room temperature obtained from Equation 17.23, one obtains in the limit $L \gg \lambda$ [40] κ_{ph} = 3960 W m^{-1} K^{-1}. This value is in very good agreement with experiments on suspended graphene, with $\kappa_{exp} \sim$ 2000–5000 W m^{-1} K^{-1} [50,52] for exfoliated, and ~1500–5000 W m^{-1} K^{-1} for CVD graphene [53,54,56,61], respectively. A comparison in a broader context is displayed in Table 17.3, where estimates for the phonon thermal conductivity of single-layer graphene arising from alternative theoretical methods to the one described along this section is also shown. Most theoretical approaches suggest a strong size dependence of the thermal conductivity of graphene structures at the nanoscale [34,40,43,60,62–66].

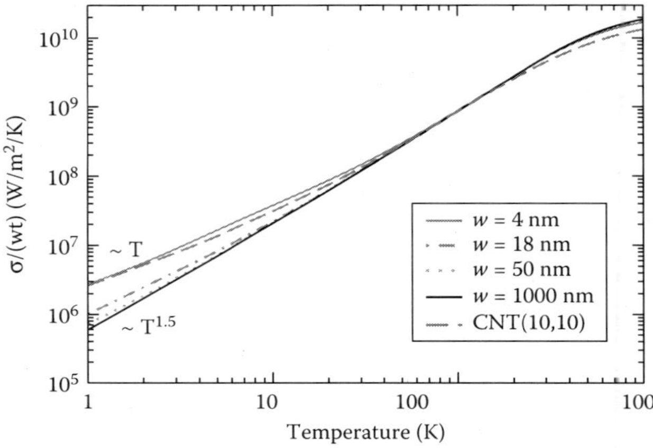

FIGURE 17.3 Dependence of the low-temperature exponents for the ballistic phonon thermal conductance on the ribbon width w, calculated from Equation 17.23.

17.3 ELECTRONIC SPECTRUM OF SINGLE-LAYER GRAPHENE

The honeycomb lattice of single-layer graphene can be described as a Bravais triangular lattice with a basis of two carbon atoms (A and B) per unit cell (Figure 17.4). The corresponding lattice vectors are $\boldsymbol{a}_1 = a(3/2, \sqrt{3}/2)$ and $\boldsymbol{a}_2 = a(3/2, -\sqrt{3}/2)$, with a = 1.42 Å the carbon–carbon distance. The two triangular sublattices A and B are connected by the nearest-neighbor vectors (see Figure 17.4)

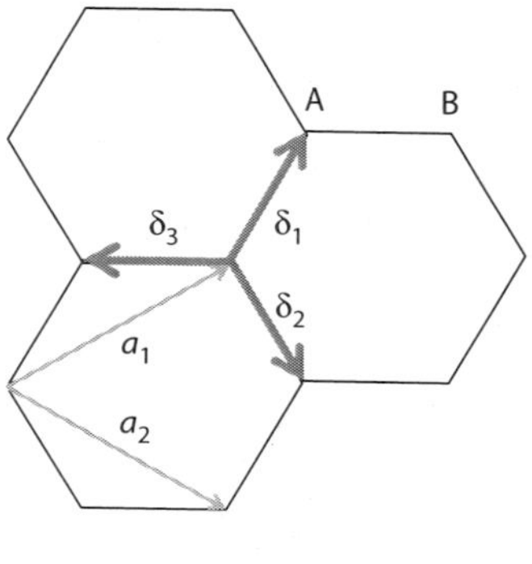

FIGURE 17.4 Graphene primitive lattice vectors $\boldsymbol{a}_1$, $\boldsymbol{a}_2$. Also shown are the nearest-neighbor vectors δ_i (i = 1, 2, 3).

$\delta_1 = (a/2, a\sqrt{3}/2)$, $\delta_2 = (a/2, -a\sqrt{3}/2)$ and $\delta_3 = (-a, 0)$. The reciprocal lattice is also triangular with a basis, with reciprocal lattice vectors $\boldsymbol{b}_1 = (2\pi/3a)(1, \sqrt{3})$ and $\boldsymbol{b}_2 = (2\pi/3a)(1, -\sqrt{3})$. The tight-binding Hamiltonian for single-layer graphene, including nearest-neighbor $\langle i,j \rangle$ and next-nearest-neighbor $\langle\langle i,j \rangle\rangle$ electron hopping, is expressed by [25,39,46]

$$H = -\gamma_0 \sum_{\langle i,j\rangle,\sigma} (a_{\sigma i}^\dagger b_{\sigma j} + H.c.) - s \sum_{\langle\langle i,j\rangle\rangle,\sigma} (a_{\sigma i}^\dagger a_{\sigma j} + b_{\sigma i}^\dagger b_{\sigma j} + H.c.). \tag{17.26}$$

Here, $a_{\sigma j}$ and $b_{\sigma j}$ are two independent $[a_{\sigma i}, b_{\sigma' j}]_+ = 0$ sets of Fermionic operators $[a_{\sigma i}, a_{\sigma' j}^\dagger]_+ = [b_{\sigma i}, b_{\sigma' j}^\dagger]_+ = \delta_{i,j}\delta_{\sigma,\sigma'}$ defined on the A and B sublattice sites, respectively. The subindex $\sigma = \{\uparrow, \downarrow\}$ represents electronic spin. In Equation 17.26, $\gamma_0 \sim 3$ eV from first-principles calculations while $s \sim 0.13$ [39,67]. By direct diagonalization of this tight-binding Hamiltonian, one obtains the π, π^* energy bands [1,25]

$$E^{(\pm)}(k,s) = \pm\gamma_0 \sqrt{1 + 4\cos^2\left(\frac{\sqrt{3}k_y a}{2}\right) + 4\cos\left(\frac{\sqrt{3}k_y a}{2}\right)\cos\left(\frac{3k_x a}{2}\right)} - 2s\left[\cos(\sqrt{3}k_x a) + 2\cos\left(\frac{\sqrt{3}k_y a}{2}\right)\cos\left(\frac{3k_x a}{2}\right)\right] \tag{17.27}$$

Neglecting the next-nearest-neighbor hopping term ($s \ll \gamma_0$), the resulting tight-binding Hamiltonian equation (17.26) can be conveniently expressed by the operator

$$H = -\gamma_0 \sum_{\boldsymbol{R}_A, j, \sigma} (a_\sigma^\dagger(\boldsymbol{R}_A) b_\sigma(\boldsymbol{R}_A + \delta_j) + H.c.) \tag{17.28}$$

Here, $\boldsymbol{R}_A = n_1\boldsymbol{a}_1 + n_2\boldsymbol{a}_2$ are the sites on the A sublattice, $\boldsymbol{R}_B = \boldsymbol{R}_A + \delta_j$ are the sites on the B sublattice (see Figure 17.4), and $\sigma = \{\uparrow, \downarrow\}$ is the electronic intrinsic spin. Expanding the local operators in momentum space

$$a_\sigma(\boldsymbol{R}_A) = \frac{1}{\sqrt{N}} \sum_{\boldsymbol{k}} e^{i\boldsymbol{k}\cdot\boldsymbol{R}_A} a_\sigma(\boldsymbol{k}) \tag{17.29}$$

and, similarly for $b_\sigma(\boldsymbol{R}_A + \delta_j)$, upon substitution into Equation 17.28 we obtain

$$H = \sum_{\boldsymbol{k},\sigma} (a_\sigma^\dagger(\boldsymbol{k}), b_\sigma^\dagger(\boldsymbol{k})) \begin{bmatrix} 0 & -\gamma(\boldsymbol{k}) \\ -\gamma^*(\boldsymbol{k}) & 0 \end{bmatrix} \begin{pmatrix} a_\sigma(\boldsymbol{k}) \\ b_\sigma(\boldsymbol{k}) \end{pmatrix} \tag{17.30}$$

with

$$\gamma(\boldsymbol{k}) = \gamma_0 \sum_{j=1}^{3} e^{i\boldsymbol{k}\cdot\sigma_j} \tag{17.31}$$

The eigenvalues of the effective tight-binding Hamiltonian equation (17.30) are clearly

$$E^{(\pm)}(\boldsymbol{k}, s=0) = \pm|\gamma(\boldsymbol{k})| \tag{17.32}$$

a result consistent with neglecting the next-nearest-neighbor parameter ($s \to 0$) in Equation 17.27. There are six points, corresponding to the corners of the hexagonal Brillouin zone, which satisfy the condition $\gamma(\boldsymbol{k}) = 0$, where the π, π^* energy bands touch each other. In particular, we choose among them two nonequivalent points, corresponding to $\boldsymbol{K}_+ = ((2\pi/3a), (2\pi/3\sqrt{3}a))$ and $\boldsymbol{K}_- = ((2\pi/3a), -(2\pi/3\sqrt{3}a))$. By considering $\boldsymbol{k} = \boldsymbol{K}_+ + \boldsymbol{q}$, or $\boldsymbol{k} = \boldsymbol{K}_- + \boldsymbol{q}$ in the vicinity of the $\boldsymbol{K}_\pm$ points, respectively, we can expand $\gamma(\boldsymbol{k})$ up to first order in $\boldsymbol{q}$ [68]

$$\begin{aligned} \gamma(\boldsymbol{K}_\pm + \boldsymbol{q}) &\approx \gamma_0 \sum_j e^{i\boldsymbol{K}_{1,2}\cdot\delta_j}(1 + i\boldsymbol{q}\cdot\delta_j) \\ &= i\frac{3}{2}\gamma_0 a e^{i\pi/3}(q_x \pm iq_y) \end{aligned} \tag{17.33}$$

Substituting these linear approximations into Equation 17.30, and going to the continuum limit in $\boldsymbol{q}$, an effective Hamiltonian in momentum space is obtained [25,68]

$$H_{eff} = H_{\boldsymbol{K}_+} + H_{\boldsymbol{K}_-} = \hbar v_F \int \frac{d^2q}{(2\pi)^2} \sum_\sigma \left(\hat{\Psi}_{+,\sigma}^\dagger(\boldsymbol{q}), \hat{\Psi}_{-,\sigma}^\dagger(\boldsymbol{q})\right) \times \begin{bmatrix} \sigma\cdot\mathbf{q} & 0 \\ 0 & -\sigma\cdot\boldsymbol{q} \end{bmatrix} \begin{pmatrix} \hat{\Psi}_{+,\sigma}(\boldsymbol{q}) \\ \hat{\Psi}_{-,\sigma}(\boldsymbol{q}) \end{pmatrix} \tag{17.34}$$

Here, we have defined the two-component operators in the vicinity of the $\boldsymbol{K}_\pm$ points, respectively [68].

$$\begin{aligned} \hat{\Psi}_{+,\sigma}(\boldsymbol{q}) &= e^{-i(\pi/3)\hat{\sigma}_z} \begin{pmatrix} a_\sigma(\boldsymbol{K}_+ + \boldsymbol{q}) \\ b_\sigma(\boldsymbol{K}_- + \boldsymbol{q}) \end{pmatrix}, \\ \hat{\Psi}_{-,\sigma}(\boldsymbol{q}) &= e^{-i(\pi/3)\hat{\sigma}_z} \hat{\sigma}_x \begin{pmatrix} a_\sigma(\boldsymbol{K}_+ + \boldsymbol{q}) \\ b_\sigma(\boldsymbol{K}_- + \boldsymbol{q}) \end{pmatrix} \end{aligned} \tag{17.35}$$

representing a noninteracting pair of massless Dirac fermions. The Fermi velocity in the vicinity of each Dirac point is thus given by $\hbar v_F = 3\gamma_0 a/2 \approx 10^6$ m/s, while $\sigma = (\sigma_x, \sigma_y)$ represents the pseudospin operator expressed in terms of the Pauli matrices. In direct space (continuum limit), and in second quantization, the effective Dirac Hamiltonian equation (17.34) becomes

$$H = \hbar v_F \int d^2r \sum_\sigma \left(\hat{\Psi}_{+,\sigma}^\dagger(\boldsymbol{r}), \hat{\Psi}_{-,\sigma}^\dagger(\boldsymbol{r})\right) \times \begin{bmatrix} \sigma\cdot(-i\nabla) & 0 \\ 0 & -\sigma\cdot(-i\nabla) \end{bmatrix} \begin{pmatrix} \hat{\Psi}_{+,\sigma}(\boldsymbol{r}) \\ \hat{\Psi}_{-,\sigma}(\boldsymbol{r}) \end{pmatrix} \tag{17.36}$$

Therefore, in first quantization the associated pseudo-spinors $\psi^{(\xi)}_{\eta,\sigma}(\boldsymbol{r})$ are obtained as solutions of the two-dimensional Dirac equation [46]

$$-i\eta\hbar v_F \sigma\cdot\nabla\psi^{(\xi)}_{\eta,\sigma}(\boldsymbol{r}) = E^{(\xi)}_{\eta}\psi^{(\xi)}_{\eta,\sigma}(\boldsymbol{r}) \tag{17.37}$$

Note that in our notation, $\eta = \pm$ represents the "valley" $\boldsymbol{K}_\eta$ index while $\xi = \pm$ represents the π and π^* energy bands, respectively. The analytical plane-wave solutions to the eigensystem equation (17.37), normalized to the area of sample A, are given by

$$\psi^{(\xi)}_{\eta,\sigma}(\boldsymbol{k},\boldsymbol{r}) = \frac{1}{\sqrt{2A}} e^{i(k_x x + k_y y)} \begin{pmatrix} e^{-i\theta_k/2} \\ \eta\xi e^{i\theta_k/2} \end{pmatrix}, \quad \text{for } \boldsymbol{k} \text{ near } K_{\eta=\pm} \tag{17.38}$$

with eigenvalues

$$E^{(\xi)}_{\eta}(\boldsymbol{k}) = \xi\hbar v_F k \tag{17.39}$$

for $\xi = \pm$ representing the π and π^* bands, respectively. Note the valley ($\eta = \pm$) degeneracy of the energy solutions in Equation 17.39. The polar angle $\theta_k = \tan^{-1}(k_y/k_x)$ defines the direction of the two-dimensional wave vector $\boldsymbol{k} = (k_x, k_y)$ on the graphene plane.

It is remarkable that the *helicity operator* $\hat{h} = \sigma\cdot\boldsymbol{p}/|\boldsymbol{p}|$, whose eigenvalues represent the relative direction of momentum and pseudospin σ (+1, parallel and −1, antiparallel) commutes with the Hamiltonian equation (17.36). Therefore, it is straightforward to check that the pseudo-spinors in Equation 17.38 are also eigenfunctions of the helicity operator, $\hat{h}\hat{\Psi}^{(\xi)}_{\eta,\sigma}(\boldsymbol{k},\boldsymbol{r}) = h_k \hat{\Psi}^{(\xi)}_{\eta,\sigma}(\boldsymbol{k},\boldsymbol{r})$, since

$$\begin{aligned}\hat{h}\hat{\Psi}^{(\xi)}_{\eta,\sigma}(\boldsymbol{k},\boldsymbol{r}) &= \frac{1}{\sqrt{2A}} e^{i(k_x x + k_y y)} \begin{pmatrix} 0 & e^{-i\theta_k} \\ e^{i\theta_k} & 0 \end{pmatrix}\begin{pmatrix} e^{-i\theta_k/2} \\ \xi\eta e^{i\theta_k/2} \end{pmatrix} \\ &= \xi\eta\hat{\Psi}^{(\xi)}_{\eta,\sigma}(\boldsymbol{k},\boldsymbol{r})\end{aligned} \tag{17.40}$$

Moreover, Equation 17.40 shows that the helicity eigenvalues $h_k = \{+1, -1\}$ are directly related to the valley ($\eta = \pm$) and band ($\xi = \pm$) indexes through the identity $h_k = \xi\eta$.

The second quantized operators (fields) in Equation 17.36 are expressed in terms of the spinor eigenfunctions equation (17.38), and a set of fermionic creation $\hat{c}^{\dagger}_{k\eta\xi\sigma}$ and annihilation $\hat{c}_{k\eta\xi\sigma}$ operators in momentum space, $[\hat{c}_{k'\eta'\xi'\sigma'}, \hat{c}^{\dagger}_{k\eta\xi\sigma}]_+ = \delta_{k,k'}\delta_{\eta,\eta'}\delta_{\xi,\xi'}\delta_{\sigma,\sigma'}$

$$\hat{\Psi}_{\eta\sigma}(\boldsymbol{r}) = \sum_{\boldsymbol{k},\xi} \psi^{(\xi)}_{\eta,\sigma}(\boldsymbol{k},\boldsymbol{r})\hat{c}_{k\eta\xi\sigma} \tag{17.41}$$

Here, $\eta = \pm 1$ represents the valley index, $\xi = \pm$ the band index, and $\sigma = \{\uparrow, \downarrow\}$ the intrinsic electronic spin (not to be confused with the pseudospin defined before). Despite one may get the wrong impression that it is only a mathematical artifact of the effective Dirac theory described along this section, the pseudospin property of charge carriers in graphene has important physical consequences that can be exploited in the emerging field of spintronics [69].

17.4 ELECTRON–PHONON INTERACTIONS AS GAUGE FIELDS IN SINGLE-LAYER GRAPHENE

At finite temperatures, the lattice phonon modes described in the previous section are thermally excited and, therefore, interact with the conduction electrons. Mechanical strain, at the atomistic level, modifies the strength of the tight-binding parameters, particularly the nearest-neighbor hopping coefficient γ_0 introduced in the previous section. For weak deformations described by small displacement vectors $\boldsymbol{u}$, the hopping parameters to each of the three nearest neighbors from every carbon atom become [70]

$$\gamma_i = \gamma_0 + \frac{\beta\gamma_0}{a^2}\delta_i\cdot(\boldsymbol{u}_i - \boldsymbol{u}_0) \tag{17.42}$$

Here, δ_i are the nearest-neighbor lattice vectors defined in Section 17.3 (see Figure 17.4), $\boldsymbol{u}_0$ is the displacement of the central atom, whereas $\boldsymbol{u}_i$ is the displacement of each nearest neighbor (i = 1, 2, 3). The electron Grüneisen parameter [70, 71] $\beta = -\partial \ln(\gamma_0)/\partial \ln(a) \approx 2$ for graphene.

In the continuum limit, where the atomic displacements are described as a continuous vector field $\boldsymbol{u}(\boldsymbol{r})$, the relative displacement in Equation 17.42 reduces to [70]

$$(\boldsymbol{u}_i - \boldsymbol{u}_0) \sim (\delta_i\cdot\nabla)\boldsymbol{u}(\boldsymbol{r}) \tag{17.43}$$

The effect of the deformation expressed by Equation 17.43, in the continuum limit, reduces to the inclusion of an effective "vector potential" into the two-dimensional Dirac equation. As in the case of an external magnetic vector potential, this strain-generated gauge field is included into the canonical momenta via the minimal coupling prescription [32,33,72]

$$\boldsymbol{p} \to \Pi = -i\hbar\nabla - \mathbf{A} \tag{17.44}$$

For the case of a pure shear strain, with no lattice dilation $\nabla\cdot\boldsymbol{u} = 0$, the components of the corresponding vector potential are [71]

$$\begin{aligned} A_x &= \frac{\sqrt{3}}{2}(t_3 - t_2) \sim c\frac{\beta\gamma_0}{a}(u_{xx} - u_{yy}), \\ A_y &= \frac{1}{2}(t_2 + t_3 - 2t_1) \sim -c\frac{2\beta\gamma_0}{a}u_{xy}. \end{aligned} \tag{17.45}$$

The electronic interaction with ripples, and particularly with the flexural out-of-plane phonon mode (ZA), is expressed in terms of the field $\phi(\boldsymbol{r})$ introduced in Section 17.2 and Equation 17.10. This represents the local vertical deformation of the graphene sheet with respect to a horizontal reference plane. The effect of this type of deformation on the nearest-neighbor hopping parameters is given, in the continuum limit, by [73]

$$\gamma_i \approx \gamma_0 - \frac{\gamma_0}{2}[(\delta_i \cdot \nabla)\nabla\phi(\boldsymbol{r})]^2 \tag{17.46}$$

This type of deformation also leads to a gauge field term in the effective Dirac equation, with components [73]

$$\begin{aligned} A_x &= -\frac{3}{8}\eta a^2[(\partial_x^2\phi)^2 - (\partial_y^2\phi)^2], \\ A_y &= \frac{3}{4}\eta a^2(\partial_x^2\phi + \partial_y^2\phi)\partial_x\phi\partial_y\phi \end{aligned} \tag{17.47}$$

Here, the constant $\eta = 2.89$ eV depends on microscopic details [73].

Remarkably, significant recent progress has been recently made to experimentally control crumpling and unfolding of graphene [23,24,30], thus creating the real possibility to design transport properties in graphene via "strain engineering" techniques. If the strain is such as to induce changes in the unit cell area, $\delta A/A = \nabla \cdot \boldsymbol{u}$, a deformation potential $V_{dp}(\boldsymbol{r})$ diagonal in the sublattice indexes (A, B) arises

$$V_{dp}(\boldsymbol{r}) = \tilde{D}\nabla \cdot \boldsymbol{u} \equiv \tilde{D}\hat{\Delta}(\boldsymbol{r}) \tag{17.48}$$

A simple phenomenological interpretation for the deformation potential is generally applicable to three-dimensional (3D) systems when the Fermi surface is approximately spherical [37,74]. In the case of extrinsic graphene, the Fermi level is raised above the Dirac point, and hence the Fermi surface corresponds to a circle whose radius is the modulus of the Fermi wave vector $k_F = \sqrt{\pi n}$, with n the surface density of extrinsic charges (see Figure 17.5). The deformation potential

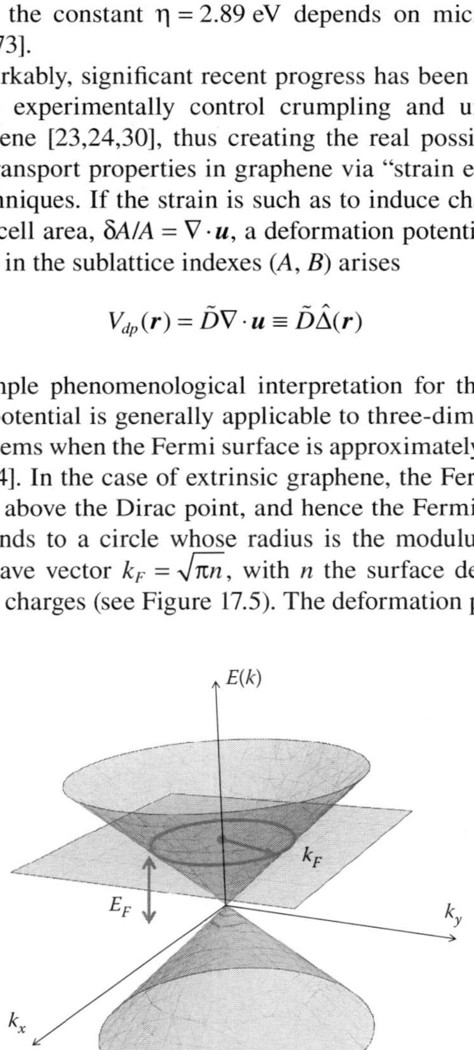

FIGURE 17.5 Extrinsic graphene: Pictorial description of the Fermi energy and Fermi "surface," in this case a circle formed by the intersection between the plane $E(\boldsymbol{k}) = E_F = \eta k_F$, and the Dirac cone. At finite extrinsic carrier surface density n, the Fermi wavevector determines the radius of the circle, $k_F = \sqrt{\pi n}$.

then captures the energy change due to the isotropic lattice dilation, in the form [37,74]

$$\delta E(k) = E(k) - E_0(k) \sim \left(A\frac{\partial E(0)}{\partial A}\right)\frac{\delta A}{A} = \tilde{D}\Delta \tag{17.49}$$

Here, we have defined $\tilde{D} = A\partial E(0)/\partial A$ as the (two-dimensional) deformation potential constant, and $\Delta = \delta A/A = \partial_x u_x + \partial_y u_y$ is the lattice dilation.

In second quantization, the lattice displacement field [37,71]

$$\boldsymbol{u}(\boldsymbol{r}) = \sum_{\mu,q}\sqrt{\frac{\hbar}{2A\rho_s\omega_{\mu q}}}(a_{\mu q} + a_{\mu q}^{\dagger})\hat{e}_{\mu q}e^{iq.r} \tag{17.50}$$

is written in terms of a set of bosonic operators representing phonon modes $[a_{\mu q}, a_{\mu' q'}^{\dagger}] = \delta_{\mu,\mu'}\delta_{q,q'}$, with μ labeling each of the in-plane phonon modes. Therefore, the lattice dilation is expressed by the operator [34]

$$\hat{\Delta}(\boldsymbol{r}) = i\sum_{q,\mu}(2A\rho_s\omega_{\mu q})^{-1/2}(\boldsymbol{q}\cdot\hat{e}_{q\mu})(a_{\mu q}^{\dagger}e^{iq\cdot r} - a_{\mu q}e^{-iq\cdot r}) \tag{17.51}$$

Here, $a_{\mu q}$ are bosonic operators, satisfying the commutation relation $[a_{\mu q}, a_{\mu' q'}^{\dagger}]_{-} = \delta_{\mu,\mu'}\delta_{q,q'}$. In this approach, only the LA phonons couple to the conduction electrons. The electron–phonon interaction Hamiltonian obtained from this approach is given by the operator [34]

$$\begin{aligned} H_{e-ph} &= \int d^2r\sum_{i=1,2}\hat{\Psi}_i^{\dagger}(\boldsymbol{r})\tilde{D}\hat{\Delta}(\boldsymbol{r})\Psi_i(\boldsymbol{r}) \\ &= i\sum_{k,q,\eta,\sigma}\sqrt{\frac{\hbar q^2\tilde{D}^2}{2\rho_s\omega_q}}\cos\left(\frac{\theta}{2}\right)(a_q - a_{-q}^{\dagger})c_{k+q,\eta,\sigma}^{\dagger}c_{k,\eta,\sigma} \end{aligned} \tag{17.52}$$

The subindex μ = LA corresponding to the LA mode, which is the only one contributing in this mechanism, has been omitted for simplicity. Here, $\cos(\theta/2) = \cos(\theta_k - \theta_{k+q})/2$ represents the electron–phonon scattering angle. This factor comes into place due to the inner product of the two spinors (defined in Equation 17.38) $\psi_{k',1}^{(+)^{\dagger}} \cdot \psi_{k,1}^{(+)}$, and it clearly forbids the possibility of backscattering ($\theta = \pi$). This phenomenon, a direct consequence of the conservation of chirality upon scattering (see Section 17.3), has been experimentally observed in graphene [46,75,76] and referred to as Klein tunneling for its historical roots in the context of relativistic quantum field theory.

17.5 BOLTZMANN EQUATION AND TRANSPORT COEFFICIENTS

In the presence of external fields and temperature gradients, the Bloch–Boltzmann transport equation [77–80]

$$\frac{\partial f_k}{\partial t} - e(\boldsymbol{E} + \boldsymbol{v}_k \times \boldsymbol{B})\cdot\nabla_k f_k + \boldsymbol{v}_k \cdot \nabla T\frac{\partial f_k}{\partial T} = (\dot{f}_k)_{scatt} \tag{17.53}$$

provides a semiclassical picture to describe the time evolution of the distribution f_k of "wave-packets" with quasi-momentum $\hbar\boldsymbol{k}$, drifted by a deterministic (Lorentz) force $\hbar\dot{\boldsymbol{k}} = -e(\boldsymbol{E} + \boldsymbol{v}_k \times \boldsymbol{B})$ and a temperature gradient ∇T. A diffusive component in the dynamics is described by the scattering term $(\dot{f}_k)_{scatt}$, involving microscopic dissipation processes, such as electron–electron Coulomb scattering, electron–phonon interactions, and so on. This approximation assumes that the scattering processes are local and instantaneous events and, therefore, its validity is restricted to the existence of a clear separation of time scales $\delta t \ll \tau$, with τ the relaxation time (inverse of the average scattering rate) and δt the time span of a collision. In semiconducting and insulating materials where typically $\delta t \sim \hbar/k_B T$, the condition of validity corresponds to the Peierls criterion [37,77,78] $\hbar/\tau \ll k_B T$, which is also commonly discussed in the context of phonon thermal transport [51]. On the other hand, in metallic systems the characteristic energy scale is the Fermi energy E_F, and hence the criteria of applicability of the Bloch–Boltzmann transport equation (17.53) is [37,77,78] $\hbar/\tau(E_F) \ll E_F$. Remarkably, this criterion is also satisfied in some semimetallic systems, particularly in extrinsic graphene [34,80], where the presence of extrinsic charges with surface density n lifts the Fermi level above the Dirac neutrality points, as seen by the relation [34,46,80] $E_F = \hbar v_F k_F$, with $k_F = \hbar v_F(\pi n)^{1/2}$ the Fermi wavevector (see Figure 17.5).

One of the standard methods to obtain approximated analytical solutions for the Bloch–Boltzmann equation (17.53), and hence for the transport coefficients, is the relaxation time approximation [37,74,81,82]. This corresponds to the substitution of the scattering term in Equation 17.53 by the expression

$$(\dot{f}_k)_{scatt} \to -\frac{f_k - f_k^0}{\tau_k} \tag{17.54}$$

Moreover, the deviation of the electronic distribution f_k form the equilibrium Fermi distribution f_k^0 is expressed, in the absence of thermal gradients and external magnetic fields, as a linear function of the applied electric field

$$f_k = f_k^0 - \frac{\partial f_k^0}{\partial E_k} e\tau_k \boldsymbol{v}_k \cdot \boldsymbol{E} = f_k^0 - \chi_k \partial f_k^0 / \partial E_k \tag{17.55}$$

The electrical conductance is then obtained from the expression for the electric current

$$\boldsymbol{J} = 4\sum_k (-e)\boldsymbol{v}_k f_k = 4\sum_k (-e)\boldsymbol{v}_k \chi_k \frac{\partial f_k^0}{\partial E_k} = \sigma \boldsymbol{E} \tag{17.56}$$

When the scattering mechanism involved is strictly *elastic*, that is, $E_{k'} = E_k$, then the relaxation time approach provides an exact solution to the *linearized* form of the Boltzmann equation [37,74,78].

17.5.1 Transport and Scattering by Static Impurities

An important example of an elastic process is the scattering of electrons by static impurities, where the relaxation time can be expressed by [81,82]

$$\frac{1}{\tau_k} = n_{imp} \sum_{k'} \mathcal{P}_k^{k'} (1 - \cos(\theta_{k,k'})) \tag{17.57}$$

Here, n_{imp} is the (surface) concentration of impurities and $\mathcal{P}_k^{k'}$ is the transition rate for the process $\boldsymbol{k} \to \boldsymbol{k}'$, which is approximated by Fermi's "golden rule"

$$\mathcal{P}_k^{k'} = \frac{2\pi}{\hbar} \left|\langle \boldsymbol{k}' | V_{scatt} | \boldsymbol{k} \rangle\right|^2 \delta\left(E_{k'} - E_k\right) \tag{17.58}$$

The scattering matrix element in Equation 17.58 is calculated at first-order in perturbation theory, considering the spinor wave-functions for a pristine graphene sheet defined in Equation 17.38

$$\langle \boldsymbol{k}' | V_{scatt} | \boldsymbol{k} \rangle = \int d^2 r (\psi_{k,\eta}^{(\xi)}(\boldsymbol{r}))^\dagger V_{scatt}(\boldsymbol{r}) \psi_{k,\eta}^{(\xi)}(\boldsymbol{r}) \tag{17.59}$$

Let us first consider the limit of a very short-range interaction, in the limit of a "contact" potential

$$V_{scatt}(\boldsymbol{r}) = V_{short}(\boldsymbol{r}) = V_0 \delta(\boldsymbol{r}). \tag{17.60}$$

The relaxation time for this contact potential is [81,82]

$$\tau_k^{short} = \frac{4\hbar^2 v_F}{n_{imp} V_0^2} \frac{1}{k}, \tag{17.61}$$

and, the corresponding expression for the electrical conductivity is [81,82]

$$\sigma^{short} = [\rho_{short}]^{-1} = \frac{8e^2}{h} \frac{(\hbar v_F)^2}{n_{imp} V_0^2}. \tag{17.62}$$

For static impurities of charge $Q_{imp} = Ze$, generating a long-range screened Coulomb potential of the form [81]

$$V_{Coulomb}(r) = \frac{Ze^2}{4\pi\epsilon\epsilon_0 r} e^{-r/L}, \tag{17.63}$$

the relaxation time is given by the expression [81]

$$\left[\tau_k^{Coul}\right]^{-1} = \frac{u_0^2}{v_F \hbar^2 k} \left(1 - \frac{\sqrt{1 + 4k^2 L^2} - 1}{2k^2 L^2}\right), \tag{17.64}$$

where $u_0 = n_{imp}Ze^2/(16\epsilon^2\epsilon_0^2)$. In the limit of an infinite screening length $L \to \infty$, Equation 17.64 reduces to

$$\tau_k^{Coul} \to \frac{v_F \hbar^2 k}{u_0^2} \quad (17.65)$$

The electrical conductivity, in this case, corresponds to [81,82]

$$\sigma_{Coulomb} = [\rho_{Coulomb}]^{-1} = 2\frac{e^2}{h}\frac{\pi(\hbar v_F)^2}{u_0^2}n \quad (17.66)$$

A similar analysis can be carried out to calculate other transport coefficients within the relaxation time approximation. In particular, to obtain the thermal conductivity, the function χ_k defined in Equation 17.55 includes a linear dependence in both the electric field and temperature gradient [37,78]. In the absence of external magnetic fields

$$\chi_k = -\frac{\partial f_k^0}{\partial E_k}\tau_k \boldsymbol{v}_k \cdot \left[e\boldsymbol{E} - \frac{E_k - E_F}{T}\nabla T\right] \quad (17.67)$$

The transport coefficients are obtained by writing down expressions for the charge and energy currents, similar to Equation 17.56

$$\boldsymbol{J} = 4\sum_k (-e)\boldsymbol{v}_k \chi_k \frac{\partial f_k^0}{\partial E_k} = e^2 L_0 \boldsymbol{E} + \frac{e}{T}L_1(-\nabla T),$$

$$\boldsymbol{U} = 4\sum_k (E_k - E_F)\boldsymbol{v}_k \chi_k \frac{\partial f_k^0}{\partial E_k} = eL_1\boldsymbol{E} + \frac{1}{T}L_2(-\nabla T). \quad (17.68)$$

In this case, the tensors L_i are diagonal. To keep the notation simple, we shall use the same symbol for the associated scalar diagonal elements. In particular, for static impurities under a long-range Coulomb potential as described in Equations 17.63 through 17.65, the coefficients are given by [81]

$$L_0 = \frac{2E_F^2}{hu_0^2}, \quad L_1 = \frac{4\pi^2}{3h}(k_BT)^2\frac{E_F}{u_0^2}, \quad L_2 = \frac{2}{3}\frac{\pi^2}{h}(k_BT)^2\frac{E_F^2}{u_0^2} \quad (17.69)$$

The thermal conductivity is defined under conditions where the electric current vanishes $\boldsymbol{J} = 0$, and hence after the first equation (17.68) the electric field and the thermal gradient satisfy the linear relation $\boldsymbol{E} = -(L_1/eTL_0)(-\nabla T) = Q(T)\nabla T$, where the proportionality factor is the Seebeck coefficient

$$Q(T) = \frac{2\pi^2}{3eE_F}k_B^2 T \quad (17.70)$$

Substitution of this linear relation into the second equation (17.68) for the energy current, the resulting expression $\boldsymbol{U} = \kappa_{imp}(-\nabla T)$ provides an explicit definition of the thermal conductivity due to scattering by static Coulomb impurities

$$\kappa_{imp} = \frac{1}{T}\left(\frac{2}{3}\frac{\pi^2}{h}(k_BT)^2\frac{E_F^2}{u_0^2} - \frac{8}{9}\frac{\pi^4}{hu_0^2}(k_BT)^4\right) \quad (17.71)$$

17.5.2 Transport and Electron–Phonon Scattering

For electron–phonon scattering, energy conservation imposes the relation

$$E_{k'} = E_k \pm \hbar\omega_q. \quad (17.72)$$

Therefore, from the perspective of the electron system solely, these are not strictly elastic processes, since the initial and final electron energies differ precisely by the amount of energy involved in the creation/absorption of the scattered phonon.

If one neglects umklapp processes, then the crystal quasi-momentum is conserved upon scattering, that is, $\boldsymbol{k}' = \boldsymbol{k} \pm \boldsymbol{q}$. This last equation implies the relation

$$q^2 = k^2 + k'^2 \pm 2kk'\cos\theta \quad (17.73)$$

where θ is the scattering angle. Since both momenta $\boldsymbol{k}$, $\boldsymbol{k}'$ lie within a thin layer near the Fermi surface, then one has $k \sim k' \sim k_F$, with $k_F = \sqrt{\pi n}$ the Fermi wavevector and n the extrinsic charge surface density. Therefore, Equation 17.73 becomes

$$\left|\sin\left(\frac{\theta}{2}\right)\right| = \frac{q}{2k_F} \quad (17.74)$$

Within a Debye model approximation, this condition restricts the range of phonon momenta that effectively contribute to electron–phonon scattering by the condition

$$q \le \min\{q_D, 2k_F\} \quad (17.75)$$

with q_D the Debye cutoff. Let us now consider the deformation potential model described in the previous Section 17.4 by Equation 17.52. Then, for LA phonons with dispersion $\omega_q = c_{LA}q$, the condition stated in Equation 17.75 can be expressed in terms of the energy scales involved

$$\hbar\omega_q/k_B \le \min\{\Theta_D, \Theta_{BG}\} \quad (17.76)$$

Here, we have defined the Bloch–Grüneisen temperature

$$\Theta_{BG} = 2\hbar c_{LA}k_F/k_B \quad (17.77)$$

with $k_F = \sqrt{\pi n}$ the Fermi wavevector and n the extrinsic charge surface density. In graphitic nanomaterials, such as graphene and carbon nanotubes, the Debye temperature is extremely high, exceeding 1000°C. Therefore, even at high carrier densities in graphene, the relevant phonon momenta involved in scattering processes are determined by the Bloch–Grüneisen rather than the Debye temperature.

Let us now specialize the Boltzmann transport equation (17.53) to the case when an external electric field $\boldsymbol{E}$, and a temperature gradient ∇T are present. The nonequilibrium distribution function is expressed, as usual, in the form

$$f_k = f_k^0 - \chi_k \partial f_k^0 / \partial E_k \tag{17.78}$$

with f_k^0 the equilibrium Fermi distribution function and χ_k representing the deviation from equilibrium. In steady state, the linearized form of the Boltzmann equation (17.78) is [34]

$$-\boldsymbol{v}_k \cdot \nabla T \frac{\partial f_k^0}{\partial T} - \boldsymbol{v}_k \cdot \boldsymbol{E}(-\mathrm{e}) \frac{\partial f_k^0}{\partial E_k} = \frac{4}{k_B T} \sum_{k',q} [\{\chi_k - \chi_{k'}\} \mathcal{P}_{kq}^{k'} - \{\chi_{k'} - \chi_k\} \mathcal{P}_{k'}^{kq}] \tag{17.79}$$

Here, $\mathcal{P}_{kq}^{k'}$ and $\mathcal{P}_{k'}^{kq}$ are the probability rates for electron–phonon scattering processes of the form $\boldsymbol{k} + \boldsymbol{q} \rightleftharpoons \boldsymbol{k}'$. These rates are calculated via Fermi's "golden rule" by the expression

$$\mathcal{P}_{kq}^{k'} = \frac{2\pi}{\hbar} |\langle n_{k'} + 1; n_k - 1; N_q - 1 | H_{e-ph} | n_{k'}; n_k; N_q \rangle|^2 \times \delta(E_k - E_{k'} + \hbar\omega_q) \delta_{k',k+q} \tag{17.80}$$

Calculating the transition matrix element directly from Equation 17.52, the probability rate is given by the expression [34]

$$\mathcal{P}_{kq}^{k'} = \left(\frac{2\pi}{\hbar}\right) \delta_{G,k'-k-q} |M_q|^2 \, \delta(E_k - E_{k'} + \hbar\omega_q) N_q^0 f_k^0 (1 - f_{k'}^0) \tag{17.81}$$

Here, $N_q^0 = (\exp(\hbar\omega_q / k_B T) - 1)^{-1}$ is the Bose–Einstein distribution for phonons with momentum $\boldsymbol{q}$. Note that here we have applied Bloch's approximation [37], by decoupling the relaxation time scales of electrons and phonons, thus allowing us to treat the phonon bath as if it was in quasi-equilibrium from the perspective of the electronic distribution. The scattering matrix element that follows from Equation 17.81 in this approximation is [34]

$$M_q = (\hbar / 2\rho_m \omega_q)^{1/2} \tilde{D} q (1 - 2(q/2k_F)^2)^{1/2} \tag{17.82}$$

The square root factor appearing in Equation 17.82 is a characteristic feature of the pseudo-relativistic behavior of conduction electrons in graphene, since it prevents from backscattering $\theta = \pi \Leftrightarrow q = 0$, in agreement with the experimentally observed Klein-tunneling effect in graphene [46,76].

By Onsager's theorem [83], within the linear response regime, the fluxes, and their conjugated affinities, that is, the gradients in temperature and electric potential, are linearly coupled

$$\boldsymbol{J} = L_{EE} \boldsymbol{E} + L_{ET} \nabla T, \tag{17.83}$$

$$\boldsymbol{U} = L_{TE} \boldsymbol{E} + L_{TT} \nabla T. \tag{17.84}$$

Here, $\boldsymbol{J}$ and $\boldsymbol{U}$ are the macroscopic charge and energy fluxes, respectively. The transport coefficients are defined in terms of the tensors in Equations 17.83 and 17.84. In the absence of a thermal gradient, Equation 17.83 reduces to $\boldsymbol{J} = L_{EE}\boldsymbol{E}$, which allows us to identify the electrical resistivity by

$$\rho = L_{EE}^{-1} \tag{17.85}$$

On the other hand, under conditions where the electric current vanishes, $\boldsymbol{J} = 0$ and after Equation 17.83 one concludes that the electric field and thermal gradient are linearly related $\boldsymbol{E} = -L_{EE}^{-1} L_{ET} \nabla T$. The proportionality factor is then identified as the Seebeck coefficient

$$Q = -L_{EE}^{-1} L_{ET} \tag{17.86}$$

From Equation 17.84, one obtains that under the aforementioned condition of vanishing electric current, the heat flux is given by the expression $\boldsymbol{U} = -(L_{TE} L_{EE}^{-1} L_{ET} - L_{TT}) \nabla T$. Therefore, we conclude that the thermal conductivity is given by the expression

$$\kappa = L_{TE} L_{EE}^{-1} L_{ET} - L_{TT} \tag{17.87}$$

The macroscopic energy and charge currents are expressed in terms of the nonequilibrium distribution function

$$\boldsymbol{J} = 4 \sum_k (-e) \boldsymbol{v}_k f_k = 4 \sum_k (-e) \boldsymbol{v}_k \chi_k \frac{\partial f_k^0}{\partial E_k}, \tag{17.88}$$

$$\boldsymbol{U} = 4 \sum_k \boldsymbol{v}_k \left(E_k - E_F\right) f_k = 4 \sum_k \boldsymbol{v}_k \left(E_k - E_F\right) \chi_k \frac{\partial f_k^0}{\partial E_k} \tag{17.89}$$

where the factor of 4 accounts for the spin (±1/2) and valley ($\boldsymbol{K}_{1,2}$) degeneracies, respectively.

17.5.3 Variational Solution to the Boltzmann Transport Equation

A standard approach to solving the linearized Boltzmann equation, Equation 17.79, and then to calculate the transport coefficients by Equations 17.85 through 17.89, is to apply the relaxation time approximation [37,80]. This provides an exact solution, under conditions such that the scattering mechanism is strictly elastic, that is, if $E_{k'} = E_k$. This is not the case, however, in electron–phonon scattering, since the energy of the scattered electron state $E_{k'} = E_k \pm \hbar\omega_q$ differs from the initial one precisely on the energy $\hbar\omega_q$ of the phonon that is being absorbed (emitted). It is a plausibly good approximation though in graphene, since $\hbar\omega_q \ll E_k$ for the energy scales involved [80]. An alternative treatment of the problem, which is not subjected to this restriction, is provided by the variational method [34]. In this context, following Onsager [83], in steady-state conditions one should look for a maximum in the macroscopic entropy production rate

$$\partial_t S = \frac{\boldsymbol{J}\cdot\boldsymbol{E}}{T} + \boldsymbol{U}\cdot\nabla\left(\frac{1}{T}\right) \tag{17.90}$$

subjected to the constraint that it equals the entropy production rate expressed in terms of the microscopic dissipative mechanisms involved, $\partial_t S = \dot{S}_{scatt}$. After appropriate linearization using Equation 17.78, one finds [34]

$$\dot{S}_{scatt} = \frac{4}{k_B T^2}\sum_{k,q,k'}\{\chi_k - \chi_{k'}\}^2 \mathcal{P}_{kq}^{k'} \tag{17.91}$$

The technique is then based on expanding the deviation in the distribution function from equilibrium

$$\chi_k = \sum_i \alpha_i \phi_i(\boldsymbol{k}) \tag{17.92}$$

in terms of a variational set of functions of the form $\phi_i(\boldsymbol{k}) = (\epsilon_k - E_F)^{i-1}(\hat{u}\cdot\boldsymbol{k})$, with $\hat{u}$ a unit vector along the direction of the external field. Using this variational set, let us define the symmetric matrix $\hat{P} = [P_{ij}]$, with coefficients $P_{ij} = P_{ji}$ given by the expressions

$$P_{ij} = \frac{16}{k_B T}\sum_{k,k',q}\{\phi_i(k) - \phi_i(k')\}\{\phi_j(k) - \phi_j(k')\}\mathcal{P}_{kq}^{k'} \tag{17.93}$$

In terms of the variational solution expressed by Equation 17.92, Equations 17.88 and 17.89 for the electric and heat currents become

$$\boldsymbol{J} = \sum_i \alpha_i \boldsymbol{J}_i, \qquad \boldsymbol{U} = \sum_i \alpha_i \boldsymbol{U}_i. \tag{17.94}$$

Here, we have defined

$$\boldsymbol{J}_i = 4(-e)\sum_k \boldsymbol{v}_k \phi_i(\boldsymbol{k})\frac{\partial f_k^0}{\partial \epsilon_k}, \qquad \boldsymbol{U}_i = 4\sum_k \boldsymbol{v}_k(\epsilon_k - E_F)\phi_i(\boldsymbol{k})\frac{\partial f_k^0}{\partial \epsilon_k}. \tag{17.95}$$

In terms of the currents in Equation 17.95, the microscopic and macroscopic entropy production rates defined in Equations 17.90 and 17.91 become

$$\partial_t S = \sum_i \alpha_i [T^{-1}\boldsymbol{J}_i\cdot\boldsymbol{E} - T^{-2}\boldsymbol{U}_i\cdot\nabla T], \tag{17.96}$$

$$\dot{S}_{scatt} = T^{-1}\sum_{i,j}\alpha_i\alpha_j P_{ij}. \tag{17.97}$$

The variational problem then reduces to find the set of coefficients $\{\alpha_j\}$ that maximizes $\dot{S}_{scatt}$, subject to the constraint $\partial_t S = \dot{S}_{scatt}$, which is enforced by a Lagrange multiplier λ

$$\frac{\delta}{\delta\alpha_n}\left[\sum_{i,j}\alpha_i\alpha_j\frac{P_{ij}}{T} - \lambda\sum_i \alpha_i(T^{-1}\boldsymbol{J}_i\cdot\boldsymbol{E} - T^{-2}\boldsymbol{U}_i\cdot\nabla T)\right] = 0 \tag{17.98}$$

The solution to the constrained variational problem is found to be [34] $\lambda = 1/2$ and

$$\alpha_i = \sum_j [\hat{P}^{-1}]_{ij}\left(\boldsymbol{J}_i\cdot\boldsymbol{E} - \boldsymbol{U}_i\cdot\frac{\nabla T}{T}\right) \tag{17.99}$$

Here, $\hat{P}^{-1}$ is the inverse of the matrix $\hat{P} = [P_{ij}]$ whose elements were defined in Equation 17.93.

Upon direct substitution of this solution into the expressions for the electric and heat currents equation (17.94), we obtain explicit analytical formulae for the tensors defined in Equations 17.83 and 17.84

$$L_{EE} = \sum_{ij}\boldsymbol{J}_i[\hat{P}^{-1}]_{ij}\boldsymbol{J}_j, \tag{17.100}$$

$$L_{ET} = -T^{-1}\sum_{ij}\boldsymbol{J}_i\,[\hat{P}^{-1}]_{ij}\boldsymbol{U}_j, \tag{17.101}$$

$$L_{TE} = \sum_{ij}\boldsymbol{U}_i[\hat{P}^{-1}]_{ij}\boldsymbol{J}_j, \tag{17.102}$$

$$L_{TT} = -T^{-1}\sum_{ij}\boldsymbol{U}_i[\hat{P}^{-1}]_{ij}\boldsymbol{U}_j. \tag{17.103}$$

17.5.4 Phonon-Limited Electrical Resistivity

The phonon-limited electrical resistivity is obtained from this approach by Equation 17.85, $\rho_{e-ph} = L_{EE}^{-1}$. By direct substitution of the coefficients and currents in Equations 17.93 and 17.95 into Equation 17.100, the analytical expression for the leading contribution to the electrical resistivity due to electron–phonon scattering is obtained [34]

$$\rho_{e-ph}(T) = \rho_0\left[\left(\frac{T}{\Theta_{BG}}\right)^4 \mathcal{J}_4(T/\Theta_{BG}) + \frac{4\pi^2}{3}\left(\frac{c_{LA}}{v_F}\right)^2\left(\frac{T}{\Theta_{BG}}\right)^6\right.$$
$$\left.\times\left\{\mathcal{J}_4\left(\frac{T}{\Theta_{BG}}\right) + \frac{3}{2\pi^2}\mathcal{J}_6(T/\Theta_{BG})\right\}\right] \qquad (17.104)$$

Here, the coefficient

$$\rho_0 = \frac{8\tilde{D}^2 k_F}{e^2\rho_s c_{LA} v_F^2} \qquad (17.105)$$

has dimensions of a two-dimensional resistivity (Ω), D is the deformation potential constant, ρ_s the surface mass density of the single-layer graphene (see Table 17.1), and c_{LA} the speed of sound for LA phonons. Θ_{BG} is the Bloch–Grüneisen temperature defined in Equation 17.77. For $p > 0$ an integer, we have defined the special functions

$$\mathcal{J}_p(z) = \int_0^z \frac{x^p\sqrt{1-(x/z)^2}}{(1-\exp(-x))(\exp(x)-1)} \qquad (17.106)$$

An important asymptotic property of these functions is $\mathcal{J}_p(\infty) = p!\zeta(p)$, with $\zeta(p)$ the Riemann zeta function. In particular, for $p = 4$, 6 it has the values $\zeta(4) = \pi^4/90$ and $\zeta(6) = \pi^6/945$. Therefore, the low-temperature behavior ($T \ll \Theta_{BG}$) of the phonon contribution to the electrical resistivity is

$$\rho_{e-ph}(T) = \rho_0 4!\zeta(4)\left(\frac{T}{\Theta_{BG}}\right)^4 + o\left(\frac{T}{\Theta_{BG}}\right)^6 \qquad (17.107)$$

In the high-temperature limit ($T \gg \Theta_{BG}$), from Equation 17.105 and the asymptotic properties of the functions $\mathcal{J}_p(z)$, one finds

$$\rho_{e-ph}(T) = \left\{\frac{1}{3} - \frac{1}{10}\cdots\right\}\left(\frac{T}{\Theta_{BG}}\right)\rho_0 = \frac{\pi}{8}\left(\frac{T}{\Theta_{BG}}\right)\rho_0 \qquad (17.108)$$

Here, the constant ρ_0 was defined in Equation 17.105. This behavior is in excellent agreement with transport experiments at high carrier densities [35], and the asymptotic limits are also in agreement with an earlier theory proposed by Hwang and Das Sarma [84] based on the relaxation time approximation. Comparison of Equation 17.104 with experiments is displayed in Figure 17.6. The experimental data for the total electrical resistivity $\rho(T)$ were fitted to the expression

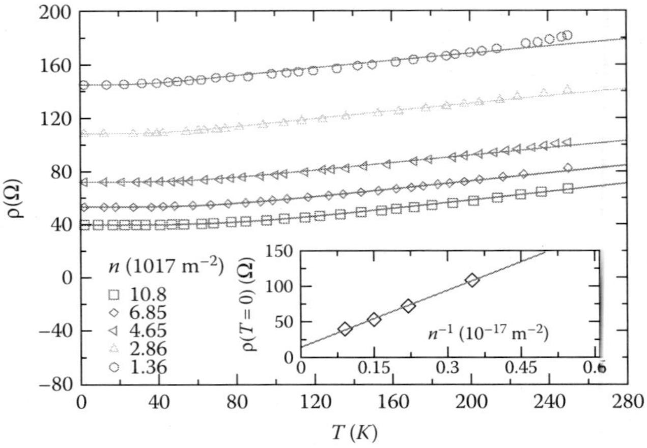

FIGURE 17.6 Electrical resistivity calculated from Equation 17.104 (colored lines), compared with experimental data (symbols) from Reference 35, at different carrier densities n (10^{17} m^{-2}). Inset: Zero-temperature contribution to the total resistivity at different carrier densities n. Also shown (solid line in inset) is the linear fit Equation 17.110.

$$\rho(T) = \rho_{T=0} + \rho_{e-ph}(T) \qquad (17.109)$$

where $\rho_{T=0}$ is the temperature independent component of the total electric resistivity, and $\rho_{e-ph}(T)$ the temperature-dependent electron–phonon scattering contribution calculated [34] from Equation 17.104. The temperature-independent contribution can be attributed to the presence of impurities and defects in the sample. As discussed in Section 17.5.1, for a surface concentration of impurities n_{imp}, the impurity contribution to the electrical resistivity due to a short-range scattering potential is given after Equation 17.62 by $\rho_{short} \sim n_{imp}V_0^2$. On the other hand, as shown in Equation 17.66, for a long-range Coulomb potential the impurity contribution to the resistivity depends on the inverse of the surface charge concentration $\rho_{Coulomb} \sim n_{imp}n^{-1}$. By taking both mechanisms into account, the zero temperature contribution to the total resistivity in Equation 17.109 is expected to follow the parametric form

$$\rho_{T=0} = a + bn^{-1} \qquad (17.110)$$

where the coefficients a and b depend on details of the sample, as the concentration and distribution of impurities and defects.

As shown in the inset of Figure 17.6, the experimental data are in excellent agreement with Equation 17.110, for numerical values $a = 14.276$ Ω and $b = 266.17$ Ω m^2 giving a linear regression coefficient $r = 0.999$. The total electrical resistivity equation (17.109) is also in excellent agreement with experimental data, as seen in Figure 17.6. The temperature-dependent part in Equation 17.109, calculated from Equation 17.104, requires just two independent parameters to be adjusted to the experimental data: The deformation potential parameter

$\tilde{D} = 23.5 \pm 0.5\,\text{eV}$ and the speed of sound for longitudinal phonons $c_{LA} = 24.0 \pm 0.6$ km/s. It is remarkable that the deviations from the mean values in both parameters among the different sets of data are quite small, on the order of 2%. Moreover, the fitted value for the speed of sound of acoustic phonons is in excellent agreement with the theoretical estimates discussed in Section 17.2 (see Table 17.2). There is some discrepancy in the literature concerning the values for the displacement potential energy parameter, being in the range $20\,\text{eV} < \tilde{D} < 30\,\text{eV}$ [70], with 16 eV for bulk graphite as a theoretical reference [85]. The value $\tilde{D} = 23.5 \pm 0.5\,\text{eV}$ obtained by fitting the experimental data to Equation 17.104 is then clearly within the range reported in the literature.

Note that the ratio $(c_{LA}/v_F)^2 \sim 10^{-4}$ is extremely small, and the terms proportional to this factor can in practice be neglected in the expression obtained for the electrical resistivity, as well as in other transport coefficients to be discussed next. This explains the quantitative success in the application of the relaxation time approximation to model electron–phonon scattering in previous theoretical work [84], even though this scattering mechanism is not strictly elastic.

17.5.5 Phonon-Limited Thermal Conductivity

As discussed in Section 17.5.2, by setting the condition $\boldsymbol{J} = 0$ one obtains Equation 17.87 for the thermal conductivity $\kappa = L_{TE}L_{EE}^{-1}L_{ET} - L_{TT}$. Upon direct substitution of the coefficients calculated in Equations 17.100 through 17.103, and neglecting terms of $o(c_{LA}/v_F)^2$ and higher, the phonon-limited thermal conductivity obtained from this approach is given by the analytical equation

$$[\kappa_{e-ph}(T)]^{-1} = \frac{\rho_0}{\mathcal{L}_0 T}\left\{\left[\left(\frac{T}{\Theta_{BG}}\right)^4 + \frac{3}{4\pi^2}\left(\frac{T}{\Theta_{BG}}\right)^2\right]\mathcal{J}_4\left(\frac{T}{\Theta_{BG}}\right) - \frac{1}{2\pi^2}\left(\frac{T}{\Theta_{BG}}\right)^4 \mathcal{J}_6\left(\frac{T}{\Theta_{BG}}\right)\right\} \quad (17.111)$$

Here, $\mathcal{L}_0 = \pi^2 k_B^2/(3e^2)$ is the Lorenz number for the free electron gas, the constant ρ_0 was defined in Equation 17.105, and the Bloch–Grüneisen temperature Θ_{BG} was defined in Equation 17.77. From the asymptotic properties of the $\mathcal{J}_p(z)$ functions defined in Equation 17.106, it follows after Equation 17.111 that at very low temperatures ($T \ll \Theta_{BG}$)

$$[\kappa_{e-ph}(T)]^{-1} = \frac{\rho_0}{\mathcal{L}_0\Theta_{BG}}\frac{3}{4\pi^2}4!\zeta(4)\left(\frac{T}{\Theta_{BG}}\right) + o\left(\frac{T}{\Theta_{BG}}\right)^3 \quad (17.112)$$

This linear in T behavior at low temperatures is clearly different from the typical thermal resistivity in normal metals, which scales as $\sim T^2$ [37,74].

The electronic thermal conductivity can be estimated, using Mathiesen's rule, by adding the thermal resistivity due to static Coulomb impurities. As discussed in Section 17.5.1 and Equation 17.71, this corresponds to $[\kappa_{imp}(T)]^{-1} = 3hT^{-1}u_0^2/(2\pi^2 k_B^2 E_F^2)$, with $u_0 = n_{imp}Ze^2/(16\epsilon^2\epsilon_0^2)$. Then, the total electronic thermal conductivity is estimated by Matthiessen's rule as

$$[\kappa(T)]^{-1} \sim [\kappa_{e-ph}(T)]^{-1} + [\kappa_{imp}(T)]^{-1} \quad (17.113)$$

To obtain an estimation of the numerical values of the electronic thermal conductivity, let us consider the concentration of impurities extracted [34] from the fit to the experimental data in Reference 35, $n_{imp} = 1.3 \times 10^{15}\ \text{m}^{-1}$. Similarly, for the experimental system described in Reference 35, we estimate a dielectric constant $\varepsilon = 3.1$ representing an average between the poly(ethylene oxide) polymer electrolyte and the SiO_2 substrate. The two-dimensional conductivity defined by Equation 17.113 is further converted to a 3D basis by considering the nominal "packing thickness" of $t = 3.4$ Å as in graphite, $\tilde{\kappa}(T) = \kappa(T)/t$. The numerical values of the total thermal conductivity as a function of temperature are presented, at different extrinsic carrier densities, in Figure 17.7. It is clear that even at very high carrier densities, the total electronic thermal conductivity at room temperature does not exceed ~400 W m^{-1} K^{-1}, a value that represents on the order of 10% of the thermal conductivity due to phonons [50–52], $\kappa_{ph} \sim 4000$ W m^{-1} K^{-1}, as discussed in Section 17.2.1 (see also Table 17.3 for a comparison between different estimations). These results are in quantitative agreement with the commonly accepted belief that it is phonons rather than electrons that dominate the total thermal conductivity in graphene and, therefore, phonons are indeed responsible for the extremely high thermal conductivity of this material [40,50,51,54,56,86].

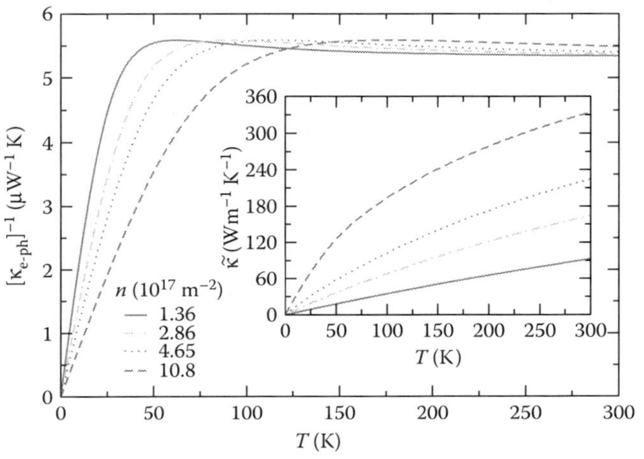

FIGURE 17.7 Phonon-limited electronic thermal resistivity $[\kappa_{e\text{-}ph}]^{-1}$, calculated from Equation 17.111, at different extrinsic carrier densities n (10^{17} m^{-2}). Inset: 3D thermal conductivity $\tilde{\kappa}(T) = \kappa(T)/t$ calculated from Equation 17.113, with $t = 3.4$ Å the nominal "packing thickness" of a single graphene layer.

17.5.6 Phonon-Limited Seebeck Coefficient

The Seebeck coefficient is obtained by setting $\boldsymbol{J} = 0$ in Equation 17.86, which yields $Q(T) = -L_{ET}/L_{EE}$. Substituting the expressions obtained for the coefficients from Equations 17.100 through 17.103 into Equation 17.86, the Seebeck coefficient predicted by this theory is

$$Q(T) = -\frac{\pi^2}{3e}\frac{k_B^2 T}{E_F}\left\{\left[1+\frac{4\pi^2}{3}\left(\frac{T}{\Theta_{BG}}\right)^2\right]\mathcal{J}_4\left(\frac{\Theta_{BG}}{T}\right) + \frac{v_s}{v_F}\left(\frac{T}{\Theta_{BG}}\right)\mathcal{J}_5\left(\frac{\Theta_{BG}}{T}\right) - 2\left(\frac{T}{\Theta_{BG}}\right)^2 \mathcal{J}_6\left(\frac{\Theta_{BG}}{T}\right)\right\} \times \left\{\left[1+\frac{4\pi^2}{3}\left(\frac{T}{\Theta_{BG}}\right)^2\right]\mathcal{J}_4\left(\frac{\Theta_{BG}}{T}\right) - \frac{2}{3}\left(\frac{T}{\Theta_{BG}}\right)^2 \mathcal{J}_6\left(\frac{\Theta_{BG}}{T}\right)\right\}^{-1}. \quad (17.114)$$

Applying the asymptotic properties of the functions $\mathcal{J}_p(z)$ defined in Equation 17.106, we obtain the low-temperature ($T \ll \Theta_{BG}$) behavior for the Seebeck coefficient defined in Equation 17.114

$$Q(T) \sim -\frac{\pi^2}{3e}\frac{k_B^2 T}{E_F}\left[1+\frac{v_s}{v_F}\frac{5!\zeta(5)}{4!\zeta(4)}\left(\frac{T}{\Theta_{BG}}\right)\right] + o(T^3) \quad (17.115)$$

The analytical expressions obtained for the Seebeck coefficient are in quantitative agreement with experiments [36], as shown in Figure 17.8 inset and in Table 17.4. It has been experimentally observed [36,61,87–89] that the Seebeck coefficient in graphene depends on the extrinsic carrier density as $n^{-1/2}$, and exhibits a linear in T dependence at very low temperatures [36]. Since the Fermi energy $E_F = \hbar v_F\sqrt{\pi n}$, Equation 17.114 clearly reproduces correctly the experimental behavior, as shown in the inset of Figure 17.8, where Equation 17.114 is compared with experimental data [36]. In Reference 90, the contribution to the Seebeck coefficient due to phonon drag effects was analyzed and shown to scale as $Q_{drag} \sim T^3$ at low temperatures. Therefore, the phonon drag represents a negligible contribution to the total Seebeck coefficient as compared to the diffusive contribution calculated from the theory in Equation 17.114. A precise assessment of the Seebeck coefficient is important not only for energy harvesting in thermoelectric applications, but also as a sensing method to characterize structural disorder, since this parameter is very sensitive to local fluctuations in the charge density even at the atomic scale [58,66,87]. Estimated values of the Seebeck coefficient in graphene, in the context of different experimental and theoretical methods other than the one described along this section, are presented for comparison in Table 17.4.

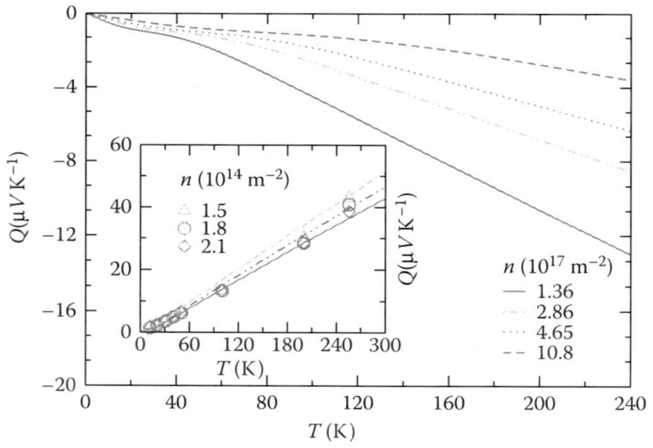

FIGURE 17.8 Seebeck coefficient at different extrinsic charge surface densities n (10^{17} m^{-2}), calculated from Equation 17.114 with $c_{LA} = 24$ km/s as inferred from experimental data in [35]. Inset: Experimental data [36] for the Seebeck coefficient due to extrinsic holes in graphene, compared with Equation 17.114 at the corresponding hole surface densities (no fitting parameters except $c_{LA} = 24$ km/s are used).

17.6 CONCLUSIONS

Along this chapter, we have discussed some of the remarkable electronic and mechanical properties of single-layer graphene. In particular, we have analyzed the effects of electron–phonon

TABLE 17.4
Seebeck Coefficient (Thermopower) of Single-Layer Graphene and Related Nanostructures Near Room Temperature

Experimental Method	Comment	Seebeck (Absolute Value) (μV/K)
STM tip temperature gradient	Atomic resolution. Sensitivity to local electronic density	~40–80 [66]
		~10–30 [87]
	$\lvert Q\rvert \sim 1/n^{1/2}$	~42 [58]
Temperature gradient imposed by metallic leads	Magnetic field and electronic density dependence	~50–100 [59]
		<100 [91]
	$\lvert Q\rvert \sim 1/n^{1/2}$	<80 [36,88]
	Controlled hydrogenation	~10–60 [89]
Theoretical Method		
Ballistic quantum transport	Local spin–orbit interaction, at very low temperature	~20 [92]
	Nanoribbons with antidots; Giant spin-Seebeck	Up to ~500 [93]
Relaxation time approximation	Scattering by Coulomb screened impurities, electronic density dependence $\lvert Q\rvert \sim 1/n^{1/2}$	<80 [34]
Boltzmann transport equation, semiclassical	Electron–phonon scattering; scattering with impurities. Electronic density dependence $\lvert Q\rvert \sim 1/n^{1/2}$	Up to ~150 [94]
Ab initio	Graphene-BN superlattice; Giant Seebeck effect	Up to ~3000 [95]

coupling, which in the continuum description of graphene adopts the form of a gauge field in the Dirac equation, over the different transport coefficients. The theoretical estimations presented here for these coefficients were obtained from a variational solution of the Boltzmann transport equation, whose limits of applicability in this context were also discussed. The theoretical calculations show an excellent agreement with independent experiments that measure electrical resistivity and the Seebeck coefficient, respectively, in extrinsic single-layer graphene. By comparing the theoretical and experimental estimates for the thermal conductivity, it is apparent that the extremely high value of the thermal conductivity in graphene is due to phonon transport. In particular, the theoretical estimations presented along this chapter suggest that the electronic contribution is less than 10% of the total thermal conductivity at room temperature, and that phonons dominate thermal transport even at relatively low temperatures. On the other hand, the strong effects of electron–phonon interactions on the values of the different transport coefficients in extrinsic graphene, suggest the possibility of controlling electrical, thermal, and thermoelectric transport in graphene via engineering the phonon properties. From the perspective of potential applications in thermoelectric energy conversion, a high total thermal conductivity due to the phonon contribution limits the value of the figure of merit $ZT = Q^2\rho^{-1}T/(\kappa_e + \kappa_{ph})$. Possible strategies to control the phonon thermal conductivity could be to enhance edge roughness scattering [62] or to isotopically engineer the graphene lattice [86,91,92].

ACKNOWLEDGMENTS

The author wishes to thank the direction of the Max Planck Institute for the Physics of Complex Systems for their hospitality during the preparation of this work, and Fondecyt Grant No. 1141146.

REFERENCES

1. Wallace, P.R., The band theory of graphite. *Physical Review*, 1947. **71**: p. 622–634.
2. Tudorovskiy, T., Reijnders, K.J.A., and Katsnelson, M.I., Chiral tunneling in single-layer and bilayer graphene. *Physica Scripta*, 2012. **T146**: p. 014010 (17pp).
3. Novoselov, K.S., Electric field effect in atomically thin carbon films. *Science*, 2004. **306**: p. 666–669.
4. Kodepelly, S.R., Senthilnathan, J., Liu, Y.-F., and Yoshimura, M., Role of peroxide ions in formation of graphene nanosheets by electrochemical exfoliation of graphite. *Scientific Reports*, 2014. **4**: p. 4237.
5. Huang, G., Hou, C., Shao, Y., Wang, H., Zhang, Q., Li, Y., and Zhu, M., Highly strong and elastic graphene fibres prepared from universal graphene oxide precursors. *Scientific Reports*, 2014. **4**: p. 4248.
6. Wang, S.M., Gong, Q., Li, Y.Y., Cao, C.F., Zhou, H.F., Yan, J.Y., Liu, Q.B. et al., A novel semiconductor compatible path for nano-graphene synthesis using CBr_4 precursor and Ga catalyst. *Scientific Reports*, 2014. **4**: p. 4653.
7. Chuang, C.-H., Wang, Y.-F., Shao, Y.-C., Yeh, Y.-c., Wang, D.-Y., Chen, C.-W., Chiou, J.W. et al., The effect of thermal reduction on the photoluminiscence and electronic structures of graphene oxides. *Scientific Reports*, 2014. **4**: p. 4525.
8. Wang, C., Chen, W., Han, C., Wang, G., Tang, B., Tang, C., Wang, Y. et al., Growth of milimeter-size single crystal graphene on Cu foils by circumfluence chemical vapor deposition. *Scientific Reports*, 2014. **4**: p. 4537.
9. Zhang, J., Wang, Z., Niu, T., Wang, S., Li, Z., and Chen, W., Elementary process for CVD graphene on Cu(110): Size-selective carbon clusters. *Scientific Reports*, 2014. **4**: p. 4431.
10. Torres, J.A., and Kaner, R.B., Graphene closer to fruition. *Nature Materials*, 2014. **13**: p. 328.
11. Zhang, H., Fonseca, A.G., and Cho, K., Tailoring thermal transport property of graphene through oxygen functionalization. *Journal of Physical Chemistry C*, 2013. **118**: p. 1436–1442.
12. Gupta, P., Dongare, P.D., Grover, S., Dubey, S., Mamgain, H., Bhattachayra, A., and Deshmukh, M.M., A facile process for soak-and-peel delamination of CVD graphene from substrates using water. *Scientific Reports*, 2014. **4**: p. 3882.
13. Novoselov, K.S., Geim, A.K., Morozov, S.V., Jiang, D., Katsnelson, M.I., Grigorieva, I.V., Dubonos, S.V., and Firsov, A.A., Two-dimensional gas of massless Dirac fermions in graphene. *Nature*, 2005. **438**: p. 197–200.
14. Gierz, I., Petersen, J.C., Mitrano, M., Cacho, C., Turcu, I.C.E., Springate, E., Stohr, A., Kohler, A., Starke, U., and Cavalleri, A., Snapshots of non-equilibrium Dirac carrier distributions in graphene. *Nature Materials*, 2013. **12**: p. 1119–1124.
15. Novoselov, K.S., Fal'ko, V.I., Colombo, L., Gellert, P.R., Schwab, M.G., and Kim, P., A roadmap for graphene. *Nature*, 2012. **490**: p. 192–200.
16. Liu, C.-H., Chang, Y.-C., Norris, T.B., and Zhong, Z., Graphene photodetectors with ultra-broadband and high responsivity at room temperature. *Nature Nanotechnology*, 2014. **9**: p. 273–277.
17. Lee, S., Hong, J.-Y., and Jang, J., Multifunctional graphene sheets embedded in silicone encapsulant for superior performance of light-emitting diodes. *ACS Nano*, 2013. **7**: p. 5784–5790.
18. Kim, J.Y., and Kim, S.O., Electric fields line up graphene oxide. *Nature Materials*, 2014. **13**: p. 325.
19. Allain, A., Han, Z., and Bouchiat, V., Electrical control of the superconducting-to-insulating transition in graphene-metal hybrids. *Nature Materials*, 2012. **11**: p. 590–594.
20. Shih, M.-H., Li, L.-J., Yang, Y.-C., Chou, H.-Y., Lin, C.-T., and Su, C.-Y., Efficient heat dissipation of photonic crystal microcavity by monolayer graphene. *ACS Nano*, 2013. **7**: p. 10818–10824.
21. Srisonphan, S., Kim, M., and Kim, H.K., Space charge neutralization by electron-transparent suspended graphene. *Scientific Reports*, 2014. **4**: p. 3764.
22. Russo, C.J., and Passmore, L.A., Controlling protein adsorption on graphene for cryo-EM using low-energy hydrogen plasmas. *Nature Methods*, 2014. **11**: p. 649–652.
23. Baringhaus, J., Ruan, M., Edler, F., Tejeda, A., Sicot, M., Taleb-Ibrahimi, A., Li, A.-P. et al., Exceptional ballistic transport in epitaxial graphene nanoribbons. *Nature*, 2014. **506**: p. 349–353.
24. Hicks, J., Tejeda, A., Taleb-Ibrahimi, A., Nevius, M.S., Wang, F., Shepperd, K., Palmer, J. et al., A wide-bandgap metal–semiconductor–metal nanostructure made entirely from graphene. *Nature Physics*, 2013. **9**: p. 49–54.
25. Goerbig, O.M., Electronic properties of graphene in a strong magnetic field. *Reviews of Modern Physics*, 2011. **83**: p. 1216–1243.
26. Wendler, F., Knorr, A., and Malic, E., Carrier multiplication in graphene under Landau quantization. *Nature Communications*, 2014. **5**: p. 3703.

27. Gusynin, V.P., and Sharapov, S.G., Unconventional integer quantum Hall effect in graphene. *Physical Review Letters*, 2005. **95**: p. 146801.
28. Booth, T.J., Blake, P., Nair, R., Jiang, D., Hill, E.W., Bangert, U., Blecloch, A. et al., Macroscopic graphene membranes and their extraordinary stiffness. *Nano Letters*, 2008. **8**: p. 2442–2446.
29. Zandiatashbar, A., Lee, G.-H., An, S.J., Lee, S., Mathew, N., Terrones, M., Hayashi, T., Picu, C.R., Hone, J., and Koratkar, N., Effect of defects on the intrinsic strength and stiffness of graphene. *Nature Communications*, 2014. **5**: p. 3186.
30. Zang, J., Ryu, S., Pugno, N., Wang, Q., Tu, Qiming, Buehler, M.J., and Zhao, X., Multifunctionality and control of the crumpling and unfolding of a large-area Graphene. *Nature Materials*, 2013. **12**: p. 321–325.
31. Pumarol, M.E., Rosamond, M.C., Tovee, P., Petty, M.C., Zeze, D.A., Falko, V., and Kolosov, O.V., Direct nanoscale imaging of ballistic and diffusive thermal transport in Graphene nanostructures. *Nano Letters*, 2012. **12**: p. 2906–2911.
32. De Juan, F., Manez, J.L., and Vozmediano, M.A.H., Gauge fields from strain in graphene. *Physical Review B*, 2013. **87**: p. 165131.
33. Vozmediano, M.A.H., Katsnelson, M.I., and Guinea, F., Gauge fields in graphene. *Physics Reports*, 2010. **496**: p. 109–148.
34. Muñoz, E., Phonon-limited transport coefficients in extrinsic graphene. *Journal of Physics: Condensed Matter*, 2012. **24**: p. 195302 (8pp).
35. Efetov, D.K., and Kim, P., Controlling electron-phonon interactions in graphene at ultrahigh carrier densities. *Physical Review Letters*, 2010. **105**: p. 256805.
36. Wei, P., Bao, W., Pu, Y., Lau, C.N., and Shi, J., Anomalous thermoelectric transport of Dirac particles in graphene. *Physical Review Letters*, 2009. **102**: p. 166808.
37. Madelung, O., *Introduction to Solid-State Theory*. 1st ed. *Springer Series in Solid-State Sciences*, ed. M. Cardona, Fulde, P., von Klitzing, K., and Queisser, H.-J. 1978, New York, USA.: Springer. 488.
38. Jishi, R.A., Venkataraman, L., Dresselhaus, M.S., and Dresselhaus, G., Phonon modes in carbon nanotubules. *Chemical Physics Letters*, 1993. **209**(1–2): p. 77–82.
39. Saito, R., Dresselhaus, G., and Dresselhaus, M.H., *Physical Properties of Carbon Nanotubes*. 1998, Imperial College, London: Imperial College Press.
40. Muñoz, E., Lu, J., and Yakobson, B.I., Ballistic thermal conductance of graphene ribbons. *Nano Letters*, 2010. **10**: p. 1652–1656.
41. Muñoz, E., Lu, J., and Yakobson, B.I., Supporting information: Ballistic thermal conductance of graphene ribbons. *Nano Letters*, 2010. **10**(5): p. 1652–1656.
42. Landau, L.D., and Lifshitz, E.M., *Theory of Elasticity*. 3rd ed. *Course of Theoretical Physics*. 1986, Oxford, Great Britain: Butterworth-Heinemann.
43. Muñoz, E., Singh, A.K., Ribas, M.A., Penev, E.S., and Yakobson, B.I., The ultimate diamond slab: GraphAne versus graphEne. *Diamond and Related Materials*, 2010. **19**: p. 368–373.
44. Kudin, K.N., Scuseria, G.E. and Yakobson, B.I., C2F, BN, and C nanoshell elasticity from ab initio computations. *Physical Review B*, 2001. **64**: p. 235406 (10pp).
45. Chaikin, P.M., and Lubensky, T.C., *Principles of Condensed Matter Physics*. 1995, Cambridge, Great Britain: Cambridge University Press.
46. Castro Neto, A.H., Guinea, F., Peres, N.M.R., Novoselov, K.S., and Geim, A.K., The electronic properties of graphene. *Reviews of Modern Physics*, 2009. **81**: p. 109–162.
47. Nilsson, J., Castro Neto, A.H., Guinea, F., and Peres, N.M.R., Electronic properties of bilayer and multilayer graphene. *Physical Review B*, 2008. **78**(4): p. 045405.
48. Nelson, D.R., Piran, D.R., and Weinberg, S., *Statistical Mechanics of Membranes and Surfaces*. 2004, Singapore: World Scientific.
49. Meyer, J.C., Geim, A.K., Katsnelson, M.I., Novoselov, K.S., Booth, T.J., and Roth, S., The structure of suspended graphene sheets. *Nature*, 2007. **446**: p. 60–63.
50. Balandin, A.A., Ghosh, S., Bao, W., Calizo, I., Teweldebrhan, D., Miao, F., and Lau, C.N., Superior thermal conductivity of single-layer graphene. *Nano Letters*, 2008. **8**(3): p. 902–907.
51. Balandin, A.A., Thermal properties of graphene and nanostructured carbon materials. *Nature Materials*, 2011. **10**: p. 569–581.
52. Ghosh, S., Calizo, I., Teweldebrhan, D., Pokatilov, E.P., and Nika, D.L., Extremely high thermal conductivity of graphene: Prospects for thermal management applications in nanoelectronic circuits. *Applied Physics Letters*, 2008. **92**: p. 151911.
53. Cai, W., Moore, A.L., Zhu, Y. Li, X., Chen, S., Shi, L., and Ruoff, R.S., Thermal transport in suspended and supported monolayer graphene grown by chemical vapor deposition. *Nano Letters*, 2010. **10**: p. 1645–1651.
54. Jauregui, L.A., Yue, Y., Sidorov, A.N., Hu, J., Yu, Q., Lopez, G., Jalilian, R. et al., Thermal transport in graphene nanostructures: Experiments and simulations. *ECS Transactions*, 2010. **28**: p. 73–83.
55. Faugeras, C., Faugeras, B., Orlita, M., Potemski, M., Nair, R.R., and Geim, A.K., Thermal conductivity of graphene in corbino membrane geometry. *ACS Nano*, 2010. **4**: p. 1889–1892.
56. Dorgan, V.E., Behnam, A., Conley, H.J., Bolotin, K.I., and Pop, E., High-field electric and thermal transport in suspended graphene. *Nano Letters*, 2013. **13**: p. 4581–4586.
57. Seol, J.H., Jo, I., Moore, A.L., Lindsay, L., Aitken, Z.H., Pettes, M.T., Li, X. et al., Two-dimensional phonon transport in supported graphene. *Science*, 2010. **328**: p. 213–216.
58. Jang, W., Chen, Z., Bao, W., Lau, C.N., and Dames, C., Thickness-dependent thermal conductivity of encased graphene and ultrathin graphite. *Nano Letters*, 2010. **10**: p. 3909–3913.
59. Wang, Z., Xie, R., Bui, C.T., Liu, D., Ni, X., Li, B., and Thong, J.T.L., Thermal transport in suspended and supported few-layer graphene. *Nano Letters*, 2011. **11**: p. 113–118.
60. Nika, D.L., Ghosh, S., Pokatilov, E.P., and Balandin, A.A., Lattice thermal conductivity of graphene flakes: Comparison with bulk graphite. *Applied Physics Letters*, 2009. **94**: p. 203103.
61. Chen, S., Moore, A.L., Cai, W., Suk, J.W., An, J., Mishra, C., Amos, C. et al., Raman measurements of thermal transport in suspended monolayer graphene of variable sizes in vacuum and gaseous environments. *ACS Nano*, 2011. **5**: p. 321–328.
62. Nika, D.L., Pokatilov, E.P., Askerov, A.S., and Balandin, A.A., Phonon thermal conduction in graphene: Role of Umklapp and edge roughness scattering. *Physical Review B*, 2009. **79**: p. 155413.
63. Lindsay, L., Broido, D.A., and Mingo, N., Diameter dependence of carbon nanotube thermal conductivity and extension to the graphene limit. *Physical Review B*, 2010. **82**: p. 161402(R).
64. Nika, D.L., Askerov, A.S., and Balandin, A.A., Anomalous size dependence of the thermal conductivity of graphene ribbons. *Nano Letters*, 2012. **12**: p. 3238–3244.
65. Evans, W.J., Hu, L., and Keblinski, P., Thermal conductivity of graphene ribbons from equilibrium molecular dynamics: Effect of ribbon width, edge roughness, and hydrogen termination. *Applied Physics Letters*, 2010. **96**: p. 203112.

66. Mu, X., Wu, X., Zhang, T., Go, D.B., and Luo, T., Thermal transport in graphene oxide—From ballistic extreme to amorphous limit. *Physics Reports*, 2014. **4**: p. 3909.
67. Painter, G.S., and Ellis, D.E., Electronic band structure and optical properties of graphite from a variational approach. *Physical Review B*, 1970. **1**: p. 4747–4752.
68. Semenoff, G.W., Condensed-matter simulation of a three-dimensional anomaly. *Physical Review Letters*, 1984. **53**(26): p. 2449–2452.
69. Pesin, D., and MacDonald, A., Spintronics and pseudospintronics in graphene and topological insulators. *Nature Materials*, 2012. **11**: p. 409–415.
70. Suzuura, H., and Ando, T., Phonons and electron–phonon scattering in carbon nanotubes. *Physical Review B*, 2002. **65**: p. 235412.
71. Ando, T., Anomaly of optical phonon in monolayer graphene. *Journal of the Physical Society of Japan*, 2006. **75**(12): p. 124701.
72. Guinea, F., Katsnelson, M.I., and Vozmediano, M.A.H., Midgap states and charge inhomogeneities in corrugated graphene. *Physical Review B*, 2008. **77**: p. 075422.
73. Kim, E.-A., and Castro Neto, A.H., Graphene as an electronic membrane. *Europhysics Letters*, 2008. **84**: p. 57007.
74. Ziman, J.M., *Electrons and Phonons*. 1960, Oxford: Oxford University Press.
75. Peres, N.M.R., Colloquium: The transport properties of graphene: An introduction. *Reviews of Modern Physics*, 2010. **82**: p. 2673–2700.
76. Beenaker, C.W.J., Colloquium: Andreev reflection and Klein tunneling in graphene. *Reviews of Modern Physics*, 2008. **80**: p. 1337–1354.
77. Pottier, N., *Nonequilibrium Statistical Physics*. 2010, Oxford: Oxford University Press.
78. Lifshitz, E.M., and Pitaevskii, L.P., *Physical Kinetics*. Vol. 9. *Landau Course of Theoretical Physics*. 1997, Oxford, Great Britain: Butterworth-Heinemann.
79. Ashcroft, N.W., and Mermin, D.N., *Solid State Physics*. 1976, United States: Harcourt College Publishers.
80. Das Sarma, S., Adam, S, Hwang, E.H., and Rossi, E., Electronic transport in two-dimensional graphene. *Reviews of Modern Physics*, 2011. **83**: p. 407–470.
81. Peres, N.M.R., Lopes dos Santos, J.M.B., and Stauber, T., Phenomenological study of the electronic transport coefficients of graphene. *Physical Review B*, 2007. **76**(7): p. 073412 (4 p.).
82. Stauber, T., Peres, N.M.R., and Guinea, F., Electronic transport in graphene: A semiclassical approach including midgap states. *Physical Review B*, 2007. **76**(20): p. 205423 (10pp.).
83. Onsager, L., Reciprocal relations in irreversible processes. I. *Physical Review* 1931. **37**: p. 405–426.
84. Hwang, E.H., and Das Sarma, S., Acoustic phonon scattering limited carrier mobility in two-dimensional extrinsic graphene. *Physical Review B*, 2008. **77**(11): p. 115449 (6pp.).
85. Sughihara, K., Thermoelectric power of graphite intercalation compounds. *Physical Review B*, 1983. **28**(4): p. 2157–2165.
86. Chen, S., Wu, Q., Mishra, C., Kang, J., Zhang, H., Cho, K., Cai, W., Balandin, A.A., and Ruoff, R.S., Thermal conductivity of isotopically modified graphene. *Nature Materials*, 2012. **11**: p. 203–207.
87. Cho, S., Kang, S.D., Kim, W., Lee, E.-S., Woo, S.J., Kong, K.-J., Kim, I. et al., Thermoelectric imaging of structural disorder in epitaxial graphene. *Nature Materials*, 2013. **12**: p. 913–917.
88. Xiao, N., Dong, X., Song, L., Liu, D., Tay, Y., Wu, S., Li, L.-J. et al., Enhanced thermopower of graphene films with oxygen plasma treatment. *ACS Nano*, 2011. **5**: p. 2749–2755.
89. Jayasingha, R., Sherehiy, A., Wu, S.-Y., and Sumanasekera, G.U., In situ study of hydrogenation of graphene and new phases of localization between metal-insulator transitions. *Nano Letters*, 2013. **13**: p. 5098–5105.
90. Kubakaddi, S.S., Interaction of massless Dirac electrons with acoustic phonons in graphene at low temperatures. *Physical Review B*, 2009. **79**(7): p. 075417 (6pp.).
91. Kim, J.Y., Lee, J.-H., and Grossman, J.C., Thermal transport in functionalized graphene. *ACS Nano*, 2012. **6**: p. 9050–9057.
92. Alomar, M.I., and Sánchez, D., Thermoelectric effects in graphene with local spin-orbit interaction. *Physical Review B*, 2014. **89**: p. 115422.
93. Wierzbicki, M., Swirkowicz, R., and Barnas, J., Giant spin thermoelectric efficiency in ferromagnetic graphene nanoribbons with antidots. *Physical Review B*, 2013. **88**: p. 235434.
94. Zuev, Y.M., Chang, W., and Kim, P., Thermoelectric and magnetothermoelectric transport measurements of graphene. *Physical Review Letters*, 2009. **102**: p. 096807.
95. Yokomizo, Y., and Nakamura, J., Giant Seebeck coefficient of the graphene/h-BN superlattices. *Applied Physics Letters*, 2013. **103**: p. 113901.

18 Thermoelectric Effects in Graphene

N. S. Sankeshwar, S. S. Kubakaddi, and B. G. Mulimani

CONTENTS

ABSTRACT

Graphene, owing to its unique electronic properties, has become one of the active areas of condensed matter research with promising applications in future efficient thermoelectric (TE) and energy storage devices. The present work reviews the status of thermoelectric power (TEP) of graphene systems, including single-layer, bilayer, and nanoribbons. The theory of TEP, based on the Boltzmann transport formalism in 2D systems, is given. An analysis of the experimental data, in terms of the diffusion and the phonon-drag contributions to TEP, with regard to the various scattering mechanisms operative in graphene systems, is presented. The outlook on TEP for better understanding of the TE properties of graphene is discussed.

18.1 INTRODUCTION

A useful measure of the potential of a material for thermoelectric (TE) applications is its TE figure of merit, $Z = S^2\sigma/\kappa$, where S is the thermoelectric power (TEP) (also called Seebeck coefficient), σ the electrical conductivity, and κ the thermal conductivity of the material. Materials, therefore, that have an enhanced power factor ($S^2\sigma$) and reduced κ, are suitable candidates for efficient TE devices [1,2]. A desirable combination of the quantities, S, σ, and κ, are generally possessed by semiconducting materials [1–3]. Bi_2Te_3 is found to have a high room temperature value of $ZT \sim 1$. Search is on for nanostructured materials for use in smaller and more efficient TE devices, necessary for cleaner, more efficient cooling, and power generation [1,2,4,5]. In recent years, graphene, which exhibits unique properties such as high thermal conductivity and high electron mobility, has generated renewed interest in the search [6–9]. Although there is no well-defined theoretical limit to ZT, even a modest increase in value of ZT would provide opportunities for applications [10]. Recent studies indicate that ZT could be enhanced nearly fourfold by optimizing the potential of graphene systems [11].

The TEP of a material is an important and interesting transport property for study because of its sensitivity to the composition and structure of a system and to the external fields. It has been able to shed much light on the interaction of electrons with phonons, impurities, and other defects, be it metals [12], bulk semiconductors [3], or conventional 2D semiconductor systems [13–15]. It is also known to provide information complementary to that of resistivity (or conductivity), which alone is inadequate, say, in distinguishing different scattering mechanisms operative in a system [12,16]. An optimization of the Seebeck coefficient for any material, therefore, involves understanding and appropriately modifying its electronic properties and provides a challenge for theoreticians and experimentalists alike to search for ways to increase the value of TE figure of merit. Graphene, a monolayer of graphite with the carbon atoms arranged in a honeycomb crystal lattice, exhibits interesting TE effects. For instance, compared to elemental semiconductors, it has higher TEP and can be made to change sign by varying the gate bias [17–19].

Technical advances have now made possible the realization of tailor-made 2D graphene systems, such as single-layer graphene (SLG), bilayer graphene (BLG), graphene nanoribbon (GNR), graphene dots, graphene superlattices, and defected graphene. The electronic properties of graphene systems have been studied and reviewed [7,8]. In the recent past, literature has also accumulated on the TE properties of these systems.

In the present work, we summarize the status of the efforts to understand TEP in graphene systems. After a brief description of the electronic structures of SLG, BLG, and GNR systems, the basic theory of TEP, based on the semiclassical Boltzmann transport formalism in 2D systems, is given. An analysis of the experimental data, in terms of the diffusion and the phonon-drag contributions is presented.

18.1.1 Graphene Systems: Electronic Structure

One, generally, distinguishes between three types of graphenes: SLG, BLG, and few-layer graphene (FLG), with number of layers <10 [20]. In SLG, commonly referred to simply as graphene, the carriers are in a 2D layer consisting of carbon atoms in the sp^2 hybridization state, with the three nearest-neighbor carbon atoms in the honeycomb lattice forming σ-bonds [6–8]. A BLG consists of two graphene monolayers weakly coupled by interlayer carbon hopping, which depends on the manner of stacking of the two layers with respect to each other; typically they are arranged in A–B stacking arrangement. A GNR is a quasi-one-dimensional (Q1D) system that confines the electrons in a thin strip of large length and a finite but very small (a few nm) width.

18.1.1.1 Single Layer Graphene

The 2D honeycomb structure of graphene lattice with two equivalent lattice sites, A and B, is equivalent to a triangular lattice with a basis of two atoms per unit cell. The inequivalent points **K** and **K′** of the Brillouin zone, at which the Dirac cones for electrons and holes touch each other are called Dirac points (Figure 18.1). A Dirac point, in gapless graphene, where the character of graphene changes from being electron-like to being hole-like, is termed as charge neutrality point (CNP). A system with no free carriers at $T = 0$ K and with Fermi level at the Dirac point has a completely filled valence band and an empty conduction band and is called intrinsic graphene. At any finite temperature and with any doping, electrons present in the conduction band make the system "extrinsic." Experimentally, it is possible to tune the system from being electron-like to being hole-like by varying the external gate voltage, V_g (Figure 18.2a) [6].

The transport characteristics of a material are intimately related to the energy band structure. In graphene, the carrier (electron/hole) transport, close to the Dirac points, is described by a Dirac-like equation for massless particles [6–8], whose solution has been calculated in the tight-binding model up to the next-nearest-neighbor approximation [21]. Being interested mostly in understanding electron transport for small energies, only the low-$\boldsymbol{k}$, linear dispersion aspects of the band structure are considered. An SLG is thus a zero band-gap semiconductor with the conduction and valence bands intersecting at $\boldsymbol{k} = 0$. In the long-wavelength ($\boldsymbol{k} \ll 2\pi/a$; a being the C–C distance) limit, the energy dispersion for

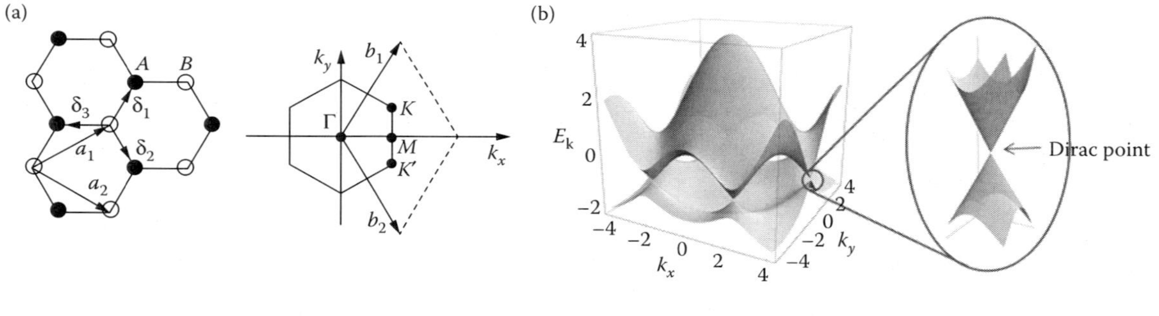

FIGURE 18.1 (a) Graphene lattice and Brillouin zone. (b) Graphene band structure. An enlargement close to the K and K' point shows Dirac cones. (From Castro Neto, A.H. et al. 2009. *Rev Mod Phys* 81:109–62. With permission.)

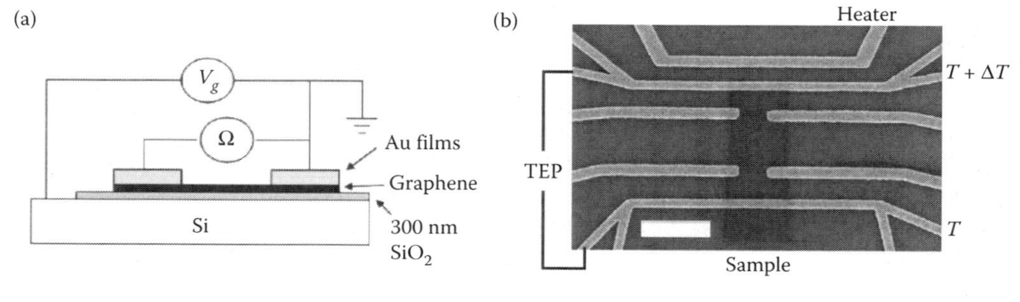

FIGURE 18.2 (a) A typical configuration for gated graphene. (b) SEM image of a typical device; scale bar is 2 μm. (From Zuev, Y.M., Chang, W., and Kim, P. 2009. *Phys Rev Lett* 102:096807-1–4. With permission.)

both electrons (in the conduction band) and holes (in the valence band) [6–8] is

$$E_k = s\hbar v_F\,|\boldsymbol{k}|. \tag{18.1}$$

Here, $s = +1$ (−1) corresponds to the conduction (valence) band, $\boldsymbol{k} = \boldsymbol{i}k_x + \boldsymbol{j}k_y$ denotes the carrier wavevector measured from the relevant Dirac point, and v_F (=$3ta/2\hbar$; t being nearest-neighbor hopping amplitude), a constant, is called the graphene (Fermi) velocity; with $t \sim 2.5$ eV, $v_F \sim 10^6$ m s^{-1}. The linearity of the dispersion relation signifies that graphene can be modeled as a 2D gas of massless Dirac fermions [6–8].

18.1.1.2 Bilayer Graphene

In a BLG with A–B stacking of the two layers, the low energy, long wavelength electronic structure is parabolic [22] (Figure 18.3). Consisting of two branches, the BLG electronic structure depends on the electrostatic potential, V, between the two layers. It controls the effective band-gap opening near Dirac point. Unbiased ($V = 0$ V) BLG is gapless semiconductor with electron effective mass $m = 0.033m_e$. A gap can be induced by applying an external voltage [8]. The low-field electron transport is discussed with respect to the lowest branch.

18.1.1.3 Graphene Nanoribbon

A graphene sheet can be patterned into nanoribbons. The spectrum of GNRs depends on the nature of their edges (Figure 18.4). The low-energy electronic states of GNRs near the two nonequivalent Dirac points (**K** and **K′**) can be described by the Dirac equation and the appropriate boundary conditions. The confinement of electrons gives rise to the subband structure with an energy gap at the Dirac point [23]. The resulting confinement gap, E_g, depends on the width, W, of the ribbon and the chirality (armchair or zigzag) of the edges. A GNR with armchair terminated edges along its length gives an armchair graphene nanoribbon (AGNR). Choosing a GNR with zig-zag terminations along its length gives a zigzag graphene nanoribbon (ZGNR). A ZGNR is metallic in nature, whereas an AGNR can be metallic or semiconducting, with E_g inversely proportional to W [8,24].

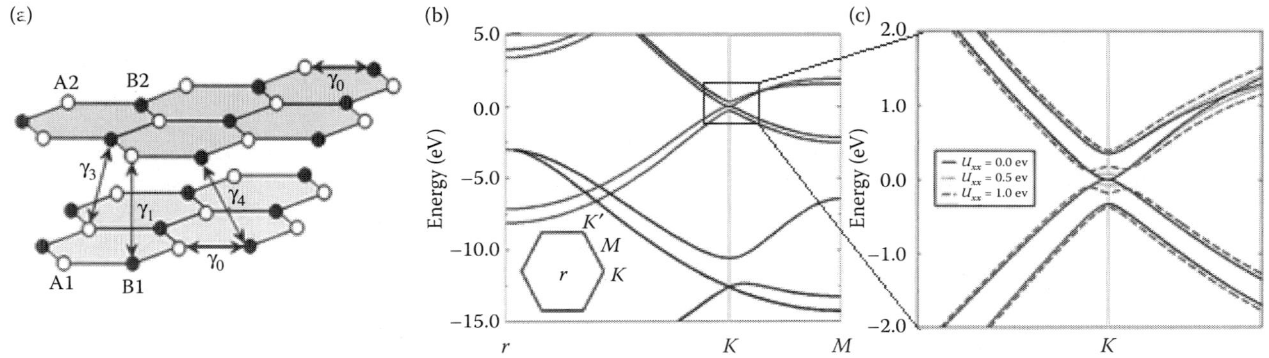

FIGURE 18.3 (a) BLG lattice. (b) BLG band structure. (c) An enlargement close to the K point. (From Mucha-Kruczynski, M., McCann, E., and Fal'ko, V.I. 2010. *Semicond Sci Technol* 25:33001-1–6. With permission.; Min, H. et al. 2009. *Phys Rev B* 75:155115-1–7. With permission.)

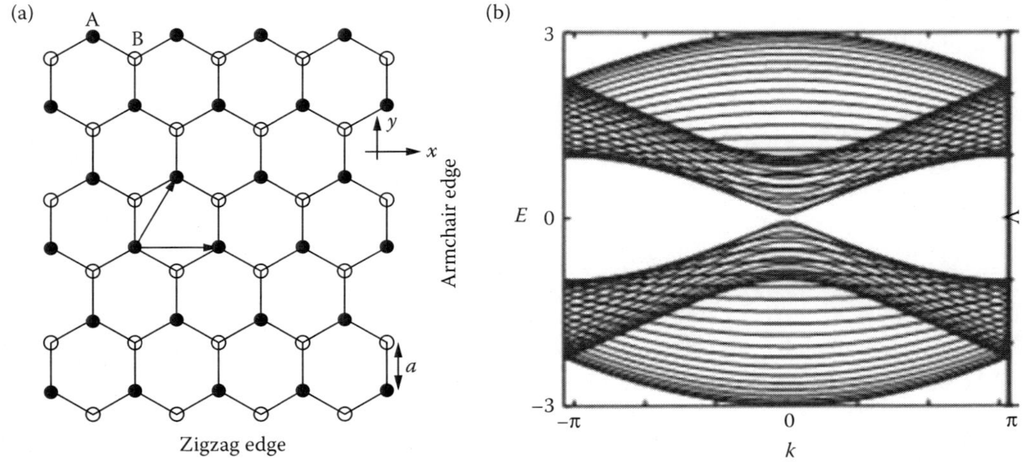

FIGURE 18.4 (a) Graphene nanoribbon lattice. (b) AGNR band structure. (From Castro Neto, A.H. et al. 2009. *Rev Mod Phys* 81:109–62. With permission.; Nakada, K. et al. 1996. *Phys Rev B* 54:17954–61. With permission.)

TABLE 18.1
Electron Wavefunctions, Energy Eigenvalues, and Density of States of Graphene Systems

System	Wavefunctions	Eigenvalues	Density of States
SLG	$\psi_{\pm,K}(r) = \frac{1}{\sqrt{2LW}}\begin{pmatrix} e^{-i\theta_k} \\ \pm 1 \end{pmatrix} e^{ik\cdot r}$	$E_k = \hbar v_F \lvert k \rvert$	$\rho(E_k) = \frac{2\lvert E_k \rvert}{\pi(\hbar v_F)^2}$
BLG	$\psi_{\pm,K}(r) = \frac{1}{\sqrt{2LW}}\begin{pmatrix} e^{-2i\theta_k} \\ \pm 1 \end{pmatrix} e^{ik\cdot r}$	$E_k = \frac{\hbar^2 k^2}{2m}$	$\rho(E_k) = \frac{2m}{\pi\hbar^2}$
AGNR	$\Psi_{n,k_y}(\mathbf{r}) = [\psi_{n,k_y}(\mathbf{r}) - \psi'_{n,k_y}(\mathbf{r})]/\sqrt{2}$ with $\psi_{n,k_y}(r) = \sqrt{1/2LW}\; e^{i[(\Delta K/2)-k_n]x}\phi_{n,k_y}(y)$, $\psi'_{n,k_y}(r) = \sqrt{1/2LW}\; e^{-i[(\Delta K/2)-k_n]x}\phi_{n,k_y}(y)$, $\phi_{n,k_y}(y) = e^{ik_y y}\begin{pmatrix} 1 \\ -e^{-i\theta_{n,k_y}} \end{pmatrix}$	$E_{n,k_y} = \hbar v_F \sqrt{k_n^2 + k_y^2}$	$\rho(E_{n,k_y}) = \frac{2}{\pi\hbar v_F}\left[\frac{E_{n,k_y}}{\{E_{n,k_y}^2 - (E_n)^2\}^{1/2}}\right]$

Source: Castro Neto, A.H. et al. 2009. *Rev Mod Phys* 81:109–62; Min, H. et al. 2009. *Phys Rev* B 75:155115-1–7; Fang, T. et al. 2008. *Phys Rev* B 78:205403-1–8.

Note: L, W: Length and width of graphene system; $\boldsymbol{r} \equiv (x,y)$; $\boldsymbol{k} \equiv (k_x,k_y)$; v_F: Fermi velocity; m: effective mass; $\theta_k = \tan^{-1}(k_x/k_y)$; AGNR subband energy, $E_n = \pm n\pi\hbar v_F/3W$, with subband index, $n = 1,2,4,5,7,8\ldots$; electron transverse wave vector $k_n = \pm n\pi/3W$; $\Delta K = 4\pi/3a$, a: lattice constant; $\theta_{n,k} = \tan^{-1}(k_n/k_y)$.

Expressions for electron wavefunctions, energy eigenvalues, and the density of states for SLG [6–8] and BLG [25] and for AGNR (satisfying hard wall boundary conditions) [26] are given in Table 18.1.

In the following, we review the TE property of TEP with regard to three graphene systems, SLG, BLG, and AGNR.

18.2 THEORY OF TE EFFECTS IN GRAPHENE SYSTEMS

18.2.1 Thermoelectric Power

The TE effect is due to the interdependence of electric potential and temperature gradient in a system where no electric current flows. The absolute TEP, S, which is a unique physical property of a material, is defined by the relation [13]

$$\mathbf{E} = S\nabla T \tag{18.2}$$

under open-circuit conditions, where $\mathbf{E}$ is the effective electric field produced by the temperature gradient ∇T.

There are, in general, two contributions to the TEP of the system: the diffusion TEP, S_d, and the phonon-drag TEP, S_g. In the presence of a temperature gradient, the electrons diffuse through the specimen interacting with a random distribution of scattering centers. The heat flux, $\mathbf{U}_d$, carried by the electrons yields S_d. In addition, the flow of phonons carries a momentum current, a fraction of which is transferred to electrons via phonon–electron interaction, giving rise to S_g. The phonon current thus "drags" electrons with it and extra electrons tend to pile up at the cold end over and above those electrons which are there as a result of the diffusion process. The heat flux, $\mathbf{U}_g$, carried by the phonon system yields S_g. The total heat current density $\mathbf{U}$, therefore, consists of two parts: $\mathbf{U} = \mathbf{U}_d + \mathbf{U}_g$, and, correlatively, the total TEP, S, can be expressed as $S = S_d + S_g$. The treatment presented here is quite general and is applicable to graphene systems.

18.2.2 Diffusion Thermopower: Boltzmann Approach

TEP is determined by the band structure and scattering mechanisms operative in a system. Low field transport in many of the systems is often described by the Boltzmann transport equation (BTE) [27,28]. The conventional theory of charge carrier transport in 2D semiconductors is based on this semiclassical formalism [13]. Adopting the Boltzmann approach, found to be robust especially for transport in graphene away from the Dirac point [6,29], we give below, in brief, the basic theory of TEP and the expressions used in the analysis of S_d in graphene systems.

We consider a graphene system of length L, and width W as a 2D homogeneous material. An external gate bias V_g applied across the system (Figure 18.2) is assumed to induce charge carriers of density, n_s. The electric and thermal current densities $\mathbf{J}$ and $\mathbf{U}$ can be expressed as

$$\mathbf{J} = \left(\frac{g_s g_v}{Ad}\right)\sum_k e\mathbf{v}_\mathbf{k} f(E_k) \tag{18.3}$$

and

$$\mathbf{U} = \left(\frac{g_s g_v}{Ad}\right)\sum_k (E_k - E_F)\mathbf{v}_\mathbf{k} f(E_k), \tag{18.4}$$

where g_v and g_s denote, respectively, the valley and spin degeneracies, $\mathbf{v}_k$ is the velocity of the electrons, $A = LW$ is the area of the surface, d is the layer thickness, e is the charge of the carriers, and E_F is the Fermi energy of the graphene system.

Assuming the electric field to be weak and the displacement of the distribution function $f(E_k)$ from its equilibrium value $f^0(E_k)$ to be small, $\mathbf{J}$ and $\mathbf{U}$ can be expressed, using Equations 18.3 and 18.4, as [30]

$$\begin{pmatrix} \mathbf{J} \\ \mathbf{U} \end{pmatrix} = \begin{pmatrix} e^2K_{11} & -\frac{e}{T}K_{12} \\ eK_{21} & -\frac{e}{T}K_{22} \end{pmatrix} \begin{pmatrix} E \\ \nabla T \end{pmatrix}. \quad (18.5)$$

The expressions for the coefficients K_{rs}, depend on the graphene system [31].

Under open circuit conditions ($\mathbf{J} = 0$), one obtains, from Equation 18.5, the expression for the diffusion contribution to thermopower, S_d as

$$S_d = \frac{1}{eT} K_{11}^{-1} K_{21}. \quad (18.6)$$

Equation 18.6 may be expressed as [32–36]

$$S_d = \frac{1}{eT}\left(\frac{\langle E_k \tau(E_k) \rangle}{\langle \tau(E_k) \rangle} - E_F \right) \quad (18.7)$$

with

$$\langle F(E_k) \rangle = \frac{\int F(E_k) E_k (-\partial f^0/\partial E_k) dE_k}{\int E_k (-\partial f^0/\partial E_k) dE_k}. \quad (18.8)$$

Here, $f^0(E_k)$ is the Fermi–Dirac distribution function. Equations 18.6 and 18.7 show that evaluation of S_d requires a knowledge of the electron momentum relaxation time(s), $\tau(E_k)$, in the system considered [31].

Often, in literature, limiting forms of S_d are used in the analysis of data. In the degenerate limit a good approximation to Equation 18.6 is the well-known Mott expression [13]:

$$S_d = \frac{\pi^2 k_B^2 T}{3e} \left[\frac{d \ln \sigma(E_k)}{dE_k} \right]_{E_k = E_F}, \quad (18.9)$$

where $\sigma(E_k)$ is the energy dependent electrical conductivity. In the case of AGNR, E_k is replaced with $E_{n\mathbf{k}}$ (see Table 18.1) [35]. If the energy dependence of relaxation time is taken as $\tau(E) \sim E^p$, Equation 18.9 can be more simply expressed for the three graphene systems, as [32–37]

$$\text{SLG, BLG:} \quad S_d = \frac{\pi^2 k_B^2 T}{3eE_F}[p-1] \quad (18.10)$$

$$\text{AGNR:} \quad S_d = \frac{\pi^2 k_B^2 T}{3eE_F}\left[p + \left(\left(\frac{E_F}{E_n} \right)^2 - 1 \right)^{-1} \right], \quad (18.11)$$

where the first term reflects the scattering mechanisms. The parameter p can also be expressed in the form

$$p = \frac{E_F}{\tau(E_F)} \left[\frac{d\tau(E_k)}{dE_k} \right]_{E_k = E_F}. \quad (18.12)$$

It may be noted that provided the energy dependence of $\tau(E_k)$ does not vary with temperature (though its magnitude will), S_d will be a linear function of temperature.

18.2.2.1 Electron Scattering Mechanisms in Graphene Systems

Any deviation from the perfect periodicity of the graphene lattice will scatter electrons. Carrier scattering results from both intrinsic and extrinsic sources. The extrinsic sources may be vacancies, surface roughness arising from rippling of the graphene sheet, disorder, which can create electron–hole puddles, and charged impurities, known to be the main scattering mechanism in graphene. Phonons constitute an intrinsic source of scattering in a system. Phonon scattering may be due to intravalley acoustic and optical phonons which induce the electronic transitions within a single valley, and intervalley phonon scattering [6–8]. Apart from the graphene layers, the substrates may also influence transport. They could be a source of impurities and remote interface polar optical phonons. In principle, the limitation due to the extrinsic scattering mechanisms can be reduced by improved growth/fabrication techniques.

One of the strategies adopted to reduce carrier scattering and achieve higher mobility in graphene layers is to suspend it over a trench thus avoiding the influence of the substrate [38]. In the case of suspended graphene (SG) layers, the intrinsic scattering mechanisms limiting electron transport are due not only to in-plane but also out-of-plane (flexural) acoustic phonons. Carrier transport in SG can be affected by strain [39].

At low temperatures the major contribution to intrinsic phonon-limited resistivity is from acoustic phonons [8]. Further, as recent investigations [40] show, there is a change in temperature dependence of resistivity from $\rho(T) \sim T$, in the high temperature limit, to $\rho(T) \sim T^4$, at low temperatures, reflecting the 2D nature of the electrons and the acoustic phonons in graphene. The crossover between the two distinct regimes is shown to be described by the characteristic Bloch–Gruneisen (BG) temperature defined as $T_{BG} = 2\hbar v_s k_F / k_B$, where v_s is the velocity of sound and $k_F = (\pi n_s)^{1/2}$ is the Fermi wavevector. The electrostatic tunability of E_F ($=\hbar v_F k_F$) allows for a wide range of control of T_{BG}. For instance, for a sample with $n_s = 1 \times 10^{16}$ m^{-2} [17], $T_{BG} \sim 54$ K, for LA phonons. The transition into BG regime reflects a change in the character of electron scattering by acoustic phonons [40]. This change influences strongly the energy dependence of, and results in a sharp decrease in, the relaxation rates $\tau^{-1}(E_k)$ of the acoustic phonon scattering and gets reflected in the behavior of S_d [41].

18.2.3 Phonon-Drag Thermopower

Unlike the diffusion contribution, S_g depends only on the electron–acoustic phonon coupling strength. It has been extensively studied in conventional semiconductor 2DEG systems [13–15]. The very large thermal conductivity observed [42] in graphene suggests a large phonon mean free path, Λ, decided by the various phonon-scattering mechanisms operative. S_g, which depends on Λ is also expected to be large.

Solving the coupled Boltzmann equations for electrons and phonons, Kubakaddi and Bhargavi [33,36,43] have developed the theory of S_g for graphene systems. Assuming the 2D electrons to interact with the in-plane 2D LA phonons via unscreened deformation potential coupling, S_g can be expressed as

$$S_g = \frac{2e}{A\sigma k_B T^2}\sum_{\mathbf{k},\mathbf{k}',\mathbf{q}} \hbar\omega_{\mathbf{q}} f(E_{\mathbf{k}})[(1-f(E_{\mathbf{k}'})] \times P_{\mathbf{q}}^{a}(\mathbf{k},\mathbf{k}')\tau_{\mathbf{p}}\mathbf{v}_p\cdot(\mathbf{v}_{\mathbf{k}}\tau_{\mathbf{k}}-\mathbf{v}_{\mathbf{k}'}\tau_{\mathbf{k}'}), \tag{18.13}$$

where $\mathbf{v}_p$ is the phonon group velocity, $\tau_k(\tau_{\mathbf{q}})$ is the electron (phonon) momentum relaxation time, and $P_{\mathbf{q}}^{a}(\mathbf{k},\mathbf{k}')$ is the transition rate at which the electron in state **k** makes transition to state **k′** by absorbing a phonon of energy $\hbar\omega_{\mathbf{q}}$. The phonon relaxation processes may be due to the sample boundaries, impurities, defects, and other phonons. At low temperatures, wherein the phonon scattering is dominated by boundary scattering, one may assume $\tau_{\mathbf{p}} = \Lambda/v_s$, with Λ corresponding to smallest linear dimension of the sample.

One may obtain, from Equation 18.13, the expressions for S_g in SLG [33], BLG [36], and AGNR (with appropriate modifications) [43].

18.3 ANALYSIS AND DISCUSSION OF THERMOPOWER DATA

Measurements of the TEP of graphene systems have helped elucidate some details of the unique electronic structure of the ambipolar nature of graphene, which cannot be probed by conductivity measurements alone. Tables 18.2, 18.3, and 18.4, respectively, list the recent experimental investigations made with regard to the TEP of SLG, BLG, and other multilayer graphene (MLG) systems.

In a typical set up (Figure 18.2), a controlled temperature difference ΔT is maintained along the plane of the sample and the resulting thermally induced voltage ΔV is measured. The acquired TEP, $S = -\Delta V/\Delta T$ of graphene is studied as a function of temperature and carrier density by modulating the gate voltage, V_g. The nonexistence of a gap in the carrier dispersion of graphene (SLG, BLG) leads to a direct transition between electron-like transport to hole-like transport as the gate voltage is tuned through the charge neutral Dirac point, V_D.

Here, we discuss the measurements made on, and the theory of, TEP in SLG, BLG, and AGNR in the absence of an applied magnetic field. An analysis of measured TEP is usually done by separating the two contributions, S_d and S_g, by making use of their characteristic temperature dependences at low temperatures [13]. The Mott relation (Equation 18.9), for the diffusion component, S_d, is expressible in terms of V_g as [17,18]

$$S_d = \frac{\pi^2 k_B^2 T}{3eE_F}\left[\frac{1}{G}\frac{dG}{dV_g}\frac{dV_g}{dE_{\mathbf{k}}}\right]_{E_{\mathbf{k}}=E_F}, \tag{18.14}$$

where G is the conductance and $E_F \,\alpha\, |V_g - V_D|^{1/2}$. Often, in literature, as suggested by Mott relation S_d, for a degenerate system, is assumed to vary linearly with temperature. S_d reflects the energy dependence of a scattering mechanism and is determined not only by the magnitude of scattering, but also by details concerning the distribution of the scatterers and their type. The phonon-drag component, S_g, on the other hand, unlike S_d, depends only on the electron–acoustic phonon coupling strength. Its low-temperature ($T < \sim 10$ K) dependence is known to play an important role in bulk and conventional 2D semiconductor systems [12–15,44] and displays a characteristic peak, exhibiting the role of phonon scattering mechanisms. Since the effect of electron–phonon coupling in determining the resistivity of graphene is observed to be weak, the phonon-drag component is expected to be absent for temperatures $10 < T < 300$ K [45]. In graphene, the existing TEP data has been explained based on the diffusion component.

18.3.1 Single-Layer Graphene

The theoretical investigations of TEP, some of which preceded the measurements [17–19,46–52] (see Table 18.2), in graphene systems have been devoted mostly to SLG at not-too-low temperatures ($10 < T < 300$ K).

Peres and coworkers [53,54], using a phenomenological theory for transport in SLG based on the semiclassical Boltzmann approach, studied, close to Dirac point, the carrier density dependence of S_d considering scattering mechanisms involving midgap states, arising from local point defects in the form of vacancies, cracks, boundaries, impurities in the substrate, or in corrugated graphene. They found that the dependence is different from that of the conventional 2DEG and of graphene with only charged impurities in the substrate.

Lofwander and Fogelstrom [55], have studied the linear response to electrical and thermal forces in graphene for the case of strong impurity scattering in the self-consistent t-matrix approximation. At low temperatures ($T < \sim 50$ K), the electronic contribution to TEP is found to be linear in T with slope proportional to the inverse of the impurity density and the impurity strength, suggesting TEP could provide information about impurities in graphene. Further, for moderately large impurity strengths, a nonlinear temperature dependence is obtained.

TABLE 18.2
Measurements of Thermoelectric Power in SLG Systems

Sample	Value (μV/K)	Reference
Mechanical exfoliation on 300 nm SiO_2 substrate; $\mu \sim 1$–7×10^3 cm²/Vs	~80 @ RT	[17]
Mechanical exfoliation on 300 nm SiO_2 substrate; $\mu \sim 3 \times 10^3$ cm²/Vs	$S_{xx} \sim 39$ @ 255 K	[18]
Exfoliation on 300 nm SiO_2/Si substrate	~100 @ RT	[19]
Exfoliated and supported on SiO_2; W: 1.5–3.2; L: 9.5–12.5 μm; G1: 3.2 μm parallel to 1.5 μm; G2: 2.4 μm $\mu = 20 \times 10^3$ cm²/Vs	G1,G2: ~80 @ RT	[46]
Exfoliated from Kish graphite/HOPG; $\mu \sim 1.5$–13×10^3 cm²/Vs; $n_i = 3.3 \times 10^{16}$ m^{-2}	~75 @ 295 K	[47]
Epitaxial on C-face of SiC; hole-doped: $n_s \sim 10^{12}$ cm^{-2}; $\mu \sim 20 \times 10^3$ cm²/Vs @ 4 K	~55 @ 230 K	[48]
Exfoliated on SiO_2/Si using e-beam lithography; $\mu \sim 4.56$–12.9×10^3 cm²/Vs	~50–75 @ 150 K	[49]
Fabricated on SiO_2/Si with e-beam lithography; $\mu \sim 12.9 \times 10^3$ cm²/Vs	$S_{xx}^{peak} \sim 10$ @ 20 K	[50]
Suspended Cu-CVD SLG	9 @ 300 K	[51]
Large-area (10 μm × 50 μm) Cu-CVD graphene on Si substrate covered by 300 nm SiO_2; $\mu < \sim 3 \times 10^3$ cm²/Vs	~22 @ 150 K	[52]

Kubakaddi [32,33] predicted the $n_s^{-1/2}$ and linear-in-T dependences for S_d. S_d was found to be relatively dominant, in relation to S_g, for $T < \sim 4$ K (see Figure 18.9).

The experimental investigations of the TE effect of Dirac electrons have been initially made on graphene samples mechanically exfoliated on ~300 nm SiO_2/Si substrates [17–19]. Zuev et al. [17] have measured both the conductivity and TEP of different SLG samples with mobilities in the range 1–7 × 10³ cm²/Vs. Figure 18.5 shows the measurements as a function of applied gate voltage V_g over a temperature range of 10–300 K. The conductivity becomes minimum at the CNP, corresponding to $V_g = 0$ V. Across the CNP, as the majority carrier density changed from electrons to holes, a change in sign of the TEP is observed. The linear temperature dependence of TEP seen (inset of Figure 18.5) for two values of V_g, far away from the CNP, suggests that the mechanism for TE generation is diffusive.

Wei et al. [18] have reported similar experimental results (see Figure 18.6) and showed that TE transport is sensitive to the electronic band structure. Away from the Dirac point, the magnitude of the thermovoltage, V_{th}, decreases, scaling approximately with $|V_g - V_D|^{-1/2}$; the dependence is more noticeable in the linear dependence of V_{th}^{-2} on V_g (Figure 18.6b). The divergence of the Seebeck coefficient S_{xx}

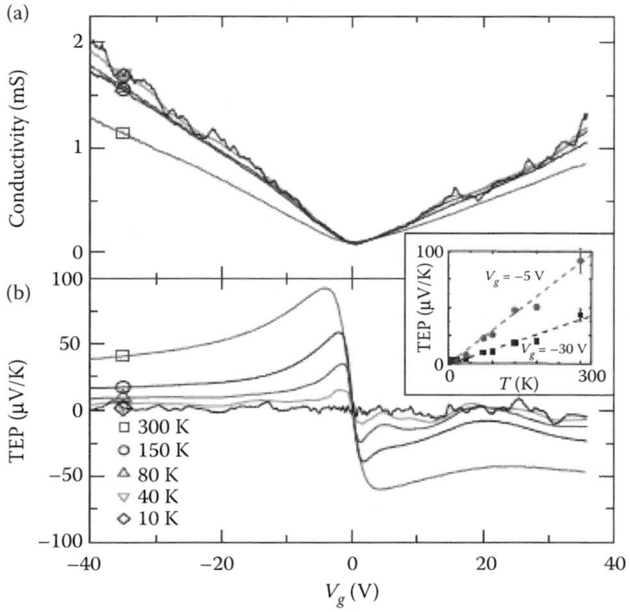

FIGURE 18.5 Measured (a) conductivity and (b) TEP of an SLG sample as function of V_g [17]. Inset: TEP values taken at $V_g = -30$ V (square) and −5 V (circle). Dashed lines are linear fits to data. (From Zuev, Y.M., Chang, W., and Kim, P. 2009. *Phys Rev Lett* 102:096807-1–4. With permission.)

as $n_s^{-1/2}$ is a direct manifestation of the linear dispersion of Dirac particles in graphene. For the highly doped 2D system, the Mott relation [13] yields $S_{xx} \sim -|V_g - V_D|^{-1/2}$. This is in contrast to conventional 2D systems with a quadratic dispersion relation, for which $S_{xx} \sim n_s^{-1}$. Measurements of S_{xx} on a device with $V_D \sim 33$ V, have indicated (see Figure 18.6c) electron–hole asymmetry. On the hole side S_{xx} decreases with decreasing V_g, whereas on the electron side S_{xx} remains constant. Further, S_{xx} is found to follow different T dependence for different V_g; S_{xx} is nearly linear on the hole side, whereas, on the electron side, it remains nonlinear in T except at very low temperatures (see Figure 18.6d). The departure from the linear T dependence on the electron side is attributed to the asymmetric nature of the band of impurity states, which in the impurity scattering model, can be highly asymmetric near Dirac point [55]. This behavior of TEP may be used as a sensitive probe for impurity bands near the Dirac point.

Checkelsky and Ong [19] have also reported measurements of S in graphene. S is found to change sign with V_g, with the TEP curves antisymmetric about the Dirac point, at the offset voltage $V_0 = 15.5$ V (see Figure 18.7). A nominally linear-in-T dependence of the peak value S_m, from ~20 K to 300 K, is also observed.

The above three pioneering experimental measurements [17–19] of TEP in SLG in the temperature range 10 < T < 300 K have brought out the following characteristic features of TEP. The measured TEP reaches a value ~100 μV/K at room temperature. The sign of TEP changes across the CNP as the majority carriers change from electrons to holes. Away from the CNP, the TEP shows a $n_s^{-1/2}$ dependence. At

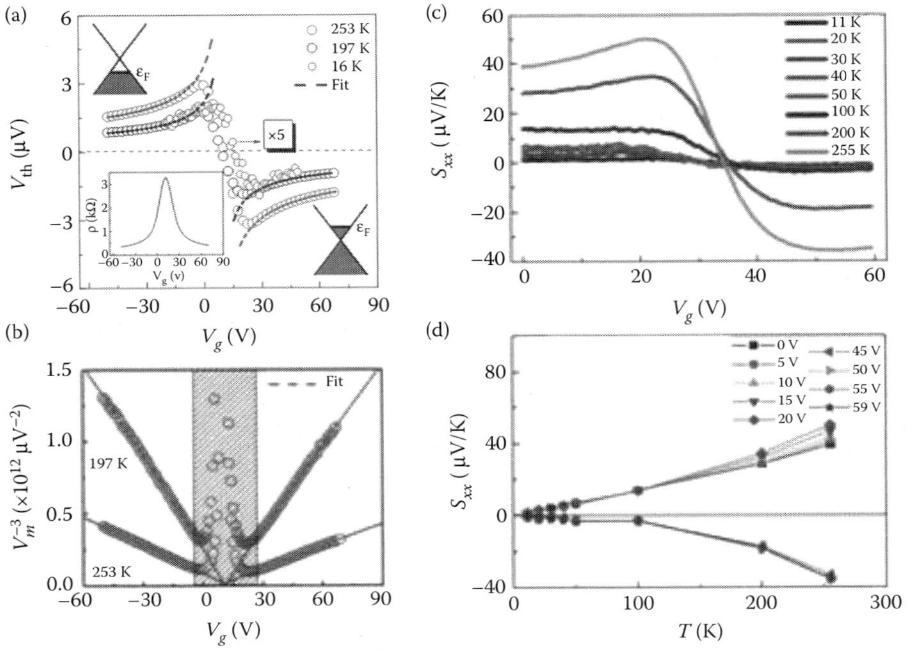

FIGURE 18.6 (a) Thermovoltage (V_{th}) of SLG as function of V_g, for three different temperatures [18]. Dashed curves: fits described by $|S_{xx}| \sim |V_g - V_D|^{-1/2}$. (b) V_{th}^{-2} versus V_g plot for the same data shown in (a). Shaded area is for $|V_g - V_D| < 10$ V. Dashed lines: best power-law fits with exponent ~0.95. (c) V_g dependence of S_{xx} at different temperatures. (d) T dependence of S_{xx} at different gate voltages. (From Wei, F. et al. 2009. *Phys Rev Lett* 102:166808-1–4. With permission.)

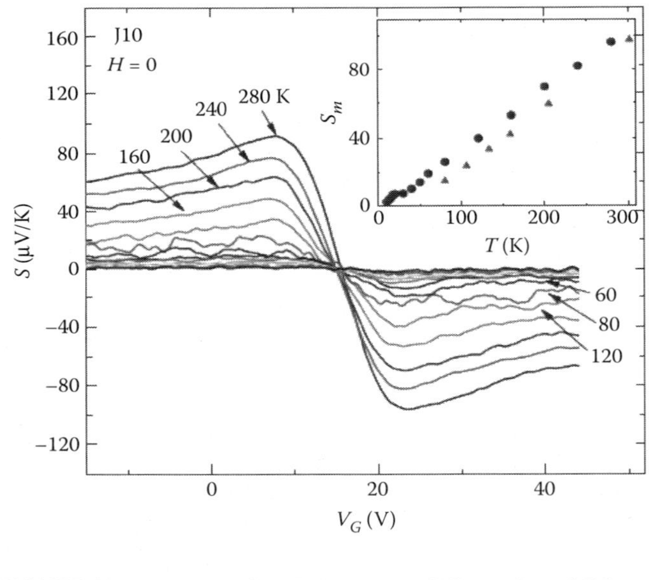

FIGURE 18.7 Measured TEP, S versus V_g [19] at selected T. Inset: T dependence of peak value S_m. (From Checkelsky, J.G. and Ong, N.P. 2009. *Phys Rev B* 80:081413(R)-1–4. With permission.)

low temperatures, it exhibits a linear temperature dependence, in consonance with the Mott formula Equation 18.10.

The experiments of Zuev et al. [17], Wei et al. [18], and Chekelsky and Ong [19], motivated further theoretical efforts. Hwang et al. [34] developed a theory for S_d of graphene, in the linear response approximation. Considering the energy dependence of the transport times corresponding to various

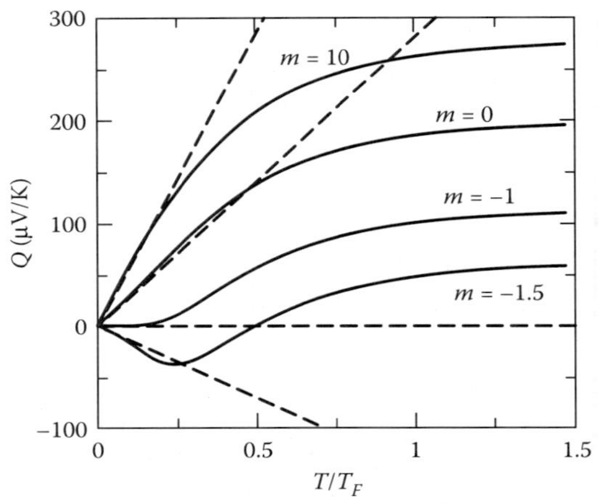

FIGURE 18.8 Hole TEP [34] with $\tau \sim E^m$. Dashed lines show corresponding Mott variations. (From Hwang, E.H., Rossi, E., and Das Sarma, S. 2009. *Phys Rev B* 80:235415-1–5. With permission.)

scattering mechanisms, they elucidate their importance using Mott formula. Figure 18.8 shows their calculations of S_d for different values of the energy exponent "p" (symbol "m," used in [34]); note S_d can change sign if $p < -1$. The Mott formula is found to apply well, for $T \le 0.2T_F$ (T_F is Fermi temperature), but fails in the low-density limit, where electron–hole puddles may dominate. At high temperatures, the TEP is independent of T and approaches a limiting value. A quadratic temperature dependence arising from the temperature-dependence of screened charged impurity scattering is predicted. At high

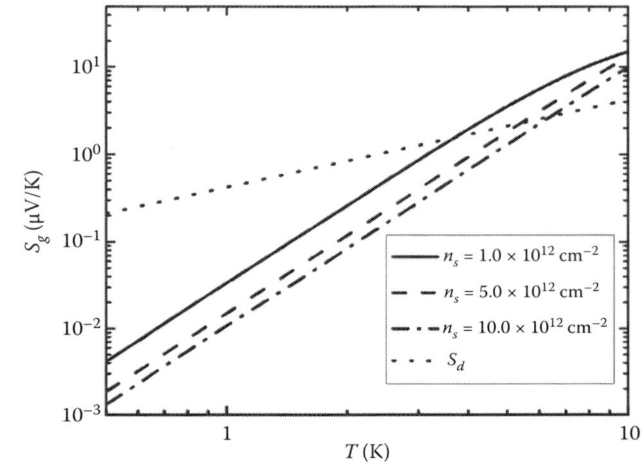

FIGURE 18.9 S_g versus T [33] for $n_s = 1.0 \times 10^{12}$ cm^{-2} (solid curve), 5.0×10^{12} cm^{-2} (dashed curve), and 10.0×10^{12} cm^{-2} (dot-dashed curve). Dotted curve represents S_d with $p = 1$ for $n_s = 1.0 \times 10^{12}$ cm^{-2}. (From Kubakaddi, S.S. 2009. *Phys Rev B* 79:075417-1–6. With permission.)

carrier densities, their calculations show the $n_s^{-1/2}$ behaviour observed in experiments [17–19]. The main features of TEP near CNP are accounted for using a simple two-component model.

Kubakaddi [33] made a theoretical study of the low-temperature behavior of S_g in SLG. Considering the phonon relaxation process to be due to only boundary scattering, S_g is shown to increase with an increase of temperature. Its magnitude near 10 K becomes comparable with S_d, described by Mott's formula (see Figure 18.9). The relative importance of S_g and S_d is shown to depend upon the values of D_{ac}, Λ, n_s, T, and p. Contrary to the earlier qualitative remarks [34], Kubakaddi shows S_g to be important for $T < \sim 10$ K. For SLG in the BG regime, $S_g \sim T^3$, a characteristic of 2D nature of phonons in contrast to the $S_g \sim T^4$ dependence of unscreened deformation potential scattering in conventional 2DEG [13,14,56]. Further, $S_g \sim n_s^{-1/2}$ in contrast to the $n_s^{-3/2}$ dependence in conventional 2DEG [13,14]. The Herring's formula: $S_g\mu_pT^{-1}$ (μ_p being phonon-limited mobility), first given for bulk semiconductors [57], is shown to be valid in 2D graphene.

Bao et al. [58] have compared their calculations of S with the data of Zuev et al. [17] for two values of gate voltage: $V_g = -30$ V and -5 V. Covering a larger range of temperatures, and considering the scattering of electrons by both impurities and phonons, they study the TE effect in SLG away from CNP. Following the balance equation approach, they obtain an expression for S_d, which in the limit of large concentrations is shown to reduce to Mott formula. To estimate the phonon-drag contribution, they take account of boundary scattering as well as phonon–phonon interaction in the phonon relaxation processes. Figure 18.10 depicts their analysis. They find that, as indicated in earlier theoretical studies, S_d, for $T > 10$ K, plays an essential role and S $(= S_d + S_g)$ shows an approximately linear-in-T and $n_s^{-1/2}$ dependences, as observed [17]. For $T > 10$ K, S_g is limited by the phonon–phonon interaction, leading to a peak in its T dependence.

Vaidya et al. [35] have investigated the T and n_s dependences of S_d considering the electrons to be scattered not only by acoustic phonons and impurities but also optical phonons and other disorder-related scattering mechanisms, namely, vacancies and surface roughness (arising from the rippling of graphene sheet deposited on SiO_2 substrate). They obtain a fit with the data of Zuev et al. [17] with S_d for $n_s = 2 \times 10^{16}$ m^{-2} for $30 < T < 300$ K (see Figure 18.11). Although a nonlinear T-dependence due to optical phonons is noticed, the overall S_d is determined mainly by vacancy and impurity scatterings and shows linear T-dependence for $T < \sim 50$ K. As a function of n_s, they see a change in sign of S_d.

With regard to the data of Wei et al. [18], the low-temperature, nonlinear-in-T dependence of TEP exhibited on

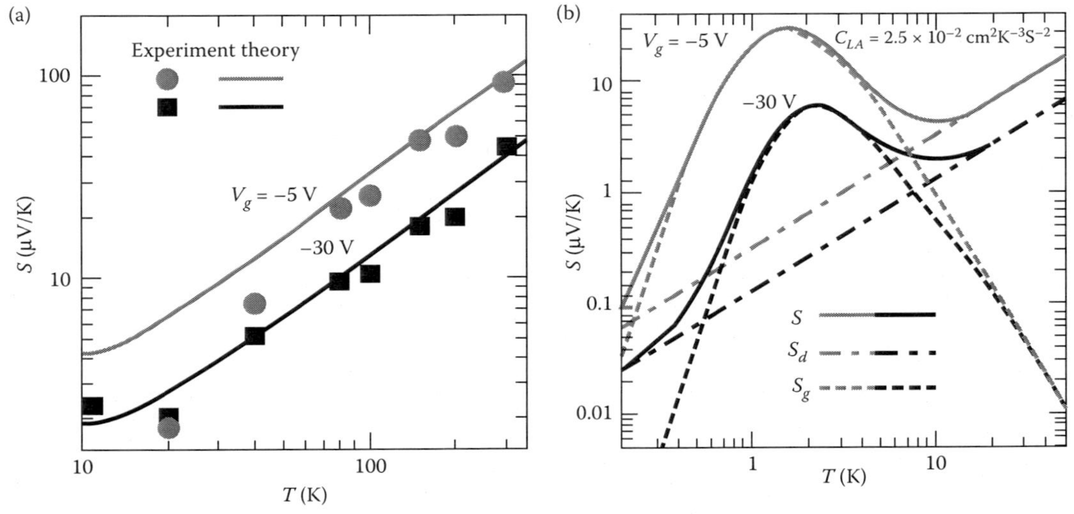

FIGURE 18.10 (a) Comparison of calculations of S [58] (curves) with experimental data [17] (dots) on SLG. (b) T dependences of individual contributions to S $(= S_g + S_d)$. (From Bao, W.S., Liu, S.Y., and Lei, X.L. 2010. *J Phys: Condens Matter* 22:315502-1–7. With permission.)

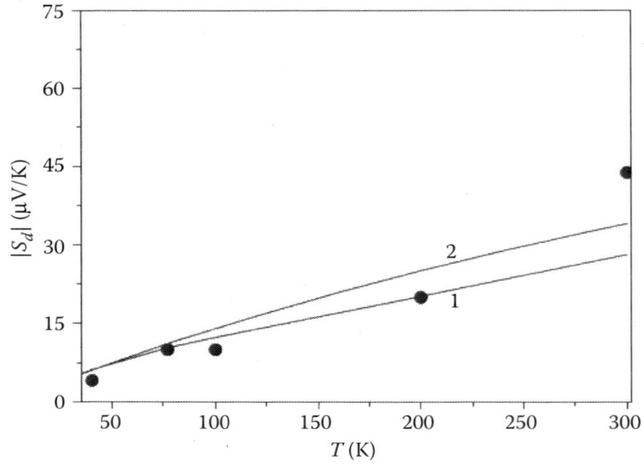

FIGURE 18.11 Comparison of S calculations [35] (curves) with the experimental data (dots) of Reference 17. Curves 1 and 2 represent calculations using Equation 18.7 and the Mott formula, respectively. (From Vaidya, R.G. et al. 2010. *Semicond Sci Technol* 25:092001-1–6. With permission.)

the electron side, has been examined by Sankeshwar et al. [41]. They show that the distinctive feature (a dip at $E = E_F$) observed in the energy dependence of the relaxation times $\tau(E_k)$ of the intrinsic acoustic phonon scattering in graphene in the BG regime, results in S_d exhibiting a nonlinear temperature dependence, before becoming linear at very low temperatures. The influence that the nonlinear behavior of S_d can have on the total TEP, S, is shown to depend on the S_g contribution. With S_g calculated in phonon-boundary scattering limit, with $\Lambda < \sim 500$ nm, S is found to reveal the non-linear structure of S_d. Figure 18.12a shows comparison of the calculations ($\Lambda = 200$ nm, comparable with sample dimensions) with the data. For $\Lambda > 500$ nm, S_g masks the nonlinear behavior of S_d. More detailed investigations, considering all scattering mechanisms operative in real systems [55,59], are called for.

To explain the data of Wei et al. [18], Munoz [60] obtains, from a variational solution of the Boltzmann equation, an analytical expression for the leading contribution to phonon-limited TEP. Figure 18.12b represents his calculations for different hole densities, n. For $T < \sim 60$ K, calculations are found to reproduce the linear-in-T and the $n_s^{-1/2}$ dependences.

Seol et al. [46] have reported measurements of TEP (shown in Figure 18.13) as well as electrical and thermal conductivities of supported and suspended SLG flakes. They obtain room-temperature values of $S \sim -80$ μV/K and explain TEP data on supported SLG with calculations based on the semiclassical theoretical model of Hwang et al. [34].

Wang and Shi [47] have measured, over wide-temperature-range ($4 < T < 300$ K), TEP and electrical conductivity of SLG samples with varying degree of disorder, characterized by a range of mobilities ($1.5–13.0 \times 10^3$ cm^2/Vs). In low-mobility samples, the high charged impurity concentration could induce relatively high residual carrier density near CNP so that $T_F \gg T$ and the Mott relation holds at all gate voltages (see Figure 18.14). On the other hand, for high-mobility graphene, the carrier density near the CNP can be so low that the condition $k_BT \ll E_F$ no longer holds, and Mott relation fails; the deviation is insignificant below 100 K. Taking account of high-temperature effects, the Boltzmann formalism is shown to explain their measured data for low-density systems.

The carrier mobility-dependence of S and σ of SLG near CNP has been further investigated by Liu et al. [49] (see Figure 18.15). With increase in mobility, the maximum value of S_{xx} is found to increase and exhibit an increasingly diverging trend accompanied by a sharper peak-to-dip transition around Dirac point. The peak-to-dip width is related to the width of the minimum conductivity plateau, which is broader for the low-mobility state, associated with disorder in graphene. At high gate voltages on either side of CNP, S_{xx} is found to converge to the same values, for all mobility values. This suggests that charge density fluctuations induced by charged impurities is much smaller than the effective carrier density.

The characteristics of TEP have also been investigated in graphene systems produced by other methods as well. (For a recent review of graphene production techniques, see [20,61].)

The TE effect in relatively high-mobility (~20×10^3 cm^2/Vs at 4 K) SLG, grown epitaxially on SiC substrates, has been

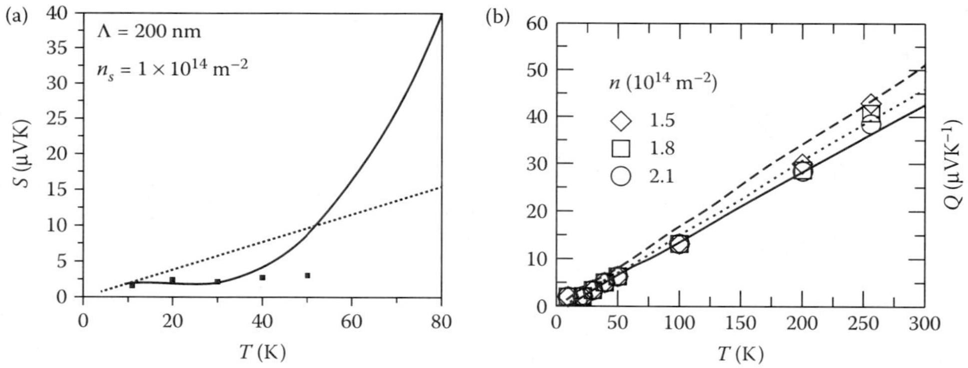

FIGURE 18.12 Comparison of two calculations [41,60] of acoustic phonon-limited TEP as function of temperature with experimental data [18] on SLG. (a) Dots: data, full curve: S, dashed curve: Mott S_d [41]. (b) Symbols: data, curves: S_d at corresponding hole densities, n [60]. (From Sankeshwar, N.S., Vaidya, R.G., and Mulimani, B.G. 2013. *Phys Stat Sol B* 250:1356–62. With permission.; Munoz, E. 2012. *J Phys Condens Matter* 24:195302-1–8. With permission.)

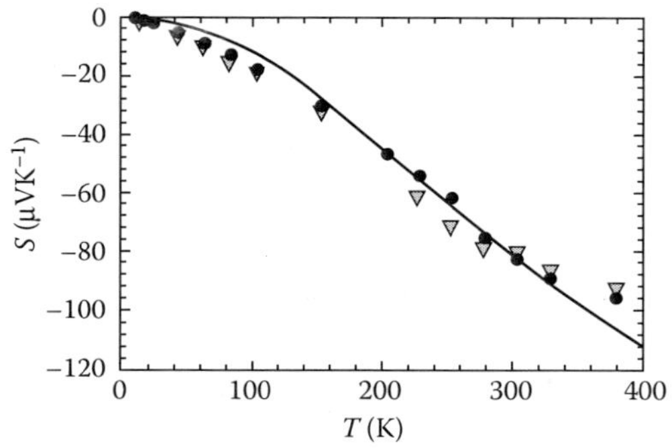

FIGURE 18.13 Measured TEP [46] for two SLG samples G1 (triangles) and G2 (circles). Curve represents calculations using model of [34]. (From Seol, J.H. et al. 2010. *Science* 328:213–6. With permission.)

studied by Wu et al. [48]. The temperature dependence of TEP, away from the Dirac point with hole density 1×10^{12} cm^{-2}, shown in Figure 18.16, displays a deviation from the Mott relation. The data is found to fit to $AT + BT^2$, where A and B are temperature independent constants. The additional quadratic dependence, instead of the otherwise linear dependence, reflects the importance of the temperature-dependent screening effect [34], expected to be stronger in epitaxial graphene. Recently, Wierzbowska and Dominiak studied the effect of C-face 4H-SiC (0001) deposition on TEP of SLG and MLG within density-functional theory using the semiclassical Boltzmann equations [62,63]

Gharari and coworkers [64,65] report measurements of TEP on samples deposited on hexa-boron nitride (hBN) substrates where drastic suppression of disorder is achieved. At high temperatures where the inelastic scattering rate due to electron–electron interactions is higher than the elastic scattering rate by disorders, the measured TEP exhibits an enhancement compared to the expected TEP from the Mott relation. In their recent study [65] the deviation at high temperatures is shown to be greater for higher mobility samples, and the BTE is used to explain the deviation.

Measurements of TEP of chemical vapour deposition (CVD)–grown graphene (see Figure 18.17) have also been reported [51,52,66]. The values of S (~+ 9 μV/K at 300 K) reported by Xu et al. [51] in suspended Cu-CVD SLG for $60 < T < 300$ K suggest hole-like majority carriers (Figure 18.17a). The measurements of TEP by Nam et al. [52] on large-area inhomogeneous Cu-CVD graphene as a function of charge carrier density for $T < 150$ K show that, the electron–hole point asymmetry in TEP (not seen in resistance data), near

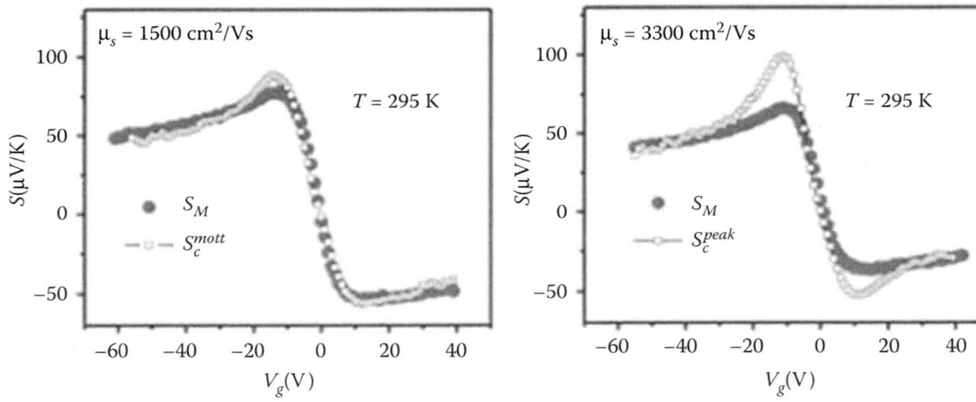

FIGURE 18.14 Comparison of measured (S_M, solid circles) and calculated (S_C^{Mott}, open squares) TEP for an SLG sample with two mobility values [47]. (From Wang, D. and Shi, J. 2011. *Phys Rev B* 83:113403-1–4. With permission.)

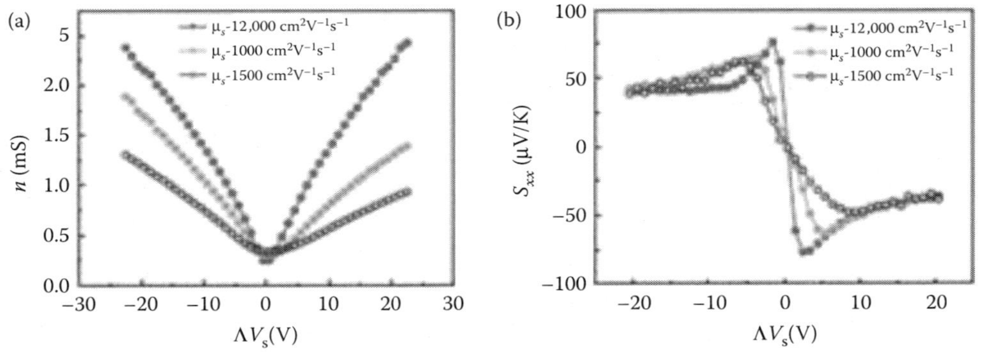

FIGURE 18.15 Gate voltage dependence of (a) electrical conductivity, σ, and (b) S_{xx}, of device A [49] at 150 K with three hole mobility values. (From Liu, X. et al. 2012. *Phys Rev B* 86:155414-1–7. With permission.)

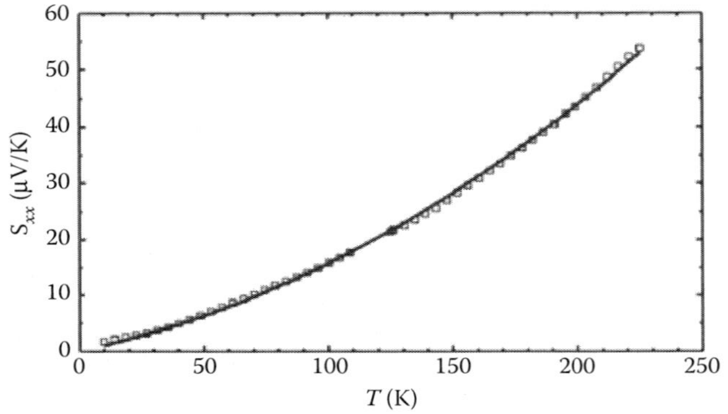

FIGURE 18.16 Temperature dependence of measured (open squares) TEP, S_{xx}, of epitaxial SLG, and least square fit to $AT + BT^2$ (curve) [48]. (From Wu, X. et al. 2011. *Appl Phys Lett* 99:133102:1–3. With permission.)

the Dirac point, increases with decreasing temperature (see Figure 18.17b). They attribute it to individual graphene regions having different TEPs. Babichev et al. [66] have studied TEP of approximately four-atomic-thick graphene samples grown on $Si/SiO_2/Ni$ substrates by CVD (Figure 18.17c). Other investigations have demonstrated the TEP of CVD-grown graphene to be a sensitive probe to the surface charge doping from the environment promising use in gas/chemical sensing [67].

The calculations of S_d by Vaidya et al. [68] obtain a fit (see Figure 18.18a) for the data (dots) of Xu et al. [51]. Although the major contribution to TEP for $T < 170$ K may be due to S_d, as observed by Xu et al. [51], the deviation from experimental data for higher temperatures suggests the role of additional scattering mechanisms.

Mariani and Oppen [69] and Ochoa et al. [39], have investigated the effect of strain on the resistivity of SG, and shown that, in the absence of strain, the flexural phonons (FPs) dominate the phonon contribution to resistivity, whereas in the presence of strain, the contribution due to FPs is suppressed and

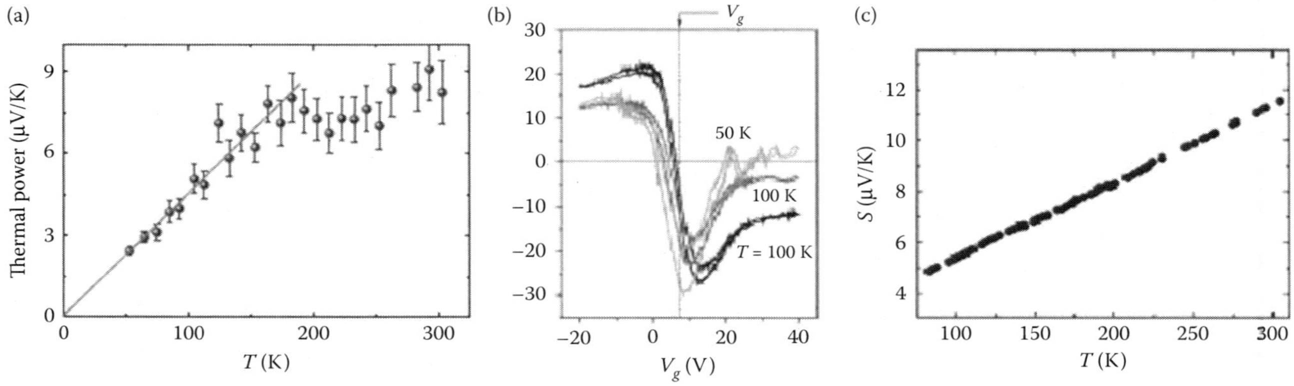

FIGURE 18.17 Measured TEP of CVD graphene samples (a) of Xu et al. [51] as a function of T, (b) of Nam et al. [52] as a function of V_g, at selected T, and (c) of Babichev et al. [66] as a function of T. Curve in (a) is a linear fit. (From Xu, X. et al. 2010. *arXiv*:1012.2937v1 [cond-mat.mes-hall] Dec 14, 2010. With permission.; Nam, Y. et al. 2014. *Appl Phys Lett* 104:021902-1–5. With permission.; and Babichev, A.V., Gasumyants, V.E., and Butko, V.Y. 2013. *J Appl Phys* 113:076101-1–3. With permission.)

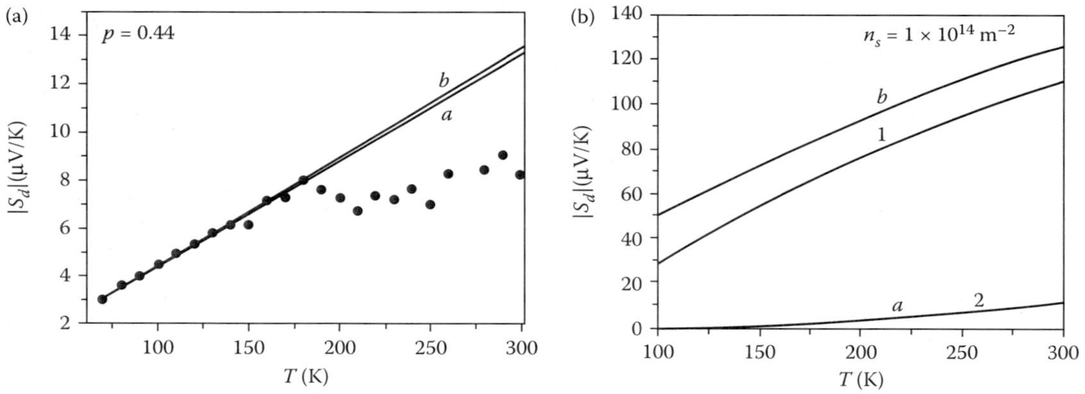

FIGURE 18.18 Temperature dependence of acoustic phonon-limited TEP of suspended SLG. (a) Comparison of calculations [68] of S_d (curves) with data (dots) of [51]. Curves a and b depict S_d due to FPs calculated using Equations 18.7 and 18.10, respectively. (b) Calculations of S_d (curve 1) with contributions from in-plane phonons (curve a) and FPs (curve b). Curve 2 shows S_d in strained SG. (From Vaidya, R.G., Sankeshwar, N.S., and Mulimani, B.G. 2012. Washington: *SPIE* p 85491U-1-6, doi: 10.1117/12.925343. With permission.; Vaidya, R.G., Sankeshwar, N.S., and Mulimani, B.G. 2012. *J Appl Phys* 112:093711-1–6. With permission.)

the in-plane phonon modes become dominant. These features seen in the scattering rates get reflected in the TE properties [68,70]. Vaidya et al. [68] have investigated the influence of strain on the acoustic phonon-limited S_d considering contributions from both in-plane and flexural acoustic phonons (see Figure 18.18b). With the analysis made for SG with high electron densities in the equipartition regime (phonon energy, $\hbar\omega_q \ll k_BT$), it is found that the dominant contribution to S, for $T < 90$ K, is from S_g in the absence and presence of strain. However, for higher temperatures ($T > 150$ K) the contribution from S_d becomes important. The effect of strain, which is known to suppress the electron–FP interaction [39], is found to suppress S_d and to alter its behavior, the effect being larger at higher temperatures.

There have been theoretical studies looking into the different aspects of TEP. In the case of FLG samples, Li et al. [71] find that TEP depends on the number of layers, increasing with increase in thickness and reaching a peak value for six layers. Sharapov and Varlamov [72] and Patel and Mukerjee [73], have studied the effect of opening a gap in the graphene spectrum on TEP and find that the TEP is proportional to the gap. It may be noted that, recently, Woods et al. [74], have reported a commensurate–incommensurate transition in graphene-on-hBN and demonstrated experimentally the opening of a gap in the spectrum. It would be interesting to have measurements of TEP on systems with varying band gaps.

18.3.2 Bilayer Graphene

In the last few years, TEP in BLG has been studied both theoretically [36,75,76] and experimentally [77–79] (see Table 18.3). Hao and Lee [75] have studied the T and n_s dependence of TEP of gapped, relatively clean BLG employing Kubo formula, with impurity scattering treated in the self-consistent Born approximation. They find that introducing a gap enhances the TEP. Wang et al. [77] and recently, Chien et al. [80], have reported enhancement in the TEP associated with bandgap opening in a BLG device, with perpendicular electric field applied.

Nam et al. [78] have measured (see Figure 18.19) the TEP, S_{xx}, in the temperature range $30 < T < 250$ K for different

TABLE 18.3
Measurements of TEP in BLG Systems

Sample	Value (μV/K)	Reference
Mechanical exfoliation on 300 nm SiO_2/Si substrate	$\|S_m\|$ ~95 @ 300 K	[77]
Mechanical exfoliation on 300 nm SiO_2/Si substrate; μ ~ 2–4 × 10^3 cm^{-2}/Vs	~100 @ 250 K	[78]
Mechanical exfoliation on 300 nm SiO_2/Si substrate; μ ~ 2–3 × 10^3 cm^{-2}/Vs	S_m^{peak} ~ 180 @ 100 K	[79]

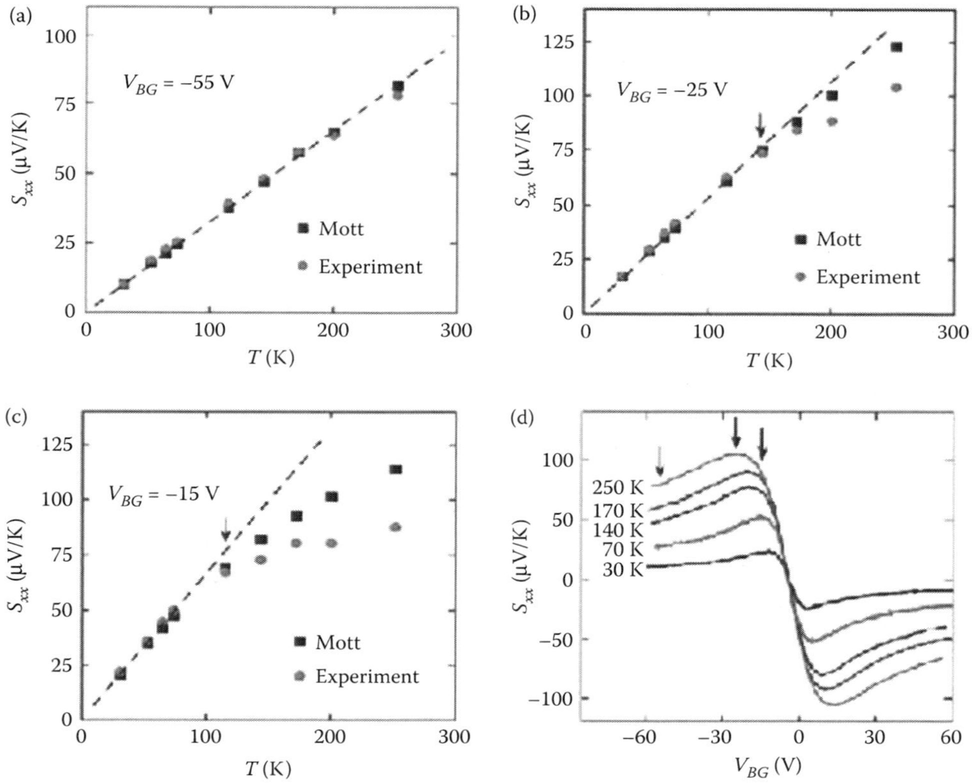

FIGURE 18.19 Measured TEP [78] of BLG as function of T, for three values of V_{BG}: (a) −55 V, (b) −25 V, and (c) −15 V. Arrows denote T^* (see text). (d) TEP as a function of V_{BG} for different temperatures. (From Nam, S.G., Ki, D.K., and Lee, H.J. 2010. *Phys Rev B* 82:245416-1–5. With permission.)

values of n_s (i.e., back-gate voltages, V_{BG}). As in SLG [17–19], the ambipolar nature of the carriers manifests itself as the sign change at the CNP (see Figure 18.19d). As V_{BG} approaches the CNP, S_{xx} is found to deviate from Mott relation at a certain onset temperature, T^*. At lower n_s, only low-T data follow the Mott formula. For high n_s, TEP shows a linear-in-T dependence implying a weak electron–phonon interaction and negligible phonon-drag effect. Further, for lower values of n_s, a deviation from the Mott relation along with a saturating tendency of TEP, is observed at higher temperatures (Figure 18.19c); this is attributed to the low T_F. Such a saturating tendency in SLG was discussed with regard to different scattering mechanisms [34].

Kubakaddi and Bhargavi [36,76] have compared their calculations of S_d with the data of Nam et al. [78]. Taking into account the contributions from acoustic phonons, surface polar phonons, charged impurity, and short range disorder, they obtain a qualitative agreement with data [78] (see Figure 18.20a). The inverse dependence of S_d on n_s, is unlike $\sim n_s^{-1/2}$ dependence observed for SLG, but is similar to that for degenerate conventional 2DEG [13]. Kubakaddi and Bhargavi [36] have also developed the theory for low-T S_g in the phonon-boundary scattering regime. With increase in temperature, S_g shows a T^3 dependence at very low T, gradually changes to sublinear dependence, and then flattens, producing a knee. They find the influence of the chirality of electrons on S_g is to reduce its magnitude, as compared to that of ideal 2DEG and the reduction is T dependent. Further, they show (their Figure 18.20b) that the low-T ($30 < T < 70$ K) data of Nam et al. [78] for $n_s = 3.67 \times 10^{12}$ cm^{-2} can be explained by S_g alone by varying D_{ac}. In BG regime, as in SLG, $S_g \sim T^3$ is a manifestation of the 2D phonons. This is in contrast with the T^4 dependence in conventional 2D systems [81]. In BLG, since $\mu_p \sim T^{-4}$ [82], Herring's formula $S_g\mu_p \sim T^{-1}$ is shown to be valid [36]. With regard to n_s dependence, $S_g \sim n_s^{-3/2}$ [36], as in conventional 2DEG [81]. This is in contrast with $S_g \sim n_s^{-1/2}$ in SLG [33]. This difference can be attributed to the different dispersion relation of the electron spectrum in BLG compared to that in SLG. It may be inferred that, the temperature (density) dependence of S_g is determined by both the dimensionality and the dispersion relation of the phonons (electron gas). The n_s and Λ dependence of S_g suggests the possibility of tuning the magnitude of S_g in BLG.

Detailed investigations of TEP in BLG over a wider temperature range are required to better understand the contribution of S_d relative to that of S_g.

18.3.3 Armchair Graphene Nanoribbons

There have been a few theoretical efforts, to understand the diffusion and drag contributions to TEP of GNR, a Q1D system. Divari and Kliros [83] have studied S_d of graphene ribbons with aspect ratio $(W/L) \geq 3$, using linear response theory and the Landauer formalism. Xing et al. [84], using nonequilibrium Green's function, show that TEP depends on the chirality of the GNR. The TE properties have also been investigated by solving the electron and phonon transport equations in the nonequilibrium Green's function formalism [11,85–87]. Mazzamoto et al. [88], in their study of the TE properties, have proposed optimized patterning of the GNRs with regard to their width and edge orientations, to achieve a high ZT.

In their systematic study of S_d of semiconducting AGNR systems, Nissimagoudar and Sankeshwar [37] employ Boltzmann formalism to study the relative importance of various electron scattering mechanisms, and the influence of GNR width and subband structure on the behavior of S_d. Considering the electrons to be scattered by edge roughness, impurities, and acoustic and optical phonons, they demonstrate that, unlike as in the case of SLG, owing to the peculiar nature of the density of states, the energy dependences of the momentum relaxation times, and hence S_d, exhibit distinctive features.

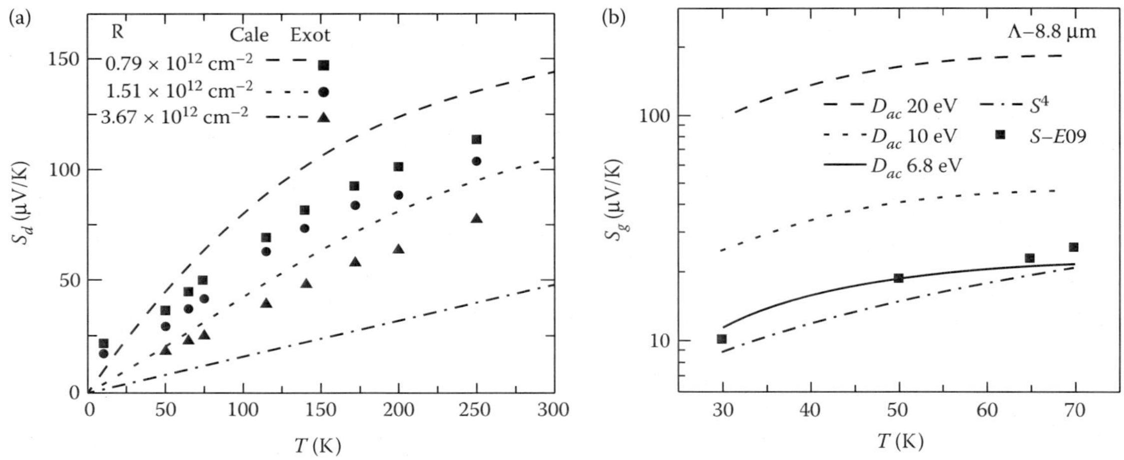

FIGURE 18.20 Comparison of measured TEP of BLG samples of Reference 78 with calculations [36,76]. (a) S_d (curves) for samples with three values of n_s [76], and (b) S_d (dotted curve) for sample with $n_s = 3.67 \times 10^{12}$ cm^{-2}. Other curves denote S_g, for three values of D_{ac}: 6.8 eV (solid curve), 20 eV (dashed curve), and 10 eV (dash-dotted curve) [36]. (From Kubakaddi, S.S. and Bhargavi, K.S. 2010. *Phys Rev B* 82:155410-1–7. With permission.; Bharaghavi, K.S. and Kubakaddi, S.S. 2013. *Physica E* 52:116–21. With permission.)

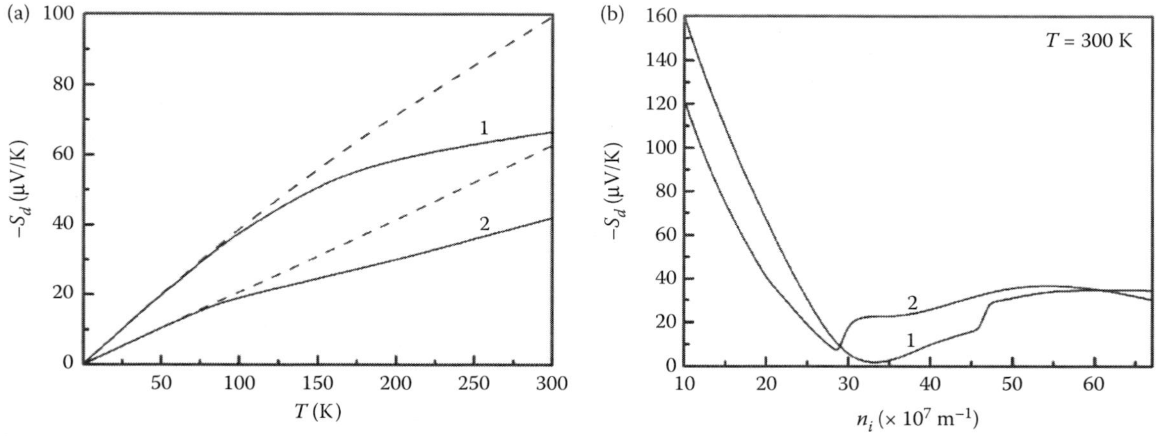

FIGURE 18.21 Calculation of S_d for AGNRs [37] with $W = 3$ nm (curve 1) and 5 nm (curve 2) as function of (a) temperature for $n_l = 2 \times 10^8$ m^{-1}, and (b) linear carrier density, n_l, at $T = 300$ K. Dashed curves in (a) show S_d according to Mott expression. (From Nissimagoudar, A.S. and Sankeshwar, N.S. 2013. *Carbon* 52:201–8. With permission.)

For $T > 100$ K, S_d shows a deviation from linear-in-T dependence (Figure 18.21a). A decrease in GNR width is found to enhance the magnitude of S_d and alter its behavior. Figure 18.21b illustrates the influence of subband structure on room temperature S_d. Reflecting the singular nature of the AGNR density of states, S_d, as a function of linear carrier density, n_l, shows a step-like behavior, increasing in magnitude, when Fermi energy moves into the second subband. With a proper choice of parameters characterizing the extrinsic scattering mechanisms, and the possibility of modulating the Fermi level with a control on gate and bias voltages [89], the behavior of S_d in AGNRs could be tuned and the subband structure in S_d be detected. These features portend promising applications in TE devices.

The role of Q1D, temperature, Fermi energy, and ribbon width on boundary scattering limited S_g of an n-type AGNR is investigated by Bhargavi and Kubakaddi [43]. Assuming the electrons to interact via unscreened deformation potential coupling, with the 1D acoustic phonons of small energy, an expression for S_g is obtained for intrasubband ($n = n'$) transitions. Figure 18.22 shows S_g versus T in the 1D quantum limit ($n = 1$). At very low T (< ~2 K), S_g is found to be exponentially suppressed, a new feature, which is attributed to the peculiar nature of the 1D Fermi-surface consisting of discrete points. Such suppression is similar to that observed in semiconducting single wall carbon nanotubes (SWCNTs) [15,90], but is in contrast to the well-known power laws: $S_g \sim T^3$ in SLG [33] and BLG [36], and $S_g \sim T^4$ in conventional 1DEG [91]. At higher temperatures, S_g deviates from exponential behavior and finally levels off as observed in SWCNT [92,93]. Also, S_g depends strongly on E_F and W. Since the ribbon edge geometry of a GNR determines the electronic structure of the system, S_d and S_g in ZGNR are expected to be different from those in AGNR.

Detailed experiments on TEP in GNRs are needed to better understand the behavior of both S_d and S_g contributions.

In literature, for transport studies in all the three graphene systems considered here, the value of D_{ac} is taken in the range 3–50 eV [39,82]. A value of 19 eV for D_{ac} is found to explain

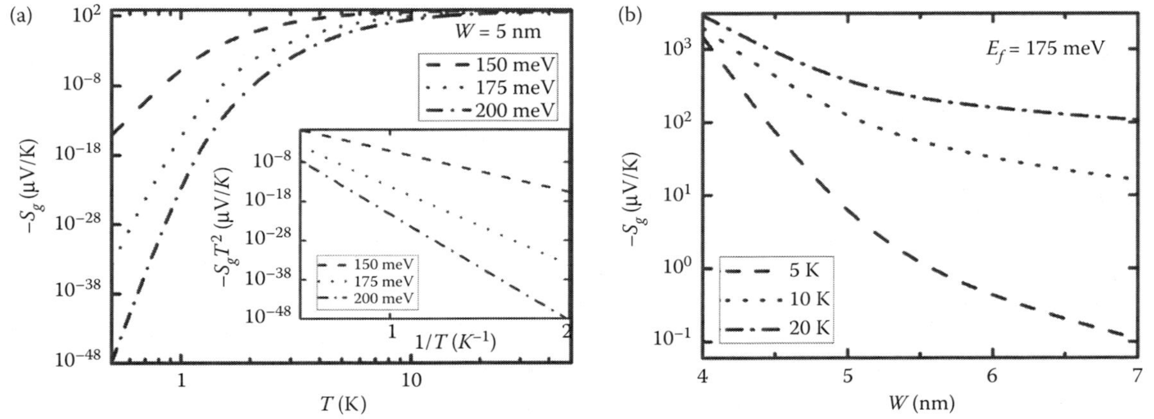

FIGURE 18.22 Calculation of S_g for AGNR [43] as function of (a) T, for $E_F = 150$, 175, and 200 meV, (b) W, for $T = 5$, 10, and 20 K. Inset: S_gT^2 versus T^{-1} showing activated behavior. (From Bhargavi, K.S. and Kubakaddi, S.S. 2011. *J Phys: Condens Matter* 23:275303-1–5. With permission.)

TABLE 18.4
Measurements of TEP in MLG Systems

Sample	Value (μV/K)	Reference
Few atomic layer thick, cm size sample CVD grown on Si/SiO_2/Ni substrates; $\rho \sim 3 \times 10^{-5}$ Ω cm	10 @ 300 K	[66]
SLG–MLG : CVD on Cu	~30 (SLG) –54 (HLG) @ RT	[71]
Pristine: on SiO_2/Si substrate, $t \sim 5$ nm with possible structural defects; treated: ACN, TPA attachments	Pristine: ~40 Treated: 180; $300 < T < 575$ K	[97]
On SiO_2/Si substrate; SLG and rGO; SLG: p @low T and n @ high T	Pristine: ~80 Treated: ~700 @ 575 K; SLG: ~ –40 to 50	[98]
SLG–BLG transistor: Mechanical exfoliation of graphene sheets onto 90 nm SiO_2/Si wafer; SLG/BLG identified by optical contrast and Raman	~6 @ 12 K	[99]
SLG–TLG: epitaxial on 6H-SiC	10 (p); –20 (n) @ 500 K	[100]

well the low-T energy-loss rates in SLG [94]. As S_g ($\sim D_{ac}^2$) depends only upon electron–acoustic phonon coupling, a systematic analysis of low-T data, where S_g is expected to manifest in graphene systems, may enable determination of D_{ac}.

For a better understanding of TE effects in graphene systems, detailed theoretical studies of S_d and S_g are required, for instance, with the inclusion of screening effects of the electron–acoustic phonon interaction via deformation potential and vector potential couplings. It is interesting to note that recent measurements of resistivity [40] and hot electron energy loss rates [94] in SLG indicate the electron–acoustic phonon interaction to be unscreened. Further, in all the three graphene systems considered, the T dependence of TEP is unlike that in graphite [95] where a large negative peak due to S_g appears at low temperatures. In this context, it may be mentioned here that recent measurements on FLG/PVDF composites [96] show TEP to display a peak near 30 K. It would be informative to make a systematic study of how the behavior of TEP evolves as the number of graphene layers increases [71]. FLG and MLG systems have also been investigated for their potential use in TE applications [66,71,97–100] (see Table 18.4).

There has been continuing interest in the TEP of graphene systems with regard to various aspects, such as characterizing epitaxial graphene samples using thermovoltage maps generated using STM [101], designing high-ZT GNR systems using first principles studies on strain engineering [102], including quantum corrections to conductivity and TEP [103], and nanostructuring [104–106].

18.4 SUMMARY

In this chapter, we have reviewed the current status of the studies on TEP, an important and interesting transport property, in graphene systems, namely, SLG, BLG, and AGNR. Measurements of TEP of SLG reveal characteristic features, some of which not observed in metals [12] and conventional 2D semiconductor systems [13–15]: (a) a range of values up to 100 μV/K, at room temperature, (b) a change in sign across the CNP as the gate bias is varied, (c) $n_s^{-1/2}$ dependence away from the CNP, and (d) linear temperature dependence, at low temperatures, in consonance with the Mott relation. However, in BLG a deviation from Mott formula is observed for higher temperatures [78]. In FLG samples, TEP depends on the number of layers, increasing with increase in thickness [71]. There are not many reports of measurements of TEP in BLG, and measurements of TEP of GNRs are also awaited.

The theoretical calculations of diffusion and phonon-drag components of TEP, employing the conventional Boltzmann formalism in the relaxation time approximation, give a basic understanding of TEP in graphene systems. They give a description of the dependences of TEP on temperature and gate bias. The effect of physical modifications, such as structure patterning (say, width and edge chirality) of GNRs [37,43], and suspending and/or straining of the graphene [70,72–74], is found to alter the magnitude as well as the behavior of TEP. This understanding is expected to provide a useful guideline for improvement and optimization of performances of graphene-based TE modules.

One of the challenges for TE materials revolves around the synthesis and fabrication of structures that do not appreciably compromise the innate high power factor of the system, but significantly reduce lattice thermal conductivity to enhance the overall figure of merit, ZT. Efforts are on to understand the TE effects in graphene for its utilization in TE applications. Future experimental endeavors may aid not only in improving applicability in TE devices but also in understanding better the TE processes in graphene. Detailed investigations of low-temperature TEP will not only enable a better analysis of the relative contributions of S_d and S_g but also provide a better understanding of the scattering mechanisms operative. Although, recently, measurements on SLG and BLG samples to probe thermal conductance due to electron–phonon coupling at sub-Kelvin temperatures have been reported [107], a systematic investigation of TEP as well as thermal conductivity of graphene systems is needed to better describe TE transport in graphene systems. Further, it would be interesting to compare results of measurements of σ, κ, and S on the same sample over a wide temperature range with calculations for parameters characteristic of the sample. This will provide a better understanding of ways to achieve enhanced ZT.

ACKNOWLEDGMENT

We thank Mr. A.S. Nissimagoudar for his help in the preparation of the manuscript.

REFERENCES

1. Rowe, D.M. (ed.) 2006. *CRC Thermoelectrics Handbook—Macro to Nano*. Boca Raton: Taylor & Francis.

2. Tritt, T.M. (vol. ed.) 2001. Recent trends in thermoelectric materials research III. In *Semiconductors and Semimetals, Vol. 71*. Willardson, R.K. and Weber, E.R. (eds.) San Diego: Academic Press.
3. Ure, R.W. 1972. Thermoelectric effects in III–V compounds. In *Semiconductors and Semimetals, vol 8,* Willardson, R.K. and Beer, A.C. (eds.) New York; Academic Press, p. 67–102.
4. Dresselhaus, M.S., Lin, Y.-M., Cronin. S.B., Rabin, O., Black, M.R., Dressehaus, G., and Koga, T. 2001. Quantum wells and quantum wires for potential thermoelectric applications. In *Semiconductors and Semimetals*, Vol. 71. Willardson, R.K. and Weber, E.R. (eds.) New York: Academic Press, p. 1.
5. Chen, G. 2001. Phonon transport in low-dimensional structures. In *Semiconductors and Semimetals*, Vol. 71. Willardson, R.K. and Weber, E.R. (eds.) New York: Academic Press, p. 203.
6. Katsnelson, M.I. 2012. *Graphene: Carbon in Two Dimensions*. Cambridge: Cambridge University Press.
7. Castro Neto, A.H., Guinea, F., Peres, N.M.R., Novoselov, K.S., and Geim, A.K. 2009. The electronic properties of graphene. *Rev Mod Phys* 81:109–62.
8. Das Sarma, S., Adam, S., Hwang, E.H., and Rossi, E. 2011. Electronic transport in two-dimensional graphene. *Rev Mod Phys* 83:407–70.
9. Balandin, A.A. 2011. Thermal properties of graphene and nanostructured carbon materials. *Nat Mater* 10:569–81.
10. Mahan, G.D. 1998. Good thermoelectrics. In: *Solid State Physics*, Vol. 51. Ehrenreich, H. and Spaefen F. (eds.) New York: Academic Press, p. 81.
11. Sevincli, H. and Cuniberti, G. 2010. Enhanced thermoelectric of merit in edge-disordered zigzag graphene nanoribbons. *Phys Rev B* 81:113401-1–4.
12. Blatt, F.J., Schroeder, P.A., Foiles, C.L., and Greig, D. 1976. *Thermoelectric Power of Metals*. New York: Plenum Press.
13. Gallagher, B.L. and Butcher, P.N. 1992. Classical transport and thermoelectric effects in low dimensional and mesoscopic semiconductor structures. In *Handbook on Semiconductors*, Vol. 1. Landsberg, P.T. (ed.) Amsterdam: Elsevier, p. 721–816.
14. Fletcher, R. Zaremba, E., and Zeitler, U. 2003. Phonon drag thermopower of low dimensional systems. In *Electron-Phonon Interactions in Low Dimensional Structures*. Challis, L. (ed.) Oxford: Oxford Science; p. 149.
15. Tsaousidou, M. 2010. Thermopower of low dimensional structures: The effect of electron-phonon coupling. In *Frontiers in Nanoscience and Nanotechnology*, Vol. 2. Narlikar, A.V. and Fu, Y.Y. (eds.) Oxford: Oxford University Press, p. 477.
16. Barnard, R.D. 1972. *Thermoelectricity in Metals and Alloys*. London: Taylor & Francis.
17. Zuev, Y.M., Chang, W., and Kim, P. 2009. Thermoelectric and magnetothermoelectric transport measurements of graphene. *Phys Rev Lett* 102:096807-1–4.
18. Wei, P., Bao, W., Pu, Y., Lau, C.N., and Shi, J. 2009. Anomalous thermoelectric transport of Dirac particles in graphene. *Phys Rev Lett* 102:166808-1–4.
19. Checkelsky, J.G. and Ong, N.P. 2009. Thermopower and Nernst effect in graphene in a magnetic field. *Phys Rev B* 80:081413(R)-1–4.
20. Rao, C.N.R., Sood, A.K., and Subrahmanyam, G.A. 2009. Graphene: The new two-dimensional nanomaterial. *Angew Chem Int Ed* 48:7752–77.
21. Wallace, P.R. 1947. The band theory of graphite. *Phys Rev* 71:622–34.
22. Mucha-Kruczynski, M., McCann, E., and Fal'ko, V.I. 2010. Electron–hole asymmetry and energy gaps in bilayer graphene. *Semicond Sci Technol* 25:33001-1–6.
23. Nakada, K. et al. 1996. Edge state in graphene ribbons: Nanometer size effect and edge shape dependence. *Phys Rev B* 54:17954–61.
24. Wakabayashi, K., Takane, Y., Yamamoto, M., and Sigrist, M. 2009. Electronic transport properties of graphene nanoribbons. *New J Phys* 11:095016-1–21.
25. Min, H., Sahu, B., Banerjee, S.K., and MacDonald, A.H. 2009. *Ab initio* theory of gate induced gaps in graphene bilayers. *Phys Rev B* 75:155115-1–7.
26. Fang, T., Konar, A., Xing, H., and Jena, D. 2008. Mobility in semiconducting graphene nanoribbons: Phonon, impurity, and edge roughness scattering. *Phys Rev B* 78:205403-1–8.
27. Nag, B.R. 1980. *Electron Transport in Compound Semiconductors*. Berlin: Springer-Verlag.
28. Ferry, D.K., Goodnick, S.M., and Bird, J. 2009. *Transport in Nanostructures*. Cambridge: Cambridge University Press.
29. Shaffique, A., Hwang, E.H., Galitski, V.M., and Das Sarma, S. 2007. A self consistent theory for graphene transport. *PNAS* 104:18392–7.
30. Ziman, J.M., 1960. *Electrons and Phonons*. Oxford: Claredon Press.
31. Sankeshwar, N.S., Kubakaddi, S.S., and Mulimani, B.G. 2013. Thermoelectric power in graphene. In *Advances in Graphene Science*. Aliofkhazraei, M. (ed.), Rijeka, Croatia InTech, DOI: 10.5772/56720. Available from: http://www.intechopen.com/books/advances-in-graphene-science/thermoelectric-power-in-graphene.
32. Kubakaddi, S.S. and Katti, V.S. 2008. Diffusion thermopower in graphene. In *Proceedings of International Conference* on 'Recent trends in nanostructured materials and their applications'. Reddy, K.N. (ed.) New Delhi: Exel India, p. 60.
33. Kubakaddi, S.S. 2009. Interaction of massless Dirac electrons with acoustic phonons in graphene at low temperatures. *Phys Rev B* 79:075417-1–6.
34. Hwang. E.H., Rossi, E., and Das Sarma, S. 2009. Theory of thermopower in two-dimensional graphene. *Phys Rev B* 80:235415-1–5.
35. Vaidya, R.G., Kamatagi, M.D., Sankeshwar, N.S., and Mulimani, B.G. 2010. Diffusion thermopower in graphene. *Semicond Sci Technol* 25:092001-1–6.
36. Kubakaddi, S.S. and Bhargavi, K.S. 2010. Enhancement of phonon-drag thermopower in bilayer graphene. *Phys Rev B* 82:155410-1–7.
37. Nissimagoudar, A.S. and Sankeshwar, N.S. 2013. Electronic thermal conductivity and thermopower of armchair graphene nanoribbons. *Carbon* 52:201–8.
38. Bolotin, K.I., Sikes, K.J., Hone, J., Stormer, H.L., and Kim, P. 2009. Temperature dependent transport in suspended graphene. *Phys Rev Lett* 101:096802-1–4.
39. Ochoa, H., Castro, E., Katsnelson, M.I., and Guinea, F. 2012. Scattering by flexural phonons in suspended graphene under back gate induced strain. *Physica E* 44:963–6.
40. Efetov, D.K. and Kim, P. 2010. Controlling electron-phonon interactions in graphene at ultrahigh carrier densities. *Phys Rev Lett* 105:256805-1–4.
41. Sankeshwar, N.S., Vaidya, R.G., and Mulimani, B.G. 2013. Behavior of thermopower of graphene in Bloch–Gruneisen regime. *Phys Stat Sol B* 250:1356–62.
42. Ghosh, S., Nika, D.L., Pokatilov, E.P., and Balandin, A.A. 2009. Heat conduction in graphene: Experimental study and theoretical interpretation. *New J Phys* 11:095012-1–19.
43. Bhargavi, K.S. and Kubakaddi, S.S. 2011. Phonon-drag thermopower in an armchair graphene nanoribbon. *J Phys: Condens Matter* 23:275303-1–5.

44. Kamatagi, M.D., Sankeshwar, N.S., and Mulimani, B.G. 2009. Wide temperature range thermopower in GaAs/AlGaAs heterojunctions. *AIP Conf Proc* 1147:514–20.
45. Hwang, E.H. and Das Sarma, S. 2008. Acoustic phonon scattering limited carrier mobility in two-dimensional extrinsic graphene. *Phys Rev B* 77:115449-1–6.
46. Seol, J.H., Jo, I., Moore, A.L., Lindsay, L., Aitken, Z.H., Pettes, M.T., Li, X. et al. 2010. Two dimensional phonon transport in supported graphene. *Science* 328:213–6.
47. Wang, D. and Shi, J. 2011. Effect of charged impurities on the thermoelectric power of graphene near the Dirac point. *Phys Rev B* 83:113403-1–4.
48. Wu, X., Hu, Y., Ruan, M., Madiomanana, N.K., Berger, C., and de Heer, W.A. 2011. Thermoelectric effect in high mobility single layer epitaxial graphene. *Appl Phys Lett* 99:133102:1–3.
49. Liu, X., Wang, D., Wei, P., Zhu, L., and Shi, J. 2012. Effect of carrier mobility on magnetothermoelectric transport properties of graphene. *Phys Rev B* 86:155414-1–7.
50. Liu, X., Ma, Z., and Shi, J. 2012 Derivative relations between electrical and thermoelectric quantum transport coefficients in graphene. *Solid State Commun* 152:469–72.
51. Xu, X., Wang, Y., Zhang, K., Zhao, X., Bae, S., Heinrich, M., Bui, C.T. et al. 2010. Phonon transport in suspended single layer graphene. http://arxiv.org/pdf/1012.2937v1.pdf.
52. Nam, Y., Sun, J., Lindvall, N., Yang, S. J., Park, C.R., Park, Y.W., and Yurgens, A. 2014. Unusual thermopower of inhomogeneous graphene grown by chemical vapor deposition. *Appl Phys Lett* 104:021902-1–5.
53. Stauber, T., Peres, N.M.R., and Guinea, F. 2007. Electronic transport in graphene: A semiclassical approach including midgap states. *Phys Rev B* 76:205423-1–10.
54. Peres, N.M.R., Lopes dos Santos, J.M.B., and Stauber, T. 2007. Phenomenological study of the electronic transport coefficients of graphene. *Phys Rev B* 76:073412-1–4.
55. Löfwander, T. and Fogelström, M. 2007. Impurity scattering and Mott's formula in graphene. *Phys Rev B* 76:193401-1–4.
56. Cantrell, D.G. and Butcher, P.N. 1987. A calculation of the phonon-drag contribution to the thermopower of quasi-2D electrons coupled to 3D phonons. *J Phys C: Solid State Phys* 20:1985–2003.
57. Herring, C. 1954. Theory of thermoelectric power of semiconductors. *Phys Rev* 96:1163-1–25.
58. Bao, W.S., Liu, S.Y., and Lei, X.L. 2010. Thermoelectric power in graphene. *J Phys: Condens Matter* 22:315502-1–7.
59. Chen, W. and Clerk, A.A. 2012. Electron-phonon mediated heat flow in disordered graphene. *Phys Rev B* 86:125443-1–14.
60. Munoz, E. 2012. Phonon-limited transport coefficients in extrinsic graphene. *J Phys Condens Matter* 24:195302-1–8.
61. Bonaccorso, F., Lombardo, A., Hasan, T., Sun, Z., Colombo, L., and Ferrari, A.C. 2012. Production and processing of graphene and 2D crystals. *Mater Today* 15:564–89.
62. Wierzbowska, M., Dominiak, A., Pizzi, G. Effect of C-face 4H-SiC (0001) deposition on thermopower of single and multilayer graphene in AA, AB and ABC stacking. 2D Mater 1:035002:1–11.
63. Wierzbowska, M. and Dominiak, A. Longitudinal spin-Seebeck effect and huge growth of thermoelectric parameters at band edges in H- and F-doped graphene, free-standing and deposited on 4H-SiC (0001) C–face. *Carbon* 80:255–267.
64. Ghahari, F., Zuev, Y., Watanabe, K., Taniguchi, T., and Kim, P. 2012. Effect of electron-electron interactions in thermoelectric power in graphene. http://meetings.aps.org/link/BAPS.2012.MAR.B6.11.
65. Ghahari, F., Taniguchi, T., Watanabe, K. and Kim, P. 2014. Thermopower of graphene and the validity of Mott's formula. http://meetings.aps.org/link/BAPS.2014.MAR.L51.2.
66. Babichev, A.V., Gasumyants, V.E., and Butko, V.Y. 2013. Resistivity and thermopower of graphene made by chemical vapor deposition technique. *J Appl Phys* 113:076101-1–3
67. Sidorov, A.N., Sherehiy, A., Jayasinghe, R., Stallard, R., Benjamin, D.K., Yu, Q., Liu, Z. et al. 2011. Thermoelectric power of graphene as surface charge doping indicator. *Appl Phys Lett* 99:013115:1–3.
68. Vaidya, R.G., Sankeshwar, N.S., and Mulimani, B.G. 2012. Diffusion thermopower in suspended graphene. In *SPIE Proceedings* Vol. 8549. Singh, V., Katiyar, M., Mazhari, B., Iyer, S.S., Das, U., Dutta, A., Sodhi, R., and Anantharamakrishna, S. (eds.) Washington: SPIE; p. 85491U-1-6, doi: 10.1117/12.925343
69. Mariani, E. and Oppen, F. 2010. Temperature-dependent resistivity of suspended graphene. *Phys Rev B* 82:195403-1–11.
70. Vaidya, R.G., Sankeshwar, N.S., and Mulimani, B.G. 2012. Diffusion thermopower in suspended graphene: Effect of strain. *J Appl Phys* 112:093711-1–6.
71. Li, X., Yin, J., Zhou, J., Wang, Q., and Guo, W. 2012. Exceptional high Seebeck coefficient and gas-flow-induced voltage in multilayer graphene. *Appl Phys Lett* 100:83108:1–3.
72. Sharapov, S.G. and Varlamov, A.A. 2012. Anomalous growth of thermoelectric power in gapped graphene. *Phys Rev B* 86(3):035430-1–5.
73. Patel, A.A. and Mukerjee, S. 2012. Thermoelectricity in graphene: Effects of a gap and magnetic fields. *Phys Rev B* 86:075411-1–5.
74. Woods, C.R., Britnell, L., Eckmann, A., Ma, R.S., Lu, J.C., Guo, H.M., Lin, X. et al. 2014. Commensurate–incommensurate transition in graphene on hexagonal boron nitride. *Nat Phys* 10:451–456.
75. Hao, L. and Lee, T.K. 2010. Thermopower of gapped bilayer graphene. *Phys Rev B* 81:165445-1–8.
76. Bharaghavi, K.S. and Kubakaddi, S.S. 2013. Scattering mechanisms and diffusion thermopower in a bilayer graphene. *Physica E* 52:116–21.
77. Wang, C.R., Lu, W.S., and Lee, W.L. 2010. Transverse thermoelectric conductivity of bilayer graphene in the quantum Hall regime. *Phys Rev B* 82:121406 (R)-1–4.
78. Nam, S.G., Ki, D.K., and Lee, H.J. 2010. Thermoelectric transport of massive Dirac fermions in bilayer graphene. *Phys Rev B* 82:245416-1–5.
79. Wang, C.R., Lu, W.S., Hao, L., Lee, W.L., Lee, T.K., Lin, F., Cheng, I.C., and Chen, J.Z. 2011. Enhanced thermoelectric power in dual-gated bilayer graphene. *Phys Rev Lett* 107:186602-1–4.
80. Chien, Y-Y., Yuan, H., Wang, C-R., Lin, C.-H., and Lee, W-L. 2014. Thermoelectric power in a bilayer graphene device. http://meetings.aps.org/link/BAPS.2014.MAR.C1.55.
81. Fletcher, R., Pudalov, V.M., Feng, Y., Tsaousidou, M., and Butcher, P.N. 1997. Thermoelectric and hot-electron properties of a silicon inversion layer. *Phys Rev B* 56:12422–3.
82. Viljas, J.K. and Heikkila, T.T. 2010. Electron-phonon heat transfer in monolayer and bilayer graphene. *Phys Rev B* 81:245404-1–9.
83. Divari, P.C. and Kliros, G.S. 2010. Modeling the thermopower of ballistic graphene ribbons. *Physica E* 42:2431–5.
84. Xing, Y., Sun, Q., and Wang, J. 2009. Nernst and Seebeck effects in a graphene nanoribbon. *Phys Rev B* 80:235411-1–8.
85. Ouyang, Y. and Guo, J. 2009. A theoretical study on thermoelectric properties of graphene nanoribbons. *Appl Phys Lett* 94:263107-1–3.

86. Karamitaheri, H., Neophytou, N., Pourfath, M., Faez, R., and Kosina, H. 2012. Engineering enhanced thermoelectric properties in zigzag graphene nanoribbons. *J Appl Phys* 111:054501-1–9.
87. Zheng, H., Liu, H.J., Tan, X.J., Lv, H.Y., Pan, L., and Shi, J. 2012. Enhanced thermoelectric performance of graphene nanoribbons. *Appl Phys Lett* 100:093104-1–5.
88. Mazzamuto, F., Hung-Nguyen, V., Apertet, Y., Caer, C., Chassat, C., Saint-Martin, J., and Dollfus, P. 2011. Enhanced thermoelectric properties in graphene nanoribbons by resonant tunneling of electrons. *Phys Rev B* 83:235426-1–7.
89. Lin, Y., Perebeinos, V., Chen, Z., and Avouris, P. 2008. Electrical observation of subband formation in graphene nanoribbons. *Phys Rev B* 78:161409(R)-1–4.
90. Tsaousidou, M. 2010. Theory of phonon-drag thermopower of extrinsic semiconducting single-wall carbon nanotubes and comparison with previous experimental data. *Phys Rev B* 81:235425-1–9.
91. Kubakaddi, S.S. 2007. Electron–phonon interaction in a quantum wire in the Bloch-Gruneisen regime. *Phys Rev B* 75:075309-1–7.
92. Vavro, J., Llaguno, M.C., Fischer, J.E., Ramesh, S., Saini, R.K., Ericson, L.M., Davis, V.A., Hauge, R.H., Pasquali, M., and Smalley, R.E. 2003.Thermoelectric power of p-doped single-wall carbon nanotubes and the role of phonon drag. *Phys Rev Lett* 90: 065503-1–4.
93. Scarola, V.W. and Mahan, G.D. 2002. Phonon drag effect in single-walled carbon nanotubes. *Phys Rev B* 66:205405-1–7.
94. Baker, A.M.R., Alexander-Webber, J.A., Altebaeumer, T., McMullan, S.D., Janssen, T.J. B.M., Tzalenchuk, A., Lara-Avila, S. et al. 2013. Energy loss rates of hot Dirac fermions in epitaxial, exfoliated and CVD graphene. *Phys Rev B* 87:0454014-1–6.
95. Sugihara, K., Hishiyama, Y., and Ono, A. 1986. Low-temperature anomalies in the thermoelectric power of highly oriented graphite. *Phys Rev B* 34:4298–303.
96. Hewitt, C.A., Kaiser, A.B., Craps, M., Czerw, R., Roth, S., and Carroll, D.L. 2013. Temperature dependent thermoelectric properties of free standing few layer graphene/polyvinylidene fluoride composite thin films. *Synth Metals* 165:56–9.
97. Sim, D., Liu, D., Dong, X., Xiao, N., Li, S., Zhao, Y., Li, L., Yan, Q., and Hng, H.H. 2011. Power factor enhancement for few-layered graphene films by molecular attachments. *J Phys Chem C* 115:1780–5.
98. Xiao, N., Dong, X., Song, L., Liu, D., Tay, Y.Y., Wu, S., Li, L. et al. 2011. Enhanced thermopower of graphene films with oxygen plasma treatment. *ACS Nano* 5:2749–55.
99. Xu, X., Gabor, N.M., Alden, J.S., van der Zande, A.M., and McEuen, P.L. 2010. Photo-thermoelectric effect at a graphene interface junction. *Nano Lett* 10:562–6.
100. Sidorov, A.N., Gaskill, K., Nardelli, M.B., Tedesco, J.L., Myers-Ward, R.L., Eddy, C.R Jr., Jayasekera, T. et al. 2012. Charge transfer equilibria in ambient-exposed epitaxial graphene on (0001) 6H-SiC. *J Appl Phys* 111:113706-1–6.
101. Park, J., He, G., Feenstra, R., and Li, A-P. 2014. Spatially-resolved thermopower of graphene: Role of boundary and defects. http://meetings.aps.org/link/BAPS.2014.MAR.L37.14.
102. Yeo, P.S.E., Sullivan, M.B., Loh, K.P., and Gan, C.K. 2013. First-principles study of the thermoelectric properties of strained graphene nanoribbons. *J Mater Chem A* 1:10762–7
103. Hinz, A.P., Kettemann, S., and Mucciolo, E.R. 2014. Quantum corrections to thermopower and conductivity in graphene. *Phys Rev* B 89:075411-1–21.
104. Li, K.M., Xie, Z.-H., Su, K.-L., Luo, W.-H., and Zhang, Y. 2014. Ballistic thermoelectric properties in double-bend graphene nanoribbons. *Phys Lett A* 378:1383–7.
105. Yokomizo, Y. and Nakamura, J. 2013. Giant Seebeck coefficient of the graphene/h-BN superlattices. *Appl Phys Lett* 103:113901-1–4.
106. Chang, P.H., Bahramy, M.S., Nagaosa, N. and Nikolic, B.K. 2014. Giant thermoelectric effect in graphene-based topological insulators with nanopores. *Nano Lett* 14:3779–3784.
107. McKitterick, C., Vora, H., Du, X., Rooks, M., and Prober, D. 2014. Electron-phonon coupling in bilayer and single-layer graphene at sub-Kelvin temperatures. http://meetings.aps.org/link/BAPS.2014.MAR.A30.10.

Section II

Optical Properties

19 Optical Properties of Graphene

Adam Mock

CONTENTS

ABSTRACT

This chapter presents a numerical method for simulating the electromagnetic field interaction with single-layer graphene. The numerical approach is the finite-difference time-domain (FDTD) method. The linear electric material response of graphene is dispersive and demands special care for implementation using a time-domain method. Graphene's electric conductivity arises from two physical mechanisms: (1) intraband electronic transitions and (2) interband electronic transitions. The intraband electronic transitions are well described by a Drude-type frequency model that is straightforward to implement in FDTD using the auxiliary differential equation (ADE) method. The second (interband) contribution does not have a form directly amenable to the ADE technique. This chapter presents a fitting approach based on Padé interpolation that facilitates incorporation of the interband conductivity term into the FDTD method using ADE. The method is used to model the surface plasmon polariton modes of finite-width graphene waveguides. Highly doped graphene is modeled, so that the surface plasmon modes propagate at near-infrared wavelengths. Both the propagation loss and waveguide dispersion are provided.

19.1 INTRODUCTION

Graphene is a two-dimensional crystalline solid composed of a planar single-atom-thick hexagonal arrangement of carbon atoms [1,2]. Graphene has a conical electronic band structure with zero band gap. The electrons in graphene behave as massless Dirac fermions with ballistic transport lengths of hundreds of nanometers even at room temperature. Its electron mobility is weakly dependent on temperature and is still limited by impurity scattering at room temperature. The massless Dirac fermions are relativistic with velocities on the order of $0.01c$ [3].

In addition to its large mechanical strength [4] and thermal conductivity [5], graphene's electronic properties give rise to novel light–matter interactions that have important implications for optical technologies [6–8]. Graphene has the potential to impact photonic signal modulation, transmission, and detection technologies. Graphene has a high optical damage threshold of 3×10^6 MW/cm^2 [9], and many of graphene's optical properties are constant over large-wavelength ranges [10,11]. Hybrid Si–graphene electroabsorption near-infrared modulators with modulation rates as high as 1 GHz and a 25-μm^2 footprint have been demonstrated [12]. Quasi-transverse electromagnetic modes of graphene-based parallel-plate waveguides that operate at terahertz frequencies [13] have been described theoretically. Graphene-based photodetectors capable of operating frequencies as high as 40 GHz at a wavelength of 1.55 μm have been demonstrated, and the authors estimate that the peak bandwidth of a graphene photodetector could be as large as 500 GHz [14]. Subsequent work demonstrated error-free detection at 10 Gbps [15]. Theoretical description [16] and experimental demonstration [17] of microcavity-enhanced graphene photodetection has been carried out. Optical bistability, regenerative oscillations, and four-wave mixing have been experimentally observed in a silicon photonic crystal microcavity with a single-layer graphene top coating [18]. The effect of a graphene layer on a photonic metamaterial has been investigated; a single layer of graphene over a split-ring resonator metamaterial resulted in resonance shifts and transmission enhancement when compared to devices without graphene [19,20]. Bao et al. [21] demonstrated graphene-based in-line fiber polarizers with 27-dB extinction ratio. The transverse electric (TE)-pass functionality of their polarizer was enabled by the conical electronic band structure of graphene. In addition to a unique conical electronic band structure, the chemical potential of graphene is tunable via external direct current (DC) electric or magnetic fields and via chemical doping [22,23]. The control of the chemical potential allows one to frequency tune the optical absorption and spectral features of the permittivity of graphene. The nonlinear-saturable absorption properties of graphene have been intensely studied recently in demonstrations of passively mode-locked lasers emitting around 1.5 μm with pulse durations shorter than 1 ps [24–36]. The advantages that graphene-saturable absorbers have over semiconductor-saturable absorber mirrors include high damage threshold, broadband operation, and low cost.

This chapter presents advances in time-domain numerical modeling of surface plasmon polariton (SPP) modes [37] on single-layer graphene using the finite-difference time-domain (FDTD) method [38,39]. Utilization of graphene as a plasmonic material has received significant attention recently [40–42]. Vakil and Engheta [43,44] discussed metamaterials and transformation optics in which the SPP properties of highly doped graphene play a key role. Noble metals such as gold or silver have dominated plasmonics research [45,46] due to their delocalized electron distributions that behave like a plasma. The electron plasma gives rise to a negative real part of the electric permittivity for frequencies below the plasma frequency. Under these conditions, a collective electron oscillation can couple to an external electromagnetic wave to form a highly localized surface wave. Using graphene for SPP devices instead of noble metals is advantageous because the material properties of graphene are tunable via an external electric or magnetic field, and it supports SPPs with longer decay times. The longer decay time is due to the ability to produce defect-free graphene layers yielding ballistic electron transport. The tunability of graphene via external electric fields could facilitate the "transfer" of a metal electronic circuit pattern to a graphene-based plasmonic circuit.

Progress has been made on the analytical and numerical treatment of the SPP modes of graphene [47,48]. Hanson analyzed the modes using dyadic Green's functions in terms of Sommerfeld integrals. Jablan et al. [49] calculated the SPP dispersion and loss using the random-phase approximation to describe the electric polarizability of graphene. Nikitin et al. [50] elucidated the connection between infinite sheet modes and edge modes in graphene ribbon SPP waveguides. Christensen et al. [51] discussed hybridized modes associated with multiple graphene sheets. Theoretical proposals have been published for graphene SPP modulators operating at 10 μm whose maximum speed is limited by a 0.1-ps carrier relaxation time [52]. Recently, graphene-resonant filtering devices have been theoretically investigated for compact optical signal processing [53–55]. Ren et al. [56] proposed tunable nanoantennas based on graphene.

Kim and Choi [57] experimentally demonstrated transverse magnetic (TM) SPP propagation in single-layer graphene strips. Yan et al. [58] fabricated graphene–insulator stacks for tunable far-infrared SPP notch filters. Yeung et al. [59] numerically and experimentally explored a periodic nanopatterned graphene for plasmonic band engineering.

Clearly, significant research effort has resulted in impressive advances in a variety of graphene-based photonic technologies. Whether the light–graphene interaction comes in the form of metamaterials, plasmonics, modulators, or detectors, the tight submicrometer field confinement and complicated vectorial field distribution will demand accurate full-wave modeling tools. One of the most popular numerical techniques is the FDTD method [38,39] which is an explicit discretization of Maxwell's equations in space and time. FDTD is attractive due to its generality, ease of implementation, linear scaling in execution time with problem size, straightforward parallelizability, favorable speedup with parallelization, and ability to handle dispersive and nonlinear materials. It is often the default method when modeling large three-dimensional irregular geometries.

The unique electronic properties of graphene give rise to a highly dispersive electromagnetic frequency response [60–62]. The frequency response is related to the energy-dependent electron conductivity. In the frequency domain, the relationship between the electric field and current density for isotropic materials is given by $\vec{J}(\vec{r},\omega) = \sigma(\vec{r},\omega)\vec{E}(\vec{r},\omega)$ where $\sigma(\vec{r},\omega)$ is the spatially varying volume conductivity with units of siemens per meter. Because FDTD solves Maxwell's equations in the time domain, the relationship between $\vec{J}(\vec{r},t)$ and $\vec{E}(\vec{r},t)$ is a convolution

$$\vec{J}(\vec{r},t) = \int_0^t \sigma(\vec{r},t-t')E(t')\,dt' \qquad (19.1)$$

While piecewise linear recursive convolution approaches [63] incorporate dispersive conductivity starting from Equation 19.1, the auxiliary differential equation (ADE) method is typically more efficient and simpler to implement [39,64,65]. An attractive feature of the ADE method is that it incorporates the conductivity spectrum into the time-domain simulation without having to find the inverse Fourier transform $\sigma(\vec{r},t) = F^{-1}[\sigma(\vec{r},\omega)]$ where F^{-1} denotes the inverse Fourier transform. A limitation of the ADE method is that it requires that the dispersive conductivity be represented as a ratio of polynomials in ω. Popular forms include the Drude [66], Debye [64], Lorentz [65], and Cole [67] models.

This chapter considers the frequency-dependent conductivity as arising from two components: intraband relaxation due to scattering from phonons, electrons, and impurities and interband relaxation due to electron–hole recombination. While the intraband relaxation contribution to graphene's conductivity can be represented using the Drude model alone, the interband transition component results in a dispersive conductivity function that cannot be directly incorporated into FDTD using the ADE method. In this chapter, a Padé approximant [39,68–70] is used to represent the interband term in a form that can be incorporated into FDTD. The general recipe is described for performing the fit, fitting parameters are given, and modeling results of SPPs in graphene are presented.

Lin et al. [71] fitted a similar function to the interband conductivity term but used vector-fitting techniques instead of the recipes outlined in this chapter. Other than the mechanics of the fitting, these techniques are similar in nature.

Other recent work has modeled an infinite graphene sheet using FDTD [72–76]; however, only the intraband (Drude) conductivity was used. While neglecting the interband term may provide good results for long wavelengths, it must be included when considering optical behavior with optical energies near the chemical potential. Ahmed et al. [77] also modeled graphene using only the intraband conductivity component, but they use a locally one-dimensional FDTD method to resolve the small atomic-size thickness of graphene while also maintaining a reasonably large FDTD time step.

This chapter presents a form for the interband conductivity term that is especially important for near-infrared SPP modeling in doped graphene. FDTD is an extremely powerful tool for nanophotonic modeling [78–81], and the results presented here will further enable the discovery and design of photonic devices that exploit the exciting properties of graphene.

19.2 GRAPHENE CONDUCTIVITY

The frequency-dependent electric surface conductivity spectrum of graphene is complex valued and displayed in Figure 19.1a for values of the chemical potential ranging from 0.2 to 0.6 eV. The conductivity in Figure 19.1 is a surface conductivity with units of Siemens. The model for graphene's electric conductivity arises from the Kubo formalism [60–62,82,83] and, in the absence of an applied electric or magnetic field, is given by

$$\sigma(\omega,\mu_c,\tau,T)=\frac{je^2/\pi\hbar^2}{\omega-j2/\tau}\int_0^\infty \epsilon\left(\frac{\partial f_d(\epsilon)}{\partial\epsilon}-\frac{\partial f_d(-\epsilon)}{\partial\epsilon}\right)d\epsilon + je^2/\pi\hbar^2\left(\omega-j(2/\tau)\right)\int_0^\infty \frac{f_d(\epsilon)-f_d(-\epsilon)}{(\omega-j2/\tau)^2-4(\epsilon/\hbar)^2}d\epsilon \tag{19.2}$$

where μ_c is the chemical potential, $1/\tau$ is the carrier-scattering rate, T is temperature, $f_d(\epsilon)=1/(e^{(\epsilon-\mu_c)/k_BT}+1)$ is the Fermi–Dirac distribution function, and k_B is the Boltzmann constant. The spatial dependence has been suppressed for brevity but arises through spatial variation of $\mu_c(\vec{r})$ and $\tau(\vec{r})$. The first integral in Equation 19.2 describes the intraband carrier relaxation contribution, and the second integral describes the interband carrier transition contribution. The first integral can be evaluated resulting in

$$\sigma_{\text{intra}}(\omega,\mu_c,\tau,T)=-j\frac{8\sigma_0 k_BT/h}{\omega-j2/\tau}\left[\frac{\mu_c}{k_BT}+2\ln(e^{-\mu_c/k_BT}+1)\right] \tag{19.3}$$

where $\sigma_0=(\pi e^2/2h)$. For the case $k_BT \ll |\mu_c|,\hbar\omega$, the second integral may be approximated by

$$\sigma_{\text{inter}}(\omega,\mu_c,\tau,T)\approx -j\frac{\sigma_0}{\pi}\ln\left[\frac{2|\mu_c|-(\omega-j2/\tau)\hbar}{2|\mu_c|+(\omega-j2/\tau)\hbar}\right] \tag{19.4}$$

Experimental measurements of room-temperature, undoped graphene have confirmed that in the frequency regime dominated by interband transitions, the real part of the optical conductivity is approximately constant with a value of $e^2/4\hbar$ [10,84]. Mak et al. [11] were able to fit Equations 19.3 and 19.4 to frequency-dependent optical absorbance measurements using μ_c in the range of 0.1–0.2 eV and τ in the range of 6.6–66 fs.

19.3 GRAPHENE ELECTRIC PERMITTIVITY

From the time-harmonic Maxwell equations, the relationship between volume conductivity and electric permittivity is given by

$$\epsilon(\omega)=\epsilon_r(\omega)-i(\sigma(\omega)/\omega) \tag{19.5}$$

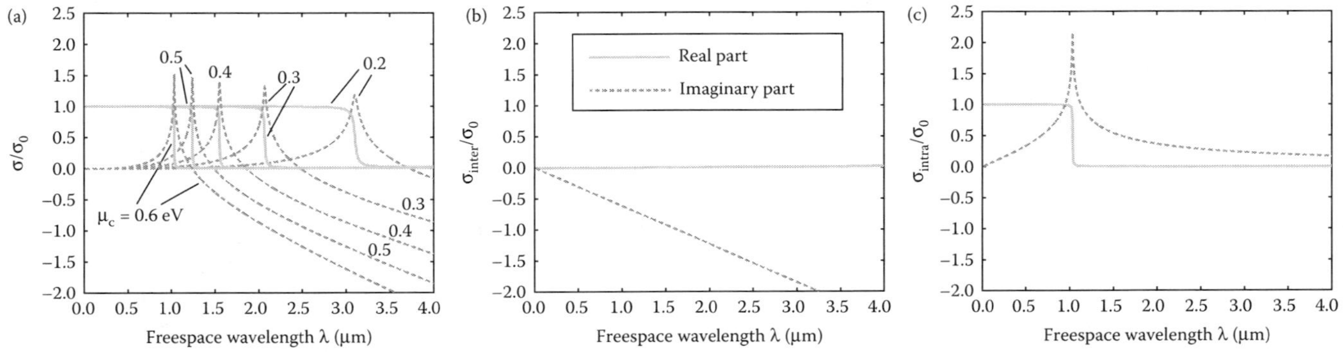

FIGURE 19.1 (a) Real and imaginary parts of the total surface electric conductivity of graphene versus free-space wavelength normalized to $\sigma_0=(\pi e^2/2h)$ for values of the chemical potential ranging from 0.2 to 0.6 eV. (b) Real and imaginary parts of intraband surface electric conductivity of graphene versus free-space wavelength normalized to $\sigma_0=(\pi e^2/2h)$ for the chemical potential of 0.6 eV. (c) Real and imaginary parts of interband surface electric conductivity of graphene versus free-space wavelength normalized to $\sigma_0=(\pi e^2/2h)$ for the chemical potential of 0.6 eV. All plots use $T=300$ K and $\tau=0.5$ ps.

where $\epsilon_r(\omega)$ is the real part of the electric permittivity and $-\sigma(\omega)/\omega$ is the imaginary part [85]. However, if the conductivity $\sigma(\omega) = \sigma_r(\omega) + i\sigma_i(\omega)$ is complex, then the real and imaginary parts of the electric permittivity are given by

$$\epsilon(\omega) = \epsilon_r(\omega) + \frac{\sigma_i(\omega)}{\omega} - i\left(\frac{\sigma_r(\omega)}{\omega}\right) \tag{19.6}$$

This demonstrates that a negative imaginary part of the conductivity $\sigma_i(\omega)$ will result in a negative real part of the electric permittivity when $\epsilon_r(\omega) + \sigma_i(\omega)/\omega < 0$. In this case, the graphene layer behaves like a thin metal capable of supporting SPP waves.

The conductivity of graphene is expressed as a surface conductivity which is appropriate due to the two-dimensional crystalline structure of graphene. However, graphene does have a volume with a thickness on the order of atomic dimensions, and its volume conductivity is given by $\sigma_v = \sigma_s/\delta$ where δ is the atom-size graphene thickness, σ_v is the volume conductivity that appears in Equations 19.5 and 19.6, and σ_s is the surface conductivity discussed in the previous section and displayed in Figure 19.1. Because a volume conductivity will be used in FDTD simulations, an appropriate value of δ must be determined. For electromagnetic simulations, the appropriate value for δ depends on whether the field distribution can "see" the thickness of graphene. The practical implication is that δ should be chosen smaller than the smallest spatial electromagnetic field variation. For SPP, the electromagnetic field is extremely localized and can vary over tens of nanometers. Previous FDTD simulation of graphene has used $\delta = \Delta$ on the order of 1–2 nm where Δ is the spatial discretization used in the FDTD simulation [20,43,86].

As displayed in Figure 19.1, graphene's surface conductivity is on the order of $\sigma_0 = (\pi e^2/2h) = 6.08 \times 10^{-5}$ S. Using this value to (incorrectly) determine whether $\epsilon_r(\omega) + \sigma_i(\omega)/\omega < 0$ at a wavelength of 1.55 μm, results in $\sigma_i(\omega)/\omega = \sigma_0(1.55\ \mu)\text{m}/(2\pi c) = 5.00 \times 10^{-20}$ F where c is the vacuum speed of light. This number is much smaller than $\epsilon_r = \epsilon_0 = 8.854 \times 10^{-12}$ F/m. The correct calculation incorporates δ; for $\delta = 1$ nm, $\sigma_i(\omega)/\omega = \sigma_0\ (1.55\ \mu)\text{m}/(2\pi c)/\delta = 5.00 \times 10^{-11}$ F/m which is significantly larger than ϵ_0. It should be noted that $\delta = 1$ nm is chosen larger than typical atomic dimensions to accommodate reasonable FDTD simulations. A true value for δ would be an order of magnitude smaller that would yield an even larger value for graphene's volume conductivity. Because of this, the $\epsilon_r(\omega)$ in $\epsilon_r(\omega) + \sigma_i(\omega)/\omega < 0$ can be ignored, and the condition for TM SPP waves is $\sigma_i(\omega) < 0$.

From this chapter, it may appear that the value for volume conductivity of graphene is not well defined since it depends on the arbitrary choice of δ. However, the micrometer-scale variation of the electromagnetic waves see only macroscopic averages of microscopic charge and current distributions [87]. In FDTD, each spatially discretized cell represents a spatial average of the material properties within that cell. Therefore, using σ_s/δ may be interpreted as assigning a nanometer-length FDTD cell a volume conductivity that is a smeared-out version of the surface conductivity. The net effect of this subwavelength material distribution on the properties of the electromagnetic wave is then weighted by the size of the cell similar to the confinement factor in semiconductor lasers [88]. If only a single spatial grid cell is used to represent the graphene region, then, the net effect of graphene's material properties on the electromagnetic wave is $(\sigma_s/\delta) \times \delta = \sigma_s$, and we see that δ cancels. This shows that the choice of δ does not affect the accuracy of FDTD simulations of graphene so long as δ is smaller than the spatial variation of the electromagnetic wave. Alternatively, the surface conductivity of graphene can be directly incorporated into FDTD simulations using conformal techniques [39,72]. This approach is potentially more accurate since it handles the boundary condition between graphene and the surrounding medium more carefully; however, it introduces complexity of implementation for irregular geometries.

19.4 DOPED GRAPHENE

In this chapter, highly doped graphene is considered with $\mu_c = 0.6$ eV. With this chemical potential, the condition $k_B T \ll |\mu_c|, \hbar\omega$ is satisfied, so that the assumption leading to Equation 19.4 holds even at room temperature $T = 300$ K. A useful property of graphene is the tunability of its chemical potential with an external DC electric or magnetic field. However, chemical doping can permanently tune graphene for long-term device operation assuming that once graphene is chemically prepared and doped, it is isolated from the environment. The relationship between the carrier density n and chemical potential is given by

$$n = \frac{2}{\pi\hbar^2 v_F^2} \int_0^\infty \epsilon[f_d(\epsilon) - f_d(\epsilon + 2\mu_c)]d\epsilon \tag{19.7}$$

where the Fermi velocity v_F has a value around 10^6 m/s. Previous discussions of graphene plasmonics dealt with chemical potential values less than or equal to 0.2 eV [49,83,89]. These small values of the chemical potential are consistent with low-to-moderate levels of chemical doping in graphene. Because the imaginary part of the conductivity is negative for optical excitations less than 2 times the chemical potential, these SPP devices are expected to function only for wavelengths longer than 3 μm. The FDTD simulations presented in this chapter are envisioned to be applicable to densely integrated photonic circuits for use in high-capacity digital communication systems. The most common wavelength used in these systems is 1.55 μm due to the absorption minimum in standard fiber-optic cables. For SPP waves to exist at 1.55 μm, the chemical potential of graphene will need to be at least 0.5 eV.

Zhao et al. [90] reported one example of how such a large doping level may be achieved in graphene. Graphene flakes were submerged in solutions of 14 M concentration of H_2SO_4 and the doping concentration was monitored *in situ* using

Raman spectroscopy. They achieved carrier concentrations of 0.2–0.4 × 10^{13} cm^{-2} in bilayer graphene which is the predicted carrier concentration necessary for near-infrared graphene plasmonics [49]. Once graphene is doped to support SPP waves near 1.55 μm, external electric and magnetic fields can still be used for dynamic tuning in switching or modulating applications [43].

Figure 19.1a displays the real and imaginary parts of the conductivity of graphene normalized to σ_0 for different values of the chemical potential (μ_c) ranging from 0.2 to 0.6 eV. The imaginary part of the conductivity passes from positive to negative when $\hbar\omega < 2\mu_c$. As μ_c is increased to values exceeding 0.5 eV, the imaginary part of σ becomes negative for wavelengths longer than 1.55 μm. Figure 19.1b and c display the real and imaginary parts of the intraband and interband contributions to graphene's conductivity for $\mu_c = 0.6$ eV. The real part of the intraband conductivity is small throughout the wavelength range of 0–4 μm and gradually increases for large wavelengths. The imaginary part of σ_{intra} is negative and significant in magnitude throughout the wavelength range of 0–4 μm. This term is the primary contributor to making the real part of the permittivity negative. The real part of the interband conductivity exhibits a step-function-like behavior with a transition at a wavelength corresponding to 2 μ_c. In Figure 19.1c, it is significant for $\lambda < 1$ μm, but becomes unimportant for larger wavelengths. The imaginary part of the interband conductivity has the most complicated behavior exhibiting a singularity-like peak at $\lambda = hc/2/\mu_c$. This term also contributes to the real part of the electric permittivity and is important for modeling SPP wave propagation. From these observations, it is clear that incorporation of both the intraband and interband conductivities is crucial for the accurate representation of graphene in FDTD simulations.

19.5 SCATTERING RATE AND TEMPERATURE DEPENDENCE

Figure 19.2 displays the real and imaginary parts of the conductivity of graphene with $\mu_c = 0.6$ eV and $T = 300$ K for scattering lifetimes ranging from 0.01 to 5 ps. One sees that the qualitative features of the conductivity spectrum remain intact; however, the spectrum is smeared as τ decreases. Figure 19.2c and d show zoomed-in views of Figure 19.2a and b where the smoothing of sharp features is clear. Also, one notices that the real part of the conductivity increases as τ decreases for wavelengths longer than 1.1 μm. The real part of the conductivity causes heating (ohmic) losses and results in attenuation of SPP waves. Table 19.1 displays the range of carrier-scattering lifetimes reported in the literature. In this chapter, a value of 0.5 ps will be used.

Assuming that the condition leading to Equation 19.4, $k_BT \ll |\mu_c|, \hbar\omega$, holds, the only temperature dependence in

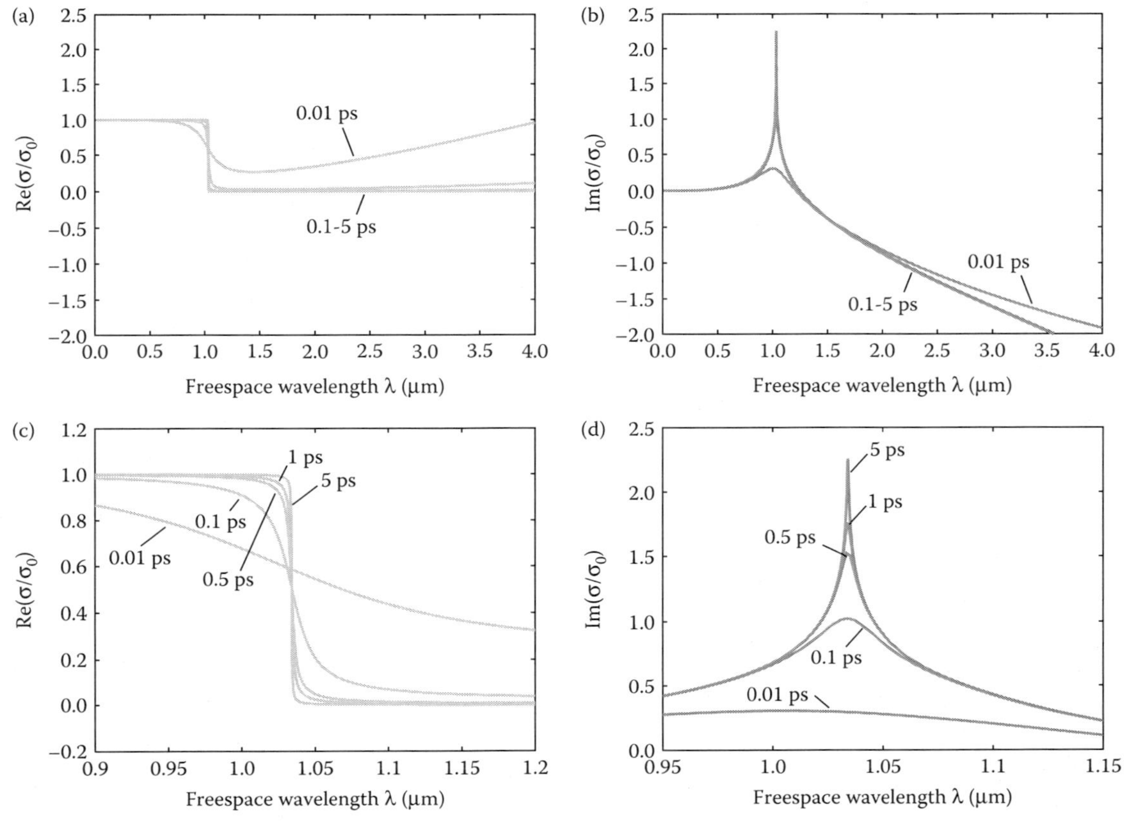

FIGURE 19.2 (a) Real and (b) imaginary parts of the total (intraband plus interband) conductivity for different values of the scattering lifetime (c), (d) is a zoom in of (a), (b), respectively.

TABLE 19.1
Reported Values for the Scattering Lifetime

Reference	τ (ps)	Comments
[103]	1.1	Calculated from DC conductivity measurement
[104]	0.35	Obtained from optical transmission measurement
[105]	0.33	Obtained from optical transmission measurement
[106]	0.5	Used in theoretical treatment
[86]	0.38	Used in theoretical treatment
[71]	0.5	Used in theoretical treatment
[49]	0.64	Used in theoretical treatment
[11]	0.2–0.4	Used in theoretical treatment

this model appears in the intraband term (Equation 19.3). Under the condition $k_BT \ll |\mu_c|$, $\ln(e^{-\mu_c/k_BT}+1) \approx \ln(1) = 0$ and

$$\sigma_{\text{intra}}(\omega,\mu_c,\tau,T) \approx -j\frac{8\sigma_0/h\mu_c}{\omega - j2/\tau} \tag{19.8}$$

This shows that the frequency dispersion of graphene's conductivity is stable under temperature variations over a wide range of temperatures. This is consistent with the extremely high electron mobility in graphene even at room temperature [41].

19.6 PADÉ FIT TO INTERBAND CONDUCTIVITY TERM

The primary objective of this chapter is to incorporate the linear electromagnetic properties of graphene into FDTD simulations. As discussed in Section 19.1, the ADE method is a popular and efficient way of modeling frequency-dispersive materials in the time domain. However, the ADE method is based on the Fourier transform property $d/dt \leftrightarrow j\omega$, so that a frequency dispersion expressed in terms of a ratio of polynomials in ω can be inversely Fourier transformed to the time domain yielding time derivatives of various electromagnetic field quantities that can be discretized. The intraband conductivity term is already expressed as a ratio of polynomials in ω

$$\sigma_{\text{intra}}(\omega,\mu_c,\tau,T) = -j\frac{8\sigma_0 k_BT/h}{\omega - j2/\tau}\left[\frac{\mu_c}{k_BT} + 2\ln(e^{-\mu_c/k_BT}+1)\right]$$
$$= \frac{K}{j\omega + 2/\tau} \tag{19.9}$$

where K is the frequency-independent factor

$$K = \frac{8\sigma_0 k_BT}{h}\left[\frac{\mu_c}{k_BT} + 2\ln(e^{-\mu_c/k_BT}+1)\right] \tag{19.10}$$

To illustrate the ADE method, consider for the moment just the intraband term (Equation 19.9) in the relation $\vec{J}(\vec{r},\omega) = \sigma_{\text{intra}}(\vec{r},\omega)\vec{E}(\vec{r},\omega)$

$$\vec{J}(\vec{r},\omega) = \frac{K}{j\omega + 2/\tau}\vec{E}(\vec{r},\omega) \tag{19.11}$$

$$\left(j\omega + \frac{2}{\tau}\right)\vec{J}(\vec{r},\omega) = K\vec{E}(\vec{r},\omega) \tag{19.12}$$

Inverse Fourier transforming Equation 19.12 involves substituting $\partial/\partial t$ for $j\omega$

$$\left(\frac{\partial}{\partial t} + \frac{2}{\tau}\right)\vec{J}(\vec{r},t) = K\vec{E}(\vec{r},t) \tag{19.13}$$

Considering only the x component of $\vec{J}$ and $\vec{E}$ along the z direction, the discretized version of Equation 19.13 is

$$\frac{J_x(i\Delta z,(n+1/2)\Delta t) - J_x(i\Delta z,(n-1/2)\Delta t)}{\Delta t}$$
$$+\frac{2J_x(i\Delta z,n\Delta t)}{\tau} = KE_x(i\Delta z,n\Delta t) \tag{19.14}$$

Because J_x and E_x are sampled at the same time, the 1/2-shift time step in Equation 19.14 is handled with the average $J_x(i\Delta z, (n+1/2)\Delta t) = (J_x(i\Delta z, (n+1)\Delta t) + J_x(i\Delta z, n\Delta t))/2$. The resulting update equation for $J_x(i\Delta z, (n+1)\Delta t)$ is

$$J_x(i\Delta z,(n+1)\Delta t) = -4\frac{\Delta t}{\tau}J_x(i\Delta z,n\Delta t)$$
$$+J_x(i\Delta z,(n-1)\Delta t) + 2\Delta t K E_x(i\Delta z,n\Delta t) \tag{19.15}$$

The discretized equation for J_x is time stepped in parallel with the usual discretized Maxwell equations. Because of its supporting role to the Maxwell equations, Equation 19.15 is referred to as an ADE. This chapter illustrates how the intraband conductivity term (in a Drude form) is incorporated into FDTD. Performing a similar procedure for the interband conductivity term is difficult because Equation 19.4 is not a ratio of polynomials in $j\omega$. To address this issue, a Padé approximant is fitted to the frequency spectrum $\sigma_{\text{inter}}(\omega)$. A Padé approximant fits a rational polynomial of the form

$$\frac{a_0 + a_1\omega + \cdots + a_M\omega^M}{1 + b_1\omega + \cdots + b_N\omega^N} = \sigma_{\text{inter}}(\omega) \tag{19.16}$$

to a set of data points or samples [39,68–70]. In this chapter, a polynomial of the form

$$\frac{a_0 + a_1 j\omega + a_2(j\omega)^2}{1 + b_1 j\omega + b_2(j\omega)^2} = \sigma_{\text{inter}}(\omega) \tag{19.17}$$

is used. Substituting $j\omega$ for ω preserves the required conjugate symmetry ($\sigma(\omega) = \sigma^*(-\omega)$) in the Padé fit assuming that a_i and b_i are real. The conjugate symmetry condition ensures a purely real impulse response. Polynomials of order $M = N = 2$ in the numerator and denominator lead to a straightforward and efficient FDTD implementation. The use of higher-order polynomials did not result in significantly improved fits. The fitting procedure consists of moving the denominator polynomial to the right side and then moving all terms containing ω back to the left as shown below

$$a_0 + a_1 j\omega + a_2(j\omega)^2 - b_1 j\omega\sigma_{\text{inter}}(\omega) - b_2(j\omega)^2\sigma_{\text{inter}}(\omega) = \sigma_{\text{inter}}(\omega) \tag{19.18}$$

This procedure results in a linear equation in the expansion coefficients a_i and b_i. To enforce purely real expansion coefficients, Equation 19.18 must be separated into its real and imaginary parts according to

$$a_0 - a_2\omega^2 + b_1\omega\text{Im}[\sigma_{\text{inter}}(\omega)] + b_2\omega^2\text{Re}[\sigma_{\text{inter}}(\omega)] = \text{Re}[\sigma_{\text{inter}}(\omega)] \tag{19.19}$$

and

$$a_1\omega - b_1\omega\text{Re}[\sigma_{\text{inter}}(\omega)] + b_2\omega^2\text{Im}[\sigma_{\text{inter}}(\omega)] = \text{Im}[\sigma_{\text{inter}}(\omega)] \tag{19.20}$$

Letting $\gamma(\omega) = \text{Re}[\sigma_{\text{inter}}(\omega)]$ and $\eta(\omega) = \text{Im}[\sigma_{\text{inter}}(\omega)]$, the resulting matrix equation is

$$\begin{bmatrix} 1 & 0 & -\omega_1^2 & \omega_1\eta(\omega_1) & \omega_1^2\gamma(\omega_1) \\ 0 & \omega_1 & 0 & -\omega_1\gamma(\omega_1) & \omega_1^2\eta(\omega_1) \\ 1 & 0 & -\omega_2^2 & \omega_2\eta(\omega_2) & \omega_2^2\gamma(\omega_2) \\ 0 & \omega_2 & 0 & -\omega_2\gamma(\omega_2) & \omega_2^2\eta(\omega_2) \\ 1 & 0 & -\omega_3^2 & \omega_3\eta(\omega_3) & \omega_3^2\gamma(\omega_3) \end{bmatrix} \begin{bmatrix} a_0 \\ a_1 \\ a_2 \\ b_1 \\ b_2 \end{bmatrix} = \begin{bmatrix} \gamma(\omega_1) \\ \eta(\omega_1) \\ \gamma(\omega_2) \\ \eta(\omega_2) \\ \gamma(\omega_3) \end{bmatrix}$$

where ω_1, ω_2, and ω_3 are three distinct frequency values at which $\sigma_{\text{inter}}(\omega)$ is sampled. Figure 19.3 displays the real and imaginary parts of the resulting Padé fit to the interband conductivity term associated with $\mu_c = 0.6$ eV. In Figure 19.3a and b, the three wavelengths used for the fit were 1.1, 1.45, and 1.8 μm. The match between the Padé fit and the exact formulas is excellent for wavelengths longer than 1.1 μm. The real

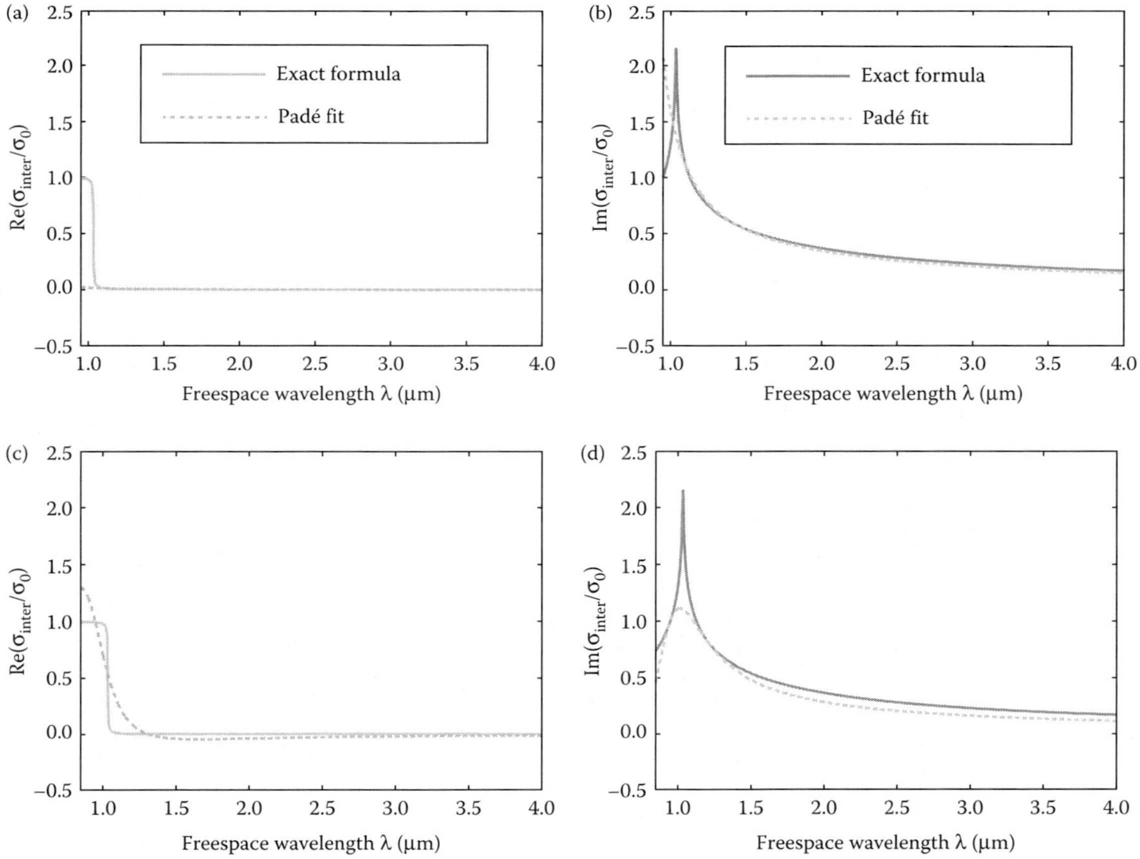

FIGURE 19.3 Comparison between Padé fit and exact formulas for the interband conductivity of graphene. Real (a), (c) and imaginary (b), (d) are shown separately. (a) and (b) show a fit that is accurate for wavelengths longer than 1.1 μm but is poor for shorter wavelengths. The three wavelengths used for the fit were 1.1, 1.45, and 1.8 μm. (c) and (d) show a fit that attempts to follow the sharp detail at 1.0 μm but is of less accuracy for longer wavelengths. The three wavelengths used in this fit were 0.95, 1.275, and 1.6 μm.

TABLE 19.2
Padé Expansion Coefficients for the Fit Displayed in Figure 19.2a and b

a_0	a_1	a_2	b_1	b_2
2.077e-37	1.852e-20	−1.104e-37	−4.421e-27	1.800e-31

part of the interband conductivity is essentially zero in this wavelength range whereas the imaginary part monotonically decreases to zero. Note that this particular Padé fit does not capture the sharp spectral features for wavelengths shorter than 1.1 μm. Figure 19.3c and d show a Padé fit using 0.95, 1.275, and 1.6 μm. Some of the complicated features of the exact formulas are evident in the Padé fit, but a second-order Padé function is clearly limited in its ability to capture all the details. Furthermore, there is a trade-off between fitting to the complicated spectral features and the overall bandwidth of the fit. The fit in Figure 19.3a and b is valid for wavelengths up to 10 μm. In the remainder of this chapter, the fit displayed in Figure 19.3a and b will be used. Table 19.2 displays the coefficients used in the fit.

19.7 SPP WAVEGUIDE

Previous work has determined that the dispersion relation for a TM SPP wave on a graphene layer infinitely extended in two dimensions is given by $\beta = k_0[1 + 2/\eta_0\sigma(\omega)]^{1/2}$ where k_0 is the free-space wave number, and η_0 is the impedance of free space [43,89]. Figure 19.4 displays the dispersion relation of a TM SPP wave using the exact formulas (Equations 19.3 and 19.4) for $\sigma = \sigma_{intra} + \sigma_{inter}$ as well as the Padé approximation to σ_{inter}. For comparison, the dispersion relation using only the intraband term $\sigma = \sigma_{intra}$ is shown to illustrate the importance of including the interband term in the near-infrared region. The gray-shaded region shows the frequency region where a significantly improved fit is obtained. The gray-shaded region also corresponds to the targeted frequency region for near-infrared graphene plasmonics that makes this fitting procedure highly effective for this wavelength region. As expected, for shorter wavelengths above the gray-shaded region, there is less agreement between the frequency dispersion using the Padé fit and the exact formulas.

19.8 COMPACT FDTD WITH ADE METHOD

The SPP waveguide dispersion and loss can be calculated using the FDTD method. For comparison to the results in Figure 19.4, an FDTD algorithm optimized to model infinitely long straight waveguides will be used. The approach is known as compact FDTD (CFDTD) because it compresses an infinitely long domain into a single two-dimensional computational domain representing the cross section of the waveguide geometry [91–99]. CFDTD retains the three-dimensional and fully vectorial nature of the field solutions.

For waveguide geometries uniform along the z direction and assuming a time-harmonic sign convention of $\exp(j\omega t)$, the spatial variation of electric and magnetic fields for modes propagating along the positive z direction is given by

$$F_i^{\beta}(x,y,z) = f_i^{\beta}(x,y)\exp(-j\beta z) \tag{19.21}$$

where $F_i^{\beta}(x,y,z)$ refers to the ith vector component of the electric or magnetic field, and β is the propagation constant. It is further noted that the results in Figure 19.4 apply to

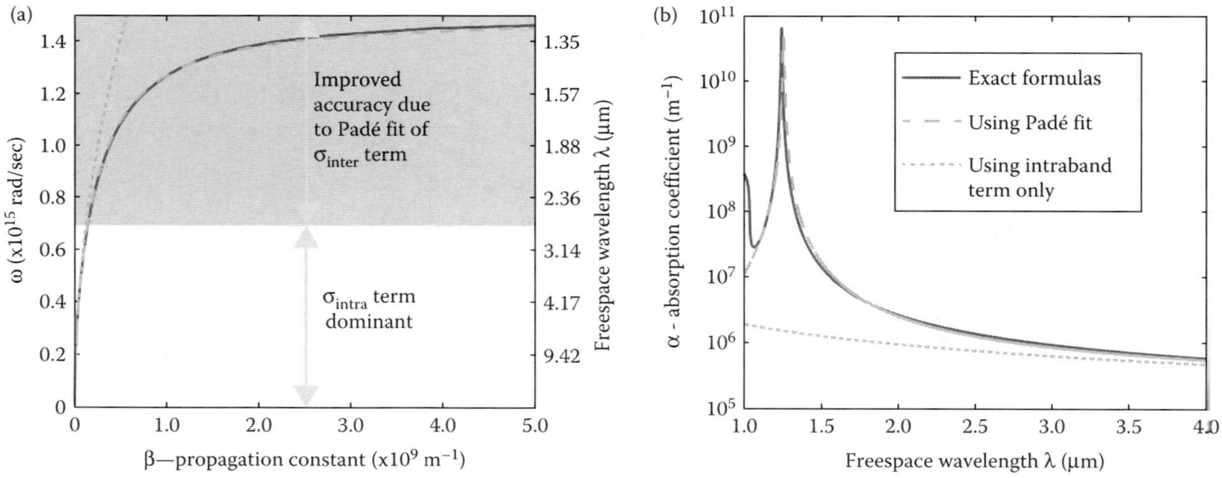

FIGURE 19.4 (a) TM SPP waveguide dispersion for a graphene layer of an infinite two-dimensional extent. The waveguide dispersion using the exact formulas matches well to the dispersion using the graphene fit. The dispersion using only the intraband conductivity term is plotted to show the importance of including the interband conductivity term as well. The gray-shaded region shows where the Padé-fitted interband conductivity is most important. (b) The propagation loss versus wavelength of a TM SPP mode on a graphene layer of an infinite two-dimensional extent. The propagation loss using the exact formulas matches well to the dispersion using the graphene fit. The loss using only the intraband conductivity term is plotted to show the importance of including the interband conductivity term as well.

a waveguide of uniform and infinite extent in two dimensions. The y direction will be assigned to the direction perpendicular to the graphene layer. z is the propagation direction as noted in Equation 19.21. And, we assume that the field is uniform along the x direction. This last property implies

$$\frac{\partial F_i^{\beta}(x,y,z)}{\partial x} = 0 \tag{19.22}$$

In this case, the field components in Maxwell's equations divide into two independent sets of modes distinguished by their transversality to the x direction: transverse magnetic to x (TM_x) consisting of E_x, H_y, and H_z and transverse electric to x (TE_x) consisting of H_x, E_y, and E_z. Using this coordinate system, the SPP waveguide mode in Figure 19.4 is TM to z which corresponds to the TE_x set of field components

$$\frac{\partial D_y}{\partial t} + J_y = \epsilon_y \frac{\partial E_y}{\partial t} + \sigma_y E_y = \frac{H_x}{\partial z} \tag{19.23}$$

$$\frac{\partial D_z}{\partial t} + J_z = \epsilon_z \frac{\partial E_z}{\partial t} + \sigma_z E_z = -\frac{H_x}{\partial y} \tag{19.24}$$

$$\frac{\partial B_x}{\partial t} = \mu_0 \frac{\partial H_x}{\partial t} = \frac{E_y}{\partial z} - \frac{E_z}{\partial y} \tag{19.25}$$

In Equation 19.25, the magnetic permeability is explicitly set to the vacuum permeability. Because the time-domain Maxwell equations are purely real quantities, a purely real version of the ansatz Equation 19.21 is needed to avoid complex numbers. The following decomposition into in-phase (I) and quadrature (Q) terms is used:

$$E_y(y,z,t) = e_{y,I}(y,t)\cos(\beta z) + e_{y,Q}(y,t)\sin(\beta z) \tag{19.26}$$

$$E_z(y,z,t) = e_{z,I}(y,t)\cos(\beta z) + e_{z,Q}(y,t)\sin(\beta z) \tag{19.27}$$

$$H_x(y,z,t) = h_{x,I}(y,t)\cos(\beta z) + h_{x,Q}(y,t)\sin(\beta z) \tag{19.28}$$

Inserting into Equations 19.23 through 19.25 results in

$$\left(\epsilon_y \frac{\partial}{\partial t} + \sigma_y\right)[e_{y,I}(y,t)\cos(\beta z) + e_{y,Q}(y,t)\sin(\beta z)]$$
$$= -\beta h_{x,I}(y,t)\sin(\beta z) + \beta h_{x,Q}(y,t)\cos(\beta z) \tag{19.29}$$

$$\left(\epsilon_z \frac{\partial}{\partial t} + \sigma_z\right)[e_{z,I}(y,t)\cos(\beta z) + e_{z,Q}(y,t)\sin(\beta z)]$$
$$= -\frac{\partial}{\partial y}[h_{x,I}(y,t)\cos(\beta z) + h_{x,Q}(y,t)\sin(\beta z)] \tag{19.30}$$

$$\mu_0 \frac{\partial}{\partial t}[h_{x,I}(y,t)\cos(\beta z) + h_{x,Q}(y,t)\sin(\beta z)]$$
$$= -\beta e_{y,I}(y,t)\sin(\beta z) + \beta e_{y,Q}(y,t)\cos(\beta z)$$
$$-\frac{\partial}{\partial y}[e_{z,I}(y,t)\cos(\beta z) + e_{z,Q}(y,t)\sin(\beta z)] \tag{19.31}$$

Using the orthogonality of $\sin(\beta z)$ and $\cos(\beta z)$, Equations 19.29 through 19.31 result in the following sets of equations:

$$\left(\epsilon_y \frac{\partial}{\partial t} + \sigma_y\right)e_{y,Q}(y,t) = -\beta h_{x,I}(y,t) \tag{19.32}$$

$$\left(\epsilon_z \frac{\partial}{\partial t} + \sigma_z\right)e_{z,I}(y,t) = -\frac{\partial}{\partial y}h_{x,I}(y,t) \tag{19.33}$$

$$\mu_0 \frac{\partial}{\partial t}h_{x,I}(y,t) = \beta e_{y,Q}(y,t) - \frac{\partial}{\partial y}e_{z,I}(y,t) \tag{19.34}$$

and

$$\left(\epsilon_y \frac{\partial}{\partial t} + \sigma_y\right)e_{y,I}(y,t) = \beta h_{x,Q}(y,t) \tag{19.35}$$

$$\left(\epsilon_z \frac{\partial}{\partial t} + \sigma_z\right)e_{z,Q}(y,t) = -\frac{\partial}{\partial y}h_{x,Q}(y,t) \tag{19.36}$$

$$\mu_0 \frac{\partial}{\partial t}h_{x,Q}(y,t) = -\beta e_{y,I}(y,t) - \frac{\partial}{\partial y}e_{z,Q}(y,t) \tag{19.37}$$

The only significant difference between Equations 19.32 through 19.34 and Equations 19.35 through 19.37 is the sign of β. However, this difference is inconsequential as the user can choose the sign of β during simulation runs. Therefore, the two sets of Equations 19.32 through 19.34 and Equations 19.35 through 19.37 are redundant. In this chapter, Equations 19.32 through 19.34 will be used. In this case, the Maxwell equations take on the following form:

$$\epsilon_y \frac{\partial E_y}{\partial t} + \sigma_y E_y = -\beta H_x \tag{19.38}$$

$$\epsilon_z \frac{\partial E_z}{\partial t} + \sigma_z E_z = -\frac{H_x}{\partial y} \tag{19.39}$$

$$\mu_0 \frac{\partial H_x}{\partial t} = \beta E_y - \frac{E_z}{\partial y} \tag{19.40}$$

In this algorithm, the user specifies a value for β, and then time steps Equations 19.38 through 19.40 to obtain the spatiotemporal behavior of the field components. More details on how the waveguide dispersion and field profiles are determined are given below. The next step is including the frequency dispersion of σ using the ADE method. The auxiliary equation for the intraband conductivity term is already given in Equation 19.15. The auxiliary equation for the interband term starts with Equation 19.17

$$J_y(\omega) = \sigma(\omega)E_y(\omega) = \frac{a_0 + a_1 j\omega + a_2(j\omega)^2}{1 + b_1 j\omega + b_2(j\omega)^2} E_y(\omega) \quad (19.41)$$

$$[1 + b_1 j\omega + b_2(j\omega)^2]J_y(\omega) = [a_0 + a_1 j\omega + a_2(j\omega)^2]E_y(\omega) \quad (19.42)$$

which after inverse Fourier transforming becomes

$$\left[1 + b_1\frac{\partial}{\partial t} + b_2\frac{\partial^2}{\partial t^2}\right]J_y(t) = \left[a_0 + a_1\frac{\partial}{\partial t} + a_2\frac{\partial^2}{\partial t^2}\right]E_y(t) \quad (19.43)$$

Discretization is performed in a similar manner as before

$$\frac{\partial^2}{\partial t^2}J_y(t) = \frac{J_y[(n+1)\Delta t] + J_y[(n-1)\Delta t] - 2J_y(n\Delta t)}{\Delta t^2} \quad (19.44)$$

with a similar equation for $E_y(t)$. Plugging into Equation 19.43 and solving for $J_y((n + 1)\Delta t)$ yields

$$J_y[(n+1)\Delta t] = -F_0 J_y(n\Delta t) - F_1 J_y[(n-1)\Delta t] + F_2 E_y[(n+1)\Delta t] + F_3 E_y(n\Delta t) + F_4 E_y[(n-1)\Delta t] \quad (19.45)$$

$$F_0 = \frac{2\Delta t^2 - 4b_2}{b_1\Delta t + 2b_2} \quad (19.46)$$

$$F_1 = \frac{2b_2 - b_1\Delta t}{b_1\Delta t + 2b_2} \quad (19.47)$$

$$F_2 = \frac{a_1\Delta t + 2a_2}{b_1\Delta t + 2b_2} \quad (19.48)$$

$$F_3 = \frac{2a_0\Delta t^2 - 4a_2}{b_1\Delta t + 2b_2} \quad (19.49)$$

$$F_4 = \frac{2a_2 - a_1\Delta t}{b_1\Delta t + 2b_2} \quad (19.50)$$

Equations 19.15 and 19.45 comprise the two auxiliary equations updated in parallel with the usual FDTD time-stepping algorithm. Equations 19.15 and 19.45 update the current density based on the electric field. To update the electric field based on the current density, we return to Equations 19.38 and 19.40 and write J for σE

$$\epsilon_y\frac{\partial E_y}{\partial t} = -\beta H_x - J_y \quad (19.51)$$

$$\epsilon_z\frac{\partial E_z}{\partial t} = -\frac{H_x}{\partial y} - J_z \quad (19.52)$$

$$\mu_0\frac{\partial H_x}{\partial t} = \beta E_y - \frac{E_z}{\partial y} \quad (19.53)$$

The discretized version of Equations 19.38 and 19.40 are

$$\epsilon_y\left(j + \frac{1}{2}\right)\frac{E_y(j+1/2, n+1) - E_y(j+1/2, n)}{\Delta t} = -\beta H_x\left(\frac{j+1}{2}, \frac{n+1}{2}\right) - J_y\left(\frac{j+1}{2}, \frac{n+1}{2}\right) \quad (19.54)$$

$$\epsilon_z(j)\frac{E_z(j, n+1) - E_z(j, n)}{\Delta t} = -\frac{H_x(j+1/2, n+1/2) - H_x(j+1/2, n+1/2)}{\Delta y} - J_z\left(j, \frac{n+1}{2}\right) \quad (19.55)$$

$$\mu_0\frac{H_x(j+1/2, n+3/2) - H_x(j+1/2, n+1/2)}{\Delta t} = \beta E_y\left(\frac{j+1}{2}, n+1\right) - \frac{E_z(j+1, n+1) - E_z(j, n+1)}{\Delta y} \quad (19.56)$$

Note that in Equations 19.54 and 19.55, the current density at time $(n + 1/2)\Delta t$ is needed, but Equations 19.15 and 19.45 provide the current density at time $(n + 1)\Delta t$. To address this, an average value is used $J[(n + 1/2)\Delta t] = J[(n + 1)\Delta t] + J(n\Delta t)]/2$. The final time-stepping equations are found by inserting $J[(n + 1/2)\Delta t]$ into Equations 19.15 and 19.45 and collecting electric fields sampled at time $(n + 1)\Delta$ on the left. The result is lengthy but straightforward to obtain.

19.9 SIMULATION METHODOLOGY

Once the CFDTD and ADE method is implemented, the time stepping is initiated with a broadband initial condition. Setting one or more field components at one or more discrete low-symmetry positions in the domain equal to a nonzero value has the effect of exciting the system with an impulse in both space and time. The simulation is run for 10^5 time steps, and the temporal variation of one or more field components at one or more discrete locations is recorded and saved. A discrete Fourier transform (typically using the fast Fourier transform or FFT) of the resulting time sequence provides information about the mode frequencies and absorption loss. The waveguide modes have Lorentzian peaks centered at the mode frequency and a linewidth related to the propagation loss. A similar Padé interpolation algorithm is used to fit a continuous spectrum to the discrete Fourier transform that results in improved accuracy in the determination of center frequency and linewidth [70,100]. This process is repeated for all β values of interest.

Figure 19.5 shows a comparison between the FDTD simulation and exact analytical results for the TM SPP waveguide. The spatial discretization was set to $\Delta y = 0.5$ nm, and a time step 0.8 times the Courant stability limit is used. The initial condition consisted of point excitations of all field components at points near the graphene region. The spatial domain included 2000 discretization points along y. The boundary is truncated with 15 layers of perfectly matched layers (PMLs) [39,101]. A single grid point at the center of the domain $y = 1000\Delta y$ is used to represent the graphene sheet. The computational domain and FDTD gridding scheme is shown in Figure 19.6. An inherent feature of the traditional Yee algorithm is the spatially offset field components. In this simulation, the E_z components are located at the integer multiples of Δy whereas E_y and H_x are located at the midpoints between the integer multiples of Δy. The E_y and H_x are collocated in the CFDTD algorithm, but in a standard three-dimensional FDTD simulation, they would be separated by $\Delta z/2$ along the z direction.

The graphene sheet is located at $i = 1000$ which is centered on the $E_z[i\Delta y]$ field value. The $E_y[(i + 1/2)\Delta y]$ and $E_y[(i - 1/2)\Delta y]$ fields partially overlap the ith grid point. A common approach to address this issue is spatial averaging at the interface between disparate media. According to Mohammadi et al. [102], for fields perpendicular to a material interface, the appropriate averaging procedure is $1/\epsilon_{ave} = 1/\epsilon_1 + 1/\epsilon_2$ where ϵ_1 and ϵ_2 are the dielectric permittivities on either side of the interface. As discussed in Section 19.3, the real part of the permittivity is a negative number much larger than the vacuum permittivity, so that $1/\epsilon_{ave} = 1/\epsilon_0 + 1/(\epsilon_0 - \sigma_i/\omega) \approx 1/\epsilon_0$. This argument suggests that the graphene material should be inserted at grid point $i = 1000$ for the E_z component but it should *not* be included at $i \pm 1/2$ for E_y. Another argument

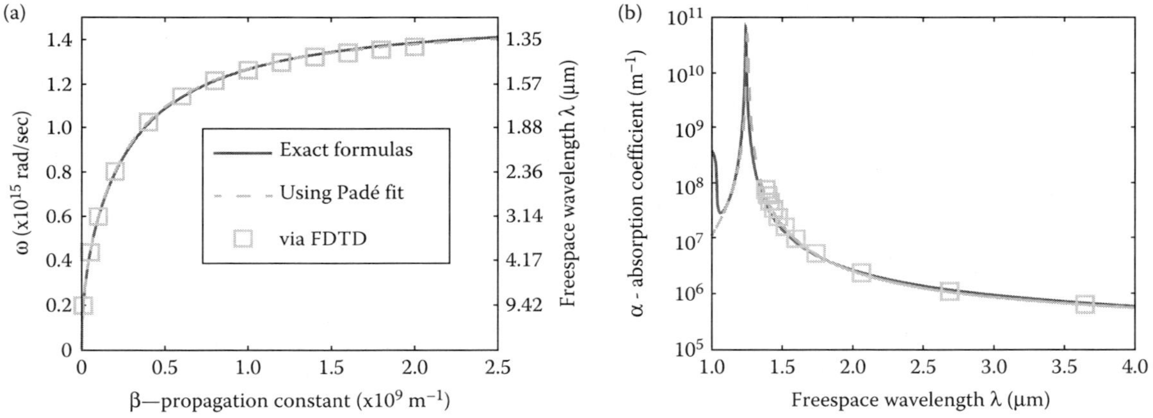

FIGURE 19.5 (a) TM SPP waveguide dispersion and (b) absorption coefficient for a graphene layer of an infinite two-dimensional extent calculated using CFDTD and compared to exact formulas.

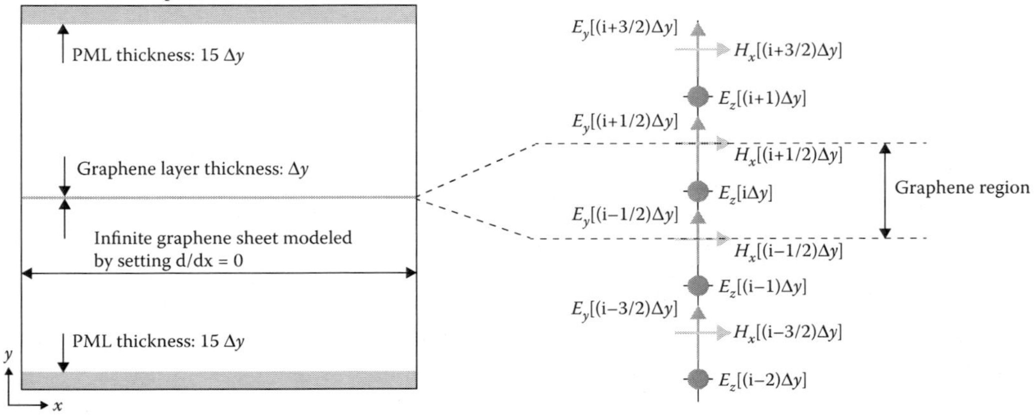

FIGURE 19.6 Computational domain for simulating TM SPP modes. The fields are assumed to be uniform along x; so, the computational domain is actually one dimensional. On the right, the staggered arrangement of the field components is shown in relation to the graphene layer.

in favor of not inserting graphene material values at $i \pm 1/2$ is based on the field behavior of E_y for the TM SPP mode. The E_y field is odd in the direction perpendicular to the graphene sheet (y direction); so, the field has a spatial zero collocated with the graphene layer. Therefore, the E_y component does not "feel" the graphene sheet strongly. Because the field is peaked and localized on either side of the graphene layer, including an average material value at $i \pm 1/2$ actually introduces error as the E_y is forced to "feel" the graphene sheet much more heavily than it would a truly monoatomic graphene layer. Therefore, in the results presented in this chapter, an anisotropic material model is used for the graphene layer.

The agreement between the FDTD and exact results shown in Figure 19.5 is good. For the waveguide dispersion, the maximum error is less than 1% and occurs at the right-most point in the dispersion (at $\beta = 2 \times 10^9$). Error is expected to increase with β because the field becomes more localized at higher β values so that the spatial variation of the field is more rapid relative to the FDTD spatial discretization size. The absorption loss is calculated from the linewidth of the mode resonance $\Delta\omega$ according to $\alpha = \Delta\omega/2v_g$ where $v_g = \partial\omega/\partial\beta$ is the group velocity of the waveguide, and the factor of 2 is due to α being a field loss coefficient (as opposed to a power loss coefficient). For the data in Figure 19.5b, the group velocity was estimated by calculating the slope between adjacent FDTD data points according to

$$v_g = \frac{\omega}{\beta} \approx \frac{\omega_{i+1} - \omega_i}{\beta_{i+1} - \beta_i} \qquad (19.57)$$

The agreement between the CFDTD results and exact results is excellent and holds over a two-order-of magnitude range of variation. Figure 19.7 shows the spatial field profiles for the three nonzero field components comprising the TM SPP wave. The nanometer-scale localization of the fields is clear. While the z component of the electric field is even with

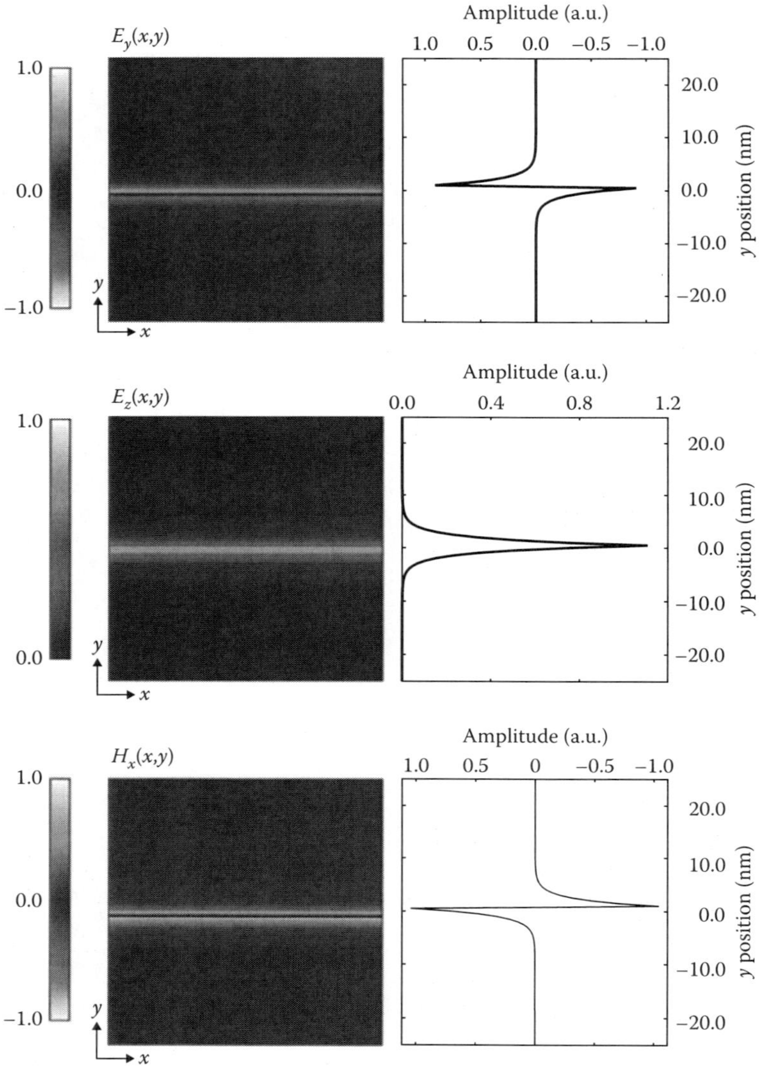

FIGURE 19.7 Three nonzero field components associated with the TM SPP wave in a graphene layer of an infinite two-dimensional extent. The fields are uniform along the x direction. Fields are calculated using the CFDTD method. Fields correspond to the mode at $\beta = 0.8 \times 10^9\ \text{m}^{-1}$ and $\lambda_0 = 1.55\ \mu\text{m}$.

respect to the direction normal to the graphene layer, both the x component of the magnetic field and the y component of the electric field are odd and opposite to each other in sign. The symmetry of the field components is consistent with the application of electromagnetic boundary conditions. The y component of the electric field is equal to $\rho_s\hat{y}$ at the top of the graphene–air boundary and $-\rho_s\hat{y}$ at the bottom of the graphene–air boundary (where ρ_s is the graphene surface charge density). The z component of the electric field is expected to be continuous across the graphene layer and, thus, possesses even symmetry. The x component of the magnetic field on the two sides of the graphene layer differs by an amount equal to the surface current along the z direction. Because of the planar nature of the graphene layer, the magnetic field on either side of it must be equal in magnitude because it is on two sides of the same surface. But, it is opposite in sign because if it was equal in sign (i.e., even), then that would imply zero surface current which is obviously not the case.

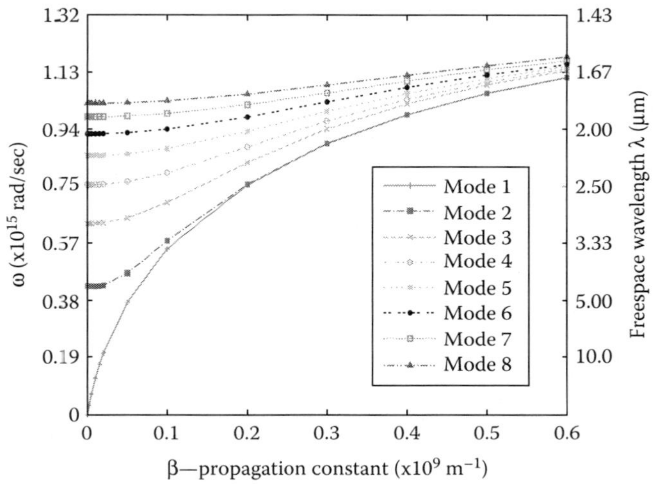

FIGURE 19.8 SPP waveguide dispersion diagram for a 50-nm-wide single-layer graphene waveguide.

19.10 FINITE-WIDTH GRAPHENE SPP WAVEGUIDE

This chapter concludes with analysis of the SPP modes of a uniform graphene waveguide with a width of 50 nm. While a finite-width waveguide is relatively simple, it represents a geometry without closed-form solutions for the SPP modes. This makes a numerical approach necessary for clear elucidation of the field properties. Figure 19.8 shows the SPP mode dispersion for the first eight modes supported by the finite-width graphene waveguide. The simulation parameters are the same as before, but $\Delta y = 1$ nm and $\Delta_x = 2.5$ nm, and the domain extends to 100 points along x and 400 points along y. The spatial profiles of the electric field along with the transverse vector fields are shown in Figure 19.9. Figures 19.10 through 19.17 show the individual spatial distribution of all six vector field components of the modes whose dispersion is shown in Figure 19.8. The dispersion of mode 1 is qualitatively similar (though slightly lower in frequency) to the TM SPP mode dispersion for the infinitely wide graphene layer. However, the spatial mode profile of mode 1 differs from the infinitely wide graphene layer. First, all six vector field components are nonzero. Second, the fields along the x direction are clearly nonuniform with an enhanced field amplitude at the edges of the finite-width graphene layer. Nikitin et al. [50] have discussed the SPP modes of finite-width graphene waveguides and noted the role of edge modes in which the electromagnetic fields are entirely localized at the edges of

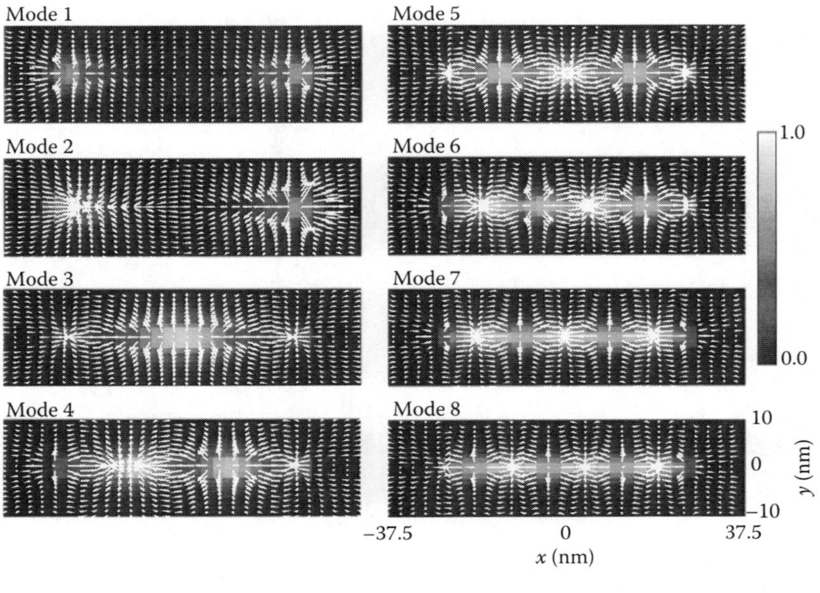

FIGURE 19.9 Three nonzero field components associated with the SPP wave in a graphene layer of an infinite two-dimensional extent. The fields are uniform along the x direction. Fields are calculated using the CFDTD method. Fields correspond to the mode at $\beta = 0.8 \times 10^9$ m^{-1} and $\lambda_0 = 1.55$ μm.

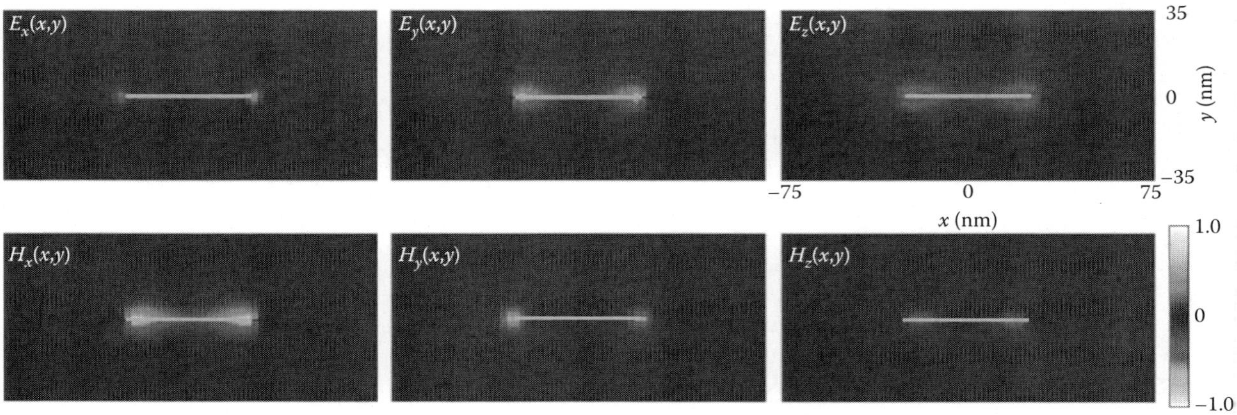

FIGURE 19.10 Spatial distribution of individual vector components associated with mode 1 in Figure 19.8. The spatial scale shown at the top right applies to all figures. All components associated with the electric field are normalized to the same color scale. All components associated with the magnetic field are normalized (separately from the electric field) to the same color scale. Fields correspond to the mode at $\beta = 0.2 \times 10^9$ m^{-1} and $\lambda_0 = 2.49$ µm.

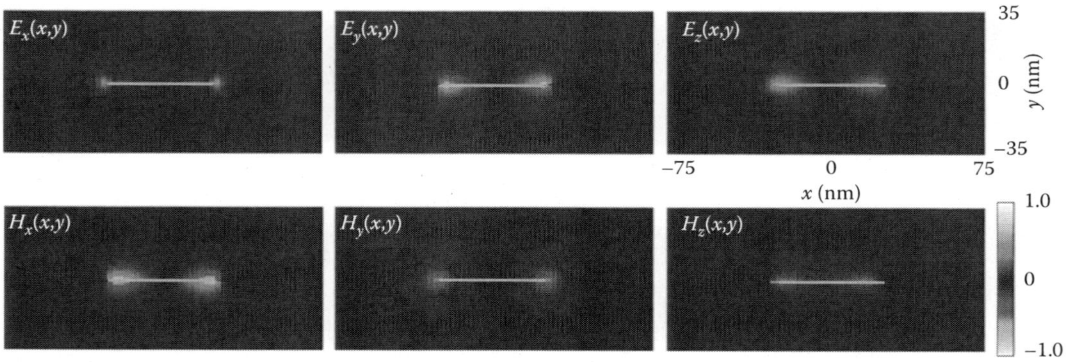

FIGURE 19.11 Spatial distribution of individual vector components associated with mode 2 in Figure 19.8. The spatial scale shown at the top right applies to all figures. All components associated with the electric field are normalized to the same color scale. All components associated with the magnetic field are normalized (separately from the electric field) to the same color scale. Fields correspond to the mode at $\beta = 0.2 \times 10^9$ m^{-1} and $\lambda_0 = 2.48$ µm.

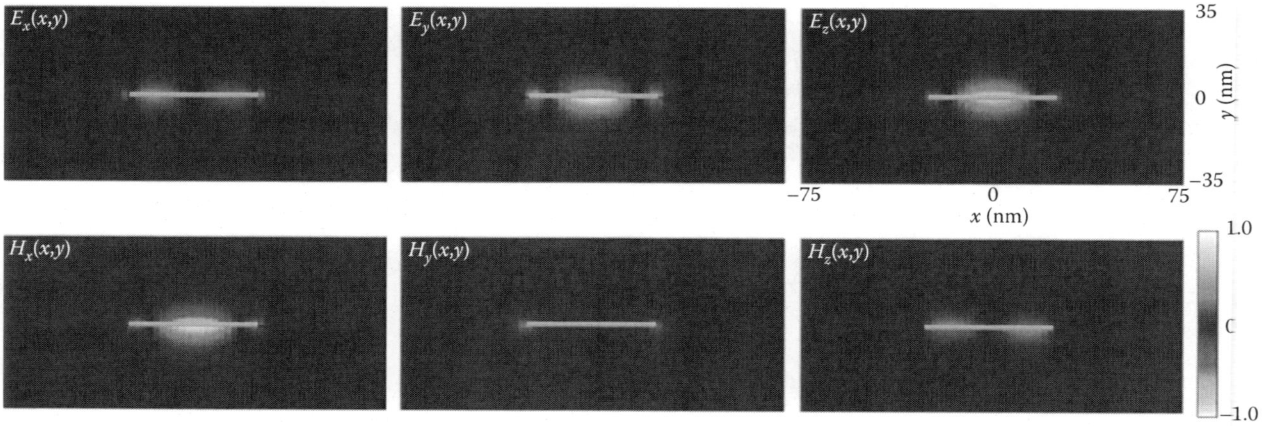

FIGURE 19.12 Spatial distribution of individual vector components associated with mode 3 in Figure 19.8. The spatial scale shown at the top right applies to all figures. All components associated with the electric field are normalized to the same color scale. All components associated with the magnetic field are normalized (separately from the electric field) to the same color scale. Fields correspond to the mode at $\beta = 0.2 \times 10^9$ m^{-1} and $\lambda_0 = 2.27$ µm.

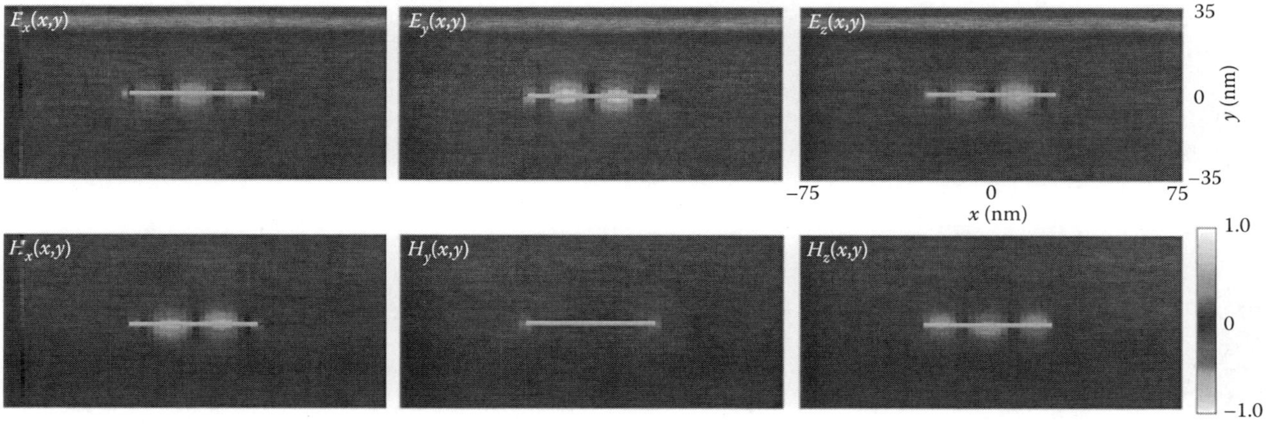

FIGURE 19.13 Spatial distribution of individual vector components associated with mode 4 in Figure 19.8. The spatial scale shown at the top right applies to all figures. All components associated with the electric field are normalized to the same color scale. All components associated with the magnetic field are normalized (separately from the electric field) to the same color scale. Fields correspond to the mode at $\beta = 0.2 \times 10^9$ m^{-1} and $\lambda_0 = 2.13$ μm.

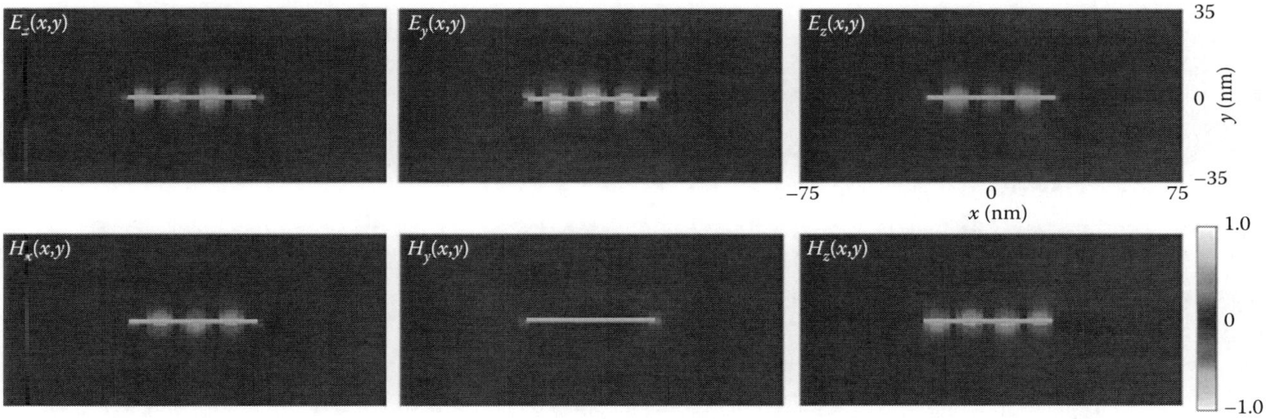

FIGURE 19.14 Spatial distribution of individual vector components associated with mode 5 in Figure 19.8. The spatial scale shown at the top right applies to all figures. All components associated with the electric field are normalized to the same color scale. All components associated with the magnetic field are normalized (separately from the electric field) to the same color scale. Fields correspond to the mode at $\beta = 0.2 \times 10^9$ m^{-1} and $\lambda_0 = 2.02$ μm.

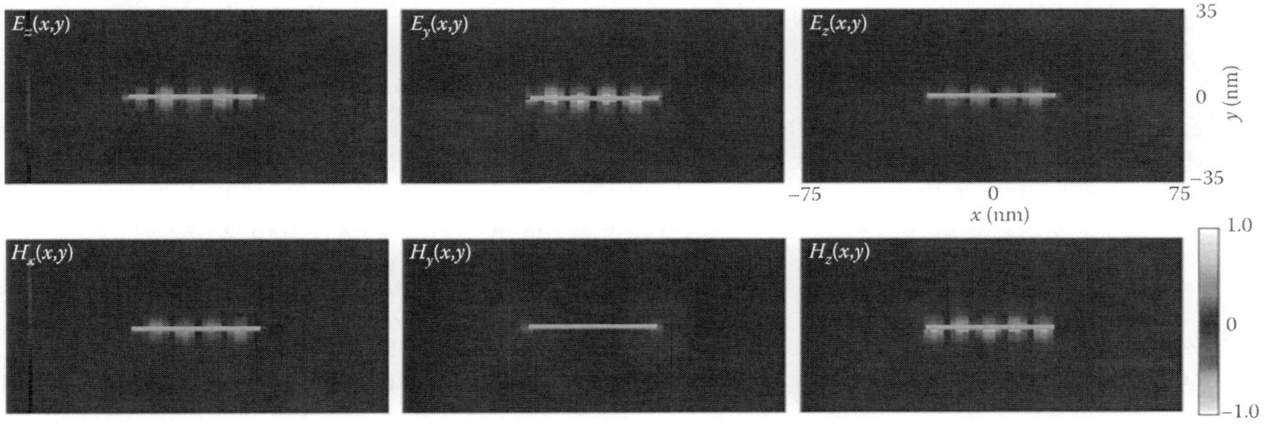

FIGURE 19.15 Spatial distribution of individual vector components associated with mode 6 in Figure 19.8. The spatial scale shown at the top right applies to all figures. All components associated with the electric field are normalized to the same color scale. All components associated with the magnetic field are normalized (separately from the electric field) to the same color scale. Fields correspond to the mode at $\beta = 0.2 \times 10^9$ m^{-1} and $\lambda_0 = 1.92$ μm.

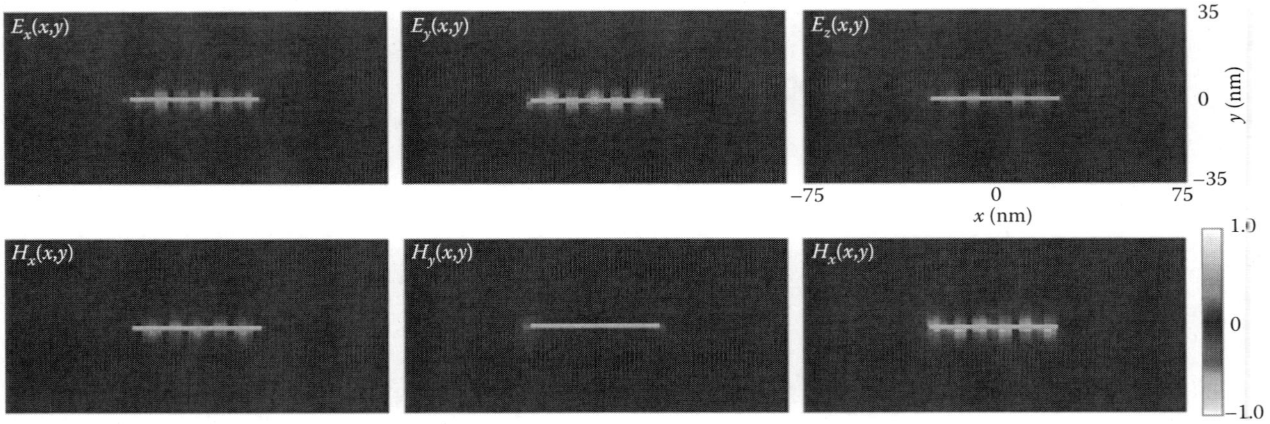

FIGURE 19.16 Spatial distribution of individual vector components associated with mode 7 in Figure 19.8. The spatial scale shown at the top right applies to all figures. All components associated with the electric field are normalized to the same color scale. All components associated with the magnetic field are normalized (separately from the electric field) to the same color scale. Fields correspond to the mode at $\beta = 0.2 \times 10^9$ m^{-1} and $\lambda_0 = 1.84$ μm.

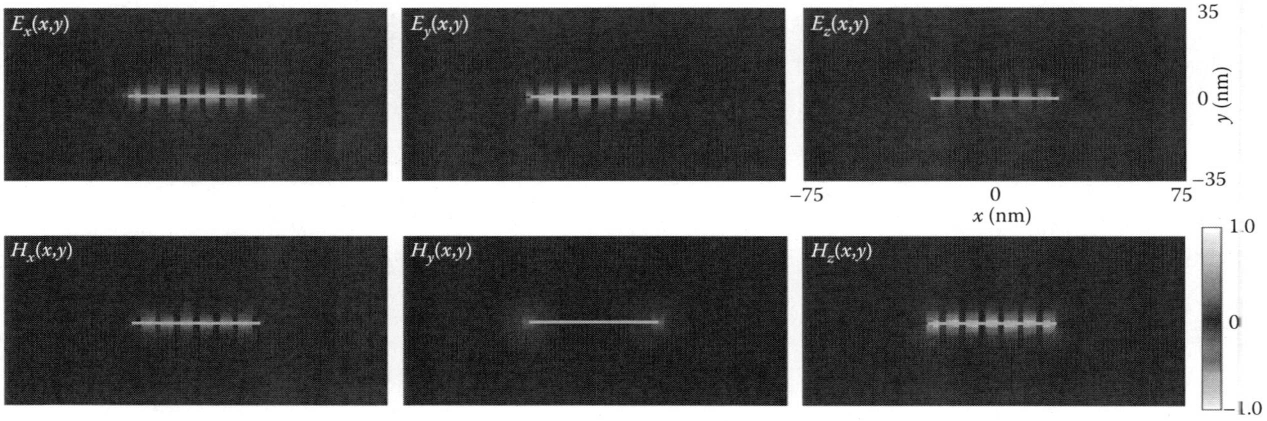

FIGURE 19.17 Spatial distribution of individual vector components associated with mode 8 in Figure 19.8. The spatial scale shown at the top right applies to all figures. All components associated with the electric field are normalized to the same color scale. All components associated with the magnetic field are normalized (separately from the electric field) to the same color scale. Fields correspond to the mode at $\beta = 0.2 \times 10^9$ m^{-1} and $\lambda_0 = 1.78$ μm.

the graphene layer. This is related to the well-known field enhancement in conductors near sharp edges or corners [87]. The fields shown in this section may be qualitatively interpreted as hybridized modes with contributions from edge modes and distributed SPP waveguide modes.

The field distribution of mode 2 shown in Figure 19.11 is similar to that of mode 1 but the E_z field is odd along the x direction. This suggests that an alternative classification for the modes is in terms of the field symmetry and number of field nulls along the x direction. First, we point out that the parity along the y direction of E_y, E_z, and H_x is the same as that shown in Figure 19.7, which suggests that these modes are derived from the TM SPP mode of the infinitely wide graphene waveguide. Focussing on the parity along x of the E_z component shows that it alternates between even (modes 1, 3, 5, and 7) and odd (modes 2, 4, 6, and 8). Furthermore, the number of field nulls along x is zero for mode 1, one for mode 2, two for mode 3, and so on. While the spatial mode profiles for modes 1 and 2 show a strong edge effect, the spatial field profiles for modes 3–8 are concentrated toward the middle of the waveguide with smaller contributions from the edge effects. In addition, modes 3–8 show transverse-standing wave patterns along the x direction, which is similar to that found in higher-order modes of dielectric waveguides.

Returning to the dispersion diagram in Figure 19.8, another interesting difference between mode 1 and modes 2–8 is the behavior near $\beta = 0$. Mode 1 tends to zero frequency as β goes to zero (similar to the infinitely wide graphene waveguide) whereas the other modes intersect the vertical axis at finite frequency values. Because the higher-order modes are associated with standing waves along the x direction, at $\beta = 0$, the x-directed standing waves simply become stationary standing waves with no propagation along z. At $\beta = 0$, the finite-width graphene waveguide becomes a Fabry–Perot resonator with x-directed waves reflected by the termination of the finite-width graphene layer. Noting that the light line is nearly coincident with the vertical axis, one may conclude that radiative coupling to modes 2–8 is possible for dispersion points above the light line. The linewidths of the modes of the finite-width waveguide are similar to the graphene layer of an infinite extent. However, because the group velocity tends to zero near $\beta = 0$ for modes 2–8, these modes have significantly larger propagation losses in this region.

19.11 CONCLUSIONS

This chapter presents a method for numerical modeling of the electrodynamics of graphene in the time domain. A particular emphasis is the incorporation of the complex conductivity frequency dispersion into the FDTD method. The CFDTD method is used to analyze the SPP modes of longitudinally uniform graphene waveguides. The agreement between numerical results and closed-form solutions is excellent for both waveguide dispersion and propagation loss. The more complicated SPP modes of a finite-width graphene waveguide are analyzed. The results show a strong influence of the graphene edges on the SPP modes as well as transverse-standing wave patterns for higher-order modes. The numerical tools presented here will enable the design and simulation of devices with applications in compact optical signal processing. While the numerical methods presented here enable the design and simulation of a multitude of useful and complicated graphene-based device geometries, a number of improvements are likely to be achieved in the coming years. These limitations are due to the ongoing characterization of the graphene material itself and include better knowledge of frequency-dependent scattering, deviations from linearity of the electronic band structure, many-body effects, and conductivity properties at the boundary of a finite-width graphene layer.

REFERENCES

1. K. S. Novoselov, A. K. Geim, S. V. Morozov, D. Jiang, Y. Zhang, S. V. Dubonos, I. V. Grigorieva, and A. A. Firsov. Electric field effect in atomically thin carbon films. *Science*, 306:666–669, 2004.
2. A. K. Geim. Graphene: Status and propects. *Science*, 324:1530–1534, 2009.
3. A. K. Geim and K. S. Novoselov. The rise of graphene. *Nature Materials*, 6:183–191, 2007.
4. C. Lee, X. Wei, J. W. Kysar, and J. Hone. Measurement of the elastic properties and intrinsic strength of monolayer graphene. *Science*, 321:385–388, 2008.
5. A. A. Balandin, S. Ghosh, W. Bao, I. Calizo, D. Teweldebrhan, F. Miao, and C. N. Lau. Superior thermal conductivity of single-layer graphene. *Nano Letters*, 8(3):902–907, 2008.
6. P. Avouris. Graphene: Electronic and photonic properties and devices. *Nano Letters*, 10(11):4285–4294, 2010.
7. F. Bonaccorso, Z. Sun, T. Hasan, and A. C. Ferrari. Graphene photonics and optoelectronics. *Nature Photonics*, 4:611–622, 2010.
8. Q. Bao and K. P. Loh. Graphene photonics, plasmonics, and broadband optoelectronic devices. *ACS Nano*, 6(5):3677–3694, 2012.
9. M. Currie, J. D. Caldwell, F. J. Bezares, J. Robinson, T. Anderson, H. Chun, and M. Tadjer. Quantifying pulsed laser induced damage to graphene. *Applied Physics Letters*, 99(21):211909, 2011.
10. R. R. Nair, P. Blake, A. N. Grigorenko, K. S. Novoselov, T. J. Booth, T. Stauber, N. M. R. Peres, and A. K. Geim. Fine structure constant defines visual transparency of graphene. *Science*, 320:1308, 2008.
11. F. Mak, M. Y. Sfeir, Y. Wu, C. H. Lui, J. A. Misewich, and T. F. Heinz. Measurement of the optical conductivity of graphene. *Physical Review Letters*, 101(19):196405, 2008.
12. M. Liu, X. Yin, E. Ulin-Avila, B. Geng, T. Zentgraf, L. Ju, F. Wang, and X. Zhang. A graphene-based broadband optical modulator. *Nature*, 474(21):64–67, 2011.
13. G. W. Hanson. Quasi-transverse electromagnetic modes supported by a graphene parallel-plate waveguide. *Journal of Applied Physics*, 104(8):084314, 2008.
14. F. Xia, T. Mueller, Y.-M. Lin, A. Valdes-Garcia, and P. Avouris. Ultrafast graphene photodetector. *Nature Nanotechnology*, 4(12):839–843, 2009.
15. T. Mueller, F. Xia, and P. Avouris. Graphene photodetectors for high-speed optical communications. *Nature Photonics*, 4(5):297–301, 2010.
16. A. Ferreira, N. M. R. Peres, R. M. Ribeiro, and T. Stauber. Graphene-based photodetector with two cavities. *Physical Review B*, 85(11):115438, 2012.
17. M. Engel, M. Steiner, A. Lombardo, A. C. Ferrari, H. v. Löhneysen, P. Avouris, and R. Krupke. Light–matter interaction in a microcavity-controlled graphene transistor. *Nature Communications*, 3(6):1–6, 2012.
18. T. Gu, N. Petrone, J. F. McMillan, A. van der Zande, M. Yu, G. Q. Lo, D. L. Kwong, J. Hone, and C. W. Wong. Regenerative oscillation and four-wave mixing in graphene optoelectronics. *Nature Photonics*, 6(11):544–549, 2012.
19. E. D. Fabrizio, A. E. Nikolaenko, and N. I. Zheludev. Graphene in a photonic metamaterial. *Optics Express*, 18(8):8353–8359, 2010.
20. Y. Zou, P. Tassin, T. Koschny, and C. M. Soukoulis. Interaction between graphene and metamaterials: Split rings vs. wire pairs. *Optics Express*, 20(11):12198–12204, 2012.
21. Q. Bao, H. Zhang, B. Wang, Z. Ni, C. Haley, Y. X. Lim, Y. Wang, D. Y. Tang, and K. P. Loh. Broadband graphene polarizer. *Nature Photonics*, 5:411–415, 2011.
22. V. P. Gusynin and S. G. Sharapov. Transport of Dirac quasiparticles in graphene: Hall and optical conductivities. *Physical Review B*, 73(24):245411, 2006.
23. G. W. Hanson. Dyadic Green's functions for an anisotropic non-local model of biased graphene. *IEEE Transactions on Antennas and Propagation*, 56(3):747–757, 2008.
24. Q. Bao, H. Zhang, Y. Wang, Z. Ni, Y. Yan, Z. X. Shen, K. P. Loh, and D. Y. Tang. Atomic-layer graphene as a saturable absorber for ultrafast pulsed lasers. *Advanced Functional Materials*, 19(19):3077–3083, 2009.

25. H. Zhang, Q. Bao, D. Tang, L. Zhao, and K. Loh. Large energy soliton erbium-doped fiber laser with a graphene–polymer composite mode locker. *Applied Physics Letters*, 95(14):141103, 2009.
26. H. Zhang, D. Y. Tang, L. M. Zhao, Q. L. Bao, and K. P. Loh. Large energy mode locking of an erbium-doped fiber laser with atomic layer graphene. *Optics Express*, 17(20):17630–17635, 2009.
27. Z. Sun, T. Hasan, F. Torrisi, D. Popa, G. Privitera, F. Wang, F. Bonaccorso, D. M. Basko, and A. C. Ferrari. Graphene mode-locked ultrafast laser. *ACS Nano*, 4(2):803–810, 2010.
28. Y.-W. Song, S.-Y. Jang, W.-S. Han, and M.-K. Bae. Graphene mode-lockers for fiber lasers functioned with evanescent field interaction. *Applied Physics Letters*, 96(5):051122, 2010.
29. H. Zhang, D. Tang, R. J. Knize, L. Zhao, Q. Bao, and K. P. Loh. Graphene mode locked, wavelength-tunable, dissipative soliton fiber laser. *Applied Physics Letters*, 96(11):111112, 2010.
30. Y. M. Chang, H. Kim, J. H. Lee, and Y.-W. Song. Multilayered graphene efficiently formed by mechanical exfoliation for nonlinear saturable absorbers in fiber mode-locked lasers. *Applied Physics Letters*, 97(21):212202, 2010.
31. G. Xing, H. Guo, X. Zhang, T. C. Sum, C. Hon, and A. Huan. The physics of ultrafast saturable absorption in graphene. *Optics Express*, 18(5):4564–4573, 2010.
32. W. D. Tan, C. Y. Su, R. J. Knize, G. Q. Zie, L. J. Li, and D. Y. Tang. Mode locking of ceramic Nd:yttrium aluminum garnet with graphene as a saturable absorber. *Applied Physics Letters*, 96(3):031106, 2010.
33. S. Yamashita. A tutorial on nonlinear photonic applications of carbon nanotube and graphene. *Journal of Lightwave Technology*, 30(4):427–447, 2012.
34. A. Mock. Dynamic material response of graphene for FDTD simulation of passive modelocking. In *13th International Conference on Numerical Simulation of Optoelectronic Devices*, p. TuB3, Vancouver, British Columbia, Canada, August 19–August 22, 2013.
35. A. Mock. FDTD simulation of passively mode-locked lasers using a dynamic saturable absorption model for graphene. In *Integrated Photonics Research, Silicon and Nanophotonics*, p. IM2B.5, Rio Grande, Puerto Rico, July 14–July 19, 2013.
36. A. Mock. Electrodynamic modeling of graphene photonic devices. In *Energy, Materials and Nanotechnology Fall Meeting*, Orlando, FL, USA, December 6–December 10, 2013.
37. S. A. Maier. *Plasmonics: Fundamentals and Applications*, Springer, New York, 2007.
38. K. S. Yee. Numerical solution of initial boundary value problems involving Maxwells equations in isotropic media. *IEEE Transactions on Antennas and Propagation*, 14(3):302–307, 1966.
39. A. Taflove and S. C. Hagness. *Computational Electrodynamics*, Artech House, Massachusetts, 2000.
40. F. H. L. Koppens, D. E. Chang, and J. García de Abajo. Graphene plasmonics: A platform for strong light–matter interactions. *Nano Letters*, 11:3370–3377, 2011.
41. A. N. Grigorenko, M. Polini, and K. S. Novoselov. Graphene plasmonics. *Nature Photonics*, 6:479–488, 2012.
42. N. Kumada, S. Tanabe, H. Hibino, H. Kamata, M. Hashisaka, K. Muraki, and T. Fujisawa. Plasmon transport in graphene investigated by time-resolved electrical measurements. *Nature Communications*, 19:1–6, 2013.
43. A. Vakil and N. Engheta. Transformation optics using graphene. *Science*, 332:1291–1294, 2011.
44. A. Vakil and N. Engheta. Fourier optics on graphene. *Physical Review B*, 85(7):075434, 2012.
45. W. A. Murray and W. L. Barnes. Plasmonic materials. *Advanced Materials*, 19:3771–3782, 2007.
46. M. Pelton, J. Aizpurua, and G. Bryant. Metal–nanoparticle plasmonics. *Laser and Photonics Reviews*, 2(3):136–159, 2008.
47. K. Ziegler. Minimal conductivity of graphene: Nonuniversal values from the Kubo formula. *Physical Review B*, 75(23):233407, 2007.
48. L. A. Falkovsky and S. S. Pershoguba. Optical far-infrared properties of a graphene monolayer and multilayer. *Physical Review B*, 76(15):153410, 2007.
49. M. Jablan, H. Buljan, and M. Soljačić. Plasmonics in graphene at infrared frequencies. *Physical Review B*, 80(24):245435, 2009.
50. A. Y. Nikitin, F. Guinea, F. J. García-Vidal, and L. Martín-Moreno. Edge and waveguide terahertz surface plasmon modes in graphene microribbons. *Physical Review B*, 84(16):161407, 2011.
51. J. Christensen, A. Manjavacas, S. Thongrattanasiri, F. H. L. Koppens, and F. J. García de Abajo. Graphene plasmon wave guiding and hybridization in individual and paired nanoribbons. *ACS Nano*, 6(1):431–440, 2012.
52. D. R. Andersen. Graphene-based long-wave infrared TM surface plasmon modulator. *Journal of the Optical Society of America B*, 27(4):818–823, 2010.
53. X. Zhu, W. Yan, N. A. Mortensen, and S. Xiao. Bends and splitters in graphene nano ribbon waveguides. *Optics Express*, 21(3):3486–3491, 2013.
54. G. Rosolen and B. Maes. Graphene ribbons for tunable coupling with plasmonic sub wavelength cavities. *Journal of the Optical Society of America B*, 31(5):1096–1102, 2014.
55. J. Wang, L. W. Bing, X. B. Li, Z. H. Ni, and T. Qui. Graphene plasmon guided along a nano ribbon coupled with a nanoring. *Journal of Physics D: Applied Physics*, 47:135106, 2014.
56. X. Ren, W. E. I. Sha, and W. C. H. Choy. Tuning optical responses of metallic dipole nanoantenna using graphene. *Optics Express*, 21(26):31824–31829, 2014.
57. J. T. Kim and S.-Y. Choi. Graphene-based plasmonic waveguides for photonic integrated circuits. *Optics Express*, 19(24):24557–24562, 2011.
58. H. Yan, X. Li, B. Chandra, G. Tulevski, Y. Wu, M. Freitag, W. Zhu, P. Avouris, and F. Zia. Tunable infrared plasmonic devices using graphene/insulator stacks. *Nature Nanotechnology*, 7:330–334, 2012.
59. K. Y. M. Yeung, J. Chee, H. Yoon, Y. Song, J. Kong, and D. Ham. Far-infrared graphene plasmonic crystals for plasmonic band engineering. *Nano Letters*, 14:2479, 2014.
60. E. H. Hwang and S. D. Sarma. Dielectric function, screening, and plasmons in two-dimensional graphene. *Physical Review B*, 75(20):205418, 2007.
61. S. A. Mikhailov and K. Ziegler. New electromagnetic mode in graphene. *Physical Review Letters*, 99(1):016803, 2007.
62. V. P. Gusynin, S. G. Sharapov, and J. P. Carbotte. Magnet-optical conductivity in graphene. *Journal of Physics: Condensed Matter*, 19:026222, 2007.
63. D. F. Kelley and R. J. Luebbers. Piecewise linear recursive convolution for dispersive media using FDTD. *IEEE Transactions on Antennas and Propagation*, 44(6):792–797, 1996.
64. M. Okoniewski, M. Mrozowski, and M. A. Stuchly. Simple treatment of multi-term dispersion in FDTD. *IEEE Microwave and Guided Wave Letters*, 7(5):121–123, 1997.
65. F. Hao and P. Nordlander. Efficient dielectric function for FDTD simulation of the optical properties of silver and gold nanoparticles. *Chemical Physics Letters*, 446:115–118, 2007.

66. A. Vial, A.-S. Grimault, D. Macías, D. Barchiesi, and M. Lamy de la Chapelle. Improved analytical fit of gold dispersion: Application to the modeling of extinction spectra with a finite-difference time-domain method. *Physical Review B*, 71(8):085416, 2005.
67. I. T. Rekanos and T. G. Papadopoulos. FDTD modeling of wave propagation in Cole–Cole media with multiple relaxation times. *IEEE Antennas and Wireless Propagation Letters*, 9:67–69, 2010.
68. G. A. Baker and P. Graves-Morris. *Padé Approximants*, Cambridge University Press, New York, 1996.
69. S. Dey and R. Mittra. Efficient computation of resonant frequencies and quality factors of cavities via a combination of the finite-difference time-domain technique and the Padé approximation. *IEEE Microwave and Guided Wave Letters*, 8(12):415–417, 1998.
70. A. Mock and J. D. O'Brien. Direct extraction of large quality factors and resonant frequencies from Padé interpolated resonance spectra. *Optical and Quantum Electronics*, 40(14):1187–1192, 2008.
71. H. Lin, M. F. Pantoja, L. D. Angulo, J. Alvarez, R. G. Martin, and S. G. Garcia. FDTD modeling of graphene devices using complex conjugate dispersion material model. *IEEE Microwave and Wireless Components Letters*, 22(12):612–614, 2012.
72. G. D. Bouzianas, N. V. Kantartzis, C. S. Antonopoulos, and T. D. Tsiboukis. Optimal modeling of infinite graphene sheets via a class of generalized FDTD schemes. *IEEE Transactions on Magnetics*, 48(2):379–382, 2012.
73. G. D. Bouzianas, N. V. Kantartzis, and T. D. Tsiboukis. Plasmon mode excitation on graphene layers via obliquely-incident focused sideband pulses in rigorous time-domain algorithms. *IEEE Transactions on Magnetics*, 49(5):1773–1776, 2013.
74. V. Nayyeri, M. Soleimani, and O. M. Ramahi. Time-domain method using a surface boundary condition. *IEEE Transactions on Antennas and Propagation*, 61(8):4176, 2013.
75. V. Nayyeri, M. Soleimani, and O. M. Ramahi. Wideband modeling of graphene using the finite-difference time-domain method. *IEEE Transactions on Antennas and Propagation*, 61(12):6107–6114, 2013.
76. G. D. Bouzianas, N. V. Kantartzis, T. V. Yioultsis, and T. D. Tsiboukis. Consistent study of graphene structures through the direct incorporation of surface conductivity. *IEEE Transactions on Magnetics*, 50(2):7003804, 2014.
77. I. Ahmed, E. H. Khoo, and E. P. Li. Efficient modeling and simulation of graphene devices with the LOD–FDTD method. *IEEE Microwave and Wireless Components Letters*, 23(6):306–308, 2013.
78. H. Wang, Y. Wu, B. Lassiter, C. L. Nehl, J. H. Hafner, P. Nordlander, and H. J. Halas. Symmetry breaking in individual plasmonic nanoparticles. *Proceedings of the National Academy of Sciences*, 103(29):10856–10860, 2006.
79. M. T. Hill, Y.-S. Oei, B. Smalbrugge, Y. Zhu, T. de Vries, P. J. van Veldhoven, F. W. M. Van Otten et al. Lasing in metal-coated nanocavities. *Nature Photonics*, 1:589–594, 2007.
80. I. Ahmed, E. H. Khoo, O. Kurniawan, and E. P. Li. Modeling and simulation of active plasmonics with the FDTD method by using solid state and Lorentz–Drude dispersive model. *Journal of the Optical Society of America B*, 28(3):352–359, 2011.
81. A. Mock. Modal analysis of nanoplasmonic multilayer spherical resonators. *IEEE Photonics Journal*, 3(4):765–776, 2011.
82. A. B. Kuzmenko, E. van Heumen, F. Carbone, and D. van der Marel. Universal optical conductance of graphite. *Physical Review Letters*, 100(11):117401, 2008.
83. T. Stauber, N. M. Peres, and A. K. Geim. Optical conductivity of graphene in the visible region of the spectrum. *Physical Review B*, 78(8):085432, 2008.
84. J. M. Dawlaty, S. Shivaraman, J. Strait, P. George, M. V. S. Chandrashekar, F. Rana, M. G. Spencer, D. Veksler, and Y. Chen. Measurement of the optical absorption spectra of epitaxial graphene from terahertz to visible. *Applied Physics Letters*, 93(13):131905, 2008.
85. D. K. Cheng. *Field and Wave Electromagnetics*, Addison Wesley, New York, 1992.
86. A. Mock. Padé approximate spectral fit for FDTD simulation of graphene in the near infrared. *Optical Materials Express*, 2(6):771–781, 2012.
87. J. D. Jackson. *Classical Electrodynamics,* 3rd ed., John Wiley and Sons, New Jersey, 1999.
88. A. Mock. First principles derivation of microcavity semiconductor laser threshold condition and its application to FDTD active cavity modeling. *Journal of the Optical Society of America B*, 27(11):2262–2272, 2010.
89. G. W. Hanson. Dyadic Green's functions and guided surface waves for a surface conductivity model of graphene. *Journal of Applied Physics*, 103(6):064302, 2008.
90. W. Zhao, P. H. Tan, J. Zhang, and J. Liu. Charge transfer and optical phonon mixing in few-layer graphene chemically doped with sulfuric acid. *Physical Review B*, 82(24):245423, 2010.
91. A. Asi and L. Shafai. Dispersion analysis of anisotropic inhomogeneous waveguides using compact 2D-FDTD. *IEEE Electronics Letters*, 28(15):1451–1452, 1992.
92. S. Xiao, R. Vahldieck, and H. Jin. Full-wave analysis of guided wave structures using a novel 2-D FDTD. *IEEE Microwave and Guided Wave Letters*, 2(5):165–167, 1992.
93. S. Xiao and R. Vahldieck. An efficient 2-D FDTD algorithm using real variables [guided wavestructure analysis]. *IEEE Microwave and Guided Wave Letters*, 3(5):127–129, 1993.
94. Y. Chen and R. Mittra. A highly efficient finite-difference time domain algorithm for analyzing axisymmetric waveguides. *Microwave and Optical Technology Letters*, 15(4):201–203, 1997.
95. M. Qiu. Analysis of guided modes in photonic crystal fibers using the finite-difference time-domain method. *Microwave and Optical Technology Letters*, 30(5):327–330, 2001.
96. S. F. Hadi and M. F. Mahmoud. Optimizing the compact-FDTD algorithm for electrically large wave guiding structures. *Progress in Electromagnetics Research*, 75:253–269, 2007.
97. A. Mock and J. D. O'Brien. Dependence of silicon-on-insulator waveguide loss on lower oxide cladding thickness. In *Integrated Photonics and Nanophotonics Research and Applications Topical Meeting*, p. IWG4, Boston, MA, USA, July, 2008. Optical Society of America.
98. Q. Lu, W. Guo, D. C. Byrne, and J. F. Donegan. Compact 2-D FDTD method combined with Padé approximation transform for leaky mode analysis. *Journal of Lightwave Technology*, 28(11):1638–1645, 2010.
99. A. Mock and P. Trader. Photonic crystal fiber analysis using cylindrical FDTD with Bloch boundary conditions. *PIERS Online*, 6(8):783–787, 2010.
100. W. Kuang, W. J. Kim, A. Mock, and J. D. O'Brien. Propagation loss of line-defect photonic crystal slab waveguides. *IEEE Journal of Selected Topics in Quantum Electronics*, 12(6):1183–1195, 2006.

101. J.-P. Berenger. A perfectly matched layer for the absorption of electromagnetic waves. *Journal of Computational Physics*, 114(2):185–200, 1994.
102. A. Mohammadi, H. Nadgaran, and M. Agio. Contour-path effective permittivities for the two-dimensional finite-difference time-domain method. *Optics Express*, 13(25):10367–10381, 2007.
103. Z. Q. Li, E. A. Henriksen, Z. Jiang, Z. Hao, M. C. Martin, P. Kim, H. L. Stormer, and D. N. Basov. Dirac charge dynamics in graphene by infrared spectroscopy. *Nature Physics*, 4(7):532–535, 2008.
104. C. Lee, J. Y. Kim, S. Bae, K. S. Kim, B. H. Hong, and E. J. Choi. Optical response of large scale single layer graphene. *Applied Physics Letters*, 98(7):071905, 2011.
105. J. Y. Kim, C. Lee, S. Bae, K. S. Kim, B. H. Hong, and E. J. Choi. Far-infrared study of substrate-effect on large scale graphene. *Applied Physics Letters*, 98(20):201907, 2011.
106. G. W. Hanson, A. B. Yakovlev, and A. Mafi. Excitation of discrete and continuous spectrum for a surface conductivity model of graphene. *Journal of Applied Physics*, 110(11):114305, 2011.

20 Visible Optical Extinction and Dispersion of Graphene in Water

John Texter

CONTENTS

ABSTRACT

Aqueous dispersion of graphene is of interest as it affords environmentally safe handling of graphene for coating, composite, electronic, and other materials applications. The dispersion of graphene in water and certain other solvents using surfactants, polymers, and other dispersants is reviewed, and results show that almost completely exfoliated graphene may be obtained at concentrations varying from 0.001 to over 10% by weight in water. The molecular features promoting good dispersion are reviewed and demonstrated. A critical review of optical extinction shows that the visible absorption coefficients of graphene have been reported over the ranges of 7–66 cm²/mg at various wavelengths. We critically present data that prove the higher among these estimates to be good lower bounds to the completely exfoliated dispersion state, with an extinction coefficient of about 100 cm²/mg at 500 nm (80–86 cm²/mg at 660 nm). An inescapable conclusion is that the practice of energetically activating graphene in various solvents with various stabilizers, followed by centrifugation to isolate the "good" dispersion components is fine for producing samples amenable to transmission electron microscopy (TEM) analysis and quantification, but cannot be expected to drive value-added productions of products on the kilogram or higher scale. We also show that nearly complete exfoliation at concentrations of 1%–5% by weight in water is feasible without centrifugation.

20.1 INTRODUCTION

It has been pointed out that graphene has low solubility in water and organic solvents,[1] and it must be borne in mind that graphene is *insoluble* in water. Single-sheet graphene constitutes a separate phase from any solvent, and no finite solubility of graphene in any solvent has been credibly reported; very dilute and metastable dispersions of graphene may exist in any solvent. The primary motivation for preparing *dispersions* of graphene in water and other solvents, as with multiphase dispersions of almost anything else, is to produce a highly processable material form that can be pumped and metered without risk of respiratory distress or exposure.

The production of graphene on various length scales and in subgram to kg quantities has been reviewed,[1–11] and chemical vapor deposition (CVD) appears satisfactory for producing defect-free platelets with area dimensions of order μm². Much larger quantities may be obtained from graphite, but defects, grain boundaries, and polycrystallinity remain problematic. Applications requiring large and contiguous graphene sheets with widths of dimension greater than tens of μm at present will not benefit from graphene dispersion, though advances in our ability to process graphene micro and nanoplatelets may obviate this restriction in the future.[8] A summary of various ways to produce graphene sheets and platelets on different length scales is given in Table 20.1.

The focus of water-based dispersions is driven by now long-standing efforts to decrease volatile organic components (VOC) in industrial processing and in our environment. As a practical matter, nonaqueous solvents continue in use as long as they provide a significant cost or feature advantage. Nonaqueous graphene dispersions are favored for ease of exfoliation and other reasons for some select solvents. But some of these, such as chlorosulfonic acid (CSA), may exhibit significant toxicity. In the sequel we will show that an evolution to certain stimuli-responsive stabilizers make it possible to disperse graphene nicely in water, and to adjust the dispersion stability of the stabilizer to facilitate phase transfer to other solvents of choice at concentrations far in excess of most reports to date.

TABLE 20.1
Graphene Production Methods

Process	Method/Amounts/Dimensions	References
Graphite exfoliation	Mechanical (scotch tape)/prototype/cm^2	9,10
	Sonication in liquids/g to kg/0.5 to 10 μm^2	3,11
	Electrochemical activation in liquids/mg to g	12
Chemical vapor deposition	Carbon dissolution and phase segregation/0.5 mm to nm scale with polycrystallinity/up to 76 cm wide roll to roll	13–16
Graphite oxidation and reduction	Wet chemical oxidation, dispersion, and reduction/scalable to large quantities/10 cm^2 flakes possible	17–19

The basic options available for dispersing graphene and graphene oxide (GO) in water and other solvents and in exfoliating graphene have been reviewed by various groups.[3,6,7,20] The general dispersion process is to mix powdered graphene with a solvent and to apply some sort of shear field or activating energy to promote graphene exfoliation into a more refined dispersion form. The primary source of graphene sheets is either graphite powder or graphene powder in which individual graphene sheets are stacked. This type of stacking makes the dispersion of graphene very different from most dispersion processes applied in the dispersion of inorganics and organics, although the basic problem is topologically equivalent to the exfoliation of clays into polymers and monomers. Such inorganics and organics are usually agglomerated masses of spheroidal or lathe-like primary particles,[21,22] and are often highly susceptible to activated separation when subjected to strong shear fields as produced in small media comminution systems highly suitable for large volume throughput. In graphene it has become clear that ultrasonic activation appears to be the most effective type of activation in getting single- and few-layer platelets to exfoliate. As in most dispersion processing, this kind of exfoliation or transformation of large aggregates into primary particles is greatly facilitated by the presence of stabilizers or dispersing aids that physically bind to the newly generated sheet or particle surfaces while also providing strong interactions with the solvent, thereby sterically or electrostatically stabilizing the newly dispersed material. This type of direct stabilization by adsorbed dispersing aids (surfactants, polymers) is the predominant mode of dispersion, and is generally found to be the preferred route for high volume and lower-cost dispersions.

A more expensive approach, but arguably a more elegant and effective one, is to covalently attach stabilizers to the surfaces of graphene. The chemistry that produces surface-reactive centers on graphene surfaces is generally not also effective at exfoliating graphene, so conventional activation such as sonication must be part of the dispersion process. These methods have been more extensively applied to the indirect production of graphene dispersions, wherein graphene is first chemically transformed into graphene oxide (GO), a material with much higher surface energy than normal graphene, as a result of extensive oxidation of interior and edge sites. The attachment of many different types of stabilizing brushes, polymers, etc. to GO edge and interior sites is relatively straightforward using the tools of surface polymer chemistry, and such stabilized GO dispersions may then be chemically reduced to produce highly stabilized reduced graphene (rGO) dispersions.

An intriguing form of dispersion is that of solubilization by a number of "strong" solvents, including CSA and *N*-methylpyrrole. This kind of dispersion has been termed solvation, and the extent to which actual solvation is achieved is arguable (see sequel). However, it is an application area that has stimulated some interesting surface energy argumentation and some particularly interesting solubility parameter modeling (well and ill founded).

The outer planar surfaces of a graphene few to multilayer platelet and the surfaces of graphene sheets are generally recognized as being hydrophobic. By hydrophobic we mean that water does not spontaneously wet graphene surfaces, and this lack of wetting will occur whenever the spreading coefficient, $S_{w/g}$ for water on graphene is negative. This spreading coefficient is rigorously thermodynamically defined as[23]

$$S_{w/g} = \gamma_g - \gamma_w - \gamma_{wg}$$

where γ_w is the surface tension (surface energy) of liquid water, γ_g is the surface energy of graphene, and γ_{wg} is the interfacial free energy of the water/graphene interface. Each of these terms is a Gibbs free energy, and should not be confused with internal energy, adhesion energies, or cohesive energies. This spreading coefficient is simply the negative of the Gibbs free energy for water spreading on graphene, so a negative spreading coefficient corresponds to a positive free energy change that cannot occur spontaneously. The surface tension of water at 25°C is about 72.8 mN/m (erg/cm^2). The surface energy of graphene is purported to be 46.7 mN/m,[24] and a recently estimated interfacial energy is 77 mN/m for the water/graphene interface.[25] The spreading coefficient of water on graphene, therefore, sums to about −103 mN/m, and spontaneous wetting by water is theoretically precluded. A similar magnitude interfacial energy of 90.5 mN/m can be derived from the aqueous wetting data of Wang et al.[24] This value yields a spreading coefficient of −117 mN/m. One caution to be borne in mind is that the films examined by Wang and coworkers were not contiguous atomically smooth films, but films obtained by filtration of graphene suspensions. The inherent roughness from platelet edges would be expected to present higher surface energy than an atomically smooth layer. Another caution is that the fitting procedure used an equation of state proposed by Neuman.[26,27] Zisman's method of estimating critical surface energies would provide useful comparative data.[28,29]

The large contact angle of 127° reported by Wang et al.[24] for water on graphene and its deviation from smaller values observed for graphene were questioned by extensive molecular dynamics (MD) and theoretical studies of Shih et al.[25] and of Taherian et al.[30] The first of these reports predicts a largest observable contact angle of 96°,[25] and the other predicts an angle in the range of 95–100°.[30] These studies fix the difference between graphene's surface free energy and the graphene/water interfacial free energy, $\gamma_g - \gamma_{wg}$, at $\gamma_w \cos \theta$, or −6.3 mN/m to −12.6 mN/m. These estimates also ensure a negative spreading coefficient of −79.1 mN/m to −89.4 mN/m for water on graphene.

These graphene/water interfacial free energy terms can vary significantly with solvent and perhaps with underlying substrates (depending on the graphene thickness), and so spontaneous wetting can be obtained with some solvents while being precluded with others. This spontaneous wetting is key to liquid exfoliation in the absence of surfactants and polymers, and is examined in the next section.

However, naturally occurring graphite and graphene subjected to plasma, chemical, or high-frequency sonication may have different types of surface defects or surface functionalities. In this case, the surface energy may be substantially higher than it would be in the absence of such functionalities. For graphene derived from naturally occurring graphite, a surface energy of about 57 mN/m has been estimated.[24] The above-mentioned graphene surface energy estimate of 46.7 mN/m was obtained for rGO obtained by reducing graphite oxide (GO). The source of this GO was the same graphite studied, after being oxidized by a combination of nitric acid and sodium oxychloride.[24]

Essentially three scenarios have been pursued in the direct dispersion of graphene. In the first, graphene is exfoliated in a given solvent without any added surface modifying additive. The processing method may induce chemical modifications of the graphene surface that promote dispersion. In the second approach, chemical functionality is produced on the graphene surface, and this functionality promotes solvation, dissolution, and exfoliation. The third approach utilizes surfactants, polymers, and other agents that physically adsorb onto the graphene surface and stabilize graphene surfaces in a given solvent.[31] One must largely rely on thermal activation (e.g., sonication) to produce exfoliating fluctuations (de-adhesions) and diffusion of stabilizers to freshly formed surfaces to "trap" and stabilize exfoliation.

The characterization of graphene dispersion has involved a lot of centrifugation and sedimentation. These methods are quite commonly applied in the broader field of dispersion analysis. Transmission electron microscopy (TEM) and scanning electron microscopy (SEM) have been heavily applied to dispersions, and in combination with atomic force microscopy (AFM), have been used to produce some interesting thickness and lateral size distributions. It seems Raman spectroscopy has been applied more than any other technique in characterizing graphene dispersions. Its main attribute has been to affirm the existence of graphene, or that the material being examined has remained graphene. A secondary application area is the use of Raman bandwidths to estimate platelet thicknesses.[32–34] Overall, one has to conclude that these thickness rules need to be scaled for use in dispersions. Optical absorption in the visible and UV has also been used to estimate the quality of dispersion. It has been generally recognized that the visible optical absorption of a given amount of dispersed graphene increases with the extent of its dispersion or exfoliation. An inescapable conclusion is that optical absorption is the most useful characterization method for estimating the overall extent of graphene dispersion. Other sizing methods such as photon correlation spectroscopy appear to offer promise. Conductivity methods, optical absorption, and diffraction techniques are useful in examining graphene coatings, as also is SEM.

Here we attempt to provide a cohesive overview of graphene dispersions. We first examine attempts to solubilize or disperse graphene nanosheets in various solvents in the absence of interfacial stabilizers of any kind. This process is represented by part "d" of Figure 20.1.[35–40] These efforts are useful in helping us focus on some of the thermodynamic factors that are critically important in solubilization and wetting. We next focus on rGO dispersions derived from graphitic oxide (GO) in different contexts. Such an oxidative approach, illustrated by process "f" in Figure 20.1, involves the creation of many oxygen-containing surface defects and the subsequent partial to full removal of such defects using various reducing chemistries. These methods offer possible advantages as well as disadvantages.

The drive for aqueous dispersions and minimizing the use of VOC is well established in every arena where processes and products work well with "solvents," but eliminating or reducing atmospheric emissions of VOC and reducing handling of such solvents is an ongoing environmental and workplace goal. Aqueous surfactant- and polymer-stabilized dispersions are each more exhaustively reviewed, and these processes are represented by part "e" of Figure 20.1. We apologize for any significant omissions. We then summarize a variety of alternative dispersion approaches, before focusing on the optical extinction of aqueous graphene dispersions. These alternative approaches include electrochemically assisted exfoliation (part "c" of Figure 20.1), exfoliation of graphite intercalation compounds (part "a" of Figure 20.1), and exfoliation of thermally expanded graphite (part "b" of Figure 20.1). Eventually, graphene dispersions in liquids will probably be produced by small media milling (comminute ion) technologies.[40] These, processing methods, generally at room temperature, are highly scalable, though whether the degrees of exfoliation obtained by sonication can be matched is yet to be shown.

One of the simplest experiments to perform in a chemical or analytical laboratory, the optical extinction of graphene dispersions has received attention only from a few groups. It will become apparent that this extinction is about the most significant property that one can measure when addressing gradual exfoliation.

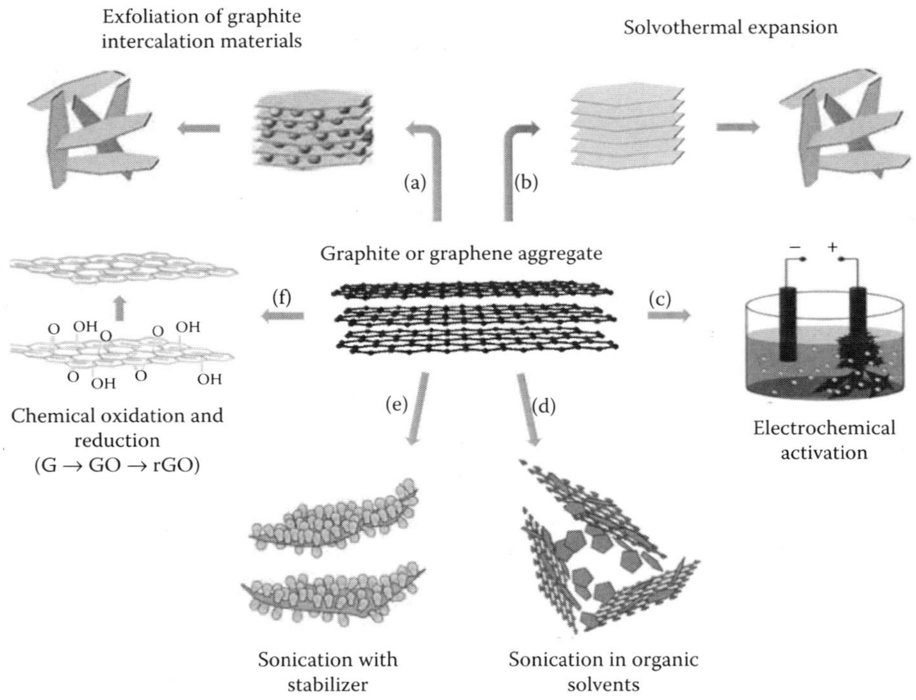

FIGURE 20.1 Cartoon illustrating processing approaches to obtaining graphene dispersions: (a) exfoliation of graphite intercalation materials involves activation of graphite intercalation compounds (GIC); (b) solvothermal exfoliation is applied to graphite that has first been thermally expanded by a high-temperature heat treatment; (c) electrochemical activation uses an applied potential to drive exfoliation; (d) sonication in organic solvents uses direct sonication in various solvents, but without auxiliary stabilizer present; (e) sonication with stabilizer, such as with surfactants or polymers, uses specific stabilizers to mitigate against re-aggregation; (f) chemical oxidation to graphite oxide followed by chemical reduction back to reduced single-sheet graphite oxide (graphene) can be done with or without added stabilizers. Most of these methods use sonication as an activation process tool, and centrifugation is used in some of these processes to remove poorly dispersed aggregates and particulates. (Adapted with permission from: (a) and (b) Neha, B. et al. *Open Journal of Organic Polymer Materials* 2:75–79. Copyright 2012, SciRes; (c) Loh, K. P. et al. 2010. The chemistry of graphene. *J. Mater. Chem.* 20:2277–2289. Reproduced by permission of The Royal Society of Chemistry; (d) Gong, P. W. et al.: One-pot sonochemical preparation of fluorographene and selective tuning of its fluorine coverage. *J. Mater. Chem.* 22:16950–16956. Copyright 2012, Wiley-VCH Verlag GmbH & Co. KGaA. Reproduced with permission; (e) Laaksonen, P. et al.: Interfacial engineering by proteins: Exfoliation and functionalization of graphene by hydrophobins. *Angew. Chemie Int. Ed.* 49:4946–4949. Copyright 2010, Wiley-VCH Verlag GmbH & Co. KGaA. Reproduced with permission; (f) Singh, K., Ohlan, A., Dhawan, S. K. Polymer-graphene nanocomposites: Preparation, characterization, properties, and applications. In *Nanocomposites—New Trends and Developments*. Ebrahimi F., editor. InTech, Rijeka, Croatia, pp. 37–71. Copyright 2012, InTech.)

20.2 GRAPHENE IN SOLVENT DISPERSIONS WITHOUT STABILIZERS

An important result that emerges from the several attempts to disperse graphene in various solvents with and without stabilizers is that the extent of exfoliation increases with sonication time. Sonication provides high energy activation, and this activation is necessary in order to overcome graphene interlayer adhesion. Diffusion of a suitable stabilizer to the new interfaces created is required to stabilize the exfoliation process against re-aggregation. The Coleman group[41,42] recently has provided detailed studies of mild sonication time on graphene exfoliation in *N*-methyl pyrrolidone (NMP), wherein concentrations up to 1 mg/mL (~0.1% w/w) can be obtained. A scaling analysis of their experiments is illustrated in Figure 20.2, where the dispersed concentrations, C_G, were estimated assuming an extinction coefficient of 24.6 cm^2/mg at 660 nm. The slope of 0.5 provides the following significant result[41]: $C_G \sim t^{1/2}$.

It was recognized that more useful coatings and films could be obtained if the single- and few-layer graphene sheets obtained during exfoliation were larger in greatest dimension.[43] The Coleman group addressed this problem by developing a centrifugation selection method that generated successively larger single and few sheet graphene dispersions. After exfoliation into NMP and centrifugation, the sediment was redispersed using less-activated processing and centrifuged at a slower speed. This process was repeated using slower and slower speeds, and sediments having successively larger dimensions were obtained.[43] An alternative process based on a very crude form of gravity-driven size exclusion chromatography has also been developed and applied to aqueous graphene dispersions stabilized by surfactants.[44]

Unfortunate confusion about the meaning of adhesion energy, cohesion energy density, and surface energy have resulted in ill-founded justifications of solvent exfoliations in terms referred to as "matching" the surface energy of graphene

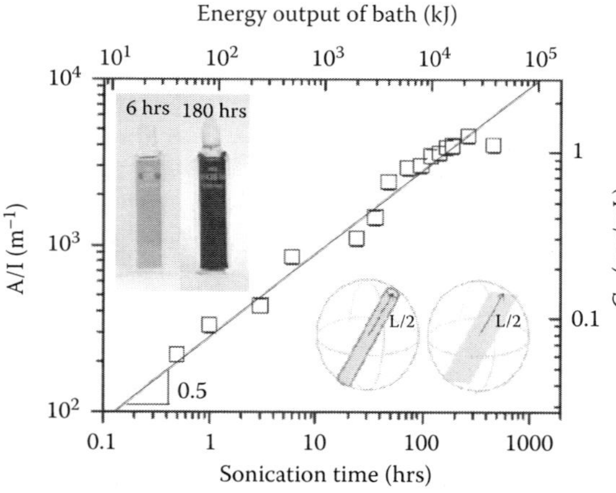

FIGURE 20.2 Scaling analysis of graphene dispersion (A/l and C_G) with sonication time and bath energy output. (Khan, U. et al.: High concentration solvent-exfoliation of graphene. *Small* 6:864–871. Copyright 2010, Wiley-VCH Verlag GmbH & Co. KGaA. Reproduced with permission.)

with the surface (energy) tension of various solvents.[6] Similarly, rationalizations[45,46] of why some solvents are better exfoliating agents than others, put forward in terms of Hildebrand[47] and Hansen[48] solubility parameters, are not entirely well founded. A cohesive energy density of 131 cal/cm^3, about half of experimental estimates of 267 cal/cm^3 (61 meV),[49] was assumed.[46] Dispersed graphene quantities obtained in various solvents were correlated with solvent Hildebrand parameters resulting in a functionally multivalued "scatter-plot" with an apparent peak around the assumed value.[45,46] This questionable fitting was explained by expressing a need to further fit this overall Hildebrand parameter to a vectorial basis in terms of three-dimensional Hansen parameters. These parameters partition the cohesive energy density (δ^2) into van der Waals dispersion (δ_d^2), polar (δ_p^2), and hydrogen bonding (δ_h^2) components. Unfortunately, this Hansen parameter fitting[45] was not done in terms of three-dimensional surfaces as is accepted,[50,51] but in terms of one-dimensional weighting of scatter plot data for each vectorial component.

MD simulations of C_{60} fullerene hydration were extended to carbon nanotubes (CNTs) and to graphene platelets (5 nm × 5 nm) by Smith and coworkers, and these studies produced interesting qualitative results.[52] The graphene surface is attractive to water molecules because of dispersion forces, and thus water was found to adsorb to these surfaces independently of curvature. Water density profiles varied significantly for different radii of curvature when a Weeks–Chandler–Anderson water–carbon purely repulsive potential was used, and graphene appeared to be the most hydrophobic of the nanocarbons studied. However, when a more realistic Lennard–Jones potential was used, the resulting density profiles were hardly distinguishable.[52] Water-induced repulsions between nanocarbon particles were found to be less significant with decreasing curvature, and in this sense these MD calculations indicate graphene to be more hydrophobic than CNT and C_{60}.

An elegant model with firm thermodynamic basis for understanding why various polar solvents appear to be better graphene sheet stabilizers than others has been developed by Blankschtein and coworkers.[53] Potentials of mean force per unit area graphene (PMF) were derived from MD simulations for each of the solvents studied. These potentials were expressed as a function of separation distance of parallel sheets, and then combined with a colloid model of slow kinetic aggregation to estimate the relative resistance to aggregation provided by different solvents. The solvents studied were NMP, dimethyl sulfoxide (DMSO), dimethylformamide (DMF), γ-butyrolactone (GBL), and water. These studies predicted the following relative stabilizing power of these solvents in the following order: NMP ≈ DMSO > DMF > GBL > H_2O.[53] This ranking appears to be consistent with experimental experience. Example results for NMP over a three-month aggregation period are illustrated in Figure 20.3, where the initial conditions were taken from a Hernandez and coworkers TEM analysis. This model was subsequently extended (see sequel) to analyze surfactant stabilization of graphene sheets in water.[54]

Following the work of Aida and Fukushima and coworkers[55–57] who had great success in exfoliating CNTs in various ionic liquids, a variety of groups used imidazolium-based ionic liquids as alternative polar solvents for graphene exfoliation. Nuvoli et al.[58] successfully exfoliated graphene in the ionic liquid, 1-hexyl-3-methylimidazolium hexafluorophosphate ($C_6C_1ImPF_6$) at a weight concentration of 0.53%. This extent of exfoliation is one of the highest values reported. However, 90% of the source graphite was discarded following the centrifugation process for isolating this dispersion. Luo and coworkers[59] examined electrochemical exfoliation in $C_8C_1ImPF_6$ (1-octyl-3-methyl-imidazolium hexafluorophosphate) and an equal volume of water was used along with

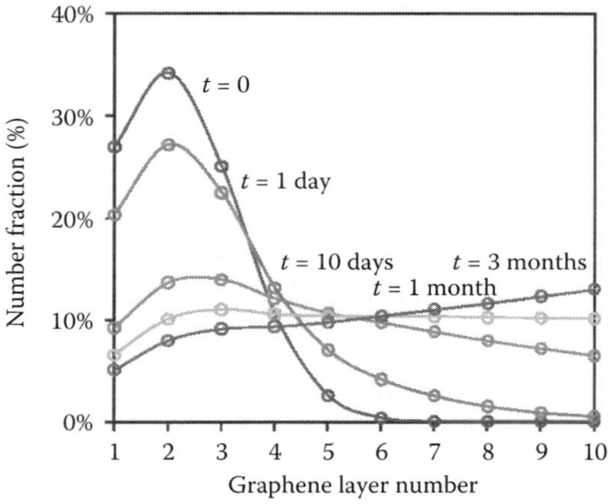

FIGURE 20.3 Predicted graphene sheet layer thickness distribution in NMP as a function of sheet layer number over a three-month aggregation period. (Reprinted with permission from Shih, C.-J. et al. Understanding the stabilization of liquid-phase-exfoliated graphene in polar solvents: Molecular dynamics simulations and kinetic theory of colloid aggregation. *J. Am. Chem. Soc.* 132:14638–14648. Copyright 2010 American Chemical Society.)

graphite electrodes; a 15 V applied potential was used to effect exfoliation. The exfoliated material was not stable in water, but it was somewhat stable in DMF, DMSO, and NMP for appreciable periods. Similar results were obtained for the chloride and tetrafluoroborate homologs (C_8C_1ImCl; $C_8C_1ImBF_4$) and for $C_4C_1ImPF_6$ (1-butyl-3-methylimidazolium hexafluorophosphate).[59] Loh and coworkers in a similar electrochemical study used potentials in the 1.5–15 V range to expand and disperse highly oriented pyrolytic graphite (at the anode) in a mixture of water and $C_4C_1ImBF_4$ (1-butyl-3-methylimidazolium tetrafluoroborate).[60] Dispersions of almost 0.1% graphene in 1-butyl-3-methylimidazolium ditriflateimide ($C_4C_1ImTf_2$) and in 1-butyl-1-methylpyrrolidinium ditriflateimide (C_4mpyTf_2) were reported by Wang et al.[61] These dispersions were prepared by sonication of a crude dispersion followed by centrifugation and disposal of >90% of the original graphite. These and other results are summarized in Table 20.2.

Khan et al.[5] succeeded in dispersing graphite in *N*-methyl-2-pyrrolidone (NMP) at up to about 6% by weight. These dispersions were noted to be unstable with respect to sedimentation and agglomeration, but the authors made the point that certain applications could be gainfully pursued by using such unstable dispersions prior to significant sedimentation. They also illustrated the collection of treated but unexfoliated particulates for subsequent dispersion.

O'Neill et al.[62] examined the dispersion of graphene in low boiling solvents. The motivation for such dispersions is that graphene films can easily be formed by spraying if the solvent is sufficiently volatile. Acetone (bp 56°C), chloroform (bp 61°C), and isopropanol (bp 82°C) were examined, and the best result was obtained with chloroform, with a final dispersion of 0.05% graphene by weight.

A process called aqueous counter collision (ACC), and somewhat analogous to jet milling, in which liquid water replaces the high-velocity gas carrier of jet milling, has been used to disperse fullerene, multiwall carbon nanotubes (MWCNT), and graphite.[66] In this process, opposing nozzles direct aqueous suspension streams into one another, and shear forces emanating from the accompanying collisions serve to exfoliate the substrate. Graphite particles initially greater than 100 μm in equivalent spheroidal greatest dimension were reduced to particulates of a few tens of μm's in diameter. Centrifugation of such suspensions produced supernatants composed of graphitic particles having equivalent diameters less than 1 μm. Visible spectra of such supernatants suggest graphene concentrations less than 0.002 mg/mL. The simple sonication of graphite powder in water without added stabilizer was shown to produce platelets 300–500 nm in lateral dimension and 20–50 nm thick.[67] While such dispersions were inherently unstable and did not exhibit lengthy lifetimes, and added stabilizers greatly lengthened such lifetimes, such stabilizer-free dispersions may find profitable application wherein the dispersions are used very shortly after production. More analytical details about the extent of exfoliation were not given. A more recent effort at sonication in water without stabilizer has produced graphene in water dispersions at graphene concentration up to 1 mg/mL by using much higher power sonication than usual (1200 W).[68] The resulting dispersions appeared stable with respect to sedimentation for up to 25 days. These dispersions were prepared as part of a study of "high thermal conductivity nanofluids," attempting to capitalize on the very high lateral thermal conductivity of graphene (3 kW/m/K). The stability obtained in this last case, no doubt, arises from oxidative chemical defects imparted by the high sonication power used. In any case, such processing is noteworthy, and coated properties of the graphene produced should be studied. Optical absorption of a 1 mg/mL dispersions at 500 nm produced an absorbance of about 2. Dividing by the concentration, 1 mg/mL, and assuming a 1 cm pathlength, we obtain an effective absorption coefficient of 2 mg/cm². This is considerably lower than the single sheet limit (see discussion in last section), and suggests that the resulting platelets are quite thick and far from the few-sheet platelets (most likely closer

TABLE 20.2
Graphene Dispersed in Solvent without Added Stabilizer

Solvent	Process/Character	Dispersed Graphene (μg/mL)	Reference
NMP	Sonication/unstable	60,000	5
NMP	Sonication/centrifugation	≤1000	4 ,42
$CHCl_3$	Sonication/centrifugation	500	62
Acetone	Sonication/centrifugation	<500	62
Isopropanol	Sonication/centrifugation	<500	62
$C_6C_1ImPF_6$	Sonication	5300	58
$C_4C_1ImTf_2$	Sonication/centrifugation	100	61
$C_8C_1ImPF_6$/water	Electrochemical expansion/ centrifugation	–	59
$C_4C_1ImBF_4$/water	Electrochemical expansion/ centrifugation	–	60
DMF	Sonication	150	63
DMF/water or AMP	Sonication/centrifugation	300	63
DMF/water	Sonication/centrifugation	1000	64
Alcohol/water	Sonication/centrifugation	10–20	65
Water	ACC	<2	66
Water	Sonication without stabilizer/centrifugation; 200–500 nm in lateral dimension and 20–50 nm thick	–	67
Water	Sonication without stabilizer	1000	68
Water	Sonication/centrifugation; N-doped graphene	160	69
Hypophosphorous acid	Graphene nanoribbons derived from MWCNT by splitting them	200	70
CSA	Sonication of graphene "bricks"	–	41
	Cotreatment with H_2O_2	–	71

to 20–50 nm thick). Such dispersions may be excellent for some applications, and good starting materials for processes that utilize stabilizers.

The last "water only" application we discuss is that where chitosan is thermally transformed into N-doped graphene by a hydrothermal process.[69] Chitosan was treated by a sequence ending in moderate (200°C) to high temperature (900°C) carbonization in supercritical CO_2. Sonication in water followed by centrifugation produced supernatant suspensions 0.16 mg/mL in graphene.

Zhang et al.[63] found that additives could increase the dispersion of graphene in DMF. They found that both water and 2-amino-2-methyl-1-propanol (AMP) in combination doubled (0.03% by weight) the amount of graphene that could be dispersed relative to DMF alone. Zhu et al.[64] examined various water/DMF ratios and found such mixtures with 10% water to yield about 0.1% by weight dispersed graphene. Water/alcohol mixtures produced dispersions only 10–20 μg/mL for ethanol and isopropanol when applied to graphene exfoliation from graphite.[65]

A synergistic effect of combining CSA and hydrogen peroxide (H_2O_2) to exfoliate and stabilize graphene few-layer sheets in CSA at up to 0.3% by weight (3 mg/mL) was reported by Lu et al.[50] Various control experiments provided support for a mechanism involving intercalation of CSA between graphene sheets, diffusion of H_2O_2 to the intercalated CSA, conversion of CSA by H_2O_2 to Caro's acid with concomitant heat generation, graphene expansion, and exfoliation following heat generation. Such dispersions may foreseeably exhibit practical applications, although removal of the CSA may arise as a processing issue. Hypophosphorous acid (H_3PO_2) has been used to stabilize graphene nanoribbons (GNR) in water, wherein the GNR are derived from MWCNT by splitting open MWCNT.[70] The GNR are sonicated in the presence of H_3PO_2, and 0.2 mg/mL concentrations were obtained.

Mohanty et al.[72] used "nanotomy" to cut graphene multilayers into well-defined rectangular, square, and ribbon kinds of "bricks." These bricks were then exfoliated in CSA to produce single- and few-sheet nanostructures. These and other studies discussed above are summarized in Table 20.2.

20.3 GRAPHITE OXIDE-BASED GRAPHENE DISPERSIONS

Graphite oxide (GO) production and various routes to its conversion to single- and few-sheet graphene have been reviewed.[17–19,73] The essential steps are summarized in Figure 20.4, where strong acid is used to oxidize graphite to graphite oxide (GO), the GO is dispersed in a particular solvent with physically or covalently attached stabilizer, followed by chemical or thermal reduction of stabilized GO to stabilized rGO.[74] Because GO has a much higher surface energy, it is intrinsically easier to disperse in water. In addition, since this higher surface energy emanates from the various types of oxygen species introduced during oxidation, including $-CO_2^-$, $-OH$, $>C=O$, and glycidyl ether (epoxy), covalent functionalization with reagents that couple with such species is facile, as is covalently linking many other types of functional and reactive groups. The select studies discussed below are summarized in Table 20.3. After functionalization of GO, chemical or thermal reduction to rGO is also usually facile. The resulting rGO is generally then stable in solvents that solvate the functional groups attached. Because these chemistries have been well established, more work has evolved in the production and stabilization of rGO types of graphene, than the direct dispersion of graphite or graphene aggregates in solvents. Uses for GO and stabilized GO, such as in fillers for various composites, have evolved for GO that are orthogonal to applications for graphene. We therefore omit review of such studies and applications herein.

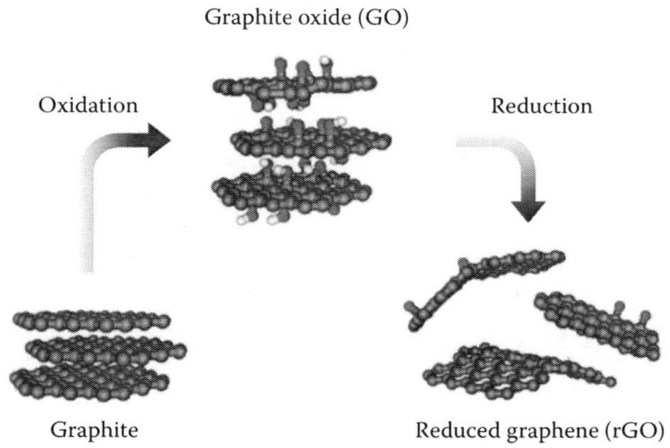

FIGURE 20.4 Chemical cycle for the production of graphene from graphite. The oxidation and reduction steps may be electrochemical or wet chemical. (Adapted with permission from McAllister, M. J. et al. *Chem. Mater.* 19:4396–4404. Copyright 2007 American Chemical Society.)

Graphite oxide (GO) sheets at 1 mg/mL in water were stabilized with a series of Pluronic triblock surfactants, F68 ($EO_{76}PO_{30}EO_{76}$), P123 ($EO_{20}PO_{70}EO_{20}$), and F127 ($EO_{106}PO_{70}EO_{106}$) at a Pluronic to GO weight ratio of 27:1.[78] The graphitic oxide was then reduced to graphene using hydrazine under sonication and extremely stable dispersions of graphene were obtained at a final concentration of graphene of about 0.05% by weight. The stabilization mechanism was the standard one,[104] where the propylene oxide (PO) blocks stick to the graphitic and graphene surfaces and the ethylene oxide (EO) arms extend into the aqueous domain to provide steric stabilization. Concentration limits were not explored, but interesting gels were prepared using these highly stable and dilute dipsersions.[78] Similar experiments using TX-100 as surfactant failed to produce a stable graphene dispersion.[75]

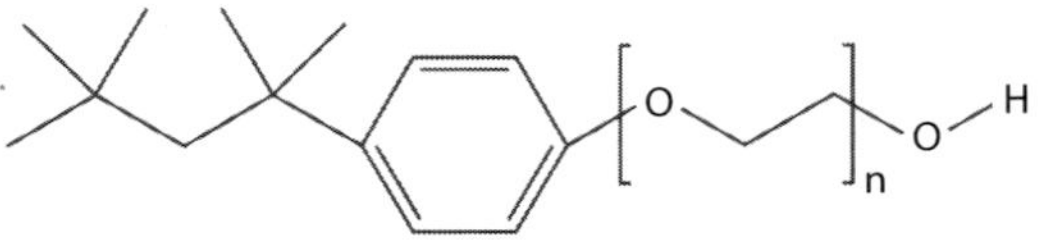

TX-100 (n ~ 9–10)

A similar synthetic approach stabilizing GO platelets with tryptophan and then using hydrazine to reduce the

TABLE 20.3
Stabilized rGO Dispersions

Stabilizer[a]/Solvent	Chief Result	Reference
TX-100/H_2O	Unstable	75
Tryptophan/H_2O	≤0.1 mg rGO/mL	76
RD/H_2O	0.1–0.5 mg/mL; photoluminescent sensor	77
SDBS/H_2O	Used to displace RD	77
LPC/H_2O	Used to displace RD	77
F68/H_2O	~0.5 mg rGO/mL; 1:27 rGO:F68; gelation	78
P123/H_2O	~0.5 mg rGO/mL; 1:27 rGO:P123; gelation	78
F127/H_2O	~0.5 mg rGO/mL; 1:27 rGO:F127; gelation	78
—CO_2^-/H_2O	0.65 mg rGO/mL; incomplete reduction with HMTA	79
—CO_2^-/aqueous NaOH	15 mg rGO/mL; incomplete thermal reduction to rGO	80
—CO_2^-/basic acetone and IPA	Incomplete thermal reduction to rGO	80
—CO_2H and /NMP, DMA	Incomplete thermal reduction to rGO	81
—NHC_3C_1ImBr/H_2O, DMF, and DMSO	≤0.5 mg rGO/mL	82
Poly(ε-caprolactone) —$C_{11}C_1ImTFSI$/H_2O	GO surface modified with $HOC_{11}C_1ImTFSI$; poly(ε-caprolactone) grown from these surface hydroxyl groups	83
Poly(VBImBr)/H_2O	Phase transfer to PC induced by Br^- to Tf_2N^- anion exchange	84
pVC_2ImBr/H_2O	1.5 mg/mL; cyclical water to PC (or NMP, NM) phase transfer induced by Br^- to Tf_2N^- anion exchange and reversed by anion exchange back to Br^-	84
PGMIC/H_2O	Dilute dispersion after reduction/sonication/centrifugation	85
Poly(Bu_4P^+ 4-styrenesulfonate)	1.3 mg rGO/mL; LCST transition at 61°C	86
HPC/H_2O	<1mg rGO/mL; 1:50 rGO/HPC composite films formed above and below LCST	87
SPEEK/H_2O	≤2.4 mg rGO/mL; ~500 F/g double-layer capacitance	88
MMT/H_2O	Thermally robust freestanding films at 10/90 and 20/80 rGO/MMT	89
Poly(norepinephrine)	Simultaneous surface functionalization and reduction of GO to rGO/p(nor)	90
PMMA	GO grafted with PMMA followed by reduction to rGO	91
PEI/H_2O	PEI adsorbed onto exfoliated GO followed by reduction to rGO; used for double-layer capacitors	92
Poly(ε-caprolactone)–p(nor)	SIP of ε-caprolactone by ring opening polymerization	90
–Poly(ε-caprolactone)	GO modified with 2-azidoethanol followed by ring opening polymerization with ε-caprolactone and reduction to rGO	92,93
PVC	Physically adsorbed PVC/rGO composite	94
–PVA/H_2O	PVA condensation onto GO followed by reduction to rGO	95
–PVC	PVC modified with 4-hydroxythiophenolate esterified onto GO followed by reduction to rGO	94
–hexadecyl	GO modified with 3-azidopropan-1-amine followed by amidation with palmitoyl chloride	92
–PS	SIP of ATRP from GO; thermal reduction to form porous rGO films	96,97
–PS	SIP ATRP of styrene from diazonium anchored surface initiator; tensile mechanical properties increased for 0.9% rGO relative to bulk PS	98
–PS	ATRP of styrene after GO modified with 2-azidoethyl-2-bromo-2-methylpropanoate	92
–PEMA/DCB	SIP of ATRP of EMA from rGO modified with hydroxyarene radical cation	99
–PCN/DMF	Azo coupling of diazonium salt of 4-aminobenzo-nitrile with rGO–PEMA to produce azo brush; 0.5 mg/mL; surface relief gratings photoproduced	99
–PU/H_2O	Condensation of polyurethane onto GO, reduction to rGO, improved mechanical properties of rGO/PU composites relative to PU alone	95
–Poly(L-lysine)/H_2O	GO amidated with poly(L-lysine) followed by reduction to rGO	100
81PAM/H_2O	LCST switching of aggregation	91
–PNIPAM/H_2O	LCST switching of aggregation	101
–PNIPAM/H_2O	GO was esterified with HTEMPO and thermally reduced to rGO-TEMPO, and then bromide terminated PNIPAM was grafted onto the rGO-TEMPO	102
PDMAEMA	SIP via ATRP from surface alkyl bromides; LCST	103
PAA	SIP via ATRP from surface alkyl bromides; LCST	103

[a] Denotes covalent attachment to GO, then rGO surface.

GO to graphene produced stable graphene dispersions.[76] The graphene dispersion concentration was a bit less than 0.1 mg/mL (0.01% w/w), and a 10:1 weight ratio of tryptophan to graphene was used.

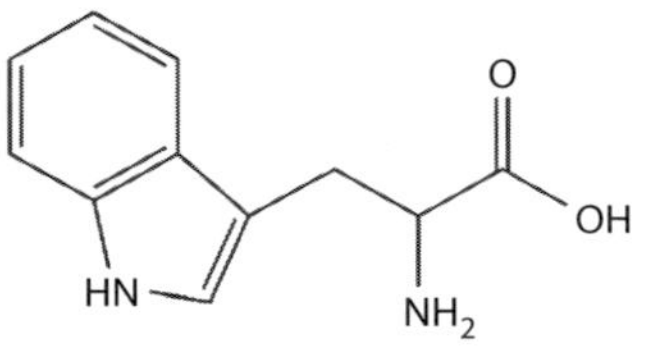

Tryptophan

GO was prepared similarly by a modified Hummers method,[105] and then poly-L-lysine (M_w ~ 30–70 kDa) was attached to surface epoxide sites by alkaline incubation for 24 h at 70°C. This was followed by sodium borohydride reduction to eliminate any remaining oxidation defects to return the GO to rGO.[100,106]

Hydrazine has been the predominant treatment for reducing GO to rGO in aqueous dispersions. A "green" approach was put forward by Shen et al.[79] who used hexamethylenetetramine (HTMA) as a reducing agent and also claimed it acted postreduction as a stabilizer. The reduction mechanism was ascribed to slow hydrolysis of HTMA to produce formaldehyde and ammonia, with the formaldehyde then acting as a chemical-reducing agent of GO oxygen-rich sites. Sufficient reduction was achieved to produce black-appearing rGO dispersions, but the resulting stability of the rGO points to incomplete reduction and retention of stabilizing carboxylate groups. An extent of dispersion of only 0.65 mg/mL (0.065% w/w) was obtained, and while green (if generating formaldehyde can be considered green), this extent of dispersion cannot be considered useful.

GO was favorably stabilized in water by 1-(11-hydroxyundecyl)-3-methyl-imidazolium-*N*,*N*-bis(trifluoromethane) sulfonamide (HO-$C_{11}C_1$ImTFSI).[83] The imidazolium π system provides strong binding overlap to the GO, and the hydroxyl groups were then used to anchor poly(ε-caprolactone) oligomers grown by ring opening polymerization of ε-caprolactone. These GO sheets were then chemically reduced to produce stable rGO dispersions. A polymerized ionic liquid (PIL)[107] based on 1-vinyl-3-butyl imidazolium bromide (VBImBr) was used[84] to stabilize GO in water and during chemical reduction of the GO to rGO. Following a phase transfer demonstration invented by Mecerreyes and coworkers[108] for single wall carbon nanotubes (SWCNT), Kim et al.[84] added lithium ditriflate imide ($LiNTf_2$) to destabilize the aqueous dispersion, collected the precipitate, and redispersed the PIL-stabilized rGO in propylene carbonate. This phase transfer aspect arises from the stimuli responsiveness of imidazolium ionic liquid-based PIL and is illustrated in the sequel for pristine exfoliated graphene sheets using a nanolatex[109] based on a PIL.

Tölle et al.[80] have developed a highly performing dispersion process based on GO (from graphite by Hummer's method[105]), wherein rGO (TrGO; Tr for thermal reduction) concentrations as high as 1.5% (w/w) are obtained by thermal reduction. Surfactants or polymeric dispersing aids are not used, but in aqueous dispersion it is required to use an alkaline (NaOH) treatment in order to ionize surface carboxyl groups. Such aqueous dispersions are claimed to be stable for months. These dispersions are created after alkali treatment by a high-pressure homogenization process, and are described as high viscosity pastes.[80] Such high viscosity may be viewed as a major contributor to the apparent dispersion stabilization. Agglomeration processes are severely impeded when particle and sheet diffusion are retarded by high viscosity; it may also be the case that such pastes are effective networks or gels, physical states that inherently retard diffusion and agglomeration.

The thermal reduction process was very elegantly correlated with the oxygen content remaining after heating at a given temperature (400°C, 500°C, 600°C, 800°C, 900°C, 1000°C). For stable dispersions or pastes in water it was found that oxygen contents of 13%–16% by weight were needed. Dispersion in nonaqueous solvents such as isopropanol and acetone required less retained oxygen (5%–10% w/w) and higher temperature reductions could be used while maintaining apparent dispersion stability.[80]

Printing of TrGO films with and without binder (polyvinylpyrrolidone) showed that conductivity decreased with binder addition. Binder-free images on paper, foil, and by vacuum-assisted flow produced electrical conductivities of 8–16 S/cm. The test objects printed with binder had conductivities of about 0.5 S/cm. This dispersion approach is particularly interesting because it uses an intermediate form of rGO, TrGO, wherein the oxygen content is controlled and utilized to effect dispersion stability. In water, the electrostatic stabilization is lost when the dispersions are acidified, and the repulsion provided by oxygen anions is lost. The 1.5% (w/w) level of TrGO is also highly significant, and is the highest concentration reported for any GO-based rGO dispersion. It is also highly significant that complete reduction of GO to rGO is not needed in order to obtain significant levels of conductivity in printed test objects. It has been shown by Zhang et al.[81] that rGO with sufficient sodium carboxy functionality could be precipitated by adding excess salt, filtered, and redispersed in organic media, such as DMF and NMP.

Chang et al.[77] demonstrated the stabilization of graphene derived by reduction of GO to rGO in water using an apparently delocalized multiring dendrimer (RD). The phenyl and benzothiadiazole rings as well as the alkyne groups would presumably provide strong adsorption to the graphenic surfaces and the pendent EO oligomers would be highly solvated in water, providing thereby amphiphilic stabilization. However, the strength of this dendrimeric adsorption has to be questioned by the facile replacement of RD by sodium dodecylbenzenesulfonate (SDBS) and by lysophospholipid (LPC). This limitation is attributable to the degrees of conformational freedom the RD possesses. We conjecture that the separate phenyl-alkyne-benzothiadiazole-phenyl chains would be better dispersing aids than the tertiary amine. Following similar earlier work,[110] stable rGO dispersions (~1 mg/mL) wrapped with SDBS were prepared[111] from GO by standard hydrazine reduction followed by arene radical cation coupling[112,113] with remaining rGO surface nucleophiles following denitrogenation of various diazonium salts. The choice or *p*-substituents on the ring allowed for stabilization in aqueous and other solvents.

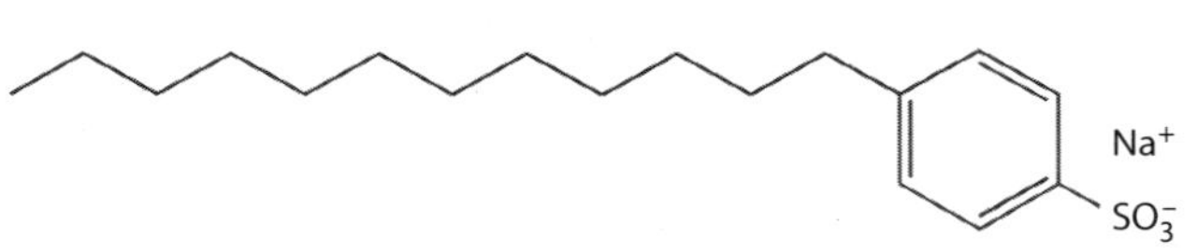

SDBS

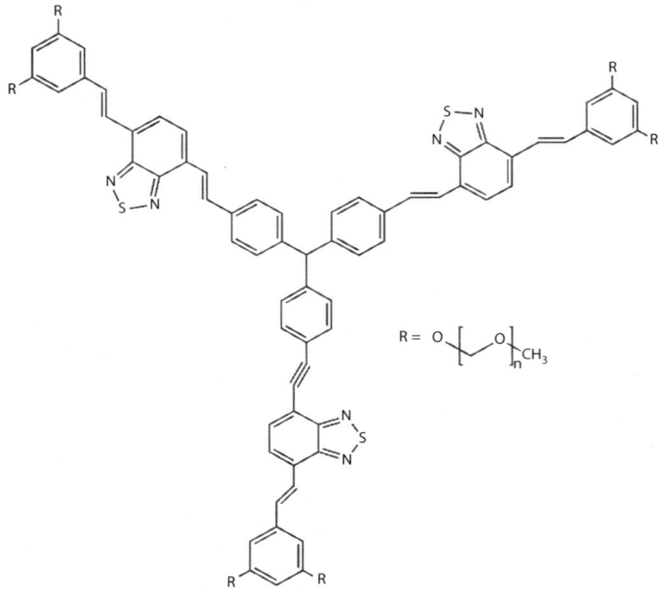

RD

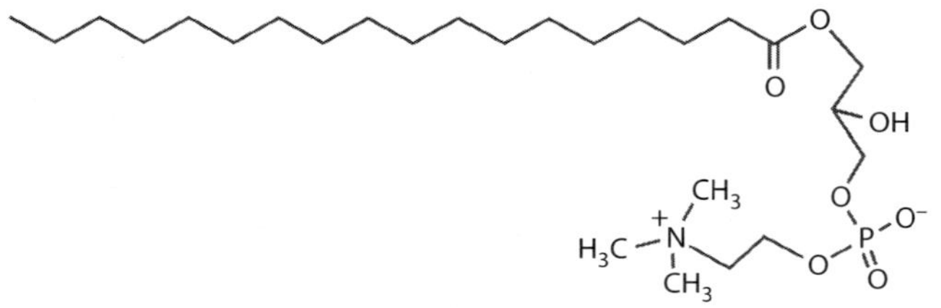

LPC

The stabilization of GO and rGO by various surface-initiated polymerization (SIP) or grafting from methods and grafting to methods has been recently reviewed.[114] SIP or grafting from methods include atom transfer radical polymerization (ATRP),[96–99] ring opening,[90] and condensation[115] methods.[114] Grafting to methods include esterification,[95,94] amidation,[92] nitrene cycloaddition,[116] nucleophilic epoxy ring opening,[100] atom transfer nitroxide radical coupling (ATNRC),[102] radical grafting,[91] and condensation[93] methods.[114] Grafting from methods usually are more expensive to implement than grafting to methods; grafting from methods generally provide greater control over areal surface density.

Reduced graphene (rGO) was stabilized in water by Lee et al.[117] using poly(N-isopropyl acrylamide) (PNIPAM). The PNIPAM backbone was presumed to provide anchoring to the rGO plate surfaces and solubilization and stabilization came from the solubility of hydrated substituted amide groups in the aqueous phase. While the specious argument that stabilization ensued from a matching of graphenic surface energy with aqueous PNIPAM surface tensions[117] is lacking in thermodynamic rigor or basis (when the interfacial surface energy vanishes), this paper illustrated a very interesting effect arising from lower critical solution temperature (LCST) shifts of PNIPAM because of its adsorption to graphene. Perhaps one of the most interesting effects demonstrated was the thermoreversible transition from a nonactinic dispersion at T < LCST, where one of the dispersions was sufficiently transparent to see through a given thickness, to an opaque form at T > LCST. Here we may conclude that for T < LCST the individually stabilized sheets were oriented randomly, producing a net optical extinction that could be "seen through." Further, upon raising the temperature to T > LCST, the contracted PNIPAM strands collapsed on themselves, leading to limited graphene sheet stacking parallel to the substrate and a concomitant increase in effective visible optical absorption.

An SIP process was applied by Ren et al.[101] but a modified Hummer's method[105] was used to produce GO. In this study, N-isopropylacrylamide (NIPAM) was used to grow oligomers by SIP using ATRP. Three batches of SIP-modified functional graphene sheets (FGS) were prepared by starting the polymerization at monomer to surface initiator ratios of 676:1, 1081:1, and 1516:1, and these ratios produced surface modified graphene having 7.5%, 32.5%, and 37.5% PNIPAM on a weight basis, respectively. The LCST for PNIPAM is 32°C, and these dispersions exhibited thermoreversible destabilization on heating to temperatures T > LCST; the same dispersion redispersed on cooling to room temperature, T < LCST. The mechanism in this system is similar to that discussed for the PDMEAMA system later.[101] Below the LCST, the PNIPAM oligomers are water loving and serve as excellent steric stabilizers for the graphene, rGO, sheets. Above the LCST the PNIPAM oligomers collapse and become hydrophobic, leading to aggregation of the graphene. On cooling this process is reversed.

Zhou et al.[118] used a PIL, poly(1-vinyl-3-butylimidazolium chloride), to stabilize rGO in water and IL. An initially aqueous GO dispersion was sonicated and then centrifuged. The resulting dilute GO dispersion was then mixed with PIL, stirred, and then the GO was reduced to rGO by addition of hydrazine. This aqueous rGO dispersion was then mixed with an equal volume of IL, 1-butyl-3-methylimidazolium hexafluorophosphate ($C_4C_1ImPF_6$), and the water was removed by evaporation yielding a PIL-stabilized rGO dispersion in IL, $C_4C_1ImPF_6$.

Yang et al.[82] used an interesting variant in covalently attaching IL moieties to graphenic surfaces. Their rational was that graphitic oxide as usually produced is believed to be replete with epoxide surface groups in addition to other oxygen species. By reacting such epoxide groups (glycidyl ether groups) with a primary amine, one should achieve facile coupling to the GO surface. A suitable surface modifying group was prepared by quaternizing methyl imidazole with 1-bromo-3-amino propane, to produce 1-(3-aminopropyl)-3-methyl imidazolium bromide ($H_2NC_3C_1ImBr$). This species reacts spontaneously with available glycidyl ether groups on GO to produce $—NHC_3C_1ImBr$ surface species and an accompanying hydroxyl group. While explicit reducing agents were not mentioned, it appeared that the lengthy surface treatment resulted in reducing the GO to rGO with a significant change in dispersions color from yellowish-orange to gray/black. This surface modified material was soluble in water, DMF, and DMSO at the dilute level of 0.05% by weight.

Dilute aqueous GO dispersions were mixed with each of several ILs, $C_4C_1ImBF_4$ (3-butyl-1-methyl imidazolium tetrafluoroborate), CA_3C_1ImCl (3-allyl-1-methyl imidazolium chloride), and C_4PyBF_4 (*N*-butyl pyridinium chloride).[119] These mixtures were then subjected to rotoevaporation to remove the water and to yield stable dispersions of GO in IL. These dispersions were then mixed with hydrazine to chemically reduce the GO to IL-stabilized rGO. Poly(1-viny-3-ethylimidazolium bromide) (pVC_2ImBr) was used by Suh and coworkers[84] to stabilize reduced graphite oxide sheets and to demonstrate phase transfer with this stabilizing polymer by stimuli responsive anion exchange. The homopolymer pVC_2ImBr had an $M_w \sim 170$ kDa. It was used to stabilize graphite oxide sheets in water, after which these GO sheets were reduced to graphene, rGO. Stabilized rGO sheets in water were at a concentration of about 1.5 mg/mL (0.15 wt%). These aqueous dispersions were very stable (lifetime greater than 6 months). On mixing with a water-immiscible solvent, propylene carbonate (PC), the stabilized rGO remained in the aqueous phase and the lower PC phase remained clear. However, on addition of an equivalent of lithium bistrifluoromethylsulfonimide ($LiNTf_2$), the hygroscopic imidazolium bromide underwent exchange to become a hydrophobic imidazolium ditriflateimide, and the aqueous graphene sheets aggregated. These aggregates could then be fully and stably dispersed in PC and in other solvents such as DMF, acetonitrile (AN), tetrahydrofuran (THF), NMP, and nitromethane (NM). These organic solvent dispersions could then be destabilized by addition of tetrabutylammonium bromide (TBAB) or tetrabutylphosphonium bromide (TBPB) and redispersed in water. This achievement follows an essentially identical one reported by the Mecerreyes group 4 years earlier using the same homopolymer in stabilizing SWCNT and in using lithium bispentafluoroimide for anion exchange processes to affect water/organic solvent phase transfer.[108]

Following the exposition of Lu et al.[107] in detailing new applications of PILs, particularly those based on imidazolium-based monomers, and the efficacy demonstrated[120,121] in stabilizing nanocarbons in water with such materials, Gao et al.[85] demonstrated the stabilization of graphene using an interesting polymer, poly(1-glycidyl-3-methylimidazolium chloride-*co*-epichlorohydrin) (PGMIC), derived from poly(epichlorohydrin) by refluxing with 1-methyl-imidazole in water.

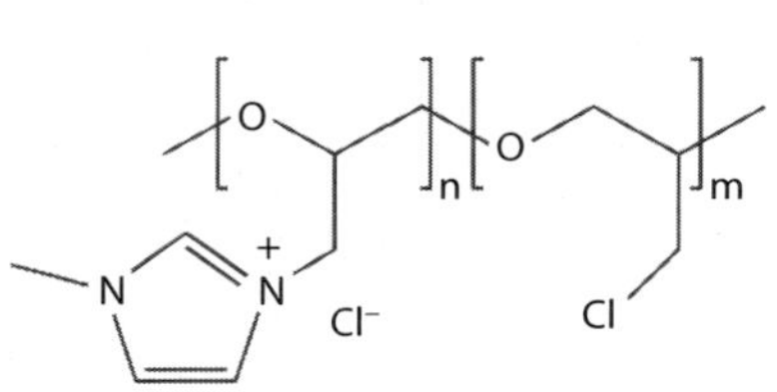

PGMIC

GO (5 mg) and PGMIC (200 mg) were mixed along with hydrazine. The GO was reduced to rGO, and the rGO/PGMIC mixture was sonicated and then centrifuged to produce a dilute and small amount of stable graphene in water.

A novel PIL based on the monomer tetrabutylphosphonium ($^+P(Bu)_4$) 4-styrenesulfonate has been reported by Yuan and coworkers[86] and by Kohno and Ohno.[122] Its efficacy in stabilizing rGO was demonstrated, and the thermal responsiveness of the PIL stabilizer, poly[p-$CH_2{=}CH{-}\phi{-}SO_3^-\ {}^+P(Bu)_4$], was shown to be useful in controlling coagulation and redispersion by triggering the LCST transition of this PIL at 61°C.[86] Below this temperature, the rGO dispersion was completely stable. Upon raising the temperature to 61°C, the PIL undergoes a volume phase transition while becoming incompatible with water, and the dispersion coagulates and sediments.

Hydroxypropylcellulose (HPC) stabilized GO/rGO dispersions were made by focusing on LCST properties of the HPC.[87] A dilute suspension of GO (1 mg/mL) was sonicated in the presence of HPC (~5% w/w) and then reduced by addition of hydrazine dropwise. Dispersions were dried and compressed to make solid composites. Higher conductivity composites were obtained when the temperature of the GO/HPC dispersions was raised above the LCST prior to hydrazine reduction. The stabilization of GO and rGO aqueous dispersions by sulfonated poly(ether-ether-ketone) (SPEEK) was demonstrated up to 2.4 mg/mL.[88] The mechanism of adsorption of the SPEEK to the GO and rGO surfaces was by π–π overlap. More significant than these dilute dispersions was their use in making double layer capacitor electrodes exhibiting specific capacitances of 476 F/g in 1 N H_2SO_4.

Zhang et al.[89] used montmorillonite (MMT) sheets to produce graphene/MMT freestanding films. Portions of stable suspensions of partially exfoliated MMT were mixed with GO dispersions to produce hybrid colayered GO–MMT suspensions. Both the GO and MMT surfaces attract water and produce broadened hydroxyl region FTIR spectra. These layered platelets were then chemically reduced with hydrazine to produce suspensions of rGO–MMT, that formed aligned and layered freestanding films upon vacuum filtration. These films were thermally very robust, and produced electrical conductivities of 3.8 and 290 mS/cm at rGO/MMT weight ratios, respectively, of 10/90 and 20/80.

A more extensive exhibition of thermoreversible responsiveness was demonstrated by Bak et al.[103] when they surface functionalized the faces of graphene sheets with alkyl bromides and used such surface initiators to grow poly(2-(dimethylamino)ethyl methacrylate) (PDMAEMA) by ATRP methods. The so-called FGS had been prepared by the Brodie method[123] and consequently were decorated with a population of hydroxyl and carboxyl groups to an extent of about 28% by weight. 4-Aminophenethyl alcohol was condensed onto these FGS surfaces in the presence of isopentyl nitrate to produce covalently anchored 4-aminophenethyl alcohol coupling sites. The nitrate converts the amine into a diazo group, and this group undergoes a first-order denitrogenation to produce a highly reactive arene radical cation. This radical couples in a diffusion controlled limit with nucleophilic

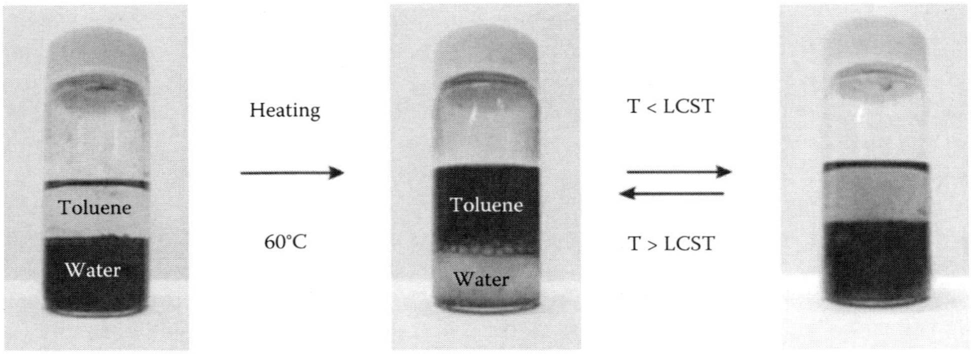

FIGURE 20.5 Phase transfer of FGS-PDMAEMA-16h between aqueous and organic phase upon temperature control. (Reprinted from *Polymer*, 53, Bak, J. M. et al. Thermoresponsive graphene nanosheets by functionalization with polymer brushes. 316–323, Copyright 2012, with permission from Elsevier.)

surface sites (and with other nucleophiles like water and halides). α-Bromoisobutyryl bromide was then condensed onto these 4-hydroxyethylphenyl sites to provide FGS decorated with alkylbromo surface initiator. This modified FGS-Br was about 12% by weight elemental bromine. These initiation sites were used to grow 2,2-(dimethylamino)ethyl methacrylate (DMAEMA) oligomers by ATRP using CuBr as catalyst and *N,N,N′,N″,N″*-pentamethyldiethylenetriamine (PMDETA) as ligand. Free oligomers were grown using ethyl α-bromoisobutyrate (EBiB) as auxiliary initiator to promote surface brush growth. Thermal gravimetric analysis (TGA) and gel permeation chromatography (GPC) methods were used to analyze the resulting oligomers formed. Brushes with M_n (PDI) of 7400 (1.09), 11700 (1.08), 17300 (1.06), 22000 (1.08), 26700 (1.09), and 59200 (1.22) were grown, respectively, in 0.5, 1, 2, 3, 4, and 16 h. The 1, 3, and 16 h polymerized samples had contents, respectively, of 24.9%, 40.1%, and 48.8% by weight of PDMAEMA brushes.

Bak and Lee[124] produced FGS–PAA (polyacrylic acid) by similar SIP methods starting with FGS-Br ATRP initiators. Poly(*t*-butyl acrylate) brushes were first grown, and then these *t*-butyl groups were hydrolyzed to produce PAA brushes. These brushes were effectively destabilized by lowering pH to 1.

The resulting thermoreversibility behavior produced interesting aqueous dispersion properties.[103] In a series of dispersions, 0.03% by weight with small to large M_n (7400 to 59,200 Da) oligomers of PDMAEMA, it was observed that a clear LCST effect existed in the larger M_n samples, but not so in the shorter length modified FGS samples. The effect observed when heating to 60°C (above the LCST for PDMAEMA) was dispersions destabilization (aggregation). This effect was first noticed for samples polymerized for 3 h (M_n ~ 22,000 Da). A possibly practical property enabled by these SIP brushes is thermoreversible phase transfer as illustrated in Figure 20.5. There we see that the nicely aqueous stabilized dispersion with M_n ~ 59,200 does not partition much into a proximal toluene phase when T < LCST. On raising temperature to T > LCST we see that the resulting hydrophobicity of the transformed graphene sheets drives these sheets into the toluene phase. When temperature is then lowered to T < LCST, most of the graphene is restabilized in the aqueous phase. Somewhat similar switching was observed in the surface energies of thin films cast from the same dispersion upon a silicon wafer.[103] At room temperature, a water droplet exhibited a contact angle of 23.2°. Raising the temperature to 60°C resulted in a contact angle of 98.1°, a transition from hydrophilic to hydrophobic.

20.4 SURFACTANT-STABILIZED GRAPHENE EXFOLIATION

Quite significantly, practical results have been achieved in using various types of surfactants to disperse graphene in water. Our discussion here begins with some of the efforts of the Coleman group. We then review quantitative results obtained with diverse types of surfactants. It will be shown that the most significant aspect of surfactant-mediated exfoliation is a strong interaction of the surfactant with the graphene surface. In this section we limit our attention to physical processing that starts with solid particulate graphene or graphite. Quantitative results are briefly summarized in Table 20.4. Not included in Table 20.4 are results from a recent comprehensive review[139] for small charged and nonionic surfactants and results for polymer modified pyrenes.[140]

The successful aqueous dispersion of graphene using surfactants follows many studies examining the use of diverse surfactants to exfoliate and to stabilize aqueous CNT dispersions. Several reviews are available,[141,142] and the "best" surfactants for dispersing CNTs are most likely very effective for dispersing graphene in view of the homologous surfaces involved. The first detailed study of surfactant-stabilized graphene from the Coleman group[125] provided very good physical characterizations of the dispersed materials. The surfactant used, SDBS was once the primary component in low-density (powdered) laundry detergent products. Lotya et al.[125] showed that a distribution of graphene platelets stable to gravitational sedimentation and to flocculation could be generated using low concentrations of graphene in water, very mild sonication,

TABLE 20.4
Aqueous Surfactant-Stabilized Graphene Dispersions

Surfactant or Stabilizer	Chief Results	Reference
SDBS	0.002–0.05 mg/mL (SC[a])	125
SDS	<0.02 mg/mL; formed into PS composite material	126
Sodium cholate	0.3 mg/mL (SC)	127
Sodium taurodeoxycholate	2.6 mg/mL (SC)	128
C_4C_1ImCholate	<5 mg/mL (SC)	129
$PySO_3Na$	0.028 mg/mL (SC)	130,131
$Py(SO_3Na)_2(OH)_2$	~0.05 mg/mL (sonication)	131
$Py(SO_3Na)_3OH$	~0.01 mg/mL (sonication)	131
$Py(SO_3Na)_4$	<0.01 mg/mL (sonication)	131
$PyCH_2NH_2$	<0.1 mg/mL (SC)	131
$PyCO_2H$	0.1–0.4 mg mL	132
PyBA	0.1–0.4 mg mL	132
$PySO_3Na$	0.1–0.7 mg/mL	132
ATS	1.5 mg/mL (SC)	133
C10	0.1–0.8 mg/mL (SC)	134
Humic acid	0.04 mg/mL (SC)	135
Perylene bisimides	Qualitative discussion	136
TB	10–11 mg/mL; sonication without centrifugation; anion responsive phase transfer	137,138

[a] SC—sonication followed by centrifugation.

and forceful centrifugation to remove larger and less colloidally stable platelets. This approach followed that adopted in the study of CNTs and some highly significant results were obtained. Detailed AFM analyses showed that the dimensions and thicknesses of the stably dispersed graphene flakes could be measured, and flake thickness distributions constructed. When flakes thicker than 12 nm were explored they found that the thickness number-frequency distribution peaked at about 1.5 nm; this thickness corresponds to slightly less than five graphene layers, assuming the bulk layer thickness of 0.34 nm applies. The experimental distribution function tailed to much higher thicknesses (6–10 nm), and a weight or volume normalized thickness distribution would peak at a much higher thickness. An early estimate of graphene's optical extinction at 660 nm was made, yielding a value of 13.6 cm²/mg. That this value was a rather low estimate (and lower bound; see sequel for more detailed optical extinction discussion) follows from the fact that these early exfoliations had mass average thicknesses much greater than five layers. These workers also showed that rather uniform graphene films could be deposited from dispersion by vacuum filtration.[125] This is a very significant demonstration, and will be key to the practical use of such dispersions in the future. This seminal article also contained an approximate Derjaguin, Landau, Verwey, and Overbeek (DLVO), colloidal stability treatment, and for charged surfactants such as SDBS it was qualitatively reasonable. The actual dispersions obtained using 0.5–10 mg SDBS/mL resulted in graphene dispersions about 0.002–0.05 mg/mL in carbon, or weight fractions of 2×10^{-6}–50×10^{-6} (2×10^{-4}–5×10^{-3} wt%). These results were obtained after centrifugation of "crude" dispersions, 0.1–10 mg/mL graphene (0.01–1 wt%). Despite these very low levels of dispersion, too low for practical application, a framework for further progress was provided.

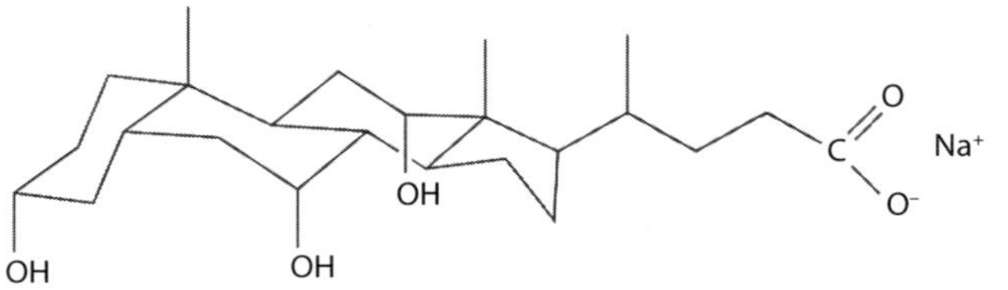

Sodium Cholate

In a continuation study,[127] colloidally stable dispersions up to 0.3 mg/mL (0.03 wt%) in graphene were obtained using a very different surfactant, sodium cholate (SC). One can see sterochemically that this surfactant may prefer to lie "flat" on a graphene surface with the hydroxyl groups and carboxyl group oriented toward the aqueous phase. SDBS has a hydrated sulfonate group attached to the benzene ring, and this arrangement prevents the π system of SDBS from a more favorable overlapping interaction with the graphene surface; the dodecyl group has unfettered availability to adsorb to the surface.

Similar methods to those of their initial efforts[125] involving centrifugation, TEM, optical absorption, and AFM were used.[127] The most significant result obtained was a showing that the effective optical absorption of their dispersions, and the amount of graphene remaining dispersed after a given amount of centrifugation increased more or less steadily with increasing sonication time. In this study a quite different estimate of 66 cm²/mg for the absorption coefficient at 660 nm was obtained. Green and Hersham showed that density gradient centrifugation of polydisperse aqueous graphene dispersions stabilized with SC could be used to fractionate such dispersions into fractions having uniform flake thicknesses.[143] Due to the intervening adsorbed surfactant onto graphene flake surfaces, buoyant densities were found to vary with the overall flake thicknesses. It has subsequently been shown without approximations of any sort that this extinction coefficient is considerably higher than the initial 13.6 cm²/mg value.[137]

A cholate-based stabilizer derived from the ionic liquid 1-butyl-3-methylimidazolium chloride (C_4C_1ImCl) by ion exchange with SC to yield C_4C_1ImCholate was claimed to be an IL, but corroborating physical properties were not reported.[129] Graphene powder was sonicated in an aqueous solution of 5 mg/mL C_4C_1ImCholate. The resulting centrifuged aqueous dispersion was used to support metal nanoparticles for catalytic applications. A subsequent study by Sun et al.[144] used sodium taurodeoxycholate to disperse graphene starting with graphite powder. They employed similar centrifugation treatments but the sonifier had a much wider tip, and they sonicated for only 24 h, but kept the sonication cool by conducting the treatment in an ice bath (similar to that adopted by Ager and Texter[137] and by Regev et al.[145]). A stable dispersion of 0.71 wt% was claimed, and it was further stated that the concentration could be increased to 1.2% by simple evaporation (of water). A difficulty in this study was that an extinction coefficient of only 29.2 cm^2/mg graphene was measured at 660 nm, and this low value indicates that the degree of exfoliation obtained was significantly less than that achieved by Lotya et al.[127] using SC and their processing scheme. There is no reasonable argument challenging the concentration of graphene in the dispersion. However, the degree of exfoliation and effective extinction are unequivocally linked.

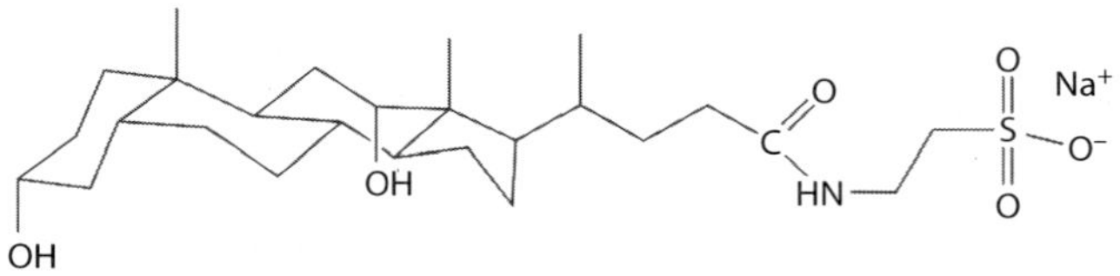

Sodium Taurodeoxycholate

A comparative study of SC and sodium deoxycholate confirmed the results reported by Lotya et al.[127] for SC and showed that deoxycholate was fivefold more effective.[128] The highest level of graphene dispersion with deoxycholate at 5 mg/mL used a 7 g graphite/10 mL crude dispersion and produced, after centrifugation, a dispersed graphene level of 2.6 mg/mL (0.26% w/w), producing almost 99% waste sediment.

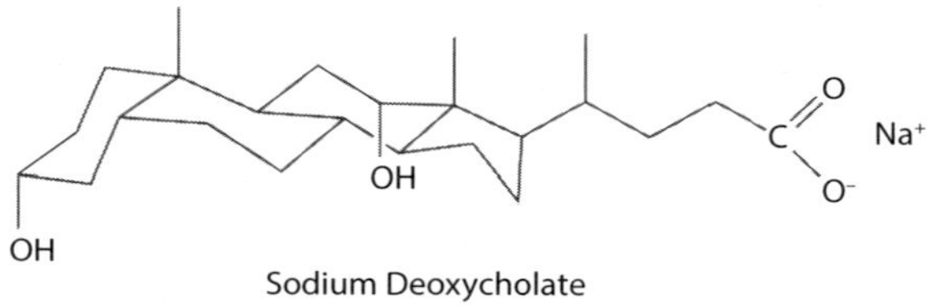

Sodium Deoxycholate

Anionic (AOT, SDBS) and cationic surfactants (ODA, EDMB) at 3% (w/w) in 1% graphite crude dispersions were sonicated at high power and then centrifuged to produce dispersions 0.02%–0.05% (w/w) in graphene that were stable for "some time."[146] The solubility of AOT [bis(2-ethylhexyl) sulfosuccinate, sodium salt] at room temperature is about 1% by weight, so those dispersions were most likely saturated in lamellar AOT vesicles. Octadecylamine (ODA) was protonated, and EDMB is ethylhexadecyldimethylammonium bromide.

Sodium 1-pyrenesulfonate was used to disperse graphite originally suspended at 3 mg/mL with the surfactant at 0.1 mg/mL.[130] After mild sonication and repetitive centrifugation to attempt to get rid of excess surfactant, a final graphene concentration of 0.074 mg/mL (0.0074% w/w) was estimated using an extinction coefficient of 24.6 cm^2/mg at 660 nm. Using the extinction determined more recently by Lotya et al.,[127] the more likely concentration obtained was 0.028 mg/mL.

The relative adsorptivity to graphene and exfoliation effectiveness for sodium 1-pyrenesulfonate ($PySO_3Na$) and three increasingly polar derivatives, $Py(SO_3Na)_2(OH)_2$, $Py(SO_3Na)_3OH$, and PS4, were analyzed in dilute dispersion.[131] The adsorptivity onto graphene (graphite) surfaces was found to be inversely proportional to the strength of the polar groups and followed the ranking $PySO_3Na > Py(SO_3Na)_2(OH)_2 > Py(SO_3Na)_3OH > Py(SO_3Na)_4$. Interestingly, $Py(SO_3Na)_2(OH)_2$ was slightly more effective in dispersing graphene (under identical processing conditions) than was $PySO_3Na$, an this effect correlated with $Py(SO_3Na)_2(OH)_2$ having the largest dipole moment.[131] $Py(SO_3Na)_3OH$ was slightly better than $Py(SO_3Na)_4$, but both were much less effective than $PySO_3Na$ and $Py(SO_3Na)_2(OH)_2$.

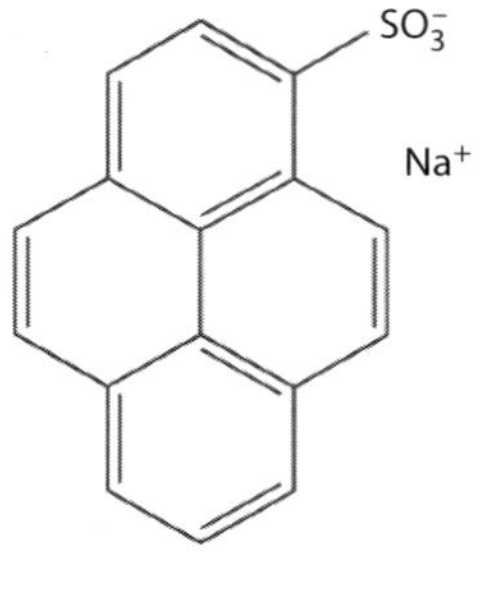

$PySO_3Na$

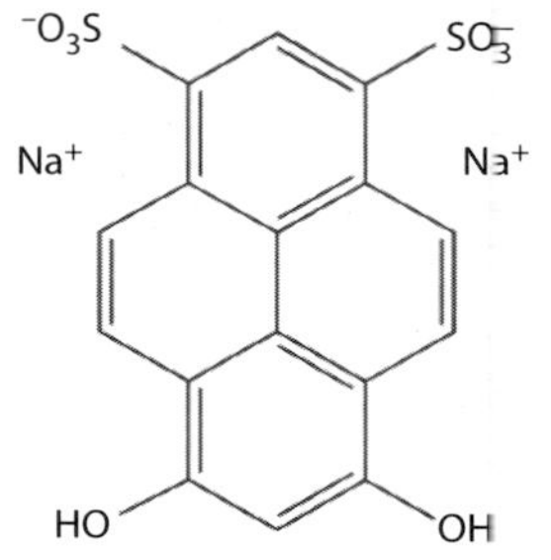

$Py(SO_3Na)_2(OH)_2$

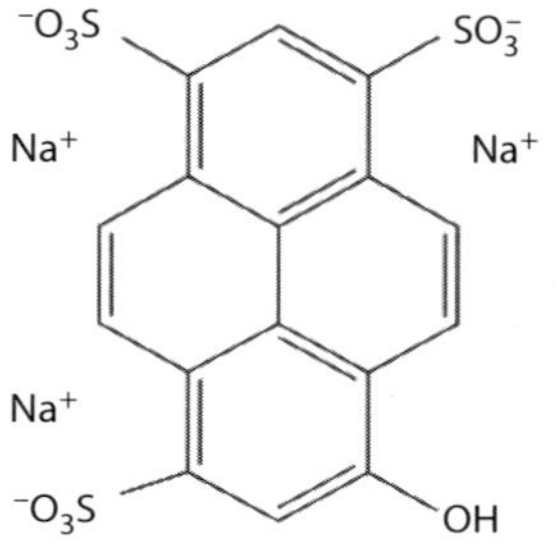

$Py(SO_3Na)_3OH$

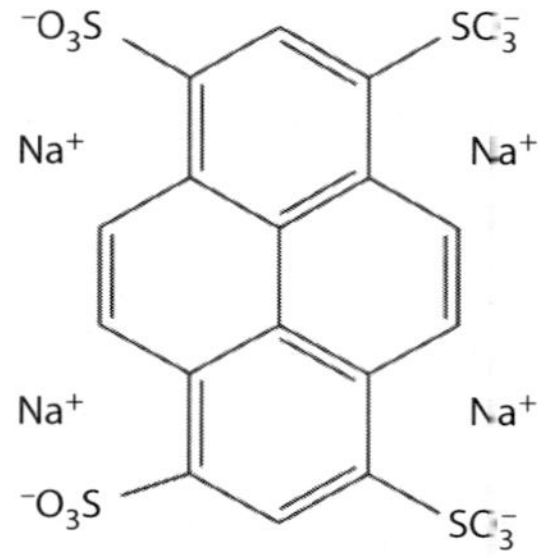

$Py(SO_3Na)_4$

Parvez et al.[132] attempted to determine the minimal dispersant required to disperse a given amount of graphene. They focused on a series of pyrene derivatives including $PyCH_2NH_2$ (1-aminomethyl pyrene), $PyCO_2H$ (1-pyrenecarboxylic acid), PyBA (1-pyrenebutyric acid), $PySO_3H$ (1-pyrenesulfonic acid), $PySO_3H$ (1-pyrenesulfonic acid sodium salt), and $Py(SO_3)_4$ (1,3,6,8-pyrenetetrasulfonic acid tetra sodium salt). The tetrasulfonate moiety was the least effective dispersant in the series, consistent with the above-discussed results of Schlierf et al.;[131] the aminomethyl derivative was also fairly ineffective and dispersed graphene at 0.1 mg/mL at most. The other moieties produced results of 0.1–0.7 mg/mL. The sodium 1-pyrenesulfonate yielded the best results, consistent with Schlierf et al., given that $Py(SO_3Na)_2(OH)_2$ was not considered. $PySO_3Na$ was compared to SDBS and to polyvinylpyrrole (PVP). A fivefold higher concentration of graphene was obtained relative to SDBS, and the respective graphene to stabilizer ratios were 0.33 and 0.036. Py-SASS produced 1 mg/mL graphene dispersed the same as PVP, but with 1/3 the weight of stabilizer.

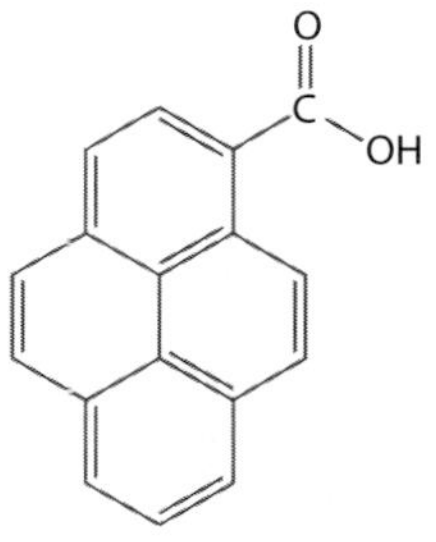

$PyCO_2H$

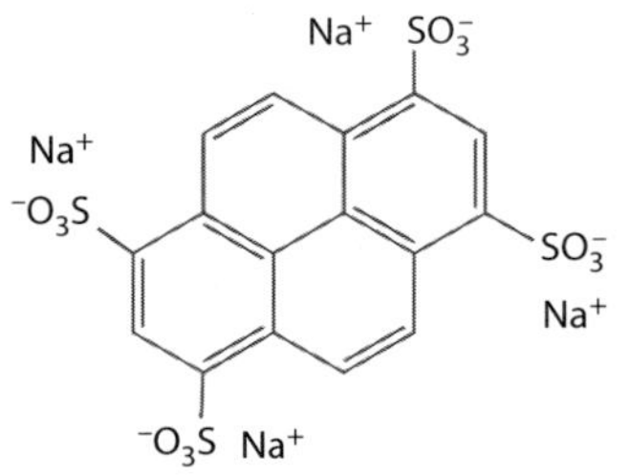

$Py(SO_3)_4$

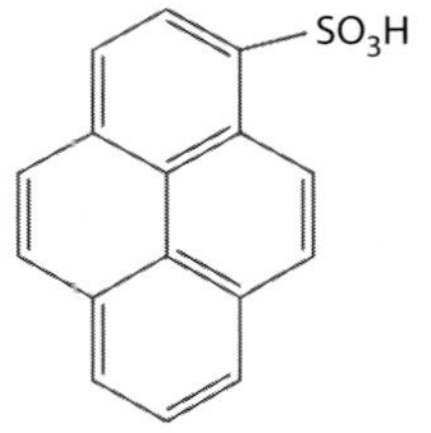

$PySO_3H$

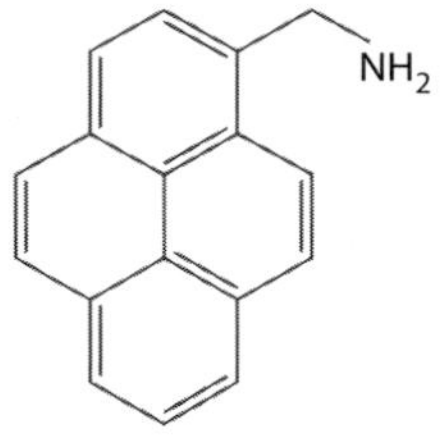

$PyCH_2NH_2$

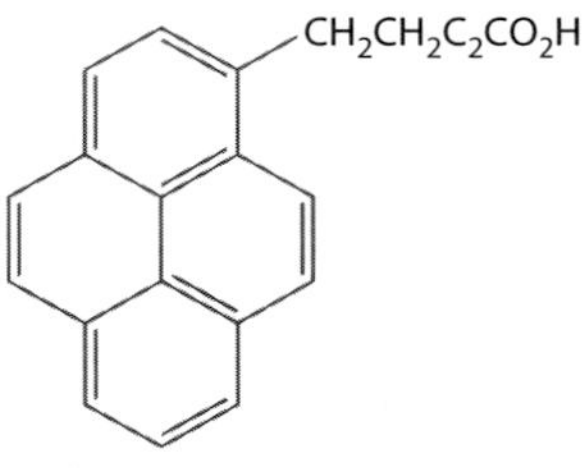

PyBA

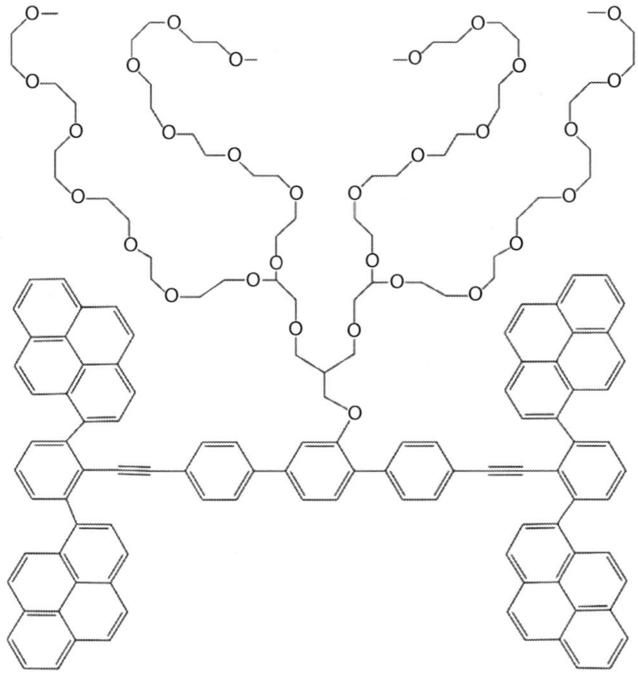

Amphiphilic tetrapyrene sheet (ATS)

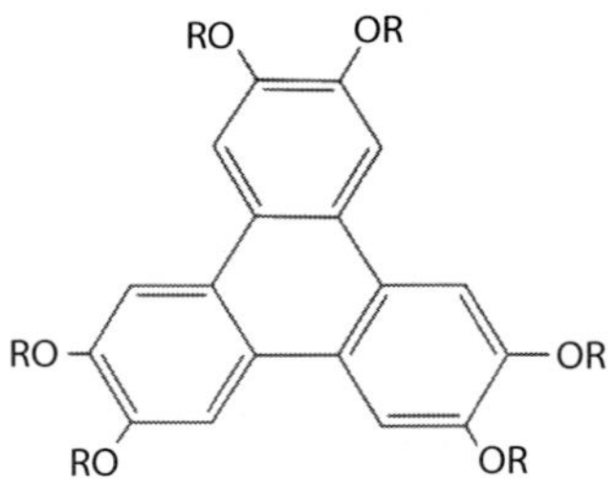

"10" ($R{=}(CH_2)10CO_2H$)

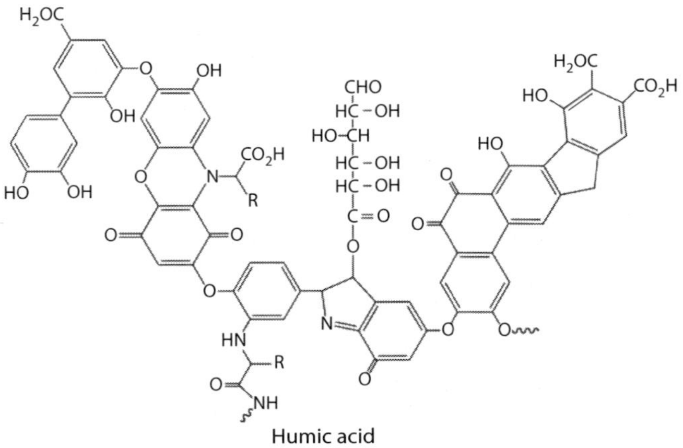

Humic acid

Lee et al.[133] devised an interesting nonionic amphiphile based on pyrene, amphiphilic tetrapyrene sheet (ATS). This stabilizer provided strong π–π overlap with the graphene surface, and the four hepta(EO) arms provide aqueous solubilization. This stabilizer was found to be ineffective for stabilizing SWCNT because of the high curvature of the SWCNT, and the extensive delocalization of the ATS, making it suitable only for zero to low curvature surfaces. Graphene concentrations of 1.5 mg/mL (0.15 wt%) were obtained with ATS.

Another graphene-like stabilizer that has been used to stabilize SWCNT[147] and graphene[134] is the hexa(carboxydecylether) triphenylene stabilizer, C10. Very dilute aqueous dispersions of graphene in the range of 0.1–0.8 mg/mL in which the C10 concentration was fivefold higher were obtained by sonication and centrifugation. This stabilizer is pH responsive because of the carboxylic functionality. Optical absorption measurements at 660 nm yielded an estimate of 15 cm^2/mg as an extinction coefficient, about fourfold smaller than the most recent estimate from the Coleman group. Humic acid and other "natural organic matter" were used by Ion et al.[135] to disperse graphene in water at levels not exceeding 0.04% w/w graphene using humic acid at about 0.029% w/w. The adsorptivity can be understood by examining the multiring structure that is a good match for sp^2 graphenic surfaces.

Perylene bisimides have been proposed as possible stabilizers by Hirsch and coworkers in aqueous and nonaqueous solvent systems.[136] The suitability for various solvents can be tuned by selecting the outer-most amide substituents. Symmetrical and asymmetrical examples can similarly be designed. The perylene component, because of its delocalized π system, offers the potential for strong binding to graphenic sp^2 surfaces.

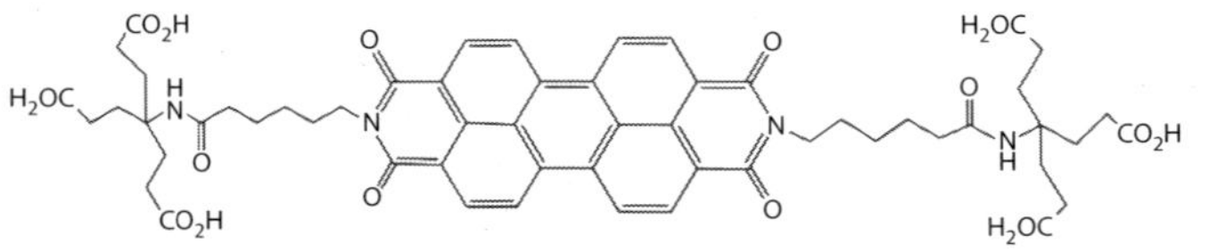

Perylene Bisimides

An interesting exfoliation mechanism was recently proposed by Notley[148] where continuous surfactant addition was claimed to produce more highly concentrated graphene dispersions. The crude dispersions used were much more concentrated than most other workers use, and ranged up to 15% w/w. Cationic (dodecyltrimethylammonium bromide, tetradecyltrimethylammonium bromide, and hexadecyltrimethylammonium bromide), anionic (sodium dodecylsulfate), and nonionic (Pluronics F127 and F108) were used and produced dispersed graphene concentrations after centrifugation of 0.5%–1.0% w/w. Of course the "waste" graphene sediment in this process was of the order of 90%. A dispersion of 1.5% w/w was claimed but without supporting data. Perhaps of most interest was the interesting mechanism put forward to justify the process. It was claimed that an optimal aqueous/air interfacial tension of 41 mN/m was needed, and that continuous surfactant addition made maintenance of this interfacial tension possible. It was not explained how such an interfacial tension meaningfully couples to the exfoliation/stabilization process or why simply adding much more surfactant at the process initiation might not result in an equivalent result.

Other "critical parameters" for sonication/centrifugation processing were discussed recently by Buzaglo et al.[145] using a series of nonionic surfactants, TX-100 and the Pluronics P65, P84, P103, D127, and P123, the anionics SC and SDS, and the cationics didecyldethyl ammonium bromide (DDAB) and (CTAB). Dispersed graphene concentrations ranged from about 0.03 to 0.18 mg/mL. Mixed sequences of low-power bath sonication and higher power tip sonication were used as well as keeping the dispersion in an ice bath while sonicating. All these dispersions were derived from crude dispersions initially, 1% w/w in graphene and surfactant levels of 0.5%–1% w/w. Again the waste graphene sediment was 98% and greater. Similarly, "scant" concentrations were reported by Seo et al.[149] using Pluronic and Tetronic block copolymers. "High concentration" dispersions containing ~0.07 mg/mL were obtained. Pluronics L64 and L68 were judged to be the least effective. Pluronic F77 and Tetronic 1107 were found to be much more effective, but with such low "high concentrations," no significant advance was obtained over the results reported by Buzaglo et al.[145] Very significant results have been obtained using block copolymer surfactants, however.

All these surfactant and some polymeric dispersion studies seem apparently interesting, and if one assumes the process of weak sonication with various amounts of surfactant or other stabilizer followed by centrifugation and examination of the dilute supernatant is meaningful, one can generate a perspective such as that illustrated in Figure 20.6.[139] One may then conclude that particular stabilizers are better than others, but contradictions to the meaningfulness of such collective results

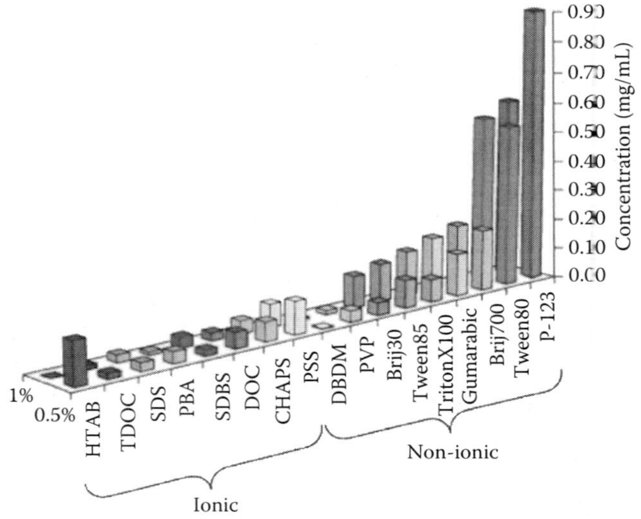

FIGURE 20.6 Comparison of supernatant dispersed graphene concentrations in aqueous stabilizers at 0.5% and 1% levels after centrifugation. (Reprinted from *Carbon*, 49, Guardia, L. et al High-throughput production of pristine graphene in an aqueous dispersion assisted by non-ionic surfactants. 1653–1662, Copyright 2011, with permission from Elsevier.)

are apparent. We see some stabilizers are over 10-fold more effective at half the concentration (HTAB), more active at half the concentration (DOC, CHAPS, PSS, PVP, P123), and some are insensitive to concentration (Brij700 and Tween80). All these results are for supernatants obtained after discarding almost 99% of the graphene treated to yield carbon weight fractions of less than 0.001. It is really hard to say which of these stabilizers is better than another, solely because the evaluation methods adopted by so many investigators fail to really address practical utility. A useful alternative approach is illustrated in the sequel.

An attempt was made by Smith et al.[151] to correlate zeta potentials with the concentration of graphene dispersion in extremely dilute aqueous dispersions produced by sonication and followed by centrifugation with concomitant discarding of 99% of the graphene originally added to the crude dispersion. While it was claimed that the degree of dispersion linearly correlated with zeta potential, the actual data represented a scatter plot, and no significant correlation was presented.

Blankschtein and coworkers[54] extended the colloidal slow aggregation model that they developed using a potential of mean force (PMF) between parallel graphene sheets for dispersion stabilization by polar solvents,[53] in examining dispersion stabilization by surfactants. They focused on aqueous dispersions stabilized by SC, and their MD simulations indicated a compact monolayer of SC forms that covers 60% of the surface, with the remainder covered by water and sodium counter ions. The PMF provided very significant insight into the stabilization mechanism. For separations less than 1.7 nm the PMF derived deviates significantly from DLVO theory and illustrates a very significant repulsive barrier (up to 8 kJ/mol nm^{-2}) in excess of the DLVO repulsion over 1.1–1.7 nm separations, and attractive weak potential minimum at 1.05 nm (−3 kJ/mol nm^{-2}). Thus, the essence of SC stabilization in water is based on steric stabilization rather than upon electrostatic repulsion.[54]

A series of C_nE_m [$C_nH_{2n+1}(OCH_2CH_2)_m$] nonionic surfactants were studied by Wu and Yang[152] using a coarse grain MD simulation procedure on finite graphene slabs. Several very interesting results were obtained in examining effects of surfactant concentration, tail and head group structure (length), and graphene sheet dimensions (finite size effects). Adsorption modes on sheet surfaces were distinguished from modes involving edge adsorption. Such effects, eventually, will provide important mechanistic insight into exfoliation chemomechanics. At low surfactant concentrations these oligoethoxyalkanes adsorbed in parallel arrangements at monolayer coverages with segment density distributions of both tail groups and head groups peaking in the surface layers. At higher concentrations various hemimicellar structures were observed, including hemispheres, hemicylinders, and U-shape hemicylinders (graphene sheet size effect).[152] Spherical micellar structures and hemispheres were observed enveloping edges under various conditions.

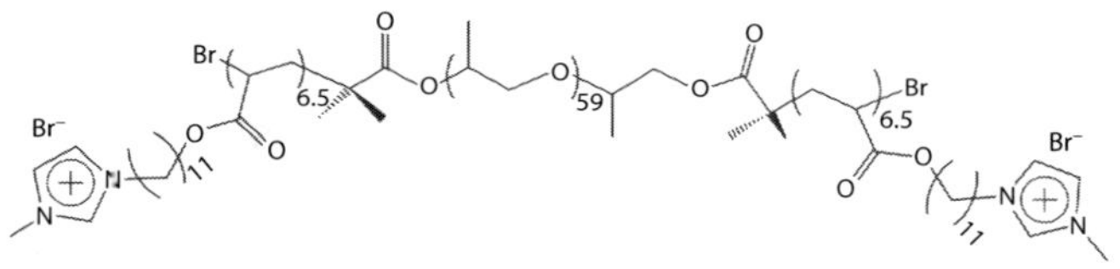

Poly(ILBr-*b*-PO-*b*-ILBr) (TB)

The triblock copolymeric surfactant TB provided almost completely exfoliated graphene dispersions at 10–11 mg/mL dispersed graphene.[138,153–156] This alternative top-down approach may obviate the need to disperse graphene oxide sheets followed by chemical reduction in certain applications. Such processing is also highly scalable from a practical manufacturing point of view, especially since centrifugation processing is not used. These studies provided an increased estimate of the extinction of graphene, although it was obtained at 500 nm rather than 660 nm. These data indicated that $\varepsilon_\lambda > 40$ cm^2/mg. This number appears significantly larger than the 12–24 cm^2/mg (at 660 nm) obtained by centrifugation methods that limit extinction analysis to a very minor fraction of the graphene dispersed.[41,42,58,125]

Contemplation of these various data force several conclusions. The most significant conclusion is that the practice of very mild sonication followed by centrifugation does not provide a very significant or meaningful basis for comparing the efficacy of one surfactant relative to another. Consider that sonication and aqueous collision processing have been used to disperse graphene in water at levels from 0.002 mg/mL to greater than 1 mg/mL without added stabilizer. Certainly, adding stabilizers such as surfactants and polymers protract and extend the stability of such dispersions, but when processes exist that produce larger degrees of dispersion than those processes used to evaluate various stabilizers, one needs to question analyses based on ineffective processes if maximizing the dispersion of graphene is one's objective. The most recent results relying on TB as a stabilizer involves no centrifugation, and essentially no significant sediment is produced, while the optical quality and extinction is 50% greater than that typically obtained with dilute "few-sheet dispersions" (see discussion of optical extinction in a later section). Another significant conclusion is that dispersing aids with groups having delocalized π bonding, such as phenyl, pyrene, and imidazolium groups, have enhanced attractiveness for graphenic sp^2 surfaces, over simple van der Waals interactions. Many of these surfactants deserve to be re-evaluated without bothering to centrifuge, to see how well they can really perform. Centrifugation as a processing step is simply not compatible with any kind of high-volume production scheme. Perhaps for some specialized application, the extra expense can be justified. In the meantime, more effort is needed to better understand the chemical and physical types of defects introduced by processing. The very large body of work associated with the GO to rGO production of graphene dispersions, suggests that some processing defects can be removed chemically.

20.5 AQUEOUS POLYMER-STABILIZED GRAPHENE EXFOLIATION

The basic results obtained for surfactant and small molecule stabilization of graphene dispersions, summarized in the last section and in Table 20.4, are similar to those obtained in polymer stabilization, discussed here. A primary difference between these two types of stabilization is that when polymers adsorb to a surface they usually do so nearly irreversibly, when many adsorption points are obtained. One can readily see that short-time detachment or desorption of a single polymer segment adsorbing moiety is much more probable than the desorption simultaneously of a dozen such moieties. We have discussed moderate molecular weight block copolymeric surfactants in the surfactant section above. More conventional polymers and copolymers are discussed in this section, and key results are summarized in Table 20.5. The use of DNA and especially single-stranded DNA, along with oligonucleotides (OGN) in stabilizing graphene has recently been reviewed by Premkumar and Geckeler.[140] This review covers direct dispersion of graphite and aggregates of graphene, in addition to rGO dispersions derived from GO.

The dispersion of graphene in water by sonicating graphite powder in the presence of polyvinylpyrrolidone (PVP), albumin, and sodium carboxymethylcellulose was investigated by Bourlinos et al.[157] After centrifugation, concentrations of 0.1–0.2 mg carbon/mL (0.01–0.02 wt%) were obtained. These dilute dispersions were ideal for optical, Raman, and AFM characterizations, and it was concluded that greater than 3% of the resulting dispersed flakes were single sheet in nature. Graphite was exfoliated by sonication with 2/3 weight equivalents of PVP and PVA, and gravitationally sedimented overnight. The grayish supernatant, about 0.25 mg/mL, was formed into a hydrogel, and tin nanoparticles were formed within the hydrogel by reduction of $SnCl_2$.[158]

TABLE 20.5
Aqueous Polymer-Stabilized Graphene Dispersions

Polymeric Stabilizer	Chief Results	Reference
PVP (polyvinylpyrrolidone)	0.1–0.2 mg/mL (SC[a])	157
PVP/PVA	0.25 mg/mL; sonication followed by overnight sedimentation; dilute supernatant formed into hydrogel	158
SDS/PANI	Anilinium dodecylsulfate reactive surfactant oxidized to form PANI coating during exfoliation (SC)	126
Sodium carboxymethyl celluose	0.1–0.2 mg/mL (SC)	157
Ethyl cellulose	0.06 mg/mL (SC)	159
Cellulose nanocrystals	0.3–1.1 mg mL (SC)	160
6,6-Nylon and 6,10-Nylon	Produced by interfacial polymerization upon SDBS-stabilized aqueous G	161
PEI	Adsorbed onto CTAB-stabilized sheets; decreasing pH increased stability	162
Poly(isoprene-*b*-acrylic acid)	pH-sensitive phase transfer	163
Albumin	0.1–0.2 mg/mL (SC)	157
Protein Hydrophobins—$(NCysHFBI)_2$ and HFBI-ZE	~0.12 mg/mL (SC); mixtures of single- and few-sheet aggregates claimed.	37
G844	4.7 mg/mL; concentrated to 7–11 mg/mL by evaporation; redox polymer for enzymatic biofuel cell anodes	164
PVAc latex	0.5–20 mg/mL; sonicated; extent of exfoliation not monitored; pronounced graphene concentration effect on viscosity and on adhesion failure (lap joint)	165
ILBr/MMA nanolatex	10–50 mg/mL; exhaustive sonication without centrifugation; anion tunable phase transfer	137

[a] SC—Sonication followed by centrifugation.

Cellulose nanocrystals have also been used to stabilize small quantities of exfoliated graphite, 0.3–1.1 mg/mL.[160] The extraction process of the cellulose nanocrystals introduced sulfate groups that provide charge stabilization in water. The 199 nm × 11 nm cylindrical dimensions suggest a specific projected cylindrical surface area of about 7.8×10^4 cm^2/g cellulose, assuming a nanocrystal density of 1.5 g/cm^3. Single-layer graphene has an areal density of 7.62×10^{-5} g/cm^2, and saturating both sides would require coverage of 2.63×10^4 cm^2/g graphene. Full saturation of both sides of a graphene sheet by such nanocrystals would place an upper bound on the graphene/cellulose nanocrystal weight ratio, 3 g graphene per g cellulose. This limit appears to have been reflected in the data reported,[160] with the exception of a 3.8:1 ratio that can be explained by the presence of multisheet graphene layers rather than single sheet.

Expanded graphite was stabilized in water using sodium dodecylbenzene sulfonate using low-power sonication to activate exfoliation and centrifugation to isolate few-layer graphene sheets.[161] Nylons were then step-polymerized using interfacial polymerization by dispersing aqueous graphene dispersion in carbon tetrachloride, 0.5 M in sebacoyl chloride, to make 6,6-nylon. 6,10-Nylon was made by using adipoyl chloride instead. After phase separation the aqueous continuous phase containing surfactant-stabilized graphene multilayers with surfactant swollen with CCl_4 solutions of diacid chloride was separated from the CCl_4-continuous phase. Hexamethylenediamine was then injected into this aqueous dispersion, and the respective nylons condensed at the water/CCl_4 interface to form a protective layer on either side of the graphene layer sheets, while the HCl condensation by-product partitions into the water.[161] The motivation for this type of encapsulation was that applications in graphene/polymer composites would ensue.

In an interesting application of a reactive surfactant, anilinium dodecylsulfate (AniDS), expanded graphite was stabilized in water using AniDS and sonication, and then the emaraldine form of polyaniline was formed by addition of ammonium persulfate, followed by additional sonication, and centrifugation.[126] While the efficacy of the exfoliation processing was not characterized or quantified, the resulting composite materials were evaluated for morphology, thermal degradation, and electrical conductivity. The PANI/graphene composite materials exhibited conductivities of about 180 S/cm, about half that of the expanded graphite alone.

Liang and Hershem[159] used ethyl cellulose to disperse graphene in ethanol as an alternative to aqueous dispersion. They achieved a postsedimentation concentration of about 0.06 mg/mL. Such a concentration is in the same range as achieved by many other polymers and surfactants in water, but such low concentration offer few practical opportunities, despite the claim of unarticulated outstanding coating properties.

Skaltsas et al.[163] found that both *o*-dichlorobenzene and NMP could be used to disperse graphene via sonication. A particularly interesting subsequent process step was that they introduced the diblock, poly(isoprene-*b*-acrylic acid), and found that it facilitated the phase transfer of graphene sheets from solvent into water. In this system, presumably, the isoprene blocks adsorb to the graphene surface and the water-soluble acrylic acid, solvated by water, provides steric stabilization and a driving force for phase transfer.

Polyethyleneimines (PEI) of 600 Da and 70 kDa were used to stabilize exfoliated graphene, and classical stability studies as a function of NaCl were done by Griffith and Notley.[162] The graphene was first stabilized with the cationic surfactant CTAB, centrifuged, dialyzed to constant conductivity, and then equilibrated with PEI after additional sonication. Apparent zeta potentials varied from about 70 mV at pH 3.5 to about zero at pH 10 for stabilization with 600 Da PEI.

Similarly, for a 70 kDa PEI, zeta potentials varied from about 55 mV at pH 3.5 to about 10 mV at pH 8–10. Stability ratios were determined classically by measuring coagulation rates using turbidimetry. Decreasing pH increased stability ratios, and increasing PEI molecular weight increased stability. The 70 kDa PEI provided both charge and steric stabilization, while the 600 Da material appeared to only provide charge stabilization.[162]

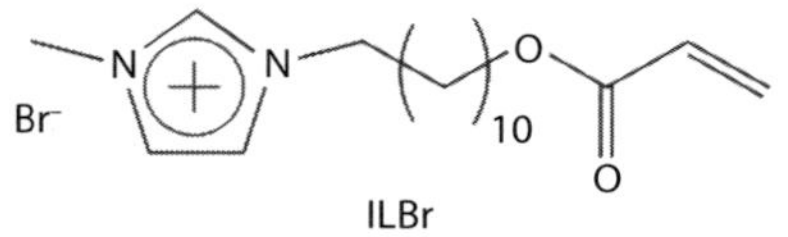

ILBr

The most efficient polymeric (or any other kind of dispersing aid) stabilizers reported to date for graphene in water are nanolatexes derived from a reactive imidazolium bromide acrylate, 1-(11-acryloyloxyundecyl)-3-methyl imidazolium bromide (ILBr). Such nanolatxes[109,166] have been shown to be excellent aqueous stabilizers for SWCNT,[120] MWCNT,[138,167,168] tungsten carbide nanoparticles,[169] hydrothermal carbon,[109,121,169] and graphene.[137,168,170] The use of such nanolatexes in dispersing graphene in water is illustrated in Figure 20.7 where the overall and final graphene weight concentration is 5%. Such a dispersion is the most concentrated to date (many orders of magnitude more concentrated than most of the studies discussed herein) claiming extensive single-sheet exfoliation. In addition, the largest effective optical densities illustrated in Figure 20.7 indicate a lower bound to the extinction of graphene at 500 nm: $\varepsilon_\lambda \geq 50$ cm^2/mg.

An interesting type of protein, hydrophobins, was studied by Linder and coworkers[38] in aqueous graphene exfoliation. Two variants of an HFBI family of proteins, a dimer $(NCysHFBI)_2$, and HFBI–ZE were examined at concentrations up to 3 mg/mL. These types of proteins are amphiphilic (as are most proteins), and have surfaces regarded as being particularly hydrophobic and other facets that are hydrophilic. The amounts of graphene dispersed as single and few sheet aggregates were about 0.12 mg/mL.

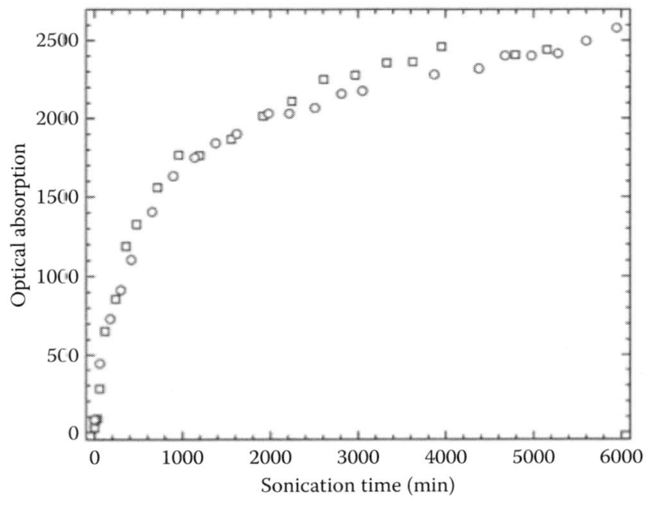

FIGURE 20.7 Effective optical absorption at 500 nm of 5% (w/w) graphene in water.

Dispersion of graphene by sonicating graphite in a polyvinylacetate (PVAc) latex has been reported[165] using a latex wherein the particles were about 300–400 nm in diameter. The dispersions were made by mixing graphite with latex (40% solids) and then sonicating the mixture. Graphite levels of 0.1%–2% by weight were introduced, and the apparent neutral density observed by reflectance steadily increased with graphene content. Sedimentation was observed in mixtures with the higher loadings. Viscosity was observed to increase at shear rates of 2–20 s^{-1} with added graphite, as expected from the Einstein equation. It would have been useful to examine viscosity as a function of sonication time. The tensile stress needed to cause lap joints to fail increased markedly (by 50%) on addition of 0.1% graphite, but the stress needed to cause joint failure subsequently decreased with increasing added graphite, but at 2% this failure (adhesive) stress was still 17% greater than that of the PVAc control. SEM suggested that the mechanism of failure may have involved multisheet platelets splitting.[165]

An interesting family of redox acrylate copolymers have been reported to serve as graphene stabilizers in water and as electron transfer materials for use in bioanodes in enzymatically catalyzed fuel cells.[164] An example is the copolymer G844 with a pendant Os(VI) moiety. Using the "conventional" sonication-centrifugation dispersion approach with graphite/G844 mixtures, after centrifugation graphene concentrations as high as 4.7 mg/mL were obtained at low centrifugation speeds (1000 rpm). This concentration was increased by evaporation of water from 7 to 11 mg/mL, albeit with some loss of stability.

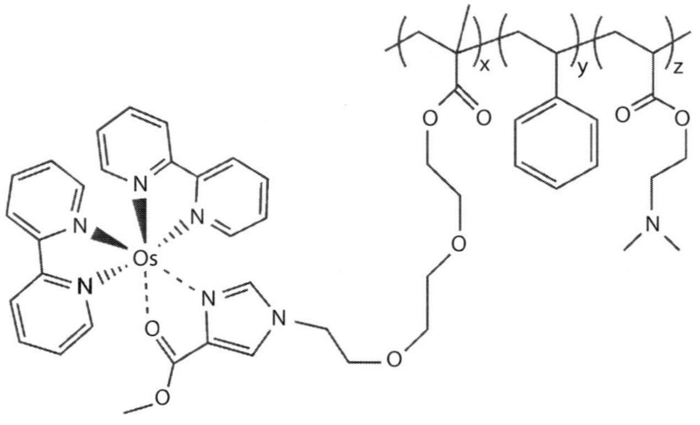

G844

The role of solubility parameters in understanding polymeric stabilization of graphene sheets in dispersion has been discussed by May et al.[171] Unfortunately, their starting assumption of a Hildebrand solubility parameter of 21.3 MPa$^{1/2}$/cm$^{3/2}$ (11.5 cal$^{1/2}$/cm$^{3/2}$), corresponding to a cohesive energy density of single-sheet graphene (131 cal/cm^3), which is about twofold too small relative to the value of 267 cal/cm^3 estimated from experiment (~61 meV).[49] The vectorial magnitude of the Hansen solubility parameters must equal the Hildebrand parameter, but the values put forward fall far short of these requirements. This disparity is a simple result of the thermodynamically ill-founded derivation of the empirically fitted Hansen parameters as discussed earlier in the exfoliation in solution section. This "ill foundedness" resides first in the fact that the concentrations of dispersed graphene used in fitting are not equilibrium or thermodynamic values. Second, the parameter fittings were not done in the accepted way of producing three-dimensional surfaces that evolve smoothly to increased solubility or swelling.

One needs to understand three things to appreciate polymeric steric stabilization: (1) the polymer is anchored, covalently or physically, to the surface it is protecting; (2) the portion of polymer interfacing with solvent is significantly solvated by this solvent; (3) on collision, this solvated polymer does not exhibit strongly attractive interactions with itself.

20.6 ALTERNATIVE EXFOLIATION AND DISPERSION METHODS

Covalent functionalization of edges and surfaces with solubilizing species has been reviewed.[1,114,172] These methods, when pursued too extensively, can disrupt the black visible extinction of graphene, and when done extensively, as in the production of graphene oxide, result in a yellowish material.

Carbon nanosheets (CNS) were produced by a carbonization of bulk betaine;[106] this material has the form of nanoscopic graphene aggregates (Figure 20.8). Activation with *N*-methylglycine and 3,4-dihydroxybenzaldehyde in DMF produces *N*-methyl pyrrole rings on the edges of the CNS with 3,4-dihydroxyphenyl substituents. These hydroxyl groups provide diverse functionalization opportunities for any stabilizing reagent exhibiting facile coupling with hydroxyl groups, such as isocyanates, acylhalides, and alkoxysilanes. Coupling with hexadecyltrichlorosilane was illustrated[106] to produce a functionalized CNS with hexadecyl groups, suitable for dispersing such $CNS{-}C_{16}$ sheets in hydrocarbon solvents. It was argued that silation was through adjacent $-Si(OH)(-O-)_2$ bonds.

Li and coworkers demonstrated an electrochemical exfoliation process that produced quite large dimensioned (3–30 μm) graphene sheets.[173] This DC-biased electrochemical approach used aqueous sulfuric acid as a solvent. AFM analysis of sheet thicknesses ranged from 0.5 to 3 nm with a distribution peak in the 1.5–2 nm range. Coatings of these sheets produced very highly conducting films with sheet resistances <1 kΩ/square. This method was touted as a practical approach to graphene ink, but the sulfuric acid may limit the applicability of such inks. The very large aspect ratios obtained with concomitant excellent conducting properties provide motivation for overcoming any acid-related issues.

20.7 VISIBLE OPTICAL EXTINCTION OF GRAPHENE

Quite a few experimentalists have noted that the optical absorption in the visible apparently increases as more sonication is applied in dispersions of various concentrations as well as in various solvents. It is then empirically established that the optical extinction of graphene becomes more pronounced as exfoliation proceeds, and it would seem logical that this extinction would become a maximum when few layer platelets become completely exfoliated into single sheets. However, such an expectation remains to be proven. Nevertheless, the interaction of adjacent sheets in contact with one another serves, somehow, to dampen the optical absorption. If this effect derives from selection rules, it will be useful to transform some of the high-quality theoretical treatments into molecular terms understandable in terms of electronic dipole moments. However, a significant gap persists between the optical absorption measurements performed on graphene dispersions and theoretical and experimental optical studies of pristine graphene under various conditions. It has also been shown that some optical properties can be tuned using biasing potentials;[174] these aspects remain outside the scope of our discussion.

A model for optical absorption throughout the visible in graphene, derived from the fine structure constant,[175,176] α ($= e^2/\hbar c$; e is electronic charge, ħ is Planck's constant divided by 2π, and c is the speed of light) that describes how white light interacts with relativistic electrons, gives $1 - \pi\alpha$ (~0.977)

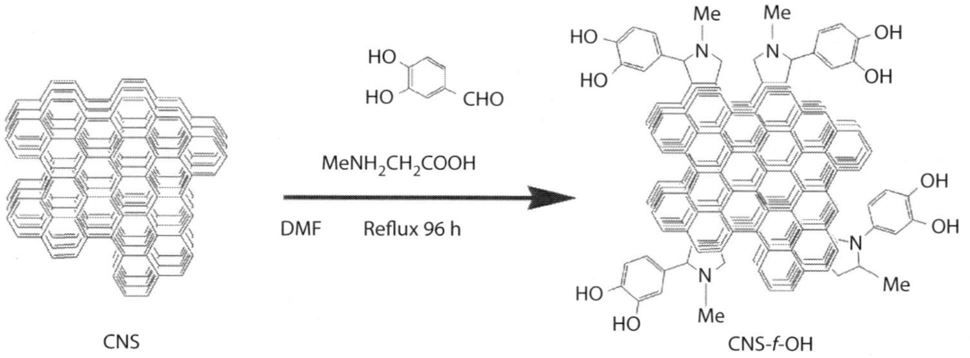

FIGURE 20.8 Scheme for chemically functionalizing CNS with accessible hydroxyl groups. (Reproduced with permission from Georgakilas, V. 2010. Chemical functionalization of ultrathin carbon nanosheets, *Fullerenes Nanotubes Carbon Nanostructures* 18:87–95, Copyright 2010, Taylor & Francis, Ltd.)

as the transmittance.[177] The corresponding fraction of light absorbed is 0.023, and corresponds to a single-layer absorbance ($-\log[1 - \pi\alpha]$) of 0.01007 at all visible wavelengths. This quantity corresponds to an extinction coefficient ε_α (cm^2/mg) ~ 0.01007/(0.34 nm × 10^{-7} cm/nm × $\rho_{graphene}$ g/cm^3 × 10^3 mg/g) = 296/$\rho_{graphene}$ (cm^2/mg) where density is given in g/cm^3. If graphene density is approximated by a value of 2.260 g/cm^3, we obtain an extinction coefficient of about 131 cm^2/mg. That single- and few-layer graphene conforms to this model was the main experimental conclusion of Nair et al.[177] when suspended in air. In spite of having a nonnegligible extinction, single-layer to few-layer stacks of graphene can appear quite transparent, and this property supports many active discussions and efforts aimed at using thin graphene layers to replace ITO electrodes in various display devices.

Complete optical absorption in periodically patterned graphene[178] and other methods of increasing absorption in graphene are discussed by Ye et al.[179] Depolarization effects in optical absorption measurements of one- and two-dimensional nanostructures have been experimentally studied by Herman et al.[180] They point out that the universal absorption discussed above was shown by Mak et al.[181] to occur only for light polarized parallel to the graphene surface, and further that Saito and Kataura[182] showed that light polarized perpendicularly to the surface is not absorbed. Their experiments and theory indicate the universal extinction of 148 cm^2/mg for light polarized parallel to the graphene surface is attenuated to about 108 cm^2/mg, and that this extinction gradually disappears as polarization shifts to perpendicular (normal). The extinction exhibited by any particular single-layer graphene flake in suspension will, therefore, depend on the local polarization of the incident light to the graphene surface. The data of Nair et al.[177] also show that the extinction per layer in few layer flakes deviates from the "universal" theory for four layers and above, becoming smaller with increasing thickness. This aspect is consonant with experimental graphene exfoliation studies that have repeatedly shown increasing optical extinction with progressively greater exfoliation. These single- and few-layer studies have not been done in suspension, and the impact of various solvents on the experimental extinction has yet to be examined experimentally or theoretically.

Experimental visible extinction coefficients for exfoliated graphene, mostly in water, are summarized in Table 20.6. Results range from 12 to 66 cm^2/mg at 660 nm, and studies at 500 nm produce lower bounds, with 50 cm^2/mg as a greater lower bound reported. None of the above-cited theoretical studies account quantitatively for variations in extinction seen in dispersions as a slowly increasing quantity through the visible with sharper increases in the UV. Based on a variety of published spectra[125,131] we expect that the extinction at 500 nm is 14.5% larger than that at 660 nm. At present it seems that the 24.6 cm^2/mg[41] value is the most widely accepted value at 660 nm, although it post dates a 66.0 cm^2/mg[127] estimate. All the 660 nm extinctions were obtained for very dilute supernatant suspensions obtained by centrifugation believed to sediment out "more than few-layer" flakes. The statistical impact and uncertainty of such an assumption has yet to be articulated. The gravimetric uncertainties in concentration determinations in such very dilute suspensions are not negligible. The 500 nm lower bounds, on the other hand, involved no centrifugation or discarding of any of the originally composed graphene. While centrifugation times were lengthy, the ice bath environment used during sonication appeared to avoid noticeable thermal degradation, insofar as the apparent dispersion optical density at 500 nm was concerned.

TABLE 20.6
Visible Optical Extinction of Graphene in Water, ε_λ (cm^2/mg)

λ (nm)	ε_λ (cm^2/mg)	Reference
660	13.6	125
660	24.6	41,42
660	25.0	164
660	11.7	58
660	66.0	127
500	>43	170
500	≥50	137

While a variety of experimental issues need to be resolved, the greatest lower bound of 50 cm^2/g at 500 nm appears as the most credible of reported extinctions in aqueous dispersion. These bounds were obtained at weight concentrations of 1%–5% by weight of carbon, orders of magnitude higher than the dilute suspensions analyzed at 660 nm. Since flake dimensions in such sonicated systems usually fall beneath 1 μm × 1 μm, and as low as 100 nm × 100 nm, theoretical analyses of optical cross sections, polarization effects, and periphery chemical effects need to be examined. Because of randomization of graphene sheets and flakes in dispersion, the extinction we would expect to measure in dispersion is 2/3 × 148 cm^2/mg (at 500 nm), or about 100 cm^2/mg.[137]

20.8 SUMMARY

Graphene dispersions prepared in the absence of added stabilizer have shown some limited promise in a few solvents, but the concentrations appear too low for large volume applications. The examination of a plethora of small molecule surfactants to high molecular weight polymers yields the conclusion that just about anything can promote the dispersion of graphene to a small extent. Unfortunately, the true potential of many of the stabilizers examined to date has not been realized due to the unfortunate practice of centrifuging. On the other hand, this practice has produced a few dozen studies that show very similar results by TEM, Raman, and visible absorption.

The extant data support a conclusion that molecules and polymers exhibited delocalized π-systems emanating from single- to multiple-ring structures have favorable binding affinities for the sp^2 graphene surface. Imidazolium-based polymeric surfactants and nanolatexes have demonstrated the highest degree of graphene dispersion at 1%–5% by weight. Those studies were not exhaustive and higher concentrations may be expected to be obtained in the near future. An outstanding need exists to molecularly engineer such stabilizers to be electrically and thermally conducting. At present, the best conductivities are obtained with dispersions essentially devoid of stabilizer, since many stabilizers are effectively insulating rather than conducting. However, the barriers to such advances are likely not overly daunting, considering the advances made in the past 20 years in high-performance polymeric materials. The very practical dispersion concentrations that have been demonstrated provide great promise for many low-cost and high-volume processes for producing double layer supercapacitors, highly electrically conducting inks and thin layers, and highly thermally conducting coatings.

REFERENCES

1. Yan, L., Zheng, Y. B., Zhao, F., Li, S., Gao, X., Bingqian X. B., Weiss, P. S., Zhao, Y. 2012. Chemistry and physics of a single atomic layer: Strategies and challenges for functionalization of graphene and graphene-based materials. *Chem. Soc. Rev.* 41:97–114.
2. Rao, C. N. R., Sood, A. K., Subrahmanyam, K. S., Govindaraj, A. 2009. Graphene: The new two-dimensional nanomaterial. *Angew. Chem. Int. Ed.* 48:7752–7777.
3. Cai, M., Thorpe, D., Adamson, D. H., Schniepp, H. C. 2012. Methods of graphite exfoliation. *J. Mater. Chem.* 22:24992–25002.
4. Edwards, R. S., Coleman, K. S. 2013. Graphene synthesis: Relationship to applications. *Nanoscale* 5:38–51.
5. Khan, U., Porwal, H., O'Neill, A., Nawaz, K., May, P., Coleman, J. N. 2011. Solvent-exfoliated graphene at extremely high concentration. *Langmuir* 27:9077–9082.
6. Texter, J. 2014. Graphene dispersions. *Curr. Opinion Colloid Interface Sci.* 19:168–174.
7. Penicaud, A., Drummond, C. 2013. Deconstructing graphite: Graphenide solutions. *Acc. Chem. Res.* 46:129–137.
8. Ruoff, R. S. 2008. Graphene—Calling all chemists. *Nature Nanotech.* 3:10–11.
9. Lin, T. Q., Chen, J., Bi, H., Wan, D. Y., Huang, F. Q., Xie, X. M. 2013. Facile and economical exfoliation of graphite for mass production of high-quality graphene sheets. *J. Mater. Chem. A* 1:500–504.
10. Chen, J. F., Duan, M., Chen, G. H. 2012. Continuous mechanical exfoliation of graphene sheets via three-roll mill. *J. Mater. Chem.* 22:19625–19628.
11. Ciesielski, A., Samori, P. 2014. Graphene via sonication assisted liquid-phase exfoliation. *Chem. Soc. Rev.* 43:381–398.
12. Low, C. T. W., Walsh, F. C., Chakrabarti, M. H., Hashim, M. A., Hussain, M. A. 2013. Electrochemical approaches to the production of graphene flakes and their potential applications. *Carbon* 54:1–21.
13. Li, X. S., Cai, W. W., An, J. H., Kim, S., Nah, J., Yang, D. X., Piner, R. D. et al. 2009. Large-area synthesis of high-quality and uniform graphene films on copper foils. *Science* 324:1312–1314.
14. Bae, S., Kim, H., Lee, Y., Xu, X., Park, J.-S., Zheng, Y., Balakrishnan, J. et al. 2010. Roll-to-roll production of 30-inch graphene films for transparent electrodes. *Nat. Nanotechnol.* 5:574–578.
15. Seah, C.-M., Cai, S.-P., Mohamed, A. R. 2014. Mechanisms of graphene growth by chemical vapour deposition on transition metals. *Carbon* 70:1–21.
16. Zhang, Y., Zhang, L. Y., Zhou, C. W. 2013. Review of chemical vapor deposition of graphene and related applications. *Acc. Res.* 46:2329–2339.
17. Park, S. J., Ruoff, R. S. 2009. Chemical methods for the production of graphenes. *Nat. Nanotech.* 4:217–224.
18. Mao, S., Pu, H. H., Chen, J. H. 2012. Graphene oxide and its reduction: Modeling and experimental progress. *RSC Adv.* 2:2643–2662.
19. Zhang, C., Lu, W., Xie, X. Y., Tang, D. M., Liu, C., Yang, Q.-H. 2013. Towards low temperature thermal exfoliation of graphite oxide for graphene production. *Carbon* 62:11–24.
20. Paredes, J. I., Villar-Rodil, S., Solís-Fernández, P., Fernández-Merino, M. J., Guardia, L., Martínez-Alonso, A., Tascón, J. M. D. 2012. Preparation, characterization and fundamental studies on graphenes by liquid-phase processing of graphite. *J. Alloys Compounds* 536S:S450–S455.
21. Texter, J. 1995. Multiphase coupling with nanocrystalline coupler dispersions. *Imaging Sci. Technol.* 39:340–354.
22. Texter, J. 2001. Precipitation and condensation of organic particles. *J. Disp. Sci. Technol.* 22:497–522.
23. Adamson, A. W., Gast, A. P. 1997. *Physical Chemistry of Surfaces*. Wiley, New York.
24. Wang, S., Zhang, Y., Abidi, N., Cabrales, L. 2009. Wettability and surface free energy of graphene films. *Langmuir* 25:11078–11081.
25. Shih, C.-J., Wang, Q. H., Lin, S., Park, K.-C., Jin, Z., Strano, M. S., Blankschtein, D. 2012. Breakdown in the wetting transparency of graphene. *Phys. Rev. Lett.* 109:176101.
26. Li, D., Neumann, A. W. 1992. Contact angles on hydrophobic solid surfaces and their interpretation. *J. Colloid Interface Sci.* 148:190–200.
27. Li, D., Neumann, A. W. 1993. Equilibrium of capillary systems with an elastic liquid vapor interface. *Langmuir* 9:50–54.
28. Batchelor, T., Cunder, J., Fadeev, A. Y. 2009. Wetting study of imidazolium ionic liquids. *J. Colloid Interface Sci.* 330:415–420.
29. Dahbi, M., Violleau, D., Ghamouss, F., Jacquemin, J., Tran-Van, F., Lemordant, D., Anouti, M. 2012. Interfacial properties of LiTFSI and $LiPF_6$-based electrolytes in binary and ternary mixtures of alkylcarbonates on graphite electrodes and Celgard separator. *Ind. Eng. Chem. Res.* 51:5240–5245.
30. Taherian, F., Marcon, V., van der Vegt, N. F. A., Leroy, F. 2013. What is the contact angle of water on graphene? *Langmuir* 29:1457–1465.
31. Texter, J. 2014. Graphene dispersions. *Curr. Opin. Coll. Interface Sci.* 19:163–174.
32. Graf, D., Molitor, F., Ensslin, K., Stampfer, C., Jungen, A., Hierold, C., Sirtz, L. 2007. Spatially resolved Raman spectroscopy of single- and few-layer graphene. *Nano Lett.* 7:238–242.
33. Wu, Y. H., Yu, T., Shen, Z. X. 2010. Two-dimensional carbon nanostructures: Fundamental properties, synthesis, characterization, and potential applications. *J. Appl. Phys.* 108:071301 (38 pp).
34. Zhu, Y. W., Murali, S., Cai, W. W., Li, X. S., Suk, J. W., Potts, J. R., Ruoff, R. S. 2010. Graphene and graphene oxide: Synthesis, properties, and applications. *Adv. Mater.* 22:3906–3924.
35. Neha, B., Manjula, K., Srinivasulu, B., Subhas, S. 2012. Synthesis and characterization of exfoliated graphite/ABS composites. *Open J. Org. Polym. Mater.* 2:75–79.
36. Loh, K. P., Bao, Q. L., Ang, P. K., Yang, J. X. 2010. The chemistry of graphene. *J. Mater. Chem.* 20:2277–2289.

37. Gong, P. W., Wang, Z. F., Wang, J. Q., Wang, H. Q., Li, Z. Q., Fan, Z. Q., Xu, Y., Han, X. X., Yang, S. R. 2012. One-pot sonochemical preparation of fluorographene and selective tuning of its fluorine coverage. *J. Mater. Chem.* 22:16950–16956.
38. Laaksonen, P., Kainlauri, M., Laaksonen, T., Shchepetov, A., Jiang, H., Ahopelto, J., Linder, M. B. 2010. Interfacial engineering by proteins: Exfoliation and functionalization of graphene by hydrophobins. *Angew. Chemie Int. Ed.* 49:4946–4949.
39. Singh, K., Ohlan, A., Dhawan, S. K. 2012. Polymer-graphene nanocomposites: Preparation, characterization, properties, and applications. In *Nanocomposites—New Trends and Developments*. Ebrahimi, F., editor. InTech, Rijeka, Croatia, pp. 37–71.
40. Lin, T. Q., Chen, J., Bi, H., Wan, D. Y., Huang, F. Q., Xie, X. M., Jiang, M. H. 2013. Facile and economical exfoliation of graphite for mass production of high-quality graphene sheets. *J. Mater. Chem. A* 1:500–503.
41. Khan, U., O'Neill, A., Loyta, M., De, S., Coleman, J. N. 2010. High concentration solvent-exfoliation of graphene. *Small* 6:864–871.
42. Hernandez, Y., Nicolosi, V., Lotya, M., Blighe, F. M., Sun, Z., De, S., McGovern, I. T. et al. 2008. High-yield production of graphene by liquid-phase exfoliation of graphite. *Nat. Nanotechnol.* 3:563–568.
43. Khan, U., Porwal, H., O'Neill, A., Nawaz, K., May, P., Coleman, J. N. 2012. Size selection of dispersed, exfoliated graphene flakes by controlled centrifugation. *Carbon* 50:470–475.
44. Smith, R. J., King, P. J., Wirtz, C., Duesberg, G. S., Coleman, J. N. 2012. Lateral size selection of surfactant-stabilised graphene flakes using size exclusion chromatography, *Chem. Phys. Lett.* 531:169–172.
45. Hernandez, Y., Lotya, M., Rickard, D., Bergin, S. D., Coleman, J. N. 2010. Measurement of multicomponent solubility parameters for graphene facilitates solvent discovery. *Langmuir* 26:3208–3213.
46. Coleman, J. N., 2013. Liquid exfoliation of defect-free graphene, *Acc. Chem. Res.* 46:14–22.
47. Barton, A. F. M. 1991. *Handbook of Solubility Parameters and Other Cohesion Parameters*, Second Edition. Boca Raton, FL: CRC Press.
48. Hansen, C. M. 2007. *Hansen Solubility Parameters—A User's Handbook*. Boca Raton, FL: CRC Press.
49. Zacharia, R., Ulbricht, H., Hertel, T. 2004. Interlayer cohesive energy of graphite from thermal desorption of polyaromatic hydrocarbons. *Phys. Rev. B* 69: 155405 (7 pp).
50. Wisniewski, R., Smieszek, E., Kaminska, E. 1995. Three-dimensional solubility parameters: Simple and effective determination of compatibility regions. *Prog. Org. Coatings* 26:265–274.
51. Zellers, E. T., Anna, D. H., Sulewski, R., Wei, X.-R. 1996. Critical analysis of the graphical determination of Hansen's solubility parameters for lightly crosslinked polymers. *J. Appl. Polym. Sci.* 62:2069–2080.
52. Li, L., Bedrov, D., Smith, G. D. 2006. Water-induced interactions between carbon nanoparticles. *J. Phys. Chem. B* 110:10509–10513.
53. Shih, C.-J., Lin, S., Strano, M. S., Blankschtein, D. 2010. Understanding the stabilization of liquid-phase-exfoliated graphene in polar solvents: Molecular dynamics simulations and kinetic theory of colloid aggregation. *J. Am. Chem. Soc.* 132:14638–14648.
54. Lin, S., Shih, C.-J., Strano, M. S., Blankschtein, D. 2011. Molecular insights into the surface morphology, layering structure, and aggregation kinetics of surfactant-stabilized graphene dispersions. *J. Am. Chem. Soc.* 133:12810–12823.
55. Fukushima, T., Kosaka, A., Ishimura, Y., Yamamoto, T., Takigawa, T., Ishii N., Aida, T. 2003. Molecular ordering of organic molten salts triggered by single-walled carbon nanotubes. *Science* 300:2072–2074.
56. Fukushima, T., Aida, T. 2007. Ionic liquids for soft functional materials with carbon nanotubes. *Chem. Eur. J.* 13:5048–5058.
57. Sekitani, T., Noguchi, Y., Hata, K., Fukushima, T., Aida, T., Someya, T. 2008. A rubberlike stretchable active matrix using elastic conductors. *Science* 321:1468–1472.
58. Nuvoli, D., Valentini, L., Alzari, V., Scognamillo, S., Bon, S. B., Piccinini, M., Illescas, J., Mariani, A. 2011. High concentration few-layer graphene sheets obtained by liquid phase exfoliation of graphite in ionic liquid. *J. Mater. Chem.* 21:3428–3431.
59. Liu, N., Luo, F., Wu, H., Liu, Y., Zhang, C., Chen, J. 2008. One-step ionic-liquid-assisted electrochemical synthesis of ionic-liquid-functionalized graphene sheets directly from graphite. *Adv. Funct. Mater.* 18:1518–1525.
60. Lu, J., Yang, J.-X., Wang, J., Lim, A., Wang, S., Loh, K. P. 2009. One-pot synthesis of fluorescent carbon nanoribbons, nanoparticles, and graphene by the exfoliation of graphite in ionic liquids. *ACS Nano* 8:2367–2375.
61. Wang, X., Fulvio, P. F., Baker, G. A., Veith, G. M., Unocic, R. R., Mahurin, S. M., Chi, M., Dai, S. 2010. Direct exfoliation of natural graphite into micrometre size few layers graphene sheets using ionic liquids. *Chem. Commun.* 46:4487–4489.
62. O'Neill, A., Khan, U., Nirmalraj, P. N., Boland, J., Coleman, J. N. 2011. Graphene dispersion and exfoliation in low boiling point solvents. *J. Phys. Chem. C* 115:5422–5428.
63. Zhang, H., Wen, J., Meng, X., Yao, Y., Yin, G., Liao, X., Huang, Z. 2012. An improved method to increase the concentration of graphene in organic solvent. *Chem. Lett.* 41:747–749.
64. Zhu, L., Zhao, X., Li, Y., Yu, X., Li, C., Zhang, Q. 2013. High-quality production of graphene by liquid-phase exfoliation of expanded graphite. *Mater. Chem. Phys.* 137:984–990.
65. Yi, M., Shen, Z., Ma, S., Zhang, X. 2012. A mixed-solvent strategy for facile and green preparation of graphene by liquid-phase exfoliation of graphite. *J. Nanopart. Res.* 14:1003–1009.
66. Kawano, Y., Kondo, T. 2014. Preparation of aqueous carbon material suspensions by aqueous counter collision. *Chem. Lett.* 43:483–485.
67. Ioni, Yu. V., Tkachev, S. V., Bulychev, N. A., Gubin, S. P. 2011. Preparation of finely dispersed nanographite. *Inorg. Mater.* 47:597–602; *Neorganich. Materialy* 47:671–677.
68. Mehrali, Md., Sadeghinezhad, E., Latibari, S. T., Kazi, S. N., Mehrali, M., Zubir, M. N. B. M., Metselaar, H. S. C. 2014. Investigation of thermal conductivity and rheological properties of nanofluids containing graphene nanoplatelets. *Nanoscale Res. Lett.* 9:15 (12 pp).
69. Primo, A., Sanchez, E., Delgado, J. M., Garcia, H. 2014. High-yield production of N-doped graphitic platelets by aqueous exfoliation of pyrolyzed chitosan. *Carbon* 68:777–783.
70. Dimiev, A. M., Gizzatov, A., Wilson, L. J., Tour, J. M. 2013. Stable aqueous colloidal solutions of intact surfactant-free graphene nanoribbons and related graphitic nanostructures. *Chem. Commun.* 49:2613–2615.
71. Lu, W., Liu, S., Qin, X., Wang, L., Tian, J., Yonglan Luo, Y., Asiri, A. M., Al-Youbic, A. O., Sun, X. 2012. High-yield, large-scale production of few-layer graphene flakes within seconds: Using chlorosulfonic acid and H_2O_2 as exfoliating agents. *J. Mater. Chem.* 22:8775–8777.

72. Mohanty, N., Moore, D., Xu, Z., Sreeprasad, T. S., Ashvin Nagaraja, A., Rodriguez, A. A., Vikas Berry, V. 2012. Nanotomy-based production of transferable and dispersible graphene nanostructures of controlled shape and size. *Nature Commun.* 3:834–839.
73. Dreyer, D. R., Park, S. J., Bielawski, C. W., Ruoff, R. S. 2010. The chemistry of graphene oxide. *Chem. Soc. Rev.* 39:228–240.
74. McAllister, M. J., Li, J.-L., Adamson, D. H., Schniepp, H. C., Abdala, A. A., Liu, J., Herrera-Alonso, M. et al. 2007. Single sheet functionalized graphene by oxidation and thermal expansion of graphite. *Chem. Mater.* 19:4396–4404.
75. Stankovich, S., Piner, R. D., Chen, X. Q., Wu, N. Q., Nguyen, S. T., Ruoff, R. S. 2006. Stable aqueous dispersions of graphitic nanoplatelets via the reduction of exfoliated graphite oxide in the presence of poly(sodium 4-styrenesulfonate). *J. Mater. Chem.* 16:155–158.
76. Guo, J., Ren, L., Wang, R., Zhang, C., Yang, Y., Liu, T. 2011. Water dispersible graphene noncovalently functionalized with tryptophan and its poly(vinyl alcohol) nanocomposite. *Composites B* 42:2130–2135.
77. Chang, D. W., Sohn, G.-J., Dai, L., Baek, J.-B., 2011. Reversible adsorption of conjugated amphiphilic dendrimers onto reduced graphene oxide (rGO) for fluorescence sensing. *Soft Matter* 7:8352–8357.
78. Zu, S.-Z., Han, B. H. 2009. Aqueous dispersion of graphene sheets stabilized by Pluronic copolymers: Formation of supramolecular hydrogel. *J. Phys. Chem. C* 113:13651–13657.
79. Shen, X., Jiang, L., Ji, Z., Wua, J., Zhou, H., Zhu, G. 2011. Stable aqueous dispersions of graphene prepared with hexamethylenetetramine as a reductant. *J. Colloid Interface Sci.* 354:493–497.
80. Tölle, F. J., Fabritius, M., Mülhaupt, R. 2012. Emulsifier-free graphene dispersions with high graphene content for printed electronics and freestanding graphene films. *Adv. Funct. Mater.* 22:1136–1144.
81. Zhang, C., Tjiu, W. W., Fan, W., Huang, S., Liu, T. 2012. A novel approach for transferring water-dispersible graphene nanosheets into organic media. *J. Mater. Chem.* 22:11748–11754.
82. Yang, H. F., Shan, C. S., Li, F. H., Han, D. X., Zhang, Q. X., Niu, L. 2009. Covalent functionalization of polydisperse chemically-converted graphene sheets with amine-terminated ionic liquid. *Chem. Commun.* 45:3880–3882.
83. Lonkar, S. P., Bobenrieth, A., De Winter, J., Gerbaux, P., Raquez, J.-M., Philippe Dubois, P. 2012. A supramolecular approach toward organo-dispersible graphene and its straightforward polymer nanocomposites. *J. Mater. Chem.* 22:18124–18126.
84. Kim, T. Y., Lee, H. W., Kim, J. E., Suh, K. S. 2010. Synthesis of phase transferable graphene sheets using ionic liquid polymers. *ACS Nano* 4:1612–1618.
85. Gao, H., Zhang, S., Fei Lu, F., Han Jia, H., Zheng, L. 2012. Aqueous dispersion of graphene sheets stabilized by ionic liquid-based polyether. *Colloid Polym. Sci.* 290:1785–1791.
86. Men, Y. J., Li, X.-H., Antonietti, M., Yuan, J. Y. 2012. Poly(tetrabutylphosphonium 4-styrenesulfonate): A poly(ionic liquid) stabilizer for graphene being multi-responsive. *Polym. Chem.* 3:871–873.
87. Liao, R., Lei, Y., Wan, J., Tang, Z., Guo, B., Zhang, L. 2012. Dispersing graphene in hydroxypropyl cellulose by utilizing its LCST behavior. *Macromol. Chem. Phys.* 213:1370–1377.
88. Kuila, T., Mishra, A. K., Khanra, P., Kim, N. H., Uddin, M. E., Lee, J. H. 2012. Facile method for the preparation of water dispersible graphene using sulfonated poly(ether–ether–ketone) and its application as energy storage materials. *Langmuir* 28:9825–9833.
89. Zhang, C., Tjiu, W. W., Fan, W., Yang, Z., Huang, S., Liu, T. 2011. Aqueous stabilization of graphene sheets using exfoliated montmorillonite nanoplatelets for multifunctional free-standing hybrid films via vacuum-assisted self-assembly. *J. Mater. Chem.* 21:18011–18017.
90. Kang, S. M., Park, S., Kim, D., Park, S. Y., Ruoff, R. S., Lee, H. 2011. Simultaneous reduction and surface functionalization of graphene oxide by mussel-inspired chemistry. *Adv. Funct. Mater.* 21:108–112.
91. Vuluga, D., Thomassin, J. M., Molenberg, I., Huynen. I., Gilbert, B., Jerome, C., Alexandre, M., Detrembleur, C. 2011. Straightforward synthesis of conductive graphene/polymer nanocomposites from graphite oxide. *Chem. Commun.* 47:2544–2546.
92. Yu, D., Dai, L. 2010. Self-assembled graphene/carbon nanotube hybrid films for supercapacitors. *J. Phys. Chem. Lett.* 1:467–470.
93. Xu, Z., Gao, C. 2010. *In situ* polymerization approach to graphene-reinforced Nylon-6 composites. *Macromolecules* 43:6716–6723.
94. Salavagione, H. J., Martínez, G. 2011. Importance of covalent linkages in the preparation of effective reduced graphene oxide-poly(vinyl chloride) nanocomposites. *Macromolecules* 44:2685–2692.
95. Salavagione, H. J., Gómez, M. A., Martínez, G. 2009. Polymeric modification of graphene through esterification of graphite oxide and poly(vinyl alcohol). *Macromolecules* 42:6331–6334.
96. Lee, S. H., Dreyer, D. R., An, J., Velamakanni, A., Piner, R. D., Park, S., Zhu, Y., Kim, S. O., Bielawski, C. W., Ruoff, R. S. 2010. Polymer brushes via controlled, surface-initiated atom transfer radical polymerization (ATRP) from graphene oxide. *Macromol. Rapid Commun.* 31:281–288.
97. Lee, S. H., Kim, H. W., Hwang, J. O., Lee, W. J., Kwon, J., Bielawski, C. W., Ruoff, R. S., Kim, S. O. 2010. Three-dimensional self-assembly of graphene oxide platelets into mechanically flexible macroporous carbon films. *Angew. Chem. Int. Ed.* 49:10084–10088.
98. Fang, M., Wang, K., Lu, H., Yang, Y., Nutt, S. 2009. Covalent polymer functionalization of graphene nanosheets and mechanical properties of composites. *J. Mater. Chem.* 19:7098–7105.
99. Wang, D., Ye, G., Wang, X., Wang, X. 2011. Graphene functionalized with azo polymer brushes: Surface-initiated polymerization and photoresponsive properties. *Adv. Mater.* 23:1122–1125.
100. Shan, C., Yang, H., Han, D., Zhang, Q., Ivaska, A., Niu, L. 2009. Water-soluble graphene covalently functionalized by biocompatible poly-L-lysine. *Langmuir* 25:12030–12033.
101. Ren, L., Huang, S., Zhang, C., Wang, R., Tjiu, W. W., Liu, T. 2012. Functionalization of graphene and grafting of temperature responsive surfaces from graphene by ATRP "on water". *J. Nanopart. Res.* 14:940–949.
102. Deng, Y., Li, Y., Dai, J., Lang. M., Huang, X. 2011. An efficient way to functionalize graphene sheets with presynthesized polymer via ATNRC chemistry. *J. Polym. Sci. A Polym. Chem.* 49:1582–1590.
103. Bak, J. M., Lee, T., Seo, E., Lee, Y., Jeong, H. M., Kim, B.-S., Lee, H.-I. 2012. Thermoresponsive graphene nanosheets by functionalization with polymer brushes. *Polymer* 53:316–323.
104. Bishop, J., Mourey, T., Texter, J. 1995. Adsorption of triblock copolymers on nanoparticulate pharmaceutical imaging agents. In *Surfactant Adsorption and Surface Solubilization*. Sharma, R., editor. ACS Symposium Series, No. 615. ACS Books, Washington, DC, 1995, pp. 205–216.

105. Hummers, W. S., Offeman, R. E. 1958. Preparation of graphitic oxide. *J. Am. Chem. Soc.* 80:1339.
106. Georgakilas, V. 2010. Chemical functionalization of ultra-thin carbon nanosheets, *Fullerenes, Nanotubes, Carbon Nanostruct.* 18:87–95.
107. Lu, J., Yan, F., Texter, J. 2009. Advanced applications of ionic liquids in polymer science. *Prog. Poly. Sci.* 34:431–448.
108. Marcilla, R., Curri, M. L., Cozzoli, P. D., Martínez, M. T., Loinaz, I., Grande, H., Pomposo, J. A., Mecerreyes, D. 2006. Nano-objects on a round trip from water to organics in a polymeric ionic liquid vehicle. *Small* 2:507–512.
109. Texter, J. 2012. Osmotic nanospheres—Ideal and switchable dispersants for waterborne dispersions and advanced coatings of nanocarbon, *Polym. Prepr.* 53:143–144.
110. Li, D., Mueller, M. B., Gilje, S., Kaner, R. B., Wallace, G. G. 2008. Processible aqueous dispersions of graphene nanosheets. *Nat. Nanotechnol.* 3:101–105.
111. Lomeda, J. R., Doyle, C. D., Kosynkin, D. V., Hwang, W.-F., Tour, J. M. 2008. Diazonium functionalization of surfactant-wrapped chemically converted graphene sheets. *J. Am. Chem. Soc.* 130:16201–16206.
112. Romsted, L. S. 2001. Interfacial compositions of surfactant assemblies by chemical trapping with arenediazonium ions: Method and applications. In *Reactions and Synthesis in Surfactant Systems.* Texter, J., editor. Surfactant Sci. Ser. Vol. 100, Marcel Dekker, New York, pp. 265–294.
113. Paulus, G. L. C., Wang, Q. H., Strano, M. S. 2013. Covalent electron transfer chemistry of graphene with diazonium salts. *Acc. Chem. Res.* 46:160–170.
114. Salavagione, H. J., Martínez, G., Ellis, G. 2011. Recent advances in the covalent modification of graphene with polymers. *Macromol. Rapid Commun.* 32:1771–1789.
115. Wang, X., Hu, Y., Song, L., Yang, H., Xing, W., Lu, H. 2011. *In situ* polymerization of graphene nanosheets and polyurethane with enhanced mechanical and thermal properties. *J. Mater. Chem.* 21:4222–4227.
116. He, H., Gao, C. 2010. General approach to individually dispersed, highly soluble, and conductive graphene nanosheets functionalized by nitrene chemistry. *Chem. Mater.* 22:5054–5064.
117. Lee, D. Y., Yoon, S., Oh, Y. J., Park, S. Y., In, I. 2011. Thermo-responsive assembly of chemically reduced graphene and poly-(*N*-isopropylacrylamide). *Macromol. Chem. Phys.* 212:336–341.
118. Zhou, X., Wu, T., Ding, K., Hu, B., Hou, M., Han, B. 2010. Dispersion of graphene sheets in ionic liquid [bmim][PF6] stabilized by an ionic liquid polymer. *Chem. Commun.* 46:386–388.
119. Zhang, B., Ning, W., Zhang, J., Qiao, X., Zang, J., He, J., Liu, C.-Y. 2010. Stable dispersions of reduced graphene oxide in ionic liquids. *J. Mater. Chem.* 20:5401–5403.
120. Antonietti, M., Shen, Y., Nakanishi, T., Manuelian, M., Campbell, R., Gwee, L., Elabd, Y., Tambe, N., Crombez, R., Texter, J. 2010. Single-wall carbon nanotube latexes. *ACS Appl. Mater. Interfaces* 2:649–653.
121. Zhao, L., Crombez, R., Antonietti, M., Texter, J., Titirici, M.-M. 2010. Sustainable nitrogen-doped carbon latexes with high electrical and thermal conductivity. *Polymer* 51:4540–4546.
122. Kohno, Y., Ohno, H. 2012. Key factors to prepare polyelectrolytes showing temperature-sensitive LCST-type phase transition in water. *Aust. J. Chem.* 65:91–94.
123. Kudin, K. N., Ozbas, B., Schniepp, H. C., Prud'homme, R. K., Aksay, I. A., Car, R. 2008. Raman spectra of graphite oxide and functionalized graphite sheets. *Nano Lett.* 8:36–41.
124. Bak, J. M., Lee, H.-I. 2012. pH-tunable aqueous dispersion of graphene nanocomposites functionalized with poly(acrylic acid) brushes. *Polymer* 53:4955–4960.
125. Lotya, M., Hernandez, Y., King, P. J., Smith, R. J., Nicolosi, V., Karlsson, L. S., Blighe, F. N. et al. 2009. Liquid phase production of graphene by exfoliation of graphite in surfactant/water solutions. *J. Am. Chem. Soc.* 131:3611–3620.
126. Vega-Rios, A., Renteria-Baltierrez, F. Y., Hernandez-Escobar, C. A., Zaragoza-Contreras, E. A. 2013. A new route toward graphene nanosheet/polyaniline composites using a reactive surfactant as polyaniline precursor. *Syn. Metals* 184:52–60.
127. Lotya, M., King, P. J., Khan, U., De, S., Coleman, J. N. 2010. High-concentration, surfactant-stabilized graphene dispersions. *ACS Nano* 4:3155–3162.
128. Ramalingam, P., Pusuluri, S. T., Periasamy, S., Veerabahu, R., Kulandaivel, J. 2013. Role of deoxy group on the high concentration of graphene in surfactant/water media. *RSC Adv.* 3:2369–2378.
129. Xiao, W., Sun, Z., Chen, S., Zhang, H., Zhao, Y., Huang. C., Zhimin Liu, Z. 2012. Ionic liquid-stabilized graphene and its use in immobilizing a metal nanocatalyst. *RSC Adv.* 2:8189–8193.
130. Yang, H., Hernandez, Y., Schlierf, A., Felten, A., Eckmann, A., S. Johal, S., P. Louette, P. et al. 2013. A simple method for graphene production based on exfoliation of graphite in water using 1-pyrenesulfonic acid sodium salt. *Carbon* 53:357–365.
131. Schlierf, A., Yang, H., Gebremedhn, E., Treossi, E., Ortolani, L., Chen, L., Minoia, A. et al. 2013. Nanoscale insight into the exfoliation mechanism of graphene with organic dyes: Effect of charge, dipole and molecular structure. *Nanoscale* 5:4205–4216.
132. Parviz, D., Sriya Das, S., Ahmed, H. S. T., Irin, F., Bhattacharia, S., Green, M. J. 2012. Dispersions of non-covalently functionalized graphene with minimal stabilizer. *ACS Nano* 6:8857–8867.
133. Lee, D.-W., Taehoon Kim, T., Myongsoo Lee, M. 2011. An amphiphilic pyrene sheet for selective functionalization of graphene. *Chem. Commun.* 47:8259–8261.
134. Das, S., Irin, F., Ahmed, H. S. T., Cortinas, A. B., Wajid, A. S., Parviz, D., Jankowski, A. F., Kato, M., Green, M. J. 2012. Non-covalent functionalization of pristine few-layer graphene using triphenylene derivatives for conductive poly (vinyl alcohol) composites. *Polymer* 53:2485–2494.
135. Ion, A. C., Alpatova, A., Ion, I., Culetu, A. 2011. Study on phenol adsorption from aqueous solutions on exfoliated graphitic nanoplatelets. *Mater. Sci. Eng. B* 176:588–595.
136. Backes, C., Frank Hauke, F., Hirsch, A. 2011. The potential of perylene bisimide derivatives for the solubilization of carbon nanotubes and graphene. *Adv. Mater.* 23:2588–2601.
137. Ager, D., Texter, J. 2014. Aqueous graphene dispersions—Optical properties and stimuli-responsive phase transfer. *ACS Nano* 8:11191–11205 doi: 10.1021/nn502946f.
138. Texter, J., Ager, D., Arjunan Vasantha, V., Crombez, R., England, D., Ma, X., Maniglia, R., Tambe, N. 2012. Advanced nanocarbon materials facilitated by novel stimuli-responsive stabilizers. *Chem. Lett.* 3:1377–1379.
139. Rastogi, R., Kaushal, R., Tripathi, S. K., Sharma, A. L., Kaur, I., Bharadwaj, L. M. 2008. Comparative study of carbon nanotube dispersion using surfactants. *J. Colloid Interface Sci.* 328:421–428.
140. Premkumar, T., Geckeler, K. E. 2012. Graphene–DNA hybrid materials: Assembly, applications, and prospects. *Prog. Polym. Sci.* 37:515–529.

141. Uddin, M. E., Kuila, T., Nayak, G. C., Kim, N. H., Ku, B. C., Lee, J. H. 2013. Effects of various surfactants on the dispersion stability and electrical conductivity of surface modified graphene. *J. Alloys Comp.* 562:134–142.
142. Wang, H. 2009. Dispersing carbon nanotubes using surfactants. *Curr. Opinion Colloid Interface Sci.* 14:364–371.
143. Green, A. A., Hersam, M. C. 2009. Solution phase production of graphene with controlled thickness via density differentiation. *Nano Lett.* 9:4031–4036.
144. Sun, Z., Masa, J., Liu, Z., Schuhmann, W., Muhler, M. 2012. Highly concentrated aqueous dispersions of graphene exfoliated by sodium taurodeoxycholate: Dispersion behavior and potential application as a catalyst support for the oxygen-reduction reaction. *Chem. Eur. J.* 18:6972–6978.
145. Buzaglo, M., Shtein, M., Kober, S., Lovrincic, R., Vilan, A., Oren Regev, O. 2013. Critical parameters in exfoliating graphite into graphene. *Phys. Chem. Chem. Phys.* 15:4428–4435.
146. Ioni, Yu. V., Tkachev, S. V., Bulychev, N. A., Gubin, S. P. 2011. Preparation of finely dispersed nanographite. *Inorg. Mater.* 47:597–602.
147. Yamamoto, T., Miyauchi, Y., Motoyanagi, J., Fukushima, T., Aida, T., Kato, M., Maruyama, S. 2008. Improved bath sonication method for dispersion of individual single-walled carbon nanotubes using new triphenylene-based surfactant. *Japan. J. Appl. Phys.* 47:2000–2004.
148. Notley, S. M. 2012. Highly concentrated aqueous suspensions of graphene through ultrasonic exfoliation with continuous surfactant addition. *Langmuir* 28:14110–14113.
149. Seo, J.-W. T., Green, A. A., Antaris, A. L., Hersam, M. C. 2011. High-concentration aqueous dispersions of graphene using nonionic, biocompatible block copolymers. *J. Phys. Chem. Lett.* 2:1004–1008.
150. Guardia, L., Fernandez-Merino, M. J., Paredes, J. I., Solis-Fernandez, P., Villar-Rodil, S., Martinez-Alonso, A., Tascon, J. M. D. 2011. High-throughput production of pristine graphene in an aqueous dispersion assisted by non-ionic surfactants. *Carbon*, 49:1653–1662.
151. Smith, R. J., Lotya, M., Coleman, J. N. 2010. The importance of repulsive potential barriers for the dispersion of graphene using surfactants. *New J. Phys.* 12:125008–125011.
152. Wu, D., Yang, X. 2012. Coarse-grained molecular simulation of self-assembly for nonionic surfactants on graphene nanostructures. *J. Phys. Chem. B* 116:12048–12056.
153. Texter, J. 2011. Nanoparticle dispersions with ionic liquid-based stabilizers, US Patent Application Publication, US 2011/0233458 A1.
154. Texter, J., Arjunan Vasantha, V., Bian, K., Ma, X., Slater, L., Mourey, T., Slater, G. 2011. Stimuli-responsive triblock copolymers—Synthesis, characterization, and application, Ch. 9. In *Non-Conventional Functional Block Copolymers*. Coughlin, B., Theato, P., Kilbinger, A. editors. American Chemical Society, Washington, DC, pp. 117–130; *ACS Symp. Ser.* 1066:117–130.
155. Texter, J., Arjunan Vasantha, V., Maniglia, R., Slater, L., Mourey, T. 2012. Triblock copolymer based on poly(propylene oxide) and poly(1-[11-acryloylundecyl]-3-methyl-imidazolium bromide). *Macromol. Rap. Com.* 33:69–74.
156. Texter, J., Arjunan Vasantha, V., Maniglia, R., Slater, L., Mourey, T. 2012. Ionic liquid-based anion and temperature responsive triblock copolymers. *Poly. Mater. Sci. Eng.* 106:786–787.
157. Bourlinos, A. B., Georgakilas, V., Radek Zboril, R., Steriotis, T. A., Stubos, A. K., Trapalis, C. 2009. Aqueous-phase exfoliation of graphite in the presence of polyvinylpyrrolidone for the production of water-soluble graphenes. *Sol. State Commun.* 149:2172–2176.
158. Delbecq, F., Kono, F., Kawai, T. 2013. Preparation of PVP–PVA–exfoliated graphite cross-linked composite hydrogels for the incorporation of small tin nanoparticles. *Eur. Polym. J.* 49:2654–2659.
159. Liang, Y. T., Hersam, M. C. 2010. Highly Concentrated graphene solutions via polymer enhanced solvent exfoliation and iterative solvent exchange. *J. Am. Chem Soc.* 132:17661–17663.
160. Carrasco, P. M., Montes, S., Garcia, I., Borghei, M., Jiang, H., Odriozola, I., Cabanero, G., Ruiz, V. 2014. High-concentration aqueous dispersions of graphene produced by exfoliation of graphite using cellulose nanocrystals. *Carbon* 70:157–163.
161. Das, S., Wajid, A. S., Shelburne, J. L., Liao, Y.-C., Green, M. J. 2011. Localized *in situ* polymerization on graphene surfaces for stabilized graphene dispersions. *ACS Appl. Mater. Interfaces* 3:1844–1851.
162. Griffith, A., Notley, S. M. 2012. pH dependent stability of aqueous suspensions of graphene with adsorbed weakly ionisable cationic polyelectrolyte. *J. Colloid Interface Sci.* 369:210–215.
163. Skaltsas, T., Karousis, N., Hui-Juan Yan, H.-J., Wang, C.-R., Pispas, S., Tagmatarchis, N. 2012. Graphene exfoliation in organic solvents and switching solubility in aqueous media with the aid of amphiphilic block copolymers. *J. Mater. Chem.* 22:21507–21512.
164. Sun, Z. Y., Vivekananthan, J., Guschin, D. A., Huang, X., Kuznetsov, V. Ebbinghaus, P., Sarfraz, A., Muhler, M., Schuhmann, W. 2014. High-concentration graphene dispersions with minimal stabilizer: A scaffold for enzyme immobilization for glucose oxidation. *Chem. Eur. J.* 20:5752–5761.
165. Pinto, A. M., Martins, J., Moreira, J. A., Mendes, A. M., Magalhaes, F. D. 2013. Dispersion of graphene nanoplatelets in poly(vinyl acetate) latex and effect on adhesive bond strength. *Polym. Int.* 62:928–935.
166. England, D., Tambe, N., Texter, J. 2012. Stimuli-responsive nanolatexes—Porating films. *ACS Macro Lett.* 1:310–314.
167. Texter, J. 2012. Anion responsive imidazolium-based polymers. *Macromol. Rap. Com.* 33:1996–2014.
168. Texter, J., Crombez, R., Ma, X., Zhao, L., Perez-Caballero, F., Titirici, M.-M., Antonietti, M. 2011. Waterborne nanocarbon dispersions for electronic and fuel applications. *Prepr. Symp.-Am. Chem. Soc., Div. Fuel Chem.* 56:388–389.
169. Giordano, C., Yang, W., Lindemann, A., Crombez, R., Texter, J., 2011. Waterborne WC nanodispersions. *Colloids Surfaces A* 374:84–87.
170. Texter, J., Ager, D. 2012. Waterborne graphene dispersions and advanced coatings. *Prepr. Symp.-Am. Chem. Soc., Div. Fuel Chem.* 57:90–92.
171. May, P., Khan, U., Hughes, J. M., Coleman, J. N. 2012. Role of solubility parameters in understanding the steric stabilization of exfoliated two-dimensional nanosheets by adsorbed polymers. *J. Phys. Chem. C* 116:11393–11400.
172. Kuila, T., Bose, S., Mishra, A. K. Khanra, P., Kim, N. H., Lee, J. H. 2012. Chemical functionalization of graphene and its applications. *Prog. Mater. Sci.* 57:1061–1105.
173. Su, C.-Y., Lu, A.-Y., Xu, Y., Chen, F. R., Khlobystov, A. N., Li, L. J. 2011. High-quality thin graphene films from fast electrochemical exfoliation. *ACS Nano* 5:2332–2339.
174. Coraux, J., Marty, L., Bendiab, N., Bouchiat, V. 2013. Functional hybrid systems based on large-area high-quality graphene. *Acc. Res.* 46:2193–2201.
175. Geim, A. K., Novoselov, K. S. 2007. The rise of graphene. *Nat. Mater.* 6:183–191.

176. Mak, K. F., Sfeir, M. Y., Wu, Y., Lui, C. H., Misewich, J. A., Heinz, T. F. 2008. Measurement of the optical conductivity of graphene. *Phys. Rev. Lett.* 101:196405.
177. Nair, R. R., Blake, P., Grigorenko, A. N., Novoselov, K. S., Booth, T. J., Stauber, T., Peres, N. M. R., Geim, A. K. 2008. Fine structure constant defines visual transparency of graphene. *Science* 320:1308.
178. Thongrattanasiri, S., Koppens, F. H. L., Garcia de Abajo, F. J. 2012. Complete optical absorption in periodically patterned graphene. *Phys. Rev. Lett.* 108:047401 (5 pp).
179. Ye, Q., Wang, J., Liu, Z., Deng, Z.-C., Kong, X.-T., Xing, F., Chen, X.-D., Zhou, W.-Y., Zhang, C.-P., Tian, J.-G. 2013. Polarization-dependent optical absorption of graphene under total internal reflection. *Appl. Phys. Lett.* 102:021912 (4 pp).
180. Herman, L. H., Kim, C.-J., Wang, Z., Jo, M.-H., Park, J. 2012. Depolarization effect in optical absorption measurements of one- and two-dimensional nanostructures. *Appl. Phys. Lett.* 101:123101 (4 pp).
181. Mak, K. F., Shan, J., Heinz, T. F. 2011. Seeing many-body effects in single- and few-layer graphene: Observation of two-dimensional saddle-point excitons. *Phys. Rev. Lett.* 106:046401.
182. Saito, R., Kataura, H. 2001. Optical properties and Raman spectroscopy of carbon nanotubes. In. *Carbon Nanotubes, Topics in Applied Physics*, Vol. 80. Dresselhaus, M. S., Dresselhaus, G., Avouris, P., editors. Springer, Berlin, pp. 213–247.

21 Graphene Applications for Photoelectrochemical Systems

Rui Cruz, José Maçaira, Luísa Andrade, and Adélio Mendes

CONTENTS

ABSTRACT

Graphene is a very versatile nanomaterial and can be used in a wide range of applications. This chapter focuses on the use of graphene in photoelectrochemical systems, in particular for photocatalysis, hydrogen generation, and in dye-sensitized solar cells.

Currently, titanium dioxide (TiO_2)–anatase is the most used photocatalyst for environmental applications. Very recently, the TiO_2/graphene composite proved to be an interesting material for photocatalytic purposes, exhibiting enhanced energy-harvesting properties and consequently improved photocatalytic activity. When graphene is bounded to semiconductors such as TiO_2, the semiconductor band gap becomes smaller and makes the rate of hole/electron recombination after light excitation decrease, acting as an efficient electron acceptor. Hydrogen production via photoelectrochemical water splitting is a thriving alternative that combines photovoltaic cells with an electrolysis system. The major advantage is that solar harvesting, conversion, and storage are combined in a single integrated system. Dye-sensitized solar cells (DSCs) are among the leading third-generation photovoltaic devices, beginning to position themselves as a reliable low-cost alternative to traditional silicon devices. Graphene is the ideal candidate for incorporation in DSCs due to its tunable surface properties and high electrical conductivity. Its versatile nature enables it to be used in different parts of the cell, namely as a counterelectrode or incorporated in the photoelectrode to increase the electronic conductivity.

21.1 PHOTOCATALYTIC DEGRADATION

The photocatalysis principle is the activation of a semiconductor (SC) by the sun or artificial light. When exposed to sunlight, the SC absorbs photons with sufficient energy to inject electrons from the valence band to its conduction band (CB), creating electron–hole pairs. These holes have a potential that is sufficiently positive to generate $OH^{\bullet}$ radicals from water molecules adsorbed onto the SC surface, which can then oxidize organic contaminants. The electrons react with oxygen molecules to form the superoxide anion, $O_2^{\bullet-}$—Figure 21.1. The photocatalytic efficiency depends on the competition between the process in which the electron reacts with a chemical species on the SC surface and the electron–hole recombination process, which results in heat or radiation release.

Photocatalysis is governed by the kinetics of charge carriers and redox reactions that take place at the surface of the photocatalyst. Therefore, the understanding of the electronic processes occurring at the SC nanoparticles (NPs) level, as well as the dynamics of charge separation/transport and reactive mechanisms in the different interfaces are of great importance. In the SC context, the photocatalysis process is usually interpreted to a band model, where at least two reactions

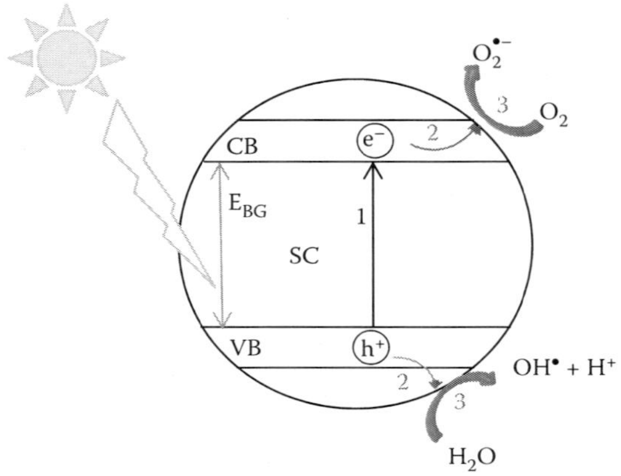

FIGURE 21.1 Mechanism of photocatalysis with the main reactions: electron–hole pair generation (1), charge separation and migration to surface reaction sites (2), and surface chemical reaction on active sites (3).

may occur simultaneously: (i) oxidation from photogenerated holes and (ii) reduction from photogenerated electrons. These processes should occur at the same rate to maintain the photocatalyst mostly electrically neutral. Figure 21.1 shows the following sequence of charge transfer processes responsible for the photocatalytic process:

1. Photons with energy higher than the band gap of the SC generate electron–hole pairs in the conduction and valence bands, respectively:

$$SC \xrightarrow{h\nu} e^-_{CB} + h^+_{VB} \tag{21.1}$$

2. Holes react with water molecules adsorbed on the SC surface, resulting in the formation of hydroxyl radicals:

$$SC(h^+_{VB}) + H_2O_{ads} \longrightarrow SC + HO^\bullet_{ads} + H^+ \tag{21.2}$$

3. Electrons in the CB reduce the adsorbed oxygen to superoxide:

$$SC(e^-_{CB}) + O_{2ads} \longrightarrow SC + O_2^{\bullet-} \tag{21.3}$$

For an SC to be photochemically active, the potential of the photogenerated valence band hole must be sufficiently positive to generate $OH^\bullet$ radicals, which can subsequently oxidize the organic pollutant. The redox potential of the photogenerated CB electrons must be sufficiently negative to be able to reduce adsorbed oxygen to superoxide. The surface-adsorbed water on the photocatalyst particles is essential for photocatalysis since it provides the electrolyte support needed for the redox reaction promoted by the e^-/h^+ (electron/hole) pairs generated by the excited SC. Nevertheless, the possibility of the occurring electron–hole recombination greatly limits the photocatalytic activity and thus, several efforts are envisaged to allow a more efficient charge carrier separation [1,2]. During the past few decades, several strategies have been followed to improve the photocatalytic performance of SC photocatalysts [3]. In particular, combining graphene with SC photocatalysts is an approach to improve photocatalyst performance that has captured the interest of researchers [4].

From the available SCs that can be used as photocatalysts, titanium dioxide (TiO_2) is generally considered to be the best material [1,5,6]. In fact, TiO_2 is close to being an ideal photocatalyst, exhibiting almost all the required properties for an efficient photocatalytic process, with only the important drawback of not absorbing visible light. Furthermore, TiO_2 is nontoxic, thermally stable, chemically inert, photostable (i.e., not prone to photocorrosion), readily available, and relatively cheap. It shows band edges well positioned, exhibiting a strong oxidizing power at ambient temperature and pressure (3.0 V) and the photogenerated electrons are able to reduce oxygen to superoxide (–0.2 V). Thus, TiO_2/graphene composites are widely used for photocatalytic applications for their potential in environmental and energy-related applications.

21.1.1 Working Principle of Photocatalysis

In TiO_2/graphene composites, the electron–hole pairs are generated upon TiO_2 excitation under ultraviolet (UV) light irradiation. These photogenerated electrons are then injected into graphene due to its more positive Fermi level [7]. Graphene platelets are an excellent platform for scavenging photogenerated electrons by dissolved oxygen, thus facilitating the hole–electron separation. Moreover, the high carrier mobility of graphene accelerates electron transport that enhances the photocatalytic performance [8]. The dissolved oxygen is normally present in the adsorbed water at the surface of the TiO_2 particles. In photocatalysis, as in all photoelectrochemical systems, an electrolyte medium must exist since during redox reactions, there is a transfer of ions between oxidation and reduction sites in the photocatalyst surface. In heterogeneous photocatalysis, the electrolyte is water favored by the hydrophilicity of TiO_2 and the abundance of molecular water in different environments. The holes can either react with adsorbed water to form hydroxyl radicals or directly oxidize various organic compounds. The common reaction steps in photocatalytic degradation under UV-light irradiation are summarized as follows [4]:

$$TiO_2 + h\nu(UV) \rightarrow TiO_2(e^- + h^+) \tag{21.4}$$

$$H_2O \leftrightarrow H^+ + OH^- \tag{21.5}$$

$$TiO_2(e^-) + \text{Graphene} \rightarrow TiO_2 + \text{Graphene}(e^-) \tag{21.6}$$

$$\text{Graphene}(e^-) + O_2 \rightarrow \text{Graphene} + O_2^{\bullet-} \tag{21.7}$$

$$TiO_2(h^+) + OH^- \rightarrow TiO_2 + OH^\bullet \tag{21.8}$$

$$OH^\bullet + \text{Pollutants} \rightarrow \text{Degradation Products} \tag{21.9}$$

Simultaneously, Ti–O–C bonds formed in the TiO_2/graphene photocatalyst originate a red shift of a few dozens of nanometers in the solar spectrum, reducing its band gap and making it sensitive to longer-wavelength light [9,10]. The resulted photocatalyst then presents an extended photoresponse of up to ca. 440 nm [10]. In particular, Lee et al. [11] synthesized highly photoactive graphene-wrapped TiO_2 composites by one-step hydrothermal graphene oxide (GO) reduction and TiO_2 crystallization from GO-wrapped amorphous TiO_2 NPs. The new material presented a red shift of the solar spectrum with consequent reduction of the band gap to 2.80 eV. This produced an excellent photocatalytic activity in the visible range with a rate constant of 3.41×10^{-2} min^{-1}.

TiO_2 photooxidation is normally intermediated with free radicals of $OH^{\bullet}$ (oxidation potential of 2.8 V [12]) and $O_2^{\bullet-}$ (reduction potential of −0.137 V [13]), making a thermodynamic minimum band gap of 2.94 eV necessary for generating both radicals. Since most of the band gap-shortening approaches consider the creation of intermediate energy levels making the electron energy gain a stepwise process, the lowest and highest energy levels are still available. This means that, despite the band gap shortening below, for example, 2.8 eV, the photocatalyst is still active toward $OH^{\bullet}$ and $O_2^{\bullet-}$ generation [11].

Nevertheless, the visible light activity of the TiO_2/graphene composites is not fully understood [14,15]. The general explanation for visible light harvesting is that GO acts as a sensitizer in the composite and this conclusion has been based on the improvement of the degradation of dyes [10,16,17]. This way, graphene absorbs visible light and injects excited electrons into the CB (d-orbital) of the TiO_2 due to the d–π interaction. Then, the excited electrons are transferred to the TiO_2 surface where they will react with oxygen, producing superoxide radicals, as follows:

$$\text{Graphene} + h\nu(\text{Visible}) \rightarrow \text{Graphene}(e^- + h^+) \quad (21.10)$$

$$H_2O \leftrightarrow H^+ + OH^- \quad (21.11)$$

$$\text{Graphene}(e^-) + TiO_2 \rightarrow TiO_2(e^-) + \text{Graphene} \quad (21.12)$$

$$TiO_2(e^-) + O_2 \rightarrow TiO_2 + O_2^{\bullet-} \quad (21.13)$$

$$\text{Graphene}(h^+) + OH^- \rightarrow \text{Graphene} + OH^{\bullet} \quad (21.14)$$

$$OH^{\bullet} + \text{Pollutants} \rightarrow \text{Degradation Products} \quad (21.15)$$

When graphene is bounded to SCs such as TiO_2, ZnO, or SiO_2, the overall photocatalytic performance is largely improved when compared with bare TiO_2. This is mainly attributed to three effects: (i) efficient charge separation and transportation; (ii) extended light absorption range; and (iii) enhanced adsorptivity.

TiO_2/graphene composites slow the rate of h^+/e^- recombination after light excitation [18], increasing the charge transfer rate of electrons and surface-adsorbed chemical species through π–π interactions. Williams et al. [18] studied the electron transfer between excited TiO_2 and GO by a nanosecond flash photolysis technique using a 308-nm laser pulse excitation; the absorption of the trapped electron was monitored at 650 nm. It was verified that, in the presence of GO, the magnitude of absorption decreased by approximately 50% indicating that the electrons transfer to GO occurs within the pulse duration, confirming the transfer of electrons to GO in the submicrosecond time scale—Figure 21.2. Moreover, Zhou et al. [8] verified that graphene acts as an electron-acceptor material to effectively hinder the electron–hole pair recombination of TiO_2. The composite photocatalyst tested exhibited excellent photocatalysis to methylene blue (MB) under irradiation of simulated sunlight. Finally, Zou et al. [19] analyzed TiO_2/graphene nanocomposites by transmission electron microscopy, which showed that TiO_2 NPs were uniformly dispersed on the graphene plane. Fluorescence emission spectra and photoelectrical experiments showed that TiO_2/graphene nanocomposites have better charge separation capability than pure TiO_2. The photocurrent varies with the amount of graphene in the TiO_2/graphene nanocomposite and it obtained a saturated photocurrent density about 3 times higher than that obtained on a TiO_2/fluorine-doped tin oxide (FTO) electrode.

Finally, a good adsorptivity of the contaminant molecules is a prerequisite for an efficient photocatalysis process and TiO_2/graphene composites exhibit enhanced adsorptivity. This is mainly attributed to its very large π-conjugation system and two-dimensional (2D) planar structure [10,20]. Nguyen-Phan et al. [15] reported the development of TiO_2/graphene composites with higher adsorption and photocatalysis performance under both UV and visible radiation. The increase in graphene content facilitated the photodegradation

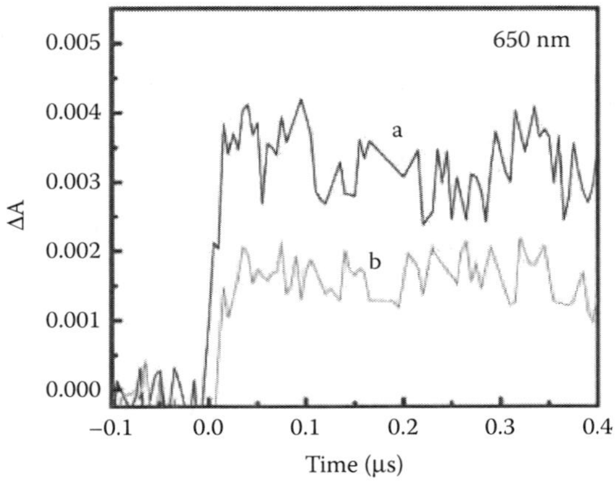

FIGURE 21.2 (a) TiO_2 without GO and (b) TiO_2/GO composite. (Reprinted with permission from Williams, G. et al., TiO_2–graphene nanocomposites. UV-assisted photocatalytic reduction of graphene oxide. *ACS Nano*, **2**(7): 1487–1491. Copyright 2008 American Chemical Society.)

rate of MB. The greater photocatalytic performance of the TiO_2/graphene composite was ascribed to the formation of both π–π conjugations between dye molecules and aromatic rings and the ionic interactions between MB and oxygen-containing functional groups at the edges or on the surfaces of carbon-based nanosheets. GO worked as the adsorbent, electron acceptor, and photosensitizer to efficiently enhance the dye photodecomposition. Furthermore, Khoa et al. [20] verified that TiO_2/graphene films present a higher contact area, higher photon scattering on the film surface, and enhanced probing dye adsorption than simple TiO_2 films. It was concluded that the π-conjugation system and the 2D planar structure of GO improve the adsorption of the dye that explains the better properties of the composite to photooxidize the dye.

21.1.2 Preparation Methods for Graphene-SC Composites

The preparation methods of graphene-SC composites can be divided into (i) *ex situ* hybridization and (ii) *in situ* crystallization. The *ex situ* hybridization process involves the mixture of graphene nanosheets (GNS) with presynthesized or commercially available SC nanocrystals in suspension [10,20–24]. The *in situ* crystallization can be conducted through different methods, such as electroless deposition, electrochemical deposition, thermal evaporation, chemical reduction, sol–gel, or hydrothermal methods [25,26]. These *in situ* crystallization methods take advantage of GO/reduced graphene oxide (RGO) nucleation sites to originate a continuous film of nanocrystals on the graphene surface.

Tanaka and Mendes [27] disclosed a graphene composite material based on bonding TiO_2 NPs to graphene platelets. A TiO_2/graphene composite was prepared by mixing a colloidal solution of GO in water and a solution of a titanium precursor. GO was reduced using hydrazine to recover graphene's conductive properties by restoring the π-network. During calcination, titanium hydroxide clusters dehydrated, being converted into TiO_2. Metal NPs can be added to the final composite via thermal evaporation of the desired salt solution followed by reduction via a thermal, chemical, or photochemical method. The authors claim that the composite photocatalyst has a higher photocatalytic activity than the reference TiO_2 photocatalyst (P25 from Evonik) and also higher activity than photocatalysts vlp7000 from Kronos. Moreover, in this newly developed composite photocatalyst, TiO_2 NPs are attached to microsize graphene platelets and thus, the risk of harm to health due to absorption or inhalation of these particles is substantially smaller [28].

21.1.3 Metal Decoration of TiO_2/Graphene Composites

In the case of TiO_2/graphene composite photocatalysts, the decoration of TiO_2 with metals such as Ag, Cu, platinum (Pt), and Au decreases charge recombination, shows a plasmonic effect, and reduces redox overvoltages [14,29,30]. In fact, the RGO platelet is an ideal substrate for metal deposition since it serves both as a nucleation site and as an inhibitor of metal NPs aggregation. In metal-decorated TiO_2/graphene composites, the main function of metals is to serve as a migration center for the excited electrons. These electrons are trapped in the metal minimizing electron–hole recombination in this way. Besides, some metals, mainly noble metals, present surface plasmon resonance that normally results in strong and broad absorption bands in the visible light region, which can promote the visible light activity of the photocatalyst.

Neppolian et al. [30] reported the development of a novel nanosized Pt/TiO_2/graphene photocatalyst. The photocatalytic activity of the composite was assessed concerning the degradation rate of dodecylbenzenesulfonate (DBS). The photocatalyst degraded DBS at a higher rate than commercial P25 and TiO_2/graphene composites, increasing DBS mineralization by a factor of 3 compared to P25. The photogenerated electrons in TiO_2 travel to the noble metal that is attached to the photocatalyst surface. Then, Pt captures the electron, which facilitates the interfacial charge transfer. Furthermore, electrons are easily transported through the GO considerably increasing the oxidation rate of DBS. Moreover, the authors also justify the improved photocatalytic activity by a significant decrease of the e^-/h^+ recombination, although no experimental evidences were shown.

Li et al. [14] prepared and characterized an Ag/TiO_2/graphene composite film. The photocatalytic activity of this composite was evaluated photoelectrochemically and compared to the performance of bare TiO_2 and the TiO_2/graphene composite. It was observed that Ag/TiO_2/graphene composite film produced the highest photocurrent under both UV and visible light illumination. The presence of graphene and silver, which facilitate the transfer of electrons from the SC promoted a decrease in e^-/h^+ recombination. Besides, silver particles showed surface plasmon resonance, facilitating the interfacial charge transfer; silver also acted as a sensitizer promoting a downward bending of TiO_2 energy bands.

For photocatalytic indoor applications, such as for photoinactivation of microorganisms, a very promising photocatalyst is Au/TiO_2/graphene. The use of gold NPs is expected to promote increased values of photoactivity due to the high surface plasmon resonance effect verified with these NPs [31,32].

21.2 PHOTOCATALYTIC HYDROGEN GENERATION

Hydrogen production via photoelectrochemical water splitting is a thriving alternative that combines photovoltaic cells with an electrolysis system [33,34]. The major advantage is that solar harvesting, conversion, and storage are combined in a single integrated system [35]. The hydrogen generated by this process has the potential to be a sustainable carbon-neutral fuel since it is produced from a renewable source and it can be stored or transformed into other chemicals such as methanol or methane [36].

The single-photon photon emission-computed (PEC) system for water splitting is composed of anSC photoelectrode that absorbs photons with sufficient energy to inject electrons from the valence to the CB, creating electron–hole pairs. The excited electrons percolate through the SC layer reaching the counterelectrode (CE), via the external circuit, to promote water reduction at its surface, while holes oxidize water in the SC surface [37,38]. The cycle is closed when the electrolyte anions generated at the CE diffuse back to the surface of the SC to react with holes, thus oxidizing water—Figure 21.3.

Several SC photocatalysts have been used for the reduction of water to hydrogen. Nevertheless, practical applications of this very promising technology are limited due to the inability of using visible light, low quantum efficiency, and photodegradation of the catalyst. In particular, the fast recombination of photogenerated electrons and holes within the photocatalyst strongly contributes to the low efficiency of these systems. Indeed, graphene with its very high electron mobility and high specific surface area may be used as an efficient electron acceptor enhancing the photoinduced charge transfer and as an inhibitor agent of the recombination pathway by separating the evolution sites of hydrogen and oxygen [39].

The first material recognized to split water under UV light was TiO_2, reported by Fujishima and Honda [40] in 1971. Thenceforward, extensive efforts have been made to turn this material into an efficient photoelectrode. Zhang et al. [41] studied the feasibility of producing solar hydrogen by using a composite photocatalyst of TiO_2/graphene sheets (GS) prepared by a sol–gel method. The as-prepared photocatalysts showed to have an enhanced photocatalytic activity than bare P25 and the maximum activity was obtained for a 5 wt% sample of graphene. It was observed that when the graphene content exceeds this value, the photocatalytic activity starts decreasing since electron–hole recombination centers are introduced into the composite. For a 5 wt% content of GS, the amount of hydrogen evolution reaches ca. 17.2 μmol during 2 h at a rate of 8.6 μmol h^{-1}. This is attributed to the good electronic conductivity and a large specific surface area of graphene, resulting in a fast-generated electrons transport to the photocatalyst surface, thus inhibiting electron/hole recombination. Moreover, an improvement in the light absorption in the visible part of the solar spectrum is also noticed, probably attributed to the presence of carbon in the composites. More recently, other TiO_2/graphene composites were synthesized by three different methods: UV-assisted photoreduction, chemical reduction using hydrazine, and a hydrothermal method. All the composites prepared by the mentioned methods exhibited better photocatalytic activity for hydrogen evolution from a methanol aqueous solution than simple P25. Moreover, the hydrothermal method allowed the preparation of the photocatalyst with the best performance for hydrogen evolution with an optimum mass ratio of 1/0.2 (TiO_2/RGO). In all the methods, GO can be reduced to GS, which enhances the photocatalytic activity for H_2 evolution; the best performance in the hydrothermal method of preparation may be explained by a stronger interaction between P25 and GS in the composite. This accelerates the charge transfer, thus suppressing the surface recombination, as is being argued by several authors working in this field. This TiO_2/graphene composite also showed better photocatalytic performance than a TiO_2/carbon nanotubes (CNTs) composite and higher hydrogen production rate under pure water [42]. The same conclusions were arrived at by other researchers who have worked on the association of GO to TiO_2 SCs and they also showed an enhanced photocatalytic activity with this kind of composite. Park et al. [43] used a spontaneous exfoliation method of bulky graphite oxide and they also justified the photocatalytic improvement by the ability of these systems to retard the undesired back reaction of recombination.

TiO_2 is actually a very interesting photocatalyst for photo-water splitting and it has been widely studied as shown above.

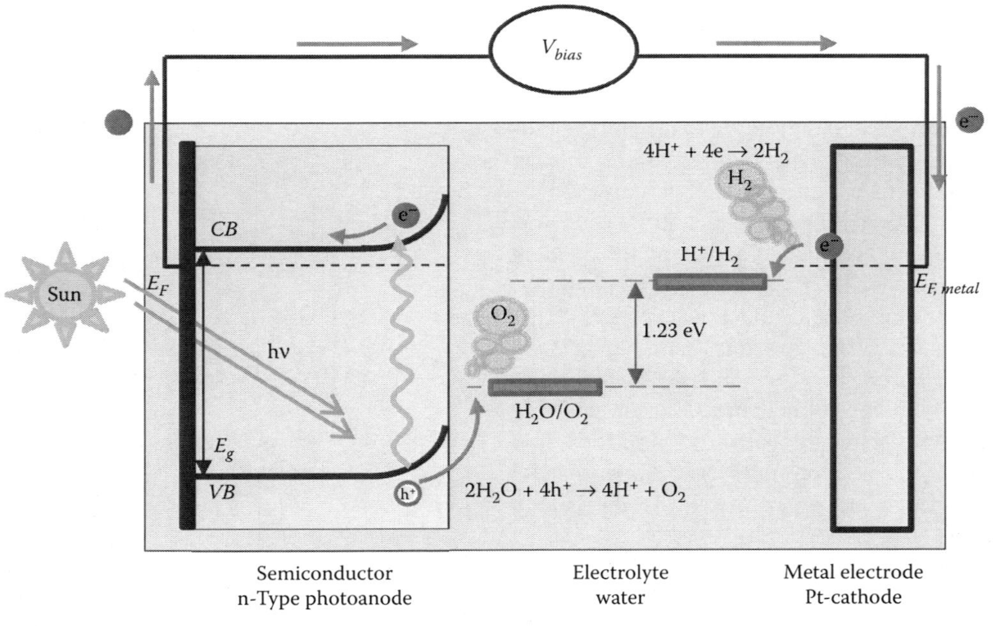

FIGURE 21.3 PEC cell working principles.

Nevertheless, this material still presents a low rate of hydrogen production mainly due to its wide band gap, making it mainly active under UV light radiation. Since the percentage of UV light is less than 5% of the total solar spectrum incident on earth, this photocatalyst absorbs a very low amount of solar energy, thus reflecting a very low hydrogen production rate. During the last three decades, different types of SCs have been studied such as metal oxide (e.g., Fe_2O_3, $SrTiO_3$, TiO_2, WO_3, $BiVO_4$, Cu_2O, etc.) and non-metal oxide SCs (e.g., GaAs, CdS, InP, etc.) [44]. Many efforts have been made to prepare appropriate photocatalysts active in the visible light region with a band gap of less than 3 eV, and recently a layered-perovskite tantalate SC ($Sr_2Ta_2O_7$) that doped with nitrogen ($Sr_2Ta_2O_{7-x}N_x$) exhibits a significant increase in the visible light absorption. The N-doping resulted in the shifting of the absorption edge from 290 to 550 nm, meaning a photocatalytic hydrogen production of 87% increase. Moreover, this material also presents a strong ability of photoinduced reduction of exfoliated GO to GS. By using graphene as a support for the Pt cocatalyst, the system graphene–Pt (5 wt% of graphene) and $Sr_2Ta_2O_{7-x}N_x$ exhibited a hydrogen production rate of 293 μmol h^{-1}, against the 194 μmol h^{-1} of the Pt cocatalyst/$Sr_2Ta_2O_{7-x}N_x$ system without graphene. Moreover, the quantum efficiency of the newly developed tantalate system in the wavelength range of 280–550 nm was increased to 6.45% (pristine undoped $Sr_2Ta_2O_7$ presented 2.33%) [45].

Following the same strategy and aiming to enhance the visible light absorption, CdS SC appears to be a very promising photocatalyst with a band gap of 2.4 eV. However, prolonged irradiation of CdS suspensions leads to decomposition and therefore the photochemical instability of these materials is improved by the association of CdS with other SCs. Still, the prepared composites present a low separation efficiency of electron–hole pairs. Jia et al. [46] used graphene/CdS composites to enhance the photocatalytic activity due to the ability of graphene to reduce the surface recombination. This was further improved by doping graphene with nitrogen, helping to tailor the electronic properties, and thus increasing the final photocatalytic activity.

$BiVO_4$ was also incorporated with RGO to promote photoelectrochemical water splitting. The obtained results showed the feasibility of using this SC to photocatalytically reduce GO in the visible light range. The final quantum yield was improved in one order of magnitude as shown by the incident photon-to-current efficiency (IPCE) results and the photocurrent density was increased 10-fold under visible light when compared with pristine $BiVO_4$; this was even higher than the photocurrent density generated by TiO_2 under UV light. Again, the enhanced photoelectrochemical water splitting using GO/$BiVO_4$ composites is attributed to the introduction of GS, which facilitate the electron transport between the SC and the electrode through graphene, increasing the charge transport and charge collection [47].

Interestingly, graphite oxide itself can be used as a photocatalyst for solar water splitting if the oxidized level is appropriately controlled. A graphite oxide SC photocatalyst with a band gap of 2.4–4.3 eV under UV radiation or a visible light can catalyze hydrogen generation from an aqueous methanol solution and pure water. A cocatalyst for promoting charge generation is not needed [48].

21.3 DYE-SENSITIZED SOLAR CELLS

21.3.1 Introduction

The dye-sensitized solar cell (DSC) is a potentially low-cost photovoltaic technology that has recently achieved 12% efficiency [49,50]. A certified world record for DSCs of 14% has been announced by M. Graztel and has already been updated in the National Renewable Energy Lab (NREL) chart of best-research cell efficiencies. This result is the promise of a future change in the photovoltaic market. DSCs mimic natural photosynthesis and differ from conventional p–n junction devices because light collection and charge transport are separated in the cell [51]—Figure 21.4. Light absorption occurs in the chemisorbed sensitizer molecule, while electron transport occurs in the SC. The photoconversion efficiency (η) of the solar cell is determined by its current–voltage characteristics, specifically the open-circuit photo voltage (V_{oc}), the photogenerated current density measured under short-circuit conditions (J_{sc}), the intensity of incident light (I_s), and the fill factor (FF) of the cell.

The operation of DSCs, illustrated in Figure 21.1, has been thoroughly discussed and results from a balance of several electron transfer kinetic reactions are exemplified in Figure 21.5. The operation of these solar cells can be summarized as follows: the photosensitizer, adsorbed on the surface

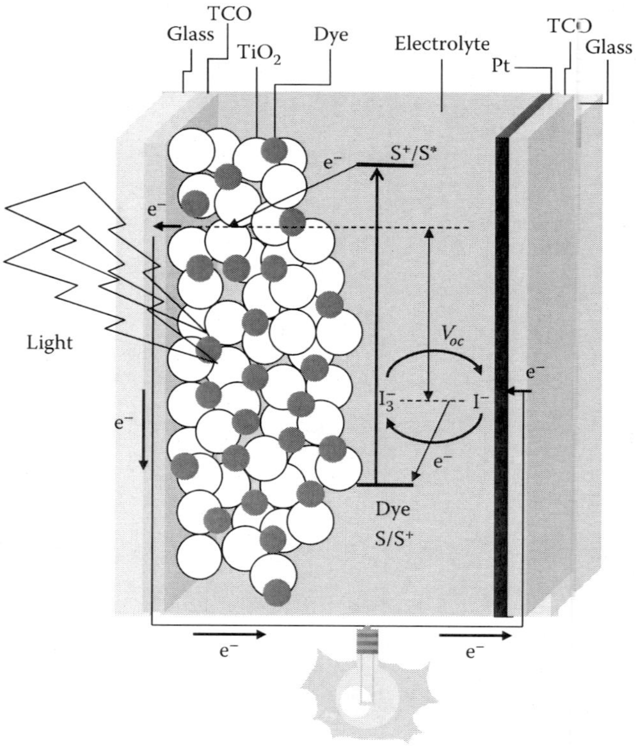

FIGURE 21.4 Schematic diagram of the operating principle of a DSC.

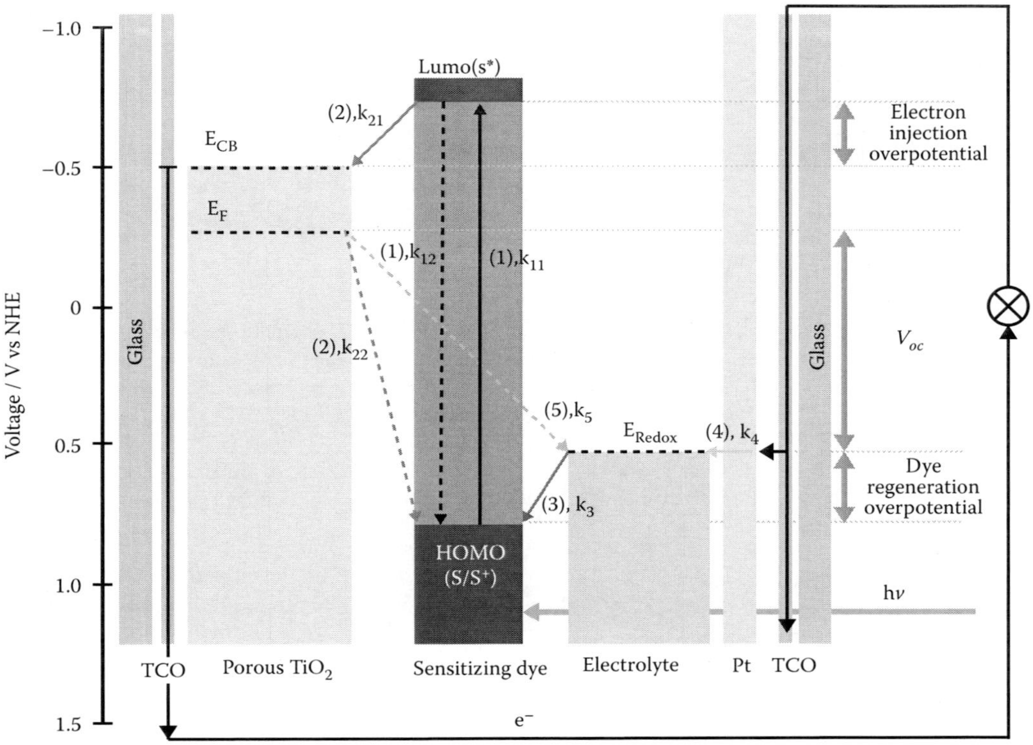

FIGURE 21.5 Schematic diagram of the DSCs kinetics. The solid arrows represent the forward electron transfer reactions; the dashed arrows represent the electron losses possible reactions.

of the porous SC (typically TiO_2), absorbs incident radiation and its electrons are excited from the ground state (S) to the excited state (S^*, k_{11}). There is also the possibility of decay from the excited state of the dye to its ground state (k_{12}) before injecting any electron in the CB of the SC; the excited electrons are injected into the CB of the SC (k_{21}), originating the oxidation of the sensitizer (S^+); reversibly, the reduction reaction of the dye cations (S^+) with electrons present in the CB of the SC is also possible (k_{22}); the oxidized sensitizer is regenerated to its original state by electron donation from the redox species present in the liquid electrolyte–iodide, producing triiodide: the triiodide species diffuse toward the CE where the Pt catalyst reduces it back to iodide by reaction with an electron from the external circuit. Instead of being collected to the external circuit, the CB electrons can also react with the electrolyte species (triiodide or iodide), reaction usually called the recombination reaction. This reaction is obviously undesirable and instead of producing electrical current, it generates heat. Because the electrolyte is present throughout the porous structure of the SC, the recombination reaction is affected by the iodide concentration, electrolyte viscosity, and dye structure.

In sum, the successful operation of DSCs relies on the countercurrents of photogenerated electrons and holes, which flow in their respective conductors (nanostructured metal oxide and liquid or solid electrolyte [or hole transport medium], respectively) with minimal losses to recombination. From the different possible strategies to slow down recombination, increasing charge mobility in the photoanode seems to be one of the most promising [52,53].

In the last couple of years, graphene has attracted enormous attention because of its unique properties and an array of applications [54–57]. With a theoretical surface area of $2630\ m^2 \cdot g^{-1}$, graphene presents itself as an ideal support material with enhanced interfacial contact even when used in small amounts [58]. Its electron mobility of $10^4\ cm^2 \cdot V^{-1}$ [54] at room temperature means that graphene has an excellent ability to transfer electrons. These facts have made it very attractive particularly for the incorporation in photocatalytic mesoporous films [39,59–61] and in the DSC's CE [62–65], where its electrocatalytic properties have been put to use to substitute the expensive Pt catalyst.

21.3.2 Photoanode

Recently, graphene has also been incorporated in DSC's photoelectrodes—Figure 21.6 [66–71]. Fourier transform infrared spectroscopy (FT-IR) experimental results of pure TiO_2 and graphene–TiO_2 hybrid films showed that Ti–O–C bonds are formed, indicating the chemical interaction between surface hydroxyl groups of TiO_2 and functional groups of GO [15]. Experimental data also show that the presence of graphene in the TiO_2 photoelectrode also increases the total amount of dye adsorbed in the film [15,67,72]. This fact is explained by these authors based on the huge surface area of graphene that provides more anchoring sites for TiO_2. But in fact, and similarly to what happens in photocatalytic studies where there are π–π conjugations between MB molecules and the aromatic rings of GO sheets, a chemical interaction between the sensitizing

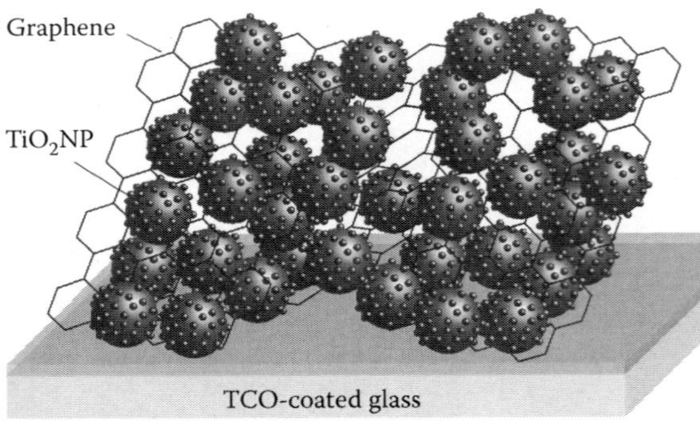

FIGURE 21.6 Schematic representation of a TiO_2/graphene hybrid photoanode. (Reprinted from *Renewable and Sustainable Energy Reviews*, **27**(0), Macaira, J. et al., Review on nanostructured photoelectrodes for next generation dye-sensitized solar cells, 334–349, Copyright 2013, with permission from Elsevier.)

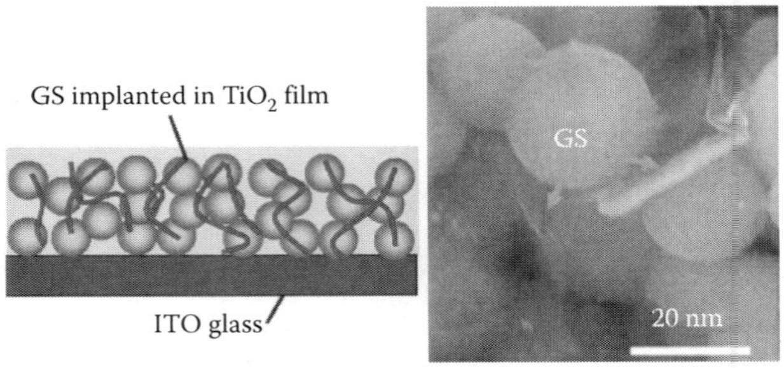

FIGURE 21.7 Schematic representation of a TiO_2 film with GS attached (left) and SEM micrographs of the prepared film with detail to the graphene sheet attached to the TiO_2. (Reprinted with permission from Tang, Y.-B. et al., Incorporation of graphenes in nanostructured TiO_2 films via molecular grafting for dye-sensitized solar cell application. *ACS Nano*, **4**(6): 3482–3488. Copyright 2010 American Chemical Society.)

dye and graphene might exist, helping to explain the higher amount of adsorbed dye in hybrid films [15,69,73].

Ideally, pristine graphene should be used in DSC photoelectrodes due to its absolute lack of electrocatalytic properties. However, this unaltered form of graphene is seldom used due to the difficulty associated with its manipulation, dispersion, and incorporation in metal oxides. Because of this fact, graphene is mainly produced by the chemical oxidation method resulting in GO [74,75], which can be then reduced either chemically or thermally. However, RGO contains oxygen functional groups (–OH and epoxides) on the planes and –OH and –C = O groups in the periphery of the planes [62,63]. This, along with the lattice surface defects created during the exfoliation process, is believed to be responsible for the electrocatalytic behavior of graphene [76–78]. Because of GO's electrocatalytic properties and its high extinction coefficient [79], RGO should be handled carefully when used in the photoelectrode of DSCs; if not properly reduced, the remaining oxygen-containing groups could promote recombination of electrons with the electrolyte. In high amounts, graphene can also compete with the sensitizer molecules on light absorption and thus decrease the performance of DSCs. Therefore, the successful introduction of graphene in the photoelectrodes of DSCs depends on a careful balance between its conductive ability and electrocatalytic behavior.

Tang et al. [67] reported the development of a photoanode where exfoliated GS were attached to a TiO_2 NPs matrix—Figure 21.7. The authors used a molecular-grafting method where the GS were chemically exfoliated and chemisorbed in the TiO_2 matrix. By controlling the oxidation time, it was possible to achieve a highly efficient electronic conductive film and thus a good attachment between the GO and the NPs. In fact, the determined resistivity of a GO/TiO_2 decreased by more than two orders of magnitude from $2.1 \pm 0.9 \times 10^5$ to $3.6 \pm 0.9 \times 10^2$ Ω [67]. These authors determined that the GO provided additional and more efficient electronic transport paths but also increased the dye loading of the film, leading to a photocurrent increase.

Sun et al. [66] developed a method where graphene was dispersed using Nafion and then incorporated in TiO_2 particles by a heterogeneous coagulation method. Because of the opposite zeta potentials of P25 particles (~15 mV) and graphene (–42 mV), there is a strong electrostatic attractive force that binds the TiO_2 NP to the surface of graphene. The authors managed to coat GS with TiO_2 resulting in a composite structure with a P25 to graphene ratio of 200:1 (w/w). The DSC with P25 NPs showed a J_{sc} of 5.04 mA cm^{-2} and an η of 2.70%. In the presence of 0.5 wt% of graphene, the J_{sc} increased by 66% to 8.38 mA cm^{-2}, resulting in an efficiency of 4.28%. The authors ascribe this enhanced performance to an increase in dye adsorption due to the creation of surface morphologies with more sites available and an extended electron lifetime since electrons travel through long mean-free paths without recombining.

Yang et al. [68] successfully incorporated graphene in a TiO_2 nanostructure to form 2D graphene bridges in DSCs. The authors reported an optimum result for GO content of 0.6 wt% that originated a DSC with a J_{sc} of 16.29 mA · cm^{-2}, a V_{oc} of 690 mV, an *FF* of 0.62, and an η of 6.97%. This result represents an increase of 45% in the short-circuit current density and 39% in the conversion efficiency, when compared to a standard TiO_2 film-based DSC, shown in Figure 21.8. Comparing this result to DSCs equipped with a photoanode containing CNTs prepared by the same method and weight percentage—0.4%—a huge difference emerges since CNT devices perform significantly worse: J_{sc} of 3.35 mA · cm^{-2}, V_{oc} of 420 mV, *FF* of 0.41, and η of 0.58%. The Fermi level of a CNT is between its CB (–4.5 eV versus vacuum) and its valence band. Besides, its CB is below the CB of TiO_2 (–4 eV versus vacuum) resulting in a decrease of the V_{oc}. In opposition, graphene is a zero-band material [80] and its work function is calculated to be higher than the CNT value (4.42–4.5 eV versus vacuum) [66,81,82]. This makes graphene perfect for introduction in the TiO_2 structure because the apparent Fermi level is not decreased. This fact explains why in the Yang et al. results, the V_{oc} was not affected by the introduction of graphene (up to 0.4 wt%) [68], as can be seen in Figure 21.8. The unaffected V_{oc} and enhanced efficiency

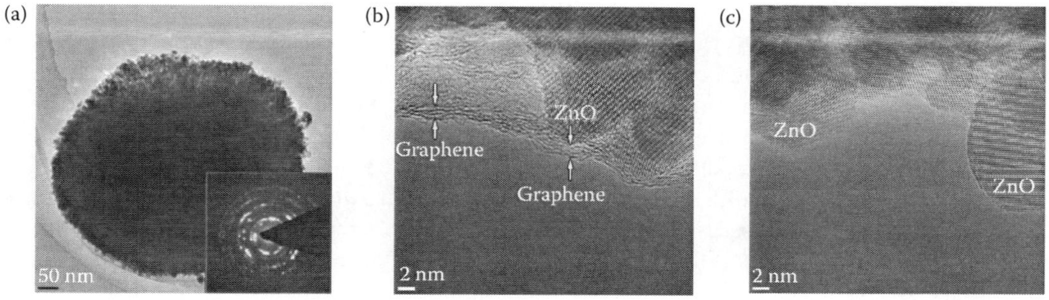

FIGURE 21.8 (a) TEM image, SAED pattern, (b) HRTEM image of an individual ZnO HSN with graphene loading, and (c) HRTEM image of as-prepared ZnO HSN without graphene loading. (Reprinted with permission from Xu, F. et al., Graphene scaffolds enhanced photogenerated electron transport in ZnO photoanodes for high-efficiency dye-sensitized solar cells. *The Journal of Physical Chemistry C*, **117**(17): 8619–8627. Copyright 2013 American Chemical Society.)

mean that graphene increased charge transport and partially suppressed electron recombination with the electrolyte. In higher concentrations, graphene starts to compete with the sensitized TiO_2 NP for light absorption and becomes a recombination center for electrons; consequently, DSC performance is affected—Figure 21.8.

Recently, graphene has also been incorporated with ZnO hierarchically structured nanoparticles (HSNs) [71]—Figure 21.9. Benefiting from the inherent higher electrical conductivity of ZnO relative to TiO_2, graphene has been used as a high conductive scaffold to enhance electron transport in the photoanode. Interestingly, in these ZnO photoanodes, the graphene bridges increased the J_{sc} and η by 43% and 38% respectively, relative to ZnO photoanodes without the graphene scaffolds. The authors found that the graphene scaffold, produced by the *in situ* reduction of GO, significantly prolonged the electron lifetime and effective diffusion length. This fact allowed the increase of the photoanode length up to 9 μm, resulting in a power conversion efficiency of 5.9% with a graphene loading of 1.2 wt%, which could be the highest efficiency obtained for this type of photoanode with 9-μm thickness.

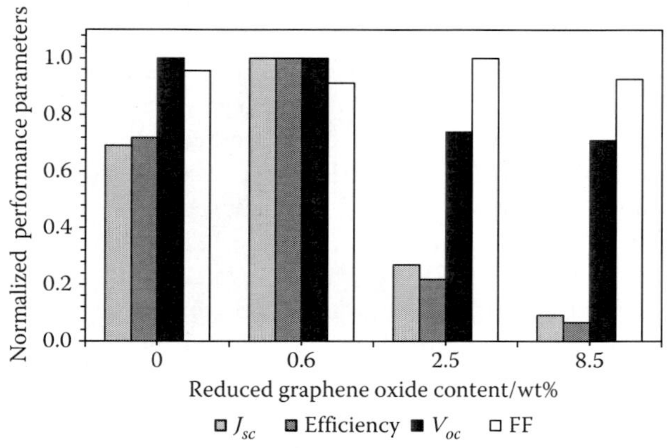

FIGURE 21.9 TiO_2/graphene hybrid DSC normalized performance values versus different graphene oxide contents present in the TiO_2 photoanode. (Reprinted from *Renewable and Sustainable Energy Reviews*, **27**(0), Macaira, J. et al., Review on nanostructured photoelectrodes for next generation dye-sensitized solar cells, 334–349, Copyright 2013, with permission from Elsevier.)

The reports concerning the use of graphene in combination with TiO_2 or ZnO photoelectrodes show that graphene enhances the DSC performance when used in very low amounts (~0.5 and 1.2 wt% for TiO_2 and ZnO respectively). This material promises to bring DSC photoelectrodes to the next level of development. The presence of graphene bridges in a DSC photoelectrode provides (i) a higher electron mobility; (ii) higher amount of adsorbed dye; and (iii) decreases the electron recombination with the electrolyte, improving the overall solar cell efficiency. Owing to its electrocatalytic properties and high extinction coefficient, when graphene is used in higher concentrations (>0.5 wt%) and is not properly reduced, it competes with the sensitizer on light absorption and acts as a recombination center. Graphene structures are very promising for increasing the overall kinetics of DSCs. Their influence and interaction with different metal oxide particles should be deeply understood and put to use for effectively increasing the efficiency of DSCs.

21.3.3 Electrolytes

In a DSC, the electrolyte is composed of a redox species suspended in a medium. The redox species are responsible for regenerating the oxidized sensitizer (after the formed electron had been injected into the SC) and carrying the formed holes (positive charges) toward the CE. The electrolyte has a great influence in the maximum current density delivered by a DSC since it is limited by the reduction kinetics of the oxidized species at the CE. It also strongly influences the dark current reaction kinetics and determines the charge transport rate in the electrolyte medium and in the pores of the SC. Typically, a liquid electrolyte is used so that it is able to penetrate the porous network of the photoanode (SC). The most commonly used are iodine-based electrolytes, composed of a redox couple of triiodide/iodide (I_3^-/I^-), dissolved in nonprotonic solvents (e.g., propionitrile, acetonitrile, and methoxypropionitrile).

Up until now, graphene has had a very small input in the redesign/modification of DSC's electrolytes. Gun et al. used GO to produce a quasi-solid electrolyte. An organogel composed of an iodide/triiodide redox couple, acetonitrile, and 1% GO yielded a DSC with an energy conversion efficiency of 7.5% versus a 6.9% DSC reference [83]. Neo and Ouyang also fabricated iodine-based organogels using GO as the gelator but with 3-methotxypropionitrile (MPN) as the solvent. The authors reported its use as a quasi-solid electrolyte in a DSC yielding efficiencies of 6.7% similar to the 7.2% control DSC [84]. Jung et al. [85] developed a highly conductive polymer gel electrolyte based on the functionalization of ODI (1-octyl-2,3-dimethylimidazolium iodide) by graphene. The authors demonstrated that the presence of graphene facilitated the electron transfer from the CE to the SC. The DSC fabricated with such an electrolyte yielded an efficiency of 5.84% [85]. Nguyen et al. [86] used few-layer graphene nanosheets (FLGs) and poly(3,4-ethylenedioxythiophene):poly(styrenesulfonate) (PEDOT:PSS) to produce a composite layer that could be effectively used in a solid-state device as a hole-transporting layer. Velten et al. [87] used few-layer RGO nanoribbons suspended by ultrasonication in the electrolyte to create a semitransparent electrolyte that can be used advantageously for structurally inverted DSCs or tandem cells.

21.3.4 Counterelectrodes

The CE plays a key role in a DSC. For DSCs fabricated with liquid-based electrolytes, the CE is responsible for reducing the redox species that are used for regenerating the sensitizer (dye) after electron injection; in the case of solid-state DSCs, the task of the CE is to collect holes coming from the hole-conducting medium placed between the CE and the photoanode. The type of the redox mediator species influences the reactions that occur at the interface of the CE. Typically, a triiodide/iodide redox couple is employed for dye regeneration with the overall redox catalytic reaction being $I_3^- + 2e^- \rightarrow 3I^-$. The preferred CE material is the one that has the lowest electrical resistance and the highest electrocatalytic activity possible toward the redox species. Additionally, it should also have high chemical stability and low cost. Transparency is not mandatory but offers added value to numerous applications (facades, windows, roof panels, etc.). Moreover, it enables its use in other applications, such as tandem cells [88,89].

Pt is currently the most used material for the catalysis reaction at the CE, due to its high electrocatalytic activity and high corrosion stability against iodine [90]; CEs thermally platinized are typically very transparent [91–93]. However, due to its scarcity and high cost [94], it is preferable to find alternatives that are simultaneously abundant, nontoxic, and capable of yielding DSCs of relatively high conversion efficiency. For that reason, other materials have been suggested as potential alternatives to Pt: transition metals such as carbides [95,96], sulfides [97–99] and nitrides [100,101], polymers [102], Pt composites [103,104], and carbonaceous materials [91,105,106]. Carbon materials have been widely reported due to their electronic conductivity, corrosion resistance toward I_2, high reactivity for triiodide reduction, and low cost [91,105,106]. The active sites responsible for catalysis are located at the edges of the carbon crystals [91]. For instance, Kay and Grätzel [107] fabricated a DSC with a relatively high efficiency using a mixture of graphite and carbon black (CB). Since then, CB (stand alone or with graphite powder) [91,105,106,108], activated carbon [77], single- [109], or multiwall carbon nanotubes (MWCNTs) [33,110,111] have been tested. As the performance of these materials is also strongly affected by the available surface area for reaction [77,106], it is necessary to have a large amount of carbon making the electrodes opaque and bulky [91,106], and undermining the DSC's transparency properties as delivered by the Pt CE. Among the carbon materials, graphene has recently emerged as one of the frontrunner candidates as a potential replacement for Pt as the catalyst material for DSCs—Table 21.1.

21.3.4.1 Graphene-Based CEs

Graphene is seen as a promising material for use as a CE in a DSC due to its inherent high conductivity that should decrease the charge transfer resistance, R_{ct} (a measure of the catalytic activity of the CE [112]) at the CE/electrolyte interface, making it more electrocatalytic [114]. Kaniyoor and Ramaprabhu discovered that thermally exfoliated graphite (TEG) CEs had an R_{ct} of 11.7 Ω cm^2, close to that of Pt (6.5 Ω cm^2). The respective DSC had an efficiency of 2.8% (versus 3.4% for Pt). Zhang et al. [62] obtained similar results for graphene dispersed in a mixture of terpineol and ethyl cellulose. Their graphene CE displays an R_{ct} of 1.20 Ω cm^2 and an efficiency of 6.81%, compared to the R_{ct} of 0.76 Ω cm^2 and 7.59% efficiency of the Pt CE. Additionally, graphene has an exceptional specific high surface area and because it has a higher oxidation potential than Pt, it should be more resistant to electrocorrosion [115]. The difference in efficiency in several graphene CEs in the literature is derived from the different techniques used in the preparation of graphene films. Some of them include oxidative exfoliation of graphite followed by hydrazine reduction [62], thermal exfoliation from graphite oxide [114], chemical reduction of GO colloids under microwave irradiation [118,144], and electrophoretic deposition (EPD) followed by annealing treatment [145,146] among many others.

The most promising path for producing graphene-related compounds, from both scale-up and cost perspectives, is starting from GO obtained by a chemical oxidation method and later reducing it either chemically or thermally [74,75]. This RGO possesses some oxygen-containing functional groups: epoxides and hydroxyls on both sides of the basal planes and hydroxyl and carbonyl groups decorating the periphery of the planes [147], which, along with lattice surface defects produced during the oxidation of the GS, are believed to be responsible for the electrocatalytic activity [34,106,115,148–152]. Owing to this, its use as a CE material is favored in opposition to perfect, fully reduced, and defect-free graphene. However, at present, it is still not clear what specific sites are mainly responsible for electrocatalysis. Likewise, the role of the size and morphology of the graphene films is not fully understood [153,154]. It is also important to bear in mind that this specificity could also change according to the type of redox mediators used. Furthermore, there must be a compromise between the electrical conductivity and the electrocatalytic ability of the film for the graphene CE to be as efficient as possible.

The influence of the active sites was demonstrated by Kavan et al. [78] who used commercially available graphene

TABLE 21.1
Selected Examples of Graphene-Based CEs (and Redox Mediators) for DSCs

Cathode Material	Redox Mediator	R_{ct}/ Ωcm^2	J_{sc}/ $mA \cdot cm^{-2}$	V_{oc}/V	*FF*	η/% (η:Pt)	References	Optical Transmittance/Note
FTO pure	I_3^-/I^-	2.5×10^7	–	–	–	–	[112]	–
Pt	I_3^-/I^-	1.3	–	–	–	–	[112]	–
Pt	I_3^-/I^- (Z946)	0.4	13.1	0.711	0.74	6.89 (–)	[78]	$T_{550} \approx 0.96$
Pt	I_3^-/I^-	–	16.8	0.804	0.71	9.53 (–)	[113]	–
PProDOT	I_3^-/I^-	–	17.0	0.761	0.71	9.25 (9.53)	[113]	Electropolymerization deposition
TEG	I_3^-/I^-	11.7	7.7	0.68	0.54	2.82 (3.37)	[114]	Opaque film (several μm)
GNs	I_3^-/I^-	1.20	16.988	0.747	0.5362	6.81 (7.59)	[62]	Opaque film, in a mixture of terpineol and EC
GNP	I_3^-/I^- (Z946)	308	13.1	0.724	0.52	5.00 (6.89)	[78]	$T_{550} = 0.87$
FGS	I_3^-/I^- (AN-50)	9.4	13.16	0.64	0.60	4.99 (5.48)	[115]	Opaque film
FGS w/EC	I_3^-/I^- (AN-50)	≈1	13.4	0.737	0.69	6.8 (6.8)	[116]	Opaque film, 6-μm-thick; N719 dye
Hemin–RGO	I_3^-/I^-	9	5.75	0.65	0.31	2.45 (3.18)	[117]	–
AGO	I_3^-/I^- (EL-HSE)	258.5	8.11	0.72	0.46	2.64 (3.5)	[63]	$T_{550} > 0.80$
RGO	I_3^-/I^-	38	14.3	0.64	0.653	5.69	[118]	$T_{550} \approx 0.60$; EPD
GNs	I_3^-/I^-	2.37	12.22	0.812	0.697	6.93 (7.23)	[119]	Opaque film; e-spray deposition
RGO–PEG	I_3^-/I^-	1.5	12.82	0.78	0.72	7.19 (7.76)	[120]	15 μm thick
NGR	I_3^-/I^-	–	10.55	0.82	0.55	4.75 (5.03)	[121]	–
N-GF	I_3^-/I^- (AN-50)	5.6	15.84	0.77	0.58	7.07 (7.44)	[122]	Opaque film, 30 μm thick
mGOM5/Ni	I_3^-/I^- (EL-HPE)	2.30	15.7	0.74	0.64	7.54 (7.45)	[123]	$T_{550} > 0.90$
Graphene/Metal Composites								
Pt–GO	I_3^-/I^-	–	14.1	0.72	0.669	6.77 (6.29)	[124]	$T_{550} = 0.65$; low-temperature synthesis
Graphene/Pt	I_3^-/I^-	7.7	11.42	0.73	0.73	6.09 (6.23)	[125]	ELSA of PDDA/graphene/PDDA/H_2PtCl_6
Graphene/Pt	I_3^-/I^-	2.36	6.67	0.74	0.59	2.91 (2.11)	[126]	27.43 wt% of Pt particles
PtNP/NG	I_3^-/I^-	0.48	17.57	0.63	0.64	7.07 (6.65)	[127]	10-mM Pt NP
PtNP/GR	I_3^-/I^-	0.67	12.06	0.79	0.67	6.35 (5.27)	[128]	–
ITO–PG	I_3^-/I^-	1.9	15.84	0.733	0.652	7.57 (7.03)	[129]	–
Graphene–$Ni_{12}P_5$	I_3^-/I^-	4.93	12.86	0.727	0.61	5.70 (6.08)	[130]	–
NiS_2@RGO	I_3^-/I^-	2.9	16.55	0.749	0.69	8.55 (8.15)	[131]	–
Graphene/Carbon Composites								
GMWNT	I_3^-/I^-	–	5.6	0.76	0.70	3.0	[132]	CVD deposition
RG-CNT	I_3^-/I^-	9.8	12.86	0.78	0.613	6.17 (7.88)	[133]	Opaque film; 60 wt% of CNT; EPD
MWCNT/GNS	I_3^-/I^-	–	8.8	0.77	0.58	4.0 (5.0)	[134]	Opaque film; 60 wt% of MWCNT
MWCNT and graphene	I_3^-/I^-	1.4	16.05	0.75	0.627	7.55 (8.8)	[135]	–
R-GONR/CNT	I_3^-/I^-	–	16.73	0.734	0.67	8.23 (7.61)	[136]	500-nm-thick film; 16 wt% of R-GONR
Graphene/Polymer Composites								
Graphene/ PEDOT–PSS	I_3^-/I^-	–	12.96	0.72	0.48	4.5 (6.3)	[34]	$T_{550} > 0.80$; 60-nm-thick film; 1 wt% of graphene
PDDA@ERGO	I_3^-/I^- (Z946)	≈1	18.77	0.692	0.74	9.54 (9.11)	[65]	$T_{550} \approx 0.90$; LBL deposition; C106TBA dye
GN/Nafion	I_3^-/I^-	5.75	19.50	0.65	0.64	8.19 (8.89)	[64]	0.75 v/v of Nafion; CYC-B11 dye
Graphene/PANI	I_3^-/I^-	–	13.28	0.685	0.67	6.09 (6.88)	[137]	–
Graphene/PEDOT	I_3^-/I^-	6.15	12.5	0.77	0.63	6.26 (6.68)	[138]	$T_{550} > 0.90$; TCO-free flexible CE
PDDA–HEG	I_3^-/I^-	484	9.1	0.71	0.47	3.1 (4.5)	[139]	$T_{550} \approx 0.85$
FTO pure	$(Co[bpy\text{-}pz]_2)^{3+/2+}$	2.5×10^4	–	–	–	–	[140]	–
Pt	$([Co(bpy)_3])^{3+/2+}$	5.5	14.0	0.904	0.65	8.2	[141]	$T_{550} \approx 0.96$, Y123 dye
Pt	$(Co[bpy\text{-}pz]_2)^{3+/2+}$	≈50	13.45	1.015	0.697	9.52	[142]	Unstable, Y123 dye
PProDOT	$(Co[bpy\text{-}pz]_2)^{3+/2+}$	≈2.5	13.06	0.998	0.774	10.08 (9.52)	[142]	160-nm-thick film;Y123 dye
FGS w/EC	$([Co(bpy)_3])^{3+/2+}$	<1	8.49	0.813	0.65	4.51 (4.39)	[116]	Opaque film, 6-μm-thick; D35 dye

(Continued)

TABLE 21.1 (*Continued*)
Selected Examples of Graphene-Based CEs (and Redox Mediators) for DSCs

Cathode Material	Redox Mediator	R_{ct}/ Ωcm²	J_{sc}/ mA·cm⁻²	V_{oc}/V	FF	η/% (η:Pt)	References	Optical Transmittance/Note
GNP	$([Co(bpy)_3])^{3+/2+}$	≈0.08	14.8	0.878	0.72	9.4 (8.2)	[141]	T_{550} = 0.66, Y123 dye
GNP	$(Co[bpy\text{-}pz]_2)^{3+/2+}$	0.70	–	–	–	8–10 (8–10)	[140]	T_{550} = 0.66, Y123 dye
GNP–PAN	$([Co(bpy)_3])^{3+/2+}$	≈1	14.3	0.865	0.74	9.11 (8.61)	[143]	Pyrolized; 20% of GNP
Pt	T_2/T^-	≈55	9.62	0.650	0.31	1.97	[116]	–
FGS w/EC	T_2/T^-	<1	9.50	0.663	0.55	3.45 (1.97)	[116]	Opaque film, 6-μm-thick; D35 dye

Note: R_{ct}—Charge transfer resistance; J_{sc}—short-circuit current density; V_{oc}—open-circuit voltage; *FF*—fill factor; η—overall energy conversion efficiency. AN50: Commercial I_3^-/I^- electrolyte solution from Solaronix; EL-HSE: commercial I_3^-/I^- high-stability electrolyte solution from Dyesol; EL-HPE: commercial I_3^-/I^- high-performance electrolyte solution from Dyesol. TEG, thermally exfoliated graphite; GNs, graphene nanosheets; GNPs, graphene nanoplatelets (commercial product from CheapTubes); FGS, functionalized graphene sheets (commercial product from Vorbeck Materials, GO modified by thermal treatment); EC, ethyl cellulose; Hemin-RGO, hemin-functionalized reduced graphene oxide; AGO, thermally annealed graphene oxide; RGO, reduced graphene oxide; EPD, electrophoretic deposition; PEG, polyethylene glycol; NGR, nitrogen-doped graphene; N-GF, N-doped graphene foam; mGOM5/Ni, oxidized commercial graphene nanoplatelets spray deposited on top of Ni particles electrophoretically deposited onto covered-glass fluorine-doped tin oxide (FTO) substrate; PProDOT, poly(3,4-propylenedioxythiophene); PtNP/NG: platinum nanoparticle/nanographite; PtNP/GR, platinum nanoparticle/graphene; ITO–PG, graphene/platinum-coated indium-tin oxide; Graphene–$Ni_{12}P_5$, nickel phosphide- embedded graphene; NiS_2@RGO, nanocomposite of nickel sulfide and reduced graphene oxide; GMWNT, graphene-based multiwalled carbon nanotubes structure; RG–CNT, reduced graphene–carbon nanotube; MWCNTs/GNS, multiwalled carbon nanotubes/graphene nanosheets; GNP–PAN, graphene nanoplatelets (commercial product from CheapTubes)–poly(acrylonitrile); Graphene/PEDOT–PSS, graphene/(polystyrenesulfonate-doped poly[3,4-ethylenedioxythiophene]); PDDA@ERGO, electrochemically reduced graphene oxide decorated with poly(diallyldimethylammonium chloride); GNs/Nafion, graphene nanosheets/Nafion; Graphene/PANI, graphene/polyaniline; Graphene/PEDOT, graphene/poly(3,4-ethylenedioxythiophene); PDDA-HEG, graphene sheets synthesized by a hydrogen-induced simultaneous exfoliation–reduction method (HEG) and postfunctionalized with poly(diallyldimethylammonium chloride).

to fabricate a CE by drop casting, without any annealing. The produced CE exhibited a transmittance of ~85% (λ = 550 nm) and an efficiency of 5%, a 72.5% relative efficiency compared to a Pt CE. Roy-Mayhew et al. [115,116] demonstrated a dependence of oxygen-containing functional groups with electrocatalytic activity. The use of functionalized GS (produced from a thermal exfoliation method) yielded opaque CEs with efficiencies similar to Pt—6.8% [116]. Xu et al. [117] further suggested that —NHCO-groups could increase the catalytic activity. Cruz et al. [63] demonstrated the crucial role that thermal annealing under an inert atmosphere has on graphene films as CEs in DSCs, as it permits increasing the electrical conductivity while preserving a sufficient number of active sites, thus greatly increasing catalytic activity. They were able to use GO without it being prereduced, as a relative semitransparent (>80% at λ = 550 nm) and efficient CE (75% comparatively to the reference Pt CE) [63]. Choi et al. [118] have used an electrophoretic technique to chemically deposit RGO, followed by annealing at 600°C, to fabricate a semitransparent CE (~60% at λ = 550 nm [146]) with an ~5.7% efficiency. Jang et al. [119] used a different technique—electrospray method (e-spray), for deposition of chemically reduced GS—Figure 21.10. Upon thermal annealing, the developed opaque CE displayed an efficiency of 6.93 % (versus 7.23% for the thermolytically prepared Pt CE), with a very high FF of 69.7% [119]. The importance of the morphology of graphene films was studied by Zheng et al. [120]. They reported the fabrication of 15-μm-thick RGO films with liquid polyethylene glycol (PEG) of low molecular weigth (which was removed through heating), with an efficiency of 7.19% (versus 7.76% for the control DSC with the Pt CE) [120]. It was observed that the preparation of RGO–PEG using ultrasonication yielded lower efficiencies (6.15%) than when mechanical grinding was used. The authors attributed the difference in performance due to different morphologies caused by the preparation methods [120]. Ultrasonication led to a reduction in the size of the RGO sheets; upon the removal of PEG, these sheets aggregate easily, leading to a decrease in pore size.

Graphene films can also have their surface modified, using fluorine ions [155] or nitrogen doping [121,122,156]. It is believed that nitrogen-doped graphene films can have several advantages, specifically high selectivity toward the reduction of redox species and introduction of electrocatalytic active sites [121,122,156]. Yen et al. [121] used a hydrothermal method to prepare a nitrogen-doped graphene CE with an efficiency of 4.75% (versus 5.03 for Pt). Xue et al. [122] reported a three-dimensional (3D) nitrogen-doped graphene foam (N-GF) by annealing freeze-dried GO foams in ammonia. The produced CE yields an efficiency of 7.07%, close to that of the Pt reference CE (7.44%) and greatly higher than undoped graphene films (4.84%) [122]. The authors attributed this performance to a combination of doping-induced high electrical conductivity and electrocatalytic activity, and large surface area, good surface hydrophilicity, and well-defined porosity created by the 3D structure that enhanced the electrolyte diffusion and contact area [122].

Cruz et al. [123] were able to create a CE that was as efficient and transparent as the reference Pt CE for iodine-based liquid-junction electrolytes. The structured CE (mGOM5/Ni) was composed of oxidized commercial GNPs (mGOM5) spray

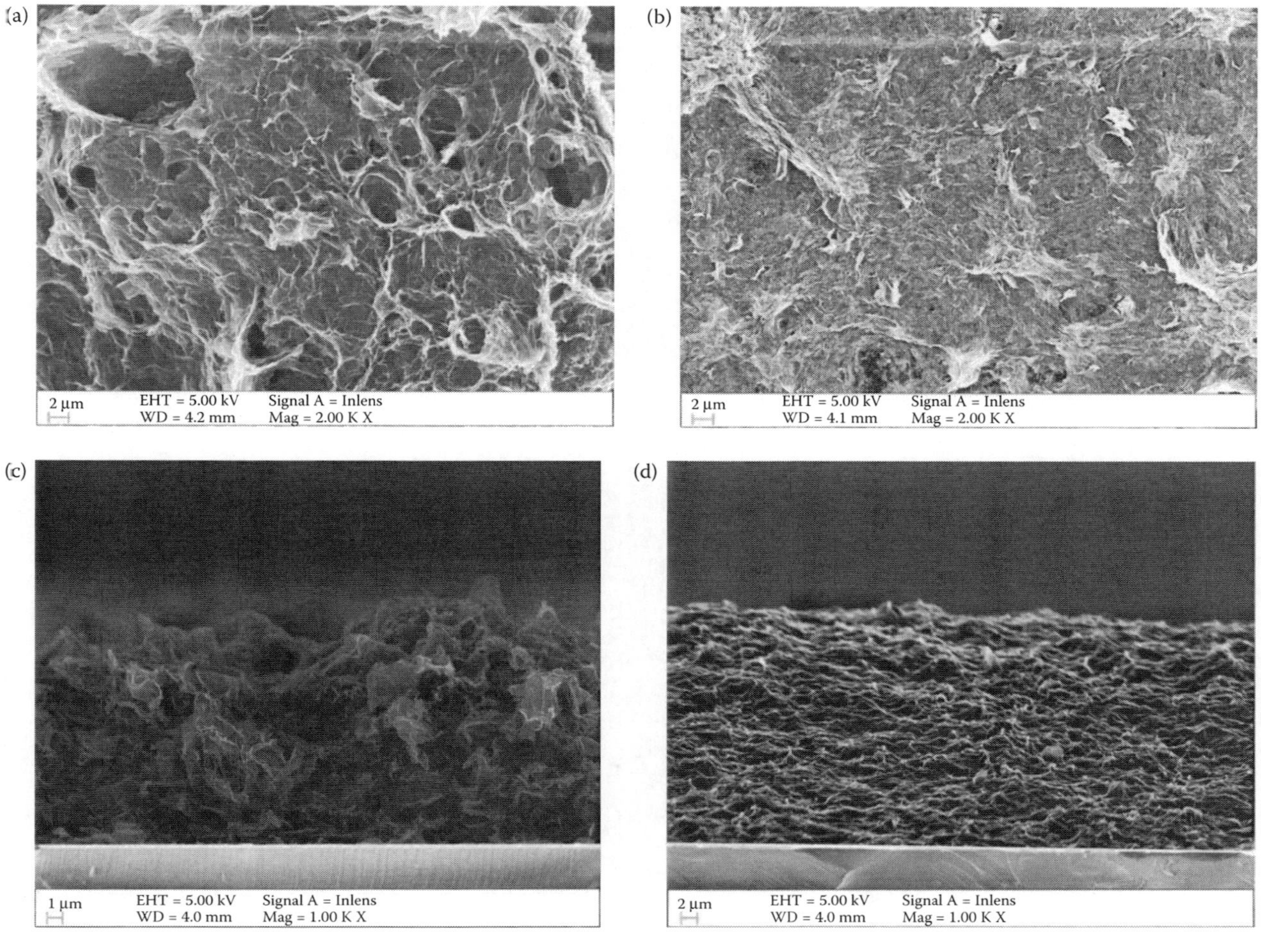

FIGURE 21.10 Surface SEM images of (a) grinded rGO films and (b) ultrasonicated rGO films; cross-sectional SEM films of (c) grinded rGO films and (d) ultrasonicated rGO films. (Reprinted from Zheng, H.Q. et al., *Journal of Materials Chemistry*, 2012, **22**(29): 14465–14474. With permission.)

deposited on top of a covered-glass FTO substrate where it first deposited nickel NPs through an EPD technique. For the mGOM5/Ni to become efficient, it was necessary to submit it to thermal annealing under an inert atmosphere [123]. As a result, the mGOM5/Ni CE displayed an efficiency of 7.54% and transparency of 91.8% at $\lambda = 550$ nm, comparatively to the Pt reference with an efficiency of 7.45% and transparency of 92.0% at $\lambda = 550$ nm [123].

Graphene-based CEs have been proven to be better options than the reference Pt CE for cobalt-based and sulfur-based mediators as they exhibit equal and even greater electrocatalytic activity than Pt [116,140]. Roy-Mayhew et al. [116] used their functionalized GS-based CE with cobalt-based and sulfur-based electrolytes yielding efficiencies of 4.5% and 3.5% respectively, greater than those of the Pt CE (4.4% and 2.0% respectively). Kavan et al. [140,141] fabricated a CE from commercial GNS annealing them under an inert atmosphere. For two types of cobalt-based mediators ($[Co(bpy)_3]^{3+/2+}$ and $[Co(bpy\text{-}pz)_2]^{3+/2+}$) in a low volatile electrolyte medium coupled with a Y123 sensitizer, their fabricated graphene semi-transparent CE (>65% at $\lambda = 550$ nm) outperformed the Pt CEs—Figure 21.11.

Graphene also exhibits excellent flexibility properties. Tung et al. [157] found that graphene films can sustain successive bending (60°, 10 times) without loss of electrical conductivity, contrary to indium tin oxide (ITO) films. Nguyen et al. [86]

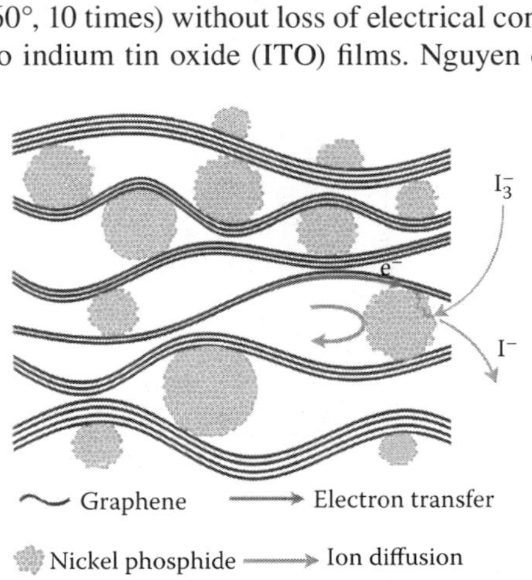

FIGURE 21.11 Electrochemical reaction scheme using a metal ($Ni_{12}P_5$) and graphene composite. (Reprinted from Dou, Y.Y. et al., *Physical Chemistry Chemical Physics*, 2012, **14**(4): 1339–1342. With permission.)

discovered that graphene films can withstand cycling/bending at angles up to 140° for more than 100 cycles without deviation in its sheet resistance. Lee et al. [138] performed bending tests on solid-state DSCs with flexible substrates and PEDOT:PSS as hole–medium-layer DSCs. The DSCs with graphene CE were able to retain the same performance after bending, contrary to a Pt-/ITO-based CE [138].

21.3.4.2 Graphene-Based Composites for CEs

21.3.4.2.1 Graphene/Metal Composites

When combined with metal particles, graphene will primarily act as a scaffold for better dispersions of these particles, functioning as a support material. Graphene should increase electron transfer as the metal particles become more accessible [128,130] to the redox species due to an increase in the available surface area in contact with the electrolyte. Concurrently, the GS should cause a decrease in the electron injection resistance between the substrate and the catalytic particles ensuring a fast electron transfer at the interfaces. Alternatively, besides functioning primarily as reaction centers, the metal NPs can also act as spacers between the GS, thus helping the diffusion of the electrolyte.

Generally, the most employed metal NP used in these kinds of composites is Pt. Pt/graphene composites use a lower load of Pt than the traditional used sputtered or thermally platinized CEs, thus contributing to a decrease in the overall cost of the DSC. Tjoa et al. [124] were able to synthesize a graphene/Pt composite via a low-temperature route by light-assisted spontaneous coreduction of GO and chloroplatinic acid without the reducing agent . This CE had a 6.77% efficiency, higher than the Pt reference [124]. Compatibility with a flexible plastic substrate was also demonstrated. Another facile approach for the construction of a graphene/Pt composite was performed by Gong et al. [125], through electrostatic layer-by-layer self-assembly (ELSA) based on poly(diallyldimethylammonium chloride) (PDDA), graphene, and H_2PtCl_6. A comparable performance with the reference-sputtered Pt CE (~6%) was achieved while reducing the load of Pt by 1000 × [125]. The effect of Pt loading was studied by Bajpai et al. [126]. Using a pulsed laser ablation method, an optimal load of 27.43 wt% of Pt NPs were deposited directly on few-layered GS, yielding a 2.9%-efficiency DSC [126]. Liu et al. [127] prepared a CE using aniline (ANI) monomers as a dispersing medium, followed by spin-coating and annealing processes, to study the combined effect of graphene with different Pt loads. It was discovered that a (10 mM Pt)/graphene CE yielded a maximum efficiency of 7.07% (versus 6.65%) as a result of increased surface roughness [127]. All the previous mentioned results could be explained by the increase of defects in the GS caused by the Pt particles as shown by Yen et al. [128]. Fabricated Pt/graphene composite CEs were also characterized as having a much smoother surface, resulting in a lower resistance to diffusion, which improves the total redox reaction rate that occurs at the CE [128]. Their Pt/graphene CE had an efficiency of 6.35%, 20% higher than the Pt reference [128].

The importance of the available surface area for electrocatalysis was demonstrated by Guai et al. [129]—Figure 21.12.

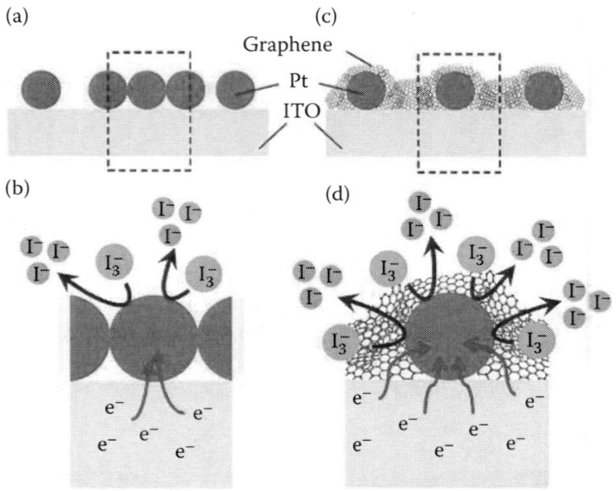

FIGURE 21.12 (a and b) Pt electrode–individual Pt particle have a small contact interface with the substrate restricting the charge transfer. (c and d) Pt/graphene electrode–the thin porous graphene film covers the Pt NPs and increases the contact surface area with ITO for fast charge transfer. (Reprinted from *Solar Energy*, **86**(7), Guai, G.H. et al., Graphene–Pt/ITO counter electrode to significantly reduce Pt loading and enhance charge transfer for high performance dye-sensitized solar cell, 2041–2048, Copyright 2012, with permission from Elsevier.)

A low-loaded Pt-sputtered/transparent conductive oxide (TCO) substrate was covered by a porous GO film deposited electrophoretically. Afterward, the GO was electrochemically reduced to become electrocatalytically active. It was shown that similar performances (~7% efficiency) can be achieved while reducing the Pt load by more than 60% [129].

Transition metals have also been used in the incorporation of graphene composites. Graphene/$Ni_{12}P_5$ (nickel phosphide) CEs were fabricated by Dou et al. [130] using a hydrothermal reaction of red phosphorus, nickel chloride, and GO in a mixture of ethylene glycol/water. The produced CE exhibited the $Ni_{12}P_5$ particles embedded into the GS. Its efficiency was 5.70%, similar to that of Pt (6.08%) [130]. Nickel sulfide (NiS_2) particles also displayed a synergetic effect similar to that of Pt and $Ni_{12}P_5$ when combining with GS [131]. Graphene/NiS_2 CEs prepared via a facile hydrothermal reaction of nickel ions and sulfur source in the presence of GO yielded efficiencies of 8.55%, higher than the Pt CE (8.15%) [131]. Other metals used include titanium nitride (TiN) [101] and nickel (placed on top of the graphene platelets) [158].

21.3.4.2.2 Graphene/Carbon Composites

Recently, interest has been sparked in the combination of CNTs with graphene. This is because CNTs have shown tunable and large areas and promising electrical conductivity; in these, CNTs act as a support material. Furthermore, its associated edges could provide a boost in the electron reduction of the redox species in the electrolyte.

Choi et al. constructed a composite CE made of MWCNT and graphene, in which the CNTs were vertically grown by chemical vapor deposition (CVD) on top of also-grown CVD GS. This caused an increase in the interface reaction area available for catalysis, yielding a DSC with an efficiency of 4.46% [132].

Zhu et al. [133] fabricated MWCNTs—(microwave-assisted) reduced graphene composite CE, using an EPD technique. It was found that the concentration of GNS and CNTs in an MWCNT/graphene composite CE strongly affects its performance, bringing about a compromise between the specific area surface of the CNT in comparison to the GNS and the type of structure formed by the MWCNT/GNS network [133]. The optimal content of CNT was found to be 60 wt%, with these CEs yielding a DSC with an efficiency of 6.17% (versus 7.88% for the Pt CE) [133]. The optimized content of CNTs was further confirmed by Battumur et al. [134] who prepared an MWCNT/GNS composite CE with 60 wt% of MWCNTs through a simple doctor blade method, with an efficiency of 4.0% .

MWCNTs were dry spun with graphene flakes to produce another composite CE [135]. The resulting CE had an efficiency of 7.55%, higher than using the MWCNT alone (6.62%) [135]. According to the authors, this was due to increased electrical conductivity between the MWCNT bundles and the graphene flakes and increased catalytic activity from the edges of the graphene flakes and the defect sites in the MWCNT [135]. Yang et al. [136] mixed MWCNT with graphene oxide nanoribbons (GONRs) chemically unzipped from MWCNT and after deoxygenation by hydroiodic acid, they were able to yield (reduced) R-GONR/CNT composite CEs—Figure 21.13. The composite R-GONR/CNT CE with an optimal 16 wt% of R-GONR had an efficiency of 8.23%, higher than a pristine MWCNT CE (6.09%) and the Pt CE (7.61%) [136]. The authors were also able to produce a semi-transparent and flexible R-GONR (16%)/MWCNT composite CE, albeit being less efficient due to the lower thickness of the produced film.

Stefik et al. [143] fabricated a carbon–graphene nanocomposite but for cobalt-based electrolytes and Y123 dye, instead of the traditional iodine-based electrolytes and ruthenium dyes. These nanocomposites were synthesized by carbonizing mixtures of GNPs with a carbon source, poly(acrylonitrile) (PAN), combining the high catalytic performance of GNS with the conductive carbon matrix. An optimal 20 wt% of GNS in the carbon matrix yielded a 9.11%-efficiency DSC (versus 8.61% for Pt) [143]. The authors underlined that this method allowed for an increased mechanical adhesion.

21.3.4.2.3 Graphene/Polymer Composites

The first composite materials of conductive polymers and graphene were done by Stankovich et al. [159] in 2006, immediately attracting attention. Soon after, the authors began using this setup as CEs for DSCs. In a possible configuration, the polymer acts as the conductive support with graphene being responsible for the electrocatalytic reaction. Polymer matrixes are used to avoid agglomeration of powder-reduced GS, but sometimes, they prevent the free flowing of electrons, as shown by Wang et al. [150]. Hong et al. [34] used 1-pyrenebutyrate (PB^-)-functionalized graphene dispersed in an organic matrix composite of PEDOT:PSS (polystyrenesulfonate-doped poly[3,4-ethylenedioxythiophene]) yielding a transparent CE (efficiency of 4.5%) with a relative difference in efficiency of ca. less than 30% than that of a Pt CE. Xu et al. [65] prepared CEs composed of thin films of a cationic polymer–PDDA (shortened as polyDADMAC or polyDDA [PDDA] decorated with GO, prepared by a layer-by-layer [LBL]) assembling technique—Figure 21.14. After electrochemical reduction of GO to reduced graphene—ERGO, the respective semitransparent (>80% at $\lambda = 550$ nm) CE (PDDA@ERGO) yielded extraordinary efficiencies of 9.5% and 7.6% for a low-volatility electrolyte and solvent-free ionic liquid electrolyte, respectively. The reference Pt CE exhibited an efficiency of 9.14% in the presence of the low-volatility electrolyte. However, only a reasonable stability was achieved and only with the ionic liquid electrolyte. Nafion can also be used as a dispersant of GNS. Composite films of GNS/Nafion coupled with a TBA(Ru[(4-carboxylic acid-4-carboxylate-2,2-bipyridine)(4,4-bis(5-(hexylthio)-2,2-bithien-5-yl)-2,2-bipyridine)(NCS)2]), that is, CYC-B11 dye, yielded DSCs with an efficiency of 8.19% [64], versus an 8.89%-sputtered Pt film. TEG mixed with Nafion should also ensure a good substrate adhesion [114].

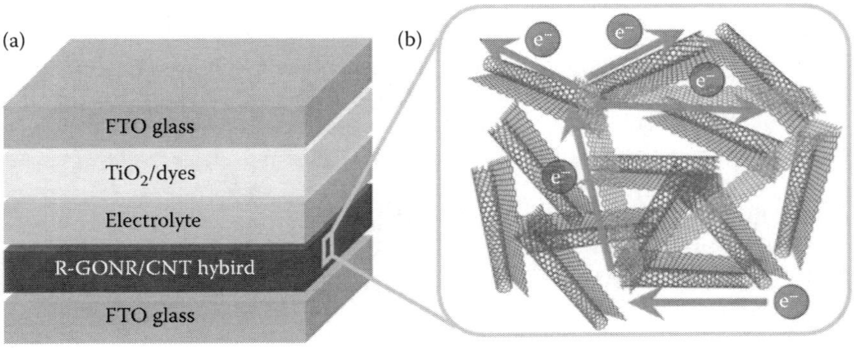

FIGURE 21.13 (a) Structure of DSCs based on R-GONR/MWCNTs as CE. (b) The mechanism of rapid electron transport in the CE. (Yang, Z. et al.: Carbon nanotubes bridged with graphene nanoribbons and their use in high-efficiency dye-sensitized solar cells. *Angewandte Chemie International Edition*, 2013, **52**: 11603–11606. Copyright 2013, Wiley-VCH Verlag GmbH & Co. KGaA. Reproduced with permission.)

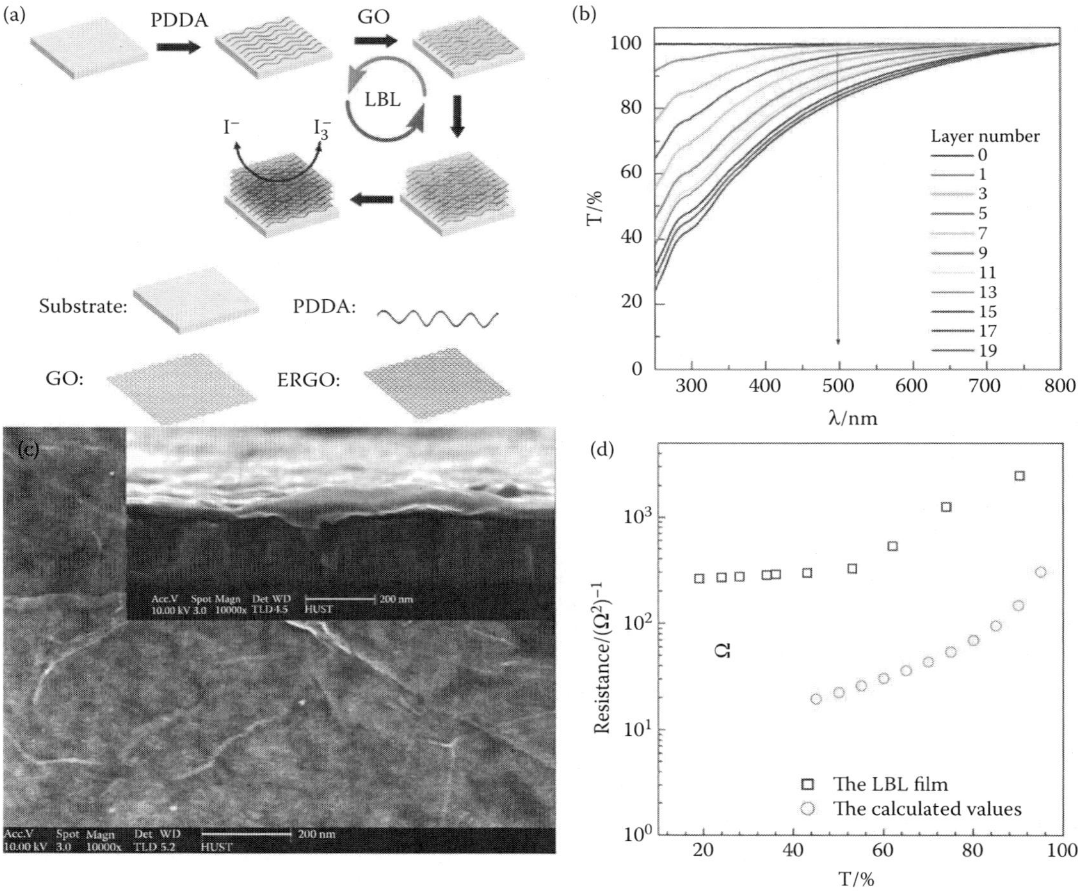

FIGURE 21.14 Preparation method and characteristics of the PDDA@ERGO films. (Reprinted by permission from Macmillan Publishers Ltd. *Science Reports*, Xu, X. et al., Electrochemically reduced graphene oxide multilayer films as efficient counter electrode for dye-sensitized solar cells, copyright 2013.)

In an alternative configuration, graphene is also employed as a support material. Wang et al. synthesized polyaniline (PANI) on GS using an *in situ* polymerization technique. As a result, the GS were homogeneously coated on the surface with PANI NPs, resulting in a DSC with an efficiency of 6.09% close to that of a Pt CE (6.88%) [137]. Lee et al. managed to create a TCO-free CE by incorporating graphene in PEDOT—Figure 21.15. This allowed for an increase in electron transport and a decrease in surface resistance. The fabricated CE was applied to polymeric, flexible substrates yielding an efficiency of 6.26% [138]. Several other examples exist where PANI was coupled with graphene to create a composite CE [160,161].

Kaniyoor and Ramaprabhu [139] tried a different approach: GS were synthesized by a hydrogen-induced simultaneous exfoliation–reduction method (HEG) and then postfunctionalized with several polyelectrolytes. They discovered that the presence of charges affects the electrochemical behavior toward triiodide reduction, with the semitransparent PDDA–HEG CE (~85% at $\lambda = 550$ nm) yielding a DSC with an efficiency of 3.1%. The PDDA induced a positive charge on the graphene surface.

21.3.4.3 Summary and Outlook

The typical CE made of Pt is bound to be replaced by graphene as it exhibits excellent conductivity and electrocatalytic properties; additionally, besides being a low-cost material, it is an easy tunable material and has a high-film flexibility. The tuning of the number and type of active sites (surface defects and oxygen-containing groups) present in GS is critical to provide simultaneously high electrocatalytic activity and electrical conductivity. The improvement of future graphene-based CEs could also be accomplished through the use of new composite materials or by using graphene in combination with such materials in the form of structured films.

It is also important to fully understand the electrocatalytic mechanism of graphene as well as the role of the morphology of its films in the electrochemistry phenomena. This understanding could be achieved using modeling tools along with new experimental data. Such information could enable the creation of prescriptively custom-made tunable graphene structures, opening the door to the fabrication of low-temperature-processing graphene films as well as expanding their application to other devices (such as sensors and fuel cells).

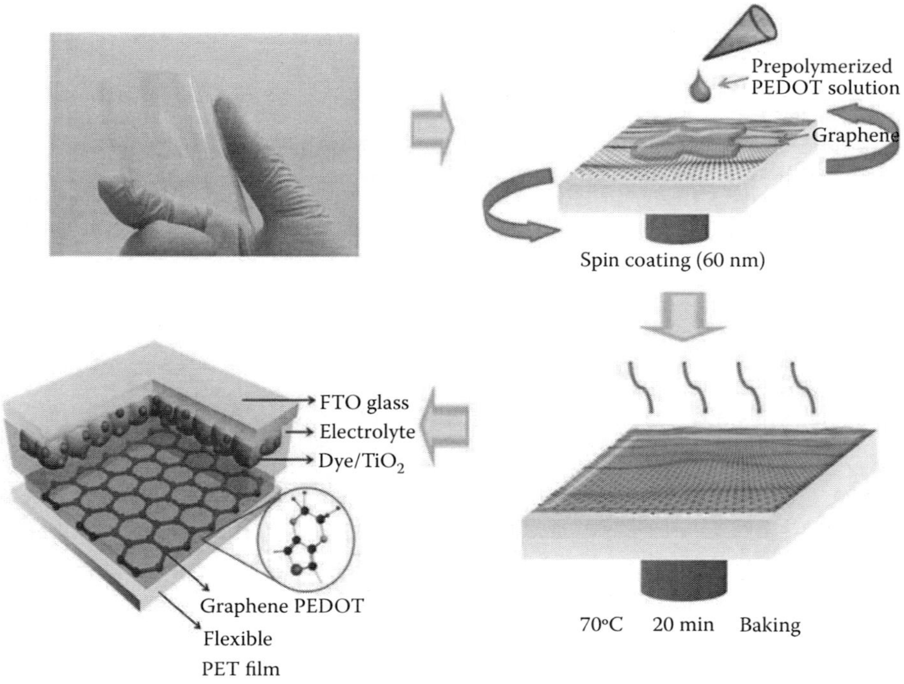

FIGURE 21.15 Fabrication of TCO-free CE with a PEDOT/graphene film. (Lee, K.S. et al.: Flexible and platinum-free dye-sensitized solar cells with conducting-polymer-coated graphene counter electrodes. *Chemistry Sustainable Chemistry*, **5**(2): 379–382, Copyright 2012, Wiley-VCH Verlag GmbH & Co. KGaA. Reproduced with permission.)

ACKNOWLEDGMENTS

Rui Cruz is grateful to the Portuguese Foundation for Science and Technology (FCT) and Efacec—Engenharia e Sistemas S.A. for his PhD grant (SFRH/BDE/33439/2008). This chapter was accomplished in the framework of the project WinDSC SI&IDT (Reference: 21539/2011), cofinanced by the European Regional Development Fund and the Portuguese Government through ADI—Agência de Inovação, under the framework of the QREN Initiative, through the Operational Programme for Competitiveness Factors. R. Cruz and Luisa Andrade also acknowledge the project SolarConcept (PTDC/EQU-EQU/120064/2010) for funding. Professor Adélio Mendes would like to acknowledge the BDI-DSC grant. José Miguel Lopes Maçaira Nogueira is grateful to FCT for his PhD Grant (Reference: SFRH/BD/80449/2011). Financial support by FCT through the project SolarConcept (Reference: PTDC/EQU-EQU/120064/2010) is also acknowledged.

REFERENCES

1. Teh, C.M. and A.R. Mohamed, Roles of titanium dioxide and ion-doped titanium dioxide on photocatalytic degradation of organic pollutants (phenolic compounds and dyes) in aqueous solutions: A review. *Journal of Alloys and Compounds*, 2011, **509**(5): 1648–1660.
2. Hager, S. and R. Bauer, Heterogeneous photocatalytic oxidation of organics for air purification by near UV irradiated titanium dioxide. *Chemosphere*, 1999, **38**(7): 1549–1559.
3. Herrmann, J.-M., Photocatalysis fundamentals revisited to avoid several misconceptions. *Applied Catalysis B: Environmental*, 2010, **99**(3–4): 461–468.
4. Xiang, Q. et al., Graphene-based semiconductor photocatalysts. *Chemical Society Reviews*, 2012, **41**(2): 782–796.
5. Águia, C., *Study and Development of Paints for the Photocatalytic Oxidation of NO_x, PhD Thesis*. 2011, FEUP: Porto.
6. Fujishima, A. et al., Titanium dioxide photocatalysis. *Journal of Photochemistry and Photobiology C: Photochemistry Reviews*, 2000, **1**(1): 1–21.
7. Hu, G. and B. Tang, Photocatalytic mechanism of graphene/titanate nanotubes photocatalyst under visible-light irradiation. *Materials Chemistry and Physics*, 2013, **138**(2–3): 608–614.
8. Zhou, K. et al., Preparation of graphene–TiO_2 composites with enhanced photocatalytic activity. *New Journal of Chemistry*, 2011, **35**(2): 353–359.
9. Liu, S. et al., Graphene facilitated visible light photodegradation of methylene blue over titanium dioxide photocatalysts. *Chemical Engineering Journal*, 2013, **214**(0): 298–303.
10. Zhang, H. et al., P25–graphene composite as a high performance photocatalyst. *ACS Nano*, 2010, **4**(1): 380–386.
11. Lee, J.S. et al., Highly photoactive, low bandgap TiO_2 nanoparticles wrapped by graphene. *Advanced Materials*, 2012, **24**(8): 1084–1088.
12. Malato, S. et al., Photocatalysis with solar energy at a pilot-plant scale: An overview. *Applied Catalysis B: Environmental*, 2002, **37**(1): 1–15.
13. Petlicki, J. and T.G.M. van de Ven, The equilibrium between the oxidation of hydrogen peroxide by oxygen and the dismutation of peroxyl or superoxide radicals in aqueous solutions

in contact with oxygen. *Journal of the Chemical Society, Faraday Transactions*, 1998, **94**(18): 2763–2767.

14. Li, G. et al., Preparation and photoelectrochemical performance of Ag/graphene/TiO_2 composite film. *Applied Surface Science*, 2011, **257**(15): 6568–6572.
15. Nguyen-Phan, T.-D. et al., The role of graphene oxide content on the adsorption-enhanced photocatalysis of titanium dioxide/graphene oxide composites. *Chemical Engineering Journal*, 2011, **170**(1): 226–232.
16. Zhang, L.-W. et al., Efficient TiO_2 photocatalysts from surface hybridization of TiO_2 particles with graphite-like carbon. *Advanced Functional Materials*, 2008, **18**(15): 2180–2189.
17. Ren, L. et al., Upconversion-P25–graphene composite as an advanced sunlight driven photocatalytic hybrid material. *Journal of Materials Chemistry*, 2012, **22**(23): 11765–11771.
18. Williams, G. et al., TiO_2–graphene nanocomposites. UV-assisted photocatalytic reduction of graphene oxide. *ACS Nano*, 2008, **2**(7): 1487–1491.
19. Zou, F. et al., A novel approach for synthesis of TiO_2–graphene nanocomposites and their photoelectrical properties. *Scripta Materialia*, 2011, **64**(7): 621–624.
20. Khoa, N.T. et al., Photodecomposition effects of graphene oxide coated on TiO_2 thin film prepared by electron-beam evaporation method. *Thin Solid Films*, 2012, **520**(16): 5417–5420.
21. Akhavan, O. et al., Photodegradation of graphene oxide sheets by TiO_2 nanoparticles after a photocatalytic reduction. *The Journal of Physical Chemistry C*, 2010, **114**(30): 12955–12959.
22. Liu, J. et al., Self-assembling TiO_2 nanorods on large graphene oxide sheets at a two-phase interface and their anti-recombination in photocatalytic applications. *Advanced Functional Materials*, 2010, **20**(23): 4175–4181.
23. Zhu, G. et al., Cascade structure of TiO_2/ZnO/CdS film for quantum dot sensitized solar cells. *Journal of Alloys and Compounds*, 2011, **509**(29): 7814–7818.
24. Yoo, D.-H. et al., Enhanced photocatalytic activity of graphene oxide decorated on TiO_2 films under UV and visible irradiation. *Current Applied Physics*, 2011, **11**(3): 805–808.
25. Zhang, Y. et al., TiO_2 – graphene nanocomposites for gas-phase photocatalytic degradation of volatile aromatic pollutant: Is TiO_2 – graphene truly different from other TiO_2 – carbon composite materials? *ACS Nano*, 2010, **4**(12): 7303–7314.
26. Lu, J. et al., Facile synthesis of TiO_2/graphene composites for selective enrichment of phosphopeptides. *Nanoscale*, 2012, **4**(5): 1577–1580.
27. Tanaka, D. and A. Mendes, *Composite grapheno-metal oxide platelet catalyst and the method of preparation*. EP2560754 A1. 2011, Universidade do Porto.
28. Long, T.C. et al., Titanium dioxide (P25) produces reactive oxygen species in immortalized brain microglia (BV2): Implications for nanoparticle neurotoxicity. *Environmental Science and Technology*, 2006, **40**(14): 4346–4352.
29. Lv, X.-J. et al., Synergetic effect of Cu and graphene as cocatalyst on TiO_2 for enhanced photocatalytic hydrogen evolution from solar water splitting. *Journal of Materials Chemistry*, 2012, **22**(35): 18542–18549.
30. Neppolian, B. et al., Graphene oxide based Pt–TiO_2 photocatalyst: Ultrasound assisted synthesis, characterization and catalytic efficiency. *Ultrasonics Sonochemistry*, 2012, **19**(1): 9–15.
31. Kochuveedu, S.T. et al., Surface-plasmon-induced visible light photocatalytic activity of TiO_2 nanospheres decorated by Au nanoparticles with controlled configuration. *The Journal of Physical Chemistry C*, 2011, **116**(3): 2500–2506.
32. Tanaka, A. et al., Preparation of Au/TiO_2 with metal cocatalysts exhibiting strong surface plasmon resonance effective for photoinduced hydrogen formation under irradiation of visible light. *ACS Catalysis*, 2012, **3**(1): 79–85.
33. Cha, S.I. et al., Pt-free transparent counter electrodes for dye-sensitized solar cells prepared from carbon nanotube micro-balls. *Journal of Materials Chemistry*, 2009, **20**(4): 659–662.
34. Hong, W.J. et al., Transparent graphene/PEDOT–PSS composite films as counter electrodes of dye-sensitized solar cells. *Electrochemistry Communications*, 2008, **10**(10): 1555–1558.
35. Tian, L. et al., Graphene oxides for homogeneous dispersion of carbon nanotubes. *ACS Applied Materials and Interfaces*, 2010, **2**(11): 3217–3222.
36. Vadukumpully, S. et al., Functionalization of surfactant wrapped graphene nanosheets with alkylazides for enhanced dispersibility. *Nanoscale*, 2011, **3**(1): 303–308.
37. Krol, R.v.d. et al., Solar hydrogen production with nanostructured metal oxides. *Journal of Materials Chemistry*, 2008, **18**(20): 2311–2320.
38. Hoshikawa, T. et al., Impedance analysis of internal resistance affecting the photoelectrochemical performance of dye-sensitized solar cells. *Journal of the Electrochemical Society*, 2005, **152**(2): E68–E73.
39. Ng, Y.H. et al., To what extent do graphene scaffolds improve the photovoltaic and photocatalytic response of TiO_2 nanostructured films? *The Journal of Physical Chemistry Letters*, 2010, **1**(15): 2222–2227.
40. Fujishima, A. and K. Honda, Electrochemical photolysis of water at a semiconductor electrode. *Nature*, 1972, **238**(5358): 37–38.
41. Zhang, X.-Y. et al., Graphene/TiO_2 nanocomposites: Synthesis, characterization and application in hydrogen evolution from water photocatalytic splitting. *Journal of Materials Chemistry*, 2010, **20**(14): 2801–2806.
42. Fan, W. et al., Nanocomposites of TiO_2 and reduced graphene oxide as efficient photocatalysts for hydrogen evolution. *The Journal of Physical Chemistry C*, 2011, **115**(21): 10694–10701.
43. Park, Y. et al., Exfoliated and reorganized graphite oxide on titania nanoparticles as an auxiliary co-catalyst for photocatalytic solar conversion. *Physical Chemistry Chemical Physics*, 2011, **13**(20): 9425–9431.
44. Shin, H.-J. et al., *Reduced Graphene Oxide Doped with Dopant, Thin Layer and Transparent Electrode*. US 2009/0146111 A1, 2009, Samsung Electronics Co., Ltd.
45. Mukherji, A. et al., Nitrogen doped $Sr_2Ta_2O_7$ coupled with graphene sheets as photocatalysts for increased photocatalytic hydrogen production. *ACS Nano*, 2011, **5**(5): 3483–3492.
46. Jia, L. et al., Highly durable N-doped graphene/CdS nanocomposites with enhanced photocatalytic hydrogen evolution from water under visible light irradiation. *The Journal of Physical Chemistry C*, 2011, **115**(23): 11466–11473.
47. Ng, Y.H. et al., Reducing graphene oxide on a visible-light $BiVO_4$ photocatalyst for an enhanced photoelectrochemical water splitting. *The Journal of Physical Chemistry Letters*, 2010, **1**(17): 2607–2612.
48. Yeh, T.-F. et al., Graphite oxide as a photocatalyst for hydrogen production from water. *Advanced Functional Materials*, 2010, **20**(14): 2255–2262.
49. Heo, J.H. et al., Efficient inorganic–organic hybrid heterojunction solar cells containing perovskite compound and polymeric hole conductors. *Nature Photonics*, 2013, **7**(6): 486–491.
50. Yella, A. et al., Porphyrin-sensitized solar cells with cobalt (II/III)-based redox electrolyte exceed 12 percent efficiency. *Science*, 2011, **334**(6056): 629–634.

51. O'Regan, B. and M. Grätzel, A low-cost, high-efficiency solar-cell based on dye-sensitized colloidal TiO_2 films. *Nature*, 1991, **353**(6346): 737–740.
52. Docampo, P. et al., Charge transport limitations in self-assembled TiO_2 photoanodes for dye-sensitized solar cells. *The Journal of Physical Chemistry Letters*, 2013, **4**(5): 698–703.
53. Crossland, E.J.W. et al., Mesoporous TiO_2 single crystals delivering enhanced mobility and optoelectronic device performance. *Nature*, 2013, **495**(7440): 215–219.
54. Novoselov, K.S. et al., Electric field effect in atomically thin carbon films. *Science*, 2004, **306**(5696): 666–669.
55. Wu, Z.-S. et al., Graphene/metal oxide composite electrode materials for energy storage. *Nano Energy*, 2012, **1**(1): 107–131.
56. Yang, S. et al., Sandwich-like, graphene-based titania nanosheets with high surface area for fast lithium storage. *Advanced Materials*, 2011, **23**(31): 3575–3579.
57. Craciun, M.F. et al., Tuneable electronic properties in graphene. *Nano Today*, 2011, **6**(1): 42–60.
58. Peigney, A. et al., Specific surface area of carbon nanotubes and bundles of carbon nanotubes. *Carbon*, 2001, **39**(4): 507–514.
59. Robin, I.C. et al., Doping in homoepitaxial ZnO films: Effect of intentional and unintentional impurities on the optical and the electrical properties. *Journal of the Korean Physical Society*, 2008, **53**(5): 2888–2892.
60. Bell, N.J. et al., Understanding the enhancement in photoelectrochemical properties of photocatalytically prepared TiO_2-reduced graphene oxide composite. *The Journal of Physical Chemistry C*, 2011, **115**(13): 6004–6009.
61. Du, J. et al., Hierarchically ordered macro – mesoporous TiO_2 – graphene composite films: Improved mass transfer, reduced charge recombination, and their enhanced photocatalytic activities. *ACS Nano*, 2010, **5**(1): 590–596.
62. Zhang, D.W. et al., Graphene-based counter electrode for dye-sensitized solar cells. *Carbon*, 2011, **49**(15): 5382–5388.
63. Cruz, R. et al., Reduced graphene oxide films as transparent counter-electrodes for dye-sensitized solar cells. *Solar Energy*, 2012, **86**(2): 716–724.
64. Yeh, M.-H. et al., A low-cost counter electrode of ITO glass coated with a graphene/Nafion® composite film for use in dye-sensitized solar cells. *Carbon*, 2012, **50**(11): 4192–4202.
65. Xu, X. et al., Electrochemically reduced graphene oxide multilayer films as efficient counter electrode for dye-sensitized solar cells. *Science Reports*, 2013, **3**: Article number 1489.
66. Sun, S. et al., Enhanced dye-sensitized solar cell using graphene–TiO[sub 2] photoanode prepared by heterogeneous coagulation. *Applied Physics Letters*, 2010, **96**(8): 083113.
67. Tang, Y.-B. et al., Incorporation of graphenes in nanostructured TiO_2 films via molecular grafting for dye-sensitized solar cell application. *ACS Nano*, 2010, **4**(6): 3482–3488.
68. Yang, N. et al., Two-dimensional graphene bridges enhanced photoinduced charge transport in dye-sensitized solar cells. *ACS Nano*, 2010, **4**(2): 887–894.
69. Yen, M.-Y. et al., Preparation of graphene/multi-walled carbon nanotube hybrid and its use as photoanodes of dye-sensitized solar cells. *Carbon*, 2011, **49**(11): 3597–3606.
70. Kim, S.R. et al., UV-reduction of graphene oxide and its application as an interfacial layer to reduce the back-transport reactions in dye-sensitized solar cells. *Chemical Physics Letters*, 2009, **483**(1–3): 124–127.
71. Xu, F. et al., Graphene scaffolds enhanced photogenerated electron transport in ZnO photoanodes for high-efficiency dye-sensitized solar cells. *The Journal of Physical Chemistry C*, 2013, **117**(17): 8619–8627.
72. Tang, B. and G. Hu, Two kinds of graphene-based composites for photoanode applying in dye-sensitized solar cell. *Journal of Power Sources*, 2012, **220**: 95–102.
73. Liu, Z. et al., PEGylated nanographene oxide for delivery of water-insoluble cancer drugs. *Journal of the American Chemical Society*, 2008, **130**(33): 10876–10877.
74. Li, D. et al., Processable aqueous dispersions of graphene nanosheets. *Nature Nano*, 2008, **3**(2): 101–105.
75. Titelman, G.I. et al., Characteristics and microstructure of aqueous colloidal dispersions of graphite oxide. *Carbon*, 2005, **43**(3): 641–649.
76. Murakami, T.N. and M. Grätzel, Counter electrodes for DSC: Application of functional materials as catalysts. *Inorganica Chimica Acta*, 2008, **361**(3): 572–580.
77. Imoto, K. et al., High-performance carbon counter electrode for dye-sensitized solar cells. *Solar Energy Materials and Solar Cells*, 2003, **79**(4): 459–469.
78. Kavan, L. et al., Optically transparent cathode for dye-sensitized solar cells based on graphene nanoplatelets. *ACS Nano*, 2010, **5**(1): 165–172.
79. Gray, A. et al., Optical detection and characterization of graphene by broadband spectrophotometry. *Journal of Applied Physics*, 2008, **104**(5): 053109.
80. Freitag, M., Graphene: Nanoelectronics goes flat out. *Nature Nano*, 2008, **3**(8): 455–457.
81. Czerw, R. et al., Substrate–interface interactions between carbon nanotubes and the supporting substrate. *Physical Review B*, 2002, **66**(3): 033408.
82. Wilder, J.W.G. et al., Electronic structure of atomically resolved carbon nanotubes. *Nature*, 1998, **391**(6662): 59–62.
83. Gun, J. et al., Graphene oxide organogel electrolyte for quasi solid dye sensitized solar cells. *Electrochemistry Communications*, 2012, **19**(1): 108–110.
84. Neo, C.Y. and J.Y. Ouyang, The production of organogels using graphene oxide as the gelator for use in high-performance quasi-solid state dye-sensitized solar cells. *Carbon*, 2013, **54**: 48–57.
85. Jung, M.-H. et al., Iodide-functionalized graphene electrolyte for highly efficient dye-sensitized solar cells. *Journal of Materials Chemistry*, 2012, **22**(32): 16477–16483.
86. Nguyen, D.D. et al., Synthesis of ethanol-soluble few-layer graphene nanosheets for flexible and transparent conducting composite films. *Nanotechnology*, 2011, **22**(29): 295606.
87. Velten, J.A. et al., Photoinduced optical transparency in dye-sensitized solar cells containing graphene nanoribbons. *Journal of Physical Chemistry C*, 2011, **115**(50): 25125–25131.
88. Grätzel, M., The artificial leaf, bio-mimetic photocatalysis. *Cattech*, 1999, **3**: 3–17.
89. Yamaguchi, T. et al., Series-connected tandem dye-sensitized solar cell for improving efficiency to more than 10%. *Solar Energy Materials and Solar Cells*, 2009, **93**(6–7): 733–736.
90. Papageorgiou, N. et al., An iodine/triiodide reduction electrocatalyst for aqueous and organic media. *Journal of the Electrochemical Society*, 1997, **144**(3): 876–884.
91. Murakami, T.N. et al., Highly efficient dye-sensitized solar cells based on carbon black counter electrodes. *Journal of the Electrochemical Society*, 2006, **153**(12): A2255–A2261.
92. Lan, Z. et al., Morphology controllable fabrication of Pt counter electrodes for highly efficient dye-sensitized solar cells. *Journal of Materials Chemistry*, 2012, **22**(9): 3948–3954.
93. Kaniyoor, A. and S. Ramaprabhu, Enhanced efficiency in dye sensitized solar cells with nanostructured Pt decorated multiwalled carbon nanotube based counter electrode. *Electrochimica Acta*, 2012, **72**: 199–206.
94. Smestad, G. et al., Testing of dye sensitized TiO_2 solar cells I: Experimental photocurrent output and conversion

efficiencies. *Solar Energy Materials and Solar Cells*, 1994, **32**(3): 259–272.
95. Yun, S.N. et al., A new type of low-cost counter electrode catalyst based on platinum nanoparticles loaded onto silicon carbide (Pt/SiC) for dye-sensitized solar cells. *Physical Chemistry Chemical Physics*, 2013, **15**(12): 4286–4290.
96. Wu, M. et al., Low-cost molybdenum carbide and tungsten carbide counter electrodes for dye-sensitized solar cells. *Angewandte Chemie—International Edition*, 2011, **50**(Compendex): 3520–3524.
97. Sun, H.C. et al., Dye-sensitized solar cells with NiS counter electrodes electrodeposited by a potential reversal technique. *Energy and Environmental Science*, 2011, **4**(8): 2630–2637.
98. Xin, X. et al., Low-cost copper zinc tin sulfide counter electrodes for high-efficiency dye-sensitized solar cells. *Angewandte Chemie—International Edition*, 2011, **50**(49): 11739–11742.
99. Zhang, L.L. et al., Metal/metal sulfide functionalized single-walled carbon nanotubes: FTO-free counter electrodes for dye sensitized solar cells. *Physical Chemistry Chemical Physics*, 2012, **14**(28): 9906–9911.
100. Zhang, X.Y. et al., Transition-metal nitride nanoparticles embedded in N-doped reduced graphene oxide: Superior synergistic electrocatalytic materials for the counter electrodes of dye-sensitized solar cells. *Journal of Materials Chemistry A*, 2013, **1**(10): 3340–3346.
101. Wen, Z. et al., Metal nitride/graphene nanohybrids: General synthesis and multifunctional titanium nitride/graphene electrocatalyst. *Advanced Materials*, 2011, **23**(45): 5445–5450.
102. Lee, B. et al., An all carbon counter electrode for dye sensitized solar cells. *Energy and Environmental Science*, 2012, **5**(5): 6941–6952.
103. Chen, H.Y. et al., Highly catalytic carbon nanotube/Pt nanohybrid-based transparent counter electrode for efficient dye-sensitized solar cells. *Chemistry—An Asian Journal*, 2012, **7**(8): 1795–1802.
104. Dao, V.D. et al., Pt–NP–MWNT nanohybrid as a robust and low-cost counter electrode material for dye-sensitized solar cells. *Journal of Materials Chemistry*, 2012, **22**(28): 14023–14029.
105. Kitamura, T. et al., Improved solid-state dye solar cells with polypyrrole using a carbon-based counter electrode. *Chemistry Letters*, 2001, **30**(10): 1054–1055.
106. Murakami, T. and M. Grätzel, Counter electrodes for DSC: Application of functional materials as catalysts. *Inorganica Chimica Acta*, 2008, **361**(3): 572–580.
107. Kay, A. and M. Grätzel, Low cost photovoltaic modules based on dye sensitized nanocrystalline titanium dioxide and carbon powder. *Solar Energy Materials and Solar Cells*, 1996, **44**(1): 99–117.
108. Denaro, T. et al., Investigation of low cost carbonaceous materials for application as counter electrode in dye-sensitized solar cells. *Journal of Applied Electrochemistry*, 2009, **39**(11): 2173–2179.
109. Suzuki, K. et al., Application of carbon nanotubes to counter electrodes of dye-sensitized solar cells. *Chemistry Letters*, 2003, **32**(1): 28–29.
110. Chou, C.-S. et al., The applicability of SWCNT on the counter electrode for the dye-sensitized solar cell. *Advanced Powder Technology*, 2009, **20**(4): 310–317.
111. Ramasamy, E. et al., Spray coated multi-wall carbon nanotube counter electrode for tri-iodide (I_3^-) reduction in dye-sensitized solar cells. *Electrochemistry Communications*, 2008,**10**(7): 1087–1089.
112. Hauch, A. and A. Georg, Diffusion in the electrolyte and charge-transfer reaction at the platinum electrode in dye-sensitized solar cells. *Electrochimica Acta*, 2001, **46**(22): 3457–3466.
113. Ahmad, S. et al., Efficient platinum-free counter electrodes for dye-sensitized solar cell applications. *Chemistry Physics Chemistry*, 2010, **11**(13): 2814–2819.
114. Kaniyoor, A. and S. Ramaprabhu, Thermally exfoliated graphene based counter electrode for low cost dye sensitized solar cells. *Journal of Applied Physics*, 2011, **109**(12): 124308.
115. Roy-Mayhew, J.D. et al., Functionalized graphene as a catalytic counter electrode in dye-sensitized solar cells. *ACS Nano*, 2010, **4**(10): 6203–6211.
116. Roy-Mayhew, J.D. et al., Functionalized graphene sheets as a versatile replacement for platinum in dye-sensitized solar cells. *ACS Applied Materials and Interfaces*, 2012, **4**(5): 2794–2800.
117. Xu, C. et al., Synthesis of hemin functionalized graphene and its application as a counter electrode in dye-sensitized solar cells. *Materials Chemistry and Physics*, 2012, **132**(2–3): 858–864.
118. Choi, H. et al., Graphene counter electrodes for dye-sensitized solar cells prepared by electrophoretic deposition. *Journal of Materials Chemistry*, 2011, **21**(21): 7548–7551.
119. Jang, S.-Y. et al., Electrodynamically sprayed thin films of aqueous dispersible graphene nanosheets: Highly efficient cathodes for dye-sensitized solar cells. *ACS Applied Materials and Interfaces*, 2012, **4**(7): 3500–3507.
120. Zheng, H.Q. et al., Reduced graphene oxide films fabricated by gel coating and their application as platinum-free counter electrodes of highly efficient iodide/triiodide dye-sensitized solar cells. *Journal of Materials Chemistry*, 2012, **22**(29): 14465–14474.
121. Yen, M.-Y. et al., Metal-free, nitrogen-doped graphene used as a novel catalyst for dye-sensitized solar cell counter electrodes. *RSC Advances*, 2012, **2**(7): 2725–2728.
122. Xue, Y. et al., Nitrogen-doped graphene foams as metal-free counter electrodes in high-performance dye-sensitized solar cells. *Angewandte Chemie International Edition*, 2012, **51**(48): 12124–12127.
123. Cruz, R. et al., Transparent graphene-based counter-electrodes for iodide/triiodide mediated dye-sensitized solar cells. *Journal of Materials Chemistry A*, 2014, **2**(7): 2028–2032.
124. Tjoa, V. et al., Facile photochemical synthesis of graphene–Pt nanoparticle composite for counter electrode in dye sensitized solar cell. *ACS Applied Materials and Interfaces*, 2012, **4**(7): 3447–3452.
125. Gong, F. et al., Self-assembled monolayer of graphene/Pt as counter electrode for efficient dye-sensitized solar cell. *Physical Chemistry Chemical Physics*, 2011, **13**(39): 17676–17682.
126. Bajpai, R. et al., Graphene supported platinum nanoparticle counter-electrode for enhanced performance of dye-sensitized solar cells. *ACS Applied Materials and Interfaces*, 2011, **3**(10): 3884–3889.
127. Liu, C.Y. et al., Enhanced efficiency of dye-sensitized solar cells with counter electrodes consisting of platinum nanoparticles and nanographites. *Electrochimica Acta*, 2012, **59**: 128–134.
128. Yen, M.Y. et al., Platinum nanoparticles/graphene composite catalyst as a novel composite counter electrode for high performance dye-sensitized solar cells. *Journal of Materials Chemistry*, 2011, **21**(34): 12880–12888.
129. Guai, G.H. et al., Graphene–Pt/ITO counter electrode to significantly reduce Pt loading and enhance charge transfer for high performance dye-sensitized solar cell. *Solar Energy*, 2012, **86**(7): 2041–2048.
130. Dou, Y.Y. et al., Nickel phosphide-embedded graphene as counter electrode for dye-sensitized solar cells. *Physical Chemistry Chemical Physics*, 2012, **14**(4): 1339–1342.

131 Li, Z. et al., NiS_2/reduced graphene oxide nanocomposites for efficient dye-sensitized solar cells. *The Journal of Physical Chemistry C*, 2013, **117**(13): 6561–6566.
132. Choi, H. et al., Dye-sensitized solar cells using graphene-based carbon nano composite as counter electrode. *Solar Energy Materials and Solar Cells*, 2011, **95**(1): 323–325.
133. Zhu, G. et al., Electrophoretic deposition of reduced graphene–carbon nanotubes composite films as counter electrodes of dye-sensitized solar cells. *Journal of Materials Chemistry*, 2011, **21**(38): 14869–14875.
134. Battumur, T. et al., Graphene/carbon nanotubes composites as a counter electrode for dye-sensitized solar cells. *Current Applied Physics*, 2012, **12**: E49–E53.
135. Velten, J. et al., Carbon nanotube/graphene nanocomposite as efficient counter electrodes in dye-sensitized solar cells. *Nanotechnology*, 2012, **23**(8): 085201.
136. Yang, Z. et al., Carbon nanotubes bridged with graphene nanoribbons and their use in high-efficiency dye-sensitized solar cells. *Angewandte Chemie International Edition*, 2013, **52**: 11603–11606.
137. Wang, G. et al., Graphene/polyaniline nanocomposite as counter electrode of dye-sensitized solar cells. *Materials Letters*, 2012, **69**: 27–29.
138. Lee, K.S. et al., Flexible and platinum-free dye-sensitized solar cells with conducting-polymer-coated graphene counter electrodes. *Chemistry Sustainable Chemistry*, 2012, **5**(2): 379–382.
139. Kaniyoor, A. and S. Ramaprabhu, Soft functionalization of graphene for enhanced tri-iodide reduction in dye sensitized solar cells. *Journal of Materials Chemistry*, 2012, **22**(17): 8377–8384.
140. Kavan, L. et al., Graphene nanoplatelet cathode for Co(III)/(II) mediated dye-sensitized solar cells. *ACS Nano*, 2011, **5**(11): 9171–9178.
141. Kavan, L. et al., Graphene nanoplatelets outperforming platinum as the electrocatalyst in co-bipyridine-mediated dye-sensitized solar cells. *Nano Letters*, 2011, **11**(12): 5501–5506.
142. Yum, J.H. et al., A cobalt complex redox shuttle for dye-sensitized solar cells with high open-circuit potentials. *Nature Communications*, 2012, **3**: Article number 631.
143. Stefik, M. et al., Carbon–graphene nanocomposite cathodes for improved Co(ii/iii) mediated dye-sensitized solar cells. *Journal of Materials Chemistry A*, 2013, **1**(16): 4982–4987.
144. Hsieh, C.-T. et al., One- and two-dimensional carbon nanomaterials as counter electrodes for dye-sensitized solar cells. *Carbon*, 2011, **49**(9): 3092–3097.
145. Choi, H. et al., Electrophoretic graphene for transparent counter electrodes in dye-sensitised solar cells. *Electronics Letters*, 2011, **47**: 281–283.
146. Kim, H. et al., Fabrication and characterization of carbon-based counter electrodes prepared by electrophoretic deposition for dye-sensitized solar cells. *Nanoscale Research Letters*, 2012, **7**: 1–12.
147. Si, Y. and E.T. Samulski, Synthesis of water soluble graphene. *Nano Letters*, 2008, **8**(6): 1679–1682.
148. Trancik, J.E. et al., Transparent and catalytic carbon nanotube films. *Nano Letters*, 2008, **8**(4): 982–987.
149. Zhang, D.W. et al., Graphene nanosheet counter-electrodes for dye-sensitized solar cells. In *Nanoelectronics Conference (INEC), 2010 3rd International*. 2010. New York.
150. Wang, X. et al., Transparent, conductive graphene electrodes for dye-sensitized solar cells. *Nano Letters*, 2008, **8**(1): 323–327.
151. Xu, Y. et al., Flexible graphene films via the filtration of water-soluble noncovalent functionalized graphene sheets. *Journal of the American Chemical Society*, 2008, **130**(18): 5856–5857.
152. Wan, L. et al., Room-temperature fabrication of graphene films on variable substrates and its use as counter electrodes for dye-sensitized solar cells. *Solid State Sciences*, 2011, **13**(2): 468–475.
153. Xu, S. et al., Investigation of catalytic activity of glassy carbon with controlled crystallinity for counter electrode in dye-sensitized solar cells. *Solar Energy*, 2011, **85**(11): 2826–2832.
154. Veerappan, G. et al., Sub-micrometer-sized graphite as a conducting and catalytic counter electrode for dye-sensitized solar cells. *ACS Applied Materials and Interfaces*, 2011, **3**(3): 857–862.
155. Das, S. et al., Amplifying charge-transfer characteristics of graphene for triiodide reduction in dye-sensitized solar cells. *Advanced Functional Materials*, 2011, **21**(19): 3729–3736.
156. Yang, S.-Y. et al., A powerful approach to fabricate nitrogen-doped graphene sheets with high specific surface area. *Electrochemistry Communications*, 2012, **14**(1): 39–42.
157. Tung, V.C. et al., Low-temperature solution processing of graphene–carbon nanotube hybrid materials for high-performance transparent conductors. *Nano Letters*, 2009, **9**(5): 1949–1955.
158. Bajpai, R. et al., Graphene supported nickel nanoparticle as a viable replacement for platinum in dye sensitized solar cells. *Nanoscale*, 2012, **4**(3): 926–930.
159. Stankovich, S. et al., Graphene-based composite materials. *Nature*, 2006, **442**(7100): 282–6.
160. Huang, K.-C. et al., Nanographite/polyaniline composite films as the counter electrodes for dye-sensitized solar cells. *Journal of Materials Chemistry*, 2011, **21**(Compendex): 10384–10389.
161. Bourdo, S.E. and T. Viswanathan, Graphite/polyaniline (GP) composites: Synthesis and characterization. *Carbon*, 2005, **43**(14): 2983–2988.
162. Maçaira, J. et al., Review on nanostructured photoelectrodes for next generation dye-sensitized solar cells. *Renewable and Sustainable Energy Reviews*, 2013, **27**(0): 334–349.

22 Direct Threat of UV–Ozone-Treated Indium-Tin Oxide in Organic Optoelectronics and Stability Enhancement Using Graphene Oxide as Anode Buffer Layer

Tsz-Wai Ng, Ming-Fai Lo, Qing-Dan Yang, and Chun-Sing Lee

CONTENTS

ABSTRACT

Ultraviolet (UV)–ozone treatment on indium-tin-oxide (ITO) glass substrates has been widely used in the field of organic optoelectronics for over several decades. ITO glass substrates are currently recognized as standard substrates for high-efficiency organic electronics devices. In the first part of this chapter, however, we present evidence for rapid decay of common organic films (e.g., NPB, Alq_3, and rubrene) when they are in direct contact with UV–ozone-treated ITO (UV–ITO) substrates. Photoluminescence (PL), x-ray, and ultraviolet photoemission spectroscopies (XPS, UPS) are used to characterize the reliability of UV–ITO substrates. We conclude that the degradation of organic thin films on UV–ITO substrates are mainly attributed to the active oxygen species generated upon UV–ozone treatments. The oxygen species behave as an oxygen reservoir that interacts with the adsorbed molecules by forming a gap state within its original bandgap and cause charge recombination.

In the second part, we demonstrate the application of a high work function graphene oxide (GO) film as an anode buffer layer for inhibiting those surface-active species on UV–ITO substrates. With the introduction of a GO anode buffer layer, significant enhancements in both the power conversion efficiency (PCE) and stabilities of organic photovoltaic (OPV) devices are observed. This chapter clearly demonstrates the use of a GO buffer layer as a simple and effective approach to improve the performance of OPV devices.

22.1 INTRODUCTION

OPVs are recognized as a new generation of solar devices due to their low cost, promising energy conversion efficiency, and potential for making flexible photovoltaics, and easy fabrication over a large area.[1–4] In the past couple of decades, up to ~10% of PCE has been achieved with a simple donor–acceptor bilayer structure and blended bulk heterostructures.[3,5–11] While many attempts have been made on improving device

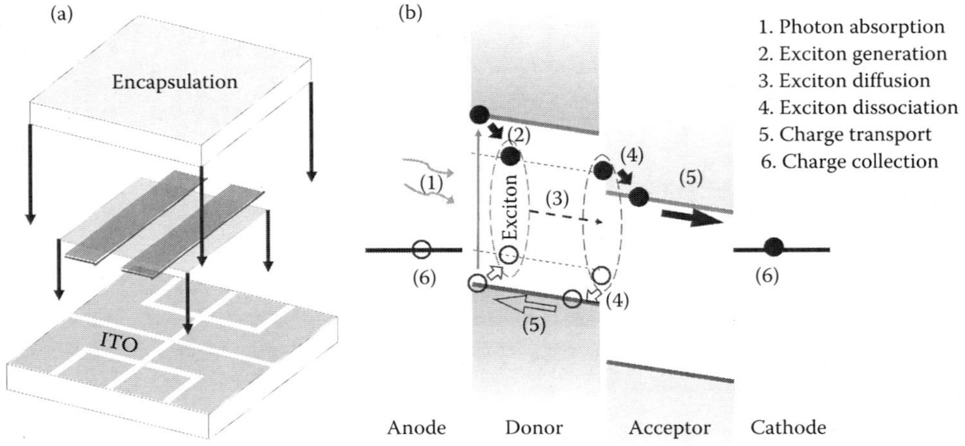

FIGURE 22.1 (a) Typical device structure and (b) schematic diagram of photovoltaic processes in OPV.

efficiencies, issues of device reliabilities and operation stabilities however are still major hurdles for the applications of OPV devices.[6,12–16] Nowadays, works focusing on stability enhancement are rarely reported.[13,17–19]

22.1.1 Operation Mechanism of OPV Devices

Figure 22.1 shows the device structure of a typical OPV device. It consists of donor and acceptor materials with staggered energy levels, that is, a donor material with a low ionization potential (IP) forms a heterojunction with an acceptor material with high electron affinity (EA). This organic donor/acceptor interface is sandwiched between the ITO anode and Al cathode. The fundamental working mechanism of an OPV device can be summarized as the following steps: the absorption of an incident photon leads to the photogenerated excitons (i.e., a bounded electron–hole pair). Owing to the concentration gradient, these generated excitons would diffuse within the organic layers until they reach the donor/acceptor interface where they encounter a strong E-field for subsequent dissociation as free charge carriers. These forming holes and electrons would be transported and finally be collected by the UV–ITO anode and Al cathode, respectively, yielding a photocurrent.

22.1.2 Organic Film Deterioration When in Contact with UV–ITO Substrates

As shown in Figure 22.2, the ambient gases diffusion from the top Al electrodes are recognized as a major cause of the degradation of organic active materials in OPV (and other organic) devices.[20] To protect the device from environmental effects, several encapsulation schemes have been proposed.[4,6,12,13,16,20] Nevertheless, degradation is still observed even when the devices were stored in vacuum where the effects from ambient gases are isolated.[12,13,20] All these studies suggest that in

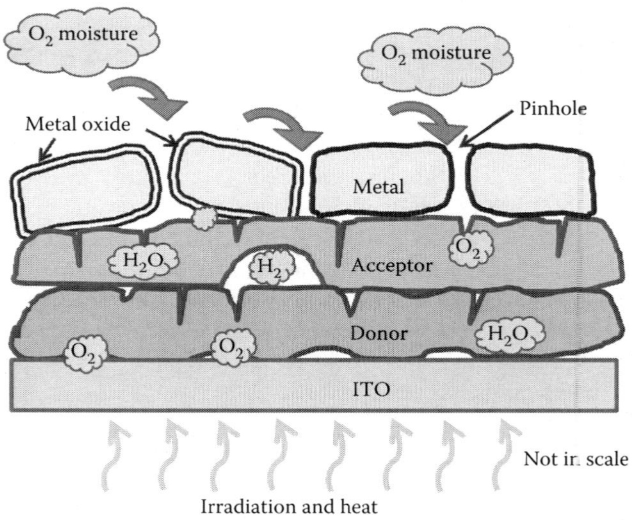

FIGURE 22.2 Schematic diagram of degradation pathways in an OPV device.

addition to the working environment, there are still other causes for device degradation.

To date, an ITO substrate is the most typical transparent conductive electrode used in organic optoelectronic devices.[1,21–28] Although early concerns for device degradation were the In diffusion from ITO into the organic layers that causes a subsequent stability problem,[29–33] Lee et al.[29] later showed that In diffusion is indeed not obvious before a prolonged operation for more than 2000 h. Nowadays, ITO is extensively used with little alert.[25,26,29–35]

Over the past two decades, many surface modification techniques of the ITO substrate have been developed to enhance the performance and stabilities of organic electronic devices.[34–38] Among these surface modification methods, UV–ozone and oxygen plasma are well recognized as

TABLE 22.1
Summary of Degradation Factor in an OPV Device

Environmental Factor	Intrinsic Factor
1. Oxygen	1. Metal diffusion (indium, Al)
2. Moisture	2. Impurities
3. Heat	3. Subsequent decrease in work functions of the organic film
4. UV irradiation	

the standard treatments for ITO substrates, which are able to enhance the work function of ITO via increasing the surface oxygen content.[34,38,39] However, the effects of such treatments drop over time.[40–42] The subsequent decrease in work functions over time would cause carriers unbalance, thus affecting device reliability.[31,40,43] Up till now, the implications and possible threats of UV–ITO to other common organic materials have not been systematically investigated. The above possible degradation factors are summarized in Table 22.1.

The chemical studies and potential threats of UV–ozone on the ITO substrate are reviewed in Section 22.2. We will illustrate the effects of ITO substrates on the stability of three commonly used organic semiconductors, namely including *N,N′*-bis(naphthalen-1-yl)-*N,N′*-bis(phenyl)-benzidine (NPB), tris(8-hydroxy-quinolinato) aluminum (Alq_3), and 5-, 6-, 11-, 12-tetraphenylnaphthacene (rubrene). While mild decay was observed in the PLs of all three materials when they are in contact with as-cleaned ITO (AC-ITO), the degradation is rapid when the same films are deposited on UV–ITO. The effects of UV–ITO on the degradation of rubrene were further investigated via UPS and XPS. Furthermore, we show that surface modification of UV–ITO substrates prior to organic film deposition is an important procedure to maintain a high stability of organic devices. This chapter points out that the most widely used UV–ITO substrate indeed causes a potential threat to stability issues of many common organic semiconducting films. Our studies show that oxygen radicals generated upon UV–ozone treatments may be another cause for degradation in OPV devices fabricated on ITO substrates.[17,44,45]

22.1.3 Improved Organic Film Stability Using GO Anode Buffer Layer

Owing to the high electrical/thermal conductivities and transparency, graphene is generally considered as a potential candidate to replace ITO such that the reliability problem related to ITO can be solved. However, it is problematic to apply graphene directly in general optoelectronic devices due to the difficulties in its fine patterning and its high sensitivity to contamination such as residuals.[46–48] Nevertheless, upon oxidation, cyclohexane-like units decorated with hydroxyl, epoxy, ether, diol, and ketone groups can be attached to graphene sheets. These functional groups impart water dispensability to graphene sheets, which allow them to be simply spin coated on the patterned ITO surface.

In Section 22.3, we explore the possible applications of solution-processed GO thin film as an anode buffer layer in standard OPV devices. A GO layer was found not only to enhance the efficiency, but also to substantially enhance the storage and operation stabilities of organic devices. Mechanisms for the efficiency and stability enhancements are discussed in detail. In this chapter, we demonstrate a high work function GO film as an anode buffer layer on a UV–ITO substrate for inhibiting the surface-active species.

22.2 Potential Threat of UV–Ozone-Treated ITO Substrate

22.2.1 PL Quenching of Organic Materials on UV–ITO Substrate

Figure 22.3a compares PL spectra of NPB, Alq_3, and rubrene films (i.e., materials generally used in organic light-emitting diodes) prepared on quartz, AC-ITO, and UV–ITO glass substrates before (black solid line) and after (symbol) illumination. Significant PL deteriorations are clearly observed

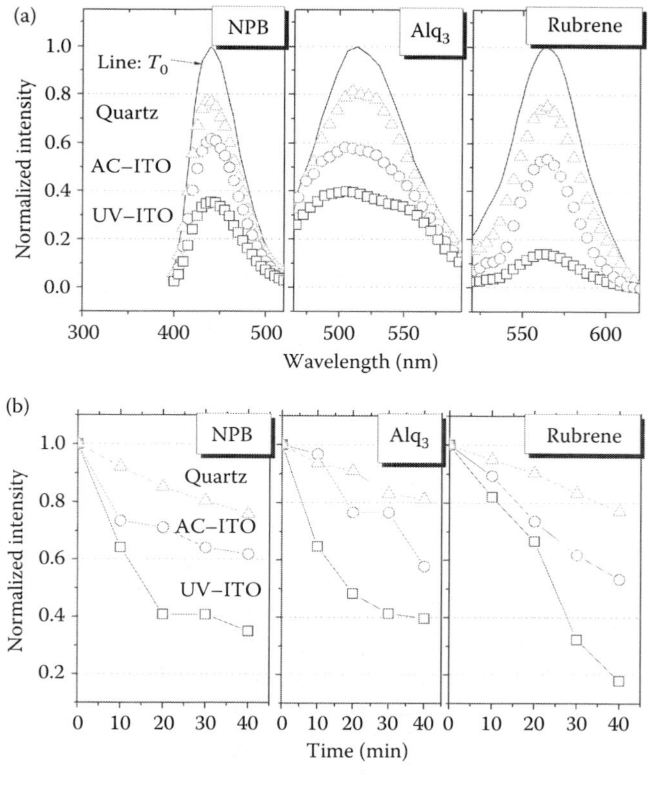

FIGURE 22.3 (a) PL spectra of NPB, Alq_3, and rubrene films prepared on quartz (Δ), AC–ITO (○), and UV–ITO (□) substrates. All PL signals are normalized with respect to the initial PL intensity. The solid line shows the initial PL signal measured at room condition (T_0) while the symbols show the PL signal after 40-min illumination under 1 Sun ($T_{40,1\ sun}$). (b) PL deterioration in NPB, Alq_3, and rubrene films prepared on quartz (Δ), AC–ITO (○), and UV–ITO (□) substrates over a continuous illumination for 40 min under 1 Sun.

in all films upon illumination.[49] However, it is interesting to note that the extent of organic PL deterioration was found to be dependent on substrates where these organic films were deposited. Figure 22.3b compares the degradation rates of the above film samples. Rubrene showed an obvious difference in PL characteristics before and after illumination. For rubrene films prepared on quartz and AC–ITO substrates, there are still considerable PL responses (>50%) even after illumination. However, PL of rubrene prepared on UV–ITO substrates is almost completely quenched over the same illumination time. The above observations show that the UV–ITO substrate does play an important role in accelerating the organic degradation.

To clearly distinguish the effects of oxygen species from the ITO surface and those from the atmosphere, a PL degradation experiment was carried out using a specially patterned UV–ITO substrate. Some parts of the ITO at the central region were etched in a way with a capital letter "A" as shown in Figure 22.4a. A rubrene film was subsequently deposited onto the specially treated substrate (shaded with "\" line). The sample was held by a sample holder covering the bottom region (shaded with "/" line). The sample was then irradiated with 1 Sun AM 1.5G illuminations for 0, 15, and 30 min. It can be seen from Figure 22.4c that the rubrene film shows uniform PL before illumination.

After 15 min of illumination, the region where the rubrene film is directly in contact with ITO shows a much weaker PL intensity than the region with no direct contact to ITO ("A"). After 30 min of illumination, PL from the rubrene region in contact with ITO can hardly be observed as shown in Figure 22.4e. These results clearly demonstrate that both the ITO substrate and the simulated solar illumination would accelerate degradation of the rubrene film.

From the above experiment, it is substantiated that the most widely used UV–ITO substrates can indeed result in significant deterioration of common organic semiconductors. We continue our discussion by examining the degradation mechanisms caused by the UV–ITO substrate using photoemission spectroscopic techniques.

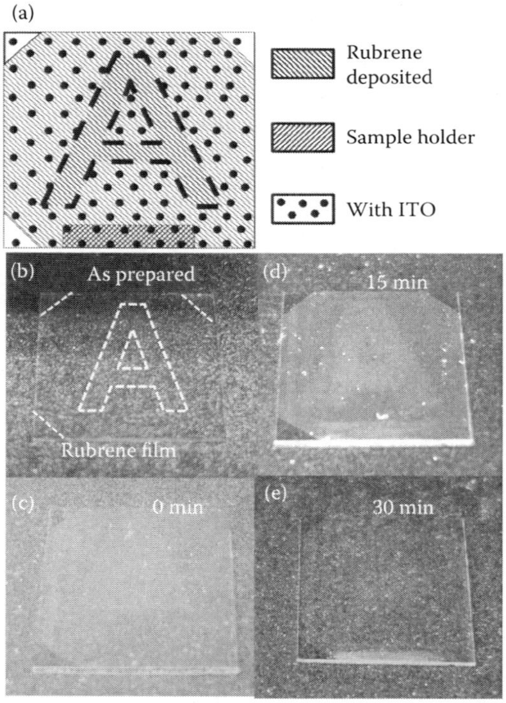

FIGURE 22.4 (a) Schematic diagram of a patterned UV–ITO substrate. The area inside the letter "A" is ITO free. Photographs of an as-prepared rubrene sample taken under (b) normal-light and (c) UV-illuminated condition. Photographs of rubrene-coated sample taken under UV-illuminated condition after (d) 15- and (e) 30-min irradiation under 1 Sun.

22.2.2 XPS Studies of Rubrene Degradation on UV–ITO Substrate

With the observed PL degradation of organic films, in this section, we focus on the chemical interaction between the UV–ITO substrate and the addlayer organic films. Here, a 5-nm rubrene film was first thermally evaporated onto a UV–ITO substrate in ultrahigh vacuum (UHV) and then transferred without vacuum break to an analysis chamber for photoemission studies using both XPS and UPS.

Figure 22.5a compares the XPS C1s and O1s core levels spectra of a freshly prepared rubrene film measured under UHV without aging (T_0) and subjected to storage aging for 840 min in UHV ($T_{840,UHV}$). The different aging conditions hereafter are referred to as $T_{time,\ condition}$. The pristine rubrene ($C_{42}H_{28}$) film shows only a single C peak at ~284 eV (and no

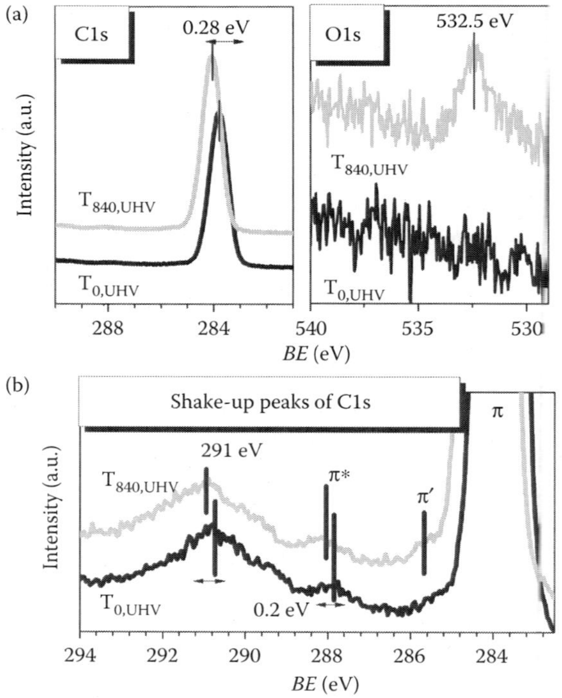

FIGURE 22.5 XPS core-level spectra of (a) C1s and O1s of rubrene film prepared on UV–ITO substrate at $T_{0,UHV}$ and $T_{840,UHV}$. (b) Enlarged shake-up peaks of C1s at $T_{0,UHV}$ and $T_{840,UHV}$.

O1s peak) at initial measurement. However, after 840 min of aging in UHV, the C peak of rubrene obviously shifts toward a higher binding energy (BE) by 0.28 eV. More importantly, a small O peak at ~532.5 eV emerges. These phenomena suggest that there is a possible oxidation of rubrene, resulting in the formation of a rubrene$^{\delta+}$–O_x^{δ} complex.

The oxidation of rubrene film may be caused by residual oxygen in the UHV analysis chamber. To rule out this possibility, the same experiment was repeated by depositing the same thickness of rubrene film on a sputter-cleaned Au substrate. The corresponding XPS results are presented in Figure 22.6. The XPS results are significantly different from those in Figure 22.5. For rubrene film prepared on an Au substrate, neither peak shift of C1s nor appearance of O1s is observed even after aging for the same time in a UHV environment. The distinct difference in chemistry as observed in Figures 22.5 and 22.6 further confirms that the oxidation of rubrene is not due to environmental oxygen in the UHV analysis chamber. On the contrary, it is attributed to the oxygen species from the UV–ITO substrate.

The chemical interaction between the UV–ITO and rubrene also changes the in-electronic structures of the rubrene film. Figure 22.5b shows a magnified portion of Figure 22.5a for a BE from 294 to 283 eV. The broad peaks found at ~291 eV are the shake-up peaks commonly observed in molecules with aromatic structures. The peaks at around 288 eV correspond to shake-up peaks corresponding to the π–π* electron transition in rubrene film. Thus, the difference of BE between the shake-up peak at ~288 eV and the main peak at ~284 eV can roughly estimate the lowest-unoccupied molecular orbital (LUMO) position of the rubrene films.[50] Upon 840 min of aging in UHV, it can be seen that both the shake-up peaks at 288 and 291 eV shift toward the higher BE by ~0.2 eV. In the meantime, a new gap-state peak, marked as π′ emerges at 285.7 eV. The emergence of this new peak is probably due to oxygen doping that generates an oxygen-induced gap state within the forbidden energy gap of rubrene film. The XPS results suggest that the UV–ITO substrate would alter the electronic structures of the overlying rubrene film even if it is aged in an UHV environment.

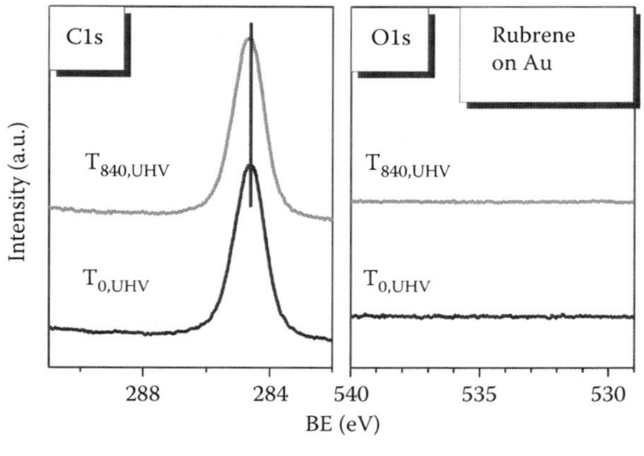

FIGURE 22.6 XPS core-level spectra of (a) C1s and O1s of rubrene film prepared on Au substrate at $T_{0,UHV}$ and $T_{840,UHV}$.

TABLE 22.2
Valence Parameters of Rubrene Prepared on Different Substrates Obtained by UPS Studies before and after 840-min Aging

	Work Function (eV)		HOMO Edge with Respect to the Fermi Level (eV)	
Devices	T_0	$T_{840,UHV}$	T_0	$T_{840,UHV}$
UV–ITO/rubrene	4.7	4.5	0.6	0.8
Au/rubrene	4.2	4.1	1.3	1.3
CHF–ITO/rubrene	4.8	4.7	0.6	0.6

22.2.3 Valence Band Structures of Rubrene Degraded on Different Substrates

To probe the subsequent impact of the oxygen-induced gap state on the valence electronic structures and charge-transporting properties of the organic film on a UV–ITO substrate, UPS was utilized to characterize the aging of rubrene films in UHV. Table 22.2 summarizes the information obtained from the UPS studies, including the measured work functions and the highest-occupied molecular orbital (HOMO) energies of rubrene prepared on various substrates. All the reported values presented are with respect to the Fermi levels (E_f), that is, $E_f = 0$.

Figure 22.7b illustrates the change in the HOMO levels of rubrene film prepared on various substrates such as *UV–ITO* and *Au* for more than 840 min. It is noted that for rubrene film prepared on the *UV–ITO* substrate, the HOMO level gradually shifts away from the Fermi level by more than 0.2 eV. However, the same films prepared on *Au* substrates show negligible change in valence electronic and chemical structures over an aging time of 840 min.

All the above photoemission results ascertain that the UV–ITO behaves as an intrinsic oxygen reservoir for the oxidation of rubrene. The surface oxygen induced during the UV–ozone treatment of ITO substrate acts as an active reactant that initiates and accelerates the degradation of the overlying organic film. The direct interaction between the surface reactive oxygen species on the ITO surface and rubrene initiates a surface charge exchange process at the ITO/rubrene junction, leading to the formation of the rubrene$^{\delta+}$–O_x^{δ} complex. This process seriously affects the valence electronic structures of rubrene film. The above observations have far-reaching implications in device degradation, suggesting that avoiding the use of UV–ozone-treated ITO or isolation of the UV zone-treated ITO substrate from organic films should be considered to enhance device stability.

22.2.4 Stability Enhancement in Rubrene-Based OPV Device

To further consolidate our photoemission spectroscopic results, rubrene-based OPV devices were fabricated on UV–ITO *(UV–ITO device)*, ITO/Au (Au–ITO device), and ITO/polymerized

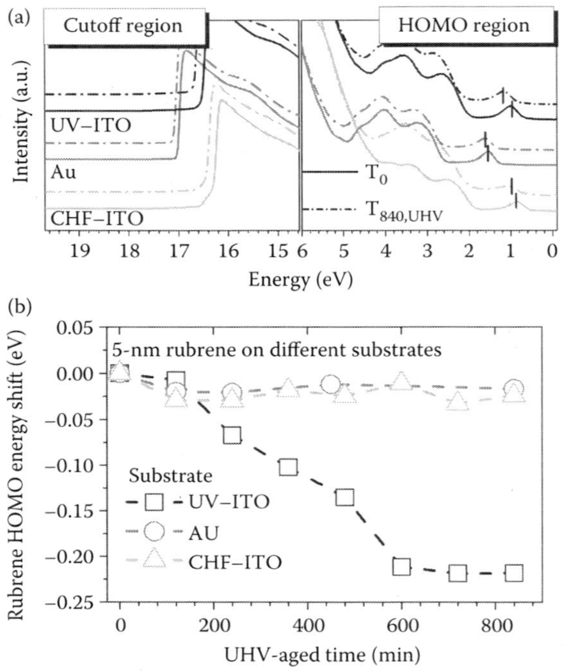

FIGURE 22.7 (a) UPS He-Iα spectra of rubrene films prepared on UV–ITO (top), Au (middle), and CHF–ITO (bottom) substrates before (solid line) and after 840-min aging (solid dotted) in UHV condition. (b) Energy shifts in the HOMO level of rubrene prepared on UV–ITO (□), Au (○), and CHF–ITO (Δ) substrates aging in UHV condition.

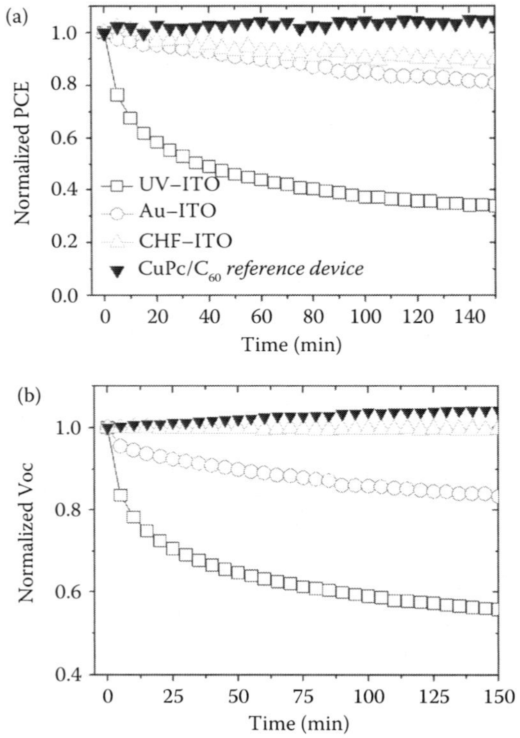

FIGURE 22.8 Photovoltaic responses showing the normalized (a) PCE and (b) Voc of an encapsulated *UV–ITO device* (□), *Au–ITO device* (○), *CHF–ITO device* (Δ), and *CuPc/*C_{60} *reference device* (▼) during the 150-min continuous 1 Sun AM1.5G illumination.

fluorocarbon trifluoromethane (CHF_3) *(CHF–ITO device)* substrates. Deposition of the CHF layer was carried out in a preparation chamber connected to the organic deposition chamber by plasma polymerization of CHF_3 gas. The device configurations are anode substrate/rubrene (35 nm)/fullerene (C_{60}) (45 nm)/bathocuproine (BCP) (5 nm)/Al. After encapsulation, devices were illuminated with 100 mW/cm^2 solar-simulated irradiation continuously for 150 min.

Figure 22.8 compares degradation rates of the three devices. The PCE and open-circuit voltage (Voc) are normalized to their corresponding initial values for better comparison while their absolute values are summarized in Table 22.3. The performance of a *CuPc/*C_{60} *reference device* (▼) with a configuration of UV–ITO/CuPc (35 nm)/C_{60} (45 nm)/BCP (5 nm)/Al was also included for comparison. The *reference device* shows a negligible change in performance, that is, no degradation over the testing period. This observation suggests that the current encapsulation scheme does provide enough protection to the device against degradation caused by the atmospheric surroundings.[13,20]

Consistent with the photoemission results, the PCE and Voc of the *UV–ITO device* rapidly dropped to 40% and 50% of their initial values, respectively after 150 min of continuous illumination. It is noteworthy that with an effective isolation of the ITO interface, much less losses in PCE were observed in both the *Au–ITO* and the *CHF–ITO devices.* More importantly, a very sustainable high Voc of 0.92 V is maintained in the *CHF–ITO device* throughout the prolonged degradation test. The device parameters during a continuous illumination test are illustrated in Table 22.3. These OPV device performances show a concrete evidence of device degradation if a UV–ITO substrate is used for device fabrication. Upon the ITO-induced aging, carrier trap states are induced by oxygen near the HOMO of the rubrene, as shown in Figure 22.5. These new trap states behave as recombination centers for photoinduced excitons, which degrade the charge generation at the rubrene/C_{60} interface, thus resulting in a deteriorated photovoltaic response.

It should be pointed out that the thickness of the CuPc and C_{60} in the *CHF–ITO device* as shown in Figure 22.8 has not been fully optimized. The experimental results for optimized devices are shown in Figure 22.9 and Table 22.4. A similar stability enhancement was observed for the optimized devices after 150 min of continuous illumination. More importantly, the PCE of the optimized *CHF–ITO device* can reach as high as 2.9% for CuPc and C_{60} with a thickness of 45 nm respectively. This performance is much higher than those reported recently in standard CuPc/C_{60} devices.[51–53]

22.2.5 Surface Modification of ITO Substrate upon CHF Plasma Treatment

The above device results suggest that surface modification of the UV–ITO surface such as CHF treatment is important for stability enhancement before the deposition of organic films.

TABLE 22.3
Photoresponse of OPV Devices with Rubrene (35 nm)/C_{60} (45 nm)/BCP (5 nm)/Al Deposited on (i) UV–ITO (UV–ITO Device), (ii) ITO/Au (5 nm) (Au–ITO Device), and (iii) ITO/CHF (1.5 nm) (CHF–ITO Devices) Substrates before and after 150-min Continuous Illumination

	Voc (V)		*Jsc* (mA/cm²)		*FF*		*PCE* (%)	
Devices	T_0	$T_{150,1\ sun}$	T_0	$T_{150,1\ sun}$	T_0	$T_{150,1\ sun}$	T_0	$T_{150,1\ sun}$
i. UV–ITO device	0.88	0.49	3.6	2.8	0.46	0.35	1.4	0.5
ii. Au–ITO device	0.76	0.63	0.8	0.8	0.35	0.34	0.2	0.2
iii. CHF–ITO device	0.92	0.92	4.2	3.9	0.6	0.56	2.3	2.0
iv. CuPc/C_{60} reference device	0.46	0.48	5.9	5.8	0.55	0.57	1.5	1.6

Note: The photoresponse of the reference device with CuPc (35 nm)/C_{60} (45 nm)/BCP (5 nm)/Al deposited on UV–ITO substrate *(CuPc/C_{60} reference device)* is also included.

Here, we examine the difference in oxygen species on the UV–ITO and CHF–ITO substrates by XPS. Figure 22.10a is the XPS spectra comparing the O1s core level of untreated ITO, UV–ITO, and CHF–ITO. To better illustrate the difference of UV–ozone and CHF plasma treatments, the normalized spectrum of UV–ITO minus that of the untreated ITO and the spectrum of CHF–ITO minus that of the UV–ITO are shown in Figure 22.10b.

It can be seen that the main difference of peaks between the above two cases are located at around the BE of 532.2 and 530 eV, which are attributed to O–H bond and O-radicals, respectively.[38] It can be seen that upon UV–ozone treatment, the intensity is diminished at 532.2 eV and increased at 530 eV. This difference is attributed to the dehydroxylation of Sn^{IV}–OH groups and the generation of "stannoxyl" Sn^{IV}–O$^•$ surface components (i.e., Sn^{IV}–OH → Sn^{IV}–O$^•$).[38] In contrast, a further CHF plasma treatment leads to a reverse effect. The peak at 530 eV corresponds to the –OH group and a peak depletion at 532.2 eV is related to –O$^•$, implying that there is a hydroxylation process of the reactive –O$^•$ component during CHF plasma (i.e., Sn^{IV}–O$^•$ → Sn^{IV}–OH). The hydrogen atom in CHF_3 gas readily reacts with Sn^{IV}–O$^•$ on the ITO surface.

Our results here clearly show that the introduction of a CHF anode buffer layer can eventually prevent the possible

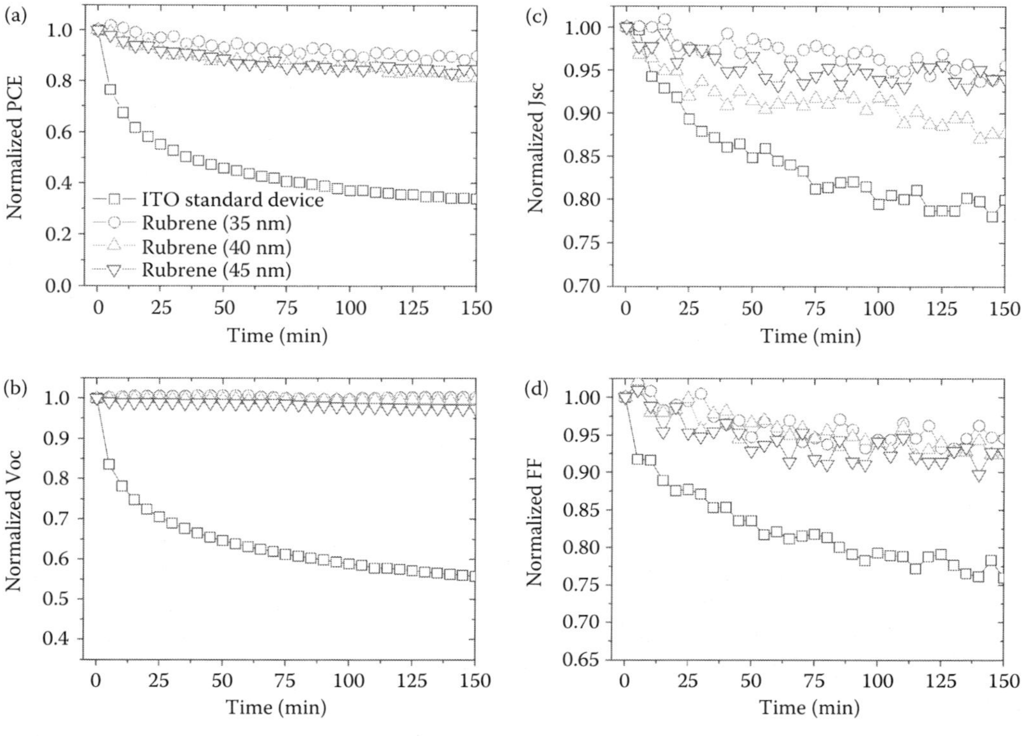

FIGURE 22.9 Photovoltaic responses showing the normalized (a) PCE, (b) Voc, (c) Jsc, and (d) FF of encapsulated rubrene-based OPV devices prepared on CHF–ITO substrate with different rubrene thickness during the 150-min continuous illumination with light intensity of 100 mW/cm². The device configuration is CHF–ITO/rubrene (x nm)/C_{60} (45 nm)/BCP (5 nm)/Al.

TABLE 22.4
Photoresponses of OPV Devices with Different Rubrene Thickness Prepared on CHF–ITO Substrates before and after 150-min Continuous Illumination

	Voc (V)		*Jsc* (mA/cm²)		*FF*		*PCE* (%)	
Devices Structure	T_0	$T_{150,1\ sun}$	T_0	$T_{150,1\ sun}$	T_0	$T_{150,1\ sun}$	T_0	$T_{150,1\ sun}$
Rubrene (35 nm)/C_{60} (45 nm)	0.92	0.92	4.2	3.9	0.60	0.56	2.3	2.0
Rubrene (40 nm)/C_{60} (45 nm)	0.93	0.93	4.7	4.1	0.59	0.55	2.6	2.1
Rubrene (45 nm)/C_{60} (45 nm)	0.95	0.93	4.9	4.6	0.62	0.57	2.9	2.46

Note: The device configuration is CHF–ITO/rubrene (x nm)/C_{60} (45 nm)/BCP (5 nm)/Al.

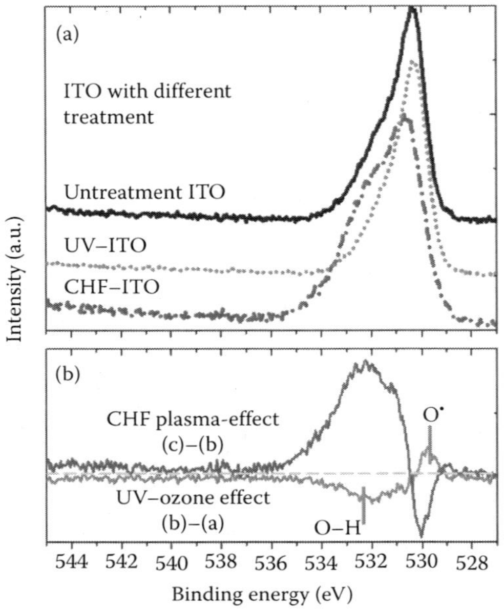

FIGURE 22.10 (a) XPS spectra showing the normalized O1s core peak of a pristine ITO, UV–ozone-treated ITO, and CHF_3 plasma-treated ITO. (b) Differences of O1s spectrum highlighting the effect of UV, ozone, and CHF_3 plasma treatment.

interaction of the –O• components with the organic active materials in the OPV devices. After all, both the rubrene and C_{60} are oxygen sensitive and unstable materials in the device operation, and therefore, the existence of oxygen will greatly diminish the OPV performance. As shown in our XPS data above in Figure 22.5a, the oxygen-induced gap states have been identified in rubrene during a prolonged aging on a UV–ITO surface.[54] The gap states act as a recombination center that would limit the lifetime of charge carriers. The trapped charge carriers would eventually recombine within the organic active layers, leading to charge quenching and deterioration in the overall device performance, for instance, low short-circuit current density (Jsc), shunt resistance (R_{shunt}), and small fill factor (FF).[6] Therefore, elimination of any side interactions due to the substrate and environment are important to enhance OPV device stability. Our presented results clearly showed that the introduction of CHF anode buffer layer can eventually prevent the possible interaction of the –O• components with the organic active materials in OPV devices.

In summary, while UV–ITOs have been widely used in organic electronic devices, here, we showed that they can cause PL degradation of the common organic semiconductors such as NPB, Alq_3, and rubrene. The degradation was further accelerated with illumination. UPS and XPS studies show that the reactive surface oxygen species generated upon UV–ozone treatment are the major cause for the degradation of rubrene film deposited on ITO substrates. The reactive oxygen species on the UV–ITO induce a gap state in the energy gap of rubrene and shift its HOMO away from the Fermi level. The observations were further supported by the performance data of rubrene-based OPV devices. This chapter suggests that the commonly used ITO-coated glass with UV–ozone treatment has far-reaching implications to device stability.

22.3 OPV DEVICES STABILITY ENHANCEMENT USING GO AS ITO ANODE BUFFER LAYER

In the previous section, we demonstrated the stability enhancement of OPV devices by using ITO substrates with a polymeric CHF buffer layer prior to organic film deposition.[17] However, the CHF treatment based on the plasma-enhanced chemical vapor deposition (CVD) implement is rather impractical for large-scale manufacturing. More importantly, the use of CVD would inevitably increase the fabrication cost of the OPV device. A simple and effective alternate approach is thus highly desirable.

Graphene is recently a hot topic due to its high electrical and thermal conductivities, transparency, and flexibility.[55–58] With controllable oxidation, it might be used as a high work function buffer layer that can modify the contact surface of the ITO/organic interface. In this section, we explore the possibility of GO as an anode buffer layer that modifies the surface of the UV–ITO surface. GO was first synthesized in the nineteenth century, and it could be fabricated using the Brodie,[59] Staudenmaier,[60] or Hummers' methods.[61] These methods presented the same starting route by oxidizing graphite by using strong acids and oxidants. In this chapter, the GO is prepared based on Hummers' method as shown in Figure 22.11.[61,62]

Graphite powders are first oxidized using potassium permanganate ($KMnO_4$), sulfuric acid (H_2SO_4), and hydrogen

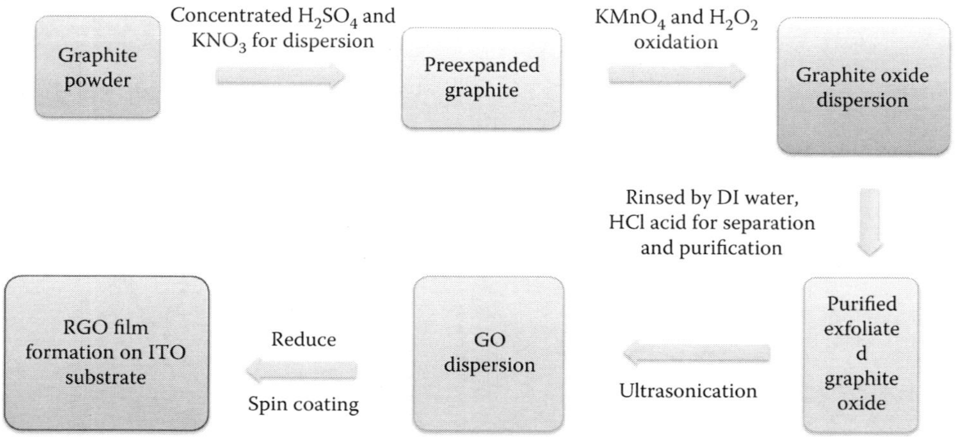

FIGURE 22.11 Schematic diagram showing the synthesis process of GO sheets.

peroxide (H_2O_2), etc. The obtained products are thoroughly washed using hydrochloric acid (HCl) and deionized (DI) water until its pH reaches neutral. After removing all the possible residuals, the forming products are then freeze dried and are dispersed into a water:ethanol solution with 1:1 ratio. The solution is then taken for ultrasonic treatment for 25 min until a clear GO suspension is obtained with a concentration of 1.0 mg/mL. Figure 22.12a schematically describes the chemical structure of the final GO products. The GO products obtained should contain ^{13}C-labeled GO as characterized

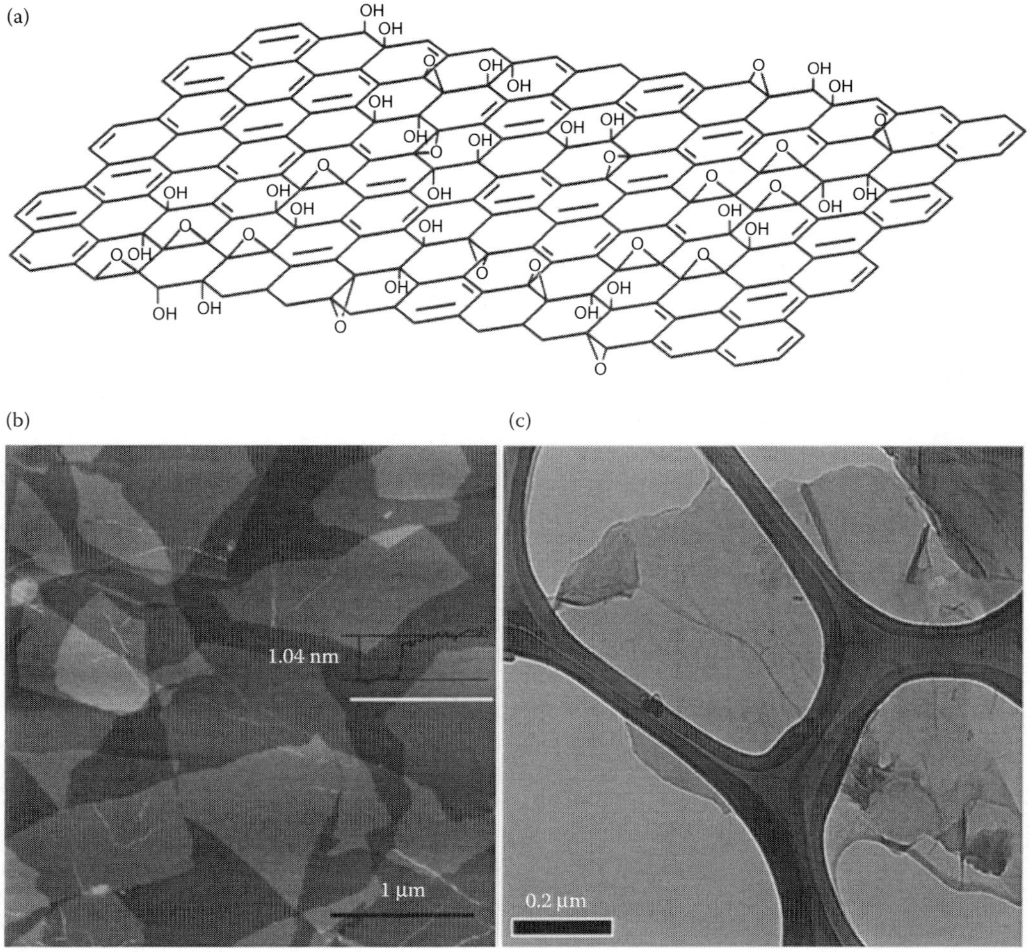

FIGURE 22.12 (a) Chemical structure of GO, (b) tapping-mode AFM image of GO sheets with a height profile (black curve), and (c) TEM image of GO sheets.

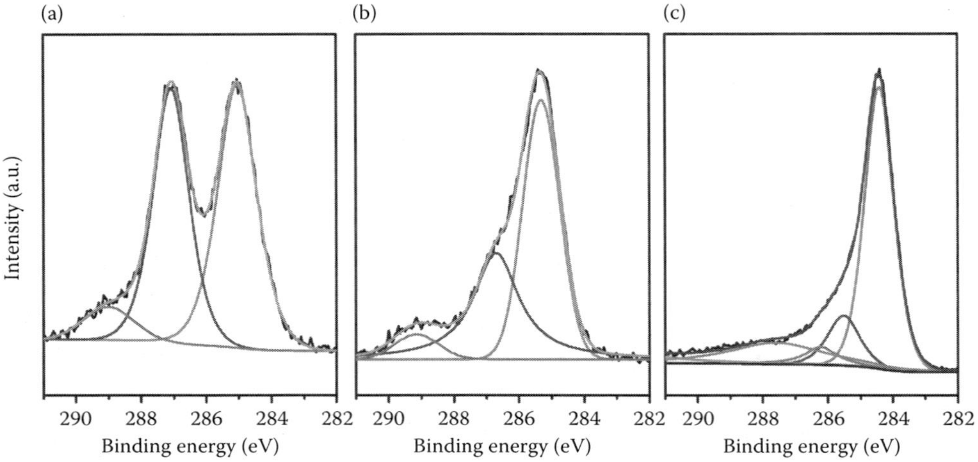

FIGURE 22.13 XPS C1s core-level spectra of (a) GO, (b) hydrazine hydrate-reduced GO, and (c) hydrazine hydrate-reduced GO after thermal annealing.

previously by Cai et al.[63] using solid-state nuclear magnetic resonance (SSNMR). After oxidation, the sp^2-bonded carbon networks are strongly distorted by connecting with these polar oxygen functional groups such as hydroxyl, epoxide, carboxylic, and carbonyl groups. The chemical structure of graphene obtained by such a chemical process is considerably different compared to those fabricated using the CVD method. Figure 22.12b shows the atomic force microscope (AFM) image of the experimentally formed GO sheet. The height profile of the graphene sheet edge is also included as an inset. Figure 22.12c is the transmission electron microscope (TEM) image of the exfoliated GO sheets.

GO prepared by this method is electrically insulated due to the distorted sp^2-bonding networks. Fortunately, the electrical conductivities of the forming GO can be recovered by restoring the π-networks. The obtained GO then underwent a reduction process by either (1) chemical, or (2) electrochemical reduction or thermal treatment. The final products were then used for a spin-coated thin film on ITO substrates. Figure 22.13 compares the XPS C1s core-level spectra of (a) as-prepared GO, hydrazine hydrate-reduced GO before (b), and after (c) thermal annealing.

22.3.1 Device Performance Using UV–ITO and GO-Buffered Substrate

In this section, the as-prepared GO solutions were spin coated to form a thin anode buffer film on top of UV–ITO substrates. The performances of standard CuPc/C_{60} OPV devices forming on various substrates were examined. The devices configurations are ITO/GO (0 or 2 nm)/CuPc (35 nm)/C_{60} (40 nm)/bathophenanthroline (BPhen) (10 nm)/Al. Figure 22.14 shows the absorption spectra of 35-nm CuPc films deposited on a UV–ITO and GO-buffered ITO (GO-ITO) substrates, respectively. It can be seen that both the absorption peaks at ~620 and 690 nm are enhanced when the CuPc film is deposited on the GO-ITO substrate compared to that on the UV–ITO substrate.

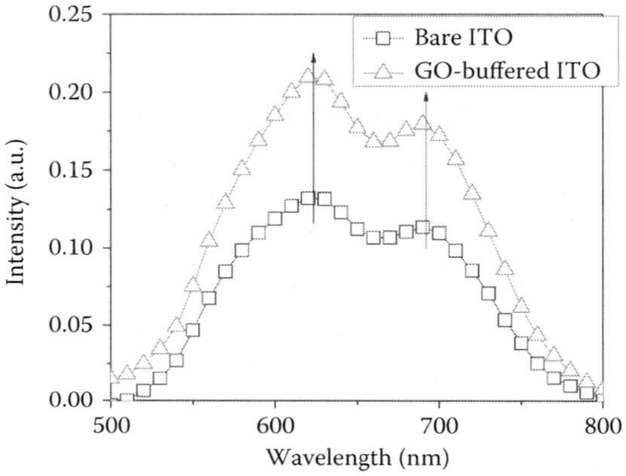

FIGURE 22.14 UV–Vis absorption spectra of 35-nm CuPc films evaporated on (square) ITO glass substrate and (triangle) on GO-buffered substrate.

Current density–voltage (J–V) characteristic of the OPV devices prepared on the UV–ITO and GO-ITO substrates is shown in Figure 22.15. The standard UV–ITO device shows an initial Voc, Jsc, FF, and PCE of 0.48 V, 5.9 mA/cm, and 0.54% and 1.5%, respectively. These values are comparable with those recently reported for standard CuPc/C_{60} OPV devices.[52,64,65]

With a 2-nm-thick GO buffer layer added on top of the UV–ITO substrate (Δ in Figure 22.15), the photovoltaic responses of the device show much improvement with Voc, Jsc, FF, and PCE that reach 0.52 V, 7.3 mA/cm, and 0.51% and 1.9%, respectively. However, the photovoltaic performance decreases with the increasing thickness of the GO. The PCE of the 8-nm-thick GO-buffered device decreased to 1.0%. As shown in Table 22.5, the decrease in PCE is mainly attributed to the decrease in Jsc and FF. The observed performance deteriorating with increasing GO buffer layer thickness could be easily understood because of the moderate conductivity of GO films.[66] Here, the optimized thickness for the GO anode buffer layer is 2 nm.

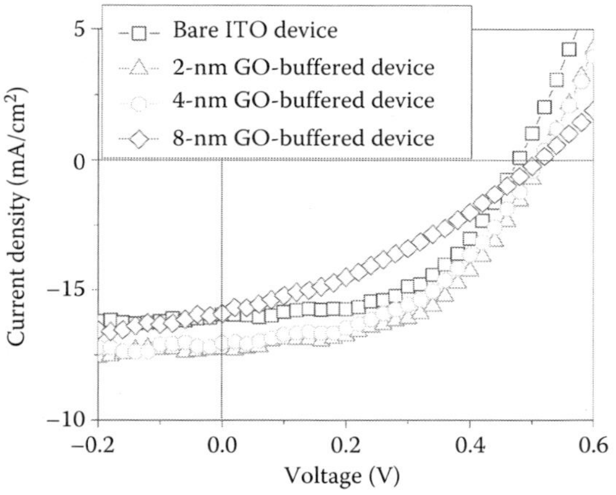

FIGURE 22.15 J–V characteristics of an encapsulated ITO device and GO-buffered devices with different GO thickness of 2, 4, and 8 nm, respectively.

TABLE 22.5
Photovoltaic Performances of ITO Device and GO-Buffered Devices with Different GO Thicknesses

Devices	Voc (V)	Jsc (mA/cm^2)	FF	PCE (%)
ITO device	0.48	5.9	0.54	1.53
2-nm GO-buffered device	0.52	7.3	0.51	1.91
4-nm GO-buffered device	0.51	7.03	0.48	1.71
6-nm GO-buffered device	0.51	5.91	0.33	1.01

To investigate the effects of the GO buffer layer, the physical morphologies and crystallinity of the CuPc films prepared on UV–ITO and GO-ITO substrates were examined using AFM and x-ray diffraction (XRD) analysis. Figure 22.16 shows the physical morphology of the 5-nm-thick CuPc films deposited on UV–ITO and GO-ITO substrates. A significant difference in the grain size of CuPc film is observed. The CuPc film prepared on the GO-ITO substrate has a much larger grain size of ~80 nm, while those prepared on the UV–ITO substrate only have ~30 nm. The larger CuPc grain size on the GO buffer layer is definitely beneficial to charge transport since it contains less grain boundaries. The formation of grain boundaries in organic films is well recognized as a trapping site for charge carrier transport. It will also cause charge carrier scattering.[67–69] It is noteworthy that according to the previous report on organic field-effect transistor (OFET) devices, the increase of CuPc grain size may result in hole mobility enhancement.[70] Thus, it is believed that the increased grain size of CuPc observed in Figure 22.16 leads to an increase of charge mobility when it is deposited on the GO-ITO substrate.

Other than physical morphologies, crystallinity also directly influences the optical properties of thin films,[71] for instance, the annealed poly (3-hexylthiophene) (P3HT) has been reported to have better absorption due to better crystallinity.[72] To study the crystallinity of CuPc films deposited on UV–ITO and GO-ITO substrates, Figure 22.17 shows XRD measurements for a 35-nm-thick CuPc thin film. The prominent peak located at $2\theta = 6.9°$ matches

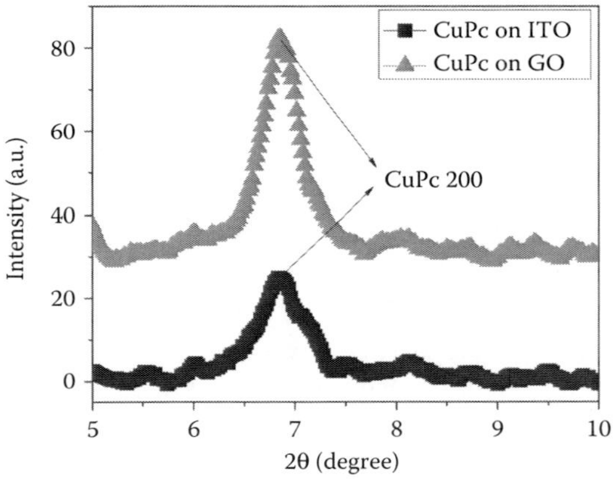

FIGURE 22.17 XRD pattern of 35-nm-thick CuPc films grown on (square) ITO glass and (triangle) GO-buffered substrates.

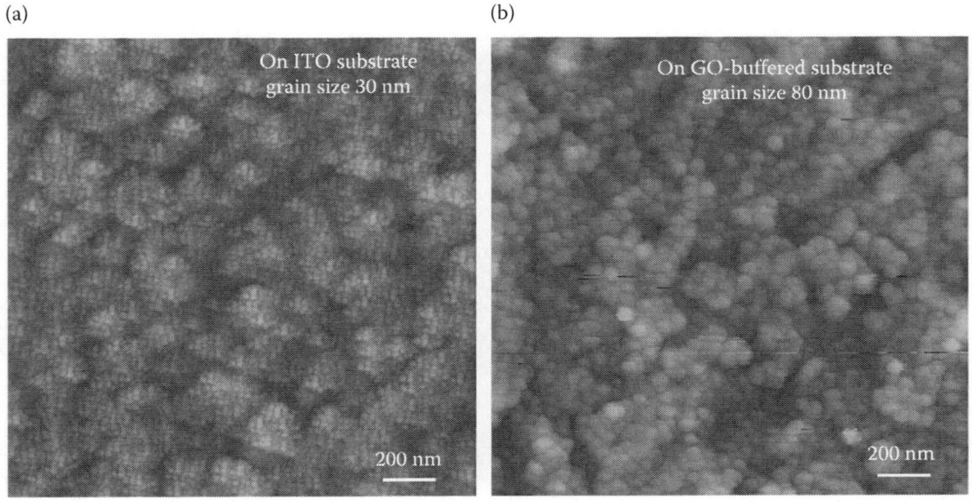

FIGURE 22.16 AFM images of 5-nm-thick CuPc films grown on (a) UV–ITO glass and (b) 2-nm GO-buffered substrates.

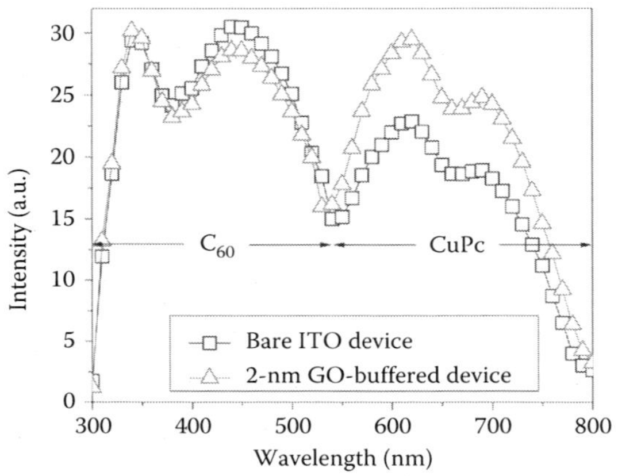

FIGURE 22.18 EQE spectra of OPV devices on (square) ITO glass substrate and (triangle) 2-nm GO-buffered substrate.

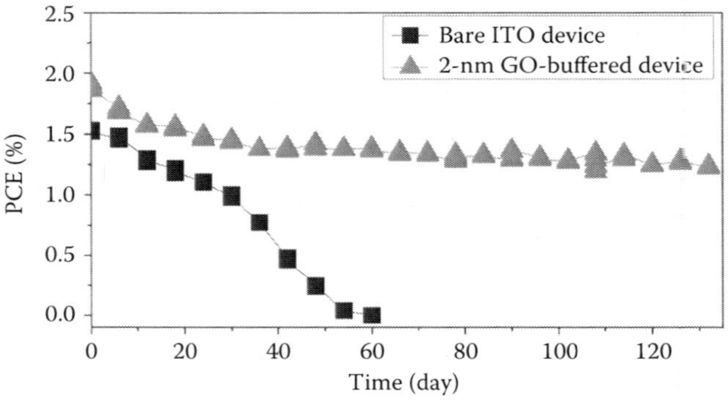

FIGURE 22.19 Summarized PCE response of an encapsulated (square) ITO device and (triangle) GO-buffered device at each circle measurement.

well to the (200) peak of α-CuPc.[67] For CuPc film prepared on the GO-ITO substrate, there is a significant increase in diffraction peak height as well as narrowing of full-width-half-maximum (FWHM). This result suggests a higher crystallinity of the CuPc film.[73]

Figure 22.18 shows external quantum efficiencies (EQEs) of the devices prepared on UV–ITO and GO-ITO substrates. While the EQEs of the UV–ITO device (■) at wavelengths of 620 and 690 nm are 23% and 18%, respectively, those of GO-ITO device (○) at the same wavelengths are 30% and 25%, respectively. The EQE enhancement in the GO-ITO device observed in Figure 22.18 is consistent with the enhanced CuPc absorption shown in Figure 22.14. It is also noted that the EQE in the wavelength region from 300 to 500 nm, which corresponds to C_{60} absorption, indeed remains unchanged. Therefore, the observed performance enhancement presented in Table 22.5 is mainly due to the difference in CuPc-forming properties as presented above.

To further investigate the effects of using the GO anode buffer layer, the storage/operation stability of the CuPc/C_{60} devices fabricated on UV–ITO and GO-ITO substrates were also investigated. The freshly formed devices were encapsulated in a glove box immediately after fabrication. After the initial J–V measurement, the devices were continuously illuminated for 1 h using an Oriel 150-W solar simulator with AM 1.5G filters at 100 mW/cm and then stored in the dark under standard laboratory conditions. After 6 days of storage, the J–V characteristics of the devices were measured again, and then followed by 1 h of continuous illumination as before. The "J–V measurement + 1 hr illumination + 6 days storage" cycle was repeated until the PCE of the device fell below 1/1000 of its original value.

Figure 22.19 summarizes PCEs of the two devices over prolonged storage–operation cycles. From the figure, it can be seen that the PCE of the standard UV–ITO device rapidly falls below 1/1000 of the initial value after 60 days of storage and operation. On the contrary, the GO-ITO device shows a much better long-term stability. Its PCE decreases by only 16% after 132 days of storage–operation. Such a remarkable difference suggests that the GO buffer layer can efficaciously enhance OPV device stability.

To understand the origin of degradation, EQEs of the two devices were measured after different durations of operation–storage, and the results are shown in Figure 22.20. After 30 days of aging, the EQEs that correspond to C_{60} (~300–550 nm) and CuPc (~550–750 nm) decrease for those prepared on the UV–ITO device. One can see that the decay in EQE is much slower when the CuPc/C_{60} device is fabricated on the GO-ITO substrate. In particular, the EQEs related to CuPc show a negligible decrease after 132 days of storage–operations. While the EQEs of C_{60} show a mild decay in the first 30 days, further degradation from the 30th to the 132nd day is negligible.

In contrast to CuPc, C_{60} is highly sensitive to an oxygen environment; p-type doping effect and reduction in conductivity have been observed in previous studies, leading to a decrease in Jsc and FF.[44] More importantly, some reactive oxygen species that generated after the UV–ozone treatment of the ITO substrate might diffuse from the ITO into the active layer.[17] All these observations suggest the different degradation behaviors of CuPc and C_{60} films in OPV devices. For the EQE spectrum as shown in Figure 22.20, we successfully demonstrate that the use of the GO buffer layer effectively prolongs the lifetime of the OPV device mainly by improving the stability of CuPc.

The J–V characteristics of the two devices are shown in Figure 22.21. It can be seen that while the J–V characteristic of the standard UV–ITO device decays continuously, the J–V characteristic of the GO-ITO device shows little changes after the first 30 days of aging. In fact, the PCE of the GO-ITO device still maintains 78% of its original value after 205 days of storage–operations.

The two devices studied here have the identical structure and encapsulation except for the GO buffer layer. The efficiency and stability enhancement should therefore be mainly caused by the GO anode buffer layer. As shown from the AFM and XRD results, the CuPc layer grown on the GO layer has a much larger crystal size and better crystallinity. It results in enhancement in the optical absorption of the CuPc film and subsequently contributes to the EQE of the OPV device.

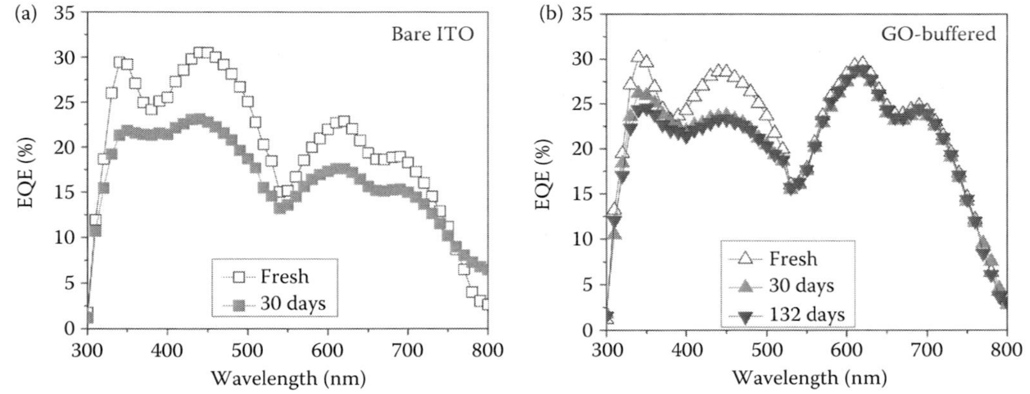

FIGURE 22.20 EQE spectra of (a) ITO device before (empty) and after (solid) 30-days storage; and (b) GO-buffered device before (empty) and after 30- (up triangle) and 132-days (down triangle) storage, respectively.

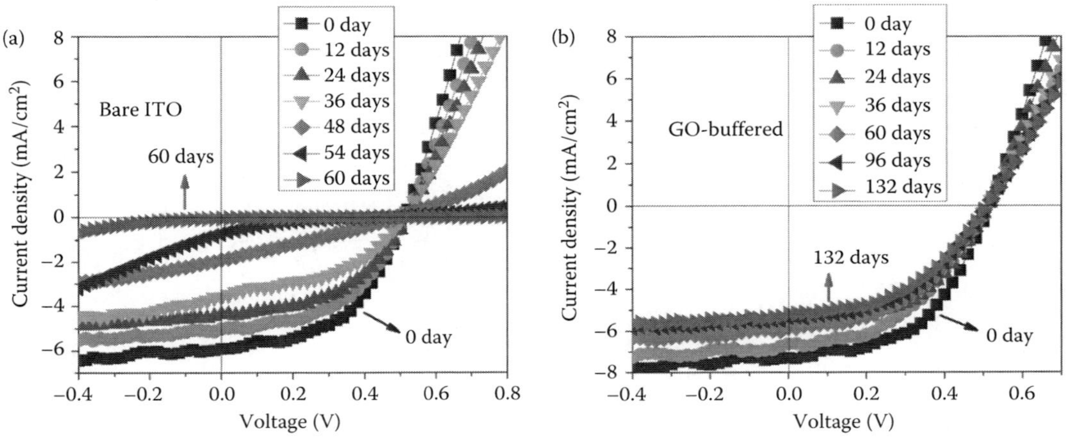

FIGURE 22.21 J–V characteristics of an encapsulated (a) ITO device at 0-, 12-, 24-, 36-, 48-, 54-, and 60-days storage and measurements; and (b) GO-buffered device at 0-, 12-, 24-, 36-, 60-, 96-, and 132-days storage and measurements.

It is believed that the GO layer used here acts as a diffusion barrier isolating the photoactive layers from the ITO substrate. On the other hand, the larger grain size and higher crystallite quality of CuPc may also render it less vulnerable to the oxygen species. While the stability enhancement effect of the GO layer is obvious and significant, detailed enhancement mechanisms are not entirely clear. In particular, the negligible decay in the CuPc layer and the much-decelerated decay of C_{60} after the first 30 days are interesting but not fully understood. It requires further systematic investigations.

It is well recognized that operation stability is an important factor that has to be addressed before OPV devices can be commercialized. This chapter demonstrates a substantial enhancement in the stability of OPV device by simply spin coating a thin GO layer on the ITO substrate without any further treatment. It is expected that this simple, low-cost, and solution-processed process should be useful and contribute towards the future commercialization of OPV devices.

In the second part of this chapter, we showed that oxygen species from the ITO substrate can cause degradation of the organic semiconductors. It is believed that the GO layer used here plays a role of the diffusion barrier that prevents the direct contact of the photoactive layers with the underlying UV–ITO substrate. On the other hand, the larger grain size and higher crystallite quality of CuPc may also render it less vulnerable to the oxygen species. While the stability enhancement effect of the GO layer is obvious and significant, detailed enhancement mechanisms are not entirely clear. In particular, the negligible decay in the CuPc layer and the much-decelerated decay of C_{60} after the first 30 days are interesting and not fully understood and require further systematic investigations.

22.4 CONCLUSIONS

It is well recognized that operational stability is an important factor that has to be addressed before the commercialization of OPV devices application. This chapter demonstrates a substantial enhancement in the stabilities of OPV devices by simply spin coating a thin GO layer on the ITO substrate without any further treatment. It is expected that this simple, low-cost, and solution-oriented process should be useful and contribute toward the realistic application of OPV devices.

ACKNOWLEDGMENTS

The work described in this chapter was supported by the Research Grants Council of the Hong Kong Special Administrative Region, China (Project No. T23-713/11); and the National Natural Science Foundation of China (No. 21303150).

REFERENCES

1. C. W. Tang, *Appl Phys Lett* 1986, *48*, 183.
2. P. Peumans, S. R. Forrest, *Appl Phys Lett* 2001, *79*, 126.
3. M. Y. Chan, S. L. Lai, M. K. Fung, C. S. Lee, S. T. Lee, *Appl Phys Lett* 2007, *90*, 203504.
4. M. F. Lo, T. W. Ng, T. Z. Liu, V. A. L. Roy, S. L. Lai, M. K. Fung, C. S. Lee, S. T. Lee, *Appl Phys Lett* 2010, *96*, 113303.
5. H. Y. Chen, J. H. Hou, S. Q. Zhang, Y. Y. Liang, G. W. Yang, Y. Yang, L. P. Yu, Y. Wu, G. Li, *Nat Photon* 2009, *3*, 649.
6. A. Moliton, J. M. Nunzi, *Polym Int* 2006, *55*, 583.
7. T. W. Ng, M. F. Lo, M. K. Fung, S. L. Lai, Z. T. Liu, C. S. Lee, S. T. Lee, *Appl Phys Lett* 2009, *95*, 203303.
8. Y. Y. Liang, D. Q. Feng, Y. Wu, S. T. Tsai, G. Li, C. Ray, L. P. Yu, *J Am Chem Soc* 2009, *131*, 7792.
9. Y. Y. Liang, Z. Xu, J. B. Xia, S. T. Tsai, Y. Wu, G. Li, C. Ray, L. P. Yu, *Adv Mater* 2010, *22*, E135.
10. R. F. Service, *Science* 2011, *332*, 293.
11. Q. Q. Gan, F. J. Bartoli, Z. H. Kafafi, *Adv Mater* 2013, *25*, 2385.
12. H. R. Wu, Q. L. Song, M. L. Wang, F. Y. Li, H. Yang, Y. Wu, C. H. Huang, X. M. Ding, X. Y. Hou, *Thin Solid Films* 2007, *515*, 8050.
13. M. F. Lo, T. W. Ng, S. L. Lai, F. L. Wong, M. K. Fung, S. T. Lee, C. S. Lee, *Appl Phys Lett* 2010, *97*, 143304.
14. M. T. Lloyd, D. C. Olson, P. Lu, E. Fang, D. L. Moore, M. S. White, M. O. Reese, D. S. Ginley, J. W. P. Hsu, *J Mater Chem* 2009, *19*, 7638.
15. S. A. Gevorgyan, M. Jorgensen, F. C. Krebs, *Sol Energ Mat Sol C* 2008, *92*, 736.
16. F. C. Krebs, J. E. Carle, N. Cruys-Bagger, M. Andersen, M. R. Lilliedal, M. A. Hammond, S. Hvidt, *Sol Energ Mat Sol C* 2005, *86*, 499.
17. M. F. Lo, T. W. Ng, S. L. Lai, M. K. Fung, S. T. Lee, C. S. Lee, *Appl Phys Lett* 2011, *99*, 033302.
18. M. F. Lo, T. W. Ng, H. W. Mo, C. S. Lee, *Adv Funct Mater* 2013, *23*, 1718.
19. D. Gao, M. G. Helander, Z. B. Wang, D. P. Puzzo, M. T. Greiner, Z. H. Lu, *Adv Mater* 2010, *22*, 5404.
20. M. Jorgensen, K. Norrman, F. C. Krebs, *Sol Energ Mat Sol C* 2008, *92*, 686.
21. J. C. Scott, J. H. Kaufman, P. J. Brock, R. DiPietro, J. Salem, J. A. Goitia, *J Appl Phys* 1996, *79*, 2745.
22. M. P. de Jong, L. J. van IJzendoorn, M. J. A. de Voigt, *Appl Phys Lett* 2000, *77*, 2255.
23. Z. L. Shen, P. E. Burrows, V. Bulovic, S. R. Forrest, M. E. Thompson, *Science* 1997, *276*, 2009.
24. Y. Kuwabara, H. Ogawa, H. Inada, N. Noma, Y. Shirota, *Adv Mater* 1994, *6*, 677.
25. Y. Shirota, *J Mater Chem* 2000, *10*, 1.
26. H. J. Kim, J. W. Kim, H. H. Lee, B. Lee, J. J. Kim, *Adv Funct Mater* 2012, *22*, 4244.
27. T. W. Ng, M. F. Lo, Q. D. Yang, M. K. Fung, C. S. Lee, *Adv Funct Mater* 2012, *22*, 3035.
28. T. W. Ng, M. F. Lo, S. T. Lee, C. S. Lee, *Org Electron* 2012, *13*, 1641.
29. S. T. Lee, Z. Q. Gao, L. S. Hung, *Appl Phys Lett* 1999, *75*, 1404.
30. Y. Qiu, Y. D. Gao, L. D. Wang, D. Q. Zhang, *Synth Met* 2002, *130*, 235.
31. C. C. Wu, C. I. Wu, J. C. Sturm, A. Kahn, *Appl Phys Lett* 1997, *70*, 1348.
32. H. Aziz, Z. D. Popovic, *Chem Mater* 2004, *16*, 4522.
33. F. So, D. Kondakov, *Adv Mater* 2010, *22*, 3762.
34. L. S. Hung, C. H. Chen, *Mat Sci Eng R* 2002, *39*, 143.
35. M. Jorgensen, K. Norrman, S. A. Gevorgyan, T. Tromholt, B. Andreasen, F. C. Krebs, *Adv Mater* 2012, *24*, 580.
36. J. X. Tang, Y. Q. Li, X. Dong, S. D. Wang, C. S. Lee, L. S. Hung, S. T. Lee, *Appl Surf Sci* 2004, *239*, 117.
37. H. T. Lu, M. Yokoyama, *J Cryst Growth* 2004, *260*, 186.
38. D. J. Milliron, I. G. Hill, C. Shen, A. Kahn, J. Schwartz, *J Appl Phys* 2000, *87*, 572.
39. J. M. Moon, J. H. Bae, J. A. Jeong, S. W. Jeong, N. J. Park, H. K. Kim, *Appl Phys Lett* 2007, *90*, 163516.
40. M. G. Helander, Z. B. Wang, J. Qiu, M. T. Greiner, D. P. Puzzo, Z. W. Liu, Z. H. Lu, *Science* 2011, *332*, 944.
41. J. Olivier, B. Servet, M. Vergnolle, M. Mosca, G. Garry, *Synth Met* 2001, *122*, 87.
42. Z. H. Huang, X. T. Zeng, X. Y. Sun, E. T. Kang, J. Y. H. Fuh, L. Lu, *Org Electron* 2008, *9*, 51.
43. H. Aziz, Z. D. Popovic, N. X. Hu, A. M. Hor, G. Xu, *Science* 1999, *283*, 1900.
44. T. W. Ng, M. F. Lo, Y. C. Zhou, Z. T. Liu, C. S. Lee, O. Kwon, S. T. Lee, *Appl Phys Lett* 2009, *94*, 193304.
45. Y. Tanaka, K. Kanai, Y. Ouchi, K. Seki, *Chem Phys Lett* 2007, *441*, 63.
46. J. Y. Fan, J. M. Michalik, L. Casado, S. Roddaro, M. R. Ibarra, J. M. De Teresa, *Solid State Commun* 2011, *151*, 1574.
47. D. Teweldebrhan, A. A. Balandin, *Appl Phys Lett* 2009, *94*, 013101.
48. M. S. Xu, D. Fujita, N. Hanagata, *Nanotechnology* 2010, *21*, 265705, 131911.
49. M. Kytka, A. Gerlach, F. Schreiber, J. Kovac, *Appl Phys Lett* 2007, *90*, 131911.
50. L. S. Liao, L. F. Cheng, M. K. Fung, C. S. Lee, S. T. Lee, M. Inbasekaran, E. P. Woo, W. W. Wu, *Phys Rev B* 2000, *62*, 10004.
51. M. D. Perez, C. Borek, P. I. Djurovich, E. I. Mayo, R. R. Lunt, S. R. Forrest, M. E. Thompson, *Adv Mater* 2009, *21*, 1517.
52. K. S. Yook, B. D. Chin, J. Y. Lee, B. E. Lassiter, S. R. Forrest, *Appl Phys Lett* 2011, *99*, 043308.
53. W. I. Jeong, Y. E. Lee, H. S. Shim, T. M. Kim, S. Y. Kim, J. J. Kim, *Adv Funct Mater* 2012, *22*, 3089.
54. A. Hamed, Y. Y. Sun, Y. K. Tao, R. L. Meng, P. H. Hor, *Phys Rev B* 1993, *47*, 10873.
55. T. Li, J. R. Hauptmann, Z. M. Wei, S. Petersen, N. Bovet, T. Vosch, J. Nygard et al., *Adv Mater* 2012, *24*, 1333.
56. S. Lee, S. J. Kang, G. Jo, M. Choe, W. Park, J. Yoon, T. Kwon et al., *Appl Phys Lett* 2011, *99*, 083306.
57. W. H. Lee, J. Park, S. H. Sim, S. Lim, K. S. Kim, B. H. Hong, K. Cho, *J Am Chem Soc* 2011, *133*, 4447.
58. G. Wang, Y. Kim, M. Choe, T. W. Kim, T. Lee, *Adv Mater* 2011, *23*, 755.
59. B. C. Brodie, *Philos Trans R Soc Lond* 1859, *149*, 249
60. L. Staudenmaier, *Ber Dtsch Bot Ges* 1898, 31, 1481.
61. W. S. Hummers, R. E. Offeman, *J Am Chem Soc* 1958, 80, 1339.

62. S. Y. Bae, I. Y. Jeon, J. Yang, N. Park, H. S. Shin, S. Park, R. S. Ruoff, L. M. Dai, J. B. Baek, *ACS Nano* 2011, *5*, 4974.
63. W. W. Cai, R. D. Piner, F. J. Stadermann, S. Park, M. A. Shaibat, Y. Ishii, D. X. Yang et al., *Science* 2008, *321*, 1815.
64. N. Li, S. R. Forrest, *Appl Phys Lett* 2009, *95*, 123309.
65. K. L. Mutolo, E. I. Mayo, B. P. Rand, S. R. Forrest, M. E. Thompson, *J Am Chem Soc* 2006, *128*, 8108.
66. M. L. Tang, S. C. B. Mannsfeld, Y. S. Sun, H. A. Becerril, Z. N. Bao, *J Am Chem Soc* 2009, *131*, 882.
67. C. H. Cheng, J. Wang, G. T. Du, S. H. Shi, Z. J. Du, Z. Q. Fan, J. M. Bian, M. S. Wang, *Appl Phys Lett* 2010, *97*, 083305.
68. J. Q. Zhong, H. Y. Mao, R. Wang, D. C. Qi, L. Cao, Y. Z. Wang, W. Chen, *J Phys Chem C* 2011, *115*, 23922.
69. W. Chen, D. C. Qi, H. Huang, X. Y. Gao, A. T. S. Wee, *Adv Funct Mater* 2011, *21*, 410.
70. K. Xiao, Y. Liu, G. Yu, D. Zhu, *Appl Phys A-Mater* 2003, *77*, 367.
71. K. S. Lee, I. Kim, S. Gullapalli, M. S. Wong, G. E. Jabbour, *Appl Phys Lett* 2011, *99*, 223515.
72. C. W. Chu, H. C. Yang, W. J. Hou, J. S. Huang, G. Li, Y. Yang, *Appl Phys Lett* 2008, *92*, 103306.
73. S. Tatemichi, M. Ichikawa, T. Koyama, Y. Taniguchi, *Appl Phys Lett* 2006, *89*, 112108.

23 Chemical and Optical Aspects of Supported Graphene

D. Tasis, C. Galiotis, and K. Papagelis

CONTENTS

ABSTRACT

Recent advances in the production, manipulation, and properties of graphene have established this material as a promising candidate for the next-generation building block for the electronics industry. While the development of integrated graphene electronics requires the utilization of traditional processes, such as lithography, it is vital to complement the existing methods with specific chemical strategies that can modulate selectively the optical and electronic properties of graphene-based devices. In this chapter, the application of various *in-situ* functionalization approaches on graphene deposited onto different types of substrates (polymeric, metallic, insulating, or semiconducting) is discussed. In addition, the effect of various environmental conditions (atmosphere, temperature) is analyzed in detail. Emphasis is given on the assessment of doping effects and structural changes of graphene lattice by optical spectroscopy means.

23.1 INTRODUCTION

In recent years, conducting oxides, such as indium tin oxide (ITO), have been widely used as transparent electrode components in photovoltaic devices. The main disadvantages of such materials are its high cost, owing to the limited resources of metal precursors, as well as the lack of bendability. Thus, the development of ITO-based optoelectronic devices in large scale remains a difficult task in commercial terms. With the emergence of novel electronics, there is a growing demand for flexible electrodes. In recent times, graphene has been recognized as an important material for transparent electrodes in optoelectronic devices. Due to its high transparency, good conductivity, and enhanced flexibility, it is anticipated that graphene will be an efficient substitute for ITO. Recently, there has been tremendous research in the development of graphene-based assemblies as transparent/conducting electrodes [1].

Graphene has the smallest mechanical thickness $t_{eff} = (12\kappa/E_{2D})^{1/2}$ ($\kappa \approx 1$ eV is the bending rigidity and $E_{2D} \approx 2.12 \times 10^3$ eV/nm is the tensile rigidity) for any known material. This is due to the inextensible but bendable nature of the sp^2 carbon bonds in graphene [2]. This extremely small thickness makes graphene more susceptible to out-of-plane deformations (bending) resulting in morphological transitions, such as wrinkling [3], and governs the ability of graphene to integrate or adhere to various substrates. The mechanical interaction between the substrate and the graphene membrane affects not only the electronic and mechanical properties of graphene but also its handling and integration into various types of devices. For instance, the engineering of the peeling and stamping depends on the adhesion of the graphene to both the carrier and target substrates.

There are several recent experimental and theoretical reports on how the morphology of graphene membrane affects its properties when deposited on various substrates. Apart from corrugations observed in the suspended graphene sheets, a comparable degree of height variation has also been reported in several studies of graphene monolayers deposited on insulating substrates [4,5]. These rippling or bending fluctuations have been invoked to explain many phenomena observed in graphene, such as the formation of electron–hole puddles [6], the suppression of weak localization, decreased carrier mobility, and enhanced chemical reactivity. Moreover, morphological features such as wrinkles and conical singularities may create a nonuniform strain in graphene [7], which produces both scalar and vector potentials and mimics the effect of a magnetic field on graphene's electronic structure [8]. Although graphene adheres conformally to smooth nanoscale features with high fidelity, graphene wrinkling has been observed under compressive or tensile stress caused by thermal cycling. Also, graphene on periodically corrugated elastic and metal substrates exhibits transitions from adhesion to delamination [8].

The graphene adhesion/interaction with different materials and surfaces have been extensively studied, mainly theoretically or computationally. Only limited number of works present direct experimental evidence with emphasis on silicon substrates [9,10]. For other widely used materials as components in micro/nanoelectronics, such as SiN, Au, or even mica, the adhesion properties with graphene are still unknown. Recently, scientists obtained atomically flat graphene on mica substrates [11] and also visualized the structures of water molecules using a sheet of graphene on mica [12]. The interaction of polymers with graphene sheets is also important in various technological fields (nanocomposites, flexible displays, etc.) but no experimental data exists for adhesion properties and graphene morphology on polymeric surfaces.

Besides the issue of interaction/adhesion of supported graphene with any substrate material, an equally important research question is to what extent the chemical environment plays a role in the optical/electrical properties of carbon nanostructures? Both theoretical and experimental results have demonstrated that the Fermi energy shift of single-layer graphene (SLG) can be controlled through deliberate doping by aromatic molecules, gas species, or electrostatic field tuning [13]. These results suggest that controlling the work function of graphene electrodes is possible. The tuning of graphene's Fermi level is an important factor in determining the successful operation of the electronic devices based on the two-dimensional nanostructure (field effect transistors, light emitting diodes, solar cells, etc.). However, it is necessary to tune the Fermi level of graphene by controlled doping without significantly changing the unique optical/electrical properties of graphene. The vibrational properties of graphene as well as its doping in various chemical environments may be monitored by nondestructive optical characterization techniques, such as Raman spectroscopy.

Raman spectroscopy, as a noninvasive probing technique, has been extensively employed in graphene providing valuable information about the number of layers, domain grain size, doping levels, structure of graphene edges, anharmonic processes, thermal conductivity, etc. [14]. This has been possible through a combined investigation of the Raman peaks D, G, and 2D in graphene films of various thicknesses and morphologies [15]. The Raman active modes of graphene respond quite sensitively to stress/strain, doping, and temperature. Thus, monitoring phonons is often the clearest and simplest way to quantify the external perturbations (substrate nature and morphology, thermal effects, stress/strain, excess charges) imparted to the graphene sheets by establishing certain calibration relations [16–24]. More specifically, the strain sensitivity of the Raman frequencies of G and 2D modes has been determined for graphene under uniaxial or biaxial stress by several groups in agreement with the theoretical predictions [18,21,22,25–27]. Both ω(G) and ω(2D) are also strongly dependent on the extra charges induced by either physical or chemical means [28,29] owing to the static effects on the bond lengths and nonadiabatic electron–phonon coupling [30]. It is important to mention that graphene is also doped under environmental conditions such as trapped molecular layers at graphene–substrate interface [31]. For example, water molecules adsorbed to the substrate act as electron acceptor and thus would lead to hole doping [32]. Therefore, the aforementioned sensitivity of ω(G) and ω(2D) complicates independent determination of either strain or charge density and decoupling of both contributions is required [17,33]. This is indeed important since most of the pristine graphene sheets, deposited onto various substrates by the mechanical exfoliation method, were shown to be under in-plane stress either tensile or compressive in the range of (−0.2)%–0.4% [33]. Moreover, the graphene produced by chemical vapor deposition (CVD) appears to be under compressive stress when cooling down to ambient temperature from high production temperatures (800–1000°C), due to the significant mismatch in thermal expansion coefficients between graphene and metal substrates.

In this chapter, various functionalization approaches on graphene supported onto different types of substrates such as polymeric, metallic/insulating/semiconducting will be discussed. Special emphasis will be provided on the assessment of doping effects and structural changes of graphene lattice by means of optical spectroscopy.

23.2 SUBSTRATE EFFECTS

The interaction of two-dimensional graphene membranes with substrates determines their morphological state (wrinkles, crumpling, conformation on the surface roughness, local stress concentrations, etc.) as well as the electric charge density induced by the surface. These, in turn, tune the mechanical, electronic, and optical properties of graphene and, thus, its functionality in a plethora of applications and devices.

As already mentioned, a possible substrate effect is the accumulation of charge carriers as a result of the difference of the work functions between graphene and the substrate [34]. When graphene is placed on the top of certain substrate, having a higher work function, electrons can be transferred to the substrate. On the contrary, a substrate with a lower work function results to electron accumulation in the graphene lattice. It should be stressed that assessment of the work function of graphene can be directly determined using Kelvin probe force microscopy [35].

Because graphene is the ultimate thin membrane, it can conform more closely to a substrate than any other material. Therefore, direct measurement of the adhesion energy of various types of graphene is important to better understand its morphology and bonding mechanism as well as to control its mechanical release from the substrates. Koening et al. [10] used a pressurized blister test to measure the adhesion energy of graphene sheets on SiO_2. An adhesion energy of 0.45(2) J/m for monolayer graphene and 0.31(3) J/m for samples

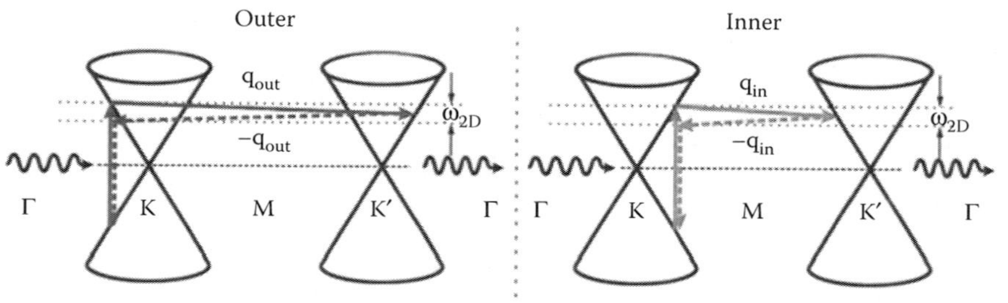

FIGURE 23.1 Incoming DR mechanisms (left, outer process, and right, inner process) plotted into the band structure scheme of graphene along the high symmetry KMK′ path. For simplicity we omit the equivalent process for hole–phonon scattering. (Adapted from O. Frank et al., 2011. *ACS Nano* 5:2231.)

containing two to five graphene sheets was extracted. These large values are attributed to the extreme flexibility of graphene, which allow it to conform to the topography of even the smoothest substrates. Also, Yoon et al. [36] measured a double cantilever beam fracture mechanics testing the adhesion energy of large-area monolayer graphene synthesized on Cu foil. An adhesion energy of 0.72(7) J/m was found.

Berciaud et al. [37a] investigated the intrinsic properties of exfoliated graphene monolayers suspended over micrometer-sized trenches using Raman spectroscopy. Freestanding samples provide an excellent reference system free from doping induced by the substrate or any other relevant perturbation (stress, strain, wrinkling, thermal). According to the authors [37a], in the areas of SiO_2-supported graphene, a spatially inhomogeneous hole doping in the order of 8×10^{12} cm^{-2} was exhibited. The ratio of the 2D and G peak intensities was systematically lower in doped supported regions compared to the suspended ones [37b]. This was ascribed to the presence of larger amount of charged impurities in supported graphene. Also, the 2D band of freestanding graphene for 514.5-nm excitation was found to exhibit an asymmetric spectral line shape, in sharp contrast with the relatively broad and symmetric feature observed in supported monolayers. In this line, a very recent work motivated from the aforementioned experiments showed that the 2D-mode exhibits a bimodal behavior for laser excitation energies between 1.53 and 2.71 eV [38].

It is well documented that the 2D band in graphene is activated from a double resonance (DR) Raman scattering mechanism [39–41]. This band consists of an intervalley DR process, which involves an electron or a hole in the vicinity of the Dirac K-point and two in-plane transverse optical (TO) phonons around the K-point having zero total phonon pseudo-momentum. The contributing TO-derived phonon branch is the one with the highest energy along the K – M direction of the Brillouin zone (BZ). Recently, it has been shown theoretically [42] that both inner and outer processes contribute to the Raman cross section of the 2D peak. In the outer (inner) process the participating phonons satisfy DR conditions coming from K – M (Γ – K) with wavevectors $\mathbf{q}_{out} > \mathbf{KK'}$ ($\mathbf{q}_{in} < \mathbf{KK'}$). Since both electron and phonon dispersions are anisotropic displaying the trigonal warping effect, it is expected that the selected "inner" and "outer" phonon frequencies will slightly differ, giving rise to two distinct DR scattering processes (inner and outer) in the Raman signal (Figure 23.1) and a complex line-shape for the 2D mode [39,43]. Nevertheless, the 2D band in supported graphene samples is assigned to the dominant contribution of the "outer" scattering channels [15,44] and is usually fitted by a single Lorentzian peak. On the other hand, Luo et al. [45] reported in freestanding graphene samples the splitting of 2D band providing a direct support for the coexistence of both outer process and inner process in the DR Raman scattering. Moreover, uniaxial strain measurements [17] on graphene samples embedded into polymer matrix showed a progressive increase of the splitting between the "inner" and "outer" components as a function of the strain level. Very recently, Berciaud et al. [38] showed that for laser excitations in the range 1.53–2.71 eV, the 2D-mode of freestanding samples appears as bimodal, in sharp contrast with the behavior of supported samples (Figure 23.2). The transition between the freestanding and supported behavior is attributed to the enhanced broadening of both components due to an increased electronic relaxation rate that takes place at relatively low levels of doping ($<2 \times 10^{11}$ cm^{-2}). This electronic effect can obscure the intrinsically bimodal 2D-mode line shape when graphene is placed on the top of a substrate [38].

Das et al. [23] extracted the charge carrier density effect on the spectral characteristics of graphene G and the 2D bands from electrochemical n- and p-doping experiments. They concluded that, compared to the neutrality point, the G mode shifts to higher wave numbers, the intensity ratio I(2D)/I(G) decreases for both electron and hole doping and the 2D mode responds differently to electron (red shift) or hole (blue shift) charge injection (Figure 23.3). This work is a benchmark to quantify the amount of doping induced in graphene structure by a certain substrate.

Bukowska et al. [34] investigated the deposition of mechanically exfoliated single-, bi-, and few-layer graphene on top of SiO_2 as well as on single-crystal surfaces of Al_2O_3, $SrTiO_3$, and TiO_2. They observed changes in the shift of 2D and G peaks induced by the presence of the underlying substrate. Based on above-mentioned analysis of Das et al. [23], they concluded that in the case of Al_2O_3 and TiO_2 an accumulation of holes occurs while for $SrTiO_3$ substrates the graphene sheets are doped with electrons.

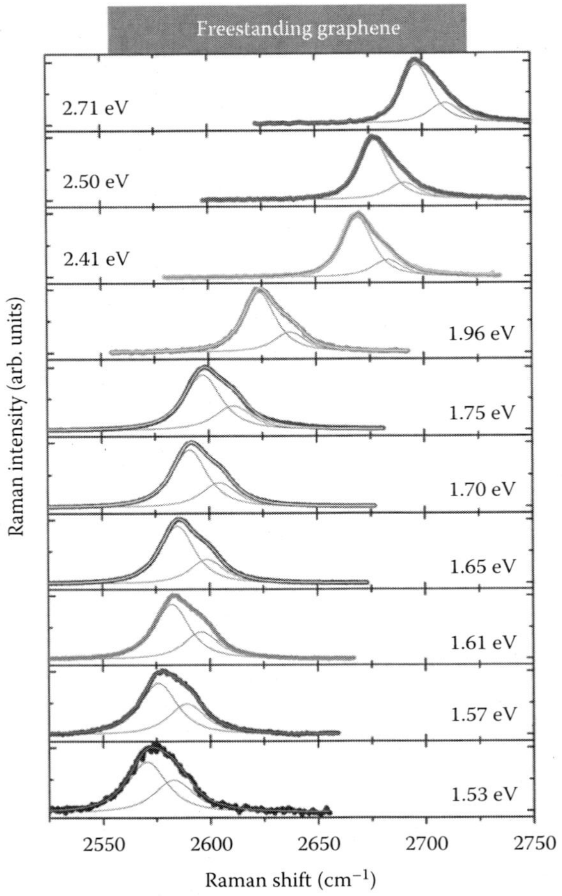

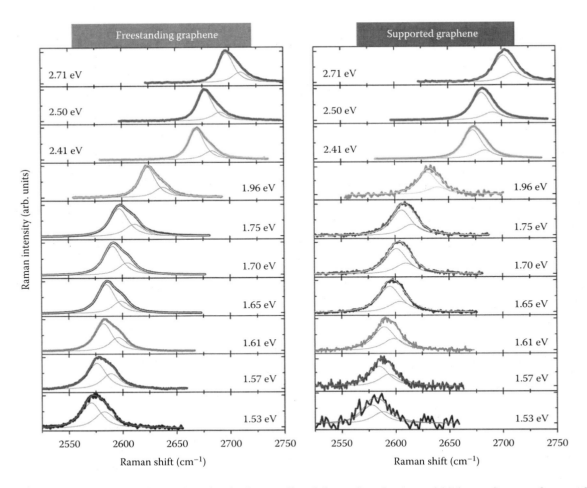

FIGURE 23.2 Raman spectra in the 2D mode region for freestanding (left panel) and supported (right panel) areas of a same SLG measured for incident photon energies in the range of 1.53–2.71 eV. (Reprinted with permission from S. Berciaud et al., 2013. Intrinsic line shape of the Raman 2D-mode in freestanding graphene monolayers. *Nano Lett.* 13, 3517–23. Copyright 2013 American Chemical Society.)

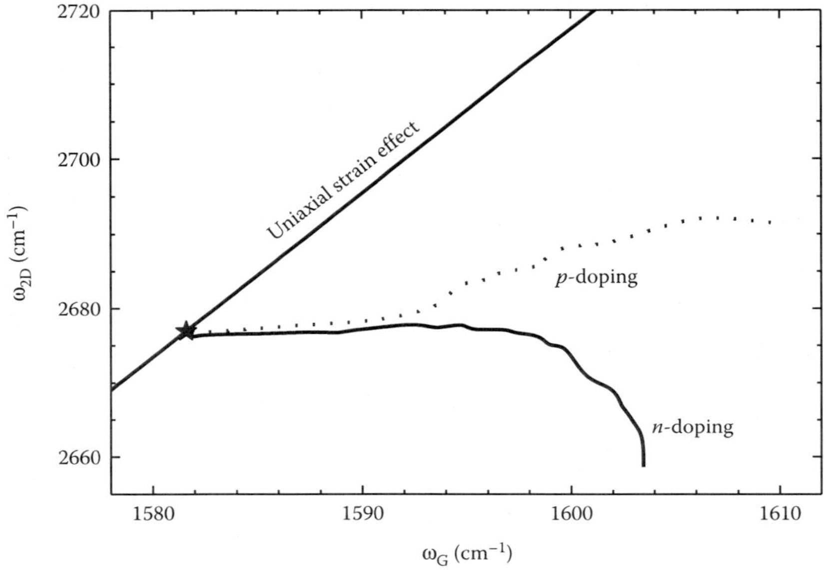

FIGURE 23.3 ω_{2D}–ω_G correlation for electrochemically electron- and hole-doped graphene. The star symbol corresponds to undoped strain-free graphene. Also, the pure uniaxial strain effect is presented. (The data are taken from A. Das et al., 2008. *Nat. Nanotechnol.* 3:210.)

In another work, Chen et al. [46] observed significant blue shifts of the G (24 cm^{-1}) and 2D (23 cm^{-1}) bands after thermal cycling of supported SLG to 700 K. The observed alterations have been attributed to the mechanical compression induced by the SiO_2 substrate and from doping effects from the surrounding air molecules and the surface charges trapped in the underlying substrate. In a similar work for suspended graphene [47], the same shift for the G band after thermal cycling between 300 and 700 K was observed. It is important to mention the formation of ripples in suspended samples after thermal cycling due to the large mismatch in compression between the substrate and the suspended regions. Also, the authors observed a broadening of the G-peak width in the suspended region, which is attributed to an inhomogeneous broadening due to the local strain variation caused by the ripples inside the laser spot [47].

Furthermore, Allard and Wirtz [48] studied the graphene/metallic substrate interaction strength and calculated the influence of different metallic substrates such as Ni, Pt, Ru on the phonon dispersion and the Raman response of the absorbed graphene layer. As stated from Reference [48], charge transfer from the substrate to graphene and a hybridization of the π electronic bands with the d electrons from the metal substrate will reduce (i) the Kohn anomalies in the highest optical branch at the Γ and κ-points of the BZ, and (ii) the energy dispersion of the 2D mode as a function of the laser frequency.

23.3 EFFECTS OF TEMPERATURE AND CHEMICAL ENVIRONMENT

23.3.1 Thermal Processes under Vacuum Conditions or in Air

Being, in essence, a one-atom-thick material, graphene, and its electrical/optical properties proves extremely sensitive to the specific conditions of its environment (temperature, atmosphere, etc.). Although there have been some recent works dealing with the effect of environmental conditions on the optical/electrical properties of supported graphene (see below), their exact role needs to be elucidated further. Charge transfer interactions between graphene and chemical species can create negative or positive net charge density. Different behaviors are observed between samples prepared by various processes, such as CVD and mechanical exfoliation. The electronic properties of mechanically exfoliated graphene supported onto SiO_2/Si substrate under high vacuum conditions was studied in the seminal work of Eklund and coworkers [49]. Assemblies exposed to open air for prolonged time were found to be initially p-type (positive net charge density). It was suggested that p-doping stems from adsorption of ambient gas substances that remove electrons from graphene surface. After a thermal treatment at 200°C under high vacuum, the device slowly evolved to n-type behavior with an excess electron density of about 4×10^{12} cm^{-2}. When the temperature was lowered to 25°C under continuous vacuum conditions, a partial reduction of the graphene n-doping was observed. As a first attempt to elucidate the temperature effect, the authors

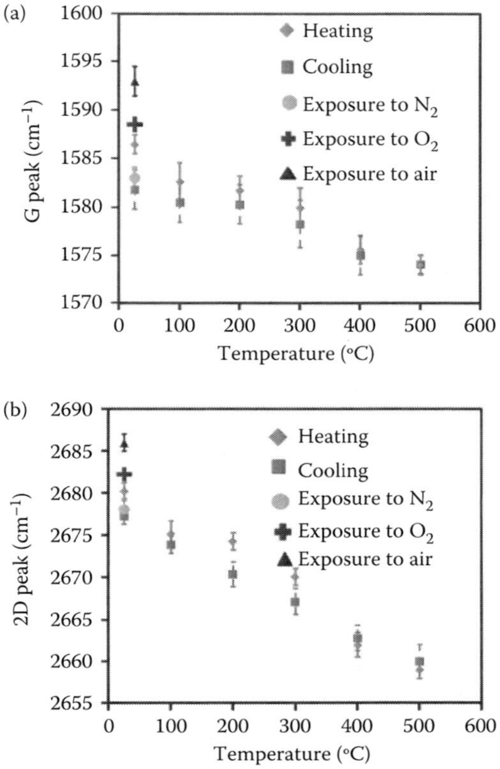

FIGURE 23.4 (a) Temperature dependence of G peak with about 12 cm^{-1} blue shift due to vacuum annealing and reexposure to air. (b) Temperature dependence of 2D peak with about 9 cm^{-1} blue shift due to vacuum annealing and reexposure to air. (Reprinted with permission from H. Sojoudi et al., 2012. Impact of post growth thermal annealing and environmental exposure on the unintentional doping of CVD graphene films. *J. Vac. Sci. Technol. B* 30:041213. Copyright 2012. American Vacuum Society.)

assumed a model where the charge density is adjusted by temperature variations. These results were strongly supported by *in situ* Raman spectroscopy measurements onto graphene/SiO_2/Si devices during heating up under vacuum [50].

It was observed that G and 2D band positions shifted to lower wavenumbers as the temperature increased (Figure 23.4). This may be partially explained due to desorption of oxygen and moisture, which results in dedoping of graphene.

To examine the post-effects of thermal treatment, graphene/SiO_2/Si devices were sequentially annealed from 100°C up to 500°C in vacuum for a specific time [51a]. It was found that the two-probe mobilities of graphene devices started to degrade, when heated at temperatures above 300°C. Raman spectroscopy was used in order to assess the structural and doping characteristics of annealed graphene. First, the absence of any defect-related D-band at about 1300 cm^{-1} clearly showed the lack of damage to the graphitic network. In addition, G and 2D bands blue shifted by 12 and 10 cm^{-1}, respectively, after annealing at 400°C. This phenomenon was attributed to hole doping in the graphene, whereas the hole concentration was estimated to be about 1×10^{13}/cm^2. The mechanism of hole doping in annealed devices could be explained due to enhanced contact between graphene and SiO_2 substrate or even an increase of adsorption of gas dopants (water and/or oxygen). Similar level of p-doping ($1.2 \times 10^{13}/cm^2$) could be achieved by microwave-induced annealing in mechanically exfoliated graphene deposited onto SiO_2/Si wafers [51b]. Relatively high temperatures of about 400°C could be reached by short time (5 min) microwave irradiation. The authors demonstrated that the dielectric heating of the silicon substrate was the origin of the elevation in temperature.

23.3.2 Doping of Graphene in the Presence of Various Substances in the Gas Phase

Oxygen gas at high temperatures has been found to be an efficient oxidant for graphitic nanostructures. Single- and multiple-layer graphene sheets deposited onto SiO_2/Si wafers were treated in an O_2/Ar gas flow at temperature range between 200°C and 600°C [52]. Atomic force microscopy (AFM) imaging showed that oxidative etching was observed to proceed faster in graphene monolayers than in multiple ones. Raman scattering was employed to directly probe the density of defects on the graphene surface. The intensity ratio of the D mode to that of the G mode (I(D)/I(G)) for various graphene layers showed that both triple- and bilayer sheets are etched efficiently at temperatures above 500°C (Figure 23.5). In single layers, the ratio increases appreciably after oxidation at temperatures approximately above 350°C. Significant blue shift of both G and 2D bands was observed after thermal oxidation.

Most of this shift occurs after oxidizing the samples in the temperature range between 200°C and 300°C and is not linked with the etching process. The authors suggested that bound oxygen at such temperatures creates strong hole doping in graphene via electron transfer mechanism, with a hole density of about $2.3 \times 10^{13}/cm^2$. In a subsequent study, Yamamoto et al. [53] have deposited mechanically exfoliated graphenes onto various substrates (hexagonal boron nitride, mica, SiO_2/Si, and SiO_2 nanoparticle thin films) and evaluated the oxygen reactivity in elevated temperatures (from 350°C up to 600°C). In contrast to SiO_2-supported monolayer graphene, the hole doping of boron nitride-supported sheets is significantly suppressed. This was extracted from the minor blue shift of the G band after oxidation at 500°C. Furthermore, the defect-induced D band is absent, indicating that monolayer graphene on boron nitride substrate is not etched at such temperature. The observed reduced reactivity of monolayer graphene onto boron nitride could be as well ascribed to the flatness of substrate or its reduced charged inhomogeneity. By probing the oxidative reactivity of monolayer graphene supported on a corrugated substrate (SiO_2 nanoparticle thin film), the authors concluded that roughness plays some role only at temperatures above 400°C.

The effects of thermal annealing in inert Ar gas atmosphere of SiO_2-supported monolayer graphene was investigated by Nourbakhsh et al. [54]. Both graphene signatures in Raman spectra (G and 2D band) collected after each annealing treatment up to 400°C show a progressive upshift by increasing temperature. This trend was explained as a gradual accumulation of holes in the monolayer graphene samples as a consequence of annealing process. The hole doping in graphene is believed to originate by charge transfer interactions from the SiO_2 support. Concerning the effect of thermal annealing to the crystalline quality of deposited monolayer graphene, the absence of D band in Raman spectra revealed that heating at inert atmosphere preserves the conjugated network.

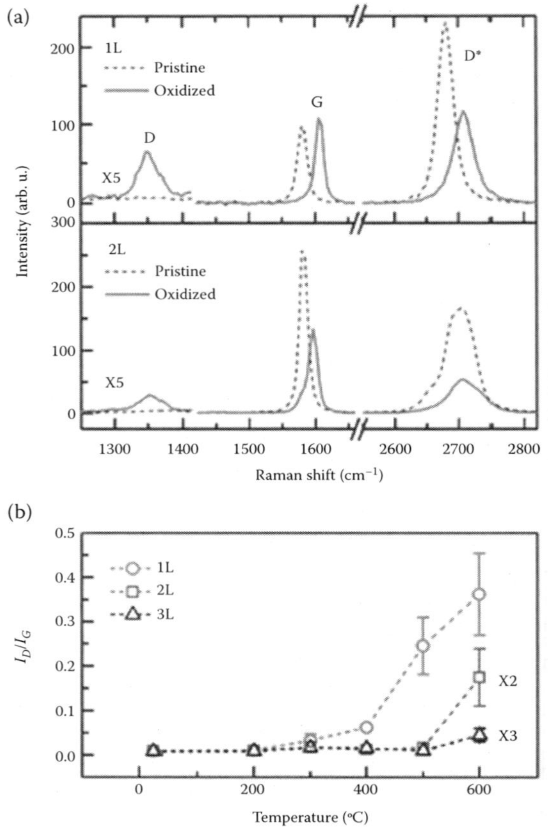

FIGURE 23.5 Raman spectra of pristine and oxidized graphene. (a) (Upper) Raman spectra of single-layer (1L) graphene: pristine (dotted), and oxidized at 500°C for 2 h ($P[O_2] = 350$ Torr) (solid). (Lower) Raman spectra of double-layer (2L) graphene: pristine (dotted), and oxidized at 600°C for 2 h ($P[O_2] = 350$ Torr) (solid). Spectra near D-mode were enlarged for clarity. (b) D-mode to G-mode integrated intensity ratio (I(D)/I(G)) as a function of oxidation temperature: 1L (circle), 2L (square), and triple-layer (3L, triangle) graphene. (Reprinted with permission from L. Liu et al., 2008. Graphene oxidation: Thickness dependent etching and strong chemical doping. *Nano Lett.* 8, 1965–70. Copyright 2008 American Chemical Society.)

In an analogous work, NO_2 gas at high pressure was used for hole doping of deposited graphenes (one up to 10 layers) [55] (Figure 23.6).

The adsorbant is paramagnetic and has a high electron affinity (2.3 eV). Raman spectra for 1–10-layer graphenes exposed to 60-Torr NO_2 showed no D peaks, meaning that the gas species is physisorbed. The G band maxima of mono- and bilayer graphene are upshifted by about 32 and 25 cm^{-1}, respectively. The three-layer graphene spectrum has a double peak with the higher energy one more intense, whereas, for layer number ≥4 the intensity of the lower energy peak increases with graphene thickness (Figure 23.6). This Raman

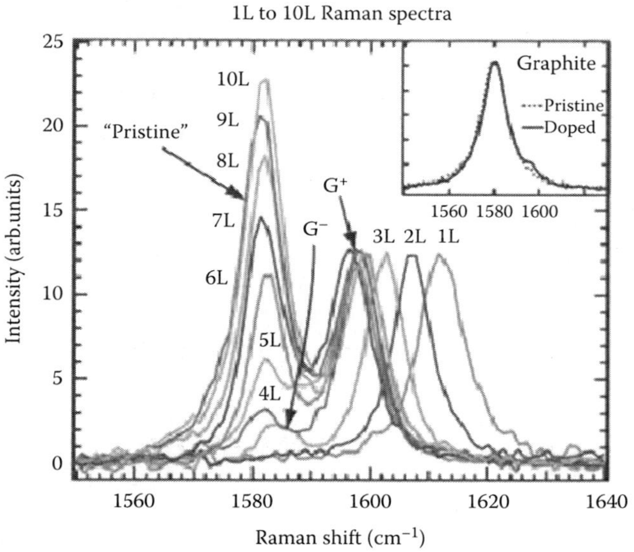

FIGURE 23.6 Raman spectra for 1L to 10L graphene on quartz exposed to 60-Torr NO_2. The spectra are normalized to the higher energy peak. The inset is the graphite spectrum before and after exposure to 60-Torr NO_2. (Reprinted with permission from A. Crowther et al., 2012. Strong charge-transfer doping of 1 to 10 layer graphene by NO_2. *ACS Nano* 6, 1865–75. Copyright 2012 American Chemical Society.)

spectral profile demonstrates that NO_2 does not intercalate, but only adsorbs on the top and bottom surfaces. Such blueshift for monolayer graphene corresponds to a hole density of $4.5 \times 10^{13}/cm^2$.

Charge transfer through adsorption of elemental species in the vapor phase is an alternative strategy to achieve graphene doping. Exposure to alkali vapors, such as K and Rb, was shown as an efficient method toward the preparation of high level n-doped graphene [56–58]. In the work of Parret et al. [58], the evolution of Raman spectra with doping time was followed *in situ* for the case of graphene/rubidium system. In the case of deposited SLG, typical Raman spectra recorded *in situ* and involve n-type doping. The G band exhibits first a gradual upshift, which saturates after 1 h of doping. As expected, the 2D band is found to continuously downshift and its relative intensity to decrease.

The authors confirmed that alkali doping allows reaching n-doping levels inaccessible by other techniques and evaluate that the maximum electron concentration attained here is ~12×10^{13} electrons/cm^2. In the case of bilayer graphene, analysis of the G-band profile evolution allowed to follow the charge repartition among the two layers. A G-band splitting in the two components G^+ and G^- was evidenced for early doping times. The G^- component was also shown to vanish for longer doping times. The authors concluded that at first only the top layer is doped, but the system gradually evolves with time toward a more symmetric repartition of the added electrons in both layers.

Molecular adsorption/intercalation of electronegative halogens onto either mono- or few-layer deposited graphenes was investigated by Brus and coworkers [59]. In the work of Jung et al. [59a], halogen (Br_2 or I_2) gas exposure was performed within a glass tube, preevacuated under high vacuum. The authors did not observe any D mode formation to mono- up to four-layer graphenes, after halogen gas exposure, meaning that no thermal or photoinduced halogenation took place under specific conditions. Raman spectra of bromine-exposed few-layer graphenes showed a peak at 238 cm^{-1}, which is attributed to the intercalated stretching mode of halogen. The latter peak was observed only in graphene samples with layer number ≥3. This is the same anionic bromine mode seen in graphite intercalation compounds (GICs). Physisorption of bromine species was evident in the profiles of G band. After bromine exposure, the G band of monolayer graphene experienced a 44 cm^{-1} energy upshift, which is unprecedented. Due to the asymmetric intercalation of bromine within a three-layer graphene, a double G band was recorded for the hybrid structure. The G peak upshift of I_2-exposed graphene is less than the corresponding of Br_2 due to two reasons. A comparison of the molecular redox potentials indicates that iodine is a weaker oxidizing agent than bromine and, in addition, iodine has a lower vapor pressure than bromine at a given temperature. Furthermore, Raman spectra showed that iodine does not intercalate between graphene few-layer samples. Similar vapor-phase process has been adopted for the fabrication of graphene/$FeCl_3$ intercalation structures by independent groups [60,61]. Due to the electron acceptor character of $FeCl_3$, the graphenes are considered as hole doped.

Modification of SiO_2-supported graphene with gold nanoparticles has been achieved by the gas aggregation technique [62]. The latter process is based on the vapor phase condensation of sputtered atoms within a magnetron sputtering system. The profiles of Raman spectra were investigated as a function of sputtering time and DC power. The evolution of G band maxima toward lower wavenumber, the increase of the full width at half maximum (G band), as well as an overall decrease of I(2D)/I(G) ratio in the Au/graphene samples is a strong indication of the presence of charge transfer interaction between the components. Specifically, the deposition of Au nanoparticles causes n-type doping in deposited graphenes.

Either sublimation or evaporation of organic compounds has been adopted as a simple method toward the fabrication of doped supported graphenes. Doping by adsorption of organic substances depends on the adsorbate orientation and the nature of graphene regions (basal plane, edges, defects, and strained spots). The presence of graphene surface defects can effectively facilitate charge transfer interactions with the adsorbate. Chiu and coworkers [63a] have studied the doping of mechanically exfoliated graphene by adsorption of melamine. The graphene samples were inserted into a furnace at 500°C, whereas melamine vapors were produced at 350°C. To demonstrate the difference between basal plane and edges adsorption capabilities, Raman spectra were recorded in the corresponding regions after the doping process. Only little change was seen in the G peak (1 cm^{-1}), whereas a pronounced blue shift of the G peak (11 cm^{-1}) was observed at the edges. This result indicates that molecular doping can be controlled by the population of graphene defect sites.

Further support to the previous conclusion was given by the fact that doping of supported graphene, which was previously treated within an Ar plasma chamber (production of extra lattice disorders), resulted in G-peak stiffening and decrease of the I(2D)/I(G) ratio to all over the graphitic surface. These plasma-induced lattice disorders favor melamine adsorption and increase charge transfer interactions. Deposition of adsorbate was found to be saturated after a period of about 7 min, under the specific conditions used, and resulted in graphene hole doping.

In an analogous approach, Hong and coworkers [63b] have studied the vapor-phase molecular doping of graphene by organic amines. The n-doping efficiency and stability was found to depend on the number of electron-donating nitrogen atoms and vapor pressure of the adsorbed molecules. In particular, the triethylene tetramine-doped graphene showed very high carrier concentration (1.4×10^{13} cm^{-2}) as well as the lowest monolayer sheet resistance (~98 Ω/sq) with excellent stability.

23.3.3 Wet Chemistry on Supported Graphene

Chemical doping of deposited graphene has also been achieved by liquid phase adsorption of either p- or n-dopants. In most cases, the latter substances involve polymers, dyes, as well as fused aromatic systems. In the independent studies of the groups of Avouris and Cho [64], the doping of graphene devices by an electron-donating macromolecule (poly[ethylene imine], PEI) was demonstrated through either soaking the supported graphenes into an alcoholic solution of the polymer [64a] or by stamping process [64b]. Measurements of conductance versus gate voltage before and after soaking process showed that adsorption of PEI on graphene results in n-type behavior.

Transfer of CVD-grown graphene onto a target substrate usually takes place by spin coating a thin polymeric film (in most cases, PMMA) onto the graphene/metal catalyst foil, followed by etching of the metal, polymer/graphene pickup, and polymer removal. In most cases, transfer agent residuals remain adsorbed onto the graphene surface, even after extensive solvent rinsing. Using a simple concept, the Ruoff group developed a method to simultaneously transfer and dope CVD-grown graphene onto various substrates, such as SiO_2/Si, SiN, or polyethylene terephthalate (PET) membranes [65] (Figure 23.7).

The protocol involves polymer coating (either a fluoropolymer [CYTOP] or a reference polymer [PMMA]), thermal annealing at 140°C, and removal in the liquid phase. Even after extensive rinsing of both graphene/polymer systems, AFM imaging showed clearly "islands" of polymeric residuals onto the graphene surface (Figure 23.7). The doping properties of graphene samples were compared through Raman spectroscopy measurements. In the case of CYTOP as transfer agent, a decrease of I(2D)/I(G) ratio and the observation of blue shifts of both G and 2D bands confirmed the graphene doping. These changes arise from p-doping of the graphene, due to its interaction with the electronegative fluorine atoms of the polymer CYTOP. On the contrary, PMMA residues have not yielded any appreciable doping effect.

Beside polymers, even low molecular weight aromatic substances have been used for graphene doping [66]. The aromatic molecules 1,5-naphthalenediamine, 9,10-dimethylanthracene, 9,10-dibromoanthracene, and tetrasodium 1,3,6,8-pyrenetetrasulfonic acid were dissolved in ethanol,

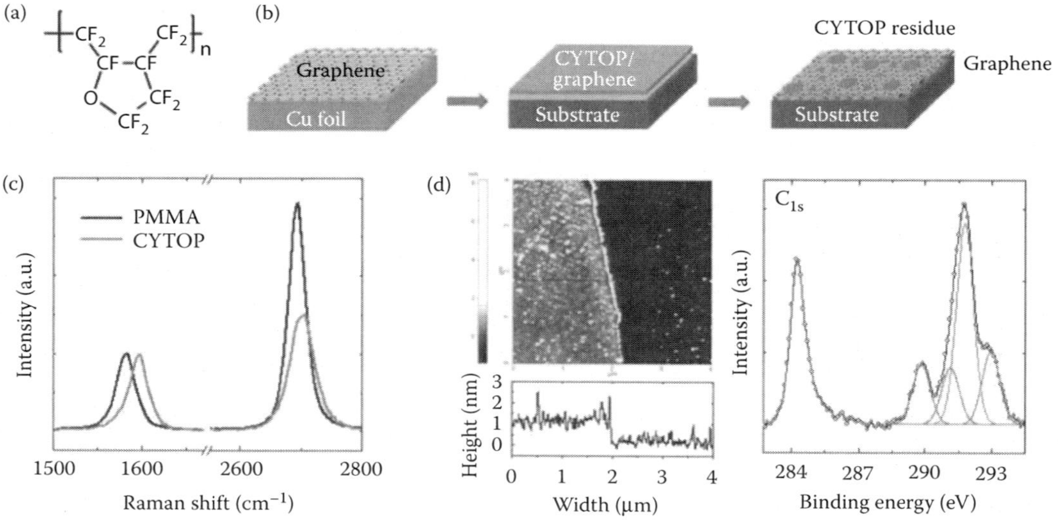

FIGURE 23.7 (a) Chemical structure of CYTOP. (b) Schematic of the graphene transfer process with CYTOP supporting layer. (c) Raman spectra of the graphene films transferred onto SiO_2/Si substrates with PMMA (black) or CYTOP (red) as the supporting layer. (d) AFM image (4 × 4 μm) (left) and XPS C1s spectrum (right) of the graphene film having an ultrathin CYTOP residue layer on the top surface. Bottom inset of the AFM image shows the cross-sectional profile. (Reprinted with permission from W. Lee et al., 2012. Simultaneous transfer and doping of CVD-grown graphene by fluoropolymer for transparent conductive films on plastic. *ACS Nano* 6, 1284–90. Copyright 2012 American Chemical Society.)

toluene, or in deionized water, depending on solubility. Such solutions (0.05 mol/L) were carefully drop cast onto the surface of SLG sheets. After evaporation of the solution at room temperature, the supported graphenes were rinsed with copious amounts of solvent. It was observed that after rinsing, the graphene surfaces were still fully covered with a thin layer of organic molecules, due to strong π–π interactions between their aromatic rings and the carbon nanostructure. Raman as well as electrical characterization suggested that substances with electron-donating groups (1,5-naphthalenediamine or 9,10-dimethylanthracene) resulted in n-doping of graphene, whereas 9,10-dibromoanthracene and tetrasodium 1,3,6,8-pyrene tetrasulfonic acid yielded p-doping.

Similar protocol involving combination of spin-coating and rinsing has been adopted for the p-doping of deposited graphenes with methyl orange (MO) [67]. As illustrated by Raman imaging, MO/graphenes showed obvious doping effects, depending on graphene layer number. After introducing the MO molecule onto the graphenes, the G band positions shift toward higher frequency (blue shift) but the blue shift decreases with the thickness of graphenes. One possible reason of thickness-dependent doping may be the so-called screening effect, that is, the holes from the surface graphene layers are transferred to the interior ones.

Potential applications of such systems involve the development of optoelectronic devices with enhanced photoefficiency. Hebard and coworkers [68] have demonstrated SLG/n-Si Schottky junction solar cells that exhibit a native (undoped) power conversion efficiency of 1.9%, under one sun AM1.5G illumination. Upon chemical charge transfer doping with bis(trifluoromethanesulfonyl)-amide ($[CF_3SO_2]_2NH$) (TFSA) the device photoefficiency was increased to 8.6%. Doping of CVD-grown graphene/Si devices was accomplished by spin coating a TFSA solution.

Analogous applications in which such graphene-based hybrids could be integrated into, include circuit systems, sensors, etc. To this goal, it is highly desirable to change the transport behavior of the graphitic nanostructures from p-type to ambipolar and n-type. In a recent work, Bao and coworkers [69] described the use of 2-(2-methoxyphenyl)-1,3-dimethyl-2,3-dihydro-1H-benzoimidazole (o-MeO-DMBI) in order to tune the electrical properties of CVD-grown graphene by simple solution processing (Figure 23.8).

It was observed that the CVD-grown graphene transistors can be controllably tuned from p-type to ambipolar and finally n-type transport behaviors, by simply varying the amount of o-MeO-DMBI spin coated onto graphene surface. o-MeO-DMBI in chlorobenzene solutions with different concentrations were used for spin-coating step.

Upon o-MeO-DMBI doping, the electrical characteristics (Dirac points) of doped graphene transistors shifted toward negative voltages, even when a very low solution concentration (0.01 mg/mL) of o-MeO-DMBI was used, indicating the strong n-type doping effect of o-MeO-DMBI on graphene sheets (Figure 23.8). At low solution concentrations of 0.01 and 0.1 mg/mL, the o-MeO-DMBI-doped graphene transistors showed ambipolar character. With higher concentrations of o-MeO-DMBI solutions, all the doped graphene transistors exhibited n-type transport behaviors. The electrical data were fully corroborated by Raman measurements. It was shown that G band maxima upshifted whereas 2D peak positions downshifted, by increasing the dopant concentration.

23.3.4 Graphene Doping by Addition of Transient Species

Addition of transient species onto the graphitic lattice has been widely used for the band gap engineering of carbon nanostructures. Precursors of radical species, such as

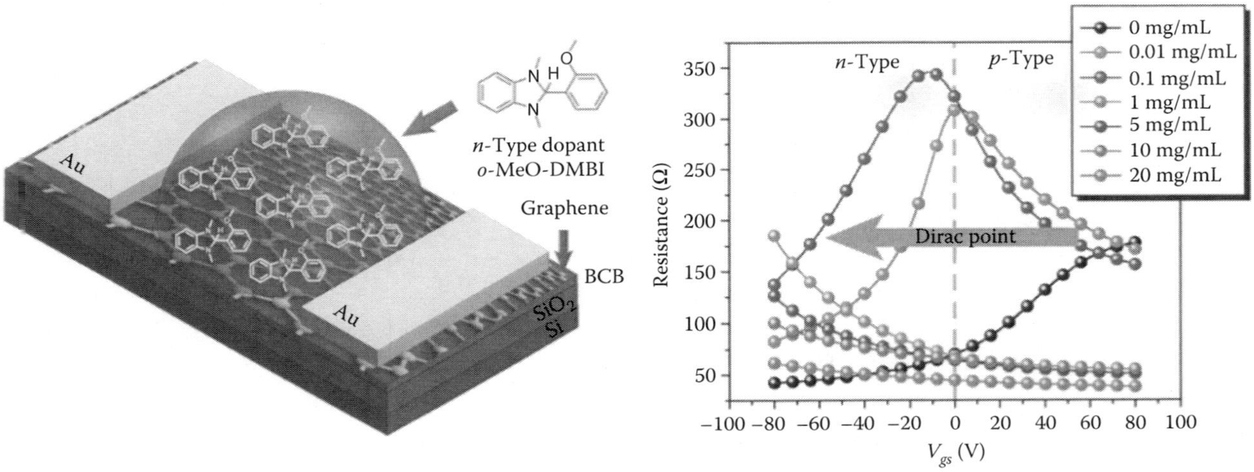

FIGURE 23.8 Left: Chemical structure of o-MeO-DMBI and the schematic illustration of o-MeO-DMBI n-type doped CVD-grown graphene transistor by solution process. Right: Resistance of the as-transferred and doped CVD graphene transistors with various concentrations of o-MeO-DMBI solutions as a function of applied bottom gate voltages (V_{gs} = 0.1 V). (Reprinted with permission from P. Wei et al., 2013. Tuning the Dirac point in CVD-grown graphene through solution processed n-type doping with 2-(2-methoxyphenyl)-1,3-dimethyl-2,3-dihydro-1H-benzoimidazole. *Nano Lett.* 13:1890–7. Copyright 2013 American Chemical Society.)

diazonium salts, have been utilized for the covalent functionalization of supported graphene, which results in specific doping effect. Graphene devices, fabricated by conventional mechanical exfoliation and subsequent deposition onto SiO_2/Si substrate, were exposed to a solution of aryl diazonium salt at room temperature [64a]. As far as the case of carbon nanotubes is concerned, the mechanism involves electron transfer from graphitic surface to the as-formed aryl cations, followed by addition of transient aryl radicals onto the double bonds of the carbon nanostructure. In the case of deposited graphene, conductivity measurements showed that alteration of conjugated network was minimum, under the specific conditions of reaction. Furthermore, the decrease of the I(2D)/I(G) ratio in Raman spectra suggested that diazonium species were mostly physically adsorbed onto graphene, resulting in p-doping. In another work, Stark and coworkers [70] tried to differentiate between the paths of covalent attachment and physical adsorption of aryldiazonium species by Raman measurements. To this goal, nitrobenzene was adsorbed to mechanically exfoliated graphene flakes and analyzed in comparison with the chemically functionalized samples (Figure 23.9).

While nitrobenzene adsorption induced a minor blue shift in the G band (3 cm^{-1}), the reaction of diazonium salt resulted in an appreciable D band increase within just 10-min time. The latter was interpreted as a sign of sp^3 carbon formation. The covalent attachment of nitrobenzene moieties was found to be favorable at the edges than at the bulk of a single layer [70,71]. In addition, Raman spectra highlighted the absence of D peak in modified bilayer graphene. The different reactivity could be explained due to the more crystalline nature of bilayer graphenes, which leads to reduced flexibility to accommodate local sp^3 geometry. Furthermore, noncovalently adsorbed diazonium species were recorded in Raman spectra at reaction times over 30 min. These data indicated that covalent attachment of diazonium moieties takes place at defect sites during the early times of reaction, whereas physical adsorption of excess reagent follows.

An important issue of deposited graphene flakes on various substrates is the accumulation of charge puddles induced by charged impurities introduced during transfer processes and by ionized impurities existing in polar substrates (e.g., SiO_2/Si wafers). The impurities introduced during the experimental processes might exist in the form of adsorbed molecules from ambient air, or residues of resist used during the fabrication process of graphene devices. The effect of charge puddles on the reactivity of supported graphene toward electron transfer chemistries has been studied independently by the groups of Nouchi [72] and Strano [28]. The electronic properties of deposited graphenes in SiO_2/Si are strongly influenced by adsorbates, such as water and sodium ions. It was found that coupling of such polar adsorbates and the consequent generation of charge puddles could be effectively altered through the selection of appropriate substrate. These substrate-induced electron–hole fluctuations in graphene influence its chemical reactivity. Graphene onto either SiO_2 or Al_2O_3 substrates is highly reactive toward diazonium chemistry, whereas graphene onto boron nitride and aliphatic monolayers is much less reactive [28,72]. These conclusions were extracted after

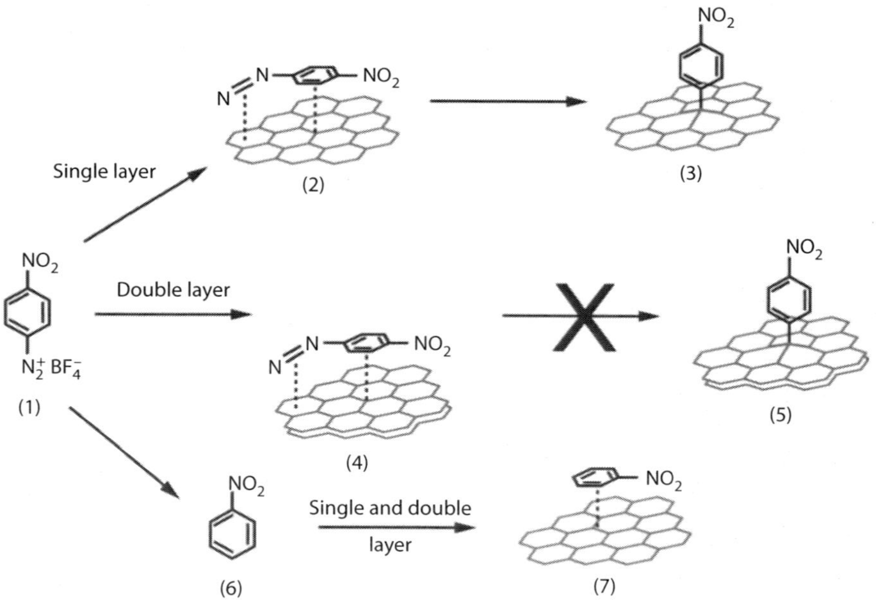

FIGURE 23.9 Schematic representation of the proposed reaction pathways on single and bilayer graphene. The diazonium reagent (1) presumably adsorbs on single- (2) or bilayer graphene (4) and decomposes in a second step to form a covalently bound nitrobenzene moiety on the graphene surface (3). On bilayer this decomposition of the diazonium group does not take place (no formation of adduct 5). (1) Can additionally decay to nitrobenzene (6), which can only adsorb on the graphene surfaces (7), however no reaction with the graphene takes place. (F. Koehler et al.: Selective chemical modification of graphene surfaces: Distinction between single and bilayer graphene. *Small*. 2010. 6. 1125–30. Copyright Wiley-VCH Verlag GmbH & Co. KGaA. Reproduced with permission.)

calculation of the I(D)/I(G) ratios in the Raman spectra of the chemically modified graphenes. This trend could be attributed to the reduction of polar adsorbates (charged impurities) in the case where the latter group of substrates is used.

Apart from using aryl radical species for the functionalization/doping of supported graphene, researchers have demonstrated the decoration of carbon nanostructures with nitrogen atoms through the combination of N^+-ion irradiation at room temperature and NH_3 annealing at 1100°C [73]. Controllable N-doping was achieved by the aforementioned process. The SiO_2-supported graphene was bombarded with 30 keV N^+-ions at various fluences. After irradiation, Raman spectra showed the typical D band, which increased gradually upon irradiation dose. In parallel, the I(2D)/I(G) ratio decayed with the increasing dose. For N-doping of irradiated graphene, NH_3 annealing was performed at 1100°C. It was suggested that the vacancy defects of the irradiated graphene were repaired with "graphitic" N atoms after annealing in NH_3, since the C–N bond formation occurs predominantly on the defect sites in the plane, where the C atoms are much more chemically reactive than the ones in the plane of the perfect graphene. It has to be noted, however, that the conductance of NH_3-annealed graphene after irradiation was a few times lower compared to that of the pristine graphene, which might be caused by the N atoms doping.

Similar substitution of carbon atoms by nitrogen could be achieved by bombardment of N ions of energies up to 35 eV at 850°C [74]. By using a tunable plasma source, the authors managed to selectively decorate the graphitic lattice with "pyridinic" N atoms after explosion of graphene to a stream of thermalized nitrogen atoms.

23.4 SUM UP RESULTS: CONCLUSIONS

Substrate effects influence considerably the structural and electronic properties of graphene governing its potential to be integrated into various types of devices. Optical spectroscopy is a sensitive tool to investigate the above-mentioned effects by establishing different calibration relations for the optical response under stress/strain, doping, and temperature. Concerning the effect of chemical environment to the structural and electronic properties of supported graphene, it was shown that carbon nanostructures may be doped, either by electrons or holes. In some cases, such doping processes lead to appreciable generation of lattice defects, introducing disorder. The utilization of dopants which, in parallel, introduce

TABLE 23.1
Examples of Doping of Deposited Graphene in Various Chemical Environments

Graphene Type (Substrate)	Production Method	Chemical Treatment	I_D/I_G	G Band Shift (cm^{-1})	2D Band Shift (cm^{-1})	Doping Type[a]—Density (cm^{-2})	Reference
Monolayer (SiO_2)	Mechanical exfoliation	Thermal annealing in Ar/O_2	0.35 (600°C)	+23 (600°C)	–	H—2.3×10^{13} (600°C)	52
Bilayer (SiO_2)			0.20 (600°C)	+13 (600°C)	–	–	
Trilayer (SiO_2)			0.05 (600°C)	+7 (600°C)	–	–	
Monolayer (SiO_2)	Mechanical exfoliation	Thermal annealing in Ar	<0.01 (400°C)	+20 (400°C)	+20 (400°C)	H—1.9×10^{13} (400°C)	54
Monolayer (SiO_2)	Mechanical exfoliation	Gas NO_2 at r.t.	<0.01	+32	Quenched	H—4.5×10^{13}	55
Monolayer (SiO_2)	Mechanical exfoliation	Rb vapor	<0.01	+25	–35	E—7.0×10^{13}	58
Monolayer (SiO_2)	Mechanical exfoliation	Br_2 gas	<0.01	+44	Quenched	H	59a
		I_2 gas	<0.01	+28	Quenched	H	
Monolayer (SiO_2)	Mechanical exfoliation	$FeCl_3$ gas	<0.01	+47	+28	H—58×10^{13}	61
Monolayer (SiO_2)	Mechanical exfoliation	Melamine vapor at 500°C	0.45	+10	+10	H—1.4×10^{13}	63a
Monolayer (SiO_2)	CVD	Vapors of ethylamine at 70°C	<0.01	+11	+1	E—1.4×10^{13}	63b
Monolayer (SiO_2)	CVD	Spincoating/annealing/removal of fluoropolymer	<0.01	+18	+20	H	65
Monolayer (SiO_2)	Mechanical exfoliation	Electron-donating organics	0.50	–2	+17	E	66
		Electron withdraw organics		+3	+3	H	
Monolayer (SiO_2)	Mechanical exfoliation	MO ads.	<0.10	+11	+8	H	67
Monolayer (SiO_2)	CVD	o-Meo-DMBI ads.	<0.01	+12	–4	E—1.2×10^{13}	69
Monolayer (SiO_2)	Mechanical exfoliation	Diazonium chemistry	0.417 (bulk)	–	–	H	71
			0.764 (edges)	–	–		

[a] H, hole doping; E, electron doping.

structural defects onto the graphene surface could be used to further tailor the carrier concentration, so there is a benefit between the nanomaterial properties and the degradation that needs to be balanced and optimized. Raman spectroscopy coupled with other characterization tools, such as scanning tunneling microscopy (STM), has been considered as a powerful probe of graphene-based materials. Data that can be extracted by analyzing a Raman spectrum of doped graphene samples include the extent of disordered sites and the type of doping. In Table 23.1, various data from the literature corresponding to the previous part of the chapter are given. These involve the graphene type and the substrate, the isolation approach, the chemical protocol, the I_D/I_G ratio of doped material, the shifts of G and 2D bands after doping, the carrier density, and the type of doping. From these data, it can be concluded that gas-phase doping of deposited graphene sheets may be the preferable approach to use in order to achieve enhanced values of carrier density. Concerning the covalent functionalization approaches used for chemical doping, further work should be done toward the control of defect location and defect concentration onto graphene basal plane and edges. In this way, we will be able to fabricate devices which will revolutionalize the field of nanotechnology.

ACKNOWLEDGMENTS

This chapter has been cofinanced by the European Union (European Social Fund-ESF) and Greek national funds through the Operational Program "Education and Lifelong Learning" of the National Strategic Reference Framework (NSRF)—Research Funding Program: ERC-10 "Deformation, Yield and Failure of Graphene and Graphene-based Nanocomposites". The financial support of the European Research Council through the projects ERC AdG 2013 ("Tailor Graphene") and Graphene FET Flagship (Graphene-Based Revolutions in ICT and Beyond, Grant No. 604391) is also acknowledged.

REFERENCES

1. H. Chang, and H. Wu, 2013. Graphene-based nanomaterials: Synthesis, properties, and optical and optoelectronic applications. *Adv. Funct. Mater.* 23:1984–97.
2. Y. Huang, J. Wu, and K.C. Hwang, 2006. Thickness of graphene and single-wall carbon nanotubes. *Phys. Rev. B* 74:245413.
3. J. Huang, M. Juszkiewicz, W.H. de Jeu, E. Cerda, T. Emrick, N. Menon, and T.P. Russell, 2007. Capillary wrinkling of floating thin films. *Science* 317:650.
4. V. Geringer, M. Liebmann, T. Echtermeyer, S. Runte, M. Schmidt, R. Rückamp, M.C. Lemme, and M. Morgenstern, 2009. Intrinsic and extrinsic corrugation of monolayer graphene deposited on SiO_2. *Phys. Rev. Lett.* 102:076102.
5. M. Ishigami, J.H. Chen, W.G. Cullen, M.S. Fuhrer, and E.D. Williams, 2007. Atomic structure of graphene on SiO_2. *Nano Lett.* 7:1643.
6. J. Martin, N. Akerman, G. Ulbricht, T. Lohmann, J.H. Smet, K. von Klitzing, and A. Yacoby, 2008. Observation of electron–hole puddles in graphene using a scanning single-electron transistor. *Nat. Phys.* 4:144.
7. V.M. Pereira, A.H. Castro Neto, H.Y. Liang, and L. Mahadevan, 2010. Geometry, mechanics and electronics of singular structures and wrinkles in graphene. *Phys. Rev. Lett.* 105:156603.
8. N. Levy, S.A. Burke, K.L. Meaker, M. Panlasigui, A. Zettl, F. Guinea, A.H.C. Neto, and M.F. Crommie, 2010. Strain-induced pseudo-magnetic fields greater than 300 tesla in graphene nanobubbles. *Science* 329:544.
9. X. Liu, N.G. Boddeti, M.R. Szpunar, L. Wang, M.A. Rodriguez, R. Long, J. Xiao, M.L. Dunn, and J.S. Bunch, 2013. Observation of pull-in instability in graphene membranes under interfacial forces. *Nano Lett.* 13:2309.
10. S.P. Koenig, N.G. Boddeti, M.L. Dunn, and J.S. Bunch, 2011. Ultrastrong adhesion of graphene membranes. *Nat. Nanotechnol.* 6:543.
11. C.H. Lui, L. Liu, K.F. Mak, G.W. Flynn, and T.F. Heinz, 2009. Ultraflat graphene. *Nature* 462:339.
12. K. Xu, P. Cao, and J.R. Heath, 2010. Graphene visualizes the first water adlayers on mica at ambient conditions. *Science* 329:1188.
13. H. Terrones, R. Lv, M. Terrones, and M.S. Dresselhaus, 2012. The role of defects and doping in 2D graphene sheets and 1D nanoribbons. *Rep. Progr. Phys.* 75:062501.
14. A.C. Ferrari, and D.M. Basko, 2013. Raman spectroscopy as a versatile tool for studying the properties of graphene. *Nat. Nanotechnol.* 8:235.
15. A.C. Ferrari, J.C. Meyer, V. Scardaci, C. Casiraghi, M. Lazzeri, F. Mauri, S. Piscanec et al., 2006. Raman spectrum of graphene and graphene layers. *Phys. Rev. Lett.* 97:187401.
16. O. Frank, G. Tsoukleri, I. Riaz, K. Papagelis, J. Parthenios, A.C. Ferrari, A.K. Geim, K.S. Novoselov, and C. Galiotis, 2011. Development of a universal stress sensor for graphene and carbon fibres. *Nat. Commun.* 2:255.
17. O. Frank, G. Tsoukleri, J. Parthenios, K. Papagelis, I. Riaz, R. Jalil, K.S. Novoselov, and C. Galiotis, 2010. Compression behavior of single-layer graphenes. *ACS Nano* 4:3131–8.
18. O. Frank, M. Mohr, J. Maultzsch, C. Thomsen, I. Riaz, R. Jalil, K.S. Novoselov et al., 2011. Raman 2-D splitting in graphene: Theory and experiment. *ACS Nano* 5:2231.
19. O. Frank, M. Bouša, I. Riaz, R. Jalil, K.S. Novoselov, G. Tsoukleri, J. Parthenios, L. Kavan, K. Papagelis, and C. Galiotis, 2012. Phonon and structural changes in deformed bernal stacked bilayer graphene. *Nano Lett.* 12:687.
20. K. Papagelis, O. Frank, G. Tsoukleri, J. Parthenios, K. Novoselov, and C. Galiotis, 2012. Axial deformation of monolayer graphene under tension and compression In *Graph. 2011*, edited by L. Ottaviano, and V. Morandi (Springer, Berlin, Heidelberg), pp. 87–97.
21. G. Tsoukleri, J. Parthenios, K. Papagelis, R. Jalil, A.C. Ferrari, A.K. Geim, K.S. Novoselov, and C. Galiotis, 2009. Subjecting a graphene monolayer to tension and compression. *Small* 5:2397.
22. T.M.G. Mohiuddin, A. Lombardo, R.R. Nair, A. Bonetti, G. Savini, R. Jalil, N. Bonini et al., 2009. Uniaxial strain in graphene by Raman spectroscopy: G peak splitting, Gruneisen parameters and sample orientation. *Phys. Rev. B* 79:205433.
23. A. Das, S. Pisana, B. Chakraborty, S. Piscanec, S.K. Saha, U.V. Waghmare, K.S. Novoselov et al., 2008. Monitoring dopants by Raman scattering in an electrochemically top-gated graphene transistor. *Nat. Nanotechnol.* 3:210.
24. J. Zabel, R.R. Nair, A. Ott, T. Georgiou, A.K. Geim, K.S. Novoselov, and C. Casiraghi, 2012. Raman spectroscopy of graphene and bilayer under biaxial strain: Bubbles and balloons. *Nano Lett.* 12:617.

25. M. Huang, H. Yan, C. Chen, D. Song, T.F. Heinz, and J. Hone, 2009. Spectroscopy of graphene under uniaxial stress: Phonon softening and determination of the crystallographic orientation. *Proc. Natl. Acad. Sci.* 106:7304.
26. F. Ding, H. Ji, Y. Chen, A. Herklotz, K. Do¨rr, Y. Mei, A. Rastelli, and O.G. Schmidt, 2010. Stretchable graphene: A close look at fundamental parameters through biaxial straining. *Nano Lett.* 10:3453.
27. C. Metzger, S. Rémi, M. Liu, S.V. Kusminskiy, A.H. Castro Neto, A.K. Swan, and B.B. Goldberg, 2010. Biaxial strain in graphene adhered to shallow depressions. *Nano Lett.* 10:6.
28. Q. Wang, Z. Jin, K. Kim, A. Hilmer, G. Paulus, C. Shih, M. Ham et al., 2012. Understanding and controlling the substrate effect on graphene electron-transfer chemistry via reactivity imprint lithography. *Nat. Chem.* 4:724–32.
29. S. Bae, H. Kim, Y. Lee, X. Xu, J.-S. Park, Y. Zheng, J. Balakrishnan et al., 2010. Roll-to-roll production of 30-inch graphene films for transparent electrodes. *Nat. Nanotechnol.* 5:574.
30. M. Lazzeri, and F. Mauri, 2006. Nonadiabatic Kohn anomaly in a doped graphene monolayer. *Phys. Rev. Lett.* 97:266407.
31. T. Tsukamoto, K. Yamazaki, H. Komurasaki, and T. Ogino, 2012. Effects of surface chemistry of substrates on Raman spectra in graphene. *J. Phys. Chem. C* 116:4732–7.
32. Z. Zhang, and J.T. Yates, 2010. Effect of adsorbed donor and acceptor molecules on electron stimulated desorption: O_2/TiO_2(110). *J. Phys. Chem. Lett.* 1:2185.
33. J.E. Lee, G. Ahn, J. Shim, Y.S. Lee, and S. Ryu, 2012. Optical separation of mechanical strain from charge doping in graphene. *Nat. Commun.* 3:1024.
34. H. Bukowska, F. Meinerzhagen, S. Akcöltekin, O. Ochedowski, M. Neubert, V. Buck, and M. Schleberger, 2011. Raman spectra of graphene exfoliated on insulating crystalline substrates. *New J. Phys.* 13:063018.
35. D. Ziegler, P. Gava, J. Güttinger, F. Molitor, L. Wirtz, M. Lazzeri, A.M. Saitta, A. Stemmer, F. Mauri, and C. Stampfer, 2011. Variations in the work function of doped single- and few-layer graphene assessed by Kelvin probe force microscopy and density functional theory. *Phys. Rev. B* 83:235434.
36. T. Yoon, W. Cheol, S. Taek, Y. Kim, J.H. Mun, T. Kim, and B.J. Cho, 2012. Direct measurement of adhesion energy of monolayer graphene as-grown on copper and its application to renewable transfer process. *Nano Lett.* 12:1448 – 52.
37. (a) S. Berciaud, S. Ryu, L.E. Brus, and T.F. Heinz, 2009. Probing the intrinsic properties of exfoliated graphene: Raman spectroscopy of free-standing monolayers. *Nano Lett.* 9:346–52; (b) C.-H. Huang, H.-Y. Lin, C.-W. Huang, Y.-M. Liu, F.-Y. Shih, W.-H. Wang, and H.-C. Chui, 2104. Probing substrate influence on graphene by analyzing Raman lineshapes. *Nanoscale Res. Lett.* 9:64.
38. S. Berciaud, X. Li, H. Htoon, L.E. Brus, S.K. Doorn, and T. F. Heinz, 2013. Intrinsic line shape of the Raman 2D-mode in freestanding graphene monolayers. *Nano Lett.* 13:3517–23.
39. J. Maultzsch, S. Reich, and C. Thomsen, 2004. Double resonant Raman scattering in graphite: Interference effects, selection rules and phonon dispersion. *Phys. Rev. B* 70:155403.
40. R. Saito, A. Jorio, A.G. Souza Filho, G. Dresselhaus, M.S. Dresselhaus, and M.A. Pimenta, 2002. Probing phonon dispersion relations of graphite by double resonance Raman scattering. *Phys. Rev. Lett.* 88:027401.
41. C. Thomsen, and S. Reich, 2000. Double resonant Raman scattering in graphite. *Phys. Rev. Lett.* 85:5214–7.
42. M. Mohr, J. Maultzsch, and C. Thomsen, 2010. Splitting of the Raman 2D band of graphene subjected to strain. *Phys. Rev. B* 82:201409.
43. P. Venezuela, M. Lazzeri, and F. Mauri, 2011. Theory of double resonant Raman spectra in graphene: Intensity and line shape of defect-induced and two-phonon bands. *Phys. Rev. B* 84:035433.
44. J. Kürti, V. Zólyomi, A. Grüneis, and H. Kuzmany, 2002. Double resonance Raman phenomena enhanced by van Hove singularities in single wall carbon nanotubes. *Phys. Rev. B* 65:165433.
45. Z. Luo, C. Cong, J. Zhang, Q. Xiong, and T. Yu, 2012. Direct observation of inner and outer G′ band double resonance Raman scattering in free standing graphene. *Appl. Phys. Lett.* 100:243107.
46. C. Chen, W. Bao, C. Chang, Z. Zhao, C.N. Lau, and S.B. Cronin, 2012. Raman spectroscopy of substrate induced compression and substrate doping in thermally cycled graphene. *Phys. Rev. B* 85:035431.
47. C. Chen, W. Bao, J. Theiss, C. Dames, C.N. Lau, and S.B. Cronin, 2009. Raman spectroscopy of ripple formation in suspended graphene. *Nano Lett.* 9:4172–6.
48. A. Allard, and L. Wirtz, 2010. Graphene on metallic substrates: Suppression of the Kohn anomalies in the phonon dispersion. *Nano Lett.* 10:4335–40.
49. H. Romero, N. Shen, P. Joshi, H. Gutierrez, S. Tadigadapa, J. Sofo, and P. Eklund, 2008. n-Type behavior of graphene supported on Si/SiO_2 substrates. *ACS Nano* 2:2037–44.
50. H. Sojoudi, J. Baltazar, C. Henderson, and S. Graham, 2012. Impact of post growth thermal annealing and environmental exposure on the unintentional doping of CVD graphene films. *J. Vac. Sci. Technol. B* 30:041213.
51. (a) Z. Cheng, Q. Zhou, C. Wang, Q. Li, C. Wang, and Y. Fang, 2011. Toward intrinsic graphene surfaces: A systematic study on thermal annealing and wet-chemical treatment of SiO_2-supported graphene devices. *Nano Lett.* 11:767–71; (b) Y. Kim, D. Cho, S. Ryu, and C. Lee, 2014. Tuning doping and strain in graphene by microwave-induced annealing. *Carbon* 67:673–9.
52. L. Liu, S. Ryu, M. Tomasik, E. Stolyarova, N. Jung, M. Hybertsen, M. Steigerwald, L. Brus, and G. Flynn, 2008. Graphene oxidation: Thickness dependent etching and strong chemical doping. *Nano Lett.* 8:1965–70.
53. M. Yamamoto, T. Einstein, M. Fuhrer, and W. Cullen, 2012. Charge inhomogeneity determines oxidative reactivity of graphene on substrates. *ACS Nano* 6:8335–41.
54. A. Nourbakhsh, M. Cantoro, A. Klekachev, F. Clemente, B. Soree, M. van der Veen, T. Vosch, A. Stesmans, B. Sels, and S. De Gendt, 2010. Tuning the Fermi level of SiO_2-supported single-layer graphene by thermal annealing. *J. Phys. Chem. C* 114:6894–900.
55. A. Crowther, A. Ghassaei, N. Jung, and L. Brus, 2012. Strong charge-transfer doping of 1 to 10 layer graphene by NO_2. *ACS Nano* 6:1865–75.
56. C.A. Howard, M.P.M. Dean, and F. Withers, 2011. Phonons in potassium-doped graphene: The effects of electron–phonon interactions, dimensionality, and adatom ordering. *Phys. Rev. B* 84:241404.
57. N. Jung, B. Kim, A.C. Crowther, N. Kim, C. Nuckolls, and L. Brus, 2011. Optical reflectivity and Raman scattering in few-layer-thick graphene highly doped by K and Rb. *ACS Nano* 5:5708–16.
58. R. Parret, M. Paillet, J. Huntzinger, D. Nakabayashi, T. Michel, A. Tiberj, J. Sauvajol, and A. Zahab, 2013. *In situ* Raman probing of graphene over a broad doping range upon rubidium vapor exposure. *ACS Nano* 7:165–73.
59. (a) N. Jung, N. Kim, S. Jockusch, N. Turro, P. Kim, and L. Brus, 2009. Charge transfer chemical doping of few layer graphenes: Charge distribution and band gap formation. *Nano Lett.*

9:4133–7; (b) J. Chen, P. Darancet, L. Wang, A.C. Crowther, Y. Gao, C.R. Dean, T. Taniguchi et al., 2014. Physical adsorption and charge transfer of molecular Br_2 on graphene. *ACS Nano* 8:2943–50.

60. D. Zhan, L. Sun, Z. Ni, L. Liu, X. Fan, Y. Wang, T. Yu, Y. Ming, W. Huang, and Z. Shen, 2010. $FeCl_3$ based few-layer graphene intercalation compounds: Single linear dispersion electronic band structure and strong charge transfer doping. *Adv. Funct. Mater.* 20:3504–9.
61. W. Zhao, P. Tan, J. Liu, and A. Ferrari, 2011. Intercalation of few-layer graphite flakes with $FeCl_3$: Raman determination of Fermi level, layer decoupling and stability. *J. Am. Chem. Soc.* 133:5941–6.
62. L. Armas, G. Landi, M. Huila, A. Champi, M. Pojar, A. Seabra, A. Santos, K. Araki, and H. Toma, 2012. Graphene modification with gold nanoparticles using the gas aggregation technique. *Diam. Rel. Mater.* 23:18–22.
63. (a) H. Medina, Y. Lin, D. Obergfell, and P. Chiu, 2011. Tuning of charge densities in graphene by molecule doping. *Adv. Funct. Mater.* 21:2687–92; (b) Y. Kim, J. Ryu, M. Park, E.S. Kim, J.M. Yoo, J. Park, J.H. Kang, and B.H. Hong, 2014. Vapor-phase molecular doping of graphene for high-performance transparent electrodes. *ACS Nano* 8:868–74.
64. (a) D. Farmer, R. Golizadeh-Mojarad, V. Perebeinos, Y. Lin, G. Tulevski, J. Tsang, and P. Avouris, 2009. Chemical doping and electron–hole conduction asymmetry in graphene devices. *Nano Lett.* 9:388–92. (b) Y. Choi, Q. Sun, E. Hwang, Y. Lee, S. Lee, J. Cho, 2015. On-demand doping of graphene by stamping with a chemically functionalized rubber lens. *ACS Nano* 9:4354–61.
65. W. Lee, J. Suk, Y. Hao, J. Park, J. Yang, H. Ha, S. Murali et al., 2012. Simultaneous transfer and doping of CVD-grown graphene by fluoropolymer for transparent conductive films on plastic. *ACS Nano* 6:1284–90.
66. X. Dong, D. Fu, W. Fang, Y. Shi, P. Chen, and L. Li, 2009. Doping single layer graphene with aromatic molecules. *Small* 5:1422–6.
67. N. Peimyoo, T. Yu, J. Shang, C. Cong, and H. Yang, 2012. Thickness-dependent azobenzene doping in mono- and few-layer graphene. *Carbon* 50:201–8.
68. X. Miao, S. Tongay, M. Petterson, K. Berke, A. Rinzler, B. Appleton, and A. Hebard, 2012. High efficiency graphene solar cells by chemical doping *Nano Lett.* 12:2745–50.
69. P. Wei, N. Liu, H. Lee, E. Adijanto, L. Ci, B. Naab, J. Zhong et al., 2013. Tuning the Dirac point in CVD-grown graphene through solution processed n-type doping with 2-(2-methoxyphenyl)-1,3-dimethyl-2,3-dihydro-1H-benzoimidazole. *Nano Lett.* 13:1890–7.
70. F. Koehler, A. Jacobsen, K. Ensslin, C. Stampfer, and W. Stark, 2010. Selective chemical modification of graphene surfaces: Distinction between single and bilayer graphene. *Small* 6:1125–30.
71. R. Sharma, J. Baik, C. Perera, and M. Strano, 2010. Anomalously large reactivity of single graphene layers and edges toward electron transfer chemistries. *Nano Lett.* 10:398–405.
72. X. Fan, R. Nouchi, and K. Tanigaki, 2011. Effect of charge puddles and ripples on the chemical reactivity of single layer graphene supported by SiO_2/Si substrate. *J. Phys. Chem. C* 115:12960–4.
73. B. Guo, Q. Liu, E. Chen, H. Zhu, L. Fang, and J. Gong, 2010. Controllable N-doping of graphene. *Nano Lett.* 10:4975–80.
74. Y.-P. Lin, Y. Ksari, J. Prakash, L. Giovanelli, J.-C. Valmalette, and J.-M. Themlin, 2014. Nitrogen-doping processes of graphene by a versatile plasma-based method. *Carbon* 73: 216–24.

24 Developments of Cavity-Controlled Devices with Graphene and Graphene Nanoribbon for Optoelectronic Applications

G. C. Shan, C. H. Shek, and M. J. Hu

CONTENTS

ABSTRACT

Recent years have witnessed many exciting breakthroughs in graphene as a promising material in photonics and optoelectronics. The wonderful optical properties of graphene afford multiple functions of signal emitting, transmitting, modulating, and detection to be realized in one material. The use of cavity to manipulate photon emission and absorption of graphene has opened unprecedented opportunities for realizing functional optoelectronic devices and also quantum photonic devices. This chapter provides an introduction to physical properties of graphene photonics, and then reviews the latest experimental and theoretical progresses on the implementation of graphene into cavity-controlled graphene phototransistors and graphene nanoribbon photodetectors as well. Particular emphasis is placed on cavity-controlled graphene nanoribbon optoelectronic devices, to integrate graphene nanoribbon photonics onto cavities to realize multiple functions of light creation, routing, modulation, computing, and detection enabled by the tunable quasiparticle bandgap and dynamical conductivity of graphene nanoribbon. These recent pioneering developments open up a route toward both the integration of graphene in hybrid silicon photonic circuits to embrace the use of monolithic electronic silicon integrated circuits and system in a "digital" optical communication network, and the implementation of subterahertz wireless communication system as well, in order to maximize system functionality, improve service flexibility, and simplify network operations.

24.1 INTRODUCTION

Low-dimensional graphene and graphene nanostructures have attracted much attention due to their importance in fundamental scientific research and potential industrial applications [1–15]. Graphene as a monolayer of carbon atoms tightly packed into a two-dimensional (2D) honeycomb lattice is a basic building block for graphitic materials of all other dimensionalities (Figure 24.1). Compared with carbon nanotubes, graphenes are ideal systems for investigating the relationship between optical, electrical transport, mechanical properties, and dimensionality [1–9]. Electrons propagating through the 2D structure of graphene have a linear relation between energy and momentum and thus, behave as massless Dirac fermions [1–6]. Consequently, graphene exhibits electronic properties of a 2D gas of charged particles governed by the relativistic Dirac equation, rather than the nonrelativistic Schrödinger

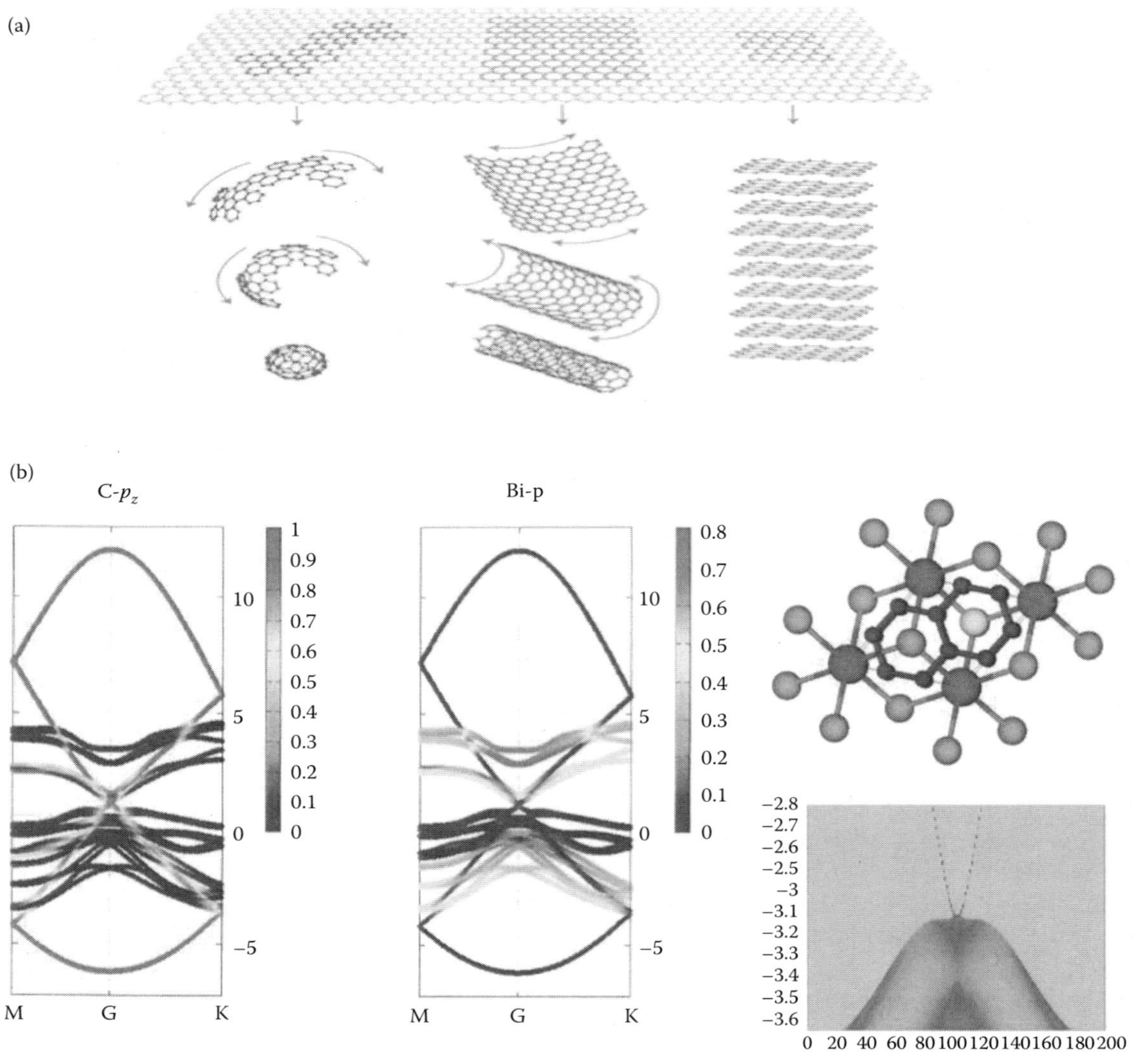

FIGURE 24.1 (a) Schematic structures of all carbon graphitic forms. Graphene is a 2D building block for other carbon allotrope materials. It can be wrapped up into 0D buckyballs C_{60}, rolled into 1D nanotubes or stacked into 3D graphite. (Reprinted by permission from Macmillan Publishers Ltd. *Nat. Mater.* Geim, A. K. and Novoselov, K. S. The rise of graphene, 6, 183–191, copyright 2007.) (b) Hybrid sandwich layer-structure of graphene–BiTeI as a topological insulator with wannier projection band structure and Dirac cone surface states.

equation with an effective mass [1–3], with carriers mimicking particles with zero mass and an effective "speed of light" of around 10^6 m/s. Graphene also exhibits a variety of transport phenomena that are characteristic of 2D Dirac fermions such as specific integer and fractional quantum Hall effects [1–4], a "minimum" conductivity of ~$4e^2/h$ even when the carrier concentration tends to zero, and Shubnikov–de Haas oscillations with a π phase shift due to Berry's phase [1]. In combination with the near-ballistic transport property at room temperature, very high mobilities (μ) of up to 10^6 cm^2/V s are observed in suspended samples, rendering graphene a potential material for nanoelectronics [4–10], particularly for high-frequency applications [11]. Graphene also shows excellent optical properties and magnetic properties [11–25]. For example, it can be optically visualized, despite being only one single atomic layer thick. Its transmittance can be expressed in terms of the fine-structure constant. The linear dispersion of the Dirac electrons makes broadband applications possible. Saturable absorption is also observed as a consequence of Pauli blocking [14,15] and nonequilibrium carriers lead to hot luminescence [16–20]. These properties make it an ideal photonic and optoelectronic material. Besides, given recent progress in graphene-based terahertz (THz) and infrared (IR) emitters and detectors, graphene may offer some interesting solutions for future THz technologies [26–39], which promise myriad applications including imaging, spectroscopy, and communications.

Furthermore, when graphene is etched or patterned along one specific direction, a novel quasi-one-dimensional structure, a stripe of graphene of nanometer in width, can be obtained, which is referred to as a graphene nanoribbon (GNR) [4,10]. The GNRs are predicted to exhibit various remarkable physical properties. Moreover, since GNRs have the large surface–volume ratio and special edge states, their properties can be modified by many methods, such as doping and adsorption. Due to their various edge structures, GNRs present different electronic and magnetic properties ranging from normal semiconductors to spin-polarized half metals, which open up the possibility of GNRs as a promising material for high-frequency electronic device application, as well as nanoscale photonic and spintronic devices. Notably, one hybrid sandwich graphene–BiTeI structure has recently been developed to be a nontrivial topological insulator with one Dirac-cone topological surface state [21], making it viable for room-temperature applications in spintronic devices.

To review recent progress in utilizing graphene to realize the graphene photonic nanodevices, we begin Section 24.2 with a discussion of the basic properties and concepts necessary to understand the properties and characteristics of graphene photonic nanodevices. Section 24.3 discusses recent experiment demonstration of several new prototypical graphene photonic devices, in which a theoretical model for the microcavity-controlled GNR photodetector (GNR-PD) consisting of an array of GNRs sandwiched between two highly reflecting mirrors of a planar microcavity working in the frequency band from THz to IR range is mainly described. Furthermore, in Section 24.4, particular emphasis is the development of graphene nanodevices for the sub-THz wireless system. Finally, Section 24.5 concludes by briefly outlining some prospective issues in the graphene photonic integrated circuit for optical communication.

24.2 BASIC PHYSICAL PROPERTIES

24.2.1 Electronic Structure of Graphene

The electronic structure of single-layer graphene (SLG) or monolayer graphene can be described using a tight-binding Hamiltonian [5,6]. Because the bonding and antibonding σ-bands are well separated in energy (>10 eV at the Brillouin zone center Γ), they can be neglected in semiempirical calculations, retaining only the two remaining π-bands. The electronic wavefunctions from different atoms on the hexagonal lattice overlap. And, the p_z electrons, which form the π-bonds, can generally be treated independently from the other valence electrons. Within this π-band approximation, it is easy to describe the electronic spectrum of the total Hamiltonian and to obtain the dispersion relations $E_\pm(k_x, k_y)$ restricted to first-nearest-neighbor interactions only

$$E_\pm(k_x,k_y) = \pm\gamma_0\sqrt{1+4\cos\frac{\sqrt{3}k_xa}{2}\cos\frac{\sqrt{3}k_ya}{2}+4\cos^2\frac{\sqrt{3}k_ya}{2}} \tag{24.1}$$

where $a = \sqrt{3}a_{cc}$ (with a_{cc} = 1.42 Å being the carbon–carbon bond length) and γ_0 is the transfer integral between the nearest-neighbor π-orbitals (typical values for γ_0 are 2.9–3.1 eV). The $\mathbf{k} = (k_x, k_y)$ vectors in the first Brillouin zone constitute the ensemble of available electronic momenta. With one p_z electron per atom in the π–π^* model, the (−) band (negative energy branch) in Equation 24.1 is fully occupied, whereas the (+) branch is totally empty. These occupied and unoccupied bands touch at the **K** points. Note that the Fermi level E_F is the zero-energy reference, and the Fermi surface is defined by **K** and **K′**. Moreover, by expanding Equation 24.1 at **K** (**K′**), the linear π- and π^*-bands of the low-energy band structures for Dirac fermions are given by

$$E_\pm(\kappa) = \pm\hbar v_F\,|\kappa| \tag{24.2}$$

where $\kappa = \mathbf{k} - \mathbf{K}$ and v_F is the electronic group velocity, which is given by $v_F = \sqrt{3}\gamma_0 a/(2\hbar) \approx 10^6$ m/s.

24.2.2 Optical Transitions and Photoconductivity of Graphene

The optical properties, inherited from its singular conical band structure, are noteworthy. These properties would suit a variety of controllable broadband photonic devices, ranging from the THz to IR and visible region (Figure 24.2 and Table 24.1) [26].

The imaginary part of the permittivity of a material is associated with loss or absorption of electromagnetic waves, which is in turn proportional to the real part of its optical conductivity. To understand the tunability of optical absorption in graphene, we need to look first at what determines its optical conductivity. Note that optical transitions (thus absorption) in graphene include intraband and interband transitions. Based on linearization of the tight-binding Hamiltonian of graphene near the **K** (**K′**) points (i.e., the Dirac points) of the first Brillouin zone, its optical conductivity can be expressed as the sum of the two contributions [6,7]

$$\sigma(\omega) = \sigma_{\text{intra}}(\omega) + \sigma_{\text{inter}}(\omega) = \frac{ie^2E_f}{\pi\hbar^2(\omega+i/\tau)} + \frac{ie^2\omega}{\pi}\int_0^\infty \frac{f(\varepsilon-E_f)-f(-\varepsilon-E_f)}{(2\varepsilon)^2-(\hbar\omega+i\Gamma)^2}d\varepsilon \tag{24.3}$$

Here, e is the electron charge, $\hbar$ the reduced Planck constant, E_f the Fermi level, f the Fermi distribution function, τ momentum relaxation time, and Γ a parameter describing the broadening of interband transitions. At low frequencies, thus low photon energies, the graphene optical conductivity is mainly determined by intraband transitions while, at high frequencies, the contribution of interband transitions becomes dominant, as shown in Figure 24.2. In the limit of $\omega \to \infty$, it follows that, at very high frequencies including the visible range, its optical conductivity reduces to a constant value of $e^2/4h$, thus, a universal absorption per graphene layer of ~2.3% at normal incidence [7,12,13]. Interestingly, the optical image contrast can be used to identify graphene on top of a Si/SiO_2 substrate [12]. This scales with the number of layers and is the result of interference, with SiO_2 acting as a spacer. The transmittance of a freestanding SLG can be derived by applying

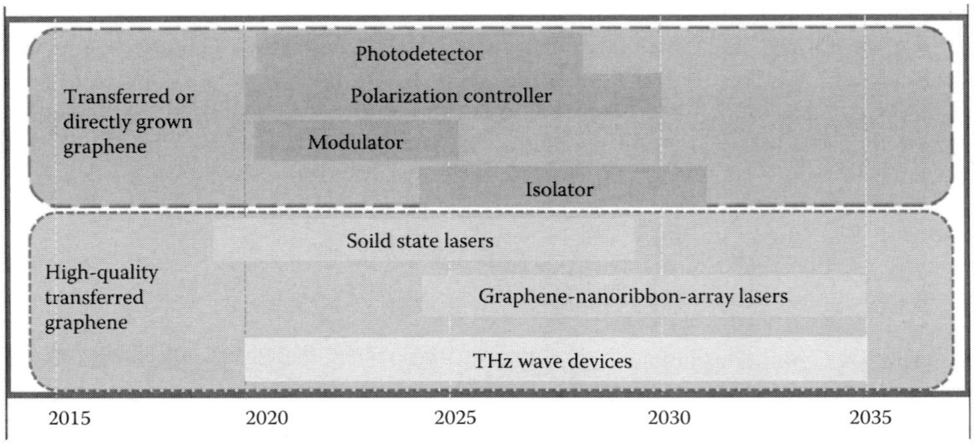

FIGURE 24.2 Graphene-based photonics applications with possible application timeline, enabled by continued advances in graphene technologies, based on projections of products requiring advanced materials such as graphene. The figure gives an indication of when a functional device prototype could be expected based on device roadmaps and the development schedules of industry leaders.

TABLE 24.1
Photonic Device Applications Made of Graphene

Device Application	Advantageous Properties	References
Solid-state mode-locked laser and GNR-array lasers	It is easy to integrate graphene-saturable absorber into the laser system and utilize the wavelength tunability of different GNRs.	[15,16] [26]
Photodetectors	Graphene can supply bandwidth per wavelength of 640 GHz for chip-to-chip communications, which is not possible with IV or III–V detectors.	[30–38]
Polarization controller	Landau quantization in 2D graphene results in a giant rotation with fast response and broadband tenability.	[9]
Optical modulator	Core-to-core and core-to-memory bandwidth increase require a dense wavelength-division-multiplexing (WDM) optical interconnect with over 50 wavelengths.	[3]
Isolator	Graphene can provide integrated miniaturized isolators on a Si substrate; note that decreasing magnetic field strength and optimization of process architecture are important for the products.	[29]

the Fresnel equations in the thin-film limit for a material with a fixed universal optical conductance $G_0 = e^2/(4\hbar)$ to give $T = (1 + 0.5\pi\alpha)^{-2} \approx 1 - \pi\alpha \approx 97.7\%$, where $\alpha = e^2/(4\pi\varepsilon_0\hbar c) = G_0/(\pi\varepsilon_0 c) \approx 1/137$ is the fine structure constant [12,13]. In a few-layer graphene (FLG) sample, each sheet can be seen as a 2D electron gas, with little perturbation from the adjacent layers, making it optically equivalent to a superposition of almost noninteracting SLG. Thus, the optical absorption of graphene layers could be taken to be proportional to the number of layers, each absorbing $A \approx 1 - T \approx \pi\alpha \approx 2.3\%$ over the visible spectrum. Graphene only reflects <0.1% of the incident light in the visible region, rising to ~2% for 10 layers [12,13]. Based on this property, graphene has been actively pursued as a material for transparent flexible electrodes [9].

The dependence of optical conductivity with Fermi level indicates that the graphene optical conductivity can be modified by controlling the Fermi level, that is, carrier concentration. For linear dispersions near the Dirac point, pair–carrier collisions cannot lead to interband relaxation, thereby conserving the total number of electrons and holes separately [13]. Interband relaxation by phonon emission can occur only if the electron and hole energies are close to the Dirac point (within the phonon energy). Radiative recombination of the hot electron–hole population has also been suggested [14–17].

In the far-infrared (FIR) and THz spectral region, the intraband response in graphene becomes pronounced, leading to the possibility of strong extinction in SLG [11,21,22], as well as of plasmon excitation through appropriate coupling [17,21–23]. Understanding the dynamics of this optical response to photoexcitation is a topic of both fundamental interest and of critical importance for the development of high-performance optoelectronic devices. Heinz et al. [23] have measured the THz frequency-dependent sheet conductivity and its transient response following femtosecond optical excitation for SLG samples grown by chemical vapor deposition. The optical excitation of the graphene is implemented by a femtosecond pulse and probing of the THz response using a time-domain spectroscopy approach. The THz field was detected by free-space electro-optic sampling (EOS). The sampling beam was scanned in time using an optical delay stage. The signal was collected with a lock-in amplifier phaselocked to an optical chopper that modulated either the THz generation or the pump beam at a frequency of 500 Hz. The conductivity of the unexcited graphene sheet, which was spontaneously doped, showed a strong free-carrier response. The THz conductivity matched a Drude model over the available THz spectral range and yielded an average carrier scattering time of 70 fs. Upon photoexcitation, a transient decrease in graphene conductivity was observed [21,23]. Characterization of the THz conductivity of the unpumped graphene sample was obtained from THz time-domain spectroscopy measurements performed in a transmission configuration. A significant attenuation of THz transmission can be observed immediately, arising from the strong response of charge carriers in the graphene. By recording the transmitted THz electric field at different fixed delay times between the pump pulse and THz probe pulse, the differential waveform results show that rather than a simple change in the amplitude of the transmitted field with pump excitation, the change in THz waveform indicates the role of the finite carrier scattering time in graphene samples. To investigate the relaxation dynamics of the graphene sample, the peak of the pump-induced differential signal $\Delta E_\tau(t)$ as a function of probe delay τ was also recorded. The physical origin of the observed effect can be understood by considering the Drude conductivity of graphene with a real conductivity of $\sigma = D/\pi\Gamma$ in the low-frequency limit. Thus, photoinduced changes in both the Drude weight D and scattering rate Γ can affect the conductivity. Under excitation, D may increase through a rise in the carrier concentration while Γ may increase through a rise in the effective temperature of the system. As a result, depending on the initial Drude weight, the scattering rate of the unpumped graphene, and the pump fluence, the ratio of D/Γ and the conductivity can either increase or decrease after photoexcitation.

Interestingly, graphene can exhibit a different type of behaviors for the transient response, depending on the detailed conditions. When both the initial doping level and scattering time are relatively high, the main effect is that of change in the scattering rate, rather than the change in the Drude weight, that is, a response resembling more than that of a metal. This is the case observed in Heinz's measurements [23]. It should be noted that the dominant change is an increase in the carrier scattering rate, rather than an increase in the Drude weight. This explains the observed negative THz photoconductivity response. The observed temporal dynamics of the THz response of the photoexcited graphene and its dependence on laser fluence is summarized in Figure 24.3. Figure 24.3a shows the increase in the THz waveform induced by the pump beam, measured at the maximum of the time-domain waveform, as a function of the delay time. Neglecting the change in the spectral characteristics, this can be considered as yielding the temporal evolution of the decrease in the graphene conductivity. The sharp rise in the response corresponds to heating of the electrons by the pump pulse, resulting in an enhanced scattering rate and a decrease of conductivity. The relaxation to a value close to the initial response occurs as the electronic system cools down by phonon emission. Several experimental results show that photoinduced switching of graphene on the time scale of 1 ps can be readily achieved [17,21]. At relatively high pump fluence, the increase of the Drude weight becomes more and more important. This gives rise to an increase in the graphene conductivity, which compensates the decrease in the conductivity caused by an increased electron scattering rate. The balance of these factors provides a natural explanation for the apparent saturation behavior in the photoresponse with increasing pump fluence (Figure 24.3b). In addition to this factor related to the amplitude of the response, there is a modest increase in the decay time [23]. The reduced sensitivity of the response of the conductivity to the electronic temperature at high fluences can explain this trend since the initial rise in the THz response will then be diminished and the apparent decay time lengthened.

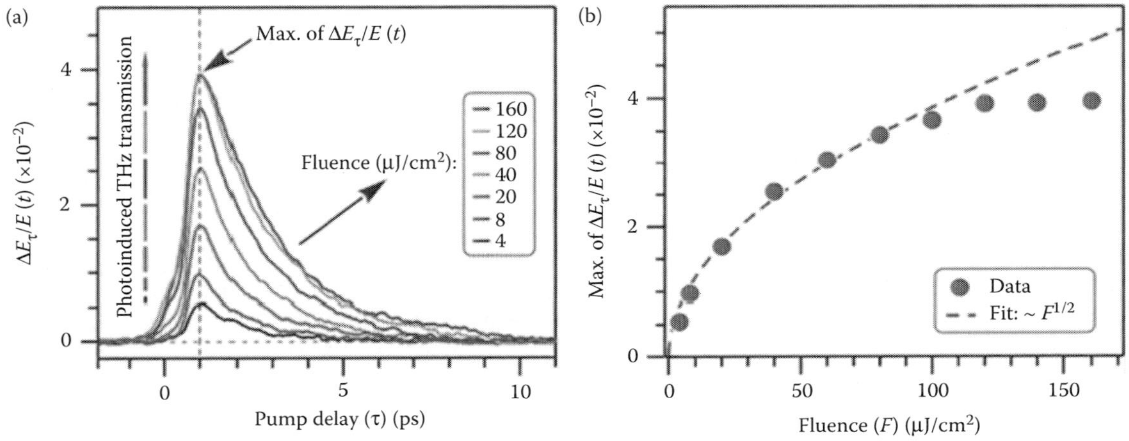

FIGURE 24.3 Transient THz transmission response of photoexcited graphene layer: (a) temporal evolution of the change in the maxima of the transmitted THz waveforms normalized by the THz probe field, that is, $\Delta E_\tau(t)/E(t)$, for different applied fluences of optical pump pulse. (b) Maxima of $\Delta E_\tau(t)/E(t)$ from (a) as a function of applied fluence, which saturates at higher fluences. The dashed line varies as the square root of the fluence, as a guide to the eye. (Adapted with permission from Jnawali, G., Rao, Y. et al. Observation of a transient decrease in terahertz conductivity of single-layer graphene induced by ultrafast optical excitation, *Nano Lett.*, 13, 524–530. Copyright 2013 American Chemical Society.)

24.2.3 Luminescent Properties

Graphene could be made luminescent by inducing a bandgap, following two main routes. One is by cutting it into ribbons and quantum dots, the other is by chemical or physical treatments, to reduce the connectivity of the π-electron network. Although GNRs have been produced with varying bandgaps [3], no photoluminescence has been reported from them thus far. However, bulk graphene oxide dispersions and solids do show a broad photoluminescence [7,24]. The combination of photoluminescent and conductive layers could be used in sandwich light-emitting diodes covering the IR, visible, and blue spectral ranges [23–26].

Even though some groups have ascribed photoluminescence in graphene oxide to the bandgap, emission from electron-confined sp^2 islands is dominant [24]. Whatever the origin, fluorescent organic compounds are of importance to the development of low-cost optoelectronic devices. Although widely used for biolabeling and bioimaging, the nanotoxicity and potential environmental hazard of luminescent quantum dots limit the widespread use and *in vivo* applications. Fluorescent biocompatible carbon-based nanomaterials may be a more suitable alternative. Fluorescent species in IR and near-infrared ranges (NIRs) are useful for biological applications because cells and tissues show little auto-fluorescence in this region.

Wang et al. have reported a gate-controlled, tunable gap up to 250 meV in bilayer graphene [20]. This may make new photonic devices possible for an FIR light generation, amplification, and detection. Broadband nonlinear photoluminescence is also possible following nonequilibrium excitation of untreated graphene layers, as recently reported by several groups [17–20]. Emission occurs throughout the visible spectrum, for energies both higher and lower than the exciting one, in contrast with conventional photoluminescence processes [7,17–20]. This broadband nonlinear photoluminescence is thought to result from the radiative recombination of a distribution of hot electrons and holes, generated by rapid scattering between photoexcited carriers after the optical excitation [17–20]. In addition, electroluminescence was also reported in pristine graphene [7].

24.3 GRAPHENE-BASED PHOTODETECTORS

Graphene-based photonic devices exhibit broadband performance, which is directly or indirectly correlated with the unique electronic structure of graphene. The broadband operation of the graphene-based devices has been used as THz, NIR, visible, and even ultraviolet (UV) light detectors. Here, we will provide a brief review of the latest development in graphene-based photodetectors.

24.3.1 Graphene Photodetectors

Photodetector gain for ultrahigh sensitivity has been exploited in avalanche photodiodes and photomultipliers, but these devices require the application of a high electric bias and are also bulky, which means they are not compatible with current circuit technologies such as complementary metal-oxide semiconductor (CMOS) electronics. The phototransistor (a type of photodetector consisting of a transistor channel with an optically controlled gate) has been reported to provide ultrahigh sensitivity at visible wavelengths or in the short-wave infrared (SWIR) using III–V epitaxially grown quantum dots in a two-dimensional electron gas (2DEG); however, this was achieved while operating at the impractically low temperature of 4 K. Heterojunction phototransistors [27], as well as photo-field-effect transistors (FETs) based on III–V semiconductors [28], have also been reported with gains of the order of 1×10^2 to 1×10^3 and gain–bandwidth products typically in the 1×10^8 Hz range. These devices rely on epitaxially grown III–V semiconductors, which curtail their monolithic integration onto CMOS electronics and impedes their utilization in high-sensitivity sensing and imaging systems.

The rapid development of graphene-based photodetection has been made to achieve the enhancement of the absorption of light in graphene, for example, by exploiting double-layer graphene heterostructure [30], graphene plasmons [31,32], or microcavities [33,34]. Graphene photodetectors may find a wide range of photonic applications including high-speed optical communications, THz detection, imaging, remote sensing, surveillance, and spectroscopy [35–45]. However, the key to ultrasensitive graphene-based photodetection is the implementation of photoconductive gain—the ability to provide multiple electrical carriers per single incident photon.

The combination of the exceptional transport and optical properties of graphene suggests novel photonic devices fundamentally different from conventional arrangements. In 2009, Xia et al. explored the use of zero-bandgap, large-area single or few layers of graphene FETs as ultrafast photodetectors [35]. On light absorption, the generated electron–hole pairs in graphene would normally recombine on a timescale of tens of picoseconds, depending on the quality and carrier concentration of the graphene [35–37]. If an external field is applied, the pairs can be separated, and a photocurrent is generated. The same happens in the presence of an internal field. It has been demonstrated that the internal fields that are formed near the metal electrode–graphene interfaces can be used to produce an ultrafast photocurrent response in graphene [35,38–40]. Owing to the high carrier transport velocity existing even under a moderate E-field, no direct bias voltage between source and drain (photocurrent generation path) is needed to ensure ultrafast and efficient (6%–16% internal quantum efficiency within the photodetection region) photocurrent generation. Such a zero-bias operation has been demonstrated previously in special unitravelling carrier photodiodes (UTC-PD) with much higher built-in E-field for high-bandwidth operation. However, in most conventional photodiodes, large external bias directly applied to the photodetecting area, and its complete depletion is essential for fast and efficient photodetection.

To generate a photocurrent in an external circuit, the photogenerated carriers must exit the photogeneration region before they recombine. Although the short photocarrier lifetime in graphene is unfavorable for efficient photocurrent generation, it is desirable for other photonic applications such as optical switches, in which quick annihilation of photogenerated carriers is essential for high-speed operation. The transit-time-limited bandwidth of the photodetector is given by

$$f_t = 3.5/2\pi t_{tr} \quad (24.4)$$

where t_{tr} is the transit time through the photodetection region. Thus, a value of f_t of 1.5 THz is obtained. If 200-nm-thick $In_{0.53}Ga_{0.47}As$ is used as the photodetection layer, the transit-time-limited bandwidth is around 150 GHz, which is 10 times smaller than that in graphene and mainly limited by the saturation velocity of the holes.

A scanning electron microscopy (SEM) image and the optical microscope image at a lower magnification of the graphene devices with two to three layers of graphene are shown in Figure 24.4a [35]. The number of graphene layers was first estimated from the color of the optical images and then confirmed by Raman spectroscopy. The SiO_2 film underneath the graphene is 300 nm thick, and the substrate is highly resistive silicon (1–10 kV cm), to minimize the parasitic capacitance. Two types of high-frequency coplanar waveguide wirings are shown: ground–signal (G–S) and ground–signal–ground (G–S–G). The layout of the devices in the G–S wiring scheme are depicted in Figure 24.4b [35].

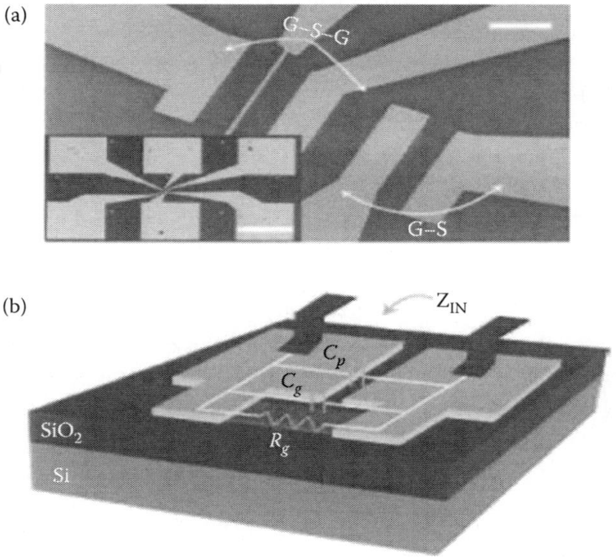

FIGURE 24.4 Electrical characterizations of the graphene photodetector: (a) SEM and optical (inset) images of the high-bandwidth graphene photodetectors. The graphene shown here has two to three layers. Two types of wirings are shown: ground (G)–signal (S) and G–S–G. The high-frequency results are from devices with G–S wirings. Scale bars: main panel, 2 mm; inset, 80 mm. (b) Device schematics and electrical model in the high-frequency domain. The high-frequency reflection coefficient GIN is measured using a network analyzer, and ZIN is inferred from GIN. The symbols, from top to bottom, represent C_p, C_g, and R_g, respectively. (Reprinted by permission from Macmillan Publishers Ltd. *Nat. Nanotechnol.* Konstantatos, G. et al. Hybrid graphene–quantum dot phototransistors with ultrahigh gain, 7, 363, copyright 2012.)

Nevertheless, it is until very recently that gain has been observed in graphene-based photodetectors. Recently, a novel hybrid graphene–quantum dot phototransistor that exhibits ultrahigh photodetection gain and high quantum efficiency, enabling high-sensitivity and gate-tunable photodetection has been developed [45]. The device consists of a graphene sheet that is sensitized with colloidal quantum dots. Graphene is the carrier transport channel, and the quantum dots are used as the photon absorbing material (Figure 24.5a). The channel of the phototransistor, consisting of a monolayer or bilayer graphene sheet decorated with PbS quantum dots, is placed on top of a silicon/SiO_2 wafer, as shown in the inset of Figure 24.5b [45]. The spatially resolved photocurrent response of the designated structure on illumination, measured with a focused laser beam is shown in Figure 24.5b [45]. The key functionality of this light-activated transistor is provided by a layer of strongly light-absorbing and spectrally tunable colloidal quantum dots, from which photogenerated charges can transfer to graphene while oppositely charged carriers remain trapped in the quantum-dot layer, where the presence of these charges changes the graphene sheet resistance through capacitive coupling. The main feature of the device is its ultrahigh gain, which originates from the high carrier mobility of the graphene sheet (1×10^3 cm^2/V s) and the recirculation of charge carriers during the lifetime of the carriers that remain trapped in the PbS quantum dots [45]. The magnitude of the photodetection gain can be quantified based on a simple physical picture incorporating the typical lifetimes of electrons and holes.

Photoexcited holes in the PbS quantum dots are transferred to the graphene layer and drift by means of a voltage bias V_{DS} to the drain, with a typical timescale of $t_{transit}$, which is inversely proportional to the carrier mobility. Electrons remain trapped (with a typical timescale of $t_{lifetime}$) in the PbS quantum dots as a result of the built-in field at the quantum dot/graphene interface as well as the electron traps in the PbS quantum dots [45]. Charge conservation in the graphene channel leads to hole replenishment from the source as soon as a hole reaches the drain. Accordingly, multiple holes circulate in the graphene channel following a single electron–hole photogeneration, leading to photoconductive gain. The ability to maximize the photoconductive gain or to fully reduce it to zero is useful for pixelated imaging applications, where the implementation of nanoscale local gates enables a locally tunable photoresponse. Furthermore, it has also been demonstrated that a short voltage pulse applied to the gate can be used to purge actively the charge carriers from the quantum dots to reset the device and thereby increase the operating speed.

24.3.2 Cavity-Controlled Graphene Nanoribbon Photodetector

Photodetectors for THz and NIR regimes of the spectrum have been demonstrated using intersubband transitions in semiconductor quantum-well structures and narrow-gap semiconductors [41,47–49]. The quantum well IR photodetector based on

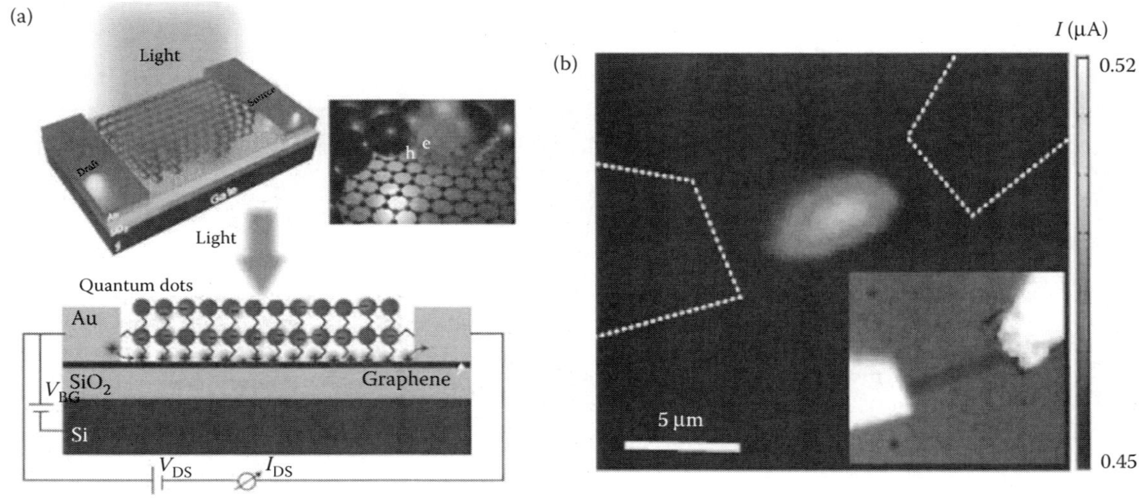

FIGURE 24.5 Hybrid graphene–quantum dot phototransistor: (a) schematic of the graphene–quantum dot hybrid phototransistor, in which a graphene flake is deposited onto a Si/SiO_2 structure and coated with PbS quantum dots. Incident photons create electron–hole pairs in the PbS quantum dots. Holes are then transferred to the graphene channel and drift toward the drain, but electrons remain in the PbS quantum dots, leading through capacitive coupling to a prolonged time during which (recirculated) carriers are present in the graphene channel. (b) Spatial photocurrent profile using a laser beam focused at 532 nm with a power of 1.7 pW. The photocurrent was recorded as the laser beam was scanned across the surface of the detector. Inset: optical image of the graphene flake used in this study in contact with the gold electrodes, forming the phototransistor. (Reprinted by permission from Macmillan Publishers Ltd. *Nat. Nanotechnol.* Konstantatos, G. et al. Hybrid graphene–quantum dot phototransistors with ultrahigh gain, *Nature Nanotech.*, 7, 363, copyright 2012.)

intersubband transitions in quantum wells was first developed for the NIR spectrum, and has become a mature technology for the wavelength region of 3–140 μm [45–50]. NIR and THz radiation and detection within the frequency between 0.1 and 400 THz has many potential applications in areas such as genetic research, drug development, environment monitoring, medical imaging, security screening, etc. [41–53]. The recent progress in THz source and detectors, as well as the blooming of application proposals, fostered the necessity to develop devices capable of actively manipulating THz waves, including switches, modulators, filters, polarizers, etc. [44–54]. Moreover, the NIR and THz photodetectors have the potential of very fast time response due to the nature of intersubband transition, which makes them very attractive for applications such as THz heterodyne detection, THz free space communications [44,45,54], high-speed THz imaging [55], etc.

As we have mentioned before, graphene photodetectors may find a wide range of photonic applications including high-speed optical communications [44,56], THz detection [43], imaging, remote sensing, surveillance, and spectroscopy [43,53]. Moreover, a GNR, for example, a strip of graphene in nanoscale width, which can be obtained by etching or patterning, is a promising candidate for optoelectronic and THz photonic nanodevices [34,57–67], because the interband transitions can be utilized to absorb photons of low energy in the NIR/THz regime. The optical response at selective frequencies can be enhanced with the use of GNRs [58–62]. Optical conductance exhibits strong peaks at frequencies corresponding to resonant interband absorption in given GNRs. In these GNRs, however, no optical response was detected and the nearly universal conductance in the visible range is totally suppressed making them transparent in this frequency range. More importantly, it was recently demonstrated that graphene can be monolithically integrated with other more established materials to form novel highly complex devices [34,62–66], which opens up real application prospects in low-cost photodetectors, nanolasers, and transmitter modules by releasing the need for temperature control, isolators, and external modulators.

We have recently proposed a GNR-PD consisting of an array of zigzag GNRs sandwiched between the highly reflecting mirrors of a planar microcavity and the theoretical model is further presented to assess its performance in the important frequency band of THz/NIR regime [58]. The structure of GNR-PD under consideration is schematically shown in Figure 24.6. The device consists of an array of GNRs as the quantum medium of mesoscopic p–n junctions, which is embedded into a planar λ/2 microcavity. Because of its 2D geometry, graphene nanostructures is ideally suited for enclosure within a photonic structure of a planar λ/2 microcavity, in which the optical cavity structure confines optical fields between two highly reflecting mirrors with a spacing of only one-half wavelength of light and is used to enhance the photodetection. The chosen thickness of the intracavity dielectrics determines the resonance wavelength of the optical microcavity. To simulate the device structure, FULLWAVE and FEM package of RSOFT software were used to calculate the transmission spectra of the planar microcavity and the electromagnetic field modes [58,68]. Final optimized structure (with all parameters and the GNR

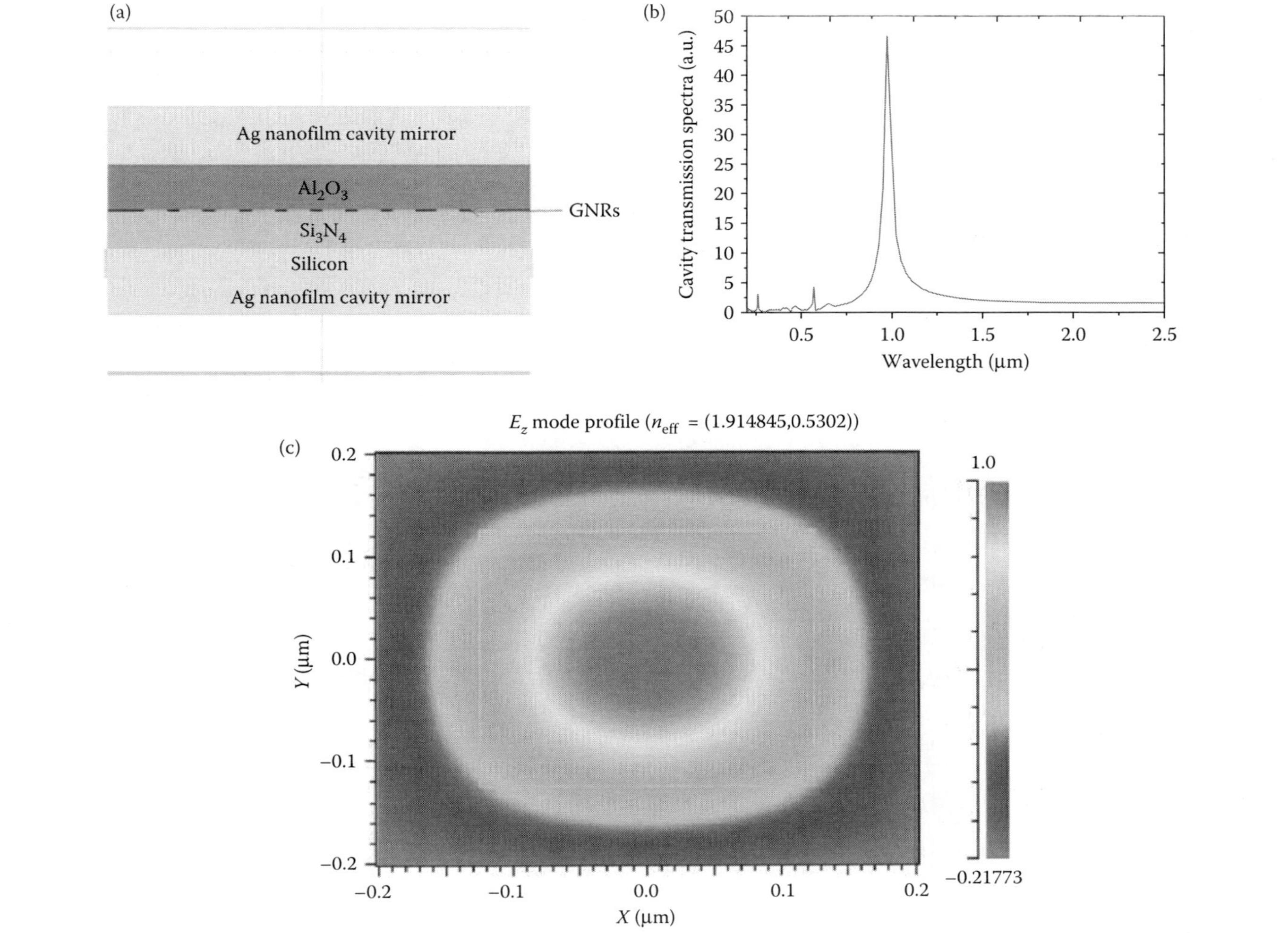

FIGURE 24.6 (a) Schematic structure of a GNR-PD. (b) Simulated transmission spectra of the planar microcavity in the GNR-PD presented here. (c) Simulated electromagnetic field mode distribution for such a GNR-PD.

array thickness $t = 0.5$ nm) and the transmission spectra of the planar microcavity and the electromagnetic field modes are shown in Figure 24.6.

To achieve a strong photoconductance or optical response in a GNR in the FIR and THz frequency range, it is necessary to lift the strict selection rules that forbid the dipole interband transitions at lower energies. For intrinsic graphene, the selection rules can be modified to allow THz/FIR transitions by applying a perpendicular magnetic field. When a magnetic field is introduced

$$E\psi_i = \sum_j t_{ij} e^{i\gamma_{ij}} \psi_j \tag{24.5}$$

where $\gamma_{ij} = (2\pi/\phi_0)\int_i^j \mathbf{A}\cdot d\mathbf{l}$ is the magnetic phase factor, with $\Phi_0 = hc/e$ being the magnetic flux quantum. The Schrödinger equation for the A and B sublattice can be obtained [58,69,70]. The B-field modifies the energy dispersions and changes the size of the bandgap, induces the semiconductor–metal transition and generates the partial flat bands [58]. As the magnetic field increases such that the cyclotron radius is smaller compared to the ribbon width, the Landau levels are developed. This allows absorption by direct transitions between the magnetic subbands. Thus, we have the following Kubo formula to calculate the conductance of GNRs:

$$J_x = e\frac{\partial H}{\partial(\hbar k_x)},\quad J_y = e\frac{\partial H}{\partial(\hbar k_y)} \tag{24.6}$$

By introducing the field operator $\hat{\psi}(x,y) = \sum_{k,j} \xi_j e^{iky} c_{kj}$, the current operator can be expressed in the second quantization notation

$$\hat{J}_\mu = \int dy \hat{\psi}^+(x,y) J_\mu \hat{\psi}(x,y) = \sum_{kjj'} J^\mu_{jj'} c^+_{kj} c_{kj'} \tag{24.7}$$

with $J_{jj} = J_j$ and $\mu = x, y$. According to the Kubo formula, the optical conductance is found as

$$G_{\mu\nu}(\omega) = -\frac{1}{i\omega}\sum_{kjj'} J^{\mu}_{jj'} J^{\nu}_{j'j} \frac{f_{kj} - f_{kj'}}{\hbar\omega + \varepsilon_{kj} - \varepsilon_{kj'} + i\delta} \quad (24.8)$$

where f_{kj} and $f_{kj'}$ are Fermi distribution functions and δ is positive infinitesimal. As the transverse average of the conductance, the optical conductivity is given by

$$\sigma_{\mu\nu}(\omega) = G_{\mu\nu}(\omega)/W \quad (24.9)$$

where W is the physical width of the nanoribbon. For armchair GNRs, $W = (\sqrt{3}a/2)(p+2)$, and for zigzag GNRs, $W = (3p + 1)a/2$.

The optical conductance of typical zigzag GNRs under zero magnetic field is shown in Figure 24.7a. A series of van Hove-type absorption peaks as the main feature exist at frequencies corresponding to the allowed interband transitions. Note that the edge states of GNRs, which play an important role in the optical absorption, are involved in many of the absorption peaks within the optical range and have no contribution to the absorption peaks beyond the optical range. For all GNRs, there exists a threshold frequency below which the optical conductance is zero. The threshold frequency decreases as the width of GNRs increases. In general, the optical absorption in the low-frequency region decreases and the threshold frequency increases as the chemical potential increases. Note that, for the armchair GNRs, the threshold frequency corresponds to the allowed transition from the second valence band to the second conduction band.

To achieve a strong optical response in a GNR in the THz and FIR range, it is necessary to lift the strict selection rules that forbid the dipole interband transitions at lower energies. The B-field modifies the energy dispersions, changes the bandgap, and generates the partial flat bands as well [58,71]. As the magnetic field increases such that the cyclotron radius is smaller than the width of nanoribbons, the Landau levels are developed. This allows absorption by direct transitions between the magnetic subbands.

Figure 24.8 shows the optical conductance of different GNRs under various magnetic fields. As shown in Figure 24.8, we can find that for a given GNR, the position and intensity of the first conductance resonance can be tuned by varying the magnetic field. The variation of the threshold frequency with the field is shown clearly. At $f = 0.001$ for the weak magnetic field, where f is the magnetic flux through the hexagon in units of Φ_0, the ω_0 does not change. The reason

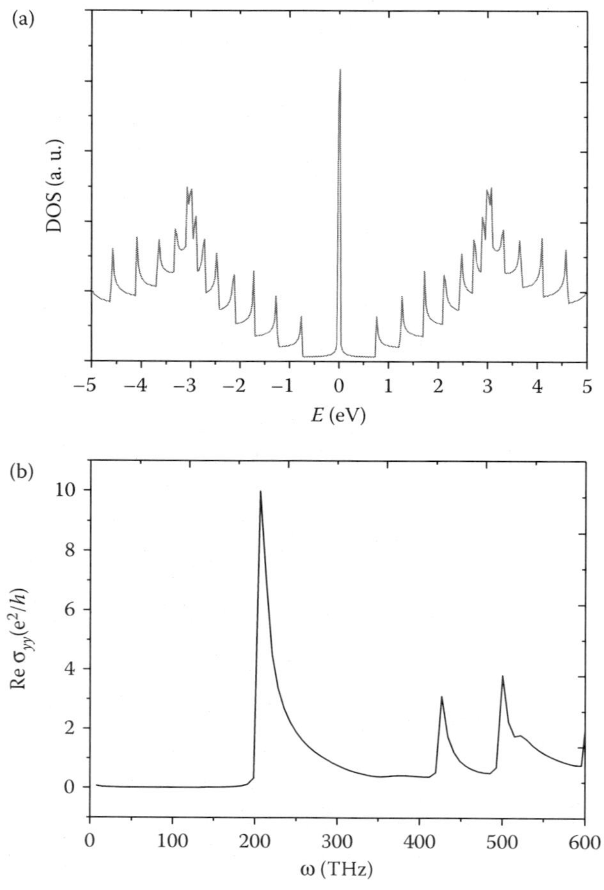

FIGURE 24.7 DOS and optical conductance for zigzag GNRs without an external magnetic field for (16, 0) zigzag GNR.

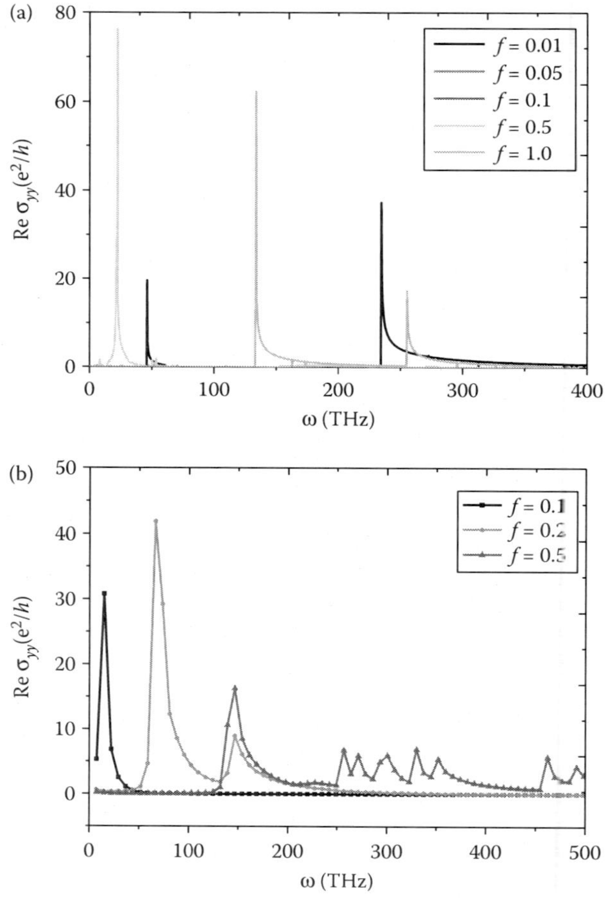

FIGURE 24.8 Optical conductance for different GNRs in a magnetic flux. (a) For (16,1) armchair GNRs, and (b) for (18,0) zigzag GNRs.

is that since the magnetic field is so weak that the cyclotron radius is still large compared to the width of nanoribbons and thereby the Landau levels are not formed. If f increases further, for example, increase to 0.1 (the solid line [with square]), the threshold frequency ω_0 decreases by a factor of about 38 for the (18, 0) zigzag GNRs. The first conductance peak now occurs around the 35 THz in THz regime. At very high frequencies, the optical conductance for a zigzag GNR tends to be the same. This is because the Landau levels are well developed in the high-energy region, but not in the low-energy region where the Landau levels are strongly influenced by the edge. The Landau levels can be seen clearly in the density of states (DOS) curves. The absorption peaks are attributed to the transition between two of these Landau levels. This is similar to the case of a graphene sheet [58]. The threshold frequencies are dramatically reduced, and a relatively strong optical response can occur at low frequencies. From Figure 24.8b, for the (16, 1) armchair GNR, the conductance resonance can occur at $\omega_0 = 46$ THz under $f = 0.01$. When the B field is increased to $f = 1.0$, ω_0 increases to about 138 THz while the resonance intensity decreases by a factor of three times in NIR regime.

The maximum responsivity of GNR-PDs can also exceed the responsivity of the customary photodetectors made of narrow gap semiconductors (e.g., PbSnTe and CdHgTe), the responsivity of which is about a few A/W [46–48], because the former can exhibit a rather high quantum efficiency at resonance arising due to the lateral quantization in GNRs. Besides, the utilization of a zigzag GNR may be expected for the realization of nanolasers with the working frequencies of about 10 THz at room temperature due to the higher efficiency of pumping, compared to the optically pumped multilayer-graphene structures [61,62,64].

Mueller and coworkers [34] have demonstrated that by monolithically integrating graphene with a Fabry–Pérot microcavity, in which the optical absorption is 26-fold enhanced and the demonstrated performance, can be significantly increased by a resonant cavity. Moreover, Englund and coworkers [69] have recently demonstrated a large enhancement in the interaction of light with graphene through coupling with localized modes in a photonic crystal nanocavity, which indicates significantly increased light absorption in the graphene layer. In light of the results above, the total optical conductance or photoconductivity for such cavity-controlled GNR-PDs for (16, 1) armchair GNR and (18, 0) zigzag GNR with different cavity enhancement factor $F_p0 = 10$ and 20 are shown in Figure 24.9. The overall performance of the photodetector is also dependent on the spectral matching of GNR absorption and cavity transmission. Though the spectral resonance peaks are not at same frequency, the resonance peak of GNR optical conductance can be tuned by applying magnetic field as shown in Figure 24.8. Moreover, the microcavity transmission spectra can also be adjusted properly through controlling the cavity length. Although practical implementation of this device is difficult to fabricate, the problem can be overcome with the use of one mirror layer of a Fabry–Pérot microcavity on one cavity mirror side and one mirror layer of planar photonic crystal nanocavity on the other cavity mirror side. In fact, it is not difficult to fabricate a photonic crystal slab for THz and/or NIR frequency [68,71,72], and by making the cavity asymmetric critical coupling can be achieved, though that needs substantial simulations. For detector applications, the light should be absorbed efficiently in the active GNR array layer. The cavity is thus intentionally made asymmetric. For on-chip applications, a waveguide can be used to guide the light and efficient couplers will be implemented to couple the light to the cavity [73].

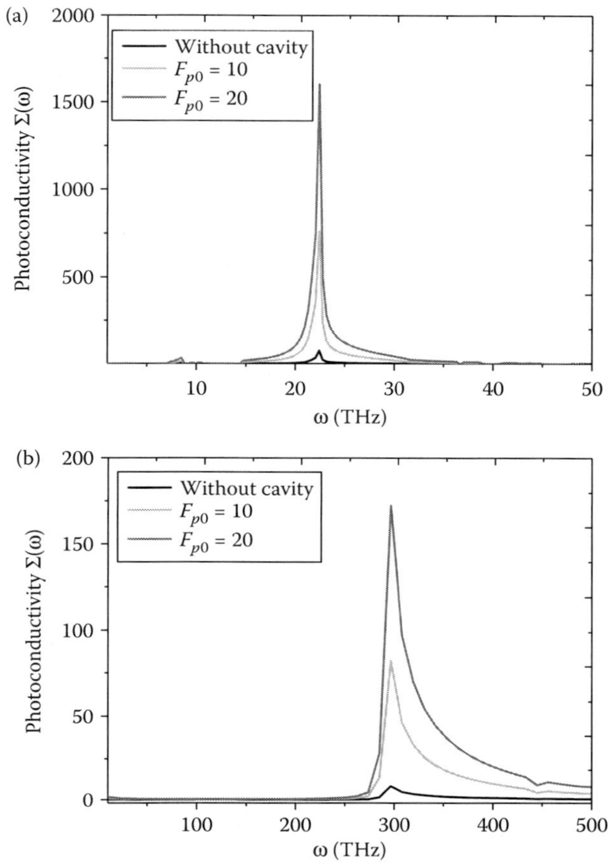

FIGURE 24.9 The total optical conductance or photoconductivity by different enhancement factor of the microcavity $F_{p0} = 10$ and 20 for (a) (16, 1) armchair GNR and (b) (18, 0) zigzag GNR.

24.4 GRAPHENE AND GRAPHENE NANORIBBON DEVICES FOR SUB-THz WIRELESS TECHNOLOGY APPLICATION

In the previous section, we have discussed several important developments of graphene-based photodetectors. In this section, a great deal of attention is paid to the design of a proof-of-concept high-speed sub-THz wireless communication technology at G-band based on advanced modulation technology and microelectromechanical system (MEMS) technology and application of graphene-based reconfigurable THz devices. Moreover, particular emphasis are placed on several key components including high-speed graphene THz modulator, high-performance 175-GHz rectangular waveguides, and sub-THz receiver with graphene transistors operating at THz frequency, which have recently been designed and demonstrated.

Nowadays, the interest in sub-THz wireless communication technology is based on the promise of a new world of information society in which data bits race through optical fiber networks that span our globe 24 h a day, 365 days a year. THz and sub-THz wireless communication refer to the use of the electromagnetic wave between 100 GHz and 3 THz as a carrier for broadband short-distance wireless transmission, high-speed wireless data transmission, and space satellite communications. Compared to conventional microwave optical communications, THz communication technology has advantages of broadband, good directionality, strong anti-interference ability, privileged communication confidentiality, and high reliability [43,44,55,74–76]. However, due to the limitation of the challenging implementation and performance of THz source and signal processing device, it is rather difficult to modulate THz-wave signals and realize a THz wireless system. By using graphene THz modulators enabled by intraband transitions and advanced optical modulation formats, high-speed modulation of THz waves can be achieved, where the carrier frequency and bit rate can be over 150 GHz and 100 Gb/s, respectively.

In the context of high-speed modulation of the sub-THz wave for sub-THz wireless technology, the 175-GHz electromagnetic-wave band is used as a carrier to achieve 100-Gb/s transmission speeds. The schematic of the 175-GHz-band wireless link is shown in Figure 24.10a. An optical sub-THz source generates optical subcarrier signals whose intensity is modulated at a frequency of 175 GHz. A Mach–Zehnder modulator (MZM) modulates the optical subcarrier signal using data signals. The modulated subcarrier signal is amplified by an optical amplifier and input into the photonic sub-THz emitter. In the emitter, a photodiode converts the optical signals into sub-THz signals, which are radiated toward the receiver via an antenna. The received sub-THz signals are input into a sub-THz photodetector, which employs a preamplifier and a demodulator by an envelope detection scheme using the InP-based high electron mobility transisor (InP-HEMT) or the newly developed graphene transistors operating at THz frequencies with sub-10-nm gate length [77,78].

24.4.1 High-Speed Sub-THz Transmission Using Advanced Modulator

Based on the technique of vector modulation, which is commonly used in wireless communication systems, the carrier frequency in radio systems should be typically lower than 100 GHz, while that should be higher than 100 THz in lightwave systems. On the basis of their peculiar band structures, several graphene photonic devices based on interband transitions have been demonstrated, including ultrafast graphene photodetectors [30,37,77], and broadband optical

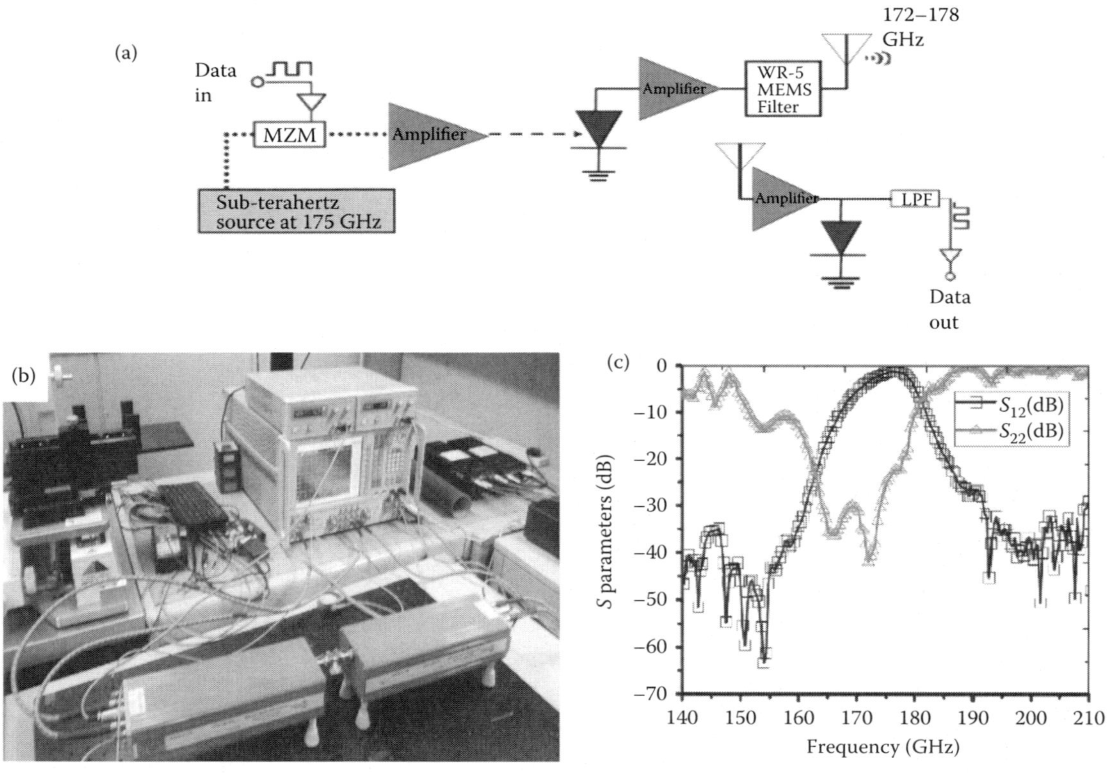

FIGURE 24.10 (a) Schematic system setup of 175 GHz band wireless link, (b) agilent millimeter-wave test setup for sub-THz waveguide filter, and (c) measured, and simulated results of rectangular waveguide with six-pole cavity.

modulators [3]. Given recent demonstrations of broadband graphene THz modulators, various types of advanced modulation formats can be built utilized with graphene modulators. Mach–Zehnder interferometer-based modulators (MZMs) consisting of two high-speed optical phase modulators can control lightwave amplitude, precisely. A quad-parallel MZM (QPMZM) can synthesize four-level signals optically both in the real and imaginary components [77], so that a 16-QAM signal can be generated from four electric binary data streams.

24.4.2 Sub-THz Waveguide Filters

Nowadays, MEMS waveguides are largely required for their low power losses and high power handling capability though planar transmission lines and coplanar waveguides are fabricated in many microwave circuits [43,44,79]. The advantages of the rectangular waveguides include wide bandwidths of operation for single-mode transmission, low attenuation, and good mode stability for the fundamental mode of propagation. The University of California first proposed the idea of implementing air-filled metal-pipe rectangular waveguides, using integrated circuit technology, back in 1980 [80]. In 2011, we have developed high-performance silicon micromachined 140-GHz band-pass MEMS waveguide filters with central frequency of 140 GHz and insertion loss of 0.4 dB by using inductively coupled plasma (ICP) dry-etching technology [43,44,81].

The sub-THz rectangular waveguide filter was designed using six inductively coupled cavities that are separated by irises and the initial geometrical structure parameters was calculated by the 0.5-dB ripple six-order Chebyscheff filter model, and then simulated with a standard cross section of 1.295 mm × 0.648 mm (WR-5) using Ansoft's HFSS. The key MEMS fabrication technologies include the silicon deep trench etching technology, the metallization process, and bonding process. Obviously, this method requires a precise alignment at bonding. The 175-GHz G-band MEMS filter irises were fabricated by the ICP reactive ion etcher (AMS100SE, Alcatel Co.). The iris pattern was then etched through the full thickness of the thick polished silicon wafer following the Bosch process. The filter circuit structures were bonded together with the metallized glass covers to form the waveguide cavity. Finally, the Au–Si eutectic bonding technology is used to bond one silicon cover coated by Au layer to form a complete sub-THz MEMS metal waveguide. The *S*-parameter tests have been performed by the Agilent millimeter-wave N5260A/V05VNA2 testing system shown in Figure 24.10b, in which the filter is installed in a standard flange to connect the test flange. The frequency response results of a prototype sub-THz waveguide filter in G-band are depicted in Figure 24.10c. The measured results show that the center frequency of ~174 GHz with a 5.6% bandwidth. The insertion loss is about 1.6 dB and the isolation is larger than 15 dB. The further characterization results indicated that the insertion loss is mainly ascribed to nonuniform electroplated coating gold layer.

24.4.3 Sub-THz Receiver

The sub-THz receiver in the wireless link is composed of all-electronic technologies. The receiver for long-distance applications, consists of a low-noise preamplifier and a demodulator, which are integrated into one chip [76,82] using the InP-HEMT or newly developed high-speed graphene transistor [77,83–85]. The device typically has a unity current gain frequency f_t of 175 GHz or even 300 GHz for recently developed graphene transistors and a maximum oscillation frequency f_{max} of 350 GHz.

24.5 CONCLUSIONS AND OUTLOOKS

Together with its ease of fabrication and integration, the unique and unconventional properties of graphene enable us to construct a variety of efficient and novel optical and THz devices, and new THz technologies. Interestingly, several demonstrations that graphene has been monolithically integrated with other more established materials to form novel highly complex devices open up real application prospects in low-cost photodetectors, nanolasers, and transmitter modules by releasing the need for temperature control, isolators, and external modulators. Implementation of graphene or GNR-based photonic devices is especially promising for optical communication system and/or THz wireless communication system. In the future, large-scale integrated circuits with all-optical and electro-optical devices on one single silicon chip will require capabilities of controlling and manipulating the direction, phase, polarization, and amplitude of optical signal beams. The graphene-based configurable THz devices can be easily integrated with other THz devices to increase functionality or to realize single chip THz system. And, such advanced large-scale monolithic photonic integrated circuits (MPhICs) represent a significant technology innovation that simplifies optical communication system design, reduces space and power consumption, and improves reliability. Hopefully, recent breakthroughs in the direct deposition of graphene on silicon can pave the way for the integration of graphene in hybrid silicon photonic circuits. This chapter discussed here has established a deep constructive understanding and prospective guideline for graphene-based nanodevices with their implications for future applications in quantum photonics and THz photonics.

ACKNOWLEDGMENTS

This chapter is supported by General Research Fund (GRF) Project No. 9041909 and Project No. CityU 119212 from RGC, Hong Kong. Moreover, G.C. would gratefully acknowledge the research activity fund support from Hong Kong to carry out research work on graphene nanostructures as a visiting scholar at Columbia University. We thank Professor Wei Huang from Nanjing Tech University, Professor Marvin Cohen from UC Berkeley, and Professor Xi Chen from Columbia University for valuable comments and fruitful discussions. G.C. also acknowledges the collaboration work from Professor Xue

Quan, Ms Jovi Mok, and Mr Xinghai Zhao at the State Key Laboratory of Millimetre Waves, City University of Hong Kong for helpful measurements of THz waveguide filter using Agilent test system of N5260A/V05VNA2.

REFERENCES

1. Novoselov, K. S., Geim, A. K., Morozov, S. V., Jiang, D., Zhang, Y., Dubonos, S. V., Grigorieva, I. V., and Firsov, A. A. Electric field effect in atomically thin carbon films, *Science*, 306, 666, 2004.
2. Zhang, Y., Tan, Y.-W., Stormer, H. L., and Kim, P. Experimental observation of quantum Hall effect and Berry's phase in graphene, *Nature*, 438, 201, 2005.
3. Liu, M., Yin, X. B., Ulin-Avila, E., Geng, B. S., Zentgraf, T., Ju, L., Wang, F., and Zhang, X. A graphene-based broadband optical modulator, *Nature*, 474, 64–67, 2011.
4. Geim, A. K. and Novoselov, K. S. The rise of graphene, *Nat. Mater.*, 6, 183–191, 2007.
5. Charlier, J. C., Eklund, P. C., Zhu, J., and Ferrari, A. C. Electron and phonon properties of graphene: Their relationship with carbon nanotubes, *Top. Appl. Phys.*, 111, 673–709, 2008.
6. Wallace, P. R. The band theory of graphite, *Phys. Rev.*, 71, 622–634, 1947.
7. Bonaccorso, F., Sun, Z., Hasan, T., and Ferrari, A. C. Graphene photonics and optoelectronics, *Nat. Photonics*, 4, 611–622, 2010.
8. Du, X., Skachko, I., Duerr, F., Luican, A., and Andrei, E. Y. Fractional quantum Hall effect and insulating phase of Dirac electrons in graphene, *Nature*, 462, 192, 2009.
9. Novoselov, K. S., Fal'ko, V. I., Colombo, L., Gellert, P. R., Schwab, M. G., and Kim, K. A roadmap for graphene, *Nature*, 490, 192–200, 2012.
10. Han, M. Y., Ozyilmaz, B., Zhang, Y., and Kim, P. Energy band-gap engineering of graphene nanoribbons, *Phys. Rev. Lett.*, 98, 206805, 2007.
11. Lin, Y.-M., Dimitrakopoulos, C., Jenkins, K. A., Farmer, D. B., Chiu, H.-Y., Grill, A., and Avouris, Ph. 100-GHz transistors from wafer-scale epitaxial graphene, *Science*, 327, 662, 2010.
12. Casiraghi, C., Hartschuh, A., Lidorikis, E., Qian, H., Harutyunyan, H., Gokus, T., Novoselov, K. S., and Ferrari, A. C. Rayleigh imaging of graphene and graphene layers, *Nano Lett.*, 7, 2711–2717, 2007.
13. Blake, P., Hill, E. W., Castro Neto, A. H., Novoselov, K. S., Jiang, D., Yang, R., Booth, T. J., and Geim, A. K. Making graphene visible, *Appl. Phys. Lett.*, 91, 063124, 2007.
14. Katsnelson, M. I., Novoselov, K. S., and Geim, A. K. Chiral tunnelling and the Klein paradox in graphene, *Nat. Phys.*, 2, 620–625, 2006.
15. Sun, Z., Hasan, T., Torrisi, F., Popa, D., Privitera, G., Wang, F., Bonaccorso, F., Basko, D. M., and Ferrari, A. C. Graphene mode-locked ultrafast laser, *ACS Nano*, 4, 803–810, 2010.
16. Hasan, T., Sun, Z., Wang, F., Bonaccorso, F., Tan, P. H., Rozhin, A. G., and Ferrari, A. C. Nanotube–polymer composites for ultrafast photonics, *Adv. Mater.*, 21, 3874–3899, 2009.
17. Nair, R. R., Blake, P., Grigorenko, A. N., Novoselov, K. S., Booth, T. J., Stauber, T., Peres, N. M., and Geim, A. K. Fine structure constant defines transparency of graphene, *Science*, 320, 1308–1308, 2008.
18. Stoehr, R. J., Kolesov, R., Pflaum, J., and Wrachtrup, J. Fluorescence of laser created electron–hole plasma in graphene, *Phys. Rev. B*, 82, 121408(R), 2010.
19. Liu, C. H., Mak, K. F., Shan, J., and Heinz, T. F. Ultrafast photoluminescence from graphene, *Phys. Rev. Lett.*, 105, 127404, 2010.
20. Wu, S. *Nonlinear photoluminescence from graphene*. Abstract number: BAPS.2010.MAR.Z22.11, APS March Meeting, Portland, Oregon, 2010.
21. Kou, L., Wu, S., Yan, B., Felser, C., and Shan, G. Dirac cone topological surface state and energy bands in graphene-BiTeI sandwiched structures, *Nanoscale,* revised, 2015.
22. Wang, F., Zhang, Y., Tian, C., Girit, C., Zettl, A., Crommie, M., and Shen, Y. R. Gate-variable optical transitions in graphene, *Science*, 320, 206–209, 2008.
23. Jnawali, G., Rao, Y., Yan, H., and Heinz, T. F. Observation of a transient decrease in terahertz conductivity of single-layer graphene induced by ultrafast optical excitation, *Nano Lett.*, 13, 524–530, 2013.
24. Gokus, T., Nair, R. R., Bonetti, A., Böhmler, M., Lombardo, A., Novoselov, K. S., Geim, A. K., Ferrari, A. C., and Hartschuh, A. Making graphene luminescent by oxygen plasma treatment, *ACS Nano*, 3, 3963–3968, 2009.
25. Maeng, I., Lim, S., Chae, S. J., Lee, Y. H., Choi, H., and Son, J.-H. Gate-controlled nonlinear conductivity of Dirac fermion in graphene field-effect transistors measured by terahertz time-domain spectroscopy, *Nano Lett*, 12, 551, 2012.
26. Zhao, X. H., Shan, G. C., and Huang, W. Steady-state property and dynamics in graphene-nanoribbon-array lasers, *Front. Phys.*, 7(5), 527–532, 2012.
27. Gansen, E. J., Rowe, M. A., Greene, M. B., Rosenberg, D., Harvey, T. E., Su, M. Y., Hadfield, R. H., Nam, S. W., and Mirin, R. P. Photon-number-discriminating detection using a quantum dot, optically gated, field-effect transistor, *Nat. Photonics*, 1, 585–588, 2007.
28. Leu, L. Y., Gardner, J. T., and Forrest, S. R. A high gain, high bandwidth $In_{0.53}Ga_{0.47}$ As/InP heterojunction phototransistor for optical communications, *J. Appl. Phys.*, 69, 1052–1062, 1991.
29. Zhou, Y., Xu, X. L., Fan, H., Ren, Z., Bai, J., and Wang, L. Tunable magnetoplasmons for efficient terahertz modulator and isolator by gated monolayer graphene, *Phys. Chem. Chem. Phys.*, 15, 5084–5090, 2013.
30. Liu, C.-H., Chang, Y.-C., Norris, T. B., and Zhong, Z. Graphene photodetectors with ultra-broadband and high responsivity at room temperature, *Nat. Nanotechnol.*, 9, 273–278, 2014.
31. Echtermeyer, T. J., Britnell, L., Jasnos, P. K., Lombardo, A., Gorbachev, R. V., Grigorenko, A. N., Geim, A. K., Ferrari, A. C., and Novoselov, K. S. Strong plasmonic enhancement of photovoltage in graphene, *Nat. Commun.*, 2, 458, 2011.
32. Koppens, F. H. L., Chang, D. E., and García de Abajo, F. J. Graphene plasmonics: A platform for strong light–matter interactions, *Nano Lett.*, 11, 3370–3377, 2011.
33. Thongrattanasiri, S., Koppens, F. H. L., and Garcia de Abajo, F. J. Complete optical absorption in periodically patterned graphene, *Phys. Rev. Lett.*, 108, 047401, 2012.
34. Furchi, M., Urich, A., Pospischil, A., Lilley, G., Unterrainer, K., Detz, H., Klang, P. et al. Microcavity-integrated graphene photodetector, *Nano Lett.*, 12(6), 2773–2777, 2012.
35. Engel, M., Steiner, M., Lombardo, A., Ferrari, A. C., Löhneysen, H. V., Avouris, P., and Krupke, R. Light–matter interaction in a microcavity-controlled graphene transistor, *Nat. Commun.*, 3, 906, 2012.
36. Xia, F., Mueller, T., Lin, Y. M., Valdes-Garcia, A., and Avouris, P. Ultrafast graphene photodetector, *Nat. Nanotechnol.*, 4, 839–843, 2009.

37. Vasko, F. T. and Ryzhii, V. Voltage and temperature dependencies of conductivity in gated graphene, *Phys. Rev. B*, 76, 233404, 2007.
38. George P. A., Strait, J., Dawlaty, J., Shivaraman, S., Chandrashekhar, M., Rana, F., and Spencer, M. G. Ultrafast optical-pump terahertz-probe spectroscopy of the carrier relaxation and recombination dynamics in epitaxial graphene, *Nano Lett.*, 8, 4248–4251, 2008.
39. Lee, E. J. H., Balasubramanian, K., Weitz, R. T., Burghard, M., and Kern, K. Contact and edge effects in graphene devices, *Nat. Nanotechnol.*, 3, 486–490, 2008.
40. Xia, F. N., Mueller, T., Golizadeh-Mojarad, R., Freitag, M., Lin, Y. M., Tsang, J., Perebeinos, V., and Avouris, P. Photocurrent imaging and efficient photon detection in a graphene transistor, *Nano Lett.*, 9, 1039–1044, 2009.
41. Mueller, T., Xia, F., Freitag, M., Tsang, J., and Avouris, Ph. Role of contacts in graphene transistors: A scanning photocurrent study, *Phys. Rev. B*, 79, 245430, 2009.
42. Ito, H., Furuta, T., Kodama, S., and Ishibashi, T. InP/InGaAs uni-travelling-carrier photodiodes with a 340 GHz bandwidth, *Electron. Lett.*, 36, 1809–1810, 2000.
43. Shan, G. C., Zhao, X., Zhu, H. S., and Shek, C.-H. Critical components in 140 GHz communication systems, *IEEE 2012 International Workshop on Microwave and Millimeter Wave Circuits and System Technology*, Vol. 1, pp. 1–4, 2012.
44. Shan, G. C., Zhao, X., Shek, C.-H., and Zhu, H. Sub-THz wireless communication technology at G-band, *2013 IEEE International Workshop on Electromagnetics*, Vol. 1, pp. 57–60, 2013.
45. Konstantatos, G., Badioli, M., Gaudreau, L., Osmond, J., Bernechea, M., de Arquer, F. P. G., Gatti, F., and Koppens, F. H. L. Hybrid graphene–quantum dot phototransistors with ultrahigh gain, *Nat. Nanotechnol.*, 7, 363, 2012.
46. Schneider, H. and Liu, H. C. *Quantum Well Infrared Photodetectors: Physics and Applications*, Springer, Berlin, pp. 100–210, 2006.
47. Ryzhii, V. Characteristics of quantum well infrared photodetectors, *J. Appl. Phys.*, 81, 6442–6448, 1997.
48. Rogalski, A. Infrared detectors: An overview, *Infrared Phys. Technol.*, 43, 187–210, 2002.
49. Ryzhii, V. *Intersubband Infrared Photodetectors*, World Scientific, Singapore, pp. 120–180, 2003.
50. Schneider, H., Drachenko, O., Winnerl, S., Helm, M., and Walther, M. Quadratic autocorrelation of free-electron laser radiation and photocurrent saturation in two-photon quantum well infrared photodetectors, *Appl. Phys. Lett.*, 89, 133508, 2006.
51. Graf, M., Scalari, G., Hofstetter, D., Faist, J., Beere, H., Linfield, E., Ritchie, D., and Davies, G. Description of transport mechanisms in a very long wave infrared quantum cascade detector under strong magnetic field, *Appl. Phys. Lett.*, 84, 475, 2004.
52. Guo, X. G., Zhang, R., Liu, H. C., SpringThorpe, A. J., and Cao, J. C. Photocurrent spectra of heavily doped terahertz quantum well photodetectors, *Appl. Phys. Lett.*, 97, 021114, 2010.
53. Bhowmick, S., Huang, G., Guo, W., Lee, C. S., Bhattacharya, P., Ariyawansa, G., and Perera, A. G. U., High-performance quantum ring detector for the 1–3 terahertz range, *Appl. Phys. Lett.*, 96, 231103, 2010.
54. Jastrow, C., Munter, K., and Piesiewicz, R. 300 GHz transmission system, *Electron. Lett.*, 44(3), 1–2, 2008.
55. Tan, Z. Y., Chen, Z., Han, Y. J., Zhang, R., Li, H., Guo, X. G., and Cao, J. C. Experimental realization of wireless transmission based on terahertz quantumcascade laser, *Acta Phys. Sin.*, 61, 098701, 2012.
56. Hattori, T., Sakamoto, M., and Rungsawang, R. Terahertz real-time imaging using spatial filtering, *Int. Quantum Electron. Conf.*, 1, 1242, 2005.
57. Shan, G. C., Zhao, X. H., and Huang, W. Nanolaser with a single-graphene-nanoribbon in a microcavity, *J. Nanoelectron. Optoelectron.*, 6, 138, 2011.
58. Tong, J., Zhao, X. H., and Shek, C. H., Antenna enhanced graphene THz detector, *J. Theor. Comput. Nanosci.*, revised.
59. Liu, J., Ma, Z., Wright, A. R., and Zhang, C. Orbital magnetization of graphene and graphene nanoribbons, *J. Appl. Phys.*, 103, 103711, 2008.
60. Wright, A. R., Xu, X. G., Cao, J. C., and Zhang, C. Strong nonlinear optical response of graphene in the terahertz regime, *Appl. Phys. Lett.*, 95, 072101, 2009.
61. Ryzhii, V., Ryzhii, M., Mitin, V., and Otsuji, T. Terahertz and infrared photodetection using p-i-n multiple-graphene-layer structures, *J. Appl. Phys.*, 107, 054512, 2010.
62. Ryzhii, V., Ryzhii, M., Mitin, V., and Otsuji, T. Toward the creation of terahertz graphene injection laser, *J. Appl. Phys.*, 110, 094503, 2011.
63. Huard, B., Sulpizio, J. A., Stander, N., Todd, K., Yang, B., and Goldhaber-Gordon, D. Transport measurements across a tunable potential barrier in gaphene, *Phys. Rev. Lett.*, 98, 236803, 2007.
64. Ryzhii, V., Ryzhii, M., Satou, A., and Otsuji, T. Current–voltage characteristics of a graphene-nanoribbon field-effect transistor, *J. Appl. Phys.*, 103, 094510, 2008.
65. Ryzhii, V., Ryzhii, M., and Otsuji, T. Tunneling current–voltage characteristics of graphene field-effect transistor, *Appl. Phys. Express*, 1, 013001, 2008.
66. Zhang, C., Chen, L., and Ma, Z. S. Orientation dependence of the optical spectra in graphene at high frequencies, *Phys. Rev. B*, 77(R), 241402, 2008.
67. Furchi, M., Urich, A., Pospischil, A., Lilley, G., Unterrainer, K., Detz, H., Klang, P., Andrews, A. M., Schrenk, W., and Strasser, G. Microcavity-integrated graphene photodetector, *Nano Lett.*, 12, 2773–2777, 2012.
68. Rsoft FemSIM software, RSOFT Design Inc., "RSOFT," (RSOFT, 2011). http://www.rsoftdesign.com/products.php?sub=Component+Design&itm=FemSIM.
69. Gan, X. T., Mak, K. F., Gao, Y. D., You, Y. M., Hatami, F., Hone, J., Heinz, T. F., and Englund, D. Strong enhancement of light–matter interaction in graphene coupled to a photonic crystal nanocavity, *Nano Lett.*, 12, 5626, 2012.
70. Motohiko, E. Peculiar width dependence of the electronic properties of carbon nanoribbons, *Phys. Rev. B*, 73, 045432, 2006.
71. Zhang, C., Cao, J. C., and Guo, X. G. Impurity mediated absorption continuum in single-walled carbon nanotubes, *Appl. Phys. Lett.*, 90, 023106, 2007.
72. Brey, L. and Fertig, H. A. Electronic states of graphene nanoribbons studied with the Dirac equation, *Phys. Rev. B*, 73, 235411, 2006.
73. Jukam, N., Yee, C., Sherwin, M. S., Fushman, I., and Vučković, J. Patterned femtosecond laser excitation of terahertz leaky modes in GaAs photonic crystals, *Appl. Phys. Lett.*, 89, 241112, 2006.
74. Cooper, K. B., Dengler, R. J., Chattopadhyay, G., Schlecht, E., Gill, J., Mehdi, I., and Siegel, P. H. A high-resolution imaging radar at 580 GHz, *IEEE Microwave Wireless Compon. Lett.*, 18(1), 64–66, 2008.
75. Song, H.-J., Ajito, K., Muramoto, Y., Wakatsuki, A., Nagatsuma, T., and Kukutsu, N. 24 Gbit/s data transmission in 300 GHz band for terahertz communications, *Electron. Lett.*, 48, 953–954, 2012.

76. Kanno, A., Inagaki, K., Morohashi, I., Sakamoto, T., Kuri, T., Hosako, I., Kawanishi, T., Yoshida, Y., and Kitayama, K. 40 Gb/s W-band (75–110 GHz) 16-QAM radio-over-fiber signal generation and its wireless transmission, *Opt. Express*, 19, B56–B63, 2011.
77. Zheng, J., Wang, L., Quhe, R., Liu, Q. H., Li, H., Yu, D. P., Mei, W. N., Shi, J. J., Gao, Z. X., and Lu, J. Sub-10 nm gate length graphene transistors: Operating at terahertz frequencies with current saturation, *Sci. Rep.*, 3, 1314, 2013.
78. Mueller, T., Xia, F., and Avouris, P. Graphene photo-detectors for high speed optical communications, *Nat. Photonics*, 4, 297–301, 2010.
79. Zhao, X. H., Shan, G. C., Du, Y. J., Chen, Y. H., and Shek, C. H. G-band MEMS rectangular waveguide filter fabricated using deep reactive ion etching and bonding processes, *Micro Nano Lett.*, 7(12), 1237–1240, 2012.
80. Rutledge, D. B., Schwarz, S. E., and Hwang, T. L. Antennas and waveguides for far-infrared integrated circuits, *IEEE J. Quantum Electron.*, 16(5), 508–516, 1980.
81. Zhao, X. H., Bao, J., Shan, G. C., Du, Y., Zheng, Y. B., Wen, Y., and Shek, C. H. D-band micromachined silicon rectangular waveguide filter, *IEEE Microwave Wireless Compon. Lett.*, 22(5), 230–232, 2012.
82. Hirata, A., Kosugi, T., Meisl, N., Shibata, T., and Nagatsuma, T. High-directivity photonic emitter using photodiode module integrated with HEMT amplifier for 10-Gbit/s wireless link, *IEEE Trans. Microwave Theory Tech.*, 52, 1843, 2004.
83. Low, T. and Avouris, P. Graphene plasmonics for terahertz to mid-infrared applications, *ACS Nano*, 8(2), 1086–1101, 2014.
84. Tassin, P., Koschny, T., and Soukoulis, C. M. Graphene for terahertz applications, *Science*, 341, 620–621, 2013.
85. Tredicucci, A. and Vitiello, M. S. Device concepts for graphene-based terahertz photonics, *IEEE J. Sel. Top. Quantum Electron.*, 20(1), 8500109, 2014.

25 On-Chip Graphene Optoelectronic Devices

Xuetao Gan, Ren-Jye Shiue, and Dirk Englund

CONTENTS

ABSTRACT

There has been increasing interest in developing optoelectronic devices based on graphene, whose tunable optical and electronic properties make it applicable for both high-speed and broadband modulators and photodetectors. The flexibility and robustness of graphene make it compatible with the geometries of various devices. In this chapter, we describe chip-integrated graphene optoelectronics with nanophotonic cavities and waveguides. By depositing graphene onto photonic crystal nanocavities, it is possible to reach near-unity absorption into graphene. The graphene-photonic crystal cavity system enables a high-contrast electro-optic modulation by electrically tuning the Fermi level of graphene. In addition, high-speed operation is possible by using capacitive gating, such as through a double-layer graphene stack. The cavity also enables dramatically enhanced and spectrally selective photodetection on graphene. A similar scheme of graphene integration on nanophotonic waveguides greatly increases the interaction time of graphene with light and enables efficient, broadband, and high-speed photodetectors. Such graphene photodetectors are compatible with photonic integrated circuits and promise an alternative to traditional on-chip photodetectors. Thus, the combination of graphene optoelectronic devices with photonic chips promises a new route for ultrafast and broadband on-chip optical interconnects.

25.1 INTRODUCTION

The unique electronic band structure of graphene results in remarkable optical properties, including a gate-tunable absorbance of 2.3% over a broad spectral range from the visible to the near-infrared [1], broadband saturable absorption [2,3], and strong third-order nonlinearity [4,5]. These optical phenomena have attracted great interest toward developing graphene as a new material for high-performance optoelectronic devices [6–11]. Already, graphene devices have been demonstrated in a range of optoelectronic systems, including optical modulators [12–14], photodetectors [15–17], plasmonic devices [18–20], and mode-locked lasers [21,22].

Graphene's absorption can be controlled by the application of a gate voltage to change the carrier concentration [23,24], as schematically shown in Figure 25.1a. For an incident photon at frequency ω, "Pauli blocking" of interband transitions within the graphene layer reduces the absorption coefficient when the Fermi energy (E_F) is tuned away from the Dirac point by more than half of the photon energy $\hbar\omega/2$. This gate-controllable absorption makes graphene attractive for a variety of applications such as electro-optic modulators [12]. Recently, a graphene modulator was demonstrated by the integration of a single-layer graphene on a silicon-on-oxide bus waveguide. A modulation with a contrast exceeding 4 dB and a dynamic response up to 1.2 GHz was observed, covering a bandwidth from 1350 to 1600 nm [12]. Improved device designs based on a double-gated graphene [25,26] or the insertion of graphene in the center of a waveguide slab promise to increase the material's overlap with the maximum mode field, which could lead to reduced device dimensions and greater modulation contrast [7,27]. A free-space-coupled broadband graphene modulator was also reported using a back reflector, although the modulation depth is below 1 dB [28]. By integrating a graphene field effect transistor (FET) on a planar photonic crystal (PPC) cavity, we recently demonstrated a graphene modulator with a modulation depth in

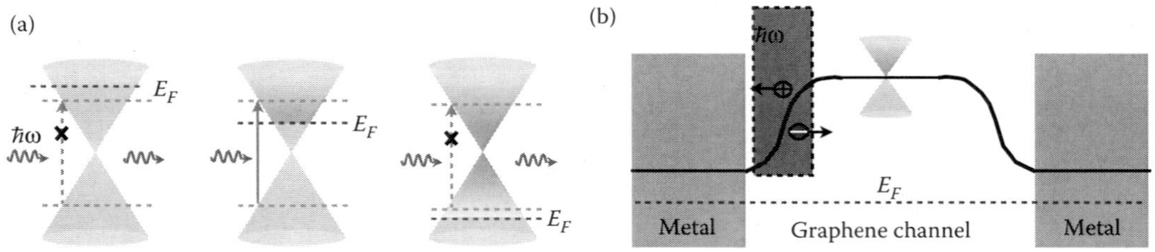

FIGURE 25.1 Schematic illustrations of graphene modulation and photodetection. (a) Band structures of graphene at different doping levels; graphene becomes more transparent when interband transitions are Pauli blocked, as shown in the left and right panels. (b) Potential profile of graphene near the metal contacts, leading to the separation of photoexcited electron–hole pairs and photocurrent generation. The dashed line represents the Fermi level.

excess of 10 dB [29]; a similar experiment reached a modulation depth up to 6 dB [30]. These PPC cavity-based graphene modulator designs have a small footprint given by the micron-scale cavity dimensions.

Much interest has also been focused on the use of graphene for high-speed photodetectors owing to the material's high carrier mobilities and strong electron–electron interactions [31–33]. In addition, a carrier multiplication process in graphene may provide inherent gain in photodetection even in the absence of external bias. Unlike conventional semiconductor photodiodes, the separation of photoexcited electron–hole pairs in a graphene photodetector require an external bias voltage to provide an electrical field [34]; however, this bias also leads to high dark current. Recently, it was shown that graphene's ultrafast carrier dynamics for electrons and holes effectively enable the separation of photoexcited carriers with a moderate internal electrical field, allowing an operation with zero externally applied bias voltage [15,16,35]. This is illustrated in Figure 25.1b. The different doping levels between the graphene beneath the metal contacts and the graphene channel lead to band bending at the graphene–electrode interface. When photons are absorbed at the narrow region adjacent to the metal/graphene interface, the photogenerated carriers can be separated effectively by the built-in electrical field across the junction. Graphene FET structures were employed to demonstrate photodetections with operation speeds up to 40 GHz [16], but the low optical absorption of graphene sheets limited their responsivity below 10 mA/W. To enhance the absorption and hence the responsivity, graphene layers have been integrated with nanocavities [36,37], microcavities [34], and plasmon resonators [38,39], though these resonant-field enhancement approaches reduce the operating bandwidth. On the other hand, it has been demonstrated that hybrid graphene–quantum dot devices can dramatically improve the responsivity [40], though at the cost of response speed. Alternatively, by integrating a graphene photodetector on a silicon (Si) bus waveguide, it is possible to greatly enhance the photodetection without sacrificing high speed and broad spectral bandwidth [41–43].

In this chapter, we review on-chip graphene optoelectronic devices integrated on nanophotonic cavities and waveguides. In Sections 25.2.1 and 25.3.1, we describe the enhanced light–graphene interaction that is achieved by coupling graphene with localized resonant modes in PPC nanocavities and the evanescent field in nanowire waveguides. Integrating graphene FETs or capacitors on PPC cavities enables modulation of the cavity reflection with a contrast exceeding 10 dB and with a high-speed response up to 0.57 GHz (Section 25.2.2). A PPC cavity-integrated photodetector was also fabricated, which strongly enhanced the photoresponsivity for resonant modes (Section 25.2.3). In Section 25.3.2, we then discuss the waveguide-integrated graphene photodetector.

25.2 GRAPHENE OPTOELECTRONIC DEVICES INTEGRATED ON PLANAR PHOTONIC CRYSTAL CAVITIES

PPC cavities can have resonant modes with high-quality (Q) factors (exceeding 10^6) and ultrasmall mode volume (V_{mode}, on the order of a cubic wavelength) [44]. When coupled to graphene, PPC cavities therefore strongly enhance the absorption into the graphene sheet. Recently, spectroscopic studies of different PPC cavity–graphene devices have been reported, showing strongly enhanced graphene absorption, Raman scattering, and hot photoluminescence via the cavity modes [36,11]. These observations can be understood from a cavity–graphene coupling model [36].

Here, we will review the design of the cavity–graphene system. The discussion will focus on the use of the system for high-contrast electro-optic modulators and high-responsivity photodetectors. In these studies, the PPC cavities were fabricated in Si or gallium phosphide (GaP) membranes using a combination of electron beam lithography and dry/wet etching processes. Graphene sheets prepared by mechanical exfoliation or chemical vapor deposition (CVD) were transferred onto the air-suspended PPC cavities using a precision transfer technique [45] and a dry transfer technique [46], respectively. The sizes and numbers of graphene layers were imaged using mirco-Raman spectroscopy. Metal electrodes of the optoelectronic devices were designed on the cavity-integrated graphene layer using electron beam lithography followed by liftoff of electron-beam-evaporated titanium/gold (Ti/Au, 1/40 nm).

25.2.1 Cavity-Enhanced Light–Matter Interaction in Graphene

Figure 25.2a shows a top view of one of the fabricated devices in which several GaP PPC cavities are covered by

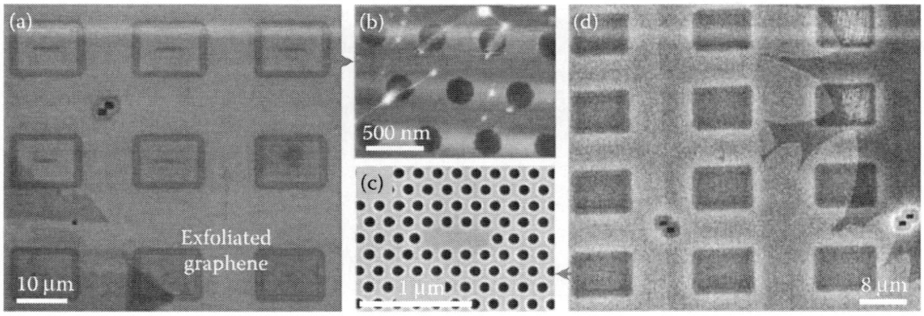

FIGURE 25.2 Integrated PPC cavity–graphene devices. (a) Optical microscope image of the PPC cavities integrated with an exfoliated graphene monolayer. (b) Atomic force microscope image of the device. (c) SEM image of an L3 PPC cavity. (d) SEM image of a device integrated with CVD-grown graphene. (Reprinted with permission from X Gan et al. Strong enhancement of light–matter interaction in graphene coupled to a photonic crystal nanocavity. *Nano Letters*, 12(11):5626–5631. Copyright 2012 American Chemical Society.)

an exfoliated graphene monolayer. A close-up atomic force microscope image (Figure 25.2b) shows that the graphene layer covers the air holes without rupturing, which is important for graphene to maintain its high-quality electronic properties. Figure 25.2c shows a scanning electron microscope (SEM) image of a typical three-hole linear defect (L3) PPC cavity employed in these studies. In addition to exfoliated graphene, wafer-scale CVD-grown graphene sheets were also integrated onto the PPC chip, as shown in Figure 25.2d, where the leaf-like shapes mark the edges of the graphene monolayer.

Figure 25.2a shows an L3 cavity used for characterizing the enhanced light–matter interaction. The device has a PPC lattice spacing of $a = 480$ nm, a hole radius of $r = 0.24a$, and a slab thickness of $d = 0.29a$. A broadband (supercontinuum laser) source was used to illuminate the device using cross-polarization dark-field microscopy, as described elsewhere [47,48]. The characterization results are displayed in Figure 25.3a. The left dotted line shows the reflection spectrum (R_0) of the unloaded cavity (before the integration of graphene), indicating a fundamental mode at 1447.3 nm with a Q factor of 2,640 (estimated by fitting to a Lorentzian). After the graphene was deposited, the cavity reflection spectrum (R_g) shows that the Q factor of the fundamental mode dropped sharply to only 360, while the resonance red shifts by 1.8 nm (the right-dotted line in Figure 25.3a). The inset of Figure 25.3a shows the spectrally resolved attenuation, calculated by [$10\text{Log}_{10}(R_g/R_0)$] from the two reflection

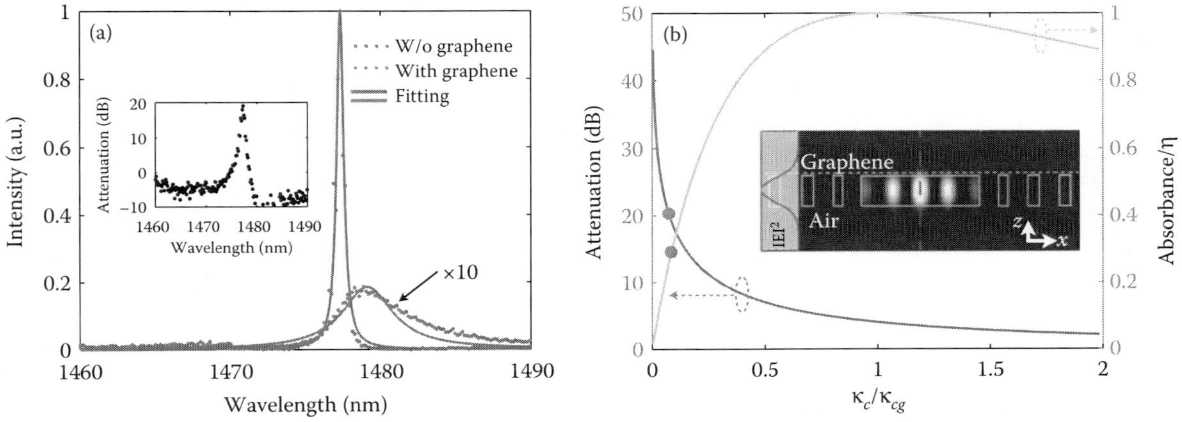

FIGURE 25.3 Cavity-enhanced light–graphene interaction. (a) Reflection spectra of the PPC cavity before and after the deposition of an exfoliated graphene monolayer (magnified 10 times). Inset: Spectrum of the relative attenuation of the cavity with graphene compared to the same cavity without graphene. At the peak of reflection of the unloaded cavity, the attenuation increases by 20 dB by inserting a single layer of graphene. (b) The calculated relative attenuation for the cavity loaded with graphene and the empty cavity (left scale) and the calculated graphene absorbance (right scale) for the graphene–cavity system versus the ratio of the cavity decay rates κ_c/κ_{cg}. The relative attenuation is calculated at the frequency of the unloaded cavity resonance, while the graphene absorbance is calculated for excitation at the frequency of the loaded cavity. The dots are the measured values for the device in (a), shown in the figure for a ratio of $\kappa_c/\kappa_{cg} \approx 0.15$. Inset: Simulated energy distribution of fundamental resonant mode of the L3 cavity shown in cross section, showing the graphene layer interacts with the evanescent field. (Reprinted with permission from X Gan et al. Strong enhancement of light–matter interaction in graphene coupled to a photonic crystal nanocavity. *Nano Letters*, 12(11):5626–5231. Copyright 2012 American Chemical Society.)

spectra. It is clear that the atomically thin graphene layer renders the cavity nearly opaque: the relative cavity attenuation increases by 20 dB at the resonance of the unloaded cavity at 1477.3 nm.

These experimental observations can be explained with a cavity–graphene-coupling model described in Reference 36. Because the GaP semiconductor has a large indirect electronic bandgap (~2.26 eV) and an absorption coefficient below 1 cm^{-1} [49], we ignored the cavity loss arising from the GaP material absorption. Hence, the loss of the unloaded PPC cavity was primarily due to scattering loss with an energy decay rate of $\kappa_c = \omega_0/Q$, where ω_0 is the angular frequency of the cavity resonance. The deposition of the graphene layer caused an additional cavity loss characterized by an energy decay rate κ_{cg}, together with a frequency shift $\Delta\omega$ in the cavity resonance. Since the graphene layer is very thin, any scattering loss of the cavity caused by the graphene deposition was also neglected, as verified by finite difference time domain (FDTD) simulations with a small planar perturbation of the cavity to model the graphene sheet. The excitation mode and collection mode of the microscope were approximated to have a Gaussian spatial distribution. We assumed a coupling efficiency η between the microscope modes and the PPC cavity out-of-plane radiation field; η typically ranges from 10% to 30% for cavity designs with integrated grating couples [50]. Therefore, we assume that the cavity mode couples to the excitation and collection modes at the same rate of $\eta\kappa_c$. Solving the steady-state solution of the coupled mode equations [36], the frequency-dependent reflection $R_g(\omega)$ and absorption $A_g(\omega)$ coefficients of the graphene-coupled cavity can then be solved to be

$$R_g(\omega) = \frac{\eta^2\kappa_c^2}{(\omega_0 + \Delta\omega - \omega)^2 + (\kappa_c/2 + \kappa_{cg}/2)^2} \tag{25.1}$$

$$A_g(\omega) = \frac{\eta\kappa_c\kappa_{cg}}{(\omega_0 + \Delta\omega - \omega)^2 + (\kappa_c/2 + \kappa_{cg}/2)^2} \tag{25.2}$$

The same expressions apply for the unloaded cavity reflection $R_0(\omega)$ and absorption $A_0(\omega)$ coefficients, but with κ_{cg} and $\Delta\omega$ set to zero. Fitting the experimental reflection spectra shown in Figure 25.3a to Equation 25.1, we obtained $\omega_0 = 1.28 \times 10^3$ THz, $\kappa_c = 1.9 \times 10^{-4}\ \omega_0$, $\Delta\omega = 1.24 \times 10^{-3}(1 \pm 0.016)\omega_0$, and $\kappa_{cg} = 2.4 \times 10^{-3}(1 \pm 0.016)\omega_0$ for the loaded cavity, where the estimated uncertainties were near those expected for shot noise from cooled charge-coupled device (CCD) camera. Thus, the resonance frequency shift $\Delta\omega$ and additional energy decay rate κ_{cg} are comparable in magnitude and far greater than κ_c. It follows that the graphene layer dominated the photon loss and accounts for approximately $\kappa_{cg}/(\kappa_{cg} + \kappa_c) \approx 92\%$ of the total loss inside the nanocavity.

The cavity-enhanced graphene absorption was confirmed further from a numerical simulation of the optical field–graphene interaction, as shown in the inset of Figure 25.3b. From perturbation theory applied to the graphene–cavity mode interaction [51], the graphene-induced resonance frequency shift $\Delta\omega$ and absorption rate κ_{cg} can be estimated as

$$\Delta\omega = -\frac{1}{2}\omega_0 \frac{\int d^3\mathbf{r}((\varepsilon_{g1\parallel}(\mathbf{r}) - 1)|\mathbf{E}_\parallel(\mathbf{r})|^2 + (\varepsilon_{g1\perp}(\mathbf{r}) - 1)|\mathbf{E}_\perp(\mathbf{r})|^2)}{\int d^3\mathbf{r}\varepsilon_s(\mathbf{r})|\mathbf{E}(\mathbf{r})|^2} \tag{25.3}$$

$$\kappa_{cg} = \omega_0 \frac{\int d^3\mathbf{r}\varepsilon_{g2\parallel}(\mathbf{r})|\mathbf{E}_\parallel(\mathbf{r})|^2}{\int d^3\mathbf{r}\varepsilon_s(\mathbf{r})|\mathbf{E}(\mathbf{r})|^2} \tag{25.4}$$

Here, the graphene layer has an anisotropic complex dielectric function with in-plane and perpendicular components given by $(\varepsilon_{g1\parallel} + i\varepsilon_{g2\parallel})$ and $(\varepsilon_{g1\perp} + i\varepsilon_{g2\perp})$, where $\varepsilon_{g2\perp} \approx 0$ for near infrared radiation [52]; $\mathbf{E}_\parallel(\mathbf{r})$ and $\mathbf{E}_\perp(\mathbf{r})$ are the in-plane and perpendicular components of the complex resonant field $\mathbf{E}(\mathbf{r})$, and $\varepsilon_s = 9.36$ denotes the dielectric constant of the GaP substrate near ω_0. Using the reported value of $\varepsilon_{g1\parallel} = 4.64$, $\varepsilon_{g1\perp} = 2.79$, and $\varepsilon_{g2\parallel} = 4.62$ [52], we found the graphene-related absorption $\kappa_{cg} = 2.31 \times 10^{-3}\ \omega_0$ and frequency shift $\Delta\omega = -1.22 \times 10^{-3}\ \omega_0$ from the simulated energy distribution of the resonant modes, in good agreement with the experimental observations.

25.2.2 Cavity-Integrated Graphene Electro-Optic Modulators

The cavity-enhanced optical absorption can be used to amplify the contrast of graphene-based optical modulators relying on gate-controlled Pauli blocking. To optimize the cavity–graphene design, it is useful to consider the modulation in the cavity reflection as a function of the intrinsic cavity loss rate κ_c and the graphene-induced loss rate κ_{cg}. Based on the theoretical model of Equation 25.1, which adequately describes the experimental observations shown in Figure 25.3a, Figure 25.3b plots the predicted attenuation in the cavity reflectivity at the frequency of the original (without graphene) resonance, given by $-10\ \mathrm{Log}_{10}(R_g/R_0)$, as a function of the ratio between the cavity intrinsic loss and the graphene loss rate, κ_c/κ_{cg}. It is apparent that, for a cavity with high Q, the additional attenuation due to the graphene absorber can exceed 40 dB. Such high-contrast modulation should be possible with existing PPC cavities ($Q \sim 10^6$). Even when κ_c is similar as $\kappa_{cg} = 1$, the reflection still drops by about 5 dB. Besides changing the intrinsic cavity absorption, it is also possible to control the graphene-related loss rate κ_{cg} by tuning the graphene-mode overlap via the thickness of the PPC membrane, as shown in the inset of Figure 25.3b.

In experiments, integrating graphene FETs or capacitors onto Si PPC cavities enabled high-contrast optical modulation near the telecom wavelength band. For the device shown in Figure 25.4a, a graphene FET is gated with a top electrolyte layer (PEO plus $LiClO_4$) [53,54], providing a strong electrical field and high carrier densities in graphene. A

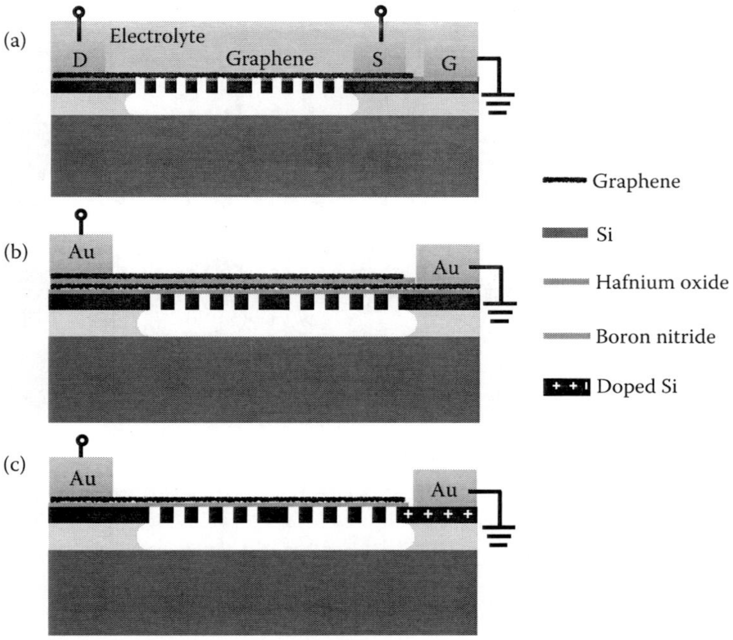

FIGURE 25.4 Schematic of different integrated PPC cavity–graphene modulators. (a) Integration of a PPC cavity and a graphene FET gated by a top electrolyte layer, enabling high doping level of the graphene layer. (b) and (c) High-speed modulators integrated PPC cavities with graphene capacitors consisting of back-gated highly doped Si membrane or dual-gated graphene layers. BN layers are integrated into the capacitors as the dielectric layer.

variation of the graphene Fermi energy up to 0.85 eV was achieved. The corresponding Pauli blocking of the interband transitions enabled high-contrast modulation (>10 dB) of the cavity reflectivity [29]. However, the low ionic mobility of the solid electrolyte limits the response of this modulator design to tens of Hz. To increase the modulation speed into the GHz regime, modified designs with high-speed contacting strategies can be employed, as shown in Figure 25.4b and c [11]. In these two strategies, the top graphene sheet is gated, while the bottom contact consists of either a doped Si PPC cavity or a second graphene layer. Boron nitride (BN) layers act as the dielectric layer between the bottom and top gates, while maintaining a high carrier mobility in the graphene layer(s) [45]. To minimize carrier diffusion from the graphene into the Si substrate, the bottom graphene layer is electrically separated from the Si slab by a hafnium oxide (HfO_2) layer, produced using atomic layer deposition, or a mechanically transferred BN layer.

Figure 25.5a shows a graphene-on-PPC cavity modulator based on the design in Figure 25.4a, where the graphene layer is demarkated by the dashed line. The PPC cavity was covered with a 10-nm-thick HfO_2 layer before the graphene transfer. After the depositions of drain, source, and gate electrodes onto the graphene layer, the electrolyte was spin coated on the entire wafer. The gate electrode was located about 15 μm from the graphene sheet for doping. This modulator employed an air-slot PPC cavity design [55] (see SEM image in Figure 25.5b) to enhance the cavity mode interaction with the graphene sheet, see Figure 25.5c. In particular, FDTD simulations indicate that this slot cavity design improves the graphene–optical mode coupling rate κ_{cg} roughly threefold compared to the previously discussed L3 cavity.

The modulator was characterized via the cavity reflectivity as a function of gate voltage (V_g). Simultaneously, the electrical resistance through source and drain electrodes was monitored to record the doping level of graphene. Figure 25.6a plots the reflection spectra. When the gate voltage is set to zero ($V_g = 0$), the cavity reflection exhibits two resonant peaks at wavelengths of 1571.7 (mode I) and 1593 nm (mode II) with Q factors of 340 and 410, as shown in the top spectrum. The expectedly low Q factors are due to the strong graphene absorption and the extra scattering loss from the electrolyte. Applying a gate voltage along the negative direction to −1 V, we observed the resonant peaks on the cavity reflection become narrower and higher, accompanied by a red shift of the central wavelength, as shown in the second panel of Figure 25.6a. At higher gate voltage, the intensities and Q factors of the resonant peaks were increased further, as shown in the third spectrum of Figure 25.6a with $V_g = -7$ V. The simultaneous increase of the reflection peaks and Q factors are consistent with the attenuation curve shown in Figure 25.3b. The two resonances also show clear blue shifts at high gate voltages. The increased reflectivity and Q factors indicate the suppression of the graphene absorption due to Pauli blocking. However, because of phonon-assisted loss processes and charge inhomogeneity in the graphene layer [24,56], the Pauli blocking is not complete even under high gate voltage. The residual absorption is responsible for the overall lower Q factors of the resonant modes compared to the devices before the graphene transfer [29].

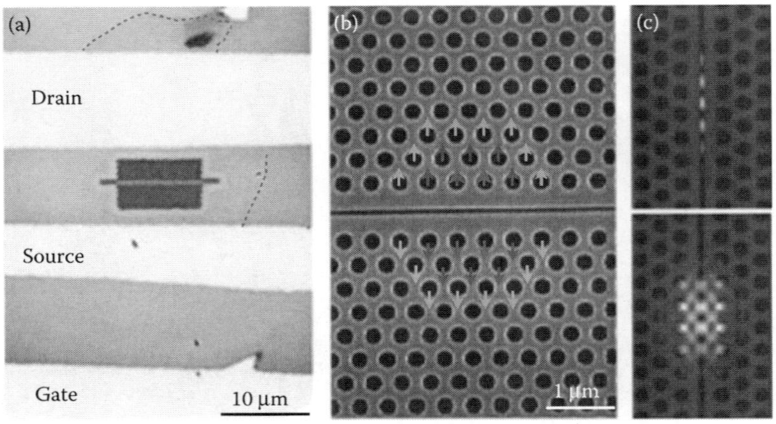

FIGURE 25.5 (a) Optical microscope image of a PPC cavity–graphene modulator gated by solid electrolyte. (b) SEM image of the employed air-slot cavity. (c) Simulated energy distributions of the resonant modes of the slot cavity showing strongly confined mode in the air gap, where the top and bottom correspond to modes I and II in Figure 25.6c. (Reprinted with permission from X Gan et al. High-contrast electrooptic modulation of a photonic crystal nanocavity by electrical gating of graphene. *Nano Letters*, 13(2):691–696. Copyright 2013 American Chemical Society.)

Figure 25.6c plots the cavity reflectivity as the gate voltage is swept between −7 and 6 V, as indicated in Figure 25.6b. Symmetric modulation of the cavity reflection was observed with negative (hole-doped) and positive (electron-doped) gate voltages with respect to the charge neutrality point of the graphene layer at $V_g = 1$ V, which coincides with the maximum electrical resistance of graphene [29]. Moreover, abrupt changes of the resonant peaks occurred when V_g changed between −1 and −2.5 V and 3 to 4.5 V, corresponding to the voltages when the Fermi energy shifts past $E_F = \hbar\omega/2$. From these reflectivity measurements, we inferred a modulation depth exceeding 10 dB for a swing gate voltage between −1 and −2.5 V.

To increase the modulation rate beyond what is possible with the electrolyte gate, we developed a device using two graphene monolayers separated by a BN dielectric layer, as

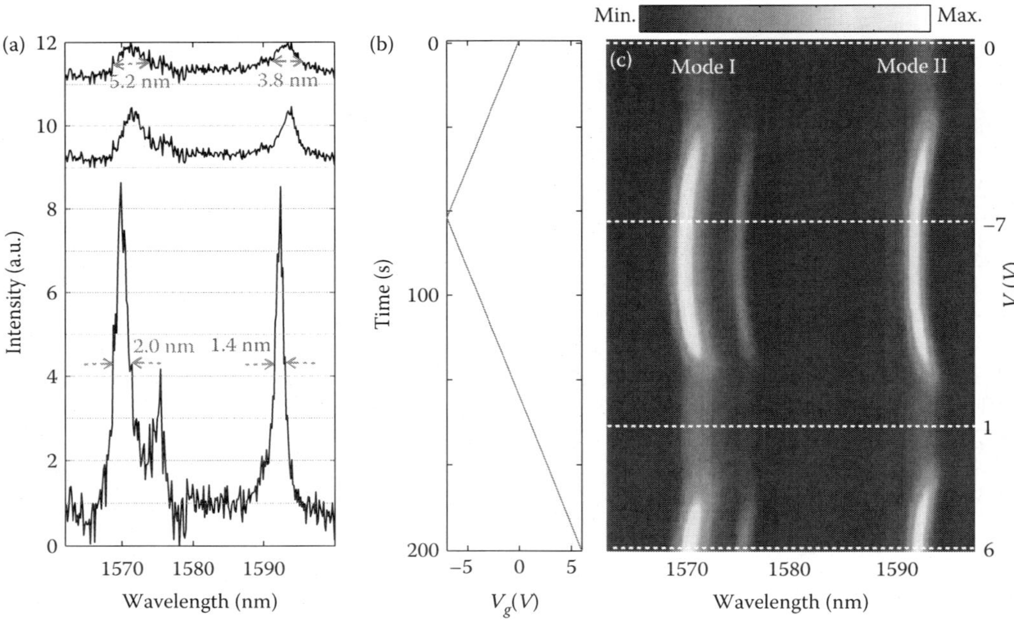

FIGURE 25.6 Characterization of the electrolyte-gated PPC cavity–graphene modulator. (a) Spectra of the cavity reflection for $V_g = 0$, −1, and −7 V (from top to bottom), which are normalized by the reflection peak at $V_g = 0$. (b) Gated voltage V_g modulated in a saw-tooth pattern at a rate of 0.1 V/s between −7 and 6 V. (c) Reflection spectra of the cavity as V_g is modulated. The resonance peaks show clear wavelength shifts, and modulations of Q factors and intensities under different gate voltages. (Reprinted with permission from X Gan et al. High-contrast electrooptic modulation of a photonic crystal nanocavity by electrical gating of graphene. *Nano Letters*, 13(2):691–696. Copyright 2013 American Chemical Society.)

indicated in Figure 25.4b. When a bias voltage is applied across this graphene capacitor, the two graphene layers reach optical transparency nearly simultaneously [25]. Alternatively, a doped Si PPC cavity could replace one of the graphene layers as an electrical contact, as shown in Figure 25.4c, though the doping in the Si layer may cause additional cavity loss compared to the "semitransparent" graphene capacitor.

A fabricated device consisting of a Si L3 PPC cavity and a dual-layer graphene capacitor is shown in Figure 25.7a. After the growth of a 10-nm HfO_2 layer on the PPC cavity, two graphene monolayers sandwiching a 7.6-nm-thick BN spacer layer were transferred onto the cavity layer by layer. Two Ti/Au electrodes were separately deposited on the top and the bottom graphene layers. The modulator was characterized via the cavity reflectivity at different gate voltages, as shown in Figure 25.7b. The curves from top to bottom correspond to the reflection spectra at gate voltages of 0 and 2.5 V, and the spectrally resolved modulation between the two spectra. As for the electrolyte-modulated device, the cavity peak narrows and rises in intensity, while the wavelength shifts when the gate voltage is applied. The modulation spectrum indicates a maximum modulation depth near 1.5 dB at the resonant wavelength of 1518.1 nm. In the electrolyte-modulated device, when the graphene Fermi energy was tuned past $E_F = \hbar\omega/2$ to reach the Pauli-blocking regime, the resonant wavelength transitioned through a red-shift region and was already strongly blue shifted (Figure 25.6c). In the graphene–BN–graphene dual-gated device, the spectra in Figure 25.7b show that the graphene doping only red shifts the cavity resonances, indicating that the doping is insufficient for the Pauli-blocking regime. This is because the voltage was limited to ~2.5 V by the current leaking of the capacitor, which we attributed to defects during the transfer processes of multiple layers, as shown in the atomic force microscope image of Figure 25.7c. Based on the gating results shown in Figure 25.6c and the wavelength red-shift of this device, one may expect a modulation depth higher than 10 dB for higher bias voltage.

We measured the temporal response by modulating a continuous-wave (CW) probe laser at the wavelength of 1518.1 nm. An alternate current (ac) electrical signal from a network analyzer (NA) was applied to the graphene capacitor via a radio frequency (RF) microwave probe. A direct current (dc) offset biased this signal near the highest slope in the reflection modulation of the graphene–cavity modulator. The cavity reflection was detected via a high-speed InGaAs photodiode connected to the same NA. The modulation response was then quantified through the $S21$ parameter in the frequency range from 20 MHz to 2 GHz, as shown in Figure 25.7d. The 3 dB cutoff is about 0.57 GHz, which agrees with the estimated RC time constant. The $S11$ parameter indicated a capacitance C of about 1.1 pF, which is consistent with a calculation from the parallel-capacitance model, taking the dielectric constant of BN to be 3.9 and the area of the capacitor as 250 μm². The series resistance, R, measured from $S11$ is about 200 Ω, which is mainly due to the contact resistance considering the low expected sheet resistance of the graphene layer. The calculated RC-limited

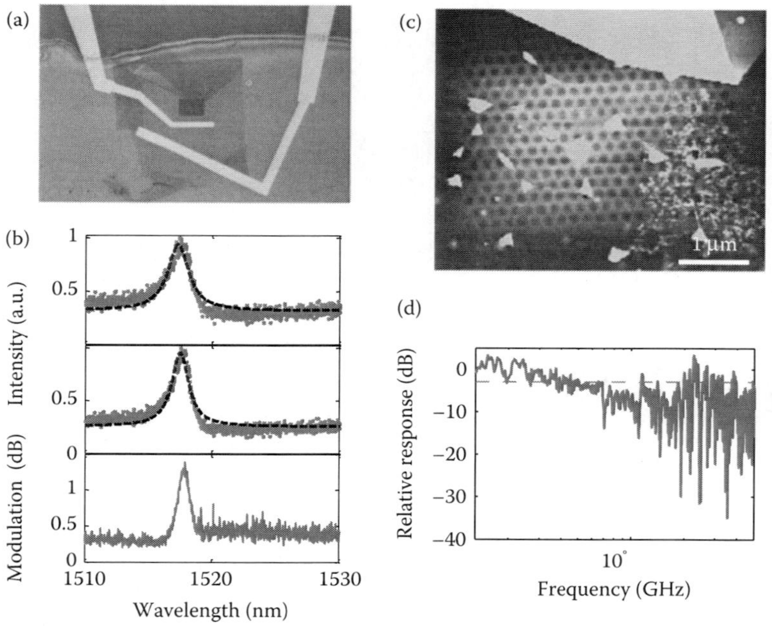

FIGURE 25.7 Characterization of the high-speed cavity–graphene electro-optic modulator with dual-gated graphene layers. (a) Optical microscope image of a fabricated device. (b) Reflection spectra of the cavity with the gated voltage of 0 V (top), 2.5 V (middle), and the spectrally resolved modulation between the two gate voltages (bottom). (c) Atomic force microscope image of the device. (d) Dynamics response of the modulator. (X Gan et al. Controlled light–matter interaction in graphene electro-optic devices using nanophotonic cavities and waveguides. *Selected Topics in Quantum Electronics, IEEE Journal of,* 20(1):6000311, © 2014 IEEE.)

bandwidth is therefore $1/(2\pi RC) = 0.72$ GHz, which is close to the measured result. An estimate of the energy consumption E per bit, given by $E = CV_g^2/2$ [57], indicates 3 pJ/bit for a swing voltage of $V_g = 2.5$ V. Note that the large spectral bandwidth of the graphene modulator would make it less sensitive to temperature changes than most ring-based modulators employing free carrier modulation [58].

25.2.3 Cavity-Enhanced Graphene Photodetector

On the cavity resonance, the graphene absorption is strongly enhanced by the PPC cavity–graphene system, which is governed by $\eta\kappa_c\kappa_{cg}/(\kappa_c/2 + \kappa_{cg}/2)^2$ derived from Equation 25.2. The enhanced on-resonance absorption versus the ratio κ_c/κ_{cg} is plotted in the light color curve in Figure 25.3b, which was normalized by the coupling efficiency η. The plot shows that the graphene absorption reaches its maximum value of η when $\kappa_c/\kappa_{cg} = 1$. For PPC cavities, the coupling efficiency η can exceed 45% using on-chip side coupling into the cavity from a waveguide [59], or tapered fiber coupling [60]. By employing efficient coupling and choosing $\kappa_c \approx \kappa_{cg}$, it therefore appears possible to reach 45% absorption into a monolayer of graphene in such a two-sided cavity.

We fabricated a PPC cavity-enhanced graphene photodetector by integrating an exfoliated graphene layer on a Si PPC cavity and depositing two metal electrodes (bias/ground) on the opposite sides of the cavity, as shown in Figure 25.8a. The graphene sheet covered the whole PPC

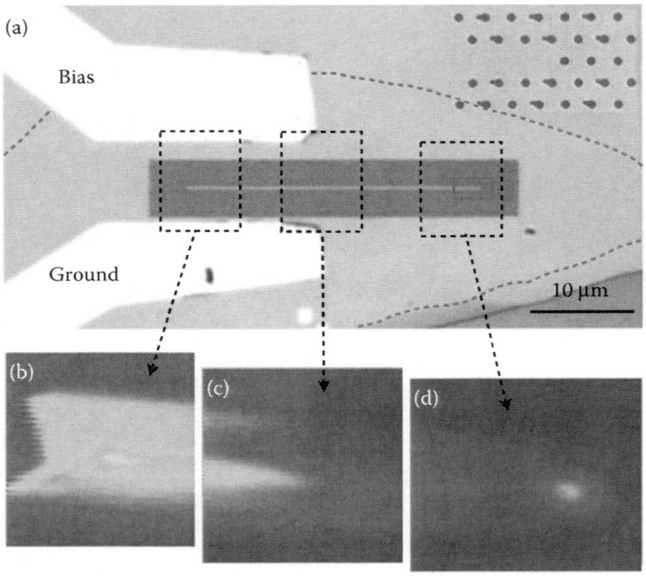

FIGURE 25.8 Cavity-integrated graphene photodetector. (a) Optical microscope image of a fabricated device. Inset: SEM image of the end of the line defect, where the small perturbation holes are shown. (b) and (c) Scanning photocurrent imagings of the corresponding regions marked in (a). (Reprinted with permission from R-J Shiue et al. Enhanced photodetection in graphene-integrated photonic crystal cavity. *Applied Physics Letters*, 103(24):241109. Copyright 2013, American Institute of Physics.)

cavity, as indicated by the light-dashed line. The PPC lattice was designed with a period of $a = 450$ nm and a hole radius of $r = 0.29a$. The cavity defect was formed by terminating a line-defect waveguide with periodic air holes, presenting several bound modes on the waveguiding band (Figure 25.9b). At the ends of the line defect, small air holes were fabricated adjacent to the PPC air holes with a period of $2a$, as shown in the inset SEM image of Figure 25.8a. This perturbation design served as a grating for vertical coupling to the cavity modes [61]. In addition, the grating couplers cause extra scattering loss of the cavity modes, which makes it possible to match the intrinsic cavity loss rate with the loss rate induced by graphene. As indicated in the light color curve of Figure 25.3b, this optimized cavity structure enabled high absorption into the graphene layer.

The fabricated device was tested on a cross-polarization microscope using a tunable CW laser [37]. Figure 25.8b shows spatial scanning photocurrent images of the corresponding regions marked in Figure 25.8a, where the input wavelength was 1535 nm and the bias voltage was 0.2 V. When light excited the cavity's grating coupler, the photocurrent was sharply enhanced. In Figure 25.8c, the photocurrent becomes invisible on the section without metal contact, illustrating the need for an external electrical field for photocurrent generation. By contrast, in the area to the right (Figure 25.8d), strong photocurrent was generated even without metal contacts when the input light was directed on the grating couplers. This observation shows that the input light coupled into the PPC cavity via the coupler and generated photocurrent in the graphene channel.

To demonstrate the cavity-enhanced photodetection, the input laser was focused on the right grating coupler of the device and scanned the input wavelength from 1514 to 1558 nm. The measured photocurrent spectrum is shown in Figure 25.9a. For wavelengths below 1550 nm, the photocurrent is magnified and shows multiple peaks. This increase follows the expected cavity-enhanced absorption into the graphene sheet, as described by Equation 25.2. Figure 25.9b plots the cavity reflection spectra before (dark color) and after (light color) the graphene deposition. The unloaded PPC cavity showed a reflection spectrum with multiple peaks in the range of 1510–1560 nm, corresponding to the cavity resonances as confirmed by FDTD simulations. After the graphene deposition, the cavity modes exhibit broadened peaks with low visibility due to the graphene absorption. Fitting the cavity peaks in the wavelength range of 1522–1541 nm (Figure 25.9c) predicted an intrinsic cavity loss rates of κ_c, a graphene-induced cavity loss rate of κ_{cg}, and the resonant frequencies ω of the cavity modes [36]. Using these parameters and Equation 25.2, the cavity-enhanced absorption coefficients of graphene were estimated, as plotted in Figure 25.9d. The fits show good agreement with the measured photocurrent spectrum, confirming the cavity enhancement of photocurrent in Figure 25.9a. Comparing the enhanced photocurrent peaks with the photocurrent obtained with off-resonance excitation, we estimated an eightfold enhancement via the cavity modes. In addition, due to the overlapping of cavity modes near

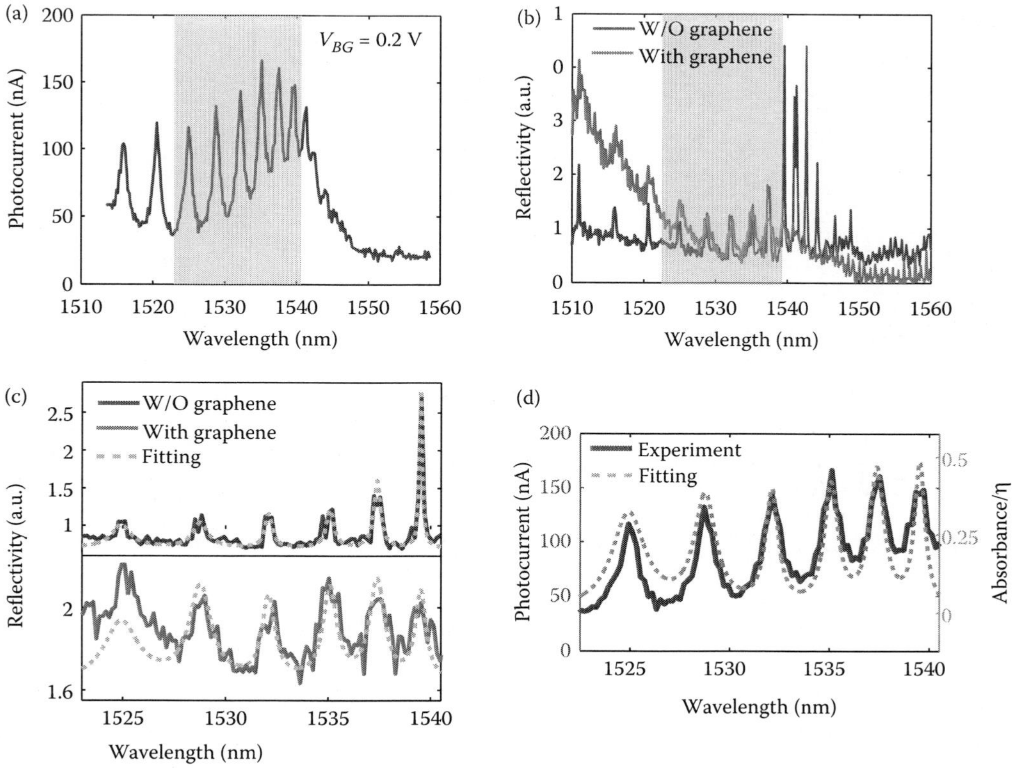

FIGURE 25.9 Cavity-enhanced photodetection in graphene. (a) Photocurrent of the graphene detector as a function of the excitation wavelength. (b) Cavity reflection spectra before and after the integration of the graphene layer. (c) Lorentizan fittings of the cavity reflection peaks with and without graphene. (d) Theoretical fitting of the cavity-enhanced photodetection. (Reprinted with permission from R-J Shiue et al. Enhanced photodetection in graphene-integrated photonic crystal cavity. *Applied Physics Letters*, 103(24):241109. Copyright 2013, American Institute of Physics.)

the photonic band edge with the narrow free spectral range, this long PPC cavity-integrated graphene photodetector also enables a broadband photocurrent enhancement in the wavelength range of 1535–1545 nm.

25.3 GRAPHENE OPTOELECTRONIC DEVICES INTEGRATED ON BUS WAVEGUIDES

So far, we have discussed the use of cavity modes to enhance the absorption in graphene photodetectors and modulators. While cavity-based devices enable wavelength-scale dimensions of the active region, the operating spectrum is inherently narrow. In this section, we show that coupling graphene with a single-mode Si bus waveguide can enhance the light–graphene interaction across a broad spectrum due to the large extension of the interaction length. Based on the waveguide-integrated architecture, a graphene photodetector with high responsivity, high speed, and operating across a broadband spectrum was demonstrated [41].

25.3.1 Waveguide-Enhanced Graphene Absorption

The Si waveguides used to integrate graphene were fabricated on a silicon-on-insulator (SOI) wafer with a 220-nm-thick Si membrane using shallow trench isolation (STI). The waveguide width was 520 nm to ensure a single transverse electrical (TE) mode with low transmission loss. To prevent the graphene from fracturing at the edges of waveguides, the chip was planarized by back filling with a thick SiO_2 layer and chemical mechanical polishing reaching the top Si layer. A ~10-nm-thick SiO_2 layer was deposited onto the chip to ensure the electrical isolation of the graphene layer from the Si membrane. Figure 25.10a shows an optical microscope image of a finished device, where a ~70-μm-long mechanically exfoliated graphene bilayer was transferred onto a waveguide.

By measuring the waveguide transmission of a CW-tunable narrowband laser in the telecom band before and after the graphene deposition, a 70-μm-long graphene bilayer was found to result in a flat transmission loss of approximately 6 dB over a wavelength range of 1500–1580 nm, as shown in Figure 25.10b. To confirm the enhanced absorption into graphene on the waveguide, the guided mode of device was simulated using a finite element method (COMSOL). The thicknesses of the bilayer graphene was chosen as 1.4 nm and the refractive indices of SiO_2, Si, and graphene were 1.48, 3.4, and 2.38 + 1.68i, respectively, for light at 1550 nm [52]. Figure 25.10c displays the simulated TE field distribution of the guided mode at 1550 nm, showing the evanescent field of the waveguide mode coupled with the graphene layer. The intensity profile across the middle of the waveguide indicates

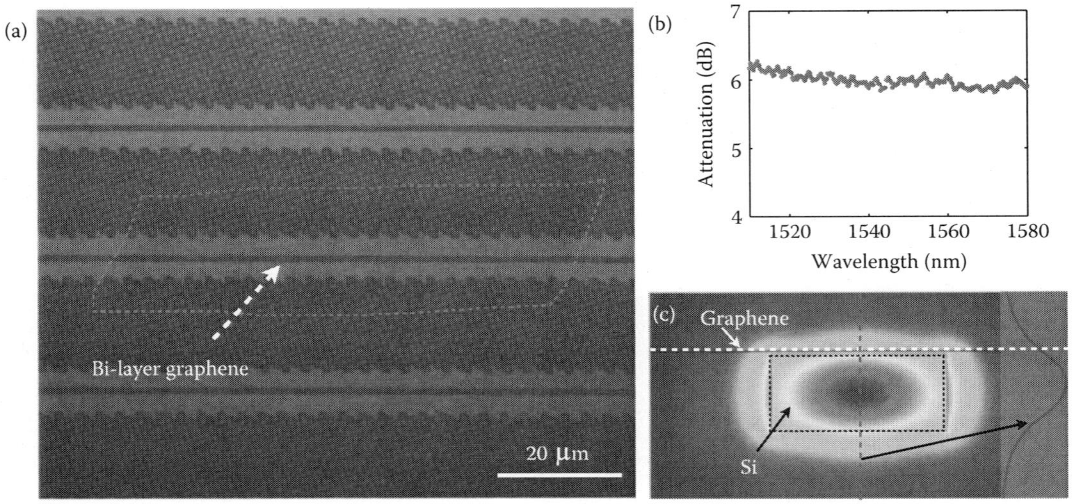

FIGURE 25.10 (a) Optical microscope image of the waveguide-integrated graphene device; the bilayer graphene is marked by the red-dashed line. (b) Graphene-induced transmission attenuations of the waveguide as a function of the input wavelength. (c) Simulated TE mode distribution of the Si waveguide integrated with bilayer graphene. (X Gan et al. Controlled light–matter interaction in graphene electro-optic devices using nanophotonic cavities and waveguides. *Selected Topics in Quantum Electronics, IEEE Journal of,* 20(1):6000311, © 2014 IEEE.)

that the electric field located at the graphene location is about 40% of the maximum. From the complex effective index of the simulated guided mode, the absorption coefficient of a graphene bilayer is determined to be 0.085 dB/μm, in good agreement with the absorption coefficient of 0.089 dB/μm calculated from the experimental results.

25.3.2 Waveguide-Integrated Photodetector with High Responsivity

The waveguide-coupled graphene photodetector [41] is schematically shown in Figure 25.11a. To eliminate the dark current due to an external bias voltage, the two contact electrodes were placed in an asymmetric configuration on opposite sides of the waveguide. One of these electrodes was positioned close to the edge of waveguide to create a lateral metal-doped junction that overlaps with the waveguide mode, as indicated by the potential profile across the graphene channel and the simulated waveguided mode shown in Figure 25.11b and c. The graphene potential gradient around the metal electrode produces an internal electrical field [62] that is close enough to the waveguide to separate the photogenerated electron–hole pairs produced over the waveguide. In addition, the absence of an overlap between the optical field and the field produced by the other electrode ensures the acceleration of electrons (or holes) in only one direction without canceling the net photocurrent [15,16].

Figure 25.12a displays a SEM image of the fabricated device integrated with a 53-μm-long graphene bilayer. The two electrodes were evaporated at separations of ~100 nm and ~3.5 μm from the edges of the waveguide. To estimate the absorption of the metal electrode on the waveguide transmission, the TE waveguided mode of the device was simulated with the parameters used in Figure 25.10 and the dielectric constant of gold chosen as $0.55 + 11.5i$ [63] for light around 1550 nm. The complex refractive index of the guided mode shown in Figure 25.11c indicates that the 40 nm thick metal contact results in an absorption coefficient of ~0.009 dB/μm on the waveguide transmission. Comparing with the value obtained from Figure 25.10c indicates that the graphene layer

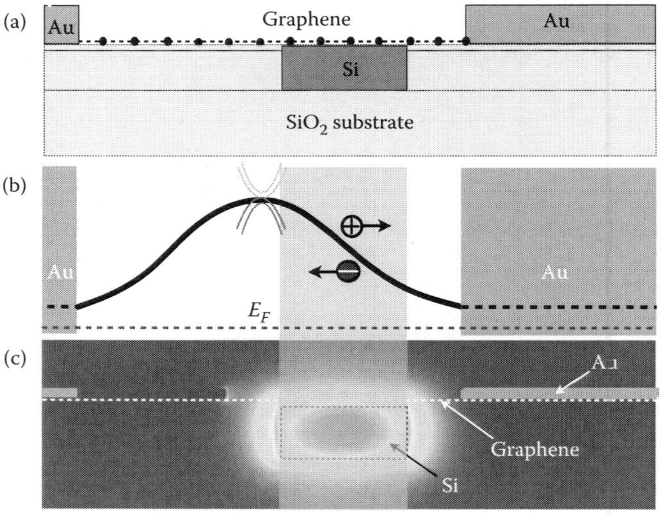

FIGURE 25.11 Waveguide-integrated graphene photodetector operating at zero bias. (a) Schematic of the waveguide-integrated graphene photodetector. (b) Calculated potential profile across the graphene channel. (c) Simulated electrical field of the TE-guided mode for the waveguide integrated with a graphene bilayer and metal contacts, showing overlap between the mode field and potential gradient on graphene. (Reprinted by permission from Macmillan Publishers Ltd. Nature Photonics, X Gan et al. Chip-integrated ultrafast graphene photodetector with high responsivity. 7(11):883–887, copyright 2013.)

was responsible for approximately 90% of the waveguide's transmission loss.

To test the performance of the photodetector, a probe laser at 1550 nm from a tunable CW laser was coupled into the device, which was modulated at a low frequency of 1 kHz by the internal modulator, and the short-circuit photocurrent signal was detected with a current preamplifier and a lock-in amplifier. Figure 25.12b plots the measured photocurrent (I_{photo}) versus incident power (P_{input}) at zero bias voltage. Here, P_{input} is the power reaching the waveguide-graphene detector, estimated by considering the input facet coupling loss of edge couplers and the Si waveguide transmission loss. This measurement indicates an external responsivity (I_{photo}/P_{input}) of 15.7 mA/W—two orders of magnitude higher than that obtained for normal incidence. This strong responsivity improvement is due to the longer light–graphene interaction and the efficient separation of the photoexcited electron–hole pairs resulting from the strong local electric field across the metal-doped junction. Moreover, the photocurrent approaches zero under low-power optical excitation.

A spectrally resolved photocurrent measurement at fixed input power (Figure 25.12c) indicated broadband operation. The flat response from 1450 to 1590 nm suggests the possibility of carrier multiplication in monolayer graphene, as reported in vertical incidence experiments [33], though further studies are needed on this effect in the waveguide configuration. Because the detector functioned in a waveguide-integrated configuration, the long absorption length of the graphene sheet enabled a large power range. There was no observable saturation of photocurrent under CW laser excitation at launching powers up to 10 dBm. Photocurrent measurements were also performed with the excitation of a pulsed optical parametric oscillator (OPO) source at a wavelength of 2000 nm and providing 250 fs pulses at a repetition rate of 78 MHz. The inset in Figure 25.12b shows the photocurrent on the device as a function of the average incident power of the OPO pulsed source, indicating a saturation of the photocurrent for an incident power near 760 μW. We estimated that under these conditions, the graphene layer will experience a peak intensity of 6.1 GW/cm^2, corresponding to the threshold of the saturable absorption in graphene due to Pauli blocking [2].

Unlike conventional semiconductors, both electrons and holes in graphene have very high mobility, and a moderate internal electric field allows an ultrafast and efficient

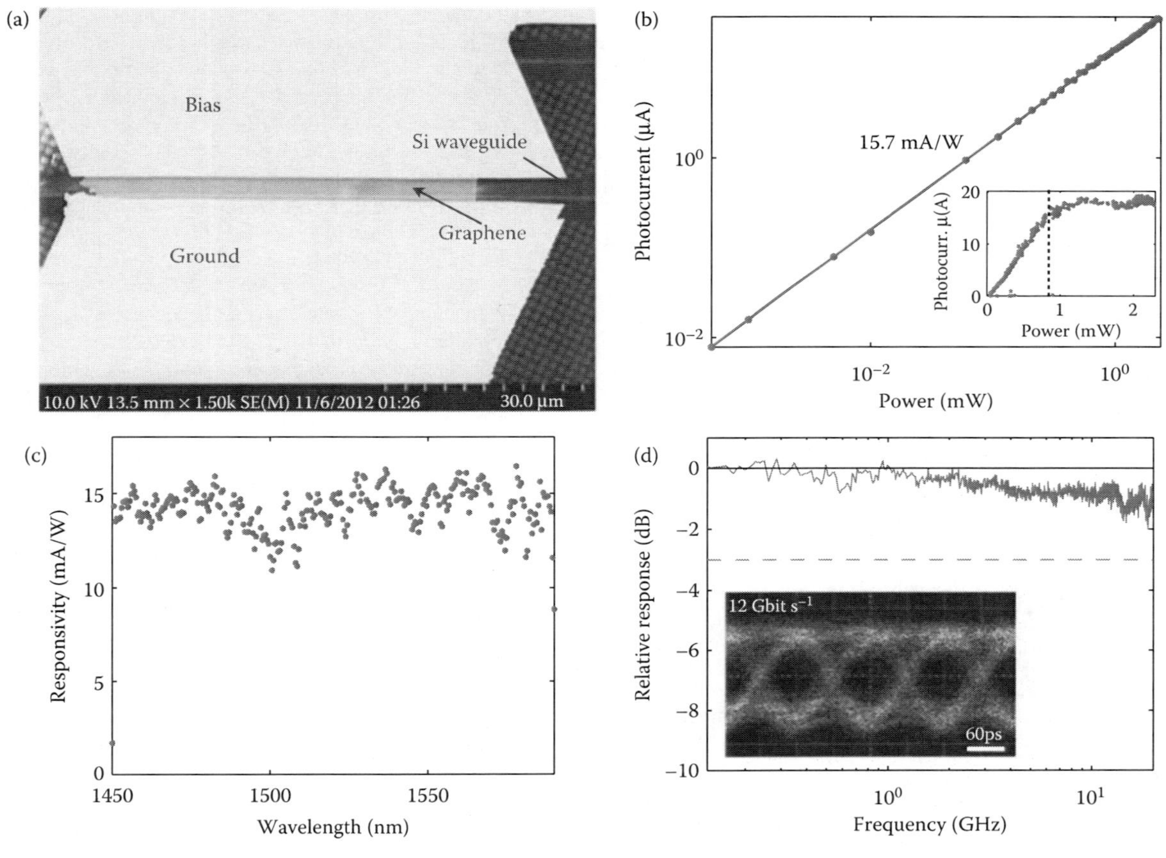

FIGURE 25.12 Characterization of the waveguide-integrated graphene photodetector. (a) SEM image of the fabricated device. (b) Photocurrent of the graphene detector with respect to the power of a CW input laser at the wavelength of 1550 nm with zero bias. Inset: Photocurrent as a function of the excited power from a pulsed OPO laser at 2000 nm. (c) Wavelength dependence of the detector with a broadband flat responsivity from 1450 to 1590 nm measured. (d) High-speed dynamic response of the detector. Inset: 12 Gbit/s optical data link test of the device, showing a clear eye opening. (Reprinted by permission from Macmillan Publishers Ltd. *Nature Photonics,* X Gan et al. Chip-integrated ultrafast graphene photodetector with high responsivity. 7(11):883–887, copyright 2013.)

photocarrier separation. The high-speed response of the device was estimated using a commercial lightwave component analyzer (LCA) with an internal laser source and NA, with a frequency range from 0.13 to 20 GHz. A modulated optical signal at a wavelength of 1550 nm with an average power of 1 mW emitted from the LCA is coupled into the device and the electrical output is measured through a ground-source RF microwave probe and fed back into the input port of the NA. The frequency response of the device is analyzed as the *S*21 parameter of the NA. Figure 25.12d displays the ac photoresponse of the device at zero bias, showing about 1 dB degradation of the signal at 20 GHz. Graphene's extremely high carrier mobility enables an intrinsic photoresponse faster than 640 GHz [16]; the observed degradation of the dynamic response can be attributed to a large capacitance from the relatively large device area. To estimate the viability of the waveguide-integrated graphene photodetector in realistic optical applications, we tested optical data transmission. A pulsed pattern generator with a maximum 12 Gbit/s internal electrical bit stream modulated a 1550-nm CW laser via an electro-optic modulator, which was launched into the waveguide–graphene detector. The output electrical data stream from the graphene detector was amplified and sent to a wideband oscilloscope to obtain an eye diagram. As shown in the inset of Figure 25.12d, a clear eye-opening diagram at 12 Gbit/s was obtained.

25.4 CONCLUSIONS

Using nanophotonic techniques to enhance the interaction of light with graphene in waveguides and resonators, a range of exciting on-chip graphene optoelectronic devices are now becoming possible. Compared with many traditional active materials that have been integrated into Si photonics, graphene promises relative ease of fabrication, compact footprint, high speed, and efficient carrier transport.

For future on-chip integration, it is important to note that the photodetector and modulator devices described in this chapter are potentially CMOS compatible. Although the current PPC cavity-integrated modulators and photodetectors were tested using a vertical coupling microscope, these devices may be coupled to on-chip waveguides. The integrated PPC cavity–waveguide chips or the planarized bus waveguide chips should enable reliable transfer of wafer-scale graphene sheets with low probability of rupture. We also fabricated photodetectors on waveguide chips with silicon nitride (SiN) couplers were also fabricated, showing 3-dB fiber-to-waveguide coupling loss. These SiN couplers enable high-temperature processing steps that are required in the CMOS process; high-temperature annealing is also compatible with graphene transfer processes. Therefore, CMOS-processing compatibility of the demonstrated device appears possible in the near-term through (1) the use of CVD-grown graphene, either transferred or selectively grown on the integrated PPC chip or bus waveguide chip [64], and (2) deposition of CMOS-compatible metal to replace gold in the Ti/Au contacts.

REFERENCES

1. R R Nair, P Blake, A N Grigorenko, K S Novoselov, T J Booth, T Stauber, N M R Peres, and A K Geim. Fine structure constant defines visual transparency of graphene. *Science*, 320(5881):1308, 2008.
2. G Xing, H Guo, X Zhang, T C Sum, C Hon, and A Huan. The physics of ultrafast saturable absorption in graphene. *Optics Express*, 18(5):4564–4573, 2010.
3. A Martinez and Z Sun. Nanotube and graphene saturable absorbers for fibre lasers. *Nature Photonics*, 7:842–845, 2013.
4. E Hendry, P Hale, J Moger, A Savchenko, and S Mikhailov. Coherent nonlinear optical response of graphene. *Physical Review Letters*, 105(9):97401, 2010.
5. M A Vincenti, D de Ceglia, M Grande, A D'Orazio, and M Scalora. Third-harmonic generation in one-dimensional photonic crystal with graphene-based defect. *Physical Review B*, 89(16):165139, 2014.
6. F Bonaccorso, Z Sun, T Hasan, and A C Ferrari. Graphene photonics and optoelectronics. *Nature Photonics*, 4(9):611–622, 2010.
7. K Kim, J-Y Choi, T Kim, S-H Cho, and H-J Chung. A role for graphene in silicon-based semiconductor devices. *Nature*, 479(7373):338–344, 2011.
8. Q Bao and K P Loh. Graphene photonics, plasmonics, and broadband optoelectronic devices. *ACS Nano*, 6(5):3677–3694, 2012.
9. P Avouris. Graphene: Electronic and photonic properties and devices. *Nano Letters*, 10: 4285–4294, 2010.
10. P Avouris and M Freitag. Graphene photonics, plasmonics and optoelectronics. *Selected Topics in Quantum Electronics, IEEE Journal of*, 20:6000112, 2014.
11. X Gan, R-J Shiue, Y Gao, S Assefa, J Hone, and D Englund. Controlled light–matter interaction in graphene electro-optic devices using nanophotonic cavities and waveguides. *Selected Topics in Quantum Electronics, IEEE Journal of*, 20(1):6000311, 2014.
12. M Liu, X Yin, E Ulin-Avila, B Geng, T Zentgraf, L Ju, F Wang, and X Zhang. A graphene-based broadband optical modulator. *Nature*, 474(7349):64–67, 2011.
13. E O Polat and C Kocabas. Broadband optical modulators based on graphene supercapacitors. *Nano Letters*, 13(12):5851–5857, 2013.
14. W Li, B Chen, C Meng, W Fang, Y Xiao, X Li, Z Hu et al. Ultrafast all-optical graphene modulator. *Nano Letters*, 14(2):955–959, 2014.
15. T Mueller, F Xia, and P Avouris. Graphene photodetectors for high-speed optical communications. *Nature Photonics*, 4(5):297–301, 2010.
16. F Xia, T Mueller, Y-M Lin, A Valdes-Garcia, and P Avouris. Ultrafast graphene photodetector. *Nature Nanotechnology*, 4(12):839–843, 2009.
17. C O Kim, S Kim, D H Shin, S S Kang, J M Kim, C W Jang, S S Joo et al. High photoresponsivity in an all-graphene p–n vertical junction photodetector. *Nature Communications*, 5:3249, 2014.
18. A N Grigorenko, M Polini, and K S Novoselov. Graphene plasmonics. *Nature Photonics*, 6:749–758, 2012.
19. J T Kim, Y-J Yu, H Choi, and C-G Choi. Graphene-based plasmonic photodetector for photonic integrated circuits. *Optics Express*, 22(1):803, 2014.
20. F H L Koppens, D E Chang, and F J G de Abajo. Graphene plasmonics: A platform for strong light–matter interactions. *Nano Letters*, 11(8):3370–3377, 2011.

21. Z Sun, T Hasan, F Torrisi, D Popa, G Privitera, F Wang, F Bonaccorso, D M Basko, and A C Ferrari. Graphene mode-locked ultrafast laser. *ACS Nano*, 4(2):803–810, 2010.
22. L Li, X Zheng, X Chen, M Qi, Z Ren, J Bai, and Z Sun. High-power diode-side-pumped Nd:YAG solid laser mode-locked by CVD graphene. *Optics Communications*, 315:204–207, 2014.
23. F Wang, Y Zhang, C Tian, C Girit, A Zettl, M Crommie, and Y R Shen. Gate-variable optical transitions in graphene. *Science (New York, NY)*, 320(5873):206–209, 2008.
24. Z Q Li, E A Henriksen, Z Jiang, Z Hao, M C Martin, P Kim, H L Stormer, and D N Basov. Dirac charge dynamics in graphene by infrared spectroscopy. *Nature Physics*, 4(7):532–535, 2008.
25. M Liu, X Yin, and X Zhang. Double-layer graphene optical modulator. *Nano Letters*, 12(3):1482–1485, 2012.
26. S J Koester and M Li. High-speed waveguide-coupled graphene-on-graphene optical modulators. *Applied Physics Letters*, 100(17):171107, 2012.
27. F Lu, T-T D Tran, W S Ko, K W Ng, R Chen, and C Chang-Hasnain. Nanolasers grown on silicon-based MOSFETs. *Optics Express*, 20(11):12171, 2012.
28. C-C Lee, S Suzuki, W Xie, and T R Schibli. Broadband graphene electro-optic modulators with sub-wavelength thickness. *Optics Express*, 20(5):5264–5269, 2012.
29. X Gan, R-J Shiue, Y Gao, K F Mak, X Yao, L Li, A Szep et al. High-contrast electrooptic modulation of a photonic crystal nanocavity by electrical gating of graphene. *Nano Letters*, 13(2):691–696, 2013.
30. A Majumdar, J Kim, J Vuckovic, and F Wang. Electrical control of silicon photonic crystal cavity by graphene. *Nano Letters*, 13(2):515–518, 2013.
31. J C W Song, M S Rudner, C M Marcus, and L S Levitov. Hot carrier transport and photocurrent response in graphene. *Nano Letters*, 11(11):4688–4692, 2011.
32. M Freitag, T Low, F Xia, and P Avouris. Photoconductivity of biased graphene. *Nature Photonics*, 7(December):53–59, 2012.
33. K J Tielrooij, J C W Song, S A Jensen, A Centeno, A Pesquera, A Zurutuza Elorza, M Bonn, L S Levitov, and F H L Koppens. Photoexcitation cascade and multiple hot-carrier generation in graphene. *Nature Physics*, 9(4):248–252, 2013.
34. M Furchi, A Urich, A Pospischil, G Lilley, K Unterrainer, H Detz, P Klang et al. Microcavity-integrated graphene photodetector. *Nano Letters*, 12(6):2773–2777, 2012.
35. A Urich, K Unterrainer, and T Mueller. Intrinsic response time of graphene photodetectors. *Nano Letters*, 11(7):2804–2808, 2011.
36. X Gan, K F Mak, Y Gao, Y You, F Hatami, J Hone, T F Heinz, and D Englund. Strong enhancement of light–matter interaction in graphene coupled to a photonic crystal nanocavity. *Nano Letters*, 12(11):5626–5631, 2012.
37. R-J Shiue, X Gan, Y Gao, L Li, X Yao, A Szep, D Walker, J Hone, and D Englund. Enhanced photodetection in graphene-integrated photonic crystal cavity. *Applied Physics Letters*, 103(24):241109, 2013.
38. T J Echtermeyer, L Britnell, P K Jasnos, A Lombardo, R V Gorbachev, A Grigorenko, A K Geim, A C Ferrai, and K S Novoselov. Strong plasmonic enhancement of photovoltage in graphene. *Nature Communications*, 2:455–458, 2011.
39. Y Liu, R Cheng, L Liao, H Zhou, J Bai, G Liu, L Liu, Y Huang, and X Duan. Plasmon resonance enhanced multicolour photodetection by graphene. *Nature Communications*, 2:579, 2011.
40. G Konstantatos, M Badioli, L Gaudreau, J Osmond, M Bernechea, F P G de Arquer, F Gatti, and F H L Koppens. Hybrid graphene quantum dot phototransistors with ultrahigh gain. *Nature Nanotechnology*, 7(June):363–368, 2012.
41. X Gan, R-J Shiue, Y Gao, I Meric, T F Heinz, K Shepard, J Hone, S Assefa, and D Englund. Chip-integrated ultrafast graphene photodetector with high responsivity. *Nature Photonics*, 7(11):883–887, 2013.
42. X Wang, Z Cheng, K Xu, H K Tsang, and J-B Xu. High-responsivity graphene/silicon-heterostructure waveguide photodetectors. *Nature Photonics*, 7(11):888–891, 2013.
43. A Pospischil, M Humer, M M Furchi, D Bachmann, R Guider, T Fromherz, and T Mueller. CMOS-compatible graphene photodetector covering all optical communication bands. *Nature Photonics*, 7(11):892–896, 2013.
44. H Sekoguchi, Y Takahashi, T Asano, and S Noda. Photonic crystal nanocavity with a q-factor of 9 million. *Optics Express*, 22(1):916–924, 2014.
45. C R Dean, A F Young, I Meric, C Lee, L Wang, S Sorgenfrei, K Watanabe et al. Boron nitride substrates for high-quality graphene electronics. *Nature Nanotechnology*, 5:722–726, 2010.
46. J W Suk, A Kitt, C W Magnuson, Y Hao, S Ahmed, J An, A K Swan, B B Goldberg, and R S Ruoff. Transfer of CVD-grown monolayer graphene onto arbitrary substrates. *ACS Nano*, 5(9):6916–6924, 2011.
47. D Englund, A Faraon, I Fushman, N Stoltz, P Petroff, and J Vučković. Controlling cavity reflectivity with a single quantum dot. *Nature*, 450(6):857–861, 2007.
48. H Altug and J Vučković. Two-dimensional coupled photonic crystal resonator arrays. *Applied Physics Letters*, 84(2):161, 2004.
49. S D Lacey. The absorption coefficient of gallium phosphide in the wavelength region 530 to 1100 nm. *Solid State Communications*, 8:1115–1118, 1970.
50. M Toishi, D Englund, A Faraon, and J Vučković. High-brightness single photon source from a quantum dot in a directional-emission nanocavity. *Optics Express*, 17(17): 14618–14626, 2009.
51. J D Joannapolous, S G Johnson, J N Winn, and R D Meade. *Photonic Crystals: Molding the Flow of Light*. Princeton University Press, Princeton, New Jersey, 2008.
52. V G Kravets, A N Grigorenko, R R Nair, P Blake, S Anissimova, K S Novoselov, and A K Geim. Spectroscopic ellipsometry of graphene and an exciton-shifted van Hove peak in absorption. *Physical Review B*, 81(15):155413, 2010.
53. K Mak, C Lui, J Shan, and T Heinz. Observation of an electric-field-induced band gap in bilayer graphene by infrared spectroscopy. *Physical Review Letters*, 102(25):100–103, 2009.
54. C Lu, Q Fu, S Huang, and J Liu. Polymer electrolyte-gated carbon nanotube field-effect transistor. *Nano Letters*, 4(4):623–627, 2004.
55. J Gao, J F McMillan, M-C Wu, J Zheng, S Assefa, and C W Wong. Demonstration of an air-slot mode-gap confined photonic crystal slab nanocavity with ultrasmall mode volumes. *Applied Physics Letters*, 96(5):51123, 2010.
56. N M R Peres, T Stauber, and A H Castro Neto. The infrared conductivity of graphene on top of silicon oxide. *Europhysics Letters*, 84(3):38002, 2008.
57. D A B Miller. Device requirements for optical interconnects to silicon chips. *Proceedings of the IEEE*, 97:1166–1185, 2009.
58. K Padmaraju, D F Logan, X Zhu, J J Ackert, A P Knights, and K Bergman. Integrated thermal stabilization of a microring modulator. *Optics Express*, 21(12):14342, 2013.

59. Y Akahane, T Asano, B-S Song, and S Noda. Fine-tuned high-Q photonic-crystal nanocavity. *Optics Express*, 13(4):1202–1214, 2005.
60. K Srinivasan, P Barclay, M Borselli, and O Painter. Optical-fiber-based measurement of an ultrasmall volume high-Q photonic crystal microcavity. *Physical Review B*, 70(8):081306, 2004.
61. C-C Tsai, J Mower, and D Englund. Directional free-space coupling from photonic crystal waveguides. *Optics Express*, 19(21):20586–20596, 2011.
62. F Xia, T Mueller, R Golizadeh-Mojarad, M Freitag, Y-M Lin, J Tsang, V Perebeinos, and P Avouris. Photocurrent imaging and efficient photon detection in a graphene transistor. *Nano Letters*, 9(3):1039–1044, 2009.
63. P B Johnson and R W Christy. Optical constants of the noble metals. *Physical Review B*, 6(12):4370–4379, 1972.
64. W Liu, B L Jackson, J Zhu, C-Q Miao, Y Park, K Sun, J Woo, and Y-H Xie. Large scale pattern graphene electrode for high performance in transparent organic single crystal field-effect transistors. *ACS Nano*, 4(7):3927–3932, 2010.

26 Photonics of Shungite Quantum Dots

B. S. Razbirin, N. N. Rozhkova, and E. F. Sheka

CONTENTS

ABSTRACT

Shungite quantum dots are associated with nanosize fragments of reduced graphene oxide similarly to synthetic graphene quantum dots thus forming a common class of graphene quantum dots (GQDs). Colloidal dispersions of powdered shungite in water, carbon tetrachloride, and toluene form the ground for the GQD photonic peculiarities manifestation. Morphological study shows a steady trend of GQDs to form fractals and a drastic change in the colloids fractal structure caused by solvent was reliably established. Spectral study reveals a dual character of emitting centers: individual GQDs are responsible for the spectra position while the fractal structure of GQD colloids provides high broadening of the spectra due to structural inhomogeneity of the colloidal dispersions and a peculiar dependence on excitation wavelength. For the first time, photoluminescence spectra of individual GQDs were observed in frozen toluene dispersions which pave the way for a theoretical treatment of GQD photonics.

26.1 INTRODUCTION

Originally, the term "graphene quantum dot" (GQD) appeared in theoretical researches and was attributed to fragments limited in size, or domains, of a single-layer two-dimensional graphene crystal. The subject of the investigations concerned the quantum size effects, manifested in the spin [1,2], electronic [3], and optical [4–9] properties of the fragments. GQDs turned out to be quite efficient fluorescent nanocarbons. Due to their luminescence stability, nanosecond lifetime, biocompatibility, low toxicity, and high water solubility, GQDs are considered excellent probes for high contrast bioimaging and biosensing applications. The latter stimulated the growth of interest in GQD, which caused a particular attention to their production. In response to this demand, a lot of synthetic methods to produce GQDs appeared, both "top-down" and "bottom-up." The former concerned such techniques as electrochemical ablation of graphite rod electrodes, chemical exfoliation from graphite nanoparticles, chemical oxidation of candle soot, intensive cavitation field in a pressurized ultrasonic batch reactor for obtaining nanosize graphite further subjected to oxidation and reduction. Laser ablation of graphite and microwave assisted small molecule carbonization present "bottom-up" techniques [10]. These and other methods are widely used in the current studies concerning GQD photoluminescence (PL) (see References 10–13 and references therein) and are described in a number of reviews [14,15]. However, GQD technology is still costly and time consuming due to which when developing efficient techniques for getting GQDs in grams-mass quantities and searching for cheap raw materials, chemists have turned to natural sources of carbon in the form of carbon fibers [16] and coal [17]. Sometimes GQDs of "natural origin" are presented as carbon quantum dots (CQDs) [10].

In spite of a large variety of techniques as well as difference in the starting materials, numerous studies exhibit a common nature of both GQDs and CQDs. The performed analysis of structure and chemical composition shows that in all cases GQDs are a few layer stacks of reduced graphene oxide (rGO) sheets of 1–10 nm in size. There is only one observation when single-layer rGO domains were synthesized in a liquid medium using trialkylphenyl polymers that form three-dimensional pores, inside which is a synthesis of carbon condensed polycyclic molecules [18]. The studied GQD stacks differ by the number of layers and linear dimension of rGO sheets as well by chemical composition of the latter. Thus, quasi planar basal plane of the sheets without chemical addends are framed quite differently with respect to chemical units that terminate the sheets' edge atoms:

C=O, C—C(COOH), C—OH, and C—H units are considered as the main terminators. As shown, a specific distribution of the latter over the sheets' circumference depends on which synthetic method of dot producing was used. In its turn, it determines the solubility of GQDs in water and other solvents as well as a strong dependence of the GQDs PL properties on the latter. Therefore, the size- and chemical-composition dependences are the main features of the GQDs PL spectra.

Due to the lack of a band gap, no optical luminescence is observed in pristine graphene. A band gap, however, can be engineered into GQDs due to quantum confinement [19,20] and chemical modification of the graphene edge [21]. Since the band gap depends on size [22], shape [23], and fraction of the sp^2 domains [24], PL emission greatly depends on the nature and size of the extended sp^2 sites [25]. This explains a large dispersion of PL properties that are affected by the synthetic method in use. The latter results in a large polydispersity of synthetic GQDs' solution and accounts for excitation-dependent PL shape and position. Standardization of both size and chemical framing of GQDs is a very complicated problem and not so many successful results are known. Controlled synthesis was realized in the case of GDQs derived from carbon fibers [16], encapsulated and stabilized in zeolitic imidazolate framework nanocrystals [26], and under a particularly scrupulous maintenance of the protocol of chemical exfoliation of graphite [11].

As mentioned earlier, the greatest hopes in the production of a grams-mass GQDs' material are pinned on the use of a natural source of carbon. The impressive results of the carbon fibers- [16] and coal [17]-based studies look quite promising. A particular feature of the studies concerned the nongraphitic origin of both carbons while the produced GQDs do not differ from those of graphite origin. Nevertheless, the natural carbons in the present case serve as raw material for the complex chemical synthetic technology of the GQDs production. At the same time, the nature, infinitely generous, as if anticipating the need for quantum dots, prepared a particular natural carbon known as shungite. Deposited exclusively in the Karelia area of Russia, shungite was attributed to one of the carbon allotropes quite long ago. Its bright individuality, as expressed, in particular, in the total absence of any similarity to other natural carbon allotropes, put shungite in a special position and formed the basis of more than half a century of careful study of its properties. The presence of sp^2 domains was one the first observations. However, it took many years to establish the domains' nature. Stimulated by current graphene chemistry, both highly extensive empirically and deeply understood theoretically, the shungite study received a new impetus which resulted in a new vision of this carbon [27,28]. At present, shungite is considered as a multilevel fractal structure which is based on nanoscale rGO fragments with an average linear dimension of ~1 nm. The fragments are grouped in 3–5 layer stacks, thus forming the second level stacked structure; the stacks form the third level globular structures of 5–7 nm in size while the latter aggregate forming large nanoparticles of 20–100 nm. Experimental evidence of the structure has

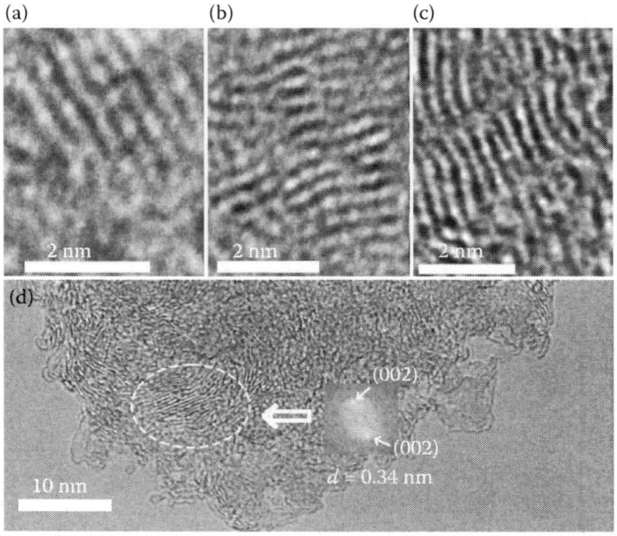

FIGURE 26.1 Darkfield HRTEM images of fragments of the atomic structure of schungite: a stack of flat (a, b) and curved rGO layers (c); (d) a general view of shungite particles: an oval marks the area subjected to the diffraction analysis, the results of which are shown on the right. (Adapted from Ye.A. Golubev and E.F. Sheka. *Techn Phys*. 2016, 61, 811.)

been recently proved by a detailed high-resolution transmission electron microscopy (HRTEM) study [29] and one of the obtained images is shown in Figure 26.1. As can be seen in Figure 26.1a through c, strips of length from a fraction to a few nanometers are clearly visible on the high-resolution images. The strips are the projections of carbon atomic planes of basic nanoscale rGO. The figure clearly shows that these basic structural fragments are grouped into stacks. The distance between the planes of the fragments in the stacks were evaluated by the Fourier diffraction patterns using a HRTEM image processing program. For a selected area of the HRTEM image in Figure 26.1d the Fourier diffraction pattern contains the spots corresponding to the interplanar distance of disordered graphite-like material $d_{002} = 0.34$ nm. Figure 26.1a through c clearly shows grouping of atomic planes in 4–7 layers stacks, which evidently are the dominant structural motif of shungite carbon in the size range 1–10 nm. The results are well consistent with the vision of the third level of the shungite structure as globules of ~7 nm in size, consisting of stacks of rGO. Provided by X-ray nanoanalysis, the averaged chemical composition of the rGOs fragments involves 95.5 ± 0.6 wt% of carbon, 3.3 ± 0.4 wt% of oxygen and 0.7 ± 0.2 wt% of hydrogen [29], which was confirmed by the recent neutron scattering study [32].

Coming back to GQDs, it should be stated that Figure 26.1 presents a picturesque image of the bodies related to shungite. Once stacked, globulized, and aggregated in solids, rGO fragments are readily dispersed in water and other solvents forming colloidal solutions with their own multilevel structures. From this viewpoint, it is quite uncertain to speak about GQDs as a few-layer stacks, which is a nowadays representation of synthetic dots. Obviously, not the stacks themselves but the rGO fragments determine PL properties of GQDs.

Apparently, it would be better to attribute GQDs to these very fragments which will help to avoid problems caused by fragment aggregation. On the other hand, such an approach opens a possibility to trace the influence of the fragment aggregation on PL properties by comparing spectra behavior of GQDs under different conditions. When applying to shungite GQDs, the approach turned out to be quite efficient which will be shown in the next sections. Besides, a common nature of shungite and synthetic GQDs will greatly expand our knowledge on the PL properties of GQDs.

26.2 PRELUDE FOR PHOTONICS OF SHUNGITE GQDs

Optical spectroscopy and PL, in particular, has become the primary method of studying the spectral properties of the GQDs. The review [15] presents a synopsis of the general picture based on the study of synthetic GQDs, which can be presented by the following features:

- The position and intensity of the GQDs PL spectrum depend on the solvent; GQDs are readily soluble in water and many polar organic solvents; upon transition from tetrafuran to acetone, dimethyl formamide, dimethyl sulfide, and water, the maximum of the PL spectrum is gradually shifted from 475 to 515 nm, which evidently manifests the GQDs interaction with the solvent.
- The PL intensity depends on the pH of the solution: being very weak at low pH, it increases rapidly when the pH is from 2 to 12, wherein the shape of the spectrum does not change.
- An important feature of the GQDs PL is a variation in a wide range (from 2% to 46%) of its quantum yield; furthermore, this variation is associated not only with different ways of the GQD producing, but is typical of the samples prepared by the same procedure—the PL quantum yield varies with time after synthesis.
- Both GQDs absorption and PL spectrum shows the size dependence and shifts to longer wavelengths with increasing particle size.
- The PL spectrum depends on the excitation wavelength λ_{exc}, which is the result of inhomogeneous broadening of the absorption spectrum caused by a mixture of GQDs of different size and chemical composition; with increasing λ_{exc}, the PL spectrum is shifted to longer wavelengths and its intensity is significantly reduced.
- The PL mechanism is still unclear, despite a large number of proposed models and intense experimental studies.

Detailed description of these features with the presentation of their possible explanations and links to the relevant publications is given in Reference 15.

As seen from the synopsis, optical spectroscopy of GQDs exhibits a complicated picture with many features. However, in spite of this diversity, common patterns can be identified that can be the basis of the GQDs spectral analysis, regardless of the method of their production. These general characteristics GQDs include the following: (1) structural inhomogeneity of GQDs solutions, better called dispersions; (2) low concentration limit that provides surveillance of the PL spectra; (3) dependence of the GQD PL spectrum on the solvent; (4) dependence of the GQD PL spectrum on the excitation light wavelength. It is these four circumstances that determine usual conditions under which the spectral analysis of complex polyatomic molecules is performed. Optimization of conditions, including primarily the choice of solvent and the experiments performance at low temperature, in many cases, led to good results, based on structural PL spectra (see, e.g., the relevant research of fullerenes solutions [33–35]). In this chapter, we will show that implementation of this optimization for spectral analysis of the GQDs is quite successful.

26.3 FRACTAL NATURE OF THE OBJECT UNDER STUDY

The GQD concept evidently implies a dispersed state of a number of nanosize rGO fragments. Empirically, the state is provided by the fragments dissolution in a solvent. Once dissolved, the sheets unavoidably aggregate forming colloidal dispersions in water or other solvents. So far only aqueous dispersions of synthetic GQDs have been studied [14,15]. In the case of shungite GQDs, two molecular solvents, namely, carbon tetrachloride (CTC) and toluene were used as well when replacing water in the pristine dispersions [36,37]. In each of these cases, the colloidal aggregates are the main object of the study. In spite of the fact that so far there has not been any direct confirmation of their fractal structure, there are serious reasons to suppose that it is an obvious reality. Actually, first, the sheets' formation occurred under conditions that unavoidably involve elements of randomness in the course of both laboratory chemical reactions and natural graphitization [28]; the latter concerns their size and shape and is clearly seen in Figure 26.1. Second, the sheets' structure certainly bears the stamp of polymers, for which fractal structure of aggregates in dilute dispersions has been convincingly proven (see Reference 38 and references therein).

As shown in Reference 38, the structure of colloidal aggregates is highly sensitive to the solvent around, the temperature of the aggregates formation, as well as other external actions such as mechanical stress, magnetic and electric fields. In addition to that previously discussed, this fact imposes extra restrictions on the definition of quantum dots of colloidal dispersions at the structural level and strengthens the previous suggestion of attributing GQDs to rGO fragments. In our case of the GQDs of different origin, the situation is even more complicated since the aggregation of synthetic (Sy) and shungite (Sh) rGO fragments occurred under different external conditions. In view of this, it must be assumed that rGO-Sy and rGO-Sh aggregates of not only different, but the same solvent dispersions, are quite different. Addressing spectral behavior of the dispersions, we should expect an obvious generality provided by the common nature of GQDs, but simultaneously complicated by the difference in packing of the dots in the different-solvent dispersions. The latter study concerned mainly the rGO-Sh dispersions [36,37] that will be considered in detail below.

26.4 rGO-Sh AQUEOUS DISPERSIONS

rGO-Sh aqueous dispersions were obtained by sonication of the pristine shungite powder [39]. The size-distribution characteristic profile of rGO-Sh aggregates is shown in Figure 26.2a. As can be seen, the average size of the aggregates is 54 nm, whereas the distribution is quite broad and

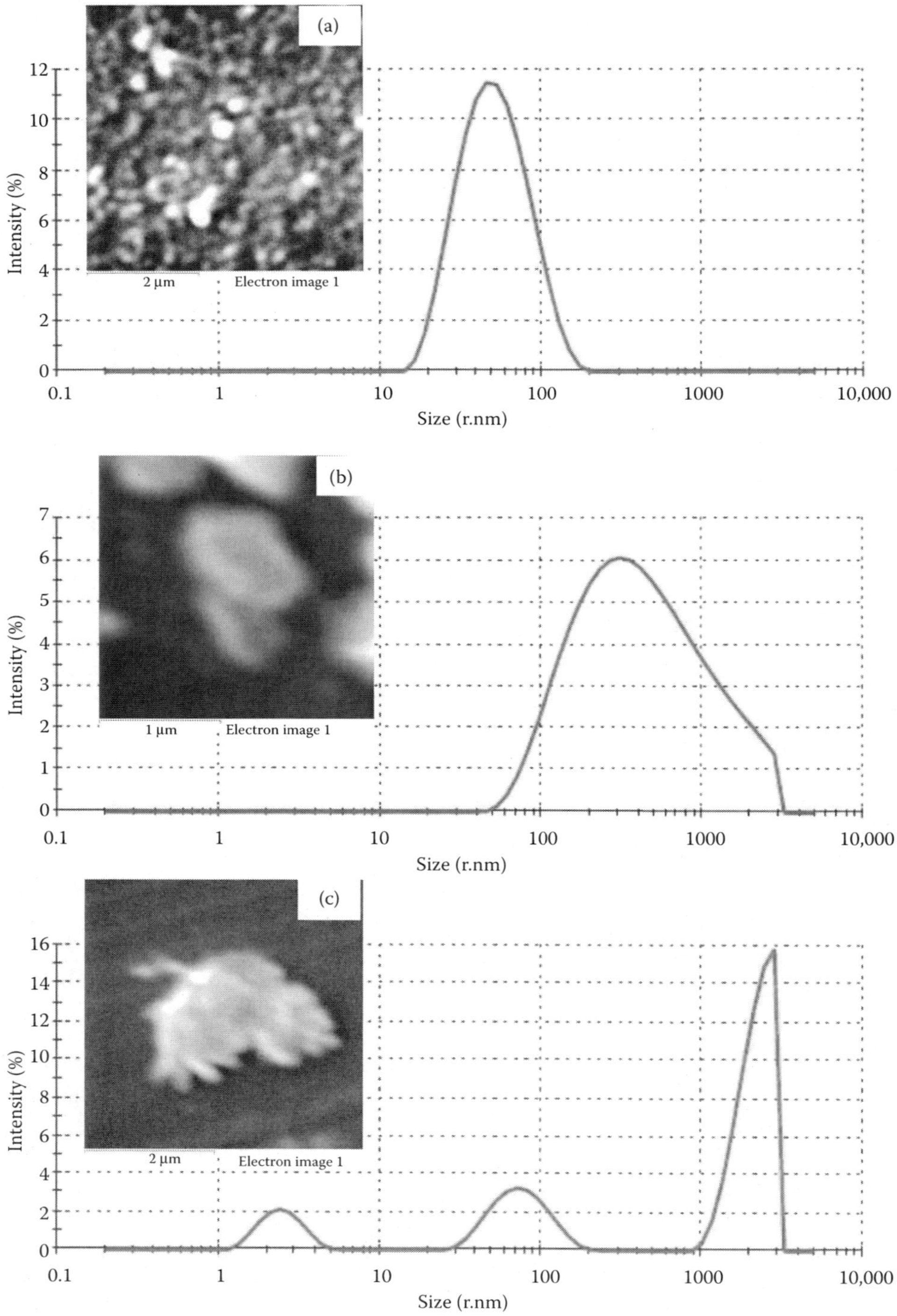

FIGURE 26.2 Size-distribution profile of colloidal aggregates of shungite in different dispersions: (a) water, (b) CTC, and (c) toluene. Carbon concentrations ~0.1 mg/mL. Insets are SEM images of the dispersion condensate on glass substrate: scale bar 2 μm. (a,c) and 1 μm (b).

characterized by full-half-width-maximum (FHWM) of 26 nm. Thus, the resulting colloids are significantly inhomogeneous. The inhomogeneity obviously concerns both size and shape (and, consequently, chemical composition) of the basic rGO-Sh fragments and, consequently, GQDs. The structure of the carbon condensate formed after water evaporation from the dispersion droplets on a glass substrate is shown in the inset. As seen in the figure, the condensate is of fractal structure formed by large aggregates, the shape of which is close to spherical.

Emission spectra of rGO-Sh aqueous dispersions consist of PL and Raman spectrum (RS) of water at the fundamental frequency of the O–H stretching vibrations of ~3400 cm^{-1} [36]. Figure 26.3 shows the PL spectra excited at λ_{exc} 405 and 457 nm after RS subtraction. The spectra present blue and green components of the GQD PL, both broad and bell shaped, that are characteristic for the PL spectra of rGO-Sy aqueous dispersions (see Reference 15). In spite of large width of the PL spectra, their position in the same spectral region for both rGO-Sy and rGO-Sh aqueous dispersions evidences a common nature of emitting GQDs. The spectra are significantly overlapped with the absorption ones, which usually points to the presence of the inhomogeneous broadening. As a consequence, different λ_{exc} provide a selective excitation of different sets of emitting centers. In the case of rGO-Sy dispersions, this feature was directly demonstrated by different fluorescence images (Fluoromax 4 [Horiba Scientific]) of the rGO-Sy colloids isolated in polymer film in the PL light excited by different λ_{exc} showing the fluorescence originates from separate particles [12]. Figure 26.4 shows a similar picture for a drop of aqueous rGO-Sh colloids deposited on a glass at room temperature [40]. As seen in the figure, at increasing λ_{exc} blue emitting centers are complemented by green emitting ones while substituted with different red emitters. A variety of rGO-Sh fragments is clearly vivid in Figure 26.1. Unfortunately, the large width of the PL spectra does not allow exhibiting those spectral details that might speak about the aggregated structure of GQDs.

26.5 rGO-Sh DISPERSIONS IN ORGANIC SOLVENTS

Traditionally, the best way to overcome difficulties caused by inhomogeneous broadening of optical spectra of complex molecules is the use of their dispersions in frozen crystalline matrices. The choice of solvent is highly important. As known, water is a "bad" solvent since the absorption and emission spectra of dissolved large organic molecules usually are broadband and unstructured. In contrast, frozen solutions of complex organic molecules, including, say, fullerenes [33–35], in CTC or toluene, in some cases provide a reliable monitoring of fine-structured spectra of individual molecules (Shpolskii's effect [41]). The detection of PL structural spectra or structural components of broad PL spectra not only simplify spectral analysis but indicate the dispersing of emitting centers into individual molecules. It is this fact that was the basis of the solvent choice when studying the spectral properties of shungite GQDs [36,37].

Organic rGO-Sh dispersions were prepared from the pristine aqueous dispersions in the course of sequential replacement of water by isopropyl alcohol first and then by CTC or toluene [30]. The morphology and spectral properties of these dispersions turned out to be quite different.

26.5.1 rGO-Sh Dispersions in CTC

When analyzing CTC-dispersions morphology, a drastic change in the size-distribution profiles of the dispersions aggregates in comparison with that one of the aqueous dispersions was the first highly important feature. The second feature concerns the high incertitude in the structure of the

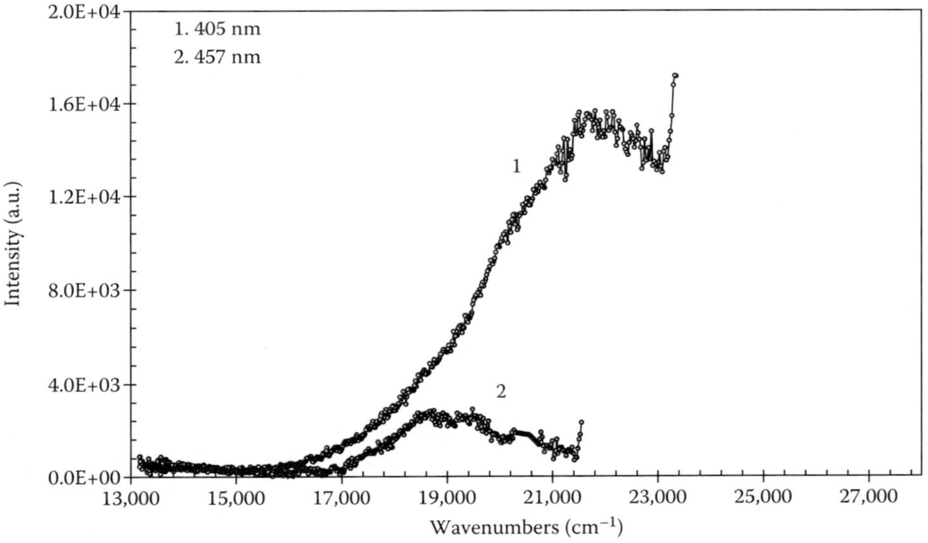

FIGURE 26.3 PL spectra of shungite water dispersions at 290 K after subtraction of Raman scattering of water. Numerals mark excitation wavelengths.

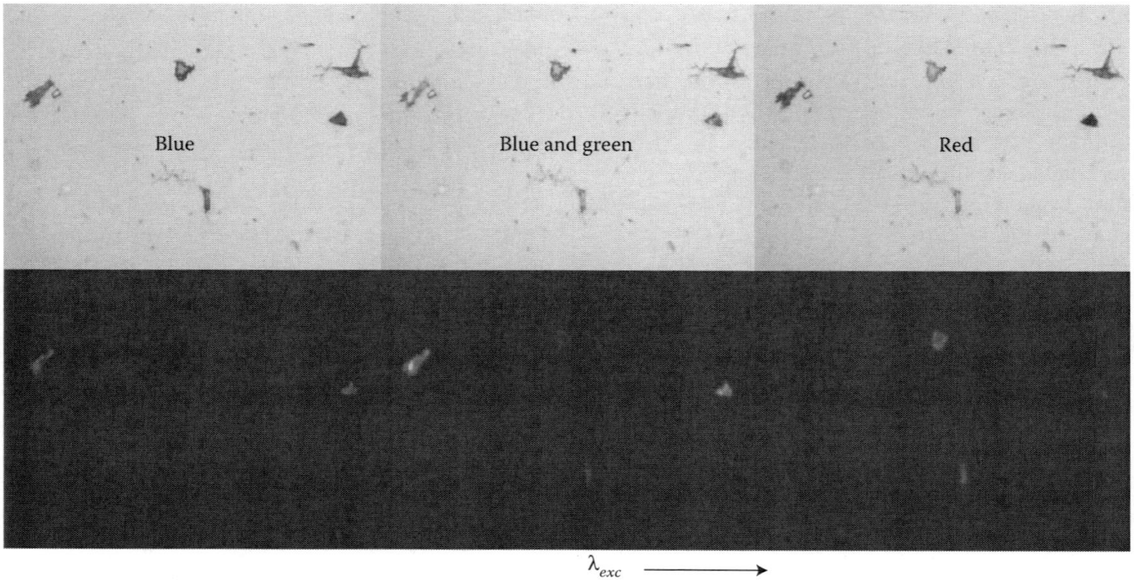

FIGURE 26.4 Fluorescence bright-field (top) and dark-field (bottom) images (Fluoromax 4 [Horiba Scientific]) of shungite aggregates deposited from water dispersion on a glass substrate excited by different wavelengths.

latter. Thus, Figure 26.2b presents a size-distribution profile related to one of the CTC-dispersions alongside the TEM image of the agglomerates of the film obtained when drying the CTC-dispersion droplets on glass. Typical for the dispersion is to increase the average size of colloidal aggregate when water is substituted with CTC. Simultaneously, becomes larger the scatter of sizes that is comparable with the size itself. The nearly spherical shape of aggregates in Figure 26.2a is replaced by lamellar faceting, mostly characteristic of microcrystals. It is necessary to note the absence of small aggregates, which indicates a complete absence of individual GQDs in the dispersions. Therefore, the change in both size-distribution profiles and shape of the aggregates of the condensate evidences a strong influence of solvent on the aggregates' structure.

The dispersion spectral features are well consistent with these findings. Figure 26.5 shows the PL spectra of the CTC-dispersion, the morphological properties of which are similar to those shown in Figure 26.2b. The dispersion is characterized by a wide absorption spectrum shown in Figure 26.5a. Arrows in the figure indicate wave numbers λ_{exc}^{-1} corresponding to laser lines. The UV excited PL spectrum in Figure 26.5a is very broad and covers the region from 27,000 to 15,000 cm^{-1} overlapping with the absorption spectrum over the entire spectral range. Such a large overlapping evidences the formation of an ensemble of emitting centers, which differ in the probability of emission (absorption) at a given wavelength. Indeed, successive PL excitation by laser lines 1, 3, 4, and 5 (see Figure 26.5a) causes a significant modification of the PL spectra (Figure 26.5b). The width of the spectra decreases as λ_{exc} increases, the PL band maximum is shifted to longer wavelengths, and the spectrum intensity decreases. This is due to the selective excitation of a certain group of centers which is typical for structurally disordered systems. To simplify further comparative analysis of the spectra obtained at different λ_{exc}, we shall denote them according to the excitation wavelength, namely: 405-, 476-, 496-spectrum, etc

Comparing the discussed PL spectra at different excitations, note the following features:

- PL spectra obtained when excited in the region of overlapping of absorption and emission spectra in Figure 26.5a, have more distinct structure than the 337-one but still evidencing a superpositioning character of the spectra.
- Intensity of the 405-spectrum is almost an order of magnitude higher than the intensity of the rest of the spectra.

As shown in Reference 36, the features are typical for a wide range of dispersions obtained at different time. However, a comparative analysis of the PL spectra of different dispersions shows that the above-mentioned spectral regularities are sensitive to the CTC-dispersions structure and are directly related to the degree of structural inhomogeneity. Thus, the narrowing of the size-distribution profile undoubtedly causes the narrowing of inhomogeneously broadened absorption and emission spectra. Unchanged in all the spectra is the predominant intensity of the 405-spectrum. The difference in the structural inhomogeneity of dispersions raises the question of their temporal stability. Spectral analysis showed that the spectra changed in the course of a long storage.

Summarizing, the following conclusions can be made on the basis of spectral features of the rGO-Sh CTC-dispersions [38]:

1. None of fine-structured spectra similar to Shpolskii's spectra of organic molecules was observed in the low-temperature PL spectra of crystalline CTC-dispersions.

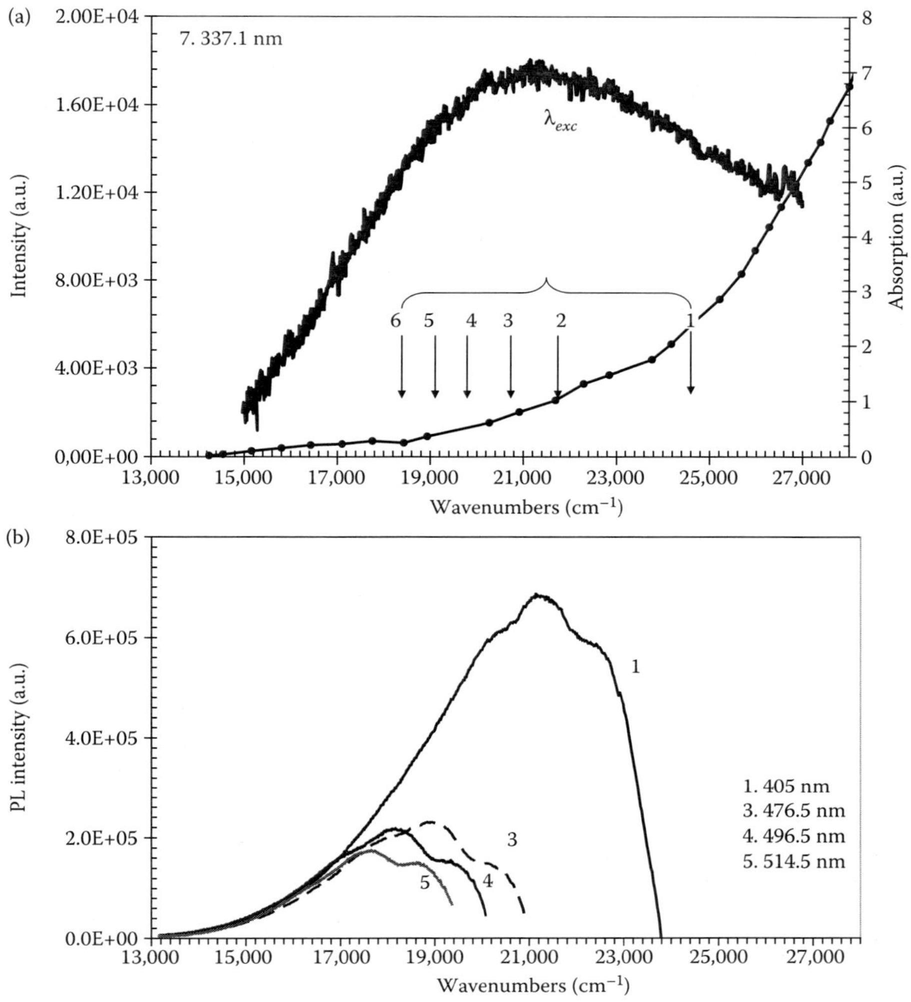

FIGURE 26.5 PL (a,b) and absorption (b) spectra of shungite dispersion in CTC at 80 K after background emission substraction. Numerals mark excitation band lengths.

This is consistent with the absence of small-size components in the size-distribution profiles of the relevant colloidal aggregates.

2. The PL spectra are broad and overlapping with the absorption spectrum over a wide spectral range. This fact testifies to the inhomogeneous broadening of the spectra, which is the result of nonuniform distribution of the dispersions colloidal aggregates, confirmed by morphological measurements.
3. The selective excitation of emission spectra by different laser lines allows decomposing the total spectrum into components corresponding to the excitation of different groups of emitting centers. In this case, common to all the studied dispersions is the high intensity of the emission spectra excited at $\lambda_{exc} = 405$ and 457 nm.
4. The observed high sensitivity of PL spectra to the structural inhomogeneity of dispersions allows the use of fluorescent spectral analysis as a method of tracking the process of the formation of primary dispersions and their aging over time.
5. The division of the GQD water dispersion spectra onto blue and green ones is not applicable to the spectra of the CTC dispersions. PL covers a large range from blue to red when λ_{exc} increases.

26.5.2 rGO-Sh Dispersion in Toluene

The behavior of toluene rGO-Sh dispersions is more intricate from both the morphological and spectral viewpoints. Basic GQDs of aqueous dispersions are very little soluble in toluene, thereby resulting toluene dispersions are essentially colorless due to low concentration of the solute. In addition, the low concentration makes the dispersion very sensitive to any change in both the content and structure of dispersions. This causes structural instability of dispersions which is manifested, in particular, in the time dependence of the relevant size-distribution profiles. Thus, the three-peak distribution of the initial toluene dispersion shown in Figure 26.2c, is gradually replaced by a single-peaked at ~1 nm for 1–2 h. The last distribution does not change with time and represents the distribution of the solute in the supernatant.

By analogy with CTC, toluene causes a drastic change in the colloidal aggregates structure. However, if the CTC action can be attributed to the consolidation of the pristine colloids, the toluene results in quite the opposite effect leading to their dispersing. The three-peak structure in Figure 26.2c shows that, at the initial stage of water replacement by toluene, there are three kinds of particles with average linear dimensions of about 2.5, 70, and 1100 nm in the resulting liquid medium. All three sets are characterized by a wide dispersion. Large particles are seen in the electron microscope (see inset in Figure 26.2c) as freaky sprawled fragments. Over time, these three entities are replaced by one with an average size of ~1 nm. Thus, freshly produced dispersions containing GQD aggregates of varying complexity, turns into the dispersion of individual GQDs. This value is well consistent with the empirical value of ~1 nm for the average size of GQDs in shungite accepted in Reference 28 as seen in Figure 26.1. The conversion of aqueous dispersion of aggregated GQDs into the colloidal dispersion of individual GQDs in toluene is a peculiar manifestation of the interaction of solvents with rGO. As for the graphene photonics, the obtained toluene dispersion has provided investigation of individual GQDs for the first time.

Figure 26.6 shows the PL spectra of colloidal dispersions of individual GQDs in toluene. The *brutto* experimental spectra, each of which is a superposition of the RS of toluene and PL spectrum of the dispersion, are presented in Figure 26.6a. Note the clearly visible enhancement of the Raman scattering of toluene in the 20,000–17,000 cm^{-1} region. Figure 26.6b shows the PL spectra after subtracting the Raman spectra. The spectra presented in the figure can be divided into three groups. The first group includes the 337-spectrum (7) that in the UV region is the PL spectrum, similar in shape to the UV PL spectrum of toluene, but shifted to longer wavelengths. This part of the spectrum should apparently be attributed to the PL of some impurities in toluene. The main contribution into the PL 337-spectrum in the region of 24,000–17,000 cm^{-1} is associated with the emission of all GQDs available in the dispersion. This spectrum is broad and structureless, which apparently indicates the structural inhomogeneity of the GQD colloids.

PL 405- and 476-spectra (1 and 3) in the region of 23,000–17,000 cm^{-1} should be attributed to the second group. Both spectra have clearly defined structure that is most clearly expressed in the 405-spectrum. The spectrum is characteristic of a complex molecule with allowed electronic transitions. Assuming that the maximum frequency at 22,910 cm^{-1} determines the position of pure electronic transition, the longer wavelength doublet at ~21,560–21,330 cm^{-1} can be interpreted as vibronic satellites. The distance between the doublet peaks and the pure electronic band constitutes 1350–1580 cm^{-1} that is consistent with the frequencies of totally symmetric vibrations of C–C graphene skeleton, commonly observed in the Raman spectra. Similarly, two peaks of the much less intensive 476-spectrum, which are wider than in the previous case, are divided by the average frequency of 1490 cm^{-1}. PL 457-spectrum, shown in Figure 26.6c (curve 2) is similar to spectra 1 and 3, in intensity closer to the 405-spectrum. All three spectra are related to individual rGO fragments albeit of different size that increase when going from 405-spectrum to 457- and 476-spectrum. All the spectra are positioned in blue–green region.

The shape of 496- and 514-spectrum substantially differs from that of the second group spectra. Instead of the two peaks observed there is a broad band observed in both cases. This feature makes these spectra attribute to the third group (red spectra) and to associate them with the appearance of not individual frozen GQDs but with their possible clusters (such as, say, dimeric homo- (GQD + GQD) and hetero- (GQD + toluene) structured charge transfer complexes and so forth) [37].

The conducted spectral studies of the rGO-Sh toluene dispersions confirmed once again the status of toluene as a good solvent and a good crystalline matrix, which allows for obtaining the fine-structured spectra of individual complex molecules in conditions under which in other solvents the molecules form fractals. This ability of toluene allowed recording the spectra of both individual GQDs and their small clusters for the first time. The finding represents the first reliable empirical basis for a comprehensive theoretical treatment of the spectra observed.

26.6 DISCUSSION

As follows from the results presented above, rGO-Sh dispersions are colloidal regardless of the solvent, whether water, CTC, or toluene. The dispersion colloids structure depends on the solvent and thereafter is substantially different. This issue deserves a special investigation. Thus, the replacement of water with CTC leads to multiple growth of the pristine colloids which promotes the formation of a quasi-crystalline image of the condensate structure. At present, the colloid detailed structure remains unclear. In contrast to CTC, toluene causes the decomposition of pristine colloids into individual rGO fragments. The last facts cast doubt on the possible direct link between the structure of the dispersions fractals and the elements of the fractal structure of solid shungite or its post-treated condensate. The observed solvent-stimulated structural transformation is a consequence of the geometric peculiarities of fractals behavior in liquids [38]. The resulting spectral data can be the basis for further study of this effect.

The spectral behavior of the aqueous and CTC-dispersions with large colloids is quite similar, despite the significant difference in size and structure of the latter. Moreover, the features of the PL spectra of these dispersions practically replicate patterns that are typical for the aqueous rGO-Sy dispersions discussed in detail in Section 26.1. This allows to conclude that one and the same structural element of the colloidal aggregates of both rGO-Sh dispersions and rGO-Sy one is responsible for the emission in spite of the pronounced morphological differences of its packing in all these cases. According to the modern view on shungite structure [23] and

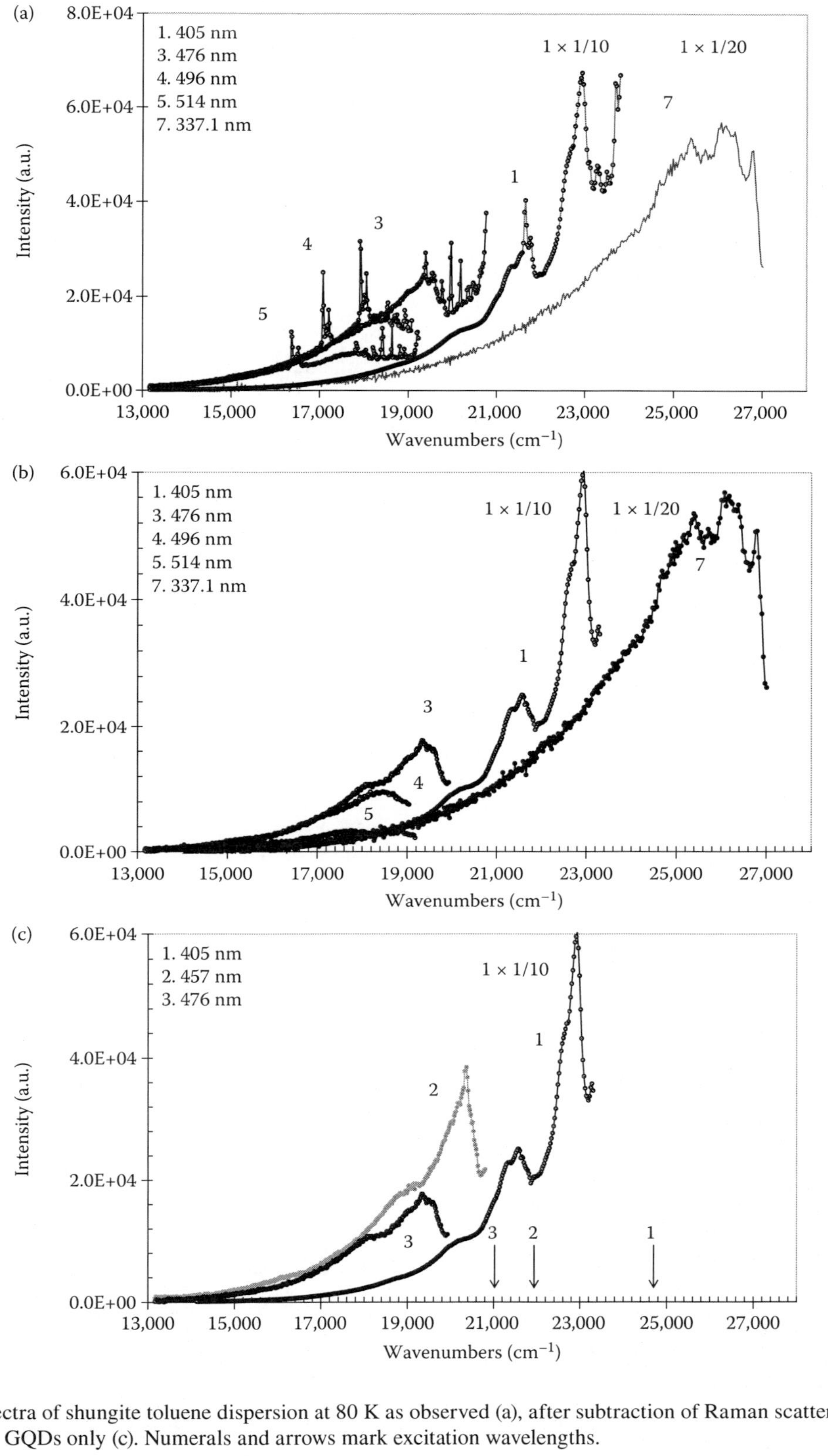

FIGURE 26.6 PL spectra of shungite toluene dispersion at 80 K as observed (a), after subtraction of Raman scattering of toluene (b), and attributed to individual GQDs only (c). Numerals and arrows mark excitation wavelengths.

a common opinion on the origin of synthetic GQDs [14,15], rGO sheets play the role representing GQDs of the rGO colloidal dispersions in all cases.

The specific effect of toluene, which caused the decomposition of pristine particles into individual rGO fragments, successfully embedding them into the toluene crystalline matrix, allowed obtaining the PL spectrum of individual rGO fragments for the first time. Obviously, the resulting fragments are of different size and shape, which determines the structural inhomogeneity of toluene dispersions. This feature of toluene dispersions is common to the other dispersions and explains the dependence of the PL spectra on λ_{exc} that is the main spectral feature of GQDs, both synthetic [14,15] and of shungite origin.

The structural inhomogeneity of GQDs colloidal dispersion is caused by two main reasons, internal and external. The internal reason concerns the uncertainty in the structure (size and shape) of basic rGO fragments. Nanosize rGO basic structural elements of solid shungite are formed under conditions of serious competition of different processes [28], among which the most valuable are (1) the natural graphitization of carbon sediments, accompanied by a simultaneous oxidation of the graphene fragments and their reduction in water vapor, (2) the retention of water molecules in the space between fragments and the exit of the water molecules from the space into the environment, and (3) the multilevel aggregation of rGO fragments providing the formation of a monolithic fractal structure of shungite. Naturally, that achieved balance between the kinetically different-factor processes is significantly influenced by random effects, so that the rGO fragments of natural shungite, which survived natural selection, are statistically averaged over a wide range of fragments that differ in size, shape, and chemical composition.

Obviously, the reverse procedure of the shungite dispersing in water is statistically also nonuniform with respect to colloidal aggregates so that there is a strong dependence of the dispersions on the technological protocol, which results in a change in the dispersion composition caused by slight protocol violations. Such a kinetic instability of dispersing, in a sense, is the reason that the composition of colloidal aggregates can vary when water is displaced by another solvent. The discussed spectral features confirm these assumptions.

The external reason is due to fractal structure of colloidal aggregates. The fractals themselves are highly inhomogeneous, moreover, they strongly depend on the solvent. The two reasons determine the feature of the GQD spectra in aqueous and CTC-dispersions while the first one dominates in the case of toluene dispersions. In view of this, the photonics of GQDs has two faces, one of which is of rGO nature while the other concerns fractal packing of the rGO fragments. As follows from what is presented in this chapter, the spectra study is quite efficient in exhibiting this duality.

Thus, the structural PL spectra allow raising the question of identifying the interaction effect of dissolved rGO fragments with each other and with the solvent. Nanosize rGO fragments have high donor-and-acceptor properties (low ionization potential and high electron affinity) and can exhibit both donor and acceptor properties so that clusters of fragments (dimers, trimers, and so forth) are typical charge transfer complexes. Besides this, toluene is a good electron donor due to which it can form a charge transfer complex with any rGO fragment, acting as an electron acceptor. The spectrum of electron–hole states of the complex, which depends on the distance between the molecules as well as on the initial parameters, is similar to the electron–hole spectrum of clusters of fullerenes C_{60} themselves and with toluene [33–35], positioned by the energy in the region of 20,000–17,000 cm^{-1}. By analogy with the nanophotonics of fullerene C_{60} solutions, the enhancement of the RS of toluene is due to its superposition over the spectrum of electron–hole states, which follows from the theory of light amplification caused by nonlinear optical phenomena [42]. Additionally, the formation of rGO–toluene charge transfer complexes may promote the formation of stable chemical composites in the course of photochemical reactions [43] that might be responsible for the PL third-group spectra observed in toluene dispersions. Certainly, this assumption requires further theoretical and experimental investigation.

26.7 CONCLUSION

The photonics of shungite colloidal dispersions faces the problem that a large statistical inhomogeneity inherent in the quantum dot ensemble makes it difficult to interpret the results in detail. Consequently, common patterns that are observed on the background of this inhomogeneity become most important. In the case of considered dispersions, the common patterns include, primarily, the dispersion PL in the visible region, which is characteristic for large molecules consisting of fused benzenoid rings. This made it possible to confirm the earlier findings that graphene-like structures of limited size, namely, rGO fragments, are the basic structural elements for all the dispersions. The second feature concerns the dependence of the position and intensity of selective PL spectra on the exciting light wavelength λ_{exc}. This feature lies in the fact that regardless of the composition and solvent of dispersions the PL excitation at λ_{exc} = 405 and 457 nm provides the highest PL intensity while excitation at either longer or shorter wavelengths produces a much lesser intensity of emission. The answer to this question must be sought in the calculated absorption and PL spectra of graphene quantum dots, which are attributed to nanoscale fragments of rGO.

ACKNOWLEDGMENTS

The financial support provided by the Ministry of Science and High Education of the Russian Federation grant 2.8223.2013 and the Basic Research Program, RAS, Earth Sciences Section-5, is gratefully acknowledged. The authors are thankful to D.K. Nelson and A.N. Starukhin for their valuable assistance in performing spectral experiments.

REFERENCES

1. B. Trauzettel, D.V. Bulaev, D. Loss, and G. Burkard. *Nat. Phys.* 2007, 3, 192.
2. A. Güçlü, P. Potasz, and P. Hawrylak. *Phys. Rev. B* 2011, 84, 035425.
3. K.A. Ritter and J.W. Lyding. *Nat. Mater.* 2009, 8, 235.
4. D. Pan et al. *Adv. Mater.* 2010, 22, 734.
5. J. Shen et al. *Chem. Commun.* 2011, 47, 2580.
6. Z.Z. Zhang and K. Chang. *Phys. Rev. B* 2008, 77, 235411.
7. V. Gupta et al. *J. Am. Chem. Soc.* 2011, 133, 9960.
8. R. Liu et al. *J. Am. Chem. Soc.* 2011, 133, 15221.
9. Y. Li et al. *Adv. Mater.* 2011, 23, 776.
10. L. Wang et al. *ACS Nano* 2014, 8, 2541.
11. F. Liu et al. *Adv. Mater.* 2013, 25, 3657.

12. S.K. Das et al. *Nano Lett.* 2014, 14, 620.
13. V. Stengl et al. *Carbon* 2013, 63, 537.
14. L. Tang et al. *ACS Nano* 2012, 6, 5102.
15. L. Li et al. *Nanoscale* 2013, 5, 4015.
16. J. Peng et al. *Nano Lett.* 2012, 12, 844.
17. R. Ye et al. *Nat. Commun.* 2013, 4, 2943.
18. X. Yan, B. Li, and L-S. Li. *Acc. Chem. Res.* 2013, 46, 2254.
19. L.A. Ponomarenko et al. *Science* 2008, 320, 356.
20. L.S. Li and X. Yan. *J. Phys. Chem. Lett.* 2010, 1, 2572.
21. S.J. Zhu et al. *Chem. Commun.* 2011, 47, 6858.
22. H. Li et al. *Angew. Chem., Int. Ed.* 2010, 49, 4430.
23. X. Yan et al. *J. Phys. Chem. Lett.* 2011, 2, 1119.
24. G. Eda et al. *Adv. Mater.* 2010, 22, 505.
25. X. Yan, X. Cui, and L.S. Li. L. *J. Am. Chem. Soc.* 2010, 132, 5944.
26. B.P. Biswal et al. *Nanoscale*, 2013, 5, 10556.
27. N.N. Rozhkova and E.F. Sheka. *arXiv*:1308.1189 [cond-mat. mtrl-sci] 2013.
28. E.F. Sheka and N.N. Rozhkova. *Int. J. Smart Nano Mater.* 2014, 5, 1.
29. Ye.A. Golubev and E.F. Sheka. *Techn Phys.* 2016, 61, 811.
30. N.N. Rozhkova. *Shungite Nanocarbon.* Karelian Research Centre of RAS, Petrozavodsk, 2011 (in Russian).
31. E.F. Sheka et al. *JETP Lett.* 2014, 8, 735.
32. E.F. Sheka et al. *Nanosyst. Phys. Chem Mat* 2014, 5, 65g.
33. B.S. Razbirin et al. *JETP Lett.* 2008, 87, 133.
34. E.F. Sheka et al. *J. Exp. Theor. Phys.* 2009, 108, 738.
35. E.F. Sheka et al. *J. Nanophotonics* 2009, 3, 033501.
36. B.S. Razbirin et al. *J. Exp. Theor. Phys.* 2014, 118, 735.
37. B.S. Razbirin et al. *Nanosyst. Phys. Chem. Math.* 2014, 5, 217.
38. T.A. Witten. In *Soft Matter Physics.* M. Daoud and C.E. Williams (Eds.). Springer-Verlag, Berlin, 1999, p. 261.
39. N.N. Rozhkova et al. *Smart Nanocompos.* 2010, 1, 71.
40. N.N. Pozhkova. Unpublished data. 2009.
41. E.V. Shpol'skii. *Phys.-Usp.* 1963, 6, 411.
42. P. Heritage, A.M. Glass. In *Surface Enhanced Raman Scattering.* R.K. Chang and T.E. Furtak (Eds.). Plenum Press, New York, 1982, p. 391.
43. E.F. Sheka. *Nanosci. Nanotechnol. Lett.* 2011, 3, 28.

27 Open-Shell Character and Nonlinear Optical Properties of Nanographenes

Kyohei Yoneda and Masayoshi Nakano

CONTENTS

ABSTRACT

Low-dimensional nanographenes such as graphene nanoflakes (GNFs) are known to display the unique spin states with nonbonding molecular orbitals localized on their zigzag edges, that is, "open-shell singlet" electronic state. The open-shell singlet nature presents a key concept for designing a new class of functional materials such as highly efficient nonlinear optical (NLO) and singlet fission materials. As an example, we have theoretically investigated the open-shell natures of GNFs with various architectures and sizes using the diradical characters, defined by quantum chemical calculations, and have revealed several structure—open-shell property relationships for the GNFs. It has been found that several zigzag-edged GNFs exhibit intermediate and large diradical characters, while GNFs having only armchair edges are closed-shell systems. Also, large GNFs are shown to display multiradical characters beyond the diradical as increasing the zigzag-edge lengths. We have further found unique structural dependences of multiradical characters in antidot hexagonal GNFs and linear GNFs composed of trigonal fused-ring units. These GNFs are expected to be possible candidates for highly efficient and tunable open-shell singlet NLO materials, which exhibit strong diradical character dependences of NLO properties and give large NLO properties as compared to conventional closed-shell NLO systems. The mechanism and rational design principle of various GNFs for highly efficient NLO materials are presented in this chapter.

27.1 INTRODUCTION

Graphene, which is a one-atom-thick layer of graphite, displays very fascinating properties such as massless quasiparticles, room-temperature quantum Hall effect, and half-metallicity due to the unique electronic state [1–20], major previous studies of which are listed in Table 27.1. Since a method for isolating graphene was discovered and its electronic properties were reported by Novoselov et al. [21], graphene has become one of the hottest topics in the fields of both chemistry and physics. For low-dimensional nanographenes such as one-dimensional (1D) graphene nanoribbons (GNRs) and zero-dimensional (0D) graphene nanoflakes (GNFs), which have two types of edge structures, that is, armchair and zigzag edges, the electronic structures exhibit a strong dependence on their edge shapes. In particular, nanographenes with zigzag edges are known to display the unique spin states with nonbonding molecular orbitals (MOs) localized on their edges [22]. In chemistry, such an electronic structure is referred to as "open-shell singlet state," which has recently been the focus of attention as a new key concept for designing highly efficient novel functional materials in photonics, optoelectronics, and spintronics [23–25].

In this chapter, on the basis of quantum chemistry, we clarify the relationships between molecular structures and open-shell singlet natures in various nanographenes, which can provide a novel molecular design strategy toward highly efficient nonlinear optical (NLO) systems composed of nanographenes. This is based on a strong correlation between

TABLE 27.1
Major Fascinating Properties of Graphene as Well as the Primary References

Properties	References
High electronic mobility	[2,3]
Massless quasiparticle	[4,5]
Anomalous quantum Hall effect	[6,7]
Klein tunneling effect	[8]
Spin transport property	[9]
Half-metallicity	[10]
High mechanical strength	[11]
NLO property	[12–19]

open-shell character (see Figure 27.1) and NLO property for open-shell singlet molecules [26–30]. This chapter is organized as follows. In Section 27.2, we expound the open-shell (diradical) character of GNFs. At first, we provide detailed explanations for the origin and physical/chemical importance of open-shell singlet states of GNFs (Section 27.2.1). Next, we introduce "diradical character," which is a theoretical quantity well defined in quantum chemistry, and subsequently evaluate the open-shell natures for several molecules (Section 27.2.2). We also describe the molecular structural and the size dependences of the diradical characters for GNFs, including rectangular GNFs, hexagonal GNFs (HGNFs), and linear GNFs composed of trigonal fused-ring units (Section 27.2.3). In Section 27.3, the NLO properties of several GNFs are described from the viewpoint of the diradical character dependence. We present a general theory of the relationship between diradical character and NLO property—in particular, third-order NLO property—as well as its mechanism using a simple diradical model (Section 27.3.1). Next, the third-order NLO properties of various GNFs shown in Section 27.2 are discussed based on their diradical characters, and the relationships between structure and third-order NLO property are presented (Section 27.3.2). In addition, we clarify unique donor (D)/acceptor (A) substitution effects on the open-shell characters and third-order NLO properties of oligoacenes, which contribute to building a new guideline for tuning and maximizing the third-order NLO properties of D/A substituted open-shell GNFs (Section 27.3.3). Several perspectives on novel NLO materials composed of open-shell nanographenes are provided in Section 27.4.

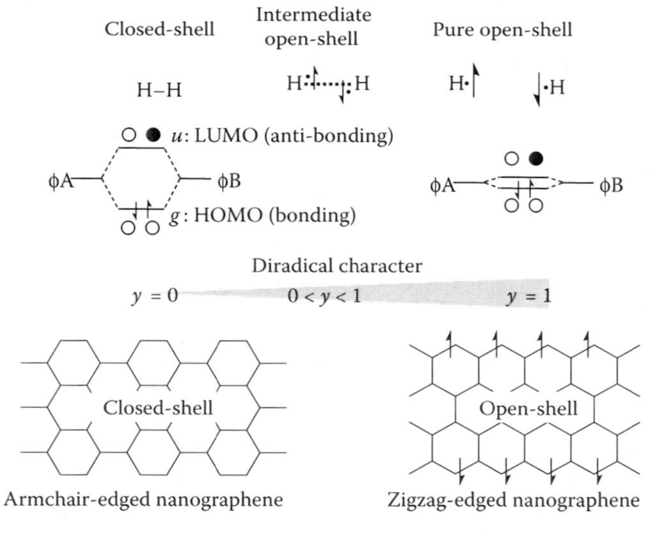

FIGURE 27.1 Open-shell character y ($0 \le y \le 1$), which is defined by twice the weight of doubly excited electron configuration from HOMO to LUMO in multiconfiguration self-consistent-field (MC-SCF) theory. As examples, dissociation of H_2 molecule and graphenes with two types of edges exhibit a wide range of diradical characters y.

27.2 OPEN-SHELL CHARACTER

27.2.1 Open-Shell Singlet States of GNFs

In 1996, Fujita et al. [22] predicted that zigzag-edged GNR (ZGNR) exhibits a peculiar electronic structure: there are degenerate two flat bands near the Fermi energy, which are derived from nonbonding orbitals localized on the zigzag edges (see Figure 27.2a). Due to the localized edge state, a strong peak of the density of states at its Fermi energy, which causes a large instability, appears in ZGNR. In systems with such instability, the degenerate flat bands are generally split to stabilize the system by the lattice distortion due to the electron–lattice interaction or by the electron spin polarization due to the electron–electron interaction. In the case of ZGNR, because the nonbonding orbitals, causing the instability, are well localized on the both zigzag edges, the lattice distortion hardly occurs, while the strong spin polarization in a direction from one zigzag edge to the other one is observed (see Figure 27.2b).

Also, finite-size GNFs are found to have similar electronic states, that is, open-shell singlet states, on account of their near-degenerate the highest occupied molecular orbital (HOMO) and the lowest unoccupied molecular orbital (LUMO) [31–36]. The open-shell singlet states of GNFs can be understood by their resonance structures based on Clar's sextet rule [37], which states that resonance structures having a larger number of six-membered rings with benzenoid structures (Clar's sextet) exhibit larger aromatic stabilities and tend to make more significant contributions to their ground electronic states. For example, for pentacene, which is regarded as one of rectangular GNFs, the open-shell resonance form with radicals located on the middle region of the zigzag edges

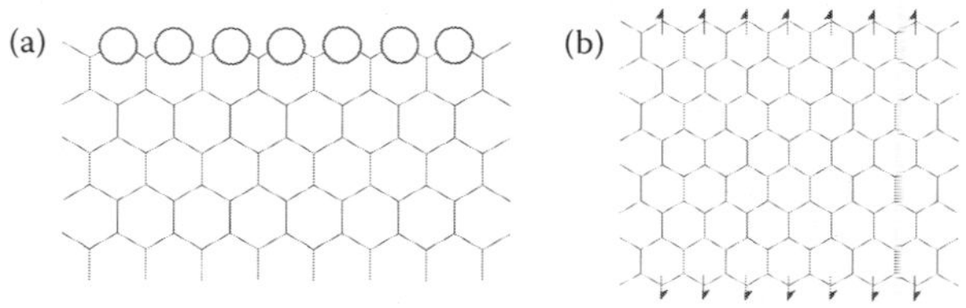

FIGURE 27.2 Localized edge state (a) and spin polarization (indicated by arrows) (b) for ZGNR.

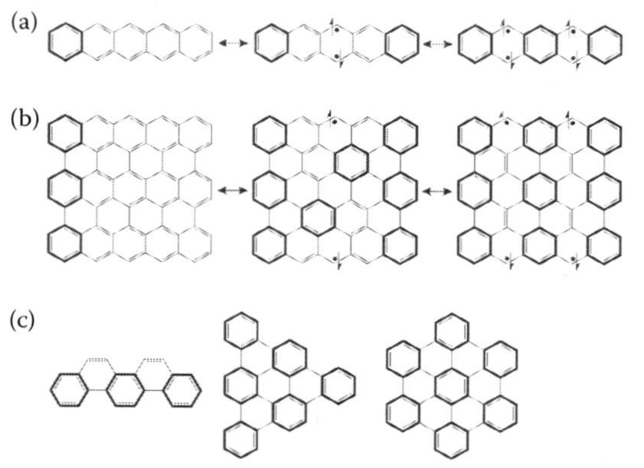

FIGURE 27.3 Resonance structures for pentacene (a) and a larger rectangular GNF (b), and linear, trigonal, and HGNFs having only armchair edges (c).

shows more Clar's sextet than the closed-shell form (see Figure 27.3a). The similar feature can also be seen for larger GNFs as shown in Figure 27.3b. However, for GNFs having only armchair edges, their closed-shell resonance forms show maximal numbers of Clar's sextet regardless of their molecular architectures including linear, trigonal, and hexagonal forms (see Figure 27.3c). As a result, the GNFs having zigzag edges tend to possess open-shell singlet ground states, where primary spin polarization is observed between the both-end zigzag edges, while those having only armchair edges tend to possess closed-shell ground states.

Note that the spin density and spin polarization observed in the open-shell singlet molecule do not represent real spins but artifacts in the broken-symmetry (BS) approaches [38,39]. To obtain an accurate description of the electronic structures of open-shell singlet molecules with small HOMO–LUMO energy gaps, it is necessary in general to take into account static electron correlations based on the multireference theory, in which the wave functions are described by multiple electron configurations, and the spin densities do not appear. On the other hand, the BS, that is, spin-unrestricted approaches, for example, the unrestricted Hartree–Fock (UHF) and unrestricted density functional theory (UDFT) methods, suffer from spin contaminations (mixing high spin states into singlet states), which cause the spin polarization in the singlet state and sometimes give incorrect singlet-triplet energy gap. The spin contamination is a deficiency of the BS approach, while its easy inclusion of electron correlation (multiple excitation) within a single determinantal scheme is an advantage. Indeed, the dissociation of a diatomic molecule into two radicals can be correctly described by the BS single-determinant approach, which gives the correct electron density and spin density of the dissociated fractions. Namely, the spin density in the singlet state can be viewed as an approximate description of the spatial spin correlations [39,40], and in particular, several UDFT methods with relevant exchange-correlation functionals are found to well reproduce the results obtained by the strong electron-correlated calculation methods with much smaller computational costs [27,41]. They are thus widely employed for obtaining approximate description of open-shell singlet molecules.

27.2.2 Quantitative Evaluation of Open-Shell Character

The degree of open-shell nature in a singlet diradical molecule can be quantified by diradical character y [38–40,42,43], which takes a value ranging from 0 (closed-shell) to 1 (pure open-shell). The diradical character y is originally defined in the multiconfigurational self-consistent field (MC-SCF) theory as twice the weight of the doubly excited configuration from the HOMO to LUMO in the singlet ground-state wave function:

$$|\psi\rangle = C_G|g\bar{g}\rangle + C_D|u\bar{u}\rangle, \tag{27.1}$$

where $y = 2|C_D|^2$, and $|g\bar{g}\rangle$ and $|u\bar{u}\rangle$ denote the ground and doubly excited configurations, respectively. On the other hand, in the case of UHF and UDFT, that is, single-determinant BS methods, the diradical character is formally evaluated by the occupation number (n_k) of the natural orbitals (NOs). In addition, not only diradical but also multiradical systems are characterized by multiple diradical character y_i ($i = 0, 1, \ldots$). The y_i values, which are related to the HONO–i and LUNO + i, are expressed by [39,40,43]

$$y_i = n_{\text{LUNO}+i} = 2 - n_{\text{HONO}-i} \tag{27.2}$$

As mentioned before, however, the diradical characters obtained by the single-determinant method tend to be overestimated due to the spin contamination, the degree of which depends on the calculation methods. For the BS approaches, we can use the approximate spin projection (ASP) scheme [39,44] to efficiently remove the spin contamination contributions in the occupation numbers. The perfect-pairing type spin-projected occupation number n_k^{ASP} is obtained as follows [40]:

$$n_{\text{HONO}-i}^{\text{ASP}} = \frac{n_{\text{HONO}-i}^2}{1+S_i^2} = 2 - y_i^{\text{ASP}}, \quad \text{and}$$
$$n_{\text{LUNO}-i}^{\text{ASP}} = \frac{n_{\text{LUNO}-i}^2}{1+S_i^2} = y_i^{\text{ASP}}, \tag{27.3}$$

where S_i is the overlap between the corresponding orbital pair [45]:

$$S_i = \frac{n_{\text{HONO}-i} - n_{\text{LUNO}+i}}{2} \tag{27.4}$$

In particular, the UHF method combined with the ASP scheme is found to well reproduce the diradical character calculated by strongly electron-correlated methods including the full configuration interaction (CI) method [39]. In the case of UDFT methods, the effectiveness of the ASP scheme depends strongly on the used functional, because the UDFT methods originally include the electron correlation and tend to suppress the spin contamination effect. For example, in the long-range corrected UDFT, LC-UBLYP, method [46], the effectiveness varies depending on the parameter within the functional. For the LC-UBLYP functional, the electron repulsion operator ($1/r_{12}$) is split into two terms, that is, short- and long-range electron–electron interactions, with a range separating parameter μ in the standard error function as [46]

$$\frac{1}{r_{12}} = \frac{1-\mathrm{erf}(\mu r_{12})}{r_{12}} + \frac{\mathrm{erf}(\mu r_{12})}{r_{12}} \tag{27.5}$$

where the first term is applied to the pure DFT exchange functional as the short-range interaction, while the second term is used for the exact HF exchange integral as the long-range interactions. It is found that the LC-UBLYP with $\mu = 0.33$ can reproduce results calculated by high accuracy calculations in the wide range of diradical character with no use of the ASP scheme, while that with $\mu = 0.47$ can be used connecting the ASP scheme [47,48].

In physics, the diradical character y is considered to represent the degree of electron correlation between a pair of radicals with up and down spins: the system with a small y value is a weakly electron-correlated system, while that with a large y value is a strongly electron-correlated system. On the other hand, from the viewpoint of chemistry, the diradical character indicates the "instability of a chemical bond" [39]. For the dissociation process of a H_2 molecule (Figure 27.4a), which is one of the simplest diradical models, the equilibrium bond distance region corresponds to a closed-shell state ($y = 0$), while the bond dissociation region to a pure open-shell (diradical) state ($y = 1$). Namely, the y value gradually increases from $y = 0$ to 1 with the bond distance. Figure 27.4b shows the variation in the spin-projected y values calculated using the UHF/6-31G** + *sp* method for H_2 models with different bond distances [28].

To deepen our understanding of the relationship between the electronic structure and diradical character, we perform the analysis using the valence configuration interaction (VCI) diradical model with two electrons in two orbitals [26]. We here consider a symmetric two-site diradical system with the localized NOs (LNO) $a(\boldsymbol{r})$ and $b(\boldsymbol{r})$ [49], which are mainly localized on each site, respectively, and have generally small tails on the other site to satisfy the orthogonal condition $\langle a|b\rangle = 0$. Then, the CI matrix using the LNO basis $\left\{|a\bar{b}\rangle, |b\bar{a}\rangle, |a\bar{a}\rangle, |b\bar{b}\rangle\right\}$ (the upper-bar [nonbar] indicates the $\beta(\alpha)$ spin) is given by

$$\begin{pmatrix} 0 & K & t & t \\ K & 0 & t & t \\ t & t & U & K \\ t & t & K & U \end{pmatrix} \tag{27.6}$$

where t, U, and K indicate a transfer integral, a difference between on- and intersite Coulomb integrals, and an exchange integral, respectively, defined as follows:

$$t \equiv \langle a|f|b\rangle, \tag{27.7}$$

$$K \equiv \left\langle a\bar{b}\left|\frac{1}{r_{12}}\right|b\bar{a}\right\rangle \tag{27.8}$$

and

$$U \equiv \left\langle a\bar{a}\left|\frac{1}{r_{12}}\right|a\bar{a}\right\rangle - \left\langle a\bar{b}\left|\frac{1}{r_{12}}\right|a\bar{b}\right\rangle, \tag{27.9}$$

where f is the Fock operator in the LNO basis representation [50]. By diagonalizing this matrix, three singlet states (S_{1g}, S_{1u}, and S_{2g}) and one triplet state (T_{1u}) are obtained. Using these solutions, the diradical character y of this system is obtained from Equation 27.1 [26]:

$$y = 1 - \frac{4|t|}{\sqrt{U^2 + 16t^2}} = 1 - \frac{4r_t}{\sqrt{1+16r_t^2}} \tag{27.10}$$

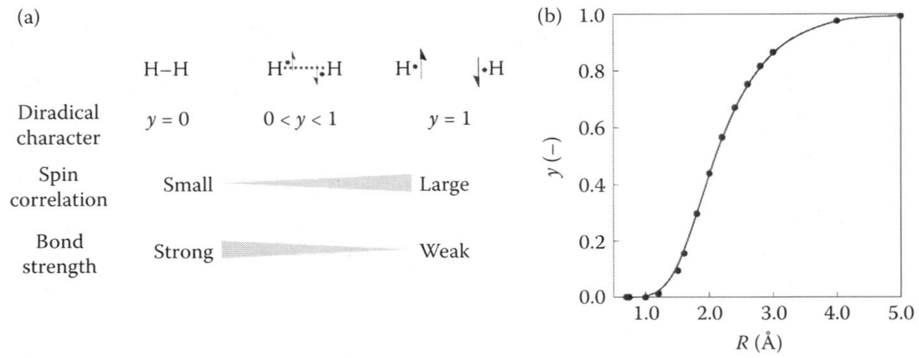

FIGURE 27.4 Dissociation process of a H_2 molecule (a), and the bond distance dependence (R) of its diradical character (y) calculated by UHF/6-31G** + *sp* with ASP scheme (b).

where $r_t = |t|/U$. Because U means the difficulty of electron transfer from one to the other site, while $|t|$ does the easiness of that, the case with much smaller $|t|$ than U ($r_t \to 0$) indicates the limit of the localization of an electron on each site, implying the pure diradical state ($y \to 1$). However, the case with $|t| \sim U$ ($r_t \sim 1$) represents a sufficient delocalization of electrons over two sites, implying a stable bonding state ($y \to 0$).

Based on the VCI scheme, we present a quantitative evaluation method for the diradical character using experimentally measurable quantities [51]. Because t and U can be described by using the total energy E_i (for state i)

$$t = -\frac{1}{4}\sqrt{(E_{S_{2g}} - E_{S_{1g}})^2 - (E_{S_{1u}} - E_{T_{1u}})^2} \quad \text{and} \quad U = E_{S_{1u}} - E_{T_{1u}}, \tag{27.11}$$

we straightforwardly obtain an expression of the diradical character from Equation 27.10 [51]:

$$y = 1 - \sqrt{1 - \left(\frac{E_{S_{1u}} - E_{T_{1u}}}{E_{S_{2g}} - E_{S_{1g}}}\right)^2} = 1 - \sqrt{1 - \left(\frac{E_{S_{1u},S_{1g}} - E_{T_{1u},S_{1g}}}{E_{S_{2g},S_{1g}}}\right)^2}, \tag{27.12}$$

where E_{ij} indicates the excitation energy between states i and j. Experimentally, $E_{S_{1u},S_{1g}}$ and $E_{S_{2g},S_{1g}}$ correspond to the lowest energy peaks of the one- and two-photon absorption (TPA) spectra, respectively, while $E_{T_{1u},S_{1g}}$ can be obtained from phosphorescence or electron spin resonance (ESR) measurement. For several realistic PAHs as shown in Figure 27.5a, including polycyclic aromatic diphenalenyl diradicaloids, IDPL (**1**), and

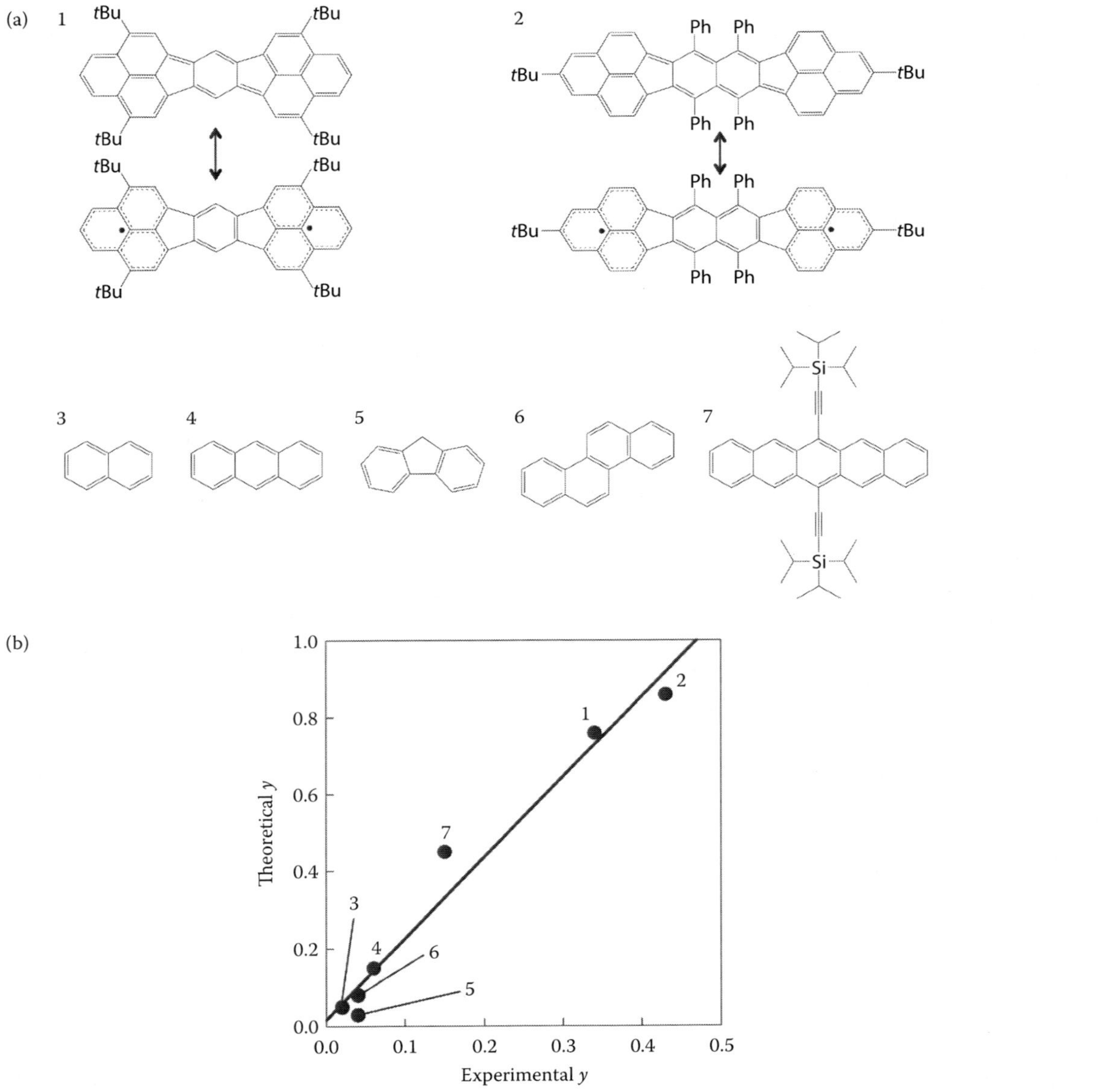

FIGURE 27.5 Molecular structures for several realistic PAHs (a), and a relationship between the experimentally deduced and theoretically calculated (using the ASP-UHF/6-31G** method) diradical characters (b).

NDPL (**2**), which are synthesized by Kubo et al. [52,53], we compare the diradical characters obtained from experimental values by using Equation 27.12 with those theoretically calculated using the ASP–UHF/6-31G** method. Figure 27.5b displays a strong correlation between the experimentally deduced diradical characters and theoretically calculated ones. Although there is a scaling factor of about 2, the strong correlation between the experimental and theoretical y values indicates that the diradical character is one of essential chemical indices for describing the ground and excited electronic structures for closed-shell and open-shell singlet systems.

Furthermore, the spatial contribution of diradical character y_i for a molecule can be described by the odd electron density, $D_{y_i}^{odd}(\boldsymbol{r})$ [40]:

$$D_{y_i}^{odd}(\boldsymbol{r}) = D_{HONO-i}(\boldsymbol{r}) + D_{LUNO+i}(\boldsymbol{r}), \qquad (27.13)$$

where each of the contributions is defined as [42]

$$D_k^{odd}(\boldsymbol{r}) = \min(2 - n_k, n_k)\phi_k^*(\boldsymbol{r})\phi_k(\boldsymbol{r}). \qquad (27.14)$$

The $\min(2 - n_k, n_k)$ can thus be regarded as the probability for the electron of being unpaired in kth NO $\phi_k(\boldsymbol{r})$. The odd electron density in the ASP scheme can also be obtained by using spin-projected n_k^{ASP} instead of n_k in the above formula. The y_i value is obtained by [40]

$$y_i = \frac{1}{2}\int\left[D_{y_i}^{odd}(\boldsymbol{r})\right]d^3\boldsymbol{r}, \qquad (27.15)$$

which is useful for clarifying the spatial contributions of odd electrons to y_i.

27.2.3 Molecular Structural and Size Dependences of Open-Shell Character for GNFs

Let us begin by considering the size dependences of the diradical characters for rectangular GNFs. The y_i $(i = 0, 1)$ values for rectangular GNFs (Figure 27.6), referred to as polycyclic aromatic hydrocarbons PAH[X,Y] (where X and Y, respectively, denote the number of fused rings in the zigzag and armchair edges), are listed in Table 27.2 [17]. The y_0 and y_1 values are shown to increase both with X when keeping Y constant and with Y when keeping X constant. The results show that the increase of y_0 value precedes that in y_1 value, and that the y_1 begins to significantly increase after the y_0 becomes close to 1. These facts imply that for zigzag-edged GNFs, the increase of π-conjugation size enhances the open-shell character and realizes a multiradical state beyond the pure diradical state. Indeed, PAH[7,7] presents the remarkable multiple diradical characters, $y_0 = 1.000$ and $y_1 = 0.899$, which indicates that this system possesses a nearly pure tetraradical singlet nature.

Figure 27.7 shows the odd electron density distributions, $D_{y_0}^{odd}(\boldsymbol{r})$ and $D_{y_1}^{odd}(\boldsymbol{r})$, of PAH[3,3] and PAH[6,7] together with their y_i values calculated by the LC-UBLYP method with

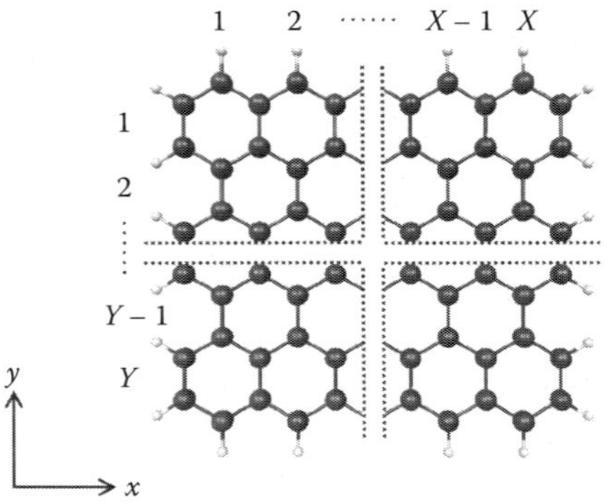

FIGURE 27.6 Definition of rectangular GNF, polycyclic aromatic hydrocarbons, PAH[X,Y], where X and Y indicate the lengths of zigzag and armchair edges, respectively. Carbon (C) and hydrogen (H) atoms are shown in black and white, respectively.

$\mu = 0.47$ and 6-31G* basis set [40]. It is found that PAH[3,3] and PAH[6,7] are classified, respectively, as an intermediate diradical system ($y_0 = 0.510$ and $y_1 = 0.053$) and as a tetraradical system having pure the first diradical character ($y_0 = 1.000$) and intermediate the second diradical character ($y_1 = 0.763$). For PAH[3,3], the $D_{y_0}^{odd}(\boldsymbol{r})$ (Figure 27.7a) is primarily distributed on the middle region of the zigzag edges, the amplitude of which rapidly decreases while going from the edge to the central region of the molecule, whereas the $D_{y_1}^{odd}(\boldsymbol{r})$ (Figure 27.7b) has negligible amplitude, which reflects $y_1 \sim 0$. The $D_{y_0}^{odd}(\boldsymbol{r})$ of PAH[6,7] (Figure 27.7c) is well localized on the zigzag edges similar to the case of PAH[3,3], the feature of which is generally observed in zigzag-edged GNFs with open-shell singlet characters. Because of the relatively large y_1 value of PAH[6,7] in contrast to that of PAH[3,3], there are some distributions of $D_{y_1}^{odd}(\boldsymbol{r})$ in the end region, though not in the middle region, of the zigzag edges in PAH[6,7] (Figure 27.7d), which are also observed for tetraradical or more multiradical rectangular GNFs.

TABLE 27.2
Diradical Characters y_0 and y_1 [–] for PAH[X,Y] ($1 \le X \le 7$, $1 \le Y \le 7$) Calculated Using the UHF/6-31G* Connecting the ASP Scheme

System	X	y_0	y_1	System	X	y_0	y_1
PAH[X,1]	1	0.000	0.000	PAH[X,5]	1	0.069	0.014
	2	0.050	0.016		2	0.372	0.057
	3	0.149	0.022		3	0.808	0.100
	4	0.282	0.034		4	0.953	0.233
	5	0.419	0.068		5	0.989	0.420
	6	0.559	0.118		6	0.999	0.623
	7	0.696	0.183		7	0.999	0.790
PAH[X,3]	1	0.037	0.012	PAH[X,7]	1	0.094	0.026
	2	0.217	0.022		2	0.487	0.114
	3	0.510	0.053		3	0.921	0.152
	4	0.806	0.137		4	0.999	0.304
	5	0.922	0.259		5	0.999	0.534
	6	0.972	0.407		6	1.000	0.763
	7	0.995	0.560		7	1.000	0.899

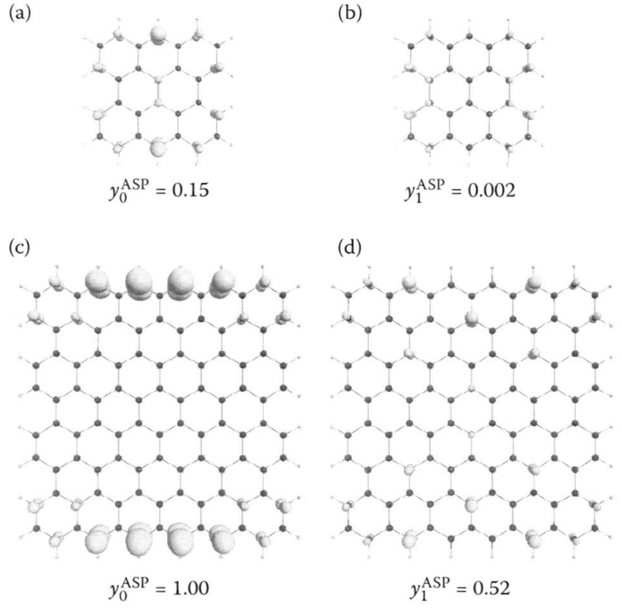

FIGURE 27.7 Odd electron densities [$D_{y_0}^{odd}(\mathbf{r})$ and $D_{y_1}^{odd}(\mathbf{r})$] for PAH[3,3], (a) and (b), and for PAH[6,7], (c) and (d), in the singlet states calculated by the LC-UBLYP (μ = 0.47)/6-31G* method with the ASP scheme. The surfaces represent the densities with isosurfaces of 0.0015 a.u. for (a), 0.0003 a.u. for (b), and 0.0025 a.u. for (c) and (d). The corresponding y_i values calculated by the LC-UBLYP (μ = 0.47)/6-31G* method with the ASP scheme are also shown.

We also introduce other structural dependences of open-shell character for several GNFs [16]. Figure 27.8 shows the molecular structures for two types of GNFs composed of two trigonal blocks, rhombic (a) and bow-tie (b) GNFs, of which trigonal blocks are connected via their sides and vertices, respectively. Their diradical characters calculated using the ASP–UHF method with 6-31G* basis set are found as $y_0 = 0.418$ (rhombic) and $y_0 = 0.970$ (bow-tie), respectively. The significant difference between their y_0 values implies that there are unique structural dependences of open-shell characters of GNFs, which cannot be predicted by the simple π-conjugation size dependence. To elucidate the origin of the linked-form dependence, we compare the resonance structures of these GNFs (see Figure 27.9). The rhombic GNF displays both closed-shell and diradical resonance forms, while there are no closed-shell forms for the bow-tie GNFs. It is shown that the number of Clar's sextets in the diradical form of rhombic GNF (3) is the same as that in the closed-shell form, while that it is less than that of diradical bow-tie GNF (4). This predicts that the thermal stabilities of closed-shell and diradical forms are similar for rhombic GNFs.

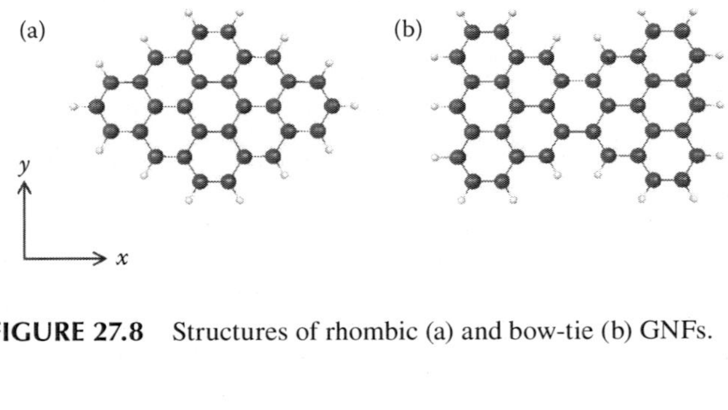

FIGURE 27.8 Structures of rhombic (a) and bow-tie (b) GNFs.

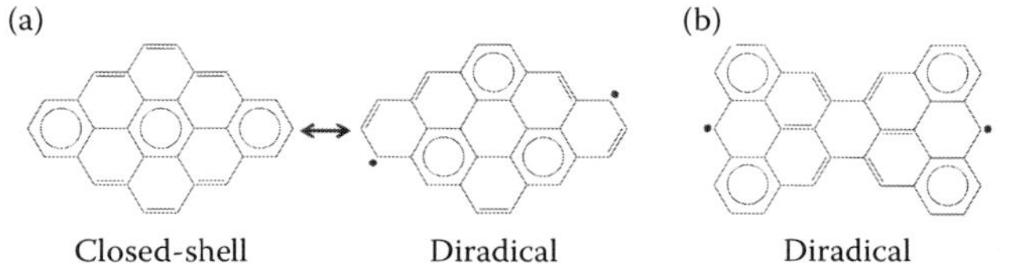

FIGURE 27.9 Resonance forms of rhombic (a) and bow-tie (b) GNFs.

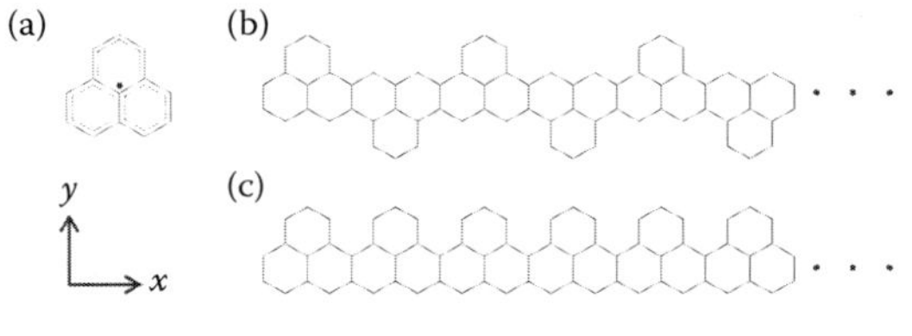

FIGURE 27.10 Molecular frameworks of the trigonal GNF unit (a), and of the two types of 1D GNFs: AL (b) and NAL (c) systems.

At the next stage, we examine the 1D GNFs composed of the smallest trigonal units, called phenalenyl [18]. Figure 27.10 shows the molecular frameworks for the trigonal GNF unit (a) and two types of 1D GNFs: the alternately linked (AL) and nonalternately linked (NAL) systems, in which the lateral directions of the linked units alternately change and are all the same, respectively. Because the trigonal GNF, which is the building block, is a mono-π-radical system (doublet), the N-units system is formally regarded as an N-radical system. Recently, a lot of fabrication methods for the nanographenes based on bottom-up approaches have been reported [54,55], so that it is meaningful to consider such multiunit systems. Table 27.3 and Figure 27.11 represent the length dependences of y_i (i = 0–4) in AL and NAL systems (N = 2, 4, 6, 8, 10) obtained using the LC-UBLYP (μ = 0.33)/6-31G* method without the ASP scheme. The diradical characters of AL systems show slower increases with N than NAL systems (see Figure 27.11a), for example, even the AL system composed of

TABLE 27.3
Diradical Characters y_i ($0 \le i \le 4$) [–] for AL and NAL Systems (N = 2, 4, 6, 8, 10) Calculated Using the LC-UBLYP (μ = 0.33)/6-31G* Method

System	N	y_0	y_1	y_2	y_3	y_4
AL	2	0.101	0.009	0.007	0.004	0.003
	4	0.302	0.118	0.021	0.015	0.013
	6	0.418	0.196	0.114	0.027	0.018
	8	0.487	0.276	0.159	0.112	0.030
	10	0.630	0.383	0.236	0.153	0.122
NAL	2	0.919	0.018	0.017	0.014	0.013
	4	0.999	0.947	0.033	0.032	0.026
	6	1.000	0.999	0.947	0.042	0.042
	8	1.000	1.000	0.999	0.943	0.044
	10	1.000	1.000	1.000	0.999	0.947

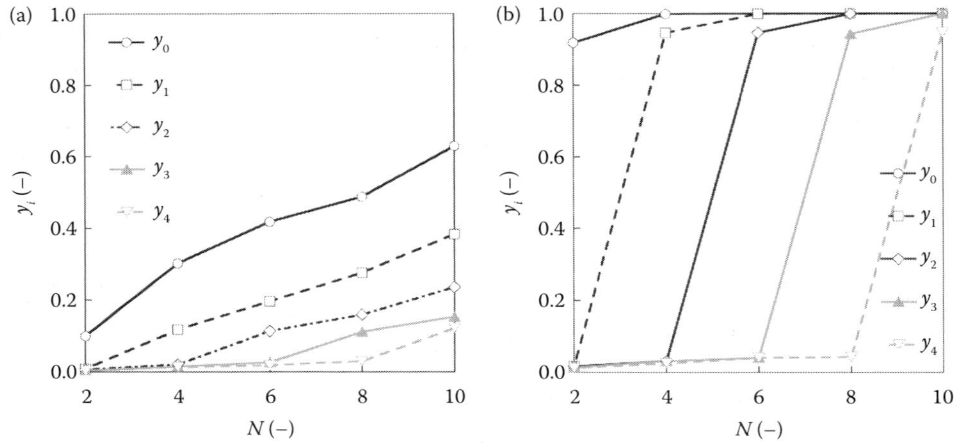

FIGURE 27.11 Size dependences of the y_i values ($0 \le i \le 4$) in AL (a) and NAL (b) systems ($N = 2, 4, 6, 8, 10$). The y_i values are calculated from the LC-UBLYP ($\mu = 0.33$)/6-31G* method.

10 units still exhibits intermediate diradical characters despite its large size. On the basis of the obtained increase rate of y_i, the AL systems are expected to keep the intermediate multiradical states up to $N \sim 16$. In contrast, the y_i values in NAL systems (Figure 27.11b) present abrupt transitions from ~0 to ~1 at every addition of a pair of the trigonal units, which indicate that the NAL systems are pure multiradical systems irrespective of the number of units. Figure 27.12 shows the resonance structures of the two-units systems with only one linked point. As shown in Figure 27.12a, the two-units AL (2-AL) system, that is, zethrene, possesses closed-shell and diradical resonance forms, both of which have the same number of Clar's sextet (2), while the two-units NAL system does not possess a closed-shell form (see Figure 27.12b). This explains the features of diradical characters for these two systems: 2-AL ($y_0 = 0.101$) and NAL ($y_0 = 0.919$) systems, which concludes that the AL form leads to a relatively strong interaction between the units, while the NAL form does a negligible interaction; indeed, the NAL (Figure 27.12b) has a triplet ground state. As a result, we observe a remarkable linked form dependence of y_i for these 1D GNFs.

Furthermore, we examine two HGNFs having only zigzag and armchair edges, respectively (Figure 27.13a and b) [15]. Of course, the armchair-edged HGNF (A-HGNF) is predicted to be a closed-shell system owing to the absence of zigzag edges. On the other hand, zigzag-edged HGNF (Z-HGNF) exhibits an intermediate tetraradical character ($y_0 = y_1 = 0.410$), in which completely the same values of y_0 and y_1 are obtained due to the respective degenerate (HONO and HONO-1) and (LUNO and LUNO + 1) orbital pairs because of D_{6h} symmetry of the molecular structure, though PAH[4,7] of similar size possesses the nearly pure diradical y_0 (= 0.999) and intermediate y_1 (= 0.304) values. The small open-shell character of Z-HGNF as compared to that of rectangular one can also be qualitatively explained based on the resonance structures together with Clar's sextet rule. For Z-HGNF (Figure 27.14a), the number of Clar's sextets in the closed-shell form is equal to those in the both diradical and tetraradical forms (12 sextets) unlike the

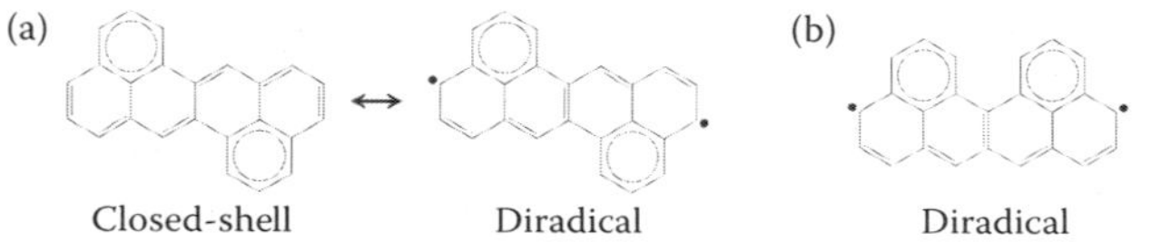

FIGURE 27.12 Resonance forms of 2-AL (a) and two-units NAL (b) systems.

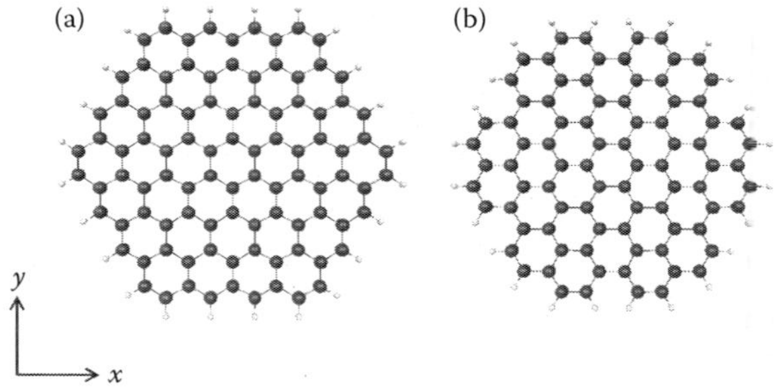

FIGURE 27.13 Structures of zigzag-edged hexagonal GNF (Z-HGNF) (a) and armchair-edged hexagonal GNF (A-HGNF) (b).

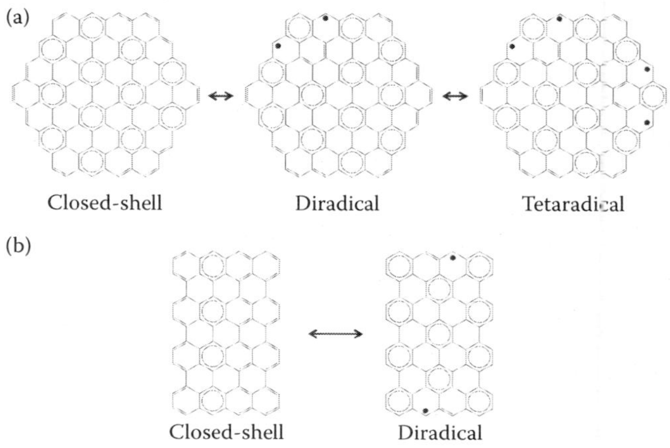

FIGURE 27.14 Resonance forms of Z-GNF (a) and PAH[4,7] (b).

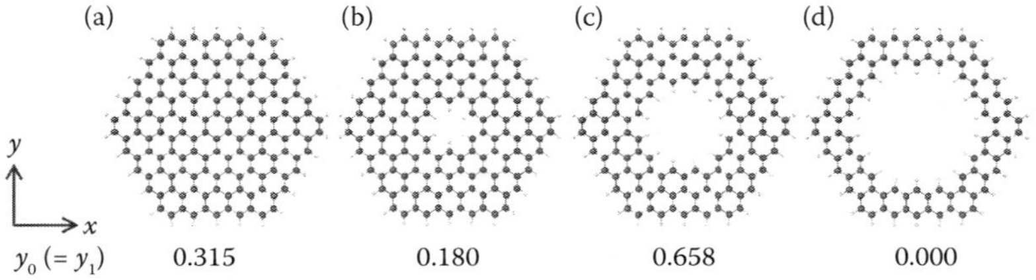

FIGURE 27.15 Molecular structures of four types of HGNFs with/without antidot structures. The y_0 (= y_1) values are also shown.

case of PAH[4,7], for which the number of Clar's sextets in the closed-shell form (4) is significantly smaller than that in the diradical form (11), as shown in Figure 27.14b.

More recently, we have revealed the relationship between the molecular structure and open-shell character for ZHGNF with a fixed HGNF size and with varying the size of the geometrical hole, which is referred to as an "antidot" structure [19]. Figure 27.15 shows the molecular structures for a perfect HGNF (**a**), in which each outer zigzag-edge side is composed of five fused rings unlike the above Z-HGNF (Figure 27.14a), and three antidot HGNFs (**b**–**d**), as well as their diradical characters y_0 (= y_1) calculated using the LC-UBLYP ($\mu = 0.33$)/6-31G* method. Note that the unnatural fact that the diradical character of system **a** ($y_0 = y_1 = 0.315$) is smaller than that of Z-HGNF in Figure 27.14a despite its larger π-conjugation size is caused by an artifact originating from the difference in the employed calculation methods. Indeed, the LC-UBLYP ($\mu = 0.33$) method tends to give smaller diradical character than the ASP–UHF method. The diradical characters of systems **a**–**d** exhibit a strange dependence on the size of their antidot structures: the diradical character shows an oscillatory behavior as a function of increasing the size of the antidot. As compared to system **a**, system **b** with a benzene-shape antidot displays small diradical characters (0.180), whereas system **c**, having a larger antidot structure, exhibits larger diradical characters (0.658), and system **d**, having the largest antidot structure, shows a zero diradical character, that is, closed-shell state. This interesting antidot effect on the open-shell character of HGNFs is rationalized by their HOMO–LUMO energy gaps and the topology of the MOs. Figure 27.16 displays the spatial distributions of the HOMO and LUMO for HGNFs **a**–**d** with the orbital energies and HOMO–LUMO gaps calculated using the LC-RBLYP ($\mu = 0.33$)/6-31G* method. Generally, the system with a smaller HOMO–LUMO gap (which can be obtained by the spin-restricted method) tends to exhibit a larger diradical character as is clear from the definition of diradical character. This tendency can also be observed in the case of these HGNFs: system **c** with the largest open-shell character possesses the smallest HOMO–LUMO gap (2.735 eV), while system **d**, which is closed-shell, possesses the largest HOMO–LUMO gap (6.221 eV). The variation of their HOMO–LUMO gaps results from the fact that the bonding and antibonding interactions between the central antidot-shape HGNFs and the surroundings alternate

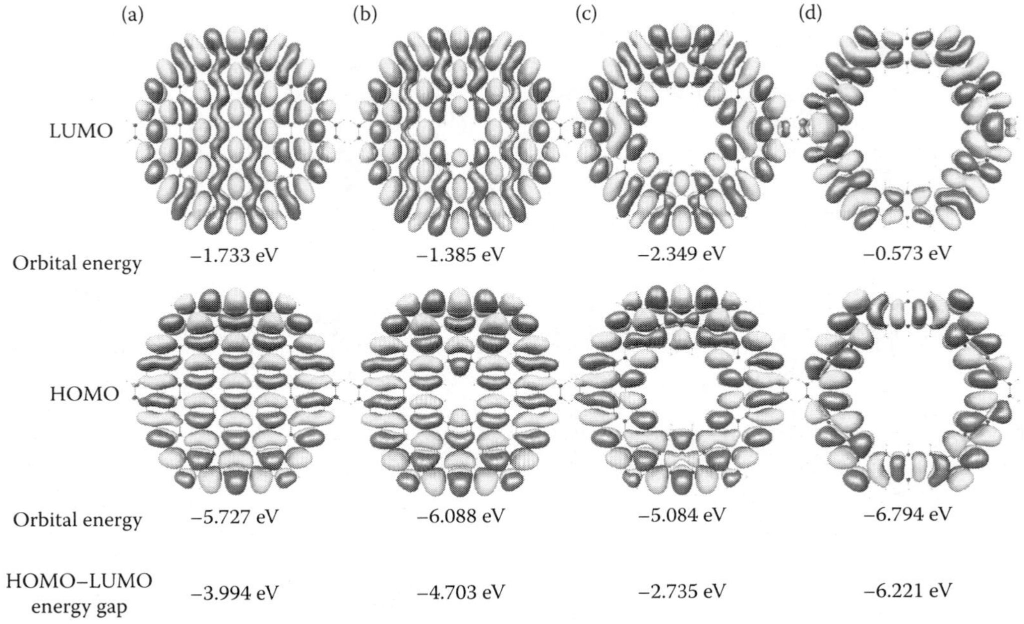

FIGURE 27.16 MOs, orbital energies, and the HOMO–LUMO energy gaps of HGNFs with/without antidot structures a–d calculated using the LC-RBLYP ($\mu = 0.33$)/6-31G* method. The light and dark gray surfaces represent positive and negative MOs with ±0.007 a.u. isosurfaces, respectively.

in both the HOMO and the LUMO as going from the center of the molecule to the peripheral region. For example, there is the bonding (antibonding) interaction between the central benzene-shape and its surrounding in the LUMO (HOMO) of system **a**. Thus, as going from **a** to **b**, having a benzene-shaped antidot in the center, the HOMO and LUMO exhibit a stabilization and a destabilization, respectively, which leads to the increase of the HOMO–LUMO energy gap. In contrast, the orbital distributions in **a** between the central coronene structure and its surrounding are found to give bonding (antibonding) interactions in the HOMO (LUMO). Thus, the removal of this central coronene from **a** is predicted to destabilize (stabilize) the HOMO (LUMO), and thus to decrease the HOMO–LUMO gap of **c** (2.735 eV). However, the HOMO and LUMO of system **d** are obviously different from those of the three other molecules. In fact, the orbitals corresponding to the HOMO (LUMO) of the other molecules are much stabilized (destabilized) in **d** due to cutting away a large central moiety, and thus become HOMO-2 (LUMO + 2). Additionally, the same feature is seen for their HOMO-1 and LUMO + 1 orbital pairs, which are related to their y_1 values.

As observed above, GNFs show a wide range of variations in their open-shell characters, which strongly depend on the edge-shape, molecular architecture, and size. It is thus concluded that the GNFs are regarded as promising candidates for a novel class of open-shell singlet molecules, the open-shell characters of which are tunable over a wide range by adjusting the molecular structures and size.

27.3 NONLINEAR OPTICAL PROPERTY

27.3.1 Diradical Character Dependence of Second Hyperpolarizability γ

In the preceding section, we demonstrated that GNFs possess open-shell singlet natures using diradical character and odd electron density, and that their open-shell singlet characters exhibit strong structural and size dependences. The next question is "In what physico-chemical phenomena, do the effects of open-shell character appear?" This question is also important for considering the applications of open-shell singlet systems in novel functional materials. As an example of theoretical proactive molecular design, we here introduce the "open-shell nonlinear optical (NLO) system," which has been theoretically proposed in our previous papers [26–28,56,57] based on a strong relationship between the open-shell character and NLO property.

We begin with a brief explanation of NLO phenomena as well as of their microscopic origins. Due to the electromagnetic interaction between material and electronic field $\boldsymbol{F}_0$ of light, the macroscopic polarization $\boldsymbol{P}$ is induced in the material, which can be expanded by the power series of $\boldsymbol{F}_0$: [58]

$$\boldsymbol{P} = \chi^{(1)}\boldsymbol{F}_0 + \chi^{(2)}\boldsymbol{F}_0\boldsymbol{F}_0 + \chi^{(3)}\boldsymbol{F}_0\boldsymbol{F}_0\boldsymbol{F}_0 + \cdots, \tag{27.16}$$

where, $\chi^{(n)}$ depicts the nth-order electric susceptibility. Under weak light such as sunlight, the polarization is dominated by only the first linear term and the higher-order nonlinear terms are negligible, because the linear susceptibility $\chi^{(1)}$ is generally much larger than other NLO susceptibilities $\chi^{(n)}$ $(n \geq 2)$. However, when the incident light possesses sufficiently strong intensity, the nonlinear components cannot be neglected. In such a case, we observe several NLO phenomenon originating from the nonlinear polarizations, such as the second harmonic generation (SHG), which is a typical second-order NLO response [59], third-harmonic generation (THG), and TPA, which are third-order NLO responses [60,61]. The macroscopic polarization $\boldsymbol{P}$ stems from the microscopic molecular polarization $\boldsymbol{p}$. The molecular polarization is expressed by [58]

$$\boldsymbol{p} = \alpha\boldsymbol{F} + \beta\boldsymbol{F}\boldsymbol{F} + \gamma\boldsymbol{F}\boldsymbol{F}\boldsymbol{F} + \cdots, \tag{27.17}$$

where $\boldsymbol{F}$ is the local electric field in the material, α, β, and γ represent the polarizability, the first and second hyperpolarizabilities, respectively. The α, β, and γ are the microscopic origins of linear optical, the second-order NLO, and third-order NLO properties, respectively.

Because the highly efficient NLO molecules are expected to be fundamental materials in future photonics and optoelectronics such as optical switching, three-dimensional memory, optical limiting, and photodynamic therapy [62–64], lots of experimental and theoretical studies [65–67] have been performed for the last decades to reveal various structure—NLO property relationships together with their tuning parameters such as π-conjugation length, the intensity of D/A substituents, and the charging effects. However, most of these studies have been limited to closed-shell systems. In recent years, we have investigated several open-shell singlet systems as a novel class of NLO molecules, and have theoretically found that the NLO properties of open-shell singlet systems have high possibilities of being far superior to those of closed-shell systems [26–28]. Indeed, we have theoretically found that the second hyperpolarizability γ exhibits a remarkable diradical character dependence: the systems with intermediate diradical characters tend to exhibit a large enhancement of γ as compared to the corresponding closed-shell and pure diradical systems of similar size [26–28,56,57]. Figure 27.17 presents the bond distance (R) dependence of γ value (longitudinal component γ_{xxxx} along the bond axis) for H_2 molecule calculated by the full-CI/6-31G** + *sp* as well as that of its diradical character at the UHF level of approximation [28]. The relationship between y and γ has also been exemplified using highly accurate *ab initio* MO calculations and UDFT methods on various model and real open-shell systems such as the diphenalenyl diradicaloids [56], transition metal complexes [57], and several GNFs, the details of which are described in the following sections. Subsequently, these theoretical predictions have been confirmed experimentally: polycyclic aromatic diphenalenyl diradicaloids, IDPL, and NDPL (**1** and **2** in Figure 27.5a), which are predicted to possess intermediate y values and large γ values [56], exhibit incredible huge TPA cross sections [29]. In particular, NDPL established a world record of TPA cross section (beyond 8000 GM at 1055 nm irradiation by the Z-scan open-aperture method) in pure hydrocarbons.

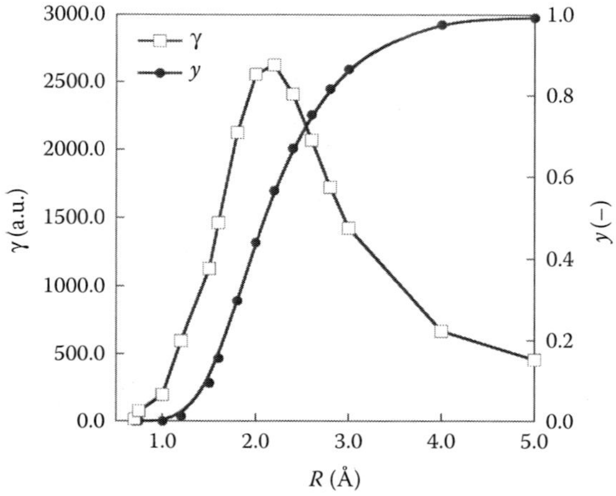

FIGURE 27.17 Bond distance (R) dependences of γ (longitudinal component γ_{xxxx} along the bond axis) calculated by the full-CI/6-31G** + sp method as well as of diradical character y calculated by the UHF/6-31G** + sp method with the ASP scheme for H_2 molecule.

The origin of diradical character dependence of γ has been clarified theoretically by the VCI diradical model with two electrons in two orbitals, as described in Section 27.2.2. Let us represent the second hyperpolarizability γ as a function of the diradical character y. Based on the perturbation theory, the static γ for the symmetric two-site model is expressed as [26]

$$\gamma = \gamma^{\mathrm{II}} + \gamma^{\mathrm{III}-2}$$

$$= -4\frac{(\mu_{S_{1g},S_{1u}})^4}{(E_{S_{1u},S_{1g}})^3} + 4\frac{(\mu_{S_{1g},S_{1u}})^2(\mu_{S_{1u},S_{2g}})^2}{(E_{S_{1u},S_{1g}})^2 E_{S_{2g},S_{1g}}}, \tag{27.18}$$

where $\mu_{i,j}$ and $E_{i,j}$ denotes the transition moment and excitation energy between states i and j, respectively. In addition, the first term represents a contribution of the virtual excitation process $S_{1g} \to S_{1u} \to S_{1g} \to S_{1u} \to S_{1g}$, while the second term does that of the virtual excitation process $S_{1g} \to S_{1u} \to S_{2g} \to S_{1u} \to S_{1g}$, which are classified to type II and type III-2, respectively [68]. These excitation properties can be described using t, U, K, and R, which indicates the intersite distance defined by

$$R \equiv \langle b|r|b\rangle - \langle a|r|a\rangle. \tag{27.19}$$

Subsequently, the dimensionless γ [$= \gamma/(R^4/U^3)$] is given as a function of two variables, y and r_K ($\equiv 2K/U$): [26]

$$\frac{\gamma}{\left(R^4/U^3\right)} = -\frac{8(1-y)^4}{\left\{1+\sqrt{1-(1-y)^2}\right\}^2\left\{1-2r_K+1\Big/\sqrt{1-(1-y)^2}\right\}^3} + \frac{4(1-y)^2}{\left\{1\Big/\sqrt{1-(1-y)^2}\right\}\left\{1-2r_K+1\Big/\sqrt{1-(1-y)^2}\right\}^2}. \tag{27.20}$$

Figure 27.18 shows the y dependence of the dimensionless γ for $r_K = 0$ as well as the dimensionless γ^{II} and $\gamma^{\mathrm{III}\text{-}2}$, which indicates that an intermediate y value (0.3586) gives the maximum of the dimensionless γ. Furthermore, it is found that the total γ is dominated by the type III-2 contribution, while the type II process gives a small negative contribution. Although the above discussion is limited to static γ, we have also examined the dynamical response by extending the VCI scheme and have substantiated that the TPA intensity for symmetric diradical model shows a similar enhancement in the intermediate y region [69].

Although the analysis for γ based on the perturbation theory gives valuable information, the perturbative approach can hardly be applied to large-size systems because the calculation of higher excited states is needed. In the following sections, to calculate the static γ for GNFs, we alternatively employ the finite-field (FF) approach [70], which can calculate γ only from the ground-state energies under applied electric fields and enable us to avoid the calculation of excited states. The FF approach consists of the fourth-order differentiation of the ground state energy (E) with respect to the applied external electric field ($\boldsymbol{F}$). The tensor component of the static γ, γ_{ijkl} is expressed by

$$\gamma_{ijkl} = \left.\frac{\partial^4 E}{\partial F_i \partial F_j \partial F_k \partial F_l}\right|_{\boldsymbol{F}=\boldsymbol{0}}, \tag{27.21}$$

where F_i denotes the i component of the field vector $\boldsymbol{F}$. We calculate the diagonal component by using the numerical differentiation with the seven-point calculations [70]:

$$\gamma_{iiii} = \frac{1}{36F_i^4}\{E(3F_i) - 12E(2F_i) + 39E(F_i) - 56E(0) + 39E(-F_i) - 12E(-2F_i) + E(-3F_i)\} \tag{27.22}$$

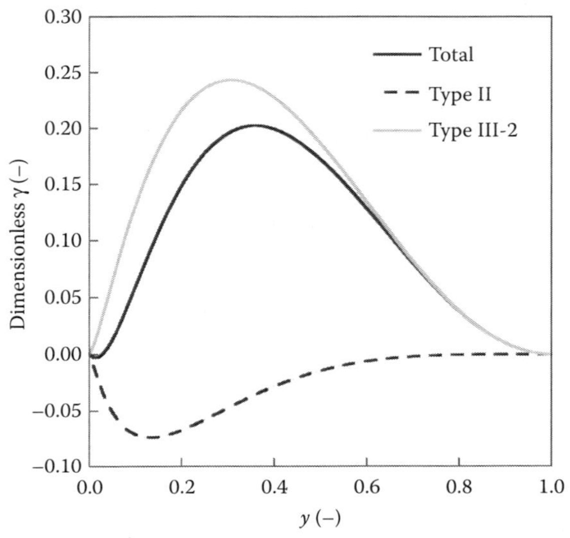

FIGURE 27.18 Diradical character dependences of dimensionless γ, $\gamma/(R^4/U^3)$, $\gamma^{\mathrm{II}}/(R^4/U^3)$, and $\gamma^{\mathrm{III}\text{-}2}/(R^4/U^3)$ for $r_K = 0$ in the VCI diradical model.

In order to elucidate the spatial electronic contributions of the electrons to γ_{iiii}, we perform γ density $[\rho_{iii}^{(3)}(\boldsymbol{r})]$ analysis [71], which utilizes the third-order derivative of the electron density with respect to the applied field:

$$\rho_{iii}^{(3)}(\boldsymbol{r}) = \left.\frac{\partial^3 \rho}{\partial F_i \partial F_i \partial F_i}\right|_{F=0}. \tag{27.23}$$

Using this density, γ is obtained by [71]

$$\gamma_{iiii} = -\frac{1}{3!}\int r_i \rho_{iii}^{(3)}(\boldsymbol{r}) d\boldsymbol{r}, \tag{27.24}$$

where r_i is the i component of the position vector. The positive (negative) value of $\rho_{iii}^{(3)}(\boldsymbol{r})$ multiplied by F_i^3 represents the field-induced increase (decrease) in the charge density in proportion to F_i^3. For the simplest example, that is, a pair of localized γ_{iiii} densities with positive and negative values, the contribution to γ_{iiii} is positive when the direction from positive to negative γ_{iiii} densities coincides with the positive direction of the i-axis. The amplitude of the contribution associated with this pair of $\rho_{iii}^{(3)}(\boldsymbol{r})$ is also proportional to the distance between them.

27.3.2 Second Hyperpolarizabilities of Open-Shell Singlet GNFs

We begin by rectangular GNFs [14,17]. Table 27.4 lists the γ_{xxxx} and γ_{yyyy} values, which are the components along its zigzag and armchair edges, respectively, of PAH[3,3] with intermediate diradical character in the singlet and triplet states calculated at the UBHandHLYP/6-31G* level of approximation. For PAH[3,3] in the singlet state, the γ_{yyyy} (14.45×10^4 a.u.) is about 4 times as large as γ_{yyyy} (3.412×10^4 a.u.). Because γ_{iiii} is generally expected to increase nonlinearly with increasing the π-conjugation length in the i-direction, the significant deference between γ_{xxxx} and γ_{yyyy} for singlet PAH[3,3] despite its square molecular structure is not usual. The exceptional enhancement of γ_{yyyy} is predicted to stem from the intermediate y_0 value (0.510). In general, the γ enhancement effect originating from intermediate diradical character is observed in the component of direction from one radical site to the other, which corresponds to the direction from one zigzag edge to the opposite one, that is, y-direction, in the case of rectangular GNFs. Moreover, there is another evidence of the open-shell NLO systems for singlet PAH[3,3], that is, significant spin state dependence of γ_{yyyy}. It is predicted that the intermediate diradical systems exhibit a significant reduction of γ amplitude by going from the singlet to the triplet state due to the Pauli effect in the triplet state [72]. Indeed, the γ_{yyyy} value exhibits a significant reduction by changing from the singlet ($\gamma_{yyyy} = 14.45 \times 10^4$ a.u.) to the triplet state ($\gamma_{yyyy} = 3.501 \times 10^4$ a.u.), while there is a little difference in the γ_{xxxx} value between the singlet ($\gamma_{xxxx} = 3.412 \times 10^4$ a.u.) and triplet states ($\gamma_{xxxx} = 3.412 \times 10^4$ a.u.). As a result, the large γ_{yyyy} for singlet PAH[3,3] is found to be caused by the intermediate diradical character.

TABLE 27.4
γ_{xxxx} and γ_{yyyy} Values [$\times 10^4$ a.u.] of PAH[3,3] in the Singlet and Triplet States Calculated at the UBHandHLYP/6-31G* Level of Approximation

Spin State	γ_{xxxx}	γ_{yyyy}
Singlet	3.412	14.45
Triplet	3.486	3.501

Table 27.5 lists the γ_{xxxx} and γ_{yyyy} values for PAH[X,Y] [$1 \le X \le 7$, $1 \le Y$ (odd number) ≤ 7] in their singlet states calculated using the UBHandHLYP/6-31G* method [17], and Figure 27.19 shows the variation in these quantities as functions of X [Y = 1,3,5,7] [17]. It is found that the γ_{xxxx} monotonically increases with X, while the γ_{yyyy} exhibits nonmonotonic X dependence, which is predicted to be caused by variations in the diradical character. PAH[6,1] ($y_0 = 0.559$), PAH[3,3] ($y_0 = 0.510$), PAH[2,5] ($y_0 = 0.372$), and PAH[2,7] ($y_0 = 0.487$), which possess intermediate y_0 values, exhibit the highest γ_{yyyy} responses among the PAH[X,1], PAH[X,3], PAH[X,5], and PAH[X,7] ($1 \le X \le 7$) families, respectively. Interestingly, the result indicates not only these γ enhancements in the intermediate diradical character (y_0) region, which are in good agreement with our previous studies, but also a multiradical character dependence of γ in the intermediate y_1 region. The PAHs with $Y \ge 3$ display the second peak in the evolution of γ_{yyyy} with X, which occurs for PAH[6,5] ($y_0 \sim 1$ and $y_1 = 0.623$) and for PAH[6,7] ($y_0 \sim 1$ and $y_1 = 0.763$) and for PAH[7,3] ($y_0 \sim 1$ and $y_1 = 0.560$). The presence of this second

TABLE 27.5
γ_{xxxx} and γ_{yyyy} Values [$\times 10^4$ a.u.] for PAH[X,Y] ($1 \le X \le 7$, $1 \le Y \le 7$) in the Singlet States Calculated Using the UBHandHLYP/6-31G* Method

System	X	γ_{xxxx}	γ_{yyyy}	System	X	γ_{xxxx}	γ_{yyyy}
PAH[X,1]	1	0.032	0.032	PAH[X,5]	1	0.068	11.5
	2	0.305	0.090		2	0.928	80.3
	3	1.52	0.160		3	5.36	50.0
	4	7.48	0.561		4	21.9	25.2
	5	19.8	0.730		5	49.9	40.5
	6	47.3	0.748		6	82.5	57.2
	7	99.3	0.667		7	152	56.0
PAH[X,3]	1	0.051	1.73	PAH[X,7]	1	0.086	37.3
	2	0.409	4.30		2	1.13	316
	3	3.41	14.5		3	7.08	102
	4	15.0	7.33		4	27.5	88.8
	5	41.2	5.51		5	53.1	197
	6	79.3	7.33		6	100	215
	7	123	9.29		7	220	147

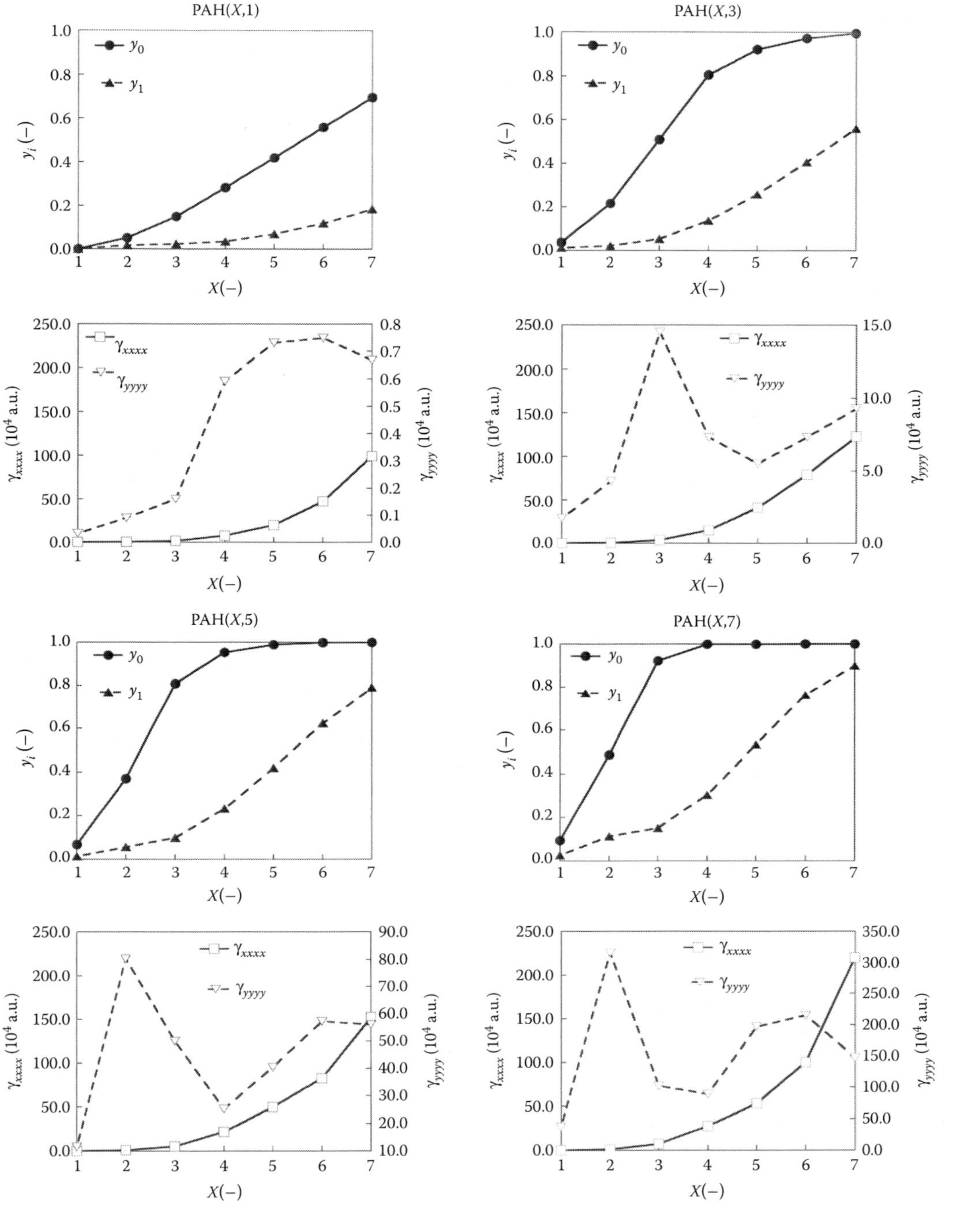

FIGURE 27.19 Variation in diradical character y_i ($i = 0, 1$) and γ (γ_{xxxx} and γ_{yyyy}) calculated using the ASP–UHF and UBHandHLYP methods, respectively, with the 6-31G* basis set for PAH[X,Y] as the function of X (Y = 1, 3, 5, 7).

peak in γ_{yyyy} as a function of X is related to y_1 and it corresponds to a region with intermediate y_1 values. In addition, the γ_{yyyy} peak for the systems with intermediate y_1 is smaller than that with intermediate y_0 value. This reduced exaltation for y_1 can be associated with the corresponding larger excitation energy, which accounts for smaller γ values. The multiradical character effects on γ can be further addressed by the analysis of their spatial electronic contributions using the γ density and odd electron density distributions. Figure 27.20 shows the γ_{yyyy} density [$\rho^{(3)}_{yyy}(\boldsymbol{r})$] distributions of PAH[3,3] (a) and PAH[6,7] (b), the γ_{yyyy} values of which are enhanced by the intermediate y_0 and y_1 values, respectively, calculated using the UBHandHLYP/6-31G* method. For both systems, π-electrons provide dominant positive and negative γ_{yyyy} densities, which are well separated to around the bottom- and top edges, respectively, and both densities rapidly decrease in amplitude toward the center region. For PAH[3,3], the $\rho^{(3)}_{yyy}(\boldsymbol{r})$ are dominantly distributed in the middle region of the both

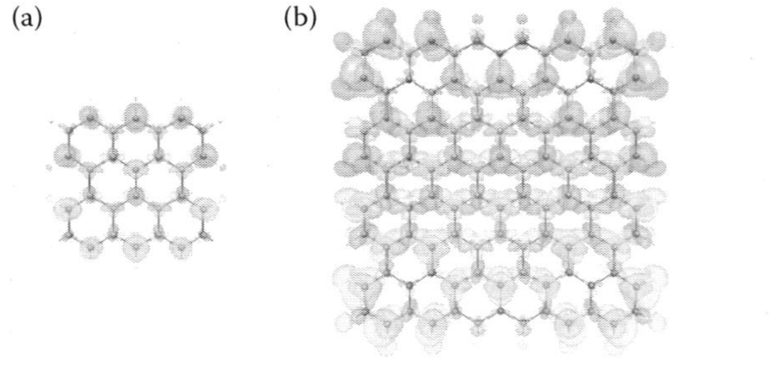

FIGURE 27.20 γ_{yyyy} density $[\rho^{(3)}_{yyy}(\boldsymbol{r})]$ distributions of PAH[3,3] (a) and PAH[6,7] (b) calculated using the UBHandHLYP/6-31G* method. The light and dark gray meshes represent positive and negative densities with isosurface ±500 a.u., respectively.

zigzag edges, which coincide with the region with large odd electron density $[D^{\text{odd}}_{y_0}(\boldsymbol{r})]$ distributions (see Figure 27.7a). In contrast, the primary amplitudes of $\rho^{(3)}_{yyy}(\boldsymbol{r})$ for PAH[6,7] are located in the end (not in the middle) zigzag-edge region, that is, four-corner phenalenyl blocks, which is similar to the odd electron density distribution $[D^{\text{odd}}_{y_1}(\boldsymbol{r})]$ related to y_1 (Figure 27.7d). These features well correspond to the fact that γ_{yyyy} of PAH[3,3] is determined by the intermediate y_0 value, while that of PAH[6,7] by the intermediate y_1 value since the $y_0 \sim 1$ reduces the HOMO–LUMO contribution. This substantiates the fact that the second γ_{yyyy} peak for PAH[X,Y] as a function of X originates from the intermediate diradical character y_1. As a result, the first and the second γ_{yyyy} peaks, which appear at intermediate y_0 and y_1 values, respectively, are evidences of the multiradical effect (tetraradical in this case) on γ.

Next, we show the linked form dependence of γ values for the GNFs composed of trigonal units [16]. The longitudinal component of γ, γ_{xxxx}, for rhombic GNF with intermediate diradical character ($y_0 = 0.418$, $\gamma_{xxxx} = 45.2 \times 10^4$ a.u.) is more than twice as large as that of bow-tie GNF with nearly pure diradical character ($y_0 = 0.970$, $\gamma_{xxxx} = 19.9 \times 10^4$ a.u.), where the γ values are calculated at the UBHandHLYP/6-31G* level of approximation. Additionally, the rhombic GNF exhibits a significant reduction of γ_{xxxx} when switching from the singlet to the triplet state [45.2 (singlet) → 11.5 (triplet)], while the γ_{xxxx} of bow-tie GNF hardly exhibits such a spin state dependence [19.9 (singlet) → 19.1 (triplet)], the feature of which confirms that the γ of singlet rhombic GNF is governed by the intermediate diradical character.

Table 27.6 lists the γ_{xxxx} values for AL and NAL systems [$2 \le N$ (even number) ≤ 10] in singlet (lowest spin) and $(N + 1)$–multiplet (highest spin) states calculated by the LC-UBLYP ($\mu = 0.33$)/6-31G* level of theory [18]. We first focus on the two-units systems. In the singlet states, the γ_{xxxx} of the 2-AL system (30.7×10^4 a.u.) is more than twice as large as that of the two-units NAL (2-NAL) system (13.4×10^4 a.u.). Then, when changing from singlet to triplet, the both amplitudes of γ are significantly reduced [γ_{xxxx}(triplet)/ γ_{xxxx}(singlet) = 0.33 (2-AL) and 0.57 (2-NAL)], and the ratio of γ_{xxxx} amplitude of 2-AL to that of 2-NAL system is also reduced [2.29 (singlet) → 1.33 (triplet)]. Nevertheless, the fact that γ_{xxxx} of the 2-AL system is also larger than 2-NAL for the triplet states suggests that the

TABLE 27.6
γ_{xxxx} Values [$\times 10^4$ a.u.] for AL and NAL Systems ($2 \le N \le 10$) in the Singlet and $(N + 1)$–Multiplet States Calculated at the LC-UBLYP ($\mu = 0.33$)/6-31G* Level of Theory

System	N	γ_{xxxx} (Singlet)	γ_{xxxx} [$(N + 1)$–Multiplet]
AL	2	30.7	10.2
	4	570	147
	6	2610	489
	8	6440	974
	10	13,300	1540
NAL	2	13.4	7.64
	4	72.0	61.7
	6	177	158
	8	295	266
	10	392	363

difference in γ_{xxxx} for their singlet states is caused by not only the diradical character dependence, but also the molecular architecture. Judging from the larger reduction of γ_{xxxx} value for 2-AL ($y_0 = 0.101$) than for 2-NAL ($y_0 = 0.919$) as changing from singlet to triplet, the diradical character giving the maximum γ_{xxxx} on the y-γ_{xxxx} curve may tend to be located in small y value region for these systems ($y < 0.5$).

Like two-units systems, the AL systems, which are intermediate open-shell singlet systems, show larger γ_{xxxx} values than the corresponding NAL systems, which always possess nearly pure open-shell characters, irrespective of the number of units N. In particular, the γ_{xxxx} value of the 10-units AL (10-AL) system in the singlet state ($13{,}300 \times 10^4$ a.u.) is more than 30 times as large as that of the analogous NAL (10-NAL) system (392×10^4 a.u.). Furthermore, the enhancement rate of γ_{xxxx} with N for AL systems is significantly large as compared to that for NAL systems, the feature of which is illustrated as the size dependences for their γ_{xxxx} values in Figure 27.21. Indeed, the ratio $\gamma_{xxxx}(N = 10)/\gamma_{xxxx}(N = 2)$ for the singlet AL systems attains 433, which is more than 10 times larger than that for the singlet NAL systems (38.4). However, all AL systems exhibit a significant spin-state dependence: the γ_{xxxx} values are significantly reduced by changing the spin states from singlet to the highest spin states, for example, 88% reduction at 10-AL system, whereas the reduction for NAL systems is negligible, for example, 7% reduction at 10-NAL. In the highest spin multiplicity state, the $\gamma_{xxxx}(N = 10)/\gamma_{xxxx}(N = 2)$ attains 151 for AL versus 48 for NAL. The spin-state dependence substantiates that the large γ_{xxxx} enhancement rate with N in singlet AL systems originates in their intermediate multiple diradical characters.

We also address the γ values of HGNFs [15]. The orthogonal components of γ ($\gamma_{xxxx} = \gamma_{yyyy}$ due to the symmetry of their structures) for A-HGNF and Z-HGNF calculated using the UBHandHLYP/6-31G* method are 139×10^4 and 41.7×10^4 a.u., respectively. The γ value of Z-HGNF, which gives an intermediate tetraradical character ($y_0 = y_1 = 0.410$), exhibits about 3.3 times larger than that of A-HGNF, which is a closed-shell system, and about 1.7 larger than γ_{yyyy} of

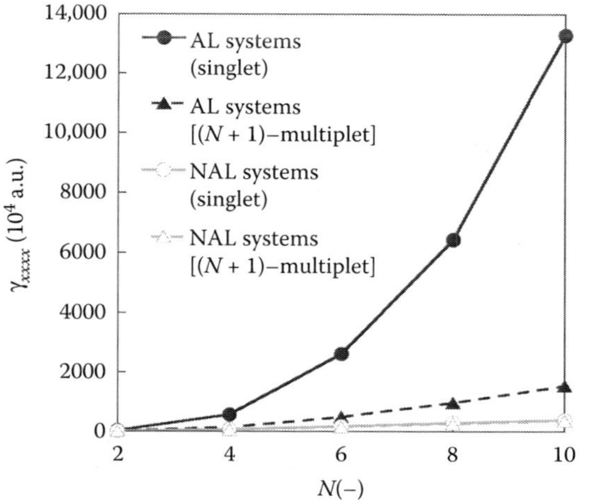

FIGURE 27.21 Size dependences of the γ_{xxxx} values (longitudinal component) for AL and NAL systems ($2 \leq N \leq 10$) in their singlet and ($N + 1$)–multiplet states calculated using the LC-UBLYP ($\mu = 0.33$)/6-31G* method.

TABLE 27.7
Diradical Characters ($y_0 = y_1$) [–] and the Orthogonal Components of γ ($\gamma_{xxxx} = \gamma_{yyyy}$) [$\times 10^4$ a.u.] Calculated Using the LC-UBLYP/6-31G* Method for HGNFs with/without Antidot Structures (Systems a–d in Figure 27.15)

System	$y_0 = y_1$	$\gamma_{xxxx} = \gamma_{yyyy}$
a	0.315	286
b	0.180	205
c	0.658	487
d	0.000	144

PAH[4,7] (88.8×10^4 a.u.), which is a pure diradical system (see Table 27.5).

Finally, we show the impact of antidot structure on the γ values for HGNF [19]. Table 27.7 lists the γ ($\gamma_{xxxx} = \gamma_{yyyy}$) for HGNFs **a**–**d** (in Figure 27.15) calculated using the LC-UBLYP ($\mu = 0.33$)/6-31G* method as well as the diradical characters ($y_0 = y_1$). The γ value of system **c** (487×10^4 a.u.) with intermediate diradical characters is about 1.7 and 2.4 times as large as those of systems **a** and **b** (286×10^4 a.u. and 205×10^4 a.u., respectively), possessing smaller diradical characters but larger π-conjugation sizes. The γ enhancement by intermediate open-shell character is also observed for these systems.

27.3.3 D/A Substitution Effects on the Diradical Character and γ

In this section, we consider the effect of asymmetric electron distribution on the NLO properties of open-shell singlet systems. At first, we revealed the effects of a static external electric field on the γ value of symmetric singlet diradical molecules using an extended VCI scheme [73]. As a result, further enhancement and control scheme of γ in the open-shell singlet systems was proposed by the application of an external field. Indeed, it is found that under the static electric field, the component of γ along the axis joining the two radical sites can be gigantically (approximately two to three orders) enhanced for symmetric diradicals having intermediate diradical characters as compared to those of closed-shell and pure diradical molecules in the absence of a field. This prediction has been confirmed for simple model diradical systems such as the H_2 molecule under dissociation and the twisted ethylene. From the viewpoint of molecular design, such an application of an external static field to molecules is partially reproduced by a substitution of D/A groups, which is thus expected to give a similar effect on γ for the open-shell singlet molecules. More recently, we have presented the general VCI theory for an asymmetric diradical model, and have clarified the origin of the asymmetricity effects on excitation energies and properties as well as the asymmetricity dependences of the first and the second hyperpolarizabilities [74]. On the basis of these results, we also examined the D/A substitution effect using realistic PAHs, that is, intermediate diradical IDPL as well as a similar-size closed-shell analog [75]. We found that the NH_2-IDPL-NO_2 exhibits a gigantic γ enhancement as compared to nonsubstituted IDPL, the enhancement ratio of which is much larger than the closed-shell one. From these results, we can conclude that the asymmetrization of open-shell system with intermediate diradical character is a promising design guide toward highly efficient NLO molecules.

As an example of asymmetric open-shell system, we introduce the D/A-substitution effect on the diradical character and γ for hexacene (Figure 27.22a) [20], which is one of rectangular GNFs, that is, PAH[6,1], having an intermediate diradical character. Table 27.8 lists the y_0 and γ_{yyyy} values for a nonsubstituted hexacene (i) and three types of D/A-tetrasubstituted hexacenes, which possess D (OH) and A (CN) groups in the middle region of the zigzag edges (ii), in the both-end region (iii), and the intermediate region (iv), calculated using the

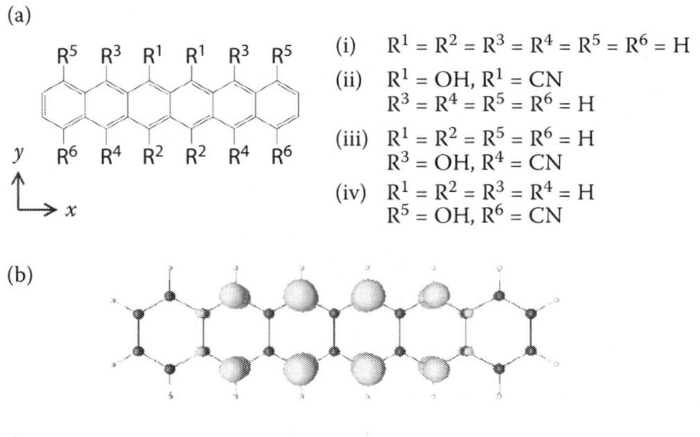

FIGURE 27.22 Structure of hexacenes with/without D (NH_2) and A (NO_2) substituents (a), and odd electron density [$D_{y_0}^{odd}(\boldsymbol{r})$] distribution calculated by the LC-UBLYP ($\mu = 0.33$)/6-31G* method (b). The surfaces represent the densities with isosurfaces of 0.002 a.u.

TABLE 27.8
y_0[–] and γ_{yyyy} [×10^4 a.u.] Values for a Nonsubstituted and Three Types of OH/CN-Tetrasubstituted (Middle-, Intermediate-, and End-Substituted) Hexacenes Calculated Using the LC-UBLYP (μ = 0.33)/6-31 + G* Method

System	y_0	γ_{xxxx}
Non	0.313	1.79
Middle	0.252	107
Intermediate	0.269	12.5
End	0.293	3.30

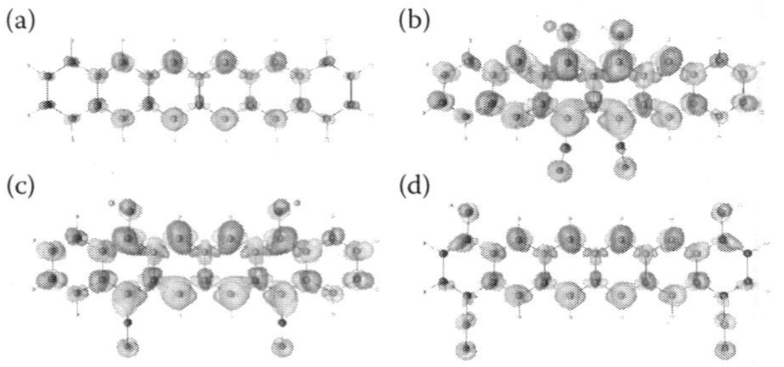

FIGURE 27.23 γ_{yyyy} density $[\rho^{(3)}_{yyy}(\boldsymbol{r})]$ distributions of nonsubstituted (a), middle-substituted (b), intermediate-substituted (c), and end-substituted (d) hexacenes calculated using the LC-UBLYP (μ = 0.33)/6-31G* method. The light (dark) gray meshes represent positive (negative) densities with isosurfaces of ±150 a.u. for (a) and (d), ±4000 a.u. for (b), and ±500 a.u. for (c), respectively.

LC-UBLYP (μ = 0.33)/6-31 + G* method. The γ_{yyyy} of these substituted hexacenes are enhanced as compared to that of the nonsubstituted system ($\gamma_{yyyy} = 1.79 \times 10^4$ a.u.), which can be easily predicted from our previous studies. The obtained data demonstrates not only a simple substituent effect but also an interesting substituted position dependence: the γ_{yyyy} values for the middle- and end-substituted systems are 107×10^4 and 3.30×10^4 a.u., respectively, which indicate a remarkable difference in their enhancement ratios of γ to the nonsubstituted systems, that is, γ_{yyyy}(middle)/γ_{yyyy}(non) = 59.8 versus γ_{yyyy}(end)/γ_{yyyy}(non) = 1.84. The ratio for intermediate-substituted system (6.98) is a value between those for other two systems. Such a remarkable substituted position dependence of γ enhancement ratios in these hexacenes is also expected to be closely related to the degree of diradical character change. In general, the external field application and D/A-substitution are found to reduce the diradical character, and thus the difference in the y_0 changes for different substitutions is expected to reflect the difference of their substituent effects. To confirm this prediction, the spatial contribution of diradical character of nonsubstituted hexacene is investigated using the odd electron density $[D^{\mathrm{odd}}_{y_0}(\boldsymbol{r})]$ distribution (Figure 27.22b), which exhibits primary amplitudes in the middle zigzag-edge regions, decreases as moving to the end regions, and have negligible amplitudes in the end regions. It is, therefore, predicted that the D/A groups substituted in the middle region cause significant effects on the unpaired electrons, that is, diradical character, and then enhance the γ_{yyyy} value, whereas those in the end regions have negligible effects on them.

In order to clarify the spatial contribution of electrons to γ_{yyyy}, we investigate the γ_{yyyy} density, $\rho^{(3)}_{yyy}(\boldsymbol{r})$, distributions of all hexacenes shown in Figure 27.23. For all the molecules, π-electron contributions are found to be dominant. It is found that hexacenes exhibit the primary $\rho^{(3)}_{yyy}(\boldsymbol{r})$ amplitudes in the middle zigzag-edge regions, which coincide with the regions with primary odd electron densities. The substituted position dependences are also illuminated in the $\rho^{(3)}_{yyy}(\boldsymbol{r})$ distributions. It is indeed found that the middle substitution in hexacene significantly enhances the $\rho^{(3)}_{yyy}(\boldsymbol{r})$ amplitudes in the middle region as compared to the nonsubstituted hexacene, while that the end substitution does not enhance the $\rho^{(3)}_{yyy}(\boldsymbol{r})$ amplitudes so much in the end region. In general, for typical closed-shell molecule with D/A groups, the γ enhancement by D/A-substitution is primarily caused by the charge transfer (CT) between D and A substituents. In sharp contrast to these features, middle-substituted hexacene exhibits a more amplitude of $\rho^{(3)}_{yyy}(\boldsymbol{r})$ in the middle zigzag edge regions than the OH and CN groups. It is thus predicted that middle-substituted hexacene remarkably increases the CT between the unpaired electrons distributed on the zigzag edge regions by the push–pull effect of the substituents attached to the middle zigzag edges having primary odd electron density distributions, while the substituents attached to the end regions hardly affect the CT between the unpaired electrons due to the absence of the odd electron density distributions on the end regions.

The enhancement of γ in D/A substituted open-shell systems are also expected to be further intensified by increasing the D–A strength. We therefore examine a hexacene middle substituted by stronger D/A substituents (NH_2 and NO_2 groups) as shown in Figure 27.24. It turns out that this molecule exhibits a smaller diradical character ($y_0 = 0.079$) than the OH/CN-middle-substituted system ($y_0 = 0.252$), which indicates a larger substituent effect, and extraordinarily large enhancement of γ_{yyyy} (2220×10^4 a.u.). Additionally, the γ density amplitudes on the zigzag edges of this molecule enhanced by the D/A-substitutions clarifies that the dominant contribution to γ originates from the CT between the unpaired electrons distributed on the zigzag edges.

27.4 CONCLUDING REMARKS AND PERSPECTIVES ON NOVEL NLO MATERIALS COMPOSED OF OPEN-SHELL NANOGRAPHENES

This chapter sheds light on the theoretical investigations of unique electronic structures of open-shell singlet states of GNFs with various types of size and architectures as well as on the molecular design guidelines for highly efficient NLO systems based on the GNFs. It has been found that zigzag-edged GNFs exhibit intermediate and large diradical characters (y),

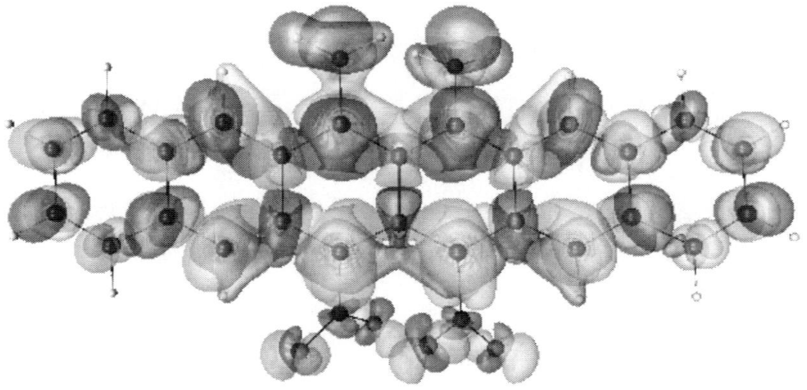

FIGURE 27.24 γ_{yyyy} density $[\rho^{(3)}_{yyy}(\boldsymbol{r})]$ distributions of NH_2/NO_2-middle-substituted hexacenes calculated using the LC-UBLYP ($\mu = 0.33$)/6-31G* method. The light (dark) gray meshes represent positive (negative) densities with isosurfaces of ±150,000 a.u., respectively.

whereas that armchair-edged GNFs present negligible y values, which means that they are nearly closed-shell systems. In addition, the zigzag-edged GNFs are expected to exhibit not only diradical but also multiradical characters with increasing π-conjugation size. We have demonstrated that zigzag-edged rectangular GNFs display significant γ enhancements and a unique size dependence of γ due to their multidiradical characters, which are emerged in the large zigzag-edge size. It has also been found that linear GNFs composed of trigonal units show a wide range of multiradical characters as well as unique size dependences of their γ values depending on their linked forms, the results of which indicate that the diradical character can be controlled by tuning the architectures except for the edge shape and size. Indeed, the introduction of antidot structures and the asymmetrization by D/A substitutions into the open-shell singlet GNFs have been found to significantly change the open-shell characters and thus to enhance and/or to control their NLO properties. These theoretical predictions and molecular design guidelines for highly efficient open-shell singlet NLO systems based on the GNFs will be realized in the near future judging from recent rapid progress in organic syntheses and physical fabrication of zigzag-edged GNFs with several shapes and sizes.

Finally, we provide some future perspectives on the NLO materials built from open-shell nanographenes. Our previous studies on the open-shell nanographenes have almost focused on the properties of single molecules except for a few studies on the dimers and trimers of slipped-stack IDPL [76] and 1D cluster models of hydrogen atoms [77,78]. These preliminary studies have clarified that the longitudinal γ values are more enhanced and show a nonlinear size-dependent γ evolution as compared to usual closed-shell clusters due to the strong covalent-like intermolecular interactions of open-shell singlet clusters [76]. Indeed, a real slipped-stack type crystal of IDPL is found to have much shorter average π–π intermolecular distance (3.137 Å) than the van der Waals radius of carbon atom due to the strong covalent-like intermolecular interaction through their radical sites [52], which shows the possibility of a variety of multiradical states in the solid states composed of open-shell monomers. On the basis of these facts, the investigations of the impact of intermolecular interaction on the open-shell characters and NLO properties for the zigzag-edged GNF clusters are expected to provide very interesting structure–property relationships toward a novel design strategy for highly efficient NLO materials beyond single molecular design. In particular, an extension of the targets from finite-sized GNFs to infinite periodic systems will provide NLO susceptibilities, which can be directly compared to the experimental values of NLO crystals. Considering a report that the spin polarization in ZGNRs are suppressed by edge defects and/or boron/nitrogen impurities [79], we can speculate a realization of infinite-size GNRs with a wide range of open-shell characters, which is a very challenging theme from the viewpoint of real open-shell NLO crystals. Furthermore, theoretical investigations of the correlations of NLO properties with a wide variety of spin states in such meso-/macroscopic open-shell NLO systems will present a basis of multifunctional materials concerning, for example, magnetism and NLO properties, which are mutually controllable by adjusting chemical and physical perturbations on the open-shell characters, charged and spin states of nanographenes. These fascinating investigations will increasingly stimulate the experimentalists and researchers in other fields to focus on the fertile science and engineering fields of open-shell nanographenes.

REFERENCES

1. Neto, A. H. C., Guinea, F., Peres, N. M. R., Novoselov, K. S., and Geim, A. K. 2009. *Reviews of Modern Physics* 81:109–162.
2. Wallace, P. R. 1947. *Physical Review* 71:622–634.
3. Semenoff, G. W. 1984. *Physical Review Letters* 53:2449–2452.
4. Geim, A. K., and Novoselov, K. S. 2007. *Nature Materials* 6:183–191.
5. Bolotin, K. I., Sikes, K. J., Jiang, Z., Klima, M., Fudenberg, G., Hone, J., Kim, P., and Stormer, H. L. 2008. *Solid State Communications* 146:351–355.
6. Novoselov, K. S., Geim, A. K., Morozov, S. V., Jiang, D., Katsnelson, M. I., Grigorieva, I. V., Dubonos, S. V., and Firsov, A. A. 2005. *Nature* 438:197–200.
7. Novoselov, K. S., Jiang, Z., Zhang, Y., Morozov, S. V., Stormer, H. L., Zeitler, U., Maan, J. C., Boebinger, G. S., Kim, P., and Geim, A. K. 2007. *Science* 315:1379.

8. Katsnelson, M. I., Novoselov, K. S., and Geim, A. K. 2006. *Nature Physics* 2:620–625.
9. Tombros, N., Jozsa, C., Popinciuc, M., Jonkman, H. T., and van Wees, B. J. 2007. *Nature* 448:571–574.
10. Son, Y. W., Cohen, M. L., and Louie, S. G. 2006. *Nature* 444:347–349.
11. Lee, C., Wei, X., Kysar, J. W., and Hone, J. 2008. *Science* 321:385–388.
12. Bao, Q., Zhang, H., Wang, Y., Ni, Z., Yan, Y., Shen, Z. X., Loh, K. P., and Tang, D. Y. 2009. *Advanced Functional Materials* 19:3077–3083.
13. Zhang, H., Virally, S., Bao, Q., Ping, L. K., Massar, S., Godbout, N., and Kockaert, P. 2012. *Optics Letters* 37:1856–1858.
14. Nakano, M., Nagai, H., Fukui, H., Yoneda, K., Kishi, R., Takahashi, H., Shimizu, A. et al. 2008. *Chemical Physics Letters* 467:120–125.
15. Nagai, H., Nakano, M., Yoneda, K., Fukui, H., Minami, T., Bonness, S., Kishi, R. et al. 2009. *Chemical Physics Letters* 477:355–359.
16. Yoneda, K., Nakano, M., Kishi, R., Takahashi, H., Shimizu, A., Kubo, T., Kamada, K., Ohta, K., Champagne, B., and Botek, E. 2009. *Chemical Physics Letters* 480:278–283.
17. Nagai, H., Nakano, M., Yoneda, K., Kishi, R., Takahashi, H., Shimizu, A., Kubo, T. et al. 2010. *Chemical Physics Letters* 489:212–218.
18. Yoneda, K., Nakano, M., Fukui, H., Minami, T., Shigeta, Y., Kubo, T., Ohta, K., Botek, E., and Champagne, B. 2011. *Chemistry Physics Chemistry* 12:1697–1707.
19. Yoneda, K., Nakano, M., Inoue, Y., Inui, T., Fukuda, K., Shigeta, Y., Kubo, T., and Champagne, B. 2012. *The Journal of Physical Chemistry C* 116:17787–17795.
20. Yoneda, K., Nakano, M., Fukuda, K., and Champagne, B. 2012. *The Journal of Physical Chemistry Letters* 3:3338–3342.
21. Novoselov, K. S., Geim, A. K., Morozov, S. V., Jiang, D., Zhang, Y., Dubonos, S. V., Grigorieva, V., and Firsov, A. A. 2004. *Science* 306:666–669.
22. Fujita, M., Wakabayashi, K., Nakada, K., and Kusakabe, K. 1996. *Journal of the Physical Society of Japan* 65:1920–1923.
23. Lambert, C. 2011. *Angewandte Chemie International Edition* 50:1756–1758.
24. Chikamatsu, M., Mikami, T., Chisaka, J., Yoshida, Y., Azumi, R., Yase, K., Shimizu, A., Kubo, T., Morita, Y., and Nakasuji, K. 2007. *Applied Physics Letters* 91:043506.
25. Minami, T., and Nakano, M. 2012. *The Journal of Physical Chemistry Letters* 3:145–150.
26. Nakano, M., Kishi, R., Ohta, S., Takahashi, H., Kubo, T., Kamada, K., Ohta, K., Botek, E., and Champagne, B. 2007. *Physical Review Letters* 99:033001.
27. Nakano, M., Kishi, R., Nitta, T., Kubo, T., Nakasuji, K., Kamada, K., Ohta, K., Champagne, B., Botek, E., and Yamaguchi, K. 2005. *The Journal of Physical Chemistry A* 109:885–891.
28. Nakano, M., Kishi, R., Ohta, S., Takebe, A., Takahashi, H., Furukawa, S., Kubo, T. et al. 2006. *The Journal of Chemical Physics* 125:074113.
29. Kamada, K., Ohta, K., Kubo, T., Shimizu, A., Morita, Y., Nakasuji, K., Kishi, R. et al. 2007. *Angewandte Chemie International Edition* 46:3544–3546.
30. Cho, S., Lim, J. M., Hiroto, S., Kim, P., Shinokubo, H., Osuka, A., and Kim, D. 2009. *Journal of the American Chemical Society* 131:6412–6420.
31. Bendikov, M., Duong, H. M., Starkey, K., Houk, K. N., Carter, E. A., and Wudl, F. 2004. *Journal of the American Chemical Society* 126:7416–7417.
32. Bettinger, H. F. 2010. *Pure and Applied Chemistry* 82:905–915.
33. Dias, J. R. 2008. *Chemical Physics Letters* 467:200–203.
34. Hachmann, J., Dorando, J. J., Avilés, M., and Chan, G. K.-L. 2007. *The Journal of Chemical Physics* 127:134309.
35. Konishi, A., Hirao, Y., Nakano, M., Shimizu, A., Botek, E., Champagne, B., Shiomi, D. et al. 2010. *Journal of the American Chemical Society* 132:11021–11023.
36. Li, Y., Heng, W.-K., Lee, B. S., Aratani, N., Zafra, J. L., Bao, N., Lee, R. et al. 2012. *Journal of the American Chemical Society* 134:14913–14922.
37. Clar, E. 1964. *Polycyclic Hydrocarbons*. London: Academic Press.
38. Salem, L., and Rowland, C. 1972. *Angewandte Chemie International Edition* 11:92–111.
39. Yamaguchi, K. 1990. Instability in chemical bonds. In *Self-Consistent Field: Theory and Applications*, eds. Carbo, R. and Klobukowski, M., 727–828. Amsterdam: Elsevier.
40. Nakano, M., Fukui, H., Minami, T., Yoneda, K., Shigeta, Y., Kishi, R., Champagne, B. et al. 2011. *Theoretical Chemistry Accounts* 130:711–724; *Erratum* 725–726.
41. Kishi, R., Bonness, S., Yoneda, K., Takahashi, H., Nakano, M., Botek, E., Champagne, B. et al. 2010. *The Journal of Chemical Physics* 132:094107.
42. Hayes, E. F. and Siu, A. K. Q. 1971. *Journal of the American Chemical Society* 93:2090.
43. Head-Gordon, M. 2003. *Chemical Physics Letters* 372: 508–511.
44. Yamaguchi, K. 1975. *Chemical Physics Letters* 33:330–335.
45. Amos, A. T., and Hall, G. G. 1961. *Proceedings of the Royal Society A* 263:483–493.
46. Iikura, H., Tsuneda, T., Yanai, T., and Hirao, K. 2001. *The Journal of Chemical Physics* 115:3540–3544.
47. Bonness, S., Fukui, H., Yoneda, K., Kishi, R., Champagne, B., Botek, E., and Nakano, M. 2010. *Chemical Physics Letters* 493:195–199.
48. Nakano, M., Minami, T., Fukui, H., Yoneda, K., Shigeta, Y., Kishi, R., Champagne, B., and Botek, E. 2010. *Chemical Physics Letters* 501:140–145.
49. Carmen, J., Calzado, C. J., Cabrero, J., Malrieu, J. P., and Caballol, R. 2002. *The Journal of Chemical Physics* 116:2728–2747.
50. Minami, T., Ito, S., and Nakano, M. 2013. *The Journal of Physical Chemistry A* 117: 2000–2006.
51. Kamada, K., Ohta, K., Shimizu, A., Kubo, T., Kishi, R., Takahashi, H., Botek, E., Champagne, B., and Nakano, M. 2010. *The Journal of Physical Chemistry Letters* 1:937–940.
52. Kubo, T., Shimizu, A., Sakamoto, M., Uruichi, M., Yakushi, K., Nakano, M., Shiomi, D. et al. 2005. *Angewandte Chemie International Edition* 44:6564–6568.
53. Kubo, T., Shimizu, A., Uruichi, M., Yakushi, K., Nakano, M., Shiomi, D., Sato, K., Takui, T., Morita, Y., and Nakasuji, K. 2007. *Organic Letters* 9:81–84.
54. Yang, X., Dou, X., Rouhanipour, A., Zhi, L., Rader, H., and Mu¨llen, K. 2008. *Journal of the American Chemical Society* 130:4216–4217.
55. Cai, J., Ruffieux, P., Jaafar, R., Bieri, M., Braun, T., Blankenburg, S., Muoth, M. et al. 2010. *Nature* 466:470–473.
56. Nakano, M., Kubo, T., Kamada, K., Ohta, K., Kishi, R., Ohta, S., Nakagawa, N. et al. 2006. *Chemical Physics Letters* 418:142–147.
57. Fukui, H., Kishi, R., Minami, T., Nagai, H., Takahashi, H., Kubo, T., Kamada, K. et al. 2008. *The Journal of Physical Chemistry A* 112:8423–8429.
58. Boyd, R. W. 2008. *Nonlinear Optics*, 3rd ed. New York: Academic Press.
59. Franken, P. A., Hill, A. E., Peters, C. W., and Weinreich, G. 1961. *Physical Review Letters* 7:118–119.

60. Göppert-Mayer, M. 1931. *Annalen der Physik* 401:273–294.
61. Kaiser, W. and Garrett, C. G. B. 1961. *Physical Review Letters* 7:229–232.
62. Pathenopoulos, D. A. and Rentzepis, P. M. 1989. *Science* 245:843–845.
63. Zhou, W., Kuebler, S. M., Braun, K. L., Yu, T., Cammack, J. K., Ober, C. K., Perry, J. W., and Marder, S. R. 2002. *Science* 296:1106–1109.
64. Frederiksen, P. K., Jørgensen, M., and Ogilby, P. R. 2001. *Journal of the American Chemical Society* 123:1215–1221.
65. Terenziani, F., Katan, C., Badaeva, E., Tretiak, S., and Blanchard-Desce, M. 2008. *Advanced Materials* 20: 4641–4678.
66. Toto, J. L., Toto, T. T., de Melo, C. P., Kirtman, B., and Robins, K. 1996. *The Journal of Chemical Physics* 104:8586–8592.
67. Limacher, P. A., Mikkelsen, K. V., and Lüthi, H. P. 2009. *The Journal of Chemical Physics* 130:194114.
68. Nakano, M., and Yamaguchi, K. 1993. *Chemical Physics Letters* 206:285–292.
69. Nakano, M., Yoneda, K., Kishi, R., Takahashi, H., Kubo, T., Kamada, K., Ohta, K., Botek, E., and Champagne, B. 2009. *The Journal of Chemical Physics* 131:114316.
70. Cohen, H. D. and Roothaan, C. C. J. 1965. *The Journal of Chemical Physics* 43:S34–S39.
71. Nakano, M., Shigemoto, I., Yamada, S., and Yamaguchi, K. 1995. *The Journal of Chemical Physics* 103:4175–4191.
72. Nakano, M., Nitta, T., Yamaguchi, K., Champagne, B., and Botek, E. 2004. *The Journal of Physical Chemistry A* 108:4105–4111.
73. Nakano, M., Champagne, B., Botek, E., Ohta, K., Kamada, K., and Kubo, T. 2010. *The Journal of Chemical Physics* 133:154302.
74. Nakano, M. and Champagne, B. 2013. *The Journal of Chemical Physics* 138:244306.
75. Nakano, M., Minami, T., Yoneda, K., Muhammad, S., Kishi, R., Shigeta, Y., Kubo, T. et al. 2011. *The Journal of Physical Chemistry Letters* 2:1094–1098.
76. Nakano, M., Takebe, A., Kishi, R., Fukui, H., Minami, T., Kubota, K., Takahashi, H. et al. 2008. *Chemical Physics Letters* 454:97–104.
77. Nakano, M., Takebe, A., Kishi, R., Ohta, S., Nate, M., Kubo, T., Kamada, K. et al. 2006. *Chemical Physics Letters* 432:473–479.
78. Nakano, M., Minami, T., Fukui, H., Kishi, R., Shigeta, Y., and Champagne, B. 2012. *The Journal of Chemical Physics* 136:0243151.
79. Huang, B., Liu, F., Wu, J., Gu, B.-L., and Duan, W. 2008. *Physical Review B* 77:153411.

28 Optical Coupling of Graphene Sheets

Bing Wang and Xiang Zhang

CONTENTS

ABSTRACT

As the chemical potential of graphene is larger than half the energy of exciting photons, graphene could support strongly confined surface plasmon polaritons (SPPs), that is, graphene plasmons (GPs), which are surface waves manifesting an antisymmetric electric field distribution on graphene. Graphene has only one-atom thickness and the properties of GPs are significantly different from that of SPPs observed in metallic thin films. GPs have many significant advantages such as shorter wavelength, lower loss, and more flexible tunability. In this chapter, we focus on the optical coupling of GPs in-between graphene sheets. The double-layer graphene system is considered first, and such a system by means of GP coupling may find important applications in optical splitters, switches, and interferometers. We further investigate the coupling of GPs in monolayer graphene sheet arrays (MGSAs) composed of periodically stacked graphene sheets. As the period of the array decreases to below a critical value, strong coupling of GPs occurs. Otherwise, weak coupling should be observed. The critical period is quantitatively determined by the plasmonic thickness of graphene, referring to the effective mode width of GPs in an individual graphene sheet. Several interesting phenomena such as beam self-splitting and collimating of GPs in MGSAs have been numerically demonstrated and theoretically analyzed. The study provides a new platform to manipulate light propagation on the scale of a deep sub-diffraction limit and could find many promising applications in optoelectronic devices and circuits.

28.1 INTRODUCTION

Owing to the two-dimensional configuration, the band structure of graphene has Dirac points at the Brillouin zone (BZ) corners where the electrons are massless and have very high mobility (Castro Neto et al. 2009). The unique features render graphene a promising material that provides an open platform for demonstrating many phenomena in graphene with great scientific significance and potential applications (Castro Neto et al. 2009; Geim 2009; Das Sarma et al. 2011). Very recently, graphene has stimulated great interest among the nanophotonics community since graphene is found to be able to support surface plasmon polaritons (SPPs) (Hanson 2008a,b; Chen et al. 2012; Fei et al. 2012). The graphene-supported SPPs, conventionally known as graphene plasmons (GPs), are guided modes that can propagate along graphene surface with an extremely strong field confinement (Grigorenko et al. 2012). As the chemical potential (Fermi energy) of graphene is larger than half the energy of exciting photons, GPs could be physically excited by metallic tips or quantum dots. Usually, the chemical potential is up to 1 ~ 2 eV (Efetov and Kim 2010; Chen et al. 2011) and hence, GPs could only be observed in the far-infrared and terahertz range. The GPs have many unique features by comparing with SPPs on metals. First, the effective index of GPs is very high. The value could be approaching to more than 100, nearly two orders larger than that of SPPs on metals. The high effective index also suggests the strong confinement of the electric field of GPs. For example, the GPs in the infrared region ($\lambda = 10\ \mu\text{m}$) have an effective mode width of ~50 nm. That means it can be manipulated deeply below the diffraction limit. Second, the loss of GPs is relatively low since graphene has Dirac electrons with very high mobility at room temperature. The propagation length could reach dozens of wavelengths of the GPs. Finally, the optical properties of graphene can be rapidly tuned by using an external static, electric, or magnetic field and gate voltage through changing the chemical potential of graphene. As a result, the optical configurations composed of graphene could be manipulated in real time other than by chemically doping graphene in advance. Owing to the fantastic characteristics, graphene has been reported to find many great applications in optical modulators (Wang et al. 2014), couplers (Wang et al. 2012; Smirnova et al. 2013), logic gates (Ooi et al. 2014), and Talbot imaging (Fan et al. 2014).

In this chapter, we focus on a very fundamental issue on the optical coupling of graphene. Coupling between waveguides is a key mechanism that contributes to the construction of many important optical elements in integrated optics, such as directional couplers, ring resonators, and optical interferometers. Adjacent waveguides arranged closely could exchange energy within the coupling length in terms of the tunneling effect of photons. Differing from coupling of symmetric modes between dielectric waveguides, the GPs in graphene are antisymmetric. The antisymmetric mode leads to negative coupling, which suggests that the propagation direction of the phase is opposite to that of the energy (Efremidis et al. 2010). In addition, the coupling of GPs in monolayer graphene sheet arrays (MGSAs) is also investigated, which follows the study of dielectric and metallic waveguide arrays. Owing to the unique properties of graphene, we find very interesting features that are quite distinct from that of dielectric and metallic waveguide arrays. Strong coupling of GPs is observed in the MGSAs and they experience self-splitting and collimating effect. The splitting angle could be modulated by electronically or magnetically tuning graphene. We confirm that the strong coupling is readily available in graphene-constituting waveguide arrays.

The chapter is arranged as follows. In Section 28.2, we discuss the fundamental plasmonic properties of graphene, especially the GPs in single-layer graphene. We also introduce the numerical approach to model the propagation of GPs in graphene. In Section 28.3, the coupling of GPs in double-layer graphene is considered. We numerically demonstrate some practical applications in optical nanodevices based on graphene coupling. Then, we focus on the coupling properties of GPs in MGSAs in Section 28.4. Both strong and weak couplings are concerned in our study. In what follows, we compare the graphene sheet arrays and the thin metal film arrays in the same section. The conclusions are finally drawn in Section 28.5.

28.2 PLASMONS IN A SINGLE LAYER OF GRAPHENE SHEET

As a two-dimensional material, the band structure of graphene seems like two cones connected tip to tip at the corners of the BZ. The connection points are known as Dirac points. In the vicinity of Dirac points, the dispersion relation of graphene is linear and the electrons turn to Dirac electrons with their effective masses approaching zero. Dirac electrons have high direct current (DC) mobility and transport quite fast on graphene surface. When the chemical potential is above the Dirac point and larger than half the energy of incident photons, graphene behaves like a metal. In this case, the intraband transitions dominate. The chemical potential is determined by the electron density in graphene, which can be changed by chemical doping or an external field. Analogous to a metal thin film, the transverse magnetic (TM) polarized SPPs could be supported by graphene, that is, the GPs. It should be mentioned that there are two kinds of SPPs in a metal thin film, the symmetric long-range SPPs and the antisymmetric short-range SPPs. Owing to the ultrasmall thickness of graphene, the corresponding long-range SPPs are nearly free plane waves and cannot be concentrated on graphene. Only the antisymmetric SPPs could be well confined. The latter are exactly the GPs that we would intensively study in this chapter.

28.2.1 Surface Conductivity of Graphene

The surface conductivity of graphene, through which the optical response is characterized, is theoretically given by the Kubo formula (Falkovsky and Pershoguba 2007; Hanson 2008a,b; Chen and Alù 2011)

$$\sigma_g = -\frac{ie^2(\omega + i\tau^{-1})}{\pi\hbar^2}\left[\int_{-\infty}^{\infty} \frac{|E|}{(\omega + i\tau^{-1})^2}\frac{\partial f_d(E)}{\partial E}dE - \int_{0}^{\infty} \frac{f_d(-E) - f_d(E)}{(\omega + i\tau^{-1})^2 - 4(E/\hbar)^2}dE\right], \tag{28.1}$$

where $f_d(E) = 1/(1 - \exp[(E - \mu_c)/(k_B T)])$ is the Fermi–Dirac distribution, E is the energy, e is the unit charge of electrons, $\hbar$ and k_B are the reduced Planck constant and Boltzmann constant, μ_c is the chemical potential, and τ is the momentum relaxation time of electrons. The formula is composed of two items, $\sigma_g = \sigma_{intra} + \sigma_{inter}$, corresponding to the intraband and interband transitions of electrons, respectively. On the condition that $\hbar\omega$, $|\mu_c| \gg k_B T$, the intraband and interband transition contributions can be approximately given by (Hanson 2008a,b)

$$\sigma_{intra} = i\frac{e^2 k_B T}{\pi\hbar^2(\omega + i\tau^{-1})}\left[\frac{\mu_c}{k_B T} + 2\ln\left(\exp\left(-\frac{\mu_c}{k_B T}\right)\right) + 1\right], \tag{28.2}$$

and

$$\sigma_{inter} = i\frac{e^2}{4\pi\hbar}\ln\left[\frac{2|\mu_c| - \hbar(\omega + i\tau^{-1})}{2|\mu_c| + \hbar(\omega + i\tau^{-1})}\right]. \tag{28.3}$$

In the infrared region, we have $\omega \gg \tau^{-1}$. Additionally, the interband contribution is effectively suppressed when $|\mu_c| \gg \hbar\omega/2$, while the intraband transition dominates. The surface conductivity is therefore governed by Equation 28.2. Furthermore, as $|\mu_c| \gg k_B T$, the surface conductivity simplifies to a Drude-like model (Hanson 2008a,b)

$$\sigma_g = \frac{ie^2 \mu_c}{\pi \hbar^2 (\omega + i\tau^{-1})}. \quad (28.4)$$

The model suggests that the surface conductivity is nearly a pure imaginary number at a low temperature and high chemical potential. The real part of σ_g, determined by τ, tends to yield Ohm loss in graphene. At room temperature ($T = 300$ K), if not mentioned especially in this chapter, the momentum relaxation time is assumed as $\tau = 0.5$ ps. The value corresponds to a mean-free distance of 500 nm, which complies with the experimental data and could fairly feature the transport loss of graphene (Chen and Alù 2011; Wang et al. 2012).

28.2.2 Plasmons in a Monolayer Graphene

We consider a single layer of graphene embedded in the dielectric with a permittivity of ε_d. For TM waves, the magnetic field around graphene could be assumed as

$$H_y = \begin{cases} A^+ \exp(-\kappa x) & x > 0 \\ A^- \exp(\kappa x) & x < 0 \end{cases}, \quad (28.5)$$

where $\kappa = (k_{GP}^2 - \varepsilon_d k_0^2)^{-1/2}$ with k_{GP} being the wave vector of GPs and $k_0 = 2\pi/\lambda$. λ is the wavelength of light in air. According to Maxwell's equations, we have $E_z = i\eta_0/(\varepsilon_d k_0)\partial H_y/\partial x$. Thus, the tangential electric field can be written as

$$E_z = \begin{cases} -i\kappa\eta_0/(\varepsilon_d k_0) A^+ \exp(-\kappa x) & x > 0 \\ ik\eta_0/(\varepsilon_d k_0) A^- \exp(\kappa x) & x < 0 \end{cases}. \quad (28.6)$$

In terms of the boundary conditions (Hanson 2008a,b), $E_z^+ = E_z^-$ and $H_y^+ - H_y^- = \sigma_g E_z$ at $x = 0$, we can figure out the dispersion relation of GPs in a single-layer graphene, where the plus and minus denote the two sides of graphene. The wave vector of the GPs is finally given by

$$k_{GP} = k_0 \sqrt{\varepsilon_d - \left(\frac{2\varepsilon_d}{\eta_0 \sigma_g}\right)^2}. \quad (28.7)$$

The decay length of GPs on both sides of the dielectric is given by $\delta = \kappa^{-1} = \eta_0\sigma_g/(2i\varepsilon_d k_0)$. In consequence, we could define a plasmonic thickness of graphene (Wang et al. 2012)

$$\xi \equiv 2\delta = \eta_0 \sigma_g/(i\varepsilon_d k_0). \quad (28.8)$$

The quantity of ξ is exactly the effective mode width of GPs in graphene. We will show later that it plays an important role in optical coupling of graphene. By using ξ, we can rewrite Equation 28.7 as

$$k_{GP} = k_0 \sqrt{\varepsilon_d + \left(\frac{2}{k_0 \xi}\right)^2}. \quad (28.9)$$

As k_{GP} is much larger than k_0, we approximately have $k_{GP} \approx 2/\xi$. Thus, the wavelength of the GPs is given by $\lambda_{GP} = 2\pi/k_{GP} \approx \pi\xi$. We would provide some typical parameters that give one an explicit recognition of GPs. At $\lambda = 10$ μm and room temperature, the surface conductivity of graphene $\sigma_g = 0.0012 + i0.0768$ mS, which is obtained by using Equations 28.2 and 28.3. Then, we can figure out the wave number of GPs $k_{GP} = 43.43 + 0.68i$ μm^{-1}. The effective index of GPs $n_{GP} \equiv k_{GP}/k_0 = 69.12 + 1.08i$. The plasmonic thickness of graphene is $\xi = 46$ nm. The propagation distance of GPs is given by $L_{GP} = [2\text{Im}(k_{GP})]^{-1}$. Figure 28.1a depicts the effective index of GPs and the propagation distance in the unit of GP wavelength (L_{GP}/λ_{GP}) as functions of the wavelength of exiting

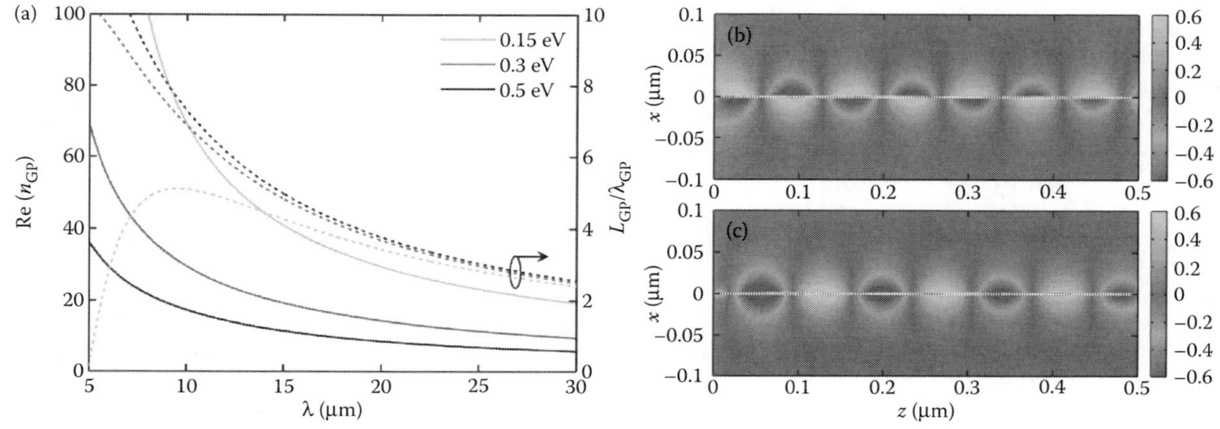

FIGURE 28.1 GPs in a single-layer graphene. (a) An effective index of GPs (solid) and propagation distance (dashed) as functions of the incident wavelength in air for $\mu_c = 0.15$, 0.3, and 0.5 eV. (b) E_x and (c) E_z distributions of GPs in a single-layer graphene.

photons in air. Generally, as the wavelength decreases, the effective index decreases, resulting in a weaker field confinement in graphene. The field confinement could be improved by increasing the chemical potential of graphene. Also, a higher chemical potential will benefit to suppressing the propagation loss and GPs can propagate a longer distance.

In most of the numerical methods based on the first principle such as finite-difference time-domain (FDTD) (Taflove and Hagness 2000) and finite-difference frequency-domain (FDFD) (Brongersma and Kik 2007; Zhang and Li 2009) methods, the objects should be composed of bulk materials. With regard to graphene, the thickness approaching to zero tends to arouse numerical divergence. Although special treatment by modifying the boundary conditions can avoid this problem, the coding becomes complicated and it is not applied for packaged commercial software. To model the propagation of GPs, one usually defines an equivalent bulk permittivity for graphene. In this manner, graphene is equivalent to a very thin metal film with a thickness of Δ, the conductivity of which is given by σ_g/Δ. Thus, the equivalent permittivity of graphene could be written as (Vakil and Engheta 2011)

$$\varepsilon_{g,eq} = 1 + i\sigma_g\eta_0/(k_0\Delta). \tag{28.10}$$

It should be mentioned that the equivalent model is only valid when Δ is very small, typically on the order of ~1 nm (Vakil and Engheta 2011). In comparison, the real thickness of graphene is 0.33 nm. Such a small length scale is not suited for time-domain simulations because the time step has to be very small, resulting in very time-consuming computation. Figure 28.1b and c illustrate the electric field distribution of GPs propagating in a single-layer graphene by using the FDFD method. One sees that the distributions of E_x are antisymmetric about the graphene surface, while E_z is symmetric. The field distribution is quite different for long-range SPPs on metals, for which E_x is symmetric and E_z is antisymmetric (Berini 2009). Since the loss is determined by the longitudinal electric field, the antisymmetric distribution of E_z could benefit to the mode propagation. That is the reason why long-range SPPs on metals have a longer propagation length than GPs (Tassin et al. 2012). However, the GPs beat SPPs on metals at the mode confinement. It should be mentioned that the transverse electric (TE) waves can also be guided in monolayer graphene (Vakil and Engheta 2011). However, the field confinement of the TE modes is very weak and they are out of the scope of this study.

28.3 OPTICAL COUPLING OF TWO GRAPHENE SHEETS

Given a coupling system consisting of two graphene sheets separated by a dielectric layer as shown in Figure 28.2a, we first investigate the eigenmodes supported in the system. According to the traditional waveguide theory, the magnetic field in different regions of the double-layer graphene system can be written as

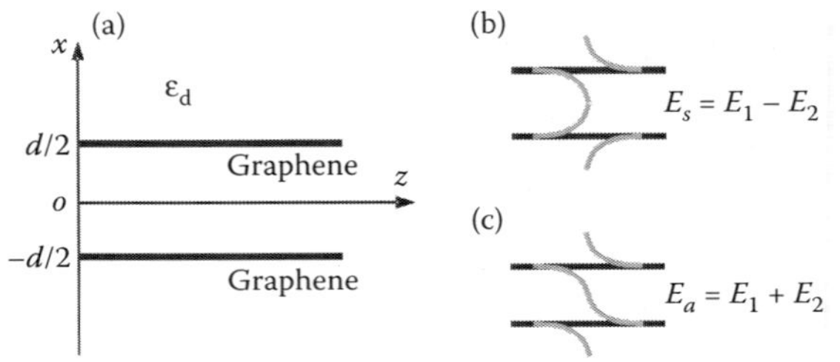

FIGURE 28.2 Coupling of GPs in the double-layer graphene. (a) Schematic diagram of the coupled double-layer graphene system. (b) Symmetric mode formed by an odd superposition and (c) antisymmetric mode formed by an even superposition of individual GPs in the single-layer graphene.

$$H_y = \begin{cases} A\exp[-\kappa(x-d/2)] & x > d/2 \\ B\exp[\kappa(x-d/2)] + C\exp[-\kappa(x+d/2)] & |x| < d/2 \\ D\exp[\kappa(x+d/2)] & x < -d/2 \end{cases}, \tag{28.11}$$

where $\kappa = (\beta^2 - \varepsilon_d k_0^2)^{1/2}$ with β being the propagation constant of the eigenmodes. A–D are mode amplitudes in different regions to be determined. The electric field along graphene is thus given by

$$E_z = \begin{cases} -i\eta_0\kappa(\varepsilon_d k_0)^{-1} A\exp[-\kappa(x-d/2)] & x > d/2 \\ i\eta_0\kappa(\varepsilon_d k_0)^{-1}\{B\exp[\kappa(x-d/2)] - C\exp[-\kappa(x+d/2)]\} & |x| < d/2 \\ i\eta_0\kappa(\varepsilon_d k_0)^{-1} D\exp[\kappa(x+d/2)] & x < -d/2 \end{cases}. \tag{28.12}$$

By matching the boundary conditions and evaluating the limit as Δ approaches to zero, we obtain the dispersion relation of the eigenmodes

$$[1 \pm \exp(-\kappa d)]\kappa\xi = 2, \tag{28.13}$$

This equation has two solutions for κ and thus for β, corresponding respectively to the symmetric and antisymmetric modes in the double-layer graphene system, as shown in Figure 28.2b and c. We denote the propagation constant of the two modes by β_s and β_a, respectively. In case the space between the two graphene sheets is large enough, the propagation constants of the symmetric and antisymmetric modes converge at the wave vector of GPs in a single-layer graphene given by Equation 28.7.

As d is large, the coupling between the graphene sheets is weak. Assuming $\beta_s = k_{GP} + \Delta\beta_s$ and $\beta_a = k_{GP} + \Delta\beta_a$, where $\Delta\beta_s$ and $\Delta\beta_a$ are small quantities with respect to k_{GP}, we approximately have

$$\beta_s \approx k_{GP} + \frac{2u}{2du - \xi(u-1)}, \quad \beta_a \approx k_{GP} + \frac{2u}{2du - \xi(u+1)}, \tag{28.14}$$

where $k_{GP} \approx 2/\xi$ and $u = \exp(-2d/\xi)$. The graphene sheets are simply assumed to be standing in air and $\varepsilon_d = 1$. As d increases, both β_s and β_a converge at k_{GP}. Because the field of GPs supported by a single-layer graphene is antisymmetric about the graphene plane, the symmetric (E_s) and antisymmetric (E_a) modes in the double-layer graphene system originate from the odd and even superpositions of the eigenmodes (E_1 and E_2) in individual graphene sheets, as shown in Figure 28.2b and c. As a result, we can derive the coupling coefficient between individual GP modes as (Okamoto 2006)

$$C_g = (\beta_a - \beta_s)/2. \quad (28.15)$$

It should be mentioned that the coupling coefficient equals $(\beta_s - \beta_a)/2$ for conventional dielectric waveguide coupling (Okamoto 2006). This is because the fundamental mode in the individual dielectric waveguide is symmetric. As the space between graphene sheets is much bigger than the wavelength of GPs, according to Equation 28.14, the coupling coefficient can further simplify to

$$C_g \approx -2/\xi \exp(-\kappa d). \quad (28.16)$$

Considering that $\beta_s > \beta_a$, the coupling coefficient has to be negative.

To directly observe the coupling process, we also perform numerical simulations to visualize the field distribution. Here, we employ FDFD method to take over the simulation. By comparing with FDTD method, which is time consuming because of requiring a very small space grid for ultrathin films and therefore an extremely short time step, FDFD method only needs spatial discretization. Graphene is treated as a metal with a permittivity of $\varepsilon_{g,eq}(\Delta)$, which is thickness dependent. The equivalent thickness of graphene is set as $\Delta = 1$ nm, at which the equivalent permittivity $\varepsilon_{g,eq} = -45.05 + 0.72i$. The minimum mesh size is set to be $\Delta/10$. The electric fields for the symmetric and antisymmetric modes are illustrated in Figure 28.3a. By extracting the spatial oscillation period of the field and the damping rate, we can obtain the propagation constant and loss of GPs in the coupled double-layer graphene. The dispersion relations calculated by FDFD method are also shown in Figure 28.3b. Both the wave vectors of the modes and coupling coefficient are shown in this figure. They agree well with the analytical results obtained by Equations 28.14 and 28.16. We ensure the numerical accuracy by resetting $\Delta = 0.2$ nm and a deviation of the propagation constant up to 2% is obtained.

The coupling of graphene sheets may find significant applications in optoelectronic devices. A typical example is the directional coupler. As two graphene sheets stand together, the energy initially input from one graphene sheet will totally couple to the other after a certain distance. The shortest distance for the total energy transfer is known as the coupling length. For a dual-waveguide system, the coupling length is given by $L_c = \pi/(2|C_g|)$ (Wang and Wang 2004). At $d = 50$ nm, for example, $L_c = 313.1$ nm. This value is much smaller than the wavelength of $\lambda = 10$ μm, which means such a coupler consisting of graphene

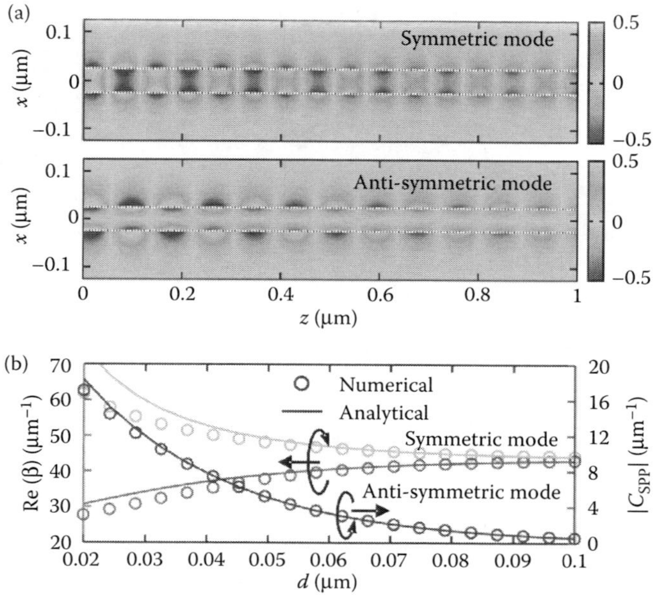

FIGURE 28.3 (a) Electric field (E_x) distribution of symmetric and antisymmetric modes in the double-layer graphene. (b) Propagation constants of the symmetric and antisymmetric modes in the double-layer graphene system, together with the coupling coefficient of GPs between the graphene sheets. (Reprinted with permission from Wang, B. et al. Optical coupling of surface plasmons between graphene sheets. *Applied Physics Letters* 100 (13):131111. Copyright 2012, American Institute of Physics.)

could work deeply below the diffraction limit. If inserting a graphene sheet into the center of two parallel graphenes with an overlapping distance of L_c, a coupling-type GP splitter is constructed. In such a three-waveguide system, the coupling length should be $L_c = \pi/(2\sqrt{2}|C_g|) = 221.4$ nm. Figure 28.4a shows the field intensity in the splitter structure and the field is evenly split into two output graphene sheets. There is no interference at the input, showing that the insertion loss could be completely avoided. The field distribution is also shown in Figure 28.4b. Heeding the coupling region between adjacent graphene sheets, the phase in the arms is always advanced to the input graphene, revealing the negative coupling between graphene. As for negative coupling, the propagation direction of energy is opposite to that of the phase (Verhagen et al. 2010).

Considering that graphene is tunable by an electric or magnetic field, the coupling effect is easily controlled by an external static, electric, or magnetic field. By using the three-waveguide system, we can construct an optical switch. Figure 28.4c and d illustrate the field distribution in the switch structure. As the chemical potential of graphene is turned to be lower than half the photon energy, graphene could not support GPs and the coupling will not happen. By tuning the chemical potential to be 0.05 eV at either graphene arms, the signal carried by GPs will come out at the other graphene sheets. Actually, in this case, the GPs coupling occurs in-between two graphene sheets and the coupling length is given by $L_c = 313.1$ nm.

Another interesting application is the Mach–Zehnder (M–Z) interferometer. The interferometer is composed of the

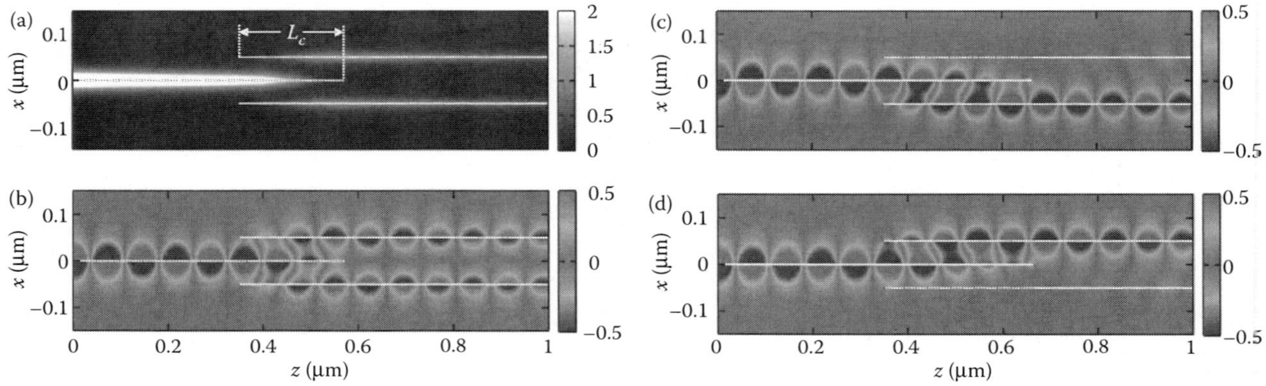

FIGURE 28.4 Optical waveguide splitter based on coupling of GPs between graphene. (a) Intensity ($|E|^2$) distributions of GPs in the splitter. (b) Electric field (E_x) distribution in the splitter. (c) Electric field (E_x) distribution of GPs when the gate voltage is applied on the upper arm of the splitter. (d) Electric field (E_x) distribution of GPs when the bottom arm is biased. (Reprinted with permission from Wang, B. et al. Optical coupling of surface plasmons between graphene sheets. *Applied Physics Letters* 100 (13):131111. Copyright 2012, American Institute of Physics.)

input and the output graphene sheets and two graphene arms, as shown in Figure 28.5a. The arms have an identical overlap distance of 221.4 nm with the input and output graphene sheets. That means the input GPs can be evenly coupled to the arms. The length of the arms except for the overlap region is given by $L_a = 200$ nm. By modifying the chemical potential of either graphene arm, the output field intensity can be manipulated accordingly. As the chemical potential of one graphene arm varies from μ_c to μ_c', the wave vector of GPs changes from k_{GP} to k_{GP}'. The phase difference due to the variation of the chemical potential can be written as

$$\Delta\varphi = \mathrm{Re}(\Delta k_{GP})L_a, \tag{28.17}$$

where $\Delta k_{GP} = k_{GP}' - k_{GP}$. The electric field distribution of GPs in the interferometer for $\mu_c' = 0.125$ eV is illustrated in Figure 28.5b. The output power is zero. That means the GPs in the two arms are out of phase, that is, $\Delta\varphi = \pi$. The transmission and phase difference as functions of the chemical potential are plotted in Figure 28.5c. The maximum of transmission locates at $\mu_c' = 0.112$, 0.15, and 0.4 eV, corresponding to $\Delta\varphi = 2m\pi$ where $m = -1$, 0, and 1 when GPs are in phase in the two arms. The minimum of transmission locates at $\mu_c' = 0.125$ and 0.2 eV, corresponding to $\Delta\varphi = (2m + 1)\pi$ with $m = -1$, 0 when GPs are out of phase in the two arms. Note that only ~25% of the total energy can be transmitted through the structure due to the intrinsic loss of graphene. By using the Drude model, we approximately have $k_{GP} \approx 2/\xi$, which is inversely proportional to μ_c. Therefore, the propagation loss of GPs can be reduced by increasing the chemical potential of graphene. It is reflected from the increasing transmission maximum as μ_c' increases. In addition, a remarkable wide band of high transmission appears from $\mu_c' = 0.25$ to 0.45 eV, where the phase difference of GPs in the two arms is less sensitive to the chemical potential and almost fixed at -2π.

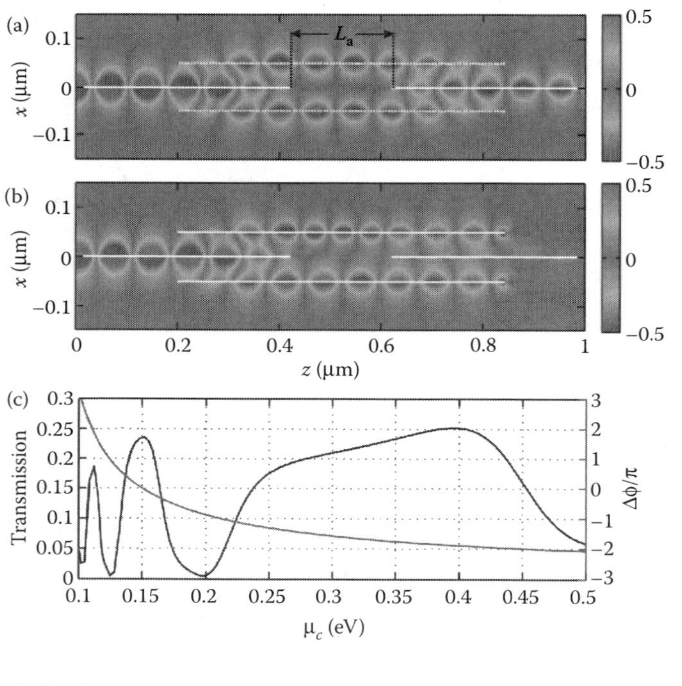

FIGURE 28.5 Optical M–Z interferometer based on coupling of GPs between graphene. (a) Electric field (E_x) distribution of GPs in a graphene interferometer. (b) Electric field (E_x) distribution in the interferometer when a voltage is applied on the upper arm. (c) Optical transmission of the GPs passing through the graphene interferometer as the chemical potential of graphene varies. The gray curve represents the phase difference between the two arms. (Reprinted with permission from Wang, B. et al. Optical coupling of surface plasmons between graphene sheets. *Applied Physics Letters* 100 (13):131111. Copyright 2012, American Institute of Physics.)

28.4 OPTICAL COUPLING OF GPs IN GRAPHENE SHEET ARRAYS

A waveguide array constituting a discrete optical system manifests interesting features in managing light diffraction (Christodoulides et al. 2003). The dielectric waveguide arrays show many interesting effects in diffraction-free optical transmission (Pertsch et al. 2002), optical solitons (Eisenberg et al. 1998; Ablowitz and Musslimani 2001), and diffraction management (Eisenberg et al. 2000). On the other hand, the metallic waveguide arrays also attract much attention because they can realize negative refraction (Fan et al. 2006; Liu et al. 2007; Verslegers et al. 2009; Verhagen et al. 2010). Following the array structures composed of dielectrics and metals, here, we consider MGSAs (Wang et al. 2012). Differing from dielectric and metallic waveguide arrays, graphene is a two-dimensional material and the eigenmode in single-layer graphene is antisymmetric. The graphene array structures have unique properties that could not be observed in conventional dielectric and metallic waveguide arrays.

28.4.1 Dispersion Relation

Let us consider an MGSA composed of periodically stacked graphene in a dielectric medium, as shown in Figure 28.6a. The period of graphene sheets is d and the dielectric constant of the medium is denoted by ε_d. In the region $-d < x < d$, the magnetic field of TM waves can be written as

$$H_y = \begin{cases} A^+ \exp[-\kappa(x+d)] + A^- \exp(\kappa x) & -d < x < 0 \\ B^+ \exp(-\kappa x) + B^- \exp[\kappa(x-d)] & 0 < x < d \end{cases}, \quad (28.18)$$

where $\kappa = \sqrt{k_z^2 - \varepsilon_d k_0^2}$ with k_z being the wave vector of SPPs in the z direction, and $A^\pm$ and $B^\pm$ represent the amplitudes of GPs in-between graphene. They abide by the Bloch theorem (Kittel 2005) in the periodic structure, that is, $[B^+, B^-]^T = [A^+, A^-]^T \exp(ik_x d)$, where k_x is the Bloch wave vector along the x direction. The tangential electric field E_z can be derived from Maxwell's equations

$$E_z = \begin{cases} \eta_0 \kappa (ik_0)^{-1}\{-A^+ \exp[-\kappa(x+d)] + A^- \exp(\kappa x)\} \\ -d < x < 0 \\ \eta_0 \kappa (ik_0)^{-1}\{-B^+ \exp(-\kappa x) + B^- \exp[\kappa(x-d)]\} \\ 0 < x < d \end{cases}. \quad (28.19)$$

The boundary conditions are $E_z^+ = E_z^-$ and $H_y^+ - H_y^- = \sigma_g E_z$ at $x = 0$. Finally, we arrive at the dispersion relation of the collective GPs in the array

$$\cos(\varphi) = \cosh(\kappa d) - \frac{\kappa \xi}{2} \sinh(\kappa d), \quad (28.20)$$

where $\varphi = k_x d$ is known as the Bloch momentum (Pertsch et al. 2002; Christodoulides et al. 2003) and ξ is the plasmonic thickness of graphene as defined previously. If d is large enough, k_z reduces to the propagation constant of SPPs in a single-layer graphene, indicating there is no coupling between graphene sheets. As the frequency is fixed, we can deduce the diffraction relation $k_z(\varphi)$ from Equation 28.20, referring to the isofrequency counter at a specific frequency (Verhagen et al. 2010). Figure 28.6b depicts the diffraction curve of the collective GP as the period of the MGSA varies. The Bloch vectors only in the first BZ are considered. Surprisingly, there is an evident distinction of the curves between the regions of $d < \xi$ and $d \gg \xi$. When $d \gg \xi$, the diffraction curves are hyperbolic and the wave vectors of the collective GPs are close to that of GPs in a single-layer graphene, indicating the weak interaction between individual GPs. As far as the region of $d < \xi$ is concerned, the wave vectors of the collective GPs vary dramatically with the Bloch momentum and could be very different from the individual GPs. That means the interplay of individual GPs is very strong and the region of $d < \xi$ refers to a strong coupling region.

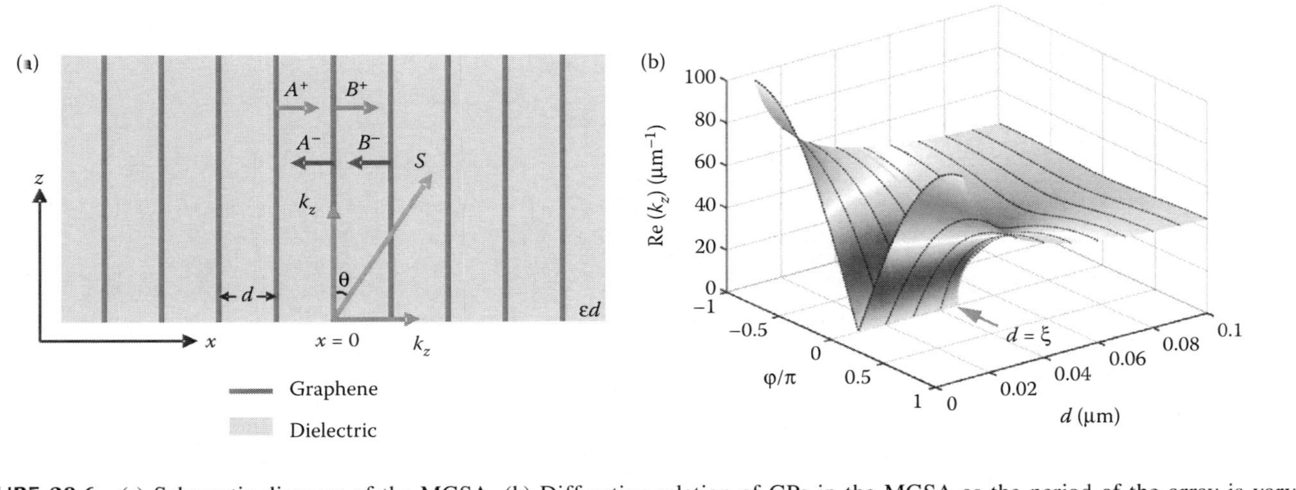

FIGURE 28.6 (a) Schematic diagram of the MGSA. (b) Diffraction relation of GPs in the MGSA as the period of the array is varying. (Reprinted with permission from Wang, B. et al. Strong coupling of surface plasmon polaritons in monolayer graphene sheet arrays. *Physical Review Letters* 109 (7):073901. Copyright 2012 by the American Physical Society.)

28.4.2 Weak Coupling

We first concern the weak coupling of GPs in the MGSA. When the period of the MGSA is large ($d \gg \xi$), the coupling between adjacent graphene sheets is weak. By using the tight-binding approximation (TBA), the dispersion relation of GPs in the MGSA can be written as (Verslegers et al. 2009)

$$k_z = \frac{\beta_e + \beta_o}{2} + \frac{\beta_e - \beta_o}{2}\cos(\varphi), \tag{28.21}$$

where β_e and β_o are the wave number of the even and odd collective GP modes, corresponding to the solutions of k_z in Equation 28.20 for $\varphi = 0$ and $\varphi = \pi$, respectively. Assuming that $\beta_e = k_{GP} + \Delta\beta_e$ and $\beta_o = k_{GP} + \Delta\beta_o$, we can obtain

$$\beta_e \approx \frac{2}{\xi} + \frac{4u}{4du - \xi}, \quad \beta_o \approx \frac{2}{\xi} + \frac{4u}{4du + \xi}, \tag{28.22}$$

where $u = \exp(-2d/\xi)$. Owing to the antisymmetric field distribution of GPs in a single-layer graphene, the even (odd) overlapping of GPs yields the antisymmetric (symmetric) modes in the array. If the period d keeps increasing, both β_e and β_o are approaching $k_{GP} \approx 2/\xi$. In this case, the diffraction relation can be simplified to (Christodoulides and Joseph 1988)

$$k_z = k_{GP} + 2C_g\cos(\varphi), \tag{28.23}$$

where $C_g = (\beta_e - \beta_o)/4$ being the coupling coefficient of GPs in-between adjacent graphene sheets. Note that the symmetric mode has a larger propagation constant than the antisymmetric mode. As a result, the coupling coefficient is negative. The coefficient can also be derived from Equation 28.15. It should be mentioned that even and odd modes here in the array system are completely different from those in dual-waveguide coupling system. The diffraction relation for weak coupling of SPPs in the MGSA is plotted in Figure 28.7a at different periods of the array, where both the rigorous data obtained through Equation 28.20 and the TBA results of Equation 28.21 are shown. As the period changes from $d = 80$ to 100 nm and 120 nm, the variation of the wave vector of collective GPs with φ becomes slight gradually. The wave vector k_z eventually shows independence on φ and converges to k_{GP}. As the period increases, moreover, the coincidence of the rigorous data and the TBA results is improved. The reason is because the TBA has its accuracy much higher in the weaker coupling region.

Now, we explore the propagation behaviors of GPs in the MGSA. The intensity distributions of GPs under in-phase ($\varphi = 0$) and out-of-phase ($\varphi = \pi/2$) illuminations are illustrated in Figure 28.7b and c for $d = 80$ nm. At $\varphi = 0$, the beam tends to spread out in a wide direction range. At $\varphi = \pi/2$, the beam of GPs tilts toward the left with an angle $\theta = \text{atan}(2C_g d)$, which is determined by the direction perpendicular to the diffraction curve (Eisenberg et al. 2000). The negative angle is due to $C_g < 0$. In addition, although the loss is large, the beam experiences good collimation due to the diffraction coefficient $\delta = 0$ at $\varphi = \pi/2$. In Figure 28.8, we plot the output field profile at a distance of $L_z = 1$ μm as the incident phase difference of GPs between adjacent graphene sheets varies. The center of the beam spot at the output is determined by

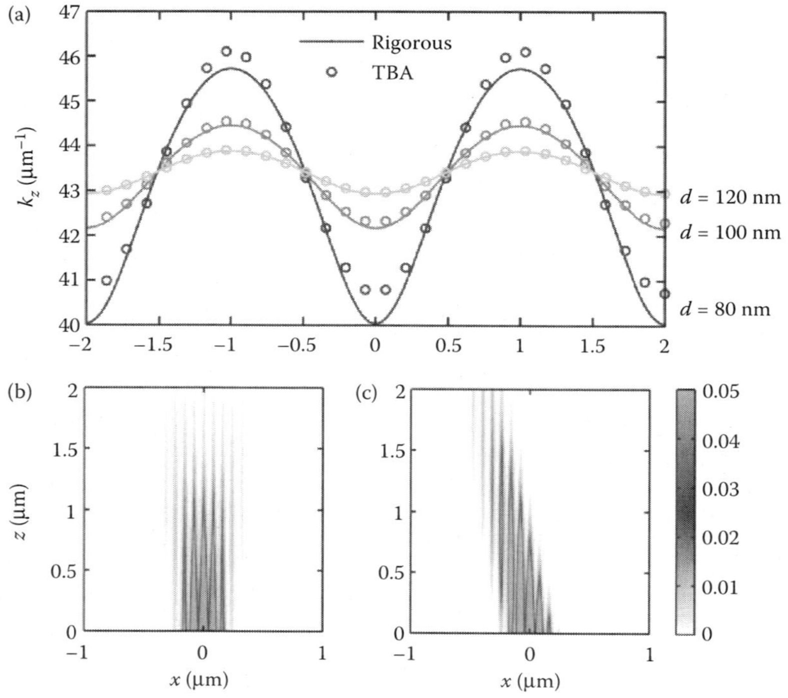

FIGURE 28.7 Weak coupling of GPs in the MGSA. (a) Diffraction relation for a different period in the weak-coupling region. (b) Intensity ($|E|^2$) distribution of GPs in the MGSA for in-phase illumination ($\varphi_0 = 0$). (c) Intensity distribution for the initial phase difference $\varphi_0 = \pi/2$.

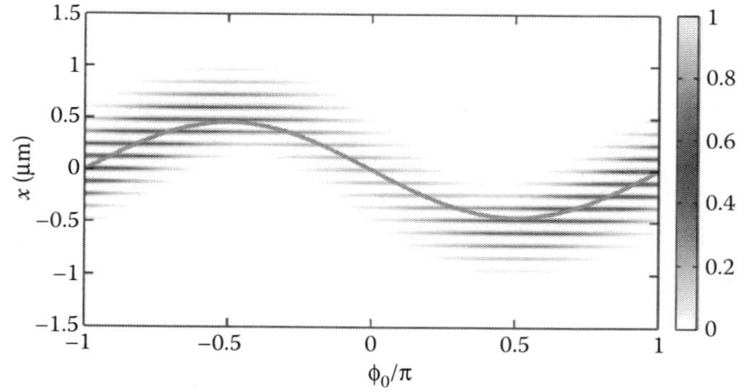

FIGURE 28.8 Profile of a normalized output intensity in the MGSA with $d = 80$ nm when the initial phase difference of incident GPs between adjacent graphene is varying. The curve of analytical prediction represents the center position of the output spot.

$x = 2C_g dL_z \sin(\varphi)$. Apart from the negative coupling, the diffraction behaviors are analogous to that in dielectric optical lattices (Pertsch et al. 2002). The conception of diffraction management initially developed for dielectric optical lattices (Eisenberg et al. 2000) can also be applied for MGSAs. The latter might be more interesting due to the flexible modulation of graphene optical properties.

28.4.3 Strong Coupling

There should be new physics unexposed in the region of $d < \xi$. The diffraction curve in this region is thoroughly different from the hyperbolic curve in the weak-coupling region. We plot the diffraction curve for $d = 20$ nm, as shown in Figure 28.9a. At the BZ center, the curve is almost linear and the wave vector of the collective GPs is very small. According to Equation 28.20, at the BZ center, the diffraction relation reduces to

$$k_z^2 - \varphi^2[d(\xi - d)]^{-1} = \varepsilon_d k_0^2. \quad (28.24)$$

Note that $k_z = k_0$ at $\varphi = 0$. In case $\varphi \gg k_0[\varepsilon_d d(\xi - d)]^{1/2}$ but φ still locates near the BZ center, the relation becomes linear as $k_z = [d(\xi - d)]^{-1/2}|\varphi|$, which is exactly the asymptotes of the hyperbolic curve governed by Equation 28.24, as depicted in Figure 28.9a. As the individual GPs in several graphene sheets are excited at the input, they will interplay to form the collective GPs in the MGSA. The excitation of individual GPs can be realized by putting quantum dots at the end of the graphene sheets (Nikitin et al. 2011). The scattered field by the quantum dots may couple to GPs in the graphene sheets. The amplitude of incident GPs follows a Gaussian distribution with a width of W_0. The phase difference of GPs between

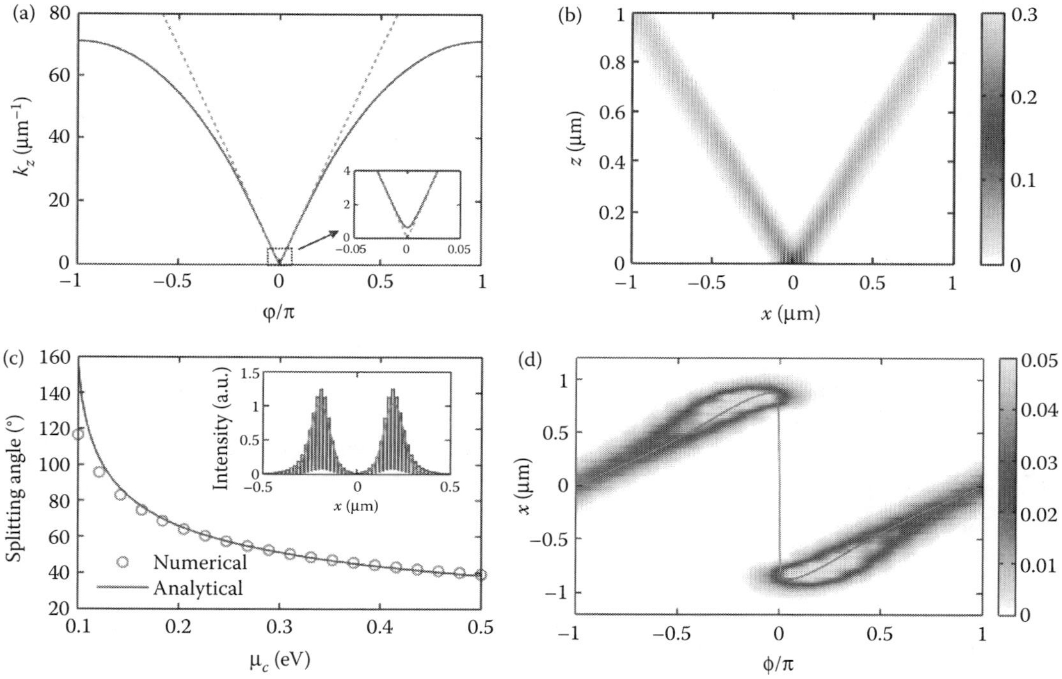

FIGURE 28.9 Strong coupling of GPs in the MGSA. (a) Diffraction curve of GPs in the MGSA for $d = 20$ nm. The red-dotted lines indicate the asymptotes of the curve in the vicinity of the BZ center. (b) Intensity distribution of GPs for in-phase illumination. (c) Splitting angle as a function of the chemical potential of graphene. The inset shows the intensity distribution of GPs at a distance of 0.5 μm in the MGSA. (d) Intensity distribution of GPs at the output of MGSA as the phase difference of incident GPs between adjacent graphene is changing. The curve represents the analytical prediction for the positions of the output spot. (Reprinted with permission from Wang, B. et al. Strong coupling of surface plasmon polaritons in monolayer graphene sheet arrays. *Physical Review Letters* 109 (7):073901. Copyright 2012 by the American Physical Society.)

adjacent graphene sheets is φ_0. The electric field distribution in the MGSA for $\varphi_0 = 0$ is illustrated in Figure 28.9b. One sees clearly that the field is split into two main directions and each direction experiences a good beam collimation. The self-splitting and collimating of the beam can be explained by the linear diffraction curve. The nonlinear region of the curve lies in $|\varphi| < \varphi_c$, where $\varphi_c = k_0[\varepsilon_d d(\xi - d)]^{1/2}$. Because of the narrow beam width, the wave vector of the incident field has a series of possible values, which corresponds to a small region of φ rather than a single value at the BZ center. As φ_c is very small ($\varphi_c = 0.0046\pi$), as shown in the inset of Figure 28.9a, the wave vectors may have both positive and negative values. The direction of the Poynting flux is determined by the group velocity, which is given by $\boldsymbol{v}_g = \nabla_k \omega(\boldsymbol{k})$. Thus, at a certain frequency, the diffraction angle of energy could be arrived at $\theta(\varphi_0) = -\text{atan}(d\partial k_z/\partial\varphi)|_{\varphi=\varphi_0}$(Pertsch et al. 2002). At a propagation distance of L_z in the z direction, the center of the output beam locates at $x_s(\varphi_0) = L_z \tan(\theta) = -L_z d\partial k_z/\partial\varphi|_{\varphi=\varphi_0}$. The diffraction of the beam is determined by $\partial^2 k_z/\partial\varphi^2$. That means there is no diffraction and the beam will be highly collimated if the $k_z(\varphi)$ curve is linear. For in-phase illumination ($\varphi_0 = 0$), the diffraction angle is given by $\theta = \pm\text{atan}([d/(\xi - d)]^{1/2})$. The splitting angle is given by $2|\theta|$. Both the numerical and analytical results are depicted in Figure 28.9c as the chemical potential of graphene is varying. They agree well with each other. The inset in Figure 28.9c shows the electric field distribution in the MGSA. At $\mu_c = 0.15$ eV, the splitting angle is 79.6° in our simulation, and the theoretically predicted value is 82.4°. For a smaller μ_c, a bigger splitting angle could be achieved. In Figure 28.9d, we also depict the output field as a function of the initial phase difference. The pattern is quite different from the weak-coupling one as shown in Figure 28.8.

Another interesting issue in the strong-coupling region is the wavelength shrinking of GPs at the BZ edges ($\varphi = \pm\pi$). In these regions, the dispersion relation becomes $\coth(\kappa d/2) = \kappa\xi/2$. Note that $k_z \approx \kappa$ and the wavelength of GPs in the MGSA can be approximately given by $\lambda_p \approx 2\pi/\kappa$. In the condition that $d \ll \xi$, we can arrive at $\kappa = 2(d\xi)^{-1/2}$ and thus $\lambda_p = \pi(d\xi)^{1/2}$. Recalling the wavelength of GPs in a single-layer graphene, we have $\lambda_{GP}/\lambda_p = (\xi/d)^{1/2}$. For example, if the period of the array is set as $d = 5$ nm, the wavelength of GPs in the array could be reduced to be ~3 times smaller than that in a single-layer graphene.

As we know, there is a critical value for the period below which the strong coupling could occur. The critical value quantitatively equals to the plasmonic thickness ξ of graphene. As the period increases from ξ, the coupling of individual GPs is gradually weakened. The plasmonic thickness of graphene also refers to the effective mode width of individual GPs in single-layer graphene. At $d = \xi$, the effective boundaries of the GP modes nicely contact each other. As the period decreases further, the overlap of the GPs in individual graphene sheets becomes stronger, resulting in the strong interplay of individual GPs. On the contrary, the separation of graphene sheets arouses a slighter overlap of the mode field, leading to the weak coupling of GPs.

28.4.4 Comparison with Thin Metal Film Arrays

It is worth comparing the array of thin layers of metal with MGSAs. Most of the previous studies focused on the weak coupling of SPPs in the array of metal layers. The diffraction relation for weak coupling is given by $k_z = k_{SP} + 2C\cos(\varphi)$, where k_{SP} is the wave vector of SPPs in a single metallic waveguide and C is the coupling coefficient of two neighboring waveguides (Fan et al. 2006). The diffraction curve used to be nonlinear around the BZ center. The phenomenon observed in graphene cannot be achieved in such metallic waveguide arrays.

However, for thinner metal layers, the coupling in the array structure becomes stronger. The rigorous dispersion relation of the array of the metal thin layer can be found in Verslegers et al. (2009). By using the formula, we calculate the diffraction curve of the array of metal thin layers. In Figure 28.10, we show some examples on the diffraction curves of the array of metal thin layers suspended in air. In the computations, the wavelengths of $\lambda = 10$ and 1 μm are considered, t_m is the thickness of the metal layer, and the period of the metal array is $d = 20$ nm. The metal is set as silver with permittivity taken from Lide (2004). For simplicity, here, we do not consider the influence of the metal thickness on the bulk permittivity. From the figures, we can see that only the diffraction curves of ultrathin metal layers with the thickness $t_m < 10$ nm are similar to those of graphene, which have deep drops at the BZ center of the diffraction curve. As the thickness increases, the variation of the diffraction curve is gradually weakened and the nonlinear region at the BZ center is also broadened. At $\lambda = 10$ μm, the modal wavelength of SPPs in the array of the thin-metal layer is much smaller than that in graphene, albeit the loss remains at the same level. At the near-infrared region of $\lambda = 1$ μm, the modal wavelength is comparable with that of graphene, but the loss increases dramatically, especially at the BZ edges.

Graphene has big advantages in realizing the beam splitting and modal wavelength reducing based on the calculations. First, the array of thin layers of metal may find similar phenomena with the graphene sheet arrays only for the metal thickness in several nanometers. In practice, the thin-metal film tends to be granular like with a large surface roughness. It is challenging to deposit the metal thin film in a large area with high quality (Nagpal et al. 2009; Liu et al. 2012), while graphene is naturally a monolayer and can be formed in a large area by chemical vapor deposition (CVD) technique. Second, the propagation distance of SPPs in graphene sheet array can be much larger than that in metal-layer array with respect to their modal wavelength. For example, $L_z/\lambda_p = 1.95$ for SPPs in the metal-layer array for $\varphi = \pi$ and $\lambda = 1$ μm at $t_m = 1$ nm, comparing to the value of $L_z/\lambda_p = 8.8$ in the MGSA (Wang et al. 2012). Finally, graphene has a big advantage over a metal on tunability. It is convenient to control the propagation of GPs in MGSAs by applying the gate voltage or an electric or magnetic field (Vakil and Engheta 2011).

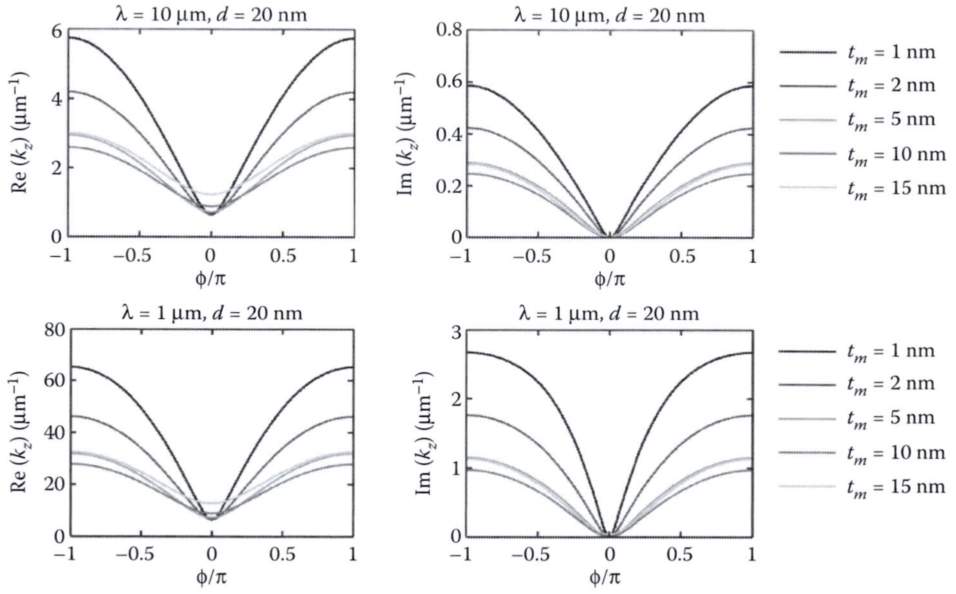

FIGURE 28.10 Diffraction relations of SPPs in an array of metal thin layers. Left: Real part of the wave vector. Right: Imaginary part of the wave vector. The excitation wavelength is 10 μm in the upper two figures and 1 μm in the lower figures.

28.5 CONCLUSION

To sum up, we have studied the optical coupling of GPs in double-layer graphene and graphene sheet arrays. Because of the unique properties of graphene, the coupling of graphene manifests different features from dielectric and metallic waveguides. First, the field distribution of GPs is antisymmetric, which leads to the negative coupling between graphene. The extremely short wavelength of GPs opens a window to design optical nanodevices that can operate deeply below the diffraction limit. Because of the one-atom thickness, the MGSA may experience a strong coupling, which leads to self-splitting and collimating of GPs in the array structure. We also compare the MGSAs with metallic waveguide arrays. The strong coupling cannot take place in the latter structures. The study provides a new platform for studying the coupling in the discrete optical system and will find a great theoretical interest and practical applications.

REFERENCES

Ablowitz, M. and Z. Musslimani. 2001. Discrete diffraction managed spatial solitons. *Physical Review Letters* 87 (25):254102.

Berini, P. 2009. Long-range surface plasmon polaritons. *Advances in Optics and Photonics* 1 (3):484–588.

Brongersma, M. L. and P. G. Kik. 2007. *Surface Plasmon Nanophotonics*. Dordrecht: Springer.

Castro Neto, A. H., N. M. R. Peres, K. S. Novoselov, and A. K. Geim. 2009. The electronic properties of graphene. *Reviews of Modern Physics* 81 (1):109–162.

Chen, C. F., C. H. Park, B. W. Boudouris, J. Horng, B. Geng, C. Girit, A. Zettl et al. 2011. Controlling inelastic light scattering quantum pathways in graphene. *Nature* 471 (7340):617–620.

Chen, J., M. Badioli, P. Alonso-González, S. Thongrattanasiri, F. Huth, J. Osmond, M. Spasenović et al. 2012. Optical nano-imaging of gate-tunable graphene plasmons. *Nature (London)* 487:77–81.

Chen, P. Y. and A. Alù. 2011. Atomically-thin surface cloak using graphene monolayers. *ACS Nano* 5 (7):5855–5863.

Christodoulides, D. N. and R. I. Joseph. 1988. Discrete self-focusing in nonlinear arrays of coupled waveguides. *Optics Letters* 13 (9):794–796.

Christodoulides, D. N., F. Lederer, and Y. Silberberg. 2003. Discretizing light behaviour in linear and nonlinear waveguide lattices. *Nature (London)* 424 (6950):817–823.

Das Sarma, S., S. Adam, E. Hwang, and E. Rossi. 2011. Electronic transport in two-dimensional graphene. *Reviews of Modern Physics* 83 (2):407–470.

Efetov, D. K. and P. Kim. 2010. Controlling electron–phonon interactions in graphene at ultrahigh carrier densities. *Physical Review Letters* 105 (25):256805.

Efremidis, N. K., P. Zhang, Z. Chen, D. N. Christodoulides, C. E. Rüter, and D. Kip. 2010. Wave propagation in waveguide arrays with alternating positive and negative couplings. *Physical Review A* 81 (5):053817.

Eisenberg, H. S., Y. Silberberg, R. Morandotti, and J. S. Aitchison. 2000. Diffraction management. *Physical Review Letters* 85 (9):1863–1866.

Eisenberg, H. S., Y. Silberberg, R. Morandotti, A. R. Boyd, and J. S. Aitchison. 1998. Discrete spatial optical solitons in waveguide arrays. *Physical Review Letters* 81 (16):3383–3386.

Falkovsky, L. A., and S. S. Pershoguba. 2007. Optical far-infrared properties of a graphene monolayer and multilayer. *Physical Review B* 76 (15):153410.

Fan, X., G. P. Wang, J. C. W. Lee, and C. T. Chan. 2006. All-angle broadband negative refraction of metal waveguide arrays in the visible range: Theoretical analysis and numerical demonstration. *Physical Review Letters* 97 (7):073901.

Fan, Y., B. Wang, K. Wang, H. Long, and P. Lu. 2014. Talbot effect in weakly coupled monolayer graphene sheet arrays. *Optics Letters* 39 (12):3371–3373.

Fei, Z., A. S. Rodin, G. O. Andreev, W. Bao, A. S. McLeod, M. Wagner, L. M. Zhang et al. 2012. Gate-tuning of graphene plasmons revealed by infrared nano-imaging. *Nature (London)* 487:82–85.

Geim, A. K. 2009. Graphene: Status and prospects. *Science* 324 (5934):1530.

Grigorenko, A. N., M. Polini, and K. S. Novoselov. 2012. Graphene plasmonics. *Nature Photonics* 6:749–758.

Hanson, G. W. 2008a. Dyadic Green's functions and guided surface waves for a surface conductivity model of graphene. *Journal of Applied Physics* 103 (6):064302.

Hanson, G. W. 2008b. Quasi-transverse electromagnetic modes supported by a graphene parallel-plate waveguide. *Journal of Applied Physics* 104 (8):084314.

Kittel, C. 2005. *Introduction to Solid State Physics.* 8th ed. New York: Wiley.

Lide, D. R. 2004. *CRC Handbook of Chemistry and Physics.* 84th ed. Boca Raton, FL: CRC Press.

Liu, H., B. Wang, L. Ke, J. Deng, C. C. Chum, S. L. Teo, L. Shen, S. A. Maier, and J. Teng. 2012. High aspect sub-diffraction-limit photolithography via a silver superlens. *Nano Letters* 12:1549–1554.

Liu, Y., G. Bartal, D. A. Genov, and X. Zhang. 2007. Subwavelength discrete solitons in nonlinear metamaterials. *Physical Review Letters* 99 (15):153901.

Nagpal, P., N. C. Lindquist, S. H. Oh, and D. J. Norris. 2009. Ultrasmooth patterned metals for plasmonics and metamaterials. *Science* 325 (5940):594–597.

Nikitin, A. Y., F. Guinea, F. J. García-Vidal, and L. Martín-Moreno. 2011. Fields radiated by a nanoemitter in a graphene sheet. *Physical Review B* 84:195446.

Okamoto, K. 2006. *Fundamentals of Optical Waveguides.* San Diego: Academic Press.

Ooi, K. J. A., H. S. Chu, P. Bai, and L. K. Ang. 2014. Electro-optical graphene plasmonic logic gates. *Optics Letters* 39 (6):1629–1632.

Pertsch, T., T. Zentgraf, U. Peschel, A. Bräuer, and F. Lederer. 2002. Anomalous refraction and diffraction in discrete optical systems. *Physical Review Letters* 88 (9):093901.

Smirnova, D. A., A. V. Gorbach, I. V. Iorsh, I. V. Shadrivov, and Y. S. Kivshar. 2013. Nonlinear switching with a graphene coupler. *Physical Review B* 88 (4):045443.

Taflove, A. and S. C. Hagness. 2000. *Computational Electrodynamics: The Finite-Difference Time-Domain Method.* Boston: Artech House.

Tassin, P., T. Koschny, M. Kafesaki, and C. M. Soukoulis. 2012. A comparison of graphene, superconductors and metals as conductors for metamaterials and plasmonics. *Nature Photonics* 6:259–264.

Vakil, A. and N. Engheta. 2011. Transformation optics using graphene. *Science* 332 (6035):1291.

Verhagen, E., R. De Waele, L. Kuipers, and A. Polman. 2010. Three-dimensional negative index of refraction at optical frequencies by coupling plasmonic waveguides. *Physical Review Letters* 105 (22):223901.

Verslegers, L., P. B. Catrysse, Z. Yu, and S. Fan. 2009. Deep-subwavelength focusing and steering of light in an aperiodic metallic waveguide array. *Physical Review Letters* 103 (3):033902.

Wang, B. and G. P. Wang. 2004. Surface plasmon polariton propagation in nanoscale metal gap waveguides. *Optics Letters* 29 (17):1992–1994.

Wang, B., X. Zhang, F. J. García-Vidal, X. Yuan, and J. Teng. 2012. Strong coupling of surface plasmon polaritons in monolayer graphene sheet arrays. *Physical Review Letters* 109 (7):073901.

Wang, B., X. Zhang, K. P. Loh, and J. Teng. 2014. Tunable broadband transmission and phase modulation of light through graphene multilayers. *Journal of Applied Physics* 115 (22):213102.

Wang, B., X. Zhang, X. Yuan, and J. Teng. 2012. Optical coupling of surface plasmons between graphene sheets. *Applied Physics Letters* 100 (13):131111.

Zhang, X. and C. Li. 2009. TE-polarized beam shaping by a subwavelength metal slit: Finite-difference at frequency-domain simulations. *Physica E: Low-Dimensional Systems and Nanostructures* 41 (8):1445–1450.

29 Optical Properties of Graphene in External Fields

Y. H. Chiu, Y. C. Ou, and M. F. Lin

CONTENTS

ABSTRACT

The generalized tight-binding model, with the exact diagonalization method, is developed to investigate optical properties of graphene in five kinds of external fields. The quite large Hamiltonian matrix is transferred into the band-like one by the rearrangement of many basis functions; furthermore, the spatial distributions of wave functions on distinct sublattices are utilized to largely reduce the numerical computation time. The external fields have a strong influence on the number, intensity, frequency, structure, and the selection rule of absorption peaks. The optical spectra in a uniform magnetic field exhibit plentiful symmetric absorption peaks and obey a specific selection rule. However, there are many asymmetric peaks and extra selection rules under the modulated electric field, the modulated magnetic field, the composite electric and magnetic fields, and the composite magnetic fields.

29.1 INTRODUCTION

Monolayer graphene (MG), constructed from a single layer of carbon atoms densely packed in hexagonal lattice, was successfully produced by mechanical exfoliation [1,2]. This particular material offers an excellent system for studying two-dimensional (2D) physical properties, such as the quantum Hall effects [3–15], and these properties could be preliminarily comprehended by the energy dispersion (or called energy band structure), which can directly reflect the main features of electronic properties. In the low-energy region of $|E^{c,v}| \leq 1$ eV, MG possesses isotropic linear bands crossing at the **K** (**K′**) point and is regarded as a 2D zero-gap semiconductor, where $c(v)$ indicates the conduction (valence) bands [16]. The linear bands are symmetric about the Fermi level ($E_F = 0$) and become nonlinear and anisotropic with $|E^{c,v}| > 1$ eV [16]. Most importantly, the quasiparticles related to the linear bands can be described by a Dirac-like Hamiltonian [17], which is associated with relativistic particles and dominates the low-energy physical properties [15,18,19]. Such a special electronic structure has been verified by experimental measurements [2,20].

MG has become a potential candidate of nano-devices due to its exotic electronic properties. Well understanding the behavior of MG under external fields is useful for improving the characteristics of graphene-based nano-devices. Five cases of external fields (see table below), which can be experimentally produced [21–27], are often applied to investigate the physical properties of few-layer graphenes (FLGs). In the presence of a uniform magnetic field (UM), the electronic states corresponding to the linear bands change into Landau levels (LLs) which obey a specific relationship $E^{c,v} \propto \sqrt{n^{c,v}B_{UM}}$, where $n^c(n^v)$ is the quantum number of the conduction (valence) states and B_{UM} the magnetic field strength. The related anomalous quantum Hall effects and particular optical excitations have been verified experimentally [14,15]. For a modulated magnetic field (MM), quasi-Landau levels (QLLs) possessing anisotropic behavior and the related optical absorption spectra with specified selection rules were shown [28,29]. Furthermore, Haldane predicted that MG in the MM could reveal quantum Hall effects even without any net magnetic flux through the whole space [13]. Concerning a modulated electric potential (ME), the linear dispersions become oscillatory and extra Dirac cones are induced by the potential [30–35]. Such a potential changes MG from a zero-gap semiconductor into a semimetal [33,35] and makes MG exhibit Klein paradox effect associated with the Dirac cones [17]. For two cases of composite fields, a uniform magnetic field combined with a modulated magnetic field (UM-MM) and a uniform magnetic field combined with a modulated electric potential (UM-ME), the LL properties are drastically changed by the modulated fields. For both composite field cases, an unusual oscillation [36–40] of the density of states (DOS) similar to the Weiss oscillation obtained in 2D electron gas (2DEG) were shown. Furthermore, the broken symmetry, displacement of the center location, and alteration of the amplitude strength of the LL wave functions were also obtained [41,42].

Graphene-related systems are predicted to exhibit rich optical absorption spectra. The spectral intensity of MG is proportional to the frequency, but no prominent peak exists at $\omega < 5$ eV [43]. In FLGs, the interlayer atomic interactions drastically alter the two linear energy bands intersecting at $E_F = 0$ [44–54]. As a result, conspicuous absorption peaks arise in optical spectra [55], where the peak structure, intensity, and frequency are dominated by the layer number and the stacking configuration. Furthermore, under an external perpendicular electric field or a uniform perpendicular magnetic field, the main features of the optical properties of the FLGs are strongly modified [49,56,57]. For theoretical studies, the complexity of calculating the optical absorption spectra is solved by the gradient approximation based on the generalized tight-binding (TB) model with exact-diagonalization method or effective-mass approximation. The way in which one can control the absorption peaks and selection rules is worthy to be reviewed in detail.

On the other hand, there has been a considerable amount of experimental research on graphene-related systems under a uniform perpendicular magnetic field. From the measured results, the features of MG and bilayer graphene are reflected in the magneto-optical spectra [58,59]. That is to say, the LL energies are proportional to $\sqrt{n^{c,v}B_0}$ or $n^{c,v}B_0$. For any graphene system, the selection rule coming from the LLs close to $E_F = 0$ is $\Delta n = |n^c - n^v| = 1$ [60–62]. Moreover, similar results may also be found in AB-stacked graphite [62]. However, experimental measurements on optical properties under a non-uniform or composite field are not available so far.

In this chapter, we would like to focus on the optical absorption spectra of MG under the five cases of external fields, UM, MM, ME, UM-MM, and UM-ME cases. The TB model with exact diagonalization method is introduced to solve the energy dispersions, and then the gradient approximation is applied to obtain the optical absorption spectra. The main features of electronic properties, which include energy dispersions and wave functions, will be shown to comprehend the optical absorption properties, where the dependence of absorption frequency on external fields, optical selection rules, and anisotropic behavior will be discussed in detail. In Section 29.1, the TB model corresponding to the five cases of external fields is shown. In Sections 29.2 through 29.6, the optical absorption spectra of MG under the UM, MM, ME, UM-MM, and UM-ME cases will be reviewed, respectively. Finally, concluding remarks are presented in Section 29.7.

Physical Properties of Graphene under External Fields

External Fields	Related Physical Properties
Uniform magnetic field	Landau level and abnormal quantum Hall effect [3–15], magneto-optical selection rule [58,59]
Modulated magnetic field	Quasi-Landau level [28,29], quantum Hall effect without Landau level [13]
Modulated electric potential	Number increment of Dirac cone [30–35], semiconductor–metal transition [33,35], Klein tunneling [17]
Uniform magnetic field + modulated magnetic field, uniform magnetic field + modulated electric potential	Weiss oscillation [36–40], destruction of Landau-level wave function [41,42]

29.2 TIGHT-BINDING MODEL WITH EXACT DIAGONALIZATION

The low-frequency optical properties of graphene are determined by the π-electronic structure due to the $2p_z$ orbitals of carbon atoms. The generalized TB model with exact diagonalization method is developed to characterize the electronic properties and then the gradient approximation is applied to obtain the optical-absorption spectra. In the absence of external fields, there are two carbon atoms, the a and b atoms, in

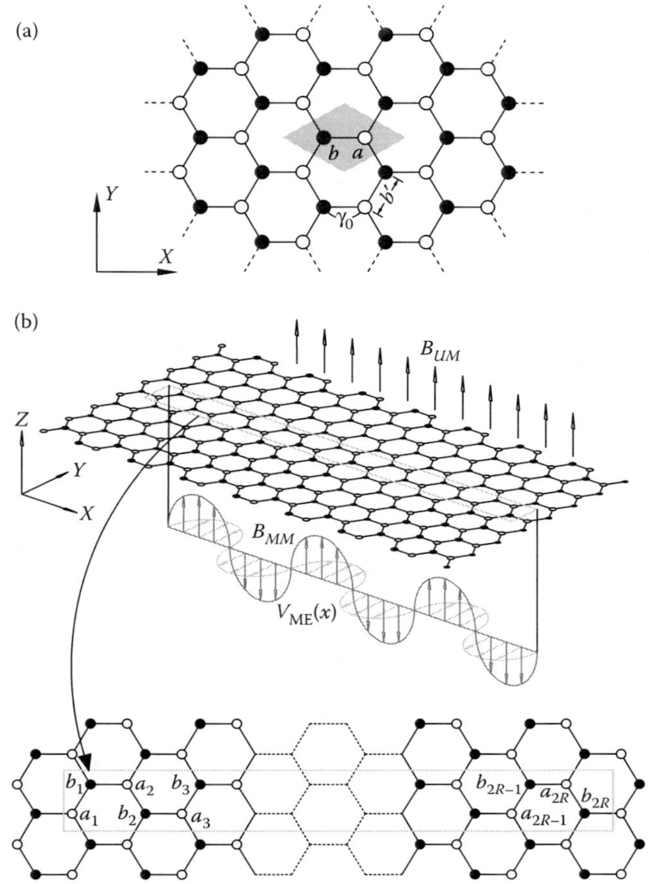

FIGURE 29.1 The primitive unit cell of monolayer graphene (a) in the absence and (b) in the presence of external fields.

a primitive unit cell of MG, as shown in Figure 29.1a by the rhombus shadow, where the x- and y-direction are, respectively, the armchair and zigzag directions of MG. This indicates that the Bloch wave function Ψ is a linear superposition of two TB functions associated with the $2p_z$ orbitals and is expressed as $\Psi = \varphi_a \pm \varphi_b$, where φ_a and φ_b, respectively, stand for the TB functions of the a and b atoms and are represented as [16]

$$\varphi_a = \sum_a \exp(i\mathbf{k}\cdot\mathbf{R}_a)\chi(\mathbf{r}-\mathbf{R}_a), \tag{29.1}$$

$$\varphi_b = \sum_b \exp(i\mathbf{k}\cdot\mathbf{R}_b)\chi(\mathbf{r}-\mathbf{R}_b). \tag{29.2}$$

$\chi(\mathbf{r})$ is the normalized orbital $2p_z$ wave function for an isolated atom. Moreover, the symbols γ_0 (=2.5 eV) and b' (=1.42 Å) shown in Figure 29.1a represent the nearest-neighbor atomic interaction (or called hopping integral) and the C–C bond length, respectively [16]. Throughout this chapter, only γ_0 is taken into account and other atomic interactions are neglected.

In the presence of an external field, the primitive unit cell is no longer the one shown in Figure 29.1a since the external field leads to a new periodic condition. Here, we choose the rectangular unit cell marked by the rectangle in Figure 29.1b as the primitive unit cell of graphene under the five kinds of external fields, where $R = R_{UM}$, R_{MM}, R_{ME}, and R_C (defined in the following) describe, respectively, the periods resulting from UM, MM, ME, and composite field. The major discussions are focused on R along the armchair direction. Consequently, an enlarged rectangular unit cell induced by an external field encompasses $2R$ a atoms and $2R$ b atoms. This implies that R determines the dimension of the Hamiltonian matrix, which is a $4R \times 4R$ Hermitian matrix spanned by $4R$ TB functions associated with the $2R$ a atoms and $2R$ b atoms. Based on the arrangement of odd and even atoms in the primitive cell, the Bloch wave function $|\Psi_k\rangle$ can have the expression:

$$|\Psi_{\mathbf{k}}\rangle = \sum_{m=1}^{2R-1} (A_{\mathbf{o}}^{c,v}|a_{m\mathbf{k}}\rangle + B_{\mathbf{o}}^{c,v}|b_{m\mathbf{k}}\rangle) + \sum_{m=2}^{2R} (A_{\mathbf{e}}^{c,v}|a_{m\mathbf{k}}\rangle + B_{\mathbf{e}}^{c,v}|b_{m\mathbf{k}}\rangle). \tag{29.3}$$

$|a_{m\mathbf{k}}\rangle$ ($|b_{m\mathbf{k}}\rangle$) is the TB function corresponding to the $2p_z$ orbital of the mth a (b) atom. $A_{\mathbf{o}}^{c,v}(A_{\mathbf{e}}^{c,v})$ and $B_{\mathbf{o}}^{c,v}(B_{\mathbf{e}}^{c,v})$ are the subenvelope functions standing for the amplitudes of the wave functions of the a- and b-atoms, respectively, where o (e) represents an odd (even) integer. Since the features of $A_o^{c,v}$ ($B_o^{c,v}$) and $A_e^{c,v}(B_e^{c,v})$ are similar, choosing only the amplitudes $A_{\mathbf{o}}^{c,v}$ and $B_{\mathbf{o}}^{c,v}$ is sufficient to comprehend the electronic and optical properties we would like to discuss in this chapter. The $4R \times 4R$ Hamiltonian matrix, which determines the magneto-electronic properties, is a giant Hermitian matrix for the external fields actually used in experiments. To make the calculations more efficient, the matrix is transformed into an $M \times 4R$ band-like matrix by a suitable rearrangement of the TB functions, where M is much smaller than $4R$. For example, one can arrange the basis functions as the sequence: $|a_{1\mathbf{k}}\rangle$, $|b_{2R\mathbf{k}}\rangle$, $|b_{1\mathbf{k}}\rangle$, $|a_{2R\mathbf{k}}\rangle$, $|a_{2\mathbf{k}}\rangle$, $|b_{2R-1\mathbf{k}}\rangle$, $|b_{2\mathbf{k}}\rangle$, $|a_{2R-1\mathbf{k}}\rangle$,…, $|a_{R-1\mathbf{k}}\rangle$, $|b_{R+2\mathbf{k}}\rangle$, $|b_{R-1\mathbf{k}}\rangle$, $|a_{R+2\mathbf{k}}\rangle$, $|a_{R\mathbf{k}}\rangle$, $|b_{R+1\mathbf{k}}\rangle$, $|b_{R\mathbf{k}}\rangle$; $|a_{R+1\mathbf{k}}\rangle$. Furthermore, distributions of the subenvelope functions are used to reduce the numerical computation time. The exact diagonalization method for numerical calculations is applicable to many kinds of magnetic, electric, and composite fields.

For the UM case, $\mathbf{B}_{UM} = B_{UM}\hat{z}$, a Peierls phase [28,63–65] related to the vector potential $\mathbf{A}_{UM} = B_{UM}x\hat{y}$ is introduced in the TB functions. The phase difference between two lattice vectors ($\mathbf{R}_m$ and $\mathbf{R}_{m'}$) is defined as $G_{UM} \equiv (2\pi/\phi_0)\int_{\mathbf{R}_{m'}}^{\mathbf{R}_m} \mathbf{A}_{UM}\cdot d\mathbf{r}$, where $\phi_0 = hc/e = 4.1356 \times 10^{-15}$ [T m^2] is the flux quantum. The Peierls phase periodic along the armchair direction provides a specific period set as $R_{UM} = (\phi_0/(3\sqrt{3}b'^2/2))/B_{UM}$ and the related Hamiltonian is a $4R_{UM} \times 4R_{UM}$ Hermitian matrix. The site energies, the diagonal matrix elements $\langle a_{m\mathbf{k}}|H|a_{m\mathbf{k}}\rangle$ and $\langle b_{m\mathbf{k}}|H|b_{m\mathbf{k}}\rangle$, are set to zero and the nonzero matrix elements related to γ_0 can be formulated as

$$\langle b_{m\mathbf{k}}|H|a_{m'\mathbf{k}}\rangle = \gamma_0 \exp i[\mathbf{k}\cdot(\mathbf{R}_m - \mathbf{R}_{m'}) + G_{UM}]. \tag{29.4}$$

Two kinds of periodic modulation fields along the armchair direction, the MM and ME cases, which can drastically change the physical properties of MG, are often selected for a study. For the MM case, $\mathbf{B}_{MM} = B_{MM}\sin(2\pi x/l_{MM})\ \hat{z}$ is exerted on MG along the armchair direction, where B_{MM} is the field strength and l_{MM} is the period length with the modulation period $R_{MM} = l_{MM}/3b'$. The vector potential is chosen as $\mathbf{A}_{MM} = (-B_{MM}(l_B/2\pi)\cos(2\pi x/l_{MM}))\hat{y}$ and the corresponding Peierls phase is $G_{MM} \equiv (2\pi/\phi_0)\int_{\mathbf{R}_{m'}}^{\mathbf{R}_m} \mathbf{A}_{MM}\cdot d\mathbf{r}$. Thus, the Hamiltonian matrix elements, which are similar to those in Equation 29.4, are represented as

$$\langle b_{m\mathbf{k}}|H|a_{m'\mathbf{k}}\rangle = \gamma_0 \exp i[\mathbf{k}\cdot(\mathbf{R}_m - \mathbf{R}_{m'}) + G_{MM}]. \quad (29.5)$$

For the ME case, $V_{ME}(x) = V_{ME}\cos(2\pi x/l_{ME})$ along the armchair direction with the potential strength V_{ME} and the period length l_{ME} is taken into account. As the period is sufficiently large, the electric potential affects only the site energies but not the nearest-neighbor hopping integral. As a result, the site energies become

$$\langle a_{m\mathbf{k}}|H|a_{m\mathbf{k}}\rangle = V_{ME}\cos\left[\frac{(m-1)\pi}{R_{ME}}\right] \equiv V_m, \quad (29.6)$$

$$\langle b_{m\mathbf{k}}|H|b_{m\mathbf{k}}\rangle = V_{ME}\cos\left[\frac{(m-2/3)\pi}{R_{ME}}\right] \equiv V_{m+1/3}, \quad (29.7)$$

where $R_{ME} = l_{ME}/3b'$ is the modulation period. The Hamiltonian matrices for the MM and the ME are $4R_{MM}\times 4R_{MM}$ and $4R_{ME}\times 4R_{ME}$ Hermitian matrices, respectively.

For a composite field case, a new periodicity, which is associated with periods induced by a UM and a modulated field, has to be defined. The rectangular unit cell is enlarged along the x-direction and the dimensionality of the Hamiltonian matrix has to agree with the least common multiple of R_{UM} and R_{MM} (R_{UM} and R_{ME}) for the UM-MM (UM-ME) case, namely R_C. The rectangular unit cell corresponding to each composite field contains $4R_C$ atoms ($2R_C$ a atoms and $2R_C$ b atoms), and the magneto-electronic wave functions are linear combinations of the $4R_C$ TB functions. In a composite field case, the matrix elements are superposed by the elements associated with each combined external field. For the sake of convenience, we put the matrix elements in Equations 29.4 through 29.7 together as a common case and the elements are rewritten as

$$\langle b_{m\mathbf{k}}|H|a_{m'\mathbf{k}}\rangle = \gamma_0 \exp i[\mathbf{k}\cdot(\mathbf{R}_m - \mathbf{R}_{m'}) + G_{UM} + G_{MM}], \quad (29.8)$$

$$\langle a_{m\mathbf{k}}|H|a_{m\mathbf{k}}\rangle = V_m, \quad (29.9)$$

$$\langle b_{m\mathbf{k}}|H|b_{m\mathbf{k}}\rangle = V_{m+1/3}. \quad (29.10)$$

The off-diagonal elements are associated with the Peierls phases induced by the magnetic fields and the diagonal elements are related to the site energies induced by the modulated electric field. By diagonalizing the matrix, the energy dispersion $E^{c,v}$ and the wave function $\Psi^{c,v}$ are obtained. It should be noted that the k_x-dependent dispersions can be ignored when the period R is sufficiently large and thus only k_y-dependent dispersions are shown for the following discussions.

When an MG is excited from the occupied valence to the unoccupied conduction bands by an electromagnetic field, only inter-π-band excitations exist at zero temperature. Based on the Fermi's golden rule, the optical absorption function results in the following form:

$$A(\omega) \propto \sum_{c,v,\tilde{n},\tilde{n}'} \int_{1stBZ} \frac{d\mathbf{k}}{(2\pi)^2} \left| \langle \Psi^c(\mathbf{k},n)| \frac{\hat{\mathbf{E}}\cdot\mathbf{P}}{m_e} |\Psi^v(\mathbf{k},n')\rangle \right|^2 \times \mathrm{Im}\left[\frac{f(E^c(\mathbf{k},n)) - f(E^v(\mathbf{k},n'))}{E^c(\mathbf{k},n) - E^v(\mathbf{k},n') - \omega - i\Gamma}\right], \quad (29.11)$$

where $f(E(\mathbf{k},\tilde{n}))$ is the Fermi–Dirac distribution function, and Γ ($=2\times 10^{-4}\gamma_0$) is the broadening parameter. The electric polarization $\hat{\mathbf{E}}$ is the unit vector of an electric polarization. Results for $\hat{\mathbf{E}}$ along the armchair and zigzag directions are taken into account for discussions. Within the gradient approximation [66–68], the velocity matrix element $M^{cv} = \langle \Psi^c(\mathbf{k},\tilde{n})|\hat{\mathbf{E}}\cdot\mathbf{P}/m_e|\Psi^v(\mathbf{k},\tilde{n}')\rangle$ is formulated as

$$\sum_{m,m'=1}^{2R_C} (A^{c*}\times B^v)\nabla_{\mathbf{k}}\langle a_{m\mathbf{k}}|H|b_{m'\mathbf{k}}\rangle + hc. \quad (29.12)$$

Equation 29.12 implies that the main features of the wave functions are major factors in determining the selection rules and the absorption rate of the optical excitations. Similar gradient approximations have been successfully applied to explain optical spectra of carbon-related systems, for example, graphite [69], graphite intercalation compounds [70], carbon nanotubes [71], FLGs [72], and graphene nanoribbons [22].

29.3 UNIFORM MAGNETIC FIELD

29.3.1 Landau Level Spectra

In this section, we mainly focus on drastic changes of the Dirac cone as the result of a uniform perpendicular magnetic field. The magnetic field causes the states to congregate and induces dispersionless LLs, as shown in Figure 29.2a for $B_{UM} = 5$ T at $k_x = 0$. The unoccupied LLs and occupied LLs are symmetric about the Fermi level ($E_F = 0$). Each LL is characterized by the quantum number $n^{c,v}$, which corresponds to the number of zeros in the eigenvectors of harmonic oscillator [50,73]. Each LL is fourfold degenerate without considering the spin degeneracy. Its energy may be approximated by a simple square-root relationship $|E_n^{c,v}| \propto \sqrt{n^{c,v}B_{UM}}$ [12,74], which is valid only for the range of $|E_n^{c,v}| \le \pm 1\,\mathrm{eV}$ [12].

29.3.2 Landau Level Wave Functions

The LL wave functions, as shown in Figure 29.2b and c, exhibit the versatility of spatial symmetry and can be described by

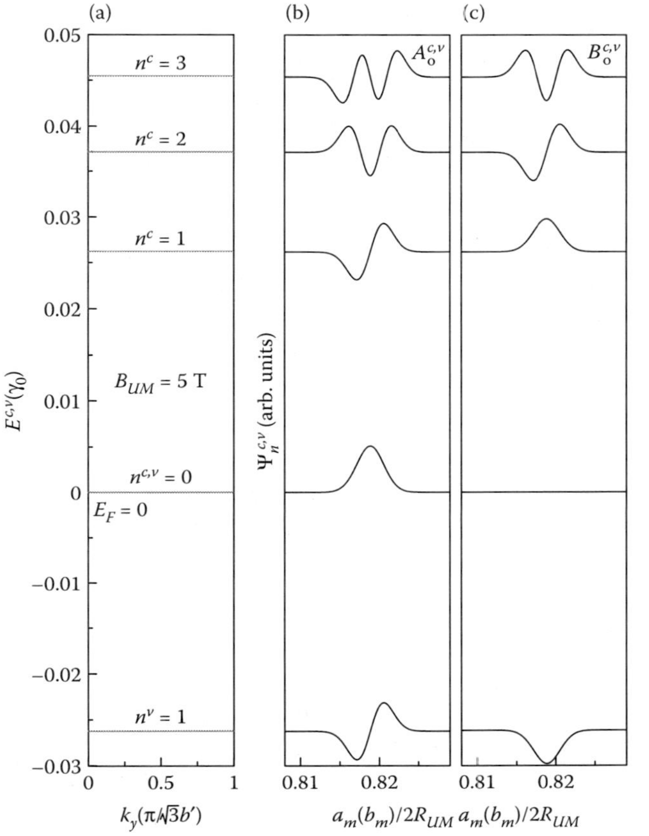

FIGURE 29.2 The Landau level spectrum for (a) the uniform magnetic field $B_{UM} = 5$ T. The Landau level wave functions corresponding to (b) the a- and (c) the b-atoms.

the eigenvectors ($\varphi_n(x)$) of harmonic oscillator, which obey the relationships, $\langle\varphi_n(x)|\varphi_{n'}(x)\rangle = \delta_{n,n'}$ and $\varphi_n(x) = 0$ for $n < 0$. The wave functions are distributed around the localization center, that is at the 5/6 position of the enlarged unit cell. Similar localization centers corresponding to the other degenerate states occur at the 1/6, 2/6, and 4/6 positions. The subenvelope functions can be expressed as

$$A_{o,e}^{c,v} \propto \varphi_{n^{c,v}}(x_1) \pm \varphi_{n^{c,v}-1}(x_2), \quad B_{o,e}^{c,v} \propto \varphi_{n^{c,v}-1}(x_1) \mp \varphi_{n^{c,v}}(x_2),$$

$$A_{o,e}^{c,v} \propto \varphi_{n^{c,v}-1}(x_3) \pm \varphi_{n^{c,v}}(x_4), \quad B_{o,e}^{c,v} \propto \varphi_{n^{c,v}}(x_3) \mp \varphi_{n^{c,v}-1}(x_4),$$

$$\text{for } x_1 = \frac{1}{6}, \quad x_2 = \frac{5}{6}, \quad x_3 = \frac{2}{6}, \quad \text{and} \quad x_4 = \frac{4}{6}. \tag{29.13}$$

However, it is adequate to only consider any one center in evaluating the absorption spectra due to their identical optical responses.

29.3.3 Optical Absorption Spectra of Landau Levels

The low-frequency optical absorption spectrum of the LLs presents many interesting features as shown in Figure 29.3a for $B_{UM} = 5$ T. The spectrum exhibits many delta-function-like symmetric peaks with a uniform intensity. Such peaks

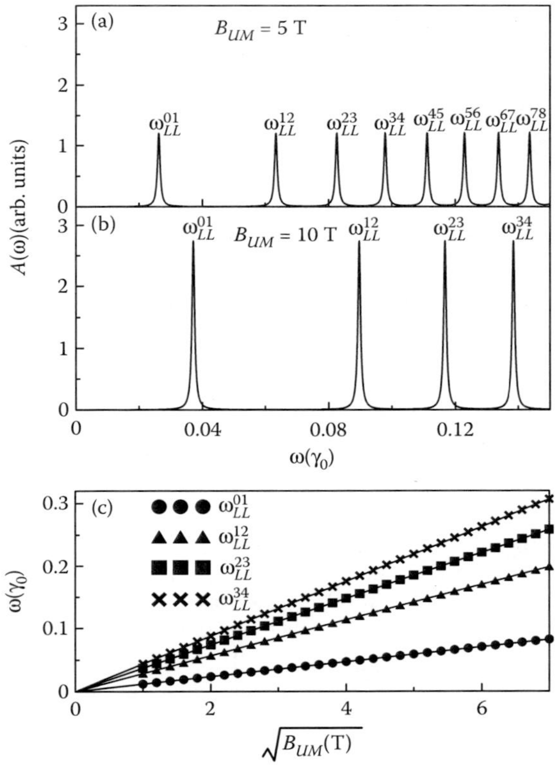

FIGURE 29.3 The optical absorption spectra for (a) $B_{UM} = 5$ T and (b) $B_{UM} = 10$ T. (c) The dependence of the absorption frequency on the square root of field strength B_{UM}.

suggest that LLs possess a zero-dimensional (0D) band structure or DOS. The optical transition channel with respect to each absorption peak can be clearly identified. A single peak $\omega_{LL}^{nn'}$ is generated by two transition channels n'LL $\rightarrow n$LL and nLL $\rightarrow n'$LL, where the symbol $n' \rightarrow n$ is used, for the sake of convenience, to represent the transition from the valence states with n' to the conduction states with n throughout this chapter. The quantum numbers related to the LL transitions must satisfy a specific selection rule, that is, $\Delta n = |n^c - n^v| = 1$. The selection rule is established by the main features of the wave functions. The velocity matrix M^{cv}, a dominant factor for the excitations of the prominent peaks, strongly depends on the number of zeros of $A_o^{c,v}$ and $B_o^{c,v}$. It has non-zero values only when $A_o^{c(v)}$ and $B_o^{v(c)}$, expressed in orthogonality of $\varphi_n(x)$, possess the same number of zeros. Moreover, examining all the transitions reveals the following relationship: $A_o^{c,v}(n^{c,v}) \propto B_o^{c,v}(n^{c,v}+1)$, with $A_o^c = A_o^v$ and $B_o^c = -B_o^v$. In other words, the quantum numbers of the conduction and valence LLs differ by one when $A_o^{c(v)}$ and $B_o^{v(c)}$ have the same $\varphi_n(x)$.

In addition to the optical selection rules, the peak intensity and absorption frequency also deserve a discussion. In Figure 29.3b, the peak intensity is strengthened, whereas the peak number is reduced as the field strength increases. This is a result of the high degree of degeneracy in the first Brillouin zone and the expanded energy spacing between the LLs. The field-dependent absorption frequencies of the first four peaks ω_{LL}^{01}, ω_{LL}^{12}, ω_{LL}^{23}, and ω_{LL}^{34} are shown in Figure 29.3c. The frequencies become much higher in a stronger field. There exists a special square-root relation between $\omega_{LL}^{nn'}$ and B_0, that is, $\omega_{LL}^{nn'} \propto \sqrt{B_{UM}}$, which has been confirmed by magneto-optical spectroscopy methods, such as experimental measurements of the absorption coefficient [75,76], cyclotron resonance [77–79], and quantum Hall conductivity [2,80–82]. This square-root relation only exists in the lower frequency range $\omega < 0.4\gamma_0$ (~1 eV). In the higher frequency range, LLs are too densely packed to be separated from one another [12]. This leads to the disappearance of the relation between $\omega_{LL}^{nn'}$ and B_{UM}.

29.4 SPATIALLY MODULATED MAGNETIC FIELD

29.4.1 Quasi-Landau Level Spectra

Compared with the uniform case, an MM has a different impact on the electronic properties and leads to the diverse features observed in the optical absorption spectra. The presence of a modulated field has multiple effects on the energy bands, as shown in Figure 29.4 for $B_{MM} = 10$ T and $R_{MM} = 500$. In the lower energy region, parabolic subbands appear around $k_y = k_1 = 2/3$. The conduction and valence subbands are symmetric about the Fermi level ($E_F = 0$). The subbands nearest to $E_F = 0$ are partially flat and nondegenerate. The other parabolic subbands characterized by weak energy dispersions have double degeneracy and one original band-edge state at k_1. The modulation effects on parabolic energy subbands result in four extra band-edge states at the sites on both sides of k_1. They demonstrate the strongest dispersion and destruction of the double degeneracy. The low-energy subbands are regarded as QLLs, which exhibit similar features of the LLs generated from a UM. Moreover, the k_y range with respect to the weak dispersion and partial flat bands grows with increasing field strength and a longer modulation period. On the contrary, when the influence of the modulation field become much weaker with increasing energy, the parabolic subbands in the higher energy region are similar to the twofold degenerate subbands directly obtained from the zone folding of MG in the $B_{MM} = 0$ case (not shown).

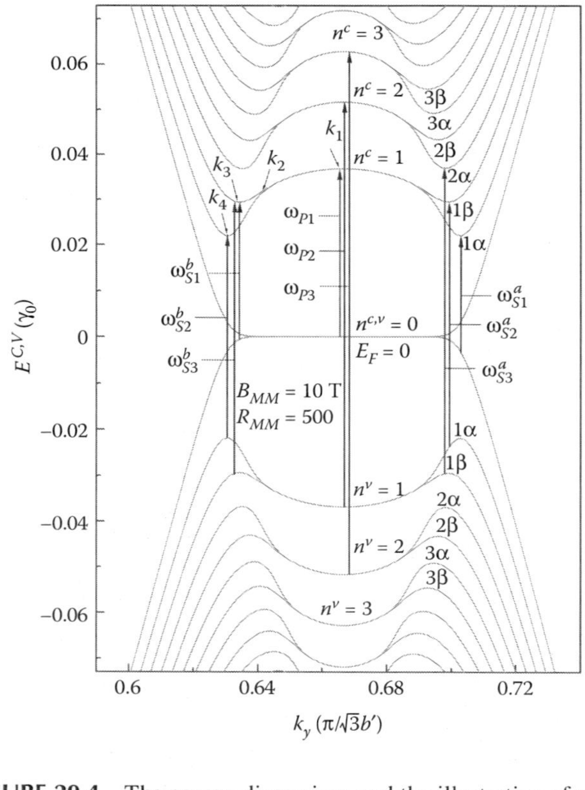

FIGURE 29.4 The energy dispersions and the illustration of optical excitation channels for the modulated magnetic field along the armchair direction with $R_{MM} = 500$ and $B_{MM} = 10$ T.

29.4.2 Quasi-Landau Level Wave Functions

In the presence of an MM, the alterations of the wave functions are rather drastic. First, the QLL wave functions corresponding to k_1 are shown in Figure 29.5a through f. The wave functions are composed of two TB functions centered at x_1 and x_2. $A_o^c(B_o^c)$ has two subenvelope functions $A_o^c(x_1)$ ($B_o^c(x_1)$) and $A_o^c(x_2)(B_o^c(x_2))$ centered at $x_1 = 1/4$ and $x_2 = 3/4$ of the primitive unit cell, respectively. The positions x_1 and x_2 are located at where the field strength is at a maximum. The number of zeros of $A_o^c(x_2)(B_o^c(x_1))$ is higher than that of $A_o^c(x_1)(B_o^c(x_2))$ by one at each QLL. A similar behavior is also shown by the valence wave function, where only the sign is flipped in either A_o^v or B_o^v. The effective quantum number $n^{c,v}$ is defined by the larger number of zeros of the subenvelope functions. In addition, the twofold degenerate QLLs have similar wave functions (solid curves and dashed curves), with the only difference in terms of the sign change in the subenvelope functions. The wave functions at k_1 can be expressed as

$$A_{o,e}^c \propto \Psi_{n^c-1}(x_1) \pm \Psi_{n^c}(x_2), \quad A_{o,e}^v \propto \Psi_{n^v-1}(x_1) \mp \Psi_{n^v}(x_2),$$

$$B_{o,e}^c \propto \Psi_{n^c}(x_1) \mp \Psi_{n^c-1}(x_2), \quad B_{o,e}^v \propto \Psi_{n^v}(x_1) \pm \Psi_{n^v-1}(x_2),$$

$$\text{for } x_1 = \frac{1}{4} \quad \text{and} \quad x_2 = \frac{3}{4}. \tag{29.14}$$

The wave functions would be strongly modified as the wave vectors gradually move away from k_1. Secondly, the wave functions at several special k points are illustrated to examine the effects caused by the MM field. As the wave vector moves to $k_y = k_2$, the doubly degenerate QLL starts to separate into two subbands. The two subenvelope functions $A_o^c(x_1)$ ($B_o^c(x_1)$) and $A_o^c(x_2)(B_o^c(x_2))$ move toward each other and shift to the center of the primitive unit cell with nearly overlapping, as shown in Figure 29.5g and h. At k_3 and k_4, the higher and lower subbands have the extra band-edge states 1α and 1β, respectively. The subenvelope functions of the 1α state, as shown in Figure 29.5i and j, exhibit a strong overlapping behavior compared to those at $k_y = k_2$ (dashed curves in Figure 29.5g and h). Similar behavior can also be found in the wave

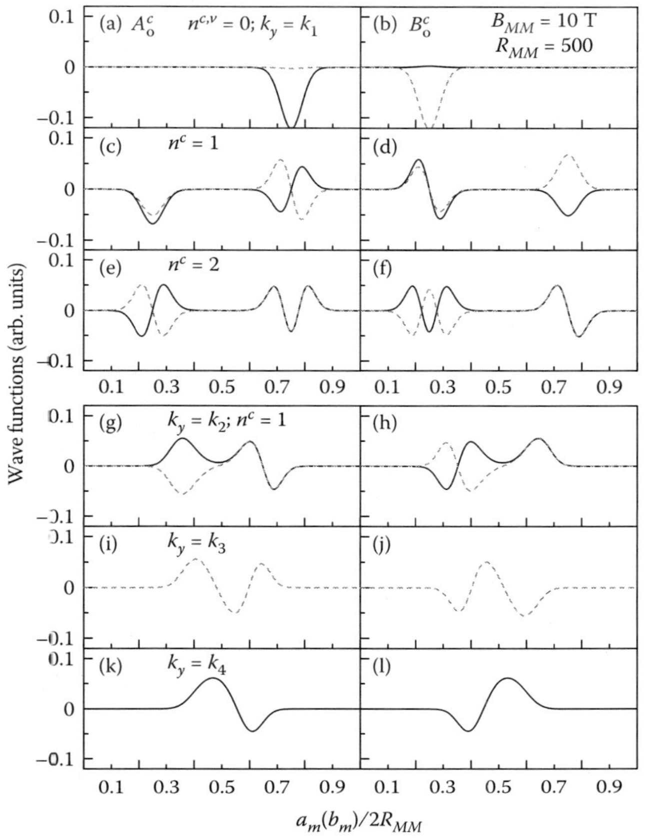

FIGURE 29.5 The wave functions of quasi-Landau levels at (a)–(f) the original band-edge state k_1 with the quantum numbers $n^{c,v} = 0$, $n^c = 1$ and $n^c = 2$, (g) and (h) the split point k_2 with $n^c = 1$, and (i)–(l) two extra band-edge states k_3 and k_4 with $n^c = 1$.

functions at 1β. This implies that there is a higher degree of overlap in the subenvelope functions at the extra band-edge states $n^{c,v}\alpha$ and $n^{c,v}\beta$. Moreover, the two states associated with the different linear combinations of $A_o^c(x_1)(B_o^c(x_1))$ and $A_o^c(x_2)$ $(B_o^c(x_2))$ are represented as

$$A_{o,e}^c \propto \Psi_{n^c-1}(x_1) + \Psi_{n^c}(x_2)$$

$$\text{for } n^c\alpha \quad \text{and} \quad \Psi_{n^c-1}(x_1) - \Psi_{n^c}(x_2) \quad \text{for } n^c\beta,$$

$$B_{o,e}^c \propto \Psi_{n^c}(x_1) - \Psi_{n^c-1}(x_2)$$

$$\text{for } n^c\alpha \quad \text{and} \quad \Psi_{n^c}(x_1) + \Psi_{n^c-1}(x_2) \quad \text{for } n^c\beta,$$

$$A_{o,e}^v \propto \Psi_{n^v-1}(x_1) - \Psi_{n^v}(x_2)$$

$$\text{for } n^v\alpha \quad \text{and} \quad \Psi_{n^v-1}(x_1) + \Psi_{n^v}(x_2) \quad \text{for } n^v\beta,$$

$$B_{o,e}^v \propto \Psi_{n^v}(x_1) + \Psi_{n^v-1}(x_2)$$

$$\text{for } n^v\alpha \quad \text{and} \quad \Psi_{n^v}(x_1) - \Psi_{n^v-1}(x_2) \quad \text{for } n^v\beta,$$

$$\text{for } x_1 \approx x_2 \simeq \frac{1}{2}. \tag{29.15}$$

29.4.3 Optical Absorption Spectra of Quasi-Landau Levels

Under the MM, the parabolic energy bands possess several band-edge states. A wave function composed of two TB functions presents a complex overlapping behavior. The above-mentioned main features of the electronic properties are expected to be directly reflected in optical excitations. The low-frequency optical absorption spectra for $R_{MM} = 500$ and $B_{MM} = 10$ T, as shown in Figure 29.6a by the black and gray solid curves for $\hat{\mathbf{E}} \perp \hat{x}$ and $\hat{\mathbf{E}} \parallel \hat{x}$, respectively, exhibit rich asymmetric peaks in the square-root divergent form. These peaks can be divided into the principal peaks ω_P's and the subpeaks ω_S's based on the optical excitations resulting from the original band-edge and extra band-edge states, respectively. ω_S's can be further classified into two subgroups, ω_S^a's and ω_S^b's, which primarily come from the excitations of extra band-edge states $\alpha \to \beta$ ($\beta \to \alpha$) and $\alpha \to \alpha$ ($\beta \to \beta$), respectively. What is worth mentioning is that the spectra for $\hat{\mathbf{E}} \perp \hat{x}$ and $\hat{\mathbf{E}} \parallel \hat{x}$ are distinct, especially for the subpeaks ω_S's. The former is mainly composed of the subgroup ω_S^a, while the latter mainly consists of the subgroup ω_S^b. This implies that the optical absorption spectra reflect

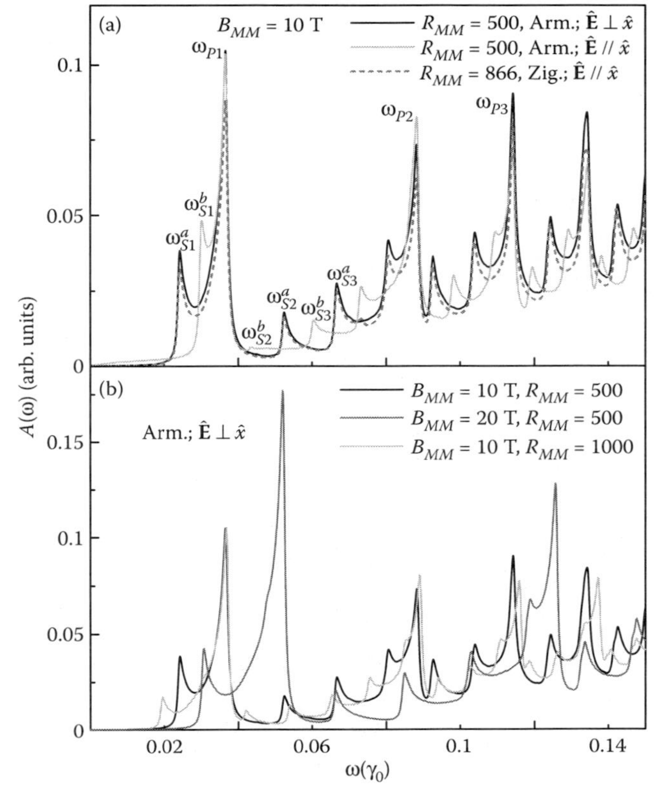

FIGURE 29.6 The optical absorption spectra for (a) $B_{MM} = 10$ T at a fixed periodic length with modulation and polarization along the armchair and zigzag directions and (b) different modulation periods and field strengths with both the modulation and polarization along the armchair direction.

the anisotropy of the polarization direction. For the modulation along the zigzag direction at $R_{MM} = 866$ and $B_{MM} = 10$ T, the absorption spectrum (dashed curve in Figure 29.6a) shows features similar to those of the spectrum corresponding to the armchair direction at $R_{MM} = 500$ and $B_{MM} = 10$ T. $R_{MM} = 866$ for the zigzag direction and $R_{MM} = 500$ for the armchair direction possess the same period length based on the definitions $R_{MM} = l_{MM}/3b'$ and $R_{MM} = l_{MM}/\sqrt{3}b'$ associated with the zigzag and armchair directions, respectively. Moreover, the anisotropic features of the modulation directions will be revealed in the higher frequency region or the smaller modulation length.

As the field strength rises, the peak height and frequency of the principal peaks increase, and the peak number decreases, as shown in Figure 29.6b by the dark gray curve for $R_{MM} = 500$ and $B_{MM} = 20$ T. These results mean that the congregation of electronic states is more pronounced as the field strength grows. In addition to the field strength, the optical-absorption spectrum is also influenced by the modulation period. In Figure 29.6b, the gray curve shows the optical spectra of $B_{MM} = 10$ T for $R_{MM} = 1000$. The subpeaks strongly depend on the period, that is, they represent different peak heights and frequencies with the variation of R_{MM}. However, the opposite is true for the principal peaks.

The peaks in the low-frequency absorption spectra can arise from the different selection rules. Figure 29.4 illustrates the transition channels of the principal peaks resulting from the original band-edge states denoted as ω_{Pn}'s in Figure 29.6a. Each ω_{Pn} corresponds to the transition channels from QLLs $n \rightarrow n+1$ and $n \rightarrow n+1$ at the original band-edge state, and the selection rule is represented by $\Delta n = |n^c - n^v| = 1$, which is same as that related to LLs. The main reason for this is that the subenvelope functions $A_o^{c(v)}(x_1)(A_o^{c(v)}(x_2))$ and $B_o^{v(c)}(x_1)$ $(B_o^{v(c)}(x_2))$ associated with the effective quantum numbers $n+1$ (n) and $n(n+1)$ have the same number of zeros, respectively. As discussed in the former section, peaks arise in the optical absorption spectra when the number of zeros is the same for $A_o^{c(v)}$ and $B_o^{v(c)}$ in Equation 29.14. The subpeaks originating from the extra band-edge states display a more complex behavior. The excitation channels for the subpeaks ω_{Sn}^a and ω_{Sn}^b in Figure 29.6a are shown in Figure 29.4. The subpeaks of different selection rules, $\Delta n = 0$ and 1, come into existence simultaneously. For example, ω_{S2}^a comes from the excitation channel $1\alpha \rightarrow 1\beta$ ($1\beta \rightarrow 1\alpha$) and ω_{S3}^a comes from the excitation channel $1\beta \rightarrow 2\alpha$ ($2\alpha \rightarrow 1\beta$). The extra selection rule $\Delta n = 0$ reflects the overlap of subenvelope functions $A_o^c(x_1)(B_o^c(x_1))$ and $A_o^c(x_2)(B_o^c(x_2))$ located around $x_1 \approx x_2 \approx 1/2$. The subenvelope functions $A_o^{c(v)}(x_1)(A_o^{c(v)}(x_2))$ and $B_o^{v(c)}(x_2)(B_o^{v(c)}(x_1))$ of the effective quantum number n also have the same number of zeros at the identical position, a cause leading to the extra selection rule $\Delta n = 0$.

The frequency of principal peaks in the optical absorption spectra is worth a closer investigation. The relation between the frequencies of the first four principal peaks and the modulation period is shown in Figure 29.7a. The ω_P's present a very weak dependence on the period as R_{MM} becomes sufficiently large, whereas they exhibit a strong dependence on the field strength. The frequencies grow with increased B_{MM}, as shown in Figure 29.7b. The dependence of ω_P's on B_{MM} is similar to what is seen in the case of a uniform perpendicular magnetic field, that is, ω_P's $\propto \sqrt{B_{MM}}$, as indicated by the black solid lines. The predicted results could be verified by optical spectroscopy [14,20,79].

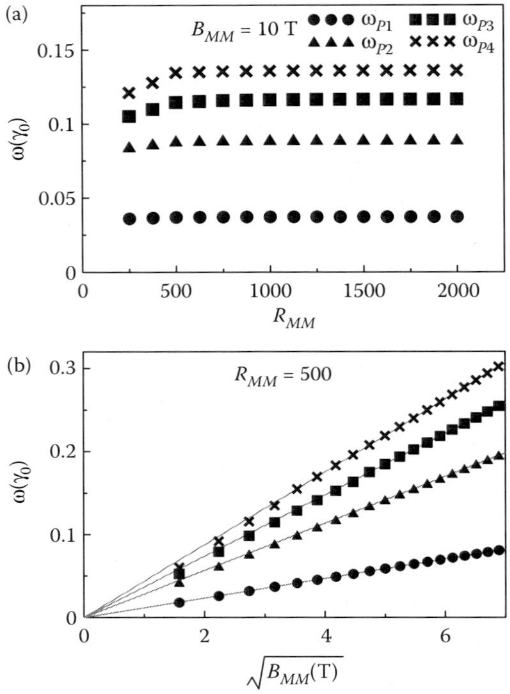

FIGURE 29.7 The dependence of the absorption frequency on (a) the period R_{MM} and (b) the square root of field strength B_{MM}.

29.5 SPATIALLY MODULATED ELECTRIC POTENTIAL

29.5.1 Oscillation Energy Subbands

Besides the spatially modulated magnetic field, the low-energy physical properties can also be strongly tuned by a spatially modulated electric potential. The energy bands for $V_{ME} = 0.05$ γ_0 and $R_{ME} = 500$ are shown in Figure 29.8. The unoccupied conduction subbands are symmetric to the occupied valence subbands about E_F. The parabolic subbands are nondegenerate and oscillate near $k_y = 2/3$. There exists an intersection where two parabolic subbands cross each other at E_F. Each subband has several band-edge states, which lead to the prominent peaks in the DOS and optical absorption spectra. For convenience, these band-edge states are further divided into two categories called μ and ν states, as indicated in Figure 29.8. The two μ (ν) states at the left- and right-hand sites of $k_y = 2/3$ might have a small difference in energies; that is, parabolic bands might be bilaterally asymmetric about $k_y = 2/3$. Not far away from $k_y = 2/3$, the energy subbands with linear dispersions intersect at E_F, preserving more Fermi-momentum states and forming several Dirac cones. Moreover, the number

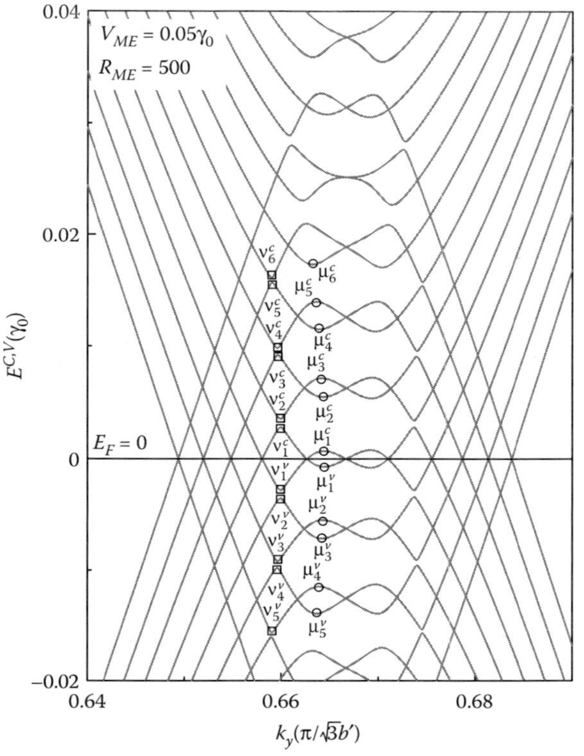

FIGURE 29.8 The energy dispersions for the modulated electric potential along the armchair direction with $R_{ME} = 500$ and $V_{ME} = 0.05\gamma_0$.

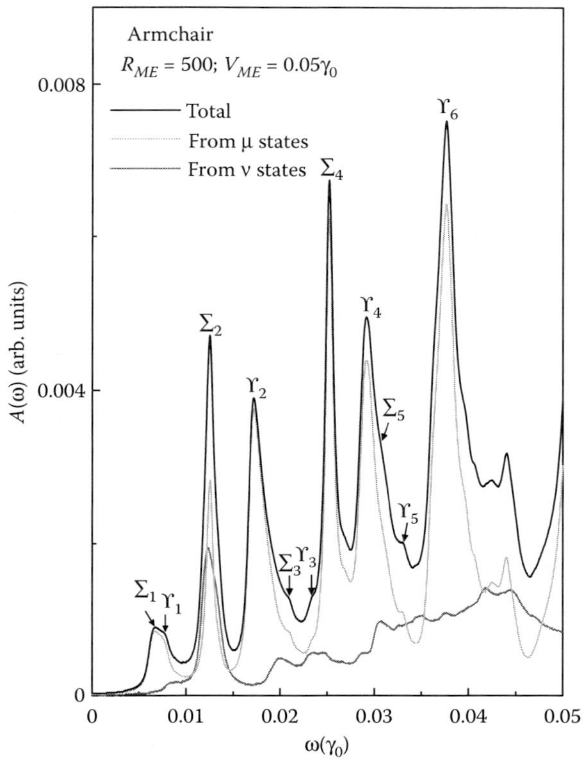

FIGURE 29.9 The optical absorption spectra corresponding to Figure 29.8, which includes the contributions from the μ and ν states, respectively.

of Fermi-momentum states or Dirac cones increases with the potential strength and modulation period.

The optical absorption spectrum for $R_{ME} = 500$ and $V_{ME} = 0.05\ \gamma_0$ along the armchair direction, as shown in Figure 29.9 by the black solid curve, exhibits two groups of prominent peaks, Σ_n's and Υ_n's. They are mainly due to the optical excitations from μ_n^v to μ_{n+1}^c (μ_{n+1}^v to μ_n^c) and μ_n^v to μ_{n+2}^c (μ_{n+2}^v to μ_n^c), respectively. Moreover, with regard to the peak intensity, the peaks Σ_n's (Υ_n's) can be further divided into two subgroups. For example, the peak heights of Σ_1, Σ_3; Σ_5, respectively, resulting from the transitions of μ_1^v to μ_2^c (μ_2^v to μ_1^c), μ_3^v to μ_4^c (μ_4^v to μ_3^c); μ_5^v to μ_6^c (μ_6^v to μ_5^c) are very low, while the peaks Σ_2, Σ_4; Σ_6 originating from the excitations μ_2^v to μ_3^c (μ_3^v to μ_2^c), μ_4^v to μ_5^c (μ_5^v to μ_4^c); μ_6^v to μ_7^c (μ_7^v to μ_6^c) present much stronger intensities than the peaks Σ_1, Σ_3, and Σ_5. That is to say, the peak of Σ_{2n}'s are higher than those of Σ_{2n-1}'s in the group Σ_n. The peaks of Υ_n's exhibit similar features to those of Σ_n's. For instance, peaks Υ_1, Υ_3; Υ_5, respectively, arising from the transitions of μ_1^v to μ_3^c (μ_3^v to μ_1^c), μ_3^v to μ_5^c (μ_5^v to μ_3^c); μ_5^v to μ_7^c (μ_7^v to μ_5^c), own the peaks with very weak intensities. In contrast to Υ_1, Υ_3; Υ_5, the peak intensities of Υ_2 and Υ_4 resulting from the excitations μ_2^v to μ_4^c (μ_4^v to μ_2^c) and μ_4^v to μ_6^c (μ_6^v to μ_4^c) are relatively stronger. Furthermore, the μ and ν states lead to different contributions to the two kinds of optical absorption peaks. Most peaks originating from the two different band-edge states have nearly the same frequencies, while the peak intensities are not the same. The gray and dark gray curves correspond to the optical absorption spectra which contain only the excitations of μ and ν states, respectively. Except for the peak Σ_2 with comparable contributions which are attributed to the transitions of μ and ν states, the other peaks with different contributions from the two states have nearly the same frequency. The peaks from the μ states exhibit much stronger intensities than those from the ν states. In other words, peaks in the optical absorption spectrum mainly result from excitations of the μ states.

29.5.2 Anisotropic Optical Absorption Spectra

The polarization direction and the strength, period, and direction of the modulating electric field strongly affect the features of the optical absorption spectrum. The spectra associated with $\hat{\mathbf{E}} \perp \hat{x}$ (black solid curve) and $\hat{\mathbf{E}} \parallel \hat{x}$ (dark gray solid curve) for $R_{ME} = 500$ and $V_{ME} = 0.05\ \gamma_0$ along the armchair direction and $R_{ME} = 866$ and $V_{ME} = 0.05\gamma_0$ along the zigzag direction (gray solid curve) are shown in Figure 29.10a for a comparison. Compared with the results of $\hat{\mathbf{E}} \perp \hat{x}$ and $\hat{\mathbf{E}} \parallel \hat{x}$, the peak structures related to the two polarization directions are totally different, which reflect the anisotropic behavior of the polarization direction. Similarly, the anisotropy of the modulation directions is reflected by that the absorption spectra corresponding to the armchair and zigzag directions display distinct features,

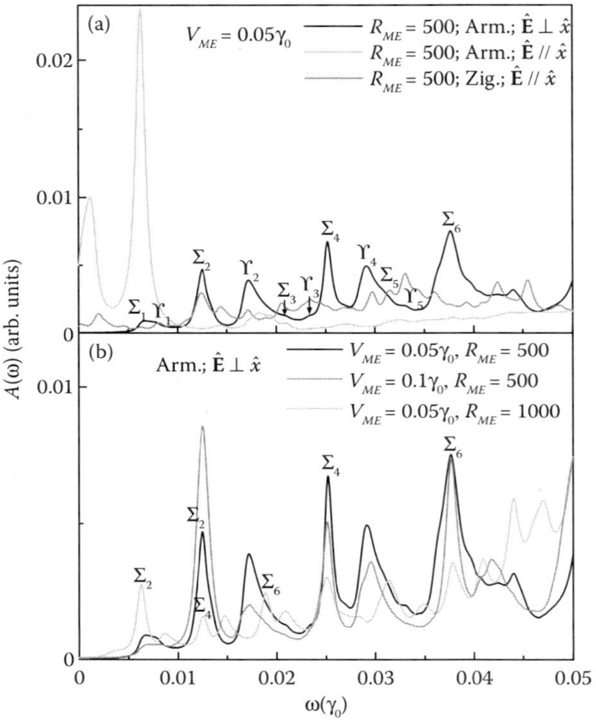

FIGURE 29.10 The optical absorption for (a) $V_{ME} = 0.05\gamma_0$ at a fixed periodic length with modulation and polarization along the armchair and zigzag directions and (b) different modulation periods and field strengths with both the modulation and polarization along the armchair direction.

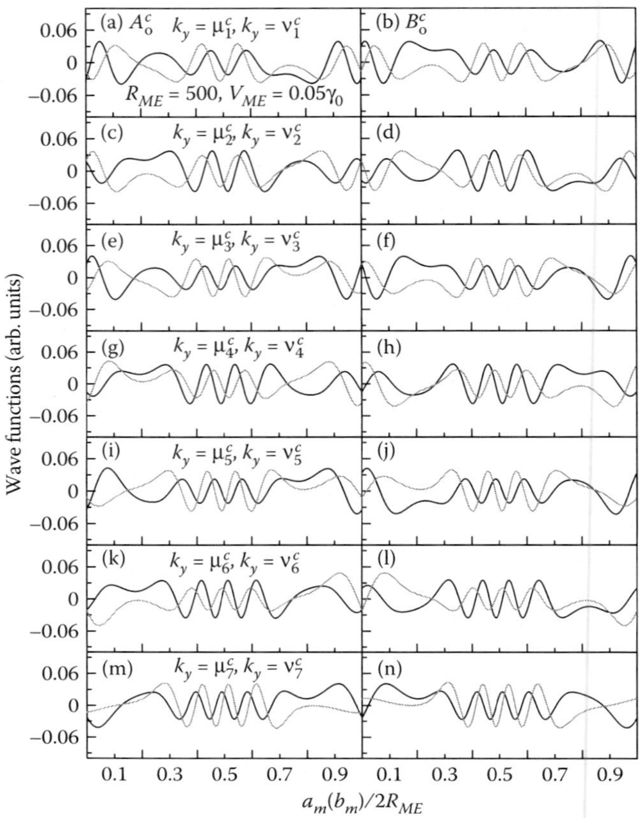

FIGURE 29.11 The wave functions at different band-edge states, the μ_i^c and ν_i^c states for i = 1–7.

that is, the anisotropic behavior of the polarization directions are more obvious than that in the MM case. With increasing the modulation strength to $V_{ME} = 0.1\gamma_0$ (dark gray solid curve in Figure 29.10b), the results show that the peak intensity strongly depends on V_{ME}, but their relationship is not straight forward. For the modulation period, the spectra at a larger $R_{ME} = 1000$ (light gray solid curve) along the armchair direction present features diverse to those in the spectra at $R_{ME} = 1000$. The peak number grows and the peak intensities decay with an increase of the period. A redshift occurs in longer periods. For example, the peak frequencies Σ_1, Σ_3; Σ_5, as indicated in black and gray curves, are almost reduced to half of the original ones when the modulation period is enlarged from 500 to 1000.

The optical absorption spectra in the ME case do not reveal certain selection rules. This is due to the fact that the amplitudes $A_o^{c,v}$ and $B_o^{v,c}$ of the wave functions do not exist as simple relationship similar to that in the UM and MM cases. The wave functions in the ME are no longer distributed around the center location; rather, they display standing-wave-like features in the primitive unit cell and are distributed over the entire primitive cell, as shown in Figure 29.11. However, the wave functions of the edge-states μ and ν exhibit irregular behavior such as disordered numbers of zero points, asymmetric spatial distributions, and random oscillations. These irregular waveforms might result from different site energies for the carbon atoms in the ME.

29.6 UNIFORM MAGNETIC FIELD COMBINED WITH MODULATED MAGNETIC FIELD

29.6.1 Landau Level Spectra Broken by Modulated Magnetic Fields

A further discussion of graphene in a composite field, the UM-MM case, is presented in this section. The main characteristics of the LLs at B_{UM} = 5 T are affected by the MM (B_{MM} = 1 T and R_{MM} = 395), as shown in Figure 29.12a by the black curves. The LL with $n^{c,v} = 0$ at $E_F = 0$ remains the feature same to that of the UM case shown in Figure 29.2a. On the other hand, each dispersionless LL with $n^{c,v} \geq 1$ splits into two parabolic subbands with double degeneracy. The subbands possess two kinds of band-edge states, $n^{c,v}\zeta$ and $n^{c,v}\eta$, which correspond to the minimum field strength $B_{UM} - B_{MM}$ and maximum field strength $B_{UM} + B_{MM}$, respectively. The surrounding electronic states at $n^{c,v}\eta$ congregate more easily, which results in the smaller band curvature. Comparably fewer states congregate at $n^{c,v}\zeta$, and the resulting band curvature is larger. Increasing B_{MM} induces more complex energy spectra, as shown in Figure 29.12b for R_{MM} = 395 and B_{MM} = 5 T. The parabolic subbands with $n^{c,v} \geq 1$ display wider oscillation

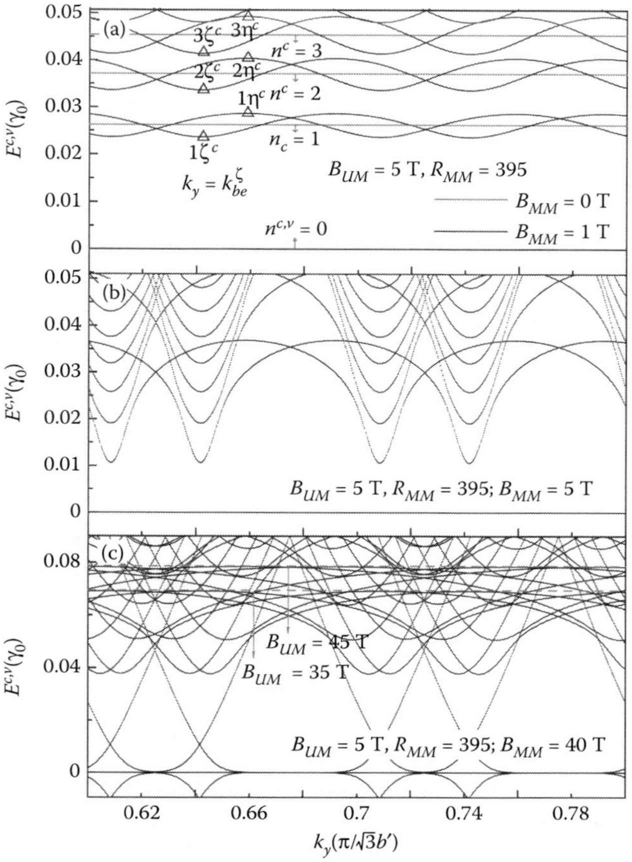

FIGURE 29.12 The energy dispersions for (a) the uniform magnetic field $B_{UM} = 5$ T by the dark gray curves and the composite field $B_{UM} = 5$ T combined with $R_{MM} = 395$ and $B_{MM} = 1$ T by the black curves, (b) $B_{UM} = 5$ T combined with $R_{MM} = 395$ and $B_{MM} = 5$ T, and (c) $B_{UM} = 5$ T combined with $R_{MM} = 395$ and $B_{MM} = 40$ T. All modulated fields are applied along the armchair direction.

amplitudes, stronger energy dispersions, and greater band curvatures. The largest and smallest band curvatures occur at the local minima $n^{c,v}\zeta$ and local maxima $n^{c,v}\eta$ states, respectively. The subband amplitudes are nearly linearly magnified by B_{MM} as $B_{MM} \leq B_{UM}$. It is noticeable that neither the minima of the conduction bands nor the maxima of the valence bands exceed $E_F = 0$ even for B_{MM} much larger than B_{UM}, as shown in Figure 29.12c for $R_{MM} = 395$ and $B_{MM} = 40$ T. Thus, no overlap exists between the conduction and valence bands, regardless of the modulation strength. With further increasing modulated field strength as $B_{MM} \gg B_{UM}$, the electronic structures are expected to approach to those in the MM case.

29.6.2 Symmetry Broken of Landau Level Wave Functions

The LL wave functions modified by the MM are shown in Figure 29.13 and the wave functions of the pure UM case are displayed by the dark gray curves. The spatial distributions corresponding to k_{be}^{ζ}, labeled in Figure 29.12, exhibit slightly broadened and reduced amplitudes, as indicated by

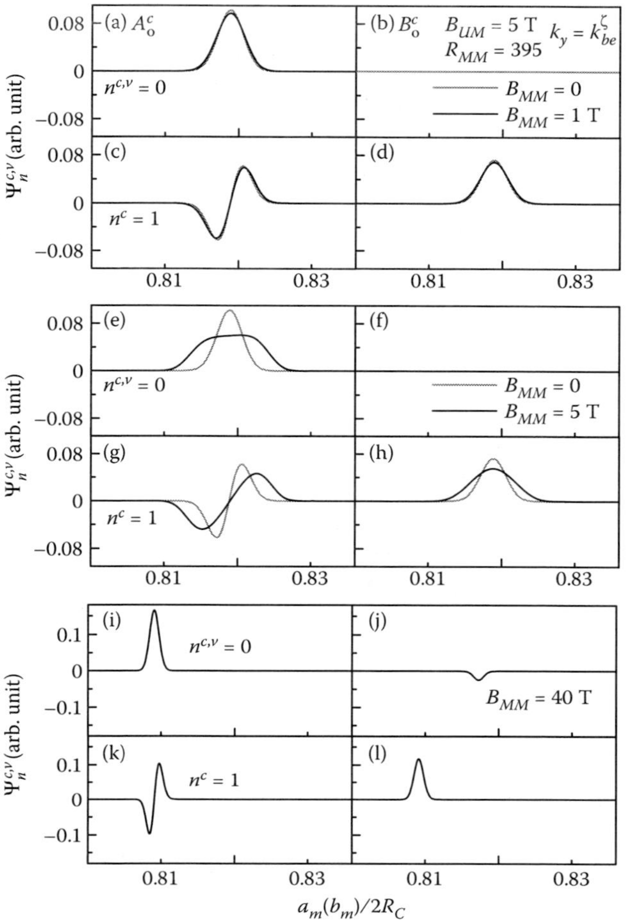

FIGURE 29.13 The wave functions with $n^{c,v} = 0$ and $n^c = 1$ at k_{be}^{ζ} for (a)–(d) the uniform magnetic field $B_{UM} = 5$ T by the dark gray curves and the composite field $B_{UM} = 5$ T combined with $R_{MM} = 395$ and $B_{MM} = 1$ T by the black curves, (e)–(h) $B_{UM} = 5$ T by the dark gray curves and $B_{UM} = 5$ T combined with $R_{MM} = 395$ and $B_{MM} = 5$ T by the black curves, and (i)–(l) $B_{UM} = 5$ T combined with $R_{MM} = 395$ and $B_{MM} = 40$ T.

the black curves in Figure 29.13a through d for $B_{MM} = 1$ T. However, the spatial symmetry and the location centers of the wave functions remain unchanged. Under the influence of a small B_{MM}, the simple relation between $A_o^{c,v}$ and $B_o^{c,v}$ of the wave functions is almost preserved. However, a stronger modulation strength results in greater spatial changes of the wave functions, as shown in Figure 29.13e through f for $B_{UM} = B_{MM} = 5$ T. The increased broadening and asymmetry of the spatial distributions of the wave functions at $n^{c,v} = 0$ are revealed. However, the spatial distributions with $n^{c,v} \geq 1$ are only widened (i.e., $n^c = 1$ in Figure 29.12g and h), but the spatial symmetry is retained. With increasing B_{MM}, as shown in Figure 29.13i through l for $B_{MM} = 40$ T, the symmetry of the wave functions with $n^{c,v} = 0$ is recovered, and one can expect that the main features of the wave functions will become similar to those in the MM case. Obviously, the electronic properties show critical changes as B_{MM} equals B_{UM}, which should be reflected to the optical properties.

29.6.3 Magneto-Optical Absorption Spectra with Extra Selection Rules

The optical absorption spectra corresponding to Figure 29.13 are shown in Figures 29.14 and 29.15. In Figure 29.14a, the absorption spectra corresponding to the UM-MM case at $B_{UM} = 5$ T with $R_{MM} = 395$ and $B_{MM} = 1$ T and the UM case at $B_{UM} = 5$ T are shown together for a comparison. The dark gray curves coming from the LLs at $B_{UM} = 5$ T display delta-function-like peaks $\omega_{LL}^{nn'}$ with the selection rule $\Delta n = 1$. However, the MM modifies each delta-function-like peak into two split square-root-divergent peaks, $\omega_{\zeta}^{nn'}$ and $\omega_{\eta}^{nn'}$, as shown by the black curves. Each $\omega_{\zeta}^{nn'}$ ($\omega_{\eta}^{nn'}$) originates from the transitions of $n\zeta \rightarrow n+1\zeta$ and $n+1\zeta \rightarrow n\zeta$ ($n\eta \rightarrow n+1\eta$ and $n+1\eta \rightarrow n\eta$) and its absorption frequency is same as that generated from the LLs at $B_{UM} - B_{MM} = 4$ T ($B_{UM} - B_{MM} = 6$ T). These absorption peaks obey a selection rule, $\Delta n = 1$, similar to that in the UM case.

By increasing the modulated field strength as $B_{MM} = B_{UM} = 5$ T, the absorption spectrum has evident variety, as shown in Figure 29.14b. In addition to the peaks $\omega_{\zeta}^{nn'}$ and $\omega_{\eta}^{nn'}$ with the selection rule $\Delta n = 1$, two extra peaks with $\Delta n = 2$ and 3, ω_{ζ}^{02} and ω_{ζ}^{03}, are generated. These two peaks do reflect the fact that the wave functions of the LLs with $n^{c,v} = 0$ are destroyed by the MM. As the modulated field strength further raises to $B_{MM} = 40$ T (dashed curves Figure 29.15), the spectrum displays some features similar to those of the spectrum in the MM case at $R_{MM} = 395$ and $B_{MM} = 40$ T (black solid curves in Figure 29.15), that is, the principal peaks ω_P's and the subpeaks ω_S's in the MM case are also shown in the UM-MM case as $B_{MM} > B_{UM}$. Moreover, the subpeaks features, which are associated with the positions at the net field strength equal to zero, are almost the same in both the MM and UM-MM cases. The principal peaks, however, possess a pair structure with ω_{Pn}^{-} and ω_{Pn}^{+}, which, respectively, correspond to two different field strengths, $|B_{UM} - B_{MM}| = 35$ T and $|B_{UM} + B_{MM}| = 45$ T, and thus the difference between two field strengths lead to distinct absorption frequencies. For $B_{MM} \gg B_{UM}$, one can anticipate that the frequency discrepancy between the pair ω_{Pn}^{-} and ω_{Pn}^{+} becomes very small, and then they will merge into one single peak, ω_{Pn}, that is, the absorption spectrum restores to that in the pure MM case.

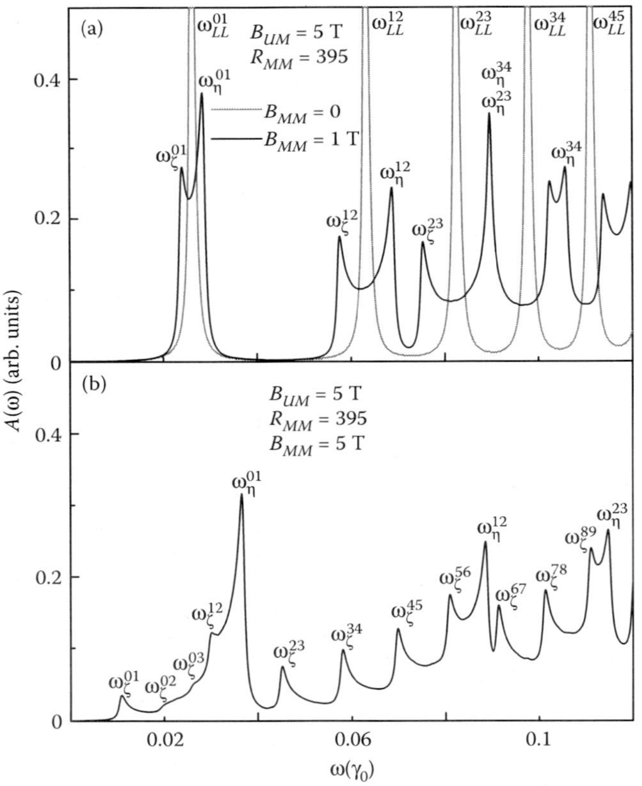

FIGURE 29.14 The optical absorption spectra corresponding to (a) the uniform magnetic field $B_{UM} = 5$ T by the dark gray curve and the composite field $B_{UM} = 5$ T combined with $R_{MM} = 395$ and $B_{MM} = 1$ T by the black curve and (b) the composite field $B_{UM} = 5$ T combined with $R_{MM} = 395$ and $B_{MM} = 5$ T.

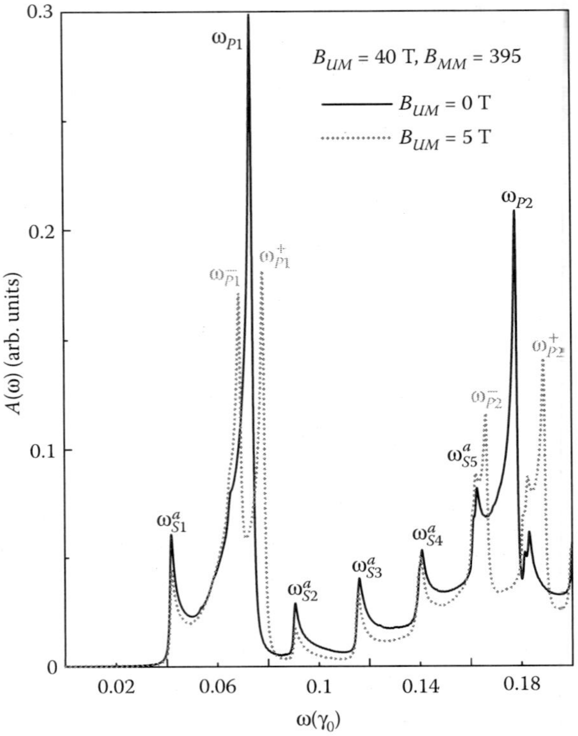

FIGURE 29.15 The optical absorption spectra corresponding to the composite field $B_{UM} = 5$ T combined with $R_{MM} = 395$ and $B_{MM} = 40$ T by the dashed curve and the pure modulated magnetic field $R_{MM} = 395$ and $B_{MM} = 40$ T by the black curve.

The dependence of the absorption frequency on the modulated field strength is shown in Figure 29.16 for $B_{MM} \leq 5$ T. In the range of $B_{MM} \leq B_{UM}$, each of absorption peaks $\omega_{\zeta}^{nn'}$ and $\omega_{\eta}^{nn'}$ is linearly dependent on B_{MM}. This reflects the fact that the subband amplitudes are nearly linearly magnified by B_{MM} within the range. However, in the higher absorption frequency

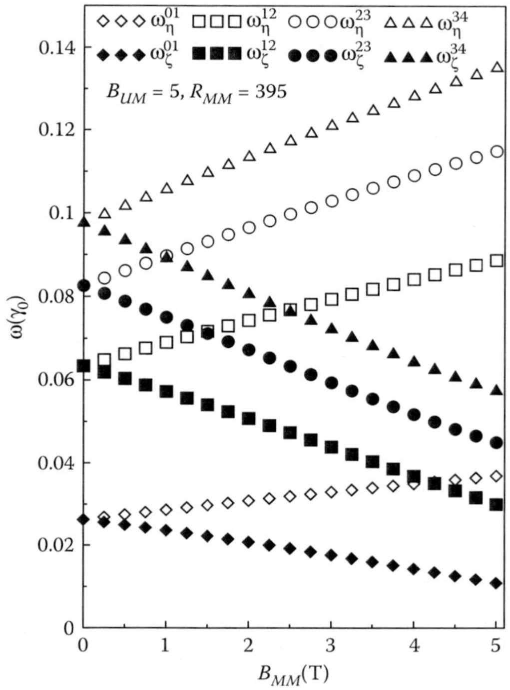

FIGURE 29.16 The dependence of absorption frequencies, $\omega_\eta^{nn'}$ and $\omega_\zeta^{nn'}$ with $|\Delta n| = |n - n'| = 1$, on the modulated strength B_{MM}.

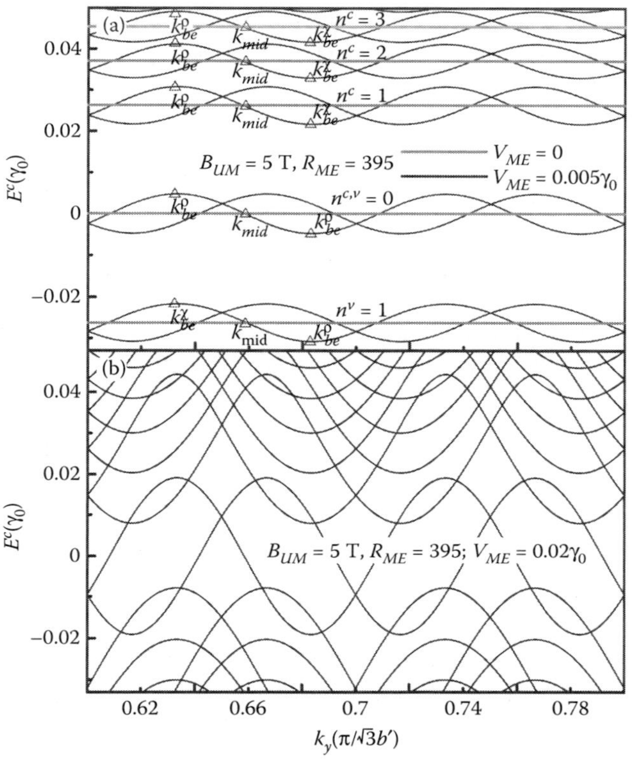

FIGURE 29.17 The energy dispersions for (a) the uniform magnetic field $B_{UM} = 5$ T by the dark gray curves and the composite field $B_{UM} = 5$ T combined with $R_{MM} = 395$ and $V_{ME} = 0.005\gamma_0$ by the black curves, (b) $B_{UM} = 5$ T combined with $R_{MM} = 395$ and $V_{ME} = 0.02\gamma_0$. All modulated fields are applied along the armchair direction.

region or in the field range of $B_{MM} > B_{UM}$, the linear-dependence relationship will be broken since the subbands become overlapping and the subband amplitudes are not linearly magnified by B_{MM} anymore.

29.7 UNIFORM MAGNETIC FIELD COMBINED WITH MODULATED ELECTRIC POTENTIAL

29.7.1 Landau Level Spectra Broken by Modulated Electric Potentials

Compared with the situation in an MM, an ME creates distinct effects on the LLs, as shown in Figure 29.17a by the black curves. The 0D LLs become the 1D sinusoidal energy subbands when electronic states are affected by the periodic electric potential. Each LL with fourfold degeneracy is split into two doubly degenerate Landau subbands (LSs), as shown in Figure 29.17a by the black curves. Each LS owns two types of extra band-edge states, $k_{be}^{\varkappa}$ and k_{be}^{ϱ}. For the conduction (valence) LSs, the band-edge states of $k_{be}^{\varkappa}$ and k_{be}^{ϱ} (k_{be}^{ϱ} and $k_{be}^{\varkappa}$) are, respectively, related to the wave functions, which possess a localization center at the minimum and maximum electric potentials (discussed in the ME case). It should be noted that the two LSs of $n^{c,v} = 0$ only have the ϱ-type band-edge states. Under a small modulation strength, the energy spacings (E_s's) or band curvatures for both $k_{be}^{\varkappa}$ and k_{be}^{ϱ} are almost the same, where E_s is the spacing between a LS and a LL at $k_{be}^{\varkappa}$ or k_{be}^{ϱ}. On the other hand, when the modulation strength is sufficiently large (e.g., $V_{ME} = 0.02\ \gamma_0$ in Figure 29.17b), the energy dispersions of LSs are relatively strong and E_s decreases with increasing state energies, that is, the oscillations of LSs decline with an increase of $n^{c,v}$. In comparison to the k_{be}^{ϱ} state, the $k_{be}^{\varkappa}$ state owns the smaller energy spacing and band curvature. Such differences are associated with the localization of the wave function within the potential well.

29.7.2 Landau Level Wave Functions Broken by Modulated Electric Potentials

The main features of the LL wave functions are altered by the ME. The illustrated wave functions at the band-edge states and the midpoint (k_{mid}, indicated in Figure 29.17a) between two band-edge states are used to examine the modulation effects. The wave functions at k_{mid} are modified by $V_{ME}(x)$, as shown in Figure 29.18a through h for $B_{UM} = 5$ T and $R_{ME} = 395$ at $V_{ME} = 0$, 0.005, and 0.02 γ_0. $A_{\mathbf{o}}^{c,v}$ of $n^{c,v} = 0$ is slightly reduced, while $B_{\mathbf{o}}^{c,v}$ of $n^{c,v} = 0$ is slightly increased (Figure 29.18a and b) after V_{ME} is introduced. This means that carriers are transferred between the a- and b-sublattices. With an increasing $n^{c,v}$, the spatial distribution symmetry of the LL wave functions is broken. The conduction and valence wave functions are, respectively, shifted toward the $+\hat{x}$ and $-\hat{x}$ directions, as shown in Figure 29.18e through h for example. The proportionality relationship between $A_{\mathbf{o}}^{c,v}$ of $n^{c,v}$ and $B_{\mathbf{o}}^{c,v}$ of $n^{c,v} + 1$

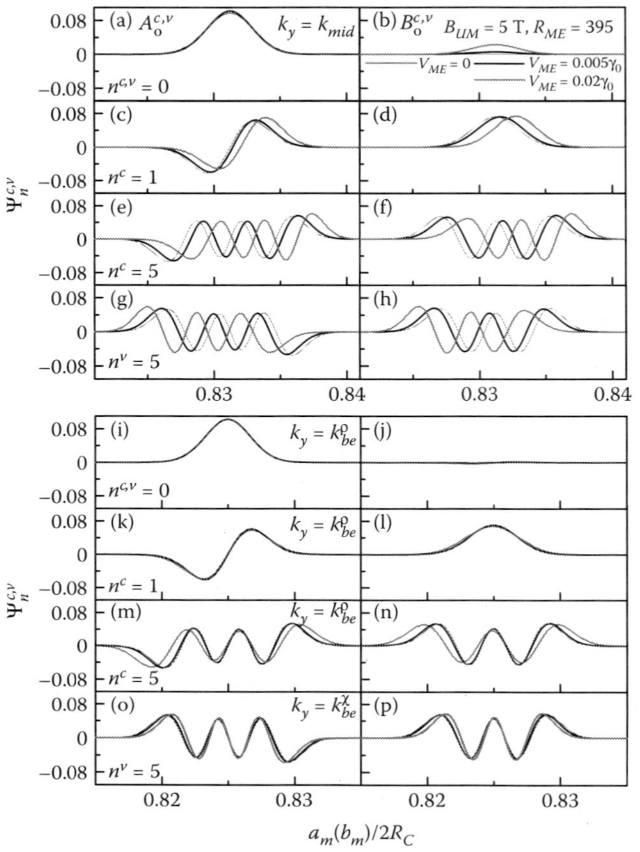

FIGURE 29.18 The wave functions for (a)–(h) $n^{c,v} = 0$, $n^c = 1$, $n^c = 5$, and $n^v = 5$ at k_{mid}, (i)–(n) $n^{c,v} = 0$, $n^c = 1$, and $n^c = 5$ at k_{be}^{ϱ}, and (o), (p) $n^v = 5$ at $k_{be}^{\varkappa}$. The results corresponding to the uniform magnetic field $B_{UM} = 5$ T, the composite field $B_{UM} = 5$ T combined with $R_{MM} = 395$ and $V_{ME} = 0.005\gamma_0$, and $B_{UM} = 5$ T combined with $R_{MM} = 395$ and $V_{ME} = 0.02\gamma_0$ are indicated by the dark gray, black, and gray curves, respectively.

no longer exists, and neither do the relationships $A_o^c = A_o^v$ and $B_o^c = -B_o^v$. Moreover, the stronger V_{ME} leads to greater changes in the spatial distributions of the wave functions. The spatial distributions of the wave functions strongly depend on k_y. As for the band-edge states, the aforementioned relationships of the wave functions are absent when $n^{c,v}$'s are sufficiently large enough. For small $n^{c,v}$'s (Figure 29.18i through l), wave functions are less influenced by the ME. However, the spatial distribution of LS with a larger $n^{c,v}$ becomes wider (narrower) for the k_{be}^{ϱ} ($k_{be}^{\varkappa}$) state, as shown in Figure 29.18m through p. Moreover, the localization center of the band-edge states is hardly affected by $V_{ME}(x)$.

29.7.3 Magneto-Optical Absorption Spectra Destroyed by Modulated Electric Potentials

Under an ME, the changes in the electronic properties of LLs are manifested in the optical absorption spectra. Each LL is split into two kinds of sinusoidal subbands, with one leading the other by 1/6 of a period (Figure 29.17a). The spatial localization region of the wave function is completely different between two kinds of subbands (not shown), but identical in corresponding to the same kind of LSs with different quantum numbers at the same k_y. This indicates that the optical transition between two different kinds of subbands is forbidden. The absorption spectrum for $R_{ME} = 395$ at $V_{ME} = 0.005\gamma_0$ is shown in Figure 29.19a by the black line. Each absorption peak, $\omega^{nn'}$, originates from a transition between two LSs with quantum numbers n and n' for the same kind of subbands. In addition to the original peaks, which correspond to the selection rule $\Delta n = 1$ similar to that of LLs, there are extra peaks not characterized by the same selection rule. In the frequency range $\omega < 0.1\ \gamma_0$, the peak intensity of ω_{LL}^{nn+1}, significantly reduced by V_{ME}, declines as the frequency increases. The extra peaks with $\Delta n \neq 1$ behave the opposite way. For a small V_{ME}, the peaks with $\Delta n = 1$ are much stronger than those with $\Delta n \neq 1$.

The original and extra peaks can be explained by the subband transitions associated with a certain set of k_y points. For the $k_{be}^{\varkappa(\varrho)} \to k_{be}^{\varrho(\varkappa)}$ transitions, the corresponding band-edge states have a high DOS and the symmetry of wave function is little changed. Therefore, they can promote the prominent peaks consistent with the $\Delta n = 1$ selection rule. As shown in Figure 29.19b by the thin dashed lines, $\omega_{\varkappa\varrho}^{nn+1}$ and $\omega_{\varrho\varkappa}^{nn+1}$ represent the absorption peaks from the transition channels $\left[nk_{be}^{\varkappa} \to (n+1)k_{be}^{\varrho}, (n+1)k_{be}^{\varrho} \to nk_{be}^{\varkappa}\right]$ and $\left[nk_{be}^{\varrho} \to (n+1)k_{be}^{\varkappa}, (n+1)k_{be}^{\varkappa} \to nk_{be}^{\varrho}\right]$, respectively. These two peaks are close to each other, and almost overlap with the original peaks. But in cases where either n or n' is zero, only the peak $\omega_{\varrho\varkappa}^{01}$ can be created by two transition channels: $\left[0k_{be}^{\varrho} \to 1k_{be}^{\varkappa}, 1k_{be}^{\varkappa} \to 0k_{be}^{\varrho}\right]$. The reduction of the transition channels is due to the fact that the $n^{c,v} = 0$ subbands oscillate between the conduction and valence bands. For the $k_{mid} \to k_{mid}$ transitions, the significant change to the symmetry of the wave function results in different selection rules, that is, $\Delta n \neq 1$. As shown in Figure 29.19b by the thick dashed lines, the absorption peak $\omega_{mid}^{nn'}$ corresponds to the transitions between two LSs of $n^v(^c)$ and $n'^c(^v)$ from the k_{mid} states. These types of peaks are responsible for the extra peaks in Figure 29.19a, that is, the $\Delta n \neq 1$ peaks primarily arise from the middle states at lower modulation strength. The $k_{mid} \to k_{mid}$ transitions also contribute to the original peaks. Such contributions decline with an increased frequency. However, the $\Delta n = 1$ absorption peaks are dominated by both the band-edge and middle states.

The peak intensities from different selection rules change dramatically with respect to the variation in frequency. The intensities of the $\Delta n = 1$ peaks gradually rise for a further increase in frequency, while the opposite is true for those of the $\Delta n \neq 1$ peaks (Figure 29.19c). Within the frequency range of $0.1\ \gamma_0 < \omega < 0.2\ \gamma_0$, the former is lower than the latter. The main reason for this is the fact that the symmetry violation of the wave functions is enhanced for LSs with larger $n^{c,v}$'s. It is deduced that the extra absorption peaks of $\Delta n \neq 1$ are relatively easily observed for experimental measurements at

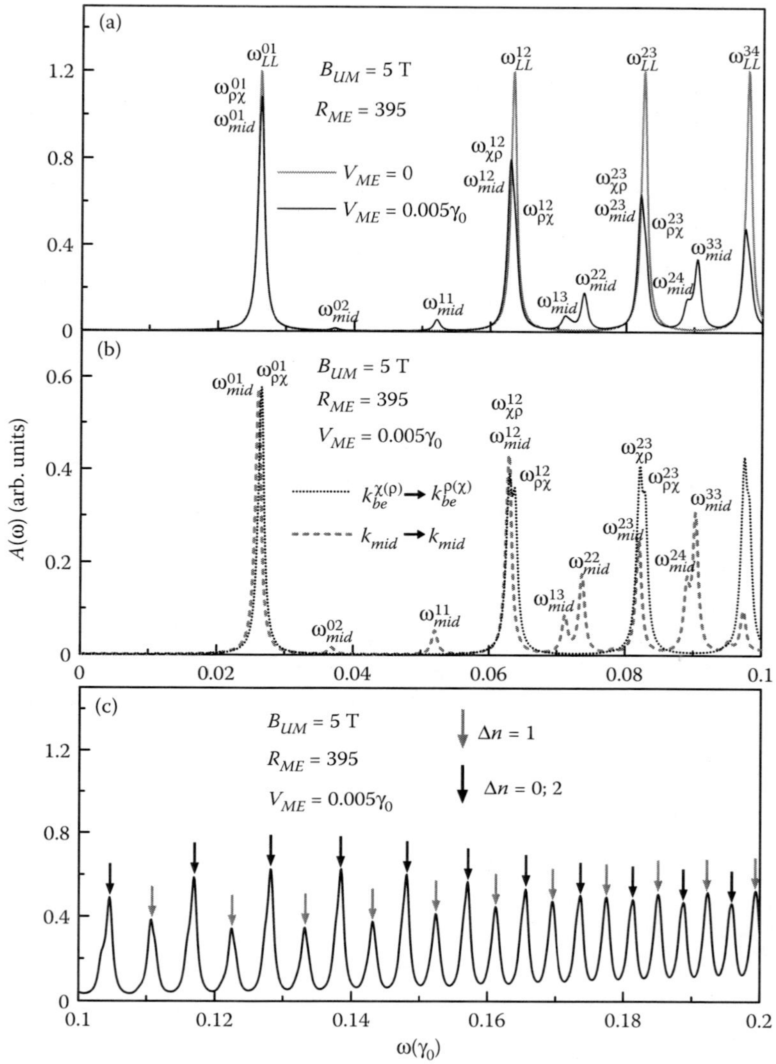

FIGURE 29.19 The optical absorption spectra corresponding to (a) the uniform magnetic field $B_{UM} = 5$ T by the dark gray curve and the composite field $B_{UM} = 5$ T combined with $R_{ME} = 395$ and $V_{ME} = 0.005\gamma_0$ by the black curve, (b) the transitions of $k_{be}^{\chi(\rho)} \rightarrow k_{be}^{\rho(\chi)}$ and $k_{mid} \rightarrow k_{mid}$ for $B_{UM} = 5$ T combined with $R_{ME} = 395$ and $V_{ME} = 0.005\gamma_0$ by the dotted and dashed curves, respectively, and (c) the higher frequency region of 0.1–0.2γ_0.

higher frequencies. Each of them is composed of two peaks, ω_{mid}^{nn} and ω_{mid}^{n-1n+1}, which satisfy $\Delta n = 0$ and $\Delta n = 2$ with nearly the same frequency.

The absorption spectrum exhibits more features for a higher modulated potential, as shown in Figure 29.20 for $R_{ME} = 395$ at $V_{ME} = 0$, $0.02\gamma_0$. In Figure 29.20a, the intensity of the extra peaks becomes comparable to that of the original peaks. Unlike in the lower V_{ME} case (Figure 29.19a), the peak intensities, regardless of their types, vary irregularly with the frequency. The oscillations of the subband structure are obviously augmented and cause the subband transitions to change greatly with respect to k_y. As a result, the absorption peaks grow much wider and split more evenly. Peaks generated by different selection rules are likely to appear, owing to the severe breakdown of the spatial symmetry of the wave functions. It should be noted that it is difficult to distinguish the original from the extra peaks solely based on the peak heights. Besides the aforementioned spectrum analysis, further discussions are made regarding two specific k_y points. For the $k_{be}^{\varkappa(\varrho)} \rightarrow k_{be}^{\varrho(\varkappa)}$ transitions, $\varkappa$- and ϱ-type band-edge states show different behavior in terms of energy spacing and curvature such that the two transition channels $[nk_{be}^{\varkappa} \rightarrow (n+1)k_{be}^{\varrho}, (n+1)k_{be}^{\varrho} \rightarrow nk_{be}^{\varkappa}]$ and $[nk_{be}^{\varrho} \rightarrow (n+1)k_{be}^{\varkappa}, (n+1)k_{be}^{\varkappa} \rightarrow nk_{be}^{\varrho}]$ possess distinct frequencies (thin dashed line in Figure 29.20b). Therefore, the initially coinciding peaks split up, and the original intensities are divided into fractions. Moreover, the band-edge states can induce the extra peaks of $\Delta n = 0$, for example, $\omega_{\varrho\varkappa}^{11}$ and $\omega_{\varrho\varkappa}^{22}$. When $k_y = k_{mid}$, the symmetry of the wave functions is destroyed. Consequently, the extra peaks are strengthened, and the original peaks are weakened or even disappear (thick dashed line in Figure 29.20b). The corruption of the

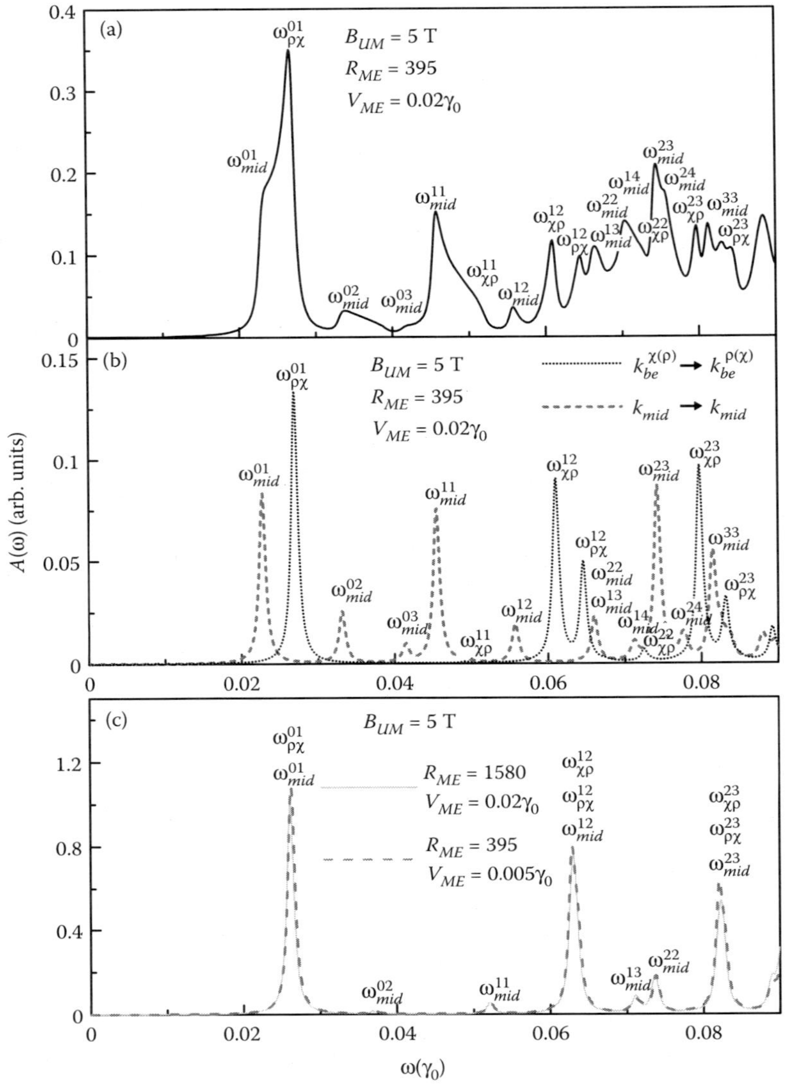

FIGURE 29.20 The optical absorption spectra for the composite field $B_{UM} = 5$ T combined with $R_{ME} = 395$ and $V_{ME} = 0.02\gamma_0$, where the contributions from the transitions $k_{be}^{\chi(\rho)} \rightarrow k_{be}^{\rho(\chi)}$ and $k_{mid} \rightarrow k_{mid}$ are shown in (b) by the dotted and dashed curves, respectively. (c) A comparison between the absorption spectra for $B_{UM} = 5$ T combined with $R_{ME} = 395$ and $V_{ME} = 0.005\gamma_0$, and $B_{UM} = 5$ T combined with $R_{ME} = 1580$ and $V_{ME} = 0.02\gamma_0$.

orthogonality of the sublattices $A_o^{c(v)}$ and $B_o^{v(c)}$ enables the subband transitions to occur from $\Delta n = 3$; such examples are seen in ω_{mid}^{03} and ω_{mid}^{14}. The very strong dispersions of LSs also lead to the splitting of the original peaks associated with k_{mid}. This can account for the $(\omega_{mid}^{01}, \omega_{\varrho\varkappa}^{01})$ peaks and the $(\omega_{mid}^{23}, \omega_{\varkappa\varrho}^{23}, \omega_{\varrho\varkappa}^{23})$ peaks in Figure 29.20a.

The modulation period strongly affects the magneto-optical spectrum of the LSs, while the modulation effect is less significant with sufficiently larger period. The absorption peaks from $\Delta n = 1$ are much higher than those from $\Delta n \neq 1$, as shown by the gray solid line in Figure 29.20c for $V_{ME} = 0.02\gamma_0$ at a larger period $R = 1580$. The transition energies between the valence and conduction LSs at different k_y values are almost the same. Thus, the frequencies $\omega_{\varkappa\varrho}^{nn+1}$, $\omega_{\varrho\varkappa}^{nn+1}$, and ω_{mid}^{nn+1} associated with these LS transitions are all the same. Since the symmetry of the wave function does not degrade much, the absorption peaks of $\Delta n = 3$ no longer exist. The modulated electric field $\mathbf{E} = -\nabla_x V_{ME}(x) = (2\pi V_{ME}/3b'R_{ME})\sin(2\pi x/3b'R_{ME})\hat{x}$ implies that the same value of V_{ME}/R_{ME} produces the same modulation effect. For instance, the absorption spectrum for $R_{ME} = 395$ at $V_{ME} = 0.005\gamma_0$ (the dashed curve in Figure 29.20c) is almost same as that for $R_{ME} = 395$ at $V_{ME} = 0$, $0.005\gamma_0$ owing to $V_{ME}/R_{ME} = 0.005/395 = 0.02/1580$.

We look at the relationship between the absorption frequency and the potential strength more closely. At $V_{ME} = 0$, the peaks denoted by the gray symbols in Figure 29.21 can only appear if they obey the selection rule $\Delta n = 1$. After an external modulation potential is applied, peaks from other selection rules ($\Delta n = 0$ and $\Delta n = 2$) appear as shown by the black symbols. Under a weak modulated potential ($V_{ME} < 0.005\ \gamma_0$),

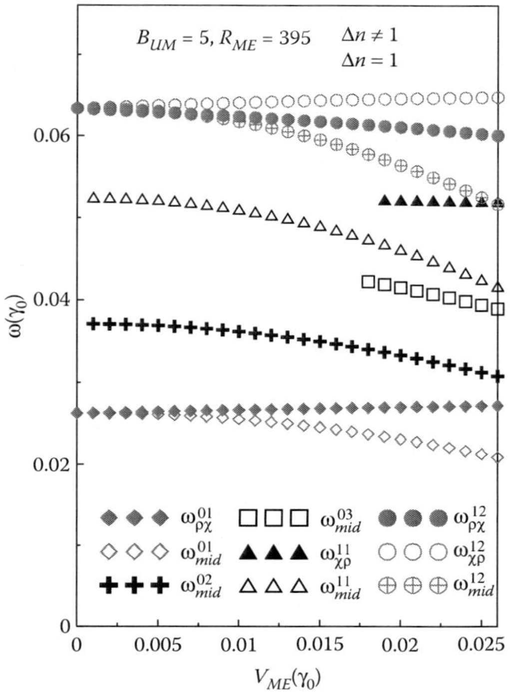

FIGURE 29.21 The dependence of absorption frequencies, $\omega^{nn'}_{\chi\rho}$, $\omega^{nn'}_{\rho\chi}$, and $\omega^{nn'}_{mid}$, on the modulated strength V_{ME}. The absorption frequencies with the selection rules $|\Delta n| = |n - n'| = 1$ and $|\Delta n| \neq 1$ are indicated by the gray and black colors, respectively.

the peak frequency is hardly affected. If the electric potential continues to grow, some peaks start to split, for example, $\omega^{nn+1}_{\chi\varrho}$, $\omega^{nn+1}_{\varrho\chi}$, and ω^{nn+1}_{mid}. Under a strong potential ($V_{ME} > 0.02$ γ_0), more peaks are created by other selection rules, such as $\Delta n = 3$ (square markers). Even for the transitions between band-edge states, peaks can be developed by the rule $\Delta n \neq 1$ (solid triangles). Finally, the absorption frequencies related to the k_{mid} states decrease rapidly with respect to the increment of V_{ME}. This is due to the fact that the conduction and valence LSs at k_{mid} move closer to the Fermi level. These theoretical predictions could be examined by the optical-absorption spectroscopy methods [76,83–86].

29.8 CONCLUSION

The results show that MG exhibits the rich optical absorption spectra, an effect being controlled by the external fields. Such fields have a strong influence on the number, intensity, frequency, and structure of absorption peaks. Moreover, there would exist the dissimilar selection rules for different external fields. In the presence of the UM, the magneto-optical excitations obey the specific selection rule $\Delta n = 1$, since the simple relationship exists between the two sublattices of *a*- and *b*-atoms. As to an MM, an extra selection rule $\Delta n = 0$ is obtained, due to the complex overlapping behavior from two subenvelope functions in the wave function. However, the wave functions exhibit irregular behaviors under the ME. As a result, it is difficult to single out a particular selection rule. In the composite fields, the symmetry of the LL wave functions is broken by introducing of two kinds of modulation fields, the modulated magnetic and electric fields, a cause resulting in the altered selection rules in the absorption excitations. One case is the UM-MM. The extra selection rules, for example, $\Delta n = 2$ and 3, exist when B_{MM} is comparable to B_{UM}. Another case is the UM-ME. The extra selection rules, for example, $\Delta n = 0$, 2, and 3, would be generated by the increase of V_{ME}.

For the other graphene systems, the magneto-optical properties corresponding to a perpendicular uniform magnetic field deserve a closer investigation. For example, the AA-stacked bilayer graphene is predicted to exhibit two groups of absorption peaks [72]; however, the selection rule $\Delta n = 1$ is same as that of MG. As for the AB-stacked bilayer graphene, there exist four groups of absorption peaks and two extra selection rules ($\Delta n = 0$ and 2). FLGs are expected to display more complex magneto-optical absorption spectra, mainly owing to the number of layers and the stacking configuration.

In this chapter, the generalized TB model is introduced to discuss MG under five kinds of external fields. The Hamiltonian, which determines the magneto-electronic properties, is a giant Hermitian matrix for the experimental fields. It is transformed into a band-like matrix by rearranging the TB functions; furthermore, the characteristics of wave function distributions in the sublattices are used to reduce the numerical computation time. In the generalized TB model, the π-electronic structure of MG is solved in the wide energy range of ± 6 eV, a solution proving valid even if the magnetic, electric, or composite field is applied. Moreover, the important interlayer atomic interactions, not just treated as the perturbations, could be simultaneous included in the calculations. The generalized model can also be extensible to other layer stacked systems, that is, AA-, AB-, ABC-stacked FLGs [87–91] and bulk graphite [57,92–94].

REFERENCES

1. K. S. Novoselov, A. K. Geim, S. V. Morozov, D. Jiang, Y. Zhang, and S. V. Dubonos, Electric field effect in atomically thin carbon films. *Science* **306**, 2004: 666.
2. K. S. Novoselov, A. K. Geim, S. V. Morozov, D. Jiang, M. I. Katsnelson, I. V. Grigorieva, S. V. Dubonos, and A. A. Firsov, Two-dimensional gas of massless Dirac fermions in graphene. *Nature (London)* **438**, 2005: 197.
3. C. R. Dean et al., Hofstadter's butterfly and the fractal quantum Hall effect in moire superlattices. *Nature* **497**, 2013: 598–602.
4. G. S. Diniz, M. R. Guassi, and F. Qu, Engineering the quantum anomalous Hall effect in graphene with uniaxial strains. *J. Appl. Phys.* **114**, 2013: 243701.
5. S. Gattenloehner, W. R. Hannes, P. M. Ostrovsky, I. V. Gornyi, A. D. Mirlin, and M. Titov, Quantum Hall criticality and localization in graphene with short-range impurities at the Dirac point. *Phys. Rev. Lett.* **112**, 2014: 026802.
6. M. Golor, T. C. Lang, and S. Wessel, Quantum Monte Carlo studies of edge magnetism in chiral graphene nanoribbons. *Phys. Rev. B* **87**, 2013: 155441.

7. G. Gumbs, A. Iurov, H. Danhong, P. Fekete, and L. Zhemchuzhna, Effects of periodic scattering potential on Landau quantization and ballistic transport of electrons in graphene. *AIP Conf. Proc.* **1590**, 2014: 134–142.
8. B. Jabakhanji, C. Consejo, N. Camara, W. Desrat, P. Godignon, and B. Jouault, Quantum Hall effect of self-organized graphene monolayers on the C-face of 6H-SiC. *J. Phys. D: Appl. Phys.* **47**, 2014: 094009.
9. Y. Kim, K. Choi, J. Ihm, and H. Jin, Topological domain walls and quantum valley Hall effects in silicene. *Phys. Rev. B* **89**, 2014: 085429.
10. Y.-X. Wang, F.-X. Li, and Y.-M. Wu, Quantum Hall effect of Haldane model under magnetic field. *EPL* **105**, 2014: 17002.
11. Q. Zhenhua, R. Wei, C. Hua, L. Bellaiche, Z. Zhenyu, A. H. MacDonald, and N. Qian, Quantum anomalous Hall effect in graphene proximity coupled to an antiferromagnetic insulator. *Phys. Rev. Lett.* **112**, 2014: 116404.
12. J. H. Ho, Y. H. Lai, Y. H. Chiu, and M. F. Lin, Landau levels in graphene. *Physica E* **40**, 2008: 1722.
13. F. D. M. Haldane, Model for a quantum Hall effect without Landau levels: Condensed-matter realization of the "parity anomaly". *Phys. Rev. Lett.* **61**, 1988: 2015.
14. Z. Jiang, E. A. Henriksen, L. C. Tung, Y. J. Wang, M. E. Schwartz, M. Y. Hun, P. Kim, and H. L. Stormer, Infrared spectroscopy of Landau levels of graphene. *Phys. Rev. Lett.* **98**, 2007: 197403.
15. Y. Zhang, Y. W. Tan, H. L. Stormer, and P. Kim, Experimental observation of the quantum Hall effect and Berry's phase in graphene. *Nature* **438**, 2005: 201.
16. P. R. Wallace, The band theory of graphite. *Phys. Rev.* **71**, 1947: 622.
17. M. I. Katsnelson, K. S. Novoselov, and A. K. Geim, Chiral tunnelling and the Klein paradox in graphene. *Nat. Phys.* **2**, 2006: 620.
18. K. I. Bolotin, F. Ghahari, M. D. Shulman, H. L. Stormer, and P. Kim, Observation of the fractional quantum Hall effect in graphene. *Nature* **462**, 2009: 196.
19. K. S. Novoselov et al., Room-temperature quantum Hall effect in graphene. *Science* **315**, 2007: 1379.
20. R. S. Deacon, K. C. Chuang, R. J. Nicholas, K. S. Novoselov, and A. K. Geim, Cyclotron resonance study of the electron and hole velocity in graphene monolayers. *Phys. Rev. B* **76**, 2007: 081406.
21. H. A. Carmona et al., Two dimensional electrons in a lateral magnetic superlattice. *Phys. Rev. Lett.* **74**, 1995: 3009.
22. M. Kato, A. Endo, S. Katsumoto, and Y. Iye, Two-dimensional electron gas under a spatially modulated magnetic field: A test ground for electron–electron scattering in a controlled environment. *Phys. Rev. B* **58**, 1998: 4876.
23. A. Messica, A. Soibel, U. Meirav, A. Stern, H. Shtrikman, V. Umansky, and D. Mahalu, Suppression of conductance in surface superlattices by temperature and electric field. *Phys. Rev. Lett.* **78**, 1997: 705.
24. C. G. Smith et al., Fabrication and physics of lateral superlattices with 40 nm pitch on high-mobility GaAs GaAlAs heterostructures. *J. Vac. Sci. Technol. B* **10**, 1992: 2904.
25. A. Soibel, U. Meirav, D. Mahalu, and H. Shtrikman, Magnetoresistance in a back-gated surface superlattice. *Phys. Rev. B* **55**, 1996: 4482.
26. S. Goswami et al., Transport through an electrostatically defined quantum dot lattice in a two-dimensional electron gas. *Phys. Rev. B* **85**, 2012: 075427.
27. M. Kato, A. Endo, M. Sakairi, S. Katsumoto, and Y. Iye, Electron–electron Umklapp process in two-dimensional electron gas under a spatially alternating magnetic field. *J. Phys. Soc. Jpn.* **68**, 1999: 1492.
28. Y. H. Chiu, J. H. Ho, C. P. Chang, D. S. Chuu, and M. F. Lin, Low-frequency magneto-optical excitations of a graphene monolayer: Peierls tight-binding model and gradient approximation calculation. *Phys. Rev. B* **78**, 2008: 245411.
29. Y. H. Chiu, Y. C. Ou, Y. Y. Liao, and M. F. Lin, Optical-absorption spectra of single-layer graphene in a periodic magnetic field. *J. Vac. Sci. Technol. B* **28**, 2010 386–390.
30. F. Sattari, and E. Faizabadi, Spin transport through electric field modulated graphene periodic ferromagnetic barriers. *Physica B* **434**, 2014: 69–73.
31. H. Yan et al., Superlattice Dirac points and space-dependent Fermi velocity in a corrugated graphene monolayer. *Phys. Rev. B* **87**, 2013: 075405.
32. L. Zheng-Fang, W. Qing-Ping, X. Xian-Bo, and L. Nian-Hua, Enhanced magnetoresistance in graphene nanostructure modulated by effective exchange field and Fermi velocity. *J. Appl. Phys.* **113**, 2013: 183704.
33. J. H. Ho, Y. H. Lai, C. L. Lu, J. S. Hwang, C. P. Chang, and M. F. Lin, Electronic structure of a monolayer graphite layer in a modulated electric field. *Phys. Lett. A* **359**, 2006: 70–75.
34. Y. H. Chiu, J. H. Ho, Y. H. Ho, D. S. Chuu, and M. F. Lin, Effects of a modulated electric field on the optical absorption spectra in a single-layer graphene. *J. Nanosci. Nanotechnol.* **9**, 2009: 6579–6586.
35. J. H. Ho, Y. H. Chiu, S. J. Tsai, and M. F. Lin, Semimetallic graphene in a modulated electric potential. *Phys. Rev. B* **79**, 2009: 115427.
36. S. K. Firoz Islam, N. K. Singh, and T. K. Ghosh, Thermodynamic properties of a magnetically modulated graphene monolayer. *J. Phys.: Condens. Matter* **23**, 2011: 445502.
37. M. Tahir, K. Sabeeh, and A. MacKinnon, Temperature effects on the magnetoplasmon spectrum of a weakly modulated graphene monolayer. *J. Phys.: Condens. Matter* **23**, 2011: 425304.
38. M. Tahir, K. Sabeeh, and A. MacKinnon, Weiss oscillations in the electronic structure of modulated graphene. *J. Phys.: Condens. Matter* **19**, 2007: 406226.
39. M. Tahir, and K. Sabeeh, Theory of Weiss oscillations in the magnetoplasmon spectrum of Dirac electrons in graphene. *Phys. Rev. B* **76.19**, 2007: 195416.
40. A. Matulis, and F. M. Peeters, Appearance of enhanced Weiss oscillations in graphene: Theory. *Phys. Rev. B* **75**, 2007: 125429.
41. Y. C. Ou, J. K. Sheu, Y. H. Chiu, R. B. Chen, and M. F. Lin, Influence of modulated fields on the Landau level properties of graphene. *Phys. Rev. B* **83**, 2011: 195405.
42. Y. C. Ou, Y. H. Chiu, J. M. Lu, W. P. Su, and M. F. Lin, Electric modulation effect on magneto-optical spectrum of monolayer graphene. *Comput. Phys. Commun.* **184**, 2013: 1821–1826.
43. C. P. Chang, C. L. Lu, F. L. Shyu, R. B. Chen, Y. K. Fang, and M. F. Lin, Magnetoelectronic properties of a graphite sheet. *Carbon* **42**, 2004: 2975.
44. A. Gruüneis et al., Tight-binding description of the quasiparticle dispersion of graphite and few-layer graphene. *Phys. Rev. B* **78**, 2008: 205425.
45. J. C. Slonczewski, and P. R. Weiss, Band structure of graphite. *Phys. Rev.* **109**, 1958: 272–279.
46. J. C. Charlier, J. P. Michenaud, and X. Gonze, First-principles study of the electronic properties of simple hexagonal graphite. *Phys. Rev. B* **46**, 1992: 4531–4539.
47. S. Latil, and L. Henrard, Charge carriers in few-layer graphene films. *Phys. Rev. Lett.* **97**, 2006: 036803.

48. B. Partoens, and F. M. Peeters, From graphene to graphite: Electronic structure around the K point. *Phys. Rev. B* **74**, 2006: 075404.
49. C. L. Lu, C. P. Chang, Y. C. Huang, R. B. Chen, and M. L. Lin, Influence of an electric field on the optical properties of few-layer graphene with AB stacking. *Phys. Rev. B* **73**, 2006: 144427.
50. Y. H. Lai, J. H. Ho, C. P. Chang, and M. F. Lin, Magnetoelectronic properties of bilayer Bernal graphene. *Phys. Rev. B* **77**, 2008: 085426.
51. E. McCann, and V. I. Fal'ko, Landau-level degeneracy and quantum hall effect in a graphite bilayer. *Phys. Rev. Lett.* **96**, 2006: 086805.
52. D. S. L. Abergel, and V. I. Fal'ko, Optical and magneto-optical far-infrared properties of bilayer graphene. *Phys. Rev. B* **75**, 2007: 155430.
53. C. L. Lu, C. P. Chang, Y. C. Huang, J. M. Lu, C. C. Hwang, and M. F. Lin, Low-energy electronic properties of the AB-stacked few-layer graphites. *J. Phys.: Condens. Matter* **18**, 2006: 5849–5859.
54. C. L. Lu, C. P. Chang, Y. C. Huang, J. H. Ho, C. C. Hwang, and M. F. Lin, Electronic properties of AA- and ABC-stacked few-layer graphites. *J. Phys. Soc. Jpn.* **76**, 2007: 024701.
55. C. L. Lu, H. L. Lin, C. C. Hwang, J. Wang, C. P. Chang, and M. F. Lin, Absorption spectra of trilayer rhombohedral graphite. *Appl. Phys. Lett.* **89**, 2006: 221910.
56. M. Koshino, and T. Ando, Magneto-optical properties of multilayer graphene. *Phys. Rev. B* **77**, 2008: 115313.
57. R. B. Chen, Y. H. Chiu, and M. F. Lin, A theoretical evaluation of the magneto-optical properties of AA-stacked graphite. *Carbon* **54**, 2012: 248–276.
58. M. Koshino, and E. McCann, Landau level spectra and the quantum Hall effect of multilayer graphene. *Phy. Rev. B* **83**, 2011: 165443.
59. T. Taychatanapat, K. Watanabe, T. Taniguchi, and P. Jarillo-Herrero, Quantum Hall effect and Landau-level crossing of Dirac fermions in trilayer graphene. *Nat. Phys.* **7**, 2011: 621–625.
60. C. Faugeras et al., Probing the band structure of quadri-layer graphene with magneto-phonon resonance. *New J. Phys.* **14**, 2012: 095007.
61. M. Orlita et al., Magneto-optics of bilayer inclusions in multilayered epitaxial graphene on the carbon face of SiC. *Phys. Rev. B* **83**, 2011: 125302.
62. N. A. Goncharuk et al., Infrared magnetospectroscopy of graphite in tilted fields. *Phys. Rev. B* **86**, 2012: 155409.
63. Y. H. Chiu, Y. H. Lai, J. H, Ho, D. S. Chuu, and M. F. Lin, Electronic structure of a two-dimensional graphene monolayer in a spatially modulated magnetic field: Peierls tight-binding model. *Phys. Rev. B* **77**, 2008: 045407.
64. N. Nemec, and G. Cuniberti, Hofstadter butterflies of bilayer graphene. *Phys. Rev. B* **75**, 2007: 201404(R).
65. T. G. Pedersen, Tight-binding theory of Faraday rotation in graphite. *Phys. Rev. B* **68**, 2003: 245104.
66. G. Dresselhaus, and M. S. Dresselhaus, Fourier expansion for the electronic energy bands in silicon and germanium. *Phys. Rev.* **160**, 1967: 649.
67. N. V. Smith, Photoemission spectra and band structures of d-band metals. VII. Extensions of the combined interpolation scheme. *Phys. Rev. B* **19**, 1979: 5019–5027.
68. L. C. Lew Yan Voon, and L. R. Ram-Mohan, Tight-binding representation of the optical matrix elements: Theory and applications. *Phys. Rev. B* **47**, 1993: 15500.
69. L. G. Johnson, and G. Dresselhaus, Optical properies of graphite. *Phys. Rev. B* **7**, 1973: 2275.
70. J. Blinowski et al., Band structure model and dynamical dielectric function in lowest stages of graphite acceptor compounds. *J. Phys. (Paris)* **41**, 1980: 47–58.
71. M. F. Lin, and K. W. K. Shung, Plasmons and optical properties of carbon nanotubes. *Phys. Rev. B* **50**, 1994: 17744.
72. Y. H. Ho, Y. H. Chiu, D. H. Lin, C. P. Chang, and M. F. Lin, Magneto-optical selection rules in bilayer Bernal graphene. *ACS Nano* **4**, 2010: 1465–1472.
73. Y. H. Ho, J. Y. Wu, R. B. Chen, Y. H. Chiu, and M. F. Lin, Optical transitions between Landau levels: AA-stacked bilayer graphene. *Appl. Phys. Lett.* **97**, 2010: 101905.
74. Y. Zheng, and T. Ando, Hall conductivity of a two-dimensional graphite system. *Phys. Rev. B* **65**, 2002: 245420.
75. M. Orltia et al., Graphite from the viewpoint of Landau level spectroscopy: An effective graphene bilayer and monolayer. *Phys. Rev. Lett.* **102**, 2009: 166401.
76. J. M. Dawlaty et al., Measurement of the optical absorption spectra of epitaxial graphene from terahertz to visible. *Appl. Phys. Lett.* **93**, 2008: 131905.
77. K.-C. Chuang, A. M. R. Baker, and R. J. Nicholas, Magnetoabsorption study of Landau levels in graphite. *Phys. Rev. B* **80**, 2009: 161410(R).
78. M. Orlita, C. Faugeras, G. Martinez, D. K. Maude, M. L. Sadowski, and M. Potemski, Dirac fermions at the H point of graphite: Magnetotransmission studies. *Phys. Rev. Lett.* **100**, 2008: 136403.
79. M. L. Sadowski, G. Martinez, M. Potemski, C. Berger, and W. A. de Heer, Landau level spectroscopy of ultrathin graphite layers. *Phys. Rev. Lett.* **97**, 2006: 266405.
80. K. S. Novoselov et al., Unconventional quantum Hall effect and Berry's phase of 2π in bilayer graphene. *Nat. Phys.* **2**, 2006: 177.
81. A. J. M. Giesbers, U. Zeitler, M. I. Katsnelson, L. A. Ponomarenko, T. M. Mohiuddin, and J. C. Maan, Quantum-Hall activation gaps in graphene. *Phys. Rev. Lett.* **99**, 2007: 206803.
82. J. G. Checkelsky, L. Li, and N. P. Ong, Zero-energy state in graphene in a high magnetic field. *Phys. Rev. Lett.* **100**, 2008: 206801.
83. S. Thongrattanasiri, F. H. L. Koppens, and F. J. García de Abajo, Complete optical absorption in periodically patterned graphene. *Phys. Rev. Lett.* **108**, 2012: 047401.
84. V. G. Kravets et al., Spectroscopic ellipsometry of graphene and an exciton-shifted van Hove peak in absorption. *Phys. Rev. B* **81**, 2010: 155413.
85. M. Orlita, and M. Potemski, Dirac electronic states in graphene systems: Optical spectroscopy studies. *Semicond. Sci. Technol.* **25**, 2010: 063001.
86. F. Wang et al., Gate-variable optical transitions in graphene. *Science* **320**, 2008: 206.
87. Y. C. Chuang, J. Y. Wu, and M. F. Lin, Electric field dependence of excitation spectra in AB-stacked bilayer graphene. *Sci. Rep.* **3**, 2013: 1368.
88. Y. H. Ho, S. J. Tsai, M. F. Lin, and W. P. Su, Unusual Landau levels in biased bilayer Bernal graphene. *Phys. Rev. B* **87**, 2013: 075417.
89. H. C. Chung, W. P. Su, and M. F. Lin, Electric-field-induced destruction of quasi-Landau levels in bilayergraphene nanoribbons. *Phys. Chem. Chem. Phys.* **15**, 2012: 868–875.
90. M. F. Lin, Y. C. Chuang, and J. Y. Wu, Electrically tunable plasma excitations in AA-stacked multilayer graphene. *Phys. Rev. B* **86**, 2012: 125434.

91. C. W. Chiu, Y. C. Huang, F. L. Shyu, and M. F. Lin, Optical absorption spectra in ABC-stacked graphene superlattice. *Synthetic Metals* **162**, 2012: 800–804.
92. C. H. Ho, C. P. Chang, W. P. Su, and M. F. Lin, Processing anisotropic Dirac cone and Landau subbands along a nodal spiral. *New J. Phys.* **15**, 2013: 053032.
93. C. W. Chiu, F. L. Shyu, M. F. Lin, G. Gumbs, and O. Roslyak, Anisotropy of π-plasmon dispersion relation of AA-stacked graphite. *J. Phys. Soc. Jpn.* **81**, 2012: 104703.
94. C. H. Ho, Y. H. Ho, Y. Y. Liao, Y. H. Chiu, C. P. Chang, and M. F. Lin, Diagonalization of landau level spectra in Rhombohedral graphite. *J. Phys. Soc. Jpn.* **81**, 2012: 024701.

30 Optoelectronic and Transport Properties of Gapped Graphene

Godfrey Gumbs, Danhong Huang, Andrii Iurov, and Bo Gao

CONTENTS

ABSTRACT

We review the transmission of Dirac electrons through a potential barrier in the presence of circularly polarized light. A different type of transmission is demonstrated and explained. Perfect transmission for nearly head-on collision in infinite graphene is suppressed in gapped dressed states of electrons. We also present our results on enhanced mobility of hot Dirac electrons in nanoribbons and magnetoplasmons in graphene in the presence of the energy gap. The calculated carrier mobility for a graphene nanoribbon as a function of the bias field possesses a high threshold for entering the nonlinear transport regime. This threshold is a function of both extrinsic and intrinsic properties, such as lattice temperature, linear density, impurity scattering strength, ribbon width, and correlation length for the line-edge roughness. Analysis of nonequilibrium carrier distribution function confirms that the difference between linear and nonlinear transport is due to sweeping electrons from the right to left Fermi one through elastic scattering as well as moving electrons from low- to high-energy ones through field-induced heating. The plasmons, as well as the electron–hole continuum are determined by both energy gap and the magnetic field, showing very specific features, which have been studied and discussed in detail.

30.1 INTRODUCTION

A considerable amount of interest in basic research and device development has been generated for both the electronic and optical properties of two-dimensional (2D) graphene material [1–5]. This began with the first successful isolation of single graphene layers and the related transport and Raman experiments for such layers [6]. It was found that the major difference between a graphene sheet and a conventional 2D electron gas (EG) in a quantum well (QW) is the band structure, where the energy dispersions of electrons and holes in the former are linear in momentum space, but quadratic for the latter. Consequently, particles in graphene behave like massless Dirac fermions and display many unexpected phenomena in electron transport and optical response, including the anomalous quantum Hall effect [7,8], bare and dressed-state Klein tunneling [9–12] and plasmon excitation [13–15], a universal absorption constant [16,17], tunable intraband [18] and interband [19,20] optical transitions, broadband p-polarization effect [21], photoexcited hot-carrier thermalization [22] and transport [23], electrically and magnetically tunable band structure for ballistic transport [24], field-enhanced mobility in a doped graphene nanoribbon (GNR) [25], and electron-energy loss in gapped graphene layers [26].

Most of the unusual electronic properties of graphene may be explained by single-particle excitation of electrons. The Kubo linear-response theory [27] and Hartree–Fock theory combined with the self-consistent Born approximation [28] were applied to diffusion-limited electron transport in doped graphene. Additionally, the semiclassical Boltzmann theory was employed for studying transport in both linear [29] and nonlinear [25] regimes. For the plasmon excitation in graphene, its important role in the dynamical screening of the electron–electron interaction [13–15,26,30] has been reported. However, relatively less attention has been received for the electromagnetic (EM) response of graphene materials, especially for low-energy intraband optical transions [18,31,32].

Gapped graphene has marked an important milestone in the study of graphene's electronic and transport properties from both a theoretical and experimental point of view as well as in practical device applications. The reason for this is that gapped graphene has applications such as a field-effect transistor where a gap is essential as well as graphene interconnects. The effective band gap may be generated by spin–orbit interaction, or when monolayer graphene is placed on a substrate such as ceramic silicon carbide or graphite. The gap may also arise dynamically when graphene is irradiated with circularly polarized light. Depending on the nature of the substrate on which graphene is placed or the intensity or amplitude of the light, the gap may be a few millielectron volts or as large as 1 eV [33]. In general, the energy gap is attributed to a breakdown in symmetry between the sublattices caused by external perturbing fields from the substrate or photons coupled to the atoms in the *A* and *B* sublattices.

This chapter will be divided into two parts covering the electrical and optical properties of electrons in gapped graphene materials. The first part will deal with electrical transport. Topics will include (1) unimpeded tunneling of chiral electrons in GNRs; (2) anomalous photon-assisted tunneling of Dirac electrons in graphene; and (3) field-enhanced mobility by nonlinear phonon scattering of Dirac electrons. The second part will focus on the collective plasmon (charge density) excitations for intraband and interband plasmons.

30.2 DIRAC FERMIONS, CHIRALITY, AND TUNABLE GAP IN GRAPHENE

Graphene, an allotrope of carbon, which has been recently discovered through experiments [7], has become one of the most important and extensively studied materials in modern condensed matter physics, mainly because of its extraordinary electronic and transport properties [9]. According to Reference 34, graphene can be described as a single atomic plane of graphite, which is sufficiently isolated from its environment and considered to be freestanding. The typical carbon–carbon distance in a graphene layer is 0.142 nm, and the interlayer distance in a graphene stack is 0.335 nm. Any graphene sample with less than 2.4×10^4 carbon atoms or less than 20 nm of length is unstable [35], tending to convert to other fullerenes or carbon structures. Now, we will briefly discuss the most crucial electronic properties of graphene, relevant to the electron tunneling phenomena. A good description of the principal electronic structure and properties may be found in References 1,36. Surprisingly, the first theoretical study of "graphene" was performed more than 60 years ago [37,38]. The most striking difference in comparison with conventional semiconductors or metals is the fact that low-energy electronic excitations in graphene are massless with Dirac cones for energies. The quite complicated energy band structure of graphene may be approximated as a cone (*Dirac cone*) in the vicinity of the two inequivalent corners, that is, *K* and *K′* points, of the Brillouin zone. In summary, the electronic properties of graphene may be approximately described by the *Dirac equation*, corresponding to the linear energy dispersion next to the *K* and *K′* points:

$$-i\hbar v_F \sigma \cdot \nabla\Psi(\mathbf{r}) = \varepsilon\Psi(\mathbf{r}) \tag{30.1}$$

where $v_F = c/300$ is the Fermi velocity. This form is similar to the high-energy quantum electrodynamics (QED) Dirac equation. In momentum (**k**) space, the Hamiltonian is simplified as follows:

$$\hat{\mathcal{H}} = \hbar v_F \sigma \cdot k = \begin{bmatrix} \mathcal{V}_0 & k_x + ik_y \\ k_x - ik_y & \mathcal{V}_0 \end{bmatrix} \tag{30.2}$$

Here, $\mathcal{V}_0$ is a uniform external potential. We will also adapt the following notation: $k_\pm = k_x \pm ik_y$. The energy dispersions are simply $\varepsilon(k) = \beta\hbar v_F k = \beta\hbar v_F \sqrt{k_x^2 + k_y^2}$ and gapless. In addition, β is the electron–hole parity index,* so that $\beta = 1$ for electrons and $\beta = -1$ for holes.

The goal of this section is to compare the electronic and transport properties of gapped graphene modeled by finite electron effective mass with those obtained in the case of massless Dirac fermions. There have been many studies [39,40], where the mass term is added to the Dirac Hamiltonian of infinite graphene. An energy gap may appear as a result of a number of physical reasons, such as by a boron nitride substrate. However, the most interesting cases are when the gap becomes *tunable* and may be varied throughout the experiment, resulting in practical applications. We will focus on the so-called *electron–photon-dressed states*, which result from the interaction between Dirac electrons in graphene and circularly polarized photons. The first complete quantum description of such a system was presented in Reference 33, and the transport properties were discussed in Reference 41. The quantum descriptions for both graphene [11] and three-dimensional topological insulators were further developed in Reference 42.

30.3 ELECTRON TUNNELING IN GAPPED GRAPHENE

The so-called *Klein paradox* is related to complete unimpeded transmission of Dirac fermions through square potential barriers of arbitrary height and width in the case of head-on collision. It has been demonstrated that certain aspects of the Klein paradox may also be observed in bilayer graphene [9], carbon nanotubes, topological insulators [42], and zigzag nanoribbons [11]. The trademark of the Klein paradox exists even in the case of electron–photon-dressed states for massive electrons [11].

* The quantity $\beta = \pm 1$ is often called *pseudospin* because of its formal resemblance to the spin index in a spinor wavefunction.

30.3.1 Square Barrier Tunneling and Klein Paradox

In order to demonstrate Klein tunneling and investigate its unique properties, we will consider a sharp potential barrier (like a *p*–*n*–*p* junction), given by $V(x) = V_0[\theta(x) - \theta(x - w)]$ and infinite in the *y*-direction specified by the Heaviside step function $\theta(x)$. Here, we will use the notations introduced in Figure 30.1b.

Klein tunneling may be explained based on a specific form of the Dirac fermion wavefunction as well as a special type of *chiral* symmetry of the Dirac Hamiltonian. The eigenvalue wavefunctions obtained from Equation 30.1 are $\varepsilon(k) = \hbar v_F k$ and

$$\Psi_\beta(k_x, k_y) = \frac{1}{\sqrt{2}}\begin{bmatrix} 1 \\ \beta e^{i(k_x x + k_y y)} \end{bmatrix} \tag{30.3}$$

Electrons are said to be chiral if their wavefunctions are eigenstates of the chirality operator $\hat{h} = \sigma \cdot \hat{\mathbf{p}}/(2p)$, where $\sigma = \{\sigma_x, \sigma_y\}$ is the Pauli vector consisting of the Pauli matrices and $\hat{\mathbf{p}} = \{\hat{p}_x, \hat{p}_y\}$ is the electron momentum operator in graphene layers. Electrons become chiral in graphene due to the fact that the chirality operator is proportional to the Dirac Hamiltonian, which automatically makes *chirality* a good quantum number. One can easily verify that the wavefunction in Equation 30.3 satisfies the chirality property.

As an example, we consider the situation of a very high potential barrier, that is, $V_0 \gg \varepsilon$, which would not allow any finite transmission amplitude possible for a conventional Schrödinger particle. For Dirac electrons, however, we have

$$T(k_{x,1}, k_{x,2}) = \frac{\cos^2\phi}{\cos^2(k_{x,2}w)\cos^2\phi + \sin^2(k_{x,2}w)} \tag{30.4}$$

where ϕ is the angle that $\mathbf{k}$ makes with the *x*-axis. Incidentally, for the case of head-on collision with $\phi = 0$, we find complete unimpeded tunneling with $T = 1$, which is a direct consequence of a special form of the electron wavefunction, resulting from the Dirac cone energy dispersion.

30.3.2 Electron–Photon-Dressed States with Tunable Gap

The peaks of transmission mainly belong to two different species. Namely, the *Klein paradox* for head-on collision where $\phi = 0$ and the so-called "transmission resonances" correspond to specific values of the electron longitudinal momenta in the barrier region.

It was shown recently [33,41] that when Dirac electrons in a single graphene layer interact with an intense circularly polarized light beam, electron states will be *dressed* by photons. Here, we investigate the transmission properties of such dressed electrons for the case of single-layer graphene.

We begin with the electron–photon interaction Hamiltonian

$$\hat{\mathcal{H}} = v_F \sigma \cdot (\hat{\mathbf{p}} - e\mathbf{A}_{circ}) \tag{30.5}$$

where the vector potential for circularly polarized light of frequency ω_0 may be expressed as

$$\mathbf{A}_{circ} = \sqrt{\frac{\hbar}{\epsilon_0 \omega_0 \mathcal{V}}}(\mathbf{e}_+\hat{a} + \mathbf{e}_-\hat{a}^\dagger) \tag{30.6}$$

in terms of photon creation and destruction operators $\hat{a}^\dagger$ and $\hat{a}^\dagger$, respectively. Here, $\mathcal{V}$ is the mode volume of an optical field. In order to study the complete electron–photon interacting system, we must add the field energy term $\hbar\omega_0\hat{a}^\dagger\hat{a}$ to the Hamiltonian Equation 30.5.

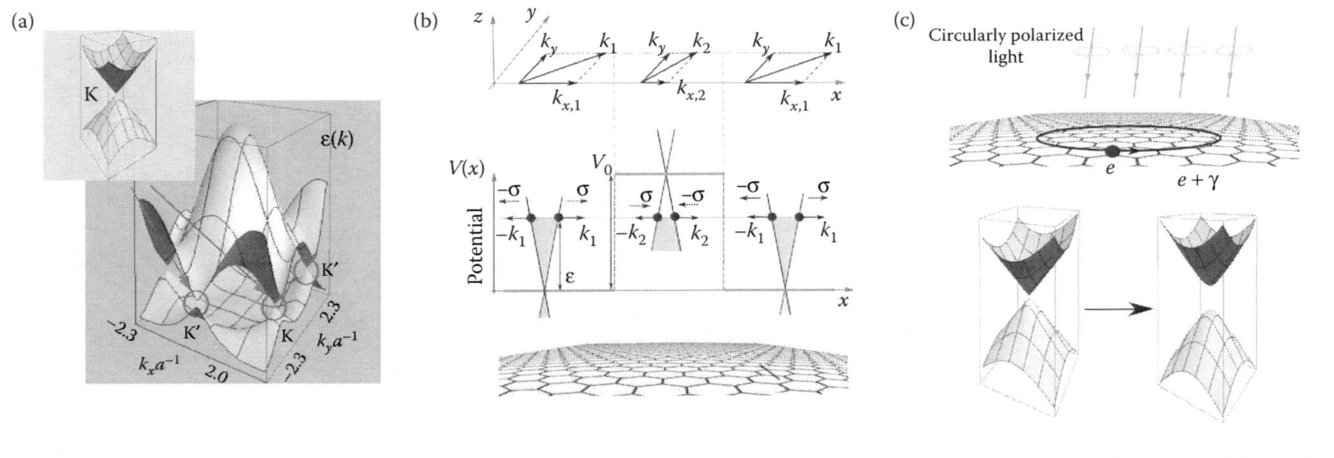

FIGURE 30.1 Schematics of Dirac cone in graphene, square barrier transmission, and electron–photon-dressed states. (a) Shows the energy bands $\varepsilon(k)$ as well as the Dirac cone as a good approximation in the vicinity of certain points in the **k**-plane. (b) Illustrates the square barrier tunneling with conserved pseudo-spin, and introduces all the notations used in the discussion of tunneling. (c) Features the electron–photon-dressed states, which appear as a result of Dirac electrons interacting with circularly polarized photons. It also demonstrates schematically that the energy dispersion of the electron-dressed states has a gap, proportional to the intensity of the light.

In terms of the energy dispersion $\varepsilon(k) = \pm\sqrt{(\hbar v_F k)^2 + \Delta^2}$, our system is formally similar to the eigenvalue equations for the case of the effective mass σ_3 Dirac Hamiltonian:

$$\hat{\mathcal{H}} = \hbar v_F \sigma \cdot \mathbf{k} + V(x)\begin{bmatrix} 1 & 0 \\ 0 & 1 \end{bmatrix} + \Delta\sigma_3 \tag{30.7}$$

where σ_3 is a Pauli matrix and $V(x)$ is a one-dimensional potential. The electron dispersion and transmission properties for both a single and multiple square potential barrier have been studied [43,44] for monolayer and bilayer graphene. It was also shown that a one-dimensional periodic array of potential barriers leads to multiple Dirac points [43]. Several published works have introduced an effective mass term into the Dirac Hamiltonian for infinite graphene, which may be justified based on different physical reasons [40,45]. For example, it has been shown [39] that an energy band gap in graphene can be created by boron nitride substrate, resulting in a finite electron effective mass. However, we would like to emphasize that the analogy between the Hamiltonian Equation 30.7 and that for irradiated graphene is not complete since the former requires $\Delta < 0$. Although this difference does not result in any modification of the energy dispersion term containing Δ^2, it certainly changes the corresponding wavefunction. Additionally, there have been a number of studies using laser radiation on single layer [46] and bilayer [47] graphene as well as GNRs [48] reporting the gap opening as a result of strong electron–photon interaction. From these aspects, it seems that topological insulators as well as gapped graphene may have potential device applications where spin plays a role. In the presence of a strong radiation field, we have the dressed-state wavefunction

$$\Phi_{dr}(k) = \begin{bmatrix} \mathcal{C}_1(k) \\ \beta\mathcal{C}_2(k)e^{i\phi} \end{bmatrix} \tag{30.8}$$

where $\mathcal{C}_1(k) \neq \mathcal{C}_2(k)$ and are given by

$$\mathcal{C}_1^{\pm}(k) = \frac{1}{\sqrt{2(1+\gamma^2) \mp 2\gamma\sqrt{1+\gamma^2}}} \tag{30.9}$$

$$\mathcal{C}_2^{\pm}(k) = \pm\frac{\sqrt{1+\gamma^2} \mp \gamma}{\sqrt{2(1+\gamma^2) \mp 2\gamma\sqrt{1+\gamma^2}}} \tag{30.10}$$

Here, $\gamma = \Delta/(\hbar v_F k)$. Consequently, the chiral symmetry is broken for electron-dressed states. The interaction between Dirac electrons in graphene and circularly polarized light has been considered in the classical limit in Reference 49. In this limit, a gap in the Dirac cone opens up due to nonlinear effects. The dressed-state wavefunction has the chirality

$$\hat{h}\Phi_{dr}(k) = \frac{1}{2}\frac{\sigma \cdot \hat{\mathbf{p}}}{p}\begin{bmatrix} \mathcal{C}_1(k) \\ \beta\mathcal{C}_2(k)\mathbf{e}^{i\phi} \end{bmatrix} = \frac{1}{2}\begin{bmatrix} \beta\mathcal{C}_2(k) \\ \mathcal{C}_1(k)\mathbf{e}^{i\phi} \end{bmatrix} \tag{30.11}$$

TABLE 30.1
Energy Gap in Graphene, Corresponding to the Different Values of the Laser Power P and Its Wavelength λ

	$\lambda = 10^{-7}$ m	$\lambda = 10^{-6}$ m	$\lambda = 10^{-5}$ m
$P = 0.1$ W	0.027	0.274	2.71
$P = 1$ W	0.264	2.738	25.03
$P = 100$ W	27.39	249.14	709.213

Note: The gap obviously increases with the increasing power, similar dependence on λ is observed, since the field amplitude is inversely proportional to the corresponding frequency. The energy gap dependence is approximately a linear function of the laser power for short wavelength.

As it appears, nonchirality of the dressed electron states becomes significant if the electron–photon interaction (the leading γ term) is increased. This affects electron tunneling and transport properties. We now turn to an investigation of the transmission of electron states through a potential barrier when graphene is irradiated with circularly polarized light.

Summarizing the description of the electron-dressed states, we would like to estimate the actual values of the energy gap in graphene and describe its dependence on the power and frequency of the applied laser beam, which obeys the equation

$$\Delta = \sqrt{\mathbb{W}_0^2 + (\hbar\omega_0)^2} - \hbar\omega_0 \tag{30.12}$$

where $\mathbb{W}_0 = 2v_F\varepsilon_0/\omega_0$, so that the gap is proportional to the square of amplitude of the electric field ε_0 for $\varepsilon_0 \to 0$. The amplitude of the EM field is determined by the laser power and its wavelength $\varepsilon_0 \simeq P^{1/2}\lambda^{-1}$. Consequently, the energy gap in graphene depends on both the power of the applied laser and its wavelength. The results are summarized in Table 30.1.

30.3.3 Tunneling and Optical Properties of Electron-Dressed States

We consider a square potential barrier given by $V(x) = V_0[\theta(x) - \theta(x - W_0)]$. The current component is $j_x = \Phi^{\dagger}\,\sigma_x\Phi$, thus we only require the wave-function continuity at the potential boundaries. There are two specific simplifications to consider here—a nearly head-on collision $k_{x,i} \ll k_{y,i}$ and the case of a high potential barrier when $\varepsilon \ll V_0$. We should also mention another relevant study [50], investigating the tunneling of Dirac electrons with a finite effective mass through a similar potential barrier.

For a nearly head-on collision with $k_y \ll k_{x,1} \ll k_{x,2}$ for high potential as well as infinite graphene ($\Delta \to 0$), the transmission coefficient has the following simplified form:

$$T = 1 - \sin^2(k_{x,2}W_0)(\theta^2 - 2\beta\theta\phi + \phi^2) \tag{30.13}$$

where we assume $V_0 \gg \varepsilon$, $\theta \ll \phi \ll 1$ and $\beta = \pm 1$.

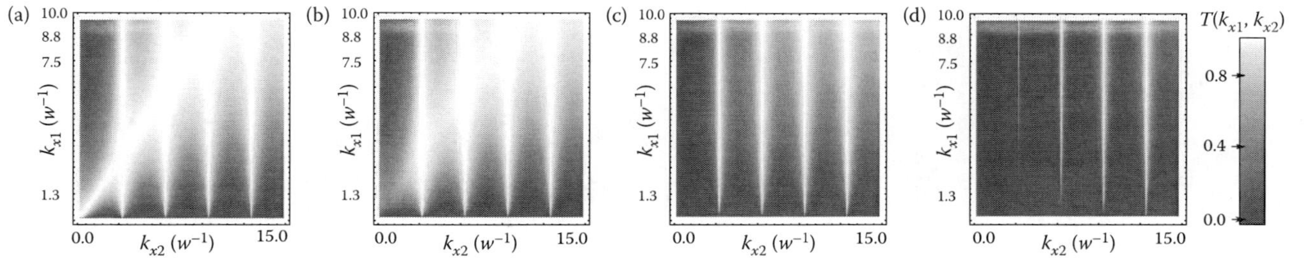

FIGURE 30.2 Density plots of the transmission probability T for electron–photon-dressed states as a function of the electron longitudinal momenta $k_{x,1}$ and $k_{x,2}$ in both *barrier* $V(x) = V_0$ and *no-barrier* $V(x) = 0$ regions, respectively. The electron momenta are given in the units of the inverse barrier width w. Here, plots (a), (b), (c), and (d) correspond to $\Delta/V_0 = 0$, 0.001, 0.008, and 0.015, respectively.

Figure 30.2 presents the calculated transmission probability T as a function of the longitudinal momenta $k_{x,1}$ (in front of the barrier) and $k_{x,2}$ (in the barrier region). We find from Figure 30.2 that the intensity of the transmission peaks in (b) through (d) are gradually distorted with increasing gap compared to infinite graphene in (a). The diagonal $k_{x,1} = k_{x,2}$ corresponds to the absence of a potential barrier and should yield a complete transmission for $\Delta = 0$ as seen in (a). However, for a finite Δ, the requirement ($\sqrt{(\varepsilon - V_0)^2 - \Delta^2} > \hbar v_F k_y$) must be satisfied, which makes the diagonal transmission incomplete, for example, missing diagonal for small $k_{x,1}$ and $k_{x,2}$ in (b), due to the occurrence of an energy gap. As Δ is further increased in (c) and (d), this diagonal distortion becomes more and more severe, which is accompanied by strongly reduced intensity of transmission peaks at small $k_{x,2}$.

Figure 30.3 displays the effect of the energy gap on T in terms of incoming particle energy ε and angle of incidence ϕ. From the upper panel of Figure 30.3a1 through a4, we see the Klein paradox as well as other resonant tunneling peaks in T for regular infinite graphene with $\Delta = 0$. The dark "pockets" on both sides of $\varepsilon = V_0$ demonstrate zero transmission for the case of $|\varepsilon - V_0| \ll \varepsilon$, which results in imaginary longitudinal momentum $k_{x,2}$ for most of the incident angles and produces a fully attenuated wavefunction.

Figure 30.4 shows T as a function of ε in (a1) and of ϕ in all other plots. From (a1), we clearly see that dressing destroys the Klein paradox for head-on collision with $\phi = 0$. Additionally, in the absence of electron–photon interaction, we display the effects on T for various values of W_0 in (a2) and V_0 in (a3). The resonant peaks are found to be shifted toward other incoming angles in both (a2) and (a3) and this effect becomes stronger for small incident angles. The effect of dressed electron states, on the other hand, is demonstrated in each plot in (b1) through (b4) for various values of Δ, as well as in individual plots of (b1) through (b4) for different barrier heights V_0 and widths W_0.

It is very helpful to compare the results obtained here with bilayer graphene having quadratic dispersion. For

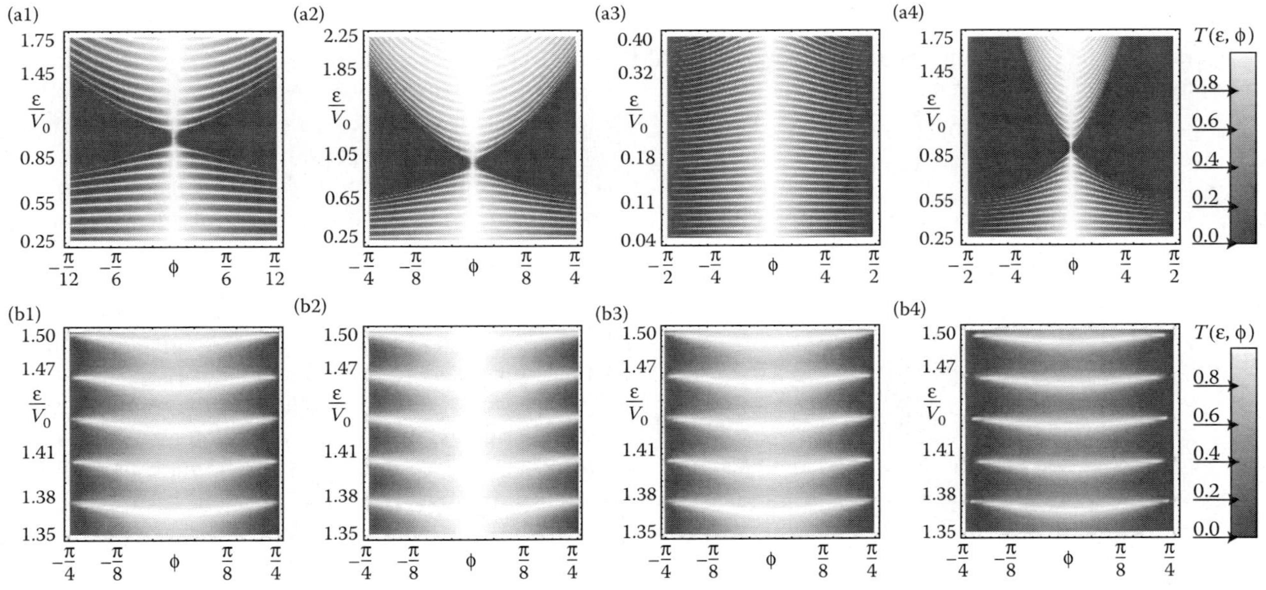

FIGURE 30.3 Density plots of T for electron–photon-dressed states as a function of the incoming electron energy ε and the angle of incidence ϕ. The upper panel displays T under $\Delta = 0$ in (a1) as well as its blowout views in (a2) through (a4). The lower panel is associated with various induced gaps $\Delta/V_0 = 0.01$, 0.05, 0.07, and 0.1 in (b1) through (b4). The significant difference in tunneling behavior at $\phi = 0$ can be seen with various Δ values.

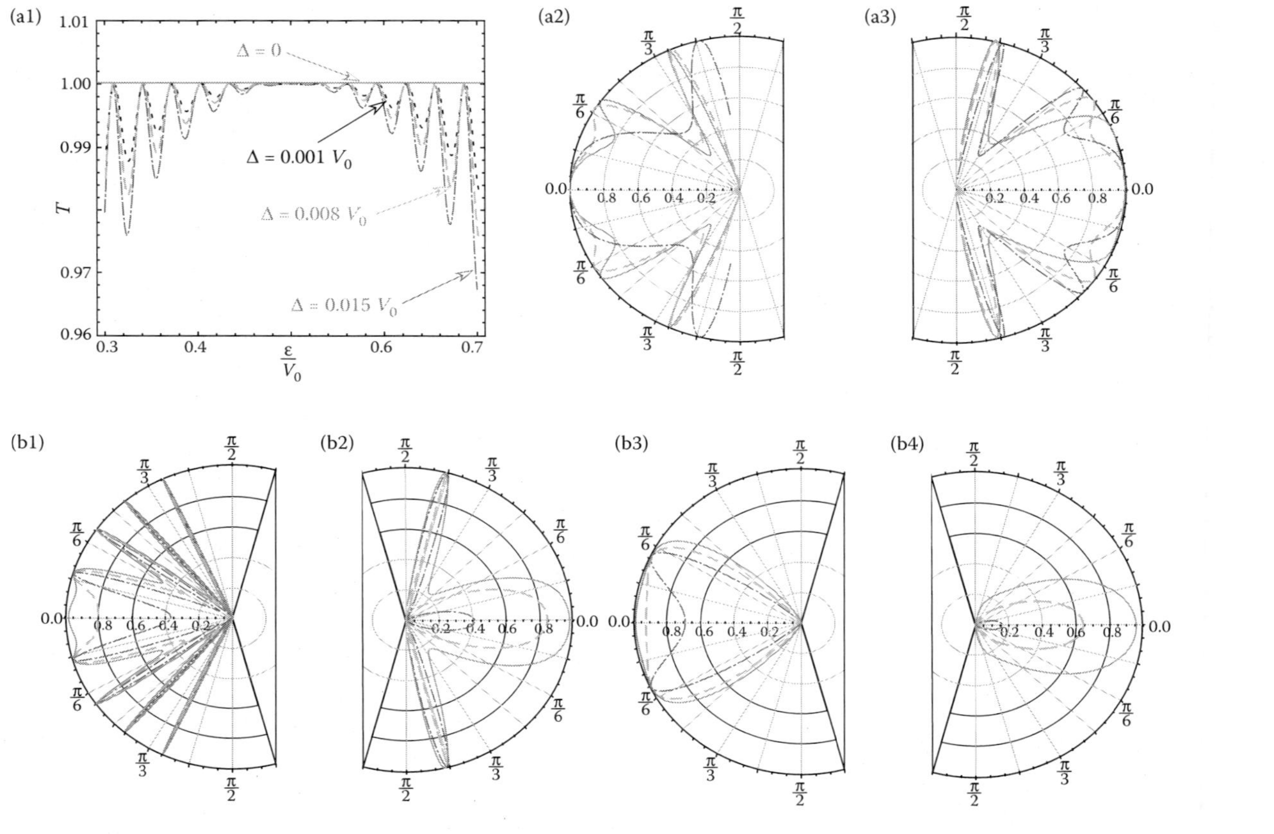

FIGURE 30.4 The upper panel presents T as functions of ε in (a1) and ϕ in (a2) and (a3). In (a1), T with $\phi = 0$ for various Δ values are exhibited with $\Delta = 0$ (solid), $\Delta/V_0 = 0.001$ (short-dashed), $\Delta/V_0 = 0.008$ (dashed), and $\Delta/V_0 = 0.015$ (dash-dotted). In (a2) and (a3), the electron–photon interaction is excluded and $\varepsilon = V_0/6$. Additionally, we assume in (a2) $V_0 = 100$ meV and various W_0 values with $W_0 = 5$ nm (solid), $W_0 = 10$ nm (dashed), and $W_0 = 20$ nm (dash-dotted). Plot (a3) shows T with $W_0 = 20$ nm for $V_0 = 150$ meV (solid), $V_0 = 200$ meV (dashed), and $V_0 = 250$ meV (dash-dotted). The lower panel presents dressed-state ϕ dependence of T with $\varepsilon = V_0/6$ for $V_0 = 0.5$ eV and $W_0 = 100$ nm in (b1); $V_0 = 2$ eV and $W_0 = 100$ nm in (b2); $V_0 = 4$ eV and $W_0 = 150$ nm in (b3); and $V_0 = 5$ eV and $W_0 = 150$ nm in (b4). Moreover, we set in (b1) through (b4) $\Delta = 0.099$ eV (dash-dotted), $\Delta = 0.050$ eV (dashed), and $\Delta = 0.010$ eV (solid).

bilayer graphene, its lowest energy states are described by the Hamiltonian [9]

$$\hat{\mathcal{H}}_{blg} = \frac{\hbar^2}{2m_b}(k_-^2\sigma_+ + k_+^2\sigma_-) \tag{30.14}$$

where m_b is the effective mass of electrons in the barrier region. In this case, the longitudinal wave vector component in the barrier region is given by $k_{x,2} = \beta'\sqrt{2m_b\beta(\varepsilon - V_0) - k_y^2}$ with $\beta, \beta' = \pm 1$.

An evanescent wave having a decay rate κ_b may coexist with a propagating wave having a wave vector $k_{x,2}$ such that $k_y^2 + k_{x,2}^2 = k_y^2 - k_b^2 = 2m_b\beta(\varepsilon - V_0)$. This implies that the evanescent modes should be taken into account simultaneously. The Klein paradox persists in bilayer graphene for chiral but massive particles. However, one finds a complete reflection, instead of a complete transmission, in this case.

In the simplest approximation to include only the two nearest subbands, our model is formally similar to the so-called σ_3 Hamiltonian used for describing the particles in a single layer of graphene with parabolic energy dispersion (nonzero effective mass). By including more than two independent pairs, this effect is expected to be weaker since perfect transmission will occur only for the two wavefunction terms. However, this effect is insensitive to the barrier width, and therefore, may be considered as reminiscent of the Klein paradox.

In spite of the fact that a significant number of papers on electron tunneling in the presence of the electron–photon interaction have already been published, this topic is receiving considerable attention at the present time. Potential barrier transmission is studied using the Floquet approach [51,52]. The effect of periodic potential is similar to the combination of the uniform magnetic and electric fields [53]. Also, we would like to mention Reference 54, where the transport properties were investigated in the framework of unitary-transformation scheme together with the nonequilibrium Green's function formalism. Similar case of Chiral tunneling modulated by a time-periodic potential on the surface states of a topological insulator was addressed in Reference 55.

In conclusion, we would like to mention *topological insulators*. Due to their band structure and a nontrivial topological order, topological insulators are insulating in the bulk, but support gapless conducting surface states [56]. The surface states are represented by a spin-polarized Dirac cone, creating an analogy with graphene. However, there is a peculiar property of the topological insulator, that is, a so-called geometrical gap in the energy dispersion. This gap appears if the sample is finite in the z-direction, which has not been observed in conventional insulators. The electron-dressed states in three-dimensional topological insulators have been obtained and their tunneling properties have been discussed [42,57].

30.4 ENHANCED MOBILITY OF HOT DIRAC ELECTRONS IN NANORIBBONS

Early studies [29,58] on transport in GNRs were restricted to the low-field limit [59], where a linearized Boltzmann equation with a relaxation-time approximation was solved. Recently, the nonequilibrium distribution of electrons is calculated by solving the Boltzmann equation beyond the relaxation-time approximation for nonlinear transport in semiconducting GNRs [60]. Enhanced mobility from field-heated electrons in high energy states is predicted. An anomalous enhancement in the line-edge roughness (LER) scattering under high fields is found with decreasing roughness correlation length due to the population of high-energy states by field-heated electrons [25].

30.4.1 Nonlinear Boltzmann Theory

We limit ourselves to single subband transport with low electron densities, moderate temperatures, ionized impurities, and LER [29,61,62]. Consequently, negligible influences from electron–electron [63], optical phonon [62], intervalley, and volume-distributed impurity scattering [29] are expected. The discrete energy dispersion for armchair nanoribbons (ANRs) is written as [1]

$$\varepsilon_j = \hbar v_F \times \begin{cases} k_j, & \text{metallic} \\ \sqrt{k_j^2 + (\pi/3W)^2}, & \text{semiconducting} \end{cases} \tag{30.15}$$

The discrete wave numbers are $k_j = [j - (N+1)/2]\delta k$ with $j = 1, 2, \ldots, N$ for a large odd integer N, and $\delta k = 2\text{ker}n1ptk_{\max}/(N-1)$ is mesh spacing. The central point is $j = M = (N+1)/2$ for the minimum of the energy. $W = (\mathcal{N}+1)a_0/2$ is the GNR width, $a_0 = 2.6$ Å the size of the graphene unit cell, and $\mathcal{N}$ the number of carbon atoms across GNRs. From Equation 30.15, we get v_F for the group velocity v_j for metallic nanoribbons, while for semiconducting ANRs, it is $v_j = v_F(\hbar v_F k_j/\varepsilon_j)$. We know the band will be filled up to $|k_j| = k_F$ at zero temperature ($T = 0$ K) with the Fermi wave number and Fermi energy given, respectively, by $k_F = \pi n_{1D}/2$ and $\varepsilon_F = \varepsilon(k_F)$. For a chosen T and chemical potential μ_0, the linear density in ANRs is $n_{1D} = \delta k/\pi \sum_{j=1}^{N} [\exp((\varepsilon_j - \mu_0)/k_B T) + 1]^{-1}$.

We assume that the wavefunction $\Psi_j(x, \text{ker}n1pty)$ corresponding to Equation 30.15 satisfies hard-wall boundary conditions [64] $\Psi_j(0, y) = \Psi_j(W, y) = 0$. This can be fulfilled by selecting the wavefunction as a mixture of ones at $\mathbf{K} = \left((2\pi/3a_0),(2\pi/\sqrt{3}a_0)\right)$ and $\mathbf{K}' = \left(-(2\pi/3a_0),(2\pi/\sqrt{3}a_0)\right)$ points as [1]

$$\Psi_j(x,y) = \frac{1}{\sqrt{2}}\left[\psi_j(x,y) - \psi'_j(x,y)\right] \tag{30.16}$$

$$\begin{cases} \psi_j(x,y) = \sqrt{\dfrac{1}{2LW}} e^{ik_j y} \begin{bmatrix} 1 \\ e^{i\phi_j} \end{bmatrix} e^{i(2\pi/3a_0 - \kappa)x} & \text{at } \mathbf{K} \text{ point} \\ \psi'_j(x,y) = \sqrt{\dfrac{1}{2LW}} e^{ik_j y} \begin{bmatrix} 1 \\ -e^{-i\phi_j} \end{bmatrix} e^{-i(2\pi/3a_0 - \kappa)x} & \text{at } \mathbf{K}' \text{ point} \end{cases} \tag{30.17}$$

Here, L is the ribbon length. For semiconducting ANRs $\kappa = \pi/3W \ll 2\pi/3a_0$ is the quantum of the transverse wave vector, and $\phi_j = \tan^{-1}(k_j/\kappa)$ the phase separation between the two graphene sublattices. For a metallic-type ribbon, we set $\kappa = 0$ and the phase assumes only $\pm\pi/2$ values.

With the wavefunction in Equation 30.16 and neglecting intervalley scattering* one may calculate the scattering from any potential $V(x, y)$. The impurity and phonon-induced intervalley scattering is neglected because the relevant phonon energy for momentum transfer is large at low temperatures and the effective scattering cross section of both volume and surface impurities is suppressed for large value of $|\mathbf{K} - \mathbf{K}'|$. Therefore, the interaction matrix elements become

$$\begin{aligned} V_{i,j} &= \int_0^W dx \int_{-\infty}^{\infty} dy \Psi_i^*(x,y) V(x,y) \Psi_j(x,y) \\ &= \frac{1}{2}\int_0^W dx \int_{-\infty}^{\infty} dy (\psi_i^* V \psi_j + \psi_i'^* V \psi'_j) \end{aligned} \tag{30.18}$$

We consider scattering potential made from the three contributions to be $V = V^{AL} + V^{LER} + V^{imp}$.

Since the longitudinal phonons induce higher deformation potential than the out-of-plane flexural ones, we neglect the flexural modes here. Under this approximation, the phonon-scattering potential is written as [29,65]

$$V^{AL}(y) = \sqrt{\frac{n^{\pm}\hbar}{2\rho LW\omega_{AL}}} D_{AL} q_y e^{iq_y y} \tag{30.19}$$

where $n^- = \left[\exp(\hbar\omega_{AL}/k_B T) - 1\right]^{-1}$ and $n^+ = 1 + n^-$ are the equilibrium phonon distributions, $\omega_{AL} = c_s q_y$ the phonon frequency, $D_{AL} \sim 16$ eV the deformation potential, $\rho \sim 7.6 \times 10^{-8}$ g/cm^2, and $c_s \sim 2 \times 10^6$ cm/s the mass density and sound velocity. Moreover, the momentum conservation gives $q_y = k_i - k_j$.

Elastic scattering is attributed to the roughness of the ribbon edges and in-plane charged impurities. For the former, we assume the width of the ribbon as $W(y) = W + \delta W(y)$ and the edge roughness to satisfy the Gaussian correlation function $\langle \delta W(y) \delta W(y + \Delta y) \rangle = \delta b^2 \exp[-(\Delta y/\Lambda_0)^2]$ with $\delta b \sim 5$ Å being

* Such intervalley scattering would require momentum transfer comparable with the distance between **K** and **K′** points.

the amplitude and $\Lambda_0 \sim 50 - 200$ Å the correlation length. The latter is related to impurities located at $(x_0, 0, 0)$ and distributed with a sheet density n_{2D}. Each point impurity produces a scattering potential. Since the momentum difference between two valleys is very large, only short-range impurities can contribute. Here, we neglect the short-range scatterers; thus making the scattering matrix diagonal. Corresponding perturbations for roughness and impurity are

$$V^{\text{LER}}(y) = \frac{\delta W(y)}{3W^2}\pi\hbar v_F \tag{30.20}$$

$$V^{\text{imp}}(x,y) = \frac{e^2}{4\pi\epsilon_0\epsilon_r\sqrt{(x-x_0)^2+y^2}} \tag{30.21}$$

where ϵ_r is the average host dielectric constant. These potentials provide the net elastic scattering rate $\hbar/\tau_j$ through

$$\frac{1}{\tau_j} = \frac{1}{\tau_j^{\text{imp}}} + \frac{1}{\tau_j^{\text{LER}}} \tag{30.22}$$

$$\begin{cases} (\tau_j^{\text{imp}})^{-1} = \gamma_0\left(\dfrac{v_F}{|v_j|}\right)\left[1+\cos(2\phi_j)\right] \\ (\tau_j^{\text{LER}})^{-1} = \gamma_1\left(\dfrac{v_F}{|v_j|}\right)\dfrac{1}{1+4k_j^2\Lambda_0^2}\left[1+\cos(2\phi_j)\right] \end{cases} \tag{30.23}$$

where γ_0 is the scattering rate of impurities at the Fermi edge and $\gamma_1 = 2(\pi v_F \delta b/3W^2)^2(\Lambda_0/v_F)$ the scattering rate from edge roughness. Here, we assume the impurities distribute within the layer. Moreover, the scattering potential is screened by electrons. Since we limit ourselves to a single subband, the screening is just a scalar Thomas–Fermi dielectric function [29,66] $\epsilon_{TF}(|k_{j'} - k_j|)$, which is calculated as $\epsilon_{TF} \approx 1 + (e^2/\pi^2\epsilon_0\epsilon_r\hbar v_F)$ in the metallic limit $(2k_F W \gg 1)$ with $\epsilon_r \approx 3.9$. We apply the static screening to both impurity and phonon scattering with a relatively large damping rate and small Fermi energy.

The deviation from the Fermi distribution under a strong field is described by the nonlinear Boltzmann equations [61,62]

$$\frac{dg'_j(t)}{dt} = b_j - \sum_{j' \neq M} a'_{j,j'}(t) g'_{j'}(t) \tag{30.24}$$

where $g'_j(t) = g_j(t) - g_M(t)$ is the reduced form of the nonequilibrium part* of the distribution. This reduced form ensures conservation of particle numbers, that is, $\sum_{j=1}^{N} g_j(t) = 0$. Equation 30.24 was used for investigating the dynamics in quantum wires [61] and quantum-dot superlattices [62]. In Equation 30.24, we introduced notation for the matrix elements $a'_{j,j'}(t) = a_{j,j'}(t) - a_{j,M}(t)$ via its components

$$a_{j,j'}(t) = \delta_{j,j'}\left[\mathcal{W}_j + \mathcal{W}_j^g(t) + \frac{1-\delta_{j,(N+1)/2}}{2\tau_j}\right] - \delta_{j+j',N+1}\left[\frac{1-\delta_{j,(N+1)/2}}{2\tau_j}\right] - \mathcal{W}_{j,j'} - \frac{e\mathcal{F}_0}{2\hbar\delta k}(\delta_{j,j'-1} - \delta_{j,j'+1}) \tag{30.25}$$

The total inelastic rate is $\mathcal{W}_j = \frac{1}{\tau_j^{AL}} = \sum_{j'} \mathcal{W}_{j,j'}$ with the scattering matrix

$$\mathcal{W}_{j,j'} = \frac{L}{2\pi}\delta k \sum_{\pm} W_{j,j'}^{\pm}(n_{j,j'} + f_{j'}^{\pm}) \tag{30.26}$$

$$W_{j,j'}^{\pm} = \theta(\pm\varepsilon_{j'} \mp \varepsilon_j)\left[\frac{D_{AL}^2|\varepsilon_{j'} - \varepsilon_j|}{2\hbar^2 c_s^3 \rho L W \epsilon_{\text{TF}}^2(|k_{j'} - k_j|)}\right]\left[1+\cos(\phi_{j'} - \phi_j)\right] \tag{30.27}$$

Here, we employ the notations $f_j^- = f_j^{(0)}$, $f_j^+ = 1 - f_j^{(0)}$, $n_{j,j'} = N_0(|\varepsilon_{j'} - \varepsilon_j|/\hbar)$, and $N_0(\omega_q) = [\exp(\hbar\omega_q/k_B T) - 1]^{-1}$ is the Bose function for phonons. Moreover, the nonlinear phonon-scattering rate $\mathcal{W}_j^g(t)$, associated with the heating of electrons, is

$$\mathcal{W}_j^g(t) = \frac{L}{2\pi}\delta k \sum_{j' \neq M} g'_{j'}(t)\left[\mathcal{W}_{j,j'}^+ - \mathcal{W}_{j,j'}^- - \left(\mathcal{W}_{j,M}^+ - \mathcal{W}_{j,M}^-\right)\right] \tag{30.28}$$

Once $g'_j(t)$ is obtained from Equation 30.24, the drift velocity $v_c(t)$ is found from

$$v_c(t) = \left[\sum_{j=1}^{N} f_j^{(0)}\right]^{-1} \times \begin{cases} \sum_{j \neq M} (v_j - v_M) g'_j(t), & \text{semiconducting} \\ 2v_F \sum_{j=1}^{M} g'_j(t), & \text{metallic} \end{cases} \tag{30.29}$$

One knows that the equilibrium part of the distribution does not contribute to the drift velocity. The steady-state drift velocity v_d is obtained from $v_c(t)$ in the limit $t \to \infty$. Furthermore, the steady-state current is given by $I = en_{1D}v_d$. The differential mobility is defined as $\mu_e = \partial v_d/\partial\mathcal{F}_0$.

* A distinct line must be drawn between equilibrium distribution function $f_j^{(0)}$ in the absence of applied electric field and stationary solution of the transport equation $\lim_{t\to 0} f_j(t)$.

30.4.2 Numerical Results for Enhanced Hot-Electron Mobility

Figure 30.5a displays the mobilities μ_e as a function of $\mathcal{F}_0$ at $T = 10$ K and $T = 6$ K. We see from Figure 30.5a a strong $\mathcal{F}_0$-dependence at a lower value of $\mathcal{F}_0$ and at higher temperature T. The observed feature comes from the electron–phonon scattering rate $\mathcal{W}_j^g(t)$ in Equation 30.28. We find a lower threshold field $\mathcal{F}^*$ is needed for entering into a nonlinear regime ($\mathcal{F} > \mathcal{F}^*$) because of enhanced phonon scattering at $T = 10$ K. $\mathcal{F}^*$ strongly depends on T, n_{1D}, γ_0, and Λ_0. As $\mathcal{F}_0 \to 0$, μ_e is larger at $T = 10$ K than at $T = 6$ K since there exists an additional thermal population of high-energy states with a high group velocity. The initial reduction of μ_e is connected to the increased frictional force with $\mathcal{F}_0$ from phonon scattering. We can see μ_e is independent of $\mathcal{F}_0$ below 0.75 kV/cm (linear regime) at $T = 10$ K. However, μ_e goes up significantly above 0.75 kV/cm (nonlinear regime). Finally, μ_e drops with $\mathcal{F}_0$ beyond 1.5 kV/cm (heating regime), giving rise to a saturation of the drift velocity. From Equation 30.15 we understand the group velocity $|v_j|$ increases with $|k_j|$. However, this increase becomes slower as approaching to v_F. Physically, the rise of μ_e with $\mathcal{F}_0$ in the nonlinear regime comes from the initially heated electrons in high-energy states with a larger group velocity, while the successive drop of μ_e in the heating regime connects to the combination of the upper limit $v_j \leq v_F$ and the significantly enhanced phonon scattering. In Figure 30.5b, v_d is presented as a function of T at $\mathcal{F}_0 = 2$ kV/cm and $\mathcal{F}_0 = 1$ kV/cm. μ_e increases with T monotonically in both cases, proving that the scattering is not dominated by phonons but by impurities and LER. Different features in the increase of μ_e are attributed to the linear and nonlinear regimes. For $\mathcal{F}_0 = 2$ kV/cm in the nonlinear regime, v_d (or μ_e) increases with T sublinearly, while v_d goes up superlinearly with T in the linear regime at $\mathcal{F}_0 = 1$ kV/cm. Different T dependence in μ_e directly connects to g_j, displayed in Figure 30.5c and d. g_j at $T = 10$ K, as well as the total distribution function f_j, are exhibited in Figure 30.5c as functions of k_j at $\mathcal{F}_0 = 2$ kV/cm and $\mathcal{F}_0 = 1$ kV/cm. The electron heating is found from Figure 30.5c at $\mathcal{F}_0 = 2$ kV/cm by moving thermally driven electrons from low- to high-energy states with heat resulting from the

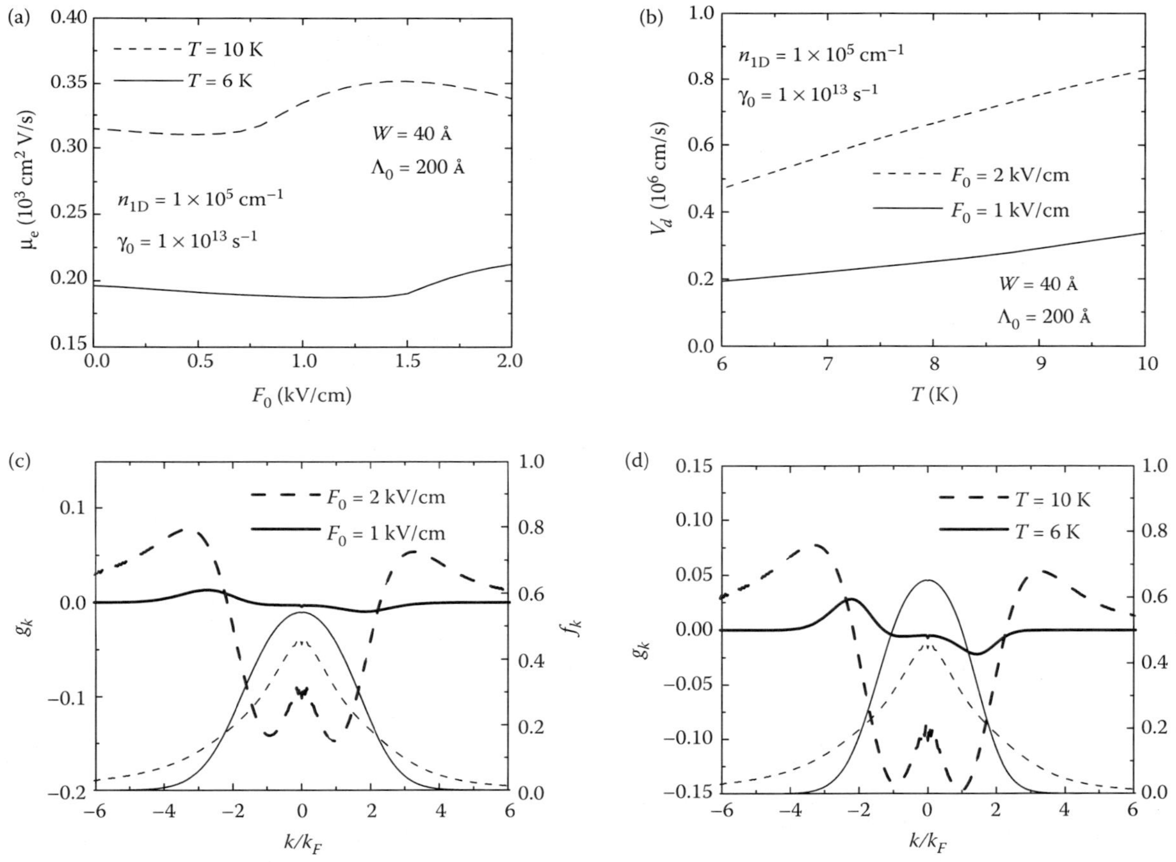

FIGURE 30.5 (a) Calculated electron mobilities μ_e as a function of applied electric field $\mathcal{F}_0$ at $T = 10$ K (dashed curve) and $T = 6$ K (solid curve); (b) electron drift velocities v_d as a function of temperature T at $\mathcal{F}_0 = 2$ kV/cm (dashed curve) and $F_0 = 1$ kV/cm (solid curve); (c) the nonequilibrium part of (g_k, left-hand-scaled thick curves) and total (f_k, right-hand-scaled thin curves) electron distribution functions at $T = 10$ K as functions of electron wave number k along the ribbon with $F_0 = 2$ kV/cm (dashed curves) and $F_0 = 1$ kV/cm (solid curves); and (d) g_k (left-hand-scaled thick curves) and f_k (right-hand-scaled thin curves) with $F_0 = 2$ kV/cm as functions of k at $T = 10$ K (dashed curves) and $T = 6$ K (solid curves). The other parameters are indicated directly in (a) and (b).

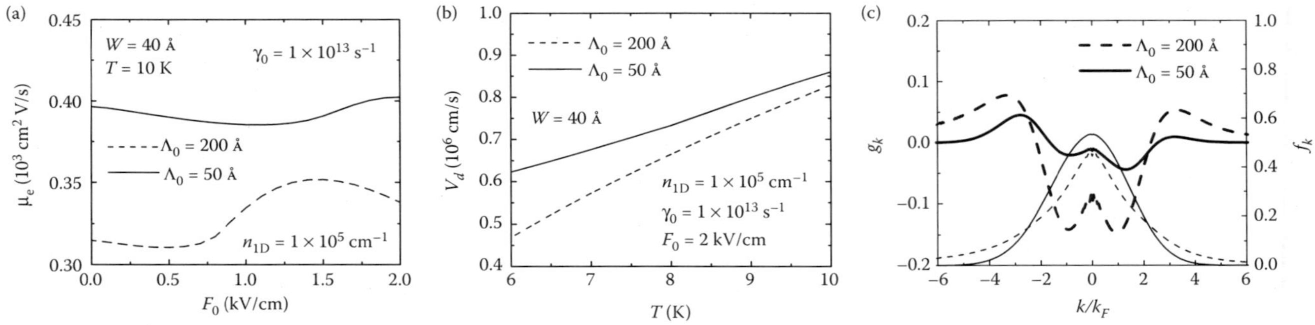

FIGURE 30.6 (a) μ_e as a function of $\mathcal{F}_0$ at T = 10 K with Λ_0 = 200 Å (dashed curve) and Λ_0 = 50 Å (solid curve); (b) v_d as a function of T with $\mathcal{F}_0$ = 2 kV/cm for Λ_0 = 200 Å (dashed curve) and Λ_0 = 50 Å (solid curve); and (c) g_k (left-hand-scaled thick curves) and f_k (right-hand-scaled thin curves) as a function of k. Here, the cases with Λ_0 = 200 Å and Λ_0 = 50 Å are represented by dashed and solid curves, respectively. The other parameters are indicated in (a) and (b).

work done by a frictional force [65] due to phonon scattering. At $\mathcal{F}_0$ = 1 kV/cm, we find electrons swept by elastic scattering from the right Fermi edge to the left one in the linear regime. Figure 30.5d demonstrates a comparison between g_j and f_j at T = 10 K and T = 6 K under $\mathcal{F}_0$ = 2 kV/cm, showing that phonon scattering is important at T = 10 K, while the elastic scattering of electrons dominates at T = 6 K, which agrees with the observation in Figure 30.5c.

The correlation-length effect for the LER is shown in Figure 30.6a through c by setting W = 50 Å and varying Λ_0 from 200 Å through 50 Å. We know from Equation 30.23 that the LER scattering may either decrease or increase with Λ_0, depending on $|k_j| \ll 1/2\Lambda_0$ or $|k_j| \gg 1/2\Lambda_0$. For $n_{1D} = 1.0\times10^5$ cm^{-1}, we know $|k_j| \ll 1/2\Lambda_0$ is met in the low-field limit ($|k_j| \sim k_F$), while we have $|k_j| \gg 1/2\Lambda_0$ for the high-field limit due to electron heating. Consequently, we see from Figure 30.6a that μ_e increases as $\mathcal{F}_0 \to 0$ when Λ_0 is 50 Å in the low-field regime. However, $\mathcal{F}^*$ increases in the high-field regime for Λ_0 = 50 Å. This is accompanied by a reduction in the enhancement of μ_e. The anomalous feature with Λ_0 gives a profound impact on the T-dependence of μ_e as shown in Figure 30.6b, where the rising rate of μ_e with T in the high-field regime is much smaller with Λ_0 = 50 Å than for Λ_0 = 200 Å. Additionally, g_j in Figure 30.6c exhibits an anomalous cooling behavior, that is, with a smaller spreading of f_j in the k_j space, in the high-field regime as Λ_0 drops to 50 Å.

30.5 MAGNETOPLASMONS IN GAPPED GRAPHENE

Electronic and transport properties of Dirac electrons in the presence of a uniform perpendicular magnetic field have been reported in a few studies [67–69]. The wavefunction, obtained in Reference 70, has a number of novel and intriguing properties. Additionally, electron–electron Coulomb interaction is found to have effects on both the quasiparticle effective mass and collective excitations [71]. Using the same method as for the case of circularly polarized light radiation in Equation 30.1, we further take into account the vector potential for a uniform perpendicular magnetic field, giving

$$\hat{\mathcal{H}} = v_F\sigma\cdot(\mathbf{p}-e\mathbf{A}) \tag{30.30}$$

$$\mathbf{B} = \nabla\times\mathbf{A} = \{0,0,B\}$$

In the case of two-dimensional electron gas (2DEG), on the other hand, the Hamiltonian contains the effective electron mass m^*:

$$\hat{\mathcal{H}}_{2\text{DEG}} = \frac{1}{2m^*}(\mathbf{p}-e\mathbf{A})^2 \tag{30.31}$$

with similar canonical momentum substitution.

Although the energy levels are equidistant for 2DEG with (n = 0, 1, 2, ...), $\varepsilon_n = \omega_c(n+1/2)$ and ω_c the cyclotron frequency, in graphene the difference between two consecutive energy levels decreases with the index n of the level as $\simeq 1/\sqrt{n}$, that is, $\varepsilon_n = (\hbar v_F/\ell_B)\sqrt{2n}$. Here $\ell_B = \sqrt{\hbar/(eB)}$. Also, $\omega_0 = \hbar v_F/\ell_B$ will be used as a unit of frequency for both the 2DEG and graphene. These eigenenergies apparently depend on the applied magnetic field B and are presented in Figure 30.7. The B dependence becomes the strongest for the case of zero gap in (c) and negligible with a large gap in (b).

In order to obtain the one-loop polarization function $\Pi^0(q,\omega)$, we perform a summation over all possible transitions between the states on both sides of the Fermi energy, so that both occupied and unoccupied states will be included. The coefficients in the summations, representing the weight of each term, are called the *oscillator strengths* or form factors. Obviously, for two well-separated eigenstates with $|n_1-n_2| \gg 1$, their coefficients become infinitesimal, which in leading order may be expressed as a factorial function $(|n_1-n_2|)!/(n_1+n_2)!$. The general definition of the form factor is the wavefunction overlap, that is,

$$\mathcal{F}\left(\frac{q^2\ell_B^2}{2}\right) = \left|\left\langle n_2,\beta_2\left|\mathbf{e}^{i\mathbf{q}\cdot\mathbf{r}}\right|n_1,\beta_1\right\rangle\right|^2 \tag{30.32}$$

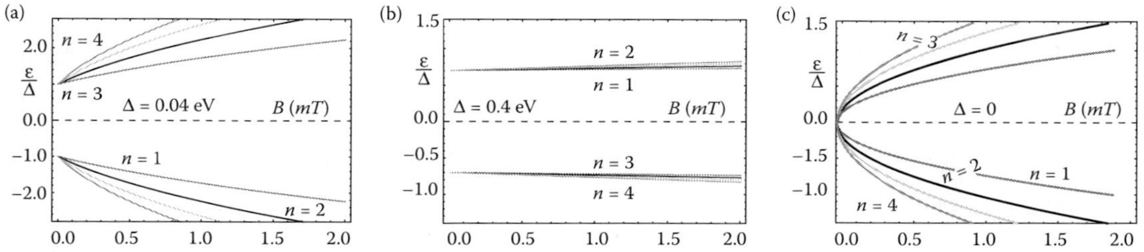

FIGURE 30.7 Dirac electron energy eigenvalues (Landau levels) as a function of the perpendicular magnetic field B. The first four levels for both electrons and holes are displayed. Plots (a) through (c) correspond to various energy gaps, that is, $\Delta = 0.04$ eV, $\Delta = 0.4$ eV, and $\Delta = 0$, respectively.

where $\beta_1, \beta_2 = \pm 1$. We will directly write down the expressions for form factors of both 2DEG and graphene. In the case of 2DEG, the form factor is calculated as

$$\mathcal{F}_{n_1,n_2}^{2\mathrm{DEG}}(q) = e^{-\xi}\xi^{|n_1-n_2|}\frac{n_<!}{n_>!}\left[\mathcal{L}_{n_<-1}^{|n_1-n_2|}(\xi)\right]^2 \quad (30.33)$$

$$\xi = \frac{q^2\ell_B^2}{2}$$

where $n_<$ is the lesser of two integers n_1 and n_2 while $n_>$ is the greater. In the case of *gapped* graphene with $\Delta > 0$, the form factor becomes

$$\mathcal{F}_{\beta_1,\beta_2}^{n_1,n_2}(q) = \mathrm{e}^{-\xi}\xi^{|n_1-n_2|}\left[\left(1+\frac{\beta_1\beta_2\Delta^2}{|\varepsilon_{n_1}\varepsilon_{n_2}|}\right)\mathcal{P}_1 + \mathcal{P}_2\right] \quad (30.34)$$

$$\mathcal{P}_1 = \frac{n_<!}{n_>!}\left[\mathcal{L}_{n_<}^{|n_1-n_2|}(\xi)\right]^2 + (1-\delta_{0n_<})\frac{(n_<-1)!}{(n_>-1)!}\left[\mathcal{L}_{n_<-1}^{|n_1-n_2|}(\xi)\right]^2$$

$$\mathcal{P}_2 = \frac{4\beta_1\beta_2 v_F^2}{\ell_B^2|\varepsilon_{n_1}\varepsilon_{n_2}|}\frac{n_<!}{(n_>-1)!}\mathcal{L}_{n_<}^{|n_1-n_2|}(\xi)\mathcal{L}_{n_<-1}^{|n_1-n_2|}(\xi)$$

$$|\varepsilon_n| = \sqrt{\frac{2\hbar^2 v_F^2}{\ell_B^2}n + \Delta^2}$$

For the case of zero energy gap, $\Delta \to 0$, one may easily obtain the standard expression for the form factor in graphene, similar to the results in References 72–74. The function $\Pi^0(q, \omega)$ contains two principal parts, that is, the vacuum polarization corresponding to *interband* transitions in *undoped* graphene, as well as another part involving the summation over all the occupied Landau levels (N_F is the highest occupied Landau level), taking into consideration the *intraband* transitions.

Our goal is to explore both incoherent (particle–hole modes) and coherent (plasmons) excitations in the system. The imaginary part of $\Pi^0(q, \omega)$ defines the structure factor $\mathcal{D}(q,\omega)$, that is,

$$\mathcal{D}(q,\omega) = -\frac{1}{\pi}\mathrm{Im}\Pi^0(q,\omega) \quad (30.35)$$

The particle–hole mode region is obtained from the condition $\mathrm{Im}\Pi^0(q, \omega) > 0$. Additionally, the plasmons become *damping free* only if they stay away from the particle–hole mode region, or equivalently, $\mathcal{D}(q,\omega) = 0$. On the other hand, the plasmon frequencies are defined by the zeros of the dielectric function $\epsilon(q,\omega) = 1 - v_c(q)\Pi^0(q,\omega)$. Here, $v_c(q) = (e^2/2\epsilon_0\epsilon_b q)$ is the Fourier-transformed unscreened Coulomb potential and ϵ_b is the background dielectric constant. These plasmons may also be obtained from the peak of the renormalized polarization $\Pi^{\mathrm{RPA}}(q, \omega)$ for interacting electrons in the random-phase approximation (RPA), giving

$$\Pi^{\mathrm{RPA}}(q,\omega) = \frac{\Pi^0(q,\omega)}{1 - v_c(q)\Pi^0(q,\omega)} \quad (30.36)$$

We note that a metal–insulator transition may also occur under a magnetic field [75], similar to the effect of circularly polarized light, along with other effects [76–78].

30.5.1 Single-Particle Excitations and Magnetoplasmons

We now present and discuss our numerical results of both $\Pi^0(q, \omega)$ and $\Pi^{\mathrm{RPA}}(q, \omega)$ for various values of chemical potentials μ (or equivalently, number of occupied Landau levels N_F) and energy gap Δ. The disorder broadening is set to be $\eta = 0.05\hbar v_F/\ell_B$ for all our numerical calculations. One of the major effects found in our numerical results is the magnetoplasmon for various energy gaps, compared with those for gapless graphene (Dirac cone). We also compare $\Pi^{\mathrm{RPA}}(q, \omega)$ for interacting electrons with noninteracting $\Pi^0(q, \omega)$ with various interaction parameters $r_s = (2m^*e^2/\epsilon_0\epsilon_b\hbar^2 k_F)$ for 2DEG and $r_s = (e^2/\epsilon_0\epsilon_b\hbar v_F)$ for graphene as well.

It is interesting to compare the results in the presence of an external magnetic field with those at zero magnetic field [13,14,57,79] for both 2DEG and gapped graphene. We emphasize that the boundaries of the particle–hole continuum are drastically different for the 2DEG and graphene. Whereas the edges of the particle–hole excitation region of 2DEG are parabolic, as seen from Figure 30.8, these boundaries

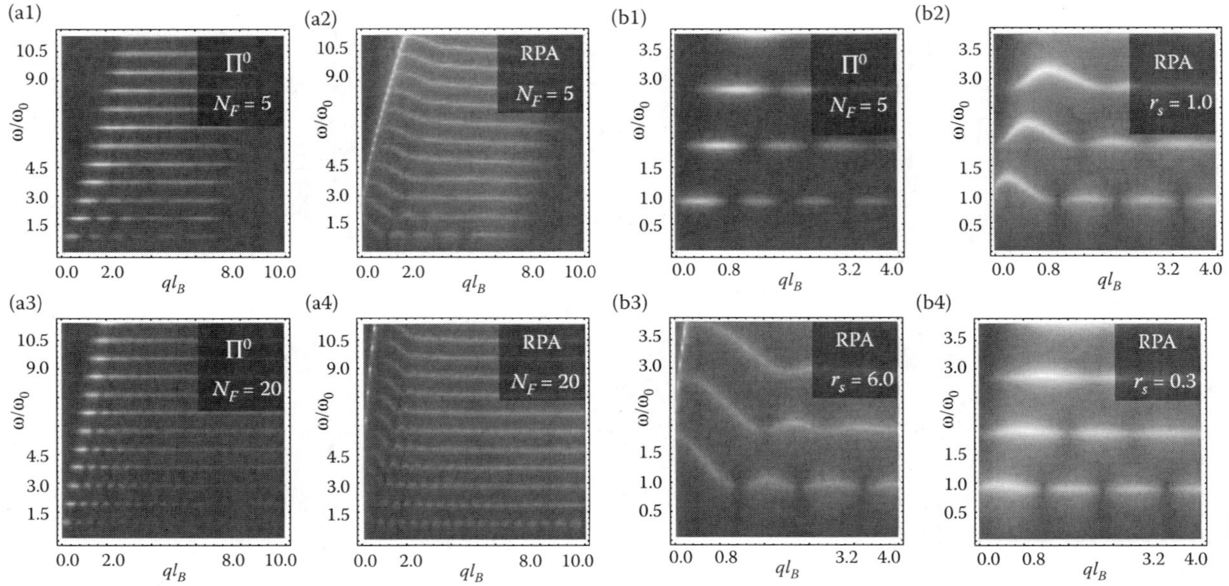

FIGURE 30.8 Density plots of Im[$\Pi^0(q, \omega)$] and Im[$\Pi^{RPA}(q, \omega)$] for 2DEG. The left panel (a1) through (a4) presents noninteracting polarizations in (a1) and (a3) as well as the renormalized polarizations in (a2) and (a4) for various values of N_F. Plots (a1) and (a2) compare the effects due to $N_F = 5$, whereas plots (a3) and (a4) are for the case of $N_F = 20$. The left panel (b1) through (b4) demonstrates Im[$\Pi^0(q, \omega)$] in (b1) and Im[$\Pi^{RPA}(q, \omega)$] in (b2) through (b4) for $N_F = 5$. Each plot in (b2) through (b4) corresponds to chosen interaction parameters $r_s = 6.0$, 1.0, and 0.3, respectively.

change to either straight lines in graphene (zero gap) or to be slightly modified by the gap. Consequently, the plasmon region defined by $\mathcal{D}(q, \omega) = 0$ in graphene is significantly increased even for a small energy gap Δ, in comparison with that for 2DEG. This demonstrates a large influence by a finite magnetic field on both the particle–hole excitation spectrum and the boundaries of the particle–hole mode regions at the same time.

Comparing Figure 30.8 with Figure 30.9, we conclude that there exist two competing types of particle–hole excitation spectral behavior. These are the horizontal lines with embedded islands for 2DEG and the inclined lines for graphene. All the lines are well aligned with the boundaries of particle–hole mode regions at zero field, including parabolic curves for 2DEG, straight lines for gapless graphene, or nearly straight lines for gapped graphene. We note that the modulation for

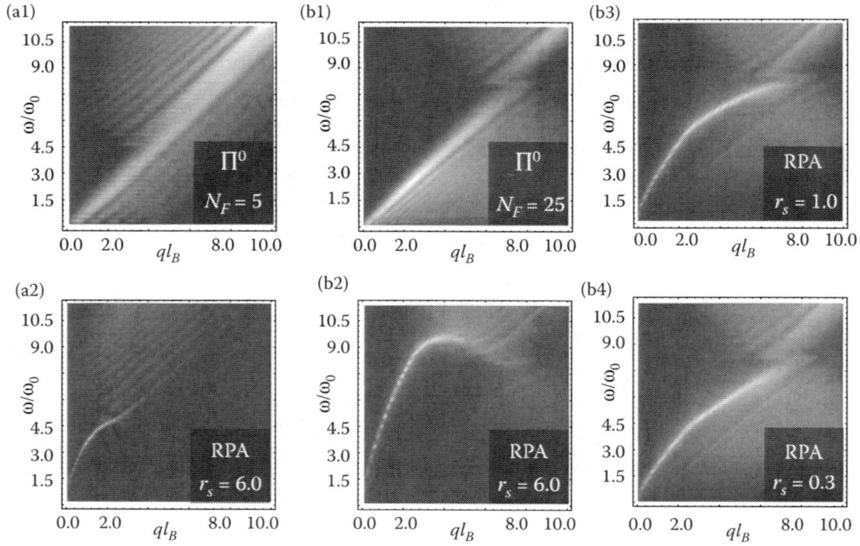

FIGURE 30.9 Density plots of Im[$\Pi^0(q, \omega)$] and Im[$\Pi^{RPA}(q, \omega)$] for *gapless graphene*. The left panel (a1) and (a2) represents the noninteracting and RPA renormalized polarization functions with $N_F = 5$. Plot (b1) corresponds to the noninteracting polarization function with $N_F = 25$, while plots (b2) through (b4) are associated with the RPA renormalized ones. Additionally, plots (b2) through (b4) present results for $r_s = 6.0$, 1.0, and 0.3, respectively.

horizontal lines in 2DEG becomes strongest, in comparison with the inclined line in graphene. This difference is attributed to the fact that the Landau levels in graphene are *not equidistant* in contrast to those in 2DEG. Additionally, horizontal modulation prevails for a small broadening parameter η, which has been verified experimentally for the range of η within $0.05-0.5\bar{v}_F/\ell_B$.

The particle–hole modes for gapless graphene with Dirac cone dispersion relation are presented in Figure 30.9. The excitation regions are basically determined by the inclined straight lines obeying $\omega = v_F q$. A dark triangular region, located near the origin, indicates the undamped plasmons. Although nondispersive particle–hole modes are not clearly seen, they still alter the particle–hole excitation spectrum at $B = 0$. The damping-free plasmon curve occurring in $\mathrm{Im}[\Pi^{RPA}(q, \omega)]$ follows the $\sqrt{q}$-dispersion in the long-wavelength limit, switching to a straight line to become damped plasmons. Similar features are found from Figure 30.9b1 and b2 for higher doping. However, the increased μ enlarges the region for undamped plasmons. Moreover, the RPA renormalization displayed in Figure 30.9b2 through b4 is most significant for a larger value of interaction parameter r_s. From Figures 30.9 and 30.8, we also compare the RPA effects ($\mathrm{Im}[\Pi^{RPA}(q, \omega)]$) with various values of the interaction parameter r_s for graphene and 2DEG. The Coulomb interaction in 2DEG leads to the existence of dispersive magnetoexcitons. In the limit of vanishing interaction $r_s \to 0$, however, the renormalized RPA polarization becomes similar to that of noninteracting electrons, as seen from the comparisons of (b1) and (b4) in Figure 30.8 as well as in Figure 30.9. Finally, Figure 30.10 shows the effect on the plasmons in graphene, where the undamped plasmon curve does not follow the $v_F q$ line anymore but follows that for $B = 0$ instead, leading to a suppression of magnetic field effects. Collective excitations in AA- and AB-stacked graphene in the presence of magnetic field have been investigated recently by Wu et al. [80]. The groupings of the Landau levels lead to considerable differences in the plasmon excitation energies, which are determined by the way in which the layers are stacked. In general, plasmonics in carbon-based nanostructures has recently received considerable development. First, plasmons modes were caclulated and studied in AA-stacked bilayer graphene [81]. The above-mentioned Klein tunneling in this type of bilayer graphene was addressed in Reference 82. Plasmon excitations of a single C_{60} molecule, induced by an external, fast moving electron, were theoretically studied based on quantum hydrodynamical model [83]. This study resulted in finding the differential cross sections. The localization of charged particles by the image potential of spherical shells, such as fullerene buckyballs, has been addressed in Reference 84. Systems without spherical symmetry demonstrate anisotropy and dimerization of the plasmon excitations [85,86]. Plasmon frequencies of such systems depend not only on the angular momentum L, but its projection M on the axis of quantization. Collective excitations in the processes of photoionization and electron inelastic scattering were addressed in Reference 87.

30.6 CONCLUDING REMARKS

In summary, we reviewed the effect of an energy gap on the electronic, transport, and many-body properties of graphene. The gap may be generated by a number of means in graphene and topological insulators. This includes a substrate or finite width. However, we paid attention to the gap that appears as a result of electron–photon interaction between Dirac electrons in graphene and circularly polarized photons. This type

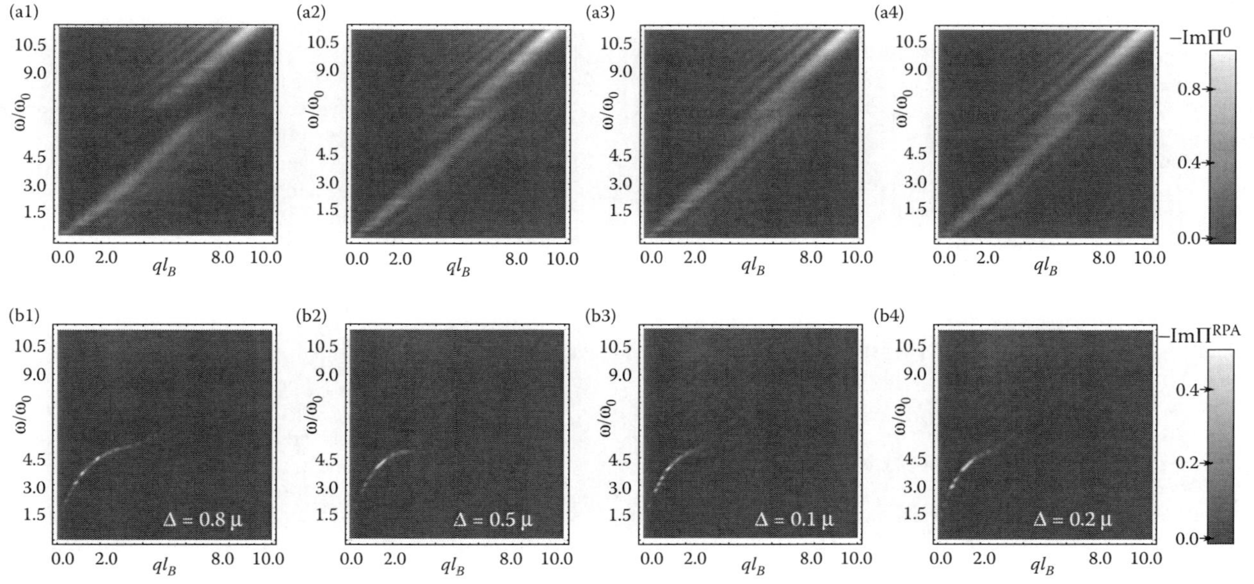

FIGURE 30.10 Density plots of $\mathrm{Im}[\Pi^0(q, \omega)]$ and $\mathrm{Im}[\Pi^{RPA}(q, \omega)]$ for *gapped graphene*. The upper panel (a1) through (a4) presents results for the noninteracting polarization function, while the lower panel (b1) through (b4) presents the RPA renormalized ones. Each plot in 1 through 4 corresponds to $\Delta/\mu = 0.8$, 0.5, 0.1, and 0.02.

of energy gap is tunable and may be varied in experiment with the intensity of the imposed light.

We note that the presence of the gap in the energy dispersion leads to the breaking of chiral symmetry. Consequently, this leads to significant changes in the electron transmission. There is a decrease of the electron transmission amplitude for a head-on collision and almost head-on incidence as a result of the fact that the Klein paradox no longer exists. The transmission resonances (the peaks for finite angles of incidence) are slightly shifted although the general structure of the peaks remains unaffected. However, if two or more pairs of subbands are taken into consideration, the transmission may even increase compared to the case of only one photon.

We showed a minimum mobility before a field threshold for entering into the nonlinear-transport regime due to buildup of a frictional force. We demonstrated a mobility enhancement after this threshold value because of heated electrons in high-energy states. We also obtained a maximum mobility enhancement due to balance between simultaneously increasing group velocity and phonon scattering. Additionally, we proved an increased field threshold by a small correlation length for the LER.

Also, we considered magnetoplasmons in gapped graphene and concluded that similar to the case of zero magnetic field, the presence of the gap increases the region free from particle–hole excitations. Consequently, the region where undamped plasmons exist is expanded. We obtained a new type of plasmon dispersion.

ACKNOWLEDGMENTS

This chapter was supported by a grant from the Air Force Office of Scientific Research (AFOSR) and a contract from the Air Force Research Laboratory (AFRL). We are grateful to Mary Michelle Easter for a critical reading of the chapter and helpful comments.

REFERENCES

1. A. H. C. Neto, F. Guinea, N. M. R. Peres, K. S. Nonoselov, and A. K. Geim, The electronic properties of graphene, *Rev. Mod. Phys.* **81**, 109–162, 2009.
2. G. Gumbs, D. H. Huang, and O. Roslyak, Electronic and photonic properties of graphene layers and carbon nanoribbons, *Phil. Trans. R. Soc. A* **138**, 1932, 2010.
3. D. S. L. Abergel, V. Apalkov, J. Berashevich, K. Ziegler, and T. Chakraborty, Properties of graphene: A theoretical perspective, *Adv. Phys.* **59**, 261–482, 2010.
4. M. Orlita and M. Potemski, Dirac electronic states in graphene systems: Optical spectroscopy studies, *Semicond. Sci. Technol.* **25** (22), 063001, 2010.
5. S. Das Sarma, S. Adam, E. H. Hwang, and E. Rossi, Electronic transport in two-dimensional graphene, *Rev. Mod. Phys.* **83**, 407–470, 2011.
6. K. S. Novoselov, A. K. Geim, S. V. Morozov, D. Jiang, M. I. Katsnelson, I. V. Grigorieva, S. V. Dubonos, and A. A. Firsov, Electric field effect in atomically thin carbon films, *Science,* **306**, 666–669, 2004.
7. K. S. Novosolev, A. K. Geim, S. V. Morozov, D. Jiang, M. I. Katsnelson, I. V. Grigorieva, S. V. Dubonos, and A. A. Firsov, Two-dimensional gas of massless Dirac fermions in graphene, *Nature (London)*, **438**, 197–200, 2005.
8. Y. Zhang, Y.-W. Tan, H. L. Störmer, and P. Kim, Experimental observation of the quantum Hall effect and Berry's phase in graphene, *Nature (London)*, **438**, 201–204, 2005.
9. M. I. Katsnelson, K. S. Novoselov, and A. K. Geim, Chiral tunnelling and the Klein paradox in graphene, *Nat. Phys.* **2**, 620–625, 2006.
10. A. F. Young and P. Kim, Quantum interference and Klein tunnelling in graphene heterojunctions, *Nat. Phys.* **5**, 222–226, 2009.
11. O. Roslyak, A. Iurov, G. Gumbs, and D. H. Huang, Unimpeded tunneling in graphene nanoribbons, *J. Phys.: Condens. Matt.* **22** (6), 165301, 2010.
12. A. Iurov, G. Gumbs, O. Roslyak, and D. H. Huang, Anomalous photon-assisted tunneling in graphene, *J. Phys.: Condens. Matt.* **24** (8), 015303, 2012.
13. B. Wunsch, T. Stauber, F. Sols, and F. Guinea, Dynamical polarization of graphene at finite doping, *New J. Phys.* **8** (15), 318, 2006.
14. E. H. Wang and S. Das Sarma, Dielectric function, screening, and plasmons in two-dimensional graphene, *Phys. Rev. B* **75** (6), 205418, 2007.
15. O. Roslyak, G. Gumbs, and D. H. Huang, Plasma excitations of dressed Dirac electrons in graphene layers, *J. Appl. Phys.* **109** (8), 113721, 2011.
16. K. F. Mak, M. Y. Sfeir, Y. Wu, C. H. Lui, J. A. Misewich, and T. F. Heinz, Measurement of the optical conductivity of graphene, *Phys. Rev. Lett.* **101** (4), 196405, 2008.
17. R. R. Nair, P. Blake, A. N. Grigorenko, K. S. Novoselov, T. J. Booth, T. Stauber, N. M. R. Peres, and A. K. Geim, Fine structure constant defines visual transparency of graphene, *Science,* **320**, 1308, 2008.
18. L. Ju, B. Geng, J. Horng, C. Girit, M. Martin, Z. Hao, H. A. Bechtel et al., Graphene plasmonics for tunable terahertz metamaterials, *Nat. Nanotechnol.* **6**, 630–634, 2011.
19. Z. Q. Li, E. A. Henriksen, Z. Jiang, Z. Hao, M. C. Martin, P. Kim, H. L. Störmer, and D. N. Basov, Dirac charge dynamics in graphene by infrared spectroscopy, *Nat. Phys.* **4**, 532–535, 2008.
20. F. Wang, Y. Zhang, C. Tian, C. Girit, A. Zettl, M. Crommie, and Y. R. Shen, Gate-variable optical transitions in graphene, *Science,* **320**, 206–209, 2008.
21. Q. Bao, H. Zhang, B. Wang, Z. Ni, C. Haley, Y. X. Lim, Y. Wang, D. Y. Tang, and K. P. Loh, Broadband graphene polarizer, *Nat. Photon.* **5**, 411–415, 2011.
22. J. H. Strait, H. Wang, S. Shivaraman, V. Shields, M. Spencer, and F. Rana, Very slow cooling dynamics of photoexcited carriers in graphene observed by optical-pump terahertz-probe spectroscopy, *Nano Lett.* **11**, 4688–4692, 2011.
23. J. C. W. Song, M. S. Rudner, C. M. Marcus, and L. S. Levitov, Hot carrier transport and photocurrent response in graphene, *Nano Lett.* **11**, 4902–4906, 2011.
24. O. Roslyak, G. Gumbs, and D. H. Huang, Tunable band structure effects on ballistic transport in graphene nanoribbons, *Phys. Lett. A* **374**, 4061–4064, 2010.
25. D. H. Huang, G. Gumbs, and O. Roslyak, Field-enhanced electron mobility by nonlinear phonon scattering of Dirac electrons in semiconducting graphene nanoribbons, *Phys. Rev. B* **83** (9), 115405, 2011.
26. O. Roslyak, G. Gumbs, and D. H. Huang, Energy loss spectroscopy of epitaxial versus free-standing multilayer graphene, *Phys. E* **44**, 1874–1884, 2012.

27. J. Z. Bernád, M. Jääskeläinen, and U. Zülicke, Effects of a quantum measurement on the electric conductivity: Application to graphene, *Phys. Rev. B* **81** (4), 173403, 2010.
28. K. Nomura and A. H. MacDonald, Quantum Hall ferromagnetism in graphene, *Phys. Rev. Lett.* **96** (4), 256602, 2006.
29. T. Fang, A. Konar, H. Xing, and D. Jena, Mobility in semiconducting graphene nanoribbons: Phonon, impurity, and edge roughness scattering, *Phys. Rev. B* **78** (8), 205403, 2008.
30. A. Bostwick, F. Speck, T. Seyller, K. Horn, M. Polini, R. Asgari, A. H. MacDonald, and E. Rotenberg, Observation of plasmarons in quasi-freestanding doped graphene, *Science*, **328**, 994–999, 2010.
31. S. A. Mikhailov and K. Ziegler, New electromagnetic mode in graphene, *Phys. Rev. Lett.* **99** (4), 016803, 2007.
32. M. Jablan, H. Buljan, and M. Soljačić, Plasmonics in graphene at infrared frequencies, *Phys. Rev. B* **80** (7), 245435, 2009.
33. O. V. Kibis, Metal-insulator transition in graphene induced by circularly polarized photons, *Phys. Rev. B* **81** (5), 165433, 2010.
34. A. K. Geim, Graphene: Status and prospects, *Science,* **324**, 1530–1534, 2009.
35. O. Shenderova, V. Zhirnov, and D. Brenner, Carbon nanostructures, *Crit. Rev. Solid State Mater. Scie.* **27**, 227–356, 2002.
36. S. Das Sarma, S. Adam, E. H. Hwang, and E. Rossi, Electronic transport in two-dimensional graphene, (arXiv:1003.4731v2).
37. P. R. Wallace, The band theory of graphite, *Phys. Rev.* **71**, 622–634, 1947.
38. J. W. McClure, Diamagnetism of graphite, *Phys. Rev.* **104**, 666–671, 1956.
39. G. Giovannetti, P. A. Khomyakov, G. Brocks, P. J. Kelly, and J. van den Brink, Substrate-induced band gap in graphene on hexagonal boron nitride: Ab initio density functional calculations, *Phys. Rev. B* **76** (4), 073103, 2007.
40. T. Low, F. Guinea, and M. I. Katsnelson, Gaps tunable by electrostatic gates in strained graphene, *Phys. Rev. B* **83** (7), 195436, 2011.
41. O. V. Kibis, Dissipationless Electron transport in photon-dressed nanostructures, *Phys. Rev. Lett.* **107** (5), 106802, 2011.
42. A. Iurov, G. Gumbs, O. Roslyak, and D. H. Huang, *J.* Photon dressed electronic states in topological insulators: Tunneling and conductance, *Phys.: Condens. Matt.* **25** (13), 135502, 2013.
43. M. Barbier, P. Vasilopoulos, and F. M. Peeters, Single-layer and bilayer graphene superlattices: Collimation, additional Dirac points and Dirac lines, *Phil. Trans. R. Soc. A* **386**, 5499–5524, 2010.
44. M. Barbier, P. Vasilopoulos, and F. M. Peeters, Dirac electrons in a Kronig-Penney potential: Dispersion relation and transmission periodic in the strength of the barriers, *Phys. Rev. B* **80** (5), 205415, 2009.
45. V. V. Chevianov and V. I. Fal'ko, Selective transmission of Dirac electrons and ballistic magnetoresistance of n-p junctions in graphene, *Phys. Rev. B.* **74** (4), 141403(R), 2006.
46. S. Roche and L. E. F. Foa Torres, *Graphene, Carbon Nanotubes, and Nanostructures* (edited by J. E. Morris and K. Iniewski, CRC Press, Taylor & Francis Group, LLC, USA, 2013), Chapter 3.
47. E. Suárez Morell and L. E. F. Foa Torres, Radiation effects on the electronic properties of bilayer graphene, *Phys. Rev. B* **86** (5), 125449, 2012.
48. H. L. Calvo, P. M. Perez-Piskunow, H. M. Pastawski, S. Roche, and L. E. F. Foa Torres, Non-perturbative effects of laser illumination on the electrical properties of graphene nanoribbons, *J. Phys.: Condens. Matt.* **25** (8), 144202, 2013.
49. T. Oka and H. Aoki, Photovoltaic Hall effect in graphene, *Phys. Rev. B* **79** (4), 081406(R), 2009.
50. J. V. Gomes and N. M. R. Peres, Tunneling of Dirac electrons through spatial regions of finite mass, *J. Phys.: Condens. Matt.* **20** (12), 325221, 2008.
51. R. Biswas and C. Sinha, Photon induced tunneling of electron through a graphene electrostatic barrier, *J. Appl. Phys.* **114**, 18, 2013.
52. L. Z. Szabó, M. G. Benedict, A. Czirják, and P. Földi, Relativistic electron transport through an oscillating barrier: Wave-packet generation and Fano-type resonances, *Phys. Rev. B* **88**, 075438, 2013.
53. G. Gumbs, A. Iurov, D. Huang, P. Fekete, and L. Zhemchuzhna, Effects of periodic scattering potential on Landau quantization and ballistic transport of electrons in graphene, *AIP Conf. Proc.* **1590**, 134, 2014.
54. M. Modarresia, A. Mogulkocb, M. R. Roknabadia, and M. Behdania, Study of the electron¨Cphoton interaction on the spin-dependent transport in nano-structures, *Phys. E: Low-Dimens. Syst. Nanostruct.* **57**, 76, 2014.
55. L. Yuan , B. A. J. Mansoor, S. G. Tan, W. Zhao, R. Bai, and G. H. Zhou, Chiral tunneling modulated by a time-periodic potential on the surface states of a topological insulator, *Nature: Sci. Rep.* **4**, 4624, 2014.
56. M. Z. Hasan and C. L. Kane, Colloquium: Topological insulators, *Rev. Mod. Phys.* **82**, 3045–3067, 2010.
57. A. Iurov and G. Gumbs, Theory for entanglement of electrons dressed with circularly polarized light in graphene and three-dimensional topological insulators, *Proc. SPIE* **8749** (11), 874903, 2013.
58. W. Xu, F. M. Peeters, and T. C. Lu, Dependence of resistivity on electron density and temperature in graphene, *Phys. Rev. B* **79** (4), 073403, 2009.
59. G. Gumbs and D. H. Huang, *Properties of Interacting Low-Dimensional Systems* (Wiley-VCH Verlag GmbH & Co. kGaA, Weinheim, Germany, 2011), Chapter 13.
60. G. Gumbs and D. H. Huang, *Properties of Interacting Low-Dimensional Systems* (Wiley-VCH Verlag GmbH & Co. kGaA, Weinheim, Germany, 2011), Chapter 14.
61. D. H. Huang and G. Gumbs, Comparison of inelastic and quasielastic scattering effects on nonlinear electron transport in quantum wires, *J. Appl. Phys.* **107** (8), 103710, 2010.
62. D. H. Huang, S. K. Lyo, and G. Gumbs, Bloch oscillation, dynamical localization, and optical probing of electron gases in quantum-dot superlattices in high electric fields, *Phys. Rev. B* **79** (19), 155308, 2009.
63. S. K. Lyo and D. H. Huang, Effect of electron-electron scattering on the conductance of a quantum wire studied with the Boltzman transport equation, *Phys. Rev. B* **73** (10), 205336, 2006.
64. L. Brey and H. A. Fertig, Electronic states of graphene nanoribbons studied with the Dirac equation, *Phys. Rev. B* **73**(5), 235411, 2006.
65. D. H. Huang, T. Apostolova, P. M. Alsing, and D. A. Cardimona, High-field transport of electrons and radiative effects using coupled force-balance and Fokker-Planck equations beyond the relaxation-time approximation, *Phys. Rev. B* **69**(12), 075214, 2004.
66. L. Brey and H. A. Fertig, Elementary electronic excitations in graphene nanoribbons, *Phys. Rev. B* **75** (6), 125434, 2007.
67. M. O. Goerbig, Electronic properties of graphene in a strong magnetic field, *Rev. Mod. Phys.* **83**, 1193–1243, 2011.
68. V. N. Kotov, B. Uchoa, V. M. Pereira, F. Guinea, and A. H. Castro Neto, Electron-electron interactions in graphene: current status and perspectives, *Rev. Mod. Phys.* **84**, 1067–1125, 2012.

69. V. P. Gusynin and S. G. Sharapov, Unconventional integer Quantum Hall effect in graphene, *Phys. Rev. Lett.* **95** (4), 146801, 2002.
70. M. Koshino and T. Ando, Splitting of the quantum Hall transition in disordered graphenes, *Phys. Rev. B* **75** (4), 033412, 2007.
71. A. P. Smith, A. H. MacDonald, and G. Gumbs, Quasiparticle effective mass and enhanced g factor for a two-dimensional electron gas at intermediate magnetic fields, *Phys. Rev. B* **45**, 8829–8832, 1992.
72. R. Roldan, J.-N. Fuchs, and M. O. Goerbig, Collective modes of doped graphene and a standard two-dimensional electron gas in a strong magnetic field: Linear magnetoplasmons versus magnetoexcitons, *Phys. Rev. B* **80** (6), 085408, 2009.
73. R. Roldán, M. O. Goerbig, and J.-N. Fuchs, The magnetic field particle¨Chole excitation spectrum in doped graphene and in a standard two-dimensional electron gas, *Semicond. Sci. Technol.* **25**(11), 034005, 2010.
74. P. K. Pyatkovskiy and V. P. Gusynin, Dynamical polarization of graphene in a magnetic field, *Phys. Rev. B* **83** (12), 075422, 2011.
75. E. V. Gorbar, V. P. Gusynin, V. A. Miransky, and I. A. Shovkovy, Magnetic field driven metal-insulator phase transition in planar systems, *Phys. Rev. B* **66** (22), 045108, 2002.
76. A. Iyengar, J. Wang, H. A. Fertig, and L. Brey, Excitations from filled Landau levels in graphene, *Phys. Rev. B* **75** (14), 125430, 2007.
77. H. A. Fertig and L. Brey, Luttinger liquid at the edge of undoped graphene in a strong magnetic field, *Phys. Rev. Lett.* **97** (4), 116805, 2006.
78. Y. Zheng and T. Ando, Hall conductivity of a two-dimensional graphite system, *Phys. Rev. B* **65** (11), 245420, 2002.
79. P. K. Pyatkovskiy, Dynamical polarization, screening, and plasmons in gapped graphene, *J. Phys.: Condens. Matt.* **21** (8), 025506, 2009.
80. J.-Y. Wu, G. Gumbs, and M.-F. Lin, Combined effect of stacking and magnetic field on plasmon excitations in bilayer graphene, *Phys. Rev. B* **89**, 165407, 2014.
81. R. Roldán and L. Brey, Dielectric screening and plasmons in AA-stacked bilayer graphene, *Phys. Rev. B* **88** (11), 115420, 2013.
82. M. Sanderson, Y. S. Ang, and C. Zhang, Klein tunneling and cone transport in AA-stacked bilayer graphene, *Phys. Rev. B* **88**, 245404, 2013.
83. C. Z. Li, Z. L. Miskovic, F. O. Goodman, and Y. N. Wang, Plasmon excitations in C60 by fast charged particle beams, *J. Appl. Phys.* **113**, 18, 2013.
84. G. Gumbs, A. Balassis, A. Iurov, and P. Fekete, Strongly localized image states of spherical graphitic particles, *Sci. World J.* **2014**, 726303, 2014.
85. G. Gumbs, A. Iurov, A. Balassis, and D. Huang, Anisotropic plasmon-coupling dimerization of a pair of spherical electron gases, *J. Phys.: Condens. Matter*, **26**, 135601, 2014.
86. A. Iurov, G. Gumbs, B. Gao, and D. Huang, Modeling anisotropic plasmon excitations in self-assembled fullerenes, *Appl. Phys. Lett.* **104**, 203103, 2014.
87. A. V. Verkhovtsev, A. V. Korol, and A. V. Solovýov, Sub-GeV electron and positron channeling in straight, bent and periodically bent silicon crystals, *J. Phys.: Conf. Ser.* **438**, 1, 2013.

Section III

Nanocomposites and Applications

31 Graphene-Based Nanocomposites with Tailored Electrical, Electromagnetic, and Electromechanical Properties

M. S. Sarto, G. De Bellis, A. Tamburrano, A. G. D'Aloia, and F. Marra

CONTENTS

ABSTRACT

Graphene nanoplatelets (GNPs) are tiny stacks of graphene layers with thicknesses ranging from 1 to several nanometers and lateral size from hundreds of nanometers to several micrometers. Depending on the fabrication route, GNPs can be produced starting from different precursors, such as graphene oxide (GO) or graphite intercalation compounds (GICs): both routes are compatible with mass production. Polymeric nanocomposites are drawing ever-growing attention thanks to their light weight and the possibility to dramatically increase the matrix properties using very low filler loadings. The main challenge in the practical application of GNP and GNP-based nanocomposites consists of tailoring their functional properties through the control of the synthesis process, using a suitable modeling tool that allows correlating the micro/nanostructure of the material to its functional properties at macroscale. This chapter focuses on the development of GNPs and GNP-based polymer composites with tailored electrical, electromagnetic, and electromechanical properties for applications such as shielding or radar-absorbing materials (RAMs), or piezoresistive strain sensors. We demonstrate the proper control of the morphological characteristics of GNPs and the final functional properties of the nanocomposite (namely, the effective complex permittivity, shielding effectiveness or reflection coefficient, the piezoresistive response) through the proper setting of the synthesis route and of the type of polymeric matrix used. Different parameters are investigated, such as the precursor expansion time and temperature, the exfoliated graphite sonication cycle, the type of solvent used, and the suspension temperature during sonication. A novel simulation model is developed to predict the effective electromagnetic properties of GNP nanocomposites.

The presented model is validated by comparison with experimental data, and represents a useful tool for the design by simulation of RAMs.

31.1 INTRODUCTION

Since it was first obtained as a single sheet in 2004 by Geim et al. [1], graphene has attracted tremendous interest in the scientific community thanks to its outstanding mechanical [2–4], thermal [5,6], and especially electrical and transport properties [7]. The interest in graphene has been particularly accelerated by the difficulties in exploiting and controlling the properties of carbon nanotubes (CNTs) throughout the last decade, as well as the high cost of the latter structures. Along with the interest in carbon-based nanostructures, recently, composites filled with CNTs [8–14], graphene nanoplatelets (GNPs) [15–22], or combined systems [23–30] have also drawn much attention from the scientific community. These new composites are continuously finding new applications in various fields such as electromagnetic (EM) shielding, aeronautics, photovoltaics, electronics, actuators, and sensors. The concern from the research and industrial community is also due to the multifunctional nature of such composites. An important upcoming application of these novel carbon-based composites, exploiting their multifunctional nature, is their use in aerospace as radar-absorbing materials (RAMs) [28–30], which are capable of absorbing the incident radiation and dissipating it as heat, that find considerable applications in both civil and military fields. As such, EM properties of composites filled with carbon nanostructure are being investigated using both thermoplastics [16–20] and thermosets matrices [12,13,17,27–33].

Yasmin et al. [15] investigated the effect on mechanical properties of different processing routes used to synthesize composites obtained by embedding GNPs into an epoxy matrix. The thermal expansion of the graphite intercalation compound (GIC) was carried out at 600°C and the effect of bath sonication time was also investigated.

Other groups [17,18] carried out an extensive investigation on thermoplastic polymers reinforced with both micrometer-sized fillers and GNPs, with a specific focus on thermal and spectroscopic properties. GNPs were synthesized using a GIC expansion step similar to the one reported in the present study. Chen et al. [18] made a direct comparison on the effect of size on the DC electrical conductivity, reporting a value of 0.31% vol for the percolation threshold of PMMA/GNP composites, which is a good result compared with the value around 3% vol reported for the same type of polymer filled with micrometer-sized conventional graphite.

Song et al. [19] expanded GICs using a microwave oven, instead of a furnace, determining the effect of different expansion ratios and filler concentrations on the electrical properties of composites based on GNPs and polysulfide, although without investigation of the percolation threshold because of the high filler loadings used (up to 70% wt).

Raza et al. [21] developed chemically conducting and highly compliant composites by dispersing GNPs into a silicone matrix for thermal interface applications. GNPs were also employed to simultaneously enhance the thermal and electrical conductivity of organic form-stable phase change materials [22].

In previous works [27–33], the authors have investigated the use of composites filled with CNTs and/or GNPs for application in RAMs. The results of these studies have highlighted that for optimum RAM design, the real and imaginary parts of the complex effective permittivity of the composite should be tailored separately. To this purpose, the use of different conventional micro- and nanofillers was combined [29,30]. In fact, it was observed that the effective electrical conductivity at radio frequency (RF) of the composite increases when the filler is characterized by a high aspect ratio (e.g., greater than 100–500). Therefore, nanofillers like CNTs or microfillers like short carbon fibers, having diameter around 7 μm and length of a few millimeters, are particularly suitable for this scope. The real part of the effective permittivity, which is representative of the polarizability of the material, rises as the volume fraction of the filler increases, and its trend is constant over the frequency if the filler is characterized by a low aspect ratio (i.e., less than 100). Therefore, among conventional carbon-based fillers, commercially available short carbon fibers with length in the range 50–400 μm are suitable for this application.

In this context GNPs can represent a valid alternative to the combined use of conventional micro- and nano-carbon fillers. In fact, due to their bidimensional shape, GNPs can play the role of the nanofiller with a high aspect ratio (intended as the ratio between the lateral dimension of the platelet, which is in the micrometer range, and the thickness, which is in the nanometer range). At the same time, owing to the fact that the platelet area lies in the range of several tens up to hundreds square micrometers, GNPs can efficiently replace the microfiller having the function of increasing the real part of the effective permittivity of the composite.

For all the aforementioned reasons, an extensive study concerning the characterization in the entire X and Ku frequency bands (8–18 GHz) of the effective complex permittivity of polymeric composites filled with graphenes or GNPs has been carried out [31]. Moreover, a novel simulation model has been developed to predict the complex dielectric permittivity of such composites at radio frequency (RF) [32,33]. GNP-based nanocomposites have also been developed for application in piezoresistive strain sensor technology [34,35] and as multifunctional thermal interface materials [36]. An *in vivo* study has been also developed in order to investigate the nanotoxicology of GNPs [37].

Table 31.1 summarizes the main advancements and achievements of previous works reported in the literature.

In the following sections, we first describe the morphological and electrical properties of GNPs produced by thermochemical expansion of GIC, and we discuss the effects of the process parameters on the DC electrical conductivity and morphology of GNP papers. Next, the fabrication routes of GNP-based nanocomposites are presented and their EM properties at RF are discussed, considering their applications in

TABLE 31.1
Main Advancements and Achievements of Recent Literature

Topic	Main Advancement and Achievements
Graphene and graphene sheet properties	• Mechanical [2–4], thermal [5,6], and electrical [7] properties and characterizations
CNT-based composites	• Composites with well-dispersed CNTs for electromagnetic shielding [8–13]
	• Effect of CNT length on mechanical, thermal, and electrical properties of CNT-based nanocomposites [14]
GNP-based composites	• Effect of different processing routes on the mechanical properties of GNP-filled composites in epoxy matrix [15]
	• Electromagnetic shielding properties of GNP-filled epoxy composites [16]
	• Extensive investigation on thermoplastics polymers reinforced with both carbon fibers and GNPs [17,18]
	• GNP-based composites obtained expanding GICs in microwave oven instead of a furnace [19]
	• Dielectric and electrical conductive properties of GNP-based composites [20]
GNP–CNT-based composites	• Thermal properties and characterizations of GNP–CNT-based epoxy composites [23,24]
	• Mechanical properties and characterizations of GNP–CNT-based epoxy composites [24,25]
	• Electromagnetic shielding properties of polymer-based composites filled with CNFs, CNTs, and GNPs [26–30]
GNP-based composites for electromagnetic shielding applications	• Composites with GNPs were prepared and characterized for EMI shielding and for RAM applications [31–33]
	• An electromagnetic model able to predict the dielectric properties of GNP-based composites was developed [32,33]
GNP-based composites for strain sensor applications	• Piezoresistive properties of GNP–polymer composites, fabrication process, and experimental characterizations [34,35,38,39]
	• Electromechanical model for GNP–polymer composites for strain sensor applications [36,39]
Nanotoxicology of GNP	• *In vivo* study aimed at investigating nanotoxicology of GNPs [37]

EM shielding, RAMs, and piezoresistive strain sensors with high gauge factor (GF).

31.2 GRAPHENE NANOPLATELETS: SYNTHESIS AND CHARACTERIZATION

GNPs are carbon nanostructures consisting of small stacks of graphene sheets, having thickness in the range from 1 up to a few tens of nanometers, and lateral linear dimensions varying from a few micrometers up to several tens of micrometers. GNPs were produced at Sapienza Nanotechnology and Nanoscience Laboratory (SNN-Lab) by thermal exfoliation. As in previous works [31–33,35,39], GNPs were synthesized starting from commercially available GIC in a muffle furnace, operating in air, at temperatures ranging from 250°C to 1350°C and dwelling times in the range 5–60 s. This step led to an increase in the volume of the original material (Figure 31.1a through d), and to the formation of worm-like graphitic structures known as thermally expanded graphite oxide (TEGO).

It was observed that no expansion occurred for GIC undergoing thermal shocks at 250°C and 500°C for times up to 60 s and at 750°C for 5 s exposure. For higher temperatures, a volume expansion by 200–400 times was observed (Figure 31.2a through d). Furthermore, GIC exposed to 1250°C for 60 s reacted completely with atmospheric oxygen, giving off carbon oxides, leaving no residue: for this reason, thermal expansion at 1350°C was carried out only for a duration of 5 s.

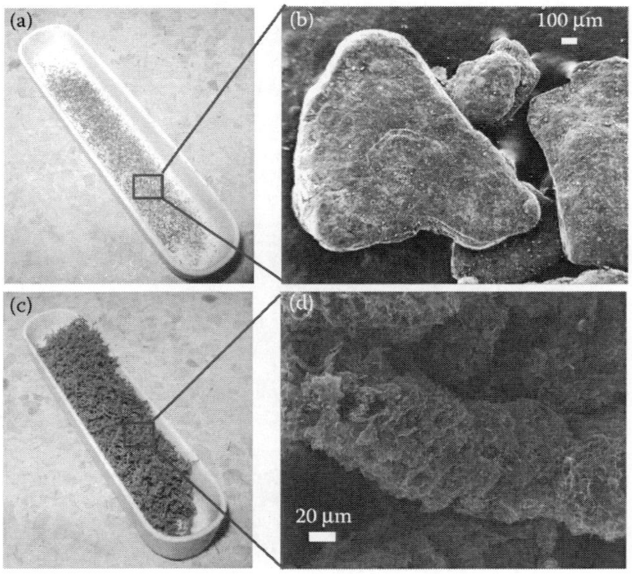

FIGURE 31.1 Photographs and corresponding SEM magnifications of GIC in the crucible before and after thermal expansion: (a), (b) GIC; (c), (d) TEGO obtained at 1150°C, 5 s.

High-resolution scanning electron microscopy (SEM) images of the produced TEGO are shown in Figure 31.3a through d, in which stacked graphene sheets can be observed.

After thermal treatment, the TEGOs is dispersed in a suitable solvent, and the obtained suspension is tip sonicated

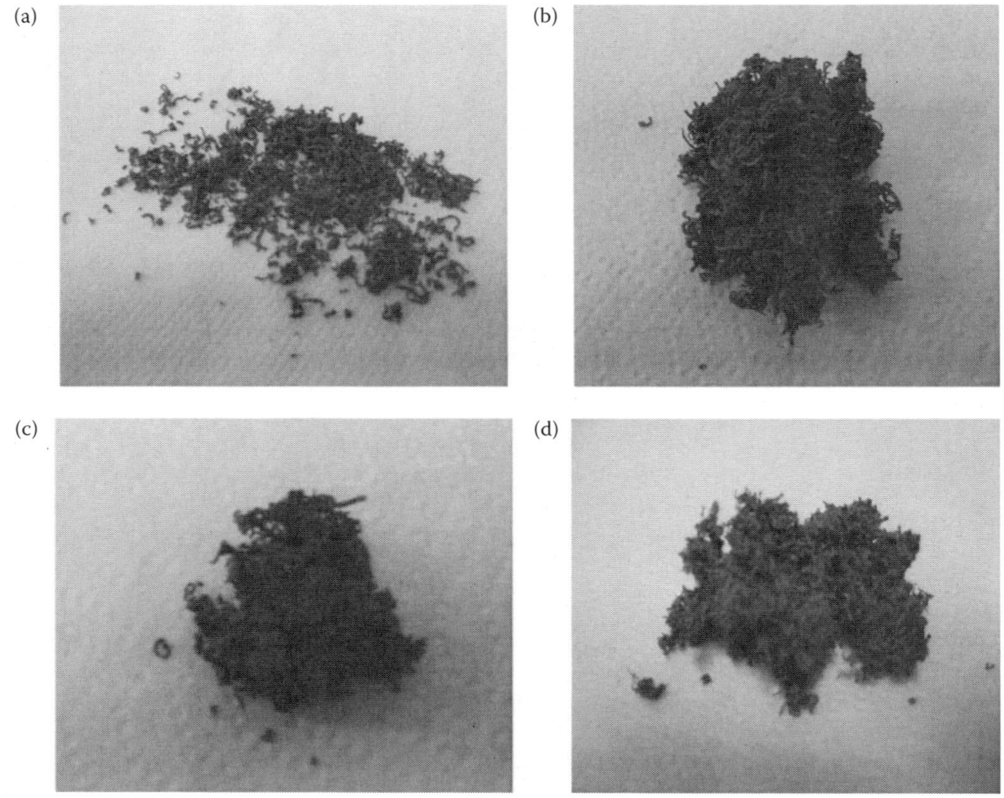

FIGURE 31.2 Photographs of TEGO obtained from thermal exfoliation of GIC in different conditions of expansion temperature and time: (a) 500°C, 60 s; (b) 1050°C, 5 s; (c) 1150°C, 60 s; (d) 1250°C, 30 s.

FIGURE 31.3 SEM micrographs of TEGO obtained from thermal exfoliation of GIC, at different magnifications: (a) 120X; (b) 500X; (c) 5000X; (d) 15000X.

using an ultrasonic processor, thus obtaining GNPs. This step of the synthesis process is particularly critical, because it has been observed that both the lateral dimensions and the thickness of GNPs are affected by the type of solvent used, the sonication cycle, and the suspension temperature control during sonication.

As regards the first point, two different solvent mixtures have been used, namely, acetone:*N*,*N*-dimethylformamide (DMF) and acetone:*N*-methyl-2-pyrrolidone (NMP), both at 9:1 volume ratio. The use of DMF and NMP is considered a standard for liquid-phase exfoliation of graphene, due to their suitable Hansen solubility parameters. According to Hernandez et al. [40], the Hansen parameters requested for a good exfoliation of graphene-based materials are $\delta D \sim 18$ MPa$^{1/2}$, $\delta P \sim 9.3$ MPa$^{1/2}$, and $\delta H \sim 7.7$ MPa$^{1/2}$, where δD, δP, and δH are the contributions to the solubility parameter related to dispersion forces, intermolecular dipoles, and hydrogen bonding, respectively. Compton and Nguyen showed that only solvents having $(\delta P + \delta H)$ in the range of 13–29 yield stable graphene suspensions [41]. Using the Hansen parameter solvent mixtures equation [42], the Hansen parameters of our solvent mixtures were calculated. The obtained results are close to the values requested to have a good solubility of graphene-based materials, as shown in Table 31.2 for the acetone–DMF mixture.

As regards the sonication cycle, most of the research groups working on liquid-phase exfoliation to obtain graphene sheets use standard sonication conditions (namely, pulsed 1–1 on–off cycles without temperature control). In our study, the GNP solution was tip sonicated using an ultrasonic processor, Sonics and Materials VC505, operated at 70% amplitude [43]. A pulsed-mode sonication was used because of the system relaxation role for the off phase, allowing a higher cavitation intensity and lower heat generation to be reached [42]. The total on-time was fixed at 20′ in order to achieve a homogeneous dispersion of GNPs. Samples were produced varying the on- and off-times (t_{on}, t_{off}) of the ultrasonic processor pulsation cycle, between 1 and 4 s. Three different conditions were experimented: (i) at room temperature (RT), that is, without the use of any refrigerator system to control the suspension temperature during sonication; (ii) using an ice bath (IB) to maintain the solution temperature during sonication in the range 14–20°C; (iii) using a thermo cryostat (TC) set at 15°C and a jacketed beaker to finely control the temperature. The final energy provided to the suspension (as indicated by the instrument display) ranged from 6.5 to 52 kJ (for uncooled and cooled cycles, respectively) with suspension final temperatures at 46°C, 17°C, and 15°C, respectively, in the uncooled condition, with the IB and with the TC.

Typical thicknesses of the produced GNPs, as observed at atomic force microscopy (AFM) (Dimension ICON manufactured by Bruker), range between 2–3 nm and 10–15 nm depending on the process parameter setting, whereas the lateral dimensions of the flakes vary from a few micrometers up to 20–25 μm (Figure 31.4a through g).

In order to investigate the effect of the expansion temperature and time on the produced GNPs, Fourier transform infrared spectroscopy (FTIR) was used. To this purpose, suspensions of TEGO expanded under different conditions, sonicated in acetone in pulsed mode with 1–1 on–off times and at RT, were vacuum-filtrated on anodic aluminum oxide (AAO) nanopore membranes to obtain paper-like GNPs, suitable for spectroscopic investigations. In particular, the effect of expansion temperature was studied in the range 500–1050°C (with expansion time fixed at 5 s), whereas the effect of expansion time from 5 to 60 s was investigated setting the temperature at 750°C. The corresponding FTIR spectra are shown in Figure 31.5a and b, respectively.

The reduction of the intensity of peaks associated with oxygen-containing functional groups, such as hydroxylic and carboxylic is clear, as the exfoliation temperature is increased (Figure 31.6a). A similar effect is observed as the dwell time is increased for a given temperature (Figure 31.6b). Nevertheless, such effects are not detectable from FTIR spectra of GIC treated at temperatures exceeding 1050°C, as shown in Figure 31.6. This suggests that the higher temperature of expansion does not significantly affect the structural characteristics of GNPs, but affects the flake topology. As an example, Figure 31.7a and b shows the effect observed by scanning electron microscopy (SEM) of the highest expansion temperatures (i.e., 1250°C for 15 s and 1350°C for 5 s) on TEGO: both samples show the presence of pits on the corrugated surfaces, due to the disruption of the conjugated sp^2 lattice. In addition, the sample treated at 1350°C shows severe edge damages (Figure 31.7b).

TABLE 31.2
Hansen Solubility Parameters for Acetone, DMF, NMP, and Acetone–DMF Mixture

Solvent Type	Surface Tension (mJ/m^2)	Hansen Solubility Parameters (MPa$^{1/2}$)		
		ä$_D$	ä$_P$	ä$_H$
Acetone	25.20	15.5	10.4	7.0
DMF	37.10	17.4	13.7	11.3
NMP	41	18	12.3	7.2
Acetone–DMF mixture	–	15.7	10.7	7.4

31.3 DC ELECTRICAL CONDUCTIVITY OF GNP PAPERS: EFFECT OF PROCESS PARAMETERS

Process parameters greatly affect the topological properties of GNPs (i.e., flake thickness and lateral dimension), and consequently their electrical and transport properties. The assessment of such influence through microscopy investigations is time consuming and not feasible for mass production applications, whereas FTIR analysis is not adequate as discussed above. On the contrary, the DC electrical conductivity of GNP papers obtained by vacuum filtration of GNP suspensions produced as described in the previous paragraph can be strongly dependent on the process parameters. In fact the

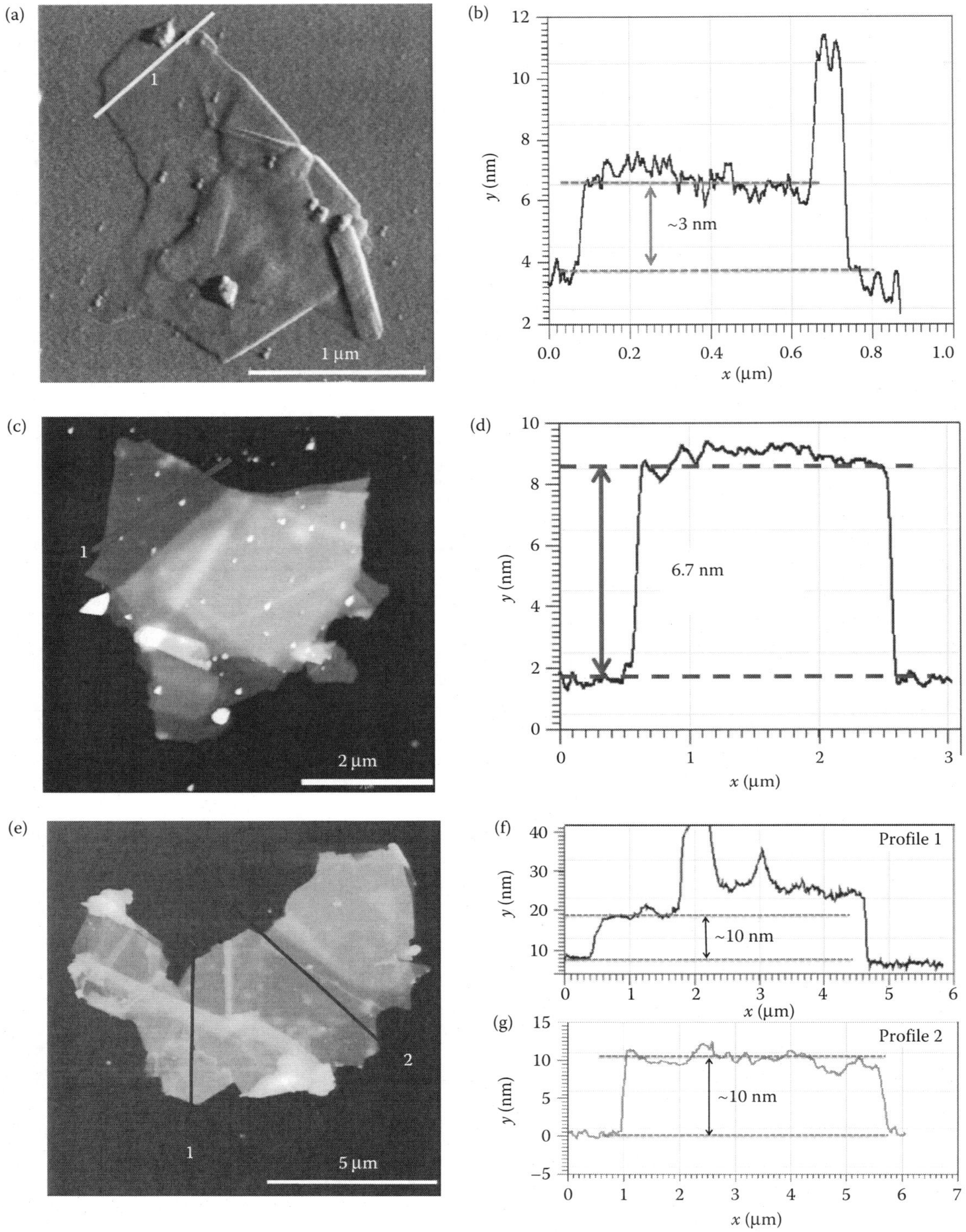

FIGURE 31.4 AFM images of produced GNPs (a), (c), (e) with lateral dimensions in the range 1–20 μm, thickness in the range 2–10 nm and corresponding line profiles (b), (d), (f), (g).

GNP morphological properties are expected to strongly affect the porosity and stratification level of GNP papers, and consequently their DC electrical conductivity [40,44].

The DC conductivity of the produced GNP papers was measured with the four-point probe method [44] at RT using a Keithley 6221 DC/AC current source connected to a Keithley 2182a nano-voltmeter. The measurement is repeated by positioning the test probes in six different locations over the sample and the resulting measurements are averaged. By taking into account the sample geometry, proper correction factors are applied to the measured average resistance in order to obtain the film conductivity value.

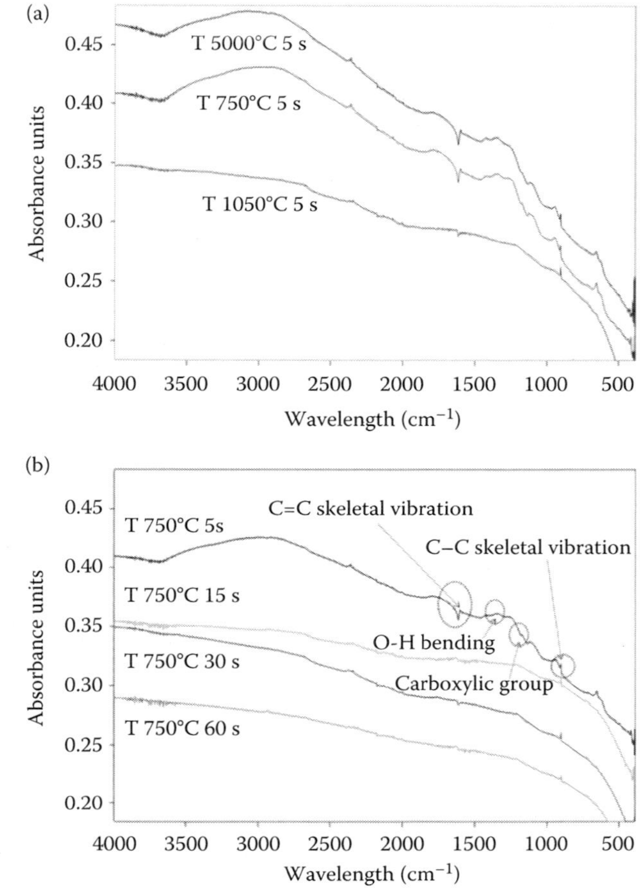

FIGURE 31.5 FTIR showing the effect of exfoliation temperature (a) and time (b) on residual oxygen-containing functional groups of GNP-based paper.

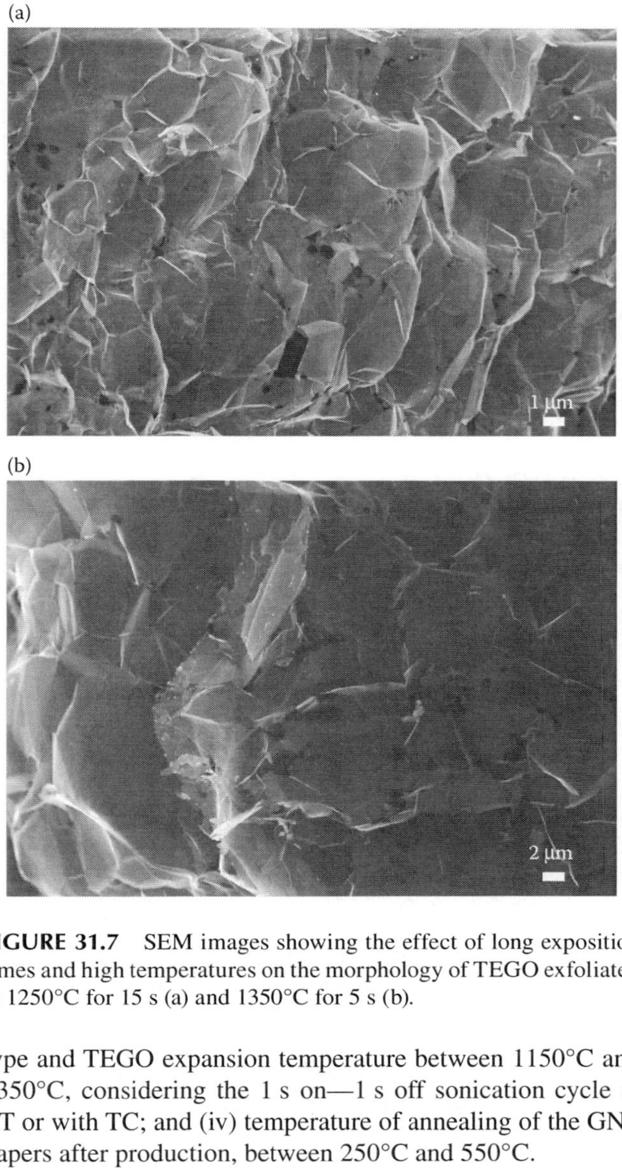

FIGURE 31.7 SEM images showing the effect of long exposition times and high temperatures on the morphology of TEGO exfoliated at 1250°C for 15 s (a) and 1350°C for 5 s (b).

In the following, we discuss the effect on the GNP paper conductivity of (i) TEGO expansion temperature and time, in the case of the RT sonication cycle, with 1 s–1 s on–off intervals, using acetone–DMF mixture as solvent; (ii) on–off phase duration of the sonication cycle, at RT or IB; (iii) solvent type and TEGO expansion temperature between 1150°C and 1350°C, considering the 1 s on—1 s off sonication cycle at RT or with TC; and (iv) temperature of annealing of the GNP papers after production, between 250°C and 550°C.

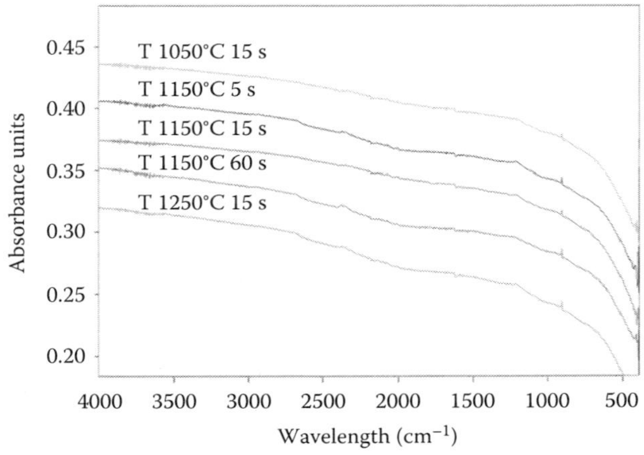

FIGURE 31.6 Comparison of FTIR spectra of GNP papers obtained from exfoliation at high temperatures (above 1050°C) and various treatment durations.

31.3.1 Influence of TEGO Expansion Conditions

GNP papers are produced in acetone–DMF solution at RT, that is, without temperature control during sonication, using TEGO expanded under different time–temperature conditions. The measured values of DC electrical conductivity are reported in Figure 31.8. The DC conductivity is shown to increase almost linearly with the exfoliation temperature for shorter exposure times. On the contrary, for longer treatment (i.e., 30 s), there is an increase of GNP paper conductivity only up to 1050°C: at higher temperatures, an increase in expansion time has a detrimental effect on the DC conductivity, probably due to the degradation of graphene basal planes and loss of material. The higher conductivity value of the paper produced by TEGO expanded at 1250°C for 5 s with respect to the one obtained with TEGO expanded at 750°C for 30 s is also due to the better exfoliation of GNPs in the former type, as shown by the AFM thickness profiles in Figure 31.9a through d.

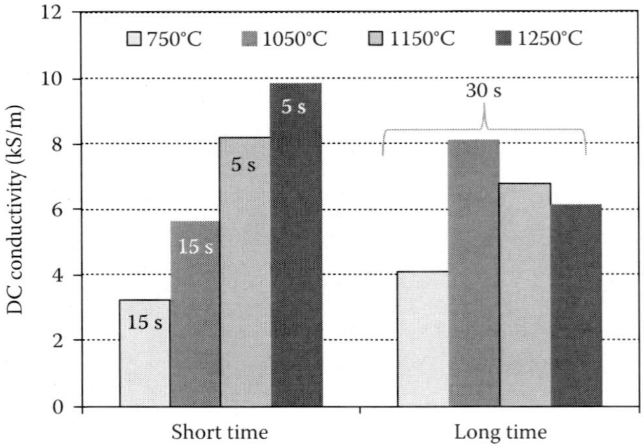

FIGURE 31.8 Measured DC conductivity of GNP papers produced at temperatures ranging from 750°C to 1250°C, and for exposition time between 5 and 30 s. Sonication in acetone–DMF solution without temperature control (uncooled condition).

31.3.2 Influence of Sonication Conditions: Duty Cycle and Suspension Temperature

The effect of different sonication duty cycles (duration of the on and off phases of the pulsed sonication process) on the DC electrical conductivity of GNP papers produced in acetone–DMF has been evaluated modulating the on and off durations (t_{on} and t_{off}) between 1 and 4 s [43] (Figure 31.10).

The measured thickness of the produced GNP films is reported in Figure 31.10, for both series produced at RT or with IB, as a function of increasing t_{on} for t_{off} = 1 s, and as a function of increasing t_{off} for t_{on} = 1 s. It is observed that thanks to the presence of the cooling system, the thickness of the GNP films is reduced by even more than 50% in some cases. This effect is attributed to the better GNP layering of the samples produced with the IB, as can be seen from the SEM images of Figure 31.11a and b, representing the fracture section of GNP films produced with sonication cycle t_{on} = 3 s, t_{off} = 1 s, using or not using the IB.

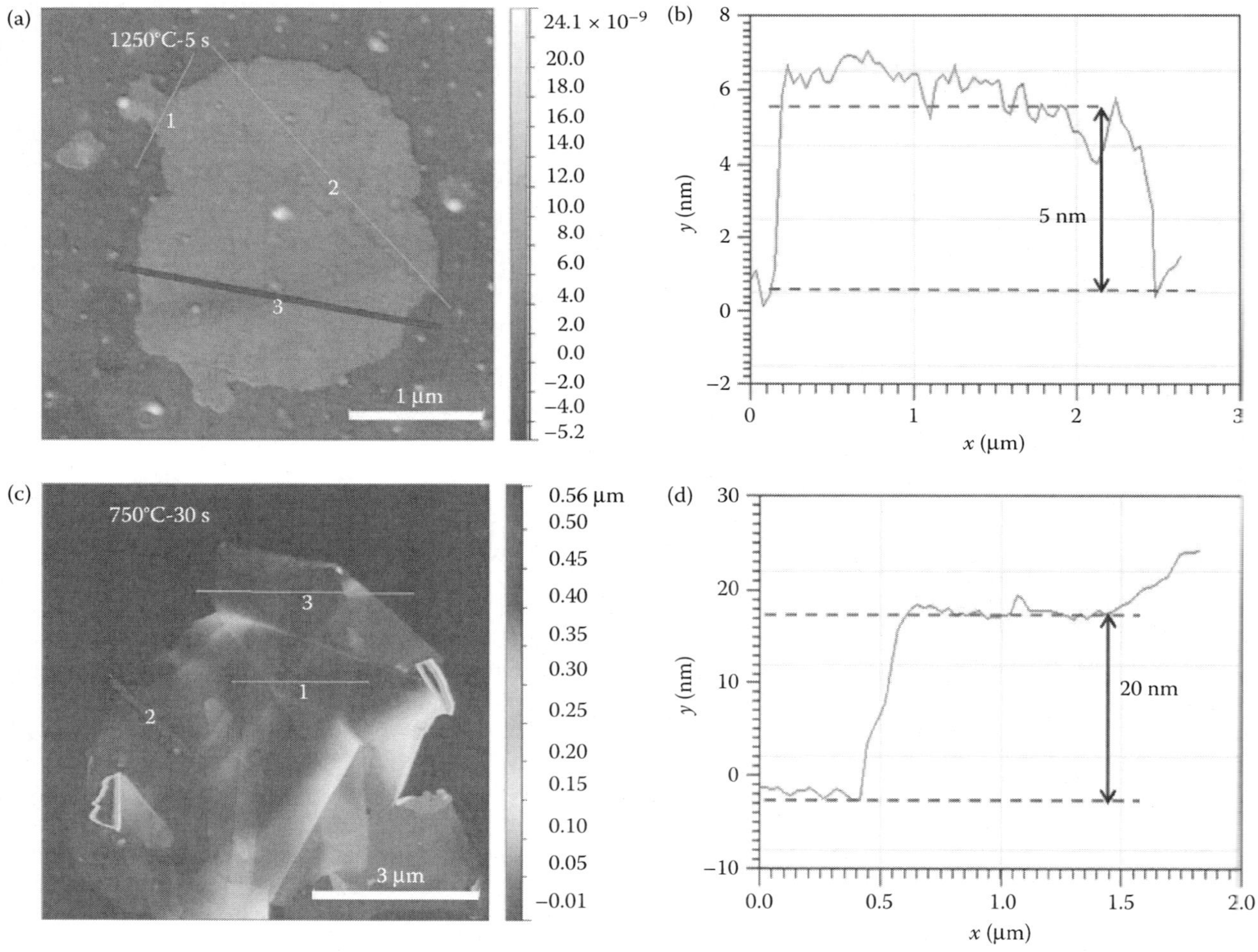

FIGURE 31.9 AFM images and related thickness profiles of GNP flakes produced at 750°C, 30 s (a,b) and at 1250°C, 5 s (c,d). (Adapted from De Bellis, G. et al. Effect of process parameters on the effective DC conductivity of GNP thick films. *Nanotechnology (IEEE-NANO), 12th IEEE Conference on*. © 2012. IEEE.)

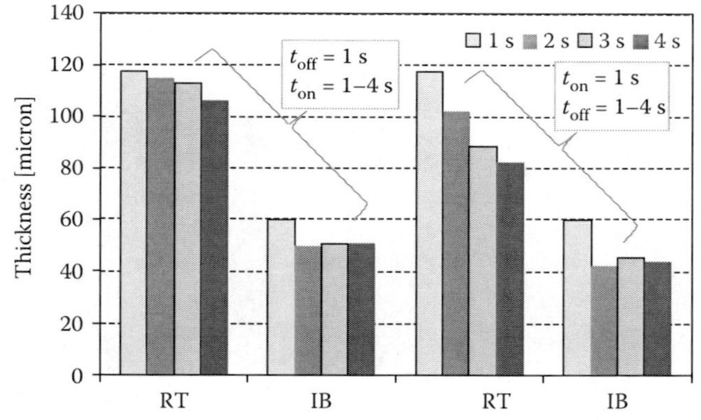

FIGURE 31.10 Measured thickness of GNP papers produced with sonication at RT or in IB, and applying different sonication cycles: t_{off} = 1 s and increasing values of t_{on} from 1 s to 4 s; t_{on} = 1 s and increasing values of t_{off} from 1 s to 4 s.

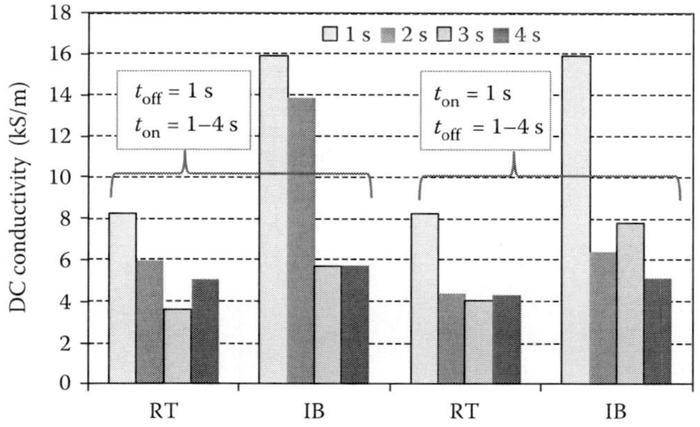

FIGURE 31.12 Measured DC electrical conductivity of GNP papers produced at RT or with IB, and applying different sonication cycles: t_{off} = 1 s and increasing values of t_{on} from 1 s to 4 s; t_{on} = 1 s and increasing values of t_{off} from 1 s to 4 s.

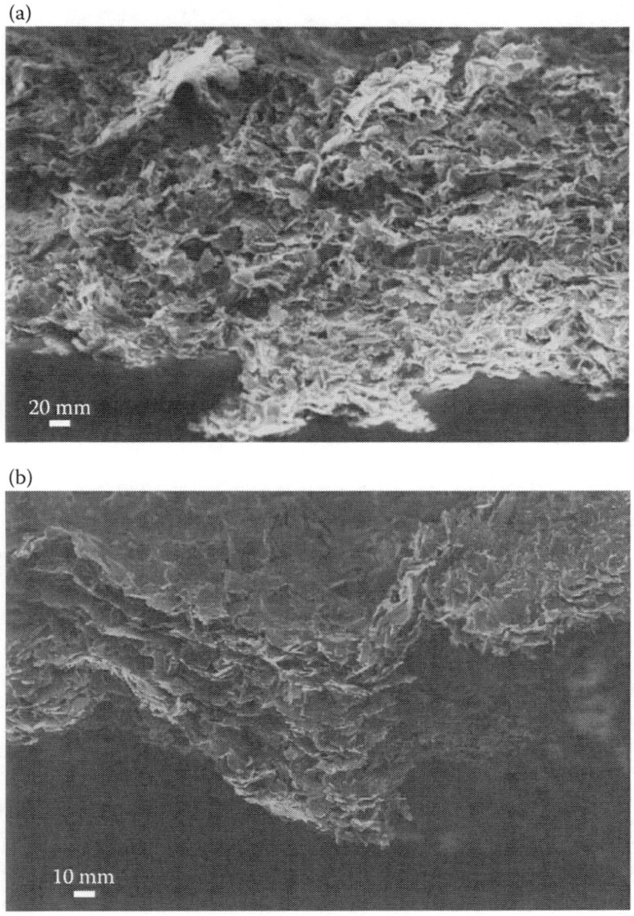

FIGURE 31.11 SEM images of the fracture borders of GNP films produced with sonication cycle t_{on} = 3 s, t_{off} = 1 s: sonication (a) at RT and (b) with IB. (Adapted from De Bellis, G. et al. Effect of sonication on morphology and dc electrical conductivity of graphene nanoplatelets-thick films. *Nanotechnology (IEEE-NANO), 13th IEEE Conference on*. © 2013. IEEE.)

Thickness reduction with increasing t_{off} is evident for RT cycles: this is probably due to the fact that the cooling down of the GNP solution during the off time causes a rapid shrinking of the collapsing bubbles produced by the cavitation activity during the on phase, thus contributing to the reagglomeration of flakes.

The corresponding measured DC conductivity values are reported in Figure 31.12. It is evident that for all the different cycles the conductivity increases significantly the suspension temperature is controlled (IB condition). This effect is more evident for the 1–1 and 2–1 (on–off duration) cycles: in fact for these duty cycles the average conductivity values for samples produced at RT are less than 50% lower than the ones of the corresponding samples produced with the IB. In particular, fixing the off duration to 1 s, the conductivity decreases for increasing t_{on}, with a reduction of more than 50% when t_{on} > 2 s. This suggests the existence of an optimal tip activation period, beyond which the effect of flakes degradation/size reduction dominates over the exfoliation degree.

31.3.3 Influence of Solvent Type and Thermal Annealing

The choice of the solvent used for the dispersion of GNPs greatly affects the morphology and DC electrical properties of GNP-based papers. This effect is observed on GNP foils produced both at RT and with a TC, using TEGO expanded at 1150°C, 1250°C, and 1350°C, for 5 s. The two different solvents used are mixtures of acetone–DMF and acetone–NMP at 9:1 ratio in volume.

NMP is known as being one of the best solvents for efficient graphene exfoliation. Nevertheless, owing to its high compatibility, NMP is entrapped more easily than DMF between the graphene layers, leading to an increase of thickness (Figure 31.13a) and to a fall of DC conductivity (Figure 31.13b) of the GNP papers as compared to DMF-based foils. According to SEM analysis, it also results that GNP foils produced using acetone-DMF solutions are characterized by an improved

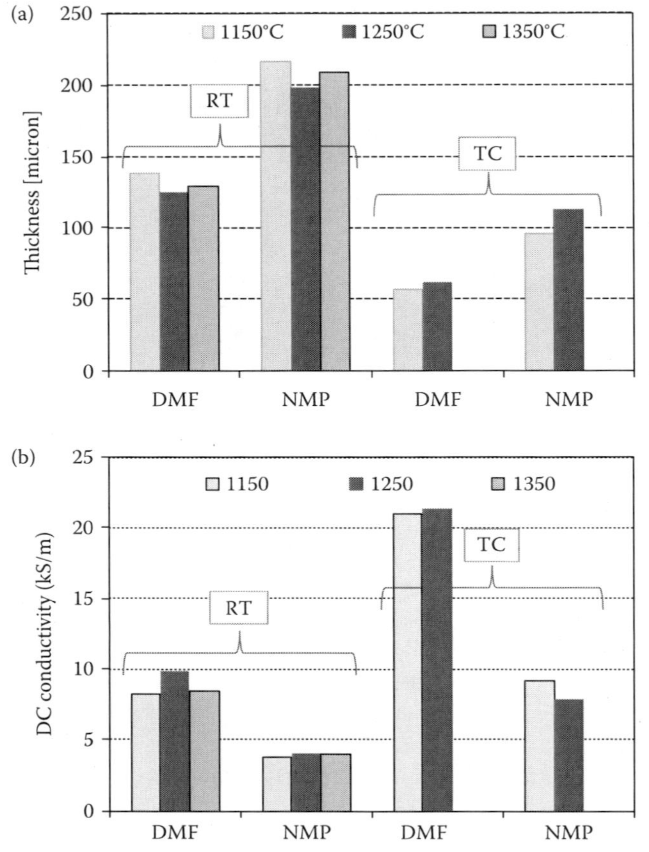

FIGURE 31.13 (a) Thickness and (b) DC electrical conductivity of GNP papers produced using as solvent acetone–DMF or acetone–NMP mixture, at RT and with TC, using TEGO expanded at temperatures of 1150°C, 1250°C, 1350°C for 5 s.

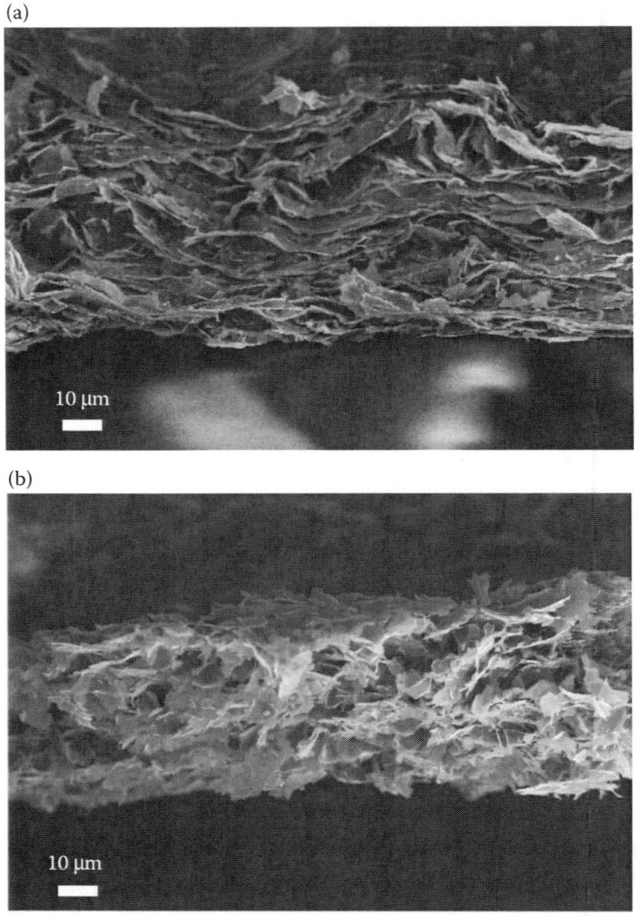

FIGURE 31.14 SEM micrographs showing the fracture sections of GNP papers produced with thermoregulated sonication in solution of (a) acetone–DMF and (b) acetone–NMP. GNPs are obtained from TEGO expanded at 1150°C for 5 s.

stratification than the ones obtained from acetone–NMP mixtures, as shown in Figure 31.14a and b.

Indeed, the highest electrical conductivity is obtained for both cooled and uncooled cycles for DMF-based GNP papers, reaching a maximum in excess of 22 kS/m for the films obtained from TEGO dispersed in DMF using thermoregulation.

Moreover, the decrease in thickness of all types of GNP papers when using the TC confirms the results obtained with the IB in Section 31.3.2, and it is explained through the formation of a clearly layered morphology.

31.3.4 Influence of the Annealing Temperature

Finally, the effect of the annealing temperature on the DC electrical conductivity of the produced GNP foils has been investigated. Annealing temperatures have been chosen according to thermogravimetric analysis, in order to avoid degradation of the sp^2-conjugated graphene lattice. To this purpose, thermal treatment has been carried out in air in the 250–550°C range, increasing the temperature in 100°C steps. Figure 31.15a and b shows the measured DC conductivity of GNP papers annealed at different temperatures and produced in acetone–DMF or acetone–NMP solvent mixtures, at RT (Figure 31.15a) and under thermally regulated sonication using a TC (Figure 31.15b). GNP papers were produced from TEGO exfoliated at temperatures of 1150°C, 1250°C, and 1350°C, fixing the expansion duration to 5 s.

The obtained results show that the DC conductivity is greatly affected by the annealing temperature, especially for GNP papers obtained in DMF-based solvents, reaching a maximum of nearly 40 kS/m for samples produced from TEGO exfoliated at 1150°C for 5 s in acetone–DMF, after annealing at 550°C for 10′ in air. On the other hand, samples obtained in NMP-based solvents show a milder increase of the DC conductivity with the annealing temperature.

The observed improvement of DC conductivity with thermal annealing is probably due to the evaporation of solvent residuals. In fact, SEM analysis shows that the stratification of GNP papers produced in acetone–DMF is not significantly affected by thermal annealing (Figure 31.16).

31.4 GNP-Based Nanocomposites

Polymer-based nanocomposites developed in the last two decades have gained considerable attention because of the possibility of strongly improving several properties through

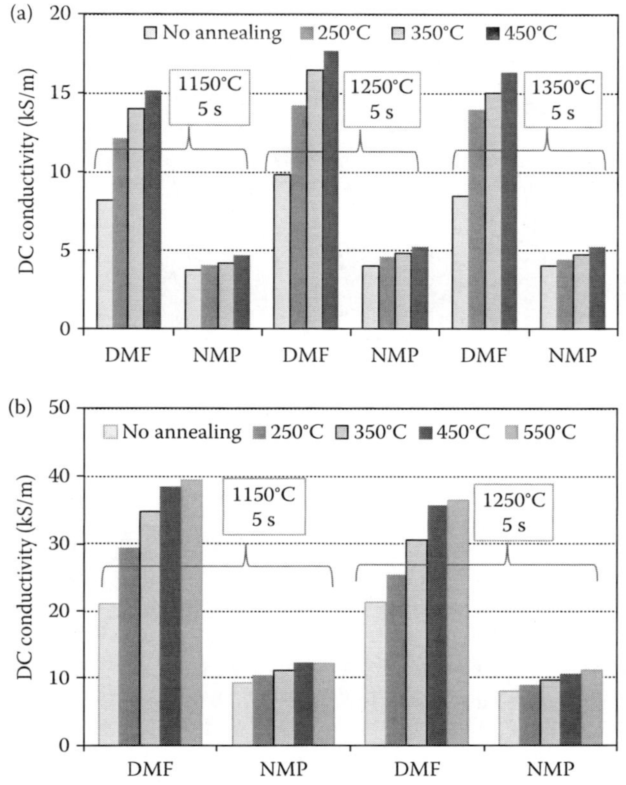

FIGURE 31.15 Influence of the annealing temperature (from 250°C to 550°C) on the DC conductivity of films produced using different TEGOs expanded at 1150°C, 1250°C, and 1350°C, in acetone–DMF or acetone–NMP (a) at RT or (b) with TC.

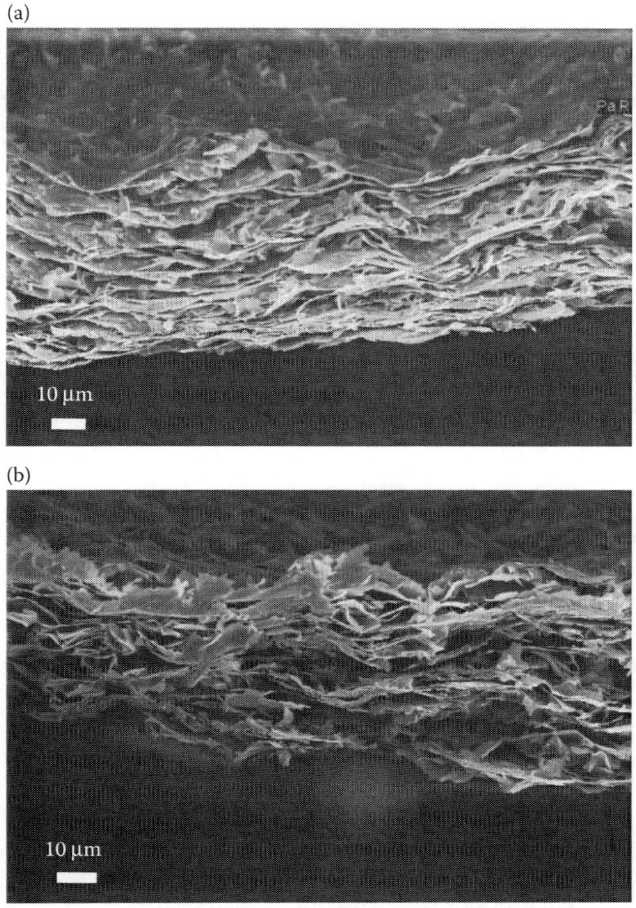

FIGURE 31.16 Fracture section SEM micrographs of GNP papers produced with thermoregulated-sonication in solution of acetone–DMF, after thermal annealing at (a) 250°C and (b) 550°C. GNP are obtained from TEGO expanded at 1150°C for 5 s.

the addition of small amounts of the nanofiller. Polymer nanocomposites are obtained mixing a given polymer (a thermoset, thermoplastic, or elastomer) with a filler, having at least one dimension below 100 nm. A wide range of fillers with different shapes can be used: the filler can be plate-like, high-aspect-ratio nanotubes or nanofibers, and lower-aspect-ratio or equiaxally shaped nanoparticles. The dramatic effects on the final properties of the resulting nanocomposite are mostly driven by the huge surface area of the nanofiller, which in turn means a very large interfacial surface with the polymer itself.

Graphene is potentially an ideal filler for nanocomposites with enhanced mechanical and electrical/EM properties. A crucial parameter in succeeding to incorporate graphene into a polymer as an efficient reinforcement is the length-to-thickness ratio, known as the aspect ratio. On the other hand, as the size of the filler particles is greatly reduced, they tend to agglomerate and become difficult to disperse in a polymer matrix.

Furthermore, the dispersion process is markedly affected by a second parameter, namely, the surface tension of graphene, which has been reported to be 46.7 mJ/m^2 [45]. As a direct consequence of their relative surface energies, graphene can be poorly or efficiently dispersed in a given solvent or polymer. The main challenge is then to disperse graphene into individual sheets, minimizing the risk for restacking of the layers, which may result in an increase in the thickness of the flakes, leading to the loss of their peculiar size-dependent properties.

This section is focused on the preparation, optimization, and characterization of GNP-based polymer nanocomposites, using different matrices. The aim is to understand how the synthesis parameters of GNPs and the processing route of the resulting nanocomposite affects the final properties of the latter, in terms of both electrical (DC properties used for the strain sensor application) and EM properties.

31.4.1 Fabrication

GNP nanocomposites were produced using two different thermosetting polymers, in order to address the influence of polymer adhesion to the nanostructure on the final composite properties: a commercially available bisphenol A-based (DGEBA type) epoxy resin (Gurit SP 106); a vinyl-ester resin (DION 9102 gently provided by Reichhold).

Epoxy resins have been widely used with carbon-based nanostructures since they have been shown to be fully

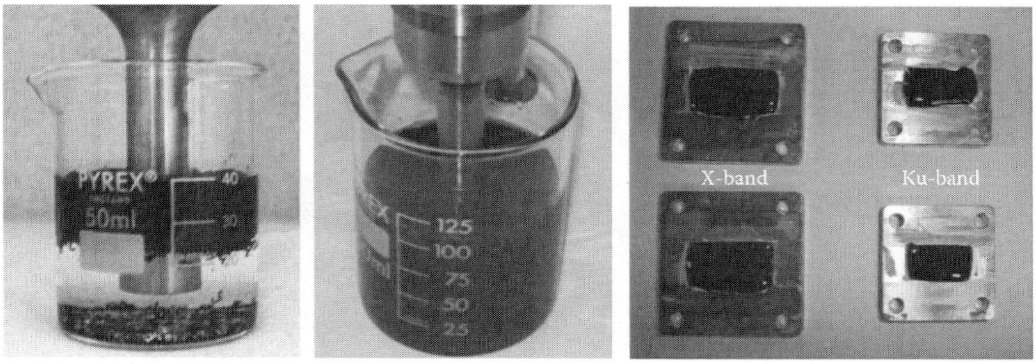

FIGURE 31.17 Steps of nanocomposite fabrication route: GNP suspension (a) before and (b) after the sonication process; (c) GNP nanocomposite after pouring into the rectangular flanges.

compatible with conventional carbon fibers. Vinyl esters are unsaturated resins made from the reaction of unsaturated carboxylic acids (principally methacrylic acid) with an epoxy, such as a bisphenol A epoxy resin. The main difference between the epoxy and the vinyl ester systems used lies in the terminal groups used to perform the cross-linking reaction. Vinyl esters were developed to combine the advantages of epoxy resins with better handling/faster cure, which are typical for unsaturated polyester resins. These resins are produced by reacting epoxy resin with acrylic or methacrylic acid. This provides an unsaturated site. The resulting material is dissolved in styrene to yield a liquid that is similar to polyester resin. Vinyl esters are also cured with conventional organic peroxides used with polyester resins.

The fabrication process of nanocomposites is based on the so-called solution processing technique. This method requires the dispersion of the filler in a suitable solvent, followed by mixing of the solution with the polymeric matrix [46].

For both vinyl-ester- and epoxy-based composites, the implemented composite synthesis procedure was as follows: first, the filler was dispersed in the organic solvent (acetone) through an ultrasonic probe (Sonics and Materials Vibracell VC505) using an amplitude of 70% at 20 kHz frequency, running for a time of 20 min, set in pulse mode (1 s on and 1 s off), in order to avoid overheating of the suspension (this step allowed the obtaining of a kinetically stable suspension): the effect of the ultrasonication process is shown in Figure 31.17a and b.

The resulting suspension was then poured into a beaker containing the liquid vinyl ester or epoxy resin, which, in the case of the vinyl ester matrix, had been previously mixed with 0.2% wt of a Co-based accelerator (Akzo Nobel Accelerator NL-51P, Cobalt (II) 2-ethylhexanoate, 6% Co in solvent mixture) and further sonicated for 90 s using an amplitude of 20% in order to minimize bubble formation. The solution was then magnetically stirred at 200–250 rpm with the aim of removing the excess of acetone. Upon complete removal of the solvent, the hardener was added at 2% wt ratio (Akzo Nobel Butanox LPT, methyl ethyl ketone peroxide, solution in diisobutyl phthalate).

The liquid formulation was then again stirred at 250 rpm until it gained a viscosity suitable for casting and it was poured directly in two sets of rectangular flanges for EM testing in the X and Ku-bands, using them as molds (Figure 31.17c). The composite obtained was then cured in air for 24 h and postcured for another 24 h at 70°C [31–33].

After the postcuring at 70°C, samples were machined through a lapping machine to reach the desired thickness. The final step required, before the electrical and EM tests, is a conditioning stage, operated in a desiccator for 24 h under controlled humidity and temperature conditions.

31.4.2 Morphological and Chemico-Structural Characterization

In order to correlate the DC and EM properties of the manufactured nanocomposites, several morphological–structural analyses, including microscopy, spectroscopy, and diffractometry were carried out.

31.4.2.1 Scanning Electron Microscopy

SEM was performed on the fracture section of the produced samples that, for imaging purposes, were broken in liquid nitrogen. Figure 31.18a through f shows micrographs of composites containing increasing weight concentrations of GNP, that is, from 0.1% wt up to 4% wt. As can be seen from the images depicting the fracture surfaces, an increase in the GNP concentration does not affect the quality of the dispersion, which seems to be very good for all the samples under investigation, showing no agglomerates.

31.4.2.2 Viscosimetry

The viscosity of two different matrix–filler mixtures has been investigated through a Haake VT500 rotational viscometer, equipped with a cone and plate sensor system. Viscosity measurements have been performed in controlled shear rate mode (CR mode), with a shear rate of 100 s^{-1}. The temperature was set at 23.5 ± 0.1°C for all the measurements. The viscosity of 0.5% wt GNP–polymer mixtures was measured over a period of time of 120 s. A constant increase in viscosity over time for a fixed value of the shear rate, probably due to both partial

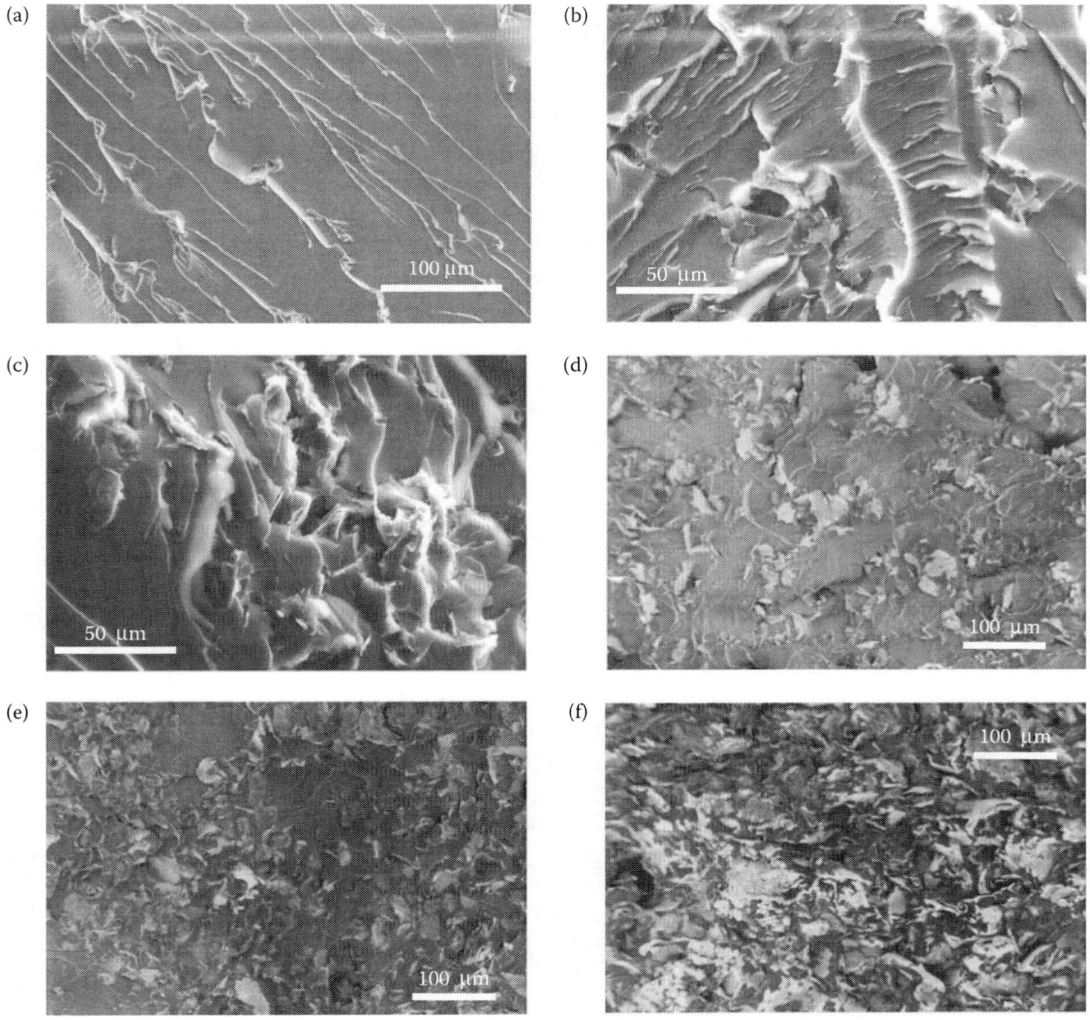

FIGURE 31.18 SEM micrographs of the fracture border of vinyl ester-based nanocomposites filled with GNP at increasing concentrations: (a) 0.1% wt; (b) 0.25% wt; (c) 0.5% wt; (d) 1% wt; (e) 2% wt; (f) 4% wt.

evaporation of the styrene contained in the vinyl ester system and reorientation of the nanofiller along the direction of the applied shear has been observed. Nevertheless, it was possible to extract an average value of dynamic viscosity of 139.1 mPa.s. The relative low viscosity value measured for the GNP-based composite gives account for easy processability during the different synthesis steps even for concentrations up to 1% wt or 2% wt.

31.4.2.3 FTIR Spectroscopy

In order to better understand the influence of the polymer used in the manufacturing process of the nanocomposite, FTIR spectroscopy has been carried out on samples produced using the two different thermosetting matrices (namely, the epoxy resin Gurit SP 106, and the vinyl ester resin, DION 9102) and GNPs produced by exfoliation of TEGO expanded at 1150°C for 5 s.

The FTIR spectra of the different nanocomposites show the appearance of phonon modes not present in the neat resin, only for composites based on the vinyl ester matrix, thanks to the formation of new bonds, not detectable in the samples based on the epoxy resin [47] (Figure 31.19).

31.4.3 Effective Relative Permittivity of GNP Nanocomposites

The fabricated nanocomposites have been characterized in terms of their EM properties by means of measurements of DC electrical conductivity and complex effective permittivity at RF.

To this purpose, after the composite curing cycle, the surfaces of the filled rectangular aluminum flanges were accurately finished through a polishing machine. Successively, all samples were dried for 24 h at controlled temperature and humidity.

The scattering parameters of the filled flanges were measured with a vector network analyzer (Anritsu Vector Star MS4647A) using two different sets of waveguides, in the X- and Ku-bands respectively, in order to cover the frequency range 8–18 GHz. In particular, for each composite type, a batch consisting of three X-band and three Ku-band flanges was prepared. A total

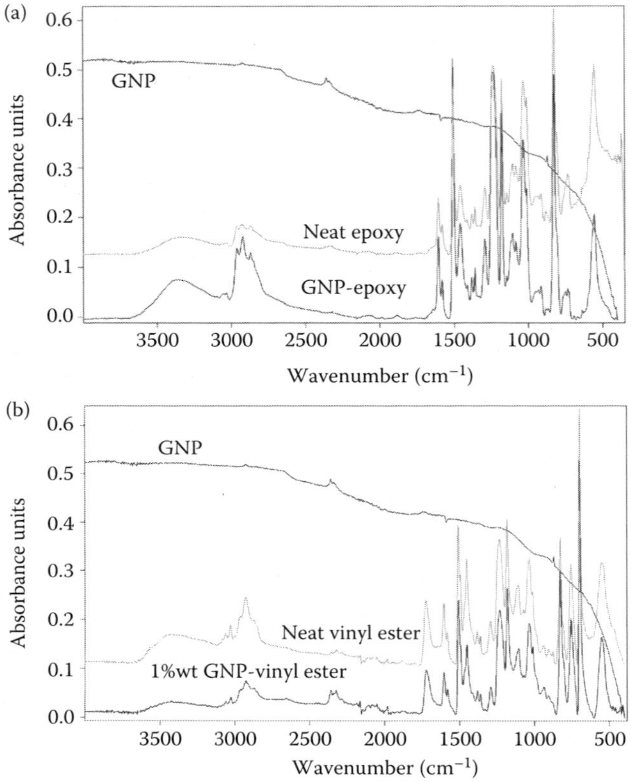

FIGURE 31.19 Measured FTIR spectra of GNP, neat resin, and nanocomposite filled at 1% wt: (a) epoxy; (b) vinyl ester.

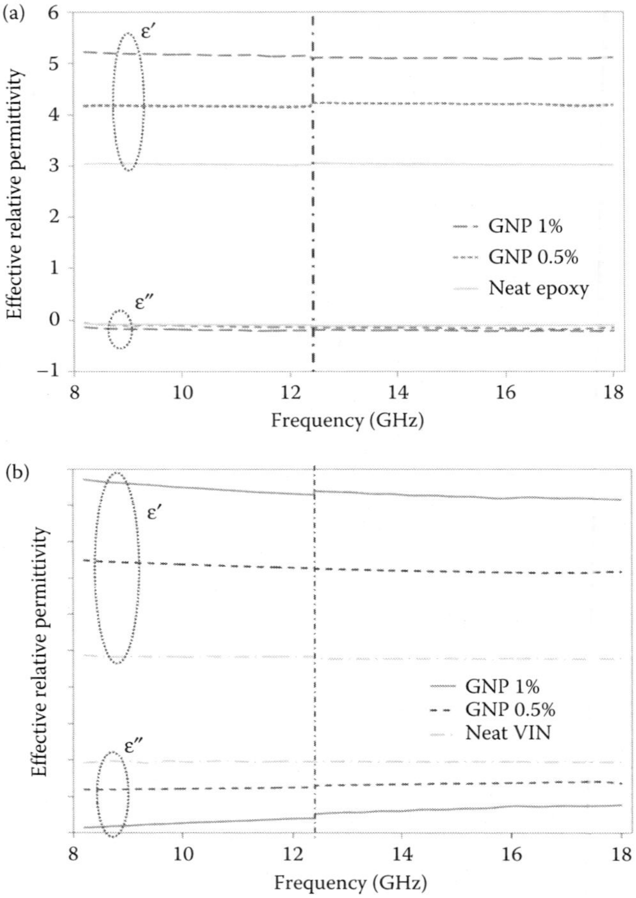

FIGURE 31.20 Measured frequency spectra of the effective relative permittivity of nanocomposites based on (a) epoxy or (b) vinyl ester matrix, filled with GNP exfoliated at 1150°C for 5 s, at 0.5% and 1% wt.

of six different measurements were performed on each batch, and data were averaged in order to reduce uncertainty. The complex effective permittivity of the composites ($\varepsilon_r = \varepsilon_r' + j\varepsilon_r''$) was finally extracted from the measured parameters following the standard method described in Reference [48].

31.4.3.1 Influence of Polymer Matrix

With the aim of comparing the effect of the chemical compatibility and adhesion of the two polymeric matrices to the GNP, the fabricated nanocomposites samples have been characterized in terms of effective complex dielectric permittivity in the frequency range from 8 to 18 GHz (X- and Ku-bands), applying the procedure described above.

The real and imaginary parts of the relative effective permittivity are reported in Figure 31.20a, b as a function of the frequency. The obtained results are in full agreement with the FTIR analysis reported in the previous section. In fact it results in the epoxy-based nanocomposite being characterized by lower values of ε_r' and $|\varepsilon_r''|$ in the whole frequency range, with respect to the vinyl ester-based one. This is due to the presence of the additional phonon modes in the FTIR spectrum of the vinyl ester nanocomposite, due to the formation of chemical bonds between the matrix and the filler, revealing an enhancement of the electron transport between the GNPs and the matrix, and consequently an increase of effective conductivity, which is related to the imaginary part of the complex effective relative permittivity by the expression $\sigma = \omega\varepsilon_0|\varepsilon_r''|$.

In general, it is observed that all nanocomposites based on the epoxy resin are characterized by values of the loss tangent much lower than the corresponding composites based on the vinyl ester resin. The obtained results demonstrate that a crucial role is played by the interaction between the nanofiller and the matrix. GNPs are better integrated into the vinyl ester rather than the epoxy resin.

The percolation curve of the vinyl ester-based nanocomposite is reported in Figure 31.21. It can be noted that the nanocomposite enters the percolation regime for loadings around 0.5% wt, in good agreement with the lowest percolation thresholds reported in the literature for GNP–thermosets composites [49].

31.4.3.2 Influence of TEGO Expansion Parameters

The effect of the synthesis parameters of TEGO on the effective complex permittivity of the corresponding nanocomposite has been investigated choosing seven different sets of expansion conditions (temperature and time). The polymer matrix used for all the composites under investigation is the vinyl ester resin (DION 9102) described in the previous chapter. For each set of expansion conditions, ranging between 750°C and

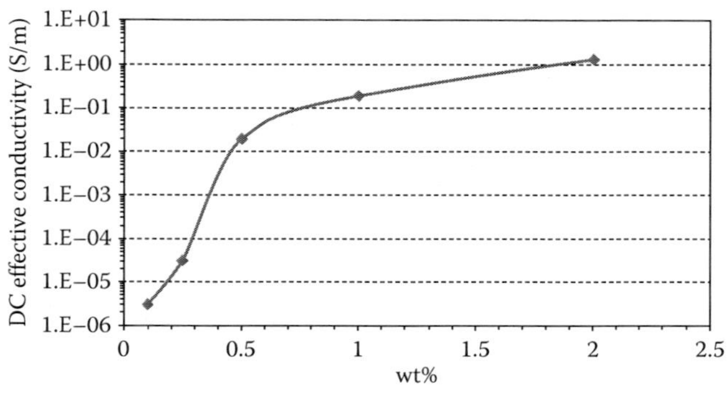

FIGURE 31.21 DC effective electrical conductivity of nanocomposites filled with GNPs at different weight concentrations.

1250°C in temperature and between 5 and 30 s in time, GNP-filled nanocomposites at 0.1% wt, 0.5% wt, and 1% wt have been fabricated, according to the procedure described above.

A comparison of the effective permittivity of composites produced with GNPs 0.1% wt, obtained from TEGO exfoliated at increasing temperatures for 30 s, is shown in Figure 31.22. As expected for the same TEGO thermal expansion time (30 s) and filler content (0.1% wt), nanocomposites filled with GNPs 1250°C–30 s show the highest values of both the real and imaginary parts of the effective complex permittivity. For the remaining composites, no significant difference is detectable in both the real and imaginary parts of the permittivity.

Successively, the analysis of the influence of TEGO exfoliation temperature and time is performed on the vinyl ester nanocomposite loaded with GNPs 0.5% wt. The obtained measured effective permittivity, in real and imaginary parts, is shown in Figure 31.23a and b.

The obtained spectra show that the highest values (absolute values) of both the real and imaginary parts of the complex effective permittivity are reached for the composite loaded with GNPs 1250°C–30 s, as in the case of the nanocomposite loaded at 0.1% wt. Nonetheless, slight variations between the different permittivities are now detectable for the different

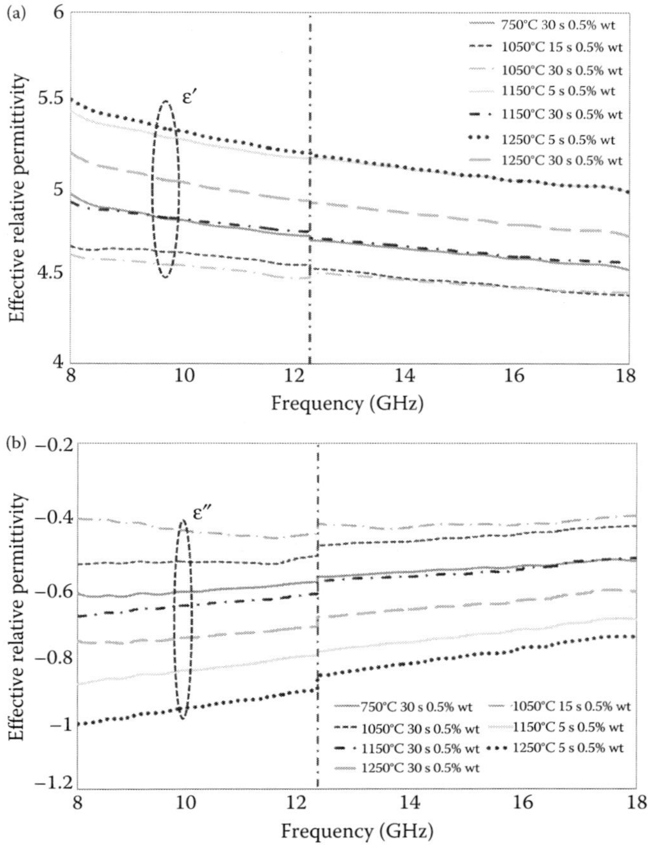

FIGURE 31.23 Measured effective permittivity of nanocomposites filled with GNPs 0.5% wt GNP, obtained from TEGO exfoliated in different conditions: (a) real and (b) imaginary parts.

composites. Moreover, though the effect of GNPs 1250°C–5 s and 1150°C–5 s on ε'_r is nearly the same, the same effect is much larger on ε''_r for GNPs 1250°C–5 s, indicating higher aspect ratio of the corresponding GNP type.

Finally, Figure 31.24 shows the real and imaginary parts of the relative effective permittivity measured for vinyl ester

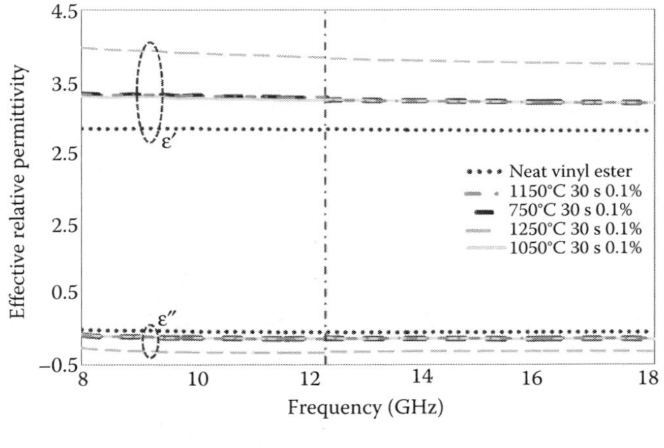

FIGURE 31.22 Effective permittivity of vinyl ester nanocomposites filled with GNPs at 0.1% wt, obtained from TEGO exfoliated at different temperatures for 30 s.

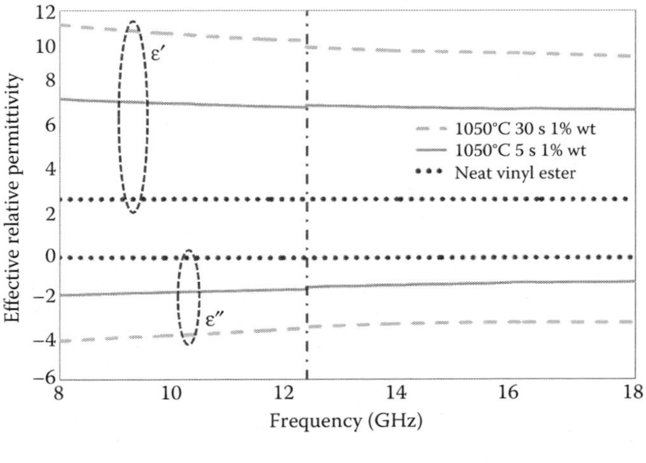

FIGURE 31.24 Effective permittivity of composites filled with 1% wt GNPs, obtained from TEGO exfoliated at 1050°C, 30 s and 1150°C, 5 s.

nanocomposites loaded with GNPs 1% wt, obtained from TEGO exfoliated at 1050°C for 30 s and at 1150°C for 5 s.

The reported spectra show a marked difference on the effect played by the different TEGO expansion conditions on the final effective permittivity of the resulting composites. Surprisingly enough, the nanocomposite filled with GNPs 1050°C–30 s provides a much higher enhancement of both ε_r' and ε_r'' with respect to the one loaded with GNPs 1150°C–5 s: all over the investigated frequency range, the value of ε_r'' measured for the former composite even doubles the one obtained for the latter. Such results are consistent with the measured DC conductivities of GNPs papers, reported in the previous section, and it can be probably addressed to a more uniform expansion of the TEGO for a longer time at 1050°C, with respect to 1150°C and 5 s.

Finally, the effective electrical conductivities of samples filled with 0.1% and 0.5% wt GNPs have been evaluated from the imaginary part of ε_r'', according to the relation $\sigma = -\omega\varepsilon_0\varepsilon_r''$ and the obtained results are reported in Figure 31.25a, b.

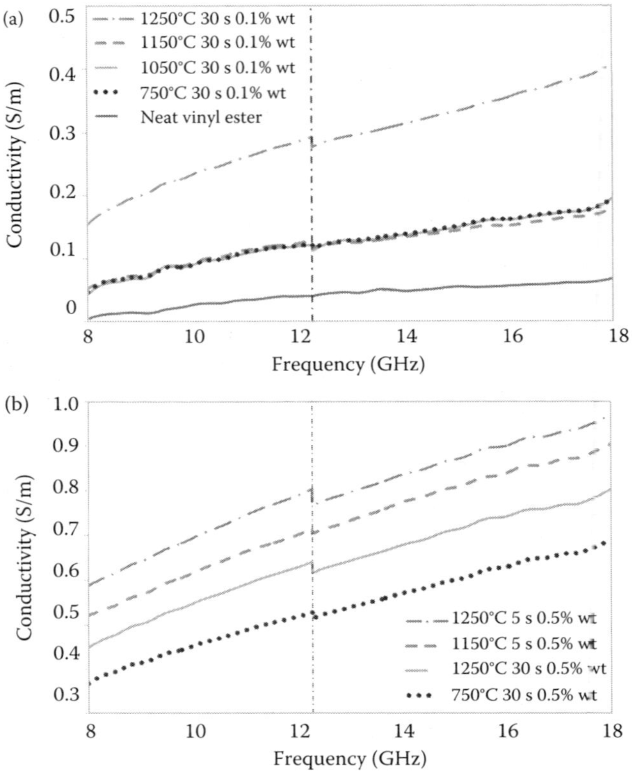

FIGURE 31.25 Effective electrical conductivity of vinyl ester nanocomposites as a function of the GNPs synthesis conditions for different concentrations: (a) 0.1% wt and (b) 0.5% wt.

31.4.3.3 Nanocomposites with High GNP Concentration

Viscosimetry tests described in the previous section demonstrated the high workability of the GNP-filled nanocomposite. In this section, we report the fabrication of nanocomposites with GNP loading at 1% wt and up to 4% wt, for possible application as EM shielding materials [50]. To this purpose, nanocomposites were prepared using the vinyl ester resin DION 9102, having demonstrated the best chemical compatibility with GNPs produced at 1150°C for 5 s.

At first, GNPs were dispersed into acetone through an ultrasonic probe working at the frequency of 20 kHz, running for 20 min, and set in pulse mode (1 s on and 1 s off). The resulting suspension was then poured into a beaker containing the liquid vinyl ester resin, which had been previously mixed with 0.2% wt of a Co-based accelerator (Akzo Nobel Accelerator NL-51P, cobalt (II) 2-ethylhexanoate, 6% wt Co in solvent mixture), and further sonicated for 30 s using a low ultrasound amplitude in order to minimize bubble formation.

Next, the solution was magnetically stirred at 200–250 rpm with the aim of removing solvent excess. Such a process step was done initially under ambient conditions, using the setup shown in Figure 31.26a. It should be noted that this phase is very critical. The amount of evaporated solvent is assessed by monitoring the mixture weight loss over time. It results in the time needed for complete solvent evaporation increasing with the filler concentration, due to the higher mixture viscosity, as

(a) (b)

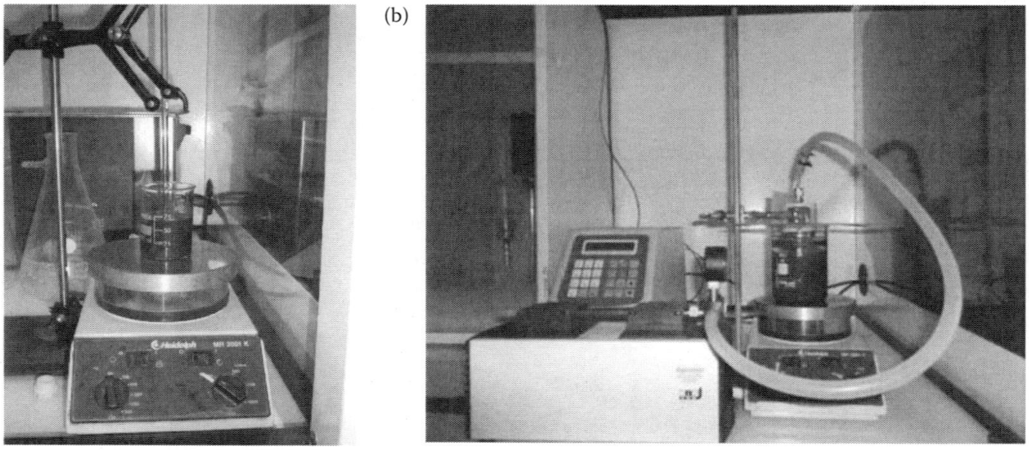

FIGURE 31.26 Solvent evaporation step: experimental setup in case of (a) ambient condition and (b) vacuum-assisted process. (Adapted from D'Aloia, A. G. et al. Synthesis and characterization of graphene-based nanocomposites for EM shielding applications. *Electromagnetic Compatibility (EMC EUROPE), 2013 International Symposium on*. © 2013. IEEE.)

TABLE 31.3
Solvent Evaporation Time for Different Filler Concentration and in Case of Ambient Condition Process or Vacuum-Assisted

GNP Concentration (% wt)	Processing Time (Vacuum-Assisted) (h)	Processing Time (Ambient Condition) (h)
1	2	6
2	3	12
4	6	n.a.

Source: D'Aloia, A. G. et al. Synthesis and characterization of graphene-based nanocomposites for EM shielding applications. *Electromagnetic Compatibility (EMC EUROPE), 2013 International Symposium on*. © 2013. IEEE.
Note: n.a.: not available.

TABLE 31.4
DC Electrical Conductivity of Different GNP Nanocomposites Fabricated with Vacuum and No-Vacuum Solvent Evaporation

	DC Conductivity (S/m)	
GNP Concentration	Ambient Condition	Vacuum-Assisted
1% wt	0.19	<10–5
2% wt	1.3	<10–5
4% wt	n.a.	2.7

Source: D'Aloia, A. G. et al. Synthesis and characterization of graphene-based nanocomposites for EM shielding applications. *Electromagnetic Compatibility (EMC EUROPE), 2013 International Symposium on*. © 2013. IEEE.
Note: n.a.: not available.

shown in Table 31.3. Therefore, an alternative processing route has been investigated in order to speed up the solvent evaporation. Figure 31.26b shows the experimental setup implementing the vacuum-assisted solvent evaporation. The beaker containing the mixture is subjected to vacuum using a diaphragm pump. Thanks to this improvement, the total processing time of the nanocomposite was limited to a few hours, as shown in Table 31.3, reaching the maximum of 6 h for the 4% wt mixture.

Upon complete evaporation of the solvent, the hardener was added at 2% wt ratio (Akzo Nobel Butanox LPT, methyl ethyl ketone peroxide, solution in diisobutyl phthalate). The liquid formulation was finally again stirred at 250 rpm until it gained a viscosity suitable for casting, and then it was poured directly in two sets of X and Ku flanges to be used for effective permittivity measurements, and in a rectangular mold for shielding and sheet-resistance characterizations. The composite obtained was cured in air for 24 h and postcured for another 24 h at 70°C.

SEM images of the fracture surfaces of the produced nanocomposites are shown in Figure 31.27a through c for different concentrations. It can be noted that GNPs are well dispersed at all concentrations (1%, 2%, and 4% wt).

The DC electrical conductivity of the samples was measured applying the four-wire volt-amperometric method using a Keithley 6221 DC/AC current source and a Keithley 2182a nanovoltmeter under controlled environmental conditions (23°C, 40% humidity). The obtained results are reported in Table 31.4.

It is noted that nanocomposites at 1% and 2% wt GNP fabricated under ambient conditions exhibit an increasing DC electrical conductivity with the GNP concentration. Moreover, it results in the specimens containing the same GNP weight fraction, but obtained applying the vacuum-assisted solvent evaporation process, being much more resistive. In fact, the measured conductivity is so low to be out of the range of the used measurement system. Only the nanocomposite filled with 4% wt GNP shows a DC conductivity of a few S/m. This demonstrates that the percolation threshold of nanocomposites synthesized under ambient conditions is well below the one of the nanocomposite fabricated with the alternative processing route, that is, below 4% wt. This effect can be explained considering that the vacuum-assisted process speeds up and forces solvent evaporation, thus producing a partial reagglomeration of the GNPs. This means that the average aspect ratio of the nanofiller is reduced, with a consequent increase of the percolation threshold of the composite.

The complex effective permittivity of the nanocomposite was extracted from scattering parameter measurements,

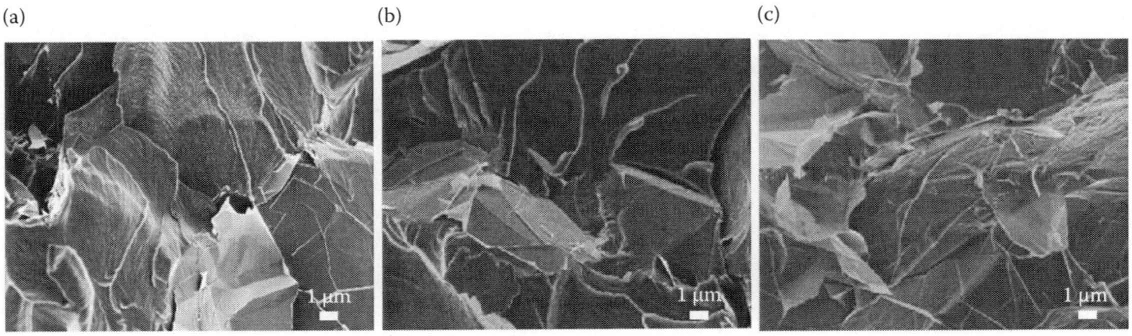

FIGURE 31.27 Low-magnification fracture surfaces of the fabricated nanocomposite loaded with increasing concentrations of GNPs: (a) 1% wt; (b) 2% wt; (c) 4% wt. (Adapted from D'Aloia, A. G. et al. Synthesis and characterization of graphene-based nanocomposites for EM shielding applications. *Electromagnetic Compatibility (EMC EUROPE), 2013 International Symposium on*. © 2013. IEEE.)

as described above, in the frequency range 8–18 GHz. For each composite type, a batch consisting of three X-band and three Ku-band flanges was prepared. A total of six different measurements were performed on each batch, and data were averaged, in order to reduce uncertainty. The complex effective permittivity of the composites ($\varepsilon_r = \varepsilon_r' + j\varepsilon_r''$) was finally extracted from the measured parameters.

The frequency spectra of the real and imaginary parts of the complex effective permittivity of the GNP nanocomposites are shown in Figure 31.28a for samples at 1% and 2% wt GNP, and in Figure 31.28b for the one loaded at 4% wt GNP. The continuous lines in Figure 31.28a refer to samples produced under ambient conditions, whereas the dotted lines refer to the ones fabricated by vacuum-assisted evaporation. It can be noted that in the case of composites filled with 1% wt GNP, there is no significant difference between the samples obtained with vacuum and no-vacuum-assisted solvent evaporation. This is reasonably due to the relatively low filler concentration dispersed inside the matrix. On the other hand, a clear difference is noted between the measured ε_r' and ε_r'' of the samples filled at 2% wt GNP and produced under ambient or vacuum-assisted conditions. Samples obtained through vacuum-assisted solvent removal are characterized by lower values in modulus of both the real and imaginary parts of the complex effective permittivity, in comparison with the samples produced under ambient conditions. However, the values of ε_r' and ε_r'' of the 2% wt GNP samples obtained with vacuum-assisted solvent evaporation are still greater in modulus than the ones measured for the 1% wt GNP samples. The observed effect confirms the hypothesis of a partial reagglomeration of the GNPs produced by the forced solved evaporation during the vacuum-assisted process, and are in line with the raising of the percolation threshold highlighted by the DC electrical conductivity measurements.

The real and imaginary parts of the complex effective permittivity of the sample filled with 4% wt GNP are reported in Figure 31.28b. This batch was prepared only through the vacuum-assisted solvent removal procedure. It is noted that the values of ε_r' and ε_r'' increase with the filler concentration. In particular, the real and imaginary parts of the complex effective permittivity of the sample filled with 4% GNP are much greater in modulus than the ones predicted in Reference [30]. This is due to the fact that the 4% wt loaded composite is well beyond the percolation threshold, and therefore it behaves definitely as a conductive material instead of a lossy dielectric.

This suggests that even if the filler concentration is so high, there is no formation of macroscopic GNP agglomerates inside the polymeric matrix.

The estimated AC electrical conductivities of the samples loaded with 1%, 2%, and 4% wt GNP are reported in Figure 31.29. As expected, there is no significant difference between the conductivities of the samples filled at 1% wt GNP with vacuum and nonvacuum solvent evaporation. On the other hand, a clear difference is noted in the case of composites filled at 2% wt GNP. Finally, it can be seen that the conductivity of the sample filled at 4% wt GNP is much higher than the other ones.

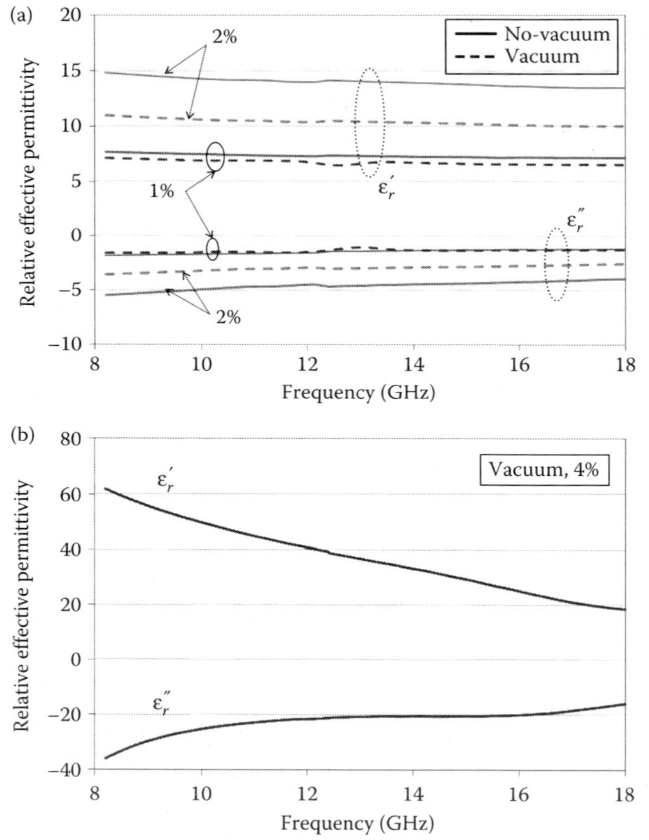

FIGURE 31.28 Measured real and imaginary parts of the relative effective permittivity of GNP nanocomposites. (a) 1% and 2% wt GNP samples produced with and without vacuum-assisted solvent evaporation. (b) 4% wt GNP sample produced with vacuum-assisted process. (Adapted from D'Aloia, A. G. et al. Synthesis and characterization of graphene-based nanocomposites for EM shielding applications. *Electromagnetic Compatibility (EMC EUROPE), 2013 International Symposium on.* © 2013. IEEE.)

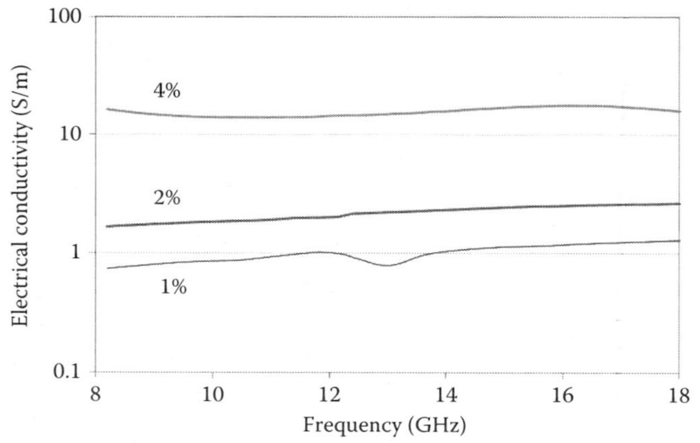

FIGURE 31.29 Electrical conductivity of samples loaded at 1%, 2%, and 4% wt GNP. (Adapted from D'Aloia, A. G. et al. Synthesis and characterization of graphene-based nanocomposites for EM shielding applications. *Electromagnetic Compatibility (EMC EUROPE), 2013 International Symposium on.* © 2013. IEEE.)

31.4.3.3.1 Shielding Effectiveness Measurements

The shielding effectiveness (SE) of the samples with 2% and 4% wt GNP produced by vacuum-assisted solvent evaporation was measured in the frequency range 8–18 GHz by the use of the double flanged coaxial sample holder (FCSH) shown in Figure 31.30, connected to a vector network analyzer (Anritsu Vector Star MS4647A). The FCSH is a modified version of the coaxial cell described in References 51, 52. It consists of an enlarged 50 Ω coaxial transmission line with an interrupted inner conductor and a flanged outer one. The cell is characterized by a central conductor with a diameter of 3 mm; the inner and outer diameters of the outer conductor are 7 and 15 mm, respectively. The material under test is inserted between the two halves of the cell that are tightly fastened together through screws placed on a polyvinyl chloride (PVC) ring.

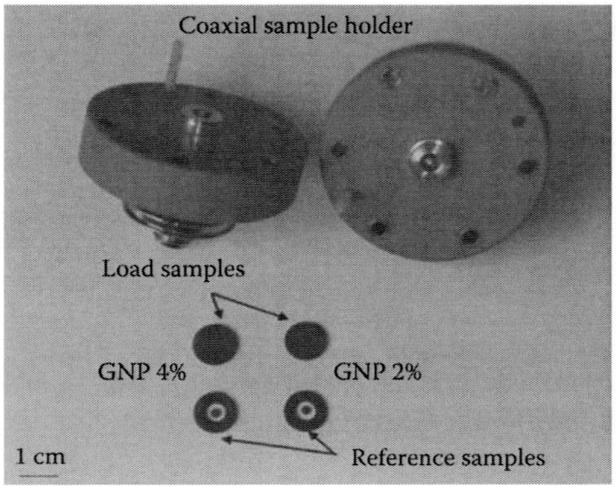

FIGURE 31.30 Picture of the coaxial sample holder, and the load and reference samples at 2% and 4% wt of GNP. (Adapted from D'Aloia, A. G. et al. Synthesis and characterization of graphene-based nanocomposites for EM shielding applications. *Electromagnetic Compatibility (EMC EUROPE), 2013 International Symposium on.* © 2013. IEEE.)

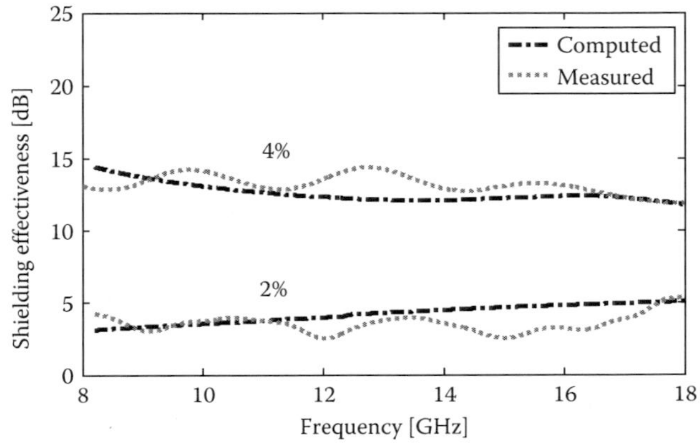

FIGURE 31.31 Measured and computed and SE of the samples loaded at 2% and 4% wt GNP. (Adapted from D'Aloia, A. G. et al. Synthesis and characterization of graphene-based nanocomposites for EM shielding applications. *Electromagnetic Compatibility (EMC EUROPE), 2013 International Symposium on.* © 2013. IEEE.)

Load and reference samples of the produced nanocomposites with 2% and 4% wt of GNP, having thickness of 0.86 and 1.15 mm, respectively, have been fabricated. The surfaces of the samples were polished carefully in order to enhance the direct electrical contact between the coaxial guide and the nanocomposite. The tested samples are reported in Figure 31.30. The measured spectra of the SE are shown in Figure 31.31. Note that the 4% wt GNP nanocomposite, having thickness of nearly 1 mm, provides an EM shielding in the range of 12–15 dB.

31.4.4 Electromagnetic Modeling of GNP Nanocomposites

The complex effective permittivity of GNP nanocomposites at RF can be predicted using the Maxwell–Garnett model [53], which has been reformulated in order to account for the shape of GNPs.

The MG formula is an algebraic expression through which the effective complex permittivity of a mixture ε_{eff} can be calculated as a function of the matrix permittivity, of the filler permittivity and of the filler volume fraction [54]. For a composite loaded with M different filler types, all having a size defined in the same dimensional scale, the MG formula reads

$$\varepsilon_{\text{eff}} = \varepsilon_m + \frac{\varepsilon_m \sum_{j=1}^{M} \theta_{Vj}(\varepsilon_{fj} - \varepsilon_m) \sum_{k=1}^{3} \frac{1}{\varepsilon_m + N_{jk}(\varepsilon_{fj} - \varepsilon_m)}}{3 - \sum_{j=1}^{M} \theta_{Vj}(\varepsilon_{fj} - \varepsilon_m) \sum_{k=1}^{3} \frac{N_{jk}}{\varepsilon_m + N_{jk}(\varepsilon_{fj} - \varepsilon_m)}} \tag{31.1}$$

where ε_m is the complex permittivity of the matrix, θ_{Vj} and ε_{fj} are the volume fraction and complex permittivity of the jth filler, and N_{jk} are the corresponding depolarization factors, which depend on the shape of the jth type of inclusion. These factors can of course be calculated for any shape, but in general this requires numerical effort. The only geometry for which simple analytical solutions can be found is the spheroidal or cylindrical shape [55].

The morphological analysis shown in the previous sections has demonstrated that GNPs cannot be modeled as spheroid or ellipsoidal particles. Owing to their flake-like geometry, it is very difficult to define depolarization factors for GNPs. Nevertheless, on the basis of the topographical characterization performed through AFM, it is observed that GNP flakes can be considered as bidimensional (2D) particles having an area in the range of a few square micrometers, but at the same time, due to the small thickness (in the range of a few nanometers), they behave as high-aspect-ratio fillers.

This suggests that GNP-filled composites can be modeled as an effective two-filler composite [32]: the first filler is constituted by oblate ellipsoids, whose presence mainly affects the dielectric permittivity of the composite; the second filler is constituted by nanorods affecting mainly the conducting properties of the material. Therefore, the MG formula is applied recursively [39]. At first, we compute the effective permittivity ε_{obl} of the polymeric matrix filled with oblate ellipsoids having axes L and t, L being the average lateral dimension of the flake and t its average thickness. This results in

$$\varepsilon_{\text{obl}} = \varepsilon_m + \frac{\varepsilon_m \theta_{\text{GNP}} \left[1 + \sigma_{\text{GNP}} / (j\omega\varepsilon_0) - \varepsilon_m\right] \sum_{j=1}^{3} \frac{1}{\varepsilon_m + N_{\text{obl}\,j}\left[1 + \sigma_{\text{GNP}} / (j\omega\varepsilon_0) - \varepsilon_m\right]}}{3 - \theta_{\text{GNP}} \left[1 + \sigma_{\text{GNP}} / (j\omega\varepsilon_0) - \varepsilon_m\right] \sum_{j=1}^{3} \frac{N_j}{\varepsilon_m + N_{\text{obl}\,j}\left[1 + \sigma_{\text{GNP}} / (j\omega\varepsilon_0) - \varepsilon_m\right]}} \tag{31.2}$$

where ε_r is the relative dielectric constant of the matrix, σ_{GNP} is the electrical conductivity, and θ_{GNP} is the GNP volume fraction. $N_{\text{obl}j}$ (j = 1,3) are the depolarization factors, given by

$$N_{\text{obl}3} = \frac{(1+e^2)(e - \tan^{-1} e)}{e^3} \tag{31.3}$$

$$N_{\text{obl}1} = N_{\text{obl}2} = \frac{1 - N_{\text{obl}3}}{2} \tag{31.4}$$

e being the eccentricity, defined as

$$e = \sqrt{\left(\frac{L}{t}\right)^2 - 1} \tag{31.5}$$

Successively, we apply again the MG expression, considering that rod-shaped inclusions are dispersed in the effective medium constituted by the matrix loaded with the oblate ellipsoids having effective permittivity ε_{obl}. It yields

$$\varepsilon_{\text{eff}} = \varepsilon_{\text{obl}} + \frac{\varepsilon_{\text{obl}} \alpha \theta_{\text{GNP}} \left[1 + \beta\sigma_{\text{GNP}} / (j\omega\varepsilon_0) - \varepsilon_{\text{obl}}\right] \sum_{k=1}^{3} \frac{1}{\varepsilon_{\text{obl}} + N_{\text{rod}\,k}\left[1 + \beta\sigma_{\text{GNP}} / (j\omega\varepsilon_0) - \varepsilon_{\text{obl}}\right]}}{3 - \alpha\theta_{\text{GNP}} \left[1 + \beta\sigma_{\text{GNP}} / (j\omega\varepsilon_0) - \varepsilon_{\text{obl}}\right] \sum_{k=1}^{3} \frac{N_{\text{rod}\,k}}{\varepsilon_{\text{obl}} + N_{\text{rod}\,k}\left[1 + \beta\sigma_{\text{GNP}} / (j\omega\varepsilon_0) - \varepsilon_{\text{obl}}\right]}} \tag{31.6}$$

where α and β are fitting parameters, which are representative of the volume concentration of the scattering edges modeled as rod-shaped fillers, and of their effective electric conductivity. The depolarization factors $N_{\text{rod}k}$ appearing in Equation 31.6 are given by

$$N_{\text{rod}1} = N_{\text{rod}2} = \frac{1}{2} \tag{31.7}$$

$$N_{\text{rod}3} = \left(\frac{\gamma t}{L}\right)^2 \ln\left[\frac{L}{\gamma t}\right] \tag{31.8}$$

where γ is a fitting parameter, which is representative of the scattering edges aspect ratio. It should be noted that α, β, and γ are strongly dependent on the GNP and composite fabrication processes.

In general, σ_{GNP} depends on the frequency due to (i) Drude's effect in the material; (ii) nonuniform current density distribution over the platelet surface due to magnetic self-inductance (e.g., skin effect); and (iii) EM field scattering. The latter two effects can be neglected in the hypothesis that, in the considered frequency range up to few tens of GHz, the GNP linear dimensions are smaller than the skin depth and wavelength. Drude's effect can be taken into account considering the following expression for σ_{GNP}:

$$\sigma_{\text{GNP}}(\omega) = \frac{\sigma_{0\text{GNP}}}{1 + j\omega\tau_{\text{GNP}}} \tag{31.9}$$

where $\sigma_{0\text{GNP}}$ is the DC electrical conductivity of the GNPs, and τ_{GNP} is the relaxation time of GNPs. According to recent studies, it results that $1/\tau_{\text{GNP}}$ is about 5 THz [56]. For EMC applications considered involving frequencies up to a few tens of GHz, it results $\omega\tau_{\text{GNP}} \ll 1$, and therefore the GNP conductivity can be considered as frequency independent.

31.4.4.1 Model Validation

GNP-filled nanocomposites at different concentration, fabricated as described in the previous section, are simulated as a two-filler composite, applying the procedure described above. GNPs are produced from TEGO expanded at 1150°C for 5 s, in acetone, with 1–1 sonication cycle, at room temperature and pressure, without thermal annealing.

The density of the vinyl ester matrix (i.e., including hardner) is set to $\rho_m = 1.2$ g/cm^3. From SEM analysis, it results that the average lateral dimension of GNPs and its average thickness are $L = 2.5$ μm and $t = 9$ nm, respectively. $\sigma_{0\text{GNP}}$ is set equal to 10.5 kS/m and $\tau_{\text{GNP}} = 0.189$ ps, as obtained in Reference [56]. Moreover, the measured relative permittivity of the plain vinyl ester resin is 3.

The fitting parameters α, β, and ξ are set equal to 0.13, 0.38, and 2.33, respectively [33].

Figure 31.32a and b shows the comparison between the measured and computed real and imaginary parts of the effective relative permittivity of composites filled with GNPs for increasing weight concentration wt%$_{\text{GNP}}$, from 0.25% wt up to 2% wt. Notice the good agreement between the measured and simulated results, with respect to both the real and imaginary parts.

The performed calculation confirmed that in the considered range, the frequency dependence of the GNP conductivity is negligible, as expected.

The effective conductivity of the nanocomposite is computed from the imaginary part of the effective permittivity. The results are reported in Figure 31.33 as a function of the frequency.

31.5 APPLICATIONS

The GNP nanocomposites described in the previous sections can find key applications as RAMs and as strain gauge

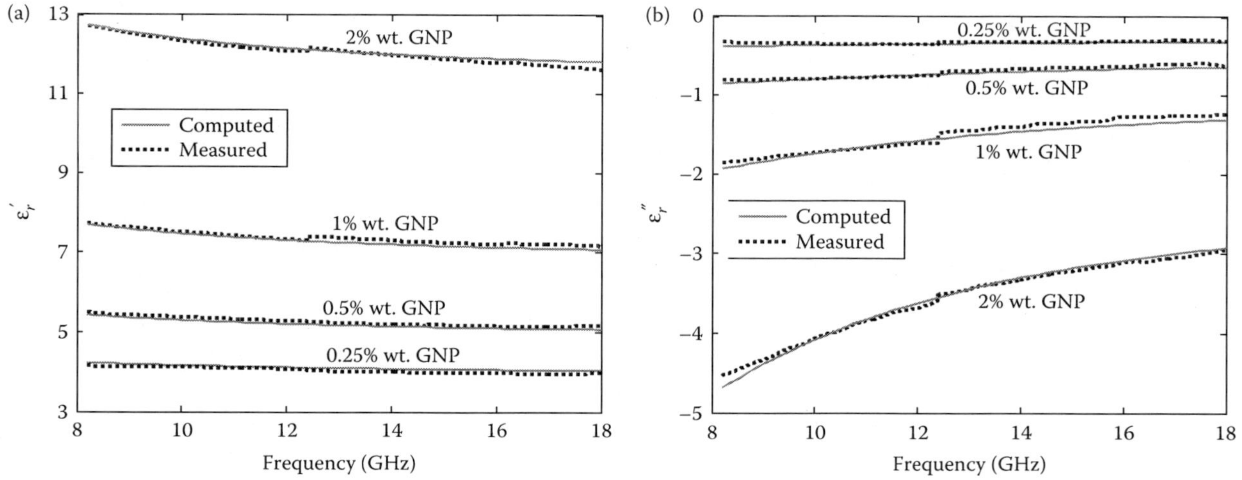

FIGURE 31.32 Measured and computed relative effective permittivity of the GNP composites: (a) real part; (b) imaginary part. (Adapted from Sarto, M.S. et al. 2012. Synthesis, modeling, and experimental characterization of graphite nanoplatelet-based composites for EMC applications. *IEEE Trans EMC* 54(1):17–27.)

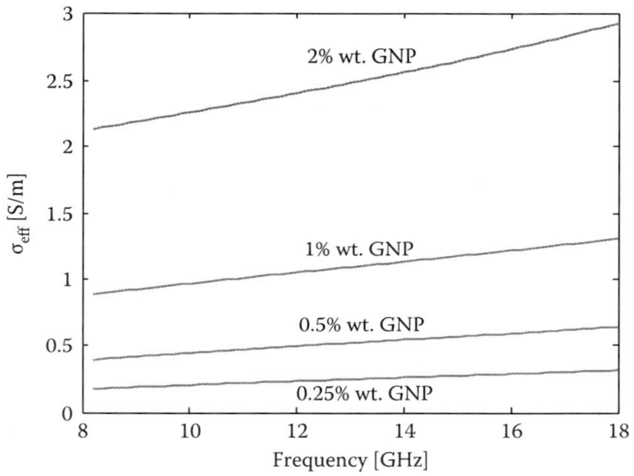

FIGURE 31.33 Computed effective electrical conductivity of the GNP composites with different concentrations. (Adapted from Sarto, M.S. et al. 2012. Synthesis, modeling, and experimental characterization of graphite nanoplatelet-based composites for EMC applications. *IEEE Trans EMC* 54(1):17–27.)

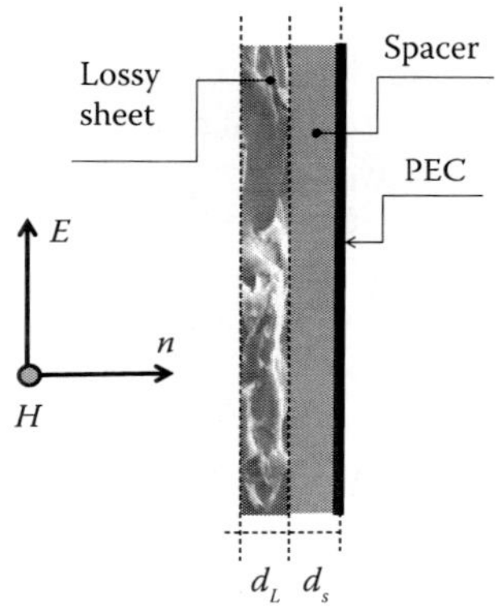

FIGURE 31.34 Schematic configuration of a dielectric Salisbury screen.

with high sensitivity, due to their unique characteristics, and to the possibility of tailoring their electrical and EM properties through the proper control of the fabrication process parameters.

31.5.1 Graphene-Based RAMs

Let us consider the dielectric Salisbury screen sketched in Figure 31.34. The lossy layer of thickness d_L is made of GNP nanocomposite and it is designed in order to absorb the energy associated with the incoming EM field. The spacer of thickness d_s is made of a lossless dielectric layer and has the function of supporting the lossy sheet and of positioning it at the desired distance from the metal backing, which is assumed to be realized with a perfect electric conducting (PEC) panel.

The fabricated panel has the characteristics reported in Table 31.5, and it is designed in order to have a minimum reflection at about 12 GHz [33].

The lossy sheet is fabricated by pouring the GNP–resin mixture (including curing agent), after solvent evaporation,

TABLE 31.5
Characteristics of the Fabricated Salisbury Screen

Layer	Material	Thickness (mm)
Lossy sheet	Nanocomposite 2% wt GNP	1
Spacer	Rohacell I51	0.6

Source: Koo, J.H. 2006. *Polymer Nanocomposites Processing, Characterization, and Applications.* McGraw-Hill.

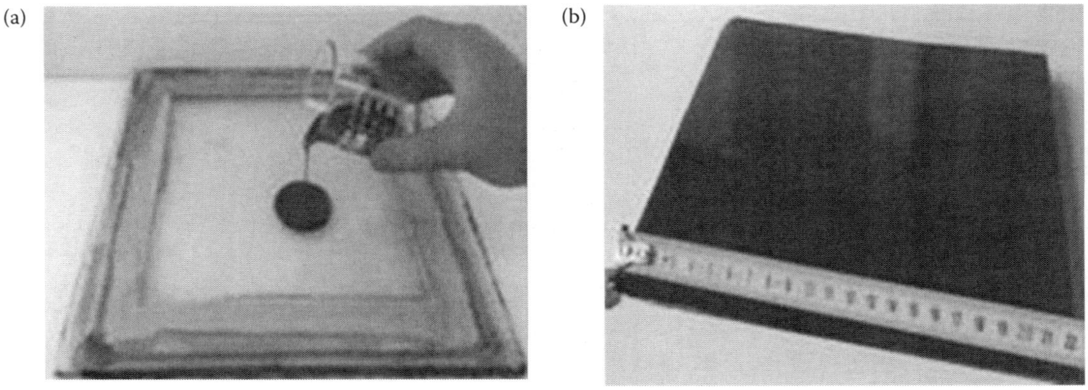

FIGURE 31.35 (a) Squared mold and (b) fabricated nanocomposite lossy sheet. (Adapted from D'Aloia, A.G. et al. 2014. *Electromagnetic absorbing properties of graphene–polymer composite shields. Carbon* 73:175–84.)

in a squared mold having dimensions 25 cm × 25 cm (Figure 31.35a). The composite is then cured in air for 24 h and postcured for another 24 h at 70°C, thus obtaining the panel shown in Figure 31.35b, which is assembled over the back metal plate with the Rohacell spacer.

The reflection coefficient of the fabricated Salisbury screen is measured in the X and Ku bands, in free-space, against a plane wave with normal incidence in the far field (Fraunhofer) region. The measured reflection coefficient is reported in Figure 31.36, and is compared with the one calculated using the effective material model described in the previous section.

The good agreement between measured and simulated results confirms the validity of the developed approach. The main differences between experimental and calculated data are due to the nonuniform thickness of the manually fabricated lossy sheet.

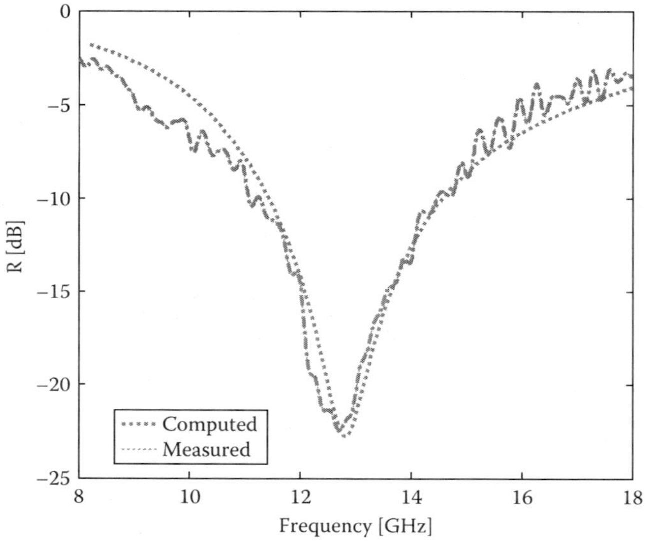

FIGURE 31.36 Measured and computed reflection coefficients of the Salisbury screen having the configuration shown in Table 31.5.

31.5.2 Graphene-Based Strain Sensors

The electromechanical properties of the fabricated GNP nanocomposites were investigated preparing samples suitable for three-point bending tests according to the standard ASTM D 790 [35,36,39].

The GNP-based polymeric composites were first cut in thin rectangular laminae (16 mm × 8 mm × 150 μm) with a Buehler® IsoMet 4000 high-precision rotating saw. Then, through the use of a mask, thin silver layers were deposited on the two end areas (1.5 mm × 5 mm) of the laminae designed for electrical contacts. Next, each lamina was bonded onto the center of the top face of a polycarbonate (PC) rectangular beam (120 mm × 24.5 mm × 6 mm) using a cyanoacrylate-based adhesive. In order to prevent any deterioration of the contact quality during the mechanical tests, the end regions of the nanocomposite laminae were not glued intentionally to the PC substrate. Finally, tin-coated copper wires 0.2 mm in diameter were attached to the contacts with a silver-based epoxy adhesive. A picture of a completed sample for flexural tests is reported in Figure 31.37.

The fabricated specimens were subjected to three-point bending tests using a Zwick/Roell Z010 universal machine equipped with a 10-kN load cell. The flexural tests were performed with a span-to-depth ratio of 16:1, setting the crosshead speed at 1 mm/min. The load range (0–390 N) applied to the specimens was specifically chosen to be in the linear

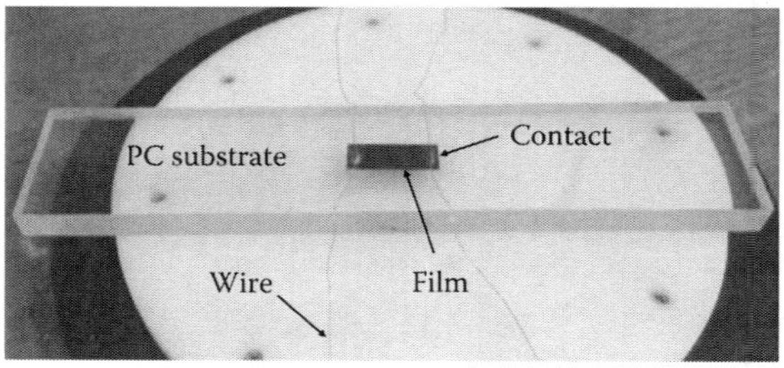

FIGURE 31.37 Picture of GNP-based polymeric nanocomposite lamina attached on a PC beam used for elecromechanical tests.

elastic portion of the response of the PC substrate. During the quasi-static monotonic loading of the PC substrate, the DC electrical resistance variation of the nanocomposite lamina was monitored and recorded as a function of the induced strain using a Keithley 6221 DC/AC current source connected to a Keithley 2182a nano-voltmeter.

The test results demonstrated a significant piezoresistive response of the fabricated GNP-based nanocomposites. In fact, the electrical resistance of the laminae showed a constant nonlinear increase with the applied strain. In general, this effect can be attributed to the disruption of the conducting GNP percolation network and to the variation of the tunneling resistance between adjacent fillers due to the loss of electrical contact and the increased interparticle distances during loading.

The measured resistance values were used to calculate the piezoresistive sensitivity (i.e., GF) of the GNP nanocomposite laminae. The GF is a dimensionless parameter defined as the relative change in electrical resistance (ΔR) due to an applied mechanical strain ε:

$$GF = \frac{\Delta R / R_0}{\varepsilon} \tag{31.10}$$

where R_0 is the initial resistance at zero load.

The results for a nanocomposite lamina with a GNP content of 2% wt are reported in Figure 31.38 up to an applied strain of nearly 2.35%. In the same figure, the GF of a lamina obtained from a polymeric nanocomposite filled with 0.5% wt multiwall carbon nanotubes (MWCNTs) is shown for comparison. It clearly shows that the GF of the GNP nanocomposite is 1–2 orders of magnitude higher (especially at large strain) than those achievable with MWCNT-based composites or conventional metallic foil strain gauges that are characterized by a constant GF of 2–2.1.

Therefore, the electromechanical properties of these novel materials seem to be promising for the realization of next-generation high-performance strain sensors.

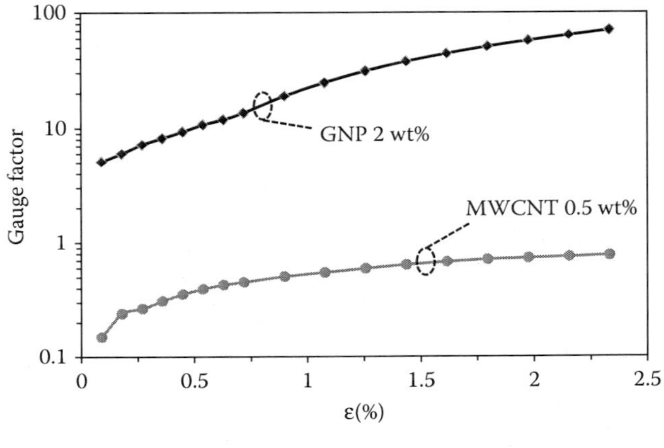

FIGURE 31.38 Gauge factor as function of strain of nanocomposite laminae characterized by a vinyl ester matrix filled with GNPs at 2% wt or MWCNT at 0.5% wt.

31.6 CONCLUSIONS

In this chapter, the synthesis of GNPs, GNP foils, and GNP nanocomposites is reported. A systematic investigation is also delivered, showing the effect of several synthesis parameters of GNPs on the resulting electrical properties of GNP-based foils and electrical/EM properties of the resulting nanocomposites. In particular, the DC conductivity of GNP foils is correlated to various process route parameters, including (i) TEGO expansion temperature and time; (ii) on–off phase duration of the sonication cycle, at RT or IB; (iii) solvent type and TEGO expansion temperature between 1150°C and 1350°C, considering the 1 s on–1 s off sonication cycle at RT or with TC; and (iv) the temperature of annealing of the GNP papers after production, between 250°C and 550°C. The systematic study has allowed the obtaining of a maximum DC conductivity of 40 kS/m for GNP foils obtained from TEGO 1150°C–5 s sonicated using a TC in an acetone: DMF mixture, after being subjected to thermal annealing in air at 550°C for 10 min. Next, a comparison of the effect of the use of different polymeric matrices and TEGO types for the manufacturing of GNP-loaded nanocomposite is investigated, measuring their DC electrical conductivity and EM properties, highlighting the crucial role played by the interaction between the nanofiller and the matrix. The percolation curve of the vinyl ester-based nanocomposite shows that the nanocomposite enters the percolation regime for loadings around 0.5% wt. The characterization of the EM properties of vinyl ester–graphene nanocomposites shows a linear increase of both ε_r' and ε_r'' with increasing GNP loadings. Only for the sample filled with 4% GNP are ε_r' and ε_r'' much greater in modulus than the ones predicted, due to the fact that the 4% wt loaded composite is well beyond the percolation threshold, and therefore it behaves definitely as a conductive material rather than a lossy dielectric. EM modeling of the fabricated nanocomposites demonstrated that, when dispersed uniformly in a polymeric matrix, GNPs behave as a combination of two different fillers: the first one consists of prolate ellipsoids that contribute mainly to the control of the real part of the complex effective permittivity of the composite; the second one consists of cylindrical rods with a high aspect ratio that affects mainly the effective electrical conductivity of the material.

Finally, two key applications of GNP nanocomposites are successfully reported, namely, a RAM and a strain gauge. The rational design and manufacture of the Salisbury screen, from a GNP nanocomposite, allowed the obtainment of a reflection coefficient of nearly 23–24 dB in the 12–13 GHz region. Based on the same composites, the proposed piezoresistive strain sensor has a measured GF of almost 100, which is two orders of magnitude higher than those achievable with MWCNT-based composites or conventional metallic foil strain gauges. Although this chapter gives only an overview of two possible GNP nanocomposite applications, it shows the potential for a variety of other possible fields where GNPs and related composites can act as a real breakthrough, when properly integrated.

REFERENCES

1. Novoselov, K.S., Geim, A.K., Morozov, S.V., Jiang, D., Zhang, Y., Dubonos, S.V. et al. 2004. Electric field effect in atomically thin carbon films. *Science* 306:666–9.
2. Schniepp, H.C., Kudin, K.N., Li, J.L., Prud'homme, R.K., Car, R., Saville, D.A., and Aksay, I.A. 2008. Bending properties of single functionalized graphene sheets probed by atomic force microscopy. *ACS Nano* 12:2577–84.
3. Lee, C., Wei, X., Kysar, J.W., Hone, J. 2008. Measurement of the elastic properties and intrinsic strength of monolayer graphene. *Science* 321:385–8.
4. Frank, I.W., Tanenbaum, D.M., van der Zande, A.M., and McEuen P.L. 2007. Mechanical properties of suspended graphene sheets. *J Vac Sci Technol B* 25(6):2558–61.
5. Balandin, A.A., Ghosh, S., Bao, W., Calizo, I., Teweldebrhan, D., Miao, F. et al. 2008. Superior thermal conductivity of single-layer graphene. *Nano Lett* 8(3):902–7.
6. Balandin, A.A. 2011. Thermal properties of graphene and nanostructured carbon materials. *Nat Mater* 10(8):569–81.
7. Du, X., Skachko, I., Barker, A., Andrei, E.Y. 2008. Approaching ballistic transport in suspended graphene. *Nat Nanotechnol* 3:491–5.
8. Liu, Z., Bai, G., Huang, Y., Ma, Y., Du, F., Li, F. et al. 2007. Reflection and absorption contributions to the electromagnetic interference shielding of single-walled carbon nanotube/polyurethane composites. *Carbon* 45:821–7.
9. Park, S.H., Thielemann, P., Asbeck, P., Bandaru, P.R. 2009. Enhanced dielectric constants and shielding effectiveness of, uniformly dispersed, functionalized carbon nanotube composites. *Appl Phys Lett* 94:243111–3.
10. Wang, L.L., Tay, B.K., See, K.Y., Sun, Z., Tan, L.K., Lua, D. 2009. Electromagnetic interference shielding effectiveness of carbon-based materials prepared by screen printing. *Carbon* 47:1905–10.
11. Al-Saleh, M.H., Sundararaj, U. 2009. Electromagnetic interference shielding mechanisms of CNT/polymer composites. *Carbon* 47:1738–46.
12. Huanga, Y., Li, N., Ma, Y., Du, F., Li, F., He, X. et al. 2007. The influence of single-walled carbon nanotube structure on the electromagnetic interference shielding efficiency of its epoxy composites. *Carbon* 45:1614–21.
13. Li, N., Huang, Y., Du, F., He, X., Lin, X., Gao, H. et al. 2006. Electromagnetic interference (EMI) shielding of single-walled carbon nanotube epoxy composites. *Nano Lett* 6:1141–5.
14. Wang, X., Jiang, Q., Xu, W., Cai, W., Inoue, Y., and Zhu, Y. 2013. Effect of carbon nanotube length on thermal, electrical and mechanical properties of CNT/bismaleimide composites. *Carbon* 53:145–52.
15. Yasmin, A., Luo, J.J., Daniel, I.M. 2006. Processing of expanded graphite reinforced polymer nanocomposites. *Comp Sci Technol* 66:1179–86.
16. Liang, J., Wang, Y., Huang, Y., Yanfeng, M., Liu, Z., Cai, J. et al. 2009. Electromagnetic interference shielding of graphene/epoxy composites. *Carbon* 47:922–5.
17. Mo, Z., Sun, Y., Chen, H., Zhang, P., Zuo, D., Liu, Y. et al. 2005. Preparation and characterization of a PMMA/Ce(OH)3, Pr2O3/graphite nanosheet composite. *Polymer* 46:12670–6.
18. Chen, G., Wenig, W., Wu, D., Wu, C. 2003. PMMA/graphite nanosheets composite and its conducting properties. *Eur Polym J* 39:2329–35.
19. Song, L.N., Xiao, M., Meng, Y.Z. 2006. Electrically conductive nanocomposites of aromatic polydisulfide/expanded graphite. *Comp Sci Technol* 66:2156–62.
20. He, B.F., Lau, S., Chan, H.L., Fan, J. 2009. High dielectric permittivity and low percolation threshold in nanocomposites based on poly(vinylidene fluoride) and exfoliated graphite nanoplates. *Adv Mater* 21:710–5.
21. Raza, M.A., Westwood, A., Brown, A., Hondow, N., Stirling, C. 2011. Characterisation of graphite nanoplatelets and the physical properties of graphite nanoplatelet/silicone composites for thermal interface applications. *Carbon* 49(13):4269–79.
22. Zhang, L., Zhu, J., Zhou, W., Wang, J., and Wang, Y. 2012. Thermal and electrical conductivity enhancement of graphite nanoplatelets on form-stable polyethylene glycol/polymethyl methacrylate composite phase change materials. *Energy* 39(1):294–302.
23. Yu, A., Ramesh, P., Sun, X., Bekyarova, E., Itkis, M.E., Haddon, R.C. 2008. Enhanced thermal conductivity in a hybrid graphite nanoplatelet–carbon nanotube filler for epoxy composites. *Adv Mater* 20(24):4740–4.
24. Yang, S.Y., Lin, W.N., Huang, Y.L., Tien, H.W., Wang, J.Y., Ma, C.C.M. et al. 2011. Synergetic effects of graphene platelets and carbon nanotubes on the mechanical and thermal properties of epoxy composites. *Carbon* 49(3):793–803.
25. Li, W., Dichiara, A., and Bai, J. 2013. Carbon nanotube–graphene nanoplatelet hybrids as high-performance multifunctional reinforcements in epoxy composites. *Comp Sci Technol* 74:221–7.
26. Kim, M.S., Yan, J., Joo, K.H., Pandey, J.K., Kang, Y.J., Ahn, S.H. 2013. Synergistic effects of carbon nanotubes and exfoliated graphite nanoplatelets for electromagnetic interference shielding and soundproofing. *J Appl Polym Sci* 130(6):3947–51.
27. De Bellis, G., De Rosa, I.M., Dinescu, A., Sarto, M.S., Tamburrano, A. 2010. Electromagnetic properties of carbon-based nanocomposites: The effect of filler and resin characteristics. In *Proc. Int. Symp. IEEE NANO 2010*, Seoul, Korea, August 17–20, 2010.
28. De Rosa, I.M., Sarasini, F., Sarto, M.S., Tamburrano, A. 2008. EMC impact of advanced carbon fiber/carbon nanotube reinforced composites for next-generation aerospace applications. *IEEE Trans EMC* 50(3):556–63.
29. De Rosa, I.M., Dinescu, A., Sarasini, F., Sarto, M.S., Tamburrano, A. 2010. Effect of short carbon fibers and MWCNTs on microwave absorbing properties of polyester composites containing nickel-coated carbon fibers. *Comp Sci Technol* 70(1):102–9.
30. De Bellis, G., De Rosa, I.M., Dinescu, A., Sarto, M.S., Tamburrano, A. 2010. Electromagnetic absorbing nanocomposites including carbon fibers, nanotubes and graphene nanoplatelets. In *Proc. IEEE Int. Symp EMC*, Fort Lauderdale, FL, USA, July 25–30, 2010.
31. De Bellis, G., Tamburrano, A., Dinescu, A., Santarelli, M.L., and Sarto, M.S. 2011. Electromagnetic properties of composites containing graphite nanoplatelets at radio frequency. *Carbon* 49(13):4291–300.
32. Sarto, M.S., D'Aloia, A.G., Tamburrano, A., Bellis, G.D. 2012. Synthesis, modeling, and experimental characterization of graphite nanoplatelet-based composites for EMC applications. *IEEE Trans EMC* 54(1):17–27.
33. D'Aloia, A.G., Marra, F., Tamburrano, A., De Bellis, G., and Sarto, M.S. 2014. Electromagnetic absorbing properties of graphene–polymer composite shields. *Carbon* 73:175–84.
34. Hou, Y., Wang, D., Zhang, X.M., Zhao, H., Zha, J.W., and Dang, Z.M. 2013. Positive piezoresistive behavior of electrically conductive alkyl-functionalized graphene/polydimethylsilicone nanocomposites. *J Mater Chem C* 1(3):515–21.

35. Tamburrano, A., Sarasini, F., De Bellis, G., D'Aloia, A.G., Sarto, M.S. 2013. The piezoresistive effect in graphene-based polymeric composites. *Nanotechnology* 24(46):465702.
36. D'Aloia, A.G., Tamburrano, A., De Bellis, G., Tirillò, J., Sarasini, F., Sarto, M.S. 2013. Electromechanical modeling of GNP nanocomposites for integrated stress monitoring of electronic devices. In *Nanoelectronic Device Applications Handbook*, J.E. Morris and K. Iniewski (editors), CRC Press (Taylor & Francis Group).
37. Zanni, E., De Bellis, G., Bracciale, M., Broggi, A., Santarelli, M.L., Sarto, M.S., Palleschi, C., Uccelletti, D. 2012. Graphite nanoplatelets and Caenorhabditis elegans: Insights from an *in vivo* model. *Nano Lett* 12(6):2740–4.
38. Chiacchiarelli, L.M., Rallini, M., Monti, M., Puglia, D., Kenny, J.M., and Torre, L. 2013. The role of irreversible and reversible phenomena in the piezoresistive behavior of graphene epoxy nanocomposites applied to structural health monitoring. *Comp Sci Technol* 80:73–9.
39. D'Aloia, A.G., Sarto, M.S., De Bellis, G., Tamburrano, A. 2011. Electromechanical modeling of GNP nanocomposites for stress sensors applications. In *IEEE NANO 2011*, Portland, OR, August 15–19, 2011.
40. Hernandez, Y., Nicolosi, V., Lotya, M., Blighe, F.M., Sun, Z.Y., De, S. et al. 2008. High-yield production of graphene by liquid-phase exfoliation of graphite. *Nat Nanotechnol* 3:563–8.
41. Compton, O.C., Nguyen, S.T. 2010. Graphene oxide, highly reduced graphene oxide, and graphene: Versatile building blocks for carbon-based materials. *Small* 6:711–23.
42. Mason, T.J., Lorimer, J.P. 2002. *Applied Sonochemistry*. Wiley-VCH.
43. De Bellis, G., Bregnocchi, A., Di Ciò, S.,Tamburrano, A., Sarto, M.S. 2013. Effect of sonication on morphology and DC electrical conductivity of graphene nanoplatelets-thick films. In *IEEE NANO 2013*, Beijing, August 5–8, 2013.
44. De Bellis, G., Tamburrano, A., Mulattieri, M., Sarto, M.S. 2012. Effect of process parameters on the effective DC conductivity of GNP thick films. In *IEEE NANO 2012*, Birmingham, UK, August 20–23, 2012.
45. Wang, S., Zhang, Y., Abidi, N., Cabrales, L. 2009. Wettability and surface free energy of graphene films. *Langmuir* 25(18):11078–81.
46. Koo, J.H. 2006. *Polymer Nanocomposites Processing, Characterization, and Applications*. McGraw-Hill.
47. De Bellis, G., De Rosa, I.M., Dinescu, A., Sarto, M.S., Tamburrano, A. 2010. Electromagnetic properties of carbon-based nanocomposites: The effect of filler and resin characteristics. In *IEEE NANO 2010*, Singapore, August 17–23, 2010.
48. ASTM D5568-0.8 2008. Standard test method for measuring relative complex permittivity and relative magnetic permeability of solid materials at microwave frequencies. *ASTM*.
49. Schlenoff, J.B. 2002. *Multilayer Thin Films Sequential Assembly of Nanocomposite Materials*, G. Decher (editor), Wiley-VCH.
50. D'Aloia, A.G., Marra, F., Tamburrano, A., De Bellis, G., Sarto, M.S. 2013. Synthesis and characterization of graphene-based nanocomposites for EM shielding applications. In *EMC EUROPE 2013*, Brugge, September 2–6, 2013.
51. Sarto, M.S., Tamburrano, A. 2006. An innovative test method for the shielding effectiveness measurement of conductive thin films in a wide frequency range. *IEEE Trans EMC*, 48(2), 331–41.
52. Tamburrano, A., Desideri, D., Maschio, A., Sarto, M.S. 2013. Coaxial waveguide methods for shielding effectiveness measurement of planar materials up to 18 GHz. *Electromagnetic Compatibility, IEEE Transactions on*, 56(6):1386–95.
53. Landau, L.D., Lifshitz, E.M. 1960. *Electrodynamics of Continuous Media*. Pergamon Press, Oxford.
54. Shivola, A. 2000. Mixing rules with complex dielectric coefficients. *SSTA* 1(4):293–415.
55. Osborn, J.A. 1945. Demagnetizing factors of the general ellipsoid. *Phys Rev* 67, 351–7.
56. Liu, H.L., Carr, G.L., Worsley, K.A., Itkis, M.E., Haddon, R.C., Caruso, A.N., Tung, L.-C., Wang, Y.J. 2010. Exploring the charge dynamics in graphite nanoplatelets by THz and infrared spectroscopy. *New J Phys* 12:113012.

32 Electronic Transport and Optical Properties of Graphene

Klaus Ziegler

CONTENTS

ABSTRACT

A brief survey is provided to discuss the transport properties of graphene in the absence of an external magnetic field. Electronic properties of mono- and bilayer graphene are strongly related to the existence of a quasiparticle spectrum, which consists of two bands that touch each other at two Dirac nodes. This structure is associated with a number of interesting features, such as Klein tunneling and electron–hole pair creation. This leads to robust transport properties, including a minimal conductivity and a constant optical conductivity, clearly indicating a deviation from conventional Drude-type transport. Moreover, recent experimental studies have revealed that the sublattice symmetry of the honeycomb lattice can be broken by chemical doping or by external gates in the case of bilayers. The broken symmetry opens a random gap in the quasiparticle spectrum which may lead to an insulating behavior. We discuss diffusion in graphene, which implies a characteristic metallic behavior in mono- and bilayer graphene, and the alternate current conductivity. The properties of the latter deviate substantially from that of a conventional metal. The behavior of plasmons in graphene, on the other hand, is similar to that of plasmons in a conventional two-dimensional electron gas.

32.1 INTRODUCTION

The enormous list of publications on transport measurements in graphene starts with the seminal papers by groups from Manchester and Columbia [1]. These studies indicate a very robust transport behavior, which is characterized by a V-shape conductivity with respect to charge density n and a minimal conductivity $\sigma_{\min} \approx 4e^2/h$ at the charge neutrality point $n = 0$. In the presence of a magnetic field there are Shubnikov–de Haas oscillations for the longitudinal conductivity σ_{xx} and quantum Hall plateaux for the Hall conductivity σ_{xy} at a sufficiently strong magnetic field. These properties have been confirmed subsequently by various experimental groups in more detail and measurements at various conditions and for different types of samples. Many of those results have been collected and discussed in a number of extensive reviews [2–4].

Optical properties of graphene for light with frequency ω are (directly) related to the optical (or AC) conductivity $\sigma_{xx}^{AC}(\omega)$. The imaginary part of the dielectric constant is related to the real part of the AC conductivity and, therefore, to the optical reflectivity and transmittance [5].

The aim of the following survey is to explain how the transport properties are related to fundamental physical principles, where we will focus on transport in the absence of a magnetic field. Transport in metals is based on the assumption that the charge carriers are fermionic quasiparticles. The quasiparticles scatter on each other and on impurities or defects of the underlying lattice structure. This represents a complex dynamical system that can be treated in practice only by some simplifying assumptions. First, we consider only independent quasiparticle of the system and average overall possible scattering effects. For the latter, we introduce a static distribution by assuming that the relevant scattering processes happen only on time scales that are large in comparison to the tunneling process of the quasiparticle in the lattice. In other words, the probability for the quasiparticle to move from site $\mathbf{r}'$ to site $\mathbf{r}$ during the time t is $P_{\mathbf{rr}'}(t) = |\langle \mathbf{r}|\exp(-iHt)|\mathbf{r}'\rangle|^2$, where H is the hopping Hamiltonian. Second, if we assume that $P_{\mathbf{rr}'}(t)$ describes diffusion, we can obtain the mean square displacement with respect to $\mathbf{r}' = 0$ from the diffusion equation

$$\langle r_k^2 \rangle = \sum_{\mathbf{r}} r_k^2 P_{\mathbf{r},0}(t) = Dt \tag{32.1}$$

Using the Green's function $G_{\mathbf{rr}'}(z) = (H - z)^{-1}_{\mathbf{rr}'}$, we obtain for large distances $|\mathbf{r} - \mathbf{r}'|$ and $\epsilon \sim 0$

$$\int_0^\infty P_{\mathbf{rr}'}(t)e^{-\epsilon t}dt \sim \int_{E_0}^{E_F} \langle G_{\mathbf{rr}'}(E+i\epsilon)G_{\mathbf{r}'\mathbf{r}}(E-i\epsilon)\rangle_d \, dE \tag{32.2}$$

where $\langle \ldots \rangle_d$ is the average with respect to disorder that is causing scattering and E_0 is the lower band edge. Then, we get from Equation 32.2 for the diffusion coefficient at energy E

$$D(E) \sim \lim_{\epsilon \to 0} \epsilon^2 \sum_{\mathbf{r}} r_k^2 \langle G_{\mathbf{r}0}(E+i\epsilon)G_{0\mathbf{r}}(E-i\epsilon)\rangle_d \tag{32.3}$$

with $D = \int_{E_0}^{E_F} D(E)dE$ in Equation 32.1. For transport in graphene at low temperatures we need the diffusion coefficient only at the Fermi energy E_F.

A quantum approach to transport starts from a Hamiltonian H (here for independent electrons) and the corresponding current operator, expressed by the commutator

$$j_k = -ie[H, r_k] \tag{32.4}$$

where r_k is a component of the position operator of the electron. The average current, induced by a weak external electric field E, is obtained in terms of linear response as Ohm's law

$$\langle j_k \rangle = \sigma_{kl} E_l \tag{32.5}$$

with conductivity σ_{kl}. The general form of the conductivity in the Kubo formalism can be expressed as a product of one-particle Green's functions $G(z)$ at different energies z [6]. In the following, we exclude an external magnetic field. This leads to a vanishing Hall conductivity $\sigma_{kl} = 0$ for $k \neq l$. Then we can distinguish transport in a field, which is constant in time, described by the DC conductivity σ^{DC}, and transport in an oscillating field $E(t) = E_0 \cos \omega t$, described by the optical or AC conductivity σ^{AC}. These two types of conductivities are discussed briefly in the remainder of this chapter.

32.1.1 DC Conductivity

The DC conductivity σ at temperature $T \sim 0$ can be calculated either from D via the Einstein relation as $\sigma \propto \rho(E_F)D(E_F)$ with the density of states ρ, or from linear response theory via the Kubo formula [3,7]

$$\sigma_{DC} = -\frac{e^2}{h} \lim_{\omega,\delta \to 0} \omega^2 \times \sum_{\mathbf{r}} r_k^2 Tr_2 \langle G_{\mathbf{r}0}(E_F + \hbar w/2 + i\delta) G_{0\mathbf{r}}(E_F - \hbar\omega/2 - i\delta)\rangle_d \tag{32.6}$$

where Tr_2 is the trace with respect to the Pauli matrices. The latter expression is obviously related to the diffusion equation in Equation 32.3 by the analytic continuation $i\varepsilon \to \hbar\omega/2 + i\delta$. Thus, the main goal for the DC transport calculation is to evaluate the average product of the Green's functions in Equations 32.3 and 32.6.

32.1.2 AC Conductivity

In contrast to the DC conductivity, where the diffusion on arbitrarily large scales dominates the conductivity, the AC conductivity at frequency ω has the maximal length scale $L_\omega = v_F/\omega$, which is the wavelength with respect to the Fermi velocity v_F. This means that all physical processes appear as if the system were restricted to a finite length L_ω. This fact simplifies the transport calculations substantially.

The drawback of the finite cut-off L_ω, however, is that we cannot approximate the AC transport by the asymptotic behavior for large scales, as in Equation 32.2, but we need to include the details on finite scales. Then, the real part of the optical conductivity at temperature $T = 1/k_B\beta$ reads [6]

$$\sigma_{AC,\mu\mu}(\omega) = \frac{i}{\hbar} \int_{-\infty}^{\infty}\int_{-\infty}^{\infty} \langle Tr(j_\mu \delta(H - E - \hbar\omega) j_\mu \delta(H - E))\rangle_d \times \frac{1}{E - E' + \hbar\omega + i0^+} \times \frac{f_\beta(E') - f_\beta(E)}{E - E'} dE\, dE' \tag{32.7}$$

with the Fermi–Dirac distribution $f_\beta(E) = [1 + \exp(\beta(E - E_F))]^{-1}$ and the current operator $j_\mu = -ie[H, r_\mu]$. Moreover, we have used the trace Tr with respect to real space and spinor components. The Dirac delta function can be expressed by the Green's function as

$$\delta(H-z)=\frac{-i}{2\pi}[G(z+i\delta)-G(z-i\delta)^{-1}]$$

Equations 32.6 and 32.7 are the basic formulas that will be used for the subsequent discussion of electronic transport and optical properties in graphene.

Before we start our survey on the properties of graphene, we briefly recall what is known about the transport properties of conventional metals. In a good approximation, the DC and and the AC conductivities are described by the Drude formula

$$\bar{\sigma}_{DC}=\frac{ne^2\tau}{m},\quad \bar{\sigma}_{AC}(\omega)=\frac{\bar{\sigma}_{DC}}{1-i\omega\tau} \tag{32.8}$$

where τ is the scattering time, m is the quasiparticle mass, and n the charge density. These parameters are given as model parameters for the specific material. These classical approximations are not valid in the case of graphene, though, neither for monolayer graphene (MLG) nor for bilayer graphene (BLG). Hwang et al. [8] applied the Boltzmann approach with a sophisticated approach to the effective scattering time. It reproduces the experimental results at high charge densities (far away from the Dirac node) but also gives $\sigma_{\min}=0$, like the conventional Boltzmann approach.

There are only two parameters, besides the frequency ω the particle mass m, namely the scattering time τ (or the related scattering rate $\eta=\hbar/\tau$) and the carrier density n which determine the transport properties. This is also the case for graphene, as we will explain in the following. For this discussion the expressions in Equations 32.3, 32.6, and 32.7 are the fundamental quantities for the subsequent discussion of MLG and BLG. Here it should also be mentioned that in MLG the charge density n is proportional to E_F^2, in contrast to the linear relation in conventional metals. This is a consequence of the linear density of states.

32.2 BASIC EXPERIMENTAL FACTS

Before we embark into the theory of transport in graphene, we briefly summarize the experimental observations that are relevant for the subsequent theoretical part.

Already the first experiments on graphene by Novoselov et al. [9] and Zhang et al. [10] revealed very characteristic transport properties in graphene. Graphene as well as a stack of two graphene sheets (graphene bilayer) are semimetals with remarkably good conducting properties [9–11]. These materials have been experimentally realized with external gates, which allow for a continuous change of the charge carrier density. It was found that the longitudinal conductivity changes linearly as a function of charge density with a negative slope for holes and a positive slope for electrons, showing a characteristic V-shape behavior. Moreover, there is a minimal conductivity $\sigma_{\min}$ near the charge neutrality point. The latter has attracted some attention because it is unexpected in terms of the classical Boltzmann approach, and its value seems to be quite robust with respect to sample quality and temperature [11–14]. More recent experiments by the group of E. Andrei [15] on suspended graphene, however, indicated that below $T\approx 150$ K the minimal conductivity decreases linearly with decreasing T and reaches the extrapolated value $\sigma_{\min}\approx 2e^2/h$ at $T=0$. A similar result was found by Danneau et al. [16]. This clearly indicates that the main mechanism of transport in graphene at the NP is diffusion, possibly with a crossover to ballistic transport due to a very large mean-free path $L_s=v_F\tau$ of several hundred nanometers. Away from the charge neutrality point, the linear behavior has not always been observed but a crossover to a sublinear behavior for decreasing temperatures [15].

32.2.1 Role of Disorder

Disorder plays an important role in the physics of graphene. First of all, a two-dimensional (2D) lattice is thermodynamically unstable. It is known that this is the origin of the strong corrugations in graphene in the form of ripples. Another source of disorder are impurities in the substrate, which probably affect the transport properties substantially. Recent experiments on suspended graphene and with clean substrates have been able to eliminate this type of disorder. Experimental evidence of strong effects of disorder comes from the observation of puddles of electrons and holes at the charge neutrality point [17]. Experiments with hydrogenated graphene (graphane), where disorder is added by an inhomogeneous coverage with hydrogen atoms, leads to the formation of localized states which causes a nonmetallic behavior characterized by a variable-range hopping conductivity [18].

32.2.2 Role of Electron-Electron Interaction

There is no clear evidence for a substantial effect of electron–electron interaction on transport properties. Coulomb interaction renormalized the Fermi velocity logarithmically near the Dirac point. But this only weakly affects the transport because the Fermi velocity drops out of the Kubo formula near the Dirac node. This is also supported by theoretical findings, based on perturbative renormalization group calculations [19–22], that Coulomb interaction provides only a correction of 1%–2% for the optical conductivity [23]. This is in good agreement with experiments on the optical transparency of graphene [24,25].

32.2.3 Role of Electron-Phonon Interaction

Although there is a remarkable electron–phonon interaction in graphene [26–28], its effect on transport properties has not been investigated in detail. Some experimental findings of a gap opening were associated with electron–phonon interaction [29] but in most samples the conductivity is explained by noninteracting particles. The optical conductivity might be affected by the electron–phonon interaction of gated graphene before interband scattering can dominate transport (i.e., when the frequency ω is less than $E_F/\hbar$) [25].

TABLE 32.1
Measured Values of the Scattering Time τ and the Fermi Energy in Graphene and Related Quantities

Quantity	Relation	Measured Values in Reference 32	Typical Values [31]
Scattering time τ	–	0.36–1.08 10^{-14} s	10^{-14}–10^{-12} s
Scattering rate η	$\hbar/\tau$	6–18 meV	0.7–70 meV
Scattering length Ls	$v_F\tau$	40–100 nm	10–100 nm
Diffusion coefficient D	$v_F^2\tau/2$	18–50 cm²/s	50–5000 cm²/s
Fermi energy E_F	–	−200–0 meV	−10–10 meV

An important question is how τ depends on the Fermi energy E_F. Its frequency dependence was measured as $\tau = 10^{-14}\ldots10^{-12}$ s [12] and was almost constant $\tau \approx 10^{-14}$ s in Reference 30. In Table 32.1, we have collected some measured values of the scattering time τ and the Fermi energy E_F in graphene. The corresponding values of the scattering rate and the scattering length are calculated. The diffusion coefficient is calculated from its weak-localization form. It should be noticed that this is only a rough estimate for D, as we explain in Section 32.4.

The experimentally measured DC conductivity as a function of the (electron or hole) charge density n is well explained by the empirical formula

$$\sigma_{DC}(n) = \sigma_{\min} + e\mu n \tag{32.9}$$

μ is the mobility, which is related to the scattering time by $\mu = ev_F^2\tau/E_F$. Comparing this expression with the Drude formula in Equation 32.8, we observe that in the latter $\sigma_{\min} = 0$ and the mass is replaced by $m \to E_F/v_F^2$.

32.3 MODELS FOR TRANSPORT IN GRAPHENE

In order to calculate the conductivities of Equations 32.6 and 32.7 we must specify the Hamiltonians for MLG and BLG, where we focus on the low-energy properties near the nodes of neutral graphene. An important aspect is to take into account random scattering caused by ripples and impurities. Moreover, a random gap can appear due to local impurities. For instance, in the case of MLG such local fluctuations appear in the coverage of MLG by additional non-carbon atoms [18,31]. In the case of BLG with a dual gate [32,33], the random gap is caused by the fact that the graphene sheets are not planar but create ripples [2,34,35]. As a result, electrons experience a randomly varying gap along each graphene sheet.

The two bands in MLG and the two low-energy bands in BLG represent a spinor-1/2 wave function. This allows us to expand the corresponding Hamiltonian $H = H_0 + V$ in terms of Pauli matrices σ_j as

$$H_0 = h_1\sigma_1 + h_2\sigma_2, \quad V = \sum_{j=0}^{3} v_j\sigma_j \tag{32.10}$$

Near each node the coefficients h_j read in low-energy approximation [36]

$$h_j = p_j(\text{MLG}), \quad h_1 = p_1^2 - p_2^2, \quad h_2 = -2p_1p_2(\text{BLG}) \tag{32.11}$$

with momentum p_j. This is a momentum expansion of the tight-binding Hamiltonians around the nodes K and K′ (Figure 32.1).

For the randomness it is assumed here that scattering appears only at small momentum such that intervalley scattering, which requires a large momentum at least near the nodes, is not relevant and can be treated as a perturbation. Then each valley contributes separately to the density of states and to the conductivity, and the contribution of the two valleys is additive. This allows us to consider the low-energy Hamiltonian in Equations 32.10 and 32.11 for each valley separately, even in the presence of randomness. Within this approximation the gap term $v_3 \equiv m$ is a random variable. The following analytic calculations will be based entirely on the Hamiltonian of Equations 32.10 and 32.11. In particular, the average Hamiltonian $\langle H\rangle$ can be diagonalized by Fourier transformation and becomes a 2D Dirac Hamiltonian for MLG

$$H_M \equiv \langle H\rangle = p_1\sigma_1 + p_2\sigma_2 + m\sigma_3 \tag{32.12}$$

with eigenvalues $E_p = \pm\sqrt{m^2 + p^2}$. For BLG, the average Hamiltonian is

$$H_B \equiv \langle H\rangle = \left(p_1^2 - p_2^2\right)\sigma_1 + 2p_1p_2\sigma_2 + m\sigma_3 \tag{32.13}$$

with eigenvalues $E_p = \pm\sqrt{m^2 + p^4}$. In order to apply results from these calculations to the real materials we must include a degeneracy factor $\gamma = 4$, referring to the two valleys K and K′ and the twofold spin degeneracy of the electrons.

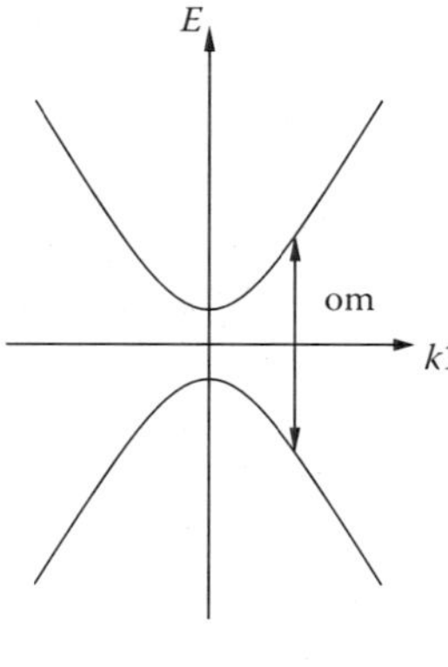

FIGURE 32.1 Schematic picture of the creation of an electron–hole pair in gapped mono- or BLG by the absorption of a photon with energy $\hbar\omega$. For this process the photon energy must be larger than the bandgap $\Delta = 2m$.

For these Hamiltonians we obtain the corresponding current matrix elements, which we need for the evaluation of the conductivity. They are commutators with respect to the position **r** and read for MLG $j_\mu = -ie[H_M, r_\mu] = e\sigma_\mu$ and for BLG

$$j_1 = -ie[H_B, r_1] = 2e(p_1\sigma_1 + p_2\sigma_2),$$
$$j_2 = -ie[H_B, r_2] = 2e(-p_2\sigma_1 + p_1\sigma_2) \tag{32.14}$$

In MLG the current is the same for all momenta, whereas it is linear in the momenta for BLG. This indicates that the low-energy spectrum reveals a distinct characterization with respect to the number of graphene layers.

32.4 DC CONDUCTIVITY

Transport in graphene, like in other materials, is based on the diffusion of quasiparticles. However, the situation in graphene is more subtle than in conventional metals. First of all, graphene is a 2D structure, where the scaling theory of Anderson localization for conventional metals predicts the localization of quantum states for any amount of disorder [37]. Surprisingly, this has not been confirmed by the experiments. Despite remarkable disorder effects from the substrate and from ripples in the graphene sheet, the metallic behavior is always dominant. Doping of graphene with hydrogen is one of the few exceptions, in which the material becomes nonmetallic. However, this is not due to Anderson localization but it is caused by sublattice symmetry breaking, which generates a small gap of a few meV up to 1 eV [38–40] (cf. Table 32.2). The reason for the absence of Anderson localization is that graphene, in contrast to a conventional metal, has two complementary bands that are connected by a particle–hole symmetry. This allows for Klein tunneling, an effect that suppresses potential scattering substantially. The particle–hole symmetry implies a chiral symmetry for the two-particle Green's function. This can be spontaneously broken by random scattering, which is indicated by a nonzero scattering rate η. Therefore, we can distinguish three different regimes: a ballistic regime with no scattering except for the sample boundaries, a diffusive regime for weak scattering, and Anderson localization for very strong scattering. Moreover, a random gap can be opened. This leads to an insulating regime for weak scattering and a metallic regime for stronger scattering and eventually to Anderson localization for very strong scattering.

TABLE 32.2
Measured Values of the Gap, Generated by Different Methods

Experimental Method	Measured Gap Value (eV)	Reference
Epitaxially grown graphene on SiC substrate	0.26	[38]
Hydrogenation of graphene	1.0	[39]
Graphene on boron nitride substrate	0.016	[40]

32.4.1 Ballistic Regime at the Dirac Node

Equation 32.3 defines the diffusion coefficient in our two-band system. In the case without disorder, the correlation of Green's functions gives for the Hamiltonian in Equation 32.12

$$D = \lim_{\epsilon \to 0} \frac{1}{4\pi}\left(1 + \frac{1+\zeta^2}{\zeta}\arctan\zeta\right)\left(\zeta = \frac{E}{\epsilon}\right) \tag{32.15}$$

This result is surprising in that it indicates diffusion at the Dirac node $E = 0$ with the diffusion coefficient $D = 1/2\pi$ even without random scattering. Away from the Dirac node ($E \neq 0$), however, D diverges, reflecting that there is no diffusion but ballistic propagation. This is accidental not only for the case for MLG, since we also get a finite D at the nodes for the BLG. This behavior reveals a characteristic transport feature at the nodes of a two-band system, which is caused by quantum fluctuations. In other words, quantum fluctuations play an important role in $2D$ transport at the points of band degeneracy.

32.4.2 Diffusive Regime with Weak Disorder Scattering

Typical disorder is due to a randomly fluctuating gap. The result of a weak scattering expansion for the correlation function in Equation 32.6 is the scaling relation [41]

$$\sum_{\mathbf{r}} r_k^2 \mathrm{Tr}_2 \langle G_{\mathbf{r}0}(E_F + i\delta) G_{0\mathbf{r}}(E_F - i\delta)\rangle_d$$
$$= \frac{1}{\delta^2}\sum_{\mathbf{r}} r_k^2 \mathrm{Tr}_2[G_{0,\mathbf{r}}(E_F + i\eta)G_{0,-\mathbf{r}}(E_F - i\eta)] \tag{32.16}$$

This relation is very important for the evaluation of the diffusion coefficient in Equation 32.3 and the DC conductivity in Equation 32.6, since it enables us to perform the averaging over disorder. The latter results in the substitution of δ by the scattering rate η and the prefactor δ^{-2}. The scattering rate is obtained from disorder distribution and can either be calculated in self-consistent Born approximation [41,42] or measured in an experiment. Then the right-hand side of the relation can be evaluated in Fourier representation and gives

$$-\frac{1}{2\pi\delta^2}\left(1 + \frac{1+\zeta^2}{\zeta}\arctan\zeta\right)\left(\zeta = \frac{E_F}{\eta}\right) \tag{32.17}$$

After inserting this expression in Equation 32.6 we obtain for the DC conductivity of MLG the simple expression

$$\sigma_{DC} = \frac{2e^2}{\pi h}\left(1 + \frac{1+\zeta^2}{\zeta}\arctan\zeta\right) \tag{32.18}$$

where the fourfold spin and valley degeneracy has been implemented. This result is remarkable, since it is in good agreement with experimental observation of a V-shape conductivity. If we compare it with the empirical formula in Equation 32.9, we can identify the minimal conductivity $\sigma_{\min} = 4e^2/\pi h$ at the Dirac node and the monotonically increasing behavior

$$\tilde{\sigma} = \frac{2e^2}{\pi h}\left(-1 + \frac{1+\zeta^2}{\zeta}\arctan\zeta\right) \sim 2\frac{e^2}{\pi h}\begin{cases}\zeta^2 & \text{for } \zeta \sim 0 \\ \pi\zeta/2 & \text{for } \zeta \sim \infty\end{cases} \tag{32.19}$$

away from the Dirac node. The minimal conductivity is independent of the scattering rate, which reflects the fact that quantum fluctuations are dominant. The behavior for small ζ can be expressed by the scattering time and leads to the expression

$$\tilde{\sigma} \sim 2\frac{e^2}{\pi h}\frac{\bar{E}_F^2\tau^2}{\hbar^2} \tag{32.20}$$

On the other hand, the linear behavior of $\tilde{\sigma}$ far away from the Dirac node agrees with the classical Boltzmann calculation. This is indicative for the fact that the scattering to the second band is irrelevant in this regime.

The calculation for the DC conductivity of MLG is also applicable to BLG. The main difference is that the minimal conductivity appears with an additional factor 2 [41], which is a consequence of the parabolic spectrum near the node. We will see later that this factor 2 is also crucial for the AC conductivity. Like for the minimal conductivity, it is not due to independent currents through the two layers but due to the spectral curvature near the nodes.

32.4.3 Anderson Localization

Is it realistic to see Anderson localization, that is, the absence of diffusion, in graphene? In a one-band system it is always present in 2D systems, according to the scaling theory [37]. In a two-band system this is less clear. For weak scattering we have seen that diffusion prevails due to spontaneous chiral symmetry breaking. For very strong scattering, when the scattering rate η exceeds the band width we have found a transition to Anderson localization [43]. However, such strong scattering rates are rather unrealistic in graphene, unless disorder is created intentionally (e.g., removing carbon atoms by a bombardment with ions). The measured scattering rates are at least two orders smaller than this value (cf. Table 32.2).

32.5 AC Conductivity for Very Weak Scattering and Thermal Fluctuations

The Kubo formula in Equation 32.7 is now employed to calculate the AC (or optical) conductivity in MLG and BLG. We assume here that $\hbar\omega \gg \eta$, such that the relevant length scale is rather the effective wavelength v_F/ω than the scattering length $v_F\tau$. This implies that disorder scattering is not important and can be neglected. We can then treat the current matrix elements of Equation 32.14 in Fourier representation with respect to energy eigenstates $|\pm E\rangle$. Of particular interest is the matrix element that describes interband scattering, for which we obtain after the integration over the circular Fermi surface with $E_F^2 \geq m^2$

$$\int_0^{2\pi} |\langle E|\sigma_1|-E\rangle|^2\, d\varphi = \pi\left(\frac{1+m^2}{E^2}\right) \tag{32.21}$$

in the case of MLG and

$$\int_0^{2\pi} |\langle E|k_x\sigma_1 + k_y\sigma_2|-E\rangle|^2\, d\varphi = \pi\sqrt{E^2-m^2}\left(\frac{1+m^2}{E^2}\right) \tag{32.22}$$

in the case of the BLG. The integrated current matrix elements behave quite differently for MLG and BLG. In particular, without gap (i.e., $m = 0$) the expression is either constant (MLG) or increases linearly with energy (BLG).

Using the Kubo formula (32.7) and the expressions of the angular integrated current matrix elements in Equations 32.21 and 32.22, the integration over E gives for $\omega^2 \geq \Delta^2$, where $\Delta = 2m$ is the gap, the expression

$$\sigma'(\omega) = \gamma\frac{\pi e^2}{8h}\left[1 + \frac{\Delta^2}{\omega^2}\right]\left[f_\beta\left(\frac{-\hbar\omega}{2}\right) - f_\beta\left(\frac{\hbar\omega}{2}\right)\right] \tag{32.23}$$

for the real part of the AC conductivity. γ is a degeneracy with $\gamma = 4$ (MLG) and $\gamma = 8$ (BLG). Thus, the conductivities of MLG and BLG agree up to a factor 2. The additive correction due to the gap parameter Δ^2 decays like ω^{-2}, which resembles the intraband scattering of the Drude behavior in Equation 32.8.

In the special case of the gapless case $m = 0$ and $T \sim 0$ we get for the AC conductivity $\sigma_{AC} = \sigma' + i\sigma''$ the real part

$$\sigma'(\omega) = \gamma\frac{\pi e^2}{8h}\Theta(\hbar\omega - 2E_F) \tag{32.24}$$

and the imaginary part

$$\sigma''(\omega) = \gamma\frac{e^2}{16h}\left[4\frac{E_F}{\hbar\omega} - \log\left(\left|\frac{2E_F + \hbar\omega}{2E_F - \hbar\omega}\right|\right)\right] \tag{32.25}$$

The first term resembles the Drude result in Equation 32.8, because this is a contribution from intraband scattering [44]. It should be noticed that $\sigma''(\omega)$ vanishes for $\omega \gg E_F$. If we take the full band structure of the honeycomb lattice into account, the AC conductivity deviates from the low-energy result of Equation 32.23. This is shown in Figure 32.2, where $\sigma'(\omega)$ versus the frequency is plotted. In particular, there is a

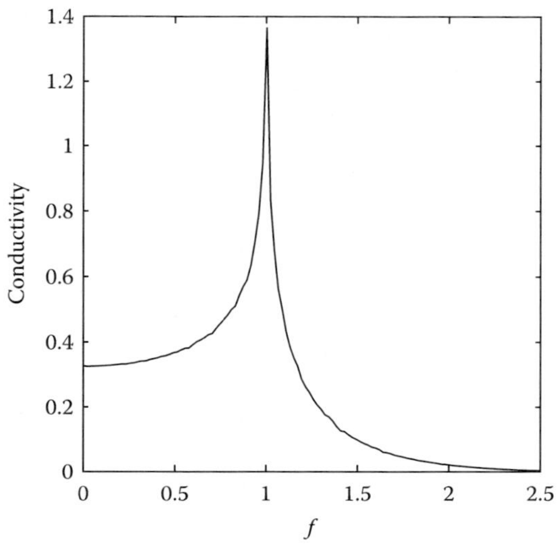

FIGURE 32.2 Real part of the AC conductivity at $T = 0$ as a function of the rescaled frequency $f = \hbar\omega/t$ for the honeycomb lattice, where $t = 2.8$ eV is the bandwidth. There is a characteristic peak due to a van Hove singularity.

characteristic conductivity maximum at the van Hove singularity, where the Fermi surfaces of the two nodes merge.

The AC conductivity provides the dielectric coefficient $\epsilon(\omega)$ via the relation [5]

$$\epsilon(\omega) = 1 + \frac{4\pi i}{\omega}\sigma(\omega) \tag{32.26}$$

such that the complex dielectric coefficient reads $\epsilon = \epsilon' + i\epsilon''$ with

$$\epsilon' = 1 - \frac{4\pi}{\omega}\sigma'', \quad \epsilon'' = \frac{4\pi}{\omega}\sigma' \tag{32.27}$$

This is the dielectric function for wavevector $\mathbf{q} = 0$: $\epsilon(\omega) = \epsilon(\mathbf{q} = 0,\omega)$. The dielectric function we need later for the description of plasmons in Section 32.6.

According to the Fresnel equations for thin layers [24,45], the optical transmittance T is directly linked to the AC conductivity through the relation

$$T \approx \frac{1}{(1 + 2\pi\sigma'(\omega)/c)^2} \tag{32.28}$$

Using the result in Equation 32.24, the transmittance becomes

$$T \approx 1 - \pi\alpha \ (MLG), \quad T \approx 1 - 2\pi\alpha \ (BLG) \tag{32.29}$$

where $\alpha = e^2/\hbar c \approx 1/137$ is the fine structure constant. This behavior was also observed in several experiments over a wide range of frequencies [24,25,45].

32.6 PLASMONS

Now we consider the electron gas in graphene, which is subject to an external potential $V_i(\mathbf{q},\omega)$. The response of the electron gas to $V_i(\mathbf{q},\omega)$ is a screening potential $V_s(\mathbf{q},\omega)$, which is created by the rearrangement of the electrons due to the external potential. Therefore, the total potential, acting on the electrons, is given by

$$V(\mathbf{q},\omega) = V_i(\mathbf{q},\omega) + V_s(\mathbf{q},\omega) \tag{32.30}$$

V_s can be evaluated self-consistently [46] and is expressed via the dielectric function $\varepsilon(\mathbf{q},\omega)$. Then the total potential reads [47]

$$V(\mathbf{q},\omega) = \frac{1}{\varepsilon(\mathbf{q},\omega)} V_i(\mathbf{q},\omega) \tag{32.31}$$

The dielectric function can be calculated from the Lindhard formula. Assuming that the wave length of the electromagnetic wave is much larger than the lattice spacing, the longitudinal component reads [46]

$$\epsilon(\mathbf{q},\omega) = 1 - \frac{2\pi e^2}{q}\chi(\mathbf{q},\omega) \tag{32.32}$$

where

$$\chi(\mathbf{q},\omega) = \lim_{\delta\to 0}\int_{\mathbf{k}} \sum_{l,l'=1,2} \frac{f_\beta(E_{\mathbf{k},l}) - f_\beta(E_{\mathbf{k+q},l'})}{E_{\mathbf{k},l} - E_{\mathbf{k+q},l'} + \hbar\omega + i\hbar\delta} |\langle \mathbf{k+q}, l'| e^{i\mathbf{q}\cdot\mathbf{r}} | k, l\rangle|^2 \tag{32.33}$$

Poles in ω of the inverse longitudinal dielectric function $1/\epsilon(\mathbf{q}, \omega)$ for a given wave vector $\mathbf{q}$ correspond to collective excitations of electrons, which are called plasmons. These poles are located either on the real axis or in the complex plane away from the real axis. The latter can be considered as damped plasmons, generated by scattering with individual electrons. An imaginary term can appear in the integral $\chi(\mathbf{q},\omega)$ of Equation 32.33 if the denominator $E_\mathbf{k} - sE_{\mathbf{k+q}} + \omega$ vanishes inside the Brillouin zone. In other words, if $(\mathbf{q}, \omega)$ is inside the band that is produced by the spectrum of the electrons, that is, where an electronic wave vector $\mathbf{k}$ exists that satisfies

$$E_{\mathbf{k+q}} - sE_\mathbf{k} = \omega \ (s = \pm 1) \tag{32.34}$$

scattering between plasmons and electrons is possible and will lead to damping of plasmons. On the other hand, outside the spectrum of electrons (i.e., when there is no electron wave vector k which solves Equation 32.34) we obtain undamped plasmons.

Using the Dirac Hamiltonian from Equation 32.12 as a low-energy approximation enables us to calculate the poles of the inverse dielectric function directly [48,49]. In this case, the plasmon dispersion follows a square root

$$\omega_P \sim cq^{1/2} \tag{32.35}$$

where the prefactor c is proportional to $\sqrt{E_F}$. The plasmon dispersion, on the other hand, depends on the spectral properties of the electrons. Therefore, deviations from Dirac cones may affect them. This can lead to a stronger damping of the electrons, since electronic excitations require lower energies on the honeycomb lattice in comparison with the linearized (Dirac) spectrum [50]. For this purpose, we plot the loss function [51]

$$Im\left(\frac{1}{(\mathbf{q},\omega)}\right) = \frac{-\epsilon''}{\epsilon'^2 + \epsilon''^2} \tag{32.36}$$

in Figure 32.3. The peak strength varies with the momentum. In particular, if the pole is away from the real axis it becomes a Lorentzian of width ϵ''. Thus, ϵ'' is a measure for damping

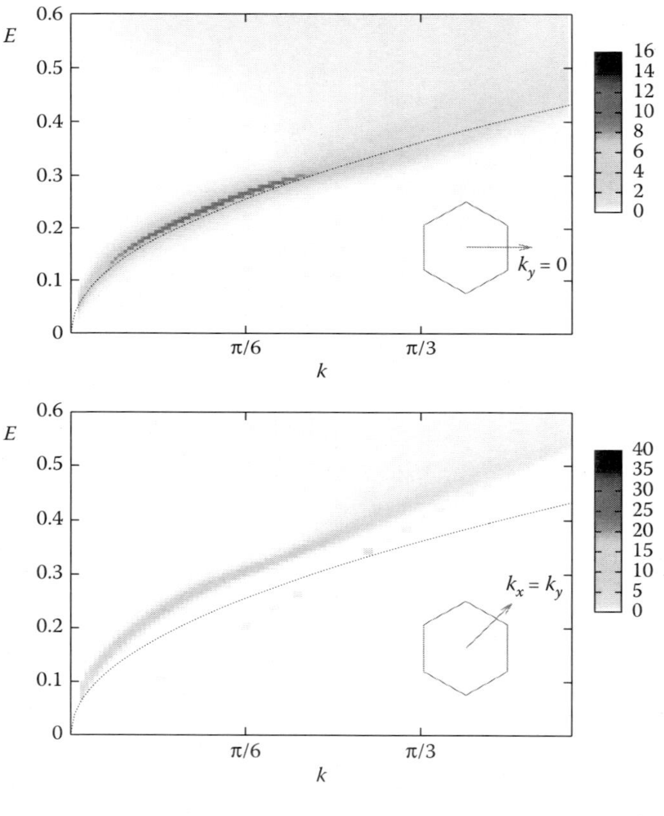

FIGURE 32.3 Plasmon dispersion $E = \hbar\omega/3t$ as a function of $k = q_x d$ at the Fermi energy $Ef = 0.25t$ on the honeycomb lattice $d = 1.42$ Å is the lattice constant of graphene. To demonstrate the anisotropy, two different directions of the **q** vector with $q_y = 0$ (upper panel) and $q_y = qx$ (lower panel) have been plotted. The isotropic square-root behavior of the Dirac case is also shown as a dashed curve. (From A. Hill, S.A. Mikhailov, and K. Ziegler, *EPL* 87, 2009, p. 27005.)

by electron scattering. Plasmons on a honeycomb lattice with an additional next-nearest neighbor hopping have also been studied [52]. Although the additional hopping break the particle–hole symmetry of the two-band system, there is no drastic effect on the plasmon dispersion.

The two panels of Figure 32.3 demonstrate the anisotropy of the plasmon dispersion for electrons on the honeycomb lattice. There is a substantial deviation from the isotropic plasmon dispersion of the Dirac Hamiltonian.

The square root behavior of the plasmon dispersion, on the other hand, is quite general for a 2D electron gas. For a conventional 2D electron gas with parabolic electron dispersion with effective electron mass m the plasmon dispersion reads [53]

$$\omega_p(q) = \frac{\sqrt{(4a + v_F^2 q)q(q^4 v_F^4 + 4q^3 v_F^2 a + 16k_F^2 a^2)}(v_F^2 q + 2a)}{4(4a + v_F^2 q)ak_F} \tag{32.37}$$

with $a = 2ne^2/m$ and the Fermi velocity v_F. Expansion for small q then gives

$$\omega_p^2 \approx aq + \frac{3}{4}v_F^2 q^2 \tag{32.38}$$

Thus, plasmons in graphene have similar properties as plasmons in a conventional 2D electron gas.

32.7 SUMMARY OF THE THEORETICAL RESULTS

Transport in graphene is remarkably different from transport in conventional metals. The main reason is the scattering between the valence and the conduction band, which leads to Klein tunneling. This has several consequences for the electronic and the optical properties in MLG as well as in BLG. First of all, it creates a minimal conductivity for the neutral system ($E_F = 0$), which can be explained by quantum fluctuations of the system without electric charges. The other distinctive feature is the constant AC conductivity over a wide range of frequencies. This was also observed in a number of experiments for frequencies ranging from infrared to visible light [24,25]. Another characteristic feature is that the AC conductivity of BLG is twice as large as the AC conductivity of MLG. This was also observed experimentally with high accuracy [24]. A third important aspect is that graphene defies Anderson localization, which is possible only for very strong scattering. Therefore, diffusion is the main transport mechanism in graphene. We can conclude that DC transport behavior of graphene becomes more conventional and Drude-like as we go deeper in the valence band (for holes) or in the conduction band (for electrons), because interband scattering becomes less important. On the other hand, exactly at the nodes of the bands the transport behavior is quite special, because the Fermi surface shrinks to a point.

REFERENCES

1. K.S. Novoselov, A.K. Geim, S.V. Morozov, D. Jiang, Y. Zhang, S.V. Dubonos, I.V. Grigorieva, and A.A. Firsov, *Science* 306, 2004, p. 666.
2. A.H. Castro Neto, F. Guinea, N.M.R. Peres, K.S. Novoselov, and A.K. Geim, *Rev. Mod. Phys.* 81, 2009, p. 109.
3. D.S.L. Abergel, V. Apalkov, J. Berashevich, K. Ziegler, and T. Chakraborty, *Adv. Phys.* 59, 2010, p. 261.
4. E.Y. Andrei, G. Li, and X. Du, *Rep. Prog. Phys.* 75, 2012, p. 056501.
5. N.W. Ashcroft and N.D. Mermin, *Solid State Physics,* Saunders College Publishing, New York, 1976.
6. K. Ziegler, *Phys. Rev. Lett.* 97, 2006, p. 266802.
7. K. Ziegler, *Phys. Rev. B* 78, 2008, p. 125401.
8. E.H. Hwang, S. Adam, and S. Das Sarma, *Phys. Rev. Lett.* 98, 2007, p. 186806.
9. K.S. Novoselov, A.K. Geim, S.V. Morozov, D. Jiang, M.I. Katsnelsnon, I.V. Grigorieva, S.V. Dubonos, and A.A. Firsov, *Nature* 438, 2005, p. 197.
10. Y. Zhang, Y.-W. Tan, H.L. Stormer, and P. Kim, *Nature* 438, 2005, p. 201.
11. A.K. Geim and K.S. Novoselov, *Nat. Mater.* 6, 2007, p. 183.
12. Y.-W. Tan, Y. Zhang, K. Bolotin, Y. Zhao, S. Adam, E.H. Hwang, S. Das Sarma, H.L. Stormer, and P. Kim, *Phys. Rev. Lett.* 99, 2007, p. 246803.
13. S.V. Morozov, K.S. Novoselov, M.I. Katsnelson, F. Schedin, D.C. Elias, J.A. Jaszczak, and A.K. Geim, *Phys. Rev. Lett.* 100, 2008, p. 016602.
14. J.H. Chen, C. Jang, S. Adam, M.S. Fuhrer, E.D. Williams, and M. Ishigami, *Nat. Phys.* 4, 2008, p. 377.
15. X. Du, I. Skachko, A. Barker, and E.Y. Andrei, *Nat. Nanotech.* 3, 2008, p. 491.
16. R. Danneau, F. Wu, M.F. Craciun, S. Russo, M.Y. Tomi, J. Salmilehto, A.F. Morpurgo, and P.J. Hakonen, *J. Low Temp. Phys.* 153, 2008, p. 374.
17. J. Martin, N. Akerman, G. Ulbricht, T. Lohmann, J.H. Smet, K. von Klitzing, and A. Yacoby, *Nat. Phys.* 4, 2008, p. 144.
18. D.C. Elias et al. *Science* 323, 2009, p. 610.
19. I.F. Herbut, V. Juricic, and O. Vafek, *Phys. Rev. Lett.* 100, 2008, p. 046403.
20. E.G. Mishchenko, *Europhys. Lett.* 83, 2008, p. 17005.
21. D.E. Sheehy and J. Schmalian, *Phys. Rev. Lett.* 99, 2007, p. 226803.
22. A. Sinner and K. Ziegler, *Phys. Rev. B* 82, 2010, p. 165453.
23. D.E. Sheehy and J. Schmalian, *Phys. Rev. B* 80, 2009, p. 193411.
24. R.R. Nair, P. Blake, A.N. Grigorenko, K.S. Novoselov, T.J. Booth, T. Stauber, N.M.R. Peres, and A.K. Geim, *Science* 320, 2008, p. 1308.
25. Z.Q. Li, E.A. Henriksen, Z. Jiang, Z. Hao, M.C. Martin, P. Kim, H.L. Stormer, and D.N. Basov, *Nat. Phys.* 4, 2008, p. 532.
26. J. Yan, Y. Zhang, P. Kim, and A. Pinczuk, *Phys. Rev. Lett.* 98, 2007, p. 166802.
27. J. Yan, E.A. Henriksen, P. Kim, and A. Pinczuk, *Phys. Rev. Lett.* 101, 2008, p. 136804.
28. T.M.G. Mohiuddin et al. *Phys. Rev. B* 79, 2009, p. 205433.
29. Y. Zhang et al., *Nat. Phys.* 4, 2008, p. 627.
30. E. Pallecchi et al., *Phys. Rev. B* 83, 2011, p. 125408.
31. A. Bostwick, J.L. McChesney, K.V. Emtsev, T. Seyller, K. Horn, S.D. Kevan, and E. Rotenberg, *Phys. Rev. Lett.* 103, 2009, p. 056404.
32. T. Ohta, A. Bostwick, T. Seyller, K. Horn, and E. Rotenberg, *Science* 313, 2006, p. 951.
33. J.B. Oostinga, H.B. Heersche, X. Liu, A.F. Morpurgo, and L.M.K. Vandersypen, *Nat. Mat.* 7, 2008, p. 151.
34. S.V. Morozov et al., *Phys. Rev. Lett.* 97, 2006, p. 016801.
35. J.C. Meyer, A.K. Geim, M.I. Katsnelson, K.S. Novoselov, T.J. Booth, and S. Roth, *Nature* 446, 2007, p. 60.
36. E. McCann et al., *Phys. Rev. Lett.* 97, 2006, p. 146805.
37. E. Abrahams, P.W. Anderson, D.C. Licciardello and T.V. Ramakrishnan, *Phys. Rev. Lett.* 42, 1979, p. 673.
38. S.Y. Zhou, G.-H. Gweon, A.V. Fedorov, P.N. First, W.A. de Heer, D.-H. Lee, F. Guinea, A.H. Castro Neto, and A. Lanzara, *Nat. Mater.* 6, 2007, p. 770.
39. D. Haberer et al., *Nano Lett.* 10, 2010, p. 3360.
40. C.R. Woods et al., *Nat. Phys.* 10, 2014, p. 451.
41. K. Ziegler, *Phys. Rev. Lett.* 102, 2009, 126802; *Phys. Rev. B* 79, 2009, 195424.
42. N.H. Shon and T. Ando, *J. Phys. Soc. Jpn.* 67, 1998, p. 2421.
43. A. Hill and K. Ziegler, arxiv:1305.6901.
44. S.A. Mikhailov and K. Ziegler, *Phys. Rev. Lett.* 99, 2007, p. 016803.
45. A.B. Kuzmenko et al., *Phys. Rev. Lett.* 100, 2008, p. 117401.
46. H. Ehrenreich and M.H. Cohen, *Phys. Rev.* 115, 1959, p. 786.
47. G. Mahan, *Many Particle Physics*, Plenum Pr., New York, 1990.
48. E.H. Hwang and S. Das Sarma, *Phys. Rev. B* 75, 2007, p. 205418.
49. B. Wunsch, T. Stauber, F. Sols, and F. Guinea, *New J. Phys.* 8, 2006, p. 318.
50. A. Hill, S.A. Mikhailov, and K. Ziegler, *EPL* 87, 2009, p. 27005.
51. M. Polini, R. Asgari, G. Borghi, Y. Barlas, T. Pereg-Barnea, and A.H. MacDonald, *Phys. Rev. B* 77, 2008, p. 081411(R).
52. V. Kadirko, K. Ziegler, and E. Kogan, *Graphene* 2, 2013, p. 97.
53. F. Stern, *Phys. Rev. Lett.* 18, 1967, p. 546.

33 Graphene Geometric Diodes and Antennas for Terahertz Applications

Zixu Zhu, Saumil Joshi, Bradley Pelz, and Garret Moddel

CONTENTS

ABSTRACT

We have developed a new ultrafast graphene diode and demonstrated its operation at 28 THz. This diode relies on geometric asymmetry and the long charge mean-free path length in graphene to provide asymmetric electrical characteristics. Fabricated geometric diodes exhibit asymmetric direct current (DC)–voltage characteristics, which agree well with Monte Carlo simulations based on the Drude model. Since an applied gate voltage can control the charge carrier type (electrons or holes) and concentration, the polarity of the diode can be reversed. The planar structure of the geometric diode provides the low resistor capacitor (RC) time constant, on the order of 10^{-15} s, necessary for operation at optical frequencies. When coupled to an antenna, the diode induces rectification of high-frequency optical signals. We formed rectennas by coupling geometric diodes with bowtie antennas and verified optical rectification at 28 THz, using both metal and graphene antennas. Applications of geometric diodes coupled to bowtie antennas include detection of terahertz and optical waves, ultra-high-speed electronics, and optical power conversion.

33.1 INTRODUCTION

Graphene's unique electrical properties provide an advantage over other materials in various high-frequency optical applications, including photodetectors (Xia et al. 2009, Zhu et al. 2011), transparent electrodes in displays and photovoltaic modules (Bae 2010), and ultrafast lasers (Sun et al. 2010). Graphene is an excellent material for terahertz electronics because of its high carrier mobility at room temperature and frequency-independent absorption (Nair et al. 2008). Furthermore, high-quality graphene samples support plasma waves that are weakly damped (Liu et al. 2008, Bostwick et al. 2010) with a gate-tunable surface plasmon frequency (Jablan et al. 2009) up to the graphene optical phonon frequency of 48.3 THz, corresponding to an energy of 0.2 eV (Park et al. 2008). When the graphene device operating frequency is above 176 THz, corresponding to a wavelength of 1.7 μm, interband loss starts affecting the device performance (Jablan et al. 2009) (Table 33.1).

Graphene devices working at terahertz frequencies have been demonstrated experimentally (Tamagnone et al. 2012, Ju et al. 2011, Vicarelli et al. 2012). Tredicucci et al. have experimentally demonstrated a graphene transistor working at terahertz frequencies (Vicarelli et al. 2012). In this antenna-coupled graphene field-effect transistor (FET), the top gate is coupled to one arm of a terahertz metal bowtie antenna while the other arm of the antenna acts as the source terminal of the transistor. A second-order nonlinear response occurs when an oscillating terahertz field is applied between the gate and source terminals. The DC photovoltage output is proportional to the derivative of the channel conductivity with respect to gate voltage (Vicarelli et al. 2012). Periodic graphene ribbon structures have shown optical resonant behavior at 3 THz, with an absorption efficiency of 13%. The resonance peak can be controlled by changing the gate voltage or the ribbon period (Ju et al. 2011).

In this chapter, we demonstrate terahertz detectors using rectennas based on graphene diodes coupled to antennas working at 28 THz. Unlike the antenna-coupled FET described above, rectennas do not rely on a gate effect in graphene. They use bowtie antennas to absorb terahertz electromagnetic (EM) waves and

TABLE 33.1
Graphene Optical Devices and Graphene Optical Properties

Graphene photodetector	Xia et al. (2009), Zhu et al. (2011)
Graphene transparent electrode	Bae (2010)
Mode-locked ultrafast laser	Sun et al. (2010)
Gate-tunable surface plasmon frequency and interband loss of graphene	Jablan et al. (2009)
Graphene optical antenna at 28 THz	Zhu et al. (2013)
Optical phonon frequency of graphene	Park et al. (2008)

rectify the alternating current (AC) using an ultrafast diode and producing a DC output. Geometric diodes are designed to be used in terahertz and optical frequency rectenna systems.

Geometric diodes rely on physical asymmetry to provide diode behavior. Other devices using a geometric effect, such as gallium arsenide-based semiconductor devices, have been demonstrated previously (Song 2002). The device uses hot electrons as charge carriers and can work as a full wave rectifier (Song 2002). A graphene diode using geometric asymmetry in the form of an oblique gate over a graphene channel was proposed and simulated (Dragoman 2010).

We coined the term "geometric diode" for a graphene device that we proposed (Moddel 2009) and demonstrated operation at DC (Zhu et al. 2011, Moddel et al. 2012) and 28 THz (Zhu et al. 2013). It makes use of an asymmetric triangular-shaped graphene layer. Using a quantum approach, Dragoman has simulated similar structures (Dragoman and Dragoman 2013) and demonstrated their DC electrical characteristics (Dragoman 2013).

33.2 DEVICE PHYSICS

The principle of operation for the geometric diode exploits an asymmetric device geometry to control the charge movement in the ballistic transport regime. Imagine a conductive thin film patterned with an asymmetric shape on the size scale of the charge carrier mean-free path length (MFPL). The electron transport within such a film can be considered ballistic and is influenced by the shape of the device. Figure 33.1 shows a schematic diagram of the geometric diode. The inverse arrowhead shape of the diode determines the preferred direction for charge flow to be from left to right. Charges can be reflected from the sloped arrowhead border, and funneled from left to right (forward bias) through the constriction, called the neck. Conversely, charges moving from right to left (reverse bias) are largely blocked by the vertical side wall of the arrowhead. The charge flows more easily in the forward direction than in the reverse direction. Such a geometric effect leads to a current–voltage [$I(V)$] asymmetry and the device behaves as a diode. Simulation results show that increasing charge MFPL and reducing neck width (d_{neck}) increases the asymmetry of the $I(V)$ curve (Zhu et al. 2013, Dragoman and Dragoman 2013), as discussed in Section 33.7.

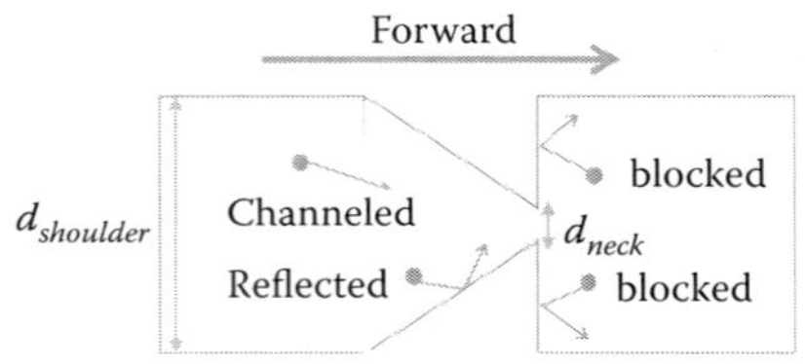

FIGURE 33.1 Inverse arrowhead geometric diode structure The width of the neck (d_{neck}) is on the order of the charge carrier MFPL in the material. Charge carriers reflect off the interior boundaries of the geometric diode. On the left side of the neck, electrons or holes can either channel directly through the neck region or reflect at the tapering edge and keep moving forward. On the right side of the neck, the vertical edge blocks a majority of the electrons or holes. We call the inverse arrowhead direction from left to right the forward direction for carrier transport.

The requirements for the neck width are determined by the material MFPL and the fabrication constraints. To obtain a large geometric effect, the carrier MFPL in the thin-film material has to be sufficiently large compared to the neck width. The MFPL of graphene is long, due to its unique atomic structure and high electron mobility (Castro Neto et al. 2009). Investigators have reported electron MFFLs in graphene of up to a few micrometers (Bolotina et al. 2008, Mayorov et al. 2011). This extremely long MFPL makes graphene an exceptional candidate for geometric diodes as the required neck width can be fabricated using available lithography (Zhu et al. 2013).

To estimate the MFPL, we measured the Dirac curve for our material prepared using the exfoliation method (Nayfeh 2011), and show typical results in Figure 33.2. By using a back gate voltage measurement (Nayfeh 2011), the MFPL of our

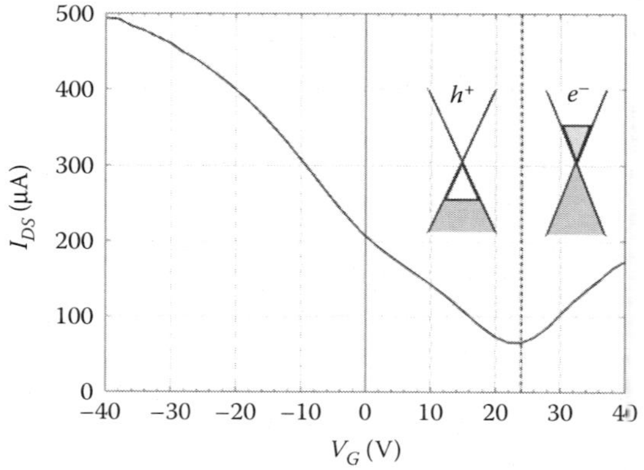

FIGURE 33.2 Dirac curve (drain–source current [I_{DS}] versus back gate voltage (V_G) of the graphene used for fabricating the graphene geometric diode. The drain–source voltage V_{DS} was 1.5 V. The dashed line at $V_G = 24$ V represents the charge neutral point voltage (V_{CNP}) of our grapheme. The graphene band diagrams in the regions for $V_G < 24$ V and $V_G > 24$ V are also shown in the figure. When V_G is less than V_{CNP}, the majority charge carriers within graphene are holes (h^+). When V_G is larger than V_{CNP}, electrons are the majority charge carriers (e^-). (From Zhu, Z. et al. 2013. *J. Phys. D: Appl. Phys.* 46: 185101. With permission.)

graphene was calculated to be between 20 and 75 nm. There are many methods to improve the MFPL of graphene and they will be discussed in Section 33.5. For the devices described here, the graphene charge neutral point (CNP) voltage V_{CNP} was ~24 V.

To operate at terahertz frequencies, the diodes in the terahertz rectenna system must be intrinsically fast and have an extremely low RC time constant. At the same time, the diode impedance must match the antenna impedance for efficient power coupling between the antenna and the diode (Grover and Moddel 2011). By taking the advantage of using graphene as the material for geometric diodes, they intrinsically are capable of operating at terahertz frequencies. More importantly, the capacitance of the graphene geometric diodes is extremely low. Based on a planar film capacitance model, the capacitance of a diode with a 100-nm neck is estimated to be between 10^{-17} and 10^{-18} F (Zhu et al. 2013). Typical antenna impedance is on the order of 100 Ω. Therefore, the RC time constant of an impedance-matched rectenna system using graphene geometric diodes is on the order of 10^{-15} s (Zhu et al. 2013).

33.3 DIODE FABRICATION

Graphene geometric diodes were fabricated using graphene exfoliation, optical and electron beam (e-beam) lithography, and oxygen plasma etching methods. We exfoliated graphite flakes onto a 300-nm-thick SiO_2 layer produced by dry thermal oxidation of a heavily doped p-type silicon wafer. After inspection under an optical microscope and verifying the graphene thickness using an atomic force microscope (AFM), we selected single-layer graphene pieces. Next, four metal contacts were patterned onto the graphene flake using optical lithography and liftoff. We used thermally evaporated 50-nm gold film on top of a 15-nm chromium adhesion layer. We used a JEOL 9300 e-beam writer at the Cornell NanoScale Science and Technology Facility (CNF) to pattern the asymmetric geometric shape in the negative tone ma-N resist. After developing the resist, O_2 plasma etching at 50 W power and 30 mTorr pressure was applied for 15 s to etch the unprotected graphene region. Figure 33.3 shows an AFM image of a fabricated graphene geometric diode. The diode neck width (d_{neck}) was measured to be 75 nm.

33.4 DC $I(V)$ MEASUREMENT OF GEOMETRIC DIODES

To obtain accurate DC $I(V)$ characteristics of the diodes, four-point measurements with pulsed bias voltage were carried out. The four-point measurement setup is shown in Figure 33.4. Channel A of a Keithley 2612 Sourcemeter provided the pulsed DC voltage to the outer two metal contacts, while the current I_{DS} was measured at the same time. The actual voltage drop V_{DS} across the diode was measured between the inner two contacts using the same channel of the source meter configured in "4-Wire" mode. Channel B was used to apply back gate voltage V_G. This four-point setup eliminated contact resistance from the measurement (Zhu et al. 2013). Additionally, to prevent the charging and hysteresis effect of graphene (Joshi et al. 2010), the applied pulsed DC voltage had a pulse width of 26 µs and followed this pattern: 0 V, $+V_1$, $-V_1$, $+V_2$, $-V_2$, ..., $+V_{end}$, $-V_{end}$.

The graphene geometric diode in Figure 33.3 exhibited a nonlinear asymmetric $I(V)$ relationship, as shown in Figure 33.5a for $V_G = 20$ V and 40 V. A figure of merit for the diodes used in rectenna systems is the responsivity, which is defined as one half of the ratio of the second derivative over the first derivative of $I(V)$ $[1/2 \times I''(V)/I'(V)]$. Responsivity represents how much DC can be generated for a given AC input power. The rectenna *system* responsivity is a function of the diode responsivity, the antenna–diode coupling efficiency, and the antenna absorption efficiency. We show the calculated responsivity curves in Figure 33.5b using the data in Figure 33.5a. For this diode, the responsivity at $V_{DS} = 0$ V and $V_G = 20$ V is 0.012 A/W.

A unique property of the graphene geometric diode is that its rectification polarity is reversible and can be controlled by the gate voltage (Moddel et al. 2012). Owing to the conical band structure of graphene, the type of the majority charge carriers and the carrier concentration can be controlled by changing V_G. In geometric diodes, electrons and holes are both subject to the same geometric effect and have the same current forward direction. Therefore, owing to the opposite charge of electrons and holes, the polarity of the diodes can be reversed by switching V_G from one side of V_{CNP} to the other (Moddel et al. 2012). An indication of this behavior

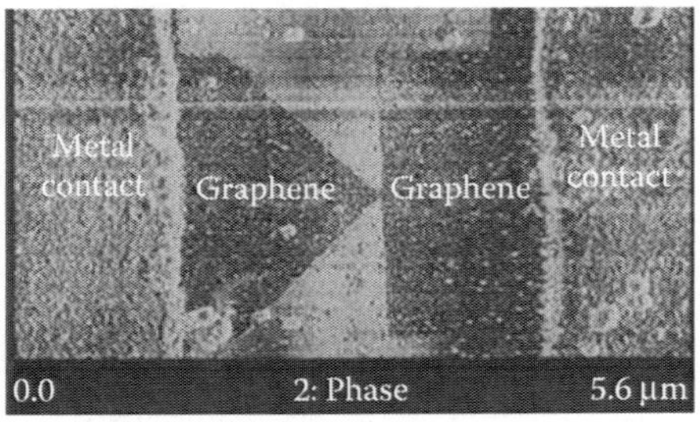

FIGURE 33.3 AFM image of a graphene geometric diode between two metal contacts. The neck width of the diode was measured to be 75 nm. (From Zhu, Z. et al. 2013. *J. Phys. D: Appl. Phys.* 46: 185101. With permission.)

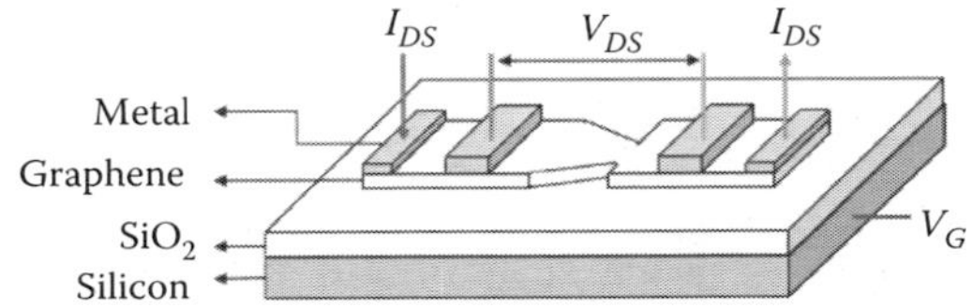

FIGURE 33.4 Four-point probe measurement setup circumvents contact resistance that distorts two-point measurements. A pulsed voltage was applied to provide drain–source current (I_{DS}) through the outer two metal contacts. The actual voltage drop across the diode (V_{DS}) was measured between the inner metal contacts. A back gate voltage (V_G) was applied directly to the silicon substrate to control the carrier type and concentration in the graphene.

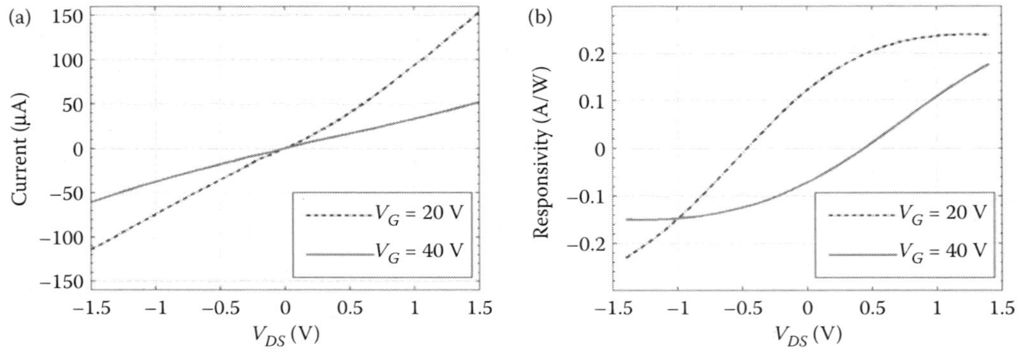

FIGURE 33.5 (a) DC $I(V)$ characteristics of the graphene geometric diode shown in Figure 33.3 at a gate voltage $V_G = 20$ V and 40 V. (b) Calculated responsivity [1/2 $|I''(V)/I'(V)|$] at $V_G = 20$ V and 40 V, using the data in (a). At $V_G = 20$ V and 0 V bias, the responsivity is ~0.12 A/W. At $V_G = 40$ V and 0 V bias, the responsivity is ~0.08 A/W.

can be seen from the diode current–voltage characteristics measured at two gate voltages on opposite sides of the CNP voltage (V_{CNP}) shown in Figure 33.5b. To observe the reversal explicitly, we varied V_G from −40 V to 40 V and measured I_{DS}, keeping V_{DS} constant. We define asymmetry A to be the absolute ratio of the current at a positive V_{DS} and the current at the negative V_{DS} with the same voltage magnitude ($|I_{DS}(+V_{DS})/I_{DS}(-V_{DS})|$). $A > 1$ means that the current flows more easily in the positive V_{DS} direction, while $A < 1$ indicates that the current flows more easily in the negative V_{DS} direction. Figure 33.6 shows A versus V_G plots for three different drain–source voltages, 0.5, 1, and 1.5 V. The reversal of diode polarity can be clearly seen at the voltage at which the asymmetry drops sharply, corresponding to where V_G sweeps through V_{CNP}.

Furthermore, by increasing the voltage difference between V_G and V_{CNP} ($|V_G - V_{CNP}|$), the charge carrier concentration n_s increases and gives a longer MFPL ($\propto \sqrt{n_s}$) (Nayfeh 2011). This increases the magnitude of A between $V_G = 12$ V and 24 V, and also between 24 V and 32 V. However, the diode asymmetry decreases as $|V_G - V_{CNP}|$ increases further because the current in the device starts to saturate (Dorgan et al. 2010). Such a current saturation effect influences the $I(V)$ characteristics and A drops back to 1. Figure 33.6 confirms the geometric effect. Because the diode asymmetry is bias voltage dependent, the relative magnitude of A increases with increase in V_{DS}. This agrees well with the simulation results in Section 33.7.

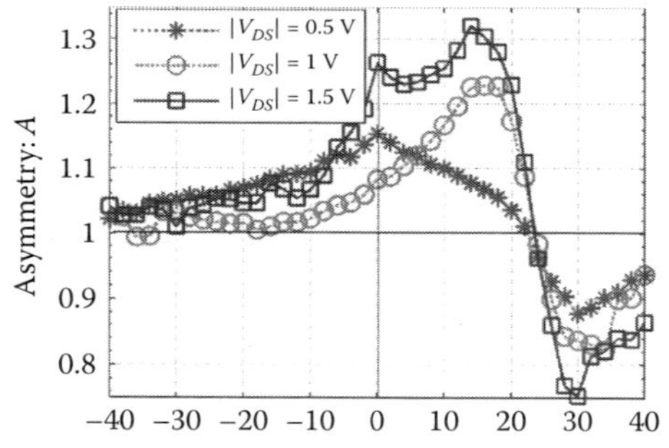

FIGURE 33.6 Measured asymmetry ($A = |I(V_{DS})/I(-V_{DS})|$) as a function of gate voltage (V_G) for a geometric diode at three magnitudes of drain–source voltages: $|V_{DS}| = 0.5$ V, $|V_{DS}| = 1$ V, and $|V_{DS}| = 1.5$ V. Diode asymmetry increases with V_{DS}. The polarity of the diode switches as the gate voltage is varied from −40 V to 40 V, due to the change of the charge carrier type from holes (h^+) to electrons (e^-) near V_{CNP} (=24 V). Current saturation effect starts playing a major role after the diode asymmetry ratio reaches its maximum as $|V_G - V_{CNP}|$ increases. (Moddel, G. et al. 2012. *Solid State Commun.* 152(19): 1842–1845. With permission.)

Asymmetry values of devices fabricated with exfoliated graphene and graphene deposited by chemical vapor deposition (CVD) are compared in Figure 33.7. The diodes made from CVD-grown graphene had approximately the same neck width as the ones made using exfoliated graphene. CVD-grown graphene diodes have lower A than the exfoliated devices, because CVD-grown graphene has a shorter MFPL compared to exfoliated graphene. To further confirm that the electrical asymmetry is due to the physical asymmetry, we also fabricated symmetric junctions using CVD-grown graphene. As predicted, symmetric junction devices show no asymmetry in the $I(V)$ behavior and A remains at one as shown in Figure 33.7.

33.5 DIODE RECTIFICATION AT 28 THz

We demonstrated the high-frequency rectifying behavior of graphene geometric diodes coupled to metal bow tie antennas. The antenna was designed to operate at 28 THz. An AFM image of the fabricated rectenna system is shown in Figure 33.8. The gold bow tie antenna consisted of two 2.3-μm long triangular arms with a 0.5-μm gap in the center (González and Boreman 2005). The graphene geometric diode was placed in the antenna gap region and electrically connected to the antenna arms. We used an edge fed configuration (Weiss et al. 2003) to lead the DC voltage and current out from the edges of the antenna to the probe contact pads. To measure the current output of the rectenna under illumination, probing the two contact pads was sufficient.

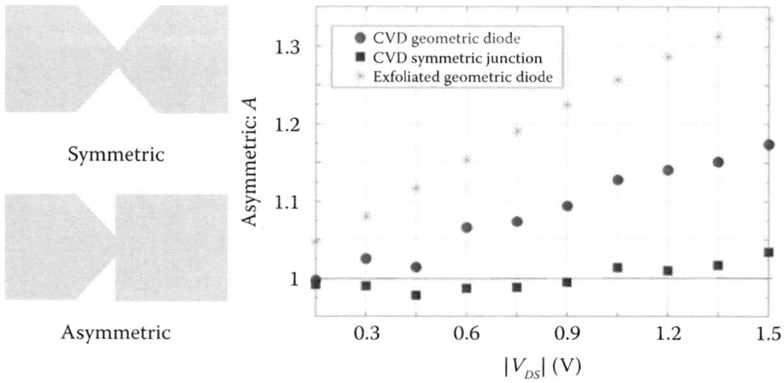

FIGURE 33.7 Asymmetry A versus drain–source voltage $|V_{DS}|$ curves for an exfoliated graphene diode (stars), a CVD graphene diode (circles), and a CVD graphene symmetric junction device (squares). The CVD graphene has a shorter charge carrier MFPL than the exfoliated graphene. This causes the CVD graphene diode to have a lower asymmetry than the exfoliated graphene diode. The CVD graphene symmetric junction device shows no asymmetry (A = 1) in its electrical behavior. (From Zhu, Z. et al. 2013. *J. Phys. D: Appl. Phys.* 46: 185101. With permission.)

To measure the optical response, we illuminated the rectenna with 28 THz radiation and used a lock-in amplifier to measure the voltage and current output. Figure 33.9 shows the setup of the optical response measurement system. A SYNRAD 48-1SWJ infrared CO_2 laser generated the 28 THz radiation. The pulse width from a pulse generator controlled the CO_2 laser's output power. A red He–Ne laser provided a visible dot for aligning the CO_2 laser with the rectenna. A half-wave plate rotated the laser's polarization relative to the antenna axis. We mechanically chopped the laser beam at 280 Hz to produce a reference for the Stanford Research Systems SR830 lock-in amplifier that measured the electrical output of the rectenna. Therefore, the lock-in amplifier was only able to detect the modulated current and voltage signal at the chopping frequency. A mercury switch short circuited the probes to ground potential during set up to avoid damage to the devices from electrostatic discharge.

Figure 33.10 shows the optical response we measured, both rectified open-circuit voltage and short-circuit current, using the setup in Figure 33.9. Owing to sufficient diode asymmetry at the zero drain–source and gate bias, no external bias voltage was applied to the rectenna. When the laser polarization was aligned with the antenna, the rectenna responded with an open-circuit voltage of 0.75 μV and a short-circuit current of 420 pA. The cosine-squared polarization dependence of the open-circuit voltage and short-circuit current confirmed that the optical response was due to radiation coupled through the bowtie antenna. This angular dependence of the optical response indicates that rectification was neither caused by diffusion of the optically generated charge carriers nor a result of thermoelectric effects due to the nonuniform illumination on the diode. Additionally, because no gate voltage was applied during the measurement and the device discharged before the

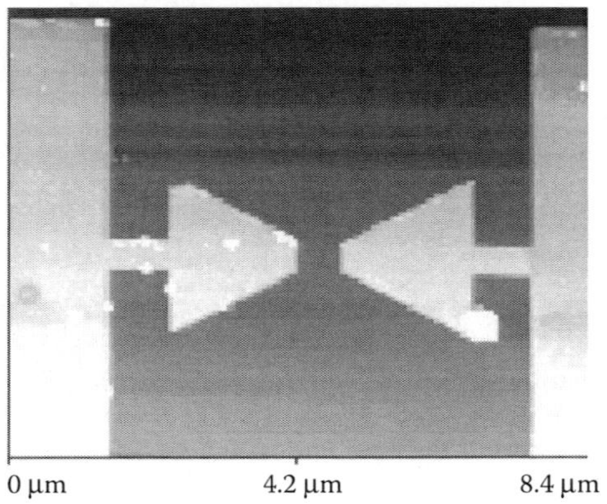

FIGURE 33.8 AFM image of a graphene geometric diode coupled to a metal bowtie antenna. Compared to the thickness of the metal antenna, the significantly thinner graphene geometric diode is not visible in the AFM image. (From Zhu, Z. et al. 2013. *J. Phys. D: Appl. Phys.* 46: 185101. With permission.)

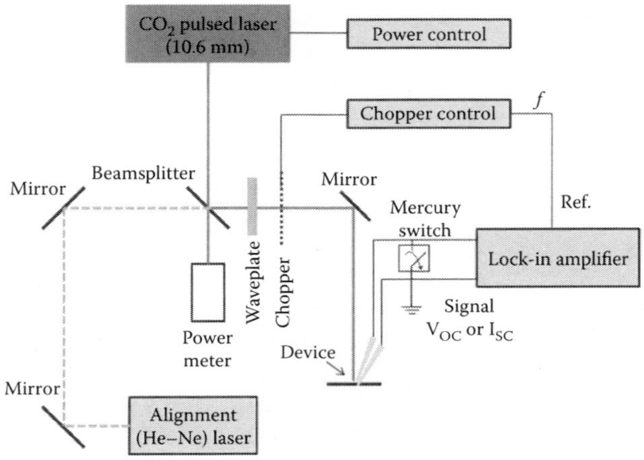

FIGURE 33.9 Optical response measurement setup of the rectenna. We used a red laser to align the CO_2 laser. The beam was guided through a set of mirrors onto the device. A two-point probe setup was used to measure the photocurrent. A chopping frequency of 280 Hz was used as a reference for the lock-in amplifier. We used a half-wave plate to study the effect of changing the angle between the antenna axis and the incident wave polarization. (From Zhu, Z. et al. 2013. *J. Phys. D: Appl. Phys.* 46: 185101. With permission.)

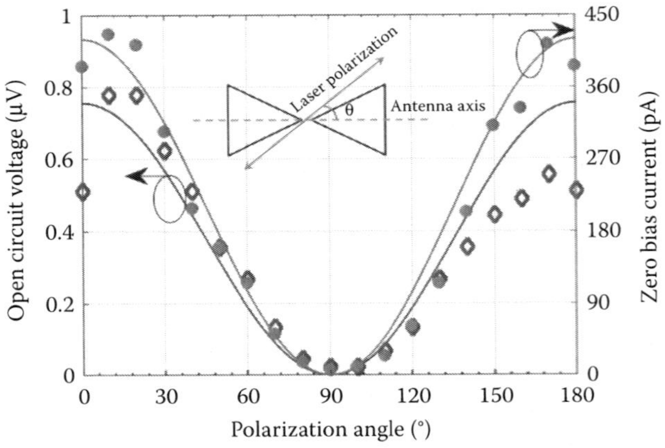

FIGURE 33.10 Metal antenna/graphene diode rectenna short-circuit current (circles) and open-circuit voltage (diamonds) as a function of polarization angle (θ), which is the difference between the laser polarization and the antenna axis. (From Zhu, Z. et al. 2013. *J. Phys. D: Appl. Phys.* 46: 185101. With permission.)

measurement, no p–n junctions could have been formed as a result of an applied field (Williams et al. 2007).

As described below, the measured maximum current of 420 pA in Figure 33.10 is of the same order of magnitude as the estimated value taking into account the following system parameters: measured laser power, diode responsivity, antenna absorption efficiency, and rectenna coupling efficiency. Our 28 THz metal bowtie antenna has an effective area approximately equal to 37.5 μm^2 (González and Boreman 2005). The illumination intensity was approximately 50 mW/mm^2. The zero-bias DC responsivity of the diode used in the optical measurements is calculated to be 0.0285 A/W. The absorption efficiency of the metal antenna configuration is estimated to be 37% (González and Boreman 2005). We chose the simplest bowtie antenna due to its ease of fabrication. This antenna absorption efficiency is not the highest reported in the literature and better antenna designs exist (González and Boreman 2005). The antenna impedance is assumed to be 100 Ω. The measured diode resistance was ~3000 Ω, and the capacitance was calculated to be ~10^{-18} F. Using a classical circuit impedance matching model (Joshi and Moddel 2013, Grover 2011), we estimate the rectenna coupling efficiency to be 12%. Combining all the above parameters, the metal antenna/graphene geometric diode rectenna gives an estimated current of 2.4 nA, which is about 6 times larger than the measured value of 0.42 nA (Table 33.2).

We performed two additional measurements to further confirm the diode rectification at 28 THz. First, we illuminated diodes without coupled antennas. As expected, diodes without antennas did not show any response. This indicates that the optical response in Figure 33.10 is not caused by *in situ* p–n doping in the graphene (Williams et al. 2007). Second, as shown in Figure 33.11, the voltage output signal increased with laser intensity when the rectenna was aligned with the laser polarization at θ = 0°. In contrast, for the misaligned case, θ = 90°, there was no change in the voltage

TABLE 33.2
Parameters for Estimating the Short-Circuit Current of the Rectenna with the Geometric Diode

Antenna effective area	37.5 μm^2 (González and Boreman 2005)
CO_2 laser illumination intensity	50 mW/mm^2
Zero-bias responsivity of this diode	0.0285 A/W
Antenna absorption efficiency	37% (González and Boreman 2005)
Antenna impedance	~100 Ω
Diode resistance	~3000 Ω
Diode capacitance	~10^{-18} F
Rectenna system coupling efficiency	12%
Estimated short circuit current	2.6 nA
Measured short circuit current	0.42 nA

output with the incident laser intensity. The graphene geometric diode genuinely rectifies the 28 THz signal.

When comparing the geometric diode rectenna with other terahertz detectors, along with responsivity, noise equivalent power (NEP) is the most important figure of merit. NEP is a measure of the system sensitivity and is the minimum incident optical power required for unity system signal-to-noise ratio (Richards 1994). Lower NEP values represent better detector sensitivity. NEP is defined as

$$\mathrm{NEP} = \frac{\sqrt{A_d}}{D^*} \tag{33.1}$$

$$\text{where } D^* = (A_d \Delta f)^{\frac{1}{2}} \frac{\beta}{I_n}, \quad \text{and} \quad I_n = \sqrt{\left(2eI_{bias} + \frac{4KT}{R_D}\right)\Delta f} \tag{33.2}$$

In the above equations, the detector area A_d is the effective antenna area (37.5 μm^2); D^* is the normalized detectivity;

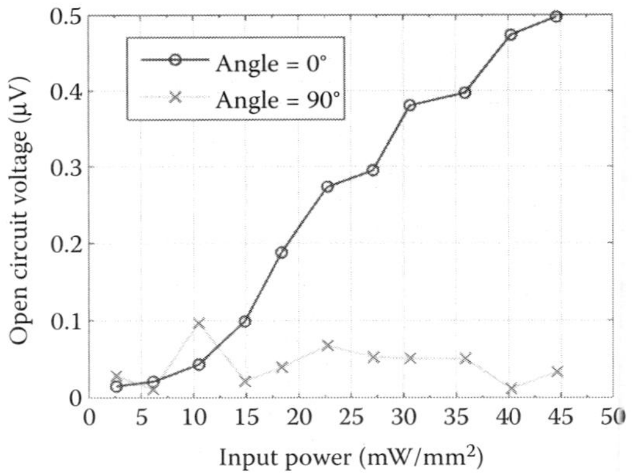

FIGURE 33.11 Open-circuit voltage versus laser input power for two polarization angles, 0° and 90°. The response at 0° (circles) indicates perfect alignment between the laser polarization and the antenna, which gives the strongest open-circuit voltage signal. At 90° (crosses), the antenna is perpendicular to the laser polarization and gives a near zero output voltage at all input intensities.

and Δf is the bandwidth of the detector (Rogalski 2003). The total system responsivity β is the product of antenna absorption efficiency (37%), antenna-to-diode coupling efficiency (12%), and the current responsivity of the diode (0.0285A/W). The noise current (I_n) is calculated as the root-mean-square value of the shot noise due to the diode DC bias current (I_{bias}) and the Johnson noise due to the finite diode resistance (R_D) (Rogalski 2003). For the graphene geometric diode operating at zero bias, I_{bias} is zero. The device operates at room temperature ($T = 300$ K). The NEP of the metal antenna/graphene geometric diode rectenna is calculated to be 10.5 nW Hz$^{-1/2}$.

We compare the graphene geometric diode rectenna to other graphene detectors (Table 33.3). At the time of writing of this chapter, there are no other graphene detectors for 28 THz detection. Therefore, we compare our rectenna to detectors operating at ~1 THz. The geometric diode rectenna is as good as the antenna-coupled graphene FETs having an NEP of 30 nW Hz$^{-1/2}$ for double-layer graphene devices, and 200 nW Hz$^{-1/2}$ for single-layer graphene devices (Vicarelli et al. 2012). The nonoptimized graphene geometric diode is comparable to the conventional semiconductor-based terahertz detectors. FET detectors based on indium arsenide nanowires achieved an optical NEP of ~1 nW Hz$^{-1/2}$ (Vitiello et al. 2012), and silicon n-MOS detectors have an NEP of ~10^{-11} W Hz$^{-1/2}$ at 0.8 THz (Knap et al. 2009). The NEP of semiconductor-based detectors increases exponentially with the operating frequency (Rogalski 2003) because their intrinsic electrical response decreases with the frequency. Our graphene geometric diode-based rectenna detector does not suffer such frequency limitations below the graphene surface plasmon frequency (Jablan et al. 2009), as discussed in Section 33.1.

There are at least three ways to increase the sensitivity of our rectenna detector. The simplest is to change the applied gate voltage V_G to the maximum diode $I(V)$ asymmetry point in Figure 33.6. This lowers the NEP of the device by half. The second approach is to improve the quality of the graphene and hence the charge MFPL. This can be achieved by improving the fabrication process and reducing chemical contamination (Fan et al. 2011). Boron nitride substrates have provided improvements in the quality of exfoliated graphene (Liu et al. 2013) (Levendorf et al. 2012). Another approach to improve the diode performance is by changing the geometric shape of the diodes. From simulation results of improved diodes (see Section 33.8), the 28 THz NEP value can be reduced to as low as 10^{-11} W Hz$^{-1/2}$.

TABLE 33.3
Comparison of Various Detectors Operating at Terahertz Frequencies and a Temperature of 300 K

Detector Type	NEP (W Hz$^{-1/2}$) at 300 K	Incident Radiation Frequency
Graphene FET detector (Vitiello et al. 2012)	2×10^{-7} (single-layer graphene) 3×10^{-8} (double-layer graphene)	1 THz
InAs FET detector (Vitiello et al. 2012)	10^{-9}	1 THz
Silicon n-MOS detector (Knap et al. 2009)	10^{-10}	0.8 THz
Rectennas using graphene geometric diodes	10^{-8}–10^{-9}	28 THz

33.6 GRAPHENE ANTENNA-COUPLED WITH GRAPHENE GEOMETRIC DIODE

In addition to fabricating graphene geometric diodes with metal antennas and measuring their response at 28 THz, we coupled the diodes to graphene antennas and repeated the optical measurements. The graphene antenna and diode were patterned during a single e-beam lithography step. Figure 33.12 shows an AFM image of the graphene rectenna. It has the same dimensions as the metal antenna/graphene diode rectenna. In retrospect, we discovered that our graphene antenna was too large, based on a recent simulation of a graphene antenna at 1.4 THz that indicates that the size of the graphene antenna must be much smaller than that of metal antennas at the same working frequency (Tamagnone et al. 2012). Nonetheless, we observed a cosine squared polarization response similar to the response of the metal antenna rectenna system, as shown in Figure 33.13. The open-circuit voltage response of the graphene rectenna device was similar to that of the graphene diode/metal antenna rectenna. However, the short-circuit current response was significantly lower due to the larger series resistance of graphene antenna compared to the metal antenna, and possibly the overly large antenna. The graphene antennas have a DC series resistance of about 1 kΩ compared to several ohms for metal antennas.

This is the first experimentally reported graphene bowtie antenna working at terahertz frequencies and it opens the possibility of using patterned graphene as an optical resonator.

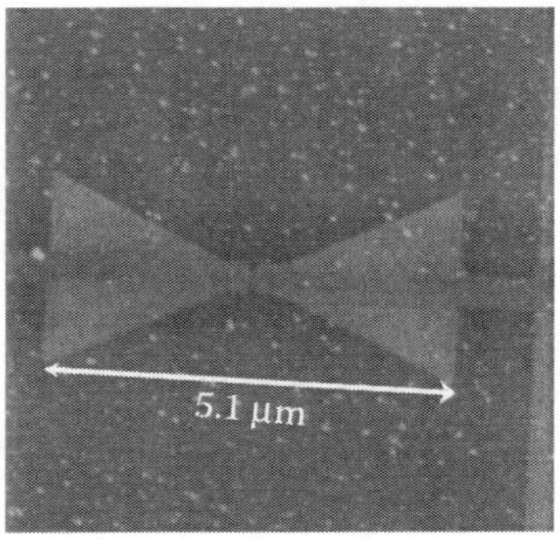

FIGURE 33.12 AFM image of a graphene rectenna system consisting of a graphene geometric diode and a graphene antenna. (From Zhu, Z. et al. 2013. *J. Phys. D: Appl. Phys.* 46: 185101. With permission.)

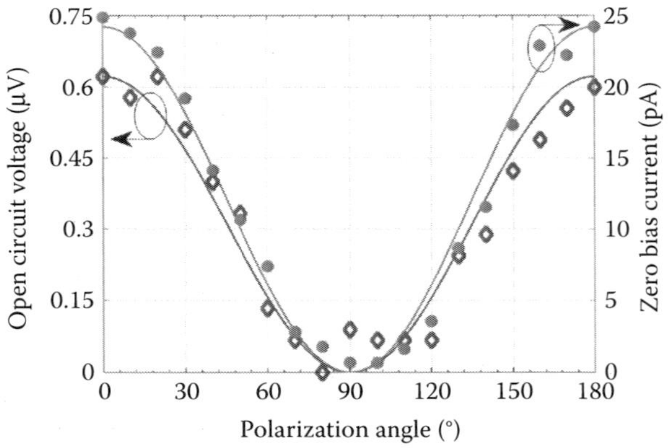

FIGURE 33.13 Graphene antenna/graphene diode rectenna zero-bias current (circles) and open-circuit voltage (diamonds) response for different polarization angles. The open-circuit voltage response is close to that of the metal antenna/graphene diode rectenna, but the lower current response in the graphene rectenna is due to a larger graphene antenna series resistance compared to the metal antenna.

Bulk graphene films on SiO_2 substrates absorb 2.3% of incident light at almost all frequencies (Nair et al. 2008). As previously discussed in Section 33.1, over 13% plasmon absorption at 3 THz has been reported (Ju et al. 2011). This indicates that patterned graphene structures could enhance the light–plasmon coupling in graphene. Graphene optical conductivity and cutoff frequency can be controlled by varying gate voltage and intrinsic doping level (Mak et al. 2008, Li et al. 2008). Hence, graphene antennas and resonators operating at terahertz frequencies can compete with conventional two-dimensional metal-based resonators.

33.7 GEOMETRIC DIODE SIMULATION

In this section, we introduce a method to simulate graphene geometric diodes, and discuss the possibility of improving the diode responsivity based on measurement and simulation results.

33.7.1 Monte Carlo Simulation Based on Drude Model

Our Monte Carlo simulation, based on the Drude model (Ashcroft and Mermin 1976), models the electron movement in geometric diodes. After defining the device shape and boundaries, the simulation places an electron randomly within the device and sets it to move with a velocity (v_{tot}) until the electron collides with defects or phonons. The MFPL is the key material parameter, which determines the average electron collision time. Between each collision, v_{tot} is the vector sum of the random Fermi velocity (v_F) in a random direction and the constant drift velocity (v_D) in response to the bias-dependent electric field. In this simple version of the simulation, the electric field is assumed to be the ratio of applied voltage and device length and is uniform within the device. In the model, the electron approaches and specularly reflects at all the boundaries while maintaining the same magnitude of v_{tot}. The motion of the electron is tracked and the Monte Carlo simulation keeps running until 10^6 collisions have occurred. Such a large number of collisions ensure a statistically stable result, which is independent of the starting position of the electron. The net current at a given bias voltage is calculated by using the electron density of graphene and counting the number of times an electron crosses a certain device cross section per unit time.

From the operating principles described in Section 33.2, we know that diode's geometric asymmetry and the constriction d_{neck} determine the $I(V)$ asymmetry of the diode. We vary d_{neck} in the simulator described above and compare the $I(V)$ curves to see the impact of changing d_{neck}. In all the simulations, we set the MFPL to be 200 nm. The $I(V)$ curves in Figure 33.14a indicate that shrinking d_{neck} increases the $I(V)$ asymmetry while keeping all other device dimensions the same. Using the $I(V)$ data of Figure 33.14a, A versus the absolute value of V_{DS} is plotted in Figure 33.14c for different d_{neck}. Besides showing that diodes with smaller d_{neck} have higher A levels, Figure 33.14c shows that increasing V_{DS} also increases A. However, the electron flow becomes insensitive to the physical structure with large V_{DS} and A starts to saturate. This is because a large electric field makes the electrons move much farther than d_{neck} during one collision time, with the result that the arrowhead-shaped boundaries start blocking the electron movement nearly as well as the vertical boundaries. At this point, the electrons will have a similar probability of transmission through the neck region for both polarities of the applied voltage. Thus, with increasing V_{DS}, the device asymmetry disappears.

Two more factors influencing asymmetry are the shoulder width and the slope of the arrowhead. We show the $I(V)$ curves with varying shoulder width in Figure 33.14b. Comparing Figure 33.14a and b, increasing the shoulder size gives a higher forward current while reducing d_{neck} has a stronger impact on lowering the reverse current. This is because the width of the shoulder determines the ease of electron movement after electrons funnel through the neck region and d_{neck} acts like a blocking mechanism for the electrons traveling in the reverse direction. The other geometric parameter, the neck slope, does not have as great an impact as the shoulder and neck widths. As shown in Figure 33.14d, changing the neck slopes between 30° and 45° has no obvious effect on A. On increasing the slope to 60° and 70°, A decreases.

33.8 SUMMARY

We have developed a new kind of graphene diode with an ultralow RC time constant for terahertz applications. Based on electrical measurements and Monte Carlo simulations, the diode's geometric asymmetry produces a DC $I(V)$ asymmetry. By using a gate voltage to change the dominant charge carrier type, the polarization of these graphene geometric diodes reversed. Optical response measurements of rectenna

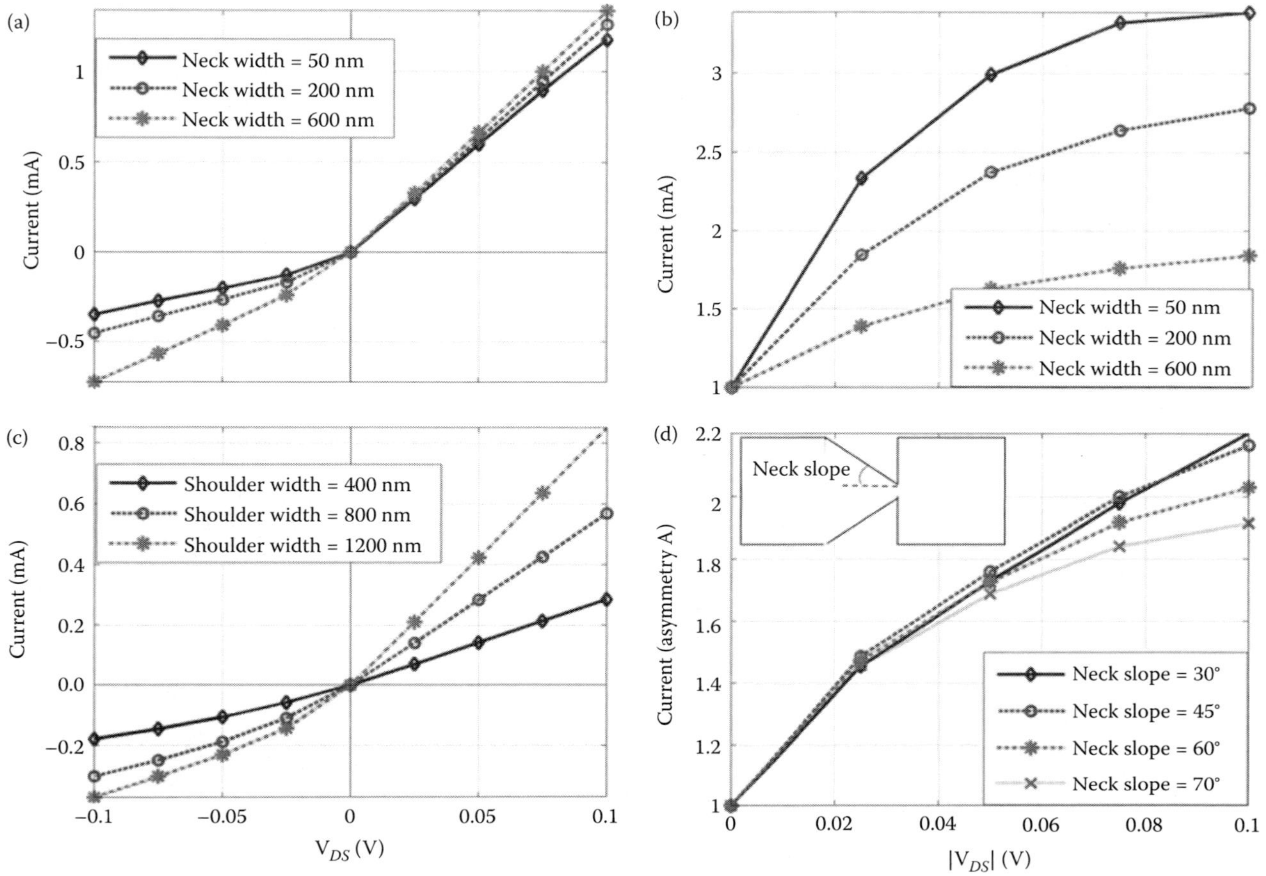

FIGURE 33.14 Simulated diode electrical characteristics based on a Monte Carlo model for charge carrier motion. (a) Simulated $I(V)$ curves for geometric diodes with different neck widths: 50, 200, and 600 nm. The shoulder width of all the simulated devices is fixed at 1 μm. A smaller neck restricts the reverse current more efficiently and has a strong effect on the diode reverse current. (b) Simulated $I(V)$ curves for geometric diodes with different shoulder widths: 400, 800, and 1200 nm. The neck width is fixed at 50 nm in all the devices. The change in forward current with increasing shoulder width is larger than that for reverse current. (c) Calculated asymmetry A versus neck width using the data from (a). Increasing V_{DS} and reducing neck width leads to higher asymmetry. (d) Calculated A versus $|V_{DS}|$ with arrowhead neck slope varying from 30° to 70°. Within this range, the slope angle does not have a great impact on A. The MFPL in all the simulations above is fixed to be 200 nm.

devices formed by coupling graphene geometric diodes to metal bowtie antennas showed diode rectification at a wavelength of 10.5 μm, corresponding to 28 THz. The measured short-circuit current compares well with the value estimated using the diode and antenna parameters. We calculated the NEP of the detector to be 43 nW $Hz^{-1/2}$ at 28 THz. In addition to using the geometric diode in rectennas with metal bowtie antennas, using the geometric diode as the rectifier, we experimentally demonstrated the optical response of graphene bowtie antennas working at 10.6 μm wavelength.

ACKNOWLEDGMENTS

We gratefully acknowledge assistance in device preparation from Kendra Krueger and David Doroski. This work was carried out under a contract from Abengoa Solar, with initial support from Hub Lab. Device processing was carried out in part at the Colorado Nanofabrication Laboratory, and in part at the Cornell NanoScale Science and Technology Facility, both members of the National Nanotechnology Infrastructure Network, which is supported by the National Science Foundation (Grant ECS-0335765). We also thank Jonathan Alden in Professor Paul McEuen's group in Cornell University for providing the CVD graphene sample.

REFERENCES

Ashcroft, N. W. and Mermin, N. D. 1976. *Solid State Physics*. New York: Holt, Rinehart and Winston.

Bae, S., Kim, H., Lee, Y., Xu, X., Park, J. S., Zheng, Y., Balakrishnan, J. et al. 2010. Roll-to-roll production of 30-inch graphene films for transparent electrodes. *Nat. Nanotech.* 5: 574–578.

Bolotina, K. I., Sikes, K. J., Jiang, Z., Klima, M., Fudenberg, G., Hone, J., Kim, P., and Stormer, H. L. 2008. Ultrahigh electron mobility in suspended graphene. *Solid State Commun.* 146: 351–355.

Bostwick, A., Speck, F., Seyller, T., Horn, K., Polini, M., Asgari, R., MacDonald, A. H., and Rotenberg, E. 2010. Observation of plasmons in quasi-freestanding doped graphene. *Science* 328: 999–1002.

Castro Neto, A. H., Guinea, F., Peres, N. M. R., Novoselov, K. S., and Geim, A. K. 2009. The electronic properties of graphene. *Rev. Mod. Phys.* 81: 109–162.

Dorgan, V. E., Bae, M.-H., and Pop, E. 2010. Mobility and saturation velocity in graphene on SiO_2. *Appl. Phys. Lett.* 97: 082112.

Dragoman, D. and Dragoman, M. 2013. Geometrically induced rectification in two-dimensional ballistic nanodevices. *J. Phys. D: Appl. Phys.* 46: 055306.

Dragoman, D., Dragoman, M., and Plana, R. 2010. Graphene-based ultrafast diode. *J. Appl. Phys.* 108: 084316.

Dragoman, M. 2013. Graphene nanoelectronics for high-frequency applications. *Proc. SPIE Nanotechnology VI.* 87660C.

Fan, J., Michalik, J. M., Casado, L., Roddaro, S., Ibarra, M. R., and Teresa, J. M. De. 2011. Investigation of the influence on graphene by using electron-beam and photo-lithography. *Solid State Commun.* 151: 1574–1578.

González, F. J. and Boreman, G. D. 2005. Comparison of dipole, bowtie, spiral and log-periodic IR antennas. *Infrared Phys. Technol.* 46(5): 418–428.

Grover, S. 2011. *Diodes for Optical Rectennas.* University of Colorado, Boulder, PhD thesis.

Grover, S. and Moddel, G. 2011. Applicability of metal/insulator/metal (MIM) diodes to solar rectennas. *IEEE J. Photovoltaics* 1(1): 78–83.

Jablan, M., Buljan, H., and Soljačić, M. 2009. Plasmonics in graphene at infrared frequencies. *Phys. Rev. B* 80: 245435.

Joshi, P., Romero, H. E., Neal, A. T., Toutam, V. K., and Tadigadapa, S. A. 2010. Intrinsic doping and gate hysteresis in graphene field effect devices fabricated on SiO_2 substrates. *J. Phys.: Condens. Matter* 22: 334214.

Joshi, S. and Moddel, G. 2013. Efficiency limits of rectenna solar cells: Theory of broadband photon-assisted tunneling. *Appl. Phys. Lett.* 102: 083901.

Ju, L., Geng, B., Horng, J., Girit, C., Martin, M., Hao, Z., Bechtel, H. A. et al. 2011. Graphene plasmonics for tunable terahertz. *Nat. Nanotech.* 6(1): 630–634.

Knap, W., Dyakonov, M., Coquillat, D., Teppe, F., Dyakonova, N., Łusakowski, J., Karpierz, K. et al. 2009. Field effect transistors for terahertz detection: Physics and first imaging applications. *J. Infrared Millim. TeraHz Waves* 30: 1319–1337.

Levendorf, M. P., Kim, C. J., Brown, L., Huang, P. Y., Havener, R. W., Muller, D. A., and Park, J. 2012. Graphene and boron nitride lateral heterostructures for atomically thin circuitry. *Nature* 488: 627–632.

Li, Z. Q., Henriksen, E. A., Jiang, Z., Hao, Z., Martin, M. C., Kim, P., Stormer, H. L., and Basov, D. N. 2008. Dirac charge dynamics in graphene by infrared spectroscopy. *Nat. Phys.* 4: 532–535.

Liu, Y., Willis, R. F., Emtsev, K. V., and Seyller, T. 2008. Plasmon dispersion and damping in electrically isolated two-dimensional charge sheets. *Phys. Rev. B* 78: 201403.

Liu, Z., Ma, L., Shi, G., Zhou, W., Gong, Y., Lei, S., Yang, X. et al. 2013. In-plane heterostructures of graphene and hexagonal boron nitride with controlled domain sizes. *Nat. Nanotech.* 8: 119–124.

Mak, K. F., Sfeir, M. Y., Wu, Y., Lui, C. H., Misewich, J. A., and Heinz, T. F. 2008. Measurement of the optical conductivity of graphene. *Phys. Rev. Lett.* 101: 196405.

Mayorov, A. S., Gorbachev, R. V., Morozov, S. V., Britnell, L., Jalil, R., Ponomarenko, L. A., Blake, P. et al. 2011. Micrometer-scale ballistic transport in encapsulated graphene at room temperature. *Nano Lett.* 11(6): 2396–2399.

Moddel, G. 2009. Geometric diode, applications and method. US Patent Application No. 20110017284, filed July 17, 2009 (provisional submitted July 18, 2008).

Moddel, G., Zhu, Z., Grover, S., and Joshi, S. 2012. Ultrahigh speed graphene diode with reversible polarity. *Solid State Commun.* 152(19): 1842–1845.

Nair, R. R., Blake, P., Grigorenko, A. N., Novoselov, K. S., Booth, T. J., Stauber, T., Peres, N. M. R., and Geim, A. K. 2008. Fine structure constant defines visual transparency of graphene. *Science* 320: 1308.

Nayfeh, O. M. 2011. Radio-frequency transistors using chemical-vapor-deposited monolayer graphene: Performance, doping, and transport effects. *IEEE Trans. Electron Devices* 58(9): 2847–2853.

Park, C. H., Giustino, F., Cohen, M. L., and Louie, S. G. 2008. Electron–phonon interactions in graphene, bilayer graphene, and graphite. *Nano Lett.* 8(12): 4229–4233.

Richards, P. L. 1994. Bolometers for infrared and millimeter waves. *J. Appl. Phys.* 76(1): 9.

Rogalski, A. 2003. Infrared detectors: Status and trends. *Progr. Quantum Electron.* 27: 69–157.

Song, A. M. 2002. Electron ratchet effect in semiconductor devices and artificial materials with broken centrosymmetry. *Appl. Phys. A* 75: 229–235.

Sun, Z., Hasan, T., Torrisi, F., Popa, D., Privitera, G., Wang, F., Bonaccorso, F., Basko, D. M., and Ferrari, A. C. 2010. Graphene mode-locked ultrafast laser. *ACS Nano* 4: 803–810.

Tamagnone, M., Gómez-Díaz, J. S., Mosig, J. R., and Perruisseau-Carrier, J. 2012. Analysis and design of terahertz antennas based on plasmonic resonant graphene sheets. *J. Appl. Phys.* 112: 114915.

Vicarelli, L., Vitiello, M. S., Coquillat, D., Lombardo, A., Ferrari, A. C., Knap, W., Polini, M., Pellegrini, V., and Tredicucci, A. 2012. Graphene field-effect transistors as room-temperature terahertz detectors. *Nat. Mater.* 6(1): 865–871.

Vitiello, M. S., Coquillat, D., Viti, L., Ercolani, D., Teppe, F., Pitanti, A., Beltram, F. Sorba, L., Knap, W., and Tredicucci, A. 2012. Room-temperature terahertz detectors based on semiconductor nanowire field-effect transistors. *Nano Lett.* 12: 96–101.

Weiss, M. D., Eliasson, B. J., and Moddel, G. 2003. Terahertz device integrated antenna for use in resonant and non-resonant modes and method. Patent 6664562.

Williams, J. R., DiCarlo, L., and Marcus, C. M. 2007. Quantum hall effect in a gate-controlled p-n junction of gaphene. *Science* 317: 638–641.

Xia, F., Mueller, T., Lin, Y. M., Valdes-Garcia1, A., and Avouris, P. 2009. Ultrafast graphene photodetector. *Nat. Nanotech.* 4: 839–843.

Zhu, Z., Grover, S., Krueger, K., and Moddel, G. 2011. Optical rectenna solar cells using graphene geometric diodes. *IEEE Photovoltaic Specialists Conference.* Seattle, WA. 002120–002122.

Zhu, Z., Joshi, S., Grover, S., and Moddel, G. 2013. Graphene geometric diodes for terahertz rectennas. *J. Phys. D: Appl. Phys.* 46: 185101.

34 Polymer Composites with Graphene
Dielectric and Microwave Properties

Vitaliy G. Shevchenko, Polina M. Nedorezova, and Alexander N. Ozerin

CONTENTS

ABSTRACT

Polymer composites with graphene or graphene oxide can be even more promising compared to polymer composites with carbon nanotubes. Different strategies to disperse graphene in polymers are discussed. They include solvent- and melt-based methods as well as *in situ* polymerization. The properties of composites strongly depend on the method of their synthesis, as this defines the dispersion of graphene in polymer. The methods facilitating graphene dispersion, such as ultrasonic treatment or chemical functionalization, are discussed. Analytical techniques for characterizing particle dimensions, surface characteristics, and dispersion in matrix polymers are reviewed. We review the electrical properties of graphene/polymer nanocomposites as a function of the method of synthesis and the type of polymer matrix (amorphous, crystalline, or semicrystalline). Percolation and electrical behavior at different frequencies are discussed, including direct current conductivity, low frequency, and microwave properties. We also review current challenges with the processing of graphene composites and future perspectives and applications for electrical properties of this new class of nanocomposites.

34.1 INTRODUCTION

Physical properties, especially electrical conductivity of polymer nanocomposites reinforced with conducting carbon nanofillers, are considered of technical importance (Hussain 2006; Qin and Brosseau 2012; Sengupta et al. 2011; Young et al. 2012). Polymer nanocomposites are highly promising for numerous applications as functional materials for electromagnetic interference (EMI) shielding, antistatic dissipation, capacitors, mobile phones, sensors, and actuators. Polymer nanocomposites with high electrical conductivity are the first candidates that can meet the requirements for those applications (Kim et al. 2010). High electrical conductivity can be achieved by proper selection of the polymer matrix and composite processing conditions and the creation of the necessary dispersion of nanofillers in the polymer matrix (Cai and Song 2010; Galpaya 2012; Terrones et al. 2011; Young et al. 2012). Polymers are mostly dielectrics with very low electrical conductivity. The incorporation of small amounts of carbon nanofillers into insulating polymers can improve their electrical conductivity significantly (Verdejo et al. 2011). When the filler content reaches the percolation threshold, the electrical conductivity of polymer nanocomposites increases by many orders of magnitude. The factors that can markedly affect the percolation threshold of conducting polymer nanocomposites are the aspect ratio and type of fillers, functionalization and dispersion of fillers, as well as the processing conditions. The addition of reduced graphene oxide (rGO) or thermally reduced graphene (TRG) nanofillers to polymers can yield a very low percolation threshold of ~0.07–0.1 vol% (Tjong 2014). This was often attributed to exfoliation of layered graphite into thin graphene sheets of large aspect ratios. Low-cost graphene-like fillers (graphene oxide [GO], rGO, graphene nanoplatelets [GNP], etc.) are considered as a feasible substitute for expensive carbon nanotubes (CNTs) in forming functional polymer composites (Choi and Lee 2012; Pati et al. 2011; Katsnelson 2012). Useful tables, summarizing the electrical properties of polymer composites with graphene, can be found in Kim et al. (2010) and Sengupta et al. (2011).

34.2 DC CONDUCTIVITY AND PERCOLATION IN POLYMER COMPOSITES WITH GRAPHENE, DEPENDENCE ON THE METHOD OF SYNTHESIS

Pristine graphene is not compatible with organic polymers and does not form homogeneous composites, which results in inferior properties. In order to improve the properties, the dispersion of graphene in polymer matrices and the graphene–polymer interaction needs to be improved, which is achieved by the surface modification of graphene (Kuilla et al. 2010). Surface modifications of graphene using a range of techniques have been carried out by different research groups with various organic modifying agents. Among these techniques, the nucleophilic addition of organic molecules to the graphene surface is the best way to achieve the bulk production of surface-modified graphene. This method is very advantageous in many aspects. For example, water can be used as a solvent, lower-cost amine compounds can be used as surface-modifying agents, the reaction can be carried in air, and surface-modified graphene can be dispersed easily in different organic media. The solvent dispersible properties of surface-modified graphene will assist in the preparation of graphene-based polymer nanocomposites. Most of the properties of the polymer/graphene nanocomposites were superior to the base polymer matrix as well as other carbon filler (CNT, carbon nanofibers [CNF], and graphite)-based composites. These improved properties of nanocomposites were obtained at very low graphene contents (≤2 wt%). In certain cases, higher graphene loading (~15 wt%) is needed to achieve the target value but is comparable to other conventional fillers.

Most graphene-based composites have binary components, although several multicomponent composites have also been fabricated to fulfill the requirements of special applications. Usually, the components of the composites are polymers (insulating and conducting), small organic compounds, metals, metal compounds, or carbon nanofillers (CNT or fullerene). Although a variety of components have been involved, the architectures of the composites can be simply classified into three types as illustrated in Figure 34.1 (Bai et al. 2011a). In type I composites, graphene sheets form a continuous phase and act as a substrate for supporting the second components, which adhere to the graphene sheets as nanoparticles. The second components are usually inorganic nanostructures, such as metal or metal compound nanoparticles, and CNTs. Occasionally, polymer nanostructures have also been induced onto the graphene surface, forming type I composites. In type III composites, graphene sheets play the role of nanofiller, distributing in the continuous matrix of the second component. In general, the graphene content in a type III composite is relatively low (<10 wt%). The architectures of type II composites are between those of type I and III composites. In these composites, both graphene and the second component are continuous phases. Composite films prepared by layer-by-layer self-assembly belong to this type. Type I composites are usually synthesized for catalysis, or other applications using their surfaces or interfaces, such as adsorption and sensing. This is mainly due to the fact that these materials usually have large specific surface areas inherited from ultrathin graphene sheets, as well as various active sites contributed by nanoparticles. Type II composites have large interfacial areas, which is favorable to chemical or electrochemical reactions, charge generation, and separation. Thus, they have potential applications in energy conversion or storage. On the contrary, specific surface area is usually not the main point in type III composites. Good mechanical properties and high electrical conductance are what one expects for these types of materials. Graphene sheets as a high-aspect-ratio and conductive nanofiller can strongly improve the mechanical, electrical, and thermal properties of the composites.

Several methods have been used to prepare graphene-based composites. They can be classified into three general strategies according to the processes of synthesis: (1) pre-graphenization strategy—graphene was synthesized before the second component was introduced; (2) post-graphenization strategy—the composite consists of the graphene precursor (usually graphene oxide) and the second component is pre-prepared, followed by converting the precursor into

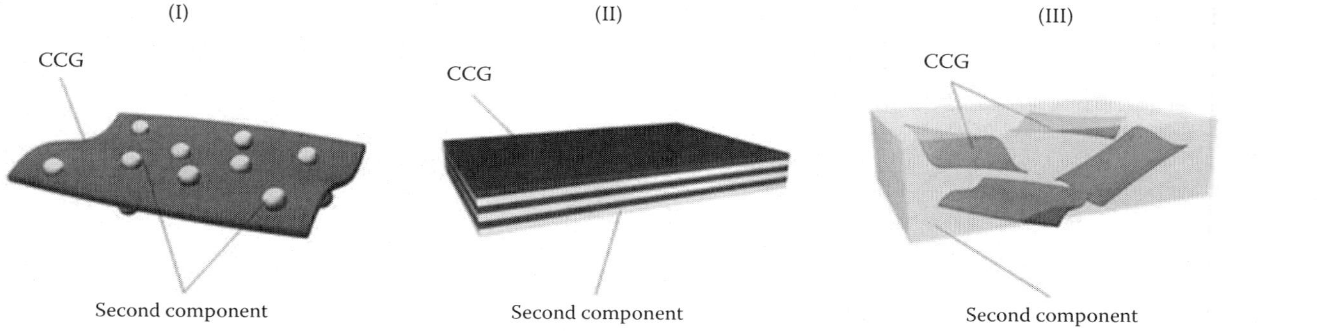

FIGURE 34.1 Three types of graphene-based composite. (Bai, H., C. Li, and G. Shi: Functional composite materials based on chemically converted graphene. *Advanced Materials* 2011a. 23 (9):1089–1115. Copyright Wiley-VCH Verlag GmbH & Co. KGaA. Reproduced with permission.)

graphene; and (3) syn-graphenization strategy: graphene and the second component are synthesized simultaneously and the two components are blended during composite formation. Two methods have been widely applied for introducing the second component to a graphene-based composite: mixing and *in situ* synthesis. Mixing includes solution mixing and melt compounding, and the latter is usually used for preparing graphene–polymer composites. This method allows precise control of the structure of the second component. *In situ* synthesis involves preparing the second component in the system containing graphene. Therefore, although *in situ* synthesis is a more convenient procedure, the control of the structure and content of the second component is more complicated.

The appreciable improvement in electrical conductivity of graphene-based composites is due to the formation of a conducting network by graphene sheets in the polymer matrix. Significant changes in electrical properties are observed according to the polymer matrix, processing method, and filler type. Figure 34.2 illustrates the variation of the percolation threshold according to the processing method.

The correlations and general dependencies between the above-mentioned parameters are difficult to establish. However, the maximum or very high electrical conductivity was obtained using a very low graphene loading in different polymer matrices compared to other carbon fillers.

The nanofiller aspect ratio has a major impact on the electrical conductivity, thermal stability, and melt rheological behavior of the nanocomposites (Li et al. 2011b). Composites were fabricated and further processed by hot compression molding. Both electrical conductivity and real permittivity increase with increasing nanofiller loading. A strongly aspect-ratio-dependent electrical conductivity percolation is observed at loadings of 15.0, 12.0, 5.0, and 3.0 wt% for carbon black (CB), GNP, CNF, and CNT, respectively. The nanofiller aspect ratio has a very significant impact on the electrical conductivity, thermal stability, and melt rheological behavior of the nanocomposites (Figures 34.3 and 34.4).

Improvement in the mechanical, thermal, and electrical properties of graphene-filled polymer nanocomposites is not only dependent on the properties of graphene but also on the properties of the host polymer matrixes. As a result, the degree of improvement in all these properties varies for different nanocomposites. The processing methods also play a significant role toward the improvement of these properties (Potts et al. 2011).

It has been said that a high degree of dispersion may not necessarily yield the lowest onset of electrical percolation (Rafiee et al. 2009), as a sheath of polymer may coat the surfaces of a well-dispersed filler and prevent direct interparticle contact. Indeed, the lowest percolation threshold achieved thus far for a graphene-based polymer nanocomposite (approximately 0.15 wt%) was observed when the filler was not homogeneously dispersed in the polymer matrix, but rather segregated from the matrix to form a conductive network. In this work, poly(ethylene) particles were mixed with GO in a water/ethanol mixture and were reduced using hydrazine, causing agglomeration of rGO and subsequent deposition onto the poly(ethylene) particles. This heterogeneous

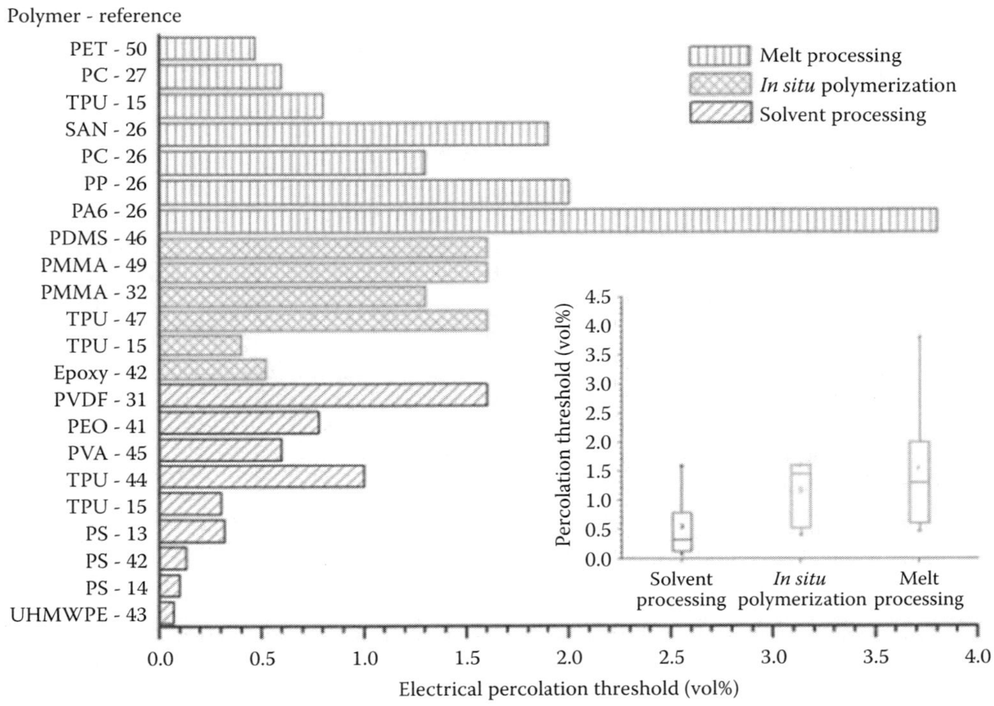

FIGURE 34.2 Electrical percolation thresholds of graphene/polymer nanocomposites according to processing strategy. (Verdejo, R. et al. 2011. Graphene filled polymer nanocomposites. *Journal of Materials Chemistry* 21 (10):3301. Reproduced by permission of The Royal Society of Chemistry.)

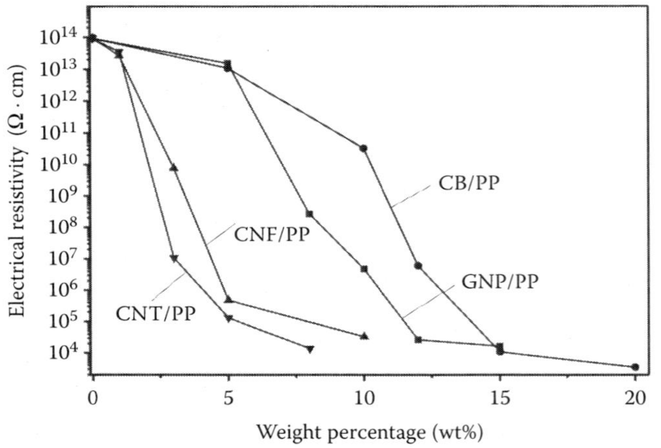

FIGURE 34.3 Volume resistivity of PP nanocomposites filled with different carbon nanostructures as a function of carbon nanofiller loading. (Li, Y. et al.: Poly(propylene) nanocomposites containing various carbon nanostructures. *Macromolecular Chemistry and Physics*. 2011b. 212:2429–2438. Copyright Wiley-VCH Verlag GmbH & Co. KGaA. Reproduced with permission.)

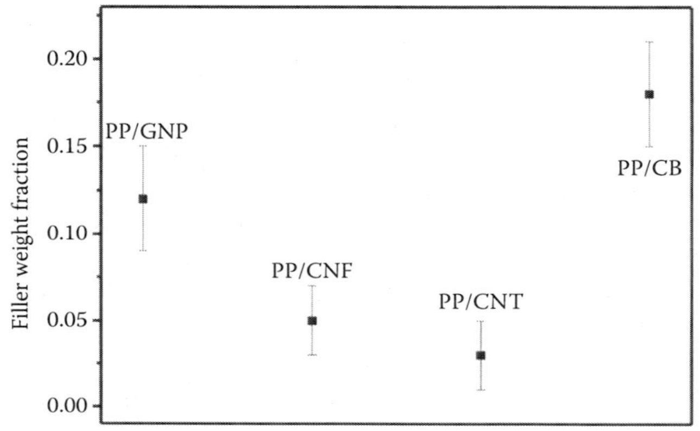

FIGURE 34.4 Percolation threshold of PP nanocomposites with different carbon materials. (Li, Y. et al.: Poly(propylene) nanocomposites containing various carbon nanostructures. *Macromolecular Chemistry and Physics*. 2011b. 212 (22):2429–2438. Copyright Wiley-VCH Verlag GmbH & Co. KGaA. Reproduced with permission.)

system was then hot-pressed to generate a composite with a segregated, highly conducting network of rGO filler (Pang et al. 2010); however, such a morphology could compromise the composite's mechanical properties due to the agglomeration of filler (Thostenson et al. 2005). Another example of low-threshold nanocomposite is the pioneering work of Stankovich (Stankovich et al. 2006), who prepared graphene–polymer composites via complete exfoliation of graphite and molecular-level dispersion of individual, chemically modified graphene sheets within polymer hosts (Figure 34.5). A polystyrene (PS)–graphene composite formed by this route exhibited a percolation threshold of ~0.1 vol%.

Alignment of the filler also plays a major role in the onset of electrical percolation: when the platelets are aligned in the matrix, there are, at least at relatively low concentrations, fewer contacts between them, and thus the percolation threshold would be expected to increase (Haggenmueller et al. 2000). The electrical percolation threshold also depends on the intrinsic filler properties, and both theoretical models (Hicks et al. 2009; Li and Kim 2007) and experiments (Eda and Chhowalla 2009) suggest that the electrical conductivity of a graphene/polymer nanocomposite depends strongly on the aspect ratio of the platelets, with a higher aspect ratio translating to a higher conductivity.

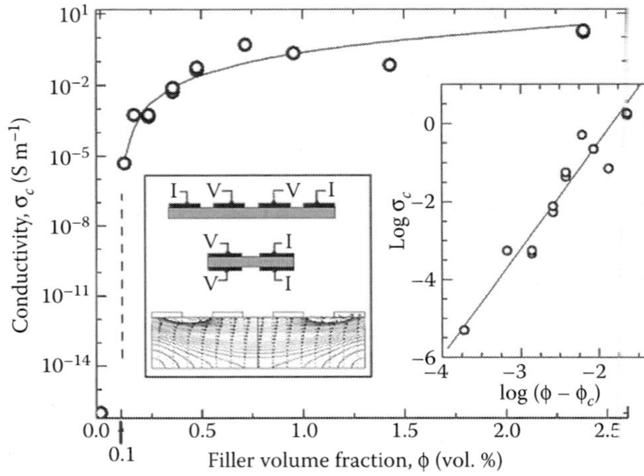

FIGURE 34.5 Composite conductivity σ_c, plotted against filler volume fraction ϕ. Right inset: log σ_c plotted against log($\phi - \phi_c$), where ϕ_c is the percolation threshold. Fitted parameters are: $t = 2.74$, $\sigma_f = 10^{4.92}$ S/m, and $\phi_c = 0.1$ vol%. Left inset shows details of measurement method. (Reprinted by permission from Macmillan Publishers Ltd. *Nature*, Stankovich, S. et al. Graphene-based composite materials, 442 (7100):282–286, copyright 2006.)

Graphene, depending on the exfoliation procedure, can be dispersed with the aid of a range of surfactants. Compared to water-soluble systems, thus far, only limited research work has been carried out with surfactant-assisted dispersions in organic solvents (Tkalya et al. 2012).

Interestingly, the recent work of Zhang et al. (2012) studied the effect of the surface chemistry of graphene (oxygen content of graphene sheets) on the electrical property of graphene–poly(methyl methacrylate) (PMMA) nanocomposites. The electrical percolation threshold increases with increasing the oxygen content of graphene sheets. PMMA composites with the lowest oxygen content in graphene show a dramatic increase in electrical conductivity of over 12 orders of magnitude, from 3.33×10^{-14} S/m with 0.4 vol% of graphene to 2.38×10^{-2} S/m with 0.8 vol% of graphene. The conductivity reaches up to 10 S/m at 2.67 vol%.

Not many theoretical models were proposed to explain the variation of the threshold with filler properties. Li and Kim (2007) introduced an analytical model based on the average interparticle distance (iPD) approach to predict the percolation threshold of conducting polymer nanocomposites containing three-dimensional (3D) randomly distributed GNPs. The individual GNP particle was assumed to be a thin and

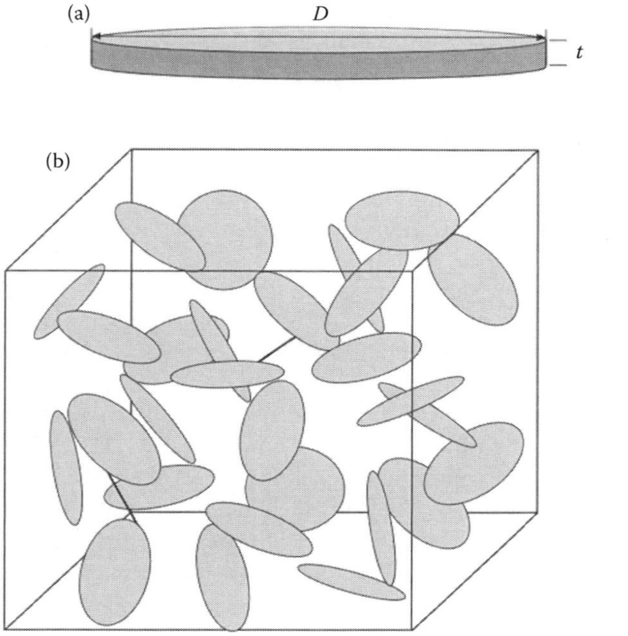

FIGURE 34.6 A model of GNP (a) and schematic model of conducting network of GNPs in 3D random distribution (b). (Reprinted from *Composites Science and Technology*, 67, Li, J., and J-K. Kim, Percolation threshold of conducting polymer composites containing 3D randomly distributed graphite nanoplatelets, 2114–2120, Copyright 2007, with permission from Elsevier.)

round platelet, and Figure 34.6 shows the model of a 3D conducting network. The formation of conducting networks is controlled by many factors, including the volume fraction, aspect ratio, and distribution of the conducting fillers.

To calculate the iPD between adjacent conductive particles, the particles were assumed to be homogeneously distributed within the matrix and perfectly bonded with the matrix (Figure 34.7). The composites were divided into cubic elements, each containing one particle in the center, and thus the total number of cubic elements was equal to the total number of particles. The orientation of the platelets was described in terms of three eulerian angles.

The expression for percolation threshold becomes the following:

$$v_f = \frac{27\pi D^2 t}{4(D + \mathrm{iPD})},$$

where D is the diameter and t is the thickness of platelet. When the iPD is equal to or less than 10 nm, electron hopping occurs, resulting in a rapid increase in the electrical conductivity of the composite, according to the quantum mechanical tunneling mechanism (Ruschau et al. 1992).

The experimental data are compared with the theoretical predictions calculated based on several models, including the present iPD model as shown in Figure 34.8. The percolation threshold increased linearly with increasing platelet

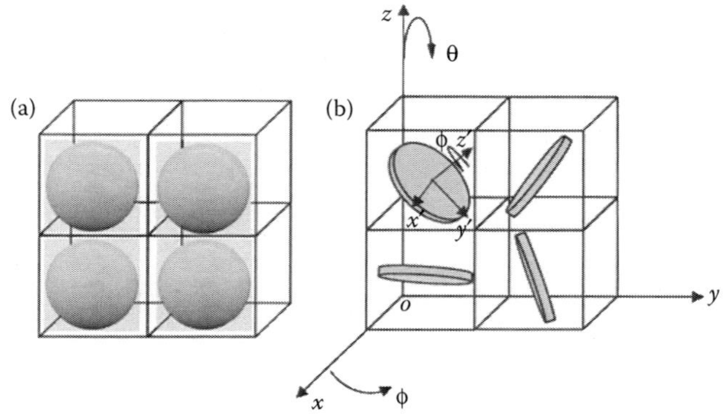

FIGURE 34.7 Schematics of the iPD model for (a) spherical fillers and (b) 3D distribution of fillers with high aspect ratio. (Reprinted from *Composites Science and Technology*, 67, Li, J., and J-K. Kim, Percolation threshold of conducting polymer composites containing 3D randomly distributed graphite nanoplatelets, 2114–2120, Copyright 2007, with permission from Elsevier.)

thickness, the rate of increase being higher for fillers with a smaller diameter (Figure 34.9).

The distribution of graphene particles was studied in several works. The conductivity behavior and volume organization of polymer composites composed of PS or polypropylene (PP) as the matrix filled with different loadings of graphene was studied (Syurik et al. 2012). During sample processing, the different melt flow behavior of the polymer matrix and the crystallization of the PP (low viscosity) cause predominantly isotropic and nonisotropic orientation of the graphene nanofillers in the bulk of graphene/PS (high viscosity) and graphene/PP composites, respectively (Figure 34.10). Scanning electron microscopy (SEM) charge contrast imaging is able

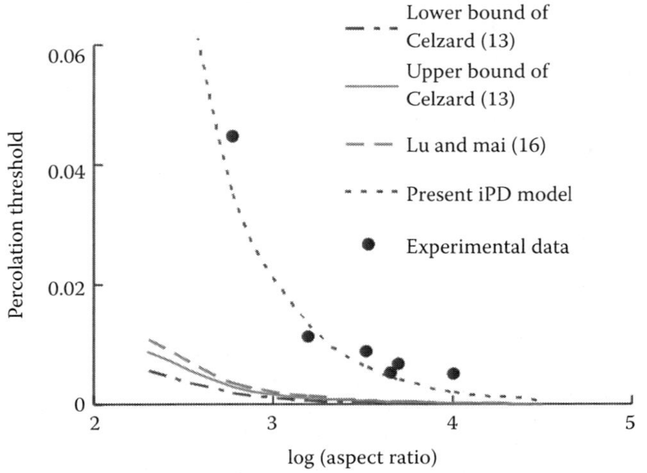

FIGURE 34.8 Comparison of experimental results and theoretical perditions for various GNP-reinforced polymer nanocomposites. (Reprinted from *Composites Science and Technology*, 67, Li, J., and J-K. Kim, Percolation threshold of conducting polymer composites containing 3D randomly distributed graphite nanoplatelets, 2114–2120, Copyright 2007, with permission from Elsevier.)

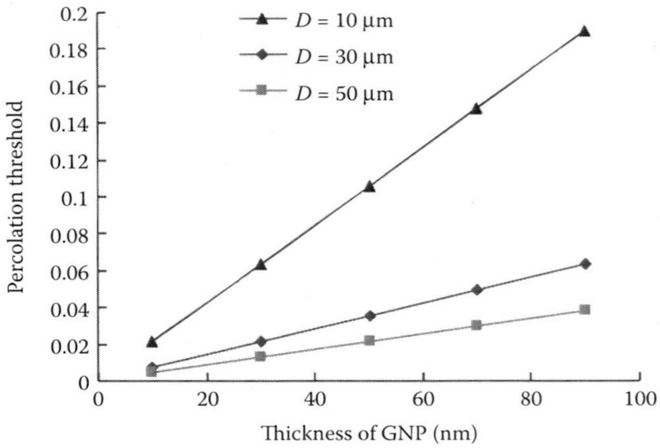

FIGURE 34.9 Effect of thickness of platelets with different diameter on percolation threshold. (Reprinted from *Composites Science and Technology*, 67, Li, J., and J-K. Kim, Percolation threshold of conducting polymer composites containing 3D randomly distributed graphite nanoplatelets, 2114–2120, Copyright 2007, with permission from Elsevier.)

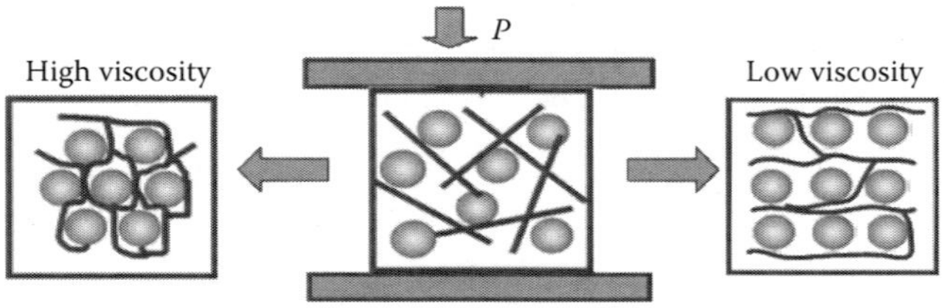

FIGURE 34.10 Illustration of graphene organization in the polymer matrixes potentially caused by the different viscosity and thus flow behavior of the polymer during compression molding. (Reprinted from *Macromolecular Chemistry and Physics*, 213, Syurik, Y. V. et al. Graphene network organisation in conductive polymer composites, 1251–1258, Copyright 2012, with permission from Elsevier.)

to visualize graphene nanoparticles and shows that for graphene/PS composites, the graphene sheets have no preferred orientation in the bulk of the sample, whereas for graphene/PP composites, most graphene sheets are aligned parallel to the sample top surface. Corresponding conductivity measurements for different setups in the direction parallel and perpendicular to the sample top surface confirm the isotropic 3D network organization of graphene in PS composites. On the other hand, for graphene/PP samples, substantial different conductivities are measured for the various measurement setups. Moreover, the broad conductivity percolation transition zone of graphene/PP ranging from about 0.2 wt% to 2.0 wt% is another indication for the nonisotropic and predominantly two-dimensional (2D) volume organization of the graphene network in PP composites. Such broad transition zone can be explained by the presence of conductive subnetworks, which gradually connect with each other with increasing graphene loading and ultimately form a single network in the bulk of the composite sample.

The formation of a segregated network of filler was observed in ultra-high-molecular-weight polyethylene (UHMWPE)/multiwalled carbon nanotubes (MWCNT), and UHMWPE/GNP composites (Ren et al. 2012). Transmission electron microscopy (TEM) and SEM images show that the filler is distributed on the surface of UHMWPE particles and forms a conducting network of segregated particles (Figure 34.11). The percolation threshold of UHMWPE/GNP composites is 0.25 wt% and that of UHMWPE/MWCNT composites is 0.20 wt%. The electrical conductivity of UHMWPE/GNP composites is almost four orders of magnitude lower than that of UHMWPE/MWCNT composites. For equivalent concentrations of GNP and MWCNT, the composites with hybrid fillers exhibit a lower percolation threshold and higher conductivity than that with GNP or MWCNT alone. This case is similar to the historical work of Malliaris, who was the first to study such segregated systems (Malliaris and Turner 1971).

Graphene particles connected and unconnected to the conductive network in graphene-based conductive composites can be clearly distinguished by measuring the same sample area with C-AFM and electric force microscopy (EFM). The different EFM signals obtained at the location of connected and unconnected particles are explained by the additional voltage drop between unconnected particles, charging of some unconnected particles, and, in some cases, by detecting particles below the surface. EFM is able to detect structure under the surface, for example, the contrast obtained at tip–sample distance of 25 nm provides information about the particles distribution down to ~60–90 nm from the surface. 3D reconstruction of the individual particles in the polymer

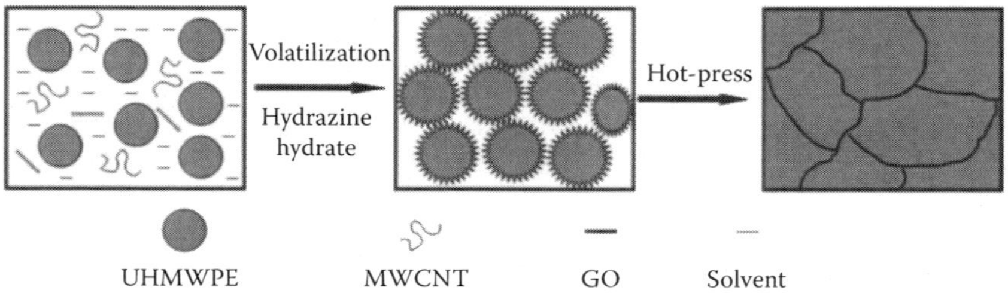

FIGURE 34.11 Schematic for preparation of UHMWPE/MWCNT and/or UHMWPE/GNS composites with segregated conductive networks. (Ren, P-G. et al.: Composites of ultrahigh-molecular-weight polyethylene with graphene sheets and/or MWCNTs with segregated network structure: Preparation and properties. *Macromolecular Materials and Engineering*. 2012. 297, 437–443. Copyright Wiley-VCH Verlag GmbH & Co. KGaA. Reproduced with permission.)

matrix with a correction procedure allows for a more precise volume reconstruction of properties and study of local organization of the graphene network on the nanoscale (Alekseev et al. 2012).

34.3 ELECTRICAL PROPERTIES OF GRAPHENE: POLYMER COMPOSITES WITH DIFFERENT MATRICES

Data for electrical properties of graphene–polymer composites with different matrices can be found in Sengupta et al. (2011).

34.3.1 Linear Crystalline Polymers

Many linear crystallizable polymers were used as nanocomposite matrices. Jiang selected polyphenylene sulfide (PPS) as the polymer matrix because of its high thermal and chemical tolerance (Jiang and Drzal 2012a). Composites with GNP were prepared by solid-state ball milling (SSBM) followed by compression molding. Although electrical conductivity exhibits a dramatic increase with increasing GNP content, it was still short of the target (>100 S/cm) even at 60 wt% GNP loading, which implies that using GNP alone as the single conductive filler may not be sufficient to provide enough conductive pathways for a desired electrical conductivity. A combination of large (20 μm) and small (5 μm) GNP particles was used. Electrical conductivity was tremendously enhanced with the increasing fraction of large GNPs. At a sample composition of PPS/CB/GNP(s)/GNP(l) (40:20:10:30 wt%), where the total filler loading is still kept at 60 wt%, the in-plane conductivity reaches 114 S/cm, which already exceeded the target. Higher electrical conductivity in the nanocomposites with large GNP was attributed to the fact that the aspect ratio of large GNP is much higher, which makes their interconnection with each other much easier to form conductive networks in the polymer matrix.

Composites PE–GNP were prepared by *in situ* polymerization (Fim et al. 2013). Nanocomposites had a percolation threshold of about 3.8 vol% (8.4 wt%). Conductivity at 8.9 vol% was 1.3×10^{-4} S/cm. This result was explained by the good dispersion of nanoparticles obtained by *in situ* polymerization, in which PE grew around the filler, isolating the filler particles with a layer of insulating material. This was unfavorable for the creation of a path for the transport of electrons and, hence, for the electrical network.

High-density polyethylene (HDPE)/exfoliated GNP nanocomposites fabricated with melt extrusion followed by injection molding had a relatively high percolation threshold between 10 and 15 vol% GNP loading. To lower the percolation threshold of injection-molded HDPE/GNP nanocomposites, two special processing methods were investigated: SSBM and solid-state shear pulverization (SSSP) (Jiang and Drzal 2012b). Results confirmed that the percolation threshold of HDPE/GNP nanocomposites could be reduced to between 3 and 5 vol% GNP loading by these two approaches. The mechanism by which SSBM and SSSP are capable of producing lower percolation is to coat the polymer surface with GNP, which facilitates the formation of conductive networks during injection molding. However, it was found that HDPE/GNP nanocomposites obtained from these two techniques exhibited lower mechanical properties at high GNP loadings.

Dodecyl amine-modified graphene (DA-G)/linear was used with low-density polyethylene (LDPE) and nanocomposites were prepared through solution mixing (Kuila et al. 2012). The functionalization of GNP facilitated dispersion of the filler in the polymer matrix. It was noted that crystallinity decreased in the nanocomposites due to the formation of random interfaces. Differential scanning calorimetry (DSC) analysis showed that the crystallization temperature of the nanocomposites was affected by the presence of DA-G due to its nucleation effects.

Composite materials of GNPs and PP were prepared by *in situ* polymerization (Polschikov et al. 2013; Pol'shchikov et al. 2013). Sonication of GNP powder was found to affect dielectric permittivity, and its dependence on the concentration of GNP dielectric permittivity of composites with sonicated GNP is much lower than composites with pristine GNP. In composites of PP and pristine GNP, appreciable electrical conductivity of 1.9×10^{-7} (S/cm) was found at filler concentration 2.4 vol%. Composites with sonicated GNP have electrical conductivity 1.5×10^{-10} and 1×10^{-6} (S/cm) at filler content 2.9 and 5.6 vol%, accordingly. This means that the value of percolation threshold is below 2.3 vol%. Calculations have shown that the aspect ratio for GNP particles in composite is equal to 77; sonication is found to decrease the aspect ratio of these aggregates in composite to 31.

A Grignard reagent, *n*-BuMgCl, was used to reduce graphene oxide and form loosely aggregated graphene sheets immobilized with few Mg–Cl species. Further complexing with $TiCl_4$ leads to a graphene-based supported catalyst that was ready for *in situ* olefin polymerization to make electrically conductive polyolefin/graphene nanocomposites (Huang et al. 2012). PP/graphene nanocomposites thus prepared possess a rather low electrical percolation threshold (approximately 0.2 vol%) and high conductivities (e.g., 3.92 S/m at 1.2 vol%, 28.5 S/m at 4.1 vol%, and 163.1 S/m at 10.2 vol%).

GNP-filled PP nanocomposites were also prepared through a facile solution dispersion method (Li et al. 2011a). SEM images show that the GNPs are bended and slightly aggregated in the polymer matrix. The GNP network formed at filler loading of around 12.0 wt% was verified by the volume resistivity and dielectric property measurements. The rheological results also verified that the critical percentage of the GNP network formation is around 12.0 wt%.

Nanocomposites made from PP and as-received graphite were prepared by SSSP as a function of graphite loading (0.3–8.4 wt%) (Wakabayashi et al. 2010). X-ray diffraction indicates that SSSP employing harsh pulverization conditions yields substantial graphite exfoliation at 0.3–2.7 wt% graphite content with less exfoliation being achieved at higher graphite content. Electrical conductivity measurements indicate percolation of graphite at 2.7 wt%. Significant tunability of graphite exfoliation and property enhancements is demonstrated as a function of SSSP processing.

PP/exfoliated graphite nanoplatelet nanocomposites with various aspect ratios of graphite nanoplatelets (large and small in diameter) were prepared by a melt-mixing procedure (Kuvardina et al. 2012). All composites exhibited typical percolation behavior: the percolation threshold value for the composites based on a smaller-size filler was 6 vol%, and decreased to 4 vol% for composites with a larger aspect ratio.

A novel process was developed to prepare electrically conducting maleic anhydride grafted polypropylene (gPP)/expanded graphite (EG) nanocomposites by solution intercalation (Jing-Wei Shen et al. 2003). The percolation threshold in these nanocomposites was 0.67 vol%, much lower than that of the conventional composites prepared by melt mixing (2.96 vol%) (Figure 34.12). When the EG content was 3.90 vol%, the electrical conductivity of the former reached 2.49×10^{-3} S/cm, whereas that of the latter was only 6.85×10^{-9} S/cm.

Poly(vinylidene fluoride) (PVDF) was widely used to prepare composites with graphene. Advanced layered GNS (graphene sheets)/PVDF nanocomposite films were prepared by solution blending and reduction, solution casting, and hot pressing (Shang et al. 2012). The process produced graphene sheets that were sufficiently exfoliated and aligned well in the PVDF matrix. The nanocomposites exhibited a much lower percolation threshold (1.29 vol%) than materials filled with CNTs. A high dielectric constant of 63 was obtained at 100 Hz when the concentration of GNS is 1.27 vol%, 9 times higher than that of pure PVDF. The dielectric performance of the nanocomposite films was ascribed mainly to the homogeneous dispersion and orientation of graphene in the PVDF matrix and strong interactions between the two components. Additionally, the breakdown strength reduction was found and attributed to the conductive graphene nanosheets and the presence of defects and impurities. The orientation of GNS in the PVDF matrix is mainly due to gravitational forces during solution flow and thickness confinement effect during solvent evaporation process. The GNS/PVDF composites constitute a new type of graphene-based composite materials to prepare high-k polymeric dielectric materials.

PVDF/GSs nanocomposites with various GSs loadings were also prepared by a solution method (Yu et al. 2011). Morphological studies revealed good dispersion of GSs in the PVDF matrix. The TGA results showed that even an extremely small amount GSs fillers can significantly enhance the thermal stability of PVDF/GSs nanocomposites.

Functionalized graphene sheets (FGSs) and exfoliated graphite (EG) were solution processed and compression molded to make PVDF composites (Ansari and Giannelis 2009). A lower percolation threshold (2 wt%) was obtained for FGS–PVDF composites compared to EG–PVDF composites (above 5 wt%). FGS–PVDF composites show an unusual resistance/temperature behavior. The resistance decreases with temperature, indicating an NTC (negative temperature coefficient of resistance) behavior while EG–PVDF composites show a PTC (positive temperature coefficient of resistance) behavior (e.g., the resistance increases with temperature). The NTC behavior of the FGS-based composites was attributed to the higher aspect ratio of FGS, which leads to contact resistance predominating over tunneling resistance.

PVDF/GNP composites were fabricated using solution mixing followed by compression molding (Li et al. 2010). Both the conductivity (Figure 34.13) and dielectric constant (Figure 34.14) grow quickly at the percolation threshold.

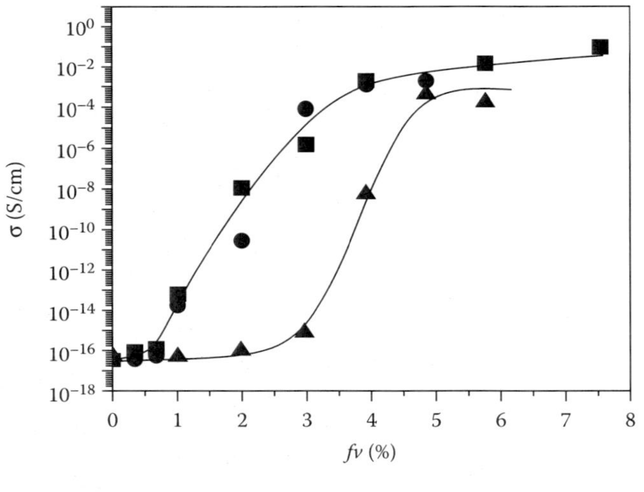

FIGURE 34.12 Electrical conductivity versus volume fraction of EG for gPP/EG composites prepared by two methods: Melt mixing (▲); solution intercalation (■); solution intercalation, repeated experiment (●). (Jing-Wei, S., X-M. Chen, and W-Y. Huang: Structure and electrical properties of grafted polypropylene graphite nanocomposites prepared by solution intercalation. *Journal of Applied Polymer Science*. 2003. 88, 1864–1869. Copyright Wiley-VCH Verlag GmbH & Co. KGaA. Reproduced with permission.)

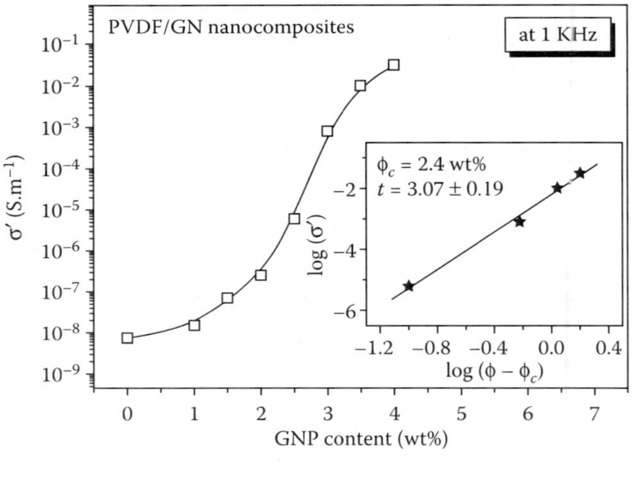

FIGURE 34.13 Electrical conductivity as a function of GNP content for PVDF/GNP composites. (Reprinted from *Synthetic Metals*, 160, Li, Y. C., S. C. Tjong, and R. K. Y. Li, Electrical conductivity and dielectric response of poly(vinylidene fluoride)–graphite nanoplatelet composites, 1912–1919, Copyright 2010, with permission from Elsevier.)

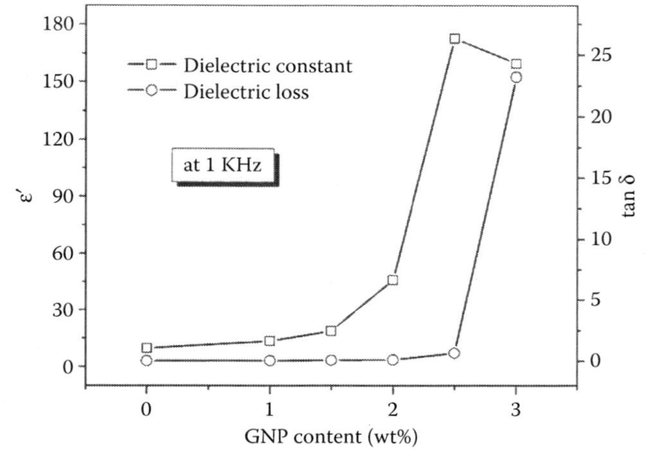

FIGURE 34.14 Dielectric constant and dielectric loss as a function of GNP content for PVDF/GNP composites. (Reprinted from *Synthetic Metals*, 160, Li, Y. C., S. C. Tjong, and R. K. Y. Li, Electrical conductivity and dielectric response of poly(vinylidene fluoride)–graphite nanoplatelet composites, 1912–1919, Copyright 2010, with permission from Elsevier.)

A large dielectric constant of 173 and low loss tangent of 0.65 were observed in the PVDF/2.5 wt% GNP nanocomposite at 1 kHz.

The best fit of experimental data yields a percolation threshold of 2.4 wt% and a critical exponent of 3.07 (inset of Figure 34.14). The critical exponent is larger than the theoretical value ($t = 2$) for the three disordered system. In general, several conducting polymer systems filled with large-aspect-ratio fillers exhibit large experimental values. For example, Panwar and Mehra (2008) obtained $t = 2$ for the SAN (styrene–acrylonitrile resin)/GNP composites. Weng et al. (2004) reported a value of 2.32 for the nylon 6/foliated graphite nanocomposites.

Graphene/PVDF composites were prepared by the *in situ* solvothermal reduction of graphene oxide in the PVDF solution (He and Tjong 2013). The percolation threshold of such a composite was determined to be 0.31 vol%, being much smaller than that of the composites prepared via blending reduced graphene sheets with a polymer matrix. This is attributed to the large aspect ratio of the reduced graphene sheets and their uniform dispersion in the polymer matrix. The dielectric constant of PVDF showed a marked increase from 7 to about 105 with only 0.5 vol% loading of reduced graphene content.

The nanocomposite films of a FGS and poly(ethylene oxide) (PEO) were cast from the physical blend of an aqueous FGS dispersion assisted by sodium dodecyl sulfate and an aqueous PEO solution (Lee et al. 2010). The percolation threshold was about 1.5 wt% for these nanocomposites.

34.3.2 Linear Amorphous Polymers

Amorphous polymers are by far the most used type of polymer matrices for graphene nanocomposites. Poly(vinylalcohol) (Jiang et al. 2010; Songet al. 2012; Tantis 2012; Vickery et al. 2009; Yang et al. 2010), thermoplastic polyurethane (TPU) (Scognamillo et al. 2012), PMMA (Yuan 2012), poly(dimethyl siloxane) (Wang et al. 2012), acrylonitrile butadiene rubber (Al-solamy et al. 2012), poly(tetraethylene glycol diacrylate) (Alzari et al. 2011), nylon-6 (Xu and Gao 2010), polyimide (Arlen et al. 2008), and other matrices were used (Sengupta et al. 2011).

The nanocomposites based on reduced graphene oxide/hydroxylated styrene–butadiene–styrene triblock copolymer were prepared via a solution blending method (Xiong et al. 2012). The percolation threshold appears at the filler concentration of 0.09–0.23 vol% (0.2–0.5 wt%). Such a low percolation threshold is evidently due to excellent dispersion of rGO and the formation of a continuous network in matrix. An initial steep increase of conductivity is observed when the filler content is lower than 0.5 wt%. Beyond this filler content, there is only a slight increase in electrical conductivity, and it seems to level off at 1.3 S/m even when the filler content is up to 5.4 vol% (12 wt%).

The construction of a 3D, compactly interconnected graphene network can offer a huge increase in the electrical conductivity of polymer composites. However, it is still a great challenge to achieve desirable 3D architectures in the polymer matrix. Highly conductive polymer nanocomposites with 3D compactly interconnected graphene networks were obtained using a self-assembly process in PS and ethylene vinyl acetate (EVA) polymer matrices (Wu et al. 2013). The procedure was as follows. First, PS nanospheres were synthesized via emulsion polymerization and modified by surface grafting of amine groups, which render the sphere surface positively charged. Then, the modified PS nanospheres were assembled with negatively charged GO by electrostatic interactions, resulting in the demulsification of PS spheres and the formation of a coagulation. GO was uniformly distributed in the gap between the spheres. Under the optimal assembly conditions, almost all the spheres were encapsulated and connected by GO in the coagulation, leaving a transparent aqueous solution. Finally, the resulting coagulation was made into films by filtering and reduced *in situ* in hydrogen iodide solution, and then hot-pressed to disk-like plates, resulting in a 3D graphene architecture in polymer matrix. The obtained PS composite film with 4.8 vol% graphene shows a high electrical conductivity of 1083.3 S/m, which is superior to that of the graphene composite prepared by a solvent mixing method.

GO synthesized by Hummers' method was chemically reduced with various reducing agents to produce reduced GOs (Ha et al. 2012). Chemically modified graphene sheets were dispersed in a high-performance polyimide (PI) matrix using polyamic acid (PAA)/graphene nanocomposite as a precursor. PI nanocomposite films with different loadings of graphene sheets were prepared by thermal imidization of the as-prepared PAA/graphene nanocomposites. The electrical conductivity of the nanocomposites containing 5 wt% rGO_L was increased by 14 orders of magnitude, compared to that of pure PI, whereas the conductivity of the GO-reinforced nanocomposite was 10 orders of magnitude higher than that of PI. rGO_L showed the maximum improvement in the electrical conductivity of PI, suggesting the efficient deoxygenation of GO with L-ascorbic acid.

Processing, morphology, and properties of TPU reinforced with exfoliated graphite were investigated (Kim et al. 2010). They compared carbon sheets exfoliated from graphene oxide (GO) via two different processes: chemical modification (isocyanate-treated GO—iGO) and thermal exfoliation (thermally reduced GO—TRG), and three different methods of dispersion: solvent blending, *in situ* polymerization, and melt compounding. Incorporation of as low as 0.5 wt% of TRG produced electrically conductive TPU. Both graphite and TRG effectively increased the electrical conductivity, but they differ greatly in the onset concentration for electrical percolation (>2.7 vol% for untreated graphite versus <0.5 vol% for TRG). Conductivity measurements indicate that the average particle aspect ratio (ratio of diameter to thickness) of TRG is considerably greater than that of graphite since for randomly oriented disks, the percolation threshold is inversely proportional to this parameter.

Polyethylene terephthalate (PET)/graphene nanocomposites were prepared by melt compounding with graphene nanosheets, prepared by complete oxidation of pristine graphite followed by thermal exfoliation and reduction (Zhang et al. 2010). The incorporation of graphene greatly improved the electrical conductivity of PET, resulting in a sharp transition from electrical insulator to semiconductor with a low percolation threshold of 0.47 vol%. A high electrical conductivity of 2.11 S/m was achieved with only 3.0 vol% of graphene. The low percolation threshold and superior electrical conductivity are attributed to the high aspect ratio, large specific surface area, and uniform dispersion of the graphene nanosheets in the PET matrix.

GNS–PMMA nanocomposites were prepared by *in situ* suspension polymerization and reduction of graphene oxide using hydrazine hydrate and ammonia (Wang et al. 2011). PMMA microspheres with a mean diameter of 2 μm are mainly covalently linked to the surface of GNS. The obtained GNS–PMMA composites not only have high electrical conductivity but also enhanced mechanical properties and thermal stability at low loadings of graphene.

Using TRG and *in situ* polymerization, Liao et al. (2012) created nanocomposites of polyurethane acrylate (PUA) with an ultralow percolation concentration of 0.15 wt% (0.07 vol%) graphene. TRG was solvent-blended in UA to form uncured TRG/UA liquids and was polymerized by free radical polymerization. Percolation concentrations of polymerized TRG/PUA nanocomposites occurred at 0.15 wt% (0.07 vol%), as determined by surface resistance measurements, bulk electrical conductivity, and modulus. The percolation concentration of TRG/UA uncured liquid (0.5 wt%) yields an aspect ratio of 210, which is lower than the aspect ratio of 750 of freestanding TRG. This lower aspect ratio may indicate aggregation of TRG in uncured liquid UA. For polymerized TRG/PUA nanocomposites, an aspect ratio of 750 was obtained by calculation from 0.15 wt% percolation concentration. This result indicates that TRG is nearly single-layer-dispersed in PUA. The latex technology concept is applied for the preparation of graphene/PS nanocomposites (Tkalya et al. 2010). Aqueous dispersions of graphene are obtained via oxidation and exfoliation of graphite and subsequent reduction in the presence of surfactant. Different amounts of aqueous graphene dispersions are then mixed with PS latex, and composites are prepared by freeze-drying and subsequent compression molding. The morphology of conductive nanocomposites was studied with SEM charge contrast imaging. The percolation threshold for conduction is below 1 wt% of graphene in the composites, and a maximum conductivity of about 15 S/m can be achieved for 1.6–2 wt% nanofiller.

Graphene oxide (GO) synthesized by Hummers' method was chemically modified with ethyl isocyanate to give ethyl isocyanate-treated graphene oxide (iGO), which is readily dispersed in dimethylformamide (DMF) (Luong et al. 2011). The iGO dispersion in DMF was then used as media for the synthesis of polyimide/functionalized graphene composites (PI/FGS) by *in situ* polymerization. It was shown that with the addition of only 0.38 wt%, the electrical conductivity of the PI/FGS was increased by more than eight orders of magnitude to 1.7×10^{-5} S/m.

A catalyst was prepared by loading TiO_2/SiO_2 nanoparticles onto the surface of organically modified GO sheets in order to promote the exfoliation of graphene nanosheets by *in situ* intercalative polymerization (Feng et al. 2011). The synthesis of PET/graphene composites was carried out using the as-made catalyst, and nanoscale dispersion and thermal reduction of graphene oxide were achieved in one step. Since the thermal reduction was performed under a mild condition, the desired chemical modification of GO was maintained, which is important for the uniform dispersion of graphene in PET. Graphene sheets were well exfoliated and homogeneously dispersed in PET, as confirmed by SEM and TEM observations. The electrical percolation threshold of the PET/graphene composites is as low as about 0.1 wt%.

Amorphous thermoplastic (PET copolymer) matrix nanocomposites based on graphite nanoparticles were produced by melt mixing, using different types of graphite (Greco et al. 2012). The PET nanocomposite with graphite showed an improvement in electrical conductivity. The nanocomposite obtained with coarser particles (graphite and graphite intercalation compound) showed a moderate increase in electrical conductivity, whereas nanocomposites produced with finely dispersed particles showed a significant improvement of electrical conductivity.

34.3.3 Network Polymers

The most widely used 3D polymer matrix for graphene nanocomposites was epoxy resin. Epoxy matrix nanocomposites reinforced with graphene stacks, ranging from 1 wt% to 3 wt% were prepared and characterized (Corcione et al. 2013). The comparison of mechanical and rheological properties with simple mathematical models indicated that the aspect ratio of EG is in the order of 1000, that is, a dispersion of nanoscale graphene stacks was obtained. This result suggests that the measurement of engineering properties of nanocomposites not only represents an objective but can also provide information about the average degree of dispersion.

Graphene platelets were successfully dispersed in an epoxy resin and cured by UV irradiation (Martin-Gallego et al. 2011). The UV-shielding effect of the graphene filler induced a slight decrease in epoxy group conversion when lower irradiation intensity was used. Nevertheless, a fully crosslinked hybrid network was obtained by increasing irradiation time and light intensity. The cured materials were characterized by high gel content, high final conversion, and good mechanical performance. The addition of the graphene platelets induced an increase in stiffness of the cured epoxy, and a marked increase of T_g, as well as enhanced storage modulus at higher temperature. Further discussion of electrical properties of network polymer nanocomposites with graphene can be found in Section 34.5.

34.3.4 Intrinsically Conducting Polymers

Graphene is added to conducting polymers in order to increase conductivity, add charge transfer sites, and enhance the mobility of charge carriers.

A novel route has been developed to synthesize polypyrrole (PPy)/graphene oxide (GO) nanocomposites via liquid/liquid interfacial polymerization where GO and initiator were dispersed in the aqueous phase and the monomer was dissolved in the organic phase (Bora and Dolui 2012). An increase in conductivity (0.360–0.507 S/cm) as compared to pure PPy was observed from electrical measurements. The composite shows reversible electrochemical response and a good cycling stability even up to 100 cycles. The PPy/GO composite bears a great application potential in energy storage devices, rechargeable batteries, biosensors, and other fields.

Nanocomposites of poly-(anthraquinonyl sulfide) (PAQS)-FGS and polyimide (PI)-FGS were successfully synthesized through a simple *in situ* polymerization process in methylpyrrolidinone solvent (Song et al. 2012). In the nanocomposite, the FGS surface is uniformly coated by the polymer, and FGSs are well dispersed in the polymer matrix due to the thorough dispersion of FGSs in N-Methyl-2-pyrrolidone (NMP) and the noncovalent interaction between the graphene surface and the polymers. The surface area and electronic conductivity of the nanocomposite materials are significantly enhanced, leading to great improvement in the battery performance. Compared to the pure polymer, the nanocomposites possess much higher active material utilization ratios and very high charge and discharge ability; for example, PAQS-FGS-b can deliver 100 mAh/g within just 16 s. The high-rate performance is explained by fast redox kinetics, fast electron transfer, and enhanced Li-ion conduction due to an increased surface area in the nanocomposites.

PPy/GR composites were prepared by *in situ* synthesis of GO and pyrrole monomer followed by chemical reduction using hydrazine monohydrate (Bose et al. 2010). The PPy/GR composites exhibited a high value of conductivity, while pure PPy shows a conductivity of 0.19 S/cm; in nanocomposites, this value reaches 1.64 S/cm. Increase in the magnitude of conductivity as compared to pure PPy was attributed to the π–π stacking between the GO layers and PPy.

Graphene oxide-doped polyaniline (PANI) nanocomposites were synthesized by a soft chemical method (Wang et al. 2010). They exhibit excellent capacitance as high as twice that of pure PANI. The size of graphite particles and their concentration have a pronounced effect on the electrochemical performance of nanocomposites. The product shows a good application potential in supercapacitors or other power source systems.

A layered PPy/graphene oxide (PPy/GO) composite was synthesized by *in situ* emulsion polymerization in the presence of cationic surfactant as emulsifier (Gu et al. 2010). The characterization of the bonding structure of the PPy/GO composite shows the π–π electron stacking effect between the PPy and the GO. PPy is decorated on the edge and surface of GO, and GO sheets disperse uniformly in the PPy/GO composite. The conductivity of the PPy/GO composite is as large as 5 S/cm, which is much larger than the usual value of pure PPy and pristine GO.

In situ polymerization was used to synthesize high-electrical-conductivity PPy/GO composites without doping materials (Gu et al. 2009). GO sheets intercalated and coated by PPy were 10 nm thick. The conductivity of the composite was improved by four orders of magnitude compared to that of GO. The PPy/GO composites show improved thermal stability compared to pure PPy, especially in the temperature range 430–700°C.

34.3.5 Biocompatible Polymers

Biopolymers and biocompatible polymers were also used as matrices for graphene nanocomposites (Lu et al. 2012).

Transparent conducting graphene–RNA nanocomposite thin films were prepared by surfactant-assisted exfoliation of graphite in the presence of RNA, a biocompatible nonionic surfactant (Sharifi et al. 2012). Parameters such as the type of RNA used and the size of starting graphite flakes are demonstrated to be essential for obtaining RNA–graphene thin films of good quality. Post- and predeposition treatments (including thermal annealing, functionalization of the films, and the preoxidation of graphite) are critical to improve the performance of graphene–RNA nanocomposites as transparent conductors. Thin films with $R \sim 200\ \Omega/\upsilon$ at 50% transmittance, or $R \sim 2.3\ \mathrm{M}\Omega/\upsilon$ at 85% transmittance are reported.

Biodegradable poly(3-hydroxybutyrate-*co*-4-hydroxybutyrate) (PHBV)/graphene nanocomposites were prepared by solution casting (Sridhar 2013). Mechanical properties of the composites were substantially improved as evident from dynamic mechanical and static tensile tests. Differential thermal analysis showed an increase in the temperature of maximum degradation. Soil degradation tests of PHBV/graphene nanocomposites showed that the presence of graphene does not interfere in its biodegradability.

A series of nanocomposite films that consisted of chitosan (CS), polyvinyl alcohol (PVA), and GO were prepared by a solution casting method (Jun Ma et al. 2012). The results indicated that the GO improved the miscibility between CS and PVA due to hydrogen bonding interaction between GO and CS/PVA matrix, and that the GO does play the role of physical cross-linking points. The results from tensile testing and thermal analyses indicated that the incorporation of GO favored the polymer mechanical properties and thermal stability of nanocomposite films.

34.4 AC DIELECTRIC PROPERTIES OF POLYMER COMPOSITES WITH GRAPHENE

Apart from DC conductivity, AC electrical properties of graphene nanocomposites are both interesting and important.

PVA polymer was blended with either graphene oxide or FGSs (Tantis 2012). A surfactant-like block copolymer was employed as an exfoliation/dispersion agent of the GNP within the PVA matrix through solution processing and compression molding. Dielectric measurements revealed that the graphene/copolymer/PVA system displayed a tendency for diminished values of permittivity with filler content. On the contrary, in the spectra of GO/PVA samples, a systematic increase of permittivity with filler content was recorded. This could be attributed to the formation of an insulating coating between the graphite inclusions and PVA because of the presence of the copolymer.

Graphene nanosheet–bisphenol A polycarbonate nanocomposites (0.027–2.2 vol%) prepared by both emulsion mixing and solution blending methods, followed by compression molding, exhibited DC electrical percolation thresholds of 0.14 and 0.38 vol%, respectively (Yoonessi and Gaier 2010). The conductivities of 2.2 vol% graphene nanocomposites were 0.512 and 0.226 S/cm for emulsion and solution mixing. The 1.1 and 2.2 vol% graphene nanocomposites exhibited frequency-independent behavior (Figure 34.15).

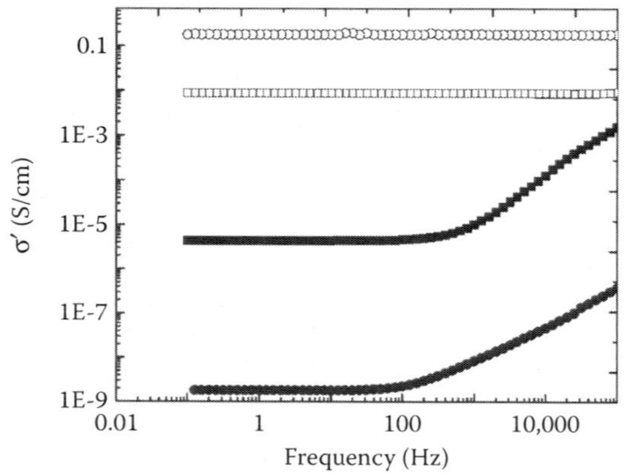

FIGURE 34.15 Frequency-dependent conductivities for solution-blended nanocomposites with graphene; volume fractions of 0.41 (●), 0.55 (■), 1.1 (□), and 2.35 (O). (Reprinted with permission from Yoonessi, M., and J. R. Gaier. Highly conductive multifunctional graphene polycarbonate nanocomposites. *ACS Nano*, 4 (12): 7211–7220. Copyright 2010 American Chemical Society.)

In general, the conductivity measured at frequency $f \to 0$ (f = 0 Hz) is often taken as the DC value. Thus the variation of conductivity with f is described by the equation $\sigma(f) = \sigma_{dc} + Af^s$, where A is a temperature-dependent constant. This equation is often referred to as AC universal dynamic response (Jonscher 1977) or AC universality law (Dyre and Schrøder 2000) because many materials display such a characteristic. Therefore, AC conductivity can be recognized as the combined effect of DC conductivity (f = 0 Hz) caused by migrating charge carriers and frequency-induced dielectric dispersion. Composites exhibiting frequency dependence of AC conductivity are below the percolation threshold, while in composites with flat conductivity–frequency dependence, the filler concentration is above the threshold.

A stacked disk model was used to obtain the average particle radius, average number of graphene layers per stack, and stack spacing by simulation of the experimental SANS data. Morphology studies indicated the presence of well-dispersed graphene and small graphene stacking with infusion of polycarbonate within the stacks. Simulation of the SANS data with the stacked disk model indicated that the average radius of the graphene particles ranged from 1.7 to 2.7 mm, and as the concentration increased from 0.1 to 2.2 vol%, the number of graphene sheets in a stack increased from 7 to 14, where the spacing within a stack decreased from 27 to 17 nm.

A new polymer poly[2,2-bis((4-acetylbenzoyloxy) methyl) propane-1,3-diyl bis(4-acetylbenzoate)] (PAPA) was used to prepare composites with graphene by ultrasonic mixing of PAPA and PVA with different GR percentages (Latif et al. 2013). Composite films were made by a conventional solvent casting technique. At low graphene weight content (≤2.50%), the electrical conductivity of graphene composites increases with increasing frequency. These specimens show a typical insulating behavior with a frequency-dependent conductivity, but when the graphene concentration reaches 4.165 wt%, the curve showed a weak slope, indicating that DC conductivity increased continuously with an increase in GR content, and conductivity became independent of frequency.

In aiming to obtain highly flexible polymer composites with high dielectric performance, graphene/PVDF composites with a multilayered structure were proposed and prepared (Fan et al. 2012). Graphene sheets were prepared by reducing graphene oxide using phenylhydrazine, which could effectively alleviate aggregation of the graphene sheets. A two-step method, including solution casting and compression molding, was employed to fabricate the graphene/PVDF composites. The composites showed an alternative multilayered structure of graphene sheets and PVDF. Owing to their unique structure, the composites had an extremely low percolation threshold (0.0018 volume fraction of graphene), which was the lowest percolation threshold ever reported among PVDF-based polymer composites. A high dielectric constant of more than 340 at 100 Hz was obtained within the vicinity of the percolation threshold when the graphene volume fraction was 0.00177. Above the percolation threshold, the dielectric constant continued to increase and a maximum value as high as 7940 at 100 Hz was observed when the graphene volume fraction was 0.0177. After the solution was cast and the solvent evaporated, the graphene sheets with large aspect ratios tended to preferentially orient in the plane of the film. The hot-press process further strengthened the preferred orientation of graphene sheets.

Composite materials of isotactic polypropylene and nanocarbon filler, GNP were synthesized by *in situ* polymerization in liquid propylene with the use of a metallocene catalyst system (Shevchenko et al. 2012). The real and imaginary part of conductivity and permittivity of PP nanocomposites with GNP particles was measured in the frequency range 10^{-2}–10^{7} Hz. From these data, the value of the percolation threshold for composites with pristine GNP particles is approximately 0.25 vol%. This compares with the available data on the low value of percolation thresholds in polymer composites with graphene. In composites with sonicated GNP particles, the percolation threshold is approximately 2–3 vol%, or much higher than in the case of particles that were not subjected to ultrasonic treatment. This fact seems to indicate (1) reduction in the size of filler particles aggregates and (2) much more uniform distribution of particles in the polymer matrix after ultrasonic treatment. Dependences of ε' on filler concentration were measured in the microwave region and analyzed with a mathematical model, developed by the authors (Koval'chuk et al. 2008). Calculations show that in composites with pristine GNP particles, their aspect ratio is 110, whereas ultrasonic processing reduces the aspect ratio to 40–50. This suggests that the particles of pristine GNP in composite form small-sized anisotropic aggregates. Ultrasonic processing provides, apparently, a more uniform distribution of GNP particles, which are largely individualized in nanocomposites. The composites have low electromagnetic wave reflection properties and are promising materials for creating electromagnetic radiation screens of absorbing type.

34.5 MICROWAVE PROPERTIES OF POLYMER COMPOSITES WITH GRAPHENE

Graphene composites can be useful materials for radio and microwave frequency applications (Liao and Duan 2012), and several research groups investigated the properties of these composites at high frequencies.

Nanocomposites of PP and graphene nanoplatelets were synthesized by *in situ* polymerization in liquid monomer in the presence of a highly effective isospecific homogeneous metallocene catalyst (Polschikov et al. 2013). Composites of isotactic PP—IPP/GNP showed high dielectric permittivity and losses in the microwave range (Shevchenko et al. 2012). Dielectric permittivity (ε') increases sharply with increasing filler content, which is apparently caused by high interface surface in composites. The values of permittivity are much higher, compared to composites with MWCNT at similar concentration. The important factor is that the percolation threshold is rather high for *in situ* polymerized composites, so that high local electrical conductivity is combined with no appreciable bulk conductivity. This results in a considerable increase of dielectric losses, while permittivity remains much less, than could be in the presence of bulk conductivity. The combination of these two factors is favorable for reducing the reflection of electromagnetic radiation and increasing its absorption.

NR-based nanocomposites containing a constant amount of standard furnace carbon black and GNP in concentrations from 1 to 5 phr have been prepared (Al-Hartomy et al. 2012). Adding small amounts of GNP to a constant amount of carbon black can be used as a way to control and primarily to improve the dielectric and microwave properties of composites based on natural rubber in the high frequency range 1–12 GHz. The results achieved allowed the recommendation of graphene as the second filler for natural rubber-based composites to afford specific absorbing properties.

34.5.1 EMI Shielding Polymer Composites with Graphene

Four interesting features of polymer composites filled with carbonaceous particles are their lightweight structure, their moldability, their thin matching thickness, and their broad bandwidth; however, most of the experimental results point out that the carbonaceous material needs to be doped or surface modified, for example, CNTs decorated by magnetic particles, in order to have strong microwave absorbers. Even if these smart materials are promising electromagnetic wave absorbers in the laboratory, it remains a challenge to develop suitable techniques for mass production of these polymer-filled nanocomposites. Of particular interest is that researchers have found applications of graphene in microwave absorption (Choudhary et al. 2012; Singh et al. 2012). For example, Bai et al. (2011b) reported a 38.8 dB absorption at 16 GHz with 2.6 vol% graphene content. Another, more recent, batch of microwave experiments were conducted on nickel-coated graphene (Fang et al. 2011). Microwave absorption is due to dielectric losses, and nickel coating makes particles ferromagnetic, improving the maximum absorption from 6.5 dB to 16.5 dB at 1–18 GHz. Zhang et al. (2011) fabricated PMMA/graphene composites and found that when the matching thickness is 1.5 mm and the graphene content is 1.8 vol%, the maximum absorption is 18 dB at 9.25 GHz and the frequency region in which the maximum reflection loss is more than 10 dB is 9.5–14.6 GHz. These nanocomposites not only exhibited a high conductivity of 3.11 S/m, but also a good EMI shielding efficiency of −13 to −19 dB at the frequency range of 8–12 GHz. The EMI shielding efficiency was mainly attributed to the absorption rather than the reflection in the investigated frequency range.

In another study, Liang et al. (2009) fabricated composites based on graphene sheets by incorporating solution-processable functionalized graphene into the epoxy matrix and their EMI shielding was studied. The composites show a low percolation threshold of 0.52 vol%. EMI shielding effectiveness was tested over a frequency range of 8.2–12.4 GHz (X-band) and −21 dB shielding efficiency was obtained for

15 wt% (8.8 vol%) loading, indicating that they may be used as lightweight, effective EMI shielding materials. Eswaraiah et al. (2011) prepared functionalized graphene (f-G)–PVDF nanocomposites and investigated their electrical conductivity and EMI shielding efficiency as a function of the mass fractions of graphene. A dramatic change in the conductivity was observed from 10^{-16} S/m for insulating PVDF to 10^{-4} S/m for 0.5 wt% composites, which can be attributed to the high-aspect-ratio and highly conducting nature of nanofiller. An EMI shielding effectiveness of ~– 20 dB was obtained in X-band (8–12 GHz) region and ~18 dB in broadband (1–8 GHz) region for 5 wt% composite. The authors reported coefficients of reflection (*R*), transmission (*T*), and absorption (*A*) of 0.78, 0.01, and 0.21, respectively, for the PVDF composite containing 5 wt% of f-G and the reflectivity increases from 10% to 80% with the increase in mass fraction of f-G from 1 to 5 wt%. It was also concluded that such f-G/PVDF composites were more reflective and less absorptive to electromagnetic radiation in both X-band and broadband frequencies, that is, the primary EMI shielding mechanism of such composites was reflection rather than absorption. The higher EMI shielding efficiency at lower loadings of f-G over other CNT-based polymer composites was attributed to the high aspect ratio and high electrical conductivity of f-G nanofiller.

Nanocomposites (polyester and epoxy) filled with different carbon nanostructures, like MWCNTs and GNPs, and nickel-coated carbon fibers were fabricated and experimentally characterized in the X-band and Ku-band (De Bellis et al. 2010). It turned out that GNPs were particularly suitable for the production of low-loss dielectrics, to be used as spacer in wideband Salisbury screens, with thickness not exceeding 2 mm in the X- and Ku-bands. The best-performing screens in the Ku-band had thicknesses around 1.5 mm, minimum reflection between −58 dB and −61 dB and bandwidth at −10 dB larger than 4.5 GHz. For the X-band, the best-performing screen had a thickness of 2 mm, a minimum reflection around −71 dB, and a bandwidth of 2.81 GHz. It should be noted that the use of the nanocomposite loaded with 0.5% with GNPs produces an improvement of the absorber performance especially in the Ku-band, whereas the spacer, filled with higher percentage of GNPs is preferable in the X-band.

Composites based on graphene-based sheets have been fabricated by incorporating solution-processable functionalized graphene into an epoxy matrix, and their EMI shielding studies were studied (Liang et al. 2009). The composites showed a low percolation threshold of 0.52 vol%. EMI shielding effectiveness was tested over a frequency range of 8.2–12.4 GHz (X-band), and 21 dB shielding efficiency was obtained for 15 wt% (8.8 vol%) loading, indicating that they may be used as lightweight, effective EMI shielding materials (Figure 34.16).

The microwave absorption ability of graphene and MWCNT was compared in the TPU matrix. Results showed that graphene has better absorption capability than MWCNT (Das et al. 2012).

EMI shielding was investigated in ICP/oxidized graphene-based nanocomposites–polyaniline/gold nanoparticles/graphene oxide in the 2.0–12.0 GHz frequency range (Saini and Aror 2012). According to the authors, the SE values observed for GO and PANI–GNP and PANI–GNP–GO composites were in the ranges −20 to −33 dB, −45 to −69 dB, and −90 to −120 dB, respectively (Basavaraja et al. 2011).

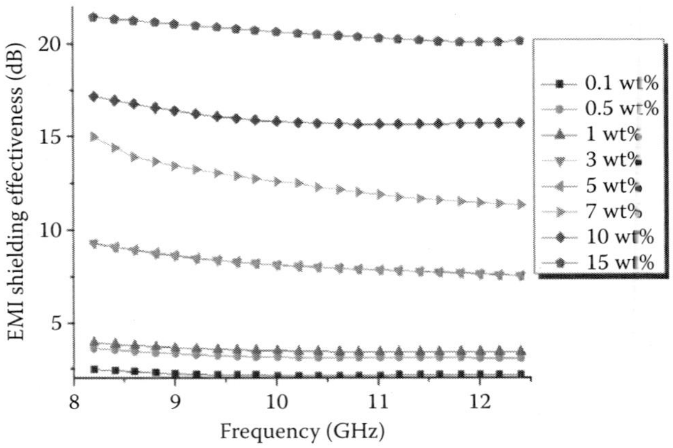

FIGURE 34.16 EMI shielding effectiveness of graphene/epoxy composites with various solution-processable functionalized graphene loadings as a function of frequency in the X-band. (Reprinted from *Carbon*, 47, Liang, J. et al. Electromagnetic interference shielding of graphene/epoxy composites, 922–925, Copyright 2009, with permission from Elsevier.)

34.6 NONLINEAR EFFECTS AND CURRENT–VOLTAGE CHARACTERISTICS

Polymer composites with graphene were found to possess other interesting and important electrical properties, apart from simple DC conductivity.

Graphene–PVA nanocomposite films with a thickness of 120 μm were synthesized by solidification of PVA in a solution with dispersed graphene nanosheets (Mitra et al. 2011). Electrical conductivity data were explained as arising due to hopping of carriers between localized states, formed at the graphene–PVA interface. Dielectric permittivity data as a function of frequency indicated the occurrence of Debye-type relaxation mechanism. The nanocomposites showed a magnetodielectric effect with dielectric constant changing by 1.8% as the magnetic field was increased to 1 T. The effect was explained as arising because of Maxwell–Wagner polarization as applied to an inhomogeneous 2D, two-component composite model. This type of nanocomposite may be suitable for applications involving nanogenerators.

Ag–graphene/epoxy nanocomposites were prepared by an *in situ* method using Ag–graphene nanocomposites as additives (Liu et al. 2012). The electrical properties of the Ag–graphene/epoxy nanocomposites were investigated by current–voltage tests. The result showed that the nanocomposites with filler contents between 0.8 and 3.0 wt% exhibited a varistor effect. This reversible electrical nonlinearity resulted from the defects in the GO-derived graphene structure and the presence of the Ag nanoparticles.

Composites of poly(dimethyl siloxane) filled with graphene oxide (GO) or slightly thermally reduced GO exhibit unusual nonlinear electrical conductivity at a small filler loading of about 3 vol% (Wang et al. 2012): At low field strength, the composite showed a lower conductivity compared to the neat polymer, probably due to the blockage of ion transport by the GO network. At a field near the switching field, the composite exhibited nonlinear conductivity with a large nonlinear coefficient. The conductivity at higher electric field was limited by electron transport along the GO network, thus potentially providing a controllable method to avoid thermal problems. The switching field and maximum conductivity can be tailored by varying the oxidation state of the GO and the volume fraction. The dielectric constant also increased at small filler loadings, and the loss factor remains low because of the insulating nature of GO at low field.

Electrically conductive acrylonitrile butadiene rubber compounds filled with different concentrations of graphite nanoplatelets (10 nm thickness) were mixed, compression molded, and vulcanized (Al-solamy et al. 2012). The percolation concentration of the investigated nanocomposites was found to be 0.5 phr and the sample in the region of the percolation transition found to be most sensitive to compressive strain. The electrical conductivity of this sample was changed by more than five orders upon a 60% compression and more than two orders for 6 MPa pressure. A model description of the microstructure providing extremely strong piezoresistive effects is proposed based on the conductive surface network of the nanocomposite. The experimental data for compressive strain are in good agreement with theoretical equations derived from a model based on the change of particle separation under applied stress. The increase of electrical resistance with uniaxial pressure may be explained on account of the destruction of the structure of the graphite electroconductive nanoplatelets network. At small compression deformations, the experimental data are consistent with the tunneling conductance model. To describe the experimental results at large deformation, destruction of the conductive network and decrease of the number of conducting pathways have to be taken into account.

High-aspect-ratio fillers are predicted to increase the dielectric constant of polymer composites more efficiently than spherical fillers according to the rule of mixtures. Using high-aspect-ratio fillers is a promising route for creating high-dielectric-constant, low-loss materials at a low filler volume fraction, for use as capacitor and electric field grading materials. A combination of barium titanate and graphene platelets yielded the highest dielectric constant when used in a polydimethyl siloxane matrix (Wang et al. 2012)—up to 300 at just over 0.006 volume fraction of graphene. The increase in dielectric loss over the pure matrix was small when the volume fraction was below the percolation threshold of graphene platelets. Electric flux density–electric field measurements showed a linear dielectric constant in barium titanate-filled composites and higher loss when graphene was added. The AC breakdown strength was reduced compared to the neat polymer and was affected by the filler aspect ratio.

The nonlinear conduction behavior of composite materials of foliated graphite nanosheets and nylon-6 subjected to a variable direct current electric field has been studied (Chen et al. 2004). The reversible nonlinear conduction behavior occurs beyond a certain value of E. Because the graphite particles are ohmic, the macroscopic nonlinearity must originate from the intrinsically nonlinear behaviors of the interfacial interactions between graphite particles (or clusters) and the new additional conduction channels generated on the insulating thin polymer gaps present on the conducting backbone. Thus, a nonlinear random dynamic resistor network model combining these two processes is used semiqualitatively to describe the nonlinear conduction in these nanocomposites. The nonlinearity is more pronounced for the nanocomposites, with their filler content approaching percolation threshold (Figure 34.17).

Magnetic graphene nanocomposites based on two different-sized graphene substrates with various nanoparticle loading levels were synthesized by a facile thermal decomposition

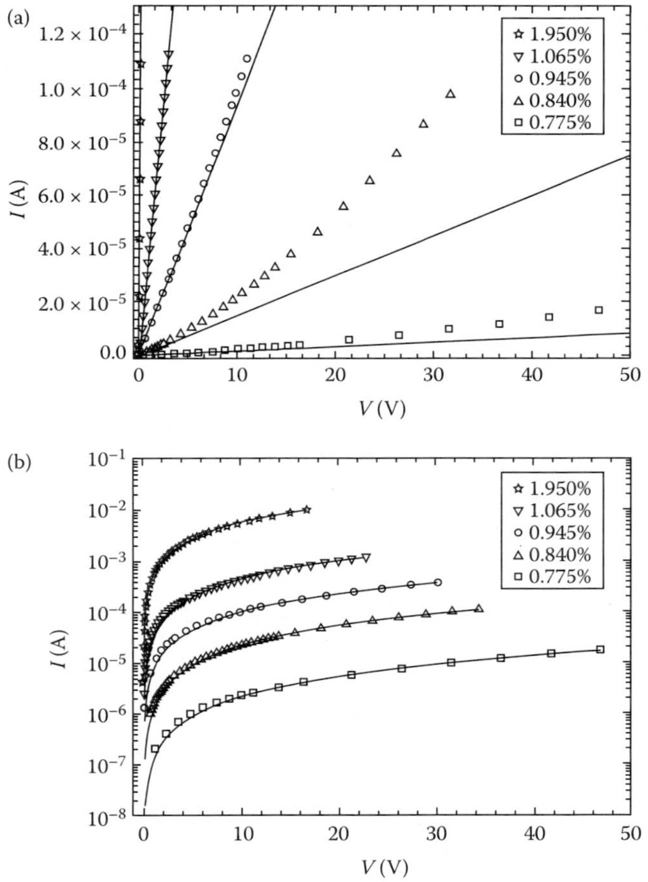

FIGURE 34.17 (a) Linear and (b) log scale of I–E responses of PA6/GNP nanocomposites filled with various GN contents. The straight line in (a) corresponds to linear portion of the curve. The GNP volume contents are given in insets. (Chen, G. H. et al.: Nonlinear conduction in nylon-6/foliated graphite nanocomposites above the percolation threshold. *Journal of Polymer Science Part B–Polymer Physics*. 2004. 42 (1):155–167. Copyright Wiley-VCH Verlag GmbH & Co. KGaA. Reproduced with permission.)

method (Zhu et al. 2012). Iron pentacarbonyl was used as the iron precursor and all the formed NPs were adhered on the graphene sheet with a uniform dispersion and narrow size distribution. The particle size and particle–particle interaction are strongly related to the iron precursor concentration and graphene size. The effects of the particle loading levels and graphene size on the electrical conductivity, dielectric permittivity, GMR, and magnetic properties of the nanocomposites were investigated and detailed. The MR was observed to vary from 38% to 64% at 130 K, and an even higher MR of about 46%–72% is observed at 290 K. More interestingly, the dielectric permittivity can be tuned from negative to positive at high frequency with increasing particle loading.

34.7 CONCLUSION

The key to the full utilization of graphene properties in composites is to control their dispersion in the matrix and the characteristics of filler–polymer interface. These are essentially the common requirements for carbon nanofillers. Much effort has been made on meeting these requirements for CNTs; the results of these efforts can no doubt be used to improve the properties of graphene composites (Nicolais et al. 2011). Both experimental and model investigations are extensively required.

New applications are emerging for graphene-based polymer nanocomposites. They have been considered as electrode materials for supercapacitors aimed at high specific capacitance coupled with high power and energy density (Bose et al. 2012). The underlying reason behind the improvement of performance of carbon-based electrode materials is due to the unique combination of chemical and physical properties.

Significant results have been achieved for the actuation by electrical, chemical, light, and other stimulus (Huang et al. 2012). This is even more obvious if we consider the combination and collective advantages of graphene in mechanical, electrical, chemical stability, and optical properties. There are many exciting opportunities ahead in this field.

REFERENCES

Al-Hartomy, O. A., A. A. Al-Ghamdi, F. Al-Salamy, N. Dishovsky, R. Shtarkova, V. Iliev, and F. El-Tantawy. 2012. Dielectric and microwave properties of graphene nanoplatelets/carbon black filled natural rubber composites. *International Journal of Materials and Chemistry* 2 (3):116–122.

Al-solamy, F. R., A. A. Al-Ghamdi, and W. E. Mahmoud. 2012. Piezoresistive behavior of graphite nanoplatelets based rubber nanocomposites. *Polymers for Advanced Technologies* 23 (3):478–482.

Alekseev, A., D. Chen, E. E. Tkalya, M. G. Ghislandi, Y. Syurik, O. Ageev, J. Loos, and G. de With. 2012. Local organization of graphene network inside graphene/polymer composites. *Advanced Functional Materials* 22 (6):1311–1318.

Alzari, V., D. Nuvoli, R. Sanna, S. Scognamillo, M. Piccinini, J. M. Kenny, G. Malucelli, and A. Mariani. 2011. *In situ* production of high filler content graphene-based polymer nanocomposites by reactive processing. *Journal of Materials Chemistry* 21 (41):16544.

Ansari, S., and E. P. Giannelis. 2009. Functionalized graphene sheet-poly(vinylidene fluoride) conductive nanocomposites. *Journal of Polymer Science Part B—Polymer Physics* 47 (9):888–897.

Arlen, M. J., D. Wang, J. D. Jacobs, R. Justice, A. Trionfi, J. W. P. Hsu, D. Schaffer, L. S. Tan, and R. A. Vaia. 2008. Thermal-electrical character of *in situ* synthesized polyimide-grafted carbon nanofiber composites. *Macromolecules* 41 (21):8053–8062.

Bai, H., C. Li, and G. Shi. 2011a. Functional composite materials based on chemically converted graphene. *Advanced Materials* 23 (9):1089–1115.

Bai, X., Y. Zhai, and Y. Zhang. 2011b. Green approach to prepare graphene-based composites with high microwave absorption capacity. *The Journal of Physical Chemistry C* 115 (23):11673–11677.

Basavaraja, C., W. J. Kim, Y. D. Kim, and D. S. Huh. 2011. Synthesis of polyaniline-gold/graphene oxide composite and microwave absorption characteristics of the composite films. *Materials Letters* 65 (19–20):3120–3123.

Bora, C., and S. K. Dolui. 2012. Fabrication of polypyrrole/graphene oxide nanocomposites by liquid/liquid interfacial polymerization and evaluation of their optical, electrical and electrochemical properties. *Polymer* 53 (4):923–932.

Bose, S., T. Kuila, A. K. Mishra, R. Rajasekar, N. H. Kim, and J. H. Lee. 2012. Carbon-based nanostructured materials and their composites as supercapacitor electrodes. *Journal of Materials Chemistry* 22 (3):767.

Bose, S., T. Kuila, Md. E. Uddin, N. H. Kim, A. K. T. Lau, and J. H. Lee. 2010. In-situ synthesis and characterization of electrically conductive polypyrrole/graphene nanocomposites. *Polymer* 51 (25):5921–5928.

Cai, D., and M. Song. 2010. Recent advance in functionalized graphene/polymer nanocomposites. *Journal of Materials Chemistry* 20 (37):7906.

Chen, G. H., W. G. Weng, D. J. Wu, and C. L. Wu. 2004. Nonlinear conduction in nylon-6/foliated graphite nanocomposites above the percolation threshold. *Journal of Polymer Science Part B—Polymer Physics* 42 (1):155–167.

Choi, W., and J.-W. Lee. 2012. *Graphene Synthesis and Applications*. CRC Press, Boca Raton, FL.

Choudhary, V., S. K. Dhawan, and P. Saini. 2012. Polymer based nanocomposites for electromagnetic interference EMI shielding. In *EM Shielding—Theory and Development of New Materials*, M. Jaroszewski and J. Ziaja, eds.. InTech, Trivandrum, Kerala, India.

Corcione, C. E., F. Freuli, and A. Maffezzoli. 2013. The aspect ratio of epoxy matrix nanocomposites reinforced with graphene stacks. *Polymer Engineering and Science* 53 (3):531–539.

Das, C. K., P. Bhattacharya, and S. S. Kalra. 2012. Graphene and MWCNT: Potential candidate for microwave absorbing materials. *Journal of Materials Science Research* 1 (2):126–132.

De Bellis, G., I. M. De Rosa, A. Dinescu, M. S. Sarto, and A. Tamburrano. 2010. Electromagnetic absorbing nanocomposites including carbon fibers, nanotubes and graphene nanoplatelets. *2010 IEEE International Symposium on Electromagnetic Compatibility (EMC 2010)*: 202–207, Fort Lauderdale, FL.

Dyre, J., and T. Schrøder. 2000. Universality of ac conduction in disordered solids. *Reviews of Modern Physics* 72 (3):873–892.

Eda, G., and M. Chhowalla. 2009. Graphene-based composite thin films for electronics. *Nano Letters* 9 (2):814–818.

Eswaraiah, V., V. Sankaranarayanan, and S. Ramaprabhu. 2011. Functionalized graphene-PVDF foam composites for EMI shielding. *Macromolecular Materials and Engineering* 296 (10):894–898.

Fan, P., L. Wang, J. Yang, F. Chen, and M. Zhong. 2012. Graphene/poly(vinylidene fluoride) composites with high dielectric constant and low percolation threshold. *Nanotechnology* 23 (36):365702.

Fang, J-J., S-F. Li, W-K. Zha, H-Y. Cong, J-F. Chen, and Z-Z. Chen. 2011. Microwave absorbing properties of Nickel-coated graphene. *Journal of Inorganic Materials* 26 (5):467–471.

Feng, R., G. Guan, W. Zhou, C. Li, D. Zhang, and Y. Xiao. 2011. *In situ* synthesis of poly(ethylene terephthalate)/graphene composites using a catalyst supported on graphite oxide. *Journal of Materials Chemistry* 21 (11):3931.

Fim, F. de C., N. R. S. Basso, A. P. Graebin, D. S. Azambuja, and G. B. Galland. 2013. Thermal, electrical, and mechanical properties of polyethylene-graphene nanocomposites obtained by *in situ* polymerization. *Journal of Applied Polymer Science* 128 (5):2630–2637.

Galpaya, D. 2012. Recent advances in fabrication and characterization of graphene-polymer nanocomposites. *Graphene* 01 (02):30–49.

Greco, A., A. Timo, and A. Maffezzoli. 2012. Development and characterization of amorphous thermoplastic matrix graphene nanocomposites. *Materials* 5 (12):1972–1985.

Gu, Z. M., C. Z. Li, G. C. Wang, L. Zhang, X. H. Li, W. D. Wang, and S. L. Jin. 2010. Synthesis and characterization of polypyrrole/graphite oxide composite by *in situ* emulsion polymerization. *Journal of Polymer Science Part B—Polymer Physics* 48 (12):1329–1335.

Gu, Z., L. Zhang, and C. Li. 2009. Preparation of highly conductive polypyrrole/graphite oxide composites via *in situ* polymerization. *Journal of Macromolecular Science, Part B* 48 (6):1093–1102.

Ha, H. W., A. Choudhury, T. Kamal, D. H. Kim, and S. Y. Park. 2012. Effect of chemical modification of graphene on mechanical, electrical, and thermal properties of polyimide/graphene nanocomposites. *ACS Applied Materials and Interfaces* 4 (9):4623–4630.

Haggenmueller, R., H. H. Gommans, A. G. Rinzler, J. E. Fischer, and K. I. Winey. 2000. Aligned single-wall carbon nanotubes in composites by melt processing methods. *Chemical Physics Letters* 330 (3–4):219–225.

He, L., and S. C. Tjong. 2013. Low percolation threshold of graphene/polymer composites prepared by solvothermal reduction of graphene oxide in the polymer solution. *Nanoscale Research Letters* 8 (1):132.

Hicks, J., A. Behnam, and A. Ural. 2009. A computational study of tunneling-percolation electrical transport in graphene-based nanocomposites. *Applied Physics Letters* 95 (21).

Huang, Y., J. Liang, and Y. Chen. 2012. The application of graphene based materials for actuators. *Journal of Materials Chemistry* 22 (9):3671.

Huang, Y., Y. Qin, N. Wang, Y. Zhou, H. Niu, J-Y. Dong, J. Hu, and Y. Wang. 2012. Reduction of graphite oxide with a grignard reagent for facile *in situ* preparation of electrically conductive polyolefin/graphene nanocomposites. *Macromolecular Chemistry and Physics* 213 (7):720–728.

Hussain, F. 2006. Review article: Polymer-matrix nanocomposites, processing, manufacturing, and application: An overview. *Journal of Composite Materials* 40 (17):1511–1575.

Jiang, L., X-P. Shen, J-L. Wu, and K-C. Shen. 2010. Preparation and characterization of graphene/poly(vinyl alcohol) nanocomposites. *Journal of Applied Polymer Science* 118 (1):275–279.

Jiang, X., and L. T. Drzal. 2012a. Exploring the potential of exfoliated graphene nanoplatelets as the conductive filler in polymeric nanocomposites for bipolar plates. *Journal of Power Sources* 218:297–306.

Jiang, X., and L. T. Drzal. 2012b. Reduction in percolation threshold of injection molded high-density polyethylene/exfoliated graphene nanoplatelets composites by solid state ball milling and solid state shear pulverization. *Journal of Applied Polymer Science* 124 (1):525–535.

Jing-Wei Shen, Xiao-Mei Chen, and Wen-Yi Huang. 2003. Structure and electrical properties of grafted polypropylene graphite nanocomposites prepared by solution intercalation. *Journal of Applied Polymer Science* 88:1864–1869.

Jonscher, A. K. 1977. The 'universal' dielectric response. *Nature* 267 (5613):673–679.

Jun Ma, C. Liu, and J. W. R. Li. 2012. Properties and structural characterization of chitosan/poly(vinylalcohol)/graphene oxide nanocomposites. *e-Polymers* (033), www.e-polymers.org.

Katsnelson, M. 2012. *Graphene Carbon in Two Dimensions*. Cambridge University Press.

Kim, H., A. A. Abdala, and C. W. Macosko. 2010. Graphene/polymer nanocomposites. *Macromolecules* 43 (16):6515–6530.

Kim, H., Y. Miura, and C. W. Macosko. 2010. Graphene/polyurethane nanocomposites for improved gas barrier and electrical conductivity. *Chemistry of Materials* 22 (11):3441–3450.

Koval'chuk, A. A., A. N. Shchegolikhin, V. G. Shevchenko, P. M. Nedorezova, A. N. Klyamkina, and A. M. Aladyshev. 2008. Synthesis and properties of polypropylene/multiwall carbon nanotube composites. *Macromolecules* 41 (9):3149–3156.

Kuila, T., S. Bose, A. K. Mishra, P. Khanra, N. H. Kim, and J. H. Lee. 2012. Effect of functionalized graphene on the physical properties of linear low density polyethylene nanocomposites. *Polymer Testing* 31 (1):31–38.

Kuilla, T., S. Bhadra, D. Yao, N. H. Kim, S. Bose, and J. H. Lee. 2010. Recent advances in graphene based polymer composites. *Progress in Polymer Science* 35 (11):1350–1375.

Kuvardina, E. V., L. A. Novokshonova, S. M. Lomakin, S. A. Timan, and I. A. Tchmutin. 2012. Effect of the graphite nanoplatelet size on the mechanical, thermal, and electrical properties of polypropylene/exfoliated graphite nanocomposites. *Journal of Applied Polymer Science* 128 (3):1417–1424.

Latif, I., T. B. Alwan, and A. H. Al-Dujaili. 2013. Low frequency dielectric study of PAPA-PVA-GR nanocomposites. *Nanoscience and Nanotechnology* 2 (6):190–200.

Lee, H. B., A. Raghu, K. S. Yoon, and H. M. Jeong. 2010. Preparation and characterization of poly(ethylene oxide)/graphene nanocomposites from an aqueous medium. *Journal of Macromolecular Science, Part B* 49 (4):802–809.

Li, J., and J-K. Kim. 2007. Percolation threshold of conducting polymer composites containing 3D randomly distributed graphite nanoplatelets. *Composites Science and Technology* 67 (10):2114–2120.

Li, Y. C., S. C. Tjong, and R. K. Y. Li. 2010. Electrical conductivity and dielectric response of poly(vinylidene fluoride)–graphite nanoplatelet composites. *Synthetic Metals* 160 (17–18):1912–1919.

Li, Y., J. Zhu, S. Wei, J. Ryu, L. Sun, and Z. Guo. 2011a. Poly(propylene)/graphene nanoplatelet nanocomposites: Melt rheological behavior and thermal, electrical, and electronic properties. *Macromolecular Chemistry and Physics* 212 (18):1951–1959.

Li, Y., J. Zhu, S. Wei, J. Ryu, Q. Wang, L. Sun, and Z. Guo. 2011b. Poly(propylene) nanocomposites containing various carbon nanostructures. *Macromolecular Chemistry and Physics* 212 (22):2429–2438.

Liang, J., Y. Wang, Y. Huang, Y. Ma, Z. Liu, J. Cai, C. Zhang, H. Gao, and Y. Chen. 2009. Electromagnetic interference shielding of graphene/epoxy composites. *Carbon* 47 (3):922–925.

Liao, K-H., Y. Qian, and C. W. Macosko. 2012. Ultralow percolation graphene/polyurethane acrylate nanocomposites. *Polymer* 53 (17):3756–3761.

Liao, L., and X. F. Duan. 2012. Graphene for radio frequency electronics. *Materials Today* 15 (7–8):328–338.

Liu, Q., X. Yao, X. Zhou, Z. Qin, and Z. Liu. 2012. Varistor effect in Ag–graphene/epoxy resin nanocomposites. *Scripta Materialia* 66 (2):113–116.

Lu, H., Z. Chen, and C. Ma. 2012. Bioinspired approaches for optimizing the strength and toughness of graphene-based polymer nanocomposites. *Journal of Materials Chemistry* 22 (32):16182.

Luong, N. D., U. Hippi, J. T. Korhonen, A. J. Soininen, J. Ruokolainen, L-S. Johansson, J-D. Nam, L. H. Sinh, and J. Seppälä. 2011. Enhanced mechanical and electrical properties of polyimide film by graphene sheets via *in situ* polymerization. *Polymer* 52 (23):5237–5242.

Malliaris, A., and D. T. Turner. 1971. Influence of particle size on the electrical resistivity of compacted mixtures of polymeric and metallic powders. *Journal of Applied Physics* 42 (2):614.

Martin-Gallego, M., R. Verdejo, M. A. Lopez-Manchado, and M. Sangermano. 2011. Epoxy-graphene UV-cured nanocomposites. *Polymer* 52 (21):4664–4669.

Mitra, S., O. Mondal, D. R. Saha, A. Datta, S. Banerjee, and D. Chakravorty. 2011. Magnetodielectric effect in graphene-PVA nanocomposites. *The Journal of Physical Chemistry C* 115 (29):14285–14289.

Nicolais, L., M. Meo, and E. Milella, eds. 2011. *Composite Materials*. Springer, Springer-Verlag, London.

Pang, H., T. Chen, G. Zhang, B. Zeng, and Z-M. Li. 2010. An electrically conducting polymer/graphene composite with a very low percolation threshold. *Materials Letters* 64 (20):2226–2229.

Panwar, V., and R. M. Mehra. 2008. Study of electrical and dielectric properties of styrene-acrylonitrile/graphite sheets composites. *European Polymer Journal* 44 (7):2367–2375.

Pati, S. K., T. Enoki, and C. N. R. Rao, eds. 2011. *Graphene and Its Fascinating Attributes*. World Scientific, Singapore.

Pol'shchikov, S. V., P. M. Nedorezova, A. N. Klyamkina, V. G. Krashenninikov, A. M. Aladyshev, A. N. Shchegolikhin, V. G. Shevchenko, E. A. Sinevich, T. V. Monakhova, and V. E. Muradyan. 2013. Composite materials based on graphene nanoplatelets and polypropylene derived via *in situ* polymerization. *Nanotechnologies in Russia* 8 (1–2):69–80.

Polschikov, S. V., P. M. Nedorezova, A. N. Klyamkina, A. A. Kovalchuk, A. M. Aladyshev, A. N. Shchegolikhin, V. G. Shevchenko, and V. E. Muradyan. 2013. Composite materials of graphene nanoplatelets and polypropylene, prepared by *in situ* polymerization. *Journal of Applied Polymer Science* 127 (2):904–911.

Potts, J. R., D. R. Dreyer, C. W. Bielawski, and R. S. Ruoff. 2011. Graphene-based polymer nanocomposites. *Polymer* 52 (1):5–25.

Qin, F., and C. Brosseau. 2012. A review and analysis of microwave absorption in polymer composites filled with carbonaceous particles. *Journal of Applied Physics* 111 (6):061301.

Rafiee, M. A., J. Rafiee, Z. Wang, H. Song, Z. Z. Yu, and N. Koratkar. 2009. Enhanced mechanical properties of nanocomposites at low graphene content. *ACS Nano* 3 (12):3884–3890.

Ren, P-G., Y-Y. Di, Q. Zhang, L. Li, H. Pang, and Z-M. Li. 2012. Composites of ultrahigh-molecular-weight polyethylene with graphene sheets and/or MWCNTs with segregated network structure: Preparation and properties. *Macromolecular Materials and Engineering* 297 (5):437–443.

Ruschau, G. R., S. Yoshikawa, and R. E. Newnham. 1992. Resistivities of conductive composites. *Journal of Applied Physics* 72 (3):953.

Saini, P., and M. Aror. 2012. Microwave absorption and EMI shielding behavior of nanocomposites based on intrinsically conducting polymers, graphene and carbon nanotubes, new polymers for special applications. Dr. Ailton De Souza Gomes (Ed.), ISBN: 978-953-51-0744-6, InTech, DOI: 10.5772/48779. Available from: http://www.intechopen.com/books/new-polymers-for-special-applications/microwave-absorption-and-emi-shielding-behavior-of-nanocomposites-based-on-intrinsically-conducting-.

Scognamillo, S., E. Gioffredi, M. Piccinini, M. Lazzari, V. Alzari, D. Nuvoli, R. Sanna, D. Piga, G. Malucelli, and A. Mariani. 2012. Synthesis and characterization of nanocomposites of thermoplastic polyurethane with both graphene and graphene nanoribbon fillers. *Polymer* 53 (19):4019–4024.

Sengupta, R., M. Bhattacharya, S. Bandyopadhyay, and A. K. Bhowmick. 2011. A review on the mechanical and electrical properties of graphite and modified graphite reinforced polymer composites. *Progress in Polymer Science* 36 (5):638–670.

Shang, J., Y. Zhang, L. Yu, B. Shen, F. Lv, and P. K. Chu. 2012. Fabrication and dielectric properties of oriented polyvinylidene fluoride nanocomposites incorporated with graphene nanosheets. *Materials Chemistry and Physics* 134 (2–3):867–874.

Sharifi, F., R. Bauld, M. S. Ahmed, and G. Fanchini. 2012. Transparent and conducting graphene-RNA-based nanocomposites. *Small* 8 (5):699–706.

Shevchenko, V. G., S. V. Polschikov, P. M. Nedorezova, A. N. Klyamkina, A. N. Shchegolikhin, A. M. Aladyshev, and V. E. Muradyan. 2012. *In situ* polymerized poly(propylene)/graphene nanoplatelets nanocomposites: Dielectric and microwave properties. *Polymer* 53 (23):5330–5335.

Singh, K., A. Ohlan, and S. K. Dhaw. 2012. Polymer-graphene nanocomposites: Preparation, characterization, properties, and applications, nanocomposites—New trends and developments. Dr. F. Ebrahimi (Ed.), ISBN: 978-953-51-0762-0, InTech, DOI: 10.5772/50408. Available from: http://www.intechopen.com/books/nanocomposites-new-trends-and-developments/polymer-graphene-nanocomposites-preparation-characterization-properties-and-applications.

Song, W-L., W. Wang, L. M. Veca, C. Y. Kong, M-S. Cao, P. Wang et al. 2012. Polymer/carbon nanocomposites for enhanced thermal transport properties—Carbon nanotubes versus graphene sheets as nanoscale fillers. *Journal of Materials Chemistry* 22 (33):17133.

Song, Z., T. Xu, M. L. Gordin, Y. B. Jiang, I. T. Bae, Q. Xiao, H. Zhan, J. Liu, and D. Wang. 2012. Polymer-graphene nanocomposites as ultrafast-charge and -discharge cathodes for rechargeable lithium batteries. *Nano Letters* 12 (5):2205–2211.

Sridhar, V. 2013. Graphene reinforced biodegradable poly(3-hydroxybutyrate-co-4-hydroxybutyrate) nano-composites. *Express Polymer Letters* 7 (4):320–328.

Stankovich, S., D. A. Dikin, G. H. Dommett, K. M. Kohlhaas, E. J. Zimney, E. A. Stach, R. D. Piner, S. T. Nguyen, and R. S. Ruoff. 2006. Graphene-based composite materials. *Nature* 442 (7100):282–286.

Syurik, Y. V., M. G. Ghislandi, E. E. Tkalya, G. Paterson, D. McGrouther, O. A. Ageev, and J. Loos. 2012. Graphene network organisation in conductive polymer composites. *Macromolecular Chemistry and Physics* 213 (12):1251–1258.

Tantis, I. 2012. Functionalized graphene—poly(vinyl alcohol) nanocomposites: Physical and dielectric properties. *Express Polymer Letters* 6 (4):283–292.

Terrones, M., O. Martín, M. González, J. Pozuelo, B. Serrano, J. C. Cabanelas, S. M. Vega-Díaz, and J. Baselga. 2011. Interphases in graphene polymer-based nanocomposites: Achievements and challenges. *Advanced Materials* 23 (44):5302–5310.

Thostenson, E., C. Li, and T. Chou. 2005. Nanocomposites in context. *Composites Science and Technology* 65 (3–4):491–516.

Tjong, S. C. 2014. Polymer composites with graphene nanofillers: Electrical properties and applications. *Journal of Nanoscience and Nanotechnology* 14 (2):1154–1168.

Tkalya, E. E., M. Ghislandi, G. de With, and C. E. Koning. 2012. The use of surfactants for dispersing carbon nanotubes and graphene to make conductive nanocomposites. *Current Opinion in Colloid and Interface Science* 17 (4):225–232.

Tkalya, E., M. Ghislandi, A. Alekseev, C. Koning, and J. Loos. 2010. Latex-based concept for the preparation of graphene-based polymer nanocomposites. *Journal of Materials Chemistry* 20 (15):3035.

Verdejo, R., M. M. Bernal, L. J. Romasanta, and M. A. Lopez-Manchado. 2011. Graphene filled polymer nanocomposites. *Journal of Materials Chemistry* 21 (10):3301.

Vickery, J. L., A. J. Patil, and S. Mann. 2009. Fabrication of graphene-polymer nanocomposites with higher-order three-dimensional architectures. *Advanced Materials* 21 (21):2180–2184.

Wakabayashi, K., P. J. Brunner, J. Masuda, S. A. Hewlett, and J. M. Torkelson. 2010. Polypropylene-graphite nanocomposites made by solid-state shear pulverization: Effects of significantly exfoliated, unmodified graphite content on physical, mechanical and electrical properties. *Polymer* 51 (23):5525–5531.

Wang, H., Q. Hao, X. Yang, L. Lu, and X. Wang. 2010. Effect of graphene oxide on the properties of its composite with polyaniline. *ACS Applied Materials and Interfaces* 2 (3):821–828.

Wang, J., H. Hu, X. Wang, C. Xu, M. Zhang, and X. Shang. 2011. Preparation and mechanical and electrical properties of graphene nanosheets-poly(methyl methacrylate) nanocomposites via *in situ* suspension polymerization. *Journal of Applied Polymer Science* 122 (3):1866–1871.

Wang, Z., J. K. Nelson, H. Hillborg, S. Zhao, and L. S. Schadler. 2012. Graphene oxide filled nanocomposite with novel electrical and dielectric properties. *Advanced Materials* 24 (23):3134–3137.

Wang, Z. P., J. K. Nelson, J. J. Miao, R. J. Linhardt, L. S. Schadler, H. Hillborg, and S. Zhao. 2012. Effect of high aspect ratio filler on dielectric properties of polymer composites: A study on Barium Titanate fibers and graphene platelets. *IEEE Transactions on Dielectrics and Electrical Insulation* 19 (3):960–967.

Weng, W. G., G. H. Chen, D. J. Wu, X. F. Chen, J. R. Lu, and P. P. Wang. 2004. Fabrication and characterization of nylon 6/foliated graphite electrically conducting nanocomposite. *Journal of Polymer Science Part B—Polymer Physics* 42 (15):2844–2856.

Wu, C., X. Huang, G. Wang, L. Lv, G. Chen, G. Li, and P. Jiang. 2013. Highly conductive nanocomposites with three-dimensional, compactly interconnected graphene networks via a self-assembly process. *Advanced Functional Materials* 23 (4):506–513.

Xiong, Y., Y. Xie, F. Zhang, E. Ou, Z. Jiang, L. Ke, D. Hu, and W. Xu. 2012. Reduced graphene oxide/hydroxylated styrene–butadiene–styrene tri-block copolymer electroconductive nanocomposites: Preparation and properties. *Materials Science and Engineering: B* 177 (14):1163–1169.

Xu, Z., and C. Gao. 2010. *In situ* polymerization approach to graphene-reinforced nylon-6 composites. *Macromolecules* 43 (16):6716–6723.

Yang, X., L. Li, S. Shang, and X-m. Tao. 2010. Synthesis and characterization of layer-aligned poly(vinyl alcohol)/graphene nanocomposites. *Polymer* 51 (15):3431–3435.

Yoonessi, M., and J. R. Gaier. 2010. Highly conductive multifunctional graphene polycarbonate nanocomposites. *ACS Nano* 4 (12):7211–7220.

Young, R. J., I. A. Kinloch, L. Gong, and K. S. Novoselov. 2012. The mechanics of graphene nanocomposites: A review. *Composites Science and Technology* 72 (12):1459–1476.

Yu, J., P. Jiang, C. Wu, L. Wang, and X. Wu. 2011. Graphene nanocomposites based on poly(vinylidene fluoride): Structure and properties. *Polymer Composites* 32 (10):1483–1491.

Yuan, X. Y. 2012. Improved properties of chemically modified graphene/poly(methyl methacrylate) nanocomposites via a facile in-situ bulk polymerization. *Express Polymer Letters* 6 (10):847–858.

Zhang, H. B., Q. Yan, W. G. Zheng, Z. He, and Z. Z. Yu. 2011. Tough graphene-polymer microcellular foams for electromagnetic interference shielding. *ACS Applied Materials and Interfaces* 3 (3):918–924.

Zhang, H-B., W.-G. Zheng, Q. Yan, Z.-G. Jiang, and Z.-Z. Yu. 2012. The effect of surface chemistry of graphene on rheological and electrical properties of polymethylmethacrylate composites. *Carbon* 50 (14):5117–5125.

Zhang, H-B., W.-G. Zheng, Q. Yan, Y. Yang, J.-W. Wang, Z.-H. Lu, G-Y. Ji, and Z-Z. Yu. 2010. Electrically conductive polyethylene terephthalate/graphene nanocomposites prepared by melt compounding. *Polymer* 51 (5):1191–1196.

Zhu, J., Z. Luo, S. Wu, N. Haldolaarachchige, D. P. Young, S. Wei, and Z. Guo. 2012. Magnetic graphene nanocomposites: Electron conduction, giant magnetoresistance and tunable negative permittivity. *Journal of Materials Chemistry* 22 (3):835.

35 Probing Collective Excitations in Graphene/Metal Interfaces by High-Resolution Electron Energy Loss Spectroscopy Measurements

Antonio Politano and Gennaro Chiarello

CONTENTS

ABSTRACT

Many of the peculiar properties of graphene are related to its collective excitations, even if their understanding is still lacking. Plasmons are collective longitudinal modes of charge fluctuation in metal samples excited by an external electric field. Plasmons find applications in magneto-optic data storage, optics, microscopy, and catalysis. Plasmons in graphene have unusual properties and offer promising prospects for plasmonic applications covering a wide frequency range, going from terahertz up to the visible. On the other hand, lattice dynamics play an important role in many chemical and physical processes. The investigation of phonons in materials provides information on numerous properties, such as sound velocity, thermal expansion, magnetic forces, heat capacity, and thermal conductivity. High-resolution electron energy loss spectroscopy is the main experimental technique for investigating collective excitations (both plasmons and phonons), with adequate resolution in both the energy and momentum domains to investigate surface electronic excitations. This chapter discusses the status and the prospect of research on collective excitations (plasmons and phonons) in graphene epitaxially grown on metals.

35.1 INTRODUCTION

The epitaxial growth of large, highly perfect graphene monolayers is a prerequisite for most practical applications of this "wonder" material [1–14] (Figure 35.1). Most systems of epitaxial graphene are spontaneously nanostructured in a periodic array of ripples by the Moiré patterns caused by the difference in lattice parameter with the different substrates such as Ru(0001) [15], Ir(111) [16], or Pt(111) [17–21]. The careful characterization of these superlattices is important because nanostructuring graphene (in superlattices, stripes, or dots), in turn, may reveal new physical phenomena and fascinating applications. In addition, it is crucial to understand the interaction of graphene with the surfaces of substrates of different nature (oxides, semiconductors, or metals), as well as with adsorbed molecules, in view of the relevance of metallic contacts, and the sensitivity of the conduction properties of graphene to gating materials and doping by adsorbed molecules [18,22].

All these topics can be characterized in detail in what has become one of the benchmarks for epitaxial graphene: a self-organized, millimeter-large, periodically "rippled" epitaxial monolayer of graphene grown by soft chemical vapor deposition

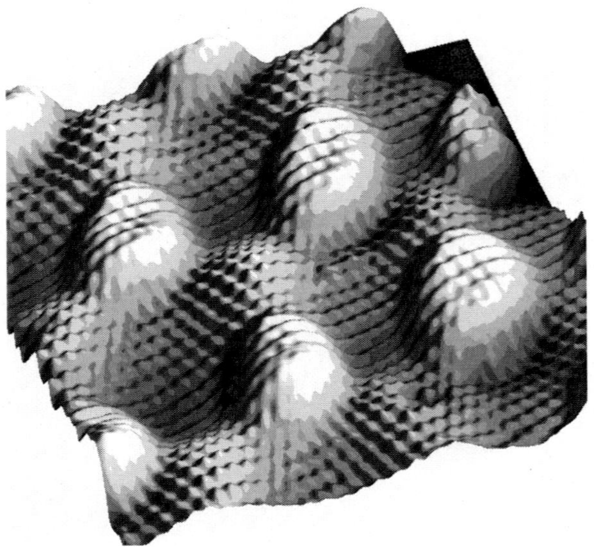

FIGURE 35.1 Periodically rippled graphene.

(CVD) under ultra-high vacuum (UHV) conditions on single-crystal metal substrates with hexagonal symmetry. The superb control allowed by the UHV environment facilitates the characterization of the system down to the atomic scale.

In this chapter, we report on high-resolution electron energy loss spectroscopy (HREELS) investigations on both the vibrational and electronic properties of graphene grown on Pt(111).

In Section 35.2, we introduce our measurement techniques, and in Sections 35.3 and 35.4, low-energy electron diffraction (LEED) and HREEL spectroscopy has been used for studying monolayer graphene (MLG) epitaxially grown on Pt(111) surface.

Our studies have clarified that acoustic-like plasmons exist in epitaxial graphene. We have also found that graphene/Pt(111) is characterized by phonon modes that are very similar to that of graphite. In particular, Kohn anomalies (KAs) are found in this system, in contrast with results for graphene/Ni(111). This finding is ascribed to the quasi-freestanding nature of the graphene sheet on Pt(111).

35.2 HREELS: A SUITABLE TOOL FOR STUDYING SURFACE COLLECTIVE MODES

HREELS is an experimental technique that permits the study of materials through the analysis of their electronic and vibrational excitations. In contrast to infrared spectroscopy, HREELS is not limited by strict dipole selection rules, which often hinder observation of important modes and adsorbates. In HREELS, both long-range *dipole* and short-range *impact* scattering mechanisms (see References 23,24 for a review) are operable and they may be effectively studied as a function of scattering angle and impact energy. Information obtained from HREELS ideally complements data obtained with other surface spectroscopies, and offers ease of interpretation for the experimentalist.

In this spectroscopy, a monochromatic electron beam is sent to the surface of a solid material. These primary electrons are partially inelastically scattered and their kinetic energy distribution is analyzed using an electron energy analyzer [24]. The energy losses are due to the excitation of electronic and vibrational transition of the investigated materials and thus provide a tool for analyzing them. The HREELS technique can be used in various configurations by changing the geometrical and physical parameters. For example, different spectra and information can be obtained by changing the electronic energy or the incidence angle of the impinging primary beam, or the angle of analysis of diffused electrons [24].

The various excitations in a HREELS spectrum cover a wide energy range, which extends from some meV, as for phonons and vibrations of atoms or molecules adsorbed onto the surface, some eV, as for interband transitions and plasmonic excitations, up to hundreds of eV, as for the excitations of core electrons [25] (Figure 35.2).

Energy losses are due mainly to three processes:

1. Excitations of network vibrations of atoms on the clean surface (optical phonons surface, acoustic phonons surface) and/or vibrations of atomic and molecular species adsorbed on the surface
2. Excitations of valence band transitions that can be divided into single particle electronic excitations (interband and intraband excitations) and collective excitations (surface, volume, and interface plasmons)
3. Excitations of core-level electrons to conduction band levels

The mentioned excitations hold a very large energy range, which extends from around tens of meV (for phonons) up to a few thousands of eV needed to excite core electrons to states above the Fermi level.

Thus, HREELS plays an important role in the investigation of surface chemical reactions [26] and dynamic screening properties [20] of metal systems.

The great diffusion of this technique in the last two decades is mainly due to the development of a new generation of high-resolution HREELS spectrometers by Harald Ibach [27,28],

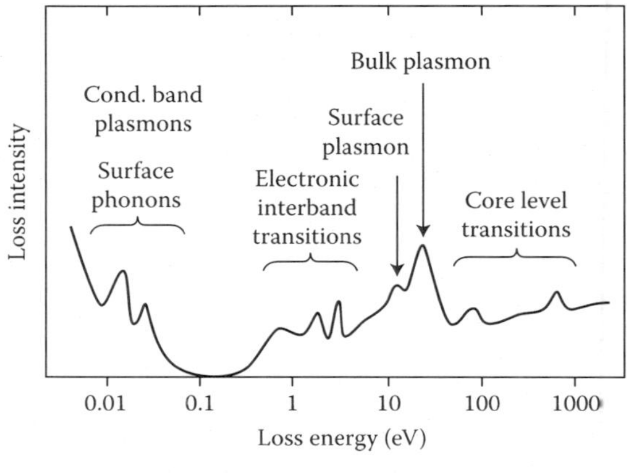

FIGURE 35.2 Regions of characteristic losses.

whose resolution ranges from 0.5 to 5 meV. This allows HREELS to have an energy resolution similar to inelastic helium atom scattering [29] and optical techniques such as infrared absorption spectroscopy [30] or second-harmonic generation [31]. The recent development of novel spectrometers for spin-polarized HREELS [32] is expected to further contribute to the diffusion of this spectroscopy.

The analysis of kinetic energy of inelastically scattered electrons provides information about excited modes at the surface. Electrons scattered from the sample are analyzed as a function of $E_{loss} = \Delta E = E_p - E_s$, where E_{loss} is the energy lost by electrons, E_p is the primary electron beam energy, and E_s is the energy of scattered electrons. If the electron excites a surface mode $\hbar\omega_0$, after the interaction with the sample, its energy will be $E_s = E_p - \hbar\omega_0$. Hence, if electrons lost E^*, they will give rise to loss peaks at energy $E_p - E^*$.

In inelastic processes, the energy change should equal the quantum energy of an electronic or vibrational surface mode, by respecting conservation laws of energy and wave vector parallel to the surface:

$$\begin{cases} E_{loss} = E_p - E_S & (35.1) \\ \hbar\vec{q}_{\parallel} = \hbar(\vec{k}_i \sin\theta_i - \vec{k}_S \sin\theta_S) & (35.2) \end{cases}$$

where $q_{\parallel}$ is the parallel momentum transfer, $\vec{k}_i$ is the wave vector of incident electrons, $\vec{k}_S$ is the wave vector of scattered electrons, and θ_i and θ_S are the angles formed with the normal to the surface by incident and scattered electrons, respectively (Figure 35.3).

Thus, an expression linking $q_{\parallel}$ with E_p, E_{loss}, and θ_i, θ_S could be obtained:

$$q_{\parallel} = \frac{\sqrt{2mE_p}}{\hbar}\left(\sin\theta_i - \sqrt{1 - \frac{E_{loss}}{E_p}}\sin\theta_S\right) \qquad (35.3)$$

Likewise, it is possible to obtain the indeterminacy on $q_{\parallel}$, that is, the window in the reciprocal space that also depends on the angular acceptance of the apparatus α [33], usually ranging between 0.5° and 1.0°:

$$\Delta q_{\parallel} = \frac{\sqrt{2mE_p}}{\hbar}\left(\cos\theta_i - \sqrt{1 - \frac{E_{loss}}{E_p}}\cos\theta_S\right)\cdot\alpha \qquad (35.4)$$

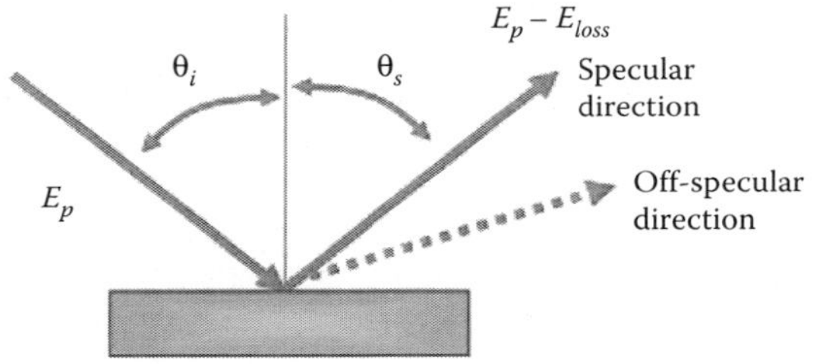

FIGURE 35.3 Scattering geometry in HREELS experiments.

Thus, $\Delta q_{\parallel}$ is minimized for low impinging energies and for grazing scattering conditions.

35.3 PHONON MODES OF EPITAXIAL GRAPHENE

35.3.1 Structure of Graphene on Pt(111)

The graphene/metal interface is a model system where the interaction between the graphene π-bands and the metal bands can be investigated. This has relevance to contacting of graphene with metal electrodes. The carbon hybridization and also the epitaxial relationship with the metal substrate were proposed to influence the contact transmittance [34] and cause local doping of graphene [35]. A variety of situations regarding the interplay of graphene and its metal support is realized depending on the support material. One type is with strong bonding between metal and graphene and a typical metal–graphene distance smaller than a metal–metal bond (around 1.4–2.2 Å) and one with weak bonds and a typical metal–graphene distance, which is much larger than the bonds between the substrate atoms (around 3.2–3.7 Å).

The difference in bonding also reflects in the phonon modes of graphene. A softening of the phonon modes indicating weakened C—C bonds is observed in the case of strong binding, whereas in the case of weak binding, the phonon frequencies stay close to the bulk values known from highly oriented pyrolytic graphite (HOPG) [36]. The electronic structure ranges from almost no interaction in the case of Pt(111) [18] to deep a modification in the case of a graphene monolayer on Ni(111) [37] or Ru(0001) [15,38].

For platinum, the lattice constant is $a = 3.92$ Å and the nearest-neighbor distance between atoms is $d = 2.77$ Å. For graphene, the lattice constant is $a = 2.46$ Å and the nearest-neighbor distance is $d = 1.42$ Å. As can be seen from a comparison of the close-packed layer of atoms of the Pt(111) surface and the honeycomb structure of graphene, they both have hexagonal symmetry. This similarity should make Pt(111) a natural fit for growing well-ordered graphene on the surface. However, the presence of a difference in lengths of the lattice constants results in a mismatch of approximately 11%.

Such a system is ideal for a study of phonon modes and elastic properties due to the absence of corrugation of the graphene overlayer found on other substrates [15,38], which has been demonstrated to be caused by the hybridization with the substrate. The growing strength of hybridization is accompanied by a gradual change in graphene morphology from nearly flat for MLG/Pt(111) to strongly corrugated in MLG on other substrates [39]. Thus, MLG on Pt(111) behaves as nearly flat freestanding graphene, as also confirmed by angle-resolved photoemission spectroscopy (ARPES) experiments [40].

Graphene was obtained by dosing ethylene onto the clean Pt(111) substrate held at 1150 K. At a certain temperature, the hydrocarbon is catalytically dissociated and hydrogen desorbs, and the remaining carbon adsorbed species can then form graphene. Moreover, the high temperature of the sample during depositions favors the increase of the size of MLG islands [41] and allows maintaining the substrate clean so as to avoid any contaminant-induced effect on graphene growth. The completion of the first layer was reached upon an exposure of 3×10^{-8} mbar for 10 min (24 L, 1 L = 1.33×10^{-6} mbar · s). After removing the C_2H_4 gas from the chamber, the temperature was held at 1150 K for further 60 s.

The attained LEED pattern (shown in Figures 35.4 and 35.5) is essentially similar to that reported in Reference 42. The ring pattern indicates the existence of different domains. Nonetheless, preferred orientations aligned with the substrate (R_0) are clearly distinguished. The presence of well-resolved spots in the LEED pattern is a clear fingerprint of the order of the MLG overstructure, also evidenced by the high electron reflectivity of the obtained surface (even higher with respect to the bare Pt substrate).

Despite the presence of other domains, the predominance of R_0 in the whole sample has been clearly inferred by the analysis of phonon dispersion measurements performed along the $\bar{\Gamma}-\bar{K}$ and the $\bar{\Gamma}-\bar{M}$ directions (inset of Figure 35.4).

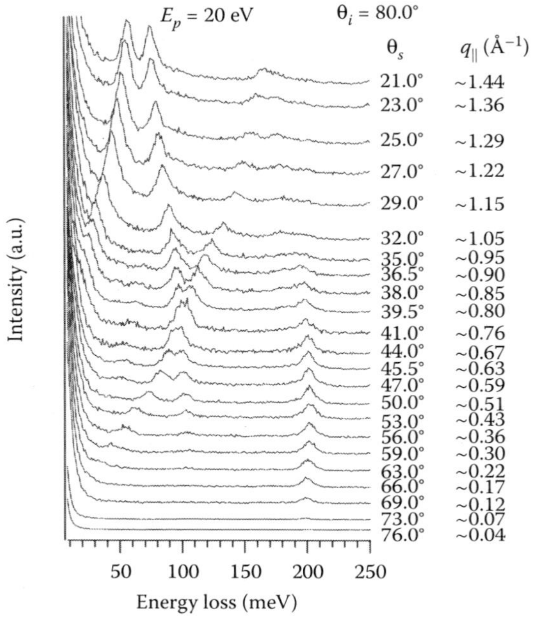

FIGURE 35.4 HREEL spectra for MLG/Pt(111) in the $\bar{\Gamma}-\bar{K}$ direction as a function of the scattering angle. The incidence angle is 80.0° and the impinging energy is 20 eV.

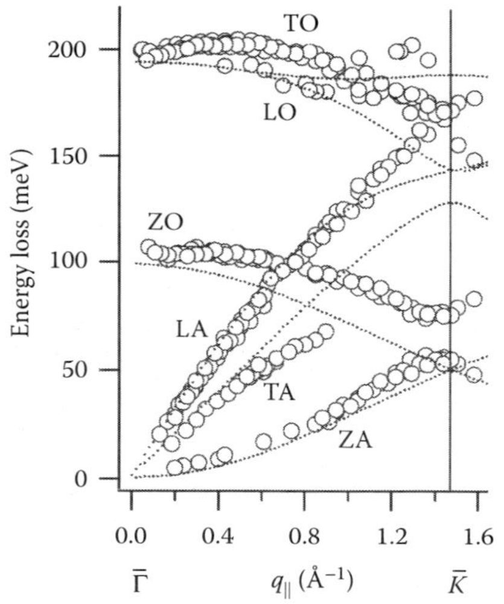

FIGURE 35.5 Dispersion relation for phonon modes in MLG/Pt(111) in the $\bar{\Gamma}-\bar{K}$ direction (empty circles). The dotted line represents the calculated phonon dispersion.

35.3.2 Phonon Dispersion

The dynamics of atoms at surfaces plays an important role in many chemical and physical processes. In particular, lattice vibrations can provide essential information on many physical properties, such as thermal expansion, heat capacity, sound velocity, magnetic forces, and thermal conductivity [43]. Recently, phonon modes of graphene sheets have been attracting much attention [6] as they influence many of the novel and unusual properties of graphene [44].

Owing to the high energy of phonon modes in graphitic materials, several techniques such as inelastic helium atom scattering cannot be used. Infrared or Raman spectroscopy cannot probe the whole Brillouin zone, while neutron or inelastic ion scattering are not always applicable. Among various detection methods for phonon dispersion, HREELS is a powerful tool to gain knowledge on phonons in graphene.

Present HREELS measurements show the dispersion relation of phonon modes in MLG/Pt(111). Results have been compared with recent calculations in Reference 45.

The energy resolution of the spectrometer was degraded to 4 meV so as to increase the signal-to-noise ratio of loss peaks. Dispersion of the loss peaks, that is, $E_{loss}(q_{||})$, was measured by moving the analyzer while keeping the sample and the monochromator in a fixed position (see Chapter 2). To measure the dispersion relation, values for the parameters E_p, impinging energy, and θ_i, the incident angle, were chosen so as to obtain the highest signal-to-noise ratio. The primary beam energy used for the dispersion, E_p = 20 eV, provided, in fact, the best compromise among surface sensitivity, the highest cross section for mode excitation, and $q_{||}$ resolution.

All measurements were made at room temperature.

The phonon dispersion was measured along the [$\bar{2}$11] direction of the Pt substrate, which corresponds to the $\bar{\Gamma}-\bar{K}$ direction of the MLG.

Loss measurements of MLG/Pt(111) recorded at E_p = 20 eV as a function of the scattering angle θ_s are reported in Figure 35.4, while the dispersion relation $E_{loss}(q_{||})$ is shown in Figure 35.5.

Measurements were repeated for several preparations of the MLG overstructure also by using different impinging energies and incident angles. All these experiments provided the same phonon dispersion. Such reproducibility further supports the occurrence of a predominant graphene domain.

HREEL spectra show several dispersing features such as a function of the scattering angle, all assigned to phonon excitations (Figure 35.6). The energy and the dispersion of phonon modes indicate a negligible interaction between MLG and the underlying Pt substrate, in agreement with previous works. Accordingly, MLG may be considered as a quasi-freestanding sheet physorbed on the underlying Pt substrate.

As for graphite [46], vibrations of the graphene lattice are characterized by two types of phonons: those vibrating in the plane of the sheet with transverse and longitudinal acoustic (TA and LA) and optical (TO and LO) branches, and those with vibrations out of the plane of the layer—the so-called flexural phonons (ZA and ZO). Modes classified with "T" are shear in-plane phonon excitations; "L" modes are longitudinal in-plane vibrations; while "Z" indicates out-of-plane polarization. In general, the ZO mode is significantly softened with respect to the other two optical modes, that is, TO and LO. This is due to the higher freedom for atom motion perpendicularly to the plane with respect to the in-plane motion. All observed phonons are similar to those of bulk graphite.

This is quite expected for modes associated with vibrations of carbon atoms in the direction of the σ bonds, that is, the LA and LO phonons (except a stiffening of the LO mode at $\bar{\Gamma}$ by 5 meV). In particular, a careful comparative analysis with respect to graphite showed a softening of the TA mode (with a maximum difference of 18 meV). On the other hand, by comparing the phonon dispersion relation of MLG/Pt(111) with that recorded for MLG/Ni(111) [47,48], we notice a significant energy shift by 10–15 meV at the $\bar{\Gamma}$ point for the ZA and ZO modes, which are connected to perpendicular vibrations of carbon atoms with respect to the surface. In particular, in MLG/Ni(111), the ZA phonon is stiffened while the ZO mode is softened with respect to MLG/Pt(111). This is caused by the orbital mixing of the π-states of the MLG with Ni d-bands [49].

Particular attention should be devoted to the ZA phonon, as it was recently found that the ZA modes in suspended graphene carry most of the heat [50]. It is a bending mode in which the two atoms in the unit cell are involved in an in-phase motion in the out-of-plane direction. At long wavelengths, it bends the MLG sheet so as to induce rippling in graphene. While the TA and the LA phonons are characterized by a linear dispersion, the ZA mode has a quadratic dispersion near the $\bar{\Gamma}$ point (Figure 35.7), as also in layered crystals [51]. Its dispersion depends on the bending rigidity τ, which is an important parameter for mechanical properties of membranes:

$$\omega_{ZA}(q_{\parallel}) = \sqrt{\frac{\tau}{\rho_{2D}}}\,|\vec{q}_{\parallel}|^2$$

where $\rho_{2D} = 4m_C/(3\sqrt{3}a^2)$ is the two-dimensional (2D) mass density (m_C is the atomic mass of carbon atoms; a is the in-plane lattice parameter). We found that the bending rigidity τ could be estimated to be about 2 eV, in agreement with results in Reference 52.

The dispersion relation of phonons in MLG/Pt(111) have been compared with recent calculations for freestanding graphene derived from long-range carbon bond order potential [45] (dotted line in Figure 35.7). In particular, we note an excellent agreement with calculations for the LA mode. The recorded dispersion for the flexural phonon ZA apparently does not match the behavior predicted by theory. However, it is worth remembering that the bending rigidity strongly depends on the temperature and an increase by about 40%

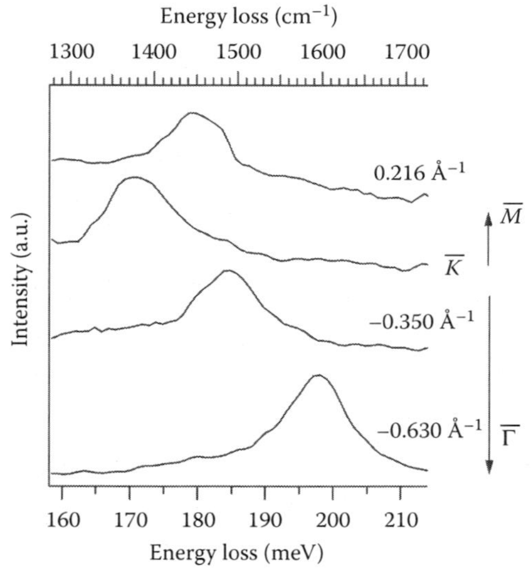

FIGURE 35.6 HREEL spectra for MLG/Pt(111) as a function of the scattering angle around the high-symmetry point $\bar{K}$. The incident angle is fixed to 86.0° with respect to the sample normal. The impinging energy is 20 eV. All measurements have been carried out at room temperature.

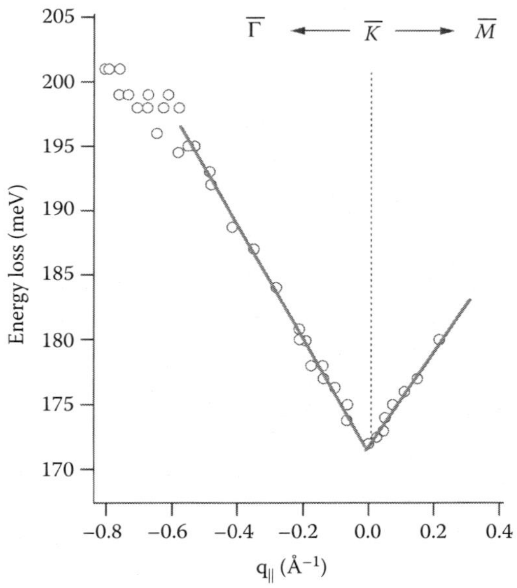

FIGURE 35.7 Graphene HOB in the nearness of the $\bar{K}$ of the Brillouin zone.

was found in the range 0–300 K [53]. The temperature dependence of the bending rigidity implies that the ZA mode should also depend on the temperature. Hence, comparison of the calculated dispersion of ZA phonons at 0 K with experiments at 300 K is not straightforward.

Interestingly, the ZA/ZO degeneracy at $\bar{K}$, predicted by theoretical calculations for freestanding graphene [45] and bulk graphite [54], is lifted and a "gap" of 20 meV in MLG/Pt(111) appears. Allard and Witz demonstrated in Reference 6 that this is a direct consequence of symmetry reduction in supported graphene systems. In fact, for freestanding graphene, all carbon atoms of the graphene unit cell are equivalent, while for adsorbed graphene, both atop and threefold adsorption sites exist. Thus, some atoms are directly connected to the substrate while others are unconnected. In the ZA mode, the unconnected atoms are performing a perpendicular motion while the bonded atoms are at rest. For the ZO branch, the opposite occurs. This implies a higher energy for the ZA phonon. Thus, the ZA/ZO gap can be taken as a fingerprint for graphene adsorption on solid substrates; in fact, it also occurs for MLG/Ni(111) [48,55]. By decoupling the MLG from the substrate through Ag intercalation, the gap can be removed [47,56].

By contrast, the degeneracy of LA and LO phonons at $\bar{K}$ was observed, even if the energy of such degenerate mode at $\bar{K}$ is blue-shifted by 25 meV with respect to the calculated value in References 45,47.

35.3.3 KAs

Atomic vibrations could be screened by electrons and, moreover, screening can change rapidly for vibrations associated with high-symmetry points of the Brillouin zone. This phenomenon leads to an anomalous behavior of the phonon dispersion around such points, which is called KA [58]. Their occurrence is completely determined by the shape of the Fermi surface.

Graphite exhibits two KAs for the $\bar{\Gamma} - E_{2g}$ and $\bar{K} - A_1'$ modes in inelastic x-ray scattering experiments [59]. In detail, only the highest optical branches (HOB) show KAs, which are evidenced in the phonon dispersion curve by two sharp kinks. Their existence is intimately related to the dispersion of the π-bands around the high-symmetry point $\bar{K}$.

With regard to graphene, several theoretical studies predicted the existence of KAs [60]. However, recently, it has been demonstrated [6] that the electron–phonon coupling in epitaxial graphene systems can be strongly modified by the interaction with the underlying metal substrate. Allard and Wirtz [6], analyzing previous phonon measurements [48] performed on MLG grown on a Ni(111) surface, suggested a complete suppression of KAs for such system. Therein, it was suggested that the absence of KAs in MLG/Ni(111) is caused by the hybridization of the graphene π-bands with the Ni d-bands which lifts the linear crossing of the π-bands at $\bar{K}$.

Experiments on phonon dispersion showed that, in contrast with the case of MLG/Ni(111), KAs could be detected in MLG/Pt(111). We ascribe this finding to the nearly quasi-freestanding behavior of π-bands in this system. Such results could

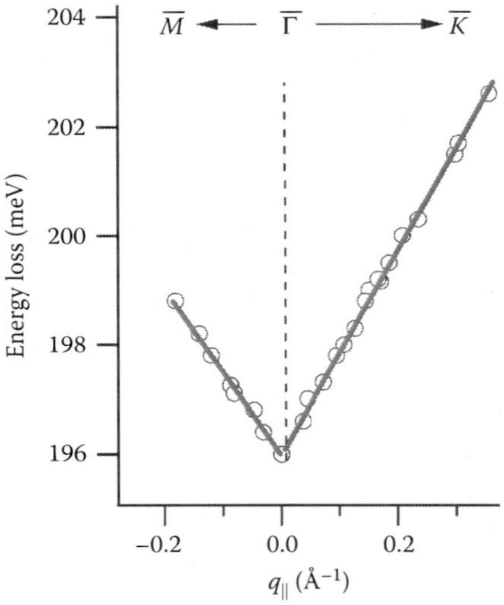

FIGURE 35.8 Graphene HOB in the nearness of the $\bar{\Gamma}$ of the Brillouin zone.

be important to evaluate the interaction strength between the graphene layer and the underlying metallic substrate.

MLG/Pt(111) loss spectra around the high-symmetry point $\bar{K}$ show (Figure 35.6) a softening by about 30 meV of the TO mode (characterized by A_1' symmetry), which reaches a minimal loss energy at $\bar{K}$ (172 meV).

Figures 35.7 and 35.8 report the dispersion relation of the HOB as a function of $q_{\|}$ around the $\bar{K}$ and $\bar{\Gamma}$ symmetry points, respectively. The most striking feature of these dispersions is the discontinuity in the derivative of the HOB (A_1' and E_{2g}, respectively) at $\bar{K}$ and $\bar{\Gamma}$, which is a direct evidence of the existence of KAs. This should be put in relationship with the abrupt change in the screening of lattice vibrations by conduction electrons. On the other hand, for MLG/Ni(111), the dispersion of the HOB is almost flat at both $\bar{K}$ and $\bar{\Gamma}$ symmetry points. This means that the interaction with the substrate leads to a complete suppression of KAs, as a consequence of the strong hybridization of the graphene π-bands with Ni d-bands [35]. In fact, the hybridization induces around $\bar{K}$ the appearance of a "gap" of almost 4 eV between unoccupied and occupied π-bands.

35.4 ELECTRONIC COLLECTIVE EXCITATIONS IN EPITAXIAL GRAPHENE

35.4.1 General Consideration on Plasmons in Graphene

Low-energy collective excitations in graphene are attracting much interest in recent years [61] as they influence many of the peculiar properties of graphene samples. In particular, the dispersion and damping of plasmons in epitaxial graphene have recently been studied for the case of graphene deposited on SiC(0001) [62,63] and Ir(111) [64]. The understanding of plasmonic excitations of graphene plays a key role in tailoring the properties of novel graphene-based devices [65].

Indeed, many of the peculiar properties of graphene are related to its collective electronic excitations [66,67], even if their understanding is still lacking. In particular, it is essential to shed light on plasmon modes in graphene/metal interfaces to understand dynamical processes and screening in such systems.

The electronic structure of MLG on Pt(111) resembles that of isolated graphene [40]. In particular, the linear dispersion of π-bands in the so-called Dirac cones, which gives rise to many manifestations of massless Dirac fermions, is preserved. ARPES experiments do not show any remarkable hybridization of graphene π-states with metal *d*-states. They just represent a superposition of graphene and metal-derived states, with minimal interaction between them. The MLG on Pt(111) is hole doped by charge transfer to the Pt substrate [68]. The Fermi energy E_F of the graphene layer shifts 0.30 ± 0.15 eV below the Dirac energy crossing point of the bands, with the Fermi wave vector $k_F = 0.09$ Å^{-1}. Epitaxial graphene on Pt(111) thus behaves as an ideal 2D system, sustaining a purely two-dimensional electron gas (2DEG) system whose collective excitations (plasmon modes) are able to propagate along the sheet. The dielectric response of the 2DEG system is determined by plasmon dispersion, which could be measured by HREELS.

The 2D plasmon, characterized by its square-root-like dispersion, has been predicted [69] and observed in metal layers on semiconductors [70,71]. On the other hand, the acoustic surface plasmon (ASP) with a linear dispersion was demonstrated to exist on semiconductor quantum wells with two interacting quantum well minibands [72]. Successively, ASP has been experimentally revealed on Be(0001) [73] and on noble metal surfaces [74,75]. The acoustic-like dispersion is a consequence of the combination of the nonlocality of the three-dimensional (3D) response and the spill-out of the 3D electron density into the vacuum, both providing incomplete screening of the 2D electron density oscillations.

Previous measurements on MLG/SiC(0001) showed a nonlinear dispersion for the sheet plasmon in MLG. Such behavior could be described by Stern's model [69]. It is interesting to study the behavior of collective excitations of MLG grown on a metal substrate in order to shed light on the screening mechanisms of the sheet plasmon in the presence of an underlying metal substrate. Present measurements by HREELS show a linear dispersion for the sheet plasmon in MLG/Pt(111). Our results indicate that the sheet plasmon of MLG survives up to a high energy, that is, 3 eV. This is a consequence of the fact that intraband excitations have negligible influence on the propagation of the plasmon mode. On the other hand, the dispersion curve of the sheet plasmon overlaps with the continuum of interband transitions above the Fermi wave vector. This broadens the plasmon peak but does not cause its disappearance, in contrast with the behavior found for ordinary sheet plasmons in 2DEG and ASP.

35.4.2 Evidence for Acoustic-Like Plasmons

To measure plasmon dispersion (see Section 35.2), values for the parameters E_p, impinging energy, and θ_i, the incident angle, were chosen so as to obtain the highest signal-to-noise ratio. The primary beam energy use for the dispersion, E_p = 7–12 eV, provided, in fact, the best compromise among surface sensitivity, the highest cross section for mode excitation and momentum resolution.

To obtain energies of loss peaks, a polynomial background was subtracted from each spectrum. The resulting spectra were fitted by a Gaussian line shape (not shown herein). All measurements were made at room temperature.

Measurements were performed for both symmetry directions (Γ–*K* and Γ–*M*), but no remarkable differences were recorded as a consequence of the existence of differently oriented domains on the sample, as observed in previous low-energy electron microscopy experiments [40]. Loss measurements of MLG/Pt(111) recorded as a function of the scattering angle θ_s are reported in Figure 35.9. HREEL spectra show a low-energy feature that develops and disperses up to 3 eV as a function of the scattering angle. This resonance exhibits a clear linear dispersion and its frequency approaches zero in the long-wavelength limit. We assign it to the sheet plasmon of MLG, in agreement with theoretical [76–80] and experimental [62–64,81–83] results.

The dispersion of the sheet plasmon for MLG on SiC(0001) well agrees with Stern's [69] prediction ($\omega \propto \sqrt{q_\parallel}$). However, the plasmon dispersion recorded in our experiments (Figure 35.10) is well described by a linear relationship, as in the case of ASP on bare metal surfaces [73–75]:

$$\hbar\omega_{2D} = Aq_\parallel$$

where $A = 7.4 \pm 0.1$ eV Å.

The sheet plasmon with a linear dispersion owes its existence to the interplay of the underlying metal substrate with the π-charge density in the MLG in the same region of space. It resembles the ASP in metal surfaces that support a partially

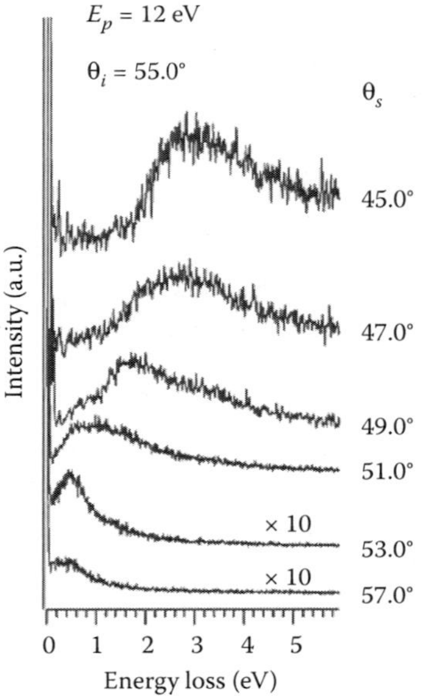

FIGURE 35.9 HREEL spectra of MLG/Pt(111) acquired as a function of the scattering angle. The incident angle is 54.0°. The impinging energy E_p is 12 eV.

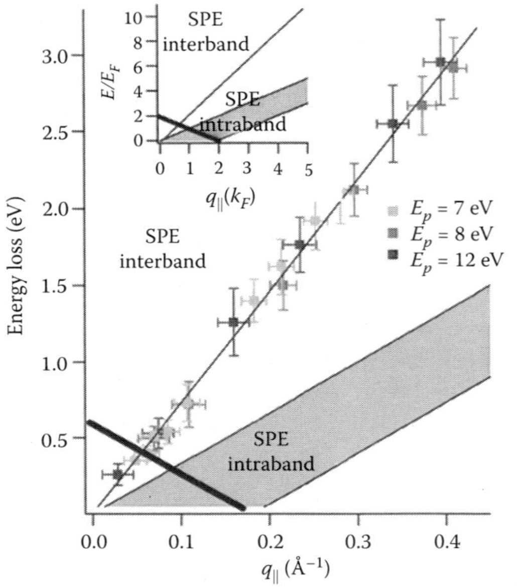

FIGURE 35.10 Plasmon dispersion in MLG/Pt(111). Data have been acquired for three different impinging energies. The thin solid line represents the best fit for experimental points. The dashed area indicates the continuum of intraband SPEs. The thick solid line represents the boundary for the continuum of interband SPEs. The plasmon mode enters the Landau damping regime by interband electron–hole excitations when its dispersion relation intercepts the boundary for the continuum of interband SPEs. In the inset, the curves are reported with respect to the dimensionless units *E*/*EF* and $q_{||}$/*kF*.

occupied surface state band within a wide bulk energy gap [84,85].

The nonlocal character of the dielectric function [86] and the screening processes in graphene [67,87] prevent the sheet plasmon from being screened out by the 3D bulk states of Pt(111).

Recently, Horing [77] predicted that the linear plasmon in graphene systems may arise from the Coulombian interaction between the native sheet plasmon (($\omega \propto \sqrt{q_{||}}$)) in MLG and the surface plasmon of a nearby thick substrate hosting a semi-infinite plasma. Calculations taking into account the electronic response of the Pt substrate could in principle put this effect in evidence, but this is not trivial due to the existence of a Moiré reconstruction in the MLG lattice on top of the Pt(111) substrate. The slope of the dispersion relation of the sheet plasmon in MLG/Pt(111) and that of acoustic-like excitations provide information about group velocities of the plasmon mode. We found that the group velocity of the sheet plasmon in MLG/Pt(111) ($1.1 \pm 0.2 \times 10^6$ m/s) is similar to that calculated for ASP [85]. The group velocity of the sheet plasmon in MLG/Pt(111) is about two orders of magnitude lower than the speed of light; thus, its direct excitation by light is not possible. However, nanometer-size objects at surfaces, such as atomic steps or molecular structures, can allow coupling between sheet plasmon and light.

The linear behavior of its dispersion implies that both phase and group velocities of the collective excitation are the same, so signals can be transmitted undistorted along the surface. Hence, this finding could be of significant importance in future graphene-based nano-optical devices, especially if we have in mind that the Moiré pattern of MLG on metal substrates offers a naturally nanostructured system [15].

In Figure 35.10, we also show the electron–hole continuum or single-particle excitation (SPE) region, which determines the absorption (Landau damping) of the external field at a given frequency.

It was calculated on the basis of results in References 78,80,88 by substituting the values of E_F and k_F for MLG/Pt(111) obtained by ARPES [40]. For a normal 2D system, only indirect transition is possible within the band. However, for graphene, both intraband and interband transitions are possible, and the boundaries are given in Figure 35.10. Owing to the phase-space restriction, the interband SPE continuum has a gap at small momenta.

For $q_{||} = 0$, the transition is not allowed at $0 < E < 2E_F$. If the collective mode enters the SPE continuum, the plasmon mode can be damped. The plasmon lies inside the interband SPE continuum, thus decaying into electron–hole pairs, above the Fermi wave vector. Plasmon can propagate without damping only in the region (see Figure 35.10) not included in the continuum of SPE (interband and intraband). Such considerations are fully confirmed by the analysis of the full-width at half maximum (FWHM) of the plasmon peak as a function of both $q_{||}$ (Figure 35.11a) and the plasmon energy (Figure 35.11b).

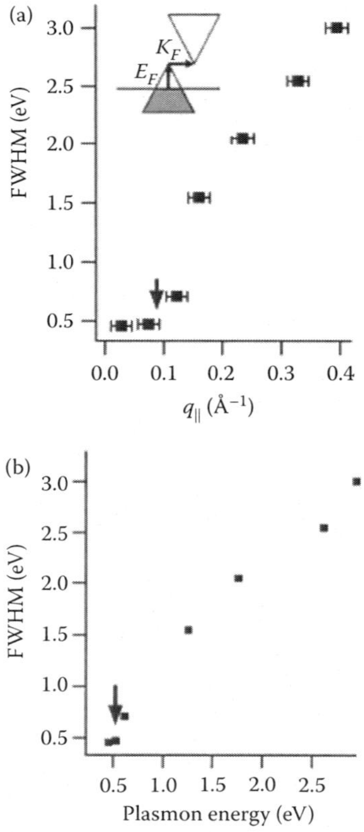

FIGURE 35.11 FWHM of the plasmon peak as a function of (a) the parallel momentum transfer $q_{||}$ and of (b) the plasmon energy. The inset in the top panel shows the origin of interband SPEs from π to π* bands. The Fermi wave vector represents the onset where plasmon enters the damping region.

Landau damping for the MLG sheet plasmon occurs for momenta above the Fermi wave vector (about 0.09 Å^{-1}) and for energies above 0.5 eV, as revealed by the sudden increase of the FWHM. Interestingly, the sheet plasmon does not enter into the intraband SPE continuum and it exists for all wave vectors. By contrast, for ASP, Landau damping occurs via intraband transitions and the plasmon mode exists only up to a few hundreds meV [73–75]. On the other hand, for MLG on SiC(0001), the FWHM continuously increases with the momentum [62,89]. For such system, it has been shown [62] that the existence of steps or grain boundaries is a source of strong damping, while the dispersion is rather insensitive to defects.

35.4.3 Dispersion and Damping Processes of π Plasmon

The electronic response of graphene systems is related to the collective excitations of the electrons, which combine in-plane and interplane interactions. Herein, we want to investigate the nature and the dispersion of π plasmon in MLG grown on a metal substrate. The π plasmon is a sensitive probe of the graphene band structure near the Fermi level [90]. Its physical origin is the electric dipole transition between the π energy bands ($\pi \rightarrow \pi^*$) mainly in the region of the M point of the Brillouin zone of graphitic systems. Recently, it has been demonstrated [67] that in the presence of a free-electron substrate, the plasmonic excitation of the graphene sheet are nearly completely quenched, as a consequence of the dynamical Coulomb interaction between induced charges in the substrate and graphene. However, in the case of graphene on Ni(111), the π plasmon was found to exist [91]. It is thus interesting to characterize the π plasmon mode also in the case of a graphene sheet weakly bonded to the metal substrate, as occurs for MLG/Pt(111).

HREELS measurements show a quadratic dispersion for the π plasmon in MLG/Pt(111), in contrast with results obtained for MLG/6H-SiC(0001) [82] and in agreement with very recent findings for MLG/Ni(111) [91]. However, the quadratic coefficient of the dispersion relation of π plasmon in MLG/Pt(111) is higher by a factor ~8 with respect to the case of MLG/Ni(111) while it is similar to the value reported for graphite.

Moreover, we found that the plasmon peak is blue-shifted by about 1.5 eV with respect to freestanding graphene and MLG/6H-SiC(0001). The presence of the metal substrate also decreases the lifetime of the plasmonic excitation, as evidenced by a careful analysis of its damping processes.

To measure the dispersion relation, primary beam energies, E_p = 30–70 eV, were used. Spectra recorded for MLG on Pt(111) with the substrate oriented in the $\bar{\Gamma}-\bar{M}$ direction are reported in Figures 35.12a (for a primary energy E_p of 70 eV) and 35.12b (for E_p = 30 eV).

A peak showing clear dispersion from 6.2 to 8.2 eV has been recorded as a function of the parallel momentum transfer $q_{\parallel}$. It has been assigned to π plasmon of graphene, in agreement with previous [82,91–93] works.

The intensity of the backscattering yield around the π plasmon energy versus the off-specular angle clearly demonstrates that the plasmon mode has a dipolar nature because it is nearly peaked in the specular direction [33,94].

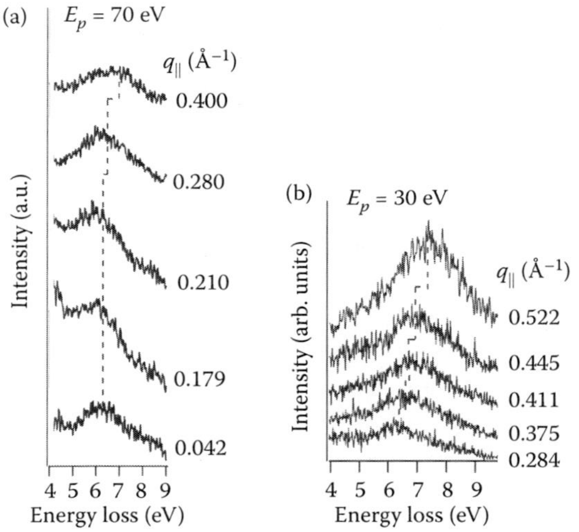

FIGURE 35.12 HREEL spectra for MLG/Pt(111) as a function of the parallel momentum transfer $q_{\parallel}$. (a) Loss spectra acquired by using an impinging energy of 70 eV and an incidence angle of 65°. (b) Spectra acquired with an impinging energy of 30 eV and an incidence angle of 55°. It is worth mentioning that, due to the very weak intensity of loss peaks (≈10^{-4} with respect to the intensity of the elastic peak), an acquisition time of several hours has been required for each spectrum to reach a sufficient signal-to-noise ratio. All measurements were made at room temperature.

The measured dispersion curve $E_{loss}(q_{\parallel})$ in Figure 35.13 has been fitted by a second-order polynomial given by

$$E_{loss}(q_{\parallel}) = E_{loss}(0) + Aq_{\parallel}^2 = E_{loss}(0) + \alpha\frac{\hbar^2}{m}q_{\parallel}^2$$

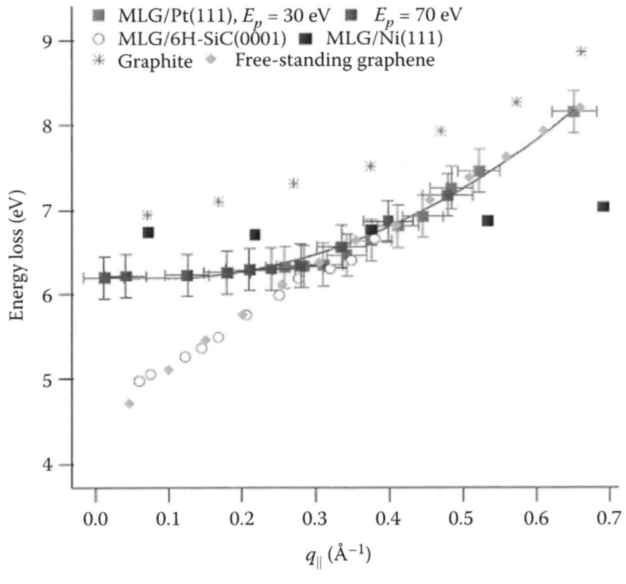

FIGURE 35.13 Dispersion relation of the π plasmon for MLG/Pt(111) (our data, acquired for two different scattering geometries), MLG/Ni(111), MLG/6H-SiC(0001), graphite, and calculations for freestanding graphene.

where $E_{loss}(0) = (6.2 \pm 0.1)$ eV, $A = (4.1 \pm 0.2)$ eV · A², and $\alpha = 0.53$.

The same measurements were repeated for the $\overline{\Gamma} - \overline{K}$ direction, giving a similar dispersion. This is a consequence of the existence of differently oriented domains on the sample with a preferential orientation aligned with respect to the substrate (Section 35.3 [42]). These findings agree with the conclusions of previous low-energy electron microscopy experiments [40].

Important information on graphene systems is provided by the analysis of the frequency of the π plasmon in the long-wavelength limit ($q_{||} \approx 0$), reported in Table 35.1 for various carbon-based systems. It can be noted that it ranges between 4.7 eV (freestanding graphene) [92] and 6.5–7.0 eV (graphite).

As regards graphite, it is important to note that π plasmon energy in the long-wavelength limit (small momenta) strongly depends on the modality of the HREELS experiment. In fact, HREELS measurements in the reflection mode provide an energy of 6.5 eV [96,97], while transmission measurements yield approximately 7 eV [98,99].

Intermediate values have been recorded for a vertically aligned single-walled carbon nanotube (VA-SWCNT) [92] and MLG on 6H-SiC(0001) [82] (about 5 eV) and magnetically aligned bundles of SWCNT [95] (about 6 eV).

A HREELS investigation on graphene grown on 6H-SiC(0001) showed the existence of a blue-shift of the π plasmon energy [82] as a function of the number of graphene layers. In fact, it shifted from 4.9 (MLG) to 4.3 eV for bilayer graphene and to 6.2 eV for 3–4 layers of graphene.

TABLE 35.1
Energy and Line-Width of the π Plasmon in the Long-Wavelength Limit (Small Momenta) for Different Systems

	E_{loss} ($q_{\|} = 0$) in eV	FHWM ($q_{\|} = 0$) in eV
Freestanding graphene	4.7 [92,95] ~6 [66]	0.45
MLG/6H-SiC(0001), data taken from Reference 82	4.9	0.95
VA-SWCNT, data taken from Reference 92	5.1	1.00
Bilayer graphene on SiC(0001), data taken from Reference 82	5.3	1.10
Magnetically aligned bundled SWCNT, data taken from Reference 95	6.0	1.25
MLG/Pt(111) (our data)	6.2	1.40
3–4 layer graphene on SiC(0001), data taken from Reference 82	6.3	1.70
Graphite, data taken from Reference 96	6.5 [96,97] ~7 [98,99]	2.90
MLG/Ni(111), data taken from References 91,93.	6.7 [91] 7.5 [93]	~3

The red-shift of the plasmon energy (at small momenta) when going from bulk graphite to quasi-2D graphene is caused by a decrease of the screening and of the interlayer coupling. This also influences the dispersion relation of the plasmon frequency. A linear dispersion was found for both VA-SWCNT [92] and MLG on 6H-SiC(0001) [82]. On the other hand, a quadratic dispersion has been recorded for bulk graphite [97], stage-1 ferric-chloride-intercalated graphite [100] and multilayer graphene on 6H-SiC(0001) [82]. For the latter case, it is clear that the realistic band structure of the system changes the dispersion of the π plasmon from linear to quadratic as a function of the number of graphene layers.

The dispersion relations that we obtained for MLG on Pt(111) (Figure 35.13) indicate a quadratic dispersion for the π plasmon. Even if a negligible hybridization between Pt and graphene states has been observed by ARPES measurements [40], the dispersion relation of the π plasmon is quadratic already for the MLG, as a consequence of the screening of the collective mode by the metal substrate. The screening in MLG on metal substrates is clearly more effective with respect to the case of graphene layers grown on the semiconductor silicon carbide substrate. This should explain the quadratic dispersion recorded in MLG/Ni(111) [91] and MLG/Pt(111) (our data), in spite of the very different band structure of two such graphene/metal interfaces. We also recall that the plasmon dispersion in the long-wavelength limit is predicted to be quadratic with respect to momentum for the interacting electron gas [101].

Another issue to be considered is the interlayer coupling. Concerning supported graphene, the interlayer interaction varies as a function of the electron density in the layers. At higher electron density, the overlap between orbitals of adjacent layers increases, thus increasing the interlayer coupling [102]. Hence, in principle, the fact that MLG is hole-doped by charge transfer from the Pt substrate (Fermi level below Dirac point) [68] could influence the electronic response of the interface. As regards graphite, it is worth mentioning that its interlayer coupling is still under debate. Band structure calculations predicted an interlayer coupling much larger than the values deduced by the c-axis conductivity measurements [103,104].

A key factor in the propagation of the plasmonic excitation is its lifetime, which is limited by the decay into electron–hole pairs (Landau damping) [105]. The damping of the plasmon peak is clearly revealed by the trend of the FWHM versus $q_{||}$, reported in Figure 35.14. The width of the plasmon rapidly increases with $q_{||}$ due to the occurrence of Landau damping. It is worth remembering that, in contrast with the low-energy sheet plasmon [20,62], the π plasmon is a mode that lies inside the continuum of particle–hole excitations and therefore it will be damped even at $q_{||} \to 0$ [66].

On the other hand, the width of the π plasmon in MLG/6H-SiC(0001) initially decreased up to 0.1 Å^{-1}, followed by a steep increase as a function of $q_{||}$. A similar behavior has been recorded on graphite [97], where the turning point has been found at 0.3–0.4 Å^{-1}. The absence of a turning point in MLG/Pt(111) could be related to the nearly linear dispersion of π-bands in the Dirac cones [40]. Instead, substrate interactions

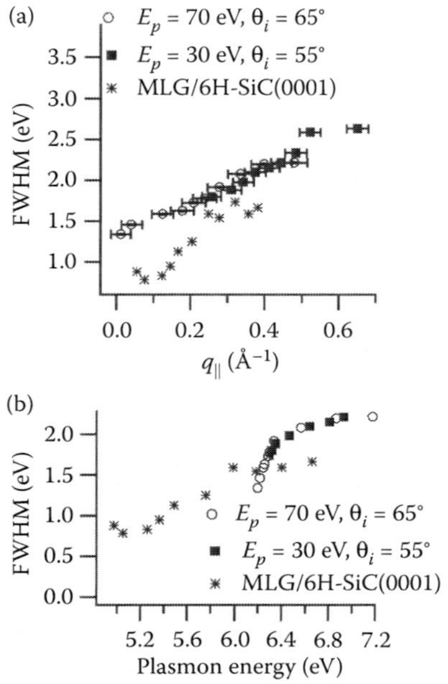

FIGURE 35.14 Behavior of the FWHM of the π plasmon of MLG/Pt(111) acquired for two different scattering conditions as a function of (a) the parallel momentum transfer and (b) the plasmon energy. Data for MLG/6H-SiC(0001) are shown for a comparison.

in graphene on silicon carbide are known to distort the linear dispersion near the Dirac point in the first graphene layer [106]. They cause the appearance of a 260 meV energy gap and enhanced electron–phonon coupling [106]. This gap decreases as the sample thickness increases and eventually approaches zero for multilayer graphene. The behavior of the FWHM as a function of the plasmon energy (Figure 35.14b) showed that for MLG/Pt(111) there is an enhanced broadening of the plasmon peak around 6.3 eV. These findings indicate that the Landau damping processes of the π plasmon in MLG/Pt(111) are mainly due to π–π* interband transitions centered around 6.3 eV.

By comparing the FWHM in the long-wavelength limit for various graphene systems (Table 35.1), it is quite evident that the presence of out-of-plane decay channels reflects into a wider line-width of the plasmon peak, that is, a shorter lifetime of the plasmon mode. As demonstrated for graphite [107], they cause additional damping of plasmons, which result in a more diffuse shape for the loss spectrum. As a matter of fact, the FWHM increases by a factor 6 from freestanding graphene to graphite.

Results for graphene on 6H-SiC(0001) showed that the π plasmon peak becomes broader and blue-shifted as the thickness of the epitaxial graphene increases. In fact, the π plasmon of 3–4 layer epitaxial graphene includes spectral contribution from both the out-of-plane and in-plane excitations of graphitic origin. This may be due to the 3D band structure of graphite, which allows interlayer coupling and out-of-plane excitation.

Moreover, it is worth noticing that the π plasmon in MLG/Pt(111) (our data) and MLG/Ni(111) [91] has a shorter lifetime (higher FWHM) than in MLG/6H-SiC(0001) and freestanding graphene as a consequence of enhanced screening by the metal substrate. This implies a broadening of the plasmon peak due to Landau damping via the creation of electron–hole pairs.

35.5 CONCLUSIONS AND OUTLOOK

Electron energy loss spectroscopy is a powerful tool for investigating collective modes on solid surfaces. In this chapter, we have reported on experiments on collective excitations in graphene/metal interfaces.

We have found a linear dispersion of the sheet plasmon in metals and such behavior is attributed to the nonlocal screening of the plasmon mode of graphene caused by the underlying metal substrate. Owing to its low energy and its linear dispersion, the sheet plasmon is expected to play an important role in graphene dynamics. This could be especially relevant for future graphene-based nano-optical devices, since the rippled, nanostructured surface of graphene on metal substrates provides an interesting scenario to couple ASPs and light. Measurements on plasmon modes also indicate that π electrons in graphene/metal interfaces behave as an interacting electron gas. We also report on collective modes due to the lattice dynamics of the graphene sheet. From experimental phonon dispersions, therefore, it is possible to draw conclusions about the interaction strength in graphene/metal interfaces. Such investigation is of fundamental importance for graphene-based devices as both electronic and optical excitations can be scattered by phonon states or decay into vibrational excitations.

While the investigation of phonon modes in graphene/metal interfaces can provide important fundamental information, graphene plasmons show very promising properties that can be used for applications: reasonably large lifetimes, tunability of the plasmon frequency, electric controllability, plasmon confinement, and strong coupling with phonons and light.

Graphene could represent an ideal playground for applications of 2D electromagnetic waves [108], so as to facilitate the design and miniaturization of nanophotonic devices [109].

Recently, plasmonics has had a groundbreaking impact in photonics [110]. Plasmonics has opened the way for the realization of high-speed and transparent photosensitive systems, which could be further functionalized to enable chemical sensing [111]. The combination of graphene with conventional plasmonic elements will allow the realization of THz plasmonic lasers [112–114], plasmonic antennas [115], plasmonic waveguides [116–118], Luneburg lenses [119], ultrasensitive biosensors [120–124] and so on. The flexibility of graphene [125] may permit the realization of graphene-based flexible plasmonic devices [126]. Moreover, the largely nonlinear optical response of graphene [127] can allow innovative experiments and applications of graphene-based nonlinear plasmonics [128,129].

However, there are still some limitations and open problems for graphene-based plasmonics, as for example the lack of reliable THz light sources.

This overall encouraging viewpoint for applications is also accompanied by the possibility to carry out many other fascinating fundamental studies. As an example, accurate experimental studies on plasmons in bilayer graphene on metals would be essential to verify and improve current theoretical models for both plasmon dispersion [130–133] and plasmaron formation [134,135]. Screened electron-electron interactions in a graphene sheet supported by a metal substrate are responsible for intriguing many-body effects. Despite the strong screening exerted by the metal, the 2DEG in graphene/metals contacts shows self-sustaining, long-living oscillations whose phase velocity coincides with the group velocity (ASP). Concerning high-energy plasmons (e.g., the π plasmons at 5–7 eV), they are suitable candidate for potential applications in UV regions in composite structures with materials able to emit or absorb UV light. However, accurate theoretical models for plasmons graphene/metal interfaces are still missing due to the difficulty in the theoretical description of the screening by the underlying metal substrate. The out-of-plane charge transfer between graphene and the metal is determined by the difference between the work function of graphene and the metal surface and, in addition, by the metal–graphene chemical interaction that creates an interface dipole that lowers the metal work function. The induced electrostatic potential decays weakly with the distance from the metal contact as $V(x) \approx x^{-1/2}$ and $\approx x^{-1}$ for undoped and doped graphene, respectively [136]. Instead, current models overestimate the screening by the metal substrate. Likely, the experimental study of plasmons in graphene deposited on jellium surfaces could help theoreticians to improve our understanding of screening processes of graphene/metals. Unfortunately, such experimental study is complicated by the difficult preparation of graphene on jellium surfaces.

REFERENCES

1. Nienhaus, H. 2002. Electronic excitations by chemical reactions on metal surfaces. *Surface Science Reports* 45:1–78.
2. Kroes, G.-J. 2012. Towards chemically accurate simulation of molecule-surface reactions. *Physical Chemistry Chemical Physics* 14:14966–81.
3. Trenhaile, B. R., Antonov, V. N., Xu, G. J., Agrawal, A., Signor, A. W., Butera, R. E., Nakayama, K. S., and Weaver, J. H. 2006. Phonon-activated electron-stimulated desorption of halogens from Si(100)-(2 × 1). *Physical Review B* 73:125318.
4. Castro Neto, A. H. 2007. Graphene: Phonons behaving badly. *Nature Materials* 6:176–7.
5. Benedek, G., Hofmann, F., Ruggerone, P., Onida, G., and Miglio, L. 1994. Surface phonons in layered crystals—Theoretical aspects. *Surface Science Reports* 20:3–43.
6. Allard, A. and Wirtz, L. 2010. Graphene on metallic substrates: Suppression of the Kohn anomalies in the phonon dispersion. *Nano Letters* 10:4335–40.
7. Aynajian, P., Keller, T., Boeri, L., Shapiro, S. M., Habicht, K., and Keimer, B. 2008. Energy gaps and Kohn anomalies in elemental superconductors. *Science* 319:1509–12.
8. Hwang, E. H. and Das Sarma, S. 2008. Screening, Kohn anomaly, Friedel oscillation, and RKKY interaction in bilayer graphene. *Physical Review Letters* 101:156802.
9. Politano, A., Marino, A. R., Campi, D., Farías, D., Miranda, R., and Chiarello, G. 2012. Elastic properties of a macroscopic graphene sample from phonon dispersion measurements. *Carbon* 50:4903–10.
10. Mante, P.-A., Wu, Y.-C., Lin, Y.-T., Ho, C.-Y., Tu, L.-W., and Sun, C.-K. 2013. Gigahertz coherent guided acoustic phonons in AlN/GaN nanowire superlattices. *Nano Letters* 13:1139–44.
11. Talwar, D. N., Vandevyver, M., Kunc, K., and Zigone, M. 1981. Lattice dynamics of zinc chalcogenides under compression: Phonon dispersion, mode Grüneisen, and thermal expansion. *Physical Review B* 24:741–53.
12. Benedek, G., Hulpke, E., and Steinhogl, W. 2001. Probing the magnetic forces in fcc-Fe(001) films by means of surface phonon spectroscopy. *Physical Review Letters* 87:027201.
13. Lin, Z., Zhigilei, L. V., and Celli, V. 2008. Electron–phonon coupling and electron heat capacity of metals under conditions of strong electron–phonon nonequilibrium. *Physical Review B* 77:075133.
14. Balandin, A. and Wang, K. L. 1998. Significant decrease of the lattice thermal conductivity due to phonon confinement in a free-standing semiconductor quantum well. *Physical Review B* 58:1544–9.
15. Borca, B., Barja, S., Garnica, M., Minniti, M., Politano, A., Rodriguez-García, J. M., Hinarejos, J. J., Farías, D., Vázquez de Parga, A. L., and Miranda, R. 2010. Electronic and geometric corrugation of periodically rippled, self-nanostructured graphene epitaxially grown on Ru(0001). *New Journal of Physics* 12:093018.
16. Müller, F., Grandthyll, S., Zeitz, C., Jacobs, K., Hüfner, S., Gsell, S., and Schreck, M. 2011. Epitaxial growth of graphene on Ir(111) by liquid precursor deposition. *Physical Review B* 84:075472.
17. Politano, A., Marino, A. R., and Chiarello, G. 2012. Phonon dispersion of quasi-freestanding graphene on Pt(111). *Journal of Physics: Condensed Matter* 24:104025.
18. Politano, A., Marino, A. R., Formoso, V., and Chiarello, G. 2011. Water adsorption on graphene/Pt(111) at room temperature: A vibrational investigation. *AIP Advances* 1:042130.
19. Politano, A., Marino, A. R., Formoso, V., and Chiarello, G. 2012. Evidence of Kohn anomalies in quasi-freestanding graphene on Pt(111). *Carbon* 50:734–6.
20. Politano, A., Marino, A. R., Formoso, V., Farías, D., Miranda, R., and Chiarello, G. 2011. Evidence for acoustic-like plasmons on epitaxial graphene on Pt(111). *Physical Review B* 84:033401.
21. Politano, A., Marino, A. R., Formoso, V., Farías, D., Miranda, R., and Chiarello, G. 2012. Quadratic dispersion and damping processes of π plasmon in monolayer graphene on Pt(111). *Plasmonics* 7:369–76.
22. Politano, A., Marino, A. R., Formoso, V., and Chiarello, G. 2011. Hydrogen bonding at the water/quasi-freestanding graphene interface. *Carbon* 49:5180–4.
23. Politano, A., Chiarello, G., Benedek, G., Chulkov, E. V., and Echenique, P. M. 2013. Vibrational measurements on alkali coadsorption systems: Experiments and theory. *Surface Science Reports* 68:305–89.
24. Ibach, H. and Mills, D. L. 1982. *Electron Energy Loss Spectroscopy and Surface Vibrations*. Academic Press, San Francisco.
25. Lüth, H. 1995. *Surfaces and Interfaces of Solid Materials*. Springer, Berlin.
26. Politano, A., Formoso, V., and Chiarello, G. 2008. Alkali-promoted CO dissociation on Cu(111) and Ni(111) at room temperature. *Journal of Chemical Physics* 129:164703.

27. Ibach, H., Balden, M., and Lehwald, S. 1996. Recent advances in electron energy loss spectroscopy of surface vibrations. *Journal of the Chemical Society—Faraday Transactions* 92:4771–4.
28. Ibach, H. 1993. Electron energy loss spectroscopy with resolution below 1 meV. *Journal of Electron Spectroscopy and Related Phenomena* 64–65:819–23.
29. Benedek, G., Brusdeylins, G., Senz, V., Skofronick, J. G., Toennies, J. P., Traeger, F., and Vollmer, R. 2001. Helium atom scattering study of the surface structure and dynamics of *in situ* cleaved MgO(001) single crystals. *Physical Review B* 64:125421.
30. Hoffmann, F. M. and Weisel, M. D. 1992. Characterization of potassium promoter states under CO hydrogenation conditions on Ru(001): An *in situ* study with FT-IRAS. *Surface Science* 269–270:495–9.
31. Balzer, F. and Rubahn, H. G. 2000. Interference effects in the optical second harmonic generation from ultrathin alkali films. *Optics Communications* 185:493–9.
32. Ibach, H., Etzkorn, M., and Kirschner, J. 2006. Electron spectrometers for inelastic scattering from magnetic surface excitations. *Surface and Interface Analysis* 38:1615–7.
33. Rocca, M. 1995. Low-energy HREELSS investigation of surface electronic excitations on metals. *Surface Science Reports* 22:1–71.
34. Nemec, N., Tománek, D., and Cuniberti, G. 2006. Contact dependence of carrier injection in carbon nanotubes: An ab initio study. *Physical Review Letters* 96:076802.
35. Giovannetti, G., Khomyakov, P. A., Brocks, G., Karpan, V. M., van den Brink, J., and Kelly, P. J. 2008. Doping graphene with metal contacts. *Physical Review Letters* 101:026803.
36. Benedek, G. and Onida, G. 1993. Bulk and surface dynamics of graphite with the bond charge model. *Physical Review B* 47:16471–6.
37. Wang, Y., Page, A. J., Nishimoto, Y., Qian, H.-J., Morokuma, K., and Irle, S. 2011. Template effect in the competition between Haeckelite and graphene growth on Ni(111): Quantum chemical molecular dynamics simulations. *Journal of the American Chemical Society* 133:18837–42.
38. Borca, B., Calleja, F., Hinarejos, J. J., Vázquez de Parga, A. L., and Miranda, R. 2009. Reactivity of periodically rippled graphene grown on Ru(0001). *Journal of Physics—Condensed Matter* 21:134002.
39. Preobrajenski, A. B., Ng, M. L., Vinogradov, A. S., and Mårtensson, N. 2008. Controlling graphene corrugation on lattice-mismatched substrates. *Physical Review B* 78:073401.
40. Sutter, P., Sadowski, J. T., and Sutter, E. 2009. Graphene on Pt(111): Growth and substrate interaction. *Physical Review B* 80:245411.
41. Zhang, H., Fu, Q., Cui, Y., Tan, D., and Bao, X. 2009. Growth mechanism of graphene on Ru(0001) and O_2 adsorption on the graphene/Ru(0001) surface. *Journal of Physical Chemistry C* 113:8296–301.
42. Gao, M., Pan, Y., Huang, L., Hu, H., Zhang, L. Z., Guo, H. M., Du, S. X., and Gao, H. J. 2011. Epitaxial growth and structural property of graphene on Pt(111). *Applied Physics Letters* 98:033101.
43. Chis, V. and Benedek, G. 2011. Phonon-induced surface charge density oscillations in quantum wells: A first-principles study of the (2×2)-K overlayer on Be(0001). *The Journal of Physical Chemistry A* 115:7242–8.
44. Efetov, D. K. and Kim, P. 2010. Controlling electron-phonon interactions in graphene at ultrahigh carrier densities. *Physical Review Letters* 105:256805.
45. Karssemeijer, L. J. and Fasolino, A. 2011. Phonons of graphene and graphitic materials derived from the empirical potential LCBOPII. *Surface Science* 605:1611–5.
46. Wirtz, L. and Rubio, A. 2004. The phonon dispersion of graphite revisited. *Solid State Communications* 131:141–52.
47. Farías, D., Rieder, K. H., Shikin, A. M., Adamchuk, V. K., Tanaka, T., and Oshima, C. 2000. Modification of the surface phonon dispersion of a graphite monolayer adsorbed on Ni(111) caused by intercalation of Yb, Cu and Ag. *Surface Science* 454:437–41.
48. Shikin, A. M., Farías, D., Adamchuk, V. K., and Rieder, K. H. 1999. Surface phonon dispersion of a graphite monolayer adsorbed on Ni(111) and its modification caused by intercalation of Yb, La and Cu layers. *Surface Science* 424:155–67.
49. Mittendorfer, F., Garhofer, A., Redinger, J., Klimeš, J., Harl, J., and Kresse, G. 2011. Graphene on Ni(111): Strong interaction and weak adsorption. *Physical Review B* 84:201401.
50. Seol, J. H., Jo, I., Moore, A. L., Lindsay, L., Aitken, Z. H., Pettes, M. T., Li, X. et al. 2010. Two-dimensional phonon transport in supported graphene. *Science* 328:213–6.
51. Hartmut, Z. 2001. Phonons in layered compounds. *Journal of Physics: Condensed Matter* 13:7679.
52. Perebeinos, V. and Tersoff, J. 2009. Valence force model for phonons in graphene and carbon nanotubes. *Physical Review B* 79:241409.
53. Fasolino, A., Los, J. H., and Katsnelson, M. I. 2007. Intrinsic ripples in graphene. *Nature Materials* 6:858–61.
54. Maultzsch, J., Reich, S., Thomsen, C., Requardt, H., and Ordejón, P. 2004. Phonon dispersion in graphite. *Physical Review Letters* 92:075501.
55. Aizawa, T., Souda, R., Ishizawa, Y., Hirano, H., Yamada, T., Tanaka, K.-I., and Oshima, C. 1990. Phonon dispersion in monolayer graphite formed on Ni(111) and Ni(001). *Surface Science* 237:194–202.
56. Farías, D., Shikin, A. M., Rieder, K. H., and Dedkov, Y. S. 1999. Synthesis of a weakly bonded graphite monolayer on Ni(111) by intercalation of silver. *Journal of Physics-Condensed Matter* 11:8453–8.
57. Viola Kusminskiy, S., Campbell, D. K., and Castro Neto, A. H. 2009. Lenosky's energy and the phonon dispersion of graphene. *Physical Review B* 80:035401.
58. Kohn, W. 1959. Image of the Fermi surface in the vibration spectrum of a metal. *Physical Review Letters* 2:393–4.
59. Piscanec, S., Lazzeri, M., Mauri, F., Ferrari, A. C., and Robertson, J. 2004. Kohn anomalies and electron–phonon interactions in graphite. *Physical Review Letters* 93:185503.
60. Lazzeri, M. and Mauri, F. 2006. Nonadiabatic Kohn anomaly in a doped graphene monolayer. *Physical Review Letters* 97:266407.
61. Apalkov, V., Wang, X. F., and Chakraborty, T. 2007. Collective excitations of Dirac electrons in graphene. *International Journal of Modern Physics B* 21:1165–79.
62. Langer, T., Baringhaus, J., Pfnür, H., Schumacher, H. W., and Tegenkamp, C. 2010. Plasmon damping below the Landau regime: The role of defects in epitaxial graphene. *New Journal of Physics* 12:033017.
63. Tegenkamp, C., Pfnür, H., Langer, T., Baringhaus, J., and Schumacher, H. W. 2011. Plasmon electron–hole resonance in epitaxial graphene. *Journal of Physics: Condensed Matter* 23:012001.
64. Langer, T., Förster, D. F., Busse, C., Michely, T., Pfnür, H., and Tegenkamp, C. 2011. Sheet plasmons in modulated graphene on Ir(111). *New Journal of Physics* 13:053006.

65. Bostwick, A., Speck, F., Seyller, T., Horn, K., Polini, M., Asgari, R., MacDonald, A. H., and Rotenberg, E. 2010. Observation of plasmarons in quasi-freestanding doped graphene. *Science* 328:999–1002.
66. Yuan, S., Roldán, R., and Katsnelson, M. I. 2011. Excitation spectrum and high-energy plasmons in single-layer and multilayer graphene. *Physical Review B* 84:035439.
67. Yan, J., Thygesen, K. S., and Jacobsen, K. W. 2011. Nonlocal screening of plasmons in graphene by semiconducting and metallic substrates: First-principles calculations. *Physical Review Letters* 106:146803.
68. Gao, M., Pan, Y., Zhang, C., Hu, H., Yang, R., Lu, H., Cai, J., Du, S., Liu, F., and Gao, H. J. 2010. Tunable interfacial properties of epitaxial graphene on metal substrates. *Applied Physics Letters* 96:053109.
69. Stern, F. 1967. Polarizability of a two-dimensional electron gas. *Physical Review Letters* 18:546–8.
70. Nagao, T., Hildebrandt, T., Henzler, M., and Hasegawa, S. 2001. Two-dimensional plasmon in a surface-state band. *Surface Science* 493:680–6.
71. Nagao, T., Hildebrandt, T., Henzler, M., and Hasegawa, S. 2001. Dispersion and damping of a two-dimensional plasmon in a metallic surface-state band. *Physical Review Letters* 86:5747–50.
72. Chen, Y., Hermanson, J. C., and Lapeyre, G. J. 1989. Coupled plasmon and phonon in the accumulation layer of InAs(110) cleaved surfaces. *Physical Review B* 39:12682.
73. Diaconescu, B., Pohl, K., Vattuone, L., Savio, L., Hofmann, P., Silkin, V. M., Pitarke, J. M. et al. 2007. Low-energy acoustic plasmons at metal surfaces. *Nature* 448:57–9.
74. Park, S. J. and Palmer, R. E. 2010. Acoustic Plasmon on the Au(111) Surface. *Physical Review Letters* 105:016801.
75. Pohl, K., Diaconescu, B., Vercelli, G., Vattuone, L., Silkin, V. M., Chulkov, E. V., Echenique, P. M., and Rocca, M. 2010. Acoustic surface plasmon on Cu(111). *EPL (Europhysics Letters)* 90:57006.
76. Hill, A., Mikhailov, S. A., and Ziegler, K. 2009. Dielectric function and plasmons in graphene. *Europhysics Letters* 87:27005.
77. Horing, N. J. M. 2010. Linear Graphene Plasmons. *IEEE Transactions on Nanotechnology* 9:679–81.
78. Hwang, E. H. and Das Sarma, S. 2007. Dielectric function, screening, and plasmons in two-dimensional graphene. *Physical Review B* 75:205418.
79. Hwang, E. H. and Das Sarma, S. 2009. Plasmon modes of spatially separated double-layer graphene. *Physical Review B* 80:205405.
80. Hwang, E. H., Sensarma, R., and Das Sarma, S. 2010. Plasmon-phonon coupling in graphene. *Physical Review B* 82:195406.
81. Liu, Y. and Willis, R. F. 2010. Plasmon-phonon strongly coupled mode in epitaxial graphene. *Physical Review B* 81:081406.
82. Lu, J., Loh, K. P., Huang, H., Chen, W., and Wee, A. T. S. 2009. Plasmon dispersion on epitaxial graphene studied using high-resolution electron energy-loss spectroscopy. *Physical Review B—Condensed Matter and Materials Physics* 80:113410.
83. Politano, A. and Chiarello, G. 2014. Emergence of a nonlinear plasmon in the electronic response of doped graphene. *Carbon* 71:176–80.
84. Silkin, V. M., Garcia-Lekue, A., Pitarke, J. M., Chulkov, E. V., Zaremba, E., and Echenique, P. M. 2004. Novel low-energy collective excitation at metal surfaces. *Europhysics Letters* 66:260–4.
85. Silkin, V. M., Pitarke, J. M., Chulkov, E. V., and Echenique, P. M. 2005. Acoustic surface plasmons in the noble metals Cu, Ag, and Au. *Physical Review B* 72:115435.
86. Horing, N. J. M. 2010. Aspects of the theory of graphene. *Philosophical Transactions of the Royal Society A: Mathematical, Physical and Engineering Sciences* 368:5525–56.
87. van Schilfgaarde, M. and Katsnelson, M. I. 2011. First-principles theory of nonlocal screening in graphene. *Physical Review B—Condensed Matter and Materials Physics* 83:081409.
88. Polini, M., Asgari, R., Borghi, G., Barlas, Y., Pereg-Barnea, T., and MacDonald, A. H. 2008. Plasmons and the spectral function of graphene. *Physical Review B* 77:081411.
89. Liu, Y., Willis, R. F., Emtsev, K. V., and Seyller, T. 2008. Plasmon dispersion and damping in electrically isolated two-dimensional charge sheets. *Physical Review B* 78:201403.
90. Sun, J., Hannon, J. B., Tromp, R. M., Johari, P., Bol, A. A., Shenoy, V. B., and Pohl, K. 2010. Spatially-resolved structure and electronic properties of graphene on polycrystalline Ni. *ACS Nano* 4:7073–7.
91. Generalov, A. V. and Dedkov, Y. S. 2012. HREELSS study of the epitaxial graphene/Ni(111) and graphene/Au/Ni(111) systems. *Carbon* 50:183–91.
92. Kramberger, C., Hambach, R., Giorgetti, C., Rümmeli, M. H., Knupfer, M., Fink, J., Büchner, B. et al. 2008. Linear plasmon dispersion in single-wall carbon nanotubes and the collective excitation spectrum of graphene. *Physical Review Letters* 100:196803.
93. Rosei, R., Modesti, S., Sette, F., Quaresima, C., Savoia, A., and Perfetti, P. 1984. Electronic structure of carbidic and graphitic carbon on Ni(111). *Physical Review B* 29:3416–22.
94. Politano, A., Agostino, R. G., Colavita, E., Formoso, V., and Chiarello, G. 2009. Collective excitations in nanoscale thin alkali films: Na/Cu(111). *Journal of Nanoscience and Nanotechnology* 9:3932–7.
95. Liu, X., Pichler, T., Knupfer, M., Golden, M. S., Fink, J., Walters, D. A., Casavant, M. J., Schmidt, J., and Smalley, R. E. 2001. An electron energy-loss study of the structural and electronic properties of magnetically aligned single wall carbon nanotubes. *Synthetic Metals* 121:1183–6.
96. Diebold, U., Preisinger, A., Schattschneider, P., and Varga, P. 1988. Angle resolved electron energy loss spectroscopy on graphite. *Surface Science* 197:430–43.
97. Papageorgiou, N., Portail, M., and Layet, J. M. 2000. Dispersion of the interband π electronic excitation of highly oriented pyrolytic graphite measured by high resolution electron energy loss spectroscopy. *Surface Science* 454–456:462–6.
98. Büchner, U. 1977. Wave-vector dependence of the electron energy losses of boron nitride and graphite. *Physica Status Solidi (B)* 81:227–34.
99. Zeppenfeld, K. 1969. Wavelength dependence and spatial dispersion of the dielectric constant in graphite by electron spectroscopy. *Optics Communications* 1:119–22.
100. Ritsko, J. J. and Rice, M. J. 1979. Plasmon spectra of ferric-chloride-intercalated graphite. *Physical Review Letters* 42:666–9.
101. Pines, D. 1964. *Elementary Excitations in Solids*. Benjamin, New York.
102. López-Sancho, M. P., Vozmediano, M. A. H., and Guinea, F. 2007. Transverse transport in graphite. *The European Physical Journal—Special Topics* 148:73–81.
103. Dion, M., Rydberg, H., Schröder, E., Langreth, D. C., and Lundqvist, B. I. 2004. Van der Waals density functional for general geometries. *Physical Review Letters* 92:246401.
104. Nilsson, J., Castro Neto, A. H., Peres, N. M. R., and Guinea, F. 2006. Electron–electron interactions and the phase diagram of a graphene bilayer. *Physical Review B* 73:214418.

105. Yuan, Z. and Gao, S. 2008. Landau damping and lifetime oscillation of surface plasmons in metallic thin films studied in a jellium slab model. *Surface Science* 602:460–4.
106. Zhou, S. Y., Gweon, G. H., Fedorov, A. V., First, P. N., de Heer, W. A., Lee, D. H., Guinea, F., Castro Neto, A. H., and Lanzara, A. 2007. Substrate-induced bandgap opening in epitaxial graphene. *Nature Materials* 6:770–5.
107. Marinopoulos, A. G., Reining, L., Rubio, A., and Olevano, V. 2004. Ab initio study of the optical absorption and wave-vector-dependent dielectric response of graphite. *Physical Review B* 69:245419.
108. Maier, S. A. 2012. Graphene plasmonics: All eyes on flatland. *Nature Physics* 8:581–2.
109. Christensen, J., Manjavacas, A., Thongrattanasiri, S., Koppens, F. H. L., and García de Abajo, F. J. 2011. Graphene plasmon waveguiding and hybridization in individual and paired nanoribbons. *ACS Nano* 6:431–40.
110. Wang, Y., Plummer, E. W., and Kempa, K. 2011. Foundations of plasmonics. *Advances in Physics* 60:799–898.
111. Bonaccorso, F., Sun, Z., Hasan, T., and Ferrari, A. C. 2010. Graphene photonics and optoelectronics. *Nature Photonics* 4:611–22.
112. Ryzhii, V., Dubinov, A. A., Otsuji, T., Mitin, V., and Shur, M. S. 2010. Terahertz lasers based on optically pumped multiple graphene structures with slot-line and dielectric waveguides. *Journal of Applied Physics* 107.
113. Davoyan, A. R., Morozov, M. Y., Popov, V. V., Satou, A., and Otsuji, T. 2013. Graphene surface emitting terahertz laser: Diffusion pumping concept. *Applied Physics Letters* 103:251102.
114. Ryzhii, V., Dubinov, A. A., Otsuji, T., Aleshkin, V. Y., Ryzhii, M., and Shur, M. 2013. Double-graphene-layer terahertz laser: Concept, characteristics, and comparison. *Optics Express* 21:31567–77.
115. Fang, Z., Liu, Z., Wang, Y., Ajayan, P. M., Nordlander, P., and Halas, N. J. 2012. Graphene-antenna sandwich photodetector. *Nano Letters* 12:3808–13.
116. Hanson, G. W. 2008. Quasi-transverse electromagnetic modes supported by a graphene parallel-plate waveguide. *Journal of Applied Physics* 104:084314–5.
117. Kim, J. T. and Choi, S.-Y. 2011. Graphene-based plasmonic waveguides for photonic integrated circuits. *Optics Express* 19:24557–62.
118. Liu, P., Zhang, X., Ma, Z., Cai, W., Wang, L., and Xu, J. 2013. Surface plasmon modes in graphene wedge and groove waveguides. *Optics Express* 21:32432–40.
119. Vakil, A. and Engheta, N. 2011. Transformation optics using graphene. *Science* 332:1291–4.
120. Salihoglu, O., Balci, S., and Kocabas, C. 2012. Plasmon-polaritons on graphene-metal surface and their use in biosensors. *Applied Physics Letters* 100:213110–5.
121. Shao, Y., Wang, J., Wu, H., Liu, J., Aksay, I. A., and Lin, Y. 2010. Graphene based electrochemical sensors and biosensors: A review. *Electroanalysis* 22:1027–36.
122. Wu, L., Chu, H. S., Koh, W. S., and Li, E. P. 2010. Highly sensitive graphene biosensors based on surface plasmon resonance. *Optics Express* 18:14395–400.
123. Yang, W. R., Ratinac, K. R., Ringer, S. P., Thordarson, P., Gooding, J. J., and Braet, F. 2010. Carbon nanomaterials in biosensors: Should you use nanotubes or graphene? *Angewandte Chemie—International Edition* 49:2114–38.
124. Zuppella, P., Tosatto, S., Corso, A. J., Zuccon, S., and Pelizzo, M. G. 2013. Graphene–noble metal bilayers for inverted surface plasmon resonance biosensors. *Journal of Optics* 15:055010.
125. Stöberl, U., Wurstbauer, U., Wegscheider, W., Weiss, D., and Eroms, J. 2008. Morphology and flexibility of graphene and few-layer graphene on various substrates. *Applied Physics Letters* 93:051906.
126. Lu, W. B., Zhu, W., Xu, H. J., Ni, Z. H., Dong, Z. G., and Cui, T. J. 2013. Flexible transformation plasmonics using graphene. *Optics Express* 21:10475–82.
127. Hendry, E., Hale, P. J., Moger, J., Savchenko, A. K., and Mikhailov, S. A. 2010. Coherent nonlinear optical response of graphene. *Physical Review Letters* 105:097401.
128. Kauranen, M. and Zayats, A. V. 2012. Nonlinear plasmonics. *Nature Photonics* 6:737–48.
129. Gullans, M., Chang, D. E., Koppens, F. H. L., de Abajo, F. J. G., and Lukin, M. D. 2013. Single-photon nonlinear optics with graphene plasmons. *Physical Review Letters* 111:247401.
130. Roldán, R. and Brey, L. 2013. Dielectric screening and plasmons in AA-stacked bilayer graphene. *Physical Review B* 88:115420.
131. Sensarma, R., Hwang, E. H., and Das Sarma, S. 2010. Dynamic screening and low-energy collective modes in bilayer graphene. *Physical Review B* 82:195428.
132. Wang, X.-F. and Chakraborty, T. 2007. Coulomb screening and collective excitations in a graphene bilayer. *Physical Review B* 75:041404.
133. Stauber, T. and Gómez-Santos, G. 2012. Plasmons and near-field amplification in double-layer graphene. *Physical Review B* 85:075410.
134. Krstajić, P. M. and Peeters, F. M. 2013. Energy-momentum dispersion relation of plasmarons in bilayer graphene. *Physical Review B* 88:165420.
135. Van-Nham, P. and Holger, F. 2012. Coulomb interaction effects in graphene bilayers: Electron–hole pairing and plasmaron formation. *New Journal of Physics* 14:075007.
136. Khomyakov, P. A., Starikov, A. A., Brocks, G., and Kelly, P. J. 2010. Nonlinear screening of charges induced in graphene by metal contacts. *Physical Review B* 82:115437.

36 Graphene/Polymer Nanocomposites for Electrical and Electronic Applications

Linxiang He and Sie Chin Tjong

CONTENTS

ABSTRACT

Graphene, a monolayer of sp^2-bonded carbon atoms with remarkable physical and mechanical properties, is regarded as an excellent filler material for polymers. Graphene/polymer nanocomposites can be fabricated by solution mixing, *in situ* polymerization, or melt compounding. They hold promise for a variety of electrical and electronic applications. Owing to the large aspect ratio of the graphene sheets, they render the polymer electrically conductive at a very low filler content. The resultant polymer composites may find applications in electrostatic discharge (ESD) materials or electromagnetic shielding coatings. Owing to the enhanced dielectric performance of the graphene/polymer composites below the percolation threshold, they can be used as the dielectric materials for electrical charge storage capacitors. In addition, graphene/polymer composite films may be utilized as the electrical transport layer in field-effect transistors (FETs), which offers the advantages of ease of processing and mechanical flexibility. Because the resistivity of graphene/polymer nanocomposites is highly sensitive to various stimuli, including gaseous environment, electrical voltage, pressure, and mechanical strain deformation, they may be used as gas, temperature, or piezoresistive sensors. Moreover, the excellent thermal conductivity of graphene renders graphene/polymer nanocomposites an ideal material for heat dissipating and thermal management in electronics.

36.1 INTRODUCTION

Graphene is a two-dimensional, one-atom-thick layer of sp^2-bonded carbon atoms with remarkable physical properties. The enormous interest in using graphene for forming polymer nanocomposites is due to its intrinsic properties, as listed in Table 36.1. Owing to its zero band gap and high carrier mobility and concentration, graphene shows nearly ballistic transport at room temperature. Electron mobility in graphene can reach 20,000 cm^2/V s, an order of magnitude higher than that of a silicon transistor [1]. After improvement in sample preparation, the mobility could even exceed 25,000 cm^2/V s [2]. A single, defect-free graphene sheet exhibits an exceptionally high Young's modulus of ~1 TPa and tensile strength of 130 GPa [3]. A single-layer graphene exhibits thermal conductivity of 4840–5300 W/m K [4], which is several times higher than that of copper. This means that graphene is capable of dissipating heat readily, thus acting as an effective heat sink material for microelectronic devices. The electrical conductivity of graphene is much higher than that of copper, but its density is almost four times lower. Even the electrical conductivity of chemically reduced graphene oxide paper reached 6.87×10^2 S/m and that of thermally reduced graphene oxide flakes is in the range of $6.2 \times 10^2 \sim 6.2 \times 10^3$ S/m [5–8]. Another beneficial property of graphene is its very high specific surface area (2630 m^2/g) [6] compared to that of a carbon

TABLE 36.1
Electrical, Mechanical, and Thermal Properties of Graphene

Properties		Reference
Electron mobility	20,000 cm^2/V s	[1,2]
Modulus	1 TPa	[3]
Strength	130 GPa	[3]
Thermal conductivity	5000 W/(m K)	[4]
Electrical conductivity	6000 S/cm	[5]
Surface area	2630 m^2/g	[6]

nanotube (CNT) (1315 m^2/g), making graphene an attractive filler material for improving the mechanical, electrical, and thermal properties of the polymers.

The routes for the preparation of graphene can be classified into two categories [9], as illustrated in Figure 36.1. One route involves the chemical synthesis of graphene on some substrates. It has been demonstrated that graphene can be grown on a specific substrate (transition metals) by chemical vapor deposition (CVD) or on silicon carbide by epitaxial growth [10,11]. It has also been reported that graphene can be synthesized from molecular assembly [12–14]. These routes cannot be used to prepare graphene sheets on a large scale to fulfill the requirements of graphene/polymer nanocomposites. The other route relates to the cleavage of graphite flakes (Table 36.2). Owing to the scalable production of graphite nowadays, this is more viable for producing a large amount of graphene sheets. Figure 36.2 shows the typical approaches to exfoliate graphite into graphene or its derivatives. The reliable technique to produce high-quality graphene is mechanical cleavage [15,22]. However, low throughput and time consumption make this route impractical for scalable production of graphene sheets. Solution-phase exfoliation of graphite to individual or few-layered graphene sheets in a certain solvent such as *N*-methylpyrrolidone (NMP) or *N*,*N*-dimethylformamide (DMF) under ultrasonication has been recently reported [16,17,23,24]. Nevertheless, the reported yield of monolayer graphene is relatively small (<1 wt%).

As an alternative, graphene can be prepared by chemically exfoliating graphite oxide (GO) into monolayer graphene oxide and subsequently reducing it. GO can be prepared by Hummers' method [25]. The advantage of this approach is that it enables high-yield production and, hence, is a cost-effective and scalable process. The monolayer graphene oxide

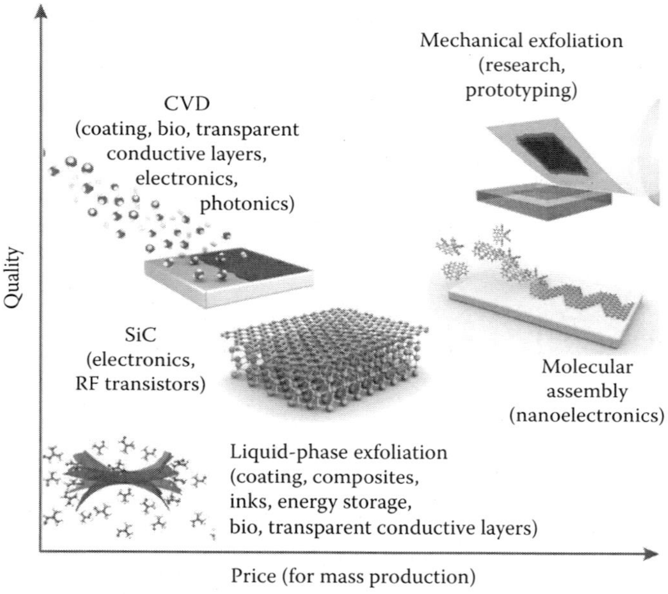

FIGURE 36.1 Fabrication routes of graphene. (Reproduced with permission from Macmillan Publishers Ltd. *Nature*, Novosolov, K. S. et al., A roadmap for graphene. 490:192–200, copyright 2012.)

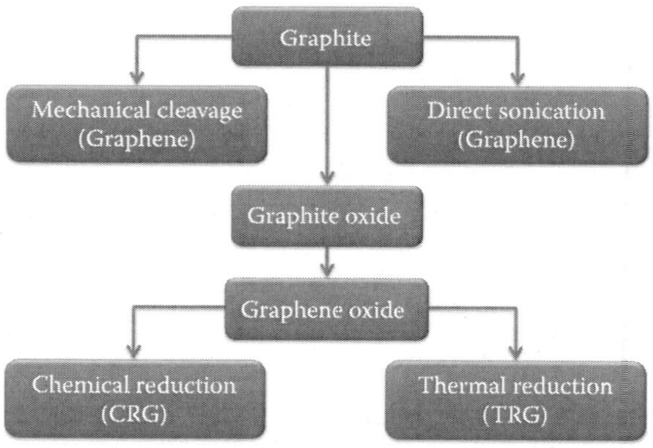

FIGURE 36.2 Different ways of cleaving graphite flakes to produce graphene or its derivatives.

TABLE 36.2
Typical Approaches of Producing Graphene or CRG/TRG Sheets Starting from Graphite

Method	Typical Dimension: Thickness	Typical Dimension: Lateral	Advantage	Disadvantage	Reference
Mechanical cleavage	Single layer	μm to cm	Large size; high quality	Low throughput and yield	[15]
Direct sonication	Single/multiple layers	μm or sub-μm	Unmodified sheets; facile operation	Small production scale	[16,17]
Chemical reduction of graphene oxide	Mostly single layer	μm or sub-μm	Larger sheet size; lower production cost; scalability	Defects; degraded properties	[7,3,18,19]
Thermal reduction of graphene oxide	Mostly single layer	μm or sub-μm	Larger sheet size; lower production cost; scalability	Defects; degraded properties	[20,21]

forms over a wide range of C:O ratios, with the oxygen bound to the carbon in the basal plane in the form of hydroxyl and epoxy functional groups, and carbonyl and carboxyl groups at the sheet edges. These functional groups make graphene oxide strongly hydrophilic and decrease the interaction energy between the layers (the interlayer distance increases from 0.34 nm in graphite to 0.7 nm in GO) [18]. Hence, GO can be readily exfoliated to graphene oxide sheets. However, graphene oxide is an insulator and the oxygen functionalities must be removed to resume its electrical conductivity. In this regard, graphene oxide can be reduced to graphene either by chemical reduction or by thermal treatment. Treating graphene oxide sheets with reducing agents such as hydrazine, dimethylhydrazine, sodium borohydride, or ascorbic acid can yield the so-called chemically reduced graphene oxide (CRG) [7,8,18,19]. Alternatively, rapid thermal heating graphene oxide in a furnace can transform it to "thermally reduced graphene oxide" (TRG) [20,21]. However, some oxygen functional groups are still retained in the TRGs despite high-temperature annealing [26]. Both CRG and TRG sheets are cost-effective filler materials for polymers.

36.2 GRAPHENE/POLYMER NANOCOMPOSITES

A critical step of forming graphene/polymer nanocomposites with desired physical properties is the attainment of homogeneous dispersion of graphene sheets in the polymer matrix. A well-dispersed state ensures a maximized reinforced surface area, thereby affecting the neighboring polymer chains and hence the properties of resultant composites.

To yield a desired state of dispersion in the polymer matrix, stable graphene or functionalized graphene sheet (FGS) dispersion is usually necessary. This can be realized in two ways, as Figure 36.3 demonstrates. One way involves the functionalization of graphene, either in covalent or in noncovalent approach [26–28]. Since graphene oxide can be readily exfoliated in water or other organic aprotic solvents, they can be readily modified with chemical agents such as long alkyl chain molecules, isocyanates, and polyallylamine. These agents can establish covalent linkages, yielding composites with well-dispersed graphene. For example, isocyanates reduce the hydrophilic character of graphene oxide sheets by forming amide and carbonate ester bonds with the carboxyl and hydroxyl groups, respectively [26].

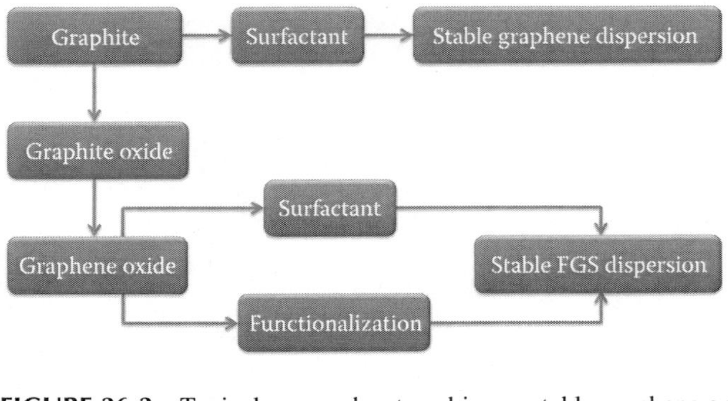

FIGURE 36.3 Typical approaches to achieve a stable graphene or FGS dispersion.

For pristine graphene sheets or CRG/TRG, surfactants are often used to disperse them well in water or other organic solvents [29–33]. Surfactants can also assist the direct liquid-phase exfoliation of pristine graphite [23,24]. In this regard, experiences of CNT dispersions might be an important guide. It should be noted, however, that the use of surfactants may adversely affect the electrical properties of the resultant composites.

Recently, it has been reported that *in situ* reduction of graphene oxide embedded in the polymer matrix leads to a uniform filler dispersion within the polymer [34,35]. This approach is expected to be applicable to most polar polymers, and thus is more attractive for graphene/polymer composite production.

36.2.1 Fabrication Processes

Graphene/polymer composites with functional properties can be prepared using three processing strategies: (a) solution mixing, (b) *in situ* polymerization, and (c) melt processing. The desired physical properties of the resulting nanocomposites are achieved at very low filler loadings by selecting the proper composite fabrication process.

36.2.1.1 Solution Mixing

Solution mixing is the most widely used wet chemical blending process for making polymer nanocomposites due to its simplicity. Figure 36.4 illustrates the process to prepare the graphene/polymer nanocomposite. It involves the dissolution of the polymer in a selected solvent followed by mixing with the graphene suspension in the solvent under stirring and/or sonication. Consequently, polymer molecules coat the graphene sheets effectively. By removing the solvent via evaporation, graphene sheets reassemble, sandwiching the polymers to form nanocomposites. The disadvantages of

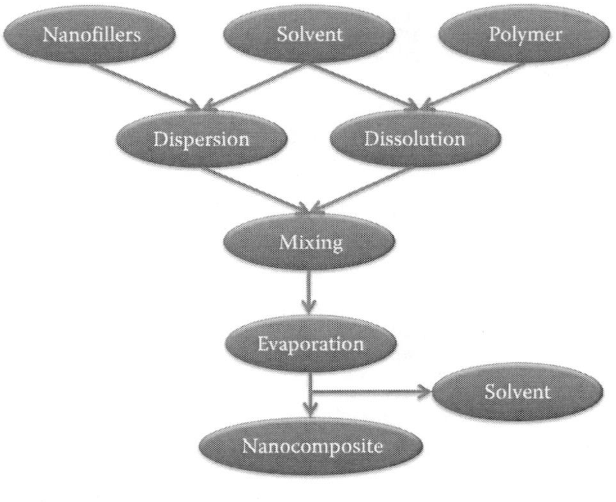

FIGURE 36.4 Schematic illustration of the solution mixing process.

this process are the necessity of using solvent and sonication for blending graphene fillers with the polymer suspension. Selected organic solvents are limited, and the use of solvent in large quantities has environmental implications. This strategy can be employed to synthesize graphene/polymer composites with a wide range of polymers such as polystyrene (PS) [36,37], poly(methyl methacrylate) (PMMA) [38], poly(vinylidene fluoride) (PVDF) [39,40], polypropylene (PP) [41], ultra-high-molecular-weight polyethylene (UHMWPE) [42], linear low-density polyethylene (LLDPE) [43], epoxy [44], etc. Very recently, He and Tjong fabricated graphene/PVDF nanocomposites by a one-step process via *in situ* solvothermal reduction of graphene oxide in the DMF solution of PVDF [45]. In this respect, DMF acted as a solvent for dissolving PVDF and also served as a medium to transmit heat and pressure for reducing graphene oxide. The reducing temperature is considerably lower than the typical thermal reduction temperature.

36.2.1.2 *In Situ* Polymerization

In this processing route, graphene sheets are mixed with the monomers, sometimes in the presence of a solvent. A suitable initiator is employed and the polymerization reaction proceeds by monitoring the parameters such as temperature and time. Figure 36.5 illustrates the *in situ* polymerization process for preparing graphene/polymer nanocomposites. Several graphene/polymer systems have been prepared using the following polymers, that is, PMMA [46], polybutylene terephthalate (PBT) [47], poly(l-lactide) [48], and polyimide [49]. A large amount of work has been done on the preparation of graphene/epoxy composites [50–59]. This involves typically first dispersing the graphene sheets in the resin followed by curing the resin with the hardener. The *in situ* polymerization route facilitates stronger interactions between graphene sheets and the polymer phase; thus, the resultant composites often exhibit better mechanical properties and lower electrical percolation threshold. The shortcomings are that the monomers are often expensive and high-temperature-reaction reactors are usually employed in the fabrication process.

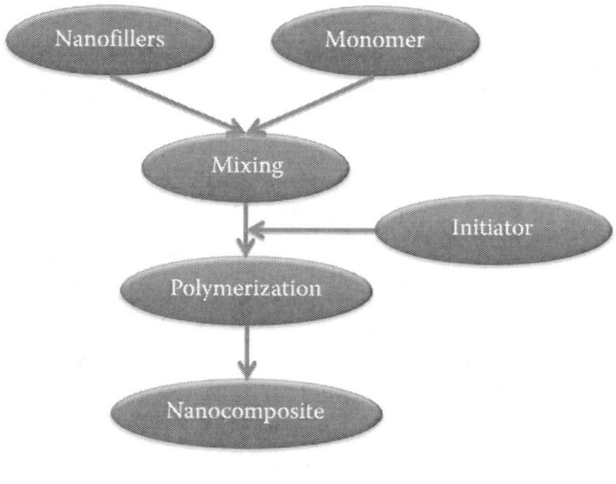

FIGURE 36.5 Schematic illustration of the *in situ* polymerization process.

36.2.1.3 Melt Compounding

Melt compounding is a versatile and commercially attractive process for large-scale production of polymer nanocomposites. As Figure 36.6 illustrates, the process involves the direct incorporation of graphene sheets into molten polymer using a twin-screw extruder and/or injection molder by adjusting parameters such as screw speed, temperature, and time. High temperature softens the polymer pellets and allows easy dispersion of graphene sheets in the polymer. Melt blending is generally less effective in dispersing graphene sheets in molten polymer, particularly at higher filler loadings due to an increase of the melt viscosity during the processing. Figure 36.7a,b shows the transmission electron microscope (TEM) micrographs of the 1 wt% TRG/LLDPE nanocomposite specimens prepared by the solution blending and melt-compounding processes, respectively. TRGs are mainly segregated into loose clusters in the polymer matrix of the melt-compounded composite. In contrast, TRGs are fully separated into independent sheets and dispersed uniformly in the LLDPE matrix of solution-mixed nanocomposite. It is noted that high-shear forces during melt mixing can reduce the aspect ratio of graphene, leading to poorer physical and mechanical properties of the nanocomposites. Many melt-processed graphene/polymer nanocomposites have been prepared using the following polymers: LLDPE [43], polycarbonate (PC) [60], polyethylene oxide (PEO) [61], and polyethylene terephthalate (PET) [62]. Compared with solution mixing and *in situ* polymerization, the melt-compounding process generally leads to poor dispersion of graphene sheets in the polymer matrix.

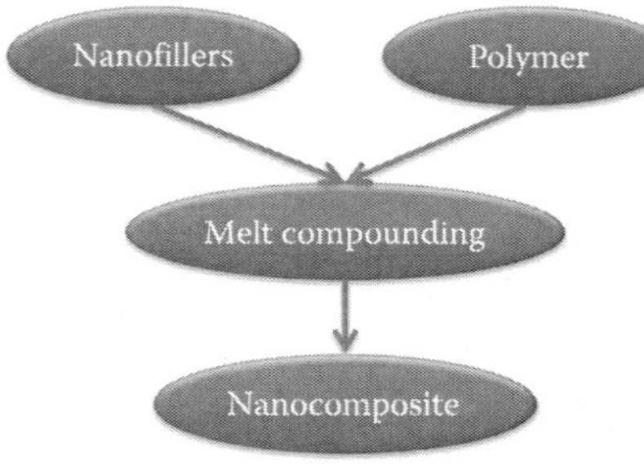

FIGURE 36.6 Schematic illustration of the melt-compounding process.

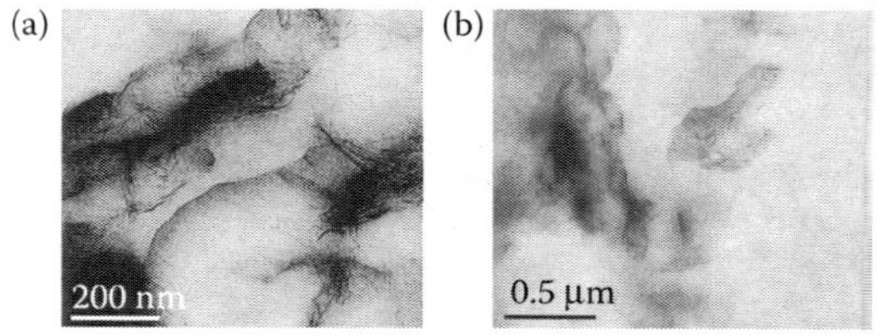

FIGURE 36.7 TEM micrographs of 1 wt% TRG/LLDPE nanocomposite specimens prepared by (a) solution mixing and (b) melt compounding. (Reprinted from *Polymer*, 52, Kim, H. W. et al., Graphene/polyethylene nanocomposites: Effect of polyethylene functionalization and blending methods, 1837–1846, Copyright 2011, with permission from Elsevier.)

36.2.2 Electrical Conductivity and Dielectric Permittivity

Graphene/polymer nanocomposites emerge as an important group of conductive materials exhibiting a typical percolation behavior. For such a system, percolation theory is usually employed to give a phenomenological description of its electrical properties. According to the percolation theory, the electrical conductivity and the dielectric permittivity follow the power law relationship near the percolation threshold (p_c) [63–66]:

$$\sigma \sim (p_c - p)^s \quad \text{for } p < p_c \tag{36.1}$$

$$\sigma \sim (p - p_c)^t \quad \text{for } p > p_c \tag{36.2}$$

$$\varepsilon \sim (p - p_c)^{s'} \quad \text{for } p < p_c \text{ or } p > p_c \tag{36.3}$$

where σ is the conductivity, ε the dielectric permittivity (i.e., dielectric constant), and p the volume fraction of the graphene sheets. The critical exponents s, s', and t are assumed to be universal and depend only on the dimensions of the percolating system. The accepted values of these exponents are $t \approx 1.3$ for two dimensions and $s \approx 0.73$, $t \approx 2.0$ for three dimensions. Nevertheless, deviations in experimental values of s and t are frequently reported [67–69].

It should be emphasized that the percolation theory is limited to cases in which either the insulating medium has zero conductance or the conducting filler has zero resistance. In addition, the classical percolation theory for describing conductive filler/polymer systems must meet certain criteria. The theoretically predicted value only agrees reasonably with the experimental result provided the following criteria are fulfilled: the particles are spherical, are monodispersed, and have an isotropic conductivity. If one or all these conditions are not fulfilled, the theoretical value can deviate significantly from the practical one. As an example, numerical calculations for the hard-core circles give $p_c \approx 16$ vol% [70]. The percolation model based on the hard-core circles has its own limitation for predicting the p_c of the polymer composites with conducting nanofillers of large aspect ratios. CRGs and TRGs exhibit very large aspect ratios, and thus facilitate the formation of conducting path network at filler concentrations well below a theoretical value of 16 vol%.

Stankovich et al. fabricated PS nanocomposites with graphene by mixing isocyanate-treated graphene oxide with PS in DMF followed by hydrazine reduction [36]. The nanocomposites exhibit a low p_c of 0.1 vol% at room temperature (Figure 36.8). Such a low p_c results from the homogeneous dispersion and extremely large aspect ratio of graphene sheets. At 0.15 vol% CRG, the conductivity of the composites satisfies the antistatic criterion (10^{-6} S/m) for thin films. The value increases rapidly over a 0.4 vol% range. By increasing the CRG content to 1 vol%, a conductivity of ~0.1 S/m can be achieved, being sufficient for electronic device applications.

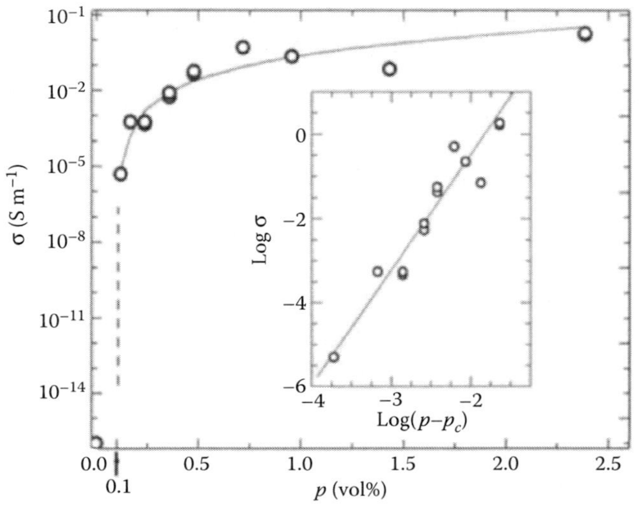

FIGURE 36.8 Electrical conductivity of the graphene/polystyrene composites as a function of the filler volume content. The conductivity is determined using a standard four-probe technique. Inset shows the plot of log σ_c versus log(p–p_c), where p_c is the percolation threshold. Best fitting Equation 36.2 with the data yields $t = 2.74 \pm 0.20$, and $p_c = 0.1$ vol%. (Reprinted by permission from Macmillan Publishers Ltd. *Nature*, Stankovich, S. et al., Graphene-based composite materials. 442:282–286, copyright 2006.)

Similarly, Fan et al. reported a low p_c of 0.18 vol% for CRG/PVDF nanocomposites [40]. The lowest p_c of 0.076 vol% is found in CRG/UHMWPE nanocomposites [42]. In contrast, melt-compounded TRG/PET nanocomposites exhibit a large p_c of 4.22 vol% [62]. Table 36.3 summarizes the typical percolating behavior of a variety of graphene/polymer nanocomposites processed by different routes.

Very recently, He and Tjong reported that the dielectric permittivity of PVDF increases from 7 to 105 by adding 0.5 vol% solvothermal reduced graphene oxide [45]. Wang et al. demonstrated that the dielectric permittivity of CRG/PP nanocomposites follows the percolation scaling law, yielding a low p_c of 0.033 vol% and a critical exponent of 0.74 [79]. Near p_c, the dielectric permittivity of the nanocomposite is about three orders of magnitude higher than that of PP due to the large aspect ratio of CRG (Figure 36.9). These CRG/PP composites with ultralow p_c can be potentially applied as novel dielectric materials for making energy storage capacitors.

36.3 APPLICATIONS

Functional graphene/polymer composites are being developed currently; thus, their potential applications in technological fields are still in the embryonic stage and subject to further research. Near the percolation threshold, novel graphene-based polymer nanocomposites exhibit a several order increase in electrical conductivity and permittivity compared with the neat polymer. The appreciable improvement in electrical conductivity is associated with the formation of a conducting network of graphene sheets in the polymer matrix.

TABLE 36.3
Percolation Threshold and Electrical Conductivity of Graphene/Polymer Composites Prepared through Different Routes

	Filler	p_c	σ (S/m) (Filler Content)	Processing Method	Reference
PS	CRG	0.1 vol%	1 (2.5 vol%)	Solution mixing	[36]
PS	TRG	0.33 vol%	3.5 (1.1 vol%)	Solution mixing	[37]
PMMA	TRG	0.62 vol%	2.4×10^{-2} (0.8 vol%)	Solution mixing	[38]
PVDF	CRG	0.18 vol%	10^{-3} (3.5 vol%)	Solution mixing	[40]
PVDF	CRG	0.31 vol%	3×10^{-4} (1.4 vol%)	Solution mixing	[45]
PVDF	CRG	0.87 vol%	1×10^{-3} (3.55 vol%)	Solution mixing	[71]
UHMWPE	CRG	0.07 vol%	7×10^{-2} (0.6 vol%)	Solution mixing	[42]
PUA	TRG	0.07 vol%	—	*In situ* polymerization	[72]
PU	TRG	2 wt%	4.9×10^{-2} (7 wt%)	*In situ* polymerization	[73]
Epoxy	CRG	0.52 vol%	8 (8.8 vol%)	*In situ* polymerization	[74]
PC	TRG	2.5 wt%	0.1 (12 wt%)	Melt compounding	[75]
PA6	TRG	7.5 wt%	7.1×10^{-3} (12 wt%)	Melt compounding	[75]
PU	TRG	2 wt%	2.75×10^{-2} (6 wt%)	Melt compounding	[76]
PET	TRG	0.47 vol%	2.1 (3 vol%)	Melt compounding	[62]
PVDF	TRG	0.1 vol%	2×10^{-2} (0.76 vol%)	Suspension blending	[77]
HDPE	TRG	1 vol%	1×10^{-2} (3.4 vol%)	Suspension blending	[78]

Note: The weight percentage loading of graphene can be converted to volume fractions using a graphite bulk density of 2.2 g/cm^3.
PUA: polyurethane acrylate; PU: polyurethane; HDPE: high-density polyethylene.

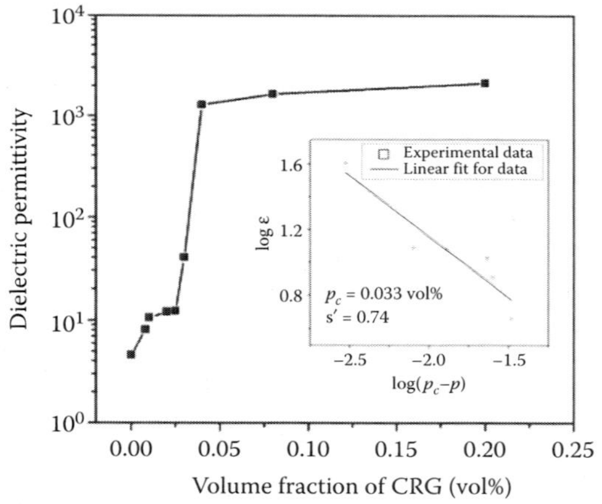

FIGURE 36.9 Dielectric permittivity of the CRG/PP composites as a function of CRG volume fraction, measured at 10^3 Hz and room temperature. The inset shows the best fits of experimental data to Equation 36.3. (Reprinted from *Polymer*, 54, Wang, D. et al., Dielectric properties of reduced graphene oxide/polypropylene composites with ultralow percolation threshold. 1916–1922, Copyright 2013, with permission from Elsevier.)

Such properties render the graphene/polymer nanocomposites favorable for a variety of applications, including antistatic coatings, ESD materials, electromagnetic interference shielding layers, dielectric materials, sensors, energy storage devices, and thermal management materials.

36.3.1 Conductive Graphene/Polymer Composites

36.3.1.1 Electrostatic Discharge

ESD is a transfer of electrostatic charges between bodies at different potentials caused by direct contact. Static charges are often created by the triboelectric effect, simply by rubbing or separating surfaces, resulting in the transfer of electrons. Electrostatic dissipation has become an important issue in the electronics industry, especially in electronic components such as data storage devices, chip carriers, and computer internals. Because of miniaturization, electronics are very sensitive to particulate contamination and must be protected from lower level of static charge. Conductive polymer composites filled with carbon black (CB) find useful application for ESD protection. CB-loaded static controlling products usually contain 15–20 wt% by weight. Such high CB contents have a negative effect on the processability and mechanical properties of the composites. Contamination is also an important issue since the carbon powder tends to slough in highly filled CB composites and thus contaminates the environment [80]. Obviously, CNTs of remarkable mechanical strength and excellent electrical conductivity have attracted considerable attention for forming polymer composites for ESD protection [81]. However, CNTs are expensive and have a high tendency to form clusters due to their strong van der Waal interactions. In this regard, graphene acts as an ideal filler material for polymers, since it can be readily mass produced at a relatively low cost. As mentioned previously, the conductivity of thin-film materials for antistatic application should be at least 10^{-6} S/m.

This conductivity value can be attained readily for graphene/polymer nanocomposites with filler content above p_c (Table 36.1). Thus, the graphene/polymer nanocomposites offer greater design flexibility, lighter weight, and cost effectiveness for antistatic application and ESD protection.

36.3.1.2 Electromagnetic Interference Shielding

The extensive demand for fast telecommunication networks and power lines in our surroundings create serious concerns about electromagnetic pollution and human safety from radiation. Electromagnetic (EM) waves with frequencies in the range of 8.2–12.4 GHz (X-band) are generated from weather radar, TV picture transmission, and telephone microwave relay systems. EM waves generated at microwave and radio frequencies can interfere with most electrical equipment and electronic devices, leading to deterioration in their electrical performances. Electromagnetic interference (EMI) shielding of a material is associated with its ability to absorb and reflect EM waves at high frequencies, thereby acting as an effective shield against the penetration of radiation. The shield materials are electrically conductive such that their mobile charge carriers can interact with EM waves.

An incident electromagnetic wave (E_0) is absorbed by a shield with a thickness t, reflection (E_R) and transmission (E_1) as shown in Figure 36.10. The corresponding reflectivity (R), absorptivity (A), and transmissivity (T) follow the relation:

$$A + R + T = 1 \tag{36.4}$$

The shielding effectiveness (SE) of a material expressed in decibels (dB) is defined as the ratio of transmitted power (P_1) to incident power (P_0) of an EM wave [44]:

$$SE = -10\log\left(\frac{P_1}{P_0}\right) = -20\log\left[\frac{E_1}{E_0}\right] \tag{36.5}$$

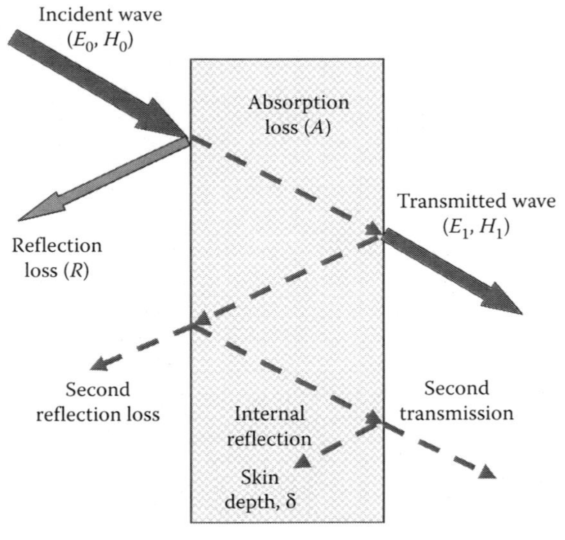

FIGURE 36.10 Schematic showing EMI shielding in a conductive plate.

The total EMI SE of a shielding material with a thickness t is the summation of the SE due to absorption (SE_A), reflection (SE_R), and multiple reflection (SE_M), that is,

$$\text{EMI SE} = SE_A + SE_R + SE_M \tag{36.6}$$

Shielding by absorption is enhanced when the electric and/or magnetic dipoles of a material interact with the EM field. The electric dipoles may derive from the materials having high dielectric constants or magnetic permeability. The multiple reflection mechanism is associated with reflections at various surfaces or interfaces in the shield. This requires the presence of large surface or interface areas in the shield such as a foamed material.

Conventional EMI shielding materials are metals and alloys, including copper, aluminum, and nickel. They are heavy and often suffer from degradation when exposed to corrosive environments. Alternatively, conducting polymer composites filled with carbon fibers have been used as EMI shielding material due to their flexibility, light weight, good moldability, and processability. Carbon fibers of high volume content (e.g., 35 vol%) have been used as the reinforcing material for polymers designed for EMI shielding [82,83]. The addition of large filler contents of microparticles to polymers generally impair processability and increase the weight of resulting composites. The usage of high carbon fiber content in the polymer composites also increases the production cost of polymer composites. Although CNTs can also serve as conductive fillers for polymers to enhance EMI shielding properties [84], their high price makes them less competitive than graphene. Owing to its high aspect ratio, a low graphene loading can lead to electrically conductive polymer composites appropriate for EMI shielding applications. Very recently, Hong et al. reported that graphene has an excellent EMI shielding performance, and the shielding mechanism is adsorption rather than reflection [85]. The monolayer CVD graphene has an average SE value of 2.27 dB, corresponding to about 40% shielding of incident waves.

The EMI shielding efficiency of a nanocomposite material is rather complicated due to the presence of conducting nanofillers of large surface areas for reflection and multiple reflections. Therefore, EMI SE depends on several factors, including the filler's intrinsic conductivity, dielectric constant, and aspect ratio. In practice, the EMI SE target value needed for commercial applications is ~20 dB. The studies on the application of graphene/polymer composites for EMI shielding were started very recently [38,74,86]. Liang et al. tested the EMI shielding performance of CRG/epoxy composites in the X-band (8–12 GHz) and reported an SE of 21 dB at 8.8 vol% (15 wt%) CRG (Figure 36.11) [74].

Very recently, Zhang et al. prepared TRGs of different C/O ratios of 13.2, 9.6, and 5.0 by rapidly heating graphene oxide sheets at 1050°C, 145°C, and 135°C, respectively [38]. The TRGs obtained were denoted as TRG-13.2, TRG-9.6, and TRG-5, respectively. They then prepared TRG/PMMA nanocomposites by solution blending. Figures 36.12a shows a typical percolating behavior of the TRG/PMMA nanocomposites.

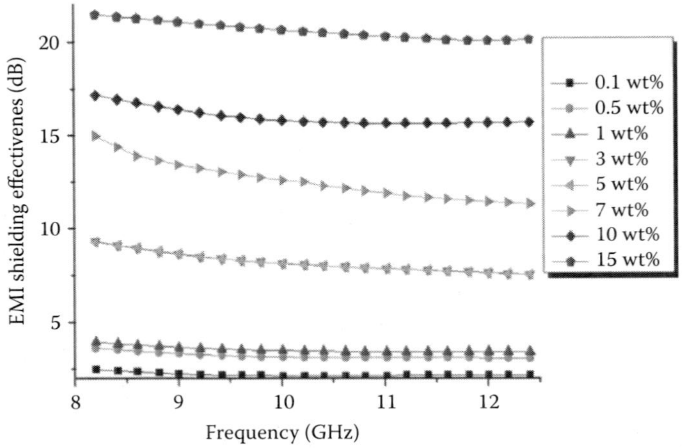

FIGURE 36.11 EMI SE versus frequency of solvent cast CRG/epoxy composites of different CRG contents. (Reprinted from *Carbon*, 47, Liang, J. et al., Electromagnetic interference shielding of graphene/epoxy composites, 922–925, Copyright 2009, with permission from Elsevier.)

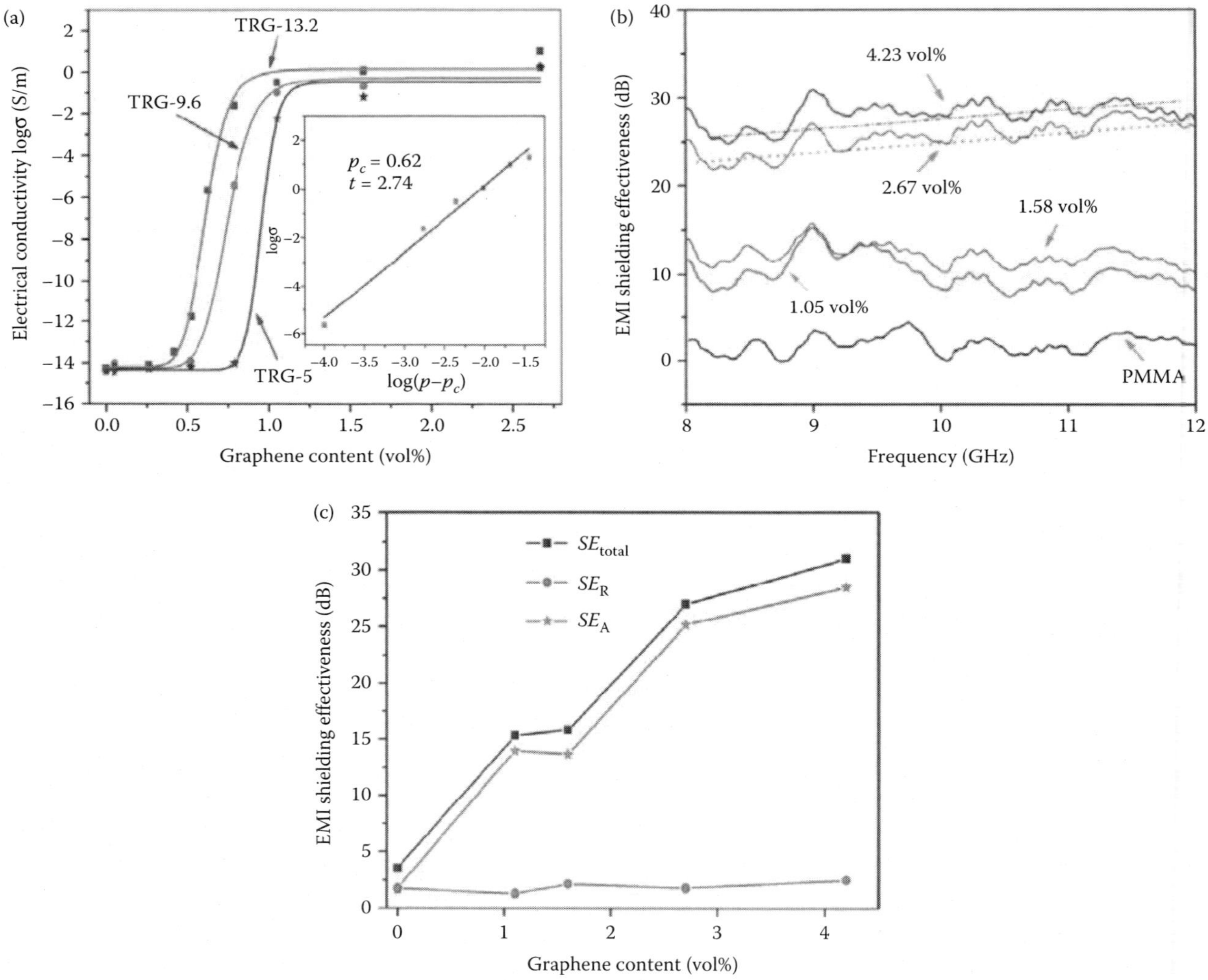

FIGURE 36.12 (a) Plots of electrical conductivity of TRG/PMMA composites as a function of graphene content, (b) EMI SE of TRG-13.2/PMMA composites, and (c) contribution of SE_A and SE_R to SE_{total} of the TRG-13.2/PMMA composites at 9 GHz as a function of graphene content. (Reprinted from *Carbon*, 50, Zhang, H. B. et al., The effect of surface chemistry of graphene on rheological and electrical properties of polymethylmethacrylate composites, 5117–5125, Copyright 2012, with permission from Elsevier.)

Apparently, the TRG-13.2/PMMA system exhibits the lowest percolation threshold of 0.62 vol% and the highest conductivity above the percolation threshold due to the highest C/O ratio of TRG fillers. In other words, the p_c of TRG/PMMA nanocomposites decreases with decreasing oxygen content in TRGs. The presence of oxygen-containing groups in graphene disrupts its carbon sp^2 network, thereby decreasing its intrinsic conductivity. Figure 36.12b shows the EMI SE of the TRG-13.2/PMMA system as a function of frequency. Apparently, the total EMI SE (SE_{total}) increases rapidly with the increase of TRG loading, as the case in electrical conductivity. The SE_{total} of the composite is about 10 dB at 1.1 vol% TRG and reaches above 25 dB at 2.67 vol% TRG. Further increasing the TRG loading to 4.23 vol%, a high SE_{total} value of about 30 dB is achieved over the frequency range of 8.8–12 GHz. This is higher than the value of 21 dB reported for the 8.8 vol% graphene/epoxy composite [74]. From Equation 36.6, SE_{total} is the summation of SE_A, SE_R, and SE_M. The SE_M contribution can be ignored when the $SE_A \geq 8$ dB. Figure 36.12c shows the plots of SE_{total}, SE_A, and SE_R of the TRG-13.2/PMMA system at 9 GHz as a function of graphene content. Apparently, the EMI shielding mechanism of the composite system is absorption dominant.

In general, foams can be introduced into conducting polymer composites to increase their *SE* via multiple internal reflections [87,88]. The beneficial effect of multiple reflections is that the material attenuates and absorbs EM radiation rather than reflects it. Therefore, the effect of foaming formation of the EMI SE of graphene/polymer nanocomposites has been reported very recently in the literature [89–92]. For example, Esmaraai et al. foamed the graphene/PVDF composites with an agent, that is, 2,2′-azobisisobutyronitrile (AIBN) [88]. They obtained an EMI SE of ~20 dB in the X-band region for the composite with 5 wt% graphene. The foams generated from the AIBN agent are heterogeneous, so the dominant EMI shielding in these foams is reflection rather than absorption. In contrast, subcritical carbon dioxide can be used to yield homogeneous cellular foams in the TRG/PMMA nanocomposites. Therefore, the EMI SE of nanocomposite foams is mainly contributed by the absorption rather than the reflection mechanism [91].

36.3.2 Dielectric Materials

High-permittivity polymeric materials show attractive applications as energy storage layers for high-performance capacitors. Most polymers are ineffective for making embedded passive capacitors because of their intrinsic low permittivity, for example, $\varepsilon < 5$. To enhance the dielectric permittivity, ferroelectric/piezoelectric ceramic nanofillers with high permittivity such as barium titanate (BT) are usually incorporated into the polymers [93]. The additions of nanoparticles of large surface area can result in an increase of the filler–matrix interfacial region, thus enhancing the interfacial polarization that would substantially increase the dielectric permittivity. In addition to inorganic BT nanoparticles, conducting carbonaceous nanofillers such as CNT and graphene were

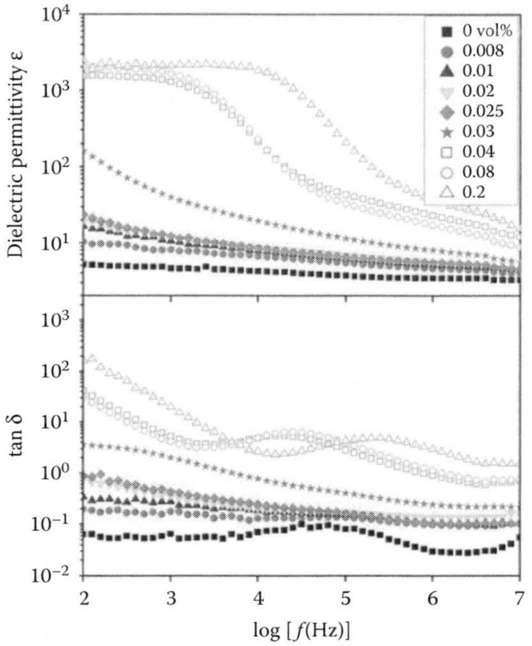

FIGURE 36.13 Dielectric permittivity (ε) and loss factor (tan δ) of the CRG/PP composites with different CRG contents as a function of frequency at room temperature. (Reprinted from *Polymer*, 54, Wang, D. et al., Dielectric properties of reduced graphene oxide/polypropylene composites with ultralow percolation threshold, 1916–1922, Copyright 2003, with permission from Elsevier.)

also utilized to enhance the dielectric performance of the polymers [63,79]. A typical example for this is the CRG/PP composites as shown in Figure 36.13 [79]. It should be noted that the percolated CRG/PP composites tend to exhibit larger dielectric loss compared with pure polymer (Figure 36.13). This is mainly caused by the leakage current associated with the formation of the conductive path network. Thus, higher filler contents facilitate the establishment of more conductive pathways, resulting in more significant leakage current and dielectric loss. From Figure 36.13, the dielectric loss of 0.04 vol% CRG/PP composite at 1 kHz is about 6, being two orders of magnitude higher than that of PP having a loss value of 0.07 at the same frequency.

More recently, low dielectric loss of about 0.03 at 1 kHz at room temperature has been observed in polydimethylsiloxane (PDMS) filled with 2 wt% TRG (Figure 36.14) [94]. The dielectric loss of this nanocomposite is nearly frequency independent in the range of 10^2 to 10^7 Hz. As mentioned previously, TRGs still contain some remnant oxygenated functionality. The oxygenated groups of TRGs interrupt the π-conjugation in the graphene layers, diminishing the surface electrical conductivity. However, such remnant oxygen functionalities are beneficial in achieving low dielectric loss while maintaining high conductivity and large dielectric permittivity of 0.2 wt% TRG/PDMS composite. It seems that the remnant oxygen groups can screen charge mobility to shut off leakage current. This low loss value together with high permittivity renders it very attractive as dielectric layers for energy storage

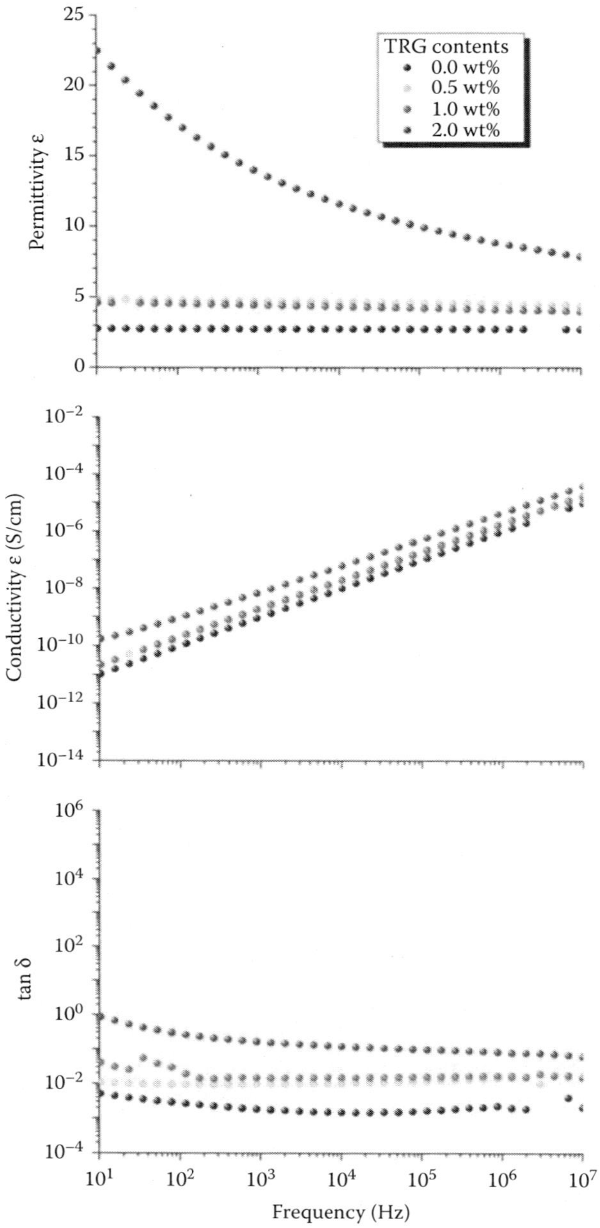

FIGURE 36.14 Dielectric permittivity, conductivity, and loss tangent as a function of frequency at room temperature for TRG/PDMS composites. (With kind permission from Springer Science+Business Media: *Nanoscale Res Lett*, Functionalized graphene sheets as effective high dielectric constant fillers, 6, 2011, 508, Romasanta, L. J. et al.)

applications. It is worth noting that the 0.2 wt% CNT/PDMS composite exhibits a very high dielectric loss value of 2×10^4. The percolating CNT/polymer system generally exhibits high dielectric loss since the leakage current readily converts electrical energy into heat [95,96].

36.3.3 Field-Effect Transistors

Owing to its high charge carrier mobility, graphene is an ideal candidate for fabricating FET. However, pure graphene sheet

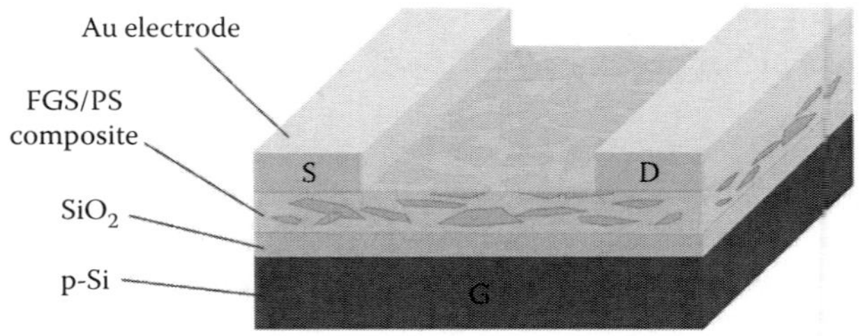

FIGURE 36.15 Schematic of FGS/PS composite thin-film field effect devices. (Reprinted with permission from Eda, G. et al., Graphene-based composite thin films for electronics. *Nano Lett* 9:814–818. Copyright 2009 American Chemical Society.)

is difficult to prepare as mentioned previously. As an alternative, graphene oxide/polymer nanocomposites offer the advantages of ease of processing and mechanical flexibility. Recently, Eda and Chhowalla prepared FGS/PS composite thin-film field effect device as illustrated in Figure 36.15 [97]. They reported that the transistors derived from graphene-based composite thin films exhibit ambipolar field effect characteristics. The transport via percolation among FGS in the PS matrix was believed to yield such behaviors. In the process, the nanocomposite was first prepared by functionalized graphene oxide with phenyl isocyanate [26] and dispersed in DMF followed by adding PS. To reduce functionalized graphene oxide, dimethylhydrazine was added to the solution and spin-coated onto a SiO_2/Si substrate subsequently. Gold source (*S*) and drain (*D*) electrodes were thermally evaporated on the composite thin film while the Si substrate was used as the gate (*G*) electrode. The transport measurements were conducted at 4.2–370 K. Figure 36.16a,b shows the typical transfer characteristics of the FGS/PS composite thin film in air and vacuum, as well as at 90 K in vacuum, respectively. The minima in air and vacuum are located at threshold voltages of ~20 V and +16 V, respectively (Figure 36.16a). The temperature-dependent transfer characteristics of the nanocomposite film are displayed in Figure 36.16c. Apparently, the composite film exhibits typical V-shaped transfer characteristic curves, being similar to those of graphene-based devices with ambipolar behaviors [98]. The *I–V* characteristics of the device measured at 4.2 K with different gate voltages are shown in Figure 36.16d.

Organic semiconductors are attractive materials for fabricating FETs due to their low cost and flexibility. Solution-processed organic FETs have high on/off ratios. However, they usually suffer from low mobility that limits their applications. In contrast, transistors based on graphene usually exhibit very high mobility. So it is expected that incorporation of graphene sheets into organic semiconductors can improve the effective mobility while keeping the on/off ratio sufficiently high. In this respect, Huang et al. [99] developed hybrid FETs by doping graphene sheets into an organic semiconductor, that is, poly(3,3-didodecylquaterthiophene) (PQT-12), via solution blending followed by spin coating onto a SiO_2/Si substrate (Figure 36.17) [99]. Compared with the FETs with only pure organic semiconductors, hybrid FETs exhibit up to 20 times

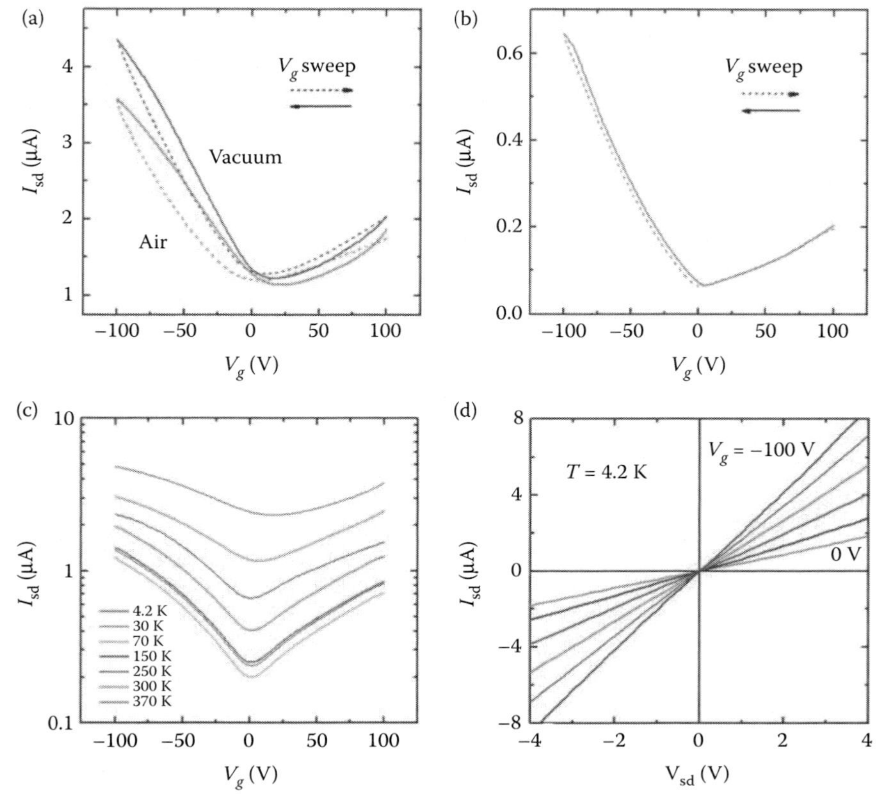

FIGURE 36.16 (a) Transfer characteristics of an FGS/PS composite thin film in air and vacuum. Hysteresis from different sweep directions of V_g (V_{sd} = 1 V) are shown. Transfer characteristics (b) at 90 K in linear scale and (c) at various temperatures in logarithmic scale (V_{sd} = 1 V). (d) I_{sd}–V_{sd} characteristics with different V_g measured at 4.2 K. (Reprinted with permission from Eda, G. et al., Graphene-based composite thin films for electronics. *Nano Lett* 9:814–818. Copyright 2009 American Chemical Society.)

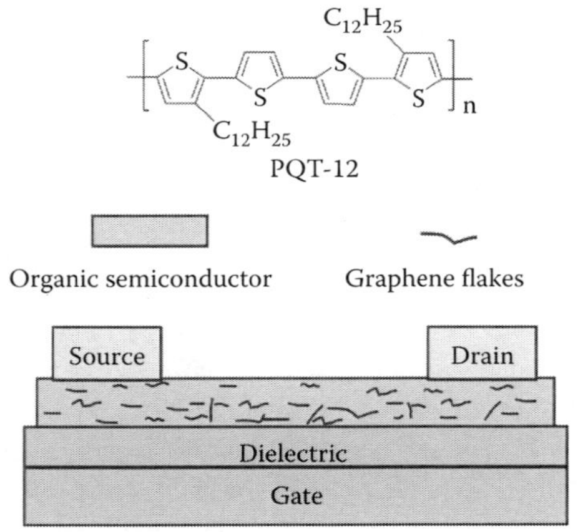

FIGURE 36.17 Structure of organic semiconductor/graphene hybrid FET. (Reprinted from *Org Electron*, 12, Huang, J. et al., Polymeric semiconductor/graphene hybrid field-effect transistors, 1471–1476, Copyright 2011, with permission from Elsevier.)

higher effective mobility, and yet they keep the on/off ratios comparable or better. Figure 36.18 shows the *I*–*V* curve of a PQT-12 FET and a PQT-12/graphene hybrid FET. From these curves, the effective mobility of PQT-12 FET is determined to be 0.15 cm²/V s while that of the PQT-12/graphene hybrid FET is 0.55 cm²/V s. Thus, the composite strategy provides a low-cost yet effective route to enhance the performance of organic semiconductor-based FETs.

36.3.4 Sensors

Sensors can be made from materials with a wide range of conductivities, from electrically conducting polymers [100] to percolating polymer composites. The electrical resistivity of conductive graphene/polymer nanocomposites is highly sensitive to applied stimuli, including gaseous environment, electrical voltage, pressure, and mechanical strain deformation, thus making such composites potentially useful as gas [101,102], temperature [57], and piezoresistive [103,104] sensors.

Two-dimensional graphene can detect the adsorption of parts per million (ppm) of gases on its surface, causing it to show concentration-dependent changes in resistivity [105]. The gas-induced change in resistivity varies in magnitude with different gases. The sign of the change indicates whether the gas is an electron acceptor (e.g., nitrogen dioxide, moisture) or an electron donor (e.g., carbon monoxide, ethanol, and ammonia) [106–109].

The response of the sensors generally derives from a change in electrical resistance due to the thermal expansion of the polymer matrix. The expansion causes increases in

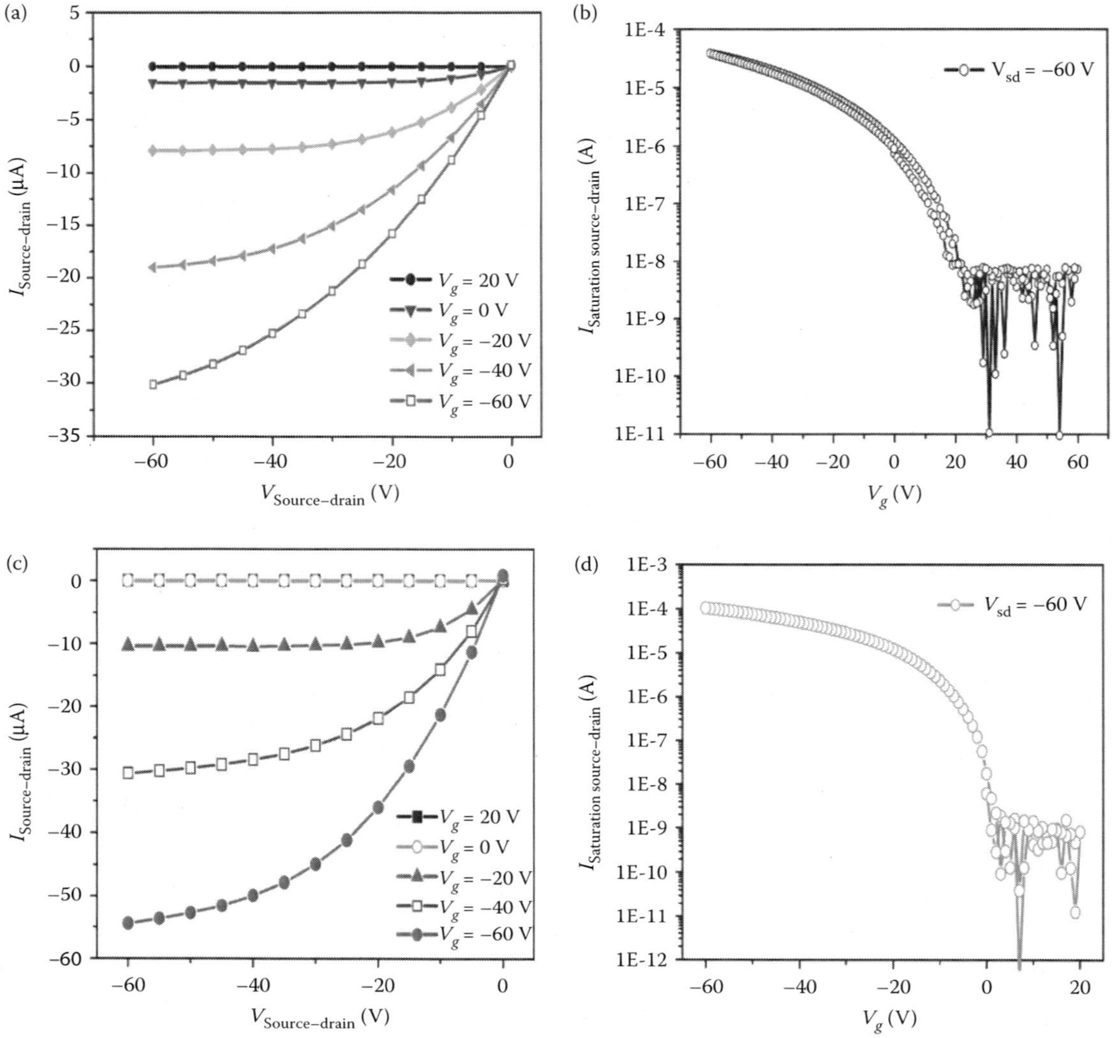

FIGURE 36.18 Source–drain *I–V* curves (a) and transfer curves (b) of a PQT-12 FET (c) and a PQT-12/graphene hybrid FET (d) with octyltrichlorosilane surface treatment and vacuum annealing. (Reprinted from *Org Electron*, 12, Huang, J. et al., Polymeric semiconductor/graphene hybrid field-effect transistors, 1471–1476, Copyright 2011, with permission from Elsevier.)

distances between conducting fillers, leading to an increase in resistance. Very recently, Arizadeh and Soltanini reported that the CRG/PMMA composite can be used as a chemiresistor sensor for the sensitive and selective determination of formaldehyde vapor [99]. Wu et al. demonstrated that the graphene/polyaniline thin-film sensor shows fast response and good reproducibility for ammonia gas [102].

The performance of graphene/polymer nanocomposites acting as the heating element is recently considered. An and Jeong examined the electric heating behavior of solution-cast TRG/epoxy nanocomposites under an applied voltage of 1–100 V [95]. Figure 36.19 shows electrical resistivity versus TRG content for the TRG/epoxy nanocomposite films. The percolation threshold lies between 1 and 2 wt%. Figure 36.20a through f shows time-dependent temperature changes of the TRG/epoxy composite films at various applied voltages. For pure epoxy resin and its composite films with 0.3–1.0 wt% TRG, there exists no temperature change for applied voltages of 1–100 V (Figure 36.20a). Thus, these specimens exhibit no electric heating behavior. For the 2 wt% TRG/epoxy composite, the temperature also shows no changes with time at relatively low applied voltages of 1–15 V (Figure 36.20b). However, the temperature of the composite film increases rapidly in ~5 s when the voltage reaches 30 V and above. The temperature reaches a maximum value in ~25 s. When the applied voltage is turned off at 180 s, the temperature falls rapidly to room temperature in a few seconds. The TRG/epoxy nanocomposites with filler contents ≥3 wt% also exhibit a similar heating behavior, with the exception that higher maximum temperature can be attained at a given applied voltage (Figure 36.20c through f). From these, the maximum temperatures of the composite films can be readily monitored by varying TRG content and the applied voltage. Furthermore, the 10 wt% TRG/epoxy film shows excellent reproducibility and operational stability when subjected to cyclic heating–cooling treatment under a cyclic voltage step of 30 V (Figure 36.21). Despite cyclic voltage

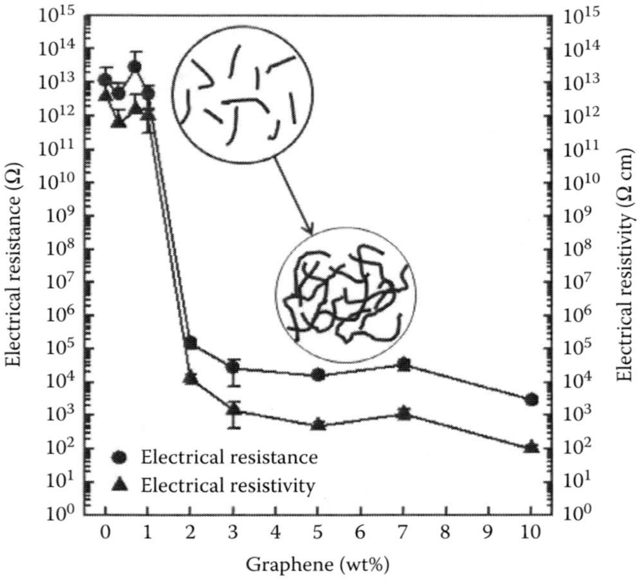

FIGURE 36.19 Electrical resistance and resistivity of TRG/epoxy composite films as a function of TRG content. (Reprinted from *Eur Polym J*, 49, An, J. E. et al., Structure and electric heating performance of graphene/epoxy composite films, 1322–1330 Copyright 2013, with permission from Elsevier.)

changes, the maximum temperature of 126°C is always maintained. This result is believed to be caused by the presence of highly conductive graphene path network in the thermosetting epoxy matrix. Therefore, it appears that the TRG/epoxy composite films can be employed as high performance electrical heating elements for advanced industrial applications.

36.3.5 Thermal Management

With increasing power levels in modern microelectronic devices, and miniaturization of personal computers, the premature failure of those devices due to overheating becomes more serious. Accordingly, thermal management is a key challenge for materials scientists to prevent thermal damage of electronic systems. Only efficient heat dissipation from microchips through the heat spreader and heat sink can ensure their fast and reliable operation [110]. As recognized, thermal conductivity (κ) of a material is governed by the lattice vibration (phonon). The excellent thermal conductivity of graphene renders it an ideal material for heat dissipating and thermal management in electronics. From confocal micro-Raman spectroscopic measurements, the thermal conductivity of a single-layer graphene prepared by mechanical cleavage at room temperature is extremely

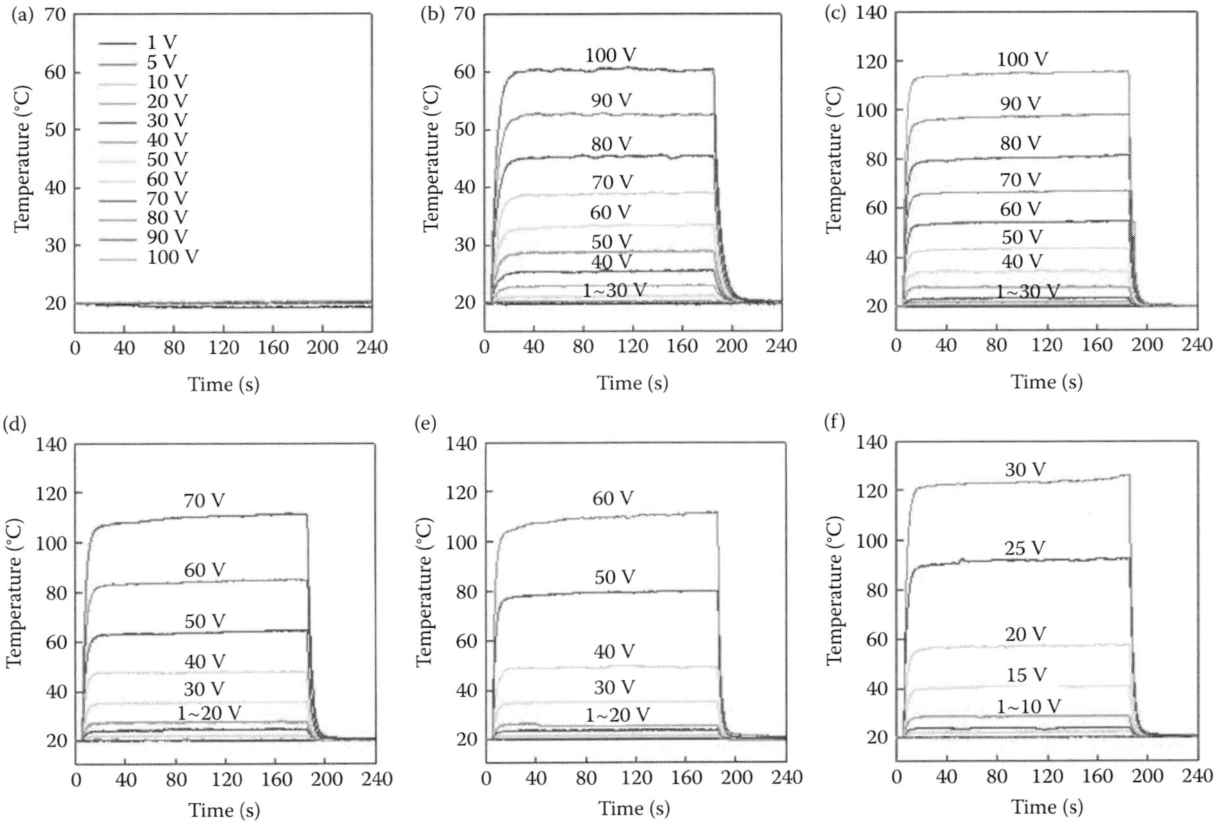

FIGURE 36.20 Time-dependent temperature changes at various applied voltages of 1–100 V for TRG/epoxy composite films with different TRG contents: (a) 1 wt%; (b) 2 wt%; (c) 3 wt%; (d) 5 wt%; (e) 7 wt%; and (f) 10 wt%. (Reprinted from *Eur Polym J*, 49, An, J. E. et al., Structure and electric heating performance of graphene/epoxy composite films, 49:1322–1330, Copyright 2013, with permission from Elsevier.)

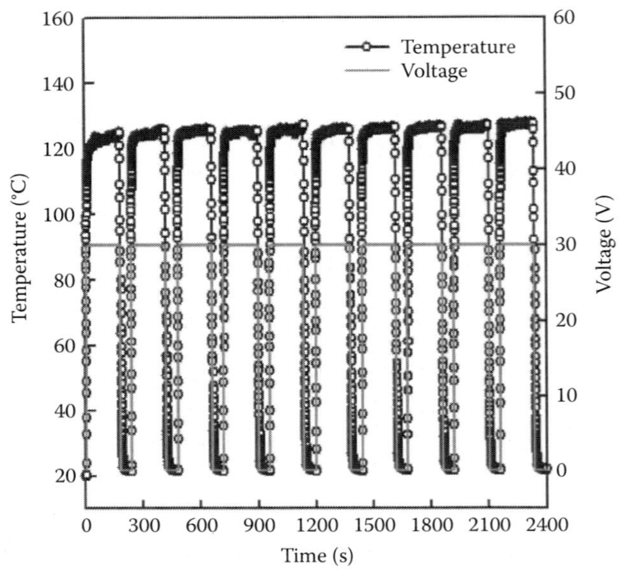

FIGURE 36.21 Variation of temperature with time of 10 wt% TRG/epoxy composite film under a cyclic voltage step. (Reprinted from *Eur Polym J*, 49, An, J. E. et al., Structure and electric heating performance of graphene/epoxy composite films, 1322–1330, Copyright 2013, with permission from Elsevier.)

high, that is, 4840–5300 W/m K [4]. Disappointingly, the measured thermal conductivity of graphene/polymer nanocomposites is far smaller than that of graphene [56]. Figure 36.22 shows thermal conducting behavior of the 1 wt% graphene oxide/epoxy and 5 wt% graphene oxide/epoxy nanocomposites. It is evident that the κ value of the 1 wt% graphene oxide/epoxy nanocomposite is comparable to that of the 1 wt% single-walled carbon nanotube (SWNT)/epoxy nanocomposite. The κ value reaches ~0.83 W/m K by adding 5 wt% graphene oxide, being a fourfold increase over the neat epoxy with a κ value of ~0.195 W/m K. The presence of thermal interface resistance is generally considered to be responsible for low experimental κ values of the polymer composites since it acts as the phonon scattering site [111,112]. More research studies are needed in

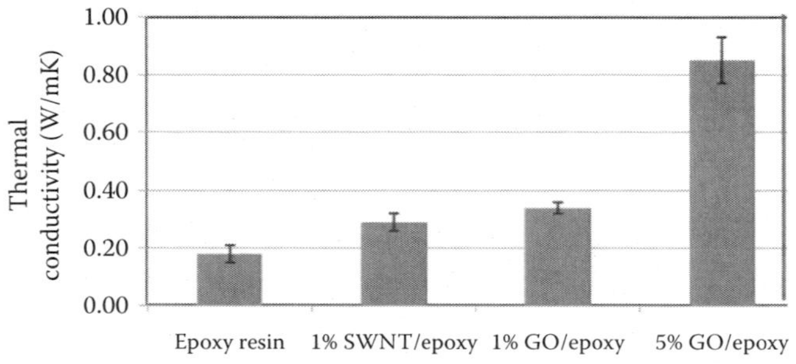

FIGURE 36.22 Thermal conductivity of epoxy filled with 1 wt% SWNT, 1 wt% GO, and 5 wt% GO. (Reprinted with permission from Wang, S. et al., Thermal expansion of graphene composites, *Macromolecules* 42:5251–5255. Copyright 2009 American Chemical Society.)

the near future to eliminate or minimize thermal interface resistance of graphene/polymer nanocomposites for their effective use as heat sinks in microelectronics.

36.4 CONCLUSIONS

Few materials have been as intensively studied as graphene. As a result of the enormous amount of work carried out on it since its discovery in 2004, we now have a wealth of information on its mechanical, electrical, thermal, optical, and other physical properties. One important route to exploit these properties is to incorporate graphene into a polymer matrix. At present, however, there are still relatively few graphene-containing plastic products on the market. Even laboratory research is far from what is desired for many electrical and electronic applications. Many technical issues, including structural control, filler dispersion, interfacial interaction, etc., need to be addressed. Therefore, it would be fair to say that graphene has not yet fulfilled its potential. The main reason for this is quite simple: it is still far too expensive.

The scalable production of pure graphene sheets is crucial to its application. However, only wet chemical approaches can be used to produce a relatively large amount of graphene sheets. The CRG and TRG obtained usually exhibit more degraded physical properties than pristine graphene sheets. The intrinsic properties of graphene are the main factors governing the electrical properties of the resultant composites. Clearly, there is still much work to be done in improving the quality and yield of graphene production.

REFERENCES

1. Katsnelson, M. I. 2007. Graphene: Carbon in two dimensions. *Mater Today* 10:20–27.
2. Soldano, C., A. Mahmood et al. 2010. Production, properties and potential of graphene. *Carbon* 48:2127–2150.
3. Lee, C. G., X. D. Wei et al. 2008. Measurement of the elastic properties and intrinsic strength of monolayer graphene. *Science* 321:385–388.
4. Balandin, A. A., S. Ghosh et al. 2008. Superior thermal conductivity of single-layer graphene. *Nano Lett* 8:902–907.
5. Du, X., I. Skachko et al. 2008. Approaching ballistic transport in suspended graphene. *Nat Nanotechnol* 3:491–495.
6. Stoller, M. D., S. Park et al. 2008. Graphene-based ultracapacitors. *Nano Lett* 8:3498–3502.
7. Park, S. J., J. H. An et al. 2008. Aqueous suspension and characterization of chemically modified graphene sheets. *Chem Mater* 20:6592–6594.
8. Schwamb, T., B. R. Burg et al. 2009. An electrical method for the measurement of the thermal and electrical conductivity of reduced graphene oxide nanostructures. *Nanotechnology* 20:405704.
9. Novoselov, K. S., V. I. Fal'ko et al. 2012. A roadmap for graphene. *Nature* 490:192–200.
10. Zhou, S. Y., G. H. Gweo et al. 2007. Substrate-induced band gap opening in epitaxial graphene. *Nat Mater* 6:770–775.
11. Berger, C., Z. Song et al. 2006. Electronic confinement and coherence in patterned epitaxial graphene. *Science* 312:1191–1196.

12. Rader, H. J., A. Rouhanipour et al. 2006. Processing of giant graphene molecules by soft-landing mass spectrometry. *Nat Mater* 5(4):276–280.
13. Yang, X. Y., X. Dou et al. 2008. Two-dimensional graphene nanoribbons. *J Am Chem Soc* 130(13):4216–4217.
14. Cai, J. M., P. Ruffieux et al. 2010. Atomically precise bottom-up fabrication of graphene nanoribbons. *Nature* 466(7305):470–473.
15. Novoselov, K. S., A. K. Geim et al. 2004. Electric field effect in atomically thin carbon films. *Science* 306 (5296):666–669.
16. Hernandez, Y., V. Nicolosi et al. 2008. High-yield production of graphene by liquid-phase exfoliation of graphite. *Nat Nanotechnol* 3:563–568.
17. Blake, P., P. D. Brimicombe et al. 2008. Graphene-based liquid crystal device. *Nano Lett* 8(6):1704–1708.
18. Stankovich, S., D. A. Dikin et al. 2007. Synthesis of graphene-based nanosheets via chemical reduction of exfoliated graphite oxide. *Carbon* 45:1558–1565.
19. Stankovich, S., R. D. Piner et al. 2006. Synthesis and exfoliation of isocyanate-treated graphene oxide nanoplatelets. *Carbon* 44:3342–3347.
20. Schniepp, H. C., J. L. Li et al. 2006. Functionalized single graphene sheets derived from splitting graphite oxide. *J Phys Chem B* 110:8535–8539.
21. McAllister, M. J., J. L. Li et al.2007. Single sheet functionalized graphene by oxidation and thermal expansion of graphite. *Chem Mater* 19:4396–4404.
22. Novoselov, K. S., D. Jiang et al. 2004. Two-dimensional atomic crystals. *Proc Natl Acad Sci* 102:10451–10453.
23. Ciesielski, A. and P. Samori. 2014. Graphene via sonication assisted liquid-phase exfoliation. *Chem Soc Rev* 43:381–398.
24. Du, W., X. Jiang, and L. Zhu. 2013. From graphite to graphene: Direct liquid-phase exfoliation of graphite to produce single- and few-layered pristine graphene. *J Mater Chem A* 1:10592–10606.
25. William, S., Jr. Hummers et al. 1958. Preparation of graphitic oxide. *J Am Chem Soc* 80(6):1339–1339.
26. Ramanathan, T., A. A. Abdala et al. 2008. Functionalized graphene sheets for polymer nanocomposites. *Nat Nanotechnol* 3:327–331.
27. Georgakilas, V., M. Otyepka et al. 2012. Functionalization of graphene: Covalent and non-covalent approaches, derivatives and applications. *Chem Rev* 112(11):6156–6214.
28. Chen, D., H. Feng, and J. Li. 2012. Graphene oxide: Preparation, functionalization, and electrochemical applications. *Chem Rev* 112(11):6027–6053.
29. Lotya, M., P. J. King et al. 2010. High-concentration, surfactant-stabilized graphene dispersions. *ACS Nano* 4(6):3155–3162.
30. Notley, S. M. 2012. Highly concentrated aqueous suspensions of graphene through ultrasonic exfoliation with continuous surfactant addition. *Langmuir* 28(40):14110–14113.
31. Smith, R. J., P. J. King et al. 2012. Lateral size selection of surfactant-stabilised graphene flakes using size exclusion chromatography. 531:169–172.
32. Hsieh, A. G., C. Punckt et al. 2013. Adsorption of sodium dodecyl sulfate on functionalized graphene measured by conductometric titration. *J Phys Chem B* 117:7950–7958.
33. Vadukumpully, S., J. Paul, and S. Valiyaveettil. 2009. Cationic surfactant mediated exfoliation of graphite into graphene flakes. 47(14):3288–3294.
34. He, L. and S. C. Tjong. 2013. A graphene oxide–polyvinylidene fluoride mixture as a precursor for fabricating thermally reduced graphene oxide–polyvinylidene fluoride composites. *RSC Adv* 3:22981–22987.
35. Tang, H., G. J. Ehlert et al. 2012. Highly efficient synthesis of graphene nanocomposites. *Nano Lett* 12(1):84–90.
36. Stankovich, S., D. A. Dikin et al. 2006. Graphene-based composite materials. *Nature* 442:282–286.
37. Qi, X. Y., D. Yan et al. 2011. Enhanced electrical conductivity in polystyrene nanocomposites at ultra-low graphene content. *ACS Appl Mater Interf* 3:3130–3133.
38. Zhang, H. B., W. G. Zheng et al. 2012. The effect of surface chemistry of graphene on rheological and electrical properties of polymethylmethacrylate composites. *Carbon* 50:5117–5125.
39. Jin, Y., X. Huang et al. 2011. Permittivity, thermal conductivity and thermal stability of poly(vinylidene fluoride)/graphene nanocomposites. *IEEE Trans Dielect Insul* 18:478–484.
40. Fan, P., L. Wanj et al. 2012. Graphene/poly(vinylidene fluoride) composites with high dielectric constant and low percolation threshold. *Nanotechnology* 23:365702.
41. Yun, Y. S., Y. H. Bae et al. 2011. Reinforcing effects of adding alkylated graphene oxide to polypropylene. *Carbon* 49:3553–3559.
42. Pang, H., T. Chen et al. 2010. An electrically conducting polymer/graphene composite with a very low percolation threshold. *Mater Lett* 64:2226–2229.
43. Kim, H. W., S. Kobayashi et al. 2011. Graphene/polyethylene nanocomposites: Effect of polyethylene functionalization and blending methods. *Polymer* 52:1837–1846.
44. Gudarzi, M. and F. Sharif. 2012. Enhancement of dispersion and bonding of graphene-polymer through wet transfer of functionalized graphene oxide. *Express Polym Lett* 6:1017–1031.
45. He, L., S. C. Tjong. 2013. Low percolation threshold of graphene/polymer composites prepared by solvothermal reduction of graphene oxide in the polymer solution. *Nanoscale Res Lett* 8:132.
46. Potts, J. R., S. H. Lee et al. 2011. Thermomechanical properties of chemically modified graphene/poly(methyl methacrylate) composites made by *in situ* polymerization. *Carbon* 49:2615–2623.
47. Fabbri, P., E. Bassoli et al. 2012. Preparation and characterization of poly(butylene terephthalate)/graphene composites by in-situ polymerization of cyclic butylene terephthalate. *Polymer* 53:897–902.
48. Yang, J. H., S. H. Lin et al. 2012. Preparation and characterization of poly(L-lactide)–graphene composites using the *in situ* ring-opening polymerization of PLLA with graphene as the initiator. *J Mater Chem* 22:10805–10815.
49. Luong, N. D., U. Hippi et al. 2011. Enhanced mechanical and electrical properties of polyimide film by graphene sheets via *in situ* polymerization. *Polymer* 52:5237–5242.
50. Tang, L. C., Y. J. Wan et al. 2013. The effect of graphene dispersion on the mechanical properties of graphene/epoxy composites. *Carbon* 60:16–27.
51. Naebe, M., J. Wang et al. 2014. Mechanical property and structure of covalent functionalised graphene/epoxy nanocomposites. *Sci Rep* 4:4375.
52. Teng, C. C., C. C. Ma et al. 2011. Thermal conductivity and structure of non-covalent functionalized graphene/epoxy composites. *Carbon* 49:5107–5116.
53. Wajid, A. S., H. S. Tanvir Ahmed et al. 2012. High-performance pristine graphene/epoxy composites with enhanced mechanical and electrical properties. *Macromol Mater Eng* 298(3):339–347.
54. Ribeiro, H., W. M. Silva et al. 2013. Glass transition improvement in epoxy/graphene composites. *J Mater Sci* 48(22):7883–7892.
55. Rafiee, M. A., J. Rafiee et al. 2010. Fracture and fatigue in graphene nanocomposites. *Small* 6(2):179–183.

56. Wang, S., M. Tambraparni et al. 2009. Thermal expansion of graphene composites. *Macromolecules* 42:5251–5255.
57. An, J. E. and Y. G. Jeong. 2013. Structure and electric heating performance of graphene/epoxy composite films. *Euro Polymer J* 49(6):1322–1330.
58. Bao, C., Y. Guo et al. 2011. *In situ* preparation of functionalized graphene oxide/epoxy nanocomposites with effective reinforcements. *J Mater Chem* 21:13290–13298.
59. Bortz, D. R., E. G. Heras et al. 2012. Impressive fatigue life and fracture toughness improvements in graphene oxide/epoxy composites. *Macromolecules* 45:238–245.
60. Kim, H. and C. W. Macosko. 2009. Processing-property relationships of polycarbonate/graphene composites. *Polymer* 50:3787–3809.
61. Mahmoud, W. E. 2011. Morphology and physical properties of poly(ethylene oxide) loaded graphene nanocomposites prepared by two different techniques. *Eur Polym J* 47:1534–1540.
62. Zhang, H. B., W. G. Zheng et al. 2010. Electrically conductive polyethylene terephthalate/graphene nanocomposites prepared by melt compounding. *Polymer* 51:1191–1196.
63. Kirkpatrick, S. 1973. Percolation and conduction. *Rev Mod Phys* 45:574–588.
64. Bergman, D. J. and Y. Imry. 1977. Critical behavior of the complex dielectric constant near the percolation threshold of a heterogeneous material. *Phys Rev Lett* 39:1222–1225.
65. Stephen, M. J. 1978. Mean-field theory and critical exponents for a random resistor network. *Phys Rev B* 17:4444–4453.
66. Nan, C. W. 1993. Physics of inhomogeneous inorganic materials. *Prog Mater Sci* 37:1–116.
67. Reghu, M., C. O. Yoon et al. 1994. Transport in polyaniline networks near the percolation threshold. *Phys Rev B* 50:13931–13941.
68. Wu, J. J. and D. S. McLachlan. 1998. Scaling behavior of the complex conductivity of graphite-boron nitride percolation systems. *Phys Rev B* 58:14880–14887.
69. Fraysse, J. and J. Planè. 2000. Interplay of hopping and percolation in organic conducting blends. *Phys Stat Sol B* 218:273–277.
70. Scher, H. and R. Zallen. 1970. Critical density in percolation processes. *J Chem Phys* 53:3759–3761.
71. Fan, Ping, L. Wang et al. 2012. Graphene/poly(vinylidene fluoride) composites with high dielectric constant and low percolation threshold. *Nanotechnology* 23:365702.
72. Liao, K. H., Y. Qian et al. 2012. Ultralow percolation graphene/polyurethane acrylate nanocomposites. *Polymer* 53:3756–3761.
73. Nguyen, D. A., Y. R. Lee et al. 2009. Morphological and physical properties of a thermoplastic polyurethane reinforced with functionalized graphene sheet. *Polym Int* 58(4):412–417.
74. Liang, J., Y. Wang et al. 2009. Electromagnetic interference shielding of graphene/epoxy composites. *Carbon* 47:922–925.
75. Steurer, P., R. Wissert et al. 2009. Functionalized graphenes and thermoplastic nanocomposites based upon expanded graphite oxide. *Macromol Rapid Commun* 30:316–327.
76. Raghu, A. V., Y. R. Lee et al. 2008. Preparation and physical properties of waterborne polyurethane/functionalized graphene sheet nanocomposites. *Macromol Chem Phys* 209:2487–2493.
77. Li, M., C. Gao et al. 2013. Electrical conductivity of thermally reduced graphene oxide/polymer composites with a segregated structure. *Carbon* 65:371–373.
78. Du, J., L. Zhao et al. 2011. Comparison of electrical properties between multi-walled carbon nanotube and graphene nanosheet/high density polyethylene composites with a segregated network structure. *Carbon* 49:1094–1100.
79. Wang, D., X. Zhang et al. 2013. Dielectric properties of reduced graphene oxide/polypropylene composites with ultralow percolation threshold. *Polymer* 54:1916–1922.
80. Narkis, M., G. Lidor et al. 1999. New injection moldable electrostatic dissipative (ESD) composites based on very low carbon black loadings. *J Electrostatics* 47:201–214.
81. Lee, J. I., S. M. Yang et al. 2009. Carbon nanotube polypropylene nanocomposites for electrostatic discharge applications. *Macromolecules* 42:8328–8334.
82. Wong, K. H., S. J. Pickerin et al. 2010. Recycled carbon fibre reinforced polymer composite for electromagnetic interference shielding. *Composites Part A* 41:693–702.
83. Wu, J. and D. D. L. Chung. 2002. Increasing the electromagnetic interference shielding effectiveness of carbon fiber polymer–matrix composite by using activated carbon fibers. *Carbon* 40:445–467.
84. Al-Saleh, M. H. and U. Sundararaj. 2009. Electromagnetic shielding mechanism of CNT/polymer composites. *Carbon* 47:1738–1746.
85. Hong, S. K., K. Y. Kim et al. 2012. Electromagnetic interference shielding effectiveness of monolayer graphene. *Nanotechnology* 23:455704.
86. Singh, A. P., P. Garg et al. 2012. Phenolic resin-based composite sheets filled with mixtures of reduced graphene oxide, c-Fe_2O_3 and carbon fibers for excellent electromagnetic interference shielding in the X-band. *Carbon* 50:3868–3875.
87. Thomassin, J. M., C. Pagnoulle et al. 2008. Foams of polycaprolactone/MWNT nanocomposites for efficient EMI reduction. *J Mater Chem* 18:792–796.
88. Chen, L. M., R. Ozisik et al. 2010. The influence of carbon nanotube aspect ratio on the foam morphology of MWNT/PMMA nanocomposite foams. *Polymer* 51:2368–2375.
89. Esmaraai, V., S. Sankaranarayanan et al. 2011. Functionalized Graphene–PVDF Foam Composites for EMI Shielding. *Macromol Mater Eng* 296:894–898.
90. Chen, Z., C. Xu et al. 2013. Lightweight and flexible graphene foam composites for high-performance electromagnetic interference shielding. *Adv Mater* 25:1296–1300.
91. Zhang, H. B., Q. Yan et al. 2011. Tough graphene-polymer microcellular foams for electromagnetic interference shielding. *ACS Appl Mater Interf* 3:918–924.
92. Yuan, B., L. Yu et al. 2012. Comparison of electromagnetic interference shielding properties between single-wall carbon nanotube and graphene sheet/polyaniline composites. *J Phys D: Appl Phys* 45:235108.
93. He, L. and S. C. Tjong. 2010. Polymer/ceramic composite hybrids containing multi-walled carbon nanotubes with high dielectric permittivity. *Current Nanosci* 6:40–44.
94. Romasanta, L. J., M. Hernández et al. 2011. Functionalized graphene sheets as effective high dielectric constant fillers. *Nanoscale Res Lett* 6:508.
95. Li, Q., Q. Xue et al. 2008. Large dielectric constant of the chemically functionalized carbon nanotube/polymer composites. *Compos Sci Technol* 68:2290–2296.
96. Dang, Z. M., L. Wang et al. 2007. Giant dielectric permittivities in functionalized carbon nanotube/electroactive polymer composites. *Adv Mater* 19:852–857
97. Eda, G. and M. Chhowalla. 2009. Graphene-based composite thin films for electronics. *Nano Lett* 9:814–818.
98. Lin, M. W., C. Ling et al. 2011. Room-temperature high on/off ratio in suspended graphene nanoribbon field-effect transistors. *Nanotechnology* 22:265201.
99. Huang, J., R. H. Daniel et al. 2011. Polymeric semiconductor/graphene hybrid field-effect transistors. *Org Electron* 12:1471–1476.

100. Carlos, M., C. M. Hangarter et al. 2013. Hybridized conducting polymer chemiresistive nano-sensors. *Nano Today* 8:39–55.
101. Arizadeh, T. and L. H. Soltani. 2013. Graphene/poly(methyl methacrylate) chemiresistor sensor for formaldehyde odor sensing. *J Hazard Mater* 248–249:401–406.
102. Wu, Z., X. Chen et al. 2013. Enhanced sensitivity of ammonia sensor using graphene/polyaniline nanocomposite. *Sens Actuators B: Chem* 178:485–493.
103. Chen, L., G. Chen et al. 2007. Piezoresistive behavior study on finger-sensing silicone rubber/graphite nanosheet nanocomposites. *Adv Funct Mater* 17:898–904.
104. Chiacchiarelli, L. M., M. Rallini et al. 2013. The role of irreversible and reversible phenomena in the piezoresistive behavior of graphene epoxy nanocomposites applied to structural health monitoring. *Compos Sci Technol* 80:73–79.
105. Schedin, F., A. K. Geim et al. 2007. Detection of individual gas molecules adsorbed on graphene. *Nat Mater* 6:652–655.
106. Leenaerts, O., B. Pasrtoens et al. 2008. Adsorption of H_2O, NH_3, CO, NO_2, and NO on graphene: A first-principles study. *Phys Rev B* 77:125416.
107. Lu, G. H., L. E. Ocola et al. 2009. Reduced graphene oxide for room-temperature gas sensors. *Nanotechnology* 20:445502.
108. Yoon, H. J., D. H. Jun et al. 2011. Carbon dioxide gas sensor using a graphene sheet. *Sens Actuators B: Chem* 157:310–313.
109. Llobet, E. 2013. Gas sensors using carbon nanomaterials: A review. *Sens Actuators B: Chem*179:32–45.
110. Schelling, P. K., L. Shi et al. 2005. Managing heat for electronics. *Mater Today* 8:30–35
111. Nan, C. W., G. Liu et al. 2004. Interface effect on thermal conductivity of carbon nanotube composites. *Appl Phys Lett* 85:3549–3551.
112. Bagchi, A. and S. Nomura. 2006. On the effective thermal conductivity of carbon nanotube reinforced polymer composites. *Compos Sci Technol* 66:1703–1712.

37 Chemical Vapor Deposition of Graphene for Electronic Device Application

Golap Kalita, Masayoshi Umeno, and Masaki Tanemura

CONTENTS

ABSTRACT

The discovery of a single atomic layer of carbon called graphene brought tremendous attention as a next-generation electronic material. Graphene has been derived or synthesized through various approaches, such as exfoliation of graphitic materials, epitaxial growth on silicon carbide (SiC), and the chemical vapor deposition (CVD) technique. Among these processes, the CVD approach has several advantages for the controlled synthesis of high-quality and large-area graphene in electronic device applications. The growth and nucleation process of single-crystal graphene domain and the formation of continuous large-area graphene film in a CVD process are discussed in this chapter. CVD graphene shows an excellent potential as the flexible transparent conducting optical window for various electronic devices. The fabrication process and application of transparent conductor using CVD graphene are discussed in detail.

37.1 BACKGROUND

Prior to the discovery of graphene, Peierls and Landau predicted that two-dimensional (2D) crystals are thermodynamically unstable and difficult to find in nature [1–6]. However, the finding of atomically thin graphene revolutionized 2D materials in physics, chemistry, and engineering fields, with enormous application possibilities [7–13]. The 2010 Nobel Prize in Physics was awarded to Andre Geim and Konstantin Novoselov for their groundbreaking work on graphene. Although the planar structure of graphene has been isolated recently, several other carbon allotropes have been discovered previously. For many decades, it has been well known that graphite is a naturally occurring crystalline form of carbon. If we look carefully at the structure of graphite, it is a stack of millions of layers held together by van der Waals forces [14]. Similarly, another form of crystalline carbon is the buckminsterfullerene, also known as the buckyball or fullerene (C_{60}), which was discovered in 1986 by R.F. Curl, H.W. Kroto, and R.E. Smalley. The soccer ball look-alike buckyball has a diameter of only 7.1 Å and consists of 12 pentagon and 20 hexagon carbon rings [15]. The zero-dimensional (0D) buckyball can resemble a graphene sheet by making a planar structure. Similarly, carbon nanotube (CNT), another form of carbon, attracted significant research interest for its fascinating properties as a one-dimensional (1D) structure [16]. Basically, CNT can be considered as a rolled sheet of graphene forming a 1D tubular structure. Single-walled and multiwalled CNTs consist of a single layer and few layers of graphene, respectively [17]. Recent studies also demonstrated the unzipping of CNTs to open up in sheet-like structures to fabricate graphene nanoribbons [18]. Graphene is the main building block of these forms of carbon and can be considered as the mother of all graphitic forms as demonstrated in Figure 37.1. These various forms of crystalline carbon in different dimensions (i.e., zero, one, two, and three dimensions) show contrasting physical and chemical properties and are prominent materials for a wide range of applications. Various physical phenomena and structure-related properties have been explored or discovered, while graphene can be an ideal candidate to understand the other forms. The development of 2D graphene crystal is summarized in Table 37.1 to provide a better understanding.

37.2 PROPERTIES OF GRAPHENE

The unit cell of graphene contains two carbon atoms and the graphene lattice can be considered to be made of two sublattices, A and B, as shown in Figure 37.2a [19]. Graphene is known to be a zero-band-gap semiconductor, where the electrons can only

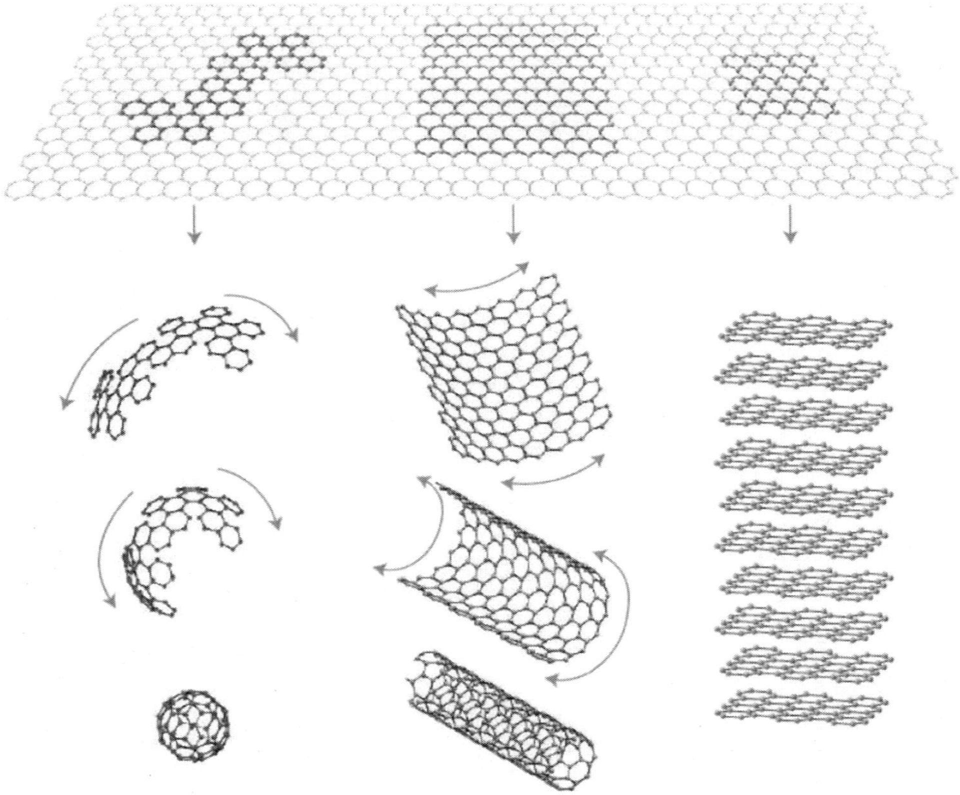

FIGURE 37.1 Graphene with a 2D structure is the main building block of other forms of sp^2-hybridized carbon in all other dimensions. Graphene is the mother of these forms of carbon as it can be wrapped up to form a 0D fullerene, rolled into 1D nanotubes, or stacked into 3D graphite. (Reprinted from Geim, A.K. and K.S. Novoselov. 2007. *Nat. Mater.* 6: 183–91. With permission.)

move between carbon atoms in the 2D lattice structure. The energy band structure of graphene involves π electrons of sp^2-hybridized carbon atoms. The first band structure calculations were performed by P.R. Wallace in 1947 and the band structure can be presented as shown in Figure 37.2b [20]. The valence band is formed by bonding π states, while the conduction band is formed by the antibonding π^* states.

The quantum confinement of electrons in the absence of a third dimension provides graphene with various exciting properties. Electrons in graphene behave as massless relativistic fermions at low temperatures, which is an unusual behavior for a condensed-matter system. Graphene shows an unusual (relativistic) quantum Hall effect with an applied perpendicular magnetic field at a temperature as high as room temperature. The massless Dirac fermion (i.e., charge carrier) of graphene moves at ballistic speed in submicron length, close to relativistic speeds. It has been estimated that the intrinsic carrier mobility of graphene is as high as 200,000 $cm^2/V \cdot s$ [21]. The 2D graphene sheet is itself an excellent current conductor; the sustainability of the current density is six orders higher than that of normal copper [22]. Graphene also shows exceptional mechanical strength, with a breaking strength of ~40 N/m and a Young's modulus of ~1.0 TPa [10]. The thermal conductivity of a suspended graphene sheet at room temperature has been measured as in the range of $4.84 \times 10^3 \sim 5.3 \times 10^3$ W/mK [13]. It has also been revealed that the absorbance and transparency of a single layer of graphene does not depend on the wavelength, which is measured from the fine structure constant ($\alpha = e^2/\hbar c$). A monolayer graphene absorbs $2.3 \pm 0.1\%$ of the incident light with a negligible reflectance of <0.1% [9]. Graphene with only one atomic layer has a high surface-area-to-volume ratio without affecting much of the mechanical properties. From the chemistry point of view, graphene sheets can be functionalized with other elements to achieve heterogeneous chemical and electronic structures. The properties of graphene-based material are summarized in Table 37.2.

TABLE 37.1
Development of Graphitic or sp^2-Hybridized Carbon-Related Materials

Year	Carbon-Related Material	Reference
1789	Graphite	Named by Abraham Gottlob Werner (Wikipedia) [14]
1986	Buckyball or C_{60}	Kroto et al. [15]
1991	Carbon nanotube	Iijima et al. [16]
2004	Graphene sheet	Novoselov et al. [1]
2009	Graphene nanoribbon by unzipping of a CNT	Kosynkin et al. [18]

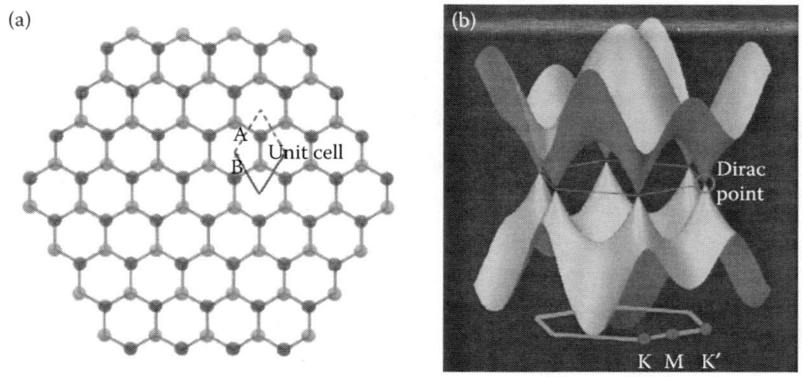

FIGURE 37.2 (a) Two graphene sublattices (A and B) and unit cell. (b) Electronic band structure and presentation of Dirac point of 2D graphene crystal. (Reprinted from Avouris, P. and C. Dimitrakopoulos. 2012. *Mater. Today* 15: 86–97. With permission.)

TABLE 37.2
Properties of Graphene 2D Crystals

Property	Graphene	Reference
Electrical	• Band structure for graphite	P.R. Wallace in 1947 [20]
	• Carrier mobility ~200,000 $cm^2/V \cdot s$	Morozov et al. [21]
	• Unusual (relativistic) quantum Hall effect	Zhang et al. [8]
	• Excellent conductor (can sustain 6 orders higher current than Cu)	Geim et al. [11]
Optical	• Graphene absorbs $2.3 \pm 0.1\%$ of light (from fine structure constant), independent of wavelength	Nair et al. [9]
Thermal	Thermal conductivity ~4.84×10^3 to 5.3×10^3 W/mK	Balandin et al. [13]
Mechanical	Breaking strength of ~40 N/m and Young's modulus of ~1.0 TPa	Lee et al. [10]
Chemical	Functionalization with various functional groups	Loh et al. [23]

37.3 DERIVATION AND SYNTHESIS OF GRAPHENE

With a view to studying basic properties as well as device integration, graphene has been derived or synthesized by micromechanical exfoliation of graphite [24], epitaxial growth on SiC [25], reduction of chemically exfoliated graphite oxide [26], and the CVD technique [27–33]. Mechanical exfoliation, also known as the scotch tape method, has been widely studied to investigate the properties of graphene and electronic device fabrication. In this approach, bulk pyrolytic graphite is placed on common sticky adhesive tape. The tape is pressed on a desired substrate and then peeled away. Repeating the peel-off process, graphene flakes can be obtained on the desired substrate, along with some graphitic portions and adhesive residues. Graphene flakes can be identified on the substrate by optical microscopy as the thin-film interference shows variation in color. However, the technique is time consuming and basically depends on a trial-and-error approach with difficulties to locate with repeatability. Again, the mechanically exfoliated graphene flakes are of the order of only a few microns, scattered randomly on the substrate, which makes it difficult for many applications.

Reduction of chemically exfoliated graphene oxide (GO) is another approach to derive large quantity of graphene flakes. GO has been derived by a modified Hummers' method that involves rigorous oxidation of pure graphitic materials [34,35]. GO is an insulating graphene sheet, where several oxygen-containing functional groups were covalently bonded with carbon atoms of the basal plane and edges [36–38]. Thus, the oxygen-containing functional groups of GO sheet determine the chemical and electrical properties, which can be tuned by defect-related structures and oxygen content. It can be considered that the GO-based materials are electronically hybrid structures, where both π state of sp^2 and σ state of sp^3 bonded carbon are present. The optical and electronic properties can be controlled by changing the ratio of sp^2 and sp^3 bonded carbon atoms [23]. In this aspect, high-temperature annealing and chemical treatments of GO are unavoidable to obtain a conducting graphene sheet by reducing the oxygen-containing functional groups. Considerable structural defects remain in the graphene sheet derived by the reduction of GO, which has a significant effect on the mechanical and electrical properties. Similarly, graphene has been synthesized on a SiC surface by a high-temperature thermal processing approach. Graphitization on the SiC crystal with annealing at a high temperature was reported by Badami in 1961 [39]. In the annealing process, the top layer of SiC undergoes thermal decomposition, where Si atoms desorb and carbon atoms remain on the surface, rearranging and bonding to form a graphene structure [25,40,41]. However, the main disadvantage of graphene growth on SiC is small domain structures with the presence of steps and terrace edges. Graphene grown on SiC can only be used in *in situ* device fabrication as transferring to another substrate is quite difficult. The most promising and simple approach to the synthesis of large-area

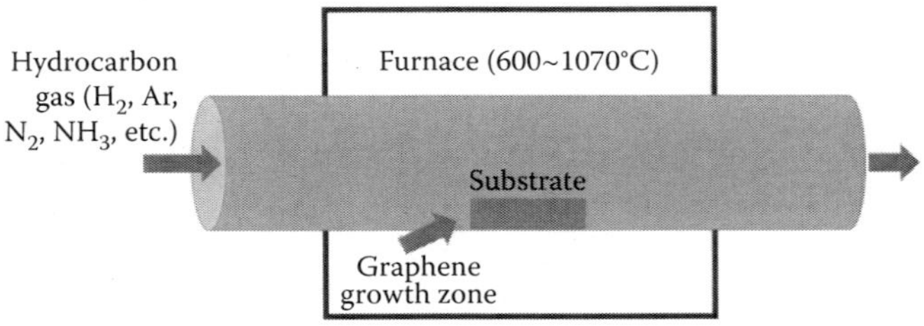

FIGURE 37.3 A simple schematic diagram of the CVD process for graphene synthesis.

continuous graphene is the CVD technique on various metal substrates. Recent significant development of high-quality and single-crystal graphene synthesis by the CVD technique on metal substrates opened up new possibilities for applications [31,32,42]. In the following, we discuss in detail about the CVD synthesis of graphene-based materials. Figure 37.3 presents a simple schematic diagram of the CVD process, generally used for graphene synthesis.

37.4 GRAPHENE-LIKE STRUCTURES BY CVD

In the earlier stage, the synthesis of a transparent graphene-constructed film as electrodes has been explored by the thermal fusion of organic molecules by a CVD technique without using any catalytic layer. In this regard, Wang et al. demonstrated the synthesis of graphene-constructed structure from superphenalene derivative, a large aromatic core with good solubility [43]. This has been achieved by the thermal reaction of synthetic nanographene molecules of giant polycyclic aromatic hydrocarbons (PAHs), which were fused to form a large graphene sheet. Figure 37.4 shows the molecular structure of the superphenalene molecule used to synthesize such a graphene-constructed film. Highly transparent graphene-constructed films on glass substrates were obtained by this process (Figure 37.4). Transparency of the synthesized material has been controlled by tuning the film thickness in the demonstrated process. The developed materials show sheet resistance and conductivity of 1.6 kΩ/Sq and 206 S/cm, respectively, which can be suitable for various transparent conductor applications [43].

Similarly, transparent graphene-constructed carbon film has been synthesized from camphor ($C_{10}H_{16}O$) molecules by the CVD process [44]. The solid camphor powder derived from trees has a molecular structure consisting of hexagonal and pentagonal carbon rings, and methyl carbons (Figure 37.5). Similar to the previous approach, these molecules can also be pyrolyzed to obtain graphene-constructed carbon film. The thickness and transparency of the graphene-constructed film can be controlled with the amount of pyrolyzed camphor and growth duration. Transmission electron microscopy (TEM) analysis shows layer structures with an interplanar spacing of about 0.34 nm, corresponding to the graphite d(002) spacing. Similarly, Raman study showed the graphitic peak (1598 cm^{-1}) and a low-intensity second order, 2D peak Raman speak

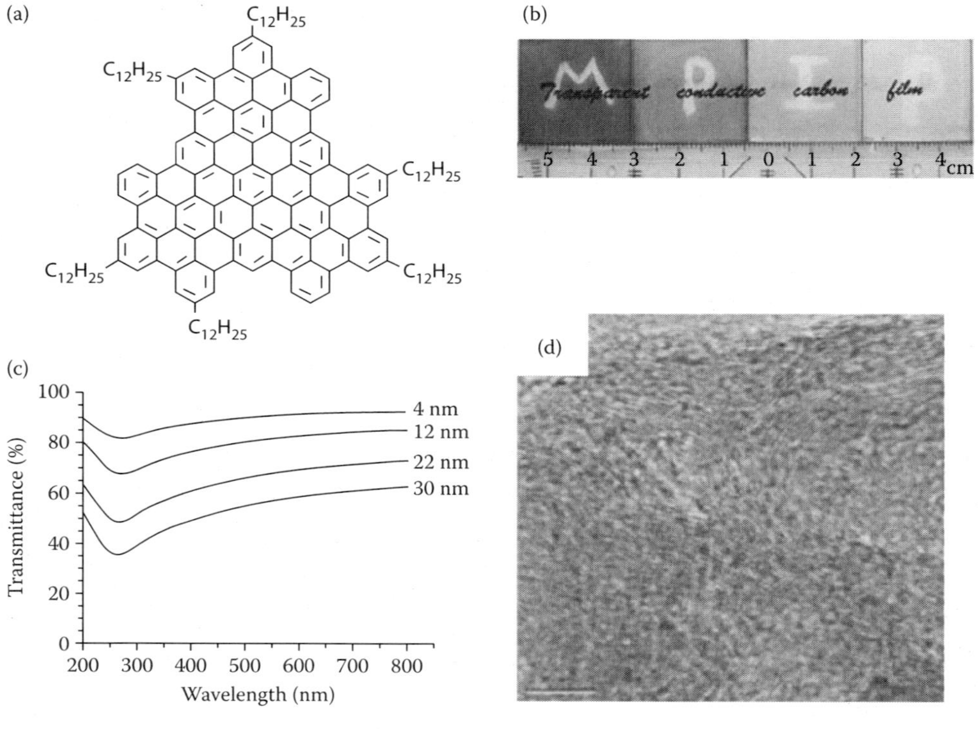

FIGURE 37.4 (a) Molecular structure of hexadodecyl-substituted superphenalene C96-C12. (b) (c) Photographs and transmittance spectra of the fabricated graphene-constructed film on quartz substrates. (d) TEM image of a graphene-constructed film (scale bar: 5 nm). (Reprinted from Wang, X. et al. 2008. *Angew. Chem. Int. Ed.* 47: 2990–2. With permission.)

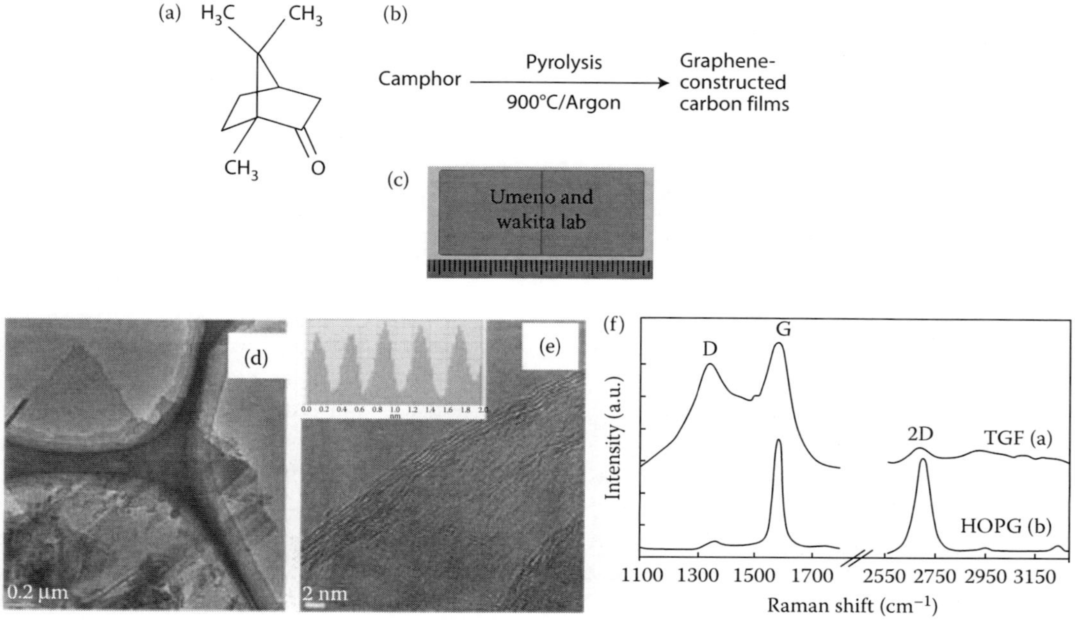

FIGURE 37.5 (a) Molecular structure of solid camphor, (b) pyrolysis process to obtain graphene-constructed carbon film, and (c) photograph of deposited film on quartz substrate. (d), (e) TEM image of synthesized graphene-constructed carbon film and (f) Raman spectra of the as-synthesized carbon film by camphor pyrolysis in comparison to HOPG. (Kalita, G. et al. 2010. Graphene constructed carbon thin films as transparent electrodes for solar cell applications. *J. Mater. Chem.* 20: 9713–7. Reproduced by permission of The Royal Society of Chemistry.)

(2700 cm^{-1}). Thus, TEM and Raman studies of the synthesized materials indicated the features of graphene-like structure [45,46]. The above discussed process and materials are significant for stiff coating as they remain unaffected after exposure to organic solvents and a harsh chemical atmosphere.

Now, we discuss some of the applications of such graphene-based thin film. The transparent and highly conducting film on a glass substrate can be an ideal candidate for optoelectronic device applications. The transparent conductor can be integrated into solar cell, light-emitting diode, photodiode, and flat-panel displays as an alternative to oxide-based transparent conductors (such as indium tin oxide (ITO), fluorine-doped tin oxide (FTO), etc.). Organic solar cells based on poly(3-hexylthiophene) (P3HT) and [6,6]-phenyl C_{61} butyric acid methyl ester (PCBM) blends have been successfully fabricated on graphene-constructed films. A solar cell fabricated with the structure graphene/PEDOT:PSS/P3HT:PCBM/Al can be compared with an ITO-based solar cell. Solar cells with graphene and ITO electrodes showed almost similar open circuit voltage (V_{oc}). However, the short circuit current density (J_{sc}) and conversion efficiency of the solar cell with graphene film were affected with high sheet resistance (Figure 37.6).

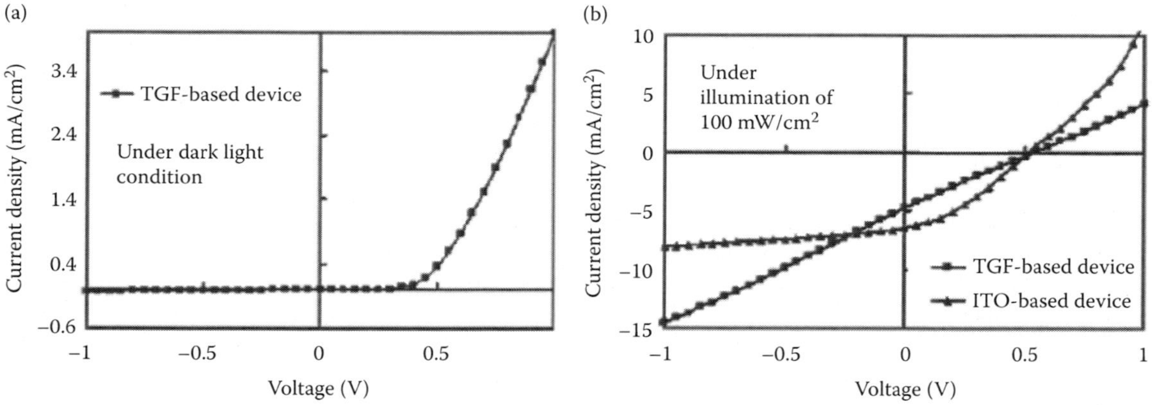

FIGURE 37.6 J–V characteristics of a device fabricated with P3HT:PCBM composite on a graphene-constructed carbon film: (a) dark conditions (without illumination) and (b) light conditions (under illumination of 100 mW/cm^2). (Kalita, G. et al. 2010. Graphene constructed carbon thin films as transparent electrodes for solar cell applications. *J. Mater. Chem.* 20: 9713–7. Reproduced by permission of The Royal Society of Chemistry.)

The work function of graphene (i.e., 4.4–4.65 eV) is almost similar to that of ITO (i.e., 4.5–5.1 eV), while the work function is tunable with chemical treatments [44–51]. This will enable the application of graphene-based transparent conductors in a wide range of applications. As we progress in the chapter, we further discuss applications of highly conducting graphene film as flexible transparent conductors for various modern electronic devices.

37.5 GRAPHENE ON TRANSITION METALS BY CVD

As discussed above, low-crystalline-quality graphene-like structures can be directly synthesized on various substrates at a high temperature using organic molecules as the precursor. To achieve high-quality graphene growth, a catalytic substrate is almost unavoidable. In this regard, the formation of graphitic structures on transition metal surfaces has been known for many years [52]. It has been understood that the dissociation of hydrocarbon-based materials from other materials to form graphite was first proposed in 1896 [53,54]. The growth of pyrolytic carbon films or layers of graphite had been first observed on the Ni surface with the introduction of hydrocarbon or evaporated carbon materials [53,55]. Almost at the same time, the formation of graphitic layers on single-crystal Pt was observed in catalysis experiments [56,57]. The formation of graphitic layers can be explained by the diffusion and segregation of carbon impurities on the metal surface during the annealing and cooling process. In the last few years, significant interest in 2D crystalline materials has led to significant development in the CVD graphene growth process. Graphene-based materials have been synthesized on various transition and poor metal surfaces such as Ru, Ir, Co, Ni, Pt, Pd, Au, Ag, and Ge. At an elevated temperature, hydrocarbons were thermally decomposed and surface absorption or segregation of carbon occurred depending on carbon–metal solubility for graphene growth [58–68]. The carbon solubility on the metal surface and the growth conditions determine the growth mechanism, consequently morphology and thickness of graphene. The binary phase diagrams of transition metals and carbon such as Ni–C Co–C Fe–C is presented in Figure 37.7. Ni, Co, and Fe exhibited significant carbon solubility (~0.5–1.0 wt.% C) in the solid solution within the temperature ranges of 800–900°C. Metastable carbides such as Co_3C, Co_2C, and Ni_3C form immediately on the Co and Ni surface following saturation of carbon in the solid solution, and additional carbon diffuses into the catalyst layer thereby forming graphene layers [69,70].

37.6 GRAPHENE ON NICKEL BY CVD

As discussed above, graphitic structure formation on the Ni surface has been well known and observed many years ago. However, controlled growth of monolayer graphene or few-layer graphene on the Ni surface has been recently realized [32,33]. Polycrystalline Ni thin film deposited on SiO_2/Si and foils have been used as the substrate material for graphene growth in a CVD process. Various hydrocarbons in solid (camphor, poly(methyl methacrylate) [PMMA], etc.), liquid (alcohol, toluene, etc.), and gaseous (methane, acetylene, etc.) forms have been investigated for graphene synthesis [28–33]. The high carbon solubility on the Ni surface allows the segregation of carbon to form graphene structures. The graphene growth mechanism on polycrystalline and single-crystal Ni surfaces will be further explained as we progress. The synthesis of planar graphene structure on the Ni surface has been demonstrated by Somani et al. in 2006 by a CVD process using solid camphor as the carbon source [27]. The process was further optimized to achieve controlled growth of monolayer, bilayer, and few-layer graphene on Ni-deposited SiO_2/Si and Ni foil [29,71,72]. In an unoptimized condition, a large number of graphene layer formation can be observed on Ni foil with higher carbon segregation (Figure 37.8a). Synthesized material showed a sharp and intense x-ray diffraction (XRD) peak at about $2\theta = 26.68°$, indicating good graphitic nature of the film (Figure 37.8b). The composition of the carrier gas along with the hydrocarbon quantity plays a critical role in high-quality graphene growth. Controlling the growth conditions such as the nature of the Ni surface, growth temperature, carrier gas composition, and pressure in the CVD chamber, high-quality graphene were grown as shown in Figure 37.8c through e [71–73].

Similarly, methane, alcohol, and various other hydrocarbon precursors have been explored to achieve a controlled growth of graphene on Ni thin film and foils. In the graphene growth process, hydrocarbon decomposes at an elevated temperature and carbon atoms dissolve on the Ni surface to form a solid solution. Ni has relatively high carbon solubility at a high temperature (800–1000°C) and the solubility decreases as the temperature goes down. During the cooling process, carbon atoms diffuse to the top surface of Ni and precipitate from the Ni and C solid solution to form graphene layers [28]. The carbon dissolution on the Ni surface and the effect of the cooling process are significant in graphene formation (Figure 37.9). Furthermore, the graphene growth on polycrystalline Ni and single-crystal Ni (111) has been reported by Zhang et al. giving an insight of the growth process. It is observed that highly uniform monolayer/bilayer graphene forms on the Ni (111) plane by the CVD process. However, the main drawback of the polycrystalline Ni catalyst layer is the formation of a few-layer structure at the grain boundaries with a higher amount of carbon segregation. This has been explained with diffusion–segregation model for carbon precipitation on the Ni surface. A single-crystal Ni (111) surface shows much more uniform growth of graphene, while the rough surface of polycrystalline Ni with a large amount of grain boundaries facilitates multilayer formation [74]. The graphene growth mechanism on single-crystal Ni (111) with a smoother surface and polycrystalline Ni with grain boundaries is illustrated in Figure 37.9. The presence of grain boundaries in polycrystalline Ni can facilitate the segregation of a large amount of carbon atoms during the CVD process, resulting in the formation of multilayer graphene.

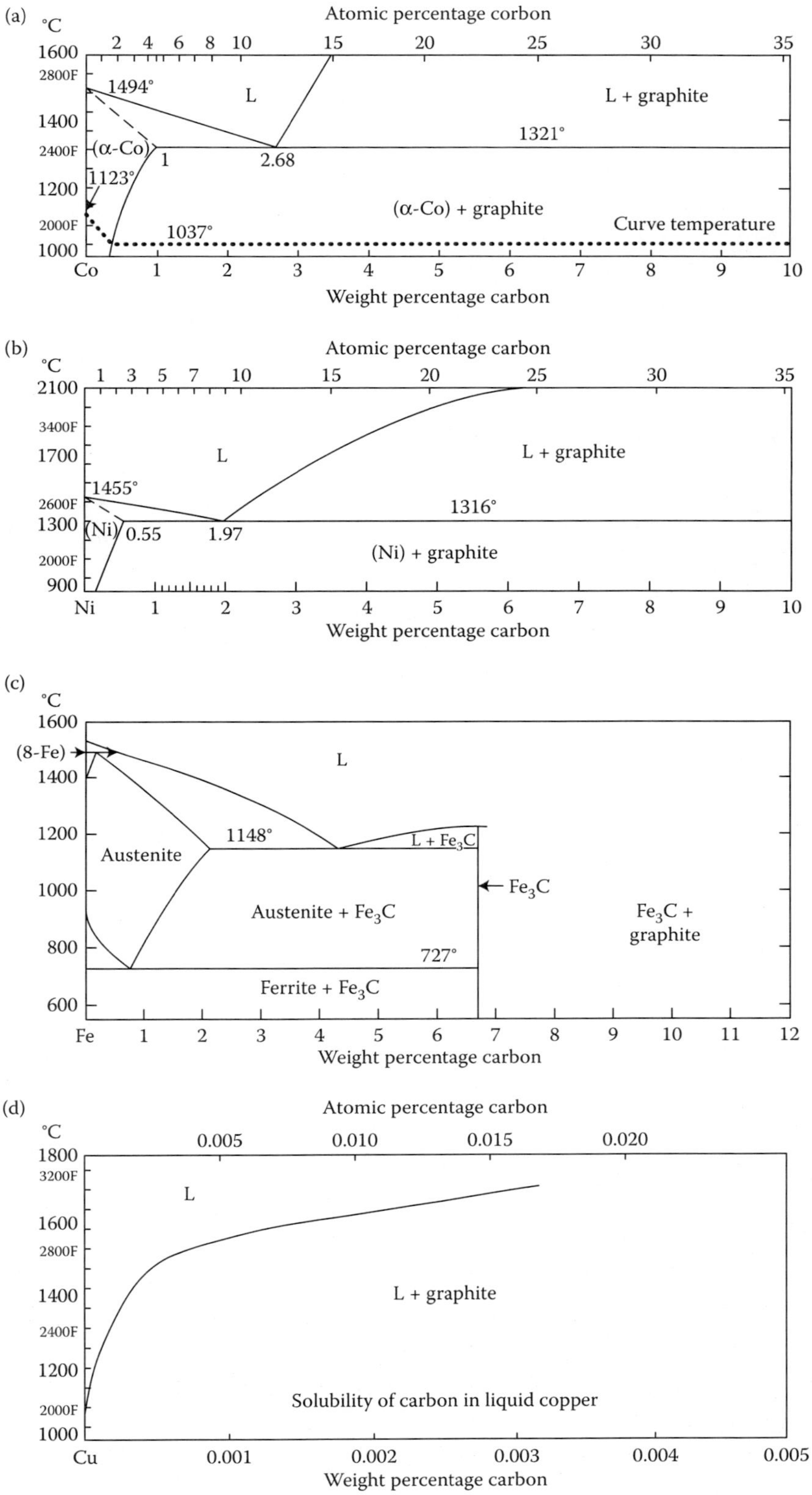

FIGURE 37.7 Binary phase diagrams of (a) cobalt–carbon, (b) nickel–carbon, (c) iron–carbon, and (d) copper–carbon. (Reprinted with permission from Hawkins D.T. et al. 1973. *Metal Handbook, 8th Edition, Metallography, Structures and Phase Diagram*, ASM International.)

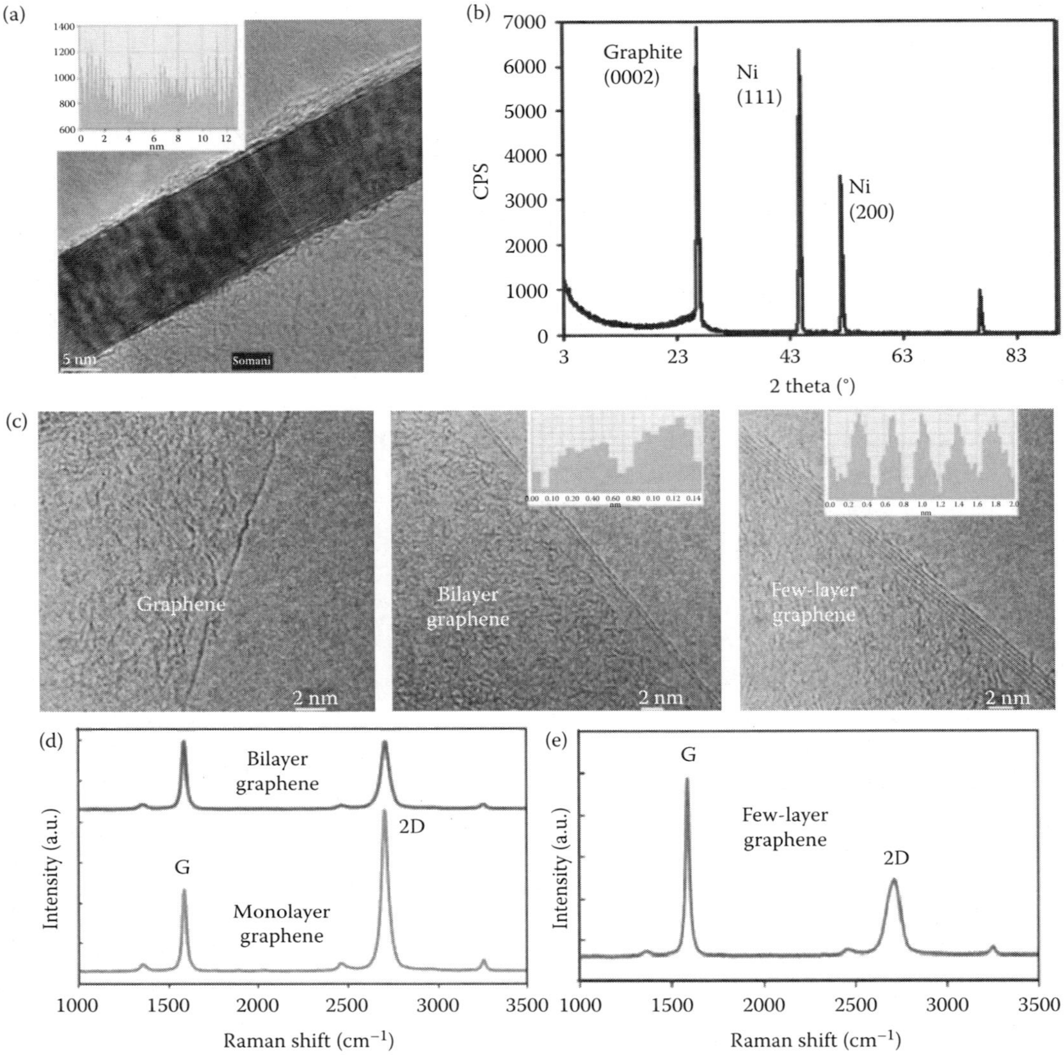

FIGURE 37.8 (a) TEM image of the few-layer graphene. (b) XRD pattern of the graphite film synthesized on Ni foil. ((a, b) Reprinted from Somani, P.R. et al. 2006. *Chem. Phys. Lett.* 430: 56–9. With permission.) (c) TEM images of the single, bilayer, and few-layer graphene synthesized by CVD process on Ni. Raman spectra of the (d) single and bilayer graphene identified at various points and (e) few-layer graphene on Ni. ((c, d, e) Reprinted from Sharma, S. et al. 2013. *Mater. Lett.* 93: 258–62. With permission.)

The most important feature of CVD graphene is the possibility of transferring it to an arbitrary substrate without disturbing the intrinsic properties. The transfer of a CVD graphene can be achieved by wet or dry etching of the catalytic layer and placing on a desirable substrate surface. A significantly investigated Ni catalytic layer can be etched with a diluted acid solution such as HNO_3. However, bubbling with H_2 generation during the etching process can damage the graphene and hinders large-area transfer. To overcome this problem, other metal chlorides and nitrates such as iron (III) chloride ($FeCl_3$) and iron nitride $Fe(NO_3)_3$ are used as an oxidizing etchant to remove the Ni layer, as presented in Figure 37.10b. The redox process slowly etches the Ni layer effectively within a mild pH range without forming any gas and precipitant. Subsequently, the graphene film floated on the solution after separating from the substrate surface and was ready to transfer on the other substrate. Similarly, the dry-transfer process for the graphene film using a soft substrate such as polydimethylsiloxane (PDMS) stamp can be effectively used to transfer on an insulting substrate as demonstrated by Kim et al. [31]. In this approach, the PDMS mold was directly attached to the graphene film on the Ni substrate synthesized by the methane-based CVD technique (Figure 37.10). The Ni substrate can be etched away using $FeCl_3$ as mentioned above, leaving the adhered graphene film on the PDMS. Now, the graphene film on PDMS can be easily stamped on a selected substrate considering the suitability of application (Figure 37.10). The transfer process opened up a huge interest in CVD graphene as it can be applied to various kinds of electronics devices [30–32]. CVD-grown

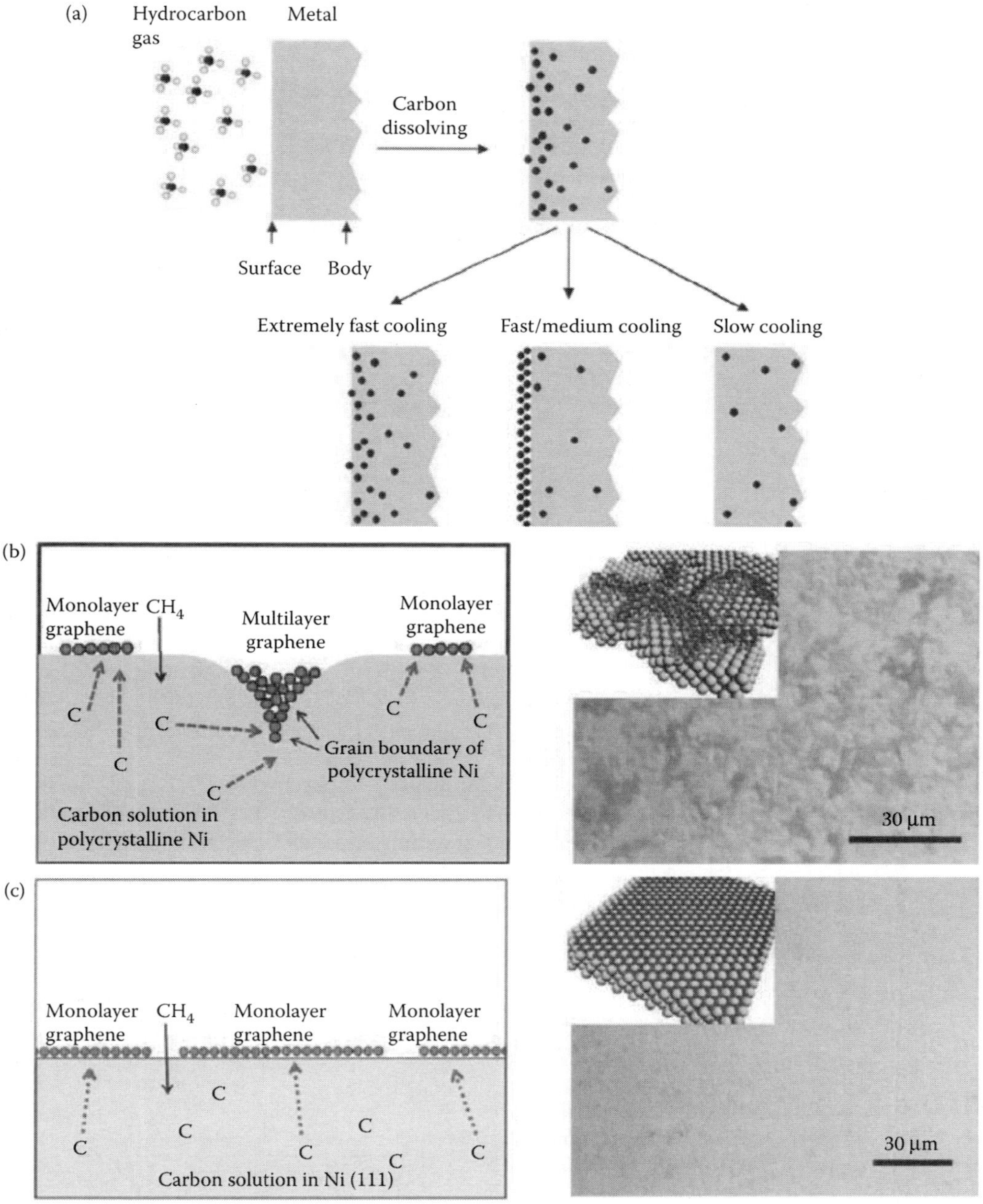

FIGURE 37.9 (a) Schematic diagram of the graphene growth process on the Ni surface. (Reprinted with permission from Yu, Q., Lian, J., S. Siriponglert et al. 2008. Graphene segregated on Ni surfaces and transferred to insulators. *Appl. Phys. Lett.* 93: 113103.) Graphene growth mechanism (b) on polycrystalline Ni surface and optical image of graphene/Ni surface. (c) Graphene growth on a Ni (111) single crystal and optical image of a graphene/Ni (111) surface after CVD process. (Reprinted from Zhang, Y. et al. 2010. *J. Phys. Chem. Lett.* 1: 3101–7. With permission.)

graphene film can be sufficiently thin to be transparent over the relevant range of wavelengths. The transparency of the graphene film on glass or a plastic substrate can be tuned by controlling the number of layers, similar to the above discussions. A graphene film transferred to an arbitrary substrate shows excellent adhesion with strong van der Waals interaction at the interface (Figure 37.11a). Rather than using the PDMS stamp, a polymer layer can be coated on the graphene film for the transfer process. In a standard process, a PMMA layer is coated by spin or drop casting to hold up the graphene film. The PMMA-coated graphene film can be transferred to any arbitrary substrate for several applications. The PMMA layer on graphene can be removed by dissolving in acetone and other organic solvents. The transferred graphene films have been used for transparent conductors for touch panel display electronics, organic solar cells, photodiodes, field-effect transistors (FET), etc. Reina et al. showed a highly transparent (~90%) graphene as well as the fabrication of

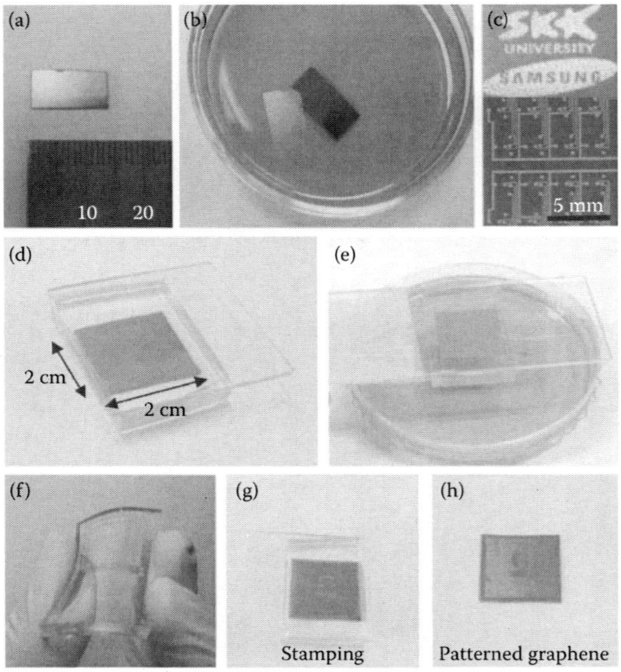

FIGURE 37.10 (a) CVD graphene growth on Ni(300 nm)/SiO_2(300 nm)/Si substrate. (b) Floated graphene film after etching the Ni layer. (c) Demonstration of patterned graphene growth on top of patterned Ni layer. (d), (e) Etching of Ni layer for dry-transfer process using PDMS stamp. (f), (g), and (h) PDMS stamping and peeling process. (Reprinted from Kim, K.S. et al. 2009. *Nature* 457: 706–10. With permission.)

FET devices (Figure 37.11). A Si back-gated device fabricated with 300 nm SiO_2 layer showed the characteristic ambipolar nature of graphene-based devices. The mobility calculated from the conductivity and gate voltage characteristic was in the range of 100–2000 cm^2/V · s [30]. This is an explicit demonstration of CVD-derived graphene for the fabrication of high-performance transistors similar to mechanically exfoliated graphene. The electrical properties of the graphene film strongly depend on the quality and number of layers. The sheet resistance of graphene film for application as transparent conductor is the critical factor. The graphene-constructed carbon with random graphene structure shows high sheet resistance, resulting in poorer device performance than that of the ITO electrode. High-quality ordered graphene film grown on a catalytic metal substrate shows significantly lower sheet resistance. The intrinsic mobility of a graphene film grown on a Ni substrate has been achieved as high as 3750 cm^2/V · s. The mobility of the presented graphene film by Kim et al. is higher than many other reports for graphene grown on a Ni substrate [31].

In Section 37.4, we discussed the fabrication of solar cells on directly obtained graphene-constructed film; however, the device performance was significantly influenced by the high sheet resistance of graphene film. Here, we explain the possibility of fabricating flexible organic solar cells using high-quality transferred graphene film. Arco et al. demonstrated such an approach as shown in Figure 37.12a [75]. CVD graphene film was transferred onto a plastic substrate and organic layers were deposited with conventional device structure for flexible

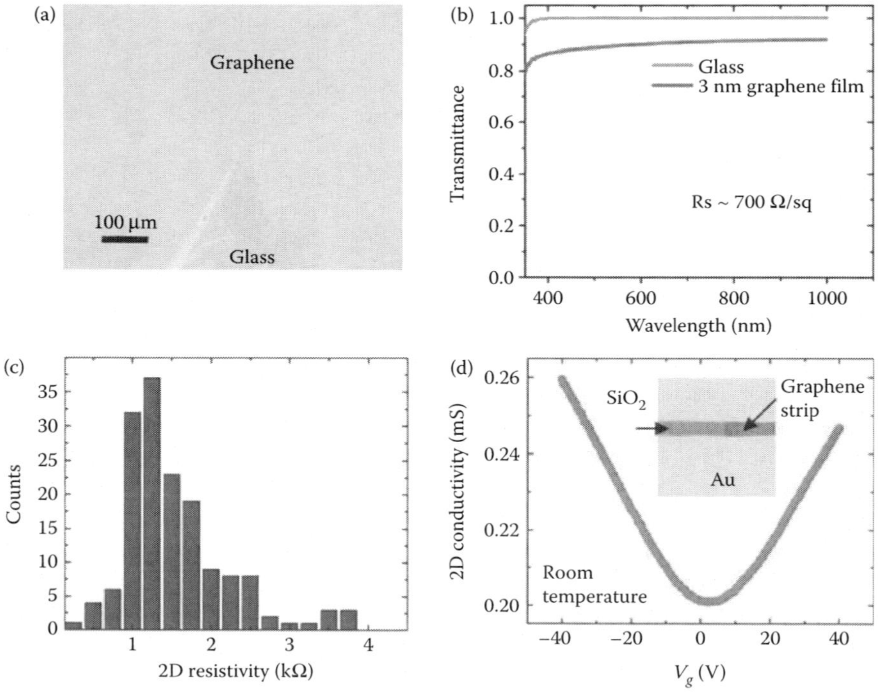

FIGURE 37.11 (a) Optical image of CVD graphene transferred to a glass substrate. (b) Transmittance of the graphene film (thickness ~3 nm). (c) 2D resistivity histogram (kΩ) for 100 graphene devices. (d) 2D conductivity versus gate voltage of a graphene transistor (inset shows optical image of a graphene strip for the device). (Reprinted with permission from Reina, A. et al. 2009. Large area, few-layer graphene films on arbitrary substrates by chemical vapor deposition. *Nano Lett.* 9, 30–5. Copyright 2009 American Chemical Society.)

solar cell fabrication. Figure 37.12b presents a schematic of the graphene transfer process for CVD graphene films, which is also discussed elaborately. The important factor in organic solar cells is the surface roughness of the electrode materials, as the organic layers are extremely thin (<100 nm). The higher roughness may lead to shorting of the device and high leakage current and thereby poor solar cell efficiency. These studies have also revealed that the transferred CVD graphene film on a plastic substrate has almost comparable surface roughness to that of ITO [75]. Performing the bending experiment, it has been observed that the conductance of the graphene/PET film remained almost unchanged. Interestingly, there has been a steady decrease in the conductance of the ITO electrode with bending cycle. This has been explained with the appearance of cracks in the ITO film, which significantly influence the conductivity [75–77]. The J–V characteristics of a typical photovoltaic device fabricated on CVD graphene compared to a device fabricated on a normal ITO anode has been investigated [66]. Figure 37.12d presents the J–V characteristics for the CVD graphene and ITO-based solar cells. The two fabricated devices showed almost similar device characteristics with identical V_{oc} as observed for graphene-constructed film

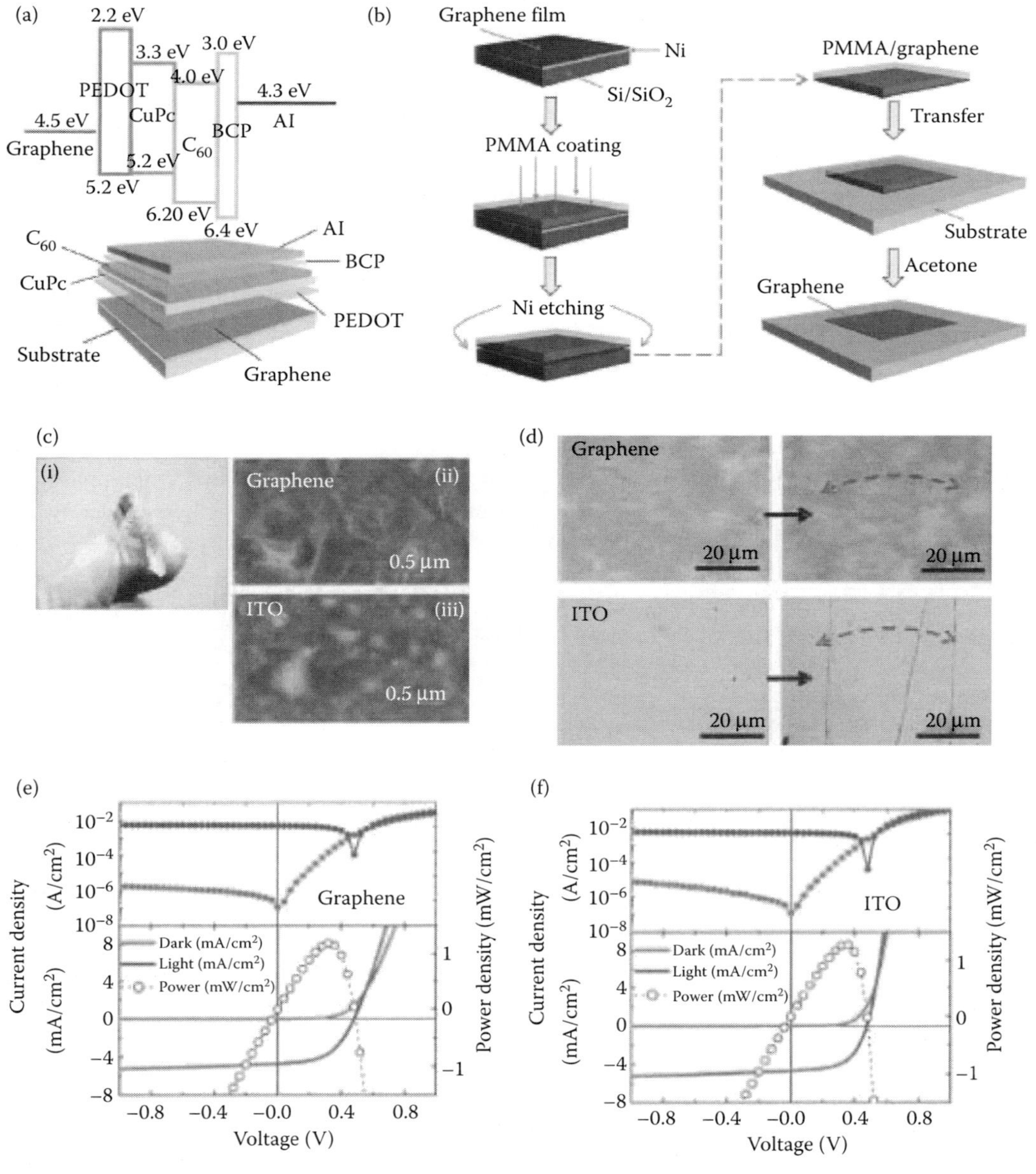

FIGURE 37.12 (a) Schematic presentation of heterojunction organic solar cell fabricated on CVD graphene and corresponding energy level diagram. (b) Graphene transfer process for solar cell fabrication. (c) Atomic force microscopic (AFM) images of flexible CVD graphene surface compared with ITO. (d) Optical images of CVD graphene (top) and ITO (bottom) films on PET before and after bending test. Characteristics of the PEDOT/CuPc/C_{60}/BCP/Al solar cell fabricated on (e) graphene and (f) ITO electrodes. (Reprinted with permission from Arco de, L.G. et al. 2010. Continuous, highly flexible, and transparent graphene films by chemical vapor deposition for organic photovoltaics. *ACS Nano* 4, 2865–73. Copyright 2010 American Chemical Society.)

[43,44]. The device performance and conversion efficiency of the two devices fabricated with CVD graphene and conventional ITO electrode were almost comparable [75]. The efficiency of CVD graphene-based solar cell performance can be enhanced much further with reduced sheet resistance and contact resistance with the deposited polymer layers [78,79]. The greater flexibility of a graphene transparent electrode and suitable transparency for a wide range of wavelength than that of an ITO can be ideal for solar cells and other device applications.

37.7 GRAPHENE ON COPPER BY CVD

The synthesis of graphene on a Cu surface by the CVD process has been significantly explored as it gives a much better opportunity to grow large-area monolayer graphene. Graphene growth on a Cu surface occurs through absorption of carbon atoms with the decomposition of various gaseous, liquid, and solid hydrocarbons at an elevated temperature (~1000°C). The growth of a graphene-like structure on a Cu surface was unintentionally achieved in 1991 [80,81]. However, controlled growth of monolayer graphene on a Cu foil has been demonstrated recently by Li et al. [32]. The growth of a continuous graphene film on the Cu surface has been achieved using CH_4 and H_2 gas mixture at a temperature higher than 1000°C by the CVD technique (Figure 37.13). As the graphene film is significantly thin, the Cu surface steps are clearly visible after the growth process. At the same time, the Cu grain boundaries also appear after the high-temperature growth process, as the grain size is enhanced with annealing. The grain structure of the Cu surface significantly influences the quality of graphene, depending on the growth conditions. Again, formation of wrinkles in the graphene can be observed after the growth process due to the thermal expansion coefficient difference between Cu and graphene. It is well known that Cu has a positive thermal expansion coefficient, whereas graphene has a negative coefficient. Thus, the post-growth cooling process has a significant influence on the formation of wrinkles. Graphene synthesized on the Cu surface can be transferred on an arbitrary substrate by a chemical etching process as discussed in the previous section. The number of graphene layers on the Cu foil can be controlled during the growth process, which provides a much better opportunity for device applications. Further, we discussed the graphene growth mechanism on the Cu surface in a CVD technique in detail to provide a better understanding.

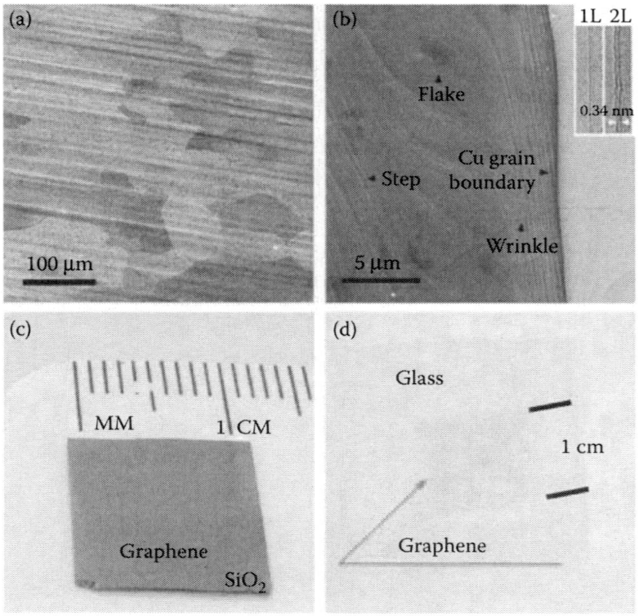

FIGURE 37.13 (a) SEM image of synthesized graphene on Cu foil. (b) High-resolution SEM image showing the Cu grain boundary and steps (inset TEM images of folded graphene edges). Graphene films transferred onto a (c) SiO_2/Si and (d) glass substrate. (Reprinted from Li, X.S. et al. 2009. *Science* 324: 1312–4. With permission.)

The formation of graphene film on the Cu surface cannot be identified by the naked eye, and required spectroscopy and microscopy analysis techniques. The synthesized graphene on the Cu surface can be identified with Raman and TEM studies alike on the Ni surface [31]. Raman spectra of graphene synthesized on a copper foil present graphitic G and second-order 2D peaks. The layer number can be identified with the intensity ratio of 2D and G peaks (Figure 37.14). It has been well established that the Raman 2D-to-G peak intensity ratio >3 corresponds to a monolayer graphene. The 2D peak intensity decreases with an increase in layer numbers. The number of graphene layers can also be identified by the high-resolution transmission electron microscopy (HRTEM) study at the folded edge of a graphene sheet [82]. In Figure 37.14, monolayer, bilayer, and trilayer graphene sheets synthesized on the Cu foil by the CVD synthesis process are shown. Graphene growth by the CVD technique has been achieved using various solid carbon source materials [83,84]. There are several aspects of controlled graphene growth on Cu, depending on precursor materials, gas composition, and pressure (atmospheric and low pressure), which determine the quality of graphene.

As mentioned above, gas composition plays a significant role in the graphene growth process. It has been observed that hydrogen has a significant role in graphene growth kinetics on the Cu substrate [85,86]. In the growth process, hydrocarbons chemisorb on the Cu surface to form active carbon species, such as (CH_3)s, (CH_2)s, (CH)s, and Cs. However, experiments and density functional theory (DFT) calculation studies show that the dehydrogenation process is not thermodynamically favorable on the Cu surface [86–88]. In this aspect, H_2 can play the catalytic role in the activation of physisorbed hydrocarbons. The catalytic role of H_2 in activating carbon is illustrated by reactions 1 and 4 as shown in Scheme 37.1. Hydrogen can more readily dissociate on the Cu surface and form active hydrogen atoms (1) [86,89,90]. These hydrogen atoms can promote the activation of physisorbed hydrocarbons as presented in reaction 4, and lead to the formation of the (CH_3)s radical. Subsequently, dehydrogenation leads to the formation of more active surface-bound (CH_2)s and (CH)

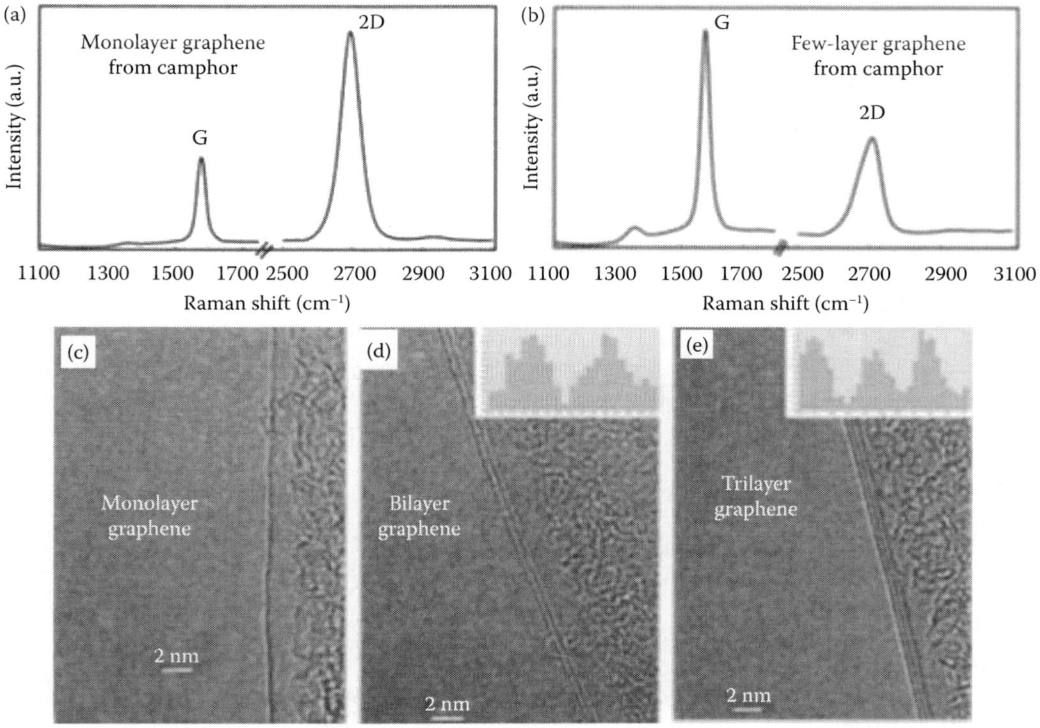

FIGURE 37.14 Raman spectra of (a) monolayer and (b) few-layer graphene synthesized from solid carbon source (camphor) on Cu foil. TEM images of (c) monolayer, (d) bilayer, and (e) trilayer graphene. (Reprinted from Kalita, G., Wakita, K., and M. Umeno. 2011. *Physica E* 43: 1490–3. With permission.)

$$Cu + H_2 \Leftrightarrow 2H_s \quad (1)$$

$$Cu + CH_4 \rightarrow (CH_3)_s + H_s - \text{slow} \quad (2)$$

$$Cu + CH_4 \Leftrightarrow (CH_4)_s \quad (3)$$

$$(CH_4)_s + H_s \Leftrightarrow (CH_3)_s + H_2 \quad (4)$$

$$(CH_4)_s + \text{graphene} \Leftrightarrow (\text{graphene} + C) + H_2 \quad (5)$$

$$H_s + \text{graphene} \Leftrightarrow (\text{graphene} - C) + (CH_x)_s \quad (6)$$

SCHEME 37.1 Reaction process of hydrogen dissociation and carbon radical formation for nucleation and growth of graphene on Cu substrate. (Reprinted with permission from Vlassiouk, I. et al. 2011. Role of hydrogen in chemical vapor deposition growth of large single-crystal graphene. *ACS Nano* 5: 6069–76. Copyright 2011 American Chemical Society.)

s species, which enable nucleation and growth of graphene on the Cu surface [86].

Graphene growth kinetics on the Cu substrate by a mixture of hydrocarbon (CH_4) and H_2 in comparison to the growth on the Ni substrate has been explained by Losurdo et al. [85]. As discussed above, the hydrocarbon is absorbed (1) in the surface of Cu and subsequent dehydrogenation occurs through the H_2 reactions (2). The dehydrogenation and radical formation on the Cu surface initiate graphene growth. On the other hand, it is important to note that dehydrogenation on Cu is an endothermic reaction, whereas metals such as Ni, Pd, and Ru show exothermic reaction [85,91,92]. The well-understood solubility of carbon in Cu and Ni define the atomic carbon diffusion and graphene growth process.

Bae et al. demonstrated the roll-to-roll synthesis and transfer process of large-area graphene film for touch panel application [93]. In the developed CVD process, a 30-inch large Cu foil is used in an 8-inch quartz tube reactor. Such a large graphene film was transferred without deformation to a plastic substrate using a thermal release tape (Figure 37.15b). Putting the thermal release tape on the graphene film, basal Cu foil was etched out as discussed in the previous sections. Subsequently, the graphene with thermal release tape was inserted to a roller with the target substrate and the graphene was realized from the tape. The developed technique has the possibility of the roll-to-roll transfer of a CVD graphene to an arbitrary substrate, which can enable continuous large-scale production. Figure 37.15d through f showed the fabrication process of graphene touch screen for touch panel device application [93]. This shows the suitability of graphene-based material for display and touch panel device application replacing oxide-based transparent conductors. The electrical properties of synthesized large-area graphene have also been demonstrated by the layer-by-layer transfer process. The sheet resistance of a graphene film with a transparency of ~97% has been achieved as ~125 Ω/Sq. Again, stacking the graphene layer-by-layer, the sheet resistance can be reduced much further (Figure 37.16). Electrical properties of such graphene conductors can be further enhanced with

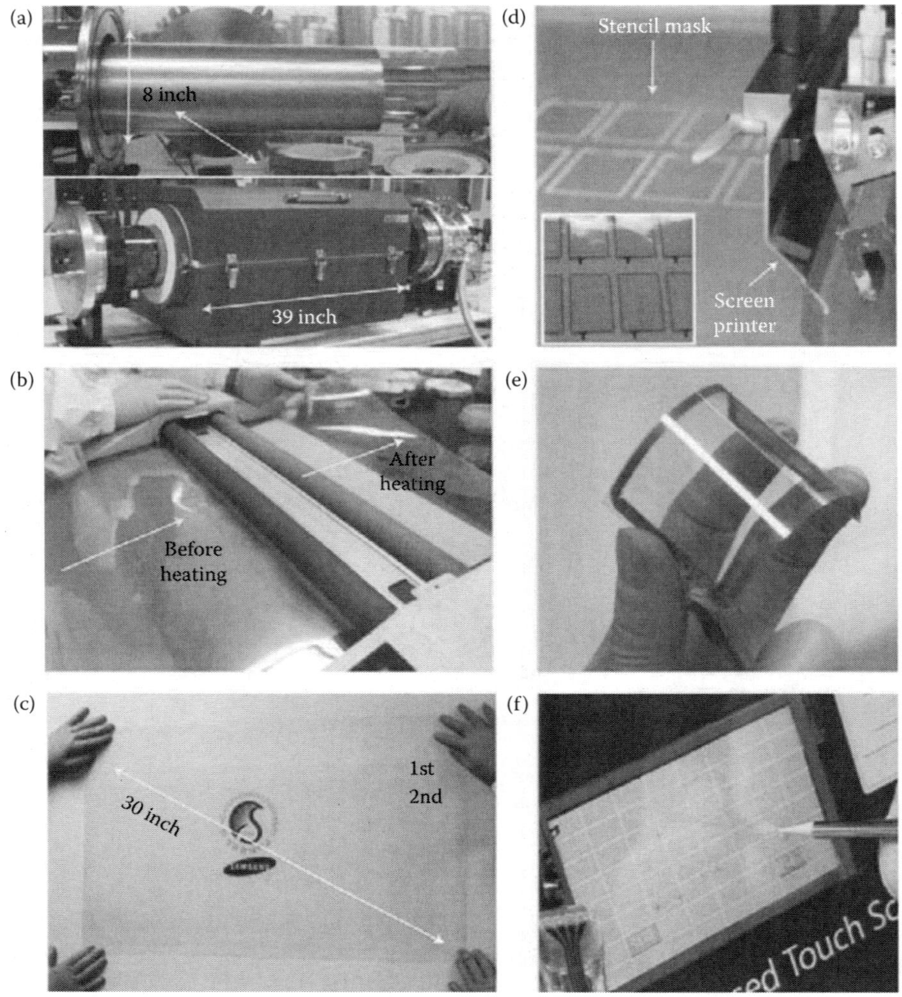

FIGURE 37.15 Photographs of roll-to-roll graphene transparent electrode fabrication process. (a) Large CVD system for graphene synthesis. (b) Roll-to-roll transfer of graphene film with a thermal release tape. (c) Large transparent graphene film on a 35-inch PET sheet. (d) Screen printing of silver paste electrodes on graphene/PET film for device fabrication. (e) Flexible assembled graphene/PET touch panel. (f) Demonstration of a graphene-based touch screen panel application. (Reprinted from Bae, S., Kim, H.K., and Y. Lee. 2010. *Nat. Nanotechnol.* 5: 574–8. With permission.)

chemical doping (such as HNO_3, $AuCl_3$, $FeCl_3$ treatments) [93,94]. A Hall effect mobility as high as $7350\ cm^2/V \cdot s$ has been obtained for synthesized graphene film, which is comparable to exfoliated graphene. A touch screen has been developed with the transferred graphene film, which has been integrated for device fabrication. The fabricated graphene-based touch panel showed much better stability under higher strain than that of an ITO-based electrode [95]. This is quite important point in achieving the full potential of a graphene-based electrode as an alternative to conventional oxide-based electrode material.

The synthesis of large-area and high-quality graphene on polycrystalline Cu foil or substrate has been achieved by several research groups. However, the Cu grain boundaries and small size of graphene domain significantly influence the overall physical properties of a CVD synthesized graphene [96–101]. One of the solutions to overcome the problem of smaller domain structures and grain boundaries is to grow graphene on a single-crystal metal substrate [102,103]. The growth of a single-crystal metal layer on sapphire or other single-crystal substrate significantly enhances the cost of the growth process and may be not possible for large-area growth. With this in view, graphene growth on polycrystalline Cu foil has been investigated significantly to enhance the domain structure [97,104,105]. Enhanced domain structure of graphene has been obtained with an increase in grain size of the Cu foil by annealing [104]. Growth of hexagonal graphene crystals across the grain, grain boundaries, and twin boundaries of polycrystalline Cu foil have also been observed [42,106–114]. Figure 37.17a presents the hexagonal-shaped large graphene domain synthesized by CVD on polycrystalline Cu foil as demonstrated by Yan et al. [94]. Microscopy analysis has confirmed the formation of homogeneous monolayer graphene with hexagonal edges (Figure 37.17b and c).

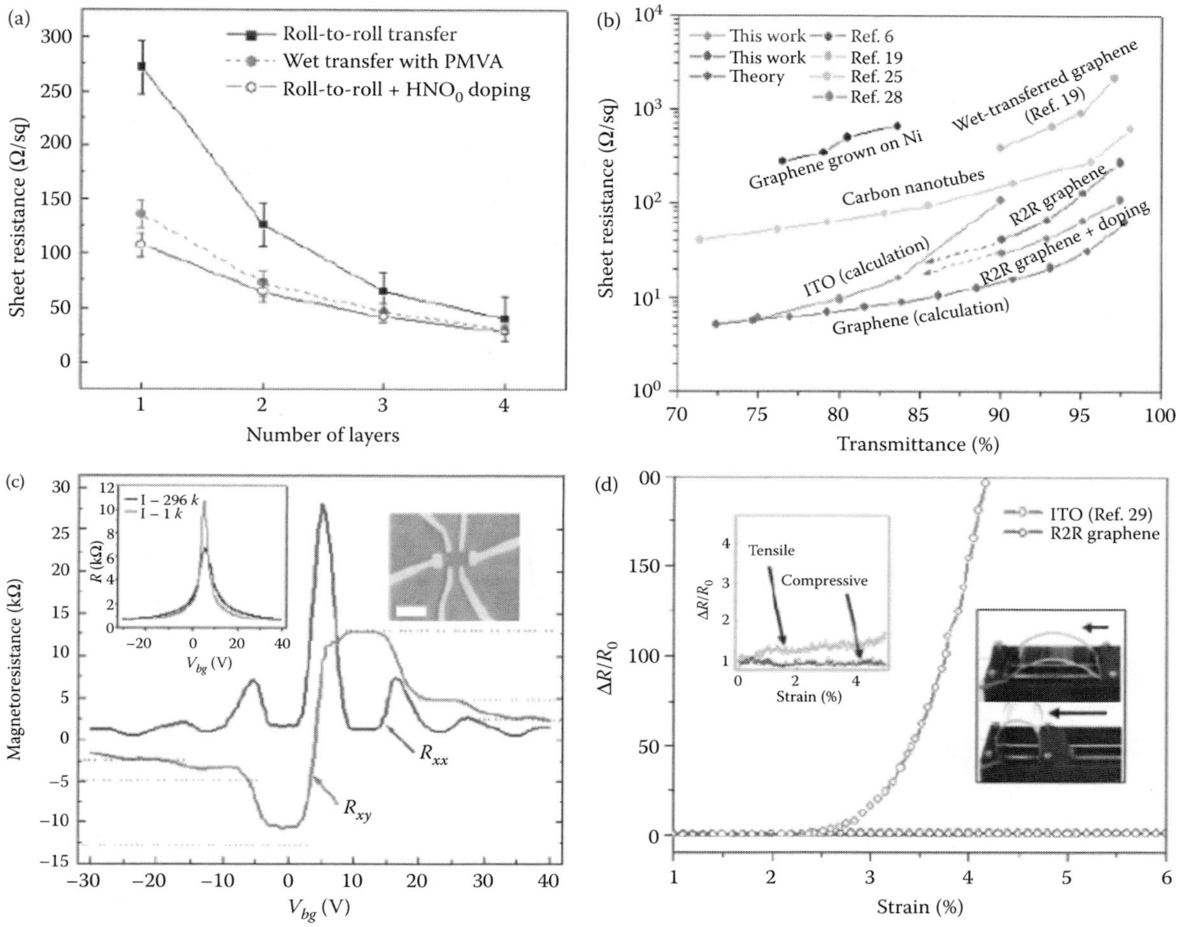

FIGURE 37.16 Electrical characterizations of roll-to-roll transferred graphene films. (a) Sheet resistances of graphene films with different number of layers. (b) Sheet resistance versus transmittance of the transferred graphene comparing various references. (c) Electrical properties of a monolayer graphene as investigated by a Hall bar device. (d) Electromechanical properties of graphene-based touch screen device compared with ITO/PET electrode. (Reprinted from Bae, S., Kim, H.K., and Y. Lee. 2010. *Nat. Nanotechnol.* 5: 574–8. With permission.)

Similarly, Raman spectra measured at different points and mapping analysis shows uniform monolayer graphene. The electrical characteristic of a hexagonal graphene crystal has been investigated with a back-gated Hall bar FET device [109]. A maximum charge carrier mobility of 11,000 $cm^2/V \cdot s$ has been obtained for individual graphene crystals [109]. The overall performance of a FET device fabricated with large hexagonal CVD graphene showed much better characteristics than that of many other CVD graphene. At present, much effort has been made to synthesize such high-quality large single-crystal graphene on Cu foil. The important aspect of achieving large single crystals on the Cu foil is controlling the nucleation density and growth conditions. The nucleation points on Cu foil have to be controlled preciously such that the lateral growth of graphene is unperturbed. Achieving synthesis of large-area single-crystal graphene has a lot of significance in improving the conductivity of large-area transparent conductor as well as achieving high-performance devices with CVD graphene.

37.8 SUMMARY

In the very outset of the chapter, we gave a brief introduction of 2D graphene structure and properties. Derivation or synthesis of graphene by various processes were introduced. Among the various synthesis processes, the CVD approach has several advantages to achieve the controlled growth of high-quality and large-area graphene for electronic device applications. Graphene-constructed carbon film on an arbitrary substrate using organic molecules has been synthesized by the CVD process. With high transparency and comparable sheet resistance, graphene-constructed films have been explored as an electrode for the fabrication of organic solar cells. Then, we discussed the control of graphene synthesis on transition metal substrates. The growth of graphitic material on a Ni surface has been known for a long time. However, recent developments in the controlled growth of graphene layer number and their transfer process to an arbitrary substrate enabled application in various electronic devices. The growth of graphene on

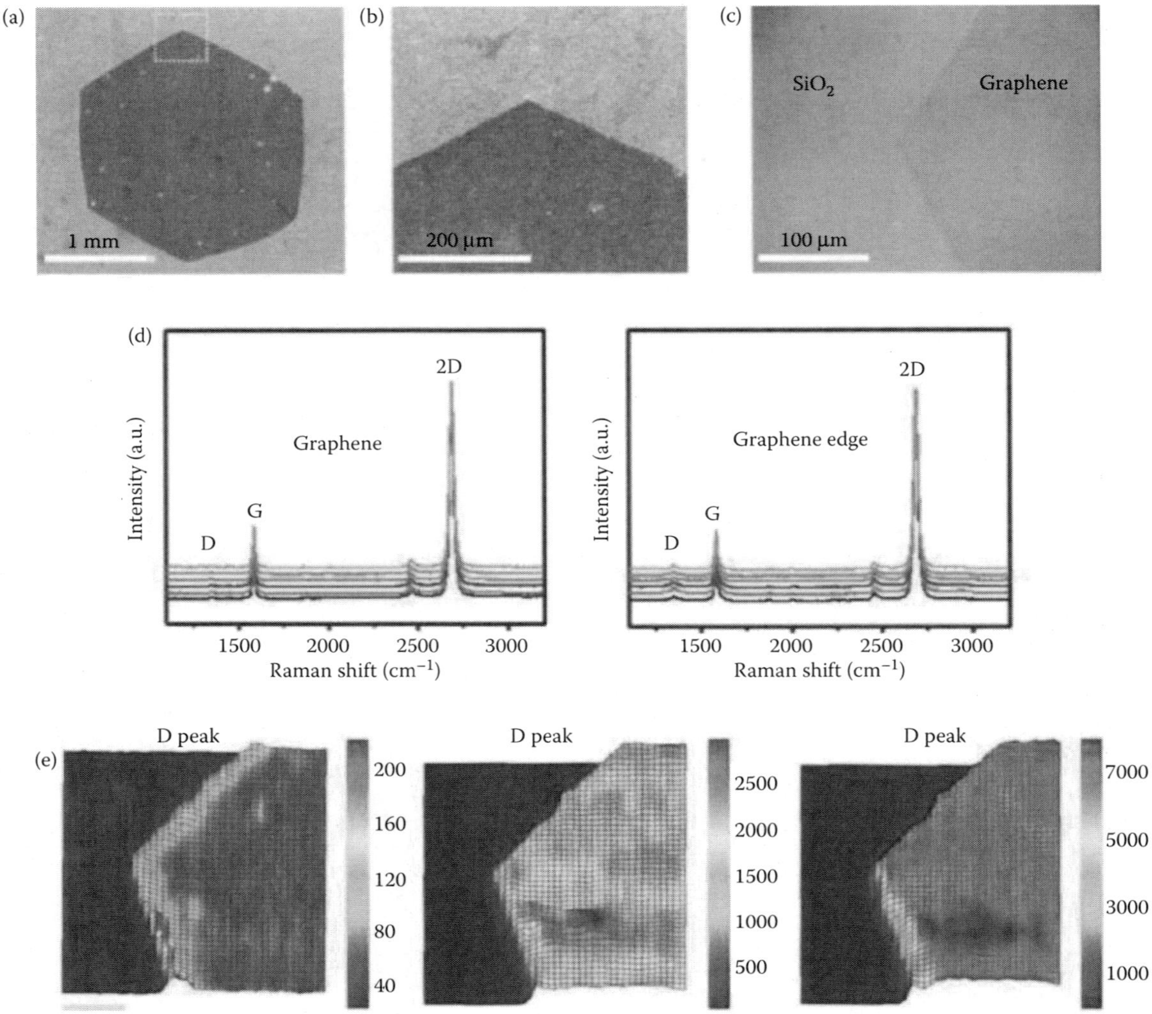

FIGURE 37.17 (a) and (b) SEM images of large hexagonal graphene after transferring to a SiO_2/Si substrate. (c) Optical image at the edge of the graphene crystal. (d) Raman spectra of the graphene domain. (e) Three-dimensional Raman mapping of the D, G, and 2D peak (scale bar 5 μm). (Reprinted with permission from Yan, Z. et al. 2012. Toward the synthesis of wafer-scale single-crystal graphene on copper foils. *ACS Nano* 6: 9110–7. Copyright 2012 American Chemical Society.)

a Ni surface with high carbon dissolution and their subsequent segregation and precipitation has been discussed in detail. The high carbon solubility of the Ni substrate makes it difficult to control the number of graphene layers. Therefore, Cu is an ideal transition metal substrate for graphene growth with low carbon solubility. The graphene growth process on Cu differs significantly from that on Ni, as graphene growth on Cu occurs through surface adsorption of carbon radicals. The graphene nucleation and growth process on Cu foil with hydrogen reaction are discussed thoroughly. Graphene domains nucleate on the Cu surface, and with enhanced lateral growth, the domain constitutes a continuous film. However, the grain boundaries in polycrystalline Cu foil lead to graphene domain boundaries, which significantly affect the transport properties. To overcome the grain boundary effect in CVD graphene, growth of graphene on single-crystal metal substrate can be ideal. Recent studies show that large single-crystal graphene in the order of few millimeters can be grown on Cu foil by the CVD process. The growth of high-quality large hexagonal single crystals on the Cu foil is discussed in detail. As discussed, the progress of CVD graphene has been significant in the last few years and is a potential candidate for realistic electronic device applications.

REFERENCES

1. Novoselov, K.S., Geim, A.K., S.V. Morozov et al. 2004. Electric field effect in atomically thin carbon films. *Science* 306: 666–9.
2. Peierls, R.E. 1935. Quelques propriétés typiques des corpses solides. *Ann. Inst. H. Poincaré* 5: 177–222.
3. Landau, L.D. 1937. Zur Theorie der Phasenumwandlungen II. *Phys. Z. Sowjetunion* 11: 26–35.
4. Landau, L.D. and E.M. Lifshitz. 1980. *Statistical Physics, Part I.* Pergamon Press: Oxford, Sections 137 and 138.

5. Mermin, N.D. and H. Wagner. 1966. Absence of ferromagnetism or antiferromagnetism in one- or two-dimensional isotropic Heisenberg models. *Phys. Rev. Lett.* 17: 1133–6.
6. Mermin, N.D. 1968. Crystalline order in two dimensions. *Phys. Rev.* 176: 250–4.
7. Novoselov, K.S., Geim, A.K., S.V. Morozov et al. 2005. Two-dimensional gas of massless Dirac fermions in graphene. *Nature* 438: 197–200.
8. Zhang, Y., Tan, Y.W., H.L. Stormer et al. 2005. Experimental observation of the quantum Hall effect and Berry's phase in graphene. *Nature* 438: 201–4.
9. Nair, R.R., Blake, P., A.N. Grigorenko et al. 2008. Fine structure constant defines visual transparency of graphene. *Science* 320: 1308.
10. Lee, C., Wei, X., J.W. Kysar et al. 2008. Measurement of the elastic properties and intrinsic strength. *Science* 321: 385–8.
11. Geim, A.K. and K.S. Novoselov. 2007. The rise of graphene. *Nat. Mater.* 6: 183–91.
12. Berger, C., Song, Z., X. Li et al. 2006. Electronic confinement and coherence in patterned epitaxial graphene. *Science* 312: 1191–6.
13. Balandin, A.A., Ghosh, S., W. Bao et al. 2008. Superior thermal conductivity of single-layer graphene. *Nano Lett.* 8: 902–7.
14. Wikipedia, http://en.wikipedia.org/wiki/Graphite.
15. Kroto, H.W., Heath, J.R., S.C. Obrien et al. 1985. C60: Buckminsterfullerene. *Nature* 318: 162–3.
16. Iijima, S. 1991. Helical microtubules of graphitic carbon. *Nature* 354: 56–8.
17. Martel, R., Schmidt, T., H.R. Shea et al. 1998. Single- and multi-wall carbon nanotube field-effect transistors. *Appl. Phys. Lett.* 73: 2447.
18. Kosynkin, D.V., Higginbotham, A.L., A. Sinitskii et al. 2009. Longitudinal unzipping of carbon nanotubes to form graphene nanoribbons. *Nature* 458: 872–6.
19. Avouris, P. and C. Dimitrakopoulos. 2012. Graphene: Synthesis and applications—Review article. *Mater. Today* 15: 86–97.
20. Wallace, P.R. 1947. The band theory of graphite. *Phys. Rev.* 71: 622.
21. Morozov, S.V., Novoselov, K.S., M.I. Katsnelson et al. 2008. Giant intrinsic carrier mobilities in graphene and its bilayer. *Phys. Rev. Lett.* 100: 016602.
22. Geim, A.K. 2009. Graphene: Status and prospects. *Science* 324: 1530–4.
23. Loh, K.P., Bao, Q., G. Eda et al. 2010. Graphene oxide as a chemically tunable platform for optical applications. *Nat. Chem.* 2: 1015–24.
24. Novoselov, K.S., Jiang, D., F. Schedin et al. 2005. Two-dimensional atomic crystals. *Proc. Natl. Acad. Sci.* 102: 10451–3.
25. Stankovich, S., Dikin, D.A., G.H.B. Dommett et al. 2006. Graphene-based composite materials. *Nature* 442: 282–6.
26. Berger, C., Song, Z., T. Li et al. 2004. Ultrathin epitaxial graphite: 2D electron gas properties and a route toward graphene-based nanoelectronics. *J. Phys. Chem. B* 108: 19912–6.
27. Somani, P.R., Somani, S.P. and M. Umeno. 2006. Planar nano-graphenes from camphor by CVD. *Chem. Phys. Lett.* 430: 56–9.
28. Yu, Q., Lian, J., S. Siriponglert et al. 2008. Graphene segregated on Ni surfaces and transferred to insulators. *Appl. Phys. Lett.* 93: 113103.
29. Kalita, G., Masahiro, M., H. Uchida et al. 2010. Few layers of graphene as transparent electrode from botanical derivative camphor. *Mater. Lett.* 64: 2180–3.
30. Reina, A., Jia, X., H. John et al. 2009. Large area, few-layer graphene films on arbitrary substrates by chemical vapor deposition. *Nano Lett.* 9: 30–5.
31. Kim, K.S., Zhao, Y., H. Jang et al. 2009. Large-scale pattern growth of graphene films for stretchable transparent electrodes. *Nature* 457: 706–10.
32. Li, X.S., Cai, W.W., J.H. An et al. 2009. Large-area synthesis of high-quality and uniform graphene films on copper foils. *Science* 324: 1312–4.
33. Sun, J., Hannon, J.B., R.M. Tromp et al. 2010. Spatially-resolved structure and electronic properties of graphene on polycrystalline Ni. *ACS Nano* 4: 7073–7.
34. Hummers, W.S. and R.E. Offeman. 1958. Preparation of graphitic oxide. *J. Am. Chem. Soc.* 80: 1339.
35. Kalita, G., Sharma, S., K. Wakita et al. 2013. A photoinduced charge transfer composite of graphene oxide and ferrocene. *Phys. Chem. Chem. Phys.* 15: 1271–4.
36. Gilje, S., Han, S., M. Wang et al. 2007. A chemical route to graphene for device applications. *Nano Lett.* 7: 3394–8.
37. Eda, G., Fanchini, G. and M. Chhowalla. 2008. Large-area ultrathin films of reduced graphene oxide as a transparent and flexible electronic material. *Nat. Nanotechnol.* 3: 270–4.
38. Kalita, G., Wakita, K., M. Umeno et al. 2013. Fabrication and characteristics of solution-processed graphene oxide–silicon heterojunction. *Phys. Stat. Sol. RRL* 7: 340–3.
39. Badami, D.V. 1962. Graphitization of α-silicon carbide. *Nature* 193: 569–70.
40. Bommel van, A.J., Crombeen, J.E. and A. van Tooren. 1975. LEED and Auger electron observations of the SiC(0001) surface. *Surf. Sci.* 48: 463–72.
41. Forbeaux, I., Themlin, J.M., A. Charrier et al. 2000. Solid-state graphitization mechanisms of silicon carbide 6H-S1C polar faces. *Appl. Surf. Sci.* 162: 406–12.
42. Yan, Z., Lin, J., Z. Peng et al. 2012. Toward the synthesis of wafer-scale single-crystal graphene on copper foils. *ACS Nano* 6: 9110–7.
43. Wang, X., Zhi, L., N. Tsao et al. 2008. Transparent carbon films as electrodes in organic solar cells. *Angew. Chem. Int. Ed.* 47: 2990–2.
44. Kalita, G., Matsushima, M., H. Uchida et al. 2010. Graphene constructed carbon thin films as transparent electrodes for solar cell applications. *J. Mater. Chem.* 20: 9713–7.
45. Ferrari, A.C., Meyer, J.C., V. Scardaci et al. 2006. Raman spectrum of graphene and graphene layers. *Phys. Rev. Lett.* 97: 187401.
46. Klar, P., Lidorikis, E., A. Eckmann et al. 2013. Raman scattering efficiency of graphene. *Phys. Rev. B* 87: 205435.
47. Huang, J.H., Fang, J.H., C.C. Liu et al. 2011. Effective work function modulation of graphene/carbon nanotube composite films as transparent cathodes for organic optoelectronics. *ACS Nano* 5: 6262–71.
48. Li, N., Oida, S., G.S. Tulevski et al. 2013. Efficient and bright organic light-emitting diodes on single-layer graphene electrodes. *Nat. Commun.* 4: 2294.
49. Bao, Q. and K.P. Loh. 2012. Graphene photonics, plasmonics, and broadband optoelectronic devices. *ACS Nano* 6: 3677–94.
50. Sun, Z. and H. Chang. 2014. Graphene and graphene-like two-dimensional materials in photodetection: Mechanisms and methodology. *ACS Nano* 8:4133–56.
51. de Abajo, F.J.G. 2014. Graphene plasmonics: Challenges and opportunities. *ACS Photonics* 1: 135–52.
52. Banerjee, B.C., Hirt, T.J. and P.L. Walker. 1961. Pyrolytic carbon formation from carbon suboxide. *Nature* 192: 450–1.
53. Acheson, E.G. 1896. United States Patent, 568323.

54. Arsem, W.C. 1911. Transformation of other forms of carbon into graphite. *Ind. Eng. Chem.* 3: 799–804.
55. Karu, A.E. and M.J. Beer. 1966. Pyrolytic formation of highly crystalline graphite films. *J. Appl. Phys.* 37: 2179.
56. Morgan, A.E. and G.A. Samorjai. 1968. Low energy electron diffraction studies of gas adsorption on the platinum (100) single crystal surface. *Surf. Sci.* 12: 405–25.
57. Hagstrom, S., Lyon, H.B. and G.A. Somorjai. 1965. Surface structures on the clean platinum (100) surface. *Phys. Rev. Lett.* 15: 491–3.
58. Hawkins, D.T. et al. 1973. *Metal Handbook, 8th Edition, Metallography, Structures and Phase Diagram*, ASM International.
59. Himpsel, F.J., Christmann, K., P. Heimann et al. 1982. Adsorbate band dispersions for C on Ru(0001). *Surf. Sci.* 115: L159.
60. Sutter, E., Albrecht, P. and P. Sutter. 2009. Graphene growth on polycrystalline Ru thin films. *Appl. Phys. Lett.* 95: 133109.
61. Kholin, A., Rut'kov, E.V. and A.Y. Tontegode. 1985. *Sov. Phys. Solid State* 27: 155.
62. Coraux, J., N'Diaye, A.T., C. Busse et al. 2008. Structural coherency of graphene on Ir(111). *Nano Lett.* 8: 565–70.
63. Hamilton, J.C. and J.M. Blakely. 1980. Carbon segregation to single crystal surfaces of Pt, Pd and Co. *Surf. Sci.* 91: 199–217.
64. Eizenberg, M. and J.M. Blakely. 1979. Carbon monolayer phase condensation on Ni(111). *Surf. Sci.* 82: 228–36.
65. Obraztsov, A.N., Obraztsova, E.A., A.V. Tyurnina et al. 2007. Chemical vapor deposition of thin graphite films of nanometer thickness. *Carbon* 45: 2017–21.
66. Wang, G., Zhang, M., Y. Zhu et al. 2013. Direct growth of graphene film on germanium substrate. *Sci. Rep.* 3: 2465.
67. Wang, G., Chen, D., Z. Lu et al. 2014. Growth of homogeneous single-layer graphene on Ni-Ge binary substrate. *Appl. Phys. Lett.* 104: 062103.
68. Lee, J.H., Lee, E.K., W.J. Joo et al. 2014. Wafer-scale growth of single-crystal monolayer graphene on reusable hydrogen-terminated germanium. *Science* 344: 286–9.
69. *Metallography, Structures and Phase Diagrams*. 1973. Metals Park, OH: American Society for Metals.
70. Deck, C.P. and K. Vecchio. 2006. Prediction of carbon nanotube growth success by the analysis of carbon–catalyst binary phase diagrams. *Carbon* 44: 267–75.
71. Sharma, S., Kalita, G., R. Hirano et al. 2013. Influence of gas composition on the formation of graphene domain synthesized from camphor. *Mater. Lett.* 93: 258–62.
72. Kalita, G., Wakita, K. and M. Umeno. 2011. Structural analysis and direct imaging of rotational stacking faults in few-layer graphene synthesized from solid botanical precursor. *Jap. J. Appl. Phys.* 50: 070106.
73. Kalita, G., Wakita, K., M. Takahashi et al. 2011. Iodine doping in solid precursor-based CVD growth graphene film. *J. Mater. Chem.* 21: 15209–13.
74. Zhang, Y., Gomez, L., F.N. Ishikawa et al. 2010. Comparison of graphene growth on single-crystalline and polycrystalline Ni by chemical vapor deposition. *J. Phys. Chem. Lett.* 1: 3101–7.
75. Arco de, L.G., Zhang, Y., C.W. Schlenker et al. 2010. Continuous, highly flexible, and transparent graphene films by chemical vapor deposition for organic photovoltaics. *ACS Nano* 4: 2865–73.
76. Tong, S.W. and K.P. Loh. 2014. Graphene properties and application. In Sarin, V.K. (Ed.), *Comprehensive Hard Materials 3: Super Hard Materials*, Burlington, MA: Elsevier, pp. 565–83.
77. Bult, J.B., Crisp, R. and C.L. Perkins. 2013. Role of dopants in long-range charge carrier transport for p-type and n-type graphene transparent conducting thin films. *ACS Nano* 7: 7251–61.
78. Wang, Y., Tong, S.W., X.F. Xu et al. 2011. Interface engineering of layer-by-layer stacked graphene anodes for high-performance organic solar cells. *Adv. Mater.* 23: 1514–8.
79. Liu, Z., Li, J., Z.H. Sun et al. 2012. The application of highly doped single-layer graphene as the top electrodes of semitransparent organic solar cells. *ACS Nano* 6: 810–18.
80. Lee, S.T., Chen, S., G. Braunstein et al. 1991. Heteroepitaxy of carbon on copper by high-temperature ion implantation. *Appl. Phys. Lett.* 59: 785–7.
81. Alstrup, I., Chorkendorff, I. and S. Ullmann. 1992. The interaction of CH_4 at high temperatures with clean and oxygen precovered Cu(100). *Surf. Sci.* 264: 95–102.
82. Kalita, G., Wakita, K. and M. Umeno. 2011. Monolayer graphene from a green solid precursor. *Physica E* 43: 1490–3.
83. Ruan, G., Sun, Z., Z. Peng et al. 2011. Growth of graphene from food, insects, and waste. *ACS Nano* 5: 7601–7.
84. Wang, H., Zhou, Y., D. Wu et al. 2013. Synthesis of boron-doped graphene monolayers using the sole solid feedstock by chemical vapor deposition. *Small* 9: 1316–20.
85. Losurdo, M., Giangregorio, M.M., P. Capezzuto et al. 2011. Graphene CVD growth on copper and nickel: Role of hydrogen in kinetics and structure. *Phys. Chem. Chem. Phys.* 13: 20836–43.
86. Vlassiouk, I., Regmi, M., P. Fulvio et al. 2011. Role of hydrogen in chemical vapor deposition growth of large single-crystal graphene. *ACS Nano* 5: 6069–76.
87. Zhang, W., Wu, P., Z. Li et al. 2011. First-principles thermodynamics of graphene growth on Cu surfaces. *J. Phys. Chem. C* 115: 17782–7.
88. Galea, N.M., Knapp, D. and T. Ziegler. 2007. Density functional theory studies of methane dissociation on anode catalysts in solid-oxide fuel cells: Suggestions for coke reduction *J. Catal.* 247: 20–33.
89. Gelb, A. and M. Cardillo. 1976. Classical trajectory studies of hydrogen dissociation on a Cu(100) surface. *Surf. Sci.* 59: 128–40.
90. Gelb, A. and M. Cardillo. 1977. Classical trajectory study of the dissociation of hydrogen on copper single crystals: II. Cu(100) and Cu(110). *Surf. Sci.* 64: 197–208.
91. Zhang, C.J. and P. Hu. 2002. Methane transformation to carbon and hydrogen on Pd(100): Pathways and energetics from density functional theory calculations. *J. Chem. Phys.* 116: 322.
92. Ciobica, I.M., Frechard, F., R.A. van Senten et al. 2000. A DFT study of transition states for C–H activation on the Ru(0001) surface. *J. Phys. Chem. B* 104: 3364–9.
93. Bae, S., Kim, H.K. and Y. Lee. 2010. Roll-to-roll production of 30-inch graphene films for transparent electrodes. *Nat. Nanotechnol.* 5: 574–8.
94. Geng, H.Z., Kim, K.K., K.P. So et al. 2007. Effect of acid treatment on carbon nanotube-based flexible transparent conducting films. *J. Am. Chem. Soc.* 129: 7758–9.
95. Lee, J.Y., Connor, S.T., Y. Cui et al. 2008. Solution-processed metal nanowire mesh transparent electrodes. *Nano Lett.* 8: 689–92.
96. Li, X.S., Magnuson, C.W., A. Venugopal et al. 2010. Graphene films with large domain size by a two-step chemical vapor deposition process. *Nano Lett.* 10: 4328–34.
97. Yu, Q.K., Jauregui, L.A., W. Wu et al. 2011. Control and characterization of individual grains and grain boundaries in graphene grown by chemical vapour deposition. *Nat. Mater.* 10: 443–9.

98. Mattevi, C., Kim, H. and M. Chhowalla. 2011. A review of chemical vapour deposition of graphene on copper. *J. Mater. Chem.* 21: 3324–34.
99. Schriver, M., Regan, W., W.J. Gannett et al. 2013. Graphene as a long-term metal oxidation barrier: Worse than nothing. *ACS Nano* 7: 5763–8.
100. Hwang, C., Yoo, K., S.J. Kim et al. 2011. Initial stage of graphene growth on a Cu substrate. *J. Phys. Chem. C* 115: 22369–74.
101. Wood, J.D., Schmucker, S.W., A.S. Lyons et al. 2011. Effects of polycrystalline cu substrate on graphene growth by chemical vapor deposition. *Nano Lett.* 11: 4547–54.
102. Ago, H., Ito, Y., N. Mizuta et al. Epitaxial chemical vapor deposition growth of single-layer graphene over cobalt film crystallized on sapphire. *ACS Nano* 4: 7407–14.
103. Sutter, P.W., Flege, J.I. and E.A. Sutter. 2008. Epitaxial graphene on ruthenium. *Nat. Mater.* 7: 406–11.
104. Wang, H., Wang, G.Z., P.F. Bao et al. 2012. Controllable synthesis of submillimeter single-crystal monolayer graphene domains on copper foils by suppressing nucleation. *J. Am. Chem. Soc.* 134: 3627–30.
105. Li, X., Magnuson, C.W., A. Venugopal et al. 2011. Large-area graphene single crystals grown by low-pressure chemical vapor deposition of methane on copper. *J. Am. Chem. Soc.* 133: 2816–9.
106. Robertson, A.W. and J.H. Warner. 2011. Hexagonal single crystal domains of few-layer graphene on copper foils. *Nano Lett.* 11: 1182–9.
107. Wu, W., Jauregui, L.A., Z. Su et al. 2011. Growth of single crystal graphene arrays by locally controlling nucleation on polycrystalline Cu using chemical vapor deposition. *Adv. Mater.* 23: 4898–903.
108. Geng, D., Wu, B., Y. Guo et al. 2012, Uniform hexagonal graphene flakes and films grown on liquid copper surface. *Proc. Natl. Acad. Sci.* 109: 7992–6.
109. Wu, T., Ding, G., H. Shen et al. 2013. Triggering the continuous growth of graphene toward millimeter-sized grains. *Adv. Funct. Mater.* 23: 198–203.
110. Vlassiouk, I., Fulvio, P., H. Meyer et al. 2013. Large scale atmospheric pressure chemical vapor deposition of graphene. *Carbon* 54: 58–67.
111. Gao, L.B., Ren, W.C., H.L. Xu et al. 2012. Repeated growth and bubbling transfer of graphene with millimetre-size single-crystal grains using platinum. *Nat. Commun.* 3: 699.
112. Murdock, A.T., Koos, A., T.B. Britton et al. 2013. Controlling the orientation, edge geometry, and thickness of chemical vapor deposition graphene. *ACS Nano* 7: 1351–9.
113. Sharma, S., Kalita, G., M.E. Ayhan et al. 2013. Synthesis of hexagonal graphene on polycrystalline Cu foil from solid camphor by atmospheric pressure chemical vapor deposition. *J. Mater. Sci.* 48: 7036–41.
114. Sharma, S., Kalita, G., R. Papon et al. 2014. Synthesis of graphene crystals from solid waste plastic by chemical vapor deposition. *Carbon* 72: 66–73.

38 Chemically Converted Graphene Thin Films for Optoelectronic Applications

Farzana A. Chowdhury, Joe Otsuki, and M. Sahabul Alam

CONTENTS

ABSTRACT

In this chapter, we present large-area graphene oxide (GO) thin films as a potential transparent electrode fabricated by using a simple route. GO thin films were prepared by the solution-casting method onto glass substrates, and were rendered conductive either with hydrazine treatment or annealing. Annealed films treated without hydrazine revealed superiority over the others. Surface morphology of these films showed a uniform film texture as seen by scanning electron microscopy and atomic force microscopy, which is wrinkle-free with roughness as low as ~1.4 nm. Desired optical and electrical properties were achieved by thermal reduction at a temperature as low as 170°C without any reagent, which is an important factor in practical application fields. Microstructural perfection is quite evident from the abrupt descent around a specific energy of photons in the transmittance spectrum. In particular, we have demonstrated that device-quality GO thin films can be obtained via a low-cost scalable technique in comparison with other previous works. The nature and extent of the band gap of GO are analyzed. The Urbach energy and different electrical parameters are addressed. The findings with skills to develop such good-quality GO thin films are very significant in view of an optimized use of these films in electrical and photoelectric applications.

38.1 INTRODUCTION

Nowadays, graphene, which is a single nanosheet of hexagonally arrayed carbon atoms, has received immense scientific and technological attention due to its extraordinary properties such as unique optical behavior, combined with its remarkably high electron mobility, high chemical stability, superior mechanical strength, and flexibility [1–19]. The fabrication of thin films with graphene derivatives such as graphene oxide (GO) with smooth surface and controllable optical and electrical properties in a cost-effective route is of great interest in modern nanoscience [20–33]. Particularly with the increasing demand for energy, much research has been devoted to find a simple solution-based route to fabricate highly stable, optically transparent, and electrically conductive thin films as substrates for solar cells [9,23,34–38]. Intense research attention has been paid to the study of GO in thin-film form due to its unique structural, electronic, and electrochemical properties, and cellular imaging and drug delivery applications [39–43]. There is ongoing research to find a more straightforward and cheaper technique for the fabrication of GO thin film to be utilized as a transparent electrode [34,40–45]. The preparation of GO films on various substrates, including mica, SiO_2/Si, and quartz, has been exploited to date via various methods [46–49]. It is well known that GO produced from flake graphite can be easily dispersed in water and, furthermore, the hydrophilic nature of GO allows it to be uniformly placed onto substrates in the form of thin films [50,51]. Several efforts have been made to assemble GO platelets from their water suspensions to large-area films, such as vacuum filtration, spin coating, chemical vapor deposition [7,34,44,45,52,53], etc.

Polymer photovoltaic devices with transparent graphene electrodes were produced by the spin-casting technique. Among them, chemical vapor deposition is scalable for wrinkle-free, large-area graphene film [54–56]. However, it is found that the preparation conditions and procedure are complex and it allows growing GO only on metals [57].

GO single sheet consists of a hexagonal-ring-based carbon network having both sp^2-hybridized carbon atoms and sp^3-hybridized carbons bearing hydroxyl and epoxide functional groups on either side of the sheet, whereas the sheet edges are mostly decorated by carboxyl and carbonyl groups [38,58]. Since GO is a nonstoichiometric compound, the optical and electrical properties depend on its synthesis processing, environmental, and growth conditions. The dispersion of graphene or reduced graphene oxide (RGO) in colloidal suspensions offers favorable pathways of making large-area graphene films [59]. The dispersion of GO in water at different concentrations are presented in Figure 38.1.

We synthesized GO by an "improved" synthetic protocol [60] and prepared large-area (1.0×2.5 cm^2) GO films by simply solution-casting of the aqueous suspension to an appropriate amount onto a clean glass substrate that demonstrated well-adhered films with better uniformity and high chemical and thermal stability. No pretreatment is required to promote adhesion of GO to the substrate. This process allows us to control film thickness by simply varying the concentrations of the GO in the suspension. The nonreduced pristine GO showed insulating behavior due to the presence of intact graphitic regions interspersed with sp^3-hybridized carbons related to hydroxyl and epoxy functional groups. The introduction of such sp^3 components destroys the p-system of the sp^2 framework, which leads to the disruption of the conduction pathway. Therefore, reduction was necessary to make the layers conductive for different optoelectronic applications [34,37]. RGO films have been prepared either chemically or thermally [61–66]. Many different kinds of reducing agents such as hydrazine [61], sodium borohydride [62,63], and hydrohalic acids [64] have been used for this purpose in different studies [9,44,45,61,67,]. These chemicals introduce many difficulties in practice. For example, hydrazine reduction introduces C–N bonds [61], which can act as defect sites in some applications, and sodium borohydride incompletely reduces hydroxyl groups [62]. In addition, these reducing agents are environmentally burdensome because they are highly toxic and hazardous. Although hydrazine is most widely used as a chemical reductant, its toxicity and explosiveness pose a problem.

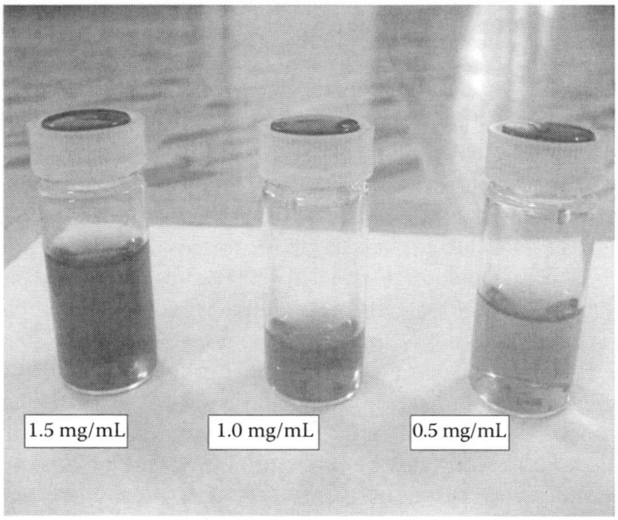

FIGURE 38.1 Photographic image of different concentrations of GO dispersion in water. It is seen that the color of the GO suspension in water changes as the concentration of the solution decreases.

Alternatively, the as-prepared electrically insulating GO films could be turned into RGO by annealing at elevated temperatures, to facilitate the carrier transport more efficiently. The film becomes conductive via a method called "percolation" inside the thin film. Increased percolation through the film causes the conductivity of GO film to be promoted by the removal of oxidized functionalities. As thermal reduction has been performed in the mean time only at high temperatures (400–1100°C) [34,68], a low processing temperature is highly desirable for making it useful for practical use [34]. In the present work, GO was reduced both by hydrazine and heat treatment. Thin films were prepared in both cases at different growth conditions, and optoelectronic properties were extensively studied and compared. Herein, we report a cost-effective practical route to grow device-quality thin films on glass at a low percolation temperature without using a reagent by simply solution-casting that has not been reported so far. Annealing at a temperature as low as 170°C for 1 h was assessed to be the optimum percolation condition in this study, for obtaining GO thin films having required transparency and conductivity. Thus, it paves an economically viable and easier way for the mass production of good-quality GO films for practical applications.

38.2 SYNTHESIS, GROWTH MECHANISM, AND CHARACTERIZATION TECHNIQUES

38.2.1 Materials

Graphite flakes and $KMnO_4$ were purchased from Alfa Aesar and Kanto Chemical, respectively. A polytetrafluoroethylene (PTFE) membrane filter with a 0.45 μm pore size was purchased from Millipore. Polyester hollow fiber (Tetoron, 90 dtex, 38 mm) was purchased from Teijin Fibers.

38.2.2 Synthesis of GO

GO was synthesized with Marcano's improved method [60].

38.2.3 Preparation of GO Thin Films

Material properties changes significantly in a thin film compared to the bulk state. In order to prepare thin films, at first the suspension was sonicated for 20 min by an ultrasonic bath to obtain a uniform solution of GO. Thin films were grown onto chemically and ultrasonically cleaned glass substrates by employing solution-casting of chemically synthesized nonreduced aqueous GO suspension of 0.5, 0.3, 0.15, and 0.075 mg/mL concentrations. A certain amount of GO suspension was first drop casted on the glass and subsequently annealed at elevated temperatures (ranging from 150°C to 200°C, for a few hours). The casting of the solution was carefully controlled to avoid overdropping that might result in thicker films. The hydrazine-treated films did not stick well to the glass and were found to be nonuniform even to the naked eye. The non-hydrazine-treated films resulted in a well-adhered uniform thin film of RGO. Here, the GO thin films were rendered conductive by annealing where thermal reduction was achieved using an oven and keeping the films inside the oven overnight after the completion of the annealing process. The films showed the optimum performance at the annealing condition of 170°C for 1 h.

38.2.4 AFM Measurements

The thicknesses (d) and root mean square (rms) surface roughness of the GO thin films were measured directly by atomic force microscopy (AFM). AFM measurements were performed by using a dynamic force-mode atomic force microscope (SII, SPI3800N-400A) with a silicon cantilever (SI-DF3; spring constant 1.3 N m^{-1}, resonance frequency 26 kHz). The frequency of the cantilever was set higher than the resonance frequency. A voltage of 1.0 V was applied to drive the vibration. The amplitude-damping factor was determined to a default value of the apparatus. Mica and glass substrates were purchased from Nisshin EM Corporation. In order to measure the thickness of GO nanoplatelets, mica substrates were used while glass substrates were used to prepare GO thin films. The mica substrates were cleaved and dried before sample preparation. Glass substrates were cleaned in an ultrasonic bath with DI water, acetone, and IPA for 15 min in each case. A sufficient amount of dispersed GO/H_2O solution was deposited onto the surface and dried.

38.2.5 Optical Characterization

A systematic study of different optical properties of thin films such as transmittance (%*T*), reflectance (%*R*), absorption coefficient (α), band gap (E_g), and refractive index (n) is of paramount importance in assessing the device performance in optoelectronic applications. Knowledge about the variation of these parameters with different growth parameters, such as concentration of the solution, annealing temperature and duration, etc., play a key role in understanding the applicability of the material. A higher transmittance value (~80%–97%) is generally required for GO thin films to be applicable as the transparent electrode in solar cells. Tunability of these optical properties at different growth conditions and thereby, optimizing the parameters is essential to be precisely investigated for suitable device applications.

38.2.5.1 Transmittance and Reflectance

The variations of optical transmittance (T) and absolute specular reflectance (R) of the films with wavelength of light incident on them were measured using a dual-beam UV–Vis–NIR recording spectrophotometer (Shimadzu, UV-3100, Japan) in the photon wavelength range between 250 and 2500 nm. Light signals coming from the samples were detected by an integrating sphere. The overall attachment of the instrument coupled with the MPC-3100 unit is shown in Figure 38.2a [3]. This attachment was composed of three components: (i) MPC-3100 main body, (ii) sample compartment unit, and (iii) accessory mirror unit. This unit leads the optical beams out of the UV-3100 main body into the large-sized sample compartment. The sample beam is reflected by a mirror, passes through the sample space, and then enters the integrating sphere. Prior to measurement, base line correction of the instrument is performed. After confirming the flatness of the baseline, we mount a sample to be measured in the sample beam. The accessory mirror unit incorporates three flat mirrors, M1, M4, and M5 to introduce the light beams to the MPC-3100 unit. Monochromatic beams irradiated into the MPC-3100 are reflected by the mirrors M2 and M3. On the other hand, sample beams were reflected by the mirror M6. The inside diameter of the integrating sphere is 60 mm and the internal surface is coated with a paint of $BaSO_4$. The PbS cell and the photo multiplier detectors were mounted at the upper and the lower sides of the integrating sphere, respectively.

A schematic diagram of the arrangement of the film position for transmittance measurements is shown in Figure 38.2b. The film was facing the beam. The beam that was transmitted through the film entered into the integrating sphere. There, the beam was multiply scattered by the $BaSO_4$ coating without appreciable loss of light and ultimately the PbS and PM detector detected the light. In this way, we measured the transmittance (T%) of light through the film as a function of the wavelength of incident light. A schematic diagram of the arrangement of the film position for reflectance measurements is shown in Figure 38.2c. The sample was placed in the position as shown in the figure. The optical path length and the incident angles to the mirrors remain the same as of those in Figure 38.2a with an additional reflection at the sample surface. Thus, the absolute reflectance, R%, of the sample surface was measured.

38.2.5.2 Absorption Coefficient and Band Gap Energy

The optical absorption coefficient was calculated with the help of a conventional software by using the measured values of transmittance (T%) and reflectance (R%). Expressions for the multiple reflected systems for transmittance (T%) at normal incidence and reflectance (R%) at near-normal incidence of light on the films has been given by Heavens [69]. Tomlin

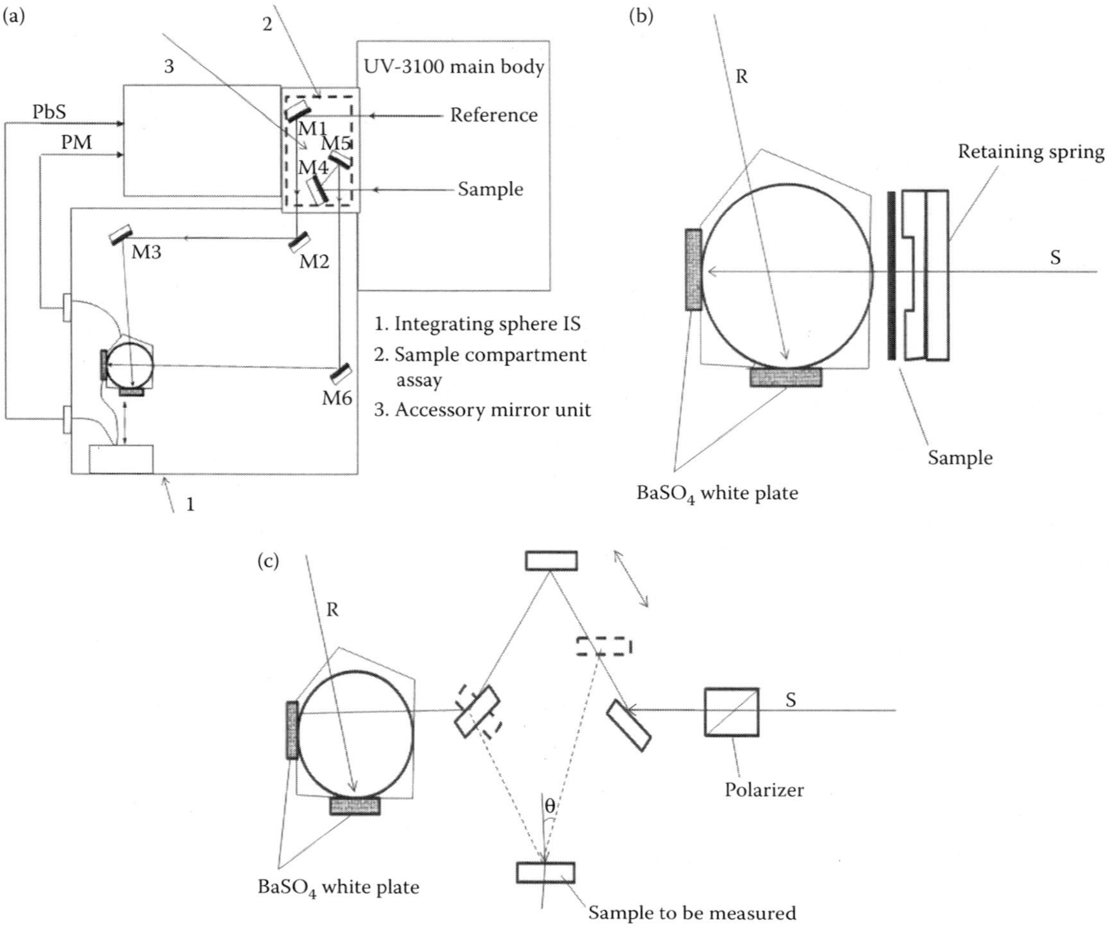

FIGURE 38.2 (a) Schematic diagram of the large-sized sample compartment unit. (b) Schematic diagram to measure the transmittance spectrum of a thin film. (c) Schematic diagram to measure the reflectance spectrum of a thin film.

[70] simplified these expressions for absorbing films on nonabsorbing substrates and expressed as

$$\frac{1-R}{T}=\frac{1}{2n_2(n_1^2+k_1^2)}\times\begin{bmatrix} n_1\{(n_1^2+n_2^2+k_1^2)\sinh 2\alpha_1+2n_1n_2\cosh 2\alpha_1\} \\ +k_1(n_1^2-n_2^2+k_1^2)\sin 2\gamma_1+2n_2k_1\cosh 2\gamma_1\} \end{bmatrix} \quad (38.1)$$

$$\frac{1-R}{T}=\frac{1}{4n_2(n_1^2+k_1^2)}\times\begin{bmatrix} (1+n_1^2+k_1^2)\{(n_1^2+n_2^2+k_1^2)\cosh 2\alpha_1 \\ +2n_1n_2\sinh 2\alpha_1 \\ +(1-n_1^2-k_1^2)\{n_1^2-n_2^2+k_1^2)\cos 2\gamma_1 \\ -2n_2k_1\sinh 2\gamma_1\} \end{bmatrix} \quad (38.2)$$

where n_1 and n_2 are the refractive indices of the film and substrate, respectively, k_1 is the extinction coefficient of the film, $n_2 = 1.45$, $\alpha_1 = (2\pi k_1 d/\lambda)$, and $\gamma_1 = (2\pi n_1 d/\lambda)$, where λ is the wavelength of light and d is the thickness of the film. Equations 38.1 and 38.2 have been solved for k_1 and n_1. The absorption coefficient, α, was then calculated using $\alpha = (4\pi k_1/\lambda)$. The dependence of absorption coefficient (α) on photon energy has been analyzed to find the nature and extent of the band gap energy.

38.3 PHYSICAL PROPERTIES OF GO THIN FILMS

38.3.1 Optical Properties

A large amount of GO was synthesized by oxidizing low-cost graphite powders using strong oxidants (H_2SO_4, H_3PO_4, and $KMnO_4$) according to Marcano's improved method [60]. It has already been demonstrated that Marcano's improved method [60] for producing GO shows significant advantages over other methods [1]. GO nanosheets are chemically modified graphene with hanging hydroxyl, epoxyl, and carboxyl functional groups. These functional groups facilitate GO to be readily dispersed in water. GO was dispersed in water, resulting in a brown colloidal suspension. The individual nanosheets were obtained via exfoliation by a moderate sonication. The exfoliation to achieve GO sheets were confirmed by thickness measurements of a single sheet (~1 nm height

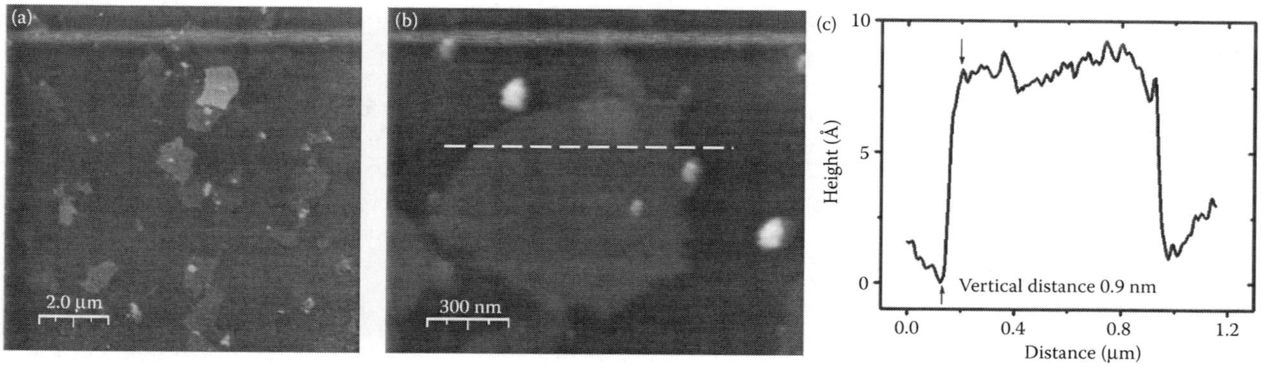

FIGURE 38.3 (a) Typical AFM images of exfoliated GO sheets deposited on a mica substrate. (b) A high-resolution AFM image of a GO sheet in (a); the sheet is ~1 nm thick. (c) 2D line profiles recorded along the dashed line in (b). (Reprinted from *Appl. Surf. Sci.*, 259, Chowdhury, F. et al., Optoelectronic properties of graphene oxide thin film processed by cost effective-route, 460–64, Copyright 2012, with permission from Elsevier.)

on mica substrate) using AFM. The typical AFM images and a two-dimensional (2D) profile are presented in Figure 38.3.

38.3.1.1 Hydrazine-Treated GO Thin Films

Initially, thin films were prepared by depositing a sufficient amount of hydrazine-treated GO aqueous suspension onto clean glass substrates. These films became conducting at a temperature of 350°C for an annealing duration of 1.5 h. The hydrazine-treated GO films exhibit very weak adhesion to the glass substrates. Consequently, on deposition to the surface, these films showed nonuniform and defective textures. A typical optical transmittance (%*T*) spectrum for these films is shown over the photon wavelength ranging between 250 and 2500 nm in Figure 38.4a. The maximum transmittance value is of 26% at 2500 nm of photon wavelength. The transmittance gradually decreases and reaches to almost zero at 250 nm of wavelength. It showed low transmittance that might occur due to the increased scattering of photons by crystal defects. The absorption coefficient (α) of this film is plotted as a function of photon wavelength (λ) in Figure 38.4b.

The hydrazine-treated film has a tendency to absorb light significantly where a steeper absorption region is observed over the entire photon wavelength regime. The optical absorption behavior of GO thin films above the fundamental absorption edge can be interpreted by considering the existence of direct allowed optical transition. The band gap was determined from Tauc's plot [71–75] of $(\alpha h\nu)^2$ versus photon energy ($h\nu$), wherein absorption coefficient varies as a function of frequency, according to $\alpha h\nu \propto (h\nu - E_g)^n$, where $h\nu$ is the photon energy and E_g is the band gap, and $n = 1/2$ for direct transitions, and $n = 2$ for indirect transitions. The variation of $(\alpha h\nu)^2$ with photon energy, $h\nu$ for the GO thin films is shown in Figure 38.4c. It has been observed that the plot of $(\alpha h\nu)^2$ versus $h\nu$ are linear over a wide range of photon energies indicating the direct type of transition. The band gap was determined by extrapolating the linear part of the curve of the incident radiation to intercept the energy axis (at $\alpha = 0$). The linear dependence of this plot indicates that GO film is a direct transition-type semiconductor. The intercepts of these plots on the energy axis give the energy band gaps. The value lies between 2.1 and 2.4 eV, which is consistent with other studies [50]. Poor transmittance and weak adhesion encourages us to concentrate on non-hydrazine-treated thin films.

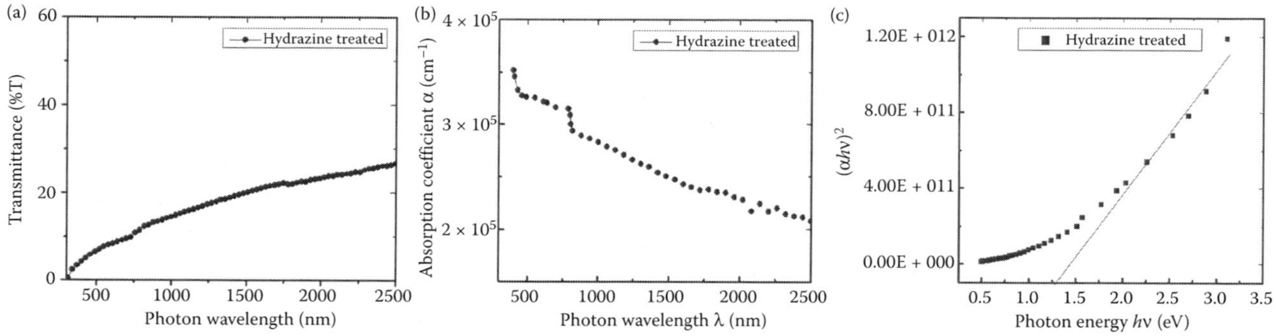

FIGURE 38.4 (a) Transmittance spectrum of GO thin films of 0.5 mg/mL concentration treated with hydrazine. (b) Absorption coefficient as a function of photon wavelength for the same film. (c) $(\alpha h\nu)^2$ versus photon energy ($h\nu$) plot. (With kind permission from Springer Science+Business Media: *Appl. Nanosci.*, Annealing effect on the optoelectronic properties of graphene oxide thin films, 3, 2013, 477–83, Chowdhury, F. et al.)

38.3.1.2 Non-Hydrazine-Treated GO Thin Films

The GO suspension not treated with hydrazine was reduced only by heat treatment at different temperatures. Two GO thin films were prepared with 0.5 mg/mL concentration at 170°C and 195°C annealing temperatures, with 45 min duration for both. In contrast to the hydrazine-treated film, the non-hydrazine-treated annealed films are uniform and more transparent. This is also confirmed by the transmittance spectra of the films as shown in Figure 38.5a. It represents the comparison where the film annealed with 170°C gives higher transmittance than the film annealed with 195°C. The high transparency is normally associated with a good structural homogeneity. Figure 38.5b shows transmittance spectra for two other GO thin films prepared with 0.15 mg/mL concentration; annealed with 160°C, 35 min and 170°C, 1 h duration. It again shows that a film produced at 170°C for 1 h annealing shows better performance than the others. By considering 170°C and 1 h duration of annealing as the optimum condition, another three films were prepared at this annealing condition with 0.3, 0.15, and 0.075 mg/mL concentrations.

Figure 38.6a shows a comparative view of transmittance spectra for GO thin films with 0.5, 0.3, 0.15, and 0.075 mg/mL concentrations for 170°C and a 1 h annealing condition. It is observed that the transmission increases significantly with decreasing solution concentration. The film with the lowest concentration (i.e., 0.075 mg/mL) gives the highest transmittance and at 550 nm the transmittance value is approximately 82%. The transmittance shows almost horizontal regime up to 500 nm, and after that decreases rapidly and reaches down toward zero. The average transmittance is about 82%–94% over a wide range of solar spectrum. Absorption coefficient (α) as a function of photon wavelength (λ) for these four GO films is shown in Figure 38.6b.

It is seen that the direct optical band gap (E_g) has increased from 1.25 to 2.24 eV with the decrease of the concentration of GO solution from 0.5 to 0.075 mg/mL. The band gap energy increases almost linearly with the increase of transmittance (Figure 38.6c). The decrease of band gap may be attributed to the presence of unstructured defects, which increases the density of localized states in the band gap and consequently decreases the energy. However, there is a shift in the absorption band edge. The shift in the absorption edge can be accounted for in terms of the increase in carrier concentration and blocking of low-energy transitions, which causes a Burstein–Moss effect, which enhances the optical band gap. Refractive index (n) versus photon wavelength curve for these GO films is represented in Figure 38.6d. The pick value increases with decreasing concentration and shifts toward higher photon wavelength in an irregular manner. The refractive index value ranges between 1 and 2.39, which is similar to those obtained by other groups [50]. All the curves show the same pattern except the film with 0.075 mg/mL concentration, which gives the highest transmittance. The variation of the refractive index with photon wavelength for this film is similar to the results of Reference 50. Different optical parameters for GO films with different maximum optical transmittance studied in this work are given in Table 38.1.

The comparison of other studies with ours is in the area of low-temperature processing of thin films that gives good quality films for optoelectronic applications. Here, GO was synthesized by a modified protocol that promotes a better percolation in the optoelectronic behavior. The work represents intensive research findings with different thin-film growth conditions, and a comparative view is addressed. Many other studies on GO report different investigations. Ours is based on thin films prepared in various solution concentrations and at different annealing conditions, such as temperature and duration, in view of an optimization of the preparation condition for optoelectronic applications. The findings obtained from this study are as follows:

1. GO was synthesized by an "improved" protocol [60].
2. This process allows us to control film thickness by simply varying the concentrations of GO in the suspension.

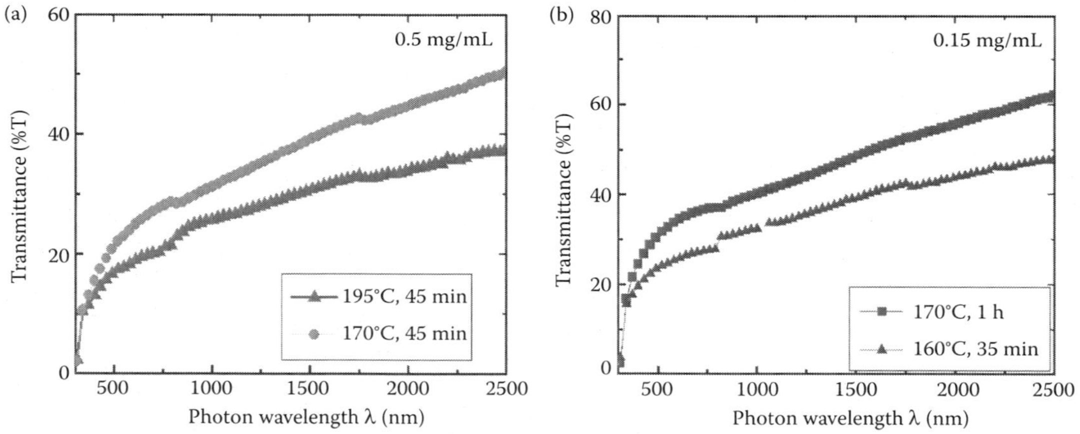

FIGURE 38.5 (a) Transmittance spectra for 0.5 mg/mL concentration of GO thin films at different annealing temperatures. (b) Transmittance spectra for 0.15 mg/mL concentration for GO thin films at different annealing conditions. (With kind permission from Springer Science+Business Media: *Appl. Nanosci.*, Annealing effect on the optoelectronic properties of graphene oxide thin films, 3, 2013, 477–83, Chowdhury, F. et al.)

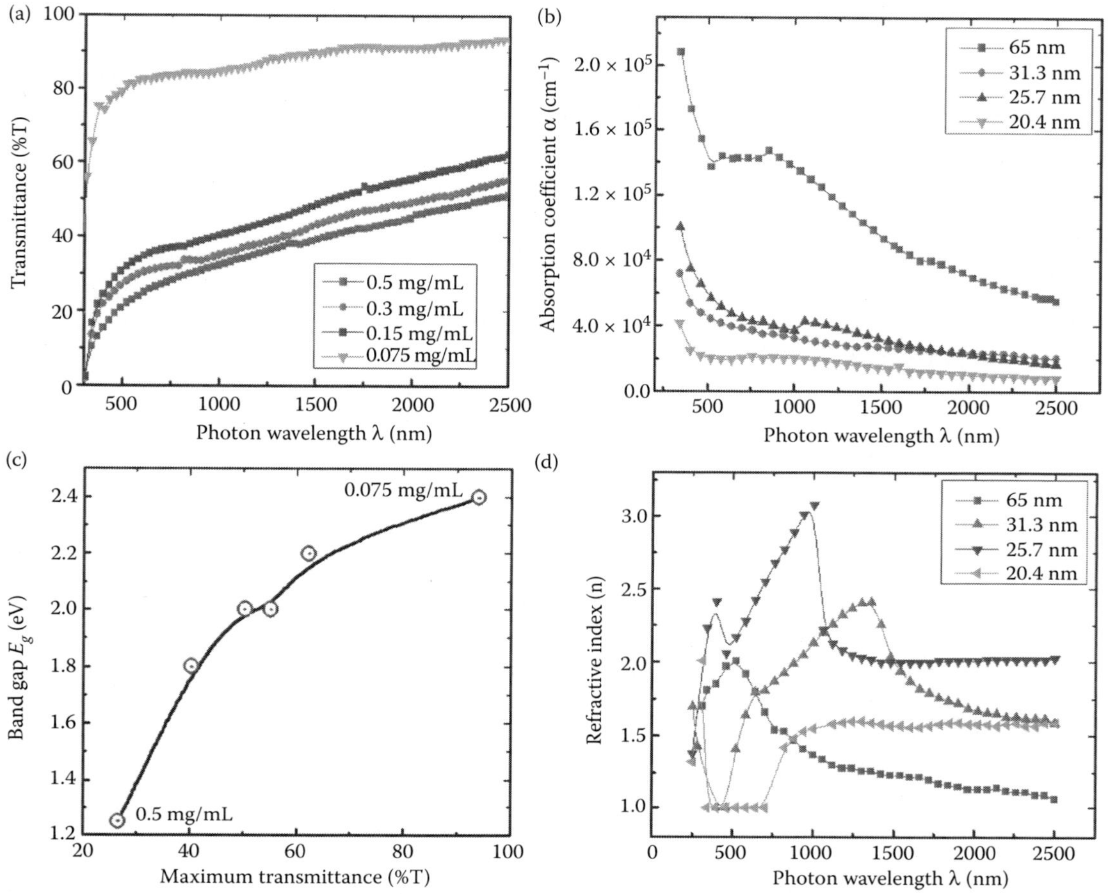

FIGURE 38.6 (a) Transmittance spectra of GO thin films for 0.5, 0.3, 0.15, and 0.075 mg/mL concentrations of GO suspension annealed at 170°C for 1 h. (b) Absorption coefficient as a function of photon wavelength for these films. (c) Band gap as a function of maximum transmittance. (d) Dependence of refractive index (n) on photon wavelength, λ, for these GO thin films. (Reprinted from *Appl. Surf. Sci.*, 259, Chowdhury, F. et al., Optoelectronic properties of graphene oxide thin film processed by cost effective-route, 460–64, Copyright 2012, with permission from Elsevier.)

TABLE 38.1
Optical Parameters of GO Thin Films

GO Thin Films	Maximum Transmittance (%T)	Absorption Coefficient at 550 nm (cm^{-1})	Band Gap, E_g (eV)	Refractive Index (n) at 550 nm
1	26.64	326,393	1.25	1
2	50.28	144,968	1.82	1.3
3	55.14	41,834	2.0	1.5
4	62.23	53,474	2.2	2.24
5	93.95	19,847	2.4	1.45

Source: With kind permission from Springer Science+Business Media: *Appl. Nanosci.*, Annealing effect on the optoelectronic properties of graphene oxide thin films, 3, 2013, 477–83, Chowdhury, F. et al.

3. Although hydrazine is most widely used as a chemical reductant, its toxicity and explosiveness pose a problem. Alternatively, the as-prepared electrically insulating GO films were turned into RGO by annealing in this study.
4. As thermal reduction has been performed in the mean time only at high temperatures (400–1100°C) [34,68], a low processing temperature is highly desirable for making it useful for practical use [34].
5. Annealing at a temperature as low as 170°C for 1 h was assessed to be the optimum percolation condition in this study, for obtaining GO thin films having required transparency and conductivity.
6. No pretreatment is required to promote the adhesion of GO to the substrate.
7. The hydrazine-treated GO films exhibit very weak adhesion to the glass substrates.
8. On deposition to the surface, these films showed nonuniform and defective textures.
9. The maximum transmittance value is of 26% at 2500 nm of photon wavelength.
10. The hydrazine-treated film has a tendency to absorb light significantly where a steeper absorption region is observed over the entire photon wavelength regime.
11. In contrast to the hydrazine-treated film, the non-hydrazine-treated annealed films are uniform and more transparent.

12. It represents the comparison where the film annealed with 170°C gives higher transmittance than the film annealed with 195°C. The high transparency is normally associated with a good structural homogeneity.
13. It is observed that transmission increases significantly with decreasing solution concentration. The film with the lowest concentration (i.e., 0.075 mg/mL) gives the highest transmittance and at 550 nm, the transmittance value is approximately 82%.
14. The average transmittance is about 82%–94% over a wide range of the solar spectrum.
15. It is seen that the direct optical band gap (E_g) increased from 1.25 to 2.24 eV with the decrease of the concentration of GO solution from 0.5 to 0.075 mg/mL.
16. The refractive index value ranges between 1 and 2.39, which is similar to those obtained by other groups [50].
17. Sheet concentration and electrical conductivity vary similarly with transmittance, that is, decreases with increase of transmittance. The conductivity, that is, sheet concentration, increases as the mobility decreases and vice versa.
18. Transmittance and sheet resistance decreased with increasing film thickness (i.e., for 0.5 mg/mL suspension), which is similar in the case of a transparent graphene electrode as was obtained by Wu et al. [76].
19. The smaller the value of Urbach energy, the smaller the compositional, topological, or structural disorder.

38.3.2 Electrical Properties of Non-Hydrazine-Treated GO Thin Films

From the above discussion, it can be concluded that non-hydrazine-treated thin films show good structural and optical properties. In this section, therefore, only electrical properties of non-hydrazine-treated or annealed GO thin films are discussed extensively. Electrical properties such as type of extent of conductivity, sheet concentration, mobility, Urbach energy (E_U), etc. are of significant concern for the investigation of the potentiality of thin films in the fabrication of optoelectronic devices. Electrical properties can be altered over several orders of magnitude with different growth conditions. Electrical properties of GO thin films prepared by different techniques were studied. Many electrical properties of GO thin films processed by the solution-casting technique are yet to be known. GO is a prospective semiconductor for fabricating optoelectronic devices. The mobility of charge carriers is affected by various sources of scattering when they move through the material. Hence, precise knowledge about the mode of conductivity in the material is very important concerning its application as a transparent conductor in the preparation of photovoltaic devices. In this study, we investigate this material as a transparent conducting electrode and optimize its electrical parameters. Electrical measurements were conducted in lateral direction by using "Ecopia Hall effect measurement system" by a four-point van der Pauw approach at ambient condition. This measurement can provide the necessary information to determine both the concentration and mobility of the carriers. The films of GO were prepared on the glass substrates. The area of the films was approximately 1.0×2.5 cm^2. Indium has been used to make ohmic contact with GO film. Four contacts, two end contacts for the current flow and two mid-contacts across the sample for the Hall voltage measurement, were made by soldering four indium dots of about 0.1 mm at the four corners on the top surface of each film as electrodes.

The average thickness of the films measured directly by AFM for 0.5, 0.3, 0.15, and 0.075 mg/mL suspensions are 31.3, 25.7, and 20.4 nm, respectively. The surface root mean square roughness are 5.06, 1.52, and 1.39 nm, for the films of concentrations 0.3, 0.15, and 0.075 mg/mL, respectively. The film thickness far exceeds the thickness of a single layer of graphene (~1 nm), indicating that the films prepared here mostly consist of multiple layers of GO.

Figure 38.7a and b represent the scanning electron microscopy (SEM) and AFM image, respectively, of the best film

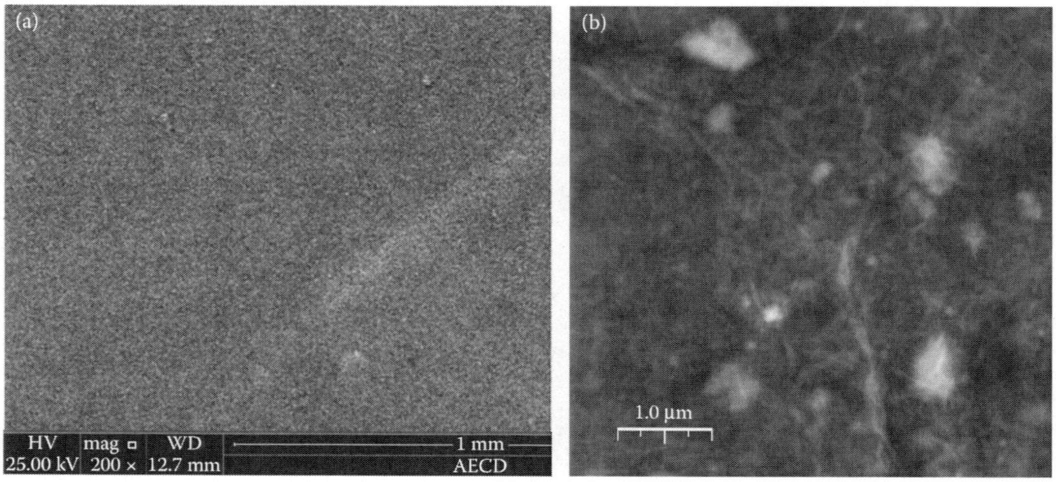

FIGURE 38.7 SEM and AFM images of solution-casted RGO films on glass substrates for concentration 0.075 mg/mL: (a) SEM image and (b) AFM image, where the films were annealed at 170°C for 1 h. (With kind permission from Springer Science+Business Media: *Appl. Nanosci.*, Annealing effect on the optoelectronic properties of graphene oxide thin films, 3, 2013, 477–83, Chowdhury, F. et al.)

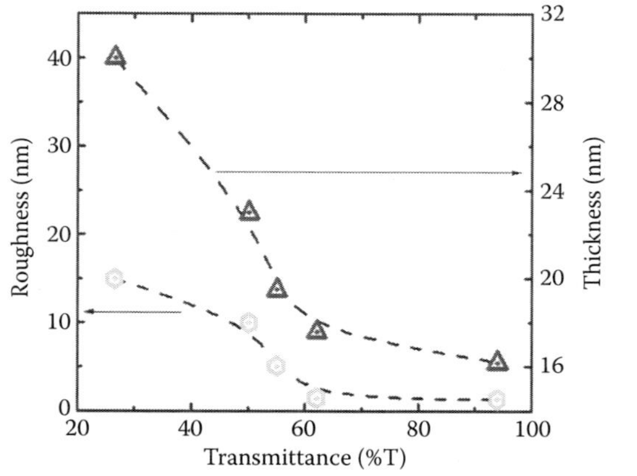

FIGURE 38.8 Surface roughness and thickness of GO thin films as a function of transmittance.

prepared in this study, where the concentration of the suspension was 0.075 mg/mL. The image reflects homogeneous deposition of the dispersed GO/H_2O solution where no wrinkles and folded regions were observed on the surface.

Surface roughness and thickness as a function of film transmittance is shown in Figure 38.8. The lowest roughness was observed for the 0.075 mg/mL suspension where optical transmittance revealed the maximum value. It is seen clearly from Figure 38.8 that the transmittance increases as the surface roughness and thickness of the films decreases. It is expected because smooth surfaces reduce the reflection of light at the surface of the films and hence the transmittance increases consequently.

The sheet concentration and electrical conductivity were obtained from Hall measurement data. Figure 38.9 shows the Hall effect measurements using the van der Pauw method that was performed in order to investigate the electrical properties of the GO thin films.

Sheet concentration and electrical conductivity vary similarly with transmittance, that is, decrease with increase of transmittance (Figure 38.9a). Figure 38.9b shows that as the conductivity, that is, sheet concentration, increases, the mobility decreases and vice versa. In the case of the thin film, it is reasoned that the volatilization of oxygen atoms for the thin film with lowest concentration would decrease the number of oxygen vacancies during annealing compared to the higher concentrated films, which results in a decrease in sheet concentration and an increase in resistance. As shown in Figure 38.9a, if we decrease the solution concentration from 0.5 to 0.075 mg/mL, the sheet concentration decreases from 9.20E + 18 to 3.74E + 18 cm^{-2}, where the conductivity decreases from 200 to 32.50 $(\Omega\text{-cm})^{-1}$. It is logical that when the solution concentration is decreased, the number of oxygen atoms adsorbed on the thin film will be increased, causing a decrease of the defect density and resulting in a decrease of the sheet concentration. These results imply that as the solution concentration is decreased, thin films come closer to the stoichiometrical composition and the Hall mobility is increased from 0.138 to 5.183 cm^2/Vs. It is rational that Hall mobility is more affected by the quality enhancement of the thin film than by the change in its resistance. Figure 38.9c represents sheet resistance as a function of transmittance. Transmittance and sheet resistance decreased with increasing film thickness (i.e., for 0.5 mg/mL suspension), which is similar in case of a transparent graphene electrode as was obtained by Gilje et al. [23]. In case of annealed films, the lowest sheet resistance of 869.99 (Ω/Sq.) was found at the annealing condition of 160°C, 35 min, where the transmittance value was low. There is a trade-off between transparency and sheet resistance, since the highest transmittance (82% at 550 nm) of GO thin film for transparent conducting applications was found at the optimum annealing condition of 170°C, 1 h in this work, where the sheet resistance increases (15 kΩ/Sq.). However, it is noticeable that this low value of optimum annealing condition is favorable in practice, as when compared with other studies where GO films were prepared by using various reagents at very high temperatures [77].

The dependence of the band gap on film thickness is depicted in Figure 38.10. The band gap varies linearly with the reciprocal square of thickness ($1/d^2$). The quantum size effect could be a plausible cause for the change in the band gap of GO thin films. This can be explained in the following

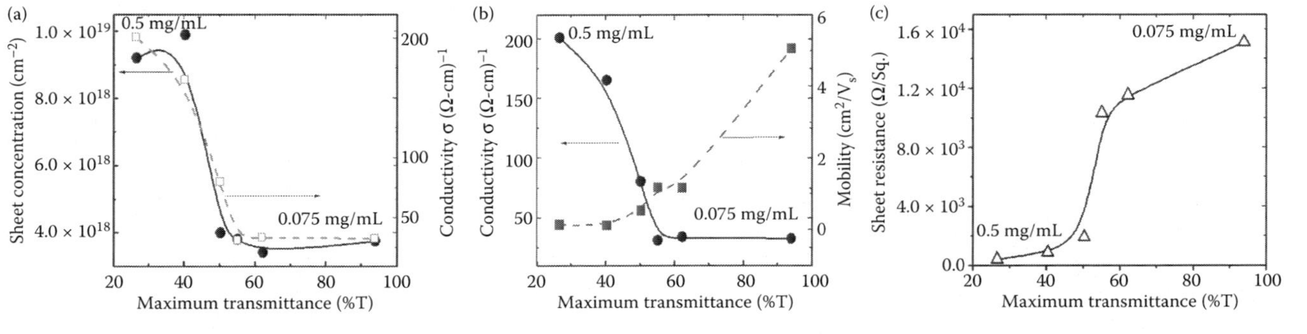

FIGURE 38.9 (a) Sheet concentration and conductivity versus transmittance curve of GO thin films. (b) Conductivity and mobility versus transmittance plot of GO thin films. (c) Sheet resistance as a function of transmittance of GO thin films. (With kind permission from Springer Science+Business Media: *Appl. Nanosci.*, Annealing effect on the optoelectronic properties of graphene oxide thin films, 3, 2013, 477–83, Chowdhury, F. et al.)

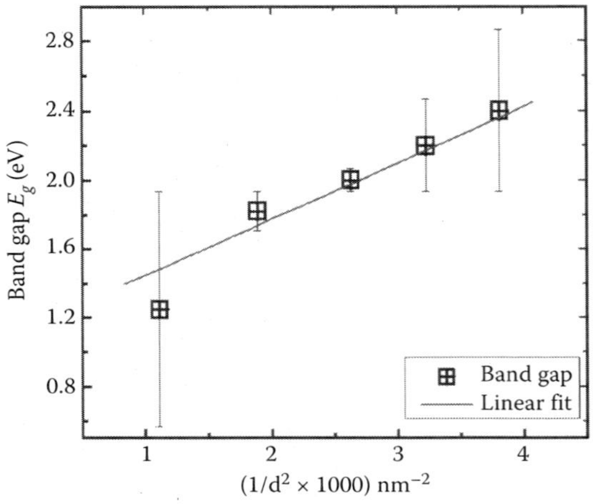

FIGURE 38.10 Band gap, E_g, as a function of film thickness, *d*. (Reprinted from *Appl. Surf.* Sci., 259, Chowdhury, F. et al., Optoelectronic properties of graphene oxide thin film processed by cost effective-route, 460–64, Copyright 2012, with permission from Elsevier.)

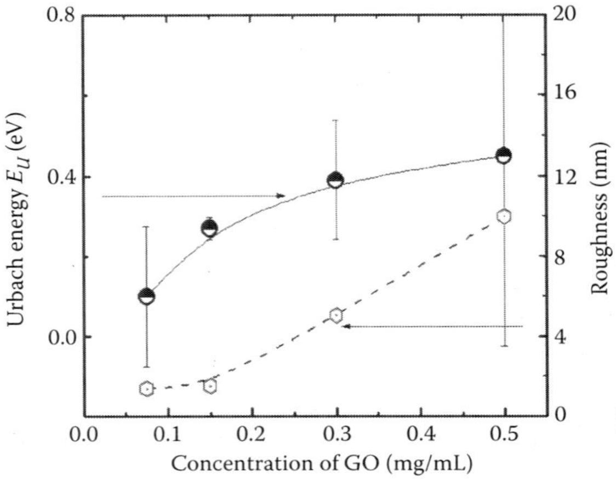

FIGURE 38.11 Urbach energy (E_U) and surface roughness as a function of GO concentration. (Reprinted from *Appl. Surf. Sci.*, 259, Chowdhury, F. et al., Optoelectronic properties of graphene oxide thin film processed by cost effective-route, 460–64, Copyright 2012, with permission from Elsevier.)

way. The thin GO films (nanoscale range) are considered as a one-dimensional (1D) confined system. In such a 2D system, elementary excitations will experience quantum confinement, resulting in restricted motion in the confinement direction (i.e., the film thickness direction) and unrestricted motion in other two directions [78,79].

It is well known that as the confining dimension decreases, typically in nanoscale, the energy spectrum twists to discrete and thus the band gap of a semiconductor becomes size dependent. For 1D confinement, the quantization energies increase when the size along the confinement direction (i.e., the film thickness in this study) reduces, as illustrated by the equation $\Delta E_g \sim (\hbar^2/2md^2)$ [80,81], where *d* is the thickness of the layer and *m* is the effective mass of electron. In general, the above equation can elucidate the bad gap increase with decrease of film thickness (or vice versa) in the nanometer scale range.

The dependence of Urbach energy and surface roughness as a function of GO concentration is presented in Figure 38.11. The disorder in GO films can be represented by the Urbach energy, E_U, which represents the width of localized state. It is the disorder that results in the difficulty in the determination of the energy gap. Urbach energy can be obtained using the relation $\alpha = \alpha_0 \exp(h\nu/E_U)$, where α_0 is a constant. "Urbach energy" is interpreted as a broadening of the intrinsic absorption edge due to disorder and related to the transitions from the extended valence band states to the localized states at the conduction band tail. The smaller the value of E_U, the smaller is the compositional, topological, or structural disorder. The lowest Urbach energy was found as 0.1 eV for the film of concentration 0.075 mg/mL. Surface roughness has a direct correlation with the Urbach energy (width of localized states), as is seen from Figure 38.11 that both of these parameters increase with the increase of GO concentration. This similarity indicates that Urbach energy is the consequence of surface roughness of the films.

38.4 CONCLUDING REMARKS

In summary, high-quality GO films were obtained by annealing at a low temperature (170°C). This is expected because GO prepared via Marcano's improved GO synthesis (that we adopted in this study) possesses fewer defects in the basal plane as compared to those prepared via conventional methods, including Hummers' method [50,82]. Furthermore, the high transmittance of the films provides them with potential applications in power-producing windows. A higher extent of graphitization of GO film is important to improve not only the transmittance but also the conductivity of the finally formed GO films. Thus, our synthesis of GO and the ability to process it into high-quality films bode well for use in high-performance applications such as solar cells. Finally, we can state that the technique employed in this work for fabrication of GO thin films is appealing particularly, because of its high yield, simple apparatus, and low-temperature thermal reduction. GO thin films prepared by simply solution-casting have properties of transparency and conductivity, which are comparable to those prepared by other methods and are thus eminently suitable for application as a low-cost transparent electrode. Different optical and electrical data provide valuable information on GO thin films under this study. It is found that GO films are direct transition-type semiconductors in nature. Obtained sheet resistance value (~15 kΩ/Sq.) at a low percolation condition without using any reagent is comparable to other studies [79]. Through this study, it was confirmed that GO thin films show device-quality performance by using this compatible technique, and such skills to develop these films with varying band gap energies are useful for next-generation optical component applications such as a solar cell.

REFERENCES

1. Eda, G., Chhowalla M. 2010. Chemically derived graphene oxide: Towards large-area thin-film electronics and optoelectronics. *Adv. Mater.* 22:2392–415.
2. Chowdhury, F., Morisaki, T., Otsuki, J. et al. 2012. Optoelectronic properties of graphene oxide thin film processed by cost effective-route. *Appl. Surf. Sci.* 259:460–64.
3. Chowdhury, F., Morisaki, T., Otsuki, J. et al. 2013. Annealing effect on the optoelectronic properties of graphene oxide thin films. *Appl. Nanosci.* 3:477–83.
4. Li, D., Kaner, R. B. 2008. Graphene-based materials. *Science* 320:1170–71.
5. Geim, A. K. 2009. Graphene: Status and prospects. *Science* 324:1530–34.
6. Compton, O. C., Jain, B., Dikin, D. A. et al. 2011. Chemically active reduced graphene oxide with tunable C/O ratios. *ACS Nano* 5:4380–91.
7. Novoselov, K. S., Geim, A. K., Morozov, S. V. et al. 2004. Electric field effect in atomically thin carbon films. *Science* 306:666–69.
8. Watcharotone, S., Dikin, D. A., Stankovich, S. et al. 2007. Graphene–silica composite thin films as transparent conductors. *Nano Lett.* 7:1888–92.
9. Gómez-Navarro, C., Weitz, R. T., Bittner, A. M. et al. 2007. Electronic transport properties of individual chemically reduced graphene oxide sheets. *Nano Lett.* 7: 3499–503.
10. Allahbakhsh, A., Sharif, F., Mazinani, S. et al. 2014. Synthesis and characterization of graphene oxide in suspension and powder forms by chemical exfoliation method. *Int. J. Nano Dimens.* 5:11–20.
11. Eigler, S., Hof, F., Enzelberger-Heim, M. et al. 2014. Statistical Raman microscopy and atomic force microscopy on heterogeneous graphene obtained after reduction of graphene oxide. *J. Phys. Chem. C* 118:7698–704.
12. Cincotto, F. H., Moraes, F. C., Machado, S. A. S. 2014. Graphene nanosheets and quantum dots: A smart material for electrochemical applications. *Chem. Eur. J.* 20:4746–53.
13. Wei, N., Lv, C., Xu, Z. 2014. Wetting of graphene oxide: A molecular dynamics study. *Langmuir* 30:3572–78.
14. Ho, N. T., Senthilkumar, V., Kim, Y. S. 2014. Impedance spectroscopy analysis of the switching mechanism of reduced graphene oxide resistive switching memory. *Solid-State Electron.* 94:61–65.
15. Zhang, D., Tong, J., College, B. X. 2014. Humidity-sensing properties of chemically reduced graphene oxide/polymer nanocomposite film sensor based on layer-by-layer nano self-assembly. *Sens. Actuators B* 197:66–72.
16. Chabot, V., Higgins, D., Yu, A. et al. 2014. A review of graphene and graphene oxide sponge: Material synthesis and applications to energy and the environment. *Energy Environ. Sci.* 7:1564–96.
17. L , W., Chen, B., Meng, C. et al. 2014. Ultrafast all-optical Graphene modulator. *Nano Lett.* 14:955–59.
18. Zhang, H., Babichev, A. V., Jacopin, G. et al. 2013. Characterization and modeling of a ZnO nanowire ultraviolet photodetector with graphene transparent contact. *J. Appl. Phys.* 114:234505–9.
19. Liu, C.-H., Chang, Y.-C., Norris, T. B. et al. 2014. Graphene photodetectors with ultra-broadband and high responsivity at room temperature. *Nat. Nanotechnol.* 9:273–78.
20. Li, X., Cai, W., An, J. et al. 2009. Large-area synthesis of high-quality and uniform graphene films on copper foils. *Science* 324:1312–14.
21. Berger, C., Song, Z. M., Li, X. B. et al. 2006. Electronic confinement and coherence in patterned epitaxial graphene. *Science* 312:1191–96.
22. Stankovich, S., Piner, R. D., Chen, X. Q. et al. 2006. Stable aqueous dispersions of graphitic nanoplatelets via the reduction of exfoliated graphite oxide in the presence of poly(sodium 4-styrenesulfonate). *J. Mater. Chem.* 16:155–58.
23. Gilje, S., Han, S., Wang, M. et al. 2007. A chemical route to graphene for device applications. *Nano Lett.* 7:3394–98.
24. Schedin, F., Geim, A. K., Morozov, S. V. et al. 2007. Detection of individual gas molecules adsorbed on graphene. *Nat. Mater.* 6:652–55.
25. Yin, Z. Y., Wu, S. X., Zhou, X. Z. et al. Electrochemical deposition of ZnO nanorods on transparent reduced graphene oxide electrodes for hybrid solar cells. *Small* 6:307–12.
26. Liu, W., Jackson, B. L., Zhu, J. et al. Large scale pattern graphene electrode for high performance in transparent organic single crystal field-effect transistors. *ACS Nano* 4:3927–32.
27. Dong, X. C., Shi, Y. M., Huang, W. et al. 2010. Electrical detection of DNA hybridization with single-base specificity using transistors based on CVD-grown graphene sheets. *Adv. Mater.* 22:1649–53.
28. Yin, Z. Y., Sun, S., Salim, T. et al. 2010. Organic photovoltaic devices using highly flexible reduced graphene oxide films as transparent electrodes. *ACS Nano* 4:5263–68.
29. He, Q. Y., Sudibya, H. G. Yin, Z. Y. et al. 2010. Centimeter-long and large-scale micropatterns of reduced graphene oxide films: Fabrication and sensing applications. *ACS Nano* 4: 3201–8.
30. Liu, J. Q., Lin, Z. Liu, T. et al. 2010. Multilayer stacked low-temperature-reduced graphene oxide films: Preparation, characterization, and application in polymer memory devices. *Small* 6:1536–42.
31. Liu, J. Q., Yin, Z. Y., Cao, X. H. et al. 2010. Bulk heterojunction polymer memory devices with reduced graphene oxide as electrodes. *ACS Nano* 4:3987–92.
32. Cao, X. H., He, Q. Y., Shi, W. H. et al. 2011. Graphene oxide as carbon source for controlled growth of carbon nanowires. *Small* 7:1199–202.
33. Sudibya, H. G., He, Q. Y., Zhang, H. et al. 2011. Electrical detection of metal ions using field-effect transistors based on micropatterned reduced graphene oxide films. *ACS Nano* 5:1990–94.
34. Becerril, H. A., Mao, J., Liu, Z. et al. 2008. Evaluation of solution-processed reduced graphene oxide films as transparent conductors. *ACS Nano* 2:463–70.
35. Stankovich, S., Dikin, D. A., Dommett, G. H. B. et al. 2006. Graphene-based composite materials. *Nature* 442:282–86.
36. Hong, W., Xu, Y., Lu, G. et al. 2008. Transparent graphene/PEDOT–PSS composite films as counter electrodes of dye-sensitized solar cells. *Electrochem. Commun.* 10:1555–58.
37. Wang, X., Zhi, L., Mullen, K. 2008. Transparent, conductive graphene electrodes for dye-sensitized solar cells. *Nano Lett.* 8:323–27.
38. Geim, A. K., Novoselov, K. S. 2007. The rise of graphene. *Nat. Mater.* 6:183–91.
39. Mkhoyan, K. A., Contryman, A. W., Silcox, J. et al. 2009. Atomic and electronic structure of graphene-oxide. *Nano Lett.* 9:1058–63.
40. Cai, W. W., Piner, R. D., Stadermann, F. J. et al. 2008. Synthesis and solid-state NMR structural characterization of c-labeled graphite oxide. *Science* 321:1815–17.

41. Yang, D., Velamakanni, A., Bozoklu, G. et al. 2009. Chemical analysis of graphene oxide films after heat and chemical treatments by x-ray photoelectron and micro-Raman spectroscopy. *Carbon* 47:145–52.
42. Eda, G., Mattevi, C., Yamaguchi, H. et al. 2009. Insulator to semimetal transition in graphene oxide. *J. Phys. Chem. C* 113:15768–71.
43. Dai, B., Fu, L., Liao, L. et al. 2011. High-quality single-layer graphene via reparative reduction of graphene oxide. *Nano Res.* 4:434–39.
44. Eda, G., Lin, Y. Y., Miller, S. et al. 2008. Transparent and conducting electrodes for organic electronics from reduced graphene oxide. *Appl. Phys. Lett.* 92:233305–7.
45. Eda, G., Fanchini, G., Chhowalla, M. et al. 2008. Large-area ultrathin films of reduced graphene oxide as a transparent and flexible electronic material. *Nat. Nanotechnol.* 3:270–74.
46. He, C. L., Zhuge, F., Zhou, X. F. et al. 2009. Nonvolatile resistive switching in graphene oxide thin films. *Appl. Phys. Lett.* 95:232101–3.
47. Zhuang, X.-D., Chen, Y., Liu, G. et al. 2009. Conjugated-polymer-functionalized graphene oxide: Synthesis and nonvolatile rewritable memory effect. *Adv. Mater.* 22:1731–35.
48. Liu, G., Zhuang, X., Chen, Y. et al. 2009. Bistable electrical switching and electronic memory effect in a solution-processable graphene oxide-donor polymer complex. *Appl. Phys. Lett.* 95:253301.
49. Han, T. H., Lee, W. J., Lee, D. et al. 2010. Peptide/graphene hybrid assembly into core/shell nanowires. *Adv. Mater.* 22:2060–64.
50. Jin, M., Jeong, H.-K., Yu, W. J. et al. 2009. Graphene oxide thin film field effect transistors without reduction. *J. Phys. D Appl. Phys.* 42:135109–13.
51. Jeong, H-K., Lee, Y. P., Jin, M. H. et al. 2009. Thermal stability of graphene oxide. *Chem. Phys. Lett.* 470:255–58.
52. Xu, Y., Long, G., Huang, L. et al. 2010. Polymer photovoltaic devices with transparent graphene electrodes produced by spin-casting. *Carbon* 48:3308–11.
53. Rollings, E., Gweon, G.-H., Zhou, S. Y. et al. 2006. Synthesis and characterization of atomically thin graphite films on a silicon carbide substrate. *J. Phys. Chem. Solids* 67:2172–77.
54. Reina, A., Jia, X., Ho, J. et al. 2009. Layer area, few-layer graphene films on arbitrary substrates by chemical vapor deposition. *Nano Lett.* 9:3087.
55. Chae, S. J., Gunes, F., Kim, K. K. et al. 2009. Synthesis of large-area graphene layers on poly-nickel substrate by chemical vapor deposition: Wrinkle formation. *Adv. Mater.* 21:2328–33.
56. Kim, K. S., Zhao, Y., Jang, H. et al. 2009. Large-scale pattern growth of graphene films for stretchable transparent electrodes. *Nature 457:706-10.*
57. Wang, Y., Gu, P., Cao, J. et al. 2012. Graphene based transparent conductive electrode. *Adv. Mat. Res.* 468:1823–26.
58. Hirata, M., Gotou, T., Horiuchi, S. et al. 2004. Thin-film particles of graphite oxide 1: High-yield synthesis and flexibility of the particles. *Carbon* 42:2929–37.
59. Park, S., Ruoff, R. S. 2009. Chemical methods for the production of graphene. *Nat. Nanotechnol.* 4:217–24
60. Marcano, D. C., Dmitry, V., Kosynkin, J. M. et al. 2010. Improved synthesis of graphene oxide. *ACS Nano* 4:4806–14.
61. Stankovich, S., Dikin, D. A., Piner, R. D. et al. 2007. Synthesis of graphene-based nanosheets via chemical reduction of exfoliated graphite oxide. *Carbon* 45:1558–65.
62. Shin, H.-J., Kim, K. K., Benayad, A. et al. 2009. Efficient reduction of graphite oxide by sodium borohydride and its effect on electrical conductance. *Adv. Funct. Mater.* 19:1987–92.
63. Pei, S., Zhao, J., Du, J. et al. 2010. Direct reduction of graphene oxide films into highly conductive and flexible graphene films by hydrohalic acids. *Carbon* 48:4466–74.
64. Gao, W., Alemany, L., Ci, B. et al. 2009. New insights into the structure and reduction of graphite oxide. *Nat. Chem.* 1:403–8.
65. Dreyer, D. R., Park, S., Bielawski, C. M. et al. 2010. The chemistry of graphene oxide. *Chem. Soc. Rev.* 39:228–40.
66. Zhu, Y., Murali, S., Cai, W. et al. 2010. Graphene and graphene oxide: Synthesis, properties, and applications. *Adv. Mater.* 22:3906–24.
67. Chua, L. L., Wang, S., Chia, P. J. et al. 2008. Deoxidation of graphene oxide nanosheets to extended graphenites by "unzipping" elimination. *J. Chem. Phys.* 129: 114702.
68. Jung, I., Dikin, D. A., Piner, R. D. et al. 2008. Tunable electrical conductivity of individual graphene oxide sheets reduced at low temperatures. *Nano Lett.* 8:323–27.
69. Heavens, O. S. 1965. *Optical Properties of Thin Solid Films*. London: Butterworth Scientific Publications.
70. Tomlin, S. G. 1968. Optical reflection and transmission formulae for thin films. *J. Phys. D Appl. Phys.* 1:1667–72.
71. Tauc, J. 1974. *Amorphous and Liquid Semiconductors*. New York: Plenum Press.
72. Narayan, R., Deepa, M., Srivastava, A. K. 2012. Nanoscale connectivity in a TiO_2/CdSe quantum dots/functionalized graphene oxide nanosheets/Au nanoparticles composite for enhanced photoelectrochemical solar cell performance. *Phys. Chem. Chem. Phys.* 14:767–78.
73. Jeon, S., Ahn, S.-E., Song, I. et al. 2012. Gated three-terminal device architecture to eliminate persistent photoconductivity in oxide semiconductor photosensor arrays. *Nat. Mater.* 11:301–5.
74. Matsuo, Y., Mimura, T., Sugie, Y. 2010. Preparation of semiconducting graphene-based carbon films from silylated graphite oxide and covalent attachment of dye molecules. *Chem. Lett.* 39:636–37.
75. Mathkar, A., Tozier, D., Cox, P. et al. 2012. Controlled, stepwise reduction and band gap manipulation of graphene oxide. *J. Phys. Chem. Lett.* 3:986–91.
76. Wu, J., Becerril, H. A., Bao, Z. et al. 2008. Organic solar cells with solution-processed graphene transparent electrodes. *Appl. Phys. Lett.* 92:263302–4.
77. Zhu, Y., Cai, W., Piner, R. D. et al. 2009. Transparent self-assembled films of reduced graphene oxide platelets. *Appl. Phys. Lett.* 95:103104–6.
78. Goh, E. S. M., Chen, T. P., Sun, C. Q. et al. 2010. Thickness effect on the band gap and optical properties of germanium thin films. *J. Appl. Phys.* 107:024305–5.
79. Mitin, V. V., Sementsov, D. I., Vagidov, N. Z. 2010. *Quantum Mechanics for Nanostructures*. Cambridge: Cambridge University Press.
80. Shik, A. 1997. *Quantum Wells: Physics and Electronics of Two-Dimensional Systems*. Singapore: World Scientific.
81. Fox, M. 2001. *Optical Properties of Solids*. Oxford: Oxford University Press.
82. Chowdhury, F., Otsuki, J., Alam, M. S. et al. 2014. Thermally reduced solution-processed graphene oxide thin film: An efficient infrared photodetector. *Chem. Phys. Lett.* 593:198–203.

39 Electrical and Thermal Conductivity of Indium–Graphene and Copper–Graphene Composites

K. Jagannadham

CONTENTS

ABSTRACT

The synthesis of graphene composites with the matrix of In, In–Ga, or Cu by low-cost methods to achieve uniform distribution of graphene platelets (GPs) is described. Characterization of the microstructure, topography, and distribution of the GPs using optical, scanning electron microscopy (SEM), and x-ray diffraction is presented, and the difficulty of evaluation of the volume fraction of GPs is emphasized. Results of measurement of electrical and thermal conductivity coupled with modeling using effective medium approximation are used to determine the volume fraction, electrical conductivity of GPs, and the interface thermal conductance between GPs in the ab plane and the matrix. The electrical conductivity of thin GPs is found to be between 2 and $3 \times 10^6\ \Omega^{-1}\ cm^{-1}$ while the average value is between 1 and $2 \times 10^6\ \Omega^{-1}\ cm^{-1}$. The thermal conductivity of composites is improved by 100% compared to that of In and by 25% compared to that of Cu when the volume fraction of GPs is near 0.20. The interface thermal conductance in the ab plane between GPs and the matrix is found to be close to 1 $GWm^{-2}\ K^{-1}$ and thus not a limiting factor in the heat spreading in the composites. Transient thermo reflectance (TTR) measurements showed that the composites are better heat spreaders than In and Cu, respectively. These results suggest that composites with In are useful as thermal interface materials and composites with Cu are useful for electrical contact brushes and heat spreaders.

39.1 INTRODUCTION

The electrical and thermal conductivity of graphene platelets (GPs) is anisotropic with two orders of magnitude higher values in the ab plane compared to that in the c-direction. The magnitude of the transport property is also strongly dependent on the thickness and size of the platelets. The sheet resistance of GPs is predicted [1] to decrease following $R_{gp} = 62.4/N\ \Omega/$sq where N is the number of layers. However, beyond thickness of 5–10 atomic layers, GP approaches graphite behavior, which has a much lower conductivity. The mobility of carriers, although predicted to be high, is reduced by scattering from impurities at low temperature and phonons at room temperature [2,3] to a value close to 15,000 $cm^2\ V^{-1}\ s^{-1}$. The mean free path of carriers is expected to be several micrometers only limited by defects such as vacancies, impurities, edges, and ripples in GP [4,5]. Injection of carriers from metal contacts such as Cu was also observed [6].

The progress in characterizing thermal conductivity of graphene is illustrated in Table 39.1.

Thermal conductivity (K) of graphene is phonon-mediated. The thermal conductivity of freestanding graphene is found [7] to be ~4800 $Wm^{-1}\ K^{-1}$ and is reduced to that of bulk graphite, 2000 $Wm^{-1}\ K^{-1}$, when the number of atomic layers reaches 8–10 [8–10]. Thermal conductivity of thinner graphene on a substrate or matrix is also reduced by phonon leakage into the substrate [11–13]. The thermal conductivity of GP on SiO_2/Si [11] is reduced drastically with the size and is found to be as small as 600 $Wm^{-1}\ K^{-1}$ both as a result of phonon leakage and scattering. The interface thermal conductance between GP and metals in the c-direction has been found to be close to 25 $MWm^{-2}K^{-1}$, which is expected to be a limiting factor in the conduction [14–16]. The interface thermal conductance in the ab plane or normal to c-direction is expected to be higher [17–20], which helps to improve the thermal conductivity of the composites.

It is evident from these above observations that the transport properties of the composites depend very much on the size and thickness of the GPs. GPs of single layer are not expected to be the choice as scattering from matrix severely reduces the magnitude. Similarly, very thick or more than 10 atomic layers of GPs approach graphite. Therefore, the benefits of graphene in the composites are achieved with GPs of intermediate thickness such as between 2 and 8 atomic layers. The volume fraction V_g of GPs is an important parameter that determines the improvement obtained over that of the matrix. The determination of V_g in the composite by image analysis has limitations because the contrast in the backscattering mode is not very strong from thinner GPs when examined by scanning electron microscopy (SEM) [18,19]. Also, thicker GPs present below the surface provide contrast that interferes with that of the thinner GPs present close to the surface. Therefore, transmission electron microscopy (TEM) of the thinned electron beam transparent samples is needed requiring special methods to prepare.

TABLE 39.1
Investigations of Thermal Conductivity (K) of Graphene in the ab Plane and Interfacial Thermal Conductance (h) on Substrates

Characteristic Measurement (Interaction between Layers)	K ($Wm^{-1}K^{-1}$)	Reference
Freestanding graphene	4800–5300	[7–10]
Graphene platelets	2000–4800	[10]
Graphene on substrate (size effect, phonon scattering, and leakage)	600	[11–13]
Highly ordered pyrolytic graphite	2000	[13]
h ($MWm^{-2}K^{-1}$)		
Interface thermal conductance (normal to ab plane)	20–60	[14–16]
Interface thermal conductance (parallel to ab plane)	1400	[17–20]

Although GPs are anisotropic, dispersion in a matrix that is isotropic enables isotropic improvement provided GPs are randomly oriented and uniformly distributed [17–19]. The processing of composites, thus, must ensure uniform distribution of larger size GPs in the matrix. Procedures using ball milling reduce the size of GPs, which is detrimental to achieve optimum properties. We have used In, In–Ga, or Cu as the matrix to prepare the composites [17–19,21]. These composites are useful for applications in electronic devices as thermal interface materials and heat spreaders, respectively. In the following, the methods of synthesis of these composites are presented briefly and the experimental results and modeling used to understand the contribution from GPs are described.

39.2 SYNTHESIS OF COMPOSITES

Preparation of GPs and combination with the matrix are two separate steps in the synthesis of the composites. Among many methods [22,23] reported in the literature on the preparation of GPs, we use chemical exfoliation [23–25] because it is suitable for large-scale industrial synthesis. The details of chemical exfoliation are provided in detail in our previous work [25]. Briefly, microcrystalline graphite is oxidized by boiling in a mixture of 50–50 by volume nitric and sulfuric acids until a brown residue of graphene oxide (GO) is obtained. Boiling in the acid mixture with addition of 50% by volume H_2O_2 is continued to further ensure complete oxidation. The residue is washed repeatedly with deionized (DI) water to remove the acidic sulfate and nitrate ions. The resulting residue is mixed with DI water and sonicated for several hours until suspension of GO is obtained without sedimentation. This is an important step that determines the size and thickness of the GO platelets from which GPs are obtained as both decrease with exfoliation time. Collection of the GO platelet suspension [26,27] from the top of the partially exfoliated medium at intermediate intervals will provide larger size GPs. Exfoliation of the remaining residue is continued before the next batch is collected. The next step of preparation of the composites follows different routes as described below.

Several methods have been employed to process composites of metals with graphene, as shown in Table 39.2.

39.2.1 Synthesis of In–Graphene and In–Ga–Graphene Composites

GO platelets placed in a ceramic (alumina) crucible are reduced to GPs by thermal treatment [17–19,21] at 400°C in

TABLE 39.2
Techniques of Synthesis Used to Make Metal–Graphene Composites

Technique	Metals	References
Electrochemical deposition	Cu and Ni	[30–32]
Powder processing and compacting	Cu, Al, Mg, Pt, Au	[33–37]
Chemical vapor deposition-graphene	Cu	[38–40]
Arc deposition-graphene	Mo, Cu	[41,42]
Chemical solution	Ag, Au, Cu, Co	[43,44]
Hydrothermal process	Ni	[45]
Physical vapor deposition	Ti, Cu, W	[46]

argon atmosphere for 1 h. The GPs are suspended in isopropyl alcohol and sonicated mildly to achieve suspension. An In foil of 1 mm thickness is rolled to a thickness of 0.1 mm and prepared in the form of a small boat so that the GP suspension collected using a 3 mm diameter hollow glass tube is deposited inside. The procedure is repeated several times after drying. The foil is folded and deformation rolled again to smaller thickness. Addition of GP suspension and rolling is repeated until the desired volume fraction is reached [18,25]. In the second method, GO platelets suspended in alcohol are deposited on In foil. The platelets are subjected to laser beam incidence using Nd-YAG laser with wave length 266 nm. The laser fluence is kept low so that the In foil does not melt while the GO is converted to GP by UV photo-induced reduction [28,29]. The volume fraction is increased by repeated suspension and laser beam incidence [28] followed by folding and rolling.

The preparation of In–Ga–graphene (In–Ga–gr) composites follows the same procedure used for In–graphene (In–gr), however, In–Ga foil is first prepared by adding Ga in liquid state to In. Ga liquid is added to the In boat in small quantities. The In boat is folded and heated so that In–Ga alloy forms. The mass of the sample is weighed to determine the weight percent of Ga added. This procedure is repeated to increase the Ga to 3.1 wt% (5 at%). The rolled foils of In–Ga are used to prepare the composite. The composite foils are finally annealed in argon for 3 h near 120°C so that recrystallization is complete and any trapped H^+ or OH^- ions are eliminated from the sample. A list of the In–gr and In–Ga–gr samples prepared for characterization is provided in Table 39.3.

TABLE 39.3
Room Temperature Resistivity and Temperature Coefficient of Resistance of the In, In–Ga, In–gr, and In–Ga–gr Samples

Sample	Resistivity (10^{-6} Ω·cm)	TCR (10^{-3} K^{-1})	V_g (SEM)	r_g (10^{-6} Ω·cm)	V_g (ρ_g)
In	9.42	4.35	–		
In–Ga	10.74	4.32	–		
In–gr-1	7.35	4.12	0.18	1.19	0.13
In–gr-2	8.85	4.26	0.24	1.73	0.11
In–gr-3	8.65	4.00	0.17	1.68	0.14
In–Ga–gr-1	9.55	4.21	0.20	1.76	0.12
In–Ga–gr-2	9.25	4.15	0.28	1.65	0.13
In–Ga–gr-3	9.32	4.26	0.24	1.68	0.13
In–Ga–gr-4	9.53	4.15	0.22	1.76	0.13
In–Ga–gr-5	9.81	4.24	0.28	1.87	0.12

Source: With kind permission from Springer Science+Business Media: *J. Elect. Mat.*, Electrical conductivity of graphene composites with In and In-Ga alloy, v. 39, 2010, pp. 1268–1276, A. Naga Sruti and K. Jaganandham. Copyright 2010.

Note: V_g (SEM) is the volume fraction determined from SEM. V_g (ρ_g) is the volume fraction determined from EMA. ρ_g is the resistivity of GPs.

39.2.2 Synthesis of Cu–Graphene (Cu–gr) Composites

Cu–gr composites were prepared by electrochemical codeposition [17–19,21] from a slightly acidic (pH = 6) 0.2 M solution of $CuSO_4$ containing GPs using a pure anode of Cu and pure Cu foil of 135 μm thickness for cathode. The GO platelets were suspended in 250 mL of DI water. The volume fraction of GPs in the composite was adjusted by varying the proportion of GO suspension in the 0.2 M solution of $CuSO_4$. A magnetic stirrer was used to achieve uniformity of GO suspension. A dc power supply was used to operate the deposition unit for longer periods of up to 2 weeks and achieve a thickness of 600 μm. The current density was kept low, near 1.75 $mAcm^{-2}$, to achieve smooth films with good surface finish. The samples consisted of the pure Cu foil in the middle with Cu–gr film, close to 250 μm, deposited on the opposite surfaces. The growth rate of the composite film was between 2 and 3 μm h^{-1}.

The samples were annealed in hydrogen atmosphere at 20 Torr for 4 h and 400°C to reduce GO to GP. GO is reduced by Cu to form CuO, and CuO was reduced to Cu with water vapor removed from the sample. Diffusion of oxygen from the sample is determined by the diffusion coefficient D = 1.16 exp (−67300/RT) cm^2s^{-1} with R = 8.31 $Jmol^{-1}$ K^{-1} so that D takes a value of 6.8×10^{-8} cm^2 s^{-1} at 400°C. The average diffusion distance is greater than 350 μm in 3 h so that GO in the composite is converted to GP. The samples were further heated in argon for 4 h so that any trapped hydrogen or moisture at interfaces and grain boundaries is removed from the composite films. Table 39.4 gives the list of Cu–gr samples prepared for characterization.

39.3 CHARACTERIZATION OF COMPOSITE

The composite samples were examined by optical and SEM in backscattering mode for evaluation of distribution of the GPs and volume fraction, f_g (SEM) shown in Table 39.5. Energy dispersive spectrometry (EDS) was used to determine the composition of the films. The images of the In–gr and the results of EDS are shown in Figure 39.1. The GP size distribution is provided in Figure 39.2. The size varies from 0.15 to 5 μm with a maximum near 0.64 μm.

SEM images of cross section and planar samples of Cu–gr taken in backscattering mode are shown in Figure 39.3a and b, respectively. A uniform distribution of GP is observed. Also, the

TABLE 39.4
Electrical Resistivity ρ_c and TCR of the Cu–gr Films Prepared from Electrochemical Solutions Containing Different Amounts of GO Are Presented

Sample	Stirrer	R_o	ρ_C	TCR_C	V_g	f_g	σ_g	σ_{gf}
Cu–gr-1	No	1.0	1.6994	3.5289	0.110	0.35	3.41	1.32
Cu–gr-2	No	0.50	1.8560	3.5306	0.125	0.28	2.82	1.46
Cu–gr-3	No	0.33	1.9020	3.5381	0.126	0.21	2.68	1.75
Cu–gr-4	No	0.25	2.0180	3.5981	0.122	0.14	2.36	2.10
Cu–gr-5	Yes	1.0	1.8020	3.4546	0.116	0.35	2.99	1.23
Cu–gr-6	Yes	0.5	2.0160	3.4603	0.133	0.28	2.37	1.31
Cu–gr-7	Yes	0.33	2.1418	3.4642	0.133	0.21	2.08	1.45
Cu–gr-8	Yes	0.25	2.2170	3.6153	0.123	0.14	1.92	1.73
Cu	No	0.0	2.0310	3.8121	0.0	–	–	–

Source: With kind permission from Springer Science+Business Media: *Met. Mat. Trans.*, Volume fraction of graphene platelets in copper-graphene composites, v. 44A, 2013, pp. 552–559, K. Jagannadham. Copyright 2013.

Note: The volume fraction V_g and conductivity σ_g of GPs obtained using EMA are shown. The average electrical conductivity σ_{gf} of GPs evaluated after taking into account the conductivity of thicker GPs is also shown. The volume fraction f_g from Table 39.5 determined from thermal conductivity measurements is also included. R_o is ratio of 250 mL of GO suspension in distilled water to volume of 0.2M solution of $CuSO_4$. Resistivity in μΩcm, TCR in 10^{-3} K^{-1}, σ_g and σ_{gf} are in 10^6 Ω^{-1} cm^{-1}.

TABLE 39.5
Thermal Conductivity of In, In–Ga and In–gr, and In–Ga–gr Composite Samples Determined by the 3ω Method and Multilayer Analysis

			Thermal Conductivity ($Wcm^{-1}K^{-1}$)			
Sample	f_g	Thickness (μm)	T = 250K	T = 275K	T = 300K	T = 350K
In	0	150		0.8	0.7	0.6
In–gr1	0.18	70		1.4	1.3	1.2
In–gr3	0.18	60	1.6		1.4	1.2
In–gr3	0.18	35		1.3	1.2	1.1
In–gr4	0.19	430	2.1		1.6	1.2
In–Ga	0	330	0.6		0.5	0.4
In–Ga–gr2	0.21	55	1.4		1.2	0.9
In–Ga–gr3	0.24	65		1.4	1.3	1.1
In–Ga–gr3	0.24	145		1.6	1.4	1.2
In–Ga–gr5	0.22	330		1.8	1.5	1.3

Source: With kind permission from Springer Science+Business Media: *J. Electr. Mat.*, Thermal conductivity of indium-graphene and indium-gallium-graphene composites, v. 40, 2011, pp. 25–34, K. Jagannadham. Copyright 2011.

Note: f_g is the volume fraction of graphene determined from SEM.

presence of GPs along grain boundaries is observed in the planar samples. EDS of the Cu–gr, pure and electrolytic Cu, and GO on Si, shown in Figure 39.3c, allows identification of the particulates as GPs. Presence of small oxygen signal in the EDS is thought to arise from the chamber environment but the carbon signal from GPs is much stronger. Carbon with lower atomic number gives a smaller signal when present in a composite.

X-ray diffraction profile from GO particulates present on Cu surface before hydrogen treatment is shown in Figure 39.4. The GO peaks are identified along with that of Cu, Ag, and Ag_2O that arise from the silver paste used to hold the sample. The peaks from GPs in Cu–gr after hydrogen treatment are not observed because of low scattering from carbon and absorption of the weak diffracted peaks by the matrix. However, electron diffraction of the GP showed the hexagonal pattern of reflections [21] from the ab planes.

39.4 ELECTRICAL CONDUCTIVITY

39.4.1 Electrical Conductivity Measurements in In–gr and In–Ga–gr

Electrical conductivity measurements at room temperature on In–gr and In–Ga–gr were made using four probe collinear method using the Signatone instrument on samples of smaller width (<2 mm) and longer length. The outer probes were used to supply current using Keithley 2400 current source and the inner probes separated by 1.66 mm were used to measure voltage using Keithley 2182 nanovoltmeter. Reverse current approach was used to cancel the thermal electromotive force (emf) at the contacts. The resistivity ρ_c of the composite calculated from $R = \rho l/A$ where R is the resistance, l the length, and A the area of cross section is provided in Table 39.3. The temperature coefficient of resistance ($TCR = dR/RdT$) is measured using the same procedure except that the sample was pressed on to four parallel and small width In contact lines created on insulating Si wafer. The outer In lines provided the source of current and the inner In lines were used to measure the voltage. The sample was held in a temperature-controlled chamber to ±0.1°. The temperature was varied

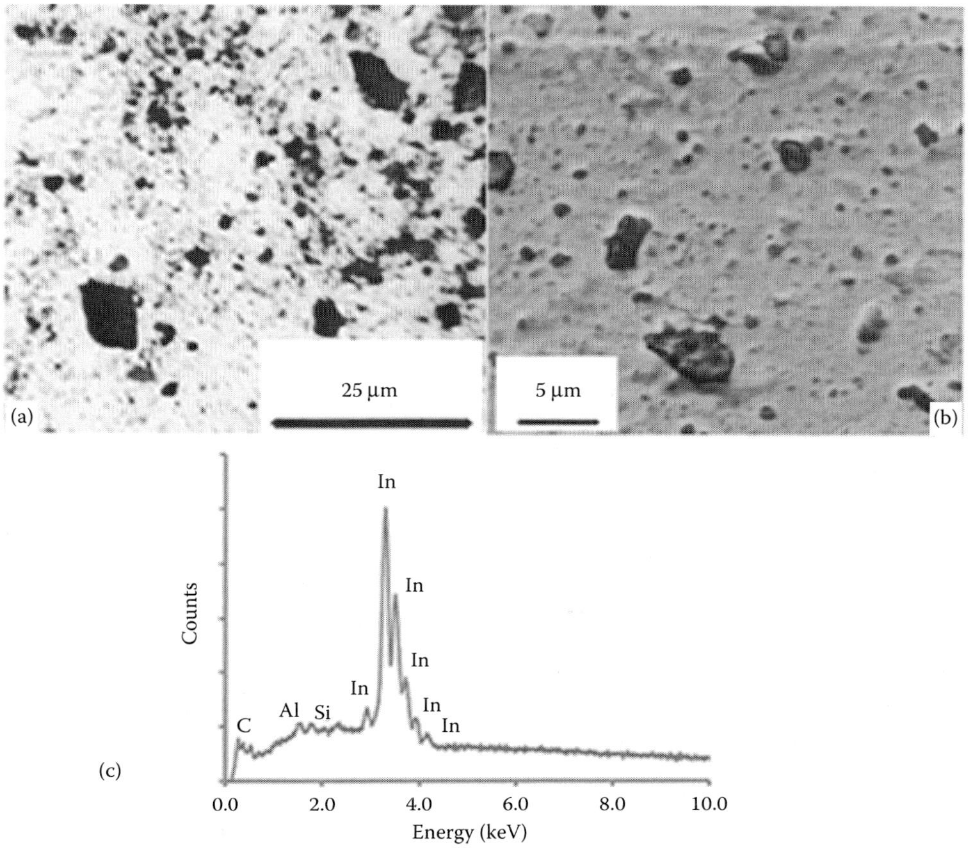

FIGURE 39.1 (a) Optical micrograph, (b) SEM image, and (c) EDS of the In–gr composite. Al and Si signals are from sample holder. (With kind permission from Springer Science+Business Media: *J. Elect. Mat.*, Electrical conductivity of graphene composites with In and In-Ga alloy, v. 39, 2010, pp. 1268–1276, A. Naga Sruti and K. Jaganandham. Copyright 2010.)

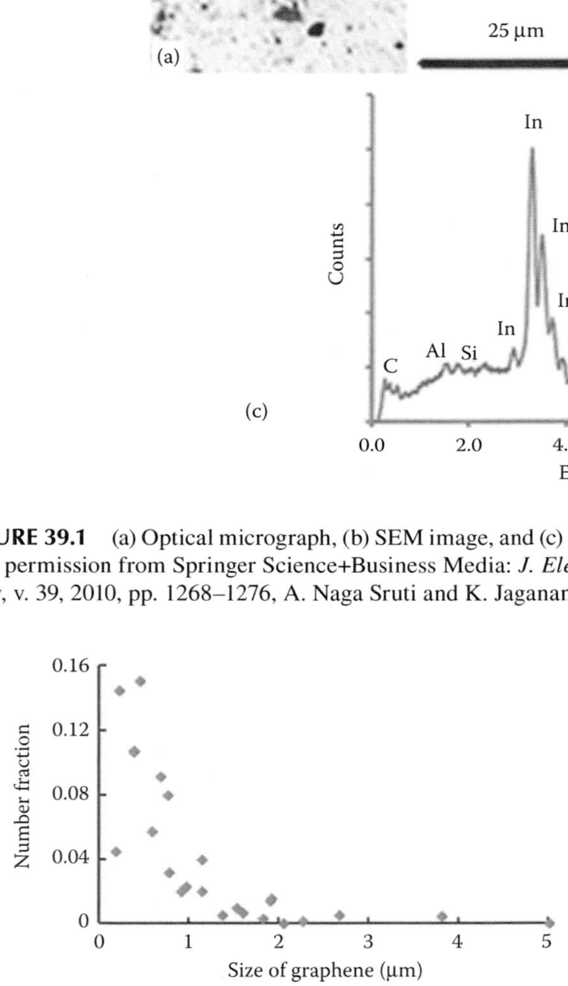

FIGURE 39.2 Number fraction shown against size of GPs in the suspension. (With kind permission from Springer Science+Business Media: *J. Elect. Mat.*, Thermal conductivity of indium-graphene and indium-gallium-graphene composites, v. 40, 2011, pp. 25–34, K. Jagannadham. Copyright 2011.)

from 285 to 315 K in steps of 5° with the resistance measured at each. The slope and the resistance were used to determine the TCR that is shown in Table 39.3 [25]. Modeling analysis using effective mean field approximation (EMA) was performed to determine the V_g and conductivity of graphene (σ_g) using the resistivity ρ and TCR of each samples, as described in the following sections.

39.4.2 Electrical Conductivity Measurements in Cu–gr

Electrical conductivity measurements of Cu–gr samples were made using the same procedure but with current passed from outer ends of the samples using conducting tapes and voltage measured at the two inner pins. This ensured a more uniform flow of current through the sample of width between 2 and 3 mm and thickness between 350 and 600 μm. The length of the samples was near 1.5 cm. Because the Cu–gr layers were on the opposite surfaces of Cu foil, we have used $1/R_s = 1/R_c + 1/R_{cu}$ to determine R_c where the subscript s, c, and cu refer to sample, Cu–gr composite, and Cu, respectively. The resistivity ρ_c of the Cu–gr composite is determined from $R_c = \rho_c l_c / t_c w_c$ where t_c and w_c are the thickness and width, respectively, and $l_c = 1.66$ mm between the inner pins as before. TCR of the samples was also measured using four contacts made with colloidal silver dots to gold wires that are connected to the Keithley instruments. TCR_c is determined from

$$TCR_s = \left(\frac{\rho_s t_c}{\rho_c t_s}\right) TCR_c + \left(\frac{\rho_s t_{cu}}{\rho_{cu} t_s}\right) TCR_{cu} \tag{39.1}$$

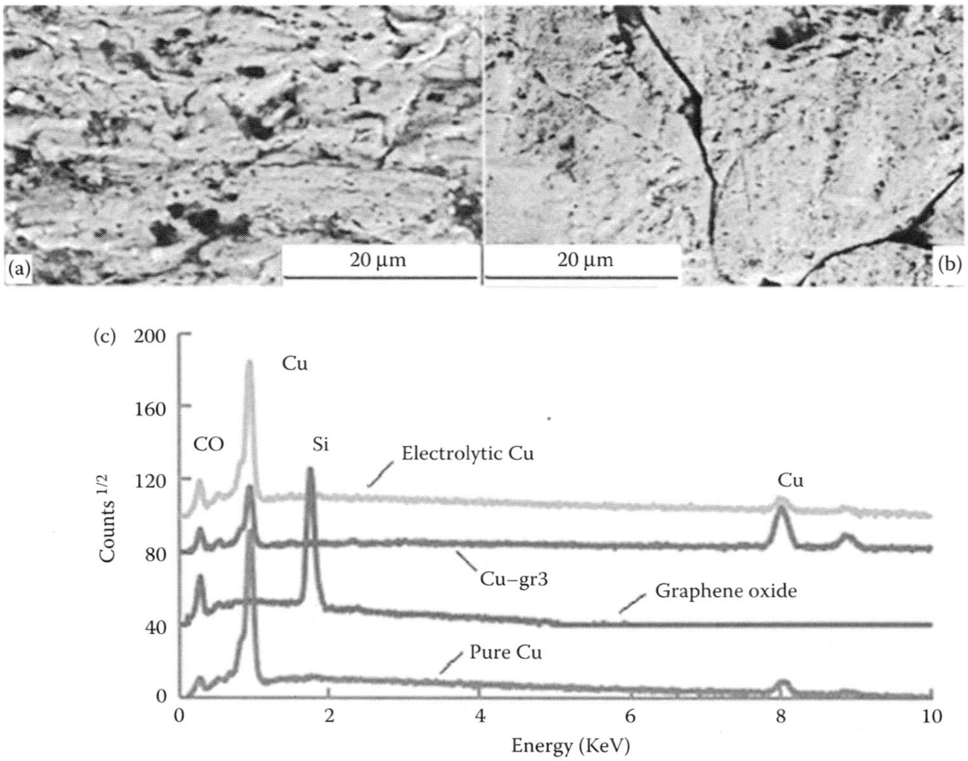

FIGURE 39.3 (a) SEM of cross section and (b) planar of Cu–gr sample. (With kind permission from Springer Science+Business Media: *Met. Mat. Trans.*, Volume fraction of graphene platelets in copper-graphene composites, v. 44A, 2013, pp. 552–559, K. Jagannadham.) (c) EDS of pure Cu, GO on Si, electrolytic Cu, and Cu–gr sample. (Reprinted with permission from K. Jagannadham, *J. Vac. Sci. Technol.*, B, v. 30, pp. 03D109-1–9, 2012. Copyright 2012, AVS.)

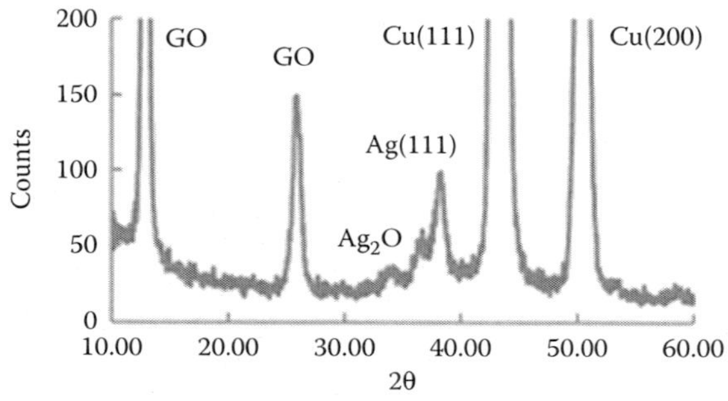

FIGURE 39.4 X-ray diffraction peak profile of GO on Cu. (With kind permission from Springer Science+Business Media: *Met. Mat. Trans.*, Volume fraction of graphene platelets in copper-graphene composites, v. 44A, 2013, pp. 552–559, K. Jagannadham. Copyright 2013.)

The values of TCR_c determined experimentally from Equation 39.1 are shown in Table 39.4 along with $\sigma_g = 1/\rho_g$ and V_g determined from analysis using EMA that is described in the following sections.

39.4.3 Modeling of Electrical Conductivity

The contribution of GPs to electrical conductivity of the composites is a function of thickness, as mentioned previously. Electrical conductivity of thinner GPs is higher but scattering of the carriers from matrix atoms reduces the value. The carrier concentration in graphene, close to 10^{12} cm^{-3}, is improved by injection of carriers from the matrix. It has been found that [6] Cu-doped graphene becomes n-type. Thicker GPs with more than 5–10 atomic layers have much lower conductivity similar to graphite. Therefore, it is important to determine the volume fractions of thinner and thicker GPs. SEM imaging does not provide a good estimation of these volume fractions due to lack of good contrast. However, thermal conductivity of GPs is less sensitive to the thickness compared to electrical conductivity. Therefore, we have evaluated the total volume fraction using thermal conductivity measurements and that of thinner GPs using electrical conductivity.

First, the contribution from GPs to electrical conductivity of the composite will be modeled using EMA. The electrical conductivity σ of the composite films from EMA is given by [21,47,48],

$$\frac{3V_m(\sigma_m-\sigma_c)}{(\sigma_m+2\sigma_c)}+\frac{(V_g/3)(2\sigma_g^a/\sigma_c-\sigma_c/\sigma_g^c)-V_g}{3}=0 \qquad (39.2)$$

where V is the volume fraction, the subscripts g, m, and c have the same meaning as before and the superscripts a and c represent the conductivity of graphene in the basal plane (ab plane) and normal to the basal plane (c-direction), respectively. The derivative of the above Equation 39.2 with respect to temperature (T) provided a second equation, which is not shown here for simplicity. Further, we use

$$\frac{(d\rho/dT)}{\rho} = \frac{(dR/dT)}{R} + \beta \quad (39.3)$$

with β the linear thermal expansion coefficient. The determination of β of the composite is described below. The value of β for In is 32.1×10^{-6} K^{-1} [25]. The above formulations are used to determine the V_g and ρ_g of GPs in In–gr composites and is shown in Table 39.3. The value of V_g determined from EMA or SEM is shown in two different columns in Table 39.3. It should be noted that ρ_g of GPs shown in Table 39.3 is the resistivity of thinner GPs obtained from modeling analysis because thicker GPs do not contribute significantly to conductivity.

The details of the results obtained for Cu–gr composites are discussed further. The value of β of Cu [49,50] is taken to be 16.5×10^{-6} K^{-1} and that of GP [48] is 6.5×10^{-6} K^{-1}. TCR of graphene [51] is -5.0×10^{-4} K^{-1}. The value of β of the composite using Turner's formula [52] is close to that of GP, which has much higher modulus than Cu. Also, σ_g^a/σ_g^c is found [47,48] to be close to 40 so that σ_g^a and V_g are the only unknowns determined from Equation 39.2 and its derivative with respect to temperature by an iterative procedure using the data provided in Table 39.4. The results of σ_g and V_g are shown in Table 39.4. It should again be noted that σ_g in Table 39.4 is the value of conductivity of thinner GPs as thicker GPs do not contribute significantly. The average value of σ_g is refined further after determining the total volume fraction f_g of GPs from thermal conductivity measurements. The value of f_g obtained from thermal conductivity measurements is also shown in Table 39.4.

39.4.4 Average Value of Electrical Conductivity of GPs

The difference between f_g and V_g determined from thermal conductivity and electrical conductivity, respectively, is the volume fraction of thicker GP. The electrical conductivity of thicker GPs, σ_{GP}, is expected to be small and we take the value to be 3.6×10^5 Ω^{-1} cm^{-1}. However, this value is higher than σ of graphite in the ab plane, which is close to 2.5×10^3 Ω^{-1} cm^{-1}. The contribution of the thicker GPs is included using the rule of mixtures in the form

$$\sigma_{gf} = \left(\frac{V_g}{f_g}\right)\sigma_g + \left(\frac{1-V_g}{f_g}\right)\sigma_{gp} \quad (39.4)$$

to determine the weighted average value of all the GPs. The value of σ_{gf} is smaller than σ_g, as shown in Table 39.4.

39.5 THERMAL CONDUCTIVITY

Thermal conductivity of the samples of size ~1.5 cm × 0.5 cm was determined using the 3ω method. A description of this method is provided in our previous work [17,18,30] and in the original reference [53]. A thin insulating polymer film was deposited on the surface of the sample so that the gold heater line deposited on the surface is isolated from the substrate electrically. In addition, thin insulating films of Si and ZrO_2 were also deposited on top so that the gold heater line is adherent to the surface.

Power was supplied to the heater at a sinusoidal voltage V_1 and frequency f (ω = 2πf) and the voltage V_3 at frequency 3f was measured using Model SR 830 DSP lock-in amplifier. The temperature increment of the surface wherein the gold heater acted both as source and measure was calculated using $dT = 2V_3/(\alpha V_1)$ where $\alpha = dR/RdT$ is the TCR of the heater measured separately. The variation of dT per unit power into the heater with frequency was curve fitted with multilayer analysis as described in our previous work [17,18,30] and reference [54]. The analysis used the width of the heater, thermal conductivity, heat capacity, and thickness of each layer.

39.5.1 In–gr and In–Ga–gr Composites and Laser Irradiation

The results from In–gr4 sample obtained after curve fitting are presented in Figure 39.5. The values of thermal conductivity in all the samples at three temperatures along with the volume fraction of graphene, f_g, determined from SEM imaging [18] are provided in Table 39.5.

The results presented in Table 39.5 illustrate that the thermal conductivity in In–gr with larger thickness (430 μm) improved by a factor of 2.3 at 300 K and by a factor of 2 at 350 K compared to that of In. Similarly, thermal conductivity of In–Ga–gr5 composite (330 μm) improved by a factor of 3 at 300 and 350 K compared to that of In–Ga. The larger improvement is associated with higher f_g in the In–Ga–gr compared to that in In–gr and the lower thermal conductivity of In–Ga.

The results also show that the thermal conductivity decreases inversely with temperature, which is expected

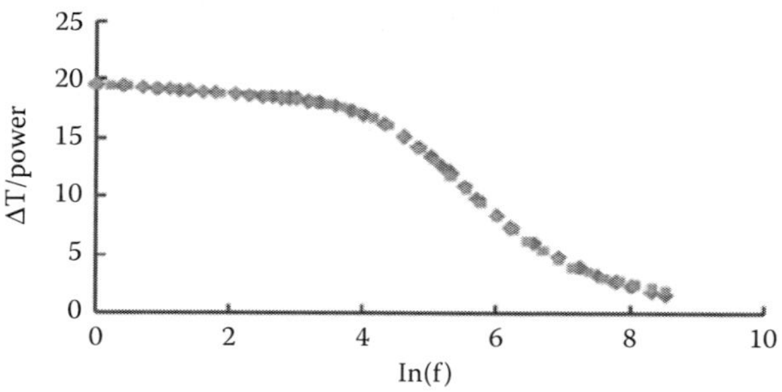

FIGURE 39.5 The temperature increment dT of the gold heater line shown as a function of frequency of power input at frequency f in In-gr-4 sample. The diamond symbols are experimental values and the square symbols are curve fitted data. (With kind permission from Springer Science+Business Media: *J. Electr. Mat.*, Thermal conductivity of indium-graphene and indium-gallium-graphene composites, v. 40, 2011, pp. 25–34, K. Jagannadham. Copyright 2011.)

from scattering of electrons by phonons in metals [55] and Umklapp scattering in graphene [12,56]. The phonon limited mean free path of electrons in In is dependent on crystallographic directions in the tetragonal structure and varies [57] between $0.769T^{-3}$ and $0.909T^{-3}$ cm. Thus, the value remains between 0.28 and 0.34 nm. The mean free path of electrons in In determined from electrical conductivity is close to 5 nm [58]. The average separation between GPs determined from SEM [25] is 0.8–0.9 μm in In–gr and 0.6–0.7 μm in In–Ga–gr. The electron mean free path is smaller than the separation between GPs. Therefore, there is no reduction in the bulk electronic thermal conductivity of In or In–Ga due to the presence of GPs. Ga atoms in In lattice are responsible for scattering of electrons and mean free path is reduced based on Matthiessen's rule [59]. Therefore, the experimentally observed thermal conductivity of 0.5 $Wcm^{-1} K^{-1}$ of In–Ga alloy is lower than 0.7 $Wcm^{-1} K^{-1}$ of In [60].

The thermal conductivity of In–gr samples prepared after laser irradiation of GPs was also determined [30]. GPs were dispersed on the In foil as before and irradiation using Nd-YAG laser beam with wave length 266 nm, pulse duration 6 ns and repetition rate of 10 Hz was carried to cover the sample area. The foil was folded and deformation rolled. This process was repeated several times to increase the volume fraction of GPs. Samples were also prepared with GPs that had not undergone laser treatment. Characterization by Raman spectroscopy showed that the D band width associated with GPs is not significantly increased at the laser fluence used [30]. Thermal conductivity measurements showed that laser irradiation has not changed the improvement brought about by GPs [30]. The results indicate that although defects are created by laser irradiation, the mean free path of phonons is not changed significantly. These results also emphasize that defects formed in GPs are either healed during annealing or helped to improve wetting between In and graphene. It has also been pointed out that septa- and penta-rings formed around defects in hexagonal carbon network of GPs do not prevent phonon propagation as large area of lattice is available around the defects [61,62]. Therefore, quenching of phonons from the substrate and the edge scattering due to finite size of graphene may be the only limiting parameters [11–13]. GPs with thickness between two and five atomic layers will be better than single layer GP so that quenching of phonons is limited.

39.5.2 Cu–gr Composites

The results of dT per unit power input along with that of multilayer analysis are shown in Figure 39.6 for the Cu–gr-5 sample [17]. The values of thermal conductivity of Cu–gr films deposited either in the presence or the absence of stirrer, shown in Table 39.6, are not different for the same value of R_o or the volume fraction of GPs. Also, the value is reduced with increasing ratio R_o indicating that dilution of the GO suspension with 0.2 M $CuSO_4$ solution reduced the volume fraction f_g of graphene. The thermal conductivity is improved by a factor of 1.25 for $f_g = 0.35$ in Cu–gr.

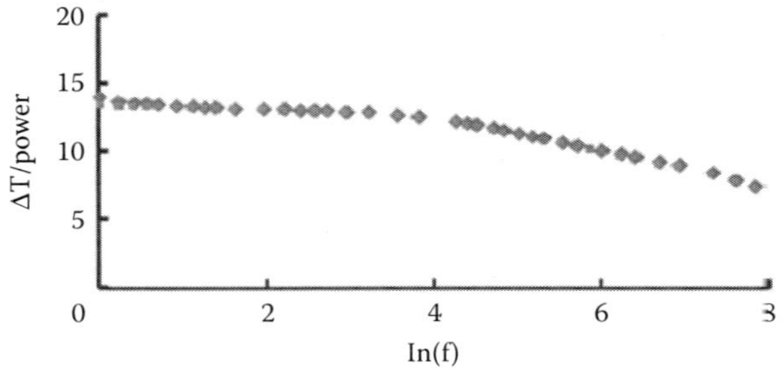

FIGURE 39.6 The temperature increment dT of the gold heater line shown as a function of frequency of power input at frequency f in Cu–gr-5 sample. The diamond symbols are experimental values and the square symbols are curve fitted data. (With kind permission from Springer Science+Business Media: *Met. Mat. Trans.*, Volume fraction of graphene platelets in copper-graphene composites, v. 44A, 2013, pp. 552–559, K. Jagannadham. Copyright 2013.)

TABLE 39.6
Thermal Conductivity of the Cu-gr Films Prepared with Different Electrolytes Containing GO Is Shown

Sample ID	Stirrer	R_o	K	f_g	h
Cu–gr-1	No	1.0	5.0	0.35	1.49
Cu–gr-2	No	0.50	4.8	0.28	1.47
Cu–gr-3	No	0.33	4.6	0.21	1.43
Cu–gr-4	No	0.25	4.4	0.14	1.40
Cu–gr-5	Yes	1.0	5.0	0.35	1.49
Cu–gr-6	Yes	0.5	4.8	0.28	1.47
Cu–gr-7	Yes	0.33	4.6	0.21	1.43
Cu–gr-8	Yes	0.25	4.4	0.14	1.40
Cu-9	No	0.0	3.8	0.0	–

Source: With kind permission from Springer Science+Business Media: *Met. Mat. Trans.*, Volume fraction of graphene platelets in copper-graphene composites, v. 44A, 2013, pp. 552–559, K. Jagannadham. Copyright 2013.

Note: R_o is the ratio of 250 mL of GO suspension in distilled water to volume of 0.2 M solution of $CuSO_4$. The volume fraction f_g of graphene and the interfacial conductance h between graphene and Cu determined from EMA are also shown. The value of K is in $Wcm^{-1} K^{-1}$ and h is in 10^5 $Wcm^{-2} K^{-1}$.

39.5.3 Modeling of Thermal Conductivity

39.5.3.1 In–gr Composites

The thermal conductivity of graphene composites is modeled using EMA [63]. The GPs are treated as flat ellipsoids that are randomly oriented in the matrix. Under this approximation, the isotropic thermal conductivity of the composite medium is given by [63],

$$K_c = K_g \frac{\{K_m(3-2f_g)+ 2f_gK_g\}}{\{f_gK_m + K_g(3-f_g+af_g)\}} \tag{39.5}$$

where f_g is the volume fraction of graphene and K stands for thermal conductivity. The parameter "a" is given by K_m/hr where h is the interfacial thermal conductance between the

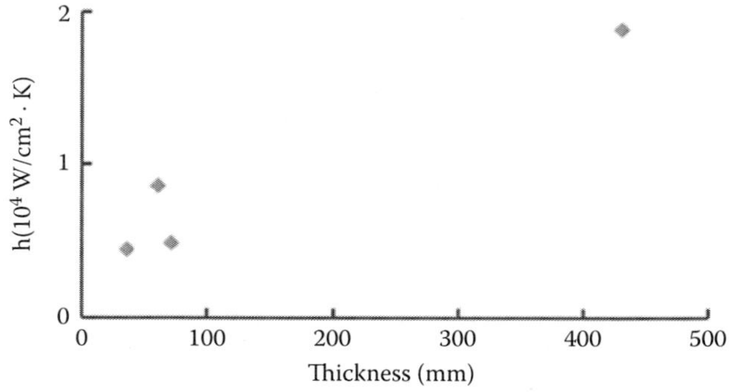

FIGURE 39.7 Interface thermal conductance, h, shown as a function of thickness of the sample in In–gr composite samples. (With kind permission from Springer Science+Business Media: *J. Electr. Mat.*, Thermal conductivity of indium-graphene and indium-gallium-graphene composites, v. 40, 2011, pp. 25–34, K. Jagannadham. Copyright 2011.)

matrix and graphene and r is the effective radius of the GPs. The value of h is different when heat flux is normal or parallel to the ab plane. The thermal conductivity of GPs is higher in the ab plane and therefore h parallel to the ab plane is important in composites. Also, we assume random orientation of GPs so that heat transfer along ab planes is effective. EMA is used to determine h and evaluate whether it is responsible for the observed improvement in the thermal conductivity of the composites.

The parameter "a" is evaluated from Equation 39.5. K_g depends on the observed size L of GPs. Experimental determinations [7,8] of K_g have shown that its value is very high when the L is bigger such as 5 μm. Modeling analysis [9,10,12] also indicated that the value decreases to 10 $Wcm^{-1}K^{-1}$ when L is 1 μm. The value of K_g follows [64] $(L)^b$ where b remains between 0.35 and 0.45. More importantly, Klemens pointed out [12,13] that thermal energy leaks into the matrix so that K_g is reduced by 20%–50% if the phonon velocity in the matrix is smaller than that in graphene, a result certainly true for In or In–Ga matrix.

The volume fraction of graphene f_g and the size distribution shown in Figure 39.2 have been evaluated from SEM images and f_g is shown in Table 39.5. The weighted average of the thermal conductivity K_g of GPs was determined to be 8 $Wcm^{-1} K^{-1}$. K_c from Table 39.5 and K_m are used to evaluate the parameter "a" in Equation 39.5. Further, the value of h between the matrix and GPs in the ab plane is evaluated using $r = L_{av}/2$ or $a = 2/L_{av}h$ where L_{av} is the average size. The value of h in In–gr increases, as shown in Figure 39.7, with thickness of the composite, which is a direct result of increase in K_c.

39.5.3.2 Cu–gr Composites

The thermal conductivity of Cu–gr films deposited either with or without the stirrer as shown in Table 39.6 was not different for two reasons. First, the average thickness of GPs in the films deposited in the presence of the stirrer is slightly larger but the average thermal conductivity of the GPs is not significantly changed. Second, the volume fraction of the GPs is also slightly larger but cannot be evaluated by SEM. Thus, the net contribution from GPs remained same within experimental accuracy. The values of f_g and h in the a–b plane determined from Equation 39.5 are shown in Table 39.6. The results illustrate that h remains between 1.4 and 1.5 $GWm^{-2} K^{-1}$, which is reasonably high. These results emphasize that heat transfer in the composites is not limited by the interface thermal conductance in the ab plane.

39.6 TRANSIENT THERMO REFLECTANCE

39.6.1 Transient Thermo Reflectance Measurements

In the experimental set up for transient thermo reflectance (TTR) measurements [65,66], Nd-YAG laser source in the second harmonic (λ = 532 nm), pulse duration >6 ns, and repetition of 10 Hz was used to heat the sample surface. A probe red laser with continuous wave length of 650 nm and variable power up to 20 mW was used to measure the TTR signal. The two laser beams were focused on the sample to a spot size close to 60 μm using a 10 × objective. The reflected laser beam was detected by a silicon photo diode with 1 ns rise time. The output from the detector was amplified and recorded using Tektronix DPO4104 B oscilloscope. The signal in the oscilloscope was triggered by output from another silicon diode used to detect the Nd-YAG laser pulse prior to incidence on the sample surface. A more detailed description is available in Panzer et al. [66].

Cu and Cu–gr substrates were polished with 3M 600 grade abrasive paper and further on rotating wheel with polishing cloth dispersed with 0.05 μm size alumina abrasive suspension in water. The substrates were etched with dilute nitric and sulfuric acid mixture to remove the abrasive particles embedded on the surface. A thin In film was pressed on to Cu and Cu–gr samples. Al film was deposited by magnetron sputtering on another set of Cu and Cu–gr samples. The TTR signals from these samples are shown in Figures 39.8 and 39.9. The wavy signal is the experimental result and the continuous line is the result obtained by modeling. The TTR signal is a measure of the surface temperature of the film on the sample surface. The results shown in Figures 39.8 and 39.9 illustrate that Cu–gr is a better heat spreader than Cu.

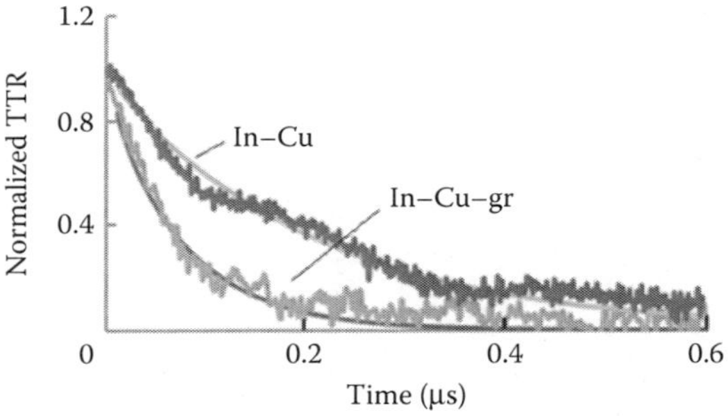

FIGURE 39.8 Normalized TTR signal shown as a function of time for In foil pressed on to polished Cu and Cu–gr composite surface. The curve fitted lines from analysis are also shown. (Reprinted with permission from H. Zheng and K. Jagannadham, *AIP Adv.*, Transient thermoreflectance from graphene composites with matrix of indium and copper, v. 3, pp. 032111-1–12, 2013. Copyright 2013, American Institute of Physics.)

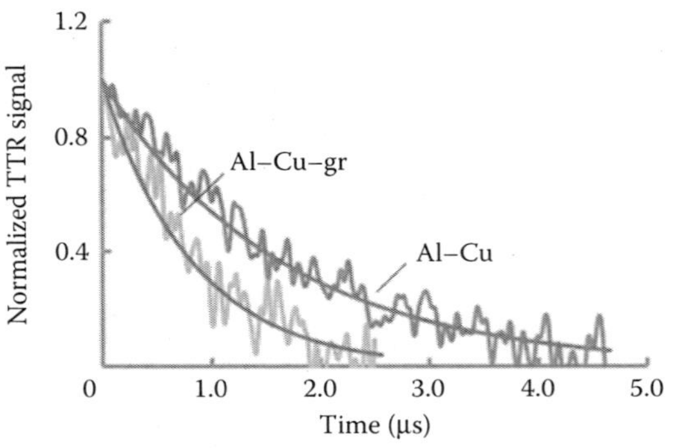

FIGURE 39.9 Normalized TTR signal shown as a function of time for Al film deposited on to Cu and Cu–gr surface by magnetron sputtering. The curve fitted lines from analysis are also shown. (Reprinted with permission from H. Zheng and K. Jagannadham, *AIP Adv.*, Transient thermoreflectance from graphene composites with matrix of indium and copper, v. 3, pp. 032111-1–12, 2013. Copyright 2013, American Institute of Physics.)

39.6.2 Analysis of the TTR Signal

The experimental TTR signal variation as a function of time was modeled to determine the interface thermal conductance between the film and the substrate. The thermal conductivity of the film and the substrate, thermal conductance of the interface and the thickness of the film on the surface are the parameters used in modeling. The thickness of the film and the substrate was measured. The thermal conductivity of the film and the substrate is taken from the values determined by 3ω method [17–19,21] so that the interface thermal conductance is the only unknown parameter evaluated. The one-dimensional heat equation was used [66] in the form

$$\frac{\partial T}{\partial t} = D\frac{\partial^2 T}{\partial x^2} \tag{39.6}$$

where $T = T(x,t)$ is the temperature as a function of depth x from the surface and time t, and D is the thermal diffusivity. Equation 39.6 was solved using finite difference method. The flux at the surface from laser incidence in the form [66]

$$J = -\frac{K_f \partial T}{\partial x} = \left(\frac{1}{\pi\tau^2}\right)^{1/2} L_f C_f dT_{max} \exp\left(\frac{-t^2}{\tau^2}\right) \tag{39.7}$$

where τ is the laser pulse width of 6 ns, dT_{max} is the peak temperature change, K, L, and C are thermal conductivity, thickness, and heat capacity, respectively, of the film represented with subscript, f. The continuity of the flux at the interface between film and substrate was applied in the form

$$\frac{K_f \partial T_f}{\partial x} = \frac{K_s \partial T_s}{\partial x} = -h(T_f - T_s) \tag{39.8}$$

with the subscript, s, used to represent the substrate. In Equation 39.8, h is the interface thermal conductance. The normalized temperature profile (T/T_{max}) was used to fit the normalized TTR signal variation with time t obtained experimentally [65].

The values of h for In film pressed on to Cu and Cu–gr were found to be 4 and 15 $MWm^{-2}K^{-1}$, respectively. These values are lower as a result of incomplete contact between In and polished surfaces of Cu or Cu–gr. However, these values are higher than that of h between Pb–Sn eutectic solder and Cu [67] bonded in liquid state, which is 0.5 $MWm^{-2}K^{-1}$. The thermal conductivity of Pb–Sn eutectic is 50 $Wm^{-1}K^{-1}$, thus, lower than that of In. The value of h between In and carbon nanotubes after metallization has been estimated to be 1 $MWm^{-2}K^{-1}$ [66] although the thermal conductivity of In and CNTs is reasonably high.

The values of h between Al film and Cu or Cu–gr with TTR shown in Figure 39.9 were found to be 1 and 2 MWm^{-2} K^{-1}, respectively. Again, two factors are thought to be responsible for these lower values. First, the real area of contact between the Al film and the polished substrate is a fraction of the ideal area [68]. Second, atomic layers of water vapor at the interface may be present in the valleys associated with surface markings [69], which are responsible for scattering of electrons that transport thermal energy.

39.7 MODELING OF INTERFACE THERMAL CONDUCTANCE IN ab PLANE BETWEEN GPs AND IN OR CU

The interface thermal conductance in the ab plane between GPs and In, from Figure 39.7, was close to 0.1–0.2 $GWm^{-2}K^{-1}$ and that between GPs and Cu was close to 1.4–1.5 GWm^{-2} K^{-1} as shown in Table 39.6. These values are reasonably high, which does not make the limiting factor in the heat spreading. The value of h normal to the ab plane between GPs and metals was found to be lower [14,15] and close to 25 MWm^{-2} K^{-1}, thus, limits the heat flux. We now provide the modeling of interface thermal conductance in ab plane using acoustic mismatch (AMM) and diffuse mismatch (DMM) models.

The electron–phonon coupling term, h_{ep}, in In is determined using [70]

$$G_{ep} = \frac{C_{em}}{\tau} \quad \text{and} \quad h_{ep} = (G_{ep}K_{pm})^{1/2} \tag{39.9}$$

where C_{em} is the electronic specific heat of In and τ is the relaxation time for electron–phonon energy transfer. The electronic specific heat is calculated using the $C_{em} = \gamma T$, where $\gamma = 0.105$ $mJcm^{-3}$ K^{-2} as provided in Reference 18. The relaxation time is assumed to be 1 ps [70] so that $G_{ep} = 31.6$ $GWcm^{-3}$. The phonon thermal conductivity of In is calculated from $K_{pm} = C_{lm}v_{dm}l_p/3$ where C_{lm} is the lattice specific heat, v_{dm} the Debye phonon velocity, and l_p is the phonon mean free path taken to be 5 nm [70] and from values provided in Reference 18, $K_{pm} = 0.074$ Wcm^{-1} K^{-1} so that h_{ep} is 4.84×10^4 Wcm^{-2} K^{-1}.

The phonon–phonon contribution, h_{pp}, is determined by AMM and DMM [71,72]. In the AMM,

$$h_{pp} = C_{lm}v_{dm}\frac{\alpha}{4} \tag{39.10}$$

where α is the transmission coefficient determined from

$$\alpha = \frac{4Z_m Z_g}{(Z_m + Z_g)^2} \quad (39.11)$$

where Z is the acoustic impedance of each medium given by the product of density and Debye velocity, the subscripts, m and g, refer to In and graphene, respectively. The value of h_{pp} from AMM determined using the values provided in Reference 18 is 5.1×10^4 Wcm^{-2} K^{-1}. The expression for h_{pp} in the DMM [71] is

$$h_{pp} = 1.02 \times 10^{10}\, T^3 \frac{\left\{\Sigma\left(1/v_m^2\right)\Sigma\left(1/v_g^2\right)\right\}}{\left\{\Sigma\left(1/v_m^2 + 1/v_g^2\right)\right\}} \quad (39.12)$$

The summation in Equation 39.12 is performed over the longitudinal and the two transverse modes in In but only over longitudinal and one transverse mode in graphene because it is a two-dimensional medium. Using the results given in reference [18], the value of h_{pp} is 1.6×10^5 Wcm^{-2} K^{-1}. The net thermal conductance h [70] given by $1/h = 1/h_{ep} + 1/h_{pp}$ is 3.7×10^4 Wcm^{-2} K^{-1} using DMM and 2.5×10^4 Wcm^{-2} K^{-1} using AMM.

An alternative expression from DMM [72] is

$$h_{pp} = \frac{C_{lm} v_{dm} C_{lg} v_{dg}}{4(C_{lm} v_{dm} + C_{lg} v_{dg})} \quad (39.13)$$

Using the results provided in Reference 18, the value of $h_{pp} = 6.7 \times 10^4$ Wcm^{-2} K^{-1}. In this case, the net value of h is 2.8×10^4 Wcm^{-2} K^{-1}.

Alternately, h is calculated based on graphene as a conducting medium. In this case, h consists of electron–electron coupling between In and graphene and electron–phonon coupling in graphene. The first part is determined from [73]

$$h_{ee} = \frac{C_{eg} v_{fg} C_{em} v_{fm}}{[4(C_{eg} v_{fg} + C_{em} v_{fm})]} \quad (39.14)$$

where v_{fg} and v_{fm} stand for Fermi velocity and C_{eg} and C_{em} represent the electronic specific heat of graphene and In, respectively. The electronic specific heat is determined as before and the value of h_{ee} at $T = 300$ K is found to be 5.9×10^4 Wcm^{-2} K^{-1}. The electron–phonon coupling term [70] in graphene, h_{ep}, is determined as before using the values provided in Jagannadham [18] and $G_{ep} = 2.38$ GWcm^{-3} K^{-1}. The phonon thermal conductivity, K_{pg}, in graphene is size-dependent with the average value shown before to be 8.0 Wcm^{-1} K^{-1} so that $h_{ep} = 1.38 \times 10^5$ Wcm^{-2} K^{-1}. The net thermal conductance h is 4.1×10^4 Wcm^{-2} K^{-1}. Thus, the value of h based on electron–electron coupling is slightly higher than when phonon–phonon coupling is used between In and graphene.

The different components and the total value of h are listed in Table 39.7. In both these models, the value of h is higher and closer to the experimentally determined value in In–gr4 with thickness 430 μm. The experimentally observed value of h in In–Ga–gr5 composites is lower although the percentage of Ga is only 5 at%. The presence of Ga introduces interface mismatch and scattering of the acoustic phonons responsible for heat transfer [74]. In addition, Ga atoms are also responsible for scattering of electrons at the interface. These results suggest that a lower interface thermal conductance is expected in In–Ga–gr composites.

Similar calculations performed using different parameters associated with GPs and Cu, provided in Reference 19, are shown in Table 39.8. These results also indicate the value of h

TABLE 39.7
Components of Thermal Conductance across In and Graphene Interface

Electron–Phonon and Phonon–Phonon Coupling		Total Thermal Conductance	Electron–Electron and Electron–Phonon Coupling		Total Thermal Conductance
h_{ep} (In)	h_{pp}(In–graphene)	h	h_{ee}(In–graphene)	h_{ep} (graphene)	h
4.8	5.1 (AMM)	2.5	5.9 (DMM)	13.8	4.1
4.8	16 (DMM)	3.7			
4.8	6.7 (DMM)	2.8			

Source: With kind permission from Springer Science+Business Media: *J. Electr. Mat.*, Thermal conductivity of indium-graphene and indium-gallium-graphene composites, v. 40, 2011, pp. 25–34, K. Jagannadham. Copyright 2011.

Note: The first model is based on phonon–phonon coupling and the second model is based on electron–electron coupling between In and graphene. All values of thermal conductance are given in units of 10^8 Wm^{-2} K^{-1}. Values obtained using AMM and DMM are listed for h_{pp}.

TABLE 39.8
Different Components of h across Cu and Graphene Interface

Electron–Phonon and Phonon–Phonon Coupling		Total Thermal Conductance	Electron–Electron and Electron–Phonon Coupling		Total Thermal Conductance
h_{ep} (Cu)	h_{pp}(Cu–graphene)	h	h_{ee}(Cu–graphene)	h_{ep} (graphene)	h
6.9	22.6 (AMM)	5.3	6.2 (DMM)	13.8	4.3
6.9	15.0 (DMM)	4.7			
6.9	18.9 (DMM)	5.0			

Source: With kind permission from Springer Science+Business Media: *Met. Mat. Trans.*, Thermal conductivity of copper-graphene composite films synthesized by electrochemical deposition with exfoliated graphene platelets, v. 43B, 2012, pp. 316–324, K. Jagannadham. Copyright 2012.

Note: The first model is based on phonon–phonon coupling and the second model is based on electron–electron coupling between Cu and graphene. All values of thermal conductance are given in units of 10^8 Wm^{-2} K^{-1}. Values obtained using AMM and DMM are listed for h_{pp}.

between GPs and Cu in the ab plane is close to 0.5 $GWm^{-2} K^{-1}$ which is fairly high so that heat transfer is not limited by the interface.

39.8 SUMMARY AND CONCLUSIONS

We have shown that In–gr and In–Ga–gr composites with uniform dispersion of GPs are achieved by dispersion on the metal foils followed by deformation rolling. Similarly, Cu–gr composites are synthesized by electrochemical codeposition of Cu and GPs. These are low-cost methods suitable for manufacturing. Higher volume fraction of GPs is also easily achieved. It has been shown that determination of volume fraction by SEM image analysis is subject to limitation of weak contrast from thin GPs. Electrical and thermal conductivity are improved by thinner GPs. However, thermal conductivity is improved by thicker GPs as well although not to the same magnitude. It is important to use GPs with thickness of 2–4 atomic layers for electrical conductivity and 2–8 atomic layers for thermal conductivity improvement.

The interface thermal conductance in the ab plane between GPs and metals is fairly high that it is not a limiting factor in heat spreading. The thermal conductivity of In–gr and In–Ga–gr is improved by more than a factor of 2 for volume fraction of 0.2 and that of Cu–gr by a factor of 1.25 for volume fraction of 0.35. These improvements enable the use of In–gr and In–Ga–gr as thermal interface materials and Cu–gr as heat spreaders in high-power electronic devices.

The electrical conductivity of Cu–gr composites is improved by a factor of 1.2 with a volume fraction of 0.1 of thin GPs. The key to improvement of transport properties remains in achieving uniform dispersion of larger size GPs. Lower coefficient of friction, improved wear resistance, and higher electrical conductivity are favorable factors for use of Cu–gr composites as electrofriction materials.

ACKNOWLEDGMENTS

This research is supported by National Science Foundation, Grant CMMI #1049751.

REFERENCES

1. D. S. Hecht, L. Hu, and G. Irvin, Emerging transparent electrodes based on thin films of carbon nanotubes, graphene, and metallic nanostructures, *Adv. Mater.*, v. 23, pp. 1482–1513, 2011.
2. A. K. Geim and K. S. Novoselov, The rise of graphene, *Nature Mat.*, v. 6, pp. 183–191, 2007.
3. A. Akturk and N. Goldsman, Electron transport and full-band electron-phonon interactions in graphene, *J. Appl. Phys.*, v. 103, pp. 053702-1-8, 2008.
4. J. Meyer, A. K. Geim, M. I. Katsnelson, K. S. Novoselov, T. J. Booth, and S. Roth, The structure of suspended graphene sheets, *Nature*, v. 446, pp. 60–63, 2007.
5. J. C. Meyer, C. Kisielowski, R. Erni, M. D. Rossell, M. F. Crommie, and A. Zettl, Direct imaging of lattice atoms and topological defects in graphene membranes, *Nano Lett.*, v. 8, pp. 3582–3586, 2008.
6. Y. Ren, S. Chen, W. Cai, Y. Zhu, C. Zhu, and R. S. Ruoff, Controlling the electrical transport properties of graphene by in situ metal deposition, *Appl. Phys. Lett.*, v. 97, pp. 053107-1-3, 2010.
7. A. A. Balandin, S. Ghosh, W. Bao, I. Calizo, D. Teweidebrhan, F. Miao, and C. N. Lau, Superior thermal conductivity of single-layer graphene, *Nano Lett.*, v. 8, pp. 902–907, 2008.
8. S. Ghosh, I. Calizo, D. Teweldebrhan, E. P. Pokatilov, D. L. Nika, A. A. Balandin, W. Bao, F. Miao, and C. N. Lau, Extremely high thermal conductivity of graphene: Prospects for thermal management applications in nanoelectronic circuits, *Appl. Phys. Lett.*, v. 92, pp. 151911-1-3, 2008.
9. D. L. Nika, E. P. Pokatilov, A. S. Askerov, and A. A. Balandin, Phonon thermal conduction in graphene: role of Umklapp and edge roughness scattering, *Phys. Rev. B*, v. 79, pp. 155413-1-12, 2009.
10. D. L. Nika, S. Ghosh, E. P. Pokatilov, and A. A. Balandin, Lattice thermal conductivity of graphene flakes: Comparison with bulk graphite, *Appl. Phys. Lett.*, v. 94, pp. 203103-1-3, 2009.
11. J. H. Seol, I. Jo, A.L. Moore, L. Lindsay, Z. H. Aitken, M. T. Pettes, X. Li et al., Two-dimensional phonon transport in supported graphene, *Science*, v. 328, pp. 213–216, 2010.
12. P. G. Klemens, Theory of thermal conduction in thin ceramic films, *Int. J. Thermophysics*, v. 22, pp. 265–275, 2001.
13. P. G. Klemens and D. F. Pedraza, Thermal conductivity of graphite in the basal plane, *Carbon*, v. 32, pp. 735–741, 1994.
14. P. E. Hopkins, M. Baraket, E. V. Barnat, T. E. Beechem, S. P. Kearney, J. C. Duda, J. T. Robinson, and S. G. Walton, Manipulating thermal conductance at metal-graphene contacts via chemical functionalization, *Nano Lett.*, v. 12, pp. 590–595, 2012.
15. Y. K. Koh, M. H. Bae, D. G. Cahill, and E. Pop, Heat conduction across monolayer and few-layer graphenes, *Nano Lett.*, v. 10, pp. 4363–4368, 2010.
16. A. J. Schmidt, K. C. Collins, A. J. Minnich, and G. Chen, Thermal conductance and phonon transmissivity of metal-graphite interfaces, *J. Appl. Phys.*, v. 107, pp. 104907-1-7, 2010.
17. K. Jagannadham, Volume fraction of graphene platelets in copper-graphene composites, *Met. Mat. Trans.*, v. 44A, pp. 552–559, 2013.
18. K. Jagannadham, Thermal conductivity of indium-graphene and indium-gallium-graphene composites, *J. Electr. Mat.*, v. 40, pp. 25–34, 2011.
19. K. Jagannadham, Thermal conductivity of copper-graphene composite films synthesized by electrochemical deposition with exfoliated graphene platelets, *Met. Mat. Trans.*, v. 43B, pp. 316–324, 2012.
20. S. W. Chang, A. K. Nair, and M. J. Buehler, Geometry and temperature effects of the interfacial thermal conductance in copper- and nickel-graphene nanocomposites, *J. Phys. Condens. Matt.*, v. 24, pp. 245301-1-6, 2012.
21. K. Jagannadham, Electrical conductivity of copper-graphene composite films synthesized by electrochemical deposition with exfoliated graphene platelets, *J. Vac. Sci. Technol., B*, v. 30, pp. 03D109-1-9, 2012.
22. M. J. McAllister, J-L Li, D. H. Adamson, H. C. Schniepp, A. A. Abdala, J. Liu, M. Herrera-Alonso, and D. L Milius, Single sheet functionalized graphene by oxidation and thermal expansion of graphite, *Chem. Mater.*, v. 19, pp. 4396–4404, 2007.

23. S. Stankovich, D. A. Dikin, R. D. Piner, K. A. Kohlhaas, A. Kleinhammes, Y. Jia, Y. Wu, S. T. Ngyyen, and R. S. Ruoff, Synthesis of graphene-based nanosheets via chemical reduction of exfoliated graphite oxide, *Carbon*, v. 45, pp. 1558–1565, 2007.
24. S. Park and R. S. Ruoff, Chemical methods for the production of graphenes, *Nature Nanotechnol*., v. 4, pp. 217–224, 2009.
25. A. N. Sruti and K, Jaganandham, Electrical conductivity of graphene composites with In and In-Ga alloy, *J. Elect. Mat.* v. 39, pp. 1268–1276, 2010.
26. A. A. Green and M. C. Hersam, Emerging methods for producing monodisperse graphene dispersions, *J. Phys. Chem. Lett*., v. 1, pp. 544–549, 2010.
27. U. Khan, A. O'Neill, H. Porwal, P. May, K. Nawaz, and J. N. Coleman, Size selection of dispersed, exfoliated graphene flakes by controlled centrifugation, *Carbon*, v. 50, p. 470–475, 2012.
28. P. Kumar, K. S. Subrahmanyam, and C. N. R. Rao, Graphene produced by radiation-induced reduction of graphene oxide, *In. J. Nanoscience*, v. 10, p. 556–559, 2011.
29. F. Ostovari, Y. Abdi, and F. Ghasemi, Controllable formation of graphene and graphene oxide sheets using photo-catalytic reduction and oxygen plasma treatment, *Eur. Phys. J. Appl. Phys*., v. 60, pp. 30401-1-6, 2012.
30. K. Jagannadham, Influence of laser and thermal treatment on the thermal conductivity of In-graphene composites, *J. Appl. Phys*., 110, 094907-1-4, 2011.
31. B. P. Singh, S. Nayak, K. K. Nanda, B. K. Jena, S. Bhattacharjee, and L. Besra, The production of a corrosion resistant graphene reinforced composite coating on copper by electrophoretic deposition, *Carbon*, v. 61, pp. 47–56, 2013.
32. D. Kuang, L. Xu, L. Liu, W. Hu, and Y. Wu, Graphene-nickel composites, *Appl. Surf. Sci*., v. 273, p. 484–490, 2013.
33. J. Fang, W. Zha, M. Kang, S. Lu, L. Cui, and S. Li, Microwave absorption response of nickel/graphene nanocomposites prepared by electrodeposition, *J. Mat. Sci*., v. 48, pp. 8060–8067, 2013.
34. M. Li, H. Che, X.Liu, S. Liang, and H. Xie, Highly enhanced mechanical properties in Cu matrix composites reinforced with graphene decorated metallic nanoparticles, *J. Mat. Sci*., v. 49, pp. 3725–3731, 2014.
35. J. Wang, Z. Li, G. Fan, H. Pan, Z. Chen, and D. Zhang, Reinforcement with graphene nanosheets in aluminum matrix composites, *Scripta Mater*., v. 66, p. 594–597, 2012.
36. L.-Y. Chen, H. Konishi, A. Fehrenbacher, C. Ma, J.-Q. Xu, H. Choi, F. E. Pfefferkorn, and X.-C. Li, Novel nanoprocessing route for bulk graphene nanoplatelets reinforced metal matrix nanocomposites, *Scripta Mater*., v. 67, pp. 29–32, 2012.
37. A. Takasaki, Y. Furuya, and M. Katayama, Mechanical alloying of graphite and magnesium powders, and their hydrogenation, *J. Alloys. Compounds*, v. 446–447, pp. 110–113, 2007.
38. H. D. Jang, S. K. Kim, H. Chang, J. W. Choi, and J. Hunag, Synthesis of graphene based noble metal composites for glucose biosensor, *Mat. Lett*., v. 106, p. 277–280, 2013.
39. H. Meng, J. Luo, W. Wang, Z. Shi, Q. Nia, L. Dai, and G. Qin, Top-emission organic light-emitting diode with a novel copper/graphene composite anode, *Adv. Funct. Mat*., v. 23, pp. 3324–3328, 2013.
40. C. Xu, X. Wang, L. Yang, and Y. Wu, Fabrication of a graphene-cuprous oxide composite, *J. Solid State Chem*., v. 182, pp. 2486–2490, 2009.
41. A.G. Nasibulin, T. Koltsova, L. I. Nasibulina, I. V. Anoshkin, A. Semencha, O. V. Tolochko, and E. I. Kauppinen, A novel approach to composite preparation by direct synthesis of carbon nanomaterial on matrix or filler particles, *Acta Mater*., v. 61, pp. 1862–1871, 2013.
42. S.-S. Li, J.-C. Lu, and M.-H. Teng, Synthesis of carbon encapsulated non-ferromagnetic metal nanoparticles, *Diam. Rel. Mat*., v. 24, pp. 88–92, 2012.
43. S. R. Sahu, M. M. Devi, P. Mukherjee, P. Sun, and K. Biswas, Optical property characterization of novel graphene-X (X = Ag, Au and Cu) nanoparticle hybrids, *J. Nano Mat*., 232409, pp. 9, 2013.
44. K. Gotoh, T. Kinumoto, E. Fujii, A. Yamamoto, H. Hashomoto, T. Ohkubo, A. Itadani, Y. Kuroda, and H. Ishida, Exfoliated graphene sheets decorated with metal/metal oxide nanoparticles: Simple preparation from cation exchanged graphite oxide, *Carbon*, v. 49, pp. 118–1125, 2011.
45. F. Yue, C. W. Jing, Y. Fei, and Z. D. Yu, Synthesis of graphene load nickel nanoparticles composites with hydrothermal process, *Adv. Mat. Res*., v. 760–762, pp. 793–796, 2013.
46. H. Zheng and K. Jagannadham, Thermal conductivity and interface thermal conductance in composites of titanium with graphene platelets, *J. Heat Transfer*, v. 136, pp. 061301-1-9, 2014.
47. A. Celzard, J. Mareche and G. Furdin, Modelling of exfoliated graphite, *Prog. Mater. Sci*., v. 50, pp. 93–179, 2005.
48. J. Helsing and A. Helte, Effective conductivity of aggregates of anisotropic grains, *J. Appl. Phys*., v. 69, pp. 3583–3588, 1991.
49. P. A. Tipler and G. Mosca, *Physics for Scientists and Engineers* (5th edition, v. 2), Freeman, New York, 2004, pp. 792.
50. *Thermal Expansion—The Physics Hyper textbook*, 1998–2010, Glenn Elert. http://physics.info/expansion/
51. J. B. Nelson and D. P. Riley, The thermal expansion of graphite from 15°C to 300°C: I. Experimental. II. Theoretical, *Proc. Phys. Soc.,* v. 57, pp. 477–95, 1945.
52. P. S. Turner, Thermal expansion stresses in reinforced plastics, *J. Res. Natl. Bur. Standards Res Rep.* 1745, v. 37, pp. 239–250, 1946.
53. D. G. Cahill, Thermal conductivity measurement from 30 to 750 K: The 3ω method, *Rev. Sci. Instrum*., v. 61, pp. 802–808, 1990.
54. J. H. Kim, A. Feldman, and D. Novotny, Application of the three omega thermal conductivity measurement method to a film on a substrate of finite thickness, *J. Appl. Phys*., v. 86, pp. 3959–3963, 1999.
55. J. Yang, in *Thermal Conductivity: Theory, Properties, and Applications*, Ed. T. M. Tritt, Kluwer Academic/Plenum publishers, New York, 2004, p. 1.
56. S. Ghosh, D. L. Nika, E. P. Pokatilov, and A. A. Balandin, Heat conduction in graphene: experimental study and theoretical interpretation, *New J. Phys*., v. 11, pp. 095012-1-19, 2009.
57. D. G. de Groot, A. B. M. Hoff, and J. H. P. va Weeren, Determination of mean free paths in indium from the radio frequency size effect and the peak near zero field, *J. Phys. F. Metal Physics,* v. 5, pp. 1559–1567, 1975.
58. N. W. Ashcroft and N. D. Mermin, *Solid State Physics*, Harcourt, San Diego, CA, 1976.
59. A. Matthiessen, On the electric conducting power of the metals, *Phil. Trans. Royal Soc.* (London), v. 148, pp. 383–387, 1858.
60. Data obtained from Indium Corporation of America, http://www.indium.com/products/indiummetal/physicalconstants.php.
61. M. Morooka, T. Yamamoto, and K. Watnabe, Defect-induced circulating thermal current in graphene with nanosized width, *Phys. Rev.B*, v. 77, pp. 033412-1-4, 2008.

62. M. T. Lusk and L. D. Carr, Nanoengineering defect structures on graphene, *Phys. Rev. Lett.*, v. 100, pp. 175503-1-4, 2008.
63. C. W. Nan, R. Birringer, D. R. Clarke, and H. Gleiter, Effective thermal conductivity of particulate composites with interfacial thermal resistance, *J. Appl. Phys.*, v. 81, pp. 6692–6695, 1997.
64. Z. Ghuo, D. Zhang, and X.G. Gong, Thermal conductivity of graphene nanoribbons, *Appl. Phys. Lett.*, v. 95, pp. 163103-1-3, 2009.
65. H. Zheng and K. Jagannadham, Transient thermoreflectance from graphene composites with matrix of indium and copper, *AIP Advances*, v. 3, pp. 032111-1-12, 2013.
66. M. A. Panzer, G. Zhang, D. Mann, X. Hu, E. Pop, H, Dai, and K. E. Goodson, Thermal properties of metal-coated vertically aligned single-wall nanotube arrays, *J. Heat Trans.*, v. 130, pp. 052401-1-9, 2008.
67. J. G. Bai, Z. Zhang, G.-Q. Lu, and D. P. H. Hasselman, Measurement of solder/copper interfacial thermal resistance by the flash technique, *Int. J. Thermophysics*, v. 26, pp. 1607–1615, 2005.
68. B.N.J. Persson, B. Lorenz, and A. I. Volokitin, Heat transfer between elastic solids with randomly rough surfaces, *Eur. Phys. J. E*, v. 31, pp. 3–24, 2010.
69. D. W. Oh, S. Kim, J. A. Rogers, D. G. Cahill, and S. Sinha, Interfacial thermal conductance of transfer-printed metal films, *Adv. Mat.*, v. 23, pp. 5028–5033, 2011.
70. A. Majumdar and P. Reddy, Role of electron-phonon coupling in thermal conductance of metal-nonmetal interfaces, *Appl. Phys. Let.*, v. 84, pp. 4768–4771, 2004.
71. E. T. Swartz and R. O. Pohl, Thermal boundary resistance, *Reviews of Modern Physics*, v. 33, pp. 605–668, 1989.
72. A. Minnich and G. Chen, Modified effective medium formulation for the thermal conductivity of nanocomposites, *Appl. Phys. Lett.*, v. 91, pp. 073105-1-3, 2007.
73. B. C. Gundrum, D. G. Cahill, and R. S. Averback, Thermal conductance of metal-metal interfaces, *Phys. Rev. B*, v. 72, pp. 245426–245435, 2005.
74. B. M. Clemens, G. L. Eesley, and C. A. Paddock, Time-resolved thermal transport in compositionally modulated metal films, *Phys. Rev.*, v. 37, pp. 1085–1096, 1988.

40 Electronic Properties of Carbon Nanotubes and Their Applications in Electrochemical Sensors and Biosensors

Xuefei Guo and Woo Hyoung Lee

CONTENTS

ABSTRACT

Carbon nanotubes (CNTs) are a one-dimensional allotrope of graphene. Sheets of graphene are "rolled" into tubes to form well-ordered, hollow graphitic CNTs. CNTs have attracted intense interest, including in sensors and biosensors, due to their important properties, such as increased electrode surface area, fast electron transfer rate, significant mechanical strength, and good chemical stability. CNTs' unique combination of mechanical, thermal, and electrical properties make CNT fibers or CNT threads and yarns even better candidates for multifunctional materials. Apart from good electrical and mechanical properties, CNT fibers inherit the advantages of high surface area and good electrocatalytic properties of CNTs, while avoiding the potential toxicity caused by individual CNTs in the form of small particles with a high aspect ratio. Thus, CNT fibers have great potential for sensing applications. In this chapter, CNT fibers are compared with carbon fibers, a common material used for biosensing, in their applications in biosening.

40.1 INTRODUCTION

Carbon materials have been widely used in both analytical and industrial electrochemistry. Low cost, wide potential windows, relatively inert electrochemistry, and electrocatalytic activity have led to the popularity of carbon materials. The well-known allotropes of carbon include graphite, diamond, and fullerenes, each of which can exist in a variety of materials with different electrochemical properties. Iijima discovered multiwalled carbon nanotubes (MWNTs), another allotrope of carbon, in carbon soot made by an arc-discharge method in 1991 (Iijima 1991), and two years later, he observed single-walled nanotubes (SWNTs). Since then, CNTs and their potential applications have attracted tremendous interest, mainly motivated by the potential to make miniaturized biosensors by incorporating CNTs, and partly motivated by a desire to develop completely new nanoscale biosensors. However, the swell of interest in graphene appears to have dwarfed CNTs' role. The capability of graphene in fluorescence resonance energy transfer (FRET) has shown that it is a promising candidate for developing fluorescent sensing methods. Graphene has also been used to construct various sensing platforms with excellent performance for molecular probing, DNA detection, enzymes, etc. (Lin et al. 2011). Recently, Yang et al. (2010) excellently compared these two carbon nanomaterials (CNTs and graphene) and concluded that there are valuable lessons that we can learn from the developments in nanotube-based biosensors to expedite developments in graphene-based biosensors. Thus, this chapter focuses on CNTs' electrochemical properties and the shared future and challenges with graphene.

40.2 ELECTROCHEMICAL SENSORS AND BIOSENSORS

Chemical sensors transform chemical information into analytically useful signals, ranging from the concentration of specific sample components to total composition analysis. According to the IUPAC definition, a biosensor is a

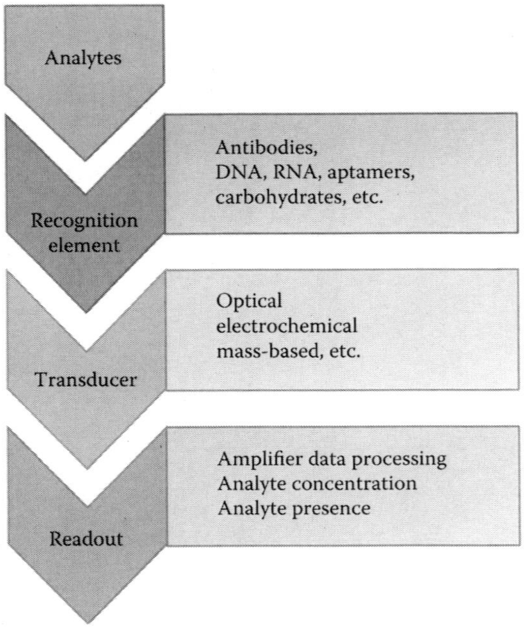

FIGURE 40.1 Basic structures and functional parts of biosensors.

self-contained integrated device that is capable of providing specific quantitative or semiquantitative analytical information by using a biological recognition element (biochemical receptor) in direct spatial contact with a transduction element. However, the terms of sensor and biosensor are often used with interchangeable meaning (Justino et al. 2010).

All biosensors contain three main functional components as shown in Figure 40.1: (1) a recognition element, (2) a signal transducer that converts analyte signals into measurable signals, and (3) a signal processor that relays and displays the results. Biosensors can be classified based on analytes, recognition elements, or transducers. Classified by transducers, electrochemical biosensors measure electrochemical signals produced during a biochemical interaction between analytes and recognition elements immobilized on the sensor. Electrochemical biosensors can be further differentiated into potentiometric, amperometric, and conductometric biosensors depending on the electrochemical property they measure (Ronkainen et al. 2010). Optical biosensors measure the light observed or emitted during a biochemical interaction between analytes and recognition elements. Optical biosensors can be further categorized based on luminescence, fluorescence, etc. Electrochemical sensors offer several advantages over conventional fluorescence measurements, such as portability, less expense, and the ability to carry out measurements in turbid samples. There have been a number of reports of CNT-based electrochemical biosensors for the detection of diverse biological analytes due to the promoted electron transfer (McCreery 2008).

Despite the existence of various sensors, the quest for better fabrication and detection methods and smaller devices continues. Ideally, a sensor should be perfectly selective, highly sensitive, stable, and rugged. Also, it should be cost effective, reproducible between successive measurements, have appropriate low detection limits, and dynamic range for the intended application, and have a reasonable response time. Unfortunately, sensors that meet all of these requirements rarely exist. The idea of a portable field device that can produce immediate and accurate results is extremely appealing and has the potential to be useful in many applications. Development of electrochemical sensors and biosensors using nanomaterials is a fast-growing research area. Nanomaterials, particularly carbon-based nanomaterials, have a significant role to play in the development of electrochemical biosensors with desirable properties discussed previously. For example, nanomaterials can make the biosensing interface more selective, achieve efficient transduction of the biorecognition event, increase sensitivity of the biosensor, and improve response times. Both optical (colorimetric, fluorescent, luminescent, and interferometric) biosensors and electrochemical (amperometric, potentiometric, and conductometric) biosensors have been reported using advanced carbon materials, including CNTs and graphene (Gong et al. 2005; Yang et al. 2010)

40.3 STRUCTURE AND ELECTRONIC PROPERTIES OF CNTs

CNTs belong to the fullerene family of carbon allotropes. CNTs are hollow graphitic nanomaterials made of cylinders of sp^2-hybridized carbon atoms. A SWNT is usually referred as a graphene sheet rolled over into a cylinder with a typical diameter on the order of 1.4 nm. Every SWNT structure can be uniquely characterized by a vector $\boldsymbol{C}$ ($\boldsymbol{C} = \mathrm{m}\boldsymbol{a}_1 + \mathrm{n}\boldsymbol{a}_2$), with $\boldsymbol{a}_1$ and $\boldsymbol{a}_2$ as graphene lattice vectors (Figure 40.2). The two integers (m, n) are commonly used to label the structures of SWNTs, providing the information on the diameters, chiral angles, and electrical properties of the CNTs. By folding a graphene sheet into a cylinder, the beginning and end of an (m, n) lattice vector in the graphene plane join together. The so-called chirality (m, m) tubes are "arm-chair" tubes, since the atoms around the circumference are in an arm-chair pattern (Figure 40.3a). The "zigzag" tubes are (m, 0) nanotubes in view of the atomic configuration along the circumference

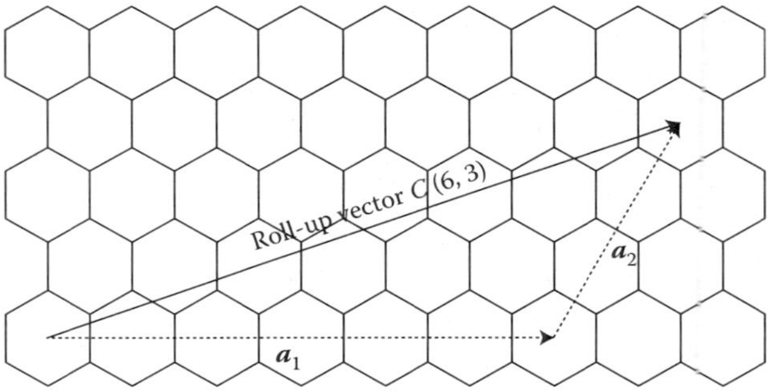

FIGURE 40.2 Roll-up vector $\boldsymbol{C}$ and graphene lattice vectors $\boldsymbol{a}_1$ and $\boldsymbol{a}_2$. (Reprinted from *TrAC-Trends in Analytical Chemistry*, 30, Huang, H. D. et al. Near-infrared fluorescence spectroscopy of single-walled carbon nanotubes and its applications, 1109–1119, Copyright 2011, with permission from Elsevier.)

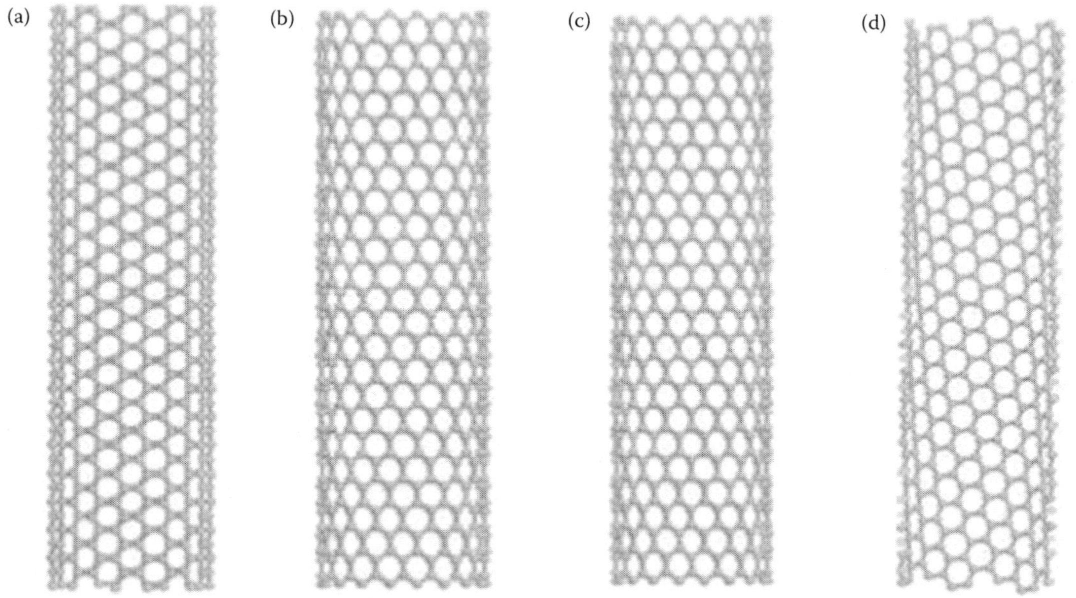

FIGURE 40.3 Schematic structures of SWNTs. (a) A (10, 10) arm-chair metallic nanotube. (b) A (12, 0) zigzag nanotube. (c) A (14, 0) zigzag semiconducting tube. (d) A (7, 16) semiconducting tube. (Reprinted from *Surface Science*, 500(1–3), Dai, H. J. Carbon nanotubes: Opportunities and challenges, 218–241, Copyright 2002, with permission from Elsevier.)

(Figure 40.3b and c). The other types of nanotubes are chiral, with the rows of hexagons spiraling along the nanotube axes (Figure 40.3d). An SWNT can be either a metal semiconductor, or small-gap semiconductor depending on the (m, n) structural parameters. For (m, m) arm-chair tubes, there are always states crossing the corner points of the first Brillouin zone; thus arm-chair tubes are always metallic. For (m, n) nanotubes with m–n ≠ 3 × integer, the nanotubes are semiconducting. For (m, n) nanotubes with m–n = 3 × integer, this type of tubes would be semimetal but become small-gap semiconductors. The sensitivity of the electronic property depending on structural parameters of CNTs poses a significant challenge to chemical synthesis in terms of controlling the nanotube diameter and chirality. Because of SWNTs' small diameter and resulting greater curvature, the overlap between the *p* orbitals on the neighboring carbon atoms involved in π bonding is reduced, suggesting increased chemical reactivity (Holloway et al. 2008). In reality, approximately two-thirds of small-diameter SWNTs are semiconductors and one-third is metallic (Ajayan 1999). To achieve uniform electrical (and optical) properties, SWNTs have to be uniform in diameter and in chirality since SWNTs with almost identical diameters can have different chirality and different electronic properties. Despite much research effort in recent years that has been seeking large-scale production of monodispersed SWNTs with uniform diameter and chirality (Hersam 2008), no one has achieved this goal yet.

An MWNT consists of concentric SWNTs and closed graphite tubules with an interlayer spacing of 3.4 Å, close to the interlayer spacing in graphite, and with a diameter typically on the order of 10–25 nm. MWNTs are usually metallic conductive and are composed of several cylinders of different helicity, which complicates any simple interpretation in terms of electronic properties (Ajayan 1999).

An important aspect of the structure of CNTs is their local anisotropy, which arises because the walls of the tubes are different from their ends. The sidewalls are a relatively inert layer of sp^2-hybridized carbon atoms, while the ends or "tips" of nanotubes have carbon atoms bonded to oxygen to give a far more reactive species, like the edge planes of pyrolytic graphite. The rate of heterogeneous electron transfer depends critically on the nanostructure of the electrode surface and particularly on the nanotube orientation and arrangement, leading to confusion and conflict in the electrochemistry of CNTs. In addition, variable levels of oxygen-containing groups or edge defects on the tips, or by trace catalytic particles from synthesis further hinder the understanding of CNTs and their applications.

40.4 ELECTROCHEMISTRY OF CNTs

Owing to the distinct structure, CNTs have shown different forms of electrochemical behavior. Generally, CNTs bear two kinds of carbon atoms: those at the tube ends and those at the sidewall. Both components show different electrochemistry. The former behaves like the edge plane of highly oriented pyrolytic graphite (HOPG) and possesses favorable electrochemical properties, while the latter resembles the basal plane of HOPG and shows very slow electron transfer kinetics.

Liu et al. (2005) modified gold electrodes with a self-assembled monolayer (SAM) of cysteamine and then shortened SWNTs. The electrochemistry of potassium ferricyanide at these aligned nanotube electrode arrays showed that SWNTs act as nanoelectrodes with the ends of the tubes more electrochemically active than the walls. The redox active center of glucose oxidase was attached to the ends of the tubes and showed efficient electron transfer (Liu et al. 2005). Their work provided strong evidence that the difference in electron transfer behavior of potassium ferricyanide was due to the amount of surface oxide species presented to the electrode (Liu et al. 2005). However, Pumera (2007) has found that oxygen-containing groups on CNTs actually slow the rate of heterogeneous electron transfer. Gong et al. (2008) explored the relative contributions of sidewalls, tips, and oxidation states in CNT electrochemistry by selectively masking the sidewalls or tips with a nonconducting polymer coating and by controlling the level of oxidation of superlong (5 mm), vertically aligned CNTs. According to this excellent work, the relative importance of these factors depends on the type of redox probe investigated and the redox reaction involved. For example, the electrochemistry of potassium ferricyanide was much enhanced at the CNT tips, especially in the presence of oxygen-containing moieties, and the electron-transfer kinetics were slower at the CNT sidewalls. In contrast, the oxidation of hydrogen peroxide favors at sidewalls rather than at tips and was relatively insensitive to the presence of oxygen-containing groups.

The compositional heterogeneity further complicated the electrochemical behavior of CNTs. CNTs always contain some impurities, such as metallic compounds or nanoparticles derived from the catalysts used in nanotube growth, which can remain trapped between the graphene sheets in nanotubes even after extensive acid washing. These metallic impurities are claimed to be responsible for the "electrocatalysis" at some nanotube-modified electrodes (Banks et al. 2006). Catalyst-free CNTs offer metal-free CNTs whose properties are not obscured by the catalyst impurities and provide an opportunity to understand which electrochemical properties are intrinsic to nanotubes and which are not (Huang et al. 2009).

Despite the complicated fundamental electrochemistry of CNTs, metal nanoparticles can be beneficial by deliberately integrating catalytic nanoparticles within CNTs to enhance the electrochemical properties of sensors. Qu et al. developed a facile yet versatile and effective substrate-enhanced electroless deposition (SEED) method for functionalizing nanotubes with a large variety of metal nanoparticles (Qu and Dai 2005). Qu et al. (2006) could also control the shape/size and site-selectively deposit these metal nanoparticles onto the outerwall, innerwall, or end-tip of CNTs. Claussen et al. (2009) decorated SWNTs with Au-coated Pd (Au/Pd) and achieved high sensitivity and low estimated detection limit (2.3 nM) amperometric sensing of hydrogen peroxide.

40.5 APPLICATIONS OF CNTs AND GRAPHENE FOR ELECTROCHEMICAL SENSORS AND BIOSENSORS

CNTs have high aspect ratios, high mechanical strength, high surface areas, excellent chemical and thermal stability, and rich electronic and optical properties. High conductivity along their length indicates that CNTs are excellent nanoscale electrode materials. These properties make CNTs important transducer materials for biosensors. For example, the combination of excellent conductivity, good electrochemical properties, and nanometer dimensions has made CNTs being plugged directly into individual redox enzymes for better transduction in electrochemical enzyme biosensors.

In 1996, Britto et al. first studied the oxidation of dopamine by using CNTs without any pretreatment (Britto et al. 1996). At a CNT electrode, a characteristic anodic peak at $E_{pa} = 0.22$ V versus a saturated calomel electrode (SCE) and a complementary cathodic peak at $E_{pc} = 0.19$ V versus SCE were shown by a cyclic voltammetric curve of dopamine in phosphate buffered saline (PBS) at pH 7.4, suggesting that dopamine was undergoing two-electron oxidation. Later, the same group reported that microelectrodes made from CNTs can electrocatalytically reduce dissolved oxygen (Britto et al. 1999). A well-defined reduction peak of oxygen was observed with a positive peak shift compared with the carbon paste electrode and the values of the current densities for nanotubes (6.3×10^{-4} A/cm^2) were at least five times higher than the paste electrode. Luo et al. (2001) studied the electrochemical behavior of acid-purified SWNTs film cast on a glassy carbon electrode. These SWNT-modified electrodes showed favorable electrocatalytic behavior toward the oxidation of dopamine, epinephrine, and ascorbic acid (Luo et al. 2001), and uric acid (Rubianes and Rivas 2003), NADH (Wang and Musameh 2003), and hydrogen peroxide (Wang et al. 2003). The monitoring of an enzyme reaction by detecting the hydrogen peroxide produced was illustrated by Wang et al. (2003). The nafion-wrapped nanotubes were deposited onto a glass carbon electrode showing a low potential for H_2O_2 oxidation at +0.2 V versus Ag/AgCl, whereas +1.0 V versus Ag/AgCl was shown on a bare glassy carbon electrode.

Functionalized CNTs have shown promises for the development of electrochemical biosensors to detect specific target compounds. For the purpose, small proteins and enzymes have been introduced into the hollow of opened nanotubes and have shown that CNTs acted as a benign host in their ability to encapsulate protein molecules with biological activities (Davis et al. 1998). Streptavidin was packed on the outer surface of MWNTs to form a helical organization (Balavoine et al. 1999). It was also reported that oligonucleotide could be immobilized on CNTs (Tsang et al. 1997). All of these reports can help CNTs to selectively sense various chemicals and biological molecules.

In addition to sense molecules, whole cells have also been sensed with CNTs. It has been shown that various cell lines can interact with CNTs by adhesion, growth, and differentiation of neuronal cells on CNT-based substrates (Hu et al. 2004). The unique mechanical, chemical, and electrical properties of CNTs make them one of the most promising materials for applications in neural biosensing (Wallace et al. 2009). Keefer et al. (2008) coated conventional tungsten and stainless-steel wire electrodes with CNTs to improve neuronal recordings. The CNT coating enhanced both recording and electrical stimulation of neurons in culture, rats, and monkeys by decreasing the electrode impedance and increasing charge transfer. Lin et al. (2009) prepared a complex electroanalytical system in which SWNTs were loaded with glucose dehydrogenase or lactate dehydrogenase to monitor glucose and lactate in rat brain tissue continuously and simultaneously. However, none of these works reported possible adverse immune responses that CNTs might generate in the long term. Kagan et al. (2005) demonstrated that aspiration of nonfunctionalized SWNTs elicited an unusual inflammatory response in the lungs of exposed mice with a very early switch from the acute inflammatory phase to fibrogenic events, accompanied by a characteristic change in the production and release of proinflammatory to anti-inflammatory profibrogenic cytokines, decline in pulmonary function, and enhanced susceptibility to infection. A possible solution to this unexplored problem is to make CNTs more biocompatible, which can be achieved by biochemical functionalization with biomolecules, including peptides, proteins, carbohydrates, or nucleic acids (Yang et al. 2007). Aptamers are oligonucleotide sequences with an affinity for a variety of specific biomolecules targets such as small molecules and proteins. They have the potential to selectively detect a wide range of molecular targets (Shamah et al. 2008). Recently, Zelada-Guillen et al. (2009) reported a novel potentiometric biosensor made with aptamer-modified SWNTs through π–π stacking interactions between the nucleic acid bases and the nanotube walls. This novel biosensor allowed specific real-time detection of a single bacterium of *Salmonella typhi* (Zelada-Guillen et al. 2009). Compared to classical microbiological tests, which take between 24 and 48 h, this early diagnosis for salmonellosis can be life-saving.

40.6 APPLICATIONS OF CNTs AND CARBON FIBER FOR BIOFILM RESEARCH

Needle-type electrochemical microelectrodes represent one of the most prominent experimental tools in biofilm research (Lee et al. 2011). Their tiny tips (5–10 μm) make them very attractive experimental tools to penetrate biofilm without destroying or disturbing their structure and to determine constituents, concentrations profiles, and functions *in situ* in biological samples (e.g., biofilm, sediments, pore water, activated sludge flocs, and biological tissues) spatially and temporally. For decades, they have been used to measure concentrations of chemical and/or biological compounds inside biofilm in aqueous systems and to determine related kinetic reaction parameters (e.g., constituent flux (J), diffusion coefficient (D), maximum reaction rate (Ks), and disinfectant penetration) (Lewandowski and Beyenal 2007). The compounds, which are measured using microelectrodes, have included pH, oxygen, nitrogen species (e.g., ammonium, nitrite, and nitrate), phosphate, free chlorine, monochloramine, redox potential, heavy metals (e.g., arsenic, cadmium), nitrous oxide, carbon dioxide, hydrogen peroxide, sulfide, and specific ions (e.g., Ca^{2+}, Mg^{2+}, Cu^{2+}) (Lewandowski and Beyenal 2007).

There are two types of sensing materials that have been widely used to fabricate electrochemical microelectrodes: solid type and liquid type. Solid-type sensing materials include noble metals (e.g., gold, platinum, and silver) and other metals (e.g., cobalt, iridium, CNTs, and carbon fibers), while liquid-type sensing material represents an ion-selective membrane, including a specific ionophore. Both types of materials require to be sealed (or encased) with a glass micropipette by heating to have a physical ability to penetrate. Several needle-type CNT sensors to measure glucose have been reported (Wang and Musameh 2003; Jia et al. 2008). But most of them are CNT composite sensors or they are unable to penetrate the biological sample. So far, needle-type CNT microelectrodes that have an ability to penetrate biofilms have not been successfully reported. Even with a configuration of needle microsensors, some electrodes have utilized a protrusion of CNT that lacks the physical strength to penetrate biofilm (YeoHeung et al. 2006). However, nanotubes have a huge surface-to-volume area for interacting with other chemical and biological substances and have outstanding electrical conductive properties. If CNTs would be well encased in a well-insulated material (e.g., glass) with a micro/nano-level exposed area, and they can thus be miniaturized for micro/nano-scale detection in biofilm, their applications to biofilm would increase extensively. On the other hand, fabrication procedures of microelectrode sensor using carbon fibers are well described elsewhere (Kawagoe et al. 1993; Yip 1996). Lee et al. (2013) developed a carbon-fiber microelectrode sensor (7 μm in tip diameter) to detect nitrite selectively in the presence of other nitrogen species (e.g., nitrate and ammonia) in a biofilm and showed well-defined nitrite concentration profiles along with wastewater biofilm thickness (~2000 μm) (Lee et al. 2013). They have also started to develop prototypes for a CNT nitrite sensor that could be carried into biofilm research (Figures 40.4 through 40.6). The fabrication method followed carbon fiber microsensor fabrication (Lee et al. 2013). Briefly, after connecting a conductive wire (e.g., silver wire or copper wire), a small bundle of CNTs (~50 μm) was inserted into a glass micropipette and was sealed with a glass by heating a glass micropipette using a puller. A bare CNT microelectrode in the test showed outstanding amperometric responses to nitrite concentration between potential +0.7 and +1.5 V versus Ag/AgCl (Figure 40.5) and showed high sensitivity (i.e., slope) at +1.0 V versus Ag/AgCl of applied potential (Figure 40.6). Other evaluation tests including ion interference and long-term exposure will also be performed to ensure complete characterization and validation of the CNT microelectrode performance. The fabrication method used for needle-type CNT sensors can also be used for creating various CNT-based biosensors with relative pretreatment (e.g., surface modification and embedded bacteria) to detect specific biological molecules or pathogens in a biofilm.

40.7 APPLICATIONS OF CNTs FOR SCANNING TUNNELING MICROSCOPY

Another important application of CNTs is their use as tips of atomic force microscopy (AFM) or scanning tunneling microscopy (STM) (Dai et al. 1996). It has been shown that SWNT tips provide significantly better resolution compared

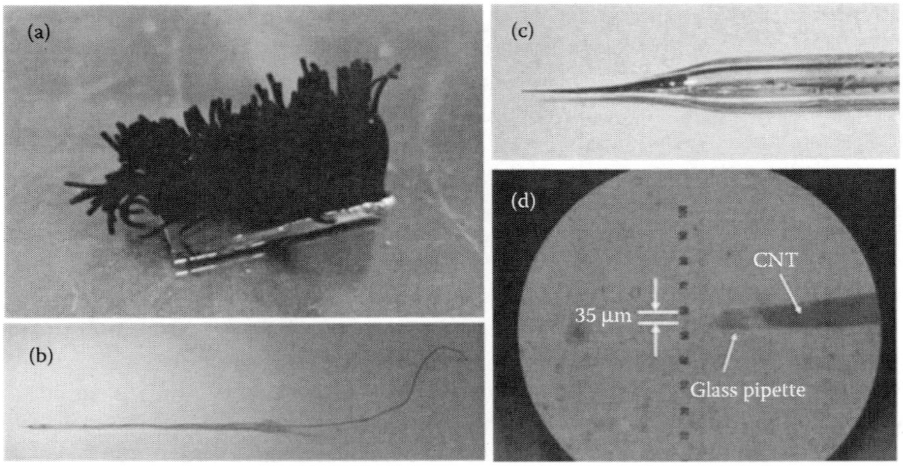

FIGURE 40.4 Carbon nanotube microsensor for nitrite detection: (a) carbon nanotubes (provided by Dr. Vasselin Sahnov in Materials Engineering at the University of Cincinnati), (b) a fabricated microelectrode sensor, (c) close-up of the microsensor tip encased with a glass micropipette, and (d) microscopic image of the microsensor tip of 35 μm diameter with a recess.

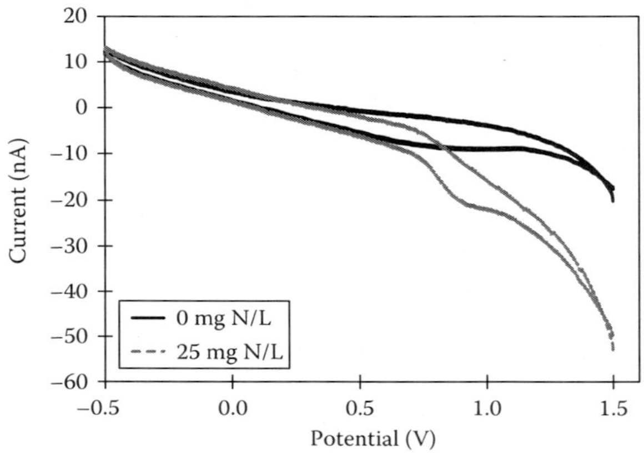

FIGURE 40.5 Cyclic voltammograms of the CNT microsensor with various nitrite concentrations (0 and 25 mg N/L) in deoxygenated buffer solution (5 mM borate buffer solution, pH 8.0, and 23°C). Scan initiated at −0.5 V versus Ag/AgCl in the positive direction at 100 mV/s. E_{final} = +1.5 V.

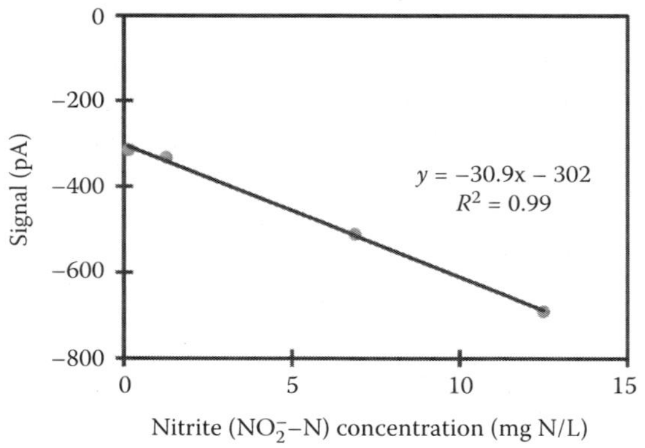

FIGURE 40.6 Representative nitrite calibration curve with various nitrite concentrations in 5 mM borate buffer solution at pH 8.0 and 23°C. Applied potential: +1.0 V versus Ag/AgCl.

with commercial Si and Si_3N_4 AFM tips. Indeed, the 0.5 nm individual SWNTs in diameter indicate that such tips could enable true molecular-resolution imaging. However, the sharp tips do not necessarily provide information about the functionality central to binding and reactivity in chemical and biological systems. Owing to the intrinsically small diameters of SWNTs, high aspect ratios, elasticity, and rich surface chemistry, functionalized CNTs have been used as AFM and STM tips, and opened up the applications of molecular recognition and chemically sensitive imaging in chemistry and biology (Wong et al. 1999). Wong et al. (1998) showed that the ends of the MWNT tip can be covalently modified to present chemically and biologically active functionality. In this work, carboxyl groups (−COOH) believed to exist at the open ends of oxidatively processed SWNT were coupled with amines to form amide-linked groups. The functionality at the SWNT tip ends was determined by measuring the adhesion force between the tip ends and hydroxyl (−OH)-terminated SAM surfaces.

40.8 APPLICATIONS OF ALIGNED CNTs FOR ELECTROCHEMICAL SENSORS AND BIOSENSORS

One of the advantages of CNTs over other carbon allotropes is their geometry and associated electrochemical properties such as small and long highly electroactive ends and relatively electroinactive walls. In addition, the ends of CNTs have shown excellent electrochemical properties, with reversible electrochemistry and low redox potentials, which are attributed to the oxygenated species at the ends of the tubes (Moore et al. 2004). However, owing to the van der Waals forces between tubes, CNTs tend to aggregate together in almost all kinds of aqueous and organic solutions. This inherent property imposes difficulties in making a homogeneous CNT composite and thus greatly hampers electrochemical studies and analytical applications of CNTs. There are some strategies to solubilize CNTs and further manipulate them (Chen et al. 1998; Georgakilas et al. 2002; Baibarac et al. 2003; Richard et al. 2003). Generally speaking, electrode configurations using CNTs are constructed by (a) mixing CNTs with a binder to form a paste for packing into an electrode, or CNT past electrodes (Valentini et al. 2003), (b) evaporating a solution of CNTs onto solid electrodes (Wang et al. 2003), (c) layer-by-layer assembly of CNT film electrodes (Zhang et al. 2004), (d) attaching individual CNTs to the end of a wire (Campbell et al. 1999), or (e) microfabricating nanoelectrode arrays (Gooding et al. 2003). However, CNT electrodes used in most of the reported electrochemical studies were simply prepared by randomly dispersed CNTs (Rubianes and Rivas 2003; Valentini et al. 2003; Wang and Musameh 2003).

Because of the random confinement of the CNTs on electrodes without orientation controlling, it is difficult to determine the contributions of the ends and the sidewall of the nanotubes to the electrochemical properties of the CNTs. These attributes provide the opportunity to make arrays of aligned nanoelectrodes in a highly controlled manner. Thus, new methods, using aligned CNTs, have been recently developed and have proven to be useful for electrochemical studies and electroanalytical applications (Gooding et al. 2003; Lin et al. 2004; Gong et al. 2008; Guo et al. 2011). Nakashima et al. (2002) dispersed shortened SWNTs using an electroactive surfactant 11-ferrocenylundecyl poly(ethylene glycol). Then, a gold electrode was held at +500 mV versus SCE in an aqueous solution containing dispersed nanotubes and gained deposited SWNTs. It was proposed that the oxidation of the ferrocene to the ferrocinium ion caused the redox surfactant to desorb from the walls of the SWNTs, leaving them to coagulate on the electrode surface. Wohlstadter et al. (2003) mixed nanotubes with a polymer of poly(vinyl acetate) (PVA) to produce an PVA–nanotube composite. The composite was extruded and SEM images suggested some alignment of the tubes. These PVA–nanotube composites were used to immobilize antibodies for a sandwich assay. Nugent et al. achieved greater control over the alignment of CNTs using the boule that forms on the electrode during arc discharge synthesis of MWNTs. The boule was broken open and a microbundle of aligned nanotubes was attached to the end of a copper electrode using conductive silver paint (Nugent et al. 2001). This microelectrode made of MWNTs showed Nernstien behavior and fast electron-transfer kinetics for electrochemical reactions of $Fe(CN)_6^{3-/4-}$ without any pretreatments for the electrodes. Liu et al. (2000) shortened SWNTs in a 3:1 mixture of concentrated sulfuric acid and concentrated nitric acid, leaving several carboxylic acid groups at each end of the SWNTs. The carboxylate ends reacted with amines on cysteamine ($NH_2CH_2CH_2SH$) and were able to self-assemble onto a gold surface. The advantages of this approach include that assembly time and cutting time gives control over the distribution of lengths of the tubes aligned on the electrode surface and a mixed SAM could give further control over the spatial distribution of the tubes. Yu et al. (2003) even showed that the CNTs can be patterned. Campbell et al. (1999) attached 80 ± 200-nm-diameter CNTs to sharpened Pt wires. This robust electrode had high length-to-diameter aspect ratio and gave a sigmoidal voltammetric response in the $Ru(NH_3)_6^{3+}$ solution.

An alternative way to obtain aligned CNTs on an electrode surface is to grow aligned nanotubes directly off a surface. Yang et al. (1999) grew MWNTs on a quartz plate by pyrolyzing iron(II) phthalocyanine under Ar/H_2 at 800–100°C followed by transferring the tubes to a gold electrode. The aligned MWNT electrode arrays can be given additional robustness by electrodepositing conducting polymer around the tubes (Gao et al. 2000). A chemical vapor deposition (CVD) method was also performed to grow aligned nanotubes directly onto a platinum electrode (Sotiropoulou and Chaniotakis 2003). For plasma-enhanced CVD, the size of the catalyst spots could be controlled to determine the number of tubes grown at a given location (Li et al. 2003). The whole surface was then encapsulated to insulate CNTs and form a robust aligned nanotube electrode array once the surface was polished to expose the ends of the nanotubes. An advantage is that controlling the distribution of the catalyst on the surface provides control over the density of the nanotubes.

The CNT tower is one kind of highly aligned CNT arrays, which shows fast electron transfer. The CNT tower was fabricated on an $Fe/Al_2O_3/SiO_2/Si$ substrate with the Fe catalyst patterned in 1 mm × 1 mm blocks with 100 μm spacing between the blocks. Patterned MWNT towers up to 4 mm high were grown and each CNT tower bundle contains approximately 25 million nanotubes (Yun et al. 2006). Guo et al. (2011) peeled off individual CNT tower bundles, connected with a conductive wire with a conductive epoxy, coated with nonconductive polymer to fabricate a sensor (Figure 40.7). This electrode can detect trace multiple heavy metals in the absence of mercury

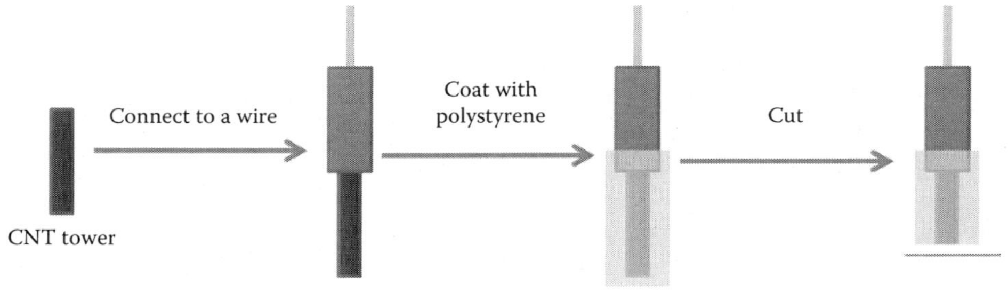

FIGURE 40.7 Fabrication of electrochemical sensors based on CNT tower.

largely because CNTs not only maintain an electrically conducting surface but also increase the electrode surface area, which results in the good detection of trace heavy metals. A CNT tower electrode has the advantage of avoiding other toxic additives such as mercury and bismuth to enhance its performance. Thus, the CNT tower electrode is suitable for *in situ* detection where toxic materials might not be allowable especially for *in vivo* and *in vitro* studies. The diameter of the CNT tower is about 400 μm. Smaller electrodes from a CNT tower can potentially expand the applications to extracellular or even intracellular studies.

CNT threads, or CNT fibers and yarns, are spun from shorter CNTs, motivated by demands of individual nanotubes with very high strength and electrical and thermal conductivities. There are two main methods for producing fibers from shorter CNTs: liquid- and solid-state spinning. Natural fibers, including wool and cotton consisting of discrete fibers, are formed by solid-state spinning, whereas most synthetic fibers are created by liquid-state spinning from a concentrated, viscous liquid, which is a melt or solution of the starting material. Both liquid- and solid-state spinning are employed for CNT-based threads. As an example of solid-state spinning, the spinning process involves two rotations about two mutually perpendicular axes, namely, spinning nanotubes into thread from an array and winding the spun thread on a spool. In the spinning process, the fibers are held together by the twisting of fibers around neighboring fibers. The twisting provides mechanical strength between individual CNTs to form long threads. As the fibers are twisted and pulled, more fibers are added to form a long strong yarn. The twist is characterized by the direction of the rotation and the twist per linear distance. While the development of a continuous and weavable pure CNT yarn remains a major challenge in the fabrication, CNT yarns so far obtained from the different processes are monolithic in structure (Vigolo et al. 2000).

CNT threads have longer length and smaller diameters (Zhang et al. 2004) than the CNT tower or array. These fibers have a promising electrical resistivity (5 Ω·cm) and tensile strength (0.8 GPa) (Behabtu et al. 2008), although not as good as those of individual CNTs. Apart from good electrical and mechanical properties, the CNT threads inherit the advantages of the high surface area and good electrocatalytic properties of CNTs, while avoiding the potential toxicity caused by individual CNTs in the form of particles with a high aspect ratio during implantation. Thus, CNT threads have great a potential for conducting and sensing applications. So far, CNT threads have been tested in potential applications for biosensors for example the detection of NADH oxidation (Viry et al. 2007) and glucose (Zhu et al. 2010).

Guo et al. (2013) connected CNT thread with a metal wire using silver conductive epoxy. Then the CNT thread was aspirated into a glass capillary. The end with the electrical wire was sealed with a hot glue gun. The other end with the CNT thread was filled entirely with epoxy. After that, the capillary section was polished to uncover the section of the fiber that becomes the active surface of the microelectrode based on the CNT thread. Cyclic voltammetry (CV) of microelectrodes

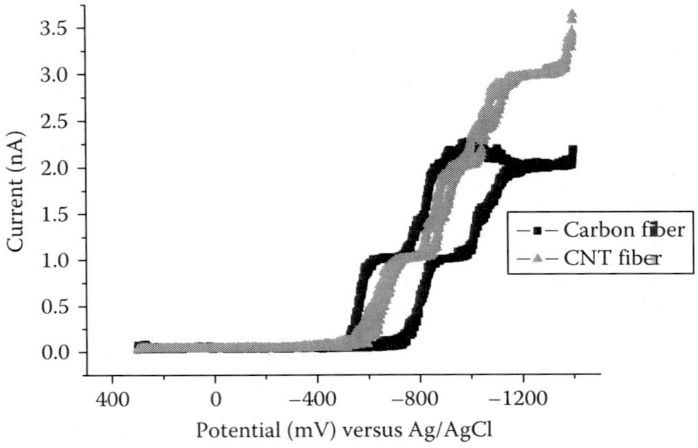

FIGURE 40.8 Cyclic voltammograms at a CNT thread electrode and carbon fiber electrode in 0.1 M acetate buffer (not deoxygenated) at pH 4.5. Scan initiated at −0.3 V versus Ag/AgCl in the negative direction at 10 mV/s. $E_{final} = -1.4$ V.

using CNT threads were compared with those using carbon fibers, since classical carbon fibers have a similar diameter to CNT threads and have been studied as *in vivo* probes for the brain for decades (Kelly and Wightman 1986; Heien et al. 2004). Figure 40.8 compares CVs of carbon fiber and CNT thread in the 0.1 M acetate buffer (pH 4.5), indicating that a microelectrode based on CNT thread has a smaller charging current and thus smaller capacitance compared with carbon fiber. Figure 40.9 shows CVs of dopamine at a CNT thread in PBS. The sigmoidal-shaped voltammogram obtained at a CNT thread indicates that the mass transport was dominated by radial diffusion or convergent diffusion, indicating that the dimensions of the diffusion layer can exceed the dimensions of the microelectrode (Wightman 1988). Radial diffusion at the CNT thread microelectrode leads to several advantages over macroelectrodes with linear diffusion (Amatore et al. 1988). For example, the magnitude of the charging current is a

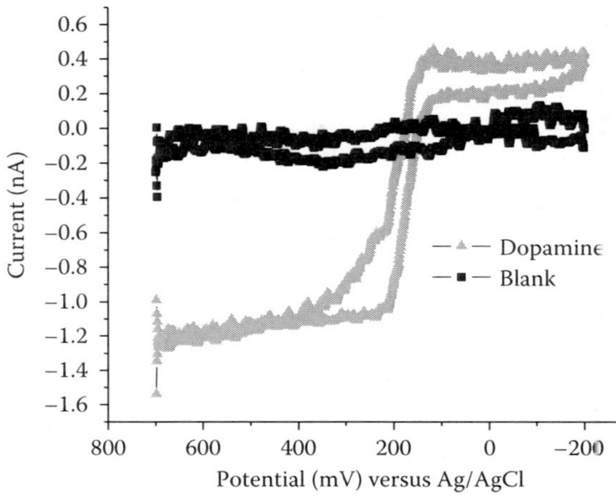

FIGURE 40.9 Cyclic voltammograms of 20 μM dopamine at a CNT thread electrode in PBS (not deoxygenated) at pH 7.4. Potential range: +700 mV to −200 mV versus Ag/AgCl. Scan rates: 100 mV/s.

concern in electrochemical experiments. The charging current of CV has a limiting value of vC_d (v: scan rate) and C_d is proportional to the electrode surface area. On the other hand, the faradaic current at the steady state is proportional to the electrode radius (Wightman 1988). Thus, the ratio of the faradaic current to the charging current is improved with a smaller electrode.

40.9 CONCLUSION

CNTs and graphene, as nanomaterials with every atom in their structures exposed on their surfaces, are highly sensitive to even small changes in surface, offering significant advantages for the development of biosensors. For example, both CNTs and graphene have shown the ability to electrochemically identify and quantify the multi-biomolecules, such as dopamine, ascorbic acid, and uric acid, which are indistinguishable with conventional glassy carbon electrodes (Zhang et al. 2005; Shang et al. 2008). Thus, both CNTs and graphene provide important diagnostic value to the distinct, high-sensitivity electrochemical peaks for these biomolecules, whose concentrations are related with a variety of medical conditions.

Compared with CNTs, graphene is not so widely and thoroughly studied given that graphene has only been available for experimental studies since 2004. But, the success of CNTs in advancing biosensors has aroused incredible interest in graphene as a material that could potentially push the boundaries of this field even farther. For example, Lu et al. dispersed thermally exfoliated graphite "nanoplatelets" in the conductive polymer nafion, similarly with electrodes fabricated with powdered graphite or CNTs. The nanocomposites of nanoplatelets and nafion exhibited significant oxidation and reduction for hydrogen peroxide (Lu et al. 2008). Development of graphene-based biosensors can be faster by learning from research in CNT-based biosensors. Yang et al. (2010) summarized that the biggest, and perhaps the most important, point is just how sensitive biosensor performance is to the exact structure and chemistry of the carbon nanomaterial. Besides the obvious differences between SWNTs and MWNTs, even supposedly similar batches of nanotubes have different lengths and structures, electronic types, purities, modifications, and levels of agglomeration. These challenges also apply to graphene. Graphene's electrical and optical properties are highly sensitive to various factors that include the number of layers, substrate, impurities and contaminants, and the edge structure and chemistry (Yang et al. 2010). To realize the bright future of CNTs and graphene for the development of biosensors, there is considerable work that needs to be done, including large-scale synthesis, optimization, consistency, the biocompatibility of carbon nanomaterials, and the possible toxicity and health risks.

REFERENCES

Ajayan, P. M. 1999. Nanotubes from carbon. *Chemical Reviews* 99 (7):1787–1799.

Amatore, C., B. Fosset, J. Bartelt, M. R. Deakin, and R. M. Wightman. 1988. Electrochemical kinetics at microelectrodes. 5. Migrational effects on steady or quasi-steady-state voltammograms. *Journal of Electroanalytical Chemistry* 256 (2):255–268.

Baibarac, M., I. Baltog, S. Lefrant, J. Y. Mevellec, and O. Chauvet. 2003. Polyaniline and carbon nanotubes based composites containing whole units and fragments of nanotubes. *Chemistry of Materials* 15 (21):4149–4156.

Balavoine, F., P. Schultz, C. Richard, V. Mallouh, T. W. Ebbesen, and C. Mioskowski. 1999. Helical crystallization of proteins on carbon nanotubes: A first step towards the development of new biosensors. *Angewandte Chemie-International Edition* 38 (13–14):1912–1915.

Banks, C. E., A. Crossley, C. Salter, S. J. Wilkins, and R. G. Compton. 2006. Carbon nanotubes contain metal impurities which are responsible for the "electrocatalysis" seen at some nanotube-modified electrodes. *Angewandte Chemie-International Edition* 45 (16):2533–2537.

Behabtu, N., M. J. Green, and M. Pasquali. 2008. Carbon nanotube-based neat fibers. *Nano Today* 3 (5–6):24–34.

Britto, P. J., K. S. V. Santhanam, and P. M. Ajayan. 1996. Carbon nanotube electrode for oxidation of dopamine. *Bioelectrochemistry and Bioenergetics* 41 (1):121–125.

Britto, P. J., K. S. V. Santhanam, A. Rubio, J. A. Alonso, and P. M. Ajayan. 1999. Improved charge transfer at carbon nanotube electrodes. *Advanced Materials* 11 (2):154–157.

Campbell, J. K., L. Sun, and R. M. Crooks. 1999. Electrochemistry using single carbon nanotubes. *Journal of the American Chemical Society* 121 (15):3779–3780.

Chen, J., M. A. Hamon, H. Hu et al. 1998. Solution properties of single-walled carbon nanotubes. *Science* 282 (5386):95–98.

Claussen, J. C., A. D. Franklin, A. ul Haque, D. M. Porterfield, and T. S. Fisher. 2009. Electrochemical biosensor of nanocube-augmented carbon nanotube networks. *ACS Nano* 3 (1):37–44.

Dai, H. J. 2002. Carbon nanotubes: Opportunities and challenges. *Surface Science* 500 (1–3):218–241.

Dai, H. J., J. H. Hafner, A. G. Rinzler, D. T. Colbert, and R. E. Smalley. 1996. Nanotubes as nanoprobes in scanning probe microscopy. *Nature* 384 (6605):147–150.

Davis, J. J., M. L. H. Green, H. A. O. Hill et al. 1998. The immobilisation of proteins in carbon nanotubes. *Inorganica Chimica Acta* 272 (1–2):261–266.

Gao, M., S. M. Huang, L. M. Dai, G. Wallace, R. P. Gao, and Z. L. Wang. 2000. Aligned coaxial nanowires of carbon nanotubes sheathed with conducting polymers. *Angewandte Chemie-International Edition* 39 (20):3664–3667.

Georgakilas, V., N. Tagmatarchis, D. Pantarotto, A. Bianco, J. P. Briand, and M. Prato. 2002. Amino acid functionalisation of water soluble carbon nanotubes. *Chemical Communications* (24):3050–3051.

Gong, K. P., S. Chakrabarti, and L. M. Dai. 2008. Electrochemistry at carbon nanotube electrodes: Is the nanotube tip more active than the sidewall? *Angewandte Chemie-International Edition* 47 (29):5446–5450.

Gong, K. P., Y. M. Yan, M. N. Zhang, L. Su, S. X. Xiong, and L. Q. Mao. 2005. Electrochemistry and electroanalytical applications of carbon nanotubes: A review. *Analytical Sciences* 21 (12):1383–1393.

Gooding, J. J., R. Wibowo, J. Q. Liu et al. 2003. Protein electrochemistry using aligned carbon nanotube arrays. *Journal of the American Chemical Society* 125 (30):9006–9007.

Guo, X., W. H. Lee, N. Alvarez, V. N. Shanov, and W. R. Heineman. 2013. Detection of trace zinc by an electrochemical microsensor based on carbon nanotube threads. *Electroanalysis* 25 (7):1599–1604.

Guo, X. F., Y. H. Yun, V. N. Shanov, H. B. Halsall, and W. R. Heineman. 2011. Determination of trace metals by anodic stripping voltammetry using a carbon nanotube tower electrode. *Electroanalysis* 23 (5):1252–1259.

Heien, M. L. A. V., M. A. Johnson, and R. M. Wightman. 2004. Resolving neurotransmitters detected by fast-scan cyclic voltammetry. *Analytical Chemistry* 76 (19):5697–5704.

Hersam, M. C. 2008. Progress towards monodisperse single-walled carbon nanotubes. *Nature Nanotechnology* 3 (7):387–394.

Holloway, A. F., K. Toghill, G. G. Wildgoose et al. 2008. Electrochemical opening of single-walled carbon nanotubes filled with metal halides and with closed ends. *Journal of Physical Chemistry C* 112 (28):10389–10397.

Hu, H., Y. C. Ni, V. Montana, R. C. Haddon, and V. Parpura. 2004. Chemically functionalized carbon nanotubes as substrates for neuronal growth. *Nano Letters* 4 (3):507–511.

Huang, H. D., M. J. Zou, X. Xu, X. Y. Wang, F. Liu, and N. Li. 2011. Near-infrared fluorescence spectroscopy of single-walled carbon nanotubes and its applications. *TrAC-Trends in Analytical Chemistry* 30 (7):1109–1119.

Huang, S. M., Q. R. Cai, J. Y. Chen, Y. Qian, and L. J. Zhang. 2009. Metal-catalyst-free growth of single-walled carbon nanotubes on substrates. *Journal of the American Chemical Society* 131 (6):2094–2095.

Iijima, S. 1991. Helical microtubules of graphitic carbon. *Nature* 354 (6348):56–58.

Jia, J., W. Guan, M. Sim, Y. Li, and H. Li. 2008. Carbon nanotubes based glucose needle-type biosensor. *Sensors* 8 (3):1712–1718.

Justino, C. I. L., T. A. Rocha-Santos, and A. C. Duarte. 2010. Review of analytical figures of merit of sensors and biosensors in clinical applications. *TrAC-Trends in Analytical Chemistry* 29 (10):1172–1183.

Kagan, V. E., H. Bayir, and A. A. Shvedova. 2005. Nanomedicine and nanotoxicology: Two sides of the same coin. *Nanomedicine* 1 (4):313–6.

Kawagoe, K. T., J. B. Zimmerman, and R. M. Wightman. 1993. Principles of voltammetry and microelectrode surface states. *Journal of Neuroscience Methods* 48 (3):225–240.

Keefer, E. W., B. R. Botterman, M. I. Romero, A. F. Rossi, and G. W. Gross. 2008. Carbon nanotube coating improves neuronal recordings. *Nature Nanotechnology* 3 (7):434–439.

Kelly, R. S., and R. M. Wightman. 1986. Beveled carbon-fiber ultramicroelectrodes. *Analytica Chimica Acta* 187:79–87.

Lee, W. H., J. H. Lee, W. H. Choi, A. A. Hosni, I. Papautsky, and P. L. Bishop. 2011. Needle-type environmental microsensors: Design, construction and uses of microelectrodes and multi-analyte MEMS sensor arrays. *Measurement Science and Technology* 22:042001 (22pp).

Lee, W. H., D. G. Wahman, and J. P. Pressman, Amperometric carbon fiber nitrite microsensor for in situ biofilm monitoring, *Sensors and Actuators B*, 2013, 188:1263–1269.

Lewandowski, Z., and H. Beyenal. 2007. *Fundamentals of Biofilm Research*. New York: CRC Press.

Li, J., H. T. Ng, A. Cassell et al. 2003. Carbon nanotube nanoelectrode array for ultrasensitive DNA detection. *Nano Letters* 3 (5):597–602.

Lin, L., Y. Liu, X. Zhao, and J. H. Li. 2011. Sensitive and rapid screening of T4 polynucleotide kinase activity and inhibition based on coupled exonuclease reaction and graphene oxide platform. *Analytical Chemistry* 83 (22):8396–8402.

Lin, Y. H., F. Lu, Y. Tu, and Z. F. Ren. 2004. Glucose biosensors based on carbon nanotube nanoelectrode ensembles. *Nano Letters* 4 (2):191–195.

Lin, Y. Q., N. N. Zhu, P. Yu, L. Su, and L. Q. Mao. 2009. Physiologically relevant online electrochemical method for continuous and simultaneous monitoring of striatum glucose and lactate following global cerebral ischemia/reperfusion. *Analytical Chemistry* 81 (6):2067–2074.

Liu, J. Q., A. Chou, W. Rahmat, M. N. Paddon-Row, and J. J. Gooding. 2005. Achieving direct electrical connection to glucose oxidase using aligned single walled carbon nanotube arrays. *Electroanalysis* 17 (1):38–46.

Liu, Z. F., Z. Y. Shen, T. Zhu et al. 2000. Organizing single-walled carbon nanotubes on gold using a wet chemical self-assembling technique. *Langmuir* 16 (8):3569–3573.

Lu, J., I. Do, L. T. Drzal, R. M. Worden, and I. Lee. 2008. Nanometal-decorated exfoliated graphite nanoplatelet based glucose biosensors with high sensitivity and fast response. *ACS Nano* 2 (9):1825–1832.

Luo, H. X., Z. J. Shi, N. Q. Li, Z. N. Gu, and Q. K. Zhuang. 2001. Investigation of the electrochemical and electrocatalytic behavior of single-wall carbon nanotube film on a glassy carbon electrode. *Analytical Chemistry* 73 (5):915–920.

McCreery, R. L. 2008. Advanced carbon electrode materials for molecular electrochemistry. *Chemical Reviews* 108 (7):2646–2687.

Moore, R. R., C. E. Banks, and R. G. Compton. 2004. Basal plane pyrolytic graphite modified electrodes: Comparison of carbon nanotubes and graphite powder as electrocatalysts. *Analytical Chemistry* 76 (10):2677–2682.

Nakashima, N., H. Kobae, T. Sagara, and H. Murakami. 2002. Formation of single-walled carbon nanotube thin films on electrodes monitored by an electrochemical quartz crystal microbalance. *ChemPhysChem* 3 (5):456–458.

Nugent, J. M., K. S. V. Santhanam, A. Rubio, and P. M. Ajayan. 2001. Fast electron transfer kinetics on multiwalled carbon nanotube microbundle electrodes. *Nano Letters* 1 (2):87–91.

Pumera, M. 2007. Electrochemical properties of double wall carbon nanotube electrodes. *Nanoscale Research Letters* 2 (2):87–93.

Qu, L. T., and L. M. Dai. 2005. Substrate-enhanced electroless deposition of metal nanoparticles on carbon nanotubes. *Journal of the American Chemical Society* 127 (31):10806–10807.

Qu, L. T., L. M. Dai, and E. Osawa. 2006. Shape/size-control led syntheses of metal nanoparticles for site-selective modification of carbon nanotubes. *Journal of the American Chemical Society* 128 (16):5523–5532.

Richard, C., F. Balavoine, P. Schultz, T. W. Ebbesen, and C. Mioskowski. 2003. Supramolecular self-assembly of lipid derivatives on carbon nanotubes. *Science* 300 (5620):775–778.

Ronkainen, N. J., H. B. Halsall, and W. R. Heineman. 2010. Electrochemical biosensors. *Chemical Society Reviews* 39 (5):1747–1763.

Rubianes, M. D., and G. A. Rivas. 2003. Carbon nanotubes paste electrode. *Electrochemistry Communications* 5 (8):689–694.

Shamah, S. M., J. M. Healy, and S. T. Cload. 2008. Complex target SELEX. *Accounts of Chemical Research* 41 (1):130–138.

Shang, N. G., P. Papakonstantinou, M. McMullan et al. 2008. Catalyst-free efficient growth, orientation and biosensing properties of multilayer graphene nanoflake films with sharp edge planes. *Advanced Functional Materials* 18 (21): 3506–3514.

Sotiropoulou, S., and N. A. Chaniotakis. 2003. Carbon nanotube array-based biosensor. *Analytical and Bioanalytical Chemistry* 375 (1):103–105.

Tsang, S. C., Z. J. Guo, Y. K. Chen et al. 1997. Immobilization of platinated and iodinated oligonucleotides on carbon nanotubes. *Angewandte Chemie-International Edition in English* 36 (20):2198–2200.

Valentini, F., A. Amine, S. Orlanducci, M. L. Terranova, and G. Palleschi. 2003. Carbon nanotube purification: Preparation and characterization of carbon nanotube paste electrodes. *Analytical Chemistry* 75 (20):5413–5421.

Vigolo, B., A. Penicaud, C. Coulon et al. 2000. Macroscopic fibers and ribbons of oriented carbon nanotubes. *Science* 290 (5495):1331–1334.

Viry, L., A. Derre, P. Garrigue, N. Sojic, P. Poulin, and A. Kuhn. 2007. Optimized carbon nanotube fiber microelectrodes as potential analytical tools. *Analytical and Bioanalytical Chemistry* 389 (2):499–505.

Wallace, G. G., S. E. Moulton, and G. M. Clark. 2009. Electrode-cellular interface. *Science* 324 (5924):185–186.

Wang, J., and M. Musameh. 2003. Carbon nanotube/teflon composite electrochemical sensors and biosensors. *Analytical Chemistry* 75 (9):2075–2079.

Wang, J., M. Musameh, and Y. H. Lin. 2003. Solubilization of carbon nanotubes by Nafion toward the preparation of amperometric biosensors. *Journal of the American Chemical Society* 125 (9):2408–2409.

Wang, J., and M. Musameh. 2003. Enzyme-dispersed carbon-nanotube electrodes: A needle microsensor for monitoring glucose. *Analyst* 128 (11):1382–1385.

Wightman, R. M. 1988. Voltammetry with microscopic electrodes in new domains. *Science* 240 (4851):415–420.

Wohlstadter, J. N., J. L. Wilbur, G. B. Sigal et al. 2003. Carbon nanotube-based biosensor. *Advanced Materials* 15 (14): 1184–1187.

Wong, S. S., A. T. Woolley, E. Joselevich, C. L. Cheung, and C. M. Lieber. 1998. Covalently-functionalized single-walled carbon nanotube probe tips for chemical force microscopy. *Journal of the American Chemical Society* 120 (33):8557–8558.

Wong, S. S., A. T. Woolley, E. Joselevich, and C. M. Lieber. 1999. Functionalization of carbon nanotube AFM probes using tip-activated gases. *Chemical Physics Letters* 306 (5–6): 219–225.

Yang, W. R., K. R. Ratinac, S. P. Ringer, P. Thordarson, J. J. Gooding, and F. Braet. 2010. Carbon nanomaterials in biosensors: Should you use nanotubes or graphene? *Angewandte Chemie-International Edition* 49 (12):2114–2138.

Yang, W. R., P. Thordarson, J. J. Gooding, S. P. Ringer, and F. Braet. 2007. Carbon nanotubes for biological and biomedical applications. *Nanotechnology* 18 (41):412001.

Yang, Y. Y., S. M. Huang, H. Z. He, A. W. H. Mau, and L. M. Dai. 1999. Patterned growth of well-aligned carbon nanotubes: A photolithographic approach. *Journal of the American Chemical Society* 121 (46):10832–10833.

YeoHeung, Y., A. Bange, V. N. Shanov et al. 2006. Fabrication and characterization of a multiwall carbon nanotube needle biosensor. *Paper read at Nanotechnology, 2006. IEEE-NANO 2006. Sixth IEEE Conference on*, 17–20 June 2006.

Yip, Tzejunn Jason. 1996. *Design, Fabrication, and Evaluation of a Carbon Fiber Polarographic Oxygen Microelectrode*, Department of Electrical Engineering and Computer Science, MIT.

Yu, X., D. Chattopadhyay, I. Galeska, F. Papadimitrakopoulos, and J. F. Rusling. 2003. Peroxidase activity of enzymes bound to the ends of single-wall carbon nanotube forest electrodes. *Electrochemistry Communications* 5 (5):408–411.

Yun, Y. H., V. Shanov, M. J. Schulz et al. 2006. High sensitivity carbon nanotube tower electrodes. *Sensors and Actuators B-Chemical* 120 (1):298–304.

Zelada-Guillen, G. A., J. Riu, A. Duzgun, and F. X. Rius. 2009. Immediate detection of living bacteria at ultralow concentrations using a carbon nanotube based potentiometric aptasensor. *Angewandte Chemie-International Edition* 48 (40):7334–7337.

Zhang, M., K. R. Atkinson, and R. H. Baughman. 2004. Multifunctional carbon nanotube yarns by downsizing an ancient technology. *Science* 306 (5700):1358–1361.

Zhang, M. N., K. P. Gong, H. W. Zhang, and L. Q. Mao. 2005. Layer-by-layer assembled carbon nanotubes for selective determination of dopamine in the presence of ascorbic acid. *Biosensors and Bioelectronics* 20 (7):1270–1276.

Zhang, M. N., Y. M. Yan, K. P. Gong, L. Q. Mao, Z. X. Guo, and Y. Chen. 2004. Electrostatic layer-by-layer assembled carbon nanotube multilayer film and its electrocatalytic activity for O-2 reduction. *Langmuir* 20 (20):8781–8785.

Zhu, Z. G., W. H. Song, K. Burugapalli, F. Moussy, Y. L. Li, and X. H. Zhong. 2010. Nano-yarn carbon nanotube fiber based enzymatic glucose biosensor. *Nanotechnology* 21 (16):165501.

41 Graphene Applications

R. M. Abdel Hameed

CONTENTS

ABSTRACT

Graphene is a two-dimensional sheet of sp^2-hybridized carbon atoms that form a flat hexagonal lattice. Since its discovery by Geim et al. in 2004, graphene has received enormous attention due to its excellent electrical conductivity, mainly originating from the delocalized π bonds above and below the basal plane, its large surface area, and low production costs. Owing to these excellent physical and chemical properties, graphene has become an interesting alternative for the development of electrical devices, sensors, and biosensors, synthesizing nanocomposites, drug delivery, catalytic and solid-phase microextraction, energy storage materials, and transparent conducting electrodes. Because of its ultrahigh specific surface area and strong π–π electrostatic stacking property, graphene is expected to be a promising stationary-phase material for open-tubular capillary electrochromatography. Graphene-based electrochemical sensors present a better performance compared to glassy carbon, graphite, and even carbon nanotube-based sensors, mainly due to sp^2-like planes and edge defects that are more exposed on the graphene nanosheets than on other carbon materials. It is believed that nanoparticles can be controllably synthesized and well dispersed on the graphene surface. Additionally, the graphene nanosheets may effectively buffer the strain from the volume change of particles during cycles and also preserve a high electrical conductivity of the electrode if highly insulating metal oxides were used as anodes (e.g., Mn_3O_4). Recently, graphene-supported metal nanocomposites as a new class of hybrid materials combining the advantages of both graphene substrate and active metal nanoparticle components have shown extensive applications in many advanced fields such as memory electronic and optoelectronic transistors.

41.1 INTRODUCTION

Graphene is a single two-dimensional (2D) layer of graphite with sp^2-hybridized carbon atoms arranged in a honeycomb lattice [1,2]. It is characterized as "the thinnest material in our universe" [3,4]. Since its discovery by Geim and Novoselov in 2004 [5], graphene has gained a lot of attention in scientific and technological fields because of its unique physical/chemical properties such as large specific surface area (2630 $m^2\ g^{-1}$), excellent electrical conductivity (10^3–10^4 S m^{-1}), mechanical strength, and potential bulk quantity production along with low cost [5–7]. The application of graphene in many fields such as energy storage materials [8,9], memory electronic, optoelectronic transistors [10,11], electrocatalysts [10,12], drug delivery [13], Li ion batteries [14,15], solid-phase microextraction [16], and biosensors [17,18] are well studied. Graphene-based electrochemical sensors present a better performance compared to glassy carbon, graphite, and even carbon nanotubes-based sensors, mainly due to sp^2-like planes and edge defects that are more exposed on the graphene nanosheets than on other carbon materials [2,19].

41.2 APPLICATION OF GRAPHENE FOR BIOSENSING PURPOSES

The combined merits of graphene and carbon fiber endow their composite with a large specific surface area and high electrical conductivity. The higher peak current intensity and lower oxidation potential of uric acid compared with that of bare glassy carbon, carbon fiber, and graphene-modified glassy carbon electrodes were observed. Further amperometric study gives a wide linear range from 0.194 to 49.68 μM and a low detection limit of 0.132 μM (S/N = 3) with fast response time [20]. Electrochemically, Ni nanoparticles deposited on Nafion/graphene film on a glassy carbon electrode (GCE) showed linear oxidative currents with ethanol concentration in the range of 0.43–88.15 mM. The detection limit was 0.12 mM (S/N = 3) that was superior to those obtained with other transition metal-based nonenzymatic sensors [21]. Figure 41.1 represented TEM image of Ni/Nafion/graphene.

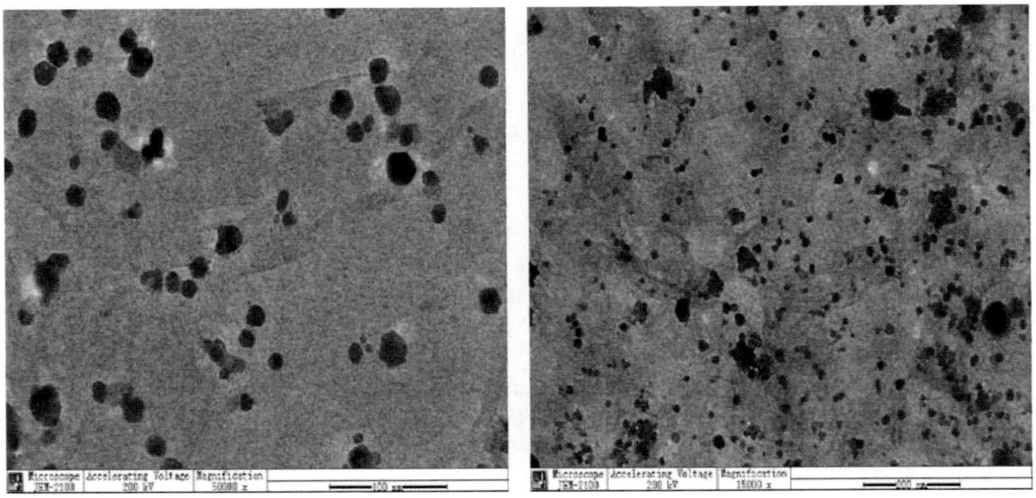

FIGURE 41.1 TEM image of Ni/Nafion/graphene. (From L.-P. Jia, H.-S. Wang, *Sens. Actuators B: Chem.* 177;2013:1035.)

A polydopamine–reduced graphene oxide (RGO) nanocomposite was prepared by a one-step procedure. This nanocomposite was applied in simultaneous determination of hydroquinone and catechol. The ΔE_p between hydroquinone and catechol was 103 mV, which was wide enough to discriminate the dihydroxy benzene isomers. The peak current of hydroquinone was linear to its concentration in the range of $1.0 \times 10^{-6} - 2.3 \times 10^{-4}$ M and the peak current of catechol was linear to its concentration in the range of $1.0 \times 10^{-6} - 2.5 \times 10^{-4}$ M [22]. The graphene oxide–manganese dioxide ($GO–MnO_2$) composite exhibited significantly decreased peak-to-peak separation of ca. 34 and 36 mV for hydroquinone and catechol, respectively, between oxidation and reduction waves in cyclic voltammetry. In differential pulse voltammetric measurements, $GO–MnO_2$-based sensor could separate the oxidation peak potentials of hydroquinone and catechol by about 115 mV, although the bare electrode gave a single broad response. This may be related to the higher surface area and catalytic ability of $GO–MnO_2$. The oxidation peak current of hydroquinone was linear over the range from 0.01 to 0.7 μM in the presence of 0.1 μM catechol, while the oxidation peak current of catechol was linear over the range from 0.03 to 1.0 μM in the presence of 0.13 μM hydroquinone. The detection limits (S/N = 3) for hydroquinone and catechol were 7.0 and 10.0 nM, respectively [23]. The differential pulse voltammograms of $GO–MnO_2$/GCE in $Na_2HPO_4–C_4H_2O_7$ buffer solution (pH 7.0) in the presence of 0.1 μM catechol with adding different concentrations of hydroquinone are shown in Figure 41.2a. Linear variation of the oxidation peak current density is observed with increasing

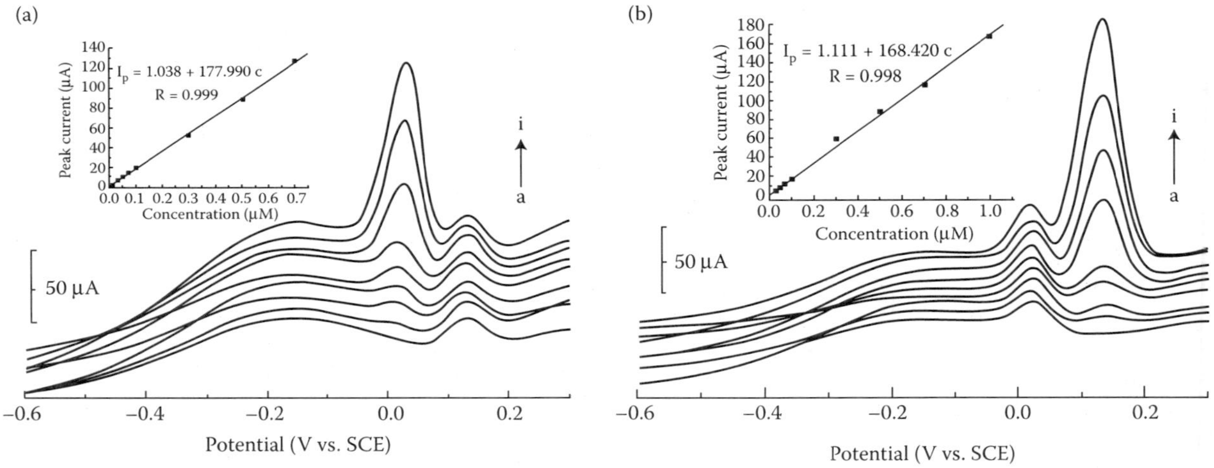

FIGURE 41.2 (a) Differential pulse voltammograms of (a) 0, (b) 0.01, (c) 0.03, (d) 0.05, (e) 0.07, (f) 0.1, (g) 0.3, (h) 0.5, and (i) 0.7 μM hydroquinone in $Na_2HPO_4–C_4H_2O_7$ buffer solution (pH 7.0) at $GO–MnO_2$/GCE in the presence of 0.1 μM catechol. Inset: linear range of the peak current versus the concentration of hydroquinone. (From T. Gan et al., *Sens. Actuators B: Chem.* 177;2013:412.) (b) Differential pulse voltammograms of (a) 0, (b) 0.03, (c) 0.05, (d) 0.07, (e) 0.1, (f) 0.3, (g) 0.5, (h) 0.7, and (i) 1.0 μM catechol in $Na_2HPO_4–C_4H_2O_7$ buffer solution (pH 7.0) at $GO–MnO_2$/GCE in the presence of 0.13 μM hydroquinone. Inset: linear range of the peak current versus the concentration of catechol. (From T. Gan et al., *Sens. Actuators B: Chem.* 177;2013:412.)

hydroquinone concentration in the inset figure. Similar trends were observed when catechol concentration was altered in the presence of hydroquinone as shown in Figure 41.2b.

CeO_2–graphene could catalyze the electrogenerated chemiluminescence of the luminol–H_2O_2 system to greatly amplify its luminal signal. CeO_2–graphene could also provide a better biocompatible microenvironment for the immobilized enzyme, which resulted in an excellent stability and long lifetime of the biosensor. A linear relation between cholesterol concentration and the current signal was extended from 12 μM to 7.2 mM with a detection limit of 4.0 μM (S/N = 3) [24]. An efficient catalytic activity was also observed at Au nanoparticles and cholesterol oxidase self-assembled to TiO_2–graphene–Pt–Pd hybrid nanocomposite. This fabricated biosensor exhibited wide linear range in the cholesterol concentration range of $5.0 \times 10^{-8} - 5.9 \times 10^{-4}$ M with a detection limit of 0.017 mM. The response time was less than 7 s with Michaelis–Menten constant of 0.21 mM. This fabricated biosensor was further tested using real food samples of eggs, meat, margarine, and fish oil, showing that this biosensor could be used as a facile cholesterol detection tool in food and supplement quality control [25].

Few-layer graphene with encased gold or silver nanoparticles were prepared in a single step by radio-frequency catalytic chemical vapor deposition (CVD) over Au_3/MgO and Ag_3/MgO systems. They were used to modify platinum substrates and subsequently employed for the electrochemical analysis of carbamazepine. Its oxidation potential decreases by 150 mV at both modified electrodes. The detection limit was found to be 2.75×10^{-5} and 2.92×10^{-5} M for silver- and gold-containing electrodes, respectively [26]. Cobalt hexacyanoferrate/RGO, synthesized by a facile precipitation route, showed an efficient electrocatalytic activity toward captopril electrooxidation. In the presence of captopril, the anodic peak current of Fe(II)/Fe(III) transition increased and the cathodic one decreased, while the peak currents of Co(II)/Co(III) transition remained almost constant. It indicates that captopril was oxidized on the modified electrode through a surface-mediated electron transfer. Captopril was determined with a detection limit of 0.331 μM [27]. The graphene–MnO_2 nanocomposite modified carbon ionic liquid electrode showed strong electrocatalytic ability to the electrochemical detection of rutin. A pair of well-defined redox peaks appeared with increased currents and decreased peak-to-peak separation when compared to other modified electrodes. Electrochemical parameters of rutin detection on this composite were calculated with the electron transfer coefficient as 0.5, the electron transfer number as 1.81, the apparent heterogeneous electron transfer rate constant as 1.61 s^{-1}, and the diffusion coefficient as 1.57×10^{-4} cm^2 s^{-1}. Under selected conditions, the reduction peak current was linear in the rutin concentration range from 0.01 to 500 μM with a detection limit of 2.73 nM (3σ) by differential pulse voltammetry. The proposed method was further applied to the rutin tablet sample determination with satisfactory results [28]. Figure 41.3a showed the cyclic voltammograms of GR–MnO_2/carbon ionic liquid electrode in phosphate buffer solution (pH 2.5) containing 100 μM rutin at different scan rates. A redox couple was observed. Its anodic and cathodic peak current density values were linearly increased with increasing the square root of scan rate in Figure 41.3b. The linear dependence of these peak potential values on the natural logarithm of scan rate is also observed in Figure 41.3c. Functionalized GO materials are prepared by the covalent reaction of GO with silver@silica–polyethylene glycol nanoparticles (~12.35 nm). The functionalized GO electrode shows a well-defined voltammetric response in phosphate-buffered saline and catalyzes the oxidation of quercetin to quinone without the need of an enzyme. Significantly, the functionalized GO modified electrode exhibited a higher sensitivity than pristine gold-printed circuit board and GO electrodes, a wide concentration range of 7.5–1040 nM and detection limit of 3.57 nM. The developed biosensor platform

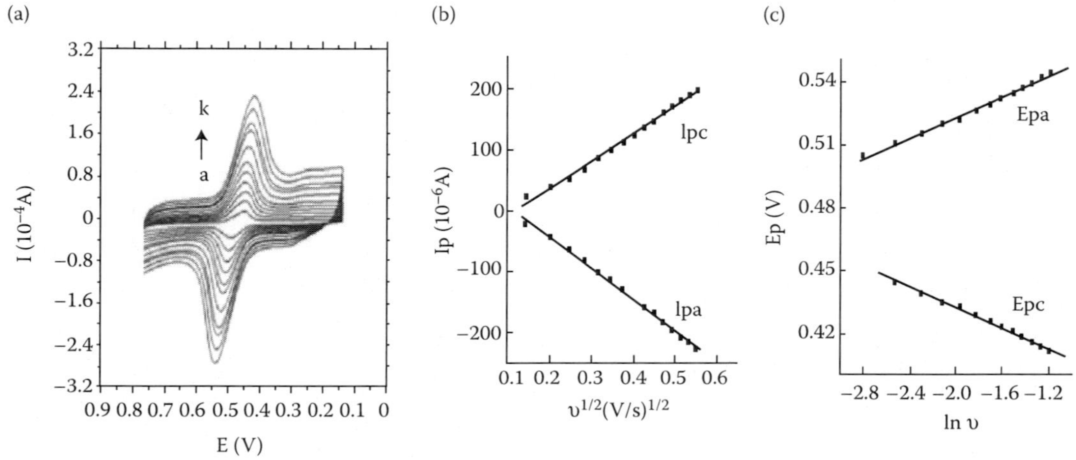

FIGURE 41.3 (a) Cyclic voltammograms of 100 μM rutin with different scan rates on GR–MnO_2/carbon ionic liquid electrode in pH 2.5 PBS (from a to k are 10, 20, 40, 60, 80, 100, 140, 180, 220, 260, and 300 mV s^{-1}, respectively). (b) Linear relationship of cathodic and anodic peak currents versus $\upsilon^{1/2}$. (c) Linear relationship between peak potentials (E_p) and ln υ. (From W. Sun et al., *Sens. Actuators B: Chem.* 178;2013:443.)

is selective toward quercetin in the presence of an interferent molecule [29]. Polystyrene–GO nanocomposite, prepared for the first time using an *in situ* polymerization method, was successfully applied to determine histamine in fish samples. It yields satisfactory results with spiked recoveries in the range of 98.2%–103.1% [30].

Cyclic voltammetry and differential pulse voltammetry were employed to evaluate the electrochemical properties of the Ag nanoparticles/GO/GCE composite toward tryptophan determination. This biosensor was prepared through a green and low-cost synthesis approach using glucose as a reducing and stabilizing agent. It exhibited a 10-fold activity increment for tryptophan oxidation than GO film. The oxidation peak currents were proportional to tryptophan concentrations over the range of 0.01–50 mM and 50–800 mM. The detection limit was 2 nM (S/N = 3). Moreover, the proposed method is free from interference by tyrosine and other coexisting species. The resulting sensor displays excellent repeatability and long-term stability. It was also successfully applied to detect tryptophan in real samples with good recoveries ranging from 99% to 103% [31]. A human serum albumin/GO/3 aminopropyltriethoxysilane modified indium tin oxide (ITO) electrode showed a clear separation between the oxidation peak potentials of D- and L-tryptophan at 0.86 and 1.26 V, respectively. The percentage of enantiomeric forms of tryptophan in a mixture can be easily measured in the presence of the other using this composite [32]. Electrochemically reduced GO displayed a greatly improved voltammetric response to tryptophan and tyrosine compared with the chemically reduced GO electrode. It could separate the voltammetric responses of ascorbic and uric acids from that of tryptophan and tyrosine. It could determine tryptophan and tyrosine with linear ranges of 0.2–40.0 μM and 0.5–80.0 μM with detection limits of 0.1 and 0.2 μM, respectively [33]. A Co_3O_4 nanoparticles/graphene/GCE showed a dramatically improved conductivity and sensitivity toward tyrosine detection. A wide linear range of tyrosine concentration was obtained from 1×10^{-8} to 4×10^{-5} M with a comparatively low detection limit of 1×10^{-9} M [34]. Multiwalled carbon nanotubes (MWCNTs)/graphene nanosheets/GCEs displayed high effective surface area, high porosity, more reactive sites and excellent electrochemical catalytic activity toward the oxidation of tyrosine and paracetamol. The peak current of their differential pulse voltammograms linearly increased with concentration ranges of 0.9–95.4 μM tyrosine and 0.8–110.0 μM paracetamol. The detection limits for tyrosine and paracetamol were 0.19 and 0.10 μM, respectively [35].

A macroporous carbon–graphene composite was synthesized by a casting method using silica nanoparticles–GO as the template and sucrose as the carbon precursor. This composite combines the porous structure of macroporous carbon with the excellent electrical and mechanical properties of graphene. The presence of macroporous carbon is beneficial for decreasing the aggregation of graphene sheets and improving the diffusion of molecules through them. A nonenzymatic amperometric sensor of glucose based on the Pt/macroporous carbon–graphene composite responds very rapidly to changes in glucose level within 3 s. Owing to the low oxidation potential, no obvious interference is observed with the addition of ascorbic acid, acetaminophen, dopamine, and uric acid [36]. Pd_1Pt_3–graphene nanomaterials, prepared by a one-pot microwave heating method, exhibited a great enhancement toward glucose oxidation in neutral phosphate buffer solution (pH 7.4) compared with unsupported Pd_1Pt_3 nanomaterials [37]. The coral-like PtAu–MnO_2 nanocomposites, prepared by a one-step template-free electrodeposition method, exhibited a unique set of structural and electrochemical properties such as better uniformity, larger active surface area, and faster electron transfer in comparison with the control electrode prepared by tandem growth of MnO_2 network and PtAu alloy in two steps. PtAu–MnO_2 decorated graphene paper has shown greatly enhanced sensing performance toward glucose such as wide linear range from 0.1 to 30 mM and high sensitivity of 58.54 mA mM^{-1} cm^{-2} and low detection limit of 0.02 mM (S/N = 3) [38]. SEM micrographs and corresponding EDX spectra of different graphene-supported nanostructures are represented in Figure 41.4. The GO/nickel oxide modified GCE showed a linear range for glucose estimation from 3.13 μM to 3.05 mM with a detection limit of 1 μM [39].

A glucose biosensor based on a polydopamine–graphene hybrid film modified glucose oxidase enzyme electrode showed a high sensitivity of 28.4 μA mM^{-1} cm^{-2} and a low detection limit of 0.1 μm. A short response period (<4 s) and a low Michaelis–Menten constant (6.77 mM) were achieved. This might be attributed to the large surface-to-volume ratio, high conductivity of graphene, and good biocompatibility of polydopamine. This could enhance the enzyme absorption and promote direct electron transfer between the redox enzymes and the electrode surface [40]. Wang et al. [41] demonstrated the immobilization of glucose oxidase on electrochemically reduced GO. A single-layer GO was adsorbed on a 3-aminopropyl-triethoxysilane modified electrode, then electrochemically reduced to electrochemically reduced graphene oxide (ERGO), electrografted with poly-*N*-succinimidyl acrylate, and finally the glucose oxidase (GO_x) molecules were cross-linked through covalent bonding. However, this sort of approach is rather time consuming and it involves tedious enzyme immobilization procedures. Unnikrishnan et al. [42] immobilized GO_x on RGO in a single step without any cross-linking agents or modifiers. A simple solution-phase approach was used to prepare exfoliated GO, followed by electrochemical reduction to obtain an ERGO–GO_x biocomposite. This ERGO–GO_x film on a GCE showed very good stability, reproducibility, and high selectivity. It exhibits excellent catalytic activity toward glucose oxidation over a wide linear range of 0.1–27 mM with a sensitivity of 1.85 mA mM^{-1} cm^{-2}. X-ray photoelectron spectroscopy analysis of an electrochemically reduced carboxyl graphene modified GCE indicated that most of the oxygen-containing groups such as epoxy/ether and hydroxyl groups in the carboxyl graphene were eliminated, while carboxylic acid groups remained. Glucose oxidase was immobilized on this composite via self-assembly. The cyclic voltammetric result of the electrode shows a pair of well-defined and quasi-reversible redox peaks with a

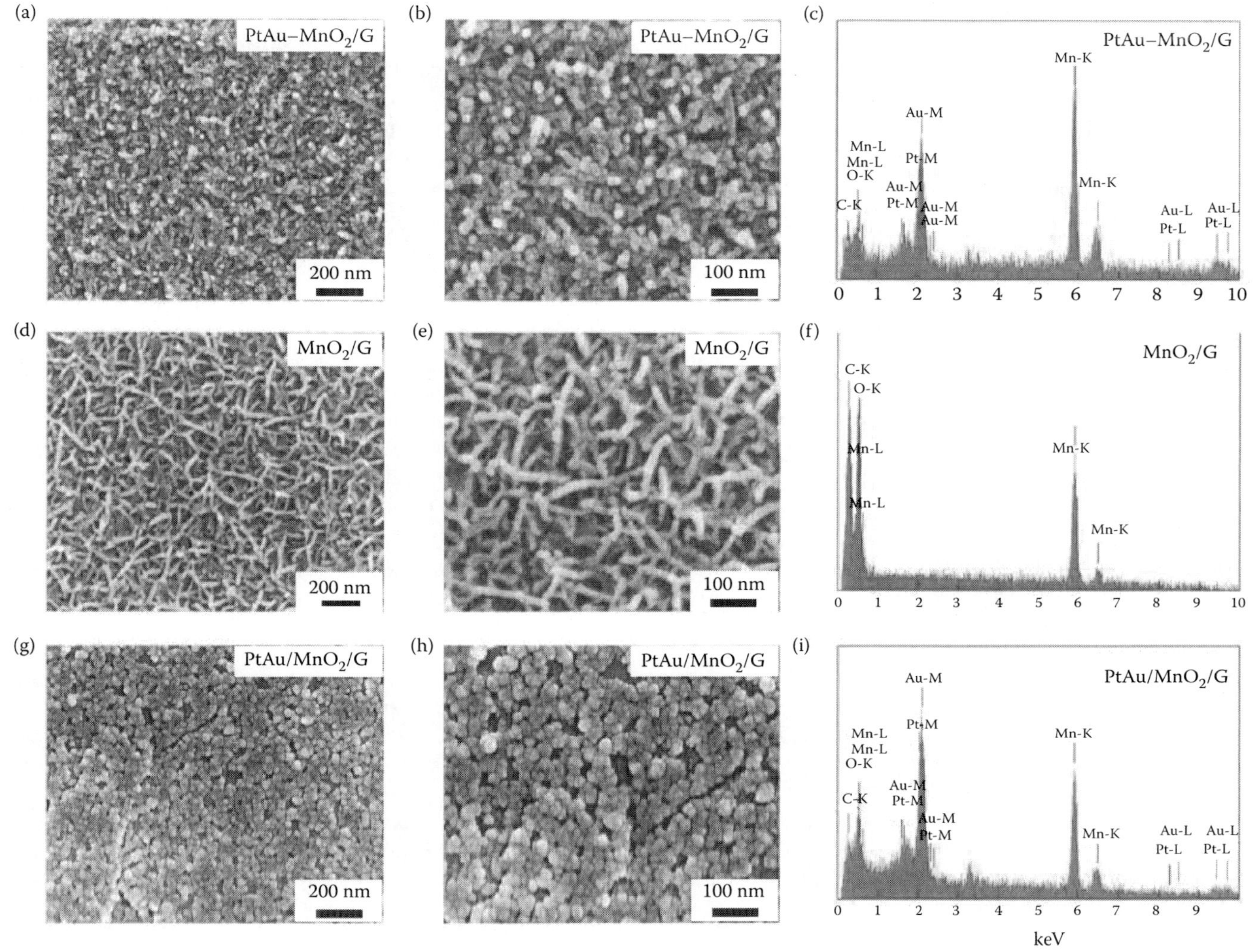

FIGURE 41.4 SEM micrographs of (a, b) PtAu–MnO_2/G, MnO_2/G (d, e) and PtAu/MnO_2/G (g, h). Their corresponding EDX spectra appear in (c), (f) and (i), respectively. (From X. Bo, L. Guo, *Electrochim. Acta* 90;2013:283.)

formal potential of −467 mV and a peak to peak separation of 49 mV, revealing direct electron transfer between glucose oxidase (GOD) and this composite. It exhibits a linear response to glucose concentrations ranging from 2 to 18 mM with a detection limit of 0.02 mM [43]. Compared with pristine MWCNTs, 2.1-fold higher peak current and very low peak-to-peak separation of 26 mV at an electrochemically reduced GO–MWCNTs hybrid modified electrode were observed, demonstrating faster electron transfer between GO_x and the modified electrode surface. This modified film exhibited high electrocatalytic activity toward glucose oxidation via the reductive detection of oxygen consumption in the presence of a mediator. A low detection limit of 4.7 mM and wide linear range of 0.01–6.5 mM were achieved [44]. An exfoliated GO sheet with carboxyl long chains was prepared in one step from primitive graphite via Friedel–Crafts acylation. It can be used as a glucose sensor by tracing the released H_2O_2 after enzymatic reaction of bound glucose oxidase. A linear response within the range from 0.01 to 1 mM H_2O_2 and 0.1 to 1.4 mM glucose was obtained with high sensitivities of 4340.6 and 1074.6 mA mM^{-1} cm^{-2}, respectively [45].

Graphene quantum dots (GQD) were introduced as an overland suitable substrate for enzyme immobilization. GO_x was immobilized on a GQD modified carbon ceramic electrode (CCE) and well-defined quasi-reversible redox peaks were observed. The electron transfer coefficient and the heterogeneous electron transfer rate constant for the redox reaction of GO_x were found to be 0.48 and 1.12 s^{-1}, respectively. The developed biosensor efficiently responds to the presence of glucose over the concentration range 5–1270 mM with a detection limit of 1.73 mM (S/N = 3) and sensitivity of 0.085 mA mM^{-1} cm^{-2}. The high value of surface coverage GO_x–GQD|CCE (1.8×10^{-9} mol cm^{-2}) and the small value of the Michaelis–Menten constant (0.76 mM) confirmed an excellent loading of the enzyme and a high affinity of the biosensor to glucose [46]. Glucose oxidase–carbon nitride dots–chemically reduced GO hybrids on a GCE showed a good electrocatalytic activity toward glucose determination with a linear concentration range from 40 μM to 20 mM and a detection limit of 40 μM (S/N = 3) [47].

A graphene–SnO_2 nanocomposite modified electrode displayed high electrocatalytic activity to the oxidation of dopamine with the increase of redox peak currents and the decrease of peak-to-peak separation. The oxidation peak current was proportional to dopamine concentration in the range from 0.5 to 500 μM with a detection limit of 0.13 μM (S/N = 3). This modified electrode showed excellent selectivity and sensitivity even in the presence of a high concentration of uric acid [48]. The current response of a GO/SiO_2–molecularly imprinted sensor toward dopamine detection was nearly 3.2 times that of the nonimprinted one. This sensor could also recognize dopamine from its relatively similar molecules of norepinephrine and epinephrine, while the sensors based on GO/SiO_2–nonimprinted and vinyl groups functionalized GO/SiO_2 did not have this ability. A wide linear range over dopamine concentration from 5.0×10^{-8} to 1.6×10^{-4} M with a detection limit of 3.0×10^{-8} M (S/N = 3) was obtained [49]. Differential pulse voltammetry was applied to determine dopamine in the presence of 500 mM ascorbic acid and 330 mM uric acid at 3,4,9,10-perylene tetracarboxylic acid functionalized graphene sheets/MWCNTs/ionic liquid composite film modified electrodes. A good linear relationship between the peak current and dopamine concentration was obtained in the range from 0.03 to 3.82 mM with a detection limit of 1.2×10^{-9} M (S/N = 3). The proposed method was successfully applied to determine dopamine in real samples with satisfactory results. This modified electrode exhibits good sensitivity, reproducibility, and long-term stability [50]. The enhanced performance of a reduced graphene oxide–multiwalled carbon nanotubes–phosphotungstic acid (RGO–MWCNTs–PTA) composite toward dopamine sensing results from (i) the enlarged surface area of electrodes due to the intercalation of MWCNTs between the stacked RGO with associated large electrochemical active sites and three-dimensional (3D) nanostructured RGO–MWCNTs covered by numerous PTA particles; (ii) the improved conductivity through the formation of a 3D network aided by MWCNTs; and (iii) the accelerated electron transfer rates, which is due to aromatic π–π stacking and electrostatic attraction between a positively charged dopamine and negatively charged RGO–MWCNTs–PTA modified electrode. A linear range of 0.5–20 μM with a detection limit of 1.14 μM (3σ) was achieved [51].

Monolithic and macroporous graphene foam grown by CVD served as the electrode scaffold to thionine molecules using *in situ* polymerized polydopamine as the linker. It can efficiently mediate H_2O_2 reduction at close proximity to the electrode surface. A wide linear range from 0.4 to 660 μM, high sensitivity of 169.7 μA mM^{-1}, and low detection limit of 80 nM were obtained [52]. The incorporation of thionine onto the GO surface resulted in more than twice an increase in the amperometric response to H_2O_2 of the thionine modified electrode. This composite also served as a biocompatible matrix for enzyme assembly and a mediator to facilitate the electron transfer between the enzyme and the electrode [53]. GO and silica nanoparticles composite showed a low H_2O_2 detection limit of 2.6 μM and a wide linear range of 5.22 μM–10.43 mM in alkaline medium [54]. Raman spectroscopy, transmission electron microscopy (TEM), atomic force microscopy, and scanning electron microscopy (SEM) all suggested that RGO in RGO/single-walled carbon nanotubes (SWCNTs) composites acted as a surfactant, covering and smoothing out the surface and that the SWCNTs acted as a conducting bridge to connect the isolated RGO sheets. The linear concentration range of H_2O_2 detection at RGO/SWCNTs modified electrode is 0.5–5 M with a sensitivity of 2732.4 mA mM^{-1} cm^{-2} and detection limit of 1.3 mM [55].

A novel nonenzymatic hydrogen peroxide sensor was developed using a graphene/Nafion/Azure I/Au nanoparticles composite modified GCE. Cyclic voltammetry demonstrated that the direct electron transfer of redox molecule, Azure I, was realized. The sensor had an excellent performance in terms of electrocatalytic reduction toward H_2O_2. It showed high sensitivity and fast response upon the addition of H_2O_2 under the conditions of pH 4.0 and potential value of −0.2 V. The time to reach the stable-state current was less than 3 s and the linear range to H_2O_2 concentration was from 30 μM to 5 mM with a detection limit of 10 μM (S/N = 3) [56]. On the other hand, H_2O_2 oxidation peak potential at monometallic Pd and Au nanoparticles supported on graphene nanoplatelets is −0.35 and 0.53 V saturated calomel electrode (SCE), respectively. The composition of Pd–Au binary catalysts affects the oxidation peak current and potential. Higher oxidation current is achieved when bimetallic Pd–Au nanoparticles with an atomic ratio of 3:1 are deposited on graphene nanoplatelets [57]. Palladium nanoparticles/aminothiophenol/electrochemically reduced GO exhibited a wide linear range from 0.1 μM to 10 mM and a low detection limit of 0.016 μM (S/N = 3) with a fast response time of less than 10 s [58].

Ag/graphene composite was prepared by surface plasmon resonance induction of Ag/GO in a two-phase (water–toluene) solvent. TEM revealed that Ag nanoparticles with a size of 5–8 nm were distributed on RGO sheets. Raman and X-ray photoelectron spectroscopy demonstrated low defect density and high deoxygenation degree of graphene in Ag/graphene composite. This excellent structure and morphology of the Ag/graphene composite contributed to superior electrical properties for H_2O_2 reduction [59]. A double pulse electrochemical method was adopted to controllably prepare silver nanocrystals on a graphene thin-film electrode. The approach relies on two potential pulses that can independently control the nucleation and subsequent growth processes of silver nanocrystals on a graphene substrate. This method also allows the observation of nanoparticles growing from a particle shape to a nanoplate form by increasing the growth time with the maximum lateral scale up to micrometer range. A proposed mechanism for silver nanoplates formation was the oriented growth of small AgNCs and 2D graphene template inducing effect. This hybrid thin-film electrode exhibits remarkable electrocatalytical activity toward H_2O_2 reduction. It displays a fast amperometric response time of less than 2 s and a good linear range from 2×10^{-5} to 1×10^{-2} M with an estimated detection limit of 3×10^{-6} M. This biosensor also exhibits good stability with relative standard deviation (RSD) value of 1.3% and high

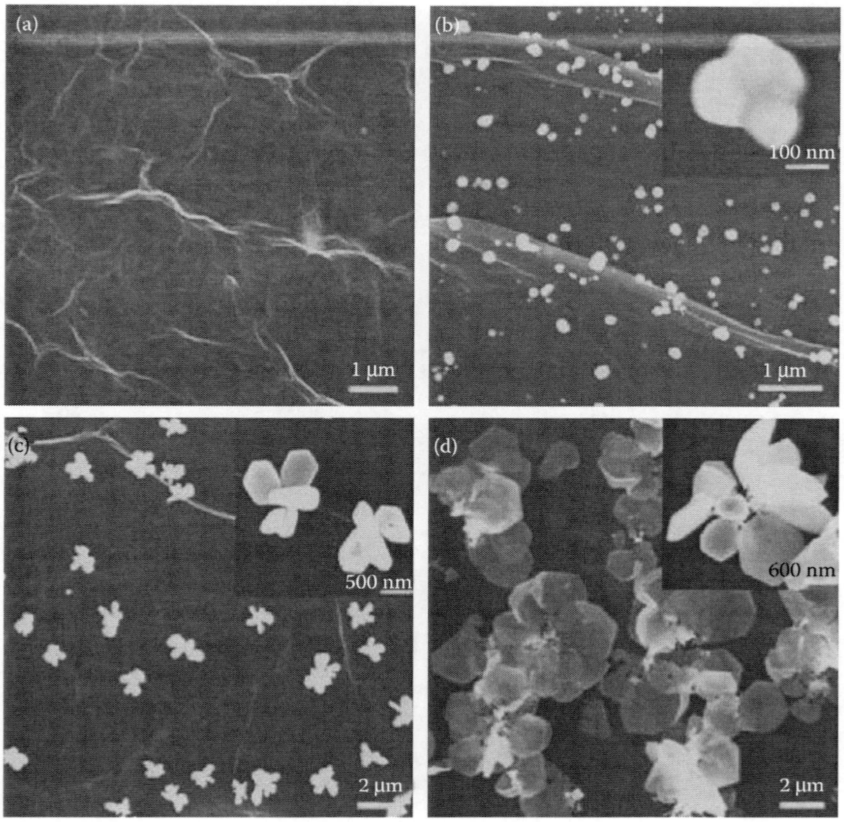

FIGURE 41.5 SEM images of ERGO–AgNCs hybrid thin films. (a) A wrinkled morphology of ERGO thin film. Parts (b)–(d) show the AgNCs electrodeposited on the surface of ERGO/GCE film electrodes under different growth times: (b) 50 s, (c) 200 s, and (d) 400 s. (From J.-M. You et al., *Sens. Actuators B: Chem.* 178;2013:450.)

sensitivity of 183.5 μA mM^{-1} cm^{-2} as well as high selectivity [60]. Figure 41.5 indicated SEM images of ERGO before and after electrodepositing AgNCs on its surface for different growth times. DNA–Ag nanoclusters/graphene supported on a GCE showed a low H_2O_2 detection limit of 3 mM and a wide linear range from 15 to 23 mM with high selectivity and good repeatability [61]. The amperometric response of bare GCE, graphene/GCE, and (DNA–Ag nanoclusters)/graphene/GCE to different concentrations of H_2O_2 in N_2-saturated phosphate buffer solution (0.1 M, pH = 7.0) is inserted in Figure 41.6a. The linear increase of H_2O_2 reduction current density at the modified graphene/GCE with its concentration is shown in Figure 41.6b. The low loading of silver nanoparticles on RGO exhibited enhanced characteristic signals for H_2O_2 detection

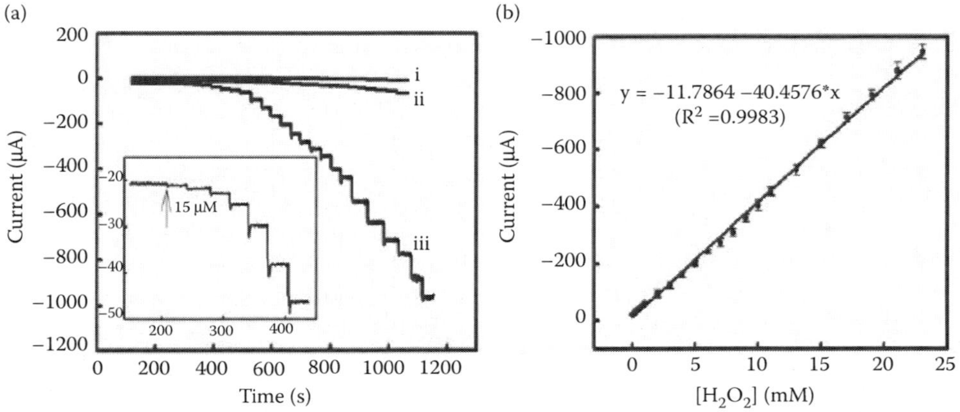

FIGURE 41.6 (a) Amperometric response to different concentrations of H_2O_2 at the bare GCE (i), graphene/GCE (ii), and (DNA–Ag nanoclusters)/graphene/GCE (iii) in N_2-saturated PBS (0.1 M, pH = 7.0). (b) Calibration plot obtained from (a(iii)). (From F.-J. Zhang et al., *Appl. Surf. Sci.* 265;2013:578.)

[62]. Ascorbic and uric acids, dopamine, glutathione, and L-cysteine did not result in any interference.

Cuprous oxide–RGO nanocomposites were prepared using three approaches, including physical adsorption, *in situ* reduction, and one pot synthesis. The composite with different morphologies and components fabricated from these three methods displayed much enhanced performance for H_2O_2 reduction than the single-component Cu_2O. Among these nanocomposites, the product prepared through the simple physical adsorption approach showed a slightly better performance than the other two composites. A wider linear range (0.03–12.8 mM), higher sensitivity (19.5 μA mM^{-1}), and better stability were achieved on this composite compared to the Cu_2O-based sensor [63]. A Fe_3O_4/RGO/GCE electrode showed an excellent electrocatalytic activity to reduce H_2O_2 at a wide and linear range from 4×10^{-6} to 1×10^{-3} M with a detection limit of 2×10^{-6} M [64]. A Nafion/exfoliated GO/Co_3O_4 nanocomposite-coated GCE showed well-defined and stable redox couples signal in both alkaline and neutral aqueous solutions with excellent electrocatalytic activity for H_2O_2 oxidation. The response of this modified electrode to H_2O_2 was examined using amperometry at 0.76 V (Ag/AgCl) in a phosphate buffer solution (pH 7.4). The detection limit was 0.3 mM with a sensitivity of 560 mA mM^{-1} cm^{-2} [65]. On the other hand, the current response of cobalt-tetraphenylporphyrin/RGO nanocomposite was linear to the H_2O_2 concentration range from 1×10^{-7} to 2.4×10^{-3} M at a potential of −0.2 V [66]. The ability of this hybrid composite for routine analyses was demonstrated by the detection of H_2O_2 present in milk samples with appreciable recovery values. Tables 41.1 through 41.3 listed the linear range and detection limit values for the determination of various compounds using graphene-based electrodes.

TABLE 41.1

Linear Range and Detection Limit Values for the Determination of Various Compounds Using Graphene-Based Electrodes

Electrode	Compound	Linear Range	Detection Limit	Reference
Ni nanoparticles electrochemically deposited on Nafion/graphene film on GCE	Ethanol	0.43–88.15 mM	0.12 mM	[21]
Polydopamine–RGO nanocomposite	Hydroquinone	1.0×10^{-6}–2.3×10^{-4} M		[22]
	Catechol	1.0×10^{-6}–2.5×10^{-4} M		
GO–manganese dioxide composite	Hydroquinone	0.01–0.7 μM	7.0 nM	[23]
	Catechol	0.03–1.0 μM	10.0 nM	
CeO_2–graphene composite	Cholesterol	12 μM–7.2 mM	4.0 μM	[24]
TiO_2–graphene–Pt–Pd hybrid nanocomposite	Cholesterol	5.0×10^{-8}–5.9×10^{-4} M	0.017 mM	[25]
Cobalt hexacyanoferrate/RGO	Captopril		0.331 μM	[27]
Graphene–MnO_2 nanocomposite modified carbon ionic liquid electrode	Rutin	0.01–500 μM	2.73 nM	[28]
Functionalized GO with silver@silica–polyethylene glycol nanoparticles	Quercetin	7.5–1040 nM	3.57 nM	[29]
Ag nanoparticles/GO/GCE composite	Tryptophan	0.01–50 mM 50–800 mM	2 nM	[31]
Electrochemically reduced GO	Tryptophan	0.2–40.0 μM	0.1 μM	[33]
	Tyrosine	0.5–80.0 μM	0.2 μM	
Co_3O_4 nanoparticles/graphene/GCE	Tyrosine	1×10^{-8}–4×10^{-5} M	1×10^{-9} M	[34]
MWCNTs/graphene nanosheets/GCE	Tyrosine	0.9–95.4 μM	0.19 μM	[35]
	Paracetamol	0.8–110.0 μM	0.10 μM	

TABLE 41.2

Linear Range, Sensitivity, and Detection Limit Values for Glucose Determination Using Graphene-Based Electrodes

Electrode	Linear Range	Sensitivity (mA mM^{-1} cm^{-2})	Detection Limit	Reference
PtAu–MnO_2 decorated graphene paper	0.1–30 mM	58.54	0.02 mM	[38]
GO/nickel oxide modified GCE	3.13 μM–3.05 mM		1 μM	[39]
GO_x on RGO	0.1–27 mM	1.85		[42]
Electrochemically reduced carboxyl graphene modified GCE	2–18 mM		0.02 mM	[43]
Electrochemically reduced GO–MWCNTs hybrid modified electrode	0.01–6.5 mM		4.7 mM	[44]
Exfoliated GO sheet with carboxyl long chains	0.1–1.4 mM	1074.6		[45]
GO_x immobilized on GQD modified carbon ceramic electrode	5–1270 mM	0.085	1.73 mM	[46]
Glucose oxidase–carbon nitride dots–chemically reduced GO hybrids on GCE	40 μM–20 mM		40 μM	[47]

TABLE 41.3
Linear Range, Sensitivity, and Detection Limit Values for Dopamine and H_2O_2 Determination Using Graphene-Based Electrodes

Electrode	Compound	Linear Range	Sensitivity ($mA\ mM^{-1}cm^{-2}$)	Detection Limit	Reference
Graphene–SnO_2 nanocomposite modified electrode	Dopamine	0.5–500 μM		0.13 μM	[48]
GO/SiO_2–molecularly imprinted sensor	Dopamine	5.0×10^{-8}–1.6×10^{-4} M		3.0×10^{-8} M	[49]
3,4,9,10-Perylene tetracarboxylic acid functionalized graphene sheets/MWCNTs/ionic liquid composite film modified electrode	Dopamine	0.03–3.82 mM		1.2×10^{-9} M	[50]
RGO–MWCNTs–phosphotungstic acid composite	Dopamine	0.5–20 μM		1.14 μM	[51]
Monolithic and macroporous graphene foam	H_2O_2	0.4–660 μM	169.7 $\mu A\ mM^{-1}$	80 nM	[52]
GO and silica nanoparticles composite	H_2O_2	5.22 μM–10.43 mM		2.6 μM	[54]
RGO/SWCNTs modified electrode	H_2O_2	0.5–5 M	2732.4	1.3 mM	[55]
Graphene/Nafion/Azure I/Au nanoparticles composite modified GCE	H_2O_2	30 μM–5 mM		10 μM	[56]
Palladium nanoparticles/aminothiophenol/electrochemically reduced GO	H_2O_2	0.1 μM–10 mM		0.016 μM	[58]
Silver nanocrystals on graphene thin-film electrode	H_2O_2	2×10^{-5}–1×10^{-2} M	183.5 $\mu A\ mM^{-1}\ cm^{-2}$	3×10^{-6} M	[60]
DNA–Ag nanoclusters/graphene supported on GCE	H_2O_2	15–23 mM		3 mM	[61]
Cuprous oxide–RGO nanocomposite	H_2O_2	0.03–12.8 mM	19.5 $\mu A\ mM^{-1}$		[63]
Fe_3O_4/RGO/GCE electrode	H_2O_2	4×10^{-6}–1×10^{-3} M		2×10^{-6} M	[64]
Nafion/exfoliated GO/Co_3O_4 nanocomposite-coated GCE	H_2O_2		560	0.3 mM	[65]

41.3 APPLICATION OF GRAPHENE IN THE ELECTROCATALYST MANUFACTURE FOR FUEL CELLS AND OXYGEN REDUCTION REACTION

Many types of carbon, including carbon black, carbon nanofibers, and carbon nanotubes have been used as support materials for catalysts in fuel cell systems. Platinum-based electrocatalysts are exclusively utilized for catalyzing oxygen reduction and methanol oxidation reactions in direct methanol fuel cells [67]. A liquid-phase reduction technique was used on nanographene oxide sheets (<100 nm), followed by a vapor-phase reduction of Pt complex to prepare Pt subnano/nanoclusters at the edge of nanographene sheets. The average sizes of Pt nanocluster were controlled from 1.1 to 2.4 nm by varying the initial concentration of the Pt complex. This composite showed superior catalytic performance for electrochemical methanol oxidation [68]. This is apparent from the catalytic activities of Pt/graphene and Pt/C electrodes as 1315 A g^{-1} Pt and 725 A g^{-1} Pt, respectively [69]. Long-term stability for methanol oxidation was achieved at Pt catalyst supported on the RGO–ordered mesoporous carbon composite [70]. The introduction of MWCNTs to the Pt/graphene electrocatalyst enhanced its electrocatalytic activity toward methanol oxidation in terms of the electroactive surface area, forward anodic peak current density, onset oxidation potential, diffusion efficiency, and the ratio of forward to backward anodic peak current density (I_f/I_b) [71]. This was estimated from the cyclic voltammograms of methanol oxidation reaction at Pt/graphene/GCE and Pt/graphene–MWCNTs/GCE in (1 M CH_3OH + 0.1 M H_2SO_4) solution at a scan rate of 50 mV s^{-1} in Figure 41.7. The SEM image of the three-dimensionally porous graphene–carbon nanotube composite-supported PtRu catalyst showed that MWCNTs act as useful nanospacers for diminishing the face-to-face aggregation of graphene sheets. The high surface area of 3D graphene–MWCNTs hybrid film due to a synergistic combination of both structural and electrical properties of 2D-graphene and 1D-MWCNTs could allow a higher dispersion of metallic nanoparticles. This may explain much higher I_f/I_b value of PtRu/graphene–MWCNTs electrocatalyst (6.33) when compared to that of a commercial catalyst, PtRu/Vulcan XC-72 (I_f/I_b = 1.33) [72]. The strong interaction between Pt and CeO_2 greatly improved methanol oxidation and oxygen reduction reactions at the Pt/graphene–CeO_2 electrocatalyst. The best electrocatalytic performance was shown at the catalyst with CeO_2 content of 7 wt.% [73]. Chen et al. [74] reported an enhanced electrochemical performance of graphene single nanosheets when

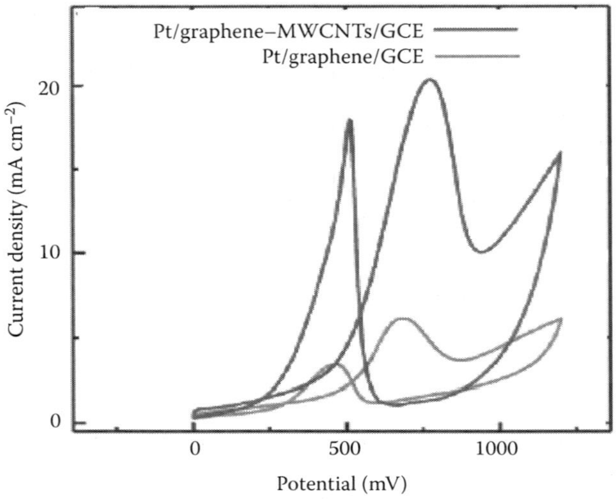

FIGURE 41.7 Electrocatalytic performance of Pt/graphene/GCE and Pt/graphene–MWCNTs/GCE in (1 M CH_3OH + 0.1 M H_2SO_4) solution at a scan rate of 50 mV s^{-1}. (From K. Kakaei, M. Zhiani, *J. Power Sources* 225;2013:356.)

compared to GO. SEM and TEM images clearly showed the lateral expansion of graphite sheet and uniform dispersion of Pt nanoparticles on graphene single nanosheets. Cyclic voltammetry showed that the electrochemical active surface areas (ECSAs) of Pt/graphene single nanosheets (Pt/GNS), commercial Pt/Vulcan XC-72 carbon black, and Pt-incorporated onto commercial carbon black (Pt/CB) are 33.1, 23.5, and 22.5 m^2 g^{-1}, respectively. Because of significant restacking of Pt/GNS sheets, carbon black with a different content is intercalated between Pt/GNS as a spacer. The ECSAs of Pt/GNS, Pt/GNS/CB20, Pt/GNS/CB30, and Pt/GNS/CB40 are evaluated to be 31.5, 28.6, 38.8, and 30.4 m^2 g^{-1}, respectively. The cell performance highly depends on the carbon black content and Pt/GNS/CB with 30 wt.% of carbon black showed the best fuel cell performance of 400 mA cm^{-2} [75]. This was clarified in Figure 41.8. A simple and environmentally-friendly method was used to prepare Pt/RGO hybrids. This approach used a redox reaction between Na_2PtCl_4 and GO nanosheets and a subsequent thermal reduction of the material at 200°C for 24 h in a vacuum oven. In contrast to other methods that use an additional reductant to prepare Pt nanoparticles, the Pt^{2+} was directly reduced to Pt^0 in the GO solution. GO was used as the reducing agent, the stabilizing agent, and the carrier. Electrochemical measurements showed that the Pt/RGO hybrids exhibit good activity as catalysts for the electrooxidation of methanol and ethanol in acid media. Interestingly, the Pt/RGO hybrids showed better electrocatalytic activity and stability for the oxidation of methanol than Pt/C and Pt/RGO hybrids made from other Pt precursors [76]. Nanometer-sized Pt islands were heterogeneously deposited on RGO subjected to oxidation prior to deposition. The functional groups can act as nucleation sites for Pt nanoparticles and homogeneous nucleation of small particles can be achieved by combining surface functionalization with diazonium chemistry and

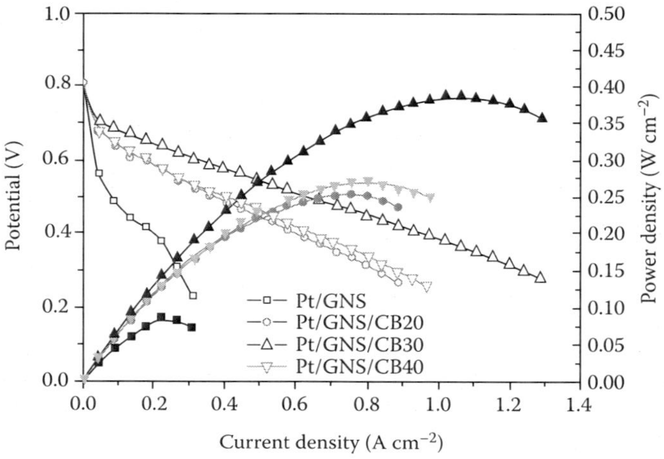

FIGURE 41.8 Cell polarization curves of Pt/GNS, Pt/GNS/CB20, Pt/GNS/CB30, and Pt/GNS/CB40. (From S. Yu et al., *Electrochim. Acta* 94;2013:245.)

appropriate stabilizers in solution. An effective electrochemical mediator is prepared using 4-aminothiophenol-modified RGO as a support decorated with Pt nanoparticles (Pt@SC_6H_4–rGO) for methanol oxidation and oxygen reduction. Results showed that Pt@SC_6H_4–rGO has an enhanced electrocatalytic activity for methanol oxidation and oxygen reduction in a solution containing 0.5 M H_2SO_4 and/or 1.0 M methanol when compared to Pt@rGO [77]. Pt–Pd hollow nanospheres supported on RGO nanosheets under mild conditions without involving any surfactant, seed, or template displayed an improved electrocatalytic activity and better stability for methanol oxidation reaction in alkaline media, compared with commercial Pt black and Pd black catalysts [78]. Uniform porous Pt–Pd nanospheres supported on RGO nanosheets were prepared under ambient temperature, where octylphenoxypolyethoxyethanol (NP-40) is used as a soft template. This nanocomposite displays an enhanced electrocatalytic activity and good stability toward methanol oxidation, compared with commercial Pd/C and Pt/C catalysts [79]. A graphene ribbon-supported Pd electrocatalyst showed an increased electrochemical surface area when compared to conventional Pd/C. This could significantly enhance its catalytic activity toward formic acid oxidation reaction. The oxidation peak current density at Pd/graphene is 1.4 A mg^{-1} Pd compared to 0.79 A mg^{-1} Pd at Pd/C with a potential shift in the negative direction by 150 mV [80]. Cyclic voltammetry and chronoamperometry indicated that higher catalytic activity and better stability for methanol oxidation were observed when MnO_2 was added to the Pd/RGO electrocatalyst [81].

In comparison to graphene, nitrogen-doped graphene (NDG) demonstrated higher electrocatalytic activity toward oxygen reduction reaction in both acidic and alkaline solutions. It can act directly as a catalyst to facilitate four-electron oxygen reduction in an alkaline solution and two-electron reduction in an acidic solution. On the other hand, when used as a catalyst support for Pt and Pt–Ru nanoparticles, NDG can contribute to a four-electron oxygen reduction in an acidic solution with much slower reaction kinetics in an alkaline solution [82]. The high activity of NDG composites may arise from the positive charge density on the adjacent carbon atoms created by the introduction of nitrogen atoms [83,84]. The bonding configurations of nitrogen atoms in graphene structure can also contribute to the enhanced activity for oxygen reduction reaction [85]. Moreover, NDG may increase the durability and catalytic activity of metal catalysts when used as a supporting material [86]. A novel strategy for fabricating NDG sheets has been developed using graphite oxide as the carbon source and urea as the nitrogen source via a hydrothermal approach. This method is allowed to obtain a high doping level of nitrogen in graphene. The doped nitrogen mainly exists as pyridinic and pyrrolic N bonding configurations. Subsequent thermal annealing will significantly transfer the pyrrolic N to graphitic N [87]. Triangular-shaped Pd nanoparticles can be then decorated over NDG by kinetically controlling the polyol reduction process. The kinetic control of nanoparticles growth and nitrogen doping of the supporting material leads to the formation of highly dispersed anisotropic nanoparticles over the graphene support. Very good electrocatalytic activity for oxygen reduction reaction with high stability and good tolerance toward methanol in acidic media was observed [88].

41.4 APPLICATION OF GRAPHENE AS ACTIVE COMPONENT IN LITHIUM ION BATTERIES

The potential application of graphene in electrochemical energy conversion and storage has rapidly drawn increasing interest [6,89,90]. The incorporation of graphene into traditional electrode materials for lithium ion batteries can improve their performance [91–93]. The introduction of graphene paper into lithium ion battery assembly offers several advantages. Fewer steps in anode fabrication and battery assembly are required with the elimination of electric conductors and polymer binders that are used in conventional powder-based anode fabrication. Graphene paper possesses equal or even higher flexibility than that of metal foils. It also has much lower mass that facilitates the fabrication of thin-film battery. Research on graphene paper as an anode material for lithium ion batteries is developing very rapidly [15,94–96]. Wang et al. [15] reported that a paper of 5 μm thickness presented a first discharge capacity of 582 mAh g^{-1}. This capacity dropped to 95 mAh g^{-1} at the second cycle. Their results were corroborated by Abouimrane et al. [96] who fabricated a nonannealed graphene paper via hydrazine reduction of prefabricated GO paper and further used it as a binder-free alternative anode. The graphene paper, about 10 μm thick, exhibited a charge–discharge profile that closely resembles those reported for polymer-bound graphene-based anodes while maintaining excellent cyclability over 70 charge–discharge cycles. Significant increases in the reversible capacity (from 84 to 214 mAh g^{-1}) were observed when the current rate was decreased from 50 to 10 mA g^{-1}. They assumed that there are possible kinetic barriers to Li ion diffusion as a result of the layered structure of the graphene paper. Gwon et al. [95] compared battery performances of graphene paper and graphene powder. T Graphene paper, about 2 μm thick, obviously outperformed graphene powder. At a rate of 1 C, the capacity of graphene paper reached 200 mAh g^{-1} and steadily increased in the following cycles. On the other hand, the capacity of graphene powder was only 160 mAh g^{-1} and decreased steadily. Hu et al. [97] studied the effect of paper thickness on the electrochemical performance of graphene paper as an anode material for lithium ion batteries. Three graphene papers, with thickness of 1.5, 3, and 10 μm, were fabricated by vacuum-assisted filtration of reduced graphene nanosheets suspended in water. They evidently deliver different lithium storage capacities. Thinner papers always outperform thicker ones. The 1.5 μm paper gives rise to initial reversible specific capacities (the first 10 cycles) of 200 mAh g^{-1} at a current density of 100 mA g^{-1}, while the 10 μm paper only presents 80 mAh g^{-1} at a current density of 50 mA g^{-1}. After 100 cycles, a specific capacity of 180 mAh g^{-1} is retained for the 1.5 μm paper; in contrast, only a specific capacity of 65 mAh g^{-1} remains for the 10 μm paper. The capacity decline with the paper thickness is associated with the dense restacking of graphene nanosheets and a large aspect ratio of the paper. The effective Li^+ diffusion distance in graphene paper is mainly controlled by the thickness of the paper and the diffusion proceeds mainly in the in-plane direction. Therefore, the effective contact of graphene nanosheets with the electrolyte is limited and the efficiency of carbon utilization is very low in the thick papers.

Bi et al. [98] adopted three kinds of graphene preparation strategies, namely, CVD, Wurtz-type reductive coupling reaction (WRC), and chemical exfoliation for improving the electrochemical properties of $LiFePO_4$ cathodes in Li ion batteries. The as-grown CVD and WRC graphene sheets have extremely high quality and their electrical conductivities are significantly superior to RGO. When 5 wt.% graphene is added, the contact resistance between $LiFePO_4$ nanoparticles is greatly reduced to enhance their electrochemical performances. When large-sized, defect-free, and highly crystalline CVD graphene sheets are used to modify $LiFePO_4$ cathodes, the capacities reach up to 132 and 80 mAh g^{-1} at discharge rates of 1 and 20 C, respectively, with corresponding capacity decay rates of 3.1% and 6.5% after 100 cycles. Fe_2O_3/graphene composite with graphene mass content of 30% exhibits outstanding cyclability with highly reversible charge capacity of 1069 mAh g^{-1} after 50 cycles at a current density of 50 mA g^{-1}. When the current density is increased to 1000 mA g^{-1}, it could still retain a charge capacity of 534 mAh g^{-1} [99]. The application of nanorod graphene as a support for Fe_3O_4 nanoparticles greatly improved their cycling stability with superior rate capacity. A charge-specific capacity of 867 mAh g^{-1} was maintained with only 5% capacity loss after the 100th cycle at 1 C. At a current density of 5 C, the charge capacity decreased to 569 mAh g^{-1} [100]. Anchored carbon-coated $Li_3V_2(PO_4)_3$ particles onto graphene sheets by a modified Pechini method showed an excellent rate performance with delivered capacities of 104, 91, and 85 mAh g^{-1} at current density values of 5, 30, and 50 C, respectively, in the voltage range of 3.0–4.3 V. The cycling performance is also improved in a higher voltage range of 3.0–4.8 V. The capacity retention is 83% with a capacity of 131 mAh g^{-1} at 10 C after 100 cycles [101].

A novel strategy for the preparation of tin and tin monoxide on graphene nanosheets by selectively using the reducing agent and precipitant was adopted. These nanoscale tin or tin monoxide particles well dispersed on graphene exhibited higher reversible capacities, better cycle performances, and higher rate capabilities [102]. The alloying reaction between SnO_2 and lithium is often associated with a huge volume variation during charge–discharge cycling. This causes pulverization and rapid capacity fading of the electrode material [103]. This problem can be overcome by fabricating composite materials based on conductive phases such as carbon-coated SnO_2 nanoparticles [104] and CNTs–SnO_2 hybrid nanocomposites [105]. A great deal of attention has been paid to prepare SnO_2–graphene nanocomposites as anode materials for lithium ion batteries [106–110]. Zhao et al. [108] prepared SnO_2–graphene composite material, which could deliver a reversible capacity of 775.3 mAh g^{-1} after 50 cycles at a current density of 100 mA g^{-1}. Wang et al. [110] reported the fabrication of SnO_2–graphene, which showed a reversible capacity of 590 mAh g^{-1} after 50 cycles at a current density of 400 mA g^{-1}. A sulfur-coated SnO_2–graphene nanocomposite exhibits a reversible capacity of 819 mAh g^{-1} after 200 cycles at 500 mA g^{-1} and an excellent rate capability of 580 mAh g^{-1} at a high current density of 4000 mA g^{-1} [111]. SnO_2–C/graphene nanosheets deliver a reversible capacity of 703 mAh g^{-1} after 80 cycles at a current density of 100 mA g^{-1} and maintain 443 mAh g^{-1} after 100 cycles at a current density of 1000 mA g^{-1}. This improved performance can be attributed to the complete confinement of SnO_2 nanoparticles between the graphene nanosheets and carbon layer, which can effectively prevent the detachment and agglomeration of SnO_2 and preserve the nanostructure's integrity during charge–discharge cycling [112]. Good results were also observed when using a Zn_2SnO_4 nanocrystals/graphene nanosheets nanohybrid as an anode material for Li ion batteries. This is a result of preventing the direct restacking of the hydrophobic graphene sheets by loading Zn_2SnO_4 nanocrystals as the spacers [113].

A highly flexible, binder-free, MoO_3 nanobelt/graphene film electrode, prepared by a two-step microwave hydrothermal method, showed an excellent rate capability, large capacity, and good cycling stability compared to pure MoO_3 film. An initial discharge capacity of 291 mAh g^{-1} can be obtained at 100 mA g^{-1} with a capacity of 172 mAh g^{-1} retained after 100 cycles [114]. Cobalt sulfides/graphene nanosheet composite delivers a reversible capacity of about 1018 mAh g^{-1}. It retains a reversible capacity of above 950 mAh g^{-1} after 50 cycles at a current density of 100 mA g^{-1} up to 1000 mA g^{-1} [115]. Graphene-wrapped SnCo nanoparticles showed a distinguished higher-than-theoretical capacity of 1117 mAh g^{-1} at 72 mA g^{-1}, which is larger than bare graphene nanosheets (727 mAh g^{-1}) or SnCo particles (599 mAh g^{-1}). A good rate performance at 720 mA g^{-1} was achieved with a retained high capacity of 922 mAh g^{-1} as a result of the complementary effect of the three elements (Sn, Co, and graphene nanosheets) in this composite. The mechanical stability of Sn upon cycling is improved by the presence of robust and electrically conductive graphene nanosheets and the alleviated function of inactive cobalt [116]. Figure 41.9 recorded the first-cycle discharge (lithium insertion) and charge (lithium extraction) curves and cycling performances of the products at 1 C.

41.5 APPLICATION OF GRAPHENE FOR ENERGY STORAGE IN SUPERCAPACITORS

The increasing demand for clean and sustainable energy has driven intensive research efforts toward the development of energy storage and delivery systems. Supercapacitors as important energy storage devices are able to provide a huge amount of energy in a short time [117,118]. Many graphene/metal oxides or hydroxides composites have been reported as good materials in supercapacitors, which exhibited remarkable capacitances (e.g., 570 F g^{-1} for graphene–RuO_2 [119], 576 F g^{-1} for graphene–NiO [120], 243.2 F g^{-1} for

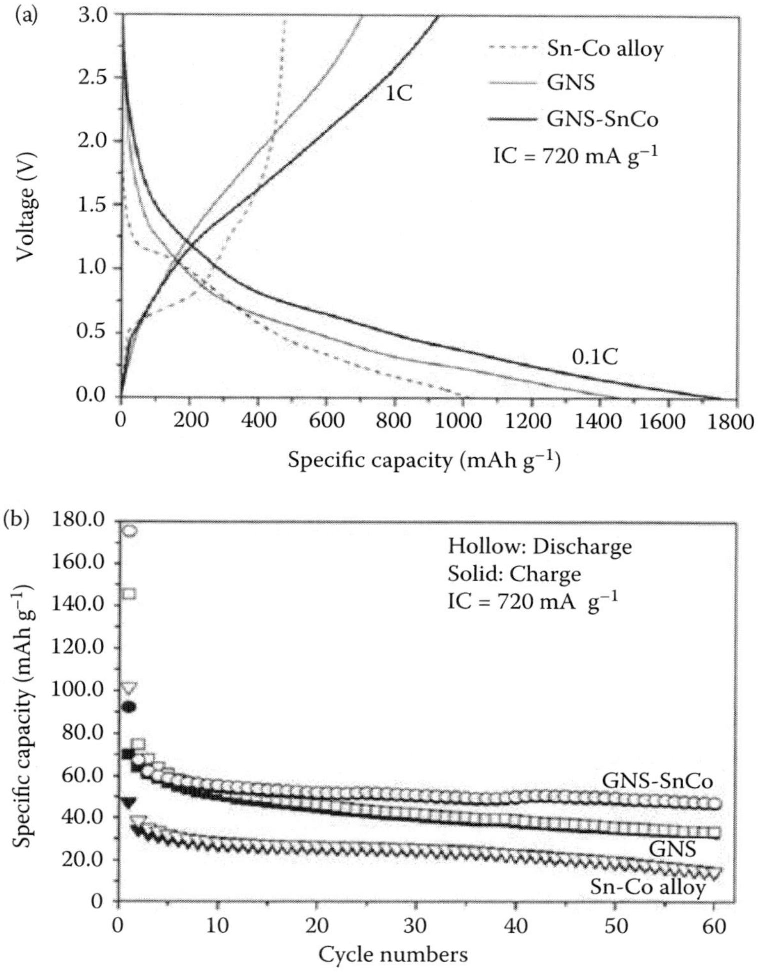

FIGURE 41.9 (a) First-cycle discharge (lithium insertion) and charge (lithium extraction) curves of the products at 1 C. (b) Cycling performances of the products at 1 C. (From X. Wang et al., *Carbon* 49;2011:133.)

graphene–Co_3O_4 [121], 256 F g^{-1} for graphene–Mn_3O_4 [122], and 935 F g^{-1} for graphene–$Ni(OH)_2$ [123]), much higher than those of pure graphene or individual metal oxides.

A new strategy to synthesize a graphene/Co_3O_4 composite in two steps is reported. The first step is the synthesis of a graphene/$Co_2(OH)_2CO_3$ composite under mild hydrothermal conditions followed by calcination to get a graphene/Co_3O_4 composite. The galvanostatic charge–discharge test of this composite showed a specific capacitance value of 443 F g^{-1} at a high current density of 5 A g^{-1}, which is higher than that of Co_3O_4. The specific capacitance retains 97.1% of its initial value after 1000 continuous charge–discharge cycles at 10 A g^{-1} indicating that this composite material has an excellent cycle life [124]. The addition of 15 wt.% carbon black to a Co_3O_4/graphene nanosheets composite resulted in a specific capacitance of 341 F g^{-1} at a scan rate of 10 mV s^{-1} in 6 M KOH solution [125]. The specific capacitances of Co_3O_4/GNS containing different carbon black contents are displayed in Figure 41.10. The uniform structure of a cobalt-doped polyaniline modified graphene composite together with high conductivity afforded high specific capacitance and good cycling stability during the charge–discharge process. This composite achieves a greater specific capacitance in comparison with HCl-doped polyaniline on graphene powder. The highest specific capacitance of 989 F g^{-1} was obtained at a scan rate of 2 mV s^{-1}. The energy density of a cobalt-doped composite could reach 352 Wh kg^{-1} at a power density of 1.581 kW kg^{-1} [126].

Mesoporous graphene/NiO composites are capable of delivering a specific capacitance of 429.7 F g^{-1} at a current density of 200 mA g^{-1}. They also offer good capacitance retention of 86.1% at 1 A g^{-1} after 2000 continuous charge–discharge cycles. This rich and evenly distributed porosity could reduce the transportation distance and offer a robust sustentation of OH^- ions due to its "ion-buffering reservoirs," ensuring that sufficient faradaic reactions can take place at the surfaces of electroactive NiO nanoparticles [127]. $Ni(OH)_2$–CNTs–RGO composite displays specific capacitances as high as 1235 and 780 F g^{-1} at current densities of 1 and 20 A g^{-1}, respectively. It also retains 80% of its original capacity after 500 cycles at a discharge current density of 10 A g^{-1}. This 3D pseudocapacitor electrode has a number of important features, such as fast

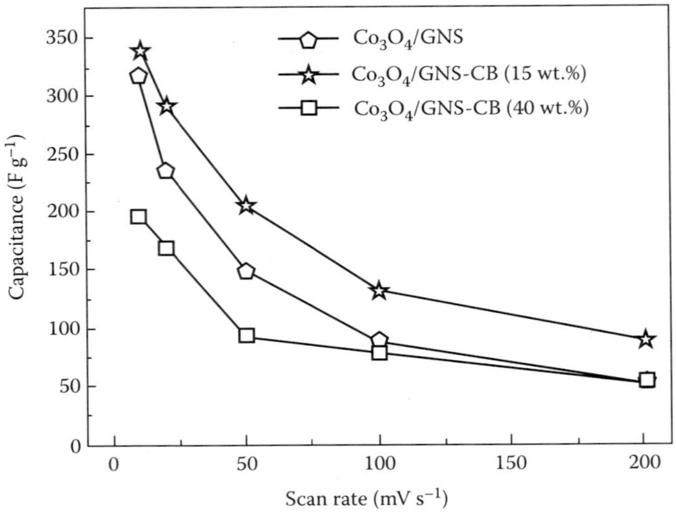

FIGURE 41.10 Specific capacitances of Co_3O_4/GNS, Co_3O_4/GNS–CB (15 wt.%), and Co_3O_4/GNS–CB (40 wt.%). (From Z.S. Wu et al., *Adv. Funct. Mater.* 20;2010:3595.)

ion and electron transfer, easy access of pseudoactive species and efficient utilization and excellent reversibility of $Ni(OH)_2$ nanoparticles [128]. Anchored copper oxide nanoparticles on GO nanosheets showed better electrochemical supercapacitive behavior and lower charge transfer resistance compared to pure components. Their specific capacitance at a current density of 0.1 A g^{-1} is 245 F g^{-1} compared to 125 F g^{-1} for CuO, 120 F g^{-1} for GO, and 155 F g^{-1} for the layer-by-layer-coated electrodes [129].

Wang et al. [122] synthesized a Mn_3O_4/graphene composite with MnO_2 organosol and graphene. It resulted in a capacitance value of 175 F g^{-1} in 1 M Na_2SO_4 solution and 256 F g^{-1} in 6 M KOH solution. But the complicated synthesis processes greatly limited its commercial application. The Mn_3O_4/graphene composite, prepared via a one-step solvothermal process, exhibited a specific capacitance of 147 F g^{-1} [130]. However, using toxic and oxidative dimethyl sulfoxide as a solvent did not make it environmentally friendly and probably could result in the incomplete reduction of graphite oxide to graphene. Mn_3O_4 nanorods deposited on graphene sheets only delivered a low capacitance of 121 F g^{-1} at 0.5 A g^{-1} [131]. This is due to the formation of the unstable and intermediate product of Mn_3O_4 [132]. The Mn_3O_4/graphene composite was prepared via a simple solvothermal process by mixing $Mn(AC)_2{\cdot}4H_2O$ and graphite oxide suspension in mixed ethanol/H_2O. Cyclic voltammetry and galvanostatic charge–discharge measurements were adopted to investigate the electrochemical properties of the Mn_3O_4/graphene composite in 1 M Na_2SO_4 solution. In a potential window of −0.2–0.8 V (SCE), it delivers an initial specific capacitance of 161 F g^{-1} at 1 A g^{-1} and increases to 230 F g^{-1} after 1000 cycles compared to significant capacitance fading for pure Mn_3O_4. At a high current density of 5 A g^{-1}, the specific capacitances of 116 and 22.5 F g^{-1} are exhibited for the Mn_3O_4/graphene composite and pure Mn_3O_4, respectively, indicating that graphene greatly enhances the electrochemical performance of Mn_3O_4 [133]. The surface area of the Mn_3O_4/GO composite, prepared by an inexpensive successive ionic layer adsorption and reaction method, is measured using Brunauer–Emmett–Teller technique as 94 m^2 g^{-1}. This increases its specific capacitance to 344 F g^{-1} at a scan rate of 5 mV s^{-1} [134]. The Mn_3O_4-anchored RGO nanocomposite has been successfully synthesized via a microwave hydrothermal technique. Its capacitance value reached 153 F g^{-1}, much higher than that of bare Mn_3O_4 (87 F g^{-1}) at a scan rate of 5 mV s^{-1} in the potential range from −0.1 to 0.8 V. X-ray photoelectron spectroscopy showed that the electrochemical activation and oxidization of Mn(II, III) to Mn(IV) during cycling at 10 mV s^{-1} increased the composite capacitance to 200% [135]. XPS spectra of the fresh RGO (31.6%)–Mn_3O_4 nanocomposite and after cyclization for 1000 cycles are represented in Figure 41.11. Through microwave-assisted hydrothermal preparation strategy, Liu et al. [136] synthesized Mn_3O_4/RGO composite. Its specific capacitance reached 193 F g^{-1} at 25 mV s^{-1} in 0.5 M Na_2SO_4 solution. By growing needle-like MnO_2 on the planes of GO in a water–isopropyl alcohol system, Chen et al. [137] obtained an MnO_2/GO composite with an enhanced capacitive performance of 216 F g^{-1} at a current density of 0.15 A g^{-1}. Yan et al. [138] also deposited MnO_2 on RGO nanosheets with a specific capacitance of 310 F g^{-1} at 2 mV s^{-1} in 1 M Na_2SO_4 solution. Using $KMnO_4$ as both the oxidant and Mn source in a modified Hummers' method, the MnO_2/GO composite was formed. By annealing at 400°C, the Mn_3O_4/graphene composite with 70.7 wt.% Mn_3O_4 nanoparticles was homogeneously anchored on each side of the graphene nanosheets. It exhibits a high specific capacitance of 271.5 F g^{-1} at a current density of 0.1 A g^{-1} as well as great cycle performance. At a higher current density of 10 A g^{-1}, the capacitance retention remains about 100% after 20,000 cycles [139]. Graphene/MWCNTs/MnO_2 hybrid material, prepared by a simple redox reaction between graphene/MWCNTs and $KMnO_4$ at room temperature, can reversibly cycle in a cell potential of 0–2.0 V. It gives a high energy density of 28.33 Wh kg^{-1},

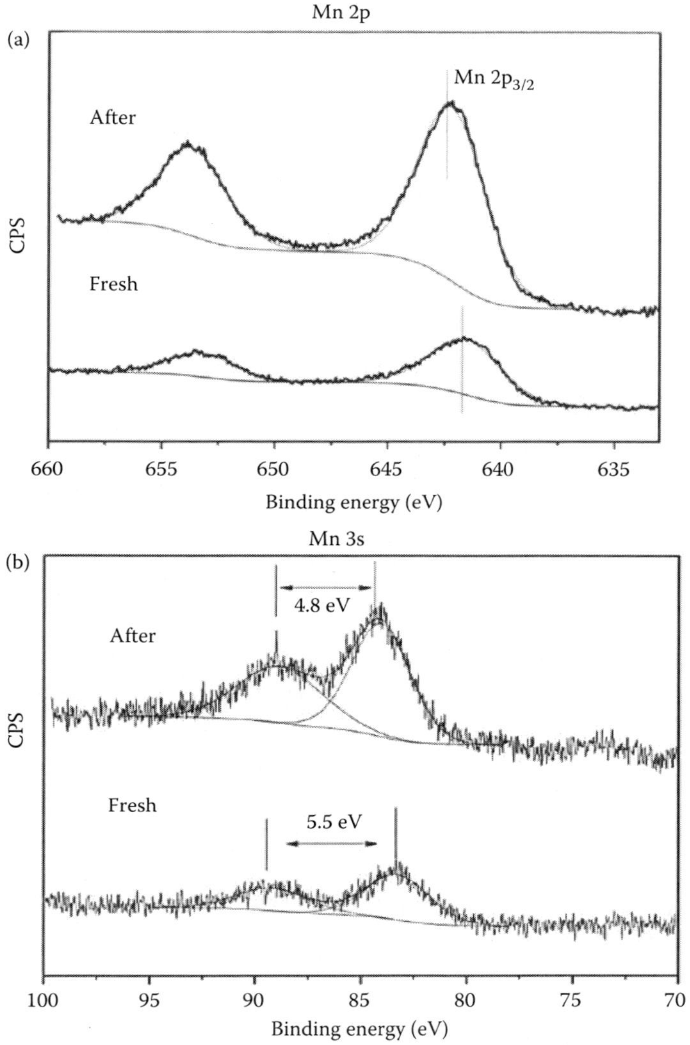

FIGURE 41.11 XPS spectra of the fresh electrode and the electrode after 1000 cycles for RGO (31.6%)–Mn_3O_4 nanocomposite in (a) the Mn 2*p* and (b) the Mn 3*s* region. Mixed Gaussian and Lorentzian component peaks are also exhibited in the spectra. (From G.M. Scheuermann et al., *J. Am. Chem. Soc.* 131;2009:8262.)

which is much higher than that based on graphene/MWCNTs (3.92 Wh kg^{-1}). This electrochemical capacitor also presents a high power density of 5 kW kg^{-1} at 13.33 Wh kg^{-1} with an excellent cycling performance of 83% retention after 2500 cycles [140]. Multilayer films of polypyrrole were prepared by the galvanostatic method with three different dopants, namely, *p*-toluenesulfonic acid, benzene sulfonic acid, and sulfuric acid ions on GO layer. The better deposition of polypyrrole films on GO is addressed by studying the influence of different electrolytes, concentrations, and current densities. These multilayer films exhibit greater capacitance compared to the single layer. Specific capacitance of a fabricated supercapacitor using multilayer electrodes is as high as 332 F g^{-1} at 10 mV s^{-1}, compared to a value of 215 F g^{-1} by a supercapacitor using a single-layer electrode [141]. Aqueous solutions of GO and nanodiamond particles were mixed in different ratios and heated at 100°C for 48 h. The electrochemical performance, including the capacitive behavior of the resultant composites, was investigated by cyclic voltammetry and galvanostatic charge–discharge curves at 1 and 2 A g^{-1} in 1 M H_2SO_4 solution. The nanocomposite matrix with 10/1 ratio displayed the best performance with a specific capacitance of 186 ± 10 F g^{-1} and excellent cycling stability [142].

41.6 APPLICATION OF GRAPHENE IN DYE-SENSITIZED SOLAR CELLS CONSTRUCTION

Since their initial discovery, dye-sensitized solar cells have attracted considerable attention due to their easy fabrication process and relatively high conversion efficiency at reasonable cost [143,144]. A dye-sensitized solar cell mainly consists of a dye-sensitized TiO_2 working electrode, a counter electrode, and an electrolyte usually containing an I^-/I^{3-} redox couple between the electrodes. The function of the counter electrode is to transfer electrons arriving from the external circuit back to the redox electrolyte and to carry the photocurrent over the width of each cell. Hence, excellent electrocatalytic activity, high surface area, and sufficient electrical conductivity are essential properties for a good counter electrode [145–147]. GO is added into a TiO_2 nanoparticles paste as an auxiliary binder. Thick mesoporous TiO_2 films free of cracks can be prepared by only single printing. GO helps binding TiO_2 nanoparticles together through the interactions between functional groups on GO and the surface species of TiO_2 nanoparticles. The presence of 0.8 wt.% GO in TiO_2 paste is sufficient to fabricate thick and crack-free TiO_2 films. This TiO_2–GO paste can act as an anode material in dye-sensitized solar cells. A power conversion efficiency of 7.70% was obtained for cells under AM1.5 illumination. It is almost the same as that of control devices with electrodes containing conventional TiO_2 paste without GO via fourfold printings [148]. On the other hand, hierarchical TiO_2 mesoporous sphere/graphene composites (HTMS/G) inherit the merits of high specific surface area derived from HTMS and graphene, as well as the high electrical conductivity of graphene. The graphene content in HTMS/G exhibited great influence on power conversion efficiencies of dye-sensitized solar cells. Lower graphene content in HTMS/G showed superior dye adsorption capacity, lower charge recombination, and thus higher photocurrent density over bare HTMS. However, excessive graphene promoted the recombination of photogenerated electrons, which deteriorated the power conversion efficiency. Owing to the high dye adsorption capacity and the prolonged electron recombination lifetime, HTMS/G containing 5.68 wt.% graphene, boosted the short circuit current density to 16.17 mA cm^{-2}. A power conversion efficiency of 7.19% was achieved that is 21.8% higher than that of bare HTMS [149]. Natural dye extracted from purple cabbage was used for fabrication of TiO_2 dye-sensitized solar cells. The effect of light intensity on the solar efficiency of the device was investigated. It was observed that its efficiency increases as light intensity increases, for example, the efficiency of the solar cell increases from $0.013 \pm 0.002\%$ to $0.150 \pm 0.020\%$ by increasing the light intensity from 30 to 100 mW cm^{-2}. The solar efficiency of the natural dye used in this research was compared with commercial dye (N 719) under similar experimental conditions and it was observed that the natural (purple cabbage) dye has higher efficiency ($0.150 \pm 0.020\%$) than N 719 ($0.078 \pm 0.002\%$). It was further evaluated that the efficiency of the fabricated solar cell could be improved by incorporating GO. The efficiency of the TiO_2 dye-sensitized solar cell was found to increase from $0.150 \pm 0.020\%$ to $0.361 \pm 0.009\%$ by incorporating GO into purple cabbage dye [150].

NDG sheets are prepared by a hydrothermal reduction of graphite oxide using ammonia as the nitrogen source and employed as the catalyst for a triiodide reduction to fabricate counter electrodes in dye-sensitized solar cells. Electrochemical impedance spectroscopy analysis reveals that the charge transfer resistance of the nitrogen-doped graphene (NDG) electrode decreases as its loading is increased. The electrode with an NDG loading of 20 μg cm^2 shows a charge transfer resistance of 0.9 Ω cm^2, which is much lower than that of a pristine graphene electrode. This enhanced electrochemical property is beneficial for improving the photovoltaic performance of dye-sensitized solar cells. Under 1 sun illumination (AM1.5), the dye-sensitized solar cell with the NDG electrode shows an energy conversion efficiency of 7.01%, which is comparable to that of the cell with the Pt electrode [151]. Pt nanoparticles/graphene nanosheets hybrid films as the counter electrode material in dye-sensitized solar cells showed higher conductivity and better electrocatalytic activity for I^{3-}/I^- redox reaction and lower charge transfer resistance on the electrolyte/electrode interface than that of the Pt electrode. A high power conversion efficiency of 7.88% was achieved under AM1.5 illumination of 100 mW cm^{-2}, which is 21.04% higher than that obtained when using the Pt electrode [152]. Dye-sensitized solar cells, prepared with Ag nanowire–graphene nanoplatelet hybrid counter electrode with graphene loading of 0.88 μg cm^{-2}, showed photoconversion efficiency of $1.61 \pm 0.13\%$, which is comparable with that made of a standard Pt-based counter electrode ($1.87 \pm 0.15\%$) [153]. A photovoltaic efficiency of 5.88% was gained from a dye-sensitized solar cell fabricated with purified and ground graphene-supported tungsten composite as a counter electrode. This result is comparable to that obtained when a standard Pt counter electrode is used (5.92%) [154].

41.7 APPLICATION OF GRAPHENE AS A STATIONARY PHASE IN CAPILLARY ELECTROCHROMATOGRAPHY

Graphene was utilized as a novel stationary phase in open-tubular capillary electrochromatography (OTCEC). The open-tubular column was easily fabricated by an electrostatic assembly of poly(diallydimethylammonium chloride) followed by self-adsorption of negatively charged graphene. The chromatographic properties of the graphene-coated open-tubular column were investigated by the separation of nitroaniline isomers through variation of the buffer pH, sodium dodecyl sulfate concentration, and separation voltage. The results demonstrate that the application of graphene in OTCEC greatly improves the separation by largely enhancing the surface area, thus increasing the stationary/mobile-phase interaction. The proposed open-tubular column exhibits good repeatability with run-to-run RSD of retention time as low as 1.5% (n = 5) and excellent stability for at least 2-week usage with a total of 200 runs. The method was successfully applied for the analysis of nitroaniline isomers in hair dye samples [155]. GO could also be immobilized on the fused-silica capillary (75 μm i.d.) derivatized by 3-aminopropyltrimethoxysilane. Compared with the bare capillary, greater separation efficiency was achieved by this capillary column as a result of the increasing interactions between the small organic molecules and the inner wall of the GO-capillary column originated from π–π electrostatic stacking. For three consecutive runs, the intraday relative standard deviations of migration time and peak areas were 0.6%–4.3% and

2.8%–9.3%, respectively. The interday relative standard deviations of migration time and peak areas were 0.2%–8.3% and 4.5%–9.6%, respectively. Additionally, one GO-capillary column could be used for more than 100 runs with no observable changes on the separation efficiency [156].

A novel organic polymer-coated stationary phase modified with inorganic nanomaterial was prepared for open-tubular capillary electrochromatography. The capillary was developed by covalent attaching graphene oxide onto the tentacle-type polymer chains grafted capillary, which was derivatized with ammonium hydroxide. This novel GO-polymer modified capillary column generated a strong and stable electroosmotic flow from anode to cathode with the running electrolytes of pH ranging from 4.0 to 9.0, and showed excellent separation ability toward neutral small molecules, amino acids, and ephedrine–pseudoephedrine isomers with resolution higher than 2.30 and column efficiency up to over 170,000 plates/m [157]. The GO column, fabricated by a new one-step coating approach for capillary gas chromatographic separations, showed average McReynolds constants of 308, suggesting the medium polar nature of the GC stationary phase. The GO stationary phase achieved good separation for analytes of different types with good peak shapes, especially for H-bonding analytes, such as alcohols and amines. Moreover, GO column showed good separation reproducibility with RSD% value less than 0.24% ($n = 5$) [158]. Graphene capillary column (0.25 mm, i.d.) resulted in column efficiency of 3100 plates/m determined by n-dodecane at 120°C. The obtained McReynolds constants suggested the weakly polar nature of graphene sheets as GC stationary phase. Graphene stationary phase differs from the conventional phase (5% phenyl polysiloxane) in its resolving ability and retention behaviors, and achieved better separation for the Grob and other mixtures. Furthermore, graphene column exhibited good repeatability with relative RSD% value in the range of 0.01%–0.07% for run-to-run and 2.5%–6.7% for column-to-column, respectively [159].

REFERENCES

1. Y. Shao, J. Wang, H. Wu, J. Liu, I.A. Aksay, Y. Lin, Graphene based electrochemical sensors and biosensors: A review, *Electroanalysis* 22;2010:1027.
2. K.R. Ratinac, W. Yang, J.J. Gooding, P. Thordarson, F. Braet, Graphene and related materials in electrochemical sensing, *Electroanalysis* 23;2011:803.
3. C.N.R. Rao, A.K. Sood, K.S. Subrahmanyam, A. Govindaraj, Graphene: The new two-dimensional nanomaterial, *Angew. Chem. Int. Ed.* 48;2009:7752.
4. M.J. Allen, V.C. Tung, R.B. Kaner, Honeycomb carbon: A review of graphene, *Chem. Rev.* 110;2010:132.
5. K.S. Novoselov, A.K. Geim, S.V. Morozov, D. Jiang, Y. Zhang, S.V. Dubonos, I.V. Grigorieva, A.A. Firsov, Electric field effect in atomically thin carbon films, *Science* 306;2004:666.
6. Y. Sun, Q. Wu, G. Shi, Graphene based new energy materials, *Energy Environ. Sci.* 4;2011:1113.
7. A.K. Geim, K.S. Novoselov, The rise of graphene, *Nat. Mater.* 6;2007:183.
8. Y. Si, E.T. Samulski, Exfoliated graphene separated by platinum nanoparticles, *Chem. Mater.* 20;2008:6792.
9. E.J. Yoo, J. Kim, E. Hosono, H.S. Zhou, T. Kudo, I. Honma, Large reversible Li storage of graphene nanosheet families for use in rechargeable lithium ion batteries, *Nano Lett.* 8;2008:2277.
10. A.N. Cao, Z. Liu, S.S. Chu, M.H. Wu, Z.M. Ye, Z.W. Cai, Y.L. Chang, S.F. Wang, Q.H. Gong, Y.F. Liu, A facile one-step method to produce graphene–CdS quantum dot nanocomposites as promising optoelectronic materials, *Adv. Mater.* 22;2010:103.
11. P. Cui, S. Seo, J. Lee, L. Wang, E. Lee, M. Min, H. Lee, Nonvolatile memory device using gold nanoparticles covalently bound to reduced graphene oxide, *ACS Nano* 5;2011:6826.
12. G.M. Scheuermann, L. Rumi, P. Steurer, W. Bannwarth, R. Mulhaupt, Palladium nanoparticles on graphite oxide and its functionalized graphene derivatives as highly active catalysts for the Suzuki–Miyaura coupling reaction, *J. Am. Chem. Soc.* 131;2009:8262.
13. X.Y. Yang, X.Y. Zhang, Y.F. Ma, Y. Huang, Y.S. Wang, Y.S. Chen, Superparamagnetic graphene oxide–Fe_3O_4 nanoparticles hybrid for controlled targeted drug carriers, *J. Mater. Chem.* 19;2009:2710.
14. G. Wang, X. Shen, J. Yao, J. Park, Graphene nanosheets for enhanced lithium storage in lithium ion batteries, *Carbon* 47;2009:2049.
15. C. Wang, D. Li, C.O. Too, G.G. Wallace, Electrochemical properties of graphene paper electrodes used in lithium batteries, *Chem. Mater.* 21;2009:2604.
16. J. Chen, J. Zou, J. Zeng, X. Song, J. Ji, Y. Wang, J. Ha, X. Chen, Preparation and evaluation of graphene-coated solid-phase microextraction fiber, *Anal. Chim. Acta* 678;2010:44.
17. C.S. Shan, H.F. Yang, J.F. Song, D.X. Han, A. Ivaska, L. Niu, Direct electrochemistry of glucose oxidase and biosensing for glucose based on graphene, *Anal. Chem.* 81;2009:2378.
18. T.T. Baby, S.S. Aravind, T. Arockiadoss, R.B. Rakhi, S. Ramaprabhu, Metal decorated graphene nanosheets as immobilization matrix for amperometric glucose biosensor, *Sens. Actuators B: Chem.* 145;2010:71.
19. F. Banhart, J. Kotakoski, A.V. Krasheninnikov, Structural defects in graphene, *ACS Nano* 5;2011:26.
20. J. Du, R. Yue, Z. Yao, F. Jiang, Y. Du, P. Yang, C. Wang, Nonenzymatic uric acid electrochemical sensor based on graphene-modified carbon fiber electrode, *Colloids Surf. A. Physicochem. Eng. Aspects* 419;2013:94.
21. L.-P. Jia, H.-S. Wang, Preparation and application of a highly sensitive nonenzymatic ethanol sensor based on nickel nanoparticles/Nafion/graphene composite film, *Sens. Actuators B: Chem.* 177;2013:1035.
22. L. Zheng, L. Xiong, Y. Li, J. Xu, X. Kang, Z. Zou, S. Yang, J. Xia, Facile preparation of polydopamine-reduced graphene oxide nanocomposite and its electrochemical application in simultaneous determination of hydroquinone and catechol, *Sens. Actuators B: Chem.* 177;2013:344.
23. T. Gan, J. Sun, K. Huang, L. Song, Y. Li, A graphene oxide–mesoporous MnO_2 nanocomposite modified glassy carbon electrode as a novel and efficient voltammetric sensor for simultaneous determination of hydroquinone and catechol, *Sens. Actuators B: Chem.* 177;2013:412.
24. M. Zhang, R. Yuan, Y. Chai, C. Wang, X. Wu, Cerium oxide–graphene as the matrix for cholesterol sensor, *Anal. Biochem.* 436;2013:69.

25. S. Cao, L. Zhang, Y. Chai, R. Yuan, An integrated sensing system for detection of cholesterol based on TiO_2–graphene–Pt–Pd hybrid nanocomposites, *Biosens. Bioelectron.* 42;2013:532.
26. S. Pruneanu, F. Pogacean, A.R. Biris, M. Coros, F. Watanabe, E. Dervishi, A.S. Biris, Electro-catalytic properties of graphene composites containing gold or silver nanoparticles, *Electrochim. Acta* 89;2013:246.
27. N. Sattarahmady, H. Heli, S.E. Moradi, Cobalt hexacyanoferrate/graphene nanocomposite—Application for the electrocatalytic oxidation and amperometric determination of captopril, *Sens. Actuators B: Chem.* 177;2013:1098.
28. W. Sun, X. Wang, H. Zhu, X. Sun, F. Shi, G. Li, Z. Sun, Graphene-MnO_2 nanocomposite modified carbon ionic liquid electrode for the sensitive electrochemical detection of rutin, *Sens. Actuators B: Chem.* 178;2013:443.
29. M. Veerapandian, Y.-T. Seo, K. Yun, M.-H. Lee, Graphene oxide functionalized with silver@silica–polyethylene glycol hybrid nanoparticles for direct electrochemical detection of quercetin, *Biosens. Bioelectron.* 58;2014:200.
30. L. Saghatforoush, M. Hasanzadeh, N. Shadjou, Polystyrene–graphene oxide modified glassy carbon electrode as a new class of polymeric nanosensors for electrochemical determination of histamine, *Chin. Chem. Lett.* 25;2014:655.
31. J. Li, D. Kuang, Y. Feng, F. Zhang, Z. Xu, M. Liu, D. Wang, Green synthesis of silver nanoparticles–graphene oxide nanocomposite and its application in electrochemical sensing of tryptophan, *Biosens. Bioelectron.* 42;2013:198.
32. E. Zor, I.H. Patir, H. Bingol, M. Ersoz, Ersoz, An electrochemical biosensor based on human serum albumin/graphene oxide/3-aminopropyltriethoxysilane modified ITO electrode for the enantioselective discrimination of D- and L-tryptophan, *Biosens. Bioelectron.* 42;2013:321.
33. K.-Q. Deng, J.-H. Zhou, X.-F. Li, Direct electrochemical reduction of graphene oxide and its application to determination of L-tryptophan and L-tyrosine, *Colloids Surf. B. Biointerfaces* 101;2013:183.
34. L. Jiang, S. Gu, Y. Ding, D. Ye, Z. Zhang, Amperometric sensor based on tricobalt tetroxide nanoparticles–graphene nanocomposite film modified glassy carbon electrode for determination of tyrosine, F. Zhang, *Colloids Surf. B. Biointerfaces* 107;2013:146.
35. M. Arvand, T.M. Gholizadeh, Simultaneous voltammetric determination of tyrosine and paracetamol using a carbon nanotube-graphene nanosheet nanocomposite modified electrode in human blood serum and pharmaceuticals, *Colloids Surf. B. Biointerfaces* 103;2013:84.
36. X. Bo, L. Guo, Simple synthesis of macroporous carbon–graphene composites and their use as a support for Pt electrocatalysts, *Electrochim. Acta* 90;2013:283.
37. H. Zhang, X. Xu, Y. Yin, P. Wu, C. Cai, Nonenzymatic electrochemical detection of glucose based on Pd_1Pt_3–graphene nanomaterials, *J. Electroanal. Chem.* 690;2013:19.
38. F. Xiao, Y. Li, H. Gao, S. Ge, H. Duan, Growth of coral-like PtAu–MnO_2 binary nanocomposites on free-standing graphene paper for flexible nonenzymatic glucose sensors, *Biosens. Bioelectron.* 41;2013:417.
39. B. Yuan, C. Xu, D. Deng, Y. Xing, L. Liu, H. Pang, D. Zhang, Graphene oxide/nickel oxide modified glassy carbon electrode for supercapacitor and nonenzymatic glucose sensor, *Electrochim. Acta* 88;2013:708.
40. C. Ruan, W. Shi, H. Jiang, Y. Sun, X. Liu, X. Zhang, Z. Sun, L. Dai, D. Ge, One-pot preparation of glucose biosensor based on polydopamine–graphene composite film modified enzyme electrode, *Sens. Actuators B: Chem.* 177;2013:826.
41. Z. Wang, X. Zhou, J. Zhang, F. Boey, H. Zhang, Direct electrochemical reduction of single-layer graphene oxide and subsequent functionalization with glucose oxidase, *J. Phys. Chem. C* 113;2009:14071.
42. B. Unnikrishnan, S. Palanisamy, S.-M. Chen, A simple electrochemical approach to fabricate a glucose biosensor based on graphene–glucose oxidase biocomposite, *Biosens. Bioelectron.* 39;2013:70.
43. B. Liang, L. Fang, G. Yang, Y. Hu, X. Guo, X. Ye, Direct electron transfer glucose biosensor based on glucose oxidase self-assembled on electrochemically reduced carboxyl graphene, *Biosens. Bioelectron.* 43;2013:131.
44. V. Mani, B. Devadas, S.-M. Chen, Direct electrochemistry of glucose oxidase at electrochemically reduced graphene oxide-multiwalled carbon nanotubes hybrid material modified electrode for glucose biosensor, *Biosens. Bioelectron.* 41;2013:309.
45. H.-W. Yang, M.-Y. Hua, S.-L. Chen, R.-Y. Tsai, Reusable sensor based on high magnetization carboxyl-modified graphene oxide with intrinsic hydrogen peroxide catalytic activity for hydrogen peroxide and glucose detection, *Biosens. Bioelectron.* 41;2013:172.
46. H. Razmi, R. Mohammad-Rezaei, Graphene quantum dots as a new substrate for immobilization and direct electrochemistry of glucose oxidase: Application to sensitive glucose determination, *Biosens. Bioelectron.* 41;2013:498.
47. X. Qin, A.M. Asiri, K.A. Alamry, A.O. Al-Youbi, X. Sun, Carbon nitride dots can serve as an effective stabilizing agent for reduced graphene oxide and help in subsequent assembly with glucose oxidase into hybrids for glucose detection application, *Electrochim. Acta* 95;2013:260.
48. W. Sun, X. Wang, Y. Wang, X. Ju, L. Xu, G. Li, Z. Sun, Application of graphene–SnO_2 nanocomposite modified electrode for the sensitive electrochemical detection of dopamine, *Electrochim. Acta* 87;2013:317.
49. Y. Zeng, Y. Zhou, L. Kong, T. Zhou, G. Shi, A novel composite of SiO_2-coated graphene oxide and molecularly imprinted polymers for electrochemical sensing dopamine, *Biosens. Bioelectron.* 45;2013:25.
50. X. Niu, W. Yang, H. Guo, J. Ren, J. Gao, Highly sensitive and selective dopamine biosensor based on 3,4,9,10-perylene tetracarboxylic acid functionalized graphene sheets/multi-wall carbon nanotubes/ionic liquid composite film modified electrode, *Biosens. Bioelectron.* 41;2013:225.
51. Y.-Y. Ling, Q.-A. Huang, M.-S. Zhu, D.-X. Feng, X.-Z. Li, Y. Wei, A facile one-step electrochemical fabrication of reduced graphene oxide–mutilwall carbon nanotubes–phospotungstic acid composite for dopamine sensing, *J. Electroanal. Chem.* 693;2013:9.
52. F. Xi, D. Zhao, X. Wang, P. Chen, Non-enzymatic detection of hydrogen peroxide using a functionalized three-dimensional graphene electrode, *Electrochem. Commun.* 26;2013:81.
53. Z. Sun, H. Fu, L. Deng, J. Wang, Redox-active thionine–graphene oxide hybrid nanosheet: One-pot, rapid synthesis, and application as a sensing platform for uric acid, *Anal. Chim. Acta* 761;2013:84.
54. Y. Huang, S.F.Y. Li, Electrocatalytic performance of silica nanoparticles on graphene oxide sheets for hydrogen peroxide sensing, *J. Electroanal. Chem.* 690;2013:8.
55. T.-Y. Huang, J.-H. Huang, H.-Y. Wei, K.-C. Ho, C.-W. Chu, rGO/SWCNT composites as novel electrode materials for electrochemical biosensing, *Biosens. Bioelectron.* 43;2013:173.

56. Y. Zhang, Y. Liu, J. He, P. Pang, Y. Gao, Q. Hu, Electrochemical behavior of graphene/Nafion/Azure I/Au nanoparticles composites modified glass carbon electrode and its application as nonenzymatic hydrogen peroxide sensor, *Electrochim. Acta* 90;2013:550.
57. Q. Wan, Y. Liu, Z. Wang, W. Wei, B. Li, J. Zou, N. Yang, Graphene nanoplatelets supported metal nanoparticles for electrochemical oxidation of hydrazine, *Electrochem. Commun.* 29;2013:29.
58. J.-M. You, D. Kim, S.K. Kim, M.-S. Kim, H.S. Han, S. Jeon, Novel determination of hydrogen peroxide by electrochemically reduced graphene oxide grafted with aminothiophenol–Pd nanoparticles, *Sens. Actuators B: Chem.* 178;2013:450.
59. F.-J. Zhang, K.-H. Zhang, F.-Z. Xie, J. Liu, H.-F. Dong, W. Zhao, Z.-D. Meng, Surface plasmon resonance induced reduction of high quality Ag/graphene composite at water/toluene phase for reduction of H_2O_2, *Appl. Surf. Sci.* 265;2013:578.
60. L. Zhong, S. Gan, X. Fu, F. Li, D. Han, L. Guo, L. Niu, Electrochemically controlled growth of silver nanocrystals on graphene thin film and applications for efficient nonenzymatic H_2O_2 biosensor, *Electrochim. Acta* 89;2013:222.
61. Y. Xia, W. Li, M. Wang, Z. Nie, C. Deng, S. Yao, A sensitive enzymeless sensor for hydrogen peroxide based on the polynucleotide-templated silver nanoclusters/graphene modified electrode, *Talanta* 107;2013:55.
62. B. Zhao, Z. Liu, W. Fu, H. Yang, Construction of 3D electrochemically reduced graphene oxide–silver nanocomposite film and application as nonenzymatic hydrogen peroxide sensor, *Electrochem. Commun.* 27;2013:1.
63. F. Xu, M. Deng, G. Li, S. Chen, L. Wang, Electrochemical behavior of cuprous oxide–reduced graphene oxide nanocomposites and their application in nonenzymatic hydrogen peroxide sensing, *Electrochim. Acta* 88;2013:59.
64. S. Zhu, J. Guo, J. Dong, Z. Cui, T. Lu, C. Zhu, D. Zhang, J. Ma, Sonochemical fabrication of Fe_3O_4 nanoparticles on reduced graphene oxide for biosensors, *Ultrason. Sonochem.* 20;2013:872.
65. A.A. Ensafi, M. Jafari–Asl, B. Rezaei, A novel enzyme-free amperometric sensor for hydrogen peroxide based on Nafion/exfoliated graphene oxide–Co_3O_4 nanocomposite, *Talanta* 103;2013:322.
66. L. Zheng, D. Ye, L. Xiong, J. Xu, K. Tao, Z. Zou, D. Huang, X. Kang, S. Yang, J. Xia, Preparation of cobalt-tetraphenylporphyrin/reduced graphene oxide nanocomposite and its application on hydrogen peroxide biosensor, *Anal. Chim. Acta* 768;2013:69.
67. B.C.H. Steele, A. Heinzel, Materials for fuel-cell technologies, *Nature* 414;2001:345.
68. T. Tomai, Y. Kawaguchi, S. Mitani, I. Honma, Pt sub-nano/nanoclusters stabilized at the edge of nanographene sheets and their catalytic performance, *Electrochim. Acta* 92;2013:421.
69. K. Kakaei, M. Zhiani, A new method for manufacturing graphene and electrochemical characteristic of graphene-supported Pt nanoparticles in methanol oxidation, *J. Power Sources* 225;2013:356.
70. J. Tang, T. Wang, X. Sun, Y. Hu, Q. Xie, Y. Guo, H. Xue, J. He, Novel synthesis of reduced graphene oxide-ordered mesoporous carbon composites and their application in electrocatalysis, *Electrochim. Acta* 90;2013:53.
71. Rajesh, R.K. Paul, A. Mulchandani, Platinum nanoflowers decorated three-dimensional graphene–carbon nanotubes hybrid with enhanced electrocatalytic activity, *J. Power Sources* 223;2013:23.
72. Y.-S. Wang, S.-Y. Yang, S.-M. Li, H.-W. Tien, S.-T. Hsiao, W.-H. Liao, C.-H. Liu, K.-H. Chang, C.-C.M. Ma, C.-C. Hu, Three-dimensionally porous graphene–carbon nanotube composite-supported PtRu catalysts with an ultrahigh electrocatalytic activity for methanol oxidation, *Electrochim. Acta* 87;2013:261.
73. S. Yu, Q. Liu, W. Yang, K. Han, Z. Wang, H. Zhu, Graphene–CeO_2 hybrid support for Pt nanoparticles as potential electrocatalyst for direct methanol fuel cells, *Electrochim. Acta* 94;2013:245.
74. Y. Chen, X. Zhang, D. Zhang, P. Yu, Y. Ma, High performance supercapacitors based on reduced graphene oxide in aqueous and ionic liquid electrolytes, *Carbon* 49;2011:573.
75. S.H. Cho, H.N. Yang, D.C. Lee, S.H. Park, W.J. Kim, Electrochemical properties of Pt/graphene intercalated by carbon black and its application in polymer electrolyte membrane fuel cell, *J. Power Sources* 225;2013:200.
76. F. Li, Y. Guo, Y. Liu, J. Yan, W. Wang, J. Gao, Excellent electrocatalytic performance of Pt nanoparticles on reduced graphene oxide nanosheets prepared by a direct redox reaction between Na2PtCl4 and graphene oxide, *Carbon* 67;2014:617.
77. A.A. Ensafi, M. Jafari-Asl, B. Rezaei, A new strategy for the synthesis of 3-D Pt nanoparticles on reduced graphene oxide through surface functionalization, application for methanol oxidation and oxygen reduction, *Electrochim. Acta* 130;2014:397.
78. S.-S. Li, J. Yu, Y.-Y. Hu, A.-J. Wang, J.-R. Chen, J.-J. Feng, Simple synthesis of hollow Pt–Pd nanospheres supported on reduced graphene oxide for enhanced methanol electrooxidation, *J. Power Sources* 254;2014:119.
79. S.-S. Li, J.-J. Lv, Y.-Y. Hu, J.-N. Zheng, J.-R. Chen, A.-J. Wang, J.-J. Feng, Facile synthesis of porous Pt–Pd nanospheres supported on reduced graphene oxide nanosheets for enhanced methanol electrooxidation, *J. Power Sources* 247;2014:213.
80. S. Wang, A. Manthiram, Graphene ribbon-supported Pd nanoparticles as highly durable, efficient electrocatalysts for formic acid oxidation, *Electrochim. Acta* 88;2013:565.
81. R. Liu, H. Zhou, J. Liu, Y. Yao, Z. Huang, C. Fu, Y. Kuang, Preparation of Pd/MnO_2-reduced graphene oxide nanocomposite for methanol electro-oxidation in alkaline media, *Electrochem. Commun.* 26;2013:63.
82. J. Bai, Q. Zhu, Z. Lv, H. Dong, J. Yu, L. Dong, Nitrogen-doped graphene as catalysts and catalyst supports for oxygen reduction in both acidic and alkaline solutions, *Int. J. Hydrogen Energy* 38;2013:1413.
83. B. Šljukić, C.E. Banks, R.G. Compton, An overview of the electrochemical reduction of oxygen at carbon-based modified electrodes, *J. Iranian Chem. Soc.* 2;2005:1.
84. L. Qu, Y. Liu, J.B. Baek, L. Dai, Nitrogen-doped graphene as efficient metal-free electrocatalyst for oxygen reduction in fuel cells, *ACS Nano* 4;2010:1321.
85. H. Wang, T. Maiyalagan, X. Wang, Review on recent progress in nitrogen-doped graphene: Synthesis, characterization, and its potential applications, *ACS Catal* 2;2012:781.
86. R.I. Jafri, N. Rajalakshmi, S. Ramaprabhu, Nitrogen doped graphene nanoplatelets as catalyst support for oxygen reduction reaction in proton exchange membrane fuel cell, *J. Mater. Chem.* 20;2010:7114.
87. B. Zheng, J. Wang, F.-B. Wang, X.-H. Xia, Synthesis of nitrogen doped graphene with high electrocatalytic activity toward oxygen reduction reaction, *Electrochem. Commun.* 28;2013:24.

88. B.P. Vinayan, K. Sethupathi, S. Ramaprabhu, Facile synthesis of triangular shaped palladium nanoparticles decorated nitrogen doped graphene and their catalytic study for renewable energy applications, *Int. J. Hydrogen Energy* 38;2013:2240.
89. D.A.C. Brownson, D.K. Kampouris, C.E. Banks, An overview of graphene in energy production and storage applications, *J. Power Sources* 196;2011:4873.
90. S.J. Guo, S.J. Dong, Graphene nanosheet: Synthesis, molecular engineering, thin film, hybrids, and energy and analytical applications, *Chem. Soc. Rev.* 40;2010:2644.
91. H.Y. Kim, S.W. Kim, J.Y. Hong, H.D. Lim, H.S. Kim, J.K. Yoo, K. Kang, Graphene-based hybrid electrode material for high-power lithium-ion batteries, *J. Electrochem. Soc.* 158;2011:A930.
92. M. Zhang, D. Lei, X.M. Yin, L.B. Chen, Q.H. Li, Y.G. Wang, T.H. Wang, Magnetite/graphene composites: Microwave irradiation synthesis and enhanced cycling and rate performances for lithium ion batteries, *J. Mater. Chem.* 20;2010:5538.
93. J. Yao, X.P. Shen, B. Wang, H.K. Liu, G.X. Wang, *In situ* chemical synthesis of SnO_2–graphene nanocomposite as anode materials for lithium-ion batteries, *Electrochem. Commun.* 11;2009:1849.
94. X. Zhao, C.M. Hayner, M.C. Kung, H.H. Kung, Flexible holey graphene paper electrodes with enhanced rate capability for energy storage applications, *ACS Nano* 5;2011:8739.
95. H. Gwon, H.S. Kim, K.U. Lee, D.H. Seo, Y.C. Park, Y.S. Lee, B.T. Ahn, S. Kang, Flexible energy storage devices based on graphene paper, *Energy Environ. Sci.* 4;2011:1277.
96. A. Abouimrane, O.C. Compton, K. Amine, S.T. Nguyen, Non-annealed graphene paper as a binder-free anode for lithium-ion batteries, *J. Phys. Chem. C* 114;2010:12800.
97. Y. Hu, X. Li, D. Geng, M. Cai, R. Li, X. Sun, Influence of paper thickness on the electrochemical performances of graphene papers as an anode for lithium ion batteries, *Electrochim. Acta* 91;2013:227.
98. H. Bi, F. Huang, Y. Tang, Z. Liu, T. Lin, J. Chen, W. Zhao, Study of $LiFePO_4$ cathode modified by graphene sheets for high-performance lithium ion batteries, *Electrochim. Acta* 88;2013:414.
99. W. Xiao, Z. Wang, H. Guo, X. Li, J. Wang, S. Huang, L. Gan, Fe_2O_3 particles enwrapped by graphene with excellent cyclability and rate capability as anode materials for lithium ion batteries, *Appl. Surf. Sci.* 266;2013:148.
100. A. Hu, X. Chen, Y. Tang, Q. Tang, L. Yang, S. Zhang, Self-assembly of Fe_3O_4 nanorods on graphene for lithium ion batteries with high rate capacity and cycle stability, *Electrochem. Commun.* 28;2013:139.
101. L. Zhang, S. Wang, D. Cai, P. Lian, X. Zhu, W. Yang, H. Wang, $Li_3V_2(PO_4)_3$@C/graphene composite with improved cycling performance as cathode material for lithium-ion batteries, *Electrochim. Acta* 91;2013:108.
102. W. Yue, S. Yang, Y. Ren, X. Yang, In situ growth of Sn, SnO on graphene nanosheets and their application as anode materials for lithium-ion batteries, *Electrochim. Acta* 92;2013:412.
103. X.W. Lou, Y. Wang, C. Yuan, J.Y. Lee, L.A. Archer, Template-free synthesis of SnO_2 hollow nanostructures with high lithium storage capacity, *Adv. Mater.* 18;2006:2325.
104. J. Chen, Y. Cheah, Y. Chen, N. Jayaprakash, S. Madhavi, Y. Yang, X. Lou, SnO_2 nanoparticles with controlled carbon nanocoating as high-capacity anode materials for lithium-ion batteries, *J. Phys. Chem.* C 113;2009:20504.
105. H. Zhang, H. Song, X. Chen, J. Zhou, H. Zhang, Preparation and electrochemical performance of SnO_2@carbon nanotube core–shell structure composites as anode material for lithium-ion batteries, *Electrochim. Acta* 59;2012:160.
106. S. Ding, D. Luan, F. Boey, J. Chen, X. Lou, SnO_2 nanosheets grown on graphene sheets with enhanced lithium storage properties, *Chem. Commun.* 47;2011:7155.
107. S.M. Paek, E. Yoo, I. Honma, Enhanced cyclic performance and lithium storage capacity of SnO_2/graphene nanoporous electrodes with three-dimensionally delaminated flexible structure, *Nano Lett.* 9;2009:72.
108. B. Zhao, G.H. Zhang, J.S. Song, Y. Jiang, H. Zhuang, P. Liu, T. Fang, Bivalent tin ion assisted reduction for preparing graphene/SnO_2 composite with good cyclic performance and lithium storage capacity, *Electrochim. Acta* 56;2011:7340.
109. Y. Li, X. Lv, J. Lu, J. Li, Preparation of SnO_2-nanocrystal/graphene-nanosheets composites and their lithium storage ability, *J. Phys. Chem.* C 114;2010:21770.
110. X. Wang, X. Zhou, K. Yao, J. Zhang, Z. Liu, A SnO_2/graphene composite as a high stability electrode for lithium ion batteries, *Carbon* 49;2011:133.
111. J. Zhu, D. Wang, L. Wang, X. Lang, W. You, Facile synthesis of sulfur coated SnO_2–graphene nanocomposites for enhanced lithium ion storage, *Electrochim. Acta* 91;2013:323.
112. J. Cheng, H. Xin, H. Zheng, B. Wang, One-pot synthesis of carbon coated-SnO_2/graphene-sheet nanocomposite with highly reversible lithium storage capability, *J. Power Sources* 232;2013:152.
113. W. Song, J. Xie, W. Hu, S. Liu, G. Cao, T. Zhu, X. Zhao, Facile synthesis of layered Zn_2SnO_4/graphene nanohybrid by a one-pot route and its application as high-performance anode for Li-ion batteries, *J. Power Sources* 229;2013:6.
114. L. Noerochim, J.-Z. Wang, D. Wexler, Z. Chao, H.-K. Liu, Rapid synthesis of free-standing MoO_3/Graphene films by the microwave hydrothermal method as cathode for bendable lithium batteries, *J. Power Sources* 228;2013:198.
115. G. Huang, T. Chen, Z. Wang, K. Chang, W. Chen, Synthesis and electrochemical performances of cobalt sulfides/graphene nanocomposite as anode material of Li-ion battery, *J. Power Sources* 235;2013:122.
116. P. Chen, L. Guo, Y. Wang, Graphene wrapped SnCo nanoparticles for high-capacity lithium ion storage, *J. Power Sources* 222;2013:526.
117. R. Kötz, M. Carlen, Principles and applications of electrochemical capacitors, *Electrochim. Acta* 45;2000:2483.
118. E. Frackowiak, F. Béguin, Carbon materials for the electrochemical storage of energy in capacitors, *Carbon* 39;2001:937.
119. Z.S. Wu, D.W. Wang, W. Ren, J. Zhao, G. Zhou, F. Li, H.M. Cheng, Anchoring hydrous RuO_2 on graphene sheets for high-performance electrochemical capacitors, *Adv. Funct. Mater.* 20;2010:3595.
120. Y.Y. Yang, Z.A. Hu, Z.Y. Zhang, F.H. Zhang, Y.J. Zhang, P.J. Liang, H.Y. Zhang, H.Y. Wu, Reduced graphene oxide–nickel oxide composites with high electrochemical capacitive performance, *Mater. Chem. Phys.* 133;2012:363.
121. J. Yan, T. Wei, W. Qiao, B. Shao, Q. Zhao, L. Zhang, Z. Fan, Rapid microwave-assisted synthesis of graphene nanosheet/Co_3O_4 composite for supercapacitors, *Electrochim. Acta* 55;2010:6973.
122. B. Wang, J. Park, C. Wang, H. Ahn, G. Wang, Mn_3O_4 nanoparticles embedded into graphene nanosheets: Preparation, characterization, and electrochemical properties for supercapacitors, *Electrochim. Acta* 55;2010:6812.
123. H. Wang, H.S. Casalongue, Y. Liang, H. Dai, $Ni(OH)_2$ nanoplates grown on graphene as advanced electrochemical pseudocapacitor materials, *J. Am. Chem. Soc.* 132;2010:7472.
124. S. Huang, Y. Jin, M. Jia, Preparation of graphene/Co_3O_4 composites by hydrothermal method and their electrochemical properties, *Electrochim. Acta* 95;2013:139.

125. S. Park, S. Kim, Effect of carbon blacks filler addition on electrochemical behaviors of Co_3O_4/graphene nanosheets as a supercapacitor electrodes, *Electrochim. Acta* 89;2013:516.
126. S. Giri, D. Ghosh, C.K. Das, In situ synthesis of cobalt doped polyaniline modified graphene composites for high performance supercapacitor electrode materials, *J. Electroanal. Chem.* 697;2013:32.
127. Y. Jiang, D. Chen, J. Song, Z. Jiao, Q. Ma, H. Zhang, L. Cheng, B. Zhao, Y. Chu, A facile hydrothermal synthesis of graphene porous NiO nanocomposite and its application in electrochemical capacitors, *Electrochim. Acta* 91;2013:173.
128. L.L. Zhang, Z. Xiong, X.S. Zhao, A composite electrode consisting of nickel hydroxide, carbon nanotubes, and reduced graphene oxide with an ultrahigh electrocapacitance, *J. Power Sources* 222;2013:326.
129. A. Pendashteh, M.F. Mousavi, M.S. Rahmanifar, Fabrication of anchored copper oxide nanoparticles on graphene oxide nanosheets via an electrostatic coprecipitation and its application as supercapacitor, *Electrochim. Acta* 88;2013:347.
130. X. Zhang, X.Z. Sun, Y. Chen, D.C. Zhang, Y.W. Ma, One-step solvothermal synthesis of graphene/Mn_3O_4 nanocomposites and their electrochemical properties for supercapacitors, *Mater. Lett.* 68;2012:336.
131. J.W. Lee, A.S. Hall, J.D. Kim, T.E. Mallouk, A facile and template-free hydrothermal synthesis of Mn_3O_4 nanorods on graphene sheets for supercapacitor electrodes with long cycle stability, *Chem. Mater.* 24;2012:1158.
132. H. Jiang, T. Zhao, C.Y. Yan, J. Ma, C.Z. Li, Hydrothermal synthesis of novel Mn_3O_4 nano-octahedrons with enhanced supercapacitors performances, *Nanoscale* 2;2010:2195.
133. Y. Wu, S. Liu, H. Wang, X. Wang, X. Zhang, G. Jin, A novel solvothermal synthesis of Mn_3O_4/graphene composites for supercapacitors, *Electrochim. Acta* 90;2013:210.
134. G.S. Gund, D.P. Dubal, B.H. Patil, S.S. Shinde, C.D. Lokhande, Lokhande, enhanced activity of chemically synthesized hybrid graphene oxide/Mn_3O_4 composite for high performance supercapacitors, *Electrochim. Acta* 92;2013:205.
135. L. Li, K.H. Seng, Z. Chen, H. Liu, I.P. Nevirkovets, Z. Guo, Synthesis of Mn_3O_4-anchored graphene sheet nanocomposites via a facile, fast microwave hydrothermal method and their supercapacitive behavior, *Electrochim. Acta* 87;2013:801.
136. C.L. Liu, K.H. Chang, C.C. Hu, W.C. Wen, Microwave-assisted hydrothermal synthesis of Mn_3O_4/reduced graphene oxide composites for high power supercapacitors, *J. Power Sources* 217;2012:184.
137. S. Chen, J. Zhu, X. Wu, Q. Han, X. Wang, Graphene oxide–MnO_2 nanocomposites for supercapacitors, *ACS Nano* 4;2010:2822.
138. J. Yan, Z. Fan, T. Wei, W. Qian, M. Zhang, F. Wei, Fast and reversible surface redox reaction of graphene–MnO_2 composites as supercapacitor electrodes, *Carbon* 48;2010:3825.
139. L. Zhu, S. Zhang, Y. Cui, H. Song, X. Chen, One step synthesis and capacitive performance of graphene nanosheets/Mn_3O_4 composite, *Electrochim. Acta* 89;2013:18.
140. L. Deng, Z. Hao, J. Wang, G. Zhu, L. Kang, Z.-H. Liu, Z. Yang, Z. Wang, Preparation and capacitance of graphene/multiwall carbon nanotubes/MnO_2 hybrid material for high-performance asymmetrical electrochemical capacitor, *Electrochim. Acta* 89;2013:191.
141. I.M. De la Fuente Salas, Y.N. Sudhakar, M. Selvakumar, High performance of symmetrical supercapacitor based on multilayer films of graphene oxide/polypyrrole electrodes, *Appl. Surf. Sci.* 296;2014:195.
142. Q. Wang, N. Plylahan, M.V. Shelke, R.R. Devarapalli, M. Li, P. Subramanian, T. Djenizian, R. Boukherroub, S. Szunerits, Nanodiamond particles/reduced graphene oxide composites as efficient supercapacitor electrodes, *Carbon* 68;2014:175.
143. A. Yella, H. Lee, H. Tsao, C. Yi, A.K. Chandiran, M.K. Nazeeruddin, E.W. Diau, S.M. Zakeeruddin, M. Grätzel, Porphyrin-sensitized solar cells with cobalt (II/III)–based redox electrolyte exceed 12 percent efficiency, *Science* 334;2011:629.
144. M. Grätzel, Photoelectrochemical cells, *Nature* 414;2001:338.
145. H. Sun, Y. Luo, Y. Zhang, D. Li, Z. Yu, K. Li, Q. Meng, In situ preparation of a flexible polyaniline/carbon composite counter electrode and its application in dye-sensitized solar cells, *J. Phys. Chem. C* 114;2010:11673.
146. W. Hong, Y. Yu, G. Lu, C. Li, G. Shi, Transparent graphene/PEDOT–PSS composite films as counter electrodes of dye-sensitized solar cells, *Electrochem. Commun.* 10;2008:1555.
147. N. Papageorgiou, Counter-electrode function in nanocrystalline photoelectrochemical cell configurations, *Coord. Chem. Rev.* 248;2004:1421.
148. C.Y. Neo, J. Ouyang, Graphene oxide as auxiliary binder for TiO_2 nanoparticle coating to more effectively fabricate dye-sensitized solar cells, *J. Power Sources* 222;2013:161.
149. J. Chang, J. Yang, P. Ma, D. Wu, L. Tian, Z. Gao, K. Jiang, L. Yang, Hierarchical titania mesoporous sphere/graphene composite, synthesis and application as photoanode in dye sensitized solar cells, *J. Colloid Interface Sci.* 394;2013:231.
150. A.A. Al-Ghamdi, R.K. Gupta, P.K. Kahol, S. Wageh, Y.A. Al-Turki, W. El Shirbeeny, F. Yakuphanoglu, Improved solar efficiency by introducing graphene oxide in purple cabbage dye sensitized TiO_2 based solar cell, *Solid State Commun.* 183;2014:56.
151. G. Wang, W. Xing, S. Zhuo, Nitrogen-doped graphene as low-cost counter electrode for high-efficiency dye-sensitized solar cells, *Electrochim. Acta* 92;2013:269.
152. G. Yue, J. Wu, Y. Xiao, M. Huang, J. Lin, L. Fan, Z. Lan, Platinum/graphene hybrid film as a counter electrode for dye-sensitized solar cells, *Electrochim. Acta* 92;2013:64.
153. M. Al-Mamun, J.-Y. Kim, Y.-E. Sung, J.-J. Lee, S.-R. Kim, Pt and TCO free hybrid bilayer silver nanowire–graphene counter electrode for dye-sensitized solar cells, *Chem. Phys. Lett.* 561–562;2013:115.
154. B. Munkhbayar, Md.J. Nine, J. Jeoun, M. Ji, H. Jeong, H. Chung, Synthesis of a graphene–tungsten composite with improved dispersibility of graphene in an ethanol solution and its use as a counter electrode for dye-sensitised solar cells, *J. Power Sources* 230;2013:207.
155. X. Liu, X. Liu, M. Li, L. Guo, L. Yang, Application of graphene as the stationary phase for open-tubular capillary electrochromatography, *J. Chromatogr. A* 1277;2013:93.
156. Y.-Y. Xu, X.-Y. Niu, Y.-L. Dong, H.-G. Zhang, X. Li, H.-L. Chen, X.-G. Chen, X.-G. Chen, Preparation and characterization of open-tubular capillary column modified with graphene oxide nanosheets for the separation of small organic molecules, *J. Chromatogr. A* 1284;2013:180.
157. X. Gao, R. Mo, Y. Ji, Preparation and characterization of tentacle-type polymer stationary phase modified with graphene oxide for open-tubular capillary electrochromatography, *J. Chromatogr. A* 1400;2015:19.
158. Y. Feng, C.-G. Hu, M.-L. Qi, R.-N. Fu, L.-T. Qu, Separation performance of graphene oxide as stationary phase for capillary gas chromatography, *Chin. Chem. Lett.* 26;2015:47.
159. J. Fan, M. Qi, R. Fu, L. Qu, Performance of graphene sheets as stationary phase for capillary gas chromatographic separations, *J. Chromatogr. A* 1399;2015:74.

42 Optical Properties of Graphene and Its Applications under Total Internal Reflection

Zhi-Bo Liu, Xiao-Qing Yan, and Jian-Guo Tian

CONTENTS

ABSTRACT

Given the unique 2D structure of graphene, its optical properties can be more fully utilized when surface waves propagate through the graphene plane. When graphene is placed on the interface of a total internal reflection (TIR) structure, its optical properties will have different performance due to longer interaction distance and difference of polarization mode supports, relative to the general structure of irradiation. Under TIR, graphene exhibits strong polarization-dependent optical absorption. Compared with universal absorbance of 2.3%, larger absorption was observed in monolayer, bilayer, and few-layer graphenes for transverse electric (TE) wave under TIR. Reflectance ratio of transverse magnetic (TM) to TE waves can easily provide the information of number of graphene layers. Furthermore, the difference of reflectivity for TE and TM modes in the context of a TIR structure is sensitive to the media in contact with the graphene. A graphene refractive index sensor can quickly and sensitively monitor changes in the local refractive index with a fast response time and broad dynamic range. The enhanced light–graphene coupling in a wide spectral range will be great potential in many applications such as photodetector, photovoltaics, and optical data storage.

42.1 INTRODUCTION

Graphene has been recognized as a revolutionary material for opto-electronic applications (Bonaccorso et al. 2010), such as optical modulator (Liu et al. 2011), ultrafast laser (Xing et al. 2013), polarizers (Bao et al. 2011), and photodetectors (Liu et al. 2009). Absorption of graphene has been experimentally observed to have a universal value $\approx\pi\alpha$ (2.3%) for light in the visible spectral range (Mak et al. 2012; Nair et al. 2008), depending on the fine structure constant α, but not on the properties of the material. Light–matter coupling, such as absorption, plays a key role in optical detectors, sensors, and photovoltaics. Note that despite graphene shows strong optical absorption, graphene photonic and optoelectronic devices (Bao and Loh 2012; Bonaccorso et al. 2010; Hecht 2012) remain affected by the problem that most of incident light does not interact with graphene for its atomic thickness. Several methods have been used to enhance light–graphene coupling, including periodically patterned graphene (Thongrattanasiri et al. 2012), coverage of plasmonic nanostructures on graphene (Echtermeyer et al. 2011), double-layer configuration (Gómez-Santos and Stauber 2012), and graphene-integrated microcavity (Engel et al. 2012). Combing graphene with silicon waveguides to increase the light–graphene interaction length, a broad, waveguide-integrated, graphene electro-optic modulator has been produced (Liu et al. 2011, 2012). These approaches lead to a large enhancement of the optical absorption in a limited range of wavelengths and angles of incidence. Increasing light–graphene coupling in a wide spectral range is kept to be a challenge.

Under total internal reflection (TIR), an evanescent wave exists on the interface between two media and exponentially attenuates in a low-refractive-index medium. The interaction between the evanescent wave and the material on the prism surface can change some of the properties of the reflected light, for example, via the surface plasma resonance effect (Homola 2008). It is shown that graphene exhibits strong polarization-dependent optical absorption under TIR (Pirruccio et al. 2013;

Ye et al. 2013). Furthermore, the enhanced light–graphene coupling in a wide spectral range will be great potential in many applications such as photodetector, photovoltaics, and optical sensor (Xing et al. 2012).

In this chapter, we present that a sandwiched graphene structure exhibits strong optical absorption for transverse electric (TE) wave under TIR. Compared with universal absorbance of 2.3%, larger absorption was observed in monolayer, bilayer, and few-layer graphenes for TE wave (s-polarized incident light) under TIR. Reflectance ratio of transverse magnetic (TM) wave (p-polarized light) to TE waves can easily provide the information of number of graphene layers. Furthermore, the difference of reflectivity for TE and TM modes in the context of a TIR structure is sensitive to the media in contact with the graphene. A graphene refractive index sensor can quickly and sensitively monitor changes in the local refractive index with a fast response time and broad dynamic range. Finally, based on polarization-dependent optical absorption, a data reading device for multilayer films was designed to read out data stored by a graphene-based multilayer-film optical data storage (ODS) medium (Xing et al. 2013).

42.2 TOTAL INTERNAL REFLECTION AND GRAPHENE MULTILAYER FILM STRUCTURE

Figure 42.1 shows the attenuated total reflection (ATR) structure of a TIR-based prism coupling without and with graphene layer. A light wave that is travelling inside a medium with refractive index n_1 is refracted when it meets the interface of this medium with a second one having a refractive index n_2. If the value of n_1 is larger than that of n_2 and the incident angle θ is larger than or equal to the critical angle, TIR occurs where the light is reflected back into the first medium. Figure 42.1 shows how the beam is reflected when the angle of incidence is larger than the critical angle.

Under TIR, a fraction of incident energy can penetrate through the interface of medium 1 and medium 2, and propagate along the interface, called as evanescent field. At $z = 0$, the intensity components of the evanescent field by TE (s-polarized) and TM (p-polarized) incident light can be written as (Axelrod et al. 1984)

$$I_x = |A_{Is}|^2 \frac{4\cos^2\theta\left(\sin^2\theta - n_{12}^2\right)}{n_{12}^4\cos^2\theta + \sin^2\theta - n_{12}^2} \tag{42.1}$$

$$I_y = |A_{Is}|^2 \frac{4\cos^2\theta\sin^2\theta}{n_{12}^4\cos^2\theta + \sin^2\theta - n_{12}^2} \tag{42.2}$$

$$I_z = |A_{Ip}|^2 \frac{4\cos^2\theta}{1 - n_{12}^2} \tag{42.3}$$

where A_{Is} and A_{Ip} are the amplitudes in the electric field for p-polarized incident waves, and $n_{12} = n_2/n_1$. Therefore, the intensities of the evanescent field for the p- and s-polarized incident light at $z = 0$ are $I_{0p} = I_{0x} + I_{0z}$ and $I_{0s} = I_{0y}$. Figure 42.2a shows the intensities of the electric field as a function of incidence angle in media 1 ($n_1 = 1.61$) and 2 ($n_2 = 1.41$), respectively. When graphene is placed between two media, the components of intensity parallel or orthogonal to graphene plane will result in different propagation behaviors for s- and p-polarized light. Furthermore, three components of intensity are also dependent on the change of n_2 when n_1 is fixed (Figure 42.2b). Thus, it is expected that the reflectance of the structure shown in Figure 42.1b will be very different for s- and p-polarized light incidence.

42.3 OPTICAL PROPERTIES OF GRAPHENE UNDER TOTAL INTERNAL REFLECTION

42.3.1 Theoretical Model

General theory of light diffraction by tri-layered structures is provided. The theory is a simplified version of that for stratified structures (Born and Wolf 1999). In particular, a tri-layered system consisting of a lossy medium sandwiched between two semi-infinite dielectrics is discussed. The lossy medium is assumed to be graphene.

For general theory of light diffraction by tri-layered structure (Figure 42.1b) consisting a loss layer (graphene) sandwiched between two semi-infinite dielectrics 1 and 2, the manner in which the transition takes place is governed by Maxwell's boundary conditions, where the tangential

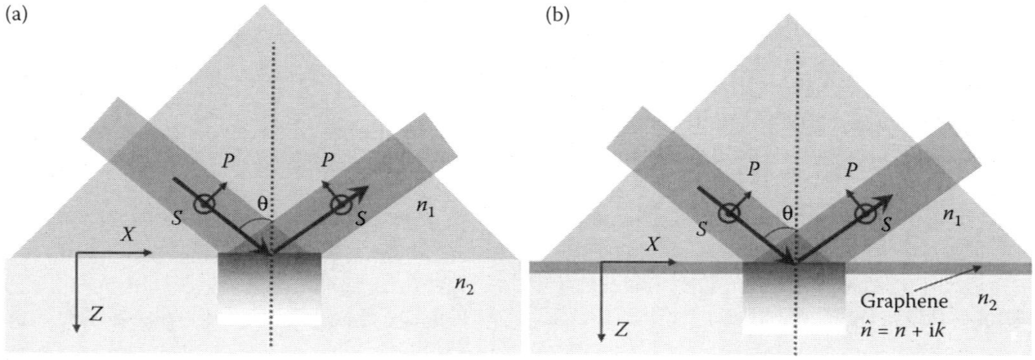

FIGURE 42.1 (a) ATR structure with the s- (S) and p-polarized (P) light. θ is the incident angle; (b) ATR structure containing a graphene layer sandwiched between the substrate and prism.

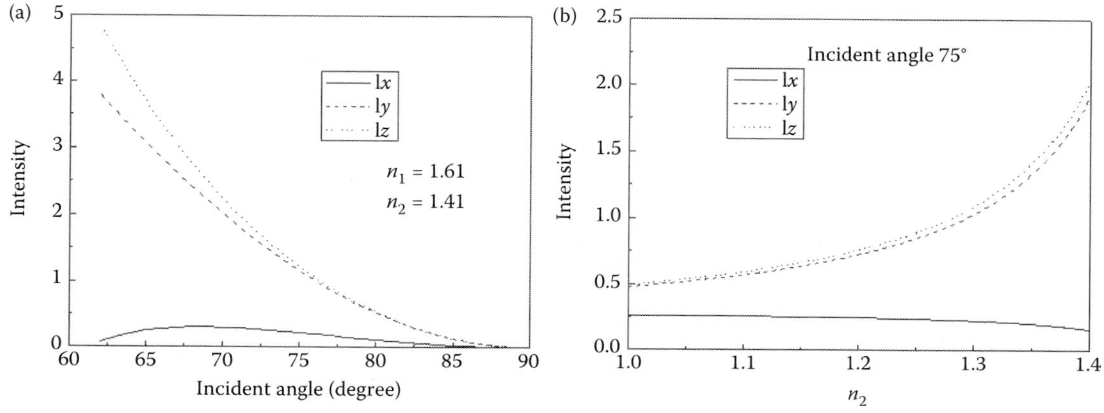

FIGURE 42.2 (a) Plot of the intensities of the evanescent waves for x and z components at p-polarization and y component at s-polarization with different incident angles. (b) Plot of the intensities of the evanescent waves for x and z components at p-polarization and y component at s-polarization with different refractive indexes n_2. The incident angle is 75°.

components of the electric (**E**) and magnetic (**H**) fields are continuous across the boundaries. The intermediate graphene layer has a complex optical constant with $\hat{n} = n + ik$, where n is the real refractive index and k is the extinction coefficient of graphene (Blake et al. 2007; Bruna and Borini 2009; Weber et al. 2010). The thickness d is relative to the number of layers of graphene (e.g., $d = 0.34$ nm for single layer graphene). Based on the boundary conditions and Snell's law, the reflectance of tri-layered structure can be obtained for s- and p-polarized light.

The concerned structure is shown in Figure 42.3. The plane of incidence is assumed to be x, z-plane, and both the boundaries are perpendicular to z axis. Maxwell equations remain unchanged when **E** and **H** and simultaneously ε and −μ are interchanged. Thus, the theorem for TM waves can be immediately deduced from the corresponding result for TE waves. For TE waves, E_x, E_z, $H_y = 0$ and $\partial/\partial y = 0$, Maxwell equations deduced to (time dependence $\exp(-i\omega t)$ being assumed)

$$\frac{\partial H_x}{\partial z} - \frac{\partial H_z}{\partial z} + i\omega\varepsilon E_y = 0,$$
$$\frac{\partial E_y}{\partial z} + i\omega\mu H_x = 0, \qquad (42.4)$$
$$\frac{\partial E_y}{\partial x} - i\omega\mu H_z = 0.$$

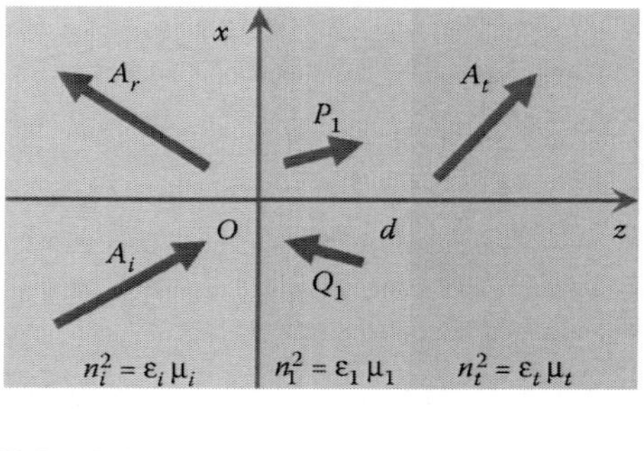

FIGURE 42.3 Schematic graph of tri-layered structure. The relative permittivity (ε_r) and permeability (μ_r) are shown in each layer.

In each layer, the fields can be written as superpositions of positive- and negative-going secondary waves. Thus, E_y is given by

$$E_y = \begin{cases} (A_i e^{ik_{iz}z} + A_r e^{-ik_{iz}z})e^{ik_x x}, & z \le 0, \\ (P_1 e^{ik_{1z}z} + Q_1 e^{-ik_{1z}(z-d)})e^{ik_x x}, & 0 \le z \le d, \\ A_t e^{ik_{tz}(z-d)} e^{ik_x x}, & z \ge d. \end{cases} \qquad (42.5)$$

Here $k_{iz} = n_i k_{0z}$, $k_{1z} = n_1 k_{0z}$, $k_{tz} = n_t k_{0z}$, and $k_x = n_i k_{0x}$, where k_0 is the free space wavevector and $n = \sqrt{\varepsilon_r \mu_r}$ is the refractive index, ε_r and μ_r being the relative permittivity and the relative permeability, respectively. A_i, A_r, and A_t are defined as incident, reflected, and transmitted amplitude coefficients. As a result of the boundary condition, x component of wavevector of all the layers is equal to each other. According to Equations 42.4 and 42.5, the magnetic components are given by

$$H_x = \begin{cases} -\dfrac{k_{iz}}{\omega\mu_0\mu_i}(A_i e^{ik_{iz}z} - A_r e^{-ik_{iz}z})e^{ik_x x}, & z \le 0, \\ -\dfrac{k_{1z}}{\omega\mu_0\mu_1}(P_1 e^{ik_{1z}z} - Q_1 e^{-ik_{1z}(z-d)})e^{ik_x x}, & 0 \le z \le d, \\ -\dfrac{k_{tz}}{\omega\mu_0\mu_t} A_t e^{ik_{tz}(z-d)} e^{ik_x x}, & z \ge d; \end{cases} \qquad (42.6)$$

and

$$H_z = \frac{kx}{\omega\mu} E_y. \qquad (42.7)$$

By applying the boundary condition of the continuities of E_y and H_x, we have

$$\begin{bmatrix} A_i \\ A_r \end{bmatrix} = \frac{1}{2} \begin{bmatrix} 1 + \dfrac{k_{1z}}{k_{iz}}\gamma_{i1} & \left(1 - \dfrac{k_{1z}}{k_{iz}}\gamma_{i1}\right) e^{ik_{1z}d} \\ 1 - \dfrac{k_{1z}}{k_{iz}}\gamma_{i1} & \left(1 + \dfrac{k_{1z}}{k_{iz}}\gamma_{i1}\right) e^{ik_{1z}d} \end{bmatrix} \begin{bmatrix} P_1 \\ Q_1 \end{bmatrix} \qquad (42.8)$$

$$\begin{bmatrix} P_1 \\ Q_1 \end{bmatrix} = \frac{1}{2} \begin{bmatrix} \left(1+\frac{k_{tz}}{k_{1z}}\gamma_{1t}\right)e^{-ik_{1z}d} \\ 1-\frac{k_{tz}}{k_{1z}}\gamma_{1t} \end{bmatrix} A_t. \quad (42.9)$$

Here, $\gamma_{i1} = \mu_i/\mu_1$ and $\gamma_{1t} = \mu_1/\mu_t$. Thus, the relation between A_i, A_r, and A_t can be written as

$$\begin{bmatrix} A_i \\ A_r \end{bmatrix} = \mathbf{M}A_t, \quad (42.10)$$

where

$$\mathbf{M} = \frac{1}{4}\begin{bmatrix} 1+\frac{k_{1z}}{k_{iz}}\gamma_{i1} & \left(1-\frac{k_{1z}}{k_{iz}}\gamma_{i1}\right)e^{ik_{1z}d} \\ 1-\frac{k_{1z}}{k_{iz}}\gamma_{i1} & \left(1+\frac{k_{1z}}{k_{iz}}\gamma_{i1}\right)e^{ik_{1z}d} \end{bmatrix} \begin{bmatrix} \left(1+\frac{k_{tz}}{k_{1z}}\gamma_{1t}\right)e^{-ik_{1z}d} \\ 1-\frac{k_{tz}}{k_{1z}}\gamma_{1t} \end{bmatrix}. \quad (42.11)$$

For TM waves, the relations between amplitude coefficients are all the same as those for TE waves, expect that the parameters γ_{i1} and γ_{1t} are defined by $\gamma_{i1} = \varepsilon_i/\varepsilon_1$ and $\gamma_{1t} = \varepsilon_1/\varepsilon_t$. For nonmagnetic materials, the relative permeability is equal to one, so for TE waves, $\gamma_{i1} = \gamma_{1t} = 1$, and for TM waves, $\gamma_{i1} = n_i^2/n_1^2$ and $\gamma_{1t} = n_1^2/n_t^2$.

In particular, if $d = 0$, the structure reduces to two semi-infinite materials bounded by $z = 0$. Thus, $P_1 = A_t$, $Q_1 = 0$. So we have

$$\mathbf{M} = \frac{1}{2}\begin{bmatrix} 1+\frac{k_{tz}}{k_{iz}}\gamma_{it} \\ 1-\frac{k_{tz}}{k_{iz}}\gamma_{it} \end{bmatrix}, \quad (42.12)$$

where γ_{it} is equal to μ_i/μ_t for TE waves, and $\varepsilon_i/\varepsilon_t$ for TM waves. It is noted that Equation 42.12 is consistent with Fresnel formulae.

M is a 2 × 1 matrix, whose elements are denoted by M_1 and M_2, respectively. Thus, we have $A_r = (M_2/M_1)A_i$ and $A_t = A_i/M_1$. The light intensity is given by the amplitude of Poynting vector

$$S = |\mathbf{E}\times\mathbf{H}| = \sqrt{\varepsilon/\mu}E^2 = \sqrt{\mu/\varepsilon}H^2. \quad (42.13)$$

For TE waves, the amount of energy that is incident on a unit area of the boundary ($z = 0$) per second is given by

$$J_i = S_{iz} = \sqrt{\frac{\varepsilon_0\varepsilon_i}{\mu_0\mu_i}}\frac{k_{iz}}{n_i k_0}|A_i|^2. \quad (42.14)$$

Similarly, the energies that are reflected and transmitted from unit areas of the boundaries ($z = 0$ and $z = d$) per second are given by

$$J_r = S_{rz} = \sqrt{\frac{\varepsilon_0\varepsilon_i}{\mu_0\mu_i}}\frac{k_{iz}}{n_i k_0}|A_r|^2, \quad (42.15)$$

$$J_t = S_{tz} = \sqrt{\frac{\varepsilon_0\varepsilon_t}{\mu_0\mu_t}}\frac{\mathrm{Re}(k_{tz})}{n_t k_0}|A_t|^2.$$

Therefore, the reflectance R and transmittance T are given by

$$R = \frac{J_r}{J_i} = \left|\frac{A_r}{A_i}\right|^2 = \frac{M_2M_2^*}{M_1M_1^*}, \quad (42.16)$$

$$T = \frac{J_t}{J_i} = \frac{\mu_i}{\mu_t}\frac{\mathrm{Re}(k_{tz})}{k_{iz}}\left|\frac{A_t}{A_i}\right| = \frac{\mu_i}{\mu_t}\frac{\mathrm{Re}(k_{tz})}{k_{iz}}\frac{1}{M_1M_1^*}. \quad (42.17)$$

For TM waves, the reflectance and transmittance can be immediately known by interchanging ε and $-\mu$ in Equations 42.16 and 42.17. We find that R for TM waves is the same as that for TE waves (Equation 42.16), whereas for TM waves,

$$T = \frac{\varepsilon_i}{\varepsilon_t}\frac{\mathrm{Re}(k_{tz})}{k_{iz}}\left|\frac{A_t}{A_i}\right| = \frac{\varepsilon_i}{\varepsilon_t}\frac{\mathrm{Re}(k_{tz})}{k_{iz}}\frac{1}{M_1M_1^*}. \quad (42.18)$$

It can be proved that $R + T = 1$ for lossless materials, which means energy conservation of the waves.

Energy density of the electromagnetic field is defined by $W = W_e + W_m$, where $W_e = (1/2)\mathbf{E}\cdot\mathbf{D}$ and $W_m = (1/2)\mathbf{H}\cdot\mathbf{B}$ are electric energy density and magnetic energy density, respectively. Note that the amplitude coefficient A_i should satisfy $|A_i| = 1/\sqrt{\varepsilon_0\varepsilon_1(1+1/\mu_1)/2}$ for TE wave, and $|A_i| = 1/\sqrt{\mu_0(1+1/\mu_1)/2}$ for TM wave in order to normalize the incident energy to unit. Strong energy density can be observed when the incident angle θ is greater than the critical angle θ_c ($\approx$61.1°). Figure 42.4a and b gives the energy density distributions of a monolayer sandwiched graphene structure for TE and TM waves with $n_1 = 1.61$, $n_2 = 1.41$, $n = 2.6$, and $\kappa = 1.3$ for monolayer graphene ($d = 0.34$ nm) (Blake et al. 2007). Strong energy density can be observed when the incident angle θ is greater than the critical angle θ_c ($\approx$61.1°). However, from the enlargement of energy density distributions in Figure 42.4c and d, a distinct higher or lower energy density in the graphene layer is observed for TE and TM waves respectively, which shows a polarization-dependence compared with those in media 1 and 2. Stronger incident energy in the graphene layer can induce more absorbance. It has been proved that the quantity of reflectance is directly related to the absorbance of the graphene film and higher accuracy can be obtained by reflection than transmission method (Mak et al. 2008). Hence, this sandwiched structure using reflectance method

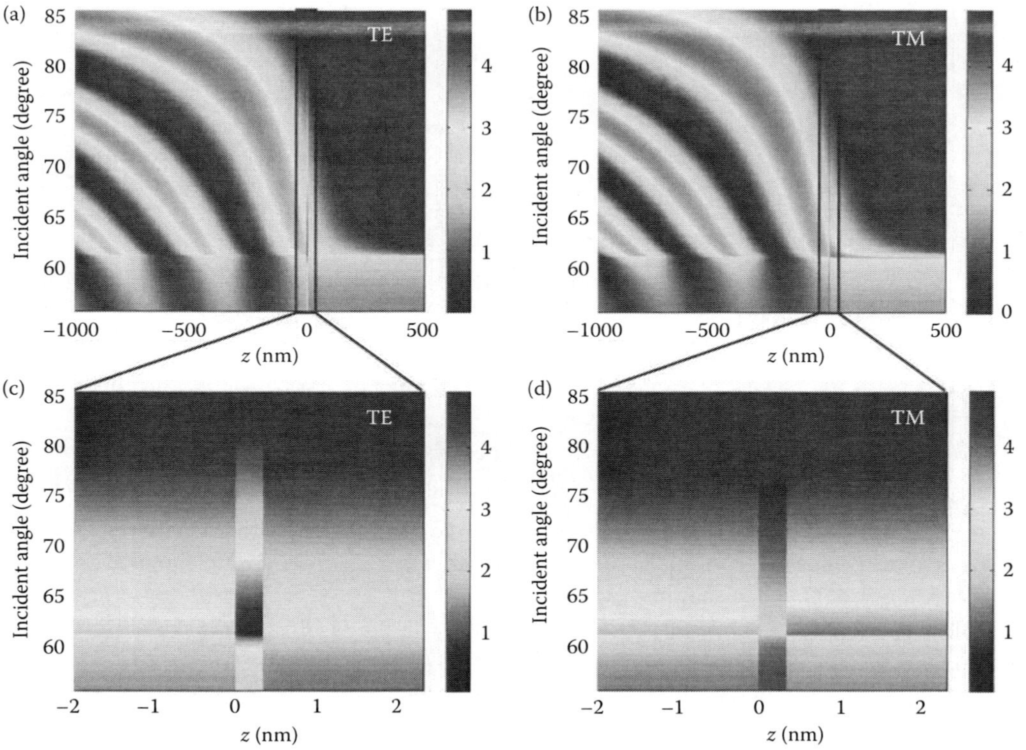

FIGURE 42.4 (a) Energy distribution for TE wave versus z-axis, and incident angle θ; (b) energy distribution for TM wave versus z-axis, and incident angle θ; (c) is the enlargement of (a); (d) is the enlargement of (b). The parameters are $n_1 = 1.61$, $n_2 = 1.41$, $n = 2.6$, $\kappa = 1.3$, and $d = 0.34$ nm. $\lambda = 632.8$ nm. The color bars in (a–d) are normalized equal. (Reprinted with permission from Ye, Q. et al. 2013. Polarization-dependent optical absorption of graphene under total internal reflection. *Applied Physics Letters* 102 (2):021912. Copyright 2013, American Institute of Physics.)

can exhibit different absorption behaviors for different polarized incident waves.

42.3.2 Polarization-Dependent Reflection and Absorption

The derivative total reflection method (DTRM) (Figure 42.5) was employed to measure the polarization-dependent absorption of graphene. Light at 632.8 nm from a He–Ne laser is split into incident and reference beams by a beam splitter. After passing through a half-wave plate and a polarizer, the incident beam reaches an equilateral prism fixed on a rotation stage. PD1 and PD2 are the two detectors of a dual-channel power meter. PD1 monitors the instantaneous fluctuations of the laser energy; PD2 measures the final light reflected by the prism. The aperture diaphragm D blocks the scattering light. The experimental setup was first calibrated using deionized water. Reflectance curves as functions of the incident angle were recorded.

Large-area monolayer ($N = 1$), bilayer ($N = 2$), and few-layer ($N = 4$) graphenes were grown by chemical vapor deposition (CVD) on copper and nickel. The graphenes were transferred to a transparent polydimethylsiloxane (PDMS, Dow Corning 184) substrate with a thickness of 1 mm. PDMS was used as the low-index medium 2 with $n_2 = 1.41$, and the prism was used as the high-index medium 1 with $n_1 = 1.61$.

Figure 42.6 shows the angle-dependent reflectance of monolayer, bilayer, and few-layer graphenes for the incidence of TE and TM waves, respectively. The solid lines represent the theoretical results, where the reflectance is given by $R = |A_{1r}/A_{1i}|^2$. If ignoring the influence of the scattering, the absorbance of

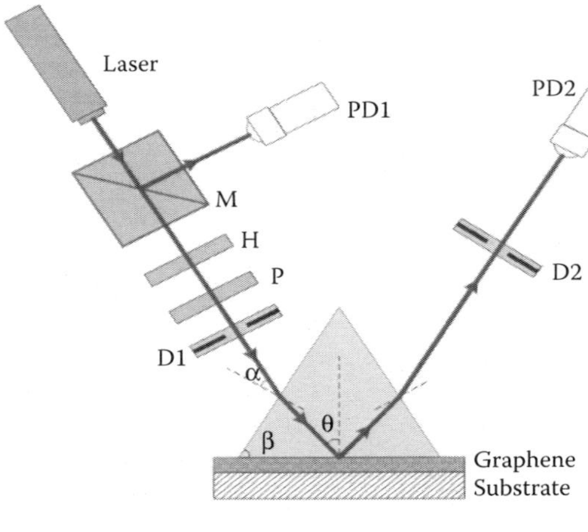

FIGURE 42.5 Schematic of the DTRM setup. M, beam splitter; H, a half-wave plate; P, a polarizer; PD1 and PD2, detectors; D1 and D2, aperture diaphragms. (Reprinted with permission from Ye, Q. et al. 2013. Polarization-dependent optical absorption of graphene under total internal reflection. *Applied Physics Letters* 102 (2):021912. Copyright 2013, American Institute of Physics.)

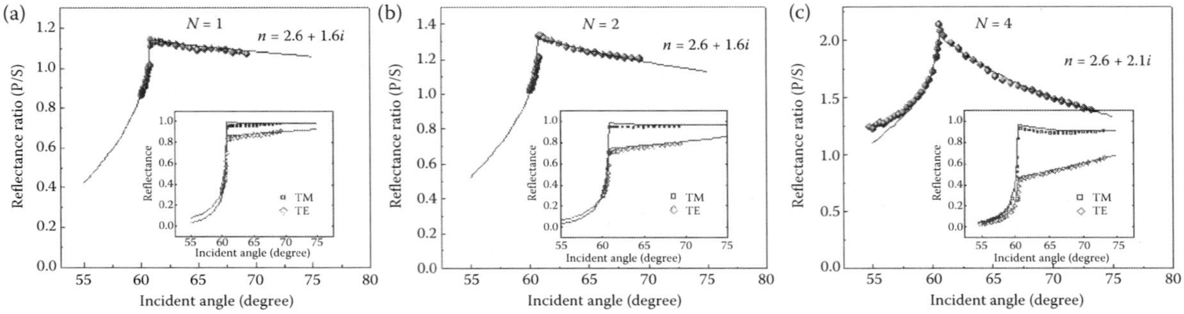

FIGURE 42.6 Experimental and calculated angular dependence of optical reflectance ratio of TM wave to TE wave. (a) Monolayer graphene ($N = 1$) with $n = 2.6$ and $\kappa = 1.6$; (b) bilayer graphene ($N = 2$) with $n = 2.6$ and $\kappa = 1.6$; (c) four-layer graphene ($N = 4$) with $n = 2.6$ and $\kappa = 2.1$. The insets are the experimental and calculated angular dependence of optical reflectance for TM and TE waves. (Reprinted with permission from Ye, Q. et al. 2013. Polarization-dependent optical absorption of graphene under total internal reflection. *Applied Physics Letters* 102 (2):021912. Copyright 2013, American Institute of Physics.)

graphene is equal to $1 - R$ because transmission is prevented by TIR. The reflectance of the sandwiched graphene structure for TE wave is considerably smaller than that for TM wave, which indicates that graphene has greater absorption for TE wave when transmission is prevented by TIR.

Compared with the results of reflectance (Figure 42.6 insets), the reflectance ratio of TM to TE wave shows superior fitting between the theoretical and experimental results. This is because some effects such as scattering may be suppressed by division of the two modes. Both the reflectance ratio and reflectance profile at the incidence of TE wave show great difference for different layers of graphene, therefore, this method can efficiently determine the thickness of the graphene layers. Several other techniques have already been applied to determine the number of graphene layers based on reflection and contrast spectroscopy (Ni et al. 2007). However, those techniques generally use a Si substrate with 300 nm SiO_2, which is not transparent in the visible range. DTRM can determine the number of graphene layers on a transparent substrate accurately and rapidly.

The optical constant $\hat{n}$ of graphene was previously measured by spectroscopic ellipsometry (Weber et al. 2010) and optical spectra (Bruna and Borini 2009). A typical value, $n = 2.6$ and $\kappa = 1.3$, of mechanically exfoliated graphene were considered in some studies (Wang et al. 2009; Yu and Hilke 2009). Using DTRM, the best fittings were obtained with $n = 2.6$, $\kappa = 1.6$ for the monolayer and bilayer graphenes, while $n = 2.6$, $\kappa = 2.1$ for the four-layer graphene. One possible reason for this discrepancy is that the graphene samples used in this analysis grown by CVD method exhibit a less even distribution than samples obtained by mechanical exfoliation. In addition, the maximum absorbance α for TE wave is considerably larger than the universal absorbance for monolayer, bilayer, and four-layer graphenes.

Spectral measurement for the four-layer graphene demonstrated a broadband polarization-dependent absorption (Ye et al. 2013). Strong absorption for TE wave can be observed in the range from 420 to 750 nm, which indicates that the polarization-dependent absorption of graphene can be employed in a wide spectral range for applications, such as determining the number of graphene layers. Using a similar TIR configuration, a remarkably large, broadband, and polarization-dependent absorption of visible light in graphene was also achieved (Pirruccio et al. 2013), as shown in Figure 42.7.

In conventional 2D electron gas, only the TM mode may exist under standard experimental conditions. This is due to the fact that the imaginary part of the conductivity is always positive in conventional 2D electron gas system such as GaAs/AlGaAs quantum-well structures. Graphene, as a 2D electron gas structure, can also only support the TM mode in the

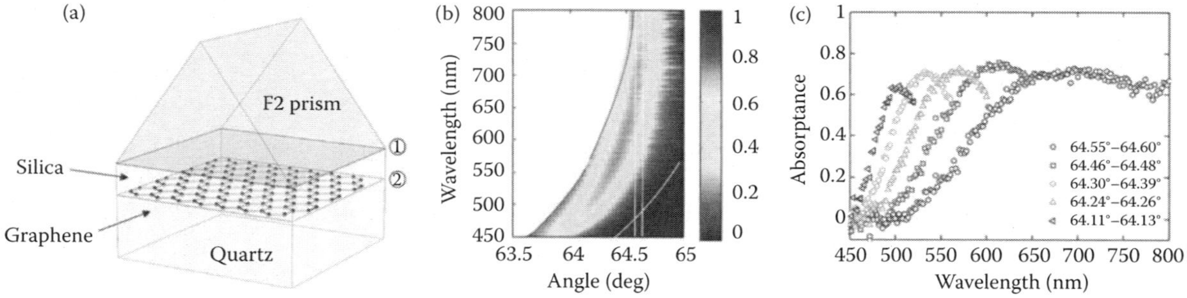

FIGURE 42.7 (a) Schematic representation of the multilayer structure formed by a quartz substrate, a graphene layer, a silica layer, and an F2 prism. (b) Measured absorbance spectra (color scale) from the structure shown in (c) in which the graphene layer is formed by five monolayers randomly stacked. The incident light is p-polarized. (Reprinted with permission from Pirruccio, G. et al. 2013. Coherent and broadband enhanced optical absorption in graphene. *ACS Nano* 7 (6), 4810–4817. Copyright 2013 American Chemical Society.)

visible spectral range (Mikhailov and Ziegler 2007). Under TIR, TE, and TM modes will propagate a short distance along the graphene layer due to Goos–Hanchen shift. Therefore, graphene has less absorption for the supported TM mode than TE mode, and large absorption of TM mode may come from that this mode is forbidden in such graphene structure.

42.4 GRAPHENE-BASED REFRACTIVE INDEX SENSOR UNDER TOTAL INTERNAL REFLECTION

For optical sensors, measurements of refractive index changes are important for a variety of applications in biosensing, drug discovery, environmental monitoring, and gas- and liquid-phase chemical sensing. Various methods for obtaining these measurements have been developed (Stanley 1989). At present, refractive-index-sensing technologies based on surface plasmon resonance (SPR) are widely used based on their ability to provide highly sensitive label-free real-time data (Homola 2006). The most common modulation approaches used in high-performance SPR sensors are based on the spectroscopy of surface plasmons in wavelength or angular domains (Homola 2008). However, when measuring a range of wavelengths or angles, optical focusing limits the sensitivity, resolution, dynamic range, and other functionalities of a spectroscopy-based SPR instrument and produces notable deficiencies in the real-time performance (>ms) of the sensor in question. An SPR sensor with amplitude modulation can produce fast, real-time monitoring and imaging functions, but its performance is typically worse than that of spectroscopy-based SPR sensors. Moreover, biomolecules adsorb poorly on gold, limiting the sensitivity of conventional SPR biosensors.

A strategy for sensing changes in refractive indexes using graphene optical sensors is based on the different reflection behaviors of TM and TE modes under TIR conditions. The differences in reflectance between the TE and TM modes are sensitive to variations in the refractive index n_2 under fixed n_1 conditions. For a given incident angle (e.g., 75°), the difference ($\Delta R = R_{TM} - R_{TE}$) in reflectance between s- and p-polarized incident light is strongly dependent on Δn, $\Delta n = n_1 - n_2$ is the difference in refractive indexes between medium 1 and medium 2. When medium 1 and the graphene layer are fixed, the change in the refractive index of medium 2 can be determined by measuring ΔR for a given incident angle.

A simple, real-time refractive index measuring system that used GRIS in combination with a microfluidic system was designed, as shown in Figure 42.8a. First, circularly polarized light obtained by adjusting a Glan–Taylor polarizer and a quarter-wave plate was focused on the GRIS at the center of its microfluid channel. The inset of Figure 42.8a shows the structure of the GRIS. Second, a multilayer graphene coating was transferred to a right-angle prism using CVD and the cast microfluid channel was bonded to the graphene layer to form

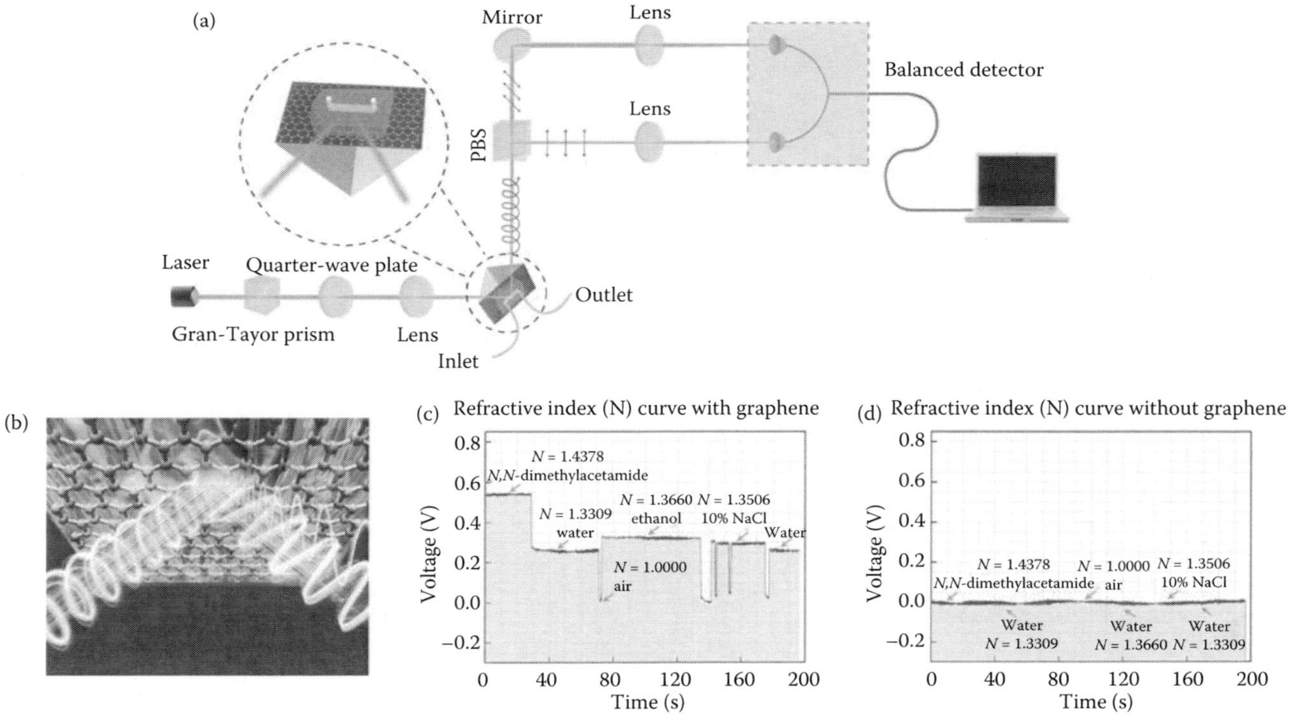

FIGURE 42.8 A sensitive, real-time microfluid refractive index measuring system. (a) A schematic illustration of the measuring system. The enlarged diagram shows the sandwiched structure of the GRIS. (b) A schematic illustration of the GRIS. (c and d) The real-time voltage signal change of microfluid with and without graphene when injecting *N,N*-dimethyl acetamide, deionized water, air, ethanol, air, a solution of 10% NaCl, air, and water successively at a flow rate of 10 μL/s. (Reprinted by permission from Macmillan Publishers Ltd. *Scientific Reports*, Xing, F. et al. 2012. Sensitive real-time monitoring of refractive indexes using a novel graphene-based optical sensor, 2:908, copyright 2012.)

a microfluid channel/graphene/prism sandwich structure. Third, the incident angle of circularly polarized light was adjusted in accordance with the refractive index of the fluid to be measured, allowing TIR to occur as incident light passed through the microfluid channel. Fourth, the reflected light was separated into *s*- and *p*-polarized light using a polarization beam splitter (PBS) and then detected by a balanced detector. Due to the polarization-sensitive absorption of graphene under TIR, the intensity of the *p*-polarized light changed little before and after microfluid injections, whereas the microfluid injection greatly altered the measured intensity of the *s*-polarized light, as shown in Figure 42.9b. The signal from the balanced detector is measured relative to the difference $\Delta R = R_{TM} - R_{TE}$ in reflectance between the *s*- and *p*-polarized incident lights. Thus, the signal from the balanced detector can be used to monitor the change in the refractive index of the fluid.

After a calibration using standard liquids known, the GRIS can be used for refractive index measuring. Furthermore, before signal measurements, a substance (generally air or water) should be chosen to calibrate the detector, and the output voltage of the balanced detector should be adjusted to 0 V for this reference substance. During the measurement process, the refractive index of the internal microfluid channel at each moment corresponds to a particular voltage calculated by subtracting the intensity of the *s*-polarized light from that of the *p*-polarized light measured by the balanced detector. The voltage varies with the microfluid's refractive index; fluids with larger refractive indexes produce higher calculated voltages.

One of the prominent features of the GRIS is its broad measuring range, as its signal depends on the value of Δn. To demonstrate this advantage, an experiment was conducted at room temperature with an incident angle of 83°; under these conditions, the refractive index of the prism was 1.51. When the incident power before the GRIS was 0.202 mW, the power of the *p*-polarization in front of the balanced detector was 10.33 μW. Subsequently, using air for calibration purposes, we injected *N*,*N*-dimethyl acetamide, deionized water, air, absolute ethyl alcohol, air, a solution of 10% NaCl, air, and water in turn at a flowing speed of 10 μL/s, generating a relationship curve indicating the real-time voltage signal change of the microfluid, as shown in Figure 42.8c. The real-time refractive index range was easily measured for refractive indexes ranging from 1 to 1.438. In fact, the dynamic range of the GRIS depends on the refractive index of the prism. If a prism with a refractive index of 1.71 is used, the refractive index of chlorobenzene (1.57) can be easily measured. The most commonly used method of measuring real-time refractive indexes employs an SPR sensor based on the modulation of incident angles. Therefore, the GRIS provides a reliable platform for measuring the refractive index of more complicated microfluids with a larger dynamic refractive index range.

The rapid response of the GRIS makes it possible to measure a fluid using a faster flow rate and to quickly reflect the changes in refractive indexes during a process of rapid fluid flow, as shown in Figure 42.9. A higher response speed facilitates the more accurate measurement of tiny changes in the refractive index; this property is particularly important for measuring refractive indexes during rapid-response processes, which are common in many biological and chemical interactions. In fact, the GRIS's response speed is theoretically dependent on the response time of the balanced detector and data acquisition equipment; the equipment used in this experiment can reach 10 ns and 340 kHz. However, the response speed of a traditional refractive index sensor, including SPR refractive index sensors, is generally in the order of milliseconds.

Moreover, the GRIS has good resolution and sensitivity. Figure 42.10 represents the voltage changes associated with

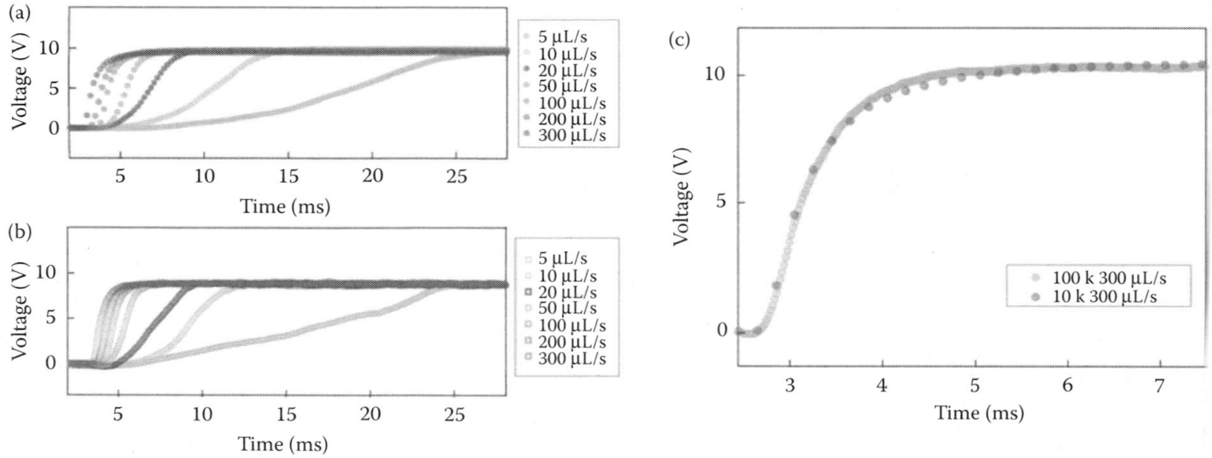

FIGURE 42.9 An evaluation of the response speed performance of the GRIS. (a and b) The real-time voltage curve at collection frequencies of 10 and 100 kHz obtained by successively injecting air and deionized water at different flow rates. The flow rate ranges from 5 to 300 μL/s. (c) The dark gray dots represent the real-time voltage curve at a collection frequency of 10 kHz and a flow rate of 300 μL/s. The light gray dots represent the real-time voltage curve at a collection frequency of 100 kHz and a flow rate of 300 μL/s. (Reprinted by permission from Macmillan Publishers Ltd. *Scientific Reports*, Xing, F. et al. 2012. Sensitive real-time monitoring of refractive indexes using a novel graphene-based optical sensor, 2:908, copyright 2012.)

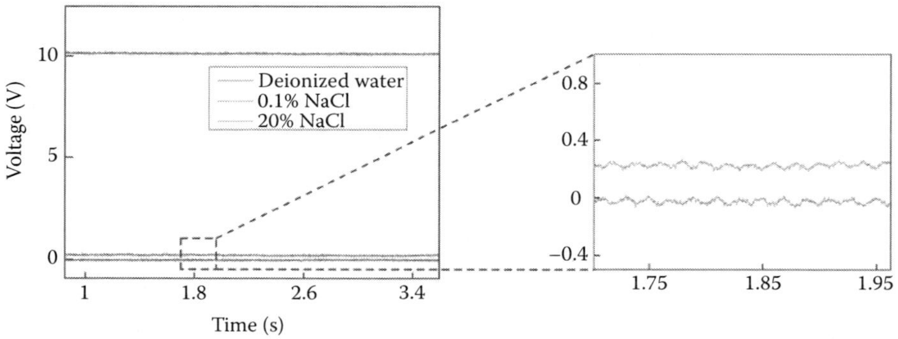

FIGURE 42.10 The resolution performance and incident power dependence of the GRIS. A curve representing the change in voltage after injecting deionized water, a solution of 0.1% NaCl and a solution of 20% NaCl into the microfluid channel, respectively. (Reprinted by permission from Macmillan Publishers Ltd. *Scientific Reports*, Xing, F. et al. 2012. Sensitive real-time monitoring of refractive indexes using a novel graphene-based optical sensor, 2:908, copyright 2012.)

injecting deionized water, a solution of 0.1% NaCl and a solution of 20% NaCl successively in the microfluid channel by shining a 13.03 mW light at the GRIS. Under the same conditions, the refractive indexes of deionized water and 0.1% NaCl solution were determined to be 1.33091 and 1.33139, respectively, which differ by 0.00048. The right of Figure 42.10 shows that the corresponding voltage difference between the deionized water and the 0.1% NaCl solution is ~245 mV and the size of the voice signal is ~45 mV. Hence, the signal-to-noise ratio (SNR) is 5. Because the relationship between the change in the refractive index and the change in voltage is considered to be roughly linear within a small range, except for the voice signal, the sensitivity of the GRIS under 13.03 mW incident light can be calculated by $245/(0.00048 * 45) \approx 11342.6$ mV/UIR. This result is comparable to the sensitivity of an SPR refractive index sensor (10^4–10^6/UIR).

As a sensing layer, graphene offers good, stable real-time performance. The evident advantages of using graphene as a sensing layer are manifold. First, with its perfect 2D structure, graphene possesses desirable mechanical properties and chemical stability, which means that a graphene sensing layer is difficult to damage. Second, the nanometer-order thickness of graphene greatly increases the propagation distance of evanescent waves in liquid, thereby producing a more accurate detection result. The velocity distribution of the fluid in a microfluid channel is known to be uneven because the fluid flows more slowly as it approaches the sensing layer; thus, if the propagation distance (in the z direction) of the evanescent waves in the fluid is short, the detected result would be subject to larger time-delay errors, resulting in poor real-time performance. For these reasons, graphene sensors provide longer service life, better real-time performance, and greater versatility in fluid measurements.

42.5 GRAPHENE-BASED OPTICAL DATA STORAGE UNDER TOTAL INTERNAL REFLECTION

Optical data storage (ODS) represents revolutionary progress for the field of information storage capacity. Given the recent significant increase in the amounts of digital information being generated, the storage capacity of ODS media must improve further. High-density ODS can be advanced in two ways; one way is to improve areal storage density (Mustroph et al. 2006). The other approach for development is to improve the volume storage density (Lott et al. 2011; Ryan et al. 2012). When the thickness of the data recording layer (DRL) is on the atomic scale, the capacity of the 3D storage layer is likely to reach its maximum limit. However, ODS media is nearly transparent on the atomic scale and difficult to be read out. In addition, the traditional process of data writing–reading largely limits the development of ODS.

The thickness of monolayer graphene is 3.4 Å, and the optical transmittance of graphene is up to 97.7% (Nair et al. 2008). The carbon-atom plane can be bent and deformed to a large degree while maintaining a stable structure. Therefore, graphene is an ideal choice as a transparent and flexible active layer for ODS media. Currently, the microstructure size of graphene can be made to be a few tens of nanometers or even several nanometers with the use of existing methods (Wang and Dai 2010), which allows for an areal storage density of up to 10^{12} byte/cm^2. The transfer technology of graphene has also been well developed. In particular, there are various methods for transferring graphene to an arbitrary flexible substrate (Lee et al. 2010; Li et al. 2009; Kim et al. 2009). Consequently, graphene-based multilayer-film ODS media can be achieved by repeatedly transferring graphene, and the volume storage density can be theoretically improved to 10^{15} byte/cm^2–10^{17} byte/cm^2. With regard to the writing and transfer of data in a graphene DRL, however, it is more difficult for graphene to be read out, especially for data that have been recorded on multilayer films using the present optical methods because graphene is nearly transparent.

Under TIR, the absorption of graphene is dependent on the polarization state of the input light. For three-layer graphene, the absorption of s-polarized light is approximately 20%, whereas the absorption of p-polarized light is very low (Ye et al. 2013). Based on this knowledge, a data reading device for multilayer films was designed to read out data stored by a graphene-based multilayer-film ODS medium. Under TIR, an evanescent wave exists on the interface between two media and exponentially attenuates in a low-refractive-index medium. As shown in Figure 42.11a, if there is no substance on the prism surface, the *s*- and *p*-polarized light intensities will not vary after the TIR. However, when there is graphene on the prism surface, the reflected *s*-polarized light intensity is much weaker than that of the *p*-polarized light after the TIR, as shown in Figure 42.11b. In fact, for monolayer graphene, the *p*-polarized light intensity shows little variation with the presence of graphene on the prism surface under TIR. However, the *s*-polarized light intensity is approximately 9% lower than before reflection, as shown in Figure 42.11c. When the light changes from *p*- to *s*-polarized, the reflectivity gradually decreases. The reflectivity of the *s*-polarized light is minimized, and the reflectivity difference between the *s*- and *p*-polarized light is maximized. For three-layer graphene, the reflectivity difference even reaches ~20%. As a result, the data recorded in the graphene are likely to be read out with a high SNR. The reflectivity difference between the *s*- and *p*-polarized light is maximized at the graphene layer, and the corresponding voltage signal is recorded as "1." The reflectivity difference between the *s*- and *p*-polarized light is minimized for the layer without graphene, and the corresponding voltage signal is recorded as "0."

Because TIR occurs between an optically dense medium and an optically thin medium, once the high-low gradient of the refractive index is built between the buffer layers, the data from the multilayer films can be read. Figure 42.11d shows the data-reading schematic diagram of the graphene-based multilayer-film ODS medium. Because the refractive indices follow the order $n_1 > n_2 > n_3$, when the incident angle is large, TIR can first occur at the interfaces of the n_1 and n_2 media. Owing to the absorption difference for graphene between the *s*- and *p*-polarized light under TIR, the data recorded on the graphene between the interfaces of n_1 and n_2 can be read out. Next, when the incident angle is reduced to the correct angle range, the incident light can be totally reflected at the interfaces of n_2 and n_3 through the interfaces of n_1 and n_2. Thus, the data recorded in the graphene between the interfaces of n_2 and n_3 are read out. Similarly, the data for the other layers can be read out.

Lithography technology can be used to write graphene microgrid structures as the data points (Tapasztó et al. 2008). Femtosecond laser direct writing technology is also a good candidate for recording data in the graphene because submicron and even nanoscale data points can be fabricated by a femtosecond pulse laser (Zhang et al. 2010).

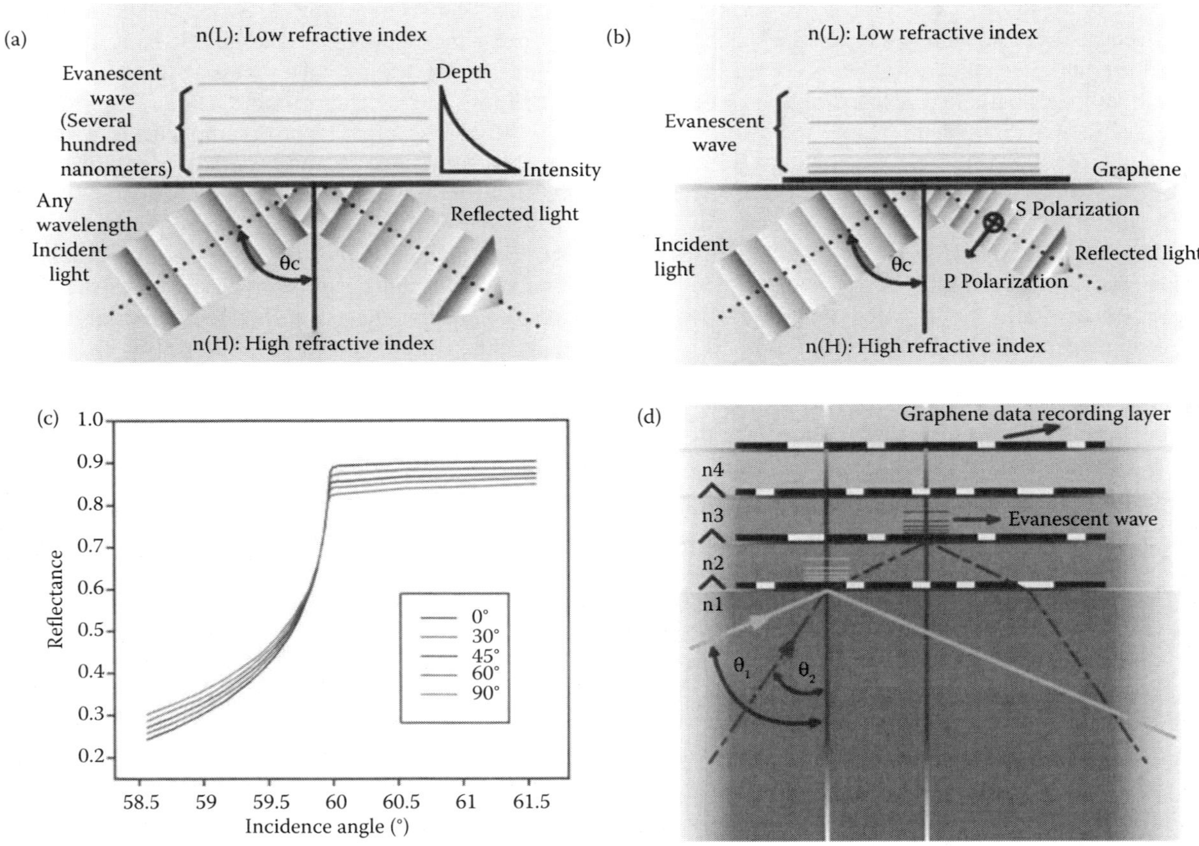

FIGURE 42.11 (a) Schematic illustration of evanescent wave propagation without graphene under TIR. (b) Schematic illustration of evanescent wave propagation with graphene on the surface of a prism under TIR. (c) Reflectivity as a function of the incident angle for different polarization directions. 0° represents *p*-polarized light, and 90° represents *s*-polarized light. (d) Data-reading schematic illustration of the graphene-based multilayer-film ODS medium.

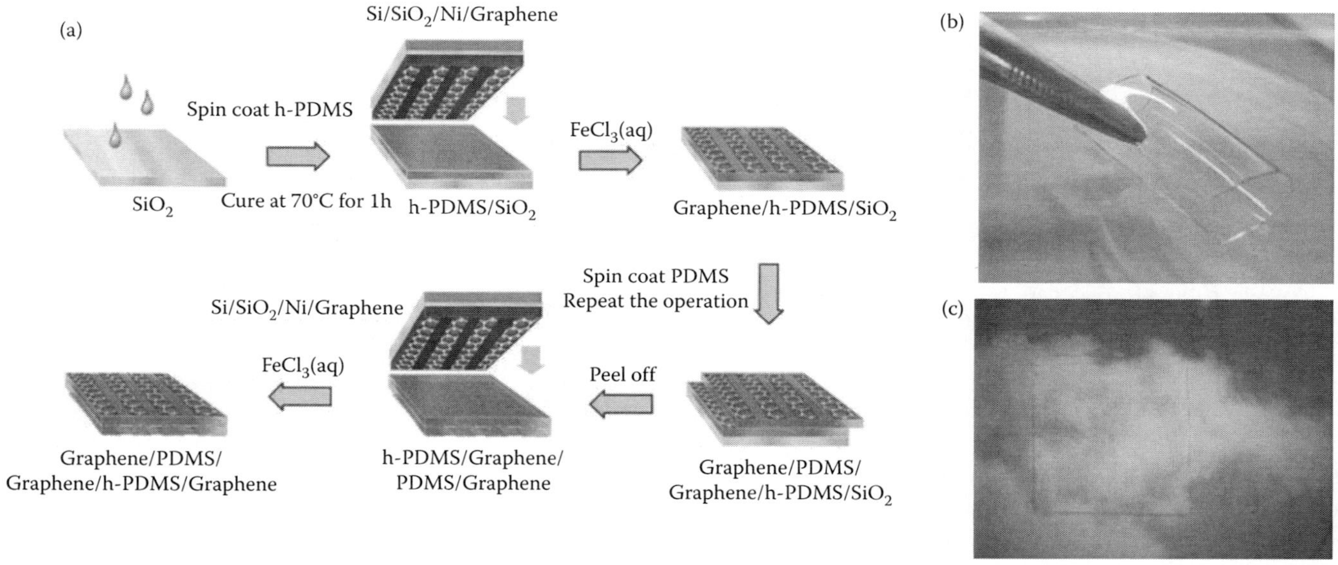

FIGURE 42.12 Preparation of flexible three-layer-film ODS by repeatedly data transferring. (a) Schematic illustration of the experimental procedure for fabricating the transparent flexible graphene-based three-layer-film ODS medium. (b and c) Photographs of the transparent flexible graphene-based three-layer-film ODS medium. (Reprinted with permission from Xing, F. et al. 2013. Transparent and flexible multi-layer films with graphene recording layers for optical data storage. *Applied Physics Letters* 102 (25):253501. Copyright 2013, American Institute of Physics.)

Owing to their excellent optical, mechanical, and surface chemical properties, PDMS and h-PDMS have been widely applied in optical fluidic chips, flexible devices, microfluidic chips, and soft lithography in recent years (Qin et al. 2010). The refractive index of PDMS is 1.4017, and the refractive index of h-PDMS is 1.4198 at room temperature. Thus, PDMS and h-PDMS can act as buffer layers with an index gradient. A graphene-based transparent and flexible multilayer-film ODS medium was fabricated through the transfer process of graphene, as shown in Figure 42.12a. First, the h-PMDS pre-polymer was poured onto a quartz substrate and spun. After curing the h-PDMS prepolymer, the h-PDMS surface was attached to Ni-based graphene with recorded data to form a structure of $Si/SiO_2/Ni$/graphene (G)/h-PDMS/quartz. The multilayer structure was then soaked in an aqueous iron (III) chloride ($FeCl_3$) solution (1 M) to remove the nickel layers. In this way, a 2D graphene-based ODS medium with h-PDMS as a buffer layer was obtained. By repeating the above process, three-layer ODS structures with h-PDMS and PDMS buffer layers can be obtained. The data transfer process breaks through the traditional process of data writing–reading and enriches the flexibility of optical data storage.

Figure 42.13 shows a sketch diagram of the data-reading process for the multilayer films. The graphene-based three-layer-film ODS medium was adsorbed on glass and formatted in the structure of glass ($n_1 = 1.5104$)/graphene/h-PDMS ($n_2 = 1.4188$)/graphene/PDMS ($n_3 = 1.4017$)/graphene/air ($n_4 = 1.0000$). The medium was placed on a prism ($n = 1.5104$) with an index-matching fluid between them. A continuous wave (CW) laser at 532 nm is transformed into circularly polarized light through a Glan–Taylor prism and a quarter-wave plate. Through objective focusing, the circularly polarized light is used to read the recorded data. The *s*- and *p*-polarized light components are separated by a polarized beam splitter. Therefore, the absorptions of graphene for the *s*- and *p*-polarized light are distinguished. The variations

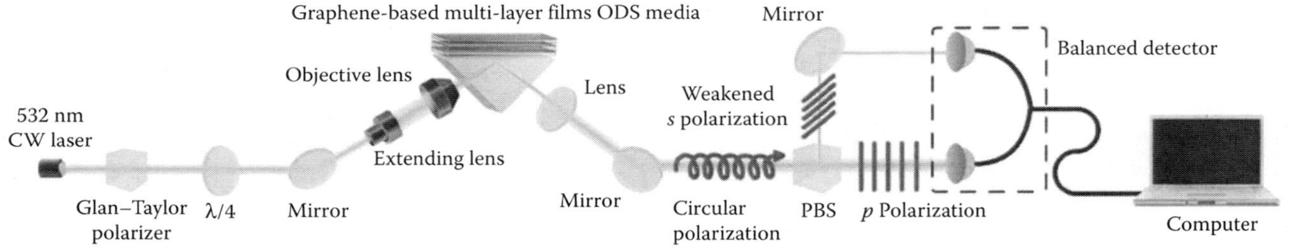

FIGURE 42.13 Sketch diagram of the data-reading process for the multilayer films. A 532-nm CW laser is used for data reading. (Reprinted with permission from Xing, F. et al. 2013. Transparent and flexible multi-layer films with graphene recording layers for optical data storage. *Applied Physics Letters* 102 (25):253501. Copyright 2013, American Institute of Physics.)

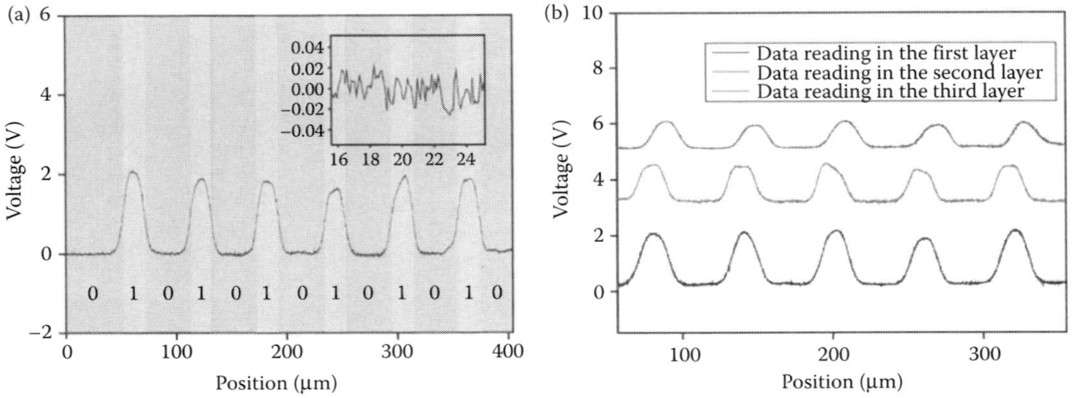

FIGURE 42.14 (a) The modulation signal of the graphene storage layer at the glass (n_1 = 1.5104) and h-PDMS (n_2 = 1.4188) interfaces. (b) The reading results for each layer of the three-layer-film ODS medium. (Reprinted with permission from Xing, F. et al. 2013. Transparent and flexible multi-layer films with graphene recording layers for optical data storage. *Applied Physics Letters* 102 (25):253501. Copyright 2013, American Institute of Physics.)

in the *s*- and *p*-polarized light components are then simultaneously detected by a balanced detector, and the difference between the *s*- and *p*-polarized light intensities is converted into a voltage signal through the detector. The data recorded on each layer of graphene can be read out in a sequence by reducing the incident angle. Figure 42.14 displays the reading results of the graphene storage layer at the glass (n_1 = 1.5104) and h-PDMS (n_2 = 1.4188) interfaces. The zone corresponding to the low-voltage value without graphene is recorded as "0", whereas the zone corresponding to the high-voltage value with graphene is recorded as "1." The reading curves show that the voltage difference is more than 2 V between reading the "0" and "1" data, which indicates an ideal readout effect and a high degree of recognizable signal. The figure shows that the curves are smooth and that the heights of the regions in the "1" zone are nearly identical, which indicates that the signal is very stable during the readout process and that the transfer process has little impact on the graphene recording layer. The SNR for this reading method is approximately 100 based on the enlarged noise/signal figure.

Crosstalk interference between layers is a significant bottleneck for multilayer-film ODS. As shown in Figure 42.11d, the data stored by the graphene at the interfaces of n_2 and n_3 are taken as an example. One aspect is that part of the incident light is reflected and absorbed when it passes though the interfaces of n_1 and n_2. Compared with the TIR, the absorption of the graphene is polarization-independent when the incident light passes though the graphene. The light intensity though the interfaces of n_1 and n_2 is weakened for both the *s*- and *p*-polarized light. Another aspect is due to the interaction between the evanescent wave and the graphene in the next layer when the evanescent wave passes though the interfaces of n_3 and n_4. This result occurs because the evanescent wave decays exponentially in the direction perpendicular to the interface, and the general penetration depth is on the order of the incident light wavelength, as shown in Figure 42.11a. Theoretically, if the thickness of the buffer layer is more than the wavelength of the incident light, the readout results cannot be influenced. Figure 42.14b shows the reading results of each layer of the three-layer-film ODS medium with a graphene recording layer. The buffer layer thickness of h-PDMS and PDMS is ~50 μm, which are orders of magnitude greater than the wavelength. The power density of the incident light is 87.5 nW/μm^2. The crosstalk interference between layers is very low because of the high SNR and the thick buffer layers. However, owing to the absorption and scattering of the graphene, the SNR decreases with an increase in the reading layers under the same reading intensity. The SNR is approximately 100 for the data on the graphene recording layer at the glass (n_1 = 1.5104) and h-PDMS (n_2 = 1.4188) interfaces, whereas the SNR reduces to 60 when the data on the graphene recording layer at the PDMS (n_3 = 1.4017) and air (n_4 = 1.0000) interfaces are read.

Figure 42.15a shows the readout signal from the graphene recording layer between the interfaces of glass (n_1 = 1.5104) and h-PDMS (n_2 = 1.4188) for a different incident light power. The figure shows that the larger the power density of the incident light, the higher the SNR of the reading results. When the power density of the incident light is 0.6 nW/μm^2, the SNR is approximately 5–6. This value corresponds to the light power density of the minimum distinguished signal in the experiment.

A flexibility test was performed by repeatedly bending and relaxing this ODS medium around a solid column (radius = 10 mm), with a tensile strain of approximately 3%. Figure 42.15b shows that as the three-layer-film ODS medium is bent 1000 times, there is no significant electrical degradation in the "0" and "1" states. In this figure, the power density of the incident light was 87.5 nW/μm^2, and the interfaces of the prism (n_1 = 1.5104) and h-PDMS (n_2 = 1.4188) were tested. Note from the figure that the recording layer's switch ratio is still stable after being bent 1000 times, and the voltage difference between the "0" and "1" states remains approximately 2 V.

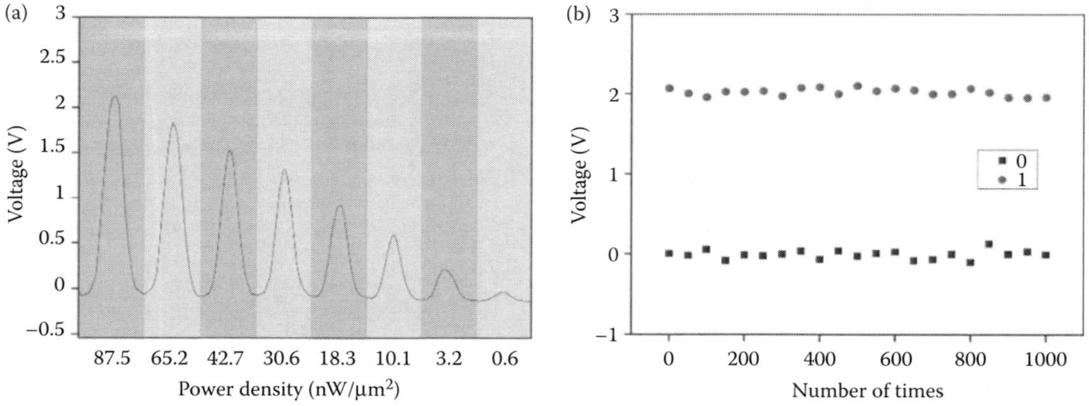

FIGURE 42.15 The SNR analysis of different incident light power and flexibility test. (a) The readout signal from the graphene recording layer between the glass (n_1 = 1.5104) and h-PDMS (n_2 = 1.4188) interfaces. (b) Bending experiment of the graphene-based three-layer-film ODS medium. (Reprinted with permission from Xing, F. et al. 2013. Transparent and flexible multi-layer films with graphene recording layers for optical data storage. *Applied Physics Letters* 102 (25):253501. Copyright 2013, American Institute of Physics.)

42.6 SUMMARY AND MORE

Given the unique 2D structure of graphene, its optical properties can be more fully utilized when surface waves propagate through the graphene plane. When graphene is placed on the interface of a TIR structure, its optical properties will have different performance due to longer interaction distance and difference of polarization mode supports, relative to the general structure of irradiation. Under TIR, graphene exhibits strong polarization-dependent optical absorption. Compared with universal absorbance of 2.3%, larger absorption was observed in monolayer, bilayer, and few-layer graphenes for TE wave under TIR. Reflectance ratio of TM to TE waves can easily provide the information of number of graphene layers.

The difference of reflectivity for TE and TM modes in the context of a TIR structure is sensitive to the media in contact with the graphene. A graphene refractive index sensor can quickly and sensitively monitor changes in the local refractive index with a fast response time and broad dynamic range. The sensitive and real-time monitoring of the refractive index was shown using a graphene-based optical sensor that has a broad dynamic refractive index range from 1.0 to 1.438, a fast response time of ~10 μs, a high sensitivity of 10^{-5} RIU, and excellent stability; these qualities are all essential for a versatile refractive index sensor. The flexibility and ease of functionalization of graphene sheets could also enable radically different refractive index sensors to be designed. For example, graphene could be integrated with flexible substrates and plastic materials. Alternatively, it could be used in novel geometries, for example, as a flexible sensor. The recent development of large-scale graphene synthesis and transfer techniques, as well as the structural modification of graphene by chemical or biological molecules, ensures that graphene will become the basis for powerful label-free monitoring methods for chemical or biomolecular interactions.

Owing to its high flexibility and transparency, a graphene-based multilayer-film ODS medium is likely to be produced as a continuous and large-scale storage medium for achieving the needed storage formats. At present, the minimum data point of graphene can be written in a few nanometers. However, this data point could not be read out in our experiment because the light spot size was ~20 μm. Further research is required to improve the reading method for reading out smaller dimensions. Two improved schemes for obtaining smaller light spot size can be considered: one is the use of TIR of the objective lens with a large numerical aperture; another is a program combining balanced detecting and imaging technique. It may be feasible to combine the breakthrough diffraction limit method and the reading method with the higher SNR. In the structure on the basis of the TIR, the smaller light spot is, the smaller data point can be read. Thus, the future development of graphene-based ODS depends on technology improvements that smaller data point can be read and the number of storage layer increases.

The interference nature of the resonance giving rise to the enhanced absorption offers the opportunity to modulate this absorption. Besides optical sensor and ODS, the enhanced light–graphene coupling in a wide spectral range will be great potential in many applications such as photodetector and photovoltaics.

REFERENCES

Axelrod, D., T. P. Burghardt, and N. L. Thompson. 1984. Total internal reflection fluorescence. *Annual Review of Biophysics and Bioengineering* 13:247–68.

Bao, Q., H. Zhang, B. Wang et al. 2011. Broadband graphene polarizer. *Nature Photonics* 5 (7):411–415.

Bao, Q., and K. P. Loh. 2012. Graphene photonics, plasmonics, and broadband optoelectronic devices. *ACS Nano* 6 (5):3677–3694.

Blake, P., E. W. Hill, A. H. C. Neto et al. 2007. Making graphene visible. *Applied Physics Letters* 91 (6):063124.

Bonaccorso, F., Z. Sun, T. Hasan, and A. C. Ferrari. 2010. Graphene photonics and optoelectronics. *Nature Photonics* 4 (9):611–622.

Born, M., and E. Wolf. 1999. *Principles of Optics*. 7th edition. Cambridge, UK: Cambridge University Press.

Bruna, M., and S. Borini. 2009. Optical constants of graphene layers in the visible range. *Applied Physics Letters* 94 (3):031901.

Echtermeyer, T. J., L. Britnell, P. K. Jasnos et al. 2011. Strong plasmonic enhancement of photovoltage in graphene. *Nature Communications* 2 (1464):458–462.

Engel, M., M. Steiner, A. Lombardo et al. 2012. Light–matter interaction in a microcavity-controlled graphene transistor. *Nature Communications* 3:906.

Gómez-Santos, G., and T. Stauber. 2012. Graphene plasmons and retardation: Strong light–matter coupling. *EPL (Europhysics Letters)* 99 (2):27006.

Hecht, J. 2012. Graphene photonics are making their way to practical use. *Laser Focus World* 48 (5):62–65.

Homola, J. 2006. *Surface Plasmon Resonance Based Sensors*. Vol. 4. Berlin and New York: Springer.

Homola, J. 2008. Surface plasmon resonance sensors for detection of chemical and biological species. *Chemical Reviews* 108 (2):462–493.

Kim, K. S., Y. Zhao, H. Jang et al. 2009. Large-scale pattern growth of graphene films for stretchable transparent electrodes. *Nature* 457 (7230):706–710.

Lee, Y., S. Bae, H. Jang et al. 2010. Wafer-scale synthesis and transfer of graphene films. *Nano Letters* 10 (2):490–493.

Li, X., Y. Zhu, W. Cai et al. 2009. Transfer of large-area graphene films for high-performance transparent conductive electrodes. *Nano Letters* 9 (12):4359–4363.

Liu, M., X. Yin, E. Ulin-Avila et al. 2011. A graphene-based broadband optical modulator. *Nature* 474 (7349):64–67.

Liu, M., X. Yin, and X. Zhang. 2012. Double-layer graphene optical modulator. *Nano Letters* 12 (3):1482–1485.

Liu, Z., Y. Wang, X. Zhang, Y. Xu, Y. Chen, and J. Tian. 2009. Nonlinear optical properties of graphene oxide in nanosecond and picosecond regimes. *Applied Physics Letters* 94 (2):021902.

Lott, J., C. Ryan, B. Valle et al. 2011. Two-photon 3D optical data storage via aggregate switching of excimer-forming dyes. *Advanced Materials* 23 (21):2425–2429.

Mak, K. F., L. Ju, F. Wang, and T. F. Heinz. 2012. Optical spectroscopy of graphene: From the far infrared to the ultraviolet. *Solid State Communications* 152 (15):1341–1349.

Mak, K. F., M. Y. Sfeir, Y. Wu, C. H. Lui, J. A. Misewich, and T. F. Heinz. 2008. Measurement of the optical conductivity of graphene. *Physical Review Letters* 101 (19):196405.

Mikhailov, S. A., and K. Ziegler. 2007. New electromagnetic mode in graphene. *Physical Review Letters* 99 (1):016803.

Mustroph, H., M. Stollenwerk, and V. Bressau. 2006. Current developments in optical data storage with organic dyes. *Angewandte Chemie International Edition* 45 (13):2016–2035.

Nair, R. R., P. Blake, A. N. Grigorenko et al. 2008. Fine structure constant defines visual transparency of graphene. *Science* 320 (5881):1308–1308.

Ni, Z. H., H. M. Wang, J. Kasim et al. 2007. Graphene thickness determination using reflection and contrast spectroscopy. *Nano Letters* 7 (9):2758–2763.

Pirruccio, G., L. M. Moreno, G. Lozano, and J. G. Rivas. 2013. Coherent and broadband enhanced optical absorption in graphene. *ACS Nano* 7 (6):4810–4817.

Qin, D., Y. Xia, and G. M Whitesides. 2010. Soft lithography for micro- and nanoscale patterning. *Nature Protocols* 5 (3):491–502.

Ryan, C., C. W. Christenson, B. Valle et al. 2012. Roll-to-roll fabrication of multilayer films for high capacity optical data storage. *Advanced Materials* 24 (38):5222–5226.

Stanley, G. F. 1989. *Refractometers: Basic Principles*. Bellingham and Stanley Ltd, UK.

Tapasztó, L., G. Dobrik, P. Lambin, and L. P. Biro. 2008. Tailoring the atomic structure of graphene nanoribbons by scanning tunnelling microscope lithography. *Nature Nanotechnology* 3 (7):397–401.

Thongrattanasiri, S., F. H. L. Koppens, and F. Javier García de Abajo. 2012. Complete optical absorption in periodically patterned graphene. *Physical Review Letters* 108 (4):047401.

Wang, X., and H. Dai. 2010. Etching and narrowing of graphene from the edges. *Nature Chemistry* 2 (8):661–665.

Wang, X., M. Zhao, and D. D. Nolte. 2009. Optical contrast and clarity of graphene on an arbitrary substrate. *Applied Physics Letters* 95 (8):081102.

Weber, J. W., V. E. Calado, and M. C. M. van de Sanden. 2010. Optical constants of graphene measured by spectroscopic ellipsometry. *Applied Physics Letters* 97 (9):091904.

Xing, F., X.-D. Chen, Z.-B. Liu et al. 2013. Transparent and flexible multi-layer films with graphene recording layers for optical data storage. *Applied Physics Letters* 102 (25):253501.

Xing, F., Z.-B. Liu, Z.-C. Deng et al. 2012. Sensitive real-time monitoring of refractive indexes using a novel graphene-based optical sensor. *Scientific Reports* 2:908.

Ye, Q., J. Wang, Z. Liu et al. 2013. Polarization-dependent optical absorption of graphene under total internal reflection. *Applied Physics Letters* 102 (2):021912.

Yu, V., and M. Hilke. 2009. Large contrast enhancement of graphene monolayers by angle detection. *Applied Physics Letters* 95 (15):151904.

Zhang, Y., L. Guo, S. Wei et al. 2010. Direct imprinting of microcircuits on graphene oxides film by femtosecond laser reduction. *Nano Today* 5 (1):15–20.

Index

F

G

H

Q

R

S

T

U